D1499765

THE COGNITIVE NEUROSCIENCES

THE COGNITIVE NEUROSCIENCES

Fifth Edition

Michael S. Gazzaniga and George R. Mangun,
Editors-in-Chief

Section Editors:

Sarah-Jayne Blakemore
James J. DiCarlo
Scott T. Grafton
Peter Hagoort
Todd F. Heatherton
Sabine Kastner
Leah Krubitzer
Elizabeth A. Phelps
Todd M. Preuss
Adina Roskies
Walter Sinnott-Armstrong
Peter L. Strick
Giulio Tononi
Anthony D. Wagner
B. A. Wandell
Rachel I. Wilson

A BRADFORD BOOK
THE MIT PRESS
CAMBRIDGE, MASSACHUSETTS
LONDON, ENGLAND

MIT Press books may be purchased at special quantity discounts for business or sales promotional use. For information, please email special_sales@mitpress.mit.edu.

This book was set in Baskerville by Toppan Best-set Premedia Limited.
Printed and bound in the United States of America.

Library of Congress Cataloging-in-Publication Data

The cognitive neurosciences / edited by
Michael S. Gazzaniga and George R. Mangun. –
[5th edition].
pages cm
"A Bradford Book"
Includes bibliographical references and index.
ISBN 978-0-262-02777-9 (hardcover : alk. paper)
1. Cognitive neuroscience. I. Gazzaniga, Michael S., editor of compilation. II. Mangun, G. R. (George Ronald), 1956– editor of compilation. III. Title: Cognitive neurosciences.
QP360.5.C63952 2014
612.8′233–dc232013048606

10 9 8 7 6 5 4 3 2 1

CONTENTS

II PLASTICITY AND LEARNING

III VISUAL ATTENTION

IV SENSATION AND PERCEPTION

V MOTOR SYSTEMS AND ACTION

VI MEMORY

VII LANGUAGE AND ABSTRACT THOUGHT

VIII SOCIAL NEUROSCIENCE AND EMOTION

IX CONSCIOUSNESS

X ADVANCES IN METHODOLOGY

PREFACE

With some amusement we note that the preface to the last edition of this book carried a hilarious error. One of us (MSG) wrote that preface and buoyantly noted that the first meeting at Lake Tahoe, which underlay that book, had occurred 20 years earlier. After all, it was the fourth edition and the book does come out every five years. Do the math, right? Upon first glance, it seems correct and yet like most everything else in our field, what seems right is not good enough. We have to think about the problems we study again and again.

Of course, it seemed right to MSG because the Summer Institute in Cognitive Neuroscience had already been up and running for the four years preceding the 1993 Tahoe meeting. It had started off at Harvard, moved to Dartmouth for two years and then to UC Davis. The fifth year of that cycle was the Tahoe meeting, which resulted in the first edition of this volume. Obviously, the volume summarized the previous four years of research presented at the Summer Institutes, not the future SI's.

So, in this 2014 edition, the result of the fifth Tahoe meeting, we can say that it really was 20 years ago that we met for the first time in Tahoe! Not only that, the fifth meeting was one of the best. Many of the speakers were former fellows, and many of the current fellows will undoubtedly be future speakers and authors. Indeed, this book and the 2013 Tahoe meeting reflect a transition. As JFK famously said, "The torch has been passed to a new generation. ..." In this case, young cognitive neuroscientists are exploring the logic of old ideas and, in some cases, finding them wanting and presenting new ideas. While some old hands were around to steady the transfer, a new era is launched with this book and is bursting with new talent.

The Institute started with a bang, a perspective on some pitfalls in the brain sciences. It was pointed out that while human brains are extremely large, neuroscientists have lacked the necessary methods both to compare humans to other animals and to determine how evolution modified the internal structure of the brain. It was noted that in the absence of such methods, neuroscientists have focused on the study of relatively few model species and have viewed those results as applicable to humans with little qualification. Fortunately, new techniques, especially MRI techniques and the methods of comparative genomics, have started to be used to study humans and other animals and have been game changers. Particularly exciting is the application of diffusion tensor imaging (DTI), a technique that has been used to compare humans to chimpanzees (the animals most closely related to humans, which makes them critical species for identifying human specializations). These studies have revealed, for the first time, differences in systems of connections between cortical areas and have provided the foundations for understanding the neural bases of human cognitive and behavioral specializations.

In the area of development, research in the past 20 years or so has generated profound insights into the infant mind and has revealed that the cognitive and social cognitive capacities of very young children are much more sophisticated than was once believed. While developmental psychology made great progress in the later part of the last century, it remained relatively removed from developmental neuroscience. In the past decade, the field of developmental cognitive neuroscience has undergone unprecedented expansion. Once again, the use of MRI and fMRI has propelled our knowledge of how the human brain develops, and the data from developmental imaging studies have in turn spurned new interest in the changing structure and function of the brain over the lifespan.

The study of sensation and perception is also being revolutionized by new experimental techniques that provide deeper insights into old problems. New tools are improving our ability to test the causal role of specific neurons and brain regions in shaping sensory perception. They are allowing us to record from increasingly large ensembles of neurons, revealing the ways that information about the state of the world is untangled from population activity evoked at the sensory epithelium. New data are also revealing the competing constraints that sculpt the architecture of sensory systems. Finally, new conceptual approaches and computational methods are beginning to reveal how the brain's limited hardware can give rise to fast, flexible, and robust perceptual categories. In short, new tools are exploding our field of view, pushing us to develop better theoretical frameworks.

People are astounding in their ability to create goal-directed behaviors that transform the world. This vast behavioral repertoire creates theoretical challenges for defining the fundamental principles of selection, planning, and control that the motor system must employ to achieve these ends. The section on motor systems and action addresses some of these challenges by first emphasizing the basics of motor system anatomy, particularly at the circuit level. This architecture includes cortical connectivity between many cortical motor areas and subcortical systems, particularly the basal ganglia and cerebellum. Functional contributions of the different motor areas are considered in terms of optimal control and the retrieval of learned motor behavior. The section then introduces some of the most important correlation methods for identifying dynamic representations of movement based on multiple neuron recordings. The section closes with two chapters on problems of human action selection. The first considers how we choose among a variety of goal-directed behavior when there are competing risks and rewards. The second evaluates the way that possible actions are planned automatically, in advance of any decision.

Approximately 20 years ago, the memory section of the first edition of *The Cognitive Neurosciences* highlighted two major themes central to theory and research at the time. First, to understand memory, one must characterize the processes governing memory encoding, storage, and retrieval. Second, memory is a nonunitary entity, as it consists of multiple forms of knowledge expression that depend on distinct neural substrates and computations. In the intervening years, technological advances in human and nonhuman neuroscience have enabled increased anatomical and computational precision in these core areas, while simultaneously sparking novel themes and challenges to earlier ideas. The memory section of this volume documents the field's progress toward establishing increasingly precise models of memory, such as specifying the distinct computations and forms of remembering supported by regions in the medial temporal lobe, delineating the multiple mechanisms that enable memory performance over short time scales, and illuminating how context is neurally represented and shapes memory behavior. The chapters in this section also document some of the newly emerging themes that have shaped the field in recent years, such as the convergence of theorizing about memory and decision making, the unexpected role of parietal cortex in memory, and the fundamental link between memory and prediction. The remarkable strides summarized in these chapters evidence past progress and predict a future rich with continued discovery.

One flourishing area in cognitive neuroscience revolves around trying to understand the affective basis of the social brain. In solving adaptive problems, the brain relies on emotional processes to solve challenges to successful adaptation. For humans, many of the most challenging adaptive problems result from the necessity of interacting with other humans, such as selecting mates, cooperating in hunting and gathering, forming alliances, competing over scarce resources, and even warring with neighboring groups. Interacting with other humans produces emotion, and these emotions serve as guidelines for successful group living. Any true understanding of human nature requires a full consideration of both the emotional and social aspects of brain functioning. There is remarkable convergence in this section that highlights the important role of the amygdala across animal species and paradigms. In addition, it is now clear from more than a decade of research that various prefrontal regions are crucial for successful functioning within social contexts. This section demonstrates that the study of emotion and social cognition has become established as a core area of cognitive neuroscience.

The session on language and abstract thought was a testimony of where the field has moved since the previous edition of this book. First, the connectivity profile of the perisylvian language areas has been mapped out in much more detail, as have the cytoarchitectonic and receptor architectonic properties of language areas such as Broca's area. It has become clear that the language network, beyond any doubt, is much more extensive than Broca's area, Wernicke's area, and the arcuate fasciculus (the classical Wernicke-Lichtheim-Geschwind model). Second, studies recording event-related brain potentials have established firmly that prediction plays an important role during language comprehension. Predictions are made at multiple levels about general event structure, a word's grammatical characteristics, and even about the upcoming word sound. Finally, we now realize that communication with language is a multimodal system in which speech is accompanied by gestures and other information. Indeed, sign language is a form of language based solely on the visual-gestural modality. How this difference in modality of expressing language and how the multimodal aspects co-determine the neural architecture of the system are starting to be understood. Moreover, understanding language is more than decoding what is written or said. In communication we try to infer the intention behind the utterance. The step from coded meaning to speaker meaning, we now know, involves the ToM network. Hence, it is clearly shown that an account of language comprehension based solely on the mirror neuron system will fail hopelessly.

The session on consciousness showed, in no uncertain terms, how far we have moved since the days when the c-word was an outcast from psychology and neuroscience (but one that had gained early admission within cognitive neurosciences, as witnessed by the previous editions of this volume). There are now bold attempts at characterizing consciousness theoretically, with respect to both quantity and quality, which do not shy away from the "hard problem" of what consciousness is and how it is generated by the brain. Impressive new work using TMS/EEG has demonstrated that the level of consciousness can be measured in individual subjects and across brain states as diverse as wake, sleep, anesthesia, and brain disorders. We are also learning a great deal about how and when consciousness is lost in different kinds of seizures—an area of study that was nearly nonexistent even 10 years ago. A strong case is now being made for the experimental and conceptual dissociation between consciousness and attention on one side and between consciousness and executive function/report on the other. Intriguingly, initial results suggest that some of the so-called neural correlates of consciousness demonstrated in past studies have to do with stimulus processing or response preparation rather than with consciousness itself. Other prominent areas of progress include the relationship of consciousness with cognitive and neural access, the conscious control of action, and the relevance of body awareness to consciousness in general.

Tackling such profoundly difficult questions will be aided by the introduction of powerful new methodologies to analyze brain function. What were leading technologies 10 years ago are not the leading technologies of today. Accordingly in this edition of *The Cognitive Neurosciences* we include a new section devoted to new methods that will shape and guide our field into the future. Moving from new, more exacting methods of measurments for fMRI to deeper probings of the massive data collected, today's cognitive neuroscientist must be well versed in both several new methodologies and concurrent research in separate related fields. Sophisticated models of causality in brain function are now beginning along with the application of network theory to massive data sets on diffusion imaging knowledge about brain connections.

Finally, other people are mysterious. We do not understand how they work, and we do not know how to treat them. Neuroscience is starting to help us unravel these mysteries. To improve our understanding of society, recent work in cognitive neuroscience has revealed some of the psychological and neural processes through which we perceive people as people, through which we determine which mental states other people have, and through which we form moral judgments about actions and choices by other people. In addition, knowledge of the brain can be used as a tool to *change* how we relate to other members of society. It can affect policies regarding addicts, psychopaths, and perhaps even "normal" criminals. It can also be used to develop drugs to enhance cognitive and emotional performance in people without illnesses and to understand and control economic transactions through neuromarketing. This power of neuroscience to illuminate and affect society can, of course, be used for good or for ill, so it poses profound moral quandaries that we are only beginning to face. The goal of the section on neuroscience and society in this volume is to raise these issues, not to resolve them.

It was a glorious three weeks of science interrupted by hikes, swimming, and pondering the lectures in the sylvan beauty of Lake Tahoe. This book and the linked videos of the lectures are our attempt to let one and all enjoy the richness and scholarship of the great scientists involved in this effort. The complex and intense planning, editing, logistics, and time put in by dozens of people should be obvious. Needless to say, the section editors and the participants make the book what it is and all deserved our profound thanks. The flawless execution of the meeting itself was due to Jayne Rosenblatt, and once again the managing editor of the book, Marin Gazzaniga, made it all seem easy. Finally, we are indebted to MIT Press, which year after year serves our field in ever so many ways. And all of this, of course, would not have been possible without the generous support of the National Institute of Mental Health.

Michael S. Gazzaniga
University of California, Santa Barbara

George R. Mangun
University of California, Davis

I

DEVELOPMENTAL AND EVOLUTIONARY COGNITIVE NEUROSCIENCE

Introduction

SARAH-JAYNE BLAKEMORE
AND TODD M. PREUSS

Fifty years ago, very little was known about how the human brain develops. Research on brain development in the second half of the twentieth century focused almost entirely on nonhuman animals and revealed a great deal about early neuronal and synaptic development (Hubel & Wiesel, 1977). These advances in animal research were paralleled by pioneering research in developmental psychology, particularly by Piaget (1977) and Vygotsky (1986). Their studies, which involved observing and analyzing children's behavior in meticulous detail, changed contemporary thinking about children's minds. Through such important behavioral studies, children were increasingly seen as more than just miniature versions of adults. Research in the past 20 years or so has generated profound insights into the infant mind and has revealed that the cognitive capacities of very young children are much more sophisticated than was once believed. Infant social cognition is the focus of chapter 1 by Baillargeon, Setoh, Sloane, Jin, and Bian in this section.

While developmental psychology made great progress in the later part of the last century, it remained relatively removed from developmental neuroscience. Research on human neural development was heavily constrained by the technical challenges of studying the living human brain and, until fairly recently, was limited to the study of post-mortem brains. Yet, in the past two decades, the field of developmental cognitive neuroscience has undergone unprecedented expansion. This can at least in part be attributed to technological advances. Electroencephalography, or EEG, has been used for several decades to study the temporal dynamics of brain function in infants and children. However, in recent years, probably more than any other advance, the increased use of magnetic resonance imaging (MRI) in children has created new opportunities to track structural and functional changes in the developing human brain. The use of MRI and functional magnetic resonance imaging (fMRI) has propelled our knowledge of how the human brain develops, and the data from developmental imaging studies has in turn spurred new interest in the changing structure and function of the brain over the life span. Who would have imagined half a century ago that scientists would be able to look inside the brains of living humans of all ages and track changes in brain structure and function across development?

Until relatively recently, it was widely believed that any changes in the brain after early development were comparably minimal. Research in the past 20 years, however, especially that emanating from the pioneering National Institutes of Health pediatric neuroimaging project, has shown that this is far from true. This large-scale longitudinal study has generated very large amounts of MRI data from childhood through adulthood and has revealed that the human brain continues to develop for many decades. Structural development in the human brain is the focus of chapter 2 by Goddings and Giedd in this section.

As developmental MRI research has expanded and flourished, so too has research on developmental changes in functional brain activity as measured by fMRI. Just like developmental MRI research, since 2000 there has been a year-on-year increase in the number of papers reporting fMRI studies on functional development in the human brain. Many labs around the world use fMRI to investigate how neural systems associated with particular cognitive processes change with age. Areas of research in this field that are fairly well established, but continue to expand, are the development of the social brain in adolescence (see chapter 4 by Blakemore and Mills, this section) and the development of decision making and executive function (see chapter 3 by Crone, this section).

Just as human developmental neuroscience has been transformed by the development of new technologies, so has human evolutionary neuroscience. Until quite recently, all that could be said about the evolution of the human brain is that it became freakishly enlarged after the human lineage separated from the branch leading to our closest relatives (chimpanzees and bonobos) 5–7 million years ago. What we have needed to know, but seemed unable to find out, was what happened to the brain's internal organization (Holloway, 1966). Lacking methods to study humans in detail, and with the field's heavy investment in model animal studies, few neuroscientists echoed the view of Crick and Jones (1993) that the field urgently needed information about the detailed structure of the human brain, particularly its connections.

This situation changed dramatically with the development of MRI and other neuroimaging modalities, techniques well suited for comparative studies, since they can be used with both humans and other primate species. What's more, cross-species application of diffusion-tensor imaging, which can track the course of fiber bundles in the brain, has for the first time made it possible to compare the cortical connectivity of humans to that of other primates, and so begin to relate human specializations of cognition and behavior to specializations of the brain. The value of these techniques is well illustrated in chapter 5 by Rilling and Stout, who review recent neuroimaging studies comparing the myeloarchitecture and cortical connectivity of humans, chimpanzees, and macaque monkeys. Consistent with traditional views of human brain evolution, evidence from magnetic resonance myeloarchitectonics indicates that the classical higher-order cortical regions—prefrontal cortex, posterior parietal cortex, and nonauditory parts of the temporal cortex—became enormously enlarged in human evolution. Comparative diffusion tensor imaging studies demonstrate that these changes in size were accompanied by changes in cortical connectivity that affected systems involved in at least three human-specialized cognitive domains: language, tool making and tool use, and social learning.

Evolutionary specializations are not, of course, restricted to the human lineage, and chapter 6 by Wise addresses an enduring issue in comparative cognitive neuroscience: the putative uniqueness of the primate prefrontal cortex. Wise maintains that the data on primate brain evolution support the view that the portions of prefrontal cortex that contain a well-developed internal granular layer are unique to primates. If this is the case, what does granular prefrontal cortex contribute to cognition? Wise contends that whereas the nongranular orbital and medial parts of prefrontal cortex (cortical regions shared widely among mammals) support classical associative-learning mechanisms that link objects and actions to expected outcomes over the course of many events, the granular regions support explicit representation of single events that enables primates to generate abstract rules, making it possible to solve novel problems and forage more efficiently.

In light of the growing body of evidence for primate and human brain specializations, Preuss and Robert, in chapter 7, examine the consequences of focusing animal research on a very few "model" species—mostly rodents—for understanding of aspects of the human brain that we cannot study directly in humans. While acknowledging the practical benefits of concentrating on a small number of species, they contend that this concentration has gone too far, with serious consequences. The central problem of the "model-animal paradigm" is its premise that differences between species are relatively minor and can generally be discounted, a view clearly at odds with the facts of comparative neuroscience and modern evolutionary biology. Adherence to the paradigm affects the interpretation of data, encouraging soft-pedaling of results when model animals are found to differ from humans, and perpetuates the view that all brains are merely minor variants of a basic brain organization. The failure to deal with diversity calls into question the integrity of neuroscience research results, and failure to accommodate modern views of evolution raises questions about the intellectual integrity of the neuroscientific enterprise generally. The authors argue that a more comparative approach would result in a more rigorous science and a better understanding of how humans resemble and differ from other animals.

New techniques derived from comparative molecular biology and genomics are beginning to illuminate molecular mechanisms at the intersection of development and evolution. It has long been appreciated that species differences in gene-expression regulation must be an important source of the differences between humans and other animals in the growth, development, and adult morphology of the brain (King & Wilson, 1975), but recent work, reviewed by Sousa, Meyer, and Šestan (chapter 8, this section) has begun to reveal just how extensive these differences are and the nature of the molecular interactions involved, as well as to identify developmental time points at which critical expression differences occur.

REFERENCES

Crick, F., & Jones, E. (1993). Backwardness of human neuroanatomy. *Nature, 361*(6408), 109–110.

Holloway, R. L. J. (1966). Cranial capacity, neural reorganization, and hominid evolution: A search for more suitable parameters. *Am Anthropol, 68,* 103–121.

Hubel, D. H., & Wiesel, T. N. (1977). Ferrier lecture: Functional architecture of macaque monkey visual cortex. *Proc R Soc Lond B Biol Sci, 198*(1130), 1–59.

King, M. C., & Wilson, A. C. (1975). Evolution at two levels in humans and chimpanzees. *Science, 188*(4184), 107–116.

Piaget, J. (1977). *Intellectual evolution from adolescence to adulthood.* Cambridge, UK: Cambridge University Press.

Vygotsky, L. S. (1986). *Thought and language* (rev. ed.). Cambridge, MA: MIT Press.

1 Infant Social Cognition: Psychological and Sociomoral Reasoning

RENÉE BAILLARGEON, PEIPEI SETOH, STEPHANIE SLOANE, KYONG-SUN JIN, AND LIN BIAN

ABSTRACT Infant social cognition depends on at least two evolved systems: the psychological- and sociomoral-reasoning systems. Each system has at its core a distinct explanatory framework of principles and concepts. The psychological-reasoning system enables infants to interpret agents' intentional actions and is constrained by a principle of rationality (with its corollaries of consistency and efficiency). When infants observe an agent act in a scene, the psychological-reasoning system infers the mental states that underlie the agent's actions; if the scene changes, infants use these mental states—together with the rationality principle—to predict the agent's likely actions. Recent evidence indicates that infants are already capable of sophisticated mentalistic reasoning and can attribute to agents not only motivational states (e.g., goals) and epistemic states (e.g., ignorance), but also counterfactual states (e.g., false beliefs). The sociomoral-reasoning system guides infants' expectations about how individuals should act toward others and is constrained by several principles, including reciprocity, fairness, and ingroup (with its corollaries of loyalty and support). When infants observe two or more individuals interact in a scene, the sociomoral-reasoning system determines what actions are obligatory, what actions are permissible, and what actions are impermissible. This chapter reviews key findings concerning each system in infancy.

Beginning in the first year of life, infants attempt to make sense of the actions of others. At least two causal-reasoning systems contribute to this process: the *psychological-* and *sociomoral-reasoning* systems (for a comprehensive review, see Baillargeon et al., 2014). Each system operates largely without explicit awareness and has at its core a distinct explanatory framework comprising key principles and concepts. The psychological-reasoning system infers agents' mental states (e.g., Leslie, 1994), and the sociomoral-reasoning system determines what is obligatory, permissible, and impermissible in social interactions (e.g., Dwyer, 2009).

To illustrate the purview of each system, imagine that infants are observing the following scene. Mary is stacking rings on a table; the last ring rolls to the floor, and Mary vainly attempts to reach it. At that point, Jane enters the scene; she observes Mary's efforts, walks over to the fallen ring, and picks it up. The psychological-reasoning system would enable infants to interpret Mary's and Jane's individual actions by inferring the mental states underlying these actions (e.g., Mary wants to stack all the rings, she noticed that the last one fell, and she is trying without success to reach it; Jane understands what Mary is trying to do, and she purposefully walks to the ring and picks it up). The sociomoral-reasoning system would allow infants to form expectations about what Jane should do next (e.g., is it obligatory for Jane to give the ring to Mary? is it permissible for Jane to drop the ring back to the floor? is it impermissible for Jane to steal the ring?) and how Mary should respond to Jane's actions.

In this chapter, we review findings on psychological and sociomoral reasoning in infancy (i.e., before age 2). These findings are important not only for developmental psychology but also for cognitive science generally, because they inform and constrain theoretical models about the nature and causal etiology of human intuitive social cognition.

Psychological reasoning

AGENTS When infants encounter a novel entity, they gather evidence about its ontological status. If the entity is capable of autonomous motion (e.g., begins to move or reverses course on its own), infants categorize it as *self-propelled* and endow it with internal energy (e.g., Luo, Kaufman, & Baillargeon, 2009). If the entity has autonomous control over its actions (e.g., uses varied means to achieve the same goal or responds purposefully to changes in its environment), infants categorize it as *agentive* and endow it with mental states (e.g., Csibra, 2008; Johnson, Shimizu, & Ok, 2007). Finally, if the entity gives evidence that it is both self-propelled and agentive, infants categorize it as an *animal* and endow it with biological properties, such as filled insides (Setoh, Wu, Baillargeon, & Gelman, 2013).

By 3 to 5 months of age, infants can already identify and reason about novel nonhuman agents (e.g., Hamlin & Wynn, 2011; Luo, 2011; Luo & Baillargeon, 2005). These findings make it unlikely that infants gradually construct (e.g., through their own experiences as agents) an abstract understanding of intentional action that is then extended to novel agents; rather, these findings support the view that psychological reasoning is an adaptation that emerges early in life and enables infants to interpret the actions of any entity they identify as an agent.

MENTAL STATES Infants who observe an agent act in a scene can attribute at least three kinds of mental states to the agent: *motivational* states, which capture the agent's motivation and include goals and attitudinal dispositions (e.g., Csibra, 2008; Woodward, 1998); *epistemic* states, which represent what the agent knows about the scene and include knowledge and ignorance states (e.g., Liszkowski, Carpenter, & Tomasello, 2008; Luo & Johnson, 2009); and *counterfactual* states, which correspond to reality-incongruent states such as false beliefs and pretense (e.g., Onishi & Baillargeon, 2005; Onishi, Baillargeon, & Leslie, 2007). A *decoupling* mechanism is recruited by the psychological-reasoning system to help represent counterfactual states (e.g., Baillargeon et al., 2014; Leslie, 1994). For example, when the agent holds a false belief about the scene, the decoupling mechanism enables infants to temporarily put aside—or decouple from—their own perspective on the scene in order to adopt the agent's perspective.

Elicited- and nonelicited-response false-belief tasks Until recently, it was generally assumed that false-belief understanding does not emerge until about age 4 and constitutes a major milestone in the development of mentalistic reasoning (e.g., Wellman, Cross, & Watson, 2001; Wimmer & Perner, 1983). This assumption was based on findings from *elicited-response* tasks. In these tasks, children are presented with a scene where an agent holds a false belief about some aspect of the scene, and they are asked a direct question about the agent's likely behavior. For example, children listen to a story enacted with props: Sally hides her toy in a basket and leaves; in her absence, Anne moves the toy to a box. Children are asked where Sally will look for her toy when she returns. At about age 4, children typically answer correctly, pointing to the basket; in contrast, most 3-year-olds point to the box, as though they do not yet understand that Sally will hold a false belief about her toy's location (e.g., Baron-Cohen, Leslie, & Frith, 1985).

The evidence that false-belief understanding is present long before age 4 comes from *nonelicited-response* tasks (e.g., Baillargeon, Scott, & He, 2010; Scott, He, Baillargeon, & Cummins, 2012). In these tasks, children are again presented with an agent who holds a false belief; instead of asking how the agent will act, however, investigators assess children's understanding of the agent's false belief using various indirect measures. For example, positive results have been obtained with infants in the second year of life using violation-of-expectation tasks (e.g., infants detect a violation when Sally searches for her toy in its current location; Onishi & Baillargeon, 2005), anticipatory-looking tasks (e.g., infants anticipate that Sally will approach the toy's original location; Senju, Southgate, Snape, Leonard, & Csibra, 2011), anticipatory-pointing tasks (e.g., infants spontaneously point to inform Sally about her toy's new location; Knudsen & Liszkowski, 2012), and prompted-action tasks (e.g., infants respond appropriately to prompts to help Sally, such as "Go on, help her!"; Buttelmann, Carpenter, & Tomasello, 2009). Evidence of false-belief understanding has been obtained using similar tasks with infants as young as 7 months of age (e.g., Kovács, Téglás, & Endress, 2010).

Processing-load account If children are capable of representing false beliefs at an early age, why do they fail at elicited-response tasks until about age 4? According to the *processing-load* account (e.g., Baillargeon et al., 2014), these tasks have considerable executive-function demands (for alternative accounts, see Butterfill & Apperly, 2013; Perner & Roessler, 2012). When children are asked the test question (e.g., "Where will Sally look for her toy?"), a *response-selection* process is activated (e.g., Mueller, Brass, Waszak, & Prinz, 2007) that inadvertently triggers a "reality bias": because agents usually look for an object where it is located, the prepotent response is that Sally will look for her toy in its current location. Thus, instead of—or in addition to—tapping their representation of Sally's false belief, children tap their own knowledge about the toy's current location. As a result, children cannot succeed unless their *inhibition* skills are sufficiently mature to suppress the prepotent response generated by the reality bias (e.g., Birch & Bloom, 2003; Leslie & Polizzi, 1998).

If young children's difficulties were entirely due to their inability to inhibit prepotent responses, we would expect them to succeed at elicited-response tasks that circumvent the reality bias. One such task is the "undisclosed-location" task: instead of moving Sally's toy from the basket to the box, Anne takes it away to an undisclosed location. As children do not know where the toy is, the reality bias should have little effect, leaving them free to answer the test question by tapping their representation of Sally's false belief. However, young children typically perform at chance in undisclosed-location tasks (Wellman et al., 2001). According to the processing-load account, the joint demands of the false-belief-representation and response-selection processes overwhelm young children's limited *working-memory* resources, leading to chance performance. This account predicts that young children should succeed at undisclosed-location tasks when response-selection demands are reduced through practice trials, and recent results confirm this prediction (Setoh, Scott, & Baillargeon, 2011).

THE RATIONALITY PRINCIPLE Adults' expectations about agents' actions are guided by a *rationality* principle: all other things being equal, adults expect agents to act rationally—indeed, this is what makes it possible to predict their actions (e.g., Dennett, 1987; Fodor, 1987). Corollaries of the rationality principle include expectations of *consistency*—agents should act in a manner consistent with their mental states—and *efficiency*—agents should expend as little effort as possible to achieve their goals (e.g., Baillargeon et al., 2014; Gergely, Nádasdy, Csibra, & Bíró, 1995). Like adults, infants expect agents to adhere to the rationality principle. When infants observe an agent act in a scene, the psychological-reasoning system attempts to build a psychological explanation for the agent's actions by positing a causally coherent set of mental states that portrays these actions as rational. If infants succeed in building such an explanation, they can then use it (along with the rationality principle) to predict how the agent will behave when the scene changes.

Consistency In a well-known consistency task (e.g., Woodward, 1998), infants first receive familiarization trials in which an agent faces two objects, object-A and object-B, and repeatedly reaches for object-A. In the test trials, the objects' locations are switched, and the agent reaches for either object-A (*old-object* event) or object-B (*new-object* event). In another version of the task (e.g., Robson & Kuhlmeier, 2013), object-B is replaced with new object-C in the test trials, and the agent reaches for either object-A (*old-object* event) or object-C (*new-object* event). In either case, infants typically look longer at the new- than at the old-object event. This result suggests that during the familiarization trials, infants notice that the agent continually chooses object-A over object-B, and, based on this systematic choice information, they attribute to the agent a particular disposition, a liking or preference for object-A. During the test trials, infants expect the agent to continue acting on this preference, and they detect a consistency violation when the agent reaches for the other object instead.

Additional findings support this interpretation. First, infants do not attribute a preference to the agent during the familiarization trials if object-A is the only object present (e.g., Luo & Baillargeon, 2005), or if the agent does not know that object-B is present (e.g., Luo & Johnson, 2009); in either case, there is no longer choice information signaling the agent's disposition toward object-A. Second, infants attribute a preference to the agent during the familiarization trials even if object-A is the only object present, as long as the agent expends effort to obtain object-A (e.g., in each familiarization trial, the agent must open a container in order to retrieve object-A; Bíró, Verschoor, & Coenen, 2011). Third, infants view preferences as attributes of individual agents: if Mary prefers object-A over object-B, infants do not expect Jane to share the same preference (e.g., Buresh & Woodward, 2007; see Egyed, Király, & Gergely, 2013, for an interesting exception involving pedagogical cues). Finally, infants attribute preferences whenever agents' choices systematically deviate from random sampling (e.g., when an agent chooses only ducks from a box containing mainly frogs; Kushnir, Xu, & Wellman, 2010).

Efficiency In a well-known efficiency task (e.g., Csibra, 2008), infants first receive familiarization trials in which an agent detours around an obstacle in order to reach a target. In the test trials, the obstacle is removed, and the agent either moves to the target in a straight line (*short-path* event) or detours as before when approaching the target (*long-path* event). Infants typically look longer at the long- than at the short-path event. This result suggests that during the familiarization trials, infants attribute to the agent the goal of reaching the target. During the test trials, infants expect the agent to maintain this goal and to pursue it efficiently: with the obstacle removed, a more efficient path to the target becomes possible, and infants detect an efficiency violation when the agent ignores this path and follows the familiar, less efficient path instead.

Additional findings support this interpretation. First, similar results have been obtained with various detour tasks, including an agent reaching over an obstacle to grasp an object (e.g., Phillips & Wellman, 2005). Second, when considering physical effort, infants evaluate not only the shortest path possible for reaching a target, but also the shortest action sequence possible for obtaining an object: infants expect an agent who has access to two identical objects to choose the one that can be retrieved with fewer actions (Scott & Baillargeon, 2013). Third, infants consider mental as well as physical effort when reasoning about efficiency: if an agent is presented with two identical objects, one under a transparent cover and one under an opaque cover, infants expect the agent to choose the visible object, which can be retrieved with less mental effort (Scott & Baillargeon, 2013).

Irrational agents The preceding results indicate that when infants are able to build a well-formed psychological explanation for an agent's actions, they use this explanation to predict how the agent will act when the scene changes. As might be expected, when infants are *unable* to build a well-formed explanation for an agent's actions, they hold no expectations about the agent's subsequent behavior. Such negative results have been obtained (1) when it is unclear why the agent is pursuing a particular goal (e.g., the agent is attempting to steal an undesirable object; Scott, Richman, & Baillargeon, 2014) or why the agent is *refraining* from pursuing a particular goal (e.g., the agent is staring at an accessible object but does not reach for it; Luo, 2010); (2) when the agent's actions are inconsistent with her knowledge about the scene (e.g., the agent expresses excitement over an empty container; Chow & Poulin-Dubois, 2009); and (3) when the agent's actions are inefficient and involve either an unnecessary action (e.g., the agent opens a container before grasping an object that stands next to the container; Bíró et al., 2011) or an unnecessary detour (e.g., the agent jumps, for no apparent reason, while approaching a target; Gergely et al., 1995).

When presented with an inefficient action on a novel object, infants occasionally give the agent the benefit of the doubt: they assume that a rational agent would not perform this action unless there was a reason for doing so. Thus, after watching a model activate a light-box by touching it with her forehead, while her hands lay idle on either side of the light-box, infants tend to imitate the model's inefficient head action (Meltzoff, 1988). Infants use their hands to activate the light-box, however, if the model's hands are occupied while she demonstrates the (now merely expedient) head action (Gergely, Bekkering, & Király, 2002).

Sociomoral reasoning

As it became clear that infants can make sense, at least in simple situations, of the actions of a single agent, researchers were naturally led to ask whether infants also possess expectations about social interactions among two or more agents, bringing about a new focus on sociomoral reasoning.

Values Because many sociomoral expectations presuppose an ability to assess the *values* of social actions, initial investigations explored this ability in infants. In general, the value of a social action is determined by its *valence* (positive, negative, or neutral) and *magnitude* (more or less) (e.g., Jackendoff, 2007; Premack, 1990). As might be expected, positive actions are those that have a beneficial effect on others, whereas negative actions are those that have a detrimental effect.

Research with infants age 10 months and older indicates that (1) they can assess the valence of both positive (e.g., helping, sharing) and negative (e.g., hindering, hitting) actions (e.g., Behne, Carpenter, Call, & Tomasello, 2005; Premack & Premack, 1997); (2) they show an affiliative preference for individuals who produce positive actions over individuals who produce negative actions (e.g., Hamlin, Wynn, & Bloom, 2007); (3) through evaluative contagion, they show affiliative preferences for individuals who behave positively toward individuals who have produced positive actions and for individuals who behave negatively toward individuals who have produced negative actions (e.g., Hamlin, Wynn, Bloom, & Mahajan, 2011); and (4) they expect others to have similar affiliative preferences (e.g., Fawcett & Liszkowski, 2012; Hamlin et al., 2007).

The Reciprocity Principle Adults expect individuals to act in accordance with a *reciprocity* principle: if A acts in some way toward B, who chooses to respond, then B's reciprocal action should match A's initial action in *value*, though it need not match in *form* (e.g., Jackendoff, 2007; Premack, 1990). Recent evidence indicates that by the second year of life, infants already possess an expectation of reciprocity (e.g., Dunfield & Kuhlmeier, 2010; He, Jin, Baillargeon, & Premack, 2014).

In one violation-of-expectation experiment, for example, 15-month-olds watched live events involving two unfamiliar women who sat at windows in the right wall (E1) and back wall (E2) of a puppet-stage apparatus (He et al., 2014). During the familiarization trials, E1 either gave a cookie to E2 (*give-cookie* condition) or stole E2's cookie (*steal-cookie* condition). During the test trial, while E2 watched, E1 stored stickers one by one in a colorful box; as she was about to store her last sticker, a bell rang, and E1 exited, leaving her last sticker on the apparatus floor. Next, either E2 stored the sticker in the box, thus helping E1 by completing her actions (*store-sticker* event), or E2 tore the sticker into four pieces and dropped them on the apparatus floor (*tear-sticker* event). In either case, after finishing her actions, E2 looked down and paused until infants looked away and the trial ended.

In the give-cookie condition, infants looked reliably longer at the final paused scene if shown the tear-sticker as opposed to the store-sticker event; in the steal-cookie condition, the opposite looking pattern was found. These results suggested two conclusions. First, infants expected E2 to follow the reciprocity principle: when E1 had acted positively toward E2, infants detected a reciprocity violation if E2 acted negatively toward E1; conversely, when E1 had acted negatively toward E2, infants detected a reciprocity violation if E2 acted positively toward E1. Second, infants could detect these violations even though E2's reciprocal actions toward E1 differed in form from E1's initial actions toward E2, pointing to an abstract expectation of reciprocity. These conclusions were supported by a second experiment identical to the first except that in the test trial E2 entered the apparatus only *after* E1 had exited. E2 found the sticker on the apparatus floor and, as before, either stored it or tore it up. Infants in both conditions looked about equally at the store-sticker and tear-sticker events. Because E2 did not know to whom the sticker belonged, her actions were not *wittingly* directed at E1. Therefore, the reciprocity principle did not apply, and infants held no expectations about E2's actions.

THE FAIRNESS PRINCIPLE According to the *fairness* principle, all other things being equal, individuals should treat others fairly when allocating windfall resources, dispensing rewards for effort or merit, and so on (e.g., Haidt & Joseph, 2007; Premack, 2007). Traditionally, investigations of fairness in 3- to 5-year-olds have used first-party tasks, where the children tested are potential recipients, and third-party tasks, where they are not. Perhaps not surprisingly given young children's pervasive difficulty in curbing their self-interest, a concern for fairness has typically been observed only in third-party tasks (e.g., Baumard, Mascaro, & Chevallier, 2012; Olson & Spelke, 2008). Extending these results, recent investigations using third-party tasks have revealed that infants in the second year of life already possess an expectation of fairness (e.g., Geraci & Surian, 2011; Schmidt & Sommerville, 2011; Sloane, Baillargeon, & Premack, 2012).

In one violation-of-expectation experiment, for example, 19-month-olds watched live events in which an experimenter divided two objects between two identical animated puppet giraffes (Sloane et al., 2012). At the start of each trial, the two giraffes protruded from openings in the back wall of the apparatus; in front of each giraffe was a small placemat. The giraffes "danced" until the experimenter entered the apparatus carrying two identical objects (e.g., edible cookies), and announced, "I have cookies!"; the giraffes then responded excitedly, "Yay, yay!" (in two distinct voices). Next, the experimenter placed either one object in front of each giraffe (*equal* event) or both objects in front of the same giraffe (*unequal* event). Finally, the experimenter left, and the two giraffes looked down at their placemats and paused until the trial ended.

Infants looked reliably longer at the final paused scene in the unequal than in the equal event, suggesting that, by 19 months, infants expect a distributor to divide resources fairly between two similar recipients. This conclusion was supported by two control conditions. In one, the giraffes were inanimate (they never moved or talked), and infants looked about equally at the two test events. In the other, instead of bringing in and distributing the two objects in each trial, the experimenter removed covers resting over the giraffes' placemats to reveal the objects; infants again looked equally at the two test events, suggesting that they did not merely expect similar individuals to have similar numbers of items.

THE INGROUP PRINCIPLE According to the *ingroup* principle, members of a social group should act in ways that sustain the group (e.g., Baillargeon et al., 2014; Brewer, 1999). The ingroup principle has two corollaries, *loyalty* and *support*, each of which carries a rich set of expectations. Ingroup loyalty dictates that in situations involving ingroup and outgroup individuals, one should (1) *prefer* and *align with* ingroup as opposed to outgroup individuals, (2) *protect* ingroup individuals who are threatened by outgroup aggressors, and (3) display *favoritism* toward ingroup over outgroup individuals (e.g., when allocating scarce resources). Ingroup support dictates that when interacting with ingroup individuals, one should (1) engage in prosocial actions such as *helping* ingroup members in need of assistance and *comforting* ingroup members in distress, and (2) limit negative interactions within the ingroup by *refraining* from unprovoked negative actions, *curbing* retaliatory actions, and engaging in *social acting*, the well-intentioned deception adults routinely practice (e.g., white lies) to support ingroup members (Baillargeon et al., 2013; Yang & Baillargeon, 2013). Although all of these expectations are being explored with infants (see Baillargeon et al., 2014), due to space constraints we focus below on the first expectation in each set.

Ingroup loyalty Infants align their toy and food choices with those endorsed by speakers of their native language (e.g., Kinzler, Dupoux, & Spelke, 2012; Shutts, Kinzler, McKee, & Spelke, 2009). In one preferential-reaching experiment, for example, 10-month-olds from English-speaking families sat at a table in front of a computer monitor and received four test trials (Kinzler et al., 2012). Each trial had a speech phase in which infants heard, in alternation, a woman who spoke English and a woman who spoke French, followed by a toy-modeling phase, in which the two women stood side by side, each silent, smiling, and holding a different toy animal; real-life replicas of the toys rested on the table below the monitor. Following the toy-modeling phase, infants were wheeled closer to the table to select one of the toys. Across trials, infants reliably chose the toy held by the English speaker. These results suggest that when infants face two unfamiliar women, one from their speech community and one from a different speech community, they extend ingroup status to the woman from their speech community and align their choices with hers, in accordance with the ingroup principle.

Ingroup support Infants *help* an experimenter in need of assistance (e.g., Warneken & Tomasello, 2006, 2007), as long as they extend ingroup status to the experimenter through either an appropriate familiarization phase (e.g., Barragan & Dweck, 2012) or affiliative primes (e.g., Over & Carpenter, 2009). For example, Warneken and Tomasello (2007) presented 14-month-olds with three out-of-reach scenarios in which a familiarized experimenter required help (e.g., he accidentally dropped a marker on the floor and unsuccessfully reached for it). Most infants helped the experimenter in at least one scenario. Infants are less likely to *comfort* a familiarized experimenter in distress, perhaps because appropriate interventions are harder to identify (e.g., Svetlova, Nichols, & Brownell, 2010). In third-party tasks, however, infants do expect an adult to comfort a crying baby (e.g., Jin et al., 2012; Johnson et al., 2010). In one violation-of-expectation experiment, for example, 12-month-olds watched videotaped *responsive* and *unresponsive* test events (Jin et al., 2012). In the responsive event, a woman folded towels on the left side of a room; at the back of the room were a chair with additional towels and a large stroller (one could not see whether there was a baby inside the stroller). Next, a baby began to cry; the woman walked to the stroller and bent over it, as though attempting to comfort the crying baby. The unresponsive event was similar except that it involved a different woman, who walked to the chair to pick up more towels, ignoring the crying baby. Infants looked reliably longer at the unresponsive than at the responsive event; this effect was eliminated when the baby laughed in the recorded soundtrack.

Conclusion

The evidence reviewed in this chapter suggests that infant social cognition involves at least two evolved systems that work together seamlessly, beginning in the first year of life. The psychological-reasoning system enables infants to interpret the intentional actions of agents and is constrained by a principle of rationality (with its corollaries of consistency and efficiency). The sociomoral-reasoning system guides infants' expectations about social interactions and is constrained by several principles, including reciprocity, fairness, and ingroup (with its corollaries of loyalty and support). Although much research is needed to understand each system and its neurological basis, the present review makes clear that key components of adult social cognition are already in place in infancy.

ACKNOWLEDGMENTS Preparation of this chapter was supported by an NICHD grant to Renée Baillargeon (HD-021104). We thank Dan Hyde for helpful comments.

REFERENCES

Baillargeon, R., He, Z., Setoh, P., Scott, R. M., Sloane, S., & Yang, D. Y.-J. (2013). False-belief understanding and why it matters: The social-acting hypothesis. In M. R. Banaji & S. A. Gelman (Eds.), *Navigating the social world: What infants, children, and other species can teach us* (pp. 88–95). New York, NY: Oxford University Press.

Baillargeon, R., Scott, R. M., & He, Z. (2010). False-belief understanding in infants. *Trends Cogn Sci, 14*, 110–118.

Baillargeon, R., Scott, R. M., He, Z., Sloane, S., Setoh, P., Jin, K., Wu, D., & Bian, L. (2014). Psychological and sociomoral reasoning in infancy. In M. Mikulincer & P. R. Shaver (Eds.) and E. Borgida & J. Bargh (Assoc. Eds.), *APA handbook of personality and social psychology: Vol. 1. Attitudes and social cognition* (pp. 79–150). Washington, DC: APA.

Baron-Cohen, S., Leslie, A. M., & Frith, U. (1985). Does the autistic child have a "theory of mind"? *Cognition, 21*, 37–46.

Barragan, R. C., & Dweck, C. S. (2012). *Is a norm of reciprocity necessary for young children to help?* Paper presented at the Biennial International Conference on Infant Studies, Minneapolis, MN.

Baumard, N., Mascaro, O., & Chevallier, C. (2012). Preschoolers are able to take merit into account when distributing goods. *Dev Psychol, 48*, 492–498.

Behne, T., Carpenter, M., Call, J., & Tomasello, M. (2005). Unwilling versus unable: Infants' understanding of intentional action. *Dev Psychol, 41*, 328–337.

Birch, S. A. J., & Bloom, P. (2003). Children are cursed: An asymmetric bias in mental-state attribution. *Psychol Sci, 14*, 283–286.

Bíró, S., Verschoor, S., & Coenen, L. (2011). Evidence for a unitary goal concept in 12-month-old infants. *Dev Sci, 14*, 1255–1260.

Brewer, M. B. (1999). The psychology of prejudice: Ingroup love and outgroup hate? *J Soc Issues, 55*, 429–444.

Buresh, J. S., & Woodward, A. L. (2007). Infants track action goals within and across agents. *Cognition, 104*, 287–314.

Buttelmann, D., Carpenter, M., & Tomasello, M. (2009). Eighteen-month-old infants show false-belief understanding in an active helping paradigm. *Cognition, 112*, 337–342.

Butterfill, S. A., & Apperly, I. A. (2013). How to construct a minimal theory of mind. *Mind Lang, 28*, 606–637.

Chow, V., & Poulin-Dubois, D. (2009). The effect of a looker's past reliability on infants' reasoning about beliefs. *Dev Psychol, 45,* 1576–1582.

Csibra, G. (2008). Goal attribution to inanimate agents by 6.5-month-old infants. *Cognition, 107,* 705–717.

Dennett, D. C. (1987). *The intentional stance.* Cambridge, MA: MIT Press.

Dunfield, K., & Kuhlmeier, V. A. (2010). Intention-mediated selective helping in infancy. *Psychol Sci, 21,* 523–527.

Dwyer, S. (2009). Moral dumbfounding and the linguistic analogy: Methodological implications for the study of moral judgment. *Mind Lang, 24,* 274–296.

Egyed, K., Király, I., & Gergely, G. (2013). Communicating shared knowledge in infancy. *Psychol Sci, 24,* 1348–1353.

Fawcett, C., & Liszkowski, U. (2012). Infants anticipate others' social preferences. *Infant Child Dev, 21,* 239–249.

Fodor, J. A. (1987). *Psychosemantics: The problem of meaning in the philosophy of mind.* Cambridge, MA: MIT Press.

Geraci, A., & Surian, L. (2011). The developmental roots of fairness: Infants' reactions to equal and unequal distributions of resources. *Dev Sci, 14,* 1012–1020.

Gergely, G., Bekkering, H., & Király, I. (2002). Rational imitation in preverbal infants. *Nature, 415,* 755.

Gergely, G., Nádasdy, Z., Csibra, G., & Bíró, S. (1995). Taking the intentional stance at 12 months of age. *Cognition, 56,* 165–193.

Haidt, J., & Joseph, C. (2007). The moral mind: How five sets of innate intuitions guide the development of many culture-specific virtues, and perhaps even modules. In P. Carruthers, S. Laurence, & S. Stich (Eds.), *The innate mind: Vol. 3. Foundations and the future* (pp. 367–391). Oxford, UK: Oxford University Press.

Hamlin, J. K., & Wynn, K. (2011). Young infants prefer prosocial to antisocial others. *Cognitive Dev, 26,* 30–39.

Hamlin, J. K., Wynn, K., & Bloom, P. (2007). Social evaluation by preverbal infants. *Nature, 450,* 557–559.

Hamlin, J. K., Wynn K., Bloom, P., & Mahajan, N. (2011). How infants and toddlers react to antisocial others. *Proc Natl Acad Sci USA, 108,* 19931–19936.

He, Z., Jin, K., Baillargeon, R., & Premack, D. (2014). Do infants understand tit-for-tat? Evidence for early expectations about reciprocation and retaliation. Manuscript under review.

Jackendoff, R. (2007). *Language, consciousness, culture: Essays on mental structure.* Cambridge, MA: MIT Press.

Jin, K., Houston, J., Baillargeon, R., Roisman, G. I., Sloane, S., & Groh, A. M. (2012). *8-month-olds expect an adult to respond to a crying but not to a laughing infant.* Paper presented at the Biennial International Conference on Infant Studies, Minneapolis, MN.

Johnson, S. C., Dweck, C. S., Chen, F. S., Stern, H. L., Ok, S.-J., & Barth, M. E. (2010). At the intersection of social and cognitive development: Internal working models of attachment in infancy. *Cognitive Sci, 34,* 807–825.

Johnson, S. C., Shimizu, Y. A., & Ok, S.-J. (2007). Actors and actions: The role of agent behavior in infants' attribution of goals. *Cognitive Dev, 22,* 310–322.

Kinzler, K. D., Dupoux, E., & Spelke, E. S. (2012). "Native" objects and collaborators: Infants' object choices and acts of giving reflect favor for native over foreign speakers. *J Cogn Dev, 13,* 67–81.

Knudsen, B., & Liszkowski, U. (2012). Eighteen- and 24-month-old infants correct others in anticipation of action mistakes. *Dev Sci, 15,* 113–122.

Kovács, Á. M., Téglás, E., & Endress, A. D. (2010). The social sense: Susceptibility to others' beliefs in human infants and adults. *Science, 330,* 1830–1834.

Kushnir, T., Xu, F., & Wellman, H. M. (2010). Young children use statistical sampling to infer the preferences of other people. *Psychol Sci, 21,* 1134–1140.

Leslie, A. M. (1994). Pretending and believing: Issues in the theory of ToMM. *Cognition, 50,* 211–238.

Leslie, A. M., & Polizzi, P. (1998). Inhibitory processing in the false belief task: Two conjectures. *Dev Sci, 1,* 247–253.

Liszkowski, U., Carpenter, M., & Tomasello, M. (2008). Twelve-month-olds communicate helpfully and appropriately for knowledgeable and ignorant partners. *Cognition, 108,* 732–739.

Luo, Y. (2010). Do 8-month-old infants consider situational constraints when interpreting others' gaze as goal-directed action? *Infancy, 15,* 392–419.

Luo, Y. (2011). Three-month-old infants attribute goals to a non-human agent. *Dev Sci, 14,* 453–460.

Luo, Y., & Baillargeon, R. (2005). Can a self-propelled box have a goal? Psychological reasoning in 5-month-old infants. *Psychol Sci, 16,* 601–608.

Luo, Y., & Johnson, S. C. (2009). Recognizing the role of perception in action at 6 months. *Dev Sci, 12,* 142–149.

Luo, Y., Kaufman, L., & Baillargeon, R. (2009). Young infants' reasoning about events involving inert and self-propelled objects. *Cognitive Psychol, 58,* 441–486.

Meltzoff, A. N. (1988). Infant imitation after a 1-week delay: Long-term memory for novel acts and multiple stimuli. *Dev Psychol, 24,* 470–476.

Mueller, V. A., Brass, M., Waszak, F., & Prinz, W. (2007). The role of the preSMA and the rostral cingulated zone in internally selected actions. *NeuroImage, 37,* 1354–1361.

Olson, K. R., & Spelke, E. S. (2008). Foundations of cooperation in young children. *Cognition, 108,* 222–231.

Onishi, K. H., & Baillargeon, R. (2005). Do 15-month-old infants understand false beliefs? *Science, 308,* 255–258.

Onishi, K. H., Baillargeon, R., & Leslie, A. M. (2007). 15-month-old infants detect violations in pretend scenarios. *Acta Psychol, 124,* 106–128.

Over, H., & Carpenter, M. (2009). Eighteen-month-old infants show increased helping following priming with affiliation. *Psychol Sci, 20,* 1189–1193.

Perner, J., & Roessler, J. (2012). From infants' to children's appreciation of belief. *Trends Cogn Sci, 16,* 519–525.

Phillips, A. T., & Wellman, H. M. (2005). Infants' understanding of object-directed action. *Cognition, 98,* 137–155.

Premack, D. (1990). The infant's theory of self-propelled objects. *Cognition, 36,* 1–16.

Premack, D. (2007). Foundations of morality in the infant. In O. Vilarroya & F. I. Argimon (Eds.), *Social brain matters: Stances on the neurobiology of social cognition* (pp. 161–167). Amsterdam: Rodopi.

Premack, D., & Premack, A. J. (1997). Infants attribute value +/− to the goal-directed actions of self-propelled objects. *J Cogn Neurosci, 9,* 848–856.

Robson, S. J., & Kuhlmeier, V. A. (2013). *Selectivity promotes 9-month-old infants to encode the goals of others.* Paper presented at the Biennial Meeting of the Society for Research in Child Development, Seattle, WA.

Schmidt, M. F. H., & Sommerville, J. A. (2011). Fairness expectations and altruistic sharing in 15-month-old human infants. *PLoS ONE, 6*, e23223.

Scott, R. M., & Baillargeon, R. (2013). Do infants really expect others to act efficiently? A critical test of the rationality principle. *Psychol Sci, 24*, 466–474.

Scott, R. M., He, Z., Baillargeon, R., & Cummins, D. (2012). False-belief understanding in 2.5-year-olds: Evidence from two novel verbal spontaneous-response tasks. *Dev Sci, 15*, 181–193.

Scott, R. M., Richman, J., & Baillargeon, R. (2014). Infants understand the art of deception: New evidence for mentalistic reasoning in infancy. Manuscript under review.

Senju, A., Southgate, V., Snape, C., Leonard, M., & Csibra, G. (2011). Do 18-month-olds really attribute mental states to others? A critical test. *Psychol Sci, 22*, 878–880.

Setoh, P., Scott, R. M., & Baillargeon, R. (2011). *False-belief reasoning in 2.5-year-olds: Evidence from an elicited-response low-inhibition task.* Paper presented at the Biennial Meeting of the Society for Research in Child Development, Montreal, Canada.

Setoh, P., Wu, D., Baillargeon, R., & Gelman, R. (2013). Young infants have biological expectations about animals. *Proc Natl Acad Sci USA, 110*(40), 15937–15942.

Shutts, K., Kinzler, K. D., McKee, C. B., & Spelke, E. S. (2009). Social information guides infants' selection of foods. *J Cogn Dev, 10*, 1–17.

Sloane, S., Baillargeon, R., & Premack, D. (2012). Do infants have a sense of fairness? *Psychol Sci, 23*, 196–204.

Svetlova, M., Nichols, S. R., & Brownell, C. (2010). Toddlers' prosocial behavior: From instrumental to empathic to altruistic helping. *Child Dev, 81*, 1814–1827.

Warneken, F., & Tomasello, M. (2006). Altruistic helping in human infants and young chimpanzees. *Science, 311*, 1301–1303.

Warneken, F., & Tomasello, M. (2007). Helping and cooperation at 14 months of age. *Infancy, 11*, 271–294.

Wellman, H. M., Cross, D., & Watson, J. (2001). Meta-analysis of theory of mind development: The truth about false belief. *Child Dev, 72*, 655–684.

Wimmer, H., & Perner, J. (1983). Beliefs about beliefs: Representation and constraining function of wrong beliefs in young children's understanding of deception. *Cognition, 13*, 103–128.

Woodward, A. L. (1998). Infants selectively encode the goal object of an actor's reach. *Cognition, 69*, 1–34.

Yang, D. Y., & Baillargeon, R. (2013). Difficulty in understanding social acting (but not false beliefs) mediates the link between autistic traits and ingroup relationships. *J Autism Dev Disord, 43*, 2199–2206.

2 Structural Brain Development During Childhood and Adolescence

ANNE-LISE GODDINGS AND JAY N. GIEDD

ABSTRACT The past 15 years have seen a major expansion in research on the structural development of the maturing human brain in childhood and adolescence. In this chapter, we summarize the major findings from histological and magnetic resonance imaging (MRI) studies looking at the structural development of the child and adolescent brain. There is substantial evidence for the ongoing development of many regions of the brain into adulthood, both microscopically and macroscopically. Histological investigations have shown that myelination continues throughout childhood and adolescence, and appears to correspond to the increasing white matter volumes seen in MRI studies. Post-mortem studies investigating synaptic density have documented a dynamic process of synaptogenesis and synaptic pruning that progresses in a region-specific manner during this developmental period. MRI studies have found that most cortical regions show evidence of increasing gray matter volume during childhood, before peaking in late childhood/early adolescence and decreasing in volume during adolescence, although there is wide variation in the timing and extent of this change. In contrast, subcortical regions show varying trajectories, which differ between structures but also across studies. Possible future directions to tackle many of the unanswered questions that remain are discussed at the end of the chapter.

The past 15 years have seen a major expansion in research on the structural development of the maturing human brain in childhood and adolescence, based largely on the results of cross-sectional and longitudinal magnetic resonance imaging (MRI) studies (Brain Development Cooperative Group [BDCG], 2012; Brown et al., 2012; Giedd et al., 1996; Lenroot et al., 2007; Østby et al., 2009; Raznahan et al., 2011a; Sowell, Trauner, Gamst, & Jernigan, 2002; Sullivan et al., 2011). Prior to the development of MRI, our understanding of the maturation of the human brain was largely limited to data from post-mortem studies. While such histological studies provide invaluable information regarding the basic processes underlying brain development, they are unable to inform us fully about how individuals change over time, how structures vary in their development both within and between individuals, and how these structural changes relate to functional correlates and behavioral patterns.

With the increased availability of MRI, a number of large-scale studies of typical structural brain development have demonstrated not only that the human brain continues to undergo substantial structural remodeling throughout childhood and adolescence and into adulthood, but have also documented region-specific differences in the timing and extent of these changes. It is important to note that common MRI techniques are limited to investigations of macroscopic changes, since the images produced are typically made up of voxels of between $1\,mm^3$ and $3\,mm^3$. By comparing MRI investigations with post-mortem histological studies, we are better able to describe the major macroscopic developmental patterns of the human brain and link these to the likely microscopic changes that underlie them.

In this chapter, we summarize the major findings from histological and MRI studies looking at the structural development of the child and adolescent brain.

Histological changes in brain structure during childhood and adolescence

Since the late 1960s, studies have shown evidence of cellular age-related differences in particular regions of the human brain in childhood and adolescence (Huttenlocher, 1990, 1979; Huttenlocher & Dabholkar, 1997; Miller et al., 2012; Petanjek et al., 2011; Yakovlev & Lecours, 1967). These developments can largely be divided into changes in myelination of axons and axonal caliber, and changes in the organization and density of synapses.

CHANGES IN MYELINATION Myelination, the process of laying down myelin by oligodendrocytes, is important for the maturing connectivity within the developing brain, allowing the synchronization of information transfer within and between brain networks (Fields & Stevens-Graham, 2002). Myelination occurs as a result of reciprocal communication between neurons and oligodendrocytes (Simons & Trajkovic, 2006). Humans are born with relatively low levels of neocortical myelin compared with other nonhuman primates such as chimpanzees (Miller et al., 2012).

Human post-mortem studies have shown that myelination continues through the first and second decade of life (Benes, Turtle, Khan, & Farol, 1994; Yakovlev & Lecours, 1967). A recent study quantifying myelinated axon fiber length density in post-mortem samples (Miller et al., 2012) showed that myelination continues until at least 28 years of age, only reaching 60% of its adult levels by adolescence/early adulthood (age 11–23 years). Myelination does not occur uniformly across the brain according to these studies, but develops in different regions of the brain at different ages (Benes et al., 1994; Yakovlev & Lecours, 1967).

Synaptic Changes Synaptogenesis in humans begins in the third trimester of fetal life and continues for a variable time depending on the brain region being examined (Huttenlocher & Dabholkar, 1997). It has, for instance, been found that the auditory cortex shows a rapid burst of synaptogenesis postnatally, peaking at age 3 months. In contrast, the prefrontal cortex continues to demonstrate synaptogenesis until 3.5 years of age (Huttenlocher & Dabholkar, 1997), and may continue for a longer period (Huttenlocher, 1990). Following this period of synaptogenesis, the cortex undergoes a period of synaptic pruning (Huttenlocher, 1979), which is again heterochronous, occurring earlier in the primary sensory cortices (e.g., visual cortex) than in the prefrontal cortex, where it continues well into the second decade of life (Huttenlocher, 1990).

This body of work by Huttenlocher and colleagues has been significantly expanded by a more recent study that investigated synaptic development in the prefrontal cortex of post-mortem brains from 32 individuals aged between 1 week and 91 years (Petanjek et al., 2011). This study also showed increasing dendritic spine density in prefrontal cortex through infancy and childhood, with a decrease in adolescence and into adulthood. The peak dendritic spine density in late childhood was approximately twice that of adult levels, and remained significantly higher than adult levels throughout adolescence. These data demonstrate that there is protracted pruning of synaptic spines through adolescence and into early adulthood (Petanjek et al., 2011).

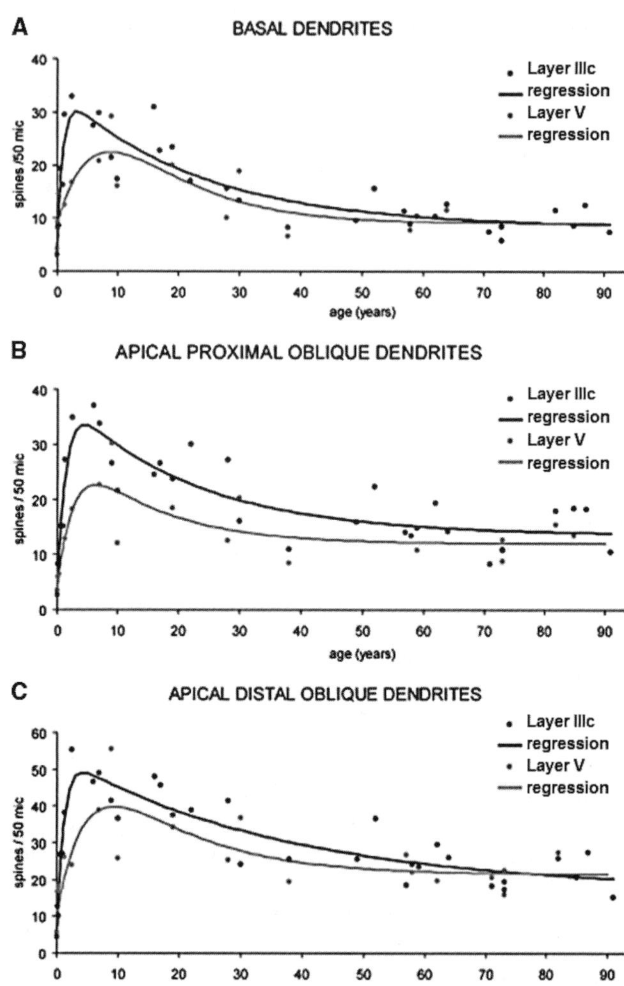

Figure 2.1 Regression curves showing the dendritic spine density from (A) basal dendrites, (B) apical proximal oblique dendrites, and (C) apical distal oblique dendrites of pyramidal cells from layers IIIc and V of the dorsolateral prefrontal cortex in 32 participants aged 1 month to 91 years. Reproduced from Petanjek et al., 2011.

Macroscopic structural changes using MRI

First used in pediatric neuroimaging studies of typical development in the late 1980s, MRI allows the collection of high-resolution anatomical brain images of large numbers of healthy children and adolescents without the use of ionizing radiation. A major benefit of MRI as an investigative technique over other forms of imaging or post-mortem studies is that it enables the collection of repeated scans in the same individuals over time. Given the large interindividual variability in structural volumes, longitudinal data have proved invaluable for documenting developmental trajectories (i.e., change in volume over time) within and between individuals.

The earliest longitudinal brain imaging project began in 1989 under the direction of Markus Krusei, M.D., at the Child Psychiatry Branch of the National Institute of Mental Health (CPB; Giedd et al., 1996). This longitudinal brain imaging project has acquired more than 6,500 structural MRI scans from thousands of individuals aged 3 years and older (see http://clinicalstudies.info.nih.gov/cgi/detail.cgi?A_1989-M-0006.html). Much of the data for the following sections regarding quantification of brain structure sizes are from subgroups of typically developing individuals within this cohort, including Giedd et al., 1996 (cross-sectional, 104 participants, 4–18 years); Giedd et al., 1999 (243 scans, 145 participants, 4–21 years); Gogtay et al., 2004 (52 scans, 13 participants, 4–21 years); Lenroot et al., 2007 (829 scans, 387 participants, 3–27 years); Raznahan et al., 2011a (376 scans, 108 participants, 9–22 years), and Raznahan et al., 2011b (1,274 scans, 647 participants, 3–30 years). We also include a number of relevant studies from other projects around the world.

Total Cerebral Volume Total brain volume has been shown to double in the first year of life and to increase by a further 15% in the second year (Knickmeyer et al., 2008). It approaches adult volumes by mid-childhood, reaching approximately 95% of peak volume by age 6 (Lenroot et al., 2007). After this time, total brain volume remains relatively stable through the second decade into adulthood, with some studies suggesting a small reduction in volume (BDCG, 2012; Koolschijn & Crone, 2013; Lenroot et al., 2007; Østby et al., 2009).

Consistent with the adult neuroimaging literature (e.g., Goldstein et al., 2001), mean total cerebral volume is approximately 10% larger in boys than girls throughout childhood and adolescence (Giedd, Raznahan, Mills, & Lenroot, 2012). This finding has often been attributed to differences in overall body size in adult populations, but the CPB study showed that the differences in brain size preceded the striking differences in body size which predominantly occurred after puberty in the sample (Lenroot et al., 2007). This study demonstrated the large interindividual variability that exists in brain volumes. Thus, healthy children at the same age may have nearly two-fold differences in total brain size, as well as large variation between volumes of various brain substructures. While the mean difference in total cerebral volume persisted between the sexes, there was significant variation within and between the sexes in terms of brain volume (Lenroot et al., 2007).

White Matter Development Cortical tissue can be broadly divided into white matter (WM) and gray matter (GM) based on its appearance on MRI. White matter is composed primarily of axons, many of which are myelinated, and associated vasculature and glia. Early cross-sectional MRI studies of WM development showed that volumes increased through childhood and adolescence (Jernigan & Tallal, 1990; Reiss, Abrams, Singer, Ross, & Denckla, 1996; Schaefer et al., 1990), complementing the anatomical studies describing ongoing myelination during childhood described above. More recent longitudinal MRI studies have tracked the trajectories of WM volume development, revealing age-dependent patterns that were broadly similar across lobes (frontal, temporal and parietal; Lenroot et al., 2007). A key WM structure in the brain is the corpus callosum, which is made up of approximately 200 million mostly myelinated axons that connect homologous areas of the left and right cerebral hemispheres. The CPB study (Lenroot et al., 2007) and other studies (Pujol, Vendrell, Junqué, Martí-Vilalta, & Capdevila, 1993; Thompson et al., 2000) have shown increasing volume of the corpus callosum in childhood and adolescence (4–20 years).

With advances in MRI technology, additional methods to characterize WM structure have been developed. One important technique is diffusion tensor imaging, which quantifies diffusion of water through different regions of the brain (Mori & Zhang, 2006). If unconstrained, water molecules will randomly diffuse in all directions. Nonrandom diffusion can be used to infer constraints placed upon the motion of water by physical features such as cell membranes or interactions with large molecules (Le Bihan et al., 2001). Fractional anisotropy is a measure used to indicate the degree of nonrandomness of the diffusion, providing both information on the microstructure of WM and the directionality of the axons contained within it. Mean diffusivity, the overall speed of diffusion, tends to be decreased by these same factors.

As would be predicted by increasing myelination, overall fractional anisotropy has been shown to increase and mean diffusivity to decrease during childhood and adolescence (Lebel & Beaulieu, 2011; Peters et al., 2012). High anisotropy indicates coherently bundled myelinated axons and axonal pruning, resulting in more efficient neuronal signaling (Suzuki, Matsuzawa, Kwee, & Nakada, 2003) and improved cognitive performance (e.g., reading ability: Beaulieu et al., 2005; IQ: Schmithorst, Wilke, Dardzinski, & Holland, 2005). This development varied between regions and specific tracts, with tracts connecting frontal and temporal regions (e.g., cingulum, uncinate fasciculus, and superior longitudinal fasciculus) showing more prolonged periods of maturation than other regions (e.g., corpus callosum and fornix; Lebel, Gee, Camicioli, Wieler, Martin, & Beaulieu, 2012), and one study showing fractional anisotropy continuing to increase beyond 23 years of age in frontal and temporal regions (Tamnes et al., 2010).

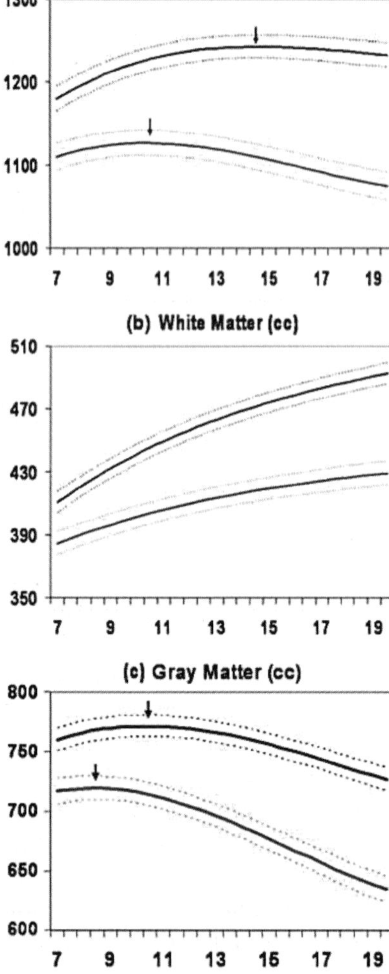

FIGURE 2.2 Mean volume (with 95% confidence intervals) by age in years for males (blue) and females (red) for (a) total brain volume, (b) white matter volume, and (c) gray matter volume. Reproduced from Lenroot et al., 2007. (See color plate 1.)

CORTICAL GRAY MATTER DEVELOPMENT Examining the developmental trajectory of total GM volume in the CPB study revealed inverted U-shaped trajectories with an increasing volume across childhood followed by a decrease in adolescence (Giedd et al., 1999; Lenroot et al., 2007; see figure 2.2). The rate of decreasing volume in adolescence reduces with age, suggesting relative stability of overall GM volume by early adulthood (Hedman, van Haren, Schnack, Kahn, & Hulshoff Pol, 2012; Lenroot et al., 2007; Østby et al., 2009). The timing at which GM volume peaks differs between lobes, with parietal lobes peaking before the frontal lobes and the temporal lobes peaking last (Giedd et al., 1999; Muftuler et al., 2011). Across the cortical surface, there is extensive heterogeneity in this developmental timing, with the earliest GM volume peak and subsequent decline occurring in the primary sensorimotor areas, and the latest in higher-order association areas that integrate those primary functions, such as the dorsolateral prefrontal cortex, inferior parietal cortex, and superior temporal gyrus (Gogtay et al., 2004; Muftuler et al., 2011). The underlying mechanisms associated with a reduction in GM volume are still debated (Carlo & Stevens, 2013; Paus, 2010), and to date there are no studies that have tested the relationship between developmental changes in underlying cellular or synaptic anatomy and structural MRI measures. Despite these limitations, it is thought that reductions in GM volume may reflect synaptic reorganization and/or increases in WM (as axons become myelinated), resulting in GM encroachment and apparent GM volume reduction.

Improving magnetic resonance image resolution and the development of new surface-based reconstruction tools (e.g., CIVET, Freesurfer) has enabled the study of distinct aspects of GM structure, such as cortical thickness and surface area, in addition to GM volume. Cortical thickness refers to the distance between the two surfaces of the GM cortex—that is, the GM-WM interface and the outer GM-pia interface. Surface area refers to the total or regional area of the cortical surface. GM volume is a product of cortical thickness and surface area. Cortical thickness and surface area are driven by distinct genetic (Panizzon et al., 2009; Winkler et al., 2010), evolutionary (Rakic, 1995), and cellular (Chenn & Walsh, 2002) processes. Differences in surface area are pronounced across species (Hill et al., 2010; Rakic, 1995), whereas cortical thickness is highly conserved in comparison. Many areas of the brain that have expanded in surface area across evolution also show relatively greater surface-area expansion between infancy and adulthood (Hill et al., 2010).

The CPB dataset has been used to characterize changes in surface area and cortical thickness (Raznahan et al., 2011b). In this study, both cortical thickness and surface area showed decreases throughout adolescence following peak measurements in childhood. In males, the changes in surface area explained two-thirds of the changes seen in cortical volume, while in females cortical thickness and surface area appeared to contribute equally (Raznahan et al., 2011b). Reductions in cortical thickness during adolescence have also been demonstrated in other studies (Brown et al., 2012; Tamnes et al., 2010; van Soelen et al., 2012). Further work is needed to elucidate the developmental trajectories of cortical thickness during childhood, since the findings across these studies are not consistent in this younger age range.

Subcortical GM Development The relatively small size and ambiguous MRI signal of the borders between subcortical structures mean that reliable automated techniques for extracting volumes on a large scale have only recently been developed, and to date there are few published large-scale studies documenting the developmental trajectories of subcortical structures.

Amygdala and hippocampus Amygdala and hippocampus volumes increase during early to mid-childhood (1 month to 8 years; Uematsu et al., 2012). The developmental trajectory in late childhood and adolescence is less clear. An early cross-sectional analysis of amygdala and hippocampus volumes using data from the CPB cohort, validated by expert rater manual tracing, showed increases in amygdala volumes in males only and increases in hippocampal volumes in females only (Giedd et al., 1996; *n* = 104, 4–18 years). A later, larger cross-sectional study of amygdala and hippocampal development between 8 and 30 years of age (*n* = 171) showed a nonlinear increase in volume with age of both the hippocampus and amygdala in both sexes (Østby et al., 2009), and a further cross-sectional study of 885 participants between 3 and 20 years of age showed nonlinear increases in hippocampus volume (Brown et al., 2012). However, a longitudinal study that incorporated the developmental trajectories of the amygdala and hippocampus described small decreases in both amygdala and hippocampus volume (*n* = 85, 8–22 years; Tamnes et al., 2013a). A decrease in amygdala volume (but not hippocampal volume) was also reported by Uematsu and colleagues (*n* = 109, 1 month–25 years; Uematsu et al., 2012). The discrepancy between studies may reflect the large interindividual variation in amygdala and hippocampus volume, where there is a twofold variation in absolute volume (Giedd et al., 1996; Østby et al., 2009; Tamnes et al., 2013a), or different study designs and methods of measurement and analysis.

Thalamus Analyses of the changes in thalamus volume across adolescence show a similar discrepancy between studies to those seen when examining the amygdala and hippocampus. One longitudinal study showed decreasing volume with age (8–22 years; Tamnes et al., 2013a) and an early cross-sectional study also reported thalamic decreases with age (*n* = 35, 7–16 years; Sowell et al., 2002). In contrast, other cross-sectional studies have reported slight increases in volume over adolescence (BDCG, 2012; Brown et al., 2012; Koolschijn & Crone, 2013; Østby et al., 2009).

Basal ganglia The basal ganglia are a collection of subcortical nuclei (caudate, putamen, globus pallidus, subthalamic nucleus, and substantia nigra). The caudate decreases in volume across adolescence (Lenroot et al., 2007; Tamnes et al., 2013a). The CPB longitudinal study found that the caudate volume followed an inverted U-shaped trajectory, increasing in childhood and peaking before decreasing in adolescence (Lenroot et al., 2007), while the longitudinal study by Tamnes et al. showed a steady decrease in volume throughout the study (Tamnes et al., 2013a). This discrepancy may reflect the differing starting ages of the studies and emphasizes the relative paucity of information regarding the childhood trajectories of many subcortical regions. The large cross-sectional samples available support the longitudinal data showing a decrease in caudate volume during adolescence (BDCG, 2012; Østby et al., 2009).

The data for the putamen, globus pallidus, and nucleus accumbens are broadly in agreement, showing decreases in volume across adolescence, although the extent of these changes varies between studies and some do not show significant changes in all regions (BDCG, 2012; Koolschijn & Crone, 2013; Østby et al., 2009; Sowell et al., 2002; Tamnes et al., 2013a).

Cerebellum Developmental curves of total cerebellum size in the CPB study followed an inverted U-shaped developmental trajectory, similar to the cerebral cortex (Tiemeier et al., 2010). The cerebellum can be subdivided into GM and WM in the same way as the cerebral cortex, and similar developmental trajectories were seen in a cross-sectional study, with increasing WM and decreasing GM volumes over late childhood and adolescence (Koolschijn & Crone, 2013; Østby et al., 2009). However, different subregions, which arise from different embryologic precursors, followed different developmental trajectories. In contrast to the evolutionarily more recent cerebellar hemispheric lobes that followed the inverted U-shaped developmental trajectory, cerebellar vermis size did not change across the age span covered in this study (5–24 years; Tiemeier et al., 2010).

Future directions

As our understanding of the ongoing development in brain structure during adolescence increases, so does the need to relate this development to the characteristic behaviors associated with the teenage years. Adolescents typically show changes in social behaviors including heightened self-consciousness, increasingly complex peer networks, and the initiation of sexual feelings and relationships, as well as an escalation in risk-taking behaviors compared with children (Spear, 2009; Steinberg & Morris, 2001; Steinberg, 2004). Decisions made during this time can have long-lasting ramifications, such as those related to relationships, education, and identity. Given that structural changes seen in MRI studies are thought to reflect changes in synaptic number and organization, it is possible that they correlate with differences in behavior and cognitive abilities. Studies have started to investigate these links in typically developing and clinical populations of children and adolescents, looking at structural correlates of, for example, hyperactivity and impulsivity traits (Shaw et al., 2007, 2011); conduct problems (Huebner et al., 2008; Walhovd, Tamnes, Østby, Due-Tønnessen, & Fjell, 2012); intelligence (Shaw et al., 2011; Sowell et al., 2004; Tamnes et al., 2011); and working memory performance (Østby et al., 2011; Tamnes et al., 2013b). Further studies are needed to extend and broaden this area of research to allow a clearer and more accurate interpretation of the structural changes seen in childhood and adolescence.

The number of studies documenting structural brain changes in childhood and adolescence is small, particularly for subcortical regions, and further investigation of these functionally important regions is crucial to understand their maturation. Ideally, these studies should be large scale and longitudinal to tackle the analysis difficulties faced by wide interindividual variation in many of the measurements used. We know little about the underlying mechanisms and potential impact of these interindividual differences between sizes of different structures within the brain, and future studies focusing on this will prove invaluable for investigating relationships with behavioral and functional differences. The majority of large-scale investigations to date have focused on changing brain structures with age. More emphasis needs to be placed on other biological markers of development during childhood and adolescence such as puberty in order to help elucidate some of the underlying mechanisms for the observed differences (e.g., Goddings, Mills, Clasen, Giedd, Viner, & Blakemore, 2014). Lastly, there are few data available from MRI of childhood brain changes (<8 years), largely due to the technical difficulties in obtaining high-quality data in this young age group. In order to understand later childhood and adolescent brain changes in context, efforts should be made to improve our knowledge and understanding of these earlier changes to complete the picture of human brain development.

Conclusion

In this chapter, we have described some of the main findings that inform us about the changes in the brain's structure through childhood and adolescence. There is ongoing development of many regions of the brain into adulthood, both microscopically and macroscopically. Myelination is ongoing throughout this developmental period and appears to correspond to increasing WM volumes seen in MRI studies. There is a dynamic process of synaptogenesis and synaptic pruning that progresses in a region-specific manner during childhood and adolescence. While this may be reflected in the changing volumes of GM detected on MRI over this period, further work is needed to confirm this. In general, most cortical regions show evidence of increasing GM volume during childhood, before peaking in late childhood/early adolescence and decreasing in volume during adolescence, although there is wide variation in the timing and extent of this change. Subcortical regions show varying trajectories that differ between structures but also across studies.

REFERENCES

Beaulieu, C., Plewes, C., Paulson, L. A., Roy, D., Snook, L., Concha, L., & Phillips, L. (2005). Imaging brain connectivity in children with diverse reading ability. *NeuroImage, 25*, 1266–1271.

Benes, F. M., Turtle, M., Khan, Y., & Farol, P. (1994). Myelination of a key relay zone in the hippocampal formation occurs in the human brain during childhood, adolescence, and adulthood. *Arch Gen Psychiatry, 51*, 477–484.

Brain Development Cooperative Group (BDCG). (2012). Total and regional brain volumes in a population-based normative sample from 4 to 18 years: The NIH MRI study of normal brain development. *Cereb Cortex, 22*, 1–12.

Brown, T. T., Kuperman, J. M., Chung, Y., Erhart, M., McCabe, C., Hagler, D. J., Jr, Venkatraman, V. K., ... Dale, A. M. (2012). Neuroanatomical assessment of biological maturity. *Curr Biol, 22*, 1693–1698.

Carlo, C. N., & Stevens, C. F. (2013). Structural uniformity of neocortex, revisited. *Proc Natl Acad Sci USA, 110*, 1488–1493.

Chenn, A., & Walsh, C. A. (2002). Regulation of cerebral cortical size by control of cell cycle exit in neural precursors. *Science, 297*, 365–369.

Fields, R. D., & Stevens-Graham, B. (2002). New insights into neuron-glia communication. *Science, 298*, 556–562.

Giedd, J. N., Blumenthal, J., Jeffries, N. O., Castellanos, F. X., Liu, H., Zijdenbos, A., Paus, T., ... Rapoport, J. L. (1999). Brain development during childhood and adolescence: A longitudinal MRI study. *Nat Neurosci, 2*, 861–863.

Giedd, J. N., Raznahan, A., Mills, K. L., & Lenroot, R. K. (2012). Review: Magnetic resonance imaging of male/female differences in human adolescent brain anatomy. *Biol Sex Differ, 3*, 19.

Giedd, J. N., Snell, J. W., Lange, N., Rajapakse, J. C., Casey, B. J., Kozuch, P. L., Vaituzis, A. C., ... Rapoport, J. L. (1996). Quantitative magnetic resonance imaging of human brain development: Ages 4–18. *Cereb Cortex, 6*, 551–560.

Goddings, A-L., Mills, K. L., Clasen, L. S., Giedd, J. N., Viner, R. M., & Blakemore, S-J. (2014). The influence of puberty on subcortical structural development. *NeuroImage, 88,* 242–251.

Gogtay, N., Giedd, J. N., Lusk, L., Hayashi, K. M., Greenstein, D., Vaituzis, A. C., Nugent, T. F., 3rd, ... Thompson, P. M. (2004). Dynamic mapping of human cortical development during childhood through early adulthood. *Proc Natl Acad Sci USA, 101,* 8174–8179.

Goldstein, J. M., Seidman, L. J., Horton, N. J., Makris, N., Kennedy, D. N., Caviness, V. S., Jr, Faraone, S. V., & Tsuang, M. T. (2001). Normal sexual dimorphism of the adult human brain assessed by in vivo magnetic resonance imaging. *Cereb Cortex, 11,* 490–497.

Hedman, A. M., van Haren, N. E. M., Schnack, H. G., Kahn, R. S., & Hulshoff Pol, H. E. (2012). Human brain changes across the life span: A review of 56 longitudinal magnetic resonance imaging studies. *Hum Brain Mapp, 33,* 1987–2002.

Hill, J., Inder, T., Neil, J., Dierker, D., Harwell, J., & Van Essen, D. (2010). Similar patterns of cortical expansion during human development and evolution. *Proc Natl Acad Sci USA, 107,* 13135–13140.

Huebner, T., Vloet, T. D., Marx, I., Konrad, K., Fink, G. R., Herpert, S. C., & Herpertz-Dahlmann, B. (2008). Morphometric brain abnormalities in boys with conduct disorder. *J Am Acad Child Psy, 47,* 540–547.

Huttenlocher, P. R. (1979). Synaptic density in human frontal cortex — Developmental changes and effects of aging. *Brain Res, 163,* 195–205.

Huttenlocher, P. R. (1990). Morphometric study of human cerebral cortex development. *Neuropsychologia, 28,* 517–527.

Huttenlocher, P. R., & Dabholkar, A. S. (1997). Regional differences in synaptogenesis in human cerebral cortex. *J Comp Neurol, 387,* 167–178.

Jernigan, T. L., & Tallal, P. (1990). Late childhood changes in brain morphology observable with MRI. *Dev Med Child Neurol, 32,* 379–385.

Knickmeyer, R. C., Gouttard, S., Kang, C., Evans, D., Wilber, K., Smith, J. K., Hamer, R. M., ... Gilmore, J. H. (2008). A structural MRI study of human brain development from birth to 2 years. *J Neurosci, 28,* 12176–12182.

Koolschijn, P. C. M. P., & Crone, E. A. (2013). Sex differences and structural brain maturation from childhood to early adulthood. *Dev Cogn Neurosci, 5C,* 106–118.

Le Bihan, D., Mangin, J. F., Poupon, C., Clark, C. A., Pappata, S., Molko, N., & Chabriat, H. (2001). Diffusion tensor imaging: Concepts and applications. *J Magn Reson Imaging, 13,* 534–546.

Lebel, C., & Beaulieu, C. (2011). Longitudinal development of human brain wiring continues from childhood into adulthood. *J Neurosci, 31,* 10937–10947.

Lebel, C., Gee, M., Camicioli, R., Wieler, M., Martin, W., & Beaulieu, C. (2012). Diffusion tensor imaging of white matter tract evolution over the lifespan. *NeuroImage, 60,* 340–352.

Lenroot, R. K., Gogtay, N., Greenstein, D. K., Wells, E. M., Wallace, G. L., Clasen, L. S., Blumenthal, J. D., ... Giedd, J. N. (2007). Sexual dimorphism of brain developmental trajectories during childhood and adolescence. *NeuroImage, 36,* 1065–1073.

Miller, D. J., Duka, T., Stimpson, C. D., Schapiro, S. J., Baze, W. B., McArthur, M. J., Fobbs, A. J., ... Sherwood, C. C. (2012). Prolonged myelination in human neocortical evolution. *Proc Natl Acad Sci USA, 109,* 16480–16485.

Mori, S., & Zhang, J. (2006). Principles of diffusion tensor imaging and its applications to basic neuroscience research. *Neuron, 51,* 527–539.

Muftuler, L. T., Davis, E. P., Buss, C., Head, K., Hasso, A. N., & Sandman, C. A. (2011). Cortical and subcortical changes in typically developing preadolescent children. *Brain Res, 1399,* 15–24.

Østby, Y., Tamnes, C. K., Fjell, A. M., & Walhovd, K. B. (2011). Morphometry and connectivity of the fronto-parietal verbal working memory network in development. *Neuropsychologia, 49,* 3854–3862.

Østby, Y., Tamnes, C. K., Fjell, A. M., Westlye, L. T., Due-Tønnessen, P., & Walhovd, K. B. (2009). Heterogeneity in subcortical brain development: A structural magnetic resonance imaging study of brain maturation from 8 to 30 years. *J Neurosci, 29,* 11772–11782.

Panizzon, M. S., Fennema-Notestine, C., Eyler, L. T., Jernigan, T. L., Prom-Wormley, E., Neale, M., Jacobson, K., ... Kremen, W. S. (2009). Distinct genetic influences on cortical surface area and cortical thickness. *Cereb Cortex, 19,* 2728–2735.

Paus, T. (2010). Growth of white matter in the adolescent brain: Myelin or axon? *Brain Cogn, 72,* 26–35.

Petanjek, Z., Judaš, M., Šimic, G., Rasin, M. R., Uylings, H. B. M., Rakic, P., & Kostovic, I. (2011). Extraordinary neoteny of synaptic spines in the human prefrontal cortex. *Proc Natl Acad Sci USA, 108,* 13281–13286.

Peters, B. D., Szeszko, P. R., Radua, J., Ikuta, T., Gruner, P., DeRosse, P., Zhang, J.-P., ... Malhotra, A. K. (2012). White matter development in adolescence: Diffusion tensor imaging and meta-analytic results. *Schizophr Bull, 38,* 1308–1317.

Pujol, J., Vendrell, P., Junqué, C., Martí-Vilalta, J. L., & Capdevila, A. (1993). When does human brain development end? Evidence of corpus callosum growth up to adulthood. *Ann Neurol, 34,* 71–75.

Rakic, P. (1995). A small step for the cell, a giant leap for mankind: A hypothesis of neocortical expansion during evolution. *Trends Neurosci, 18,* 383–388.

Raznahan, A., Lerch, J. P., Lee, N., Greenstein, D., Wallace, G. L., Stockman, M., Clasen, L., Shaw, P. W., & Giedd, J. N. (2011a). Patterns of coordinated anatomical change in human cortical development: A longitudinal neuroimaging study of maturational coupling. *Neuron, 72,* 873–884.

Raznahan, A., Shaw, P., Lalonde, F., Stockman, M., Wallace, G. L., Greenstein, D., Clasen, L., Gogtay, N., & Giedd, J. N. (2011b). How does your cortex grow? *J Neurosci, 31,* 7174–7177.

Reiss, A. L., Abrams, M. T., Singer, H. S., Ross, J. L., & Denckla, M. B. (1996). Brain development, gender and IQ in children. A volumetric imaging study. *Brain, 119* (Part 5), 1763–1774.

Schaefer, G. B., Thompson, J. N., Jr, Bodensteiner, J. B., Hamza, M., Tucker, R. R., Marks, W., Gay, C., & Wilson, D. (1990). Quantitative morphometric analysis of brain growth using magnetic resonance imaging. *J Child Neurol, 5,* 127–130.

Schmithorst, V. J., Wilke, M., Dardzinski, B. J., & Holland, S. K. (2005). Cognitive functions correlate with white matter architecture in a normal pediatric population: A diffusion tensor MRI study. *Hum Brain Mapp, 26,* 139–147.

Shaw, P., Eckstrand, K., Sharp, W., Blumenthal, J., Lerch, J. P., Greenstein, D., Clasen, L., ... Rapoport, J. L. (2007). Attention-deficit/hyperactivity disorder is characterized by a delay in cortical maturation. *Proc Natl Acad Sci USA, 104*, 19649–19654.

Shaw, P., Gilliam, M., Liverpool, M., Weddle, C., Malek, M., Sharp, W., Greenstein, D., ... Giedd, J. (2011). Cortical development in typically developing children with symptoms of hyperactivity and impulsivity: Support for a dimensional view of attention deficit hyperactivity disorder. *Am J Psychiatry, 168*, 143–151.

Simons, M., & Trajkovic, K. (2006). Neuron-glia communication in the control of oligodendrocyte function and myelin biogenesis. *J Cell Sci, 119*, 4381–4389.

Sowell, E. R., Thompson, P. M., Leonard, C. M., Welcome, S. E., Kan, E., & Toga, A. W. (2004). Longitudinal mapping of cortical thickness and brain growth in normal children. *J Neurosci, 24*, 8223–8231.

Sowell, E. R., Trauner, D. A., Gamst, A., & Jernigan, T. L. (2002). Development of cortical and subcortical brain structures in childhood and adolescence: A structural MRI study. *Dev Med Child Neurol, 44*, 4–16.

Spear, L. P. (2009). Heightened stress responsivity and emotional reactivity during pubertal maturation: Implications for psychopathology. *Dev Psychopathol, 21*, 87–97.

Steinberg, L. (2004). Risk taking in adolescence: What changes, and why? *Ann N Y Acad Sci, 1021*, 51–58.

Steinberg, L., & Morris, A. S. (2001). Adolescent development. *Annu Rev Psychol, 52*, 83–110.

Sullivan, E. V., Pfefferbaum, A., Rohlfing, T., Baker, F. C., Padilla, M. L., & Colrain, I. M. (2011). Developmental change in regional brain structure over 7 months in early adolescence: Comparison of approaches for longitudinal atlas-based parcellation. *NeuroImage, 57*, 214–224.

Suzuki, Y., Matsuzawa, H., Kwee, I. L., & Nakada, T. (2003). Absolute eigenvalue diffusion tensor analysis for human brain maturation. *NMR Biomed, 16*, 257–260.

Tamnes, C. K., Fjell, A. M., Østby, Y., Westlye, L. T., Due-Tønnessen, P., Bjørnerud, A., & Walhovd, K. B. (2011). The brain dynamics of intellectual development: Waxing and waning white and gray matter. *Neuropsychologia, 49*, 3605–3611.

Tamnes, C. K., Østby, Y., Fjell, A. M., Westlye, L. T., Due-Tønnessen, P., & Walhovd, K. B. (2010). Brain maturation in adolescence and young adulthood: Regional age-related changes in cortical thickness and white matter volume and microstructure. *Cereb Cortex, 20*, 534–548.

Tamnes, C. K., Walhovd, K. B., Dale, A. M., Østby, Y., Grydeland, H., Richardson, G., Westlye, L. T., ... Fjell, A. M. (2013a). Brain development and aging: Overlapping and unique patterns of change. *NeuroImage, 68C*, 63–74.

Tamnes, C. K., Walhovd, K. B., Grydeland, H., Holland, D., Østby, Y., Dale, A. M., & Fjell, A. M. (2013b). Longitudinal working memory development is related to structural maturation of frontal and parietal cortices. *J Cogn Neurosci, 25*, 1611–1623.

Thompson, P. M., Giedd, J. N., Woods, R. P., MacDonald, D., Evans, A. C., & Toga, A. W. (2000). Growth patterns in the developing brain detected by using continuum mechanical tensor maps. *Nature, 404*, 190–193.

Tiemeier, H., Lenroot, R. K., Greenstein, D. K., Tran, L., Pierson, R., & Giedd, J. N. (2010). Cerebellum development during childhood and adolescence: A longitudinal morphometric MRI study. *NeuroImage, 49*, 63–70.

Uematsu, A., Matsui, M., Tanaka, C., Takahashi, T., Noguchi, K., Suzuki, M., & Nishijo, H. (2012). Developmental trajectories of amygdala and hippocampus from infancy to early adulthood in healthy individuals. *PLoS ONE, 7*, e46970.

van Soelen, I. L. C., Brouwer, R. M., van Baal, G. C. M., Schnack, H. G., Peper, J. S., Collins, D. L., ... Hulshoff Pol, H. E. (2012). Genetic influences on thinning of the cerebral cortex during development. *NeuroImage, 59*, 3871–3880.

Walhovd, K. B., Tamnes, C. K., Østby, Y., Due-Tønnessen, P., & Fjell, A. M. (2012). Normal variation in behavioral adjustment relates to regional differences in cortical thickness in children. *Eur Child Adolesc Psychiatry, 21*, 133–140.

Winkler, A. M., Kochunov, P., Blangero, J., Almasy, L., Zilles, K., Fox, P. T., Duggirala, R., & Glahn, D. C. (2010). Cortical thickness or grey matter volume? The importance of selecting the phenotype for imaging genetics studies. *NeuroImage, 53*, 1135–1146.

Yakovlev, P. A., & Lecours, I. R. (1967). The myelogenetic cycles of regional maturation of the brain. In A. Minkowski (Ed.), *Regional development of the brain in early life* (pp. 3–70). Oxford, UK: Blackwell.

3 Cognitive Control and Affective Decision Making in Childhood and Adolescence

EVELINE A. CRONE

ABSTRACT One of the major steps in cognitive development during childhood and adolescence is an increase in cognitive control functions, such as working memory, response inhibition, adaptation, and complex reasoning. Yet, at the same time adolescence is characterized by a relative dominant focus on rewards and short-term outcomes at the cost of long-term consequences. This chapter reviews neurocognitive studies showing a change in recruitment of prefrontal cortex and parietal cortex over time when performing cognitive control tasks and heightened sensitivity of subcortical areas in response to emotional stimuli in mid-adolescence. Together, these findings have been interpreted in terms of different developmental trajectories for brain regions involved in making complex decisions, which creates sensitivities but also important opportunities for adaptation and explorative learning.

Cognitive and affective decision making in adolescence

FAST IMPROVEMENT IN COGNITIVE CONTROL One of the most consistent observations in the development of cognitive capacities across childhood and adolescence is the rapid increase in cognitive-control functions, otherwise referred to as "executive functions." Cognitive-control functions are capacities that allow one to keep relevant information in mind in order to obtain a future goal (Huizinga, Dolan, & van der Molen, 2006). In early development, there is a marked improvement in a wide range of cognitive-control functions, such as the ability to store and manipulate information in one's mind to inhibit responses, to filter irrelevant information, and to switch between tasks (Zelazo, 2004). This improvement continues during school-aged development, when children refine their cognitive-control functions, with adult levels achieved around mid-adolescence. These cognitive-control functions are of crucial importance for all kinds of daily activities, and for school performance. For example, working memory, a key component of cognitive control, predicts future academic performance in areas such as reading and arithmetic (St Clair-Thompson & Gathercole, 2006).

SOCIAL AFFECTIVE SENSITIVITIES At the onset of adolescence the influence of social and affective context starts to impact the decisions that adolescents make (see chapter 4, this volume). Adolescence starts with puberty around 10 to 11 years of age, although there is variance between children at the onset of puberty (Forbes & Dahl, 2010). During pubertal development there are substantial changes in hormone release; these changes propagate alterations in both bodily characteristics and in social-affective sensitivities, such as an increased tendency towards risk-taking and a greater sensitivity to peer group influence (Crone & Dahl, 2012). Most changes in social sensitivity and increases in risk-taking behavior are adaptive, to stimulate explorative learning, and to eventually obtain mature social goals. However, in some cases such changes can have serious consequences, including accidents, drug abuse, and in extreme cases suicide attempts (Dahl & Gunnar, 2009).

These developmental patterns have inspired neuroscientists to investigate how different regions in the brain work together when children, adolescents, and adults make decisions.

The neurocognitive development of cognitive control

BASIC COGNITIVE CONTROL: WORKING MEMORY AND RESPONSE INHIBITION The basic components of cognitive control consist of several processes. Most developmental cognitive-control research has focused on the processes of working memory and response inhibition.

Working memory Drawing from the adult literature (D'Esposito, 2007), several developmental neuroimaging studies have examined the role of the lateral prefrontal cortex (Brodmann area [BA] 44 and BA 9/46) and posterior parietal cortex (BA 7) in working memory performance and development. These studies have reported that when 8- to 12-year-old children and adults are performing a visuospatial working memory task, adults showed more activation in the lateral prefrontal cortex and posterior parietal cortex than children (Klingberg, Forssberg, & Westerberg, 2002; Kwon, Reiss, & Menon, 2002; Scherf, Sweeney, & Luna, 2006; Thomason et al., 2009). Age-related increases in recruitment of these areas were also found for other domains of working memory, such as verbal working memory (Thomason et al., 2009) and object working memory (Ciesielski, Lesnik, Savoy, Grant, & Ahlfors, 2006; Crone, Wendelken, Donohue, van Leijenhorst, & Bunge, 2006; Jolles, van Buchem, Rombouts, & Crone, 2011). This increase in activation in the lateral prefrontal and posterior parietal cortex correlates with performance in both adults and children (Crone et al., 2006; Finn, Sheridan, Kam, Hinshaw, & D'Esposito, 2010; Olesen, Nagy, Westerberg, & Klingberg, 2003), suggesting that both age and performance have partly independent contributions to activation levels in these areas.

However, specific task analyses also point to more complex patterns when different domains of working memory are taken under study. For example, during long-delay visuospatial memory trials, children (8–12 years) and adolescents (13–17 years) recruited a more widespread network in the lateral prefrontal cortex compared with adults (18–30 years; Geier, Garver, Terwilliger, & Luna, 2009). Furthermore, when a visuospatial task was performed with distractors, adults recruited the lateral prefrontal cortex and posterior parietal cortex more during working memory demands, while 12- to 13-year-old children recruited the lateral prefrontal cortex more than adults when distractors were present, independent of working memory demands (Olesen, Macoveanu, Tegner, & Klingberg, 2007). Finally, some studies have reported that adults are more responsive in terms of neural activity to specific task demands (e.g., load dependency or modality dependency) than are 7- to 13-year-old children (Brahmbhatt, White, & Barch, 2010; Libertus, Brannon, & Pelphrey, 2009; O'Hare, Lu, Houston, Bookheimer, & Sowell, 2008).

Taken together, a comparison among children, adolescents, and adults shows that while adults recruit the lateral prefrontal cortex and posterior parietal cortex in a way that is more specifically helpful for successful performance of a given task, children and adolescents also recruit these areas, though under different, possibly less performance-related task conditions.

Response inhibition A second basic control process contributing to cognitive control is the ability to inhibit inappropriate responses. To study this, most studies have made use of either a stop-signal task, in which an already initiated response needs to be inhibited, or a go/no-go task, which involves the inhibition of a response to a rarely presented specific stimulus (e.g., the letter *X*) that is presented in a series of other, more prevalent stimuli requiring a response (e.g., other letters of the alphabet).

Neuroimaging studies in adults and patient studies have consistently reported that the right inferior frontal gyrus (BA 45/47) is important for successful inhibitory control (Aron & Poldrack, 2005) and for the greater attention demands associated with response inhibition (Hampshire, Chamberlain, Monti, Duncan, & Owen, 2010). Developmental neuroimaging studies have shown that the right inferior frontal gyrus is recruited more in adults than in 8- to 17-year-old children and adolescents, and that adults perform better on inhibition tasks, suggesting that the right inferior frontal gyrus is important for successful inhibition (Durston et al., 2006; Rubia et al., 2006; Tamm, Menon, & Reiss, 2002). Indeed, performance on the stop-signal task correlates with activity in the right inferior frontal gyrus in children, adolescents, and adults (Cohen et al., 2010). Aside from the age-related increase in activation in the right inferior frontal gyrus, some studies have reported that 8- to 12-year-old children show more left inferior frontal gyrus activation than do adults (Booth et al., 2003; Bunge, Dudukovic, Thomason, Vaidya, & Gabrieli, 2002), while several studies have also reported more widespread activation in other parts of the lateral and medial prefrontal cortex in 8- to 12-year-old children than adults when inhibiting responses (Booth et al., 2003; Velanova, Wheeler, & Luna, 2008).

This evidence suggests that better inhibitory control is accompanied by greater activation of the right inferior frontal gyrus, and that, overall, adults recruit this region more strongly than do children. Thus, both performance and age drive activation differences. However, children recruit other areas of prefrontal cortex more than adults, suggesting that they rely on a wider network of areas to perform a task.

Adaptive Control: Feedback Monitoring Whereas working memory and response inhibition require the implementation of specific task rules, most of our cognitive control requires us to respond to changing task demands. The ability to show adaptive behavior in response to changing task demands is referred to as adaptive control, a process required when, for example, feedback cues inform us that we need to change our behavior on a subsequent occasion.

Feedback monitoring has been widely studied in neuropsychological literature, using classic tasks such as the

Wisconsin Card Sorting Task. Based on patient studies, it has been found that several regions of the lateral and medial prefrontal cortex are important for monitoring negative feedback cues that inform participants a previously applied rule (for example, sorting cards according to color) is no longer correct. This feedback cue (such as a minus sign, or the word "incorrect") informs the participant to switch to a new rule (for example, sorting cards according to shape; Barcelo & Knight, 2002).

Several tasks have been developed in which positive or negative feedback ensues from the response to certain task rules. Neuroimaging analyses reveal that in adults, receiving negative feedback results in activation in the same frontal-parietal network and medial prefrontal cortex as is activated in working memory and inhibition studies (Zanolie, van Leijenhorst, Rombouts, & Crone, 2008). The negative feedback–related activity was greater for adults and 13- to 17-year-old adolescents than for 8- to 12-year-old children, specifically in the lateral prefrontal and posterior parietal cortices (Crone, Zanolie, van Leijenhorst, Westenberg, & Rombouts, 2008; van den Bos, Güroğlu, van den Bulk, Rombouts, & Crone, 2009; van Duijvenvoorde, Zanolie, Rombouts, Raijmakers, & Crone, 2008). This activation increase correlated with successful performance independent of age, suggesting that these areas are important for updating behavior following negative feedback.

However, in children aged 8 to 10 years, the lateral prefrontal cortex and posterior parietal cortex are typically more active in the reverse contrast; that is to say, more activation is reported following positive compared to negative feedback, with a shift occurring in adolescence (van den Bos et al., 2009; van Duijvenvoorde et al., 2008). This developmental difference is specific to situations in which participants learn new rules, but not when applying rules that are already learned (van den Bos et al., 2009).

This suggests that adults show more activation in the lateral prefrontal cortex and posterior parietal cortex when updating behavior following negative feedback, while children recruit these same areas more following positive feedback, with a transition occurring in adolescence.

Complex Cognitive Control: Relational Reasoning and Creativity The ability to interpret problems from multiple perspectives, to integrate knowledge, or to infer new solutions from presently available information probably lies at the highest level of cognitive control. This type of complex reasoning often involves combining different control processes for successful performance.

Relational reasoning Previous research in adults has demonstrated that this ability to integrate information relies on the most anterior part of the prefrontal cortex, the rostrolateral prefrontal cortex (Christoff et al., 2001). In a series of developmental neuroimaging studies, an adaptation of the Raven Progressive Matrices task was used to study how neural activation differs when individuals need to integrate one dimension (e.g., follow a horizontal line of reasoning) or two dimensions (e.g., follow and integrate a horizontal and a vertical line of reasoning; see figure 3.1). The rostrolateral prefrontal cortex was more active in 8- to 12-year-old children at the onset of stimulus presentation, but failed to show sustained activation during problem solving. In contrast, in adults this region showed sustained activation throughout the problem-solving period (Crone et al., 2009). In addition, a study including children, adolescents, and adults showed that children aged 7 to 10 years recruited the rostrolateral prefrontal cortex for both one- and two-dimensional problems, whereas adolescents aged 11 to 14 years showed a small differentiation in activation patterns, and adolescents aged 15 to 18 years recruited the rostrolateral prefrontal cortex for two-dimensional, but not for one-dimensional problems (Wendelken, O'Hare, Whitaker, Ferrer, & Bunge, 2011).

Relational-reasoning problems can be solved in different ways, and strategy differences may also result in developmental differences in brain activation. For example, one study showed that adolescents (14–18 years) had relatively more activation than did children (11–13 years) and adults (22–37 years) in the rostrolateral prefrontal cortex when solving complex relational-reasoning problems. This increase in activation was accompanied by a dip in performance, which suggests that adolescents might use a qualitatively different way of solving relational-reasoning problems (Dumontheil, Houlton, Christoff, & Blakemore, 2010). This shows that relational-reasoning demands are associated with specialization of the rostrolateral prefrontal cortex and may be highly sensitive to strategy use.

Creativity Creativity requires the integration of multiple dimensions to come to new solutions that are novel and useful at the same time (Runco, 2004). When a task requires participants to generate multiple solutions to an open-ended problem, this is often referred to as divergent thinking. Originality in verbal divergent thinking develops with age (ages 10–25 years), whereas spatial divergent thinking is highest at 15 to 16 years (Kleibeuker, de Dreu, & Crone, 2013).

A neuroimaging study tested spatial divergent thinking using a Match Stick task in 15- to 17-year-old

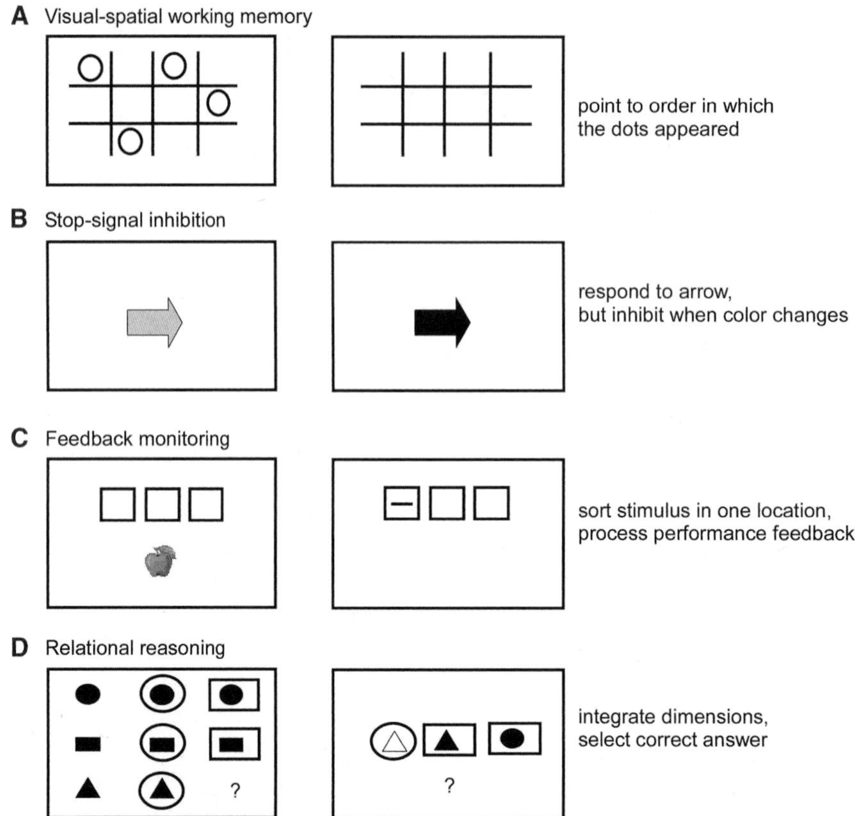

FIGURE 3.1 Examples of cognitive-control paradigms. (A) A visuospatial working memory typically involves the presentation of a grid in which dots are consecutively presented and need to be reproduced on the next trial. More dots will make the task more difficult. (B) A stop-signal paradigm involves the presentation of a stimulus that requires a left- or a right-hand response. On some trials, the arrow quickly turns color, which informs the participant that he/she should inhibit responding. (C) A feedback learning task typically involves a stimulus that needs to be sorted in a specific location. The feedback screen informs the participant whether the response was correct or incorrect. (D) Relational reasoning requires participant to integrate dimensions of a presented stimulus. One-dimensional trials are those trials where only one direction needs to be followed (e.g., a horizontal line) and two-dimensional trials are trials where more dimensions (e.g., a horizontal line and a vertical line) need to be integrated to come to the correct solution.

adolescents and 25- to 30-year-old adults (Kleibeuker et al., 2013). Participants were instructed to find solutions for Match Stick problems by removing matches to find new spatial shapes (e.g., remove six matches to make two squares). Adolescents slightly outperformed adults in finding spatial solutions and showed more activation in the dorsolateral prefrontal cortex while solving these problems. This same area correlated positively with the number of solutions found. Such findings suggest that the flexible recruitment of the lateral prefrontal cortex in adolescence is adaptive for successful performance, such as in the case of spatial creativity.

The neurocognitive development of affective decision making

RISKS AND REWARDS To understand how affective context influences the way we control our actions and make decisions, research has often focused on how children, adolescents, and adults process rewards. Reward processing has been examined in the context of risk-taking, based on the observation that adolescents are more prone than children and adults to take risks in daily life (Steinberg, 2011). Laboratory studies have demonstrated an age-related reduction in risk-taking (Crone, Bullens, van der Plas, Kijkuit, & Zelazo, 2008; van Duijvenvoorde, Jansen, Bredman, & Huizenga, 2012), but also nonlinear age effects, suggesting that adolescents take more risks than children and adults when there is a strong affective context (Burnett, Bault, Coricelli, & Blakemore, 2010; Figner, Mackinlay, Wilkening, & Weber, 2009). The specificity of these developmental differences has been studied in more detail using neuroimaging, for example, during studies that aim to differentiate between different phases of risk-taking.

Taking risks When participants are faced with a risky decision, they typically need to decide between a certain chance of getting a small reward (safe bet) and an uncertain chance of getting a high reward (risky bet). This design was used in a developmental study where it was found that, overall, a safe bet was associated with activation in the dorsolateral prefrontal cortex, whereas a risky bet was associated with activation in the ventromedial prefrontal cortex (van Leijenhorst, Moor et al., 2010). In addition, a lower part of the ventromedial prefrontal cortex/subgenual anterior cingulate cortex (the subcallosal cortex), a region previously implicated in affective processing, was more active in 12- to 17-year-old adolescents compared to 8- to 10-year-old children and 18- to 25-year-old adults when taking risks, suggesting a unique affective coding of risks in adolescence. Using a comparable design, adults showed more activation in the anterior cingulate cortex than adolescents when taking risks (Eshel, Nelson, Blair, Pine, & Ernst, 2007), suggesting that adolescents rely more on affective prefrontal cortex regions, whereas adults rely more on prefrontal cortex regions that are important for deliberative processing.

Anticipating rewards To gain more insight into reward processing independent of choice, studies used simple paradigms that were either passive or where choices had no consequence on outcome (e.g., press the button when a stimulus appears that is tagged to a certain amount of reward). These studies reported that a specific area in the limbic circuit, the ventral striatum, is active when anticipating rewards. This reward anticipation in the ventral striatum is elevated in 13- to 17-year-old adolescents, such that adolescents recruit the ventral striatum more compared to 8- to 11-year-old children and 18- to 29-year-old adults in response to the same level of reward (Galvan et al., 2006; van Leijenhorst, Moor, et al., 2010). This response in the striatum was not found in more complex or uncertain reward-anticipation paradigms (Bjork, Smith, Chen, & Hommer, 2010). Thus, when adolescents anticipate predictable rewards, they show more activation in the ventral striatum than children or adults.

Receiving rewards After taking risks, individuals typically receive a reward or loss following their choices. The rewarding outcome also results in activation in the ventral striatum. Several studies reported that this reward response in the ventral striatum is higher in 13- to 17-year-old adolescents compared to children and adults (Ernst et al., 2005; Galvan et al., 2006; Padmanabhan, Geier, Ordaz, Teslovich, & Luna, 2011; van Leijenhorst, Zanolie, et al., 2010). Recently, it was found that adolescents also show elevated activation compared with adults in the ventral striatum following an aversive event in comparison to a neutral baseline (Galvan & McGlennen, 2013). Thus, the receipt of positive and negative learning signals resulted in elevated activation in the ventral striatum in adolescents, suggesting that the ventral striatum is more sensitive to affective learning signals in adolescence.

Short-Term and Long-Term Consequences The way individuals weigh short- versus long-term consequences is often examined in delay discounting tasks where the choice for an immediate reward (e.g., 1 euro now) and a delayed reward (e.g., 2 euros later) is presented with variable delays. Individuals tend to prefer the immediate reward more when the delay for the larger reward is longer. In a delay discounting task, preference for a delayed reward was associated with activation in the lateral prefrontal cortex, whereas preference for an immediate reward was associated with activation in the ventral striatum in both adolescents and adults (Christakou, Brammer, & Rubia, 2011). In adults, the ventromedial prefrontal cortex was more active than in 11- to 17-year-old adolescents when they chose immediate rewards compared to 18- to 31-year-old adults. This region also showed age-related increases in functional connectivity within the ventral striatum, suggesting that the ventromedial prefrontal cortex works together with the striatum when selecting or inhibiting impulsive choices. Consistent with this line of reasoning, a diffusion tensor imaging study in adults reported that structural connectivity between the ventromedial prefrontal cortex and the striatum predicts fewer short-term choices, suggesting that the ventromedial prefrontal cortex plays a regulatory role when making impulsive choices (Peper et al., 2012). The development of this connectivity may underlie the developmental improvements in inhibition of impulsive choices between adolescence and adulthood.

Models of neurocognitive development

Several models have been introduced to explain how differential development of various brain regions is important for control, and thus influences decision making in adolescence. These dual-processing models (Ernst & Fudge, 2009; Somerville, Jones, & Casey, 2010; Steinberg et al., 2008) suggest that affective limbic brain regions, such as the ventral striatum, develop at a faster pace than brain regions that are important for control and regulation, such as the prefrontal cortex, the dorsal anterior cingulate cortex, and the posterior parietal cortex. This imbalance makes adolescence a

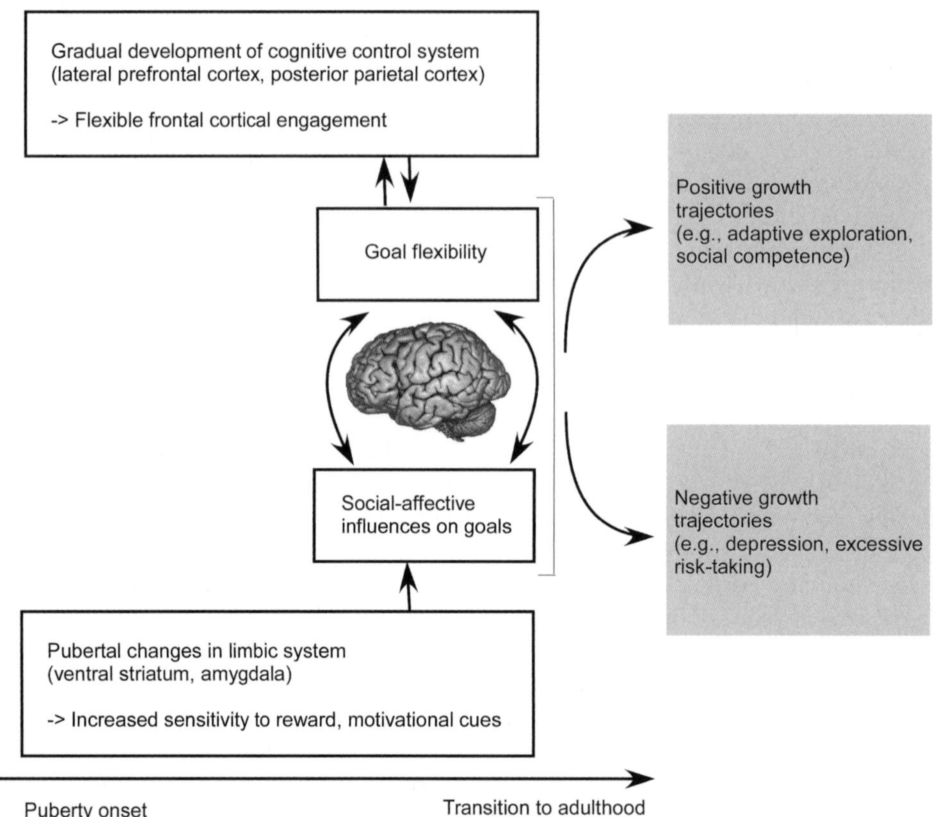

FIGURE 3.2 This model explains the slow developmental trajectory and flexible recruitment of the prefrontal and parietal cortices in adolescence, in combination with puberty-specific changes in the limbic system. This combination leads to positive growth trajectories in adolescence, as this is a natural time of exploration and social learning. However, in some cases the imbalance between these systems can cause negative growth trajectories, which can result in depression or excessive risk-taking. (Adapted from Crone & Dahl, 2012).

sensitive time for risk-taking, but also brings opportunities for exploration and adaptive learning.

Crone and Dahl suggested that puberty could be a driving force for heightened affective sensitivity in adolescence (Crone & Dahl, 2012). Extensive animal research has pinpointed the timing of puberty in rodents and reported specific effects of puberty on brain function and structure (Spear, 2011). Furthermore, pubertal hormones were found to have a steering influence on the structural development of the human brain (Ladouceur, Peper, Crone, & Dahl, 2012). Finally, puberty influences affective responses to reward in the ventral striatum, independent of age (Forbes et al., 2010; Op de Macks et al., 2011). Puberty has a strong influence on the way we process affective and social information, preparing adolescents to obtain independence and adapt quickly to changing social contexts. Therefore, pubertal development may be an important contributor to the increased sensitivity to affective information, which, together with flexibility in recruitment of prefrontal cortex, may facilitate explorative learning (figure 3.2).

Conclusion and future directions

This chapter has described the neural correlates of cognitive and affective decision making in school-aged children, adolescents, and adults. Literature relating to cognitive control shows that the development of basic to complex levels of control follows a pattern of specialization with age in the prefrontal cortex and the posterior parietal cortex, such that these areas are more selectively recruited for specific tasks. The transition from widespread to focused networks takes place during adolescence, a period of change in explorative learning. This development coincides with increased affective sensitivity in mid-adolescence to affective cues, pinpointing nonlinear contributions of control and affective brain regions in development. This integrative approach, in which the development of cognitive control and decision making are studied in combination, is expected to allow for a richer description of adolescent brain development.

ACKNOWLEDGMENTS The author of this chapter is supported by an innovative ideas grant from the European Research Council. ERC-2010-StG-263234.

REFERENCES

ARON, A. R., & POLDRACK, R. A. (2005). The cognitive neuroscience of response inhibition: Relevance for genetic research in attention-deficit/hyperactivity disorder. *Biol Psychiatry, 57*(11), 1285–1292.

BARCELO, F., & KNIGHT, R. T. (2002). Both random and perseverative errors underlie WCST deficits in prefrontal patients. *Neuropsychologia, 40*(3), 349–356.

BJORK, J. M., SMITH, A. R., CHEN, G., & HOMMER, D. W. (2010). Adolescents, adults and rewards: Comparing motivational neurocircuitry recruitment using fMRI. *PLoS One, 5*(7), e11440.

BOOTH, J. R., BURMAN, D. D., MEYER, J. R., LEI, Z., TROMMER, B. L., DAVENPORT, N. D., LI, W., ... MESULAM, M. M. (2003). Neural development of selective attention and response inhibition. *NeuroImage, 20*(2), 737–751.

BRAHMBHATT, S. B., WHITE, D. A., & BARCH, D. M. (2010). Developmental differences in sustained and transient activity underlying working memory. *Brain Res, 1354*, 140–151.

BUNGE, S. A., DUDUKOVIC, N. M., THOMASON, M. E., VAIDYA, C. J., & GABRIELI, J. D. (2002). Immature frontal lobe contributions to cognitive control in children: Evidence from fMRI. *Neuron, 33*(2), 301–311.

BURNETT, S., BAULT, N., CORICELLI, G., & BLAKEMORE, S. J. (2010). Adolescents' heightened risk-seeking in a probabilistic gambling task. *Cogn Dev, 25*(2), 183–196.

CHRISTAKOU, A., BRAMMER, M., & RUBIA, K. (2011). Maturation of limbic corticostriatal activation and connectivity associated with developmental changes in temporal discounting. *NeuroImage, 54*(2), 1344–1354.

CHRISTOFF, K., PRABHAKARAN, V., DORFMAN, J., ZHAO, Z., KROGER, J. K., HOLYOAK, K. J., & GABRIELI, J. D. (2001). Rostrolateral prefrontal cortex involvement in relational integration during reasoning. *NeuroImage, 14*(5), 1136–1149.

CIESIELSKI, K. T., LESNIK, P. G., SAVOY, R. L., GRANT, E. P., & AHLFORS, S. P. (2006). Developmental neural networks in children performing a Categorical N-Back Task. *NeuroImage, 33*(3), 980–990.

COHEN, J. R., ASARNOW, R. F., SABB, F. W., BILDER, R. M., BOOKHEIMER, S. Y., KNOWLTON, B. J., & POLDRACK, R. A. (2010). Decoding developmental differences and individual variability in response inhibition through predictive analyses across individuals. *Front Hum Neurosci, 4*, 47.

CRONE, E. A., BULLENS, L., VAN DER PLAS, E. A., KIJKUIT, E. J., & ZELAZO, P. D. (2008). Developmental changes and individual differences in risk and perspective taking in adolescence. *Dev Psychopathol, 20*(4), 1213–1229.

CRONE, E. A., & DAHL, R. E. (2012). Understanding adolescence as a period of social-affective engagement and goal flexibility. *Nat Rev Neurosci, 13*(9), 636–650.

CRONE, E. A., WENDELKEN, C., DONOHUE, S., VAN LEIJENHORST, L., & BUNGE, S. A. (2006). Neurocognitive development of the ability to manipulate information in working memory. *Proc Natl Acad Sci USA, 103*(24), 9315–9320.

CRONE, E. A., WENDELKEN, C., VAN LEIJENHORST, L., HONOMICHL, R. D., CHRISTOFF, K., & BUNGE, S. A. (2009). Neurocognitive development of relational reasoning. *Dev Sci, 12*(1), 55–66.

CRONE, E. A., ZANOLIE, K., VAN LEIJENHORST, L., WESTENBERG, P. M., & ROMBOUTS, S. A. (2008). Neural mechanisms supporting flexible performance adjustment during development. *Cogn Affect Behav Neurosci, 8*(2), 165–177.

DAHL, R. E., & GUNNAR, M. R. (2009). Heightened stress responsiveness and emotional reactivity during pubertal maturation: Implications for psychopathology. *Dev Psychopathol, 21*(1), 1–6.

D'ESPOSITO, M. (2007). From cognitive to neural models of working memory. *Proc R Soc Lond B Biol Sci, 362*(1481), 761–772.

DUMONTHEIL, I., HOULTON, R., CHRISTOFF, K., & BLAKEMORE, S. J. (2010). Development of relational reasoning during adolescence. *Dev Sci, 13*(6), F15–24.

DURSTON, S., DAVIDSON, M. C., TOTTENHAM, N., GALVAN, A., SPICER, J., FOSSELLA, J. A., & CASEY, B. J. (2006). A shift from diffuse to focal cortical activity with development. *Dev Sci, 9*(1), 1–8.

ERNST, M., & FUDGE, J. L. (2009). A developmental neurobiological model of motivated behavior: Anatomy, connectivity and ontogeny of the triadic nodes. *Neurosci Biobehav Rev, 33*(3), 367–382.

ERNST, M., NELSON, E. E., JAZBEC, S., MCCLURE, E. B., MONK, C. S., LEIBENLUFT, E., BLAIR, J., & PINE, D. S. (2005). Amygdala and nucleus accumbens in responses to receipt and omission of gains in adults and adolescents. *NeuroImage, 25*(4), 1279–1291.

ESHEL, N., NELSON, E. E., BLAIR, R. J., PINE, D. S., & ERNST, M. (2007). Neural substrates of choice selection in adults and adolescents: Development of the ventrolateral prefrontal and anterior cingulate cortices. *Neuropsychologia, 45*(6), 1270–1279.

FIGNER, B., MACKINLAY, R. J., WILKENING, F., & WEBER, E. U. (2009). Affective and deliberative processes in risky choice: Age differences in risk-taking in the Columbia Card Task. *J Exp Psychol Learn Mem Cogn, 35*(3), 709–730.

FINN, A. S., SHERIDAN, M. A., KAM, C. L., HINSHAW, S., & D'ESPOSITO, M. (2010). Longitudinal evidence for functional specialization of the neural circuit supporting working memory in the human brain. *J Neurosci, 30*(33), 11062–11067.

FORBES, E. E., & DAHL, R. E. (2010). Pubertal development and behavior: Hormonal activation of social and motivational tendencies. *Brain Cogn, 72*(1), 66–72.

FORBES, E. E., RYAN, N. D., PHILLIPS, M. L., MANUCK, S. B., WORTHMAN, C. M., MOYLES, D. L., TARR, J. A., SCIARRILLO, S. R., & DAHL, R. E. (2010). Healthy adolescents' neural response to reward: Associations with puberty, positive affect, and depressive symptoms. *J Am Acad Child Adolesc Psychiatry, 49*(2), 162–172.

GALVAN, A., HARE, T. A., PARRA, C. E., PENN, J., VOSS, H., GLOVER, G., & CASEY, B. J. (2006). Earlier development of the accumbens relative to orbitofrontal cortex might underlie risk-taking behavior in adolescents. *J Neurosci, 26*(25), 6885–6892.

GALVAN, A., & MCGLENNEN, K. M. (2013). Enhanced striatal sensitivity to aversive reinforcement in adolescents versus adults. *J Cogn Neurosci, 25*(2), 284–296.

Geier, C. F., Garver, K., Terwilliger, R., & Luna, B. (2009). Development of working memory maintenance. *J Neurophysiol, 101*(1), 84–99.

Hampshire, A., Chamberlain, S. R., Monti, M. M., Duncan, J., & Owen, A. M. (2010). The role of the right inferior frontal gyrus: Inhibition and attentional control. *NeuroImage, 50*(3), 1313–1319.

Huizinga, M., Dolan, C. V., & van der Molen, M. W. (2006). Age-related change in executive function: Developmental trends and a latent variable analysis. *Neuropsychologia, 44*(11), 2017–2036.

Jolles, D. D., van Buchem, M. A., Rombouts, S. A., & Crone, E. A. (2011). Developmental differences in prefrontal activation during working memory maintenance and manipulation for different memory loads. *Dev Sci, 14,* 713–724.

Kleibeuker, S. W., de Dreu, C. K. W., & Crone, E. A. (2013). The development of creative cognition across adolescence: Distinct trajectories for insight and divergent thinking. *Dev Sci, 16,* 2–12.

Kleibeuker, S. W., Koolschijn, P. C., Jolles, D. D., Schel, M. A., de Dreu, C. K., & Crone, E. A. (2013). Prefrontal cortex involvement in creative problem solving in middle adolescence and adulthood. *Dev Cogn Neurosci, 5,* 197–206.

Klingberg, T., Forssberg, H., & Westerberg, H. (2002). Increased brain activity in frontal and parietal cortex underlies the development of visuospatial working memory capacity during childhood. *J Cogn Neurosci, 14*(1), 1–10.

Kwon, H., Reiss, A. L., & Menon, V. (2002). Neural basis of protracted developmental changes in visuo-spatial working memory. *Proc Natl Acad Sci USA, 99*(20), 13336–13341.

Ladouceur, C. D., Peper, J. S., Crone, E. A., & Dahl, R. E. (2012). White matter development in adolescence: The influence of puberty and implications for affective disorders. *Dev Cogn Neurosci, 2,* 36–54.

Libertus, M. E., Brannon, E. M., & Pelphrey, K. A. (2009). Developmental changes in category-specific brain responses to numbers and letters in a working memory task. *NeuroImage, 44*(4), 1404–1414.

O'Hare, E. D., Lu, L. H., Houston, S. M., Bookheimer, S. Y., & Sowell, E. R. (2008). Neurodevelopmental changes in verbal working memory load-dependency: An fMRI investigation. *NeuroImage, 42*(4), 1678–1685.

Olesen, P. J., Macoveanu, J., Tegner, J., & Klingberg, T. (2007). Brain activity related to working memory and distraction in children and adults. *Cereb Cortex, 17*(5), 1047–1054.

Olesen, P. J., Nagy, Z., Westerberg, H., & Klingberg, T. (2003). Combined analysis of DTI and fMRI data reveals a joint maturation of white and grey matter in a fronto-parietal network. *Brain Res Cogn Brain Res, 18*(1), 48–57.

Op de Macks, Z., Gunther Moor, B., Overgaauw, S., Guroglu, B., Dahl, R. E., & Crone, E. A. (2011). Testosterone levels correspond with increased ventral striatum activation in response to monetary rewards in adolescents. *Dev Cogn Neurosci, 1,* 506–516.

Padmanabhan, A., Geier, C. F., Ordaz, S. J., Teslovich, T., & Luna, B. (2011). Developmental changes in brain function underlying the influence of reward processing on inhibitory control. *Dev Cogn Neurosci, 1*(4), 517–529.

Peper, J. S., Mandl, R. C., Braams, B. R., de Water, E., Heijboer, A. C., Koolschijn, P. C., & Crone, E. A. (2012). Delay discounting and frontostriatal fiber tracts: A combined DTI and MTR study on impulsive choices in healthy young adults. *Cereb Cortex, 23*(7), 1695–1702.

Rubia, K., Smith, A. B., Woolley, J., Nosarti, C., Heyman, I., Taylor, E., & Brammer, M. (2006). Progressive increase of frontostriatal brain activation from childhood to adulthood during event-related tasks of cognitive control. *Hum Brain Mapp, 27*(12), 973–993.

Runco, M. A. (2004). Creativity. *Annu Rev Psychol, 55,* 657–687.

Scherf, K. S., Sweeney, J. A., & Luna, B. (2006). Brain basis of developmental change in visuospatial working memory. *J Cogn Neurosci, 18*(7), 1045–1058.

Somerville, L. H., Jones, R. M., & Casey, B. J. (2010). A time of change: Behavioral and neural correlates of adolescent sensitivity to appetitive and aversive environmental cues. *Brain Cogn, 72*(1), 124–133.

Spear, L. P. (2011). Rewards, aversions and affect in adolescence: Emerging convergences across laboratory animal and human data. *Dev Cogn Neurosci, 1,* 390–403.

St Clair-Thompson, H. L., & Gathercole, S. E. (2006). Executive functions and achievements in school: Shifting, updating, inhibition, and working memory. *Q J Exp Psychol, 59*(4), 745–759.

Steinberg, L. (2011). *The science of adolescent risk-taking.* Washington, DC: The National Academies.

Steinberg, L., Albert, D., Cauffman, E., Banich, M., Graham, S., & Woolard, J. (2008). Age differences in sensation seeking and impulsivity as indexed by behavior and self-report: Evidence for a dual systems model. *Dev Psychol, 44*(6), 1764–1778.

Tamm, L., Menon, V., & Reiss, A. L. (2002). Maturation of brain function associated with response inhibition. *J Am Acad Child Adolesc Psych, 41*(10), 1231–1238.

Thomason, M. E., Race, E., Burrows, B., Whitfield-Gabrieli, S., Glover, G. H., & Gabrieli, J. D. (2009). Development of spatial and verbal working memory capacity in the human brain. *J Cogn Neurosci, 21*(2), 316–332.

van den Bos, W., Güroğlu, B., van den Bulk, B. G., Rombouts, S. A., & Crone, E. A. (2009). Better than expected or as bad as you thought? The neurocognitive development of probabilistic feedback processing. *Front Hum Neurosci, 3,* 52.

van Duijvenvoorde, A. C., Jansen, B. R., Bredman, J. C., & Huizenga, H. M. (2012). Age-related changes in decision making: Comparing informed and noninformed situations. *Dev Psych, 48*(1), 192–203.

van Duijvenvoorde, A. C., Zanolie, K., Rombouts, S. A., Raijmakers, M. E., & Crone, E. A. (2008). Evaluating the negative or valuing the positive? Neural mechanisms supporting feedback-based learning across development. *J Neurosci, 28*(38), 9495–9503.

van Leijenhorst, L., Moor, B. G., Op de Macks, Z. A., Rombouts, S. A., Westenberg, P. M., & Crone, E. A. (2010). Adolescent risky decision-making: Neurocognitive development of reward and control regions. *NeuroImage, 51*(1), 345–355.

van Leijenhorst, L., Zanolie, K., Van Meel, C. S., Westenberg, P. M., Rombouts, S. A., & Crone, E. A. (2010). What motivates the adolescent? Brain regions mediating reward sensitivity across adolescence. *Cereb Cortex, 20*(1), 61–69.

Velanova, K., Wheeler, M. E., & Luna, B. (2008). Maturational changes in anterior cingulate and frontoparietal recruitment support the development of error processing and inhibitory control. *Cereb Cortex, 18*(11), 2505–2522.

Wendelken, C., O'Hare, E. D., Whitaker, K. J., Ferrer, E., & Bunge, S. A. (2011). Increased functional selectivity over development in rostrolateral prefrontal cortex. *J Neurosci, 31*(47), 17260–17268.

Zanolie, K., van Leijenhorst, L., Rombouts, S. A., & Crone, E. A. (2008). Separable neural mechanisms contribute to feedback processing in a rule-learning task. *Neuropsychologia, 46*(1), 117–126.

Zelazo, P. D. (2004). The development of conscious control in childhood. *Trends Cogn Sci, 8*(1), 12–17.

4 The Social Brain in Adolescence

SARAH-JAYNE BLAKEMORE AND KATHRYN L. MILLS

ABSTRACT Adolescence is a period of formative biological and social transition. Social cognitive processes involved in navigating increasingly complex and intimate relationships continue to develop throughout adolescence. Here, we review neuroimaging studies of social cognitive processes across development during adolescence. These studies show that areas of the social brain undergo both structural changes and functional reorganization during the second decade of life, possibly reflecting a sensitive period for adapting to the social environment.

Adolescence is defined as the period between the onset of puberty and the achievement of relative self-sufficiency. The adults who emerge from adolescence must be equipped to navigate the social complexities of their community. Therefore, it has been proposed that adolescence is a time of particular sensitivity to social signals (Choudhury, 2010; Fiske, 2009) and that the impact of puberty on the brain makes adolescents particularly sensitive to their social environments (Crone & Dahl, 2012; Peper & Dahl, 2013). Knowledge about the development of the adolescent brain has significantly increased in the last 20 years, and there is now extensive evidence that the brain undergoes substantial structural and functional change during this transition from childhood to adulthood, including in networks of brain regions involved in social cognition (Blakemore, 2008; Mills, Lalonde, Clasen, Giedd, & Blakemore, 2013). This chapter reviews neuroimaging studies of social cognitive processes across development, with emphasis on the changes occurring in mentalizing—that is, the attribution of mental states to other people—during adolescence.

Social cognitive development

Social cognition refers to the ability to make sense of the world through processing signals generated by members of the same species (Frith, 2008). Social cognition includes basic perceptual processes such as face processing (Farroni et al., 2005), biological motion detection (Pelphrey & Carter, 2008), and joint attention (Carpenter, Nagell, & Tomasello, 1998)—all of which are developing rapidly from birth (see chapter 1, this volume). Other social cognitive processes are relatively complex, such as understanding others' mental states (theory of mind/mentalizing; Blakemore, den Ouden, Choudhury, & Frith, 2007), social emotion processing (Burnett, Bird, Moll, Frith, & Blakemore, 2009), and negotiating complex interpersonal decisions (Crone, 2013). In this chapter we focus on the more complex social cognitive processes. Recent behavioral and neuroimaging studies have shown that these processes continue to develop past childhood and throughout adolescence (Apperly, 2010; Blakemore, 2012).

Adolescent social environment

Adolescents go through a period of "social reorienting," during which the opinions of peers become more important than those of family members (Larson & Richards, 1991; Larson, Richards, Moneta, Holmbeck, & Duckett, 1996). This enhanced sensitivity to the social environment and the rewarding nature of peer acceptance is reflected in studies showing that adolescents are particularly susceptible to the negative consequences of social exclusion (Sebastian, Viding, Williams, & Blakemore, 2010). Adolescents aged 13–17 years reported that peer evaluations affected their feelings of social or personal worth, and that peer rejection indicated their unworthiness as an individual (O'Brien & Bierman, 1988). While the adolescents and children aged 10–13 years similarly felt that peers provided companionship, stimulation, and support, the younger group did not indicate that peer acceptance impacted self-evaluation. O'Brien and Bierman suggest that increasing abilities to form abstract representations, as well as increasing motivation for peer acceptance, might account for the influence of peer evaluations on self-evaluations in adolescence (O'Brien & Bierman, 1988). These self-reported accounts of the importance of peer acceptance are supported by the results of a behavioral study investigating the effects of social exclusion in the lab. After being excluded by other players in a laboratory ball-throwing game called "Cyberball," young and mid-adolescents (11–15 years) reported lowered overall mood and young adolescents (11–13 years) reported higher state anxiety compared with adults (Sebastian et al., 2010). Thus, it appears that complex social cognitive processes, such as the desire to be accepted by one's peers, the importance of social reward, and avoidance of social rejection, are particularly acute in adolescence, and might drive much adolescent-typical behavior. Thus, it is important to understand how social cognition and the social brain develop during adolescence.

Until recently, there was a shortage of studies looking into social cognitive abilities after childhood, as it was generally assumed that these abilities were already mature by mid-childhood in typically developing children. Most paradigms have been designed to investigate social cognition (in particular, theory of mind) in young children and result in very high levels of performance after mid-childhood (Apperly, 2010). However, recently more complex and challenging paradigms have been designed to be appropriate for older children and adolescents. One of the first studies to investigate changes in social cognitive behavior in adolescence showed that the ability to integrate the perspectives and intentions of others when making fairness considerations continues to improve throughout adolescence (Güroğlu, van den Bos, & Crone, 2009). The authors of this study suggested that adolescents perceive interactions with peers as rewarding, and that this could affect social decision-making processes (Güroğlu et al., 2009).

Another study demonstrated that online social cognitive skills improve across adolescence (Dumontheil, Apperly, & Blakemore, 2010). Participants aged 7–27 years were tested on their ability to take the perspective of another person when making decisions. Their paradigm, referred to here as the "Director Task," adapted a referential communication task in which adults frequently make mistakes (Keysar, Barr, Balin, & Brauner, 2000; Keysar, Lin, & Barr, 2003). In our computerized version of the task, participants view a computer screen showing a set of shelves containing objects, which they are instructed to move by a Director. The Director is standing behind the shelves and he cannot see all the objects the participant can see because some objects are occluded from his point of view by a gray screen (Dumontheil et al., 2010). Thus there is a "conflict" between the participant's perspective and that of the Director. In order to move the correct object, the participant needs to take into account the Director's perspective by moving only objects that the Director can see (the Director condition). In addition, we included a control condition in which the Director is absent, and instead participants are instructed to follow a rule ("ignore objects with a gray background") in order to decide which objects to move. Thus, the control condition is matched to the Director condition in terms of demands on working memory, inhibiting a prepotent response and other executive functions; the main difference is the need to take into account someone else's perspective in order to guide decisions in the Director condition.

We tested participants aged between 7 and 27 years and found that performance in both Director and Control conditions improved gradually, following the same trajectory from mid-childhood until mid-adolescence. In contrast, for the Director condition only, there was continued improvement between mid-adolescence and adulthood. These results suggest that the ability to take into account another person's perspective to guide decisions is still developing in late adolescence (Dumontheil et al., 2010).

Social brain network

Mentalizing, the ability to infer the mental states of others, has been associated with a network of brain regions including the dorsal medial prefrontal cortex (dmPFC), temporoparietal junction (TPJ), posterior superior temporal sulcus (pSTS), and anterior temporal cortex (ATC). This is referred to as the social brain network, or the mentalizing network. The mentalizing tasks that recruit this network use a wide variety of stimuli such as animated shapes (Castelli, Happé, Frith, & Frith, 2000), cartoon stories (Brunet, Sarfati, Hardy-Baylé, & Decety, 2000; Gallagher et al., 2000), and written stories (Fletcher et al., 1995) to elicit the representation of mental states.

STRUCTURAL DEVELOPMENT OF THE SOCIAL BRAIN NETWORK Cellular studies of post-mortem human brain tissue provided some of the first evidence that the brain undergoes profound changes in anatomy across the first decades of life (Petanjek et al., 2011; Webb, Monk, & Nelson, 2001; Yakovlev & Lecours, 1967). Magnetic resonance imaging (MRI) studies over the past 20 years have illuminated how and when the human brain develops. Structural MRI studies have consistently shown continuing neuroanatomical development in gray matter and white matter (Brain Development Cooperative Group, 2012; Lenroot et al., 2007; Sowell et al., 2003; Tamnes et al., 2013), with association cortices reducing in gray matter volume across adolescence and white matter increasing into adulthood (see chapter 2, this volume).

Areas within the social brain network continue to undergo neuroanatomical changes across adolescence before relatively stabilizing in the early twenties (Mills et al., 2013). In a study using MRI data from 288 individuals with at least two brain scans between ages 7 and 30 years, we examined the neuroanatomical

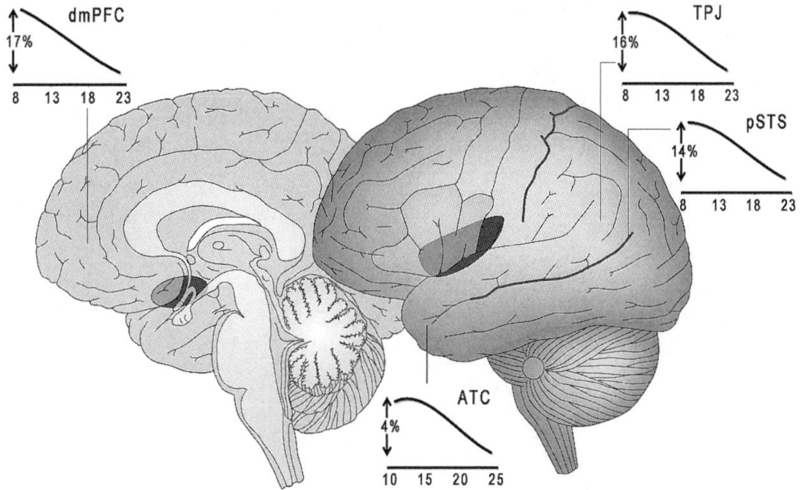

FIGURE 4.1 Social brain network: Areas of the brain that are sensitive to social cognitive processes necessary to navigate the adolescent social environment accompanied by graphs displaying the developmental trajectories of each region's gray matter volume. Regions represented in this figure include the dorsomedial prefrontal cortex (dmPFC) and temporoparietal junction (TPJ), which are involved in thinking about mental states; the posterior superior temporal sulcus (pSTS), which is involved in observing faces and biological motion; and the anterior temporal cortex (ATC), which is involved in applying social knowledge. For each graph, age is on the x-axis and gray matter volume is on y-axis, with the displayed percentage representing the decrease in gray matter volume for each region. (Data for graphs adapted from Mills et al., 2013.)

developmental trajectories of the social brain network. Gray matter volume and cortical thickness in medial Brodmann Area 10 (a proxy for dmPFC), TPJ, and pSTS showed continued reductions from childhood into the early twenties, whereas the ATC continued to increase in gray matter volume until adolescence and in cortical thickness until early adulthood. This protracted development demonstrates that areas of the brain involved in deciphering the mental states of others are still maturing in terms of their structure from late childhood to early adulthood.

FUNCTIONAL DEVELOPMENT OF THE SOCIAL BRAIN NETWORK There have been a number of functional MRI (fMRI) studies that show functional changes across adolescence in the brain networks associated with social cognition, including mentalizing, social emotion, peer evaluation, and peer influence. We discuss these studies below.

Mentalizing Several fMRI studies on mentalizing report decreases in dmPFC recruitment between adolescence and adulthood (Blakemore, 2008, 2012). These studies have used a variety of tasks that require mental state attribution, such as understanding irony (Wang, Lee, Sigman, & Dapretto, 2006), thinking about social emotions like guilt (Burnett et al., 2009), understanding intentions (Blakemore et al., 2007), understanding emotions from photographs of eyes (Gunther Moor et al., 2012), and thinking about the preferences and dispositions of oneself or a fictitious story character (Pfeifer et al., 2009). In some studies, higher activity in more posterior regions, such as the pSTS/TPJ (Blakemore et al., 2007) and the ATC (Burnett et al., 2009), was observed in adults as compared to adolescents. These changes in functional recruitment are hypothesized to reflect changes in neurocognitive strategy and/or neuroanatomy (Blakemore, 2008).

In an adapted version of the Director Task (Apperly et al., 2010; Dumontheil et al., 2010), areas of the social brain network were engaged when participants had to use social cues to select an appropriate action in a communicative context (Dumontheil, Hillebrandt, Apperly, & Blakemore, 2012). While both adults and adolescents recruited the dmPFC when the social cues were needed to accurately perform the task, adolescents also recruited the dmPFC when social cues were not needed. The authors suggest that this engagement of the dmPFC in social conditions, even when social signals are irrelevant, may reflect the use of brain regions involved in mentalizing even when they are not necessary during adolescence.

Adolescents also show developmental changes in sensitivity to the perspectives of others. In an fMRI study, young adolescents (12–14 years), older adolescents (15–17 years), and emerging adults (18–22 years) completed a social exchange game in which participants were the "second players" in an investment game (van den Bos, van Dijk, Westenberg, Rombouts, & Crone, 2011). Initially, these participants were given an amount of money by an anonymous "first player," which they could divide equally among themselves and the first player (reciprocate), or keep mostly for themselves (defect). Participants' ability to understand the intentions of the first player was also measured by comparing trials in which the first player stood to lose a large amount of money by trusting the second player with trials where the first player stood to lose a relatively small amount of money. Older adolescents and emerging adults were more likely to reciprocate when the first player stood to lose more money, whereas the younger adolescents did not differentiate, supporting the idea that the ability to understand the intentions of others increases into adulthood. The recruitment of the left TPJ when participants were shown that the first player trusted them increased with age, and this level of activation correlated with participants' sensitivity to the first player's intentions. All participants showed greater recruitment in the dmPFC when making self-oriented choices (defecting), but only young adolescents engaged this region when making reciprocal choices. This heightened activation in the dmPFC for reciprocal choices decreased between early and late adolescence and remained stable into early adulthood, possibly reflecting a shift away from engaging in social interactions from an egocentric perspective (van den Bos et al., 2011).

Social emotion Social emotions—such as guilt, embarrassment, shame, and pride—require thinking about someone else's mental states or emotions, whereas basic emotions like fear and disgust do not. As adolescence is a period of increased sensitivity to peer evaluation, there may be changes in how social emotions are processed. One fMRI study investigated changes in neural recruitment during a social-emotional task between adolescence (11–18 years) and adulthood (23–32 years; Burnett et al., 2009). Participants were instructed to read sentences describing social- or basic-emotion scenarios. Adolescents recruited the dmPFC more than adults when reading social-emotional sentences relative to basic-emotion sentences. In contrast, adults recruited the left ATC more than did adolescents when reading social-emotional sentences relative to basic-emotion sentences (Burnett et al., 2009). These results reflect a developmental shift from prefrontal to temporal cortex recruitment during the processing of social emotion.

A more recent study investigated the influence of puberty on social-emotion processing in adolescence (Goddings, Burnett Heyes, Bird, Viner, & Blakemore, 2012). In a sample of 42 female adolescents (11–13 years), level of pubertal hormones (testosterone, estradiol, and DHEA) was related to ATC recruitment during social-emotional processing. While activity in the left ATC was positively correlated with hormone levels (irrespective of age), activity in the dmPFC was negatively correlated with chronological age (irrespective of hormone levels), providing evidence for a dissociation between puberty- and age-related changes in social brain function during adolescence (Goddings et al., 2012).

Peer evaluation There are a number of fMRI investigations of experimentally manipulated social exclusion using the Cyberball task. As described above, this task involves participants playing a game of "catch" with two other players, under the guise that they are playing with real people over the Internet. However, the other players are actually preprogrammed to include or exclude the participant. In one study, recruitment of the mPFC during exclusion conditions relative to inclusion conditions was associated with greater self-reported susceptibility to peer influence in adolescents, but not in adults (Sebastian et al., 2011). This study also found age-related differences in right ventrolateral PFC (vlPFC) recruitment during exclusion conditions, with adults recruiting right vlPFC more than adolescents (Sebastian et al., 2011). Another fMRI study using the Cyberball task specifically in a group of adolescents aged 12–13 years found that the recruitment of the right vlPFC during exclusion conditions was negatively correlated with self-reported measures of distress following exclusion (Masten et al., 2009). Together these studies suggest the vlPFC plays a role in regulating distress following social exclusion, and that this region is still developing functionally between adolescence and adulthood. A recent study found that healthy adolescents who display heightened activity in the subgenual anterior cingulate cortex while being excluded from peers in Cyberball were more likely to show an increase in depressive symptoms over the following year (Masten et al., 2011).

Prompted by research linking good peer relationships to well-being, Masten and colleagues examined how 12- and 13-year-olds respond to witnessing peer rejection in an online game (Masten, Eisenberger, Pfeifer, & Dapretto, 2010). Participants first completed a self-reported measure of trait empathy before participating in an fMRI task where they witnessed peer exclusion in a game of Cyberball. Afterward, they were asked to write a letter to the rejected player as a measure of prosocial behavior. Activity in parts of the social brain network was related to observed exclusion compared to observed inclusion. While recruitment of the dmPFC and ATC appeared to be related to self-reported trait empathy, only the anterior insula showed a positive correlation with prosocial behavior, perhaps reflecting increased distress during observed exclusion. Together, these findings suggest that young adolescents recruit the social brain network more while witnessing peer rejection, in contrast to a situation where peers are being treated equally. One speculative interpretation is that adolescents might be paying special attention to situations of peer rejection in order to figure out why a peer may become excluded from the social group.

Peer influence Peer influence on conformity shows a curvilinear pattern between middle childhood and late adolescence, reaching a peak in early adolescence (Berndt, 1979). The popularity rankings of a given song influence how much adolescents like it, for example (Berns, Capra, Moore, & Noussair, 2010). In an fMRI task, adolescents aged 12–17 years listened to and rated the likeability of short music clips, first without knowing the popularity of the song, and then after receiving its popularity ranking. Adolescents' change in song evaluation correlated with increased recruitment of the anterior insula and ACC. Since these regions are involved in arousal, the authors speculate that this may reflect the anxiety of having dissimilar preferences to others (note that this is speculative because these brain regions are involved in multiple other cognitive and emotional processes).

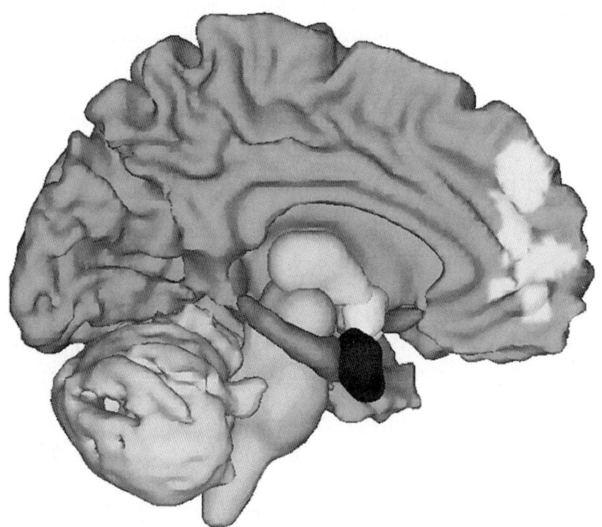

FIGURE 4.2 A qualitative meta-analysis of the region of dmPFC that consistently shows decreased activity during mentalizing tasks between late childhood and adulthood. This meta-analysis shows voxels in mPFC (in yellow) that are within 10 mm of the peak voxel found to have a significant negative relationship with age in three or more of eight published developmental fMRI studies of social cognition (Blakemore et al., 2007; Burnett et al., 2009; Gunther Moor et al., 2012; Güroğlu et al., 2009; Pfeifer et al., 2007, 2009; Sebastian et al., 2011; van den Bos et al., 2011). Meta-analysis was performed using Neurosynth software (www.neurosynth.org). (See color plate 2.)

The presence of peers affects how likely adolescents are to take risks in a driving game. Adolescents (13–16 years), youths (18–22 years), and adults (24+ years) took around the same number of driving risks when alone, whereas the adolescents took almost three times that number of risks in the presence of their friends. In contrast, peers had no impact on risk-taking in adults and had an intermediate effect on risk-taking in youths (Gardner & Steinberg, 2005). In an fMRI version of this task, in the peers-present condition, two friends communicated with the participant in the scanner over the intercom (Chein, Albert, O'Brien, Uckert, & Steinberg, 2011). Adults (24–29 years) showed higher activity in lateral PFC than did adolescents (14–18 years) or younger adults (19–22 years) when they had to make critical decisions in the driving game, both when alone and when peers were present. Relative to both groups of adults, adolescents showed increased recruitment of the ventral striatum and orbitofrontal cortex during the driving decisions with peers compared to when alone.

Social context also modulates risk attitudes adopted by adolescents (Engelmann, Moore, Monica Capra, & Berns, 2012). Relative to adults, adolescents showed greater risk-adverse behavior after receiving expert advice, and this effect is modulated by increased engagement of the dorsolateral PFC by adolescents during valuation in the presence of advice (Engelmann et al., 2012). The authors suggest enhanced inhibitory and cognitive control processes might underlie the effect of social context on risky decision making in adolescents.

Decision making and social influence

Adolescents are stereotypically known for their engagement in risky behaviors. There is experimental evidence supporting the idea that, while in laboratory settings, adolescents are more likely than children and adults to make risky decisions in "hot" (e.g., affective) contexts (Blakemore & Robbins, 2012). Experimental evidence from risky decision-making and probabilistic reward paradigms mostly support the hypothesis that adolescents are biased toward taking risks due to overactive reward-related circuitry (i.e., the ventral striatum; Ernst, Pine, & Hardin, 2006; van Leijenhorst et al., 2010; also see chapter 3, this volume). Probabilistic reward paradigms in laboratory experiments on risk-taking often involve gambling tasks. Children and adolescents showed adult levels of probability estimation and reward evaluation during one such gambling task, suggesting that heightened risky decision making in adolescents is probably not related to a change in risk perception (van Leijenhorst, Westenberg, & Crone, 2008). When asked in a laboratory setting to estimate the risks of negative outcomes to some risky behaviors, adolescents actually overestimate risks (Reyna & Farley, 2006). Adolescents also rate the potential reward to be gained as very high, which may make the perceived benefits outweigh the perceived risk.

Social and contextual cues can bias the way adolescents perceive the risks involved in certain behaviors (Reyna & Adam, 2003; Reyna & Farley, 2006; Reyna, 2008). While risky decision making during adolescence is often framed as maladaptive and unavoidable, this perspective leaves out many key features of risky decision making, including the fact that the outcome can be positive and some risky decision making is necessary in development and throughout life (see chapter 3, this volume). A recent report highlights the benefits of asking "What's in it for the adolescent?" when studying risky behavior and risky decision making in adolescence (Ellis et al., 2012). We propose that some rewards gained by risky behaviors are social in nature, such as peer acceptance or the avoidance of social exclusion, and that this is a potential major driver of risky behavior.

Conclusion

Neuroimaging and behavioral studies in humans have demonstrated that the social brain and social cognition undergo a profound period of development in adolescence. As such, adolescence might represent a sensitive period for the processing of social signals.

ACKNOWLEDGMENTS Support for KLM came from the National Institute of Mental Health Graduate Partnership Program. Support for SJB came from the Royal Society and the Leverhulme Trust. SJB has a Royal Society University Research Fellowship.

REFERENCES

Apperly, I. (2010). *Mindreaders: The cognitive basis of "theory of mind."* East Sussex, UK: Psychology Press.

Apperly, I. A., Carroll, D. J., Samson, D., Humphreys, G. W., Qureshi, A., & Moffitt, G. (2010). Why are there limits on theory of mind use? Evidence from adults' ability to follow instructions from an ignorant speaker. *Q J Exp Psychol A, 63*(6), 1201–1217.

Berns, G. S., Capra, C. M., Moore, S., & Noussair, C. (2010). Neural mechanisms of the influence of popularity on adolescent ratings of music. *NeuroImage, 49*(3), 2687–2696.

Berndt, T. J. (1979). Developmental changes in conformity to peers and parents. *Dev Psychol, 15,* 608–616.

Blakemore, S.-J. (2008). The social brain in adolescence. *Nat Rev Neurosci, 9*(4), 267–277.

Blakemore, S.-J. (2012). Development of the social brain in adolescence. *J R Soc Med, 105*(3), 111–116.

Blakemore, S.-J., den Ouden, H., Choudhury, S., & Frith, C. (2007). Adolescent development of the neural circuitry for thinking about intentions. *Soc Cogn Affect Neurosci, 2*(2), 130–139.

Blakemore, S.-J., & Robbins, T. W. (2012). Decision-making in the adolescent brain. *Nat Neurosci, 15*(9), 1184–1191.

Brain Development Cooperative Group. (2012). Total and regional brain volumes in a population-based normative sample from 4 to 18 years: The NIH MRI Study of Normal Brain Development. *Cereb Cortex, 22*(1), 1–12.

Brunet, E., Sarfati, Y., Hardy-Baylé, M. C., & Decety, J. (2000). A PET investigation of the attribution of intentions with a nonverbal task. *NeuroImage, 11*(2), 157–166.

Burnett, S., Bird, G., Moll, J., Frith, C., & Blakemore, S.-J. (2009). Development during adolescence of the neural processing of social emotion. *J Cogn Neurosci, 21*(9), 1736–1750.

Carpenter, M., Nagell, K., & Tomasello, M. (1998). Social cognition, joint attention, and communicative competence from 9 to 15 months of age. *Monogr Soc Res Child Dev, 63*(4), i–vi, 1–143.

Castelli, F., Happé, F., Frith, U., & Frith, C. (2000). Movement and mind: A functional imaging study of perception and interpretation of complex intentional movement patterns. *NeuroImage, 12*(3), 314–325.

Chein, J., Albert, D., O'Brien, L., Uckert, K., & Steinberg, L. (2011). Peers increase adolescent risk taking by enhancing activity in the brain's reward circuitry. *Dev Sci, 14*(2), F1–10.

Choudhury, S. (2010). Culturing the adolescent brain: What can neuroscience learn from anthropology? *Soc Cogn Affect Neurosci, 5*(2–3), 159–167.

Crone, E. A. (2013). Considerations of fairness in the adolescent brain. *Child Dev Perspect, 7*(2), 97–103.

Crone, E. A., & Dahl, R. E. (2012). Understanding adolescence as a period of social-affective engagement and goal flexibility. *Nat Rev Neurosci, 13*(9), 636–650.

Dumontheil, I., Apperly, I. A., & Blakemore, S.-J. (2010). Online usage of theory of mind continues to develop in late adolescence. *Dev Sci, 13*(2), 331–338.

Dumontheil, I., Hillebrandt, H., Apperly, I. A., & Blakemore, S.-J. (2012). Developmental differences in the control of action selection by social information. *J Cogn Neurosci, 24*(10), 2080–2095.

Ellis, B. J., Del Giudice, M., Dishion, T. J., Figueredo, A. J., Gray, P., Griskevicius, V., ... Wilson, D. S. (2012). The evolutionary basis of risky adolescent behavior: Implications for science, policy, and practice. *Dev Psychol, 48*(3), 598–623.

Engelmann, J. B., Moore, S., Monica Capra, C., & Berns, G. S. (2012). Differential neurobiological effects of expert advice on risky choice in adolescents and adults. *Soc Cogn Affect Neurosci, 7*(5), 557–567.

Ernst, M., Pine, D. S., & Hardin, M. (2006). Triadic model of the neurobiology of motivated behavior in adolescence. *Psychol Med, 36*(3), 299–312.

Farroni, T., Johnson, M. H., Menon, E., Zulian, L., Faraguna, D., & Csibra, G. (2005). Newborns' preference for face-relevant stimuli: Effects of contrast polarity. *Proc Natl Acad Sci USA, 102*(47), 17245–17250.

Fiske, S. T. (2009). Cultural processes. In G. G. Bernston & J. T. Cacioppo (Eds.), *Handbook of neuroscience for the behavioral sciences.* Hoboken, NJ: Wiley.

Fletcher, P. C., Happé, F., Frith, U., Baker, S. C., Dolan, R. J., Frackowiak, R. S., & Frith, C. D. (1995). Other minds in the brain: A functional imaging study of "theory of mind" in story comprehension. *Cognition, 57*(2), 109–128.

Frith, C. D. (2008). Social cognition. *Philos Trans R Soc Lond B Biol Sci, 363*(1499), 2033–2039.

Gallagher, H. L., Happé, F., Brunswick, N., Fletcher, P. C., Frith, U., & Frith, C. D. (2000). Reading the mind in cartoons and stories: An fMRI study of "theory of mind" in verbal and nonverbal tasks. *Neuropsychologia, 38*(1), 11–21.

Gardner, M., & Steinberg, L. (2005). Peer influence on risk taking, risk preference, and risky decision making in adolescence and adulthood: An experimental study. *Dev Psychol, 41*(4), 625–635.

Goddings, A.-L., Burnett Heyes, S., Bird, G., Viner, R. M., & Blakemore, S.-J. (2012). The relationship between puberty and social emotion processing. *Dev Sci, 15*(6), 801–811.

Gunther Moor, B., de Macks, Z. A. O., Güroğlu, B., Rombouts, S. A. R. B., van der Molen, M. W., & Crone, E. A. (2012). Neurodevelopmental changes of reading the mind in the eyes. *Soc Cogn Affect Neurosci, 7*(1), 44–52.

Güroğlu, B., van den Bos, W., & Crone, E. A. (2009). Fairness considerations: Increasing understanding of intentionality during adolescence. *J Exp Child Psychol, 104*(4), 398–409.

Keysar, B., Barr, D. J., Balin, J. A., & Brauner, J. S. (2000). Taking perspective in conversation: The role of mutual knowledge in comprehension. *Psychol Sci, 11*(1), 32–38.

Keysar, B., Lin, S., & Barr, D. J. (2003). Limits on theory of mind use in adults. *Cognition, 89*(1), 25–41.

Larson, R., & Richards, M. H. (1991). Daily companionship in late childhood and early adolescence: Changing developmental contexts. *Child Dev, 62*(2), 284–300.

Larson, R. W., Richards, M. H., Moneta, G., Holmbeck, G., & Duckett, E. (1996). Changes in adolescents' daily interactions with their families from ages 10 to 18: Disengagement and transformation. *Dev Psychol, 32*(4), 744–754.

Lenroot, R. K., Gogtay, N., Greenstein, D. K., Wells, E. M., Wallace, G. L., Clasen, L. S., ... Giedd, J. N. (2007). Sexual dimorphism of brain developmental trajectories during childhood and adolescence. *NeuroImage, 36*(4), 1065–1073.

Masten, C. L., Eisenberger, N. I., Borofsky, L. A., McNealy, K., Pfeifer, J. H., & Dapretto, M. (2011). Subgenual anterior cingulate responses to peer rejection: A marker of adolescents' risk for depression. *Dev Psychopathol, 23*(1), 283–292.

Masten, C. L., Eisenberger, N. I., Borofsky, L. A., Pfeifer, J. H., McNealy, K., Mazziotta, J. C., & Dapretto, M. (2009). Neural correlates of social exclusion during adolescence: Understanding the distress of peer rejection. *Soc Cogn Affect Neurosci, 4*(2), 143–157.

Masten, C. L., Eisenberger, N. I., Pfeifer, J. H., & Dapretto, M. (2010). Witnessing peer rejection during early adolescence: Neural correlates of empathy for experiences of social exclusion. *Soc Neurosci, 5*(5–6), 496–507.

Mills, K. L., Lalonde, F., Clasen, L. S., Giedd, J. N., & Blakemore, S.-J. (2013). Developmental changes in the structure of the social brain in late childhood and adolescence. *Soc Cogn Affect Neurosci, 9*(1), 123–131.

O'Brien, S. F., & Bierman, K. L. (1988). Conceptions and perceived influence of peer groups: Interviews with preadolescents and adolescents. *Child Dev, 59*(5), 1360–1365.

Pelphrey, K. A., & Carter, E. J. (2008). Charting the typical and atypical development of the social brain. *Dev Psychopathol, 20*(4), 1081–1102.

Peper, J. S., & Dahl, R. E. (2013). Surging hormones: Brain–behavior interactions during puberty. *Curr Dir Psychol Sci, 22*(2), 134–139.

Petanjek, Z., Judaš, M., Šimic, G., Rasin, M. R., Uylings, H. B. M., Rakic, P., & Kostovic, I. (2011). Extraordinary neoteny of synaptic spines in the human prefrontal cortex. *Proc Natl Acad Sci USA, 108*(32), 13281–13286.

Pfeifer, J. H., Lieberman, M. D., & Dapretto, M. (2007). "I know you are but what am I?!": Neural bases of self- and social knowledge retrieval in children and adults. *J Cogn Neurosci, 19*(8), 1323–1337.

Pfeifer, J. H., Masten, C. L., Borofsky, L. A., Dapretto, M., Fuligni, A. J., & Lieberman, M. D. (2009). Neural correlates of direct and reflected self-appraisals in adolescents and adults: When social perspective-taking informs self-perception. *Child Dev, 80*(4), 1016–1038.

Reyna, V. F. (2008). A theory of medical decision making and health: Fuzzy trace theory. *Med Decis Making, 28*(6), 850–865.

Reyna, V. F., & Adam, M. B. (2003). Fuzzy-trace theory, risk communication, and product labeling in sexually transmitted diseases. *Risk Anal, 23*(2), 325–342.

Reyna, V. F., & Farley, F. (2006). Risk and rationality in adolescent decision making: Implications for theory, practice, and public policy. *Psychol Sci Public Interest, 7*(1), 1–44.

Sebastian, C. L., Tan, G. C. Y., Roiser, J. P., Viding, E., Dumontheil, I., & Blakemore, S.-J. (2011). Developmental influences on the neural bases of responses to social rejection: Implications of social neuroscience for education. *NeuroImage, 57*(3), 686–694.

Sebastian, C., Viding, E., Williams, K. D., & Blakemore, S.-J. (2010). Social brain development and the affective consequences of ostracism in adolescence. *Brain Cogn, 72*(1), 134–145.

Sowell, E. R., Peterson, B. S., Thompson, P. M., Welcome, S. E., Henkenius, A. L., & Toga, A. W. (2003). Mapping cortical change across the human life span. *Nat Neurosci, 6*(3), 309–315.

Tamnes, C. K., Walhovd, K. B., Dale, A. M., Østby, Y., Grydeland, H., Richardson, G., ... Fjell, A. M. (2013). Brain development and aging: Overlapping and unique patterns of change. *NeuroImage, 68C*, 63–74.

van den Bos, W., van Dijk, E., Westenberg, M., Rombouts, S. A. R. B., & Crone, E. A. (2011). Changing brains, changing perspectives: The neurocognitive development of reciprocity. *Psychol Sci, 22*(1), 60–70.

van Leijenhorst, L., Westenberg, P. M., & Crone, E. A. (2008). A developmental study of risky decisions on the cake gambling task: Age and gender analyses of probability estimation and reward evaluation. *Dev Neuropsychol, 33*(2), 179–196.

van Leijenhorst, L., Zanolie, K., Van Meel, C. S., Westenberg, P. M., Rombouts, S. A. R. B., & Crone, E. A.

(2010). What motivates the adolescent? Brain regions mediating reward sensitivity across adolescence. *Cereb Cortex, 20*(1), 61–69.

Wang, A. T., Lee, S. S., Sigman, M., & Dapretto, M. (2006). Developmental changes in the neural basis of interpreting communicative intent. *Soc Cogn Affect Neurosci, 1*(2), 107–121.

Webb, S. J., Monk, C. S., & Nelson, C. A. (2001). Mechanisms of postnatal neurobiological development: Implications for human development. *Dev Neuropsychol, 19*(2), 147–171.

Yakovlev, P. A., & Lecours, I. R. (1967). The myelogenetic cycles of regional maturation of the brain. In A. Minkowski (Ed.), *Regional development of the brain in early life.* Oxford, UK: Blackwell.

5 Evolution of the Neural Bases of Higher Cognitive Function in Humans

JAMES K. RILLING AND DIETRICH STOUT

ABSTRACT Three of the most distinctive attributes of *Homo sapiens* are language, complex technology, and our degree of reliance on social learning. It is an open question whether these human faculties reflect a proliferation of distinct cognitive adaptations or the diverse behavioral expression of a smaller number of key underlying capacities. By comparing modern human brains with the brains of other living primate species, we can glean insights into the neurobiological changes that evolved to support these abilities. In particular, comparison with our closest living primate relative, the chimpanzee, is crucial for reaching conclusions about human brain evolution. The recent advent of noninvasive neuroimaging techniques has opened new possibilities for such comparative studies. These and other studies collectively suggest that language, technology, and aspects of social learning draw on overlapping regions of association cortex in prefrontal, temporal, and parietal regions that expanded and became functionally specialized during human evolution.

Like all primate species, humans have species-specific cognitive and behavioral attributes. Three of the most distinctive attributes of *Homo sapiens* are language, complex technology, and our facility with and propensity for social learning. While other species communicate, none does so by combining thousands of symbols according to a defined set of rules to generate phrases with a nearly infinite variety of meanings (Pinker, 2000). Humans are also uniquely technological. Although nonhuman primates are capable of simple tool use, using objects to implement motor to mechanical transformations that amplify movements of the upper limbs, humans are far superior at complex tool use, in which objects are used to convert movements of the hands into mechanical actions qualitatively different from manual actions (Frey, 2007). Humans also excel at tool making. Not even the most highly enculturated modern great apes approach the stone tool-making abilities of our ancestors 2.5 million years ago (Schick et al., 1999). Only in humans are complex tool use and tool making elaborated into truly "technological" systems involving socially coordinated action, intentional teaching, and the cultural accumulation of complexity. Our facility with technology thus depends in part on our capacity to learn from others through observation and imitation. Imitation involves reproducing both the end result of an observed action as well as the specific movements used to achieve it (Hecht et al., 2012; Visalberghi & Fragazy, 2002). Although chimpanzees are capable of imitation, they more commonly emulate, reproducing the end result of observed actions but not the specific movements used to achieve them (Tennie, Call, & Tomasello, 2009). Humans are exceptional imitators and will even over-imitate by reproducing movements in an observed action that do not contribute to reaching its end result (Whiten, McGuigan, Marshall-Pescini, & Hopper, 2009). This expertise in social learning is essential for the social transmission of knowledge upon which human culture is based (Boyd, Richerson, & Henrich, 2011). It is an open question to what degree human faculties for language, technology, and social learning reflect a proliferation of distinct cognitive adaptations as opposed to the diverse behavioral expression of a smaller number of key underlying capacities (Stout & Chaminade, 2012).

It is thus natural to ask how the human brain was modified throughout evolution to enable these unique faculties. The fossil record shows that brain size approximately tripled over the last 2.5 million years of human evolution (Holloway, Sherwood, Rilling, & Hof, 2008), and this may be part of the explanation. However, there may also have been important evolutionary changes in the internal organization of the brain, changes that would be difficult to identify in fossils (Holloway, 1968). To investigate internal changes, we must turn to the comparative study of the brains of living primate species. If we can identify a characteristic of the human brain that is not found in the brain of any closely related primate species, then we can infer that the trait evolved in the hominin lineage after we diverged from our common ancestor with chimpanzees some 5 to 7 million years ago. This approach renders the study of chimpanzees crucial for learning about human brain evolution: we cannot infer that a trait uniquely evolved in the human lineage unless it is absent in modern chimpanzees (Preuss, 2006).

We have extensive knowledge of rhesus macaque brain anatomy and physiology obtained through lesion studies, single-cell electrophysiology, and tracer studies. We would dearly like to have similarly detailed knowledge from humans and great apes, but these invasive methods cannot ethically be applied in humans and great apes. Fortunately, the recent advent of noninvasive neuroimaging techniques has opened new possibilities for comparative studies (Preuss, 2011).

These comparative studies have revealed multiple human brain specializations (Sherwood, Subiaul, & Zawidzki, 2008), beyond its large overall size. One such specialization that may be of particular importance to language, technology, and social learning is the larger proportion of the cortical surface dedicated to the classical association regions of the frontal, temporal, and parietal lobes compared to primary sensory or motor cortex in humans as opposed to nonhuman primates (Glasser, Goyal, Preuss, Raichle, & Van Essen, 2013; Preuss, 2011). At the risk of oversimplifying, this suggests that relatively more of the human cerebral cortex, compared with other living primate species, is dedicated to conceptual as opposed to perceptual processing. Below we provide evidence that human language, technology, and social learning each tap into similar expanded regions of association cortex in prefrontal, temporal, and parietal regions.

Language

Averaging 1330 cc, human brains are much larger than the brains of any other living primate species (Holloway et al., 2008). The brains of our closest living primate relatives, the great apes, range from an average of 405 cc in chimpanzees to 500 cc in gorillas (Holloway et al., 2008). Rhesus macaque brains average only 88 cc (Herndon, Tigges, Klumpp, & Anderson, 1998). Might our unique capacity for language simply be a product of our large brain size? Studies with high-functioning human microcephalics suggest that this is unlikely to be the whole explanation, as some have ape-sized brains but greater linguistic abilities than language-trained chimpanzees. Thus, there must be qualitative differences supporting the human propensity to acquire language (Holloway, 1968).

To identify human brain specializations related to language, we must consider the neural circuitry of language in humans and how it may differ from what is found in nonhuman primate brains. Geschwind's classic model of the functional neuroanatomy of language (Geschwind, 1970), although somewhat simplistic in light of current knowledge, is nevertheless a useful starting point for comparative analyses. The model postulates that a region in the posterior portion of the left superior temporal gyrus, Wernicke's area, is responsible for speech comprehension, while a region in the left inferior frontal cortex, Broca's area, is involved in speech production. These two regions are connected by a large white matter bundle known as the arcuate fasciculus. An obvious question is whether homologues of Broca's and Wernicke's areas exist in nonhuman primate brains, and it appears that they do, based on cytoarchitectonic similarities and shared nonlinguistic functional properties (Preuss, 2004). But are there any differences in the organization of these areas between humans and nonhuman primates?

Broca's Area Comparative evidence suggests that Broca's area (BA 44 and 45) and surrounding regions (e.g., BA 6) have had an important role in communication throughout anthropoid primate evolution (Corballis, 2009). Mirror neurons are found in macaque area F5, part of which is homologous to the posterior part of human Broca's area (area 44). These neurons presumably allow monkeys to better understand the actions of others by mapping those observed actions onto their own motor repertoire. Mirror neurons, best known for their response to reaching and grasping movements of both self and others, also fire when producing communicative mouth movements, or when observing them (Rizzolatti & Fogassi, 2007). Macaque monkeys rely very heavily on orofacial expressions for communication.

Mirror neuron regions seem to retain this function in orofacial communication in humans (Carr, Iacoboni, Dubeau, Mazziotta, & Lenzi, 2003); however, human Broca's area is additionally responsible for organizing linguistic actions, including speech. Could Broca's area homologue mediate the production of nonhuman primate vocalizations? In contrast to human speech, monkey calls are largely involuntary expressions of emotional arousal (Deacon, 1997), implying that they may not be under cortical control. Indeed, lesioning the macaque homologue of Broca's area does not interfere with vocalizations (Aitken, 1981). Monkey calls are instead mediated by the limbic system and brainstem. Chimpanzees, however, exhibit volitional calls in captivity (Hopkins, Taglialatela, & Leavens, 2007), and there is evidence that Broca's homologue could be involved in their production (Taglialatela, Russell, Schaeffer, & Hopkins, 2008). This raises the intriguing prospect that Broca's area was recruited for vocal, in addition to orofacial, communication at some point in hominoid evolution.

Still, human and chimpanzee Broca's areas have obvious functional differences, not just in the motor control of vocalizations but also in the involvement of human Broca's area with syntax (Vigneau et al., 2006), and we expect these to be supported by anatomical differences. Indeed, human Broca's area has wider cortical minicolumns than its great ape homologues (Schenker et al., 2008), whereas there is no difference between humans and great apes in the width of minicolumns in primary somatosensory, motor, or visual cortex (Semendeferi et al., 2011). Larger minicolumns have more space for connections, perhaps facilitating greater integration of information. Given the strong tendency for many aspects of language function to be lateralized to the left hemisphere, it may also be relevant that Broca's area is larger on the left than on the right in humans but not chimpanzees (Schenker et al., 2010).

Wernicke's Area In contrast to the production of vocalizations, vocalization comprehension seems to depend on similar neural substrates in humans and macaque monkeys. The left posterior superior temporal gyrus (i.e., Wernicke's area) is involved in human speech comprehension. Likewise, lesions of the left superior temporal gyrus of Japanese macaques are associated with selective impairment for discriminating species-specific vocalizations but not other types of auditory stimuli, whereas right hemisphere lesions do not produce these deficits (Heffner & Heffner, 1986). Additionally, monkey superior temporal cortex comprises three sets of auditory areas: a core region, a surrounding belt region, and an adjacent parabelt region (Hackett, Preuss, & Kaas, 2001). Single-cell electrophysiology studies reveal that whereas core areas respond best to pure tones, lateral belt areas respond best to complex sounds, including species-specific vocalizations (Rauschecker, Tian, & Hauser, 1995).

As with Broca's area, the human Wernicke's area has functional properties that distinguish it from its nonhuman homologue, such as phonological processing (Vigneau et al., 2006). Therefore, we again might expect to find underlying anatomical differences. Many years ago, it was discovered that the planum temporale, a portion of Wernicke's area, is leftwardly asymmetric in most humans (Geschwind & Levitsky, 1968). However, this same asymmetry has been found in great apes (Gannon, Holloway, Broadfield, & Braun, 1998; Hopkins, Marino, Rilling, & MacGregor, 1998). While these findings are consistent with the notion that human brain language systems evolved from homologues present in nonhuman primate ancestors, they also imply that planum temporale asymmetries do not mediate human-specific linguistic abilities. Despite this, at the microscopic level, potentially important specializations of human Wernicke's area are apparent. Similar to Broca's area, planum temporale minicolumns are wider in humans compared with chimpanzees, and also wider in the left than in the right hemisphere in humans but not chimpanzees (Buxhoeveden, Switala, Litaker, Roy, & Casanova, 2001).

Connections Between Wernicke's and Broca's Areas In addition to these structural differences in Broca's and Wernicke's areas, there are differences in the white matter connections between them that could help to account for the human capacity for language. The connectivity of the macaque homologues of Wernicke's and Broca's areas have been explored in detail using anterograde and retrograde tracer techniques (Deacon, 1992; Petrides & Pandya, 1988, 2002, 2009). Interestingly, the dominant frontal connection of Wernicke's homologue (area Tpt) is not with Broca's area homologue, but rather with dorsal prefrontal cortex (Petrides & Pandya, 1988). This pathway is involved with localizing sounds in space, as part of the auditory "where" pathway (Romanski et al., 1999). Macaques also have a second auditory pathway that is involved in auditory object recognition (Romanski et al., 1999). This pathway connects the anterior portion of Broca's area not with Wernicke's homologue (area Tpt), as might be expected, but instead with cortex of the middle superior temporal gyrus (Petrides & Pandya, 2002). This connection travels via the extreme capsule pathway, which is ventral to the auditory where pathway. A macaque homologue of the arcuate fasciculus has been described by Deacon (1992) and by Petrides and Pandya (2009). However, this is a relatively weak pathway in macaques, and its function is unknown.

The original descriptions of white matter fascicles in the human brain were based on gross dissections of post-mortem brains (Dejerine, 1895), which cannot precisely define the cortical terminations of axons. The invasive tracer methods used in macaques, which can identify cortical terminations, cannot be used in humans and great apes. Noninvasive diffusion tensor imaging (DTI), however, can be used to delineate major white matter fiber tracts in both humans and nonhuman primates. DTI reveals connections between Broca's and Wernicke's areas (or their homologues) in humans, chimpanzees, and rhesus macaques; however, there is one striking difference between humans and the other two species. In both rhesus macaques and chimpanzees, the posterior terminations of the arcuate are focused

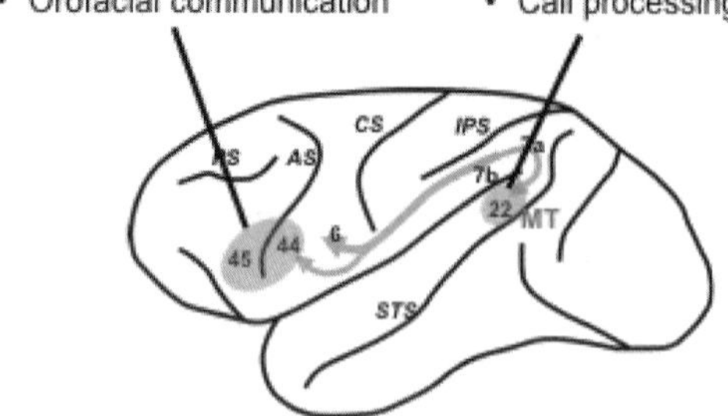

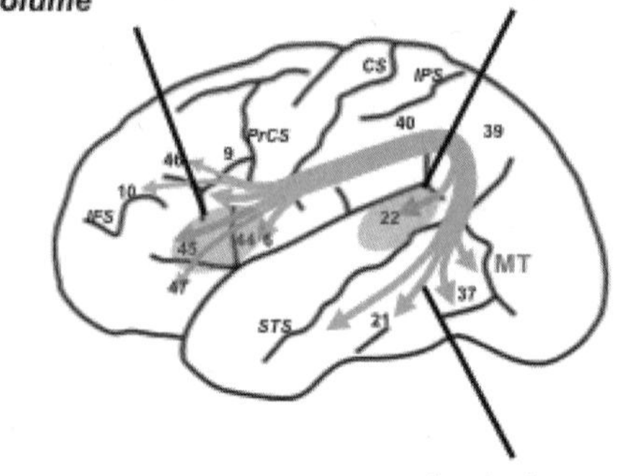

FIGURE 5.1 Schematic of language/communication systems in humans (right) and nonhuman primates (left). Neurological attributes are in italicized font. Cognitive attributes are in nonitalicized font. Human specializations are designated by bold font. MT: the middle-temporal visual motion area; PS: principal sulcus; AS: arcuate sulcus; CS: central sulcus; IPS: intraparietal sulcus; STS: superior temporal sulcus; PrCS: precentral sulcus; IFS: inferior frontal sulcus.

on the homologue of Wernicke's area in the posterior superior temporal gyrus. Humans, however, also possess a massive projection of the arcuate into the middle and inferior temporal gyri, ventral to classic Wernicke's area (Rilling et al., 2008).

These projections lie within a region of temporal association cortex that seems to have expanded in human evolution, displacing the nearby extrastriate visual cortex in the process. For example, the middle temporal visual motion area, or MT, lies within the superior temporal sulcus (STS) in chimpanzees and rhesus macaques. In humans, however, MT is located posterior to the STS, which instead contains association cortex (Glasser et al., 2013). The posterior limit of temporal-lobe arcuate fasciculus terminations in the human brain coincides very closely with the anterior limit of visual motion area MT, consistent with displacement of MT by the highly expanded arcuate pathway.

The region of expanded cortex that receives arcuate projections has been dubbed an "epicenter for lexical-semantic processing," based on lesion, functional MRI (fMRI), and structural and functional connectivity data (Turken & Dronkers, 2011). Thus, this portion of the arcuate fasciculus may carry lexical-sematic information to Broca's area. Although some have postulated that evolution recruited the extreme capsule pathway into the language system and that this pathway has played a key role in language evolution, comparative DTI data suggest that the evolutionary expansion of the arcuate fasciculus outstripped that of the extreme capsule, implicating the arcuate fasciculus as the key substrate for human language evolution. Furthermore, the human arcuate fasciculus is leftwardly asymmetric, while the extreme capsule is not (Rilling, Glasser, Jbabdi, Andersson, & Preuss, 2011).

In summary, specializations of the human brain that may be relevant to explaining our capacity for language include (1) wider cortical minicolumns in both Broca's and Wernicke's areas, (2) leftward asymmetries in Broca's area volume and PT minicolumn width, and 3) arcuate fasciculus projections beyond Wernicke's area to a region of expanded association cortex in the middle and inferior temporal cortex involved in processing word meaning (figure 5.1).

Technology

Many species make and use tools, but humans are distinguished by the extent and complexity of their technological behavior. In fact, the term "technology" properly refers to a complex assemblage of artifacts, institutions, social relations, knowledge, and skills that has no strict counterpart in nonhumans. Comparative studies are critical to unraveling the evolutionary origins of this uniquely human mode of behavior. Technological behavior may be heuristically divided into simple tool use, complex tool use, and tool making.

SIMPLE TOOL USE In humans and macaques, grasping and manipulation depend on circuits that connect the intraparietal sulcus with the inferior frontal cortex via

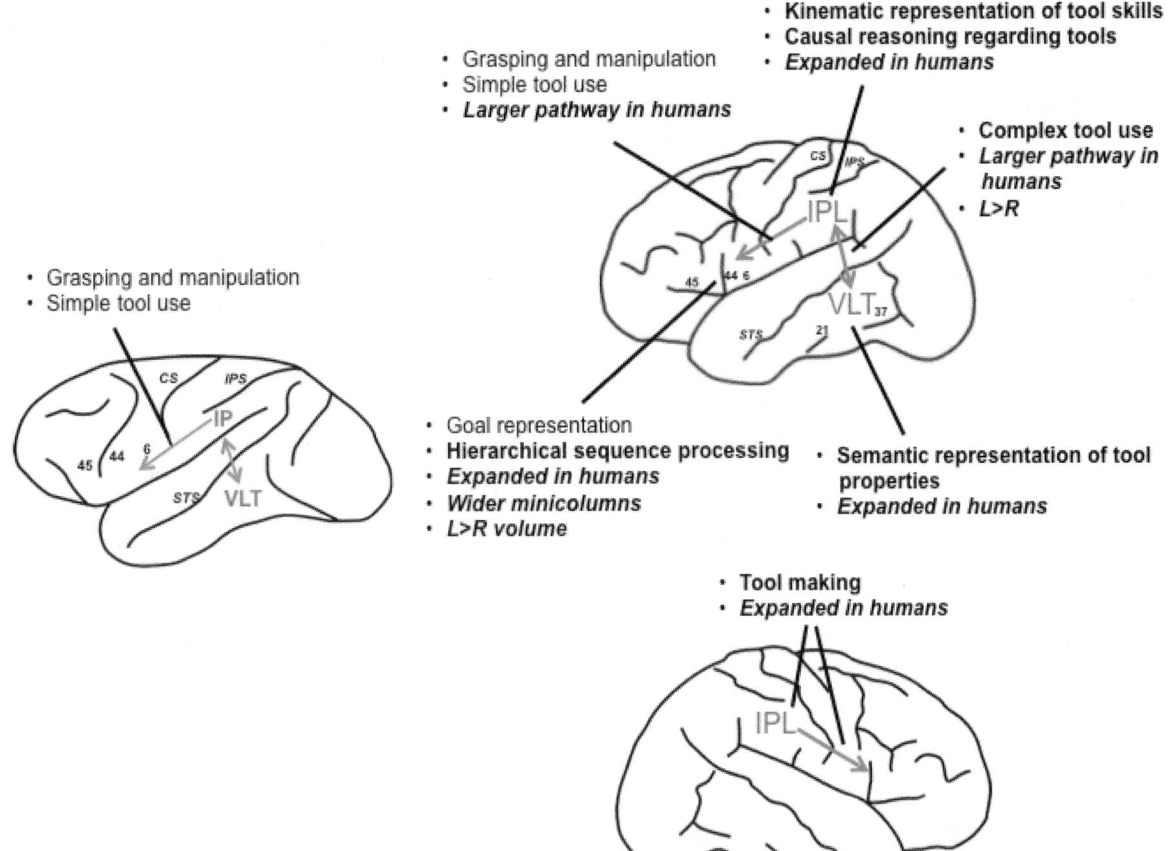

Figure 5.2 Schematic of tool-use systems in humans (right) and nonhuman primates (left). Conventions are the same as in figure 5.1. VLT: ventral lateral temporal lobe; IP: inferior parietal lobe.

the third branch of the superior longitudinal fasciculus (SLF III). As macaques learn to use rakes to extend their reach, the receptive fields of visuotactile neurons in the intraparietal sulcus increase over time (Iriki, 2006). It has been argued that this reflects an extension of the body schema to incorporate the handheld tool, and that simple tool use may be accounted for by straightforward extension of the sensorimotor mechanisms involved in prehension.

Complex Tool Use Body schema extension is insufficient to explain the use of complex tools that alter the functional properties of the hand (e.g., knives, hammers, potholders), which require an additional causal understanding of tool properties as distinct from the hand (Frey, 2007). Complex tool use is understood with reference to the "two-streams" account of visual perception (Milner & Goodale, 1995). A "dorsal stream" from extrastriate visual cortices to the posterior parietal lobe supports visuospatial-motor transformations for action, whereas a "ventral stream" from occipital to posterior temporal cortices maps visual percepts to stored semantic knowledge about tool form and function. These streams converge in the inferior parietal lobule (IPL) to integrate the action and semantic knowledge required for complex tool use. This information is communicated via SLF III to the inferior frontal cortex, which supports hierarchically structured action sequencing during production and perception of object-directed actions (Fadiga, Craighero, & D'Ausilio, 2009). Below, we present evidence that these tool-relevant regions of temporal, parietal, and frontal cortex are relatively larger and more extensively interconnected in humans than in other primates (figure 5.2).

Evidence for temporal lobe expansion is discussed above in the context of language. Human IPL is similarly expanded relative to macaques (Orban et al., 2006) and chimpanzees (Glasser et al., 2013). Finally, human Broca's area displays wider minicolumns than its ape homologue (Schenker et al., 2008), is one of the most greatly expanded cortical regions in humans compared to apes, particularly in the left hemisphere (Schenker et al., 2010), and displays greater bilateral connectivity with IPL via SLF III (Hecht et al., 2012).

The cortical control of tool use is left-lateralized in humans (Johnson-Frey, Newman-Norlund, & Grafton, 2005), further stimulating interest in possible relations between tool use, handedness, and the evolution of left hemisphere language circuits. According to Frey (2007), complex tool use is enabled by integration of semantic representations in the left middle temporal gyrus with kinematic representations in left IPL. Peeters et al. (2009) reported a region of left rostral IPL that is active when humans, but not tool-experienced monkeys, view actions performed with simple tools. This region may be involved in representing the causal properties of handheld tools. DTI evidence indicates that the direct white matter connection between middle temporal gyrus and IPL is left-lateralized in humans (Ramayya, Glasser, & Rilling, 2010), and is relatively larger in both hemispheres in humans compared with chimpanzees and macaques (Hecht et al., 2012).

Tool Making Although many species make and use tools, humans are distinguished by the greater sophistication of their tool making, including the use of tools to make other tools, the construction of multicomponent tools, and the cultural accumulation of complexity in tool design (Frey, 2007). On this view, the use of tools to transform durable objects is the most distinctive element of human tool behavior. How does the human brain represent the more abstract and temporally extended goals enabled by transformation of durable objects? Available evidence suggests that, in contrast to tool use, representation of such material outcomes may be right-lateralized in humans.

Hamilton and Grafton (2008) used repetition suppression to demonstrate that, while repeated presentation of the same grasping target objects produces suppression of left inferior frontal cortex and IPL, repetition of demonstrated material action outcomes (e.g., closing or opening a box) yields suppression of right inferior frontal cortex and IPL. Frey and Gerry (2006) found that the right intraparietal sulcus was the only brain region where activity predicted successful imitation of an object-construction sequence, and Hartmann et al. (2005) report that right-brain damage is associated with deficits in the organization of multistep technical actions. PET and fMRI investigations of Paleolithic stone tool making by modern experimental subjects (Stout, Passingham, Frith, Apel, & Chaminade, 2011; Stout, Toth, Schick, & Chaminade, 2008) also reveal right hemisphere activation. Stout et al. (2008) found greater activation of right IPL and inferior frontal cortex when making more sophisticated tools, a result mirrored by fMRI data from the observation of tool production (Stout et al., 2011). As tasks were matched for manipulative complexity (Faisal, Stout, Apel, & Bradley, 2010), this result is attributable to differences in the complexity of object transformations.

Functional studies of tool making, though sparse, thus implicate right hemisphere homologues of left hemisphere inferior frontal and parietal regions involved in tool use and may be underwritten by the bilateral expansion and increased interconnectivity of these regions. Interestingly, Ramayya et al. (2010) report DTI evidence of a strong rightward asymmetry in the human white matter tract connecting posterior IPL and inferior frontal cortex (although anterior IPL to IFG is larger on the left). The differential lateralization of tool use versus tool making parallels that of small-scale (e.g., phonology, syntax) versus large-scale (e.g., prosody, context) linguistic processing and may reflect a more general hemispheric division of labor between rapid, routine action control on the left and large-scale integration on the right (Stout & Chaminade, 2012).

Social learning

The intersubjectivity upon which human social learning depends involves simulation of others' actions, thoughts, and feelings. Here, we restrict our comments to action simulation that is believed to mediate imitation, because the underlying neurobiology overlaps that of language and technology, and because humans have a stronger propensity to imitate than do chimpanzees or macaques.

The putative ability of the mirror-neuron system to map the observed actions of others onto one's own motor repertoire makes it well suited to support imitation (Molenberghs, Cunnington, & Mattingley, 2009). The mirror-neuron system consists of the STS, inferior parietal cortex, and inferior frontal gyrus. STS processes the visual motion of observed actions and is thought to then relay this information to mirror regions in the inferior parietal cortex and inferior frontal gyrus, allowing the observer to covertly simulate the observed action. Species differences in the mirror-neuron system might explain species differences in social learning abilities (Hecht et al., 2012). For example, human but not monkey mirror regions will respond to intransitive actions such as mimed grasping movements that lack an object (Rizzolatti & Fogassi, 2007). Additionally, DTI has revealed species differences in connections among components of the mirror-neuron system. In both macaques and chimpanzees, most connections between STS and the inferior frontal gyrus travel via the ventral extreme capsule pathway mentioned earlier. In humans, by contrast, a substantial number of connections travel by a dorsal route between STS and inferior parietal

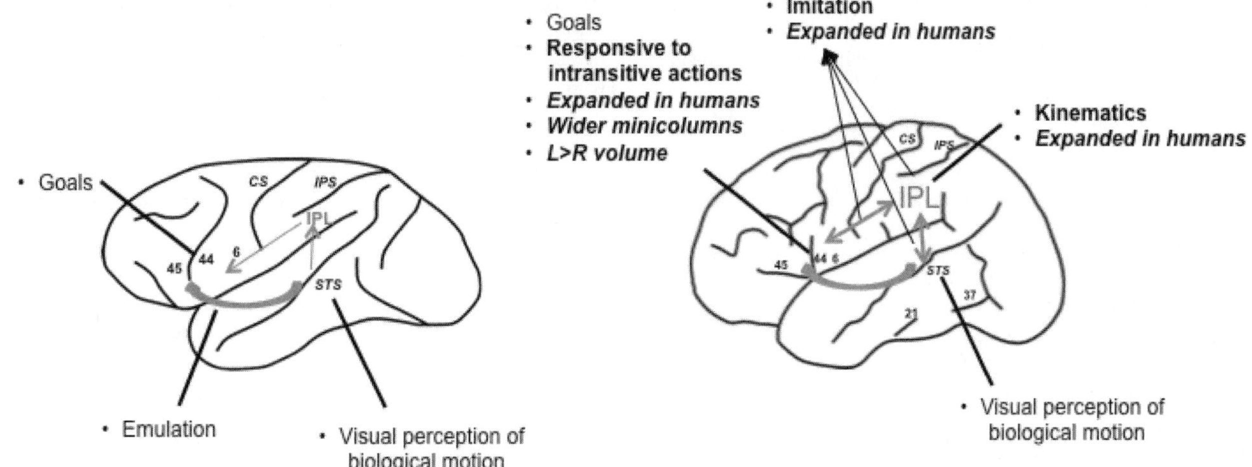

FIGURE 5.3 Schematic of social learning system in humans (right) and nonhuman primates (left). Conventions are the same as in previous figures.

cortex (via the middle/inferior longitudinal fasciculus), and between inferior parietal cortex and inferior frontal gyrus (via SLF III; figure 5.3). If IFG processes the goals of actions, then the ventral connections that dominate macaque and chimpanzee brains would be involved mainly with linking the observed action with its goal or intention. These pathways may primarily support emulation. Inferior parietal cortex is involved in the spatial mapping of movement and may be useful for extracting more kinematic details from observed actions. This pathway might support reproducing the specific actions that lead to the goal, and its greater development in humans may explain our penchant for imitation (Hecht et al., 2012). Thus, the connectivity of the inferior parietal cortex, another region of association cortex that appears to have expanded in human evolution (Bruner, 2004; Glasser et al., 2013; Orban et al., 2006), emerges as potentially crucial for the human capacity for learning through imitation.

Conclusion

The distinctive attributes of human language, complex technology, and reliance on social learning draw on overlapping regions of association cortex in prefrontal, temporal, and parietal regions that have expanded and become functionally specialized during human evolution. Specializations supporting human language include microstructural specializations of Wernicke's and Broca's areas, as well as expansion of ventral lateral temporal cortex involved in lexical-semantic processing and its connection with Broca's area via the arcuate fasciculus. Semantic knowledge of tools seems also to be represented within the expanded ventral lateral temporal cortex; however, skill is represented in inferior parietal cortex, which is functionally specialized for specific aspects of tool use. Both the inferior parietal cortex and the pathway linking it with ventral lateral temporal cortex, which presumably integrates these dorsal and ventral streams of information about tools, have also expanded disproportionately in human evolution. Whereas tool use seems to be lateralized to left hemisphere parietofrontal systems, tool making is more reliant on the right hemisphere homologues of these areas. Finally, imitation may depend on connections between the superior temporal sulcus and the inferior frontal gyrus that run via the inferior parietal lobe, a pathway that is better developed in humans than nonhuman primates. More fine-grained studies are needed to determine if the cortical substrates supporting these different human specializations are indeed the same, or merely in close proximity to one another.

REFERENCES

AITKEN, P. G. (1981). Cortical control of conditioned and spontaneous vocal behavior in rhesus monkeys. *Brain Lang, 13*(1), 171–184.

BOYD, R., RICHERSON, P. J., & HENRICH, J. (2011). The cultural niche: Why social learning is essential for human adaptation. *Proc Natl Acad Sci USA, 108*(Suppl 2), 10918–10925.

BRUNER, E. (2004). Geometric morphometrics and paleoneurology: Brain shape evolution in the genus *Homo. J Hum Evol, 47*(5), 279–303.

BUXHOEVEDEN, D. P., SWITALA, A. E., LITAKER, M., ROY, E., & CASANOVA, M. F. (2001). Lateralization of minicolumns in human planum temporale is absent in nonhuman primate cortex. *Brain Behav Evol, 57*(6), 349–358.

Carr, L., Iacoboni, M., Dubeau, M. C., Mazziotta, J. C., & Lenzi, G. L. (2003). Neural mechanisms of empathy in humans: A relay from neural systems for imitation to limbic areas. *Proc Natl Acad Sci USA, 100*(9), 5497–5502.

Corballis, M. C. (2009). The evolution of language. *Ann NY Acad Sci, 1156*, 19–43.

Deacon, T. W. (1992). Cortical connections of the inferior arcuate sulcus cortex in the macaque brain. *Brain Res, 573*(1), 8–26.

Deacon, T. W. (1997). What makes the human brain different? *Annu Rev Anthropol, 26*, 337–357.

Dejerine, J. (1895). *Anatomie des centres nerveux.* Paris: Rueff et Cie.

Fadiga, L., Craighero, L., & D'Ausilio, A. (2009). Broca's area in language, action, and music. *Ann NY Acad Sci, 1169*(1), 448–458.

Faisal, A., Stout, D., Apel, J., & Bradley, B. (2010). The manipulative complexity of lower Paleolithic stone toolmaking. *PLoS One, 5*(11), e13718.

Frey, S. H. (2007). What puts the how in where? Tool use and the divided visual streams hypothesis. *Cortex, 43*(3), 368–375.

Frey, S. H., & Gerry, V. (2006). Modulation of neural activity during observational learning of action and their sequential orders. *J Neurosci, 26*(51), 13194–13201.

Gannon, P. J., Holloway, R. L., Broadfield, D. C., & Braun, A. R. (1998). Asymmetry of chimpanzee planum temporale: Humanlike pattern of Wernicke's brain language area homolog. *Science, 279*(5348), 220–222.

Geschwind, N. (1970). The organization of language and the brain. *Science, 170*(961), 940–944.

Geschwind, N., & Levitsky, W. (1968). Human brain: Left-right asymmetries in temporal speech region. *Science, 161*, 186–187.

Glasser, M. F., Goyal, M. S., Preuss, T. M., Raichle, M. E., & Van Essen, D. C. (2013). Trends and properties of human cerebral cortex: Correlations with cortical myelin content. *NeuroImage*, in press. http://dx.doi.org/10.1016/j.neuroimage.2013.03.060

Hackett, T. A., Preuss, T. M., & Kaas, J. H. (2001). Architectonic identification of the core region in auditory cortex of macaques, chimpanzees, and humans. *J Comp Neurol, 441*(3), 197–222.

Hamilton, A. F. d. C., & Grafton, S. T. (2008). Action outcomes are represented in human inferior frontoparietal cortex. *Cereb Cortex, 18*, 1160–1168.

Hartmann, K., Goldenberg, G., Daumuller, M., & Hermsdorfer, J. (2005). It takes the whole brain to make a cup of coffee: The neuropsychology of naturalistic actions involving technical devices. *Neuropsychologia, 43*, 625–637.

Hecht, E. E., Gutman, D. A., Preuss, T. M., Sanchez, M. M., Parr, L. A., & Rilling, J. K. (2012). Process versus product in social learning: Comparative diffusion tensor imaging of neural systems for action execution-observation matching in macaques, chimpanzees, and humans. *Cereb Cortex, 23*(5), 1014–1024.

Heffner, H. E., & Heffner, R. S. (1986). Effect of unilateral and bilateral auditory cortex lesions on the discrimination of vocalizations by Japanese macaques. *J Neurophysiol, 56*(3), 683–701.

Herndon, J. G., Tigges, J., Klumpp, S. A., & Anderson, D. C. (1998). Brain weight does not decrease with age in adult rhesus monkeys. *Neurobiol Aging, 19*(3), 267–272.

Holloway, R. L. (1968). The evolution of the primate brain: Some aspects of quantitative relations. *Brain Res, 7*, 121–172.

Holloway, R. L., Sherwood, C. C., Hof, P. R., & Rilling, J. K. (2008). Evolution of the brain in humans—paleoneurology. In M. D. Binder, N. Hirokawa, & U. Windhorst (Eds.), *Encyclopedia of neuroscience* (pp. 1326–1334). Berlin: Springer-Verlag.

Hopkins, W. D., Marino, L., Rilling, J. K., & MacGregor, L. A. (1998). Planum temporale asymmetries in great apes as revealed by magnetic resonance imaging (MRI). *Neuroreport, 9*(12), 2913–2918.

Hopkins, W. D., Taglialatela, J., & Leavens, D. A. (2007). Chimpanzees differentially produce novel vocalizations to capture the attention of a human. *Anim Behav, 73*(2), 281–286.

Iriki, A. (2006). The neural origins and implications of imitation, mirror-neurons and tool use. *Curr Opin Neurobiol, 16*(6), 660–667.

Johnson-Frey, S. H., Newman-Norlund, R., & Grafton, S. T. (2005). A distributed left hemisphere network active during planning of everyday tool use skills. *Cereb Cortex, 15*(6), 681–695.

Milner, A. D., & Goodale, M. A. (1995). *The visual brain in action.* Oxford, UK: Oxford University Press.

Molenberghs, P., Cunnington, R., & Mattingley, J. B. (2009). Is the mirror-neuron system involved in imitation? A short review and meta-analysis. *Neurosci Biobehav Rev, 33*(7), 975–980.

Orban, G. A., Claeys, K., Nelissen, K., Smans, R., Sunaert, S., Todd, J. T., … Vanduffel, W. (2006). Mapping the parietal cortex of human and non-human primates. *Neuropsychologia, 44*(13), 2647–2667.

Peeters, R., Simone, L., Nelissen, K., Fabbri-Destro, M., Vanduffel, W., Rizzolatti, G., & Orban, G. A. (2009). The representation of tool use in humans and monkeys: Common and uniquely human features. *J Neurosci, 29*(37), 11523–11539.

Petrides, M., & Pandya, D. N. (1988). Association fiber pathways to the frontal cortex from the superior temporal region in the rhesus monkey. *J Comp Neurol, 273*(1), 52–66.

Petrides, M., & Pandya, D. N. (2002). Comparative cytoarchitectonic analysis of the human and the macaque ventrolateral prefrontal cortex and corticocortical connection patterns in the monkey. *Eur J Neurosci, 16*(2), 291–310.

Petrides, M., & Pandya, D. N. (2009). Distinct parietal and temporal pathways to the homologues of Broca's area in the monkey. *PLoS Biol, 7*(8), e1000170.

Pinker, S. (2000). *The language instinct: How the mind creates language.* New York, NY: Perennial Classics.

Preuss, T. M. (2004). What is it like to be a human? In M. S. Gazzaniga (Ed.), *The cognitive neurosciences* (pp. 5–22). Cambridge, MA: MIT Press.

Preuss, T. M. (2006). Who's afraid of *Homo sapiens*? *J Biomed Discov Collab, 1*, 17.

Preuss, T. M. (2011). The human brain: Rewired and running hot. *Ann NY Acad Sci, 1225*(Suppl 1), E182–191.

Ramayya, A. G., Glasser, M. F., & Rilling, J. K. (2010). A DTI investigation of neural substrates supporting tool use. *Cereb Cortex, 20*(3), 507–516.

Rauschecker, J. P., Tian, B., & Hauser, M. (1995). Processing of complex sounds in the macaque nonprimary auditory cortex. *Science, 268*(5207), 111–114.

Rilling, J. K., Glasser, M. F., Jbabdi, S., Andersson, J., & Preuss, T. M. (2011). Continuity, divergence, and the evolution of brain language pathways. *Front Evol Neurosci, 3*, 11.

Rilling, J. K., Glasser, M. F., Preuss, T. M., Ma, X., Zhao, T., Hu, X., & Behrens, T. E. (2008). The evolution of the arcuate fasciculus revealed with comparative DTI. *Nat Neurosci, 11*(4), 426–428.

Rizzolatti, G., & Fogassi, L. (2007). Mirror neurons and social cognition. In R. I. M. Dunbar & L. Barrett (Eds.), *The Oxford handbook of evolutionary psychology* (pp. 179–196). Oxford, UK: Oxford University Press.

Romanski, L. M., Tian, B., Fritz, J., Mishkin, M., Goldman-Rakic, P. S., & Rauschecker, J. P. (1999). Dual streams of auditory afferents target multiple domains in the primate prefrontal cortex. *Nat Neurosci, 2*(12), 1131–1136.

Schenker, N. M., Buxhoeveden, D. P., Blackmon, W. L., Amunts, K., Zilles, K., & Semendeferi, K. (2008). A comparative quantitative analysis of cytoarchitecture and minicolumnar organization in Broca's area in humans and great apes. *J Comp Neurol, 510*(1), 117–128.

Schenker, N. M., Hopkins, W. D., Spocter, M. A., Garrison, A. R., Stimpson, C. D., Erwin, J. M., … Sherwood, C. C. (2010). Broca's area homologue in chimpanzees (*Pan troglodytes*): Probabilistic mapping, asymmetry, and comparison to humans. *Cereb Cortex, 20*(3), 730.

Schick, K. D., Toth, N., Garufi, G., Savage-Rumbaugh, E. S., Rumbaugh, D., & Sevcik, R. (1999). Continuing investigations into the stone tool-making and tool-using capabilities of a bonobo (*Pan paniscus*). *J Archaeol Sci, 26*, 821–832.

Semendeferi, K., Teffer, K., Buxhoeveden, D. P., Park, M. S., Bludau, S., Amunts, K., … Buckwalter, J. (2011). Spatial organization of neurons in the frontal pole sets humans apart from great apes. *Cereb Cortex, 21*(7), 1485–1497.

Sherwood, C. C., Subiaul, F., & Zawidzki, T. W. (2008). A natural history of the human mind: Tracing evolutionary changes in brain and cognition. *J Anat, 212*(4), 426–454.

Stout, D., & Chaminade, T. (2012). Stone tools, language and the brain in human evolution. *Philos Trans R Soc Lond B Biol Sci, 367*(1585), 75–87.

Stout, D., Passingham, R., Frith, C., Apel, J., & Chaminade, T. (2011). Technology, expertise and social cognition in human evolution. *Eur J Neurosci, 33*(7), 1328–1338.

Stout, D., Toth, N., Schick, K. D., & Chaminade, T. (2008). Neural correlates of early Stone Age tool-making: Technology, language and cognition in human evolution. *Philos Trans R Soc Lond B Biol Sci, 363*, 1939–1949.

Taglialatela, J. P., Russell, J. L., Schaeffer, J. A., & Hopkins, W. D. (2008). Communicative signaling activates "Broca's" homolog in chimpanzees. *Curr Biol, 18*(5), 343–348.

Tennie, C., Call, J., & Tomasello, M. (2009). Ratcheting up the ratchet: On the evolution of cumulative culture. *Philos Trans R Soc Lond B Biol Sci, 364*(1528), 2405–2415.

Turken, A. U., & Dronkers, N. F. (2011). The neural architecture of the language comprehension network: Converging evidence from lesion and connectivity analyses. *Front Syst Neurosci, 5*, 1.

Vigneau, M., Beaucousin, V., Herve, P. Y., Duffau, H., Crivello, F., Houde, O., … Tzourio-Mazoyer, N. (2006). Meta-analyzing left hemisphere language areas: Phonology, semantics, and sentence processing. *Neuroimage, 30*(4), 1414–1432.

Visalberghi, E., & Fragazy, D. (2002). Do monkeys ape? Ten years after. In K. Dautenhahn & C. L. Nehaniv (Eds.), *Imitation in animals and artifacts* (pp. 471–499). Cambridge, MA: MIT Press.

Whiten, A., McGuigan, N., Marshall-Pescini, S., & Hopper, L. M. (2009). Emulation, imitation, over-imitation and the scope of culture for child and chimpanzee. *Philos Trans R Soc Lond B Biol Sci, 364*(1528), 2417–2428.

6 The Primate Prefrontal Cortex in Comparative Perspective

STEVEN P. WISE

ABSTRACT In *The Neurobiology of the Prefrontal Cortex*, Passingham and Wise build on the work of Preuss and Goldman-Rakic, who found that new prefrontal areas emerged during primate evolution. We propose that the new prefrontal areas of early primates perform search and valuation functions, and that later, in anthropoid primates, additional prefrontal areas evolved to perform related functions. The new anthropoid areas augment ancestral reinforcement-learning mechanisms, which use feedback events to update the state of memories through the cumulative adjustment of associations, averaged over many events. In contrast, the anthropoid prefrontal areas store the memory of single events and use event representations to reduce the number of unproductive, risky, or costly foraging choices. They do so by using the memory of events to speed learning and by using event representations to generate goals from abstract rules and strategies, thus bringing previous solutions to bear on novel problems.

Passingham and Wise (2012) present a proposal on the fundamental function of the primate prefrontal cortex (PF), which this chapter presents in summary form. Figure 6.1 illustrates the terminology used to explain our idea. Primates include strepsirrhines and haplorhines; haplorhines consist of tarsiers and anthropoids; anthropoids include platyrrhines (New World monkeys) and catarrhines; and catarrhines comprise Old World monkeys, humans, and apes (figure 6.1B). The term "prosimian" refers to tarsiers and strepsirrhines. Figure 6.1A recognizes seven subdivisions of PF: caudal (PFc), orbital (PFo), ventral (PFv), dorsolateral PF (PFdl), dorsal (PFd), medial (PFm), and polar (PFp).

Early primate evolution

Most primatologists agree that early primates adapted to life in the fine branches of angiosperm trees and that their innovations included grasping hands and feet, nails instead of claws, and frontally directed eyes (Fleagle, 1999; Rose, 2006). These adaptations probably supported visually guided leaping (Martin, 1990) and foraging (Cartmill, 1972), and fossil evidence suggests that the grasping specializations arose first (Bloch & Boyer, 2002). Early primates foraged nocturnally (Rose, 2006), and their visual advances probably provided improved depth perception and object detection, as well as vision below and in front of their snout (Allman, 2000; Barton, 1998, 2004; Changizi, 2009).

Among their motor specializations, early primates used their hindlimbs more in locomotion than other mammals and generated less force to move from place to place (Larson, 1998; Schmitt, 2010). The latter trait minimized predator-attracting movements of flimsy branches, and the former enabled forelimb specializations for reaching and grasping. Like many modern prosimians, early primates probably used a hand-to-mouth feeding technique (MacNeilage, Studdert-Kennedy, & Lindblom, 1987), one that required coordination among head, mouth, and hand movements (Wise, 2007). Primates also reach in a unique way (Shadmehr & Wise, 2005), which depends on a suite of new cortical areas (Preuss, 2007a, 2007b), a topic taken up later.

Anthropoid evolution

Early anthropoids engaged in diurnal foraging, as indicated by their relatively small eyes (Fleagle, 1999; Heesy & Ross, 2004). Daylight foraging increased the risk of predation, as well as competition from other diurnal animals, such as frugivorous birds. Anthropoids inherited the primate fovea, which emerged in haplorhines (Ross, 2004), and amplified several traits of early primates, such as large brains, long lives, and frontally directed eyes. Later anthropoids evolved various forms of trichromatic vision, with catarrhines developing the "routine" form that depends on three distinct photopigment genes (Williams, Kay, & Kirk, 2010).

Early anthropoids were small animals, approximately 100 grams, and tooth-wear patterns suggest that they foraged mostly on fruits and insects (Fleagle, 1999; Rose, 2006). An increase in size began around 34 million years ago, after the divergence of platyrrhines and catarrhines (Williams et al., 2010). The first of these larger anthropoids were probably arboreal quadrupeds (Fleagle, 1999; Kay, Williams, Ross, Takai, &

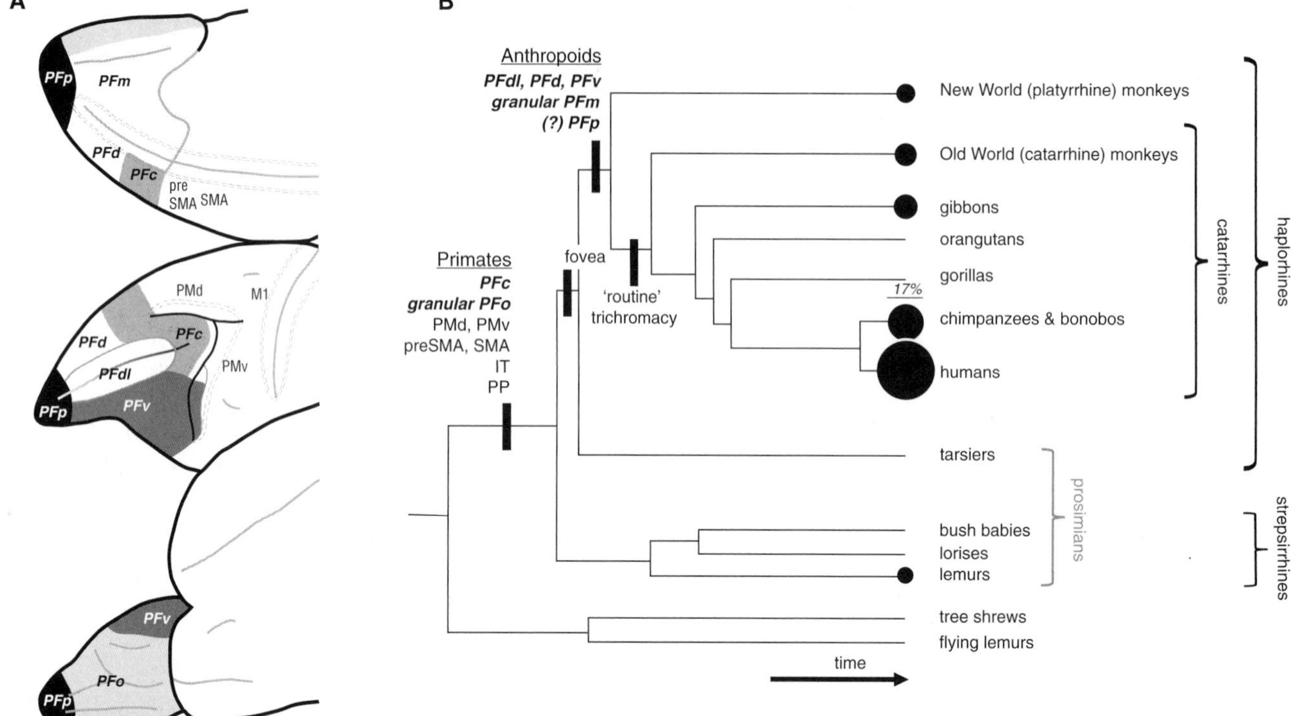

FIGURE 6.1 (A) PF regions in macaques. (B) Primate cladogram showing shared derived traits (black bars). The gray lettering and bracket in (B) indicate a paraphyletic group. Circle diameter proportional to percentage of cortex comprising granular PF. Abbreviations: IT, inferotemporal areas; PP, posterior parietal areas; PF, prefrontal cortex; PM, premotor cortex; M1, primary motor cortex; SMA, supplementary motor area; c, caudal; d, dorsal; dl, dorsolateral; o, orbital; p, polar; m, medial; v, ventral. Lines in (A) reproduced from Petrides and Pandya (2007) by permission of the Society for Neuroscience. (A) and (B) reproduced from Passingham and Wise (2012) by permission of Oxford University Press.

Shigehara, 2004), a trait that many modern anthropoids have retained (Schmitt, 2010). This kind of locomotion requires considerable energy, especially with frequent changes in elevation.

As anthropoids increased in size, they began to forage over a larger home range, an energy-intensive activity (Martin, 1981). To meet their needs, they probably obtained much of their nutrition from ripe, high-energy fruits and tender, easily digestible leaves. Like modern anthropoids, these animals foraged among thousands of angiosperm trees, few of which had ripe fruit on any given day. Among their many foraging challenges, trees of a given species generated fruit at various times and frequencies, with substantial annual variation and ripening rates that varied by location (Janmaat, Byrne, & Zuberbühler, 2006). When shortfalls in preferred nutrients occurred, competition intensified, as did the risk of predation.

Cortical evolution in primates

EARLY PRIMATES As early primates adapted to a nocturnal, fine-branch niche, they developed several new premotor and posterior parietal areas, along with a dozen or more new visual and auditory areas (Kaas, 2012). In addition, a new part of the PF cortex emerged, the main topic of this chapter. As Preuss (1995) pointed out, this part of the PF cortex has a prominent internal granular layer, and so can be called the granular PF cortex. This name, however, reflects just one of its diagnostic traits; as reviewed elsewhere, extensive additional evidence confirms that the granular PF cortex occurs uniquely in primates (Passingham & Wise, 2012; Preuss, 1995, 2007a, 2007b; Wise, 2008).

Preuss and Goldman-Rakic (1991a, 1991b) provided the key support for this idea. They concluded that agranular PF areas have homologues in all mammals and that PFc and PFo emerged as new granular areas in early primates. First, they identified the granular PF cortex in the bush baby, a strepsirrhine primate. As in anthropoids, these granular areas are situated rostral to the premotor cortex (area 6), a group of agranular areas. Second, they compared the connections, cytoarchitectonics, and myeloarchitectonics of the frontal cortex in bush babies with anthropoid monkeys (macaques) and other mammals. They concluded that one part of the bush baby PF cortex was homologous with area 8 of macaque monkeys, which Passingham and Wise (2012) call PFc. This area has a "very thick, dense" layer 4 and bundles of vertically oriented myelinated fibers in the infragranular layers, which disperse as they enter the supragranular cortex (Preuss & Goldman-Rakic, 1991b, p. 459). Preuss and Goldman-Rakic further concluded that another set of areas, the granular parts of PFo, also had homologues in both bush babies and macaques.

Contemporary support for these conclusions includes the identification of the frontal eye field within PFc in anthropoids and strepsirrhines; the topological relationship of PF areas with the allocortex; the pattern of their projections to the striatum; the lack of direct inputs to granular PF from olfactory, gustatory, and visceral systems (which characterizes agranular PF areas); and the relatively high threshold for evoking autonomic effects from granular PF with electrical stimulation (in contrast to agranular PF areas; Wise, 2008).

Passingham and Wise (2012) review the functions of PFc and PFo. The evidence comes from neuropsychological, neurophysiological, connectional, and brain-imaging studies, which indicate that PFc functions in the search for salient objects, their visual identification, and the orientation of attention toward them. The same kinds of evidence indicate that the granular parts of PFo function in estimating the current value of objects and other stimuli. For example, evidence from lesion studies in macaque monkeys indicates that the granular PFo performs two related functions: (1) it maps the choice of an object to the specific outcome apparently caused by that choice, including the taste, smell, and visual appearance of the foods or fluids that compose such outcomes (Walton, Behrens, Buckley, Rudebeck, & Rushworth, 2010), and (2) in cooperation with the amygdala, it updates the valuation of these outcomes to reflect an animal's current state, such as selective satiety (Baxter & Murray, 2002; Izquierdo, Suda, & Murray, 2004). Importantly, and unlike the orbital cortex of other mammals, the granular PF cortex of primates can establish this mapping on the basis of a single event, a capacity called credit assignment. The next section returns to this topic.

These new PF functions developed in the context of a suite of adaptations to the fine-branch niche (Preuss, 2007a). As mentioned earlier, primate innovations include many new sensory areas for specialized processing of visual, auditory, and somatosensory information, along with new posterior parietal and premotor areas that guide reaching toward visual goals, often consisting of nutritious objects. Parietal and premotor areas also control the grasping of these objects, bringing them to the mouth in coordination with head and mouth movements, and postural stabilization on a flimsy, thin-branch substrate (Wise, 2007). The adaptive advantages of PFc and PFo become clearer in this context:

- PFc contributes to selective foraging among the many objects found in the clutter of the fine-branch niche, including search and identification functions, as well as directing and maintaining attention toward valuable objects, which the animal might acquire at some future time. The oculomotor functions of the frontal eye field, part of PFc, can therefore be understood as fundamentally attentional rather than "motor."
- PFo evaluates objects in terms of current biological needs and links the sensory properties of predicted outcomes with these valuations, especially in the visual domain.

Anthropoids As anthropoids became larger animals, their brains expanded more than predicted by their increase in body size, sometimes called an "upward grade shift" in brain size. An early catarrhine, *Aegyptopithecus* (Kay et al., 2004), had relatively small frontal lobes (Radinsky, 1979) compared to modern catarrhines, but had a visual cortex (broadly defined) within the range of modern catarrhines. On this basis, Radinsky concluded that the visual cortex expanded earlier during primate evolution than did the frontal lobes, and probably prior to the upward grade shift in total brain size. Even with its large visual cortex, the *Aegyptopithecus* brain falls within the modern prosimian size range, when regressed against body size (figure 6.2A, point A). Likewise, other early anthropoids have brains within the prosimian range, including *Parapithecus* (Bush, Simons, & Allman, 2004), an early anthropoid (point P); and *Chilecebus* (Sears, Finarelli, Flynn, & Wyss, 2008), an early platyrrhine (point C). This evidence suggests that the upward grade shift in brain size occurred in parallel in catarrhines and platyrrhines (Williams et al., 2010).

As for the types of cortex that expanded, comparative studies in modern primates point to granular PF, which in anthropoids comprises 9–11% of the neocortex (figure 6.1B) and 43–50% of the frontal lobe (figure 6.2B; Elston et al., 2006). In strepsirrhines, only 7–8% of the neocortex (figure 6.1B) and 41–43% of the frontal lobe (figure 6.2B) consists of granular PF. Coupled with Radinsky's conclusion about the early expansion of visual cortex, the upward grade shift in brain size probably depended in part on expansion of granular PF. Posterior parietal areas also developed extensively in anthropoids, and they play a role in visually guided reaching and in providing the granular PF cortex with contexts concerning the order, number,

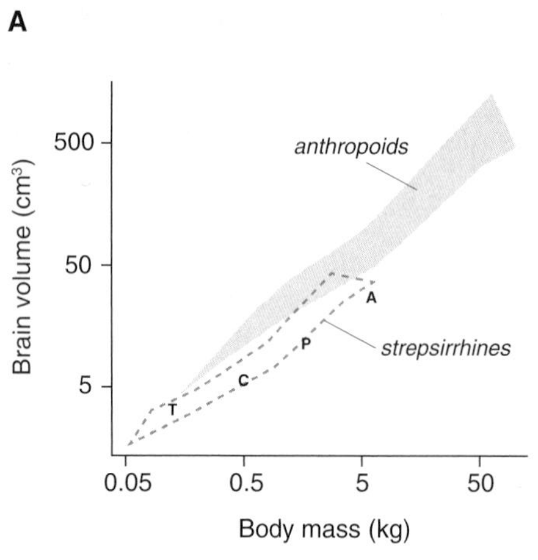

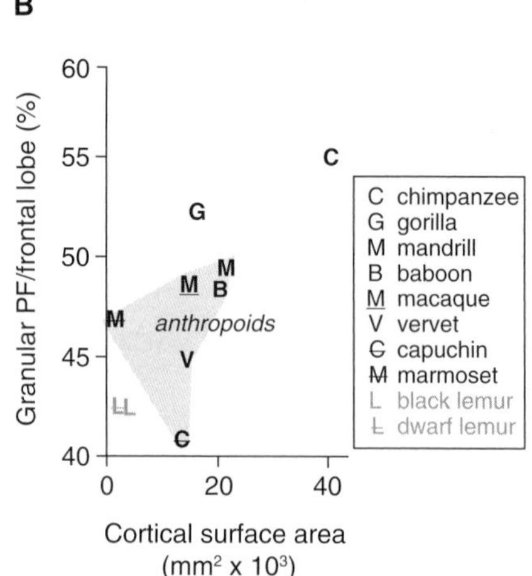

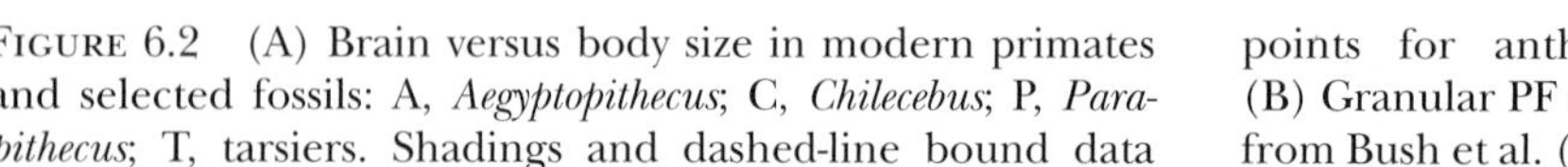

FIGURE 6.2 (A) Brain versus body size in modern primates and selected fossils: A, *Aegyptopithecus*; C, *Chilecebus*; P, *Parapithecus*; T, tarsiers. Shadings and dashed-line bound data points for anthropoids and strepsirrhines, respectively. (B) Granular PF as percentage of frontal cortex. Data for (A) from Bush et al. (2004); data for (B) from Elston et al. (2006).

durations, distances, speeds, and locations of visual events (Passingham & Wise, 2012). These areas also contribute to the control of attention, in part by orienting the fovea (Paré & Dorris, 2011).

In support of the idea that granular PF cortex expanded in anthropoids, Preuss and Goldman-Rakic (1991b) made two key observations in addition to those cited earlier: many of the granular PF areas in macaque monkeys have a paucity of myelin, and bush babies lack large myelin-poor regions within their granular PF. Chimpanzees and humans, like anthropoid monkeys, also have myelin-poor regions within their PF (Glasser, Goyal, Preuss, Raichle, & Van Essen, 2014). The myelin-poor parts of granular PF were probably, therefore, anthropoid innovations.

Proposal

Anthropoid innovations include several parts of granular PF: PFv, PFdl, PFd, the rostral, granular parts of PFm, and probably PFp. Passingham and Wise (2012) suggest specialized roles for each of these areas, but their most important contribution comes from the integrated function of granular PF as a whole (Wilson, Gaffan, Browning, & Baxter, 2010). Specifically, we propose that the primate PF cortex generates the goals appropriate to a current context and to current biological needs, and that it can sometimes do so based on a single event.

As a result, the new granular PF areas of anthropoids reduce foraging errors, which they accomplish in at least two ways: through fast learning and by applying abstract rules and strategies. Both capacities reduce errors in novel situations: the former by solving a problem quickly, the latter by transferring solutions from similar problems to novel ones. A reduction of errors results in more productive foraging choices, less wasted energy, and lower risk of predation.

Our proposal relies partly on neuroimaging and neurophysiological findings, but the most direct evidence comes from the effects of selective brain lesions, as summarized in the next section.

Using events to reduce foraging errors

CREDIT ASSIGNMENT As noted earlier, PFo functions in updating the mappings between objects and the outcomes that follow the appearance or choice of those objects. PFo receives high-level, conjunctive visual representations of objects and outcomes from inferotemporal cortex (IT); olfactory, gustatory, and visceral information about outcomes from agranular parts of orbital and insular cortex; and signals that allow it to update the value of these outcomes from the amygdala (Izquierdo et al., 2004).

Walton et al. (2010) provided evidence that monkeys use their memory of single events to choose a foraging goal. They studied the choices that monkeys made among three objects that produced rewards at a variable rate. Normal monkeys can generate appropriate choices based on a memory of which previous choice seemed to cause a particular outcome. In contrast, monkeys with lesions of the granular PFo were more affected by the mappings between objects and outcomes that accumulate across trials, as an average of several events. The choices of lesioned monkeys thus

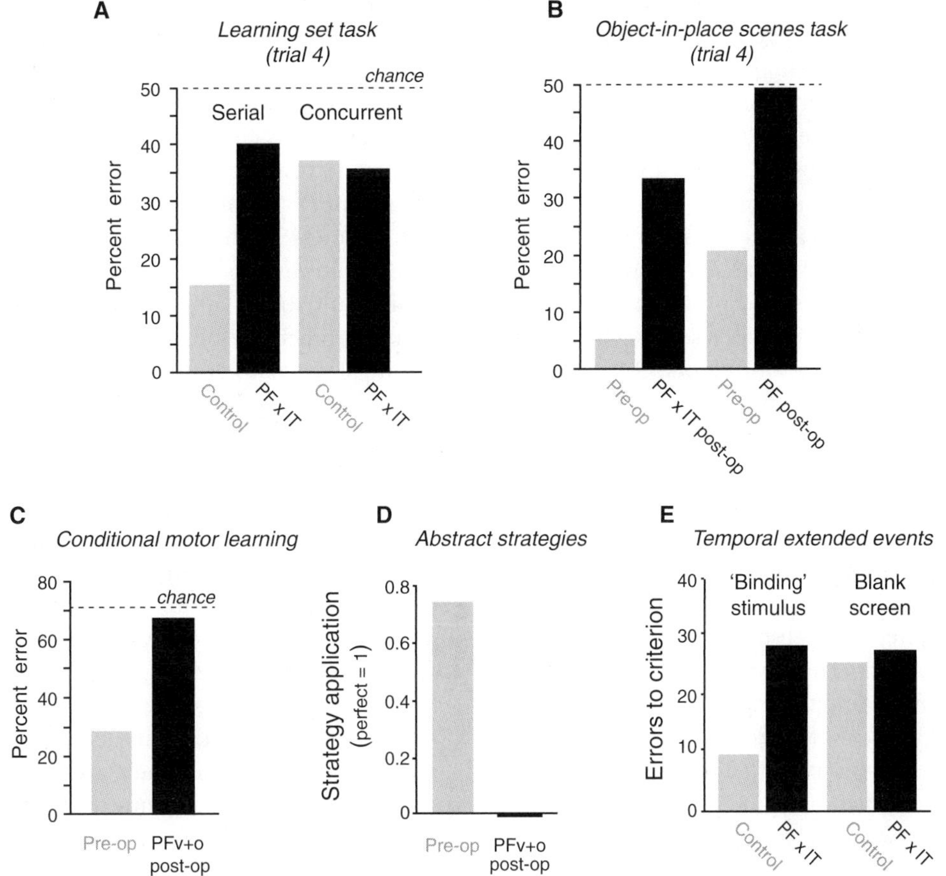

FIGURE 6.3 (A) Learning set, performance on trial 4. (B) Object-in-place scenes task, performance on trial 4. (C) Conditional motor learning, mean over 48 trials. (D) Application of the change-shift strategy, first eight change trials. (E) Temporally extended events. Data for (A) from Browning et al. (2007); for (B), from Browning et al. (2005); for (C) and (D), from Bussey et al. (2001); for (E), adapted from Wilson et al., 2010, with permission from Elsevier; reproduced from Passingham and Wise (2012), by permission of Oxford University Press.

resemble those of nonprimate mammals and other vertebrates, which learn associations between responses and outcomes through cumulative experience. The capacity for learning the outcome caused by a single event provides an advantage in generating future foraging choices.

DISCRIMINATION AND REVERSAL LEARNING SET With experience, macaque monkeys can make the correct choice between two objects on the second opportunity to make that choice, a capability called "learning set." These experiments usually involve the serial discrimination task, in which the same two stimuli appear trial after trial. In the concurrent discrimination task, by contrast, monkeys choose between two objects and then many trials with different objects intervene before monkeys face the same choice again.

Murray and Gaffan (2006) found that monkeys fail to develop a strong learning set when tested on the concurrent discrimination task. Monkeys learn much faster and make fewer errors on the serial version of the task (figure 6.3A, gray). In general, their high error rate on the concurrent task resembles that of naive monkeys on the serial task. Murray and Gaffan concluded that when experienced monkeys face the serial discrimination task, they choose their next goal based on events from the previous trial. If the choice of an object produced a reward, then the monkey will choose that object as the goal for the next trial; otherwise, it chooses the alternative object. Monkeys can then maintain the goal in short-term memory until the opportunity comes to make the next choice, a process called prospective coding. The concurrent discrimination task precludes this strategy by imposing intervening objects.

The development of a strong learning set depends on granular PF and its interactions with IT (Browning, Easton, & Gaffan, 2007). Disconnection lesions that block interactions between PF and IT slow learning and increase errors on the serial discrimination task, bringing performance into the range observed for the concurrent version of that task (figure 6.3A, black) and for naive monkeys on the serial version.

The term "reversal set" refers to an improvement in switching between two objects when experimenters change the one designated as the correct choice. After macaque monkeys develop a reversal set, PF-IT disconnections impair reversal performance (Wilson & Gaffan, 2008). This finding suggests that granular PF enables monkeys to generalize from previous reversal problems, thereby reducing the number of errors. Rats and cats generalize across reversal problems poorly, although some corvid birds show convergence with anthropoids for this trait (Bond, Kamil, & Balda, 2007).

Object-in-Place Scenes Task In the object-in-place scenes task, monkeys use a background context, called a scene, to help them choose between two object-like stimuli in that scene. The presence of the background leads to faster learning and fewer errors. Bilateral PF lesions cause a severe impairment on this task (Browning, Easton, Buckley, & Gaffan, 2005), and PF-IT disconnections do so to a lesser extent (figure 6.3B). Without granular PF or its interactions with IT, macaque monkeys perform as they do without background scenes. Browning et al. argue that this task measures the ability to remember single events, sometimes called episodic memories.

Conditional Motor Learning The term conditional motor learning refers to the ability to use arbitrary cues, usually visual, to choose an action. Naive macaques learn novel cue-action mappings relatively slowly, but with experience they can reduce errors dramatically, often learning from single events. As in learning set and reversal tasks, this error reduction requires granular PF (figure 6.3C; Bussey, Wise, & Murray, 2001). Bilateral lesions of PFo and PFv cause a reversion to error rates typical of naive monkeys and rats (Bussey, Muir, Everitt, & Robbins, 1996).

In addition to fast learning, the application of abstract rules and strategies also reduces errors. Two abstract strategies, called repeat-stay and change-shift, also depend on granular PF (figure 6.3D). These strategies depend on event memory, because what occurs on the previous trial determines whether the monkeys should stay with their previous choice or shift to an alternative.

Temporally Extended Events In a task used by Browning and Gaffan (2008), macaque monkeys see a picture during an interval between their choice and a subsequent reward instead of a blank screen. This "binding" stimulus probably fosters the linkage of a series of events into a single conjunctive representation, called a temporally extended (or complex) event. Like the other results cited here, PF-IT disconnections cause monkeys to revert to the level of performance achieved without the "binding" stimulus (figure 6.3E).

An account of PF function in terms of conjunctive representations also applies to credit assignment (choice-outcome), the object-in-place scenes task (background-choice-outcome), and conditional motor learning (cue-action-outcome).

Conclusion

Lesion studies in macaque monkeys show that an intact granular PF cortex decreases errors in laboratory tasks. According to Passingham and Wise (2012), it does so by generating goals from a current context, based on current biological needs. A goal, in this sense, refers to the objects and places that serve as targets of action. PF sometimes uses a single remembered event—the conjunction of a context, goal, action, and outcome—to generate a future goal when a context recurs. It also reduces errors by using events to choose goals based on abstract rules and strategies and by encoding goals prospectively until the opportunity arises to attain them. Connections between PF and premotor areas serve to implement the chosen goals at that time, in part by providing a bias among competing visuomotor plans (Pastor-Bernier & Cisek, 2011). Connections with other areas provide PF with high-level information about contexts, goals, actions, and outcomes.

The ecological and comparative perspective provided earlier helps us understand these laboratory findings. Granular PF first emerged in early primates as they adapted to life in the fine branches of angiosperm trees. Their new PF areas, PFc and PFo, provided advantages in the search for and valuation of foraging goals in this cluttered environment. The search function of PFc, including the frontal eye field, involves top-down attention of both the overt and covert varieties.

Additional granular areas appeared later, in anthropoids, as these animals increased in size and came to rely on rich but volatile resources distributed over a large home range. They foraged by day in a competitive environment, under threat of predation, using an energy-intensive form of locomotion. These anthropoid ancestors used visual advances, such as foveal and trichromatic vision, to improve their perception of color, shape, and visual texture. Temporal areas, including IT, provided PFv with signs of resources at a distance based on those attributes. Posterior parietal areas, in parallel, provided PFdl and PFd with the metrics of resource patches, such as relative quantities, distances, durations, and order (Genovesio, Wise, & Passingham, 2014). The foraging requirements of these animals placed a premium on making good foraging choices and learning to do so quickly based on limited experience, especially during periods of dearth. Foraging errors entail costs and dangers, and they can be fatal. Any reduction in poor foraging choices therefore provides an adaptive advantage. According to our proposal, the new PF areas of anthropoid primates evolved as an adaptation for reducing the number of such errors.

ACKNOWLEDGMENTS I thank Drs. Elisabeth A. Murray, Richard E. Passingham, and Todd M. Preuss for their comments on a previous version of this chapter.

REFERENCES

Allman, J. M. (2000). *Evolving brains*. New York, NY: Freeman.

Barton, R. A. (1998). Visual specialization and brain evolution in primates. *Proc R Soc Lond B Biol Sci, 265*, 1933–1937.

Barton, R. A. (2004). Binocularity and brain evolution in primates. *Proc Nat Acad Sci USA, 101*, 10113–10115.

Baxter, M. G., & Murray, E. A. (2002). The amygdala and reward. *Nat Rev Neurosci, 3*, 563–573.

Bloch, J. I., & Boyer, D. M. (2002). Grasping primate origins. *Science, 298*, 1606–1610.

Bond, A. B., Kamil, A. C., & Balda, R. P. (2007). Serial reversal learning and the evolution of behavioral flexibility in three species of North American corvids (*Gymnorhinus cyanocephalus, Nucifraga columbiana, Aphelocoma californica*). *J Comp Psychol, 121*, 372–379.

Browning, P. G. F., Easton, A., Buckley, M. J., & Gaffan, D. (2005). The role of prefrontal cortex in object-in-place learning in monkeys. *Eur J Neurosci, 22*, 3281–3291.

Browning, P. G. F., Easton, A., & Gaffan, D. (2007). Frontal-temporal disconnection abolishes object discrimination learning set in macaque monkeys. *Cereb Cortex, 17*, 859–864.

Browning, P. G. F., & Gaffan, D. (2008). Prefrontal cortex function in the representation of temporally complex events. *J Neurosci, 28*, 3934–3940.

Bush, E. C., Simons, E. L., & Allman, J. M. (2004). High-resolution computed tomography study of the cranium of a fossil anthropoid primate, *Parapithecus grangeri*: New insights into the evolutionary history of primate sensory systems. *Anat Rec A Discov Mol Cell Evol Biol, 281*, 1083–1087.

Bussey, T. J., Muir, J. L., Everitt, B. J., & Robbins, T. W. (1996). Dissociable effects of anterior and posterior cingulate cortex lesions on the acquisition of a conditional visual discrimination: Facilitation of early learning vs. impairment of late learning. *Behav Brain Res, 82*, 45–56.

Bussey, T. J., Wise, S. P., & Murray, E. A. (2001). The role of ventral and orbital prefrontal cortex in conditional visuomotor learning and strategy use in rhesus monkeys. *Behav Neurosci, 115*, 971–982.

Cartmill, M. (1972). Arboreal adaptations and the origin of the order Primates. In R. Tuttle (Ed.), *Functional and evolutionary biology of primates* (pp. 97–122). Chicago, IL: Aldine-Atherton.

Changizi, M. A. (2009). *The visual revolution*. Dallas, TX: Benbella.

Elston, G. N., Benavides-Piccione, R., Elston, A., Zietsch, B., DeFelipe, J., Manger, P., Casagrande, V., & Kaas, J. H. (2006). Specializations of the granular prefrontal cortex of primates: Implications for cognitive processing. *Anat Rec A Discov Mol Cell Evol Biol, 288*, 26–35.

Fleagle, J. G. (1999). *Primate adaptation and evolution* (2nd ed.). San Diego, CA: Academic Press.

Genovesio, A., Wise, S. P., & Passingham, R. E. (2014). Prefrontal–parietal function: From foraging to foresight. *Trends Cogn Sci, 18*, 72–81.

Glasser, M. F., Goyal, M. S., Preuss, T. M., Raichle, M. E., & Van Essen, D. C. (2014). Trends and properties of human cerebral cortex: Correlations with cortical myelin content. *Neuroimage, 93*, 165–175.

Heesy, C. P., & Ross, C. F. (2004). Mosaic evolution of activity patterns, diet, and color vision in haplorhine primates. In C. F. Ross & R. F. Kay (Eds.), *Anthropoid origins: New visions* (pp. 665–698). New York, NY: Academic/Plenum.

Izquierdo, A., Suda, R. K., & Murray, E. A. (2004). Bilateral orbital prefrontal cortex lesions in rhesus monkeys disrupt choices guided by both reward value and reward contingency. *J Neurosci, 24*, 7540–7548.

Janmaat, K. R. L., Byrne, R. W., & Zuberbühler, K. (2006). Evidence for a spatial memory of fruiting states of rainforest trees in wild mangabeys. *Animal Behav, 72*, 797–807.

Kaas, J. H. (2012). The evolution of neocortex in primates. *Prog Brain Res, 195*, 91–102.

Kay, R. F., Williams, B. A., Ross, C. F., Takai, M., & Shigehara, N. (2004). Anthropoid origins: A phylogenetic analysis. In C. F. Ross & R. F. Kay (Eds.), *Anthropoid origins: New visions* (pp. 91–135). New York, NY: Academic/Plenum.

Larson, S. G. (1998). Unique aspects of quadrupedal locomotion in nonhuman primates. In E. Strasser, J. G. Fleagle, A. L. Rosenberger, & H. M. McHenry (Eds.), *Primate locomotion: Recent advances* (pp. 157–173). New York, NY: Plenum.

MacNeilage, P. F., Studdert-Kennedy, M. G., & Lindblom, B. (1987). Primate handedness reconsidered. *Behav Brain Sci, 10*, 247–303.

Martin, R. D. (1981). Relative brain size and basal metabolic rate in terrestrial vertebrates. *Nature, 293*, 60.

Martin, R. D. (1990). *Primate origins and evolution: A phylogenetic reconstruction*. Princeton, NJ: Princeton University Press.

Murray, E. A., & Gaffan, D. (2006). Prospective memory in the formation of learning sets by rhesus monkeys (*Macaca mulatta*). *J Exp Psychol Anim Behav Process, 32*, 87–90.

Paré, M., & Dorris, M. C. (2011). Role of posterior parietal cortex in the regulation of saccadic eye movements. In S. P. Liversedge, I. Gilchrist, & S. Everling (Eds.), *Oxford handbook of eye movements* (pp. 257–278). Oxford, UK: Oxford University Press.

Passingham, R. E., & Wise, S. P. (2012). *The neurobiology of the prefrontal cortex: Anatomy, evolution, and the origin of insight*. Oxford, UK: Oxford University Press.

Pastor-Bernier, A., & Cisek, P. (2011). Neural correlates of biased competition in premotor cortex. *J Neurosci, 31*, 7083–7088.

Petrides, M., & Pandya, D. N. (2007). Efferent association pathways from the rostral prefrontal cortex in the macaque monkey. *J Neurosci, 27,* 11573–11586.

Preuss, T. M. (1995). Do rats have prefrontal cortex? The Rose-Woolsey-Akert program reconsidered. *J Cog Neurosci, 7,* 1–24.

Preuss, T. M. (2007a). Evolutionary specializations of primate brain systems. In M. J. Ravosa & M. Dagasto (Eds.), *Primate origins: Adaptations and evolution* (pp. 625–675). New York, NY: Springer.

Preuss, T. M. (2007b). Primate brain evolution in phylogenetic context. In J. H. Kaas & T. M. Preuss (Eds.), *Evolution of nervous systems* (Vol. 4, pp. 2–34). Oxford, UK: Elsevier.

Preuss, T. M., & Goldman-Rakic, P. S. (1991a). Ipsilateral cortical connections of granular frontal cortex in the strepsirhine primate *Galago,* with comparative comments on anthropoid primates. *J Comp Neurol, 310,* 507–549.

Preuss, T. M., & Goldman-Rakic, P. S. (1991b). Myelo- and cytoarchitecture of the granular frontal cortex and surrounding regions in the strepsirhine primate *Galago* and the anthropoid primate *Macaca. J Comp Neurol, 310,* 429–474.

Radinsky, L. (1979). *The fossil record of primate brain evolution.* 49th James Arthur lecture on the evolution of the human brain. New York, NY: American Museum of Natural History.

Rose, K. D. (2006). *The beginnings of the age of mammals.* Baltimore, MD: Johns Hopkins University Press.

Ross, C. F. (2004). The tarsier fovea: Functionless vestige or nocturnal adaptation? In C. F. Ross & R. F. Kay (Eds.), *Anthropoid origins: New visions* (pp. 477–537). New York, NY: Academic/Plenum.

Schmitt, D. (2010). Primate locomotor evolution: Biomechanical studies of primate locomotion and their implications for understanding primate neuroethology. In M. L. Platt & A. A. Ghazanfar (Eds.), *Primate neuroethology* (pp. 31–63). New York, NY: Oxford University Press.

Sears, K. E., Finarelli, J. A., Flynn, J. J., & Wyss, A. (2008). Estimating body mass in New World "monkeys" (*Platyrrhini,* Primates), with consideration of the Miocene platyrrhine, *Chilecebus carrascoensis. Am Mus Novit, 3617,* 1–29.

Shadmehr, R., & Wise, S. P. (2005). *The computational neurobiology of reaching and pointing: A foundation for motor learning.* Cambridge MA: MIT Press.

Walton, M. E., Behrens, T. E., Buckley, M. J., Rudebeck, P. H., & Rushworth, M. F. (2010). Separable learning systems in the macaque brain and the role of orbitofrontal cortex in contingent learning. *Neuron, 65,* 927–939.

Williams, B. A., Kay, R. F., & Kirk, E. C. (2010). New perspectives on anthropoid origins. *Proc Natl Acad Sci USA, 107,* 4797–4804.

Wilson, C. R., & Gaffan, D. (2008). Prefrontal-inferotemporal interaction is not always necessary for reversal learning. *J Neurosci, 28,* 5529–5538.

Wilson, C. R., Gaffan, D., Browning, P. G., & Baxter, M. G. (2010). Functional localization within the prefrontal cortex: Missing the forest for the trees? *Trends Neurosci, 33,* 533–540.

Wise, S. P. (2007). The evolution of ventral premotor cortex and the primate way of reaching. In T. M. Preuss & J. H. Kaas (Eds.), *The evolution of primate nervous systems* (pp. 157–166). Amsterdam: Elsevier.

Wise, S. P. (2008). Forward frontal fields: Phylogeny and fundamental function. *Trends Neurosci, 31,* 599–608.

7 Animal Models of the Human Brain: Repairing the Paradigm

TODD M. PREUSS AND JASON S. ROBERT

ABSTRACT Despite advances in in vivo human studies, neuroscientists still depend on studies of nonhuman species for information that cannot be obtained without invasive techniques. This research is concentrated in a very few "model" species, mostly macaque monkeys, rats, and mice. Although it is well documented that brain organization varies markedly across mammals, species differences are perceived as threatening to model-organism research and so are often ignored or, if acknowledged, are interpreted in the context of a simple-to-complex phylogenetic progression that is at odds with the modern, branching-tree conception of evolution. We argue that the failure of the model-organism paradigm to deal forthrightly with phyletic variation can compromise the integrity of research by slanting the generation, reporting, and interpretation of results, and by placing the enterprise outside the bounds of evolutionary plausibility. We propose that many of the deficiencies of the model-organism paradigm could be remedied, and the quality of inferences about humans improved, by increasing the number and phylogenetic diversity of core experimental species and embedding model research in a broader comparative framework.

Research in cognitive neuroscience primarily involves studying humans with one of several noninvasive imaging techniques. Nevertheless, the field is still very dependent on experimental studies of nonhuman species for obtaining certain kinds of information, especially small-scale features of brain organization, and information that can only be obtained with invasive techniques that are considered ethically unacceptable for use in humans. Given this dependence, what is the best strategy for studying animals in order to understand human beings? For most scientists, the answer is obvious: focus on one of the widely used "model" species. In the neurosciences, those would be rats and mice, mainly, but also macaque monkeys and other nonhuman primates (Manger et al., 2008). This approach, which we will refer to as the "model-organism paradigm" (Preuss, 2000), or MOP, is the dominant animal-research paradigm not only in the neurosciences, but in the biomedical sciences generally, as reflected in its support by funding organizations such as the National Institutes of Health (cf. Robert, 2008a, on "model organism-ism").

The purpose of this essay is to critique the MOP. We emphasize that we are *not* arguing against animal research. Rather, we identify limitations and liabilities of animal research conducted within the current framework of the MOP, and suggest ways to modify science policy so as to enhance our understanding of humans through the study of other animals.

What model organisms do for *us*

The advancement of knowledge in the biological and behavioral sciences has been pursued primarily with research on a relatively small number of model organisms. Model organisms are physical representations of phenomena of interest, and those representations are of several kinds (Bolker, 2009; Preuss, 2001): a species might serve as a disease model—that is, as a proxy or in vivo replica of a disease, symptom, or disease process (e.g., the rat model of Parkinson's disease), or as an exhibitor of some widespread phenomenon (e.g., a model for studying neurogenesis, a model of prefrontal cortex), or a general-purpose stand-in for other species. In any case, the goal is to advance research by standardizing experimental methods and materials (Robert, 2008a).

Current model organisms were established for primarily pragmatic reasons: they are lab-friendly (relatively cheap, small, and generally susceptible to husbandry), experimentally tractable (relatively easy to observe and manipulate), and standard (the ones that everyone uses). Model organisms canalize entire domains of research because they create research efficiencies, including beneficial economies of scale and reliable replication of results (Robert, 2004b, 2008a).

The "model organism" honorific is most properly applied to those animals sanctioned by the National Institutes of Health (NIH) and other funders, and around which a model "system" has grown (including research networks, literatures, databases, and so on). Thus, model organisms are more than just animals: they are the foci of human institutions (e.g., Logan, 2005b;

Rader, 2004). When scientists achieve prominence through their research on a particular model, they attract students who typically work on that model. Funding agencies take notice and direct resources to research on the model. Research know-how on the model accumulates and supporting infrastructures develop. The community develops an ideology about the best way to do research and how it applies to humans. Members of the model-organism community—researchers, administrators, students—develop vested interests in the political as well as scientific success of their model, interests that, we argue, influence their view of biological reality.

Typically, organisms destined to become models come to the attention of scientists owing to some scientifically interesting or methodologically convenient characteristic. For example, Norway rats came to the attention of researchers because they mature quickly and were perceived as being hypersexual, at a time when researchers were beginning to study the physiology of sex (Logan, 2001). However, as Logan has documented, when rats became domesticated through generations of breeding, and researchers became increasingly reliant on their use, they underwent a remarkable transformation in the minds of scientists, their rodent- and species-specific characteristics receiving progressively less attention and their status as "typical mammals" promoted (Logan, 1999, 2002, 2005a). With this change, and the availability of pure strains, the status of rats changed from biological species to laboratory reagent. This is an accurate depiction of model organisms more generally: they are standardized, quality-controlled laboratory materials useful for making experiments work. The purification of these reagents probably increases the internal validity and repeatability of experimental results, but unfortunately does not assure their external validity—that is, their generalizability to other species—and may, in fact, compromise it.

What model organisms do to *us*

While the benefits of model-organism research are widely recognized, it also has liabilities, especially in the neurosciences, where research is concentrated in so few species. Here, we consider how scientists' commitment to models affects how they understand the place of their models in the phylogenetic scheme, how they view their work in the broader context of science, and how they represent their findings.

At the heart of the MOP is the assumption that findings obtained in model species will hold for many or most other species. This was dubbed the "August Krogh principle" by Hans Krebs (1975), although Krogh would probably not have endorsed Krebs's view that species differences are "minor modifications" of a standard animal plan (Logan, 2002). It is the MOP's default assumption of generality that justifies the concentration of effort on internal validity and repeatability rather than on external validity (see also van der Staay, Arndt, & Nordquist, 2009).

This ideology promotes a kind of *species neglect.* Since species are thought not to matter in any fundamental way, titles of neuroscience papers commonly do not name them—as one can easily confirm by scanning the contents of the *Journal of Neuroscience*—and species are sometimes omitted from abstracts as well. Instead, papers focus on the structure or function of particular biological entities, such as brain areas and genes. These entities are thus presented as independent of the species that harbor them, although it is well known that the phenotypic effects of a particular gene or protein, or the functional properties of a particular brain structure, can differ dramatically between even closely related species (see, e.g., Krubitzer, Campi, & Cooke, 2011; McNamara, Namgung, & Routtenberg, 1996; Rekart, Sandoval, & Routtenberg, 2007; and, generally, Robert, 2004a, 2008b). The reality of species differences is further highlighted when models come into competition with each other, leading to sometimes rancorous disputes about the best model for a particular purpose, as for example in the debate over primate versus rodent models in Parkinson's disease research (Robert, 2008a).

Species neglect in the neurosciences is reinforced by explicit doctrine: the claim that there is a "basic uniformity" of neocortical organization, an idea most fully articulated by Rockel, Hiorns, and Powell (1980). "Basic uniformity" holds that the numbers of cells and proportions of cell types in a cortical column are the same across species and cortical areas (except the primary visual area of primates), and that the local circuitry of the cortex is likewise invariant; functional differences between areas result from differences in extrinsic connectivity. Taken seriously, basic uniformity implies that for most research purposes, any mammal will do. But basic uniformity cannot be sustained, as modern research has revealed numerous differences in intrinsic as well as large-scale cortical organization between areas and species (Hof & Sherwood, 2007; Preuss, 2010). Nevertheless, basic uniformity remains an influential concept in neuroscience.

The MOP's distortions of biology extend into the realm of evolution (Preuss, 1995a, 2009; Robert, 2004a). Modern evolutionary biology views every species as a mosaic of features, some unique or specific to the species, some shared with close relatives, and some

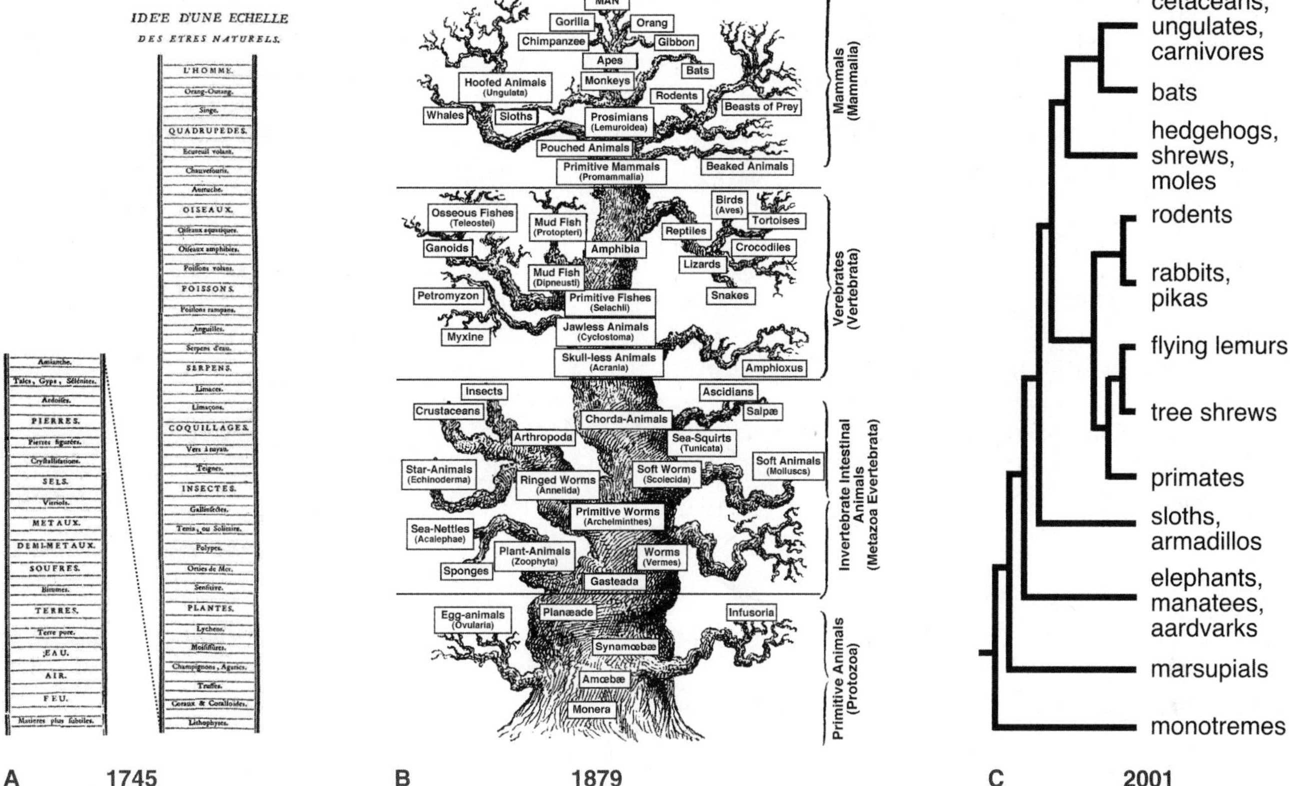

FIGURE 7.1 Evolution of biological ideas about the relationship of humans to other animals. (A) Pre-Darwinian conception of the scale of being, from Charles Bonnet (1745). (B) An early, scale-like tree of life, from Darwin's German champion, Ernst Haeckel (1879). (C) A modern phylogenetic tree, depicting the relationships among mammalian orders. From Preuss (2007), simplified from Murphy et al. (2001).

shared with more distantly related species. The totemic organism of modern evolutionary biology is the branching tree, representing both the unity and diversity of life. The rise of "tree thinking" in the latter part of the twentieth century marked a profound change in the way evolutionary biologists understand evolution. Trees drawn in the nineteenth century and first half of the twentieth century looked as much like scales as like trees (figure 7.1) and, indeed, early evolutionists (including Darwin) understood evolution as progression from undifferentiated "lower" forms to the most complex and perfect "higher" forms: humans, that is (see reviews by Fleagle & Jungers, 1980; Hodos & Campbell, 1969; Preuss, 2009; Striedter, 2007). The phylogenetic scale compresses the diversity of life onto a single, vertical axis of complexity and perfection. The bushier trees drawn by modern evolutionists (figure 7.1C) reflect the rejection of a linear, human-centered view of evolution, emphasizing instead its diversifying character, each branch of the tree possessing unique features.

The MOP is much more compatible with the old view of life than with tree thinking. Explicit references to the defunct phylogenetic scale, as well as implicit references in the use of terms such as "infrahuman" primates and "lower" mammals, remain common in the neuroscientific literature. Even NIH's animal care and use guidelines recommend using animals "lower on the phylogenetic scale" whenever possible (Striedter, 2007). The fact that the phylogenetic scale remains the common-sense, folk understanding of evolution in our culture may explain some of its continuing appeal within the MOP. Yet what the MOP has done is to promote a way of thinking about the relationship between humans and other animals incompatible with basic principles of modern biology—while still passing for modern biology.

For scientists working within the MOP, commitment to a specific model can shape the ways data are interpreted and basic concepts are framed. A case in point is prefrontal cortex. Primates possess a region of prefrontal cortex (PFC) located mainly on the dorsolateral aspect of the frontal lobe that has a well-formed granular layer IV: the granular PFC. This region, which is involved in some very high-level cognitive functions such as conscious reasoning and planning, comprises numerous subdivisions that are nodes in networks

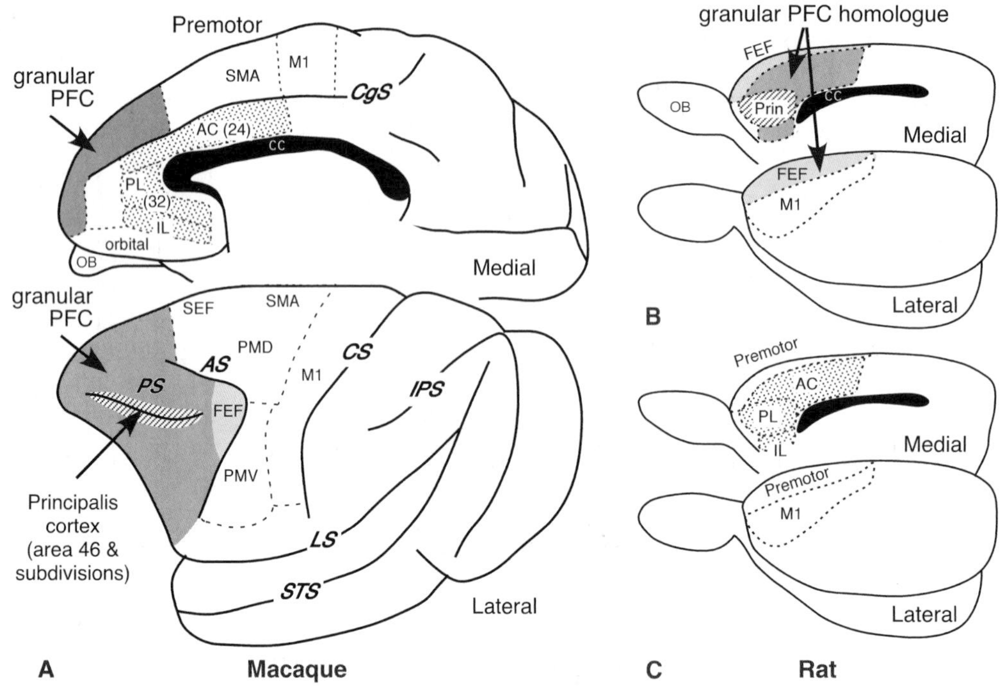

FIGURE 7.2 Alternative interpretations of frontal-lobe homologies in macaques and rats. Modified from Preuss (2007).

linking granular PFC with higher-order parietal and temporal areas (reviewed by Preuss, 1995b, and Passingham & Wise, 2012). Granular PFC presents a problem for rodent models, because obvious homologues are found only in primates (Brodmann, 1909; Preuss, 1995b). One proposed solution is to focus on other features of anatomy that are shared by rodents and primates. Since primate granular PFC receives projections from the lateral part of the mediodorsal (MD) thalamic nucleus, the part of rodent cortex that receives inputs from lateral MD would be a candidate for homology. In rodents, however, lateral MD projects not to dorsolateral frontal cortex, but rather to the medial frontal cortex, including areas 24, 32, and 25, areas originally classified as "cingulate." This has prompted the claim that rodent areas had been misidentified, and that they actually correspond to granular PFC, or perhaps to some amalgam of granular PFC and cingulate cortex (see, for example, Krettek & Price, 1977; Leonard, 1969; Uylings, Groenewegen, & Kolb, 2003; and figures 7.2A and 7.2B). Current evidence, however, clearly indicates that the medial frontal areas of rodents resemble the rostral cingulate areas of primates in virtually every feature that can be compared: in both, major inputs arise from nucleus MD, both have an agranular histology, and lesions in both produce similar behavioral effects—whereas these areas bear no diagnostic similarity to primate granular PFC (Passingham & Wise, 2012; Preuss, 1995b; figure 7.2C). In contrast to granular PFC, the medial frontal cortex in rodents, as in primates, is notable for its strong connections with subcortical limbic and autonomic structures. Rebranding rostral cingulate cortex as "prefrontal" may enhance the stature of rodent models by providing them with an enlarged region of PFC, but it amounts to little more than an accounting change, and one that obscures important differences among cortical regions and among species without advancing our understanding of brain structure and function.

Not only can models shape conceptual space, they can police access to it, influencing what data are accepted as legitimate and significant. Consider the macaque model of the visual system, which has dominated our understanding of the neurobiology of human vision. This dominance is no accident: a series of studies made a point of emphasizing the value of macaques as proxies for humans, creating the impression that there are few differences of consequence between the central visual systems of humans and macaques (e.g., de Valois, Morgan, Polson, Mead, & Hull, 1974; Harwerth & Smith, 1985; Merigan, 1980). So strong is the influence of the macaque model that one is hard-pressed to find references to human-macaque visual-system differences in the secondary literature or in textbooks. Moreover, its influence has made it seem reasonable to dismiss evidence of human specializations out of hand, on the grounds that features present in humans but not macaques must be artifacts of poor methods or

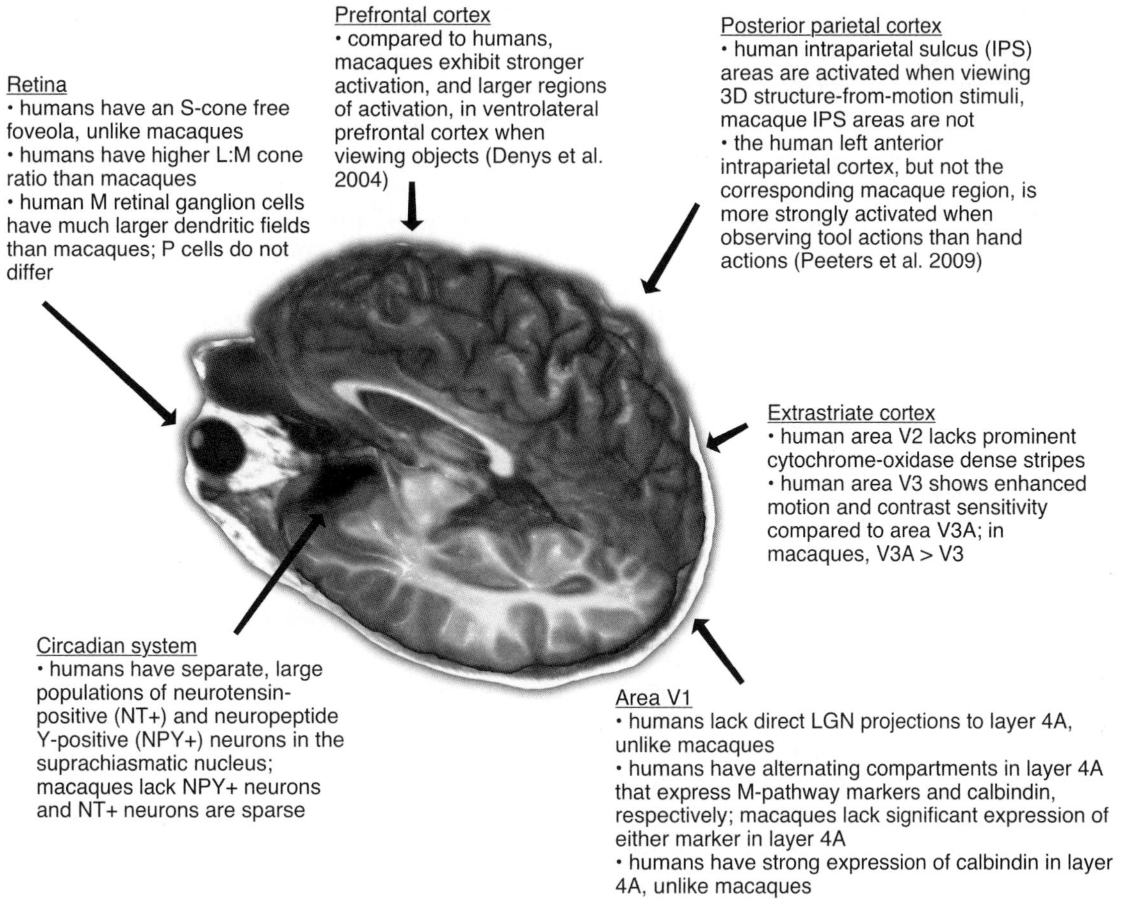

Prefrontal cortex
• compared to humans, macaques exhibit stronger activation, and larger regions of activation, in ventrolateral prefrontal cortex when viewing objects (Denys et al. 2004)
Posterior parietal cortex
• human intraparietal sulcus (IPS) areas are activated when viewing 3D structure-from-motion stimuli, macaque IPS areas are not
• the human left anterior intraparietal cortex, but not the corresponding macaque region, is more strongly activated when observing tool actions than hand actions (Peeters et al. 2009)
Retina
• humans have an S-cone free foveola, unlike macaques
• humans have higher L:M cone ratio than macaques
• human M retinal ganglion cells have much larger dendritic fields than macaques; P cells do not differ
Extrastriate cortex
• human area V2 lacks prominent cytochrome-oxidase dense stripes
• human area V3 shows enhanced motion and contrast sensitivity compared to area V3A; in macaques, V3A > V3
Circadian system
• humans have separate, large populations of neurotensin-positive (NT+) and neuropeptide Y-positive (NPY+) neurons in the suprachiasmatic nucleus; macaques lack NPY+ neurons and NT+ neurons are sparse
Area V1
• humans lack direct LGN projections to layer 4A, unlike macaques
• humans have alternating compartments in layer 4A that express M-pathway markers and calbindin, respectively; macaques lack significant expression of either marker in layer 4A
• humans have strong expression of calbindin in layer 4A, unlike macaques

FIGURE 7.3 Summary of differences in the visual-system organization of humans and macaque monkeys. Preuss (2004) reviews the findings presented here, except those of Denys et al. (2004) and Peeters et al. (2009).

materials because similar characteristics had never been reported in macaques (as an anonymous reviewer explained in critiquing the findings ultimately published by Preuss, Qi, and Kaas, 1999). Yet, while there is no question that macaque and human visual systems have much in common, the primary literature also contains abundant evidence of differences between humans and macaques, and attests that these differences are found at every level of the visual system, from the retina to association cortex (figure 7.3). Why aren't these differences better known? We suggest that acknowledging human-macaque differences is perceived as threatening to the monkey model. The lesson in this case, as in the example of prefrontal cortex, is that when the model comes into conflict with the thing modeled—humans, that is—the model wins. Alternatively, if something doesn't exist in the model, it doesn't exist.

Ethical dimensions of the MOP

We briefly note here the ethical hazards resulting from the reality-distorting effects of the model-organism paradigm. As we have seen, adopting the MOP can bias the generation of results, because so few species are studied, as well as the reporting and interpretation of results. In this way, the MOP challenges the integrity of individual researchers and of the larger research enterprise, ultimately undermining the evidence base of science. Moreover, the fundamental assumption of the MOP—that species differences are trivial—is inconsistent with the modern understanding of evolution, and thus challenges the intellectual integrity of biology.

Additionally, animal research undertaken within the MOP can undermine the credibility of animal research more broadly, and in so doing abet those who would limit or eliminate animal research altogether. Not all animal research is equally well justified; poorly justified MOP research lacking external validity threatens the moral acceptability of more epistemically legitimate research with nonhuman species. Scientists are obligated to make the best use of nonhuman species in research, and the species neglect endemic to the MOP works against this goal.

Reforming Research As we see it, the liabilities and contradictions of the model-organism paradigm stem from the way findings in model species are thought to relate to other species, including humans—specifically, from the default assumption that any given finding in a model will generalize to other animals until proven otherwise. If evolution took the form of a phylogenetic scale, and all the diversity of life could be collapsed onto a single axis of complexity, the default assumption would make sense, and the distinction between internal and external validity would be relatively unimportant. But evolution is like a tree and not a scale, and this fact requires we acknowledge that internal validity is no guarantee of external validity. Demonstrating the generalizability of results requires that neuroscientists adopt a more comparative approach, extending their research broadly and deeply to include species outside of the current stable of model animals.

Let us make clear what we are not proposing. We do not advocate eliminating any of the currently favored model animals, because we acknowledge that there is, indeed, value in working with animals that have established husbandry practices and research systems, or are particularly tractable for one purpose or another (for example, making transgenic animals, in the case of mice). Nevertheless, we believe the concentration of resources in a few highly entrenched models has passed the point of diminishing returns and that the enterprise would be better served by distributing resources over a larger number of species. We advocate the following reforms based on both epistemological and ethical considerations:

Embed research in a broader comparative framework Without broad comparative studies, it is not possible to determine how particular features of brain organization and function are distributed across the evolutionary tree, or within the human part of it. For a program of research designed to improve our understanding of humans, a reasonable approach would, at a minimum, involve identifying features of brain organization characteristic of ancestral mammals, of ancestral eutherian mammals, of the ancestors of the modern eutherian orders (including of course Rodentia and Primates), as well as the specialized features of the genera and species from which our model species are drawn. As well as identifying widely shared features, this approach would help to identify additional species that possess features similar to those present in humans that resulted from convergent evolution: adult pair bonding in voles comes to mind (Young, Wang, & Insel, 1998). Comparative studies can also highlight alternative means to similar behavioral ends: for example, certain avian species appear to have evolved cognitive capacities similar to those of primates, despite marked differences in brain structure (e.g., Avian Brain Nomenclature Consortium, 2005; Emery & Clayton, 2004). How did birds achieve that? Perhaps they share with us emergent, higher-order features of brain organization, features relevant to cognition that we do not yet appreciate even in our own species. Comparative studies can also identify extreme cases that illustrate processes or design features relevant beyond specific phylogenetic groups. For example, our appreciation of the capacity of cerebral cortex to compartmentalize processing would be diminished if we did not know that compartmental, modular organization evolved repeatedly in mammalian history, star-nosed moles providing a remarkable example of this (Catania, 2006).

Increase the number of core species In generating an appropriately comparative context for research, we recommend a strategic and systematic approach rather than a haphazard and idiosyncratic one. Determining which species are especially well suited for addressing particular biological problems will and should depend on the problem itself, not strictly on historical antecedents and canalized research programs. We cannot, of course, expect to study every species in great detail, or even in modest detail, but we would benefit from adding species for concentrated research in a way that increases the breadth and depth of phyletic coverage. This would entail adding more representatives of the rodent and primate orders, adding species from other mammalian orders, and adding selected taxa from other animal groups, especially those meeting the criteria discussed in the previous paragraph. Species to be favored might also include those for which genome sequences are available. Access to primates is a particular concern, given the reduction of primate species available at NIH's National Primate Research Centers over the past two decades, as this represents a serious obstacle to obtaining a deeper understanding of primate (including human) biology.

Foster multispecies research and training For the sake of improving external validation, funding institutions should promote the study of specific processes and phenomena in multiple species (models or otherwise). Also, studies of phenomena in a particular animal group should be favored, not disfavored, if most previous research has been concentrated in some other species. In addition, just as we accept that it is usually better for students to work with several mentors and at different institutions during their training, we should promote cross-species training at all levels of scientific education.

Privilege research that directly compares humans and nonhuman species, and study humans in as many ways as possible In the neurosciences, we now have some very powerful tools for in vivo research, including the wide array of neuroimaging techniques. These techniques, along with the development of improved methods for collecting, storing, and processing human tissue obtained post-mortem, have made human brains accessible as never before (Preuss, 2010). What is more, these techniques have made it possible to directly compare humans and other animals in ways scarcely imaginable a decade ago. Examples of this research include comparisons of gene and protein expression in the brain, of brain morphology and connectivity, and of the cellular and histological organization of the brain (reviewed by Preuss, 2011; Rilling, 2008; Sherwood, Subiaul, & Zawidzki, 2008). These studies are important, as they have, for the first time, identified features of brain organization that are uniquely human.

Organizing research along these lines would have many direct scientific benefits. It would mitigate the current malignant species neglect in animal research and encourage investigators to adopt a more rigorous, evidence-based understanding of phyletic similarities and differences. By expanding the range of species studied, it would provide more opportunity for identifying valuable new model species. Perhaps most importantly, a broader, more comparative research program would enable us to achieve a much better understanding of what humans share with other species, and what is distinctively human. This will allow us to make better judgments about how to model specific features of human organization in other species and in tissue and cell preparations, and make better inferences about development, evolution, structure, and function.

Conclusion

We are not the first to travel this path—historians of science have shown that every generation of experimental biologists has generated critiques of over-reliance on one or a few model organisms. That these critiques have not been heeded suggests that the choice of species to be studied owes considerably more to science as a human and social institution—that is, to the politics of science—than to the way nature is actually structured. Taking these critiques seriously is critical for making progress in neuroscience and other branches of experimental biology, generating findings that more accurately depict the real world of biological organisms.

ACKNOWLEDGMENTS Supported by the James S. McDonnell Foundation, NSF 0926058, and NIH Office of Research Infrastructure Programs/OD P51OD11132.

REFERENCES

Avian Brain Nomenclature Consortium. (2005). Avian brains and a new understanding of vertebrate brain evolution. *Nat Rev Neurosci, 6*(2), 151–159.

Bolker, J. A. (2009). Exemplary and surrogate models: Two modes of representation in biology. *Perspect Biol Med, 52*(4), 485–499.

Bonnet, C. (1745). *Traité d'insectologie.* Paris: Chez Durand.

Brodmann, K. (1909). *Vergleichende Lokalisationslehre der Grosshirnrhinde.* Leipzig: Barth (reprinted as *Brodmann's localisation in the cerebral cortex*; trans. and ed. L. J. Garey. London: Smith-Gordon, 1994).

Catania, K. C. (2006). Evolution of the somatosensory system: Clues from specialized species. In J. H. Kaas & L. A. Krubitzer (Eds.), *Evolution of nervous systems, Vol. 3: Mammals* (pp. 189–206): Oxford, UK: Academic Press.

de Valois, R. L., Morgan, H. C., Polson, M. C., Mead, W. R., & Hull, E. M. (1974). Psychophysical studies of monkey vision. I. Macaque luminosity and color vision tests. *Vision Res, 14*(1), 53–67.

Denys, K., Vanduffel, W., Fize, D., Nelissen, K., Sawamura, H., Georgieva, S., ... Orban, G. A. (2004). Visual activation in prefrontal cortex is stronger in monkeys than in humans. *J Cogn Neurosci, 16*(9), 1505–1516.

Emery, N. J., & Clayton, N. S. (2004). The mentality of crows: Convergent evolution of intelligence in corvids and apes. *Science, 306*(5703), 1903–1907.

Fleagle, J. G., & Jungers, W. L. (1980). Fifty years of higher primate phylogeny. In F. Spencer (Ed.), *A history of American physical anthropology, 1930–1980.* New York, NY: Academic Press.

Haeckel, E. H. P. A. (1879). *The evolution of man: A popular exposition of the principal points of human ontogeny and phylogeny, from the German of Ernst Haeckel.* New York, NY: Appleton.

Harwerth, R. S., & Smith, E. L. (1985). Rhesus monkey as a model for normal vision of humans. *Am J Optom Physiol Opt, 62*(9), 633–641.

Hodos, W., & Campbell, C. B. (1969). Scala naturae: Why there is no theory in comparative psychology. *Psychol Rev, 76*(4), 1–14.

Hof, P. R., & Sherwood, C. C. (2007). The evolution of neuron classes in the neocortex of mammals. In J. H. Kaas & L. A. Krubitzer (Eds.), *Evolution of nervous systems, Vol. 3: Mammals* (pp. 113–124). Oxford, UK: Elsevier.

Krettek, J. E., & Price, J. L. (1977). The cortical projections of the mediodorsal nucleus and adjacent thalamic nuclei in the rat. *J Comp Neurol, 171*(2), 157–191.

Krubitzer, L., Campi, K. L., & Cooke, D. F. (2011). All rodents are not the same: A modern synthesis of cortical organization. *Brain Behav Evol, 78*(1), 51–93.

Leonard, C. M. (1969). The prefrontal cortex of the rat. I. Cortical projection of the mediodorsal nucleus. II. Efferent connections. *Brain Res, 12*(2), 321–343.

Logan, C. A. (1999). The altered rationale for the choice of a standard animal in experimental psychology: Henry H.

Donaldson, Adolf Meyer, and "the" albino rat. *Hist Psychol, 2*(1), 3.

Logan, C. A. (2001). "[A]re Norway Rats ... Things?": Diversity versus generality in the use of albino rats in experiments on development and sexuality. *J Hist Biol, 34*(2), 287–314.

Logan, C. A. (2002). Before there were standards: The role of test animals in the production of empirical generality in physiology. *J Hist Biol, 35*(2), 329–363.

Logan, C. A. (2005a). The legacy of Adolf Meyer's comparative approach: Worcester rats and the strange birth of the animal model. *Integr Physiol Behav Sci, 40*(4), 169–181.

Logan, C. A. (2005b). Making mice: Standardizing animals for American biomedical research, 1900–1955 (book review). *J Hist Behav Sci, 41*(3), 293–294.

Manger, P. R., Cort, J., Ebrahim, N., Goodman, A., Henning, J., Karolia, M., ... Strkalj, G. (2008). Is 21st century neuroscience too focussed on the rat/mouse model of brain function and dysfunction? *Front Neuroanat, 2*, 5.

McNamara, R. K., Namgung, U., & Routtenberg, A. (1996). Distinctions between hippocampus of mouse and rat: Protein F1/GAP-43 gene expression, promoter activity, and spatial memory. *Brain Res Mol Brain Res, 40*(2), 177–187.

Merigan, W. H. (1980). Temporal modulation sensitivity of macaque monkeys. *Vis Res, 20*(11), 953–959.

Murphy, W. J., Eizirik, E., O'Brien, S. J., Madsen, O., Scally, M., Douady, C. J., ... Springer, M. S. (2001). Resolution of the early placental mammal radiation using Bayesian phylogenetics. *Science, 294*(5550), 2348–2351.

Passingham, R. E., & Wise, S. P. (2012). *The neurobiology of the prefrontal cortex: Anatomy, evolution, and the origin of insight.* Oxford, UK: Oxford University Press.

Peeters, R., Simone, L., Nelissen, K., Fabbri-Destro, M., Vanduffel, W., Rizzolatti, G., & Orban, G. A. (2009). The representation of tool use in humans and monkeys: Common and uniquely human features. *J Neurosci, 29*(37), 11523–11539.

Preuss, T. M. (1995a). The argument from animals to humans in cognitive neuroscience. In M. S. Gazzaniga (Ed.), *The cognitive neurosciences* (pp. 1227–1241). Cambridge, MA: MIT Press.

Preuss, T. M. (1995b). Do rats have prefrontal cortex? The Rose-Woolsey-Akert program reconsidered. *J Cogn Neurosci, 7*(1), 1–24.

Preuss, T. M. (2000). Taking the measure of diversity: Comparative alternatives to the model-animal paradigm in cortical neuroscience. *Brain Behav Evol, 55*(6), 287–299.

Preuss, T. M. (2001). The discovery of cerebral diversity: An unwelcome scientific revolution. In D. Falk & K. R Gibson (Eds.), *Evolutionary anatomy of the primate cerebral cortex* (pp. 138–164). Cambridge, UK: Cambridge University Press.

Preuss, T. M. (2004). Specializations of the human visual system: The monkey model meets human reality. In J. H. Kaas & C. E. Collins (Eds.), *The primate visual system* (pp. 231–259). Boca Raton, FL: CRC Press.

Preuss, T. M. (2007). Evolutionary specializations of primate brain systems. In M. J. Ravosa & M. Dagasto (Eds.), *Primate origins: Adaptations and evolution* (pp. 625–675). New York, NY: Springer.

Preuss, T. M. (2009). The cognitive neuroscience of human uniqueness. In M. S. Gazzaniga (Ed.), *The cognitive neurosciences*, 4th ed. (pp. 49–64). Cambridge, MA: MIT Press.

Preuss, T. M. (2010). Reinventing primate neuroscience for the twenty-first century. In M. L. Platt & A. A. Ghazanfar (Eds.), *Primate neuroethology* (Vol. 1, pp. 422–454). Oxford, UK: Oxford University Press.

Preuss, T. M. (2011). The human brain: Rewired and running hot. *Ann NY Acad Sci, 1225*(Suppl. 1), E182–E191.

Preuss, T. M., Qi, H., & Kaas, J. H. (1999). Distinctive compartmental organization of human primary visual cortex. *Proc Natl Acad Sci USA, 96*(20), 11601–11606.

Rader, K. A. (2004). *Making mice: Standardizing animals for American biomedical research, 1900–1955.* Princeton, NJ: Princeton University Press.

Rekart, J. L., Sandoval, C. J., & Routtenberg, A. (2007). Learning-induced axonal remodeling: Evolutionary divergence and conservation of two components of the mossy fiber system within Rodentia. *Neurobiol Learn Mem, 87*(2), 225–235.

Rilling, J. K. (2008). Neuroscientific approaches and applications within anthropology. *Am J Phys Anthropol*, Suppl. *47*, 2–32.

Robert, J. S. (2004a). *Embryology, epigenesis and evolution: Taking development seriously.* Cambridge, UK: Cambridge University Press.

Robert, J. S. (2004b). Model systems in stem cell biology. *BioEssays, 26*(9), 1005–1012.

Robert, J. S. (2008a). The comparative biology of human nature. *Phil Psych, 21*(3), 425–436.

Robert, J. S. (2008b). Taking old ideas seriously: Evolution, development, and human behavior. *New Ideas Psychol, 26*(3), 387–404.

Rockel, A. J., Hiorns, R. W., & Powell, T. P. (1980). The basic uniformity in structure of the neocortex. *Brain, 103*(2), 221–244.

Sherwood, C. C., Subiaul, F., & Zawidzki, T. W. (2008). A natural history of the human mind: Tracing evolutionary changes in brain and cognition. *J Anat, 212*(4), 426–454.

Striedter, G. F. (2007). A history of ideas in evolutionary neuroscience. In J. H. Kaas (Ed.), *Evolution of nervous systems: A comprehensive reference* (Vol. 1, pp. 1–5). Amsterdam: Academic Press.

Uylings, H. B. M., Groenewegen, H. J., & Kolb, B. (2003). Do rats have a prefrontal cortex? *Behav Brain Res, 146*(1–2), 3–17.

van der Staay, F. J., Arndt, S. S., & Nordquist, R. E. (2009). Evaluation of animal models of neurobehavioral disorders. *Behav Brain Funct, 5*(1), 11.

Young, L. J., Wang, Z., & Insel, T. R. (1998). Neuroendocrine bases of monogamy. *Trends Neurosci, 21*(2), 71–75.

8 Molecular and Cellular Mechanisms of Human Brain Development and Evolution

ANDRÉ M. M. SOUSA*, KYLE A. MEYER*, AND NENAD ŠESTAN

ABSTRACT The immense complexity of the human brain is reflected in its cellular organization and the vast behavioral and cognitive repertoire that it can generate. The human brain develops through a dynamic and prolonged process during which a myriad of cell types are generated and assembled into intricate synaptic circuitry. Deviations from this normal course of development can lead to a variety of pathologies, including disorders, such as autism and schizophrenia, that affect some of the most distinctly human aspects of cognition and behavior. While humans share many features with other mammals, in particular with other primates, organizational and developmental differences have allowed for the elaboration of human-specific cognition and behavior. Analyzing molecular and cellular processes involved in human brain development, along with parallel studies in nonhuman primates, is necessary for defining both ancestral and uniquely human features, but this is often difficult to do in a systematic and comprehensive manner. In this review, we summarize current knowledge about molecular and cellular processes underlying human brain development and evolution. Particular emphasis is given to studies of the cerebral cortex because of its importance in higher cognition and because it has been the focus of many comparative and developmental studies.

The human brain is composed of over eighty billion neurons and at least an equal number of glial cells (Azevedo et al., 2009). Neurons are connected with approximately 150,000 to 180,000 km of myelinated axons, and within the neocortex alone, there are about 0.15 quadrillion synaptic contacts (Pakkenberg et al., 2003). These basic facts illustrate the organizational complexity of the human brain and highlight some difficulties we face when trying to understand the molecular and cellular mechanisms of its development and evolution. In this chapter, we will first review the sequences of cellular events in the developing human brain, with a focus on the cerebral neocortex, and then highlight advances in understanding the molecular processes associated with its development and evolution.

*These authors contributed equally to this work.

Cellular mechanisms of human brain development

Human brain development involves many cellular and molecular processes that unfold over the course of almost two decades (Kang et al., 2011; Kostović & Judas, 2002; Sidman & Rakic, 1973; see figure 8.1). One of the most remarkable aspects of human development is that, by the time of birth, the general architecture of the brain has been assembled and the majority of neurons have migrated to their final positions. The organization of human neurodevelopment can be divided into three main sequences of events: generation of neuronal and glial cells types, migration of newly born cells to their final destination, and their differentiation into mature and properly functioning cells within neural circuits.

GENESIS OF NEURONAL AND GLIAL CELLS The ventricular and subventricular zones (VZ and SVZ, respectively) comprise the germinal zones of the developing telencephalon and give rise to neurons and macroglia (astrocytes and oligodendrocytes) of the cerebral cortex (figure 8.2; Caviness, Takahashi, & Nowakowski, 1995; Fishell & Kriegstein, 2003; Sidman & Rakic, 1973). VZ is the first germinal zone to form and is composed of elongated polarized neuroepithelial cells that undergo interkinetic nuclear migration. Early neuroepithelial progenitor cells each produce two daughter cells that re-enter the cell cycle. This symmetrical division doubles the number of progenitor cells each time and exponentially expands the pool of progenitor cells (Caviness et al., 1995; Fishell & Kriegstein, 2003; Rakic, 1995). Early in neurogenesis, neuroepithelial progenitor cells transform into radial glial cells that elongate along apico-basal axis and begin dividing asymmetrically to generate a new progenitor and a postmitotic neuron or glial cell. The generation of glial cells follows neurogenesis and peaks around birth in humans (Sidman & Rakic, 1973).

There are important species differences in the organization of neural progenitor cells and the generation

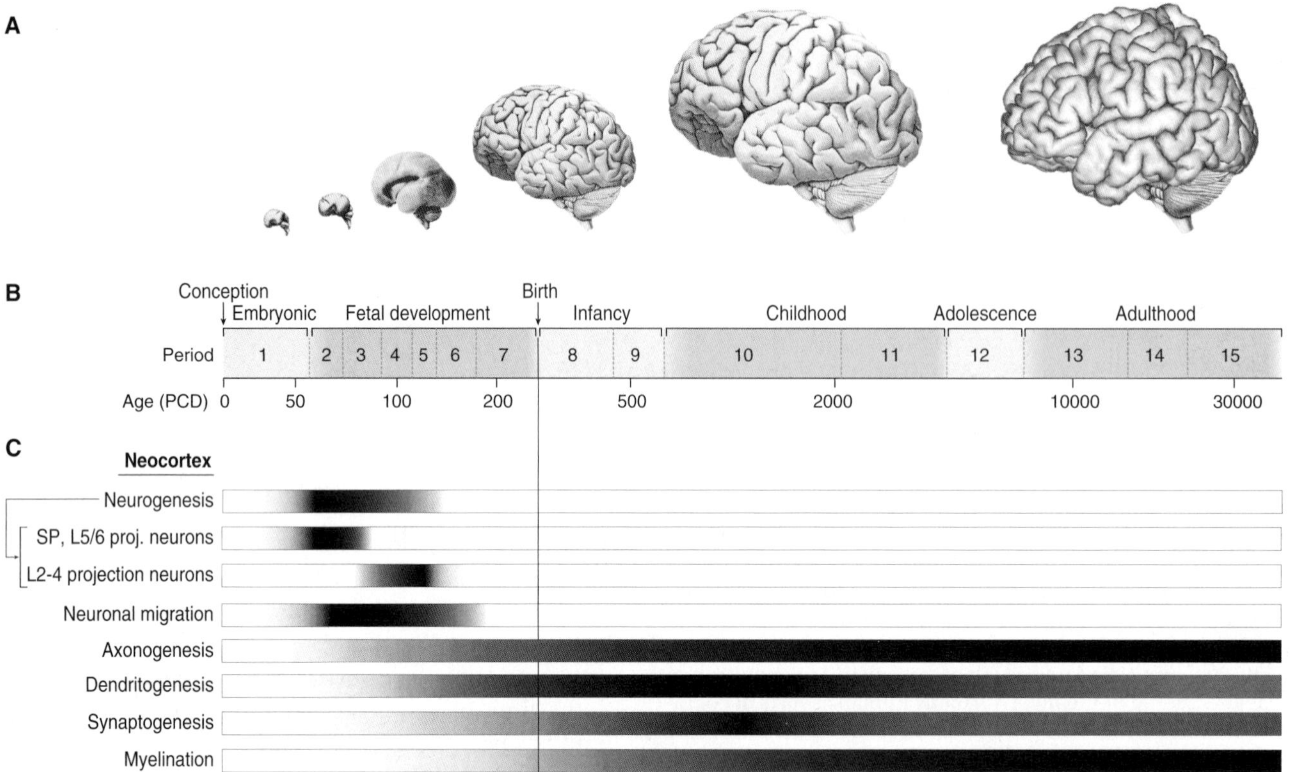

FIGURE 8.1 Timeline of major cellular events and gross morphological changes in human brain development. (A) Schematic images of developing brains from Gustaf Retzius's 1896 atlas and an adult brain generated by magnetic resonance imaging. (B) Periods of development and adulthood as previously defined (Kang et al., 2011). Age is represented in post-conception days (PCD). (C) Summaries of the occurrence and progression of major cellular events in the human neocortex. Black indicates the developmental age when the defined event reaches its peak or is indistinguishable from adult periods.

of neural cell types. One of the most prominent is the prolonged period of neuronal and glial production in humans compared to other primates and mammals (see figure 8.2), which has been postulated to play a critical role in regulating the size of the brain and maturation of neural circuits (Caviness et al., 1995; Rakic, 1995). Increases in neocortical size, particularly in primates and humans, have been linked to the expansion of progenitor cells in the outer subventricular zone (oSVZ) during development (Fietz et al., 2010; Hansen, Lui, Parker, & Kriegstein, 2010; Smart, Dehay, Giroud, Berland, & Kennedy, 2002).

There are also potential differences in the origin and migrational routes of cortical neurons, particularly between primates and rodents. The neurons of the cerebral cortex can be roughly classified into two distinct groups: excitatory and inhibitory. The excitatory neurons utilize the excitatory neurotransmitter glutamate. The great majority of them have characteristic pyramidal-shaped cell bodies and a long apical dendrite covered with spines, and project long axons to other regions of the central nervous system (Kwan, Sestan, & Anton, 2012; Leone, Srinivasan, Chen, Alcamo, & McConnell, 2008; Molyneaux, Arlotta, Menezes, & Macklis, 2007). In contrast, the inhibitory neurons or interneurons are GABAergic, form local circuit connections, and account for 15–25% of all cortical neurons (DeFelipe et al., 2013). A number of studies have shown that cortical interneurons share a common origin with striatal neurons, arising from progenitors within the ganglionic eminences of the ventral forebrain and migrating tangentially into the cortex (Marín & Rubenstein, 2003). Intriguingly, studies in humans and nonhuman primates (NHPs) have reported that certain cortical interneurons arise from dorsal, instead of ventral, pallial progenitors (Jakovcevski, Mayer, & Zecevic, 2011; Letinic, Zoncu, & Rakic, 2002; Petanjek, Kostović, & Esclapez, 2009), suggesting that the origin and migration of primate cortical interneurons are evolutionarily divergent. However, the extent of these species differences is unclear. A study of human holoprosencephaly brains, which exhibit severe ventral forebrain hypoplasia and lack a subgroup of ventral progenitors that generate distinct types of interneurons

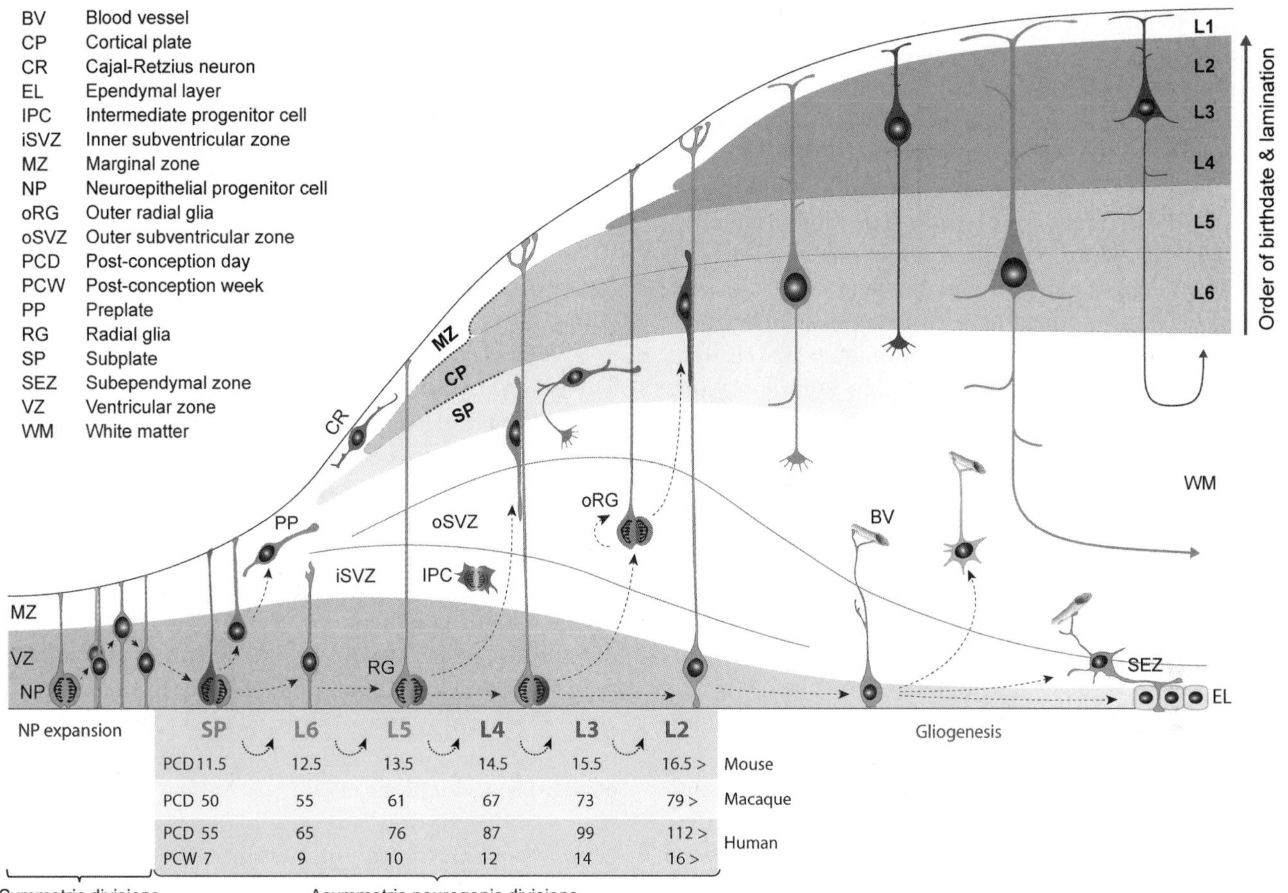

FIGURE 8.2 Schematic of generation and migration of projection neurons and glia in the neocortex. Projection neurons are generated by progenitor cells in the ventricular zone (VZ) and subventricular zone (SVZ). Their generation and migration into the cortical plate (CP) occurs in an inside-first, outside-last manner. At the end of neurogenesis, radial glial (RG) cells lose their polarity and generate glia. Adapted from Kwan et al. (2012). (See color plate 3.)

in rodents, revealed that these patients also lack the same subtypes of cortical interneurons (Fertuzinhos et al., 2009). Further support for the importance of the ventral forebrain in generating human cortical interneurons comes from a recent study (Hansen et al., 2010) demonstrating that cultured human neocortical progenitors do not generate interneurons.

NEURONAL MIGRATION Upon leaving the cell cycle, neurons migrate toward their final destination (figure 8.2). The first generated neurons settle immediately above the VZ, forming the early marginal zone MZ, which is also referred to as the primordial plexiform layer or the preplate (Marin-Padilla, 1978). The subsequent generation of neurons destined for the cortical plate (future layers 2–6) are thought to split the preplate into a superficial marginal zone (future layer 1) and a deep subplate zone. Neurons in the marginal and subplate zones are the first to be generated and achieve functional maturity, as well as the first to establish synaptic contacts with ingrowing cortical afferents. These neurons also play a key role in establishing cortical lamination and patterning of cortical connections (Allendoerfer & Shatz, 1994; Kostović & Judas, 2002; Rice & Curran, 2001).

Neurons use different modes of migration to reach their final destination. For example, newly born pyramidal neurons either undergo somal translocation (Morest & Silver, 2003) or migrate while attached to radial glial fibers (Rakic, 1971) to their final position in the cortical plate, following a precise inside-first, outside-last order. Hence, radial glia serves dual roles as a neural progenitor and a transient scaffold for neuronal migration. Cortical interneurons, on the other hand, migrate tangentially from the ventral forebrain or within the growing dorsal cerebral wall (Marín & Rubenstein, 2003).

In all mammals, including humans, at least two types of spatial information must be imprinted onto neurons of the developing cerebral cortex: (1) their position in the radial direction, corresponding to their laminar position, and (2) their position in the tangential plane, corresponding to their particular cortical area. The physical separation of layers and areas is functionally determined and maintained through their distinct composition of neuronal cell types and a unique set of afferent and efferent connections. The laminar identity of projection neurons reflects their birth order, with first-born neurons occupying the deepest layers and later-born neurons present in more superficial layers. A neuron's laminar identity is also intimately linked to its eventual function; neurons of the deepest layers (layers 5 and 6) send connections to either other cortical areas or the subcortical regions, while those in upper layers (layers 2 to 4) form exclusively intracortical connections. The upper cortical layers are overrepresented in primates, especially in humans, and have been proposed to contribute to some of the cognitive and motor abilities that are unique to humans (Marin-Padilla, 2014).

Cellular Differentiation and Neural Circuit Formation Upon arriving at the cortical plate, neurons stop migrating and continue to differentiate. Despite the fact that neurons rapidly extend their axons as they migrate, most differentiation processes, such as the extension and elaboration of dendrites and the formation of synaptic connections, take place only after neurons have assumed their final laminar position in the cortical plate. Overlaid onto this laminar specification is the parcellation of neurons into distinct areas. Work in rodents has shown that the patterning is initiated by molecular signaling centers in and around the embryonic cortex (Fukuchi-Shimogori & Grove, 2001; O'Leary & Nakagawa, 2002; Sur & Rubenstein, 2005). Extrinsic influences also affect laminar and areal fate, particularly during the ingrowth of thalamocortical fibers and elaboration of cortico-cortical projections.

The timing of synaptogenesis differs across layers and regions of the developing human cortex. Synaptogenesis begins within the subplate zone and the limbic cortical areas (Huttenlocher & Dabholkar, 1997; Kostović & Judas, 2002). The earliest evidence for intraneocortical synapses has been found within deep layers at 18 post-conception weeks (Kwan et al., 2012). In early postnatal life, exuberant dendritic growth and local elaboration of axon terminals characterize a critical period during which there is a marked increase in the number of established synaptic connections and an extended capacity for neuronal remodeling in response to environmental cues and activity-dependent mechanisms (Huttenlocher & Dabholkar, 1997).

Several lines of evidence indicate that processes such as myelination and synaptogenesis progress at different rates across human neocortical areas with the general trend of earlier maturation of the primary sensory-motor areas (Giedd & Rapoport, 2010; Huttenlocher & Dabholkar, 1997).

Molecular mechanisms of human brain development

The generation of different neural cell types in proper numbers at the right time and location, followed by their assembly into a complex network, requires precise spatial and temporal regulation of gene expression. Valuable information has been obtained over the past several years by transcriptome analysis of post-mortem human brains. Transcriptome studies of the developing human brain have included a relatively small number of samples and have predominantly focused on few regions or developmental time points (Abrahams et al., 2007; Colantuoni et al., 2011; Ip et al., 2010; Johnson et al., 2009; Sun et al., 2005). Two recent studies (Kang et al., 2011; Pletikos et al., 2014) greatly expanded spatial and temporal coverage by analyzing exon-level gene expression in multiple brain regions and cortical areas across the full course of human brain development. A high percentage of genes analyzed (86%) were expressed in at least one region of the developing or adult brain. Of these, nine out of ten genes were differentially regulated at the whole-transcript or exon level across brain regions and/or time. The bulk of these transcriptional differences occurred in prenatal development. Among brain regions, the cerebellum possesses the most distinct transcriptional profile. Among neocortical areas, strong transcriptional differences were particularly prominent during fetal development and included specific transcriptional signatures associated with prefrontal and perisylvian areas, which are involved in some of the most distinctly human aspects of cognition (Johnson et al., 2009; Kang et al., 2011; Pletikos et al., 2014). These strong neocortical transcriptional differences in prenatal development diminish during infancy and childhood, and increase again after adolescence (Pletikos et al., 2014).

Gene co-expression analyses also revealed that the human developing transcriptome is organized into distinct co-expression networks enriched for specific biological functions (Kang et al., 2011). Interestingly, genetic variation in some of the most well-connected genes in these modules has previously been linked to psychiatric or neurological disorders, including schizophrenia and autism spectrum disorders, suggesting that they may have converging functions in specific brain regions and developmental periods. The same study has also identified robust sex differences in spatiotemporal gene expression, especially prenatally (Kang et al., 2011). Some of the sex-biased genes had previously been associated with disorders that differentially affect males and females, suggesting that the risk for certain disorders may be traceable to transcriptional mechanisms.

Taken together, the above-mentioned transcriptome studies have provided unique data on the developing human brain and valuable insights into the transcriptional foundations of human neurodevelopment. As discussed in the rest of this chapter, these findings in human tissues are an important step to comparative and functional analyses aimed at elucidating transcriptional mechanisms that led to the phenotypic specializations of the human brain.

Molecular mechanisms of human brain evolution

It is often suggested that differences in expression of genes, rather than the makeup of the genes themselves, have been the major drivers of phenotypic evolution. One motivation for this claim is the argument, which is proposed on the basis of exclusion, that the protein-coding differences are too small to account for this, an idea put forward by King and Wilson (1975). However, without knowledge of how genomic changes map to phenotypic differences, it is difficult to estimate a priori what degree of divergence would be necessary in order to explain observed phenotypic differences. This difficulty is highlighted by stating the sequence differences between human and chimpanzee proteins in another way: the majority of proteins differ by at least one amino acid (Chimpanzee Sequencing and Analysis Consortium, 2005; Glazko, Veeramachaneni, Nei, & Makalowski, 2005).

An important argument for the contribution of regulatory evolution to human-specific features is that most changes are quantitative rather than qualitative, which points to changes in developmental processes and timing via regulatory evolution (Carroll, 2003; see also Hoekstra & Coyne, 2007, for a different perspective). Within the context of regulatory evolution, noncoding *cis*-regulatory evolution has been proposed as the primary source of phenotypic change, based on the idea that mutations to *cis*-regulatory regions circumvent increased selective pressure due to pleiotropic effects because regulatory elements act in a modular fashion, enabling tissue- and time-specific changes in gene-expression levels. Despite the focus on *cis*-regulatory regions, there is also evidence that transcription factors can avoid negative pleiotropy (Wagner & Lynch, 2008), making it unclear what the relative contributions of *cis* and *trans* mutations are to regulatory evolution.

Thus far, several studies have tried to globally characterize *cis*-regulatory elements that show evidence of human-specific changes compared to other NHPs. Most of these studies use conservation to gauge functional importance, as these regions have likely been preserved by purifying selection. Thousands of conserved noncoding regions show signs of accelerated evolution in the human genome (Bird et al., 2007; Pollard et al., 2006a; Prabhakar, Noonan, Paabo, & Rubin, 2006), and some regions that are highly conserved in other NHPs are deleted in the human genome (McLean et al., 2011). Global analysis of positive selection in coding and noncoding regions found that neural genes were enriched for regulatory evolution (Haygood, Babbitt, Fedrigo, & Wray, 2010). In a few cases, the ability of these regions to regulate human-specific expression has been tested using mouse transgenic assays (McLean et al., 2011; Pennacchio et al., 2006; Prabhakar et al., 2008), but the vast majority of them have unknown consequences.

As of yet, there has not been any extensive study linking human-specific *cis*-regulatory evolution to changes in human brain development. A human accelerated region (*HAR1F*) that is composed of a noncoding RNA is specifically expressed in Cajal-Retzius neurons of the cortical marginal zone in a developmentally regulated and human-specific manner, but the phenotypic consequence of this is not known (Pollard et al., 2006b). Another interesting example is *GADD45G*, a tumor suppressor gene. If a human-specific deletion neighboring this gene is introduced into the mouse, expression is no longer driven in the forebrain SVZ, leading to the speculation that this loss could have a role in human brain development (McLean et al., 2011).

A complementary approach to sequence comparisons is to compare gene expression in human and NHP brains to identify genes that are differentially expressed between species. This has the advantage of being able to identify regulatory changes regardless of the underlying regulatory mechanism. From this point, both the regulatory changes responsible for the expression change as well as the phenotypic consequences of the expression changes can be investigated.

Even though differences in gene expression during development is an area of high interest, most studies on transcriptome evolution were done in adult specimens due to the scarcity of human and NHP developmental tissue (especially from great apes) in good condition. A handful of these studies reported evidence that gene expression in the human brain has diverged more from other primate species than gene expression of other tissues examined (Enard et al., 2002) and that there was a bias for upregulated expression in the human brain that was not observed in the other tissues (Caceres et al., 2003; Gu & Gu, 2003; Khaitovich et al., 2004). However, other studies did not find higher divergence in brain gene expression (Hsieh, Chu, Wolfinger, & Gibson, 2003) or a bias for upregulated expression in the human brain (Uddin et al., 2004). Some other works focused on differences in metabolic genes, especially in aerobic metabolism (Babbitt et al., 2010; Uddin et al., 2008), groups of genes with human-specific co-expression (Konopka et al., 2012; Oldham, Horvath, & Geschwind, 2006), and also noncoding RNAs (Babbitt et al., 2010), which were surprisingly conserved in terms of expression, suggesting that they may have a functional role.

Despite the difficulties in obtaining developmental specimens and interpreting results across species due to developmental heterochronicity (Clancy, Darlington, & Finlay, 2001), several studies have focused on analyzing gene- and protein-expression data across postnatal primate brains (Liu et al., 2012; Miller et al., 2012; Somel et al., 2009; Somel et al., 2011). An interesting finding is that the human brain, compared to the brains of other primates, appears to have an increased number of genes with a delayed expression pattern (Somel et al., 2009) and that genes involved in synaptogenesis have a prolonged expression pattern (Liu et al., 2012). Similarly, Miller et al. (2012) quantified myelinated fiber length density and myelin-associated protein expression and found that myelination is protracted in humans compared to chimpanzees. It is important to note that, while the above studies provide valuable information on human-specific expression patterns, they have used mostly postnatal samples, making them unable to detect any critical differences in the transcriptional program that shapes the human brain during prenatal development.

Functional approaches and future directions

The current feasibility of sequencing is enabling the production of an enormous quantity of data on the human and NHP genomes and transcriptomes. It is important to focus on integrating the findings at various levels and to begin to work out which species-specific differences are functionally and developmentally meaningful. Functional characterization can be approached in several ways, from using transgenic mice carrying human-specific genetic variants (Enard et al., 2009), *in utero* electroporation (Charrier et al., 2012; Kwan et al., 2012; Shim, Kwan, Li, Lefebvre, & Sestan, 2012), and human neural progenitors (Konopka et al., 2009). Induced pluripotent stem cell research is another area that may provide valuable tools for exploring human-specific features of neurodevelopment. As we advance our knowledge of the differences and the similarities in mammalian brain development, we will also be better positioned to perform informative and relevant experiments in model organisms.

ACKNOWLEDGMENTS We apologize to all colleagues whose important work was not cited because of space limitations. This article is supported by grants from the Kavli Foundation, the James S. McDonnell Foundation, and the National Institutes of Health.

REFERENCES

Abrahams, B. S., Tentler, D., Perederiy, J. V., Oldham, M. C., Coppola, G., & Geschwind, D. H. (2007). Genome-wide analyses of human perisylvian cerebral cortical patterning. *Proc Natl Acad Sci USA, 104*(45), 17849–17854.

Allendoerfer, K. L., & Shatz, C. J. (1994). The subplate, a transient neocortical structure: Its role in the development of connections between thalamus and cortex. *Annu Rev Neurosci, 17*, 185–218.

Azevedo, F. A. C., Carvalho, L. R. B., Grinberg, L. T., Farfel, J. M., Ferretti, R. E. L., Leite, R. E. P., ... Herculano-Houzel, S. (2009). Equal numbers of neuronal and nonneuronal cells make the human brain an isometrically scaled-up primate brain. *J Comp Neurol, 513*(5), 532–541.

Babbitt, C. C., Fedrigo, O., Pfefferle, A. D., Boyle, A. P., Horvath, J. E., Furey, T. S., & Wray, G. A. (2010). Both noncoding and protein-coding RNAs contribute to gene expression evolution in the primate brain. *Genome Biol Evol, 2*, 67–79.

Bird, C. P., Stranger, B. E., Liu, M., Thomas, D. J., Ingle, C. E., Beazley, C., ... Dermitzakis, E. T. (2007). Fast-evolving noncoding sequences in the human genome. *Genome Biol, 8*(6), 118.

Caceres, M., Lachuer, J., Zapala, M. A., Redmond, J. C., Kudo, L., Geschwind, D. H., ... Barlow, C. (2003). Elevated gene expression levels distinguish human from non-human primate brains. *Proc Natl Acad Sci USA, 100*(22), 13030–13035.

Carroll, S. B. (2003). Genetics and the making of *Homo sapiens. Nature, 422*(6934), 849–857.

Caviness, V. S., Takahashi, T., & Nowakowski, R. S. (1995). Numbers, time and neocortical neuronogenesis: A general developmental and evolutionary model. *Trends Neurosci, 18*(9), 379–383.

Charrier, C., Joshi, K., Coutinho-Budd, J., Kim, J.-E., Lambert, N., de Marchena, J., ... Polleux, F. (2012). Inhibition of SRGAP2 function by its human-specific paralogs induces neoteny during spine maturation. *Cell, 149*(4), 923–935.

Chimpanzee Sequencing and Analysis Consortium. (2005). Initial sequence of the chimpanzee genome and comparison with the human genome. *Nature, 437*(7055), 69–87.

Clancy, B., Darlington, R. B., & Finlay, B. L. (2001). Translating developmental time across mammalian species. *Neuroscience, 105*(1), 7–17.

Colantuoni, C., Lipska, B. K., Ye, T. Z., Hyde, T. M., Tao, R., Leek, J. T., … Kleinman, J. E. (2011). Temporal dynamics and genetic control of transcription in the human prefrontal cortex. *Nature, 478*(7370), 519–523.

DeFelipe, J., Lopez-Cruz, P. L., Benavides-Piccione, R., Bielza, C., Larranaga, P., Anderson, S., … Ascoli, G. A. (2013). New insights into the classification and nomenclature of cortical GABAergic interneurons. *Nature Rev Neurosci, 14*(3), 202–216.

Enard, W., Gehre, S., Hammerschmidt, K., Hölter, S. M., Blass, T., Somel, M., … Paabo, S. (2009). A humanized version of Foxp2 affects cortico-basal ganglia circuits in mice. *Cell, 137*(5), 961–971.

Enard, W., Khaitovich, P., Klose, J., Zöllner, S., Heissig, F., Giavalisco, P., … Paabo, S. (2002). Intra- and interspecific variation in primate gene expression patterns. *Science, 296*(5566), 340–343.

Fertuzinhos, S., Krsnik, Z., Kawasawa, Y. I., Rasin, M.-R., Kwan, K. Y., Chen, J.-G., … Sestan, N. (2009). Selective depletion of molecularly defined cortical interneurons in human holoprosencephaly with severe striatal hypoplasia. *Cereb Cortex, 19*(9), 2196–2207.

Fietz, S. A., Kelava, I., Vogt, J., Wilsch-Brauninger, M., Stenzel, D., Fish, J. L., … Huttner, W. B. (2010). OSVZ progenitors of human and ferret neocortex are epithelial-like and expand by integrin signaling. *Nat Neurosci, 13*(6), 690–699.

Fishell, G., & Kriegstein, A. R. (2003). Neurons from radial glia: The consequences of asymmetric inheritance. *Curr Opin Neurobiol, 13*(1), 34–41.

Fukuchi-Shimogori, T., & Grove, E. A. (2001). Neocortex patterning by the secreted signaling molecule FGF8. *Science, 294*(5544), 1071–1074.

Giedd, J. N., & Rapoport, J. L. (2010). Structural MRI of pediatric brain development: What have we learned and where are we going? *Neuron, 67*(5), 728–734.

Glazko, G., Veeramachaneni, V., Nei, M., & Makalowski, W. (2005). Eighty percent of proteins are different between humans and chimpanzees. *Gene, 346*, 215–219.

Gu, J., & Gu, X. (2003). Induced gene expression in human brain after the split from chimpanzee. *Trends Genet, 19*(2), 63–65.

Hansen, D. V., Lui, J. H., Parker, P. R. L., & Kriegstein, A. R. (2010). Neurogenic radial glia in the outer subventricular zone of human neocortex. *Nature, 464*(7288), 554–561.

Haygood, R., Babbitt, C. C., Fedrigo, O., & Wray, G. A. (2010). Contrasts between adaptive coding and noncoding changes during human evolution. *Proc Natl Acad Sci USA, 107*(17), 7853–7857.

Hoekstra, H. E., & Coyne, J. A. (2007). The locus of evolution: Evo devo and the genetics of adaptation. *Evolution, 61*(5), 995–1016.

Hsieh, W.-P., Chu, T.-M., Wolfinger, R. D., & Gibson, G. (2003). Mixed-model reanalysis of primate data suggests tissue and species biases in oligonucleotide-based gene expression profiles. *Genetics, 165*(2), 747–757.

Huttenlocher, P. R., & Dabholkar, A. S. (1997). Regional differences in synaptogenesis in human cerebral cortex. *J Comp Neurol, 387*(2), 167–178.

Ip, B. K., Wappler, I., Peters, H., Lindsay, S., Clowry, G. J., & Bayatti, N. (2010). Investigating gradients of gene expression involved in early human cortical development. *J Anat, 217*(4), 300–311.

Jakovcevski, I., Mayer, N., & Zecevic, N. (2011). Multiple origins of human neocortical interneurons are supported by distinct expression of transcription factors. *Cereb Cortex, 21*(8), 1771–1782.

Johnson, M. B., Kawasawa, Y. I., Mason, C. E., Krsnik, Z., Coppola, G., Bogdanovic, D., … Sestan, N. (2009). Functional and evolutionary insights into human brain development through global transcriptome analysis. *Neuron, 62*(4), 494–509.

Kang, H. J., Kawasawa, Y. I., Cheng, F., Zhu, Y., Xu, X., Li, M., … Sestan, N. (2011). Spatio-temporal transcriptome of the human brain. *Nature, 478*(7370), 483–489.

Khaitovich, P., Muetzel, B., She, X., Lachmann, M., Hellmann, I., Dietzsch, J., … Paabo, S. (2004). Regional patterns of gene expression in human and chimpanzee brains. *Genome Res, 14*(8), 1462–1473.

King, M. C., & Wilson, A. C. (1975). Evolution at two levels in humans and chimpanzees. *Science, 188*(4184), 107–116.

Konopka, G., Bomar, J. M., Winden, K., Coppola, G., Jonsson, Z. O., Gao, F. Y., … Geschwind, D. H. (2009). Human-specific transcriptional regulation of CNS development genes by FOXP2. *Nature, 462*(7270), 213–217.

Konopka, G., Friedrich, T., Davis-Turak, J., Winden, K., Oldham, M. C., Gao, F., … Geschwind, D. H. (2012). Human-specific transcriptional networks in the brain. *Neuron, 75*(4), 601–617.

Kostović, I., & Judas, M. (2002). Correlation between the sequential ingrowth of afferents and transient patterns of cortical lamination in preterm infants. *Anat Rec, 267*(1), 1–6.

Kwan, K. Y., Lam, M. M., Johnson, M. B., Dube, U., Shim, S., Rašin, M. R., … Sestan, N. (2012). Species-dependent posttranscriptional regulation of NOS1 by FMRP in the developing cerebral cortex. *Cell, 149*(4), 899–911.

Kwan, K. Y., Sestan, N., & Anton, E. S. (2012). Transcriptional co-regulation of neuronal migration and laminar identity in the neocortex. *Development, 139*(9), 1535–1546.

Leone, D. P., Srinivasan, K., Chen, B., Alcamo, E., & McConnell, S. K. (2008). The determination of projection neuron identity in the developing cerebral cortex. *Current Opin Neurobiol, 18*(1), 28–35.

Letinic, K., Zoncu, R., & Rakic, P. (2002). Origin of GABAergic neurons in the human neocortex. *Nature, 417*(6889), 645–649.

Liu, X., Somel, M., Tang, L., Yan, Z., Jiang, X., Guo, S., … Khaitovich, P. (2012). Extension of cortical synaptic development distinguishes humans from chimpanzees and macaques. *Genome Res 22*(4), 611–622.

Marín, O., & Rubenstein, J. L. R. (2003). Cell migration in the forebrain. *Annu Rev Neurosci, 26*, 441–483.

Marin-Padilla, M. (1978). Dual origin of the mammalian neocortex and evolution of the cortical plate. *Anat Embryol, 152*(2), 109–126.

Marin-Padilla, M. (2014). The mammalian neocortex new pyramidal neuron: A new conception. *Front Neuroanat, 7*(51), doi:10.3389/fnana.2013.00051.

McLean, C. Y., Reno, P. L., Pollen, A. A., Bassan, A. I., Capellini, T. D., Guenther, C., … Kingsley, D. M. (2011).

Human-specific loss of regulatory DNA and the evolution of human-specific traits. *Nature, 471*(7337), 216–219.

Miller, D. J., Duka, T., Stimpson, C. D., Schapiro, S. J., Baze, W. B., McArthur, M. J., ... Sherwood, C. C. (2012). Prolonged myelination in human neocortical evolution. *Proc Natl Acad Sci USA, 109*(41), 16480–16485.

Molyneaux, B. J., Arlotta, P., Menezes, J. R. L., & Macklis, J. D. (2007). Neuronal subtype specification in the cerebral cortex. *Nat Rev Neurosci, 8*(6), 427–437.

Morest, D. K., & Silver, J. (2003). Precursors of neurons, neuroglia, and ependymal cells in the CNS: What are they? Where are they from? How do they get where they are going? *Glia, 43*(1), 6–18.

Oldham, M. C., Horvath, S., & Geschwind, D. H. (2006). Conservation and evolution of gene coexpression networks in human and chimpanzee brains. *Proc Natl Acad Sci USA, 103*(47), 17973–17978.

O'Leary, D. D., & Nakagawa, Y. (2002). Patterning centers, regulatory genes and extrinsic mechanisms controlling arealization of the neocortex. *Curr Opin Neurobiol, 12*(1), 14–25.

Pakkenberg, B., Pelvig, D., Marner, L., Bundgaard, M. J., Gundersen, H. J., Nyengaard, J. R., & Regeur, L. (2003). Aging and the human neocortex. *Exp Gerontol, 38*(1–2), 95–99.

Pennacchio, L. A., Ahituv, N., Moses, A. M., Prabhakar, S., Nobrega, M. A., Shoukry, M., ... Rubin, E. M. (2006). In vivo enhancer analysis of human conserved non-coding sequences. *Nature, 444*(7118), 499–502.

Petanjek, Z., Kostovi, I., & Esclapez, M. (2009). Primate-specific origins and migration of cortical GABAergic neurons. *Front Neuroanat, 3,* 26.

Pletikos, M., Sousa, A. M. M., Sedmak, G., Meyer, K. A., Zhu, Y., Cheng, F., ... Sestan, N. (2014). Temporal specification and bilaterality of human neocortical topographic gene expression. *Neuron, 81*(2), 321–332.

Pollard, K. S., Salama, S. R., King, B., Kern, A. D., Dreszer, T., Katzman, S., ... Haussler, D. (2006a). Forces shaping the fastest evolving regions in the human genome. *PLoS Genet, 2*(10), 168.

Pollard, K. S., Salama, S. R., Lambert, N., Lambot, M.-A., Coppens, S., Pedersen, J. S., ... Haussler, D. (2006b). An RNA gene expressed during cortical development evolved rapidly in humans. *Nature, 443*(7108), 167–172.

Prabhakar, S., Noonan, J. P., Paabo, S., & Rubin, E. M. (2006). Accelerated evolution of conserved noncoding sequences in humans. *Science, 314*(5800), 786.

Prabhakar, S., Visel, A., Akiyama, J. A., Shoukry, M., Lewis, K. D., Holt, A., ... Noonan, J. P. (2008). Human-specific gain of function in a developmental enhancer. *Science, 321*(5894), 1346–1350.

Rakic, P. (1971). Neuron-glia relationship during granule cell migration in developing cerebellar cortex. A Golgi and electronmicroscopic study in Macacus Rhesus. *J Comp Neurol, 141*(3), 283–312.

Rakic, P. (1995). A small step for the cell, a giant leap for mankind: A hypothesis of neocortical expansion during evolution. *Trends Neurosci, 18*(9), 383–388.

Rice, D. S., & Curran, T. (2001). Role of the reelin signaling pathway in central nervous system development. *Annu Rev Neurosci, 24,* 1005–1039.

Shim, S., Kwan, K. Y., Li, M., Lefebvre, V., & Sestan, N. (2012). Cis-regulatory control of corticospinal system development and evolution. *Nature, 486*(7401), 74–79.

Sidman, R. L., & Rakic, P. (1973). Neuronal migration, with special reference to developing human brain: A review. *Brain Res, 62*(1), 1–35.

Smart, I. H., Dehay, C., Giroud, P., Berland, M., & Kennedy, H. (2002). Unique morphological features of the proliferative zones and postmitotic compartments of the neural epithelium giving rise to striate and extrastriate cortex in the monkey. *Cereb Cortex, 12*(1), 37–53.

Somel, M., Franz, H., Yan, Z., Lorenc, A., Guo, S., Giger, T., ... Khaitovich, P. (2009). Transcriptional neoteny in the human brain. *Proc Natl Acad Sci USA, 106*(14), 5743–5748.

Somel, M., Liu, X., Tang, L., Yan, Z., Hu, H., Guo, S., ... Khaitovich, P. (2011). MicroRNA-driven developmental remodeling in the brain distinguishes humans from other primates. *PLoS Biol, 9*(12), e1001214.

Sun, T., Patoine, C., Abu-Khalil, A., Visvader, J., Sum, E., Cherry, T. J., ... Walsh, C. A. (2005). Early asymmetry of gene transcription in embryonic human left and right cerebral cortex. *Science, 308*(5729), 1794–1798.

Sur, M., & Rubenstein, J. L. (2005). Patterning and plasticity of the cerebral cortex. *Science, 310*(5749), 805–810.

Uddin, M., Goodman, M., Erez, O., Romero, R., Liu, G., Islam, M., ... Wildman, D. E. (2008). Distinct genomic signatures of adaptation in pre- and postnatal environments during human evolution. *Proc Natl Acad Sci USA, 105*(9), 3215–3220.

Uddin, M., Wildman, D. E., Liu, G., Xu, W., Johnson, R. M., Hof, P. R., ... Goodman, M. (2004). Sister grouping of chimpanzees and humans as revealed by genome-wide phylogenetic analysis of brain gene expression profiles. *Proc Natl Acad Sci USA, 101*(9), 2957–2962.

Wagner, G. P., & Lynch, V. J. (2008). The gene regulatory logic of transcription factor evolution. *Trends Ecol Evol, 23*(7), 377–385.

II

PLASTICITY AND LEARNING

Introduction

LEAH KRUBITZER

My first thought when Mike Gazzaniga and Ron Mangun asked me to be the editor for this section of this book was: "What do I know about plasticity and learning?" My second thought was: "Aren't they the same thing?" This motivated a venture to the dictionary, where I looked up the definition of both terms.

Learning: the acquisition of knowledge or skills through experience, study or by being taught. (*The New Shorter Oxford English Dictionary*)

Plasticity: the quality of being plastic; especially: capacity for being molded or altered. (*Merriam-Webster*)

Although a variety of definitions are founsd for both terms, the most common definitions of learning emphasize experience and repetition, while those for plasticity emphasize the capacity to change. If our underlying assumption is that behavior in all animals is mediated by the brain, and the covert behavior of acquiring new information through exposure and repetition requires the capacity to change the brain, then brain plasticity and learning are inextricably interwoven, but not the same thing. It is possible that the extent to which the brain, particularly the neocortex, is plastic distinguishes mammals from each other, and that the mechanisms that generate plasticity evolve. Thus, while the ability to be plastic may be genetically determined, the actual changes to the phenotype are context-dependent. Such changes can appear to be hereditary, masquerading as evolution, when in fact a stable developmental environment is merely triggering the same plasticity in each new generation.

I began my research career trying to understand how complex brains evolve, and quickly realized how restricted this question was. Evolution deals with heritability, and therefore only the genetic component of the phenotype. What I grew to appreciate was that the important question was how complex phenotypes emerge. This deals with both genetic, heritable traits as well as traits that are context-dependent. Every given species is a combination of both, and any individual within a species could have a multitude (or at least a broad distribution) of phenotypes of brain organization and behavior that are expressed in varying environmental contexts. Thus, evolution and plasticity are also intertwined.

The focus of this section, "Plasticity and Learning," is the neocortex, the portion of the brain associated with volitional motor control, perception, and higher cognitive processing. Collectively, this section spans many levels of organization. When considering how the neocortex works and how it changes within the lifetime of an individual and across species over time, it is important to recognize that the neocortex is only one component of the entire nervous system. It should be obvious, but it is valuable to remember that the nervous system is embedded in a body that has sensory receptor arrays and a particular morphology that is often highly derived. For example, a bat interacts with the world in a very different way than a monkey. Each have different forelimb morphologies, a wing versus a hand, and each have enhanced sensory receptor arrays that allow for extraordinary abilities associated with the auditory (sonar) versus the visual (form vision) system, respectively. The cortical areas that process inputs from these specialized effector arrays have expanded and are specified by their unique inputs, and additional fields have been added to the portions of the neocortex that ultimately receive inputs from these sensory systems. Further, cortical networks necessary for sensory motor integration for flight versus manual dexterity, and for interfacing the specialized effectors (ears/pinna versus eyes) with the object to be explored are also uniquely wired within the brain. Finally, each body faces a different and changing physical environment, areal versus terrestrial-arboreal, that constrains movement and perception.

This body/brain configuration of a given individual within a species interacts with other bodies and brains, chasing, courting, mating, attacking, fleeing, communicating, and forming social systems. These individuals interact with other species and other social systems in a variable physical environment composed of living and nonliving elements. This group of organisms and their environment generates a complex and highly dynamic collective biomass that has emergent properties that differ from, and often exceed, the individual parts of which it is composed. For example, a social system has properties that cannot be easily explained by the behavior of a single individual, or brain, or neural firing pattern.

On the other end of the spectrum, brains can be decomposed into cortical networks, made up of nodes or cortical areas, which are composed of local neural circuits. These circuits are formed by neurons, which are held together by the cytoskeleton and plasma membranes and connected to each other by synapses. These in turn include smaller elements such as synaptic receptors, which are composed of molecules encoded by genes, and the expression and transcription of genes can be controlled by environmental context. While genes are the heritable part of this complex hierarchy, it is the larger, often context-dependent components of organization, such as behavior, that are the targets of natural selection. To further complicate matters, the relationship between the targets of selection and genes is often correlative and generally indirect.

Keeping these different levels of organization in mind, the authors of this section were charged with addressing several important questions. What is changing in the neocortex, and does it directly or indirectly co-vary with the target of selection, which in this case is the ability to learn and respond appropriately in a complex and dynamic multisensory context? At what level should we be looking for these alterations in the neocortex? Will there be changes in the microcircuitry (Rodney J. Douglas and Kevan A. C. Martin) and/or the macrociruitry (Olaf Sporns)? How does enhanced or degraded input from our sensory receptor arrays affect the functional organization of the brain and our perceptions of a sensory experience (Jon H. Kaas and Charnese Bowes)? Do particular types of neocortical organizations promote plasticity (Man Chen and Michael W. Deem)? What is the impact of groups of brains (social systems) on the development of brain structure and function (Courtney Stevens and Helen Neville), and how does the brain change throughout the entire life of an individual (Gregg H. Recanzone)? What are the underlying mechanisms that generate these phenotypic alterations at all levels of organization (Tania Roth), from cellular to systems levels, and how do these changes result in variant behaviors? The objective of addressing these questions is to begin to understand the boundary conditions of each level of organization; how higher levels of organization emerge from and interrelate with lower levels; how the brain dynamically generates highly adaptive, context-dependent behaviors throughout a lifetime; and how human social and cultural evolution impact brain evolution.

9 Organizing Principles of Cortical Circuits and Their Function

RODNEY J. DOUGLAS AND KEVAN A. C. MARTIN

ABSTRACT Every percept, every concept, every memory, every decision, and every action we make arise from the activity of neurons in our brains. Of all brain structures, the neocortex, which forms over 80% of the volume of the human brain, is arguably the most critical structure that makes us what we are and allows us to create the societies in which we live. This is a paradox, because the local circuits of the neocortex in all mammals, from mouse to man, are thought to be very similar and determined by the laminar distribution of relatively few types of excitatory and inhibitory neurons organized according to common principles of connectivity. Here we trace how all these different elements contribute to the dense thickets of cortical circuits, and how evolution has organized these circuits across the two-dimensional surface of the cortical sheet to produce a map of behavior.

The most astonishing hypothesis about the human neocortex is not that it is the seat of consciousness, but that it achieves its enormous range of competence through local circuits that may bear a close similarity to those found in the brain of a mouse. Of course, the human neocortex forms a far larger percentage of the brain by volume than that of the mouse (35% vs. 80%), but the basic local circuits, areal organization, and interconnections with subcortical structures are qualitatively similar in man and mouse. The hypothesis that the neocortex of all mammals is built of a common circuit is longstanding: in the early twentieth century, Cecile and Oscar Vogt (1919) referred to neocortex as "isocortex," because of its relatively uniform six-layered appearance. Pioneers of cortical functional architecture, like Mountcastle (1997) and Hubel and Wiesel (1974), were convinced that their favorite areas were built of repeated units, "like a crystal" (Hubel & Wiesel, 1974). In 1980, Rockel, Hiorns, and Powell published their controversial classic entitled "The basic uniformity of structure of the neocortex," in which they claimed that all cortical areas, with the exception of primate area 17, had the same number of neurons lying under each unit of surface area (see Carlo & Stevens, 2013, for a confirmation of Rockel et al.'s findings and a rebuttal of the counterclaims). By the end of the same decade, Douglas, Martin, and Whitteridge (1989) expressed a functional version of this same idea in their "canonical circuit" for the neocortex, based on intracellular recordings and computer simulations of cat area 17. While there is broad agreement that neocortex is built of common components linking in a stereotyped combination, it is worth emphasizing from the start that, however beguiling this concept, it is still a working hypothesis 100 years after the Vogts proposed it.

Composition

A quick look at the composition of the neocortex is instructive. In each cubic millimeter of cortical gray matter we find up to 100,000 neurons, about 100 million synapses, and 4 km of axon (Braitenberg & Schüz, 1991). By contrast the white matter, which is generally thought to contain most of the wiring of the brain, contains only 9 m of axon per cubic millimeter. These bald numbers suggest that the processing strategy of the neocortex is to exploit direct broadband transmission for local circuits in the gray matter and to use the white matter only for long-distance, low-bandwidth transmission. The implication of this is that most of the synapses made in any area originate from neurons within that area, not from outside sources, like the thalamus or other cortical areas. Thus we should first look to the local circuits if we are to understand how the neocortex performs its multifarious tasks of perception, memory, cognition, and action. Of course, the long-distance external inputs to these local circuits are not incidental players, since some of them connect cortex to the peripheral structures, like the spinal cord, while others connect cortex to itself and to other essential machinery like the basal ganglia, claustrum, brain stem, and cerebellar nuclei. Unfortunately, the multiple roles of these external pathways are rather poorly documented. We have nothing like a comprehensive description of their synaptic connectivity and physiology, despite such iconic representations as the summary circuit diagram from Felleman and Van Essen's (1991) meta-study of the hierarchy of connections among visual cortical areas in the macaque monkey.

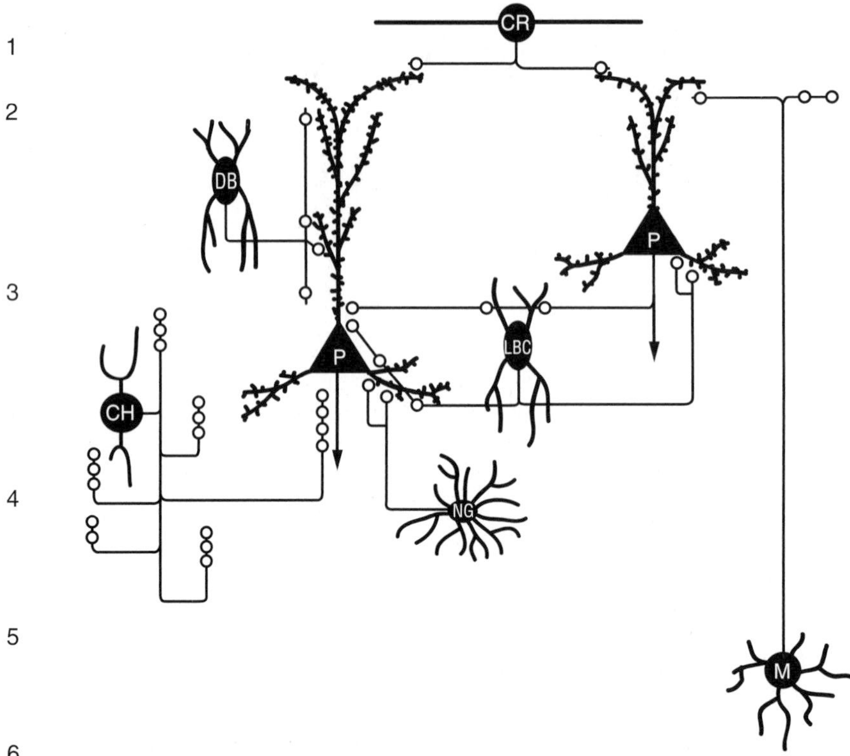

FIGURE 9.1 Pyramidal cells and their principal inputs. Most excitatory (glutamatergic) synapses are formed on spines and arise from other pyramidal cells within the same area. Inhibitory (GABAergic) synapses are provided by neurons that target different portions of the dendritic tree—only the best studied are shown here. P, pyramidal cell; CH, chandelier cell; DB, double bouquet cell; NG, neurogliaform cell; LBC, large basket cell; CR, Cajal-Retzius cell; M, Martinotti cell.

All textbooks will tell you that about 80% of the neurons that occupy the six layers of the neocortex are pyramidal cells. These neurons are spiny, have a prominent apical dendrite, use glutamate as their neurotransmitter, and are excitatory in function (figure 9.1). The remaining 20% are smooth neurons, often misleadingly called "interneurons" without further qualification. These neurons do not have densely spiny dendrites (hence "smooth"), use gamma amino-butyric acid (GABA) as their neurotransmitter, and are inhibitory in function. The confusion about cortical "interneurons" arises because they were traditionally defined as cells whose entire axon arborizes in the region around its parent cell body; that is, in contrast to "projection neurons," their axon does not project out of their own cortical area. While this is largely true of the smooth neurons, is it also true of some pyramidal cells and of most members of the second type of excitatory neuron, the "spiny stellate," which is spiny, lacks an apical dendrite, and has a rather symmetric dendritic arbor, which gives it its name. Spiny stellate neurons are not found throughout the neocortex, but only in the middle layers of certain sensory cortical areas like rodent barrel cortex and the primary visual cortex of cats and monkeys. Interestingly, they are not found in some primary sensory areas, like the auditory cortex of the mouse. Spiny stellate cells are usually lumped together with the pyramidal cells, since both are spiny and excitatory. Spiny stellate neurons are not found throughout the neocortex, but only in the middle layers of certain sensory cortical areas like rodent barrel cortex and the primary visual cortex of cats and monkeys (Lund, Lund, Hendrickson, Bunt & Fuchs, 1975).

The 4:1 ratio of the two basic morphological and physiological types of neurons is standard in most "biologically realistic" models of neocortical functions, but in two-dimensional (2D) models of cortex their relative spatial distributions are usually tweaked to ensure that monosynaptic inhibitory connections spread further than the excitatory connections. This center-surround pattern (also called a DOG [Difference of Gaussians] or "Mexican hat") is widespread, as, for example, in the structure of the receptive fields of retinal ganglion cells and thalamic relay cells. In models of cortex the pattern of center excitation–surround inhibition is also applied to ensure stability of the network. However, in reality, the smooth neurons typically have much more compact axon arbors than the pyramidal cells, but this spatial characteristic has been hard to incorporate in models without producing a "leak" of recurrent excitation that leads to instability in the network.

The yin and yang of excitation and inhibition probably arose as an early feature in the evolution of the nervous system. In his studies of spinal cord reflexes, Charles Sherrington found that excitatory and inhibitory neurons were always in tandem and that together they provided the algebra of the nervous system: "The net change which results there when the two areas are stimulated concurrently is an algebraic sum of the plus and minus effects producible separately by stimulating singly the two antagonistic nerves" (Sherrington, 1908, 578). The algebra of the cortex has proved to be more elaborate, but the principle of controlling the balance between excitation and inhibition is a fundamental property needed for cortical computation, and not simply as a means of stabilizing excitation.

Diversity

Smooth neurons take a variety of forms, and the "diversity" of the smooth neuron morphology has become a cliché. Their fascination lies in that fact that their axonal arbors arrange themselves in a variety of exotic shapes, which reminded the neuroanatomists who named them of baskets, bouquets, chandeliers, horsetails, glial forms, and the like (figure 9.1). Other smooth neurons are named after their discoverers, such as Martinotti, Cajal, and Retzius, as are some pyramidal cells, like Betz and Meynert, who gave their names to the large solitary cells that project to the spinal cord or to the midbrain, respectively. Pyramidal and spiny stellate cells distinguish themselves not so much by the shape of their axons, but by their distinct dendritic arborizations and by the precision of their axonal projections within and between layers. Adding to this, pyramidal cells are further differentiated from each other by their projections to specific target structures lying outside their local area. Thus, two neighboring pyramidal cells can have very similar local dendritic trees and axonal arborizations, yet project to different targets. When all these characteristics are included, spiny neurons clearly show at least as much diversity in their morphology as their smooth cousins.

The anatomical diversity of the axons and dendrites of cortical neurons has strong implications for connectivity. Their very shape and size generates important constraints on what connections are possible (since the vast majority of synapses are formed between dendrites and axons) and how many synapses they can support. The laminated structure of cortex has a major influence here, as does the actual form of the dendritic and axonal arbors. Axonal arborizations of the inhibitory neurons come in a few basic forms (figure 9.1): one that forms a tight local cluster (small basket cells, neurogliaform cells, chandelier cells), one that elongates horizontally and remains largely within a single lamina (large basket cells, Cajal-Retzius cells), and a third form where the axon extends radially (double bouquet cells, Martinotti cells). In all cases the axon arbors are relatively compact compared to those of the spiny neurons, whose axons sometimes form a local cloud of synapses in the region of the dendritic tree, but then project horizontally within their home layer and/or to another layer. Pyramidal cells and spiny stellate cells send their axons in rather straight trajectories before branching to form a localized terminal cluster of synapses (see Douglas & Martin, 2004).

The diversity also extends to the differences in synapses—certainly to the dynamics of the synapses and to their palette of postsynaptic receptors. However, at the ultrastructural level there is a clear binary division in the morphology of synapses, which was first noted by Gray (1959), who named the two types of synapses Type 1 and Type 2. Subsequent investigations have strengthened the binary classification, and we now know that Type 1 synapses are glutamatergic and excitatory in function, while Type 2 are GABAergic and inhibitory in function. On average, the number of synapses formed by each pyramidal cell axon is more than that formed by each smooth neuron. In addition, the vast majority of long-distance projections entering an area are excitatory, thus further diluting the aliquot of inhibitory synapses provided by local smooth neurons. Estimates vary across species and areas, but the proportion of inhibitory synapses in any cortical area lies between 8% and 18%.

Rules of engagement

Since synapses form at an approximately constant density along a dendrite, the length of dendrite that a given neuron contributes to a layer gives a good approximation to the number of synapses that can be contributed by the multiple axons that arborize within the same layer. This potential connectivity is the basis for Peters's Rule, which predicts that neurons connect to each other in proportion to the amount of dendrite or axon they contribute to a particular layer (Braitenberg & Schüz, 1991; Peters & Payne, 1993). In this form it is a convenient "poor-man's" method of establishing an average circuit of a given patch of cortex. One common interpretation of Peters's Rule is that if it holds, it is evidence that the connections are made randomly. It is hazardous to interpret Peters's Rule in this way without qualification, however, since in its statistical averaging it says nothing about any possible specificity of neuron-to-neuron connections. Indeed, there are many studies that show that very specific connections exist, but nonetheless the overall statistics follow Peters's Rule. The most prominent lesson comes from studies of the connection between thalamus and layer 4 of the primary visual cortex in the cat, where about 6% of the synapses arise from the thalamic relay cells (as predicted by Peters himself; see Binzegger, Douglas, & Martin, 2004; Peters & Payne, 1993) and thus implying a random connectivity. Physiology, however, shows that the neurons in layer 4 are the textbook "simple cells" of Hubel and Wiesel (1962): they are orientation-selective and have separate "on" and "off" fields that are most likely a consequence of a highly orderly arrangement of inputs from thalamic relay cells. This nonrandom patterning of connections is also reflected in the highly orderly functional architecture of cortex, such as the organized maps for orientation and the arrangements of common functional properties in radial columns. The specificity in connections needed to generate these properties is invisible to the average statistics generated by applying Peters's Rule.

Deviations from Peters's Rule, which have been called White's Exceptions by Braitenberg and Schüz (1991), are also important in signposting specific connections. They are increasing in number since White first reported that the thalamic projection to layer 4 of the mouse somatosensory formed 20% of the excitatory synapses on spiny neurons, but only 4% on adjacent smooth neurons (see White, 1989). This is quite different from cat visual cortex, where the pattern follows the Peters's Rule prediction that thalamic afferents will form approximately equal numbers on spiny and smooth neurons (Ahmed, Anderson, Douglas, Martin, & Nelson, 1994; Binzegger et al., 2004). In cat and monkey visual areas, pyramidal cells form between 80% and 90% of their synapses with spines. Thus the most recent example of a White's Exception is very puzzling: Bock et al. (2011) found that smooth neurons formed fully 50% of the targets of pyramidal cells in mouse visual cortex, whereas Peters's Rule predicts at best 20%. What does this say about mouse's strategy for visual processing compared to that of the cat or monkey?

Mapping on the body

The distribution of excitatory and inhibitory synapses on their targets is also nonrandom, for they concentrate their input on distinct structures or regions of their postsynaptic neuron (figure 9.1). For example, the thalamic relay cells form multiple synapses on the somata of small basket cells in layer 4 of the cat's visual cortex, but not on the somata of adjacent spiny cells. Spiny neurons concentrate their input on spines and never form synapses with the somata of other spiny neurons. However, they do form synapses with the somata of smooth neurons. This design suggests a need to drive the inhibitory cells to spike threshold as fast as possible, without the intervention or delay of any dendritic processing, which may be a sign that even minute differences in the timing of inhibition and excitation may be critical (Ohana, Portner, & Martin, 2012). Although it is relatively rare that excitatory synapses form on the dendritic shafts of excitatory cells, one type of pyramidal cell in layer 6, which projects back to the thalamus, specifically forms the majority of its synapses on the dendritic shafts of spiny cells in layer 4 (Ahmed et al., 1994).

Traditionally, inhibitory synapses were thought to locate themselves mainly on the somata and proximal dendritic tree of spiny cells, but recent evidence shows otherwise. Inhibitory synapses are actually found on all parts of the neuron, including distal dendrites, the initial segment of the axon, and everything in between (figure 9.1). This territory, from tip of dendrite to axon, is actually divided up between the different subtypes of smooth neurons. Thus, the chandelier cells form their synapses exclusively with the initial segment of the axon of pyramidal cells (mainly in the superficial layers), whereas the basket cells tend to concentrate their synapses on the proximal parts of the neuron, although they also form a substantial fraction—between 20% and 40%—of their synapses with dendritic spines. The radial axons of the double bouquet cell and Martinotti cells span the layers and form synapses on the distal shafts and spines of pyramidal cells.

The existence of inhibitory synapses on distal dendrites has long been a puzzle, particularly for dendrites entering layer 1, which itself contains only GABAergic neurons, albeit at a very low density. Standard cable theory predicts that the most efficient site for inhibitory synapses is the region around the cell body, which is indeed dominated by inhibitory synapses. However, physiological studies *in vitro* of the large pyramidal cells of layer 5 reveal reasons for distal inhibition. The distal regions of the apical dendrite of these cells contain at least two active excitatory conductances—an N-methyl D-aspartate (NMDA) and a calcium conductance—which can be quenched by activating the inhibitory synapses to the distal regions of the apical dendrite (Larkum, Nevian, Sandler, Polsky, & Schiller, 2009;

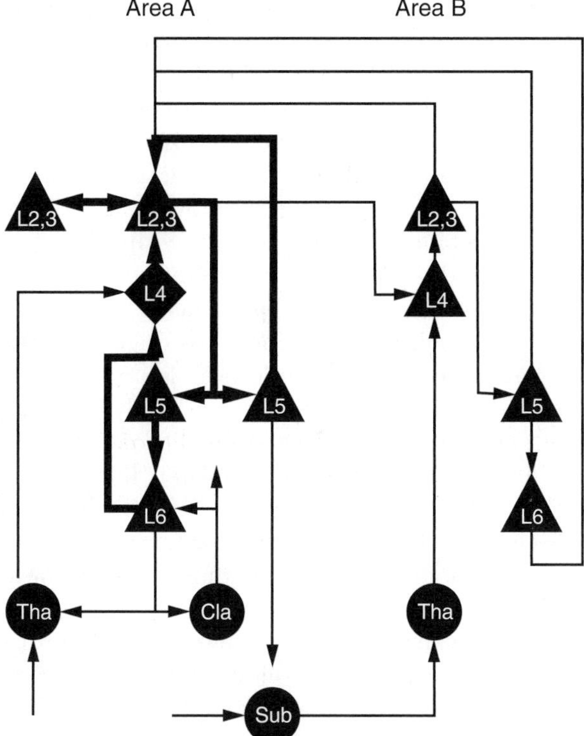

FIGURE 9.2 Main pattern of connections of the excitatory cell types found across areas of neocortex. Thick lines indicate the dominant interlaminar and recurrent intralaminar connections between excitatory neurons in the local circuit. Thin lines indicate the main long-distance connections to other cortical areas and subcortical targets. Triangle indicates pyramidal cell; diamond indicates spiny stellate. Numbers indicate layers. Source and target structure indicated by circles: Thal, thalamus; Sub, subcortical structures such as striatum, superior colliculus, or spinal cord; Cla, claustrum.

Murayama et al., 2009). The sources of the distal inhibitory synapses are the Martinotti cells of layer 5 and inhibitory neurons in layer 1, like the Cajal-Retzius cells (figure 9.2). In a revision of the conventional theories, recent theoretical work has also indicated that inhibitory synapses located on distal branches of apical dendrites can more effectively quench distal excitatory hotspots than inhibition located at sites more proximally (Gidon & Segev, 2012). Thus, both experimental and theoretical work agree that the dendritic tree, soma, and axon initial segment of pyramidal cells are all involved in the active engagement of excitation and inhibition. These new findings go beyond Sherrington's notion of an algebraic sum of plus and minus effects, and encourage the view that the soma-dendrite complex of single neurons are computationally rich devices, not simply the most convenient means of connecting A with B.

The long and the short of connections

The basic equation of cortical connectivity is simple—on average, a given neuron in the local circuit needs to make as many synapses as it receives. Estimates from the monkey are that about 70–80% of the neurons that are labeled retrogradely by a tracer injection in any area of cortex lie in the same area. Similarly, 70–80% of the synapses in any cortical area originate from neurons in that area. Thus, most of the 4 km of axonal "wire" in each cubic millimeter of the gray matter, is generated mainly by local neurons connecting with each other (figure 9.2). This strategy of making most of the connections with nearby neurons is space-saving, for lengthening the wires necessarily increases their total volume relative to the computational elements, both for neurons in brains and transistors in silicon circuits (Mead, 1990; Mitchison, 1991). The problem of scaling can be severe, too, since larger brains require longer axons to connect two areas of cortex and so occupy more white matter volume. The ratio of white matter to gray matter does increase with increasing brain size, but not explosively so (Zhang & Sejnowski, 1990). The axons in larger brains also have to increase their diameter and degree of myelination if they are to maintain reasonable speeds of conduction over long distances. Indeed, the white matter axons are myelinated, whereas most of the axons in the local circuit are finer-caliber and unmyelinated, and hence slower-conducting. Although the sheer length of wire makes the analysis of the local circuits dauntingly complex, many have been excited by the thought that a fundamentally new understanding of cortical processing will come from a detailed analysis of one cubic millimeter of gray matter, rather than a study of an entire system or an entire brain (e.g., Helmstaedter, de Kock, Feldmeyer, Bruno, & Sakmann, 2007; Markram, 2006).

Of course, there are long-distance connections to transmit information between cortical and subcortical areas (figure 9.2). The most critical of these is the input from the primary thalamic nuclei, which project principally to layers 4 and 6. The secondary thalamic nuclei project to all the remaining layers, including layer 1. Surprisingly, none of these thalamic relays form large numbers of synapses in their target layers. In the visual cortex, for example, the thalamus contributes less than 10% of all the excitatory synapses even in layer 4, its main target layer (Ahmed et al., 1994; Da Costa & Martin, 2009). On the output side, the corticospinal tract is also tiny—perhaps only a million fibers connect our primary motor cortex to output regions of the ventral spinal cord. Given the rich sensory world we seem to occupy, it is sobering to discover that the constraints of wire minimization in the brain mean that our cortex is connected to our sense organs and muscles by only the thinnest of threads. A few bytes come into a neocortex-dominated brain that dynamically interprets this sparse input and then delivers a few bytes to a final common output path that, from an evolutionary perspective, is relatively unchanged across species. Leaving aside the special modifications needed for species-specific behavior (e.g., corticospinal neurons in humans and nonhuman primates with opposable thumbs are specified for fine motor control of the hands), what has changed is not so much the overt behavior, but the conditions under which this behavior is generated, and the temporal relationship between stimulus and response (which can be very long in species with a big neocortex).

The long-distance connections between areas come in three basic motifs: feedforward, feedback, and lateral (Felleman & Van Essen, 1991; Rockland & Pandya, 1979; figure 9.2). The concept of directionality in these interareal circuits provides the major framework for understanding the hierarchy of structure-function relationships of the cortex (Crick & Koch, 1998; Felleman & Van Essen, 1991; Rockland & Pandya, 1979). Precise quantification of the laminar origins and numbers of neurons involved has revealed a second organizing principle with which to establish hierarchical relationships (Barone, Batardiere, Knoblauch, & Kennedy, 2000; Markov et al., 2012). This is achieved experimentally by tracing all the projections into a given area. This is not simply a qualitative map, for the exact proportion of superficial to deep-layer neurons labeled in a given donor area depends precisely on its hierarchical distance from its recipient target area. Thus, feedforward projections to far distant areas originate almost exclusively from superficial-layer neurons (layers 2 and 3), but as one approaches the target area there is a monotonic increase in the contribution from the deep layers (layers 5 and 6), until adjacent areas provide their input from almost equal numbers of neurons from their superficial and deep layers. Likewise for feedback projections, nearby areas show equal numbers of labeled superficial- and deep-layer neurons, and as the hierarchical distance increases there is a steady increase in the proportion of labeled cells in deep layers, so that far distant feedback projections originate almost entirely from deep-layer neurons. This regularity has been formalized as a "distance rule" (Rockland & Pandya, 1979), which has the power to define the hierarchical organization of a cortical network from the analysis of the projections to only a small number of key areas (Barone et al., 2000). In addition, the number of neurons involved also falls exponentially with distance, so that the most projection neurons are used to connect to the area next door and the fewest neurons are used for the very long-distance connections, as one would expect if wire use were being optimized. This is not good news for human tractographers, because as the studies of macaque cortex indicate (e.g., Markov et al., 2012), the number of white matter fibers involved in many area-to-area pathways falls way below the resolution of current noninvasive imaging instruments.

Although feedforward connections originate principally from neurons in the superficial layers, they form their synapses in the middle layers (mainly layer 4) of the target area. Thus they are thought to act similarly to the inputs from the primary thalamic nuclei and to provide the primary drive to the local circuits. Feedback connections, which originate predominantly from neurons in the deep layers, form their synapses outside the middle layers of their target areas, much like the secondary thalamic nuclei do. These patterns translate to functional differences: cooling feedforward neurons to reduce their activity has a much more dramatic effect on their target structures than cooling feedback neurons (Bullier, Hupé, James, & Girard, 2001). Lateral projections originate from all layers and send their axons to all layers. With the exception of a lateral projection in the monkey that carries an important attentional signal from prefrontal to extrastriate cortex (Anderson, Kennedy, & Martin, 2011; Moore & Fallah, 2001), the functional roles of lateral projections have yet to be studied.

Over 60% of cortical areas that can connect with one another in the macaque neocortex actually do (Markov et al., 2012). Thus, while "small-world" architectures (Watts & Strogatz, 1988) have been the theorists' favorite description of cortical circuits, the reality is that the interareal network is dense and highly structured. Most of the connections between two areas involve very small numbers of neurons, and only near-neighboring areas connect with each other using substantial numbers of neurons (Markov et al., 2012). This is a crucial observation, for it means that the vast majority of interareal connections in the macaque monkey involve connections to areas less than 12 mm distant in a cortical sheet of over 60 mm. A single area usually projects to multiple levels of the hierarchy, with the result that any single area receives its input from between 26 and 87 other cortical areas, and at least 10% of these are unidirectional (Markov et al., 2012).

These new connectivity studies in macaque monkeys also indicate that nearby areas typically receive their cortical inputs from many areas in common, and this raises fundamental questions as to what actually determines the functional subdivisions of cortex seen with many physiological and neurological methods. That areas share common cortical inputs, yet still differ functionally, suggests that the cortico-cortical connections are not the dominating factor in determining the functional parcellation. This leaves the thalamic input with the job of generating functional differentiation. On the other hand, the degree of convergent input from other cortical areas offers extraordinary combinatorial possibilities for context-dependent computations that lead to behavior (see, e.g., the discussion in Markov et al., 2012). Indeed, given that the thalamus maps in a highly orderly and well-conserved fashion onto the cortical sheet, variations in the precise pattern of long-distance interareal connections across the cortical sheet may be the major source of variation between species in evolution (Krubitzer, 2009). The high degree of convergence of these long-distance connections also offers a means of plasticity in response to altered environment, or even brain damage. Similarly, the robust performance of the individual in the face of ever-changing contexts is an adaptive behavior at which humans seem particularly advantaged. If best choices are based on accumulating evidence to weigh a number of alternative strategies, then the small contributions from multiple areas may be a significant factor in biasing the probability toward better decision making (Gold & Shadlen, 2002). Similarly, while the loss of some of the inputs would not be catastrophic, their loss may nevertheless alter the bias of probabilities and inject more uncertainty into the process of decision making.

Local processor

What is so special about the local circuits of the neocortex? What makes them so efficient and so adaptable to different tasks? An early insight into the cortical circuit came with the evidence of Hubel and Wiesel (1962) that input arriving from the thalamus was processed serially through the cortical layers (figure 9.2). Their physiological measurements of the receptive fields of neurons in the different cortical layers of the cat and the monkey were consistent with structural studies of the projection patterns reconstructed from intracellularly labeled spiny neurons (Gilbert & Wiesel, 1979; Martin & Whitteridge, 1984). Importantly, the pattern of excitatory connections in these very detailed reconstructions was consistent with generations of anatomical descriptions of other cortical areas, suggesting that the pattern is conserved across species and across different areas of sensory cortex. Thus the rat barrel cortex, the cat and monkey visual cortex, and the cat's auditory cortex, for example, show a very comparable "backbone" of excitatory connections between the layers (Douglas & Martin, 2004). The neurons concerned do not simply transfer information serially between layers, but also form a series of nested positive and negative feedback loops called recurrent circuits. This is a radically different means of computation than the model of serial processing offered by Hubel and Wiesel (1962).

The principle of a recurrent cortical circuit is most simply captured in the model called the "canonical circuit" (Douglas et al., 1989; Douglas & Martin, 2004), which expresses the functional relationships between the excitatory and inhibitory neurons in the different cortical layers and shows how the inputs to a local region of cortex from the sensory periphery via the thalamus, or from other cortical areas, are integrated. There are three major functional features of the circuit: first, the thalamic input is weak, which is reflected in the small proportion of thalamic synapses seen in layer 4 (Da Costa & Martin, 2009). The thalamic synapses also depress with repeated activation, thus further reducing their efficacy (Stratford, Tarczy-Hornoch, Martin, Bannister, & Jack, 1996). Second, there is strong recurrent excitation within and between layers (see figure 9.2). Third, excitation and inhibition are linked in tandem, so that they remain in balance and prevent the positive feedback from overexciting the circuit. This organization explains how it is that the relatively tiny numbers of thalamic synapses are nevertheless effective if they act in synchrony and are amplified by recurrent excitatory circuits. Explorations of this model in the visual cortex have shown how this key notion of recurrent amplification can explain the emergence of cortical properties, such as direction and velocity sensitivity, orientation selectivity, masking, contrast adaptation, and winner-take-all behavior (Ben-Yishai, Bar-Or, & Sompolinsky, 1995; Douglas et al., 1989; Douglas, Koch, Mahowald, Martin, & Suarez, 1995; Douglas & Martin, 2004; Somers, Nelson, & Sur, 1995).

Virtually "transplanted" to a very different part of the brain, a simulation of a detailed canonical circuit derived from cat visual cortex has provided a comprehensive explanation as to how the frontal eye field area of prefrontal cortex might drive saccadic eye movements in monkeys and humans (Heinzle, Hepp, & Martin, 2007). Principles captured in the canonical circuit are thus generalizable, as they should be. It should be emphasized, however, that while the notion of an "isocortex" has a long history, it is far from proven. Good evidence exists for common rules of connections at many levels across areas and species, but we still lack quantitative comparisons of even the best-studied areas, like visual cortex in cat and barrel cortex in rat, that would help convince us that these areas generate the same computations with the same canonical circuits.

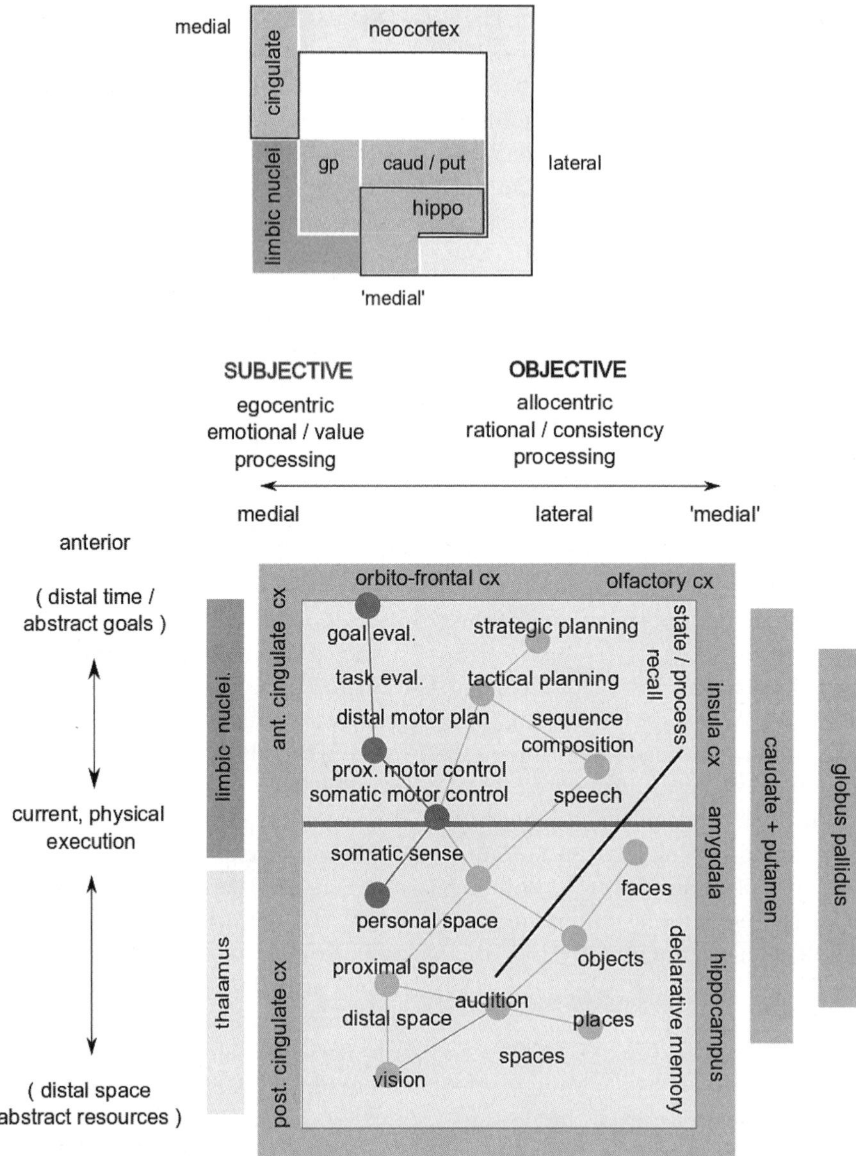

FIGURE 9.3 Organization of the cortical sheet: schematic showing functional organization of cortical sheet and some of its related structures in the right hemisphere. *Above*: Transverse section shows how limbic nuclei, basal ganglia (gp, globus pallidus; caud/put, caudate and putamen), and hippocampus (hippo) are folded and come to occupy more "medial" positions. *Below*: Plan view of the cortical sheet (cyan) with central sulcus indicated by thick red horizontal line. The sheet is surrounded by limbic cortical regions (brown; e.g., cingulate cortex, insula) and associated limbic nuclei (red; e.g., septal nuclei, amygdala). Dynamically evolving behaviors are represented schematically as "nodes," representing regions of active processing, and "edges," which represent the axonal communication channels between active nodes. The channels act directly through cortico-cortical connections, or indirectly via thalamus and basal ganglia. Multiple behaviors may evolve simultaneously (green graph; see text), while the red graph represents the various functional relations of the behavior currently being executed. (See color plate 4.)

Nonetheless, the unifying hypothesis that the neocortex is essentially a computational sheet, which assembles itself using common components and rules of connection, is not only extremely attractive, it offers new insights into the way in which the neocortex might evolve, develop, and function. Importantly, it is on this cortical sheet that the sensory, perceptual, emotive, cognitive, and motor aspects of behavior are systematically mapped and form their connections with the rest of the brain and spinal cord.

Cortical sheet

Having established some of the key principles by which local and long-distance cortical connections organize themselves, we now show that these anatomical constraints have a direct interpretation in terms of the structure and dynamics of behavior (figure 9.3). The organization of the neocortical sheet and its relation to other brain structures is a consequence of the long evolutionary history of the vertebrate forebrain, which forms from the dorsal part of the most anterior segment of the neural tube. Symmetrical telencephalic hemispheres grow out of the midline diencephalon, and the neocortex forms from the dorsal regions of these hemispheres. The dorsal diencephalon contains the thalamus, whose nuclei provide bidirectional connections to all regions of the neocortex. The margin (or limbus) of the two telencephalic hemispheres contains archaic structures that provide emotive content and valuation of current input, as well as memory. These structures include the hippocampus (spatiotemporal memory), cingulate cortex (emotion and evaluation), olfactory cortex (olfactory perception), and the amygdala (emotional memory). The ventral telencephalon gives rise to the basal ganglia, which interact with the cortex in the context of planning, initiation, and regulation of movement as well as a variety of cognitive and affective functions.

Now imagine the highly folded primate cortex laid out as a flat sheet in plan view (figure 9.3). We observe that it is organized along two axes (Douglas & Martin, 2012). Regions immediately around the central sulcus relate to ongoing sensorimotor behavior. More anteriorly, the premotor areas of frontal cortex process potential action repertoires, plans, and goals that are temporally bound to the concurrent actions. Further anteriorly lie the prefrontal areas that are not temporally bound to the immediate stimulus, but process plans and goals extended in time from the near-present to well into the distant future. The cortex posterior to the central sulcus processes space. Here, as one moves more caudally on the cortical sheet, the space to be represented expands from the body surface, as represented in the somatosensory cortex; to proximal space represented in the parietal cortex; to the more expansive distal spaces provided by the senses of hearing, represented in the posterior areas of the temporal cortex, and vision, which involves areas lying in the most posterior areas of the cortical sheet. The ventral and anterior areas of the temporal lobe also encode places, things, and faces. This lobe is greatly enlarged in primates, including humans, and is intimately related to the declarative memory functions involving the hippocampus. The emotive, subjective signals of the limbic cortices color or bias the processing of the evolving plans of the more medial areas of the cortical sheet. More lateral cortical areas are relatively isolated from these biases and process alternative action sequences and spatial structure in a more semantically and syntactically objective manner.

The mediolateral axis can be described as an evaluate/deploy axis (figure 9.3). Particularly in the prefrontal region, this axis links limbic evaluation to appropriate behavioral procedures. Thus, behaviors evolving on the cortical sheet can be viewed as coherent processing between transiently bound sets of active subregions across the sheet, the effect of each processing region being characterized by its spatiotemporal scale in the environment and the explore/exploit quality of its role. One consequence of this view is that the "hierarchical" nature of behavioral scale, as it expresses itself in action-space and action-time, is directly encoded in the cortical sheet (Koechlin, Ody, & Kouneiher, 2003). In its plan, the cortical sheet thus offers a natural syntax of permissible patterns, and a natural semantics in terms of the particular regions of the sheet that are activated. The arrangement of these fields of neurons appears crucial to the generation of behavior, not least because the fundamental organization of the forebrain is highly conserved and scales across mammals from mouse, to monkey, to man, along with increased cognitive performance (Finlay & Darlington, 1995; Jerison, 1973; Passingham, 1982; Yopak et al., 2010).

As humans, we tend to attribute our rational actions to a "self"—an individual sentient agent. Although the forebrain may support this impression in terms of its coherent external actions, we can now see that coherent behaviors arise out of neuronal processing that is at any time widely and inhomogeneously distributed across fields of neurons that are, on average, only sparsely connected with one another. This architecture is so different from the artificial ones that we have designed that it is not surprising that our very partial descriptions of it are still far richer than our ideas of how it might actually work. While uncovering the structural principles that organize the cortical circuits and constrain their interactions will not lead automatically to knowledge of their higher-order functions, it is a major step in understanding how their function leads to the intelligent behavior.

ACKNOWLEDGMENTS The authors acknowledge generous support from European Union Future and Emerging Technologies Grant, "Self Constructing Computing Systems" (SECO) (RJD & KACM), Swiss National Fund Sinergia, Grant "Neural Circuit Reconstruction," the Swiss Federal Institute of

Technology (ETH) Zurich, and Human Frontiers Science Program (KACM).

REFERENCES

AHMED, B., ANDERSON, J. C., DOUGLAS, R. J., MARTIN, K. A. C., & NELSON, J. C. (1994). The polyneuronal innervation of spiny stellate neurons in cat visual cortex. *J. Comp Neurol, 341*, 39–49.

ANDERSON, J. C., KENNEDY, H., & MARTIN, K. A. C. (2011). Pathways of attention: Synaptic relationships of frontal eye fields to V4, lateral intraparietal cortex, and area 46 in macaque monkey. *J Neurosci, 31*, 10872–10881.

BARONE, P., BATARDIERE, A., KNOBLAUCH, K., & KENNEDY, H. (2000). Laminar distribution of neurons in extrastriate areas projecting to visual areas V1 and V4 correlates with the hierarchical rank and indicates the operation of a distance rule. *J Neurosci, 20*, 3263–3281.

BEN-YISHAI, R., BAR-OR, R. L, & SOMPOLINSKY, H. (1995). Theory of orientation tuning in visual cortex. *Proc Natl Acad Sci USA, 92*, 3844–3848.

BINZEGGER, T., DOUGLAS, R. J., & MARTIN, K. A. C. (2004). A quantitative map of the circuit of cat primary visual cortex. *J Neurosci, 24*, 8441–8453.

BOCK, D. D., LEE, W. C., KERLIN, A. M., ANDERMANN, M. L., HOOD, G., WETZEL, A. W., ... REID, R. C. (2011). Network anatomy and *in vivo* physiology of visual cortical neurons. *Nature, 471*, 177–182.

BRAITENBERG, V., & SCHÜZ, A. (1991). *Anatomy of the cortex.* Berlin: Springer.

BULLIER, J., HUPÉ, J. M., JAMES, A. C., & GIRARD, P. (2001). The role of feedback connections in shaping the responses of visual cortical neurons. *Prog Brain Res, 134*, 193–204.

CARLO, C. N., & STEVENS, C. F. (2013). Structural uniformity of neocortex revisited. *Proc Natl Acad Sci USA, 110*, 1488–1493.

CRICK, F., & KOCH, C. (1998). Constraints on cortical and thalamic projections: The no-strong-loops hypothesis. *Nature, 391*, 245–250.

DA COSTA, N. M., & MARTIN, K. A. C. (2009). The proportion of synapses formed by the axons of the lateral geniculate nucleus in layer 4 of area 17 of the cat. *J Comp Neurol, 516*, 264–276.

DOUGLAS, R. J., & MARTIN, K. A. C. (2004). Neuronal circuits of the neocortex. *Ann Rev Neurosci, 27*, 419–451.

DOUGLAS, R. J., & MARTIN, K. A. C. (2012). Behavioral architecture of the cortical sheet. *Current Biol, 22*, R1–R6.

DOUGLAS, R. J., KOCH, C., MAHOWALD, M., MARTIN, K. A., & SUAREZ, H. H. (1995). Recurrent excitation in neocortical circuits. *Science, 269*, 981–985.

DOUGLAS, R. J., MARTIN, K. A. C., & WHITTERIDGE, D. (1989). A canonical microcircuit for neocortex. *Neural Comput, 1*, 480–488.

FELLEMAN, D. J., & VAN ESSEN, D. C. (1991). Distributed hierarchical processing in the primate cerebral cortex. *Cereb Cortex, 1*, 1–47.

FINLAY, B. L., & DARLINGTON, R. B. (1995). Linked regularities in the development and evolution of mammalian brains. *Science, 268*, 1578–1584.

GIDON, A., & SEGEV, I. (2012). Principles governing the operation of synaptic inhibition in dendrites. *Neuron, 75*, 330–341.

GILBERT, C. D., & WIESEL, T. N. (1979). Morphology and intracortical projections of functionally characterised neurons in the visual cortex. *Nature, 280*, 120–125.

GOLD, J. I., & SHADLEN, M. (2002). Banbarismus and the brain: Decoding the relationship between sensory stimuli, decisions and reward. *Neuron, 36*, 299–308.

GRAY, E. G. (1959). Axo-somatic and axo-dendritic synapses of the cerebral cortex: An electron microscope study. *J Anat, 93*, 420–433.

HEINZLE, J., HEPP, K., & MARTIN, K. (2007). A microcircuit model of the frontal eye fields. *J Neurosci, 27*, 9341–9353.

HELMSTAEDTER, M., DE KOCK, C. P., FELDMEYER, D., BRUNO, R. M., & SAKMANN, B. (2007). Reconstruction of an average cortical column in silico. *Brain Res Rev, 55*, 193–203.

HUBEL, D. H., & WIESEL, T. N. (1962). Receptive fields, binocular interaction and functional architecture in the cat's visual cortex. *J Physiol, 160*(1), 106–154.

HUBEL, D. H., & WIESEL, T. N. (1974). Uniformity of monkey striate cortex: A parallel relationship between field size, scatter, and magnification factor. *J Comp Neurol, 158*, 295–305.

JERISON, H. J. (1973). *Evolution of the brain and intelligence.* New York, NY: Academic.

KOECHLIN, E., ODY, C., & KOUNEIHER, F. (2003). The architecture of cognitive control in the human prefrontal cortex. *Science, 302*, 1181–1185.

KRUBITZER, L. (2009). In search of a unifying theory of complex brain evolution. *Ann NY Acad Sci, 1156*, 44–67.

LARKUM, M. E., NEVIAN, T., SANDLER, M., POLSKY, A., & SCHILLER, J. (2009). Synaptic integration in tuft dendrites of layer 5 pyramidal neurons: A new unifying principle. *Science, 325*, 756–760.

LUND, J. S., LUND, R. D., HENDRICKSON, A. E., BUNT, A. H., & FUCHS, A. F. (1975). The origin of efferent pathways from the primary visual cortex, area 17, of the macaque monkey as shown by retrograde transport of horseradish peroxidase. *J Comp Neurol, 164*, 287–304.

MARKOV, N. T., ERCSEY-RAVASZ, M. M., GOMES, A. R. R., LAMY, C., MAGROU, L., VEZOLI, J., ... GARIEL, M. A. (2012). A weighted and directed interareal connectivity matrix for macaque cerebral cortex. *Cereb Cortex, 24*(1), 17–36.

MARKRAM, H. (2006). The blue brain project. *Nat Rev Neurosci, 7*, 153–160.

MARTIN, K. A. C., & WHITTERIDGE, D. (1984). Form, function and intracortical projections of spiny neurones in the striate visual cortex of the cat. *J Physiol, 353*, 463–504.

MEAD, C. (1990). Neuromorphic electronic systems. *Proc IEEE, 78*, 1629–1636.

MITCHISON, G., (1991). Neuronal branching patterns and the economy of cortical wiring. *Proc R Soc Lond B Biol Sci, 245*, 151–158.

MOORE, T., & FALLAH, M. (2001). Control of eye movements and spatial attention. *Proc Natl Acad Sci USA, 98*, 1273–1276.

MOUNTCASTLE, V. B. (1997). The columnar organization of the neocortex. *Brain, 120*, 701–722.

MURAYAMA, M, PÉREZ-GARCI, E., NEVIAN, T., BOCK, T., SENN, W., & LARKUM, M. E. (2009). Dendritic encoding of sensory stimuli controlled by deep cortical interneurons. *Nature, 457*, 1137–1141.

OHANA, O., PORTNER, H., & MARTIN, K. A. C. (2012). Fast recruitment of recurrent inhibition in the cat visual cortex. *PLoS One, 7*(7), e40601.

Passingham, R. (1982). *The human primate.* San Francisco, CA: W. H. Freeman.

Peters, A., & Payne, B. (1993). Numerical relationships between geniculocortical afferents and pyramidal cell modules in cat primary visual cortex. *Cereb Cortex, 3,* 69–78.

Rockel, A. J., Hiorns, R. W., & Powell, T. P. (1980). The basic uniformity in structure of the neocortex. *Brain, 103,* 221–244.

Rockland, K. S., & Pandya, D. N. (1979). Laminar origins and terminations of cortical connections of the occipital lobe in the rhesus monkey. *Brain Res, 179,* 3–20.

Sherrington, C. (1908). On the reciprocal innervation of antagonistic muscles. Thirteenth note. On the antagonism between reflex inhibition and reflex excitation. *Proc R Soc Lond B Biol Sci, 80b,* 565–578 (reprinted in *Folia neurobiol,* 1908, 1, 365).

Somers, D. C., Nelson, S. B., & Sur, M. (1995). An emergent model of orientation selectivity in cat visual cortical simple cells. *J Neurosci, 15,* 5448–5465.

Stratford, K. J., Tarczy-Hornoch, K., Martin, K. A. C., Bannister, N. J., & Jack, J. J. B. (1996). Excitatory synaptic inputs to spiny stellate cells in cat visual cortex. *Nature, 382,* 258–261.

Vogt, C., & Vogt, O. (1919). Ergebnisse unserer hirnforschung. 1.–4. Mitteilung. *J Psychol Neurol, 25,* 279–461.

Watts, D. J., & Strogatz, S. H. (1988). Collective dynamics of "small-world" networks. *Nature, 393,* 440–442.

White, E. L. (1989). *Cortical circuits.* Boston, MA: Birkhäuser.

Yopak, K. E., Lisney, T. J., Darlington, R. B., Collin, S. P., Montgomery, R. C., & Finlay, B. L. (2010). A conserved pattern of brain scaling from sharks to primates. *Proc Natl Acad Sci USA, 107,* 12946–12951.

Zhang, K., & Sejnowski, T. J. (1990). A universal scaling lay between gray matter and white matter of cerebral cortex. *Proc Natl Acad Sci USA, 97,* 5621–5626.

10 Cost, Efficiency, and Economy of Brain Networks

OLAF SPORNS

ABSTRACT The architecture of brain networks has become a strong focus of recent empirical and theoretical studies, building on the availability of comprehensive data sets on anatomical brain connectivity in multiple species, including humans, as well as on the development of novel analytic techniques coming from the emerging field of network science. Numerous studies have documented a strong tendency of brain networks to minimize network cost, expressed as the physical length or volume of connections or the energy budget devoted to neural signaling. Network studies also show that cost minimization cannot fully explain all empirically observed features of network architecture. The prevalence of high-cost aspects of network organization, including long-distance projections and highly connected brain regions, suggest that these aspects confer advantages in terms of efficiency and integrative network function that outweigh their cost. This chapter explores aspects of network cost and network efficiency and suggests that the topology of brain networks represents a trade-off between these opposing driving forces.

Nervous systems are complex networks of structurally connected and dynamically interacting neural elements. Informally used, the term "network" has a long history in neuroscience. In recent years, an increasing focus on recording comprehensive data on brain connectivity has introduced a new formal and quantitative network perspective to cognitive and systems neuroscience. The important innovation lies in the application of computational and analytic tools and methods for mapping and interpreting patterns of connections and interactions among neurons and brain regions (Bullmore & Sporns, 2009; Sporns, 2014). Empirical studies of brain networks at all scales, from synaptic circuits in the mammalian retina (Briggman, Helmstaedter, & Denk, 2011) to large-scale connectivity in the monkey (Markov et al., 2011) and human cerebral cortex (Hagmann et al., 2008), continue to reveal network patterns in ever-greater detail. These patterns are characterized by distinct nonrandom structural features (Sporns, 2011a), many of which are shared across brain networks in different species. Since brain networks underpin dynamic interactions among neural populations, their structural organization is thought to be essential for efficient and adaptive brain function (Sporns, 2011b).

In formal terms, networks are collections of nodes (network elements) and edges (interconnections). Networks can be represented as connection matrices where each entry in the matrix records the presence or absence, weight, sign, and direction of connections between pairs of elements. The full set of network nodes and edges can be described as a graph, and the set of pairwise relations constitutes the graph's topology. Importantly, graph topology defines proximity or distance among elements solely in terms of the configuration of the link structure, without explicitly taking into account the spatial or geometric embedding of the graph within the real-world system from which it is derived. Thus, network topology captures the configuration of network nodes and edges in the absence of any spatial context, while spatial embedding explicitly considers the spatial relations among network elements. This distinction of network topology and spatial embedding is important for the main subject of this chapter, which focuses on the relationship between these two ways of capturing brain network architecture. Another important distinction is that between structural networks and functional networks. Structural networks are constructed from data on connectional anatomy, derived from microscopy, histology, or neuroimaging. Functional networks are based on time series data from neural recordings, capturing various aspects of correlation or dynamic coupling. Unless otherwise noted, the current chapter focuses on properties of structural brain networks and the role of these properties in brain function.

A deeper understanding of the relationship between network structure and network function is one of the central challenges for cognitive and systems neuroscience (Bassett & Gazzaniga, 2011). This chapter attempts to approach this challenge by exploring key factors that can account for and may have shaped structural brain networks, with an emphasis on "highly evolved" or complex brains (including humans). To motivate the approach taken in this chapter, let us consider the space of all possible brain network architectures. This space is truly astronomical, as it spans all of the ways in which such brain networks can be topologically connected. Yet the architectures we actually encounter in living species occupy only small regions within this vast space. The overwhelming majority of these theoretically possible architectures do not actually exist as part of any living biological organism, and have probably never been realized over the course of evolution. It is likely that a large part of the space is excluded simply because of historical contingencies. Early evolutionary choices about body and brain morphology, developmental paths, and physiological mechanisms put strong limits on future brain architectures, thus excluding many theoretically feasible networks from becoming viable options at later points in time.

But even taking into account the limiting influence of early evolution, clearly most of the remaining possible brain architectures do not exist today. One explanation for their nonexistence is based on the fact that brain networks consume limited resources as they are built and run, and many theoretically possible networks are simply too costly to be instantiated in a living and functioning brain. This explanation favors network cost as a major constraint that has shaped the spatial layout and operation of brain networks, and steered evolving brains toward the regions of network space they currently occupy. Another competing explanation is based on the idea that networks of real brains have been primarily selected to support efficient information processing and adaptive behavior. This explanation favors some form of network efficiency as the dominant criterion for evolutionary selection of brain networks.

The first part of this chapter examines to what extent network cost and network efficiency can account for structural features of brain networks, highlighting evidence for conserved cost as well as high efficiency in brain networks. The chapter then moves to a joint consideration of these two major driving forces of network organization, in the process developing the idea that extant brain networks represent the result of an economic trade-off between cost and efficiency. The chapter ends with an exploration of how the cost-efficiency trade-off may illuminate comparative analyses across species and shed light on fundamental questions of brain evolution.

Network cost

Network cost is incurred by the need to consume limited resources in the course of the construction ("wiring") or operation ("running") of a network. The realization that the nervous system's design and performance is subject to constraints imposed by its cost was already expressed over 100 years ago in Cajal's postulate that neuronal morphology is shaped by "laws of conservation for time, space and material" (Cajal, 1995). Building on pioneering work on neuronal morphology by Mitchison (1991) and Cherniak (1992), recent studies of the anatomical layout and physiology of brain networks have greatly added to our understanding of the limits imposed by wiring and running cost on network structure. The next section examines different aspects of network cost in light of their implications for network topology.

Brain Volume and Wiring Length Brain networks are spatially embedded, and this fundamental and inescapable fact immediately leads to the definition of several aspects of network cost. Limits on brain size place severe constraints on the number and density of neurons, and on the volume allocated to neural connections. Volume constraints, together with the spatial layout of neural elements, determine the physical lengths of axonal connections and place limits on axonal caliber, with important consequences for the velocity (Wang et al., 2008) and information rate (Perge, Niven, Mugnaini, Balasubramanian, & Sterling, 2012) of neural signal transmission. Brain volume generally scales with body size across species (Jerison, 1973; Roth & Dicke, 2005), and different components of the brain also exhibit allometric scaling relations across species (Changizi, 2001; Striedter, 2005) as well as across individuals within a species (Finlay, Hinz, & Darlington, 2011). An increase in the number of neurons (for example, as a result of an increase in brain volume) implies a parallel increase in the number of neuronal connections, with both neurons and connections together subject to volume constraints. Across mammalian species, the relation of gray matter volume (neurons) to white matter volume (connections) follows a robust allometric scaling law, with white matter volume growing faster than gray matter volume as brain size increases (Zhang & Sejnowski, 2000). This scaling relation has consequences for the density and layout of brain connectivity, resulting in a tendency toward sparser long-distance connections among neurons as brains increase in size (Herculano-Houzel, Mota, Wong, & Kaas, 2010), a pattern consistent with earlier theoretical predictions (Deacon, 1990; Ringo, 1991; Stevens, 1989).

Constraints imposed by the placement of neural elements and their wiring pattern have been extensively studied in the nematode *Caenorhabditis elegans*, for which the spatial arrangement and topology of all neurons and connections was mapped more than 25 years ago (White, Southgate, Thomson, & Brenner, 1986). It was noted early on that neurons that were spatially close tended to be densely and mutually interconnected, forming "triangle" motifs (White, 1985). Computational studies examined wiring cost by fixing the network topology and spatially rearranging the location of ganglia (Cherniak, 1995) and individual neurons (Chen, Hall, & Chklovskii, 2006; Kaiser & Hilgetag, 2006). These studies revealed that the placement neurons actually encountered was consistent with strongly conserved (though not strictly minimized) wiring cost. Similar cost-conserving relations between spatial layout and wiring were found in mammalian brains (Cherniak, Mokhtarzada, Rodriguez-Estaban, & Changizi, 2004; Klyachko & Stevens, 2003).

Among individual neurons, numerous studies have shown that connection probabilities are greatly reduced as physical distance increases (Hellwig, 2000; Stepanyants et al., 2008). In addition, neocortex exhibits local clustering of connections among spatially co-localized neurons, with an increased probability of inferred synaptic connections among neurons that share common neighbors (Perin, Berger, & Markram, 2011; Perin, Telefont, & Markram, 2013). Similar effects are seen at the macroscale level of brain regions and inter-regional projections. Adjoining regions or regions separated by short physical distances are much more likely to be linked by an inter-regional pathway than are pairs of regions that are spatially remote (Averbeck & Seo, 2008; Young, 1992). This decline in the projection probability with physical distance has been confirmed in recent quantitative labeling studies in mouse (Wang, Sporns, & Burkhalter, 2012) and macaque cortex (Markov et al., 2011, 2013). In addition, these studies have shown a pronounced decrease in the density of existing projections with increasing physical distance. Within single brain regions, intrinsic connection densities exhibit an exponential decrease with distance, with 95% of all connections made within a radius of 1.9 mm (Markov et al., 2011). Extrinsic projections among regions are characterized by high-density, short-distance projections, with the strongest projections among adjoining regions and much weaker long-distance projections and low-density subcortical input (Markov et al., 2011). Ordered by projection strength, regional density profiles exhibit a log-normal distribution consistent with the existence of a small number of strong (short-distance) and a larger number of weak (long-distance) projections. Log-normal scaling was found not only in macaque but also in the mouse brain (Wang et al., 2012).

Neuronal Communication and Energy Budget

Another important aspect of network cost is related to the metabolic energy that must be expended toward maintaining neuronal activity and synaptic signaling (Laughlin & Sejnowski, 2003). The brain is metabolically expensive, consuming a large fraction of the body's energy budget (approximately 20%) relative to its size of around 2% (Attwell & Laughlin, 2001; Clarke & Sokoloff, 1999). Looking beyond the brain itself, the absolute need to balance metabolic demand and energy consumption across the entire organism implies that the energetic cost of the brain is subject to the constraints imposed by the rest of the body. Increased encephalization is subject to severe energetic constraints (Isler & van Schaik, 2006) which may limit the adaptive benefits due to increased cognitive function. Increases in brain energy cost require a redistribution of energy demands across other organs and tissues, with potentially important consequences for cognition and behavior (Aiello & Wheeler, 1995).

As brains increase in size, so does the burden they place on the organism's energy budget. Global energy use of the brain has been shown to exhibit a scaling relationship with brain volume such that larger brains become increasingly metabolically expensive, imposing a disproportionate metabolic cost as brain volume expands (Karbowski, 2007). Other factors such as increases in axonal diameter and length contribute to increasing the metabolic cost of neuronal signaling, and also contribute to wiring cost. Remarkably, only a small fraction, perhaps as little as 1%, of the total energy use of the human brain is devoted to changes in neuronal activity associated with specific "functions," while the bulk is devoted to processes that occur "at rest"; that is, in the absence of any overt sensory input, attention-demanding task, or cognitive challenge (Raichle & Mintun, 2006). This observation suggests an important and evolutionarily conserved role for resting brain activity (Raichle, 2010).

It is clear from these observations that information processing in biological circuits requires considerable energy consumption, and it has been argued that this requirement has resulted in evolutionary pressure to increase energy efficiency (Laughlin, 2011). Evidence for conserved energy use while retaining high-capacity information transmission is found in the design of insect photoreceptors (Niven, Anderson, & Laughlin, 2007) and may underlie the cellular architecture of other sensory systems (Niven & Laughlin, 2008). Another way in which metabolic cost can be conserved is by neuronal signaling and coding strategies that minimize energy use associated with action potentials. The initiation and propagation of action potentials is potentially costly since it requires the restoration of ionic gradients by ATP-driven ionic pumps. An analysis of the kinetics of ionic currents engaged during action potentials in mammalian neurons suggests that these currents operate with near-minimal ATP consumption (Sengupta, Stemmler, Laughlin, & Niven, 2010). Another cost-saving strategy is sparse coding, which allows for highly specific representations of a large number of inputs while minimizing the level of neuronal activity needed to process information (Laughlin, van Steveninck, & Anderson, 1998).

Do Brain Networks Minimize Cost? These and other studies have led to the idea that network cost is a major factor in shaping patterns of brain connectivity. As summarized in the previous section, there is much evidence to support this view. Many different factors contribute to the concept of network cost, some of them determined by the network's spatial and geometric embedding, others related to the cost of neuronal signaling and coding. But there is also evidence to suggest that minimization of network cost cannot fully account for the topology of brain networks. There are other competing factors that prevent brain networks from reaching optimally low wiring or running cost, for example, the need to encode neuronal information with high precision and to transmit it at high rate and speed across the entire network. Jointly these competing factors contribute to increased efficiency of network performance.

Network efficiency

The evidence presented in the previous section demonstrated that extensive material resources are devoted to the physical infrastructure and operation of neuronal signaling networks across all scales, from local circuits to brain systems. Wiring and signaling cost are expended with the purpose of enabling efficient network communication. In theoretical network science, the concept of network efficiency has been approached from the perspective of communication processes (Boccaletti, Latora, Moreno, Chavez, & Hwang, 2006). According to one definition of global network efficiency (Latora & Marchiori, 2001), an efficient network allows pairs of network elements to interact along relatively short or direct communication paths. While this definition of global efficiency is widely used, it should be noted that it assumes that communication processes can actually access short communication paths, a capacity that generally requires some sort of global knowledge of the network topology. Alternative models of network communication rely on diffusion or spreading dynamics, which do not require such global knowledge (Goñi et al., 2013). Which model of network communication best describes neuronal signaling processes is still unknown.

Complex Network Topology The topology of brain networks has become a major focus of empirical efforts to create comprehensive maps of structural connections in a variety of species, including humans (Bullmore & Sporns, 2009). These maps can be derived from reconstructions of synaptic connections imaged at the microlevel of individual neurons, or from tract tracing or diffusion imaging and tractography studies at the macrolevel of brain regions. The relative advantages and pitfalls of various mapping techniques are the subject of much controversy, and thus it is important to keep in mind that all currently available connection maps must be continually updated and refined as empirical techniques for mapping structural connections become more sensitive and reliable. The ultimate goal of creating a comprehensive map of structural connectivity, the "connectome" (Sporns, 2012; Sporns, Tononi, & Kötter, 2005), is to deliver a structural foundation for understanding dynamic aspects of brain function.

All studies of structural brain networks carried out so far have provided evidence for several characteristic "nonrandom" features of network organization. One of the first analyses, carried out on the cellular network of the nematode *C. elegans*, revealed the existence of so-called small-world attributes, specifically the co-existence of high clustering and a short path length (Watts & Strogatz, 1998). High clustering indicates that the topological neighbors of a given node also have a tendency to be neighbors of each other, a feature that is characteristic of highly regular lattice networks. Short path length indicates that, on average, all pairs of nodes can be reached along a path with few intermediate steps, a feature that is present in random networks. The network of *C. elegans* was found to exhibit a mixture of regular and random features, resulting in a topology that was intermediate between the extremes of lattice and random networks. Such intermediate characteristics are often encountered in systems that exhibit high structural or functional complexity, defined as a mixture of order and disorder (Crutchfield, 2011).

High clustering and short path length have since been observed in virtually all connectome data sets, including several connection matrices derived from tract tracing studies in mammalian brains (Hilgetag, Burns, O'Neill, Scannell, & Young, 2000; Sporns et al., 2000; Sporns & Zwi, 2004), and diffusion imaging/tractography in humans (Gong et al., 2009; Hagmann et al., 2008). Additional features include characteristic distributions of local subgraphs or motifs (Milo et al., 2002; Sporns & Kötter, 2004), which deviate from distributions that would be expected if the brain was

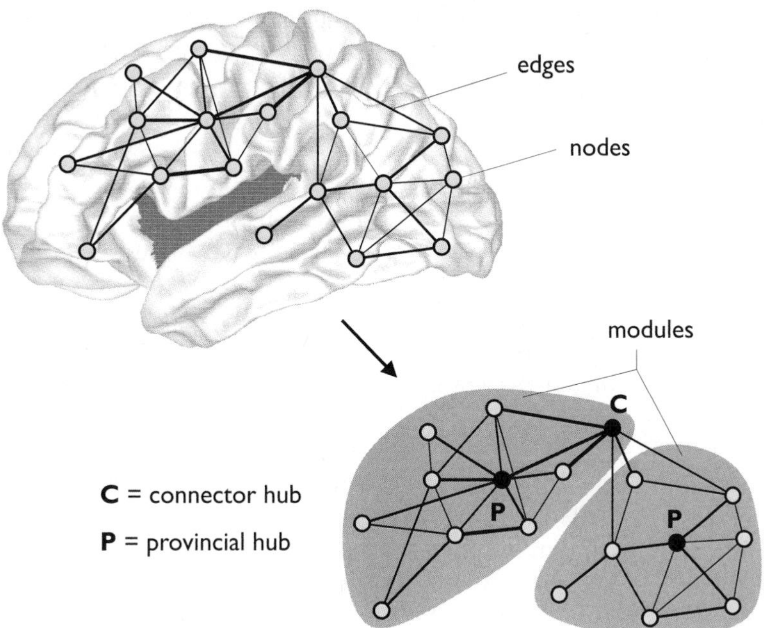

FIGURE 10.1 A schematic illustration of a distributed and modular brain network. Brain connectivity has been rendered as a network of nodes and edges (upper left), with the nodes corresponding to distinct brain regions and the edges representing estimates of weighted and undirected anatomical connections. After thresholding of weak edges, the network is partitioned into two principal modules (lower right). Both connector hubs and provincial hubs maintain a high number of connections, with connector hubs cross-linking the two modules, while provincial hubs link to numerous other nodes within their own module.

organized as a perfect lattice or random network. A statistical overabundance of some motifs is likely driven by the presence of high clustering, which tends to promote fully connected and recurrent motif configurations. The strong tendency of cortical networks to exhibit local clustering is partly due to wiring cost minimization. An abundance of short-distance connections naturally leads to high local clustering as seen in a regular lattice. Counteracting this tendency are network features that break regularity and drive brain organization in the direction of "randomness," in the process incurring higher wiring cost.

The features of network organization discussed so far, while present in many examples of structural brain networks, are also broadly, perhaps universally, shared across many other social, technological, or biological networks (Boccaletti et al., 2006; Watts & Strogatz, 1998). Hence, their presence in brain networks is not unique, which begs the question whether there are additional aspects of network organization that are compatible with these universal features but more directly indicative of specific aspects of neuronal information processing. In this context, topological features pointing to the existence of network communities or modules that are interconnected by important hub nodes are of special importance.

NETWORK MODULES AND HUBS Many real-world networks can be partitioned into a set of modules or network communities wherein each module corresponds to a set of nodes that are densely interconnected among one another, while connections between modules are relatively sparse (figure 10.1). This definition can be formalized in the form of a modularity metric that expresses the degree to which a given network exhibits modular organization, given a module partition (Newman & Girvan, 2004). An optimal partition can then be identified using an optimization algorithm (Blondel, Guillaume, Lambiotte, & Lefebvre, 2008). The optimal partition assigns each network node to exactly one module, thus decomposing the network into nonoverlapping network communities. Modularity detection is an active research area in its own right (for a review, see Fortunato, 2010), and recent methodological developments include new search algorithms, refinements of modularity measures (Fortunato & Barthelemy, 2007), and reformulations of modularity in terms of link communities (Ahn, Bagrow, & Lehmann, 2010).

Modularity is ubiquitous in brain networks. Modules of densely interconnected neurons have been identified in *C. elegans* (Sohn, Choi, Ahn, Lee, & Jeong, 2011), and modules of brain regions have been found in whole-brain analyses across multiple mammalian species (e.g., Hagmann et al., 2008; Hilgetag et al., 2000; Hilgetag & Kaiser, 2004). Some studies have demonstrated hierarchical modularity, with large-scale modules that could be further subdivided into smaller-scale modules (e.g., Bassett et al., 2010). In most cases, modules identified in structural networks comprised network elements that were located within the same spatial neighborhood, reflecting once again the high density of local or short-distance connections. This stands in contrast to modules derived from functional networks, which are often composed of more spatially distributed elements, for example, in the case of so-called resting-state networks derived from functional MRI recordings (e.g., Power et al., 2011).

Computational network studies suggest that modules confer a number of functional benefits. The dense clustering of connections within modules and the relatively weak interconnections between modules form a structural basis for functional specialization—network elements within modules are likely to share response properties, while elements across modules exhibit differences. This notion is supported by studies that have pointed to a strong relationship between membership in modules or clusters derived from structural connectivity and similarities in physiological responses (Hilgetag & Kaiser, 2004). Modular connections also shape patterns of synchronization (Müller-Linow, Hilgetag, & Hütt, 2008; Zhou, Zemanova, Zamora, Hilgetag, & Kurths, 2006) favoring local or spatially restricted synchronization behavior. Other studies have suggested that modularity supports greater adaptability of networks to changing environmental challenges (Kashtan & Alon, 2005; Kashtan, Noor, & Alon, 2007), as well as complex neuronal dynamics, especially when modules are arranged along several hierarchical levels (Kaiser & Hilgetag, 2010; Rubinov, Sporns, Thivierge, & Breakspear, 2011). Finally, the tendency of modules to limit the spread of dynamic perturbations supports the dynamic stability of the network as a whole. This effect is widely thought to be associated with evolvability, which is defined as the capacity to generate heritable, selectable phenotypes (Kirschner & Gerhart, 1998), for example, by limiting the effects of mutations and thus ensuring greater robustness. The link between modular topology and evolvability has been demonstrated in computational network models that suggest evolvability may result from selection pressure to reduce network cost (Clune, Mouret, & Lipson, 2013). Given this multitude of functional benefits, modular brain architectures are of special interest in an evolutionary context. Modular organization (columns, brain regions) becomes more prominent as brains increase in size and complexity (Kaas, 1989; Krubitzer, 2009), and the addition of new and more specialized modules may have been instrumental for the emergence of novel cognitive capacities (Chittka & Niven, 2009).

Modular brain architecture places a premium on network features that can break modularity to enable global communication and integration. Communication paths between modules largely flow through a specific subset of nodes generally referred to as network hubs (van den Heuvel & Sporns, 2013). Hubs can be defined according to a number of different criteria (Sporns, Honey, & Kötter, 2007), including high numbers of connections (degree), placement on many short paths (centrality), or a diverse connection profile. The latter criterion can be directly derived from the number and distribution of connections maintained within a single or across multiple communities. Hubs link to nodes across many modules, thus acting as points of convergence and divergence for intermodular signal traffic. Studies of mammalian cortex have identified a subset of areas in parietal, temporal, and frontal cortex as putative hub regions (Sporns et al., 2007). In human cortex, network analyses have converged on a set of regions including portions of the medial and superior parietal cortex as well as selected regions in orbitofrontal, superior frontal, and lateral prefrontal cortex (Gong et al., 2009; Hagmann et al., 2008, 2010). Many of these regions have been previously described as multimodal or transmodal association areas (Mesulam, 1998) characterized by complex physiological responses and diverse activation patterns across many tasks. Network studies suggest that the diverse responses of hub regions are strongly linked to their important role in ensuring efficient network communication. Hubs are key structures for lowering the path length of the network, as they create shortcuts among network communities and thus reduce the topological distance between many pairs of nodes, particularly nodes that belong in different structural modules.

Are Brain Networks Maximally Efficient? In the previous section, the concept of network efficiency was approached from the perspective of network features that promote the efficiency of communication processes. While this formulation of network efficiency may have some validity in the context of sensorimotor processing required for simple behavioral outputs, it should also be noted that network efficiency cannot be easily extended to considerations of adaptation or fitness at the level of the whole organism. With this limitation in mind, a communication-based definition of network efficiency does allow the construction of objective efficiency measures that can be derived from network topology. Application of such measures to brain networks suggests that these networks are not optimally efficient. On the one hand, the existence of modules and hubs offers potential adaptive advantages by creating specialized sources of information that can link to effectors and impact on behavior rather directly through connectional "off-ramps." However, on the other hand, local communities or modules can restrict long-distance brain communication, which depends heavily on hub nodes that open intermodule links. While hubs can lower the overall path length (and hence increase global network efficiency), they may also act as processing bottlenecks that can impede information flow, resulting in network congestion.

So far, we have presented evidence for conserved network cost and pointed to network features that promote network efficiency. However, neither minimization of cost nor maximization of efficiency appears sufficient to fully explain all features of brain network organization. We now turn to an alternative model, one that instead views brain architecture as resulting from an economic trade-off between the competing demands of cost and efficiency.

Network economy

In the context of economic theory, the success or profitability of a product can be measured by taking into account both the cost incurred by the producer as well as the propensity of the product to add value for its customers. Profitable businesses may be said to optimize a trade-off between the cost of production and the value added by the product itself (Bullmore & Sporns, 2012). While an explicit economic analogy can only go so far, the larger concept of an economic trade-off between competing objectives, here represented by cost minimization and efficiency maximization, appears fruitful when applied to brain architecture. Indeed, strategies for multi-objective optimization are increasingly applied in studies of complex biological systems (Shoval et al., 2012), including brain networks (Clune et al., 2013; Goñi et al., 2013; Santana et al., 2011).

A central concept first used in engineering and economics is called "Pareto optimality." It refers to the state where the construction or operation of a given system simultaneously satisfies multiple objectives. Identifying the objectives that are jointly satisfied can reveal selectional forces that drive optimal design or biological fitness. Applications of Pareto optimality to biological networks have begun to provide insight into the factors that have shaped their connectivity and function. For example, multi-objective optimization of different measures of network efficiency suggest that the topologies of real-world networks, including networks of neurons and brain regions, have been selected to favor efficient communication processes that not only rely on routing of messages along short paths but also on processes involving diffusion dynamics (Goñi et al., 2013). Pareto optimality and multi-objective strategies for maximizing performance offer promising new theoretical avenues toward understanding brain networks that merit further exploration.

The next section illustrates the trade-off between network cost and efficiency in brain networks by giving two examples of network features that violate cost-minimization in favor of gains in efficiency. One of these examples concerns increased wiring cost, and the other involves increased running cost. Both serve to enable more direct and efficient network communication.

Trade-Off Between Network Cost and Efficiency

As discussed earlier, quantitative labeling studies of inter-regional projections in macaque cortex have revealed a high density of local projections among spatially nearby regions and a lower density of projections between regions that are spatially remote, a pattern that is consistent with the principle of conservation of wiring cost. A corollary of this pattern is that long-distance projections make important contributions toward specific functional roles to individual brain regions by increasing the specificity of regional connection profiles (Markov et al., 2013). Such connection profiles are thought to be associated with functional specialization (Passingham, Stephan, & Kötter, 2002), and the added expense of long-distance projections may confer a functional benefit in allowing greater specialization of regional response profiles.

A further and perhaps even more important functional benefit of long-distance projections is due to their central role in global integrative processes and efficient network communication. Global integration must break modularity and requires information flow on connections that cross module boundaries. Such intermodular connections typically span longer physical distances than shorter connections among members of the same module (Sporns et al., 2007). While they are expensive in terms of wiring cost, network studies of the topology of several structural brain networks have shown that such long-distance connections greatly reduce the network's path length, and hence increase its global efficiency (Kaiser & Hilgetag, 2006), thus providing functional benefits, including more direct, faster, and less noisy information flow. A re-examination of the wiring diagram of *C. elegans* has shown that its wiring cost is not strictly minimized but could be reduced further by rewiring some of its connections. However, this reduction in wiring cost resulted in a network with longer path length and lower global efficiency, thus supporting the idea that the *C. elegans* connectome represents a trade-off between the competing demands imposed by cost and efficiency. Similar results were obtained for mammalian connection matrices describing the cat and macaque cortex. More recent comparisons of the spatial layout and topological complexity of brain networks and electronic circuits have shown that their advanced information-processing capabilities are associated with greater-than-minimal wiring cost (Bassett et al., 2010).

Long-distance connections tend to link high-degree nodes or brain hubs. Recent studies of brain networks including *C. elegans* (Towlson, Vértes, Ahnert, Schafer, & Bullmore, 2013), cat cortex (Zamora-López, Zhou, & Kurths, 2010), macaque cortex (Harriger, van den Heuvel, & Sporns, 2012), and the human brain (van den Heuvel & Sporns, 2011; van den Heuvel, Kahn, Goñi, & Sporns, 2012) have shown that hub nodes are interconnected to form a so-called "rich club," defined as a collective of nodes with mutual interconnections that are denser than predicted by chance (Colizza, Flammini, Serrano, & Vespignani, 2006). In both human and macaque cortex, rich club regions include a variety of multimodal association areas that are widely distributed across all major cortical lobes. The connections linking these rich club regions tend to be long distance, and hence violate the wiring minimization principle. Computational network studies strongly suggest that rich club connections are vitally important for integrative processes. For example, a large proportion of all short paths linking pairs of brain regions access rich club nodes and connections. Deletion of rich club nodes of connections results in greater disconnection than equivalent damage to more peripheral nodes and connections. Rich club regions span and interconnect all of the brain's structural and functional modules, suggestive of an essential role in integrating their functional activity. How the extent of rich-club organization may vary across brains of different sizes, for example across mammals differing in cortical size and expansion, is currently unknown.

Hub nodes and their connections are essential for global brain communication, and their existence not only incurs increased wiring but also increased running cost, expressed in their energy use and metabolic profile. A meta-analysis of the regional distribution of aerobic glycolysis (Vaishnavi et al., 2010) and regional network centrality suggests that the two measures are significantly and positively correlated, with higher centrality or hub-ness predicting a specific metabolic profile. High levels of aerobic glycolysis may index strong demand for fast, albeit relatively inefficient, supply of ATP, possibly in the service of fueling ionic pumps or cellular processes involved in biosynthesis and synaptic plasticity (Vaishnavi et al., 2010). Other aspects related to running costs were found to be associated with rich club regions (Collin et al., 2013). For example, axonal projections of these regions span long distances and show high levels of myelination, and their temporal activity tends to be more variable—all aspects of structural and functional organization that are likely to put pressure on energy and metabolic resources.

Segregation and Integration The trade-off of network cost and efficiency gives rise to an architecture that simultaneously accommodates two performance requirements—the generation of specialized information in modular or segregated circuits, as well as the need to integrate specialized information to create globally coherent brain states (Sporns, 2011b; Tononi, Sporns, & Edelman, 1994). Numerous attributes of brain networks promote these two aspects of neural information processing, segregation and integration (Sporns, 2013). This chapter has discussed several examples of such network attributes encountered in structural brain connectivity. Modularity and high clustering favor the sharing of related information within functionally specialized subsystems. Hubs and short path lengths enable the flow and integration of information across module boundaries. Rich club organization accommodates both specialized and integrated processing by providing the infrastructure for attracting and dispersing information to and from diverse sources.

Further examples are found in functional brain networks, and in this domain the trade-off between cost and efficiency may be much more dynamic and time-dependent. Several studies have shown that functional networks undergo rapid reconfigurations in response to momentary demands imposed by stimulus and task (Bassett, Meyer-Lindenberg, Achard, Duke, & Bullmore, 2006; Bassett et al., 2009; Doron, Bassett, & Gazzaniga, 2012). Over time and depending on cognitive demands, they continually iterate between functional network architectures that are characterized by more local processing and others that are more commensurate with integrated processing. Local processing may be considered less costly, since it involves signaling over smaller physical distances, while more integrated processing likely involves "expensive" long-distance pathways. Hence, it appears that the cost-efficiency trade-off not only plays out on the longer time scales of development and evolution, but may be constantly renegotiated on much faster time scales as brains respond to the ever-changing demands imposed by their environment.

Conclusion

The goal of this chapter was to examine some candidate principles that may have shaped brain networks. The phenotypic characteristics of the brain have complex origins and are best described as a "magnificent compromise" (Krubitzer, 2007; Krubitzer & Seelke, 2012) resulting from numerous constraints and selection pressures related to development, body structure, and environmental statistics. When examined in detail from the more specific perspective of the structure of brain networks, it appears that no single factor or dimension is sufficient for explaining the extant patterns of brain connectivity. Instead, concepts based on simple notions of optimization should be replaced by a framework that emphasizes the economic trade-off among multiple objectives, including network cost and efficiency. These objectives may be mutually incompatible; for example, many features of brain network organization impose increased cost while also conferring distinct advantages in terms of network efficiency. While the competition between cost and efficiency necessarily results in architectures that are suboptimal along each separate dimension, brain networks may represent optimal solutions in terms of trade-offs or compromises among conflicting goals. Comparative studies across a broader range of species are needed to further explore the driving forces that have shaped the topology of brain networks.

ACKNOWLEDGMENT The author's work was supported by the J. S. McDonnell Foundation.

REFERENCES

AHN, Y. Y., BAGROW, J. P., & LEHMANN, S. (2010). Link communities reveal multiscale complexity in networks. *Nature, 466,* 761–764.

AIELLO, L. C., & WHEELER, P. (1995). The expensive-tissue hypothesis: The brain and the digestive system in human primate evolution. *Curr Anthropol, 36,* 199–221.

ATTWELL, D., & LAUGHLIN, S. B. (2001). An energy budget for signaling in the grey matter of the brain. *J Cerebr Blood F Met, 21,* 1133–1145.

AVERBECK, B. B., & SEO, M. (2008). The statistical neuroanatomy of frontal networks in the macaque. *PLoS Comput Biol, 4,* e1000050.

BASSETT, D. S., BULLMORE, E. T., MEYER-LINDENBERG, A., APUD, J. A., WEINBERGER, D. R., & COPPOLA, R. (2009). Cognitive fitness of cost-efficient brain functional networks. *Proc Natl Acad Sci USA, 106,* 11747–11752.

BASSETT, D. S., GREENFIELD, D. L., MEYER-LINDENBERG, A., WEINBERGER, D. R., MOORE, S. W., & BULLMORE, E. T. (2010). Efficient physical embedding of topologically complex information processing networks in brains and computer circuits. *PLoS Comput Biol, 6,* e1000748.

BASSETT, D. S., & GAZZANIGA, M. S. (2011). Understanding complexity in the human brain. *Trends Cogn Sci, 15,* 200–209.

BASSETT, D. S., MEYER-LINDENBERG, A., ACHARD, S., DUKE, T., & BULLMORE, E. T. (2006). Adaptive reconfiguration of fractal small-world human brain functional networks. *Proc Natl Acad Sci USA, 103,* 19518–19523.

BLONDEL, V. D., GUILLAUME, J. L., LAMBIOTTE, R., & LEFEBVRE, E. (2008). Fast unfolding of communities in large networks. *J Stat Mech-Theory E, 10,* P10008.

BOCCALETTI, S., LATORA, V., MORENO, Y., CHAVEZ, M., & HWANG, D. U. (2006). Complex networks: Structure and dynamics. *Phys Rep, 424,* 175–308.

BRIGGMAN, K. L., HELMSTAEDTER, M., & DENK, W. (2011). Wiring specificity in the direction-selectivity circuit of the retina. *Nature, 471,* 183–188.

BULLMORE, E., & SPORNS, O. (2009). Complex brain networks: Graph theoretical analysis of structural and functional systems. *Nat Rev Neurosci, 10,* 186–198.

BULLMORE, E., & SPORNS, O. (2012). The economy of brain network organization. *Nat Rev Neurosci, 13,* 336–349.

CAJAL, S. R. (1995). *Histology of the nervous system of man and vertebrates.* New York, NY: Oxford University Press.

CHANGIZI, M. A. (2001). Principles underlying mammalian neocortical scaling. *Biol Cybern, 84,* 207–215.

CHEN, B. L., HALL, D. H., & CHKLOVSKII, D. B. (2006). Wiring optimization can relate neuronal structure and function. *Proc Natl Acad Sci USA, 103,* 4723–4728.

CHERNIAK, C. (1992). Local optimization of neuron arbors. *Biol Cybern, 66,* 503–510.

CHERNIAK, C. (1995). Neural component placement. *Trends Neurosci, 18,* 522–527.

CHERNIAK, C., MOKHTARZADA, Z., RODRIGUEZ-ESTABAN, R., & CHANGIZI, K. (2004). Global optimization of cerebral cortex layout. *Proc Natl Acad Sci USA, 101,* 1081–1086.

CHITTKA, L., & NIVEN, J. (2009). Are bigger brains better? *Curr Biol, 19,* R995–R1008.

CLARKE, D. D., & SOKOLOFF, L. (1999). Circulation and energy metabolism of the brain. In G. J. Siegel, B. W. Agranoff, R. W. Albers, S. K. Fisher, & M. D. Uhler (Eds.), *Basic neurochemistry* (6th ed., pp. 637–669). Philadelphia, PA: Lippincott-Raven.

CLUNE, J., MOURET, J. B., & LIPSON, H. (2013). The evolutionary origins of modularity. *Proc R Soc Lond B Biol Sci, 280,* 20122863.

COLIZZA, V., FLAMMINI, A., SERRANO, M. A., & VESPIGNANI, A. (2006). Detecting rich-club ordering in complex networks. *Nat Phys, 2,* 110–115.

COLLIN, G., SPORNS, O., MANDL, R. C., & VAN DEN HEUVEL, M. P. (2013). Structural and functional aspects relating to cost and benefit of rich club organization in the human cerebral cortex. *Cereb Cortex,* in press. doi:10.1093/cercor/bht064.

CRUTCHFIELD, J. P. (2011). Between order and chaos. *Nat Phys, 8,* 17–24.

DEACON, T. W. (1990). Rethinking mammalian brain evolution. *Am Zool, 30,* 629–705.

DORON, K. W., BASSETT, D. S., & GAZZANIGA, M. S. (2012). Dynamic network structure of interhemispheric coordination. *Proc Natl Acad Sci USA, 109*(46), 18661–18668.

FINLAY, B. L., HINZ, F., & DARLINGTON, R. B. (2011). Mapping behavioural evolution onto brain evolution: The strategic roles of conserved organization in individuals and species. *Philos Trans R Soc Lond B Biol Sci, 366*, 2111–2123.

Fortunato, S. (2010). Community detection in graphs. *Phys Rep, 486*, 75–174.

Fortunato, S., & Barthelemy, M. (2007). Resolution limit in community detection. *Proc Natl Acad Sci USA, 104*(1), 36–41.

Gong, G., He, Y., Concha, L., Lebel, C., Gross, D. W., Evans, A. C., & Beaulieu, C. (2009). Mapping anatomical connectivity patterns of human cerebral cortex using in vivo diffusion tensor imaging tractography. *Cereb Cortex, 19*, 524–536.

Goñi, J., Avena-Koenigsberger, A., de Mendizabal, N. V., van den Heuvel, M. P., Betzel, R. F., & Sporns, O. (2013). Exploring the morphospace of communication efficiency in complex networks. *PLoS One, 8*(3), e58070.

Hagmann, P., Cammoun, L., Gigandet, X., Meuli, R., Honey, C. J., Wedeen, V. J., & Sporns, O. (2008). Mapping the structural core of human cerebral cortex. *PLoS Biol, 6*(7), e159.

Hagmann, P., Sporns, O., Madan, N., Cammoun, L., Pienaar, R., Wedeen, V. J., ... Grant, P. E. (2010). White matter maturation reshapes structural connectivity in the late developing human brain. *Proc Natl Acad Sci USA, 107*(44), 19067–19072.

Harriger, L., van den Heuvel, M., & Sporns, O. (2012). Rich club organization of macaque cerebral cortex and its role in network communication. *PLoS One, 7*, e46497.

Hellwig, B. (2000). A quantitative analysis of the local connectivity between pyramidal neurons in layers 2/3 of the rat visual cortex. *Biol Cybern, 82*, 111–121.

Herculano-Houzel, S., Mota, B., Wong, P. Y. & Kaas, J. H. (2010). Connectivity-driven white matter scaling and folding in primate cerebral cortex. *Proc Natl Acad Sci USA, 107*, 19008–19013.

Hilgetag, C. C., Burns, G. A., O'Neill, M. A., Scannell, J. W., & Young, M. P. (2000). Anatomical connectivity defines the organization of clusters of cortical areas in the macaque monkey and the cat. *Philos Trans R Soc Lond B Biol Sci, 355*, 91–110.

Hilgetag, C. C., & Kaiser, M. (2004). Clustered organization of cortical connectivity. *Neuroinformatics, 2*, 353–360.

Isler, K., & van Schaik, C. P. (2006). Metabolic costs of brain size evolution. *Biol Lett, 2*, 557–560.

Jerison, H. J. (1973). *Evolution of the brain and intelligence.* New York, NY: Elsevier.

Kaas, J. H. (1989). The evolution of complex sensory systems in mammals. *J Exp Biol, 146*(1), 165–176.

Kaiser, M., & Hilgetag, C. C. (2006). Nonoptimal component placement, but short processing paths, due to long-distance projections in neural systems. *PLoS Comput Biol, 2*(7), e95.

Kaiser, M., & Hilgetag, C. C. (2010). Optimal hierarchical modular topologies for producing limited sustained activation of neural networks. *Front Neuroinform, 4*, 8.

Karbowski, J. (2007). Global and regional brain metabolic scaling and its functional consequences. *BMC Biol, 5*, 18.

Kashtan, N., & Alon, U. (2005). Spontaneous evolution of modularity and network motifs. *Proc Natl Acad Sci USA, 102*, 13773–13778.

Kashtan, N., Noor, E., & Alon, U. (2007). Varying environments can speed up evolution. *Proc Natl Acad Sci USA, 104*(34), 13711–13716.

Kirschner, M., & Gerhart, J. (1998). Evolvability. *Proc Natl Acad Sci USA, 95*, 8420–8427.

Klyachko, V. A., & Stevens, C. F. (2003). Connectivity optimization and the positioning of cortical areas. *Proc Natl Acad Sci USA, 100*, 7937–7941.

Krubitzer, L. (2007). The magnificent compromise: Cortical field evolution in mammals. *Neuron, 56*, 201–208.

Krubitzer, L. (2009). In search of a unifying theory of complex brain evolution. *Ann NY Acad Sci, 1156*(1), 44–67.

Krubitzer, L. A., & Seelke, A. M. (2012). Cortical evolution in mammals: The bane and beauty of phenotypic variability. *Proc Natl Acad Sci USA, 109*(Suppl. 1), 10647–10654.

Latora, V., & Marchiori, M. (2001). Efficient behavior of small-world networks. *Phys Rev Lett, 87*, 198701.

Laughlin, S. B. (2011). Energy, information, and the work of the brain. In R. Levin, S. Laughlin, C. De La Rocha, & A. Blackwell (Eds.), *Work meets life: Exploring the integrative study of work in living systems* (pp. 39–67). Cambridge, MA: MIT Press.

Laughlin, S. B., & Sejnowski, T. J. (2003). Communication in neuronal networks. *Science, 301*, 1870–1874.

Laughlin, S. B., van Steveninck, R. R. D., & Anderson, J. C. (1998). The metabolic cost of neural information. *Nat Neurosci, 1*, 36–41.

Markov, N. T., Ercsey-Ravasz, M., Lamy, C., Gomes, A. R. R., Magrou, L., Misery, P., ... Kennedy, H. (2013). The role of long-range connections on the specificity of the macaque interareal cortical network. *Proc Natl Acad Sci USA, 110*(13), 5187–5192.

Markov, N. T., Misery, P., Falchier, A., Lamy, C., Vezoli, J., Quilodran, R., ... Knoblauch, K. (2011). Weight consistency specifies regularities of macaque cortical networks. *Cereb Cortex, 21*(6), 1254–1272.

Mesulam, M. M. (1998). From sensation to cognition. *Brain, 121*, 1013–1052.

Milo, R., Shen-Orr, S., Itzkovitz, S., Kashtan, N., Chklovskii, D., & Alon, U. (2002). Network motifs: Simple building blocks of complex networks. *Science, 298*, 824–827.

Mitchison, G. (1991). Neuronal branching patterns and the economy of cortical wiring. *Proc R Soc Lond B Biol Sci, 245*, 151–158.

Müller-Linow, M., Hilgetag, C. C., & Hütt, M. T. (2008). Organization of excitable dynamics in hierarchical biological networks. *PLoS Comput Biol, 4*, e1000190.

Newman, M. E. J., & Girvan, M. (2004). Finding and evaluating community structure in networks. *Phys Rev, E69*, 026113.

Niven, J. E., Anderson, J. C., & Laughlin, S. B. (2007). Photoreceptors demonstrate energy-information trade-offs in neural coding. *PLoS Biol, 5*, e116.

Niven, J. E., & Laughlin, S. B. (2008). Energy limitation as a selective pressure on the evolution of sensory systems. *J Exp Biol, 211*, 1792–1804.

Passingham, R. E., Stephan, K. E., & Kötter, R. (2002). The anatomical basis of functional localization in the cortex. *Nat Rev Neurosci, 3*, 606–616.

Perge, J. A., Niven, J. E., Mugnaini, E., Balasubramanian, V., & Sterling, P. (2012). Why do axons differ in caliber? *J Neurosci, 32*, 626–638.

Perin, R., Berger, T. K., & Markram, H. (2011). A synaptic organizing principle for cortical neuronal groups. *Proc Natl Acad Sci USA, 108*, 5419–5424.

Perin, R., Telefont, M., & Markram, H. (2013). Computing the size and number of neuronal clusters in local circuits. *Front Neuroanat, 7*, 1.

Power, J. D., Cohen, A. L., Nelson, S. M., Wig, G. S., Barnes, K. A., Church, J. A., ... Petersen, S. E. (2011). Functional network organization of the human brain. *Neuron, 72*(4), 665–678.

Raichle, M. E. (2010). Two views of brain function. *Trends Cogn Sci, 14*(4), 180–190.

Raichle, M. E., & Mintun, M. A. (2006). Brain work and brain imaging. *Ann Rev Neurosci, 29*, 449–476.

Ringo, J. L. (1991). Neuronal interconnection as a function of brain size. *Brain Behav Evolut, 38*, 1–6.

Roth, G., & Dicke, U. (2005). Evolution of the brain and intelligence. *Trends Cogn Sci, 9*, 250–257.

Rubinov, M., Sporns, O., Thivierge, J. P., & Breakspear, M. (2011). Neurobiologically realistic determinants of self-organized criticality in networks of spiking neurons. *PLoS Comput Biol, 7*(6), e1002038.

Santana, R., Bielza, C., & Larrañaga, P. (2011). Optimizing brain networks topologies using multi-objective evolutionary computation. *Neuroinformatics, 9*(1), 3–19.

Sengupta, B., Stemmler, M., Laughlin, S. B., & Niven, J. E. (2010). Action potential energy efficiency varies among neuron types in vertebrates and invertebrates. *PLoS Comput Biol, 6*, e1000840.

Shoval, O., Sheftel, H., Shinar, G., Hart, Y., Ramote, O., Mayo, A., ... Alon, U. (2012). Evolutionary trade-offs, Pareto optimality, and the geometry of phenotype space. *Science, 336*(6085), 1157–1160.

Sohn, Y., Choi, M. K., Ahn, Y. Y., Lee, J., & Jeong, J. (2011). Topological cluster analysis reveals the systemic organization of the *Caenorhabditis elegans* connectome. *PLoS Comput Biol, 7*(5), e1001139.

Sporns, O. (2011a) The non-random brain: Efficiency, economy, and complex dynamics. *Front Comput Neurosci, 5*, 5.

Sporns, O. (2011b). *Networks of the brain.* Cambridge, MA: MIT Press.

Sporns, O. (2012). *Discovering the human connectome.* Cambridge, MA: MIT Press.

Sporns, O. (2013). Network attributes for segregation and integration in the human brain. *Curr Opin Neurobiol, 23*, 162–171.

Sporns, O. (2014). Contributions and challenges for network models in cognitive neuroscience. *Nat Neurosci.* doi:10.1038/nn.3690.

Sporns, O., Honey, C. J., & Kötter, R. (2007). Identification and classification of hubs in brain networks. *PLoS ONE, 2*, e1049.

Sporns, O., & Kötter, R. (2004). Motifs in brain networks. *PLoS Biol, 2*(11), e369.

Sporns, O., Tononi, G., & Edelman, G. M. (2000). Theoretical neuroanatomy: Relating anatomical and functional connectivity in graphs and cortical connection matrices. *Cereb Cortex, 10*, 127–141.

Sporns, O., Tononi, G., & Kötter, R. (2005). The human connectome: A structural description of the human brain. *PLoS Comput Biol, 1*, 245–251.

Sporns, O., & Zwi, J. (2004). The small world of the cerebral cortex. *Neuroinformatics, 2*, 145–162.

Stepanyants, A., Hirsch, J. A., Martinez, L. M., Kisvárday, Z. F., Ferecskó, A. S., & Chklovskii, D. B. (2008). Local potential connectivity in cat primary visual cortex. *Cereb Cortex, 18*(1), 13–28.

Stevens, C. F. (1989). How cortical interconnectedness varies with network size. *Neural Comput, 1*, 473–479.

Striedter, G. F. (2005). *Principles of brain evolution.* Sunderland, MA: Sinauer.

Tononi, G., Sporns, O., & Edelman, G. M. (1994). A measure for brain complexity: Relating functional segregation and integration in the nervous system. *Proc Natl Acad Sci USA, 91*, 5033–5037.

Towlson, E. K., Vértes, P. E., Ahnert, S. E., Schafer, W. R., & Bullmore, E. T. (2013). The rich club of the *C. elegans* neuronal connectome. *J Neurosci, 33*(15), 6380–6387.

Vaishnavi, S. N., Vlassenko, A. G., Rundle, M. M., Snyder, A. Z., Mintun, M. A., & Raichle, M. E. (2010). Regional aerobic glycolysis in the human brain. *Proc Natl Acad Sci USA, 107*(41), 17757–17762.

van den Heuvel, M. P., & Sporns, O. (2013). Network hubs in the human brain. *Trends Cogn Sci, 17*, 683–696.

van den Heuvel, M. P., Kahn, R. S., Goñi, J., & Sporns, O. (2012). A high-cost, high-capacity backbone for global brain communication. *Proc Natl Acad Sci USA, 109*, 11372–11377.

van den Heuvel, M. P., & Sporns, O. (2011). Rich-club organization of the human connectome. *J Neurosci, 31*, 15775–15786.

Wang, S. S. H., Shultz, J. R., Burish, M. J., Harrison, K. H., Hof, P. R., Towns, L. C., ... Wyatt, K. D. (2008). Functional trade-offs in white matter axonal scaling. *J Neurosci, 28*, 4047–4056.

Wang, Q., Sporns, O., & Burkhalter, A. (2012). Network analysis of corticocortical connections reveals ventral and dorsal processing streams in mouse visual cortex. *J Neurosci, 32*(13), 4386–4399.

Watts, D. J., & Strogatz, S. H. (1998). Collective dynamics of "small-world" networks. *Nature, 393*, 440–442.

White, J. G. (1985). Neuronal connectivity in *Caenorhabditis elegans. Trends Neurosci, 8*, 277–283.

White, J. G., Southgate, E., Thomson, J. N., & Brenner, S. (1986). The structure of the nervous system of the nematode *Caenorhabditis elegans. Philos Trans R Soc Lond B Biol Sci, 314*, 1–340.

Young, M. P. (1992). Objective analysis of the topological organization of the primate cortical visual system. *Nature, 358*, 152–155.

Zamora-López, G., Zhou, C., & Kurths, J. (2010). Cortical hubs form a module for multisensory integration on top of the hierarchy of cortical networks. *Front Neuroinform, 4*, 1.

Zhang, K., & Sejnowski, T. J. (2000). A universal scaling law between gray matter and white matter of cerebral cortex. *Proc Natl Acad Sci USA, 97*, 5621–5626.

Zhou, C., Zemanova, L., Zamora, G., Hilgetag, C. C., & Kurths, J. (2006). Hierarchical organization unveiled by functional connectivity in complex brain networks. *Phys Rev Lett, 97*, 238103.

11 Plasticity of Sensory and Motor Systems After Injury in Mature Primates

JON H. KAAS AND CHARNESE BOWES

ABSTRACT After damage to peripheral nerves and other parts of sensory and motor systems, adult humans often experience sensory or motor impairments that diminish over weeks to months after the injury. Studies on monkeys, rats, and other animals have helped us understand how these recoveries occur and suggest ways to promote them. While valuable information has come from a wide range of studies, those on nonhuman primates are especially informative because their nervous systems more closely resemble our own. Studies have focused on recoveries after various types of restricted damage to sensory or motor systems that produce sensory or motor deficits by deactivating parts of sensory or motor systems. These studies indicate that the growth of new connections in the central nervous system as a consequence of deactivations can be extensive and lead to reactivations that promote behavioral recovery. Importantly, several types of treatment appear to enhance adaptive growth of new nervous system connections and behavioral recovery. The mature nervous system is more plastic than previously thought.

"In the general's entourage, the discomfort José Maria Carreño experienced in the stump of his arm was reason for cordial teasing. He felt the movements of his hand, the sense of touch in his fingers, the pain bad weather caused in bones he did not have."

—Gabriel Garciá Márquez, *The General in His Labyrinth*

The developing nervous system has long been known to be quite plastic. This means that the developing nervous system has the potential to develop in different ways depending on sensory experience, other environmental factors, and nervous system damage (e.g., Kaas, Merzenich, & Killackey, 1983; Wiesel, 1982). As the developing nervous system is always changing, it seems reasonable that the course of development can be altered, sometimes in ways that compensate for nervous system damage or allow an adaptation to an abnormal environment. In contrast, early neuroscientists were often skeptical about the possibility of mature nervous systems being plastic, and even recently the concept of visual cortex in mature monkeys being plastic has been seriously challenged.

There are several reasons for this skepticism. First, early investigators such as Sperry (1959) noted that behavioral compensations did not occur after eye rotation in frogs or after a nerve was crossed from one leg to another in rats. Such observations suggested that the mature nervous system has a functional organization that is highly fixed, as functional adaptations to altered sensory inputs did not seem to occur. Second, the early experiments that demonstrated alterations in the organization and functions of the visual system after sensory deprivation in cats and monkeys found that the susceptibility to such changes was largely restricted to a short period early in postnatal development, the so-called sensitive or critical period, and the same treatments have little or no effect on the mature visual system (Daw, 1995). Third, it seems logical to conclude that a fully functional nervous system should no longer be plastic at maturity, as almost any change would impair function. Of course, early investigators recognized that at least systems for learning remain plastic throughout life, perhaps at a reduced level for language in adults, and that patients often recover some or all lost functions after some brain damage. However, evidence that the mature nervous system remains quite plastic has accumulated, and it is useful to consider this evidence and understand the nature of the nervous system changes that occur. Thus, we want to know how such changes emerge, how they relate to the recovery of abilities, and why they sometimes lead to errors in perception. We also need to evaluate treatments that have the potential to promote functional recoveries.

This review focuses on plasticity studies of sensory and motor systems in monkeys and other primates. While much has been learned about the plasticity of the human brain from clinical patients, studies in animals allow the repetition and control of variables and provide evidence that is more easily interpreted. Because monkeys and our other primate relatives have nervous systems that more closely resemble our own than those of cats, rats, and mice, most of the experimental evidence presented here comes from studies on monkeys. However, evidence from other mammals is important and relevant, and some of it is included here. In addition, the emphasis in this review is on the plasticity of the somatosensory system, since the organization of this system in monkeys is well understood and the system offers several technical advantages for studies of plasticity, including the ability for the regeneration of sensory nerves. We also consider the plasticity of visual and auditory systems, as well as the motor system, for comparison. Other systems—for example, those devoted to cognitive functions—might be even more plastic than sensory or motor systems. However, the structural and functional organizations of sensory and motor systems are much better understood, and thus changes in these systems as a result of experimental manipulation are more likely to be detected. Finally, the types of plasticity that are due to experience and learning in sensory systems (e.g., perceptual learning) and motor systems are considered elsewhere (Buomomano & Merzenich, 1998; Diamond, Armstrong-James, & Ebner, 1993; Li, Piech, & Gilbert, 2004; Nudo, Milliken, Jenkins, & Merzenich, 1996; Recanzone, Schreiner, & Merzenich, 1993).

Plasticity of the somatosensory system after sensory loss or lesions

The structural organization of the somatosensory system of primates is well known (Kaas, 2011). Thus, alterations from the normal organization as a result of plasticity can be detected (figure 11.1). Sensory afferents from the receptor sheet, the skin, largely preserve their peripheral neighborhood relationships as they terminate in the ipsilateral spinal cord and brainstem nuclei. Topographic (or somatotopic) patterns of connections are further preserved in projections to the contralateral ventroposterior nucleus of the contralateral somatosensory thalamus, and then to primary somatosensory cortex, S1 (area 3b of Brodmann, 1909). Early studies of the somatotopic organization of somatosensory cortex in primates did not have the resolution to distinguish the four systematic representations of the contralateral body in the four architectonic fields of Brodmann areas 3a, 3b, 1, and 2. The unfortunate consequence has been that the term S1 is often used to refer to all four fields combined. However, only area 3b is homologous to S1 of other mammals, and it is important to distinguish area 3b from these other fields (Kaas, 1983). Area 1 forms a mirror image representation of touch receptors just caudal to area 3b. Parallel somatotopic representations in areas 3a and 2 are merged with proprioceptive inputs from the ventroposterior superior nucleus of the thalamus. Additional representations with inputs from areas 3b, 3a, 1, and 2 include the second somatosensory area, S2, and the parietal ventral area, PV. As information about touch is distributed from area 3b to all of these areas, any plastic change in area 3b would be relayed to these higher-order areas.

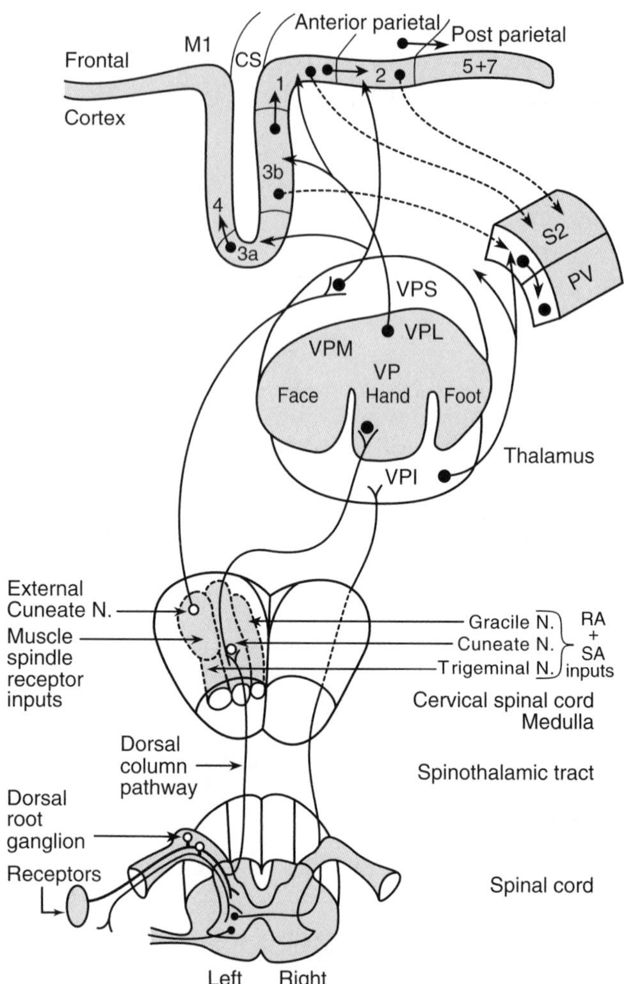

FIGURE 11.1 The somatosensory system of primates. Note that cutting the dorsal column pathway removes the direct somatosensory input from low-threshold, rapidly adapting (RA) and slowly adapting (SA) cutaneous peripheral nerve afferents from the lower body, while cutting dorsal roots of peripheral nerves removes all sensory inputs. Cutting all dorsal column branches of RA and SA afferents from the forearm deactivates neurons in the cuneate nucleus and the hand territories of the contralateral ventroposterior (VP) nucleus of the thalamus and cortical areas 3b (primary somatosensory cortex, S1) and area 1. Other areas with direct or indirect inputs from areas 3b and 1 are also deprived (e.g., area 2; the second somatosensory area, S2; the parietal ventral area, PV; posterior parietal cortex, areas 5 & 7; area 3a; and primary motor cortex, M1 or area 4.) VPM, the medial division of VP represents the face and mouth, while VPL, the lateral division, represents the forelimb, trunk, and hind limb.

Thus, we focus on changes in area 3b, although there may be further changes that are mediated in these additional processing areas.

Many studies of plasticity of the somatosensory system have demonstrated changes in the somatotopy of cortex or other parts of the system after a loss of some of the sensory inputs from the skin. Alterations or compensations likely occur throughout the system, but most studies have been on the somatotopy of area 3b, since this cortical area is accessible for recording. Any changes in area 3b would reflect those mediated within area 3b and those relayed from the thalamus, brainstem, and spinal cord. Without plasticity, any extensive loss of sensory inputs would be expected to inactivate somatotopically matched parts of the system. For example, a loss of inputs from the hand would be expected to leave the hand representation in the cuneate nucleus of the ipsilateral lower brainstem and upper spinal cord totally unresponsive to touch on the hand, as well as the hand portions of the contralateral ventroposterior nucleus and area 3b. Instead, considerable reactivation of these hand territories may occur over days to months of recovery following the sensory loss. These reactivations are typically detected and characterized with microelectrode recordings from neurons in area 3b. As time course and somatotopic features of the reactivation process depend on the type and extent of sensory loss, we describe plastic changes for four different types of sensory loss.

We are concerned with addressing several questions: How does the reactivation happen? What are the functional consequences, and are they beneficial? And can beneficial plastic changes and behavioral outcomes be enhanced by treatments that could be applied to humans? Later in this review, we describe evidence that visual, auditory, and motor systems are plastic in adult primates as well, but not always in the same ways.

Reactivation Following Damage to Peripheral Nerves Many of us have experienced some sort of accidental nerve damage where part of a peripheral nerve is cut or crushed, leaving some region of skin numb. If all goes well, sensation will return after weeks to months as the damaged nerve regenerates. However, the regeneration may be incomplete, or jumbled, leading to mislocalizations of the place of touch (Wall et al., 1986). In addition, representations of the region of numb skin in the central nervous system may have become responsive to other inputs from normally innervated skin, leading to misperceptions of where the skin was touched. However, perceptions may become more normal after the peripheral nerve regenerates and the inputs from the re-innervated skin reclaim all or most of their original central nervous system territory. These outcomes have occurred in experiments in which central nervous system representations have been evaluated at various times after section of a peripheral nerve.

In one set of experiments designed to reveal plastic changes after sensory loss, the median nerve at the level of the wrist was cut or crushed. This deprives the thumb half of the glabrous skin of the hand of sensory innervation, while the rest of the hand remains innervated. This sensory loss is so minor that monkeys behave normally and barely seem to notice. Yet about the lateral half of the hand representation in contralateral area 3b is completely deactivated, so that neurons are no longer responsive to touch on the hand. Subcortical representations that relay to area 3b are also deactivated, as is cortical area 1. Presumably, higher cortical areas are altered as well. Within days and over a few weeks, this unresponsive cortex becomes extensively or completely reactivated by preserved inputs from the hand, mainly the back of the hand and digits in a somatotopic pattern that matches the missing inputs for the glabrous hand (Merzenich et al., 1983a, 1983b). The back of the hand is sparsely innervated compared to the glabrous hand, and yet this sparse input can reactivate the large cortical territory of the glabrous hand.

Thus, sparse sensory inputs can activate much larger cortical territories than they normally do. This means that sparse brain connections can expand their functional roles. Competition between inputs for synaptic space seems to prevent this from normally happening. If a crushed medial nerve is allowed to regenerate over months of recovery, the normal map of the glabrous hand in area 3b returns as if nothing has happened (Wall, Felleman, & Kaas, 1983). However, when a cut nerve regenerates, regeneration is less precise, errors in innervation occur, and the regeneration may be incomplete. Thus, the recaptured representation of the hand in area 3b may be somewhat abnormal (Florence, Garraghty, Wall, & Kaas, 1994; Garraghty & Kaas, 1991a; Wall & Kaas, 1986; Wall et al., 1983). These results indicate that considerable plasticity of the somatosensory system can occur, cortical reactivation and reorganization can be rapid and occur over days to weeks, and the plastic effects are reversible with re-innervation. As in all types of brain damage, sensory loss not only concerns the region of direct deprivation, but also all regions that are thereby deprived of activating connections. As the competition among different sources of inputs to any nucleus or cortical area is changed, system functions widely change.

How do we explain the more rapid reactivation of cortex after nerve section? Studies of the anatomy of the somatosensory system indicate that the sparse inputs from the back of the hand terminate in the cuneate nucleus very close to the many more inputs from the glabrous skin of the hand (Florence, Wall, & Kaas, 1988, 1991; Qi & Kaas, 2006). At least some of the reactivation occurs at the level of the cuneate nucleus (Xu & Wall, 1997), and reactivations also occur at the level of the contralateral ventroposterior nucleus (Garraghty & Kaas, 1991b). Thus, with a lack of competition from the afferents of the median nerve, the sparser inputs from the radial nerve of the back of the hand likely sprout over the very short distances in the cuneate nucleus to activate deprived neurons (Florence & Kaas, 2000). Alternatively, subthreshold connections to somatotopically mismatched locations in the cuneate nucleus may already exist, and they become stronger by providing more synapses or because of other homeostatic mechanisms (Garraghty, LaChica, & Kaas, 1991; Turrigiano, 1999). As the time of regeneration of peripheral nerves depends on distance, reactivations in the central nervous system that depend on longer axon growth should take longer times. Of course, some of the cortical reactivation that follows a nerve cut may depend on axon growth at thalamic and cortical levels, perhaps occurring over longer times.

Reactivations After Damage to Dorsal Roots of Peripheral Nerves as They Enter the Spinal Cord Another type of sensory loss occurs when the sensory inputs to the spinal cord are reduced or eliminated by cutting the dorsal sensory roots of peripheral nerves as they enter the spinal cord, leaving the ventral motor root outputs to muscles intact. Years after this approach was used to evaluate arm use after a complete sensory loss from one arm, it became possible to study cortical responsiveness in some of these monkeys (Pons et al., 1991). The surprising result was that the complete arm representation in area 3b was activated by touch on the face, as well as the arm representations in areas 3a, 1, and 2. These results showed that major reactivation of deprived cortex is possible, but there was no information about the time course of the massive reactivation or mechanisms of this reactivation.

More recently, this approach has been used to only partially denervate the arm and hand by cutting only a few dorsal roots, thus removing most, but not all, of the inputs from some of the digits. As a result, only a few preserved inputs from the digits, ones that were too few to activate cortex immediately after surgery, became effective in activating most of the deprived cortical territories for those digits, with a recovery of hand use over a period of weeks (Darian-Smith & Ciferri, 2005, 2006). These experiments demonstrated that even sparse remaining inputs can reactivate their former cortical territories over short periods of time. Furthermore, the cortical reactivations and behavioral recoveries were associated with sprouting of preserved axons in the cuneate nucleus and spinal cord (Darian-Smith & Ciferri, 2006), as well as in cortex.

Lesions of Peripheral Nerve Afferents as They Travel in the Dorsal Columns of the Spinal Cord A third way of producing and studying sensory loss has been to cut the ascending branches of peripheral nerve afferents as they course in the dorsal columns of the spinal cord to the dorsal column nuclei. The major advantage of this approach is that the sensory loss is very specific, and very limited. The large axons conducting touch information from the skin of the hand and arm branch as they enter the spinal cord—with one branch entering the dorsal column of axons that project to the cuneate nucleus representing the forearm, and the other branch terminating on dorsal horn neurons of the spinal cord that preserve this touch information for use in local reflexes and motor control—and send this information to other locations, including the cuneate nucleus. All other afferents, including those that mediate crude touch, pain, and temperature, terminate normally in the dorsal horn, as they do not branch into the dorsal columns (Kaas, 2011). Thus, monkeys with dorsal column lesions appear quite normal in most behavior, including running and climbing, but do have impaired hand use for a period of 2 to 4 weeks (Qi, Gharbawie, Wynne, & Kaas, 2013). They may drop grasped objects, and even look in their hand to see if the object is present. However, they rapidly recover, and normal hand use returns. Immediately after such dorsal column lesions, the hand region of contralateral area 3b is largely or completely unresponsive to touch on the hand (figure 11.2), and this level of unresponsiveness persists for 1 to 2 weeks, after which the cortex is progressively reactivated by any of inputs in the dorsal columns from the hand that were preserved (Chen, Qi, & Kaas, 2012; Jain, Catania, & Kaas, 1997; Qi, Chen, & Kaas, 2011). Spinal cord neurons that receive the branched dorsal horn inputs from peripheral nerve axons and project to the cuneate nucleus (Rustioni, 1976; Witham & Baker, 2011) also likely contribute to the reactivation. Finally, with lesions that are extensive, and when all of the hand cortex is not reactivated by the hand, then parts of the hand representations may become activated by touch on the face (Jain et al., 1997; Jain, Qi, Collins, & Kaas, 2008). The activation by inputs from the face appears to depend on the growth of a few axons from the face past their normal targets in brainstem trigeminal nuclei to reach the cuneate nucleus for the forearm (Jain, Florence, Qi, & Kaas, 2000). Much of the hand subnucleus of the contralateral ventroposterior nucleus also becomes responsive to the face (Jain et al., 2008). The reactivation of hand cortex by inputs from the face takes 6 to 8 months to emerge, a longer time perhaps related to the longer distance the face axons need to

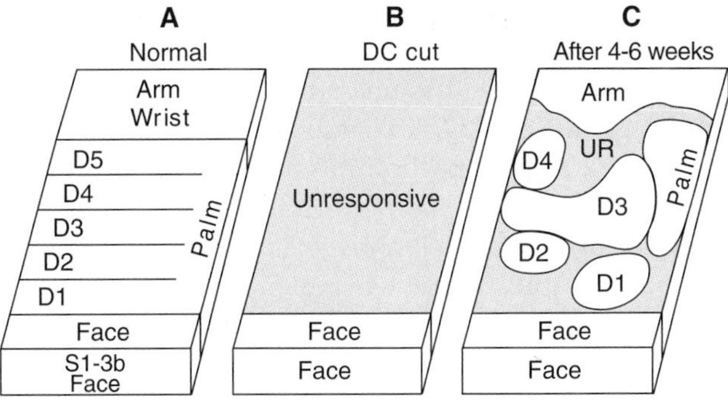

FIGURE 11.2 Effects of dorsal column (DC) section at a high cervical level on the somatotopic organization of primary somatosensory cortex, S1 (area 3b). Only the face to forearm region is shown. (A) The normal organization of area 3b in monkeys. The digits 1–5 are represented in order in the rostral two-thirds of area 3b, just medial to the representation of the face. The palm is represented caudal to the digits. (B) For days to weeks after a complete or nearly complete DC cut, neurons throughout the hand, wrist, and forearm region of area 3b are unresponsive to tactile stimulation of the forelimb. Neurons in the face region remain normally responsive. (C) After 4–6 post-lesion weeks, much of the deactivated hand and forelimb region becomes responsive to touch in the digits, palm, and forearm in a roughly normal somatotopic pattern. However, the reactivation is variable across cases, and digit territories are usually smaller than normal, but sometimes larger or discontinuous. In other regions, the neurons are unresponsive to touch. After 6–8 months of recovery, some regions of hand cortex may become responsive to touch on the face.

grow to reach the cuneate nucleus. A rapid and extensive recovery of hand use after dorsal column lesions appears to depend on the amount of cortical reactivation, as well as the somatotopy of the reactivation. If most of the fingers are represented, especially digits 1 and 2, and in their normal cortical territories, recovery is faster and better.

In summary, after a complete or nearly complete lesion of the sensory afferents in the dorsal columns of the upper spinal cord, contralateral somatosensory cortex is unresponsive to touch on the hand, and hand use is impaired. Cortical reactivation by touch on the hand is rapid, over a period of weeks, and hand use recovers. This reactivation and hand use is apparently mediated by the potentiation of the few preserved axons in the dorsal column, and by the potentiation of second-order axons that travel outside the lesion to reach the cuneate nucleus. These second-order neurons normally provide subthreshold modulating inputs. Finally, large lesions that remove more of the inputs to the cuneate nucleus leave parts of the hand representation unresponsive for a period of 6 to 8 months, after which inputs from the face may activate the cortex via the growth of axons from the face into the cuneate nucleus for the hand, while reactivations depend on the growth of preserved peripheral nerve inputs. Reactivations are likely also mediated and expanded by the growth of new connections at thalamic and cortical levels.

REORGANIZATION AFTER AMPUTATION A fourth way of studying sensory loss has been after injuries to a limb that require amputation in monkeys or humans. Such injuries are not common, but occasionally, monkeys in social groups are injured to the extent that a therapeutic amputation is necessary. The somatosensory systems of a few of these monkeys have been studied years after injury. After a loss of the lower forearm, neurons throughout the forearm portion of area 3b and adjoining areas 1 respond to touch on the stump of the amputated arm and on the lower face (Florence & Kaas, 1995). In addition, neurons in the region of the hand representation in the ventroposterior nucleus of the somatosensory thalamus became responsive to touch on the face and on the stump of the amputated arm (Florence & Kaas, 2000). As with long-standing lesions of the dorsal column afferents, tactile inputs from the face came to activate portions of the hand cortex after a major sensory loss from the entire lower arm.

We know the perceptual consequences of such major reactivations and reorganizations of the somatosensory system because similar reactivations occur in humans with long-standing amputations of an arm. Noninvasive imaging studies (fMRI) of the brain activity in humans with arm amputations show that much of the hand and arm region of contralateral somatosensory cortex (areas 3b, 1, and 2) responds to touch on the face (Flor et al., 1995; Karl, Birbaumer, Lutzenberger, Cohen, & Flor, 2001; Moore et al., 2000). Other evidence comes from a patient who was being treated for pain in the phantom limb after a long-standing amputation (Davis et al., 1998). As in many other patients with arm amputations, this patient felt that the arm was still present, a feeling called a "phantom limb." In this patient, as in some others, pain was felt in the phantom, and recordings with an electrode were necessary to locate the region producing the sensation of pain in the phantom. Recording from neurons in the hand portion of the ventroposterior nucleus contralateral to the amputation revealed receptive fields on the stump of the arm. Yet when these same neurons were electrically stimulated with the same electrode, the patient reported feeling touch on the missing hand; that is, the phantom hand. This result indicates that even though the reactivated thalamic neurons responded to the stump, they signaled touch on the hand, as if the hand were still present. This result is consistent with the reported sensations of patients with amputations of the arm. Such patients may feel touch on the missing hand (the phantom) when touched on the face, or on the stump (Ramachandran, Rogers-Ramachandran, & Stewart, 1992).

In summary, all types of sensory loss produce reactivations of deprived portions of the somatosensory system. The reactivation can be rapid, over days to weeks, with a limited sensory loss, and the reactivation from preserved afferents appears to lead to considerable functional recovery. When the sensory loss is extensive, reactivations are incomplete for periods of months, followed by reactivations based on inputs that do not contribute to any recovery of function, but instead lead to perceptual errors. This happens when inputs from the face activate cortex formerly devoted to the hand. All reactivations appear to depend on the formation of new connections in the somatosensory system, perhaps at several levels of processing.

Reactivations of visual cortex after lesions of the retina

The capacity of the mature visual system to reorganize has been studied by placing small, visuotopically matching or overlapping lesions of the retina of each eye. This procedure produced a region in primary visual cortex, V1, which no longer received its normal source of activating input from the lateral geniculate nucleus of the visual thalamus. Well before the cortical effects of such lesions were known, Eysel, Gonzalez-Aguillar, and Mayer (1980, 1981) discovered that small lesions of the retina in cats produced a permanent core of unresponsive neurons in the deprived portions of the lateral geniculate nucleus that were directly denervated by the lesions of the retina. However, the core of unresponsive neurons was bordered by a narrow zone of deprived neurons that had become activated by inputs from parts of the retina around the lesion. Thus, there was plasticity as the neurons along the margins of the deprived zones acquired new sources of activation. However, the formation of new connections from the retina in the lateral geniculate nucleus was very limited. The effects of retinal lesions on the lateral geniculate nucleus of monkeys were similar. Restricted retinal lesions produced a core of unresponsive neurons in the deprived zones of lateral geniculate nucleus layers, and plasticity was extremely limited (Darian-Smith & Gilbert, 1995). In contrast, the neurons in the cuneate nucleus with direct inputs from the hand and arm are reactivated by more extensive growth of new connections (Jain et al., 2000; Xu & Wall, 1997).

Given the limited plasticity that was observed in the lateral geniculate nucleus after retinal lesions, a surprising level of reactivation was found after in primary visual cortex. After months of recovery from small, bilateral 10^0 or less lesions of the retina, neurons throughout the deprived region of visual cortex responded to visual stimulation with receptive fields based on parts of the retina just outside the lesion. As this fringe zone of retina activated neurons in both their normal locations in V1, and in neurons across the deprived zone, the normal retinotopic organization of V1 was disrupted, as neurons in the fringe and in the deprived zone had overlapping receptive fields. This overlap was called receptive field "pile-up." Cortical reorganization after retinal lesions was first reported for visual cortex of cats (Kaas et al., 1990), and soon thereafter for V1 of monkeys (Darian-Smith & Gilbert, 1995; Gilbert & Wiesel, 1992; Heinen & Skavenski, 1991). With much larger cortical deactivation produced by bilateral lesions of the foveal region of the retina in monkeys, the reactivation was incomplete, revealing a cortical core of unresponsive cortex (Heinen & Skavenski, 1991). In a more recent study with microelectrode recording and optical imaging, the authors focused on the unresponsive core (Smirnakis et al., 2005). However, microelectrode recordings from neurons in cortex bordering the unresponsive core revealed receptive field "pile-up"—clear evidence of plasticity. The reactivation of the deprived zone of cortex, called the lesion projection zone, continued to increase, and neuron responses to visual stimuli improved for at least one year (Giannikopoulos & Eysel, 2006). Overall, the evidence for the reactivation of visual cortex is compelling. Considerable cortical plasticity does occur after retinal lesions, but when the region of deactivated primary visual cortex is large, complete reactivation may not occur.

Because of the limited reactivation of deprived neurons in the lateral geniculate nucleus, the reactivation of cortical neurons appears to be largely the result of the sprouting of cortical axons intrinsic to V1 into the deprived zone, as the extents of new axon growth match the sizes of the reactivated zones (Darian-Smith & Gilbert, 1994; Yamahachi, Mavik, McManus, Denk, & Gilbert, 2009). In addition, lesions of V1 along the margin of the reactivated cortex block the responsiveness of neurons in the lesion-projection zone (Calford, Wright, Metha, & Taglianetti, 2003). The response properties of the reactivated cortical neurons to visual stimuli appear to be nearly normal, although response thresholds may be elevated and response magnitudes reduced (Chino et al., 1995). In humans with retinal lesions, the scotoma produced by the lesion "fills in" with information from the retina around the lesion, so that the blind spot in the visual field is not perceived (Sergent, 1988). Cortical reactivation appears to occur in adult humans after damage to the retina (Baker, Dilks, Peli, & Kanwisher, 2008), as in adult monkeys. Thus, the perceptual change where the scotoma fills in is likely mediated by cortical reactivation and receptive field pile-up.

Cortical reorganization after a hearing loss

Noise-induced hearing loss is common in humans, especially in older adults. Because of differences in coding mechanisms and cochlear anatomy, the hearing loss is most pronounced for high-frequency sounds. As the high-frequency hearing loss becomes more and more pronounced, more of the auditory system becomes deprived of its normal sources of activation. It was once assumed that the mature auditory system in mammals was hard-wired and did not change as a result of hearing loss. Thus, many neurons throughout the auditory system would be unresponsive to sound after a high-frequency hearing loss. While the plasticity of the auditory system has not been extensively studied, especially in primates, there is compelling evidence at the level of primary auditory cortex that deprived cortical neurons become responsive to remaining auditory inputs, very much like the reactivation of somatosensory cortex after sensory loss. In a study of three adult macaque monkeys with a high-frequency hearing loss induced by ototoxic drugs, recording with microelectrodes after 2 to 3 months demonstrated that the tonotopic organization of primary auditory cortex, A1, had changed so that the caudal portion of A1 that is normally responsive to high-frequency tones had become responsive to middle-frequency tones (Kaas, 1996; Schwaber, Garraghty, & Kaas, 1993). Similar results have been reported for mice (Willott, Aitkin, & McFadden, 1993), cats (Irvin & Rojan, 1997), and guinea pigs (Robertson & Irvine, 1989). Although these changes have been detected in primary auditory cortex, they likely occur, to a lesser extent, in subcortical structures. Of course, changes in the activation pattern in primate auditory cortex would be relayed to a host of other auditory areas, adding substantially to the number of neurons that are responsive to middle-frequency sounds. The functional consequences of this reactivation of auditory neurons might be beneficial for some auditory tasks, but sounds could also be misinterpreted as having more high-frequency components than were present. To the extent that incompletely reactivated neurons compensate by having higher levels of spontaneous activity, they could contribute to the perception of nonexistent sounds, as in tinnitus (Rauschecker, 1999).

Plasticity of motor cortex and corticospinal projections

Primary motor cortex represents movements of body parts via connections to groups of motor and premotor neurons in the brainstem and spinal cord. Overall, the movement map in M1 is somatotopically organized, so that foot, trunk, hand, and face movements are evoked by electrical stimulation in a mediolateral sequence across M1, but locally the organization forms a mosaic so that wrist, elbow, and shoulder movement can be evoked at some sites that are mixed with sites that evoke digit movements (Gould, Cusick, Pons, & Kaas, 1986; Kaas, Stepniewska, & Gharbawie, 2012; Schieber, 2001).

A loss of sensory input to cortex, and the subsequent reorganization of sensory cortex, could alter the organization of motor cortex due to the reorganized and possibly misleading sensory relay of information from somatosensory cortex to M1. However, abnormalities in the motor map in the M1 hand region were not detected after long-standing lesions of the contralateral dorsal column in adult moneys (Qi, Stepniewska, & Kaas, 2000). In contrast, changes in M1 occur after long-standing dorsal column lesions placed in early postnatal monkeys, as a reduced representation of evoked finger movements was detected in such monkeys reared to adulthood (Qi et al., 2010). Likely, in these monkeys that were deprived of normal sensory guidance of skilled use of their fingers of the affected hand, motor cortex developed differently.

Restricted lesions of part of the hand-forelimb representation of primary motor cortex are typically followed by a period of impaired hand function, followed by a recovery of much of the lost function (Glees & Cole, 1950; Hoffman & Strick, 1995; Nudo & Milliken, 1996; see Dancause & Nudo, 2011, for review). The recovery may be enhanced by practice or training in forelimb and hand use. In addition, the motor map, revealed by electrical stimulation with microelectrodes to produce hand and digit movements, is altered over the recovery period so that cortex adjoining the region of the lesion in primary motor cortex, M1, produces some of the movements formerly evoked from the lesioned tissue (Nudo & Milliken, 1996). Some of the functional changes could be mediated by alterations in intrinsic connections within M1, so that output patterns of M1 are altered. In addition, the activating inputs to M1 may change, as M1 receives goal-directed information from posterior parietal cortex and premotor cortex (Kaas, Gharbawie, & Stepniewska, 2011; Dancause et al., 2006). But much of the recovery of function likely depends on new termination patterns of corticospinal projections of intact parts of M1 to occupy and activate deprived motor and premotor neurons that mediate muscle activity and organize movements.

Other types of nervous system damage also alter the organization of M1. Amputation of limbs not only deprives motor cortex of its normal sensory input, but also of its normal motor functions. After months to years of recovery after therapeutic amputation of the forelimb in monkeys, electrical stimulation of the deprived forelimb portion of M1 produces shoulder and arm stump movements (Qi et al., 2000; Schieber & Deuel, 1997; Wu & Kaas, 1999). In part, these changes in cortical organization were the result of the sprouting of cut peripheral-nerve motor axons that had innervated the muscles of the missing limb to newly innervate muscles of the stump (Wu & Kaas, 2000).

The functional reorganization of motor cortex is also altered by the sprouting of cut corticospinal axons to innervate new spinal cord motor and premotor neurons. Lesions of the corticospinal tract at the level of the upper cervical spinal cord are followed by impaired hand use, followed by a period of recovery. The corticospinal projections arise predominantly, but not exclusively, from M1 (Galea & Darian-Smith, 1994). Most of these projections from M1 are to the contralateral spinal cord, but a significant number are to the ipsilateral spinal cord. Experimental results so far indicate that there is little spontaneous sprouting of cut corticospinal axons past the barrier of the glial scar of the lesion site ("axons do not penetrate the lesion, cross it, or grow around it ... in significant numbers"; Steward et al., 2008, p. 6836). Instead, much of the behavioral recovery may be due to the considerable sprouting of the spared ipsilateral corticospinal axons to cross the spinal cord midline below the spinal cord lesion (Rosenzweig et al., 2010).

In summary, the functional reorganization in motor cortex may depend in part on rearrangements of intrinsic connections and cortical inputs, but the terminal arbors of corticospinal axons are plastic and capable of forming new contacts. In addition, the synaptic contacts of spinal motor neurons on muscles are plastic and can change to innervate different muscles.

All cortical areas in adult mammals are plastic

All cortical areas in adult primates and other mammals are likely plastic. The evidence is not comprehensive, but it is compelling. Partial lesions of primary visual cortex lend to reactivation of deprived neurons in higher-order visual areas, either by a potentiation of other sources of activation, as proposed for the mediation of blindsight (Cowey, 2010; Jenkins & Merzenich, 1987; Stoerig & Cowey, 1997; Weiskrantz, 1990; Weiskrantz et al., 1974), or by the potentiation of intrinsic, lateral connections from activated to deprived cortex within the higher-order visual area (Collins, Lyon, & Kaas, 2003, 2005). Similarly, a small lesion of primary somatosensory cortex, area 3b, is followed by a local reorganization of that cortex (Jenkins & Merzenich, 1987), likely induced by a sprouting of thalamocortical inputs, or the potentiation of subthreshold connections that existed before the lesions, but new connections are also likely formed in area 1, as well as in other areas with a loss of direct inputs from area 3b. Lesions of primary somatosensory cortex, and adjoining areas 1 and 2, resulted in a reactivation of deprived parts of the second somatosensory area, S2 (Pons, Garraghty, & Mishkin, 1988). Furthermore, lesions of the hand representation in M1 lead to an increase in size of the hand representation in the ventral premotor area and an increase in somatosensory connections (Dancause et al., 2005). In addition, small lesions of cortical regions in posterior parietal cortex that are involved in a specific behavior, such as reaching, are followed by a rapid recovery of the impaired behavior (Padberg et al., 2010). Thus, we propose that wherever neurons are deprived of their normal sources of activation in cortex, and likely at other levels, they are partially or totally reactivated by the potentiation of other inputs, by the growth of axons to form new connections, or both.

Treatments that promote or refine plasticity

Repairing the Injured Central Nervous System: A Focus on Spinal Cord Injury Although the mature nervous system retains considerable ability to form new connections and recover from damage, various treatments have the potential of increasing useful plasticity and promoting functional recovery. Recent studies, especially in the spinal cord injury field, have made great progress in elucidating conditions under which it can be coaxed to increase its malleability. These comprise not only alterations in synaptic strength, short-range rewiring or terminal sprouting (Darian-Smith & Ciferri, 2006; Florence & Kaas, 1995; Jain et al., 2000; Kaas et al., 2008), but also the extension of large numbers of axons over remarkable distances (Lu et al., 2012b). It is worth a reminder that increased plasticity within the central nervous system (CNS) doesn't necessarily translate to improved functional recovery (Lu et al., 2012a), and indeed may actually induce mechanical or thermal hypersensitivity, or cutaneous mislocation. Oftentimes, however, whether as a result of axonal regeneration or facilitation of intraspinal or cortical reorganization of spared circuits, researcher interventions have led to improved outcomes after the central nervous system has suffered an insult such as a spinal cord injury. The following examples are largely from studies in mice and rats due to the greater availability of molecular biology tools for rodent gene manipulation.

Neuroprotection After a central nervous system injury, the primary and most beneficial method of preserving function is via neuroprotection. Within 30 minutes of CNS injury, cells express proinflammatory cytokines that lead to the recruitment of even greater numbers of peripheral immune cells into the site of injury (Yang et al., 2004, 2005). These inflammatory cytokines (Ankeny & Popovich, 2009; Detloff et al., 2008; Donnelly & Popovich, 2008) are crucial to the processes of wound healing, and to the sequestration of an excitotoxic environment that kills neighboring cells via increases in extracellular glutamate and the ubiquitous cell-signaling calcium. The deleterious effects of inflammatory molecules, however, soon begin to hinder recovery. Therefore, various immunomodulatory drugs have been under scrutiny for reducing the secondary damage that significantly compounds an injury to the CNS (Casha et al., 2012; David, Lopez-Vales, & Wee Yong, 2012; Lord-Fontaine et al., 2008). While it is clear that the best outcome after CNS injury involves protecting the remaining susceptible infrastructure, interventions may not be timely, and this therefore requires that we maximize the potential of the remaining circuits either via the replacement of lost cells, intraspinal or cortical reorganization, or axonal sprouting or regeneration. The most commonly utilized spinal cord injury (SCI) repair strategies follow.

Overcoming Inhibitory Molecules in the Central Nervous System

Nogo Approximately 25 years ago, researchers discovered that axonal myelin actively inhibits the outgrowth of neurites, both in vitro and in vivo (Caroni, Savio, & Schwab, 1988; Caroni & Schwab, 1988; Schwab & Caroni, 1988). Subsequently, a major myelin-associated neurite growth inhibitor has been identified and named Nogo-A. Nogo-A is one of the transcripts of the Nogo gene that has been experimentally neutralized with function-blocking antibodies in efforts to enhance the regeneration and growth of injured axons in order to restore functions (see Fawcett, Schwab, Montani, Brazda, & Muller, 2012). Studies in rats demonstrated the efficacy of Nogo inhibition in stimulating the regrowth of cortical spinal tract axons up to 11 mm caudal to the lesion site by means of intracerebral implantation of Nogo-A antibody producing hybridoma grafts (Schnell & Schwab, 1990). Other promising results were seen in further studies (Liebscher et al., 2005; Simonen et al., 2003) that utilized different delivery routes and dosages, and across different SCI models, including a primate SCI study wherein neurites were observed to grow into the lesion site in four of the five treated animals (Fouad, Klusman, & Schwab, 2004). Importantly, a Nogo immunotherapy clinical trial has been underway for several years.

Chondroitin sulphate proteoglycans Chondroitin sulphate proteoglycans, or CSPGs, comprise an integral part of the extracellular matrix around neurons, as well as the glial scar that walls off the injury after SCI (Lemons, Howland, & Anderson, 1999; Rolls et al., 2008). This scar formation serves the beneficial role of protecting the unassaulted perilesion spinal cord from the waves of inflammation and excitotoxicity that otherwise exacerbate the injury (Silver, 2008). The scar is, however, a potent chemical and physical barrier to the outgrowth of regenerating axons (Plant, Bates, & Bunge, 2001; Snow, Brown, & Letourneau, 1996), and therefore is a continued target for the dissolution or modulation of its inhibitory components.

The commonly used method for digestion of CSPGs has been the application of the bacterial digestive enzyme chondrotinase ABC. The *Proteus vulgaris* product has been used in a large number of experiments and has repeatedly led to increased CNS plasticity. Initial reports measuring regeneration of dopaminergic nigrostriatal axons following axotomy and treatment with chondroitinase ABC indicated that nigral axons grew up to 4 mm through the injury site along the course of the original nigrostriatal tract back toward the ipsilateral striatum (Moon, Asher, Rhodes, & Fawcett, 2001). Subsequently, Bradbury et al. (2002) demonstrated that chABC application could also be used to promote the growth of dorsal column fibers, which were apparent 4 mm past the dorsal column lesion site. These studies also reported a correlational improvement in beam and grid locomotor tasks, but it is likely that these improvements were mediated via short-distance intraspinal sprouting. More recently, the application of chABC to the cuneate nucleus in primates promoted the effectiveness of surviving D1 afferents to activate larger portions of somatosensory cortex after most or all of the dorsal column cutaneous afferents of other digits had been cut (Bowes et al., 2012). Whereas most other studies had focused on the resulting changes at the spinal cord level, the reactivation of primary somatosensory cortex to which dorsal column afferents are relayed was evaluated. By intentionally sparing dorsal column afferents from digit 1 (thumb), it was demonstrated that months after the dorsal column lesion and cuneate nucleus treatment, the cortical territory in area 3b of somatosensory cortex activated by touch on D1 was larger in chABC-treated than in control monkeys.

Gene therapy

Growth Factors Given the discovery that secretion of exogenous growth factors has the ability to not only promote the regrowth of axons after injury, but to also serve in chemotactic guidance toward appropriate targets (Massey et al., 2008), researchers have sought methods to deliver these neurotrophic factors constitutively after injury. Additionally, growth-factor delivery has been shown to be effective not only when delivered acutely, but also after chronic injuries (Kadoya et al., 2009). However, widespread delivery of these neurotrophic factors can lead to adverse effects, and therefore local and targeted administration is required. Extended neurotrophic delivery has been achieved via infusion, ex vivo methods wherein grafts of tissue genetically modified to secrete brain-derived neurotropic factor, neurotrophin -4/5, or glial-cell-line–derived neurotropic factor have been implanted into the lesion site after SCI, or via viral vector gene delivery (Blits, Dijkhuizen, Boer, & Verhaagen, 2000; Piantino, Burdick, Goldberg, Langer, & Benowitz, 2006; Ruitenberg, Vukovic, Sarich, Busfield, & Plant, 2006; Schnell, Schneider, Kolbeck, Barde, & Schwab, 1994; Zhou & Shine, 2003). These grafts were usually successful in inducing axonal elongation, but only to a limited extent. Typically, greater elongation of sensory axons occurred (Bradbury et al., 1999; Taylor, Jones, Tuszynski, & Blesch, 2006), but corticospinal tract axons remained resistant to manipulation.

Regeneration-Associated Genes As the central nervous system matures, a down-regulation of mTOR activity occurs. Age and injury-dependent down-regulation of neuronal mTOR activity was identified as a major cause of the lack of regeneration of optic nerve axons after injury, and genetic activation of mTOR was observed to facilitate axon regrowth (Park et al., 2008). Subsequent efforts (Liu et al., 2010) revealed that considerable sprouting of the corticospinal tract also occurred after the conditional deletion of the gene for the phosphate tensor homologue protein (PTEN) in the sensorimotor cortex of adult mice. Liu and colleagues (2010) used Cre-expressing adeno-associated virus (AAV-Cre) to delete the gene encoding PTEN in homozygous conditional PTEN mutant mice. After lesions of the corticospinal tract in adults that had received neonatal AAV-Cre injections, PTEN deletion elicited considerable sprouting of the intact ipsilateral corticospinal tract into the contralateral cord. Further studies by these authors (Sun et al., 2011) showed an increased capacity for axonal regeneration after co-deletion of PTEN and suppressor of cytokine signaling 3 (SOCS3) in an optic nerve crush model, but similar observations have not yet been demonstrated in the spinal cord. Importantly, it has been shown that conditionally deleted SOCS3 expression in nestin-expressing cells have increased STAT3 activation, and that this limits the inflammatory expression after injury, thereby reducing cell death (Okada et al., 2006). Alternatively, CNS deletion of STAT3 delays formation of the glial scar, leading to greater inflammation, cell death, and functional impairment (Herrmann et al., 2008). Therefore, SOCS3 reduction may indeed be a good combinatorial component for reducing post-SCI deficits.

In 2009, a group of researchers (Moore et al., 2009) screened developmentally regulated genes in retinal ganglion cells and identified a family of Kruppel-like transcription factors (KLF) that regulated intrinsic axon regeneration ability. Of these, KLF7 was later found to have a large impact on axon growth after its developmental down-regulation was countered with overexpression in *in vitro* cortical neurons (Blackmore et al., 2012). This held true in vivo when overexpression of a KLF7 chimera with a VP16 transactivation domain in adult mice produced axonal regrowth in the CST tract.

Cell Transplantation The allure of cell transplantation is quite understandable given the many potential benefits that could be achieved with an effective transplant. Cell transplantation has the capacity to physically bridge a lesion site, replace lost neurons or supporting cells, secrete growth factors that can enhance regeneration or sprouting of damaged axons, and even aid in chemotactic signaling for guidance toward appropriate targets (Biernaskie et al., 2007; Grill, Blesch, & Tuszynski, 1997; Guest, Rao, Olson, Bunge, & Bunge, 1997; Haas & Fischer, 2013; Kajikawa, de la Mothe, Blumell, & Hackett, 2003; Kobayashi et al., 2012; Martin et al., 1996; Ruitenberg et al., 2006; Tuszynski, Grill, Jones, McKay, & Blesch, 2002; X. M. Xu, Zhang, Li, Aebischer, & Bunge, 1999). Regenerating axons will grow into a cell graft, but the greatest difficulty still lies in coaxing them through and beyond the lesion site for any considerable distance. However, this is not an insurmountable problem in terms of therapeutic potential since axons do grow around the lesion site, where they can potentially form new or re-establish functional connections that increase the remaining circuit's usefulness. Grafted Schwann cells overexpressing glial-cell-line–derived neurotrophic factor have also been used to promote regeneration of descending propriospinal axons through and beyond the lesion gap of a spinal cord hemisection where they formed new synapses (Deng et al., 2013).

Combinatorial Treatments With the promising, yet restricted, gains in individual therapies available, it is perhaps more productive to tackle the problem with multiple interventions at once. The potential of combinatorial treatments to transform the central nervous system into a more permissive state is already being realized.

Crucial to the usefulness of new synaptic connections is that they occur within a suitable target. Therefore, in addition to co-administration of neurotrophic factors, and dissolution of the CSPGs within the glial scar or around target neurons, physical rehabilitation might mediate an important component of functional recovery after injury via activity-dependent axonal guidance. With this in mind, Garcia-Alias, Barkhuysen, Buckle, and Fawcett (2009) investigated whether chondroitinase ABC–induced plasticity combined with physical rehabilitation could promote the recovery of manual dexterity in rats with cervical spinal cord injuries. After assigning animals to either a specific skilled reaching task or one that reinforced general locomotion, these investigators found that chondroitinase treatment facilitated the positive effects of rehabilitation. However, improvement was restricted to the trained task, whereas performance on the untrained task was negatively affected. The treatment was later found to be effective at the chronic stage as well (Wang et al., 2011). Recently, the regeneration of axons over incredible distances of 25 mm and beyond was reported (Lu et al., 2012a) after spinal cord transection models in rats. Two weeks after injury, the animals were grafted with human neural stem cells (or other cell types) delivered with a high-concentration growth factor cocktail into a fibrin/thrombin matrix in the lesioned spinal cord. Lu et al. reported the formation of functional synapses between host and grafted cells, and electrophysiology revealed an evoked response across the lesion site.

These observations of long-distance regrowth illuminate a promising new path via which a host of treatments that individually fall short can be incorporated into a complementary strategy to rebuild a circuit that best facilitates recovery after CNS injury. Treatment with chondroitinase ABC has also been recently combined with an older approach that had previously led to only limited success. A number of investigations had used a segment of peripheral nerve to bridge a gap in the spinal cord due to injury. However, a buildup of growth-inhibitory CSPGs prevented regenerating spinal cord axons from entering or leaving the bridge. In several studies, chondroitinase ABC has been used to counter the CSPGs, resulting in axon growth across the bridge that restored function (see Lee et al., 2013).

Conclusion

In summary, behavioral training has long been the most productive treatment for promoting behavioral recoveries after nervous system injury. The recoveries may be, in part, due to the learning of behavior compensations, but they also may promote the formation of favorable new connections as a result of activity-dependent mechanisms. In addition, a number of other treatments are promising to be effective by removing barriers to new axon growth and enhancing growth with growth factors. While the research emphasis has been on promoting formation of new connections, more research is needed to find ways that restrict new growth that could lead to misperception, motor errors, and impaired cognition.

REFERENCES

Ankeny, D. P., & Popovich, P. G. (2009). Mechanisms and implications of adaptive immune responses after traumatic spinal cord injury. *Neurosci, 158*, 1112–1121.

Baker, C. I., Dilks, D. D., Peli, E., & Kanwisher, N. (2008). Reorganization of visual processing in macular degeneration: Replication and clues about the role of foveal loss. *Vis Res, 48*, 1910–1919.

Biernaskie, J., Sparling, J. S., Liu, J., Shannon, C. P., Plemel, J. R., Xie, Y., … Tetzlaff, W. (2007). Skin-derived precursors generate myelinating Schwann cells that promote remyelination and functional recovery after contusion spinal cord injury. *J Neurosci, 27*, 9545–9559.

Blackmore, M. G., Wang, Z., Lerch, J. K., Motti, D., Zhang, Y. P., Shields, C. B., … Bixby, J. L. (2012). Kruppel-like Factor 7 engineered for transcriptional activation promotes axon regeneration in the adult corticospinal tract. *Proc Natl Acad Sci USA, 109*, 7517–7522.

Blits, B., Dijkhuizen, P. A., Boer, G. J., & Verhaagen, J. (2000). Intercostal nerve implants transduced with an adenoviral vector encoding neurotrophin-3 promote regrowth of injured rat corticospinal tract fibers and improve hindlimb function. *Exp Neurol, 164*, 25–37.

Bowes, C., Massey, J. M., Burish, M., Cerkevich, C. M., & Kaas, J. H. (2012). Chondroitinase ABC promotes selective reactivation of somatosensory cortex in squirrel monkeys after a cervical dorsal column lesion. *Proc Natl Acad Sci USA, 109,* 2595–2600.

Bradbury, E. J., Khemani, S., Von, R., King, V. R., Priestley, J. V., & McMahon, S. B. (1999). NT-3 promotes growth of lesioned adult rat sensory axons ascending in the dorsal columns of the spinal cord. *Euro J Neurosci, 11,* 3873–3883.

Bradbury, E. J., Moon, L. D., Popat, R. J., King, V. R., Bennett, G. S., Patel, P. N., ... McMahon, S. B. (2002). Chondroitinase ABC promotes functional recovery after spinal cord injury. *Nature, 416,* 636–640.

Brodmann, K. (1909). *Vergleichende Lokalisationslehre der Grosshirnrhinde.* Barth, Leipzig.

Buomomano, D. V., & Merzenich, M. M. (1998). Cortical plasticity, from synapses to maps. *Ann Rev Neurosci, 21,* 149–196.

Calford, M. B., Wright, L. I., Metha, A. B., & Taglianetti, V. (2003). Topographic plasticity in primary visual cortex is mediated by local corticocortical connections. *J Neurosci, 23,* 6434–6442.

Caroni, P., Savio, T., & Schwab, M. E. (1988). Central nervous system regeneration, oligodendrocytes and myelin as non-permissive substrates for neurite growth. *Prog Brain Res, 78,* 363–370.

Caroni, P., & Schwab, M. E. (1988). Antibody against myelin-associated inhibitor of neurite growth neutralizes nonpermissive substrate properties of CNS white matter. *Neuron, 1,* 85–96.

Casha, S., Zygun, D., McGowan, M. D., Bains, I., Yong, V. W., & Hurlbert, R. J. (2012). Results of a phase II placebo-controlled randomized trial of minocycline in acute spinal cord injury. *Brain, 135*(PT 4), 1224–1236.

Chen, L. M., Qi, H-X., & Kaas, J. H. (2012). Dynamic reorganization of digit representations in somatosensory cortex of nonhuman primates after spinal cord injury. *J Neurosci, 32,* 14649–14663.

Chino, Y. M., Smith, E. L., 3rd, Kaas, J. H., Sasaki, Y., & Cheng, H. (1995). Receptive-field properties of deafferentated visual cortical neurons after topographic map reorganization in adult cats. *J Neurosci, 15,* 2417–2433.

Collins, C. E., Lyon, D. C., & Kaas, J. H. (2003). Responses of neurons in the middle temporal visual area after long-standing lesions of the primary visual cortex in adult New World monkeys. *J Neurosci, 23,* 2251–2264.

Collins, C. E., Lyon, D. C., & Kaas, J. H. (2005). Distribution across cortical areas of neurons projecting to the superior colliculus in New World monkeys. *Anat Rec, 285*A, 619–627.

Cowey, A. (2010). The blindsight saga. *Exp Brain Res, 200,* 3–24.

Dancause, N., Barbay, S., Frost, S. B, Plautz, E. J., Chen, D., Zoubina, E. V., ... Nudo, R. J. (2005). Extensive cortical rewiring after brain injury. *J Neurosci, 25,* 10167–10179.

Dancause, N., Barbay, S., Frost, S. B, Zoubina, E. V., Plautz, E. J., Mahnken, J. D., & Nudo, R. J. (2006). Effects of small ischemic lesions in the primary motor cortex on neurophysiological organization in ventral premotor cortex. *J Neurophysiol, 96,* 3506–3511.

Dancause, N., & Nudo, R. J. (2011). Shaping plasticity to enhance recovery after injury. *Prog Brain Res, 192,* 273–291.

Darian-Smith, C., & Ciferri, M. M. (2005). Loss and recovery of voluntary hand movements in the macaque following a cervical dorsal rhizotomy. *J Comp Neurol, 491,* 27–45.

Darian-Smith, C., & Ciferri, M. M. (2006). Cuneate nucleus reorganization following cervical dorsal rhizotomy in the macaque monkey: Its role in the recovery of manual dexterity. *J Comp Neurol, 498,* 552–565.

Darian-Smith, C., & Gilbert, C. D. (1994). Axonal sprouting accompanies functional reorganization in adult cat striate cortex. *Nature, 368,* 737–740.

Darian-Smith, C., & Gilbert, C. D. (1995). Topographic reorganization in the striate cortex of the adult cat and monkey is cortically mediated. *J Neurosci, 15,* 1631–1647.

David, S., Lopez-Vales, R., & Wee Yong, V. (2012). Harmful and beneficial effects of inflammation after spinal cord injury; potential therapeutic implications. *Handb clin neurol, 109,* 485–502.

Davis, K. D., Kiss, Z. H. T., Luo, L., Tasker, R. R., Lozano, A. M., & Dostrovsky, J. O. (1998). Phantom sensations generated by thalamic microstimulation. *Nature, 391,* 385–387.

Daw, N. W. (1995). *Visual development.* New York, NY: Plenum Press.

Deng, L. X., Deng, P., Ruan, Y., Xu, Z. C., Liu, N. K., Wen, X., ... Xu, X. M. (2013). A novel growth-promoting pathway formed by GDNF-overexpressing Schwann cells promotes propriospinal axonal regeneration, synapse formation, and partial recovery of function after spinal cord injury. *J. Neurosci, 33,* 5655–5667.

Detloff, M. R., Fisher, L. C., McGaughy, V., Longbrake, E. E., Popovich, P. G., & Basso, D. M. (2008). Remote activation of microglia and pro-inflammatory cytokines predict the onset and severity of below-level neuropathic pain after spinal cord injury in rats. *Exp Neurol, 212,* 337–347.

Diamond, M. E., Armstrong-James, M., & Ebner, F. F. (1993). Experience-dependent plasticity in adult rat barrel cortex. *Proc Natl Acad Sci USA, 90*(5), 2082–2086.

Donnelly, D. J., & Popovich, P. G. (2008). Inflammation and its role in neuroprotection, axonal regeneration and functional recovery after spinal cord injury. *Exp Neurol, 209,* 378–388.

Eysel, U. T., Gonzalez-Aguillar, F., & Mayer, U. (1980). A functional sign of reorganization in the visual system of adult cats: Lateral geniculate neurons with displaced receptive fields after lesions of the nasal retina. *Brain Res, 191,* 285–300.

Eysel, U. T., Gonzalez-Aguillar, F., & Mayer, U. (1981). Time-dependent decrease in the extent of visual deafferentation in the lateral geniculate nucleus of adult cats with small retinal lesions. *Exp Brain Res, 41,* 256–263.

Fawcett, J. W., Schwab, M. E., Montani, L., Brazda, N., & Muller, H. W. (2012). Defeating inhibition of regeneration by scar and myelin components. *Handb Clin Neurol, 109,* 503–522.

Flor, H., Elbert, T., Knecht, S., Wienbruch, C., Pantev, C., Birbaumer, N., ... Taub, E. (1995). Phantom-limb pain as a perceptual correlate of cortical reorganization following arm amputation. *Nature, 375,* 482–484.

Florence, S. L., Garraghty, P. E., Wall, J. T., & Kaas, J. H. (1994). Sensory afferent projections and area 3b somatotopy following median nerve cut and repair in macaque monkeys. *Cereb Cortex, 4,* 391–407.

Florence, S. L., & Kaas, J. H. (1995). Large-scale reorganization at multiple levels of the somatosensory pathway follows therapeutic amputation of the hand in monkeys. *J Neurosci, 15*, 8083–8095.

Florence, S. L., & Kaas, J. H. (2000). Cortical plasticity: Growth of new connections can contribute to reorganization. In M. J. Rowe & Y. Iwamura (Eds.), *Somatosensory processing: From single neurons to brain imaging* (pp. 167–185). Amsterdam: Harwood Academic.

Florence, S. L., Wall, J. T., & Kaas, J. H. (1988). The somatotopic pattern of afferent projections from the digits to the spinal cord and cuneate nucleus in macaque monkeys. *Brain Res, 452*, 388–392.

Florence, S. L., Wall, J. T., & Kaas, J. H. (1991). Central projections from the skin of the hand in squirrel monkeys. *J Comp Neurol, 311*, 563–578.

Fouad, K., Klusman, I., & Schwab, M. E. (2004). Regenerating corticospinal fibers in the Marmoset (*Callitrix jacchus*) after spinal cord lesion and treatment with the anti-Nogo-A antibody IN-1. *Euro J Neurosci, 20*, 2479–2482.

Galea, M. P., & Darian-Smith, C. (1994). Multiple corticospinal neuron populations in the macaque monkey are specified by their unique cortical origins, spinal terminations, and connections. *Cereb Cortex, 4*, 166–194.

Garcia-Alias, G., Barkhuysen, S., Buckle, M., & Fawcett, J. W. (2009). Chondroitinase ABC treatment opens a window of opportunity for task-specific rehabilitation. *Nat Neurosci, 12*, 1145–1151.

Garraghty, P. E., & Kaas, J. H. (1991a). Large-scale functional reorganization in adult monkey cortex after peripheral nerve injury. *Proc Natl Acad Sci USA, 88*, 6976–6980.

Garraghty, P. E., & Kaas, J. H. (1991b). Functional reorganization in adult monkey thalamus after peripheral nerve injury. *NeuroReport, 2*, 747–750.

Garraghty, P. E., LaChica, E. A., & Kaas, J. H. (1991). Injury-induced reorganization of somatosensory cortex is accompanied by reductions in GABA staining. *Somatosens Mot Res, 8*, 347–354.

Giannikopoulos, D. V., & Eysel, U. T. (2006). Dynamics and specificity of cortical map reorganization after retinal lesions. *Proc Natl Acad Sci USA, 103*, 10805–10810.

Gilbert, C. D., & Wiesel, T. N. (1992). Receptive field dynamics in adult primary visual cortex. *Nature, 356*, 150–152.

Glees, P., & Cole, J. (1950). Recovery of skilled motor functions after small repeated lesions of motor cortex in macaque. *J Neurophysiol, 13*, 133–148.

Gould, H. J., Cusick, C. G., Pons, T. P., & Kaas, J. H. (1986). The relationship of corpus callosum connections to electrical stimulation maos of motor, supplementary motor, and the frontal eye fields in owl monkeys. *J Comp Neurol, 247*, 298–325.

Grill, R. J., Blesch, A., & Tuszynski, M. H. (1997). Robust growth of chronically injured spinal cord axons induced by grafts of genetically modified NGF-secreting cells. *Exp Neurol, 148*, 444–452.

Guest, J. D., Rao, A., Olson, L., Bunge, M. B., & Bunge, R. P. (1997). The ability of human Schwann cell grafts to promote regeneration in the transected nude rat spinal cord. *Exp Neurol, 148*, 502–522.

Haas, C., & Fischer, I. (2013). Human astrocytes derived from glial restricted progenitors support regeneration of the injured spinal cord. *J Neurotrauma, 30*(12), 1035–1052.

Heinen, S. J., & Skavenski, A. A. (1991). Recovery of visual responses in foveal V1 neurons following bilateral foveal lesions in adult monkey. *Exp Brain Res, 83*, 670–674.

Herrmann, J. E., Imura, T., Song, B., Qi, J., Ao, Y., Nguyen, T. K., ... Sofroniew, M. V. (2008). STAT3 is a critical regulator of astrogliosis and scar formation after spinal cord injury. *J Neurosci, 28*, 7231–7243.

Hoffman, D. S., & Strick, P. L. (1995). Effects of a primary motor cortex lesion on step-tracking movements of the wrist. *J Neurophysiol, 73*, 891–895.

Irvin, D. R. F., & Rojan, R. (1997). Injury-induced reorganization of frequency maps in adult auditory cortex: The role of unmasking of normally-inhibited inputs. *Acta Otolaryngol Suppl, 532*, 39–45.

Jain, N., Catania, K. C., & Kaas, J. H. (1997). Deactivation and reactivation of somatosensory cortex after dorsal spinal cord injury. *Nature, 386*, 495–498.

Jain, N., Florence, S. L., Qi, H. X., & Kaas, J. H. (2000). Growth of new brainstem connections in adult monkeys with massive sensory loss. *Proc Natl Acad Sci USA, 97*, 5546–5550.

Jain, N., Qi, H.-X., Collins, C. E., & Kaas, J. H. (2008). Large-scale reorganization of the somatosensory cortex and thalamus after sensory loss in macaque monkeys. *J Neurosci, 28*, 11042–11060.

Jenkins, W. M., & Merzenich, M. M. (1987). Reorganization of neocortical representations after brain injury: A neurophysiological model of the bases of recovery from stroke. *Prog Brain Res, 71*, 249–266.

Kaas, J. H. (1983). What, if anything, is SI? The organization of "first somatosensory area" of cortex. *Physiol Rev, 63*, 206–231.

Kaas, J. H. (1996). Plasticity of sensory representations in the auditory and other systems of adult mammals. In R. J. Salvi, D. Henderson, F. Fiorino, & J. Colletti (Eds.), *Auditory system plasticity and regeneration* (pp. 213–223). New York, NY: Thieme Medical.

Kaas, J. H. (2011). Somatosensory system. In J. K. Mai & G. Paxinos (Eds.), *The human nervous system* (3rd ed., pp. 1064–1099). London: Elsevier.

Kaas, J. H., Gharbawie, O. A., & Stepniewska, I. (2011). The organization and evolution of dorsal stream multisensory motor pathways in primates. *Front Neuroanat, 5*(34), 1–7.

Kaas, J. H., Krubitzer, L. A., Chino, Y. M., Langston, A. L., Polley, E. H., & Blair, N. (1990). Reorganization of retinotopic cortical maps in adult mammals after lesions of the retina. *Science, 248*(4952), 229–231.

Kaas, J. H., Merzenich, M. M., & Killackey, H. P. (1983). Changes in the organization of somatosensory cortex following peripheral nerve damage in adult and developing mammals. *Ann Rev Neurosci, 6*, 325–356.

Kaas, J. H., Qi, H. X., Burish, M. J., Gharbawie, O. A., Onifer, S. M., & Massey, J. M. (2008). Cortical and subcortical plasticity in the brains of humans, primates, and rats after damage to sensory afferents in the dorsal columns of the spinal cord. *Exp Neurol, 209*, 407–416.

Kaas, J. H., Stepniewska, I., & Gharbawie, O. (2012). Cortical networks subserving upper limb movements in primates. *Eur J Phys Rehabil Med, 48*, 299–306.

Kadoya, K., Tsukada, S., Lu, P., Coppola, G., Geschwind, D., Filbin, M. T., ... Tuszynski, M. H. (2009). Combined intrinsic and extrinsic neuronal mechanisms facilitate bridging axonal regeneration one year after spinal cord injury. *Neuron, 64*, 165–172.

Kajikawa, Y., de la Mothe, L., Blumell, S., & Hackett, T. A. (2003). Response properties of neurons in medial belt auditory cortex of marmoset monkeys. *Assoc Res Otolaryngol Abstr, 26*, 204.

Karl, A., Birbaumer, N., Lutzenberger, W., Cohen, L. G., & Flor, H. (2001). Reorganization of motor and somatosensory cortex in upper extremity amputees with phanton limb pain. *J Neurosci, 21*, 3609–3618.

Kobayashi, Y., Okada, Y., Itakura, G., Iwai, H., Nishimura, S., Yasuda, A., ... Okano, H. (2012). Pre-evaluated safe human iPSC-derived neural stem cells promote functional recovery after spinal cord injury in common marmoset without tumorigenicity. *PLoS ONE, 7*, e52787.

Lee, Y.-S., Lin, C.-Y., Jiang, H.-H., DePaul, M., Lin, V. W., & Silver, J. (2013). Nerve regeneration restores supraspinal control of bladder function after complete spinal cord injury. *J Neurosci, 33*, 10591–10606.

Lemons, M. L., Howland, D. R., & Anderson, D. K. (1999). Chondroitin sulfate proteoglycan immunoreactivity increases following spinal cord injury and transplantation. *Exp Neurol, 160*(1), 51–65.

Li, W., Piech, V., & Gilbert, C. D. (2004). Perceptual learning and top-down influences in primary visual cortex. *Nat Neurosci, 7*, 651–657.

Liebscher, T., Schnell, L., Schnell, D., Scholl, J., Schneider, R., Gullo, M., ... Schwab, M. E. (2005). Nogo-A antibody improves regeneration and locomotion of spinal cord-injured rats. *Ann Neurol, 58*, 706–719.

Liu, K., Lu, Y., Lee, J. K., Samara, R., Willenberg, R., Sears-Kraxberger, I., ... He, Z. (2010). PTEN deletion enhances the regenerative ability of adult corticospinal neurons. *Nat Neurosci, 13*, 1075–1081.

Lord-Fontaine, S., Yang, F., Diep, Q., Dergham, P., Munzer, S., Tremblay, P., & McKerracher, L. (2008). Local inhibition of Rho signaling by cell-permeable recombinant protein BA-210 prevents secondary damage and promotes functional recovery following acute spinal cord injury. *J Neurotrauma, 25*(25), 1309–1322.

Lu, P., Blesch, A., Graham, L., Wang, Y., Samara, R., Banos, K., ... Tuszynski, M. H. (2012a). Motor axonal regeneration after partial and complete spinal cord transection. *J Neurosci, 32*, 8208–8218.

Lu, P., Wang, Y., Graham, L., McHale, K., Gao, M., Wu, D., ... Tuszynski, M. H. (2012b). Long-distance growth and connectivity of neural stem cells after severe spinal cord injury. *Cell, 150*, 1264–1273.

Martin, D., Robe, P., Franzen, R., Delree, P., Schoenen, J., Stevenaert, A., & Moonen, G. (1996). Effects of Schwann cell transplantation in a contusion model of rat spinal cord injury. *J Neurosci Res, 45*, 588–597.

Massey, J. M., Amps, J., Viapiano, M. S., Matthews, R. T., Wagoner, M. R., Whitaker, C. M., ... Onifer, S. M. (2008). Increased chondroitin sulfate proteoglycan expression in denervated brainstem targets following spinal cord injury creates a barrier to axonal regeneration overcome by chondroitinase ABC and neurotrophin-3. *Exper Neurol, 209*, 426–445.

Merzenich, M. M., Kaas, J. H., Wall, J., Nelson, R. J., Sur, M., & Felleman, D. (1983a). Topographic reorganization of somatosensory cortical areas 3b and 1 in adult monkeys following restricted deafferentation. *Neurosci, 8*, 33–55.

Merzenich, M. M., Kaas, J. H., Wall, J. T., Sur, M., Nelson, R. J., & Felleman, D. J. (1983b). Progression of change following median nerve section in the cortical representation of the hand in areas 3b and 1 in adult owl and squirrel monkeys. *Neuroscience, 10*, 639–665.

Moon, L. D., Asher, R. A., Rhodes, K. E., & Fawcett, J. W. (2001). Regeneration of CNS axons back to their target following treatment of adult rat brain with chondroitinase ABC. *Nat Neurosci, 4*, 465–466.

Moore, C. I., Stern, C. E., Dunbar, C., Kostyk, S. K., Gehi, A., & Corkin, S. (2000). Referred phantom sensation and cortical reorganization after spinal cord injury in humans. *Proc Natl Acad Sci USA, 97*, 14703–14705.

Moore, D. L., Blackmore, M. G., Hu, Y., Kaestner, K. H., Bixby, J. L., Lemmon, V. P., & Goldberg, J. L. (2009). KLF family members regulate intrinsic axon regeneration ability. *Science, 326*, 298–301.

Nudo, R. L., & Milliken, G. W. (1996). Reorganization of movement representations in primary motor cortex following focal ischemic infarcts in adult squirrel monkeys. *J Neurophysiol, 75*, 2144–2149.

Nudo, R. L., Milliken, G. W., Jenkins, W. M., & Merzenich, M. M. (1996). Use dependent alterations of movement representations in primary motor cortex of adult squirrel monkeys. *J Neurosci, 96*, 785–807.

Okada, S., Nakamura, M., Katoh, H., Miyao, T., Shimazaki, T., Ishii, K., ... Okano, H. (2006). Conditional ablation of Stat3 or Socs3 discloses a dual role for reactive astrocytes after spinal cord injury. *Nat Med, 12*, 829–834.

Padberg, J., Recanzone, G., Engle, J., Cooke, D., Goldring, A., & Krubitzer, L. (2010). Lesions of posterior parietal area 5 in monkeys result in behavioral and cortical plasticity. *J. Neurosci, 30*, 12918–12935.

Park, K. K., Liu, K., Hu, Y., Smith, P. D., Wang, C., Cai, B., & He, Z. (2008). Promoting axon regeneration in the adult CNS by modulation of the PTEN/mTOR pathway. *Science, 322*, 963–966.

Piantino, J., Burdick, J. A., Goldberg, D., Langer, R., & Benowitz, L. I. (2006). An injectable, biodegradable hydrogel for trophic factor delivery enhances axonal rewiring and improves performance after spinal cord injury. *Exp Neurol, 201*, 359–367.

Plant, G. W., Bates, M. L., & Bunge, M. B. (2001). Inhibitory proteoglycan immunoreactivity is higher at the caudal than the rostral Schwann cell graft-transected spinal cord interface. *Mol Cell Neurosci, 17*, 471–487.

Pons, T. P., Garraghty, P. E., & Mishkin, M. (1988). Lesion-induced plasticity in the second somatosensory cortex of adult macaques. *Proc Natl Acad Sci USA, 85*, 5279–5281.

Pons, T. P., Garraghty, P. E., Ommaya, A. K., Kaas, J. H., Taub, E., & Mishkin, M. (1991). Massive cortical reorganization after sensory deafferentation in adult macaques. *Science, 252*, 1857–1860.

Qi, H. X., Chen, L. M., & Kaas, J. H. (2011). Reorganization of somatosensory cortical areas 3b and 1 after unilateral section of dorsal columns of the spinal cord in squirrel monkeys. *J Neurosci, 31*, 13662–13675.

Qi, H. X., Gharbawie, O. A., Wynne, K. W., & Kaas, J. H. (2013). Impairment and recovery of hand use after unilateral section of the dorsal columns of the spinal cord in squirrel monkeys. *Behav Brain Res, 252*C, 363–376.

Qi, H. X., Jain, N., Collins, C. E., Lyon, D. C., & Kaas, J. H. (2010). Functional organization of motor cortex of adult macaque monkeys is altered by sensory loss in infancy. *Proc Natl Acad Sci USA, 107,* 3192–3197.

Qi, H.-X., & Kaas, J. H. (2006). The organization of primary afferent projections to the gracile nucleus of the dorsal column system of primates. *J Comp Neurol, 499,* 183–217.

Qi, H. X., Stepniewska, I., & Kaas, J. H. (2000). Reorganization of primary motor cortex in adult macaque monkeys with long-standing amputations. *J Neurophysiol, 84,* 2133–2147.

Ramachandran, V. S., Rogers-Ramachandran, D., & Stewart, M. (1992). Perceptual correlates of massive cortical reorganization. *Science, 258,* 1159–1160.

Rauschecker, J. P. (1999). Auditory cortical plasticity: A comparison with other sensory systems. *Trends Neurosci, 22,* 74–80.

Recanzone, G. H., Schreiner, C. E., & Merzenich, M. M. (1993). Plasticity in the frequency representation of primary auditory cortex following discrimination training in adult owl monkeys. *J Neurosci, 1,* 87–103.

Robertson, D., & Irvine, D. R. F. (1989). Plasticity of frequency organization in auditory cortex of guinea pigs with partial unilateral deafness. *J Comp Neurol, 282,* 456–471.

Rolls, A., Shechter, R., London, A., Segev, Y. Jacob-Hirsch, J., Amariglio, N. ... Schwartz, M. (2008). Two faces of chondroitin sulfate proteoglycan in spinal cord repair: A role in microglia/macrophage activation. *PLoS Med, 5*(8), e171.

Rosenzweig, E. S., Courtine, G., Jinarich, D. L., Brock, J. H., Ferguson, A. R., Strand, S. C., ... Tuszynski, M. H. (2010). Extensive spontaneous plasticity of corticospinal projections after primate spinal cord injury. *Nat Neurosci, 13,* 1505–1510.

Ruitenberg, M. J., Vukovic, J., Sarich, J., Busfield, S. J., & Plant, G. W. (2006). Olfactory ensheathing cells: Characteristics, genetic engineering, and therapeutic potential. *J Neurotrauma, 23,* 468–478.

Rustioni, A. (1976). Spinal neurons project to the dorsal column nuclei of rhesus monkeys. *Science, 196,* 656–658.

Schieber, M. H. (2001). Constraints on somatotopic organization in the primary motor cortex. *J Neurophysiol, 86,* 2125–2143.

Schieber, M. H., & Deuel, R. K. (1997). Primary motor cortex reorganization in a long-term monkey amputee. *Somatosens Mot Res, 14,* 157–167.

Schnell, L., Schneider, R., Kolbeck, R., Barde, Y. A., & Schwab, M. E. (1994). Neurotrophin-3 enhances sprouting of corticospinal tract during development and after adult spinal cord lesion. *Nature, 367,* 170–173.

Schnell, L., & Schwab, M. E. (1990). Axonal regeneration in the rat spinal cord produced by an antibody against myelin-associated neurite growth inhibitors. *Nature, 343,* 269–272.

Schwab, M. E., & Caroni, P. (1988). Oligodendrocytes and CNS myelin are nonpermissive substrates for neurite growth and fibroblast spreading in vitro. *J Neurosci, 8,* 2381–2393.

Schwaber, M. K., Garraghty, P. E., & Kaas, J. H. (1993). Neuroplasticity of the adult primate auditory cortex following cochlear hearing loss. A*m J Otol, 14,* 252–258.

Sergent, J. (1988). An investigation into perceptual completion in blind areas of the visual field. *Brain, 111,* 347–373.

Silver, J. (2008). Special issue: Spinal cord regeneration and repair. *Exper Neurology, 209,* 293.

Simonen, M., Pedersen, V., Weinmann, O., Schnell, L., Buss, A., Ledermann, B., ... Schwab, M. E. (2003). Systemic deletion of the myelin-associated outgrowth inhibitor Nogo-A improves regenerative and plastic responses after spinal cord injury. *Neuron, 38,* 201–211.

Smirnakis, S. M., Brewer, A. H., Schmid, M. C., Tolias, A. S., Schüz, A., Augath, M., ... Logothetis, N. K. (2005). Lack of long-term cortical reorganization after macaque retinal lesions. *Nature, 435,* 300–307.

Snow, D. M., Brown, E. M., & Letourneau, P. C. (1996). Growth cone behavior in the presence of soluble chondroitin sulfate proteoglycan (CSPG), compared to behavior on CSPG bound to laminin or fibronectin. *Int J Dev Neurosci, 14,* 331–349.

Sperry, R. W. (1959). The growth of nerve circuits. *Sci Am, 201,* 68–75.

Steward, O., Zheng, B., Tessier-LaVigne, M., Hofstadter, M., Sharp, K., & Yee, K. M. (2008). Regenerative growth of corticospinal tract axons via the ventral column after spinal cord injury in mice. *J Neurosci, 28,* 6836–6847.

Stoerig, P., & Cowey, A. (1997). Blindsight in man and monkey. *Brain, 120,* 535–559.

Sun, F., Park, K. K., Belin, S., Wang, D., Lu, T., Chen, G., ... He, Z. (2011). Sustained axon regeneration induced by codeletion of PTEN and SOCS3. *Nature, 480*(7377), 372–375.

Taylor, L., Jones, L., Tuszynski, M. H., & Blesch, A. (2006). Neurotrophin-3 gradients established by lentiviral gene delivery promote short-distance axonal bridging beyond cellular grafts in the injured spinal cord. *J Neurosci, 26*(38), 9713–9721.

Turrigiano, G. G. (1999). Homostatic plasticity in neuronal networks: The more things change, the more they stay the same. *Trends Neurosci, 22,* 221–227.

Tuszynski, M. H., Grill, R., Jones, L. L., McKay, H. M., & Blesch, A. (2002). Spontaneous and augmented growth of axons in the primate spinal cord: Effects of local injury and nerve growth factor-secreting cell grafts. *J Comp Neurol, 449,* 88–101.

Wall, J. T., Felleman, D. J., & Kaas, J. H. (1983). Recovery of normal topography in the somatosensory cortex of monkeys after nerve crush and regeneration. *Science, 221,* 771–773.

Wall, J. T., & Kaas, J. H. (1986). Long-term cortical consequences of reinnervation errors after nerve regeneration in monkeys. *Brain Res, 372,* 400–404.

Wall, J. T., Kaas, J. H., Merzenich, M. M., Sur, M., Nelson, R. J., & Felleman, D. J. (1986). Functional reorganization in somatosensory cortical areas 3b and 1 of adult monkeys after median nerve repair: Possible relationships to sensory recovery in humans. *J Neurosci, 6,* 218–233.

Wang, D., Ichiyama, R. M., Zhao, R., Andrews, M. R., & Fawcett, J. W. (2011). Chondroitinase combined with rehabilitation promotes recovery of forelimb function in rats with chronic spinal cord injury. *J Neurosci, 31,* 9332–9344.

Weiskrantz, L. (1990). *Blindsight: A case study and implications.* Oxford, UK: Oxford University Press.

Weiskrantz, L., Warrington, E. K., Sanders, M. D., & Marshall, J. (1974). Visual capacity in the hemianopic field following a restricted cortical ablation. *Brain, 97,* 709–728.

Wiesel, T. N. (1982). Postnatal development of the visual cortex and the influence of environment. *Nature, 299,* 583–591.

Willott, J. F., Aitkin, L. M., & McFadden, S. L. (1993). Plasticity of auditory cortex associated with sensorineural hearing loss in adult C57BL/6J mice. *J Comp Neurol, 329,* 402–411.

Witham, C. L., & Baker, S. N. (2011). Modulation and transmission of peripheral inputs in monkey cuneate and external cuneate nuclei. *J Neurophysiol, 106,* 2764–2775.

Wu, C. W., & Kaas, J. H. (1999). Reorganization in primary motor cortex of primates with long-standing therapeutic amputations. *J Neurosci, 19,* 7679–7697.

Wu, C. W.-H., & Kaas, J. H. (2000). Spinal cord atrophy and reorganization of motor neuron connections following long-standing limb loss in primates. *Neuron, 28,* 1–20.

Xu, J., & Wall, J. T. (1997). Rapid changes in brainstem maps of adult primates after peripheral injury. *Brain Res, 774,* 211–215.

Xu, X. M., Zhang, S. X., Li, H., Aebischer, P., & Bunge, M. B. (1999). Regrowth of axons into the distal spinal cord through a Schwann-cell-seeded mini-channel implanted into hemisected adult rat spinal cord. *Euro J Neurosci, 11,* 1723–1740.

Yamahachi, H., Mavik, S. A., McManus, J. N., Denk, W., & Gilbert, C. D. (2009). Rapid axonal sprouting and pruning accompany functional reorganization in primary visual cortex. *Neuron, 64,* 719–729.

Yang, L., Blumbergs, P. C., Jones, N. R., Manavis, J., Sarvestani, G. T., & Ghabriel, M. N. (2004). Early expression and cellular localization of proinflammatory cytokines interleukin-1beta, interleukin-6, and tumor necrosis factor-alpha in human traumatic spinal cord injury. *Spine, 29,* 966–971.

Yang, L., Jones, N. R., Blumbergs, P. C., van den Heuvel, C., Moore, E. J., Manavis, J., & Ghabriel, M. N. (2005). Severity-dependent expression of pro-inflammatory cytokines in traumatic spinal cord injury in the rat. *J Clin Neurosci, Australasia, 12*(3), 276–284.

Zhou, L., & Shine, H. D. (2003). Neurotrophic factors expressed in both cortex and spinal cord induce axonal plasticity after spinal cord injury. *J Neurosci Res, 74,* 221–226.

12 Modularity, Plasticity, and Development of Brain Function

MAN CHEN AND MICHAEL W. DEEM

ABSTRACT We discuss the relationship between modularity of neural activity in the brain and cognitive ability and review observations and theories relating modularity to plasticity of brain neural activity. By analogy with evolutionary biology, we hypothesize that selection for maximum plasticity of the human brain occurs in young adulthood, which implies modularity should peak in young adults. We show that modularity of neural activity derived from functional MRI data rises from childhood, peaks in young adults, and declines in older adults. We suggest experiments to measure the impact of modularity on fast, low-level, automatic cognitive processes versus high-level, effortful, conscious cognitive processes. We suggest modularity may be a useful biomarker for disease, injury, or rehabilitation.

Biological systems are modular, and the organization of their genetic material reflects this modularity (Hartwell, Hopfield, Leibler, & Murray, 1999; Rojas, 1996; Simon, 1962; Waddington, 1942). Genes are organized into exons, and expression of different exons allows one gene to produce multiple proteins. In many cases, each exon confers a distinct function to the protein, and so an exon is a modular element. Similarly, while genes interact, each gene confers function or functions to the organism, and so a gene is a modular element. Collections of related genes clustered together on the genome occasionally form a functional unit termed an operon, which is also a module.

In this chapter, we are interested in a different sort of modularity, that of spatially and temporally correlated neural activity in the brain. A group of neurons that tends to be activated above threshold for a period of time in response to a stimulus is termed a region of interest, which is a module (Schwarz, Gozzi, & Bifone, 2008). Remarkably, these modules of neural activity are also generated spontaneously from purely subject-driven cognitive states (Shirer, Ryali, Rykhlevskaia, Menon, & Greicius, 2012).

Organization of biology into modules simultaneously restricts the possibilities for function, because the modular organization is a subset of all possible organizations, and may lead to more rapid evolution, because the evolution occurs in a vastly restricted modular subspace of all possibilities (Sun & Deem, 2007; Kashtan, Noor, & Alon, 2007). At short time, modularity improves the plasticity of a system, because the system more efficiently finds a solution in the smaller, modular subspace. At long time, the system has enough time to find the optimal organization in the space of all possibilities, which is unlikely to be in the modular subspace. There is, then, a tension between the increased ability to adapt that modularity confers upon a system at short time and the constraint that modularity imposes upon a system at long time. The amount of modularity that arises in a system reflects a balance between these two competing effects, as a function of the time scale at which selection occurs (Sun & Deem, 2007).

Modularity is an order parameter that characterizes the structure of correlations of neural activity in the brain. If a network of neural brain activity can change its architecture, it will change its modularity to a value that achieves the balance of plasticity and cognitive performance demanded of it (Lorenz, Jeng, & Deem, 2011). In particular, a system under greater pressure to respond quickly will tend to develop a more modular distribution of connections. A pressure to respond quickly selects for an effective response function of the system, at the time scale of the environmental change (Sun & Deem, 2007). More modular systems have a more rapid response at short times. The rate of growth of modularity from an initially nonmodular state is roughly proportional to the variability in selection pressure (Sun & Deem, 2007). Similarly, if there is selection pressure for a system to adapt rapidly to steady state from an initially unfit state, modularity can arise (Wagner & Altenberg, 1996). The coupling between transport and a heterogeneous spatial environment also leads to the emergence of modularity (Kashtan, Parter, Dekel, Mayo, & Alon, 2009).

The remainder of this chapter is organized as follows. In the following section, we introduce modularity and discuss a number of biological examples of modularity, culminating with a discussion of modularity in neural activity. We provide an intuitive and a mathematical definition of modularity and discuss how modularity may adapt in time. In the third section,

"Modularity and brain function," we discuss how modularity in neural activity of the brain is correlated with learning rate. We also discuss how modularity is lowered in patients with brain injury or disease. In "Modularity and plasticity," we discuss how modularity increases plasticity and how increased modularity of neural activity in the brain is expected to increase performance at task switching. In "Modularity of neural activity increases then decreases with age," we consider how modularity of neural activity of the brain develops from childhood to old age. We show that modularity rises, peaks in young adults, and declines in old adults. We also show that modularity is positively correlated with raw IQ score. In the following section, we discuss some of the details of functional magnetic resonance imaging (fMRI) analysis. We conclude in the final section.

Modularity

Biological systems have long been recognized to be modular. In 1942, Waddington presented his now classic description of a canalized landscape for development, in which minor perturbations do not disrupt the function of developmental modules (Waddington, 1942). In 1961, H. A. Simon described how biological systems are more efficiently adapted and are more stable if they are modular (Simon, 1962). A seminal paper by Hartwell, Hopfield, Leibler, and Murray (1999) firmly established the concept of modularity in cell biology. Systems biology has since provided a wealth of examples of modular cellular circuits, including metabolic circuits (Ravasz, Somera, Mongru, Oltvai, & Barabási, 2002) and modules on different scales; that is, modules of modules (da Silva, Ma, & Zeng, 2008). Protein-protein interaction networks have been observed to be modular (Gavin et al., 2006; Spirin & Mirny, 2003; von Mering et al., 2003).

Ecological food webs have been found to be modular (Krause, Frank, Mason, Ulanowicz, & Taylor, 2003). The gene regulatory network of the developmental pathway exhibits modules (Raff & Raff, 2000; Wagner, 1996), and the developmental pathway is modular (Klingenberg, 2008).

Modules have also been found in spatial correlations of brain activity (Chavez et al., 2010; Ferrarnini et al., 2009; He, Wang et al., 2009; Meunier, Achard, Morcom, & Bullmore, 2009; Meunier, Lambiotte, & Bullmore, 2010; Schwarz et al., 2008). Modularity of brain activity is computed by analyzing correlations in fMRI data (Meunier et al., 2009; Schwarz et al., 2008). Intuitively, modules are regions of correlated brain activity, and modularity is a measure of how well the brain activity clusters into these regions. The fMRI data measure brain activity with spatial and temporal resolution during a 20- to 30-minute imaging session.

The data are analyzed by extracting the activity as a function of position within the brain and may be further averaged within regions of interest—for example, all the Brodmann's areas. The correlations of neural activity among different regions of interest are then computed. A cutoff is typically applied so that only the largest of these correlations are considered. A schematic of this process is shown in figure 12.1.

Modularity is defined by rearranging and partitioning the correlation matrix, with cutoff applied, so that it is as block-diagonal as possible (Fortunato, 2010). Mathematically, modularity is the probability to have a correlation within the modules, minus the probability to have a correlation expected by random chance (Newman, 2006). In other words,

$$M = \frac{1}{2e} \sum_{\text{all modules}} \sum_{\text{nodes } ij \text{ within this module}} \left(A_{ij} - \frac{a_i a_j}{2e} \right) \qquad (1)$$

where A_{ij} is 1 if there is a link between node i and node j and 0; otherwise, $a_i = \Sigma_j A_{ij}$ is the degree of node i, and $e = \frac{1}{2}\sum_i a_i$ the total number of links. The value of modularity depends on the definition of the modules, that is, how the nodes are partitioned into modules in Eq. 1. The clustering of the nodes into modules is defined by choosing the partitioning that maximizes Eq. 1.

The modularity of biological systems changes over time. There are a number of demonstrations of the evolution of modularity in biological systems. For example, the modularity of the protein-protein interaction network significantly increases when yeast are exposed to heat shock (Mihalik & Csermely, 2011), and the modularity of the protein-protein networks in both yeast and *E. coli* appears to have increased over evolutionary time (He, Sun, et al., 2009). Additionally, food webs in low-energy, stressful environments are more modular than those in plentiful environments (Lorenz et al., 2011), arid ecologies are more modular during droughts (Rietkerk, Dekker, de Ruiter, & van de Koppel, 2004), and foraging of sea otters is more modular when food is limiting (Tinker, Bentall, & Estes, 2008). Other complex dynamical systems exhibit time-dependent modularity as well. The modularity of social networks changes over time: stock-brokers' instant messaging networks are more modular under stressful market conditions (Saavedra, Hagerty, & Uzzi, 2011), and socioeconomic community overlap decreases with increasing stress (Estrada & Hatano, 2010). Modularity of financial networks changes over time: the modularity of the world

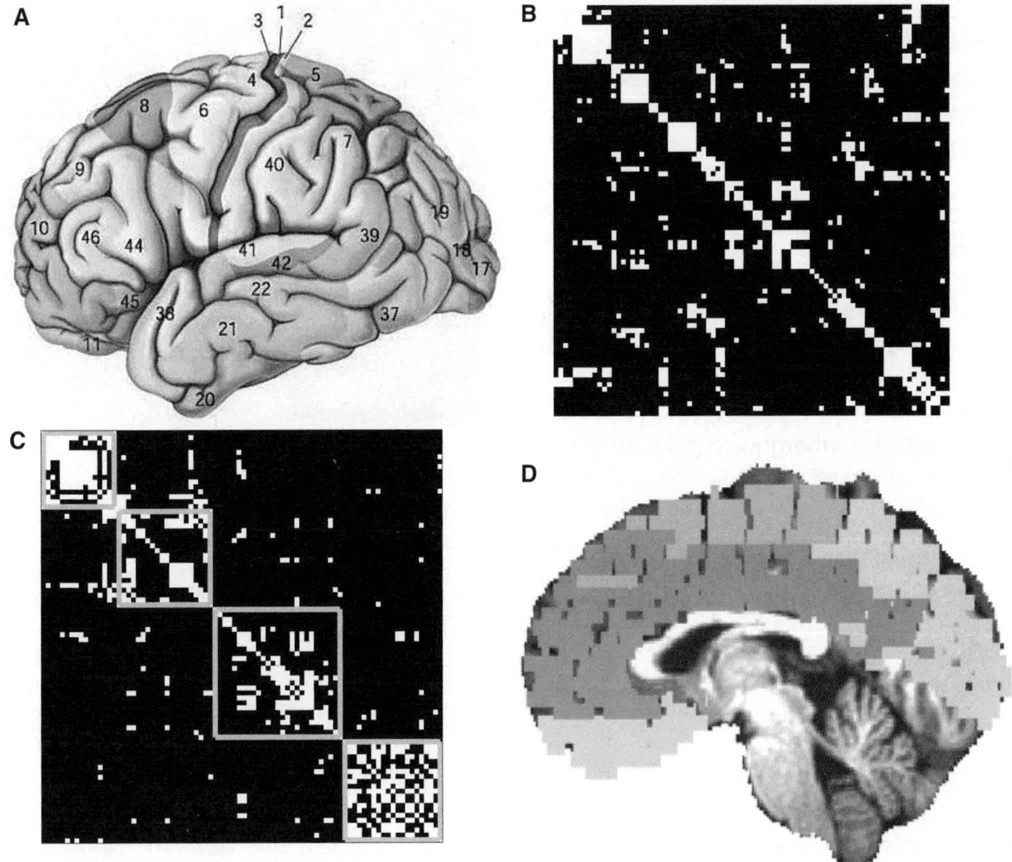

FIGURE 12.1 Clustering of brain neural activity into modules and calculation of modularity. (A) The Brodmann's areas, onto which the measured fMRI neural activity data are projected (Dubuc). (B) Correlation matrix of the neural activity between the different Brodmann's areas, with only the largest elements retained (white). (C) Reordered correlation matrix, showing the grouping of the areas into four modules. The Brodmann's areas are grouped into modules by permuting the rows and columns of the correlation matrix. The modularity calculated from Eq. 1 is 0.6441, and the contributions to M are 0.1585, 0.1310, 0.1592, and 0.1954 from the green, red, yellow, and orange modules, respectively. (D) The Brodmann's areas grouped into the four modules. In this slice, only two voxels of the second module are visible, with the majority of the voxels in a different region of the brain not visible in this slice. From Chen & Deem (2014). (See color plate 5.)

trade network has decreased over the last 40 years, leading to increased susceptibility to recessionary shocks (He & Deem, 2010), and increased modularity has been suggested as a way to increase the robustness and adaptability of the banking system (Haldane & May, 2011).

Modularity and brain function

Neural activity in the brain is modular (Fodor, 1983; Mountcastle, 1979). That is, neural activity in different regions of the brain is stimulated by different tasks. Brain function of subjects activates different regions, which can be observed in fMRI experiments (Shirer et al., 2012). Patterns of neural activity that are activated during attention- and memory-demanding tasks are correlated with structural connectivity in the brain (Hermundstad et al., 2013).

Features of modularity are correlated with learning rates. In a study of subjects learning a simple motor skill, the flexibility of neural activity was positively correlated with learning ability (Bassett et al., 2011). Flexibility was defined as the probability that voxels of brain neural activity participated in multiple modules. In other words, the ability of the brain to reconfigure neural activity into different modules was predictive of learning rate.

In some cases, modularity of neural activity can distinguish healthy subjects and patients with disease. For example, an analysis of magnetoencephalographic signals from epileptic patients showed their neural activity to be less modular than that of normal subjects (Chavez et al., 2010). Patients with epilepsy had neural activity networks with greater connectivity in addition to lower modularity. These patients also had more connections between nodes in different modules, which by the definition of Newman's modularity lowers the modularity. Anecdotal evidence suggests patients with traumatic brain injury have less modular neural activity as well (Leclercq et al., 2000).

Modularity and plasticity

A networked system that is modular adapts more easily than does a nonmodular networked system. This concept was explored in the early works of Waddington (1942) and Simon (1962). Modular networks adapt more easily because most of the rewiring of connections is done within the modules, and it is easier to rewire connections within a small module than within the entire network. Evolution of the modules can also occur in parallel if the modularity is high and the modules interact only weakly. Detailed calculations on models of molecular evolution have shown that modularity favors plasticity (Bogarad & Deem, 1999).

The cognitive ability of the brain—that is, the ability of the brain to solve a challenging problem—depends on the modularity of neural activity. Figure 12.2 shows the responses of a highly modular and a less modular system. Here the term *fitness* is used to quantify the quality of the solution to the problem. For example, fitness, f, could be the probability that a subject correctly identifies an image that is visible for a time, t. The more modular system produces a better solution at short times. Given enough time to respond, however, the nonmodular system will have a better solution. Thus, the utility of modularity depends on the time scale of the required response. The term *plasticity* is used to quantify the ability of the system to respond to the challenge. Plasticity is the increase of fitness in response to a challenge. At short times, the modular system is more plastic, because it adapts a better solution to the problem.

Figure 12.2 suggests that if a networked system is under pressure to be plastic, modularity might be expected to arise (Lipson, Pollack, & Suh, 2002). Early models showed that if a system is subjected to changing, modular goals, the system will become modular (Kashtan & Alon, 2005; Kashtan et al., 2007). These results are examples of selection for plasticity (Earl & Deem, 2004), in which modularity provides the plasticity. It was further shown that structured goals are not required, and that modularity can arise even in a randomly varying environment (Sun & Deem, 2007). Figure 12.2 is a schematic summary of these results, showing increased rates of adaptation for more modular networked systems.

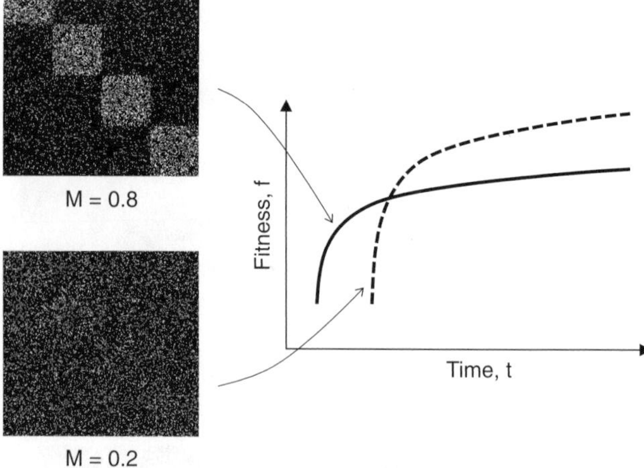

Figure 12.2 Modularity quantifies the amount of clustering in the correlations of neural brain activity. A highly modular network of neural activity provides a better response (solid curve) to a challenge than does a less modular network (dashed curve) at short times. At long times, the less modular system will produce a better response. Thus, the optimal amount of modularity depends on the time scale of the required response. Here fitness quantifies the quality of the cognitive response. For example, fitness, f, could be the probability that a subject correctly identifies an image that is visible for a time, t.

Applying this theory to the brain, one expects that more highly modular patterns of neural function should endow a brain with greater plasticity of neural function. Thus, more modular neural activity should allow the brain to more quickly switch from one type of neural activation to another. In other words, increased modularity of neural activity in the brain is expected to increase performance at task switching.

Modularity of neural activity increases then decreases with age

A natural question to ask is how modularity of neural activity in the brain changes with age. Our hypothesis is that selection for neural adaptability peaks during adulthood, adulthood being the period of peak demands on flexible brain function. We know that modularity increases adaptability, from theory described in the previous section, in particular from the spontaneous emergence of modularity under conditions of greater environmental change (Kashtan et al., 2007; Sun & Deem, 2007). Thus, modularity might peak during adulthood as well.

In other words, a part of brain development may be learning how to make modules of neural activity. These modules, while correlated with physical brain structure (Hermundstad et al., 2013), are not entirely hard-wired at birth. Interaction of the individual with a demanding environment promotes the development of modular neural activity that endows the brain with increased plasticity and ability to task switch.

The hypothesis of neural activity peaking in adulthood is borne out by experimental data. We analyzed fMRI data from 24 children aged 4 to 11 watching *Sesame Street* for 20 minutes, as well as fMRI data from 21 adults aged 18 to 26 watching the same *Sesame Street* video (Cantlon & Li, 2013). For each subject, we processed the fMRI data to make a movie of three-dimensional neural activity with 4 mm spatial resolution and 2 sec time resolution. For each subject, we despiked the data and then projected it to standard Talairach space (Talairach & Tournoux, 1988). We aligned the observed data to high-resolution deskulled anatomy data for each subject. We further aligned the data through time to correct for head motion. Two children and one adult were excluded due to excessive censor readings, inferred to be due to excessive head movement. These calculations were done using AFNI (Cox, 1996). Activity from each Brodmann area was then correlated, and a cutoff was applied to the correlation matrix. We then computed modularity of the cutoff correlation matrix.

The values of modularity calculated for children and adults are shown in figure 12.3. We see that modularity

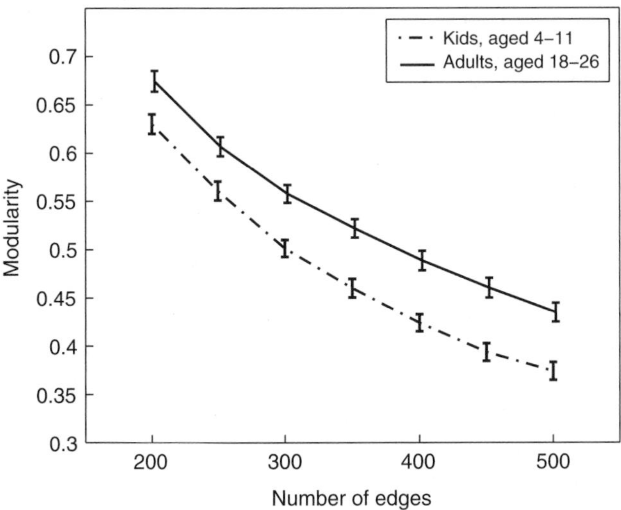

FIGURE 12.3 Modularity of neural activity for children and adults. Modularity is greater for adults than for children. Modularity is calculated from the correlation of neural activity between Brodmann's areas. The entries of the correlation matrix are projected to unity for values above a cutoff and set to zero otherwise. The number of correlation matrix entries above the cutoff value—i.e., the number of white entries shown in figure 12.1—is denoted by edges. The trend of increased modularity for adults is persistent for different cutoffs—i.e., different edges—which quantify differing sparseness of the projected correlation matrix. The error bars are one standard error. (From Chen & Deem, 2014.)

of neural activity is greater in adults than in children. This result is robust to changing the cutoff used when projecting the correlation matrix to a binary matrix.

Not only is neural activity more modular in young adults than in children, but modularity of neural activity also increases with age during childhood development. Shown in figure 12.4 is the correlation between modularity and age; the positive correlation shows that modularity increases with development. Scores on standardized test for childhood verbal and nonverbal IQ were also analyzed (Cantlon & Li, 2013). Also shown in figure 12.4 is the correlation between modularity and raw KBIT-2 overall IQ score. This raw score is not age-standardized. This positive correlation with raw IQ also supports the interpretation of modularity increasing with development.

Finally, modularity peaks in adulthood and declines in old age. Resting state fMRI data from 17 young adults aged 18 to 33 and 13 older adults aged 62 to 76 were analyzed (Meunier et al., 2009). The data were projected onto standard regions of interest (Tzourio-Mazoyer et al., 2002) and analyzed with SPM2 (Friston, Frith, Liddle, & Frackowiak, 1991). Correlations of these time series were calculated. A cutoff was applied to make a binary correlation matrix between the regions of interest.

Finally, the modularity of this correlation matrix was computed (Lorenz et al., 2011; Meunier et al., 2009). We here reprocessed these data, eliminating unnecessary wavelet filtering. The results are shown in figure 12.5. As previously (Meunier et al., 2009; Lorenz et al., 2011), we observe the modularity of adults aged 18 to 33 is greater than that of adults aged 62 to 76.

Modularity enhances plasticity, and modularity of brain neural activity peaks in adulthood. This observation may be placed in mathematical form by considering the fitness of the brain to be a function of modularity. For young children, this fitness function peaks at a lower value of modularity than it does for adults. For adults over 62, the fitness function again peaks at a lower value of modularity. Figure 12.6 shows how the fitness as a function of modularity may look for children, young adults, and older adults. The modularities measured for 400 edges are used to construct this estimation of the fitness function.

Functional MRI data analysis

Here we present some details of the calculation of modularity from fMRI data. Such fMRI data provide an indirect measure of neural activity in the brain with millimeter spatial and second temporal resolution. The fMRI signal is an approximation of blood flow to each

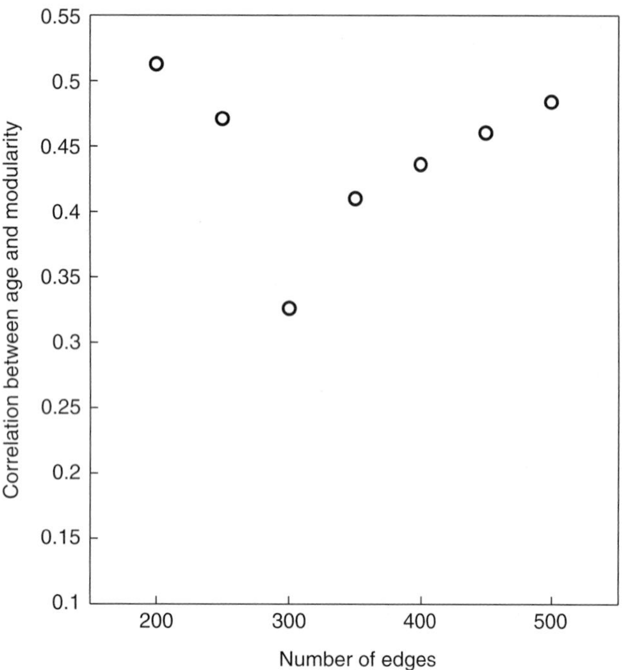

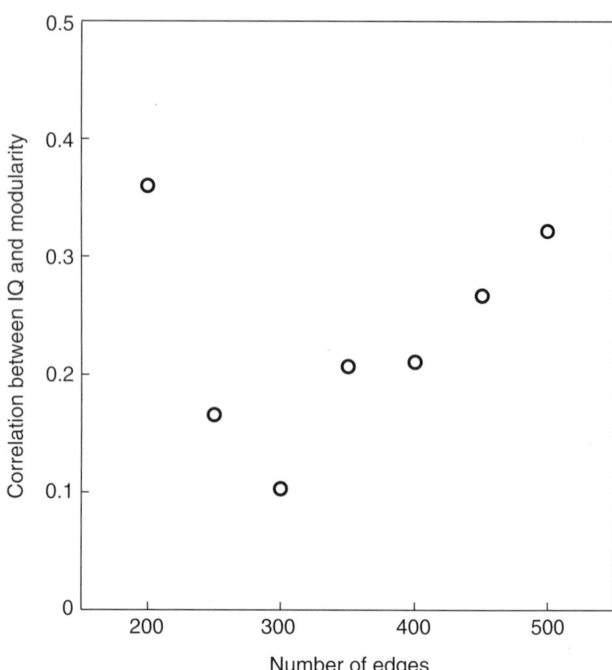

FIGURE 12.4 For each child, the age is known, the raw KBIT-2 IQ score was measured (Cantlon & Li, 2013), and the modularity of neural brain activity was calculated. The correlation between modularity and either age or IQ was calculated. Ages were in the range 4–11. *Left*: The correlation between modularity and age. *Right*: The correlation between modularity and raw KBIT-2 overall IQ score. Modularity is positively correlated with both age and IQ. These positive correlations show that modularity increases with development. (From Chen & Deem, 2014.)

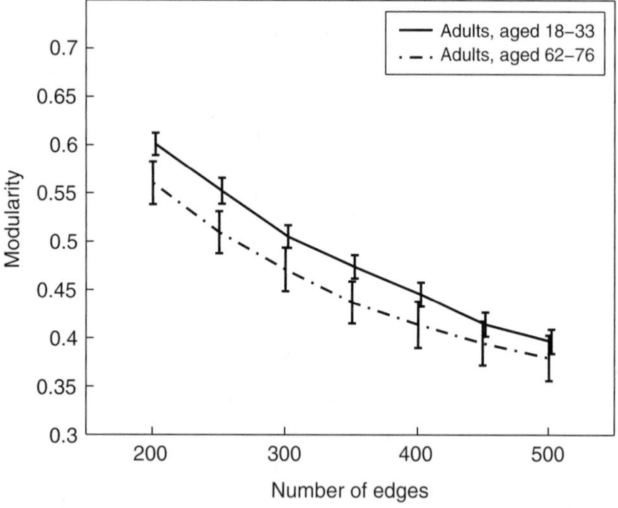

FIGURE 12.5 Modularity of neural activity for adults aged 18–33 and adults aged 62–76. Modularity is greater for younger adults than for older adults. Modularity is calculated on the correlation between Brodmann's regions. The correlation matrix entries are projected to unity for values above a cutoff and set to zero otherwise. The trend of increased modularity for adults is persistent for different cutoffs, i.e., differing sparseness of the projected correlation matrix. The error bars are one standard error. (After Lorenz et al., 2011; Meunier et al., 2009.)

region of the brain, and local blood flow is an approximation to local neural activity. Technically, fMRI measures the difference in magnetic properties of oxygenated and deoxygenated hemoglobin (Ogawa, Lee, Nayak, & Glynn, 1990). The fMRI data are collected over a time period that can be up to 30 minutes. Scans of the entire brain can be taken every 2 seconds. The spatial resolution of these data can be 1 to 4 mm. Typical data sets will be several two-dimensional slices through the brain, repeated every 2 to 3 seconds, for the duration of the experiment. In addition, for each subject, a high-resolution image of the brain at one time point will typically be collected to provide an anatomical reference.

Software tools are now available so that one need not be an imaging expert to analyze these data. AFNI and SPM2 are popular packages (Cox, 1996; Friston et al., 1991). Data from each subject are analyzed independently. The first step of the analysis is to combine the two-dimensional slices into a three-dimensional fMRI representation of the neural activity as a function of time. Due to sensor noise and subject movement, data at some of the time points may be outside the expected range for neural activity, and the values at these time points are removed and interpolated from adjacent values. The fMRI data are then registered to the

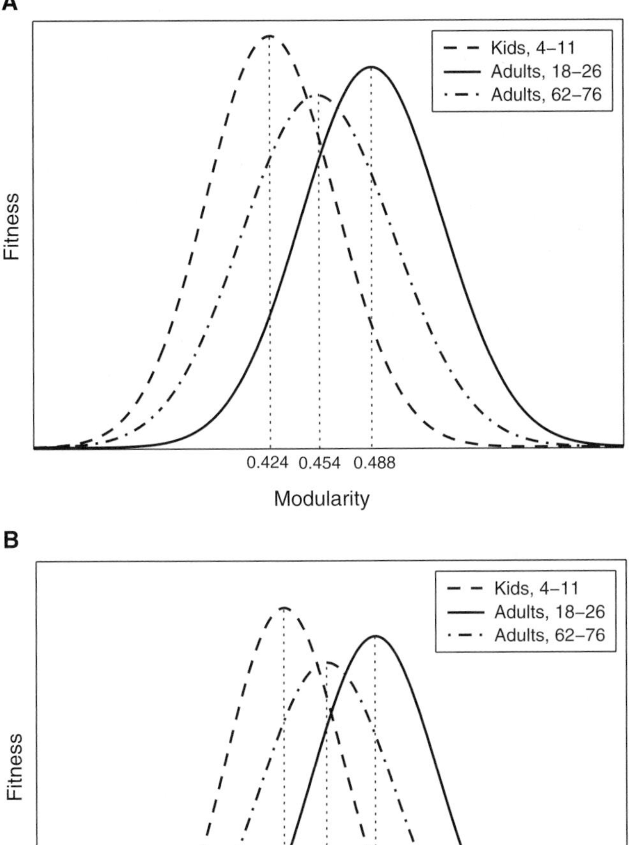

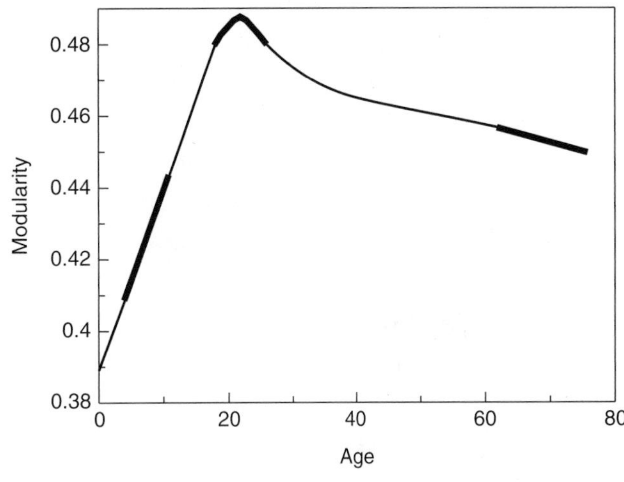

FIGURE 12.6 (A) Schematic representation of the results in figures 12.3 and 12.5. Data are from subjects with ages in the solid part of the curve. (B) Schematic figure of the fitness of brain activity as a function of modularity. Fitness is a measure of the quality of neural activity, over all the tasks demanded of the brain. To measure this average fitness, figure 12.2 is constructed for many typical challenges and typical response times, and an average is taken. The peak of fitness is set at the observed values of modularity for children, young adults, and old adults. The width of the curve is set schematically to the standard deviation of the observed subject modularity within each group.

high-resolution anatomical data, typically by registering one of the first few images of the fMRI data. Since brains of different subjects are of different sizes, the imaging data are then typically scaled into standardized Talairach brain coordinates (Talairach & Tournoux, 1988). The fMRI data are then registered during the time course to correct for frame-to-frame head motion of the subject. The outcome of these processing steps gives neural activity as a function of time in fixed, standardized spatial coordinates.

For further analysis, the fMRI data may be projected to standardized regions of interest, for example the Brodmann's areas; see figure 12.7. The average value of the processed fMRI data in each region may be calculated as a function of time. A wavelet transform may optionally be applied to these data to regress out smooth features of the time-dependent data (Meunier et al., 2009), but is unnecessary. The correlation of these data among the different regions of interest is then calculated (Meunier et al., 2009; Schwarz et al., 2008).

Only those elements above a cutoff value are retained and set to unity, with all other elements set to zero. The modularity of this cutoff correlation matrix is then calculated by Eq. 1.

Discussion and conclusion

Modularity quantifies the degree to which neural activity in the brain is modular. Modularity is a mathematical concept, expressed in Eq. 1. Physical regions of the brain were defined by early anatomical studies, and later studies identified functional regions by observing how brain lesions from disease or trauma related to loss of function. Functional MRI studies now provide detailed maps of neural activity, allowing identification of modules from correlations of neural activity (see figure 12.1).

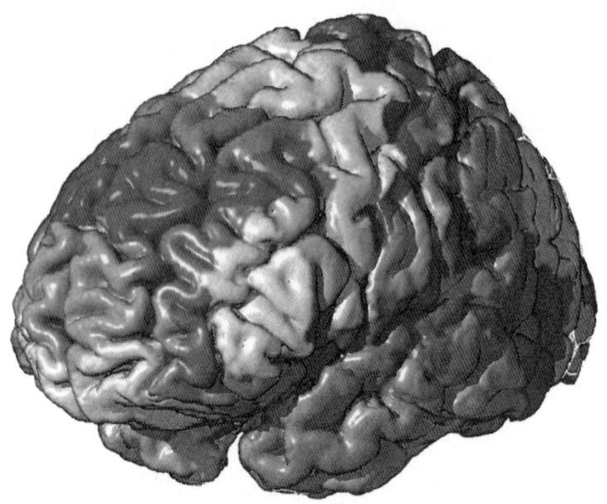

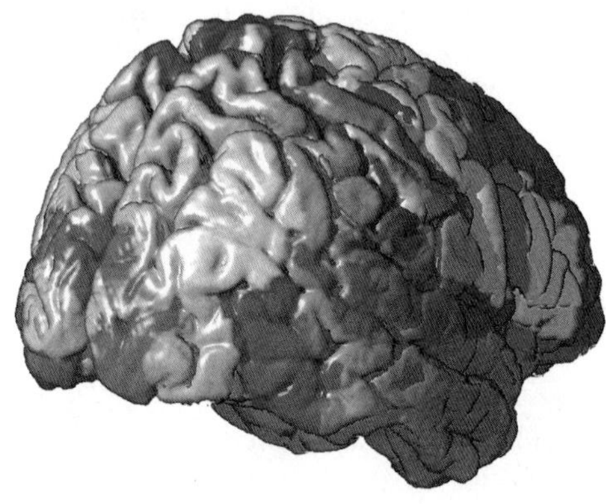

FIGURE 12.7 The Brodmann areas of the human cerebral cortex, onto which fMRI data are projected for further analysis. (From Dow.) Each Brodmann area is shaded a different color. (See color plate 6.)

We hypothesize that brains of young people are likely under more selective pressure to quickly solve challenging problems than are brains of older people. This pressure induces the emergence of modularity in human brain activity as children develop into adults (Kashtan et al., 2007; Sun & Deem, 2007). That is, this hypothesis implies neural brain activity of young adults should be more modular than that of children or older adults, as suggested schematically in figure 12.2. Figures 12.3, 12.4, and 12.5 show fMRI data that provide support for this hypothesis. In other words, modularity rises during childhood, peaks in young adults, and declines in older adults.

The ability of the brain to solve a challenge in a given amount of time is termed fitness. Fitness quantifies brain function, and fitness depends on modularity. The dependence of fitness on modularity changes with age, shown schematically in figure 12.6. The highest values of modularity are selected for in young adults.

It would be interesting to set up experiments to determine fitness as a function of modularity. For example, an experiment could challenge subjects with a cognitive task during the fMRI data collection. Among other parameters, the ability of a subject to solve the task will depend on the modularity of correlations of neural activity. Quantifying this ability as a function of modularity would provide the fitness function for this particular task. Perhaps different tasks will have fitness functions that peak at different values of modularity. It has been speculated that fast, low-level, automatic cognitive processes are modular, whereas high-level, effortful, conscious cognitive processes are less modular (Meunier et al., 2010). This speculation is related to figure 12.2: at short times, modular networks are more efficient, but at long times, nonmodular networks will find better solutions. It is possible that fast, low-level cognitive processes are to the left on this figure, and slow, high-level tasks are to the right. Measuring fitness as a function of modularity for low-level and high-level tasks would resolve this issue.

Modularity may be used as a biomarker for brain function. For example, the neural activity of epileptic patients is less modular than that of normal subjects (Chavez et al., 2010). Epileptic patients had more connections and less modular structure in their correlations of neural activity. An interesting application suggested by anecdotal evidence (Leclercq et al., 2000) would be to quantify extent of traumatic brain injury (TBI) and to quantify effectiveness of rehabilitation treatments. This diagnostic quantifies the reduction of structure in background brain activity due to TBI. The architecture of functional networks extracted from TBI patients is expected to differ from that of healthy subjects. We expect that TBI, as a debilitating condition, will reduce the natural modularity of the healthy brain. We propose modularity as a metric for evaluating extent of TBI. This metric can be used for quantification of success of therapeutic strategies such as administration of ibuprofen and as a feedback to the success of rehabilitation programs.

ACKNOWLEDGMENTS We thank Jessica Cantlon for providing the fMRI, age, and IQ data. This research was supported by the US National Institutes of Health under grant 1 R01 GM 100468–01.

REFERENCES

BASSETT, D. S., WYMBS, N. F., PORTER, M. A., MUCHA, P. J., CARLSON, J. M., & GRAFTON, S. T. (2011). Dynamic reconfiguration of human brain networks during learning. *Proc Natl Acad Sci USA, 108,* 7641–7646.

BOGARAD, L. D., & DEEM, M. W. (1999). A hierarchical approach to protein molecular evolution. *Proc Natl Acad Sci USA, 96,* 2591–2595.

Cantlon, J. F., & Li, R. (2013). Neural activity during natural viewing of Sesame Street statistically predicts test scores in early childhood. *PLoS Biol, 11*, e1001462.

Chavez, M., Valencia, M., Navarro, V., Latora, V., & Martinerie, J. (2010). Functional modularity of background activities in normal and epileptic brain networks. *Phys Rev Lett, 104*, 118701.

Chen, M., & Deem, M. W. (2014). Development of modularity in neural brain activity of children. Submitted.

Cox, R. W. (1996). AFNI: Software for analysis and visualization of functional magnetic resonance neuroimages. *Comput Biomed Res, 29*, 162–173.

Da Silva, M. R., Ma, H., & Zeng, A.-P. (2008). Centrality, network capacity, and modularity as parameters to analyze the core-periphery structure in metabolic networks. *Proc IEEE, 96*, 1411–1420.

Dow, M. http://lcni.uoregon.edu/~dow/Spaces_Software/renderings.html.

Dubuc, B. The brain from top to bottom. http://thebrain.mcgill.ca/flash/capsules/outil_jaune05.html.

Earl, D. J., & Deem, M. W. (2004). Evolvability is a selectable trait. *Proc Natl Acad Sci USA, 101*, 11531–11536.

Estrada, E., & Hatano, N. (2010). Communicability and communities in complex socio-economic networks. In M. Takayasu, T. Watanabe, & H. Takayasu (Eds.), *Econophysics approaches to large-scale business data and financial crisis* (pp. 271–288). Tokyo: Springer.

Ferrarnini, L., Veer, I. M., Baerends, E., van Tol, M.-J., Renken, R. J., van der Wee, N. J. A. ... Milles, J. (2009). Hierarchical functional modularity in the resting-state human brain. *Hum Brain Mapp, 30*, 2220–2231.

Fodor, J. (1983). *The modularity of mind.* Cambridge, MA: MIT Press.

Fortunato, S. (2010). Community detection in graphs. *Phys Rep, 486*, 75–174.

Friston, K. J., Frith, C. D., Liddle, P. F., & Frackowiak, R. S. (1991). Comparing functional (PET) images: The assessment of significant change. *J Cereb Blood Flow Metab, 11*, 690–699.

Gavin, A-C., Aloy, P., Grandi, P., Krause, R., Boesche, M., Marzioch, M., ... Superti-Furga, G. (2006). Proteome survey reveals modularity of the yeast cell machinery. *Nature, 440*, 631–636.

Haldane, A. G., & May, R. M. (2011). Systemic risk in banking ecosystems. *Nature, 469*, 351–355.

Hartwell, L. H., Hopfield, J. J., Leibler, S., & Murray, A. W. (1999). From molecular to modular cell biology. *Nature, 402*, C47–C52.

He, J., & Deem, M. W. (2010). Structure and response in the world trade network. *Phys Rev Lett, 105*, 198701.

He, J., Sun, J., & Deem, M. W. (2009). Spontaneous emergence of modularity in a model of evolving individuals and in real networks. *Phys Rev E, 79*, 031907.

He, Y., Wang, J., Wang, L., Chen, Z. J., Yan, C., Yang, H., ... Evans, A. C. (2009). Uncovering intrinsic modular organization of spontaneous brain activity in humans. *PLoS ONE, 4*, e5226.

Hermundstad, A. M., Bassett, D. S., Brown, K. S., Aminoff, E. M., Clewett, G. D., Freeman, S., ... Carlson, J. M. (2013). Structural foundations of resting-state and task-based functional connectivity in the human brain. *Proc Natl Acad Sci USA, 110*, 6169–6174.

Kashtan, N., & Alon, U. (2005). Spontaneous evolution of modularity and network motifs. *Proc Natl Acad Sci USA, 102*, 13773–13778.

Kashtan, N., Noor, E., & Alon, U. (2007). Varying environments can speed up evolution. *Proc Natl Acad Sci USA, 104*, 13711–13716.

Kashtan, N., Parter, M., Dekel, E., Mayo, A. E., & Alon, U. (2009). Extinctions in heterogeneous environments and the evolution of modularity. *Evolution, 63*, 1964–1975.

Klingenberg, C. P. (2008). Morphological integration and developmental modularity. *Annu Rev Ecol Evol S, 39*, 115–132.

Krause, A. E., Frank, K. A., Mason, D. M., Ulanowicz, R. E., & Taylor, W. W. (2003). Compartments revealed in food-web structure. *Nature, 426*, 282–285.

Leclercq, M. Couillet, J., Azouvi, P., Marlier, N., Martin, Y., Strypstein, E, & Rousseaux, M. (2000). Dual task performance after severe diffuse traumatic brain injury or vascular prefrontal damage. *J Clin Exper Neuropsychol, 22*, 339–350.

Lipson, H., Pollack, J. B., & Suh, N. P. (2002). On the origin of modular variation. *Evolution, 56*, 1549–1556.

Lorenz, D. M., Jeng, A., & Deem, M. W. (2011). The emergence of modularity in biological systems. *Phys Life Rev, 8*, 129–160.

Meunier, D., Achard, S., Morcom, A., & Bullmore, E. (2009). Age-related changes in modular organization of human brain functional networks. *NeuroImage, 44*, 715–723.

Meunier, D., Lambiotte, R., & Bullmore, E. T. (2010). Modular and hierarchically modular organization of brain networks. *Front Neurosci, 4*.

Mihalik, A., & Csermely, P. (2011). Heat shock partially dissociates the overlapping modules of the yeast protein-protein interaction network: A systems level model of adaptation. *PloS Comput. Biol, 7*, e1002187.

Mountcastle, V. B. (1979). An organizing principle for cerebral function: The unit module and the distributed system. In *The neurosciences fourth study program* (pp. 21–42). Cambridge, MA: MIT Press.

Newman, M. E. J. (2006). Modularity and community structure in networks. *Proc Natl Acad Sci USA, 103*, 8577–8582.

Ogawa, S., Lee, T. M., Nayak, A. S., & Glynn, P. (1990). Oxygenation-sensitive contrast in magnetic resonance image of rodent brain at high magnetic fields. *Magnet Reson Med, 14*, 68–78.

Raff, E. C., & Raff, R. A. (2000). Dissociability, modularity, evolvability. *Evol Dev, 2*, 235–237.

Ravasz, E., Somera, A. L., Mongru, D. A., Oltvai, Z. N., & Barabási, A. L. (2002). Hierarchical organization of modularity in metabolic networks. *Science, 297*, 1551–1555.

Rietkerk, M., Dekker, S. C., de Ruiter, P. C., & van de Koppel, J. (2004). Self-organized patchiness and catastrophic shifts in ecosystems. *Science, 305*, 1926–1929.

Rojas, R. (1996). *Neural networks: A systematic introduction.* New York, NY: Springer.

Saavedra, S., Hagerty, K., & Uzzi, B. (2011). Synchronicity, instant messaging, and performance among financial traders. *Proc Natl Acad Sci USA, 108*, 5296–5301.

Schwarz, A. J., Gozzi, A., & Bifone, A. (2008). Community structure and modularity in networks of correlated brain activity. *J Magn Reson Imaging, 26*, 914–920.

Shirer, W. R., Ryali, S., Rykhlevskaia, E., Menon, V., & Greicius, M. D. (2012). Decoding subject-driven cognitive states with whole-brain connectivity patterns. *Cereb Cortex, 22,* 158–165.

Simon, H. A. (1962). The architecture of complexity. *Proc Amer Phil Soc, 106,* 467–482.

Spirin, V., & Mirny, L. A. (2003). Protein complexes and functional modules in molecular networks. *Proc Natl Acad Sci USA, 100,* 12123–12128.

Sun, J., & Deem, M. W. (2007). Spontaneous emergence of modularity in a model of evolving individuals. *Phys Rev Lett, 99,* 228107.

Talairach, J., & Tournoux, P. (1988). *Co-planar stereotaxic atlas of the human brain.* New York, NY: Thieme.

Tinker, M. T., Bentall, G., & Estes, J. A. (2008). Food limitation leads to behavioral diversification and dietary specialization in sea otters. *Proc Natl Acad Sci USA, 105,* 560–565.

Tzourio-Mazoyer, N., Landeau, B., Papathanassiou, D., Crivello, F., Etard, O., Delcroix, N., ... Joliot, M. (2002). Automated anatomical labeling of activations in spm using a macroscopic anatomical parcellation of the MNI MRI single-subject brain. *NeuroImage, 15,* 273–289.

Von Mering, C., Zdobnov, E. M., Sophia Tsoka, S., Ciccarelli, F. D., Pereira-Leal, J. B., Ouzounis, C. A., & Bork, P. (2003). Genome evolution reveals biochemical networks and functional modules. *Proc Natl Acad Sci USA, 100*(21) 15428–15433.

Waddington, C. H. (1942). Canalization of development and the inheritance of acquired characters. *Nature, 150,* 563–565.

Wagner, G. P. (1996). Homologues, natural kinds and the evolution of modularity. *Integr Comp Biol, 36,* 36–43.

Wagner, G. P., & Altenberg, L. (1996). Complex adaptations and the evolution of evolvability. *Evolution, 50,* 967–976.

13 Specificity of Experiential Effects in Neurocognitive Development

COURTNEY STEVENS AND HELEN NEVILLE

ABSTRACT Here, we report research on the neuroplasticity of different subsystems within vision, audition, sensory integration, language, and attention. In each section, we note different profiles of plasticity observed in different subsystems within the domain, situations in which enhancements versus deficits are observed, and likely mechanisms contributing to these different profiles of plasticity. A final section describes our studies that test the hypothesis, raised by this basic research on human neuroplasticity, that interventions that target the most plastic, and thus potentially vulnerable, neurocognitive systems can protect and enhance children with, or at risk for, developmental deficits.

For several years we have employed psychophysics, electrophysiological, and magnetic resonance imaging (MRI) techniques to study the development and neuroplasticity of the human brain. We have studied deaf and blind individuals, people who learned their first or second spoken or signed language at different ages, and children of different ages and stages of cognitive development. Here, we review our research on the neuroplasticity of different brain systems and subsystems. As detailed in the sections that follow, in each of the domains examined in this research we observe the following characteristics:

- Different brain systems and subsystems and related sensory and cognitive abilities display different degrees and time periods ("profiles") of neuroplasticity.
- Neuroplasticity within a system acts as a double-edged sword, conferring the possibility for *either* enhancement or deficit.
- Multiple mechanisms both support and constrain the ability to modify different brain systems and subsystems, with the most malleable systems generally displaying longer developmental trajectories, higher levels of redundant connectivity, and a greater concentration of neurochemicals, including CAT301 and BDNF, which are both important in neuroplasticity.

In the sections that follow, we describe our research on neuroplasticity within visual, auditory, sensory integration, language, and attention systems. In each section, we note different profiles of plasticity that are system- and context-dependent and in which enhancements versus deficits are observed. We also propose likely mechanisms contributing to these different profiles of plasticity. A final section describes our studies testing the hypothesis, raised by this basic research on human neuroplasticity, that interventions that target the most plastic, and thus potentially most vulnerable, neurocognitive systems can protect and enhance children with, or at risk for, developmental deficits.

Vision

The primate visual system contains multiple distinct visual areas, with a gross segregation of cortical visual processing into dorsal and ventral streams that project respectively from V1 toward posterior parietal cortex and ventrally toward the temporal cortex (Grill-Spector & Malach, 2004; Ungerleider & Haxby, 1994). In our research, we observe that aspects of vision and attention mediated by the dorsal pathway—including motion perception, peripheral vision, and selective attention—are most modifiable with altered experience, showing both enhancements and deficits in special populations.

In one line of research, we examined changes in visual processing in congenitally, genetically deaf adults. By virtue of an increased reliance on vision, we hypothesized that deaf adults would exhibit enhancements in visual processing. Indeed, our studies revealed improvements in some—but not all—aspects of vision in these deaf adults. For example, congenitally deaf individuals have superior motion detection for peripheral visual stimuli than hearing individuals (Neville & Lawson, 1987b; Neville, Schmidt, & Kutas, 1983; Stevens & Neville, 2006). These improvements are accompanied by increases in the amplitudes of early visual event-related potentials (ERPs) for peripheral visual stimuli, as well as increased functional magnetic resonance imaging (fMRI) activation in motion-sensitive middle temporal (MT) and middle superior temporal (MST) areas (Armstrong, Hillyard, Neville, & Mitchell, 2002; Bavelier et al., 2001; Bavelier et al., 2000; Neville & Lawson, 1987b; Neville et al., 1983). In contrast, we observe no differences between deaf and hearing adults in tasks tapping ventral pathway functions, including central visual field tasks and isoluminant color processing (Armstrong et al., 2002; Bavelier et al., 2001; Bavelier et al., 2000; Neville & Lawson, 1987b; Neville et al., 1983; Stevens & Neville, 2006).

If neuroplasticity indeed acts as a double-edged sword, conferring both the potential for a system to be enhanced or to show deficits, one might predict that dorsal visual functions would—under different conditions—be selectively vulnerable to *deficit*. This is indeed the case. For example, several studies report that visual processing mediated by the dorsal (but not ventral) pathway shows deficits in many developmental disorders, including autism, Williams and Fragile X syndromes, as well as in individuals with reading or language impairments (Atkinson, 1992; Atkinson et al., 1997; Cornelissen, Richardson, Mason, Fowler, & Stein, 1995; Demb, Boynton, Best, & Heeger, 1998; Eden et al., 1996; Everatt, Bradshaw, & Hibbard, 1999; Hansen, Stein, Orde, Winter, & Talcott, 2001; Lovegrove, Martin, & Slaghuis, 1986; Sperling, Lu, Manis, & Seidenberg, 2003; Talcott, Hansen, Assoku, & Stein, 2000).

Interestingly, the literatures showing enhancements and deficits in dorsal pathway visual function have developed largely in parallel, with different experimental paradigms employed in each literature. To examine whether parallel enhancements and deficits could be observed in dorsal pathway function, we have used the same visual processing tasks in congenitally deaf adults (hypothesized to show enhancements) and dyslexic adults (hypothesized to show deficits), as shown in figure 13.1. In this study we observe the two sides of plasticity, using the same experimental paradigm to show enhancements in deaf adults and deficits in adults with dyslexia for dorsal visual pathway tasks, with no difference for either group in visual tasks relying on the ventral pathway (Stevens & Neville, 2006).

Taken together, these data suggest that the dorsal visual pathway exhibits a greater degree of neuroplasticity than the ventral visual pathway, rendering it capable of either enhancement (as is the case following congenital deafness) or deficit (as is the case in individuals with some developmental disorders). The dorsal pathway is likely more developmentally labile due to subsystem differences in rate of maturation, extent and timing of redundant connectivity, and presence of neurochemicals and receptors known to be important in plasticity (for a review, see Bavelier & Neville, 2002). For example, there is considerable, though not unequivocal, evidence indicating that the dorsal pathway matures more slowly than the ventral pathway (Coch, Skendzel, Grossi, & Neville, 2005; Hickey, 1981; Hollants-Gilhuijs,

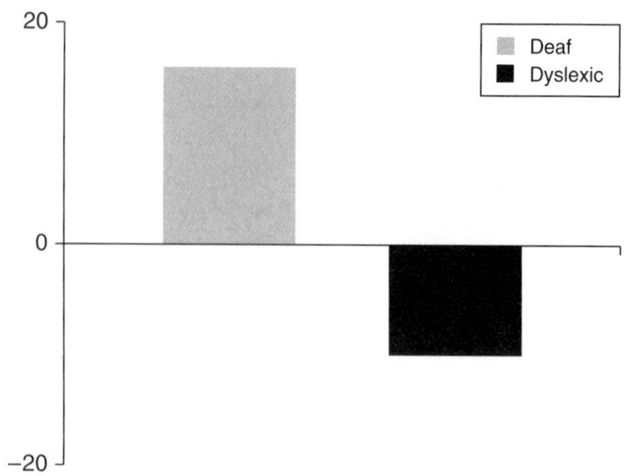

Figure 13.1 Performance on a peripheral visual field motion-detection task for deaf participants (gray bar) and dyslexic participants (black bar) relative to matched control groups. The zero line represents performance of the respective control groups. Deaf participants showed enhancements and dyslexic participants showed deficits relative to matched controls. No differences between groups were observed on a central visual field contrast-sensitivity task. (Data from Stevens & Neville, 2006.)

Ruijter, & Spekreijse, 1998a, 1998b; Mitchell & Neville, 2004; Packer, Hendrickson, & Curcio, 1990). Anatomical studies also suggest that connections within regions of the visual system that represent the central visual field are more strongly genetically specified and display fewer redundancies, whereas connections within the portions of the visual system that represent the visual periphery contain more redundant connections that can be shaped by experience over a longer developmental time course (Chalupa & Dreher, 1991). Moreover, anatomical studies in nonhuman primates (Falchier, Clavagnier, Barone, & Kennedy, 2002; Rockland & Ojima, 2003) and neuroimaging studies of humans (Eckert et al., 2008) report cross-modal connections between primary auditory cortex and the portion of primary visual cortex that represents the periphery (anterior calcarine sulcus). In our own laboratory, we observe that deaf (but not hearing) participants recruit a large, additional network of supplementary cortical areas when processing far peripheral relative to central flickering visual stimuli (Scott, Karns, Dow, Stevens, & Neville, 2014), including regions associated with auditory and visual processing (A1 and V1), multisensory integration (STS), motion processing (MT/MT+), and attention (posterior parietal and anterior cingulate regions). In sum, the anatomical, neurochemical, and developmental mechanisms mentioned above could render particular aspects of vision more modifiable by experience and likely to display either enhanced or deficient processing.

Audition

To test whether the specificity of plasticity observed in the visual system generalizes to other sensory systems, we have conducted studies on the effects of visual deprivation on the development of the auditory system. Although less is known about the organization of the auditory system, as in the visual system there are large (magno) cells in the medial geniculate nucleus that have shorter conduction times than the smaller (parvo) cells, and evidence suggests that there may be dorsal and ventral auditory processing streams with different functional specializations (Rauschecker, 1998). To determine whether similar patterns of plasticity occur following auditory and visual deprivation, we developed an auditory paradigm similar to one of the visual paradigms employed in our studies of deaf adults. Participants detected infrequent pitch changes in a series of tones that were preceded by different interstimulus intervals (Röder, Teder-Salejarvi, et al., 1999). Congenitally blind participants were faster at detecting the targets and displayed ERPs that were less refractory—that is, recovered amplitude faster—than normally sighted participants. These results parallel those of our study showing faster amplitude recovery of the visual ERP in deaf than in hearing participants (Neville et al., 1983) and suggest that rapid auditory and visual processing may show specific enhancements following sensory deprivation.

Similar to the dual nature of the plasticity observed in the dorsal visual pathway, the processing of rapidly presented acoustic information, which is enhanced in the blind, shows deficits in some developmental disorders (Bishop & McArthur, 2004; Tallal, 1975, 1976; Tallal & Piercy, 1974). Using ERPs, we have observed in two studies of children with specific language impairment (SLI) that the amplitude of auditory evoked potentials were smaller (i.e., more refractory) than in controls, but only at short interstimulus intervals (Neville, Coffey, Holcomb, & Tallal, 1993; Stevens, Paulsen, Yasen, Mitsunaga, & Neville, 2012). This suggests that in audition, as in vision, neural subsystems that display more neuroplasticity show both greater potential for enhancement and also greater vulnerability to deficit under different conditions.

The mechanisms that give rise to greater modifiability of rapid auditory processing are as yet unknown. However, as mentioned above, some changes might be greater for magnocellular divisions of the medial geniculate nucleus. For example, magno, but not parvo, cells in both the lateral and medial geniculate nucleus are smaller than normal in dyslexia (Galaburda & Livingstone, 1993; Galaburda, Menard, & Rosen, 1994). Rapid auditory processing, including the recovery cycles of neurons, might also engage aspects of attention to a greater degree than other aspects of auditory processing. In the case of congenital blindness, changes in auditory processing may be facilitated by compensatory reorganization. A number of studies confirm that primary and secondary visual areas are functionally involved in nonvisual tasks in congenitally blind adults (Burton et al., 2002; Cohen, Weeks, Celnik, & Hallett, 1999; Röder, Stock, Bien, Neville, & Rösler, 2002; Sedato et al., 1996). In addition, parallel studies in animals reveal information about mechanisms underlying cross-modal plasticity. For example, in blind mole rats, normally transient, weak connections between the ear and primary visual cortex become stabilized and more pronounced (Bavelier & Neville, 2002; Cooper, Herbin, & Nevo, 1993; Doron & Wollberg, 1994; Heil, Bronchti, Wollberg, & Scheich, 1991). Thus, portions of the auditory network that either depend upon or can recruit multimodal, attentional, or normally visual regions may show greater degrees of neuroplasticity.

Sensory integration

The above discussion suggests that, in some cases, plasticity might be related to the interactions and integration among different sensory systems. Most research examining cross-modal plasticity in congenitally deaf adults has focused exclusively on vision. In the visual domain, it is unclear whether primary auditory cortex shows cross-modal plasticity, since the studies have employed methods that poorly localize Heschl's gyrus. Only a few studies have examined the somatosensory modality (Auer, Bernstein, Sungkarat, & Singh, 2007; Lavänen, Jousmäki, & Hari, 1998). These somatosensory studies implicate Heschl's gyrus in cross-modal plasticity but are also limited in anatomical precision. A third study, using magnetoencephalography and fMRI in a single congenitally deaf individual, found neither visual nor somatosensory responses in deaf auditory cortex (Hickok et al., 1997).

A large literature on developmental disorders reports deficits in multisensory integration (e.g., see Foss-Feig et al., 2010; Hairston, Burdette, Flowers, Wood, & Wallace, 2005; Iarocci & McDonald, 2006). We recently tested the hypothesis raised by these results—i.e., that multisensory integration is a process that displays considerable neuroplasticity and may, therefore, be capable of large enhancements in the deaf. We examined whether visual, somatosensory, and bimodal processing are altered in congenitally deaf adult humans by quantifying fMRI signal change within anatomically defined Heschl's gyrus and in superior-temporal cortex in individual subjects (Karns, Dow, & Neville, 2012). We found that deaf adults did recruit Heschl's gyrus slightly for processing visual stimuli, but to a much larger extent for somatosensory processing and visual-somatosensory stimuli. Importantly, this cross-modal neuroplasticity had functional consequences, namely, altered perception in deaf individuals. Only the congenitally deaf adults in our study reported a somatosensory double-flash illusion, a visual percept induced by a somatosensory stimulus. However, this somatosensory recruitment was not constant across deaf individuals. We found that those individuals with the strongest response to somatosensory stimuli in Heschl's gyrus also saw the somatosensory-induced double flash illusion more frequently. This research shows that in congenital deafness, even primary sensory cortices can be recruited to process other sensory modalities. Interestingly, it has recently been proposed that multisensory integration and attention are tightly interconnected (Talsma, Senkowski, Soto-Faraco, & Woldorff, 2010), suggesting that attention may facilitate or enable plasticity in multisensory processing.

Language

It is reasonable to hypothesize that the same principles that characterize neuroplasticity of sensory systems—including different profiles, degrees, and mechanisms of plasticity—also characterize language. Indeed, as in the sensory systems, language exhibits a number of distinct subsystems, with nonidentical neural networks mediating the processing of, for example, semantics, syntax, and speech segmentation. As an example, when sentences contain a semantic (as opposed to syntactic) violation, ERPs to the semantic violation reveal a bilateral negative potential that is largest around 400 ms (N400; Kutas & Hillyard, 1980; Neville, Nicol, Barss, Forster, & Garrett, 1991; Newman, Ullman, Pancheva, Waligura, & Neville, 2007). In contrast, syntactic violations elicit a biphasic response consisting of an early, left-lateralized anterior negativity (LAN) followed by a later, bilateral positivity, peaking over posterior sites ~600 ms after the violation (P600; Friederici, 2002; Neville et al., 1991). The LAN is hypothesized to index more automatic aspects of the processing of syntactic structure (see below for new evidence on this) and the P600 to index later, more controlled processing of syntax associated with attempts to recover the meaning of syntactically anomalous sentences. In the case of speech segmentation, there are also distinct ERP responses to word-initial as compared to word-medial syllables. By 100 ms after word onset, syllables at the beginning of a word elicit a larger negativity than acoustically similar syllables in the middle of the word (Astheimer & Sanders, 2009; Sanders & Neville, 2003a; Sanders, Newport, & Neville, 2002). Below, we examine the degree to which these three subsystems—semantics, syntax, and speech segmentation—display neuroplasticity with altered experience.

To the extent that language is composed of distinct neural subsystems, it is possible that, as in vision and audition, these subsystems show different profiles of neuroplasticity. In support of this hypothesis, behavioral studies of language proficiency in second-language learners document that phonology and syntax are particularly vulnerable following delays in second-language acquisition (Johnson & Newport, 1989). Similarly, in studies of Chinese-English bilinguals who were first exposed to English at different ages, the neural systems associated with syntactic processing are vulnerable to differences with early delays in age of acquisition. On the other hand, the neural systems associated with semantic processing are more robust against delays in age of acquisition (Weber-Fox & Neville, 1996). We also observe atypical ERP effects of speech segmentation among late-learners when processing their second language (Sanders & Neville, 2003b). However, studies of bilingual participants suffer from two limitations: (1) the difficulty of separating out effects of age of acquisition from language proficiency and (2) the difficulty of assessing whether any observed differences in neural organization are due to delays in acquisition as compared to interference from a first-learned language. In related research, we have addressed each of these issues.

In one recent study, we examined the effects of delayed age of acquisition in a group of high-proficiency German-English bilingual participants, in which proficiency was equivalent to that of native English speakers. However, as shown in figure 13.2, even in this proficiency-matched group, later age of acquisition was associated with atypical ERP responses to syntactic violations, and in particular an absence of the early LAN effect (Pakulak & Neville, 2011). This suggests that age of acquisition can influence the neural systems recruited during language processing, independent of language proficiency. However, at the same time, we have—in separate studies—observed effects of language proficiency on the neural systems used for language processing, even among native speakers (Pakulak & Neville, 2010). Specifically, whereas high-proficiency native speakers show the typical biphasic LAN/P600 response to syntactic violations, native speakers of lower proficiency show a less spatially and temporally focal early neural response, as well as a reduced amplitude P600 (see figure 13.2). Indeed, in developmental studies as

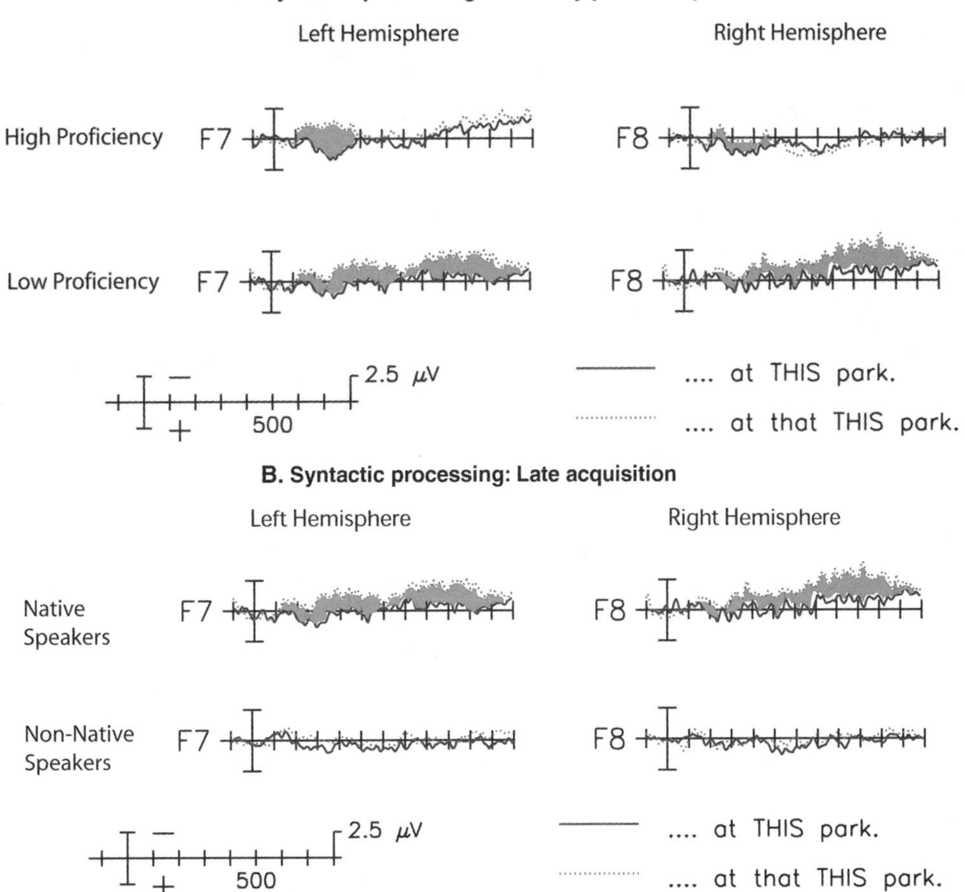

FIGURE 13.2 Event-related brain potential (ERP) to syntactic violations in spoken sentences showing independent effects of age of acquisition and proficiency on neural processing for language. (A) Comparison of high-proficiency second-language learners to native speakers of equivalent proficiency. (B) Comparison of native speakers with higher versus lower levels of English language proficiency. (Data from Pakulak & Neville, 2010, and Pakulak & Neville, 2011.) (See color plate 7.)

well, the neural response to known and unknown words and to syntactic anomalies is more strongly predicted by a child's language proficiency than by chronological age (Adamson, Mills, Appelbaum, & Neville, 1998; Adamson-Harris, Mills, & Neville, 2000; Mills, Coffey-Corina, & Neville, 1993, 1997). These data suggest that both age of acquisition and language proficiency affect the neural systems used during language processing.

To address whether altered neural organization in late bilinguals is due to age of acquisition versus interference of a first-learned language, we have studied a unique group of individuals: deaf signers who acquired sign language late in life. Many deaf children are born to hearing parents and, due to their limited access to spoken language, do not have full access to a first language until exposed to a signed language, which often occurs very late in development. This provides an opportunity to study the neural systems underlying delayed *first* language acquisition, since the neurobiology of language shows a strong degree of biological invariance, and several neural subsystems important in written and spoken language have also been observed when deaf and hearing native signers process American Sign Language (ASL), despite the fact that the eliciting sensory stimuli are very different (Capek, 2004; Capek et al., 2009; MacSweeney, Capek, Campbell, & Woll, 2009). However, it should be noted that although spoken and signed language processing share a number of modality-independent neural substrates, there is also some degree of specialization based on language modality. The processing of ASL, for example, is associated with additional and/or greater recruitment of right-hemisphere structures, perhaps owing to the use of spatial location and motion in syntactic processing in ASL (Capek et al., 2004, 2009; MacSweeney et al., 2009; Neville et al., 1998). In support of this hypothesis, we have shown that syntactic violations in ASL elicit a more bilateral anterior negativity for violations of spatial syntax, whereas a left-lateralized anterior negativity is observed for other classes of syntactic violations in ASL (Capek et al., 2009).

The effects of delayed ASL first-language acquisition have been observed in both behavioral and neuroimaging studies. Behavioral studies of deaf individuals with delayed exposure to sign language indicate that with increasing age of acquisition, proficiency in sign language decreases (Mayberry, 1993, 2003; Mayberry & Eichen, 1991; Mayberry, Lock, & Kazmi, 2002). In studies of deaf late-learners of ASL, we have examined the effects of this delayed first-language acquisition on brain organization. In one study, we demonstrated that whereas the right angular gyrus is active when native signers process ASL, it is not active in individuals who acquired ASL after puberty (Newman, Bavelier, Corina, Jezzard, & Neville, 2002). In a second study, in which we employed ERPs, we studied groups of deaf individuals who acquired ASL from birth, from 2 to 10 years, or between 11 and 21 years of age (Capek, 2004; Capek et al., 2014). All three groups display normal noncognitive skills. In all three groups of participants, the N400 index of semantic processing displays the same amplitude, latency, and cortical distribution. However, the early anterior negativity thought to index more automatic aspects of syntactic processing is only evident in those who acquired ASL before the age of 10 years. These data suggest that, in contrast to semantic processing, aspects of syntactic processing are subject to maturational constraints that render them more vulnerable following delays in either first- or second-language acquisition.

Using a novel laboratory language-learning task designed to mimic second-language immersion, we recently compared the ERP response to novel syntactic rules acquired under conditions of implicit exposure and explicit instruction. Regardless of training condition, learners who successfully acquired the novel syntactic rules showed P600 effects to syntactic violations similar to those elicited by native speakers who were tested on the same paradigm. This effect was observed after only an hour of exposure to the novel language, demonstrating that late, controlled mechanisms indexed by the P600 can be rapidly recruited (Batterink & Neville, 2013b). In a second study, we examined the neural mechanisms involved in the acquisition of novel semantic information using a task in which novel pseudowords were presented 10 times in a narrative context. During a subsequent explicit recognition task, these novel words elicited a robust N400 effect, suggesting that explicit representations of word meanings can be acquired with remarkable speed (Batterink & Neville, 2011).

We have also examined the role of awareness in semantic and syntactic processing using attentional blink manipulations. We found that the N400 was elicited only by words that were correctly reported, suggesting that this component indexes processes that are dependent upon awareness (Batterink, Karns, Yamada, & Neville, 2010). Similarly, in the domain of syntactic processing, the P600 was observed only for syntactic violations that were explicitly detected. In contrast, the LAN was elicited by both detected and undetected syntactic violations, suggesting that this response reflects automatic and implicit syntactic processing mechanisms that operate outside of conscious awareness (Batterink & Neville, 2013a).

Several mechanisms may render the neural systems important for syntax and speech segmentation more vulnerable than the neural systems important for semantics. For example, we and others have observed that the neural systems important for syntactic processing show a longer developmental time course than systems important for semantic processing (Hahne, Eckstein, & Friederici, 2004; Sabourin, Pakulak, Paulsen, Fanning, & Neville, 2007), again suggesting that systems with a longer developmental time course may be more modifiable during development. Networks involved in language processing that overlap with networks associated with selective attention may also be more vulnerable. For example, the speech segmentation ERP effect resembles the effect of temporally selective attention (Astheimer & Sanders, 2009), which allows for the preferential processing of information presented at specific time points in rapidly changing streams, and has also been shown to modulate early (100 ms) auditory ERPs (Lange & Röder, 2005; Lange, Rosler, & Röder, 2003). Thus, the neural mechanisms of speech segmentation may rely on the deployment of temporally selective attention during speech perception to aid in processing the most relevant rapid acoustic changes.

Attention

As noted above, many of the changes in vision, audition, sensory integration, and language observed during studies of neuroplasticity may depend at least in part on selective attention. The importance of selective attention for certain types of adult neuroplasticity is supported by animal research. For example, when monkeys are provided extensive exposure to auditory and tactile stimuli, experience-dependent expansions in associated auditory or somatosensory cortical areas occur, but *only* when attention is directed toward those stimuli in order to make behaviorally relevant discriminations (Recanzone, Jenkins, Hradek, & Merzenich,

1992; Recanzone, Schreiner, & Merzenich, 1993). Mere exposure is not enough. These data strongly suggest that attention is important in enabling neuroplasticity. Given this, and the central role of attention in learning more generally, we have conducted several studies on the development and neuroplasticity of attention.

In these studies, we examined the effects of sustained, selective attention on neural processing employing the Hillyard principle, that is, while keeping the physical stimuli, arousal levels, and task demands constant. For example, competing streams of stimuli are presented (e.g., two different trains of auditory stimuli delivered to different ears), with participants alternating attention to one stream at a time in order to detect rare target events. By comparing neural activity to the same physical stimuli when attended versus ignored, the effects of selective attention can be ascertained. Studies utilizing fMRI reveal that selective attention modulates the magnitude and extent of cortical activation in the relevant processing areas (Corbetta, Miezin, Dobmeyer, Shulman, & Petersen, 1990). Complementary studies using the ERP methodology have clarified the time course of attentional modulation. These studies reveal that in adults, selective attention amplifies the sensorineural response by 50–100% during the first 100 ms of processing (Hillyard, Di Russo, & Martinez, 2003; Hillyard, Hink, Schwent, & Picton, 1973; Luck, Woodman, & Vogel, 2000; Mangun & Hillyard, 1990). This early attentional modulation is in part domain-general in that it is observed across multiple sensory modalities and is based on spatial, temporal, or other stimulus attributes. Moreover, in between-group and change-over-time comparisons, ERPs can separately index processes of signal enhancement (ERP amplitude gains for attended stimuli) and distractor suppression (amplitude reductions for unattended stimuli).

In a number of studies, we document neuroplasticity in the early neural mechanisms of selective attention that, as in other neural systems, show considerable specificity. In the case of adults born deaf, employing ERPs and fMRI, we observe enhancements of attention that are specific to the peripheral, but not central, visual field (Bavelier et al., 2000, 2001; Neville & Lawson, 1987b). In parallel studies of auditory spatial attention among congenitally blind adults, we observe similar specificity in attentional enhancements to peripheral, but not central, auditory locations (Röder et al., 1999). However, there is some evidence for a sensitive period in attentional changes. For example, adults blinded later in life do not show changes in the early (N1) attention effects, though they do show changes in later (P300) attention effects (Fieger, Röder, Teder-Sälejärvi, Hillyard, & Neville, 2006). Such differences in the specific aspects of attention that are modifiable may also help to explain the apparent lifelong plasticity observed in other types of training in adulthood, including video gaming (e.g., Green & Bavelier, 2003).

If the early neural mechanisms of selective attention can be enhanced after altered experience, it is possible that, as with other systems that display a high degree of neuroplasticity, attention may be particularly vulnerable during development. To address this question, we first developed a child-friendly ERP paradigm for assessing selective auditory attention. These studies were modeled after those we and others have used with adults (Hillyard et al., 1973; Neville & Lawson, 1987a; Woods, 1990). The task was designed to be difficult enough to demand focused selective attention, while keeping the physical stimuli, arousal levels, and task demands constant. We presented two different children's stories concurrently from speakers to the left and right of the participant, and then asked participants to attend to one story and ignore the other. We superimposed probe stimuli on the stories and recorded ERPs to these identical probes both when attention was on that story and when it was on the other story. Adults tested with this paradigm show typical N1 attention effects (Coch, Sanders, & Neville, 2005). Children, who show a different ERP morphology to the probe stimuli, also show early attentional modulation within the first 100 ms of processing. This attentional modulation is an amplification of the broad positivity occurring in this time window, and we observe it in children as young as 3 years of age (Sanders, Stevens, Coch, & Neville, 2006; see also figure 13.3). These data suggest that with sufficient attentional cues, children as young as 3 years of age can attend selectively to an auditory stream and that doing so, as in adults, alters neural activity within 100 ms of

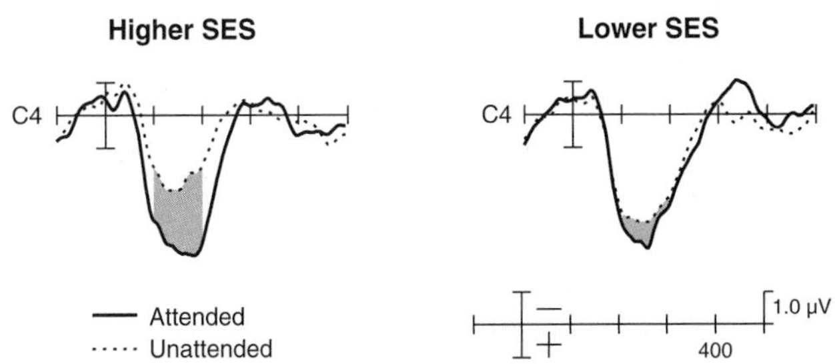

FIGURE 13.3 Effects of selective attention on neural processing in children aged 3–8 years from higher versus lower socioeconomic status (SES) backgrounds. Children from higher SES backgrounds had significantly greater effects of selective attention on neural processing than children from lower SES backgrounds. (Data from Stevens, Lauinger, & Neville, 2009.) (See color plate 8.)

processing. These data provide the baseline from which to examine possible vulnerabilities in children with or at risk for developmental disorders.

In one study, we examined children from backgrounds of differing socioeconomic status (SES; Stevens, Sanders, & Neville, 2006). Previous behavioral studies indicated that children from lower socioeconomic backgrounds experience difficulty with selective attention, particularly in tasks of executive function and in those tasks that require filtering irrelevant information or suppressing prepotent responses (Farah et al., 2006; Lupien, King, Meaney, & McEwen, 2001; Mezzacappa, 2004; Noble, McCandliss, & Farah, 2007; Noble, Norman, & Farah, 2005). We observed that children from lower SES backgrounds showed reduced effects of selective attention on neural processing (see figure 13.3) and, moreover, that these differences were related specifically to a reduced ability to filter irrelevant information (i.e., to suppress the response to ignored sounds). Other research groups report similar results (D'Angiulli, Herdman, Stapells, & Hertzman, 2008). In a second study, we examined children with specific language impairment (SLI; Stevens et al., 2006). We were interested in children with SLI, as previous behavioral studies reported deficits in aspects of attention, including filtering and noise exclusion (Atkinson, 1991; Cherry, 1981; Sperling, Lu, Manis, & Seidenberg, 2005; Ziegler, Pech-Georgel, George, Alanio, & Lorenzi, 2005). We observed that children with SLI did not show effects of selective attention on neural processing, and that this deficit was specific to a reduced ability to enhance the neural response to attended stimuli. Thus, the mechanism implicated in attention deficits in children from lower socioeconomic backgrounds (i.e., distractor suppression) was not the same as the mechanism implicated in children with SLI, who showed a deficit in signal enhancement of stimuli in the attended channel.

Taken together, these studies point to the two sides of the plasticity in early mechanisms of attention. The marked plasticity observed in attentional systems—and in particular in selective attention—may be mediated by several mechanisms. For example, sustained, selective attention shows a particularly long time course of development. Although the effects of selective attention on neural processing are quite similar in adults and young children, it may be more difficult for children to deploy selective attention successfully. In support of this, a robust literature documents that the abilities to select input for processing and successfully ignore irrelevant stimuli improve progressively with increasing age across childhood (Cherry, 1981; Geffen & Sexton, 1978; Geffen & Wale, 1979; Hiscock & Kinsbourne, 1980; Lane & Pearson, 1982; Maccoby & Konrad, 1966; Ridderinkhof & van der Stelt, 2000; Sexton & Geffen, 1979; Zukier & Hagen, 1978). Additionally, since the key sources of selective attention within the parietal and frontal lobes constitute parts of the dorsal pathway, similar neurochemical and anatomical factors noted in the section on vision may contribute to the plasticity of attention. Finally, recent evidence points to considerable genetic effects on attention (Bell et al., 2008; Fan, Fossella, Sommer, Wu, & Posner, 2003; Posner, Rothbart, & Sheese, 2007; Rueda, Rothbart, McCandliss, Saccamanno, & Posner, 2005) that may also be modified by environmental input epigenetically (Bakermans-Kranenberg, van Ijzendoorn, Pijlman, Mesman, & Femmie, 2008; Sheese, Voelker, Rothbart, & Posner, 2007). Thus, a complex interplay of genetic and experiential factors, operating across a relatively long developmental time course, contribute to the plasticity of selective attention.

Interventions

As described above, selective attention influences early sensory processing across a number of domains. In our most recent research, we are investigating the possibility that attention itself might be trainable, and that this training can impact processing in a number of different domains. We outlined this general argument in a recent review, which applies a cognitive neuroscience framework to identify the role of selective attention in the development of several foundational skills, including language, literacy, and mathematics (Stevens & Bavelier, 2012).

In one line of research, we examined whether training programs that successfully target language or literacy skills also train selective attention. We were interested in this question as several proposals suggest that some interventions designed to improve language skills might also target or train selective attention (Gillam, 1999; Gillam, Crofford, Gale, & Hoffman, 2001; Gillam, Loeb, & Friel-Patti, 2001; Hari & Renvall, 2001). We tested this hypothesis in a series of intervention studies. In this research, we document changes in the neural mechanisms of selective attention following training in typically developing children, as well as in children with language impairment or at risk for reading failure (Stevens, Coch, Sanders, & Neville, 2008; Stevens et al., 2013; Yamada, Stevens, Harn, Chard, & Neville, 2011). In all cases, increases in the effects of attention on sensorineural processing are accompanied by behavioral changes in other domains that were also targeted by the training programs, including language and preliteracy skills and/or changes in the neural systems important for literacy processing. These data suggest that modifications in behavior can arise alongside changes in the early neural mechanisms of attention, and provide a "proof of concept" for the malleability of the early mechanisms of selective attention in children.

In a second line of research, we have designed training programs that specifically target attention. Indeed, in his seminal work, *Principles of Psychology*, William James raised the idea of attention training for children, proposing that this would be "*the* education *par excellence*" (James, 1890; italics original). While James went on to say that such an education is difficult to define and bring about, attention training has recently been implemented in curricula for preschool and school-age children (Bodrova & Leong, 2007; Chenault, Thomson, Abbott, & Berninger, 2006; Diamond, Barnett, Thomas, & Munro, 2007; Rueda et al., 2005). These programs are associated with improvements in behavioral and neurophysiological indices of attention, as well as in measures of academic outcomes and nonverbal intelligence. However, these previous studies do not engage the larger context of parents and the home environment. Yet the family context plays a key role in supporting children's attention development and may specifically be targeted in intervention programs aimed at improving child outcomes. Thus, employing information from research on the neuroplasticity of selective attention and on the central role of successful parenting in child development, we have now developed and rigorously assessed an eight-week, family-based training program designed to improve brain systems for selective attention in preschool children.

The program, Parents and Children Making Connections—Highlighting Attention (PCMC-A), is unique in combining training sessions for parents, guardians, or caregivers (hereafter "parent") with attention-training exercises for children. Parents attend eight weekly, two-hour, small-group classes that occur in the evenings or on weekends, and their children participate in concurrent small-group training activities. Parents learn strategies targeting family stress regulation, contingency-based discipline, parental responsiveness and language use, and facilitation of child attention through links to child training exercises. The child component of PCMC-A consists of small-group activities (four to six children and two adults) designed to address the fundamental goal of improving regulation of attention and emotion states.

In the evaluation study (Neville et al., 2013), 141 lower SES preschoolers enrolled in a Head Start (HS) program were randomly assigned to the training program, HS alone, or an active control group. Prior to and following the eight-week intervention period, a multi-rater, multi-method assessment was conducted that included electrophysiological measures of children's selective attention using the paradigm described above. Results indicated that electrophysiological measures of children's brain functions supporting selective attention, standardized measures of cognition, and parent-reported child behaviors all favored children in the treatment program relative to both control groups. Figure 13.4 presents the ERP results from this study. Positive changes were also observed in the parents themselves, including changes in language interaction patterns with their children and reductions in parenting stress. Moreover, the effect sizes were large in magnitude, ranging from one-quarter to one-half of a standard deviation.

Importantly, this study both builds on previous research on neuroplasticity and also advances the field in several ways. Perhaps most importantly, the study demonstrates that the neural mechanism of selective attention can be improved in children from lower SES backgrounds, and this can be done in the relatively short time frame of eight weeks. Second, as the most favorable outcomes were observed in a more parent-focused (as opposed to child-focused) training model, these findings underscore the importance of engaging parents to support child development. Third, the effectiveness of PCMC-A, a short, inexpensive ($800 per family) eight-week program, supports the design of programs that efficiently build on evidence from basic research on neuroplasticity and on evidence-based practices and that can be delivered in relatively short time frames. Finally, by including multiple outcome measures, the study provides a comprehensive picture of the changes resulting from a family-based training model, including not only gains for children in a direct neural measure of selective attention, but also specific skills assessed by standardized tests (Neville, Stevens, et al., 2011a; Neville, Stevens, et al., 2011b; Neville et al., 2013).

Conclusion

The research described in this chapter illustrates the variable degrees and time periods of neuroplasticity in the human brain and likely mechanisms whereby experience influences different subsystems within perceptual and cognitive domains. Additionally, this research highlights the bidirectional nature of plasticity—those aspects of neural processing and related cognitive functioning that show the greatest capability for enhancement also display the greatest susceptibility to deficits under different conditions. Researchers are now entering a powerful frontier of neuroplasticity research that

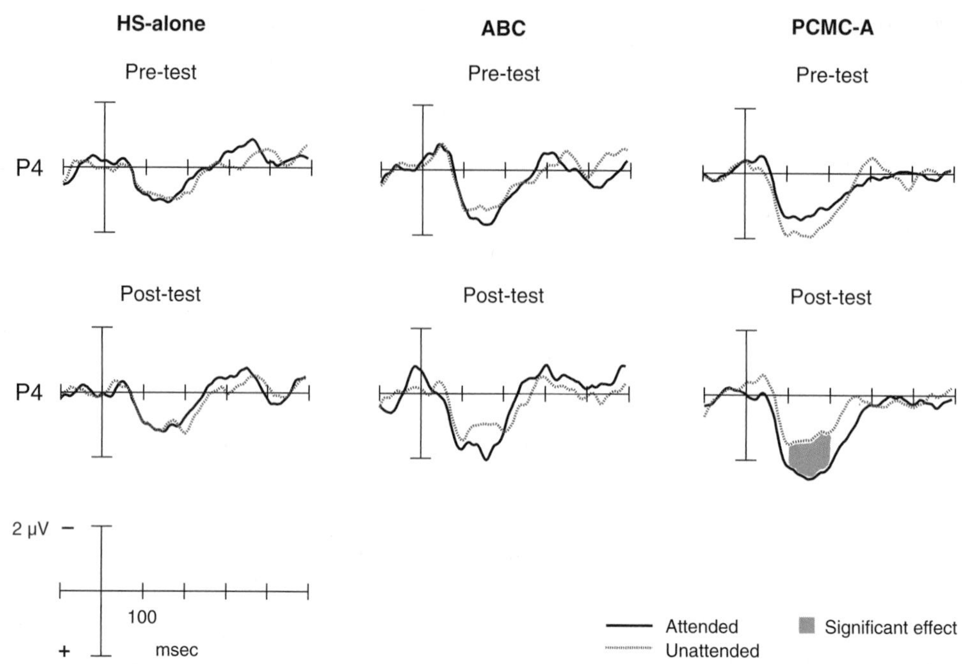

FIGURE 13.4 Effects of selective attention on neural processing in children aged 3–5 years from lower socioeconomic status backgrounds, both before and after an eight-week training period. Only children in the family-based program with a greater emphasis on parent training ("PCMC-A") showed an increase in the effects of attention on neural processing. Children in Head Start alone ("HS-alone") or in a comparison training program ("ABC") did not show significant changes in the effects of attention on neural processing from pre- to post-training. Children in the PCMC-A program also made significantly greater gains than either comparison group on standardized measures of receptive language and nonverbal IQ, as well as parent reports of child behavior. See main text for details. (Data from Neville et al., 2013.) (See color plate 9.)

transforms the results of basic research on the profiles and mechanisms of neuroplasticity into the development of training and intervention programs. Our growing understanding of the limits and mechanisms of plasticity contributes to a basic understanding of human brain development and function and can also inform and guide efforts to harness neuroplasticity both to optimize and to protect the malleable and vulnerable aspects of human development.

ACKNOWLEDGMENTS We thank our many collaborators in the research reported here, supported by grants from NIH NIDCD (R01 DC000481, R01 DC000128) and Department of Education IES (R305B070018, R305A110397, and R305A110398).

REFERENCES

ADAMSON, A., MILLS, D., APPELBAUM, G., & NEVILLE, H. (1998). Auditory sentence processing in adults and children: Evidence from ERPs. Paper presented at the Cognitive Neuroscience Society.

ADAMSON-HARRIS, A. M., MILLS, D. L., & NEVILLE, H. J. (2000). Children's processing of grammatical and semantic information within sentences: Evidence from event-related potentials. Paper presented at the Cognitive Neuroscience Society.

ARMSTRONG, B., HILLYARD, S. A., NEVILLE, H. J., & MITCHELL, T. V. (2002). Auditory deprivation affects processing of motion, but not color. *Brain Res Cogn Brain Res, 14*(3), 422–434.

ASTHEIMER, L., & SANDERS, L. (2009). Listeners modulate temporally selective attention during natural speech processing. *Biol Psych, 80*, 23–34.

ATKINSON, J. (1991). Review of human visual development: Crowding and dyslexia. In J. Cronly-Dillon & J. Stein (Eds.), *Vision and visual dysfunction* (Vol. 13, pp. 44–57). Cambridge, MA: MIT Press.

ATKINSON, J. (1992). Early visual development: Differential functioning of parvocellular and magnocellular pathways. *Eye, 6*, 129–135.

ATKINSON, J., KING, J., BRADDICK, O., NOKES, L., ANKER, S., & BRADDICK, F. (1997). A specific deficit of dorsal stream function in Williams' syndrome. *NeuroReport, 8*(8), 1919–1922.

AUER, E. T., JR, BERNSTEIN, L., SUNGKARAT, W., & SINGH, M. (2007). Vibrotactile activation of the auditory cortices in deaf versus hearing adults. *NeuroReport, 18*, 645–648.

BAKERMANS-KRANENBERG, M., VAN IJZENDOORN, M. H., PIJLMAN, F. T. A., MESMAN, J., & FEMMIE, J. (2008). Experimental evidence for differential susceptibility: Dopamine D4 receptor polymorphism (DRD4 VNTR) moderates intervention effects on toddlers' externalizing behavior in a randomized controlled trial. *Dev Psychol, 44,* 293–300.

Batterink, L., Karns, C., Yamada, Y., & Neville, H. (2010). The role of awareness in semantic and syntactic processing: An ERP attentional blink study. *J Cogn Neurosci, 22,* 2514–2529.

Batterink, L., & Neville, H. (2011). Implicit and explicit mechanisms of word learning in a narrative context: An event-related potential study. *J Cogn Neurosci, 23,* 3181–3196.

Batterink, L., & Neville, H. (2013a). The human brain processes syntax in the absence of conscious awareness. *J Neurosci, 33,* 8528–8533.

Batterink, L., & Neville, H. (2013b). Implicit and explicit second language training recruit common neural mechanisms for syntactic processing. *J Cogn Neurosci, 25,* 936–951.

Bavelier, D., Brozinsky, C., Tomann, A., Mitchell, T., Neville, H., & Liu, G. (2001). Impact of early deafness and early exposure to sign language on the cerebral organization for motion processing. *J Neurosci, 21*(22), 8931–8942.

Bavelier, D., & Neville, H. J. (2002). Cross-modal plasticity: Where and how? *Nat Rev Neurosci, 3,* 443–452.

Bavelier, D., Tomann, A., Hutton, C., Mitchell, T., Liu, G., Corina, D., & Neville, H. (2000). Visual attention to the periphery is enhanced in congenitally deaf individuals. *J Neurosci, 20*(17), 1–6.

Bell, T., Batterink, L., Currin, L., Pakulak, E., Stevens, C., & Neville, H. (2008). Genetic influences on selective auditory attention as indexed by ERPs. Paper presented at the Cognitive Neuroscience Society, San Francisco, CA.

Bishop, D., & McArthur, G. M. (2004). Immature cortical responses to auditory stimuli in specific language impairment: Evidence from ERPs to rapid tone sequences. *Dev Sci, 7,* F11–F18.

Bodrova, E., & Leong, D. (2007). *Tools of the mind: The Vygotskian approach to early childhood education* (2nd ed.). Upper Saddle River, NJ: Pearson.

Burton, H., Synder, A., Conturo, T., Akbudak, E., Ollinger, J., & Raichle, M. (2002). A fMRI study of verb generation to auditory nouns in early and late blind. Paper presented at the Society for Neuroscience.

Capek, C. (2004). The cortical organization of spoken and signed sentence processing in adults (PhD doctoral dissertation). University of Oregon, Eugene, OR.

Capek, C., Bavelier, D., Corina, D., Newman, A. J., Jezzard, P., & Neville, H. J. (2004). The cortical organization for audio-visual sentence comprehension: An fMRI study at 4 Tesla. *Brain Res Cogn Brain Res, 20*(2), 111–119.

Capek, C., Corina, D., Grossi, G., McBurney, S. L., Neville, H. J., Newman, A. J., & Röder, B. (2014). American Sign Language sentence processing: ERP evidence from adults with different ages of acquisition. Manuscript in preparation.

Capek, C., Grossi, G., Newman, A., McBurney, S., Corina, D., Röder, B., & Neville, H. (2009). Brain systems mediating semantic and syntactic processing in deaf native signers: Biological invariance and modality specificity. *Proc Natl Acad Sci USA, 106,* 8784–8789.

Chalupa, L. M., & Dreher, B. (1991). High precision systems require high precision "blueprints": A new view regarding the formation of connections in the mammalian visual system. *J Cogn Neurosci, 3*(3), 209–219.

Chenault, B., Thomson, J., Abbott, R. D., & Berninger, V. W. (2006). Effects of prior attention training on child dyslexics' response to composition instruction. *Dev Neuropsych, 29*(1), 243–260.

Cherry, R. (1981). Development of selective auditory attention skills in children. *Percept Mot Skills, 52,* 379–385.

Coch, D., Sanders, L. D., & Neville, H. J. (2005). An event-related potential study of selective auditory attention in children and adults. *J Cogn Neurosci, 17*(4), 605–622.

Coch, D., Skendzel, W., Grossi, G., & Neville, H. (2005). Motion and color processing in school-age children and adults: An ERP study. *Dev Sci, 8*(4), 372–386.

Cohen, L. G., Weeks, R. A., Celnik, P., & Hallett, M. (1999). Role of the occipital cortex during Braille reading in subjects with blindness acquired late in life. *J Neurosci, 45,* 451–460.

Cooper, H., Herbin, M., & Nevo, E. (1993). Visual system of a naturally microphthalmic mammal: The blind mole rat, *Spalax ehrenbergi*. *J Comp Neurol, 328,* 313–350.

Corbetta, M., Miezin, F., Dobmeyer, S., Shulman, G., & Petersen, S. E. (1990). Attentional modulation of neural processing of shape, color, and velocity in humans. *Science, 248,* 1556–1559.

Cornelissen, P., Richardson, A., Mason, A., Fowler, S., & Stein, J. (1995). Contrast sensitivity and coherent motion detection measured at photopic luminance levels in dyslexics and controls. *Vis Res, 35*(10), 1483–1494.

D'Angiulli, A., Herdman, A., Stapells, D., & Hertzman, C. (2008). Children's event-related potentials of auditory selective attention vary with their socioeconomic status. *Neuropsych, 22,* 293–300.

Demb, J. B., Boynton, G. M., Best, M., & Heeger, D. J. (1998). Psychophysical evidence for a magnocellular pathway deficit in dyslexia. *Vis Res, 38,* 1555–1559.

Diamond, A., Barnett, W., Thomas, J., & Munro, S. (2007). Preschool program improves cognitive control. *Science, 318,* 1387–1388.

Doron, N., & Wollberg, Z. (1994). Cross-modal neuroplasticity in the blind mole rat *Spalax ehrenbergi*: A WGA-HRP tracing study. *NeuroReport, 5,* 2697–2701.

Eckert, M. A., Kamdar, N. V., Chang, C. E., Beckmann, C. F., Greicius, M. D., & Menon, V. (2008). A cross-model system linking primary auditory and visual cortices: Evidence from intrinsic fMRI connectivity analysis. *Hum Brain Mapp, 29,* 848–857.

Eden, G. F., VanMeter, J. W., Rumsey, J. M., Maisog, J. M., Woods, R. P., & Zeffiro, T. A. (1996). Abnormal processing of visual motion in dyslexia revealed by functional brain imaging. *Nature, 382*(6586), 66–69.

Everatt, J., Bradshaw, M. F., & Hibbard, P. B. (1999). Visual processing and dyslexia. *Perception, 28,* 243–254.

Falchier, A., Clavagnier, S., Barone, P., & Kennedy, H. (2002). Anatomical evidence of multimodal integration in primate striate cortex. *J Neurosci, 22*(13), 5749–5759.

Fan, J., Fossella, J., Sommer, T., Wu, Y., & Posner, M. I. (2003). Mapping the genetic variation of executive attention onto brain activity. *Proc Natl Acad Sci USA, 100*(12), 7406–7411.

Farah, M., Shera, D., Savage, J., Betancourt, L., Giannetta, J., Brodsky, N., … Hurt, H. (2006). Childhood poverty: Specific associations with neurocognitive development. *Brain Res, 1110,* 166–174.

Fieger, A., Röder, B., Teder-Sälejärvi, W., Hillyard, S. A., & Neville, H. J. (2006). Auditory spatial tuning in late onset blind humans. *J Cogn Neurosci, 18*(2), 149–157.

Foss-Feig, J., Kwakye, L., Cascio, C., Burnette, C., Kadivar, H., Stone, W., & Wallace, M. (2010). An extended multisensory temporal binding window in autism spectrum disorders. *Exp Brain Res, 203*, 381–389.

Friederici, A. D. (2002). Towards a neural basis of auditory sentence processing. *Trends Cogn Sci, 6*(2), 78–84.

Galaburda, A., & Livingstone, M. (1993). Evidence for a magnocellular defect in developmental dyslexia. In P. Tallal, A. M. Galaburda, R. R. Llinas, & C. von Euler (Eds.), *Temporal information processing in the nervous system* (pp. 70–82). New York, NY: New York Academy of Sciences.

Galaburda, A., Menard, M., & Rosen, G. (1994). Evidence for aberrant auditory anatomy in developmental dyslexia. *Proc Natl Acad Sci USA, 91*, 8010–8013.

Geffen, G., & Sexton, M. A. (1978). The development of auditory strategies of attention. *Dev Psychol, 14*(1), 11–17.

Geffen, G., & Wale, J. (1979). Development of selective listening and hemispheric asymmetry. *Dev Psychol, 15*(2), 138–146.

Gillam, R. (1999). Computer-assisted language intervention using Fast ForWord: Theoretical and empirical considerations for clinical decision-making. *Lang Speech Hear Ser, 30*, 363–370.

Gillam, R., Crofford, J., Gale, M., & Hoffman, L. (2001). Language change following computer-assisted language instruction with Fast ForWord or Laureate Learning Systems software. *Am J Speech Lang Pathol, 10*, 231–247.

Gillam, R., Loeb, D. F., & Friel-Patti, S. (2001). Looking back: A summary of five exploratory studies of Fast ForWord. *Am J Speech Lang Pathol, 10*, 269–273.

Green, C., & Bavelier, D. (2003). Action video game modifies visual attention. *Nature, 423*, 534–537.

Grill-Spector, K., & Malach, R. (2004). The human visual cortex. *Annu Rev Neurosci, 27*, 649–677.

Hahne, A., Eckstein, K., & Friederici, A. D. (2004). Brain signatures of syntactic and semantic processes during children's language development. *J Cogn Neurosci, 16*(7), 1302–1318.

Hairston, W. D., Burdette, J., Flowers, D. L., Wood, F. B., & Wallace, M. (2005). Altered temporal profile of visual-auditory multisensory interactions in dyslexia. *Exp Brain Res, 166*, 474–480.

Hansen, P. C., Stein, J. F., Orde, S. R., Winter, J. L., & Talcott, J. B. (2001). Are dyslexics' visual deficits limited to measures of dorsal stream function? *NeuroReport, 12*(7), 1527–1530.

Hari, R., & Renvall, H. (2001). Impaired processing of rapid stimulus sequences in dyslexia. *Trends Cogn Sci, 5*(12), 525–532.

Heil, P., Bronchti, G., Wollberg, Z., & Scheich, H. (1991). Invasion of visual cortex by the auditory system in the naturally blind mole rat. *NeuroReport, 2*(12), 735–738.

Hickey, T. L. (1981). The developing visual system. *Trends Neurosci, 2*, 41–44.

Hickok, G., Poeppel, D., Clark, K., Buxton, R. B., Rowley, H. A., & Roberts, T. P. (1997). Sensory mapping in a congenitally deaf subject: MEG and fMRI studies of cross-modal non-plasticity. *Hum Brain Mapp, 5*, 437–444.

Hillyard, S., Di Russo, F., & Martinez, A. (2003). Imaging of visual attention. In N. Kanwisher & J. Duncan (Eds.), *Functional neuroimaging of visual cognition attention and performance.* Oxford, UK: Oxford University Press.

Hillyard, S., Hink, R. F., Schwent, V. L., & Picton, T. W. (1973). Electrical signals of selective attention in the human brain. *Science, 182*(4108), 177–179.

Hiscock, M., & Kinsbourne, M. (1980). Asymmetries of selective listening and attention switching in children. *Dev Psychol, 16*(1), 70–82.

Hollants-Gilhuijs, M. A. M., Ruijter, J. M., & Spekreijse, H. (1998a). Visual half-field development in children: Detection of colour-contrast-defined forms. *Vis Res, 38*(5), 645–649.

Hollants-Gilhuijs, M. A. M., Ruijter, J. M., & Spekreijse, H. (1998b). Visual half-field development in children: Detection of motion-defined forms. *Vis Res, 38*(5), 651–657.

Iarocci, G., & McDonald, J. (2006). Sensory integration and the perceptual experience of persons with autism. *J Autism Dev Disord, 36*, 77–90.

James, W. (1890). *Principles of psychology.* New York, NY: Henry Holt.

Johnson, J., & Newport, E. (1989). Critical period effects in second language learning: The influence of maturational state on the acquisition of English as a second language. *Cogn Psychol, 21*, 60–99.

Karns, C. M., Dow, M. W., & Neville, H. J. (2012). Altered cross-modal processing in primary auditory cortex of congenitally deaf adults: A visual-somatosensory fMRI study with a double-flash illusion. *J Neurosci, 32*, 9626–9638.

Kutas, M., & Hillyard, S. A. (1980). Reading senseless sentences: Brain potentials reflect semantic incongruity. *Science, 207*, 203–204.

Lane, D., & Pearson, D. (1982). The development of selective attention. *Merrill Palmer Q, 28*(3), 317–337.

Lange, K., & Röder, B. (2005). Orienting attention to points in time improves stimulus processing both within and across modalities. *J Cogn Neurosci, 18*(5), 715–729.

Lange, K., Rosler, F., & Röder, B. (2003). Early processing stages are modulated when auditory stimuli are presented at an attended moment in time: An event-related potential study. *Psychophysiology, 40*, 806–817.

Lavänen, S., Jousmäki, V., & Hari, R. (1998). Vibration-induced auditory-cortex activation in a congenitally deaf adult. *Curr Biol, 8*, 869–872.

Lovegrove, W., Martin, F., & Slaghuis, W. (1986). A theoretical and experimental case for a visual deficit in specific reading disability. *Cogn Neuropsychol, 3*, 225–267.

Luck, S. J., Woodman, G. F., & Vogel, E. K. (2000). Event-related potential studies of attention. *Trends Cogn Sci, 4*(11), 432–440.

Lupien, S. J., King, S., Meaney, M. J., & McEwen, B. S. (2001). Can poverty get under your skin? Basal cortisol levels and cognitive function in children from low and high socioeconomic status. *Dev Psychopathol, 13*, 653–676.

Maccoby, E., & Konrad, K. (1966). Age trends in selective listening. *J Exp Child Psychol, 3*, 113–122.

MacSweeney, M., Capek, C., Campbell, R., & Woll, B. (2009). The signing brain: The neurobiology of sign language. *Trends Cogn Sci, 12*, 432–440.

Mangun, G., & Hillyard, S. (1990). Electrophysiological studies of visual selective attention in humans. In A. Scheibel & A. Wechsler (Eds.), *Neurobiology of higher cognitive function* (pp. 271–295). New York, NY: Guilford.

Mayberry, R. (1993). First-language acquisition after childhood differs from second-language acquisition: The case of American Sign Language. *J Speech Hear Res, 36*(6), 1258–1270.

Mayberry, R. (2003). Age constraints on first versus second language acquisition: Evidence for linguistic plasticity and epigenesis. *Brain Lang, 87*(3), 369–384.

Mayberry, R., & Eichen, E. (1991). The long-lasting advantage of learning sign language in childhood: Another look at the critical period for language acquisition. *J Mem Lang, 30*, 486–512.

Mayberry, R., Lock, E., & Kazmi, H. (2002). Linguistic ability and early language exposure. *Nature, 417*, 38.

Mezzacappa, E. (2004). Alerting, orienting, and executive attention: Developmental properties and sociodemographic correlates in epidemiological sample of young, urban children. *Child Dev, 75*(5), 1373–1386.

Mills, D. L., Coffey-Corina, S. A., & Neville, H. J. (1993). Language acquisition and cerebral specialization in 20-month-old infants. *J Cogn Neurosci, 5*(3), 317–334.

Mills, D. L., Coffey-Corina, S. A., & Neville, H. J. (1997). Language comprehension and cerebral specialization from 13 to 20 months. *Dev Neuropsych, 13*(3), 397–445.

Mitchell, T. V., & Neville, H. J. (2004). Asynchronies in the development of electrophysiological responses to motion and color. *J Cogn Neurosci, 16*(8), 1–12.

Neville, H. J., & Lawson, D. (1987a). Attention to central and peripheral visual space in a movement detection task: An event-related potential and behavioral study. I. Normal hearing adults. *Brain Res, 405*, 253–267.

Neville, H. J., & Lawson, D. (1987b). Attention to central and peripheral visual space in a movement detection task: An event-related potential and behavioral study. II. Congenitally deaf adults. *Brain Res, 405*, 268–283.

Neville, H. J., Bavelier, D., Corina, D., Rauschecker, J., Karni, A., Lalwani, A., ... Turner, R. (1998). Cerebral organization for language in deaf and hearing subjects: Biological constraints and effects of experience. *Proc Natl Acad Sci USA, 95*(3), 922–929.

Neville, H. J., Coffey, S. A., Holcomb, P. J., & Tallal, P. (1993). The neurobiology of sensory and language processing in language-impaired children. *J Cogn Neurosci, 5*(2), 235–253.

Neville, H. J., Nicol, J., Barss, A., Forster, K., & Garrett, M. (1991). Syntactically based sentence processing classes: Evidence from event-related brain potentials. *J Cogn Neurosci, 3*, 155–170.

Neville, H. J., Schmidt, A., & Kutas, M. (1983). Altered visual-evoked potentials in congenitally deaf adults. *Brain Res, 266*, 127–132.

Neville, H., Stevens, C., Klein, S., Fanning, J., Bell, T., Cakir, E., & Pakulak, E. (2011a). Improving behavior, cognition, and neural mechanisms of attention in at-risk children. Cognitive Neuroscience Society, 18, San Francisco, CA.

Neville, H., Stevens, C., Klein, S., Fanning, J., Bell, T., Isbell, E., & Pakulak, E. (2011b). Improving behavior, cognition, and neural mechanisms of attention in lower SES children. Society for Neuroscience, 37, Washington DC.

Neville, H., Stevens, C., Pakulak, E., Bell, T., Fanning, J., Klein, S., & Isbell, E. (2013). Family-based training program improves brain function, cognition, and behavior in lower socioeconomic status preschoolers. *Proc Natl Acad Sci USA*. doi:10.1073/pnas.1304437110

Newman, A. J., Bavelier, D., Corina, D., Jezzard, P., & Neville, H. J. (2002). A critical period for right hemisphere recruitment in American Sign Language processing. *Nat Neurosci, 5*(1), 76–80.

Newman, A. J., Ullman, M. T., Pancheva, R., Waligura, D., & Neville, H. (2007). An ERP study of regular and irregular English past tense inflection. *NeuroImage, 34*(1), 435–445.

Noble, K. G., Norman, M. F., & Farah, M. J. (2005). Neurocognitive correlates of socioeconomic status in kindergarten children. *Dev Sci, 8*(1), 74–87.

Noble, K., McCandliss, B., & Farah, M. (2007). Socioeconomic gradients predict individual differences in neurocognitive abilities. *Dev Sci, 10*, 464–480.

Packer, O., Hendrickson, A., & Curcio, A. (1990). Developmental redistribution of photoreceptors across the *Macaca nemestrina* (Pigtail Macaque) retina. *J Comp Neurol, 298*, 472–493.

Pakulak, E., & Neville, H. (2010). Proficiency differences in syntactic processing of native speakers indexed by event-related potentials. *J Cogn Neurosci, 22*, 2728–2744.

Pakulak, E., & Neville, H. (2011). Maturational constraints on the recruitment of early processes for syntactic processing. *J Cogn Neurosci, 23*, 2752–2765.

Posner, M., Rothbart, M. K., & Sheese, B. E. (2007). Attention genes. *Dev Sci, 10*, 24–29.

Rauschecker, J. P. (1998). Parallel processing in the auditory cortex of primates. *Audiol Neurootol, 3*, 86–103.

Recanzone, G., Jenkins, W., Hradek, G., & Merzenich, M. (1992). Progressive improvement in discriminative abilities in adult owl monkeys performing a tactile frequency discrimination task. *J Neurophysiol, 67*, 1015–1030.

Recanzone, G., Schreiner, C., & Merzenich, M. (1993). Plasticity in the frequency representation of primary auditory cortex following discrimination training in adult owl monkeys. *J Neurosci, 12*, 87–103.

Ridderinkhof, K., & van der Stelt, O. (2000). Attention and selection in the growing child: Views derived from developmental psychophysiology. *Biol Psych, 54*, 55–106.

Rockland, K. S., & Ojima, H. (2003). Multisensory convergence in calcarine visual areas in macaque monkey. *Int J Psychophysiol, 50*, 19–26.

Röder, B., Stock, O., Bien, S., Neville, H., & Rösler, F. (2002). Speech processing activates visual cortex in congenitally blind humans. *Eur J Neurosci, 16*, 930–936.

Röder, B., Teder-Salejarvi, W., Sterr, A., Rosler, F., Hillyard, S. A., & Neville, H. (1999). Improved auditory spatial tuning in blind humans. *Nature, 400*, 162–166.

Rueda, M., Rothbart, M., McCandliss, B., Saccamanno, L., & Posner, M. (2005). Training, maturation, and genetic influences on the development of executive attention. *Proc Natl Acad Sci USA, 102*, 14931–14936.

Sabourin, L., Pakulak, E., Paulsen, D., Fanning, J. L., & Neville, H. (2007). The effects of age, language proficiency and SES on ERP indices of syntactic processing in children. Paper presented at the Cognitive Neuroscience Society, New York City, NY.

Sanders, L., & Neville, H. J. (2003a). An ERP study of continuous speech processing: I. Segmentation, semantics, and syntax in native English speakers. *Brain Res Cogn Brain Res, 15*(3), 228–240.

Sanders, L., & Neville, H. (2003b). An ERP study of continuous speech processing: II. Segmentation, semantics, and syntax in non-native speakers. *Brain Res Cogn Brain Res, 15*(3), 214–227.

Sanders, L., Newport, E. L., & Neville, H. J. (2002). Segmenting nonsense: An event-related potential index of perceived onsets in continuous speech. *Nat Neurosci, 5*(7), 700–703.

Sanders, L., Stevens, C., Coch, D., & Neville, H. J. (2006). Selective auditory attention in 3- to 5-year-old children: An event-related potential study. *Neuropsychologia, 44,* 2126–2138.

Scott, G. D., Karns, C. M., Dow, M. W., Stevens, C., & Neville, H. J. (2014). Enhanced peripheral visual processing in congenitally deaf humans is supported by multiple brain regions, including primary auditory cortex. *Front Hum Neurosci, 8,* 1–9.

Sedato, N., Pascual-Leone, A., Grafman, J., Ibanez, V., Delber, M.-P., Dold, G., & Hallett, M. (1996). Activation of the primary visual cortex by Braille reading in blind subjects. *Nature, 380*(11), 526–528.

Sexton, M. A., & Geffen, G. (1979). Development of three strategies of attention in dichotic monitoring. *Dev Psychol, 15*(3), 299–310.

Sheese, B. E., Voelker, P. M., Rothbart, M. K., & Posner, M. I. (2007). Parenting quality interacts with genetic variation in dopamine receptor D4 to influence temperament in early childhood. *Dev Psychopathol, 19,* 1039–1046.

Sperling, A., Lu, Z.-l., Manis, F. R., & Seidenberg, M. S. (2003). Selective magnocellular deficits in dyslexia: A "phantom contour" study. *Neuropsychologia, 41*(10), 1422–1429.

Sperling, A., Lu, Z., Manis, F. R., & Seidenberg, M. S. (2005). Deficits in perceptual noise exclusion in developmental dyslexia. *Nat Neurosci, 8,* 862–863.

Stevens, C., & Bavelier, D. (2012). The role of selective attention on academic foundations: A cognitive neuroscience perspective. *Dev Cogn Neurosci, 2S,* S30–S48.

Stevens, C., Coch, D., Sanders, L., & Neville, H. (2008). Neural mechanisms of selective auditory attention are enhanced by computerized training: Electrophysiological evidence from language-impaired and typically developing children. *Brain Res,* (*1205*), 55–69.

Stevens, C., Harn, B., Chard, D., Currin, J., Parisi, D., & Neville, H. (2013). Examining the role of attention and instruction in at-risk kindergarteners: Electrophysiological measures of selective auditory attention before and after an early literacy intervention. *J Learn Disabil, 46,* 73–86.

Stevens, C., & Neville, H. (2006). Neuroplasticity as a double-edged sword: Deaf enhancements and dyslexic deficits in motion processing. *J Cogn Neurosci, 18*(5), 701–704.

Stevens, C., Paulsen, D., Yasen, A., Mitsunaga, L., & Neville, H. (2012). ERP evidence for attenuated auditory recovery cycles in children with Specific Language Impairment (SLI). *Brain Res, 1438,* 35–47.

Stevens, C., Sanders, L., & Neville, H. (2006). Neurophysiological evidence for selective auditory attention deficits in children with specific language impairment. *Brain Res, 1111,* 143–152.

Talcott, J. B., Hansen, P. C., Assoku, E. L., & Stein, J. F. (2000). Visual motion sensitivity in dyslexia: Evidence for temporal and energy integration deficits. *Neuropsychologia, 38*(7), 935–943.

Tallal, P. (1975). Perceptual and linguistic factors in the language impairment of developmental dysphasics: An experimental investigation with the Token Test. *Cortex, 11,* 196–205.

Tallal, P. (1976). Rapid auditory processing in normal and disordered language development. *J Speech Hear Res, 19,* 561–571.

Tallal, P., & Piercy, M. (1974). Developmental aphasia: Rate of auditory processing and selective impairment of consonant perception. *Neuropsychologia, 12,* 83–93.

Talsma, D., Senkowski, D., Soto-Faraco, S., & Woldorff, M. (2010). The multifaceted interplay between attention and multisensory integration. *Trends Cogn Sci, 14,* 400–410.

Ungerleider, L. G., & Haxby, J. V. (1994). "What" and "where" in the human brain. *Curr Opin Neurobiol, 4,* 157–165.

Weber-Fox, C. M., & Neville, H. J. (1996). Neural systems for language processing: Effects of delays in second language exposure. *Brain Cogn, 30,* 264–265.

Woods, D. (1990). The physiological basis of selective attention: Implications of event-related potential studies. In J. Rohrbaugh, R. Parasuraman, & R. Johnson (Eds.), *Event-related brain potentials: Issues and interdisciplinary vantages.* New York, NY: Oxford University Press.

Yamada, Y., Stevens, C., Harn, B., Chard, D., & Neville, H. (2011). Emergence of the neural network for reading in five-year-old beginning readers of different levels of early literacy abilities: An fMRI study. *NeuroImage, 57,* 704–713.

Ziegler, J. C., Pech-Georgel, C., George, F., Alanio, F. X., & Lorenzi, C. (2005). Deficits in speech perception predict language learning impairment. *Proc Natl Acad Sci USA, 102*(39), 14110–14115.

Zukier, H., & Hagen, J. W. (1978). The development of selective attention under distracting conditions. *Child Dev, 49,* 870–873.

14 Auditory Cortical Plasticity as a Consequence of Experience and Aging

GREGG H. RECANZONE

ABSTRACT The auditory cortex is well known to show plasticity in its functional organization in the adult as a consequence of peripheral denervation, behavioral training, conditioning, and other manipulations. It is also well established that hearing deficits, both peripheral and central, are a common consequence of natural aging. What remains unclear is how these two phenomena are interrelated. This chapter will review the evidence for adult cortical plasticity in young adult animals and compare these functional changes that occur at younger ages with those seen at both the perceptual and neural levels in a natural aging population.

Adult mammals are continuously changing throughout their lifetime by adding new repertoires of behaviors while losing others. Nervous systems are also under a constant state of change, which should be no surprise to those of us who attribute behaviors and perceptions to the nervous system. There are a myriad of studies on neural development, from the first formation of the neural tube to the final myelination of the frontal lobes. Focus on the development of the cerebral cortex includes neurogenesis, the development of polarity, synaptogenesis, migration, and critical periods of sensory perception (e.g., see Barnes & Polleux, 2009; Clowry, Molnar, & Rakic, 2010; Kanold & Luhmann, 2010; Levelt & Hubener, 2012). These studies have given us a great perspective on how the nervous system is made and what constraints are imposed during its development. Following development, adult cortical plasticity studies have shown that injury, training, or other means of altering peripheral input will reorganize the functional representations of the sensory epithelium in the cerebral cortex (e.g., see Feldman, 2009; Irvine, Rajan, & McDermott, 2000; Pascual-Leone, Amedi, Fregni, & Merabet, 2005). These studies are largely done in relatively young adult animals, and represent changes in the brain that occur over the course of weeks or months. As with developmental studies, these studies have also provided us with great insights into the general function and constraints of the nervous system.

One constant in the life of an animal, however, is that it continuously ages once the developmental periods are over. Aging affects virtually all systems of the body, causing the central nervous system (which is also undergoing aging effects on its own) to adapt and reorganize to accommodate these changes (e.g., Davis & Leathers, 1985; Gems & Partridge, 2013). For example, changes in the flexibility of joints and hypotrophy of the muscles will necessarily influence how cortical motor circuits have to behave in order to produce the same (or similar) motor functions (Ren, Wu, Chan, & Yan, 2013). Nonneural effects of aging in sensory systems will also dramatically affect central nervous system processing. A simple example includes presbyopia, where aging causes the lens of the eye to become stiffer and less able to image near objects on the retina (Truscott, 2009). Similar affects can be noted for other sensory systems as well, all of which will require that the central nervous system in some way adapts to these changes to provide percepts that are as similar as possible to those for the same stimuli at a younger stage in life.

Neural changes also occur in aging, the most popularly attributed to aging being Alzheimer's and Parkinson's diseases. These are neurodegenerative diseases affecting the higher-order temporal and frontal cortical areas and the basal ganglion. However, there have been far fewer studies of the effects of aging on unisensory areas, for example in primary sensory cortical areas A1, V1, and S1 (David-Jürgens, Churs, Berkefeld, Zepka, & Dinse, 2008; David-Jürgens & Dinse, 2010; de Villers-Sidani et al., 2010; Engle & Recanzone, 2012; Juarez-Salinas, Engle, Navarro, & Recanzone, 2010; Mendelson & Ricketts, 2001; Recanzone, Engle, & Juarez-Salinas, 2011; Schmolesky, Wang, Pu, & Leventhal, 2000; Yang et al., 2008; Yu, Wang, Li, Zhou, & Leventhal, 2006; Zhang et al., 2008). Any age-related effects seen at this level must be at least partially due to the age-related differences in the periphery (hearing loss, cataracts, etc.) but it remains a largely open question as to the extent to which the primary sensory areas are able to compensate for these peripheral changes. What is also virtually unstudied is how natural aging influences the transfer of information throughout the cortical hierarchy. It remains unknown whether it is the case that once the compromised input is initially processed in the primary sensory area, there is relatively normal processing across cortical areas until one reaches the higher-order areas where senile plaques and neurofibrillary tangles begin to compromise this processing. Alternatively, it could be that each stage in cortical processing is compromised, leading to very poor and unrefined input to the higher cortical areas, which in itself could contribute to the formation of neurofibrillary tangles and senile plaques. This chapter will begin to address these questions in the auditory system, largely of the nonhuman primate, and draw a few examples from other systems where appropriate and available.

Historical perspective of auditory cortical plasticity

Cortical plasticity in adults is a now widely accepted phenomenon that can be induced by injury, training, and manipulations of the neuromodulatory circuits such as acetylcholine (see citations, above). In auditory cortical processing, one of the first influential reports of adult cortical plasticity was done by Robertson and Irvine (1989), who showed that damage to the high-frequency region of one cochlea in the guinea pig resulted in an expansion of the cortical representation of frequencies spared at the edge of the lesion (figure 14.1A and 14.1B). The Irvine group extended these studies in cats (Rajan, Irvine, Wise, & Heil, 1993), and subsequently demonstrated similar changes in the thalamus (Kamke, Brown, & Irvine, 2003) and inferior colliculus (Irvine, Rajan, & Smith, 2003). Similar studies using bilateral lesions in the cochleae in monkeys are consistent in these findings: namely, there is an overrepresentation of the spared frequencies at the edge of the cochlear lesion (Schwaber, Garraghty, & Kaas, 1993). This is particularly interesting as it stands in contrast to what is seen in S1 following the amputation of a single digit, which is an analogous peripheral denervation (Merzenich et al., 1984). In that case, amputation of the middle finger resulted in the expansion of the cortical representations of the ring and index fingers, but the fine topography was maintained; that is, there was no larger representation of the skin of those two fingers closest to the amputated one. Similar experiments in primary visual cortex, where a scotoma was produced in the retina either binocularly or monocularly with the other eye enucleated, showed a similar pattern as in A1 after the cochlear lesions. In this case the region that formerly represented the visual field in the scotoma came to represent the visual field locations at the spared edge of the scotoma (see Gilbert & Li, 2012), again without a continuous and orderly retinotopic representation in cortex.

One interesting feature of these now classic studies is that they were performed in young animals after relatively brief periods of time between the peripheral manipulation and the cortical mapping procedures. In natural aging, however, the "peripheral manipulation" that is experienced is quite different. Aging deficits result in a denervation that is gradual, and more similar to multiple small acute denervations. With respect to hearing, the hearing loss can occur over the span of multiple years, and even decades, as opposed to one acute event. Therefore, the amount of time that the central nervous system has to adapt to age-related changes is also very long, and has very small changes in the periphery to adjust to at any given time. Since there are very few studies that can mimic the long range of time between manipulation and effect, it is unclear how much of an impact these two factors could have on cortical reorganization and the extent to which acute manipulations can be used to interpret these aging affects. The few studies that were done in the Silver Spring monkeys, in which the time between the acute manipulation and subsequent effects was over the course of many years, indicate that cortical plasticity can occur over a much larger spatial scale than was previously predicted based on acute studies (Jones & Pons, 1998; see Jones, 2000). The question arises, then, whether natural aging is better modeled by these long-term changes as seen in the Silver Spring monkeys, which would predict that there is a more topographic re-representation of the lesioned cochlea than is seen in the more acute denervation studies. A schematic example of this is shown in figure 14.1, where it is possible that natural aging resulting in a high-frequency hearing loss would have the same cortical consequences as acute cochlear lesions (figure 14.1B), or result in a more topographic reorganization similar to that following digit amputation (figure 14.1C). Natural aging studies may be the best model system for this long-term cortical plasticity, and it may be crucial to have an animal model that ages sufficiently slowly, so that these changes have time to occur.

A second class of studies of auditory cortical plasticity has taken a more "top-down" approach. Many of these studies originated in the Merzenich laboratory and used either operant conditioning (Recanzone, Merzenich, & Schneiner, 1993) or basal forebrain microstimulation paired with acoustic stimulation (Kilgard & Merzenich, 1998a) to effectively change cortical maps. Using operant conditioning, the representation of frequencies that the animal had to discriminate from a

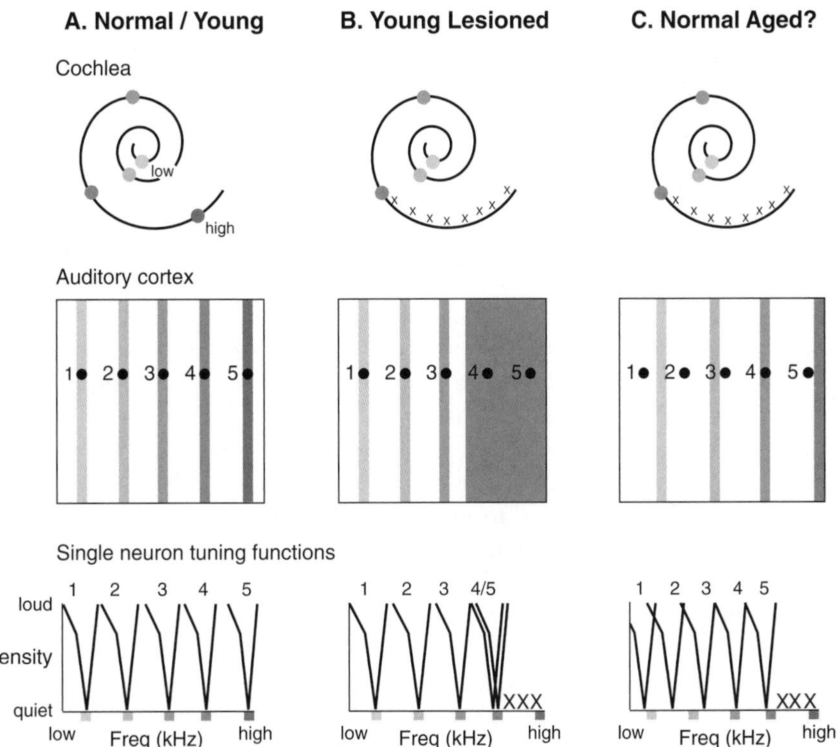

FIGURE 14.1 Cortical map changes after cochlear damage. (A) Schematic representation of the cochlea (top) with low frequencies represented at the apex (red) and progressively higher frequencies represented toward the basal end (blue). The middle panel shows a schematic of the primary auditory cortex and the isofrequency bands where neurons have similar frequency tuning corresponding to the different cochlear locations. Black dots (numbered 1–5) denote recording sites that correspond to the tuning functions of individual neurons shown in the bottom row. In the normal animal, there is an orderly and progressive shift of best frequency across A1. (B) Results from lesions of the basal turn of the cochlea (high frequencies, denoted by X) and the subsequent expansion of the representation of the frequencies at the edge of the lesion, with little change in the representation of the other frequencies. (C) One possible outcome of high-frequency hearing loss as a consequence of natural aging, where the representation of many frequencies are altered to regain an orderly, topographic representation in A1. (See color plate 10.)

standard frequency increased, and this increase was well correlated with how well the individual animals performed the perceptual task. An important finding was that the animals had to be attending to these stimuli, as attending to unrelated tactile stimulation resulted in no such cortical plasticity (Recanzone et al., 1993). Similarly in rats, basal forebrain stimulation, which causes a release of acetylcholine in the auditory cortex in much the same way that attention does, also resulted in a broad increase in the representation of the paired frequency (Kilgard & Merzenich, 1998a). Subsequent studies showed that the cortical representation of the temporal aspects of acoustic stimuli were similarly plastic (Kilgard & Merzenich, 1998b; Kilgard et al., 2001). These microstimulation and operant conditioning experiments indicate that attentional mechanisms, or "top-down" control, are critical for changes in perception and cortical plasticity, and suggest that attentional deficits will also strongly influence the magnitude and effectiveness of cortical plasticity mechanisms.

With natural aging, one of the common deficits noted is in "executive control" (Salthouse, 2012). Geriatric individuals show deficits in maintaining attention, in allocating attention to one set of stimuli in the presence of distractors, and in allocating attention as a function of the number of distractors (Li, Gratton, Fabiani, & Knight, 2013; Fabiani, 2012). Each of these deficits would undoubtedly result in both a weakened perception of the relevant stimuli, but also in a decrease in the potential for cortical plasticity mechanisms to enhance the perception of those stimuli. Thus it may be difficult to disambiguate how much of a perceptual deficit is the result of bottom-up influences from the sensory periphery and top-down influence from decreased executive control. Again, both are likely to be in operation simultaneously, and given the unique location of the auditory cortex with multiple ascending auditory nuclei processing information in the brainstem, midbrain, and thalamus, and the multiple feedback connections from the frontal and parietal lobes, the primary auditory cortex may be in an optimal position to allow us to delve into the interaction between these two classes of deficits.

Bottom-up changes in the ascending auditory pathways

One of the leading perceptual deficits as a consequence of natural aging is age-related hearing loss (Brandt & Fozard, 1990; Cruickshanks et al., 1998; Fetoni, Picciotti, Paludetti, & Troiani, 2011; Scholtz et al., 2001). This audiometric hearing loss is characterized by an upward shift in the audiogram, indicating that the individuals simply can't hear many sounds unless they are louder than necessary for unaffected individuals to hear. It had originally been proposed that certain characteristic shifts in the audiograms corresponded to specific deficits in the inner ear (Ramadan & Schuknecht, 1989; Schuknecht, 1955, 1964; Schuknecht & Gacek, 1993). This idea has been difficult to test directly, as the availability of human audiograms and cochleae for histological processing is very low (see Nelson & Hinojosa, 2006; Ohlemiller, 2004). More recently, the same style of experiment has been conducted in macaque monkeys (Engle, Tinling, & Recanzone, 2013), which age at approximately three times the rate of humans (Davis & Leathers, 1985). In this case, audiograms were measured using the auditory brainstem response (ABR), which can be derived when the animal is anesthetized. Four different metrics in the Organ of Corti were quantified and compared to the hearing deficits as well as the age of the different monkeys. These four anatomical metrics were (1) the presence of inner hair cells, (2) the presence of outer hair cells, (3) the density of spiral ganglion cells, and (4) the thickness of the stria vascularis. The main finding was that there was a great deal of individual variability in the types of inner ear histopathologies observed between the different animals. Thus, animals that had very similar hearing deficits could have very different deficits in the cochlea; for example, one could have a reduction in the number of spiral ganglion cells, while another could have fewer outer hair cells. The one thing that did effectively correlate with age was not which particular deficit existed, but rather how many different deficits were there. Thus, as aging progresses, different elements begin to degrade independently with different individuals showing different cochlear degradations over time.

These cochlear studies have shown that the input into the ascending auditory system has decreased by a variety of different mechanisms as a consequence of aging, but this decreased input is not nearly as structured as specific cochlear lesions, either mechanical or chemical, that have been used experimentally (cited above). Studies in rodent models of aging have investigated several aspects of how the auditory brainstem and midbrain potentially compensate for this decreased afferent input (Caspary, Hughes, Schatteman, & Turner, 2006; Caspary, Ling, Turner, & Hughes, 2008; Caspary, Milbrandt, & Helfert, 1995; Caspary, Schatteman, & Hughes, 2005; Palombi & Caspary, 1996). One of the key findings is that there is a decrease in GABAergic inhibition at most of the auditory relay nuclei including the cochlear nucleus, the superior olivary complex, and the inferior colliculus. This decreased inhibition predicts an increased level of both spontaneous and driven activity. This prediction has been shown physiologically in the inferior colliculus of aged rats (Simon, Frisina, & Walton, 2004; Walton, Barsz, & Wilson, 2008; Walton, Frisina, Ison, & O'Neill, 1997). In these studies, the overall neuronal activity in the inferior colliculus in anesthetized aged rats is much higher than that seen in younger animals. Importantly, while there is an increase in driven activity as well, this increase is not selective for different acoustic features. For example, activity mechanisms that can plausibly encode the presence of gaps in an otherwise continuous stimulus are severely compromised. This sort of altered neural activity presumably underlies the increased gap-detection thresholds seen in aged humans (Schneider, Pichora-Fuller, Kowalchuk, & Lamb, 1994; Schneider & Hamstra, 1999; Snell, 1997). Gap-detection ability is a key component in speech comprehension, so it is not surprising that gap-detection and speech-recognition thresholds are also tightly correlated (Snell & Frisina, 2000; Snell, Mapes, Hickman, & Frisina, 2002).

Similar studies have been conducted in macaque monkeys and show that the same basic processes also occur. These studies indicate that there is a shift in the expression of different calcium binding proteins, specifically parvalbumin, which has been implicated in the excitatory and inhibitory circuitry in these auditory nuclei (Gray, Engle, & Recanzone, 2014a, 2014b; Gray, Rudolph, Engle, & Recanzone, 2013). Thus, there appear to be some neuroanatomical compensatory mechanisms in place to help offset the decreased excitatory input to the cortex. Similarly, responses to gap stimuli in auditory cortex reflect these changes in gap detection in humans (Recanzone et al., 2011), and in increases in both spontaneous and driven activity (Juarez-Salinas et al., 2010).

Auditory cortical plasticity with aging

Very few studies have investigated the effects of natural aging on the auditory cortex (De Villers-Sidani et al., 2010; Hughes, Turner, Parrish, & Caspary, 2010;

Juarez-Salinas et al., 2010; Mendelson & Ricketts, 2001; Turner, Hughes, & Caspary, 2005). There is good evidence that aging humans show spatial-processing deficits (Abel, Giguere, Consoli, & Papsin, 2000; Brown, 1984; Kubo et al., 1998) as well as temporal-processing deficits beyond gap detection that include speech-processing deficits (Barsz, Ison, Snell & Walton, 2002; Frisina & Frisina, 1997; Gordon-Salant & Fitzgibbons, 1993, 2001; Kubo et al., 1998; Phillips, Gordon-Salant, Fitzgibbons, & Yeni-Komshian, 2000; Snell et al., 2002; Strouse, Ashmead, Ohde, & Grantham, 1998). Animal models have shown similar results, and lesion studies in animals indicate that the auditory cortex is necessary for these auditory perceptions (Harrington, Heffner, & Heffner, 2001; Heffner & Heffner, 1989, 1990). It is therefore prudent to investigate the auditory nervous system at this level in animal models, where single-neuron recordings can be made in the alert state.

One difficulty in studying aging, regardless of whether it is in humans or in an animal model, is in separating the effects of aging from the effects of hearing loss, as the two are highly correlated (references cited above). However, it is possible to find aged humans and animals that have similar audiometric hearing as younger subjects (i.e., the peripheral circuits are intact), in which case any hearing deficits must be related to changes in the central auditory processing network. Two studies on the spatial processing of auditory cortical neurons were conducted on young and aged macaque monkeys that showed similar behavioral audiograms (Engle & Recanzone, 2012; Juarez-Salinas et al., 2010). The primate auditory cortex is made up of multiple cortical fields and is organized into a series of core-belt-parabelt fields (Galaburda & Sanides, 1980; Hackett, Preuss, & Kaas, 2001; Kaas & Hackett, 2000; Pandya, 1995; Petkov, Kayser, Augath, & Logothetis, 2006). These fields are anatomically interconnected with their neighbors but are not connected beyond that; that is, core and belt fields are interconnected, but core and parabelt fields are not (Kaas & Hackett, 2000). It has been hypothesized that there are different processing streams, one for the spatial features and the other for the non-spatial features of the stimulus (Rauschecker, 1998; Rauschecker & Scott, 2009; Rauschecker & Tian, 2000; Recanzone & Cohen, 2010). Previous studies in young animals had shown the spatial receptive fields were broadly tuned in the core area A1, but much more sharply tuned in the caudal belt field known as the caudolateral field, or CL (Recanzone, Guard, Phan, & Su, 2000; Woods, Lopez, Long, Rahman, & Recanzone, 2006). This spatial tuning was the sharpest of all areas investigated, including other core and belt fields. Further studies showed that the spike rate information across the population of CL neurons, but in no other area tested, was sharp enough to account for sound localization performance in humans (Miller & Recanzone, 2009).

When the responses of neurons in the auditory cortex of aged monkeys were tested, there were several key findings that are consistent with the reduced GABAergic inhibition seen at subcortical areas and the decreased sound localization acuity common in the aged. First, the spontaneous activity in both A1 and CL was much greater in the aged animals than was seen in the young animals (Juarez-Salinas et al., 2010). Second, the amount of driven activity was also greater in both cortical areas. This increase average was several-fold higher when compared to the younger animals. Similar findings have been made in the primary visual cortex (V1) and the motion processing area MT of anesthetized animals (Yang et al., 2008; Yu et al., 2006; Zhang et al., 2008). In visual cortex, there was also a broadening in orientation and direction of motion tuning in V1 and MT, respectively, in the aged animals. The spatial tuning of auditory cortical neurons was also broader in aged monkeys compared to younger ones. One of the more salient findings was that these aging effects were relatively small in A1, in the sense that spatial tuning metrics were not greatly reduced. However, in young animals there was a considerable narrowing in the spatial tuning in CL compared to A1. In older animals, this sharpening of spatial tuning was almost completely eliminated. Indeed, across the population of neurons in the aged animals, there was very little difference in spatial tuning between A1 and CL neurons under these task conditions. This indicates that there is a breakdown in the cortical processing between hierarchically connected areas, resulting in a significant reduction in the feature extraction that normally occurs.

The next question addressed how this reduction could take place. When looking across the population responses, one feature of note was that in younger animals there was an increase in the latency of the response in area CL compared to A1 (Engle & Recanzone, 2012). This latency difference was gone in the aged animals, since the first spike latencies were the same between the population of A1 and CL neurons, and in fact were less than the first spike latencies of A1 neurons in younger animals. A second feature that was interesting was that in younger animals there was a very short latency inhibition of the response at locations beyond the best direction in CL neurons. This short latency inhibition was completely absent in older animals. Thus, consistent with a lack of inhibition in the subcortical processing areas, there was a clear lack of inhibitory drive at the level of the auditory cortex. These results indicate that spatial tuning in the normal auditory cortex is heavily dependent on, and shaped by, inhibitory mechanisms, and these inhibitory mechanisms are lost as a consequence of natural aging.

A second series of studies investigated the temporal response properties of these neurons. Hallmarks of age-related hearing deficits are noted in gap detection and the discrimination of amplitude-modulated sounds. With respect to amplitude-modulated sounds, several studies have investigated how well cortical responses in aged animals are able to follow the envelope of the stimulus. In one key study, this degradation in the ability to encode the envelope of the amplitude-modulated signal in aged animals was restored following a period of behavioral training (de Villers-Sidani et al., 2010). This neural plasticity is very much in line with that seen in younger animals following behavioral training as described above. This indicates that, while there are a myriad of changes in the ascending auditory pathways as a consequence of aging, the same neural plasticity mechanisms that are correlated with improvements in perceptual abilities can counteract these changes, at least to some degree.

Conclusion

Decades of cortical plasticity research have focused on the assumption that this plasticity underlies skill acquisition and learning in adults. While this is undoubtedly true, this type of plasticity is also likely a focal point in compensatory mechanisms to overcome peripheral changes as a consequence of natural aging. It remains unclear how well acute manipulations in animal models will reflect this natural aging process, and it may be that the time course of peripheral degradation and long-term neural changes result in qualitative differences in the reorganization of cortical representations and the consequent perceptual deficits and recoveries.

ACKNOWLEDGMENT This work was supported in part by NIH grant AG034137.

REFERENCES

Abel, S. M., Giguere, C., Consoli, A., & Papsin, B. C. (2000). The effect of aging on horizontal plane sound localization. *J Acoust Soc Am, 108,* 743–752.

Barnes, A. P., & Polleux, F. (2009). Establishment of axon-dendrite polarity in developing neurons. *Annu Rev Neurosci, 32,* 347–381.

Barsz, D., Ison, J. R., Snell, K. B., & Walton, J. P. (2002). Behavioral and neural measures of auditory temporal acuity in aging humans and mice. *Neurobiol Aging, 23,* 565–578.

Brandt, L. J., & Fozard, J. L. (1990). Age changes in pure-tone hearing thresholds in a longitudinal study of normal human aging. *J Acoust Soc Am, 88,* 813–820.

Brown, C. H. (1984). Directional hearing in aging rats. *Exp Aging Res, 10,* 35–38.

Caspary, D. M., Schatteman, T. A., & Hughes, L. F. (2005). Age-related changes in the inhibitory response properties of dorsal cochlear nucleus output neurons: Role of inhibitory inputs. *J Neurosci, 25,* 10952–10959.

Caspary, D. M., Hughes, L. F., Schatteman, T. A., & Turner, J. G. (2006). Age-related changes in the response properties of cartwheel cells in rat dorsal cochlear nucleus. *Hearing Res, 217,* 207–215.

Caspary, D. M., Ling, L., Turner, J. G., & Hughes, L. F. (2008). Inhibitory neurotransmission, plasticity and aging in the mammalian central auditory system. *J Exp Biol, 211,* 1781–1791.

Caspary, D. M., Milbrandt, J. C., & Helfert, R. H. (1995). Central auditory aging: GABA changes in the inferior colliculus. *Exp Gerontol, 30,* 349–360.

Clowry, G., Molnar, Z., & Rakic, P. (2010). Renewed focus on the developing human neocortex. *J Anat, 217,* 276–288.

Cruickshanks, K. J., Wiley, T. L., Tweed, T. S., Klein, B. E. K., Klein, R., ... Nondahl, D. M. (1998). Prevalence of hearing loss in older adults in Beaver Dam, Wisconsin. The Epidemiology of Hearing Loss study. *Am J Epidemiol, 148,* 879–886.

David-Jürgens, M., Churs, L., Berkefeld, T., Zepka, R. F., & Dinse, H. R. (2008). Differential effects of aging on fore- and hindpaw maps of rat somatosensory cortex. *PLos One, 3,* e3399.

David-Jürgens, M., & Dinse, H. R. (2010). Effects of aging on paired-pulse behavior of rat somatosensory cortical neurons. *Cereb Cortex, 20,* 1208–1216.

Davis, R. T., & Leathers, C. W. (1985). *Behavior and pathology of aging in rhesus monkeys.* New York, NY: A. R. Liss.

De Villers-Sidani, E., Alzghoul, L., Zhou, X., Simpson, K. L., Lin, R. C., & Merzenich, M. M. (2010). Recovery of functional and structural age-related changes in the rat primary auditory cortex with operant training. *Proc Natl Acad Sci USA, 107,* 13900–13905.

Engle, J. R., & Recanzone, G. H. (2012). Characterizing spatial tuning functions of neurons in the auditory cortex of young and aged monkeys: A new perspective on old data. *Front Aging Neurosci, 4,* 36.

Engle, J. R., Tinling, S., & Recanzone, G. H. (2013). Age-related hearing loss in rhesus monkeys is correlated with cochlear histopathologies. *PLoS One, 8,* e55092.

Fabiani, M. (2012). It was the best of times, it was the worst of times: A psychophysiologist's view of cognitive aging. *Psychophysiology, 49,* 283–304.

Feldman, D. E. (2009). Synaptic mechanisms for plasticity in neocortex. *Annu Rev Neurosci, 32,* 33–55.

Fetoni, A. R., Picciotti, P. M., Paludetti, G., & Troiani, D. (2011). Pathogenesis of presbycusis in animal models: A review. *Exp Gerontol, 46,* 413–425.

Frisina, D. R., & Frisina, R. D. (1997). Speech recognition in noise and presbycusis: Relations to possible neural mechanisms. *Hearing Res, 106,* 95–104.

Galaburda, A., & Sanides, F. (1980). Cytoarchitectonic organization of the human auditory cortex. *J Comp Neurol, 190,* 597–610.

Gems, D., & Partridge, L. (2013). Genetics of longevity in model organisms: Debates and paradigm shifts. *Annu Rev Physiol, 75,* 621–644.

Gilbert, C. D., & Li, W. (2012). Adult visual cortical plasticity. *Neuron, 75,* 250–264.

Gordon-Salant, S., & Fitzgibbons, P. J. (1993). Temporal factors and speech recognition performance in young and elderly listeners. *J Speech Hear Res, 36,* 1276–1285.

Gordon-Salant, S., & Fitzgibbons, P. J. (2001). Sources of age-related recognition difficulty for time-compressed speech. *J Speech Hear Res, 44,* 709–719.

Gray, D. T., Engle, J. R., & Recanzone, G. H. (2014a). Age-related neurochemical changes in the rhesus macaque superior olivary complex. *J Comp Neurol, 522,* 573–591.

Gray, D. T., Engle, J. R., & Recanzone, G. H. (2014b). Age-related neurochemical changes in the rhesus macaque cochlear nucleus. *J Comp Neurol, 522,* 1527–1541.

Gray, D. T., Rudolph, M. L., Engle, J. R., & Recanzone, G. H. (2013). Parvalbumin increases in the medial and lateral geniculate nuclei of aged rhesus monkeys. *Front Aging Res, 5,* 69.

Hackett, T. A., Preuss, T. M., & Kaas, J. H. (2001). Architectonic identification of the core region in auditory cortex of macaques, chimpanzees, and humans. *J Comp Neurol, 441,* 197–222.

Harrington, I. A., Heffner, R. S., & Heffner, H. E. (2001). An investigation of sensory deficits underlying the aphasia-like behavior of macaques with auditory cortex lesions. *NeuroReport, 12,* 1217–1221.

Heffner, H. E., & Heffner, R. S. (1989). Effect of restricted cortical lesions on absolute thresholds and aphasia-like deficits in Japanese macaques. *Behav Neurosci, 103,* 158–169.

Heffner, H. E., & Heffner, R. S. (1990). Effect of bilateral auditory cortex lesions on sound localization in Japanese macaques. *J Neurophysiol, 64,* 915–931.

Hughes L. F., Turner, J. G., Parrish, J. L., & Caspary, D. M. (2010). Processing of broadband stimuli across A1 layers in young and aged rats. *Hearing Res, 264,* 79–85.

Irvine, D. R. F., Rajan, R., & McDermott, H. J. (2000). Injury-induced reorganization in adult auditory cortex and its perceptual consequences. *Hearing Res, 147,* 188–199.

Irvine, D. R., Rajan, R., & Smith, S. (2003). Effects of restricted cochlear lesions in adult cats on the frequency organization of the inferior colliculus. *J Comp Neurol, 467,* 354–374.

Jones, E. G. (2000). Cortical and sub-cortical contributions to activity-dependent plasticity in primate somatosensory cortex. *Annu Rev Neurosci, 23,* 1–37.

Jones, E. G., & Pons, T. P. (1998). Thalamic and brainstem contributions to large-scale plasticity of primate somatosensory cortex. *Science, 282,* 1121–1125.

Juarez-Salinas, D. L, Engle, J. R., Navarro, X. O., & Recanzone, G. H. (2010). Hierarchical and serial processing in the spatial auditory cortical pathway is degraded by natural aging. *J Neurosci, 30,* 14795–14804.

Kaas, J. H., & Hackett, T. A. (2000). Subdivisions of auditory cortex and processing streams in primates. *Proc Natl Acad Sci USA, 97,* 11793–11799.

Kamke, M. R., Brown, M., & Irvine, D. R. (2003). Plasticity in the tonotopic organization of the medial geniculate body in adult cats following restricted unilateral cochlear lesions. *J Comp Neurol, 459,* 355–367.

Kanold, P. O., & Luhmann, H. J. (2010). The subplate and early cortical circuits. *Annu Rev Neurosci, 33,* 23–48.

Kilgard, M. P., & Merzenich, M. M. (1998a). Cortical map reorganization enabled by nucleus basalis stimulation. *Science, 279,* 1714–1718.

Kilgard, M. P., & Merzenich, M. M. (1998b). Plasticity of temporal information processing in the primary auditory cortex. *Nat Neurosci, 1,* 727–731.

Kilgard, M. P., Pandya, P. K., Vazquez, J., Gehi, A., Schreiner, C. E., & Merzenich, M. M. (2001). Sensory input directs spatial and temporal plasticity in primary auditory cortex. *J Neurophysiol, 86,* 326–338.

Kubo, T., Sakashita, T., Kusuki, M., Kyunai, K., Ueno, K., Hikawa, C., ... Nakai, Y. (1998). Sound lateralization and speech discrimination in patients with sensorineural hearing loss. *Acta Otolaryngol Suppl, 538,* 63–69.

Levelt, C. N., & Hubener, M. (2012). Critical-period plasticity in the visual cortex. *Annu Rev Neurosci, 35,* 309–330.

Li, L., Gratton, C., Fabiani, M., & Knight, R. T. (2013). Age-related frontoparietal changes during the control of bottom-up and top-down attention: An ERP study. *Neurobiol Aging, 34,* 477–488.

Mendelson, J. R., & Ricketts, C. (2001). Age-related temporal processing speed deterioration in auditory cortex. *Hearing Res, 158,* 84–94.

Merzenich, M. M., Nelson, R. J., Stryker, M. P., Cynader, M. S., Schoppmann, A., & Zook, J. M. (1984). Somatosensory cortical map changes following digit amputation in adult monkeys. *J Comp Neurol, 224,* 591–605.

Miller, L. M., & Recanzone, G. H. (2009). Populations of auditory cortical neurons can accurately encode acoustic space across stimulus intensity. *Proc Natl Acad Sci USA, 106,* 5931–5935.

Nelson, E. G., & Hinojosa, R. (2006). Presbycusis: A human temporal bone study of individuals with downward sloping audiometric patterns of hearing loss and review of the literature. *Larygoscope, 116*(Suppl 112), 1–12.

Ohlemiller, K. K. (2004). Age-related hearing loss: The status of Schuknecht's typology. *Curr Opin Otolaryngol Head Neck Surg, 12,* 439–443.

Palombi, P. S., & Caspary, D. M. (1996). Responses of young and aged Fischer 344 rat inferior colliculus neurons to binaural tonal stimuli. *Hearing Res, 100,* 59–67.

Pandya, D. N. (1995). Anatomy of the auditory cortex. *Rev Neurol (Paris), 151,* 486–494.

Pascual-Leone, A., Amedi, A., Fregni, F., & Merabet, L. B. (2005). The plastic human brain cortex. *Annu Rev Neurosci, 28,* 377–401.

Petkov, C. I., Kayser, C., Augath, M., & Logothetis, N. K. (2006). Functional imaging reveals numerous fields in the monkey auditory cortex. *PLoS Biol, 4,* e215.

Phillips, S. L., Gordon-Salant, S., Fitzgibbons, P. J., & Yeni-Komshian, G. (2000). Frequency and temporal resolution in elderly listeners with good and poor word recognition. *J Speech Lang Hear R, 43,* 217–228.

Rajan, R., Irvine, D. R. F., Wise, L. Z., & Heil, P. (1993). Effect of unilateral partial cochlear lesions in adult cats on the representation of lesioned and unlesioned cochleas in primary auditory cortex. *J Comp Neurol, 338,* 17–49.

Ramadan, H. H., & Schuknecht, H. F. (1989). Is there a conductive type of presbycusis? *Otolaryngol Head Neck Surg, 100,* 30–34.

Rauschecker, J. P. (1998). Parallel processing in the auditory cortex of primates. *Audiol Neurotol, 3,* 86–103.

Rauschecker, J. P., & Scott, S. K. (2009). Maps and streams in the auditory cortex: Nonhuman primates illuminate human speech processing. *Nat Neurosci, 12,* 718–724.

Rauschecker, J. P., & Tian, B. (2000). Mechanisms and streams for processing of "what" and "where" in auditory cortex. *Proc Natl Acad Sci USA, 97,* 11800–11806.

Recanzone, G. H., & Cohen, Y. E. (2010). Serial and parallel processing in the primate auditory cortex revisited. *Behav Brain Res, 206,* 1–7.

Recanzone, G. H., Engle, J. R., & Juarez-Salinas, D. L. (2011). Spatial and temporal processing of single auditory cortical neurons and populations of neurons in the macaque monkey. *Hearing Res, 271,* 115–122.

Recanzone, G. H., Guard, D. C., Phan, M. L., & Su, T. K. (2000). Correlation between the activity of single auditory cortical neurons and sound localization behavior in the macaque monkey. *J Neurophysiol, 83,* 2723–2739.

Recanzone, G. H., Merzenich, M. M., & Schreiner, C. E. (1993). Plasticity in the frequency representation of primary auditory cortex following discrimination training in adult owl monkeys. *J Neurosci, 13,* 87–103.

Ren, J., Wu, Y. D., Chan, J. S., & Yan, J. H. (2013). Cognitive aging affects motor performance and learning. *Geriatr Gerontol Int, 13,* 19–27.

Robertson, D., & Irvine, D. R. F. (1989). Plasticity of frequency organization in auditory cortex of guinea pigs with partial unilateral deafness. *J Comp Neurol, 282,* 456–471.

Salthouse, T. (2012). Consequences of age-related cognitive declines. *Annu Rev Psychol, 63,* 201–226.

Schmolesky, M. T., Wang, Y., Pu, M., & Leventhal, A. G. (2000). Degradation of stimulus selectivity of visual cortical cells in senescent rhesus monkeys. *Nat Neurosci, 3,* 384–390.

Schneider, B. A., & Hamstra, S. J. (1999). Gap detection thresholds as a function of tonal duration for younger and older listeners. *J Acoust Soc Am, 106,* 371–380.

Schneider, B. A., Pichora-Fuller, M. K., Kowalchuk, D., & Lamb, M. (1994). Gap detection and the precendence effect in young and old adults. *J Acoust Soc Am, 95,* 980–991.

Scholtz, A. W., Kammen-Jolley, K., Felder, E., Hussl, B., Rask-Andersen, H., & Schrott-Fischer, A. (2001). Selective aspects of human pathology in high-tone hearing loss of the aging inner ear. *Hearing Res, 157,* 77–86.

Schuknecht, H. F. (1955). Presbycusis. *Laryngoscope, 65,* 402–419.

Schuknecht, H. F. (1964). Further observations on the pathology of presbycusis. *Arch Otolaryngol, 80,* 369–382.

Schuknecht, H. F., & Gacek, M. R. (1993). Cochlear pathology in presbycusis. *Ann Otol Rhinol Laryngol, 102,* 1–16.

Schwaber, M. K., Garraghty, P. E., & Kaas, J. H. (1993). Neuroplasticity of the adult primate auditory cortex following cochlear hearing loss. *Am J Otolaryngol, 14,* 252–258.

Simon, H., Frisina, R. D., & Walton, J. P. (2004). Age reduces response latency of mouse inferior colliculus neurons to AM sounds. *J Acoust Soc Am, 116,* 469–477.

Snell, K. B. (1997). Age-related changes in temporal gap detection. *J Acoust Soc Am, 101,* 2214–2220.

Snell, K. B., & Frisina, D. R. (2000). Relationships among age-related differences in gap detection and word recognition. *J Acoust Soc Am, 107,* 1615–1626.

Snell, K. B., Mapes, F. M., Hickman, E. D., & Frisina, D. R. (2002). Word recognition in competing babble and the effects of age, temporal processing, and absolute sensitivity. *J Acoust Soc Am, 112,* 720–727.

Strouse, A., Ashmead, D. H., Ohde, R. N., & Grantham, D. W. (1998). Temporal processing in the aging auditory system. *J Acoust Soc Am, 104,* 2385–2399.

Truscott, R. J. (2009). Presbyopia. Emerging from a blur towards an understanding of the molecular basis for this most common eye condition. *Exp Eye Res, 88,* 241–247.

Turner, J. G., Hughes, L. F., & Caspary, D. M. (2005). Affects of aging on receptive fields in rat primary auditory cortex layer V neurons. *J Neurophysiol, 94,* 2738–2747.

Walton, J. P., Barsz, K., & Wilson, W. W. (2008). Sensorineural hearing loss and neural correlates of temporal acuity in the inferior colliculus of the C57BL/6 mouse. *J Assoc Res Otolaryngol, 9,* 90–101.

Walton, J. P., Frisina, R. D., Ison, J. R., & O'Neill, W. E. (1997). Neural correlates of behavioral gap detection in the inferior colliculus of the young CBA mouse. *J Comp Physiol A, 181,* 161–176.

Woods, T. M., Lopez, S. E., Long, J. H., Rahman, J. E., & Recanzone, G. H. (2006). Effects of stimulus azimuth and intensity on the single-neuron activity in the auditory cortex of the alert macaque monkey. *J Neurophysiol, 96,* 3323–3337.

Yang, Y., Liang, Z., Li, G., Wang, Y., Zhou, Y., & Leventhal, A. G. (2008). Aging affects contrast response functions and adaptation of middle temporal visual area neurons in rhesus monkeys. *Neuroscience, 156,* 748–757.

Yu, S., Wang, Y., Li, X., Zhou, Y., & Leventhal, A. G. (2006). Functional degradation of extrastriate visual cortex in senescent rhesus monkeys. *Neuroscience, 140,* 1023–1029.

Zhang, J., Wang, X., Wang, Y., Fu, Y., Liang, Z., Ma, Y., & Leventhal, A. G. (2008). Spatial and temporal sensitivity degradation of primary visual cortical cells in senescent rhesus monkeys. *Eur J Neurosci, 28,* 201–207.

15 The Contribution of Genetics, Epigenetics, and Early Life Experiences to Behavior

JENNIFER BLAZE, ERIC D. ROTH, AND TANIA L. ROTH

ABSTRACT A continuing challenge in neuroscience is elucidating complex relationships between nature and nurture. Both genetic polymorphisms and environmental stressors are recognized for their significant roles in producing behavioral abnormalities and psychiatric disorders. The mechanisms by which environmental factors may physically interact with genotype are likely varied, but increasing evidence argues for a prominent role of DNA methylation. This chapter will focus on the role that polymorphisms and epigenetic factors play in behavior.

For decades, psychologists and neuroscientists have realized that early life experiences have far-reaching effects on behavior. Stressful experiences and adverse social environments during sensitive periods of development have been associated with a range of negative outcomes. These include structural changes in regions such as the prefrontal cortex, hippocampus, amygdala, and cerebellum and a high prevalence of anxiety and depression (Cicchetti & Toth, 2005; Hanson, Chandra, Wolfe, & Pollak, 2011; Tottenham et al., 2010). Nature-versus-nurture questions have long plagued scientists seeking to understand mechanisms responsible for behavioral development and psychiatric disorders. Is it an individual's genetic makeup, exposure to certain environmental factors, or a gene-by-environment interaction that best predicts behavioral trajectories?

After completion of the Human Genome Project in the early 2000s, investigators realized that the human genome did not possess enough genes to fully explain the vast complexities of behavior. At the same time, studies were emerging showing that not all individuals who encounter stressful events succumb to depression or anxiety. Instead, data suggested that certain individuals may be more vulnerable to stress and adversity because of so-called genetic predispositions. Such observations led us to the realization that there is a gene-by-environment interaction influencing the development of behavior. With the birth of behavioral epigenetics research in the mid-to-late 2000s, epigenetics then emerged as a leading candidate for a biological pathway linking gene-environment interactions. Our current understanding is that epigenetic alterations, acting either separately or in conjunction with genetic polymorphisms, serve as a risk factor responsible for long-term and even multigenerational trajectories in the development of behavior following stress and social adversity encountered early in development. A central theme also explored here is that epigenetic regulation of genes is not a process exclusive to developing organisms, but is an active process that affects behavior throughout life into senescence. Such observations are thus consistent with the notion that epigenetics can provide a mechanism for gene-by-environment interaction and behavior throughout the life span.

Contribution of genetic polymorphisms to behavior

Common types of functional polymorphisms studied in the context of behavior are single nucleotide polymorphisms, or SNPs, in which a single nucleotide within the DNA sequence is altered, and short tandem repeats, or STRs, which are sections of DNA with repeating patterns of short sequences (figure 15.1A). The presence of polymorphisms often produces changes in gene transcription, conferring either higher or lower levels of gene expression. A widely studied polymorphism that codes for the serotonin transporter 5-HTT is well recognized for its capacity to influence stress responsivity and mood. Encoded by the gene *SLC6A4* in humans, 5-HTT is responsible for removal and reuptake of serotonin. 5HTTLPR (see table 15.1 for abbreviations and definitions commonly used in this chapter) is a polymorphic region that has a variable number of tandem repeats yielding either short or long alleles. The short allele has reduced transcriptional efficiency compared with the long allele, ultimately leading to reduced serotonin transporter in membranes.

Table 15.1
Abbreviations and definitions

Abbreviation	Full Term	Definition
5HTTLPR	Serotonin transporter-linked polymorphic region	Region within the *5HTT* gene containing a variable number of tandem repeats
BDNF	Brain-derived neurotrophic factor	A protein crucial for neural development and nervous system plasticity, encoded by *Bdnf* gene
CRF	*Corticotrophin-releasing factor* gene	Gene encoding the CRF hormone that is involved in stress responses
DNMTs	DNA methyltransferases	Enzymes responsible for catalyzing DNA methylation
FKBP5	*FK506 binding protein 5* gene	Gene encoding a protein that controls HPA axis activity through regulation of glucocorticoid receptors
GADD45b	Growth-arrest and DNA-damage-inducible beta protein	A protein that can actively demethylate cytosines by a DNA repair-like mechanism, encoded by *Gadd45b* gene
GR	Glucocorticoid receptor	The receptor that binds cortisol (or corticosterone in animals), encoded by the *GR* (rodent) or *Nr3C1* (human) genes
HDAC	Histone deacetylase	An enzyme that removes acetyl groups from histone tails, helping to suppress gene transcription
HPA axis	Hypothalamic-pituitary-adrenal axis	A set of pathways and feedback interactions that modulate the stress response
LG	Licking and grooming	Nurturing pup-directed behaviors displayed by rat mothers
MAN2C1	*Mannosidase alpha class 2C member 1* gene	An immune-related gene that is epigenetically regulated in PTSD
MeCP2	Methyl-CpG-binding protein	A protein that binds to methylated cytosines and recruits transcriptional repressors or activators to inhibit or activate gene transcription
OXTR	Oxytocin receptor	A receptor for oxytocin that regulates stress and social behaviors, encoded by the *OXTR* gene
PACAP	Pituitary adenylate cyclase-activating polypeptide	Protein responsive to cellular stress and implicated in neurotrophic function that is implicated in PTSD
PVN	Paraventricular nucleus	A nucleus in the hypothalamus that secretes CRF during the stress response
SLC6A4	*Solute carrier family 6, member 4* gene	The gene that encodes the serotonin transporter in humans
SNP	Single nucleotide polymorphism	The alteration of a single nucleotide within a DNA sequence
STR	Short tandem repeat	Sections of DNA with repeating patterns of short sequences
TSA	Trichostatin-A	An HDAC inhibitor known to activate gene transcription
TSST	Trier Social Stress Test	A psychosocial stress regimen in which participants perform a speech and arithmetic for a panel of "interviewers"

Neuroimaging studies have revealed that healthy subjects with at least one short allele have increased temperamental anxiety and amygdala reactivity (Pezawas et al., 2005). Work with nonhuman primates and humans provide evidence for an interaction between 5HTTLPR genotype and lifetime stressors, especially early life stress, on behavioral and emotional outcomes. For example, rhesus monkeys with the short allele that were deprived of maternal care showed deficiencies in serotonergic functioning, altered stress reactivity, and aberrant emotional behavior (Barr et al., 2004; Bennett et al., 2002). In humans, the interaction of early life adversity and 5HTTLPR genotype was pioneered by Caspi, who found that 5HTTLPR genotype moderated the influence of stressful life events on depression. Specifically, individuals with one or two copies of the short allele exhibited more depressive symptoms in relation to stressful life events, especially child maltreatment (Caspi et al., 2003).

The gene encoding the brain-derived neurotrophic factor protein (BDNF) has another commonly studied polymorphism. A polymorphism in the *Bdnf* gene causes an amino acid substitution (driven by a G/A substitution) of valine to methionine at codon 66 (Val66Met), which ultimately leads to reduced secretion of BDNF. Because BDNF is crucial for neural development and central nervous system plasticity, alterations in its secretion have been well linked to numerous cognitive deficits and neuropsychiatric disorders (Bath & Lee, 2006). Independent of environmental factors, the Met

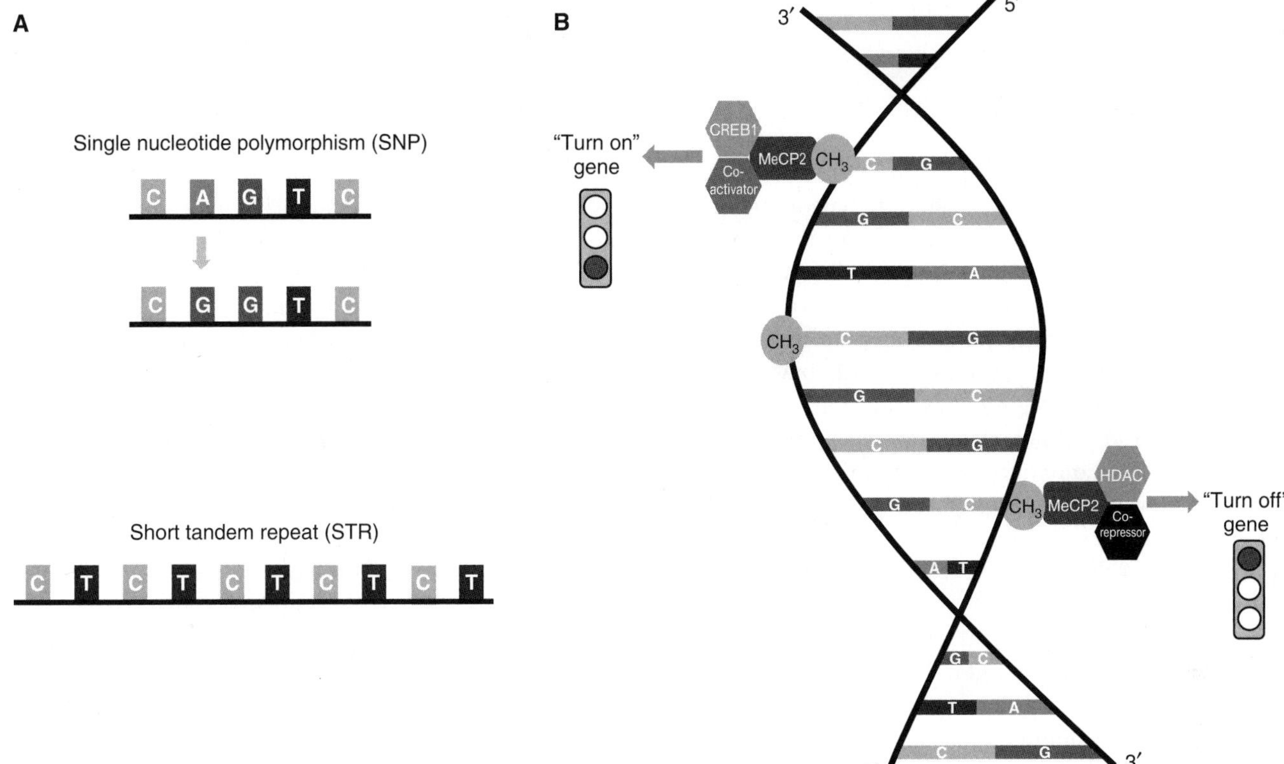

FIGURE 15.1 Schematic of common gene polymorphisms and DNA methylation. (A) Single nucleotide polymorphisms (SNPs) consist of a single nucleotide variation between individuals. Short tandem repeats (STRs) contain multiple repeats of a certain short sequence of nucleotides. (B) DNA methylation refers to the addition of methyl groups to specific cytosine residues. MeCP2 binds to methylated DNA and can either recruit HDACs and other co-repressors to suppress gene transcription or can recruit cAMP response element-binding (CREB) protein and other transcriptional co-activators to enhance gene expression. (See color plate 11.)

allele has been associated with poor episodic memory and abnormal hippocampal activation (Egan et al., 2003). An important study from B. J. Casey's lab showed that the Val66Met polymorphism is biologically and behaviorally relevant across species, since both humans and genetic knock-in mice showed abnormal neural activity and significant impairments in fear memory extinction (Soliman et al., 2010).

The *Bdnf* polymorphism also moderates the effect of early or later-life environmental factors. In carriers of the Met allele, early life stress predicts higher levels of depression and anxiety that are linked to loss of prefrontal gray matter and reduction in hippocampal volume (Gatt et al., 2009). The Trier Social Stress Test (TSST), in which participants have to perform a speech and do mathematical problems in front of a panel of "interviewers," is an effective psychosocial stress regimen commonly used in humans to explore the significance of polymorphisms. The presence of the *Bdnf* Val/Met polymorphism has been shown to modulate hypothalamic-pituitary-adrenal (HPA) axis reactivity in a sex-specific manner in this task (Shalev et al., 2009). Specifically, Val/Met females showed greater increases in salivary cortisol during the test than Val/Val homozygotes, while Val/Val males showed a higher cortisol response than heterozygotes.

Although 5HTTLPR and Val66Met polymorphisms are most commonly investigated, many other variations confer specific behavioral phenotypes either alone or in interaction with environmental factors. For example, *FKBP5* is a gene that controls HPA axis activity through regulation of glucocorticoid receptors (GR), and various *FKBP5* SNPs have been recently linked to anxiety and depression following early life stress (Binder et al., 2008; Gillespie, Phifer, Bradley, & Ressler, 2009). It is likely that having multiple polymorphisms in combination with specific environmental events confers an even greater moderating consequence in terms of neural and behavioral outcomes. Indeed, such gene-by-gene interactions (having the Met allele of the *Bdnf* gene and two short alleles for 5HTTLPR) in conveying greater vulnerability to depression have been demonstrated in maltreated children (Kaufman et al., 2006).

Introduction to epigenetics

While the genetic polymorphism literature implicates gene-by-environment interactions in behavior, the question arises: How can environmental factors interact with genes? The idea that environmental factors can physically interact with our genome and in turn produce central nervous system responses capable of driving a change in behavior is a central tenet of behavioral epigenetics. The word *epigenetics* as traditionally defined involves the study of changes in gene expression that affect phenotypic outcome of cells during differentiation and development. Epigenetics as we currently define it refers to changes in genome function that occur without a change in the DNA sequence of nucleotide bases.

Epigenetic processes such as DNA methylation (figure 15.1b) actively regulate our genome in response to environmental input and are poised to do so not just during early life development, but also throughout the life span. DNA methylation is a process in which a methyl group attaches to a cytosine of a cytosine-guanine dinucleotide. This reaction is initiated by a family of DNA methyltransferases (DNMTs), including DNMT1 and DNMT3a. Most literature is consistent with the view of DNA methylation being associated with suppressing gene transcription. This occurs through interference of the methyl groups with transcription factors or is mediated by methyl-CpG-binding protein 2 (MeCP2), which can recruit histone deacetylases (HDACs) and other co-repressors to "turn off" a gene. In some cases, MeCP2 can instead associate with transcriptional activators and promote gene transcription. Active demethylation of cytosines can occur through a DNA repair–like mechanism mediated by the growth-arrest DNA-damage-inducible beta protein, GADD45b.

Contribution of DNA methylation to behavior

Our discussion regarding genetic polymorphisms emphasized a gene-by-environment interaction that plays a prominent role in mediating neural and behavioral changes. A plausible mechanism for how environmental factors could interact with genes is through epigenetics. These mechanisms are recognized for their ability to alter gene expression in response to environmental factors, whether they are prenatal, postnatal, or occur later in life. In the next few sections, we review key studies that have helped us understand this point. For a more comprehensive review of this literature and epigenetic mechanisms in general, we refer readers to several of our recent reviews (Blaze & Roth, 2012; Roth, 2012).

Prenatal Environmental Events Some of the first clues that prenatal events could have behavioral consequences via epigenetics came from observations that exposure of pregnant women to war-related stressors (including dietary restrictions) in the 1940s was associated with an increase in prevalence of schizophrenia and anxiety disorders in their offspring (Roseboom, Painter, van Abeelen, Veenendaal, & de Rooij, 2011). Additional clues came from observations of descendants of the Dutch Famine (six decades removed), whereby prenatal exposure to famine was linked to hypomethylation of an imprinted gene regulating body growth (Heijmans et al., 2008).

Work in rodents has provided more direct evidence of epigenetic factors linking maladaptive maternal nutrition to adverse health outcomes. Offspring of mice fed a high-fat diet during gestation and lactation show persisting hypomethylation of genes that moderate reward-related and feeding behavior, including those coding for the dopamine reuptake transporter and mu-opioid receptor in the nucleus accumbens and hypothalamus (Vucetic, Kimmel, Totoki, Hollenbeck, & Reyes, 2010). The long-standing epigenetic modifications due to maternal high-fat diet are accompanied by increased gene expression and preference for sucrose and fat. In addition to nutritional abnormalities, prenatal toxin exposure is known to epigenetically modify genes and offspring behavior across multiple generations (Crews et al., 2012; Skinner, Anway, Savenkova, Gore, & Crews, 2008).

Our first hint that methylation status of the human genome is also sensitive to maternal stress and emotion came in 2008. Infants born to mothers who reported high levels of depression and anxiety during their third trimester of pregnancy exhibited increased methylation of the human *GR* gene (*Nr3c1*) promoter in cord blood cells (Oberlander et al., 2008). At 3 months old, these infants also had increased salivary cortisol in response to an information-processing task involving the presentation of novel visual stimuli. Work in a rodent model has also established a link between epigenetic mechanisms and the maladaptive effects of a mother's stress on offspring behavior (Mueller & Bale, 2008). To model a stressful early life environment, investigators applied different stressors during gestation, including exposing pregnant dams to fox odor, restraint stress, and saturated bedding. When they examined male offspring from the stressed mothers, they found decreased methylation of the *corticotrophin-releasing factor* (*CRF*) promoter in the adult hypothalamus and central nucleus of the amygdala, which corresponded to increased *CRF* mRNA levels. Also present was an increase in methylation of the *GR* promoter accompanied by a decrease in GR gene expression in the hippocampus. These epigenetic alterations were shown to be relevant to increased stress reactivity.

Early Life Environmental Events The previous section highlighted that epigenetics can provide a mechanism for gene-by-environmental interactions within the womb. We have learned, too, that epigenetics can do the same outside the womb and during early development. Work by Meaney's lab pioneered the idea that early life caregiving can alter gene methylation with consequences for later behavior. Prior to their epigenetic study, they had identified natural variations in maternal caregiving in a strain of rats, in which rat dams either displayed high levels of licking or grooming (LG) and arched-back nursing, or low levels of these maternal behaviors (high vs. low LG). Their 2004 study provided evidence that epigenetic patterns in offspring were directly attributable to differences in these maternal behaviors (Weaver et al., 2004). Specifically, they demonstrated that pups raised by low-LG mothers had increased methylation of the *GR* promoter in the hippocampus, an effect that emerged during the first week of infancy and persisted into adulthood. Consistent with the idea of DNA methylation as a suppressor of gene transcription, these effects paralleled a decrease in expression of the *GR* gene. Offspring of high-LG mothers instead showed an opposite pattern of *GR* promoter methylation and expression. They also demonstrated that patterns of methylation (higher to low) could be induced by cross-fostering infants, and that the effects in low-LG offspring were reversible with trichostatin-A, an HDAC inhibitor that eliminated the DNA methylation and attenuated corticosterone levels in response to restraint stress (Weaver et al., 2004). The ability of child maltreatment to epigenetically modify the human *GR* gene has since been demonstrated (McGowan et al., 2009; Tyrka, Price, Marsit, Walters, & Carpenter, 2012).

The broad array of phenotypes affected by caregiving would suggest the ability of caregiver-infant interactions to epigenetically program a multitude of genes and systems. Research has shown that this is indeed the case. For example, a whole-genome methylation analysis of institutionalized children (maternally separated) in comparison to children raised by their biological parents showed greater methylation of several genes involved in controlling immune responses and cell signaling (Naumova et al., 2012). Additionally, we have explored the ability of DNA methylation to link *Bdnf* gene-by-caregiver maltreatment interactions in the rat (Roth, Lubin, Funk, & Sweatt, 2009). In our model, we repeatedly expose infants to adverse caregiving conditions for 30-minute bouts each day during the first seven days of life. We have found that adults that experienced these caregiving conditions have increased methylation of DNA associated with several regions of the *Bdnf* gene within their prefrontal cortex, with a concomitant decrease in *Bdnf* mRNA levels. We have also demonstrated that these epigenetic changes appear in the next generation of infants, and are potentially reversible with chronic treatment of a DNA methylation inhibitor.

Later-Life Environmental Events There is a growing body of literature demonstrating that epigenetic mechanisms facilitate similar environmentally driven phenomena later in life. These studies make it clear that DNA methylation alterations are biological responses to psychological and social-contextual factors. Furthermore, these observations are consistent with the hypothesis that epigenetic marking of genes could underlie aspects of neuropsychiatric disorders that can be associated with stress and abnormal brain function.

Various genes have been implicated in post-traumatic stress disorder (PTSD), and a growing body of literature has begun to investigate underlying epigenetic modifications. Because a substantial trauma is necessary for this disorder to emerge, the molecular machinery governing the symptoms must be significantly affected by environmental factors, making epigenetics a viable candidate. One candidate gene shown to be epigenetically modified in individuals with PTSD is *α-mannosidase* (*MAN2C1*), an immune-related gene. Individuals with increased methylation of *MAN2C1* and greater exposure to self-reported potentially traumatic experiences showed greatest risk for PTSD (Uddin et al., 2011). Another report has shown an increase in global methylation, as well as increased methylation of specific genes associated with plasticity (*Bdnf*) and immune regulation in PTSD patients who have high self-reports of total life stress, including child abuse (Smith et al., 2011). Together, peripheral measures show strong associations between child abuse/life stressors, methylation of DNA, and the diagnosis of PTSD.

In human epigenetic studies, it is difficult to distinguish methylation patterns that developed specifically as a consequence of exposure to a traumatic event. There is evidence, however, that this can be the case, as DNA methylation is an active mechanism in the mature brain and is part of a biological response repertoire to psychological and social-contextual factors. To illustrate this point, we will first discuss some important work from animal models, and then end with recent work in humans.

David Sweatt and colleagues pioneered the idea that if DNA methylation remains an active mechanism in the mature rat brain, then it can subserve gene-environment interactions relevant to learning and memory. A contextual fear-conditioning paradigm, which pairs an aversive stimulus such as a shock with a neutral context, was used to explore this provocative notion. They showed that adult rats that had formed a fear memory had decreased methylation of DNA associated with *reelin* (Miller & Sweatt, 2007) and *Bdnf* (Lubin, Roth, & Sweatt, 2008) genes in their hippocampus. Consistent with DNA demethylation as a mechanism to facilitate gene transcription, rats also had increased *reelin* and *Bdnf* gene expression. To make more of a causal argument, they have also shown that disrupting DNA methylation impairs memory formation (reviewed in Blaze & Roth, 2012).

Spatial learning and memory in rodents is often assessed using the Morris water maze, a task in which rats are placed in a pool of opaque water with a hidden escape platform under the water surface. With successive trials, rats learn the location of the hidden platform. GADD45b is known to mediate gene-specific demethylation in the hippocampus following seizure (Ma et al., 2009), and one study has examined performance in the Morris water maze in mutant *Gadd45b* knock-out mice (Sultan, Wang, Tront, Liebermann, & Sweatt, 2012). Although these knock-out mice did not exhibit differences in latency to reach the platform, they did show an increased number of platform crossings in probe trials compared to wild-types. An increase in platform crossings is typically used as an indicator of enhanced spatial learning and memory. While the role of DNA methylation in spatial learning and memory needs further exploration, other spatial studies are suggestive of different epigenetic mechanisms. Histone acetylation, an epigenetic molecular mechanism that regulates chromatin structure to promote gene transcription, has been linked to spatial memory (Bousiges et al., 2010). During spatial memory consolidation following the Morris water maze task, increases in histone acetylation were observed in the dorsal hippocampus. Furthermore, increased histone acetylation through HDAC inhibition resulted in dendrite sprouting, synapse formation, and enhanced Morris water maze performance in a mouse model of Alzheimer's disease (Fischer, Sananbenesi, Wang, Dobbin, & Tsai, 2007). Together, these studies are consistent with the notion that gene-environment interactions play a role in spatial learning and memory.

Social stress in the form of social defeat likewise has the capacity to affect epigenetic marks in adult rodents. In this paradigm, a rodent is placed in a cage with a more dominant peer and bullied for a set amount of time, either acutely or chronically. One study has shown that susceptible mice—mice that spend less time in a social-interaction zone after social defeat—show long-term demethylation of the *CRF* gene in the PVN, which leads to an overactive HPA axis and social avoidance behaviors (Elliott, Ezra-Nevo, Regev, Neufeld-Cohen, & Chen, 2010). Resilient mice on the other hand, which spend more time in the social-interaction zone after defeat, do not show these methylation changes. We have also explored the ability of stressful experiences in adult rats to promote DNA methylation changes. We used a clinically relevant animal model of PTSD, developed by David Diamond's group, which involves subjecting rats to psychosocial stress. Adult rats are twice immobilized and subjected to a hovering cat for a period of one hour, and in conjunction experience unstable housing conditions (Zoladz, Conrad, Fleshner, & Diamond, 2008). This stress regimen produces profound learning and memory deficits, increased anxiety-like behavior, and glucocorticoid abnormalities weeks to months following the termination of the stress regimen (Zoladz et al., 2008; Zoladz, Fleshner, & Diamond, 2012). We have recently shown that these behavioral abnormalities coincide with robust alterations in hippocampal *Bdnf* DNA methylation (that vary by subregion) and gene expression. Specifically, we have found an increase in *Bdnf* exon IV DNA methylation in the dorsal DG and dorsal CA1 of stressed rats, but decreased methylation in ventral CA3 (Roth, Zoladz, Sweatt, & Diamond, 2011). We found that these methylation changes also coincided with a decrease in *Bdnf* mRNA in both dorsal and ventral CA1.

Additional work with models of defeat or chronic stress also demonstrates epigenetic consequences of stress outside of periods of development (Tsankova et al., 2006; Uchida et al., 2011). Finally, in humans the ability of stressful experiences to evoke epigenetic modifications has been investigated in a laboratory setting using the TSST, which we described earlier in this chapter. One group found in blood samples from participants following the TSST a transient increase in methylation of the *oxytocin receptor* (*OXTR*) gene that was no longer present 90 minutes post-stress (Unternaehrer et al., 2012). Response to the TSST is known to differ for male and females, and another recent study showed greater methylation of the *NR3C-1* gene after the TSST in females compared to males, which coincided with a decrease in salivary cortisol released during the TSST (Edelman et al., 2012).

Evidence of epigenetic mechanisms mediating genotype contributions to behavior

We have reviewed key studies to illustrate that DNA methylation can serve as a mechanism for modified behavior following exposure to adversity or stress. Genetic polymorphisms can also convey susceptibility to these factors. Is there evidence, then, that epigenetic factors can mediate gene-by-environment interactions? A study with infant rhesus macaques showed that maternally deprived infants showed higher peripheral blood mononuclear cell (PBMC) *5-HTT* methylation regardless of genotype, and that methylation levels were negatively associated with *5-HTT* mRNA levels (Kinnally et al., 2010). Furthermore, they showed *5-HTT* methylation was highest in carriers of the short allele, and these infants exhibited the greatest stress response during maternal/social separation. Taken together, their results suggest that *5-HTT* methylation may mediate the risk conferred by the short allele (Kinnally et al., 2010).

In humans, Koenen and colleagues found that after controlling for *SLC6A4* promoter genotype, people with more traumatic events were at an increased risk for PTSD, but only with decreased *SLC6A4* DNA methylation. Individuals exhibiting higher methylation were protected against PTSD (Koenen et al., 2011). Methylation of genes within the pituitary adenylate cyclase-activating polypeptide (PACAP) system, a system responsive to cellular stress and implicated in neurotrophic function, also appears to predict PTSD diagnosis. Specifically, females with high plasma PACAP levels and greater methylation of the PACAP receptor gene (*ADCYAP1R1*) containing a polymorphism show strongest PTSD symptoms and physiological fear responses (Ressler et al., 2011). Finally, a study has found allele-specific and childhood trauma–dependent demethylation of the *FKBP5* gene, resulting in dysregulation of stress systems and immune function (Klengel et al., 2013). Together, such studies provide evidence for epigenetic mediation of the combined effects of genotype and environmental exposure on the risk for stress-related disorders.

Conclusion

Early life experiences can profoundly impact developmental trajectories, and mechanisms by which these changes occur are likely diverse. Genetic polymorphisms that are engrained in DNA are well recognized for their role in modifying environmental effects on behavior; thus, genetic variation is one route through which life experiences can influence behavioral outcomes. Interest has certainly increased for the role of epigenetic mechanisms as a biological basis of neural and behavioral effects of gene-environment interactions, and we now understand that changes in gene activity established through epigenetic alterations occur as a consequence of exposure to environmental adversity, social stress, and traumatic experiences. Though the epigenetic focus of this chapter has been on DNA methylation, histone modifications and microRNA processing are presumed to work in parallel with DNA methylation to exert changes in neurobiological function. Studies are also consistent with the notion that there is a so-called double-hit phenomenon underlying behavioral changes and outcomes, in which a reduced functioning allele combined with altered methylation can exacerbate the effects of the environment. It is likely, then, that incorporating both genetics and epigenetics into future environmental health research will yield substantial information regarding biological determinants of central nervous system changes and behavior.

REFERENCES

Barr, C. S., Newman, T. K., Shannon, C., Parker, C., Dvoskin, R. L., Becker, M. L., ... Higley, J. D. (2004). Rearing condition and rh5-HTTLPR interact to influence limbic-hypothalamic-pituitary-adrenal axis response to stress in infant macaques. *Biol Psychiatry, 55*(7), 733–738.

Bath, K., & Lee, F. (2006). Variant BDNF (Val66Met) impact on brain structure and function. *Cogn Affect Behav Neurosci, 6*(1), 79–85.

Bennett, A., Lesch, K. P., Heils, A., Long, J., Lorenz, J., Shoaf, S., ... Higley, J. D. (2002). Early experience and serotonin transporter gene variation interact to influence primate CNS function. *Mol Psychiatry, 7*(1), 118–122.

Binder, E., Bradley, R., Liu, W., Epstein, M., Deveau, T., Mercer, K., ... Ressler, K. J. (2008). Association of FKBP5 polymorphisms and childhood abuse with risk of posttraumatic stress disorder symptoms in adults. *J Am Med Assoc, 299*(11), 1291–1305.

Blaze, J., & Roth, T. L. (2012). Epigenetic mechanisms in learning and memory. *Wiley Interdiscip Rev Cogn Sci, 4*(1), 105–115.

Bousiges, O., Vasconcelos, A. P. D., Neidl, R., Cosquer, B., Herbeaux, K., Panteleeva, I., ... Boutillier, A. L. (2010). Spatial memory consolidation is associated with induction of several lysine-acetyltransferase (histone acetyltransferase) expression levels and H2b/H4 acetylation-dependent transcriptional events in the rat hippocampus. *Neuropsychopharmacology, 35*(13), 2521–2537.

Caspi, A., Sugden, K., Moffitt, T. E., Taylor, A., Craig, I. W., Harrington, H., ... Poulton R. (2003). Influence of life stress on depression: Moderation by a polymorphism in the 5-HTT gene. *Science, 301*(5631), 386–389.

Cicchetti, D., & Toth, S. L. (2005). Child maltreatment. *Annu Rev Clin Psychol, 1*(1), 409–438.

Crews, D., Gillette, R., Scarpino, S. V., Manikkam, M., Savenkova, M. I., & Skinner, M. K. (2012). Epigenetic transgenerational inheritance of altered stress responses. *Proc Natl Acad Sci USA, 109*(23), 9143–9148.

Edelman, S., Shalev, I., Uzefovsky, F., Israel, S., Knafo, A., Kremer, I., ... Ebstein, R. P. (2012). Epigenetic and genetic factors predict women's salivary cortisol following a threat to the social self. *PLoS One, 7*(11), e48597.

Egan, M. F., Kojima, M., Callicott, J. H., Goldberg, T. E., Kolachana, B. S., Bertolino, A., ... Weinberger, D. R. (2003). The BDNF Val66Met polymorphism affects activity-dependent secretion of BDNF and human memory and hippocampal function. *Cell, 112*(2), 257–269.

Elliott, E., Ezra-Nevo, G., Regev, L., Neufeld-Cohen, A., & Chen, A. (2010). Resilience to social stress coincides with functional DNA methylation of the CRF gene in adult mice. *Nat Neurosci, 13*(11), 1351–1353.

Fischer, A., Sananbenesi, F., Wang, X., Dobbin, M., & Tsai, L.-H. (2007). Recovery of learning and memory is associated with chromatin remodelling. *Nature, 447*(7141), 178–182.

Gatt, J. M., Nemeroff, C. B., Dobson-Stone, C., Paul, R. H., Bryant, R. A., Schofield, P. R., ... Williams, L. M. (2009). Interactions between BDNF Val66Met polymorphism and early life stress predict brain and arousal pathways to syndromal depression and anxiety. *Mol Psychiatry, 14*(7), 681–695.

Gillespie, C. F., Phifer, J., Bradley, B., & Ressler, K. J. (2009). Risk and resilience: Genetic and environmental influences on development of the stress response. *Depress Anxiety, 26*(11), 984–992.

Hanson, J. L., Chandra, A., Wolfe, B. L., & Pollak, S. D. (2011). Association between income and the hippocampus. *PLoS One, 6*(5), e18712.

Heijmans, B. T., Tobi, E. W., Stein, A. D., Putter, H., Blauw, G. J., Susser, E. S., ... Lumey, L. H. (2008). Persistent epigenetic differences associated with prenatal exposure to famine in humans. *Proc Natl Acad Sci USA, 105*(44), 17046–17049.

Kaufman, J., Yang, B.-Z., Douglas-Palumberi, H., Grasso, D., Lipschitz, D., Houshyar, S., ... Gelernter, J. (2006). Brain-derived neurotrophic factor–5-HTTLPR gene interactions and environmental modifiers of depression in children. *Biol Psychiatry, 59*(8), 673–680.

Kinnally, E. L., Capitanio, J. P., Leibel, R., Deng, L., LeDuc, C., Haghighi, F., ... Mann, J. J. (2010). Epigenetic regulation of serotonin transporter expression and behavior in infant rhesus macaques. *Genes Brain Behav, 9*(6), 575–582.

Klengel, T., Mehta, D., Anacker, C., Rex-Haffner, M., Pruessner, J. C., Pariante, C. M., ... Binder, E. B. (2013). Allele-specific FKBP5 DNA demethylation mediates gene-childhood trauma interactions. *Nat Neurosci, 16*(1), 33–41.

Koenen, K. C., Uddin, M., Chang, S.-C., Aiello, A. E., Wildman, D. E., Goldmann, E., ... Galea, S. (2011). SLC6A4 methylation modifies the effect of the number of traumatic events on risk for posttraumatic stress disorder. *Depress Anxiety, 28*(8), 639–647.

Lubin, F. D., Roth, T. L., & Sweatt, J. D. (2008). Epigenetic regulation of BDNF gene transcription in the consolidation of fear memory. *J Neurosci, 28*(42), 10576–10586.

Ma, D. K., Jang, M. H., Guo, J. U., Kitabatake, Y., Chang, M. L., Pow-Anpongkul, N., ... Song, H. (2009). Neuronal activity-induced GADD45B promotes epigenetic DNA demethylation and adult neurogenesis. *Science, 323*(5917), 1074–1077.

McGowan, P. O., Sasaki, A., D'Alessio, A. C., Dymov, S., Labonte, B., Szyf, M., ... Meaney, M. J. (2009). Epigenetic regulation of the glucocorticoid receptor in human brain associates with childhood abuse. *Nat Neurosci, 12*(3), 342–348.

Miller, C. A., & Sweatt, J. D. (2007). Covalent modification of DNA regulates memory formation. *Neuron, 53*(6), 857–869.

Mueller, B. R., & Bale, T. L. (2008). Sex-specific programming of offspring emotionality after stress early in pregnancy. *J Neurosci, 28*(36), 9055–9065.

Naumova, O. Y., Lee, M., Koposov, R., Szyf, M., Dozier, M., & Grigorenko, E. L. (2012). Differential patterns of whole-genome DNA methylation in institutionalized children and children raised by their biological parents. *Dev Psychopathol, 24*(01), 143–155.

Oberlander, T. F., Weinberg, J., Papsdorf, M., Grunau, R., Misri, S., & Devlin, A. M. (2008). Prenatal exposure to maternal depression, neonatal methylation of human glucocorticoid receptor gene (NR3C1) and infant cortisol stress responses. *Epigenetics, 3*(2), 97–106.

Pezawas, L., Meyer-Lindenberg, A., Drabant, E. M., Verchinski, B. A., Munoz, K. E., Kolachana, B. S., ... Weinberger, D. R. (2005). 5-HTTLPR polymorphism impacts human cingulate-amygdala interactions: A genetic susceptibility mechanism for depression. *Nat Neurosci, 8*(6), 828–834.

Ressler, K. J., Mercer, K. B., Bradley, B., Jovanovic, T., Mahan, A., Kerley, K., ... May, V. (2011). Post-traumatic stress disorder is associated with PACAP and the PAC1 receptor. *Nature, 470*(7335), 492–497.

Roseboom, T. J., Painter, R. C., van Abeelen, A. F. M., Veenendaal, M. V. E., & de Rooij, S. R. (2011). Hungry in the womb: What are the consequences? Lessons from the Dutch famine. *Maturitas, 70*(2), 141–145.

Roth, T. L. (2012). Epigenetics of neurobiology and behavior during development and adulthood. *Dev Psychobiol, 54*(6), 590–597.

Roth, T. L., Lubin, F. D., Funk, A. J., & Sweatt, J. D. (2009). Lasting epigenetic influence of early life adversity on the BDNF gene. *Biol Psychiatry, 65*(9), 760–769.

Roth, T. L., Zoladz, P. R., Sweatt, J. D., & Diamond, D. M. (2011). Epigenetic modification of hippocampal BDNF DNA in adult rats in an animal model of post-traumatic stress disorder. *J Psychiatr Res, 45*(7), 919–926.

Shalev, I., Lerer, E., Israel, S., Uzefovsky, F., Gritsenko, I., Mankuta, D., ... Kaitz, M. (2009). BDNF Val66Met polymorphism is associated with HPA axis reactivity to psychological stress characterized by genotype and gender interactions. *Psychoneuroendocrinology, 34*(3), 382–388.

Skinner, M. K., Anway, M. D., Savenkova, M. I., Gore, A. C., & Crews, D. (2008). Transgenerational epigenetic programming of the brain transcriptome and anxiety behavior. *PLoS One, 3*(11), e3745.

Smith, A. K., Conneely, K. N., Kilaru, V., Mercer, K. B., Weiss, T. E., Bradley, B., ... Ressler, K. J. (2011). Differential immune system DNA methylation and cytokine regulation in post-traumatic stress disorder. *Am J Med Genet B Neuropsychiatr Genet, 156*(6), 700–708.

Soliman, F., Glatt, C. E., Bath, K. G., Levita, L., Jones, R. M., Pattwell, S. S., ... Casey, B. J. (2010). A genetic variant BDNF polymorphism alters extinction learning in both mouse and human. *Science, 327*(5967), 863–866.

Sultan, F. A., Wang, J., Tront, J., Liebermann, D. A., & Sweatt, J. D. (2012). Genetic deletion of GADD45B, a regulator of active DNA demethylation, enhances long-term memory and synaptic plasticity. *J Neurosci, 32*(48), 17059–17066.

Tottenham, N., Hare, T. A., Quinn, B. T., McCarry, T. W., Nurse, M., Gilhooly, T., … Casey, B. J. (2010). Prolonged institutional rearing is associated with atypically large amygdala volume and difficulties in emotion regulation. *Dev Sci, 13*(1), 46–61.

Tsankova, N. M., Berton, O., Renthal, W., Kumar, A., Neve, R. L., & Nestler, E. J. (2006). Sustained hippocampal chromatin regulation in a mouse model of depression and antidepressant action. *Nat Neurosci, 9*(4), 519–525.

Tyrka, A. R., Price, L. H., Marsit, C., Walters, O. C., & Carpenter, L. L. (2012). Childhood adversity and epigenetic modulation of the leukocyte glucocorticoid receptor: Preliminary findings in healthy adults. *PLoS One, 7*(1), e30148.

Uchida, S., Hara, K., Kobayashi, A., Otsuki, K., Yamagata, H., Hobara, T., … Watanabe, Y. (2011). Epigenetic status of Gdnf in the ventral striatum determines susceptibility and adaptation to daily stressful events. *Neuron, 69*(2), 359–372.

Uddin, M., Galea, S., Chang, S.-C., Aiello, A. E., Wildman, D. E., de los Santos, R., … Koenen, K. C. (2011). Gene expression and methylation signatures of MAN2C1 are associated with PTSD. *Dis Markers, 30*(2), 111–121.

Unternaehrer, E., Luers, P., Mill, J., Dempster, E., Meyer, A. H., Staehli, S., … Meinlschmidt, G. (2012). Dynamic changes in DNA methylation of stress-associated genes (OXTR, BDNF) after acute psychosocial stress. *Transl Psychiatry, 2*, e150.

Vucetic, Z., Kimmel, J., Totoki, K., Hollenbeck, E., & Reyes, T. M. (2010). Maternal high-fat diet alters methylation and gene expression of dopamine and opioid-related genes. *Endocrinology, 151*(10), 4756–4764.

Weaver, I. C., Cervoni, N., Champagne, F. A., D'Alessio, A. C., Sharma, S., Seckl, J. R., & Meaney, M. J. (2004). Epigenetic programming by maternal behavior. *Nat Neurosci, 7*(8), 847–854.

Zoladz, P. R., Conrad, C. D., Fleshner, M., & Diamond, D. M. (2008). Acute episodes of predator exposure in conjunction with chronic social instability as an animal model of post-traumatic stress disorder. *Stress, 11*(4), 259–281.

Zoladz, P. R., Fleshner, M., & Diamond, D. M. (2012). Psychosocial animal model of PTSD produces a long-lasting traumatic memory, an increase in general anxiety and PTSD-like glucocorticoid abnormalities. *Psychoneuroendocrinology, 37*(9), 1531–1545.

III

VISUAL ATTENTION

Introduction

SABINE KASTNER

"ATTENTION" IS a broad term that refers to a large variety of different behavioral phenomena and their underlying neural mechanisms. In the context of this section, we will define this term as the set of processes that leads to the "selection of behaviorally relevant information from our sensory environment." This definition implies a selective process and sets it apart from related, but more general processes such as vigilance that will not be considered in depth (but see chapter 19 by Corbetta and colleagues in this section). Importantly, in our consideration, attentional selection operates on the external world and is directed at our sensory environment in contrast to selection processes that are internally directed at different mental operations or thoughts.

Early behavioral studies on attention used the auditory domain to investigate fundamental problems of the selection process. For example, the well-known "cocktail party" problem illustrates our limited ability to process and deal with multiple competing inputs at the same time. In order to process the speech of the person you talk to at a party, you have to filter out the many other voices in the room. This situation exemplifies that, in typical natural environments, mechanisms are needed to route sensory information that is most relevant to ongoing behavior—such as the speech of the person you are currently interacting with at a cocktail party—preferentially through the brain and at the expense of the overwhelming majority of the sensory information available, that is, all the irrelevant, distracting stimuli.

Limitations in processing capacity are a characteristic of all sensory systems. During the last 30 years, the study of attention has shifted from the auditory to the visual domain, partly because of the rising interest in the neural basis of attention and the wealth of knowledge available for the primate visual system. For that reason, the chapters in our section focus exclusively on the visual domain. However, it is conceivable to assume that many of the principled findings from vision can be generalized to other sensory domains.

The chapters in this section provide updates on traditionally important topics in the field of attention and, at the same time, pave the path toward future challenges in gaining a deeper understanding of attentional operations in the primate brain. The first four chapters summarize state-of-the-art research directed at the human brain using complimentary approaches, including psychophysics, functional brain imaging, and electrophysiology. These are followed by three chapters that summarize monkey physiology studies and discuss different aspects of attention mechanisms as measured at the cellular and systems level. The final chapter presents a theoretical account of core problems in attention and how they can be conceptualized through computational models.

Even though much of this section is dedicated to the review of literature on the neural underpinnings of visual attention, it is important to carefully consider the behavioral literature that has provided not only many of the standard paradigms used in research on neural foundations, but also the theoretical basis for hypothesis-driven empirical studies. Jeremy Wolfe sets the stage for the section by reviewing behavioral studies on a variety of attention phenomena that illustrate our limited processing capacity in space and time.

The selection of behaviorally relevant information can be based on many different criteria, including a particular location, feature of an object, or an object itself. For example, when we look for a friend in a crowd who wears a red shirt, we scan all people with red shirts at different locations in the crowd to guide the search. Attention can even be directed at categories of objects, such as when we look for cars when crossing the street. Each of these individual selection criteria is associated with a different set of underlying neural mechanisms. John Serences and I review the neural basis of such space-, feature-, and object-based attention as evidenced by functional imaging studies in the human brain. Together, these studies reveal the large-scale network that mediates selective attention, including several areas, or functional nodes, in occipital, temporal, parietal and frontal cortex as well as the thalamus. It is a major goal of the field to characterize the functional roles of each of these nodes and their interactions across the network.

While functional brain imaging has been instrumental in identifying the large-scale organization of the attention network, little knowledge has been gained from these studies regarding the temporal dynamics across the network. Steve Luck and Steve Hillyard fill this gap by reviewing the literature on human electrophysiology. Event-related and steady-state potential studies reveal the time course of modulatory attention effects in sensory processing areas, as well as the dynamics in higher-order areas that control the routing of information in sensory systems during the selection process. Intriguingly, the steady-state potential technique that operates by evoking neural oscillations at different frequencies has been successful in teasing apart temporal dynamics associated with different components of the selection process such as space- and feature-based mechanisms to study their interactions.

The significance of attention mechanisms for successful behavior becomes most apparent when they fail. Attention deficits are associated with many psychiatric and neurological diseases and have devastating consequences for the impaired individual. Spatial neglect is a particularly dramatic example of an attention deficit, given that patients have difficulty attending to large portions of visual space contralateral to the brain lesion. Even though the neglect syndrome has been studied extensively, its neural basis is still poorly understood. Maurizio Corbetta and colleagues present an elegant recent account of the pathological basis of spatial neglect, proposing that the deficit is mediated by abnormal interactions of attention control networks. Their account reconciles many of the unresolved questions in the neglect field and constitutes an important example that our growing knowledge about normal brain function from neuroimaging can finally be applied to revealing the neural basis of brain pathology.

Stefan Treue and Julio Martinez-Trujillo revisit one of the most fundamental problems in attention research. Specifically, how do we filter out the unwanted information from our sensory environment—that is, the majority of information available to us? Biased competition accounts of attention have provided compelling evidence that attention directed to one of two competing stimuli that are present in a neuron's receptive field (RF; a computational unit that may be one of the neural correlates underlying capacity limitations in vision) leads to a response modulation that reflects the stimulus-evoked response to the same stimulus when presented alone. This and other findings led early on to the proposal that the RF may shrink around the attended stimulus. The chapter summarizes recent evidence from monkey physiology showing that RF dynamics do play an important role in filtering distractor information during attentional selection. Such dynamic RF properties are likely to play a fundamental role in many other cognitive operations.

Behrad Noudoost, Eddy Albarran, and Tirin Moore have shown in seminal studies that modulatory attention effects observed in visual cortex are generated in higher-order cortex and transmitted via feedback connections. These studies, which have provided unprecedented causal evidence for interactions between nodes of the attention network, are summarized in addition to recent advances that have used hypothesis-driven pharmacological interventions to manipulate network interactions and behavior. Together, these studies pave the way toward a thorough understanding of the mechanistic operations within the attention network.

Traditional monkey electrophysiology has focused on response properties of neurons in local areas without much consideration of the greater network that these local entities are part of. Simultaneously recording from several neural populations across a large-scale network has been technically challenging, particularly in behaving primates trained on sophisticated cognitive tasks. However, understanding how communication is set up in cognitive networks and translates into successful behavior is one of the ultimate frontiers of cognitive neuroscience. Thilo Womelsdorf, Ayelet Landau, and Pascal Fries summarize recent studies of electrocorticographic recordings from several brain regions providing first insights into the functional role of oscillatory activity that may serve large-scale communication to mediate attentional selection.

The final chapter by Laurent Itti and Ali Borji underlines the need for computational models to organize the increasing amount of empirical data into coherent frameworks and to derive hypotheses for guiding future quests. Ultimately, the integration of data across different primate brain models (e.g., human, monkey), different levels of analyses (e.g., cellular and systems methods), and different conceptual approaches (e.g., theoretical and empirical) will be necessary for future progress in understanding the neural foundations of selective attention.

16 Theoretical and Behavioral Aspects of Selective Attention

JEREMY WOLFE

ABSTRACT The nervous system is limited in its ability to process stimuli and information. *Attention* is a term covering the many ways in which the system can selectively process a useful subset of all that could be processed. This chapter focuses on the behavioral manifestations of attention, specifically in visual processing. Subsequent chapters will describe the neural basis for some of these aspects of attention.

In a text on cognitive neuroscience, there is going to be a heavy emphasis on those aspects of attention that are amenable to study with electrophysiology and neuroimaging. In focusing on that subset of the field, one can lose track of the broader sweep of attentional phenomena—the behaviors that got us interested in the neural underpinnings of attention in the first place. The purpose of this chapter is to briefly review behavioral manifestations of attention and some of the theoretical constructs that have been used to organize our understanding of those behaviors.[1]

Posner's taxonomy: Alerting, orienting, and executive function

"Attention" refers to a wide range of mechanisms in the mind/brain that help us deal with our inability to do everything at the same time. There are fundamental bottlenecks in processing, and there are attentional mechanisms that control access and navigation for many of these. At least since William James, it has been clear that attention is not A Single Thing. There have been many efforts to organize this broad category of different processes and, indeed, one way to do this has been to look for anatomical or physiological networks that seem to be devoted to one or the other class of attentional process (see chapter 19, this volume). One useful taxonomy grows out of the work of Michael Posner (Posner & Cohen, 1984). He divides the territory into three networks: alerting, orienting, and executive function.

When your teacher suggested that you "pay attention," we might say that she was primarily addressing your *alerting* mechanisms: those processes that allow you to be ready to sustain your sensitivity to incoming stimuli. Vigilance studies, in which an observer might be asked to simply push a key each time an infrequent light appears, are measures of the decline or lapse of alertness over time (Warm, Dember, & Hancock, 1996). When your friend in that class suggested that you "look here," he was *orienting* your attention: inducing you to select a subset of the currently available sensory stimuli. A vast body of research makes use of Posner's cuing paradigm to study orienting of attention. In its simplest form, observers are asked to detect a stimulus that can appear at one of two locations. Prior to the stimulus, a cue directs attention to one or the other location. After a delay, the stimulus is presented at the cued or uncued location. A benefit in speed, accuracy, or both at the cued location or a penalty, relative to a baseline, at the uncued location is evidence for the selective orienting of attention. Simple cueing and probe-detection paradigms are readily adapted to animal and neuroimaging work.

The *executive* function aspect of attention refers to those processes that serve to monitor our ongoing flow of thoughts, reactions, and responses. These processes are of particular use in resolving conflicting demands. The tumbling kittens and babies at the top of a web page are trying to summon your attention by virtue of their high perceptual salience (Yantis, 1993), but you wish to deploy top-down control in order to read the turgid scientific prose elsewhere on the page. Executive processes mediate that conflict. Researchers have developed a number of laboratory tasks to study these processes. For instance, in "stop" signal tasks, you must try to countermand an order or plan to do something when a second command tells you to stop. For instance, you might be asked to push a button when a light appears but to stop or withhold that response if that light turns red (Schall & Godlove, 2012). In anti-saccade tasks, a signal at one location tells you to make a saccadic eye movement to a different location, again

[1]This chapter is lightly referenced because of space considerations. A more fulsomely referenced version can be obtained from jwolfe@partners.org.

requiring an act of volition to overcome the reflexive tendency to move the eyes to the signal. Similarly, executive processes are presumed to be involved whenever you switch from one task to another (Kiesel et al., 2010).

The "Stroop" effect is a classic illustration of the costs of invoking executive control. In the Stroop effect, observers are asked to name the color in which a word is printed. If that word is, itself, a color word, it will interfere with the response if the color named is different than the color of the word (e.g., "RED" printed in green). The conflict between the two colors, competing to control the response, must be managed by the executive network. A whole family of similar conflict situations can be described as "Stroop-like" (MacLeod, 1992). "Flanker" effects pit the identity or response demands of a central target against the identity of flanking items. For example, you might be shown a triplet of letters and asked to press one key in response to an "A" in the central position and another in response to a "B." Given those instructions, you will be faster with an "AAA" triple than with "BAB," even though you have been told to ignore those flanking letters. Rather like the Stroop effect, there seems to be some processing of the technically irrelevant flanking stimuli that interferes or supports response to the central letter. Interestingly, if the central task is made harder, it actually becomes easier to ignore the flankers (Lavie & Tsal, 1994), as if attention were a resource of some limited size. If it is entirely consumed in dealing with the central letter, the flankers do not interfere. However, if there is leftover capacity, attention spills over to the flankers, leading to interference.

This chapter will focus on visual selective attention. Needless to say, the processes of selective attention operate in all the senses (Shamma, Elhilali, & Micheyl, 2011) and between senses (see various chapters in Stein, 2012) but the largest body of work has been done in the visual modality.

Alerting, orienting, and executive function are one way to divide up the territory of attention. Chun, Golomb, and Turk-Browne (2011) suggest that the most basic taxonomic division should distinguish between external attention and internal attention: attention to the outside world and attention to internal states and thoughts. In a sense, this is an orthogonal division to Posner's. For example, you have both internal and external processes of selective attention. Internal and external processes probably share some fundamental components. Thus, if you are looking for your coworker's child at the birthday party, you are performing a visual search, requiring visual selective attention. If you try to remember the name of that child, you are searching through memory in a manner that bears more than a passing resemblance to that visual search (Hills, Jones, & Todd, 2012).

Attention exists because capacity is limited

We have attentional processes at every level in our mental architecture because, large and powerful as our brains may be, they cannot handle all possible demands at the same time. Tsotsos, in a "back-of-the-envelope" calculation, determined that we would need brains larger than the size of the known universe to process all visual input in parallel, without selection (an almost theological thought; see Tsotsos, 1990). On the output side, the hands cannot wield knife and fork while typing at the keyboard. Again, a selection must be made. In between, many components of our mental hardware have very restrictive limits. Visual short-term (or working) memory can handle only a very few items at any one time (Cowan, 2001). Consequently, access to that limited resource must be managed intelligently.

Given that there are attentive processes, are there "preattentive" processes?

Back in the 1960s, Neisser (1967) introduced the distinction between attentive and preattentive processes. This seemed straightforward enough if you imagined a simple box diagram. Outputs from the eyes fed preattentive processes which, in turn, fed attentive processes. Attention could be seen as a gate (Broadbent, 1958) or filter (Treisman, 1969) between preattentive and attentive stages in vision. In Treisman's powerful feature integration theory (Treisman & Gelade, 1980), preattentive processes were those that operated in parallel, across the visual field. These processes would support the "pop-out" of one salient feature on a background of homogeneous distractors; for example, a red spot among green, or a vertical among horizontal lines (Egeth, Jonides, & Wall, 1972). Similarly, preattentive processes would support texture segmentation, a process by which regions in the visual field were defined and separated from neighboring regions by parallel processing of basic features (Julesz, 1984), though texture segmentation and efficient discovery of single items are not quite the same thing (Wolfe, 1992).

A substantial literature developed, seeking to identify the basic stimulus attributes that could be processed preattentively (Wolfe & Horowitz, 2004) but, at the same time, challenges arose to the preattentive-attentive dichotomy. First, as is discussed in chapters 17 through 22 of this volume, it is possible to find neural evidence of attention at every stage of visual processing except the retina (Kastner & Pinsk, 2004). More problematic from a behavioral science vantage point, what would it mean for a process to be "preattentive," if it can be clearly modulated by attention? Going back to the basic Posner cuing task, response to a single spot of light (a clear preattentive "pop-out" stimulus) is influenced by cues that direct attention. Using more subtle, psychophysical and signal detection methods, Carrasco, among others, has shown that the extraction of very basic visual properties (contrast, saturation, etc.) is influenced by attention (Carrasco & McElree, 2001).

Nevertheless, even though attention can reach the earliest, post-retinal stages of processing, and even though attention can modulate the most basic of visual functions, the concept of "preattentive" processing retains its meaning. Suppose that you are attending to one location, perhaps a spot of light in an otherwise darkened room. Now suppose the lights are turned on. You are quickly aware of visual stimuli everywhere in the visual field. If we assume that visual selective attention is still spatially selective and, thus, selecting some portion of the stimulus, then the unselected portion of the field is tautologically "preattentive"—selective attention has not gotten there yet. The crisis over the notion of the preattentive arises, in part, from something like a confusion of space and time. Preattentive has a clear meaning if it refers to processing of stimuli in some period of time before those stimuli are selected. In the older, Neisser-Broadbent era, the view of visual processing was a sequence of steps, so that earlier in time also meant earlier in the space of visual processes. Thus, one could imagine labeling some pieces of the visual pathway as preattentive and other pieces, further along this linear chain, as attentive. However, it is now clear that processing involves feedback as well as feedforward processing, and that attentive effects in early visual centers can be driven by feedback from later centers (Hochstein & Ahissar, 2002). As a consequence, the same piece of early visual hardware (e.g., V1) could be engaged in preattentive and subsequent attentive processing of the same stimuli. This raises the interesting question of the "postattentive" status. Do visual stimuli return to their preattentive state when an attended location becomes unattended (Wolfe, Klempen, & Dahlen, 2000)?

The idea of initially preattentive processing that is altered by subsequent selective attention to a locus or object goes back, at least, to the French philosopher Condillac (1715–1780). In a more modern formulation, the initial preattentive experience would include basic features and some aspects of the semantic "gist" of the scene (Oliva, 2005).

What does attention select?

Selection in Space Many aspects of selection by attention can be appreciated by inspection of figure 16.1A.

To begin, though all of these items are highly visible, some are more likely to attract your attention in a *bottom-up*, stimulus-driven manner. Odds are that your attention was directed, at first, to the picture of Michael Gazzaniga or to the white item with spikes. In both cases, these are especially different from all of their neighbors and, thus, likely to be the most "salient" items in the field (Itti & Koch, 2001). The bottom-up salience of an item would increase if all of the other items were more similar to each other. This can be seen in figure 16.1B, where the small black square "pops out" among homogeneous big black squares. Bottom-up salience increases with distractor homogeneity (Distractor-Distractor "D-D" similarity; Duncan & Humphreys, 1989). Unsurprisingly, salience *decreases* when target and distractors become more similar (T-D similarity). Thus, in 16.1B, the target would be less salient if it were a little bigger, and in 16.1A the photo would become more salient if it were in color (Duncan & Humphreys, 1989). "Distance" in feature space is not trivially measured in terms of the physical units of measurement. It is the psychophysical response to the physical stimulus that would be important and, even then, basic psychophysical measures like a "just noticeable difference" are not necessarily the correct metric for what we could call "salience space." For instance, a pink target amidst white and red distractors is found much more quickly and, thus, would be described as much more salient than a light green amidst white and saturated green distractors. This is true even if the distances between targets and distractors are carefully controlled to be identical in a just noticeable difference–based color space (Lindsey et al., 2010). In the absence of a general rule, salience and stimulus "distance" must be

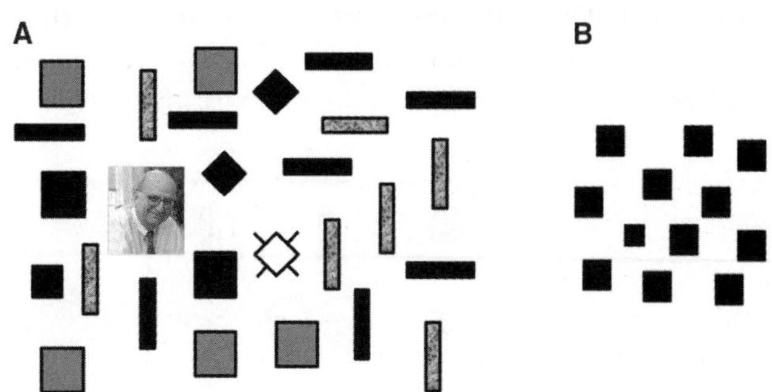

Figure 16.1 (A) Many rules of selection can be illustrated with this collection of items. (B) The small(er) square, not salient in 16.1A, attracts attention here.

determined empirically. One approach to measuring salience is to use some standard salient stimulus (e.g., brightness) as a yardstick (Nothdurft, 1993).

While you were looking for small black squares in figures 16.1A and 16.1B, you may not have noticed that the "small" black square in 16.1A is the same size as the larger squares of 16.1B. When specifying a property like size, the relative value of that attribute is important (Becker, 2009).

Attention can be guided by "top-down" user-driven factors as well (Wolfe, Cave, & Franzel, 1989). Thus, returning to figure 16.1A, if we ask if all the dark gray objects are the same, your appreciation of the figure now changes—though, of course, the physical stimulus has not changed. You may find that the gray objects actually appear to be more salient now that they are the objects of attention. Indeed, you may find that the very qualia of the stimuli change with the application of attention (Carrasco, Ling, & Read, 2004).

This attention to features seems to be applied at the same time to all locations having that feature. Notice that, if asked, you could direct your attention from one gray square to another while maintaining the sense of having highlighted the entire group of gray squares. This is, at least, introspective support for the idea that attention to a feature is separable from attention to a location or an object (Chen, 2009; but see Bundesen, 1991).

If you now look for a horizontal, textured bar, you will be able to quickly deploy your attention to this conjunction of two features (Wolfe et al., 1989). In this case, you are directing *feature* attention to two different features in order to permit *spatial* attention to quickly select one object and to enable its identification as horizontal and textured. Introspectively, you may notice that either the textured items or the horizontal items, or both, are available to you as sets, highlighted by attention to their relevant features. The single, horizontal textured item does not "pop out" in the way that the face does, even once selected. Were there multiple, conjunctively defined, textured horizontal items, they would not group as effectively as the gray squares, as will be discussed below under the heading of "binding."

Feature guidance is one way to get spatial selective attention to an object. We could also direct attention using a cue of some sort. One of the standard divisions of attentional cues is into endogenous and exogenous categories (Posner & Cohen, 1984). Endogenous cues are typically symbolic and are typically placed at fixation. Thus, the words "upper left" could be an endogenous cue, directing attention to one gray square in the upper left of figure 16.1A. Endogenous cueing takes a few hundred milliseconds to take effect (longer, if reading is required). Exogenous cues are typically peripheral, at the intended destination of attention. A flash of light at the lower right would exogenously direct attention to the vertical textured bar in that location. Exogenous cueing works more quickly with a lag of perhaps 50 msec. It is more reflexive. Flashes of light "capture" attention in ways that endogenous cues do not (Yantis, 1993). Exogenous cueing is also more transient than endogenous cueing. Interestingly, some central cues that might be thought to be obviously endogenous and symbolic can behave as though they are exogenous, fast attention-capturing cues. The clearest examples are eye gaze (Friesen & Kingstone, 1998) and arrows (Eimer, 1997). In figure 16.2, you can compare three sorts of central cues and decide if the arrows and eyes, even when very schematic, seem to do their cueing work more reflexively and automatically than a word cue.

What is cued by a cue? If, for example, we use exogenous cuing by a spot of light on one end of one of the black bars in figure 16.1A, attention would be summoned rapidly to the cue and, we would find, the effects of selective attention would spread throughout the object. A cue at one end of a bar would produce more facilitation in the detection of a subsequent probe at the other end of the bar than at an equidistant location that was not on the same object (Reppa, Schmidt, & Leek, 2012). All else being equal, spatial selective attention appears to select objects (Goldsmith, 1998).

Returning once more to figure 16.1A, attention could be directed to the item at the lower right of the figure. If you deploy your attention in response to this endogenous cue, this will lead you to become aware that this

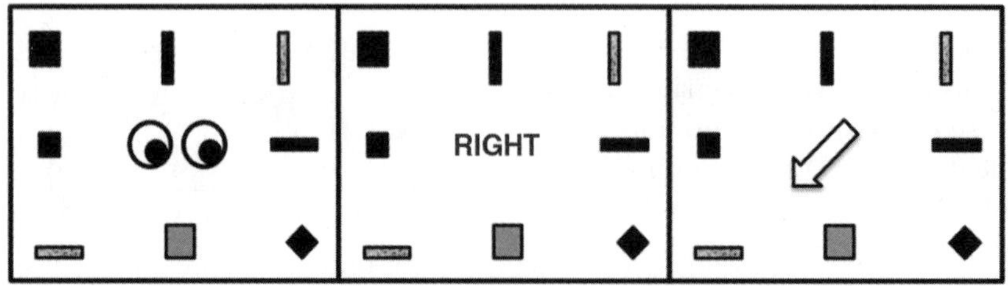

Figure 16.2 Three central cues: Eyes and arrows behave more like exogenous cues than like endogenous, symbolic cues.

item is a textured, vertical bar. The relationship between attention to features and attention to objects or locations is not a one-way street. The act of attending to that textured, vertical bar will "prime" the features "textured" and "vertical." This will make it somewhat more likely that you would subsequently direct attention to an item with either or both of those features (Maljkovic & Nakayama, 1994).

The idea of object-based attention is not as simple as it might seem. For instance, if you were asked to count the objects in front of you, you would not have a definitive answer. Is the bookshelf an object, or do we count each book or each letter on the spine of each book? Any of those could be the "object of attention." Moreover, there is a chicken-and-egg problem here. In order to direct attention to "objects," don't objects need to exist as perceptual entities ahead of time? How is that to be accomplished? Part of the solution or, at least, an acknowledgement of the problem, is to propose that the preattentive world is divided into "proto-objects" (Rensink, 2000). These would be some default approximations of objects that might be carved out of the visual scene preattentively and might be converted to perceived, recognized objects by the subsequent deployment of attention. There is currently no consensus about how to partition the visual world into proto-objects.

Feature-based selection is subject to a variety of constraints (reviewed in Wolfe & Horowitz, 2004). There appears to be a limited set of features that can be selected. That is, you can select on the basis of surface properties like color or, perhaps, shininess, but not something like "furriness," even though you are remarkably good at identifying furriness in a brief flash of an attended stimulus. Moreover, even selection by a member of that limited set of features is subject to further limits. For example, it is easy to select vertical from horizontal lines in figure 16.1A on the basis of their orientation. However, even though you can see the difference between lines tilted 10 and 15 degrees from vertical, you would not be able to guide attention to the 10-degree ones in the way that you can select on the basis of more dramatic orientation differences.

As there are limits on selection by feature, there are strong limits on the ability to select in space. These are well illustrated by "crowding" phenomena (Levi, 2011). For example, if one views a set of lines, some distance from fixation, it may be possible to see that there are lines lying between the leftmost and rightmost lines in the set, but it may not be possible to determine their individual orientations. Nevertheless, we can show that the orientation of these unattendable lines is still being processed at some loci in the visual system. For example, they produce tilt aftereffects (He, Cavanagh, & Intriligator, 1996).

There is a long-standing debate about the ability to select multiple, discrete locations in space (Jans, Peters, & De Weerd, 2010). Some of the debate is engendered by the multiplicity of attentional processes. Thus, as noted before, in figure 16.1A, if you attend to gray squares, feature-based attention would seem to be selecting multiple loci in, at least, two discrete groups. Object-based attention might be restricted to just one of these items. If one is merely keeping track of the location of objects and is not concerned with the specific features attached to those objects, then one can select and track a small number of objects (typically three to four) as they move around on the computer screen (Cavanagh & Alvarez, 2005).

The rules of selection become more complex when we think about selection in real scenes. As discussed above, it is essentially impossible to enumerate the possible objects of attention in a real scene. Moreover, the basic features like color or orientation tend to be very heterogeneous in real scenes. Such heterogeneity is known to make it harder to get attention to a desired target (Duncan & Humphreys, 1989). Nevertheless, search in real scenes is relatively easy (Wolfe et al., 2008). Part of the answer appears to be that there are forms of attentional guidance that are specifically based on our knowledge of the structure of real world scenes. Borrowing from the study of language, we can talk about *semantic* and *syntactic* guidance (Biederman, Mezzanotte, & Rabinowitz, 1982). Semantic guidance directs attention to locations based on meaning. If you were looking for your computer mouse, semantic guidance might direct your attention to the neighborhood of the keyboard and screen, even though your wireless mouse could be sitting on the floor or the shelf behind you. Syntactic guidance directs attention based on the structure and physical rules of the world. Syntactic guidance would tell you the computer mouse should be sitting on a horizontal surface. The floor would be syntactically but not semantically acceptable. Accounts of visual attention in the real world need to consider these scene-based rules as well as space-, feature-, and object-based rules.

Selection in Time Attention can be used to select locations in time as well as in space. Posner's "alerting" network was invoked earlier to describe the fluctuations of attention over time as vigilance varies. We can also select specific instants in time (Nobre, 2001). In the study of auditory attention, much of what is spatial in vision becomes temporal. Thus, attention to an auditory "object" is more usefully considered as attention over some piece of time than over some piece of space (Shinn-Cunningham, 2008).

In vision, much of the interest in temporal selection focuses on the consequences of selecting one object on subsequent selection of others. This is best seen in the attentional blink (AB) phenomenon (Raymond, Shapiro, & Arnell, 1992). In the AB, observers typically view a stream of stimuli at fixation in so-called RSVP, or rapid serial visual presentation. Observers are instructed to monitor the stream for two targets. For example, in a stream of black letters, they might be asked to name the only red letter and name the identity of the only number that appears. The red letter is the first target (T1) while the number is T2. Items appear, perhaps, every 100 msec. The key finding is that there is a period on the order of 200–500 msec after the appearance of T1 during which T2 is much harder to report. Interestingly, the observer is not completely blind to T2 when it is missed. For instance, a "blinked" word still produces semantic priming effects (Shapiro, Driver, Ward, & Sorenson, 1997) based on the meaning of the word, so the word was, on some level, read. Given that word reading almost undoubtedly requires selective attention to the word, this tells us that AB represents yet another attentional mechanism.

A metaphorical way to think about this is to imagine dipping a fishing net into a river. If one pulls out a fish (T1), there is a period of time when one might be able to see another fish (T2) but would be unable to pull that fish from the water because the net is occupied with the first fish (Chun & Potter, 1995). There are many less metaphorical models of AB and a large body of data addressing the theoretical possibilities (Dux & Marois, 2009).

Inhibitory Processes The AB might be thought of as an inhibition of attention following the deployment of attention to an object in space and time. Inhibitory processes are important by-products of many acts of attention. In a spatial array, when attention moves away from a selected item, inhibition of return (IOR) makes it harder and slower to get attention back to that item or location than to move attention to some other item or location (Posner & Cohen, 1984). IOR may serve a role in preventing observers from perseverating on salient, but irrelevant, points in the field. Once you figure out that a bright spot does not mark what you are looking for, it would be useful if attention were not continually dragged back to that spot by uninformed, bottom-up attentional guidance. One useful way to describe IOR is as a "foraging facilitator" (Klein & MacInnes, 1999). It inclines a search to move forward, not to perseverate.

There also appears to be inhibition surrounding the locus of attention. It can be harder to perform a task at a location near to a recent deployment of attention than a bit further away. Attention is sometimes metaphorically described as a "*spotlight* that enhances the efficiency of the detection of events within its beam" (Posner, Snyder, & Davidson, 1980, p. 172). It might be better to describe it as having a "Mexican hat" profile, with an enhancing center surround by an inhibitory rim (Muller, Mollenhauer, Rosler, & Kleinschmidt, 2005).

Binding

A central function of attention is to permit the binding of features into coherent objects. Different pieces of information about the same object will be represented in different brain areas. Some essentially passive combination of separate features will occur when those features co-occur in space (e.g., color and orientation; Houck & Hoffman, 1986) but the active integration of features, especially if multiple objects are present, requires a process of binding, enabled by attention. Thus, binding is a central tenet of Treisman's feature integration theory, mentioned earlier. This process of binding can be experienced in figure 16.3, if you look for the black-headed arrow, pointing left. It should be intuitively clear to you that you know that the colors in the figure are black and white and that the component shapes are triangular and rectangular. You also know that the arrows are oriented on the main vertical and horizontal axes, with no oblique arrows. However, even though it is trivial to determine if an arrow has a black head and is pointing left, you do not know which arrow that is until the bundle of black, white, triangle and rectangle features is attended and bound into a coherent object (Treisman, 1996).

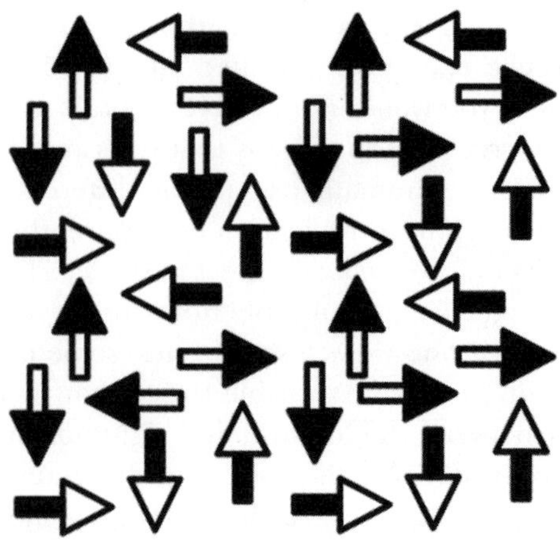

Figure 16.3 Find the black-headed arrow, pointing left.

Binding is typically discussed in the context of firmly attaching visual features like color and shape to specific visual objects. However, this may be merely a convenient arena in which to study a general role for binding attention. When you deliberately connect the dots of a thought or associate a specific feeling with a specific memory, you may well be using attentional processes to enable binding processes that are similar to basic visual feature binding. Of course, not all associations are attentionally mediated acts of binding (see chapter 50, this volume). It is the more effortful and controlled aspects of human cognition that may be analogs of visual binding.

Questions of the neural mechanisms of binding are discussed in later chapters (e.g., see chapters 20 and 21, this volume). Where does feature binding occur? Parietal structures are strongly implicated (Friedman-Hill, Robertson, & Treisman, 1995). As will be discussed, there is some evidence that binding is implemented by synchrony of neural activity (von der Malsburg, 1981). Some have argued that we don't need a locus or mechanism for binding because there is no "binding problem" to be solved (DiLollo, 2012).

Individuation, Scrutiny, and Awareness For many attention-demanding tasks, there is no obvious binding that needs to be done. If you want to appreciate the exact shade of green of a new leaf or the precise orientation of grating in an experiment, you will need to "pay attention." Attention allows finer perceptual decisions to be made (Yeshurun & Carrasco, 1998). In the presence of multiple items, whose features would tend to average together, attention can help individuate items (Intriligator & Cavanagh, 2001).

In some cases, a failure to attend to an item or a location will cause a failure of awareness of the existence of that item. The iconic example of this "inattentional blindness" is the failure to notice a gorilla wandering through a ball game (Simons & Chabris, 1999). Even if the observer is aware of the contents of a scene, she may fail to detect quite salient changes to that scene. This is known as "change blindness" (Simons & Rensink, 2005). While these phenomena have been used to argue that all conscious awareness requires attention, arguing that we are "blind" outside of the spotlight of attention seems too extreme. As Condillac noted, we see *something* everywhere before selective attention has had a chance to visit more than a spot or two. Our visual experience is made up of nonselective processing of the entire scene, selective attention and binding of one or a few objects and some memory for previously attended items (Wolfe, Vo, Evans, & Greene, 2011).

The mechanics of attention: How might attention modulate a signal?

Much of the work on the neural basis of attention, especially at the single-cell level, is devoted to asking how attention might be implemented in a brain. Behavioral research points to a variety of ways in which attention might be implemented. Figure 16.4 shows a selection of these, many of which will recur in subsequent chapters.

If we imagine a signal progressing from sense organs to more central processes, attention could act as a gate (see figure 16.4A) preventing some signals from moving beyond some point in the pathways (Broadbent, 1958). If the gate is not absolute, attention could be seen as a filter (see figure 16.4B), attenuating some signals (Treisman, 1969). A gate or filter can also been seen as a spatial allocation of attention to some portion of the visual field or to some object(s) within that field (see figure 16.4C). Behaviorally, signals outside of that spatial window are responded to less effectively (Cepeda, Cave, Bichot, & Kim, 1998). In the neural realm, this might represent a shrinking of receptive fields to include only the current object of attention (Moran & Desimone, 1985). Rather than impeding unattended signals, attention could speed attended signals (see figure 16.4D) (Carrasco & McElree, 2001). In models where items are racing for access to later limited capacity processing (Bundesen, 1990), giving some signals a temporal boost would be equivalent to blocking or attenuating other signals.

Lu and Dosher (e.g., Lu & Dosher, 2004) have pursued a program of research in which the behavioral responses to added external noise can be used to infer different roles of attention including distractor exclusion (see figure 16.4E, similar to "sharpening" in other contexts) in which a process, channel, or neuron that had responded to a wider range of stimuli comes to respond to a narrower range (here illustrated as a channel coming to respond to a narrower range of orientations). Attention could also boost the response to a signal (see figure 16.4F). That might take the form of a response gain, where attention acts to multiply all outputs of a process. Alternatively, attention might produce a contrast gain where the effect of attention would be to make a weak attended signal the equivalent of a stronger unattended signal (see figure 2 of Reynolds & Heeger, 2009). Again, these various signs of attentional effects should not be seen as mutually exclusive. For instance, a normalization process can produce several of these patterns, depending on the experimental conditions (Reynolds & Heeger, 2009).

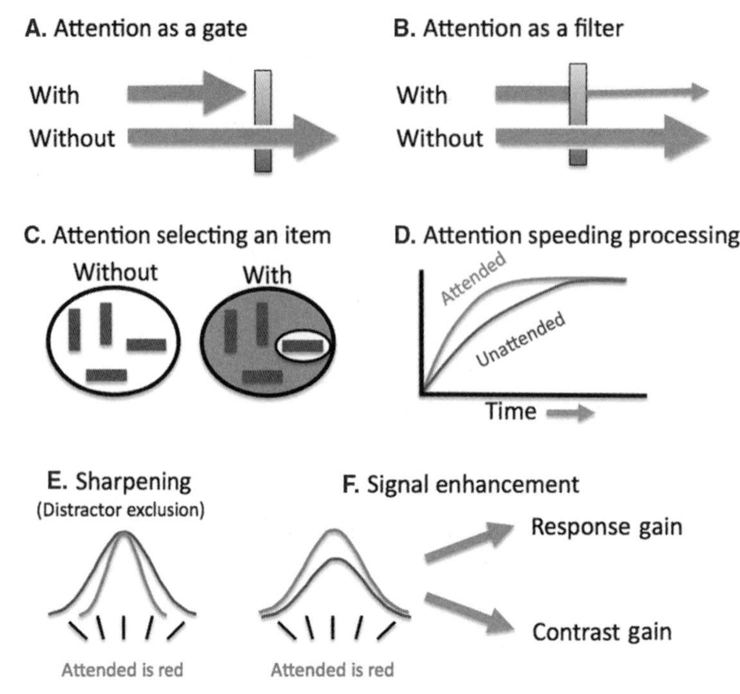

FIGURE 16.4 A sampling of the ways that attention could alter a signal. (See color plate 12.)

Conclusion

This chapter should have made it clear that it is important not to fall into the trap of reifying "attention" as a single thing with a single locus. The term should be understood to refer to a wide range of processes that make it possible for our limited minds to deal with a subsection of the overwhelming amount of information constantly flowing in through internal and external stimuli.

REFERENCES

BECKER, S. (2009). The role of target-distractor relationships in guiding attention and the eyes. *J Exp Psychol Gen, 139*(2), 247–265.

BIEDERMAN, I., MEZZANOTTE, R. J., & RABINOWITZ, J. C. (1982). Scene perception: Detecting and judging objects undergoing relational violations. *Cogn Psychol, 14,* 143–177.

BROADBENT, D. E. (1958). *Perception and communication.* London: Pergamon Press.

BUNDESEN, C. (1990). A theory of visual attention. *Psychol Rev, 97*(4), 523–547.

BUNDESEN, C. (1991). Visual selection of features and objects: Is location special? A reinterpretation of Nissen's (1985) findings. *Percept Psychophys, 50*(1), 87–89.

CARRASCO, M., LING, S., & READ, S. (2004). Attention alters appearance. *Nat Neurosci, 7*(3), 308–313.

CARRASCO, M., & MCELREE, B. (2001). Covert attention accelerates the rate of visual information processing. *Proc Natl Acad Sci USA, 98*(9), 5363–5367.

CAVANAGH, P., & ALVAREZ, G. A. (2005). Tracking multiple targets with multifocal attention. *Trends Cogn Sci, 9*(7), 349–354.

CEPEDA, N. J., CAVE, K. R., BICHOT, N., & KIM, M.-S. (1998). Spatial selection via feature-driven inhibition of distractor locations. *Percept Psychophys, 60*(5), 727–746.

CHEN, Z. (2009). Not all features are created equal: Processing asymmetries between location and object features. *Vision Res, 49*(11), 1481–1491.

CHUN, M. M., GOLOMB, J. D., & TURK-BROWNE, N. B. (2011). A taxonomy of external and internal attention. *Annu Rev Psychol, 62,* 73–101.

CHUN, M. M., & POTTER, M. C. (1995). A two-stage model for multiple target detection in rapid serial visual presentation. *J Exp Psychol Human, 21*(1), 109–127.

COWAN, N. (2001). The magical number 4 in short-term memory: A reconsideration of mental storage capacity. *Behav Brain Sci, 24*(1), 87–114; discussion 114–185.

DILOLLO, V. (2012). The feature- binding problem is an ill-posed problem. *Trends Cogn Sci, 16*(6), 317–321.

DUNCAN, J., & HUMPHREYS, G. W. (1989). Visual search and stimulus similarity. *Psychol Rev, 96,* 433–458.

DUX, P. E., & MAROIS, R. (2009). The attentional blink: A review of data and theory. *Atten Percept Psychophys, 71*(8), 1683–1700.

EGETH, H., JONIDES, J., & WALL, S. (1972). Parallel processing of multielement displays. *Cogn Psychol, 3,* 674–698.

EIMER, M. (1997). Uninformative symbolic cues may bias visual-spatial attention: Behavioral and electrophysiological evidence. *Biol Psychol, 46*(1), 67–71.

FRIEDMAN-HILL, S. R., ROBERTSON, L. C., & TREISMAN, A. (1995). Parietal contributions to visual feature binding: Evidence from a patient with bilateral lesions. *Science, 269*(11), 853–855.

Friesen, C., & Kingstone, A. (1998). The eyes have it! Reflexive orienting is triggered by nonpredictive gaze. *Psychon Bull Rev, 5*(3), 490–495.

Goldsmith, M. (1998). What's in a location? Comparing object-based and space-based models of feature integration in visual search. *J Exp Psychol Gen, 127*(2), 189–219.

He, S., Cavanagh, P., & Intriligator, J. (1996). Attentional resolution and the locus of visual awareness. *Nature, 383*, 334–337.

Hills, T. T., Jones, M. N., & Todd, P. M. (2012). Optimal foraging in semantic memory. *Psychol Rev, 119*(2), 431–440.

Hochstein, S., & Ahissar, M. (2002). View from the top: Hierarchies and reverse hierarchies in the visual system. *Neuron, 36*, 791–804.

Houck, M. R., & Hoffman, J. E. (1986). Conjunction of color and form without attention. Evidence from an orientation-contingent color aftereffect. *J Exp Psychol Human, 12*, 186–199.

Intriligator, J., & Cavanagh, P. (2001). The spatial resolution of visual attention. *Cogn Psychol, 43*(3), 171–216.

Itti, L., & Koch, C. (2001). Computational modelling of visual attention. *Nat Rev Neurosci, 2*(3), 194–203.

Jans, B., Peters, J. C., & De Weerd, P. (2010). Visual spatial attention to multiple locations at once: The jury is still out. *Psychol Rev, 117*(2), 637–682.

Julesz, B. (1984). A brief outline of the texton theory of human vision. *Trends Neurosci, 7*, 41–45.

Kastner, S., & Pinsk, M. A. (2004). Visual attention as a multilevel selection process. *Cogn Affect Behav Neurosci, 4*(4), 483–500.

Kiesel, A., Steinhauser, M., Wendt, M., Falkenstein, M., Jost, K., Philipp, A. M., & Koch, I. (2010). Control and interference in task switching: A review. *Psychol Bull, 136*(5), 849–874.

Klein, R. M., & MacInnes, W. J. (1999). Inhibition of return is a foraging facilitator in visual search. *Psychol Sci, 10*, 346–352.

Lavie, N., & Tsal, Y. (1994). Perceptual load as a major determinant of the locus of selection in visual attention. *Percept Psychophys, 56*(2), 183–197.

Levi, D. M. (2011). Visual crowding. *Curr Biol, 21*(18), R678–R679.

Lindsey, D. T., Brown, A. M., Reijnen, E., Rich, A. N., Kuzmova, Y., & Wolfe, J. M. (2010). Color channels, not color appearance or color categories, guide visual search for desaturated color targets. *Psychol Sci, 21*(9), 1208–1214.

Lu, Z. L., & Dosher, B. A. (2004). Spatial attention excludes external noise without changing the spatial frequency tuning of the perceptual template. *J Vis, 4*(10), 955–966.

MacLeod, C. M. (1992). The Stroop task: The "gold standard" of attentional measures. *J Exp Psychol Gen, 121*(1), 12–14.

Maljkovic, V., & Nakayama, K. (1994). Priming of popout: I. Role of features. *Mem Cognition, 22*(6), 657–672.

Moran, J., & Desimone, R. (1985). Selective attention gates visual processing in the extrastriate cortex. *Science, 229*, 782–784.

Muller, N. G., Mollenhauer, M., Rosler, A., & Kleinschmidt, A. (2005). The attentional field has a Mexican hat distribution. *Vision Res, 45*(9), 1129–1137.

Neisser, U. (1967). *Cognitive psychology*. New York, NY: Appleton Century Crofts.

Nobre, A. C. (2001). Orienting attention to instants in time. *Neuropsychologia, 39*(12), 1317–1328.

Nothdurft, H. C. (1993). The conspicuousness of orientation and visual motion. *Spatial Vis, 7*(4), 341–366.

Oliva, A. (2005). Gist of the scene. In L. Itti, G. Rees, & J. Tsotsos (Eds.), *Neurobiology of attention* (pp. 251–257). San Diego, CA: Academic Press.

Posner, M. I., & Cohen, Y. (1984). Components of attention. In H. Bouma & D. G. Bouwhuis (Eds.), *Attention and performance* (Vol. X, pp. 55–66). Hillside, NJ: Erlbaum.

Posner, M. I., Snyder, C. R. R., & Davidson, B. J. (1980). Attention and the detection of signals. *J Exp Psychol General, 109*, 160–174.

Raymond, J. E., Shapiro, K. L., & Arnell, K. M. (1992). Temporary suppression of visual processing in an RSVP task: An attentional blink? *J Exp Psychol Human, 18*(3), 849–860.

Rensink, R. A. (2000). Seeing, sensing, and scrutinizing. *Vision Res, 40*(10–12), 1469–1487.

Reppa, I., Schmidt, W. C., & Leek, E. C. (2012). Successes and failures in producing attentional object-based cueing effects. *Atten Percept Psychophys, 74*(1), 43–69.

Reynolds, J. H., & Heeger, D. J. (2009). The normalization model of attention. *Neuron, 61*(2), 168–185.

Schall, J. D., & Godlove, D. C. (2012). Current advances and pressing problems in studies of stopping. *Curr Opin Neurobiol, 22*(6), 1012–1021.

Shamma, S. A., Elhilali, M., & Micheyl, C. (2011). Temporal coherence and attention in auditory scene analysis. *Trends Neurosci, 34*(3), 114–123.

Shapiro, K., Driver, J., Ward, R., & Sorenson, R. E. (1997). Priming from the attentional blink. *Psychol Sci, 8*(2), 95–100.

Shinn-Cunningham, B. G. (2008). Object-based auditory and visual attention. *Trends Cogn Sci, 12*(5), 182–186.

Simons, D. J., & Chabris, C. F. (1999). Gorillas in our midst: Sustained inattentional blindness for dynamic events. *Perception, 28*, 1059–1074.

Simons, D. J., & Rensink, R. A. (2005). Change blindness: Past, present, and future. *Trends Cogn Sci, 9*(1), 16–20.

Stein, B. E. (2012). *The new handbook of multisensory processes.* Cambridge, MA: MIT Press.

Treisman, A. (1969). Strategies and models of selective attention. *Psychol Rev, 76*(3), 282–299.

Treisman, A. (1996). The binding problem. *Curr Opin Neurobiol, 6*, 171–178.

Treisman, A., & Gelade, G. (1980). A feature-integration theory of attention. *Cogn Psychol, 12*, 97–136.

Tsotsos, J. K. (1990). Analyzing vision at the complexity level. *Behav Brain Sci, 13*(3), 423–469.

Von der Malsburg, C. (1981). The correlation theory of brain function. Max-Planck-Institute for Biophysical Chemistry, Göttingen, Germany. Reprinted in E. Domany, J. L. van Hemmen, & K. Schulten (Eds.), *Models of neural networks* (Vol. 2, 1994). Berlin: Springer.

Warm, J. S., Dember, W. N., & Hancock, P. A. (1996). Vigilance and workload in automated systems. In R. Parasuraman & M. Mouloua (Eds.), *Automation and human performance* (pp. 183–200). Mahwah, NJ: Erlbaum.

Wolfe, J. M. (1992). "Effortless" texture segmentation and "parallel" visual search are *not* the same thing. *Visual Res, 32*(4), 757–763.

Wolfe, J., Alvarez, G., Rosenholtz, R., Oliva, A., Torralba, A., Kuzmova, Y., & Uhlenhuth, M. (2008). Search for arbitrary objects in natural scenes is remarkably efficient. *J Vision, 8*(6), 1103–1103.

Wolfe, J. M., Cave, K. R., & Franzel, S. L. (1989). Guided search: An alternative to the feature integration model for visual search. *J Exp Psychol Human, 15*, 419–433.

Wolfe, J. M., & Horowitz, T. S. (2004). What attributes guide the deployment of visual attention and how do they do it? *Nat Rev Neurosci, 5*(6), 495–501.

Wolfe, J. M., Klempen, N., & Dahlen, K. (2000). Post-attentive vision. *J Exp Psychol Human, 26*(2), 693–716.

Wolfe, J. M., Vo, M. L.-H., Evans, K. K., & Greene, M. R. (2011). Visual search in scenes involves selective and nonselective pathways. *Trends Cogn Sci, 15*(2), 77–84.

Yantis, S. (1993). Stimulus-driven attentional capture. *Curr Dir Psychol Sci, 2*(5), 156–161.

Yeshurun, Y., & Carrasco, M. (1998). Attention improves or impairs visual performance by enhancing spatial resolution. *Nature, 396*, 72–75.

17 Representations of Space and Feature-Based Attentional Priority in Human Visual Cortex

JOHN T. SERENCES AND SABINE KASTNER

ABSTRACT Computational theories suggest that attention shapes the topographical landscape of priority maps in visual areas extending from the lateral geniculate nucleus to prefrontal cortex. Here we review evidence suggesting that top-down deployments of attention to spatial locations and to behaviorally relevant simple (e.g., orientations, colors) and complex (e.g., categorical object information) features can jointly influence the topography of attentional priority maps. These topographic representations appear to integrate multiple sources of information to efficiently represent sensory stimuli and to guide motor interactions with important objects in the environment.

Selective attention is the mechanism by which an organism prioritizes the processing of relevant sensory inputs at the expense of irrelevant distractors. The need for selective sensory processing arises for a variety of reasons (Tsotsos, 1991). First, the neurons that encode environmental stimuli are highly variable, such that the same physical stimulus will evoke a slightly different response each time it is presented. In turn, this variability in neural response patterns limits the amount of information that neurons can encode about basic stimulus features (Pouget, Dayan, & Zemel, 2003; Seung & Sompolinsky, 1993). Second, multiple items in the visual field compete for representation, and this competition must be resolved so that the most behaviorally relevant sensory stimuli are represented robustly to guide goal-directed behavior (Desimone & Duncan, 1995). Collectively, these factors restrict the speed and accuracy of perceptual decisions and place an upper limit on our ability to interact with objects in the environment.

Computational theories propose that attention modulates the topographical landscape of spatial maps in the visual system such that the location of the most important object is associated with the highest activation level, and the activation level of irrelevant objects declines in descending order of importance. Recent studies suggest that these *attentional priority maps* of visual space are ubiquitous and span subcortical areas, as well as occipital, temporal, parietal, and frontal cortex. The activity level of neural representations that code each position in a given map can be modulated either via foreknowledge of an important spatial location (*space-based attention*) or by foreknowledge of the relevant visual features that define an object (e.g., attend to yellow objects when searching for a NYC taxi cab, termed *feature- and object-based attention*). Here we will review functional magnetic resonance (fMRI) studies that highlight the flexible and adaptive nature of attentional modulation of neural activity across the visual hierarchy (see chapters 18 and 20 by Hillyard & Luck and by Treue & Martinez-Trujillo, this volume, for related reviews that focus on monkey and human electrophysiology, respectively).

The influence of spatial attention on attentional priority maps

Although the existence of visual maps in human cortex was established nearly 100 years ago (Holmes, 1918), it took until the advent of functional magnetic resonance imaging (fMRI) in the early 1990s for researchers to reliably define maps on a whole-brain basis. Mapping the visual system also opened the door to the systematic exploration of the influence of attentional factors on the topographic representations maintained in each visual area. In the earliest investigations, a key question was: How early in the visual system does attention have an effect on spatially selective representations? In a pioneering set of studies, several groups independently established that attention to the location of a moving stimulus (Gandhi, Heeger, & Boynton, 1999; Somers, Dale, Seiffert, & Tootell, 1999) and to a stationary stimulus (Martinez et al., 1999) reliably modulated activation levels in primary visual cortex, or area V1. This result, which has subsequently been replicated many times, provided the most compelling evidence to date that attention could operate on the earliest cortical node of information processing, thus putting to rest a long-standing debate in the single-unit physiology literature (e.g., Motter, 1993).

A few years later, the effects of spatial attention were pushed even one step earlier in the processing hierarchy with the discovery of modulations in the lateral geniculate nucleus (LGN) of the thalamus, which is one synapse removed from the retina (a structure that does not have feedback connectivity) and is also the first post-retinal structure that can be modulated via feedback projections through afferent inputs from V1, the brainstem, and the thalamic reticular nucleus (O'Connor, Fukui, Pinsk, & Kastner, 2002). Moreover, attentional modulation in the LGN was shown to be even stronger than in V1, suggesting that activity in the LGN was influenced not just by feedback from cortical sources. Instead, attentional modulation in the LGN more likely reflects the joint influence of cortical, thalamic, and brainstem sources.

More recent studies have shifted focus to better understand how attention impacts topographical representations in areas of parietal and frontal cortex. Thus far, seven topographic maps have been identified in parietal cortex, each with a bias toward representing the contralateral hemifield and located along the intraparietal sulcus (IPS; termed areas IPS0-5) and the superior parietal lobule (SPL; termed area SPL1; Sereno, Pitzalis, & Martinez, 2001; Silver & Kastner, 2009; Swisher, Halko, Merabet, McMains, & Somers, 2007). In addition, two sets of maps have been identified in frontal cortex: one each in the superior and inferior branches of the precentral cortex (PCC; Kastner et al., 2007; Silver & Kastner, 2009), in the vicinity of what is typically referred to as human frontal eye fields. These frontal maps were first defined using a delayed saccade task (Kastner et al., 2007), but have subsequently also been found when observers perform an attentionally demanding peripheral task such as monitoring biological motion stimuli or performing a difficult speed discrimination task (Jerde, Merriam, Riggall, Hedges, & Curtis, 2012; Saygin & Sereno, 2008).

Silver and colleagues (2005) showed strong attentional modulation in IPS1 and 2 using a task in which subjects sequentially shifted attention around a circular stimulus array. Importantly, the physical attributes of the stimulus were matched in all attention conditions, so this task isolated the effects of attention independent from changes in sensory stimulation. In addition, Silver et al. established a gradient of sensitivity across occipital and parietal areas, such that V1 responded maximally to changes in the sensory stimulus (a large flickering checkerboard), whereas IPS1 and IPS2 showed a relatively modest response to changes in the sensory stimulus but large modulations in response to changes in the locus of attention (figure 17.1). This finding suggests that—at least with the type of simple stimulus employed by Silver et al.—retinotopic maps in higher-order areas of parietal cortex are primarily sensitive to attentional as opposed to sensory factors (see also Kastner, Pinsk, De Weerd, Desimone, & Ungerleider, 1999).

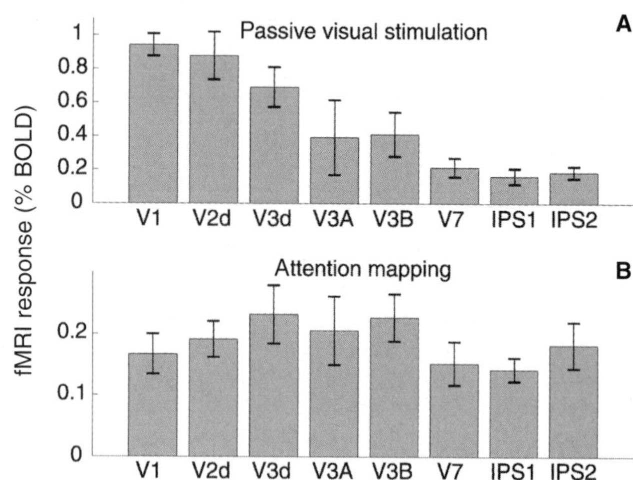

FIGURE 17.1 The relative influence of sensory and attentional factors on occipital and parietal visual areas. Responses to passive visual stimulation and attention. (A) Functional MRI response amplitudes to passive visual stimulation with a high-contrast visual stimulus. (B) Functional MRI response amplitudes during performance of the attention-mapping task in which subjects sequentially shifted covert attention around a circular annulus (with very low-contrast visual stimuli that were presented with equal probability at attended and unattended locations). For a given cortical area, the response amplitudes were computed for each subject and then averaged across subjects ($n = 4$). Error bars: standard error of the mean (SEM) across subjects. As shown in A, early visual areas show a large sensory-evoked response, even in the absence of attention. In contrast, in later stages of occipital cortex and in intraparietal sulcus (IPS) passive sensory stimulation yielded a very modest response. In contrast, B shows that almost all areas exhibited a robust and roughly equivalent modulation due purely to covert shifts of spatial attention.

The high degree of sensitivity to attentional factors suggests that maps in parietal cortex might play an important role in representing the attentional priority of competing objects in the visual field and in providing top-down biasing signals to modulate activity in earlier areas of occipital cortex. One basic prediction of this account is that attentional signals in parietal cortex should precede signals observed in occipital visual areas. Consistent with this prediction, two recent studies examined the temporal structure of blood oxygenation level dependent (BOLD) signals and found that attentional modulation occurred earlier in parietal than in occipital cortex (Lauritzen et al., 2009). In addition, Greenberg et al. (2012) measured the magnitude of attention effects in occipital areas V1–V3 using fMRI, and then measured the strength of white matter connections between IPS1 and each of the early visual areas using diffusion spectrum imaging. Consistent with many previous reports (e.g., Kastner, De Weerd, Desimone, & Ungerleider, 1998), larger spatial attention effects were observed in V3 compared to V1. Moreover, white matter connectedness was stronger between IPS1 and V3 as compared to IPS1 and V1, consistent with the notion that the strength of projections from topographic maps in IPS to occipital visual areas may play a role in determining the magnitude of attentional modulation.

In another recent study, Jerde et al. (2012) identified maps in both IPS and prefrontal cortex and tested a key prediction regarding attentional priority maps: if these IPS/frontal regions form candidate attentional priority maps of space, then they should topographically represent important locations, irrespective of why a location was initially prioritized. To test this hypothesis, subjects performed one of three tasks: they covertly attended a peripheral location, they remembered a peripheral location across a delay interval, or they planned a delayed saccade to a peripheral location. Despite differences in the type of cognitive operation that the subjects were carrying out, in all three cases a specific location in space should be prioritized, and thus the spatial representation in a putative priority map should be similar. Jerde et al. trained separate pattern classifiers to discriminate the spatial position of the prioritized location in each task condition based on the spatial distribution of the activation across voxels in each region of IPS and prefrontal cortex. First, they established that each classifier could decode the prioritized location using training data from the same task. Next, they demonstrated that the classifiers trained using data from IPS2 and superior PCC generalized across tasks, suggesting that a common spatial representation was maintained in these regions irrespective of the cognitive or motor operation being carried out (figure 17.2). This is exactly what one would predict if the topographic representations within these regions reflected the attentional priority of locations in the visual field as opposed to simply encapsulating a specific cognitive or motor facet of information processing.

How are control functions of space-based attentional priority implemented in higher-order cortex? The classic notion based on patients with visuospatial hemineglect has been that the right hemisphere controls attention across the entire visual field, while the left hemisphere exerts control only over the right hemifield. This asymmetry explains the frequently observed right-hemispheric dominance of the syndrome (Heilman & van den Abell, 1980; see also chapter 19 in this volume by Corbetta and colleagues). However, recent detailed fMRI and transcranial magnetic stimulation studies in healthy subjects do not support the notion of a biased parietal control mechanism. Instead, these studies suggest that space-based attention is controlled via an opponent processor control system in which each hemisphere directs attention toward the contralateral visual field and balance between the hemispheres is achieved through reciprocal inhibition (Kinsbourne, 1977). Consistent with this account, stronger fMRI responses were observed in each topographically mapped region of parietal cortex when attention was directed to the contralateral as compared to the ipsilateral visual field, thereby generating multiple contralateral spatial biasing signals. On average, the biasing signals were balanced across the two hemispheres, with only a few notable hemispheric asymmetries (Szczepanski, Konen, & Kastner, 2010). First, only right SPL1 generated contralateral biasing signals, and left frontal eye field (FEF) as well as left IPS1/2 generated stronger contralateral biasing signals than their counterparts in the right hemisphere. These asymmetries may be important in generating a more powerful left-hemispheric processor in the case of right-hemispheric damage, thereby accounting for the well-documented hemispheric asymmetries in spatial neglect that were described earlier. Importantly, this evidence for interhemispheric competition was further substantiated by a causal intervention using transcranial magnetic stimulation: interfering with either right or left parietal areas altered the attentional weights of individual subjects and predictably shifted spatial attention toward one visual field or the other (Szczepanski & Kastner, 2013).

While the evidence that topographically organized subregions of IPS and prefrontal cortex maintain domain independent representations of attentional priority is consistent with the notion of an attentional priority map, computational theories also predict a graded representation such that locations are continuously represented in descending order of importance (Bisley & Goldberg, 2010; Itti & Koch, 2001; Serences & Yantis, 2006). While some evidence for graded representations exists in the single unit literature (e.g., Bichot, Rossi, & Desimone, 2005; Bichot & Schall, 1999), less is known about graded representations in human visual cortex. In one line of relevant work, recent fMRI studies in humans have systematically manipulated the perceptual or cognitive load, or both, associated with a target stimulus while simultaneously assessing the response associated with unattended

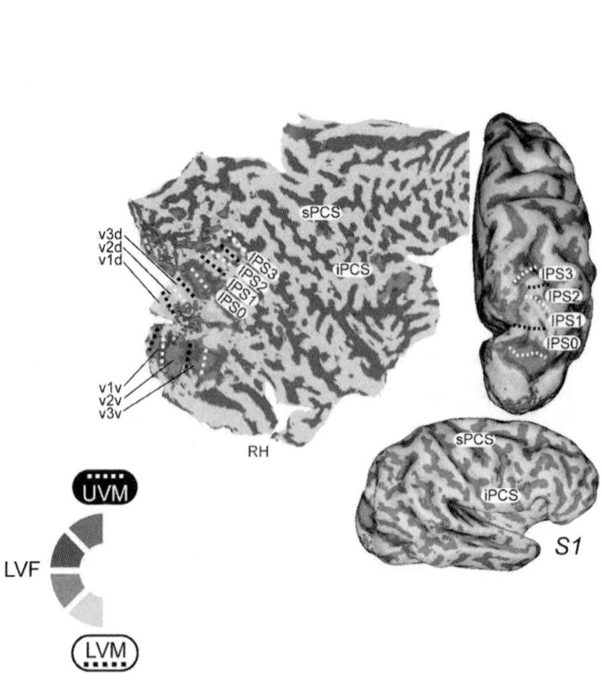

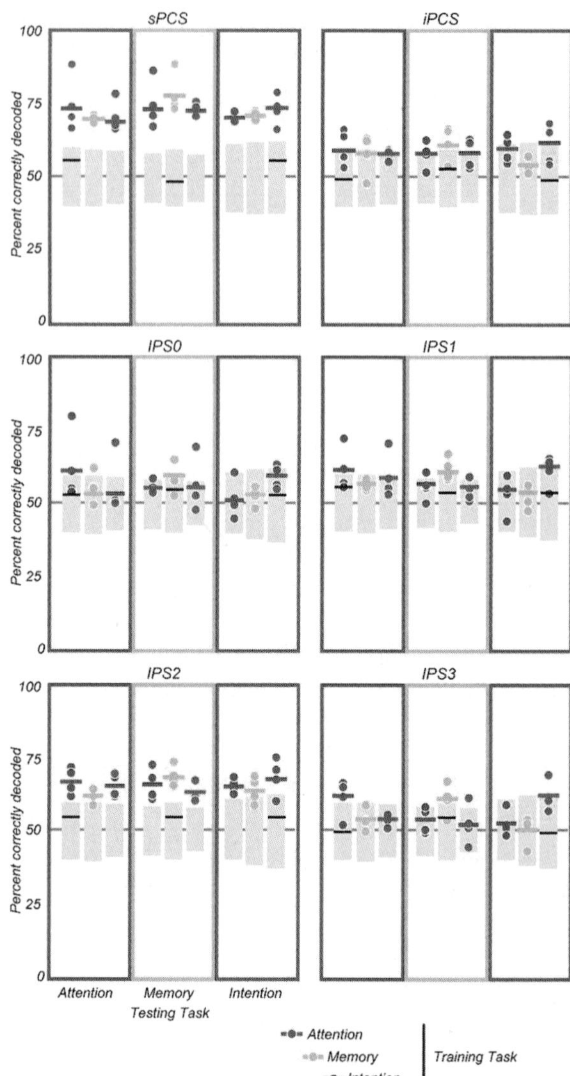

FIGURE 17.2 Task general representation of attentional priority in parietal and frontal cortex. *Left:* Topographic maps in the right hemisphere of a single subject. As indicated in the central color wheel key, warm colors reflect the position of the stimulus in the contralateral left visual field (LVF). The borders of topographic areas in occipital and parietal cortex are demarked by dotted black lines reflecting the lower visual meridian (LVM) and dotted white lines reflecting the upper visual meridian (UVM). In this subject, anterior and dorsal to early visual areas V1–V3, four topographic areas are found along the caudal-rostral intraparietal sulcus (IPS0–IPS3). In the prefrontal cortex (PFC), two topographic areas are found along the dorsal—ventral precentral sulcus (sPCS and iPCS; LH, left hemisphere; RH, right hemisphere). Note that the sPCS and the iPCS in these figures correspond to the superior and inferior precentral cortex (PCC) in the main text, respectively. *Right:* Classifier results for decoding the visual field that was currently prioritized with attention. For each topographic area, the percentage of correctly decoded trials is plotted. Each dot is an individual subject, and each horizontal line is the mean performance across subjects. The color of the dots keys the task used to train the classifier. The color of the boxes keys the task used to test the classifier. Within-task classification, dot and box colors match; across-task classification, dot and box colors do not match. The gray boxes represent the 2.5th and 97.5th percentile of the null distribution generated by random permutation analysis. Dots and bars beyond these cutoffs are significantly different from chance. The multivoxel pattern of delay period activity only in sPCS (superior PCC) and IPS2 predicts the prioritized hemifield both within and across the three spatial cognitive tasks. The black horizontal bars are the mean performance of the control analyses, in which the mean signal difference of all voxels in the left and right hemisphere topographic areas was used to predict the prioritized hemifield. (All data adapted and reprinted with permission from Jerde et al., 2012.) (See color plate 13.)

distractors. If topographic representations represent attended items in a graded manner, then the representation of unattended stimuli should be attenuated as attentional demands increase at the target location. In one study, the attentional demands associated with a task at fixation were parametrically manipulated while fMRI was used to measure neural responses associated with an unattended peripheral motion stimulus (Rees, Frith, & Lavie, 1997). The magnitude of the response in motion-selective area MT scaled inversely with task difficulty, suggesting a graded representation of unattended items as a function of their attentional priority. This same pattern of results has also been observed in the thalamus, primary visual cortex, and extrastriate cortex in other studies that employed a similar experimental logic (O'Connor et al., 2002; Pinsk, Doniger, & Kastner, 2004; Schwartz et al., 2005). In addition, the magnitude of load-dependent attenuation is larger in the LGN as compared to V1, mimicking the larger overall effect of attention on target processing that was discussed above (O'Connor et al., 2002). Thus, the data to date suggest that early subcortical and cortical regions maintain graded representations of stimuli in which the magnitude of the response corresponds to the relative attentional priority granted to each spatial location in the visual field. However, analogous experiments have not been carried out in topographically organized regions of human IPS and prefrontal cortex. Thus, future work is required to specify the functional properties of these regions. They may also maintain a graded representation, or the representations may increasingly evolve to resemble a winner-take-all mechanism in which only the most relevant location is tagged at any moment in time.

The influence of feature-based attention on attentional priority maps

In addition to being able to direct attention to specific locations in the visual field, attention can also be directed to basic visual features such as orientation, direction of motion, or color (Maunsell & Treue, 2006; Wolfe, Cave, & Franzel, 1989). In contrast to spatial attention, feature-based attention modulates the gain of neurons based on their feature-tuning profile, as opposed to the location of their spatial receptive field. As a result, feature-based attention enhances the gain of neurons that are tuned to behaviorally relevant features while simultaneously suppressing the gain of neurons that are tuned to behaviorally irrelevant features (Martinez-Trujillo & Treue, 2004; Scolari, Byers, & Serences, 2012; Serences, Saproo, Scolari, Ho, & Muftuler, 2009). Thus, a simple change in the gain (up or down) of sets of feature-selective neurons can influence the selectivity of response profiles at the population level and increase the precision of neural coding. Finally, attending to a specific feature can modulate neural activity in a spatially global manner, such that neurons tuned to an attended feature will increase their firing irrespective of their spatial receptive field (Cohen & Maunsell, 2011; Martinez-Trujillo & Treue, 2004; Sàenz, Buraĉas, & Boynton, 2002; Serences & Boynton, 2007). Thus, feature-based attention can increase the activation level of all locations in attentional priority maps that contain a potentially relevant target feature (see also chapters 20 and 22 by Treue & Martinez-Trujillo and Wolfe, this volume).

In human visual cortex, the first demonstrations of feature-based attention examined cross-category modulations as opposed to within-category modulations (e.g., attend color versus motion, as opposed to attend orientation 1 versus orientation 2). In the early 1990s, Corbetta, Miezin, Dobmeyer, Shulman, and Petersen (1991) used positron emission topography to demonstrate that attending to the color, shape, or velocity of an object differentially activated human extrastriate visual areas that were thought to be selectively involved in processing each feature dimension (e.g., MT was more active in general when subjects attended motion compared to color).

However, it was not until many years later that researchers started to examine within-category feature-based attentional modulations, which is necessary to understand how the competition between populations of similarly tuned neurons plays out to support a coherent representation of relevant features. These studies took several forms and progressed rapidly in their complexity in a relatively short period of time. Sàenz, Buraĉas, and Boynton (2002) instructed human subjects to attend to one of two overlapping moving dot fields on the left side of space; one dot field always moved up and the other down. In addition, there was always an ignored field of dots presented on the right side of space that moved either upward or downward. This design, which was similar to that employed in earlier single-unit recording studies (Martinez-Trujillo & Treue, 2004; Treue & Martinez Trujillo, 1999), allowed the researchers to fix the locus of spatial attention on the left while simultaneously examining the isolated influence of feature-based attention by measuring responses to the dot field on the right. In line with the idea of a feature-based attentional mechanism that operates in parallel across the entire visual field, the BOLD response in area MT that was associated with the ignored right stimulus was higher when it moved in the same direction as the attended dot field on the left (compared to when it mismatched the direction of the attended dot field on the left). Subsequent work further demonstrated that these modulations in areas like MT actually underwent a systematic change as a function of the attended direction, such that the pattern of activation across voxels in a given area could be used to infer the precise feature that was being attended (e.g., 45° versus 135° motion; Kamitani & Tong, 2006; Serences & Boynton, 2007). Moreover, these modulations could occur even when a region was not being directly driven by a stimulus (Serences & Boynton, 2007), suggesting that all neurons tuned to the relevant feature are modulated by attention, irrespective of their spatial receptive field (Liu & Hou, 2011; Sàenz, Buraĉas, & Boynton, 2002; White & Carrasco, 2011).

Interestingly, the speed with which feature-based attention can modulate responses in visual cortex is also consistent with a role for rapidly tagging potentially relevant locations in an attentional priority map. For example, investigators combined event-related potential and event-related field data (magnetoencephalographic signal) that was concurrently collected to determine the temporal dynamics of feature-based attentional selection in feature-selective cortical areas (Schoenfeld et al., 2007). Using these techniques, feature-based modulations were found to onset about 90–120 ms following stimulus presentation when subjects shifted between attending to colors or directions of motion. This estimate was much faster than previous estimates made while subjects shifted between features in the same dimension (e.g., Anllo-Vento, Luck, & Hillyard, 1998; Liu, Stevens, & Carrasco, 2007), suggesting that feature-based shifts between dimensions might be faster than shifts within a dimension. However, Zhang and Luck (2009) subsequently demonstrated that within-dimension shifts also occur very rapidly, at least when the stimuli were already visible before the attention-shift cue was presented.

The rapid action of feature-based shifts of attention further raised the possibility that early subcortical areas of the visual system might play a key role just as they do in the spatial domain. Schneider (2011) used fMRI to measure BOLD signals associated with between-category shifts of feature-attention within the LGN and the pulvinar, as these regions exhibit differential sensitivity to color and motion, respectively (Petersen, Robinson, & Keys, 1985). In line with predictions based on neural selectivity, the BOLD signal was larger in the LGN when subjects attended to stimulus color, and the response in the pulvinar was higher when subjects attended to motion. Thus, just as in the spatial domain, feature-based attention shaped responses at the earliest stages of visual information processing to support the efficient selection of behaviorally relevant objects.

The selection of behaviorally relevant information from the environment based on simple features such as a particular color or direction of motion has been well documented. However, we also have a remarkable ability to efficiently extract complex information related to entire objects at a categorical level in natural vision, such as when looking for cars while crossing the street. Even though exemplars from any given object category can appear in different locations, can be viewed from different viewpoints, and can contain different featural details, we are able to detect categorical information about objects in the near-absence of focused attention (Fei-Fei et al., 2002). In one recent study that explored the neural basis of such object-category based selection effects, subjects were asked to attend to briefly presented city scenes and to detect the presence of people or cars. In object-selective ventral temporal cortex, only task-relevant information was processed to the categorical level, even when the information was not spatially attended, whereas task-irrelevant information was not processed to the categorical level, even when the scenes were attended (figure 17.3B; Peelen et al., 2009). These object-category biasing signals were also found following an abstract letter cue that indicated the target category, that is, in the absence of a visual stimulus (Peelen & Kastner, 2011). These cue-related biasing signals were predictive of behavioral outcome: the probability to correctly detect a target category increased with the strength of the biasing signal. These studies indicate that neural activity in object-selective cortex is entirely dominated by task-related demands as subjects extract categorical information from natural scenes. The mechanism that mediates object category–based selection is independent from that for space-based attention and appears to operate in a spatially global fashion, similar to feature-based attention mechanisms. However, in contrast to feature-based attention, object category–based selection mechanisms are implemented at a later stage of cortical processing (i.e., in object-selective ventral temporal cortex; figure 17.3B).

While much is known about the control of spatial attention, most studies of feature- and object category–based attention have focused solely on their role in modulating activity in visual cortex. Thus, relatively little is known about the role of higher-order cortical areas in mediating these forms of attentional selection. However, two early studies implicated a subregion of the SPL in switching attention between different categories of features (color/motion) and object categories (faces/houses; Liu et al., 2003; Serences et al., 2004). More recently, Serences and Boynton (2007a, 2007b)

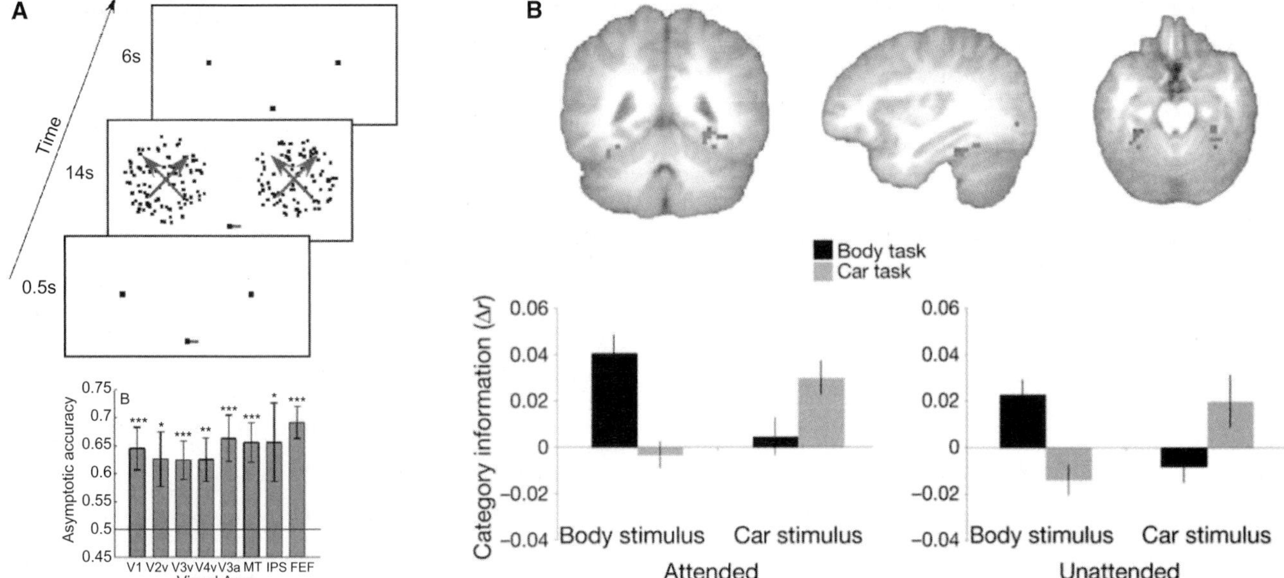

FIGURE 17.3 Feature- and object category–based attention effects. (A) *Top panel*: Sequence of events on a trial where the observer was attending to 45° motion in the right stimulus aperture. One-half of the dots in each stimulus aperture moved at 45°, and the other half moved at 135°, for the duration of the 14 sec presentation period. Targets were defined as a brief slowing of the dots at the attended location that moved in the currently attended direction (45° in this figure); distractors were defined as a brief slowing of the dots at the attended location that moved in the unattended direction (135° in this figure). *Bottom-left panel*: Asymptotic multi-voxel classification accuracy for the attended direction of motion based on fMRI activation patterns in each visual region (error bars, ±SEM across observers). (Data adapted and reprinted with permission from Serences and Boynton, 2007.) (B) *Top panel*: Clusters of object-selective activations in ventral temporal cortex (as determined by contrasting activations evoked by viewing intact vs. scrambled objects) in a group-average analysis at $P < 0.005$ (Talairach coordinates of peak: $x = 35$, $y = -41$, $z = -18$; $N = 10$). The lower panel shows category information as a function of category, task, and attention in individually defined object-selective cortex. Category information was calculated by taking the difference between within-category comparisons and between-category comparisons, and reflects the amount of category information in multi-voxel patterns of activation. Significant category information depended solely on task instruction and not on spatially attended location. Task-relevant information was processed to the categorical level, even when unattended, whereas task-irrelevant information was not represented at that level, even when attended. Error bars indicate ± SEM. (Data adapted and reprinted with permission from Peelen et al., 2009.) (See color plate 14.)

demonstrated that regions of IPS and PCC encode information about the specific direction of an attended motion stimulus, suggesting that these regions may play a role in generating the biasing signals observed in early visual areas. Liu et al. (2011) extended these findings by showing that areas of IPS and PCC also encode specific colors (as well as directions of motion). In addition, Liu et al. (2011) demonstrated that partially separable regions of IPS and PCC carry information about attended directions and colors, suggesting graded specialization or sensitivity in the putative priority maps that are thought to be maintained in these regions.

Conclusion

Together, the rapid implementation of feature- and object category–based attentional shifts and their modulatory impact on the visual system raise many parallels with the spatial attention system. Indeed, jointly directing attention to locations and to simple and complex features—to the extent supported by top-down knowledge—is a highly efficient way to rapidly hone in on the most relevant objects in the visual scene. The combined information provided by foreknowledge of relevant visual information is likely combined in higher-order topographic maps that are found in subregions of IPS and prefrontal cortex, where the landscape of topographical representations is partially divorced from the exact task or motor plan that initially induced attentional priority (Jerde & Curtis, 2013; Jerde et al., 2012).

In the future, a major challenge will be to form a more integrated understanding of the role that each region in the visual system plays in building an increasingly selective and refined representation of relevant stimuli in the environment. For instance, neurons in each area of the visual system differentially encode information about stimulus attributes (orientations, colors, etc.). Therefore, it stands to reason that selective attention implements different functions in different visual areas based on each region's innate processing capabilities. In the domain of spatial attention, this is easily seen given that early regions such as V1 are more sensitive to physical stimulus attributes, whereas later areas in IPS are increasingly sensitive to attentional demands (figure 17.1). In addition, each of these regions contains a diverse and unique set of cortical and subcortical projections. Thus, even if topographic representations in subregions of IPS and prefrontal cortex appear to be modulated in a similar manner, the modulations might have a very different impact on the ultimate behavioral response that an organism initiates in order to interact with relevant objects in the environment.

ACKNOWLEDGMENTS Supported by NIH R01-MH092345 and a James S McDonnell Foundation Scholar Award to J. T. S. and by NIH RO1-MH64043, RO1-EY017699, R21-EY0211078, and NSF BCS-1025149 to S. K.

REFERENCES

Anllo-Vento, L., Luck, S. J., & Hillyard, S. A. (1998). Spatio-temporal dynamics of attention to color: Evidence from human electrophysiology. *Hum Brain Mapp, 6*(4), 216–238.

Bichot, N. P., Rossi, A. F., & Desimone, R. (2005). Parallel and serial neural mechanisms for visual search in macaque area V4. *Science, 308*(5721), 529–534.

Bichot, N. P., & Schall, J. D. (1999). Effects of similarity and history on neural mechanisms of visual selection. *Nat Neurosci, 2*(6), 549–554.

Bisley, J. W., & Goldberg, M. E. (2010). Attention, intention, and priority in the parietal lobe. *Annu Rev Neurosci, 33*, 1–21.

Cohen, M. R., & Maunsell, J. (2011). Using neuronal populations to study the mechanisms underlying spatial and feature attention. *Neuron, 70*, 1192–1204.

Corbetta, M., Miezin, F., Dobmeyer, S., Shulman, G., & Petersen, S. (1991). Selective and divided attention during visual discriminations of shape, color, and speed: Functional anatomy by positron emission tomography. *J Neurosci, 11*(8), 2383–2402.

Desimone, R., & Duncan, J. (1995). Neural mechanisms of selective visual attention. *Annu Rev Neurosci, 18*, 193–222.

Fei-Fei, L., VanRullen, R., Koch, C., & Perona, P. (2002). Rapid natural scene categorization in the near absence of attention. *Proc Natl Acad Sci USA, 99*, 9596–9601.

Gandhi, S. P., Heeger, D. J., & Boynton, G. M. (1999). Spatial attention affects brain activity in human primary visual cortex. *Proc Natl Acad Sci USA, 96*(6), 3314–3319.

Greenberg, A. S., Verstynen, T., Chiu, Y. C., Yantis, S., Schneider, W., & Behrmann, M. (2012). Visuotopic cortical connectivity underlying attention revealed with white-matter tractography. *J Neurosci, 32*(8), 2773–2782.

Heilman, K. M., & van den Abell, T. (1980). Right hemisphere dominance for attention: The mechanism underlying hemispheric asymmetries of inattention (neglect). *Neurology, 30*, 327–330.

Holmes, G. (1918). Disturbances of vision by cerebral lesions. *Br J Ophthalmol, 2*, 353–384.

Itti, L., & Koch, C. (2001). Computational modelling of visual attention. *Nat Rev Neurosci, 2*(3), 194–203.

Jerde, T. A., & Curtis, C. E. (2013). Maps of space in human frontoparietal cortex. *J Physiol (Paris), 107*(6), 510–516.

Jerde, T. A., Merriam, E. P., Riggall, A. C., Hedges, J. H., & Curtis, C. E. (2012). Prioritized maps of space in human frontoparietal cortex. *J Neurosci, 32*(48), 17382–17390.

Kamitani, Y., & Tong, F. (2006). Decoding seen and attended motion directions from activity in the human visual cortex. *Curr Biol, 16*(11), 1096–1102.

Kastner, S., DeWeerd, P., Desimone, R., & Ungerleider, L. G. (1998). Mechanisms of directed attention in the human extrastriate cortex as revealed by functional MRI. *Science, 282*, 108–111.

Kastner, S., Pinsk, M., De Weerd, P., Desimone, R., & Ungerleider, L. G. (1999). Increased activity in human visual cortex during directed attention in the absence of visual stimulation. *Neuron, 22*(4), 751–761.

Kastner, S., DeSimone, K., Konen, C. S., Szczepanski, S. M., Weiner, K. S., & Schneider, K. A. (2007). Topographic maps in human frontal cortex revealed in memory-guided saccade and spatial working-memory tasks. *J Neurophysiol, 97*(5), 3494–3507.

Kinsbourne, M. (1977). Hemi-neglect and hemisphere rivalry. *Adv Neurol, 18*, 41–49.

Lauritzen, T. Z., D'Esposito, M., Heeger, D. J., & Silver, M. A. (2009). Top-down flow of visual spatial attention signals from parietal to occipital cortex. *J Vision, 9*(13), 1–14.

Liu, T., Hospadaruk, L., Zhu, D., & Gardner, J. L. (2011) Feature-specific attentional priority signals in human cortex. *J Neuroscience, 31*, 4484–4495.

Liu, T., Slotnick, S. D., Serences, J. S., & Yantis, S. (2003) Cortical mechanisms of feature-based attentional control. *Cereb Cortex, 13*, 1334–1343.

Liu, T., & Hou, Y. (2011). Global feature-based attention to orientation. *J Vision, 11*(10).

Liu, T., Stevens, S. T., & Carrasco, M. (2007). Comparing the time course and efficacy of spatial and feature-based attention. *Vision Res, 47*(1), 108–113.

Martinez-Trujillo, J. C., & Treue, S. (2004). Feature-based attention increases the selectivity of population responses in primate visual cortex. *Curr Biol, 14*(9), 744–751.

Martinez, A., Anllo-Vento, L., Sereno, M. I., Frank, L. R., Buxton, R. B., Dubowitz, D. J., … Hillyard, S. A. (1999). Involvement of striate and extrastriate visual cortical areas in spatial attention. *Nat Neurosci, 2*(4), 364–369.

Maunsell, J. H., & Treue, S. (2006). Feature-based attention in visual cortex. *Trends Neurosci, 29*(6), 317–322.

Motter, B. C. (1993). Focal attention produces spatially selective processing in visual cortical areas V1, V2, and V4 in the presence of competing stimuli. *J Neurophysiol, 70*(3), 909–919.

O'Connor, D. H., Fukui, M. M., Pinsk, M. A., & Kastner, S. (2002). Attention modulates responses in the human lateral geniculate nucleus. *Nat Neurosci, 5*(11), 1203–1209.

PEELEN, M. V., FEI-FEI, L., & KASTNER, S. (2009). Neural mechanisms of rapid natural scene categorization in human visual cortex. *Nature, 460,* 94–97.

PEELEN, M. V., & KASTNER, S. (2011). Is that a bathtub in your kitchen? *Nat.Neurosci, 14,* 1224–1226.

PETERSEN, S. E., ROBINSON, D. L., & KEYS, W. (1985). Pulvinar nuclei of the behaving rhesus monkey: Visual responses and their modulation. *J Neurophysiol, 54*(4), 867–886.

PINSK, M. A., DONIGER, G. M., & KASTNER, S. (2004). Push-pull mechanism of selective attention in human extrastriate cortex. *J Neurophysiol, 92*(1), 622–629.

POUGET, A., DAYAN, P., & ZEMEL, R. S. (2003). Inference and computation with population codes. *Annu Rev Neurosci, 26,* 381–410.

REES, G., FRITH, C. D., & LAVIE, N. (1997). Modulating irrelevant motion perception by varying attentional load in an unrelated task. *Science, 278*(5343), 1616–1619.

SÀENZ, M., BURAĈAS, G. T., & BOYNTON, G. M. (2002). Global effects of feature-based attention in human visual cortex. *Nat Neurosci, 5*(7), 631–632.

SAYGIN, A. P., & SERENO, M. I. (2008). Retinotopy and attention in human occipital, temporal, parietal, and frontal cortex. *Cereb Cortex, 18*(9), 2158–2168.

SCHNEIDER, K. A. (2011). Subcortical mechanisms of feature-based attention. *J Neurosci, 31*(23), 8643–8653.

SCHOENFELD, M. A., HOPF, J. M., MARTINEZ, A., MAI, H. M., SATTLER, C., GASDE, A., ... HILLYARD, S. A. (2007). Spatiotemporal analysis of feature-based attention. *Cereb Cortex, 17*(10), 2468–2477.

SCHWARTZ, S., VUILLEUMIER, P., HUTTON, C., MARAVITA, A., DOLAN, R. J., & DRIVER, J. (2005). Attentional load and sensory competition in human vision: Modulation of fMRI responses by load at fixation during task-irrelevant stimulation in the peripheral visual field. *Cereb Cortex, 15*(6), 770–786.

SCOLARI, M., BYERS, A., & SERENCES, J. T. (2012). Optimal deployment of attentional gain during fine discriminations. *J Neurosci, 32*(22), 7723–7733.

SERENCES, J. T., & BOYNTON, G. M. (2007a). Feature-based attentional modulations in the absence of direct visual stimulation. *Neuron, 55*(2), 301–312.

SERENCES, J. T., & BOYNTON, G. M. (2007b). The representation of behavioral choice for motion in human visual cortex. *J Neurosci, 27*(47), 12893–12899.

SERENCES, J. T., SCHWARZBACH, J., COURTNEY, S. M., GOLAY, X., & YANTIS, S. (2004). Control of object-based attention in human cortex. *Cereb Cortex, 14*(12), 1346–1357.

SERENCES, J. T., SAPROO, S., SCOLARI, M., HO, T., & MUFTULER, L. T. (2009). Estimating the influence of attention on population codes in human visual cortex using voxel-based tuning functions. *NeuroImage, 44*(1), 223–231.

SERENCES, J. T., & YANTIS, S. (2006). Selective visual attention and perceptual coherence. *Trends Cogn Sci, 10*(1), 38–45.

SERENO, M. I., PITZALIS, S., & MARTINEZ, A. (2001). Mapping of contralateral space in retinotopic coordinates by a parietal cortical area in humans. *Science, 294*(5545), 1350–1354.

SEUNG, H. S., & SOMPOLINSKY, H. (1993). Simple models for reading neuronal population codes. *Proc Natl Acad Sci USA, 90*(22), 10749–10753.

SILVER, M. A., & KASTNER, S. (2009). Topographic maps in human frontal and parietal cortex. *Trends Cogn Sci, 13*(11), 488–495.

SILVER, M. A., RESS, D., & HEEGER, D. J. (2005). Topographic maps of visual spatial attention in human parietal cortex. *J Neurophysiol, 94*(2), 1358–1371.

SOMERS, D. C., DALE, A. M., SEIFFERT, A. E., & TOOTELL, R. B. (1999). Functional MRI reveals spatially specific attentional modulation in human primary visual cortex. *Proc Natl Acad Sci USA, 96*(4), 1663–1668.

SWISHER, J. D., HALKO, M. A., MERABET, L. B., MCMAINS, S. A., & SOMERS, D. C. (2007). Visual topography of human intraparietal sulcus. *J Neurosci, 27*(20), 5326–5337.

SZCZEPANSKI, S. M., & KASTNER, S. (2013). Shifting attentional priorities: Control of spatial attention through hemispheric competition. *J Neurosci, 33,* 5411–5421.

SZCZEPANSKI, S. M., KONEN C. S., & KASTNER, S. (2010). Mechanisms of spatial attention control in frontal and parietal cortex. *J Neurosci, 30,* 148–160.

TREUE, S., & MARTINEZ TRUJILLO, J. C. (1999). Feature-based attention influences motion processing gain in macaque visual cortex. *Nature, 399*(6736), 575–579.

TSOTSOS, J. K. (1991). Analyzing vision at the complexity level. *Behav Brain Sci, 14*(4), 768–768.

WHITE, A. L., & CARRASCO, M. (2011). Feature-based attention involuntarily and simultaneously improves visual performance across locations. *J Vision, 11*(6).

WOLFE, J. M., CAVE, K. R., & FRANZEL, S. L. (1989). Guided search: An alternative to the feature integration model for visual search. *J Exp Psychol Human, 15*(3), 419–433.

ZHANG, W., & LUCK, S. J. (2009). Feature-based attention modulates feedforward visual processing. *Nat Neurosci, 12*(1), 24–25.

18 Electrophysiology of Visual Attention in Humans

STEVEN J. LUCK AND STEVEN A. HILLYARD

ABSTRACT Human vision unfolds rapidly—with dozens of distinct processes occurring within a half-second of the onset of a stimulus—and attention influences many of these processes. The event-related potential (ERP) technique is particularly well suited for measuring these rapidly varying attentional processes because of its millisecond-level temporal resolution. Recent ERP studies of attention have been used to study two major topics in attention, namely, the mechanisms by which attention is directed to specific types of stimuli and the consequences of attention on the processing of incoming sensory information. In particular, recent research on attentional control shows that salient stimuli are always detected, even if task-irrelevant, but can be prevented from capturing attention by top-down suppression mechanisms. Once attention has been captured by a stimulus, these same suppression mechanisms may be used to terminate attention when the perception of the attended object is complete. In addition, recent studies of the effects of attention on sensory processing have focused on a subset of ERPs called steady-state visual evoked potentials (SSVEPs), which are oscillating potentials induced by flickering stimuli. By tagging individual stimuli with different flicker frequencies, SSVEPs make it possible to assess the effects of attention on sensory processing for multiple simultaneous stimuli. Recent SSVEP studies have shown that the "spotlight" of attention can be divided among multiple discontinuous locations and have delineated the speed at which attention can be shifted.

The role of event-related potentials in studying attention

Studies of visual selective attention are typically focused on understanding two broad questions (see reviews by Luck & Gold, 2008, and Luck & Vecera, 2002). First, how does the brain determine which sources of input should be attended? This is called the *control of attention.* Second, once attention has been focused on a given source of information, how does this change the processing of the attended and unattended information? This is called the *implementation of selection.*

Event-related potentials (ERPs) have played a key role in addressing both of these questions for many years (see reviews by Hillyard, Vogel, & Luck, 1998; Luck & Kappenman, 2012). ERPs have two main advantages over behavioral measures for addressing these questions. First, ERPs provide a continuous, millisecond-resolution measure of processing between a stimulus and a response (as well as prior to the stimulus and after the response). This makes it possible to track shifts of attention as they occur and to determine the stage at which attention influences the processing of incoming sensory information. Second, ERPs make it possible to "covertly" monitor the processing of a stimulus in the absence of a behavioral response. This is especially valuable in attention research, because it is difficult to obtain a behavioral response to a stimulus that is truly unattended.

The temporal resolution of ERPs also makes it possible to answer questions that cannot be addressed by techniques that are limited by the slow time course of the hemodynamic response, such as functional magnetic resonance imaging (fMRI). Many different attention-related processes may occur within a given brain area within the first half-second after the onset of a stimulus, and it is difficult to pull these processes apart with fMRI. In addition, fMRI typically mixes together brain activity elicited by attended and unattended stimuli, making it difficult to assess the implementation of selection with fMRI. In contrast, the ERPs elicited by attended and unattended stimuli can easily be recorded and measured separately, as long as these stimuli are separated by a few hundred milliseconds.

A special class of ERPs elicited by repetitively flashing or flickering stimuli has proven to be particularly useful for studying the implementation of visual selection. These so-called *steady-state visual evoked potentials* (SSVEPs) are continuous oscillatory neural responses elicited in the visual cortex at the same fundamental frequency as the flickering stimulus. If several stimuli are presented simultaneously, but each is flickering at a different frequency, the brain activity for each stimulus can be separated by extracting oscillations at the specific frequencies of stimulation. This is called *frequency tagging* the stimuli, and it makes it possible to separately measure sensory activity to attended and ignored stimuli that are presented simultaneously but flicker at different frequencies. Frequency-tagged

SSVEPs can reveal selection mechanisms that are different from those employed when the stimuli are brief flashes (Andersen, Müller, & Hillyard, 2012). Frequency-tagged SSVEPs can also be used to study the selective processing of spatially intermingled stimuli (e.g., spatially overlapping fields of red and green dots), which is difficult to achieve with other neural recording methods. Finally, since SSVEPs provide a continuous measure of selective stimulus processing, the time course of their amplitude modulations can reveal the speed with which selection is implemented following a cue to attend (Müller et al., 1998).

ERP studies of attentional control

THE N2PC AND P_D COMPONENTS Many studies of the control of visual attention make use of the *N2pc* component, an attention-related ERP response that is useful for tracking the spatial allocation of attention millisecond by millisecond (see review by Luck, 2012). Figure 18.1 shows a prototypical paradigm for isolating the N2pc component. Each display consists of a large number of randomly located squares, most of which are black. One square is red and another is green, and these two colored squares are randomly located, with the constraint that they are always on opposite sides of the display. Subjects are told to attend to the red item in some trial blocks and the green item in other blocks, and they press a button on each trial to indicate if the attended-color item has a gap on the top versus a gap on the bottom. In this paradigm, the location of the to-be-attended object is not known in advance, and attention must be shifted to the target after an initial wave of processing that finds a likely target object in the stimulus array.

The N2pc is a negative-going ERP deflection in the N2 latency range (200–300 ms) at posterior electrode sites over the hemisphere contralateral to the location of the target item (N2pc stands for *N2-posterior-contralateral*). As shown in figure 18.1, the voltage over the left hemisphere is more negative from 200–300 ms when the target is in the right hemisphere compared to when the target is in the left hemisphere. Similarly, the voltage over the right hemisphere is more negative when the target is in the left hemisphere than when it is in the right hemisphere. The waveforms are usually collapsed across hemispheres, with one waveform for contralateral (left hemisphere, right target, averaged with right hemisphere, left target) and another for ipsilateral (left hemisphere, left target, averaged with right hemisphere, right target). The N2pc component is typically superimposed on the positive-polarity P2 and P3 components, but P2 and P3 are not lateralized, so the N2pc component can be isolated by means of a contralateral-minus-ipsilateral difference wave (see figure 18.1, right).

Many studies have shown that the N2pc component reflects the focusing of attention onto a given item (reviewed by Luck, 2012). Single-unit recording studies in monkeys using similar tasks have found analogous attention-related changes in neural firing rates in area V4 and inferotemporal cortex with approximately the same latency (Chelazzi, Miller, Duncan, & Desimone, 1993, 2001). The amplitude of the N2pc and the size of these single-unit attention effects show similar modulations in response to a variety of stimulus and task manipulations (Luck, Girelli, McDermott, & Ford, 1997). Moreover, magnetoencephalographic and fMRI experiments suggest that the N2pc is generated in the human homologues of macaque area V4 and inferotemporal cortex (Hopf et al., 2006). Thus, the N2pc component likely reflects the same mechanism of attention reflected by changes in single-unit activity in these areas.

A recent study discovered a component called P_D (distractor positivity) that is much like N2pc, but is opposite in polarity and reflects a suppression of processing rather than attentional enhancement (Hickey, di Lollo, & McDonald, 2009). As illustrated in figure 18.2A, the stimulus arrays in this study contained one object on the vertical midline and another object on the left or right side of the display. On some trials, the item on the midline was the target and the lateralized item was the distractor, and on other trials this was reversed. If a difference wave is formed by means of a contralateral-minus-ipsilateral subtraction relative to the location of the lateralized item, this difference wave will contain little or no contribution from the midline item (because the brain activity for the midline item should be largely bilateral). Thus, the difference wave will reflect target-related activity when the target is the lateralized item and distractor-related activity when the distractor is the lateralized item.

When the target was lateralized and the distractor was on the midline, Hickey et al. (2009) observed a typical N2pc component (figure 18.2B). In contrast, when the distractor was lateralized and the target was on the midline, they observed a more positive voltage over the hemisphere contralateral to this distractor (figure 18.2C). There is growing evidence that this positivity, the P_D component, reflects some kind of suppressive influence on distractor processing.

TRACKING THE TIME COURSE OF ATTENTION WITH N2PC In a well-controlled N2pc experiment, the

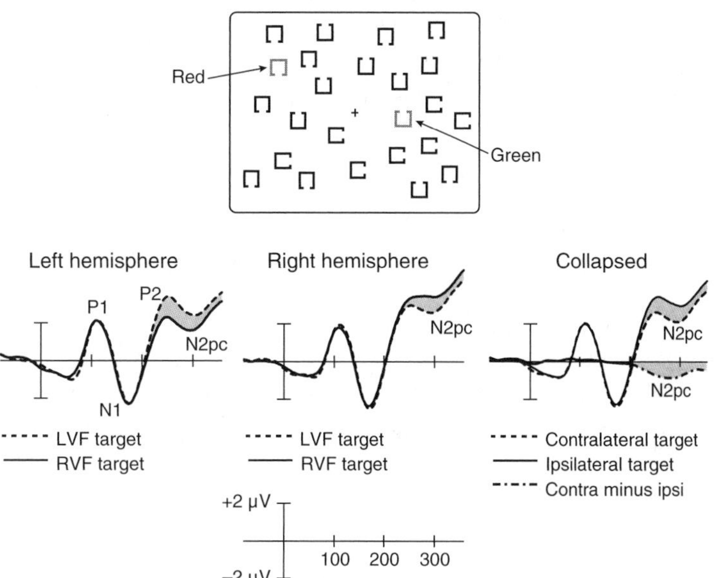

FIGURE 18.1 Typical N2pc paradigm and results. Each stimulus array contains a distinctly colored item on each side, and one of these two colors is designated as the target color in the instruction screen for each trial block. The locations of the items varied at random from trial to trial, except that the two pop-out colors are always on opposite sides. The subject is required to press one of two buttons to indicate whether the gap in the target item is on the top or the bottom of the square. The N2pc component is defined as the difference between the contralateral and ipsilateral waveforms (shown as the shaded region), which can be isolated by constructing a contralateral-minus-ipsilateral difference wave. (Adapted with permission from Luck, 2014.)

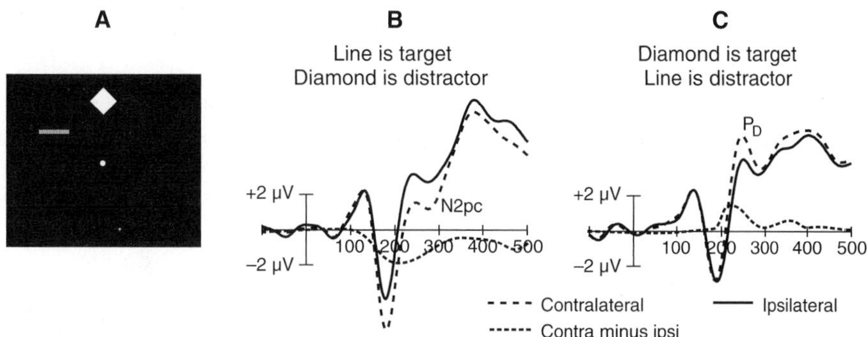

FIGURE 18.2 Example stimuli and results from the study of Hickey et al. (2009). Each array (A) contained a bright green square or diamond along with a short or long red line that was isoluminant with the background. The red stimulus shows almost no lateralized sensory activity, making it possible to attribute any later lateralized activity to top-down processing. When the red line was the target (B), a contralateral negativity (N2pc) was elicited by the red item. When the green diamond was the target (C), the ERP was more positive contralateral to the red distractor item than ipsilateral to this item. This contralateral positivity is called the distractor positivity (P_D). Waveforms courtesy of John McDonald. (Adapted with permission from Luck, 2012.)

contralateral-minus-ipsilateral difference wave cannot deviate from zero microvolts until the brain has detected and localized the to-be-attended item and begun to shift attention toward it. Thus, N2pc onset latency provides a means of tracking the time course of attentional control. Not surprisingly, N2pc onset latency varies with the salience of the target and is earlier for targets that are associated with large rewards (Kiss, Driver, & Eimer, 2009).

N2pc onset latency can also be used to study individual and group differences in the speed of attentional control. For example, N2pc onset latency is increased in older individuals (Lorenzo-Lopez, Amenedo, & Cadaveira, 2008) and in patients with hepatic encephalopathy (Schiff et al., 2006). In contrast, no change in N2pc amplitude or latency has been observed in patients with schizophrenia (Luck et al., 2006) or Parkinson's disease (Praamstra & Plat, 2001).

The N2pc component has also been used to test the hypothesis that some visual search tasks involve serial shifts of covert attention. If a search task is so difficult that the observer cannot tell if an item is a target or a distractor without focusing attention on it, some theories propose that attention is focused sequentially on each item until the target is found (Treisman, 1988). Other theories propose that attention is simply divided among all the items in parallel (Duncan, 1996). Woodman and Luck (1999, 2003) provided strong evidence in favor of serial processing by showing that, when the distractor and target are on opposite sides of the display, the N2pc can be observed to shift from the hemisphere contralateral to the distractor to the hemisphere contralateral to the target (spending about 100 ms on each item).

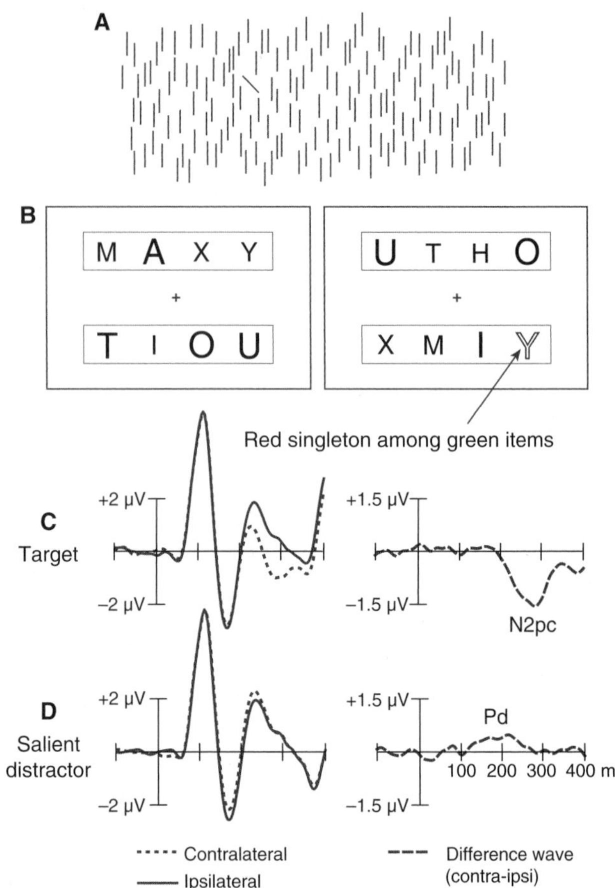

FIGURE 18.3 (A) Example of a feature singleton, which appears to "pop out" from the background. (B) Stimuli from the study of Sawaki and Luck (2010). Subjects searched for a specific target letter (e.g., the large A) and ignored color. (C) N2pc elicited by the target. (D) P_D elicited by the irrelevant color singleton.

BOTTOM-UP AND TOP-DOWN CONTROL OF ATTENTION Figure 18.3A displays a homogeneous field of vertical lines along with a single tilted line. The tilted line is called an orientation singleton, and it "pops out" from the other lines. For the last two decades, attention researchers have debated whether this "pop-out" phenomenon reflects a completely automatic, bottom-up attraction of attention to the singleton (van der Stigchel et al., 2009), or is the result of the match between the singleton and a top-down attentional set (Folk, Remington, & Johnston, 1992). Behavioral evidence has failed to resolve this debate, and recordings of N2pc yielded conflicting evidence, with an N2pc observed for salient but task-irrelevant singletons under some conditions (e.g., Hickey, McDonald, & Theeuwes, 2006) but not others (e.g., Eimer & Kiss, 2008).

In attempt to resolve this conflict, Sawaki and Luck (2010) proposed a *signal-suppression* hypothesis that combines elements of both the bottom-up and top-down hypotheses. According to the signal-suppression hypothesis, feature singletons always elicit an attentional priority signal (an *attend-to-me* signal), but this signal can be suppressed by top-down control processes before attention is actually allocated to the source of the priority signal. The suppression process is further hypothesized to produce a P_D wave.

Figure 18.3B shows the design of an experiment that was designed to test this hypothesis. Subjects performed a visual search task in which they searched for a specific combination of size and letter identity (e.g., a large A). All of the items in the display were one color (e.g., green), but a color singleton distractor was sometimes present (e.g., a red letter). The target, when present, was found to elicit an N2pc, as would be expected given that subjects were motivated to focus attention on this item (figure 18.3C). In contrast, the salient singleton distractor elicited a P_D component (figure 18.3D). Thus, although subjects had no incentive to attend to the color singleton, it was detected (i.e., it generated an attend-to-me priority signal). However, it did not elicit a shift of attention; instead, it triggered the suppression process indexed by the P_D component, thus lending strong support to the signal-suppression hypothesis.

DON'T FADE AWAY: TERMINATING SHIFTS OF ATTENTION The timing of the N2pc component indicates that attention typically dwells on a target for about 100 ms (although it may dwell longer when the target is particularly difficult to perceive). Presumably, this reflects the amount of time needed to perform perceptual processing and perhaps working memory encoding on the target once attention has been shifted to it. But what happens to attention once target processing is complete? Does attention passively fade away, or is an active process needed to terminate the episode of attention?

A recent series of experiments has demonstrated that a shift of attention to a target typically leads to an N2pc component that is immediately followed by a P_D component (Sawaki, Geng, & Luck, 2012). This suggests that the same mechanism that is used to prevent a shift of attention to a salient but irrelevant distractor item is also used to terminate a shift of attention after target processing is complete. This is analogous to the phenomenon of inhibition of return, in which a peripheral transient (e.g., the onset of a new stimulus) will trigger an automatic shift of attention to the location of the transient and then an inhibition of processing at that location (Klein, 2000). However, inhibition of return is ordinarily observed only after peripheral transients, not after voluntary shifts of attention. In addition, the P_D process appears to bring attention back to a neutral state rather than creating inhibition at the previously attended location (relative to other locations in the display). Thus, the termination process reflected by the P_D component appears to be different from the mechanisms that produce inhibition of return.

ERP studies of the implementation of selection

Spatial Attention ERPs have been used to study visual-spatial attention for several decades (see reviews by Hillyard et al., 1998; Hopfinger, Luck, & Hillyard, 2004; Luck & Kappenman, 2012). Figure 18.4 shows the most common experimental paradigm and typical results. Subjects direct gaze to a central fixation point while a rapid sequence of stimuli is presented to the left and right visual fields. Subjects are instructed to attend to one side for a block of trials and detect rare targets on that side, and eye movements are monitored to make sure that they attend to this side covertly rather than overtly. A stimulus in this paradigm typically elicits larger P1 and N1 waves when it is attended than when attention is directed to the opposite side of the display (figure 18.4B). The P1 effect typically begins within 100 ms of stimulus onset and appears to arise from lateral areas of extrastriate visual cortex (Di Russo, Martinez, & Hillyard, 2003; Martinez et al., 1999). However, attention does not reliably influence the C1 wave, which is generated in striate cortex (Mangun, Hillyard, & Luck, 1993; reviewed in Ding, Martinez, Qu, & Hillyard, 2013). Thus, spatial attention appears to modulate the flow of information through intermediate levels of visual cortex. Analogous effects of attention in extrastriate visual cortex but not in striate cortex have also been observed in monkey single-unit studies (Luck, Chelazzi et al., 1997), although attention can influence single-unit activity in striate cortex under some conditions (Briggs, Mangun, & Usrey, 2013). Attention also influences the blood-oxygen-level-dependent (BOLD) signal

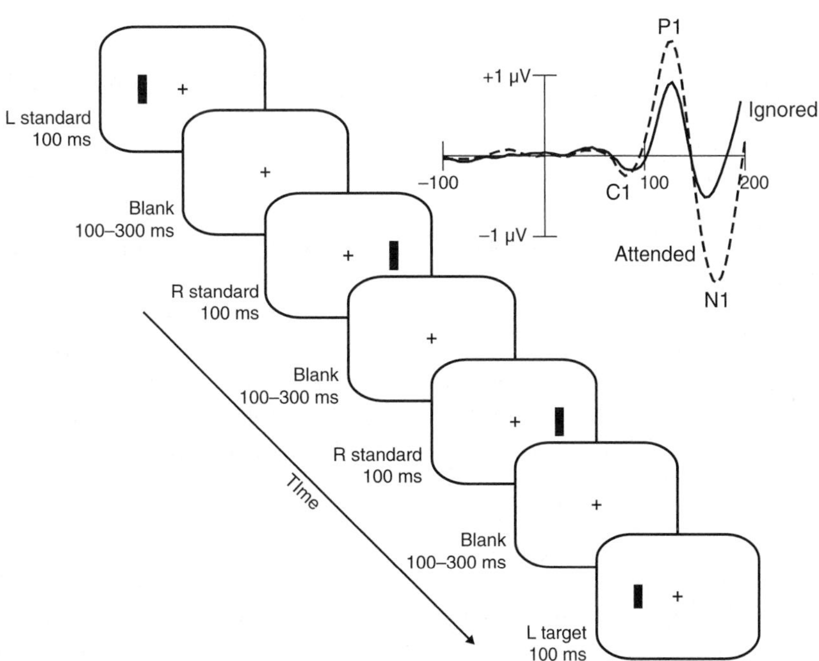

Figure 18.4 Prototypical paradigm and results for using ERPs to assess the effects of spatial attention on sensory processing. Subjects attend to the left side of the display in some trial blocks and to the right side on others, making it possible to examine the ERP elicited by a stimulus in a given location when it is attended and when it is ignored. The P1 and N1 components are typically found to be increased in amplitude for attended-location stimuli. (Adapted with permission from Luck & Kappenman, 2012.)

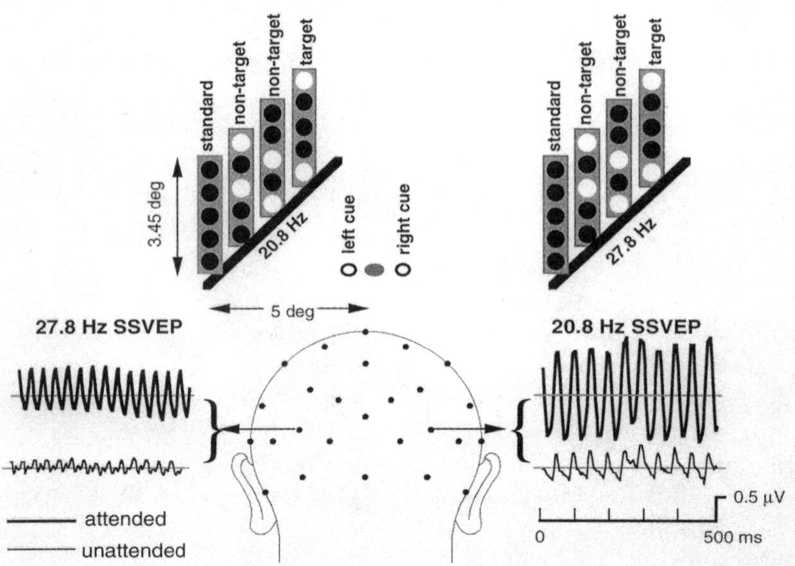

FIGURE 18.5 Schematic diagram of stimulus array and SSVEP waveforms from one subject in the study of Müller et al. (1998). On each trial, subjects were cued to attend to the left or right array of flickering LEDs and detect a target pattern of color changes. SSVEP waveforms shown were averaged in the time domain using a moving window average over the interval 1–3 sec after cue onset. (Reprinted with permission from Müller, Teder-Sälejärvi, & Hillyard, 1998.)

in striate cortex, but this likely reflects late feedback signals rather than a feedforward modulation of incoming sensory signals (Martinez et al., 1999).

SSVEPs elicited by flickering stimuli at multiple locations in the visual field are strongly modulated by the allocation of spatial attention among those locations (reviewed in Andersen et al., 2012). Typically, SSVEPs to stimuli within the spotlight of attention show enhanced amplitudes in early visual-cortical areas relative to stimuli outside the spotlight (figure 18.5). When flickering stimuli are presented at two spatially separated locations that need to be attended simultaneously, the spotlight can be effectively divided in two for periods of at least several seconds, as evidenced by SSVEPs being enhanced to attended-location stimuli but not to intermediate, unattended stimuli (Müller et al., 2003). Such a sustained splitting of the attentional spotlight is not readily achieved when stimuli are flashed transiently at the attended locations, suggesting that the human visual system can only divide attention effectively between separated locations when stimuli are continuously present (or flickering rapidly so that they appear to be continuous), as is usually the case in the natural environment. The sensitivity of SSVEPs to spatial attention has led to the development of brain-computer interfaces whereby the intentions of an immobile subject can be conveyed via amplitude modulations of the SSVEP produced by directing attention within an array of flickering stimuli (Vialatte, Maurice, Dauwels, & Cichocki, 2010). SSVEP amplitude modulations have also been used to show that improvements in visual attention seen in fast-action video game players were associated with a more effective suppression of irrelevant competing inputs (Krishnan, Kang, Sperling, & Srinivasan, 2012; Mishra, Zinni, Bavelier, & Hillyard, 2011).

FEATURE-BASED ATTENTION In a typical ERP study of feature-selective attention, a series of stimuli is presented at a single location, and the subject is asked to attend only to the stimuli that possess a specific feature (e.g., the color blue). Under these conditions, the P1 and N1 waves are not any larger for stimuli containing the attended feature than for stimuli that lack this feature (Anllo-Vento & Hillyard, 1996). This suggests that selection based on nonspatial features occurs at a later stage of processing than selection based on spatial location. However, recent studies have shown that color-based attention can influence P1 amplitude under certain conditions, operating at the same stage as spatial attention (Zhang & Luck, 2009). In these studies, the attended and ignored colors were present continuously and were spatially overlapping, which presumably maximized the competition between them, and the ERPs were elicited by transient probe events.

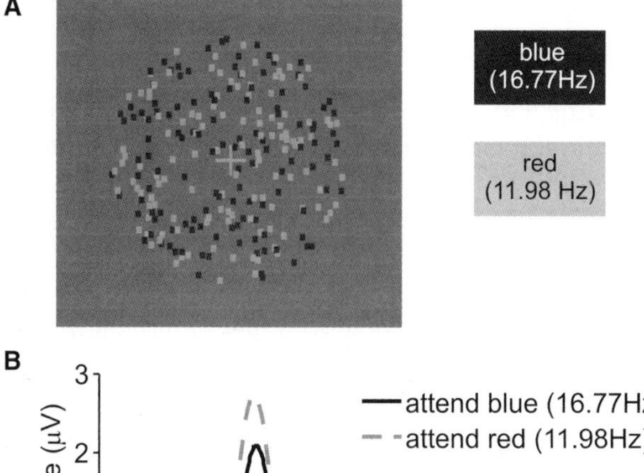

FIGURE 18.6 Stimulus display and spectral pattern of SSVEP amplitude modulations with attention in study of Andersen and Müller (2010). Each trial began with concurrent flickering of the overlapping red (11.98 Hz) and blue (16.77 Hz) dot populations, followed by a color-change cue indicating which color should be attended. Amplitude spectra of SSVEPs were obtained from Fourier transforms over the interval −300 to +1500 ms with respect to the cue, averaged over all subjects. (Reprinted with permission from Andersen & Müller, 2010.)

Feature-selective attention to overlapping fields of stimuli can also be studied with SSVEPs using the frequency tagging principle. In the experiment shown in figure 18.6 (Andersen & Müller, 2010), overlapping fields of randomly moving red and blue dots were flashed at 11.98 and 16.77 Hz, respectively. Subjects were cued to attend to either the red or blue dots on each trial, with the task of detecting occasional targets of coherent dot motion in the attended color. In this and several other similar studies (reviewed in Andersen et al., 2012), it was found that the attended-color dots elicited an enhanced SSVEP in a region that encompassed visual areas V1–V3. This facilitation of attended-color inputs at early cortical levels was shown to operate independently of spatial attention mechanisms, since the randomly moving dots could not be tracked spatially (Andersen, Müller, & Hillyard, 2009).

The combined effects of spatial- and feature-selective attention were studied by Andersen, Fuchs, & Müller (2011) in a design where overlapping fields of red and blue dots were presented to both the right and left visual fields. Each of these four randomly moving dot populations was flickered at a different rate and hence elicited a separately measurable SSVEP. Subjects were required to attend to dots of one color at one location at a time, again searching for coherent-motion targets. It was found that the SSVEP was enhanced maximally to dots having the attended color and location, and that these feature and spatial selective effects were largely additive. It thus appears that spatial and feature selective attention can independently enhance signal gain at early levels of visual-cortical processing. Moreover, the finding that stimuli of the attended color were facilitated equivalently at attended and unattended locations provides strong support for the feature-similarity gain model (Maunsell & Treue, 2006), which specifies that paying attention to a feature at one location results in a global facilitation of the processing of that feature throughout the entire visual field. Current evidence indicates this global facilitation of attended-feature inputs at an early cortical level only occurs when the attended stimuli are continuously visible and hence can be readily studied with SSVEP techniques (Andersen, Müller, & Hillyard, 2013).

TRACKING THE TIME COURSE OF ATTENTION SHIFTS WITH SSVEPs The speed with which attention can be shifted from one item or location to another has been estimated using a variety of behavioral and electrophysiological techniques. Because SSVEPs are recorded continuously over time, the speed of attentional shifting can be measured with high temporal resolution by examining the time course of SSVEP amplitude modulations following an attention-directing cue. In a study of spatial attention shifting, Müller et al., (1998) presented flickering stimulus arrays in the left (20.8 Hz) and right (27.8 Hz) visual fields as shown in figure 18.5; the stimuli began to flicker at the beginning of each trial, and after a delay a central (endogenous) cue was given directing attention to the left or right array. The SSVEP elicited by the cued array started to increase about 400–500 ms after the cue onset, and this increase was paralleled by an improvement in behavioral target detection rates, thereby confirming earlier reports that endogenously cued shifts of spatial attention require several hundreds of milliseconds to be fully implemented. Interestingly, this cued spatial attention effect appeared to involve early cortical facilitation of the attended-side input with no parallel suppression of the unattended-side input. These effects were confirmed by Kashiwase, Matsumiya, Kuriki, and Shioiri (2012), who showed further that attention enhanced the phase coherence of the SSVEP, thereby suggesting that attention increases neural-response synchronization and increases neural population activity.

A similar approach was used by Andersen and Müller (2010) to study cued shifting of color-selective attention

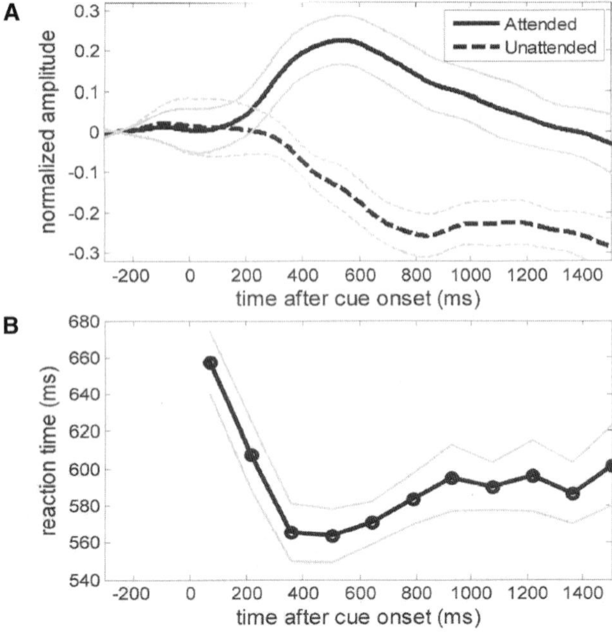

Figure 18.7 (A) Time course of SSVEP amplitude modulations with attention in study by Andersen and Müller (2010). Stimulus layout is shown in figure 18.6. SSVEP amplitudes calculated by moving Gabor filters were collapsed over the red and blue dot populations and show distinct time courses of enhancement for the attended dots and suppression for the unattended dots. (B) Time course of behavioral reaction times to targets (brief intervals of coherent motion of the attended dots).

using the stimuli shown in figure 18.6. On each trial, the red and blue dot arrays began flickering for over a second before a color change of the fixation point cued the subject to attend to the red or blue dots. The SSVEP elicited by the attended-color dots showed a significant amplitude increase 220 ms after the cue onset, which was followed after another 150 ms by an amplitude suppression of the SSVEP to the unattended dots (figure 18.7A). These amplitude modulations were paralleled by a speeding of behavioral reaction times to target detections following the cue (figure 18.7B). These results suggest that color-selective attention may be engaged more rapidly than endogenously cued spatial attention, and that color selectivity involves both a facilitation of the attended-color items and a suppression of the unattended-color items. This is consistent with many prior visual search experiments showing that, when the to-be-attended color is known in advance, the target color triggers an N2pc component beginning 175–200 ms after the onset of the visual search array. Moreover, a small probe stimulus that appears 250 ms after the onset of the search array elicits enhanced sensory-evoked ERPs when presented at the target location and suppressed sensory-evoked ERPs when presented at a distractor location (Luck & Hillyard, 1995).

Several recent studies have also capitalized on the continuous nature of SSVEPs to track different aspects of attentional selection over time. In a divided spotlight experiment akin to that of Müller et al. (2003), Itthipuripat, Garcia, and Serences (2013) examined the time course of SSVEPs elicited by spatially separated target items and by an intermediate distractor. The results showed that the spotlight of attention could be divided across noncontiguous regions of space over a temporally discrete interval prior to the behavioral response, and that the degree of amplitude separation between target and distractor SSVEPs was predictive of behavioral accuracy. In a task where subjects judged the orientation tilt of a contrast-reversing circular grating, Garcia, Srinivasan, and Serences (2013) identified scalp distribution patterns of time-varying SSVEPs that were orientation-specific; amplitude changes of these feature-selective SSVEP profiles over time were found to be predictive of tilt discrimination performance. The time course of attentional capture by a meaningful picture could also be revealed through SSVEP recordings. Hindi-Attar, Andersen, and Müller (2010) presented pictures that were overlaid with flickering dots and found that the dot-evoked SSVEP was sharply reduced within 200 ms after the picture emerged from a scrambled background. This reduction was greater for more emotionally charged pictures, indicating a more intensive focusing of attention on the picture (and away from the irrelevant dots).

Conclusion

This chapter has highlighted recent research using both transient ERPs and SSVEPs. Transient ERPs have been used for decades to study attention, and they continue to be an important tool in attention research. In particular, the N2pc component and the more recently discovered P_D component have become popular measures for assessing the control of attention. Experiments with these components have shown that it is possible to avoid shifting attention to some kinds of salient stimuli (e.g., color singletons), but that this reflects an active suppression of the salient stimuli. The N2pc component is also useful for tracking the time course of attention over the half-second following the onset of a static stimulus array. SSVEPs have been more recently harnessed for studying attention, and they have the advantage of being able to separately measure the sensory processing of multiple simultaneous stimuli. Recent SSVEP studies have been particularly important for showing that the spotlight of attention can be divided among discontinuous regions and for tracking how the sensory processing varies over time as attention is shifted to different spatial locations or nonspatial features. Although the relative poor spatial resolution of ERPs and SSVEPs makes it difficult to determine the neuroanatomical loci of the attention effects, they are extremely well suited for "unpacking" the rapidly evolving attentional processes that occur in human vision.

ACKNOWLEDGMENTS Preparation of this chapter was supported by grants from NSF (BCS-1029084) and NIMH (1P50MH86385) to S. A. H., and by NIH grant R01MH076226 to S. J. L.

REFERENCES

Andersen, S. K., Fuchs, S., & Müller, M. M. (2011). Effects of feature-selective and spatial attention at different stages of visual processing. *J Cogn Neurosci, 23*, 238–246.

Andersen, S. K., & Müller, M. M. (2010). Behavioral performance follows the time course of neural facilitation and suppression during cued shifts of feature-selective attention. *Proc Natl Acad Sci USA, 107*, 13878–13882.

Andersen, S. K., Müller, M. M., & Hillyard, S. A. (2009). Color-selective attention need not be mediated by spatial attention. *J Vision, 9*(2), 1–7.

Andersen, S. K., Müller, M. M., & Hillyard, S. A. (2012). Tracking the allocation of attention in visual scenes with steady-state evoked potentials. In M. I. Posner (Ed.), *Cognitive neuroscience of attention* (2nd ed., pp. 197–216). New York, NY: Guilford.

Andersen, S. K., Müller, M. M., & Hillyard, S. A. (2013). Global facilitation of attended features is obligatory and restricts divided attention. *J Neurosci, 33*(46), 18200–18207.

Anllo-Vento, L., & Hillyard, S. A. (1996). Selective attention to the color and direction of moving stimuli: Electrophysiological correlates of hierarchical feature selection. *Percept Psychophys, 58*, 191–206.

Briggs, F., Mangun, G. R., & Usrey, W. M. (2013). Attention enhances synaptic efficacy and the signal-to-noise ratio in neural circuits. *Nature, 499*, 476–480.

Chelazzi, L., Miller, E. K., Duncan, J., & Desimone, R. (1993). A neural basis for visual search in inferior temporal cortex. *Nature, 363*, 345–347.

Chelazzi, L., Miller, E. K., Duncan, J., & Desimone, R. (2001). Responses of neurons in macaque area V4 during memory-guided visual search. *Cereb Cortex, 11*, 761–772.

Di Russo, F., Martinez, A., & Hillyard, S. A. (2003). Source analysis of event-related cortical activity during visuo-spatial attention. *Cereb Cortex, 13*, 486–499.

Ding, Y., Martinez, A., Qu, Z., & Hillyard, S. A. (2013). The earliest stages of visual cortical processing are not modified by attentional load. *Hum Brain Mapp.* Advance online publication. doi:10.1002/hbm.22381

Duncan, J. (1996). Cooperating brain systems in selective perception and action. In T. Inui & J. L. McClelland (Eds.), *Attention and performance, Vol. XVI: Information integration in perception and communication* (pp. 549–578). Cambridge, MA: MIT Press.

Eimer, M., & Kiss, M. (2008). Involuntary attentional capture is determined by task set: Evidence from event-related brain potentials. *J Cogn Neurosci, 208*, 1423–1433.

Folk, C. L., Remington, R. W., & Johnston, J. C. (1992). Involuntary covert orienting is contingent on attentional control settings. *J Exp Psychol Human, 18*, 1030–1044.

Garcia, J. O., Srinivasan, R., & Serences, J. T. (2013). Near-real-time feature-selective modulations in human cortex. *Curr Biol, 23*, 515–522.

Hickey, C., di Lollo, V., & McDonald, J. J. (2009). Electrophysiological indices of target and distractor processing in visual search. *J Cogn Neurosci, 21*, 760–775.

Hickey, C., McDonald, J. J., & Theeuwes, J. (2006). Electrophysiological evidence of the capture of visual attention. *J Cogn Neurosci, 18*, 604–613.

Hillyard, S. A., Vogel, E. K., & Luck, S. J. (1998). Sensory gain control (amplification) as a mechanism of selective attention: Electrophysiological and neuroimaging evidence. *Philos Trans R Soc Lond B Biol Sci, 353*, 1257–1270.

Hindi-Attar, C., Andersen, S. K., & Müller, M. M. (2010). Time course of affective bias in visual attention: Convergent evidence from steady-state visual evoked potentials and behavioral data. *NeuroImage, 53*, 1326–1333.

Hopf, J.-M., Luck, S. J., Boelmans, K., Schoenfeld, M. A., Boehler, N., Rieger, J., & Heinze, H.-J. (2006). The neural site of attention matches the spatial scale of perception. *J Neurosci, 26*, 3532–3540.

Hopfinger, J. B., Luck, S. J., & Hillyard, S. A. (2004). Selective attention: Electrophysiological and neuromagnetic studies. In M. S. Gazzaniga (Ed.), *The cognitive neurosciences* (3rd ed., pp. 561–574). Cambridge, MA: MIT Press.

Itthipuripat, S., Garcia, J. O., & Serences, J. T. (2013). Temporal dynamics of divided spatial attention. *J Neurophysiol, 109*, 2364–2373.

Kashiwase, Y., Matsumiya, K., Kuriki, I., & Shioiri, S. (2012). Time courses of attentional modulation in neural amplification and synchronization measured with steady-state visual-evoked potentials. *J Cogn Neurosci, 24*, 8, 1779–1793.

Kiss, M., Driver, J., & Eimer, M. (2009). Reward priority of visual target singletons modulates event-related potential signatures of attentional selection. *Psychol Sci, 20*, 245–251.

Klein, R. (2000). Inhibition of return. *Trends Cogn Sci, 4*, 138–147.

Krishnan, L., Kang, A., Sperling, G., & Srinivasan, R. (2012). Neural strategies for selective attention distinguish fast-action video game players. *Brain Topogr, 26*, 83–97.

Lorenzo-Lopez, L., Amenedo, E., & Cadaveira, F. (2008). Feature processing during visual search in normal aging: Electrophysiological evidence. *Neurobiol Aging, 29*, 1101–1110.

Luck, S. J. (2012). Electrophysiological correlates of the focusing of attention within complex visual scenes: N2pc and related ERP components. In S. J. Luck & E. S. Kappenman (Eds.), *The Oxford handbook of ERP components* (pp. 329–360). New York, NY: Oxford University Press.

Luck, S. J. (2014). *An introduction to the event-related potential technique* (2nd ed.). Cambridge, MA: MIT Press.

Luck, S. J., Chelazzi, L., Hillyard, S. A., & Desimone, R. (1997). Neural mechanisms of spatial selective attention in areas V1, V2, and V4 of macaque visual cortex. *J Neurophysiol, 77*, 24–42.

Luck, S. J., Fuller, R. L., Braun, E. L., Robinson, B., Summerfelt, A., & Gold, J. M. (2006). The speed of visual attention in schizophrenia: Electrophysiological and behavioral evidence. *Schizophr Res, 85*, 174–195.

Luck, S. J., Girelli, M., McDermott, M. T., & Ford, M. A. (1997). Bridging the gap between monkey neurophysiology and human perception: An ambiguity resolution theory of visual selective attention. *Cognitive Psychol, 33*, 64–87.

Luck, S. J., & Gold, J. M. (2008). The construct of attention in schizophrenia. *Biol Psychiatry, 64*, 34–39.

Luck, S. J., & Hillyard, S. A. (1995). The role of attention in feature detection and conjunction discrimination: An electrophysiological analysis. *Int J Neurosci, 80*, 281–297.

Luck, S. J., & Kappenman, E. K. (2012). ERP components and selective attention. In S. J. Luck & E. S. Kappenman (Eds.), *The Oxford handbook of ERP components* (pp. 295–327). New York, NY: Oxford University Press.

Luck, S. J., & Vecera, S. P. (2002). Attention. In S. Yantis (Ed.), *Stevens' handbook of experimental psychology, Vol. 1: Sensation and perception* (3rd ed., pp. 235–286). New York, NY: Wiley.

Mangun, G. R., Hillyard, S. A., & Luck, S. J. (1993). Electrocortical substrates of visual selective attention. In D. Meyer & S. Kornblum (Eds.), *Attention and performance, Vol. XIV: Synergies in experimental psychology, artificial intelligence, and cognitive neuroscience* (pp. 219–243). Cambridge, MA: MIT Press.

Martinez, A., Anllo-Vento, L., Sereno, M. I., Frank, L. R., Buxton, R. B., Dubowitz, D. J., ... Hillyard, S. A. (1999). Involvement of striate and extrastriate visual cortical areas in spatial attention. *Nat Neurosci, 2*, 364–369.

Maunsell, J. H., & Treue, S. (2006). Feature-based attention in visual cortex. *Trends Neurosci, 29*, 317–322.

Mishra, J., Zinni, M., Bavelier, D., & Hillyard, S. A. (2011). Neural basis of superior performance of action videogame players in an attention-demanding task. *J Neurosci, 31*, 992–998.

Müller, M. M., Malinowski, P., Gruber, T., & Hillyard, S. A. (2003). Sustained division of the attentional spotlight. *Nature, 424*, 309–312.

Müller, M. M., Teder-Sälejärvi, W. A., & Hillyard, S. A. (1998). The time course of cortical facilitation during cued shifts of spatial attention. *Nat Neurosci, 1*, 631–634.

Praamstra, P., & Plat, F. M. (2001). Failed suppression of direct visuomotor activation in Parkinson's disease. *J Cogn Neurosci, 13*, 31–43.

Sawaki, R., Geng, J. J., & Luck, S. J. (2012). A common neural mechanism for preventing and terminating attention. *J Neurosci, 32*, 10725–10736.

Sawaki, R., & Luck, S. J. (2010). Capture versus suppression of attention by salient singletons: Electrophysiological evidence for an automatic attend-to-me signal. *Atten Percept Psychophys, 72*, 1455–1470.

Schiff, S., Mapelli, D., Vallesi, A., Orsato, R., Gatta, A., Umilta, C., & Amodio, P. (2006). Top-down and bottom-up processes in the extrastriate cortex of cirrhotic patients: An ERP study. *Clin Neurophysiol, 117*, 1728–1736.

Treisman, A. (1988). Features and objects: The fourteenth Bartlett memorial lecture. *J Exp Psychol Gen, 40*, 201–237.

van der Stigchel, S., Belopolsky, A. V., Peters, J. C., Wijnen, J. G., Meeter, M., & Theeuwes, J. (2009). The limits of top-down control of visual attention. *Acta Psychol, 132*, 201–212.

Vialatte, F. B., Maurice, M., Dauwels, J., & Cichocki, A. (2010). Steady-state visually evoked potentials: Focus on essential paradigms and future perspectives. *Prog Neurobiol, 90*, 418–438.

Woodman, G. F., & Luck, S. J. (1999). Electrophysiological measurement of rapid shifts of attention during visual search. *Nature, 400*, 867–869.

Woodman, G. F., & Luck, S. J. (2003). Serial deployment of attention during visual search. *J Exp Psychol Human, 29*, 121–138.

Zhang, W., & Luck, S. J. (2009). Feature-based attention modulates feedforward visual processing. *Nat Neurosci, 12*, 24–25.

19 Spatial Neglect and Attention Networks: A Cognitive Neuroscience Approach

MAURIZIO CORBETTA, ANTONELLO BALDASSARE, ALICIA CALLEJAS, LENNY RAMSEY, AND GORDON L. SHULMAN

ABSTRACT Spatial neglect is a common syndrome after brain injury, characterized by spatial and nonspatial attention, and perceptual and motor deficits. These deficits are due to the structural and functional damage of attention networks, and their interaction with sensory and motor regions of the brain. Accordingly, white matter damage is associated with more severe neglect due to the disconnection of different interacting regions. The right hemisphere lateralization of neglect is due to the hemispheric lateralization of arousal/vigilance. Spatial neglect is an outstanding model to study how large-scale neural systems mediate behavior.

Spatial neglect is a common syndrome that follows focal brain injury (stroke, tumor, etc.). It is defined as the inability to attend and report stimuli on the side opposite the lesion (or contralesional) despite apparently normal visual perception. Neglect is also frequently associated with a reduction of arousal and vigilance, a spatial bias for directing actions toward the hemi-space or hemi-body on the same side as the lesion (or ipsilesional), and several disorders of awareness, including a degree of obliviousness toward being ill and confabulation about body ownership.

About 25–30% of all stroke patients suffer from this syndrome, an estimated 250,000–300,000 patients per year. Clinically, spatial neglect negatively impacts motor recovery and outcome. Stroke patients with neglect are less likely to go back to work, drive, and be independent than stroke patients with a comparable degree of motor disability but without spatial neglect.

Spatial neglect has attracted tremendous interest as a model for understanding the neurological basis of awareness, cerebral lateralization, spatial cognition, and recovery of function. Yet its neural bases remain poorly understood.

There are several outstanding issues that are of great interest for cognitive neuroscience. First, even though strokes causing neglect occur in different locations in the brain, both cortically and subcortically, the syndrome is relatively stereotypical. This suggests that some of the deficits are supported by a common pathophysiological mechanism despite variability in lesion location. Second, although acutely both left- and right-hemisphere lesions can cause neglect, only right-hemisphere lesions cause severe and persistent deficits (Stone, Halligan, & Greenwood, 1993). This still-unexplained hemispheric asymmetry after nearly 70 years of research is one of the primary reasons for the widely held view that the right hemisphere is dominant for attention. Finally, spatial neglect is unique among the behavioral disorders resulting from focal lesions, since its severity can be modulated by behavioral interventions over very short timescales (e.g., seconds to minutes). Deficits in attending to and reporting objects in contralesional space can be lessened by (1) encouraging a patient to attend to the previously ignored stimuli using verbal cues; (2) presenting salient sensory stimuli, such as noises (Robertson, Mattingley, Rorden, & Driver, 1998); (3) asking the patient to perform hand movements controlled by the injured hemisphere (Robertson & North, 1992); or (4) training patients to increase their alertness (Robertson et al., 1995). These observations suggest that the neural mechanisms underlying the spatial deficit can be dynamically modulated by signals from other parts of the brain reflecting endogenous or exogenous attention, movement, and arousal. Moreover, in a matter of days or weeks, most patients with spatial neglect recover from the more obvious spatial impairments, which, however, continue to negatively influence their ability to return to a productive life (Denes, Semenza, Stoppa, & Lis, 1982; Paolucci, Antonucci, Grasso, & Pizzamiglio, 2001).

To reconcile these observations, we have proposed that neglect is mediated by the abnormal interaction between brain networks that control attention to the environment in the healthy brain (Corbetta & Shulman, 2011; see Heilman, Watson, & Valenstein, 1985, and Mesulam, 1999, for other network formulations). We argue for a core set of spatial and nonspatial deficits that match the physiological properties of these networks. First, the core spatial deficit, a bias in spatial attention and salience mapped in an egocentric coordinate frame, is caused by the dysfunction of a dorsal frontal-parietal network that controls attention and eye movements and represent stimulus saliency. Core nonspatial deficits of arousal, reorienting, and detection reflect structural damage to more ventral regions that partly overlap with a right-hemisphere dominant ventral frontal-parietal network recruited during reorienting and detection of novel behaviorally relevant events.

Second, ventral lesions that result in neglect alter the physiology of structurally undamaged dorsal frontoparietal regions, consistent with the fact that dorsal and ventral attention regions interact in the healthy brain. Physiological dysfunction in dorsal frontoparietal regions is empirically observed not only during task performance, but also at rest, and correlates with the severity of the egocentric spatial bias. Moreover, this dysfunction decreases the top-down modulation of visual cortex, reducing its responsiveness, which can also contribute to neglect. The highly dynamic and plastic nature of the spatial deficits in neglect strongly argues that they are mediated by parts of the brain that still function, even if abnormally. Measurements of the physiology of brain regions, not just of structural damage, are essential for understanding neglect (Deuel & Collins, 1983), and we emphasize neuroimaging methods that provide a window on physiological function.

Finally, perhaps the least understood clinical feature of spatial neglect is its right-hemisphere lateralization. Brain-imaging studies have shown that dorsal frontoparietal regions controlling spatial attention and eye movements are largely symmetrically organized, with each hemisphere predominantly representing the contralateral side of space. In contrast, ventral regions that underlie the core nonspatial deficits observed in neglect patients are strongly right-hemisphere dominant. We argue that lateralization of these latter functions, and their interaction with dorsal regions, rather than asymmetries of spatial attention per se, primarily account for the hemispheric asymmetry of neglect.

The core spatial deficit: Egocentric bias

A large body of neuropsychological research has tried to characterize the deficits in neglect that are "spatial" (i.e., involve predominantly one side of space). Factor-analytic studies of behavioral deficits have consistently isolated at least one factor associated with impairments in attending, searching, or responding to targets in contralateral space, but have yielded inconsistent conclusions with regard to other factors (Azouvi et al., 2002; Coslett, Bowers, Fitzpatrick, Haws, & Heilman, 1990; Farne et al., 2004; Halligan, Marshall, & Wade, 1989; Hamilton, Coslett, Buxbaum, Whyte, & Ferraro, 2008; Kinsella, Olver, Ng, Packer, & Stark, 1993; Rengachary, He, Shulman, & Corbetta, 2011; Verdon, Schwartz, Lovblad, Hauert, & Vuilleumier, 2010). Simple speeded responses to lateralized visual targets are highly sensitive and specific in classifying patients with neglect both subacutely (2 weeks) and chronically (9 months; Rengachary, d'Avossa, Sapir, Shulman, & Corbetta, 2009). We argue that at its core, spatial neglect represents a deficit of spatial attention and stimulus saliency.

Gradients of Spatial Attention and Stimulus Saliency Virtually all patients with spatial neglect manifest a lateralized bias in visual information processing that is evident both clinically and experimentally as a gradient across space (Behrmann, Watt, Black, & Barton, 1997; Pouget & Driver, 2000). Clinically this is evident as a tendency to look ipsilesionally when presented with stimuli in the visual field, as a deficit of exploration of the contralesional space, and as a reluctance to direct actions (eye, hand) toward the contralesional space. Low sensitivity and responsiveness to behaviorally relevant stimuli improve as one moves from contralesional to ipsilesional locations. Figure 19.1A shows the spatial bias of movie director Federico Fellini, who suffered a right-hemisphere stroke. The bias is apparent as a deviation toward the right (ipsilesionally) of line bisection, and as the missing half of Fellini's own body image.

This spatial bias does not reflect abnormal early visual mechanisms, as indicated by contrast sensitivity (Spinelli, Guariglia, Massironi, Pizzamiglio, & Zoccolotti, 1990), image segmentation based on low-level features in the neglected visual field (Driver & Mattingley, 1998), visually evoked potentials (Di Russo, Aprile, Spitoni, & Spinelli, 2008; Watson, Miller, & Heilman, 1977), and functional magnetic resonance imaging (fMRI; Rees et al., 2000).

In contrast, the "saliency" of objects in the neglected visual field is impaired. Saliency refers to the sensory distinctiveness and behavioral relevance of an object relative to other objects. In a recent study, the saliency of objects in the ipsilesional or contralesional visual field of neglect subjects was measured by indexing the patients' tendency to look at distinctive but task-irrelevant stimuli or at task-relevant stimuli (Bays, Singh-Curry, Gorgoraptis, Driver, & Husain, 2010). The probability of an eye movement increased along a spatial gradient from contralesional to ipsilesional locations that was the same for both kinds of stimuli, suggesting that exogenous (automatic) and goal-driven components of spatial attention were equally affected (figure 19.1B). The abnormally high salience of ipsilesional stimuli may prevent them from being filtered

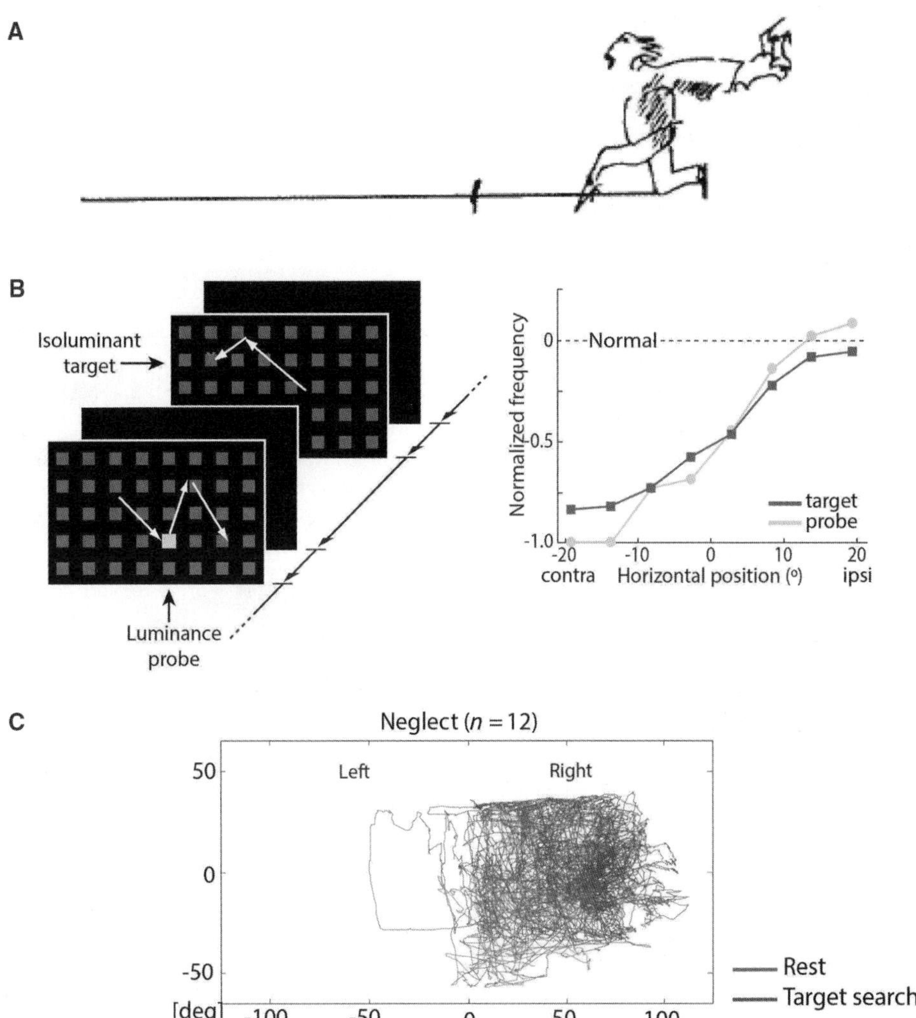

FIGURE 19.1 (A) Rightward bias in line bisection and neglect left body image. (B) Gradients of saliency for both goal-driven and stimulus-driven attention. The task requires making an eye movement to a target defined by an isoluminant color target that was cued prior to the task (goal-driven). Occasional luminance distractors are presented, and they attract eye movements automatically (stimulus-driven). The frequency of eye movements linearly normalizes according to a gradient from left to right visual field.

when they are task-irrelevant (Bays et al., 2010; Shomstein, Lee, & Behrmann, 2010; Snow & Mattingley, 2006) or lead to repeated re-fixations during search tasks (Husain et al., 2001).

Importantly, a spatially lateralized bias is observed even in the absence of a stimulus. When neglect subjects searched for a nonexistent object in complete darkness, search patterns as measured by eye or head position were strongly biased toward the ipsilesional field (Hornak, 1992). Moreover, gaze deviations are observed tonically at rest, similar to those observed during task performance (figure 19.1C; Fruhmann Berger, Johannsen, & Karnath, 2008).

Spatial biases in the dark during target search could reflect a reduced salience of contralesional spatial locations during task performance, but the biases observed at rest also suggest an indwelling imbalance in the mechanism's controlling gaze. These motor biases are likely associated with attentional biases because of the functional relationship between the corresponding neural systems (Corbetta et al., 1998; Rizzolatti, Riggio, Dascola, & Umiltá, 1987).

AN EGOCENTRIC FRAME OF REFERENCE Spatial deficits can be separated based on the reference frame in which stimuli are coded. Neglect is often *egocentric* (viewer-centered), with left and right hemi-spaces based on the observer's midline. Neglect can also be *allocentric*, where the midline is defined from the central axis of a stimulus, irrespective of its position in the environment (stimulus-centered) or of both its position and orientation (object-centered). (Chechlacz, Rotshtein,

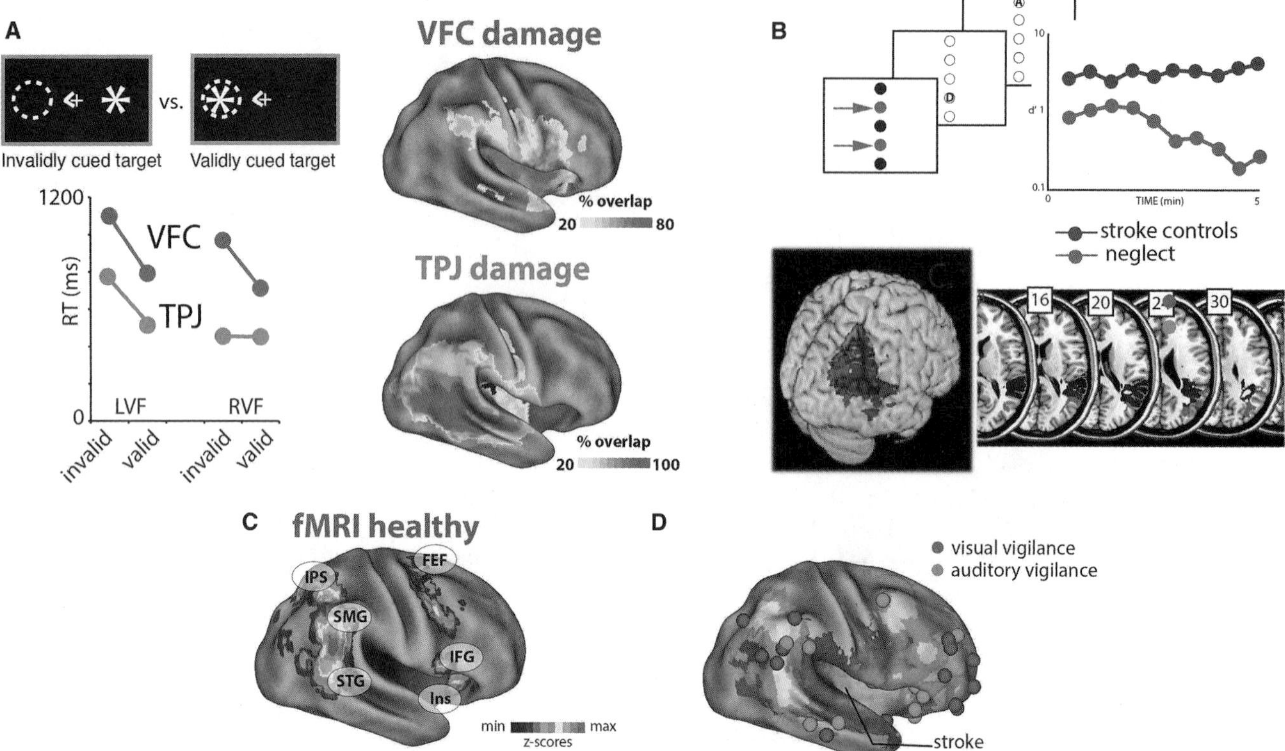

FIGURE 19.2 (A) Reorienting deficit. Slower reaction time to visual targets presented at unattended (invalidly cued targets in the Posner orienting task). Ventral frontal cortex damage causes damage in both visual fields while temporoparietal junction (TPJ)/superior temporal gyrus (STG) damage causes deficits mainly contralesionally (Friedrich et al., 1998; Rengachary et al., 2011). (B) Neglect patients show a decrement in performance as time elapses on a task in which they have to maintain attention to specific locations on the screen. The region most associated with this vigilance is the right TPJ region (Malhotra et al., 2009). (C) Reorienting response (invalidly-validly cued targets) in healthy subjects involves both dorsal and ventral frontoparietal regions (i.e., dorsal and ventral attention networks). (D) The map shows regions activated either by reorienting or target detection in ventral frontoparietal cortex. Activity is lateralized to the right hemisphere (Stevens, Calhoun, & Kiehl, 2005; Shulman et al., 2009). Plotted are foci of activation during visual and auditory vigilance experiments. Note overlap, especially in TPJ, between response to target, vigilance, and lesions that cause neglect.

& Humphreys, 2012; Marsh & Hillis, 2008). Egocentric neglect is associated with regions classically damaged in spatial neglect, i.e., inferior parietal lobule (IPL), superior temporal gyrus (STG), and inferior frontal gyrus (IFG), while allocentric neglect has been associated with damage of inferior temporal regions (figure 19.2A; but see Chechlacz et al., 2012). We suggest that egocentric neglect underlies a spatial disorder and is associated with damage of systems involved in spatial processing (see below), while allocentric neglect reflects a problem with object coding and is associated with damage to the ventral visual system (Ungerleider & Mishkin, 1982).

WHAT ABOUT REPRESENTATIONAL NEGLECT? An influential account proposes that neglect is a deficit in forming, storing, or manipulating the left side of mental images or information in VSTM, termed representational neglect (i.e., representation within VSTM; Bisiach & Luzzatti, 1978; Della Sala, Logie, Beschin, & Denis, 2004). In a famous experiment, patients with right-hemisphere stroke and neglect were asked to report by memory buildings to the left and right of the cathedral in the Piazza del Duomo in Milan while imagining they were facing the front of the cathedral. Patients with neglect tended to ignore buildings on the left side of the square. However, when asked to repeat the same task while imagining being rotated by 180 degrees, thus looking toward the square with the shoulders to the cathedral, now they reported buildings on the right side that had been previously ignored, while ignoring buildings on the left side that had been formerly reported. This experiment clearly indicates the integrity of spatial information in working memory, and the involvement of a "scanning" or "saliency" mechanism similarly to what is observed during perceptual tasks. Furthermore, mechanisms for spatial attention, VSTM, and imagery are closely related, as shown by psychological and physiological studies (Awh & Jonides, 2001; Kosslyn, Kosslyn, Ganis, & Thompson, 2001). Not surprisingly, representational neglect is nearly always observed in association with perceptual neglect (Bartolomeo, D'Erme, & Gainotti, 1994).

Summary Spatial neglect is characterized by a spatial gradient of impaired attention, saliency, and representation within an egocentric reference frame. The saliency deficit reflects both task and sensory factors and is linked with indwelling motor imbalances that produce resting ipsilesional deviations in eye, head, and body movements. Clinical observations indicate that the gradient fluctuates depending on arousal and task instructions, suggesting that the underlying neural mechanisms are modulated by signals from other parts of the brain, and are dysfunctional rather than obliterated by structural damage.

Core nonspatial deficits

Patients with spatial neglect also suffer from a host of nonspatial deficits, including problems with arousal/vigilance, reorienting, and spatial-temporal capacity. Clinically, especially in the first few days after stroke, patients with right-hemisphere injuries and neglect show an overall decrement in their ability to maintain a normal vigilance state. They tend to be sleepier, less responsive, and more prone to fluctuations of arousal than left-hemisphere damaged patients, alternating moments of responsiveness with periods in which they appear comatose.

Reorienting of Attention Michael Posner and colleagues reported that neglect patients are impaired in reorienting to unexpected events (Posner, Walker, Friedrich, & Rafal, 1984). Patients showed especially large deficits in detecting contralesional targets when they were expecting an ipsilesional target, suggesting a deficit in *disengaging* attention from the ipsilesional field. Several studies have localized disengagement/reorienting deficits to right temporoparietal junction (TPJ)/STG (Friedrich, Egly, Rafal, & Beck, 1998; Rengachary et al., 2011), but also right inferior frontal cortex (Rengachary et al., 2011; figure 19.2A).

Detection of Behaviorally Relevant Stimuli Right hemisphere and neglect patients show deficits in target detection in even the simplest paradigms. Simple auditory reaction time, for example, is much slower following right- than left-hemisphere damage (Howes & Boller, 1975). Related differences have been reported for auditory stimuli presented at ipsilesional locations in right-hemisphere patients with neglect versus without neglect (Samuelsson, Hjelmquist, Jensen, Ekholm, & Blomstrand, 1998), indicating that the reaction-time slowing may reflect damage to right-hemisphere brain regions specifically associated with neglect (figure 19.2B).

Arousal and Vigilance Arousal refers to the combination of autonomic, electrophysiological, and behavioral activity that is associated with an alert state, while vigilance refers to the ability to sustain this state over time. Kenneth Heilman and colleagues have argued that neglect patients have decreased arousal due to hypoactivation of the right hemisphere (Heilman, Watson, Valenstein, & Goldberg, 1987). For instance, patients with right- as opposed to left-hemisphere damage do not show the typical slowing of heart rate following a cue that signals a subsequent target and show reduced galvanic skin response to electrical stimulation. Lesion studies have associated right TPJ damage with decreased sustained attention to spatial locations (Malhotra et al., 2009; figure19.2B), and right frontal damage with a decrement over time in sustaining attention and detecting targets (vigilance decrement; Rueckert & Grafman, 1996; Wilkins et al., 1987).

Interaction between spatial and nonspatial deficits

Recent behavioral evidence in healthy adults indicates that arousal interacts with spatial attention, the primary spatial function impaired in neglect patients. Healthy adults show a slight tendency to attend to the left side of an object (Nicholls, Bradshaw, & Mattingley, 1999), but this bias is reduced or shifted to the right under conditions of low arousal (Bellgrove, Dockree, Aimola, & Robertson, 2004; Manly, Dobler, Dodds, & George, 2005; Matthias et al., 2009).

Importantly, there is also evidence in neglect patients for an interaction between arousal/vigilance and spatial deficits. For instance, a study reported a specific association between damage to right TPJ cortex and vigilance decrements, but only when attention had to be sustained to a spatial location, not for targets presented at random locations (figure 19.2B; Malhotra et al., 2009).

The specific form of the interaction between arousal and spatial attention predicts that increases in arousal should bias attention to the left visual field, ameliorating left field neglect. Ian Robertson and colleagues observed this result in two important studies. Increases in either phasic (Robertson et al., 1998) or sustained (Robertson, Tegner, Tham, Lo, & Nimmo-Smith, 1995) arousal decreased neglect of the contralesional field,

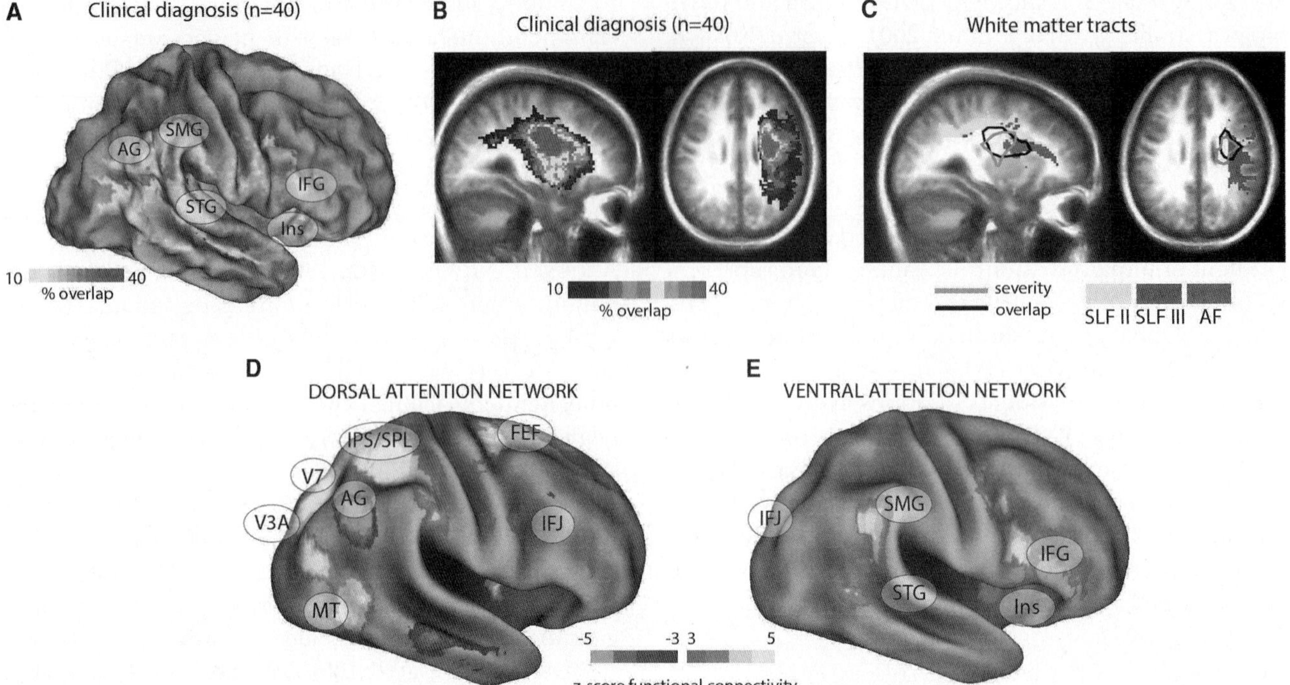

FIGURE 19.3 (A) Anatomy of neglect defined based on clinical diagnosis by treating therapist (N = 40 patients). Note ventral distribution of lesions with involvement especially of superior temporal gyrus (STG), supramarginal gyrus (SMG), angular gyrus (AG), inferior frontal gyrus (IFG), and insula (Ins). (B) Lesions extend in the periventricular white matter both ventrally and dorsally. White matter damage is more common in severe neglect. (C) The region of maximal damage in the white matter overlaps with common white matter pathways: superior longitudinal fasciculus (SLF) and arcuate (AF), which connect frontal and parietal regions, and ventral frontal to dorsal parietal cortex. (D) Functional connectivity maps of dorsal attention network with core regions identified. (E) Functional connectivity maps of ventral attention network with core regions identified. (See color plate 15.)

consistent with a direct effect of the activation of ventral right-hemisphere mechanisms on the dorsal attention network.

The interaction between mechanisms underlying nonspatial and spatial deficits is likely critical for the pathogenesis and right-hemisphere lateralization of spatial neglect. We argue that this is due to the interaction of structural damage and secondary physiological dysfunction of networks involved in the control of attention (see below).

Anatomy of spatial neglect

Lesions causing spatial neglect are heterogeneous in location, since they can occur in frontal, temporoparietal, superior temporal, insular cortices, and even subcortical structures (thalamus, basal ganglia). Often lesions involve the underlying white matter that connects anterior to posterior cortex. These observations raise the question: How does a relatively homogenous behavioral syndrome arise from a heterogeneous set of structural lesions?

STRUCTURAL DAMAGE DOES NOT EXPLAIN THE EGOCENTRIC SPATIAL BIAS Spatial neglect is traditionally associated with damage to parietal cortex (Critchley, 1953), especially IPL (Mort et al., 2003; Vallar & Perani, 1987). However, subsequent studies have also emphasized STG (Karnath, Ferber, & Himmelbach, 2001) and IFG (Husain & Kennard, 1996), and convincing evidence exists that damage to other regions, including anterior insula and middle frontal gyrus, sometimes produces neglect (figure 19.3A). Importantly, neglect patients, especially in severe cases, have white matter fiber damage, which can disconnect frontal, temporal, and parietal cortex (Bartolomeo, Thiebaut de Schotten, & Doricchi, 2007; Gaffan & Hornak, 1997; He et al., 2007; Thiebaut de Schotten et al., 2005; figure 19.3B–C). White matter damage most commonly involves a dorsal region lateral to the ventricle where arcuate and superior longitudinal fasciculi (II and III) run parallel in an anterior-to-posterior direction (Doricchi & Tomaiuolo, 2003; figure 19.3C). Finally, neglect can be also caused by damage to subcortical nuclei (pulvinar, caudate, putamen; Karnath et al., 2002; Vallar & Perani, 1987) that cause cortical hypo-activation of regions important for the genesis of neglect (Karnath et al., 2005; Perani, Vallar, Cappa, Messa, & Fazio, 1987).

While the detailed profile of behavioral deficits undoubtedly depends on the site of the lesion (Medina et al., 2008; Verdon et al., 2010), and some lesion sites are more likely than others to produce neglect, the striking fact remains that neglect of the left field can be caused by many different right-hemisphere lesions. Attempts to identify a critical region based on structural damage alone inevitably must explain away a large number of reported lesions that produce neglect and yet do not involve the supposedly critical region (Friedrich et al., 1998; Malhotra et al., 2009; Rengachary et al., 2011).

Importantly, lesions that cause neglect tend to occur in ventral aspects of cortex that are associated with the nonspatial deficits discussed above, but do not contain the physiological signals that mediate the egocentric spatial bias and core spatial processing deficits characteristic of neglect. Instead, as discussed next, these physiological signals are housed in dorsal regions that often are not structurally damaged in neglect patients.

Cognitive neuroscience of attention and relationship to spatial neglect

A Dorsal Frontoparietal Network for Spatial Attention, Stimulus Salience, and Eye Movements Does Not Match the Anatomy of Neglect In healthy subjects, regions in dorsal frontal-parietal cortex form a network (dorsal attention network) that contains physiological signals for spatial attention and eye movements (figure 19.3D). For example, bilateral medial intraparietal sulcus, SPL, precuneus, supplementary eye field and frontal eye field regions respond to symbolic cues to shift attention voluntarily to a location (e.g., Corbetta, Kincade, Ollinger, McAvoy, & Shulman, 2000; Hopfinger, Buonocore, & Mangun, 2000; Kastner, Pinsk, De Weerd, Desimone, & Ungerleider, 1999). These regions are also recruited when attention is shifted to salient objects based on task relevance and sensory distinctiveness (Shulman et al., 2009). Regions of the dorsal attention network are recruited not only by "attention" signals, but also by visually and memory-guided saccadic signals, with almost complete overlap of attention and eye movement activations (Corbetta et al., 1998). These regions are well suited to code for the location of objects in space, since they contain topographic maps (Hagler & Sereno, 2006; Sereno, Pitzalis, & Martinez, 2001; Swisher, Halko, Merabet, McMains, & Somers, 2007) that precisely represent locations for attention, eye movement, and saliency signals (Koch & Ullman, 1985).

Physiological studies in both human and nonhuman primates indicate that a possible mechanism for coding the location of relevant stimuli is a computation that calculates differences in activity between attended versus nonattended parts of the topographic map. The idea that the locus of attention may be efficiently coded by the two hemispheres is not novel (Kinsbourne, 1987), but the notion that this could be represented by a differencing signal coded either through the corpus callosum (Innocenti, 2009) or subcortical routes is supported by a growing literature (Bisley & Goldberg, 2003; Sylvester et al., 2007; Thut, Nietzel, Brandt, & Pascual-Leone, 2006).

A Ventral Frontoparietal Network for Target Detection, Reorienting, and Vigilance/Arousal Matches the Anatomy of Neglect In contrast to the poor correlation between the anatomy of neglect and the physiology of spatial attention, eye movements, and saliency, the anatomy of nonspatial deficits matches the underlying physiology much more closely (figure 19.3E). Neuroimaging studies of healthy adults have shown that reorienting to stimuli in either visual field that are presented outside the focus of attention ("stimulus-driven" reorienting) recruits a right-lateralized "ventral attention" network in TPJ (including separate foci in the supramarginal gyrus and STG; Shulman et al., 2010) and VFC (insula, IFG, middle frontal gyrus), in conjunction with the dorsal network (figure 19.3C; Corbetta & Shulman, 2002).

The ventral (and dorsal) attention networks are also activated by detection of behaviorally relevant stimuli. Regions of activation in the ventral network that respond to targets in both visual fields are lateralized to the right hemisphere (Shulman et al., 2010; Stevens, Calhoun, & Kiehl, 2005; figure 19.3D). These studies are consistent with lesion studies noted above in which damage of STG/TPJ and ventral frontal cortex leads to reorienting deficits and problems with target detection (Husain, Shapiro, Martin, & Kennard, 1997; Rueckert & Grafman, 1996; Rueckert & Grafman, 1998; Wilkins, Shallice, & McCarthy, 1987).

Neuroimaging studies of arousal and vigilance have qualitatively reported right-hemisphere dominance, usually in lateral prefrontal, insula/frontal operculum, and TPJ regions (Coull, Frackowiak, & Frith, 1998; Foucher, Otzenberger, & Gounot, 2004; Pardo, Fox, & Raichle, 1991; Paus et al., 1997; Sturm et al., 1999; Sturm et al., 2004; figure 19.3D). Arousal-related activity has been recorded more frequently in ventral cortex of the right hemisphere than left hemisphere.

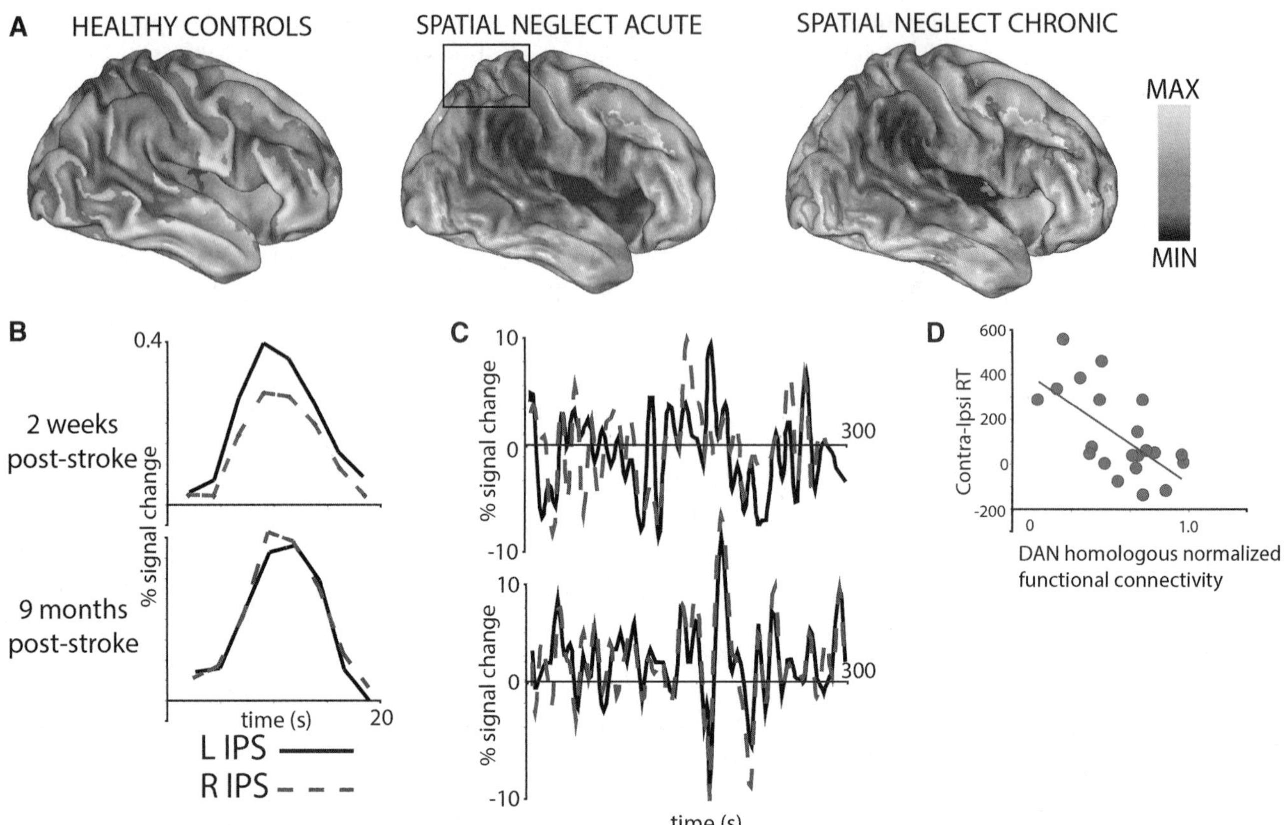

FIGURE 19.4 (A) Maps of activity in healthy subjects and neglect patients at 3 weeks (acute) and 9 months (chronic) during the Posner orienting task. Note lesions and paucity of response at acute stage. (B) Task imbalance in dorsal network at acute stage. The response in L IPS is larger than response in R IPS. Both regions are structurally intact. The imbalance is also present in visual cortex (not shown). (C) Abnormal interhemispheric temporal correlation in spontaneous blood-oxygen-level-dependent (BOLD) activity at acute stage and recovery at chronic stage between the same dorsal areas. (D) Interhemispheric coherence inversely correlates to spatial neglect, measured as the difference in response to visual targets in the left and right visual field.

Importantly, arousal-related activations in the TPJ and insula/frontal operculum overlap with regions that are damaged in spatial neglect and recruited during reorienting and target detection (Heilman, Bowers, Valenstein, & Watson, 1987; Malhotra, Coulthard, & Husain, 2009).

Therefore, a consistent set of right-hemisphere-dominant regions is recruited during tasks that involve reorienting, target detection, and arousal/vigilance. Damage to these regions acutely or chronically causes nonspatial deficits similar to those described in patients with neglect.

Pathogenesis of spatial neglect

PHYSIOLOGICAL CORRELATES OF THE EGOCENTRIC SPATIAL BIAS IN NEGLECT The model we propose aims at reconciling the seeming contradiction that signals for saliency, attention, and eye movement; the primary spatial deficits in neglect are housed in dorsal regions, while the lesions that cause neglect occur in ventral regions. The main insight comes from neuroimaging studies in neglect patients showing that ventral lesions cause physiological abnormalities in structurally intact dorsal frontoparietal cortex corresponding to the dorsal attention network. First, at 3 weeks post-stroke, a widespread cortical hypo-activation during a spatial attention task was observed more strongly in the right than left hemispheres (figure 19.4A). The cortical hypo-activation was associated with a large interhemispheric imbalance of activity in dorsal parietal cortex (figure 19.4B). The interhemispheric imbalance normalized at the chronic stage (9 months post-stroke) in parallel with an overall improvement of cortical activity and spatial neglect (figure 19.4B; Corbetta, Kincade, Lewis, Snyder, & Sapir, 2005). Interestingly, activity in ipsilesional occipital visual cortex was also abnormal, showing reductions in magnitude and spatial selectivity

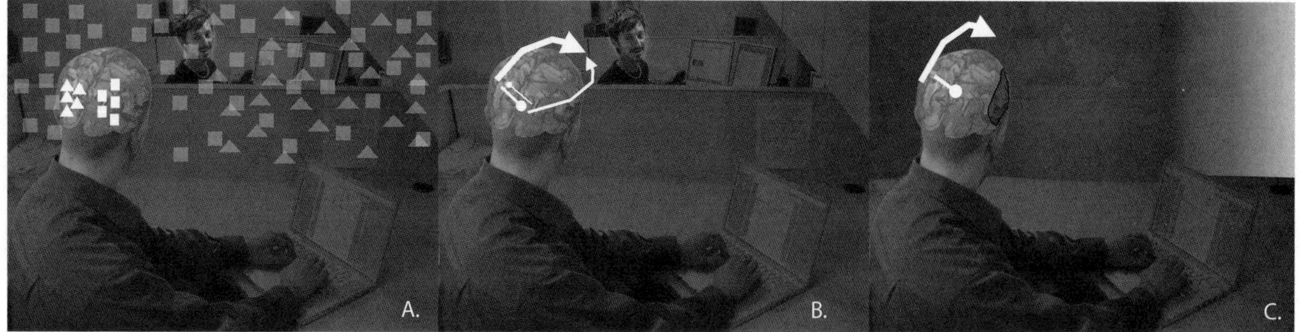

FIGURE 19.5 Models of neglect. (A) Mesulam/Heilman's model. The right hemisphere contains neurons representing space bilaterally (ipsi, contra), while the left hemisphere contains only neurons coding for contralateral locations. Accordingly, ipsilateral right-hemisphere maps compensate for lesions in the left hemisphere. (B) Kinsbourne's model. Each hemisphere contains an opponent mechanism to orient contralaterally. Each hemisphere's orienting mechanism cross-inhibits the other, and the locus of attention depends on a balance of activity between hemispheres. The right-lateralization of neglect comes from the stronger unopposed rightward bias generated by the left hemisphere in the presence of a right lesion. A left lesion does not produce the same amount of right neglect, given the relatively weak orienting bias of the right hemisphere. (C) Corbetta and Shulman model. Neglect involves a bilateral decrement of detection due to structural damage of the ventral attention network and associated deficits in vigilance/arousal and target detection. This problem is further exacerbated by a rightward spatial and sensory-motor bias induced by an interhemispheric imbalance caused by the disconnection of the right ventral network with the right dorsal network. Left lesions do not cause neglect because the ventral network is lateralized to the right hemisphere.

(Corbetta et al., 2005). These impairments in sensory-evoked activity, possibly reflecting abnormal top-down control (Bressler, Tang, Sylvester, Shulman, & Corbetta, 2008), may further lessen the saliency of contralesional stimuli. Secondly, neglect patients show anomalies in the pattern of spontaneous activity fluctuations within the dorsal attention network (figure 19.4C). Coherence between left and right parietal regions was disrupted 3 weeks post-stroke and improved over time in parallel with the improvement of neglect (He et al., 2007).

Both task-evoked activity and resting coherence are functionally significant. Left parietal activity is stronger in subjects with more severe neglect, as indexed by the difference in response times to contralesional versus ipsilesional visual targets (Corbetta et al., 2005). Stronger left- than right-hemisphere activation of parietal and occipital regions may reflect a biased representation of stimulus salience and the locus of spatial attention (Bisley & Goldberg, 2003; Sylvester, Shulman, Jack, & Corbetta, 2007). This interpretation is consistent with transcranial magnetic studies, in which inactivation of left posterior parietal cortex reduced left-field neglect (Brighina et al., 2003; Koch et al., 2008). Secondly, at rest, significant correlations are found throughout the dorsal attention network between reductions in interhemispheric coherence and the magnitude of the spatial bias; for example, the difference in reaction times between contralesional and ipsilesional targets (figure 19.4D; Carter et al., 2010; He et al., 2007). The anomalies in spontaneous activity may underlie the lateral rotation of the eyes, head, and body in neglect patients at rest, and contribute to the observed abnormalities in task-evoked responses. Overall, then, in structurally intact regions of the dorsal attention network, ventral lesions induce changes in blood-oxygen-level-dependent (BOLD) resting functional connectivity and task-evoked activity that reflect abnormal interhemispheric interactions and response balances (Kinsbourne, 1987), which plausibly explain the egocentric spatial bias in neglect.

RIGHT HEMISPHERE LATERALIZATION OF SPATIAL DEFICITS Previous models have explained the right-lateralization of neglect, and hence attention in the human brain, by proposing that the right hemisphere contains a bilateral representation of space while the left hemisphere contains predominantly a contralateral space representation (Heilman, Bowers et al., 1987; Mesulam, 1999; figure 19.5A). Damage to the right hemisphere impairs attention to the left hemifield, while damage to the left hemisphere can be compensated. A second, "opponent-process" theory proposes that each hemisphere promotes orienting in a contralateral direction, but the strength of this bias is stronger in the left than right hemisphere (Kinsbourne, 1987; figure 19.5B). Left hemisphere lesions cause only mild right spatial neglect, because the unopposed orienting bias generated by the right hemisphere is relatively weak.

According to our model, right-hemisphere dominance of neglect reflects an asymmetrical interaction between bilaterally represented spatial attention and visuomotor systems in dorsal frontal and parietal cortex, and right-hemisphere-dominant systems in ventral frontoparietal cortex for arousal/vigilance, target detection, and reorienting. Lesions in the right hemisphere cause more contralateral (left) spatial neglect for two reasons. First, they lead to more severe nonspatial deficits, given the lateralization of the underlying neural mechanisms. Second, they lead to a greater disruption of spatial processes in the right dorsal attention network. This effect, combined with an imbalance of the normal push-pull relationship between spatial-orienting mechanisms in dorsal cortex, leads to a rightward bias in saliency and eye movements that corresponds well with clinical and experimental evidence (figure 19.5C). A lateralization of sustained attention/vigilance processes has been also proposed by Husain and Rorden (2003).

There is currently only limited evidence of the anatomy and physiology underlying the interactions between nonspatial and spatial mechanisms observed in behavioral studies of healthy adults, and the effects of ventral lesions on dorsal physiology observed in neglect patients. Both may involve similar pathways. For example, stimulating a white matter tract that connects right frontal and parietal cortex produced rightward deviations in bisection performance (Thiebaut de Schotten et al., 2005).

Conclusion

Our review suggests the following account of spatial neglect. Lesions of ventral cortex (frontal, temporoparietal, and basal ganglia, with secondary cortical functional disconnection of the same regions) directly affect nonspatial processes of reorienting, target detection, and vigilance/arousal and cause a relative global hypoactivation of the right hemisphere. The relatively stronger effect of right- versus left-hemisphere lesions depends according to our model on physiological asymmetries in the cortical input of subcortical projection systems regulating arousal/vigilance. Evidence on this critical point is scarce. Old anatomical evidence showed a preferential catecholaminergic input to the inferior parietal lobe, or posterior core of the ventral attention network (Morrison & Foote, 1986), and some right-left catecholamine input asymmetries in the thalamus (Oke, Keller, Mefford, & Adams, 1978).

Lesions in the underlying white matter tend to produce more severe neglect (Gaffan & Hornak, 1997; Karnath, Rennig, Johannsen, & Rorden, 2011). This is related to disconnection of ventral cortex from dorsal cortex, and the relative functional de-synchronization and response imbalance in dorsal cortex and mirror regions in the opposite hemisphere. Because the locus of attention is coded by mechanisms that take into account activity from both sides of the brain, this interhemispheric functional disruption drives spatial attention and eye movements to the right visual field.

ACKNOWLEDGMENTS We acknowledge the help of many colleagues who generously made their data available for this review, and Drs. Bays and Driver, University College London, for figure 19.1B; Drs. Karnath and Fruhmann-Berger, University of Tübingen, for figure 19.1C; Drs. Medina, Pavlak, and Hillis, Johns Hopkins and University of Pennsylvania, for figure 19.2; and Drs. Malhotra and Husain, University College London, for figure 19.4B. This work was supported by the National Institute of Child Health and Human Development RO1 HD061117-05A2 and the National Institute of Mental Health R01 MH096482-01.

REFERENCES

AWH, E., & JONIDES, J. (2001). Overlapping mechanisms of attention and spatial working memory. *Trends Cogn Sci, 5,* 119–126.

AZOUVI, P., SAMUEL, C., LOUIS-DREYFUS, A., BERNATI, T., BARTOLOMEO, P., BEIS, J. M., … FRENCH COLLABORATIVE STUDY GROUP ON ASSESSMENT OF UNILATERAL NEGLECT (GEREN/GRECO). (2002). Sensitivity of clinical and behavioural tests of spatial neglect after right hemisphere stroke. *J Neurol Neurosurg Psychiatry, 73,* 160–166.

BARTOLOMEO, P., D'ERME, P., & GAINOTTI, G. (1994). The relationship between visuospatial and representational neglect. *Neurology, 44,* 1710–1714.

BARTOLOMEO, P., THIEBAUT DE SCHOTTEN, M., & DORICCHI, F. (2007). Left unilateral neglect as a disconnection syndrome. *Cereb Cortex, 17,* 2479–2490.

BAYS, P. M., SINGH-CURRY, V., GORGORAPTIS, N., DRIVER, J., & HUSAIN, M. (2010). Integration of goal- and stimulus-related visual signals revealed by damage to human parietal cortex. *J Neurosci, 30,* 5968–5978.

BEHRMANN, M., WATT, S., BLACK, S., & BARTON, J. (1997). Impaired visual search in patients with unilateral neglect: An oculographic analysis. *Neuropsychologia, 35,* 1445–1458.

BELLGROVE, M. A., DOCKREE, P. M., AIMOLA, L., & ROBERTSON, I. H. (2004). Attenuation of spatial attentional asymmetries with poor sustained attention. *NeuroReport, 15,* 1065–1069.

BISIACH, E., & LUZZATTI, C. (1978). Unilateral neglect of representational space. *Cortex, 14,* 129–133.

BISLEY, J. W., & GOLDBERG, M. E. (2003). Neuronal activity in the lateral intraparietal area and spatial attention. *Science, 299,* 81–86.

BRESSLER, S. L., TANG, W., SYLVESTER, C. M., SHULMAN, G. L., & CORBETTA, M. (2008). Top-down control of human visual cortex by frontal and parietal cortex in anticipatory visual spatial attention. *J Neurosci, 28*(1005), 6–61.

BRIGHINA, F., BISIACH, E., OLIVERI, M., PIAZZA, A., LA BUA, V., DANIELE, O., & FIERRO, B. (2003). 1 Hz repetitive transcranial magnetic stimulation of the unaffected hemisphere ameliorates contralesional visuospatial neglect in humans. *Neurosci Lett, 336,* 131–133.

Carter, A. R., Astafiev, S. V., Lang, C. E., Connor, L. T., Rengachary, J., Strube, M. J. ...., & Corbetta, M. (2010). Resting interhemispheric functional magnetic resonance imaging connectivity predicts performance after stroke. *Ann Neurol, 67,* 365–375.

Chechlacz, M., Rotshtein, P., & Humphreys, G. W. (2012). Neuroanatomical dissections of unilateral visual neglect symptoms: ALE meta-analysis of lesion-symptom mapping. *Front Hum Neurosci, 6,* 230.

Corbetta, M., Akbudak, E., Conturo, T. E., Snyder, A. Z., Ollinger, J. M., Drury, H. A. .... , & Shulman, G. L. (1998). A common network of functional areas for attention and eye movements. *Neuron, 21,* 761–773.

Corbetta, M., Kincade, M. J., Lewis, C., Snyder, A. Z., & Sapir, A. (2005). Neural basis and recovery of spatial attention deficits in spatial neglect. *Nat Neurosci, 8,* 1603–1610.

Corbetta, M., Kincade, J. M., Ollinger, J. M., McAvoy, M. P., & Shulman, G. L. (2000). Voluntary orienting is dissociated from target detection in human posterior parietal cortex. *Nat Neurosci, 3,* 292–297.

Corbetta, M., & Shulman, G. L. (2002). Control of goal-directed and stimulus-driven attention in the brain. *Nat Rev Neurosci, 3,* 201–215.

Corbetta, M., & Shulman, G. L. (2011). Spatial neglect and attention networks. *Annu Rev Neurosci, 34,* 569–599.

Coslett, H. B., Bowers, D., Fitzpatrick, E., Haws, B., & Heilman, K. M. (1990). Directional hypokinesia and hemispatial inattention in neglect. *Brain, 113,* 475–486.

Coull, J. T., Frackowiak, R. S., & Frith, C. D. (1998). Monitoring for target objects: Activation of right frontal and parietal cortices with increasing time on task. *Neuropsychologia, 36,* 1325–1334.

Critchley, M. (1953). *The parietal lobes.* London: Edward Arnold.

Della Sala, S., Logie, R. H., Beschin, N., & Denis, M. (2004). Preserved visuo-spatial transformations in representational neglect. *Neuropsychologia, 42,* 1358–1364.

Denes, G., Semenza, C., Stoppa, E., & Lis, A. (1982). Unilateral spatial neglect and recovery from hemiplegia: A follow-up study. *Brain, 105,* 543–552.

Deuel, R. M., & Collins, R. C. (1983). Recovery from unilateral neglect. *Exp Neurol, 81,* 733–748.

Di Russo, F., Aprile, T., Spitoni, G., & Spinelli, D. (2008). Impaired visual processing of contralesional stimuli in neglect patients: A visual-evoked potential study. *Brain, 131,* 842–854.

Doricchi, F., & Tomaiuolo, F. (2003). The anatomy of neglect without hemianopia: A key role for parietal-frontal disconnection? *NeuroReport, 14,* 2239–2243.

Driver, J., & Mattingley, J. B. (1998). Parietal neglect and visual awareness. *Nat Neurosci, 1,* 17–22.

Farne, A., Buxbaum, L. J., Ferraro, M., Frassinetti, F., Whyte, J., Veramonti, T. .... , & Ladavas, E. (2004). Patterns of spontaneous recovery of neglect and associated disorders in acute right brain-damaged patients. *J Neurol Neurosurg Psychiatry, 75,* 1401–1410.

Foucher, J. R., Otzenberger, H., & Gounot, D. (2004). Where arousal meets attention: A simultaneous fMRI and EEG recording study. *NeuroImage, 22,* 688–697.

Friedrich, F. J., Egly, R., Rafal, R. D., & Beck, D. (1998). Spatial attention deficits in humans: A comparison of superior parietal and temporal-parietal junction lesions. *Neuropsychology, 12,* 193–207.

Fruhmann Berger, M., Johannsen, L., & Karnath, H. O. (2008). Time course of eye and head deviation in spatial neglect. *Neuropsychology, 22,* 697–702.

Gaffan, D., & Hornak, J. (1997). Visual neglect in the monkey. Representation and disconnection. *Brain, 120*(9), 1647–1657.

Hagler, D. J., Jr., & Sereno, M. I. (2006). Spatial maps in frontal and prefrontal cortex. *NeuroImage, 29,* 567–577.

Halligan, P. W., Marshall, J. C., & Wade, D. T. (1989). Visuospatial neglect: Underlying factors and test sensitivity. *Lancet, 2,* 908–911.

Hamilton, R. H., Coslett, H. B., Buxbaum, L. J., Whyte, J., & Ferraro, M. K. (2008). Inconsistency of performance on neglect subtype tests following acute right hemisphere stroke. *J Int Neuropsychol Soc, 14,* 23–32.

He, B. J., Snyder, A. Z., Vincent, J. L., Epstein, A., Shulman, G. L., & Corbetta, M. (2007). Breakdown of functional connectivity in frontoparietal networks underlies behavioral deficits in spatial neglect. *Neuron, 53,* 905–918.

Heilman, K. M., Bowers, D., Valenstein, E., & Watson, R. T. (1987). Hemispace and hemispatial neglect. In M. Jeannerod (Ed.), *Neurophysiological and neuropsychological aspects of spatial neglect* (pp. 115–150). Amsterdam: Elsevier.

Heilman, K. M., Watson, R. T., & Valenstein, E. (1985). Neglect and related disorders. In K. M. Heilman & E. Valenstein (Eds.), *Clinical neuropsychology* (pp. 243–293). New York, NY: Oxford University Press.

Heilman, K. M., Watson, R. T., Valenstein, E., & Goldberg, M. E. (1987). Attention, behaviour and neural mechanisms. In F. Plum, V. B. Mountcastle, & S. T. Geiger (Eds.), *The handbook of physiology, Vol. V: Higher functions of the brain,* Part 2 (pp. 461–481). Bethesda, MD: American Physiological Society.

Hopfinger, J. B., Buonocore, M. H., & Mangun, G. R. (2000). The neural mechanisms of top-down attentional control. *Nat Neurosci, 3,* 284–291.

Hornak, J. (1992). Ocular exploration in the dark by patients with visual neglect. *Neuropsychologia, 30,* 547–552.

Howes, D., & Boller, F. (1975). Simple reaction time: Evidence for focal impairment from lesions of the right hemisphere. *Brain, 98,* 317–332.

Husain, M., & Kennard, C. (1996). Visual neglect associated with frontal lobe infarction. *J Neurol, 243,* 652–657.

Husain, M., Mannan, S., Hodgson, T., Wojciulik, E., Driver, J., & Kennard, C. (2001). Impaired spatial working memory across saccades contributes to abnormal search in parietal neglect. *Brain, 124,* 941–952.

Husain, M., & Rorden, C. (2003). Non-spatially lateralized mechanisms in hemispatial neglect. *Nat Rev Neurosci, 4,* 26–36.

Husain, M., Shapiro, K., Martin, J., & Kennard, C. (1997). Abnormal temporal dynamics of visual attention in spatial neglect patients. *Nature, 385,* 154–156.

Innocenti, G. M. (2009). Dynamic interactions between the cerebral hemispheres. *Exp Brain Res, 192,* 417–423.

Karnath, H. O., Ferber, S., & Himmelbach, M. (2001). Spatial awareness is a function of the temporal not the posterior parietal lobe. *Nature, 411,* 950–953.

Karnath, H. O., Himmelbach, M., & Rorden, C. (2002). The subcortical anatomy of human spatial neglect: Putamen, caudate nucleus and pulvinar. *Brain, 125*(2), 350–360.

Karnath, H. O., Rennig, J., Johannsen, L., & Rorden, C. (2011). The anatomy underlying acute versus chronic spatial neglect: A longitudinal study. *Brain, 134*, 903–912.

Karnath, H. O., Zopf, R, Johannsen, L., Fruhmann Berger, M., Nèagele, T., & Klose, U. (2005). Normalized perfusion M. R. I. to identify common areas of dysfunction: Patients with basal ganglia neglect. *Brain, 128*(10), 2462–2469.

Kastner, S., Pinsk, M. A., P., De Weerd, P., Desimone, R., & Ungerleider, L. G. (1999). Increased activity in human visual cortex during directed attention in the absence of visual stimulation. *Neuron, 22*, 751–761.

Kinsbourne, M. (1987). Mechanisms of unilateral neglect In M. Jeannerod (Ed.), *Neurophysiological and neuropsychological aspects of spatial neglect* (pp. 69–86). Amsterdam: Elsevier.

Kinsella, G., Olver, J., Ng, K., Packer, S., & Stark, R. (1993). Analysis of the syndrome of unilateral neglect. *Cortex, 29*, 135–140.

Koch, C., & Ullman, S. (1985). Shifts in visual attention: Towards the underlying circuitry. *Hum Neurobiol, 4*, 219–227.

Koch, G., Oliveri, M., Cheeran, B., Ruge, D., Lo Gerfo, E., Salerno, S., ... Caltagirone, C. (2008). Hyperexcitability of parietal-motor functional connections in the intact left-hemisphere of patients with neglect. *Brain, 131*, 3147–3155.

Kosslyn, S. M., Ganis, G., & Thompson, W. L. (2001). Neural foundations of imagery. *Nat Rev Neurosci, 2*, 635–642.

Malhotra, P., Coulthard, E. J., & Husain, M. (2009). Role of right posterior parietal cortex in maintaining attention to spatial locations over time. *Brain, 132*, 645–660.

Manly, T., Dobler, V. B., Dodds, C. M., & George, M. A. (2005). Rightward shift in spatial awareness with declining alertness. *Neuropsychologia, 43*, 1721–1728.

Marsh, E. B., & Hillis, A. E. (2008). Dissociation between egocentric and allocentric visuospatial and tactile neglect in acute stroke. *Cortex, 44*, 1215–1220.

Matthias, E., Bublak, P., Costa, A., Muller, H. J., Schneider, W. X., & Finke, K. (2009). Attentional and sensory effects of lowered levels of intrinsic alertness. *Neuropsychologia, 47*, 3255–3264.

Medina, J., Kannan, V., Pawlak, M. A., Kleinman, J. T., Newhart, M., Davis, C., ... Hillis, A. E. (2008). Neural substrates of visuospatial processing in distinct reference frames: Evidence from unilateral spatial neglect. *J Cogn Neurosci, 21*, 2073–2084.

Mesulam, M. M. (1999). Spatial attention and neglect: Parietal, frontal and cingulate contributions to the mental representation and attentional targeting of salient extrapersonal events. *Philos Trans R Soc Lond B Biol Sci, 354*, 1325–1346.

Morrison, J. H., & Foote, S. L. (1986). Noradrenergic and serotoninergic innervation of cortical, thalamic and tectal visual structures in old and new world monkeys. *J Comp Neurol, 243*, 117–128.

Mort, D. J., Malhotra, P., Mannan, S. K., Rorden, C., Pambakian, A., Kennard, C., & Husain, M. (2003). The anatomy of visual neglect. *Brain, 126*(1986).

Nicholls, M. E., Bradshaw, J. L., & Mattingley, J. B. (1999). Free-viewing perceptual asymmetries for the judgement of brightness, numerosity and size. *Neuropsychologia, 37*, 307–314.

Oke, A., Keller, R., Mefford, I., & Adams, R. N. (1978). Lateralization of norepinephrine in human thalamus. *Science, 200*, 1411–1413.

Paolucci, S., Antonucci, G., Grasso, M. G., & Pizzamiglio, L. (2001). The role of unilateral spatial neglect in rehabilitation of right brain-damaged ischemic stroke patients: A matched comparison. *Arch Phys Med Rehabil, 82*, 743–749.

Pardo, J. V., Fox, P. T., & Raichle, M. E. (1991). Localization of a human system for sustained attention by positron emission tomography. *Nature, 349*, 61–64.

Paus, T., Zatorre, R. J., Hofle, N., Zografos, C., Gotman, J., Petrides, M., & Evans, A. C. (1997). Time-related changes in neural systems underlying attention and arousal during the performance of an auditory vigilance task. *J Cogn Neurosci, 9*, 392–408.

Perani, D., Vallar, G., Cappa, S., Messa, C., & Fazio, F. (1987). Aphasia and neglect after subcortical stroke: A clinical/cerebral perfusion correlation study. *Brain, 110*, 1211–1229.

Posner, M. I., Walker, J. A., Friedrich, F. J., & Rafal, R. D. (1984). Effects of parietal injury on covert orienting of attention. *J Neurosci, 4*, 1863–1874.

Pouget, A., & Driver, J. (2000). Relating unilateral neglect to the neural coding of space. *Curr Opin Neurobiol, 10*, 242–249.

Rees, G., Wojciulik, E., Clarke, K., Husain, M., Frith, C., & Driver, J. (2000). Unconscious activation of visual cortex in the damaged right hemisphere of a parietal patient with extinction. *Brain, 123*(8), 1624–1633.

Rengachary, J., d'Avossa, G., Sapir, A., Shulman, G. L., & Corbetta, M. (2009). Is the Posner reaction time test more accurate than clinical tests in detecting left neglect in acute and chronic stroke? *Arch Phys Med Rehabil, 90*(12), 2081–2088.

Rengachary, J., He, B. J., Shulman, G. L., & Corbetta, M. (2011). A behavioral analysis of spatial neglect and its recovery after stroke. *Front Hum Neurosci, 5*, 29.

Rizzolatti, G., Riggio, L., Dascola, I., & Umiltá, C. (1987). Reorienting attention across the horizontal and vertical meridians: Evidence in favor of a premotor theory of attention. *Neuropsychologia, 25*, 31–40.

Robertson, I. H., Mattingley, J. B., Rorden, C., & Driver, J. (1998). Phasic alerting of neglect patients overcomes their spatial deficit in visual awareness. *Nature, 395*, 169–172.

Robertson, I., & North, N. (1992). Spatio-motor cueing in unilateral left neglect: The role of hemispace, hand and motor activation. *Neuropsychologia, 30*, 553–563.

Robertson, I., Tegner, R., Tham, K., Lo, A., & Nimmo-Smith, I. (1995). Sustained attention training for unilateral neglect: Theoretical and rehabilitation implications. *J Clin Exp Neuropsychol, 17*, 416–430.

Rueckert, L., & Grafman, J. (1996). Sustained attention deficits in patients with right frontal lesions. *Neuropsychologia, 34*, 953–963.

Rueckert, L., & Grafman, J. (1998). Sustained attention deficits in patients with lesions of posterior cortex. *Neuropsychologia, 36*, 653–660.

Samuelsson, H., Hjelmquist, E. K., Jensen, C., Ekholm, S., & Blomstrand, C. (1998). Nonlateralized attentional deficits: An important component behind persisting visuospatial neglect? *J Clin Exp Neuropsychol, 20*, 73–88.

Sereno, M. I., Pitzalis, S., & Martinez, A. (2001). Mapping of contralateral space in retinotopic coordinates by a parietal cortical area in humans. *Science, 294,* 1350–1354.

Shomstein, S., Lee, J., & Behrmann, M. (2010). Top-down and bottom-up attentional guidance: Investigating the role of the dorsal and ventral parietal cortices. *Exp Brain Res, 206*(2), 197–208.

Shulman, G. L., Astafiev, S. V., Franke, D., Pope, D. L., Snyder, A. Z., McAvoy, M. P., & Corbetta, M. (2009). Interaction of stimulus-driven reorienting and expectation in ventral and dorsal frontoparietal and basal ganglia-cortical networks. *J Neurosci, 29,* 4392–4407.

Shulman, G. L., Pope, D. L., Astafiev, S. V., McAvoy, M. P., Snyder, A. Z., & Corbetta, M. (2010). Right hemisphere dominance during spatial selective attention and target detection occurs outside the dorsal frontoparietal network. *J Neurosci, 30,* 3640–3651.

Snow, J. C., & Mattingley, J. B. (2006). Goal-driven selective attention in patients with right hemisphere lesions: How intact is the ipsilesional field? *Brain, 129,* 168–181.

Spinelli, D., Guariglia, C., Massironi, M., Pizzamiglio, L., & Zoccolotti, P. (1990). Contrast sensitivity and low spatial frequency discrimination in hemi-neglect patients. *Neuropsychologia, 28,* 727–732.

Stevens, M. C., Calhoun, V. D., & Kiehl, K. A. (2005). Hemispheric differences in hemodynamics elicited by auditory oddball stimuli. *NeuroImage, 26,* 782–792.

Stone, S. P., Halligan, P. W., & Greenwood, R. J. (1993). The incidence of neglect phenomena and related disorders in patients with an acute right or left hemisphere stroke. *Age Ageing, 22,* 46–52.

Sturm, W., de Simone, A., Krause, B. J., Specht, K., Hesselmann, V., Radermacher, I., ... Willmes, K. (1999). Functional anatomy of intrinsic alertness: Evidence for a fronto-parietal-thalamic-brainstem network in the right hemisphere. *Neuropsychologia, 37,* 797–805.

Sturm, W., Longoni, F., Fimm, B., Dietrich, T., Weis, S., Kemna, S. ... Willmes, K. (2004). Network for auditory intrinsic alertness: A PET study. *Neuropsychologia, 42,* 563–568.

Swisher, J. D., Halko, M. A., Merabet, L. B., McMains, S. A., & Somers, D. C. (2007). Visual topography of human intraparietal sulcus. *J Neurosci, 27,* 5326–5337.

Sylvester, C. M., Shulman, G. L., Jack, A. I., & Corbetta, M. (2007). Asymmetry of anticipatory activity in visual cortex predicts the locus of attention and perception. *J Neurosci, 27*(52), 14424–14433.

Thiebaut de Schotten, M., Urbanski, M., Duffau, H., Volle, E., Lévy, R., Dubois, B., & Bartolomeo, P. (2005). Direct evidence for a parietal-frontal pathway subserving spatial awareness in humans. *Science, 309,* 2226–2228.

Thut, G., Nietzel, A., Brandt, S. A., & Pascual-Leone, A. (2006). Alpha-band electroencephalographic activity over occipital cortex indexes visuospatial attention bias and predicts visual target detection. *J Neurosci, 26,* 9494–9502.

Ungerleider, L. G., & Mishkin, M. (1982). Two cortical visual systems. In D. J. Ingle, M. A. Goodale, & R. J. W. Mansfield (Eds.), *Analysis of visual behavior* (pp. 549–580). Cambridge, MA: MIT Press.

Vallar, G., & Perani, D. (1987). The anatomy of spatial neglect in humans. In M. Jeannerod (Ed.), *Neurophysiological and neuropsychological aspects of spatial neglect* (pp. 235–258). Amsterdam: Elsevier.

Verdon, V., Schwartz, S., Lovblad, K. O., Hauert, C. A., & Vuilleumier, P. (2010). Neuroanatomy of hemispatial neglect and its functional components: A study using voxel-based lesion-symptom mapping. *Brain, 133,* 880–894.

Watson, R. T., Miller, B. D., & Heilman, K. M. (1977). Evoked potential in neglect. *Arch Neurol, 34,* 224–227.

Wilkins, A. J., Shallice, T., & McCarthy, R. (1987). Frontal lesions and sustained attention. *Neuropsychologia, 25,* 359–365.

20 Attentional Modulation of Receptive Field Profiles in the Primate Visual Cortex

STEFAN TREUE AND JULIO C. MARTINEZ-TRUJILLO

ABSTRACT During the allocation of spatial attention, humans and animals selectively filter information that is relevant to behavior, avoiding information-processing overload. A change in receptive field profiles may be one neural correlate of such effects of attention. A neuron's visual receptive field is the region of the visual field where the presentation of a stimulus evokes a train of action potentials. Receptive fields therefore are spatiotemporal filters representing the spatial allocation of processing resources across the visual field. Receptive fields have been seen as invariant, providing the visual system with a stable "labeled line" code for the spatial position of visual stimuli. In recent years, this view has changed; it has been proposed that under certain circumstances the spatial profile of receptive fields changes, providing the visual system with flexibility for allocating processing resources in space. Here, we focus on the brain system for encoding visual motion information and review recent findings documenting interactions between spatial attention and receptive fields in the visual cortex of primates. Such interactions create a fine balance between the benefits of receptive field invariance and those derived from the attentional modulation of receptive field profiles.

Visual receptive fields: Caught between labeled lines and dynamic resource allocation

Neurons in the primate visual system represent visual information by variations in their firing rate (frequency of action potentials) when different visual stimuli are presented within a specific, circumscribed region of visual space. This region, or *receptive field* (RF), is considered a neuron's essential physiological property; it acts as a spatial filter, shaping the encoding of retinal signals across the visual system. Since the seminal work of Hubel and Wiesel in the 1960s (e.g., Hubel & Wiesel, 1968), electrophysiological methods in nonhuman primates have been continuously refined to determine the location and shape of neuronal RFs throughout the visual cortex. Such studies have revealed a basic organization of the primate visual system in which the image of an organism's environment projected onto the retina is encoded by populations of neurons in a multitude of areas in striate and extrastriate visual cortices, each containing retinotopically organized maps emerging from the systematic tiling of the visual field by the RFs of neighboring neurons. Within a given cortical area, the size of RFs increases with eccentricity. Similarly, RF sizes increase along the processing hierarchy, from early striate to extrastriate visual cortices. In association areas of the frontal, parietal, and temporal lobes, RFs become so large that they cover most or all of the visual field captured by the eyes.

The spatial tuning embodied by a visual neuron's RF is combined with a tuning for nonspatial features, that is, a systematic selectivity for one or several nonspatial stimulus properties. For example, neurons in area V1 change their firing rate when stimuli shown inside the RFs vary their color, orientation, motion direction, or speed. In areas downstream from V1, feature selectivity becomes segregated into two main streams, the "what" and "where" pathways. In the former, reaching from V1 to inferotemporal cortex, neurons show selectivity for orientation, color, shape, and even complex objects such as faces and houses. In the latter, which comprises a chain of areas leading into parietal cortex, neurons show selectivity for motion features such as direction and speed, and for stereoscopic disparity (depth). The selectivity of a neuron's RF for a given stimulus feature can be described by the cell's tuning curve, a plot of mean response as a function of the feature value presented inside its RF.

A predominant and long-held view has been that the spatial structure of the neurons' RF, as well as their nonspatial tuning, are hard-wired properties. Such invariance ensures a stable "labeled line code" for the location and other basic aspects of sensory stimuli. This provides a powerful, simple, and easily decodable representation of the spatial layout of the visual environment. On the other hand, these advantages are counterbalanced by the inflexibility of such a system. Relevant information is not homogenously distributed in the sensory environment and is dynamically changing, providing a continuous strain on the processing resources of a nervous system aiming to achieve an efficient and powerful representation of just the relevant aspects of the sensory input at any given moment.

The "missing link" between these opposing approaches of strict "labeled line coding" and an entirely flexible resource allocation according to momentary behavioral demands is selective attention, or the ability to bias the distribution of processing resources according to momentary demands. For spatial attention, this manifests behaviorally in our ability to selectively perceive visual stimuli within our "spotlight of attention" at the expense of those stimuli outside our attentional focus.

Over recent decades there has been substantial progress in understanding how spatial attention relies on changes in the RF of neurons in the visual cortex and how spatial attention causes those changes. One major avenue of investigation is to explore the properties of visual RFs in many areas of awake macaque monkeys while manipulating the behavioral relevance of visual stimuli.

Multiple stimuli inside the receptive field of visual neurons

It has been shown that in extrastriate visual areas, RFs increase in size up to several degrees of visual angle. Thus, in a normal visual scene, stimuli located nearby will unavoidably share the same RF. Some studies have shown that when two stimuli falling inside a neuron's RF have similar behavioral relevance, the cell's response is a weighted sum of the responses to each stimulus presented alone (Recanzone, Wurtz, & Schwarz, 1997; Treue, Hol, & Rauber, 2000). This, of course, introduces ambiguity in the coding of stimulus properties. Thus, this neuron cannot unambiguously signal one of these two particular stimulus configurations inside its RF.

A solution to this problem was proposed by Moran and Desimone more than two decades ago. They positioned two oriented bars (one with the preferred and one with a less preferred orientation) inside the RF of a neuron in area V4 of macaque monkeys trained in an attentional task, and observed that when the animals directed their attention to one of the stimuli, the neurons increased their responses when attention was on the preferred stimulus (i.e., the one evoking a strong response) and reduced their response when attention was on the less-preferred stimulus (i.e., the one evoking a weak response; Moran & Desimone, 1985). They hypothesized that the mechanism underlying this nonsensory response modulation was a change in the neurons' RF profile. Essentially, the neurons responded as if the RF had shrunk around the attended stimulus, effectively excluding the unattended stimulus from the RF.

Other studies have demonstrated similar attentional modulation along the dorsal pathway. For example, Treue and Maunsell used two oppositely moving dots within the RF of direction-selective neurons in area MT and showed that directing attention to one dot modulated the MT neuron's responses as if the influence of the second, unattended dot had been reduced (Treue & Maunsell, 1996). The same result was reported by Treue and Martinez-Trujillo in 1999 using random dot patterns (RDPs; Treue & Martinez-Trujillo, 1999). They positioned two RDPs inside the RF of MT neurons and instructed the animals to direct attention to one or the other. As anticipated, when the two stimuli were unattended and the animals detected a small color change on a central fixation spot, the responses were a weighted average of the responses to the two stimuli presented alone. However, when the animals directed attention to the RDP moving in the neuron's preferred direction, responses approximated the ones to this pattern presented alone inside the RF. A similar effect was obtained when the animals directed attention to the RDP varying direction from trial to trial (tuning pattern); then responses approximated the ones to this pattern alone inside the RF (figure 20.1). Thus, directing attention to one out of two stimuli inside the RF of an MT neuron will increase the influence of that stimulus on the response and filter out the influence of the second unattended stimulus.

These and many later findings in other areas of visual cortex demonstrate that visual neurons are not invariant filters that exclusively encode their sensory input, but that their response properties are modulated by the allocation of attention. Thus, information gets effectively filtered out of the neurons' RFs at an early point during visual processing according to behavioral relevance.

Directly measuring changes in receptive field profiles with attention

To directly determine whether the mechanism of spatial-attention modulation is indeed a reshaping of RFs, as suggested originally by Moran and Desimone, it is

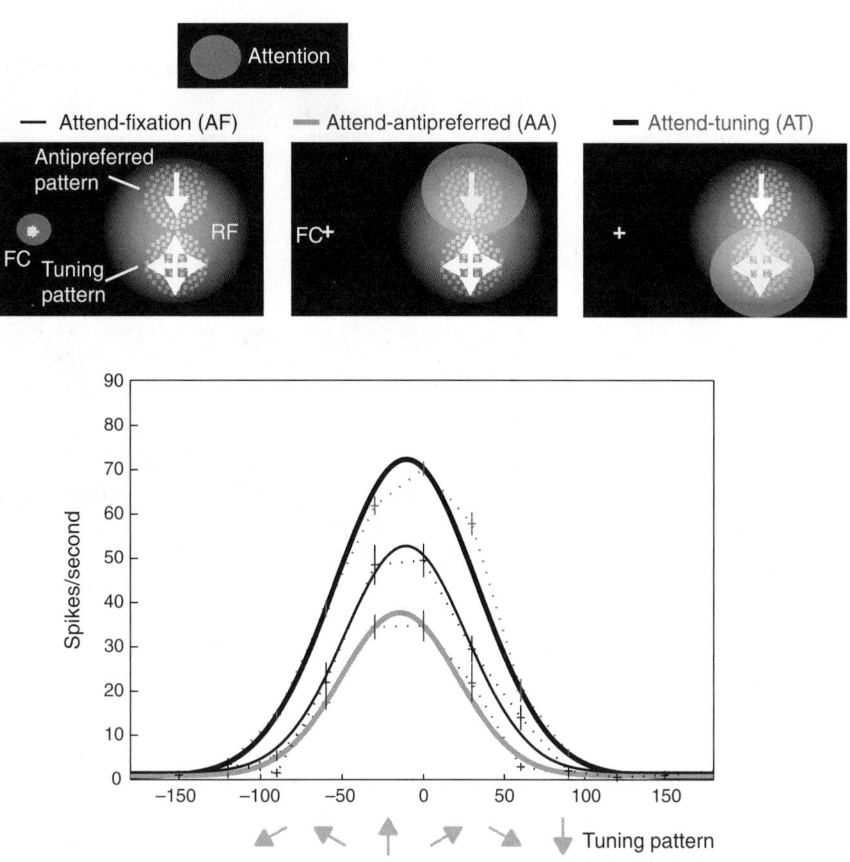

FIGURE 20.1 Switching attention between two moving stimuli inside the receptive field (RF) of MT neurons. The top panels illustrate the experimental design of Treue and Martinez-Trujillo (1999). Two random dot patterns (RDPs) were presented inside the RF of an MT neuron (white area) while the animal fixated a dot (small square). One RDP always moved in the neurons' anti-preferred direction (anti-preferred pattern) and the other could change direction from trial to trial (tuning pattern). The lower panel illustrates tuning curves (responses as a function of the tuning pattern direction) obtained when the animal attended to the fixation point (black), to the tuning pattern (red), and to the anti-preferred pattern (blue). (See color plate 16.)

necessary to map RFs while spatial attention is allocated to one or the other of two locations inside a given RF. Womelsdorf and colleagues (Womelsdorf, Anton-Erxleben, Pieper, & Treue, 2006) recorded the responses of single neurons in area MT of rhesus monkeys while the animals attended to one of two identical RDPs positioned inside the recorded neuron's RF. The task for the animals was to sustain attention on one pattern, wait for a change in its direction, and release a lever within a short time window after the change onset. Because the second, unattended pattern could also change direction, and trials were terminated without reward if the animal released the lever in response to the distractor change, they were motivated to attend to the target RDP and ignore the distractor. By flashing probes at different locations inside and around the RF while the animal was waiting for the stimulus to change, the researchers could map a neuron's RF separately for each of the two spatial attention conditions. When the animal directed spatial attention to one of the patterns, the RF shifted its center toward that attended stimulus's location. In addition, there was also a small shrinkage of the RF around that stimulus (figure 20.2A). This modulation of RF profiles indicates that early in the hierarchy of visual processing, unattended information is filtered out from the RF and only behaviorally relevant attended signals are further processed in downstream areas.

One feature of RFs in many visual areas is their center-surround structure; the center is an excitatory region where stimuli produce an increase in response, and the surround is an inhibitory region where stimuli produce response inhibition. One interesting question arising from the results of Womelsdorf and colleagues is whether both of these regions shift during the

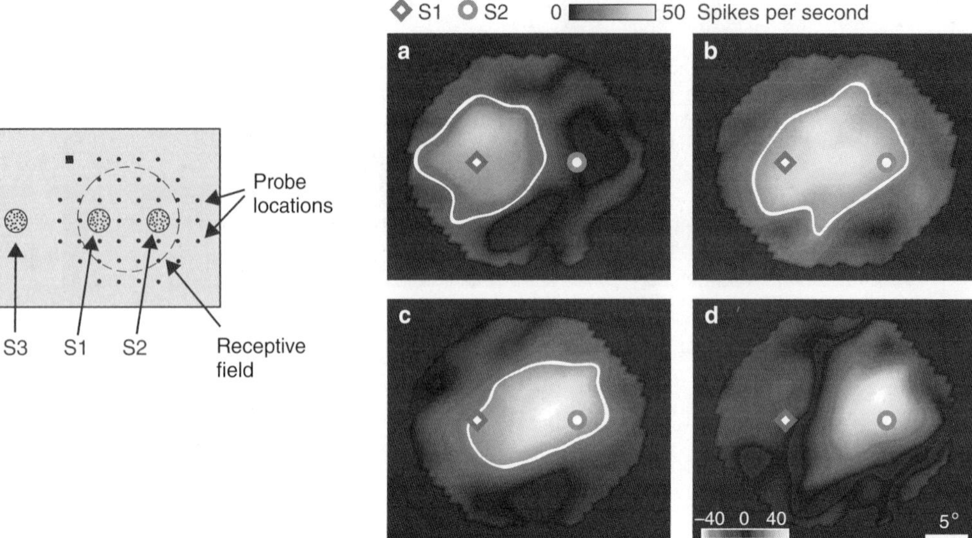

FIGURE 20.2A *Left:* Sketch of the layout in the study of Womelsdorf et al. (2006), depicting an example of the placement of the three moving random dot patterns (shown here as textured circles) that were present in every trial as well as the grid of locations at which a series of small probes could briefly appear within a trial. The single dot represents the fixation point where the animal has to maintain his gaze throughout every trial. *Right:* RF profiles of an example neuron, when attention was directed inside the RF, to stimulus S1 (panel a) or S2 (panel c), or when attention was directed outside the RF, to S3 (panel b). The surface color at each point in the plots indicates the increase in the neuron's response elicited by the presentation of a probe stimulus at that position, over the response observed in the absence of a probe (that is, when only S1 and S2 were present). Panel d depicts a difference map computed by subtracting the RF when attention was on S1 from the RF when attention was on S2. The map illustrates that shifting attention from S1 to S2 enhances responsiveness around S2 and reduces it near S1. (See color plate 17a.)

allocation of attention. Anton-Erxleben and colleagues (2009) answered this question using a similar experimental design as in figure 20.2B, with the difference that they also stimulated the RF surround with the small probes. They found that both RF regions shift with shifts in the allocation of attention (Anton-Erxleben, Stephan, & Treue, 2009).

These findings provide support for Moran and Desimone's hypothesis that spatial attention can change a neuron's RF profile. Moreover, they indicate that the sensitivity of the attended RF region increases at the expense of one of the unattended parts. Thus, the original view outlined above of neurons in visual cortex as hard-wired, invariant filters for the location and other properties of visual stimuli has to be abandoned in favor of a system where neurons receive retinal input from a restricted region of the visual field, but the sensitivity of different RF regions can dynamically change depending on the allocation of spatial attention.

When more than two stimuli share a visual receptive field: Splitting the spotlight of attention

In many visual scenes, stimuli can be cluttered over a small region of visual space (e.g., while watching a football game or staring at a menu in a restaurant). In these circumstances it is very likely that more than two stimuli will share a single visual RF. This is even more frequent in extrastriate visual areas, where RFs comprise several degrees of visual angle. Nevertheless, humans can allocate spatial attention to one stimulus within a crowd, selectively processing details for that stimulus while ignoring other stimuli nearby (He, Cavanagh, & Intriligator, 1996; Palmer & Moore, 2009).

Moreover, studies of multi-object tracking have shown that human subjects can simultaneously track (attend to) multiple, independently moving objects without moving their gaze; that is, they can split the "spotlight" of spatial attention into more than one focus (Cavanagh & Alvarez, 2005). An interesting question arising from this observation is whether the attentional mechanism that causes changes in RF profiles is flexible enough to selectively enhance processing in two or more attended regions of a given RF while suppress processing in a third region. For the particular scenario of two attended objects separated by an irrelevant distractor in between, that may mean that the RF would be virtually split by attention into two regions of enhancement, separated by a region of inhibition.

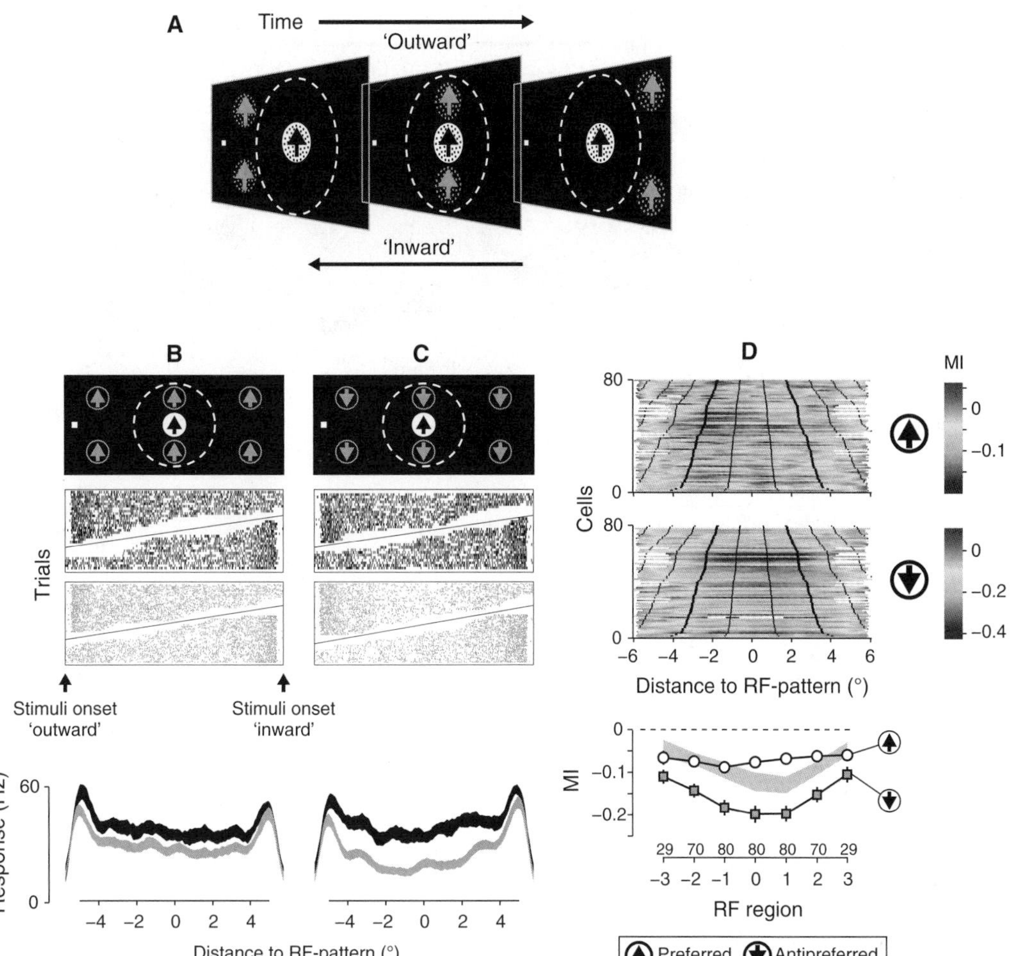

FIGURE 20.2B Splitting receptive fields with attention. (A) Sketch of the experimental paradigm of Niebergall et al. (2011b). The experiment was identical to the one described in figure 20.4, except that a third RDP (with local dot motion in the neuron's preferred direction) was placed and stayed in the center of the RF, and the animal was instructed to attend either to the translating RDPs (*tracking*) or to the RF stimulus (*attend RF*). (B) Responses for *attend RF* (green rasters) and *tracking* (red rasters) trials. Data from an example cell for the stimulus configuration with the translating RDPs dots locally moving in the preferred direction. The bottom plot shows average responses (± standard error) as a function of the translating RDP position during *attend RF* (green) and *tracking* (red) trials. (C) Responses when the translating pattern dots moved in the anti-preferred direction. (D) Response modulation for all neurons. The cells have been sorted according to their RF size, and the strength of the modulation appears in color. The arrows indicate the direction of the translating patterns (up: preferred, down: anti-preferred). The bottom graph plots the mean modulation for both directions (two lines) as well as the difference in modulation between them (gray band, representing the average difference of the two line plots ± 95% confidence interval). (See color plate 17b.)

A recent study by Niebergall, Khayat, Treue, and Martinez-Trujillo (2011b) using single-cell recordings in area MT of macaque monkeys provides a neural correlate of the ability to split the focus of spatial attention between two stimuli within a visual RF separated by an irrelevant distractor. They trained rhesus monkeys to attentively track two RDPs that translated across a computer screen along parallel paths and passed a third RDP positioned inside a recorded neuron's RF (figure 20.3A). During task trials, the animals maintained their gaze on a light spot. This allowed testing of the hypothesis that splitting the attentional spotlight across the two tracked stimuli would produce an area of suppression at the RF center, affecting the processing of the irrelevant distractor RDP.

They found that when the two tracked patterns passed alongside or entered the RF, the responses to the central stimulus in the receptive field were reduced, with the maximum suppression when the three patterns were aligned (figure 20.3B). Interestingly, they tested two stimulus configurations: in the first, the three stimuli moved in the neuron's preferred direction. In

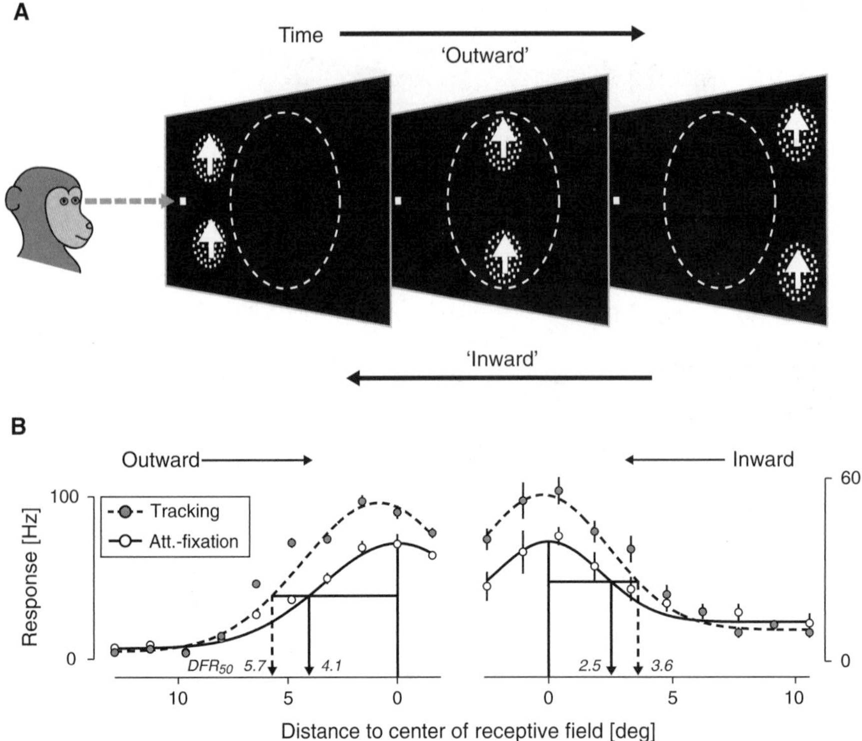

FIGURE 20.3 Expanding receptive fields with attention. (A) Sketch of the experimental paradigm of Niebergall et al. (2011a). Two RDPs moved across the screen following parallel trajectories (moving outward or inward relative to the fixation point) and moving through the peripheral parts of a neuron's receptive field (RF, dashed circle). During *attend-fixation* trials, animals ignored the RDPs and had to detect a luminance reduction in the fixation point. During *tracking* trials, animals had to detect a change in the local speed of the dots in one of the RDPs, which occurred at an unpredictable point in time. (B) Result from an example neuron. *Left:* Data from "outward" trials. *Right:* Data from "inward" trials. Data points represent the average responses evoked by RDPs with local dots moving in the neuron's preferred direction during *tracking* (gray), and *attend fixation* (white). Gaussian fit–predicted values are superimposed (dashed line, *tracking*; solid line, *attend fixation*). The DFR_{50} (downward arrows) represents the distance from the Gaussian center to the point of half-maximum response during *attend fixation* (horizontal line) as a measure of receptive field size.

the second, the translating patterns moved in the anti-preferred direction and the pattern at the center of the RF moved in the preferred direction. They found that in both configurations, directing attention to the translating patterns effectively suppressed the responses to the pattern positioned in between, relative to a condition where the latter pattern was attended and the translating patterns were irrelevant. However, in the configuration with the translating patterns moving in the neuron's anti-preferred direction, the suppression of the central pattern was larger. These results demonstrate that focusing spatial attention on the two translating patterns produced a suppressive area of inattention between them. Moreover, such suppression was dependent on the selectivity of the neurons for the attended objects. This finding matches the observation reported by functional MRI studies of spatially separated peaks of activity across the retinotopic spatial maps in visual cortex when splitting spatial attention between two distant stationary stimuli (McMains & Somers, 2004; Morawetz, Baudewig, Treue, & Dechent, 2010). It demonstrates that attention is a flexible and adaptable mechanism able to produce dramatic changes in RF profiles according to task demands.

Expansion of RFs during attentive tracking

The studies described above have demonstrated shifting and contraction of RF profiles with spatial attention in order to reduce the influence of irrelevant stimuli on neuronal responses. An additional important function of attention is to dynamically allocate processing resources to the relevant visual input. This flexible allocation is restricted across visual space toward attended locations and is constrained by the RF boundaries. This is because the RF constrains a neuron's processing resources to input from a highly restricted region of space. Whether and to what extent the RF boundaries can be reshaped by attention has only recently been investigated at the level of single neurons.

Niebergall and colleagues (2011a) trained macaque monkeys to attentively track two RDPs as they translated across a computer screen (figure 20.3A) while recording the responses of direction-selective neurons in area MT. The motion paths of the tracked stimuli were designed to approach, pass through, and exit a given neuron's RF. A comparison between neuronal responses to passing stimuli that were attended (tracked) or unattended (not tracked) revealed that in the former condition the RFs expanded toward the attended stimulus (figure 20.3B). Interestingly, this effect was stronger at the RF boundaries, as if there was a selective increase of the neurons' RF sensitivity along those boundaries when facing the tracked stimuli. The consequence of such a systematic expansion of RFs is that the spatial path traveled by an attentively tracked stimulus is represented by more neurons than an untracked path and likely during a longer time. Thus, attentively tracking a moving object dynamically allocates additional processing resources to that stimulus representation. Again, this flexibility allows the attentional system to adjust RF profiles to task demands.

Contribution of feature-based attention to shaping RFs profiles in visual cortex

Several of the studies mentioned above used behavioral tasks where spatial attention is shifted between stimuli inside the RF, and the results are consistent with an attentional modulation of RF profiles. However, a different type of attentional modulation of neuronal responses observed after the initial studies on spatial attention may provide an alternative interpretation.

In a series of studies in area MT of the dorsal visual pathway, Treue and Martinez-Trujillo showed that attending to different motion directions outside of a neuron's RF caused systematic changes in neuronal responses, even though spatial attention was far outside the RF. This phenomenon was termed feature-based attention and led to the proposal of the feature-similarity gain model. It predicts the attentional modulation of visual neurons based on the similarity between a given visual neuron's preferences (RF location and tuning for nonspatial properties) and the attended stimulus's properties. When the animal's attention is well matched to a given neuron's preferences, the neuron up-regulates its firing rate, and when the attentional match is poor, the neuron down-regulates its firing rate (Martinez-Trujillo & Treue, 2004). This feature-based attention provided an alternative or additional account for the results observed when switching spatial attention between two stimuli inside a RF, as this always involves a switching between a preferred and a nonpreferred stimulus. Thus the attentional response modulation observed could reflect this change in feature-similarity without a need to invoke a change in the RF profile.

To further explore the mechanism of feature-based attention, Patzwahl and Treue (2009) used superimposed RDPs moving in opposite directions positioned inside the RF of MT neurons, and instructed monkeys to attend to one of the RDPs while ignoring the other. In this design, a shrinking of RFs could not account for the attentional modulation of responses, since the two RDPs were fully superimposed and in the same fixation disparity plane (i.e., projected on the computer screen). They observed that the attentional modulation of responses was half of that observed when the RDPs were spatially separated inside the RF of the same neurons (see figure 20.1). This result points toward a combined effect of spatial attention (perhaps equivalent to a shrinking of the RF) and the feature-based attention effect observed by Treue and Martinez-Trujillo (1999). Thus, feature-based attention seems to act below the resolution of visual receptive fields in area MT, producing changes that interact with those produced by spatial attention in a predictable manner.

A mechanism for changes in receptive field profiles with attention

The studies discussed above document a wide variety of attention-evoked changes in the structure of extrastriate neuron RFs: changes in size and position as well as splitting RF profiles have been observed. While these changes are highly adaptive in the given attentional condition, it is unclear which mechanisms are responsible for such changes. A clue to answering this question might be provided by existing models of RFs in extrastriate visual cortical neurons. One of the most well-known RF models in extrastriate area MT was proposed by Simoncelli and colleagues (Heeger, Simoncelli, & Movshon, 1996; Simoncelli & Heeger, 1998). The response of a neuron *(R)* is computed by taking the stimulus drive *(E)*, representing the excitatory inputs into the neuron, and dividing it by a constant (σ) plus the suppressive drive *(S)*. *S* is provided by neighboring neurons with RF at the same location via interneurons. The normalized responses are then subjected to a threshold *(T)*, and the firing rate of the neuron is taken to be proportional to the amount of response exceeding the threshold (see equation 1 from Reynolds & Heeger, 2009).

$$R(x, \theta) = |E(x, \theta)/[S(x, \theta) + \sigma]_T \tag{1}$$

This model explains the sigmoid shape of the contrast response function of MT neurons, and it seems to be applicable to visual neurons in general (Heeger et al., 1996). How can we relate this model to the changes in the RF profiles observed in studies of attention? Important clues come from studies that have examined the mechanisms of attentional modulation of responses to two stimuli inside the RF of extrastriate neurons. Ghose and Maunsell (2008) recorded the responses of V4 neurons to two stimuli inside the units' RF. They also recorded the responses to each one of the stimuli alone. Then they tried to predict the responses to the stimulus pairs from the responses to the single stimuli using an input-summation model. They showed that by modulating the individual inputs into V4 neurons, they could account for the attentional modulation of responses to two stimuli in the RF and reconcile apparently contradictory observations made by previous single-cell studies of attention.

Another input model of attention was proposed by Reynolds and Heeger (2009). In their approach, attention also modulates responses of extrastriate neurons in a given area by modulating inputs into neurons. This modulation affects the intensity of the normalization mechanism that causes response saturation (Albrecht & Hamilton, 1982). This input-normalization model also explains the effect of attention on the contrast response function of extrastriate visual neurons (Reynolds, Pasternak, & Desimone, 2000; Martinez-Trujillo & Treue, 2002). As an alternative to these input models, Lee and Maunsell proposed that attention may act by controlling the intensity of the normalization step (i.e., the firing rate of neurons in the normalization pool; Lee & Maunsell, 2009). However, because the neurons in the normalization pool share inputs with the recorded neuron, any modulation of stimulus inputs would also be a modulation of the normalization pool activity. Thus, one can conclude that despite small differences between these normalization models, they all provide good accounts for most of the known modulatory effects of visual attention.

Empirical evidence in favor of changes in the strength of input signals into extrastriate visual neurons during attention tasks has been provided by Khayat and coworkers (Khayat, Niebergall, & Martinez-Trujillo, 2010a). They recorded from two macaque monkeys the responses of MT neurons to two stimuli inside their RFs, while varying the contrast and direction of one of the stimuli. By instructing the animals to switch attention between different stimuli, they observed a modulation of the neurons' firing rate that was incompatible with a modulation of responses at the level of MT neurons by a gain-control mechanism. The results were better explained by a modulation of inputs into MT neurons. Moreover, in a related study, the same authors demonstrated a modulation of local field potentials recorded in area MT compatible with a modulation of inputs into the area. They proposed that a modulation of responses of V1 neurons feeding into MT (Born & Bradley, 2005), which contribute the most to the local field potentials' high frequencies (Khayat et al., 2010b), would account for the observed pattern of modulation.

A proposed mechanism for changes in the RF profile with attention is illustrated in figure 20.4. Here a layer of neurons with small RFs (e.g., V1 neurons) feeds into an MT neuron. The connectivity strength of each neuron is determined by the weight function, which approximates a Gaussian shape (Rust, Mante, Simoncelli, & Movshon, 2006). The contribution of neurons in the middle is larger than that of neurons at both sides; thus, the MT neuron RF also approximates a Gaussian function, but it is much larger than the input neurons' RFs. Note that although this is illustrated in one dimension, this applies in the same way to a two-dimensional model of the RF. The effects of attention can be simulated by applying different gains to different neurons in the input layer (see figure 6B of in Womelsdorf, Anton-Erxleben, & Treue, 2008). In this case, by increasing the contribution of neurons on the right side and decreasing that of neurons on the left, one can obtain a shift of the RF to the right. Different transformations are possible by using different patterns of modulation (e.g., multiply the weight function by differently shaped functions that describe how attention acts on the input layer). Note that in this simplified model we have not considered the normalization step. However, it is also possible that a modulation of the normalization pool activity contributes to changing the shape of RFs, for example, to the nonlinear changes observed by Niebergall et al. (2011b) during tracking. Building a quantitative model that captures the dynamics proposed in figure 20.4 and that generates testable new predictions remains a challenge for future studies of attention.

Conclusion

Investigations of the detailed effects of spatial attention on the response properties of individual neurons in primate visual cortex have revealed a great deal of flexibility in the profile of receptive fields. Contrary to the classical view of RFs as static, hard-wired entities that form the central component of encoding incoming sensory information, it is now apparent that RFs are highly flexible entities. By being able to dynamically

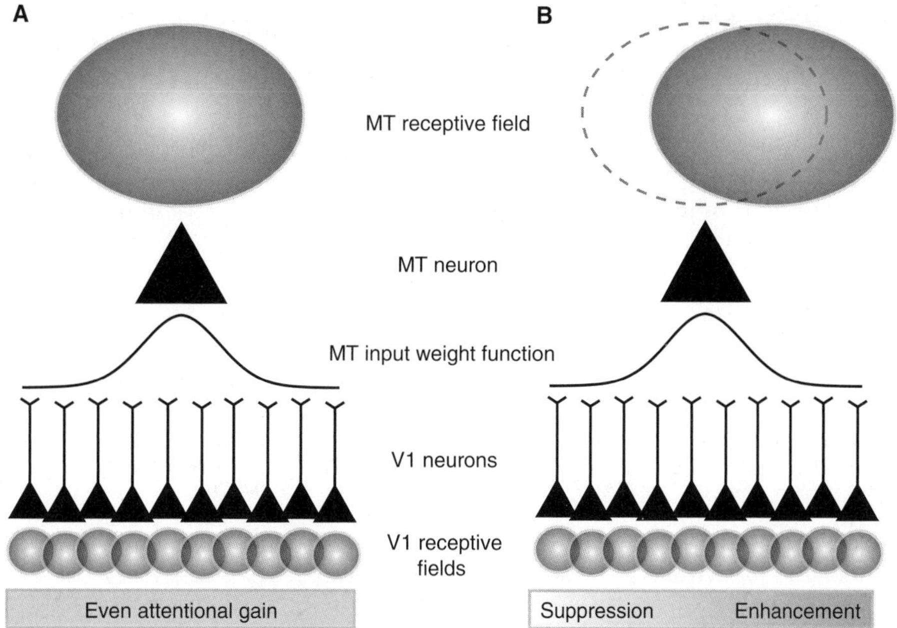

FIGURE 20.4 A possible mechanism for the attentional shift of receptive fields. The two panels are sketches of the putative V1 inputs into an MT neuron. The small circles at the bottom indicate the RFs of the V1 input neurons along a horizontal axis crossing the middle of the MT receptive field. Each small triangle represents the cell body of a V1 neuron, and the connected lines, the projection toward the MT neuron (large triangle). The bell-shaped function represents the strength of the weights from the V1 neurons into MT. (A) The large ellipse represents the MT neuron's RF. The gray band at the bottom represents the even attentional gain applied to all V1 neurons in the absence of a spatial focus of attention. (B) If attention is directed to the right, the attentional gain increases for the right V1 neurons and decreases for the left V1 neurons. This causes the MT RF profile to shift toward the right. The dashed circle represents the old RF, and the filled circle the new RF in the presence of spatial attention on the right side.

alter its RF properties, the visual system can fulfill two core goals of attentional modulation of information processing, namely, the suppression of irrelevant information and the allocation of additional resources to the processing of relevant incoming sensory information. Given the importance of a balance between the benefits and disadvantages of an invariant system and a highly adaptive and dynamic system, evolution seems to have opted for a compromise. The changes to RF profiles induced by attention seem to be highly predictable and stay within relatively well-circumscribed boundaries while still adapting to the demands of a given task. In essence, the brain combines the benefits of both a stationary and a dynamic system.

REFERENCES

ALBRECHT, D. G., & HAMILTON, D. B. (1982). Striate cortex of monkey and cat: Contrast response function. *J Neurophysiol, 48*, 217–237.

ANTON-ERXLEBEN, K., STEPHAN, V. M., & TREUE, S. (2009). Attention reshapes center-surround receptive field structure in macaque cortical area MT. *Cereb Cortex, 19*(10), 2466–2478.

BORN, R. T., & BRADLEY, D. C. (2005). Structure and function of visual area MT. *Annu Rev Neurosci, 28*, 157–189.

CAVANAGH, P., & ALVAREZ, G. A. (2005). Tracking multiple targets with multifocal attention. *Trends Cogn Sci, 9*(7), 349–354.

GHOSE, G. M., & MAUNSELL, J. H. (2008). Spatial summation can explain the attentional modulation of neuronal responses to multiple stimuli in area V4. *J Neurosci, 28*, 5115–5126.

HE, S., CAVANAGH, P., & INTRILIGATOR, J. (1996). Attentional resolution and the locus of visual awareness. *Nature, 383*, 334–337.

HEEGER, D. J., SIMONCELLI, E. P., & MOVSHON, J. A. (1996). Computational models of cortical visual processing. *Proc Natl Acad Sci USA, 93*, 623–627.

HUBEL, D. H., & WIESEL, T. N. (1968). Receptive fields and functional architecture of monkey striate cortex. *J Physiol, 195*, 215–243.

KHAYAT, P. S., NIEBERGALL, R., & MARTINEZ-TRUJILLO, J. C. (2010a). Frequency-dependent attentional modulation of local field potential signals in macaque area MT. *J Neurosci, 30*, 7037–7048.

KHAYAT, P. S., NIEBERGALL, R., & MARTINEZ-TRUJILLO, J. C. (2010b). Attention differentially modulates similar neuronal responses evoked by varying contrast and direction stimuli in area MT. *J Neurosci, 30*, 2188–2197.

Lee, J., & Maunsell, J. H. (2009). A normalization model of attentional modulation of single unit responses. *PLoS One, 4,* e4651.

Martinez-Trujillo, J., & Treue, S. (2002). Attentional modulation strength in cortical area MT depends on stimulus contrast. *Neuron, 35,* 365–370.

Martinez-Trujillo, J. C., & Treue, S. (2004). Feature-based attention increases the selectivity of population responses in primate visual cortex. *Curr Biol, 14,* 744–751.

McMains, S. A., & Somers, D. C. (2004). Multiple spotlights of attentional selection in human visual cortex. *Neuron, 42,* 677–686.

Moran, J., & Desimone, R. (1985). Selective attention gates visual processing in the extrastriate cortex. *Science, 229,* 782–784.

Morawetz, C., Baudewig, J., Treue, S., & Dechent, P. (2010). Diverting attention suppresses human amygdala responses to faces. *Front Hum Neurosci, 4,* 226.

Niebergall, R., Khayat, P. S., Treue, S., & Martinez-Trujillo, J. C. (2011a). Expansion of MT neurons excitatory receptive fields during covert attentive tracking. *J Neurosci, 31,* 15499–15510.

Niebergall, R., Khayat, P. S., Treue, S., & Martinez-Trujillo, J. C. (2011b). Multifocal attention filters targets from distracters within and beyond primate MT neurons' receptive field boundaries. *Neuron, 72,* 1067–1079.

Palmer, J., & Moore, C. M. (2009). Using a filtering task to measure the spatial extent of selective attention. *Vis Res, 49,* 1045–1064.

Patzwahl, D. R., & Treue, S. (2009). Combining spatial and feature-based attention within the receptive field of MT neurons. *Vision Res, 49,* 1188–1193.

Recanzone, G. H., Wurtz, R. H., & Schwarz, U. (1997). Responses of MT and MST neurons to one and two moving objects in the receptive field. *J Neurophysiol, 78,* 2904–2915.

Reynolds, J. H., & Heeger, D. J. (2009). The normalization model of attention. *Neuron, 61,* 168–185.

Reynolds, J. H., Pasternak, T., & Desimone, R. (2000). Attention increases sensitivity of V4 neurons. *Neuron, 26,* 703–714.

Rust, N. C., Mante, V., Simoncelli, E. P., & Movshon, J. A. (2006). How MT cells analyze the motion of visual patterns. *Nat Neurosci, 9,* 1421–1431.

Simoncelli, E. P., & Heeger, D. J. (1998). A model of neuronal responses in visual area MT. *Vision Res, 38,* 743–761.

Treue, S., Hol, K., & Rauber, H. J. (2000). Seeing multiple directions of motion: Physiology and psychophysics. *Nat Neurosci, 3,* 270–276.

Treue, S., & Maunsell, J. H. (1996). Attentional modulation of visual motion processing in cortical areas MT and MST. *Nature, 382,* 539–541.

Treue, S., & Martinez-Trujillo, J. C. (1999). Feature-based attention influences motion processing gain in macaque visual cortex. *Nature, 399,* 575–579.

Womelsdorf, T., Anton-Erxleben, K., Pieper, F., & Treue, S. (2006). Dynamic shifts of visual receptive fields in cortical area MT by spatial attention. *Nat Neurosci, 9,* 1156–1160.

Womelsdorf, T., Anton-Erxleben, K., & Treue, S. (2008). Receptive field shift and shrinkage in macaque middle temporal area through attentional gain modulation. *J Neurosci, 28,* 8934–8944.

21 Attentional Selection Through Rhythmic Synchronization at Multiple Frequencies

THILO WOMELSDORF, AYELET N. LANDAU, AND PASCAL FRIES

ABSTRACT Selective visual attention is realized in brain circuits by the dynamic restructuring of cortical information flow in favor of sensory information that is behaviorally relevant, rather than irrelevant and distracting. Electrophysiological evidence suggests that this attentional selection of neuronal information routing is implemented flexibly through selective synchronization of neuronal activities at multiple frequencies and across all spatial levels of neuronal interactions. This chapter delineates the underlying neuronal processes that could give rise to selective neuronal communication through coherence and highlights recent insights about (1) selective attentional routing through synchronized gamma activity in sensory cortices, (2) long-range reverberation of selective top-down information at beta and gamma-band frequencies, and (3) the role of low-frequency theta band rhythms to selectively sample and switch between attentionally relevant sensory information. This survey suggests that core processes underlying attentional structuring of neuronal communication employ precise phase synchronization that is selective in space, time, and frequency.

The role of selective neuronal synchronization in the implementation of attentional top-down biases

Attentional top-down biases on perceptual processes are ubiquitous, ensuring that neuronal circuitry is devoted to prioritize processing resources to the typically small subset of sensory inputs that should control behavior. In visual cortex, this identification process has to act highly selectively, because the same sensory aspects can pertain to different visual objects. Attention is thus required to functionally connect those neurons that represent information from attended visual objects, and decouple or filter out the contribution of cells encoding distracting information (figure 21.1A). For example, neurons at the highest visual processing stage in IT cortex have receptive fields that span much of a visual field and respond selectively to complex objects composed of simpler visual features. Part of this selectivity arises from their broad and convergent anatomical input at earlier processing stages from neurons that have smaller receptive fields and simpler tuning properties (Kravitz, Saleem, Baker, Ungerleider, & Mishkin, 2013). During natural vision, the large receptive field of an IT neuron will typically contain multiple objects. However, when attention is directed to only one of those objects, the IT neuron's response is biased toward the response that would be obtained if only the attended object were presented (Chelazzi, Miller, Duncan, & Desimone, 1993).

Such dynamic biasing of responses in later visual processing stages could be achieved by selective enhancement (suppression) of the impact of those afferent inputs from neurons in earlier visual areas coding for the attended (nonattended) input (Reynolds, Chelazzi, & Desimone, 1999). However, the mechanisms underlying this up-and down-modulation of neuronal input gain for subsets of converging connections are only poorly understood. They likely entail a selective increase of temporally precise and coincident inputs from those neurons activated by an attended stimulus in earlier areas. Such a role of spike timing is suggested by fine-grained attentional modulation of precise neuronal synchronization within area V4. Enhanced synchronization of the spiking output among those neuronal groups activated by attended sensory input (Fries, Womelsdorf, Oostenveld, & Desimone, 2008) translates into enhanced coincident arrival of spikes at their postsynaptic target neurons at later stages of visual processing. These temporally coincident inputs will be highly effective in driving neuronal activity (Azouz & Gray, 2003; Tiesinga, Fellous, & Sejnowski, 2008). It is therefore likely that selective synchronization within area V4 underlies attentional biasing within IT cortex, and could thus underlie effective spatial routing of information flow within visual cortex (Bosman et al., 2012).

Synchronized spike output not only leads to increased impact on postsynaptic target neurons in a feedforward manner, it also rhythmically modulates the group's ability to communicate such that rhythmic synchronization between two neuronal groups likely supports their interaction, because rhythmic inhibition within the two groups is coordinated and mutual inputs are optimally timed. We capture these implications in the framework of selective neuronal communication through neuronal coherence (Fries, 2005, 2009).

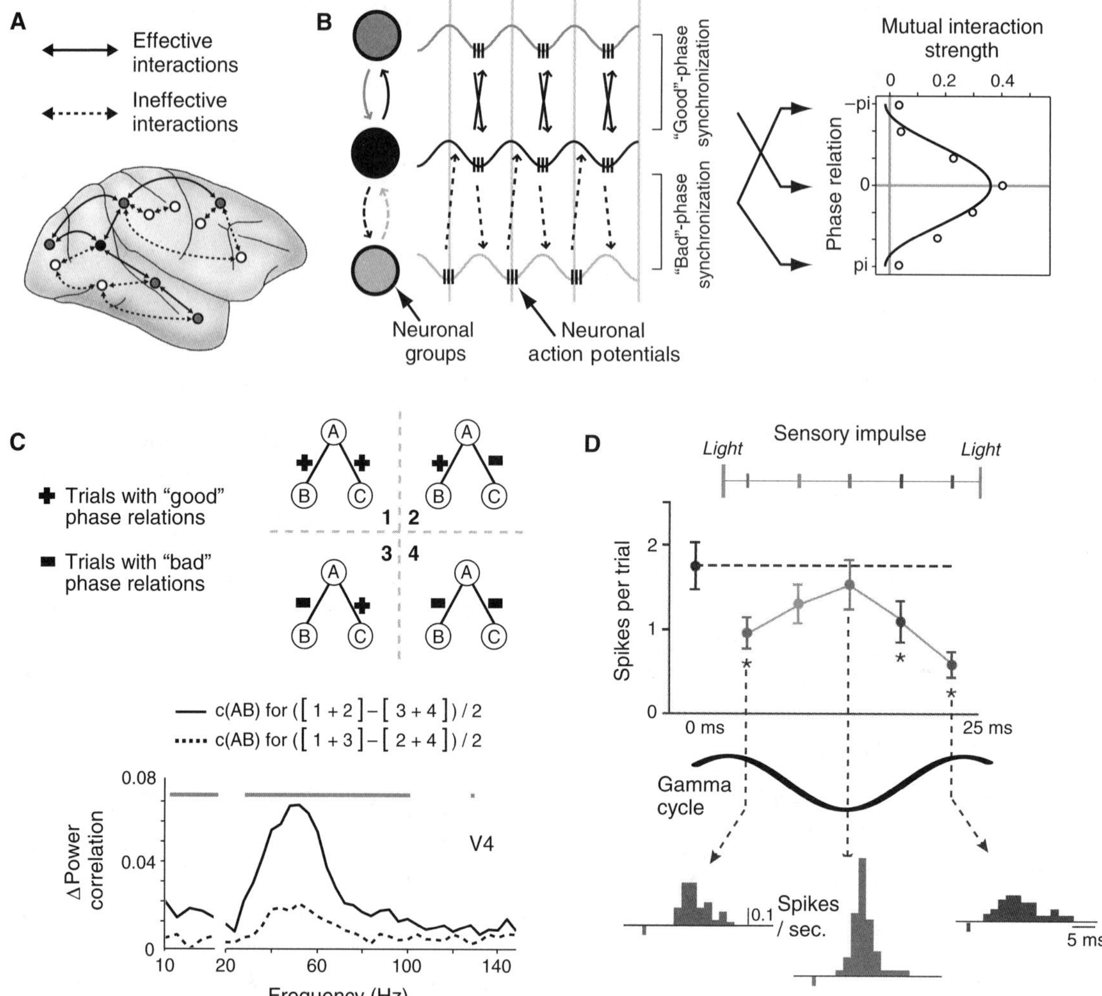

FIGURE 21.1 (A) Attention renders a subset of cortical connections effective. (B) Rhythmic activity provides briefly recurring time windows of maximum excitability (LFP troughs), which are either in phase (black and dark gray groups), or in anti-phase (black and light gray groups). Mutual interactions (right panel, upper axis) are high during periods of in-phase synchronization and lower otherwise. (C) The trial-by-trial interaction pattern between neuronal groups (A to B, and A to C) is predicted by the pattern of synchronization. (D) Evidence that rhythmic optogenetic activation in mouse somatosensory cortex modulates gain of responses to whisker stimulation. A whisker was stimulated at different phases of an optogenetic activation rhythm. The spiking responses of pyramidal cells (*y*-axis) varied as a function of the phase (*x*-axis). ((B) and (C) adapted from Womelsdorf et al., 2007; (D) adapted from Cardin et al., 2009. Reprinted by permission from Macmillan Publishers Ltd: *Nature*.)

SELECTIVE COMMUNICATION THROUGH COHERENCE Local neuronal groups frequently engage in periods of rhythmic synchronization. During activated states, rhythmic synchronization is typically evident in the gamma-frequency band (30–90 Hz; Fries, 2009). In vitro experiments and computational studies illustrate that gamma-band synchronization can emerge from the interplay of excitatory drive and rhythmic inhibition imposed by interneuron networks (Bartos, Vida, & Jonas, 2007; Börgers, Epstein, & Kopell, 2005). Interneurons impose synchronized inhibition onto the local network (Bartos et al., 2007; Cardin et al., 2009). The brief time periods between inhibition provide time windows for effective neuronal interactions with other neuronal groups, because they reflect enhanced postsynaptic sensitivity to input from other neuronal groups as well as maximal excitability for generating spiking output to other neuronal groups (Azouz & Gray, 2003; Fries, Nikolic, & Singer, 2007; Tiesinga et al., 2008). As a consequence, when two neuronal groups open their temporal windows for interaction at the same time, they will be more likely to influence each other (Womelsdorf et al., 2007). The consequences for selective neuronal communication are illustrated in figure 21.1B: if the rhythmic activities within neuronal groups are precisely synchronized between the two groups, they are maximally likely to interact. By the same token, if rhythmic activity within neuronal groups is unsynchronized between groups, or if activity synchronizes consistently out of phase, their interaction is curtailed (figure 21.1B).

This scenario entails that the pattern of synchronization between neuronal groups flexibly structures the pattern of interactions between neuronal groups (figure 21.1C). Consistent with this hypothesis, the interaction pattern of one neuronal group (group "A") with two other groups (groups "B" and "C") can be predicted by their pattern of precise synchronization (figure 21.1C). This has been demonstrated for interactions of triplets of neuronal groups from within and between visual areas in the awake cat and monkey (Womelsdorf et al., 2007). This study measured the trial-by-trial changes in correlated amplitude fluctuation and changes in precise synchronization between pair "AB" and pair "AC," using the spontaneous variation of neuronal activity during constant visual stimulation. The strength of amplitude covariation—the covariation of power in the local field potential (LFP), multiunit spiking responses, or both—was used as the measure of mutual interaction strength (figure 21.1C, bottom panel). The results showed that the interaction strength of "AB" could be inferred from the phase of gamma-band synchronization between group A and group B, being rather unaffected by the phase of synchronization of group A with group C (figure 21.1C). This finding was evident for triplets of neuronal groups spatially separated by less than 750 μm illustrating a high spatial specificity of the influence of precise phase synchronization between neuronal groups on the efficacy of neuronal interaction. Further analysis of the temporal evolution of the selective phase-dependent power correlations revealed that the phase relations at which power correlations were maximal preceded by a few milliseconds the time of maximal power correlation (Womelsdorf et al., 2007). Such a temporal precedence is consistent with a mechanistic role of phase synchronization to instantiate neuronal interactions.

Synchronization in Interneuron Networks and Their Attentional Modulation The described characteristics of selective neuronal interactions entail the major components required for selective attentional routing: attentional selection evolves at rapid time scales and with high spatial resolution by enhancing (reducing) the effective connectivity among neuronal groups conveying task relevant (irrelevant) information.

Modeling studies have shown that such selective routing of information flow can be achieved by synchronizing an inhibitory gating network with only one of multiple input streams (Börgers & Kopell, 2008; Mishra, Fellous, & Sejnowski, 2006; Tiesinga et al., 2008). Possible sources of inhibitory gating are fast spiking interneurons of the basket cell type that synapse predominantly onto perisomatic regions of principal cells and thereby impose a powerful gain control on synaptic inputs arriving at sites distal to a cell's soma (Cardin et al., 2009). The inhibitory synaptic influence is inherently rhythmic at high (gamma) frequencies due to $GABA_A$ membrane time constants (Cardin et al., 2009; Bartos et al., 2007). Selectively activating basket cells at the gamma frequency (using optogenetic methods) has been shown to gate input through sensory cortex as a function of the phase at which the input arrives (figure 21.1D). The phase-specific inhibitory gate increases the gain of pyramidal cell output, enhances the precision of pyramidal cell firing, and shortens the neuronal response latency to sensory stimuli (figure 21.1D; Cardin et al., 2009).

During attentional processes, interneuron networks could be activated by various possible sources. They may be activated by transient and spatially specific neuromodulatory inputs (Lin, Gervasoni, & Nicolelis, 2006; Rodriguez, Kallenbach, Singer, & Munk, 2004). Alternatively, selective attention could target local interneuron networks directly via top-down inputs from neurons in upstream areas (Buia & Tiesinga, 2008; Mishra et al., 2006). In these models, selective synchronization emerges either by depolarizing selective subsets of interneurons (Buia & Tiesinga, 2008), or by biasing the phase of rhythmic activity in a more global inhibitory interneuron pool (Mishra et al., 2006). In either case, rhythmic inhibition controls the spiking responses of groups of excitatory neurons, enhancing the impact of those neurons that spike synchronously during the periods of disinhibition, while actively reducing the impact of neurons spiking asynchronously to this rhythm. This suppressive influence on excitatory neurons, which are activated by distracting feedforward input, is the critical ingredient for the concept of selective communication through coherence (CTC): attention is expected not only to enhance synchronization of already more coherent activity representing attended stimuli, but to actively suppress the synchronization and impact of groups of neurons receiving strong, albeit distracting, inputs, because these inputs arrive outside the noninhibited periods in the target group (Börgers & Kopell, 2008).

The outlined framework of the interneuron-mediated gating of neuronal information flow at rhythmic gamma-band synchronization predicts that synchronization should not only selectively increase with focused attention, but that phase-dependent activity implements a selective gate on bottom-up processing of sensory information that is integrated within a large-scale attention network that encodes the higher-level, top-down information. Recent evidence supports these predictions.

Experimental evidence for the role of selective gamma-band synchronization for attentional selection

Direct evidence for the functional significance of selective synchronization within and between local neuronal groups for attentional selection has been obtained from recordings in macaque visual cortical areas V1, V2, and V4 (e.g., Bosman et al., 2012; Buffalo, Fries, Landman, Buschman, & Desimone, 2011; Fries et al., 2008; Gregoriou, Gotts, Zhou, & Desimone, 2009; Grothe et al., 2012). One consistent result across studies in V4 is that spatial attention enhances local (i.e., within-area) gamma-band synchronization of the neuronal groups that have receptive fields overlapping the attended location (Fries et al., 2008). Attention selectively modulates phase synchrony among LFPs, spike-LFP couplings, and mutual spike-train coherence, demonstrating that those V4 neurons that convey attended stimulus information send more coherent spike output to their postsynaptic projection targets (Fries et al., 2008). This conclusion is strongly supported by the recent findings that gamma-band spike-LFP synchronization is enhanced by attention to receptive field stimuli, particularly for neurons in superficial layers that host the majority of corticocortical feedforward projection neurons (Buffalo et al., 2011). A similar pattern of layer-specific attentional modulation is observed at the major input regions to area V4: within areas V1 and V2, selective attention modulates gamma-band synchronization in superficial—that is, feedforward projecting—layers (Buffalo et al., 2011).

The strength of selective increases of spike-LFP gamma-band synchronization, as well as the attentional modulation of firing rates, is lower within V1 and V2 as compared to area V4 (Buffalo et al., 2011). This progression of attentional modulation from early to extrastriate visual areas may result from the increased demand of neurons with larger receptive fields in later visual processing areas to select subsets of inputs from a broader set of converging inputs from neurons with smaller receptive fields. According to this suggestion, the larger attentional enhancement of selective gamma-band synchronization within area V4 is based on the requirement of V4 neurons to selectively gate the input from only those V1/V2 populations of neurons that convey attended (i.e., relevant) stimulus information.

The selective gating hypothesis of V1 to V4 connectivity has been verified in recent inter-areal recording studies in awake nonhuman primate cortex (Bosman et al., 2012; Grothe et al., 2012). Bosman et al. (2012) investigated the coherence of electrocorticographic activity from one V4 neuronal population with activity of two spatially separate V1 populations (figure 21.2A). The receptive fields at each of the V1 populations converged onto the larger receptive field of the V4 neuronal population. This arrangement allowed the placement of separate stimuli in each of the two non-overlapping V1 receptive fields, and at the same time within the confines of the same V4 receptive field (figures 21.2B and 21.2C). With this spatial arrangement, the activity of the V4 recording site reflected the response to the combined, convergent input from the two V1 sites. When attention is selectively focused on one of the two stimuli, then only the V1 site processing the attended stimulus should synchronize to V4 and thereby mediate the selective routing of attended information according to the CTC mechanism. Figures 21.2E and 21.2F show the local effect of spatial attention in V1, revealing a similar degree of gamma-band synchronization locally, that is, within each of the neuronal populations. When the animals focused spatial attention on one of the stimuli, gamma-band coherence emerged between V4 and the specific V1 site that conveyed the relevant, attended stimulus. At the same time, gamma-band synchronization was strongly reduced between V4 and the V1 site that conveyed the irrelevant, distracting stimulus (figures 21.2G and 21.2H). Thus, the top-down influence of selective attention determined which anatomically converging connections from primary visual cortex to extrastriate area V4 became functionally effective (Bosman et al., 2012). This emergence of selective functional connectivity was brought about by selective gamma-band coherence, providing direct support for the concept of selective communication through coherence (Fries, 2005).

Communication though neuronal coherence could be based on a symmetric mutual influence between neuronal groups, or it could arise from nonsymmetric influences. The asymmetry, or directedness, of coherence can be estimated with the Granger causality metric, which quantifies the influence of one neuronal group "A" (e.g., in V1) onto another group "B" (e.g., the V4 group) as the variance in "B" that is explained by the immediate past of "A," rather than by the past of "B"

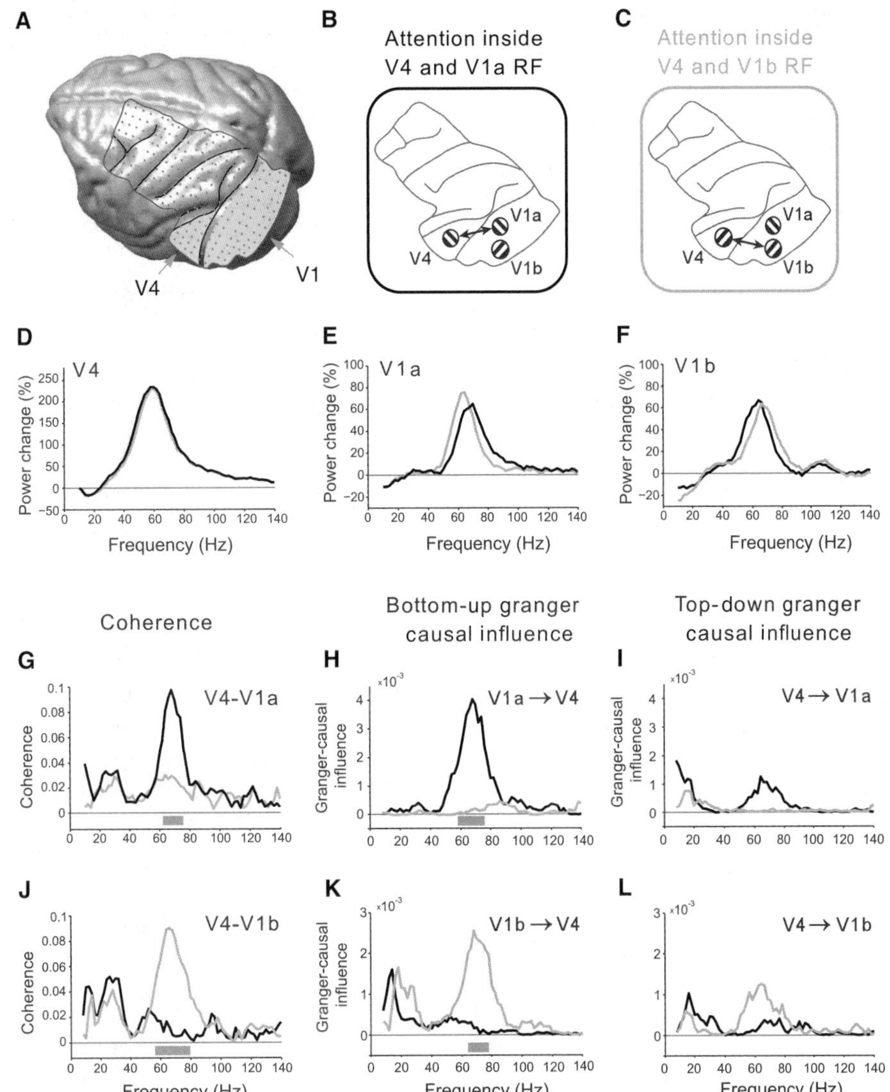

FIGURE 21.2 (A) The spatial coverage of ECoG electrode locations. Shaded regions indicate areas V1 and V4. (B, C) Attention conditions with arrows denoting that only the behaviorally relevant and therefore attended stimulus was routed from the respective V1 site to the V4 site. The shading of the box contours (black, gray) is used in the remaining panels to denote the attention conditions. (D) Spectral power change relative to pre-stimulus baseline in V4 (gray line mostly occluded by black line). (E, F) Spectral power change relative to pre-stimulus baseline in V1a (E) and V1b (F). (G, H) Coherence spectra between V4 and V1a (G) and between V4 and V1b (H). (I, J) Bottom-up Granger causal influence of V1a onto V4 (I) and of V1b onto V4 (J). (K, L) Top-down Granger causal influence of V4 onto V1a (K) and of V4 onto V1b (L). Gray bars in (G)–(L) indicate the frequencies with a significant effect. (Adapted from Bosman et al., 2012. Reprinted by permission from Macmillan Publishers Ltd: *Nature.*)

(Dhamala, Rangarajan, & Ding, 2008). Applying the Granger causal metric, Bosman et al. (2012) revealed a predominance of Granger causal influences from the attended V1 neuronal group onto the neuronal group in V4 in the gamma-frequency band (figures 21.2I and 21.2J), and only a subtle Granger causal feedback influence of V4 to V1 (figures 21.2K and 21.2L). These findings resonate well with the predominance of gamma-band synchronization in superficial versus deep visual cortical areas reported above (Buffalo et al., 2011), because feedforward connectivity originates from superficial layers in primary visual cortex.

Origins of attentional top-down biases onto visual cortex: Selective long-range coordination at gamma- and beta-frequency synchronization

The described findings illustrate that the behavioral relevance of a stimulus location or feature determines how the neuronal circuitry in V4 synchronizes its activity to earlier stages of the sensory processing hierarchy. The top-down information about which stimulus aspect is behaviorally relevant arrives in area V4 from diverse sources, including the posterior parietal cortex and the prefrontal cortex. Both of these brain regions are part of larger frontoparietal networks that implement selective attention and working memory (Corbetta & Schulman, 2011; see also Serences & Kastner and Corbetta et al., this volume). Notably, prefrontal and parietal network nodes convey top-down attention information about relevant stimulus features in local and long-range, phase-synchronized activity that critically support the outlined role of synchronization as a key mechanism for selective neuronal communication (Gregoriou et al., 2009; Gregoriou, Gotts, & Desimone, 2012; Liebe, Hoerzer, Logothetis, & Rainer, 2012; Phillips, Vinck, Everling, & Womelsdorf, 2013; Pesaran, Nelson, & Andersen, 2008; Salazar, Dotson, Bressler, & Gray, 2012; Siegel, Donner, & Engel, 2012). In particular, recent studies suggest that neuronal circuitry in area V4 receives specific top-down control signals through selective phase locking to the activity of two major networks. First, area V4 phase-couples in the classical gamma-frequency band to visually tuned neurons in the frontal eye field (Gregoriou et al., 2009, 2012). And second, V4 synchronizes to a beta-synchronous cortico-cortical network that is composed of the dorsolateral prefrontal and the posterior parietal cortex (Buschman, Denovellis, Diogo, Bullock, & Miller, 2012; Buschman & Miller, 2007; Salazar et al., 2012; Siegel et al., 2012).

The frontal eye field (FEF), a known source of attentional top-down control, contains neuronal circuitry whose electrical and pharmacological stimulation imposes a selective change of neuronal activity in visual area V4 that resembles the activity modulation of spatial attention (Noudoost & Moore, 2011; Noudoost, Albarran, & Moore, this volume). Notably, electrical microstimulation of circuitry within FEF induces 40 Hz gamma oscillations of those distantly connected V4 sites that contain neurons with receptive fields overlapping the stimulation sites (Premereur, Vanduffel, Roelfsema, & Janssen, 2012). These long-distance influences from FEF onto visual cortex likely depend on monosynaptic and excitatory long-range connections that are particularly dense and bidirectional between FEF and V4 (Anderson, Kennedy, & Martin, 2011). Empirical evidence shows that this recurrent connectivity is functionally coordinated by selective attention at a narrow gamma-frequency band (Gregoriou et al., 2009). Attention to a stimulus overlapping the receptive fields of simultaneously recorded V4 and FEF sites enhances long-range gamma-band coherence when compared to attention elsewhere. Moreover, this attentional modulation is based on a subpopulation of neurons in FEF that is tuned to visual inputs and rich in long-range connectivity to visual cortex (figure 21.3; Anderson et al., 2011; Gregoriou et al., 2012). Figures 21.3A and 21.3C illustrate that selective attention inside the receptive field of visually tuned FEF neurons synchronizes their spike-trains to the distant sites in visual cortex. In contrast, FEF neurons tuned to oculomotor planning and execution did not modulate their long-range coherence as a function of covert visual attention (figures 21.3B, 21.3D; Gregoriou et al., 2012). These findings reveal that the stimulus-specific attention information that is evident in gamma cycle–dependent firing of area V4 cells (Fries et al., 2008) is phase-synchronized to visually encoding neurons of the top-down control circuitry in FEF, but decoupled from neurons that are part of the sensorimotor network that control saccadic eye movements (figure 21.3E).

Figure 21.3 (A) Average coherence of spike-trains of visually tuned neurons in the frontal eye field (FEF) to the local field potential (LFP) when the monkeys attended inside (red) or outside (blue) the visual receptive field of the FEF cell. (B) Same as in (A), but for "motor" cells in the FEF that are tuned to saccadic target directions and are located predominantly in deeper cortical layers in the FEF. (C) Long-range coherence of visually tuned FEF cells with the LFP recorded in visual area V4 FEF. (D) Long-range coherence of motor-tuned FEF cells with the LFP recorded in visual area V4. (E) Illustration of cells in the FEF that are primarily responsive to visual stimuli and located in supra-granular layers (red), visuomotor neurons (half-red, half-green), and motor-tuned neurons (green) that are predominantly located in deeper layers. The solid/dashed arrows to V4 cells indicate selective coupling of visual FEF to visual V4 cells, but a lack of coupling from motor-related FEF cells. (F) The working memory task required encoding of a sample object or sample object location. Following a 0.8–1.2 sec delay, the macaque monkeys had to identify which of two objects or object locations matched the sample. (G) Average selectivity of coherence for object identity (*upper panel*) and location (*bottom panel*) during sample and delay epochs (separated by white vertical lines). The LFP coherence was computed between pairs of frontal and posterior parietal sites. (H) Average coherence, for each of three objects, rank-ordered for each recorded pair according to their object preference. The time course shows differences in coherence with the start of the delay period. (I) The proportion of frontoparietal LFP-LFP pairs that showed individually significant selectivity of beta band coherence, split according to their anatomical locations. (Panels (A)–(D) adapted from Gregoriou et al., 2012; panels (F)–(I) adapted from Salazar et al., 2012. Reprinted with permission from AAAS.) (See color plate 18.)

Selective attention demands not only synchronize FEF-V4 connections in a location- and feature-specific manner. Recent work has shown that the selective encoding and maintenance of location- and object-specific information is achieved by an anatomically specific pattern of beta-band synchronization spanning prefrontal and parietal cortex. The cognitive nature of this long-range coherence has been documented for active visual search (Buschman & Miller, 2007) and for dynamic working memory processes (Salazar et al., 2012). Figures 21.3F and 21.3G illustrate the temporal evolution of the content-specific synchronization between prefrontal and posterior parietal cortex of the LFPs in a 12–22 Hz beta-frequency band. The strength of this long-range phase-locking started to predict either the location or the identity of a sample object at the time when the sample stimulus was removed from the visual scene, and working memory circuitry began to sustain its information during the delay of the task, until the test stimuli were presented >0.8 seconds later (figure 21.3H). Critically, the beta-rhythmic frontoparietal communication of working memory content was widespread, but

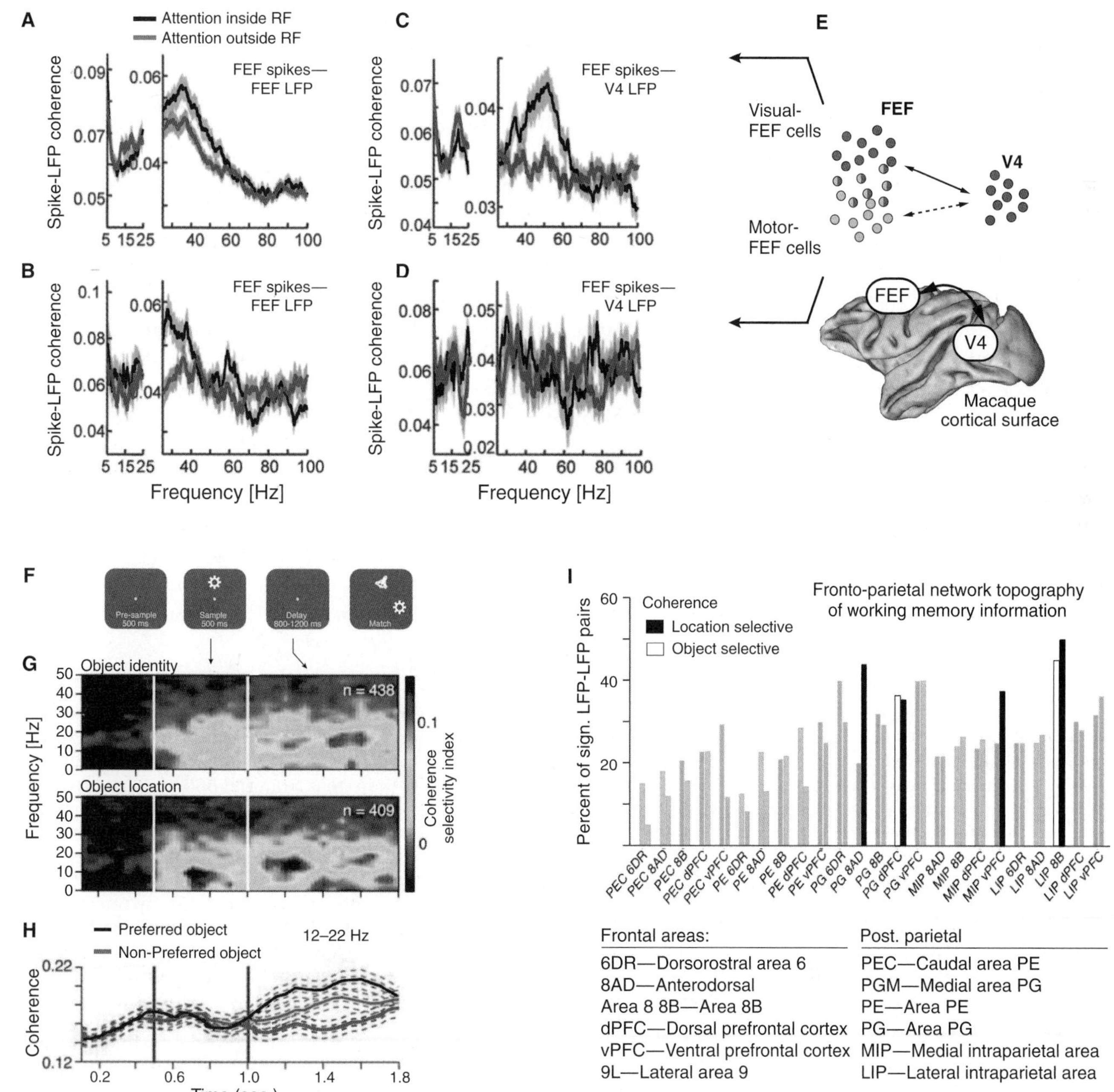

particularly pronounced for the dorsolateral prefrontal cortex (including area 8) and the lateral intraparietal cortex. More than 40% of pairs of dorsolateral prefrontal–lateral intraparietal cortex connections engaged in beta frequency–specific phase-locking that conveyed the content of the current visuospatial working memory (figure 21.3I; Salazar et al., 2012).

Taken together, the delineated evidence for selective long-range synchronization illustrates that precisely timed activity at gamma- and beta-frequency ranges (1) conveys information that is specific to behaviorally relevant stimulus dimensions, (2) emerges in anatomically clustered subareas that form selective long-range networks, and (3) is evident for selected subsets of cells of the topographic clusters that are tuned to the attended stimulus dimension.

Theta-rhythmic modulation of gamma-band synchronization

As outlined above, the CTC framework describes a mechanism for selecting relevant incoming information as well as curtailing irrelevant input from further processing (figure 21.4A). However, in continuous sensory processing there are many factors, endogenous as well as external (bottom-up), that affect the process of sensory selection. Under those circumstances, how does selective synchronization, as the likely mechanism for attentional selection, dynamically adapt to the ever-changing goals and sensory environment? Once interareal synchronization is achieved, how is it broken in order to support shifting of attention to other potentially relevant locations or objects? A possible elaboration of the CTC framework that would account for the updating of attentional selection is the incorporation of dynamic nesting of gamma rhythmic activity within slower frequencies in the theta band (<10 Hz; figure 21.4B). Previous studies suggest that gamma-band power increases are often limited to a certain phase of slower neural oscillations in the theta range. The link between gamma and theta activity was first found in rodent hippocampal recordings (Bragin et al., 1995) but later was also described in cortex (see Fries, 2009). The finding that the gamma-rhythmic synchronization is reestablished every theta cycle implies that the selection process itself is reset at a theta rhythm. This type of temporal structure gives rise to active sampling and updating of the selection process, akin to other exploration behaviors that operate at a theta rhythm. For example, sniffing and whisking in rodents occur at a theta frequency (Buszaki, 2005; Kleinfeld, Berg, & O'Connor, 1999) as does saccadic exploration in primates (Otero-Millan, Troncoso, Macknik, Serrano-Pedraza, & Martinez-Conde, 2008). Finally, attentional top-down control processes have recently been shown to proceed on large-scale theta-synchronous frontoparietal networks (Phillips et al., 2013). Figure 21.4B illustrates the proposed extension of selection through synchrony incorporating the nesting of each synchronous state within a given theta cycle. The resetting of gamma by the theta cycle as an implementation of a fundamental exploration rhythm that shapes the selection process provides a mechanistic and functional interpretation for the prevalent finding of theta-gamma cross-frequency coupling. More importantly, it also generates predictions regarding the temporal properties of brain-behavior relationships and the course of attentional selection. In what follows, we survey evidence for this theoretical possibility from three main research areas: (1) investigations of the temporal properties of overt exploration behaviors, (2) investigations of the

FIGURE 21.4 (A) Schematic illustrating the communication through coherence hypothesis in an attentional task where one location (apple) is attended while the other (pepper) is ignored. Lower visual areas represent both objects, yet gamma-band synchronization supports the communication of the attended object for further processing and curtails the communication of the unattended object from further processing in higher visual areas. (B) Incorporating the nesting of gamma-band synchronization within a theta sampling rhythm when, for example, two objects are behaviorally relevant. Within each theta cycle, a given object is selected, and this selection is implemented in the gamma-band synchronization much like illustrated in panel (A). In each sampling cycle, the selection alternates between the two objects. (C) Statistical significance of a phase bifurcation between detected and missed targets. The white rectangle highlights a time-frequency region of significance. (D) The relationship between phase at 7.1 Hz, for the last 120 ms prior to target onset and standardized performance after the phases were aligned for each participant so that the optimal phase would be lined up with the zero degrees bin. Note that from the optimal phase, performance declines and reaches a minimum at the opposite phase. (E) Schematic illustration of a task that measures the sampling rhythm using psychophysics. (F) Detection performance as a function of temporal interval between target event and flash event (attention reset event). Zero point denotes onset of flash event; negative values denote trials in which the target preceded the irrelevant flash. Black (gray) line: detection for targets appearing in the same (opposite) visual field as the flash. These time courses of performance reveal a rhythmic sampling of each location sequentially. (Panels (C) and (D) adapted from Busch, Dubois, & van Rullen, 2009, and van Rullen, Busch, Drewes, & Dubois, 2011.)

functional relationship between ongoing oscillatory phase and behavioral outcome, and (3) behavioral investigations of the temporal structure of attentional selection.

Investigations of the Temporal Properties of Overt Exploration Behaviors The proposal that attentional selection entails a theta sampling rhythm serving exploration behaviors is supported by studies investigating the temporal dynamics of saccadic eye movements (Bosman, Womelsdorf, Desimone, & Fries, 2009; Otero-Millan et al., 2008). Otero-Millan et al. (2008) have investigated the properties of saccadic eye movements in naturalistic viewing setups in humans. In addition to demonstrating inter-saccadic intervals that are consistent with a theta rhythmic sampling mechanism, they have shown that microsaccades (i.e., involuntary fixational eye movements) have velocity magnitude characteristics that are consistent with regular, exploratory saccades, suggesting that both types of saccadic eye movements lie on a single continuum. Bosman et al. (2009) extended the finding of theta rhythmic microsaccades to nonhuman primates, combining physiological measurements from V1 and V4 during the

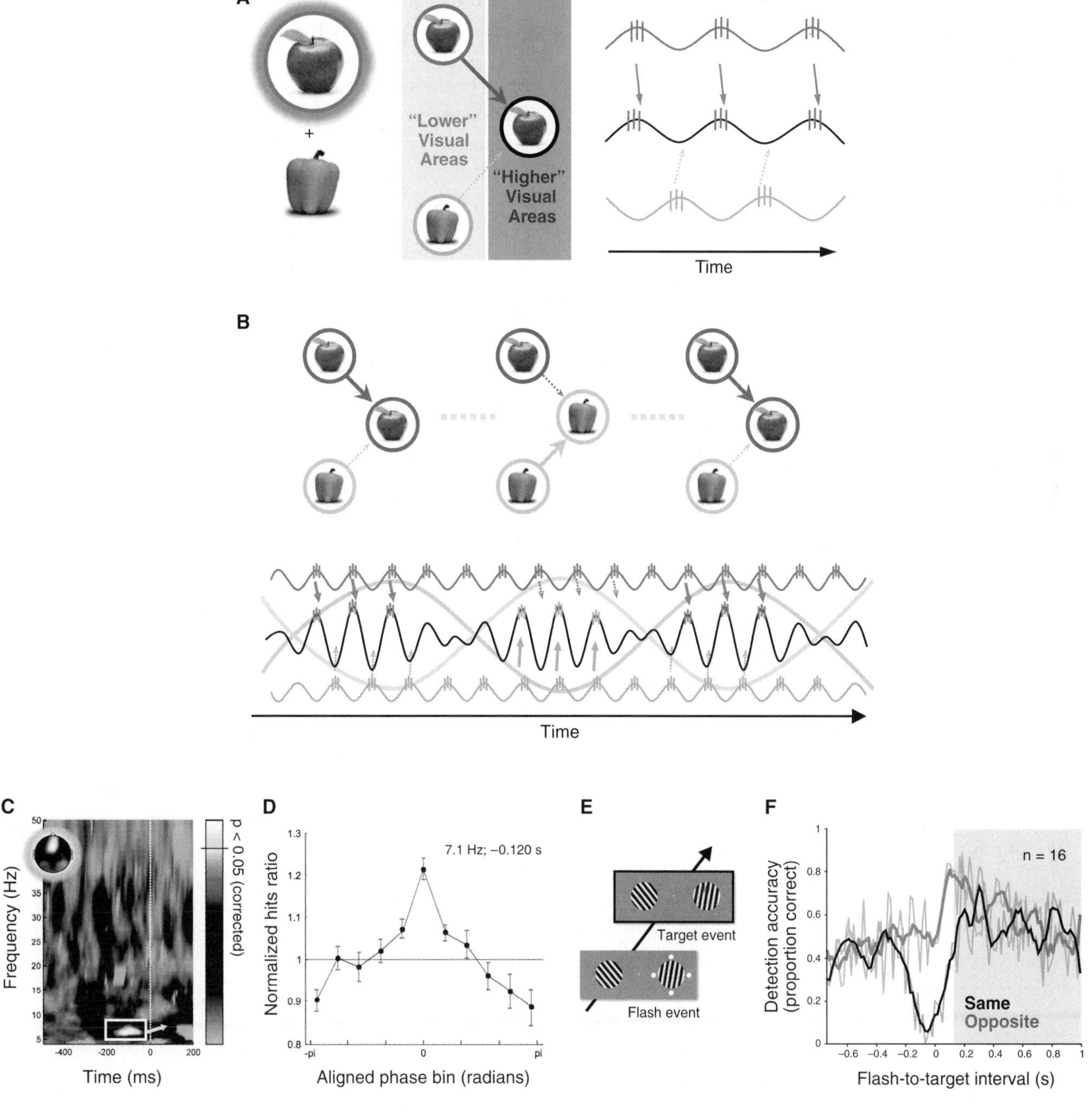

performance of an attention task while the animal was trained to maintain fixation. Gamma-band synchronization modulated around individual microsaccades in both V1 and V4, revealing a link between the overt exploration behavior and the physiological implementation of attentional selection (Bosman et al., 2009). Consistently, response times to targets were predicted by the ongoing pattern of microsaccades, establishing a relationship between miniature eye movements and attentional performance. It should be noted that, although most attention tasks engage attention in the nominal absence of eye movements (i.e., covert attention), the detailed investigation of eye movements over the past decade has highlighted that even when saccadic eye movements are suppressed by instruction and training, microsaccades are present and often biased toward behaviorally relevant locations (Hafed, Lovejoy, & Krauzlis, 2011).

Investigations of the Functional Relationship Between Ongoing Oscillatory Phase and Behavioral Outcome If a slow theta-frequency oscillation is nesting the attentional selection process that is evident in the gamma band, then measuring ongoing oscillatory activity should reveal the state of the system in a way that could explain the variance in performance during constant stimulation conditions. Studies investigating the link between ongoing oscillations and behavioral outcomes aim to predict performance variability by using the phase information of the ongoing oscillation of single trials (e.g., Busch et al., 2009). Trials would then be sorted according to behavioral responses (e.g., visible-detection target vs. invisible-detection target) and phase consistency measured within each response category. Such a procedure has by now been performed for several tasks, consistently revealing that shortly prior to the onset of a detection target, the phase of ongoing alpha or theta (5–15 Hz) predicted the behavioral outcome (figure 21.4C; Busch, Dubois, & VanRullen, 2009, threshold detection task; Mathewson, Gratton, Fabiani, Beck, & Ro, 2009, masked detection task; Busch & VanRullen, 2010, threshold detection task with and without attention; Drewes & VanRullen, 2011, saccadic reaction times). This common result provides evidence for the functional significance of the theta phase to attentional selection processes. Importantly, the findings highlight that the sampling at theta is a robust and fundamental mechanism. It will be important for future work to further test for direct links of the pre-stimulus phase effect to the physiological manifestation of the selection process at different levels of the sensory processing hierarchy.

Behavioral Investigations of the Temporal Structure of Attentional Selection If the neural implementation of attentional selection is modulated at a slow theta rhythm, the behaviors associated with attentional selection should likewise entail such periodicity. Recently, there have been a few demonstrations of how rhythmic structure can be found directly in behavioral performance (Fiebelkorn, Saalmann, & Kastner, 2013; Holcombe & Chen, 2013; Landau & Fries, 2012; Mathewson, Fabiani, Gratton, Beck, & Lleras, 2010). In a study we conducted (Landau & Fries, 2012), participants distributed attention over two locations, and performance was probed finely in time over the course of more than 1 second. Crucially, we reset attention to one of the two locations using a flash event and tested the temporal structure of detection performance thereafter. We probed performance both at the location of the reset event and at the opposite location and found that behavioral detection performance fluctuated rhythmically at 4 Hz for each location. The performance fluctuations for the two locations were in anti-phase, suggesting a sequential sampling of one location at a time. The finding of a 4-Hz sampling rhythm at each of two locations, and in anti-phase between them, is consistent with a general sampling mechanism operating at twice that frequency, or 8 Hz. Accordingly, if we had a way to examine the sampling of three locations, we should predict that the same 8-Hz sampling rhythm would be sequentially sampling over the three locations. Holcombe and Chen (2013) recently provided such evidence in the context of a motion-tracking task. Commonly, capacity limitations in multiple-object tracking tasks were quantified as a function of an object's speed. Accordingly, with increasing speed, participants' item capacity to track multiple objects decreases. Holcombe and Chen applied a crucial modification to this analysis: they quantified the motion of an object in terms of its frequency; that is, the time that elapses from the appearance of a target at a given location until it is replaced by a distractor. With this approach, they found that attentional tracking capacity remains constant for different numbers of distractors and only changes when a different number of targets are to be tracked. The temporal resolution of attentional tracking is 7 Hz for tracking one object, 4 Hz for tracking two objects, and 2.6 Hz for tracking three objects. These findings are consistent with a sampling mechanism that is sequentially sampling the targets with an overall sampling periodicity of 7 Hz.

The converging evidence, linking the physiological dynamic of theta-gamma coupling to different aspects of attentional selection, supports the notion that attention, like many other exploration behaviors, entails sequential sampling and takes a rhythmic course. The evidence summarized in this section highlights elements of active sensing through voluntary and involuntary motor exploration (saccades and microsaccades) and aspects of perceptual processing (phase-dependent perceptual windows), as well as direct delineation of the temporal aspects of attentional performance. Taken together, this diverse collection of findings illustrates the complex behavioral and physiological interactions that result in the selected content of visual perception.

REFERENCES

Anderson, J. C., Kennedy, H., & Martin, K. A. (2011). Pathways of attention: Synaptic relationships of frontal eye field to V4, lateral intraparietal cortex, and area 46 in macaque monkey. *J Neurosci, 31,* 10872–10881.

Azouz, R., & Gray, C. M. (2003). Adaptive coincidence detection and dynamic gain control in visual cortical neurons in vivo. *Neuron, 37*(3), 513–523.

Bartos, M., Vida, I., & Jonas, P. (2007). Synaptic mechanisms of synchronized gamma oscillations in inhibitory interneuron networks. *Nat Rev Neurosci, 8*(1), 45–56.

Börgers, C., Epstein, S., & Kopell, N. J. (2005). Background gamma rhythmicity and attention in cortical local circuits: A computational study. *Proc Natl Acad Sci USA, 102*(19), 7002–7007.

Börgers, C., & Kopell, N. J. (2008). Gamma oscillations and stimulus selection. *Neural Comput, 20*(2), 383–414.

Bosman, C. A., Schoffelen, J. M., Brunet, N., Oostenveld, R., Bastos, A. M., Womelsdorf, T., … Fries, P. (2012). Attentional stimulus selection through selective synchronization between monkey visual areas. *Neuron, 75,* 875–888.

Bosman, C. A., Womelsdorf, T., Desimone, R., & Fries, P. (2009). A microsaccadic rhythm modulates gamma-band synchronization and behavior. *J Neurosci, 29*(30), 9471–9480.

Bragin, A., Jandó, G., Nádasdy, Z., Hetke, J., Wise, K., & Buzsáki, G. (1995). Gamma (40–100 Hz) oscillation in the hippocampus of the behaving rat. *J Neurosci, 15*(1), 47–60.

Buffalo, E. A., Fries, P., Landman, R., Buschman, T. J., & Desimone, R. (2011). Laminar differences in gamma and alpha coherence in the ventral stream. *Proc Natl Acad Sci USA, 108,* 11262–11267.

Buia, C. I., & Tiesinga, P. H. (2008). Role of interneuron diversity in the cortical microcircuit for attention. *J Neurophysiol, 99*(5), 2158–2182.

Busch, N. A., Dubois, J., & VanRullen, R. (2009). The phase of ongoing EEG oscillations predicts visual perception. *J Neurosci, 29*(24), 7869–7876.

Busch, N. A., & VanRullen, R. (2010). Spontaneous EEG oscillations reveal periodic sampling of visual attention. *Proc Natl Acad Sci USA, 107*(37), 16048–16053.

Buschman, T. J., Denovellis, E. L., Diogo, C., Bullock, D., & Miller, E. K. (2012). Synchronous oscillatory neural ensembles for rules in the prefrontal cortex. *Neuron, 76,* 838–846.

Buschman, T. J., & Miller, E. K. (2007). Top-down versus bottom-up control of attention in the prefrontal and posterior parietal cortices. *Science, 315*(5820), 1860–1862.

Buzsaki, G. (2005). Theta rhythm of navigation: Link between path integration and landmark navigation, episodic and semantic memory. *Hippocampus, 15,* 827–840.

Cardin, J. A., Carlen, M., Meletis, K., Knoblich, U., Zhang, F., Deisseroth, K., … Moore, C. I. (2009). Driving fast-spiking cells induces gamma rhythm and controls sensory responses. *Nature, 459,* 663–667.

Chelazzi, L., Miller, E. K., Duncan, J., & Desimone, R. (1993). A neural basis for visual search in inferior temporal cortex. *Nature, 363*(6427), 345–347.

Corbetta, M., & Shulman, G. L. (2011). Spatial neglect and attention networks. *Ann Rev Neuro, 34,* 569–599.

Dhamala, M., Rangarajan, G., & Ding, M. (2008). Analyzing information flow in brain networks with nonparametric Granger causality. *NeuroImage, 41,* 354–362.

Drewes, J., & VanRullen, R. (2011). This is the rhythm of your eyes: The phase of ongoing electroencephalogram oscillations modulates saccadic reaction time. *J Neurosci, 31*(12), 4698–4708.

Fiebelkorn, I. C., Saalmann, Y. B., & Kastner, S. (2013). Rhythmic sampling with and between objects despite sustained attention at a cued location. *Curr Biol, 23*(24), 2553–2558.

Fries, P. (2005). A mechanism for cognitive dynamics: Neuronal communication through neuronal coherence. *Trends Cogn Sci, 9*(10), 474–480.

Fries, P. (2009). Neuronal gamma-band synchronization as a fundamental process in cortical computation. *Annu Rev Neurosci, 32,* 209–224.

Fries, P., Nikolic, D., & Singer, W. (2007). The gamma cycle. *Trends Neurosci, 30*(7), 309–316.

Fries, P., Womelsdorf, T., Oostenveld, R., & Desimone, R. (2008). The effects of visual stimulation and selective visual attention on rhythmic neuronal synchronization in macaque area V4. *J Neurosci, 28*(18), 4823–4835.

Gregoriou, G. G., Gotts, S. J., & Desimone, R. (2012). Cell-type-specific synchronization of neural activity in FEF with V4 during attention. *Neuron, 73,* 581–594.

Gregoriou, G. G., Gotts, S. J., Zhou, H., & Desimone, R. (2009). High-frequency, long-range coupling between prefrontal and visual cortex during attention. *Science, 324,* 1207–1210.

Grothe, I., Neitzel, S. D., Mandon, S., & Kreiter, A. K. (2012). Switching neuronal inputs by differential modulations of gamma-band phase-coherence. *J Neurosci, 32,* 16172–16180.

Hafed, Z. M., Lovejoy, L. P., & Krauzlis, R. J. (2011). Modulation of microsaccades in monkey during a covert visual attention task. *J Neurosci, 31*(43), 15219–15230.

Holcombe, A. O., & Chen, W.-Y. (2013). Splitting attention reduces temporal resolution from 7 Hz for tracking one object to <3 Hz when tracking three. *J Vision, 13*(1), 12.

Kleinfeld, D., Berg, R. W., & O'Connor, S. M. (1999). Anatomical loops and their electrical dynamics in relation to whisking by rat. *Somatosens Mot Res, 16*(2), 69–88.

Kravitz, D. J., Saleem, K. S., Baker, C. I., Ungerleider, L. G., & Mishkin, M. (2013). The ventral visual pathway: An expanded neural framework for the processing of object quality. *Trends Cogn Sci, 17,* 26–49.

Landau, A. N., & Fries, P. (2012). Attention samples stimuli rhythmically. *Curr Biol, 22*(11), 1000–1004.

Liebe, S., Hoerzer, G. M., Logothetis, N. K., & Rainer, G. (2012). Theta coupling between V4 and prefrontal cortex predicts visual short-term memory performance. *Nat Neurosci, 15*, 456–462, S451–S452.

Lin, S. C., Gervasoni, D., & Nicolelis, M. A. (2006). Fast modulation of prefrontal cortex activity by basal forebrain noncholinergic neuronal ensembles. *J Neurophysiol, 96*(6), 3209–3219.

Mathewson, K. E., Fabiani, M., Gratton, G., Beck, D. M., & Lleras, A. (2010). Rescuing stimuli from invisibility: Inducing a momentary release from visual masking with pre-target entrainment. *Cognition, 115*(1), 186–191.

Mathewson, K. E., Gratton, G., Fabiani, M., Beck, D. M., & Ro, T. (2009). To see or not to see: Prestimulus phase predicts visual awareness. *J Neurosci, 29*(9), 2725–2732.

Mishra, J., Fellous, J. M., & Sejnowski, T. J. (2006). Selective attention through phase relationship of excitatory and inhibitory input synchrony in a model cortical neuron. *Neural Netw, 19*, 1329–1346.

Noudoost, B., & Moore, T. (2011). Control of visual cortical signals by prefrontal dopamine. *Nature, 474*, 372–375.

Otero-Millan, J., Troncoso, X. G., Macknik, S. L., Serrano-Pedraza, I., & Martinez-Conde, S. (2008). Saccades and microsaccades during visual fixation, exploration, and search: Foundations for a common saccadic generator. *J Vision, 8*(14), 21–21.

Pesaran, B., Nelson, M. J., & Andersen, R. A. (2008). Free choice activates a decision circuit between frontal and parietal cortex. *Nature, 453*, 406–409.

Phillips, J. M., Vinck, M., Everling, S., & Womelsdorf, T. (2013). A long-range fronto-parietal 5- to 10-Hz network predicts "top-down" controlled guidance in a task-switch paradigm. *Cereb Cortex*. Advance online publication. doi:10.1093/cercor/bht050

Premereur, E., Vanduffel, W., Roelfsema, P. R., & Janssen, P. (2012). Frontal eye field microstimulation induces task-dependent gamma oscillations in the lateral intraparietal area. *J Neurophysiol, 108*, 1392–1402.

Reynolds, J. H., Chelazzi, L., & Desimone, R. (1999). Competitive mechanisms subserve attention in macaque areas V2 and V4. *J Neurosci, 19*(5), 1736–1753.

Rodriguez, R., Kallenbach, U., Singer, W., & Munk, M. H. (2004). Short- and long-term effects of cholinergic modulation on gamma oscillations and response synchronization in the visual cortex. *J Neurosci, 24*(46), 10369–10378.

Salazar, R. F., Dotson, N. M., Bressler, S. L., & Gray, C. M. (2012). Content-specific fronto-parietal synchronization during visual working memory. *Science, 338*, 1097–1100.

Siegel, M., Donner, T. H., & Engel, A. K. (2012). Spectral fingerprints of large-scale neuronal interactions. *Nat Rev Neurosci, 13*, 121–134.

Tiesinga, P., Fellous, J. M., & Sejnowski, T. J. (2008). Regulation of spike timing in visual cortical circuits. *Nat Rev Neurosci, 9*(2), 97–107.

Womelsdorf, T., Schoffelen, J. M., Oostenveld, R., Singer, W., Desimone, R., Engel, A. K., & Fries, P. (2007). Modulation of neuronal interactions through neuronal synchronization. *Science, 316*, 1609–1612.

22 Neural Signatures, Circuitry, and Modulators of Visual Selective Attention

BEHRAD NOUDOOST, EDDY ALBARRAN, AND TIRIN MOORE

ABSTRACT Attention is the process of selecting a subset of sensory information for further processing while ignoring the rest. A great deal of neurophysiological work has established that the dependence of visual perception on selective attention is reflected in the responses of neurons throughout the visual system via changes in the signal-to-noise ratio of visual information. However, until recently, little was known about the neural circuitry that controls the perceptual effects of attention and their neural correlates within the visual system. In this chapter, we review recent work implicating particular brain structures, both cortical and subcortical, in the control of visual spatial attention. Studies aimed at localizing neural circuits controlling selective visual attention are distinct from those that have explored the long-suspected role of particular neuromodulators in attention. We also discuss the current knowledge and major questions concerning the role that neuromodulators play in attentional control.

What is attention?

Attention is the process of selecting a subset of sensory information for further processing while ignoring the rest. The neural basis of attention is typically studied through experiments that measure neural activity (e.g., neuronal spike rate) within brain structures of animal subjects (e.g., monkeys) during attention-demanding tasks. Once obtained, these neural measures can be analyzed to establish correlates or signatures of attentional control. These signatures reflect and encode behavioral effects at the neural level. Because attention is the means by which we focus and amplify behaviorally relevant information (i.e., an important stimulus), neural signatures of attention are generally expected to reflect that as an increase in the ratio of relevant to irrelevant neural responses (i.e., signal-to-noise ratio, or SNR). In principle, increases in the SNR can be achieved through a variety of means, such as strengthening the selected visual signal, improving the efficacy of incoming inputs, or reducing the noise. Understanding these *signatures* of attention—that is, the ways in which the SNR is altered—presumably enables one to identify the mechanisms underlying both them and their corresponding behavioral effects.

Signatures of visual attention

One signature of visual attention is an increase in the spiking rates of neurons to stimuli within their receptive fields (RFs). The effects of visual attention on neuronal firing rate have largely been observed in neurophysiological experiments involving monkeys. Monkeys can be trained to perform complex visuospatial tasks such as fixating on a central fixation point while covertly attending to one of many peripheral stimuli on the same display in return for a reward. The experimenter can control which of the peripheral stimuli the monkey attends to by making the reward contingent upon the detection or discrimination of some aspect of that stimulus (e.g., change in color). While the monkey attends to the RF stimulus or stimuli outside of the RF, the spiking activity of neurons within the visual system can be measured and compared between those two conditions. Generally, one observes that the spiking activity increases when the monkey attends to the stimuli located within the RF relative to when the monkey attends to stimuli outside of the RF (Noudoost, Chang, Steinmetz, & Moore, 2010). Importantly, this difference in firing rate is observed even though the retinal stimulation is identical across conditions. Such increases in firing rate due to attention have been observed beginning at the earliest stage of processing beyond the retina, namely, in neurons within the lateral geniculate nucleus (McAlonan, Cavanaugh, & Wurtz, 2006). Furthermore, this type of neuronal modulation has been observed across multiple stages and areas of visual processing. Increase in firing rate is only one measurable signature of attention, and recent studies have revealed other, complementary ways that attention may increase the SNR (table 22.1).

Another way to potentially alter the SNR during attention is through changes in synchronous neural activity, either in the spiking rates of neurons or in the low-frequency electrical signals recorded locally (local field potentials; LFPs), or both. Increased synchrony of neuronal signals is thought to enhance the transmission

Table 22.1

Summary of signatures of top-down attention within the visual system. An overview of currently known signatures of attention and their potential implications. Abbreviations: LFP, local field potential; RF, receptive field; FEF, frontal eye field; LIP, lateral intraparietal area; IT, inferotemporal cortex; SC, superior colliculus; LGN, lateral geniculate nucleus; Pulv, pulvinar nucleus of the thalamus; PFC, prefrontal cortex.

Signature	Potential implication	Area
↑ Spiking rate	Increases output signal	LGN, TRN, Pulv, SC, V1, V2, V4, MT, IT, LIP, FEF, PFC
↑ γ-band LFP power	Increases locally synchronous synaptic activity	FEF, V4
↑ Local γ-band coherence	Potentially increases signal efficacy at postsynaptic targets	FEF, V4, LIP
↑ Cross-areal γ-band coherence	Potentially facilitates long-range interactions by providing a common temporal reference frame	FEF-V4, PFC/FEF-V4, LIP-MT
↓ Response variability	Increases reliability of encoded information	V4
↓ Low-frequency synchrony	Potentially reduces output redundancy by decreasing correlation	V4
↓ Correlated noise at low frequencies	Potentially increases the sensitivity of pooled responses	V4
Receptive field shift	Potentially increases output signal by recruiting neurons	MT, V4

of signals by reducing the phase lag of inputs to downstream structures and thus potentially increasing their postsynaptic efficacy. Spikes reflect the output of neurons, while LFPs are thought to largely reflect the inputs (excitatory and inhibitory post-synaptic potentials). Coherence in neural activity can be measured between the spiking activity within different sets of neurons, between their LFPs, or between the spikes of one set of neurons and the LFPs of the other. Specifically, it has been argued that gamma-band (25–100 Hz) synchronization of spiking from selected groups of neurons can increase the influence of these spikes on downstream areas (Salinas & Sejnowski, 2001). Synchrony in the gamma band among neurons encoding an attended representation could facilitate the integration of spikes from these populations converging on their postsynaptic targets, enhancing the attended representation. Various studies have reported increases in the local gamma band (Fries, Womelsdorf, Oostenveld, & Desimone, 2008; Gregoriou, Gotts, Zhou, & Desimone, 2009), and that synchrony has been found to correlate with behavioral performance (Womelsdorf, Fries, Mitra, & Desimone, 2006). Although these observations support the positive role of synchrony in attentional selection, there are differences in the actual reported frequency ranges in which significant effects of attention are observed, and the role of synchrony in attention remains an active area of research (see chapter 21, this volume).

Other studies have found that attention also decreases the variability of neuronal firing of individual neurons. Variability in firing patterns of neurons can be attributed to independent sources of noise within each neuron, the correlated fluctuations shared across groups of neurons, or both. For individual neurons, the Fano factor, a measure of neural firing variability quantified as the variance divided by the mean response, is reduced when attention is directed toward the respective RFs of those neurons (Mitchell, Sundberg, & Reynolds, 2007). Recent studies have found that attention reduces the correlated variability among pools of many neurons in area V4, which appears to improve the SNR more effectively than increases in firing rate or decreases in independent firing variability (Cohen & Maunsell, 2009; Mitchell, Sundberg, & Reynolds, 2009). This decorrelation arises mainly from a suppression of low-frequency (<5 Hz) rate fluctuations shared across neuronal populations and is thought to reflect the attention-dependent low-frequency desynchronization observed in area V4 (Fries et al., 2008).

Another measurable signature of attention is the number of neurons involved in processing particular visual information. The act of attention potentially amplifies and sharpens selected signals through the recruitment of additional neurons to represent selected information and, additionally, through the narrowing down of their spatial tuning (RFs). Studies that show shifting and shrinking of RFs toward attended stimuli provide evidence for this adaptive prioritization of visual representations (Connor, Preddie, Gallant, & Van, 1997)

In summary, attention increases the efficacy of visual processing via various different potential mechanisms, including (1) increasing firing rates of visual neurons, (2) increasing neural coding through synchronous activity, and (3) reducing redundancies by decorrelating the responses of populations of neurons. In principle, understanding the ways in which the SNR can be enhanced during attention should aid in revealing their underlying causal mechanisms.

Sources of attention

It is clear that visual perception depends on the direction of attention, and as is now well established, this dependence is reflected in the responses of neurons throughout the visual system via changes in the SNR of visual information (table 22.1). However, until recently little was known about the neural circuitry that controls the perceptual effects of attention and their neural correlates within the visual system. In the following section, we discuss several brain structures suspected of playing a causal role in attention (figure 22.1 and table 22.2).

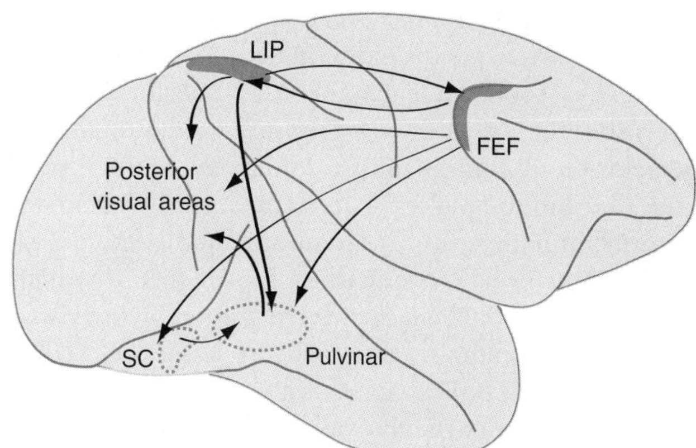

FIGURE 22.1 Macaque brain highlighting structures implicated in the control of visual selective attention and their connections, namely, the lateral intraparietal area (LIP) in posterior parietal cortex, the frontal eye field (FEF) in prefrontal cortex, the superior colliculus (SC) in the midbrain, and the pulvinar nuclei in the thalamus. Dotted outlines indicate subcortical structures. (See color plate 19.)

Frontal eye field

The frontal eye field (FEF), located within the prefrontal cortex (PFC), is an area originally known for its role in overt attention, specifically saccadic eye movements (Bruce, 1990). FEF neurons exhibit a spectrum of visual and motor properties. Most FEF neurons respond to visual stimulation within their RFs. A subset of those neurons will also respond prior to saccades of a particular direction and amplitude, most often (but not always) to locations within the visual RF. The remaining subset of FEF neurons consists of those that do not respond to visual stimulation, but only prior to saccades. In addition, like neurons within earlier stages of the visual system, FEF visual responses are modulated by covert attention (e.g., Armstrong, Chang, & Moore, 2009; Thompson, Bichot, & Sato, 2005). However, more recent studies revealed a prominent role of the FEF in covert spatial attention. FEF neurons have a rich connectivity with posterior visual areas, both feedforward and feedback, the latter being implicated in the modulation of posterior visual signals by the PFC. Electrical microstimulation of sites within the FEF (through microelectrodes) with sufficiently high currents can evoke highly reproducible saccades. More importantly, microstimulation of FEF sites with currents too weak to

TABLE 22.2
Candidate brain areas implicated in the control of visual processing

Method of causality test	Signature of attention	FEF	SC	LIP	Pulvinar
Stimulation	Neural	Increased visual response and stimulus selectivity	?	?	?
	Behavioral	Increased visual detection and visual guidance of saccades	Increased visual detection	Spatially non-specific increases in visual detection	?
Inactivation	Neural	Decreased stimulus selectivity	Decreased stimulus selectivity modulation	?	?
	Behavioral	Distractor-dependent decrease in visual detection	Distractor-dependent decrease in visual detection	Distractor-dependent decrease in visual detection (LIPv not LIPd)	Distractor-dependent decrease in visual discrimination

evoke saccades (*subthreshold* microstimulation) nonetheless improves performance in an attention-demanding task (Moore & Fallah, 2001). Specifically, FEF microstimulation improves the ability of monkeys to detect minute changes in the luminance of targets positioned within the part of space represented by neurons at the stimulation site. In addition, subthreshold FEF microstimulation also enhances the degree to which voluntarily generated saccades are guided by the visual features of saccadic targets (Schafer & Moore, 2007). Thus, activation of FEF neurons is sufficient to increase the influence of particular visual stimuli on behavior.

Subthreshold microstimulation of FEF sites also produces a brief enhancement of visually driven responses of area V4 neurons with RFs in the region of space represented by neurons at the stimulated FEF site (Moore & Armstrong, 2003). Furthermore, it has been shown that the magnitude of this enhancement increases when more effective RF stimuli are used and when non-RF (i.e., distractor) stimuli are present. Conversely, microstimulation of FEF sites that do not overlap with particular V4 RFs suppresses the responses of those V4 neurons, which reflects the effects observed in endogenous attention studies. Additionally, this enhancement of V4 responses is observed only with RF stimuli that align with the endpoint of the saccade vector represented at the FEF site. As a result of this, when two competing stimuli are present within a V4 RF, FEF microstimulation drives visual responses toward those observed when an aligned stimulus is presented alone (Armstrong, Fitzgerald, & Moore, 2006). Functional magnetic resonance imaging (fMRI) has been used to demonstrate the influence of FEF microstimulation on visual activation throughout the cortex (Ekstrom, Roelfsema, Arsenault, Bonmassar, & Vanduffel, 2008). This work shows that FEF microstimulation enhances the visual activation of retinotopically corresponding foci within multiple visual areas and increases the contrast sensitivity within multiple visual areas (Ekstrom, Roelfsema, Arsenault, Kolster, & Vanduffel, 2009).

In addition to studies using electrically induced changes in FEF activity to test its role in attention, Schafer and Moore (2011) tested the hypothesis that behaviorally conditioned, voluntary changes in FEF neuronal activity are sufficient to bring about the deployment of visual attention (figure 22.2). The authors used operant training techniques to examine the impact of voluntary control of FEF activity on visually driven behavior. Monkeys were provided with real-time auditory feedback based on the firing rate of FEF activity, and rewarded for either increasing or decreasing that activity (in alternating "UP" and "DOWN" blocks of trials), while maintaining central fixation on a blank visual display. The authors probed the behavioral and neurophysiological consequences of operant control of FEF activity. They did this by introducing probe trials in which the monkey performed a visual search task while exerting operant control over FEF activity. They found that when the search targets

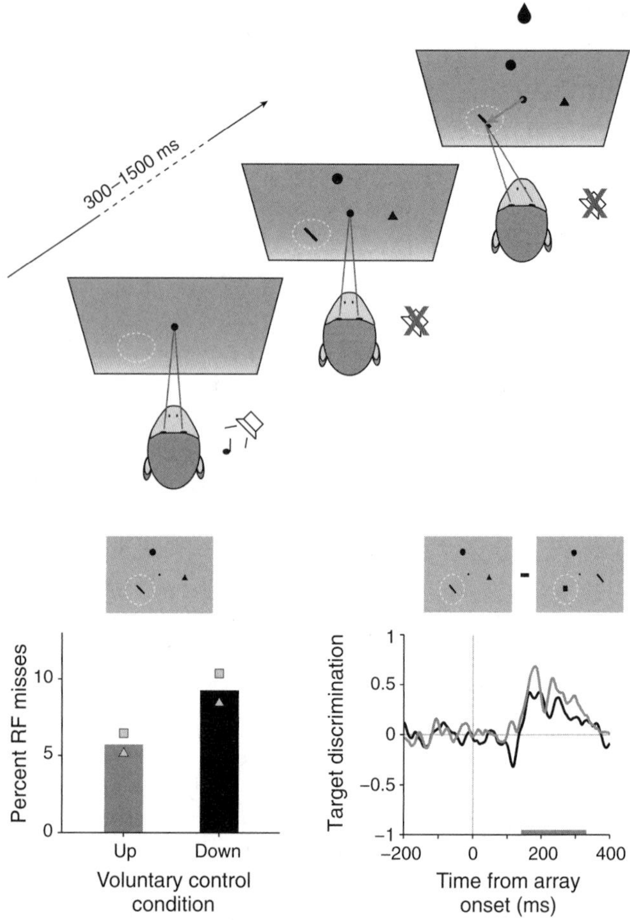

FIGURE 22.2 Operant control of FEF neurons and its effects on selective attention measured behaviorally and neurophysiologically. *Top*: The operant control task in which the monkey fixated a central spot on an otherwise blank video display and was rewarded for increasing or decreasing the firing rate of FEF neurons. Dotted circle shows the FEF RF. Speaker icon and musical note depict auditory feedback of FEF neuronal activity (spike train). Subsequent panels depict the visual-search probe trials in which a search array appeared, the auditory feedback ceased ("x" on speaker icon) and the monkey was rewarded (blue droplet) for directing a saccade toward an oriented bar target. *Bottom left*: The frequency of trials in which the monkey failed to respond to the target ("misses") opposite the RF was increased during DOWN operant control of FEF activity in both monkeys (square and triangle symbols). *Bottom right*: Target discrimination by FEF neurons was increased during upward (red) operant control relative to downward (blue). (Adapted from Schafer and Moore, 2011.) (See color plate 20.)

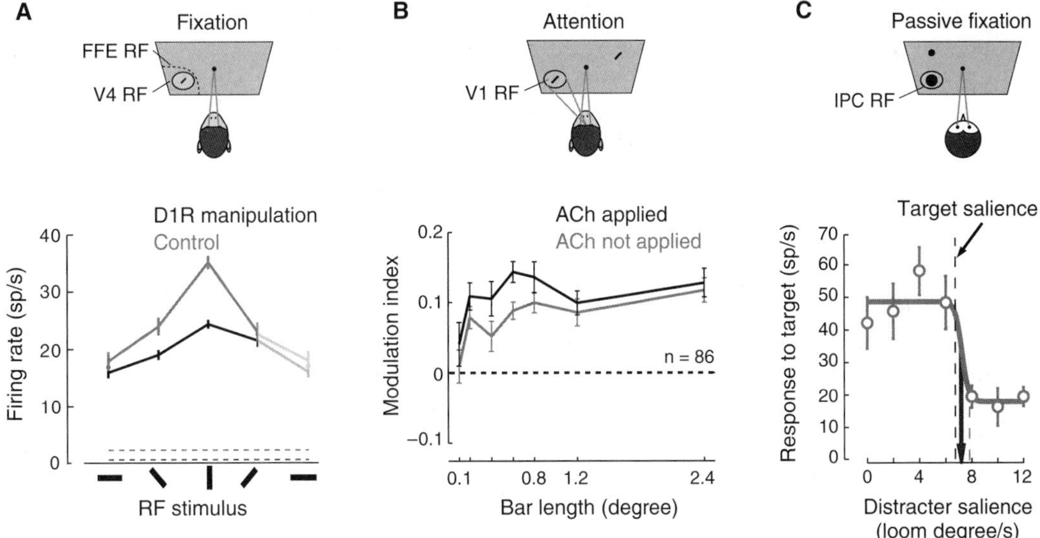

FIGURE 22.3 (A) Visual responses of a V4 neuron with a RF that overlapped the FEF RF measured during passive fixation (*top panel*). The bottom plot shows mean and standard error of the mean (SEM) visual responses to bar stimuli presented at varying orientations and the baseline-firing rate in the absence of visual stimulation (dashed lines) before (black) and after (gray) the FEF D1R manipulation. (B) Enhancement of attention effects in macaque V1 by ACh. The cartoon above depicts the behavioral task in which attention was directed covertly to a neuron's RF stimulus (V1 RF) or to a stimulus outside of the RF (not shown) during fixation of a central spot (gray lines). The effect of spatial attention on the responses of V1 neurons to visual stimuli is quantified with a modulation index for stimuli of varying lengths. Positive indices indicate greater responses when attention is directed toward the stimulus within the neuron's RF compared to when attention is directed elsewhere. Indices measured during control trials (gray) and during iontophoretic application of acetylcholine (black) are shown. (C) Cholinergic neurons in the owl's IPC nucleus signal the physical salience of stimuli by a characteristic switch response. The exemplar neuron responds almost invariantly to RF stimuli across a range of stimulus intensities as long as the stimulus is more physically salient than the other stimulus on the screen (distractor). When the RF stimulus is less salient, the neuron responds at a uniformly low rate. Salience is manipulated by varying the speed at which a given stimulus looms (target salience in this example is set at 7 degrees/sec). IPC neurons were recorded in owls during passive viewing (*top panel*).

appeared in the FEF RF, monkeys were less likely to detect the target (i.e., they had more "misses") on the DOWN-trials than UP-trials. In addition, the authors found that FEF neurons could discriminate targets from distractors better during UP trials compared to DOWN trials. This change in target discriminability was dependent on the direction of operant control and not on spontaneous fluctuations in firing rate. These results show that endogenous, voluntary changes in FEF neural activity are sufficient to bring about both the behavioral and neurophysiological effects of visual attention, and that explicit learning of an attention task is not required.

Noudoost and Moore (2011a) pharmacologically altered FEF activity by infusing small quantities of dopamine receptor agonists and antagonists at localized cortical sites. They found that, similar to FEF microstimulation, manipulation of FEF neuronal activity through dopamine D1 receptors could enhance the visual responses of area V4 neurons at corresponding spatial locations (figure 22.3A). Specifically, they found that blocking D1 receptors (1) increased the spiking responses of V4 neurons to visual stimulation, (2) increased stimulus selectivity, and (3) reduced cross-trial variability of visual responses. Note that similar changes to each of these three measures have been observed during endogenously directed attention (McAdams & Maunsell, 1999; Mitchell et al., 2007). The results of Noudoost and Moore (2011a) provide two crucial pieces of evidence. First, they demonstrate changes in FEF neuronal firing rate are sufficient to bring about changes in sensory signals in posterior visual cortex. Although the previous microstimulation results suggest this to be the case, they do not rule out other possibilities such as antidromic spread of microstimulation-evoked activity to other cortical or subcortical structures. Second, the results also demonstrate that the influence the FEF has on posterior sensory signals can itself be influenced by changes in dopaminergic activity. This latter point may establish a basis by which prefrontal dopamine contributes to attentional control, as appears to be the case in studies of attention-deficit hyperactivity disorder (Castellanos & Tannock, 2002).

In addition to the studies taking a gain-of-function approach, inactivation experiments have consistently shown that removal of FEF output disrupts behavioral measures of attention. Wardak and colleagues studied effects of FEF inactivation on behavioral performance in visually guided saccades and covert visual search tasks and found that FEF inactivation causes deficits in both saccade and attention tasks (Wardak, Ibos, Duhamel, & Olivier, 2006). Monosov and Thompson have also shown that inactivation of the FEF produces spatially selective perceptual deficits in covert attention tasks (Monosov & Thompson, 2009). These deficits are strongest on invalid cue trials that require endogenous attention shifting. Experiments using search classification tasks to record responses of inferotemporal (IT) neurons before and after FEF inactivation show that both object-selective responses of IT neurons and behavioral performance are affected by changes in FEF activity (Monosov, Sheinberg, & Thompson, 2010). Thus, not only does FEF activation (electrically or pharmacologically) enhance behavioral and neuronal measures of attention, its inactivation reduces performance on attention-based tasks, suggesting that FEF activity is not only sufficient to drive attentional deployment, but that it is also necessary for that deployment.

Superior colliculus

Another oculomotor area suspected of playing a causal role in attention is the superior colliculus (SC). As with the FEF, SC neurons exhibit a spectrum of visual and motor properties. Most SC neurons respond to visual stimulation within their RFs (Goldberg & Wurtz, 1972). A subset of those neurons will also respond prior to saccades of a particular direction and amplitude, most often to locations within the visual RF. The remaining subset of FEF neurons consists of those that do not respond to visual stimulation, but only prior to saccades. In general, neurons with largely visual properties are found in more superficial layers of the SC, while neurons with largely motor properties are found in deeper layers. In addition, like neurons within earlier stages of the visual system, SC visual responses are modulated by covert attention (Ignashchenkova, Dicke, Haarmeier, & Their, 2004). As with the FEF, microstimulation of the SC enhances performance on attention-demanding tasks. Subthreshold microstimulation of sites within the SC spatial map enhances visual performance in the corresponding region of space. This has been observed in studies in which monkeys performed a direction-of-motion discrimination task (Muller, Philiastides, & Newsome, 2005) or they detected changes in visual stimuli (Cavanaugh & Wurtz, 2004). Other studies further demonstrated that the behavioral benefits of SC stimulation on performance are not dependent on whether microstimulation occurs in the visual or motor layers of the SC (Cavanaugh, Alvarez, & Wurtz, 2006), indicating that the benefit does not result from an evoked visual experience.

The results of pharmacological inactivation of the SC also suggest a causal role of this area in visuospatial attention. Lovejoy and Krauzlis (2010) trained monkeys in a covert spatial attention task. Monkeys were required to deploy covert attention to a cued location and report the motion direction of stimuli appearing at that location. The authors found that inactivation of the SC resulted in profound deficits for targets in the corresponding part of space. This deficit was only observed when target stimuli were presented in the presence of distractors outside of the affected visual field, consistent with the idea that the observed deficits are attentional in nature, rather than merely visual. More recently, the authors tested the effects of SC inactivation on the neural signatures of attention within visual areas MT and MST. Surprisingly, they found that in spite of prominent behavioral deficits in visual attention following SC inactivation, the measured neural signatures of attention within the two visual areas were unaffected (Zenon & Krauzlis, 2012). This work suggests that although the SC may be necessary for attentionally directed behavior, it may not be a source of attention-related modulation within visual cortex.

Lateral intraparietal area

Another brain region that has been implicated as a source of attentional modulation is the lateral intraparietal area (LIP), an area located in the lateral bank of the intraparietal sulcus of the parietal lobe. As with the FEF and the SC, LIP neurons exhibit a spectrum of visual and motor properties (Mazzoni, Bracewell, Barash, & Andersen, 1996). As with the other structures, LIP visual responses are also modulated by covert attention (e.g., Bisley & Goldberg, 2003). The first evidence for a causal role of the LIP in visual attention comes from inactivation studies. Wardak et al. found that inactivation of the LIP had no direct influence on saccade mechanisms for visually and memory-guided saccades (i.e., saccades maintained the same accuracy, omission rate, latency, duration, and amplitude; Wardak, Olivier, & Duhamel, 2002). However, LIP inactivation did cause monkeys to choose targets appearing on the ipsilateral side of visual field over contralateral ones and induced a delay in searching for visual targets in the ipsilateral field. Follow-up experiments testing the monkeys' performance on covert attention tasks found that LIP inactivation slows the reaction time of the task, but that this deficit is highly dependent on the number of distractors and the task load (Wardak et al., 2004). Comparing FEF (Wardak et al., 2006) and LIP inactivation studies (Wardak et al., 2004), Wardak and colleagues concluded that whereas both FEF and LIP inactivation cause deficits in covert attention tasks, there are essential differences between their observed inactivation effects: (1) unlike FEF inactivation, which produces deficits in the presence or absence of distractors, attention deficits due to LIP inactivation are more prominent when the target stimulus is presented in competition with distractors and when the task load is high; (2) FEF inactivation results in severe deficits in memory-guided saccade execution, whereas these saccades are mostly intact after LIP inactivation (Wardak, Olivier, & Duhamel, 2002). Experiments by Liu et al. further explored the dissociation of saccade deficits from attention deficits after LIP inactivation. They found that well-localized inactivation of the dorsal region of the LIP (LIPd) specifically impairs saccades, but that attention-driven search remains intact (Liu, Yttri, & Snyder, 2010). In contrast, well-localized inactivation of the ventral region (LIPv) impairs both attention-driven search as well as saccades, suggesting a dissociation of function within area LIP.

Microstimulation techniques have also been used to test the causal role of the LIP in attention. Cutrell and Marrocco observed that LIP stimulation overrides the effects of a distracting visual cue, eliminating the "cost" of incorrectly deploying attention to the extraneous location (Cutrell & Marrocco, 2002). However, when there was no cue, such stimulation resulted in a behavioral advantage for both ipsi- and contra-stimulation fields, suggesting the presence of a general vigilance for visual stimuli, rather than a spatially selective orientation. Others have found that microstimulation of the LIP slightly biases animal subjects to saccade to visual stimuli at the stimulated location during visual search tasks (Mirpour, Ong, & Bisley, 2010). Studies comparing the FEF and LIP areas in attention tasks (Wardak et al., 2004, 2006) and in tasks requiring filtering of distractors (Suzuki & Gottlieb, 2013) suggest a difference between the roles of these two areas in visuospatial attention tasks, in which the LIP is more involved in bottom-up attention whereas the FEF is more involved in top-down attention (Buschman & Miller, 2007).

The pulvinar

The pulvinar nuclei of the thalamus comprise another set of candidate structures controlling visual attention. Based on its widespread connectivity with other attention-related structures and visual cortex, the pulvinar is thought to mediate the control of attention by those structures (Allman, Kaas, Lane, & Miezin, 1972). Generally, cortical areas connecting directly to each other are also connected via pulvinar (Sherman & Guillery, 2002), and thus the pulvinar may regulate cortico-cortical transmission via these cortico-pulvino-cortical pathways. As in the above-mentioned structures, the visual responses of pulvinar neurons are modulated by attention (Petersen, Robinson, & Keys, 1985), and there is evidence of a causal role of the pulvinar in attention. Pharmacological inactivation of the lateral pulvinar causes attention deficits in behaving monkeys (Desimone, Wessinger, Thomas, & Schneider, 1990). A more recent study by Kastner and colleagues provides evidence of a role of the pulvinar in regulating the transmission of information between cortical areas. They first showed that visual cortical areas V4 and temporo-occipital area connect to overlapping areas within the pulvinar. They then studied the visual responses of neurons in these three regions in monkeys performing an attention task. They observed an increase in coherence between pulvinar spikes and LFPs and the LFPs in areas V4 and temporo-occipital area. Furthermore, they found evidence that the coherence of activity between the pulvinar and the two visual areas originated from the pulvinar (Saalmann, Pinsk, Wang, Li, & Kastner, 2012). These results suggest that the pulvinar contributes to attentional control via its intermediary role in cortico-cortical transmission.

Neuromodulators and attention

Distinct from studies aimed at localizing neural circuits controlling selective visual attention are those that have been aimed at exploring the long-suspected role of particular neuromodulators in attentional control. These studies have focused on a variety of different species, including humans, in both clinical and nonclinical subjects (Brennan & Arnsten, 2008; Sarter, Hasselmo, Bruno, & Givens, 2005). Neuromodulators are classes of neurotransmitters that influence synaptic transmission broadly within neural circuits. They have several characteristics in common. (1) They are all released by neurons within specific brainstem or midbrain nuclei. (2) These subcortical neurons project broadly to many other subcortical and cortical structures. Projections to the cortex include both posterior sensory areas where correlates of selective attention are observed, as well as projections to the PFC, where the control of selective attention is thought to originate. (3) Each of the specific nuclei also receives projections from the PFC, suggesting a means by which PFC control can exert network-wide attentional effects. For a more detailed review of the role of neuromodulators in attention, see Noudoost and Moore (2011c). Here, we briefly discuss recent evidence of a role for two major neuromodulators, acetylcholine and dopamine.

Acetylcholine Within the past twenty years, a number of studies using human and animal subjects have yielded evidence for the role of acetylcholine (ACh) in attention (Warburton & Rusted, 1993). Systemic increases in ACh activity have been shown to enhance visual selective attention in healthy human subjects (Warburton & Rusted, 1993). Cholinergic receptors are classified into two classes: metabotropic muscarinic receptors (mAChRs) and ionotropic nicotinic receptors (nAChRs; Cooper, Bloom, & Roth, 2003). Although much of the evidence for behavioral enhancement through cholinergic stimulation involves nAChRs, there is evidence for a role of both receptors in certain aspects of attentional control. Rodent studies suggest that the processing of sensory signals within posterior areas might be influenced by the interaction of PFC with ascending cholinergic projections (Sarter et al., 2005), and the interaction appears to involve nAChRs (Guillem et al., 2011). Within posterior areas, basal forebrain stimulation enhances sensory signals within somatosensory (Tremblay, Warren, & Dykes, 1990) and visual cortex (Goard & Dan, 2009), and in all cases the effects appear to involve mAChRs. However, other studies have recently shown that within primary visual cortex (V1) it is in fact nAChRs that are involved in gain control (Disney, Aoki, & Hawken, 2007). Within V1, nAChRs are localized presynaptically at geniculocortical inputs to layer IVc neurons, where they enhance responsiveness and contrast sensitivity of thalamorecipient neurons. Recent experiments concerning the role of ACh in attentional modulation recorded visual responses of V1 neurons in monkeys performing a covert attention task (Herrero et al., 2008), and, consistent with previous results (e.g., McAdams & Reid, 2005), an increase in V1 responses was found when monkeys attended toward RF stimuli compared to when they attended to non-RF stimuli (figure 22.3B). However, iontophoretic application of ACh augmented the attentional modulation of V1 responses. Furthermore, application of scopolamine, an mAChR antagonist, reduced the attentional modulation, while application of the nAChR antagonist, mecamylamine, had no effect. These results demonstrate a robust interaction between attentional deployment and mAChR activity on the representation of stimuli within visual cortex.

Studies employing behavioral paradigms that manipulate bottom-up attentional orienting—for example, using spatial cues (Posner, 1980)—have generally found that lesions of the basal cholinergic nuclei impair such orienting (Voytko et al., 1994), while increased cholinergic activity (e.g., via nicotine agonist) increases orienting (Witte, Davidson, & Marrocco, 1997). Moreover, both systemic administration of the muscarinic antagonist scopolamine (Davidson, Cutrell, & Marrocco, 1999) and its local injection into posterior parietal cortex have been shown to slow bottom-up orienting of attention in nonhuman primates (Davidson & Marrocco, 2000), suggesting a role of ACh in the mechanism of bottom-up attention. In contrast to top-down attention, in which selection among different sensory stimuli depends solely on the relevance of those stimuli to behavioral goals, bottom-up driven selection is based solely on the (physical) salience of stimuli. As with top-down attention, bottom-up salience enhances the responses of neurons within visual cortex (Burrows & Moore, 2009). Yet it remains unclear how ACh contributes to these effects. Recent studies employing owl models reveal that neurons in a cholinergic nucleus exhibit response characteristics consistent with a potential role in the selection of visual objects based on salience (Asadollahi, Mysore, & Knudsen, 2010). Neurons within the nucleus isthmi pars parvocellularis (IPC) of owls transmit cholinergic inputs to the tectum and respond to both auditory and visual stimuli. Interestingly, the visual responses of these neurons depend heavily on whether the stimulus in their RF is more salient than stimuli outside of their RF; the magnitude of their responses decreases sharply at the boundary where the relative salience of the RF stimulus falls below that of a stimulus outside of the RF (figure 22.3C). Similar effects are observed within the owl optic tectum (Mysore & Knudsen, 2011), which is reciprocally connected with IPC in a precisely topographic manner. These results suggest that salience-driven selection may originate in part from cholinergic inputs, or at least it involves those inputs.

Dopamine Dopamine receptors are grouped into two classes, D1 and D2. The D1 family includes D1 and D5 receptors, whereas D2, D3, and D4 receptors make up the D2 family (Seamans & Yang, 2004). Within the PFC, D1Rs exhibit a bilaminar pattern of expression, while D2Rs are less abundant and appear to be expressed primarily within infragranular layers (Lidow, Goldman-Rakic, Gallager, & Rakic, 1991; Santana, Mengod, & Artiga, 2009). Recent studies have explored the impact of manipulating D1R-mediated activity within the FEF on saccadic target selection and on visual responses of extrastriate area V4 neurons (figure 22.3A; Noudoost & Moore, 2011a). Manipulation of D1R-mediated FEF activity was achieved via volume injections (Noudoost & Moore, 2011b) of a D1 antagonist (SCH23390) into sites within the FEF where neurons represented the same part of visual space as simultaneously recorded area V4 neurons. Following the D1R manipulation, visual targets presented within the affected part of space became more likely to be chosen as targets for saccades than during control trials. Thus, the D1 manipulation increased saccadic target selection. In the same experiments, area V4 neurons with RFs within the part of space affected by the D1R manipulation exhibited an enhancement of visual responses during passive fixation. Responses of V4 neurons were altered in three important ways: (1) via an enhancement in the magnitude of responses to visual stimulation. (2) The visual responses become more selective to stimulus orientation. (3) The visual responses become less variable across trials. Importantly, all three changes in V4 visual activity can also be observed in monkeys trained to covertly attend to RF stimuli (McAdams & Maunsell, 1999; Mitchell et al., 2007). Thus, manipulation of D1R-mediated FEF activity not only increased saccadic target selection, but it also increases the magnitude, selectivity, and reliability of V4 visual responses within the corresponding part of space. Interestingly, injection of a D2 agonist into FEF sites resulted in equivalent target-selection effects as the D1 antagonist. However, only the D1 manipulation produced attention-like effects within area V4. Thus, in addition to being dissociable at the level of functional subclasses of FEF neurons (Thompson et al., 2005), the control of attention and target selection appears to be dissociable at the level of dopamine receptors.

Conclusion

Attention has been one of the most active fields in cognitive neuroscience during the last few decades. It is clear from work carried out in both the human and nonhuman primate model systems that directed attention alters the representation of sensory stimuli within the brain, particularly within the visual system. In more recent years, much more effort has been devoted to identifying the neurons, neural circuits, and other neural mechanisms (e.g., neuromodulators) that can be causally implicated in the control of attention. In the above review, we have focused on neurophysiological studies of visual selective attention in nonhuman primates, since this particular model system has thus far provided the greatest insight.

REFERENCES

Allman, J. M., Kaas, J. H., Lane, R. H., & Miezin, F. M. (1972). A representation of the visual field in the inferior nucleus of the pulvinar in the owl monkey (*Aotus trivirgatus*). *Brain Res, 40*(2), 291–302.

Armstrong, K. M., Chang, M. H., & Moore, T. (2009). Selection and maintenance of spatial information by frontal eye field neurons. *J Neurosci, 29*(50), 15621–15629.

Armstrong, K. M., Fitzgerald, J. K., & Moore, T. (2006). Changes in visual receptive fields with microstimulation of frontal cortex. *Neuron, 50,* 791–798.

Asadollahi, A., Mysore, S. P., & Knudsen, E. I. (2010). Stimulus-driven competition in a cholinergic midbrain nucleus. *Nat Neurosci, 13,* 889–895.

Bisley, J. W., & Goldberg, M. E. (2003). Neuronal activity in the lateral intraparietal area and spatial attention. *Science, 299*(5603), 81–86.

Brennan, A. R., & Arnsten, A. F. T. (2008). Neuronal mechanisms underlying attention deficit hyperactivity disorder: The influence of arousal on prefrontal cortical function. *Ann NY Acad Sci, 1129,* 236–245.

Bruce, C. J. (1990). Integration of sensory and motor signals for saccadic eye movements in the primate frontal eye fields. In G. M. Edelman, W. E. Gall, & W. M. Cowan (Eds.), *Signals and sense: Local and global order in perceptual maps* (pp. 261–314). New York, NY: Wiley.

Burrows, B. E., & Moore, T. (2009). Influence and limitations of popout in the selection of salient visual stimuli by area V4 neurons. *J Neurosci, 29,* 15169–15177.

Buschman, T. J., & Miller, E. K. (2007). Top-down versus bottom-up control of attention in the prefrontal and parietal cortices. *Science, 315,* 1860-1862.

Castellanos, F. X., & Tannock, R. (2002). Neuroscience of attention-deficit/hyperactivity disorder: The search for endophenotypes. *Nature Rev Neurosci, 3,* 617–628.

Cavanaugh, J., Alvarez, B. D., & Wurtz, R. H. (2006). Enhanced performance with brain stimulation: Attentional shift or visual cue? *J Neurosci, 26*(44), 11347–11358.

Cavanaugh, J., & Wurtz, R. H. (2004). Subcortical modulation of attention counters change blindness. *J Neurosci, 24*(50), 11236–11243.

Cohen, M. R., & Maunsell, J. H. R. (2009). Attention improves performance primarily by reducing interneuronal correlations. *Nat Neurosci, 12,* 1594–1600.

Connor, C. E., Preddie, D. C., Gallant, J. L., & Van, E. (1997). DC spatial attention effects in macaque area V4. *J Neurosci, 17,* 3201–3214.

Cooper, J. R., Bloom, F. E., & Roth, R. H. (2003). *The biochemical basis of neuropharmacology.* New York, NY: Oxford University Press.

Cutrell, E. B., & Marrocco, R. T. (2002). Electrical microstimulation of primate posterior parietal cortex initiates orienting and alerting components of covert attention. *Exp Brain Res, 144*(1), 103–113.

Davidson, M. C., Cutrell, E. B., & Marrocco, R. T. (1999). Scopolamine slows the orienting of attention in primates to cued visual targets. *Psychopharmacology, 142,* 1–8.

Davidson, M. C., & Marrocco, R. T. (2000). Local infusion of scopolamine into intraparietal cortex slows covert orienting in rhesus monkeys. *J Neurophysiol, 83,* 1536–1549.

Desimone, R., Wessinger, M., Thomas, L., & Schneider, W. (1990). Attentional control of visual perception: Cortical and subcortical mechanisms. *Cold Spring Harb Symp Quant Biol, 55*, 963–971.

Disney, A. A., Aoki, C., & Hawken, M. J. (2007). Gain modulation by nicotine in macaque V1. *Neuron, 56*, 701–713.

Ekstrom, L. B., Roelfsema, P. R., Arsenault, J. T., Bonmassar, G., & Vanduffel, W. (2008). Bottom-up dependent gating of frontal signals in early visual cortex. *Science, 321*, 414–417.

Ekstrom, L. B., Roelfsema, P. R., Arsenault, J. T., Kolster, H., & Vanduffel, W. (2009). Modulation of the contrast response function by electrical microstimulation of the macaque frontal eye field. *J Neurosci, 29*(34), 10683–10694.

Fries, P., Womelsdorf, T., Oostenveld, R., & Desimone, R. (2008). The effects of visual stimulation and selective visual attention on rhythmic neuronal synchronization in macaque area V4. *J Neurosci, 28*, 4823.

Goard, M., & Dan, Y. (2009). Basal forebrain activation enhances cortical coding of natural scenes. *Nat Neurosci, 12*, 1444–1449.

Goldberg, M. E., & Wurtz, R. H. (1972). Activity of superior colliculus in behaving monkey. I. Visual receptive fields of single neurons. *J Neurophysiol, 35*(4), 542–559.

Gregoriou, G. G., Gotts, S. J., Zhou, H., & Desimone, R. (2009). High-frequency, long-range coupling between prefrontal and visual cortex during attention. *Science, 324*, 1207–1210.

Guillem, K., Bloem, B., Poorthuis, R. B., Loos, M., Smit, A. B., Maskos, U. … Mansvelder, H. D. (2011). Nicotinic acetylcholine receptor b2 subunits in the medial prefrontal cortex control attention. *Science, 333*, 888–891.

Herrero, J. L., Roberts, M. J., Delicato, L. S., Gieselmann, M. A., Dayan, P., & Thiele, A. (2008). Acetylcholine contributes through muscarinic receptors to attentional modulation in V1. *Nature, 454*, 1110–1114.

Ignashchenkova, A., Dicke, P. W., Haarmeier, T., & Thier, P. (2004). *Neuron*-specific contribution of the superior colliculus to overt and covert shifts of attention. *Nat Neurosci, 7*(1), 56–64.

Lidow, M. S., Goldman-Rakic, P. S., Gallager, D. W., & Rakic, P. (1991). Distribution of dopaminergic receptors in the primate cerebral-cortex: Quantitative autoradiographic analysis using [3H]raclopride, [3H]spiperone and [3H]Sch23390. *Neuroscience, 40*, 657–671.

Liu, Y., Yttri, E. A., & Snyder, L. H. (2010). Intention and attention: Different functional roles for LIPd and LIPv. *Nat Neurosci, 13*(4), 495–500.

Lovejoy, L. P., & Krauzlis, R. J. (2010). Inactivation of primate superior colliculus impairs covert selection of signals for perceptual judgments. *Nat Neurosci, 13*(2), 261–266.

Mazzoni, P., Bracewell, R. M., Barash, S., & Andersen, R. A. (1996). Motor intention activity in the macaque's lateral parietal area. I. Dissociation of motor plan from sensory memory. *J Neurophysiol, 76*(3), 1439–1456.

McAdams, C. J., & Maunsell, J. H. R. (1999). Effects of attention on orientation-tuning functions of single neurons in macaque cortical area V4. *J Neurosci, 19*, 431–441.

McAdams, C. J., & Reid, R. C. (2005). Attention modulates the responses of simple cells in monkey primary visual cortex. *J Neurosci, 25*, 11023–11033.

McAlonan, K., Cavanaugh, J., & Wurtz, R. H. (2006). Attentional modulation of thalamic reticular neurons. *J Neurosci, 26*, 4444–4450.

Mirpour, K., Ong, W. S., & Bisley, J. W. (2010). Microstimulation of posterior parietal cortex biases the selection of eye movement goals during search. *J Neurophysiol, 104*(6), 3021–3028.

Mitchell, J. F., Sundberg, K. A., & Reynolds, J. H. (2007). Differential attention-dependent response modulation across cell classes in macaque visual area V4. *Neuron, 55*, 131–141.

Mitchell, J. F., Sundberg, K. A., & Reynolds, J. H. (2009). Spatial attention decorrelates intrinsic activity fluctuations in macaque area V4. *Neuron, 63*, 879–888.

Monosov, I. E., Sheinberg, D. L., & Thompson, K. G. (2010). Paired neuron recordings in the prefrontal and inferotemporal cortices reveal that spatial selection precedes object identification during visual search. *Proc Natl Acad Sci USA, 107*(29), 13105–13110.

Monosov, I. E., & Thompson, K. G. (2009). Frontal eye field activity enhances object identification during covert visual search. *J Neurophysiol, 102*(6), 3656–3672.

Moore, T., & Armstrong, K. M. (2003). Selective gating of visual signals by microstimulation of frontal cortex. *Nature, 421*, 370–373.

Moore, T., & Fallah, M. (2001). Control of eye movements and spatial attention. *Proc Natl Acad Sci USA, 98*, 1273–1276.

Muller, J. R., Philiastides, M. G., & Newsome, W. T. (2005). Microstimulation of the superior colliculus focuses attention without moving eyes. *Proc Natl Acad Sci USA, 102*(3), 524–529.

Mysore, S. P., & Knudsen, E. I. (2011). Flexible categorization of relative stimulus strength by the optic tectum. *J Neurosci, 31*, 7745–7752.

Noudoost, B., Chang, M. H., Steinmetz, N. A., & Moore, T. (2010). Top-down control of visual attention. *Curr Opin Neurobiol, 20*, 183–190.

Noudoost, B., & Moore, T. (2011a). Control of visual cortical signals by prefrontal dopamine. *Nature, 474*, 372–375.

Noudoost, B., & Moore, T. (2011b). A reliable microinjectrode system for use in behaving monkeys. *J Neurosci Methods, 194*, 218–223.

Noudoost, B., & Moore, T. (2011c). The role of neuromodulators in selective attention. *Trends Cogn Sci, 15*, 585–591.

Petersen, S. E., Robinson, D. L., & Keys, W. (1985). Pulvinar nuclei of the behaving rhesus monkey: Visual responses and their modulation. *J Neurophysiol, 54*(4), 867–886.

Posner, M. I. (1980). Orienting of attention. *Q J Exp Psychol A, 32*, 3–25.

Saalmann, Y. B., Pinsk, M. A., Wang, L., Li, X., & Kastner, S. (2012). The pulvinar regulates information transmission between cortical areas based on attention demands. *Science, 337*(6095), 753–756.

Salinas, E., & Sejnowski, T. J. (2001). Correlated neuronal activity and the flow of neural information. *Nat Rev Neurosci, 2*, 539–550.

Santana, N., Mengod, G., & Artiga, F. (2009). Quantitative analysis of the expression of dopamine D1 and D2 receptors in pyramidal and GABAergic neurons of the rat prefrontal cortex. *Cereb Cortex, 19*, 849–860.

Sarter, M., Hasselmo, M. E., Bruno, J. P., & Givens, B. (2005). Unraveling the attentional functions of cortical cholinergic inputs: Interactions between signal-driven and cognitive modulation of signal detection. *Brain Res Brain Res Rev, 48*, 98–111.

Schafer, R. J., & Moore, T. (2007). Attention governs action in the primate frontal eye field. *Neuron, 56*(3), 541–551.

Schafer, R. J., & Moore, T. (2011). Selective attention from voluntary control of neurons in prefrontal cortex. *Science, 332*(6037), 1568–1571.

Seamans, J. K., & Yang, C. R. (2004). The principal features and mechanisms of dopamine modulation in the prefrontal cortex. *Prog Neurobiol, 74,* 1–57.

Sherman, S. M., & Guillery, R. W. (2002). The role of the thalamus in the flow of information to the cortex. *Philos Trans R Soc Lond B Biol Sci, 357*(1428), 1695–1708.

Suzuki, M., & Gottlieb, J. (2013). Distinct neural mechanisms of distractor suppression in the frontal and parietal lobe. *Nat Neurosci, 16*(1), 98–104.

Thompson, K. G., Bichot, N. P., & Sato, T. R. (2005). Frontal eye field activity before visual search errors reveals the integration of bottom-up and top-down salience. *J Neurophysiol, 93,* 337–351.

Tremblay, N., Warren, R. A., & Dykes, R. W. (1990). Electrophysiological studies of acetylcholine and the role of the basal forebrain in the somatosensory cortex of the cat. II. Cortical neurons excited by somatic stimuli. *J Neurophysiol, 64,* 1212–1222.

Voytko, M. L., Olton, D. S., Richardson, R. T., Gorman, L. K., Tobin, J. R., & Price, D. L. (1994). Basal forebrain lesions in monkeys disrupt attention but not learning and memory. *J Neurosci, 14,* 167–186.

Warburton, D. M., & Rusted, J. M. (1993). Cholinergic control of cognitive resources. *Neuropsychobiology, 28,* 43–46.

Wardak, C., Ibos, G., Duhamel, J. R., & Olivier, E. (2006). Contribution of the monkey frontal eye field to covert visual attention. *J Neurosci, 26*(16), 4228–4235.

Wardak, C., Olivier, E., & Duhamel, J. R. (2002). Saccadic target selection deficits after lateral intraparietal area inactivation in monkeys. *J Neurosci, 22*(22), 9877–9884.

Wardak, C., Olivier, E., & Duhamel, J. R. (2004). A deficit in covert attention after parietal cortex inactivation in the monkey. *Neuron, 42*(3), 501–508.

Witte, E. A., Davidson, M. C., & Marrocco, R. T. (1997). Effects of altering brain cholinergic activity on covert orienting of attention: Comparison of monkey and human performance. *Psychopharmacology, 132,* 324–334.

Womelsdorf, T., Fries, P., Mitra, P. P., & Desimone, R. (2006). Gamma-band synchronization in visual cortex predicts speed of change detection. *Nature, 439,* 733–736.

Zenon, A., & Krauzlis, R. J. (2012). Attention deficits without cortical neuronal deficits. *Nature, 489*(7416), 434–437.

23 Computational Models of Attention

LAURENT ITTI AND ALI BORJI

ABSTRACT This chapter reviews recent computational models of visual attention. We begin with models for the bottom-up or stimulus-driven guidance of attention to salient visual items, which we examine in seven different broad categories. We then examine more complex models, which address the top-down or goal-oriented guidance of attention toward items that are more relevant to the task at hand.

A large body of psychophysical evidence on attention can be summarized by postulating two forms of visual attention (James, 1981). The first is driven by the visual input; this so-called exogenous, bottom-up, stimulus-driven, or saliency-based form of attention is rapid, operates in parallel throughout the entire visual field, and helps mediate pop-out, the phenomenon by which some visual items tend to stand out from their surroundings and to instinctively grab our attention (Itti & Koch, 2001; Koch & Ullman, 1985; Treisman & Gelade, 1980). The second, endogenous, top-down, or task-driven form of attention depends on the exact task at hand and on subjective visual experience, takes longer to deploy, and is volitionally controlled. Normal vision employs both processes simultaneously to control both overt and covert shifts of visual attention (Itti & Koch, 2001). Covert focal attention has been described as a rapidly shiftable "spotlight" (Crick, 1984), which serves the double function of selecting particular locations and objects of interest and of enhancing visual processing at those locations and for specific attributes of those objects. Thus, attention acts as a shiftable information-processing bottleneck, allowing only objects within a circumscribed visual region to reach higher levels of processing and visual awareness (Crick & Koch, 1998).

Most computational models of attention to date have focused on bottom-up guidance of attention toward salient visual items. Many, but not all, of these models have embraced the concept of a topographic saliency map (Koch & Ullman, 1985) that highlights scene locations according to their relative conspicuity or salience. This has allowed for validation of saliency models against eye movement recordings, by measuring the extent to which human or monkey observers fixate locations that have higher predicted salience than expected by chance (Parkhurst, Law, & Niebur, 2002; Tatler, Baddeley, & Gilchrist, 2005). However, recent behavioral studies have shown that the majority of eye fixations during execution of many tasks are directed to task-relevant locations that may or may not also be salient, and fixations are coupled in a tight temporal relationship with other task-related behaviors, such as reaching and grasping (Hayhoe, Shrivastava, Mruczek, & Pelz, 2003). Several of these studies have used naturalistic interactive or immersive environments to give high-level accounts of gaze behavior in terms of objects, agents, "gist of the scene" (Potter & Levy, 1969; Torralba, 2003), and short-term memory to describe, for example, how task-relevant information guides eye movements while subjects make a sandwich (Hayhoe et al., 2003; Land & Hayhoe, 2001), or how distractions such as setting the radio or answering a phone affect eye movements while driving (Sodhi, Reimer, & Llamazares, 2002). While these more complex attentional behaviors have been more difficult to capture in computational models, we review below several recent efforts that have successfully modeled attention guidance during complex tasks, such as driving a car or running a hot-dog stand that serves many hungry customers.

While we focus on purely computational models (which autonomously process visual data without the requirement of a human operator, or of manual parsing of the data into conceptual entities), we also point the reader to several relevant previous reviews on attention theories and models more generally (Frintrop, Rome, & Christensen, 2010; Gottlieb & Balan, 2010; Itti & Koch, 2001; Paletta, Rome, & Buxton, 2005; Toet, 2011; Tsotsos & Rothenstein, 2011).

Computational models of bottom-up attention

Development of computational models of attention started with the feature integration theory of Treisman and Gelade (1980), which proposed that only simple visual features are computed in a massively parallel manner over the entire visual field. Attention is then necessary to bind those early features into a united object representation, and the selected bound representation is the only part of the visual world that passes though the attentional bottleneck (figure 23.1A). Koch and Ullman (1985) extended the theory by proposing

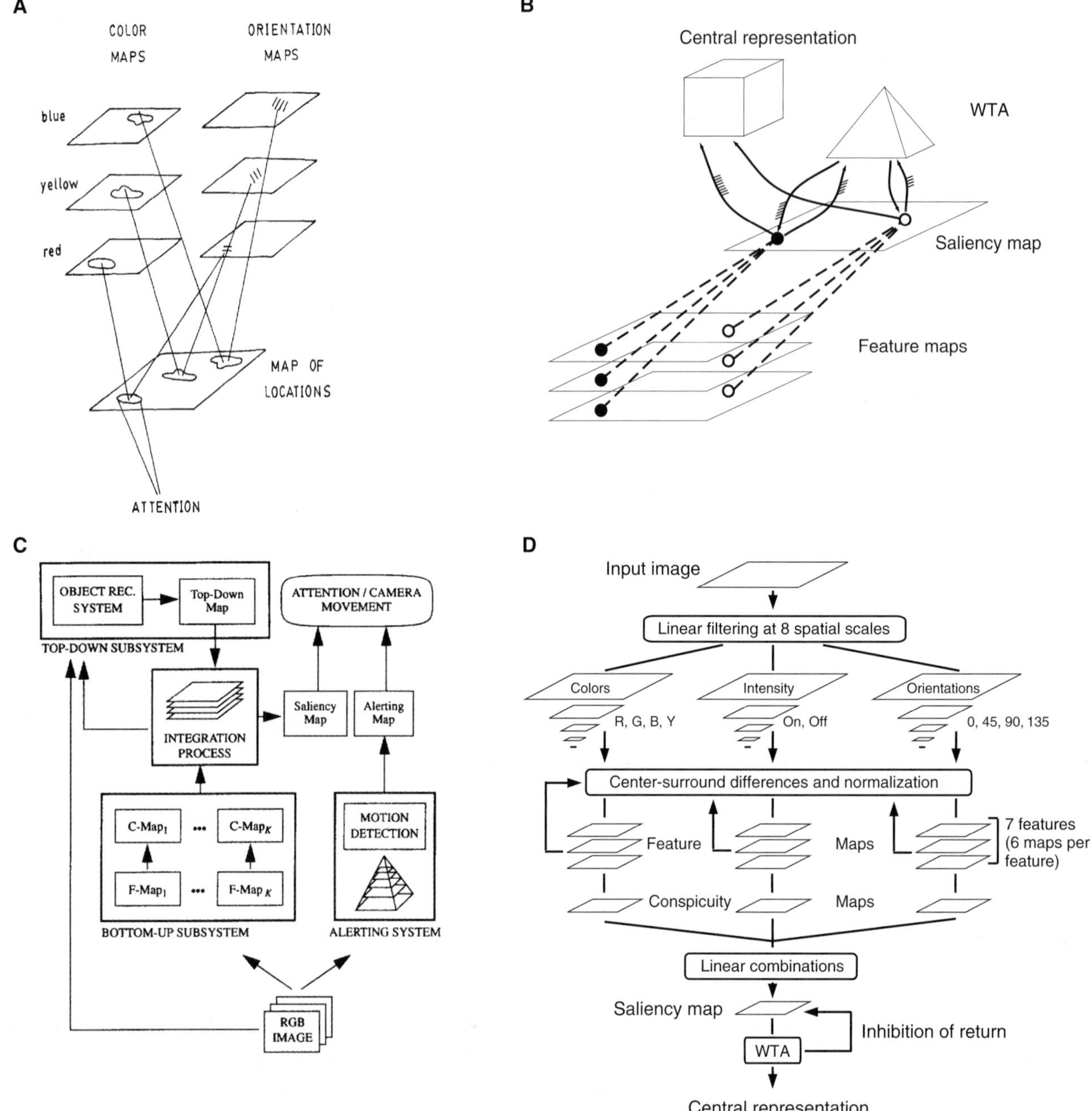

FIGURE 23.1 Early bottom-up attention theories and models. (A) Feature integration theory of Treisman and Gelade (1980) posits several feature maps, and a focus of attention that scans a map of locations and collects and binds features at the currently attended location (from Treisman & Souther, 1985). (B) Koch and Ullman (1985) introduced the concept of a saliency map receiving bottom-up inputs from all feature maps, where a winner-take-all (WTA) network selects the most salient location for further processing. (C) Milanese, Wechsler, Gill, Bost, and Pun (1994) provided one of the earliest computational models. They included many elements of the Koch and Ullman framework and added new components, such as an alerting subsystem (motion-based saliency map) and a top-down subsystem (which could modulate the saliency map based on memories of previously recognized objects). (D) Itti et al. (1998) proposed a complete computational implementation of a purely bottom-up and task-independent model based on Koch and Ullman's theory, including multiscale feature maps, saliency map, winner-take-all, and inhibition of return.

the idea of a single topographic saliency map, receiving inputs from the feature maps, as a computationally efficient representation upon which to operate the selection of where to attend next: a simple maximum-detector or winner-take-all neural network was proposed to simply pick the next most salient location as the next attended one, while an active inhibition-of-return mechanism would later inhibit that location and thereby allow attention to shift to the next most salient location (figure 23.1B). From these ideas, a number of fully computational models started to be developed (e.g., figure 23.1C and 23.1D).

Many research groups have more recently developed new models of bottom-up attention. Fifty-three bottom-up models are classified along 13 different factors in figure 23.2. Most of these models fall into one of the seven general categories described below, with some models spanning several categories (also see Tsotsos & Rothenstein, 2011, for another taxonomy). Additional details and benchmarking of these models was recently proposed by Borji, Sihite, and Itti (2012a, 2012b).

COGNITIVE MODELS Development of saliency-based models escalated after Itti, Koch, and Niebur's (1998) implementation of Koch and Ullman's (1985) computational architecture. Cognitive models were the first to approach algorithms for saliency computation that could apply to any digital image. In these models, the input image is decomposed into a set of multiscale feature maps, selective for elementary visual attributes (e.g., luminance or color contrast, or motion energy). The feature maps are combined across features and scales to form a master saliency map. An important element of this theory is the idea of center-surround operators, which define saliency as distinctiveness of an image region compared to its surroundings. Almost all saliency models are directly or indirectly inspired by cognitive concepts of visual attention (e.g., Le Meur, Le Callet, Barba, & Thoreau, 2006; Marat et al., 2009).

INFORMATION-THEORETIC MODELS Stepping back from biologically plausible implementations, models in this category are based on the premise that localized saliency computations serve to guide attention to the most informative image regions first. These models thus assign higher saliency to scene regions with rare features. Information regarding visual feature F is $I(F) = -log\ p(F)$, inversely proportional to the likelihood of observing F. By fitting a distribution $P(F)$ to features (e.g., using Gaussian mixture models or kernels), rare features can be immediately found by computing $P(F)^{-1}$ in an image. While, in theory, using any feature space is feasible, often these models (inspired by efficient coding in visual cortex) utilize a sparse set of basis functions learned from a repository of natural scenes. Examples in this category are the attention based on information maximization, or AIM, model (Bruce & Tsotsos, 2005); the rarity model (Mancas, 2007); the local + global image patch rarity model (Borji & Itti, 2012a); and incremental coding length models (Hou & Zhang, 2008).

GRAPHICAL MODELS Graphical models are generalized Bayesian models that have been employed for modeling complex attention mechanisms over space and time. Torralba (2003) proposed a Bayesian approach for modeling contextual effects on visual search which was later adopted in the saliency using natural statistics, or SUN, model (Zhang, Tong, Marks, Shan, & Cottrell, 2008) for fixation prediction in free viewing. Itti and Baldi (2005) defined surprising stimuli as those that significantly change beliefs of an observer. Harel, Koch, and Perona (2007) propagated similarity of features in a fully connected graph to build a saliency map. Avraham and Lindenbaum (2010), Li, Zhou, Yan, Niu, and Yang (2010), and Rezazadegan, Tavakoli, Rahtu, and Heikkilä (2011), have also exploited Bayesian concepts for saliency modeling.

DECISION-THEORETIC MODELS This interpretation states that attention is driven optimally with respect to the task. Gao and Vasconcelos (2004) argued that, for recognition of objects, salient features are those that best distinguish a class of objects of interest from all other classes. Given some set of features $X = \{X_1, \ldots, X_d\}$, at locations l, where each location is assigned a class label Y ($Y_l = 0$ for background, $Y_l = 1$ for objects of interest), saliency is then a measure of mutual information (usually the Kullback-Leibler divergence), computed as $I(X, Y) = \sum_{i=1}^{d} I(X_i, Y)$. Besides having good accuracy in predicting eye fixations, these models have been very successful in computer vision applications (e.g., anomaly detection and object tracking).

SPECTRAL ANALYSIS MODELS Instead of processing an image in the spatial domain, these models compute saliency in the frequency domain. Hou and Zhang (2007) derive saliency for an image by computing its Fourier transform, preserving the phase information while discarding most of the amplitude spectrum (to focus on image discontinuities), and taking the inverse Fourier transform to obtain the final saliency map. Bian and Zhang (2009) and Guo and Zhang (2010) further proposed spatiotemporal models in the spectral domain.

Group	No	Model	Year	f1	f2	f3	f4	f5	f6	f7	f8	f9	f10	f11	f12	f13
Bottom-up visual saliency models (for fixation prediction or salient region detection)	1	Itti et al.	1998	+	-	-	+	-	-	+	f	+	CIO	C	-	-
	2	Privitera & Stark	2000	+	-	-	+	-	-	+	f	+	-	O	-	Stark and Choi
	3	Salah et al.	2002	+	+	-	+	-	-	+	-	+	O	G	DR	Digit & Face
	4	Itti et al.	2003	+	-	+	+	+	+	+	f	+	CIOFM	C	-	-
	5	Torralba	2003	-	+	-	+	-	-	+	s	+	CI	B	DR	Torralba et al.
	6	Sun & Fisher	2003	+	-	-	+	-	-	+	-	-	CIO	G	-	-
	7	Gao & Vasconcelos	2004	-	+	-	+	-	-	+	s	-	DCT	D	DR	Brodatz, Caltech
	8	Ouerhani et al.	2004	+	-	-	+	-	-	+	f	+	CIO+Corner	C	CC	Ouerhani
	9	Boccignone & Ferraro	2004	+	-	+	-	+	-	+	f	-	Optical Flow	B	-	BEHAVE
	10	Frintrop	2005	+	+	+	+	+	+	+	f/s	+/-	CIOM	C	-	-
	11	Itti & Baldi	2005	+	-	+	+	+	+	-	f	+	CIOFM	B	KL, AUC	ORIG-MTV
	12	Ma et al.	2005	+	-	+	+	-	-	+	f	+	M*	O	-	-
	13	Bruce & Tsotsos	2006	+	-	-	+	-	+	+	f	+	DOG, ICA	I	KL, ROC	Bruce and Tsotsos
	14	Navalpakkam & Itti	2006	-	+	-	+	-	+	+	s	+	CIO	C	-	-
	15	Zhai & Shah	2006	+	-	+	+	+	-	+	f	+	SIFT	O	-	-
	16	Harel et al.	2006	+	-	-	+	-	-	+	f	+	IO	G	AUC	Bruce and Tsotsos
	17	Le Meur et al.	2006	+	-	-	+	-	-	+	f	+	LM*	C	CC, KL	Le Meur et al.
	18	Walther & Koch	2006	+	-	-	+	-	+	+	f	+/-	CIO	C	-	-
	19	Peters & Itti	2007	+	+	+	+	+	-	+	i	+	CIOFM	P	KL, NSS	Peters and Itti
	20	Liu et al.	2007	+	-	-	+	-	-	+	f	-	Liu*	G	F-measure	Regional
	21	Shic & Scassellati	2007	+	-	+	+	+	-	+	f	+	CIOM	C	ROC	Shic and Scassellati
	22	Hou & Zhang	2007	+	-	-	+	-	+	+	f	+	FFT, DCT	S	NSS	DB of Hou and Zhang, 2007
	23	Cerf et al.	2007	+	+	-	+	-	+	+	f/s	+	CIO :)	C	AUC	Cerf et al.
	24	Le Meur et al.	2007	+	-	+	+	+	-	+	f	+	LM*	C	CC, KL	Le Meur et al.
	25	Mancas	2007	+	-	+	+	+	+	+	f	+	CI	I	CC	Le Meur et al.
	26	Guo et al.	2008	+	-	-	+	-	-	+	f	+	CIO	D	CC	Self data
	27	Zhang et al.	2008	+	-	-	+	-	+	+	f	+	DOG, ICA	B	KL, AUC	Bruce and Tsotsos
	28	Hou & Zhang	2008	+	-	+	+	+	-	+	f	+	ICA	I	AUC, KL	Bruce and Tsotsos, ORIG
	29	Pang et al.	2008	+	+	+	+	+	-	+	f	+	CIOM	G	NSS	ORIG, Self data
	30	Kootstra et al.	2008	+	-	-	+	-	-	+	f	+	Symmetry	C	CC	Kootstra et al.
	31	Ban et al.	2008	+	-	+	+	+	-	+	f	+	CIO+SYM	I	-	-
	32	Rajashekar et al.	2008	+	-	-	+	-	-	+	f	+	R*	S	CC	Rajashekar et al.
	33	Aziz and Mertsching	2008	+	-	-	+	-	+	+	f	+	CO, size, sym	I	_	-
	34	Kienzle et al.	2009	+	-	-	+	-	-	+	f	+	I	P	K*	Kienzle et al.
	35	Marat et al.	2009	+	-	+	+	+	-	+	f	+	SM*	C	NSS	Marat et al.
	36	Judd et al.	2009	+	-	-	+	-	-	+	f	+	J*	P	AUC	Judd et al.
	37	Seo & Milanfar	2009	+	-	+	+	+	+	+	f	+	LSK	I	AUC, KL	Bruce and Tsotsos, ORIG
	38	Rosin	2009	+	-	-	+	-	-	+	f	+	C+ Edge	O	PR, F-measure	DB of Liu et al, 2007
	39	Yin Li et al.	2009	-	+	+	+	+	+	+	s	+	RGB	S	DR	DB of Hou and Zhang, 2007
	40	Bian & Zhang	2009	+	-	+	+	+	+	+	f	+	FFT	S	AUC	Bruce and Tsotsos
	41	Diaz et al.	2009	+	-	-	+	-	+	+	f	+	CIO	O	AUC	Bruce and Tsotsos
	42	Zhang et al.	2009	+	-	+	-	+	-	+	f	+	DOG, ICA	B	KL, AUC	Bruce and Tsotsos
	43	Achanta et al.	2009	+	-	-	+	-	-	+	f	+	DOG	S	PR	DB of Liu et al, 2007
	44	Gao et al.	2009	+	-	+	+	+	+	+	f	+	CIO	D	AUC	Bruce and Tsotsos
	45	Chikkerur et al.	2010	+	+	-	+	-	+	+	f/s	+/-	CIO	B	AUC	Bruce and Tsotsos, Chikkerur
	46	Mahadaven & Vasconcelos	2010	+	-	+	-	+	-	+	-	+	I	D	DR, AUC	SVCL background data
	47	Avraham & Lindenbaum	2010	+	+	-	+	-	+	+	f/s	+/-	CIO	G	DR, CC	UWGT, Ouerhani et al.
	48	Jia Li et al.	2010	-	+	+	+	+	-	+	f	+	CIO	B	AUC	RSD, MTV, ORIG, Peters and Itti
	49	Guo et al.	2010	+	-	+	+	+	+	+	f/s	+/-	FFT	S	DR	Self data
	50	Borji et al.	2010	-	+	-	+	-	+	+	s	+/-	CIO	O	DR	-
	51	Goeferman et al.	2010	+	-	-	+	-	-	+	-	+	C :)	O	AUC	DB of Hou and Zhang, 2007
	52	Murray et al.	2011	+	-	-	+	-	-	+	f	+	CIO	C	AUC, KL	Bruce and Tsotsos, Judd et al.
	53	Wang et al.	2011	+	-	-	+	-	-	+	f	+	ICA	I	AUC	Self data
Top-down (general models)	54	McCallum	1995	-	+	-	+	-	+	-	i	+	-	R	-	Self data
	55	Rao et al.	1995	-	+	-	+	-	-	+	s	+	CIO	O	-	Self data
	56	Ramstrom & Christiansen	2002	-	+	-	+	-	-	+	-	+	CI	O	-	-
	57	Sprague & Ballard	2003	-	+	+	-	+	+	+	i	-	S*	R	-	-
	58	Renninger et al.	2004	-	+	-	+	-	+	-	s	-	Edgelet	I	DR	Self data
	59	Navalpakkam & Itti	2005	-	+	-	+	-	+	+	-	+	CIO	C	-	Self data
	60	Paletta et al.	2005	-	+	-	+	-	-	+	-	-	SIFT	R	DR	COIL-20, TSG-20
	61	Jodogne & Piater	2007	-	+	-	+	-	-	+	i	-	SIFT	R	-	-
	62	Butko & Movellan	2009	-	+	+	+	+	+	+	s	-	-	R	-	-
	63	Verma & McOwan	2009	+	-	-	+	-	+	-	s	-	CIO	O	-	-
	64	Borji et al.	2010	-	+	-	+	-	-	+	i	-	CIO	R	-	Self data
	65	Borji et al.	2012	-	+	-	+	+	-	+	i	-	CIO	B	AUC, NSS	Self data

FIGURE 23.2 Survey of bottom-up and top-down computational models, classified according to 13 factors. These factors, in order: bottom-up (f_1); top-down (f_2); spatial (−)/spatiotemporal (+) (f_3); static (f_4); dynamic (f_5); synthetic (f_6) and natural (f_7) stimuli; task-type (f_8); space-based(+)/object-based(−) (f_9); features (f_{10}); model type (f_{11}); measures (f_{12}); and used data set (f_{13}). In the task type (f_8) column: free-viewing (f); target search (s); interactive (i). In the features (f_{10}) column: color, intensity, and orientation (CIO) saliency; CIOFM: CIO plus flicker and motion saliency; M* = motion saliency, static saliency, camera motion, and object (face) and aural saliency (speech-music); LM*: contrast sensitivity, perceptual decomposition, visual masking, and center-surround interactions; Liu*: center-surround histogram,

PATTERN-CLASSIFICATION MODELS Models in this category use machine learning techniques to learn stimulus-to-saliency mappings, from image features to eye fixations. They estimate saliency s by computing $p(s|f)$, where f is a feature vector which could be the contrast of a location and its surrounding neighborhood. Kienzle, Wichmann, Scholkopf, and Franz (2007), Peters and Itti (2007), and Judd, Ehinger, Durand, and Torralba (2009) used image patches, scene gist, and a vector of several features at each pixel, respectively, and used classical support vector machine (SVM) and regression classifiers for learning saliency. Rezazadegan et al. (2011) used sparse sampling and kernel density estimation to estimate the above probability in a Bayesian framework. Note that some of these models may not be purely bottom-up since they use features that guide top-down attention, such as faces or text (Cerf, Harel, Einhauser, & Koch, 2008; Judd et al., 2009).

OTHER MODELS Other models exist that do not easily fit into our categorization. For example, Seo and Milanfar (2009) proposed self-resemblance of local image structure for saliency detection. The idea of decorrelation of neural response was used for a normalization scheme in the adaptive whitening saliency model (Garcia-Diaz Fdez-Vidal, Pardo, & Dosil, 2009). Kootstra, Nederveen, and de Boer (2008) developed symmetry operators for measuring saliency, and Goferman, Zelnik-Manor, and Tal (2010) proposed a context-aware saliency detection model with successful applications in re-targeting and summarization.

In summary, modeling bottom-up visual attention is an active research field in computational neuroscience and machine vision. New theories and models are constantly proposed that keep advancing the state of the art.

Top-down attention models

Models that address top-down, task-dependent influences on attention are more complex, since some representations of goal and of task become necessary. In addition, top-down models typically involve some degree of cognitive reasoning, not only attending to but also recognizing objects and their context, to incrementally update the model's understanding of the scene and to plan the next most task-relevant shift of attention (Beuter, Lohmann, Schmidt, & Kummer, 2009; Navalpakkam & Itti, 2005; Yu, Mann, & Gosine, 2008, 2012). For example, one may consider the following information flow, aimed at rapidly extracting a task-dependent compact representation of the scene, that can be used for further reasoning and planning of top-down shifts of attention, and of action (Itti & Arbib, 2006; Navalpakkam & Itti, 2005).

- Interpret task definition: by evaluating the relevance of known entities (in long-term symbolic memory) to the task at hand and storing the few most relevant entities into symbolic working memory. For example, if the task is to drive, be alert to traffic signs, pedestrians, and other vehicles.
- Prime visual analysis: by priming spatial locations that have been learned to usually be relevant, given a set of desired entities and a rapid analysis of the "gist" and rough layout of the environment (Rensink, 2000; Torralba, 2003), and by priming the visual features (e.g., color, size) of the most relevant entities being looked for (Wolfe, 1994).
- Attend and recognize: the most salient location given the priming and biasing done at the previous step. Evaluate how the recognized entity relates to the relevant entities in working memory, using long-term knowledge of interrelationships among entities.
- Update: based on the relevance of the recognized entity, decide whether it should be dropped as uninteresting or retained in working memory (possibly creating an associated summary "object file" in working memory; see Kahneman, Treisman, & Gibbs, 1992) as a potential object and location of interest for action planning.
- Iterate: the process until sufficient information has been gathered to allow a confident decision for action.
- Act: based on the current understanding of the visual environment and the high-level goals.

An example of a top-down model that includes the above elements, although not in a very detailed implementation, was proposed by Navalpakkam and Itti

---

multiscale contrast and color spatial-distribution; R*: luminance, contrast, luminance-bandpass, contrast-bandpass; SM*: orientation and motion; J*: CIO, horizontal line, face, people detector, gist, etc.; S*: color matching, depth and lines; and :) represents the face. In the model type (f_{11}) column, R means that a model is based on reinforcement learning. In the measures (f_{12}) column, K* means the Wilcoxon-Mann-Whitney test was used (the probability that a randomly chosen target patch receives higher saliency than a randomly chosen negative one); DR means models have used a measure of detection/classification rate to determine how successful the model was. PR stands for precision recall. In the data set (f_{13}) column: self data means that authors gathered their own data. For a detailed definition of these factors, please refer to Borji and Itti (2012b).

(2005). Given a task definition as keywords, the model first determines and stores the task-relevant entities in symbolic working memory, using prior knowledge from symbolic long-term memory. The model then biases its saliency-based attention system to emphasize the learned visual features of the most relevant entity. Next, it attends to the most salient location in the scene and attempts to recognize the attended object through hierarchical matching against stored representations in visual long-term memory. The task-relevance of the recognized entity is computed and used to update the symbolic working memory. In addition, a visual working memory in the form of a topographic task-relevance map is updated with the location and relevance of the recognized entity. The implemented prototype of this model has emphasized four aspects of biological vision: determining task-relevance of an entity, biasing attention for the low-level visual features of desired targets, recognizing these targets using the same low-level features, and incrementally building a visual map of task-relevance. The model was tested on three types of tasks: single-target detection in 343 natural and synthetic images, where biasing for the target accelerated its detection over two-fold on average; sequential multiple-target detection in 28 natural images, where biasing, recognition, and working memory contributed to rapidly finding all targets; and learning a map of likely locations of cars from a video clip filmed while driving on a highway (Navalpakkam & Itti, 2005).

While the previous example model uses explicit cognitive reasoning about world entities and their relationships, a complementary trend in top-down modeling uses fuzzy or probabilistic reasoning to explore how several sources of bottom-up and top-down information may combine. For example, Ban, Kim, and Lee (2010) proposed a model where the bottom-up and top-down components interact through a fuzzy learning system (figure 23.3A). During training, a bottom-up saliency map selects locations, and their features are incrementally clustered and learned in a growing fuzzy topology adaptive resonance theory model (GFTART). During testing, top-down interest in a given object activates its features stored in the GFTART model and biases the bottom-up saliency model to become more sensitive to these features, thereby increasing the probability that the object of interest will stand out. In a related approach, Akamine, Fukuchi, Kimura, and Takagi (2012; also see Kimura, Pang, Takeuchi, Yamato, & Kashino, 2008) developed a dynamic Bayesian network that combines the following factors. First, input video frames give rise to deterministic saliency maps. These are converted into stochastic saliency maps via a random process that affects the shape of salient blobs over time (e.g., dynamic Markov random field; Kimura et al., 2008). An eye-focusing map is then created that highlights maxima in the stochastic saliency map, additionally integrating top-down influences from an eye-movement pattern (a stochastic selection between passive and active state with a learned transition probability matrix). The authors use a particle filter with Markov chain Monte Carlo sampling to estimate the parameters; this technique, often used in machine learning, allows for fast and efficient estimation of unknown probability density functions. Several additional recent related models using graphical models have been proposed (e.g., Chikkerur, Serre, Tan, & Poggio, 2010).

In a recent example, using probabilistic reasoning and inference tools, Borji et al. (2012b) introduced a framework for top-down overt visual attention based on reasoning, in a task-dependent manner, about objects present in the scene and about previous eye movements. They designed a dynamic Bayesian network (DBN) that infers future probability distributions over attended objects and spatial locations from past observed data. Briefly, the Bayesian network is defined over object variables that matter for the task. For example, in a video game where one runs a hot-dog stand and has to serve multiple customers while managing the grill, those include raw sausages, cooked sausages, buns, ketchup, etc. Then, existing objects in the scene, as well as the previous attended object, provide evidence toward the next attended object (figure 23.3B). The model also allows reading out which spatial location will be attended, thus allowing one to verify its accuracy against the next actual fixation of the human player. The parameters of the network are learned from training data in the same form as the test data (human players playing the game). This object-based model was significantly more predictive of eye fixations, compared to simpler classifier-based models, several state-of-the-art bottom-up saliency models, and control algorithms such as mean eye position (figure 23.3C). This points toward the efficacy of this class of models to capture spatiotemporal, visually guided behavior in the presence of a task.

While fully computational top-down models are more complex than their bottom-up counterparts, many recent examples thus exist that provide an inspiration for future efforts in developing models that more accurately emulate the human cognitive processes that control top-down attention.

Outlook

Our review shows that tremendous progress has been made in modeling both bottom-up and top-down

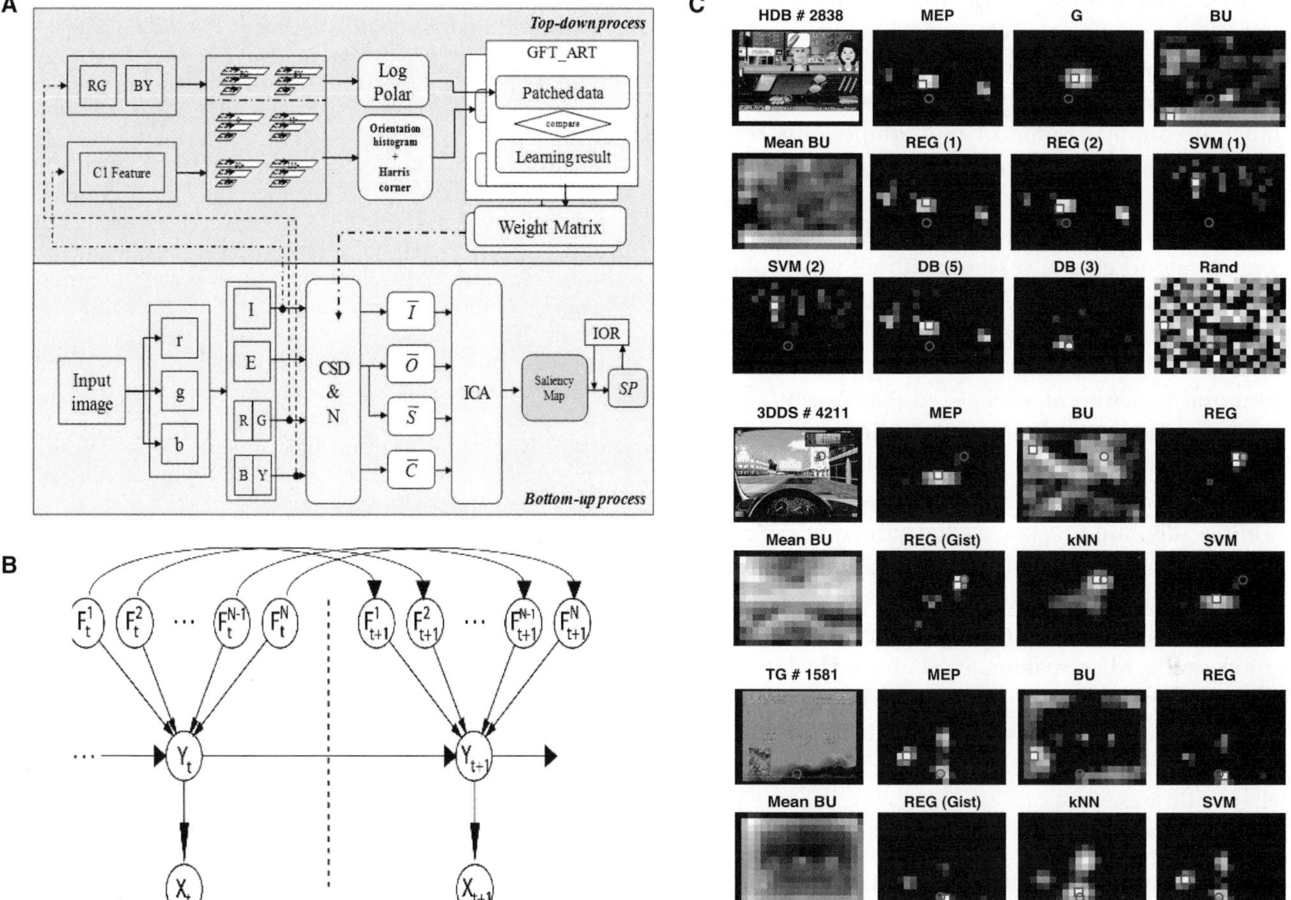

FIGURE 23.3 Examples of recent top-down models. (A) Model of Ban et al. (2010), which integrates bottom-up and top-down components; r, g, b: red, green, and blue color channels. I: intensity feature. E: edges. R, G: red-green color. B, Y: blue-yellow color. CSD&N: center-surround differences and normalization. ICA: independent component analysis. GFT_ART: growing fuzzy topology adaptive resonance theory. SP: saliency point. (B) Graphical representation of the dynamic Bayesian network (DBN) approach of Borji et al. (2012b), unrolled over two time-slices. X_t is the current saccade position, Y_t is the currently attended object, and F_t^i is the function that describes object i at the current scene. All variables are discrete. It also shows a time-series plot of probability of objects being attended and a sample frame with tagged objects and eye fixation overlaid. (C) Sample predicted saccade maps of the DBN model (shown in B) on three video games and tasks: running a hot-dog stand (HDB; top three rows), driving (3DDS; middle two rows), and flight combat (TG, bottom two rows). Each red circle indicates the observer's eye position superimposed with each map's peak location (blue squares). Smaller distance indicates better prediction. Models compared are as follows. MEP: mean eye position over all frames during the game play (control model). G: trivial Gaussian map at the image center. BU: bottom-up saliency map of the Itti model. Mean BU: average saliency maps over all video frames. REG(1): regression model that maps the previous attended object to the current attended object and fixation location. REG(2): similar to REG(1), but the input vector consists of the available objects at the scene augmented with the previously attended object. SVM(1) and SVM(2) correspond to REG(1) and REG(2) but using a support vector machine (SVM) classifier. Similarly, DB(5) and DB(3) correspond to REG(1) and REG(2), meaning that in DB(5) the network considers just one previously attended object, while in DB(3) each network slice consists of the previously attended object as well as information about the previous objects in the scene. REG(Gist): regression based only on the gist of the scene. kNN: k-nearest-neighbors classifier. Rand: white noise random map (control). Overall, DB(3) performed best at predicting where the player would look next (Borji et al., 2012b). (See color plate 21.)

aspects of attention computationally. Tens of new models have been developed, each bringing new insights into the question of what makes some stimuli more important to visual observers than other stimuli.

While many models have approached the problem of modeling top-down attention, a fully implemented cognitive system that reasons about objects, their relationships, and their locations to guide the next shift of attention remains an elusive goal to date.

Several barriers exist in building even more sophisticated visual attention models, which, we argue, depend on progress in complementary aspects of machine vision, knowledge representation, and artificial intelligence, to support some of the components required to implement attention-driven scene-understanding systems. Of prime importance is object recognition, which remains a hard problem in machine vision, but is necessary to enable reasoning about which object to look for next (using top-down strategies) given the set of objects that have been attended to and recognized so far. Also important is understanding the spatial and temporal structure of a scene, so that reasoning about objects and locations in space and time can be exploited to guide attention (e.g., understanding pointing gestures, or trajectories of objects in three dimensions). Additionally, building knowledge bases that can capture what an observer may know about different world entities and that allow reasoning over this knowledge is required to build more able top-down attention models. For example, when making tea (Land & Hayhoe, 2001), knowledge about different objects relevant to the task, where they usually are stored in a kitchen, and how to manipulate them is needed to decide where to look and what to do next.

ACKNOWLEDGMENTS Supported by the National Science Foundation (grant numbers CCF-1317433 and CMMI-1235539), the Army Research Office (W911NF-11-1-0046 and W911NF-12-1-0433), the US Army (W81XWH-10-2-0076), and Google. The authors affirm that the views expressed herein are solely their own, and do not represent the views of the United States government or any agency thereof.

REFERENCES

Akamine, K., Fukuchi, K., Kimura, A., & Takagi, S. (2012). Fully automatic extraction of salient objects from videos in near real time. *Comp J, 55*(1), 3–14.

Avraham, T., & Lindenbaum, M. (2010). Esaliency (extended saliency): Meaningful attention using stochastic image modeling. *IEEE Trans Pattern Anal, 32*(4), 693–708.

Ban, S. W., Kim, B., & Lee, M. (2010). Top-down visual selection attention model combined with bottom-up saliency map for incremental object perception. *Proc. International Joint Conference on Neural Networks* (pp. 1–8).

Beuter, N., Lohmann, O., Schmidt, J., & Kummert, F. (2009). Directed attention: A cognitive vision system for a mobile robot. *IEEE International Symposium on Robot and Human Interactive Communication* (pp. 854–860).

Bian, P., & Zhang, L. (2009). Biological plausibility of spectral domain approach for spatiotemporal visual saliency. In M. Köppen, N. Kasabov, & G. Coghill (Eds.), *Lecture notes in computer science: Vol. 5506. Advances in neuro-information processing* (pp. 251–258). Berlin: Springer.

Borji, A., & Itti, L. (2012a). Exploiting local and global patch rarities for saliency detection. *Proc IEEE Conference on Computer Vision and Pattern Recognition* (pp. 1–8).

Borji, A., & Itti, L. (2012b). State-of-the-art in visual attention modeling. *IEEE Trans Pattern Anal, 35*(1), 185–207.

Borji, A., Sihite, D. N., & Itti, L. (2012a). Quantitative analysis of human-model agreement in visual saliency modeling: A comparative study. *IEEE Trans Image Process, 22*, 55–69.

Borji, A., Sihite, D. N., & Itti, L. (2012b). What/where to look next? Modeling top-down visual attention in complex interactive environments. *IEEE Trans Syst Man Cy A, 99*, 1.

Bruce, N. D. B., & Tsotsos, J. K. (2005). Saliency based on information maximization. *Adv Neural Inf Process Syst, 18*, 155–162.

Butko, N. J., & Movellan, J. R. (2009). Optimal scanning for faster object detection. *Proc. IEEE Conference on Computer Vision and Pattern Recognition* (pp. 2751–275).

Cerf, M., Harel, J., Einhauser, W., & Koch, C. (2008). Predicting human gaze using low-level saliency combined with face detection. *Adv Neural Inf Process Syst, 20*, 241–248.

Chikkerur, S., Serre, T., Tan, C., & Poggio, T. (2010). What and where: A Bayesian inference theory of attention. *Vis Res, 50*(22), 2233–2247.

Crick, F. (1984). Function of the thalamic reticular complex: The searchlight hypothesis. *Proc Natl Acad Sci USA, 81*, 4586–4590.

Crick, F., & Koch, C. (1998). Constraints on cortical and thalamic projections: The no-strong-loops hypothesis. *Nature, 391*(6664), 245–250.

Frintrop, S., Rome, E., & Christensen, H. I. (2010). Computational visual attention systems and their cognitive foundations: A survey. *ACM Trans Appl Percept, 7*(1), 6.

Gao, D., & Vasconcelos, N. (2004). Discriminant saliency for visual recognition from cluttered scenes. *Adv Neural Inf Process Syst, 17*, 481–488.

Garcia-Diaz, A., Fdez-Vidal, X., Pardo, X., & Dosil, R. (2009). Decorrelation and distinctiveness provide with human-like saliency. In J. Blanc-Talon, W. Philips, D. Popescu, & P. Scheunders (Eds.), *Lecture notes in computer science: Vol. 5807. Advanced concepts for intelligent vision systems* (pp. 343–354). Berlin: Springer.

Goferman, S., Zelnik-Manor, L., & Tal, A. (2010). Context-aware saliency detection. *Proc IEEE Conference on Computer Vision and Pattern Recognition* (pp. 2376–2383).

Gottlieb, J., & Balan, P. (2010). Attention as a decision in information space. *Trends Cogn Sci, 14*(6), 240–248.

Guo, C., & Zhang, L. (2010). A novel multiresolution spatiotemporal saliency detection model and its applications in image and video compression. *IEEE Trans Image Process, 19*, 185–198.

Harel, J., Koch, C., & Perona, P. (2007). Graph-based visual saliency. *Adv Neural Inf Process Syst, 19*, 545.

Hayhoe, M. M., Shrivastava, A., Mruczek, R., & Pelz, J. B. (2003). Visual memory and motor planning in a natural task. *J Vision, 3*(1), 49–63.

Hou, X., & Zhang, L. (2007). Saliency detection: A spectral residual approach. *Proc IEEE Conference on Computer Vision and Pattern Recognition* (pp. 1–8).

Hou, X., & Zhang, L. (2008). Dynamic visual attention: Searching for coding length increments. *Adv Neural Inf Process Syst, 21*, 681–688.

ITTI, L., & ARBIB, M. A. (2006). Attention and the minimal subscene. In M. A. Arbib (Ed.), *Action to language via the mirror neuron system* (pp. 289–346). Cambridge, UK: Cambridge University Press.

ITTI, L., & BALDI, P. F. (2005). A principled approach to detecting surprising events in video. *Proc IEEE Conference on Computer Vision and Pattern Recognition* (pp. 631–637).

ITTI, L., & KOCH, C. (2001). Computational modelling of visual attention. *Nat Rev Neurosci, 2*(3), 194–203.

ITTI, L., KOCH, C., & NIEBUR, E. (1998). A model of saliency-based visual attention for rapid scene analysis. *IEEE Trans Pattern Anal, 20*(11), 1254–1259.

JAMES, W. (1981). *The principles of psychology*. Cambridge, MA: Harvard University Press. Original work published 1890.

JUDD, T., EHINGER, K., DURAND, F., & TORRALBA, A. (2009). Learning to predict where humans look. *Proc IEEE Conference on Computer Vision and Pattern Recognition* (pp. 2106–2113).

KAHNEMAN, D., TREISMAN, A., & GIBBS, B. J. (1992). The reviewing of object files: Object-specific integration of information. *Cogn Psychol, 24*(2), 175–219.

KIENZLE, W., WICHMANN, F., SCHOLKOPF, B., & FRANZ, M. (2007). A nonparametric approach to bottom-up visual saliency. *Proc Adv Neural Inf Process Syst, 19*, 689–696.

KIMURA, A., PANG, D., TAKEUCHI, T., YAMATO, J., & KASHINO, K. (2008). Dynamic Markov random fields for stochastic modeling of visual attention. *Proc International Conference on Pattern Recognition* (pp. 1–5).

KOCH, C., & ULLMAN, S. (1985). Shifts in selective visual attention: Towards the underlying neural circuitry. *Hum Neurobiol, 4*(4), 219–227.

KOOTSTRA, G., NEDERVEEN, A., & DE BOER, B. (2008). Paying attention to symmetry. *Proc British Machine Vision Conference* (pp. 1115–1125).

LAND, M. F., & HAYHOE, M. (2001). In what ways do eye movements contribute to everyday activities? *Vis Res, 41*(25–26), 3559–3565.

LE MEUR, O., LE CALLET, P., BARBA, D., & THOREAU, D. (2006). A coherent computational approach to model bottom-up visual attention. *IEEE Trans Pattern Anal, 28*(5), 802–817.

LI, Y., ZHOU, Y., YAN, J., NIU, Z., & YANG, J. (2010). Visual saliency based on conditional entropy. *Lecture Notes in Computer Science: Vol. 5994. Computer Vision–ACCV 2009* (pp. 246–257).

MANCAS, M. (2007). *Computational attention: Towards attentive computers.* Belgium: Presses universitaires de Louvain.

MARAT, S., HO PHUOC, T., GRANJON, L., GUYADER, N., PELLERIN, D., & GUÉRIN-DUGUÉ, A. (2009). Modelling spatio-temporal saliency to predict gaze direction for short videos. *Int J Comput Vis, 82*(3), 231–243.

MCCALLUM, A. K. (1996). *Reinforcement learning with selective perception and hidden state.* PhD thesis, University of Rochester, Rochester, NY.

MILANESE, R., WECHSLER, H., GILL, S., BOST, J. M., & PUN, T. (1994). Integration of bottom-up and top-down cues for visual attention using non-linear relaxation. *Proc IEEE Conference on Computer Vision and Pattern Recognition* (pp. 781–785).

NAVALPAKKAM, V., & ITTI, L. (2005). Modeling the influene of task on attention. *Vis Res, 45*(2), 205–231.

PALETTA, L., ROME, E., & BUXTON, H. (2005). Attention architectures for machine vision and mobile robots. In L. Itti, G. Rees, & J. Tsotsos (Eds.), *Neurobiology of attention* (pp. 642–648). New York, NY: Academic Press.

PARKHURST, D., LAW, K., & NIEBUR, E. (2002). Modeling the role of salience in the allocation of overt visual attention. *Vis Res, 42*(1), 107–123.

PETERS, R. J., & ITTI, L. (2007). (Jun). Beyond bottom-up: Incorporating task-dependent influences into a computational model of spatial attention. *Proc IEEE Conference on Computer Vision and Pattern Recognition* (pp. 1–8).

POTTER, M. C., & LEVY, E. I. (1969). Recognition memory for a rapid sequence of pictures. *J Exp Psychol Gen, 81*(1), 10.

RENSINK, R. A. (2000). The dynamic representation of scenes. *Vis Cogn, 7*, 17–42.

REZAZADEGAN TAVAKOLI, H., RAHTU, E., & HEIKKILÄ, J. (2011). Fast and efficient saliency detection using sparse sampling and kernel density estimation. In *Image Analysis, SCIA 2011 Proc, Lecture Notes in Computer Science, 6688*, 666–675.

SEO, H., & MILANFAR, P. (2009). Static and space-time visual saliency detection by self-resemblance. *J Vision, 9*(12), 1–27.

SODHI, M., REIMER, B., & LLAMAZARES, I. (2002). Glance analysis of driver eye movements to evaluate distraction. *Behav Res Meth Ins C, 34*(4), 529–538.

TATLER, B. W., BADDELEY, R. J., & GILCHRIST, I. D. (2005). Visual correlates of fixation selection: Effects of scale and time. *Vision Res, 45*(5), 643–659.

TOET, A. (2011). Computational versus psychophysical bottom-up image saliency: A comparative evaluation study. *IEEE Trans Pattern Anal, 33*(11), 2131–2146.

TORRALBA, A. (2003). Modeling global scene factors in attention. *J Opt Soc Am A, 20*(7), 1407–1418.

TREISMAN, A. M., & GELADE, G. (1980). A feature-integration theory of attention. *Cognit Psychol, 12*(1), 97–136.

TREISMAN, A., & SOUTHER, J. (1985). Search asymmetry: A diagnostic for preattentive processing of separable features. *J Exp Psychol Gen, 114*(3), 285–310.

TSOTSOS, J. K., & ROTHENSTEIN, A. (2011). Computational models of visual attention. *Scholarpedia, 6*(1), 6201.

WOLFE, J. M. (1994). Guided search 2.0 A revised model of visual search. *Psychon Bull Rev, 1*(2), 202–238.

YU, Y., MANN, G. K. I., & GOSINE, R. G. (2008). An object-based visual attention model for robots. *Proc IEEE International Conference on Robotics and Automation* (pp. 943–948).

YU, Y., MANN, G., & GOSINE, R. (2012). A goal-directed visual perception system using object-based top-down attention. *IEEE Trans Auton Ment Dev, 4*(1), 87–103.

ZHANG, L., TONG, M. H., MARKS, T. K., SHAN, H., & COTTRELL, G. W. (2008). SUN: A Bayesian framework for saliency using natural statistics. *J Vis, 8*(7), 32.

IV

SENSATION AND PERCEPTION

Introduction

JAMES J. DICARLO AND RACHEL I. WILSON

When we took on the task of editing this section, we had one aim in mind: to outline the key concepts underlying the science of sensation, both those that had stood the test of time and those that are now emerging. We selected authors whose own research has contributed to the field's understanding of these concepts. However, we asked them not to focus on their own research, or to write a conventional review article. Rather, we asked them to explain the key ideas that have motivated their research, or that raise questions for research of the future.

Inevitably, different authors ended up tackling some of the same concepts, but always with a different angle or emphasis. For us, one of the pleasures in reading and editing these chapters was the experience of viewing the same concepts refracted through these different lenses. We hope the resulting collection is more than the sum of its parts, because it invites the reader to compare how the same idea is treated by different authors.

Often, the term *sensation* is used to refer to the neural processes resulting from a sensory stimulus, whereas the term *perception* is used to refer to the conscious awareness of those processes. We would argue that it is useful to think of percepts as particular neural states that can be flexibly linked to a variety of behaviors. In scientific practice, we generally seek to understand sensory perceptions by means of behavioral reports, because conscious awareness is not directly observable. Careful quantitative measurements of sensory perceptions (in this sense of the term) are the domain of psychophysics, and these types of measurements can be applied to both human and nonhuman subjects. Indeed, one of our goals in selecting authors was to illustrate relationships between studies in a wide range of organisms. This implicitly assumes that some of the neural states that link sensory stimuli to behaviors will have characteristics that are common to a variety of different species.

This section begins with a chapter by Rachel Wilson, which outlines the *internal constraints* on sensory systems, with a particular focus on visual and olfactory systems. These constraints include limitations imposed by evolution, development, metabolism, neural noise, and the electrical limitations of single neurons. This is followed by a chapter by Fred Rieke that considers how neural systems might actually reach *optimality* in balancing internal constraints against the demands of behavioral tasks. Rieke focuses on the very first stage of visual processing, which is unusually costly in terms of energy, and thus unusually constrained.

We then turn to the question of the neural code. Several chapters describe principled ways to consider why neurons use the codes they do. Adrienne Fairhall summarizes the *statistical regularities* that are characteristic of natural stimuli, and describes how *adaptation* continually adjusts neural codes to match these statistical properties. Bruno Olshausen extends these ideas to show how it is useful to conceptualize the sensory system as implementing a *probabilistic inference* about the true state of the visual world under ill-posed conditions. He uses this approach to show how the properties of neurons in primary visual cortex can be derived from this viewpoint.

Next, Xaq Pitkow and Markus Meister ask how we should characterize the response properties of neurons that detect low-dimensional, but behaviorally relevant, aspects of the world from a very high-dimensional stream of data, highlighting the need for highly nonlinear operations to accomplish such tasks. They explore how such neurons challenge conventional ways of characterizing *receptive field properties*, and they describe how selectivity for complex features could arise through cascades of elementary *neural computations*. Elad Schneidman's chapter takes on the problem of *population coding*, and in particular the question of why neurons may carry redundant information. Schneidman proposes that neural networks might achieve near-optimal levels of redundancy, given that redundancy can be useful in combating noise and facilitating decoding. Nicole Rust's chapter links together all these ideas. Rust describes how information about complex stimulus features is only *implicit* near the sensory periphery, but is made *explicit* via a series of *hierarchical transformations* of population activity.

Finally, we turn to the relationship between neural activity and perception. Doug Ruff and Marlene Cohen discuss how one can design experiments to establish a *causal role* for specific neurons in a particular percept. This chapter builds upon classic ideas, but updates these ideas in light of new data and new tools for perturbing neural activity. Bill Geisler, Johannes Burge, Melchi Michel, and Anthony D'Antona describe *psychophysical approaches* to characterizing perception, as well as modern statistical methods to characterize the difficulty of sensory tasks based on the information available in sensory data and the goal of the task. These methods allow one to determine the efficiency of neuronal systems and to isolate the sources of variability that limit its performance. This chapter also describes how the same framework can be used to generate principled hypotheses about the neural implementation of a behavioral task.

The final chapters consider the relationship between neural activity and perception in particularly difficult tasks. Specifically, all of these chapters deal with the problem of integrating information about fine-scale sensory features (textures) with information about large-scale sensory features (objects). First, Steven Hsiao and Manuel Gomez-Ramirez describe how information from many types of sensory neurons might be combined to account for perceptual performance in complex somatosensory tasks. They point out that sensing is often an *active* process, and thus sensory signals must be integrated with information about the active behaviors involved in collecting those signals. Next, Dan Kersten and Alan Yuille tackle the problem of visual scene analysis. Their chapter describes the computational importance of hierarchical processing, and also shows how that processing can be usefully conceptualized as a form of statistical inference. Their chapter argues that feedforward ("bottom-up") processing streams convey uncertain messages that can be disambiguated by virtue of lateral and *top-down processing*. Finally, Kalanit Grill-Spector and Kevin Weiner approach the same problem from the viewpoint of human functional magnetic resonance imaging studies. They emphasize the importance of explicitly identifying the *goals* of any neural process, along with the specific computations that must be achieved to reach those goals. They also show how the *spatial maps* within visual cortical regions can be linked to these specific computations. Although these three chapters deal with specific cases of visual or tactile object recognition, they describe concepts that are likely equally relevant to processing auditory objects or olfactory objects.

Inevitably, there are gaps here, reflecting our own preoccupations, and perhaps those of the field as well. For example, we have overemphasized vision and neglected audition. And while there is an interesting dialogue in these pages between primates and flies, there is little about neural systems of intermediate size—especially rodents. In the future, we expect these gaps will be filled in, especially with the development of new tools for monitoring and perturbing neural activity. Regardless, each of these chapters contains key principles that we believe will continue to stand the test of time, and thus serve as important reading for all serious students of sensation and perception. In addition, we invite the reader to consider how future studies might test the more speculative ideas described in these chapters.

24 Constraints on Sensory Processing

RACHEL I. WILSON

ABSTRACT The evolution of sensory systems is driven by the need to survive and reproduce in a naturalistic sensory environment. However, neural architectures are also shaped by their own internal constraints. These constraints include the evolutionary-developmental constraints that restrict the possible effects of mutations, as well as stochastic developmental noise. Neural systems are further limited by metabolic costs, particularly the cost of maintaining ionic gradients. Finally, neural systems are limited by the properties of their own electronics—namely, the existence of intrinsic electrical noise, the limited speed of signal propagation, the restricted dynamic range of synapses and firing rates, and the quasi-linear nature of synaptic integration. Importantly, many of these internal constraints are fundamentally in conflict with each other, insofar as they pressure evolution in opposite directions. Thus, neural systems are shaped by the need to find a satisfactory trade-off between competing factors. Many such trade-offs are common to many sensory modalities, and they are shared by organisms as different as flies and humans. As such, they can help explain why some features of neural systems are also shared.

"The enemy of art is the absence of limitations."
—Orson Welles

The ability of neural systems to process sensory information is subject to internal constraints. A constraint (as defined here) is a limitation on the performance of a sensory system that arises from the intrinsic properties of the neural system itself. This is distinct from limitations arising from the sensory environment, or the nature of the behavioral task.

Any given internal constraint may not directly limit performance, because it may be possible to compensate for the constraint. However, compensation may push the nervous system toward a particular architecture, rather than other architectures that might hypothetically produce better performance in the absence of this constraint. Thus, constraints often limit performance indirectly, by pressuring neural systems toward compromise architectures. A compromise may be optimal, or it may simply be good enough to permit the organism to survive and reproduce (Marder & Goaillard, 2006; Rieke, 2014).

The constraints discussed here are common to all sensory systems. Moreover, they are common to very different organisms. As such, they can potentially explain why certain neural architectures occur again and again: they are good compromises.

The design of artificial systems is sometimes inspired by biological systems. This represents an additional motivation to seriously consider the constraints on biological systems. Because artificial systems are not subject to many of the constraints discussed here, an implementation that works well in biology may be suboptimal in an artificial system, and thus engineers must be careful not to draw the wrong lessons from biological systems (Stafford, 2010).

This chapter reviews the key constraints on neural systems and their implications for sensory processing. Organisms have evolved intriguing ways to cope with some of these constraints. Such cases are generally clearer in sensory neuroscience than in other branches of neuroscience, simply because the function of sensory systems is particularly transparent. For this reason, sensory systems are a good setting for investigating how neural systems evolve in the context of constraints.

Evolutionary inheritance

Neural systems in related species have a similar organization. This is true even in species that inhabit very different ecological niches. For example, the relative size of major brain divisions is remarkably constant across mammals. A meta-analysis of 131 species showed that the volumes of all major brain divisions (including the medulla, hippocampus, cerebellum, striatum, and neocortex) were highly systematically related to total brain volume, except the olfactory bulb. Different brain divisions showed different relationships to brain volume—in particular, neocortex grew particularly steeply with increasing volume—but for all brain divisions there was a systematic dependence on brain volume that extended to species with widely varying body sizes and lifestyles. Notably, this analysis included species as diverse as simians, prosimians, insectivores, and bats. This result argues that the expansion of brain volume resulting from natural selection for any behavioral trait is constrained to be a coordinated growth of the entire nonolfactory brain (Finlay & Darlington, 1995). This finding raises a provocative question: might there be excess signaling capacity in some brain divisions, as a by-product of a strong pressure to expand other divisions?

Whereas the size of an entire brain division (e.g., neocortex) appears to be highly constrained, there is relatively more flexibility in the regionalization of brain divisions (e.g., the division of neocortex into sensory regions). Even so, there is evidence that regionalization is also subject to constraints. For example, all mammals share a common set of primary and secondary sensory cortical regions. This includes visual areas V1 and V2, somatosensory areas S1 and S2, and auditory area A1, as defined by cytoarchitectonic landmarks and afferent/efferent connections. Even the relative positions of these regions are grossly conserved in all mammals. Because mammals inhabit a wide diversity of ecological niches, from treetops to fields to oceans, this conserved pattern of regionalization suggests that the neocortex is evolutionarily constrained to a particular architecture (Krubitzer & Kahn, 2003).

More evidence of constraints upon regionalization comes from studies of animals that have completely lost one sensory modality, and yet still preserve a vestige of the corresponding regions of neocortex and thalamus. For example, the subterranean mole rat *Spalax ehrenberghi* is completely blind: its eyes are entirely covered with skin and fur, and recordings from cortex show no evidence of visually evoked potentials. The only function for the retina in this species is to entrain its circadian clock, which occurs via projections to the superchiasmatic nucleus. Nevertheless, the retina still sends sparse projections to all the visual areas that normally process form and motion in other mammals, including the superior colliculus and the lateral geniculate nucleus (LGN) of the thalamus. Moreover, the LGN still sends a topographic projection to occipital cortex, where V1 is normally located (Cooper, Herbin, & Nevo, 1993). Although these regions are severely reduced in size, their persistence in this species argues that the regionalization of thalamus and cortex is constrained by evolutionary inheritance.

Despite sharing conserved features, homologous neural structures can dramatically switch functions. This idea is illustrated by the case of the blind mole rat, where large parts of the LGN and occipital cortex are taken over by auditory inputs (Bronchti et al., 2002; Heil, Bronchti, Wollberg, & Scheich, 1991). This illustrates the principle that inherited constraints are typically incorporated into functional neural systems.

Developmental programs

Canalization Because certain developmental programs are robust to genetic variation, they tend to persist across evolutionary time, and they channel neural systems into stereotyped architectures. These programs can be likened to a canal that channels the progress of a waterway along a stereotyped route; thus, this phenomenon has been termed *canalization.* The idea that neural systems are constrained by these developmental canals is closely linked to the idea of inherited evolutionary constraints (see above). Indeed, there is no real distinction between a developmental constraint of this sort and an evolutionary constraint: developmental programs are the means by which evolutionary constraints are imprinted on an individual organism.

An example of a developmental canal is the sequence of neurogenesis in different divisions of the mammalian brain. This sequence is stereotyped across species, suggesting it is difficult to alter by genetic variation, perhaps due to some robustness of the master regulatory genetic networks that control it. The sequence of neurogenesis is important because it affects the relative volume of different brain divisions. The later the onset of neurogenesis in a particular brain division, the larger the potential pool of neural precursors in that division. Delaying the onset of neurogenesis in one brain division should therefore have cascading effects on the volume of all later-developing brain divisions. Indeed, as we might expect, the brain divisions where neurogenesis occurs last are those that have enlarged disproportionately in large-brained species. The implication is that disproportionate enlargement is constrained to occur preferentially in these brain divisions, as compared to other ones (Finlay & Darlington, 1995; Finlay, Darlington, & Nicastro, 2001). If so, then the disproportionately large size of the human neocortex (and its associated sensory regions) may have arisen initially as a by-product of a constraint on brain development.

Developmental Noise In addition to being limited by developmental canals, neural systems are also limited by developmental noise. A clear example is the rodent olfactory bulb. This structure is divided into ~1,000 discrete neuropil compartments called glomeruli. Each glomerulus is uniquely associated with a single olfactory receptor neuron type, corresponding to a single odorant receptor (figure 24.1). On a coarse spatial scale, the relatively spatial location of each glomerulus is completely stereotyped. On a fine spatial scale, however, there is notable imprecision: the relative positions of adjacent glomeruli are often swapped. Notably, imprecision along the anterior-posterior axis is significantly larger than along the medial-lateral axis (Soucy, Albeanu, Fantana, Murthy, & Meister, 2009). This is particularly interesting because anterior-posterior position is specified by an unusual axon guidance mechanism that depends on the intrinsic properties of odorant receptors themselves (Imai, Suzuki, & Sakano, 2006). This suggests that

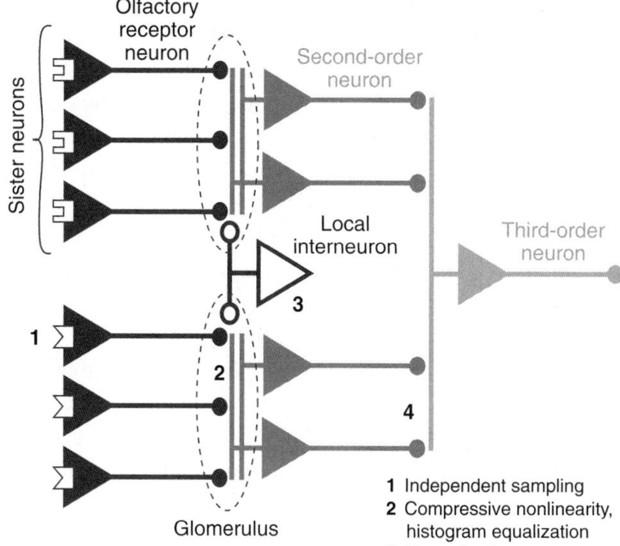

FIGURE 24.1 Circuit organization of the olfactory system. The architecture of the olfactory system provides several examples of how neural systems can respond to internal constraints. Each olfactory receptor neuron (ORN) generally expresses a single odorant receptor, and all the ORNs that express the same receptor converge on the same compartment of neuropil ("glomerulus," dashed line) in the olfactory bulb (in vertebrates) or antennal lobe (in insects). This schematic depicts only two glomeruli, but in reality there are on the order of 1,000 glomeruli (in vertebrates) and 50 in *Drosophila.* The spatial placement of each glomerulus is coarsely stereotyped but can vary on a fine spatial scale, likely reflecting stochastic developmental noise. There are on the order of 10,000 ORNs per glomerulus in vertebrates, and about 40 ORNs per glomerulus in *Drosophila.* In *Drosophila,* all sister ORNs synapse precisely on every second-order neuron, with about 3 second-order neurons per glomerulus. (Vertebrates have 10–100 second-order neurons per glomerulus.) Studies in *Drosophila* have shown that several features of this circuit can be viewed as a response to a constraint (see text): (1) sister ORNs provide independent samples of the olfactory world; (2) within a glomerulus, the transfer function from ORNs to second-order neurons takes the form of a compressive nonlinearity; (3) inhibitory local interneurons mediate lateral inhibition between glomeruli, thereby normalizing the amplitude of population activity across odor stimuli; (4) sister second-order neurons carry largely redundant signals, and they are thought to reconverge onto the same third-order neurons.

variation in glomerular position is due mainly to limitations in the precision of the hardwired developmental mechanisms that specify glomerular position, rather than variations across animals in odor-evoked neural activity. Consistent with this idea, a mutation that eliminates odor-evoked neural activity has relatively little effect on the glomerular map (Lin et al., 2000; Zheng, Feinstein, Bozza, Rodriguez, & Mombaerts, 2000).

As this example illustrates, developmental mechanisms can be noisy. The origin of developmental noise is stochasticity in signal transduction and gene expression. Stochasticity at the single-cell level reflects the low copy number of some proteins within cells (McAdams & Arkin, 1997). This type of noise places limits on the theoretical maximum rate of information transmission in biochemical signaling networks, including the biochemical signals that instruct neural development. Recent studies using information-theoretic analyses have formalized this intuition (Cheong, Rhee, Wang, Nemenman, & Levchenko, 2011).

The effect of developmental noise can be mitigated by using multiple mechanisms having partly redundant functions. For example, retinal ganglion cell axons must project in an orderly fashion to retino-recipient brain regions, forming retinotopic maps where axonal position is systematically related to retinal position. These retinotopic maps are specified by multiple ligand-receptor systems, and some of these have partly redundant functions, such that multiple mechanisms must be genetically disrupted in order to reveal any substantial phenotypic defect (Feldheim et al., 2000).

In spite of such compensatory strategies, developmental noise is still likely to limit the function of neural systems. For example, intrinsic imprecision in glomerular targeting may limit the pattern of horizontal connectivity between glomeruli, which could limit the computations performed in the olfactory bulb (Murthy, 2011). It will be interesting in future to learn whether some features of neural circuit architecture might be adaptive responses to stochasticity in single-cell developmental programs.

Metabolic constraints

ORIGINS OF METABOLIC COSTS For many species, neural systems represent a major metabolic burden. In humans, about 20% of the resting metabolic rate is consumed by the brain (Rolfe & Brown, 1997). Neurons consume more energy when they are active, but even inactive neurons impose a substantial energy burden. When neurons run out of energy, the consequences are swift: a human subject falls unconscious only 7 seconds after circulation to the neck is blocked (Ames, 2000). For these reasons, metabolic demands strongly constrain the architecture of neural systems. This idea has been explored in several comprehensive reviews (Laughlin, 2001; Niven & Laughlin, 2008), which also serve as primers on how neural systems respond to competing constraints.

The highest metabolic costs are imposed by the need to maintain steep ionic gradients across the plasma membrane. This alone accounts for about half of the energy consumed by neural systems (Ames, 2000). Ionic gradients are dissipated by transmembrane currents, and so there is a price associated with the channels that carry these currents. Among all transmembrane currents, the most costly are the currents associated with action potentials, followed by synaptic currents and leak currents. By comparison, the cost of recycling synaptic vesicles is low (Attwell & Laughlin, 2001).

Architectures That Minimize Metabolic Costs Metabolic constraints shape neural architecture in several ways. To begin with, they create an incentive to match the intrinsic properties of neurons to the signals they must carry. A nice example is provided by the potassium conductances of fly photoreceptors. Visual signals fluctuate rapidly, and, to capture these fluctuations, photoreceptors should have fast membrane time constants. However, there is a metabolic cost to the high leak conductances that would be required to create a fast membrane (Niven, Anderson, & Laughlin, 2007; Niven & Laughlin, 2008). The solution is to match the temporal bandwidth of the membrane to the characteristics of input signals. Phototransduction is slow at low light levels and fast at higher light levels; accordingly, at rest the membrane acts as a low-pass filter, but when depolarized it acts as a high-pass filter. Moreover, fast-flying flies have faster membrane time constants, whereas slow-flying flies have slower membrane time constants (Laughlin, 1994).

Moreover, metabolic costs favor architectures where connected neurons are located near each other in space. For a typical neuron, most of the metabolic cost associated with a single spike is incurred by axonal currents, with a smaller contribution from dendritic and somatic currents (Attwell & Laughlin, 2001). The cost of axonal currents grows with axon length, and so axons should be as short as possible. This means that there is a strong pressure for connected neurons to be spatially close, and indeed most connectivity in neural systems is local rather than long range. This is sometimes called the *wiring economy principle* (Chklovskii & Koulakov, 2004).

In addition, metabolic costs create an incentive to keep redundancy low (Barlow, 1961, 2001). Indeed, metabolic costs would argue that redundancy should be minimized both in space and in time (i.e., both across neurons and within neurons). There are several ways that neural systems can do this. First, redundancy can be minimized by using an array of sensors that under-samples the sensory world. This is exemplified by the photoreceptor array in both vertebrates and invertebrates, which under-samples the optical image; this may related to the very high metabolic costs incurred by these cells (Laughlin, 1994; Snyder, Bossomaier, & Hughes, 1986). Second, redundancy can also be reduced by cell-intrinsic mechanisms of adaptation or gain control, which tend to reduce redundancy over time. Cells possess a variety of these intrinsic negative feedback mechanisms (Shapley & Enroth-Cugell, 1984; Wark, Lundstrom, & Fairhall, 2007). Third, redundancy can be reduced by *lateral inhibition* among neurons that have correlated activity. This type of lateral inhibition also represents a form of negative feedback. The key feature here is that information in other neurons is used to make a prediction about what level of gain is needed. This has been called a *predictive coding* architecture. Finally, redundancy is reduced simply by virtue of the fact that most neurons have a nonlinear spike threshold, because any nonlinearity tends to reduce linear correlations (Pitkow & Meister, 2012).

Electrical noise

Origins of Noise As electrical signals propagate through neural systems, they are continually contaminated by noise that arises in neurons themselves. This represents a major intrinsic constraint on sensory processing. The origin of neural noise lies in stochastic microscopic processes. Chief among these are ion-channel gating and synaptic vesicle release. The noise created by these stochastic microscopic events can then be amplified by the nonlinear properties of neurons (Faisal, Selen, & Wolpert, 2008).

Intuitively, one might think that channel noise is not a major problem for most neurons, because noise should average out across many channels. Surprisingly, this is not true—even for neurons that contain relatively large numbers of channels (White, Rubinstein, & Kay, 2000). There are three reasons for this result. First, the signal-to-noise ratio (SNR) of total conductance grows only slowly with increasing channel number, because it is proportional to the square root of the number of channels (N). Given the metabolic costs of increasing N, this limits the ability of a cell to overcome noise by increasing N. Second, for a channel that is gated by depolarization, the probability of opening is low at hyperpolarized potentials, and so the SNR of total conductance can be relatively poor. Third, individual channels are not independent: a stochastic opening of one voltage-gated Na^+ channel will tend to depolarize the cell, thereby increasing the probability that another Na^+ channel will open.

Synaptic noise arises primarily from the fact that the release of synaptic vesicles is stochastic (del Castillo & Katz, 1954). At many synapses, the mean number of released vesicles is low, and so trial-to-trial variability in the number of released vesicles is quite high. This, together with variability in the amount of neurotransmitter per vesicle, can create large trial-to-trial fluctuations in the size of the postsynaptic response to a single presynaptic spike (Bekkers & Clements, 1999; Sargent, Saviane, Nielsen, DiGregorio, & Silver, 2005).

Together, channel noise and synaptic noise inject a substantial stochastic element into neural activity. This places a limitation on the precision of stimulus encoding, because it causes identical presentations of the same stimulus to elicit different neural responses. Moreover, the mean membrane potential of many neurons *in vivo* is just below their spike threshold; neural noise sources tend to push these neurons above their threshold, so that they fire spikes even in the absence of a stimulus. Finally, it should be kept in mind that neural noise is injected at every layer of a sensory processing circuit, and so each central neuron inherits noise from previous layers.

ARCHITECTURES THAT MINIMIZE THE EFFECTS OF NOISE Given these considerations, neural systems are under pressure to minimize the negative effects of noise. One strategy is to pool redundant signals from independent sensors. This is exemplified by the first relay in the olfactory system (figure 24.1), where each second-order neuron pools input from many olfactory receptor neurons, all of which express the same odorant receptor and all of which project their axons to the same glomerulus. In the fruit fly *Drosophila melanogaster*, all olfactory receptor neurons that express the same odorant receptor (called "sister ORNs") are known to synapse quite precisely onto each and every second-order neuron in their target glomerulus, and to make synapses of a rather uniform strength. All sister ORNs have the same noise level, and their noise is independent, so this architecture ought to maximize the SNR of second-order neurons. Indeed, the SNR of second-order neurons is better than that of their cognate ORNs, despite the fact that principal neurons are subject to additional synaptic and channel noise (Bhandawat, Olsen, Schlief, Gouwens, & Wilson, 2007; Kazama & Wilson, 2008, 2009). There are several second-order neurons in each glomerulus, and because they pool input from exactly the same sister ORNs, they carry highly redundant signals. Interestingly, there is evidence that they synapse onto some of the same third-order neurons (Marin, Jefferis, Komiyama, Zhu, & Luo, 2002; Wong, Wang, & Axel, 2002). This arrangement should allow third-order neurons to average out some of the noise that arises *de novo* in second-order neurons. Note that there is substantial redundancy in this circuit at two successive layers, in spite of its metabolic costs. Note also the peculiar architecture of this circuit (figure 24.1): signals first converge (all sister ORNs synapse onto each sister second-order neuron), then diverge (in the form of redundant sister second-order neurons), and then reconverge (as sister second-order neurons wire onto the same third-order neuron). This architecture suggests the system is under strong pressure to minimize the maladaptive effects of noise. A similar architecture has been proposed for the transmission of visual information between retina, thalamus, and visual cortex (Alonso, Usrey, & Reid, 2001).

Another strategy is to impose a filter that selectively discards noise, retaining the signal. This is possible only if the properties of signal and noise are distinctively different. For example, the phototransduction cascade downstream from rhodopsin is spontaneously active, which generates continuous voltage noise in photoreceptors. Absorption of a photon generates a predictable discrete "bump" of activity in the phototransduction cascade, which is distinctively different from continuous noise. Accordingly, the synapse between rod photoreceptors and bipolar cells is configured to impose a threshold on rod output, such that continuous noise cannot pass, but (many) single-photon responses can pass (Field & Rieke, 2002).

Yet another strategy is to distribute signals as uniformly as possible within the available coding space, ensuring that all codes are used with equal frequency. This strategy is sometimes known as *histogram equalization*, because it produces a flat histogram of response probabilities. The classic example of this phenomenon is the contrast-response function of second-order fly visual neurons, which is nicely matched to the distribution of contrasts in natural visual scenes. As a consequence, these neurons use each response level with equal probability (Laughlin, 1981). Importantly, histogram equalization cannot help combat existing noise, but it helps immunize signals from noise that is added later: when signals are well separated in coding space, adding noise has a minimal effect on their discriminability. Another example of histogram equalization occurs in the *Drosophila* olfactory system. Most odor responses of olfactory receptor neurons fall within the lower part of the dynamic range of these neurons. This might reflect a metabolic constraint on average firing rate, especially as ORNs are numerous, outnumbering second-order neurons by ~10:1. Weak ORN responses are then preferentially boosted as they are transmitted to second-order neurons. As a consequence, second-order neurons use each response level with roughly equal probability. Because second-order neurons are less numerous, the relative pressure of metabolic constraints and noise constraints may be different in these neurons.

Constraints on neural electronics

In principle, single neurons can perform a vast array of operations on their synaptic inputs. Biophysicists tend to emphasize this viewpoint. The fanciest single-neuron operations rely on complex dendritic morphologies, the specific spatial placement of synaptic inputs onto dendritic trees, and well-tuned voltage-gated conductances in dendrites. There is ample evidence that neurons can achieve these things, although their roles *in vivo* are necessarily difficult to demonstrate (London & Hausser, 2005; Silver, 2010).

Nonetheless, the operations that a neuron can perform are also constrained by the nature of cellular electronics. Many of these constraints can be mitigated, but sometimes the solution would be costly and so is not worth the price. In short, single-neuron operations are not arbitrarily flexible. In some cases, this creates incentives for neural circuits to evolve architectures that can compensate for the limitations of single neurons.

Limitations on Speed Several factors limit the speed of neural processing. As we have seen, fast membrane time constants are metabolically costly, and this may be why many neurons have relatively slow membrane time constants. In addition, dendritic cable filtering tends to slow synaptic potentials as they travel to the spike initiation zone. In particular, axonal conduction delays can be as large as 100 milliseconds in long axons, much longer than the typical delay involved in synaptic transmission (150–400 microseconds; Sabatini & Regehr, 1999). Axonal conduction speed can be increased by increasing axon diameter, but because volume grows with the square of the diameter, this strategy consumes valuable space (Swadlow, 2000).

Notably, many organisms have evolved neural subsystems with unusual cellular specializations for speed. In invertebrates, these subsystems are characterized by large-diameter axons and electrical synapses, and they mediate escape reflexes (Allen, Godenschwege, Tanouye, & Phelan, 2006; Faulkes, 2008). In the brains of many mammals, the distribution of axon diameters is right-skewed, and the largest axons are always myelinated, which further increases speed. This subpopulation of particularly large axons has been proposed to serve brain functions that require fast conduction speed (Perge, Koch, Miller, Sterling, & Balasubramanian, 2009; Wang et al., 2008). The fact that these specializations for speed are only found in a small fraction of neurons is consistent with the idea that they come at a high price.

Limitations on Dynamic Range Spike rates and vesicular release rates cannot be negative. This poses a constraint, because many sensory systems must encode fluctuations above and below some mean ambient level of a stimulus (e.g., light). In principle, neural systems might respond to this constraint by setting basal firing rates high, allowing the same neuron to encode both increases and decreases about the mean. However, high basal firing rates are metabolically costly. An alternative is to create opponent populations of neurons having opposite stimulus preferences. Classic cases are the ON-OFF neurons and color opponent neurons of the retina, but opponent neurons can also be found in mechanosensory, auditory, and thermosensory systems (Jacobs, Miller, & Aldworth, 2008; Ma, 2010; Yorozu et al., 2009).

Firing rates also cannot be arbitrarily large. Moreover, it seems that there can be strong constraints on the firing rate of a neuron averaged over long time scales. The evidence for this idea comes from the observation that firing rate distributions are exponentially distributed in visual cortical neurons. This is notable, because information theory predicts that firing-rate distributions should actually be flat, since this should maximize the rate of information transmission (see above). Exponentially distributed firing rates are consistent with the existence of a constraint on mean firing rates, together with a pressure to maximize the rate of information transmission within that constraint (Baddeley et al., 1997). Given the energy budget of the human brain, it has been estimated that the average neuron is constrained to fire at rates around 1–7 Hz (Wang et al., 2008). A constraint on firing rates will constrain the number of distinguishable messages that a neuron can send, because noise limits the number of different firing rates that can be reliably distinguished from each other.

Finally, synapses cannot be arbitrarily strong. In a passive dendrite, increasing the conductance of a synapse brings diminishing marginal returns on the postsynaptic voltage response, because a large localized synaptic conductance will simply shunt synaptic currents. Dendritic voltage-gated conductances can of course amplify synaptic potentials, but this also has a cost. Specifically, if a dendrite is endowed with active conductances, this will increase not only the efficacy of the synapse in question, but also the barrage of ongoing noise arising from spontaneous presynaptic spikes at other synapses. As a consequence, there may be no net increase in the ability of the synapse in question to control the postsynaptic cell (London, Schreibman, Hausser, Larkum, & Segev, 2002).

All these considerations limit the dynamic range of any neuron's output. These limits create an incentive for sensory neurons to implement some form of adaptation or gain control. This allows a neuron to continually adjust its sensitivity based on the current characteristics of the sensory environment, so that both large and small inputs fall within its dynamic range (Wark et al., 2007). These mechanisms are ubiquitous, but they also come at a cost: because the signals that control adaptation or gain will tend to be noisy, any feedback mechanism based on these signals will tend to amplify noise (Dunn & Rieke, 2008). Moreover, because gain control changes the relationship between input and output, it necessarily creates ambiguity about the stimulus (Fairhall, Lewen, Bialek, & de Ruyter Van Steveninck, 2001).

LIMITATIONS ARISING FROM LINEAR SUMMATION At least to a first approximation, the dendritic trees of most neurons perform a rather simple operation on their synaptic inputs. Inputs are weighted and summed quasi-linearly, and then the neuron fires roughly in proportion to the sum above some threshold. This simplistic description, of course, does not capture all that single neurons are capable of. However, it encapsulates the basic hardware that all neurons possess.

Insofar as many neurons integrate their inputs in this manner, this situation creates a constraint on what categorizations these neurons can perform. Imagine we would like to create a neuron that responds only to a particular subset of stimuli. This neuron can receive N synaptic inputs. The free parameters we have are the weights associated with each input synapse, as well as the threshold of that neuron. It is possible to find a set of parameter values that generates the desired solution if and only if one can draw a hyperplane in N-dimensional space that separates target stimuli from off-target stimuli (Rosenblatt, 1958). In other words, stimulus representations must be *linearly separable.* An idealized binary linear classifier of this type is called a *perceptron.*

Real neurons are even more constrained than classical perceptrons. This is because synaptic weights and thresholds cannot be arbitrarily large. In addition, the synaptic inputs that can be tuned to confer the desired selectivity may be constrained to be nonnegative, because they are synaptic inputs from excitatory neurons.

There are several ways that neural systems can improve linear separability under these strong constraints. First, it is helpful to begin with a high-dimensional stimulus representation. This is because the likelihood that there is a separating hyperplane increases with the number of synaptic inputs N (Cover, 1965). For example, it has been pointed out that high dimensionality is a characteristic property of olfactory encoding (Itskov & Abbott, 2008). Odors are encoded combinatorially by the activity of many odorant receptors having diverse odor tuning, where each receptor binds multiple odors, and each odor binds multiple receptors. As the number of odorant receptors (N) grows, so does the potential selectivity of higher-order olfactory neurons that linearly sum the activity of different receptors.

Second, there is a special case of the linear separation problem where *normalization* and a *compressive nonlinearity* can be helpful. This special case can be termed sparse recoding. Imagine that we want to create an array of neurons where each neuron responds to only one stimulus out of many, and each stimulus activates at least one neuron. To achieve this, each neuron needs synaptic input weights and a threshold that confers specificity for a single stimulus. The problem of finding these values amounts to the problem of drawing a line (or in high-dimensional space, a hyperplane) that separates each individual stimulus from all other stimuli (figure 24.2). Sparse codes are typical of many sensory brain regions, so this is a biologically relevant special case of the linear separability problem.

In this situation, separability can be improved by *normalization.* Normalization involves dividing activity in individual neurons by the summed activity of many neurons in the same brain region. This operation occurs in a wide range of sensory modalities and organisms (Carandini & Heeger, 2012). Normalization facilitates sparse recoding, because it tends to equalize the total population firing rates evoked by different stimuli. In geometric terms, if we imagine each stimulus as a point in N-dimensional space, then normalization tends to move all points toward the surface of a hypersphere in that space, making it is easier to find a line or a hyperplane that separates each stimulus from the rest (figure 24.2). An example of normalization has been described in the *Drosophila* olfactory system, where input from individual olfactory receptor neurons is divided by the total activity of all olfactory receptor neurons. This is accomplished via lateral inhibition from local interneurons (figure 24.1). In simulations that were tightly constrained by data, this operation makes it easier to construct linear classifiers that respond sparsely and selectively to single odors (Luo, Axel, & Abbott, 2010; Olsen, Bhandawat, & Wilson, 2010). Normalization has been proposed to serve a similar function in visual cortical areas (DiCarlo & Cox, 2007; Olshausen & Field, 2005).

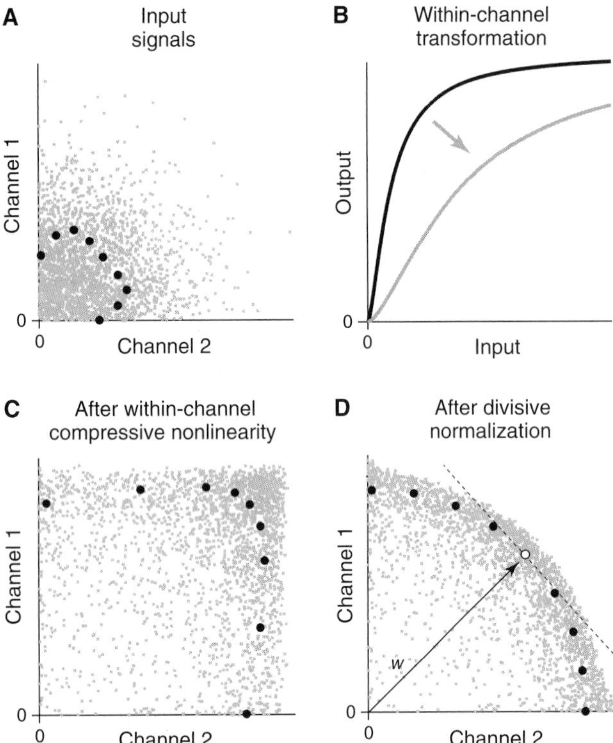

FIGURE 24.2 Compression, normalization, and linear separability. (A) Consider a sensory representation in multiple dimensions, with each dimension corresponding to a coding channel (a neuron or a chain of neurons). For concreteness, we might imagine the two-dimensional space in this schematic corresponds to two different types of olfactory receptor neurons. Every odor stimulus is a point in this space. A subset of points are highlighted for comparison across panels (black circles). (B) A compressive nonlinearity transforms signals within each channel (black curve). Divisive normalization adjusts the steepness of this nonlinearity according to the total input to both channels. Namely, when total activity is high, the nonlinearity is less steep (gray curve; Carandini & Heeger, 2012). Different stimuli will therefore fall on different curves, because they produce different amounts of total input to both channels. Curves in this schematic are fit to data recorded from *Drosophila* olfactory neurons in vivo (Olsen et al., 2010). (C) A steep compressive nonlinearity (black curve in B) preferentially boosts weak input signals (compare to A). (D) Normalization tends to equalize the distance of all representations from the origin. As a consequence, stimuli tend to lie near the surface of a circle. The vector **w** corresponds to the weights on the two input channels that define a linear separation (dashed line) between one highlighted symbol and the other symbols. Note that, in this example, we can separate each highlighted symbol from the rest using only nonnegative weights. In the *Drosophila* olfactory system, there are 50 coding channels (glomeruli), as compared to the two channels in this schematic. This system has a large capacity for generating sparse and selective representations, and compression and normalization together improve this capacity (Luo et al., 2010; Olsen et al., 2010).

In addition, sparse recoding is also facilitated by a *compressive nonlinearity* in feedforward excitation. The intuition here is similar. A compressive nonlinearity tends to make a responsive neuron fire at a fixed rate (i.e., its saturated response level). This operation produces total population firing rates that are relatively equal for different stimuli, as compared with a scenario where there is no compressive nonlinearity. Again drawing an example from the *Drosophila* olfactory system, it is notable that there is a compressive nonlinearity in the relationship between the odor responses of olfactory receptor neurons and postsynaptic second-order neurons. Implementing this nonlinearity in data-driven simulations makes it easier to construct linear classifiers that respond sparsely and selectively to single odors (figure 24.2; Luo et al., 2010; Olsen et al., 2010). It is perhaps nonintuitive that equalizing firing rates (via compressive nonlinearity and normalization) can improve separability. The key is that these operations occur in a high-dimensional coding space.

Conclusion

To be a film director, in the words of Orson Welles, is simply to "preside over accidents." This could stand as a description of how a nervous system evolves and develops—accident by accident. Neural systems are subject to constraints, and this influences which random variations are passed on to successive generations. As we have seen, the history of which accidents survive becomes another constraint on neural system architecture, in the form of the organism's evolutionary inheritance and developmental programs.

The central argument of this chapter is that internal constraints leave their imprint on the architecture of neural systems. Because these constraints are ubiquitous, they can potentially explain why some architectures are so common. Understanding how this might occur will require a comparative approach that embraces a variety of sensory modalities and organisms.

Thinking about constraints is more important than ever before, because the field of experimental neuroscience is undergoing a revolution in techniques. New techniques allow us to precisely perturb neural activity, and to test how this affects perception and behavior. In practice, one often begins with a specific element of a neural system (e.g., a cell type) and one searches for behaviors that fall apart when this element is perturbed. This search assumes we will understand the function of each element by identifying the behaviors that rely on it. However, the nature of the behavioral task is only one pressure that drives the evolution of neural systems: equally important is the pressure to cope with internal constraints. Understanding these constraints may inspire more sophisticated experiments and interpretations.

ACKNOWLEDGMENTS Larry Abbott, James Jeanne, Andreas Liu, and Wendy Liu provided helpful comments on drafts of the manuscript. The author's research is supported by the National Institutes of Health (R01 DC008174) and the Howard Hughes Medical Institute.

REFERENCES

Allen, M. J., Godenschwege, T. A., Tanouye, M. A., & Phelan, P. (2006). Making an escape: Development and function of the *Drosophila* giant fibre system. *Semin Cell Dev Biol, 17*(1), 31–41.

Alonso, J. M., Usrey, W. M., & Reid, R. C. (2001). Rules of connectivity between geniculate cells and simple cells in cat primary visual cortex. *J Neurosci, 21*(11), 4002–4015.

Ames, A., III. (2000). CNS energy metabolism as related to function. *Brain Res Rev, 34*(1–2), 42–68.

Attwell, D., & Laughlin, S. B. (2001). An energy budget for signaling in the grey matter of the brain. *J Cereb Blood Flow Metab, 21*(10), 1133–1145.

Baddeley, R., Abbott, L. F., Booth, M. C., Sengpiel, F., Freeman, T., Wakeman, E. A., & Rolls, E. T. (1997). Responses of neurons in primary and inferior temporal visual cortices to natural scenes. *Proc Biol Sci, 264*(1389), 1775–1783.

Barlow, H. (1961). Possible principles underlying the transformation of sensory messages. In W. A. Rosenblith (Ed.), *Sensory communication* (pp. 217–234). Cambridge, MA: MIT Press.

Barlow, H. (2001). Redundancy reduction revisited. *Network, 12*(3), 241–253.

Bekkers, J. M., & Clements, J. D. (1999). Quantal amplitude and quantal variance of strontium-induced asynchronous EPSCs in rat dentate granule neurons. *J Physiol, 516*, 227–248.

Bhandawat, V., Olsen, S. R., Schlief, M. L., Gouwens, N. W., & Wilson, R. I. (2007). Sensory processing in the *Drosophila* antennal lobe increases the reliability and separability of ensemble odor representations. *Nat Neurosci, 10*(11), 1474–1482.

Bronchti, G., Heil, P., Sadka, R., Hess, A., Scheich, H., & Wollberg, Z. (2002). Auditory activation of "visual" cortical areas in the blind mole rat (*Spalax ehrenbergi*). *Eur J Neurosci, 16*(2), 311–329.

Carandini, M., & Heeger, D. J. (2012). Normalization as a canonical neural computation. *Nat Rev Neurosci, 13*(1), 51–62.

Cheong, R., Rhee, A., Wang, C. J., Nemenman, I., & Levchenko, A. (2011). Information transduction capacity of noisy biochemical signaling networks. *Science, 334*(6054), 354–358.

Chklovskii, D. B., & Koulakov, A. A. (2004). Maps in the brain: What can we learn from them? *Annu Rev Neurosci, 27*, 369–392.

Cooper, H. M., Herbin, M., & Nevo, E. (1993). Visual system of a naturally microphthalmic mammal: The blind mole rat, *Spalax ehrenbergi*. *J Comp Neurol, 328*(3), 313–350.

Cover, T. M. (1965). Geometrical and statistical properties of systems of linear inequalities with applications in pattern recognition. *IEEE Trans Comput, 14*(3), 326–334.

Del Castillo, J., & Katz, B. (1954). Quantal components of the end-plate potential. *J Physiol, 124*(3), 560–573.

DiCarlo, J. J., & Cox, D. D. (2007). Untangling invariant object recognition. *Trends Cogn Sci, 11*(8), 333–341.

Dunn, F. A., & Rieke, F. (2008). Single-photon absorptions evoke synaptic depression in the retina to extend the operational range of rod vision. *Neuron, 57*(6), 894–904.

Fairhall, A. L., Lewen, G. D., Bialek, W., & de Ruyter Van Steveninck, R. R. (2001). Efficiency and ambiguity in an adaptive neural code. *Nature, 412*(6849), 787–792.

Faisal, A. A., Selen, L. P., & Wolpert, D. M. (2008). Noise in the nervous system. *Nat Rev Neurosci, 9*(4), 292–303.

Faulkes, Z. (2008). Turning loss into opportunity: The key deletion of an escape circuit in decapod crustaceans. *Brain Behav Evolut, 72*, 251–261.

Feldheim, D. A., Kim, Y. I., Bergemann, A. D., Frisen, J., Barbacid, M., & Flanagan, J. G. (2000). Genetic analysis of ephrin-A2 and ephrin-A5 shows their requirement in multiple aspects of retinocollicular mapping. *Neuron, 25*(3), 563–574.

Field, G. D., & Rieke, F. (2002). Nonlinear signal transfer from mouse rods to bipolar cells and implications for visual sensitivity. *Neuron, 34*(5), 773–785.

Finlay, B. L., & Darlington, R. B. (1995). Linked regularities in the development and evolution of mammalian brains. *Science, 268*(5217), 1578–1584.

Finlay, B. L., Darlington, R. B., & Nicastro, N. (2001). Developmental structure in brain evolution. *Behav Brain Sci, 24*(2), 263–308.

Heil, P., Bronchti, G., Wollberg, Z., & Scheich, H. (1991). Invasion of visual cortex by the auditory system in the naturally blind mole rat. *NeuroReport, 2*(12), 735–738.

Imai, T., Suzuki, M., & Sakano, H. (2006). Odorant receptor-derived cAMP signals direct axonal targeting. *Science, 314*(5799), 657–661.

Itskov, V., & Abbott, L. F. (2008). Pattern capacity of a perceptron for sparse discrimination. *Phys Rev Lett, 101*(1), 018101.

Jacobs, G. A., Miller, J. P., & Aldworth, Z. (2008). Computational mechanisms of mechanosensory processing in the cricket. *J Exp Biol, 211*(Pt 11), 1819–1828.

Kazama, H., & Wilson, R. I. (2008). Homeostatic matching and nonlinear amplification at genetically identified central synapses. *Neuron, 58*, 401–413.

Kazama, H., & Wilson, R. I. (2009). Origins of correlated activity in an olfactory circuit. *Nat Neurosci, 12*(9), 1136–1144.

Krubitzer, L., & Kahn, D. M. (2003). Nature versus nurture revisited: An old idea with a new twist. *Prog Neurobiol, 70*(1), 33–52.

Laughlin, S. B. (1981). A simple coding procedure enhances a neuron's information capacity. *Z Naturforsch, 36c*(9–10), 910–912.

Laughlin, S. B. (1994). Matching coding, circuits, cells, and molecules to signals: General principles of retinal design in the fly's eye. *Prog Retin Eye Res, 13*, 165–196.

Laughlin, S. B. (2001). Energy as a constraint on the coding and processing of sensory information. *Curr Opin Neurobiol, 11*(4), 475–480.

Lin, D. M., Wang, F., Lowe, G., Gold, G. H., Axel, R., Ngai, J., & Brunet, L. (2000). Formation of precise connections in the olfactory bulb occurs in the absence of odorant-evoked neuronal activity. *Neuron, 26*(1), 69–80.

London, M., & Hausser, M. (2005). Dendritic computation. *Annu Rev Neurosci, 28,* 503–532.

London, M., Schreibman, A., Hausser, M., Larkum, M. E., & Segev, I. (2002). The information efficacy of a synapse. *Nat Neurosci, 5*(4), 332–340.

Luo, S. X., Axel, R., & Abbott, L. F. (2010). Generating sparse and selective third-order responses in the olfactory system of the fly. *Proc Natl Acad Sci USA, 107*(23), 10713–10718.

Ma, Q. (2010). Labeled lines meet and talk: Population coding of somatic sensations. *J Clin Invest, 120*(11), 3773–3778.

Marder, E., & Goaillard, J. M. (2006). Variability, compensation and homeostasis in neuron and network function. *Nat Rev Neurosci, 7*(7), 563–574.

Marin, E. C., Jefferis, G. S., Komiyama, T., Zhu, H., & Luo, L. (2002). Representation of the glomerular olfactory map in the *Drosophila* brain. *Cell, 109*(2), 243–255.

McAdams, H. H., & Arkin, A. (1997). Stochastic mechanisms in gene expression. *Proc Natl Acad Sci USA, 94*(3), 814–819.

Murthy, V. N. (2011). Olfactory maps in the brain. *Annu Rev Neurosci, 34,* 233–258.

Niven, J. E., Anderson, J. C., & Laughlin, S. B. (2007). Fly photoreceptors demonstrate energy-information trade-offs in neural coding. *PLoS Biol, 5*(4), e116.

Niven, J. E., & Laughlin, S. B. (2008). Energy limitation as a selective pressure on the evolution of sensory systems. *J Exp Biol, 211*(Pt 11), 1792–1804.

Olsen, S. R., Bhandawat, V., & Wilson, R. I. (2010). Divisive normalization in olfactory population codes. *Neuron, 66,* 287–299.

Olshausen, B. A., & Field, D. J. (2005). How close are we to understanding V1? *Neural Comput, 17*(8), 1665–1699.

Perge, J. A., Koch, K., Miller, R., Sterling, P., & Balasubramanian, V. (2009). How the optic nerve allocates space, energy capacity, and information. *J Neurosci, 29*(24), 7917–7928.

Pitkow, X., & Meister, M. (2012). Decorrelation and efficient coding by retinal ganglion cells. *Nat Neurosci, 15*(4), 628–635.

Rieke, F. (2014). Are neural circuits the best they can be? In M. S. Gazzaniga (Ed.), *The cognitive neurosciences* (5th ed.). Cambridge, MA: MIT Press.

Rolfe, D. F., & Brown, G. C. (1997). Cellular energy utilization and molecular origin of standard metabolic rate in mammals. *Physiol Rev, 77*(3), 731–758.

Rosenblatt, F. (1958). The perceptron: A probabilistic model for information storage and organization in the brain. *Psychol Rev, 65*(6), 386–408.

Sabatini, B. L., & Regehr, W. G. (1999). Timing of synaptic transmission. *Annu Rev Physiol, 61,* 521–542.

Sargent, P. B., Saviane, C., Nielsen, T. A., DiGregorio, D. A., & Silver, R. A. (2005). Rapid vesicular release, quantal variability, and spillover contribute to the precision and reliability of transmission at a glomerular synapse. *J Neurosci, 25*(36), 8173–8187.

Shapley, R., & Enroth-Cugell, C. (1984). Visual adaptation and retinal gain controls. *Prog Retin Res, 3,* 263–346.

Silver, R. A. (2010). Neuronal arithmetic. *Nat Rev Neurosci, 11*(7), 474–489.

Snyder, A. W., Bossomaier, T. R., & Hughes, A. (1986). Optical image quality and the cone mosaic. *Science, 231*(4737), 499–501.

Soucy, E. R., Albeanu, D. F., Fantana, A. L., Murthy, V. N., & Meister, M. (2009). Precision and diversity in an odor map on the olfactory bulb. *Nat Neurosci, 12*(2), 210–220.

Stafford, R. (2010). Constraints of biological neural networks and their consideration in AI applications. *Lect Notes Artif Int, 2010,* 1–6.

Swadlow, H. A. (2000). Information flow along neocortical axons. In R. Miller (Ed.), *Time and the brain* (pp. 150–180). Amsterdam: Harwood Academic.

Wang, S. S., Shultz, J. R., Burish, M. J., Harrison, K. H., Hof, P. R., Towns, L. C., & Wyatt, K. D. (2008). Functional trade-offs in white matter axonal scaling. *J Neurosci, 28*(15), 4047–4056.

Wark, B., Lundstrom, B. N., & Fairhall, A. (2007). Sensory adaptation. *Curr Opin Neurobiol, 17*(4), 423–429.

White, J. A., Rubinstein, J. T., & Kay, A. R. (2000). Channel noise in neurons. *Trends Neurosci, 23*(3), 131–137.

Wong, A. M., Wang, J. W., & Axel, R. (2002). Spatial representation of the glomerular map in the *Drosophila protocerebrum. Cell, 109*(2), 229–241.

Yorozu, S., Wong, A., Fischer, B. J., Dankert, H., Kernan, M. J., Kamikouchi, A., & Anderson, D. J. (2009). Distinct sensory representations of wind and near-field sound in the *Drosophila* brain. *Nature, 458*(7235), 201–205.

Zheng, C., Feinstein, P., Bozza, T., Rodriguez, I., & Mombaerts, P. (2000). Peripheral olfactory projections are differentially affected in mice deficient in a cyclic nucleotide-gated channel subunit. *Neuron, 26*(1), 81–91.

25 Are Neural Circuits the Best They Can Be?

FRED RIEKE

ABSTRACT Neural computation is impressive, yet we have made limited progress in understanding the underlying algorithms. This is in large part because of the complexity of neural circuits and the difficulty in identifying their relevant computational features. One approach is to construct predictive models for relevant computational features based on the idea that computation is not just good but optimal according to some well-defined metric. Such models can provide a conceptual framework for neural computation and can guide experiment. This chapter highlights some notable successes and some of the challenges of such approaches.

Neural circuits have the potential for enormous complexity: they are dynamic across a wide range of time scales, they exhibit strong nonlinearities, and they are highly and elaborately connected. Making sense of this complexity is a central challenge in computational neuroscience.

One approach to systematizing the complexity of neural circuits is to seek conceptual frameworks that make clear and testable predictions about the relevant computational features. "Optimality"—the idea that neural circuits perform as well as they can given a set of operating conditions and hardware constraints—is one such framework. Arguments based on optimality begin by proposing a metric by which system performance is evaluated—for example, maximizing sensitivity to a particular stimulus feature or minimizing energy consumption. Optimizing this metric, often in conjunction with hardware constraints such as finite wiring length, leads to predictions about how neural circuits implement specific computations.

Predictions based on optimization could fail either because the solutions neural circuits employ are far from optimal (e.g., circuit behavior could depend strongly on the particular evolutionary trajectory a system has taken), or because capturing all the constraints that contribute to circuit design with a tractable performance metric is effectively impossible. Thus, pragmatically, a key issue is whether simple performance metrics allow for successful predictions. In some cases, admittedly specialized ones, the answer seems to be yes.

Physical limits

Physical laws set performance limits that no system—biological or engineered—can exceed (Bialek, 1987). Two basic limits are posed by statistical variations in the input signals themselves and irreducible noise that results from the operation of any physical system at finite temperature.

The division of physical stimuli into discrete units—for example, single photons of light or single molecules of a pheromone, odor, or tastant—means that the actual stimulus reaching a detector will fluctuate even if the stimulus is nominally constant. For example, the number of photons absorbed by a light detector varies over time, even for a source of nominally fixed intensity. To be more specific, consider a stimulus that results on average in n absorbed photons in some time period Δt. For conventional light sources, the actual number of photons absorbed in any given time window Δt will vary according to Poisson statistics (figure 25.1). These fluctuations become smaller relative to the mean number of absorbed photons as light levels increase, but they never disappear. Such extrinsic noise sets a fundamental limit to the sensitivity of any system—biological or man-made—designed to detect the stimulus. Put differently, statistical variations in the stimulus set a design goal for the noise intrinsic to the detector, since intrinsic detector noise rapidly becomes unimportant when it is smaller than the extrinsic noise in the inputs.

A second unavoidable source of noise is created by the fact that biological systems operate at finite temperature and hence are subject to thermal fluctuations from the environment. Thus all molecules undergo spontaneous, thermally driven changes in conformation. The rate of these spontaneous conformational changes depends on the energy barrier E that the molecule has to cross to become active, as captured by the Boltzmann factor $e^{-E/kT}$, where k is Boltzmann's constant and T is temperature. These conformational changes can introduce noise into neural signals if they cause a molecule to transition between two different signaling states (e.g., close-open transitions of an ion channel or

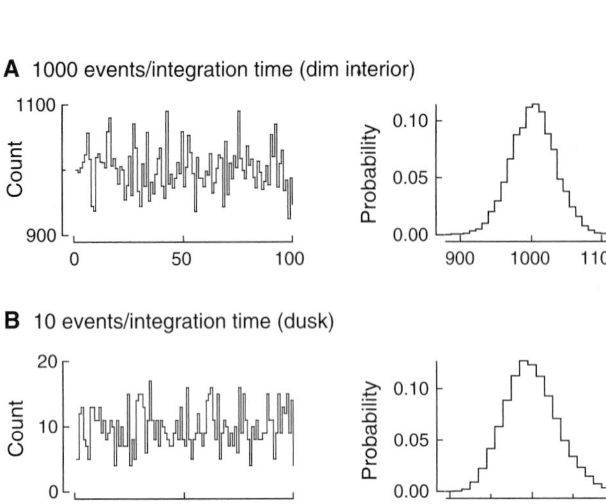

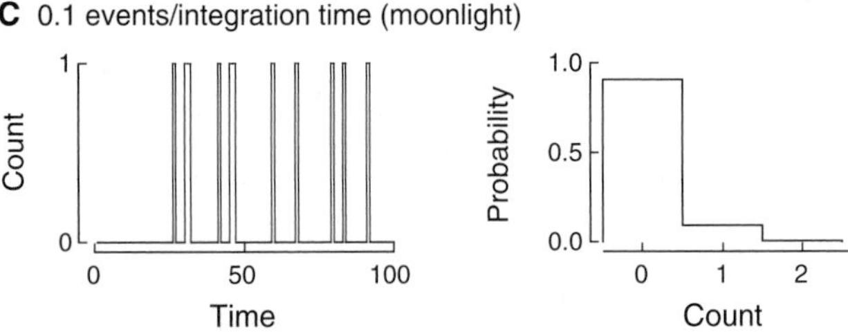

FIGURE 25.1 Poisson fluctuations in photon arrival. Left panels show simulated photon arrivals at a single detector (e.g., a photoreceptor) as counts per integration time at three mean light levels, corresponding roughly to dim room light (A), dusk (B), and moonlight (C). Right panels show histograms of photon counts in a single integration time.

spontaneous activation of an enzyme). Below we focus on one such source of noise created by spontaneous activation of the same photopigment molecules that absorb photons and initiate the visual process; spontaneous activation of these molecules generates photon-like noise events that provide an important source of intrinsic noise in the responses of rod photoreceptors.

Hardware constraints and the view of the ideal observer

Optimization approaches are closely related to ideal observer analysis (Geisler, 2003, 2011; Green & Swets, 1966), which determines the optimal performance obtainable from observing signals at one level of a system. Optimal performance should not be confused with perfect performance. Instead, noise in sensory inputs or in neural responses (e.g., those described in the previous section) will set a limit to performance—the limit identified by ideal observer analysis. Ideal observer analysis and optimization approaches share similar challenges—in particular, identifying an appropriate metric to evaluate performance and incorporating key hardware constraints. In the best case, both provide parameter-free predictions about how well a neural system can perform a given task and how it could achieve this limiting performance.

Ideal observer or optimization approaches provide key tools to compare performance across different levels of a system. Thus the limiting performance set by signals at one level of a system can be compared with the actual performance realized at a later stage. These two levels could be, for example, light inputs to the eye and behavior (reviewed by Donner, 1992; Field, Sampath, & Rieke, 2005; Geisler, 2003, 2011), or photoreceptor signals and retinal output (reviewed by Field et al., 2005), or pre- and post-synaptic signals. Ideal observer analysis provides a natural progression in our understanding of a system. Thus, initial applications often focus on overall performance (e.g., a comparison of detection sensitivity in vision with that permitted by the statistics of the light inputs). A more restrictive focus on specific components of a system can then help pinpoint key bottlenecks to performance—for example, separating preneural limitations in vision (stimulus statistics and optics) from neural factors. Construction of the ideal observer also necessarily identifies optimal algorithms for decoding neural signals. Hardware constraints—for example, wiring length, receptor noise, or optical aberrations—can also be built into such calculations.

Vision in starlight and photon detection

Vision at low light levels exemplifies how a focus on optimal design can guide investigation of the relevant neural circuits. Specifically, the approach of behavioral performance to fundamental limits has led to a set of precise questions and constraints about how the underlying computations work. Many of the issues have been reviewed in detail previously (Donner, 1992; Field et al., 2005).

PROBING BEHAVIORAL SENSITIVITY It has been known for more than 100 years that dark-adapted human observers can detect a small number of absorbed photons. The experiments described below support a picture in which visual sensitivity is limited as much by the division of light into discrete photons and the unavoidable statistical fluctuations in photon absorption as by biological noise or inefficiency. These observations about behavioral fidelity motivate making and testing predictions about the operation of key components of the neural circuits supporting night vision.

Absolute visual sensitivity has been measured in frequency-of-seeing experiments—measurements of the probability that a dim flash is seen as a function of flash strength (Hecht, Shlaer, & Pirenne, 1942; van der Velden, 1946). Initial experiments of this type were analyzed assuming that all flashes producing more than a threshold number of absorbed photons were seen, and that the number of absorbed photons was described by Poisson statistics. In this case, the probability of seeing is simply the probability that a flash of a given strength produces more than the threshold number of absorbed photons. This analysis does not incorporate any biological noise—and thus is equivalent to an ideal observer of the flux of photons at the rod photoreceptors. Fits to experimental frequency-of-seeing curves are unique or near-unique given these (strong) assumptions. The conclusion from this work was that dark-adapted human observers can detect <10 absorbed photons spread over an area of the retina containing ~500 rods.

Interpretation of early frequency-of-seeing experiments is hampered by the inability to incorporate false-positive responses—that is, trials in which the observer reported seeing a flash although no light was delivered. Observers in early experiments were trained to rarely generate false-positive responses, but this stringent requirement on detection likely elevates perceptual threshold (Barlow, 1956; Sakitt, 1972). The development of signal detection theory in the 1950s and 1960s provided a clearer view of the relationship between noise, false-positive responses and an observer's criterion for responding (Green & Swets, 1966). Signal-detection theory quantifies performance based on the probability of both detected events and false-positive responses. This allows, for example, an observer to vary his or her criterion for detection, revealing the relationship between perceptual threshold and false-positive rate (figure 25.2). Subsequent absolute-threshold experiments exploited the connection between false positives and perceptual threshold to estimate the internal noise-limiting detection sensitivity (Sakitt, 1972; Teich, Prucnal, Vannucci, Breton, & McGill, 1982). These estimates assumed that the internal noise could be equated with a rate of photon-like events, and that those events followed Poisson statistics.

A challenge in this extended analysis is that fits to the frequency-of-seeing curves are no longer unique when internal noise and Poisson fluctuations together limit sensitivity. Specifically, different combinations of key parameters—the probability of photon absorption, internal noise, and response threshold—account for the empirical data equally well (Rieke, 2008). This degeneracy in fitting can be at least partially resolved

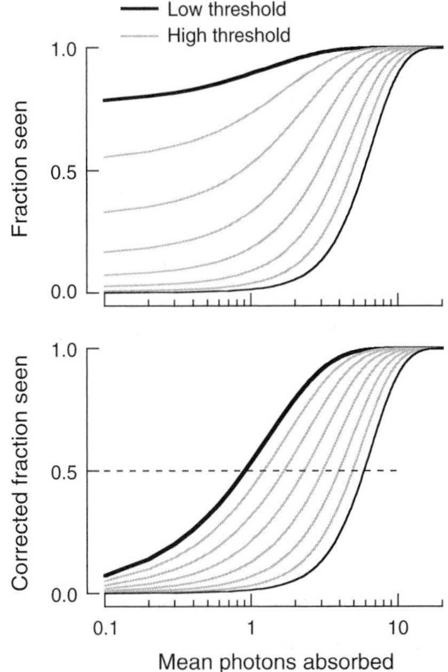

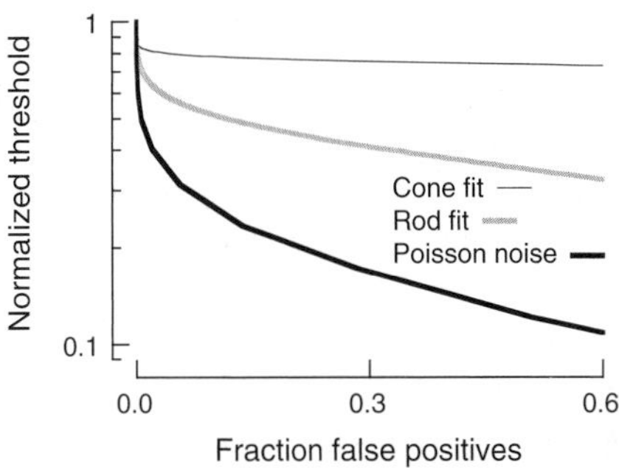

FIGURE 25.2 Trade-off between sensitivity and false-positive responses (adapted from Koenig & Hofer, 2011). (A) *Top*: simulated frequency-of-seeing curves (probability that a flash is seen as a function of the number of photons absorbed). Each curve corresponds to a different criterion or threshold for detecting the flash. *Bottom*: frequency-of-seeing curves corrected for false-positive responses. The false-positive rate (probability of seeing when no photons were absorbed) was subtracted from each curve, and the curve rescaled to a maximum of 1. (B) Dependence of sensitivity (flash strength for 0.5 corrected probability of seeing as indicated by the dashed line in bottom panel of (A) on the false-positive rate. Rod and cone curves are fits to behavioral data from Koenig and Hofer (2011), generously provided by Darren Koenig and Heidi Hofer.

by varying the observer's criterion (e.g., using different ratings that imply different degrees of certainty; Sakitt, 1972; Teich et al., 1982). Results from these experiments have been interpreted in the context of two nonexclusive models (Donner, 1992). In the first, signals from every photon-like event—whether due to noise or true photon absorption—are available to cortical circuits involved in the decision of whether the flash was seen. In the second, a thresholding nonlinearity in the visual circuitry limits information about individual photon-like events. The first model makes a clear prediction about how detection threshold and the rate of false-positive responses interact (figure 25.2B, Poisson noise curve). The actual decrease in perceptual threshold with increased false-positive rate falls short of this prediction, indicating that the "every photon" model is unlikely to be strictly correct (Koenig & Hofer, 2011). This is consistent with a thresholding nonlinearity that eliminates some responses to absorbed photons.

The ability to express the noise that limits behavior as a rate of equivalent photon-like events is convenient, but does not specify where the noise originates. Barlow suggested that internal noise in fact resulted from the thermal activation of rhodopsin molecules in the rods (Barlow, 1956). Subsequent recordings from single-rod photoreceptors revealed discrete noise events associated with spontaneous rhodopsin activation (Baylor, Matthews, & Yau, 1980; Baylor, Nunn, & Schnapf, 1984). These noise events are indistinguishable from events produced by photon absorption. They occur about once every 100–200 sec in a mammalian rod, a rate quite close to that required to account for the noise-limiting behavioral sensitivity, although the connection between the two in humans should be regarded as suggestive rather than definitive due to uncertainties in both measurements (Field et al., 2005). The strongest evidence linking these noise events to the noise-limiting behavior comes from prey-capture experiments in toads and frogs. Since toads and frogs are cold-blooded, the temperature dependence of behavioral threshold can be measured by varying the animals' temperature (Aho, Donner, Hyden, Larsen, & Reuter, 1988); behavioral sensitivity indeed improves at lower temperature. Further, the temperature dependence of behavioral threshold in these experiments agrees closely with the temperature dependence of the rate of spontaneous rhodopsin activation.

Rod Photoreceptors Act as Near-Ideal Photon Detectors The behavioral experiments summarized above indicate that rod vision reaches or approaches limits set by the division of light into discrete photons and the occasional spontaneous activation of rhodopsin. This performance motivates us to take seriously the idea that the neural circuitry responsible for rod vision is near-optimal, and to use this performance to pose specific questions about how this circuitry works. This starts with the rods themselves.

We start with a seemingly innocuous question about photon absorption: what fraction of the incident photons should a rod capture to optimize sensitivity? A reflexive answer is that maximal sensitivity is achieved by capturing as many incident photons as possible, but as we will see that answer is wrong (see Bialek, 2012, for more details). The argument is based on two properties of rhodopsin: self-screening and spontaneous activation.

To get some intuition for self-screening, approximate the outer segment of a rod photoreceptor as a cylinder filled with a uniform concentration of rhodopsin molecules (figure 25.3). As photons travel along the long axis of the cylinder, some of them are absorbed. This means that the density of photons at the location of a given rhodopsin molecule depends on its position within the cylinder; specifically, the photon density declines exponentially from the base to the tip of the cylinder. The probability of photon absorption by a rhodopsin molecule at location x depends on the photon density at the molecule's position. As a

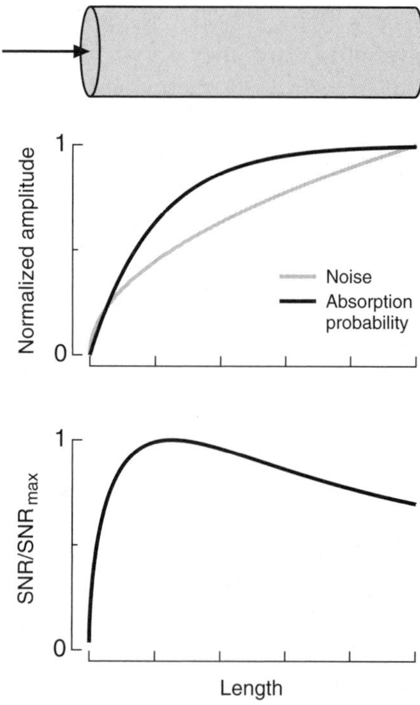

Figure 25.3 Optimizing rod outer-segment length. As photons traverse the outer segment (top) they are absorbed, causing the effective photon flux to decrease. The middle panel shows the fraction of photons absorbed as a function of outer-segment length. Also plotted is the total outer-segment noise, which increases as the square root of the outer-segment length. The bottom panel plots the signal-to-noise ratio (SNR) as a function of length. The different dependencies of photon absorption and noise on length cause the SNR to have a clear peak.

consequence, rhodopsin molecules located near the base of the outer segment see a higher density of incident photons and hence contribute more to visual signaling than those located near the tip of the outer segment. A second important property of rhodopsin is its spontaneous activation. The total rate of spontaneous activation depends on the number of rhodopsin molecules contained within the outer segment. Hence, a longer outer segment that would capture a greater fraction of the incident photons would also generate more noise.

These considerations make it clear that there is an optimal length: as the length goes to zero, the rod does not absorb any photons, while as it gets infinitely long it absorbs all incident photons but suffers from extremely high noise. The highest signal-to-noise ratio is achieved somewhere in between; formulating the problem more precisely shows that this optimum corresponds to absorbing ~70% of the incident photons (Bialek, 2012; figure 25.3). Given the concentration of rhodopsin in the outer segment and its absorption cross-section, 70% absorption is achieved for a rod ~60 microns in length. Toad and frog rods are quite close to this optimal length, while rods in primate are ~35 microns in length, and absorb ~50% of the incident photons. The imperfect agreement of the predicted and actual lengths is to be expected given the many factors this simple argument ignores.

In addition to capturing photons, our ability to see at night requires that rod photoreceptors respond reliably when they absorb a single photon. We now have a solid mechanistic understanding of how this is achieved (reviewed by Yau & Hardie, 2009; figure 25.4A), and can instantiate this understanding in the form of mathematical models that capture many aspects of rod behavior (reviewed by Rieke & Baylor, 1998). These models nicely summarize the biochemical reactions that form the rod phototransduction cascade. The behavior of the transduction cascade is dictated by a collection of key rate constants that reflect the concentrations and activities of the underlying components. We do not understand well, however, how rod behavior would change as these parameters are varied, and in particular where real rods reside within the range of possible behaviors exhibited by the general transduction scheme.

Consider a simple example in lieu of a more complete study. In addition to discrete noise due to spontaneous rhodopsin activation, rod signals are contaminated by continuous current fluctuations produced by spontaneous activation of the phosphodiesterase and resulting fluctuations in channel activity (Baylor et al., 1980; Rieke & Baylor, 1996). Spontaneous phosphodiesterase activation also produces ongoing activity in the transduction cascade that abbreviates the single-photon response by allowing the cascade to recover quickly following a perturbation such as activation of rhodopsin. Thus maintaining low noise and producing a brief response impose opposite constraints on the

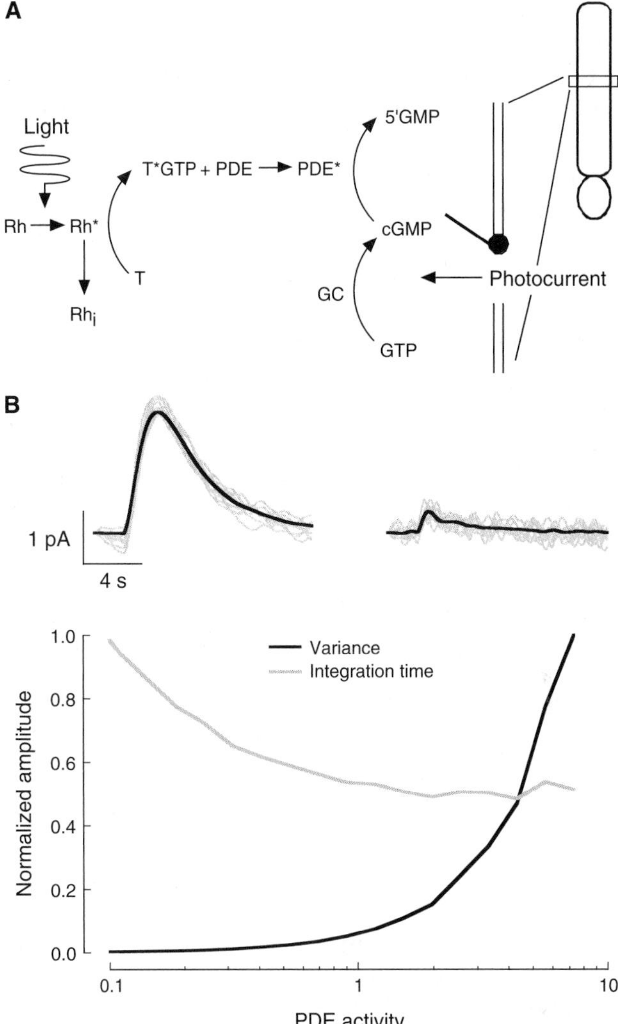

FIGURE 25.4 Trade-off of response duration and noise. (A) Schematic of phototransduction cascade that converts light into an electrical response in the rod outer segment (top right). The membrane current is controlled by cyclic GMP (cGMP) channels. Light-activated rhodopsin (Rh) activates the G-protein transducin (T), which activates the phosphodiesterase (PDE). Activated PDE hydrolyzes cGMP, allowing channels to close. Cyclic GMP is synthesized by guanylate cyclase (GC). (B) Dependence of response kinetics and noise on PDE activity. The top left panel shows simulated single-photon responses for a low level of spontaneous PDE activation. The black trace is the average response, and the gray traces are 10 superimposed individual responses. The top right shows responses for a high level of spontaneous PDE activation. The bottom panel shows how the integration time and noise depend on PDE activity.

phototransduction process, specifically on the rate of spontaneous phosphodiesterase activation.

Figure 25.4 explores these issues using a model for the phototransduction cascade based on measurements in toad rods. Figure 25.4B plots the noise variance and response integration time as the rate of spontaneous phosphodiesterase activation is increased. High phosphodiesterase activity produces a brief response, but also produces so much noise that individual single-photon responses are not identifiable (top right). Low phosphodiesterase activity produces little noise, but responses last tens of seconds (top left). The measured rate of spontaneous phosphodiesterase activation places the actual rod behavior at a location in the middle. Evaluating whether or not this is a good compromise, however, requires some means of weighing the relative costs of response duration and noise, and it is not clear how to do so. This example highlights a central challenge facing ideal observer calculations: the task facing the system is often multifaceted, and weighing different aspects of system function can be difficult or impossible.

Retinal Readout Rod-mediated signals traverse the mammalian retina through the specialized rod bipolar circuit (reviewed by Bloomfield & Dacheux, 2001; Field et al., 2005). Behavioral sensitivity requires that this circuit operate effectively when <0.1% of the rods absorb a photon within the 100–200 ms integration time of rod signals. This sensitivity poses substantial constraints. First, rod photoreceptors must respond to single photons. Second, the resulting signals must be reliably transmitted across the retina and protected from noise. Third, absorption of one or a few photons spread across many rods must noticeably impact the spike responses sent from the retina to the brain. These constraints have guided investigation of the underlying mechanisms, and the resulting focus has provided multiple insights into circuit function.

At light levels near perceptual threshold, photons arrive rarely at individual rods, but all of the rods generate noise. This noise threatens to overwhelm the sparse single-photon responses when signals from multiple rods are combined. How should downstream retinal neurons combine signals across the rods from which they receive input? This is a situation in which averaging is a disaster. Averaging is a good strategy when both signal and noise are spread uniformly across receptors. But when the signal is sparse, averaging inextricably mixes signal from a few receptors with noise from receptors that do not carry any signal. Instead, sensitivity can be dramatically improved by a nonlinear mechanism that selectively retains signals from those rods likely generating single-photon responses while rejecting noise from the remaining rods (Baylor et al., 1984; van Rossum & Smith, 1998).

Similar issues of how to integrate sparse, noisy inputs recur in many other neural circuits. Photon detection in the retina offers several key advantages for investigating this general issue. First, the signal and noise properties of the rod's single-photon responses can be thoroughly characterized. This permits construction of parameter-free predictions of the optimal strategies for separating rod signal and noise. Second, separation of single-photon responses from noise is effective only prior to mixing of rod signals. This restricts the possible anatomical locations for such signal processing to sites at or prior to the synapse from rods to rod bipolar cells. Thus we can generate parameter-free predictions about the properties, location, and functional importance of a key signal-processing step supporting night vision.

The predictions about processing of rod responses have been tested by recording from rod bipolar cells (Berntson, Smith, & Taylor, 2004; Field & Rieke, 2002). Responses of mouse rod bipolar cells indeed differ substantially from those of the rods from which they receive input. In particular, rod bipolar responses depend supralinearly on flash strength, and are more discrete than responses of the rods. These properties are consistent with the presence of a threshold-like nonlinearity at the synapse between rods and rod bipolar cells. Further, the synaptic nonlinearity has properties closely aligned to those predicted from the rod signal and noise. Surprisingly, the synapse discards many of the rod's single-photon responses. This is, however, a good strategy when photon absorptions are exceedingly rare, as the change in current produced by a small or average single-photon response is more likely to have originated from noise from spontaneous phosphodiesterase activity (figure 25.4) than from a true photon absorption.

Thresholding of the rod signals occurs through a simple synaptic mechanism (Sampath & Rieke, 2004). Rods release glutamate continuously in the dark, and decrease this release during the single-photon response. Rod bipolar cells sense glutamate released by the rods via metabotropic receptors that control ion-channel activity through a second-messenger cascade. Glutamate release from the rods saturates this signaling cascade; as a consequence, small changes in rod voltage and glutamate release fail to modulate the rod bipolar voltage. Larger changes in glutamate release—such as those associated with a large single-photon response—relieve this saturation and produce a response. An interesting and unresolved issue is how the level of synaptic saturation is controlled.

Signal transfer from rods to rod bipolar cells also serves to substantially abbreviate the rod responses. This temporal filtering can be nicely predicted in amphibian retina, which lacks the synaptic nonlinearity described above. Rod responses themselves are strikingly long-lasting—single-photon responses of mammalian rods exceed 0.5 sec at body temperature (Baylor et al., 1984). The long duration of the single-photon response, however, does not fundamentally limit temporal precision. This is because linear filtering can be undone—for example, low-pass filtering can be compensated by matched high-pass filtering. This deconvolution process is a common strategy employed in both biological and engineered systems. For example, deconvolution is a key part of several microscopy approaches aimed at achieving resolution beyond the nominal diffraction limit.

While in principle linear filtering can be undone to exactly recover the input signals, in practice the addition of noise after the initial filtering step often makes complete recovery of the input signal impossible. In this situation, modified deconvolution approaches take into account the temporal properties of the signal and noise so as not to amplify the noise in an attempt to recover a weak signal. This general argument, appropriately formalized, leads to predictions about the dynamics of signal transfer from rods to bipolar cells, on the assumption that these dynamics serve to optimally estimate the times of photon absorption from the noisy rod signals (Bialek & Owen, 1990; Rieke, Owen, & Bialek, 1991). Such predictions agree well with the kinetics of signal transfer in amphibian retina.

Daylight vision

The sensitivity of cone vision is impressive: we can see spatial structure smaller than the spacing between cone photoreceptors (Klein & Levi, 1985; Westheimer, 1981), resolve 1–2% changes in contrast (Banks, Geisler, & Bennett, 1987), and detect ~2 nm changes in the wavelength of a monochromatic light (Mollon, Estevez, & Cavonius, 1990). How close are these measures of visual performance to fundamental limits? In particular, do these observations indicate that the underlying neural circuits make effective use of all the information available in the light inputs or in the cone photoreceptor signals? Can optimal design approaches help us understand the operation of the neural circuits responsible for cone-mediated vision, much as they have for rod-mediated vision? For the most part, these questions remain unanswered, although progress has been made in several areas.

Behavior and Preneural Factors The detection sensitivity of cone-mediated vision has been quantified using frequency-of-seeing approaches like those used for rod-mediated vision. Cone signals are typically isolated by centering stimuli on the rod-free fovea at the center of the visual field. Flashes producing ~5 absorbed photons per cone in a region containing ~10 cones are just detectable (Koenig & Hofer, 2011; Marriott, 1963); for comparison, just-detectable flashes for rod-mediated vision produce ~0.005 absorbed photons per rod in a region containing ~2,000 rods (Koenig & Hofer, 2011). Estimates of the effective cone dark noise are ~20–200 photon-like noise events per cone per second, compared to estimates of ~0.01 events per rod per second for rod vision. These measures of sensitivity and noise pose important constraints on the underlying neural circuits.

Further constraints on the operation of the neural circuits responsible for cone-mediated vision come from a relative inability to trade false positives for sensitivity (Koenig & Hofer, 2011; see figure 25.2). Sensitivity is, instead, quite constant across a broad range of false-positive rates. The lack of improvement in sensitivity with increased false-positive rate is inconsistent with linear processing limited by a combination of quantal fluctuations and spontaneous pigment activation (or any other early noise source that follows Poisson statistics). Two nonexclusive models have been proposed to account for this result. The first model posits a neural threshold that eliminates small cone-mediated responses together with a post-threshold noise source that contributes to false-positive responses. The second model posits that noise is pooled nonoptimally due to an observer's uncertainty about the timing and location of the stimulus. This model also requires a neural threshold. In both models, the neural threshold limits the increase in sensitivity associated with lower behavioral criteria for detecting a flash.

In addition to detection sensitivity, several other well-characterized aspects of cone vision provide clear comparisons with fundamental limits. A striking example is spatial sensitivity, which can substantially exceed the spacing between cone photoreceptors (Klein & Levi, 1985; Westheimer, 1981). Spatial acuity is often quantified using Vernier tasks such as that depicted in figure 25.5. The behavioral task in such an experiment is to determine whether the center line is closer to the left or the right flanking line. Human observers can detect displacements of the center line in such tasks that are as small as 3–5% of the spacing between foveal cones. Interpretation of this sensitivity requires considering the quality of the optical image at the cone photoreceptors, spatial sampling by the cones, and cone noise.

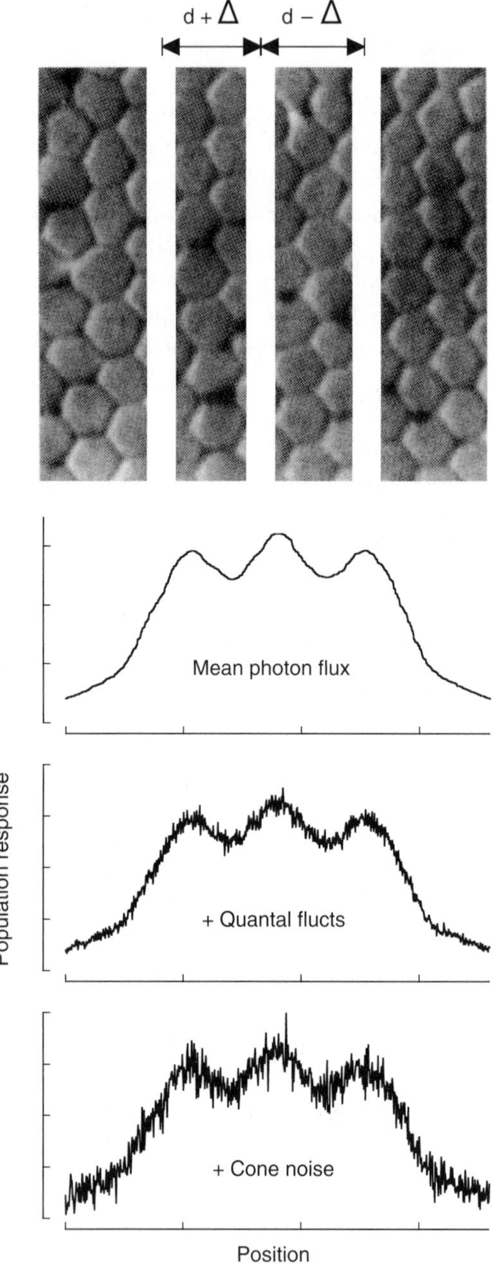

FIGURE 25.5 Vernier acuity task and cone signals. *Top*: Three-line stimulus used in many Vernier discrimination tasks superimposed on foveal cone mosaic (modified from Curcio et al., 1990). *Second from top*: Estimated distribution of mean photon catches in the cone array as a function of position relative to the lines. Photon fluxes were obtained by passing lines through an estimate of the point spread function of the optics (Westheimer, 1984) and then sampling with a simulated cone array. *Third from top*: Distribution of cone quantal catches on a single 1 sec trial of the three-line stimulus. *Bottom*: Distribution of cone responses, complete with noise, on a single 1 sec trial.

The optics of the eye blur the lines in the Vernier task of figure 25.5 so that they are spread over many cones (Westheimer, 1984). Thus the distribution of mean cone photon fluxes (without any noise) resembles that in figure 25.5 (second panel from top). The asymmetry in the distribution—particularly the different depths of the minima corresponding to the regions between the center and flanking lines—is indicative of the displacement of the central line; from this simple picture it is clear that, without noise, an ideal observer could detect exceedingly small displacements even given the finite cone sampling and optical blur. Figure 25.5 (second panel from bottom) depicts these distributions on a single 0.6 sec trial of such a task for a displacement of 1/20th of the cone spacing; noise due to quantal fluctuations is included. This illustrates the quality of signal available to an ideal observer of the photon flux at the cone photoreceptors.

Human behavioral performance in a Vernier task like that in figure 25.5 falls short of that possible given the photon flux at the cones. The difference between actual and ideal performance is a factor of 3–10 depending on details of the spatial configuration (Banks, Sekuler, & Anderson, 1991). A similar difference between performance supported by the photon flux at the cones and that measured behaviorally holds for several other discrimination tasks, including grating acuity and contrast sensitivity (reviewed by Geisler, 2003, 2011). Thus limits imposed by the preneural factors that determine the quality of the image reaching the cone photoreceptors (optics, cone sampling, and quantal fluctuations) do not fully account for the limitations to the sensitivity of cone vision.

PHYSIOLOGY: DOMINANCE OF CONE NOISE The failure of behavioral sensitivity to reach limits set by quantal fluctuations and optical properties highlights the importance of identifying the location and properties of the neural noise sources active at cone light levels. Several results indicate that much of the noise in the retinal output originates in the cones themselves.

Barlow proposed that the >10,000-fold difference in estimates of the dark noise-limiting rod and cone vision might be explained by differences in the wavelength of peak sensitivity of the rod and long-wavelength-sensitive (L) cone photoreceptors, and corresponding differences in the rate of spontaneous pigment activation (Barlow, 1972). The basic idea is that the rate of spontaneous activation is proportional to $e^{-E/kT}$, where E is the energy required for spontaneous activation. If E is equal to the energy of a photon at the wavelength of the peak photopigment absorption, a shift in peak sensitivity from 500 nm (rods) to 580 nm (L cones) would cause a ~10^5-fold increase in spontaneous activation rate. Indeed, rates of spontaneous activation of a broad range of photopigments can be quantitatively predicted from this basic idea (Luo, Yue, Ala-Laurila, & Yau, 2011). Measured noise in mammalian cones, however, shows much more rapid variations than expected from pigment activation, and thus appears to be dominated by other sources (Ala-Laurila, Greschner, Chichilnisky, & Rieke, 2011; Angueyra & Rieke, 2013; Schneeweis & Schnapf, 1999).

Evidence for the importance of cone noise in the retinal output comes from several types of experiments. The first evidence comes from comparing the sensitivity of an ideal observer of the photon flux at the cones with that of an ideal observer of the responses of horizontal cells (which receive input from cones) or of the ganglion cells (the retinal output cells) in guinea pig retina (Borghuis, Sterling, & Smith, 2009). Losses of sensitivity between two locations—for example, between the photon flux at the cones and the horizontal cells—indicate a source of noise between the two locations. This comparison indicated that noise in the output signals of the most sensitive ganglion cells contained roughly equal contributions from noise intrinsic to the cones and from post-cone noise introduced in the retinal circuitry.

A second indication of the importance of cone noise comes from correlated noise in the responses of nearby ganglion cells, particularly in mammalian retina. Correlated noise is present in the spike outputs and the excitatory synaptic inputs to pairs of ganglion cells, indicating an origin from noise in a common upstream cell in the retinal circuitry (reviewed by Mastronarde, 1989; Meister, 1996; Shlens, Rieke, & Chichilnisky, 2008). Correlated noise persists in the absence of modulated light inputs and often has a time scale substantially briefer than that associated with responses to light (Brivanlou, Warland, & Meister, 1998; DeVries, 1999; Mastronarde, 1983; Trong & Rieke, 2008). These properties indicate that it is not produced by quantal fluctuations in photon absorption, but instead originates after photopigment activation. The time course of correlated noise in primate retina is quite similar to that characterizing the total noise in the inputs to a single ganglion cell, suggesting that both correlated and uncorrelated noise originate from a common source (Trong & Rieke, 2008). Correlated noise occurs across cells of different types, and generally respects the polarity (On or Off) of the cells involved (Greschner et al., 2011). This suggests that correlated noise originates from noise introduced prior to the divergence of signals into the parallel circuits that control the responses of these different cells. The strength of correlated noise can be predicted by receptive field or dendritic overlap (Ala-Laurila et al., 2011; Greschner et al., 2011; Trong & Rieke, 2008). Together these observations support a simple model in which correlated noise originates from noise introduced by the cone photoreceptors, which then propagates across the retinal circuits that control the responses of diverse retinal output cells.

Additional evidence that cones account for much of the noise in the output signals of primate retina comes from suppressing cone inputs to On bipolar cells (Ala-Laurila et al., 2011). On bipolar cells signal via metabotropic glutamate receptors, and can be targeted quite selectively by pharmacological agents. A potential issue in such experiments is that manipulating receptor activity could lead to large changes in bipolar voltage and hence in transmitter release; such changes will confound interpretation of changes in ganglion cell signaling, since noise could be reduced simply because it can no longer be transmitted from bipolar cell to ganglion cell. An appropriate mixture of receptor agonists and antagonists, however, can suppress cone input to On bipolar cells with minimal changes in the rate of bipolar transmitter release. Suppressing cone inputs to the retina in this way reduced noise in the excitatory synaptic inputs to primate On parasol ganglion cells ~3-fold. Thus much of the noise in a ganglion cell's excitatory inputs appears to be inherited from the cones themselves.

Together, these observations support two conclusions: (1) the cones produce considerably more noise than expected from quantal fluctuations across a broad range of light levels, and (2) at least in primates, the retinal readout adds little noise to that already present in the cone signals. These conclusions suggest that much of the discrepancy between the performance of an ideal observer of the light flux at the cones and behavioral performance may be due to noise introduced by the cones themselves. In turn, this suggests that we compare the fidelity of the cone signals with that of behavior. As an example, figure 25.5 (bottom panel) shows the distribution of cone responses (including intrinsic cone noise) for the Vernier acuity task. Cone noise substantially degrades the signals available for discriminating small shifts in the line position; indeed, the limiting spatial resolution including cone noise is within a factor of 2–3 of that measured behaviorally for these stimuli.

Conclusion

Relating the mechanistic operation of neural circuits to function is challenging in large part because of the many circuit properties that can be measured and the vast array of stimulus conditions under which such measurements can be made. Further complicating matters, models that characterize responses for one set of stimulus conditions rarely generalize well to describe responses for different stimulus conditions. What, then, is a neuroscientist to do?

Predictions based on the idea that neural circuits approach ideal performance provide one way to navigate these complexities. This chapter has focused on illustrating how this approach can be applied to the operation of the retina in the context of simple detection tasks. Limits to performance in these tasks due to unavoidable stimulus variations or early sensory noise are particularly clear, and the computations required to realize such performance provide experimentally testable predictions about the relevant computational dynamics of retinal circuits.

There are many reasons such optimization approaches could fail. Calculations of the performance of an ideal observer necessarily focus on simple measures of system performance; in reality, a host of factors, many unknown, are certain to have imposed evolutionary pressures that have shaped circuit operation. Indeed, other aspects of the operation of the retina can be predicted from arguments based on minimizing energy expenditure or wiring length. Thus it is somewhat surprising that predictions based on optimizing a simple performance metric are ever successful. The fact that they can be indicates some order in the apparent complexity of neural computation.

ACKNOWLEDGMENTS I thank Juan Angueyra and Rachel Wilson for many helpful comments, Darren Koenig and Heidi Hofer for providing data for figure 25.2, and past and present members of the laboratory for helping develop and refine many of the ideas here.

REFERENCES

Aho, A. C., Donner, K., Hyden, C., Larsen, L. O., & Reuter, T. (1988). Low retinal noise in animals with low body temperature allows high visual sensitivity. *Nature, 334*(6180), 348–350.

Ala-Laurila, P., Greschner, M., Chichilnisky, E. J., & Rieke, F. (2011). Cone photoreceptor contributions to noise and correlations in the retinal output. *Nat Neurosci, 14*(10), 1309–1316.

Angueyra, J. M., & Rieke, F. (2013). Origin and effect of phototransduction noise in primate cone photoreceptors. *Nat Neurosci, 16*(11), 1692–1700.

Banks, M. S., Geisler, W. S., & Bennett, P. J. (1987). The physical limits of grating visibility. *Vis Res, 27*(11), 1915–1924.

Banks, M. S., Sekuler, A. B., & Anderson, S. J. (1991). Peripheral spatial vision: Limits imposed by optics, photoreceptors, and receptor pooling. *J Opt Soc Am A, 8*(11), 1775–1787.

Barlow, H. B. (1956). Retinal noise and absolute threshold. *J Opt Soc Am, 46*(8), 634–639.

Barlow, H. B. (1972). Dark and light adaptation: Psychophysics. *Handbook of sensory physiology, 8*(4), 1–24.

Baylor, D. A., Matthews, G., & Yau, K. W. (1980). Two components of electrical dark noise in toad retinal rod outer segments. *J Physiol, 309*, 591–621.

Baylor, D. A., Nunn, B. J., & Schnapf, J. L. (1984). The photocurrent, noise and spectral sensitivity of rods of the monkey *Macaca fascicularis*. *J Physiol, 357*, 575–607.

Berntson, A., Smith, R. G., & Taylor, W. R. (2004). Transmission of single photon signals through a binary synapse in the mammalian retina. *Vis Neurosci, 21*(5), 693–702.

Bialek, W. (1987). Physical limits to sensation and perception. *Annu Rev Biophys Biophys Chem, 16*, 455–478.

Bialek, W. S. (2012). *Biophysics: Searching for principles*. Princeton, NJ: Princeton University Press.

Bialek, W., & Owen, W. G. (1990). Temporal filtering in retinal bipolar cells. Elements of an optimal computation? *Biophys J, 58*(5), 1227–1233.

Bloomfield, S. A., & Dacheux, R. F. (2001). Rod vision: Pathways and processing in the mammalian retina. *Prog Retin Eye Res, 20*(3), 351–384.

Borghuis, B. G., Sterling, P., & Smith, R. G. (2009). Loss of sensitivity in an analog neural circuit. *J Neurosci, 29*(10), 3045–3058.

Brivanlou, I. H., Warland, D. K., & Meister, M. (1998). Mechanisms of concerted firing among retinal ganglion cells. *Neuron, 20*(3), 527–539.

Curcio, C. A., Sloan, K. R., Kalina, R. E., & Hendrickson, A. E. (1990). Human photoreceptor topography. *J Comp Neurol, 292*(4), 497–523.

DeVries, S. H. (1999). Correlated firing in rabbit retinal ganglion cells. *J Neurophysiol, 81*(2), 908–920.

Donner, K. (1992). Noise and the absolute thresholds of cone and rod vision. *Vis Res, 32*(5), 853–866.

Field, G. D., & Rieke, F. (2002). Nonlinear signal transfer from mouse rods to bipolar cells and implications for visual sensitivity. *Neuron, 34*(5), 773–785.

Field, G. D., Sampath, A. P., & Rieke, F. (2005). Retinal processing near absolute threshold: From behavior to mechanism. *Annu Rev Physiol, 67*, 491–514.

Geisler, W. S. (2003). Ideal observer analysis. In L. Chalupa & J. Werner (Eds.), *The visual neurosciences* (pp. 825–837). Cambridge, MA: MIT Press.

Geisler, W. S. (2011). Contributions of ideal observer theory to vision research. *Vis Res, 51*(7), 771–781.

Green, D. M., & Swets, J. A. (1966). *Signal detection theory and psychophysics*. New York, NY: Wiley.

Greschner, M., Shlens, J., Bakolitsa, C., Field, G. D., Gauthier, J. L., Jepson, L. H., ... Chichilnisky, E. J. (2011). Correlated firing among major ganglion cell types in primate retina. *J Physiol, 589*(Pt 1), 75–86.

Hecht, S., Shlaer, S., & Pirenne, M. H. (1942). Energy, quanta, and vision. *J Gen Physiol, 25*, 819–840.

Klein, S. A., & Levi, D. M. (1985). Hyperacuity thresholds of 1 sec: Theoretical predictions and empirical validation. *J Opt Soc Am A, 2*(7), 1170–1190.

Koenig, D., & Hofer, H. (2011). The absolute threshold of cone vision. *J Vis, 11*(1), 1–24.

Luo, D. G., Yue, W. W., Ala-Laurila, P., & Yau, K. W. (2011). Activation of visual pigments by light and heat. *Science, 332*(6035), 1307–1312.

Marriott, F. H. (1963). The foveal absolute visual threshold for short flashes and small fields. *J Physiol, 169,* 416–423.

Mastronarde, D. N. (1983). Correlated firing of cat retinal ganglion cells. I. Spontaneously active inputs to X- and Y-cells. *J Neurophysiol, 49*(2), 303–324.

Mastronarde, D. N. (1989). Correlated firing of retinal ganglion cells. *Trends Neurosci, 12*(2), 75–80.

Meister, M. (1996). Multineuronal codes in retinal signaling. *Proc Natl Acad Sci USA, 93*(2), 609–614.

Mollon, J. D., Estevez, O., & Cavonius, C. R. (1990). The two subsystems of colour vision and their roles in wavelength discrimination. In C. B. Blakemore (Ed.), *Vision: Coding and efficiency* (pp. 119–131). Cambridge, UK: Cambridge University Press.

Rieke, F. (2008). Seeing in the dark: Retinal processing and absolute visual threshold. In R. Masland & T. Albright (Eds.), *The senses: A comprehensive reference* (pp. 393–412). San Diego, CA: Academic Press.

Rieke, F., & Baylor, D. A. (1996). Molecular origin of continuous dark noise in rod photoreceptors. *Biophys J, 71*(5), 2553–2572.

Rieke, F., & Baylor, D. A. (1998). Single-photon detection by rod cells of the retina. *Rev Mod Phys, 70,* 1027–1036.

Rieke, F., Owen, W. G., & Bialek, W. (1991). Optimal filtering in the salamander retina. In R. P. Lippman, J. E. Moody, & D. S. Touretzky (Eds.), *Advances in neural information processing,* Vol. 3. San Mateo, CA: Morgan Kaufmann.

Sakitt, B. (1972). Counting every quantum. *J Physiol, 223*(1), 131–150.

Sampath, A. P., & Rieke, F. (2004). Selective transmission of single photon responses by saturation at the rod-to-rod bipolar synapse. *Neuron, 41*(3), 431–443.

Schneeweis, D. M., & Schnapf, J. L. (1999). The photovoltage of macaque cone photoreceptors: Adaptation, noise, and kinetics. *J Neurosci, 19*(4), 1203–1216.

Shlens, J., Rieke, F., & Chichilnisky, E. (2008). Synchronized firing in the retina. *Curr Opin Neurobiol, 18*(4), 396–402.

Teich, M. C., Prucnal, P. R., Vannucci, G., Breton, M. E., & McGill, W. J. (1982). Multiplication noise in the human visual system at threshold: 1. Quantum fluctuations and minimum detectable energy. *J Opt Soc Am, 72*(4), 419–431.

Trong, P. K., & Rieke, F. (2008). Origin of correlated activity between parasol retinal ganglion cells. *Nat Neurosci, 11*(11), 1343–1351.

van der Velden, H. A. (1946). The number of quanta necessary for the perception of light in the human eye. *Ophthalmologica, 111,* 321–331.

van Rossum, M. C., & Smith, R. G. (1998). Noise removal at the rod synapse of mammalian retina. *Vis Neurosci, 15*(5), 809–821.

Westheimer, G. (1981). Visual hyperacuity. *Prog Sens Physiol, 1,* 1–30.

Westheimer, G. (1984). Spatial vision. *Annu Rev Psychol, 35,* 201–226.

Yau, K. W., & Hardie, R. C. (2009). Phototransduction motifs and variations. *Cell, 139*(2), 246–264.

26 Adaptation and Natural Stimulus Statistics

ADRIENNE FAIRHALL

ABSTRACT In trying to understand the structure and dynamics of sensory neural representations, it has been productive to examine the complex structure of natural images to discover the nature of the task that the nervous system performs. Two different forms of adaptation help sensory systems to handle the large variations and multiple spatial and temporal scales of natural inputs. Slow changes in excitability can be understood as an encoding of slowly varying stimulus properties that highlights change, while adaptive changes to neural coding strategies lead to more efficient representations of stimuli drawn from time-varying distributions.

Animals must parse a sea of sensory data to extract information that informs their behavioral interactions with the world. Early sensory systems must provide sufficiently resolved data to extract, produce, and manipulate representations of meaningful structure in the world. The composition of much of the natural world in terms of discrete events and objects at varying distances imposes a strong spatial and temporal structure on sensory stimuli from all modalities. Several properties are common: a multiplicity of spatial and temporal scales, a prevalence of sharp transitions or edges, and complex correlations. It has long been hypothesized that this special structure might be reflected in the organization of sensory neural systems.

Sensory neural coding describes the mapping of a sensory input to its representation in neuronal activity. While the classic notion of a tuning curve implies that this mapping is stationary, several different processes combine to create what can appear to be a dynamic relationship between input and output. These nonstationarities span a wide range of time scales, from neuronal adaptations that are almost instantaneous to the reconfiguration of brain circuitry that underlies long-term memory.

We focus here on the shorter time-scale processes referred to as adaptation. We will discuss how properties of adaptation may be relevant to the precise and efficient coding of inputs with the special structure of natural inputs. The term *adaptation* is used in two broadly defined ways. In the first sense, it describes the transience of many neural responses: the fact that a neuron or neural system's response to a specific stimulus often reduces over time. While this is often thought of as a decay in the response, it can be interpreted as part of the dynamical coding properties of the neural system. *Adaptation* is also used to describe more general nonstationarities in the neural code: shifts in the form of neural representation over time. These shifts often appear to allow neural systems to represent inputs more efficiently.

Natural scenes and their challenges

Natural stimuli have a distinctive structure that results from the properties of the natural world (Ruderman, 1997; Simoncelli & Olshausen, 2001). These properties pose particular challenges for sensory systems, and also opportunities for selecting efficient strategies for encoding.

LARGE VARIATIONS Natural stimuli (e.g., figure 26.1A–D) exhibit modulations in amplitude over many orders of magnitude in intensity. Audible, naturally occurring sounds span a range from the threshold of hearing at around 10 dB to 120 dB, around 10 orders of magnitude, with little alteration in our ability to detect and identify sound sources (figure 26.1A). Olfactory stimuli are detectable by humans at fractions of a part per million (figure 26.1B), and odorants are also uniquely identified over a very wide range of concentrations. Light level and contrast can vary by 10 orders of magnitude, yet permit accurate viewing and stable object recognition (figure 26.1C). Given that the activity of primary sensory receptors typically increases with the intensity of the input, this tremendous dynamic range poses two questions. First, how can sensory systems represent such a wide range of inputs without saturation, and second, how do they achieve the ability to extract information about object identity that is invariant with intensity?

STRUCTURE AT MANY SCALES Because visual scenes contain structure in the form of objects and

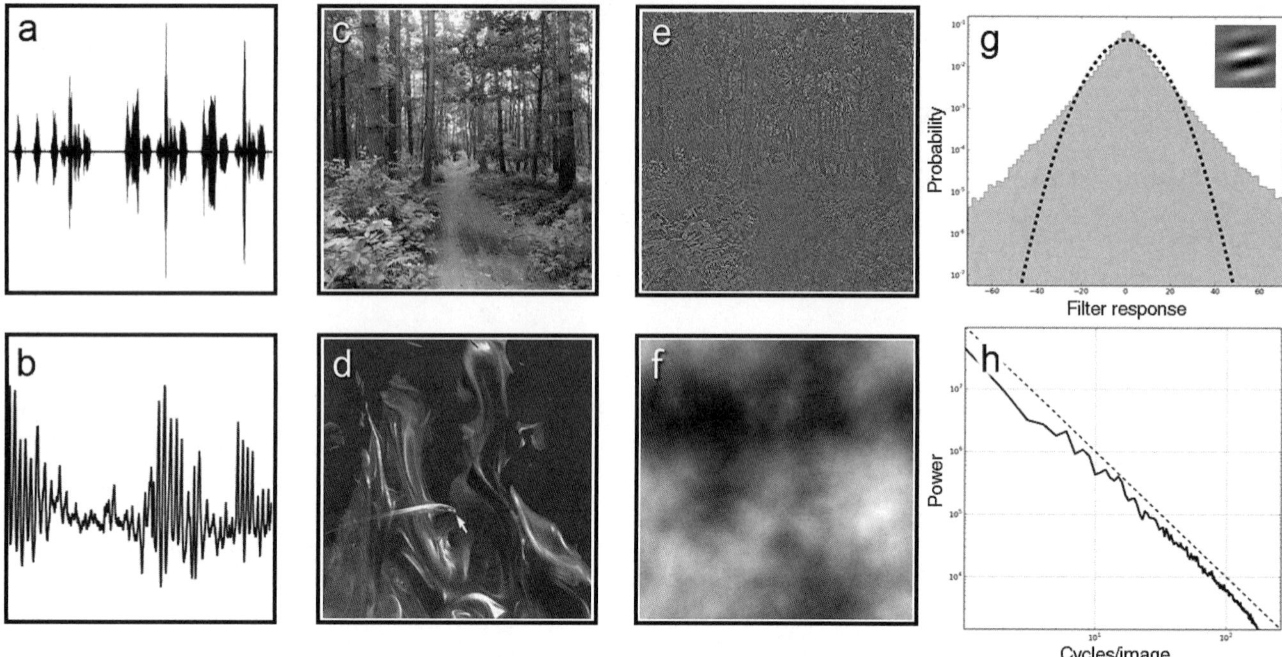

FIGURE 26.1 Examples and properties of natural stimuli. (A) A natural auditory stimulus, the sound pressure wave of a zebra finch's song (M. Fee). (B) The position of a rat's vibrissa during natural whisking (data courtesy of D. Kleinfeld). (C) A natural image (photograph by Ruben Holthuijsen). (D) An underwater olfactory plume passing a lobster antenna (photograph by M. Koehl). (E) The image in (C), whitened. (F) A Gaussian random image with spectral characteristics matching those of natural scenes. (G) The probability density, gray, of the output of a Gabor filter (inset) acting on the image in (C); differs strongly from Gaussian (dotted). (H) Natural images have a close to $1/f^2$ power spectrum. (See color plate 22.)

arrangements that can be viewed from a range of distances and that exist in a wide range of sizes (figure 26.1C), images exhibit self-similarity: as one zooms through different scales, the statistical composition of the scene is relatively invariant (Ruderman & Bialek, 1994; Simoncelli & Olshausen, 2001). For an image with intensity as a function of space given by $I(x, y)$ we can compute the image's autocorrelation function, $C(\delta x, \delta y) = I(x + \delta x, y + \delta y)I(x, y)$, where the average is over all locations x and y. As images are translation-invariant, one can equivalently compute this function in the Fourier domain to obtain the power spectrum. This two-dimensional function can be reduced to a function of spatial frequency f by averaging over phase. For natural images, this function behaves as a power law, $1/f\alpha$, where the exponent α is very close to 2 (figure 26.1H; Ruderman & Bialek, 1994; Tolhurst, Tadmor, & Chao, 1992). The power-law form of the power spectrum is likely a result of the scale invariance of natural scenes; the particular exponent of 2 may be a result of the prevalence of edges, as edges themselves have a power spectrum of $1/f^2$. This property is not unique to vision: under ideal conditions, odorants that are advected by a turbulent airflow also become self-similar and exhibit a power-law spectrum (Catrakis & Dimotakis, 1996).

Not all of the structure in a natural scene is captured by the power spectrum: one can easily construct a random image with a $1/f^2$ spectrum that has no object-like properties (figure 26.1F). The existence of higher-order correlations is apparent when one examines the distribution of two-point differences. If images were fully explained by the mean and power spectrum, these distributions would be Gaussian; however, they are typically strongly peaked at zero and have long tails (figure 26.1G), showing that higher-order moments, or equivalently, multipoint correlations, are needed to fully describe such images.

Despite the overall self-similarity, natural inputs often have a separation of meaning between large-scale and small-scale features (Attias & Schreiner, 1998; Joris, Schreiner, & Rees, 2004). Large-scale features might be termed *objects*, whereas small-scale features may be termed *textures*; features on these different scales convey different types of information. Here, "scale" can refer to either time or space. For example, the large-scale features that identify object boundaries are filled with finer-scale textural structure that conveys distinct information both in visual images and in the temporal signals created by tactile exploration (Hill, Curtis, Moore, & Kleinfeld, 2011). Speech, the auditory input with which humans are most concerned, consists of a large-scale amplitude modulation or envelope that carries most of the information about word meaning, while fine textural details of frequency composition or pitch convey information about the speaker, nuance, and emotion. This general separation between large- and small-scale features suggests that different coding schemes might be used at different time or length scales.

Given these general properties, how might neural encoding be optimized for natural scenes? One hypothesis is that the coding properties of neurons should be organized to represent natural stimuli as efficiently as possible. *Efficient coding* means that a neuron's response properties are adjusted to the distribution of possible stimuli in such a way as to maximize the neuron's representational capacity (Attneave, 1954; Barlow, 1961). This idea has been generalized to neuronal populations through the notion of *sparse coding*, which imposes an additional constraint on the number of active neurons (Barlow, 2001; Bell & Sejnowski, 1997; Olshausen & Field, 1996a, 1996b, 1997, 2004). These approaches have led to a qualitative understanding of the form of receptive fields in a number of sensory systems. However, in the course of natural behavior and through transitions in environments and natural diurnal cycles, the statistics of stimuli continuously vary in space and time. An extension of the efficient coding hypothesis suggests that neural systems should dynamically adjust their coding strategies to best represent these time-varying stimulus statistics.

Adaptation as temporal decay

The term *adaptation* is often used to describe the tendency of neural responses to decay over time. When a stimulus is presented after no input, neuronal responses often initially rapidly increase, then gradually decrease to a lower steady state (Adrian, 1928). This ubiquitous effect occurs through a variety of mechanisms. In the periphery, sensory transduction often occurs through second messenger cascades that contain several opportunities for feedback that diminishes response to constant stimuli over time (Soo, Detwiler, & Rieke, 2008); further, receptors tend to inactivate (Sato et al., 2008). In spiking neurons, this effect is known as spike-frequency adaptation and is due to a range of ion channels, including a noninactivating K^+ current known as the M current (Brown & Adams, 1980); the medium afterhyperpolarization current, mAHP, due to Ca^{2+}-dependent K^+ channels (Adelman, Maylie, & Sah, 2012); Na^+-dependent K^+ channels (Schwindt, Spain, & Crill, 1989); and slow inactivation of sodium channels (Marom, 1998). Beyond single-neuron properties, short-term synaptic plasticity (Zucker & Regehr, 2002) can also reduce input strength in an activity-dependent way (Castro-Alamancos & Connors, 1996).

How one interprets the coding role of such adaptation depends on the method used to characterize the neural code. If one presents a sequence of inputs I that are fixed for some period of time, then change, one can construct an input-output curve derived from the responses immediately following each change in input, $f_0(I)$, and a different input-output curve from the steady state part of the response, $f_\infty(I)$ (Fernandez & White, 2010; Higgs, Slee, & Spain, 2006). This leads to a characterization of schemes of neural coding as either instantaneous or steady-state relative to an onset time. Benda and Herz (2003) developed a general description of adapting dynamics whereby the firing rate f relaxes between these two coding regimes according to the dynamics of an adaptation variable A:

$$f = f_0(I - A) \tag{1}$$

$$\tau \frac{dA}{dt} = f_\infty^{-1}(f) - f_0^{-1}(f) - A. \tag{2}$$

In this model, the adaptation variable A is driven by the neuronal activity f and relaxes with a fixed time constant of τ. When the two f-I curves are linear, with slopes f_0' and f_∞', these dynamics correspond to a high-pass filter imposing gain $G(f_c)$ on the input component at f_c,

$$G(f_c) = f_\infty' \sqrt{\frac{1 + (2\pi f_c \tau_{eff} f_0'/f_\infty')^2}{1 + (2\pi f_c \tau_{eff})^2}}. \tag{3}$$

The time constant τ_{eff} is given by $\tau f_\infty'/f_0'$ and corresponds to the measured time course of decay to a step of input. In this model, Eq. (1) captures the rapid onset of the response seen experimentally, and Eq. (2) the slow relaxation. This model predicts the behavior of electroreceptors in the electric fish, which respond with high gain to chirps embedded in a more slowly varying amplitude modulation signal (Benda, Longtin, & Maler, 2005).

The continuous variation of natural stimuli suggests an alternative perspective: that the dynamics of adaptation is an essential part of the encoding of the stimulus. The transient responses are a strong response to a *change* in stimulus: that is, the firing rate can be considered as a combined function of the new stimulus value and the derivative of the stimulus. As the response $R(t)$ to a stimulus stream s at any time t depends on the

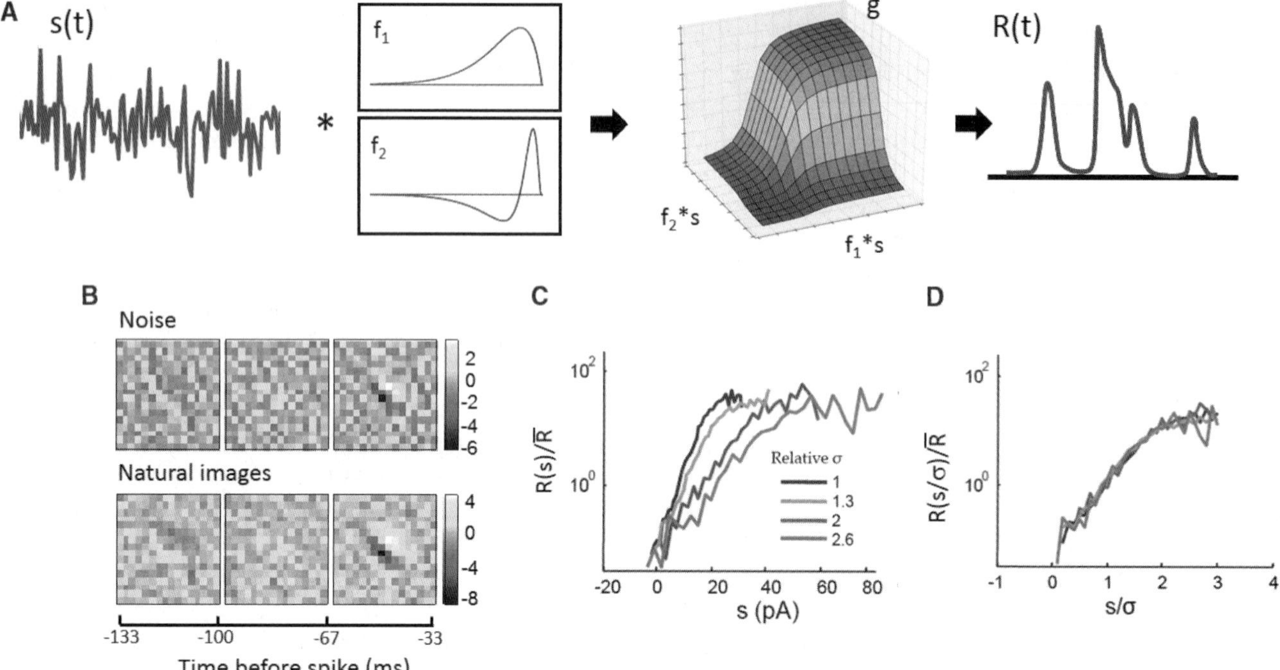

FIGURE 26.2 Linear-nonlinear models and adaptation. (A) Multidimensional linear-nonlinear cascade model: the stimulus $s(t)$ is filtered through one or more filters f_i and passed through a nonlinear function $g(f_1*s, f_2*s, \ldots)$. Here, f_1 is a leaky integrating filter while f_2 is closer to a derivative. The filters f_i and g can both change with adaptation to stimulus context. (B) Example of filter adaptation: spatiotemporal filters from V1 computed using the maximally informative dimension method (Sharpee et al., 2004) when driven either by a flickering noise input or a natural image ensemble. The filters show changes in low-frequency content (adapted from Sharpee et al., 2006). (C) The nonlinearity g can also adapt. Input-output curves from a cortical neuron driven by a filtered white noise current input (R. Mease). The stimulus s is the current filtered through the spike-triggered average. (D) The same input-output curves plotted against the stimulus normalized by the standard deviation of the current. (See color plate 23.)

recent past of the stimulus (for example, the fact that it just underwent a step, or that it has been constant for some time), one can approximate it as some function f of the recent history of the stimulus:

$$R(t) = f(s(t' \le t)).$$

In the simplest case, this response might be a linear function:

$$R(t) = f * s = \int_0^\infty d\tau f(\tau) s(t - \tau), \tag{4}$$

where here, $f(\tau)$ is a set of coefficients that weights the stimulus at times in the recent past, denoted by τ. Equivalently, $f(\tau)$ is a *filter* acting on the stimulus. One can think about $f(\tau)$ as defining a feature of the input to which the system is sensitive, and the operation performed in Eq. (4) is a projection of the stimulus onto that feature. The nature of $f(\tau)$ reveals the dependence of the output on variations in $s(t)$: for example, the response might simply be an average of the recent stimulus, in which case $f(\tau)$ will have a form like f_1 in figure 26.2A. If, as is often observed, the output activity responds transiently to changes in the stimulus, $f(\tau)$ is likely to have a differentiating nature, as in f_2 in figure 26.2A. Generally, one might expect the response to have a sensitivity to both the recent mean stimulus and its derivative, so that $f(\tau)$ might be a composition of an averaging filter and a differentiating filter. If the response is truly approximately linear, these components will simply add. It is likely, however, that the response is a nonlinear function of this weighted sum:

$$R(t) = g\left(\int_0^\infty d\tau f(\tau) s(t - \tau)\right), \tag{5}$$

where $g(\cdot)$ is a nonlinear function that relates the firing rate output to the stimulus projected onto the feature $f(\tau)$. In the case that two or more features contribute to the response nonlinearly, the response could then be approximated as

$$R(t) = g\left(\int_0^\infty d\tau f_1(\tau) s(t - \tau), \int_0^\infty d\tau f_2(\tau) s(t - \tau), \ldots\right), \tag{6}$$

where now $s_1 = \int_0^\infty d\tau f_1(\tau)s(t-\tau)$ and $s_2 = \int_0^\infty d\tau f_2(\tau)s(t-\tau)$ are the components of the entire temporal history of s, projected onto features f_1 and f_2 and $g(\cdot,\cdot)$ is a two-dimensional function of these two variables (figure 26.2A). The features to which a given neuron is sensitive, and the set of features to which a population of neurons is sensitive, define a basis set that captures the components of an input to which the neural population responds (Schwartz, Pillow, Rust, & Simoncelli, 2006).

In this formulation, the initial rapid increase followed by an exponential decay of the firing rate of the neuron in response to a constant stimulus corresponds to convolving the input with a high-pass filter like that defined in Eq. (3). It is important to realize that the decaying dynamics observed in $R(t)$ can be usefully interpreted not as a fatigue of the response, but as part of a well-defined coding property of the system in which the neuron's response is not an instantaneous input-output function but incorporates history dependence and sensitivity to derivatives (Lundstrom, Higgs, Spain, & Fairhall, 2008; Nagel & Wilson, 2011).

Power-law adaptation

While modeling spike-frequency adaptation with a single exponential has been successful in several applications (Benda & Herz, 2003), the decay observed in a variety of different neurons exhibits multiple temporal components (French & Torkkeli, 2008; La Camera et al., 2006; Schwindt, Spain, Foehring, Chubb, & Crill, 1988; Thorson & Biederman-Thorson, 1974). The response following a stimulus step can be approximately power-law:

$$R(t) = R_0 t^{-\alpha}$$

A response function that can account for both the derivative-like rapid onset to a step and also the slow power-law decay can be expressed in the frequency domain as a transfer function of the form $R(\omega) = (i\omega)\alpha$. If the exponent α is 1, this operation is differentiation: the system responds to the step with a pulse that decays immediately. When α is less than 1, this is known as fractional differentiation (Oldham & Spanier, 2006), and can be conceived of as a mixture of integration and differentiation. Such a system also responds with a sharp increase to a step, but following the step, the response exhibits a slow decay. Because of the power-law nature of the response, if such a system is driven by a switching stimulus that changes amplitude with some period T (Baccus & Meister, 2002; Kvale & Schreiner, 2004; Maravall et al., 2007; Smirnakis, Berry, Warland, Bialek, & Meister, 1997), the time scale of the response appears to scale with T (Fairhall, Lewen, Bialek, & de Ruyter van Steveninck, 2001b; Lundstrom et al., 2008; Wark, Fairhall, & Rieke, 2009; Wark, Lundstrom, & Fairhall, 2007). This response function scales each input frequency component's amplitude by a power-law function of frequency $\omega\alpha$ and rotates each component in the complex plane by a constant amount, $i\alpha = e^{i}\pi\alpha^{/2}$. This rotation corresponds to a frequency-independent phase shift of $\varphi = \pi\alpha/2$. These dynamics are approximately realized by several sensory systems (Anastasio, 1994; Fairhall et al., 2001b; Lundstrom et al., 2008; Lundstrom, Fairhall, & Maravall, 2010); however, the constant phase property is not always observed along with power-law-like decays (Drew & Abbott, 2006; Pozzorini, Naud, Mensi, & Gerstner, 2013). Note that this response is a linear function of the time-varying input amplitude. For a linear function, the response simply scales with the size of the input; more biophysical (Nagel & Wilson, 2011; Ozuysal & Baccus, 2012) or normative (Wark et al., 2009) models have been evoked to account for nonlinear effects such as a change in the time constant of the response with a change in input amplitude.

What might power-law adaptation accomplish in terms of signal processing? Since many natural inputs have a power-law spectrum, it is possible that this adaptive transformation has the effect of whitening such inputs: following this transformation, different frequencies in the input are represented more uniformly, and thus, more efficiently, by the firing rate output. Some direct evidence for this was found by Pozzorini et al. (2013), who showed that the spectrum of the input to layer V cortical pyramidal neurons *in vivo* is indeed matched to the power-law filtering induced by multiple adaptation time scales in spike-frequency adaptation in these neurons, such that natural current inputs are whitened.

There are several possible biophysical mechanisms that might underlie power-law-like responses at the single-neuron and network level, including multiple time scales in Na^+ channel inactivation (Gilboa, Chen, & Brenner, 2005; Toib, Lyakhov, & Marom, 1998), spike-frequency adaptation currents (Lundstrom et al., 2008), and synaptic dynamics (Drew & Abbott, 2006; Puccini, Sanchez-Vives, & Compte, 2007). In principle, one or more of these mechanisms could work together to increase the power-law exponent (Lundstrom et al., 2010).

Adaptive coding with changing stimulus statistics

So far, we have considered how adaptive dynamics themselves constitute a channel for encoding stimulus variations in the time domain. We will now use the descriptions of responses given in Eqs. (4–6) to consider a general approach to capturing the effects of different forms of adaptation on encoding. Eqs. (5) and (6) express the concept that a neuron's response can be approximated as a nonlinear function acting on one or more linear filters applied to the input. This is known as a linear-nonlinear cascade model (Schwartz et al., 2006). The filter can be generalized to act on arbitrary components of the stimulus: spatial, temporal, spectral, or other representations (figure 26.2). The two components of such a coding model, the linear filtering of the system and its nonlinear input-output mapping, may be particularly adapted to the statistics of the inputs. Furthermore, these components may be able to change in time to track nonstationary inputs.

Optimal Filters First, the linear filter stage of the response can cancel out the correlations in the input so that responses represent only "surprising" components of the stimulus. This is known as *predictive coding*. Based on the spatial autocorrelation function of natural scenes, Srinivasan, Laughlin, and Dubs (1982) computed the amount of inhibitory surround required to cancel out these correlations and found that this prediction matched the filtering properties of first-order cells in the fly eye. Atick and Redlich (1990; Atick, 1992) used information theory along with a model of photoreceptor noise to derive filters that both optimally reduce the noise and whiten the power spectrum in natural scenes. The predicted filters qualitatively match the response characteristics of retinal ganglion cells and behavioral contrast sensitivity. Predictive coding can remove not only spatial correlations, but also temporal ones (Dong & Atick, 1995; van Hateren, 1992a, 1992b). For example, predictive coding has been invoked to explain adaptation to stimulus input in the encoding of electrical signals in cerebellum in electric fish and in dorsal root ganglia (Roberts & Portfors, 2008).

While visual processing in the retina and lateral geniculate nucleus (Dan, Atick, & Reid, 1996; Dong & Atick, 1995) may serve to whiten images and image streams, figure 26.1 shows that significant correlational structure still remains. Olshausen and Field (1996a) used natural image patches to learn a set of spatial patterns that best represent the images, but such that only a small number is needed to reconstruct any image. Constraining the number required for representing natural images makes this a "sparse" representation: if the spatial-filtering properties of neurons in cortical area V1 corresponded to these patterns, any given stimulus would activate as few neurons in the population as possible. The filters that result from this procedure strongly resemble oriented Gabor-like receptive fields with a range of spatial scales, as are characteristic of V1 simple cells (Bell & Sejnowski, 1997). Similarly, Smith and Lewicki (2006) were able to learn filters with the same form and population distribution of filters of cochlear auditory neurons from natural sound ensembles using this sparseness constraint. Together, these studies suggest that neurons encode stimuli in such a way as to minimize correlations across neurons, and across time, with the additional condition that the population response is sparse (Olshausen & Field, 2004).

Optimal Input-Output Functions The second part of the standard coding model of figure 26.2 concerns the nonlinear mapping from filtered input to output, the function $g(\cdot)$ in Eq. (5). Information-theoretic concepts can also be applied to the form of this function. If the range of the output is limited, one should use all available output symbols (e.g., firing rates) as often as possible; under certain assumptions, this will maximize the entropy of the output and thus maximize information transmission through the coding channel (Barlow, 1961; Laughlin, 1981). In order for this to happen, the input-output relation of the system must depend on the probability distribution of the inputs; with a fixed maximum output, the input-output curve should be the cumulative integral of the input distribution. Laughlin (1981) measured the distribution of contrasts in natural scenes and showed that the input-output relations of fly large monopolar cells indeed matched this predicted form.

These results indicate that the structure of some neural codes can be understood in terms of optimizing response functions relative to the statistics of natural stimuli. However, as we have emphasized, a typical property of such inputs is their local variability in space and time. One might therefore expect that the relationship between input and output changes in time as stimulus statistics change. Indeed, many sensory neural codes display the ability to adapt to the current context of the stimulus.

A dependence of the code on statistical context appears in the calculations of Atick and Redlich (1990; Atick, 1992), which predict that the receptive fields of ganglion cells will be low-pass at low light levels, or low signal-to-noise ratio, and more band-pass as the signal-to-noise ratio increases. This tendency of neural filters to shift toward higher frequencies with higher signal-to-noise as the amplitude of the stimulus increases is typical in many systems, even in single neurons (Hong, Aguera y Arcas, & Fairhall, 2007). Filter properties can also adapt to more complex statistical properties of the input. In the retina, when stimulated with random inputs that contain spatiotemporal correlations, ganglion cell filters adjust over minutes to cancel out these correlations (Hosoya, Baccus, & Meister, 2005). These changes were proposed to be due to anti-Hebbian synaptic plasticity reweighting interneuron contributions to the receptive field. Similarly, slow adaptation of the filter characteristics of cortical V1 neurons occurs when stimuli are switched from Gaussian white noise to a natural image ensemble (Sharpee et al., 2006; figure 26.2B).

Along with changing filter properties, the nonlinear input-output function may also vary dynamically with changes in stimulus statistics. A method of maintaining high information transmission through a coding channel is to continuously match response range to a changing range of inputs (Brenner, Bialek, & de Ruyter van Steveninck, 2000). Suppose that the response function appropriately spans the dynamic range of a stimulus ensemble. If this input ensemble has a mean or variance that changes in time, the input-output curve should shift its center along the stimulus axis as the mean changes, and alter its slope to accommodate changes in the variance (Enroth-Cugell & Shapley, 1973; Shapley & Victor, 1979). In the fly visual system, the input-output curve adapts to stimuli with changing variance such that the stimulus appears to be encoded in units of its time-varying standard deviation (Brenner et al., 2000; Fairhall et al., 2001a). This rescaling of the input with standard deviation in neural response functions has been shown to occur, with different levels of precision, in visual (Baccus & Meister, 2002; Kim & Rieke, 2001), auditory (Kvale & Schreiner, 2004; Nagel & Doupe, 2006), and somatosensory neurons (Diaz-Quesada & Maravall, 2008; Maravall, Petersen, Fairhall, Arabzadeh, & Diamond, 2007), and even in developing cortical neurons responding to an injected input current (Mease, Famulare, Gjorgjieva, Moody, & Fairhall, 2013; figure 26.2C and D).

Multiple time scales of information representation

We have described two apparently distinct notions of adaptation: the decay processes that tend to encode large-scale stimulus variation with transient dynamics, and the ability of a coding scheme to adjust to the local statistics of a stimulus. While some systems appear to explicitly separate representations at different time scales into different neuronal populations, for example, in electrosensation in response to amplitude modulation versus the modulation envelope (Stamper, Fortune, & Chacron, 2013), in many cases, these different components are temporally multiplexed in the same neurons (Baccus & Meister, 2002; Fairhall et al., 2001a; Fairhall et al., 2001b; Nagel & Wilson, 2011; Panzeri, Brunel, Logothetis, & Kayser, 2010; Wen, Wang, Dean, & Delgutte, 2012).

Recall the earlier separation of natural stimulus scales into "textures" and "objects." For a time-varying stimulus amplitude, we can approximately decompose the stimulus as $s(t) = \sigma(t)\eta(t)$, where $\sigma(t)$ is the envelope and $\eta(t)$ is a fast-varying fluctuation. The slower-adapting dynamics encode the stimulus envelope, while an adapting coding scheme permits responses to stimulus details that maintain high fidelity in the face of changing stimulus statistics such as overall amplitude or correlation structure. When the input-output curve adapts to the scale σ of the stimulus, the response might be approximated as

$$R[s(t' \leq t)] = F[\sigma](t)\, g((f * \eta)(t)).$$

Here the rate factor $F[\sigma]$, a functional of the time-varying envelope, multiplies a linear-nonlinear model, Eq. (5), so that the fast fluctuations modulate the firing rate on short time scales and the stimulus is treated in units normalized to the stimulus standard deviation, while the overall gain depends on the slower-varying stimulus statistics. As the mean firing rate is a function of the amplitude—although potentially a nontrivial one—the inherent ambiguity of a normalized code is removed: information about $\sigma(t)$ is available, although precise spike timing encodes stimulus fluctuations in a variance-independent way (Fairhall et al., 2001a).

Thus, adaptive processes can play two different roles in information processing. The temporal envelope dynamics contribute to the scaling of the output, called "output gain" (Carandini & Heeger, 2012), while the normalization of the stimulus is an example of "input gain." The coexistence of fast input adaptation with slow relaxation of the firing rate has been observed in the fly visual system (Fairhall et al., 2001a), the retina (Baccus & Meister, 2002), and the somatosensory system (Lundstrom et al., 2010; Maravall et al., 2007); in the auditory system, these time scales may be similar (Wen et al., 2012). At the single-neuron level, the channel dynamics that drive spiking define the feature that triggers spikes (Hong et al., 2007) and govern encoding of stimulus components at fast time scales through precise spike timing while the slower, activity-dependent time scales of adaptation encode slower stimulus variations in the time-averaged rate or membrane dynamics (Lundstrom et al., 2008; Nagel & Wilson, 2011). These processes are not necessarily independent: a biophysical model that incorporates history-dependent changes in kinetics accounts well for both aspects of multiple time scale adaptation in the retina (Ozuysal & Baccus, 2012).

Predicting time scales of adaptation

Many of the experiments previously discussed imply that neural systems incorporate in their response properties an implicit model of time-varying stimulus statistics. Some of these coding changes occur almost instantaneously (Baccus & Meister, 2002; Fairhall et al., 2001a), suggesting that no explicit learning process is occurring and that such adaptive behavior can arise as a result of intrinsic neuronal nonlinearity, both with respect to the stimulus filter (Famulare & Fairhall, 2009) and the input-output function (Paninski, Lau, & Reyes, 2003; Rudd & Brown, 1997). This is consistent with experimental and modeling observations that the nonlinearities of ion-channel dynamics in single neurons contribute to this behavior (Kim & Rieke, 2003; Mease et al., 2013; Yu & Lee, 2003). Other changes, such as alterations in filtering properties in response to changes in spatiotemporal correlations in the stimulus, occur slowly, and without a fixed time scale that is easily mapped on to an intrinsic time constant (Hosoya et al., 2005; Smirnakis et al., 1997; Sharpee et al., 2006).

What sets the time scale for adaptation? While adaptation has clear benefits in reshaping coding strategies toward higher-fidelity representations as stimulus statistics change, there are also costs: adapting too rapidly can amplify input noise (Dunn & Rieke, 2006). Furthermore, there are statistical limits to how rapidly a system can track time-varying stimulus statistics. DeWeese and Zador (1998) considered a model of adaptation as a Bayesian update process whereby incoming samples are used to update a probability distribution over possible stimulus parameters (mean or variance). This approach predicts two experimental observations. First, it predicts that systems should adapt more rapidly to mean than to variance. Second, there should be an asymmetry between adapting to a higher variance than in adapting to a lower one: a sample from the lower-variance ensemble has a high likelihood of being drawn from the higher-variance ensemble, while the higher variance sample is likely to be an outlier in the lower-variance ensemble. This is indeed observed in the fly visual system (Snippe, Poot, & van Hateren, 2004). In this particular formulation, the time scales of adaptation were dominated by the internal model for the typical rate of change of the tracked parameter. By assuming no inherent time scale for this process, Wark et al. (2009) constructed a Bayesian framework that could account for the observation that the time scale of adaptation in retinal ganglion cell inputs depends on stimulus length: a long period of observation of samples from a particular variance ensemble strengthens one's belief about the current stimulus parameter and slows the rate of adaptation to a new state. The idea of inference has also been applied to the problem of estimating a neuron's membrane potential through observation of spikes (Pfister, Dayan, & Lengyel, 2010); short-term synaptic plasticity emerges as a consequence of optimal estimation. These examples show that optimal statistical inference can serve as a normative model to qualitatively explain the temporal dynamics of adaptation (Lochmann & Deneve, 2011; Lochmann, Ernst, & Deneve, 2012), although these approaches do not yet provide a prediction of absolute time scale on theoretical grounds.

Building neural models from natural inputs

Due to the nonlinearities of neural systems, standard methods for deriving models of neural coding from data—reverse correlation (Rieke, Warland, de Ruyter van Steveninck, & Bialek, 1997; Simoncelli, Paninski, Pillow, & Schwartz, 2004) and generalized linear models (Truccolo, Eden, Fellows, Donoghue, & Brown, 2005)—lead to an inherent dependence of the resulting model on the input statistics. Furthermore, those methods are guaranteed to be correct only for Gaussian input distributions (Paninski, 2003). However, while Gaussian white noise is convenient in that it samples responses to multiple frequencies in an unbiased way (Rieke et al., 1997; Simoncelli et al., 2004), in many cases it does not drive the brain in a way that resembles sensory function during normal behavior (Quiroga, Reddy, Kreiman, Koch, & Fried, 2005; Theunissen, Sen, & Doupe, 2000). Thus, it is necessary to develop methods to obtain reliable and unbiased estimates of coding properties that generalize across different stimulus ensembles. The method of maximally informative dimensions (Sharpee, Rust, & Bialek, 2004) searches for filter directions in stimulus space that maximize mutual information between the filtered stimulus and spikes; it does not depend on the use of a Gaussian input. In general, models that have been sampled with a given white or colored noise ensemble notoriously do not do well in predicting responses to natural scenes, presumably due to rapid adaptation to changing stimulus statistics (Rieke & Rudd, 2009) as well as the effects of nonlinearities prior to linear filtering (Schwartz & Rieke, 2011). While dynamical models can reproduce some of the characteristics of adapting systems (Ozuysal & Baccus, 2012), future coding models should aim to capture a neural system's information representation properties while incorporating these adaptive properties.

ACKNOWLEDGMENTS Thanks to Rachel Wilson, Julijana Gjorgjieva, and Blaise Agüera y Arcas for very helpful comments and assistance in preparation of the manuscript. This work was supported by NSF grant 0928251.

REFERENCES

ADELMAN, J. P., MAYLIE, J., & SAH, P. (2012). Small-conductance Ca2+-activated K+ channels: Form and function. *Annu Rev Physiol, 74,* 245–269.

ADRIAN, E. D. (1928). *The basis of sensation: The action of the sense organs.* New York, NY: Norton.

ANASTASIO, T. J. (1994). The fractional-order dynamics of brainstem vestibulo-oculomotor neurons. *Biol Cybern, 72*(1), 69–79.

ATICK, J. J. (1992). Could information theory provide an ecological theory of sensory processing? *Network, 3,* 213–251.

ATICK, J. J., & REDLICH, A. N. (1990). Towards a theory of early visual processing. *Neural Comput, 2,* 308–320.

ATTIAS, H., & SCHREINER, C. E. (1998). Blind source separation and deconvolution: The dynamic component analysis algorithm. *Neural Comput, 10*(6), 1373–1424.

ATTNEAVE, F. (1954). Some informational aspects of visual perception. *Psychol Rev, 61*(3), 183–193.

BACCUS, S. A., & MEISTER, M. (2002). Fast and slow contrast adaptation in retinal circuitry. *Neuron, 36*(5), 909–919.

BARLOW, H. (2001). Redundancy reduction revisited. *Network, 12*(3), 241–253.

BARLOW, H. B. (1961). Possible principles underlying the transformation of sensory messages. In W. Rosenblith (Ed.), *Sensory communication* (pp. 217–234). Cambridge, MA: MIT Press.

BELL, A. J., & SEJNOWSKI, T. J. (1997). The independent components of natural scenes are edge filters. *Vis Res, 37*(23), 3327–3338.

BENDA, J., & HERZ, A. V. (2003). A universal model for spike-frequency adaptation. *Neural Comput, 15*(11), 2523–2564.

BENDA, J., LONGTIN, A., & MALER, L. (2005). Spike-frequency adaptation separates transient communication signals from background oscillations. *J Neurosci, 25*(9), 2312–2321.

BRENNER, N., BIALEK, W., & DE RUYTER VAN STEVENINCK, R. (2000). Adaptive rescaling maximizes information transmission. *Neuron, 26*(3), 695–702.

BROWN, D. A., & ADAMS, P. R. (1980). Muscarinic suppression of a novel voltage-sensitive K+ current in a vertebrate neurone. *Nature, 283*(5748), 673–676.

CARANDINI, M., & HEEGER, D. J. (2012). Normalization as a canonical neural computation. *Nat Rev Neurosci, 13*(1), 51–62.

CASTRO-ALAMANCOS, M. A., & CONNORS, B. W. (1996). Short-term plasticity of a thalamocortical pathway dynamically modulated by behavioral state. *Science, 272,* 274–277.

CATRAKIS, H. J., & DIMOTAKIS, P. E. (1996). Scale distributions and fractal dimensions in turbulence. *Phys Rev Lett, 77*(18), 3795–3798.

DAN, Y., ATICK, J. J., & REID, R. C. (1996). Efficient coding of natural scenes in the lateral geniculate nucleus: Experimental test of a computational theory. *J Neurosci, 16*(10), 3351–3362.

DEWEESE, M., & ZADOR, A. (1998). Asymmetric dynamics in optimal variance adaptation. *Neural Comput, 10,* 1179–1202.

DIAZ-QUESADA, M., & MARAVALL, M. (2008). Intrinsic mechanisms for adaptive gain rescaling in barrel cortex. *J Neurosci, 28*(3), 696–710.

DONG, D. W., & ATICK, J. J. (1995). Temporal decorrelation: A theory of lagged and nonlagged responses in the lateral geniculate nucleus. *Network: Comput Neural Syst, 6,* 159–178.

DREW, P. J., & ABBOTT, L. F. (2006). Models and properties of power-law adaptation in neural systems. *J Neurophysiol, 96*(2), 826–833.

DUNN, F. A., & RIEKE, F. (2006). The impact of photoreceptor noise on retinal gain controls. *Curr Opin Neurobiol, 16*(4), 363–370.

ENROTH-CUGELL, C., & SHAPLEY, R. M. (1973). Adaptation and dynamics of cat retinal ganglion cells. *J Physiol, 233*(2), 271–309.

FAIRHALL, A. L., LEWEN, G. D., BIALEK, W., & DE RUYTER VAN STEVENINCK, R. R. (2001a). Efficiency and ambiguity in an adaptive neural code. *Nature, 412*(6849), 787–792.

FAIRHALL, A. L., LEWEN, G. D., BIALEK, W., & DE RUYTER VAN STEVENINCK, R. (2001b). Multiple time scales of adaptation in a neural code. *Advances in Neural Information Processing Systems, Vol. 13* (pp. 124–130).

FAMULARE, M., & FAIRHALL, A. L. (2009). Adaptation in simple neurons: Dependence of feature selectivity on stimulus statistics. *Neural Comput, 22*(3), 581–598.

FERNANDEZ, F. R., & WHITE, J. A. (2010). Gain control in CA1 pyramidal cells using changes in somatic conductance. *J Neurosci, 30*(1), 230–241.

FRENCH, A. S., & TORKKELI, P. H. (2008). The power law of sensory adaptation: Simulation by a model of excitability in spider mechanoreceptor neurons. *Ann Biomed Eng, 36*(1), 153–161.

GILBOA, G., CHEN, R., & BRENNER, N. (2005). History-dependent multiple-time-scale dynamics in a single-neuron model. *J Neurosci, 25,* 6479–6489.

HIGGS, M. H., SLEE, S. J., & SPAIN, W. J. (2006). Diversity of gain modulation by noise in neocortical neurons: Regulation by the slow afterhyperpolarization conductance. *J Neurosci, 26,* 8787–8799.

HILL, D. N., CURTIS, J. C., MOORE, J. D., & KLEINFELD, D. (2011). Primary motor cortex reports efferent control of vibrissa motion on multiple time scales. *Neuron, 72*(2), 344–356.

HONG, S., AGUERA Y ARCAS, B., & FAIRHALL, A. L. (2007). Single neuron computation: From dynamical system to feature detector. *Neural Comput, 19*(12), 3133–3172.

HOSOYA, T., BACCUS, S. A., & MEISTER, M. (2005). Dynamic predictive coding by the retina. *Nature, 436,* 71–77.

JORIS, P. X., SCHREINER, C. E., & REES, A. (2004). Neural proessing of amplitude-modulated sounds. *Physiol Rev, 84*(2), 541–577.

KIM, K. J., & RIEKE, F. (2001). Temporal contrast adaptation in the input and output signals of salamander retinal ganglion cells. *J Neurosci, 21,* 287–299.

KIM, K. J., & RIEKE, F. (2003). Slow Na+ inactivation and variance adaptation in salamander retinal ganglion cells. *J Neurosci, 23*(4), 1506–1516.

Kvale, M. N., & Schreiner, C. E. (2004). Short-term adaptation of auditory receptive fields to dynamic stimuli. *J Neurophysiol, 91*(2), 604–612.

La Camera, G., Rauch, A., Thurbon, D., Luscher, H. R., Senn, W., & Fusi, S. (2006). Multiple time scales of temporal response in pyramidal and fast spiking cortical neurons. *J Neurophysiol, 96*, 3448–3464.

Laughlin, S. B. (1981). A simple coding procedure enhances a neuron's information capacity. *Z Naturforsch C, 36*, 910–912.

Lochmann, T., & Deneve, S. (2011). Neural processing as causal inference. *Curr Opin Neurobiol, 21*(5), 774–781.

Lochmann, T., Ernst, U. A., & Deneve, S. (2012). Perceptual inference predicts contextual modulations of sensory responses. *J Neurosci, 32*(12), 4179–4195.

Lundstrom, B. N., Fairhall, A. L., & Maravall, M. (2010). Multiple time scale encoding of slowly varying whisker stimulus envelope in cortical and thalamic neurons in vivo. *J Neurosci, 30*(14), 5071–5077.

Lundstrom, B. N., Higgs, M. H., Spain, W. J., & Fairhall, A. (2008). Fractional differentiation by neocortical pyramidal neurons. *Nat Neurosci, 11*, 1335–1342.

Maravall, M., Petersen, R. S., Fairhall, A. L., Arabzadeh, E., & Diamond, M. E. (2007). Shifts in coding properties and maintenance of information transmission during adaptation in barrel cortex. *PLoS Biol, 5*(2), e19.

Marom, S. (1998). Slow changes in the availability of voltage-gated ion channels: Effects on the dynamics of excitable membranes. *J Membr Biol, 161*(2), 105–113.

Mease, R. A., Famulare, M., Gjorgjieva, J., Moody, W. J., & Fairhall, A. L. (2013). Emergence of adaptive computation by single neurons in the developing cortex. *J Neurosci, 33*(30), 12154–12170.

Nagel, K. I., & Doupe, A. J. (2006). Temporal processing and adaptation in the songbird auditory forebrain. *Neuron, 51*(6), 845–859.

Nagel, K. I., & Wilson, R. I. (2011). Biophysical mechanisms underlying olfactory receptor neuron dynamics. *Nat Neurosci, 14*(2), 208–216.

Oldham, K. B., & Spanier, J. (2006). *The fractional calculus: Theory and applications of differentiation and integration to arbitrary order.* Mineola, NY: Dover.

Olshausen, B. A., & Field, D. J. (1996a). Emergence of simple-cell receptive field properties by learning a sparse code for natural images. *Nature, 381*(6583), 607–609.

Olshausen, B. A., & Field, D. J. (1996b). Natural image statistics and efficient coding. *Network, 7*(2), 333–339.

Olshausen, B. A., & Field, D. J. (1997). Sparse coding with an overcomplete basis set: A strategy employed by V1? *Vis Res, 37*(23), 3311–3325.

Olshausen, B. A., & Field, D. J. (2004). Sparse coding of sensory inputs. *Curr Opin Neurobiol, 14*, 481–487.

Ozuysal, Y., & Baccus, S. A. (2012). Linking the computational structure of variance adaptation to biophysical mechanisms. *Neuron, 73*(5), 1002–1015.

Paninski, L. (2003). Convergence properties of three spike-triggered analysis techniques. *Network, 14*(3), 437–464.

Paninski, L., Lau, B., & Reyes, A. (2003). Noise-driven adaptation: In vitro and mathematical analysis. *Neurocomputing, 52–54*, 877–883.

Panzeri, S., Brunel, N., Logothetis, N. K., & Kayser, C. (2010). Sensory neural codes using multiplexed temporal scales. *Trends Neurosci, 33*(3), 111–120.

Pfister, J. P., Dayan, P., & Lengyel, M. (2010). Synapses with short-term plasticity are optimal estimators of presynaptic membrane potentials. *Nat Neurosci, 13*(10), 1271–1275.

Pozzorini, C., Naud, R., Mensi, S., & Gerstner, W. (2013). Temporal whitening by power-law adaptation in neocortical neurons. *Nat Neurosci, 16*(7), 942–948.

Puccini, G. D., Sanchez-Vives, M. V., & Compte, A. (2007). Integrated mechanisms of anticipation and rate-of-change computations in cortical circuits. *PLoS Comput Biol, 3*(5), e82.

Quiroga, R. Q., Reddy, L., Kreiman, G., Koch, C., & Fried, I. (2005). Invariant visual representation by single neurons in the human brain. *Nature, 435*(7045), 1102–1107.

Rieke, F., & Rudd, M. E. (2009). The challenges natural images pose for visual adaptation. *Neuron, 64*(5), 605–616.

Rieke, F., Warland, D., de Ruyter van Steveninck, R., & Bialek, W. (1997). *Spikes: Exploring the neural code.* Cambridge, MA: MIT Press.

Roberts, P. D., & Portfors, C. V. (2008). Design principles of sensory processing in cerebellum-like structures. Early stage processing of electrosensory and auditory objects. *Biol Cybern, 98*(6), 491–507.

Rudd, M. E., & Brown, L. G. (1997). Noise adaptation in integrate-and-fire neurons. *Neural Comput, 9*(5), 1047–1069.

Ruderman, D. L. (1997). Origins of scaling in natural images. *Vis Res, 37*(23), 3385–3398.

Ruderman, D. L., & Bialek, W. (1994). Statistics of natural images: Scaling in the woods. *Phys Rev Lett, 73*(6), 814–817.

Sato, K., Pellegrino, M., Nakagawa, T., Nakagawa, T., Vosshall, L. B., & Touhara, K. (2008). Insect olfactory receptors are heteromeric ligand-gated ion channels. *Nature, 452*(7190), 1002–1006.

Schwartz, G., & Rieke, F. (2011). Perspectives on information and coding in mammalian sensory physiology: Nonlinear spatial encoding by retinal ganglion cells: When 1 + 1 ≠ 2. *J Gen Physiol, 138*(3), 283–290.

Schwartz, O., Pillow, J. W., Rust, N. C., & Simoncelli, E. P. (2006). Spike-triggered neural characterization. *J Vis, 6*, 484–507.

Schwindt, P. C., Spain, W. J., & Crill, W. E. (1989). Long-lasting reduction of excitability by a sodium-dependent potassium current in cat neocortical neurons. *J Neurophysiol, 61*, 233–244.

Schwindt, P. C., Spain, W. J., Foehring, R. C., Chubb, M. C., & Crill, W. E. (1988). Slow conductances in neurons from cat sensorimotor cortex in vitro and their role in slow excitability changes. *J Neurophysiol, 59*(2), 450–467.

Shapley, R., & Victor, J. D. (1979). The contrast gain control of the cat retina. *Vis Res, 19*(4), 431–434.

Sharpee, T. O., Rust, N. C., & Bialek, W. S. (2004). Analyzing neural responses to natural signals: Maximally informative dimensions. *Neural Comput, 16*, 223–250.

Sharpee, T. O., Sugihara, H., Kurgansky, A. V., Rebrik, S. P., Stryker, M. P., & Miller, K. D. (2006). Adaptive filtering enhances information transmission in visual cortex. *Nature, 439*, 936–942.

Simoncelli, E. P., & Olshausen, B. A. (2001). Natural image statistics and neural representation. *Annu Rev Neurosci, 24*, 1193–1216.

Simoncelli, E. P., Paninski, L., Pillow, J., & Schwartz, O. (2004). Characterization of neural responses with stochastic stimuli. In M. Gazzaniga (Ed.), *The new cognitive neurosciences* (pp. 327–338). Cambridge, MA: MIT Press.

Smirnakis, S. M., Berry, M. J., Warland, D. K., Bialek, W., & Meister, M. (1997). Adaptation of retinal processing to image contrast and spatial scale. *Nature, 386*(6620), 69–73.

Smith, E. C., & Lewicki, M. S. (2006). Efficient auditory coding. *Nature, 439*(7079), 978–982.

Snippe, H. P., Poot, L., & van Hateren, J. H. (2004). Asymmetric dynamics of adaptation after onset and offset of flicker. *J Vis, 4*(1), 1–12.

Soo, F. S., Detwiler, P. B., & Rieke, F. (2008). Light adaptation in salamander L-cone photoreceptors. *J Neurosci, 28*(6), 1331–1342.

Srinivasan, M. V., Laughlin, S. B., & Dubs, A. (1982). Predictive coding: A fresh view of inhibition in the retina. *Proc R Soc Lond B Biol Sci, 216*(1205), 427–459.

Stamper, S. A., Fortune, E. S., & Chacron, M. J. (2013). Perception and coding of envelopes in weakly electric fishes. *J Exp Biol, 216*, 2393–2402.

Theunissen, F. E., Sen, K., & Doupe, A. J. (2000). Spectral-temporal receptive fields of nonlinear auditory neurons obtained using natural sounds. *J Neurosci, 20*(6), 2315–2331.

Thorson, J., & Biederman-Thorson, M. (1974). Distributed relaxation processes in sensory adaptation. *Science, 183*(121), 161–172.

Toib, A., Lyakhov, V., & Marom, S. (1998). Interaction between duration of activity and time course of recovery from slow inactivation in mammalian brain Na channels. *J Neurosci, 18*, 1893–1903.

Tolhurst, D. J., Tadmor, Y., & Chao, T. (1992). Amplitude spectra of natural images. *Ophthalmic Physiol Opt, 12*(2), 229–232.

Truccolo, W., Eden, U. T., Fellows, M. R., Donoghue, J. P., & Brown, E. N. (2005). A point process framework for relating neural spiking activity to spiking history, neural ensemble, and extrinsic covariate effects. *J Neurophysiol, 93*(2), 1074–1089.

van Hateren, J. H. (1992a). A theory of maximizing sensory information. *Biol Cybern, 68*, 23–29.

van Hateren, J. H. (1992b). Theoretical predictions of spatiotemporal receptive fields of fly LMCs, and experimental validation. *J Comp Physiol A, 171*, 157–170.

Wark, B., Fairhall, A. L., & Rieke, F. (2009). Time scales of inference in visual adaptation. *Neuron, 61*(5), 750–761.

Wark, B., Lundstrom, B. N., & Fairhall, A. (2007). Sensory adaptation. *Curr Opin Neurobiol, 17*(4), 423–429.

Wen, B., Wang, G. I., Dean, I., & Delgutte, B. (2012). Time course of dynamic range adaptation in the auditory nerve. *J Neurophysiol, 108*(1), 69–82.

Yu, Y., & Lee, T. S. (2003). Dynamical mechanisms underlying contrast gain control in single neurons. *Phys Rev E, 68*, 011901.

Zucker, R. S., & Regehr, W. G. (2002). Short-term synaptic plasticity. *Annu Rev Physiol, 64*, 355–405.

27 Perception as an Inference Problem

BRUNO A. OLSHAUSEN

ABSTRACT Although the idea of thinking of perception as an inference problem goes back to Helmholtz, it is only recently that we have seen the emergence of neural models of perception that embrace this idea. Here I describe why inferential computations are necessary for perception, and how they go beyond traditional computational approaches based on deductive processes such as feature detection and classification. Neural models of perceptual inference rely heavily upon recurrent computation in which information propagates both within and between levels of representation in a bidirectional manner. The inferential framework shifts us away from thinking of "receptive fields" and "tuning" of individual neurons, and instead toward how populations of neurons interact via horizontal and top-down feedback connections to perform collective computations.

One of the vexing mysteries facing neuroscientists in the study of perception is the plethora of intermediate-level sensory areas that lie between low-level and high-level representations. Why in visual cortex do we have a V2, V3, and V4, each containing a complete map of visual space, in addition to V1? Why S2 in addition to S1 in somatosensory cortex? Why the multiple belt fields surrounding A1 in auditory cortex?

A common explanation for this organization is that multiple stages of processing are needed to build progressively more complex or abstract representations of sensory input, beginning with neurons signaling patterns of activation among sensory receptors in lower levels and culminating with representations of entire objects or properties of the environment in higher areas. For example, numerous models of visual cortex propose that invariant representations of objects are built up through a hierarchical, feedforward processing architecture (Fukushima, 1980; Riesenhuber & Poggio, 1999; Wallis & Rolls, 1997). Each stage is composed of separate populations of neurons that perform feature extraction and spatial pooling, with information flowing from one stage to the next, as shown in figure 27.1. The idea here is that each successive stage learns progressively more complex features of the input that are built upon the features extracted in the previous stage. By pooling over spatial position at each stage, one also obtains progressively more tolerance to variations in the positions of features, culminating in object-selective responses at the top level that are invariant to variations in the pose of an object. Such networks now form the basis of "deep learning" models in machine learning and have achieved unprecedented success on both image- and speech-recognition benchmarks.

Might such hierarchical, feedforward processing models provide insight into what is going on in the multiple areas of sensory cortex? I shall argue here that despite the strong parallels between these models and cortical anatomy and physiology, these models are still missing something fundamental. The problem lies not just with their computational architecture, but also the class of problems they have been designed to solve. Namely, benchmark tasks such as image or speech recognition—while appearing to capture human perceptual capabilities—define the problem of perception too narrowly. Perception involves much more than a passive observer attaching labels to images or sounds. Arriving at the right computational framework for modeling perception requires that we consider the wider range of tasks that sensory systems evolved to solve.

So, what are these tasks? What do animals use their senses for? Answering these questions is a research problem in its own right. One thing we can say with certainty is that visual systems did not start out processing high-definition images, and auditory systems did not start out with well-formed cochleae providing time-frequency analysis of sound. Rather, sensory systems began with crude, coarse-grained sensors attached to organisms moving about in the world. Visual systems, for example, began with simple light detectors situated in the epithelium. Remarkably, over a relatively short period of time (estimated to be 500,000 years), they evolved into the wide variety of sophisticated eye designs we see today (Nilsson & Pelger, 1994). What was the fitness function driving this process? Presumably, it was the ability to plan useful actions and predict their outcomes in complex, three-dimensional (3D) environments. For this purpose, performance at tasks such as navigation or judging scene layout is crucially important. From an evolutionary perspective, the problem of "recognition"—especially when distilled down to one of classification—may not be as fundamental as it seems introspectively to us humans.

The greater problem faced by all animals is one of *scene analysis* (Lewicki, Olshausen, Surlykke, & Moss, 2014). It is the problem of taking incoming sensory

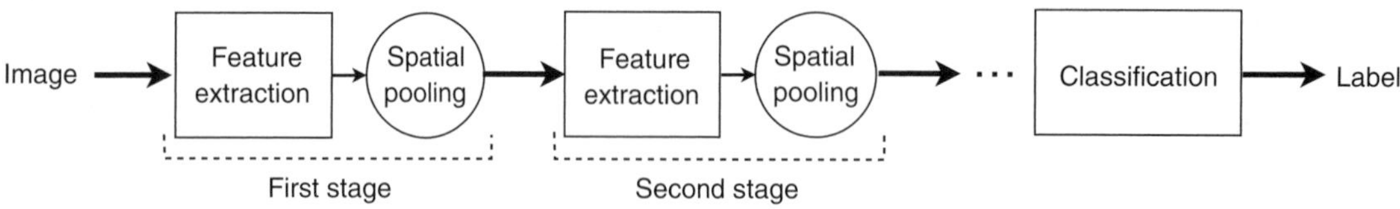

FIGURE 27.1 Hierarchical feedforward processing model of visual cortex. Each stage contains separate populations of neurons for feature extraction and pooling, resulting in object-selective responses at the top stage that are invariant to variations in pose.

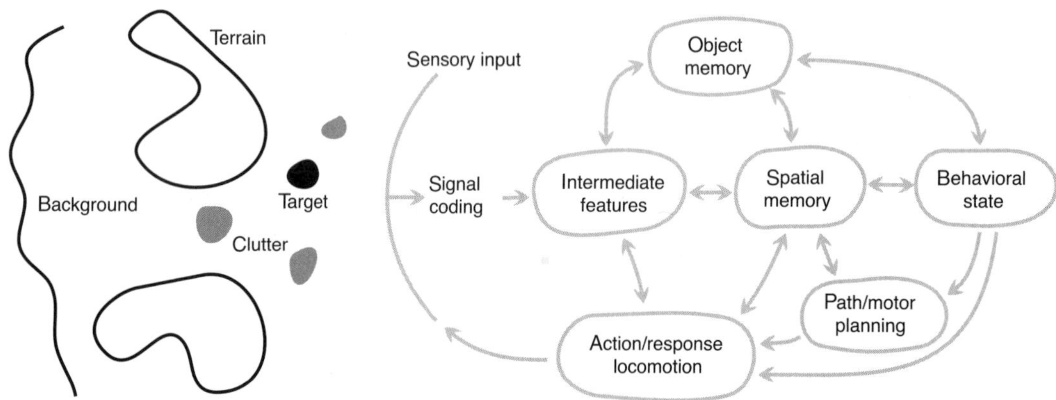

FIGURE 27.2 Components of scene analysis. The scene itself contains not just a single target object, but other objects, terrain, and background, all of which may be important for behavior. The neural structures enabling scene analysis contain multiple levels of representation and analysis. The level of "intermediate features" is where inferential processes come into play (Lewicki et al., 2014).

information and interpreting it in terms of what it conveys about the surrounding environment: terrain, obstacles, and navigable surfaces or routes, in addition to specific objects of interest and their pose and position within the scene. In contrast to a simple feedforward processing pipeline in service to the single goal of classification, scene analysis involves multiple types of representation with different functions (figure 27.2). Most actions (the interesting ones we care about) are not simply reflexive behaviors in direct response to sensory messages. Rather, they depend on goals, behavioral state, and the past history of what has occurred (memory). In other words, meaningful behavior requires having a model of the world and one's place in it. The model needn't be particularly detailed—indeed, what aspects of the environment are modeled, and to what degree of accuracy, is an important empirical question—but we may reasonably assume that it must be assimilated from diverse types of sensory input into a common format that mediates planning and execution of behavior.

The goal of intermediate levels of representation, then, is to disentangle from the raw input stream aspects of the scene appropriate for driving behavior. In contrast to classification that collapses over variability to make a discrete categorical assignment, the goal here is to *describe* the variability—such as the slope of terrain, or the pose of an object—in an analog manner that captures the components of a scene and how one might act upon them. It is at this intermediate level where inferential processes come into play.

In this chapter, I shall describe why inference provides a suitable computational framework for perception, the basic computations it entails, and specific models that have been proposed for how it is instantiated in neural systems. As we shall see, the inferential framework forces us to consider neural architectures other than the standard feedforward processing pipeline. An open problem, though, is to understand the relation between perception and action, and how inferential computations fit into the larger framework of sensorimotor systems.

For other excellent reviews of "perception as inference" from the perspective of psychophysics, see Knill and Richards (1996), Kersten, Mamassian, and Yuille (2004), Kersten and Yuille (2003), and Yuille and Kersten (2006), in addition to chapters 32 and 34 in this volume.

Why inference?

THE PROBLEM OF DISENTANGLING The properties of the world that we care about—which drive behavior—are not directly provided by sensory input. There are no sensors that measure surface shape, motion of objects, material properties, or object identity. Rather, these properties are entangled among multiple sensor values and must be *disentangled* to be made explicit (see also DiCarlo & Cox, 2007). In vision, for example, the retinal image provides a set of measurements of how much light is impinging on the eye from each direction of space. The fact that we humans can look at two-dimensional (2D) images and make sense of them unfortunately gives the misleading impression that an image tells you everything you need to know. But the image itself is simply a starting point, and its 2D format is not well suited to drive behavior in a 3D world. Similarly, the array of hair cell responses does not provide an explicit representation of sound sources, nor does the array of mechanoreceptor activities on the fingertip provide a representation of object shape. These properties are entangled in spatiotemporal patterns of sensor activities.

Importantly, the nature of these disentangling problems is that they are often *ill posed*, meaning that there is not enough information provided by the sensory data to uniquely recover the properties of interest. In other words, the various aspects of a scene that are needed to drive behavior cannot simply be *deduced* from sensory measurements. Rather, they must be *inferred* by combining sensory data together with prior knowledge. Moreover, the disentangling often requires that different aspects of scene structure be estimated simultaneously, so that the inference of one variable affects the other. Thus, it would be impossible—or at least highly inefficient—to infer these things in a purely feedforward chain of processing.

To give a concrete example, consider the simple image of a block painted in two shades of gray, as shown in figure 27.3 (Adelson, 2000). Computing a representation of the 2D edges in this image is easy, but understanding what they mean is far more difficult. Note that there are three different types of edges: (1) those due

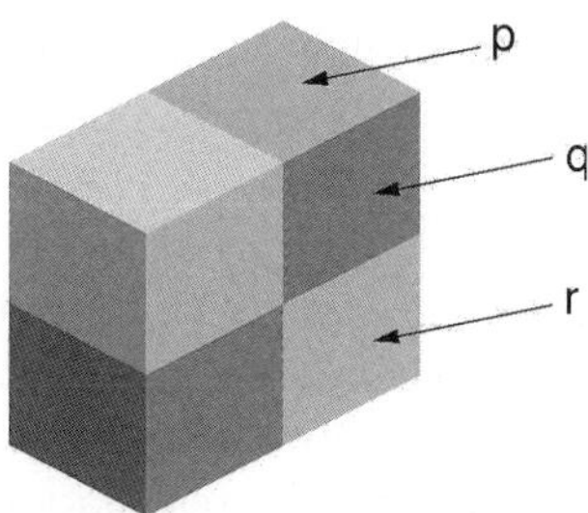

FIGURE 27.3 Image of a block painted in two shades of gray (from Adelson, 2000). The edges in this image are easy to extract, but understanding what they mean is far more difficult and cannot be discerned through local image analysis.

to a change in reflectance (the boundary between *q* and *r*); (2) those due to a change in 3D object shape (the boundary between *p* and *q*); and (3) those due to the boundary between the object and background. Obviously, it is impossible for any computation based on purely local image analysis to tell these different types of edges apart. It is the context that informs us what they mean—but how, exactly?

To interpret this image, one must understand how illumination, 3D shape, and reflectance interact, and how an object combines with its background in projecting to a 2D image (i.e., occlusion). If an edge is ascribed to a reflectance change, it cannot also be due to a shape change (an edge could be due to both, but then the contribution of each of these causes would need to be reduced, so that when combined they still match what is in the image). Thus, the computation of reflectance depends on the computation of shape, and vice versa. And both of these require prior knowledge of what shape and reflectance changes are likely in order to arrive at a plausible interpretation consistent with the data.

Early investigators such as Roberts and Waltz attempted to formally specify the logical operations needed to recover representations of 3D shape from such idealized "blocks world" scenes (Roberts, 1965; Waltz, 1975). However, their methods assumed perfect knowledge of edge segments in the image and the X and Y junctions formed at their intersections, and they utilized the constraints of geometry to then deduce 3D shape from the 2D image. Later, Marr (1982) proposed breaking this process into multiple stages: a primal sketch in which features and tokens are extracted from the image, a 2.5-D sketch that begins to make explicit aspects of depth and surface structure, and finally an object-centered, 3D model representation of objects. His model proposed a feedforward chain of processing in which features are extracted from the image and progressively built up into representations of objects through a logical series of operations in which information flows from one stage to the next. However, since these initial proposals, experience with real-world images has shown us that such bottom-up, deductive processes rarely work in practice.

Consider, for example, the simple scene of a log against a background of rocks, as in figure 27.4. It takes little conscious effort to comprehend what is going on in this scene—the boundary of the log appears obvious to most observers. But if we put ourselves in the position of a local population of neurons in V1 getting input from a local patch of this image, things are far less clear. The right panel of figure 27.4 shows the response of an array of model V1,

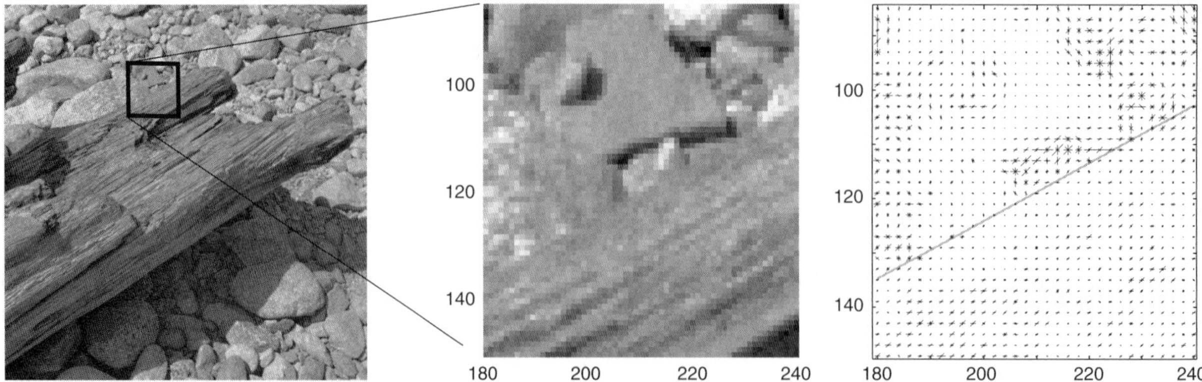

Figure 27.4 The outlined region around the boundary of the log (left panel) is shown expanded in the middle panel. The right panel shows how a hypothetical array of model V1 neurons (Gabor filters at four different orientations) would respond to the image subregion shown at left. The length of each line segment indicates the magnitude of response of a neuron whose receptive field is situated at that position and orientation. An array of such neurons provides only weak or ambiguous cues about the presence of object boundaries in natural scenes.

orientation-selective units analyzing a local patch of the image, with the boundary of the log superimposed as a faint gray line. As one can see, almost nowhere along this boundary are there neurons firing indicating the position and orientation of the boundary. Instead, one finds neurons firing at many different positions and orientations that signal structure in the background and foreground, but with little relation to the boundary itself. Thus, simply measuring oriented contrast in an image does not give us a direct measure of the boundaries or intersection of objects which can then be fed into a reasoning engine about 3D shape. The best we will get from early levels of representation is a collection of ambiguous 2D shape cues that have aspects of illumination, shape, and reflectance intermingled. These must then be aggregated and refined by higher-level processes to disambiguate these cues and what aspects of scene structure they correspond to.

Intrinsic Images One of the first attempts to grapple with the computational aspects of the disentangling problem was Barrow and Tenenbaum's work on "intrinsic images" (Barrow & Tenenbaum, 1978). They attempted to specify the rules by which scene components such as surface shape, reflectance, and illumination could be recovered from the raw-intensity image. They argued that these attributes should be separated at an early level of representation, and that by doing so it greatly facilitates the process of segmentation and object recognition. Specifically, they proposed representing these attributes as a stack of 2D maps, or intrinsic images, that are in register with the original intensity image, where each pixel location is labeled according to its shape, reflectance, or illumination properties. Importantly, the computation of each of these maps involves propagating information between maps to obey photometric constraints and within maps to obey continuity and occlusion constraints. That is, they cannot be computed in independent streams in a purely feedforward fashion, but must cooperate to reach a solution.

Although Barrow and Tenenbaum appreciated the importance of disentangling and outlined some of the computational problems that need to be solved, they stopped short of proposing a specific algorithm and testing it on real images. Somewhat surprisingly, the intervening years have seen only a handful of efforts devoted to these problems (e.g., Jojic & Frey, 2001; Wang & Adelson, 1994), and as a result there has been little progress in developing practical solutions for dealing with real-world images. Recently, however, Barron and Malik (2012) have made an important advance by using priors over shape, reflectance, and illumination to recover intrinsic images for these quantities from photographs of real objects. To date, their method obtains the best performance on this challenging problem. And, notably, it is based on inferential computation in which representations of shape, reflectance, and illumination interact in order to settle to a solution.

The intrinsic-image approach takes an important step in introducing the idea of a structured or layered representation that moves away from a flat, monolithic representation of image properties (such as an array of Gabor filters) and toward a representation of properties of the scene. But still, attributes of the scene are represented in 2D Cartesian coordinates, in a retinotopic or camera-centric frame of reference, whereas animals must act in complex 3D environments. Ultimately, then, it makes sense for scene attributes to be represented in a format that is more amenable to planning actions in the world.

SURFACE REPRESENTATION Nakayama and colleagues have argued based on psychophysical evidence that intermediate-level representations are organized around *surfaces* in the 3D environment, and that these representations serve as a basis for high-level processes such as visual search and attention (Nakayama, He, & Shimojo, 1995). This view stands in contrast to previous theories of perceptual grouping, search, and attention based on 2D maps of image features such as local orientation and motion energy (Julesz, 1981; Treisman & Gelade, 1980). Nakayama's experiments suggest that representations of 3D surface structure are formed at an early stage, and that perceptual grouping, search, and attention operate primarily on inferred surface representations rather than 2D maps of image features. For example, when colored items are arranged on surfaces in different depth planes, detection of an odd-colored target is facilitated when pre-cued to the depth plane containing the target; but if the items are arranged so as to appear attached to a common surface receding in depth, then pre-cueing to a specific depth has little effect. Thus, it would appear that attention spreads within inferred surfaces in 3D coordinates in the environment, not by 2D proximity in the image or within a common depth plane (disparity).

Nakayama's work also points to the importance of surface-occlusion relationships in determining how features group within a scene. Under natural viewing conditions, the 2D image arises from the projection of 3D surfaces in the environment. When these surfaces overlap in the projection, the one nearest the observer "over-writes," or occludes, the other. Thus, a proper grouping of features would need to take this aspect of scene composition into account in determining what goes together with what, as shown in figure 27.5. By manipulating disparity cues so as to reverse figure-ground relationships in a scene, they show that the visual system groups features in a way that obeys the rules of 3D scene composition. Features are grouped within surfaces, even when parts of the surface are not visible, but not beyond the boundary of a surface. Thus, the neural machinery mediating this grouping would seem to require an explicit representation of border ownership, such as described by Zhou, Friedman, and von der Heydt (2000), or some other variable that expresses the boundaries and ordinal relationship of surfaces.

Although the discussion above has focused mainly on visual inferential processes, the same ideas generalize

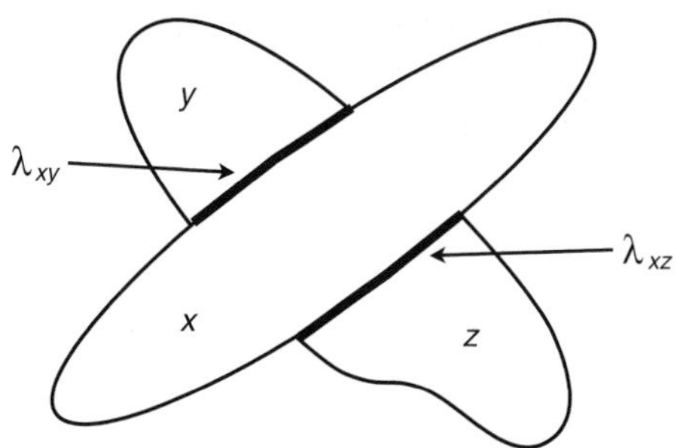

FIGURE 27.5 Occlusion and border ownership. When image regions corresponding to different surfaces meet in the projection of a scene, the region corresponding to the surface in front "owns" the border between them. A region that does not own a border is essentially unbounded and can group together with other unbounded regions. Here, surface x owns the borders λ_{xy} and λ_{xz}. Thus, regions y and z are unbounded at these borders, and they are free to group with each other but not with region x because it owns these borders and is therefore bounded by them (adapted from Nakayama et al., 1995).

to other sensory modalities such as audition or touch. The central problem we face for all of these modalities is that the properties of the world that are needed to drive behavior are not given directly by the sensory receptors, but instead must be inferred by combining sensory data together with prior knowledge. Now we turn to the question of how this is actually done by neurons.

How do neurons perform inferential computations?

Helmholtz astutely observed long ago that perception is a process of "unconscious inferences." Only recently, though, have investigators pursued this idea in a quantitative manner in order to characterize inferential computations carried out in the nervous system. Here I describe the mathematical framework for inference based on Bayes's rule, as well as neural models that have been proposed for doing perceptual inference.

BAYES'S RULE The basic mathematical framework for inference begins with Bayes's rule, which uses the laws of conditional probability to calculate the probability of a hypothesis H given the data D:

$$P(H \mid D) = \frac{P(D \mid H)P(H)}{P(D)}. \qquad (1)$$

What this equation tells us is that if we have a model that specifies how probable the data would be under a certain hypothesis—that is, the *likelihood* $P(D|H)$—in addition to the *prior* ("before data") probability of the hypothesis, $P(H)$, then we can calculate the *posterior* ("after data") probability of the hypothesis $P(H|D)$. (The term $P(D)$ often plays the role of a normalization constant, and may be ignored if we are mainly interested in the relative probability of different hypotheses for the same data.) Simply speaking, Bayes's rule provides a calculus for reasoning in the face of uncertainty. It is a powerful conceptual and mathematical framework that tells us *quantitatively* how to make inferences in the face of noisy or incomplete data. Not surprisingly, one now finds it applied to a wide variety of problems, from the control of guided missiles to spam filtering.

In perception, we are interested in estimating properties of the external environment from sensory data. For example, in vision we are given a set of photoreceptor activations or pixel intensities, I, and we wish to infer properties such as shape, s, and reflectance, r. Using Bayes's rule, we can formulate this problem as follows:

$$P(s, r \mid I) \propto P(I \mid s, r)P(s)P(r). \tag{2}$$

Here the likelihood term $P(I|s, r)$ expresses the rendering model—that is, how images are generated as a function of shape and reflectance. This is a well-posed computation that is routinely solved by computer graphics algorithms. However, the problem of going the other direction—from the image to compute shape and reflectance—is highly ill posed, because there are multiple ways to set these parameters that would result in the same image (Adelson, 2000). This degeneracy is resolved by the priors over shape and reflectance, $P(s)$ and $P(r)$, which favor certain settings of s and r over others. In the work of Barron and Malik (2012), these priors were obtained by measuring statistics of shape and reflectance on a large database of objects. The resulting posterior distribution over s and r, $P(s, r|I)$ rates the different shape and reflectance values of the image in terms of their probability of being the correct interpretation. It takes into account both how well image measurements are fit by these values and how consistent they are with prior knowledge. With strong enough priors, the posterior may be peaked around a single value of s and r, which would make these settings an obvious choice (i.e., the maximum a posteriori, or MAP, estimate).

A model that shows how the above computations may be implemented in a neurally plausible manner in the circuits of visual cortex has yet to be fully developed. In the meantime, though, we can get a feel for the nature of such a solution by looking at a simpler neural model of inference called *sparse coding*.

Sparse Coding The goal of sparse coding is to learn a set of basis patterns from the statistics of incoming sensory data and then infer a representation of the data in terms of these patterns. Although these patterns may not correspond to the actual properties of a scene, the model nevertheless illustrates the principles of inferential computation in a neural system and how it can provide new insight into the response properties of neurons.

In a visual sparse-coding model, we start with the assumption that the spatial distribution of light intensities within a local region of the image $I(\vec{x})$ may be represented in terms of a superposition of some basis patterns $\phi(\vec{x})$:

$$I(\vec{x}) = \sum_i a_i \phi_i(\vec{x}) + n(\vec{x}). \tag{3}$$

The image is then represented in terms of the coefficients a_i, which tell us which basis patterns are contained in the image (figure 27.6). The term $n(\vec{x})$ is a

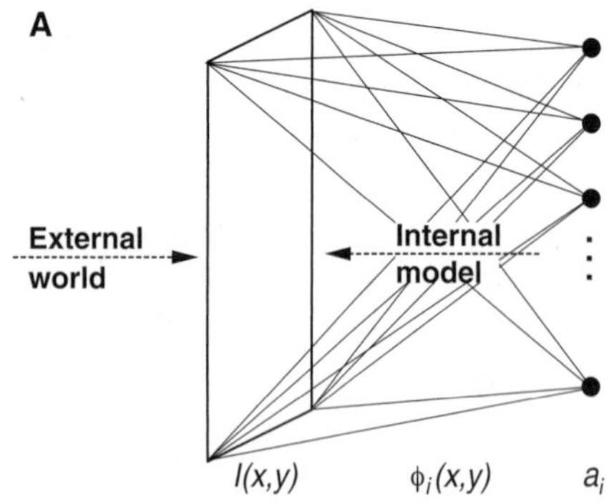

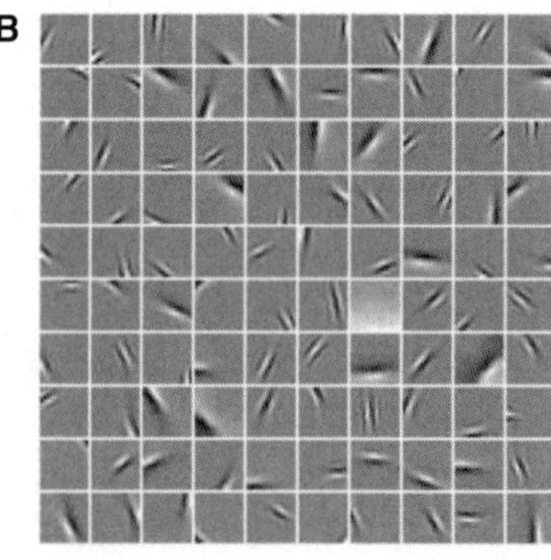

Figure 27.6 Sparse-coding model of images. (A) Images are represented in terms of a set of basis patterns $\varphi_i(\vec{x})$ that are learned from the data. Each coefficient a_i expresses how much of each basis pattern is needed to describe the image. Sparsity is imposed through a prior that encourages coefficients to be zero. (B) Basis patterns learned from image patches extracted from natural images. Each patch shows a different learned pattern $\varphi_i(\vec{x})$. The learned patterns are oriented, localized, and bandpass (selective to structure at different spatial scales), similar to the measured receptive fields of V1 neurons.

residual that is included to account for other structure, such as noise, that is not well described by the model. The basis patterns themselves are learned from the statistics of images so as to provide a sparse or compact description of the image—that is, we desire a *dictionary* of basis patterns $\{\phi_i(\vec{x})\}$ that allows us to provide a good match to any given image using the fewest (sparse) number of nonzero coefficients a_i. Sparsity is enforced by imposing a prior over the coefficients, $P(a)$, that encourages values to be zero. The coefficients themselves are then computed by maximizing the posterior distribution

$$P(a \mid I) \propto P(I \mid a)P(a). \quad (4)$$

The proposal here is that neurons in cortex (layer 4 of V1) are representing the coefficients a_i (Olshausen & Field, 1997). But how should such a population of neurons compute their responses so as to maximize the posterior $P(a|I)$? Rozell, Johnson, Baraniuk, and Olshausen (2008) have shown that a solution may be computed according to the following system of equations:

$$\tau \dot{u}_i + u_i = \sum_{\vec{x}} \phi_i(\vec{x})I(\vec{x}) - \sum_j G_{ij}a_j$$
$$a_i = g(u_i). \quad (5)$$

These equations are amenable to direct implementation in a neural network (in fact, they are the same equations as for a Hopfield network). Each neuron's membrane voltage u_i is driven by the combination of a feedforward (receptive field) term and a feedback (recurrent inhibition) term that depends on the overlap between basis patterns $G_{ij} = \sum_{\vec{x}} \phi_i(\vec{x})\phi_j(\vec{x})$. The output a_i is then computed by simply thresholding the membrane voltage u_i (the function g passes values above a specified threshold and sets values below threshold to zero). Thus, in order to arrive at a representation of the image, the population of neurons must interact. Although the receptive field provides a driving input, the neuron's actual response is determined by the context in which other neurons around it are also

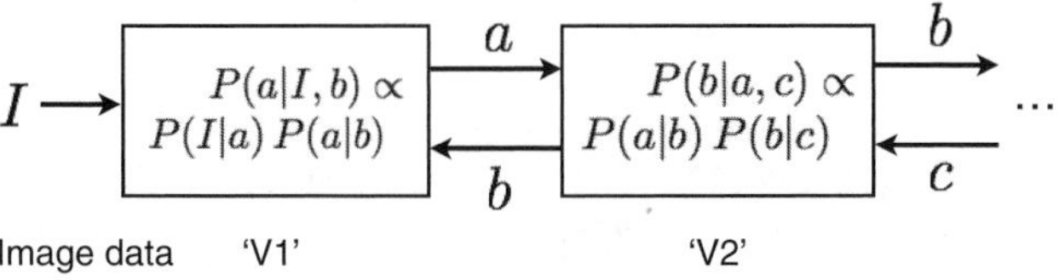

FIGURE 27.7 Hierarchical Bayesian inference. The variables represented at each level are inferred from a combination of bottom-up and top-down inputs. Bottom-up inputs enter into the likelihood, while top-down inputs enter into the prior. The two are combined to form the un-normalized posterior, which guides the inference of variables at each level.

responding. If the basis pattern of one unit is better matched to the image than another, it will attempt to cancel out or "explain away" the other unit's activity. Interestingly, Zhu and Rozell (2013) have shown that these explaining-away interactions can account for a wide variety of nonclassical receptive field effects, such as end-stopping and contrast-orientation tuning.

The sparse coding model illustrates how a variety of neural response properties found in V1—localized, oriented, bandpass receptive fields and contextual modulation—may be accounted for in terms of a model that attempts to infer a representation of the incoming sensory data in terms of its underlying features. But the most we can hope for with this approach is a direct representation of the data per se (e.g., basis patterns of the image), whereas what we ultimately desire is a representation of the properties of a scene that these data tell us about. As the simple example of the painted block in figure 27.3 shows us, this cannot be accomplished through local image analysis, but rather involves aggregating information globally across the scene in order to infer properties such as shape and reflectance. Thus, we turn now to the question we addressed at the outset: How can we build up more complex or abstract representations of sensory input through a hierarchy of multiple stages of analysis?

HIERARCHICAL REPRESENTATION Lee and Mumford (2003) have proposed a framework for hierarchical Bayesian inference that illustrates how the above inferential computations could be extended to multiple stages of representation. The general idea is illustrated in figure 27.7, which is adapted from their paper. At each stage, the variables being represented are influenced by both bottom-up and top-down inputs. At the first stage, corresponding to V1, the variables a are inferred through a combination of the likelihood and prior, as above, except now the prior over a is shaped by the variables b represented in the next-higher level ("V2"). For example, if many weak signals among the a variables suggest the presence of a contour (as in the log and rocks image of figure 27.4), then the b variables in the next stage that explicitly represent the contour would become active, in turn encouraging the elements consistent with it to increase their activity by modulating the prior over a in this direction. The variables b in turn are subject to influences from yet higher levels, such as objects or fragments of surfaces represented by variables c. Thus, the full representation of the scene involves variables at all levels, a, b, and c, and computing these variables relies upon bidirectional information flow between levels.

This is admittedly just the sketch of a theory. There are many details to be filled in here, and some efforts have already been made along these lines (Cadieu & Olshausen, 2012; Garrigues & Olshausen, 2010; Karklin & Lewicki, 2005, 2009). What is most needed now is to incorporate layered representations, such as those proposed by Barrow and Tenenbaum (1978), and Barron and Malik (2012), that separate aspects of scene structure due to surface shape, reflectance, or other scene variables. It will also be necessary to include in the generative model the ability to account for occlusion or figure-ground relationships (Le Roux, Heess, Shotton, & Winn, 2011; Lücke, Turner, Sahani, & Henniges, 2009) and geometric transformations due to variations in pose (Arathorn, 2002, 2005; Olshausen, Anderson, & Van Essen, 1993).

From a neurobiological perspective, the hierarchical inference model provides a clear role for feedback connections in the cortex, and it suggests how to design experiments to reveal what they are doing. Although there have already been numerous experimental attempts to uncover what feedback is doing—for example, by cooling or disabling neurons in a higher area and characterizing how responses in lower areas change (Andolina, Jones, Wang, & Sillito, 2007; Angelucci & Bullier, 2003; Hupé et al., 2001)—the effects to date appear rather subtle. Indeed, there is considerable doubt among neuroscientists as to whether feedback plays any role in dynamically shaping information processing (Lennie, 1998). If the hierarchical Bayesian inference model is correct, it suggests we would see the greatest effects of feedback when the system is presented with scenes containing locally ambiguous cues that can only be properly interpreted at higher levels of analysis. Indeed, functional MRI experiments along these lines have revealed evidence for strong top-down effects: when subjects perceive a collection of features as an entire 3D object as opposed to its individual parts, activity in higher levels increases while activity in lower levels decreases, consistent with disambiguation (Murray, Kersten, Olshausen, Schrater, & Woods, 2002).

Another important neurobiological consideration is the speed with which such a hierarchical inference system can settle on a solution. It has been argued, based on the latency of object-selective neural responses (Hung, Kreiman, Poggio, & DiCarlo, 2005; Oram & Perrett, 1992; Thorpe, Fize, & Marlot, 1996), that there is little time for the iterative type of processing that feedback loops would entail (Thorpe & Imbert, 1989). But conduction velocities of feedforward and feedback axons between V1 and V2 are on the order of 2–4 ms (Angelucci & Bullier, 2003). Even between thalamus and V1, the round-trip travel time can be as short as 9 ms (Briggs & Usrey, 2007). It is also not clear whether "iterative processing" is an apt analogy to describe signal flow in the cortex, since there is no clock latching each cycle of feedback. It may well be that such a system can quickly settle to a solution within physiological time constraints. More importantly, sensory processing does not work in terms of static snapshots of input that get churned away in the system one at a time. Rather, it operates as a dynamical system operating on a continuous, time-varying input stream. Thus, the consequence of feedback arriving through axonal and synaptic delays is simply that sensory information arriving at the present moment is processed in the context of past information that has gone through a higher level of processing, which undoubtedly could be quite advantageous.

Finally, it should be noted that the hierarchical Bayesian inference framework is distinct from the "predictive coding" model of Rao and Ballard (1999). Though both models advocate an important role for top-down feedback signals, the proposed effect of these signals is very different. Predictive coding proposes that feedback signals are largely inhibitory, as they carry the predictions of higher levels that attempt to cancel out signals coming from lower levels. By contrast, hierarchical Bayesian inference proposes that feedback serves to disambiguate representations in lower levels, meaning that it would facilitate, rather than cancel out, the activity of neurons at lower levels consistent with representations from higher levels, and it would suppress the activity of neurons that are inconsistent.

Conclusion

Thinking of perception as an inference problem, as opposed to a deductive computational pipeline, leads us to ask a different set of questions about what neurons are doing. Instead of asking about feature extraction, receptive fields, and tuning of individual neurons, we are led to ask how populations of neurons cooperate and interact to infer representations of scene properties. Instead of looking for organized "maps" of sensory features varying along one or a small number of feature dimensions across the cortical surface, we are led to look for different layers of representation that disentangle different scene properties and that are likely to be intermingled on a much finer scale. Instead of viewing the cortical hierarchy as a feedforward pipeline with classification as an end goal, we are led to ask how ill-posed problems are being solved at each level, and how feedback from higher levels disambiguates representations at lower levels.

While Bayes's rule provides a computational framework for perceptual inference, it leaves many questions unanswered regarding its implementation. If the brain is doing Bayesian inference, should we expect to find neurons representing probabilities and calculating Bayes's rule when we look inside? As we have seen in the case of the sparse-coding model, not necessarily. Here neurons represent hypotheses about what is contained in the scene, and Bayes's rule acts as the "invisible hand" that governs the dynamics of the circuit and leads the population to a mode of the posterior distribution. Other models have proposed ways that neurons might represent probabilities implicitly through a population code (Eliasmith & Anderson, 2004; Ma, Beck, Latham, & Pouget, 2006) or through stochastic sampling via spontaneous activity (Berkes, Orban, Lengyel, & Fiser, 2011). The advantage of these latter approaches is that the entire distribution over a set of hypotheses may be represented that allows for uncertainty or multiple probable hypotheses to be taken into account during inference.

Finally, it will be necessary to think more seriously about the link between perception and action in order to give these ideas a more solid footing. Much has been said here about representing scene properties, but which properties need to be represented and how well depends on the manner in which they are used to guide actions. As Guillery and Sherman (2011) point out, layer 5 neurons in all areas of visual cortex (or other sensory cortices) project to motor nuclei. Thus it is not just the top box of the cortical hierarchy but also representations in V1, V2, V4, and so on that are used to guide actions. Figuring out how to incorporate these aspects of cortical architecture into the hierarchical inference framework—that is, understanding how inference feeds into action—will be an important goal for future work.

ACKNOWLEDGMENTS These ideas evolved in large part through discussions with Mike Lewicki, Cindy Moss, and Anne-Marie Surlykke while on sabbatical at the Wissenschaftskolleg zu Berlin in 2009. I also thank Jim DiCarlo for encouraging me to write these things down and better explain the inferential framework of perception. Supported by NSF (IIS-1111654), NIH (EY019965), NGA (HM1582-08-1-0007), SRC STARnet (SONIC), and the Canadian Institute for Advanced Research.

REFERENCES

ADELSON, E. H. (2000). Lightness perception and lightness illusions. In M. Gazzaniga (Ed.), *The new cognitive neurosciences* (2nd ed., pp. 339–351). Cambridge, MA: MIT Press.

ANDOLINA, I. M., JONES, H. E., WANG, W., & SILLITO, A. M. (2007). Corticothalamic feedback enhances stimulus response precision in the visual system. *Proc Natl Acad Sci USA, 104*(5), 1685–1690.

ANGELUCCI, A., & BULLIER, J. (2003). Reaching beyond the classical receptive field of V1 neurons: Horizontal or feedback axons? *J Physiol (Paris), 97*(2–3), 141–154.

ARATHORN, D. W. (2002). *Map-seeking circuits in visual cognition: A computational mechanism for biological and machine vision.* Stanford, CA: Stanford University Press.

ARATHORN, D. W. (2005). Computation in the higher visual cortices: Map-seeking circuit theory and application to machine vision. *Proc. of the 33rd Applied Imagery Pattern Recognition Workshop* (pp. 73–78).

BARRON, J. T., & MALIK, J. (2012). Shape, albedo, and illumination from a single image of an unknown object. *Proc. IEEE Conference on Computer Vision and Pattern Recognition* (pp. 334–341).

BARROW, H. G., & TENENBAUM, J. M. (1978). Recovering intrinsic scene characteristics from images. In A. Hanson & E. Riseman (Eds.), *Computer vision systems* (pp. 3–26). New York, NY: Academic Press.

BERKES, P., ORBAN, G., LENGYEL, M., & FISER, J. (2011). Spontaneous cortical activity reveals hallmarks of an optimal internal model of the environment. *Science, 331*(6013), 83–87.

BRIGGS, F., & USREY, W. M. (2007). A fast, reciprocal pathway between the lateral geniculate nucleus and visual cortex in the macaque monkey. *J Neurosci, 27*(20), 5431–5436.

CADIEU, C. F., & OLSHAUSEN, B. A. (2012). Learning intermediate-level representations of form and motion from natural movies. *Neural Comput, 24*(4), 827–866.

DICARLO, J. J., & COX, D. D. (2007). Untangling invariant object recognition. *Trends Cogn Sci, 11*(8), 333–341.

ELIASMITH, C., & ANDERSON, C. (2004). *Neural engineering: Computation, representation, and dynamics in neurobiological systems.* Cambridge, MA: MIT Press.

FUKUSHIMA, K. (1980). Neocognitron: A self organizing neural network model for a mechanism of pattern recognition unaffected by shift in position. *Biol Cybern, 36*(4), 193–202.

GARRIGUES, P. J., & OLSHAUSEN, B. A. (2010). Group sparse coding with a Laplacian scale mixture prior. In *Advances in neural information processing systems*, Vol. 23 (pp. 1–9). Red Hook, NY: Curran.

GUILLERY, R. W., & SHERMAN, S. M. (2011). Branched thalamic afferents: What are the messages that they relay to the cortex? *Brain Res Rev, 66*(1–2), 205–219.

HUNG, C. P., KREIMAN, G., POGGIO, T., & DICARLO, J. J. (2005). Fast readout of object identity from macaque inferior temporal cortex. *Science, 310*(5749), 863–866.

HUPÉ, J.-M., JAMES, A. C., GIRARD, P., LOMBER, S. G., PAYNE, B. R., & BULLIER, J. (2001). Feedback connections act on the early part of the responses in monkey visual cortex. *J Neurophysiol, 85*(1), 134–145.

JOJIC, N., & FREY, B. J. (2001). Learning flexible sprites in video layers. *Proc. IEEE Conference on Computer Vision and Pattern Recognition* (pp. I-199–I-206).

JULESZ, B. (1981). Textons, the elements of texture perception, and their interactions. *Nature, 290*(5802), 91–97.

KARKLIN, Y., & LEWICKI, M. S. (2005). A hierarchical Bayesian model for learning nonlinear statistical regularities in nonstationary natural signals. *Neural Comput, 17*(2), 397–423.

KARKLIN, Y., & LEWICKI, M. S. (2009). Emergence of complex cell properties by learning to generalize in natural scenes. *Nature, 457*(7225), 83–86.

Kersten, D., Mamassian, P., & Yuille, A. (2004). Object perception as Bayesian inference. *Annu Rev Psych, 55*, 271–304.

Kersten, D., & Yuille, A. (2003). Bayesian models of object perception. *Curr Opin Neurobiol, 13*(2), 150–158.

Knill, D. C., & Richards, W. (1996). *Perception as Bayesian inference.* Cambridge, UK: Cambridge University Press.

Le Roux, N., Heess, N., Shotton, J., & Winn, J. (2011). Learning a generative model of images by factoring appearance and shape. *Neural Comput, 23*(3), 593–650.

Lee, T. S., & Mumford, D. (2003). Hierarchical Bayesian inference in the visual cortex. *J Opt Soc Am A, 20*(7), 1434–1448.

Lennie, P. (1998). Single units and visual cortical organization. *Perception, 27*(8), 889–935.

Lewicki, M. S., Olshausen, B. A., Surlykke, A., & Moss, C. F. (2014). Scene analysis in the natural environment. *Frontiers in psychology* (in press).

Lücke, J., Turner, R. E., Sahani, M., & Henniges, M. (2009). Occlusive components analysis. In *Advances in neural information processing systems*, Vol. 22 (pp. 1069–1077). Red Hook, NY: Curran.

Ma, W. J., Beck, J. M., Latham, P. E., & Pouget, A. (2006). Bayesian inference with probabilistic population codes. *Nat Neurosci, 9*(11), 1432–1438.

Marr, D. (1982). *Vision: A computational investigation into the human representation and processing of visual information.* San Francisco, CA: Freeman.

Murray, S. O., Kersten, D., Olshausen, B. A., Schrater, P., & Woods, D. L. (2002). Shape perception reduces activity in human primary visual cortex. *Proc Natl Acad Sci USA, 99*(23), 15164–15169.

Nakayama, K., He, Z., & Shimojo, S. (1995). Visual surface representation: A critical link between lower-level and higher-level vision. In D. M. Osherson & S. M. Kosslyn (Eds.), *An invitation to cognitive science*, Vol. 2: *Visual cognition* (pp. 1–70). Cambridge, MA: MIT Press.

Nilsson, D. E., & Pelger, S. (1994). A pessimistic estimate of the time required for an eye to evolve. *Proc Biol Sci, 256*(1345), 53–58. doi:10.1098/rspb.1994.0048

Olshausen, B. A., Anderson, C. H., & Van Essen, D. C. (1993). A neurobiological model of visual attention and invariant pattern recognition based on dynamic routing of information. *J Neurosci, 13*(11), 4700–4719.

Olshausen, B. A., & Field, D. J. (1997). Sparse coding with an overcomplete basis set: A strategy employed by V1? *Vis Res, 37*(23), 3311–3325.

Oram, M. W., & Perrett, D. I. (1992). Time course of neural responses discriminating different views of the face and head. *J Neurophysiol, 68*(1), 70–84.

Rao, R. P. N., & Ballard, D. H. (1999). Predictive coding in the visual cortex: A functional interpretation of some extra-classical receptive-field effects. *Nat Neurosci, 2*(1), 79–87.

Riesenhuber, M., & Poggio, T. (1999). Hierarchical models of object recognition in cortex. *Nat Neurosci, 2*(11), 1019–1025.

Roberts, L. G. (1965). Machine perception of three-dimensional solids. In J. T. Tippett (Ed.), *Optical and electro-optical information processing* (pp. 159–198). Cambridge, MA: MIT Press.

Rozell, C. J., Johnson, D. H., Baraniuk, R. G., & Olshausen, B. A. (2008). Sparse coding via thresholding and local competition in neural circuits. [Letter]. *Neural Comput, 20*(10), 2526–2563.

Thorpe, S. J., Fize, D., & Marlot, C. (1996). Speed of processing in the human visual system. *Nature, 381*(6582), 520–522.

Thorpe, S. J., & Imbert, M. (1989). Biological constraints on connectionist models. In R. Pfeifer, Z. Schreter, F. Fogelman-Soulie, & L. Steels (Eds.), *Connectionism in perspective* (pp. 63–92). Amsterdam: Elsevier.

Treisman, A. M., & Gelade, G. (1980). A feature-integration theory of attention. *Cognit Psychol, 12*(1), 97–136.

Waltz, D. (1975). Understanding line drawings of scenes with shadows. In P. H. Winston (Ed.), *The psychology of computer vision.* New York, NY: McGraw-Hill.

Wallis, G., & Rolls, E. T. (1997). Invariant face and object recognition in the visual system. *Prog Neurobiol, 51*(2), 167–194.

Wang, J. Y. A., & Adelson, E. H. (1994). Representing moving images with layers. *IEEE Trans Image Process, 3*(5), 625–638.

Yuille, A., & Kersten, D. (2006). Vision as Bayesian inference: Analysis by synthesis? *Trends Cogn Sci, 10*(7), 301–308.

Zhou, H., Friedman, H. S., & von der Heydt, R. (2000). Coding of border ownership in monkey visual cortex. *J Neurosci, 20*(17), 6594–6611.

Zhu, M., & Rozell, C. J. (2013). Visual nonclassical receptive field effects emerge from sparse coding in a dynamical system. *PLoS Comput Biol, 9*(8), e1003191.

28 Neural Computation in Sensory Systems

XAQ PITKOW AND MARKUS MEISTER

ABSTRACT An animal's brain must extract from the onslaught of raw sense data the few bits of information that actually matter for the guidance of behavior. Accordingly, one finds neurons in the upper echelons of a sensory system that report very selectively on high-level features of the stimulus while remaining invariant to many low-level perturbations. For example, "face cells" in the primate visual cortex respond selectively to one person's face regardless of the view angle, scale, or illumination. The emergence of such complex pattern detectors is one of the stunning phenomena in sensory neuroscience. In studies that trace signal flow through the associated neural circuits, the most common tool is the measurement of sensory receptive fields. Here we review the tenets and basic results of receptive field analysis and place it in a common framework with the task of selective feature detection. We show that classic receptive field measurement—with its focus on the linear summation of stimulus variables—has little to contribute on the subject of feature computation. On the other hand, a cascade of simple nonlinear operations can indeed account for high-level pattern detector neurons, and hints of such organization are found in the brains of diverse animals.

Consider a human typist copying a handwritten manuscript. His brain converts visual signals that enter the eye into motor signals that drive the fingers. How much information is involved in this task? At the input, the typist's eye can receive about 10^9 bits per second of raw visual information.[1] At the output, the typist generates about 20 bits per second of movement information.[2] This modest number is in line with other estimates for the information rate of human behavior (Eriksson, 1996; Pierce & Karlin, 1957). So the task of the typist's sensory system is to extract, every second from a data deluge of 10^9 bits, the 20 bits that really matter for visually driven behavior. Of course, those few important bits are hidden in the raw sensory input in convoluted ways. The same applies to all other cases of sensory-driven behaviors. This process of highly selective extraction is appropriately called computation; indeed, many of the tasks that we hand to our man-made computing machines involve similar challenges of pattern extraction.

How does the brain attack this problem? At the very input, it converts photons into electrical signals in nerve cells. At the output, it converts electrical signals of nerve cells into movement. In fact, all the world events with which we interact are represented by neuronal membrane potentials: sight, sound, smell, touch, speech, movement, sweating, and internal phenomena like thought, dreams, and emotions. This is a truly remarkable step of abstraction, by which phenomena of entirely different physical nature are encoded with the same symbol set of membrane voltages. From that point on the brain can use general purpose devices—namely, neurons and synapses—to create connections between these world events; for example, combining sights and sounds to cause thoughts and speech. We sometimes forget the immense power of this abstraction, but it offers another parallel to man-made electronic computers, which similarly represent all world events by voltage signals.

In this chapter, we consider how these sensory computations happen; namely, how the neurons and synapses are arranged to accomplish the selective extraction of relevant bits from the sensory stream. We start with some illustrative examples of selective computation. In tracing this process through the associated neural circuitry, the primary tool has been measurement of receptive fields, and we discuss and critique this approach. How complex neuronal-response properties come about through the actual signal flow in neural circuits presents a more daunting challenge, and we cite some examples of progress in this area. Finally, we consider several theories of brain function that purport to explain why sensory computation is organized the way it is. Throughout, we make liberal use of examples from the visual system, partly because of our own exposure bias, but the chapter focuses on principles thought to apply in the other senses as well.

Selectivity and invariance

Continuing our observation of the typist: note how his right thumb flicks downward only when he needs to type a space. So there is a motor neuron innervating his thumb muscle that has become a perfectly selective

[1] Our back of the envelope reads: Six million cone photoreceptors, each modulated at up to 25 Hz with a signal-to-noise ratio of 100.

[2] One hundred words per minute, six letters per word, two bits of entropy per letter in the English language.

visual pattern detector. It fires reliably and only in response to visual images that look like the space between two words. Of course, these images can take on many different forms, given the varieties of handwriting, types of paper and ink, and illumination. The motor neuron is invariant to this enormous variation of the raw sensory stimulus. After work, the typist may ride home on a bicycle and his right thumb is used to flick a small bell on the handlebar. Now that same motor neuron has become a selective detector of pedestrians. We draw two lessons from these simple observations. First, the brain can indeed construct circuits that make single neurons perfectly selective for a high-level abstract feature of the sensory input. Second, this mapping is flexible and changes dramatically when the brain engages a different task or context.

Given that the brain can produce motor neurons selective for high-level features, one suspects that it may do so already within the sensory system, namely, prior to the commitment to a particular motor output: once individual neurons have extracted a high-level concept, their firing could easily be mapped into different behavioral outputs, or associated with other concepts for the purpose of learning, memory retrieval, or thinking (Barlow, 1972, 2009; Quiroga, 2012). Indeed, this is the case. A compelling example can be found in the "face cells" of the macaque inferotemporal cortex (Freiwald & Tsao, 2010; Gross, 1992; Rolls, 1992; Tsao & Livingstone, 2008). When the animal is presented with a broad variety of visual images, these neurons remain silent, except on presentation of a face. Some neurons appear selective for the faces of specific individuals. Yet their response to that face is largely invariant under widely different view angles, distances, or illumination conditions that result during natural interactions. The macaque cortex has several small regions in which the great majority of neurons are specialized for faces, to the extent where these regions can even be resolved by functional MRI (fMRI) imaging (Tsao, Freiwald, Knutsen, Mandeville, & Tootell, 2003). In turn, fMRI of human subjects suggests that they have similar regions dominated by face-selective neurons (Kanwisher, McDermott, & Chun, 1997). Recordings from multimodal areas of the human brain have suggested specialist neurons with even broader invariances, such as the "Halle Berry cell" that responded selectively both to the name of the actress and her image (Quiroga, Reddy, Kreiman, Koch, & Fried, 2005).

Selective pattern detectors have been described in other species, often in the context of sensory tasks that are an essential part of an animal's behavior. For example, owls can localize prey on the ground based on sound cues alone. The information about azimuthal direction of the sound source comes from the time delay between sounds arriving at the two ears. And in the higher stations of the auditory system, one indeed finds neurons whose response is selective for a particular interaural time delay, but invariant to many other features of the sound waveform (Carr & Konishi, 1990; Konishi, 2003). Another instance arises in weakly electric fish that sense their surroundings by producing an oscillating electric field in the water. When two fish with similar oscillation frequencies approach too close, their signals threaten to interfere, and they shift the two frequencies further apart. Among the many stations of the electrosensory nervous system, one finds neurons whose response reports selectively the frequency difference of the other fish, invariant to the location or orientation of the fish or even the absolute field frequencies (Heiligenberg, 1991; Rose, Kawasaki, & Heiligenberg, 1988). These two examples are instructive, because much has been discovered about the circuits that generate the selectivity (Konishi, 2006).

We will use these high-level pattern-detector neurons as a guide to understanding sensory computation. Obviously, this is not the only purpose of a sensory system. But in exploring how a system works, it is useful to have some concrete phenomenon in mind whose explanation is likely to reveal something fundamental. Face cells—and their analogs in other species—impress us by the combination of selectivity and invariance. Their response is selective for faces, but not just one picture of a face; that could easily be accomplished by a template match to the target image. Instead, they respond equally to images of that same face rotated, translated, and illuminated in all different ways. How does that arise?

First of all, one must recognize that selectivity and invariance per se are not unusual: every neuron has them. A photoreceptor is selective for light that falls on one point in the image, and invariant to the pattern of light on all the other photoreceptors. Further into the sensory system, each neuron receives a synaptic input current that represents a single scalar function of the sensory stimulus, and that defines its selectivity. The neuron's response is invariant under all the myriad stimuli that leave this scalar function unchanged. So one is led to a more subtle assessment: the "face cell" is remarkable because its selectivity and invariance have been shaped so exquisitely to match a specific region in the space of all images that corresponds to different views of the grandmother's face. Furthermore, its response is invariant under image transformations that profoundly alter the signals of each sensory receptor and most neurons in the early visual system. To

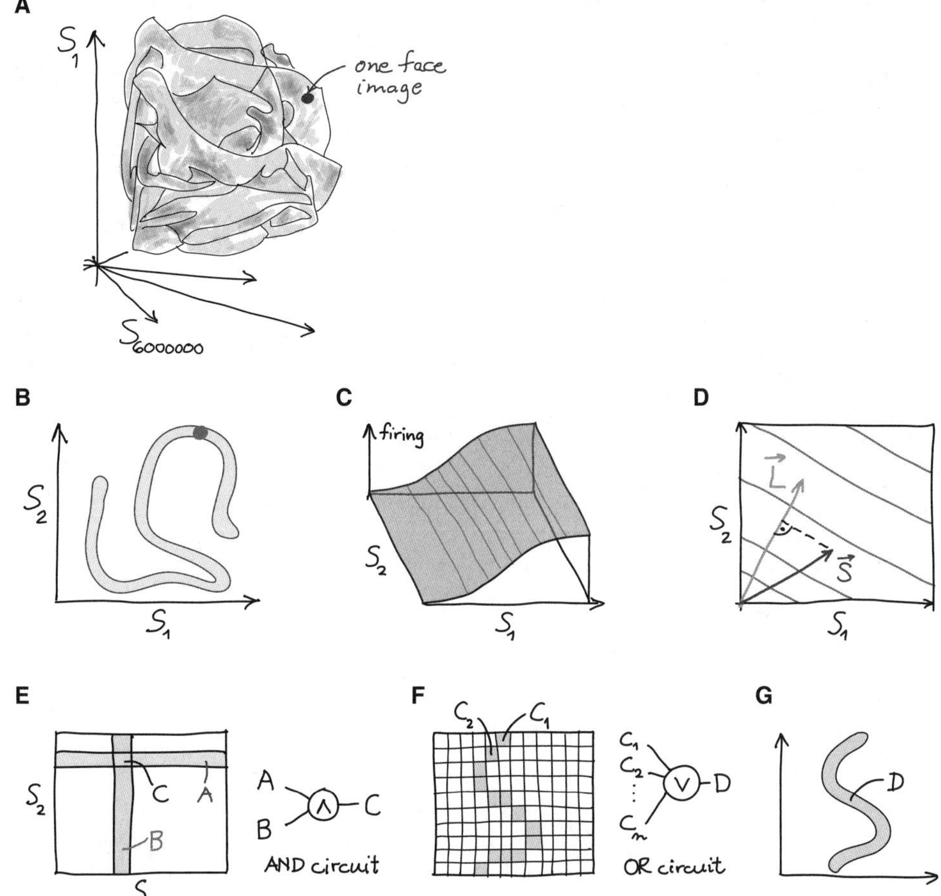

FIGURE 28.1 A geometric view of sensory computations. (A) Each stimulus is a vector in a high-dimensional space, whose axes are defined by the excitation of all the sensory receptors, for example, the intensity on the 6 million cones in a human eye. One retinal image of a face is a point in that space. The same face seen under different views and illuminations defines a high-dimensional surface (DiCarlo & Cox, 2007). (B) Two-dimensional stimulus space. A "face cell" responds only when the stimulus is in a thin, convoluted, shoelace-shaped region. (C) The response of a receptor as a function of the stimulus variables. Note the contour lines are straight and orthogonal to the receptor's axis of sensitivity (here the S_1 axis). (D) A receptive field is represented by a vector **L** in stimulus space. Integration of the stimulus **S** by this receptive field is the projection of **S** onto **L**. The response of a linear-nonlinear (LN) neuron depends only on this projection: the contour lines of its response function are straight and orthogonal to **L**. (E–G) How to create a shoelace-shaped response region in three simple steps. See text for details. (See color plate 24.)

appreciate this, it is helpful to think about the space of stimuli in geometric terms.

Your eye has about 6 million cone photoreceptors. For daylight vision, this is the number of pixels of the raw visual stimulus. Suppose the image of a face appears on the retina: like all other retinal images, it can be represented as a point in an abstract space that has 6 million dimensions (DiCarlo & Cox, 2007). Each coordinate of that space represents the intensity on one photoreceptor (figure 28.1A). Now rotate the face about one axis: this produces a series of stimulus points that trace out a curved line in the stimulus space. Rotate it also about another axis: now one has a two-dimensional surface. Add several other transformations, like scaling, translation, illumination by a point source at various angles, background illumination, diverse facial expressions. Together these define a high-dimensional and highly convoluted surface in the stimulus space (figure 28.1A). The face cell fires only if the stimulus is on or near that surface.

For those few readers with difficulty imagining a 6-million-dimensional space, let us consider a rudimentary sensory system with just two receptors. The space of all stimuli is a two-dimensional plane (figure 28.1B). Any given image is a point on that surface. The activity of those two receptors can be represented with two response functions on that surface, whose contour lines run straight and parallel to the axes (figure 28.1C). But a hypothetical "face cell" fires only in a thin stimulus region that looks like a convoluted shoelace (figure 28.1B). The challenge for the brain (or more aptly for us trying to understand the brain) is how the shoelace-shaped selectivity region can be built from the straight-line selectivities of the receptors. Of course, in the 6-million-dimensional space of real face detection, this problem is greatly magnified.

Receptive fields

How does one go about exploring these sensory circuits? The primary approach has been to record the signals of single neurons along the pathway from the primary receptors to the high-level feature detectors. At each stage, one measures the properties of neuronal responses to sensory stimuli, in the hope that they will gradually approach the selectivity and invariance of the "face cell." In characterizing the function of a sensory neuron, one would ideally like to report its response to any relevant sensory stimulus. Given limited experimental time, one can test only a small number of stimuli, and then the results must be extrapolated in a way that extends to stimuli outside that limited set.

The most common form of reporting such a summary of response properties is the "receptive field." The meaning of this term has undergone some evolution over the years, adapting to increased needs and powers of sensory physiology. Originally applied to entire reflex arcs (Sherrington, 1906), the receptive field meant the area of a dog's skin in which touch would trigger a scratch reflex. Later it was applied more specifically to characterize neural responses, but generalized to other sensory surfaces like the retina (figure 28.2A–C), and even to abstract surfaces like an auditory spectrogram (figure 28.2E). Then the concept was extended to allow for both excitation and inhibition, meaning that area of the sensory field in which point stimuli directly increase or decrease the neuron's firing rate. Finally, in today's usage, the receptive field also conveys quantitative information about how much the firing rate is altered and with what time course.

In the visual system, for example, the spatiotemporal receptive field $L(x, y, t)$ is a function of space and time (sometimes also wavelength) that spells out how much an impulse of light at location (x, y) in the visual field alters the firing a time t later. An arbitrary stimulus can be constructed from a superposition of short flashes of light at different locations and times (Rodieck & Stone, 1965). Thus a neuron's response summary should include a description of how these many point-like stimuli at different locations and times combine to generate the response of the neuron. When an author quotes just the receptive field, the best (only) guess for such integration is a linear summation: at any given time, the system sums the intensity of the point-like stimuli in the recent past, weighted by the amplitude of the receptive field. Formally, this linear summation is expressed as

$$g(t) = \int_x \int_y \int_{t'=-\infty}^{t} S(x, y, t-t') L(x, y, t') dt' dy dx, \qquad (1)$$

where $S(x, y, t)$ represents the stimulus at location (x, y) and time t and $L(x, y, t)$ is the receptive field. In practice, one always deals with a finite number of discrete locations $\vec{x}_i$ and times discretized to multiples of some small interval Δt. So the stimulus can also be represented as a vector that lists the intensities at those points in the recent past,

$$\mathbf{S} = \{S_{ij}\} = \{S(\vec{x}_i, t - j\Delta t)\}, \qquad (2)$$

and similarly the receptive field is a vector

$$\mathbf{L} = \{L_{ij}\} = \{L(\vec{x}_i, j\Delta t)\}. \qquad (3)$$

In this vector representation, the linear summation of stimuli becomes a simple dot product

$$g = \mathbf{S} \bullet \mathbf{L} \qquad (4)$$

The effect of any given stimulus is determined entirely by the projection of the stimulus vector onto the special direction indicated by the receptive field vector **L**. In this way the receptive field of a sensory neuron—at least in this modern incarnation—identifies its axis of selectivity in the abstract space of stimuli. The neuron's response varies only along that axis, and remains invariant along all orthogonal directions (figure 28.1D).

How does the response vary along the special direction? Whereas the above expression for g can take on any value, a neuron's response is generally limited to a finite range. For example, firing rates cannot be negative or exceed a biophysical limit from membrane refractoriness. Thus a more realistic model of sensory responses includes a nonlinear distortion function by which g turns into the neuron's response:

$$r = N(g) = N(\mathbf{S} \bullet \mathbf{L}). \qquad (5)$$

Generally, $N(g)$ has a sigmoid shape, with zero or constant response for very small values of g and saturation to a maximal response at very large values. This expression for a neuron's response to stimuli is often called the "LN model": a linear summation of stimuli followed by a nonlinear response function. One can measure the nonlinearity N during the same experiment that measures the receptive field L (Chichilnisky, 2001; Ringach, 2004). A complete report that includes both receptive field and nonlinearity allows the reader to predict the neuron's response to any desired stimulus.

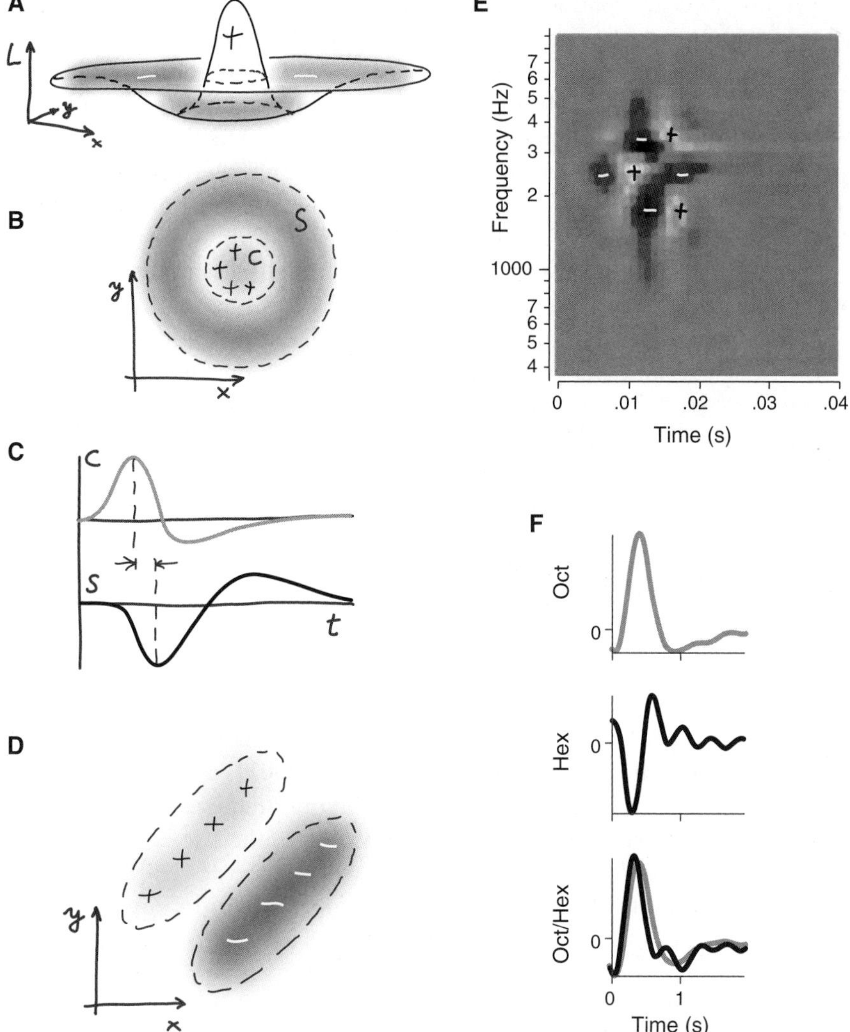

FIGURE 28.2 Receptive fields in various sensory systems. (A–C) Spatiotemporal receptive fields in the early visual system. Here the components of the stimulus vector **S** are the light intensity at different locations and time points. Any given stimulus vector consists of a short "movie" $S(x, y, t)$. The receptive field $L(x, y, t)$ can similarly be viewed as a movie. For displays on paper, one often shows a single frame $L(x, y, t_0)$ of this movie (panels A–B) along with the time course $L(x_0, y_0, t)$ at a particular location (panel C). Note that L has both positive and negative components, meaning that light is excitatory (red) at some locations and times, and inhibitory (blue) at others. For retinal bipolar cells, ganglion cells, and thalamic relay cells, the receptive fields look very similar except for some scaling along the spatial and temporal axes. The spatial profile (B) contains a center region (C) and a surround region (S) in which light has opposite action. When displayed in three dimensions (A), this profile has the appearance of a "Mexican hat." The time course of the receptive field (C) is biphasic, both in the center and the surround. Generally, the surround response is slightly delayed relative to the center (C). (D) For a simple cell in primary visual cortex, the receptive field profile typically shows elongated side-by-side regions in which light has opposite action. The time course again is biphasic, as in panel C. (E) Spectrotemporal receptive field of a neuron in the auditory area field L of the zebra finch (Nagel & Doupe, 2008). Here the stimulus $S(f, t)$ is represented as a spectrogram, plotting sound power as a function of time and frequency. The receptive field of this cell shows excitatory (red) and inhibitory (blue) regions. Along the spectral dimension one finds a "Mexican hat" profile, with a central region of excitatory frequencies and adjacent inhibitory regions. The action of the surround is delayed relative to the center. In the time domain, the receptive field is strongly biphasic or even triphasic. This is only one of many receptive field shapes encountered in this brain region. (F) Osmotemporal receptive field of a neuron in the olfactory pathway of the locust (Geffen et al., 2009). Here the stimulus $S(o, t)$ is the time-varying concentration of two discrete odors, octanone or hexanal. The receptive field $L(o, t)$ specifies the contribution to the response of each odor at different times. *Top and middle*: Receptive field derived from experiments in which only one odor was presented at a time. *Bottom*: Receptive field measured when both odors were varied simultaneously. Note the neuron's response to hexanal changes polarity under this condition. (See color plate 25.)

Figure 28.2A–D presents typical spatiotemporal receptive fields encountered along the mammalian visual pathway. Beginning in retinal bipolar cells, the spatial receptive field profile shows what is called "center-surround antagonism" (Asari & Meister, 2012; Burkhardt, Bartoletti, & Thoreson, 2011; Hare & Owen, 1990). This means that light in a small central region near the cell body has opposite effects from light in a larger surrounding region (figure 28.2A–B). The temporal part of the receptive field shows antagonism in time: light in the recent past has opposite effects from light in the more distant past (Baccus, Ölveczky, Manu, & Meister, 2008). This is reflected in the biphasic time course of the receptive field (figure 28.2C). Generally this time course is slightly different in the center and the surround, with a notable delay in the response to the surround (Baccus et al., 2008; Fahey & Burkhardt, 2003). These same features dominate the receptive fields reported for downstream neurons, like retinal ganglion cells (Benardete & Kaplan, 1997; Enroth-Cugell, Robson, Schweitzer-Tong, & Watson, 1983; Kuffler, 1953; Meister & Berry, 1999) and thalamic relay neurons (Hubel & Wiesel, 1961; Reid & Shapley, 2002). One synapse further, in the primary visual cortex, the receptive fields begin to change (DeAngelis, Ohzawa, & Freeman, 1995; Hubel & Wiesel, 1962; Martinez et al., 2005; Ringach, 2004): for so-called simple cells, the canonical shape consists of elongated side-by-side regions of opposite polarity (figure 28.2D). The temporal component of the receptive field again has a biphasic time course (figure 28.2D). The same receptive field formalism has been used for other sensory modalities, and we illustrate examples from the auditory (figure 28.2E) and olfactory systems (figure 28.2F). Note that in all these cases the dynamics look biphasic. Fundamentally, this means that the neurons emphasize changes in the sensory input.[3]

Why does the brain use neurons with these particular receptive fields? Attempts to explain the receptive field shapes have been particularly effective in the early visual system. One theory postulates that the retina is concerned primarily with transmitting visual information efficiently through the bottleneck of the optic nerve (Atick & Redlich, 1992; Doi et al., 2012; Graham, Chandler, & Field, 2006; Srinivasan, Laughlin, & Dubs, 1982; van Hateren, 1992). In particular, this means avoiding redundancy among the signals of different retinal ganglion cells. The images from the natural world contain a great deal of correlation, because nearby points in space and time tend to have very similar intensity (Field, 1987). In this view, the retinal circuitry filters the incoming movies in space and time to remove those correlations. Formalizing this principle leads to a theory that predicts image filters with center-surround antagonism in space and a biphasic time course, bearing close resemblance to actual receptive fields measured in the retina. Furthermore, the same theory appears to explain a list of performance features of the early visual system (van Hateren, 1993). This has led to widespread acceptance of the efficient coding idea, although some skeptics remain, including its earliest proponent (Barlow, 2001). For example, it appears that lateral inhibition in the early visual system performs much less redundancy reduction than should be possible (Pitkow & Meister, 2012), and there are proposals that it serves a different purpose (Balboa & Grzywacz, 2000).

Deeper into the visual system, once the coding constraints of the optic nerve have fallen away, receptive fields have been interpreted under a different perspective. Here it has been suggested that the brain seeks to represent the stimulus using a large population of neurons that are each active only rarely and are statistically independent of each other (Olshausen & Field, 2004). Such a code can help to highlight "suspicious coincidences" (Barlow, 1994) of firings in the population, which are the hallmarks of higher-level stimulus structures like objects. When analyzing natural visual scenes by this criterion, one predicts receptive fields that respond selectively to edges, similar to those seen to excite neurons in the early visual cortex (figure 28.2D; Bell & Sejnowski, 1997; Olshausen & Field, 1996). A similar correspondence has been noted in the auditory system (Smith & Lewicki, 2006). On the one hand, it is quite remarkable that an *ab initio* theory based only on concepts of signal coding and the statistics of natural stimuli can predict with some success the sensory responses in living brains. On the other hand, the particular response features that it explains, namely receptive fields, do not tell the whole story of visual processing.

[3]Incidentally, the biphasic time course of receptive fields extends even to other phylogenetic kingdoms. The dynamics of human photoreceptors (Schnapf, Nunn, Meister, & Baylor, 1990), bacterial chemosensory responses (Block, Segall, & Berg, 1982), and light responses of fungi (Lipson, 1975) all follow the same time course, except for a scaling of the time axis. It is tempting to postulate this as a universal law, presumably related to the need of all living things to detect changes in their environment. Someday, life will be discovered on other planets. While those creatures may or may not be based on carbon chemistry, it is a safe bet that they have biphasic sensory responses.

Limits of receptive field analysis

For all its sophistication, the method of receptive field measurement—even in its modern embodiment by the LN model—ultimately delivers only partial insights into sensory computations. We focus here on four commonly noted limitations.

First, the LN model for a given neuron often accounts for only a portion of its neural response. This can be tested directly by comparing the firing rate predicted from the receptive field and nonlinearity (Eq. 5) with the actual firing rate. Often this prediction works well when one uses simple artificial stimuli, with limited or tightly defined spatial and temporal structure (Chichilnisky, 2001; Keat, Reinagel, Reid, & Meister, 2001). But when natural stimuli are used as inputs, the simple receptive field models often fail dramatically (David, Mesgarani, Fritz, & Shamma, 2009; Machens, Wehr, & Zador, 2004; Mante, Bonin, & Carandini, 2008; Sharpee, Miller, & Stryker, 2008). Such complex stimuli seem to engage pathways that go beyond linear summation. As an example, consider the "object motion sensitive" ganglion cells found in the vertebrate retina (Baccus et al., 2008; Ölveczky et al., 2003; van Wyk, Taylor, & Vaney, 2006; Zhang, Kim, Sanes, & Meister, 2012). These neurons fire selectively when an object moves with a trajectory different from that of the background. This response is largely invariant to the pattern on the object. But the cells remain silent when the entire retinal image moves in concert. This complex response function is of obvious utility for detecting prey or predator within the visual scene. But it cannot be formulated in terms of an LN model.

Second, receptive fields are fickle and can change on a moment's notice. One finds very commonly that a simple change in the stimulus environment—going from dim to bright lights, or from soft to loud sounds, or from one to two odors—elicits substantial changes in the receptive field (Bair, 2005; Geffen, Broome, Laurent, & Meister, 2009; Geffen, de Vries, & Meister, 2007; Gilbert & Li, 2013; Mante et al., 2008; Nagel & Doupe, 2006; Theunissen, Sen, & Doupe, 2000; figure 28.2F). The LN model may well produce acceptable fits to the response under one stimulus environment, but changing the environment leads to entirely different results for L and N. Sometimes these changes can be interpreted as "adaptation" to the new environment (Hosoya, Baccus, & Meister, 2005; Shapley & Enroth-Cugell, 1984; Wark, Lundstrom, & Fairhall, 2007), but they are a clear indicator that the actual computation performed by the system is more intricate than the LN model. If one knew the true nature of that computation, there would be no need to invoke "adaptation" (Borst, Flanagin, & Sompolinsky, 2005; Garvert & Gollisch, 2013; Ozuysal & Baccus, 2012).

Third, the sensory computation expressed by the LN model is rather primitive. It seems implausible a priori that neurons many synapses into the brain are still concerned with a simple weighted summation of stimuli. For example, the receptive fields quoted for neurons along the mammalian visual pathway are remarkably similar. There is virtually no change from retinal bipolar cells to thalamic relay neurons: they all have spatial profiles shaped like "Mexican hats" with opposite effects from light in the center and surround, and a biphasic temporal profile (figure 28.2A–C). Some change in the spatial profile appears at the level of primary visual cortex, where the antagonistic regions tend to lie side by side rather than concentrically (figure 28.2D). This reflects a rather modest computational accomplishment, given the enormous amount of neural machinery that is engaged, including no fewer than 70 different neuron types in the retina alone. One gets the suspicion that these receptive field measurements have very little to do with what these circuits really accomplish. In the vertebrate retina, we now know that many of the approximately 20 types of retinal ganglion cell perform quite sophisticated image computations (Field & Chichilnisky, 2007; Gollisch & Meister, 2010), which become apparent under more ecologically relevant stimuli. Yet when probed with the white-noise flicker stimulus that is popular for receptive field measurements, these RGCs will always hide their true personality under a modest sombrero.

Finally, and perhaps most fundamentally, receptive field analysis of this type simply does not help much in understanding interesting computations, such as those leading to face cells. This is because the LN neuron has a trivially simple invariance. Its response varies along a single direction in stimulus space (the vector **L** in Eq. 5), and it remains invariant to all the directions orthogonal to it. This basic limitation is independent of the form of the nonlinearity *N*. In our reduced two-receptor system (figure 28.1D), the response function of any LN neuron has straight and parallel contour lines. Given such a straight-laced neuron, we are no closer to constructing the curly response space of a face cell (figure 28.1B) than we were with the original sensory receptors. At best, such a neuron can offer a local tangent to the desired response region. But within the receptive field formalism, one cannot explain why the region should bend around from one tangent to the next. So from the standpoint of stimulus geometry, the receptive field of any LN neuron has the same complexity as that of a sensory receptor, and is no more helpful in explaining the advanced stimulus selectivities one encounters at the higher echelons of sensory systems.

Beyond receptive fields

Experimentally, there have long been indications that the receptive field alone cannot account for sensory neuron responses, even early in the sensory circuits. For example, stimuli that on their own do not elicit a response—and thus fall outside the receptive field—can often modulate the responses in powerful ways. When a simple cell in primary visual cortex is probed with short line segments as stimuli, it has a compact receptive field with elongated response regions (figure 28.2D). But if one adds another line segment outside this receptive field, the response changes dramatically depending on whether the segments are collinear or not (Kapadia, Ito, Gilbert, & Westheimer, 1995). Similarly, the response of a retinal ganglion cell can be powerfully suppressed by stimulus movement in distant regions on the retina, far outside the receptive field (Baccus et al., 2008; Roska & Werblin, 2003). These and similar response components have been described as the "extraclassical" receptive field, or "surround effects," or "contextual influences" (Fitzpatrick, 2000). Presumably these effects contribute to bending a neuron's response surface. This gets us closer to explaining sensory computation, and indeed for that goal the extraclassical effects are more important than the neuron's receptive field alone.

Given the importance of characterizing a neuron's response beyond a single receptive field, how can those additional components be treated in a formal analysis?

1. One approach is to experiment in a stimulus space of reduced dimensionality, for example, with just two independent stimulus components (figure 28.1B). Within that space, one can then map out a neuron's response region in complete detail (Bolinger & Gollisch, 2012; Gollisch, Schutze, Benda, & Herz, 2002). A recent study on retinal ganglion cells showed that their response regions are often highly curved, and that different cell types produce opposite curvatures in the same part of stimulus space (Bolinger & Gollisch, 2012). These could serve as primitive parts for constructing a more complex response region.

2. Alternatively, one can maintain the full dimensionality of the stimulus space, but ask whether there exists a subset of these directions $\mathbf{L}_i$ that can affect the neuron's response. A formal framework for such responses is the multi-LN model:

$$r = N(g_1, \ldots, g_n) = N(\mathbf{S} \bullet \mathbf{L}_1, \ldots, \mathbf{S} \bullet \mathbf{L}_n). \qquad (6)$$

Here the stimulus $\mathbf{S}$ is projected onto n different vectors $\mathbf{L}_i$, producing n scalars g_i. Then the response is computed as a nonlinear function of all these variables (figure 28.3A). There are principled methods for discovering the special vectors $\mathbf{L}_i$ (Marmarelis & Orme, 1993; Schwartz, Pillow, Rust, & Simoncelli, 2006), and indeed many sensory neurons probed this way have more than one special direction in stimulus space (Fairhall et al., 2006; Maravall, Petersen, Fairhall, Arabzadeh, & Diamond, 2007; Rust, Schwartz, Movshon, & Simoncelli, 2005; Slee, Higgs, Fairhall, & Spain, 2005; Touryan, Lau, & Dan, 2002). Already the simple process of spike generation, which turns a membrane current into action potentials, involves two special directions (Agüera y Arcas, Fairhall, & Bialek, 2003). These methods can serve to identify a relevant subspace in which the neuron's computation seems to take place. However, mapping out the nonlinearity $N(\ldots)$ in the response function is a challenge when it depends on more than two variables. Moreover, if it is simple enough to characterize this way, the mathematical form of this response function is still rather restrictive, and implausible a priori for neurons deep into a sensory system.

3. A third approach is to model the neural system as a cascade of LN stages leading up to the sensory neuron in question (French & Korenberg, 1989; Mante et al., 2008; Shapley & Victor, 1981; van Hateren, Ruttiger, Sun, & Lee, 2002). The structure of such a model can be inspired by what is known about the anatomy of the circuits. For example, in modeling retinal ganglion cell responses, one would include circuit elements like bipolar cells and amacrine cells with their known synaptic relationships (Baccus et al., 2008; Chen et al., 2013; Gollisch & Meister, 2008; Greschner, Thiel, Kretzberg, & Ammermüller, 2006; figure 28.3B). Each circuit element performs a simple LN operation on its inputs. Yet a cascade of such simple units—including divergence, convergence, and feedback of signals—can in principle compute arbitrarily complex functions of the stimulus (Cybenko, 1989). This modeling approach has been successful in capturing many "extraclassical" response features (Gollisch & Meister, 2010). It also offers a useful linkage to the actual neural circuits that are the biophysical substrate of all these computations.

Expanded nonlinear representations

With these insights, we can now return to the simple two-receptor system considered above, and ask: how could one construct a neuron with a curved and convoluted response region? One very simple approach is to first construct a population of neurons that have small, local, almost point-like response regions in

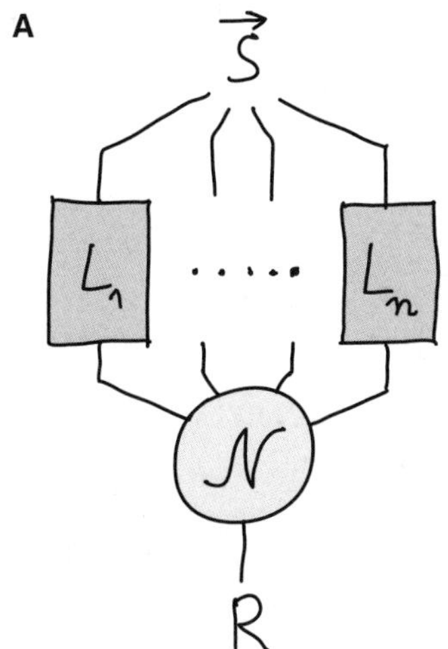

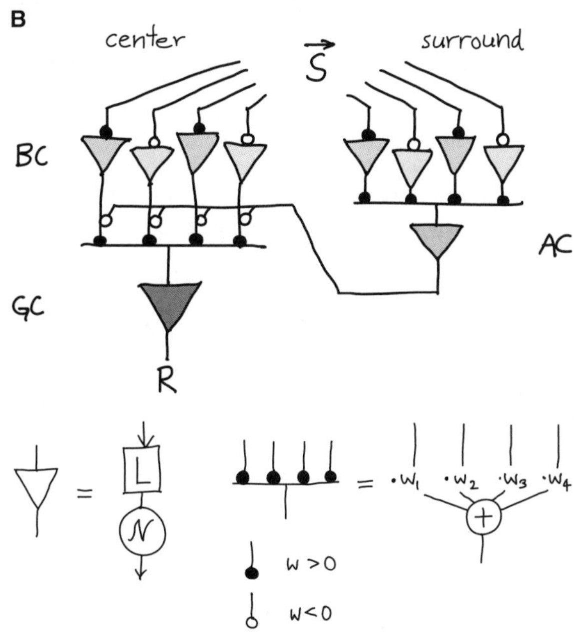

FIGURE 28.3 Response models beyond the receptive field. (A) Multi-LN model. Here the stimulus, **S**, is processed linearly through multiple parallel filters. Their outputs are then combined in a single static nonlinearity, yielding the response, *R*. (B) Circuit model with a cascade of LN stages that may be arranged in series or parallel, including feedforward and feedback pathways. This example takes inspiration from retinal circuitry, and each participating neuron type (On and Off bipolar cells, BC; amacrine cells, AC; ganglion cells, GC) is represented by a simple LN model: a weighted summation of inputs (weights w_i) followed with a temporal filter and a nonlinear response function. With appropriate choice of those parameters, the circuit makes quantitatively accurate predictions for the response of the so-called object motion sensitive ganglion cells that sense differential motion between foreground and background (Ölveczky et al., 2003). (See color plate 26.)

stimulus space. Then one can combine a suitable collection of these to create a neuron with an arbitrary response region.

In principle, this can be achieved in three steps (figure 28.1E–G):

1. For each of the two stimulus axes, create n neurons that respond to only a small range along that axis (e.g., neurons A and B in figure 28.1E).
2. Combine such neurons from each axis with a logical AND to create a large population of n^2 neurons that each respond in a tiny region of the stimulus space (neurons C_i in figure 28.1F; for example, neuron C_1 responds only when stimulus intensity S_1 is medium and intensity S_2 is maximal).
3. Combine many of these by a logical OR to create a neuron with a thin and convoluted response region (neuron D in figure 28.1G).

Note that the neurons from step 1 can still be described by an LN model—though with an unconventional hump-shaped nonlinearity—and thus have a defined receptive field, whereas this is no longer the case for the neurons in step 2. A key ingredient of this scheme is the dramatic expansion of the neural population. Starting with just two receptor neurons, after step 2 the stimuli are represented by n^2 neurons. This is followed by an equally dramatic contraction in step 3, which yields a neuron with complex response function.

Indeed, one can find such extreme expansions and contractions in several sensory systems. For example, in the retina, one cone photoreceptor connects to ~10 bipolar cells (Wässle, Puller, Muller, & Haverkamp, 2009), and each bipolar cell in turn has many synaptic terminals. Each of these terminals may receive a different set of amacrine cell inputs, resulting in different response properties (Asari & Meister, 2012; Baden, Berens, Bethge, & Euler, 2013). Finally, a retinal ganglion cell pools over a select subset of all these bipolar cell terminals to construct its specific response function (Gollisch & Meister, 2010). From the retina to primary visual cortex, one finds another expansion in the neuron number by a factor of ~100. At the same time, the activity in these cortical populations becomes more and more sparse (Barth & Poulet, 2012; Hromadka & Zador, 2009; Isaacson, 2010; Olshausen & Field, 2004; Wolfe, Houweling, & Brecht, 2010): whereas natural stimulation will drive early sensory neurons to fire much of the time, neurons in the cortex are active more rarely, with an average spike rate estimated at only one to three spikes per second (Attwell & Laughlin, 2001; Lennie, 2003). Of course, each cortical neuron pools several thousand of such inputs to construct its own response space. A similar expansion into a sparsely active population has been demonstrated in the insect olfactory system: From the second-order to third-order neurons in this circuit, the number of distinct response types in the population increases, and the activity within each type decreases dramatically (Laurent, 2002).

As illustrated above, the purpose of such a nonlinear expansion may be to allow for construction of complex response functions in the subsequent pooling stage. Interestingly, this does not require a careful design of response functions in the expanded representation. Even random combinations of features are sufficient to construct very complex response spaces (Rigotti, Ben Dayan Rubin, Wang, & Fusi, 2010). However, the nonlinearity is essential: only by bending the input into higher dimensions can linear projections produce a different outcome than on the original space. Surprisingly, only one layer of nonlinear neurons is required to generate any arbitrary function (Cybenko, 1989), although this may require a huge expansion and delicate cancellations in the subsequent convergence. A much better use of neural resources appears to be a series of nonlinear expansions and linear projections, an architecture that is generally described as a "deep network" (Bengio, Courville, & Vincent, 2013).

This kind of architecture forms the basis of an influential model of sensory processing leading to object detection (Mel, 1997; Riesenhuber & Poggio, 1999; Serre, Oliva, & Poggio, 2007). According to this model, the brain constructs complex percepts by a sequence of linear and nonlinear operations. The linear operations reduce the dimensionality of the input to encode only the presence of particular patterns, generating selectivity. The nonlinear operations compute the maximum over similar patterns that differ only by a transformation like shifting or rotation, thereby generating invariance to that transformation. A cascade of these operations can develop sensitivity to complex combinations of object features, while retaining invariance to changes in viewpoint, scale, or illumination (see chapter 30 by Nicole Rust, this volume).

Thus we have arrived at a plausible model to explain the puzzling phenomenon of "face cells" that motivated our foray, though it should be said that the computational models don't yet match human performance on such tasks (DiCarlo, Zoccolan, & Rust, 2012). Along the way we learned that the measurement of receptive fields—perhaps the single most common activity in sensory neuroscience—is of limited use for understanding how sensory computations arise. It needs to be extended with more intricate and flexible models of neural responses. Ultimately, the most effective modeling strategy will be to actually understand the neural circuits that underlie the responses: the arrangement of physiological cell types, their connectivity, synaptic integration, and the patterns of signal flow. This parallel pursuit of structural and functional information has greatly helped in understanding peripheral sensory circuits, like retina and olfactory bulb, and will ultimately be essential to cracking the mysteries of cortical circuits as well.

REFERENCES

Agüera y Arcas, B., Fairhall, A. L., & Bialek, W. (2003). Computation in a single neuron: Hodgkin and Huxley revisited. *Neural Comput, 15*, 1715–1749.

Asari, H., & Meister, M. (2012). Divergence of visual channels in the inner retina. *Nat Neurosci, 15*, 1581–1589.

Atick, J. J., & Redlich, A. N. (1992). What does the retina know about natural scenes? *Neural Comput, 4*, 196–210.

Attwell, D., & Laughlin, S. B. (2001). An energy budget for signaling in the grey matter of the brain. *J Cereb Blood Flow Metab, 21*, 1133–1145.

Baccus, S. A., Ölveczky, B. P., Manu, M., & Meister, M. (2008). A retinal circuit that computes object motion. *J Neurosci, 28*, 6807–6817.

Baden, T., Berens, P., Bethge, M., & Euler, T. (2013). Spikes in mammalian bipolar cells support temporal layering of the inner retina. *Curr Biol, 23*, 48–52.

Bair, W. (2005). Visual receptive field organization. *Curr Opin Neurobiol, 15*, 459–464.

Balboa, R. M., & Grzywacz, N. M. (2000). The role of early retinal lateral inhibition: More than maximizing luminance information. *Vis Neurosci, 17*, 77–89.

Barlow, H. B. (1972). Single units and sensation: A neuron doctrine for perceptual psychology? *Perception, 1*, 371–394.

Barlow, H. B. (1994). What is the computational goal of the neocortex? In C. Koch & J. L. Davis (Eds.), *Large-scale neuronal theories of the brain* (pp. 1–22). Cambridge, MA: MIT Press.

Barlow, H. B. (2001). Redundancy reduction revisited. *Network, 12*, 241–253.

Barlow, H. B. (2009). Single units and sensation: A neuron doctrine for perceptual psychology? *Perception, 38*, 795–798.

Barth, A. L., & Poulet, J. F. (2012). Experimental evidence for sparse firing in the neocortex. *Trends Neurosci, 35*, 345–355.

Bell, A. J., & Sejnowski, T. J. (1997). The "independent components" of natural scenes are edge filters. *Vis Res, 37*, 3327–3338.

Benardete, E. A., & Kaplan, E. (1997). The receptive field of the primate P retinal ganglion cell, I: Linear dynamics. *Vis Neurosci, 14*, 169–185.

Bengio, Y., Courville, A., & Vincent, P. (2013). Representation learning: A review and new perspectives. *IEEE Trans Pattern Anal Mach Intell, 35*, 1798–1828.

Block, S. M., Segall, J. E., & Berg, H. C. (1982). Impulse responses in bacterial chemotaxis. *Cell, 31*, 215–226.

Bolinger, D., & Gollisch, T. (2012). Closed-loop measurements of iso-response stimuli reveal dynamic nonlinear stimulus integration in the retina. *Neuron, 73*, 333–346.

Borst, A., Flanagin, V. L., & Sompolinsky, H. (2005). Adaptation without parameter change: Dynamic gain control in motion detection. *Proc Natl Acad Sci USA, 102,* 6172–6176.

Burkhardt, D. A., Bartoletti, T. M., & Thoreson, W. B. (2011). Center/surround organization of retinal bipolar cells: High correlation of fundamental responses of center and surround to sinusoidal contrasts. *Vis Neurosci, 28,* 183–192.

Carr, C. E., & Konishi, M. (1990). A circuit for detection of interaural time differences in the brain stem of the barn owl. *J Neurosci, 10,* 3227–3246.

Chen, E. Y., Marre, O., Fisher, C., Schwartz, G., Levy, J., da Silveira, R. A., & Berry, M. J. 2nd. (2013). Alert response to motion onset in the retina. *J Neurosci, 33,* 120–132.

Chichilnisky, E. J. (2001). A simple white noise analysis of neuronal light responses. *Network, 12,* 199–213.

Cybenko, G. (1989). Approximation by superpositions of a sigmoidal function. *Math Control Signal, 2,* 303–314.

David, S. V., Mesgarani, N., Fritz, J. B., & Shamma, S. A. (2009). Rapid synaptic depression explains nonlinear modulation of spectro-temporal tuning in primary auditory cortex by natural stimuli. *J Neurosci, 29,* 3374–3386.

DeAngelis, G. C., Ohzawa, I., & Freeman, R. D. (1995). Receptive-field dynamics in the central visual pathways. *Trends Neurosci, 18,* 451–458.

DiCarlo, J. J., & Cox, D. D. (2007). Untangling invariant object recognition. *Trends Cogn Sci, 11,* 333–341.

DiCarlo, J. J., Zoccolan, D., & Rust, N. C. (2012). How does the brain solve visual object recognition? *Neuron, 73,* 415–434.

Doi, E., Gauthier, J. L., Field, G. D., Shlens, J., Sher, A., Greschner, M., ... Simoncelli, E. P. (2012). Efficient coding of spatial information in the primate retina. *J Neurosci, 32,* 16256–16264.

Enroth-Cugell, C., Robson, J. G., Schweitzer-Tong, D. E., & Watson, A. B. (1983). Spatio-temporal interactions in cat retinal ganglion cells showing linear spatial summation. *J Physiol, 341,* 279–307.

Eriksson, J. T. (1996). Impact of information compression on intellectual activities in the brain. *Int J Neural Syst, 7,* 543–550.

Fahey, P. K., & Burkhardt, D. A. (2003). Center-surround organization in bipolar cells: Symmetry for opposing contrasts. *Vis Neurosci, 20,* 1–10.

Fairhall, A. L., Burlingame, C. A., Narasimhan, R., Harris, R. A., Puchalla, J. L., & Berry, M. J. 2nd. (2006). Selectivity for multiple stimulus features in retinal ganglion cells. *J Neurophysiol, 96,* 2724–2738.

Field, D. J. (1987). Relations between the statistics of natural images and the response properties of cortical cells. *J Opt Soc Am A, 4,* 2379–2394.

Field, G. D., & Chichilnisky, E. J. (2007). Information processing in the primate retina: Circuitry and coding. *Annu Rev Neurosci, 30,* 1–30.

Fitzpatrick, D. (2000). Seeing beyond the receptive field in primary visual cortex. *Curr Opin Neurobiol, 10,* 438–443.

Freiwald, W. A., & Tsao, D. Y. (2010). Functional compartmentalization and viewpoint generalization within the macaque face-processing system. *Science, 330,* 845–851.

French, A. S., & Korenberg, M. J. (1989). A nonlinear cascade model for action potential encoding in an insect sensory neuron. *Biophys J, 55,* 655–661.

Garvert, M. M., & Gollisch, T. (2013). Local and global contrast adaptation in retinal ganglion cells. *Neuron, 77,* 915–928.

Geffen, M. N., Broome, B. M., Laurent, G., & Meister, M. (2009). Neural encoding of rapidly fluctuating odors. *Neuron, 61,* 570–586.

Geffen, M. N., de Vries, S. E., & Meister, M. (2007). Retinal ganglion cells can rapidly change polarity from Off to On. *PLoS Biol, 5,* e65.

Gilbert, C. D., & Li, W. (2013). Top-down influences on visual processing. *Nat Rev Neurosci, 14,* 350–363.

Gollisch, T., & Meister, M. (2008). Rapid neural coding in the retina with relative spike latencies. *Science, 319,* 1108–1111.

Gollisch, T., & Meister, M. (2010). Eye smarter than scientists believed: Neural computations in circuits of the retina. *Neuron, 65,* 150–164.

Gollisch, T., Schutze, H., Benda, J., & Herz, A. V. (2002). Energy integration describes sound-intensity coding in an insect auditory system. *J Neurosci, 22,* 10434–10448.

Graham, D. J., Chandler, D. M., & Field, D. J. (2006). Can the theory of "whitening" explain the center-surround properties of retinal ganglion cell receptive fields? *Vis Res, 46,* 2901–2913.

Greschner, M., Thiel, A., Kretzberg, J., & Ammermüller, J. (2006). Complex spike-event pattern of transient ON-OFF retinal ganglion cells. *J Neurophysiol, 96,* 2845–2856.

Gross, C. G. (1992). Representation of visual stimuli in inferior temporal cortex. *Philos Trans R Soc Lond B Biol Sci, 335,* 3–10.

Hare, W. A., & Owen, W. G. (1990). Spatial organization of the bipolar cell's receptive field in the retina of the tiger salamander. *J Physiol, 421,* 223–245.

Heiligenberg, W. (1991). *Neural nets in electric fish.* Cambridge, MA: MIT Press.

Hosoya, T., Baccus, S. A., & Meister, M. (2005). Dynamic predictive coding by the retina. *Nature, 436,* 71–77.

Hromadka, T., & Zador, A. M. (2009). Representations in auditory cortex. *Curr Opin Neurobiol, 19,* 430–433.

Hubel, D. H., & Wiesel, T. N. (1961). Integrative action in the cat's lateral geniculate body. *J Physiol, 155,* 385–398.

Hubel, D. H., & Wiesel, T. N. (1962). Receptive fields, binocular interaction and functional architecture in the cat's visual cortex. *J Physiol, 160,* 106–154.

Isaacson, J. S. (2010). Odor representations in mammalian cortical circuits. *Curr Opin Neurobiol, 20,* 328–331.

Kanwisher, N., McDermott, J., & Chun, M. M. (1997). The fusiform face area: A module in human extrastriate cortex specialized for face perception. *J Neurosci, 17,* 4302–4311.

Kapadia, M. K., Ito, M., Gilbert, C. D., & Westheimer, G. (1995). Improvement in visual sensitivity by changes in local context: Parallel studies in human observers and in V1 of alert monkeys. *Neuron, 15,* 843–856.

Keat, J., Reinagel, P., Reid, R. C., & Meister, M. (2001). Predicting every spike: A model for the responses of visual neurons. *Neuron, 30,* 803–817.

Konishi, M. (2003). Coding of auditory space. *Annu Rev Neurosci, 26,* 31–55.

Konishi, M. (2006). Behavioral guides for sensory neurophysiology. *J Comp Physiol A, 192,* 671–676.

Kuffler, S. W. (1953). Discharge patterns and functional organization of mammalian retina. *J Neurophysiol, 16,* 37–68.

Laurent, G. (2002). Olfactory network dynamics and the coding of multidimensional signals. *Nat Rev Neurosci, 3,* 884–895.

Lennie, P. (2003). The cost of cortical computation. *Curr Biol, 13,* 493–497.

Lipson, E. D. (1975). White noise analysis of Phycomyces light growth response system. I. Normal intensity range. *Biophys J, 15,* 989–1011.

Machens, C. K., Wehr, M. S., & Zador, A. M. (2004). Linearity of cortical receptive fields measured with natural sounds. *J Neurosci, 24,* 1089–1100.

Mante, V., Bonin, V., & Carandini, M. (2008). Functional mechanisms shaping lateral geniculate responses to artificial and natural stimuli. *Neuron, 58,* 625–638.

Maravall, M., Petersen, R. S., Fairhall, A. L., Arabzadeh, E., & Diamond, M. E. (2007). Shifts in coding properties and maintenance of information transmission during adaptation in barrel cortex. *PLoS Biol, 5,* e19.

Marmarelis, V. Z., & Orme, M. E. (1993). Modeling of neural systems by use of neuronal modes. *IEEE Trans Biomed Eng, 40,* 1149–1158.

Martinez, L. M., Wang, Q., Reid, R. C., Pillai, C., Alonso, J. M., Sommer, F. T., & Hirsch, J. A. (2005). Receptive field structure varies with layer in the primary visual cortex. *Nat Neurosci, 8,* 372–379.

Meister, M., & Berry, M. J. (1999). The neural code of the retina. *Neuron, 22,* 435–450.

Mel, B. W. (1997). SEEMORE: Combining color, shape, and texture histogramming in a neurally inspired approach to visual object recognition. *Neural Comput, 9,* 777–804.

Nagel, K. I., & Doupe, A. J. (2006). Temporal processing and adaptation in the songbird auditory forebrain. *Neuron, 51,* 845–859.

Nagel, K. I., & Doupe, A. J. (2008). Organizing principles of spectro-temporal encoding in the avian primary auditory area field L. *Neuron, 58,* 938–955.

Olshausen, B. A., & Field, D. J. (1996). Emergence of simple-cell receptive field properties by learning a sparse code for natural images. *Nature, 381,* 607–609.

Olshausen, B. A., & Field, D. J. (2004). Sparse coding of sensory inputs. *Curr Opin Neurobiol, 14,* 481–487.

Ölveczky, B. P., Baccus, S. A., & Meister, M. (2003). Segregation of object and background motion in the retina. *Nature, 423,* 401–408.

Ozuysal, Y., & Baccus, S. A. (2012). Linking the computational structure of variance adaptation to biophysical mechanisms. *Neuron, 73,* 1002–1015.

Pierce, J. R., & Karlin, J. E. (1957). Reading rates and the information rate of a human channel. *Bell Syst Tech J, 36,* 497–516.

Pitkow, X., & Meister, M. (2012). Decorrelation and efficient coding by retinal ganglion cells. *Nat Neurosci, 15,* 628–635.

Quiroga, R. Q. (2012). Concept cells: The building blocks of declarative memory functions. *Nat Rev Neurosci, 13,* 587–597.

Quiroga, R. Q., Reddy, L., Kreiman, G., Koch, C., & Fried, I. (2005). Invariant visual representation by single neurons in the human brain. *Nature, 435,* 1102–1107.

Reid, R. C., & Shapley, R. M. (2002). Space and time maps of cone photoreceptor signals in macaque lateral geniculate nucleus. *J Neurosci, 22,* 6158–6175.

Riesenhuber, M., & Poggio, T. (1999). Hierarchical models of object recognition in cortex. *Nat Neurosci, 2,* 1019–1025.

Rigotti, M., Ben Dayan Rubin, D., Wang, X. J., & Fusi, S. (2010). Internal representation of task rules by recurrent dynamics: The importance of the diversity of neural responses. *Front Comput Neurosci, 4*(24).

Ringach, D. L. (2004). Mapping receptive fields in primary visual cortex. *J Physiol, 558,* 717–728.

Rodieck, R. W., & Stone, J. (1965). Analysis of receptive fields of cat retinal ganglion cells. *J Neurophysiol, 28,* 832–849.

Rolls, E. T. (1992). Neurophysiological mechanisms underlying face processing within and beyond the temporal cortical visual areas. *Philos Trans R Soc Lond B Biol Sci, 335,* 11–20.

Rose, G. J., Kawasaki, M., & Heiligenberg, W. (1988). "Recognition units" at the top of a neuronal hierarchy? Prepacemaker neurons in Eigenmannia code the sign of frequency differences unambiguously. *J Comp Physiol A, 162,* 759–772.

Roska, B., & Werblin, F. (2003). Rapid global shifts in natural scenes block spiking in specific ganglion cell types. *Nat Neurosci, 6,* 600–608.

Rust, N. C., Schwartz, O., Movshon, J. A., & Simoncelli, E. P. (2005). Spatiotemporal elements of macaque V1 receptive fields. *Neuron, 46,* 945–956.

Schnapf, J. L., Nunn, B. J., Meister, M., & Baylor, D. A. (1990). Visual transduction in cones of the monkey *Macaca fascicularis. J Physiol, 427,* 681–713.

Schwartz, O., Pillow, J. W., Rust, N. C., & Simoncelli, E. P. (2006). Spike-triggered neural characterization. *J Vis, 6,* 484–507.

Serre, T., Oliva, A., & Poggio, T. (2007). A feedforward architecture accounts for rapid categorization. *Proc Natl Acad Sci USA, 104,* 6424–6429.

Shapley, R., & Enroth-Cugell, C. (1984). Visual adaptation and retinal gain controls. *Prog Retin Eye Res, 3,* 263–346.

Shapley, R. M., & Victor, J. D. (1981). How the contrast gain control modifies the frequency responses of cat retinal ganglion cells. *J Physiol, 318,* 161–79.

Sharpee, T. O., Miller, K. D., & Stryker, M. P. (2008). On the importance of static nonlinearity in estimating spatiotemporal neural filters with natural stimuli. *J Neurophysiol, 99,* 2496–2509.

Sherrington, C. S. (1906). Observations on the scratch-reflex in the spinal dog. *J Physiol, 34,* 1–50.

Slee, S. J., Higgs, M. H., Fairhall, A. L., & Spain, W. J. (2005). Two-dimensional time coding in the auditory brainstem. *J Neurosci, 25,* 9978–9988.

Smith, E. C., & Lewicki, M. S. (2006). Efficient auditory coding. *Nature, 439,* 978–982.

Srinivasan, M. V., Laughlin, S. B., & Dubs, A. (1982). Predictive coding: A fresh view of inhibition in the retina. *Proc R Soc Lond B Biol Sci, 216,* 427–459.

Theunissen, F. E., Sen, K., & Doupe, A. J. (2000). Spectral-temporal receptive fields of nonlinear auditory neurons obtained using natural sounds. *J Neurosci, 20,* 2315–2331.

Touryan, J., Lau, B., & Dan, Y. (2002). Isolation of relevant visual features from random stimuli for cortical complex cells. *J Neurosci, 22,* 10811–10818.

Tsao, D. Y., Freiwald, W. A., Knutsen, T. A., Mandeville, J. B., & Tootell, R. B. (2003). Faces and objects in macaque cerebral cortex. *Nat Neurosci, 6,* 989–995.

TSAO, D. Y., & LIVINGSTONE, M. S. (2008). Mechanisms of face perception. *Annu Rev Neurosci, 31,* 411–437.
VAN HATEREN, J. H. (1992). A theory of maximizing sensory information. *Biol Cybern, 68,* 23–29.
VAN HATEREN, J. H. (1993). Spatiotemporal contrast sensitivity of early vision. *Vis Res, 33,* 257–267.
VAN HATEREN, J. H., RUTTIGER, L., SUN, H., & LEE, B. B. (2002). Processing of natural temporal stimuli by macaque retinal ganglion cells. *J Neurosci, 22,* 9945–9960.
VAN WYK, M., TAYLOR, W. R., & VANEY, D. I. (2006). Local edge detectors: A substrate for fine spatial vision at low temporal frequencies in rabbit retina. *J Neurosci, 26,* 13250–13263.
WARK, B., LUNDSTROM, B. N., & FAIRHALL, A. (2007). Sensory adaptation. *Curr Opin Neurobiol, 17,* 423–429.
WÄSSLE, H., PULLER, C., MULLER, F., & HAVERKAMP, S. (2009). Cone contacts, mosaics, and territories of bipolar cells in the mouse retina. *J Neurosci, 29,* 106–117.
WOLFE, J., HOUWELING, A. R., & BRECHT, M. (2010). Sparse and powerful cortical spikes. *Curr Opin Neurobiol, 20,* 306–312.
ZHANG, Y., KIM, I. J., SANES, J. R., & MEISTER, M. (2012). The most numerous ganglion cell type of the mouse retina is a selective feature detector. *Proc Natl Acad Sci USA, 109,* 2391–2398.

29 Noise, Correlations, and Information in Neural Population Codes

ELAD SCHNEIDMAN

ABSTRACT The brain relies on the spiking patterns of large groups of neurons to represent information. Uncovering the design principles of this "neural code" is fundamental to our understanding of information processing and computation in the brain. This task is made difficult by the fact that neural responses to their stimuli are highly selective and noisy, and because population activity patterns are often correlated and the number of potential population patterns is exponential in the size of the population. We present here the fundamental questions regarding the design and structure of probabilistic and correlated neural population codes and their implication for encoding and decoding. We show that correlations are neither good nor bad, but that the combination of stimulus-induced correlations and joint fluctuations of the cells balance the capacity of the code and its error-correction properties. We then show that the typically weak signal and noise correlations observed at the level of cell pairs can add up to give strong signal and noise correlations in large populations. Finally, we suggest that population codes may be designed to adapt to stimulus statistics in a way that would give collective coding that balances the information content and the code's readability or learnability.

Information is represented in the brain by the action potentials of large groups of neurons. These basic symbols form temporal sequences of multi-neuron spiking and silences that are the universal language that our brain uses. Extending the language metaphor, to decipher this code we need to understand how this alphabet is used to form the vocabulary, what the grammar is, and, ultimately, the content of this neural code.

Much of our understanding of how information is "encoded" in the brain relies on studies of single neurons. As most of what we care about is the result of the joint activity of many neurons working together, the nature of the code of populations is central to neuroscience and our understanding of the brain. Experimentally, it has been hard to record the joint activity of many neurons over long periods of time. Since this task has become more feasible in recent years using multi-electrode arrays or imaging, we will most likely have in the not-too-distant future recordings of very large populations and even recordings from full, large neural circuits. It is not clear, however, how we should interpret the code of such a large number of neurons.

Representation of information in the brain is necessarily noisy because of the stochastic nature of neurons and synapses. Overcoming this noise requires some form of error correction, but our understanding of how the brain might do this is, at best, limited. While different brain areas show distinct spiking patterns and are coding stimuli of different modalities, universal features of the code include population spiking patterns that are sparse, noisy, correlated, and adaptive (Rieke, Warland, de Ruyter van Steveninck, & Bialek, 1997). We present here some of the fundamental design principles of the code of populations of neurons and, in particular, how noise and correlations are shaping it.

We suggest that neural noise means that we must rely on probability distributions in describing and analyzing the code, and that the brain must do so as well. We argue that the combinatorial number of potential activity patterns and correlations implies that understanding the code of large populations must rely on identifying simplifying principles of the design of neural codes. We present an information theory–based framework for assessing the role and impact of correlations and noise between cells and how it shapes the nature of the code. We review the dominant theoretical idea of redundancy reduction as a design principle of neural population codes and its limitations, and suggest instead that neural population codes rely on correlated collective coding that balances the amount of conveyed information with the readability or "learnability" of the code.

The neural code

The universality of spikes as the basic symbols of neuronal activity has made it common to talk about the "neural code" (Rieke et al., 1997). But although the "letters" of the code are universal in the brain (i.e., spikes are the key carrier of information), the organization of these letters into words and their meaning differs between brain areas, between conspecifics, and presumably between species. Moreover, the mapping between stimuli and neural response is not stable even for the same cells over time due to adaptation and learning.

At the level of single cells, the study of the neural code has focused on what kind of information is carried by neurons of different types and in different brain areas, and how this information is encoded. In particular, a central question has been whether the main carrier of information is the time-dependent spiking rates of neurons (rate coding), and how much is carried by the precise timing of the spikes (temporal coding) or the relative timing between spikes within the spiking patterns (Gerstner, Kreiter, Markram, & Herz, 1997; Perkel & Bullock, 1968; Rieke et al., 1997). At the level of populations, coding has often been explored in terms of averaging over cells by combining their activity linearly, but also in terms of the magnitude and structure of correlations between cells (Ahissar et al., 1992; Georgopoulos, Schwartz, & Kettner, 1986; Gray, König, Engel, & Singer, 1989; Hatsopoulos, Ojakangas, Paninski, & Donoghue, 1998; Martignon et al., 2000; Meister, Lagnado, & Baylor, 1995; Panzeri, Schultz, Treves, & Rolls, 1999; Pillow et al., 2008; Riehle, 1997; Schneidman, Berry, Segev, & Bialek, 2006; Stopfer, Bhagavan, Smith, & Laurent, 1997; Vaadia et al., 1995).

It seems almost obvious to expect that the structure of the code of retinal neurons would be different than that of the olfactory bulb, for example, due to differences in the stimulus structure, the time scales of stimulus changes, and the nature of spatiotemporal correlations in the stimulus. Similarly, the spiking patterns of neurons in the hippocampus and those in the visual cortex would probably have different organizational principles, due to the nature of information that is represented, time scales, neural noise, communication with other areas, and so on. Thus, it might be more appropriate to replace the notion of "the neural code" with a set of neural codes that are used in the brain. Moreover, even within the same neural circuit, the structure of the code may change when the stimulus statistics change. Despite these individual differences, many aspects of the code are universal, and must be related. Sensory systems are a natural place to explore the nature of neural codes since they are close to the stimulus, and we therefore have a sense of their primary role as conveying information about the outside world or of what is being encoded.

Neurons respond to stimuli with sparse and noisy spiking patterns

Almost everywhere in the brain, neurons are silent most of the time (e.g., Shoham, O'Connor, & Segev, 2006), and respond with spikes only to a small fraction of the stimuli they are presented with. Because the mapping from stimuli to spikes at the level of a single neuron is so selective, it has been common (following Hartline, 1938) to characterize the properties of neurons in terms of the stimulus feature that a cell responds to, or "encodes." This classic notion of the "receptive field"—which describes the part of stimulus space or the properties of the stimuli that a neuron responds to—is intuitive and simple to describe, but is often a little too simplistic. The inferred mapping between stimuli and responses of a neuron depends critically on the set of the stimuli used to assess this mapping. Analysis of neural responses to simple artificial stimuli often suggests an easy way to describe the feature selectivity of a neuron. Yet the response of the same neuron to other kinds of stimuli, and in particular rich naturalistic stimuli, may differ considerably than the picture one might form based on the response to the artificial stimuli (David, Vinje, & Gallant, 2004). In particular, even stereotyped simple functions that one might infer from "standard" stimuli may have a qualitatively different feature selectivity when stimuli that are considered irrelevant are used (e.g., Geffen, de Vries, & Meister, 2007).

The common interpretation of feature selectivity of neurons is missing another aspect of neuronal encoding. The responses of neurons, even to their "preferred" stimuli, are not deterministic or reliable. Somewhat surprisingly, this "noise" in neural function and response is significant in almost every part of the nervous system, from synaptic unreliability to neural spiking, network activity, and behavior (Calvin & Stevens, 1968; De Ruyter van Steveninck, Lewen, Strong, Koberle, & Bialek, 1997; Harris & Wolpert, 1998; Körding & Wolpert, 2004; Reich, Victor, Knight, Ozaki, & Kaplan, 1997; Stevens & Zador, 1998; Warzecha & Egelhaaf, 1999). In particular, the variability of neural responses to repeated presentations of the same stimulus is apparent at different aspects of the spike trains—the number of spikes in a given window, the temporal structure of the spikes, and time delay with respect to the stimulus (figure 29.1).

Neural variability may arise from a variety of sources (Faisal, Selen, & Wolpert, 2008). One such source is sensor noise, such as bending of hair cells (Denk & Webb, 1992), fixational eye movements (Martinez-Conde, Macknik, & Hubel, 2004), and so on. These are usually associated with the mechanistic biophysical properties and the limits of the biological "hardware." Another source of variability is synaptic fluctuations, which result in unreliable response on the postsynaptic side to incoming presynaptic spikes (Destexhe,

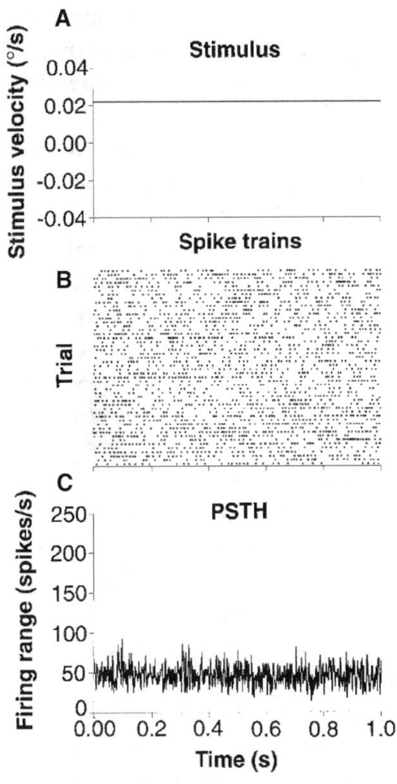

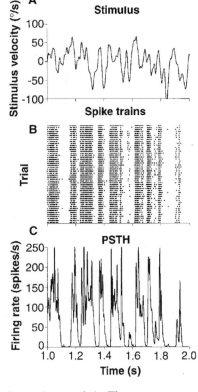

FIGURE 29.1 Reliability and reproducibility of in vivo spiking patterns in response to different stimuli. A fly (*Calliphora vicina*) viewed a pattern of random bars that was moved across the visual either with constant or dynamic velocity. The spiking activity of H1, a motion-sensitive neuron in the fly's visual system, was recorded from an immobilized fly over multiple repetitions of the same visual stimulus. *Left*: Spike-pattern reliability for constant stimuli. (A) A random bar pattern was presented across the visual field at constant speed (0.022 degrees/sec) and in the preferred direction of the H1neuron in the fly. (B) A set of 50 response traces to thestimulus in panel A, each lasting 1 sec, taken 20 sec apart. The occurrence of each spike is shown as a dot. The traces were taken from a segment of the experiment where transient responses have decayed. (C) The peristimulus time histogram (PSTH; bin width 3 ms, 96 presentations), which describes the rate at which spikes were generated in response to the stimulus shown in panel A. The responses to constant stimuli gave highly irregular and noisy spiking patterns. *Right*: Spike-pattern reliability for dynamic stimuli. (A) The fly viewed the same spatial pattern as in the left panel but now moving with a time-dependent velocity, part of which is shown. The motion approximates a random walk with diffusion constant $D \sim 14$ degrees^2/sec. For illustration, the waveform shown is low-pass filtered. In the experiment, a 10-sec waveform is presented 900 times, every 20 sec. (B) A set of 50 response traces to the repeated presentations of the stimulus waveform shown in panel A. (C) Averaged rate (PSTH) for the same segment. The rate is strongly modulated, but its time-average is very close to that in the left panel. Variability and reproducibility were much higher than for the constant stimulus, but still showed diverse, stimulus-dependent reliability. (Adapted from de Ruyter van Steveninck et al., 1997.) Such results have parallels in many experiments on sensory neurons.

Rudolph, Fellous, & Sejnowski, 2001; Seung, 2003; Deweese & Zador, 2004; Stevens & Wang, 1994). Yet another source is the stochastic nature of ion channel opening and closing (DeFelice, 1981; White, Rubinstein, & Kay, 2000), which means that even when presented with the exact same intracellular input, we can expect variability of the resulting spike train (Mainen & Sejnowski, 1995; Schneidman, Freedman, & Segev, 1998). In addition to inherent randomness or "thermal" noise, we might get variance in the response to the same exact inputs without true "randomness." Variability may result from chaotic behavior at the level of single cells, or from the dynamics of neuronal networks (see Timme, Wolf, & Geisel, 2002; van Vreeswijk & Sompolinsky, 1996). The combination of all these sources of randomness and chaos shapes the probabilistic mapping from stimuli to responses, giving a surprisingly high level of variability in many cases (see more on this below).

Thus, rather than describing neural encoding in terms of their high selectivity (which is the basis of the notion of "grandmother cells"), the mapping from stimuli to responses must be described in terms of a probability distribution over the responses to each stimulus. This implies a computational cost in terms of learning the code and processing it, which then requires accurate sampling of the encoding distribution. Therefore, understanding the design and content of the code relies on the characterization of the variability of the temporal spiking patterns of a single cell and the spatiotemporal patterns of populations.

Noise and unreliability of the neural code limit its capacity and shape the structure of the code and processing

The noise or unreliability of neural responses has a crucial role in determining how information is carried by neurons and how it may be interpreted or read by neurons "down the road." If neurons were noise-free, each stimulus would result in a single particular neural "code word." The capacity of the neural vocabulary could then be used efficiently, with every potential spiking pattern carrying distinct information. But, since the mapping from stimuli to responses is probabilistic, then for a given presentation of s, the response, r, is described by the probability distribution over the possible responses $p(r|s)$. The readout of information from the neural response then depends on which particular r was "chosen." The noise also means that different stimuli may result in the exact same neural response. Therefore, there is also a probabilistic mapping from responses to stimuli, and so observing a response r, our knowledge of the stimulus that has caused it is given by $p(s|r)$. This probabilistic relation sets the limits of the accuracy of decoding or reconstructing the stimulus, and in particular it means that it would generally be impossible to get perfect reconstruction of a stimulus. Furthermore, the noise also determines the similarity of neural responses in terms of their meaning and the similarity of stimuli: what makes stimuli similar for the brain is defined by the overlap of the distributions $p(r|s)$ for different s, and the semantic similarity of responses is then given by the overlap of $p(s|r)$ for different r. These can be considerably different from our intuition of similarity of the stimulus (Curto & Itskov, 2008; Ganmor, Segev, & Schneidman, 2011a; Kiani, Esteky, Mirpour, & Tanaka, 2007; Tkacik, Granot-Atedgi, Segev, & Schneidman, 2013).

To assess what do the neural responses convey about the stimulus and the impact of noise, we need to quantify the probabilistic nature of the relations between the stimuli $\{s\}$ and responses $\{r\}$. In particular, we seek a measure that captures the relations in both directions—both the encoding distribution $p(r|s)$ and the inverse distribution $p(s|r)$, which can be viewed as the basis for decoding a neural response; these two distributions are related through Bayes's rule $p(s|r) = p(r|s)*p(s)/p(r)$. Such a measure of the relations between s and r should also avoid arbitrary assumptions about what features of the spiking patterns matter.

Information theory gives a first-principled way to do this using a minimal set of assumptions (Cover & Thomas, 2012). It does not assume any particular kind of dependency, such as linearity of relations between stimuli and responses, nor does it make assumptions about which stimuli are similar to one another (or the similarity between responses). Instead, it measures a general form of the relation between s and r. Importantly, the mutual information does not tell us which features of the stimulus are encoded or how—but it puts a bound on what can be inferred about the stimulus from the neural response, or vice versa (Rieke et al., 1997).

The fundamental measure of this framework is the entropy of a distribution $p(x)$, which gives a mathematically unique way to quantify our uncertainty about x, or how much we do not know about it (Cover & Thomas, 2012; the basis of the logarithm sets the units for the entropy; we use basis 2, which then gives the entropy in "bits"). For a stimulus set $\{s\}$ where each particular s appears with probability $p(s)$, the entropy of $H[p(s)]$ measures how much there is to know about the set of stimuli we might encounter. After observing the neural response, our knowledge about the stimulus has changed and is now given by the posterior distribution $p(s|r)$, which we can quantify by $H[p(s|r)]$. The difference between the two is the mutual information between stimulus and response,

$$I(s;r) = H[p(s)] - \langle H[p(s|r)] \rangle_r$$

where $\langle\rangle_r$ denotes averaging over $p(r)$. The mutual information between stimuli and responses is also equal to $H[p(r)] - \langle H[p(r|s)] \rangle_s$. This implies that $I(s;r)$ can also be interpreted as the difference between the capacity or richness of the neural codebook, quantified by $H[p(r)]$ and the average noisiness of the responses to a particular stimulus, given by the average over stimuli of the stimulus-dependent conditional entropy $\langle H[p(r|s)] \rangle_s$, or "noise entropy" (Rieke et al., 1997; Strong, Koberle, de Ruyter van Steveninck, & Bialek, 1998). (We note here that estimating these entropy terms and mutual information between stimuli and responses suffers from sampling issues and bias, and usually requires large amounts of data as well as bias correction and validation techniques (Strong et al., [1998]; Treves & Panzeri, [1995]).

One can assess exactly how much of the neural response is "wasted" because of the noise by the ratio of the information that the neural response carries about the stimulus to the total response entropy, namely, $I(s;r)/H[p(r)]$. If this ratio is close to 1, then almost all the capacity is used to convey information. Estimating this fraction accurately for temporal spiking patterns is experimentally and computationally demanding, but in the few cases where it has been quantified, almost 50% of the coding capacity of single cells was lost due to noise (see, e.g., Borst & Haag, 2001; Strong et al., 1998).

The noisiness of neural machinery seems wasteful in terms of energy and computational cost, raising the question: Why wasn't noise selected against by evolution? In particular, despite the prevalence of noise at the level of receptors, synapses, and single neurons, there are examples of neural hardware that is extremely accurate, such as the low spontaneous rate of conformational change of rhodopsin molecules in photoreceptors (Rieke & Baylor, 1998). Moreover, some neural circuits are so finely tuned we can find neurons that respond with temporal accuracy of tens of microseconds (Simmons, Ferragamo, Moss, Stevenson, & Altes, 1990). At the behavioral level, we have examples of high performance despite the noise, which sometimes approaches physical limits, like photon counting (Bialek, 2012), sound localization (Simmons et al., 1990), or primate ability to follow a target at an accuracy limited by its sensory noise (Osborne, Lisberger, & Bialek, 2005).

Performing so well means that some neural circuits are very successful in overcoming their inherent variability and the noisy inputs they receive. One possibility is that these circuits have much more reliable components than other circuits, which in turn must have a cost in terms of design and maintenance. Alternatively, using many noisy cells to encode the same stimulus would allow for some form of averaging or noise reduction. In most cases, it is not immediately unclear how the brain might use population codes to that effect (Deweese & Zador, 2004), in particular since neural noise is often nonadditive and depends on the specific stimulus presented.

How are population codes built?

The brain relies on arrays of receptors and processing neurons to represent information about the outside world in almost all sensory systems. This population-based representation is inevitable given the richness of the natural sensory world that the brain is presented with, and the binary and noisy encoding capacity of single neurons. But what kind of population codes should we then expect to find in the brain? If neurons overlap in terms of the part of stimulus space they respond to, then population coding implies higher fidelity of coding, by having multiple cells covering the same part of the sensory space, and the potential for noise correction. Alternatively, every neuron might cover an almost unique part of stimulus space (with no or very little overlap with other cells), which would imply detailed coverage of the sensory space, although less potential for noise correction.

From a simple mathematical viewpoint, the potential coding capacity of populations is huge as a result of the combinatorial number of patterns they could use to encode information: If each neuron splits the stimulus space into "preferred" and "nonpreferred" parts and responds by spiking or silence when a stimulus appears in the corresponding parts of stimulus space, then with N such neurons, the space of potential responses has 2^N distinct patterns. Thus, a handful of such neurons would be able to encode any stimulus ever encountered using a unique pattern, since even just 100 cells would have $2^{100} \sim 10^{30}$ different activity patterns. However, such combinatorial codes which would efficiently utilize all potential code words would be hard to interpret and decode, because reading the meaning of a pattern would require detailed calculations to infer the contribution of each cell and the relations between them or an exponentially large lookup table. Moreover, because every population activity pattern could carry different information, a missing spike or noisy neuron could result in a very different message being represented, which would make the code even harder to interpret.

On the other extreme, to overcome noise, population codes should rely on some form of redundancy in terms of the information conveyed about their stimuli. One simple possibility is duplication of neural responses, or high overlap by noisy samples of the same stimuli by neurons with similar response properties. This form of population coding could use a very simple decoding mechanisms, such as population vector (Georgopoulos et al., 1986) and linear decoding (Bialek, Rieke, de Ruyter van Steveninck, & Warland, 1991; Dan, Atick, & Reid, 1996; Warland, Reinagel, & Meister, 1997), where the estimated or reconstructed version of the stimulus is given by $s^{est}(t) = \sum_i \mathbf{k}_i * \mathbf{r}_i$, where $s^{est}(t)$ is the value of the estimated stimulus at time t, $\mathbf{r}_i$ is a vector denoting the activity of neuron i up until time t, and $\mathbf{k}_i$ for the each of the neurons are spatio-temporal filters, and * denotes the convolution between the filters and the neurons activity patterns. The filters $\{\mathbf{k}_i\}$ are fit such that over all cells they minimize the reconstruction error between s and s^{est} over time.

However, the design of population codes may need to answer different and sometimes conflicting goals and constraints in terms of energy, error correction, cost of computation, or stimulus statistics. Thus, there is no universal optimal code. Different coding schemes might be efficient or even optimal for different combinations of stimuli, single cell properties, noise level, and connectivity between cells. In other words, to enable effective error correction and decoding of novel neural activity patterns, population codes must balance the richness of combinatorial codes with some form of overlap or dependency between the cells. This balance and therefore, the nature of the code is set by the relations between neurons.

Correlations between neurons

The relations between the spiking patterns of different cells depend on the similarity of their respective stimuli and response properties (receptive fields), the synaptic connections between cells, and their joint circuitry or cells, from which they receive inputs. These different sources of dependencies between cells may result in plain similarity of their activity, linear correlation between their spiking patterns, or in more complicated statistical dependencies. For obvious experimental and practical reasons, the dependency between cells is most readily assessed in terms of the relations between pairs of cells. The discussion bellow presents the questions and measures of correlated neural codes at the level of pairs, but can be immediately extended to more cells.

The most obvious form of dependency between cells—and the one that the brain itself observes and can assess—is reflected in the activity of one cell given that of another cell, regardless of the details of the stimulus. This is often measured at the level of individual cell pairs by comparing the joint activity pattern of both cells to what would be expected from what we know about the activity of each cell, or the prediction of independent activity of the cells (figure 29.2A). Perhaps most common is the estimation of the Pearson correlation between the firing rates or the number of spikes of the cells over some time window Δt given by, $\frac{(r_1 - \langle r_1 \rangle)(r_2 - \langle r_2 \rangle)}{\sigma_1 \sigma_2}$, where r_1 and r_2 denote the responses of each of the cells, $\langle\rangle$ denotes average over conditions, and σ_i denotes the standard deviation of r_i. However, the dependency between pairs of cells may be nonlinear, and moreover, the relation between cells may not necessarily manifest itself in terms of firing rates. Thus, in general, the dependency between cells should be measured in terms of the probability of seeing joint activity patterns of the two cells $p(r_1, r_2)$ to the independent prediction $p(r_1)p(r_2)$, where r_1 and r_2 can be any form of activity (from spiking rates to more compound spiking patterns in time). Similarly, one can estimate the temporal dependency between cells in terms of $p(r_1(t) r_2(t - \delta t))$, where δt denotes the time difference

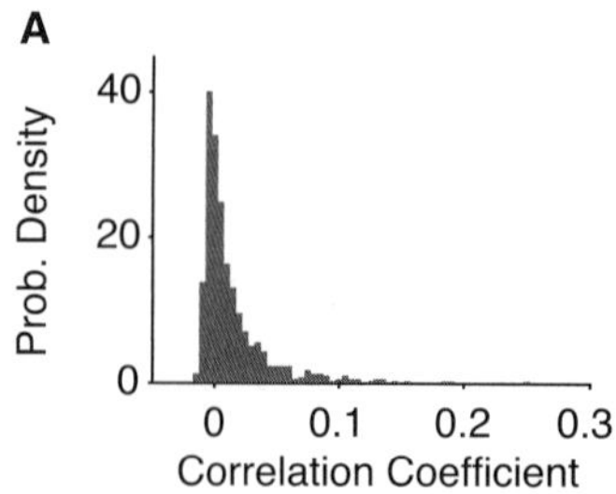

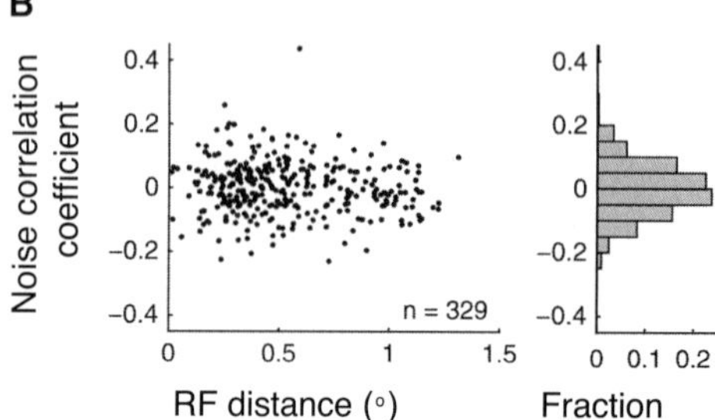

FIGURE 29.2 Distributions of pairwise signal correlations and pairwise noise correlations in two different neural systems. (A) A histogram of correlation coefficients for all pairs of 40 ganglion cells from a patch of an intact salamander retina, recorded using a multi-electrode array, which was presented with a long natural movie. These "signal correlations" are typically weak, even for cells with overlapping receptive fields. (Adapted from Schneidman et al., 2006.) (B) Example of typically weak pairwise noise correlations in the cortex. *Left*: Spike-count noise correlations of 329 pairs of neurons in the primary visual cortex (area V1) of awake monkeys were estimated using simultaneous recording by an array of chronically implanted tetrodes. Monkeys were presented with natural images that were each shown for 200 ms, and the noise correlations for each stimulus were estimated based on the first 500 ms of each response. The average noise correlations (over stimuli) are shown as a function of the receptive field distance (mapped separately) for pairs of V1 cells. *Right*: Distribution of the average noise correlations over the population. (Adapted from Ecker et al., 2010.)

between cells. This form of correlation between cells is commonly termed "signal correlation," since the probability distributions over activity patterns are the result of summing over all stimuli and conditions that the cells observe together. This measure of correlation between the cells reflects what knowing about cell 1 tells about the activity of cell 2, averaged over all stimuli, which is exactly what the brain can assess (since usually it has no knowledge of the specific stimulus that was presented). Various measures of similarity between $p(r_1, r_2)$ and $p(r_1)p(r_2)$ can be used to generalize over linear measures of correlations, but given the mathematical benefits of information theory, we can quantify the dependency between the two cells in terms of the mutual information between the cell spiking patterns, which is given by $I(r_1; r_2) = \sum_{r_1, r_2} p(r_1, r_2) \log_2 \frac{p(r_1, r_2)}{p(r_1)p(r_2)}$.

We note that this way of expressing mutual information between the two cells is mathematically equivalent to the two other ways to define information between two variables presented above.

Another measure of dependency between cells is the relation between their activity patterns with respect to a specific stimulus (or stimuli). For a pair of cells and a particular stimulus, *s*, we can compare the distribution of responses of the two neurons to that particular stimulus, given by $p(r_1,r_2|s)$, to the predicted distribution of joint responses if the cells were responding to the stimulus independently of one another, namely, $p(r_1|s)p(r_2|s)$. This comparison between the empirical distribution of joint responses to the one that assumes that the cells are conditionally independent in responding to the stimulus measures how correlated are the fluctuations (or the "noise") of the responses of the cells. These correlations are often estimated by presenting the same exact stimulus repeatedly, then shuffling the order of responses of one cell with respect to those of another. It has been common to average this measure of joint fluctuations in encoding over a range of different stimuli. This, then, measures the average nature of joint fluctuations in the responses, or the noisy nature of the joint responses, and is commonly termed "noise correlation." It is noteworthy that this name is somewhat confusing, since these correlations depend on specific stimuli and then are averaged over stimuli. Thus, these noise correlations are generally not stimulus-independent, and have a strong impact on the way stimulus information is encoded and can be read. In particular, noise correlations for one stimulus may be very different than those of another stimulus. Moreover, while it is relatively easy to assess noise correlations for the case where no external stimulus is presented (ongoing activity), this tells us very little about noise correlations for different stimuli. In terms of quantifying these correlations, we can again use information theory to quantify the information that one cells tells us about another, given a particular stimulus value, $I(r_1; r_2|s) = \sum_{r_1,r_2} p(r_1, r_2|s)\log_2 p(r_1, r_2|s)/p(r_1|s)p(r_2|s)$, to measure the noise correlations for particular *s*. The average noise correlations would be then quantified by $\langle I(r_1;r_2|s)\rangle_s$.

The two forms of dependencies between neurons we presented are coupled (the distinctions and relations between them have old roots in statistics; see, e.g., Good, 1953). At a fundamental level, the noise correlations are at the heart of the nature of the encoding of a stimulus by the population. Yet the brain has no direct access to these dependencies in general, and can only assess how cells work together over all stimuli (stimulus correlations). In other words, since our estimate of the signal correlations results from the joint probability $p(r_1,r_2)$, and this is based on the sum over stimuli of $p(r_1,r_2|s)$, then signal correlations are in a sense a marginalized reflection of the nature of the joint response and joint noise of the cells to specific stimuli or noise correlations (figure 29.2B).

How do these correlations shape the neural population code? In terms of encoding, it is clear that any form of correlation or dependency between the cells must diminish the range of possible patterns the neurons may use to convey information. Yet it is this very dependency that would allow a "listener" to the code to correct errors. Thus, signal and noise correlations set a balance of coding properties. The dependencies between cells restrict the joint vocabulary, but the nature and magnitude of the stimulus-dependent encoding noise is the key for error correction.

A population coding scheme in which the cells encode their stimuli independently of one another (i.e., without noise correlations), would have several experimental and practical benefits, which would make the neural code simple to describe and to study. First, there would be no need to use simultaneous recording of multiple neurons to study the code, and we could combine cells from different recording sessions (although to know that, one would need to record the joint activity to verify that they are indeed independent). Second, the hard problem of sampling of the population code would be reduced to sampling the individual noise of each cell, which is exponentially easier. Third, analyzing the structure of the code would then reduce to understanding how a collection of neurons are encoding their stimuli, which again would be exponentially simpler than exploring joint coding schemes. However, the lack of noise correlation would also mean that the variance of joint responses of the population is larger than in the case of any correlated response of the cells to the stimulus.

While it is common to think of error correction in terms of averaging over multiple independent identical encoders, this is not the only way and in many cases not the optimal way to overcome noise. Critically, the dependencies between cells are a *double-edged sword.* They reduce the richness or capacity of the population code, making it less diverse than it would be if the cells were encoding independently. But, this coupling between cells is also what allows for attenuation of noise or error correction by the cells themselves (Deweese & Zador, 2004). Thus, signal correlations or noise correlations are not universally "good" or "bad" (see figure 29.3B for simple toy examples of the potential effect of noise and signal correlations on the code; Averbeck,

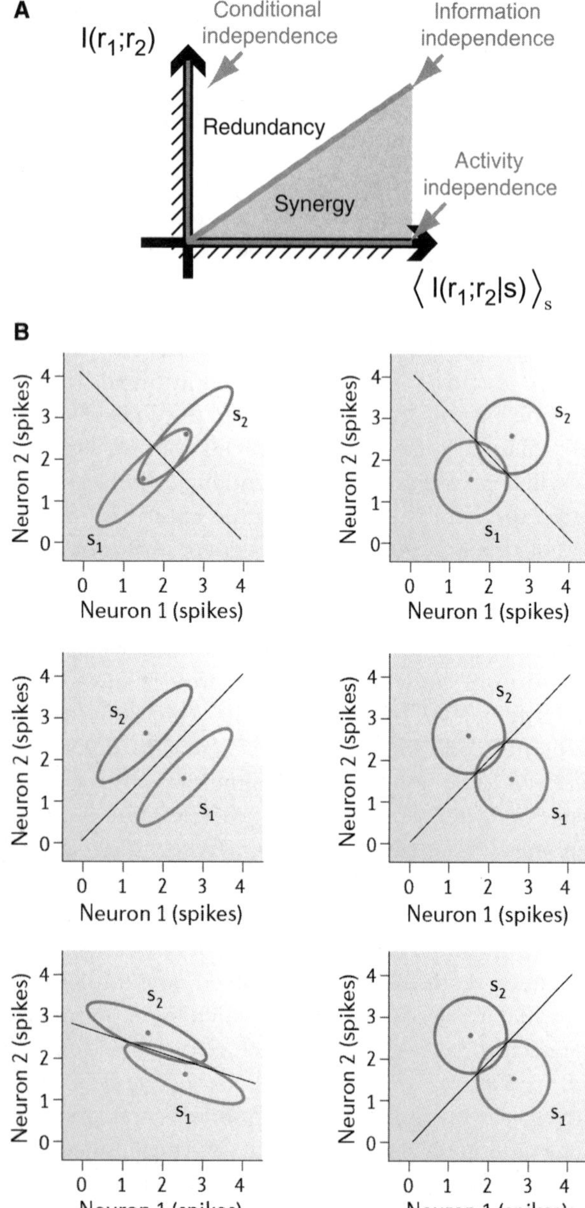

FIGURE 29.3 Graphical presentations of the complex effects of combinations of signal and noise correlations on the information content of population codes. (A) Graphical presentations of synergy as a combination of other measures of non-independence. As discussed in the text, we can represent the synergy or redundancy of a pair of cells as a point in a plane with the axes $<I(r_1;r_2|s)>_s$ and $I(r_1;r_2)$. Because both of these measures are nonnegative, only the top right quadrangle is allowed. Neuronal pairs whose activity is independent (i.e. they have no signal correlations) would be represented by points along the abscissa with $I(r_1;r_2) = 0$. Neuronal pairs whose response is independent given the stimulus, (i.e. they have no noise correlations) would result in points along the ordinate, with $I(r_1;r_2|s) = 0$. The case where the neurons carry independent information corresponds to the diagonal that separates the synergistic values from the redundant ones. (Adapted from Schneidman et al., 2003a.) (B) A toy model example of the effects of correlation on information encoding. In all three cases (rows), the response distributions for two neurons that respond to two different stimuli are shown. The panels on the left show the joint responses of the cells, whereas those on the right show the conditionally independent version of the same responses, also known as "shuffled response," which can be achieved from the joint responses by shuffling the order of responses of each cell to different presentations of the same stimulus; that is, left panels show the result of $p(r_1,r_2|s)$ and those on the right show the result of $p(r_1|s)p(r_2|s)$. Each ellipse (which appears as a circle in the uncorrelated plots) indicates the 95% confidence interval for the responses. Each diagonal line shows the optimal decision boundary—that is, responses falling above the line are classified as stimulus 2 and responses below the line are classified as stimulus 1. The *x*-axis is the response of neuron 1, the *y*-axis the response of neuron 2. (Note that in this particular simple case, signal correlations correspond to the relative positions of the mean responses and noise correlations control the shape and orientation of the ellipses, and thus whether correlations result in more or less encoded information). *Top row:* In this case the responses were chosen such that the signal and noise correlations are both positive, and this leads to larger overlap between the ellipses for the correlated rather than for the uncorrelated responses. A larger fraction of the ellipses lie on the "wrong" side of the decision boundary for the correlated responses than for the independent responses. In this case $I(r_1,r_2;s) < I(r_1;s) + I(r_2;s)$, and so the cells carry more redundant information compared to the conditional independent case. Middle row: Here the signal correlations are negative and noise correlations are positive, and there is less overlap in the correlated than uncorrelated responses. A smaller fraction of the ellipses lie on the wrong side of the decision boundary for the correlated responses, and in this case $I(r_1,r_2;s) > I(r_1;s) + I(r_2;s)$. This also means that the cells in this case carry synergistic information. *Bottom row:* The same fraction of the ellipses lies on the wrong side of the decision boundary for both the correlated and independent responses, so in this case $I(r_1,r_2;s) = I(r_1;s) + I(r_2;s)$. Thus, the presence of correlations does not guarantee an effect on the amount of information encoded—it is the detailed correlation structure of the signal and noise correlations that shapes the content and organization of the code. (Adapted from Averbeck et al., 2006.)

Latham, & Pouget, 2006). Instead, the effect of correlations depends on the nature of single cell properties, the balance of the contributions of correlations to reducing the population variability, and richness of the population code (Abbott & Dayan, 1999; Amari & Nakahara, 2006; Averbeck & Lee, 2004; Averbeck et al., 2006; Latham & Nirenberg, 2005; Quiroga & Panzeri, 2009; Romo, Hernández, Zainos, & Salinas, 2003; Schneidman, Bialek, & Berry, 2003a).

From correlations to redundancy, independence, and synergy in neural population codes

Intuitively, the design of information coding by a neural population can be described in terms of the information that is over-represented by several cells and can be read from each of these cells individually (redundant coding), information that is carried uniquely by some cells but not by others (independent coding), and information that is carried by the joint activity of cells in a way that can only be read from their joint activity (synergistic coding; Amari & Nakahara, 2006; Averbeck & Lee, 2004; Latham & Nirenberg, 2005; Schneidman et al., 2003a). It is not immediately clear, though, how to translate the correlation structure among cells, and between the cells and stimuli, into these notions of redundant, synergistic, and independent coding. First, the relation between cells and the stimuli they respond to can be quantified using various correlation measures. For example, while it is straightforward (and common) to assess the Pearson correlation between stimuli and responses, it is not clear how to use such linear measures for the correlation between the stimulus and the combined responses of two cells (or more). Second, and more importantly, it is not clear how these measures relate to one another and to these coding notions.

Information theory gives a principled and self-consistent way to combine these notions of dependency between stimuli and cells, without making arbitrary assumptions—since the mutual information between variables measures their dependency in a way that is not limited to linear correlations, or any specific metric on similarity. We can then compare the information that two cells convey together about the stimulus $I(s;r_1,r_2)$ and the information that each of the cells carries on its own, $I(s;r_1)$ and $I(s;r_2)$. The synergy or redundancy of the code can be then expressed as the difference between the two, namely,

$$SR(r_1, r_2; s) = I(s; r_1, r_2) - (I(s; r_1) + I(s; r_2)).$$

Then, the cells are *redundant* when $SR < 0$, since in this case they must be carrying redundant information. The cells are *synergistic* if $SR > 0$, since together they convey more information than taken separately. If $SR = 0$, then the cells carry independent information. This framework also allows us to make direct link to the relations between the activity of the cells and the stimulus $I(r_1,r_2;s)$, the signal correlations between cells $I(r_1;r_2)$, and the average noise correlations over stimuli $\langle I(r_1;r_2|s)\rangle_s$. The synergy/redundancy (SR) measure defined above is also equal to the difference between the correlations between neurons measured over all stimuli measured as the information between cells, and the average noise correlations, measured as the average over stimuli of the stimulus-dependent information between cells (figure 29.3A):

$$\begin{aligned} SR(r_1, r_2; s) &= I(s; r_1, r_2) - (I(s; r_1) + I(s; r_2)) \\ &= I(r_1; r_2) - \langle I(r_1; r_2 \mid s)\rangle_s \end{aligned}$$

In relative terms we can quantify the synergy or redundancy in terms of the ratio $SR(r_1,r_2;s)/\min(I(s;r_1),I(s;r_2))$, which measures what fraction of the information that is carried by the cells is redundant, and is often denoted as the fractional redundancy.

The different mutual information terms we presented and the relations between them do not tell us which stimuli are associated with which responses. Nor do they tell us what is the redundant or synergistic information that is encoded by the cells. Other kinds of measures must be used to give a breakdown of the encoded information and relations between cells or stimuli, which would have to rely on additional assumptions, such as metric approaches to the stimuli or decoding errors (Brunel & Nadal, 1998; Kang & Sompolinsky, 2001; Shamir & Sompolinsky, 2004), or measures of information carried by specific patterns of activity (Brenner, Strong, Koberle, Bialek, & de Ruyter van Steveninck, 2000; Deweese & Meister, 1999; Pola, Thiele, Hoffmann, & Panzeri, 2003; Schneidman et al., 2011). The power and benefit of using information-theory based measures here is exactly their assumption- and metric-free nature, which gives bounds on the relations between cells and stimuli.

Somewhat surprisingly, the correlations between pairs of cells are typically weak—in terms of both the signal correlations and the average of noise correlations over stimuli—in many different sensory systems (Bair, Zohary, & Newsome, 2001; Ecker et al., 2010; Nirenberg, Carcieri, Jacobs, & Latham, 2001). This has often been interpreted to suggest that we could either neglect noise correlations or signal ones. Moreover, estimates of pairwise redundancy between cells have also given relatively low values of redundancy, typically on the order of a few percent (Chechik et al., 2006; Gawne & Richmond, 1993; Narayanan, 2005; Puchalla, Schneidman, Harris, & Berry, 2005; Reich, Mechler, & Victor, 2001; see an example in figure 29.4). Yet even these typically weak correlations can shape the reliability of neural coding and computation (Cafaro & Rieke, 2010; Lee, Port, Kruse, & Georgopoulos, 1998).

The results for pairs of cells seem to be consistent with one of the more influential theoretical ideas in neuroscience, known as "redundancy reduction." Attneave (1954) and Barlow (1961, 2001) suggested that neural populations "aim" to encode information efficiently, and since redundancy is prevalent in natural sensory scenes, neurons would decorrelate their responses with respect to one another, to optimally use their joint coding capacity. This idea of a design

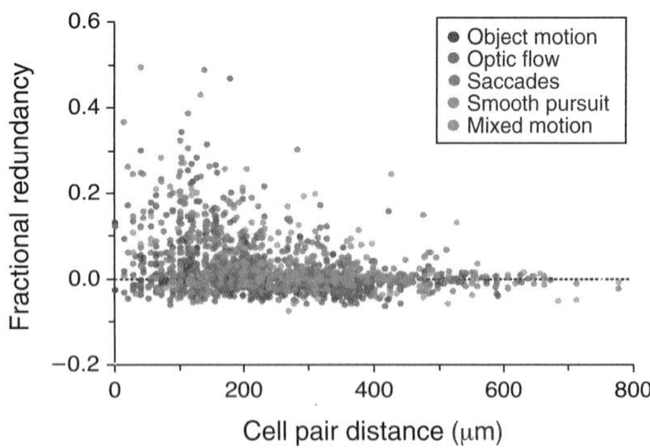

FIGURE 29.4 Pairwise redundancy in the retina, responding to natural scenes. Four salamander retina patches were presented with natural movie clips, and the ganglion cell responses were recorded extracellularly using a multi-electrode array. The fractional redundancy (see main text) for 1,838 cell pairs is shown (each dot represents one pair of cells) as a function of the distance between the centers of the cells' receptive-fields. The type of motion present in each movie clip is shown by the dot: object motion, saccades, optic flow, smooth pursuit, and combinations of motion. Most pairs show weak redundancy or information independence. (Adapted from Puchalla et al., 2005.)

principle for neural populations was the basis for exploration of possible structures of the neural code (van Hateren, 1992) and used to predict the receptive field properties of single cells in the visual system (Atick & Redlich, 1990, 1992; Dan et al., 1996), and at the population level, corresponded to the notion of independent component analysis and coding (Bell & Sejnowski, 1995, 1997; Hyvärinen & Hoyer, 2001).

However, removing redundancy and achieving decorrelation would be problematic for two reasons—as Barlow himself argued later (Barlow, 2001). First, redundancy is crucial for overcoming noise, and so reducing the redundancy, and making the cells independent encoders, would make the code highly susceptible to noise. Second, learning from examples must rely on identifying structure in patterns, which requires some form of redundancy in the patterns themselves. An efficient code in terms of its information rate would have no such redundancy, making the patterns seem almost random in structure. This would make it a hard code to learn from examples, since there would be no apparent dependencies from which one could infer the rules or structure (which is what one uses when learning a new language, for example).

What, then, is the nature of correlation and redundancy in large populations of cells? Naively, if every cell has 5% correlation with 100 of its neighbors, then every piece of information is repeated on average 5 times (Puchalla et al., 2005). Thus, even before we consider high-order dependencies, the network implications of the correlation and redundancy at the level of pairs depend on the number of pairs and the relations between the pairs.

Signal and noise correlations and the population codes of large groups of neurons

Despite the typically weak correlation and redundancy at the level of pairs, the correlation among large groups of neurons can be surprisingly strong. Even small groups of ~10 neurons in the retina already show strongly correlated activity patterns that deviate significantly from what might be expected from their weak pairwise correlations (Schneidman et al., 2006; Schnitzer & Meister, 2003; Shlens et al., 2006). These deviations are clear at the level of specific patterns, showing orders of magnitude mismatch between the rates of appearance of patterns that one observes experimentally and what would be expected from independent cells. Over all patterns, the difference between the joint distribution of activity patterns of N cells, $p(r_1,r_2,\ldots,r_N)$, and the independent model $p_{independent}(r_1,r_2,\ldots,r_N) = p(r_1)p(r_2)\ldots p(r_N)$ quantifies the strength of network correlations, implying that the population vocabulary is far less diverse than it would have been if cells were independent (Schneidman et al., 2006).

To understand the nature of the code of large neural populations, it is imperative to use tools that would enable us to dissect the nature of the different orders of dependency and correlations between cells and their coding role (Amari, 2001; Martignon et al., 2000; Nakahara & Amari, 2002; Schneidman, Still, Berry, & Bialek, 2003b). Somewhat surprisingly, strong network correlations could arise from the collective effect of the quadratic number of weak pairwise dependencies. This can be seen by building the minimal model that relies on the pairwise correlations between cells: since the entropy of a distribution measures how random or structured it is, the distribution $p(r_1,r_2,\ldots,r_N)$ with maximal entropy (Jaynes, 1957) that is consistent with the observed firing rates r_i and the pairwise correlations between cells (which is equivalent to giving the average of all products r_ir_j) is the minimal or most parsimonious model that relies only on pairwise relations between cells (Schneidman et al., 2003). If we discretize neural responses into small time bins, where in each bin a neuron either spikes $r_i = 1$ or is silent $r_i = 0$, then this minimal pairwise model is given by a distribution of the form

$$p^{(2)}(r_1,r_2,\ldots,r_N) = \frac{1}{Z}\exp\Big(\sum_i \alpha_i r_i + \sum_{i<j} \beta_{ij} r_i r_j\Big).$$

Here, α_i (for each neuron) and β_{ij} (for each pair) are Lagrange multipliers that are found to obey the required constraints, and Z is a normalization factor, or the *partition function*. This solution, which is equivalent to the Ising model from physics (Landau & Lifshitz, 1996), is mathematically unique and can be found numerically, giving the most parsimonious, or minimal pairwise model of the population codebook.

In the vertebrate retina, cultured cortical neurons, cortical slices, and in vivo cortical recordings, it was found that such pairwise maximum-entropy models give an extremely accurate description of the population activity vocabulary for both spatial and temporal population activity patterns (see figure 29.5A and Amari, Nakahara, Wu, & Sakai, 2003; Ganmor, Segev, & Schneidman, 2011b; Marre, El Boustani, Frégnac, & Destexhe, 2009; Ohiorhenuan et al., 2010; Schneidman et al., 2006; Shlens et al., 2006, 2009; Tang et al., 2008).

Going beyond the overall vocabulary of the code or signal correlation, the immediate question is what is the role of correlations in encoding specific stimuli, or what is the nature of population noise correlations? Similar to the case of the overall vocabulary, the nature of the responses of large populations of neurons to the same stimulus are often poorly described by a model that assumes no noise correlations (Granot-Atedgi, Tkacik, Segev, & Schneidman, 2013; Pillow et al., 2008; Schneidman et al., 2006). Explicitly, this means that $p(r_1,r_2,\ldots,r_N|s)$ can be significantly different from $p(r_1|s)p(r_2|s)\ldots\ p(r_N|s)$. This difference between these distributions reflects the impact of noise correlation at the level of the network for a particular stimulus (Granot-Atedgi et al., 2013), which for a population of N cells we can quantify by $I(r_1;r_2;\ldots;r_N|s) = H[p(r_1|s)p(r_2|s)\ldots p(r_N|s)] - H[p(r_1,r_2,\ldots,r_N|s)]$. Importantly, while $I(r_1;r_2;\ldots;r_N|s)$ can be very high for particular s, it is often the case that averaging over stimuli gives a relatively low value of $\langle I(r_1;r_2;\ldots;r_N|s)\rangle_s$ since for many other stimuli the noise correlations at the level of the group can be low (Granot-Atedgi et al., 2013). As it has been customary in the literature to report the average noise correlations over stimuli, and these reports often suggested weak noise correlations (see e.g., Ecker et al., 2010), one should note that these averages may indeed mask strong noise correlations for particular stimuli (see, e.g., Granot-Atedgi et al., 2013).

Two families of models have been used to show how taking the relations between cells into account gives a significant improvement in describing the correlated nature of population encoding of specific stimuli. The generalized linear model (GLM; Pillow et al., 2008) describes the response of each cell in a population to a stimulus s by its instantaneous firing rate, given by

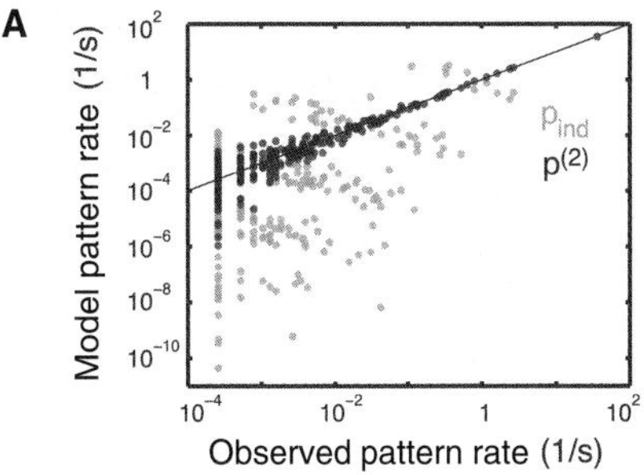

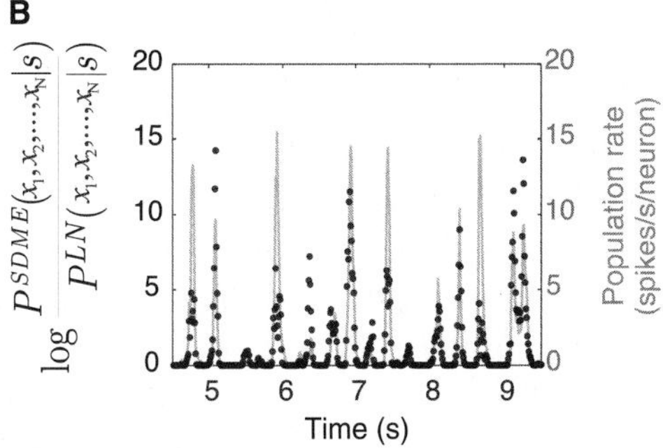

FIGURE 29.5 Strong signal and noise correlations at the level of large populations of neurons. (A) An example of strong signal correlations in the vertebrate retina and the accuracy of pairwise maximum-entropy models. The response of 10 simultaneously recorded ganglion cells in an intact patch of salamander retina, in response to a natural movie, was recorded using a multi-electrode array. Neural responses were discretized into 20 ms bins, and the activity of the group was represented in each time bin as a binary word of length 10, with $x_i = 1$ if a neuron was active, and 0 if it were silent. The rate of occurrence of each pattern predicted if all cells are independent is plotted against the measured rate shown in gray. Each dot stands for one of the 2^{10} possible binary activity patterns for the 10 cells. Black line shows equality. For the same group of cells, the rate of occurrence of each firing pattern predicted from the pairwise maximum-entropy model $p^{(2)}$ that takes into account all pairwise correlations is plotted against the measured rate (black dots). The rates of commonly occurring patterns are predicted with better than 10% accuracy, and scatter between predictions and observations is confined largely to rare events for which the measurement of rates is itself uncertain. (Adapted from Schneidman et al., 2006.) (B) The stimulus-dependent maximum-entropy (SDME) model (see main text) was fit to describe the response of 100 retinal ganglion cells to full-field flicker Gaussian stimulus. The pairwise SDME model predicts population activity patterns for $N = 100$ neurons better than a model that assumes that the cells are independent encoders, given by a set of conditionally independent LN models for each of the cells. The log-likelihood ratio of the population firing patterns under the pairwise SDME model and under the conditionally independent model is shown as a function of time (black dots, scale on left) for an example stimulus repeat (which was not used to train the model). For reference, the average firing rate of the population is shown in gray (scale on right). Thus, the noise correlations in the population are very strong at particular times along the stimulus. (Adapted from Granot-Atedgi et al., 2013.)

$\lambda_i(t) = \exp\left(\mathbf{k_i} * \mathbf{s} + \mathbf{h_i} * \mathbf{r_i} + \sum_{j \neq i} \mathbf{l_{ji}} * \mathbf{r}_j + \mu_i\right)$ where $*$ denotes the convolution between two vectors, $\mathbf{r_i}$ is a vector (over time bins) denoting the spike-train history of cell i at time t, and $\{\mathbf{r_j}\}$ are the spike-train histories of the other cells at time t, $\mathbf{k}_i$ is the stimulus filter of neuron i, $\mathbf{h}_i$ is the history filter for neuron i, $\mathbf{l_{ji}}$ are the coupling filters between neuron j to neuron i, and μ_i is the baseline log-firing rate of cell i that were found to maximize the likelihood of a spiking pattern of each of the cells. GLMs have been shown to be more accurate in capturing noise correlations among small populations of neurons (Pillow et al., 2008; Truccolo, Hochberg, & Donoghue, 2009). Maximum entropy models that are also stimulus-dependent can give the minimal models that are consistent with stimulus-dependent features of the neuronal response. A simple version of such stimulus-dependent maximum entropy models (Tkacik et al., 2010) is the minimal model that has the empirical stimulus-dependent firing rate of the cells $r_i(s)$ (where the average is over multiple presentations of the same stimulus) and the correlation between cells r_ir_j over all stimuli. This model is given by a distribution of the form

$$p^{(2)}(r_1, r_2, \ldots, r_N | s) = \frac{1}{Z(s)} \exp\left(\sum_i \alpha_i(s) r_i + \sum_{i<j} \beta_{ij} r_i r_j\right)$$

where the individual stimulus parameters for each of the cells, $\alpha_i(s)$ and the stimulus-independent pairwise interaction terms β_{ij} are set to match the measured firing rates $r_i(s)$ and the pairwise correlations between cells; $Z(s)$ is a normalization factor or partition function for each stimulus s, given by

$$Z(s) = \sum_{\{r_i\}} \exp\left(\sum_i \alpha_i(s) r_i + \sum_{i<j} \beta_{ij} r_i r_j\right).$$

These models (Granot-Atedgi et al., 2013) show orders of magnitude improvement in capturing the noise correlations for different stimuli for large populations of cells (figure 29.5B). Thus, also in the case of noise correlations, the typically weak noise correlations at the level of pairs add up to strong noise correlations at the level of the network for particular stimuli.

These modeling approaches are not only beneficial but critical for the analysis and understanding the nature of the code of large neural populations. Again, since the number of potential patterns that N cells can transmit is exponential in N, any hope of direct sampling the joint activity of even just 100 neurons, which have 2^{100} potential patterns, is doomed to fail. In fact, we often find if we take the population activity patterns of large group of cells observed in a long experiment, the model that we fit to one half of the data is a better predictor of the other half than the original data itself. Hence, even describing the data would be more accurate using a model. Ultimately, understanding the code of large groups of neurons, and in particular the relations between the correlations between cells and the stimulus, must therefore rely on models that would allow us to identify the design principles of the code, which the modeling frameworks like maximum entropy and GLMs aim to give us.

Discussion

The emerging picture of the design of population codes is one in which large groups of noisy and highly selective neurons are strongly correlated. These strong network correlations rely on a large set of typically weak dependencies that add up and result in strongly correlated collective behavior at the level of the population.

The correlations between neurons are almost inevitable, given the overlap in receptive fields, correlations in the stimuli, and synaptic connections between neurons. An overlap in the receptive field of cells (which would be reflected in signal correlations) is imperative for the ability to overcome noise, but it also means that some of the potential coding capacity of the neurons is lost due to their overlap. Weak noise correlations, or lack thereof, would suggest an obvious way to average over independent encoders of the same stimulus, but this would prevent the cells from correcting or reducing the noise through collective coding. The role and value of correlations are set by the balance of correlations in the stimulus and neural noise, while connections between the cells shape the nature of population codes. In particular, the combination of noise and signal correlations may enable neural populations to operate at a point that may give a code that has high information capacity by matching it to the statistics of stimuli and noise (figure 29.6; Tkacik, Prentice, Balasubramanian, & Schneidman, 2010). The redundancy in the code may have an additional role in creating repeating structure in the code, which is important for learning from examples (Barlow, 2001). Interestingly, recent results on the nature of correlations in large neural populations under naturalistic stimuli suggest that strong correlations result from a sparse network of high-order interactions, making the population code simple to learn (Barlow, 2001; Ganmor et al., 2011a).

Why does noise play such a significant role in shaping the neural code? One possible reason is that evolution has not optimized the neural code as well as other

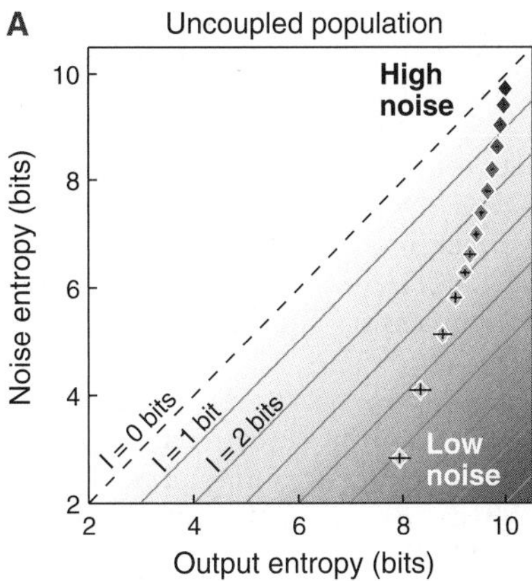

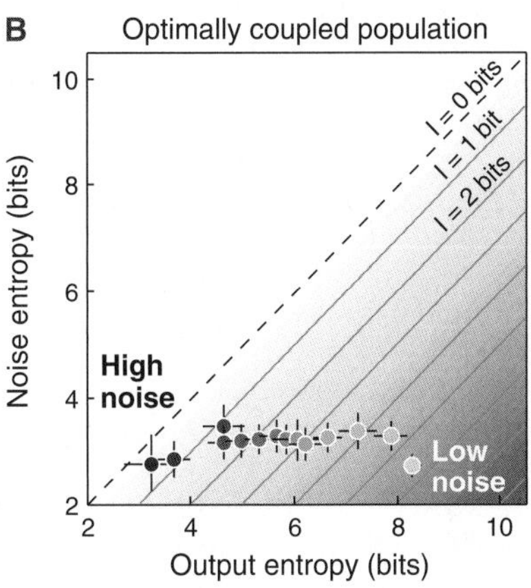

FIGURE 29.6 The effect of noise and correlation on the nature of optimal information encoding by model networks. (A and B) A toy SDME model with 10 neurons was studied under Gaussian correlated stimuli for the case of no coupling between the cells, aside from signal correlations (in panel A), and with the optimal couplings that would maximize the encoded information (in panel B). The output entropy (total population entropy) and the population noise entropy are shown for uncoupled and optimal coupled networks parametrically as neural reliability (or 1/noise) changes from high (bright symbols) to low (dark symbols), The information that the population transmits *I*, is the difference between output and noise entropies and is shown by the gray scale gradient. Networks with low reliability transmit less information and thus lie close to the diagonal, while networks that achieve high information rates lie close to the lower right corner. The optimal network uses its couplings to maintain a consistently low network noise entropy despite a 10-fold variation in neural reliability. Error bars are computed over 30 replicate optimizations for each reliability (1/noise) level. Thus, these two models show distinct information coding capability, based on optimal coupling between cells, but also distinct profiles of coding vocabulary and coding noise. (Adapted from Tkacik et al., 2010.)

features of our brains. However, with 10^{11} neurons and 10^{15} synapses in the human brain, it is clear that if this were the key aspect of optimality, the evolutionary pressure for more reliable neural circuits would be immense. The brain's performance despite its noisy hardware suggests that neural noise may be a "feature" rather than a "bug." Noisy or probabilistic-based coding and computation may enable the brain to overcome local minima in learning (Seung, 2003), and underlie innate exploration of new solutions. It may even allow the computational benefits of probabilistic algorithms, which can be better than any known deterministic ones.

The nature of noise and correlations has been suggested to imply different design principles of information processing and computation that one brain area may perform on the population-activity patterns of noisy correlated neurons it receives from another. We have presented here the notion of strongly correlated population codes that rely on typically weak pairwise relations, which give noise-tolerant and possibly learnable codes. Other, nondistinct features that have been reported include codes from which important behavioral information can be read fast and reliably, even if not perfectly (Gollisch & Meister, 2008; Haddad et al., 2010; VanRullen & Thorpe, 2001); predictive coding of stimulus features by neural populations (Hosoya, Baccus, & Meister, 2005; Palmer, Marre, & Berry, 2013; Srinivasan, Laughlin, & Dubs, 1982); sparseness and overcompleteness of the code (Olshausen & Field, 1997; Simoncelli & Olshausen, 2001); time scale of representation (Burak & Fiete, 2012); and energy efficiency (Laughlin, 2001; Laughlin & Sejnowski, 2003; Levy & Baxter, 1996). It is not unlikely that the principles would be different in different parts of the brain. In particular, the effect of noise in neural circuits goes beyond the capacity and accuracy of representing sensory information that we presented here. It also imposes limits on the nature of persistent activity in networks and propagation in the brain (London, Roth, Beeren, Häusser, & Latham, 2010), and is likely to have a crucial role in shaping the computations that underlie decision making based on population activity (Beck et al., 2008; Churchland et al., 2011; Lee, 2008) and the representation of probabilistic information in the brain (Beck et al., 2008; Berkes et al., 2011; Daw, O'Doherty, Dayan, Seymour, & Dolan, 2006; Deneve, Latham, & Pouget, 1999; Lee et al., 1998; Pouget, Dayan, & Zemel, 2000).

The ability to record large populations of neurons over long time periods would enable us in the not-so-distant future to address the design and content of very large population codes and their dynamics. Recent experimental studies suggest that network correlation structure may change during development or learning (Berkes, Orban, Lengyel, & Fiser, 2011; Tkacik et al., 2010) to match stimulus statistics. Theoretical analysis and experimental evidence suggest that very different correlation structures would be optimal for different stimuli and noise conditions (Chechik et al., 2006;

Pitkow & Meister, 2012; Tkacik et al., 2010), and that the ongoing activity of neural populations may be reflecting the result of network optimization to match stimulus statistics (Berens, Ecker, Gerwinn, Tolias, & Bethge, 2011; Berkes et al., 2011; Ganguli & Simoncelli, 2010; Tkacik et al., 2010). Similar frameworks to the one we presented here would allow us to ask how correlations change during adaptation and learning, as well as decipher the design of neural population codes and the network computations that rely on them.

ACKNOWLEDGMENTS This work was supported by grants from the European Research Council grant (No. 311238 Neuro-Popcode), the Israeli Science Foundation, the SFARI foundation, and the estate of Toby Bieber.

REFERENCES

Abbott, L. F., & Dayan, P. (1999). The effect of correlated variability on the accuracy of a population code. *Neural Comput, 11*, 91–101.

Ahissar, E., Vaadia, E., Ahissar, M., Bergman, H., Arieli, A., & Abeles, M. (1992). Dependence of cortical plasticity on correlated activity of single neurons and on behavioral context. *Science, 257*, 1412–1415.

Amari, S. I. (2001). Information geometry on hierarchy of probability distributions. *IEEE Trans Inform Theory, 47*, 1701–1711.

Amari, S.-I., & Nakahara, H. (2006). Correlation and independence in the neural code. *Neural Comput, 18*, 1259–1267.

Amari, S.-I., Nakahara, H., Wu, S., & Sakai, Y. (2003). Synchronous firing and higher-order interactions in neuron pool. *Neural Comput, 15*, 127–142.

Atick, J. J., & Redlich, A. N. (1990). Towards a theory of early visual processing. *Neural Comput, 2*, 308–320.

Atick, J. J., & Redlich, A. N. (1992). What does the retina know about natural scenes? *Neural Comput, 4*, 196–210.

Attneave, F. (1954). Some informational aspects of visual perception. *Psychol Rev, 61*, 183–193.

Averbeck, B. B., Latham, P. E., & Pouget, A. (2006). Neural correlations, population coding and computation. *Nat Rev Neurosci, 95*, 3633–3644.

Averbeck, B. B., & Lee, D. (2004). Coding and transmission of information by neural ensembles. *Trends Neurosci, 27*, 225–230.

Bair, W., Zohary, E., & Newsome, W. T. (2001). Correlated firing in macaque visual area MT: Time scales and relationship to behavior. *J Neurosci, 21*, 1676–1697.

Barlow, H. (2001). Redundancy reduction revisited. *Network, 12*, 241–253.

Barlow, H. B. (1961). Possible principles underlying the transformation of sensory messages. In W. Rosenblith (Ed.), *Sensory communication* (pp. 217–234). Cambridge, MA: MIT Press.

Beck, J. M., Ma, W. J., Kiani, R., Hanks, T., Churchland, A. K., Roitman, J., ... Pouget, A. (2008). Probabilistic population codes for Bayesian decision making. *Neuron, 60*, 1142–1152.

Bell, A. J., & Sejnowski, T. J. (1995). An Information-maximization approach to blind separation and blind deconvolution. *Neural Comput, 7*, 1129–1159.

Bell, A. J., & Sejnowski, T. J. (1997). The "independent components" of natural scenes are edge filters. *Vis Res, 37*(23), 3327–3338.

Berens, P., Ecker, A. S., Gerwinn, S., Tolias, A. S., & Bethge, M. (2011). Reassessing optimal neural population codes with neurometric functions. *Proc Natl Acad Sci USA, 108*, 4423–4428.

Berkes, P., Orban, G., Lengyel, M., & Fiser, J. (2011). Spontaneous cortical activity reveals hallmarks of an optimal internal model of the environment. *Science, 331*, 83–87.

Bialek, W. (2012). *Biophysics*. Princeton, NJ: Princeton University Press.

Bialek, W., Rieke, F., de Ruyter van Steveninck, R. R., & Warland. D. (1991). Reading a neural code. *Science, 252*, 1854–1857.

Borst, A., & Haag, J. (2001). Effects of mean firing on neural information rate. *J Comput Neurosci, 10*, 213–221.

Brenner, N., Strong, S. P., Koberle, R., Bialek, W., & de Ruyter van Steveninck, R. R. (2000). Synergy in a neural code. *Neural Comput, 12*, 1531–1552.

Brunel, N., & Nadal, J.-P. (1998). Mutual information, Fisher information, and population coding. *Neural Comput, 10*, 1731–1757.

Burak, Y., & Fiete, I. R. (2012). Fundamental limits on persistent activity in networks of noisy neurons. *Proc Natl Acad Sci USA, 109*, 17645–17650.

Cafaro, J., & Rieke, F. (2010). Noise correlations improve response fidelity and stimulus encoding. *Nature, 468*(7326), 964–967.

Calvin, W. H., & Stevens, C. F. (1968). Synaptic noise and other sources of randomness in motoneuron interspike intervals. *J Neurophysiol, 31*, 574–587.

Chechik, G., Anderson, M. J., Bar-Yosef, O., Young, E. D., Tishby, N., & Nelken, I. (2006). Reduction of information redundancy in the ascending auditory pathway. *Neuron, 51*, 359–368.

Churchland, A. K., Kiani, R., Chaudhuri, R., Wang, X.-J., Pouget, A., & Shadlen, M. N. (2011). Variance as a signature of neural computations during decision making. *Neuron, 69*, 818–831.

Cover, T. M., & Thomas, J. A. (2012). *Elements of information theory*. New York, NY: Wiley.

Curto, C., & Itskov, V. (2008). Cell groups reveal structure of stimulus space. *PLoS Comput Biol, 4*, e1000205.

Dan, Y., Atick, J. J., & Reid, R. C. (1996). Efficient coding of natural scenes in the lateral geniculate nucleus: Experimental test of a computational theory. *J Neurosci, 16*, 3351–3362.

David, S. V., Vinje, W. E., & Gallant, J. L. (2004). Natural stimulus statistics alter the receptive field structure of V1 neurons. *J Neurosci, 24*, 6991–7006.

Daw, N. D., O'Doherty, J. P., Dayan, P., Seymour, B., & Dolan, R. J. (2006). Cortical substrates for exploratory decisions in humans. *Nature, 441*, 876–879.

de Ruyter van Steveninck, R. R., Lewen, G. D., Strong, S. P., Koberle, R., & Bialek, W. (1997). Reproducibility and variability in neural spike trains. *Science, 275*, 1805–1808.

DeFelice, L. J. (1981). *Introduction to membrane noise*. New York, NY: Plenum.

Deneve, S., Latham, P. E., & Pouget, A. (1999). Reading population codes: A neural implementation of ideal observers. *Nat Neurosci, 2,* 740–745.

Denk, W., & Webb, W. W. (1992). Forward and reverse transduction at the limit of sensitivity studied by correlating electrical and mechanical fluctuations in frog saccular hair cells. *Hear Res, 60,* 89–102.

Destexhe, A., Rudolph, M., Fellous, J. M., & Sejnowski, T. J. (2001). Fluctuating synaptic conductances recreate in vivo-like activity in neocortical neurons. *Neuroscience, 107,* 13–24.

DeWeese, M. R., & Meister, M. (1999). How to measure the information gained from one symbol. *Network, 10,* 325–340.

DeWeese, M. R., & Zador, A. M. (2004). Shared and private variability in the auditory cortex. *J Neurophysiol, 92,* 1840–1855.

Ecker, A. S., Berens, P., Keliris, G. A., Bethge, M., Logothetis, N. K., & Tolias, A. S. (2010). Decorrelated neuronal firing in cortical microcircuits. *Science, 327,* 584–587.

Faisal, A. A., Selen, L. P. J., & Wolpert, D. M. (2008). Noise in the nervous system. *Nat Rev Neurosci, 9,* 292–303.

Ganguli, D., & Simoncelli, E. P. (2010). Implicit encoding of prior probabilities in optimal neural populations. *Adv Neural Inf Process Syst, 23,* 658–666.

Ganmor, E., Segev, R., & Schneidman, E. (2011a). Sparse low-order interaction network underlies a highly correlated and learnable neural population code. *Proc Natl Acad Sci USA, 108,* 9679–9684.

Ganmor, E., Segev, R., & Schneidman, E. (2011b). The architecture of functional interaction networks in the retina. *J Neurosci, 31,* 3044–3054.

Gawne, T. J., & Richmond, B. J. (1993). How independent are the messages carried by adjacent inferior temporal cortical neurons? *J Neurosci, 13,* 2758–2771.

Geffen, M. N., de Vries, S. E. J., & Meister, M. (2007). Retinal ganglion cells can rapidly change polarity from off to on. *PLoS Biol, 5,* e65.

Georgopoulos, A. P., Schwartz, A. B., & Kettner, R. E. (1986). Neuronal population coding of movement direction. *Science, 233,* 1416–1419.

Gerstner, W., Kreiter, A. K., Markram, H., & Herz, A. V. (1997). Neural codes: Firing rates and beyond. *Proc Natl Acad Sci USA, 94,* 12740–12741.

Gollisch, T., & Meister, M. (2008). Rapid neural coding in the retina with relative spike latencies. *Science, 319,* 1108–1111.

Good, I. J. (1953). The population frequencies of species and the estimation of population parameters. *Biometrika, 40,* 237–264.

Granot-Atedgi, E., Tkacik, G., Segev, R., & Schneidman, E. (2013). Stimulus-dependent maximum entropy models of neural population codes. *PLoS Comput Biol, 9,* e1002922.

Gray, C. M., König, P., Engel, A. K., & Singer, W. (1989). Oscillatory responses in cat visual cortex exhibit intercolumnar synchronization which reflects global stimulus properties. *Nature, 338,* 334–337.

Haddad, R., Weiss, T., Khan, R., Nadler, B., Mandairon, N., Bensafi, M., ... Sobel, N. (2010). Global features of neural activity in the olfactory system form a parallel code that predicts olfactory behavior and perception. *J Neurosci, 30,* 9017–9026.

Harris, C. M., & Wolpert, D. M. (1998). Signal-dependent noise determines motor planning. *Nature, 394,* 780–784.

Hartline, H. K. (1938). The response of single optic nerve fibers of the vertebrate eye to illumination of the retina. *Am J Physiol, 121,* 400–415.

Hatsopoulos, N. G., Ojakangas, C. L., Paninski, L., & Donoghue, J. P. (1998). Information about movement direction obtained from synchronous activity of motor cortical neurons. *Proc Natl Acad Sci USA, 95*(26), 15706–15711.

Hosoya, T., Baccus, S. A., & Meister, M. (2005). Dynamic predictive coding by the retina. *Nature, 436,* 71–77.

Hyvärinen, A., & Hoyer, P. O. (2001). A two-layer sparse coding model learns simple and complex cell receptive fields and topography from natural images. *Vision Res, 41,* 2413–2423.

Jaynes, E. (1957). Information theory and statistical mechanics. *Phys Rev, 106,* 620–630.

Kang, K., & Sompolinsky, H. (2001). Mutual information of population codes and distance measures in probability space. *Phys Rev Lett, 86,* 4958–4961.

Kiani, R., Esteky, H., Mirpour, K., & Tanaka, K. (2007). Object category structure in response patterns of neuronal population in monkey inferior temporal cortex. *J Neurophysiol, 97,* 4296–4309.

Körding, K. P., & Wolpert, D. M. (2004). Bayesian integration in sensorimotor learning. *Nature, 427,* 244–247.

Landau, L. D., & Lifshitz, E. M. (1996). *Statistical physics.* Oxford, UK: Elsevier.

Latham, P. E., & Nirenberg, S. (2005). Synergy, redundancy, and independence in population codes, revisited. *J Neurosci, 25,* 5195–5206.

Laughlin, S. B. (2001). Energy as a constraint on the coding and processing of sensory information. *Curr Opin Neurobiol, 11,* 475–480.

Laughlin, S. B., & Sejnowski, T. J. (2003). Communication in neuronal networks. *Science, 301,* 1870–1874.

Lee, D. (2008). Game theory and neural basis of social decision making. *Nat Neurosci, 11,* 404–409.

Lee, D., Port, N. L., Kruse, W., & Georgopoulos, A. P. (1998). Variability and correlated noise in the discharge of neurons in motor and parietal areas of the primate cortex. *J Neurosci, 18,* 1161–1170.

Levy, W. B., & Baxter, R. A. (1996). Energy efficient neural codes. *Neural Comput, 8,* 531–543.

London, M., Roth, A., Beeren, L., Häusser, M., & Latham, P. E. (2010). Sensitivity to perturbations in vivo implies high noise and suggests rate coding in cortex. *Nature, 466,* 123–127.

Mainen, Z. & Sejnowski, T. (1995). Reliability of spike timing in neocortical neurons. *Science, 268,* 1503–1506.

Marre, O., El Boustani, S., Frégnac, Y., & Destexhe, A. (2009). Prediction of spatiotemporal patterns of neural activity from pairwise correlations. *Phys Rev Lett, 102,* 138101.

Martignon, L., Deco, G., Laskey, K., Diamond, M., Freiwald, W., & Vaadia, E. (2000). Neural coding: Higher-order temporal patterns in the neurostatistics of cell assemblies. *Neural Comput, 12,* 2621–2653.

Martinez-Conde, S., Macknik, S. L., & Hubel, D. H. (2004). The role of fixational eye movements in visual perception. *Nat Rev Neurosci, 5,* 229–240.

Meister, M., Lagnado, L., & Baylor, D. A. (1995). Concerted signaling by retinal ganglion cells. *Science, 270,* 1207–1210.

Nakahara, H., & Amari, S.-I. (2002). Information-geometric measure for neural spikes. *Neural Comput, 14,* 2269–2316.

Narayanan, N. S. (2005). Redundancy and synergy of neuronal ensembles in motor cortex. *J Neurosci, 25,* 4207–4216.

Nirenberg, S., Carcieri, S. M., Jacobs, A. L., & Latham, P. E. (2001). Retinal ganglion cells act largely as independent encoders. *Nature, 411,* 698–701.

Ohiorhenuan, I. E., Mechler, F., Purpura, K. P., Schmid, A. M., Hu, Q., & Victor, J. D. (2010). Sparse coding and high-order correlations in fine-scale cortical networks. *Nature, 466,* 617–621.

Olshausen, B. A., & Field, D. J. (1997). Sparse coding with an overcomplete basis set: A strategy employed by V1? *Vision Res, 37,* 3311–3325.

Osborne, L. C., Lisberger, S. G., & Bialek, W. (2005). A sensory source for motor variation. *Nature, 437,* 412–416.

Palmer, S. E., Marre, O., & Berry II, M. J. (2013). Predictive information in a sensory population. arXiv: 1307.0225v1

Panzeri, S., Schultz, S. R., Treves, A., & Rolls, E. T. (1999). Correlations and the encoding of information in the nervous system. *Proc R Soc Lond B Biol Sci, 266,* 1001–1012.

Perkel, D. H., & Bullock, T. H. (1968). Neural coding. *Neurosci Res Prog B, 6,* 221–348.

Pillow, J. W., Shlens, J., Paninski, L., Sher, A., Litke, A. M., Chichilnisky, E. J., & Simoncelli, E. P. (2008). Spatiotemporal correlations and visual signalling in a complete neuronal population. *Nature, 454,* 995–999.

Pitkow, X., & Meister, M. (2012). Decorrelation and efficient coding by retinal ganglion cells. *Nat Neurosci, 15,* 628–635.

Pola, G., Thiele, A., Hoffmann, K. P., & Panzeri, S. (2003). An exact method to quantify the information transmitted by different mechanisms of correlational coding. *Network, 14,* 35–60.

Pouget, A., Dayan, P., & Zemel, R. (2000). Information processing with population codes. *Nat Rev Neurosci, 1,* 125–132.

Puchalla, J. L., Schneidman, E., Harris, R. A., & Berry, M. J. (2005). Redundancy in the population code of the retina. *Neuron, 46,* 493–504.

Quiroga, R. Q., & Panzeri, S. (2009). Extracting information from neuronal populations: Information theory and decoding approaches. *Nat Rev Neurosci, 10,* 173–185.

Reich, D. S., Mechler, F., & Victor, J. D. (2001). Independent and redundant information in nearby cortical neurons. *Science, 294,* 2566–2568.

Reich, D. S., Victor, J. D., Knight, B. W., Ozaki, T., & Kaplan, E. (1997). Response variability and timing precision of neuronal spike trains in vivo. *J Neurophysiol, 77,* 2836–2841.

Riehle, A. (1997). Spike synchronization and rate modulation differentially involved in motor cortical function. *Science, 278,* 1950–1953.

Rieke, F., & Baylor, D. A. (1998). Single-photon detection by rod cells of the retina. *Rev Mod Phys, 70,* 1027.

Rieke, F., Warland, D. K., de Ruyter van Steveninck, R. R., & Bialek, W. (1997). *Spikes: Exploring the neural code.* Cambridge, MA: MIT Press.

Romo, R., Hernández, A., Zainos, A., & Salinas, E. (2003). Correlated neuronal discharges that increase coding efficiency during perceptual discrimination. *Neuron, 38,* 649–657.

Schneidman, E., Berry, M. J., Segev, R., & Bialek, W. (2006). Weak pairwise correlations imply strongly correlated network states in a neural population. *Nature, 440,* 1007–1012.

Schneidman, E., Bialek, W., & Berry, M. J. (2003a). Synergy, redundancy, and independence in population codes. *J Neurosci, 23,* 11539–11553.

Schneidman, E., Freedman, B., & Segev, I. (1998). Ion channel stochasticity may be critical in determining the reliability and precision of spike timing. *Neural Comput, 10,* 1679–1703.

Schneidman, E., Puchalla, J. L., Segev, R., Harris, R. A., Bialek, W., & Berry, M. J. (2011). Synergy from silence in a combinatorial neural code. *J Neurosci, 31,* 15732–15741.

Schneidman, E., Still, S., Berry, M. J., & Bialek, W. (2003b). Network information and connected correlations. *Phys Rev Lett, 91,* 238701.

Schnitzer, M. J., & Meister, M. (2003). Multineuronal firing patterns in the signal from eye to brain. *Neuron, 37,* 499–511.

Seung, H. S. (2003). Learning in spiking neural networks by reinforcement of stochastic synaptic transmission. *Neuron, 40,* 1063–1073.

Shamir, M., & Sompolinsky, H. (2004). Nonlinear population codes. *Neural Comput, 16,* 1105–1136.

Shlens, J., Field, G. D., Gauthier, J. L., Greschner, M., Sher, A., Litke, A. M., & Chichilnisky, E. J. (2009). The structure of large-scale synchronized firing in primate retina. *J Neurosci, 29,* 5022–5031.

Shlens, J., Field, G. D., Gauthier, J. L., Grivich, M. I., Petrusca, D., Sher, A., … Chichilnisky, E. J. (2006). The structure of multi-neuron firing patterns in primate retina. *J Neurosci, 26,* 8254–8266.

Shoham, S., O'Connor, D. H., & Segev, R. (2006). How silent is the brain: Is there a "dark matter" problem in neuroscience? *J Comp Physiol A, 192,* 777–784.

Simmons, J. A., Ferragamo, M., Moss, C. F., Stevenson, S. B., & Altes, R. A. (1990). Discrimination of jittered sonar echoes by the echolocating bat, *Eptesicus fuscus*: The shape of target images in echolocation. *J Comp Physiol A, 167,* 589–616.

Simoncelli, E. P., & Olshausen, B. A. (2001). Natural image statistics and neural representation. *Annu Rev Neurosci, 24,* 1193–1216.

Srinivasan, M. V., Laughlin, S. B., & Dubs, A. (1982). Predictive coding: A fresh view of inhibition in the retina. *Proc R Soc Lond B Biol Sci, 216,* 427–459.

Stevens, C. F., & Wang, Y. (1994). Changes in reliability of synaptic function as a mechanism for plasticity. *Nature, 371,* 704–707.

Stevens, C. F., & Zador, A. M. (1998). Input synchrony and the irregular firing of cortical neurons. *Nat Neurosci, 1,* 210–217.

Stopfer, M., Bhagavan, S., Smith, B. H., & Laurent, G. (1997). Impaired odour discrimination on desynchronization of odour-encoding neural assemblies. *Nature, 390,* 70–74.

Strong, S. P., Koberle, R., de Ruyter van Steveninck, R. R., & Bialek, W. (1998). Entropy and information in neural spike trains. *Phys Rev Lett, 80,* 197.

Tang, A., Jackson, D., Hobbs, J., Chen, W., Smith, J. L., Patel, H., ... Beggs, J. M. (2008). A maximum entropy model applied to spatial and temporal correlations from cortical networks in vitro. *J Neurosci, 28,* 505–518.

Timme, M., Wolf, F., & Geisel, T. (2002). Coexistence of regular and irregular dynamics in complex networks of pulse-coupled oscillators. *Phys Rev Lett, 89,* 258701.

Tkacik, G., Granot-Atedgi, E., Segev, R., & Schneidman, E. (2013). Retinal metric: A stimulus distance measure derived from population neural responses. *Phys Rev Lett, 110,* 058104.

Tkacik, G., Prentice, J. S., Balasubramanian, V., & Schneidman, E. (2010). Optimal population coding by noisy spiking neurons. *Proc Natl Acad Sci USA, 107,* 14419–14424.

Treves, A., & Panzeri, S. (1995). The upward bias in measures of information derived from limited data samples. *Neural Comput, 7,* 399–407.

Truccolo, W., Hochberg, L. R., & Donoghue, J. P. (2009). Collective dynamics in human and monkey sensorimotor cortex: Predicting single neuron spikes. *Nat Neurosci, 13,* 105–111.

Vaadia, E., Haalman, I., Abeles, M., Bergman, H., Prut, Y., Slovin, H., & Aertsen, A. (1995). Dynamics of neuronal interactions in monkey cortex in relation to behavioural events. *Nature, 373,* 515–518.

van Hateren, J. H. (1992). A theory of maximizing sensory information. *Biol Cybernetics, 68,* 23–29.

VanRullen, R., & Thorpe, S. J. (2001). Rate coding versus temporal order coding: What the retinal ganglion cells tell the visual cortex. *Neural Comput, 13,* 1255–1283.

van Vreeswijk, C. A., & Sompolinsky, H. (1996). Chaos in neuronal networks with balanced excitatory and inhibitory activity. *Science, 274,* 1724–1726.

Warland, D. K., Reinagel, P., & Meister, M. (1997). Decoding visual information from a population of retinal ganglion cells. *J Neurophysiol, 78,* 2336–2350.

Warzecha, A. K., & Egelhaaf, M. (1999). Variability in spike trains during constant and dynamic stimulation. *Science, 283,* 1927–1930.

White, J. A., Rubinstein, J. T., & Kay, A. R. (2000). Channel noise in neurons. *Trends Neurosci, 23,* 131–137.

30 Population-Based Representations: From Implicit to Explicit

NICOLE C. RUST

ABSTRACT Many of our everyday perceptual and cognitive tasks require our brains to transform information from *implicit representations* in which task-relevant information exists but in a format that is difficult to extract, into *explicit representations* in which this information is accessible. For example, determining the identities of objects that are currently in view across naturally occurring variation, such as changes in an object's position, requires our brains to reformat the pattern of light-based representations encoded by our photoreceptors into representations that explicitly reflect object identity. The brain faces similar challenges for other perceptual tasks, such as identifying words spoken by different voices, as well as more cognitive challenges, such as determining whether a chair belongs to the category of furniture. Insight into these challenges, and the brain's solutions, can be understood using geometrical, population-based coding approaches. Once formulated, these population-based descriptions can be linked to the single- and multi-neuron mechanisms that support a successful task solution as well as provide important insights into the computations that the brain uses to process information.

Many of the computations that our brains perform can be restated as transformations of information from *implicit representations* in which task-relevant information exists but in a format that is difficult to extract, into *explicit representations* in which this information is accessible. Transformations from implicitly to explicitly formatted information are often required when our brains need to "group" the neural responses to different conditions into sets, and this requirement is ubiquitously present in the tasks that we perform every day. For example, consider the perceptual challenge of identifying the objects that are currently in view. Each of our photoreceptors encodes light intensity at a particular position, and thus the photoreceptor population represents the visual environment as patterns of light. These light patterns can differ substantially with natural variation in an object, such as changes in its position on our retina, or its retinal size as the object moves toward us, yet we have no problems identifying objects across these identity-preserving transformations. This is because our brains successfully "group" all the light patterns that contain the same object and differentiate those from the light patterns that contain different objects (DiCarlo & Cox, 2007; Hung, Kreiman, Poggio, & DiCarlo, 2005a).

The need to group neural responses into sets applies to more cognitive challenges as well. For example, signaling whether an item is a member of a particular category (e.g., whether a chair is a piece of furniture) requires our brains to group the sensory representations of different items (e.g., Freedman, Riesenhuber, Poggio, & Miller, 2001). Moreover, we know that our brains can flexibly group the same items in different ways, depending on task demands, as exemplified by the Wisconsin card sorting task in which subjects are instructed to take a deck of cards and switch between groupings of cards with the same shape, color, and number of items (Berg, 1948). In these and similar cases, our brains transform lower-level population representations in which the task solution exists but is hidden (because the neural responses to the members of different groups are intermingled in the population representation), into higher-level population representations in which the appropriate neural responses are grouped together and the task solution can be extracted.

Understanding neural transformations at multiple levels

Explicit representations tend not to emerge until higher stages of neural processing. In these high-level brain areas, neural response properties tend to be diverse, and this diversity is thought to be advantageous insofar as a population that contains a diversity of neural responses is capable of performing a diversity of tasks (Rigotti et al., 2013). However, response heterogeneity also makes these high-level brain areas difficult to understand using classical single-neuron approaches, which inherently rely on identifying regularities in the response properties of individual neurons across a population (e.g., discovering that the majority of V1 neurons are tuned for orientation).

Inspired in large part by the proposals of David Marr (1982), we and many others have converged on using a multilevel approach to understand how heterogeneous brain areas process information. Specifically, we have found it useful to begin by taking a population-based approach (Level 1), which has proven to be an effective way to investigate heterogeneous brain areas (e.g., Churchland et al., 2012; Hung, Kreiman, Poggio, & DiCarlo, 2005b; Machens, Romo, & Brody, 2010; Meyers, Freedman, Kreiman, Miller, & Poggio, 2008; Rigotti et al., 2013) as it inherently focuses on how the combined population response reflects a specific type of information. Next, we can use these population-level descriptions to constrain explanations at the response mechanism level (Level 2), where we focus on determining the single and multineuron mechanisms that give rise to the population representation. Finally, we can use both population and response mechanism level descriptions to constrain descriptions at the computational level (Level 3), which seeks to describe the computations that transform signals from one stage to the next. Below we review how we have applied this multilevel approach to understand how the brain reformats task-relevant information to make it explicitly accessible, by first describing each level in more detail, followed by its application to two specific examples.

Level 1—Population representations: Implicit, explicit, and "untangling"

To envision how a population of neurons might represent information, we begin by considering the population response to some condition (e.g., the visual response to a particular image) at one point in time as a *population response vector,* defined as the pattern of spike count responses produced by each neuron in the population on a single trial (figure 30.1A). This vector lies in a space whose dimensionality is defined by the number of neurons in the population but is most easily envisioned in a two-dimensional space that corresponds to the responses of just two neurons. Because neurons are noisy, the response vector for a given condition can

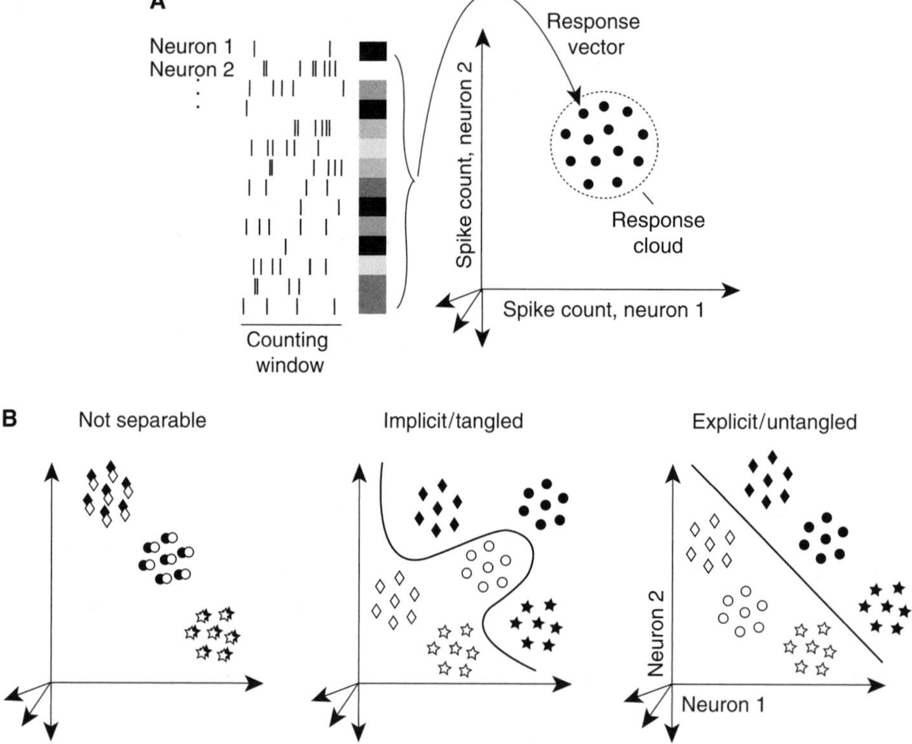

FIGURE 30.1 Population representations. (A) The spike count responses for N neurons combine to form a "response vector" of length N. This response vector exists in an N-dimensional space but is illustrated in the two-dimensional space defined by the responses of two neurons (one dot). Because neurons are noisy, repeated trials of the same condition produce slightly different response vectors, and together all trials form a "response cloud." (B) Shown are the hypothetical responses for a two-way discrimination task that requires parsing the white and black sets of response clouds. Panels show scenarios in which the information required for this task does not exist (*left*); the information required for this task is present but requires a highly nonlinear population readout (i.e., is implicit or tangled; *middle*); the information required for this task can be accessed via a linear readout of the population (i.e., is explicit or untangled; *right*).

fall at slightly shifted positions within the population space, and this distribution is called a *population response cloud* (figure 30.1A).

Tasks that require the brain to group different conditions (e.g., identify a face across changes in position, size, and pose) amount to associating the response clouds that belong to the same set and differentiating those from response clouds that belong to different sets (DiCarlo & Cox, 2007; DiCarlo, Zoccolan, & Rust, 2012). The amount of total information available in the population to perform such a task depends on the degree to which the response clouds corresponding to different sets are nonoverlapping (figure 30.1B). When information exists in a population, we envision that the brain might group responses to different members of a set by placing *decision boundaries* in this space that separate the response clouds for different sets (e.g., a boundary that parses the response clouds corresponding to one face presented at different views from the response clouds for all the views of other faces). The shape of these decision boundaries corresponds to the *population readout rule* required to discriminate these sets based on the population response. These decision boundaries can range from simple rules that correspond to lines or hyperplanes to complex, nonlinear rules that correspond to boundaries that are curved and contorted (compare figure 30.1B, center vs. right). Thus two populations could (in theory) have the same amount of total information to discriminate two sets, but that information could be formatted very differently and require differently shaped decision boundaries to parse them.

The shape of the decision boundary required to extract information determines whether a certain type of information (e.g., object identity) is represented in a manner that is implicit or explicit: "implicit" information is defined as information that requires a complex, highly nonlinear readout, whereas "explicit" information can be extracted using a simple readout rule (e.g., linear). The rationale behind this distinction is the notion that the entities that extract information from a population of neurons are higher-level neurons, and thus our models of neural machinery can be used to guide our determination about whether a representation is implicit or explicit. The simplest decision boundary, a line or a hyperplane, corresponds to the most basic model of a neuron—one that receives weighted input from a population, followed by a threshold, to produce a response that signals when a particular event occurs (e.g., when a particular object is in view). Slightly more complex decision boundaries correspond to slightly more complex models of readout neurons (e.g., a "bent hyperplane" follows from a model neuron with divisive normalization). However, highly complex decision boundaries are likely to be beyond the machinery that can be implemented by individual neurons and instead reflect scenarios in which the information must be reformatted before it becomes accessible to a neurally plausible readout. Because implicit and explicit information loosely map onto nonlinear and linear decision boundaries, respectively, the reformatting process from a nonlinear or "tangled" representation into a linear or "untangled" representation has been coined as "untangling" (DiCarlo & Cox, 2007).

Level 2—The response mechanisms that support explicit representations

As a proxy for measuring how a specific type of information is represented by a population, one can measure how well a population of neurons can solve a particular task (with a particular type of decision boundary). Population performance depends on several factors that are largely nonexclusive. First and foremost, population performance depends on the information conveyed by individual neurons. Notably, population performance also depends on population size. To gain some intuition for this relationship, we can calculate how the distance between two population response clouds ("population discriminability") depends on the discriminability of individual neurons. For a two-way classification (e.g., face 1 versus 2) and a linear decision boundary, one can calculate the commonly used measure of single neuron discriminability, d', as the difference between the mean firing rate responses to each set, divided by the pooled standard deviation of the two distributions (figure 30.2A). This single-neuron measure can be extended to a measure of the distance between two response clouds in the population space (the "normalized Euclidean distance," or NED), computed as the square root of the summed squared d' for all neurons (figure 30.2A). Consequently, the distance between two population response clouds increases as a function of the number of neurons in a population, and even small single-neuron d' can translate into large population discriminability given enough neurons.

Population performance also depends on the number of different sets that need to be parsed. Multiway classification problems (e.g., which of 100 possible objects is currently in view?) are often envisioned as multiple two-way classifications (A/not-A, B/not-B, etc.), followed by a "max" operation to determine a population's final answer (figure 30.2B). Thus one can envision that the solution to a multiway classification problem is computed by parsing the population space with multiple

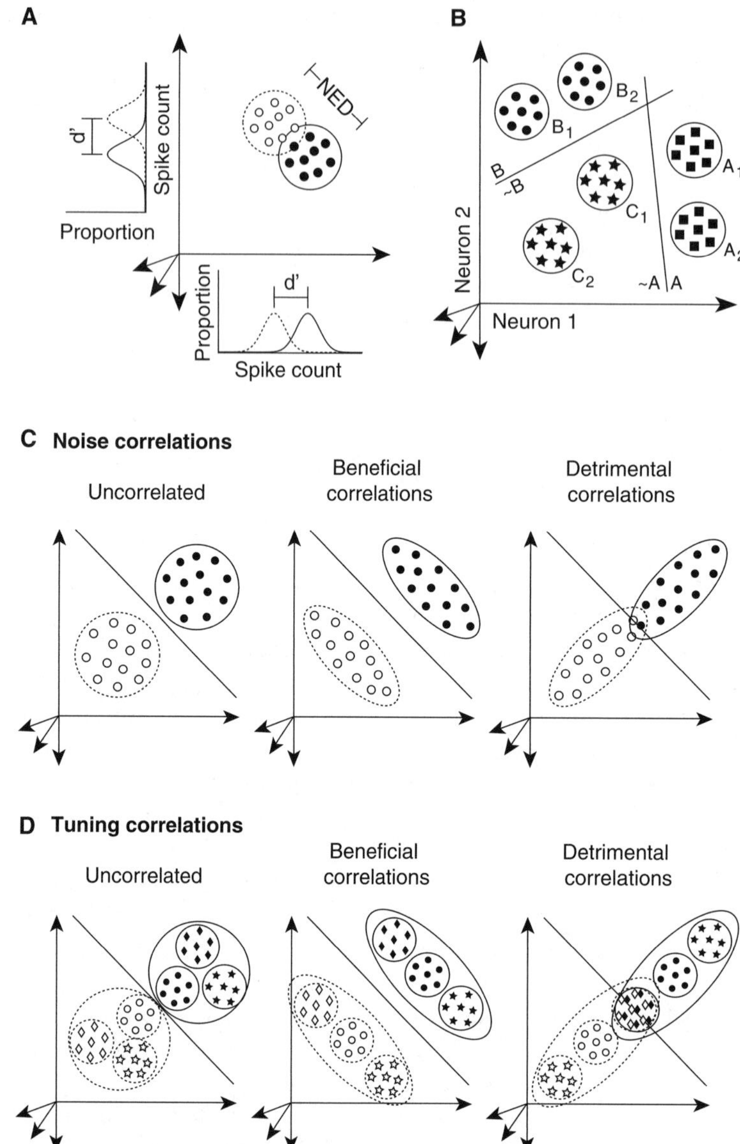

FIGURE 30.2 Factors that determine population performance. (A) The distance between population-response clouds depends on the information conveyed by single neurons and grows as a function of population size. Shown are the responses of two hypothetical neurons, each with a $d' = 3$ for two conditions. The normalized Euclidean distance (NED) for these two conditions in the 2D population space can be calculated as $NED = \sqrt{3^2 + 3^2} = 4.2$. (B) An N-way classification can be envisioned as parsing the population space into N two-way classifications (e.g., A/not A, B/not B, etc.). (C) Noise correlations are determined by the trial-by-trial variability between neurons, and they determine the shape of the individual response clouds. (D) Tuning correlations are determined by the mean firing-rate responses within and across sets, and they determine the relative positions of the different response clouds. For (C) and (D), shown are hypothetical scenarios that include no correlations, correlations that are beneficial because they are aligned to the decision boundary, and correlations that are detrimental because they are perpendicular to the decision boundary.

decision boundaries, each of which separates the response clouds corresponding to one set from all the other sets.

Population performance also depends on "correlations," which come in two varieties. "Noise correlations," defined by the correlated trial-by-trial variability between neurons, determine the shapes of the individual response clouds (i.e., uncorrelated noise has a spherical shape whereas correlated noise is oblong; figure 30.2C). Whether this type of correlation is beneficial or detrimental to population performance depends on how these correlations are aligned relative to the decision boundaries (figure 30.2C). In contrast, "tuning correlations" determine how the response clouds for the different conditions within a set are positioned relative to one another. This type of correlation can also be beneficial or detrimental to population performance, depending on how the response clouds within a set align relative to the decision boundaries (figure 30.2D). See chapter 29, this volume, for a more extensive description of noise correlations and their impact on population representation.

Finally, population performance depends not only on the shape of the decision boundary selected, but also on the details by which the decision boundary is positioned within the population space. For example, a linear decision boundary might be placed midway between the means of two sets of response clouds, or, alternatively, another position along this vector may be more appropriate (e.g., because one set of response clouds has a larger variance). Returning to the notion that higher-level neurons are responsible for reading out neural populations at an earlier stage, these decision boundaries are presumably positioned via the learning rules by which neurons are wired together (discussed in more detail below). Issues related to how classification performance depends on the manner by which decision boundaries are positioned can be informed by the rich engineering literature focused on machine learning and information processing (e.g., Manning, Raghavan, & Schutze, 2008). In practice, absolute population performance values depend on many factors, including the number of classifications, the number of neurons, the specific details about the classification scheme and its optimization, and thus it is often difficult to interpret absolute levels of performance (i.e., "75% performance" means little by itself). Thus population performance values are almost always studied as comparisons with these parameters fixed—for example, between different populations of neurons or between different conditions within the same population.

Level 3—The computations that produce explicit representations

In comparison to the other two levels, understanding the computations that the brain uses to transform information is probably the most challenging. This is because the other two levels tend to lend themselves more to pure data-analysis approaches (e.g., a comparison of classifier performance for two populations), whereas arriving at a computational description most often involves the more challenging task of finding and fitting a model that is sufficiently simple to be constrained by the data but at the same time is sufficiently complex to provide a good account of most neurons. One popular approach attempts to describe the response properties of neurons within a brain area using variants of the "linear-nonlinear" (LN) model, in which the computations performed by individual neurons are described as a weighted sum of the firing rate responses from an input brain area, followed by the application of an instantaneous nonlinearity (e.g., thresholding) to produce an output firing rate response. Due to the relative simplicity of this type of model, one can often identify regularities across the models fit to different neurons and arrive at an intuitive yet accurate description of "how" the response properties of neurons in a particular brain area are constructed from the responses of the input population (Adelson & Bergen, 1985; Carandini et al., 2005; Heeger, 1993; Rust, Mante, Simoncelli, & Movshon, 2006; Rust, Schwartz, Movshon, & Simoncelli, 2005; Simoncelli & Heeger, 1998). See chapter 29, this volume, for a more extensive description of the LN model concept.

One attractive proposal that can be regarded as an extension of the LN model concept is that each cortical brain implements the same "canonical" computation, albeit with different inputs, and thus achieves different goals (Douglas & Martin, 1991; Fukushima, 1980; Heeger, Simoncelli, & Movshon, 1996; Kouh & Poggio, 2008; Riesenhuber & Poggio, 1999). The appeal of this idea arises in part from the fact that iterative stacks of simple, LN canonical computational elements are known to be capable of powerful computations, including untangling (Fukushima, 1980; Riesenhuber & Poggio, 2000; Serre, Oliva, & Poggio, 2007), and one can envision how these computational elements might arise from relatively simple genetic programming. As a refinement of these ideas, some have proposed that we should focus on identifying whether a particular canonical quantity is optimized at each stage of processing through a learning process, such as the degree to which input responses are locally untangled within subpopulations of neighboring neurons (DiCarlo et al., 2012).

The ideas associated with each of these levels of explanation are elaborated in more detail below, where we describe two examples of how complementary descriptions at the population, response mechanism, and computational levels have been combined to understand how the brain transforms task-relevant information from an implicit format into an explicit representation.

Example 1—Explicit representations of object identity

Level 1: Population Representations of Objects

One example in which the brain is thought to untangle information is the case of invariant object recognition.

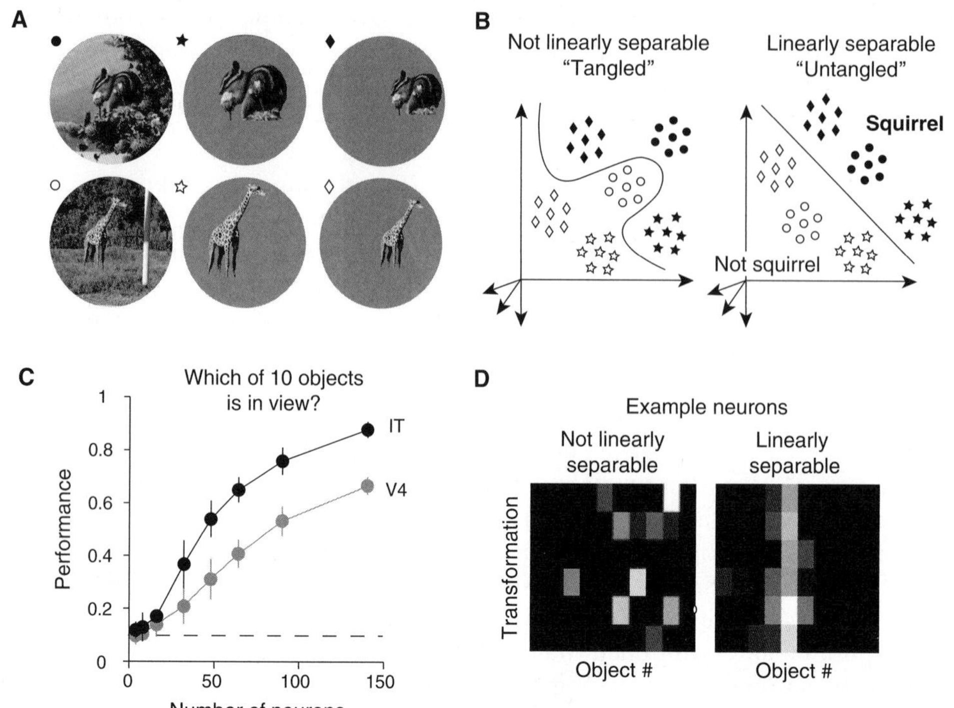

FIGURE 30.3 Explicit representations of object identity. As reported by Rust and DiCarlo (2010). (A) A subset of images used to probe the representation of object identity, across identity-preserving transformations, in V4 and the inferotemporal cortex (IT). In total, 10 objects were presented under six different transformations that included changes in the objects' position, size, and background context. (B) The problem of invariant object recognition can be formulated as multiple, two-way classifications (see also figure 30.2B). (C) Object classification performance in V4 and IT using a linear readout, reprinted from Rust and DiCarlo (2010). (D) Firing-rate response surfaces for 10 objects presented under six different transformations for one example V4 neuron that is not linearly separable, and one example IT neuron that is linearly separable. Single-neuron linear separability translates into the preservation of object preferences across these transformations, or equivalently a separable response matrix.

Information about objects is processed in the primate brain along the ventral visual pathway, a hierarchically arranged collection of neural structures that includes the retina and lateral geniculate nucleus (LGN) as well as cortical brain areas V1, V2, and V4 and inferotemporal cortex (IT; Felleman & Van Essen, 1991; reviewed by DiCarlo et al., 2012). Within each structure, the response vectors corresponding to all possible identity-preserving transformations of the same object (e.g., changes in position, scale, pose, etc.) define an *object manifold* (DiCarlo & Cox, 2007). As described above, the population of photoreceptors represents the visual scene as patterns of light, and because natural object variation produces markedly different light patterns, the manifolds for different objects are "tangled" in the photoreceptor representation. Thus, to extract object identity information, the brain must untangle these signals. To evaluate where and how the brain accomplishes this, we and others have recorded and/or simulated the responses of neurons within different ventral visual pathway brain areas to images of objects presented at different positions and sizes and within different background contexts (figure 30.3A; Hung et al., 2005a; Li, Cox, Zoccolan, & DiCarlo, 2009; Rust & DiCarlo, 2010). We then evaluated the degree to which object identity was explicitly represented at each stage by probing how well a linear decision boundary could separate the responses corresponding to one object from the others using a cross-validated linear classification scheme (figure 30.3B). These results revealed that a linear readout of object identity applied to the IT population performs robustly with only a few hundred neurons and that the IT population performs invariant object recognition tasks much better than real or simulated populations at earlier stages, such as V1 and V4 (figure 30.3C; Hung et al., 2005a; Freiwald & Tsao, 2010; Li et al., 2009; Rust & DiCarlo, 2010; figure 30.3A). These results support the notion that the ventral visual pathway reformats object manifolds to make object identity explicit at its final stage (DiCarlo & Cox, 2007).

Level 2: The Response Mechanisms Underlying Object Recognition What single-neuron responses correspond to "tangled" and "untangled" object manifolds? At the final stage of the pathway (in IT), neural responses across transformations (i.e., for changes in position) tend to change in a manner that maintains their rank-order selectivity preferences for objects across identity-preserving transformations (Ito, Tamura, Fujita, & Tanaka, 1995; Zoccolan, Kouh, Poggio, & DiCarlo, 2007). This property, coined *tolerance*, translates into single-neuron response surfaces that correspond to different object identities presented at different transformations that are linearly separable, or "untangled" (figure 30.3D; Li et al., 2009; Rust & DiCarlo, 2010). In contrast, at early stages of the ventral visual pathway, such as the retina, LGN, and V1, neurons have small receptive fields that are activated by simple light patterns, and the object identity response surfaces at earlier stages of the pathway are nonlinearly separable, or "tangled" (figure 30.3D; Li et al., 2009; Rust & DiCarlo, 2010). These nonlinearly separable responses within individual neurons translate to population object manifolds that are "tangled" together, similar to pieces of paper crumpled into a ball. Similarly, the untangled single-neuron responses observed in IT translate into object manifolds that are more "untangled," in that they are both more flat and are more separated from one another. At intermediate stages of processing, such as V4, neurons appear to have intermediate response properties, consistent with gradual untangling along the pathway.

Level 3: The Computation of Object Identity The gradual transformation from the tangled, light-pattern representation encoded by the eye into the untangled representation of objects found in IT is often envisioned as an iterative cascade of two types of computations: selectivity and invariance. Selectivity computations combine inputs (that each respond to different visual features) with what is often described as an "AND-like" operation to create a neuron that responds more robustly to the conjunction of the features (e.g., features A and B presented together) than the response predicted from the features presented individually (e.g., the response to feature A alone + the response to feature B alone). Selectivity computations are responsible for transforming the representation from its pattern-of-light based form into one based on the conjunctions of visual features that define objects. In contrast, invariance computations act with what is often described as an "OR-like" operation to create neurons that respond whenever any of the features encoded by any of its inputs are in view. Invariance computations are responsible for transforming the small receptive fields found in the eye into the larger receptive fields found at higher stages of the pathway and, similarly, for creating tolerances for the other identity-preserving transformations (e.g., size and background context).

Prototypical examples of selectivity and invariance computations are those used to describe the construction of V1 simple and complex cells (Hubel & Wiesel, 1962). In this description, V1 simple cells implement selectivity operations on the LGN inputs to produce an orientation-selective response. Next, V1 complex cells implement polarity (i.e., light/dark) invariance by combining simple cells tuned for the same orientation. Together, these two operations accomplish a small amount of "untangling" in the visual representation by transforming it from the center-surround representation found in the LGN to one based on local orientation partially invariant to polarity and position. Models of biological object recognition have extended this selectivity/invariance framework to an iterative, hierarchical cascade of these operations that loosely maps onto the processing thought to occur at different stages of the ventral visual pathway (Fukushima, 1980; Riesenhuber & Poggio, 1999; Serre et al., 2007). These relatively simple hierarchical models are largely successful at producing an "untangled" representation of object identity at their highest stage. However, they cannot completely account for human object-recognition performance, and some have speculated that doing so will require a deeper, more systematic exploration of this large class of models (DiCarlo et al., 2012).

For a complementary perspective on the neural computations underlying invariant object recognition, we can consider how the ventral visual pathway might "wire itself up" to untangle information about object identity. The scheme described above implies that neurons learn to appropriately position the decision boundaries at each stage such that object identity is ultimately untangled. How might these boundaries be determined? One proposal is that the brain might exploit the fact that the identities of the objects that are in view tend to change more slowly than the specific light patterns encoded by the eye (Foldiak, 1991; Stryker, 1992; Wallis & Rolls, 1997; Wiskott & Sejnowski, 2002). This notion incorporates both changes due to variation as objects move as well as changes in the light patterns produced by a moving observer. For example, when we move our eyes to scan a scene and across saccades, the same object is likely to fall at different positions on our retina, thus producing different light-pattern representations and adjacent points in time.

These naturally occurring temporal contiguity cues can instruct the building of tolerance to identity-preserving transformations. In the population-based description presented in figure 30.3A, the notion is that the response vectors that are produced by the retina at points close together in time tend to be those that correspond to identity-preserving image variation. Thus, a learning rule that attempts to produce similar patterns of neural responses at points close in time can achieve the larger goal of creating an untangled object representation. Experiments targeted at testing these ideas have manipulated the statistics of temporal contiguity cues within the sensory input (e.g., by changing object identity across saccades) and then evaluated the impact of these manipulations on human behavioral judgments (Cox, Meier, Oertelt, & DiCarlo, 2005) and neural responses in IT (Li & DiCarlo, 2008, 2011). The results of these experiments support the notion that invariant object recognition mechanisms are shaped via an unsupervised learning process, and they suggest that even in the adult brain, invariant object representations are constantly being recalibrated.

Example 2—Explicit representations during target search

LEVEL 1: THE POPULATION REPRESENTATION OF TARGET MATCHES The need to reformat task-relevant information to make it explicitly available is not confined to the realm of perception (i.e., determining the content of the sensory environment), but naturally extends to more cognitive tasks as well. One example is the case of finding target visual objects. Finding a target object, such as your wallet, requires your brain to not only determine the items you are looking at, but also to compare this visual representation with a working memory representation of what you are looking for. To act on the event of finding your target, your brain must transform this information into an "I found it" signal that reports when a target is in view, no matter what that target is (e.g., a neuron that fires when you view your wallet or your car keys, but only when you are looking for those items and not when you are looking for other things). Creating such a signal requires nonlinear computation (i.e., conjunctions of visual and working memory information) and, before these signals have been appropriately combined, the representation of whether a currently viewed object is a target match or not will be tangled.

While the process is not well understood, visual and target-specific information are thought to be combined at middle to higher stages of the ventral visual pathway (i.e., V4 and IT), where the firing rate responses of neurons are largely visual, but are also modulated by changing the identity of a target (e.g., Eskandar, Richmond, & Optican, 1992; Haenny, Maunsell, & Schiller, 1988; Miller & Desimone, 1994). To evaluate the degree to which target match signals are explicitly represented within and just beyond the ventral visual pathway, we recorded the responses of neurons in IT and one of its projection areas, perirhinal cortex (PRH), as monkeys performed a task in which they had to indicate when a target image appeared within a sequence of distractor images (figure 30.4A; Pagan, Urban, Wohl, & Rust, 2013). We then evaluated the degree to which each population could distinguish the same images presented as targets and as distractors with a linear readout by applying a cross-validated linear classification scheme (figure 30.4B). We found that PRH performed this task better than IT, suggesting that PRH has more untangled target match information (figure 30.4C).

More untangled target match information in PRH as compared to IT could follow from a scenario in which PRH has more total information for this task than IT because it receives information that IT does not (compare figure 30.1B, left vs. right). Alternatively, more untangled target match information in PRH could follow from a scenario in which the two areas have similar total information, but that information is more tangled in IT and more untangled in PRH (compare figure 30.1B, center vs. right). To discriminate between these alternatives, we measured the amount of total target match information using an ideal observer classifier whose performance depended on the distance between the response clouds for target matches and distractors but not on the relative positioning of response clouds within each set. We found total information to be similar in IT and PRH, consistent with a description in which visual and working memory information are combined within or before IT in the ventral visual pathway in a largely tangled fashion, and then this information is sent to PRH, which then untangles it.

LEVEL 2: THE RESPONSE MECHANISMS UNDERLYING TARGET SEARCH As described above, finding a specific target requires the brain to compare visual and working memory signals. Within both IT and PRH, we found example neurons whose responses reflected pure versions of these signals, as illustrated by firing-rate response matrices defined by each image viewed in the context of every image as a target, where visual and working memory signals are represented with vertical and horizontal matrix structures, respectively (figure 30.4D). The solution to the monkeys' task required differentiating target matches, which fall along the diagonal of these response matrices, from distractors, which fall off the diagonal, and we also

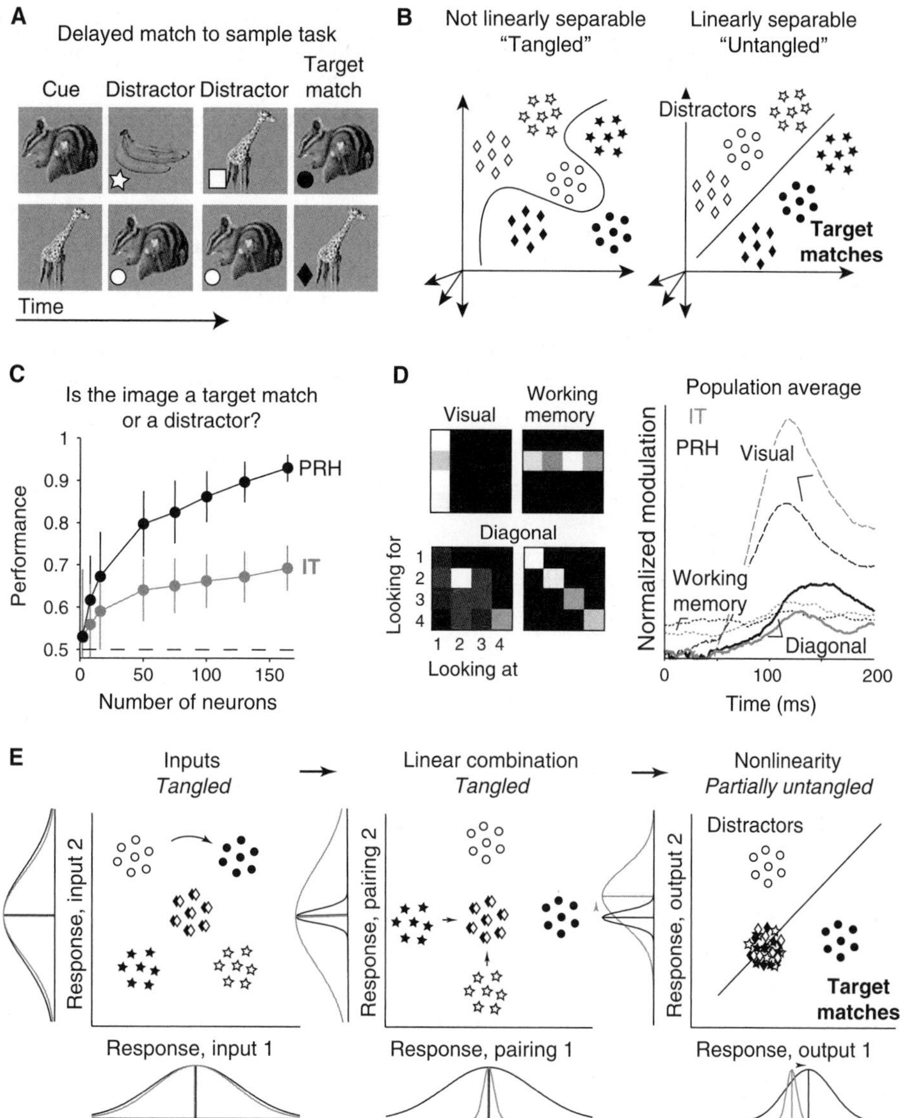

FIGURE 30.4 Explicit representations during target search. As reported by Pagan et al. (2013). (A) Monkeys performed a delayed match to sample task in which a cue image was presented, followed by a random number of distractor images and then a target match. The task required the monkeys to fixate throughout the presentation of the distractors and make a downward saccade in response to the match. The experimental design involved presenting four images in all possible combinations as a visual stimulus and a target. Throughout the figure, the different images are indicated with different shapes, with target match conditions in black and distractors in white. (B) This task can be formulated as a two-way classification of the same images presented as target matches and as distractors. For simplicity, only a subset of the conditions is depicted. (C) Target match/distractor linear classification performance measured in IT and perirhinal cortex (PRH; reprinted from Pagan et al., 2013). (D) Firing-rate response matrices for four example neurons (*left*). Amounts of each type of signal as function of time relative to stimulus onset and averaged across each population, computed via a bias-corrected, ANOVA-like decomposition (*right*; described by Pagan et al., 2013). (E) A linear-nonlinear model provided a good account of the transformation from IT to PRH. This model untangled target match information by combining IT neurons that had tuning correlations for target matches and distractors that were asymmetric (e.g., positive tuning correlations for target matches and negative correlations for distractors are shown in this hypothetical example). Pairing such neurons with appropriate weights resulted in a rotation in the population space that resulted in variance differences between the firing-rate response distributions to target matches and distractors (depicted at the bottom and left of the plot). These variance differences were then exploited to increase linear separability via an instantaneous nonlinearity that included thresholding and/or saturation.

found neurons with diagonal matrix structure. These included neurons that responded to individual images but only when they were presented as targets, which were found in both IT and PRH, as well as a handful of compelling PRH neurons that responded whenever any image was presented as a target (figure 30.4D, "diagonal"). Notably, the intuitive neurons described above were more the exception than the rule in both brain areas, as we found that most neurons were heterogeneous and difficult-to-describe mixtures of different types of signals (not shown). To quantify the magnitudes of different types of signals within each population, we developed a method to parse the responses of each neuron into different types of signals (e.g., visual, working memory, and diagonal; figure 30.4D), and we used these quantifications to constrain our models of how signals were reformatted between IT and PRH, as described below. This analysis confirmed that higher performance in PRH as compared to IT corresponded to an increased amount of diagonal structure within the response matrices of PRH neurons (figure 30.4D), consistent with other types of IT signals that were reformatted into diagonal signals in PRH. We also investigated whether noise correlations might contribute to higher PRH population performance by analyzing the responses of small, simultaneously recorded populations, but we found no evidence for that in our data. This result is likely due to the fact that performance in this task is more limited by the separation of the responses to different target match and different distractor conditions, which must be grouped together (figure 30.2D), than it is by the noise correlations, which determine the shapes of the individual response clouds (figure 30.2C).

LEVEL 3: THE COMPUTATIONS RESPONSIBLE FOR UNTANGLING TARGET MATCH SIGNALS Probing computations in high-level brain areas like PRH can be challenging, particularly in the absence of a complete understanding of all of the computations up to the brain area of interest (e.g., to describe computations in PRH that act on the inputs from IT, a model of processing up to and including IT). In the absence of such a model of IT, we circumvented these challenges by developing novel methods to "leap-frog" into the system and fit a simple linear-nonlinear model of PRH that described how our recorded IT responses were reformatted to produce a more explicit representation of target matches (despite an incomplete understanding of how these IT responses came to be). To constrain the model, we assumed that the untangling process in PRH acted optimally on its inputs arriving from IT, and we searched for the combinations of IT neurons and the model parameters that would maximally untangle information. The resulting model PRH population matched task performance of the actual PRH population for target match–distractor distinctions, and remarkably, this model of PRH also matched the PRH data in many other respects as well, including a decrease in the amount of visual modulation in PRH as compared to IT (figure 30.4D).

An investigation into the mechanism that the model used to untangle target match information revealed that the model worked by combining IT neurons with slightly offset tuning preferences (i.e., offset tuning correlations) for target matches as compared to distractors (figure 30.4E, left). Linear combinations of such neurons acted to rotate the population representation to produce differences in the variance across the responses to target matches as compared to the variance across the responses to distractors (figure 30.4E, middle). Once produced, nonlinearities could act on these response variance differences to produce a more untangled target match representation (figure 30.4E, right). We found that the offset-tuning preferences that this mechanism relied on were ubiquitously present in our data (and, by extension, any population that reflects heterogeneous mixtures of visual and target signals). This suggests that the connectivity responsible for untangling IT inputs within PRH could reasonably be determined via reinforcement learning during the natural experience of searching for targets.

Conclusion

Many tasks require our brains to group neural responses via transformations from implicit representations in which the responses to different sets are intermingled into explicit representations in which these groupings can be easily accessed. While described above for the specific problem of object recognition, the challenge of transforming the elemental representation encoded by our sensory receptors into an explicit representation of the content in our environment is a common perceptual problem. Other examples include identifying motion direction invariant to the moving pattern (Movshon, Adelson, Gizzi, & Newsome, 1985); determining the relative depths (Thomas, Cumming, & Parker, 2002; Umeda, Tanabe, & Fujita, 2007) or tilt (Nguyenkim & DeAngelis, 2003) of two surfaces independent of absolute depth; and independently determining the pitch and location of a sound (Walker, Bizley, King, & Schnupp, 2011). Similarly, target search is one instantiation of a larger class of "cognitive flexibility" problems in which the brain must flexibly switch between different states (in this case between different visual targets) by modifying its task-relevant working memory signals. To solve these tasks, the brain must combine task-relevant working memory and sensory information in a manner that ultimately produces an explicit representation that indicates when the state of the environment matches the task goals.

A complete understanding of how the brain transforms implicit information into an explicit format requires a complementary approach at multiple levels of explanation. For at least two tasks, identifying objects and flexibly switching between different visual targets, explicit representations are produced in high-level brain areas (IT and PRH), in which neural responses are heterogeneous and difficult-to-understand mixtures of different types of information. To describe how the brain solves these tasks, we have found it useful to first constrain explanations with population-level descriptions, followed by a determination of the specific single-neuron response mechanisms that give rise to those population representations. We have found that both types of data, in turn, provide useful and important constraints for computational-level descriptions of how the brain reformats and processes information.

ACKNOWLEDGMENTS NCR was supported by National Eye Institute Grant (R01EY020851), NSF CAREER 1265480 and the Alfred P. Sloan Foundation and the McKnight Endowment Fund for Neuroscience.

REFERENCES

Adelson, E. H., & Bergen, J. R. (1985). Spatiotemporal energy models for the perception of motion. *J Opt Soc Am A, 2*, 284–299.

Berg, E. A. (1948). A simple objective technique for measuring flexibility in thinking. *J Gen Psychol, 39*, 15–22.

Carandini, M., Demb, J. B., Mante, V., Tolhurst, D. J., Dan, Y., Olshausen, B. A., … Rust, N. C. (2005). Do we know what the early visual system does? *J Neurosci, 25*, 10577–10597.

Churchland, M. M., Cunningham, J. P., Kaufman, M. T., Foster, J. D., Nuyujukian, P., Ryu, S. I., & Shenoy, K. V. (2012). Neural population dynamics during reaching. *Nature, 487*, 51–56.

Cox, D. D., Meier, P., Oertelt, N., & DiCarlo, J. J. (2005). "Breaking" position-invariant object recognition. *Nat Neurosci, 8*, 1145–1147.

DiCarlo, J. J., & Cox, D. D. (2007). Untangling invariant object recognition. *Trends Cogn Sci, 11*, 333–341.

DiCarlo, J. J., Zoccolan, D., & Rust, N. C. (2012). How does the brain solve visual object recognition? *Neuron, 73*, 415–434.

Douglas, R. J., & Martin, K. A. (1991). A functional microcircuit for cat visual cortex. *J Physiol, 440*, 735–769.

Eskandar, E. N., Richmond, B. J., & Optican, L. M. (1992). Role of inferior temporal neurons in visual memory. I. Temporal encoding of information about visual images, recalled images, and behavioral context. *J Neurophysiol, 68*, 1277–1295.

Felleman, D. J., & Van Essen, D. C. (1991). Distributed hierarchical processing in the primate cerebral cortex. *Cereb Cortex, 1*, 1–47.

Foldiak, P. (1991). Learning invariance from transformation sequences. *Neural Comput, 3*, 194–200.

Freedman, D. J., Riesenhuber, M., Poggio, T., & Miller, E. K. (2001). Categorical representation of visual stimuli in the primate prefrontal cortex. *Science, 291*, 312–316.

Freiwald, W. A., & Tsao, D. Y. (2010). Functional compartmentalization and viewpoint generalization within the macaque face-processing system. *Science, 330*, 845–851.

Fukushima, K. (1980). Neocognitron: A self organizing neural network model for a mechanism of pattern recognition unaffected by shift in position. *Biol Cybern, 36*, 193–202.

Haenny, P. E., Maunsell, J. H., & Schiller, P. H. (1988). State dependent activity in monkey visual cortex. II. Retinal and extraretinal factors in V4. *Exp Brain Res, 69*, 245–259.

Heeger, D. J. (1993). Modeling simple-cell direction selectivity with normalized, half-squared, linear operators. *J Neurophysiol, 70*, 1885–1898.

Heeger, D. J., Simoncelli, E. P., & Movshon, J. A. (1996). Computational models of cortical visual processing. *Proc Natl Acad Sci USA, 93*, 623–627.

Hubel, D. H., & Wiesel, T. N. (1962). Receptive fields, binocular interaction and functional architecture in the cat's visual cortex. *J Physiol, 160*, 106–154.

Hung, C. P., Kreiman, G., Poggio, T., & DiCarlo, J. J. (2005a). Fast readout of object identity from macaque inferior temporal cortex. *Science, 310*, 863–866.

Hung, C. P., Kreiman, G., Poggio, T., & DiCarlo, J. J. (2005b). Ultra-fast object recognition from few spikes. MIT AI Memo 2005–022.

Ito, M., Tamura, H., Fujita, I., & Tanaka, K. (1995). Size and position invariance of neuronal responses in monkey inferotemporal cortex. *J Neurophysiol, 73*, 218–226.

Kouh, M., & Poggio, T. (2008). A canonical neural circuit for cortical nonlinear operations. *Neural Comput, 20*, 1427–1451.

Li, N., & DiCarlo, J. J. (2008). Unsupervised natural experience rapidly alters invariant object representation in visual cortex. *Science, 321*, 1502–1507.

Li, N., & DiCarlo, J. J. (2011). Unsupervised natural visual experience rapidly reshapes size-invariant object representation in inferior temporal cortex. *Neuron, 67*, 1062–1075.

Li, N., Cox, D. D., Zoccolan, D., & DiCarlo, J. J. (2009). What response properties do individual neurons need to underlie position and clutter "invariant" object recognition? *J Neurophysiol, 102*(1), 360–376.

Machens, C. K., Romo, R., & Brody, C. D. (2010). Functional, but not anatomical, separation of "what" and "when" in prefrontal cortex. *J Neurosci, 30*, 350–360.

Manning, C. D., Raghavan, P., & Schutze, H. (2008). *Introduction to information retrieval.* Cambridge, UK: Cambridge University Press.

Marr, D. (1982). *Vision: A computational investigation into the human representation and processing of visual information.* New York, NY: Henry Holt.

Meyers, E. M., Freedman, D. J., Kreiman, G., Miller, E. K., & Poggio, T. (2008). Dynamic population coding of category information in inferior temporal and prefrontal cortex. *J Neurophysiol, 100,* 1407–1419.

Miller, E. K., & Desimone, R. (1994). Parallel neuronal mechanisms for short-term memory. *Science, 263,* 520–522.

Movshon, J. A., Adelson, E. H., Gizzi, M. S., & Newsome, W. T. (1985). The analysis of moving patterns. *Exp Brain Res, 11*(Suppl.), 117–151.

Nguyenkim, J. D., & DeAngelis, G. C. (2003). Disparity-based coding of three-dimensional surface orientation by macaque middle temporal neurons. *J Neurosci, 23,* 7117–7128.

Pagan, M., Urban, L. S., Wohl, M. P., & Rust, N. C. (2013). Signals in inferotemporal and perirhinal cortex suggest an untangling of visual target information. *Nat Neurosci, 16,* 1132–1139.

Riesenhuber, M., & Poggio, T. (1999). Hierarchical models of object recognition in cortex. *Nat Neurosci, 2,* 1019–1025.

Riesenhuber, M., & Poggio, T. (2000). Models of object recognition. *Nat Neurosci, 3*(Suppl.), 1199–1204.

Rigotti, M., Barak, O., Warden, M. R., Wang, X. J., Daw, N. D., Miller, E. K., & Fusi, S. (2013). The importance of mixed selectivity in complex cognitive tasks. *Nature, 497,* 585–590.

Rust, N. C., & DiCarlo, J. J. (2010). Selectivity and tolerance ("invariance") both increase as visual information propagates from cortical area V4 to IT. *J Neurosci, 30,* 12978–12995.

Rust, N. C., Schwartz, O., Movshon, J. A., & Simoncelli, E. P. (2005). Spatiotemporal elements of macaque V1 receptive fields. *Neuron, 46,* 945–956.

Rust, N. C., Mante, V., Simoncelli, E. P., & Movshon, J. A. (2006). How MT cells analyze the motion of visual patterns. *Nat Neurosci, 9,* 1421–1431.

Serre, T., Oliva, A., & Poggio, T. (2007). A feedforward architecture accounts for rapid categorization. *Proc Natl Acad Sci USA, 104,* 6424–6429.

Simoncelli, E. P., & Heeger, D. J. (1998). A model of neuronal responses in visual area MT. *Vis Res, 38,* 743–761.

Stryker, M. P. (1992). Neurobiology. Elements of visual perception. *Nature, 360,* 301–302.

Thomas, O. M., Cumming, B. G., & Parker, A. J. (2002). A specialization for relative disparity in V2. *Nat Neurosci, 5,* 472–478.

Umeda, K., Tanabe, S., & Fujita, I. (2007). Representation of stereoscopic depth based on relative disparity in macaque area V4. *J Neurophysiol, 98,* 241–252.

Walker, K. M., Bizley, J. K., King, A. J., & Schnupp, J. W. (2011). Multiplexed and robust representations of sound features in auditory cortex. *J Neurosci, 31,* 14565–14576.

Wallis, G., & Rolls, E. T. (1997). Invariant face and object recognition in the visual system. *Prog Neurobiol, 51,* 167–194.

Wiskott, L., & Sejnowski, T. J. (2002). Slow feature analysis: Unsupervised learning of invariances. *Neural Comput, 14,* 715–770.

Zoccolan, D., Kouh, M., Poggio, T., & DiCarlo, J. J. (2007). Trade-off between object selectivity and tolerance in monkey inferotemporal cortex. *J Neurosci, 27,* 12292–12307.

31 Relating the Activity of Sensory Neurons to Perception

DOUGLAS A. RUFF AND MARLENE R. COHEN

ABSTRACT One of the major goals of systems neuroscience is to understand how the activity of sensory neurons gives rise to our perceptual experience. When scientists first began recording from sensory neurons while subjects performed perceptual tasks, Parker and Newsome (1998) wrote a seminal paper laying out a rubric for how to establish that the electrical impulses on the end of the electrode were actually responsible for a specific percept. The intervening years have seen an explosion of interest in this question, technical developments that have paved the way for new types of answers, and theoretical advances that provide a context for the new experimental results. In this chapter we will update the rubric of Parker and Newsome to incorporate recent work and to pose questions whose answers will provide a new level of understanding in the coming years.

Since the fourth century BC, when Aristotle claimed that the heart was the seat of the mind and the soul, scientists and philosophers have been searching for a link between biology and our internal perception of the world around us. Modern neuroscience has long recognized that the physical source of our internal experience is the brain. Over the last few decades, neuroscientists have begun to amass a body of evidence linking the activity of groups of sensory neurons in specific brain areas with individual percepts.

Although the claim that our internal experience is due solely to the activity of neurons is no longer controversial, associating a specific group of neurons with a particular perceptual experience is challenging for both conceptual and experimental reasons. First, it is necessary to measure a subject's percept, which is a fuzzy, subjective experience, in a quantifiable way. This challenge has largely been met thanks to centuries of work by psychologists and psychophysicists who have designed clever tasks to measure subjects' perceptual abilities. Still, we are left relating neuronal activity to *performance* on a task, rather than *perception.*

Determining whether the activity of a particular group of neurons is in a position to underlie performance is even trickier. Many thousands of neurons across sensory cortex as well as in subcortical areas respond every time we see, hear, touch, smell, or taste a sensory stimulus. The neurons that respond to any stimulus vary tremendously in their functional and anatomical properties. The sensory information they encode may or may not be useful for the task at hand or sufficiently sensitive to explain the detail with which a subject perceives a stimulus. Different cortical areas or other groupings of cells work as a network, so simply determining the effect of removing or activating a group of neurons might be either too crude or too subtle a manipulation to yield specific behavioral effects. To make matters worse, most neurons do more than simply provide sensory information, so dissociating their contributions to perception rather than a cognitive or motor process can be challenging. In general, establishing a link between neurons and task performance entails monitoring the activity of and manipulating specific (but often large) groups of neurons while a subject performs a task, which can be experimentally challenging.

In 1998, Parker and Newsome wrote an influential review summarizing the state of the field and providing a rubric for establishing a link between a group of candidate neurons and a specific percept (Parker & Newsome, 1998). Their rubric has become an invaluable framework for interpreting and combining the results of many studies. The nature and quantity of the experimental evidence linking sensory neurons to perception has exploded since that time, leading to many new insights. Out of technological necessity, earlier experiments typically monitored the activity of one neuron at a time and extrapolated their responses to the large groups of neurons thought to underlie any percept. The technology for activating or inactivating neurons was crude, akin to using a chain saw when a scalpel was called for. The techniques for analyzing data focused on these single neuron recordings or gross causal manipulations. In the intervening decade and a half, new experimental technology and techniques for analyzing data have revolutionized the field.

These developments have recently begun to bear fruit, and hard drives are being filled with data as quickly as engineers can increase their size. As the field adjusts to this wealth of new information, it is more important than ever to establish a conceptual framework for understanding the meaning of these new data. For those who spend their days in the lab or at the computer, such a framework will be the only way of recognizing when we have accomplished the goal. What evidence will convince us that we've found the biological basis for a percept?

Our goals in this chapter are fourfold. We aim to

1. provide an updated framework that is based largely on the rubric of Parker and Newsome but reflects new insights for linking groups of neurons to perception;
2. review the current state of the art for performing experiments and analyzing the results;
3. review the results of influential experiments using older technology and describe early experiments using the newest methods;
4. highlight current and future avenues of research aimed at understanding how neural activity gives rise to perception.

Although we will discuss evidence linking different sets of neurons with particular percepts from a variety of model systems and organisms, our goal is not to provide a comprehensive review of this field. We aim to explore the depth of knowledge necessary to present a compelling case that a particular set of neurons underlies a specific perception.

Quantifying a percept

Perception is an inherently subjective experience. We can easily quantify aspects of a physical stimulus, such as the speed of a moving object, the chemicals that give rise to an odor, or the pitch of a musical note. But the subjective experience of seeing your puppy sprint by you, smelling your mother's homemade chocolate chip cookies, or hearing the crescendo in a Beethoven symphony is not something that is readily accessible to experimentalists.

In everyday life, we try to understand the perceptual experiences of others by asking them what something looks, smells, or sounds like. Such *perceptual reports*, however, are an ineffective way of linking neuronal activity to perception for two reasons:

1. Perceptual reports are notoriously unreliable. Although it is beyond the scope of this chapter, there is abundant experimental evidence that people's reports of what they perceive (or do not perceive) are very different from what quantitative tests reveal are the limits of their perceptual capabilities. This has consequences far beyond neuroscience, including the validity of eyewitness testimony in the courtroom.
2. Most experimental methods for recording or manipulating the activity of neurons are invasive, so they cannot be used in humans except in rare cases when a patient is undergoing brain surgery for another reason. Therefore, the vast majority of data concerning the neuronal basis of perception comes from animal studies, in which the subjects cannot verbally describe their experiences.

To link neuronal activity with perception, we therefore rely on the field of *psychophysics* to quantify the relationship between physical stimuli and a subject's perceptual abilities. Psychophysical experiments can be carefully designed to minimize the ambiguity inherent in perceptual reports, often in ways that can be generalized from human subjects to nonhuman subjects. Rather than asking a subject how fast something went, for example, one could probe their ability to judge speed by asking them which of two stimuli is moving faster. This experimental design removes the ambiguity associated with the subject's internal speed calibration, because it forces the subject to evaluate one stimulus with respect to another stimulus. The subject's response on a single trial could be compared to the activity of the candidate neurons at that moment (e.g., could a subject's mistake be predicted from the activity of that group of neurons?). The subject's performance on the task could also be compared to that group of neurons (e.g., how well could a hypothetical subject do if all the information they had to go on came from the activity of that group of neurons?).

The experiments described in the rest of this chapter attempt to determine whether the responses of a group of candidate neurons are necessary and sufficient to explain performance on perceptual tasks. Therefore, if the title of this chapter were to describe the status rather than the goal of this field, it would be called the more cumbersome "relating the activity of sensory neurons to performance on perceptual tasks."

Quantifying performance

Measuring a subject's overall performance on a task is therefore critical for assessing perception. Simple measures like total percent correct are affected by factors other than the subject (such as the difficulty of the task). Psychophysicists usually measure a *psychometric curve*, which is a plot of performance (either percent correct or percent of choices in favor of a particular decision) as a function of a measure of the sensory stimulus. For example, in a speed-discrimination task, the psychometric curve might plot percent correct as a function of the difference in speed between the two

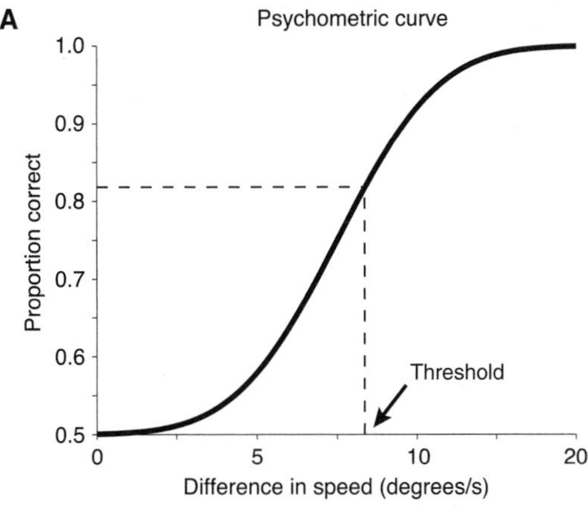

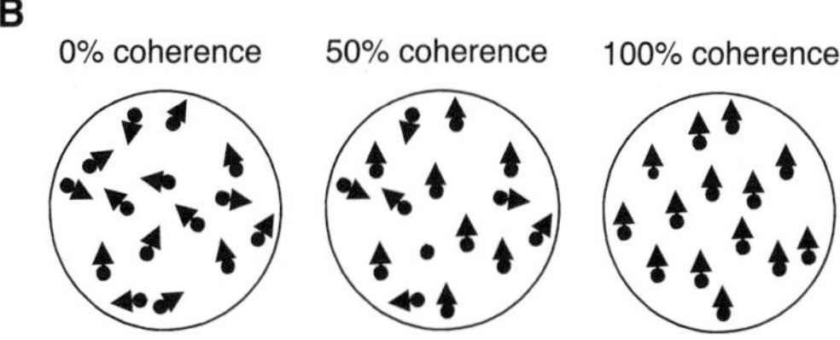

FIGURE 31.1 (A) Psychometric curves plot performance as a function of a measure of the stimulus. The dashed line represents the psychometric threshold, which is the stimulus value at which the subject can achieve a certain level of performance. In this example, the relevant stimulus parameter for the behavioral task is stimulus speed, but in principle it can be any one-dimensional analog parameter of the stimulus. (B) The stimuli in the motion-direction discrimination task as typically used by Newsome and colleagues.

stimuli (figure 31.1A). Psychometric curves usually have a sigmoidal (or "S") shape and can be characterized by two parameters: their slope (the steepness of the linear part of the curve), and the *threshold* (corresponding to the left-right position of the curve on the graph; dashed line in figure 31.1A). Because experimental manipulations do not often change the slope (for review, see Parker & Newsome, 1998), performance is often quantified as the threshold. The threshold is in units of the physical stimulus rather than in units of performance or perception. It is defined slightly differently in different studies, but it is always the value of the stimulus that is necessary to achieve a certain level of performance. For example, in a speed-discrimination task, the threshold might be the speed difference that a subject can discriminate with 82% accuracy.

COMMON PERCEPTUAL TASKS There are a multitude of psychophysical tasks designed to measure different aspects of perception. For use in comparing neurons to perception, each has a unique set of strengths and limitations. These are important to keep in mind when assessing the quality of the evidence that the results of each experiment bring to the hypothesis that a certain set of neurons is responsible for a specific percept. Much of this psychophysical work is discussed in detail elsewhere in this volume.

Two types of psychophysical tasks have been used most commonly to relate neuronal activity to perception: detection tasks and discrimination tasks.

Detection tasks In detection tasks, subjects are asked to signal that they perceive the onset or a change in a sensory stimulus. Chapter 25 (by Fred Rieke) of this volume described a classical detection task in which human observers were asked to detect weak flashes of light in a very dark room. Impressively, this task showed that humans can detect single photons of light under some conditions.

Strengths:

- Detection tasks are typically the easiest psychophysical tasks to train animals to do. Nearly all psychophysical tasks with nonprimate subjects use detection tasks for this reason.
- Because the onset of the stimulus can come at an uncertain time, subjects must remain focused on the stimulus (or expected stimulus location) for a long period of time. This is useful for physiology experiments because it gives a longer period for obtaining accurate measurements of neuronal responses.

Limitations:

- The psychophysical threshold depends on the subject's internal *criterion*, or willingness to make certain types of errors. In the light-detection task, some subjects might be willing to falsely report seeing a flash if they are unsure, while others will only report seeing the flash if they are absolutely certain. Therefore, most studies focus only on changes in threshold from experimental manipulations rather than absolute threshold values.
- It can be difficult to dissociate changes in performance or neuronal responses that are due to sensory factors from those caused by changes in the subject's cognitive state. For example, changes in alertness might increase the responses of neurons all over sensory cortex and also improve performance on psychophysical tasks (simply because increased alertness improves performance on most things). Therefore, one might observe a misleading correlation between performance and the responses of neurons, simply because both are modulated by the subject's alertness.

Discrimination tasks Discrimination tasks require the observer to choose between two or more options. A motion-direction discrimination task (figure 31.1B) was used to establish what is currently the best link between perception and sensory neurons. Newsome and his colleagues (Britten, Newsome, Shadlen, Celebrini, & Movshon, 1996; Britten, Shadlen, Newsome, & Movshon, 1992; Newsome & Paré, 1988; Salzman, Britten, & Newsome, 1990; Shadlen, Britten, Newsome, & Movshon, 1996) performed a series of experiments linking the responses of direction-selective neurons in the middle temporal area (MT) and performance on the task in figure 31.1B. In these studies, rhesus monkeys viewed a dynamic random dot display in which a percentage of the dots moved coherently in one of two opposite directions (up or down in figure 31.1B), while the rest of the dots moved randomly. The monkeys were required to indicate which of the two directions contained the coherent motion. When a large percentage of dots moved coherently, this task proved to be very easy. At low coherence, the random dots provided a masking stimulus, making the discrimination difficult.

There are two types of discrimination tasks. Some, like the direction-discrimination task (figure 31.1B), require subjects to observe a stimulus and choose between two or more options. When there are two options, these are called two-alternative forced-choice tasks (the "forced" is because the subject must pick an option rather than indicating "I don't know"). The forced choice can often reveal perceptual abilities unknown to the subject. In the direction-discrimination task, human subjects often perform better than would be expected by chance even when they report seeing no coherent motion.

In the second type of discrimination task, two-interval forced-choice tasks, subjects are required to indicate in which of two time intervals a stimulus (or a stimulus with a certain property) occurred. An example of this type of task was used by Romo and colleagues (Romo, Hernández, Zainos, & Salinas, 1998) to determine the role of somatosensory areas in the perception of vibrating tactile stimuli. Two stimuli were presented in sequence and monkeys were required to indicate which of the two vibrated at a higher frequency.

Strengths:

- Thresholds are immune to differences in criterion because a bias in favor of one option (e.g., "up" answers in the motion-direction discrimination task) can be measured by comparing performance on trials when the dots actually moved up versus down.
- While cognitive factors like alertness still improve performance and modulate neuronal responses, they do so equally for all trial types. Therefore, increases in arousal will affect "up" and "down" trials equally.

Limitations:

- For nonhuman subjects, discrimination tasks are often more difficult to train than detection tasks.
- Decisions are often made quickly, limiting the time available to record neuronal responses. If subjects are forced to observe a stimulus for a long period of time before making a decision, they may not use all of the available sensory information, making it difficult to determine when the decision actually occurred.

Linking sensory neurons to perception

We will devote the rest of the chapter to establishing a series of questions (based heavily on the rubric of Parker & Newsome, 1998) that must be answered to demonstrate that a candidate group of neurons underlies performance on a particular perceptual task. The earliest attempts to characterize sensory neurons focused on describing the general properties they encode. On average, neurons in early sensory areas are selective for simple features like the orientation of a line, the pitch of a note, or which of a rodent's whiskers has moved. In higher cortical areas, cells are tuned for more complex features like object identity. Building on this knowledge, one can make a somewhat educated guess about which neurons seem like good candidates to underlie the perception of a particular feature. The goal of the rest of this chapter is to pose a series of experimentally tractable questions that test this hypothesis.

By far, the best-established connection between sensory neurons and perception is the one studied by Newsome and colleagues (figure 31.1B) between direction-selective neurons in the middle temporal visual area (MT) and performance in a motion-direction discrimination task. To our knowledge, this is the only connection between sensory neurons and perception for which all of the questions below can be answered in the affirmative. In each section, we will describe the evidence linking MT with motion perception and also provide recent examples showing how these concepts have been applied to other systems or tested using new experimental techniques.

Do the responses of the candidate neurons encode detailed enough sensory information to support the percept? The most basic requirement for the neural underpinnings of a percept is that the candidate neurons carry enough sensory information to explain performance on a perceptual task. For example, if, as a group, MT neurons could only discriminate strong upward from downward motion with 75% accuracy but the monkey can get 90%

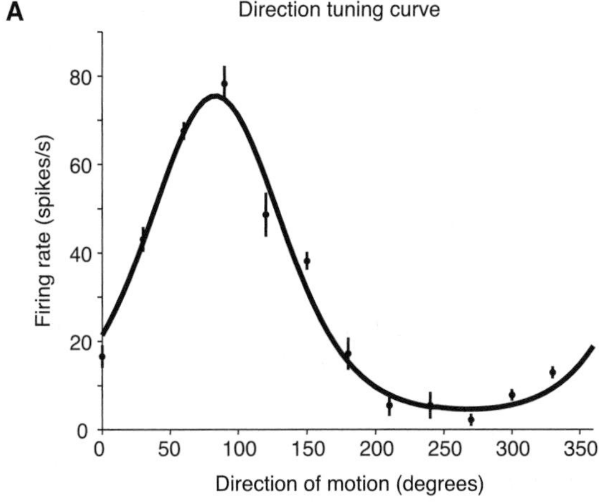

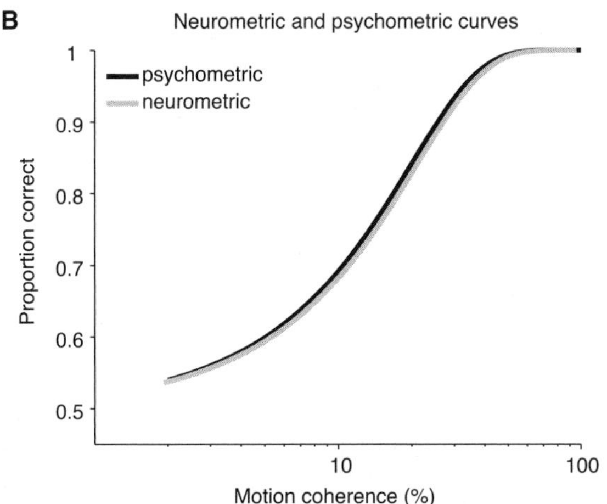

FIGURE 31.2 (A) Motion-direction tuning curve of an example MT neuron (Ruff and Cohen, unpublished data). (B) A schematic neurometric curve. The neurometric curve reflects how well an ideal observer could do a task based on the responses of a single neuron. In MT, neurometric and psychometric curves for the direction-discrimination task are similar (based on the results of Britten et al., 1992).

of the trials correct, then the monkey must be incorporating motion information from somewhere else.

Britten and colleagues (1992) quantified the amount of information about motion direction encoded by single neurons in MT using a metric that could be directly compared to the monkey's performance. They first measured each neuron's direction tuning curve (figure 31.2A) and determined the neuron's preferred direction (the direction of motion that elicits the highest response from the neuron; up, in this example). They then measured neuronal responses while the monkey discriminated motion in the preferred direction (up) from the opposite direction (down) using the direction-discrimination task in figure 31.1B.

For each stimulus strength (e.g., 12% coherence motion), they computed the accuracy with which an ideal observer could discriminate the two motion directions (e.g., up from down) based only on the responses of the one neuron on the end of the electrode. (The ideal observer's performance is equal to the area under the receiver operating characteristic curve corresponding to the responses to upward and downward motion; Green & Swets, 1966.) Using the ideal observer's performance for each stimulus strength, they constructed a *neurometric curve* that could be directly compared to the monkey's *psychometric curve* (figure 31.2B). The corresponding neurometric and psychometric thresholds from each curve (as in figure 31.1A) allowed them to compare the amount of motion information encoded by the neuron to the amount that the monkey used to perform the task.

Britten and colleagues (1992) found that MT neurons are shockingly sensitive. On average, the neurometric and psychometric thresholds were identical, meaning that the monkey behaved as if he based his decisions on a single MT neuron. Later studies suggested that some details of the original experiment tilted the results in favor of the neuron, but that the observation that MT neurons are exquisitely sensitive was true: the monkeys' behavior suggested that they used only as much motion information as was carried by two or three MT neurons (Cohen & Newsome, 2009). This begs the question, why would the monkey ignore the other 100,000 MT neurons or the motion-selective neurons in other cortical areas?

We will discuss a resolution to this question in the next section, but the answer is not that MT neurons are markedly more sensitive than neurons in other areas or that the direction-discrimination task is unusually well suited to the responses of sensory neurons. Similar experiments in a variety of species and cortical areas found that individual sensory neurons are extremely sensitive to the stimulus information they encode. For example, Prince and colleagues (Prince, Pointon, Cumming, & Parker, 2000) trained monkeys to judge the depth (or perceived distance) of ambiguous visual stimuli. They found that single neurons as early as primary visual cortex carry enough depth information to account for the monkey's performance. In rats trained to discriminate the frequency of stimuli vibrating against their whiskers, individual trigeminal ganglion neurons are similarly sensitive to vibration frequency (Gerdjikov, Bergner, Stüttgen, Waiblinger, & Schwarz, 2010).

There is no reason that single neurons alone need to be sensitive enough to account for a subject's psychophysical performance. After all, the brain contains lots of neurons! For a group of candidate neurons to underlie perception, however, they must together encode sufficient information to explain performance on psychophysical tasks. The results highlighted here show that this is not usually the most difficult requirement to fulfill in linking sensory neurons with perception.

Can the responses of the candidate neurons be used to predict the subject's choices? The idea that a group of neurons is responsible for what a subject perceives makes a strong prediction about what the subject will *do* with the sensory information encoded by those neurons. Neural responses are noisy. Say you regularly meet a friend for lunch, and you're in the habit of looking far down the street to see if she's coming. Each time you see your friend walking toward you (even if she is wearing the exact same thing and in the exact same spot), your visual neurons will respond slightly differently. If a group of neurons is responsible for your perception of your friend, the noise in their responses should affect your visual experience. Specifically, if a group of neurons happens to respond unusually strongly, you should be more likely to perceive the visual features they encode (and respond accordingly on a perceptual task).

Britten, Newsome, and colleagues tested this hypothesis by determining whether they could predict monkeys' responses on the direction-discrimination task from the fluctuations in the responses of the MT neurons the authors recorded (Britten et al., 1996). Consider a situation in which they recorded from a neuron whose preferred direction is "up" on many trials of, say, a random 0% coherence stimulus. If the responses of this neuron contribute to a monkey's decision on these trials, then it should be possible to use its responses to predict the monkey's choice on each trial. On trials in which the neuron fires more than its average, the monkey should be more likely to report seeing upward motion than on trials in which the neuron fires less than its average.

The authors of this original study coined the term *choice probability* to describe the proportion of trials on which an ideal observer could predict the monkey's choices based on the neuron under study. In the context of a detection task, a similar metric is called *detect probability*. A choice probability of 1 would mean that the neuron could be used to perfectly predict the monkey's choice (figure 31.3A). This would occur if the upward-preferring neuron fired more on every trial in which the monkey chose up than on any trial in which the monkey chose down. A choice probability of 0.5 would mean that the neuron is uninformative, so someone only looking at the responses of this neuron would have no idea which choice the monkey was about to make. This would occur if the distribution of responses on trials on which the monkey chose up were identical to the distribution when the monkey chose down.

The authors found that choice probabilities for individual MT neurons (whose tuning was well-matched to the task) were approximately 0.54, which was significantly greater than 0.5 but far from the perfect value of 1. Therefore, individual neurons carry some information about what the monkey is about to do, but not much. Neurons whose tuning does not match the stimulus to be discriminated or detected (e.g., a neuron whose preferred direction is not exactly up or down) have weak but significant detect probabilities (Bosking & Maunsell, 2011; figure 31.3B).

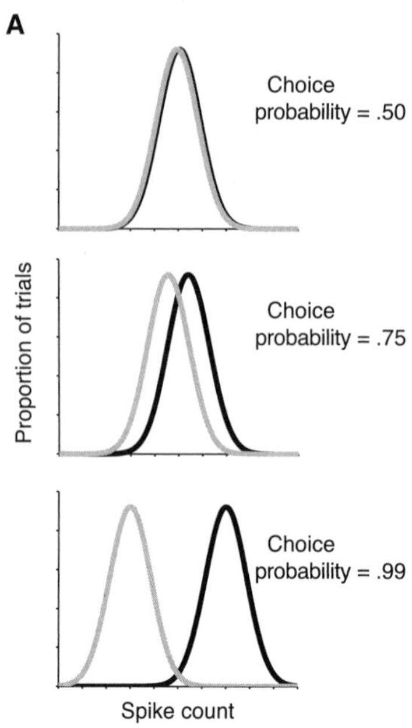

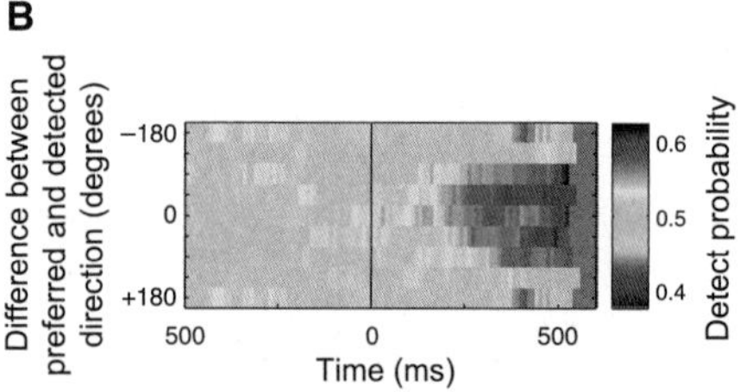

FIGURE 31.3 (A) Choice probability is calculated by comparing the distributions of neural responses on trials when the subject made each of two choices. This schematic shows pairs of distributions with different choice probabilities. (B) Choice probabilities (referred to as "detect probabilities" in this detection task) decrease for neurons whose preferred direction does not match the direction of motion being detected (adapted from Bosking & Maunsell, 2011). (See color plate 27.)

These weak choice probabilities are to be expected given how many neurons respond to any given stimulus. After all, many neurons presumably contribute to any percept, so no individual neuron should be terribly predictive of the monkey's choice.

The low but significant choice probabilities are surprising for two reasons, however. First is the observation we discussed previously: individual neurons carry nearly enough information to explain the monkey's performance on perceptual tasks. If the monkey only uses one or a few neurons to make a decision, those neurons should have very high choice probabilities. The low choice probabilities suggest that, instead, the monkey combines information from many neurons. This is a sensible strategy, but then why doesn't his performance reflect the benefit of the information encoded by all those neurons?

If the monkey indeed uses many neurons to make any decision, the second surprise is that choice probabilities are big enough to be detected at all. Trying to predict the monkey's decision from the one neuron that happens to be close to the electrode is like trying to predict the outcome of a presidential election by polling only the first person you encounter on the street. If many neurons or people vote in a decision or an election, that one person's opinion should carry very little weight in the election and should not tell you much about which way it will go.

The resolution to both of these apparent paradoxes lies in the fact that neurons (like voters ...) are not independent thinkers. If the noise in neural responses were independent, some neurons would fire more than average and some would fire less than average at any given moment. In this case, the mean of the population would remain relatively constant, so the noise would not have a big effect on the ability of the population to encode sensory information. Therefore, the monkey would have no excuse for not performing better on the task, and the average individual neuron would carry very little information about which option he would choose.

Instead, the noise in neuronal responses is shared, or *correlated*. Simultaneous recordings from multiple neurons have shown that when one neuron fires more than its average, nearby neurons are likely to be firing more than their average as well (for review, see Cohen & Kohn, 2011). If all of the neurons were doing the exact same thing, there would be no point in having more than one: the monkey would do exactly as well as any one neuron, and every neuron would perfectly predict the monkey's choices. In reality, the correlation between the noise in the responses of nearby neurons is positive but weak. A model of the monkey's decision process suggests the observed correlations can explain the monkey's performance and the observed choice probabilities pretty well (Cohen & Newsome, 2009; Shadlen et al., 1996).

This model also makes the prediction that even if monkeys use a very large number of neurons to make a decision, neither their performance nor the ability of an ideal observer to predict choices from the responses of a group of neurons should improve after about 100 neurons. This prediction appears to be true. A recent study recording from about 80 sensory neurons simultaneously showed that those neurons were enough to predict the monkey's decisions on the vast majority of trials (Cohen & Maunsell, 2010).

It is tempting to interpret the existence of choice probabilities significantly greater than 0.5 as evidence enough that a group of neurons is responsible for perception. After all, recording from a few dozen neurons is enough to predict a monkey's actions almost perfectly. The problem is that correlated noise can lead to choice probability even when the recorded neurons have nothing to do with the decision, as long as their responses are correlated with the neurons that are. If the monkey makes his decision based on neuron A's response, but neuron B responds the same way as A, both could be used to predict the decision. (The same could be true of a nonvoter whose whims reflect the feelings of the nation as a whole.) The fact that significant choice probabilities have been observed almost every time they have been measured (for review, see Nienborg, Cohen, & Cumming, 2012), including in neurons whose tuning is very poor for the particular task, suggests choice probability may sometimes be present solely due to widespread correlated noise. Further evidence is therefore necessary to make the case that a group of neurons are responsible for, rather than simply correlated with, a perceptual decision.

Does activating the candidate neurons bias perception in favor of the stimulus property they encode? If a group of neurons is responsible for a percept, modifying their activity should change the percept. Since the work of Penfield in the 1950s, experimenters have used electrical stimulation to activate small groups of neurons. Penfield found that applying large currents on the surface of cortex could elicit reliable perceptions or movements depending on the location of the injected current (Penfield & Rasmussen, 1950). This work led to the creation of well-known maps of somatosensory and motor cortex (sometimes affectionately termed "homunculus maps," because they look like a small person). As stimulation techniques were refined, experimenters realized that smaller currents delivered by fine electrodes placed inside the cortex of animals could lead to readily reproducible motor responses (for a review, see Graziano, 2006).

Newsome and colleagues used microstimulation to determine how activating small populations of MT neurons could bias the monkey's choice during the motion-direction-discrimination task (Salzman et al., 1990). This clever experiment relied on the fact that in MT, neurons located near each other tend to have similar preferred directions. Therefore, when the experimenters applied small currents, they could reasonably expect that they were modulating the activity of a group of neurons with similar tuning.

The authors put an electrode into MT, placed the motion stimulus in the receptive fields of the neurons they recorded, and had the monkey discriminate motion in the neurons' preferred direction from motion in the neurons' null direction. Unbeknownst to the monkey, the experimenters passed small amounts of current through the electrode on a subset of trials, which increased the firing rates of the neurons near the electrode tip.

The monkey behaved as if the stimulation had increased the strength of the motion in the neurons' preferred direction, which can be seen in the leftward shift in the animal's psychometric curve (figure 31.4A). That this technique works is astounding: the currents used in this experiment should have modulated the responses of only tens or hundreds of neurons (Histed, Bonin, & Reid, 2009). The fact that tickling the responses of such a small number of neurons affects the monkeys' behavior in a measurable way is strong evidence that MT neurons underlie performance in the direction-discrimination task.

In other contexts, microstimulation has an even more astounding effect: it has been used to create a percept in the absence of a physical stimulus. Romo and colleagues trained monkeys to perform a two-interval somatosensory discrimination where the monkey's job was to indicate whether the second stimulus in a pair vibrated at a higher or lower frequency than the first (Romo et al., 1998). During most trials, a mechanical vibrating flutter stimulus was applied to the monkey's fingertips for both stimuli, but on some trials, one of the mechanical stimuli was absent and instead, electrical microstimulation was applied directly in somatosensory cortex. Amazingly, the animals were able to compare electrical and mechanical stimuli after training only on mechanical stimuli. This experiment provided strong evidence that the electrically stimulated somatosensory neurons encoded vibrotactile information that the animals could use to make decisions about stimulus frequency.

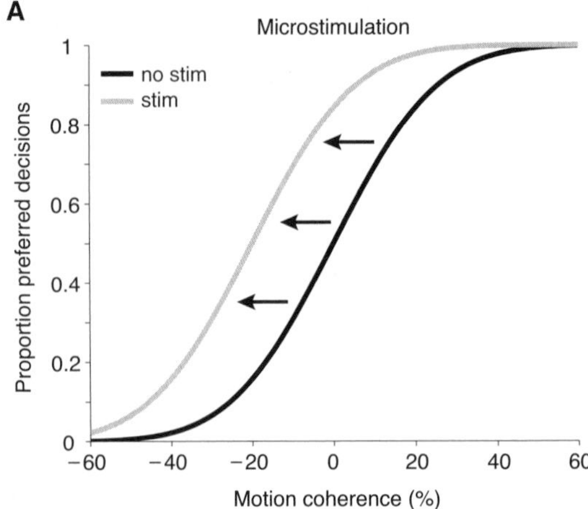

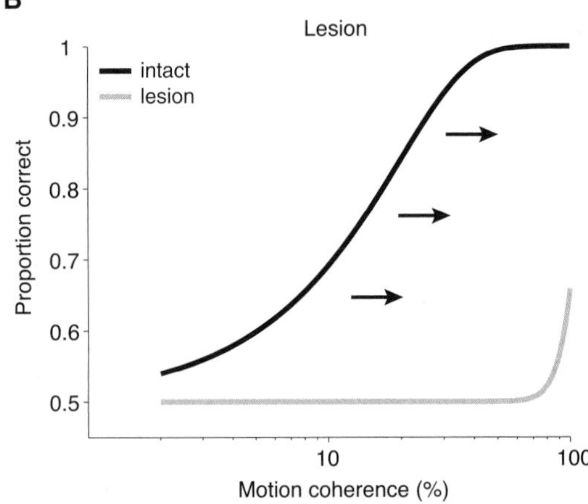

FIGURE 31.4 (A) Microstimulation biases the subject's choice toward the option that matches the preferred direction of the stimulated neurons (schematic based on the results of Salzman et al., 1990). (B) Performance on a motion-discrimination task is not much better than chance following ibotenic acid lesion of MT (schematic based on the results of Newsome & Paré, 1988).

While microstimulation is a powerful tool for evaluating the causal contribution of neurons to perception, it is a relatively coarse method both temporally and spatially, and does not allow for the precise targeting of cell types or neuronal subpopulations. New techniques make it possible to precisely target particular neurons at particular times. In particular, the field of optogenetics allows researchers to use light to activate neurons that express proteins such as channelrhodopsin (ChR2; Luo, Callaway, & Svoboda, 2008; Peron & Svoboda, 2010; Scanziani & Häusser, 2009).

The greatest promise of these techniques lies in their potential specificity: in mice and some other species, ChR2 can be genetically expressed into particular subtypes of neurons in different areas or cortical layers or that project to different areas. Making these methods usable in monkeys is a topic of active investigation, and there have been some notable early successes (Diester et al., 2011; Han et al., 2009). A recent study was the first to demonstrate the behavioral detection of opsin-mediated neuronal signals in the primate brain (Jazayeri, Lindbloom-Brown, & Horwitz, 2012).

Does inactivating the candidate neurons cause deficits in perception? If a group of neurons underlie a percept, suppressing their activity should cause a deficit in perception. The brain is remarkably flexible, and removing or inactivating large portions of cortex can have surprisingly subtle effects on behavior. Nevertheless, to make a strong statement that a group of neurons are responsible for a percept, it is necessary to show that inactivating them has a measurable effect on a subject's perceptual ability, even if it's a transient one.

To determine whether MT neurons are necessary for motion-direction discrimination, Newsome and Paré used ibotenic acid to completely remove MT in one hemisphere (Newsome & Paré, 1988). Removing MT almost completely abolished the monkeys' ability to perform the direction-discrimination task (figure 31.4B), suggesting that under normal circumstances, MT is critical for performance in this task. Interestingly, the monkeys' performance improved almost to pre-lesion levels within a few weeks, suggesting that motion signals from elsewhere in cortex can be recruited to guide behavior in the absence of MT.

While lesion studies can provide clear answers, they are a relatively crude approach for asking whether a group of neurons are necessary for a percept. It is usually not possible to measure performance on perceptual tasks for a while after the lesions have been made, and, as the MT results show, subjects may eventually be able to learn alternate strategies to make up for the missing areas. Because reversible inactivation experiments can be performed on a relatively fast time scale, such manipulations may solve this problem by not allowing for sufficient time for the animal's brain or behavioral strategy to adjust to a long-term deficit. Using muscimol, a GABA agonist, Chowdhury and Deangelis (2008) replicated the findings of Newsome and Paré, showing that the monkeys' performance returned to normal after the drug had sufficient time to wash out two days after injection. Further, because they could allow MT to recover after muscimol was applied, the authors were able to ask questions about how an area's role in a percept might change with training on different tasks. They found that MT's role in a particular type of depth judgment depended on whether monkeys had first been trained on a different, finer-grained depth discrimination.

Technological advancements are making possible even more precise inactivation and lesion methods. For example, Schlief and Wilson (2007) used a genetic approach to lesion specific, highly selective olfactory receptor neurons in the fly brain. Flies are innately attracted to certain odors, but removing these neurons made them no longer interested in the odors encoded by the removed receptor neurons.

While genetic approaches are still in their infancy in primates, researchers have begun to use rodents to combine cutting-edge genetic techniques with electrophysiology and perceptual tasks. A recent study from Znamenskiy and Zador (2013) harnessed many of the techniques we have discussed in the previous sections to provide an impressive amount of evidence linking corticostriatal neurons to decisions about the frequency of a tone stimulus. The authors used excitatory channelrhodopsin and inhibitory archaerhodopsin to demonstrate that the activity of neurons in auditory cortex that project to the striatum biases decisions either toward or away from the choice represented by the set of neurons expressing each opsin type, respectively.

Do the candidate neurons underlie perception or a planned motor response? Neurons that underlie perception should correlate with the subject's choices regardless of the way that the subject communicates that choice. The responses of neurons that encode a planned motor response could correlate with choices for a trivial reason that has nothing to do with perception. In the most common version of the direction-discrimination task, the monkey signals that he perceives upward motion by moving his eyes up and downward motion by moving his eyes downward. Imagine recording from a motor neuron responsible for upward eye movements. The responses of that neuron would be perfectly correlated with the monkey's choice; high firing rates would signal upward choices, and low rates would signal downward choices.

One straightforward way to ensure that a set of neurons is correlated with perception rather than simply a planned motor response is to ensure that their responses carry sensory information and correlate with choices regardless of the motor output used to signal the choice. MT neurons have choice-probability and signal-motion information in the motion-discrimination task regardless of whether choices are signaled with eye movements (Britten et al., 1992, 1996) or hand movements (Nichols & Newsome, 2002). In principle, the responses of the same neurons could be responsible for perception and a motor plan, so the requirement should not be for identical responses for different motor outputs. Rather, some aspects of the response (e.g., early in the behavioral trial) should not depend on the motor output used to signal the choice.

Examples of systems for which the link between sensory neurons and perceptions is most well established

To our knowledge, MT and the motion-direction discrimination task represents the only system for which all of the above questions have been answered in the affirmative. However, there is growing evidence linking other brain areas with specific percepts. Two other experimental systems have proven to be particularly fertile grounds for linking sensory neurons to perception.

Motion direction is a relatively low-level feature of a visual stimulus, but the same experimental methods used in the MT experiments have been used to link the activity of neurons in inferior temporal (IT) cortex, in both humans and monkeys, with the perception of higher-level features like faces. Early electrophysiological experiments identified neurons in primate IT that were extremely selective for faces compared to other complex objects (Desimone, Albright, Gross, & Bruce, 1984). The existence of small clusters of face-selective IT neurons was revealed by work from Doris Tsao and colleagues, who used functional imaging to guide their electrophysiological recordings and identified patches in macaque temporal cortex that were overwhelmingly selective for faces, even compared to other similar shapes (Tsao, Freiwald, Tootell, & Livingstone, 2006). Functional imaging in humans has also revealed patches of cortex in the ventral stream that are highly selective for different object categories, such as faces (Kanwisher, McDermott, & Chun, 1997). Activity in these regions may be used during perceptual decisions about complex object categories like faces and houses (Heekeren, Marrett, Bandettini, & Ungerleider, 2004), but it is currently unknown whether individual IT neurons in humans or monkeys can be used to predict choices in a face-discrimination task in a manner that is directly analogous to the choice-probability studies performed in MT.

However, the results of lesion and microstimulation studies strongly implicate IT cells in face perception. Prosopagnosia, famously documented by Oliver Sacks and others, is a well-described condition in humans thought to result from damage to the temporal lobe in which patients are completely unable to recognize or identify faces. To test whether activity of inferotemporal neurons is sufficient to lead to the percept of a face, Afraz and colleagues performed an analogous experiment in primate IT to the one performed by Newsome and colleagues in area MT. These authors trained monkeys to perform a task where they reported whether an image contained a face or not (Afraz, Kiani, & Esteky, 2006). This task was made difficult by adding various amounts of noise to the image set that degraded the clarity of the images, thus making them harder to discriminate. Similar to the results of the MT study, the authors found that microstimulation in face-selective portions of cortex led the animals to report seeing a face in a noisy stimulus more often than when stimulation was not applied.

With the advent of increasingly sophisticated molecular, genetic, and imaging techniques has come a strong interest in developing both rodent and rat models for linking the activity of neurons to behavior. Progress on this front has been made across a range of tasks and sensory modalities, including visual (Carandini & Churchland, 2013) and auditory (Brunton, Botvinick, & Brody, 2013) discriminations. Of particular interest is a series of work linking the activity across multiple brain areas in rats to olfactory discriminations.

Uchida, Mainen, and colleagues have demonstrated that rats can accurately discriminate between mixtures of odor pairs with just a single sniff (Uchida & Mainen, 2003). Neurons in the olfactory bulb, the target of olfactory receptor neurons in the nose, exhibit odor-specific selectivity and have been shown to reliably distinguish between small sets of odorants on similarly brief time scales to a sniff, as well as to correlate with reaction times during an odor-discrimination task (Cury & Uchida, 2010).

Piriform cortex, also known as olfactory cortex, receives input from the olfactory bulb and is a candidate for a cortical region that is involved in odor discriminations. A recent study recording activity in piriform cortex during an odor-mixture discrimination demonstrated that the activity of fewer than 100 neurons was sufficient to accurately predict behavioral performance and reaction time. This study also interestingly revealed that noise correlations among these neurons were extremely low, highlighting a potentially important difference between olfactory and visual processing (Miura, Mainen, & Uchida, 2012).

The systems for which the most progress has been made linking sensory neurons to perception tend to have at least two things in common. First, experimenters have designed psychophysical tasks that allow them to measure subtle changes in specific perceptual abilities. Second, the brain areas under study (in the primate visual system, at least) tend to be organized so that neurons with similar tuning tend to be located near each other in the brain. This certainly has experimental benefits, because techniques like microstimulation or chemical inactivation that are accessible in primates can be used to affect small groups of neurons with, for example, only a preference for upward (and not downward) motion. New techniques such as optogenetics may remove this technical requirement. It will be interesting to see whether anatomical organization is required for certain perceptual abilities, or whether areas with anatomical organization have been well studied simply because it is technically easier to do so.

General themes and future directions

As our knowledge of the link between sensory neurons and specific percepts has become more developed in recent years, a few conceptual themes and unanswered questions have emerged.

STUDIES OF SINGLE NEURONS MISS CRITICAL INFORMATION For technological reasons, most studies linking the responses of sensory neurons to perception focus on recordings from one neuron at a time. However, the responses of a large subset of the thousands of neurons that respond to any sensory stimulus are thought to underlie any percept. The logic is that we can learn about how big groups of neurons respond at one time (as in actual behavior) by recording how individual neurons respond over many behavioral trials and using computational models to figure out how the responses of many neurons are combined. Although these types of studies have been hugely informative, recent studies have shown that this assumption does not always hold.

New technology makes it possible to record from groups of neurons simultaneously, and early results suggest that measuring the responses of many neurons at once is very different than combining information from individual neurons recorded on separate days. For example, we discussed earlier that shared or correlated noise can make measurements of choice-probability misleading. Those same correlated responses can have big effects on the amount of information a group of neurons encodes, although different models of how the responses of many neurons are combined make different predictions about whether correlations hurt or help (Abbott & Dayan, 1999; Shadlen et al., 1996). Correlations depend on the sensory stimuli (Aertsen, Gerstein, Habib, & Palm, 1989; Ahissar, Vaadia, Ahissar, & Bergman, 1992; Espinosa & Gerstein, 1988; Kohn & Smith, 2005), learning (Ahissar et al., 1992; Gutnisky & Dragoi, 2008; Komiyama et al., 2010), and behavioral state and cognitive factors like attention (Cohen & Maunsell, 2009; Cohen & Newsome, 2008; Mitchell, Sundberg, & Reynolds, 2009; Poulet & Petersen, 2008; Vaadia et al., 1995). That so many factors affect correlations strongly suggests that they are important, but future theoretical and experimental work will be needed to determine their exact role in encoding sensory stimuli.

Recording from large groups of neurons has another advantage: it gives experimenters a snapshot of the sensory information available to a subject at a given moment rather than the average responses to many repetitions of the same sensory stimulus. It has long been known that cognitive factors and motor planning can affect the responses of sensory neurons, but the role these factors play likely differs from moment to moment and can be obscured by averaging across many trials (for review, see Desimone & Duncan, 1995; Maunsell & Cook, 2002; Maunsell & Treue, 2006). The differences in the conclusions that can be drawn from studies that record many neurons at a single moment compared to one neuron over a long period of time are an area of active investigation.

NOT ALL NEURONS ARE THE SAME Neurons come in many anatomical and physiological subtypes. Neurons differ in their pattern of connections, whether they are inhibitory and excitatory, whether they fire tonically or in bursts, and a host of other factors. Most of the studies we have discussed rely on extracellular electrophysiology. This technique makes it very difficult to determine the subtype of the neuron under study. Most models of neural circuits posit very different roles for neurons with different properties (e.g., inhibitory vs. excitatory neurons), but there is little experimental data on how they function in behaving animals. New technology, including optogenetics (Luo et al., 2008; Peron & Svoboda, 2010; Prakash et al., 2012; Scanziani & Häusser, 2009; Zeng & Madisen, 2012) and improved imaging tools (Helmchen & Denk, 2005), will make it possible to understand the role of different classes of neurons in neural computations.

THEORETICAL MODELS ARE IMPORTANT As monitoring the activity of large numbers of neurons becomes easier and easier, having a theoretical framework for interpreting all of these data becomes more and more critical. The dominant, and really only, framework for relating sensory neurons to perceptual decisions comes from work by Shadlen and colleagues (1996) modeling decision making in the motion-discrimination task. This model has significantly helped researchers to make sense of the physiological data. For example, it pointed to correlated variability as the source of choice-predictive signals in individual MT neurons. It is in many ways a high-level model, however, and does not take into account factors like cell type, dynamics, or the pattern of connections different neurons make. As data sets get more sophisticated, new theoretical frameworks will be necessary to convert data into understanding.

Tremendous progress has been made on both theoretical and experimental fronts since Parker and Newsome wrote their landmark review in 1998. As the field continues to progress, their rubric has become more important than ever. It is only through a principled application of the technological advancements discussed above that we will improve our understanding of the relationship between neural activity and perception.

ACKNOWLEDGMENTS The authors are supported by NIH grants 4R00EY020844-03 and R01 EY022930 (MRC), a training grant slot on NIH 5T32NS7391-14 (DAR), a Whitehall Fellowship (MRC), and a Klingenstein Fellowship (MRC). We thank David Montez, Regina Chang, and Trevor Stoltzfus for helpful comments on an earlier version of the chapter.

REFERENCES

Abbott, L. F., & Dayan, P. (1999). The effect of correlated variability on the accuracy of a population code. *Neural Comput, 11*(1), 91–101.

Aertsen, A. M., Gerstein, G. L., Habib, M. K., & Palm, G. (1989). Dynamics of neuronal firing correlation: Modulation of "effective connectivity." *J Neurophysiol, 61*(5), 900–917.

Afraz, S.-R., Kiani, R., & Esteky, H. (2006). Microstimulation of inferotemporal cortex influences face categorization. *Nature, 442*(7103), 692–695.

Ahissar, E., Vaadia, E., Ahissar, M., & Bergman, H. (1992). Dependence of cortical plasticity on correlated activity of single neurons and on behavioral context. *Science, 257*(5075), 1412–1415.

Bosking, W. H., & Maunsell, J. H. R. (2011). Effects of stimulus direction on the correlation between behavior and single units in area MT during a motion detection task. *J Neurosci, 31*(22), 8230–8238.

Britten, K. H., Newsome, W. T., Shadlen, M. N., Celebrini, S., & Movshon, J. A. (1996). A relationship between behavioral choice and the visual responses of neurons in macaque MT. *Vis Neurosci, 13*(1), 87–100.

Britten, K. H., Shadlen, M. N., Newsome, W. T., & Movshon, J. A. (1992). The analysis of visual motion: A comparison of neuronal and psychophysical performance. *J Neurosci, 12*(12), 4745–4765.

Brunton, B. W., Botvinick, M. M., & Brody, C. D. (2013). Rats and humans can optimally accumulate evidence for decision-making. *Science, 340*(6128), 95–98.

Carandini, M., & Churchland, A. K. (2013). Probing perceptual decisions in rodents. *Nat Neurosci, 16*(7), 824–831.

Cohen, M. R., & Kohn, A. (2011). Measuring and interpreting neuronal correlations. *Nat Neurosci, 14*(7), 811–819.

Cohen, M. R., & Maunsell, J. H. R. (2009). Attention improves performance primarily by reducing interneuronal correlations. *Nat Neurosci, 12*(12), 1594–1600.

Cohen, M. R., & Maunsell, J. H. R. (2010). A neuronal population measure of attention predicts behavioral performance on individual trials. *J Neurosci, 30*(45), 15241–15253.

Cohen, M. R., & Newsome, W. T. (2008). Context-dependent changes in functional circuitry in visual area MT. *Neuron, 60*(1), 162–173.

Cohen, M. R., & Newsome, W. T. (2009). Estimates of the contribution of single neurons to perception depend on timescale and noise correlation. *J Neurosci, 29*(20), 6635–6648.

Chowdhury, S. A., & DeAngelis, G. C. (2008). Fine discrimination training alters the causal contribution of macaque area MT to depth perception. *Neuron, 60*(2), 367–377.

Cury, K. M., & Uchida, N. (2010). Robust odor coding via inhalation-coupled transient activity in the mammalian olfactory bulb. *Neuron, 68*(3), 570–585.

Desimone, R., Albright, T., Gross, C., & Bruce, C. (1984). Stimulus-selective properties of inferior temporal neurons in the macaque. *J Neurosci, 4*(8), 2051–2062.

Desimone, R., & Duncan, J. (1995). Neural mechanisms of selective visual attention. *Annu Rev Neurosci, 18*, 193–222.

Diester, I., Kaufman, M. T., Mogri, M., Pashaie, R., Goo, W., Yizhar, O., ... Shenoy, K. V. (2011). An optogenetic toolbox designed for primates. *Nat Neurosci, 14*(3), 387–397.

Espinosa, I., & Gerstein, G. (1988). Cortical auditory neuron interactions during presentation of 3-tone sequences: Effective connectivity. *Brain Res, 450*(1), 39–50.

Gerdjikov, T. V, Bergner, C. G., Stüttgen, M. C., Waiblinger, C., & Schwarz, C. (2010). Discrimination of vibrotactile stimuli in the rat whisker system: Behavior and neurometrics. *Neuron, 65*(4), 530–540.

Graziano, M. (2006). The organization of behavioral repertoire in motor cortex. *Annu Rev Neurosci, 29*, 105–134.

Green, D. M., & Swets, J. A. (1966). *Signal detection theory and psychophysics.* New York, NY: Wiley.

Gutnisky, D. A., & Dragoi, V. (2008). Adaptive coding of visual information in neural populations. *Nature, 452*(7184), 220–224.

Han, X., Qian, X., Bernstein, J. G., Zhou, H.-H., Franzesi, G. T., Stern, P., ... Boyden, E. S. (2009). Millisecond-timescale optical control of neural dynamics in the nonhuman primate brain. *Neuron, 62*(2), 191–198.

Heekeren, H. R., Marrett, S., Bandettini, P. A., & Ungerleider, L. G. (2004). A general mechanism for perceptual decision-making in the human brain. *Nature, 431*(7010), 859–862.

Helmchen, F., & Denk, W. (2005). Deep tissue two-photon microscopy. *Nat Methods, 2*(12), 932–940.

Histed, M., Bonin, V., & Reid, R. (2009). Direct activation of sparse, distributed populations of cortical neurons by electrical microstimulation. *Neuron, 63*(4), 508–522.

Jazayeri, M., Lindbloom-Brown, Z., & Horwitz, G. D. (2012). Saccadic eye movements evoked by optogenetic activation of primate V1. *Nat Neurosci, 15*(10), 1368–1370.

Kanwisher, N., McDermott, J., & Chun, M. M. (1997). The fusiform face area: A module in human extrastriate cortex specialized for face perception. *J Neurosci, 17*(11), 4302–4311.

Kohn, A., & Smith, M. A. (2005). Stimulus dependence of neuronal correlation in primary visual cortex of the macaque. *J Neurosci, 25*(14), 3661–3673.

Komiyama, T., Sato, T. R., O'Connor, D. H., Zhang, Y.-X., Huber, D., Hooks, B. M., ... Svoboda, K. (2010). Learning-related fine-scale specificity imaged in motor cortex circuits of behaving mice. *Nature, 464*(7292), 1182–1186.

Luo, L., Callaway, E., & Svoboda, K. (2008). Genetic dissection of neural circuits. *Neuron, 57*(5), 634–660.

Maunsell, J. H. R., & Cook, E. P. (2002). The role of attention in visual processing. *Philos Trans R Soc Lond B Biol Sci, 357*(1424), 1063–1072.

Maunsell, J. H. R., & Treue, S. (2006). Feature-based attention in visual cortex. *Trends Neurosci, 29*(6), 317–322.

Mitchell, J. F., Sundberg, K. A., & Reynolds, J. H. (2009). Spatial attention decorrelates intrinsic activity fluctuations in macaque area V4. *Neuron, 63*(6), 879–888.

Miura, K., Mainen, Z. F., & Uchida, N. (2012). Odor representations in olfactory cortex: Distributed rate coding and decorrelated population activity. *Neuron, 74*(6), 1087–1098.

Newsome, W., & Paré, E. (1988). A selective impairment of motion perception following lesions of the middle temporal visual area (MT). *J Neurosci, 8*(6), 2201–2211.

Nichols, M. J., & Newsome, W. T. (2002). Middle temporal visual area microstimulation influences veridical judgments of motion direction. *J Neurosci, 22*(21), 9530–9540.

Nienborg, H., Cohen, M. R., & Cumming, B. G. (2012). Decision-related activity in sensory neurons: Correlations among neurons and with behavior. *Ann Rev Neurosci, 35*, 463–483.

Parker, A. J., & Newsome, W. T. (1998). Sense and the single neuron: Probing the physiology of perception. *Ann Rev Neurosci, 21*, 227–277.

Penfield, W., & Rasmussen, T. (1950). *The cerebral cortex of man.* New York, NY: Macmillan.

Peron, S., & Svoboda, K. (2010). From cudgel to scalpel: Toward precise neural control with optogenetics. *Nat Methods, 8*(1), 30–34.

Poulet, J. F., & Petersen, C. (2008). Internal brain state regulates membrane potential synchrony in barrel cortex of behaving mice. *Nature, 454*(7206), 881–885.

Prakash, R., Yizhar, O., Grewe, B., Ramakrishnan, C., Wang, N., Goshen, I., ... Deisseroth, K. (2012). Two-photon optogenetic toolbox for fast inhibition, excitation and bistable modulation. *Nat Methods, 9*(12), 1171–1179.

Prince, S. J., Pointon, A. D., Cumming, B. G., & Parker, A. J. (2000). The precision of single neuron responses in cortical area V1 during stereoscopic depth judgments. *J Neurosci, 20*(9), 3387–3400.

Romo, R., Hernández, A., Zainos, A., & Salinas, E. (1998). Somatosensory discrimination based on cortical microstimulation. *Nature, 392*(6674), 387–390.

Salzman, C., Britten, K., & Newsome, W. (1990). Cortical microstimulation influences perceptual judgements of motion direction. *Nature, 346*(6280), 174–177.

Scanziani, M., & Häusser, M. (2009). Electrophysiology in the age of light. *Nature, 461*(7266), 930–939.

Schlief, M. L., & Wilson, R. I. (2007). Olfactory processing and behavior downstream from highly selective receptor neurons. *Nat Neurosci, 10*(5), 623–630.

Shadlen, M. N., Britten, K. H., Newsome, W. T., & Movshon, J. A. (1996). A computational analysis of the relationship between neuronal and behavioral responses to visual motion. *J Neurosci, 16*(4), 1486–1510.

Tsao, D., Freiwald, W., Tootell, R., & Livingstone, M. (2006). A cortical region consisting entirely of face-selective cells. *Science, 311*(5761), 670–674.

Uchida, N., & Mainen, Z. F. (2003). Speed and accuracy of olfactory discrimination in the rat. *Nat Neurosci, 6*(11), 1224–1229.

Vaadia, E., Haalman, I., Abeles, M., Bergman, H., Prut, Y., Slovin, H., & Aertsen, A. (1995). Dynamics of neuronal interactions in monkey cortex in relation to behavioral events. *Nature, 373*, 515–518.

Zeng, H., & Madisen, L. (2012). Mouse transgenic approaches in optogenetics. *Prog Brain Res, 196*, 193–213.

Znamenskiy, P., & Zador, A. M. (2013). Corticostriatal neurons in auditory cortex drive decisions during auditory discrimination. *Nature, 497*(7450), 482–485.

32 Characterizing the Effects of Stimulus and Neural Variability on Perceptual Performance

WILSON S. GEISLER, JOHANNES BURGE, MELCHI M. MICHEL, AND ANTHONY D. D'ANTONA

ABSTRACT Perceptual performance is limited by both external and internal factors. External factors include the physical variability of sensory stimuli and the inherent ambiguities that exist in the mapping between the properties of the environment and the properties of stimuli at the sensory organs (natural scene statistics). Internal factors include neural noise and nonrandom computational inefficiencies. External factors have not received the study they deserve, perhaps because they are difficult to measure and because methods for characterizing them have not been standardized. This chapter describes some of the computational tools used to characterize the effects of external and internal variability on perceptual performance. These tools are based on concepts of Bayesian statistical decision theory and are illustrated for several basic natural tasks: grouping of contours across occlusions, estimation of binocular disparity, and interpolation of missing pixel-luminance values.

Evolution pushes sensory and perceptual systems to perform efficiently in those tasks necessary for the organism to survive and reproduce. Nonetheless, even in an organism's natural tasks, perceptual performance can never be perfect. Thus, to understand and predict perceptual performance it is crucial to characterize and understand the many factors that limit performance. These factors include the complexity and variability of the sensory stimuli, as well as many sources of internal variability, ranging from noise in sensory receptor responses, to noise in decision and memory circuits, to noise in motor neuron responses. The aim of this chapter is to describe some of the computational tools used to characterize and understand the effects of external (stimulus) and internal sources of variability on perceptual performance. These tools are based on principles of statistical decision and estimation theory. The computational tools described here are applicable to many perceptual systems, but the examples are drawn from the vision literature.

The most basic kinds of stimulus variability are irreducible sources of noise that occur in transmission of stimulus information from the environment to the sensory organs. For example, the quantum nature of light causes the number of photopigment molecules activated in a photoreceptor to vary according to the Poisson probability distribution, even when the stimulus is nominally the same (de Vries, 1943; Hecht, Shlaer, & Pirenne, 1942; Rose, 1948). Although this source of noise is ubiquitous, there are only a few situations where it is the primary factor limiting performance. These situations consist primarily of simple detection or discrimination tasks where brief, spatially localized targets are presented in the visual periphery under dark-adapted conditions when the rod photoreceptors are most sensitive (Hecht et al., 1942).

In some laboratory tasks, it is possible to avoid all sources of stimulus noise, other than irreducible sources such as photon noise. In such tasks, performance is usually dominated by neural variability, and by limitations in neural computations. Examples of such cases would be simple detection or discrimination tasks with fixed stimuli presented under light-adapted conditions (e.g., detection of a known pattern on a uniform gray background). In other laboratory tasks, and in most natural tasks, additional sources of stimulus variability are also major factors. Examples of such cases would be detection of targets in pixel-noise backgrounds (Burgess, Wagner, Jennings, & Barlow, 1981), or estimation of physical properties in the environment such as the depth, shape, and reflectance of object surfaces (e.g., see Geisler, 2011; Kersten, Mamassian, & Yuille, 2004).

Most perceptual tasks can be regarded as decision making in the presence of random variability, and hence an appropriate theoretical framework for analyzing perceptual performance is Bayesian statistical decision theory (e.g., see Geisler, 2011; Kersten et al., 2004; Knill & Richards, 1996). In what follows, we sketch the general Bayesian framework and then discuss several special cases, starting with a discussion of simple detection and discrimination tasks and ending with discussion of optimal estimation in natural scenes.

It is important to note that Bayesian statistical decision theory is used to analyze perceptual performance in two different ways. The first is to derive ideal-observer models, which are theoretical devices that perform a perceptual task optimally. An ideal observer usually contains no free parameters and is not meant to be a model of real observers. Rather, its purpose is (1) to help identify task-relevant stimulus properties, (2) to describe how those properties should be used to perform the task of interest, (3) to provide a rigorous benchmark against which to compare real perceptual systems, and (4) to suggest principled hypotheses and models for real performance.

The second way Bayesian statistical decision theory is used is as a framework for modeling perceptual performance. When used in modeling perception, there are generally hypothesized internal (neural or information-processing) mechanisms, which have unknown parameters that are estimated from perceptual performance data.

Bayesian statistical decision theory

Specifying the Task The first step in using Bayesian statistical decision theory is to specify the task. This includes specifying the set of possible stimuli, the set of possible responses, and the goal of the task. Specifying the set of possible stimuli typically requires specifying (1) the ground-truth (distal) stimuli, which reflect the true task-relevant state of the world $\boldsymbol{\omega}$ and (2) the proximal stimuli, which constitute the input data $\boldsymbol{s}$. For example, in a simple detection-in-noise task, the true state of the world is that a target is either absent ($\omega = a$) or present ($\omega = b$) in a noise pattern, and the proximal stimulus is the specific pattern of pixels that would be imaged on the retina. In a typical depth-estimation task, the true state of the world is the physical distance of one surface patch from another, and the proximal stimulus is the specific pattern of pixels imaged on the two retinas.

The set of possible responses can be quite complex, but in most perception experiments it is simple. For example, in the detection-in-noise task it would be one of two responses that indicate whether the observer judged the target to be absent ($r = a$) or present ($r = b$). In the depth-estimation task, the response might be an estimate of the number of centimeters in depth separating the surface patches.

Specifying the goal of a task requires specifying the costs and benefits (utility) of each possible response for each possible state of the world: $\gamma(\boldsymbol{r}, \boldsymbol{\omega})$. If the goal in the detection-in-noise task is to be as accurate as possible, then that can be represented by making the utility a positive value u when the response is correct ($\gamma(a, a) = \gamma(b, b) = u$), and $-u$ when the response is incorrect ($\gamma(a, b) = \gamma(b, a) = -u$). As another example, if the goal is to maximize accuracy, while keeping false-positive responses (saying an absent target is present) at some low rate, then that can be represented by assigning a greater cost to false-positive than false-negative responses, that is, making $\gamma(b, a) < \gamma(a, b)$. In the depth-estimation task there are many more possible combinations of response and state of the world, and hence many more possible goals (utility functions). A typical goal would be to minimize the mean squared error, which would be obtained by setting $\gamma(r, \omega) = -(r - \omega)^2$. In an ideal-observer model, the utility function (goal) is fully specified. In a perceptual-performance model, the utility function is a part of the model and may have free parameters.

Ground Truth and Input Stimuli The second step in using Bayesian statistical decision theory is to specify the statistical relationship between the states of the world and the proximal stimulus. In the most common case, this involves specifying the conditional probability of the different ground-truth states of the world given the input stimuli; this is the posterior probability distribution, $p(\boldsymbol{\omega}|\boldsymbol{s})$. In practice, it is often convenient to first specify the stimulus likelihood distribution $p(\boldsymbol{s}|\boldsymbol{\omega})$ for each possible state of the world and the prior probability distribution $p(\boldsymbol{\omega})$, and then use Bayes's rule to compute the posterior probability distribution.

Input Data In many applications of Bayesian statistical decision theory, there are properties of the perceptual system that are part of the specification of the input to the optimal Bayesian computations. These properties could be either known physical or neural properties that have no free parameters, or models of these properties that have free parameters. For example, these properties might include the optics of the eye, the sampling pattern of the photoreceptors, or the tuning and noise characteristics of retinal ganglion cells. The properties can be represented by a function g_θ that maps the input stimulus $\mathbf{s}$ onto input data,

$$\mathbf{z} = \mathbf{g}_\theta(\mathbf{s}), \tag{1}$$

where $\boldsymbol{\theta}$ represents any free parameters. In other words, this constraint function incorporates the effects of known or assumed properties, and its output is the input data to the Bayesian analysis.

Bayes Optimal Response Once the task, input data, and posterior distributions are specified, it is possible to write down an expression for the optimal response. Namely, one should pick the response that maximizes the utility (minimizes cost), averaged over the posterior probability of the possible states of the world, given the input data:[1]

$$\mathbf{r}_{opt}(\mathbf{z}) = \arg\max_{\mathbf{r}} \left[\sum_{\omega} \gamma(\mathbf{r}, \omega) p(\omega|\mathbf{z}) \right]. \qquad (2)$$

Note that $\mathbf{z}$ reduces to $\mathbf{s}$ in the case where the input data are the input stimuli. We define the "ideal observer" for a given task and constraint function to be the observer that makes responses according to Eq. 2.

In what follows, we first consider identification tasks (of which detection and discrimination are special cases), then estimation tasks, and finally make some general points about the relative importance of external and internal factors.

Identification tasks

In an identification task, the observer is required to identify which of n possible stimulus categories was presented on a trial. The special cases where there are only two possible categories of stimuli are usually referred to as *detection* or *discrimination* tasks.

In the classic yes-no task, the observer is presented on each trial with one of two randomly chosen stimuli (a or b). The observer is required to report whether the stimulus was a or b. Here we will regard a as the reference stimulus, and b as the reference plus signal. The typical goal is to maximize accuracy. In another variant, there are monetary costs and benefits associated with the different stimulus-response outcomes, and the goal is to maximize monetary gain. In the two-alternative forced choice (2AFC) task, the observer is presented both stimuli on each trial, a and b, either in two temporal intervals or two spatial locations. The temporal or spatial order is randomized, and the observer is required to report whether stimulus a or b was in the first location or interval. Although the yes-no task is more representative of real-world tasks, the 2AFC task is more common in laboratory experiments, because performance tends to be better and response biases smaller than in the yes-no task.

SIGNAL-DETECTION THEORY Signal-detection theory is a special case of Bayesian statistical decision theory that was developed to interpret the behavioral data in detection and discrimination experiments (Green & Swets, 1966; Tanner & Swets, 1954). The first key assumption of signal-detection theory is that on each trial the responses are based on the value of the decision variable ψ. On trials where the stimulus is a, the random values of ψ are described by one probability distribution, $p(\psi|a)$, whereas on trials where the stimulus is b, the random values of ψ are described by another probability distribution, $p(\psi|b)$ (see figure 32.1A). The second key assumption is that the observer's responses are selected by placing a criterion β along the decision variable axis; if the value of ψ exceeds the criterion, then the response is b; if it falls below the criterion, the response is a. Two potential criteria are shown in figure 32.1A (i.e., the solid and dashed vertical lines).

There are four possible stimulus-response outcomes in the yes-no task: responding b when the stimulus is b (hit); responding b when the stimulus is a (false alarm); responding a when the stimulus is a (correct rejection); and responding a when the stimulus is b (miss). The proportions of trials that are hits and misses must sum to 1.0, and proportions that are false alarms and correct rejections must sum to 1.0; thus, the data can be summarized by the proportions of hits and false alarms. As is clear from figure 32.1A, these two stimulus-response outcomes are interpreted in signal-detection theory as the areas under the two probability distributions to the right of the criterion. The number of standard deviations separating the means of the two distributions represents the observer's sensitivity, and is called d' (d-prime). The bigger the value of d', the greater the potential accuracy of the observer; however, actual performance will also depend on where the criterion is placed.

An important feature of signal-detection theory is that it allows estimation of both the sensitivity and the criterion from the proportion of hits and false alarms. For example, if the probability distributions are assumed to be Gaussian and of equal variance, then $d' = \Phi^{-1}(p_h) - \Phi^{-1}(p_{fa})$ and $\beta = \Phi^{-1}(1 - p_{fa}) - d'/2$, where Φ is the cumulative standard normal integral function. These formulas are useful because observers can differ in task performance due to differences in the criterion, even when they are equally sensitive, or vice versa. Once the value of d' is determined, it is also possible to calculate what would be the performance of the observer if the decision criterion were placed at some other location (e.g., the optimal location) using the formulas $p_h = \Phi(d'/2 - \beta)$ and $p_f = \Phi(-d'/2 - \beta)$.

The same logic and equations above hold for the 2AFC task; however, the colorful names hits and false alarms are reserved for the yes-no task. Also, if the responses to the two stimuli are statistically independent, then signal-detection theory predicts d' in the

[1]Note that arg max $[f(x)]$ is the value of x (the argument) for which $f(x)$ reaches its maximum value.

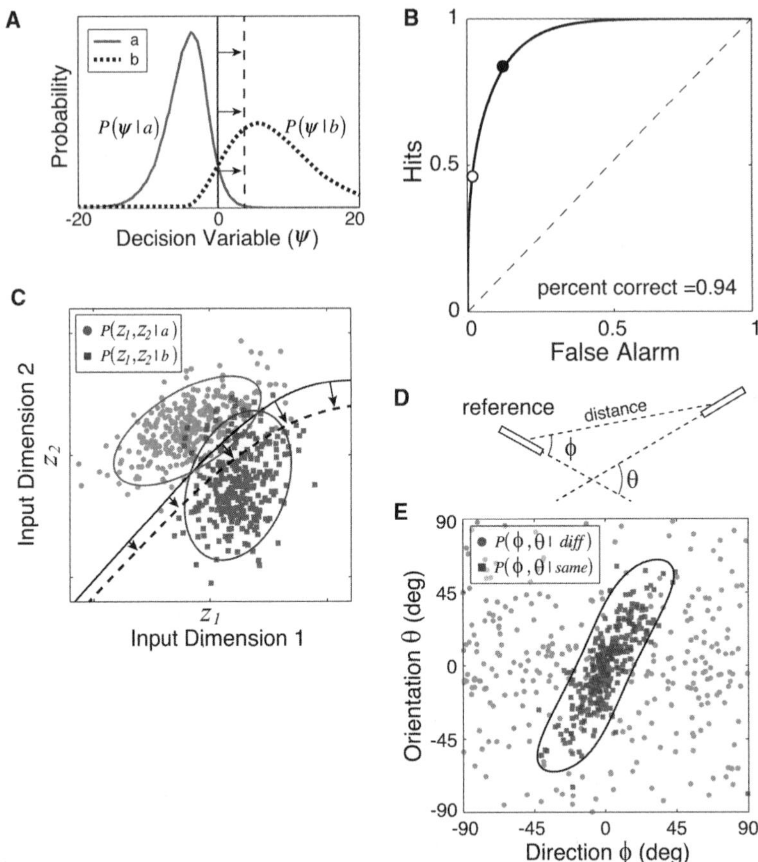

FIGURE 32.1 Detection and discrimination. (A) Distribution of decision variable values conditioned on two different states of the world (a and b). (B) The receiver operating characteristic for the distributions in A. (C) Samples from hypothetical Gaussian likelihood distributions for the two stimulus categories in a yes-no detection task. The log of the ratio of these two likelihood distributions at any point is the decision variable in A; that is, $\Psi = \log[p(z_1, z_2|b)/p(z_1, z_2|a)]$. The solid and dashed contours correspond to the two criteria in A. (D) Three parameters describing the possible geometrical relationships between two contour elements. (E) Samples of geometrical relationships for pairs of contour elements belonging to the same physical contour (blue symbols) and belonging to different physical contours (red symbols), for a particular distance between the elements. The black curve shows the optimal decision bound given equal prior probabilities. (Data from Geisler & Perry, 2009.) (See color plate 28.)

2AFC task to be $\sqrt{2}$ larger than in the yes-no task, for the same stimuli (Green & Swets, 1966).

RECEIVER OPERATING CHARACTERISTIC ANALYSIS The effect on performance of changes in the decision criterion, for a given level of sensitivity, can be represented with the receiver operating characteristic (ROC), which plots the proportion of hits as a function of the proportion of false alarms (figure 32.1B). Note that in this plot the value of the decision criterion is implicit; the ROC shows only the locus of hit and false alarm rates for all values of the criterion. The shape of the ROC depends on the shapes of the probability distributions; the ROCs plotted in figure 32.1B are for the distributions in figure 32.1A. The solid and open symbols in figure 32.1B show points on the ROC curve corresponding to the solid and dashed criteria, respectively, in figure 32.1A.

The assumptions of signal-detection theory can be tested in part by inducing the observer to adopt different decision criteria and then seeing if the observer's hit and false alarm rates fall on the ROC predicted for the measured value of d'. Typically, different decision criteria are induced by varying the relative probability of stimulus *a* and *b*, varying the monetary payoffs for the different kinds of correct and error responses, varying the instructions to observer, or by asking the observer to provide a confidence judgment along with each response (Green & Swets, 1966). Signal-detection theory predicts that in the 2AFC task the ROC curve should be symmetric about the negative diagonal, even when it is not predicted to be symmetric in the yes-no task (e.g., see Green & Swets, 1966).

If the probability distributions for the decision variable cross at just one point (as they do for the equal-variance Gaussian distributions), and if the two stimuli are equally probable, then the area under the ROC is the maximum percent correct that can be obtained in a two-interval, two-alternative forced choice task with the same stimuli (Green & Swets, 1966). In neurophysiology experiments, the area under the ROC is frequently used to quantify the "discrimination information" transmitted by individual neurons (e.g., Britten, Shadlen, Newsome, & Movshon, 1992; Tolhurst, Movshon, & Dean, 1983). The typical procedure is to record the response of a neuron to multiple presentations of two stimuli, compute the hit and false alarm rate for each value of a criterion along the response axis, and finally compute the area under the resulting ROC. This calculation provides an estimate of the maximum percent correct that could be supported by that neuron alone, assuming that all the relevant discrimination information is contained in the total spikes per trial (i.e., no information is in the temporal pattern of responses).

For an ideal observer in a yes-no task, the probability distributions for the optimal decision variable must cross at a single location (see later), and hence the area under the ROC is the maximum percent correct. However, in neurophysiology experiments the neural response (e.g., spike count or spike rate) is typically regarded as the decision variable. The neural response is not guaranteed to represent an optimal decision variable (i.e., a likelihood ratio; see below). In other words, there is no guarantee that the neural response is a monotonic with the likelihood ratio. Hence, the two probability distributions could cross at more than one location. In this case, the area under the ROC curve will not correspond to the maximum percent correct. A better procedure is to fit the ROC, assuming an appropriate family of distributions (e.g., gamma distributions), compute percent correct, and then do a statistical analysis, for the family of distributions, to correct for bias in the accuracy estimates.

Ideal Observer for Identification If the goal in an identification task is to maximize the percentage of correct identifications, then Eq. 2 reduces to the maximum *a posteriori* (MAP) rule:

$$r_{opt} = \arg\max_{a_i} \left[p(\mathbf{z}|a_i)\, p(a_i) \right]. \tag{3}$$

Further, in the case of just two categories of stimuli ($a_1 = a$, $a_2 = b$), Eq. 2 reduces to

$$\text{respond } b \text{ if } \frac{p(\mathbf{z}|b)}{p(\mathbf{z}|a)} > \frac{p(a)}{p(b)}; \text{ otherwise, respond } a. \tag{4}$$

The left side of the inequality is the likelihood ratio, and the right side is the prior-probability ratio (often called the "prior odds"). Figure 32.1C illustrates this decision rule for an example where the input data are two-dimensional, $\mathbf{z} = (z_1, z_2)$, and Gaussian, with different means and covariance matrices. The ellipses show iso-likelihood contours, and the symbols show random samples from the two Gaussians. Equation 4 says that if the two categories occur with equal probability, then the response should be b when the likelihood ratio exceeds 1.0. The solid black curve shows the locus of points where the likelihood ratio equals 1.0, and thus any input $\mathbf{z}$ below and to the right of the solid curve should be assigned response b. If the ratio of the priors is greater than 1.0, then the decision boundary should shift. For example, the dashed curve shows the locus of points where the likelihood ratio is 44. If the ratio of the priors is 44, then any input $\mathbf{z}$ below and to the right of the dashed curve should be assigned response b. The same logic applies to input data of arbitrary dimensionality and to arbitrary likelihood distributions. It also applies to arbitrary numbers of categories, except that the boundaries now define regions for each of the possible responses.

How is this analysis of optimal identification related to signal-detection theory? To see the connection, note that the ideal decision variable is the likelihood ratio (the left side of Eq. 4). Furthermore, note that the specific decision is unchanged by any strictly monotonic transformation of the two sides of the inequality in Eq. 4. In other words, for an ideal observer, the decision variable can be any monotonic transformation of the likelihood ratio. For Gaussian (and many other) distributions, a useful monotonic transformation is the logarithm (although any monotonic transformation is valid). Applying this transformation to both sides of Eq. 4, we obtain the decision variable $\psi = \log[p(z|b)/p(z|a)]$ and the criterion $\beta = \log[p(a)/p(b)]$. When the stimulus on a trial is from category a, then the decision variable will have a distribution $p(\psi|a)$ (red curve in figure 32.1A). When the actual stimulus is b, then the decision variable will have a distribution $p(\psi|b)$ (blue curve in figure 32.1A). Note that for the ideal observer's decision variable, it is impossible for the two probability distributions on the decision axis to cross at more than one point.

We see, then, that signal-detection theory is consistent with ideal-observer theory. This fact provides one rationale for the assumptions of signal-detection theory. However, signal-detection theory is more general in that an observer's decision variable need not be a monotonic transformation of the likelihood ratio, and the criterion need not be optimally placed. In other words, signal-detection theory assumes an arbitrary rather than the ideal decision variable and criterion.

To obtain quantitative predictions for the ideal observer in an identification task, it is necessary to specify the prior probability of the different stimulus categories and the likelihood of the input data for each of the categories. Specifying the likelihoods and priors can be very difficult. The most common approach is to constrain the stimuli so that it is practical to derive or compute the likelihoods and priors. One way of constraining stimuli is to create them by sampling from probability distributions specified by the experimenter. This is what is done in many perception experiments. For example, stimulus *a* might be a sample of Gaussian noise, and stimulus *b* a sample of Gaussian noise with a fixed added target. In this case, the decision variable and the criterion can be easily computed, making it straightforward to calculate or simulate ideal performance (e.g., Burgess et al., 1981).

Working with natural stimuli is more difficult, because their statistical structure is complex and generally unknown. One approach is to restrict what aspects of the natural stimuli are presented in an experiment. By considering only certain aspects of natural stimuli, it can become practical to measure the relevant probability distributions and compute ideal-observer performance. For example, consider a task where the observer is presented with just two contour elements (short line segments) at the boundary of an occluding surface, and must decide whether the elements belong to the same or different physical contours (Geisler & Perry, 2009). For a given occluder width (distance), two parameters describe the geometrical relationship between two contour elements: the direction of one element from the other, φ, and the orientation difference between the two elements, θ (see figure 32.1D). The blue symbols in figure 32.1E show samples from the actual distribution in natural images of direction and orientation difference when the contour elements belong to the same contour; the red symbols show samples when the elements belong to different contours. The solid curve shows the ideal decision bound when the goal is to maximize accuracy. Specifically, the ideal observer should report that the contour elements are on the same contour if the direction and orientation difference fall inside the boundary; otherwise, the observer should report that the elements are from different contours. The performance of this ideal observer can be determined by applying this decision rule to test stimuli that contain two contour elements (taken from natural scenes) separated by an occluder. These same test stimuli can be presented to human observers. In this case, human and ideal performance is nearly identical, implying that the human visual system accurately applies the decision boundary in figure 32.1E (Geisler & Perry, 2009).

We reiterate that ideal observers are not meant to be models of real observers. Rather, they provide a rigorous benchmark against which to compare real observers, and a principled starting point for developing models for real performance. For example, the classic signal-detection model for interpreting performance in detection and discrimination experiments is motivated by the computational principles of the ideal observer. In addition, there are many examples in the perception literature where human performance is found to parallel that of the ideal observer, showing that modest modifications of ideal observers (or heuristic approximations to the ideal observer) can serve as plausible and testable models for real performance (for a review, see Geisler, 2011).

Estimation Tasks In estimation tasks, there is some physically ordered dimension along which stimuli fall, and the observer is required to estimate the value along that dimension. The distinction between estimation and identification is not a sharp one, because one can regard estimation as identification with a large number of categories. The primary distinction is captured in the utility (cost/benefit) function. Generally, in the estimation task, the closer the estimate to the true value, the better. On the other hand, in many identification tasks all errors are equally costly; for example, if the task is to identify a criminal from a police lineup of otherwise innocent people, then all errors would be equally bad. Estimation tasks are very common under natural conditions, but in laboratory settings they are less common than identification tasks.

The typical method for measuring estimation performance is similar to that for measuring discrimination performance. On each trial, a variable test stimulus is presented, and the observer is required to respond whether it is greater or less than some standard along the stimulus dimension of interest (e.g., color, depth, size, shape, etc.). Often the standard is another stimulus, but in some tasks the standard may be an internal reference. For example, in a slant-estimation task (e.g., Burge, Girshick, & Banks, 2010), the observer may be required to respond whether a test stimulus is right-side-back from frontoparallel (an internal standard). Data are typically plotted as psychometric functions, and the estimate is taken to be the point of subjective equality (PSE)—the value of the variable stimulus where the observer reports that the test is greater than the standard with probability 0.5.

Ideal observer for estimation A typical goal in an estimation task is to minimize the mean squared error between the estimate and the true value. This is the MMSE estimate given by

$$\hat{\omega}_{opt} = \arg\min_{\hat{\omega}} \left[\sum_{\omega} (\omega - \hat{\omega})^2 p(\omega|\mathbf{z}) \right] = \sum_{\omega} \omega p(\omega|\mathbf{z}) = E(\omega|\mathbf{z}). \quad (5)$$

In other words, the optimal estimate is simply the mean of the posterior probability distribution. Using Bayes's rule to expand $p(\omega|\mathbf{z})$ in Eq. 5, the optimal estimate can also be expressed in terms of the likelihood and prior probability distributions:

$$\hat{\omega}_{opt} = \sum_{\omega} \omega \frac{p(z|\omega)\, p(\omega)}{\sum_{\mathbf{z}} p(z|\omega)\, p(\omega)}. \quad (6)$$

Although minimizing some measure of the deviation from the true value is the intuitive goal for most estimation tasks, it is not uncommon for researchers to consider the MAP estimate (which penalizes all errors equally; cf. Eq. 3):

$$\hat{\omega}_{opt} = \arg\max_{\omega} p(\omega|\mathbf{z}) = \arg\max_{\omega} [p(\mathbf{z}|\omega)\, p(\omega)]. \quad (7)$$

The reason for this choice is that sometimes the MAP estimate is easier to compute, and if the posterior distributions are unimodal and not skewed, then the MAP and MMSE estimates are the same.

As in identification experiments, the difficult step in generating ideal-observer predictions is specifying the likelihood and prior distributions (or, equivalently, the posterior distributions). For both laboratory and natural stimuli, this generally requires constraining the stimuli in some way. Below, we briefly describe two approaches that can be applied to tasks with natural stimuli.

Both approaches begin by constraining the amount of data in the input. To be concrete, suppose that the task is to estimate some state of the world (e.g., depth) at each location in the retinal image. The input $\mathbf{z}$ to the visual system is the entire image, which for natural stimuli is far too big and complex to allow specification of the likelihood or posterior distributions. Thus, it is typical to restrict the input to some small neighborhood or context, $\mathbf{c}$, over the location where the estimate is to be made. This is reasonable because, in many cases, image correlations drop rapidly with distance (e.g., Deriugin, 1956; Field, 1987). In addition, in experiments on real observers it may be possible to use stimuli restricted to the local context so that the ideal observer is appropriate for the stimuli tested on real observers.

The first approach is to make an assumption about the parametric form of the likelihood distributions. A common (but sometimes unverified) assumption is that the likelihood distributions are Gaussian,

$$p(\mathbf{c}|\omega) = gauss\left(\mathbf{c}; \boldsymbol{\mu}_{\omega}, \Sigma_{\omega}\right), \quad (8)$$

where $\boldsymbol{\mu}_{\omega}$ and Σ_{ω} are a mean vector and covariance matrix that depend on the specific state of the world (e.g., depth). This assumption implies that the probability distribution of the context vector $p(\boldsymbol{c})$ is a mixture of Gaussian distributions with weights given by the prior probabilities, a form of Gaussian mixture model (GMM): $p(\boldsymbol{c}) = \Sigma_{\omega}\, p(\boldsymbol{c}|\omega)\, p(\omega)$. In applying this approach to natural stimuli, the mean vectors and covariance matrices can be measured (learned) from a large set of contexts taken from natural images, for each state of the world. Empirical measurements of natural stimuli also allow researchers to verify whether their assumptions about the parametric form of the likelihood are valid (e.g., Burge & Geisler, 2011, 2014). If the size of the context vector is n, then the number of parameters that must be estimated for each possible state of the world is $n(n+1)/2 + n$. This number is small enough to make it practical to measure all parameters for moderate context sizes. Once the means and covariance matrices are measured, Eqs. 6 or 7 can be used to compute the ideal observer's estimates.

The second approach makes no assumptions about the parametric form of the likelihood or prior distributions, but instead makes the analysis tractable by considering only small context sizes (e.g., Geisler & Perry, 2011). One version of this approach, parallel conditional means (PCM), involves measuring separately the mean of the posterior probability distribution for all context values for two or more contexts in the input data (gray squares in figure 32.2A). These means can be measured directly by computing sample means from training data, and do not require measuring (or modeling) the posterior probability distributions. These means specify estimation functions that map context values into optimal (MMSE) estimates: $\hat{\omega}_1 = E(\omega|\boldsymbol{c}_1)$, $\hat{\omega}_2 = E(\omega|\boldsymbol{c}_2)$. Once these functions are measured, the final estimate is obtained by combining the estimates, typically by weighting the estimates by their relative reliability (e.g., Oruc, Maloney, & Landy, 2003). Another version of this approach, recursive conditional means (RCM), involves first measuring the mean of the posterior probability distribution for one context $\boldsymbol{c}_1$ in the input data $\mathbf{z}$ (gray squares in the first box in figure 32.2B). Again the optimal estimate is $\hat{\omega}_1 = E(\omega|\boldsymbol{c}_1)$. The recursive step is to define a second-level context $\boldsymbol{c}_2$ that includes one or more values of the first-level estimates (gray squares in third box in figure 32.2B), and then directly measure the mean of the posterior distribution

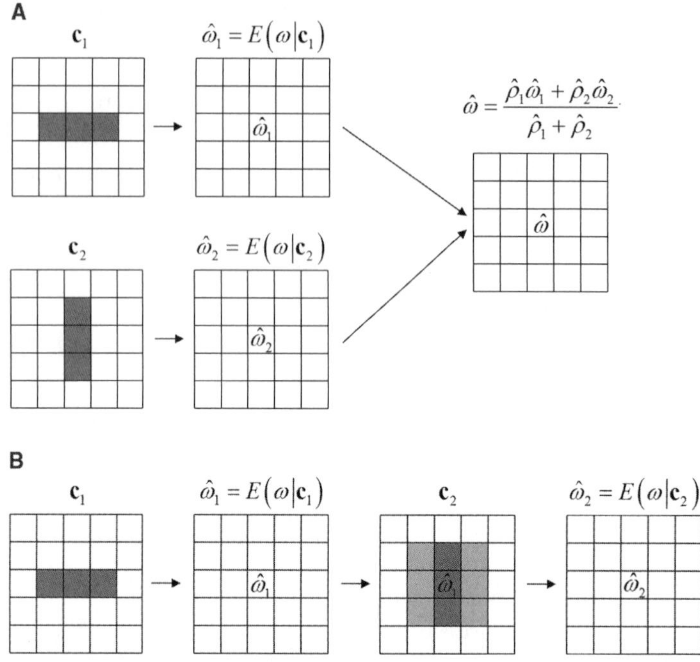

FIGURE 32.2 Simple nonparametric minimum mean squared error estimates. (A) Parallel conditional means. (B) Recursive conditional means.

for all possible values of the variables in this second-level context. The result is a second estimation function that maps second-level context values into optimal estimates: $\hat{\omega}_2 = E(\omega|\mathbf{c}_2)$. This process can be repeated to obtain a series of n estimation functions; the value of n is determined by when performance reaches asymptote. The final estimate is obtained by applying the n estimation functions sequentially.

Which version performs best depends on the particular task. Both versions require having enough training data to estimate the mean of ω or each possible pattern of context values. For example, if the values of the context variables range from 0 to 255, then the context size is limited to three or four variables, because more variables would require an impractical amount of training data. However, the context size does grow (in effect) at each step. In the recursive case, the effective context grows because the context for a higher-level estimate contains estimates that were obtained using the contexts at lower levels (in figure 32.2B, the context for the second estimate can effectively include the light gray pixels). Note that it is also possible to apply Gaussian mixture models recursively.

Another distinction between these two approaches is that the GMM observer is "generative," in the sense that the GMM parameters specify the joint distribution of the context and true values. (The term "generative" refers to the fact that if the joint distribution is specified, then it is possible to generate random samples from the distribution.) On the other hand, the RCM observer is "discriminative," in the sense that it provides optimal estimates, but does not specify the joint distribution of the context and true values (McLachlan, 1992; Vapnik, 1998). The distinction between generative and discriminative is separate from the distinction between parametric and nonparametric. For example, direct nonparametric measurements of higher-order moments beyond the conditional mean may allow generation of random samples from the joint distribution. Alternatively, parametric models such as multiple linear regression produce estimates, but cannot generate random samples from the joint distribution. In perceptual systems, a potential advantage of representing the joint distribution is that it may be possible to switch utility/cost functions without needing to learn whole new estimation functions.

We illustrate the two approaches with two examples: (1) disparity estimation, which underlies binocular depth perception, and (2) missing-pixel estimation, which is a simple form of image interpolation (amodal completion).

Disparity estimation To illustrate the first approach, consider the task of estimating horizontal disparity from the images formed in the left and right eye when binocularly viewing a small patch of natural scene. In this example, the context consists of eight variables, where each variable is the dot product of a different,

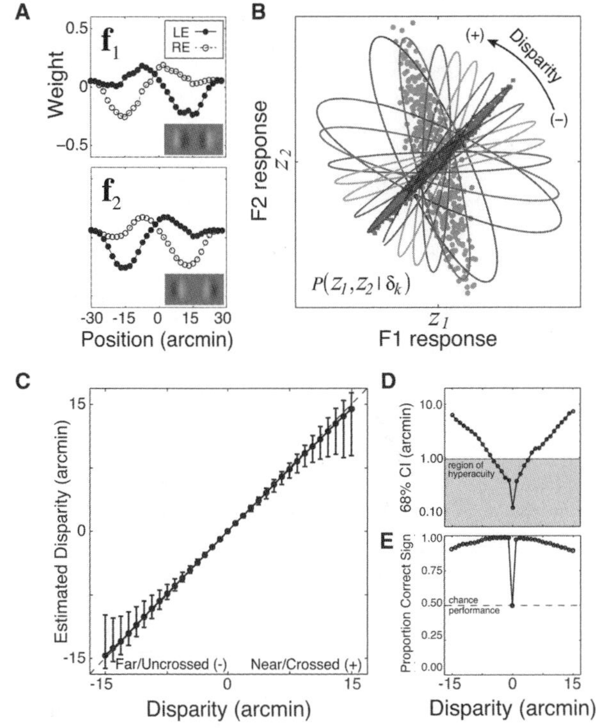

FIGURE 32.3 Disparity estimation. (A) Two of the eight vertically oriented binocular receptive fields (filters) optimal for disparity estimation. (B) Likelihood distributions of first two filter responses for disparities ranging from −15 min to 15 min. Symbols show responses to individual natural image patches for two disparities. Contours indicate 95% of the volume of Gaussian distributions fit to the joint responses. (C) Estimation performance on random natural image test patches for an ideal Gaussian mixture model using all eight optimal filter responses. Symbols are mean estimates, and error bars represent 68% confidence intervals. (D) Confidence intervals of estimates as a function of disparity. (E) Proportion correct estimation of disparity sign (crossed vs. uncrossed) as a function of disparity. (From Burge & Geisler, 2014.)

vertically oriented binocular receptive field with the retinal images in the two eyes. These eight receptive fields were found (by a separate analysis) to be the most useful vertical receptive fields for disparity estimation given the optics of human eyes and the properties of natural stereo images (Burge & Geisler, 2014). The symbols in figure 32.3B show joint responses of the first two binocular units (figure 32.3A) to randomly selected contrast-normalized natural image patches, for a range of horizontal disparities (−15 to 15 minutes of arc). As can be seen, the likelihood distributions are roughly Gaussian in shape (solid curves are 95% volume contours), with mean vectors that change little with disparity and covariance matrices that change rather dramatically. This pattern holds for all pairs of variables, and for disparities intermediate to those shown in figure 32.3B, strongly suggesting that Eqs. 7 and 8 should give near-ideal performance. Figure 32.3C shows that the optimal estimates (for a separate set of test patches) are unbiased and that the confidence intervals grow with the magnitude of disparity. Figures 32.3D and 32.3E show, in agreement with human psychophysics, that the growth in the confidence interval is approximately exponential with disparity (Blakemore, 1970; McKee, Levi, & Bowne, 1990), and that the proportion of disparity sign confusions decreases rapidly at small disparities, is minimal at intermediate disparities, and decreases gradually at large disparities (Landers & Cormack, 1997). Thus, an ideal (GMM) observer for disparity estimation in natural images shows that human performance tracks the information available in the retinal images and provides a principled starting point for developing models of human disparity estimation under natural conditions.

Missing-pixel estimation To illustrate the second approach, consider the task of estimating the gray level of missing pixels in 8-bit (0-255 gray level) calibrated natural images (D'Antona, Perry, & Geisler, 2013; task modified from Kersten, 1987). The left side of figure 32.4A shows a large patch of natural image with a missing center pixel; the right side shows an enlargement of the center 5x5 neighborhood around the missing pixel. PCM (figure 32.2A) was applied using two contexts: the four pixels left and right of the center pixel, and the four pixels above and below the center pixel. From a large set of natural image training patches (on the order 10^{10}), the mean of the center pixel is computed for each combination of the four context values. These conditional means (which are a smooth function of the context values) were used to obtain two estimates that were then combined.

For test patches of natural image, the mean squared error of the estimates of the PCM observer is approximately 90 (SD of error = 9 gray steps). Analysis shows that the most useful pixels for this task are the four neighboring horizontal and vertical pixels (other pixels provide much less information), and hence the performance of the PCM observer is likely close to the true optimum (Geisler & Perry, 2011). The mean squared error of the estimates of the GMM observer that uses simultaneously the four pixels in the horizontal direction and the four pixels in vertical directions (8-dimensional Gaussian distributions) is approximately 140. Thus, for this task the PCM observer is considerably closer to ideal.

The mean squared error of human estimates on exactly the same test patches is approximately 246 (figure 32.4C), and thus humans are well below optimal

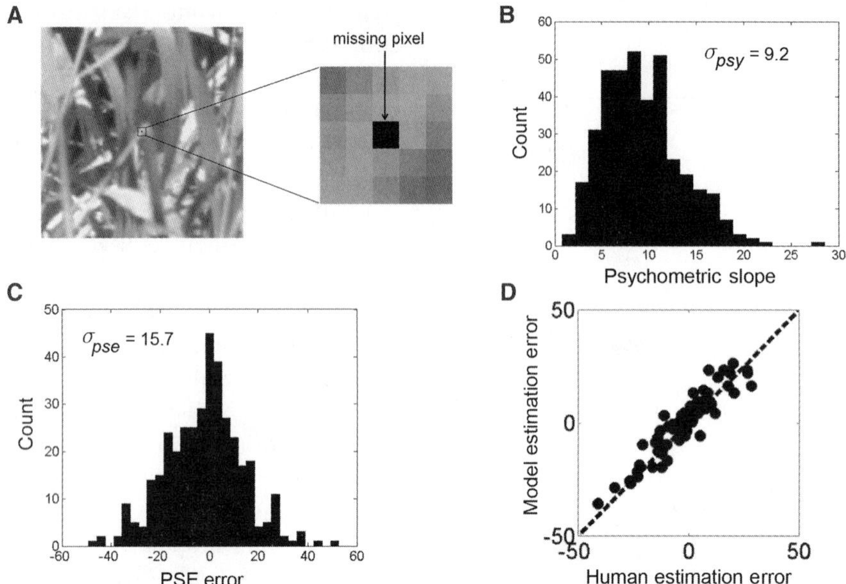

FIGURE 32.4 Missing-pixel estimation task. (A) Example test image (calibrated 8-bit gray scale). (B) Distribution of estimation errors (in gray-level) of the point of subject equality for three human observers on 62 test patches. (C) Distribution of psychometric function slopes (standard deviation values) obtained by fitting psychometric data with a cumulative Gaussian. (D) Estimation of error of a parallel conditional means (PCM) observer trained on natural images plotted as a function of the estimation error of three human observers, for 62 test patches. If PCM and human errors were identical, the points would fall on the positive diagonal.

in this task. However, because of luminance gain control and center-surround mechanisms in the retina, the output of the retina is probably better described as a contrast image rather than a luminance image (a contrast image is obtained from a luminance image by subtracting and then dividing by the local mean luminance at each pixel location). Interestingly, humans match the performance of a PCM observer trained on contrast images. Figure 32.4D shows, for 62 representative test patches, the estimation error of the contrast PCM observer plotted against the human estimation error. The contrast PCM observer does a good job of predicting the specific errors made by humans for arbitrary natural image patches.

Relative influence of external and internal factors

EFFICIENCY An observer's performance is generally limited by both external factors (variability and ambiguity of the inputs) and by internal factors (neural, decision, and motor noise, as well as nonrandom computational inefficiencies). For the purpose of estimating the relative influence of external and internal factors, the combined effect of all the internal factors can be regarded as a level of internal noise, which can be estimated by calculating how much the external (stimulus) variability must be scaled up for the performance of the ideal observer to match that of the organism. This scale factor κ is closely related to the definition of efficiency, η in signal-detection theory: $\eta = d'^2_{real}/d'^2_{ideal}$ (Tanner & Birdsall, 1958). In a detection task with a fixed signal in Gaussian noise, κ is simply the inverse of the efficiency ($\kappa = 1/\eta$). As a more general example, consider an ideal observer in an identification task, where each category is represented by a Gaussian distribution. In this case, κ is the scale factor on the covariance matrices that brings the ideal performance down to real performance. If the external variability must be scaled by a factor of κ, then the effective internal variability equals $\kappa - 1$ times the external variability. In other words, if the value of κ is near 1.0, then the internal noise is near zero and external factors dominate performance; if the value of κ is large, then internal factors dominate performance.

Whether external or internal factors dominate performance is highly task-dependent. For detection of targets in fixed backgrounds, the only external variability is photon noise (the ideal observer is limited only by the signal's energy and photon noise) and the value of κ is large (typically greater than 10), showing that internal factors dominate (e.g., Geisler, 1989). For detection in high-contrast pixel noise, the values of κ can be quite a bit smaller (sometimes less than 2), showing that external factors dominate, or at least play a major role in limiting performance (e.g., Burgess et al., 1981).

For tasks involving natural stimuli, the variability of the stimuli is often high, and hence there are likely to be many cases where external factors dominate. An example is the task described earlier, where the observer is presented with two contour elements at the boundary of an occluding surface and must decide whether the elements belong to the same or different physical contours (figures 32.1D, E). The accuracy of the ideal observer in this task is 87% correct, and is entirely due to external (stimulus) variability. Human performance under exactly the same conditions is 83% correct. The value of κ necessary to degrade ideal to real performance is 1.5, and hence in this task human performance is dominated by external factors. In other words, the human visual system uses a decision rule that closely approximates the solid curve in figure 32.1E and hence has efficiently incorporated the statistics of natural contours.

Fixed-Stimulus and Across-Stimulus Variation in Performance Another important distinction is between the variations in behavioral response that occur when an observer is presented with the same fixed stimulus repeatedly, and the variations that occur across different stimuli. Fixed-stimulus variation must be entirely due to internal factors that are varying from presentation to presentation (e.g., sensory neural noise, decision noise, or motor noise). On the other hand, variation in response across stimuli must be due either to external factors or to nonrandom internal factors.

A method for separating the two types of variation in detection-in-noise tasks is the "frozen noise" experiment, where each noise background (or natural stimulus background) is repeated occasionally in the course of the experiment. Fitting the subject's responses with standard signal-detection models allows estimation of the relative variance of the two sources of variation. Furthermore, if the pixel-based ideal observer for the task is known, then it is possible to separately estimate the effective variance due to external factors, nonrandom internal factors, and internal noise (e.g., see Swensson & Judy, 1996).

A related simple analysis for estimation experiments is to measure the fixed-stimulus variance from the slopes of the psychometric functions and the across-stimulus variance from the differences between the PSEs and the true values. For example, figure 32.4B shows the distribution of psychometric function slopes in the pixel-estimation task. These slope values are standard deviations of the cumulative Gaussian distributions fitted to the psychometric data. If the human observers had no internal variability, they would make the same decision every time the same stimulus was presented, and the psychometric functions would be step functions. Thus, the fitted standard deviations estimate all the internal variability, which in this case is equivalent to a pixel noise standard deviation of 9.2 gray steps (variance = 85). On the other hand, figure 32.4C shows the distribution of systematic errors (PSE errors). These errors are largely due to external factors or fixed (nonrandom) internal factors (the confidence intervals on the PSEs are quite small). The root mean squared PSE error is 15.7 gray steps (variance = 246), which is substantially larger than the internal variability. This result makes the important point that in many natural tasks performance is limited more by external factors and nonrandom internal inefficiencies than by neural, decision, and motor noise.

Conclusion

This chapter reviewed some tools that are useful for characterizing the external and internal factors that limit perceptual performance. These tools are based on applying the concepts of Bayesian statistical decision theory to the analysis of natural signals, neural responses, and behavioral responses. Application of the Bayesian approach to natural signals can identify task-relevant dimensions of information, provide principled hypotheses for neural mechanisms, and determine the limitations on perceptual performance imposed by external factors. Application of the Bayesian approach to neural responses can provide similar insight into task-relevant dimensions of neural information and can provide principled hypotheses for subsequent decoding. Application of Bayesian approaches to behavior can provide principled perceptual models and can be used to separate effects on performance due to sensitivity from those due to decision criteria.

acknowledgments Supported by NIH grant EY11747, NSF Grant IIS 1111328 and NIH Training Grant 1T32-EY021462.

REFERENCES

Blakemore, C. (1970). The range and scope of binocular depth discrimination in man. *J Physiol, 211*, 599–622.

Britten, K. H., Shadlen, M. N., Newsome, W. T., & Movshon, J. A. (1992). The analysis of visual motion: A comparison of neuronal and psychophysical performance. *J Neurosci, 12*(12), 4745–4765.

Burge, J., & Geisler, W. S. (2011). Optimal defocus estimation in single natural images. *Proc Natl Acad Sci USA, 108*, 16849–16854.

Burge, J., & Geisler, W. S. (2014). Optimal disparity estimation in natural stereo images. *J Vis, 14*(2), 1–18.

Burge, J., Girshick, A. R., & Banks, M. S. (2010). Visual-haptic recalibration is determined by relative reliability. *J Neurosci, 30*(22), 7714–7721.

Burgess, A. E., Wagner, R. F., Jennings, R. J., & Barlow, H. B. (1981). Efficiency of human visual signal discrimination. *Science, 214*, 93–94.

D'Antona, A. D., Perry, J. S., & Geisler, W. S. (2013). Humans make efficient use of natural image statistics when performing spatial interpolation. *J Vis, 13*(14), 1–13.

De Vries, H. L. (1943). The quantum character of light and its bearing upon threshold of vision, the differential sensitivity and visual acuity of the eye. *Physica, 10*(7), 553–564.

Deriugin, N. (1956). The power spectrum and the correlation function of the television signal. *Telecommunications, 1*(7), 1–12.

Field, D. J. (1987). Relations between the statistics of natural images and the response properties of cortical cells. *J Opt Soc Am A, 4*(12), 2379–2394.

Geisler, W. S. (1989). Sequential ideal-observer analysis of visual discrimination. *Psychol Rev, 96*, 267–314.

Geisler, W. S. (2011). Contributions of ideal observer theory to vision research. *Vis Res, 51*, 771–781.

Geisler, W. S., & Perry, J. S. (2009). Contour statistics in natural images: Grouping across occlusions. *Vis Neurosci, 26*, 109–121.

Geisler, W. S., & Perry, J. S. (2011). Statistics for optimal point prediction in natural images. *J Vis, 11*(12), 1–17.

Green, D. M., & Swets, J. A. (1966). *Signal detection theory and psychophysics.* New York, NY: Wiley.

Hecht, S., Shlaer, S., & Pirenne, M. H. (1942). Energy, quanta, and vision. *J Gen Physiol, 25*, 819–840.

Kersten, D. (1987). Predictability and redundancy of natural images. *J Opt Soc Am A, 4*(12), 2395–2400.

Kersten, D., Mamassian, P., & Yuille, A. L. (2004). Object perception as Bayesian inference. *Annu Rev Psychol, 55*, 271–304.

Knill, D. C., & Richards, W. (Eds.). (1996). *Perception as Bayesian inference.* Cambridge, UK: Cambridge University Press.

Landers, D. D., & Cormack, L. K. (1997). Asymmetries and errors in perception of depth from disparity suggest a multicomponent model of disparity processing. *Percept Psychophys, 59*, 219–231.

McKee, S. P., Levi, D. M., & Bowne, S. F. (1990). The imprecision of stereopsis. *Vis Res, 30*, 1763–1779.

McLachlan, G. J. (1992). *Discriminant analysis and statistical pattern recognition.* New York, NY: Wiley.

Oruc, I., Maloney, L. T., & Landy, M. S. (2003). Weighted linear cue combination with possibly correlated error. *Vis Res, 43*, 2451–2468.

Rose, A. (1948). The sensitivity performance of the human eye on an absolute scale. *J Opt Soc Am, 38*(2), 196–208.

Swensson, R. G., & Judy, P. F. (1996). Measuring performance efficiency and consistency in visual discriminations with noise images. *J Exp Psych, 22*(6), 1393–1415.

Tanner, W. P., & Birdsall, T. G. (1958). Definitions of d' and η as psychophysical measures. *J Acoust Soc Am, 30*(10), 922–928.

Tanner, W. P., & Swets, J. A. (1954). A decision-making theory of visual detection. *Psychol Rev, 61*(6), 401–409.

Tolhurst, D. J., Movshon, J. A., & Dean, A. F. (1983). The statistical reliability of signals in single neurons in the cat and monkey visual cortex. *Vis Res, 23*(8), 775–785.

Vapnik, V. N. (1998). *Statistical learning theory.* New York, NY: Wiley.

33 Neural Coding of Tactile Perception

STEVEN HSIAO AND MANUEL GOMEZ-RAMIREZ

ABSTRACT A fundamental problem in neuroscience is to understand how neural activity is related to behavior. This neural coding problem combines psychophysical data with neurophysiology, and it has been largely applied in two kinds of psychophysical tasks, detection and discrimination tasks, and subjective magnitude estimate (SME) tasks. The aim of the psychophysical study is to quantify the relationship between the stimulus and perception, whereas the aim of the neurophysiological study is to draw a relationship between the stimulus and neural responses. The neural code is then determined by finding the link that relates the two kinds of data. The lower-envelope principle is used to determine the neural code for detection and discrimination tasks, while the principles of consistency and falsification are used to determine the neural code for SME tasks. Throughout the chapter, we discuss neural coding studies of somatosensory function. Particularly, we will review studies of roughness, orientation, curvature, and motion, which show that the neural mechanisms for coding local cues are based on extracting two-dimensional (2D) features of the stimulus from the spatial activity across receptors. The mechanisms for spatial processing are highly similar between vision and touch. However, the mechanisms for coding global features of object size and shape are different between vision and touch.

Sensory systems are faced with the common problem of encoding and transforming inputs from the environment into patterns of action potentials that give rise to memory and perception. In touch, this process begins with the activation of a large number of receptors in the skin that have specificity for different environmental energies, such as mechanical or thermal inputs to the skin, and specificity for forces, positions, and movements of our limbs. Tactile perception of mechanical inputs alone has four different kinds of receptors located in the skin that are sensitive to different aspects of mechanical energy. Each receptor is associated with a particular afferent fiber that is thought to encode information about different features of the input. This division of labor among the peripheral nerves results in the central nervous system receiving parallel representations of the peripheral input that are processed in different regions of cortex. Along each pathway, these images are transformed by convergent and divergent synaptic circuits into central representations of objects that are matched against stored memories. The broad goals of tactile research are to determine (1) which neuronal population, or populations, at each stage throughout the pathway are linked to sensation and perception, and (2) how information is transformed and represented within those pathways.

In this chapter we discuss two concepts related to sensation and perception. The first is the concept of quantifying the neural activity with behavior, which is called the *neural coding problem*. The rationale is that embedded within the spatial and temporal patterns of action potentials is a neural representation of the sensory stimulus. The aim of neural coding studies is to determine how these representations are related to behavior. We first discuss classical neural coding studies of the tactile system and provide a theoretical framework for linking neural activity recorded from nonhuman primates with human behavior.

The second concept is that, under conditions when the form of the input representations are the same across sensory modalities, the nervous system appears to use similar coding mechanisms to solve similar problems. To illustrate this second concept, we discuss how objects are represented in the somatosensory system and draw parallels with neural mechanisms in the visual modality. We show that tactile object recognition is based on processing of the global and local features of objects. Results from a number of studies show that neural coding mechanisms for processing local features of objects, which rely on 2D arrays of receptors in the skin, are similar between touch and vision. However we will show that mechanisms for processing global features, which depend on integrating inputs across fingers, are different.

Neural coding

In the 1960s, Vernon Mountcastle initially suggested that the neural mechanisms underlying perception can be studied directly. The approach is to first select an aspect of perception and perform quantitative psychophysical experiments in humans to determine how perception changes as a function of the stimulus's parameters. Neurophysiological experiments are then performed on nonhuman primates using the same stimulus and stimulus conditions to determine how neurons are modulated by the stimulus. Given the correspondence in neural mechanisms between the two species, the final step is to "inquire which quantitative aspects of the neural response tally with psychophysical measurements" (Mountcastle, Poggio, & Werner, 1963).

This combined psychophysical-neurophysiological experimental approach has been applied to two kinds of psychophysical tasks: (1) detection or discrimination tasks, which measure the perceptual thresholds of detecting a sensory stimulus or detecting differences between two or more stimuli, respectively; and (2) subjective magnitude estimate tasks (SME; first proposed by Stevens, 1960), which provide quantitative measures of subjective behavior (e.g., the brightness of a light or the roughness of a surface) for which there is no right or wrong answer. The issues of how to link human psychophysics with neural responses are different for these two kinds of behavior.

In detection and discrimination tasks, it is assumed there are internal noise distributions of the neural activity that are known by the observer. The observer's task is to determine when the signal-plus-noise distribution is significantly different from the noise distribution alone (detection) or when the two signal-plus-noise distributions are different (discrimination). The first neural coding study was aimed at understanding which afferent fibers were responsible for coding perception of tactile vibration. In those studies, Mountcastle and his colleagues performed psychophysical studies in humans and nonhuman primates and found that the vibratory detection threshold forms a U-shaped function with the minimal threshold (i.e., bottom of the U-shaped function) being around 200 Hz for both species (LaMotte & Mountcastle, 1975; Talbot, Darian-Smith, Kornhuber, & Mountcastle, 1968; figure 33.1). They rationalized that, because the psychometric curves were similar for the two species, the neural mechanisms mediating vibration detection between the species must also be similar. Previous recordings from the peripheral nerves of nonhuman primates had revealed that there are four kinds of mechanoreceptive afferents that innervate the skin. These are called the slowly adapting type 1 (SA1), the slowly adapting type II (SA2, rarely reported in the monkey), the rapidly adapting (RA), and the Pacinian afferents (PC; see below for details). Mountcastle and colleagues set out to determine which afferent type was responsible for the perception of vibration. They recorded the responses of the SA1, RA, and PC peripheral afferents in nonhuman primates to vibratory stimuli, and found that the neural detection threshold (minimum intensity needed to activate a neuron) of the most sensitive RA and PC afferents closely matched the psychophysical measurements in humans (see figure 33.1A). They concluded that low frequencies below about 150 Hz are coded by RA afferents, while high frequencies are encoded by PC afferents. This result is now referred to as the *lower-envelope principle*, which posits that perceptual threshold for any behavior is determined by the responses of the most sensitive neurons. Since those studies, the notion that perception is bounded by neural activity of select neural populations has been applied extensively to many

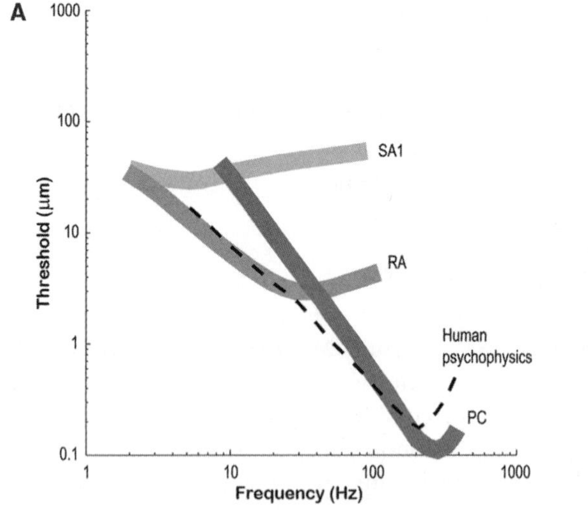

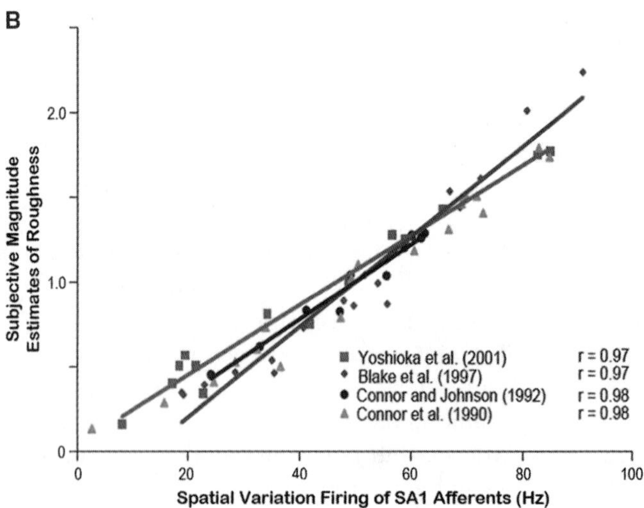

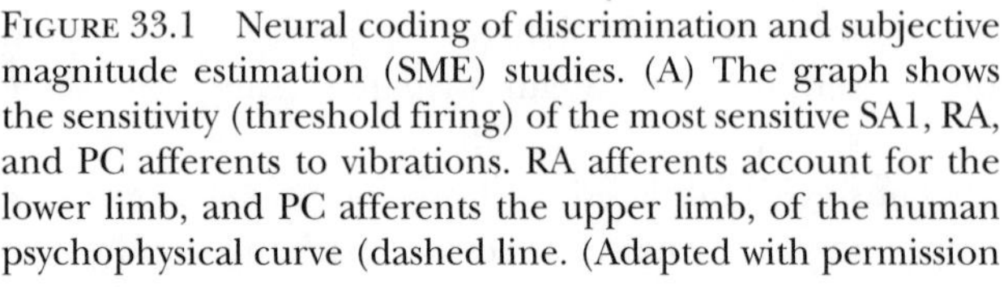
FIGURE 33.1 Neural coding of discrimination and subjective magnitude estimation (SME) studies. (A) The graph shows the sensitivity (threshold firing) of the most sensitive SA1, RA, and PC afferents to vibrations. RA afferents account for the lower limb, and PC afferents the upper limb, of the human psychophysical curve (dashed line. (Adapted with permission from Mountcastle et al., 1972, and Freeman & Johnson, 1982.) (B) Summary of four coding studies of roughness perception. *Y*-axis, SME. *X*-axis, spatial variation of firing rates of SA1 afferents (see text for details). The correlation is equal or greater than 0.97 in all studies.

studies in other sensory systems (for a review, see Parker & Newsome, 1998).

The concept that a direct link could be made between neural activity and behavior for threshold detection tasks suggests that the methods of statistical decision theory, which are widely used in many areas of science, could be applied to neural discrimination studies. Johnson addressed this issue in a series of papers in which he provided a theoretical basis for linking neural activity to sensory discrimination (Johnson, 1980a, 1980b). The theory proposes that sensory discrimination is based on two normal distributions, with the shape of these distributions known by the observer. Further, the distance between these distributions (called d') is dependent on the signal strength of the stimulus input, whereby small or weak input signals result in significantly overlapping distributions. The "observer" then decides whether the two signals are the same or different by placing a decision boundary between the two distributions. The decision boundary can change depending on the goals of the observer, with a conservative observer placing their decision boundary well into the signal distribution to ensure that they are responding to the signal and not the noise. The close fit between theory and experimental results suggests that this theory is a good model for discrimination tasks (for a review, see Parker & Newsome, 1998).

The second kind of psychophysical tasks are those in which subjects freely scale their responses to a sensory stimulus or event, such that a reported number is twice as large if a stimulus feels twice as intense. The assumption behind these studies is that the neural code underlying the aspect of perception increases monotonically as a function of the perceived magnitude of the stimulus, which could be determined using multivariate regression methods. An example of a neural coding study using SME is one by Hsiao, Johnson, and colleagues, who studied the neural coding of tactile roughness perception in four separate studies (see Johnson, Hsiao, & Yoshioka, 2002, for details on these studies). In all four studies, humans scanned their fingers across embossed spatial patterns and reported their SME of the surface's roughness. They then recorded neural activity of SA1, RA, and PC afferents in nonhuman primates and tested which afferent population and neural code underlies the perception of roughness. In each study, they systematically ruled out hypothetical neural codes when they failed to show consistent monotonic relationships with perception. In the first study, they ruled out mean rate codes and codes based on the PC afferents because there was no consistent relationship between these codes for any of the afferent fiber types and behavior (Connor, Hsiao, Phillips, & Johnson, 1990). In the second study, they designed stimuli that differentially activated neurons depending on how the spatial pattern of dots was arranged on the surface and tested whether the code was based on spatial or temporal codes (Connor & Johnson, 1992), while in the third study stimuli were designed to differentially activate SA1 and RA afferents based on their sensitivity to dot height (Blake, Hsiao, & Johnson, 1997). From these studies they rejected temporal codes and codes based on RA afferents since they failed to show consistent relationships with the behavior. The only remaining neural code that showed a consistent relationship with the SME across all of the studies was a code based on the differences in firing rate between SA1 afferents spaced about 2.0 mm apart, which they termed the SA1 spatial variation code. In a fourth study (Yoshioka, Gibb, Dorsch, Hsiao, & Johnson, 2001), they confirmed that this neural code continued to show a consistent relationship with behavior, even for fine spatial patterns that are below the spatial resolution of the SA1 afferents (see figure 33.1B for a summary of these four studies). What was remarkable about this neural code was that its relationship with perception was not only monotonic but also linear, suggesting that linearity may be a general law of perception.

The logic they used to relate the neural code to perception is based on the ideas of falsification (Popper, 1959), which proposes that codes can only be considered valid when all other possible codes have been rejected. For some percepts, such as vibratory intensity, determining the neural code is difficult because there are several potential codes that co-vary with intensity (e.g., total number of impulses evoked and total number of activated neurons). Further, it is difficult to design stimuli that can distinguish between those codes, and as such some codes cannot be falsified (Johnson, 1974).

Neural coding of tactile object perception

Determining the neural code for objects that span multiple dimensions is more difficult than the coding studies described earlier, in which only a single dimension (e.g., roughness) is the focus of investigation. The dimensions of tactile object recognition can be broken down into two broad categories of global and local features depending on how the information about the object is gathered from the hand. Global features, such as size, three-dimensional (3D) shape of the object, and its weight, rely on integrating information across the fingers and hand. The local surface features include 2D form (e.g., shape, curvature), temperature, surface texture, and direction of motion of the object. All

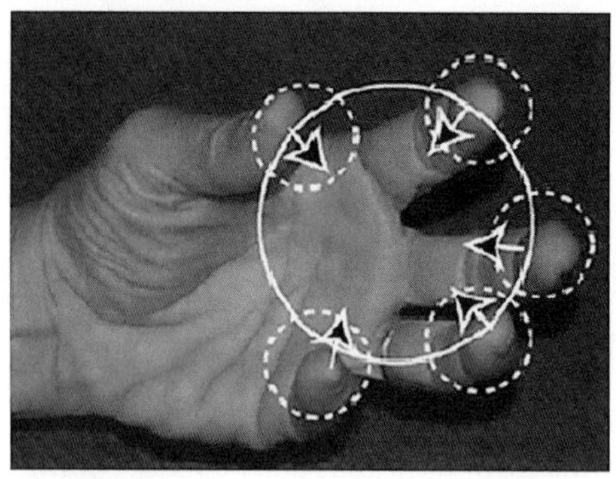

FIGURE 33.2 Working model of tactile object recognition. (A) Hand holding an object. (B) Hand in the same conformation without the object. Dashed circles represent multiple views of the object. At each view, the cutaneous input provides information about local shape, texture, and motion. (Adapted with permission from Hsiao, 2008.)

largely depend on information gathered from individual fingers and have been shown to play a role in object recognition (Klatzky & Lederman, 1995; Lederman & Klatzky, 1997). Indeed, psychophysical evidence suggests that the local and global features of objects are processed independently, with material properties processed separately from shape properties (Klatzky & Lederman, 1995). Understanding how objects are coded in cortex requires that we first understand how individual features of objects are coded.

Figure 33.2 shows a working hypothesis of how we envision the differences between global and local feature processing. We hypothesize that global properties of size and shape are derived by integrating cutaneous inputs from mechanoreceptors innervating the skin with proprioceptive receptors in the skin, muscles, and joints. Proprioception provides information about where the fingers contacting the object lie in space. The local properties of the object are derived from cutaneous receptors that provide 2D images of the spatial and temporal pattern of activity on the skin.

In the next section, we briefly describe the anatomical pathways underlying tactile perception and then describe neural coding studies of how individual features are represented in the brain.

ANATOMICAL PATHWAYS Tactile object processing begins with the activation of four types of cutaneous afferents, namely, the SA1, RA, slowly adapting type 2 (SA2), PC, and proprioceptive afferents (see Hsiao & Gomez-Ramirez, 2012, for a review). There are about 100 SA1 afferents/cm^2 at the fingertips in both human and nonhuman primates (Darian-Smith & Kenins, 1980; Johansson & Vallbo, 1979), with each afferent having circular receptive fields (RFs) of about 2–3 mm in diameter that have a Gaussian-like shape (Hsiao, Fitzgerald, Thakur, Denchev, & Yoshioka, 2006). The RA afferents also innervate the skin densely (150 afferents/cm^2), but convey a less acute spatial image of the stimulus because the receptors they innervate have large, uniform RFs (~3–5 mm in diameter). The SA2 and PC afferents have large RFs that sparsely innervate the skin, and unlike the SA1 and RA afferents, play no role in 2D form processing. The SA2 afferents provide information about skin stretch and joint angle, and are thought to play a role in 3D shape perception. The PC afferents provide information about vibration (see above) and play an important role in tool use (Johnson, 2001).

The neural mechanisms underlying proprioception are poorly understood. In addition to the SA2 afferents, there are four other types of afferents providing proprioceptive input. These are the Golgi tendon organs (1b afferents), muscle spindles (1a and II), and joint afferents. Of these, only the muscle spindles and SA2 afferents are thought to provide information about joint angle. The relative contributions of these afferents to joint-angle perception are not well understood. Psychophysical studies show that activation of either afferent can alter perception of joint angle (Collins, Refshauge, & Gandevia, 2000; Goodwin, McCloskey, & Matthews, 1972), and neurophysiological studies show that the responses of SA2 afferents to different joint-angle configurations closely match human psychophysical reports (Dimitriou & Edin, 2008; Edin & Abbs, 1991).

After leaving the arm, the main cutaneous and proprioceptive pathway of peripheral afferents project to the dorsal column nuclei in the brain stem, which in turn project to neurons in the ventroposterior lateral nucleus (VPL) of the thalamus, which itself projects to neurons in primary somatosensory cortex (S1). Neurophysiological evidence suggests that minimal processing occurs along these ascending pathways (for a review, see Hsiao & Gomez-Ramirez, 2012). However, there is new evidence indicating that neural processing of tactile inputs may begin to emerge in different columnar structures of the spinal cord (Abraira & Ginty, 2013).

The first place where significant processing occurs is S1, which is composed of four areas (3a, 3b, 1, and 2). Areas 3a and 3b receive the bulk of the feedforward thalamic input and are considered to be SI proper. Areas 1 and 2 receive both parallel inputs from VPL and serial inputs from areas 3a and 3b, suggesting that these areas serve an integrative function and lie at a higher processing stage (Felleman & Van Essen, 1991). Area 3a neurons largely encode proprioceptive input, while areas 3b and 1 process cutaneous input. Area 2 processes both proprioceptive and cutaneous inputs (Hsiao, 2008).

Neurons from the four areas of S1 cortex predominantly project to areas 5, 7b, and secondary somatosensory cortex (SII). SII cortex is thought to play an important role in object identification, since animals with SII ablated are unable to discriminate shapes of tactile objects. These deficits were not observed in animals when area 5 was ablated (Murray & Mishkin, 1984). However, the organization of SII cortex is not well understood. Like S1 cortex, it is composed of multiple areas (Burton, Fabri, & Alloway, 1995; Disbrow, Roberts, & Krubitzer, 2000; Fitzgerald, Lane, Thakur, & Hsiao, 2004; Krubitzer, Clarey, Tweedale, Elston, & Calford, 1995; Robinson & Burton, 1980; Whitsel, Petrucelli, & Werner, 1969). Hinkley et al. (2006) and Fitzgerald et al. (2004) reported that neurons in SII received both proprioception and cutaneous input, which suggests that, like area 2, it is important for processing global features of objects. These authors subdivided SII into three functional areas, SIIa, SIIc, and SIIp, with the anterior and posterior portions of SII (SIIa and SIIp) responding to proprioceptive and cutaneous inputs, while the central region (SIIc) mainly responding to cutaneous stimulation. This organization is similar to that of S1 cortex, where areas 3b and 1 are flanked by areas 3a and 2, which are the SI regions that predominantly encode proprioceptive inputs.

Neural Coding of Local Properties of Objects

Practically all of the current neural coding studies of touch have focused on understanding how 2D form, texture, and motion perception are represented in the nervous system. The perception of texture is captured by three main dimensions: rough-smooth, hard-soft, and sticky-slippery. Neural coding studies show that the dimensions of roughness-smoothness (see above) and hard-soft (not reviewed here) are also coded in the neural activity of the SA1 afferents. The neural code underlying sticky-slippery is not understood. In the next section, we discuss peripheral and central neural coding studies of 2D spatial form.

Spatial acuity Studies of spatial acuity in touch show that the discrimination threshold for gap detection, grating orientation, and letter discrimination is about 1 mm. Among the four candidate afferent types, only the SA1 and RA afferent systems have sufficient innervation density to account for this threshold. Johnson and Phillips (Johnson & Phillips, 1981; Phillips & Johnson, 1981) recorded from these afferents and showed, using the lower-envelope principle, that only the SA1 afferents have RFs with sufficient spatial acuity to account for the psychophysical performance. While SA1 afferents can resolve 1.0 mm gaps, RA afferents only resolve gaps that are 3–5 times greater. These results demonstrated that the SA1 system includes the afferent fibers and all of the ascending pathways leading to perception. This system is the tactile spatial system and is the system responsible for spatial form processing with the individual afferents playing the role of pixels (smallest processing element) in a 2D image. While the RA system also provides a dense 2D image of the central nervous system, the spatial resolution of the individual afferents is relatively poor, and as such the neural image provided by these afferents is quite blurred. The poor spatial resolution of the RA afferents is the reason why reading with an Optacon, which is a sensory substitution device for the blind that only activates RA afferents, requires spatial patterns that are five times greater in height than embossed patterns (for a review, see Johnson & Hsiao, 1992).

The results from those studies show that at the periphery, the initial coding of spatial form is an isomorphic 2D representation of the spatial pattern on the skin. This representation is identical (although much less acute) to the representation of visual images leaving the optic nerve. In the next section, we will describe the representation of form in S1 beginning with neural coding of first-order stimuli (i.e., oriented bars).

Bar orientation Psychophysical studies show that orientation-discrimination thresholds for bars is about 20 degrees (Bensmaia, Hsiao, Denchev, Killebrew, & Craig, 2008), which is less accurate in touch compared to vision (1–4 degrees in vision; Beaudot & Mullen, 2006). Although peripheral SA1 afferents have small RFs and are insensitive to orientation (Hsiao et al., 2008), many cortical neurons in areas 3b and 1 show orientation-tuned responses to both scanned and indented bars (Bensmaia, Denchev, Dammann, Craig, & Hsiao, 2008;

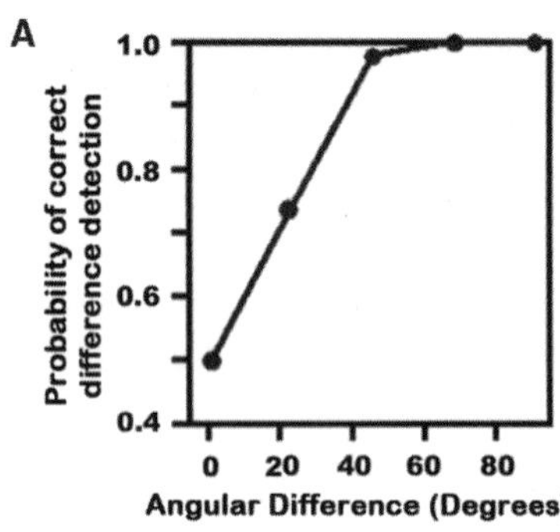

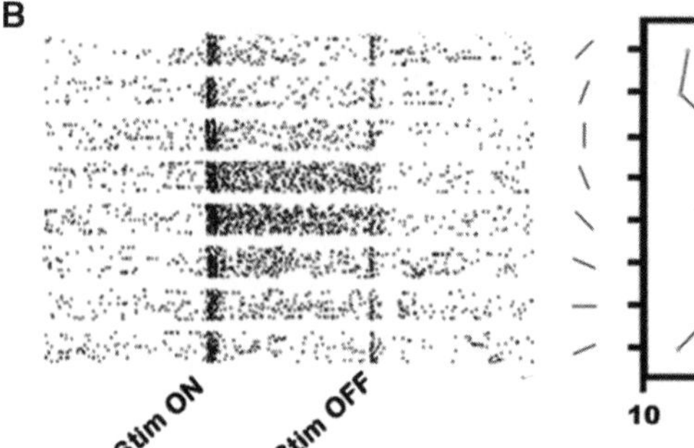

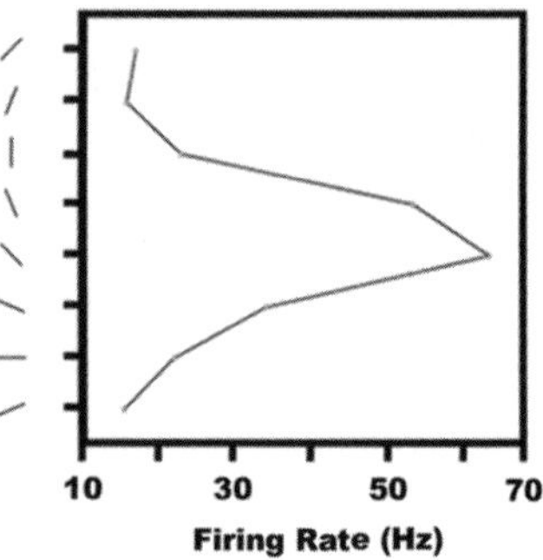

FIGURE 33.3 Neural coding of orientation tuning in area 3b. (A) Both psychometric discrimination thresholds for oriented bars indented in the skin and neurometric discrimination for neurons in area 3b predict that the threshold is about 20 degrees. Shown is the neuromatic function. Threshold is defined as 75% correct. The *x*-axis represents the angular difference between two oriented bars, while the *y*-axis indicates the probability of correct difference discrimination. (Adapted from Bensmaia, Denchev et al., 2008.) (B) Raster plot and orientation-tuning curve for an area 3b neuron. (Adapted from Hsiao et al., 2002.)

DiCarlo & Johnson, 2000; Hsiao et al., 2002; Sripati, Yoshioka, Denchev, Hsiao, & Johnson, 2006; Thakur, Fitzgerald, & Hsiao, 2012), like neurons in primary visual cortex (Hubel, 1958). This emergent property of orientation selectivity is mediated by neurons that have RFs with oriented excitatory centers and flanking inhibitory regions that also have oriented shapes (DiCarlo & Johnson, 2000; DiCarlo et al., 1998; Sripati et al., 2006; Thakur et al., 2012). The RF structures are similar to ones observed in primary visual cortex (Hubel & Wiesel, 1959). Orientation-selective neurons are also observed in SII cortex, with the majority of neurons showing position-invariant responses on a single finger (Fitzgerald et al., 2006; Thakur et al., 2006). The orientation-discrimination threshold of about 20 degrees closely matches the threshold predicted by the corresponding population response of area 3b neurons (figure 33.3; Bensmaia, Denchev et al., 2008), thus indicating (using the lower-envelope principle) that the neural code for tactile orientation resides in area 3b. Next we investigate the representation of second-order stimuli (i.e., curves) which can be either 2D and lie within the plane of the skin (e.g., the letter "C"), or 3D and indent into the skin with different depths (e.g., like a ball).

2D curvature Wheat and Goodwin (2001) showed in human psychophysics that the discrimination threshold for curvature in the plane of the skin is about $20\,m^{-1}$. In corresponding peripheral neurophysiological studies, which were also performed in humans, they showed that the neural representation of curvature is an isomorphic spatial pattern based on the responses of SA1s. The responses of the RA afferents did not match the psychophysical data, further supporting the notion that the SA1 system plays a prominent role in encoding tactile spatial information.

Studies reveal that tuning for contour features (e.g., curvature) emerges in area 2 and SII (Yau, Pasupathy, Fitzgerald, Hsiao, & Connor, 2009; Yau et al., 2013). Neurons in both areas respond to curves pointing in a specific direction, with the representations of curvature developing and peaking concurrently in both areas (Yau et al., 2013). Interestingly, the responses to tactile curved stimuli are similar to the responses in area V4 to visual curves (Yau et al., 2009).

3D curvature Similar results are observed for coding of 3D curves indented into a single finger pad, with threshold discrimination of about $5\,m^{-1}$ (Goodwin, Browning, & Wheat, 1995) and the threshold being mildly affected by indentation force. Neurophysiological observations from human peripheral nerves show that only the 3D spatial profile of the SA1 population provides a veridical representation of curvature, and that their activity accounts for human psychophysical observations (Goodwin et al., 1995; LaMotte & Srinivasan, 1993). RAs respond poorly to such stimuli and are unrelated to curvature discrimination (Goodwin et al., 1995; Khalsa et al., 1998; LaMotte, Friedman, Lu, Khalsa, & Srinivasan, 1998). The neural coding of 3D curvature in cortex has not been studied.

2D patterns composed of multiple features Further evidence that the SA1 system is essential for fine form processing, analogous to the parvocellular pathway in vision, comes from studies that use complex 2D patterns such as Braille dots or embossed letters, which are composed of combinations of curves and oriented bars (for details, see Hsiao & Gomez-Ramirez, 2012). If the

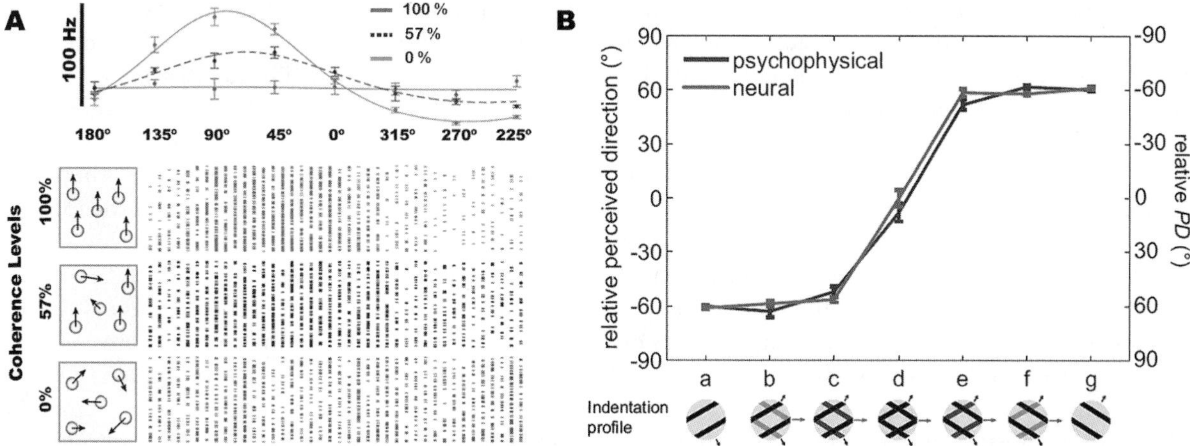

FIGURE 33.4 Neural coding of motion in area 1. (A) Response profile of a neuron tuned to ~90° moving random-dot stimulus. Tuning diminishes with lesser levels of coherent motion between the dots. (B) Human perception of moving plaid patterns and neurometric function of area 1 neurons. The *y*-axis on the left depicts human subjects' behavioral data, while the *y*-axis on the right represents the relative preferred direction of the neural population in area 1. (Adapted with permission from Pei et al., 2010, and Pei et al., 2011.)

SA1 system is devoted to processing spatial form, then can complex patterns be recognized as easily in touch as they are in vision? Phillips et al. (1983) showed that when letters are scaled in height to span the same number of receptors in touch and vision, the patterns of confusions in an identification task are nearly identical. The results showed that complex 2D patterns are perceived similarly in the two systems, suggesting that the mechanisms of 2D form processing are similar.

Subsequent studies showed that letter-recognition performance in humans is not affected by active or passive scanning (Vega-Bermudez et al., 1991). In an attention study, Hsiao trained nonhuman primates to recognize embossed letters of the alphabet scanned across their finger pads (Hsiao et al., 1993). They observed that the ability of nonhuman primates to recognize letters is nearly identical to human performance, supporting the cross-species assumption that the mechanisms of 2D form processing across the two species are similar.

Neurophysiological studies using embossed letters show that both the RA and SA1 afferents provide an isomorphic representation of the patterns (Phillips et al., 1988), with the representation being more acute for the SA1 compared to the RA system. The spatial pattern of the peripheral SA1 afferents accounts well for the psychophysical data in subjects performing letter-recognition tasks (Vega-Bermudez et al., 1991). It is not understood how spatial patterns composed of multiple features of patterns like letters are represented in cortex. In the next section, we review neural coding studies of tactile motion that is based on the activation of the RA afferents.

Motion When interacting with the world, we need to represent not only static properties of objects but also dynamic features of how and where the objects are moving. There have been a number of studies investigating tactile motion, with a series of them showing that neurons in S1 cortex are tuned to tactile motion (Costanzo & Gardner, 1980; DiCarlo & Johnson, 2000; Pei, Hsiao, Craig, & Bensmaia, 2010; Ruiz, Crespo, & Romo, 1995; Warren, Hämäläinen, & Gardner, 1986). However, the coding of stimulus motion begins with the activation of RA afferents (Gardner & Palmer, 1989). These afferents are sensitive to motion but not to motion direction (Pei et al., 2010). Further, Pei et al. reported that while a large percentage of SI neurons respond to motion signals (~60% in areas 3b and 1 and 30% in area 2), only neurons in area 1 showed motion tuning independent of the spatial form of the stimulus, and their population response closely matched humans' psychophysical data in a motion-discrimination task (figure 33.4). Pei et al. (2011) further examined the responses of area 1 neurons to tactile, plaid moving patterns (pairs of gratings moving in different directions) and showed that, similar to vision, a subset of neurons encode the individual movement of each grating composing the plaid stimulus (component neurons), while a separate set of neurons encode the combined motion across all gratings composing the plaid (pattern neurons). In corresponding neurophysiological studies, Pei and his colleagues found that neurons in area 1 responded to both component and pattern motion. These responses were accounted for by a modified vector-average model (see Pei et al., 2011), which is similar to the mechanisms used for coding visual motion. These studies suggest that area 1 is the tactile analog of the middle temporal area (MT) in the visual system, the core area of visual motion processing. These neural data, in combination with psychophysical studies, strongly support the notion that motion is processed similarly in the visual and tactile modalities.

In the next section, we review how global features of objects are represented in cortex. To the best of our knowledge, there have been no neural coding studies of global object representation because of the difficulty in determining the dimensionality of global features. We first review psychophysical studies of tactile object recognition and then, based on complementary neurophysiological data, provide a working hypothesis of how global features of tactile shapes are represented.

Neural Coding of Global Features We effortlessly recognize and manipulate objects with our hands without visual input. Klatzky, Loomis, Lederman, Wake, and Fujita (1993) had subjects identify 36 common objects that were either real objects (e.g., scissors, hammer, cups, etc.) or embossed images. The tactile performance (reaction times and accuracy) was nearly identical to visual performance when subjects freely explored the objects with their hand. Performance was mildly degraded when they scanned the objects just with their finger pads and was greatly degraded in touch but not in vision when subjects scanned the 2D embossed images. The results show that allowing the hand to be molded around the object is important for tactile object identification, indicating that recognition involves integrating both cutaneous and proprioceptive inputs originating from within and across fingers.

These results raise the question of what information is integrated across digits. Some studies suggest that there is minimal integration. Loomis, Klatzky, and Lederman (1991) found that when subjects perform 2D pattern-recognition tasks, visual performance is greatly enhanced when the aperture size is doubled, but tactile performance is not enhanced when the surface is touched by two fingers simulating the doubling of the in-aperture size in the visual condition. Similarly, Craig (1985) showed that spatial patterns presented to a single finger pad are recognized more accurately than when divided and presented to adjacent finger pads. That being said, other studies show that perception is different when scanned with multiple fingers, suggesting that there are significant interactions (Pont et al., 1999). When grasping real objects, subjects have access to both proprioception and cutaneous input. The effect of proprioception on integrating input across fingers can be exemplified by an illusion that was first described by Aristotle (1984). Normally, a single edge feels continuous when touched by two fingers. However, the same edge feels like two separate edges when touched by two fingers that are crossed. Proprioceptive influences have also been observed on the cutaneous perception of motion. Rinker and Craig (1994) reported that the direction of stimuli moving across the thumb while a nontarget stimulus moved across the index finger was affected by hand conformation. They found that the effect of the nontarget stimulus changed when the hand was in a flat position, so that stimuli moving in the opposite direction across the skin interfered with motion perception. However, when the two fingers were opposed to each other, the same cutaneous motions on the skin enhanced perception. These results suggest that the cutaneous interactions between fingers are dependent on hand conformation.

Active Sensing A fundamental issue that confronts all sensory systems is the role of active sensing. In the tactile system, the role of active sensing has received significant attention because of the inherent active role that the hand plays during object exploration. In contrast to echolocation in bats and other marine animals, in primates the receptors in the skin are completely passive and depend on encoding the stimulus input, whether it is generated by the object or by movements of the hand.

There are two issues related to active sensing. The first is the role of efference copy of the motor command. As noted above, it is clear that hand movement plays an important role in perception. The question is whether the efference copy modulates sensory input. The second issue is one of reafference, which are sensory signals that result from self movement. It was initially thought that efference copy played a significant role in sensory processing because of the active nature of hand movements (Gibson, 1962). However, recent studies suggest that it plays a minimal role (Magee & Kennedy, 1980; Yoshioka, Craig, Beck, & Hsiao, 2011). Yoshioka et al. (2011) showed that while active scanning has a significant role in tactile roughness perception, the results were the same when the experimenter moved subjects' hand compared to when subjects moved their hand independently. This result suggests that hand movements are captured by afferent proprioceptive signals as opposed to efference copy of the movement. Indeed, there is strong evidence indicating that reafference plays a major role in sensory perception. Sensory input is greatly enhanced during scanning as opposed to static touch because the primary afferents are more active (higher firing rates) during movement (Phillips & Johnson, 1985). Further hand movements appear to be tuned to gather specific kinds of sensory information, suggesting that movements are used to optimally activate the primary afferents. Lederman and Klatzky (1996) reported that subjects typically use eight exploratory procedures to extract qualities or features of objects. For example, back and forth lateral movements of the fingers are used to extract information about surface roughness, while pressure on the object is used to extract information about hardness. In addition, molding the fingers around the object and contour following (scanning) is employed to extract shape information. These studies demonstrate that hand movements are not random, but rather purposefully executed to guide the desired behavioral task of the subject. Based on studies described above, we believe that information about hand movements is conveyed to cortex by proprioceptive afferents.

Object Size Perception Studies show that tactile size perception largely depends on combining proprioceptive and cutaneous inputs. Santello and Soechting (1997) showed that the perceived size of an object is related to finger span, while Berryman, Yau, and Hsiao (2006) showed that subjects had distorted perceptions of object size when their cutaneous receptors are anesthetized. Berryman and colleagues concluded that tactile size is based on the fingers relative spread determined at the first contact point with the object. Kahrimanovic, Tiest, and Kappers (2010) showed that there are interactions between the shape of objects and their perceived size. They proposed that the perception of size is based on an estimate of the surface area of the object. The neural mechanisms of tactile size perception in cortex are unknown.

Object Shape Perception There are few studies of tactile shape perception of objects. Norman, Norman, Clayton, Lianekhammy, and Zielke (2004) asked subjects to discriminate asymmetric shapes (green pepper–like) using vision, touch, and combined vision and touch. They found that subjects are able to compare shapes within and across sensory modalities with accuracies of about 70%. This finding and others suggest that there might be functionally overlapping representations of shape in vision and touch (Amedi, Jacobson, Hendler, Malach, & Zohary, 2002; Lacey & Sathian, 2012; Lucan, Foxe, Gomez-Ramirez, Sathian, & Molholm, 2010). Perhaps the most extensively studied aspect of 3D shape is the perception of curvature (for a review, see Kappers, 2011). The main factors that affect perception of broad curvatures are location, attitude or slope of finger contact, and spread between the fingers (Pont, Kappers, & Koenderink, 1997).

Neural Mechanisms of Size and Shape Neural mechanisms of global tactile size and shape perception are thought to be carried out in neurons with RFs that span multiple fingers. Neurons in area 3b have classical RFs confined to a single finger, while neurons in areas 1, 2, and SII have RFs that span multiple fingers (Iwamura, Tanaka, Sakamoto, & Hikosaka, 1985). However, recent studies suggest that even neurons in area 3b have multidigit RFs (Reed et al., 2010; Thakur, Fitzgerald, & Hsiao, 2012), suggesting that integration across digits occurs at the earliest stages of cortical processing. The most complete multidigit RF maps are from SII cortex. Figure 33.5 shows that neurons in SII cortex have a variety of RF sizes and shapes, with mixed excitatory and inhibitory input. Many neurons (about 30%) have one or more pads that have orientation-tuned responses. There are two possible explanations for the functional role of neurons with multidigit RFs. One is spatial invariance, where neurons code for a specific feature independent of where it is located on the hand. The other is that neurons with multidigit RFs are coding for global features of objects. We rule out the first explanation as playing an important role since the RFs are not uniformly responsive to cutaneous input (figure 33.5).

The second explanation suggests that neurons with multidigit RFs are integrating information across fingers. This raises the question of how information about objects is coded when fingers move relatively independently with respect to each other during grasping and exploring objects (see the following section). During exploration there are dynamic changes in the relative positions of the cutaneous receptor sheet as the hand changes conformation. A major question facing tactile research is to understand how neurons in somatosensory cortex modulate their responses with changes in hand conformation.

Hand Synergies While we can independently move our fingers, there is strong evidence that the fingers move synergistically. That is, during grasping and exploration, there are a small set of characteristic movement patterns. The human hand has about 22 degrees of freedom. Controlling these degrees of freedom to produce smooth movements and integrating those movements with the cutaneous inputs to extract global features of objects is difficult. Reducing those dimensions by having the joints move synergistically is therefore advantageous for both motor control

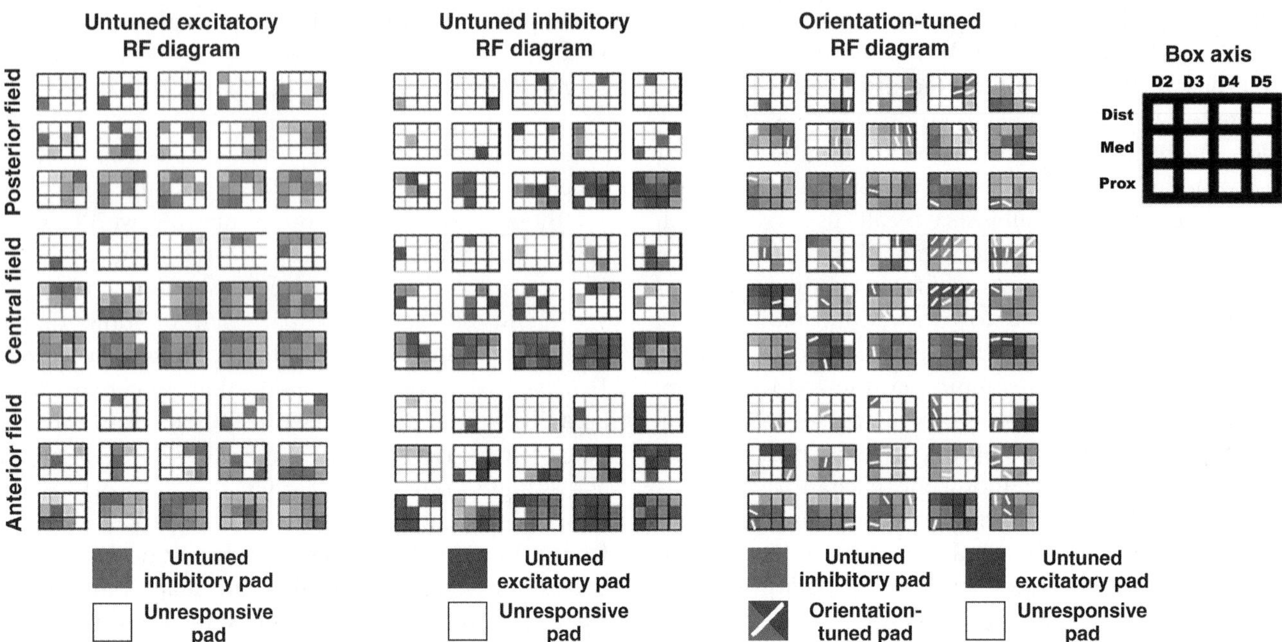

FIGURE 33.5 Receptive fields of SII neurons. Each box of twelve squares represents the RF of a single neuron, with each square representing a finger pad. Red: excitatory; blue: inhibitory. (Adapted with permission from Fitzgerald et al., 2006.) (See color plate 29.)

and sensory perception. Studies of hand movements suggest that this is in fact the case (Santello & Soechting, 2000; Thakur et al., 2008). Thakur et al. (2008) showed that only seven synergies are needed to capture more than 90% of the variance of movements during generalized object exploration. These synergies are highly conserved across subjects, suggesting that a common "library" of hand motion is used to explore objects.

WORKING MODEL Based on these results, we formulate a working model of how objects are represented in the somatosensory cortex. As the hand is molded around an object, neurons tuned for that hand conformation become active. These neurons have excitatory and inhibitory inputs from the different finger pads, and global features of objects are coded in the activity of a subset of those neurons that are selectively activated by the cutaneous input. Thus, different populations of neurons are activated for different shapes grasped with the same hand conformation (Fitzgerald et al., 2006a, 2006b; Haggard, 2006). Simultaneously, there are cutaneous neurons that code for local features of the object at the locations where the skin contacts the surface. Thus, the neural representation of objects depends conjointly on the local cutaneous properties of an object and the global properties determined by how the object is held in the hand.

Conclusion

In this chapter, we addressed the question of neural coding and how this principle has been used to understand how objects are represented in cortex. We described a set of steps that need to be implemented to determine the neural code for a particular aspect of perception. Initially, one needs to characterize the perceptual space under inquiry. Then, one needs to determine the dimensionality of the perceptual space, and the psychophysical measurements needed to characterize perception. The next step is to design a set of stimuli and psychophysical tests to explore the space. If the subjects can make greater than or less than judgments, then SME is a candidate psychophysical test. Otherwise, the classical psychophysical tests of discrimination are more appropriate. Importantly, corresponding neurophysiological studies need to be performed on either nonhuman primates or humans. Other species could be used, but the further away the species are from humans (in evolutionary terms), the harder it is to exploit the principles of the cross-species hypothesis. The lower-envelope principle is typically used when determining the correct neural code for detection and discrimination tasks, and the principle of consistency is used for SME tasks.

In the second part of this chapter, we showed how neural coding studies have been used to understand objects representation in the somatosensory system. Currently, it is not well understood how objects manipulated with our hands are represented in the central nervous system. However, evidence from a number of studies suggests that the tactile system has developed two independent strategies for representing objects. The first strategy is to take advantage of the 2D arrays of receptors on the skin and to encode individual local features of objects using similar mechanisms as those in the visual system. We propose from these studies that the nervous system has adopted common neural mechanisms for solving similar problems across sensory modalities. Using a similar coding scheme across modalities has the advantage that it greatly facilitates cross-modal processing of information. The second strategy is to extract information about global features of objects by processing cutaneous inputs with information about hand movement and conformation. The second strategy requires that the brain integrate sparse views of the object from the contact points and match this input against stored object representations. Although we do not know how this is done, we now have working hypotheses that can be tested in future neural coding studies.

REFERENCES

Abraira, V. E., & Ginty, D. D. (2013). The sensory neurons of touch. *Neuron, 79*, 618–639.

Amedi, A., Jacobson, G., Hendler, T., Malach, R., & Zohary, E. (2002). Convergence of visual and tactile shape processing in the human lateral occipital complex. *Cereb Cortex, 12*, 1202–1212.

Aristotle. (1984). On dreams. In *The complete works of Aristotle* (pp. 1–6). Princeton, NJ: Princeton University Press.

Beaudot, W. H., & Mullen, K. T. (2006). Orientation discrimination in human vision: Psychophysics and modeling. *Vis Res, 46*, 26–46.

Bensmaia, S. J., Denchev, P. V., Dammann, J. F. III, Craig, J. C., & Hsiao, S. S. (2008). The representation of stimulus orientation in the early stages of somatosensory processing. *J Neurosci, 28*, 776–786.

Bensmaia, S. J., Hsiao, S. S., Denchev, P. V., Killebrew, J. H., & Craig, J. C. (2008). The tactile perception of stimulus orientation. *Somatosens Mot Res, 25*, 49–59.

Berryman, L. J., Yau, J. M., & Hsiao, S. S. (2006). Representation of object size in the somatosensory system. *J Neurophysiol, 96*, 27–39.

Blake, D. T., Hsiao, S. S., & Johnson, K. O. (1997). Neural coding mechanisms in tactile pattern recognition: The relative contributions of slowly and rapidly adapting mechanoreceptors to perceived roughness. *J Neurosci, 17*, 7480–7489.

Burton, H., Fabri, M., & Alloway, K. D. (1995). Cortical areas within the lateral sulcus connected to cutaneous representations in areas 3b and 1: A revised interpretation of the second somatosensory area in macaque monkeys. *J Comp Neurol, 355*, 539–562.

Collins, D. F., Refshauge, K. M., & Gandevia, S. C. (2000). Sensory integration in the perception of movements at the human metacarpophalangeal joint. *J Physiol, 529*, 505–515.

Connor, C. E., Hsiao, S. S., Phillips, J. R., & Johnson, K. O. (1990). Tactile roughness: Neural codes that account for psychophysical magnitude estimates. *J Neurosci, 10*, 3823–3836.

Connor, C. E., & Johnson, K. O. (1992). Neural coding of tactile texture: Comparison of spatial and temporal mechanisms for roughness perception. *J Neurosci, 12*, 3414–3426.

Costanzo, R. M., & Gardner, E. P. (1980). A quantitative analysis of responses of direction-sensitive neurons in somatosensory cortex of awake monkeys. *J Neurophysiol, 43*, 1319–1341.

Craig, J. C. (1985). Tactile pattern perception and its perturbations. *J Acoust Soc Am, 77*, 238–246.

Darian-Smith, I., & Kenins, P. (1980). Innervation density of mechanoreceptive fibers supplying glabrous skin of the monkey's index finger. *J Physiol, 309*, 147–155.

DiCarlo, J. J., & Johnson, K. O. (2000). Spatial and temporal structure of receptive fields in primate somatosensory area 3b: Effects of stimulus scanning direction and orientation. *J Neurosci, 20*, 495–510.

DiCarlo, J. J., Johnson, K. O., & Hsiao, S. S. (1998). Structure of receptive fields in area 3b of primary somatosensory cortex in the alert monkey. *J Neurosci, 18*, 2626–2645.

Dimitriou, M., & Edin, B. B. (2008). Discharges in human muscle spindle afferents during a key-pressing task. *J Physiol, 586*, 5455–5470.

Disbrow, E., Roberts, T., & Krubitzer, L. (2000). Somatotopic organization of cortical fields in the lateral sulcus of *Homo sapiens*: Evidence for SII and PV. *J Comp Neurol, 418*, 1–21.

Edin, B. B., & Abbs, J. H. (1991). Finger movement responses of cutaneous mechanoreceptors in the dorsal skin of the human hand. *J Neurophysiol, 65*, 657–670.

Felleman, D. J., & Van Essen, D. C. (1991). Distributed hierarchical processing in the primate cerebral cortex. *Cereb Cortex, 1*, 1–47.

Fitzgerald, P. J., Lane, J. W., Thakur, P. H., & Hsiao, S. S. (2004). Receptive field properties of the macaque second somatosensory cortex: Evidence for multiple functional representations. *J Neurosci, 24*, 11193–11204.

Fitzgerald, P. J., Lane, J. W., Thakur, P. H., & Hsiao, S. S. (2006a). Receptive field properties of the macaque second somatosensory cortex: Representation of orientation on different finger pads. *J Neurosci, 26*, 6473–6484.

Fitzgerald, P. J., Lane, J. W., Thakur, P. H., & Hsiao, S. S. (2006b). Receptive field (RF) properties of the macaque second somatosensory cortex: RF size, shape, and somatotopic organization. *J Neurosci, 26*, 6485–6495.

Freeman, A. W., & Johnson K. O. (1982). Cutaneous mechanoreceptors in macaque monkey: Temporal discharge patterns evoked by vibration, and a receptor model. *J Physiol, 323*, 21–41.

Gardner, E. P., & Palmer, C. I. (1989). Simulation of motion on the skin. I. Receptive fields and temporal frequency coding by cutaneous mechanoreceptors of Optacon pulses delivered to the hand. *J Neurophysiol, 62*, 1410–1436.

Gibson, J. J. (1962). Observations on active touch. *Psychol Rev, 69*, 477–491.

Goodwin, A. W., Browning, A. S., & Wheat, H. E. (1995). Representation of curved surfaces in responses of mechanoreceptive afferent fibers innervating the monkey's fingerpad. *J Neurosci, 15*, 798–810.

Goodwin, G. M., McCloskey, D. I., & Matthews, P. B. C. (1972). Proprioceptive illusions induced by muscle vibration: Contribution by muscle spindles to perception. *Science, 175*, 1382–1384.

Haggard, P. (2006). Sensory neuroscience: From skin to object in the somatosensory cortex. *Curr Biol, 16*, 884–886.

Hinkley, L. B., Krubitzer, L., Nagarajan, S., & Disbrow, E. A. (2006). Sensorimotor integration in S2, PV, and the parietal rostroventral areas of the human Sylvian fissure. *J Neurophysiol.* doi:10.1152/jn.00733.2006

Hsiao, S. S. (2008). Central mechanisms of tactile shape perception. *Curr Opin Neurobiol, 18*, 418–424.

Hsiao, S. S., Fitzgerald, P. J., Thakur, P. H., Denchev, P., & Yoshioka, T. (2006). Receptive fields of somatosensory neurons. In L. Squire (Ed.), *The new encyclopedia of neuroscience.* Amsterdam: Elsevier.

Hsiao, S. S., & Gomez-Ramirez, M. (2012). Neural mechanisms of tactile perception. In I. B. Weiner (Ed.), *Handbook of psychology*, Vol. 3: *Behavioral neuroscience* (pp. 348–412). Hoboken, NJ: Wiley.

Hsiao, S. S., Lane, J. W., & Fitzgerald, P. (2002). Representation of orientation in the somatosensory system. *Behav Brain Res, 135*, 93–103.

Hsiao, S. S., O'Shaughnessy, D. M., & Johnson, K. O. (1993). Effects of selective attention on spatial form processing in SI and SII cortex. *J Neurophysiol, 70*, 444–447.

Hubel, D. H. (1958). Cortical unit responses to visual stimuli in nonanesthetized cats. *Am J Ophthalmol, 46*, 110–121.

Hubel, D. H., & Wiesel, T. N. (1959). Receptive fields of single neurones in the cat's striate cortex. *J Physiol, 148*, 574–591.

Iwamura, Y., Tanaka, M., Sakamoto, M., & Hikosaka, O. (1985). Comparison of the hand and finger representation in areas 3, 1, and 2 of the monkey somatosensory cortex. In M. J. Rowe & W. D. Willis (Eds.), *Development, organization, and processing in somatosensory pathways* (pp. 239–245). New York, NY: Alan R. Liss.

Johansson, R. S., & Vallbo, Å. B. (1979). Tactile sensibility in the human hand: Relative and absolute densities of four types of mechanoreceptive units in glabrous skin. *J Physiol, 286*, 283–300.

Johnson, K. O. (1974). Reconstruction of population response to a vibratory stimulus in quickly adapting mechanoreceptive afferent fiber population innervating glabrous skin of the monkey. *J Neurophysiol, 37*, 48–72.

Johnson, K. O. (1980a). Sensory discrimination: Decision process. *J Neurophysiol, 43*, 1771–1792.

Johnson, K. O. (1980b). Sensory discrimination: Neural processes preceding discrimination decision. *J Neurophysiol, 43*, 1793–1815.

Johnson, K. O. (2001). The roles and functions of cutaneous mechanoreceptors. *Curr Opin Neurobiol, 11*, 455–461.

Johnson, K. O., & Hsiao, S. S. (1992). Neural mechanisms of tactual form and texture perception. *Annu Rev Neurosci, 15*, 227–250.

Johnson, K. O., Hsiao, S. S., & Yoshioka, T. (2002). Neural coding and the basic law of psychophysics. *Neuroscientist, 8*, 111–121.

Johnson, K. O., & Phillips, J. R. (1981). Tactile spatial resolution: I. Two-point discrimination, gap detection, grating resolution, and letter recognition. *J Neurophysiol, 46*, 1177–1191.

Kahrimanovic, M., Tiest, W. M., & Kappers, A. M. (2010). Haptic perception of volume and surface area of 3-D objects. *Atten Percept Psychophys, 72*, 517–527.

Kappers, A. M. (2011). Human perception of shape from touch. *Philos Trans R Soc Lond B Biol Sci, 366*, 3106–3114.

Khalsa, P. S., Friedman, R. M., Srinivasan, M. A., & LaMotte, R. H. (1998). Encoding of shape and orientation of objects indented into the monkey finger pad by populations of slowly and rapidly adapting mechanoreceptors. *J Neurophysiol, 79*, 3238–3251.

Klatzky, R. L., & Lederman, S. J. (1995). Identifying objects from a haptic glance. *Percept Psychophys, 57*, 1111–1123.

Klatzky, R. L., Loomis, J. M., Lederman, S. J., Wake, H., & Fujita, N. (1993). Haptic identification of objects and their depictions. *Percept Psychophys, 54*, 170–178.

Krubitzer, L. A., Clarey, J., Tweedale, R., Elston, G., & Calford, M. B. (1995). A redefinition of somatosensory areas in the lateral sulcus of macaque monekeys. *J Neurosci, 15*, 3821–3839.

Lacey, S., & Sathian, K. (2012). Representation of object form in vision and touch. In M. M. Murray & M. T. Wallace (Eds.), *The neural bases of multisensory processes* (Vol. 1, chap. 10). Boca Raton, FL: CRC Press

LaMotte, R. H., Friedman, R. M., Lu, C., Khalsa, P. S., & Srinivasan, M. A. (1998). Raised object on a planar surface stroked across the finger pad: Responses of cutaneous mechanoreceptors to shape and orientation. *J Neurophysiol, 80*, 2446–2466.

LaMotte, R. H., & Mountcastle, V. B. (1975). Capacities of humans and monkeys to discriminate between vibratory stimuli of different frequency and amplitude: A correlation between neural events and psychophysical measurements. *J Neurophysiol, 38*, 539–559.

LaMotte, R. H., & Srinivasan, M. A. (1993). Responses of cutaneous mechanoreceptors to the shape of objects applied to the primate finger pad. *Acta Psychol (Amst), 84*, 41–51.

Lederman, S. J., & Klatzky, R. L. (1996). Haptic object identification II: Purposive exploration. In O. Franzén, R. S. Johansson, & L. Terenius (Eds.), *Somesthesis and the neurobiology of the somatosensory cortex* (pp. 153–162). Basel: Birkhäuser.

Lederman, S. J., & Klatzky, R. L. (1997). Relative availability of surface and object properties during early haptic processing. *J Exp Psychol Hum Percept Perform, 23*, 1680–1707.

Loomis, J. M., Klatzky, R. L., & Lederman, S. J. (1991). Similarity of tactual and visual picture recognition with limited field of view. *Perception, 20*, 167–177.

Lucan, J. N., Foxe, J. J., Gomez-Ramirez, M., Sathian, K., & Molholm, S. (2010). Tactile shape discrimination recruits human lateral occipital complex during early perceptual processing. *Hum Brain Mapp, 31*(11), 1813–1821.

Magee, L. E., & Kennedy, J. M. (1980). Exploring pictures tactually. *Nature, 283*, 287–288.

Mountcastle, V. B., Poggio, G. F., & Werner, G. (1963). The relation of thalamic cell response to peripheral stimuli varied over an intensive continuum. *J Neurophysiol, 26*, 807–834.

Mountcastle, V. B., LaMotte, R. H., & Carli, G. (1972). Detection thresholds for stimuli in humans and monkeys: Comparison with threshold events in mechanoreceptive afferent nerve fibers innervating the monkey hand. *J Neurophysiol, 35,* 122–136.

Murray, E. A., & Mishkin, M. (1984). Relative contributions of SII and area 5 to tactile discrimination in monkeys. *Behav Brain Res, 11,* 67–85.

Norman, J. F., Norman, H. F., Clayton, A. M., Lianekhammy, J., & Zielke, G. (2004). The visual and haptic perception of natural object shape. *Percept Psychophys, 66,* 342–351.

Parker, A. J., & Newsome, W. T. (1998). Sense and the single neuron: Probing the physiology of perception. *Annu Rev Neurosci, 21,* 227–277.

Pei, Y. C., Hsiao, S. S., Craig, J. C., & Bensmaia, S. J. (2010). Shape invariant coding of motion direction in somatosensory cortex. *PLoS Biol, 8,* e1000305.

Pei, Y. C., Hsiao, S. S., Craig, J. C., & Bensmaia, S. J. (2011). Neural mechanisms of tactile motion integration in somatosensory cortex. *Neuron, 69,* 536–547.

Phillips, J. R., & Johnson, K. O. (1981). Tactile spatial resolution: II. Neural representation of bars, edges, and gratings in monkey primary afferents. *J Neurophysiol, 46,* 1192–1203.

Phillips, J. R., & Johnson, K. O. (1985). Neural mechanisms of scanned and stationary touch. *J Acoust Soc Am, 77,* 220–224.

Phillips, J. R., Johnson, K. O., & Browne, H. M. (1983). A comparison of visual and two modes of tactual letter resolution. *Percept Psychophys, 34,* 243–249.

Phillips, J. R., Johnson, K. O., & Hsiao, S. S. (1988). Spatial pattern representation and transformation in monkey somatosensory cortex. *Proc Natl Acad Sci USA, 85,* 1317–1321.

Pont, S. C., Kappers, A. M. L., & Koenderink, J. J. (1997). Haptic curvature discrimination at several regions of the hand. *Percept Psychophys, 59,* 1225–1240.

Pont, S. C., Kappers, A. M. L., & Koenderink, J. J. (1999). Similar mechanisms underlie curvature comparison by static and dynamic touch. *Percept Psychophys, 61,* 874–894.

Popper, K. (1959). *The logic of scientific discovery.* New York, NY: Basic Books.

Reed, J. L., Qi, H. X., Zhou, Z., Bernard, M. R., Burish, M. J., Bonds, A. B., & Kaas, J. H. (2010). Response properties of neurons in primary somatosensory cortex of owl monkeys reflect widespread spatiotemporal integration. *J Neurophysiol, 103,* 2139–2157.

Rinker, M. A., & Craig, J. C. (1994). The effect of spatial orientation on the perception of moving tactile stimuli. *Percept Psychophys, 56,* 356–362.

Robinson, C. J., & Burton, H. (1980). Somatotopographic organization in the second somatosensory area of *M. fascicularis. J Comp Neurol, 192,* 43–67.

Ruiz, S., Crespo, P., & Romo, R. (1995). Representation of moving tactile stimuli in the somatic sensory cortex of awake monkeys. *J Neurophysiol, 73,* 525–537.

Santello, M., & Soechting, J. F. (1997). Matching object size by controlling figure span and hand shape. *Somatosens Mot Res, 14,* 203–212.

Santello, M., & Soechting, J. F. (2000). Force synergies for multifingered grasping. *Exp Brain Res, 133,* 457–467.

Sripati, A. P., Yoshioka, T., Denchev, P., Hsiao, S. S., & Johnson, K. O. (2006). Spatiotemporal receptive fields of peripheral afferents and cortical area 3b and 1 neurons in the primate somatosensory system. *J Neurosci, 26,* 2101–2114.

Stevens, S. S. (1960). Psychophysics of sensory perception. In W. A. Rosenblith (Ed.), *Sensory communication* (pp. 1–34). Cambridge, MA: MIT Press.

Talbot, W. H., Darian-Smith, I., Kornhuber, H. H., & Mountcastle, V. B. (1968). The sense of flutter-vibration: Comparison of the human capacity with response patterns of mechanoreceptive afferents from the monkey hand. *J Neurophysiol, 31,* 301–334.

Thakur, P. H., Bastian, A. J., & Hsiao, S. S. (2008). Multidigit movement synergies of the human hand in an unconstrained haptic exploration task. *J Neurosci, 28,* 1271–1281.

Thakur, P. H., Fitzgerald, P. J., & Hsiao, S. S. (2012). Second-order receptive fields reveal multidigit interactions in area 3b of the macaque monkey. *J Neurophysiol, 108,* 243–262.

Thakur, P. H., Fitzgerald, P. J., Lane, J. W., & Hsiao, S. S. (2006). Receptive field properties of the macaque second somatosensory cortex: Nonlinear mechanisms underlying the representation of orientation within a finger pad. *J Neurosci, 26,* 13567–13575.

Vega-Bermudez, F., Johnson, K. O., & Hsiao, S. S. (1991). Human tactile pattern recognition: Active versus passive touch, velocity effects, and patterns of confusion. *J Neurophysiol, 65,* 531–546.

Warren, S., Hämäläinen, H. A., & Gardner, E. P. (1986). Objective classification of motion- and direction-sensitive neurons in primary somatosensory cortex of awake monkeys. *J Neurophysiol, 56,* 598–622.

Wheat, H. E., & Goodwin, A. W. (2001). Tactile discrimination of edge shape: Limits on spatial resolution imposed by parameters of the peripheral neural population. *J Neurosci, 21,* 7751–7763.

Whitsel, B. L., Petrucelli, L. M., & Werner, G. (1969). Symmetry and connectivity in the map of the body surface in somatosensory area II of primates. *J Neurophysiol, 32,* 170–183.

Yau, J. M., Connor, C. E., & Hsiao, S. S. (2013). Representation of tactile curvature in macaque somatosensory area 2. *J Neurophysiol, 109*(12), 2999–3012.

Yau, J. M., Pasupathy, A., Fitzgerald, P. J., Hsiao, S. S., & Connor, C. E. (2009). Analogous intermediate shape coding in vision and touch. *Proc Natl Acad Sci USA, 106,* 16457–16462.

Yoshioka, T., Craig, J. C., Beck, G. C., & Hsiao, S. S. (2011). Perceptual constancy of texture roughness in the tactile system. *J Neurosci, 31,* 17603–17611.

Yoshioka, T., Gibb, B., Dorsch, A. K., Hsiao, S. S., & Johnson, K. O. (2001). Neural coding mechanisms underlying perceived roughness of finely textured surfaces. *J Neurosci, 21,* 6905–6916.

34 Inferential Models of the Visual Cortical Hierarchy

DANIEL KERSTEN AND ALAN YUILLE

ABSTRACT Human visual object decisions are believed to be based on a hierarchical organization of stages through which image information is successively transformed from a large number of local feature measurements with a small number of types (e.g., edges at many locations) to increasingly lower-dimensional representations of many types (e.g., dog, car, ...). Functional utility requires integrating a large number of local features to reduce ambiguity, while at the same time selecting task-relevant information. For example, decisions requiring object recognition involve pathways in the hierarchy in which representations become increasingly selective for specific pattern types (e.g., boundaries, textures, shapes, parts, objects), together with increased invariance to transformations such as translation, scale, and illumination. Computer vision architectures for object recognition and parsing, as well as models of the primate ventral visual stream, are consistent with this hierarchical view of visual processing. The hierarchical model has been extraordinarily fruitful, providing qualitative explanations of behavioral and neurophysiological results. However, the computational processes carried out by the visual hierarchy during object perception and recognition are not well understood. This chapter describes how a Bayesian, inferential perspective may help to explain the brain's hierarchical organization of visual knowledge, and its utilization through the feedforward and feedback flow of information.

It takes just one quick glance at the picture in figure 34.1A to see the fox, a tree trunk, some grass, and background twigs. This is a remarkable achievement in which the visual system turns a massive set of highly ambiguous local measurements (figure 34.1B) into accurate, reliable identifications. But that is just the beginning of what vision enables us to do with this picture. With a few more glances, one can see a whole lot more: the shape of the fox's legs and head, the varying properties of its fur, and so on, as well as guess what it is doing and whether it is young or old. The ability to generate an unbounded set of descriptions from a virtually limitless number of images illustrates the extraordinary versatility of human perception.

This chapter focuses on the following question: what knowledge representations and computational processes are needed to achieve reliable and versatile object vision? Although we are far from complete answers, there has been substantial progress in the overlapping fields of perceptual psychology, computer vision and robotics, and visual neuroscience.

In all three fields, theories of representation of visual knowledge and the processes acting on them are constrained by (1) functional behaviors or tasks, and their priorities; (2) the statistical structure of the visual world, and consequently in images received; (3) algorithms and knowledge structures for getting from images to behaviors; and (4) neurophysiological (or hardware) limitations on what can be computed by collections of neurons (or components and circuits).

There has been considerable growth in (4), our knowledge of the neurophysiology and anatomy of the primate visual system at the level of large-scale organization of visual areas and their connections (Kanwisher, 2010; Kourtzi & Connor, 2011), and the finer-scale level of cortical (Callaway, 1998; Lund, Angelucci, & Bressloff, 2003; Markov et al., 2013) and subcortical neurocircuitry (Guillery & Sherman, 2002). The larger picture is that visual processing involves processing within a visual area (both laterally and across laminae), and hierarchical—feedforward and feedback—processing between areas with various feature selectivities (figure 34.2).

However, despite growth in our knowledge of the visual brain, there remains a gap in our understanding of how the biology of vision enables common behaviors.[1] An immediate problem faced when beginning such an analysis is that the large-scale systems nature of the problem makes it difficult to empirically test theories of behavior at the level of neurons. One strategic solution is to temporarily ignore the details of the neurophysiology and neurocircuitry—that is, (4) above—and try to understand a narrower problem: what are the representations,

[1] Even complete knowledge of neural network connectivity and dynamics would be insufficient to explain visual function. For example, a complete description of spatial-temporal switching of the billion-plus transistors in a video game console would provide little insight into how these patterns relate to game goals, algorithms, or behavior.

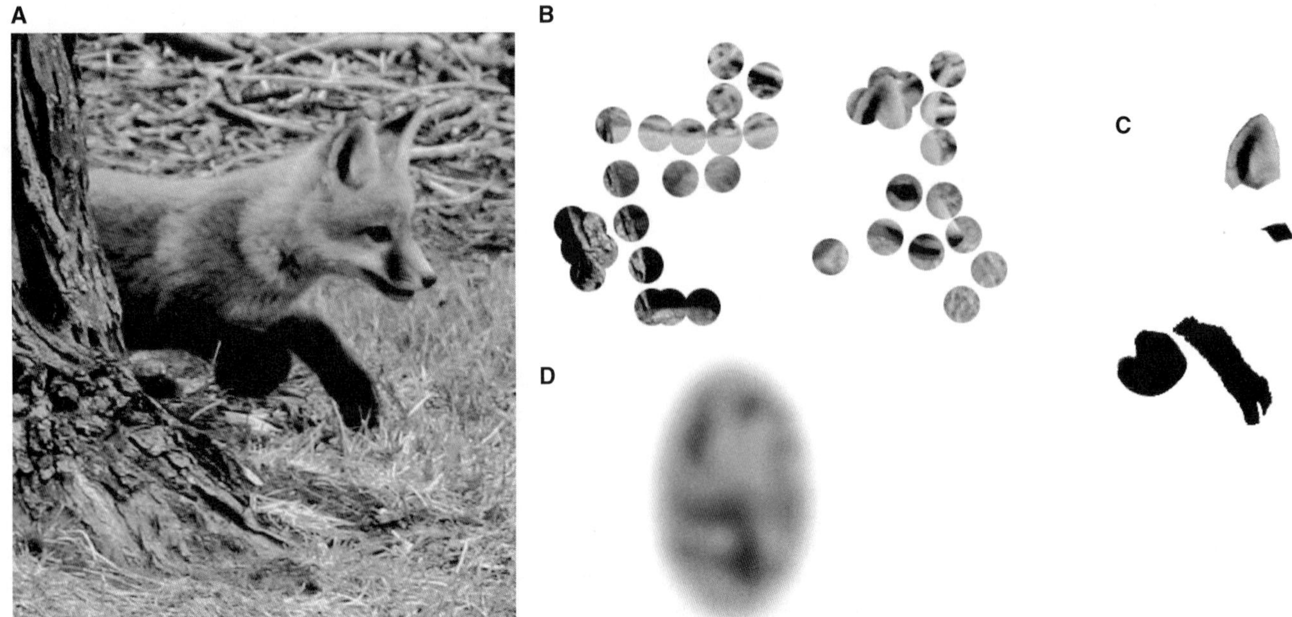

Figure 34.1 (A) This figure illustrates two problems. (1) How can local measurements made from small patches (B), using neurons with small receptive fields, be integrated to recognize objects and patterns (e.g., fox, tree trunk, grass)? (2) How does the visual system support a limitless number of descriptions of a single scene? Answers need to account for flexible access to information of various types over a range of spatial scales, such as the various edge and textural properties of local regions B, the shape of parts C, and intermediate- and higher-level concepts such as "head" D, respectively. There is a bootstrapping problem, in that the accurate interpretation of any local patch is ambiguous without knowledge of the rest. (See color plate 30.)

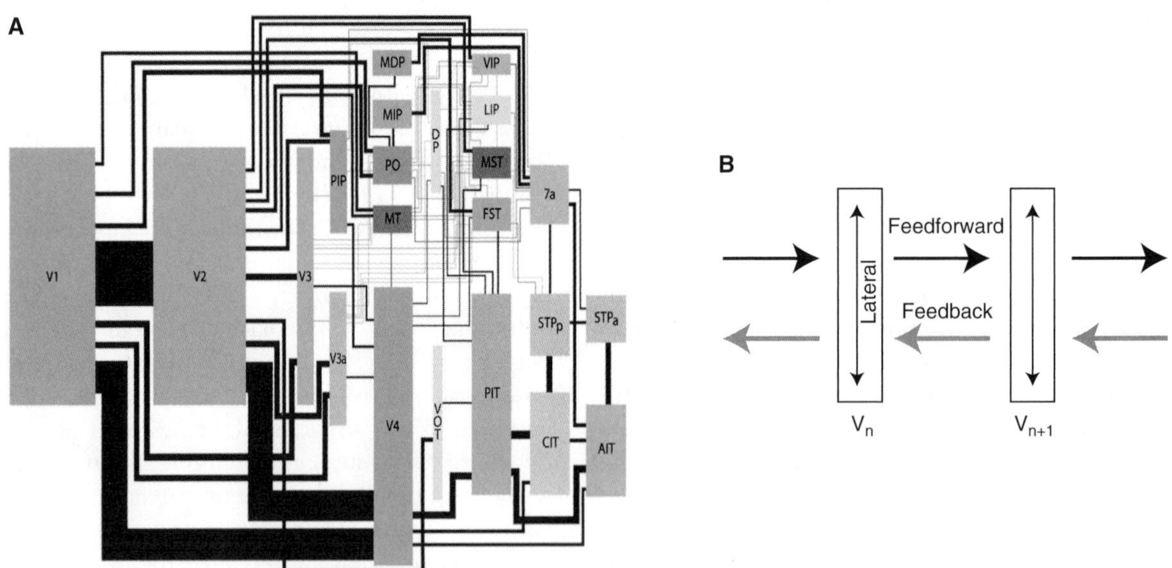

Figure 34.2 (A) Schematic of macaque monkey visual cortex (figure reprinted from Wallisch & Movshon, 2008, with permission from Elsevier; see also Lennie, 1998). The colored rectangles represent visual areas (see Felleman & Van Essen, 1991). The gray lines show the connections between areas, with the thickness proportional to estimates of the number of feedforward fibers. Areas in warm and cool tones belong to the dorsal and ventral streams, respectively. (B) Feedforward and feedback connections represent transmission of feedforward and feedback signals between visual areas. Lateral (also called "horizontal") organization within areas, representing features of similar types and level of abstraction. (See color plate 31.)

learning principles, and types of computations required for competent visual behavior?

A key idea, inspired by both computer vision research and quantitative studies of human behavior, is that vision is fundamentally inferential. More specifically, visual perception involves processes of statistical inference, which can be as simple as heuristic rules or more complex, probabilistic processes. Further, methods of statistical inference can also be applied, specifically through machine learning techniques, to understand how hierarchical representations of feature types are constrained by the statistical regularities in natural images.

In the next section, we review basic concepts of statistical inference, focusing on Bayesian decision theory. In subsequent sections, we discuss the functions of within-area (focusing on lateral representations), feedforward, and feedback visual processing from an inferential perspective, with a view toward a better understanding of how the visual cortical architecture may support human visual object perception and recognition.

Vision as statistical inference

How can one begin to model vision as inference? To begin, we need to specify the task requirements: what should be estimated, and the image information required to get there. The number of models to get from input to output can be very large, suggesting the strategy of first characterizing the requirements for optimal inference, and then interpreting actual performance in terms of approximations (see ideal-observer analysis below). Bayesian inference theory provides a well-developed set of concepts for modeling optimal inference, including discriminative and generative probability models and decision rules.

In its basic form, Bayesian theory provides mathematical tools for estimating hypotheses with potentially complex interdependencies (e.g., causal relationships), given varying degrees of uncertainty and importance. Bayesian inferences are based on knowledge represented by the joint probability distribution $p(s_1, s_2,\ldots, I_1, I_2,\ldots)$—a model of the probability of descriptions ("explanations" or "hypotheses") $s = s_1, s_2,\ldots$, together with the patterns of image measurements (or "features") $I = (I_1, I_2,\ldots)$.

The joint distribution, however, can be quite complex, reflecting causes of image patterns that are often subtle and deep. For example, the descriptions of the fox in figure 34.1 included inferences of category (which influence 3D object shape, and thus measurements available in the image projection); subcategory (baby fox, which affects the size and contours of the head); material (fur properties, together with shape and lighting produce image texture); relative depths (the tree occludes part of the fox, which in turn occludes background); and pose (the image of fox's head is to the right of the body). This suggests a causal, top-down hierarchical structure, with variables representing abstract concepts at the top to variables at the bottom representing local features shared among many objects.

Formally, the structure of images can be formulated in terms of probability distributions over structured graphs (Lauritzen & Spiegelhalter, 1988; Yuille, 2010). The graphical language helps capture the causal structures and the dependencies and independencies between causes. The nodes are random variables that represent hypotheses about events, objects, parts, features, and their relations. The links express the statistical dependencies between nodes. The links can be directed, representing causal influence, or undirected. Inference and task flexibility is achieved by fixing values of nodes based on local image measurements, or decisions made elsewhere in the system (e.g., through "priming"), together with integrating out variables that are unimportant for a given task (for a simple example, see figure 34.3A).[2]

Optimality is defined by a criterion (e.g., "minimize average error"), which determines a decision rule (e.g., "pick the values of the unknowns that maximize the posterior probability").[3]

Bayesian algorithms can be discriminative, based on a model of the posterior:

$$p(s|I) = p(s, I)/p(I),$$

the probability of a description $s = s_1, s_2,\ldots$, given a pattern of image measurements (or "features") $I = (I_1, I_2,\ldots)$. Discriminative algorithms are bottom-up, and do not incorporate explicit models of how image patterns are caused by objects. For example, in its simplest form, a discriminative algorithm could be a look-up table that

[2]Until the advent of computers, it was difficult to handle Bayesian calculations beyond a few dimensions. Today, computer vision algorithms find Bayes optimal solutions for problems involving thousands of dimensions. Optimization methods include regression, various message-passing algorithms such as expectation-maximization (EM), and belief-propagation. It is largely an open question if and how such algorithms could be implemented in a neurally plausible fashion.

[3]Bayesian decision theory generalizes "integrating out" by introducing a loss (or utility) function to allow for relative costs of imprecision in the estimation of various contributing values of s_i. Optimality is then defined as maximizing utility (or minimizing risk; Geisler & Kersten, 2002; Maloney & Zhang, 2010).

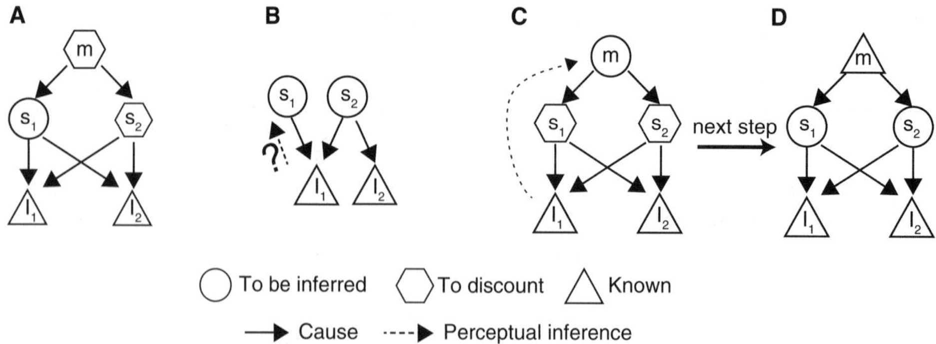

FIGURE 34.3 (A) Simple graph illustrating the generative constraints on incoming data. (See main text.) (B) More than one combination of causes, *s*, could explain local image measurement, I_1. Optimal perception seeks an explanation, i.e., values of s_1 or s_2 that give the most probable explanation for how the image measurement could have been generated. For example, Bayes optimal calculations show that without feature I_2, s_1 takes on one value, but with a measurement of I_2, it takes on a different value. Pearl (1988) calls this "explaining away." (C) Bayesian coarse-to-fine inference. Different "models," *m*, can be different functions of the parameters *s*, which in turn lead to different image features. An initial, "quick and dirty" visual inference may be at the top level (e.g., it is a "fox"), ignoring shape details (but using, for example, features from the wooded context, fur color, "features of intermediate complexity" or "fragments," that may be sufficient). Fixing the hypothesis of "fox" can be followed by reliable inferences at a lower level (e.g., "shape of the head of the fox").

maps an image pattern to the most probable hypothesis, which in neural terms is not that different from a reflex (Purves & Lotto, 2003).[4]

Bayesian algorithms can also be generative. Generative models rely on knowledge in the likelihood $p(I|s)$, which specifies how an image results from causes or explanations *s* and a prior $p(s)$. These probabilities are related to the posterior through Bayes's rule: $p(s|I) = p(I|s)p(s)/p(I)$. Generative algorithms make explicit use of top-down generative processes, in which high-level hypotheses are used to simulate the values of lower-level nodes, ultimately generating a prediction of *I* (Mumford, 1992; Yuille & Kersten, 2006). Generative models provide a number of advantages. For example, by elaborating the structure of the likelihood, computational studies have shown that a generative process can improve recognition through "explaining away," which is useful for both learning (Hinton, 2009; Zeiler, Taylor, & Fergus, 2011), and inference applied to image parsing (Tu, Chen, Yuille, & Zhu, 2005). Generative algorithms predict appearances in time (e.g., Bayes-Kalman; Burgi, Yuille, & Grzywacz, 2000), and cope more efficiently with a wider range of variability, such as the virtually unlimited ways in which objects can be composed (Chang, Jin, Zhang, Borenstein, & Geman, 2011; Yuille & Mottaghi, 2013), discussed more below.

[4]A discriminative algorithm can implement a decision rule with no explicit use of probabilities. For example, with a large number of samples, a rule to minimize empirical risk (Schölkopf & Smola, 2002) becomes equivalent to minimizing Bayes risk, as discussed in Kersten et al. (2004).

Computer vision studies have shown discriminative and generative models can be combined (Tu et al., 2005)—an algorithmic strategy similar in spirit to two-stage processing accounts of human visual recognition, in which an initial, fast decision about the "gist" of a scene narrows the space of specific objects to match to the image (Bar, 2003).

Bayesian probabilistic methods have been applied in a number of quantitative studies of human visual behavior. There is a long history to studying human perception (and neural responses) using ideal-observer analysis (Gold, Abbey, Tjan, & Kersten, 2009). Here one makes quantitative comparisons between what an ideal (Bayesian) observer can achieve with humans or neurons (Geisler, 2011; Trenti, Barraza, & Eckstein, 2010). A strategic benefit of ideal-observer analysis in studies of human behavior is that it helps to distinguish perceptual limitations inherent to the information-processing problem from limitations of the neural mechanisms (cf. Eckstein, Drescher, & Shimozaki, 2006; Weiss, Simoncelli, & Adelson, 2002).

Quantitative behavioral experiments have shown near optimality or ideal-like behavior in a variety of domains, including visual cue integration (Jacobs, 1999); visual motor control (Orban & Wolpert, 2011; Wolpert & Landy, 2012); learning (Green, Pouget, & Bavelier, 2010); and attention (Chikkerur, Serre, Tan, & Poggio, 2010; for reviews, see Geisler, 2011; Kersten & Yuille, 2013; Vilares & Körding, 2011). Findings of optimal behavior have raised the question of whether neural populations within the brain explicitly represent and compute with probabilities, for example, using information about both the mean and covariance of perceptual variables (cf. Beck, Latham, & Pouget, 2011; Knill & Pouget, 2004; Koch, Marroquin, & Yuille, 1986; Lee & Mumford, 2003; Ma, 2010, 2012; Ma, Beck, Latham, & Pouget, 2006; Ma, Beck, & Pouget, 2008; Zemel & Pouget, 1998).

Bayesian methods applied to graphical models have provided a unified framework within which to understand generative and inverse inference as well as statistical learning (Jordan & Weiss, 2002). And while it isn't always practical to develop a quantitative model for a complex visual function, the basic concepts provide a common language for describing how image representations with an area might be discovered from natural image regularities, how complexity is managed, and how reliable, flexible decisions may be made through the combination of feedforward and feedback flow of cortical information.

Representations and computations in visual hierarchies

In the following three sections, we discuss within-area, feedforward and feedback computations from an inferential perspective, with particular attention to how lateral/within-area and between-area (feedforward and feedback) processes may relate to primate vision. Because most relevant research has been on early retinotopic visual areas, our examples focus there. The computations and surface representations in early visual cortex may be more complex than traditionally thought, making V1 a good test-bed for ideas regarding hierarchical functions generally (Gilbert & Sigman, 2007; Lee, 2003; Olshausen & Field, 2005).

WITHIN-AREA REPRESENTATIONS Cortical maps are a fundamental, large-scale property of lateral, within-area cortical organization with a well-established empirical and theoretical history (Barlow, 1981; Hubel & Wiesel, 1977; Mountcastle, 1997). Specifically, the columnar organization within a visual area reflects the requirement that units representing similar image features should be nearby on the cortical surface (Durbin & Mitchison, 1990). This arrangement is believed to provide the basis for perceptual organization, for example, to group local edges into object boundaries. The presumption is that local features of a similar type can be more easily linked over cortical space. A given area represents spatially organized information of a similar type and level of abstraction (Connor, Brincat, & Pasupathy, 2007; Orban, 2008). Are there natural image regularities that support the evolution, development, and adult plasticity of lateral, within-area feature representation? If so, what theoretical learning principles might help to explain the discovery and representation of regularities? How do the task requirements of object perception constrain representations?

Insight comes from computational studies that have shown how structured image knowledge can be discovered through "unsupervised" as well as task-based learning (e.g., "supervised" learning) from collections of natural images. Such "discoveries" in an organism presumably arise through evolution and development of the visual system through exposure to natural images, as well as to their behavioral outcomes. It makes sense that early visual features would be more general purpose, involving representations shared among many objects, and thus more strongly constrained by the statistical regularities in natural images and discoverable through unsupervised learning. As one moves up the visual hierarchy, the contingencies of primary tasks become more important. This may account for multiple parallel pathways (Beauchamp, Lee, Haxby, & Martin, 2002; Freiwald & Tsao, 2010; Nassi & Callaway, 2009) and the divergence, following V1 and V2, into multiple visual areas in which different causal contributions are discounted (integrated out) based on different task requirements. Such specialization would be constrained through adaptations based on outcomes (e.g., task-based or reinforcement learning) across phylogenetic and ontogenetic time scales.

Unsupervised learning of feature representations An early idea was that, in its simplest form, N discrete levels (or areas, or layers of neural units) are required to detect Nth-order image regularities. With such a system in place, vision operates in a feedforward manner in which progressive conjunctions of features are detected, eventually leading to the detection of whole objects. Barlow (1990) suggested that mechanisms for learning Nth-order image regularities could rely on the detection of "suspicious coincidences" in the combinations of input features—that is, test whether $p(s_1, s_2) >> p(s_1)p(s_2)$, and if so recode to remove this dependency. Some coding could be "hard-wired," and modulated or built during early development. At the behavioral level, it has been shown that human adults can learn, without supervision, part combinations by detecting co-occurrence of features (Fiser, Berkes, Orbán, & Lengyel, 2010; Orbán, Fiser, Aslin, & Lengyel, 2008).

There have been a large number of computational studies aimed at explaining the neural population architecture in V1 in terms of efficient codes that exploit the regularities in natural images. Neural response properties, such as orientation and spatial frequency tuning in V1 neurons, are consistent with a sparse-coding strategy adapted to the statistics of natural images (Olshausen, 1996; Hyvärinen, 2010). In addition, neurons in primary visual cortex show nonlinear divisive-normalization behavior in which responses are inhibited by contrast variation outside the classical receptive field. Divisive normalization results in a reduction of statistical dependencies (Schwartz & Simoncelli, 2001), providing an efficient representation potentially useful for discovering (additional) suspicious coincidences. Recently, Freeman, Ziemba, Heeger, Simoncelli, and Movshon (2013) developed a texture model based on high-order statistical dependencies in natural images that could account for selectivities in both macaque and human V2. Purely bottom-up, unsupervised feature learning typically ignores task requirements (i.e., what to discount), and eventually the behavioral end-goal of a visual pathway needs to be taken into account.[5] However, some task requirements are general, suggesting that certain kinds of information can be discounted early on.

Generic task constraints on early representations It is believed that early vision involves both contour- and region-based linking (Grossberg & Mingolla, 1985; Lamme, Sup, & Spekreijse, 1998; Lee, 2003; Roe, Chen, & Lu, 2009; Roe et al., 2012). For contour features, conditional probabilities, fit with natural image statistics, predict aspects of human contour perception, such as the Gestalt property of "good continuation"—nearby contour elements tend to have similar orientations (Elder & Goldberg, 2002; Geisler & Perry, 2009). Region-based grouping relies on the prior assumption of piece-wise smoothness in low- and higher-order intensive attributes (i.e., texture; Shi & Malik, 2000). The assumed function of edge- and region-based grouping is to compute surface representations that are more reliably associated with object than image properties, providing a front end to a variety of object-based tasks, including recognition (Marr, 1982). And a first step would be to begin the process of discounting causes of image patterns that are not needed.

The accurate inference of illumination level and direction is low priority for both "what" and "how" tasks, which care primarily about objects and surfaces. This suggests that at least some components of illumination variation would be discounted early in the visual system. This is consistent with retinal lateral inhibition filtering out slow spatial gradients (presumed due to illumination), and emphasizing edges (presumed due to surface changes). However, illumination effects are complicated: slow gradients can also be caused by shape, and simple filtering neither accounts for human perception of brightness (Kingdom, 2011; Knill & Kersten, 1991), nor provides accurate reflection estimation in computer vision applied to natural images (Tappen, Freeman, & Adelson, 2005).

This problem is naturally cast in terms of Bayesian inference, where the generative knowledge is contained in the image formation model $I = f(E, R, S)$ and spatial priors on illumination (E), reflectance (R), and shape (S)—spatial maps called "intrinsic images" (Barrow & Tenenbaum, 1978). Conceptually, a Bayesian model would use a posterior proportional to the product of a likelihood function $p(I - f(R, S, E))$, and priors that characterize the spatial regularities in the natural patterns of reflectance and shape, while discounting illumination through integration (see Freeman, 1994). While computing intrinsic images from natural images can be done in special cases, it nevertheless remains a challenging problem (Barron & Malik, 2012; Grosse, Johnson, Adelson, & Freeman, 2009).

Perceptual evidence for human computation of an intrinsic image for reflectance comes from human lightness judgments, which are more strongly correlated with reflectance than image intensity or contrast. The classic Craik-O'Brien lightness illusion, shown in the upper-middle panel of figure 34.4A, illustrates this. Regions with identical physical intensities appear to have different lightnesses. The functional interpretation is that the illusion is due to a mechanism designed to produce an estimate of surface reflectance, based on the assumptions that reflectance changes are often abrupt and illumination changes tend to be gradual (figure 34.4B).

Functional MRI evidence for processes involved in computing a lightness map in human V1 and V2 is shown in figure 34.4A (Boyaci, Fang, Murray, & Kersten, 2007). Activity in localized regions of visual cortical areas V1 and V2 (distant from the central edge) responds to a perceived change in lightness in the absence of a physical change in intensity (see lower panels in figure 34.4A). While purely lateral computations have been invoked to explain this kind of "filling in," it has also been shown that human V1 response to lightness change is also sensitive to perceptual organization of occluded surfaces, suggesting that top-down feedback may be involved (Boyaci, Fang, Murray, & Kersten, 2010).

In addition to allowing for illumination variation, object recognition has the additional requirement

[5] Discounting can be achieved through unsupervised learning. For example, Cadieu and Olshausen (2012) show unsupervised learning of invariances of form by factoring out contributions from motion.

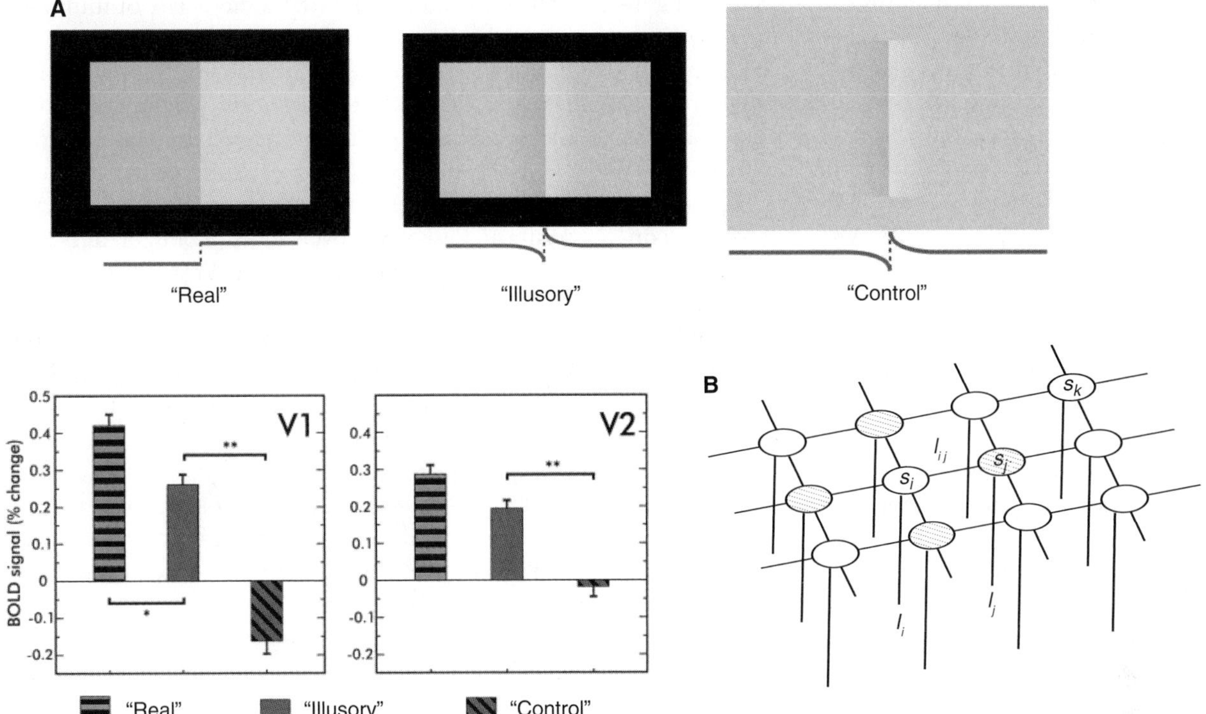

FIGURE 34.4 (A) The upper-middle panel shows a classic illusion known as the Craik-O'Brien effect. Away from the vertical border, the left and right rectangles have the same luminance, as indicated by the red line, which shows how light intensity varies from left to right. The interesting perceptual observation is that the left rectangle looks darker than the right. In fact, there is little difference between the appearance of a real intensity difference (upper left), and the illusory one. The lower graphs show that voxels in both V1 and V2 respond to apparent changes in lightness almost as strongly as real changes, as compared with a control. (Reprinted from Boyaci et al., 2007, with permission from Elsevier.) (B) An undirected graph (Markov random field) can be used to formulate prior probabilities representing lateral, spatial statistical dependencies for contours and surface properties such as reflectance (cf. Kersten, 1991; Marroquin et al., 1987). (See color plate 32.)

that variations due to position and depth need to be discounted. We discuss within-area computations supporting invariant recognition in the later section on feedforward computations.

Learning hierarchically organized area representations for recognition One can use the end goal of object classification as a constraint on learning feature hierarchies through successive, top-down categorization of intermediate-level features. Here the invariance requirements are built into the choice of what distinguishes the top-level training classes. The basic principle is to learn diagnostic features (such as "fragments" or "features of intermediate complexity") that maximize the information for distinguishing object classes (Ullman, Vidal-Naquet, & Sali, 2002). Humans and nonhuman primates seem to learn such features (Harel, Ullman, Epshtein, & Bentin, 2007; Hegdé, Bart, & Kersten, 2008; Kromrey, Maestri, Hauffen, Bart, & Hegdé, 2010; Lerner, Epshtein, Ullman, & Malach, 2008). To build a feature hierarchy, one applies this principle at the highest level to learn high-level features that optimally distinguish object classes. At the next level down, the principle is again applied to learn lower-level features that distinguish the previous features learned, and so forth (Epshtein, Lifshitz, & Ullman, 2008). The task requirement of what to discount is built into the a priori selection of the training classes to be distinguished. Simulations have shown that once the features have been learned, accurate object recognition and localization can be achieved with one forward pass followed by one backward pass through the hierarchy (Epshtein et al., 2008).

Learning object compositions to manage image complexity Compositionality refers to the human ability to construct hierarchical representations, whereby features/parts are used and shared to describe a potentially unlimited number of relational compositions (Geman, Potter, & Chi, 2002). It is argued that without such a generative structure underlying scene and object compositions, we could not account for the efficiency and versatility with which humans can acquire and generalize visual knowledge. There is also evidence that humans exploit compositionality when learning new patterns (Barenholtz & Tarr, 2011). One aspect of compositionality is the ability to represent spatial relationships between parts, an idea with an early history (Biederman, 1987; Hummel & Biederman, 1992; Marr & Nishihara, 1978; Waltz, 1972). A second aspect, consistent with current models of primate recognition, is the idea of "reusable" features or "shared" parts, where lower levels have only a few feature types (e.g., edges), but these can be combined in many ways to make compositions of parts with increasing specificity at higher levels.

An underlying compositional structure to the visual world suggests that learning should exploit that assumption, and computer vision work has demonstrated unsupervised learning of levels of reusable parts from natural image ensembles, which is then applied to multi-class recognition (Zhu, Chen, Lin, & Yuille, 2010; Zhu, Chen, Torralba, Freeman, & Yuille, 2011; see figure 34.5).

FEEDFORWARD COMPUTATIONS Invariant object recognition by the ventral stream requires discounting spatial position and size (DiCarlo, Zoccolan, & Rust, 2012; Fukushima, 1988; Riesenhuber & Poggio, 1999; Wallis, Rolls, & Foldiak, 1993). The basic feedforward computations are assumed to be the detection of conjunctions of features that belong together as part of an object, while at the same time discounting, through disjunction (which can be viewed as an approximation for "integrating out"), sources of variation, including position and scale.

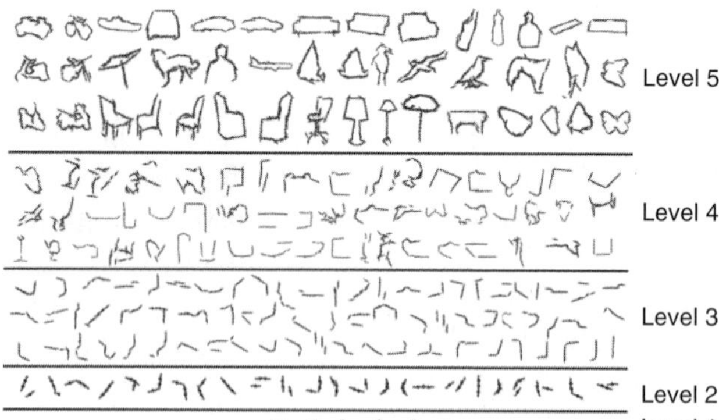

FIGURE 34.5 Examples of the mean shapes of visual concepts automatically learned for multiple objects with part sharing between objects. The specificity and the number of types of features increases as one goes up the hierarchy, consistent in general terms, with the progression of neural selectivities as one moves up the ventral stream. (Figure adapted from Zhu et al., 2011.)

It has been argued that a hierarchy of multiple areas is required to achieve functional invariance, given the biological properties of neurons and their connections (Poggio, 2011). In this account, discounting is achieved incrementally through levels of the ventral stream via the operation of AND-like (to detect feature conjunctions) and OR-like operations (to discount variations in position and size) over levels (Zhu et al., 2010) via simple- and complex-type cells, respectively (Riesenhuber & Poggio, 1999).

During the first feedforward pass, information necessarily gets left behind in the race to quickly and accurately draw from a relatively small set of high-priority, categorical hypotheses. But "no going back" requires strong a priori architectural assumptions regarding what constitutes high-priority end goals, as well as a strategic balancing of the trade-off between selectivity and invariance. Invariance is achieved at the cost of loss of information—too much loss and categories become indistinguishable; too little, and there are too many object types.

Efficiency of a compositional hierarchy for recognition Compositional arguments may help to answer the question of why a hierarchical visual architecture is desirable. Yuille and Mottaghi (2013) conjecture that the key problem of vision is complexity. The visual system needs to be organized in such a way that it can represent a very large number of objects and be able to rapidly detect which ones are present in an image. They demonstrate by mathematical analysis that this can be achieved using compositional models, which are constructed in terms of hierarchical dictionaries of parts (see figure 34.5). There are two key issues. First, this visual architecture exploits part sharing between different objects, which leads to great efficiency in representation and speed of detection. The lower-level parts are small and are shared between many objects. The high-level parts are larger (composed from lower-level parts) and are shared less because they are more specific to objects. Second, objects are represented in a distributed hierarchical manner where the positions, and other properties, of the high-level parts are specified coarsely, while the low-level parts are specified to higher precision. This "executive summary principle," combined with part-sharing, can lead to exponential gains in the number of objects that can be represented, as well as the speed of recognition. For these types of models (based on Zhu et al., 2011) recognition is performed by propagating up hypotheses about which low-level parts are present to obtain an unambiguous high-level interpretation. And, as discussed in the next section, top-down processing can be used to remove false low-level hypotheses (using a high-level context).

We noted at the beginning the extraordinary reliability and versatility of human vision, in its ability to respond both to challenging input (partially hidden objects, confusing background clutter, camouflage) and diverse task demands, such as the fox description example. What if the information for a low-level hypothesis (e.g., precise object boundary location, or the direction of movement of a local edge) is not sufficiently reliable from a single forward pass? What if a task needs information not present or easily computable within top levels of the hierarchy? Earlier we noted some of the computational advantages of generative models in resolving residual ambiguity. The next section discusses human behavioral and neuroimaging experiments, based primarily on the effects of context on local decisions, that are consistent with cortical feedback computations.

FEEDBACK COMPUTATIONS Most interpretations of top-down visual processes have focused on selective attention, which is viewed as feedback that improves sensitivity at attended locations, features, or both (Desimone & Duncan, 1995; Noudoost, Chang, Steinmetz, & Moore, 2010; Petersen & Posner, 2012). Top-down (or "endogenous") visual attention is typically interpreted as selective tuning in which information is routed through the visual processing hierarchy to amplify some features relative to others. In particular, Tsotsos et al. (1995) argue that attention acts to optimize visual search for features through a top-down hierarchy of winner-take-all processes. A Bayesian perspective emphasizes preservation of information regarding uncertainty about hypotheses, and its sequential reduction by message-passing between units and areas (Lee & Mumford, 2003). In addition, the diversity of visual descriptions suggests flexible access to hierarchically organized information. While there is no direct evidence, at this time, for neural populations representing hypotheses rather than decisions, or for probabilistic computations (as in message passing; Lochmann & Deneve, 2011), there are behavioral and neuroimaging results that are suggestive of Bayesian top-down computations down the cortical hierarchy. We briefly describe some of them.

Coarse-to-fine inferences A basic lesson learned from computer vision is that being certain about a local region of a natural image requires knowledge of the whole (figure 34.1B). Local perceptual decisions can be automatic, constrained by spatial or temporal context (as in priming or prior learning; cf. Hsieh, Vul, & Kanwisher, 2010), or be consciously task-driven and specified by a higher-level "executive."

Automatic (and executive) coarse-to-fine inference can be modeled as an initial high-level decision that "fixes" the value in the upper level of a hierarchical model, constraining subsequent lower-level decisions (figure 34.3C). An optimal decision restricted to a high level requires integrating out intermediate-level parameters. Several behavioral results are consistent with Bayesian coarse-to-fine computations over a simple hierarchical graph structure (Knill, 2003; Körding et al., 2007; Stocker & Simoncelli, 2008; Wozny, Beierholm, & Shams, 2010; Wu, Lu, & Yuille, 2008).

For example, Wu et al. (2008) have shown that human velocity-discrimination performance is consistent with an initial classification of motion type (rotation, expansion, translation).

Does feedback enhance or suppress feature representations? There are several ways in which top-down signals could change the neural representation of the probability distributions. Top-down processes may enhance or suppress low-level features consistent with a description or hypotheses at higher levels (Lee & Mumford, 2003; Mumford, 1992; Rao & Ballard, 1999; Spratling, 2012; Yuille & Kersten, 2006). Enhancement is consistent with neurophysiological and brain imaging studies that have demonstrated that perceptual grouping is correlated with the amplification of neural responses throughout the visual hierarchy (Kourtzi, Tolias, Altmann, Augath, & Logothetis, 2003; Roelfsema, 2006).

Enhancement is also consistent with the compositional models described earlier, in which information about a given object is represented and bound hierarchically. In principle, and depending on the task, feature enhancement could either be automatic or correspond to executive, top-down ("endogenous") attention. There is also evidence for suppression of lower-level features, which are consistent with a high-level hypothesis. Such a mechanism, sometimes referred to as "predictive coding," could support detecting and subsequently processing image information that does not fit with the current interpretation. Such a bottom-up signal would provide the basis for exogenous attention, but in contrast to a saliency computation (Itti & Baldi, 2009; Li, 1997; Rao & Ballard, 2013; Zhang, Tong, Marks, Shan, & Cottrell, 2008; Zhang, Zhaoping, Zhou, & Fang, 2012), which could be accomplished laterally, the signal increase is the result of a top-down prediction that fails.

Figure 34.6 shows behavioral evidence consistent with a predictive coding interpretation of "explaining away," in which occlusion cues provide an explanation for the missing vertices of the diamond (see Kersten, Mamassian, & Yuille, 2004). When the diamond is seen during an adaptation period (figure 34.6C), there was an increase in the strength of adaptation to shape (e.g., adapting to a skinny diamond results in seeing a standard comparison diamond as fatter); at the same time, there was a decrease in the strength of adaptation to the local orientation of comparison gratings. The converse was found when the occlusion cues were inconsistent with a diamond (figure 34.6D). The interpretation, consistent with other research, rests on the assumption that the sites of orientation and shape adaptation are in early and higher-level cortical areas, respectively.

There is also evidence from human fMRI studies for context-dependent suppression of neural activity in earlier areas in some cases (Alink, Schwiedrzik, Kohler, Singer, & Muckli, 2010; Cardin, Friston, & Zeki, 2011; Fang, Kersten, & Murray, 2008; Murray, Kersten, Olshausen, Schrater, & Woods, 2002; Rauss, Schwartz, & Pourtois, 2011), but not all (Mannion, Kersten, & Olman, 2013). And suppression measured using fMRI activity does not necessarily show the spatial specificity suggested by the above adaptation study or by theory (de Wit, Kubilius, Wagemans, & op de Beeck, 2012).

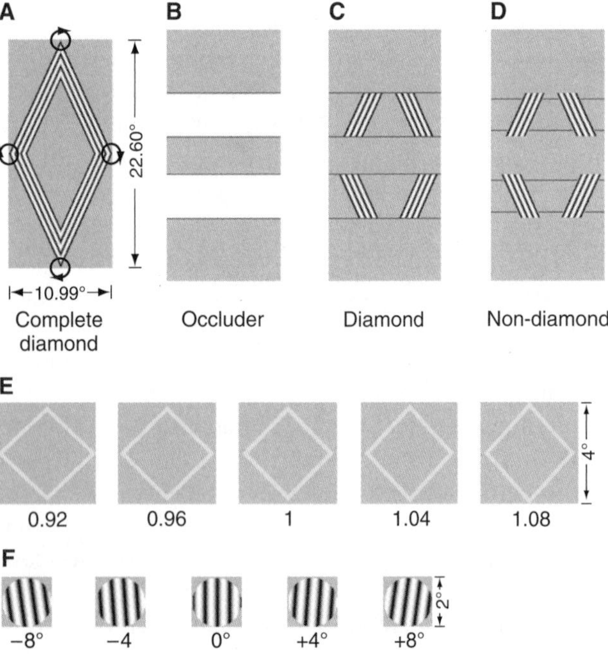

Figure 34.6 In studies with human subjects, He et al. (2012) showed that perceptual grouping amplifies the effect of adaptation to a whole shape, while reducing the strength of adaptation to local tilt. Thus, perceptual grouping is consistent with enhancement of high-level shape representation and attenuation of the low-level feature representation, possibly the result of top-down predictive coding. In panel A, the diamond corners undergo tight rotations during adaptation. When covered by an occluder, shown in B, the diamond can still be perceived, as shown in C. The diamond percept can be disrupted by the occlusion relationships shown in D. Panels E and F show test stimuli for measuring the aftereffects of shape, and tilt, respectively. (Figure adapted from He et al., 2012, with permission from Elsevier.)

In the language of signal-detection theory, the suppression of false and true positives through feedback could both be computationally useful. Suppression of false positives, enhancement of true positives in one population of neurons, or both could serve to bind object representations with parts and features at lower levels, as in the above compositional model. At the same time, increased activity in another neural population could signal false positives (i.e., inconsistent features that need to be resolved with other hypotheses; Friston, 2005; Rao & Ballard, 1999). Ultra-high field fMRI with submillimeter resolution has found stronger fMRI response in middle cortical layers of V1 during the presentation of scrambled objects as compared with intact objects (Olman et al., 2012), similar to what one might expect from prediction errors.

Hierarchically organized expertise In the race to make high-priority decisions quickly, as in "core" or basic-level recognition (DiCarlo et al., 2012), detailed information about position, size, shape, material and illumination direction is left behind, but not necessarily discarded. We know that human vision can discriminate subtle differences in shape and material and even see gradients of illumination, suggesting that it has the ability to access low-level information or recover transformations discounted earlier (Grimes & Rao, 2005; Olshausen et al., 1993; Tenenbaum & Freeman, 2000).

The ability of vision to extract information of different types and across multiple spatial scales raises the possibility that feedback signals in visual hierarchies have a richer computational function than so far discussed. Neuronal activity and receptive fields as early as primary visual cortex appear to be modulated by task requirements (Gilbert & Sigman, 2007; McManus, Li, & Gilbert, 2011). The interesting possibility is that the representation of information across levels of the visual hierarchy is accessible for a range of tasks. But for what functions, representations, and operations?

One possibility is that the optimal machinery, representations, or coordinate frames for the task exist at a lower level. Lee, Mumford, Romero, and Lamme (1998) suggested that higher-level computations involving fine-grain spatial and orientation information would necessarily involve V1. There are a number of results consistent with this idea. For example, Harrison and Tong (2009) analyzed patterns of fMRI voxel activity to show that visual areas from V4 down to V1 can retain orientation information held in working memory over

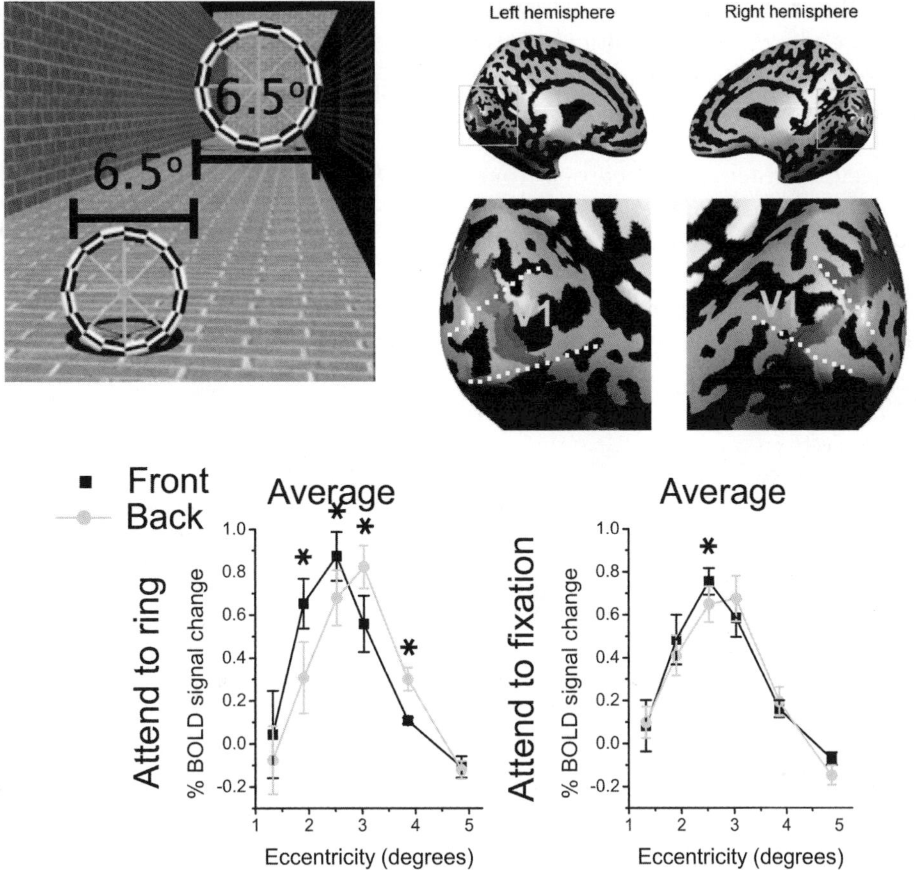

FIGURE 34.7 This figure illustrates how global, contextual information for 3D depth can shift the spatial extent of activity in human V1. *Upper left panel*: Both rings are the same size in the image, but the ring in the back of the hallway is perceived to be bigger than the one in front because it appears to be further away. *Upper right panel*: The spatial extent of activity in human cortical area V1 is larger for the back ring (red) than the front ring (green), similar to what would be found if the image of the back ring, in the absence of the 3D depth context, actually was bigger. *Lower panel*: When a person attends to the ring, the increase in spatial extent as indicated by a shift in the ring response with eccentricity, is more robust (bottom left) than when attending to the fixation mark (bottom right). (Reprinted from Fang et al., 2008, with permission from Elsevier.) (See color plate 33.)

many seconds. Variations in perceptual learning and its transfer may be understood in terms of whether the learning task requires the "expertise" of a lower versus higher level of processing (Hochstein & Ahissar, 2002). In another study, Williams et al. (2008) found that the measured patterns of fMRI activity near foveal retinotopic cortex could discriminate which object category the observers had been seeing with their peripheral vision. It has been known for some time that visual imagery involving fine spatial discrimination, and even orientation-specific tactile tasks, may activate representations in early visual areas (Kosslyn et al., 1993; Kosslyn & Thompson, 2003; Lucan, Foxe, Gomez-Ramirez, Sathian, & Molholm, 2010; Zangaladze, Epstein, Grafton, & Sathian, 1999).

Consider the everyday task of inferring an object's physical size from its image. This is a nontrivial computation with no current computer vision solution. The visual system has to decide which features form the boundary of the object's image—that is, a challenging segmentation and grouping problem that could require feedback to retinotopic areas. The locations of these features are needed to summarize the average diameter, or angular size. Then to estimate physical size from angular size, the system needs to process the larger context in order to take the object's depth into account. Further, size perception often involves comparisons with other objects, raising the question of where to make those. The complexity of the analysis suggests an interplay between high-level representations and early retinotopic areas, particularly V1 as a result of its high spatial precision. Studies by Murray et al. (2006) and Fang et al. (2008) used a classic depth illusion to show that the pattern of spatial activity in V1 activity is indeed modulated by 3D depth context (figure 34.7). When an object (a ring) appeared bigger, its "neural image" on V1 was bigger (i.e., activation shifts to a more eccentric representation of the visual field).

This effect was significantly stronger when observers attended to the object, consistent with feedback from higher-level areas that process depth in the larger context of the scene. Psychophysical data is also consistent with a top-down influence of depth on orientation-selective, and putatively early, cortical regions (Arnold, Birt, & Wallis, 2008).

The longer you look, the more you see

Not many decades ago, "perception" seemed to be not much more than a screen, admittedly with some puzzling distortions, viewed by a high-level executive agent. Then retinal and cortical studies showed that neurons were doing much more than transmitting image information: they were emphasizing certain kinds of information, such as edges, at the expense of others (smooth gradients). This led to the idea of the retina and early visual cortical areas as spatiotemporal filter banks. But still, the emphasis was on early perceptual processing as a set of filtering stages, effectively passing decisions forward from one stage to the next (Lennie, 1998).

Computer vision has provided the perspective that in order to produce useful behavioral outcomes, the human visual system is solving a decoding problem whose understanding requires concepts and a level of analysis beyond traditional neural network filtering. The past decade has seen substantial progress in both the computational and neural understanding of how vision could be solving the problems of object perception. We have discussed potential limitations on the robustness and versatility of vision with strictly feedforward processing and have reviewed arguments and results suggesting that both automatic and executive processes access built-in image knowledge at several levels of abstraction. We conjecture that the brain's ability to solve the problems of local uncertainty and task versatility rests on deep generative knowledge of the structure of images. A major challenge for the future is to better understand the way the brain represents and controls the top-down utilization of this knowledge (cf. Blanchard & Geman, 2005; Ullman, 1984), eventually explaining how the brain enables us to see so much in just one picture of a fox.

ACKNOWLEDGMENTS DK and AL were supported by the WCU (World Class University) program funded by the Ministry of Education, Science, and Technology through the National Research Foundation of Korea (R31-10008) and by ONR N000141210883.

REFERENCES

Alink, A., Schwiedrzik, C. M., Kohler, A., Singer, W., & Muckli, L. (2010). Stimulus predictability reduces responses in primary visual cortex. *J Neurosci, 30*(8), 2960–2966.

Arnold, D. H., Birt, A., & Wallis, T. S. A. (2008). Perceived size and spatial coding. *J Neurosci, 28*(23), 5954–5958.

Bar, M. (2003). A cortical mechanism for triggering top-down facilitation in visual object recognition. *J Cogn Neurosci, 15*(4), 600–609.

Barenholtz, E., & Tarr, M. J. (2011). Visual learning of statistical relations among nonadjacent features: Evidence for structural encoding. *Vis Cogn, 19*(4), 469–482.

Barlow, H. (1981). The Ferrier lecture, 1980. Critical limiting factors in the design of the eye and visual cortex. *Proc R Soc Lond B Biol Sci, 212*(1186), 1–34.

Barlow, H. (1990). Conditions for versatile learning, Helmholtz's unconscious inference, and the task of perception. *Vis Res, 30*(11), 1561–1571.

Barron, J. T., & Malik, J. (2012). Shape, albedo, and illumination from a single image of an unknown object. *Proc. IEEE Conference on Computer Vision* (pp. 1–8).

Barrow, H., & Tenenbaum, J. (1978). Recovering intrinsic scene characteristics from images. In A. Hanson & E. Riseman (Eds.), *Computer vision systems* (pp. 3–16). New York, NY: Academic Press.

Beauchamp, M. S., Lee, K. E., Haxby, J. V., & Martin, A. (2002). Parallel visual motion processing streams for manipulable objects and human movements. *Neuron, 34*(1), 149–159.

Beck, J. M., Latham, P. E., & Pouget, A. (2011). Marginalization in neural circuits with divisive normalization. *J Neurosci, 31*(43), 15310–15319.

Biederman, I. (1987). Recognition-by-components: A theory of human image understanding. *Psychol Rev, 94*(2), 115–147.

Blanchard, G., & Geman, D. (2005). Hierarchical testing designs for pattern recognition. *Ann Stat, 33*(3), 1155–1202.

Boyaci, H., Fang, F., Murray, S. O., & Kersten, D. (2007). Responses to lightness variations in early human visual cortex. *Curr Biol, 17*(11), 989–993.

Boyaci, H., Fang, F., Murray, S. O., & Kersten, D. (2010). Perceptual grouping-dependent lightness processing in human early visual cortex. *J Vis, 10*(9), 1–12.

Burgi, P. Y., Yuille, A., & Grzywacz, N. M. (2000). Probabilistic motion estimation based on temporal coherence. *Neural Comput, 12*(8), 1839–1867.

Cadieu, C. F., & Olshausen, B. A. (2012). Learning intermediate-level representations of form and motion from natural movies. *Neural Comput, 24*(4), 827–866.

Callaway, E. (1998). Local circuits in primary visual cortex of the macaque monkey. *Annu Rev Neurosci, 21*, 47–74.

Cardin, V., Friston, K. J., & Zeki, S. (2011). Top-down modulations in the visual form pathway revealed with dynamic causal modeling. *Cereb Cortex, 21*(3), 550–562.

Chang, L., Jin, Y., Zhang, W., Borenstein, E., & Geman, S. (2011). Context, computation, and optimal ROC performance in hierarchical models. *Int J Comput Vis, 93*(2), 117–140.

Chikkerur, S., Serre, T., Tan, C., & Poggio, T. (2010). What and where: A Bayesian inference theory of attention. *Vis Res, 50*(22), 2233–2247.

Connor, C. E., Brincat, S. L., & Pasupathy, A. (2007). Transformation of shape information in the ventral pathway. *Curr Opin Neurobiol, 17*(2), 140–147.

de Wit, L. H., Kubilius, J., Wagemans, J., & op de Beeck, H. P. (2012). Bistable Gestalts reduce activity in the whole of V1, not just the retinotopically predicted parts. *J Vis, 12*(11), 1–14.

Desimone, R., & Duncan, J. (1995). Neural mechanisms of selective visual attention. *Annu Rev Neurosci, 18*, 193–222.

DiCarlo, J. J., Zoccolan, D., & Rust, N. C. (2012). How does the brain solve visual object recognition? *Neuron, 73*(3), 415–434.

Durbin, R., & Mitchison, G. (1990). A dimension reduction framework for understanding cortical maps. *Nature, 343*(6259), 644–647.

Eckstein, M. P., Drescher, B., & Shimozaki, S. S. (2006). Attentional cues in real scenes, saccadic targeting, and Bayesian priors. *Psychol Sci, 17*(11), 973.

Elder, J. H., & Goldberg, R. M. (2002). Ecological statistics of Gestalt laws for the perceptual organization of contours. *J Vis, 2*(4), 324–353.

Epshtein, B., Lifshitz, I., & Ullman, S. (2008). Image interpretation by a single bottom-up top-down cycle. *Proc Natl Acad Sci USA, 105*(38), 14298.

Fang, F., Boyaci, H., Kersten, D., & Murray, S. O. (2008). Attention-dependent representation of a size illusion in human V1. *Curr Biol, 18*(21), 1707–1712.

Fang, F., Kersten, D., & Murray, S. O. (2008). Perceptual grouping and inverse fMRI activity patterns in human visual cortex. *J Vis, 8*(7), 2–9.

Felleman, D., & Van Essen, D. (1991). Distributed hierarchical processing in the primate cerebral cortex. *Cereb Cortex, 1*(1), 1–47.

Fiser, J., Berkes, P., Orbán, G., & Lengyel, M. (2010). Statistically optimal perception and learning: From behavior to neural representations. *Trends Cogn Sci, 14*(3), 119–130.

Freeman, J., Ziemba, C. M., Heeger, D. J., Simoncelli, E. P., & Movshon, J. A. (2013). A functional and perceptual signature of the second visual area in primates. *Nat Neurosci, 16*(7), 974–981.

Freeman, W. (1994). The generic viewpoint assumption in a framework for visual perception. *Nature, 368*(6471), 542–545.

Freiwald, W. A., & Tsao, D. Y. (2010). Functional compartmentalization and viewpoint generalization within the macaque face-processing system. *Science, 330*(6005), 845–851.

Friston, K. (2005). A theory of cortical responses. *Philos Trans R Soc Lond B Biol Sci, 360*(1456), 815–836.

Fukushima, K. (1988). Neocognitron: A hierarchical neural network capable of visual pattern recognition. *Neural Netw, 1*(2), 119–130.

Geisler, W. S. (2011). Contributions of ideal observer theory to vision research. *Vis Res, 51*(7), 771–781.

Geisler, W. S., & Kersten, D. (2002). Illusions, perception and Bayes. *Nat Neurosci, 5*(6), 508–510.

Geisler, W. S., & Perry, J. (2009). Contour statistics in natural images: Grouping across occlusions. *Vis Neurosci, 26*(01), 109–121.

Geman, S., Potter, D., & Chi, Z. (2002). Composition systems. *Q Appl Math, 60*(4), 707–736.

Gilbert, C. D., & Sigman, M. (2007). Brain states: Top-down influences in sensory processing. *Neuron, 54*(5), 677–696.

Gold, J. M., Abbey, C., Tjan, B. S., & Kersten, D. (2009). Ideal observers and efficiency: Commemorating 50 years of Tanner and Birdsall. *J Opt Soc Am A Opt Image Sci Vis, 26*(11), IO1–IO2.

Green, C. S., Pouget, A., & Bavelier, D. (2010). Improved probabilistic inference as a general learning mechanism with action video games. *Curr Biol, 20*(17), 1573–1579.

Grimes, D., & Rao, R. P. (2005). Bilinear sparse coding for invariant vision. *Neural Comput, 17*(1), 47–73.

Grossberg, S., & Mingolla, E. (1985). Neural dynamics of perceptual grouping: Textures, boundaries, and emergent segmentations. *Atten Percept Psychophys, 38*(2), 141–171.

Grosse, R., Johnson, M., Adelson, E., & Freeman, W. (2009). Ground truth dataset and baseline evaluations for intrinsic image algorithms. *Proc. IEEE Conference on Computer Vision* (pp. 2335–2342).

Guillery, R. W., & Sherman, S. M. (2002). Thalamic relay functions and their role in corticocortical communication: Generalizations from the visual system. *Neuron, 33*(2), 163–175.

Harel, A., Ullman, S., Epshtein, B., & Bentin, S. (2007). Mutual information of image fragments predicts categorization in humans: Electrophysiological and behavioral evidence. *Vis Res, 47*(15), 2010–2020.

Harrison, S. A., & Tong, F. (2009). Decoding reveals the contents of visual working memory in early visual areas. *Nature, 458*(7238), 632–635.

He, D., Kersten, D., & Fang, F. (2012). Opposite modulation of high- and low-level visual aftereffects by perceptual grouping. *Curr Biol, 22*(11), 1040–1045.

Hegdé, J., Bart, E., & Kersten, D. (2008). Fragment-based learning of visual object categories. *Curr Biol, 18*(8), 597–601.

Hinton, G. (2009). Learning to represent visual input. *Philos Trans R Soc Lond B Biol Sci, 365*(1537), 177–184.

Hochstein, S., & Ahissar, M. (2002). View from the top: Hierarchies and reverse hierarchies in the visual system. *Neuron, 36*(5), 791–804.

Hsieh, P. J., Vul, E., & Kanwisher, N. (2010). Recognition alters the spatial pattern of fMRI activation in early retinotopic cortex. *J Neurophysiol, 103*(3), 1501–1507.

Hubel, D., & Wiesel, T. (1977). Ferrier lecture: Functional architecture of macaque monkey visual cortex. *Proc R Soc Lond B Biol Sci, 198*(1130), 1–59.

Hummel, J. E., & Biederman, I. (1992). Dynamic binding in a neural network for shape recognition. *Psychol Rev, 99*(3), 480–517.

Hyvärinen, A. (2010). Statistical models of natural images and cortical visual representation. *Top Cogn Sci, 2*(2), 251–264.

Itti, L., & Baldi, P. (2009). Bayesian surprise attracts human attention. *Vis Res, 49*(10), 1295–1306.

Jacobs, R. (1999). Optimal integration of texture and motion cues to depth. *Vis Res, 39*(21), 3621–3629.

Jordan, M. I., & Weiss, Y. (2002). Graphical models: Probabilistic inference. In M. Arbib (Ed.), *The handbook of brain theory and neural networks* (2nd ed.). Cambridge, MA: MIT Press.

Kanwisher, N. (2010). Functional specificity in the human brain: A window into the functional architecture of the mind. *Proc Natl Acad Sci USA, 107*(25), 11163.

Kersten, D. (1991). Transparency and the cooperative computation of scene attributes. In M. S. Landy (Ed.), *Computational models of visual processing* (pp. 209–228). Cambridge, MA: MIT Press.

Kersten, D., Mamassian, P., & Yuille, A. (2004). Object perception as Bayesian inference. *Annu Rev Psychol, 55,* 271–304.

Kersten, D. J., & Yuille, A. L. (2013). Vision: Bayesian inference and beyond. In J. Werner & L. M. Chalupa (Eds.), *The new visual neurosciences* (pp. 1–16). Cambridge, MA: MIT Press.

Kingdom, F. A. A. (2011). Lightness, brightness and transparency: A quarter century of new ideas, captivating demonstrations and unrelenting controversy. *Vis Res, 51*(7), 652–673.

Knill, D. C. (2003). Mixture models and the probabilistic structure of depth cues. *Vis Res, 43*(7), 831–854.

Knill, D. C., & Kersten, D. (1991). Apparent surface curvature affects lightness perception. *Nature, 351*(6323), 228–230.

Knill, D. C., & Pouget, A. (2004). The Bayesian brain: The role of uncertainty in neural coding and computation. *Trends Neurosci, 27*(12), 712–719.

Koch, C., Marroquin, J., & Yuille, A. (1986). Analog "neuronal" networks in early vision. *Proc Natl Acad Sci USA, 83*(12), 4263–4267.

Körding, K. P., Beierholm, U., Ma, W. J., Quartz, S., Tenenbaum, J. B., & Shams, L. (2007). Casual inference in multisensory perceptions. *PloS ONE, 2*(9), e943.

Kosslyn, S. M., Alpert, N. M., Thompson, W. L., Maljkovic, V., Weise, S. B., Chabris, C. F., ... Buonanno, F. S. (1993). Visual mental imagery activates topographically organized visual cortex: PET investigations. *J Cogn Neurosci, 5*(3), 263–287.

Kosslyn, S. M., & Thompson, W. L. (2003). When is early visual cortex activated during visual mental imagery? *Psychol Bull, 129*(5), 723–746.

Kourtzi, Z., & Connor, C. E. (2011). Neural representations for object perception: Structure, category, and adaptive coding. *Annu Rev Neurosci, 34*(1), 45–67.

Kourtzi, Z., Tolias, A. S., Altmann, C. F., Augath, M., & Logothetis, N. K. (2003). Integration of local features into global shapes: Monkey and human fMRI studies. *Neuron, 37*(2), 333–346.

Kromrey, S., Maestri, M., Hauffen, K., Bart, E., & Hegdé, J. (2010). Fragment-based learning of visual object categories in non-human primates. *PLoS ONE, 5*(11), e15444.

Lamme, V. A., Sup, H., & Spekreijse, H. (1998). Feedforward, horizontal, and feedback processing cortex. *Curr Opin Neurobiol, 8,* 529–535.

Lauritzen, S., & Spiegelhalter, D. (1988). Local computations with probabilities on graphical structures and their application to expert systems. *J Roy Stat Soc B Met, 50*(2), 157–224.

Lee, T., Mumford, D., Romero, R., & Lamme, V. A. (1998). The role of the primary visual cortex in higher level vision. *Vis Res, 38*(15–16), 2429–2454.

Lee, T. S. (2003). Computations in the early visual cortex. *J Physiol (Paris), 97*(2–3), 121–139.

Lee, T. S., & Mumford, D. (2003). Hierarchical Bayesian inference in the visual cortex. *J Opt Soc Am A Opt Image Sci Vis, 20*(7), 1434–1448.

Lennie, P. (1998). Single units and visual cortical organization. *Perception, 27,* 889–936.

Lerner, Y., Epshtein, B., Ullman, S., & Malach, R. (2008). Class information predicts activation by object fragments in human object areas. *J Cogn Neurosci, 20*(7), 1189–1206.

Li, Z. (1997). Primary cortical dynamics for visual grouping. In K.-Y. M. Wong & D.-Y. Yeung (Eds.), *Theoretical aspects of neural computation.* New York, NY: Springer.

Lochmann, T., & Deneve, S. (2011). Neural processing as causal inference. *Curr Opin Neurobiol, 21*(5), 774–781.

Lucan, J. N., Foxe, J. J., Gomez-Ramirez, M., Sathian, K., & Molholm, S. (2010). Tactile shape discrimination recruits human lateral occipital complex during early perceptual processing. *Hum Brain Mapp, 31*(11), 1813–1821.

Lund, J., Angelucci, A., & Bressloff, P. C. (2003). Anatomical substrates for functional columns in macaque monkey primary visual cortex. *Cereb Cortex, 13*(1), 15–24.

Ma, W. J. (2010). Signal detection theory, uncertainty, and Poisson-like population codes. *Vis Res, 50*(22), 2308–2319.

Ma, W. J. (2012). Organizing probabilistic models of perception. *Trends Cogn Sci, 16*(10), 511–518.

Ma, W. J., Beck, J. M., Latham, P. E., & Pouget, A. (2006). Bayesian inference with probabilistic population codes. *Nat Neurosci, 9*(11), 1432–1438.

Ma, W. J., Beck, J. M., & Pouget, A. (2008). Spiking networks for Bayesian inference and choice. *Curr Opin Neurobiol, 18*(2), 217–222.

Maloney, L. T., & Zhang, H. (2010). Decision-theoretic models of visual perception and action. *Vis Res, 50*(23), 2362–2374.

Mannion, D. J., Kersten, D. J., & Olman, C. A. (2013). Consequences of polar form coherence for fMRI responses in human visual cortex. *NeuroImage, 78*(C), 152–158.

Markov, N. T., Vezoli, J., Chameau, P., Falchier, A., Quilodran, R., Huissoud, C., ... Kennedy, H. (2013). The anatomy of hierarchy: Feedforward and feedback pathways in macaque visual cortex. *J Comp Neurol, 522*(1), 225–259.

Marr, D. (1982). *Vision: A computational investigation into the human representation and processing of visual information.* New York, NY: Henry Holt.

Marr, D., & Nishihara, H. K. (1978). Representation and recognition of the spatial organization of three-dimensional shapes. *Proc R Soc Lond B Biol Sci, 200*(1140), 269–294.

Marroquin, J., Mitter, S., & Poggio, T. (1987). Probabilistic solution of ill-posed problems in computational vision. *J Am Stat Assoc, 82*(397), 76–89.

McManus, J. N. J., Li, W., & Gilbert, C. D. (2011). Adaptive shape processing in primary visual cortex. *Proc Natl Acad Sci USA, 108*(24), 9739–9746.

Mountcastle, V. B. (1997). The columnar organization of the neocortex. *Brain, 120,* 701–722.

Mumford, D. (1992). On the computational architecture of the neocortex. *Biol Cybern, 66*(3), 241–251.

Murray, S. O., Boyaci, H., & Kersten, D. (2006). The representation of perceived angular size in human primary visual cortex. *Nat Neuroscience, 9*(3), 429–434.

Murray, S. O., Kersten, D., Olshausen, B. A., Schrater, P., & Woods, D. L. (2002). Shape perception reduces activity in human primary visual cortex. *Proc Natl Acad Sci USA, 99*(23), 15164–15169.

Nassi, J. J., & Callaway, E. M. (2009). Parallel processing strategies of the primate visual system. *Nat Rev Neurosci, 10*(5), 360–372.

Noudoost, B., Chang, M. H., Steinmetz, N. A., & Moore, T. (2010). Top-down control of visual attention. *Curr Opin Neurobiol, 20*(2), 183–190.

Olman, C. A., Harel, N., Feinberg, D. A., He, S., Zhang, P., Ugurbil, K., & Yacoub, E. (2012). Layer-specific fMRI reflects different neuronal computations at different depths in human V1. *PLoS ONE, 7*(3), e32536.

Olshausen, B. A. (1996). Emergence of simple-cell receptive field properties by learning a sparse code for natural images. *Nature, 381*(6583), 607–609.

Olshausen, B. A., Anderson, C. H., & Van Essen, D. (1993). A neurobiological model of visual attention and invariant pattern recognition based on dynamic routing of information. *J Neurosci, 13*(11), 4700–4719.

Olshausen, B. A., & Field, D. J. (2005). How close are we to understanding V1? *Neural Comput, 17*(8), 1665–1699.

Orbán, G. A. (2008). Higher order visual processing in macaque extrastriate cortex. *Physiol Rev, 88*(1), 59–89.

Orbán, G., Fiser, J., Aslin, R. N., & Lengyel, M. (2008). Bayesian learning of visual chunks by human observers. *Proc Natl Acad Sci USA, 105*(7), 2745.

Orbán, G., & Wolpert, D. M. (2011). Representations of uncertainty in sensorimotor control. *Curr Opin Neurobiol, 21*(4), 629–635.

Pearl, J. (1988). *Probabilistic reasoning in intelligent systems: Networks of plausible inference.* Burlington, MA: Morgan Kaufmann.

Petersen, S. E., & Posner, M. I. (2012). The attention system of the human brain: 20 years after. *Annu Rev Neurosci, 35,* 73.

Poggio, T. (2011). The computational magic of the ventral stream: Towards a theory. *Nat Proceedings.* doi:10.1038/npre.2011.6117.1

Purves, D., & Lotto, R. (2003). *Why we see what we do: An empirical theory of vision.* Sunderland, MA: Sinauer.

Rao, R. P., & Ballard, D. (1999). Predictive coding in the visual cortex: A functional interpretation of some extra-classical receptive-field effects. *Nat Neurosci, 2,* 79–87.

Rao, R. P., & Ballard, D. H. (2013). Probabilistic models of attention based on iconic representations and predictive coding. In L. Itti, G. Rees, & J. Tsotsos (Eds.), *Neurobiology of attention* (pp. 1–16). New York, NY: Academic Press.

Rauss, K., Schwartz, S., & Pourtois, G. (2011). Top-down effects on early visual processing in humans: A predictive coding framework. *Neurosci Biobehav R, 35*(5), 1237–1253.

Riesenhuber, M., & Poggio, T. (1999). Hierarchical models of object recognition in cortex. *Nat Neurosci, 2,* 1019–1025.

Roe, A. W., Chelazzi, L., Connor, C. E., Conway, B. R., Fujita, I., Gallant, J. L., ... Vanduffel, W. (2012). Toward a unified theory of visual area V4. *Neuron, 74*(1), 12–29.

Roe, A. W., Chen, G., & Lu, H. (2009). Visual system: Functional architecture of area V2. In L. R. Squire (Ed.), *Encyclopedia of neuroscience* (pp. 331–349). Amsterdam: Elsevier.

Roelfsema, P. (2006). Cortical algorithms for perceptual grouping. *Annu Rev Neurosci, 29,* 203–227.

Schölkopf, B., & Smola, A. J. (2002). *Learning with kernels: Support vector machines, regularization, optimization, and beyond.* Cambridge, MA: MIT Press.

Schwartz, O., & Simoncelli, E. P. (2001). Natural signal statistics and sensory gain control. *Nat Neurosci, 4*(8), 819–825.

Shi, J., & Malik, J. (2000). Normalized cuts and image segmentation. *IEEE T Pattern Anal, 22*(8), 888–905.

Spratling, M. W. (2012). Unsupervised learning of generative and discriminative weights encoding elementary image components in a predictive coding model of cortical function. *Neural Comput, 24*(1), 60–103.

Stocker, A. A., & Simoncelli, E. (2008). A Bayesian model of conditioned perception. *Adv Neural Inf Process Syst, 20,* 1409–1416.

Tappen, M., Freeman, W., & Adelson, E. (2005). Recovering intrinsic images from a single image. *IEEE T Pattern Anal, 27*(9), 1459–1472.

Tenenbaum, J. B., & Freeman, W. (2000). Separating style and content with bilinear models. *Neural Comput, 12*(6), 1247–1283.

Trenti, E. J., Barraza, J. F., & Eckstein, M. P. (2010). Learning motion: Human vs. optimal Bayesian learner. *Vis Res, 50*(4), 460–472.

Tsotsos, J. K., Culhane, S. M., Kei Wai, W. Y., Lai, Y., Davis, N., & Nuflo, F. (1995). Modeling visual attention via selective tuning. *Artif Intell, 78*(1), 507–545.

Tu, Z., Chen, X., Yuille, A., & Zhu, S. (2005). Image parsing: Unifying segmentation, detection, and recognition. *Int J Comput Vis, 63*(2), 113–140.

Ullman, S. (1984). Visual routines. *Cognition, 18*(1–3), 97–159.

Ullman, S., Vidal-Naquet, M., & Sali, E. (2002). Visual features of intermediate complexity and their use in classification. *Nat Neurosci, 5*(7), 682–687.

Vilares, I., & Körding, K. P. (2011). Bayesian models: The structure of the world, uncertainty, behavior, and the brain. *Ann NY Acad Sci, 1224*(1), 22–39.

Wallis, G., Rolls, E., & Foldiak, P. (1993). Learning invariant responses to the natural transformations of objects. *Proc. IEEE International Joint Conference on Neural Networks* (pp. 1087–1090).

Wallisch, P., & Movshon, J. A. (2008). Structure and function come unglued in the visual cortex. *Neuron, 60*(2), 194–197.

Waltz, D. L. (1972). Generating semantic descriptions from drawings of scenes with shadows. Technical report, MIT Artificial Intelligence Lab. http://hdl.handle.net/1721.1/6911

Weiss, Y., Simoncelli, E. P., & Adelson, E. H. (2002). Motion illusions as optimal percepts. *Nat Neurosci, 5*(6), 598–604.

Williams, M. A., Baker, C. I., op de Beeck, H. P., Shim, W. M., Dang, S., Triantafyllou, C., & Kanwisher, N. (2008). Feedback of visual object information to foveal retinotopic cortex. *Nat Neurosci, 11*(12), 1439–1445.

Wolpert, D. M., & Landy, M. S. (2012). Motor control is decision-making. *Curr Opin Neurobiol, 22*(6), 996–1003.

Wozny, D. R., Beierholm, U. R., & Shams, L. (2010). Probability matching as a computational strategy used in perception. *PLoS Comput Biol, 6*(8), e1000871.

Wu, S., Lu, H., & Yuille, A. (2008). Model selection and velocity estimation using novel priors for motion patterns. *Adv Neural Inf Process Syst, 1793–1800.*

Yuille, A. (2010). An information theory perspective on computational vision. *Front Electr Electron Eng Chin, 5*(3), 329–346.

Yuille, A., & Kersten, D. (2006). Vision as Bayesian inference: Analysis by synthesis? *Trends Cogn Sci, 10*(7), 301–308.

Yuille, A. L., & Mottaghi, R. (2013). Complexity of representation and inference in compositional models with part sharing. arXiv:1301.3560

Zangaladze, A., Epstein, C. M., Grafton, S. T., & Sathian, K. (1999). Involvement of visual cortex in tactile discrimination of orientation. *Nature, 401*(6753), 587–590.

Zeiler, M., Taylor, G., & Fergus, R. (2011). Adaptive deconvolutional networks for mid and high level feature learning. *Proc. IEEE Conference on Computer Vision* (pp. 2018–2025).

Zemel, R. S., & Pouget, A. (1998). Probabilistic interpretation of population codes. *Neural Comput, 10*(2), 403–430.

Zhang, L., Tong, M. H., Marks, T. K., Shan, H., & Cottrell, G. W. (2008). SUN: A Bayesian framework for saliency using natural statistics. *J Vis, 8*(7), 1–20.

Zhang, X., Zhaoping, L., Zhou, T., & Fang, F. (2012). Neural activities in V1 create a bottom-up saliency map. *Neuron, 73*(1), 183–192.

Zhu, L., Chen, Y., Lin, C., & Yuille, A. (2010). Max margin learning of hierarchical configural deformable templates (HCDTs) for efficient object parsing and pose estimation. *Int J Comput Vis, 93*(1), 1–21.

Zhu, L., Chen, Y., Torralba, A., Freeman, W., & Yuille, A. (2011). Part and appearance sharing: Recursive compositional models for multi-view multi-object detection. *Proc. IEEE Conference on Computer Vision* (pp. 1919–1926).

35 The Functional Architecture of Human Ventral Temporal Cortex and Its Role in Visual Perception

KALANIT GRILL-SPECTOR AND KEVIN S. WEINER

ABSTRACT Visual recognition is amazingly rapid and requires a series of computations extending from primary visual cortex to ventral temporal cortex (VTC). Here, we examine the functional architecture of human VTC and its role in recognition. We detail the computational goals, the information computed to reach those goals (representations), and the physical layout of that information on the cortical sheet (implementation). Then, we discuss how implementational features of neural architecture may impact representations and computations in VTC. Specifically, recent discoveries detail anatomical and topological relationships among a series of fine-scale representations nested within large-scale representations in VTC, which generate predictable divergences and convergences across cortical scales. We suggest that this implementation generates a spatial hierarchy of nested functional representations that mirrors the hierarchical category information structure of VTC, where more abstract information is represented at larger spatial scales. Together these implementational, representational, and computational features enable a rapid and flexible recognition system with access to different levels of information abstraction extending from exemplars, to basic categories, to broad superordinate categorical distinctions.

Human recognition is amazingly rapid, enabling us to determine the gist of a visual scene in just one-tenth of a second (Thorpe, Fize, & Marlot, 1996). Despite this remarkable efficiency, visual recognition is nontrivial, requiring a series of computations arranged in a ventral stream processing hierarchy (Ungerleider & Haxby, 1994), ascending from primary visual cortex (V1) to high-level regions in ventral temporal cortex (VTC; also referred to as inferotemporal cortex, or IT). Human VTC consists of the posterior half of the ventral aspect of the temporal lobe. Lesions to VTC can produce specific deficits, such as the inability to recognize objects, faces, or words, while other visual faculties are preserved, indicating its critical role in recognition.

To understand how visual recognition is achieved, it is necessary to study the computational goals of the recognition system, the representations and information computed to reach those goals, and its physical layout and implementation in cortex (Marr, 1982). Thus, the goal of this chapter is to unveil the "black box" of VTC and its role in visual perception adapting Marr's framework to examine the inter-relationship among computations, representations, and neural implementation (figure 35.1). Our examination of these three factors builds upon classic cognitive neuroscience research, which provided an integrative foundation for how anatomical and functional properties of the visual system (Van Essen, Anderson, & Felleman, 1992; Zeki & Shipp, 1988)—namely, its functional architecture (Hubel & Wiesel, 1962)—contribute to information processing (figure 35.1A). We emphasize recent discoveries of implementational features of VTC, which reveal superimposition of multiple neural representations on the cortical sheet with predictable levels of anatomical convergence and divergence. This functional architecture matriculates at multiple spatial scales, generating a spatial hierarchy of nested neural representations (figure 35.1C). We propose that this spatial hierarchy generates an information hierarchy within VTC (figure 35.1B), which, in turn, enables a fast and flexible system supporting recognition at multiple levels of abstraction ranging from exemplars to broad categories.

Computational theory: What are the computational goals of ventral temporal cortex?

Theorists investigate the computational requirements necessary for achieving an efficient and rapid visual recognition system. This computational research has yielded four important insights:

1. Shape is the critical visual attribute for object recognition.
2. A visual recognition system needs to be robust to changes in object appearance, such as retinal size, position in the visual field, illumination, contrast, and view (referred to as tolerance to image transformations)

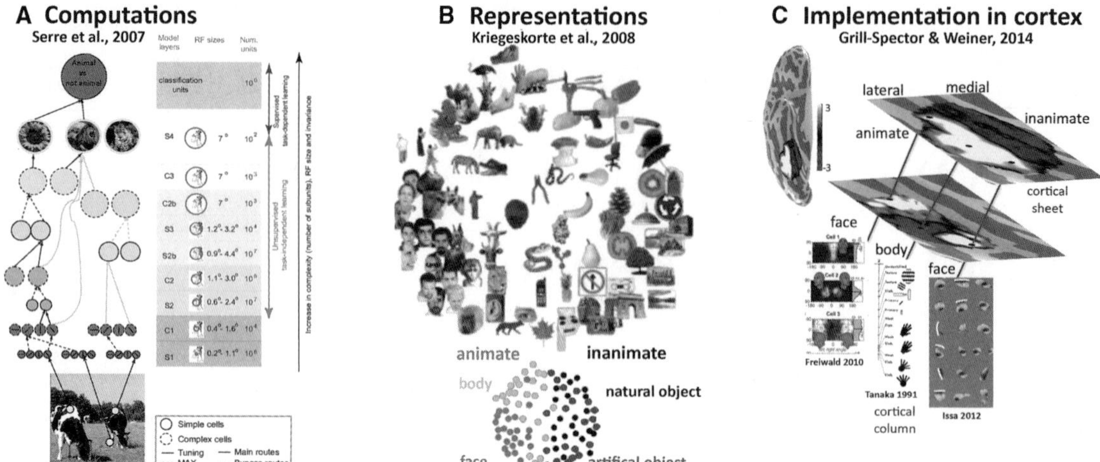

FIGURE 35.1 Marr's (1982) framework applied to ventral temporal cortex (VTC). (A) Computational models simulate the ventral stream computational hierarchy during an animate vs. inanimate task. (B) VTC representations contain a hierarchical category information structure, first separating animate from inanimate representations, then face from body representations, as well as natural from artificial objects. (C) Implementational features reveal superimposed representations at multiple spatial scales (top to bottom: large to fine) with orderly diverging (e.g., animate from inanimate) and converging (e.g., animate with face and body representations) characteristics.

without losing the ability to discriminate between similar exemplars with comparable parts and configuration (such as two different faces).

3. A visual recognition system should support recognition and perception at several levels of abstraction; for instance, differentiating among exemplars (e.g., Joey vs. Jack), basic categories (e.g., face vs. car), as well as broad superordinate categories (e.g., animate vs. inanimate).

4. There are classes of stimuli that may require specialized computations in addition to the domain-general computations mentioned in (1)–(3). For example, face perception may also require domain-specific computations to determine the gender, age, expression, or gaze of a person.

Understanding the computational goals of VTC as a recognition system has important implications for both information processing and neural implementation. In the sections below, we examine the properties of VTC representations and their relation to the computational goals of a recognition system: Are VTC representations tuned to shape? Are they tolerant to changes in appearance? How do VTC representations and their cortical implementation support different levels of information abstraction? Empirical research striving to answer these questions has led to the development of computational models that provide important insights into the computations performed by the ventral stream hierarchy (Epshtein, Lifshitz, & Ullman, 2008; Riesenhuber & Poggio, 1999; Serre, Oliva, & Poggio, 2007).

Representations: How does ventral temporal cortex represent information to support visual recognition?

VENTRAL TEMPORAL CORTEX IS TUNED TO SHAPE IRRESPECTIVE OF VISUAL CUE A general property of human VTC is that it responds more strongly to a variety of shapes and objects compared to textures, noise, or highly scrambled objects (Malach et al., 1995). Importantly, VTC responses maintain their selectivity to shape across a large spectrum of visual cues including luminance, color, motion, texture, stereo, and illusory contours (Davindenko, Remus, & Grill-Spector, 2012; Grill-Spector, Kushnir, Edelman, Itzchak, & Malach, 1998; Kourtzi & Kanwisher, 2001; Mendola, Dale, Fischl, Liu, & Tootell, 1999; Vinberg & Grill-Spector, 2008). Thus, a domain-general property of VTC is that it demonstrates perceptual constancy to shape across visual cues.

VENTRAL TEMPORAL CORTEX REPRESENTATIONS ARE SENSITIVE TO CHANGES IN EXEMPLAR IDENTITY AND EXHIBIT SOME TOLERANCE TO CHANGES IN EXEMPLAR APPEARANCE Classic theories of recognition suggest that recognizing exemplars across changes in appearance requires object-specific, invariant representations (Biederman, 1987). However, modern theories have shown that complete tolerance is unnecessary as long as representations for different exemplars are "untangled" or linearly separable from each other (DiCarlo & Cox, 2007; DiCarlo, Zoccolan, & Rust, 2012). Thus, untangling predicts that effective representations for recognition should have separable identity and transformation information (e.g., view, size, position).

Indeed, a prominent characteristic of VTC responses is that they are highly sensitive to exemplar identity. That is, neuronal responses in VTC (but not early visual areas) decrease when an exemplar is repeated (referred to as repetition suppression or adaptation; Grill-Spector, Henson, & Martin, 2006), but remain high when different exemplars are shown. Interestingly, VTC representations are highly sensitive to small changes in form that affect perceived exemplar identity (Grill-Spector et al., 1999; Jiang et al., 2007; Kourtzi & Kanwisher, 2001), but are tolerant in some degree to changes in contrast (Avidan et al., 2002) and size (Grill-Spector et al., 1999; Liu, Agam, Madsen, & Kreiman, 2009; Vuilleumier, Henson, Driver, & Dolan, 2002). Comparatively, VTC exhibits less tolerance to changes in exemplar viewpoint (Epstein, Graham, & Downing, 2003; Grill-Spector et al., 1999; Vuilleumier et al., 2002). Further, VTC responses are more tolerant to position than early visual areas (Grill-Spector, Kushnir, Hendler, et al., 1998), but not completely invariant to position as postulated by classic theories (Biederman, 1987). Indeed, VTC responses are modulated by eccentricity (Levy, Hasson, Avidan, Hendler, & Malach, 2001), are higher for contralateral than ipsilateral stimuli (Hemond, Kanwisher, & Op de Beeck, 2007; Rauschecker, Bowen, Parvizi, & Wandell, 2012; Sayres & Grill-Spector, 2008), and can be decoded to infer the category of the object across positions (Carlson, Hogendoorn, Fonteijn, & Verstraten, 2011). Together, these data provide evidence for separable object and transformation information in VTC.

Categorical Information in Ventral Temporal Cortex VTC representations also support visual categorization. Categorical information is identified at different spatial scales, from the level of distributed responses across the entire VTC (figure 35.1B, C; Haxby et al., 2001; Kriegeskorte et al., 2008), to activations in focal clusters (figure 35.2; Cohen et al., 2000; Epstein & Kanwisher, 1998; Kanwisher, McDermott, & Chun, 1997; Peelen & Downing, 2007), to responses of single neurons (Kreiman, Koch, & Fried, 2000). Regardless of the spatial scale of measurements, category information generalizes across exemplars and format (figure 35.2) and exhibits some tolerance to changes in position and size (Kanwisher, 2010), as well as a small range of views (Kietzmann, Swisher, Konig, & Tong, 2012).

Interestingly, categorical information in distributed VTC responses exhibits a hierarchical information structure. There is a broad distinction between animate versus inanimate categories, and there are additional levels of distinction within each domain (Kriegeskorte et al., 2008). For example, within the animate domain, there is separable information for faces and bodies (Kriegeskorte et al., 2008; Weiner & Grill-Spector, 2010), in addition to separable information for biological classes such as bugs, birds, and primates (Connolly et al., 2012; Haxby et al., 2011). Furthermore, within faces, there is some species-specific distinction between human and animal faces (Kriegeskorte et al., 2008). On the other hand, within the inanimate domain, there is

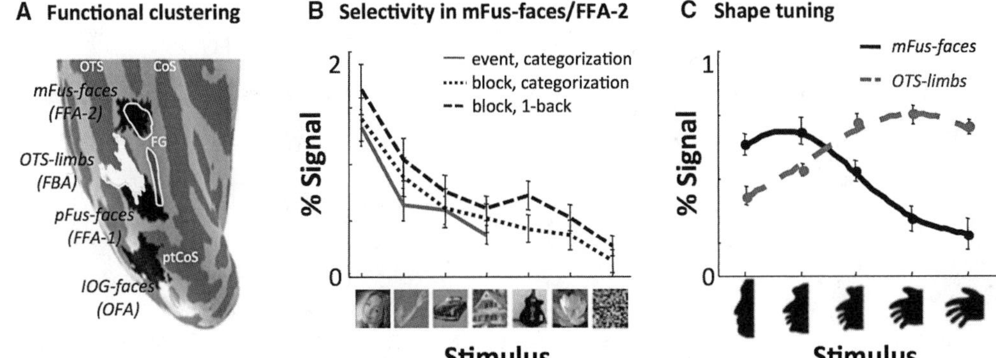

Figure 35.2 Focal categorical responses: selective and shape-tuned. (A) Example inflated right cortical surface zoomed on VTC. Face (black) and body (white) representations are arranged in focal clusters with predictable topological characteristics relative to one another and macroanatomical landmarks. CoS: collateral sulcus; FG: fusiform gyrus; OTS: occipitotemporal sulcus; MFS: mid-fusiform sulcus (white); ptCoS: posterior transverse collateral sulcus. (B) Responses (plus or minus standard error of the mean, SEM) of mFus-faces/(FFA-2) fusiform face area-2 averaged across seven subjects show reproducible preference for faces across tasks and paradigms. (Adapted from Weiner & Grill-Spector, 2010.) (C) Responses (±SEM) of mFus-faces/FFA-2 and OTS-limbs/fusiform body area (FBA) averaged across 11 subjects show differential shape tuning for a face-hand morph continuum. Selectivity is maintained across format changes. (Adapted from Davidenko et al., 2012.)

a distinction between objects and places (Kriegeskorte et al., 2008; Weiner et al., 2010), and within places, there is separable information for different types of scenes such as mountains or beaches (Walther, Caddigan, Fei-Fei, & Beck, 2009). Recent measurements using intracranial implanted electrodes in VTC of patients suggest that this categorical information is present within 150 ms after stimulus onset (Jacques et al., 2013; Liu et al., 2009), supporting the role of VTC in fast visual recognition.

Of the tens of thousands of categories in the world, VTC exhibits striking selectivity to only a handful of ecologically relevant categories: faces (Kanwisher et al., 1997), body parts (Peelen & Downing, 2007), places (Epstein & Kanwisher, 1998) and words (Cohen et al., 2000). Selective responses manifest as stronger responses to exemplars of these categories compared to exemplars of other categories within focal regions in VTC (figure 35.2), which are thought to contain a high proportion of neurons selective to these stimuli (Tsao, Freiwald, Tootell, & Livingstone, 2006). This strong selectivity profile has elicited a debate surrounding whether these activations reflect domain-specific processing of the preferred category (Kanwisher, 2010) or expertise (McGugin, Gatenby, Gore, & Gauthier, 2012), as well as a debate regarding the information in nonmaximal responses within these category-selective regions (Haxby et al., 2001; Weiner & Grill-Spector, 2010).

Ventral Temporal Cortex Responses Are Correlated with Perception By measuring brain responses when subjects viewed ambiguous, rivalrous, dichoptic, or masked stimuli, experimenters have shown that VTC responses (but not responses in early visual areas) are stronger and more reliable when subjects perceive objects, faces, and places, compared to when they are shown, but not perceived (Bar et al., 2001; Grill-Spector et al., 2000, 2004; Moutoussis & Zeki, 2002; Tong, Nakayama, Vaughan, & Kanwisher, 1998; Hasson, Hendler, Ben Bashat, & Malach, 2001). Further, activations in different VTC subregions are correlated with recognition of different categories (Grill-Spector et al., 2004; Moutoussis & Zeki, 2002; Tong et al., 1998). For example, responses in face-selective regions on the fusiform gyrus (FG) are strongest when a face is correctly identified, intermediate when a face is detected but not identified, and minimal when a face is presented, but missed (figure 35.3A). However, responses in this region do not correlate with the identification of inanimate categories such as guitars, houses, cars, or flowers. Instead, categorization and identification of these stimuli occurs in adjacent regions (figure 35.3B). Notably, the causal involvement of these face-selective FG regions in face perception has been recently demonstrated. Electrical brain stimulation of face-selective regions in a patient resulted in a specific distortion of face, but not object, perception (Parvizi et al., 2012).

Implementation: What is the relationship between the organizational and computational principles of human ventral temporal cortex?

The major implementation challenge of the brain is to arrange multidimensional representations onto the

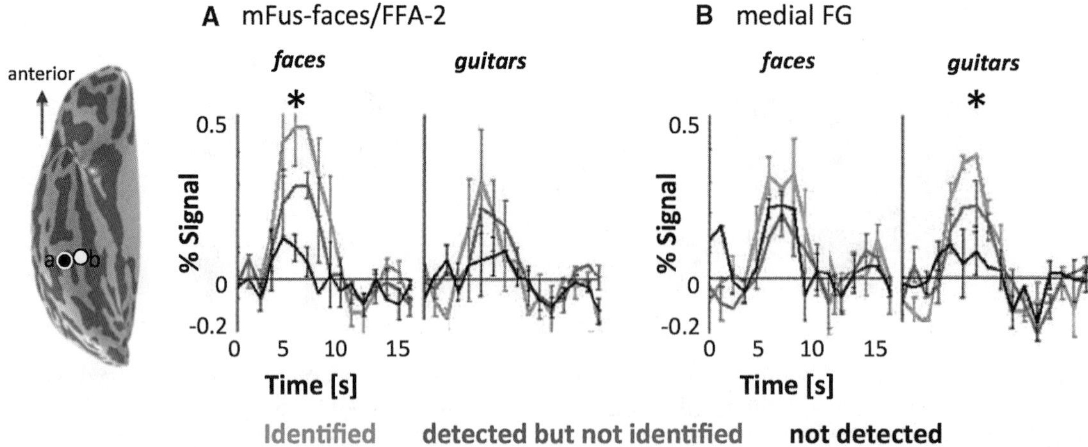

Figure 35.3 VTC responses correlate with successful identification and detection. Responses (±SEM) averaged across five subjects from (A) a 5 mm disk on mFus-faces/FFA-2 and (B) a 5 mm disk on an adjacent region on the medial FG during recognition of briefly presented and masked face or guitar images. Asterisk indicates that neural responses during identification (e.g., *this is Harrison Ford*) are significantly larger ($P < 0.01$) than during image detection. (Adapted from Grill-Spector et al., 2004.)

FIGURE 35.4 Functional architecture of VTC. (A) Several large-scale functional maps tile the cortical sheet with the same lateral-medial gradient. From left: domain specificity, animacy (the second principal component [PC2] from Haxby et al., 2011), and eccentricity bias on an example right hemisphere. (B) Fine-scale organization in a representative subject measured with 1.8 mm voxels. Functional regions are located in predictable cortical positions relative to macroanatomical landmarks and relative to one another. *White:* borders of retinotopic regions. (C) *Left:* inflated cortical surface with cytoarchitectonic regions of VTC in one representative postmortem subject showing the cytoarchitectonic transition between FG1/FG2 occurring within the MFS (black). *Right:* cytoarchitectonic characteristics of FG1 and FG2 (left) with profiles of gray level indices (right). FG1 displays a columnar arrangement of small pyramidal cells, while FG2 does not, and FG2 displays a higher neuronal density than FG1. (Adapted from Caspers et al., 2013, and Weiner et al., 2014.) (D) Correspondence between structural connectivity and functional division of VTC in a representative subject: thresholded boundaries of inferotemporal (dark red) and lingual (dark blue) connectivity are nearly identical to the faces vs. places functional map. (Adapted from Saygin et al., 2012.) (See color plate 34.)

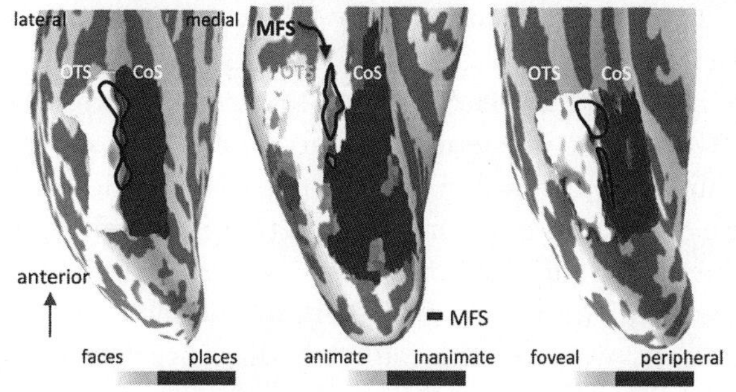

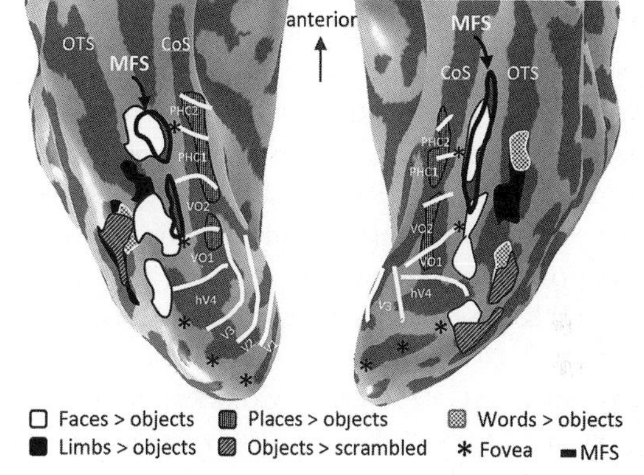

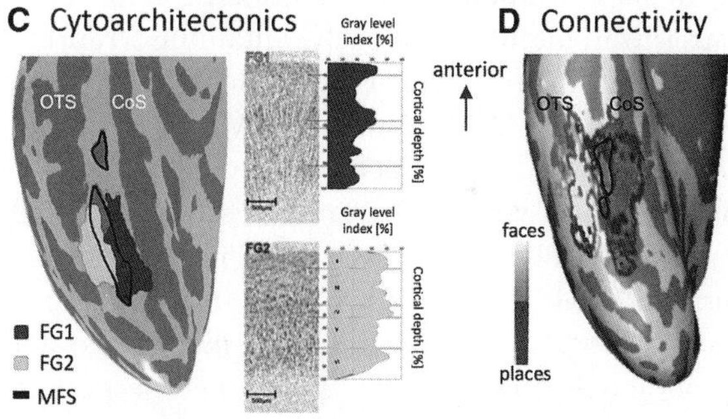

two-dimensional cortical sheet. In general, the brain employs three organizational solutions: clustering of neurons with similar properties, arranging information into large-scale maps, and superimposing multiple kinds of information on the cortical sheet. This organization can matriculate at various spatial scales: from the level of neurons (~1 μm), to columns (~0.2–1 mm), to areas (~1 cm), to the entire VTC (several cm). Next, we describe recent findings which have elucidated implementational features of VTC in the millimeter scale and larger.

LARGE-SCALE MAPS IN VENTRAL TEMPORAL CORTEX ARE ARRANGED IN A LATERAL-MEDIAL FUNCTIONAL GRADIENT TIGHTLY LINKED TO ANATOMY The organization of large-scale functional maps in VTC reveals three features (figure 35.4A): (1) several large-scale, superimposed functional maps tile the entire VTC; (2) maps display a common lateral-to-medial spatial arrangement on the cortical surface; and (3) this lateral-medial gradient is tied to anatomy in a specific way.

Large-scale functional maps in VTC range from representations differentiating animate from inanimate stimuli (Martin, 2007), faces from places (Nasr et al., 2011), and foveal from peripheral biases (Malach, Levy, & Hasson, 2002; figure 35.4A) to maps of animacy (Haxby et al., 2011; Connolly et al., 2012), semantics (Huth, Nishimoto, Vu, & Gallant, 2012), real-world object size (Konkle & Oliva, 2012), and the human body (Orlov, Makin, & Zohary, 2010). Notably, these maps have a common lateral-to-medial arrangement across VTC despite their different informational dimensions. For instance, animate, face, limb, central, and small-object representations manifest in lateral VTC, while inanimate, place, peripheral, torso, and large-object representations manifest in medial VTC (figure 35.4A). Consequently, the former representations converge on lateral VTC and diverge from those on medial VTC, and vice versa.

Not only are these maps aligned with respect to one another, but they are also aligned relative to the mid-fusiform sulcus (MFS; Weiner et al., 2014), which anatomically divides VTC into lateral and medial partitions (figure 35.4A). The positioning of the MFS in individual brains predicts the division of face-selective regions in the lateral FG from place-selective regions in the collateral sulcus (CoS; see figure 35.4A; Nasr et al., 2011;

Weiner et al., 2014), as well as the division of the eccentricity bias map, separating foveal biases in lateral VTC from peripheral biases in medial VTC (figure 35.4A; Weiner et al., 2014). Together, the large-scale maps in VTC show a consistent spatial arrangement (topology) relative to one another and the macroanatomical structure of VTC. In turn, this suggests that converging components of different functional representations may have a linking property (either computational or anatomical), whereas the lateral-medial divergence suggests that spatially distinct VTC components may have different computational goals and anatomical architecture.

Fine-Scale Clusters: Consistent Topological Arrangement with Various Levels of Convergence and Divergence Within each of the large-scale lateral and medial components of VTC, there is a series of fine-scale functional representations displaying both a tight coupling with anatomy as well as a consistent topology relative to other fine-scale clusters.

Medial Ventral Temporal Cortex: Superimposed Retinotopic Maps and Functional Clusters While VTC is traditionally described as nonretinotopic cortex, recent measurements have identified four retinotopic maps within medial VTC, each containing a representation of the contralateral visual field. These maps are arranged in two clusters in medial VTC (figure 35.4B): the ventral occipital (VO) cluster, VO-1 and VO-2 (Brewer, Liu, Wade, & Wandell, 2005; Wandell & Winawer, 2011), and the parahippocampal (PHC) cluster, PHC-1 and PHC-2 (Arcaro, McMains, Singer, & Kastner, 2009). These retinotopic maps have a tight coupling with anatomy: they are aligned perpendicular to the CoS, with their foveal representations on the medial FG, their peripheral representations on the CoS, and their posterior extent, defined by the hV4–VO1 boundary, located on the posterior transverse CoS (Winawer, Horiguchi, Sayres, Amano, & Wandell, 2010; Witthoft et al., 2013).

Likewise, researchers report a tight structural-functional link between place-selective activations (traditionally referred to as the parahippocampal place area, or PPA; Epstein & Kanwisher, 1998) and the CoS (Nasr et al., 2011; Weiner et al., 2010). While the PPA was initially described as located outside retinotopic cortex, recent data shows that place-selective activations on the CoS consistently overlap the peripheral representations of VO-2, PHC-1, and PHC-2 (Arcaro et al., 2009; figure 35.4B). Importantly, although there is not a one-to-one relationship between a retinotopic area (e.g., PHC-1) and a functional area (e.g., PPA), there is a consistent topological relationship among these functional representations, where place-selective activations converge with peripheral representations on the CoS, and diverge from foveal representations on the medial FG.

Understanding the organization of multiple fine-scale functional regions in medial VTC has important implications for information processing. First, position information in place-selective regions in medial VTC is a direct outcome of the overlap between the PPA and maps of the visual field. Second, convergence of place selectivity and retinotopy suggests that the brain may simultaneously process form and position information by superimposing them in the same cortical expanse. Third, position and place information are integrated in VO-2, PHC-1, and PHC-2, but are segregated anterior to PHC-2. Thus, there may be a functional gradient of category and retinotopic processing across what is considered to be a homogeneous region (the PPA).

Lateral Ventral Temporal Cortex: Predictable Topological Arrangement of Functional Clusters Selective to Faces, Body Parts, Words, and Objects Lateral VTC is characterized by multiple functional clusters selective to faces, body parts, words, and objects (figure 35.4B). Unlike medial VTC, retinotopic maps are yet to be identified in lateral VTC. However, there is a striking regularity in which functional clusters are arranged relative to one another and relative to anatomy. Just as the MFS predicts the large-scale division between face- and place-selective activations, the MFS predicts the positioning of two face-selective regions on the lateral FG (Weiner & Grill-Spector, 2010; Weiner et al., 2014). Thus, instead of the one homogenous Fusiform Face Area (FFA) suggested by classical work (Kanwisher et al., 1997), there are two face-selective regions in the lateral FG in predictable anatomical locations: one overlapping the anterolateral tip of the MFS (mFus-faces/FFA-2) and one 10–15 mm more posterior (pFus-faces/FFA-1) overlapping the posterolateral tip of the MFS and extending into the occipitotemporal sulcus (OTS). These face-selective clusters are separated by a body part–selective region, which is positioned between these two face-selective clusters and extends laterally into the OTS (Weiner & Grill-Spector, 2010, 2012, 2013). Additionally, object-selective and word-selective activations overlap the posterior FG and OTS and are lateral to face-selective activations, where the object-selective activation extends more posteriorly and laterally than the word-selective activation.

This organization structure has implications for information processing in lateral VTC. For instance, object and word activations tend to converge (Ben-Shachar,

Dougherty, Deutsch, & Wandell, 2007), indicating common processing (e.g., domain-general processing producing tolerance). However, face- and limb-selective activations diverge from one another, indicating segregated processing of different parts of the body. Face and limb activations are also components of a larger topographic representation of the human body extending into the lateral occipitotemporal cortex, suggesting that cortical representations may have incorporated the statistical regularity of the human body (Orlov et al., 2010; Weiner & Grill-Spector, 2013).

Discussion

Overall, this chapter links the information and computations performed by VTC to its neural implementation, which reveals two prominent characteristics: an orderly topological pattern of convergence and divergence among representations, and a spatial hierarchy of nested functional representations. *What may constrain this organization, and how does it generate an efficient perceptual system?* We hypothesize that (1) the interplay among microarchitecture, connectivity, and functional needs operate as mutual constraints underlying the functional organization of VTC, (2) the regular topological convergence and divergence of functional representations in VTC enables rapid perception, and (3) the hierarchical spatial structure of the functional organization of VTC supports its hierarchical information structure.

CYTOARCHITECTURE AND CONNECTIVITY: ANATOMICAL CONSTRAINTS WITH FUNCTIONAL BENEFITS The direct relationship among cytoarchitectonics, connectivity, and functional regions in human VTC is unknown. Nevertheless, recent breakthroughs in linking functional divisions of VTC to its cytoarchitectonic and connectivity structure may provide important insights. The MFS is not only the boundary between functional transitions in VTC (figure 35.4A, B), but it also aligns with a cytoarchitectonic division of the FG (figure 35.4C; Caspers et al., 2013; Weiner et al., 2014), as well as with a lateral-medial division of white matter connections (Saygin et al., 2012). Since cytoarchitectonic, connectivity, and large-scale functional divisions occupy the same macroanatomical expanse on either side of the MFS, it suggests the intriguing hypothesis that the ubiquitous lateral-medial functional division of VTC may be partially constrained by underlying cytoarchitectonics and white matter connections.

CONVERGENCE AND DIVERGENCE EXPEDITE CORTICAL PROCESSING BY REDUCING COMMUNICATION TIME AND PROCESSING INDEPENDENT INFORMATION IN PARALLEL We have shown that there is an orderly spatial arrangement among multiple large-scale maps and fine-scale functional regions that generates predictable convergences and divergences among neural representations in VTC. We propose that the convergent and divergent properties of this functional architecture provide an anatomical substrate for fast and efficient visual processing (Van Essen, Anderson, & Felleman, 1992; Zeki & Shipp, 1988). Convergence may reduce wiring costs and speed up neural communication by placing neurons that process related information in close spatial proximity. Thus, clustering face-selective neurons may enable fast communication among neurons coding facial features, consequently expediting facial recognition. Divergence may expedite computations by providing an anatomical substrate for parallel processing of independent information in segregated cortical circuits. Thus, the segregated cytoarchitectonic and connectivity structure of lateral and medial VTC may allow parallel processing of independent components of the visual scene such as identity and place information. For example, identity information (e.g. Kevin not Jim) may be processed in lateral VTC, and place information (e.g. beach not office) may be processed in medial VTC.

INTRIGUING HYPOTHESIS: HIERARCHICAL SPATIAL STRUCTURE OF FUNCTIONAL REPRESENTATIONS IN VENTRAL TEMPORAL CORTEX SUPPORTS A HIERARCHICAL INFORMATION STRUCTURE We propose the intriguing hypothesis that the spatial hierarchy of nested functional representations at various spatial scales in VTC may serve as the scaffolding supporting the hierarchical information structure of VTC (figure 35.1B, C). Further, the spatial structure may extend to other spatial scales beyond the ones described here (e.g., the smaller spatial scale of a cortical column; Borra, Ichinohe, Sato, Tanifuji, & Rockland, 2010; Fujita, Tanaka, Ito, & Cheng, 1992; Tanaka, Saito, Fukada, & Moriya, 1991). This proposition is consistent with hierarchical computational models of visual recognition. These models are composed of a series of layers that process increasingly more complex information, from simple features (e.g., angled lines), to intermediate complexity features (e.g., face or object fragments), to exemplars, and finally, to categories (Epshtein, Lifshitz, & Ullman, 2008; Riesenhuber & Poggio, 1999; Serre, Oliva, & Poggio, 2007). These hierarchical models are robust recognition systems that are tolerant to changes in appearance and enable extraction of different levels of information from distinct levels of the hierarchy. While researchers have traditionally mapped these hierarchies across the entire ventral visual pathway, we speculate that some aspect of this computational hierarchy may be implemented across different spatial scales of the representational hierarchy within VTC itself (figure 35.1C). Cortical columns may represent intermediate complexity features (e.g., object fragments; Tanaka, 1996; Ullman, Vidal-Naquet, & Sali, 2002); columns containing features shared by exemplars of a category may then be clustered into larger regions (Tanaka, 1996; Tsao et al., 2006) generating domain-specific representations; and these may be arranged together with other clusters to form large-scale maps containing broader categorical distinctions. Thus, the spatial hierarchy of VTC representations may provide the flexibility needed for the visual system to access multiple levels of information according to task demands, by reading out information from different spatial scales of representations in VTC.

Conclusion

We have adapted classic frameworks in vision science (Hubel & Wiesel, 1962; Marr, 1982) to examine how the functional architecture of VTC contributes to the speed and flexibility of visual perception as well as how the neural implementation generates a hierarchical representational structure of exemplar and category information that is useful for recognition. While it is debatable which part of the functional organization of VTC is the most important for perception, the complementary nature of functional representations across spatial scales is overwhelmingly clear. Future research examining the interplay among cytoarchitectonics, connectivity, functional representations, and computational theories will assuredly advance our understanding of the neural bases of visual perception.

ACKNOWLEDGMENTS We thank Julian Caspers, Andy Connolly, Nick Davidenko, Winrich Freiwald, Elias Issa, James Haxby, Niko Krigeskorte, Zeynep Saygin, and Keiji Tanaka for contributing to the figures published in this chapter. This research was funded by NSF BCS0920865 and NIH 1 RO1 EY 02231801A11 and R01 EY019279-01A.

REFERENCES

Arcaro, M. J., McMains, S. A., Singer, B. D., & Kastner, S. (2009). Retinotopic organization of human ventral visual cortex. *J Neurosci, 29*, 10638–10652.

Avidan, G., Harel, M., Hendler, T., Ben-Bashat, D., Zohary, E., & Malach, R. (2002). Contrast sensitivity in human visual areas and its relationship to object recognition. *J Neurophysiol, 87*, 3102–3116.

Bar, M., Tootell, R. B., Schacter, D. L., Greve, D. N., Fischl, B., Mendola, J. D., ... Dale, A. M. (2001). Cortical mechanisms specific to explicit visual object recognition. *Neuron, 29*, 529–535.

Ben-Shachar, M., Dougherty, R. F., Deutsch, G. K., & Wandell, B. A. (2007). Differential sensitivity to words and shapes in ventral occipito-temporal cortex. *Cereb cortex, 17*(7), 1604–1611.

Biederman, I. (1987). Recognition-by-components: A theory of human image understanding. *Psychol Rev, 94*, 115–147.

Borra, E., Ichinohe, N., Sato, T., Tanifuji, M., & Rockland, K. S. (2010). Cortical connections to area TE in monkey: Hybrid modular and distributed organization. *Cereb Cortex, 20*, 257–270.

Brewer, A. A., Liu, J., Wade, A. R., & Wandell, B. A. (2005). Visual field maps and stimulus selectivity in human ventral occipital cortex. *Nat Neurosci, 8*, 1102–1109.

Carlson, T., Hogendoorn, H., Fonteijn, H., & Verstraten, F. A. (2011). Spatial coding and invariance in object-selective cortex. *Cortex, 47*, 14–22.

Caspers, J., Zilles, K., Eickhoff, S. B., Schleicher, A., Mohlberg, H., & Amunts, K. (2013). Cytoarchitectonical analysis and probabilistic mapping of two extrastriate areas of the human posterior fusiform gyrus. *Brain Struct Funct, 218*, 511–526.

Cohen, L., Dehaene, S., Naccache, L., Lehericy, S., Dehaene-Lambertz, G., Hénaff, M. A., & Michel, F. (2000). The visual word form area: Spatial and temporal characterization of an initial stage of reading in normal subjects and posterior split-brain patients. *Brain, 123*, 291–307.

Connolly, A. C., Guntupalli, J. S., Gors, J., Hanke, M., Halchenko, Y. O., Wu, Y. C., ... Haxby, J. V. (2012). The representation of biological classes in the human brain. *J Neurosci, 32*, 2608–2618.

Davidenko, N., Remus, D. A., & Grill-Spector, K. (2012). Face-likeness and image variability drive responses in human face-selective ventral regions. *Hum Brain Mapp, 33*(10), 2334–2349.

DiCarlo, J. J., & Cox, D. D. (2007). Untangling invariant object recognition. *Trends Cogn Sci, 11*, 333–341.

DiCarlo, J. J., Zoccolan, D., & Rust, N. C. (2012). How does the brain solve visual object recognition? *Neuron, 73*, 415–434.

Epshtein, B., Lifshitz, I., & Ullman, S. (2008). Image interpretation by a single bottom-up top-down cycle. *Proc Natl Acad Sci USA, 105*, 14298–14303.

Epstein, R., Graham, K. S., & Downing, P. E. (2003). Viewpoint-specific scene representations in human parahippocampal cortex. *Neuron, 37*, 865–876.

Epstein, R., & Kanwisher, N. (1998). A cortical representation of the local visual environment. *Nature, 392*, 598–601.

Freiwald, W. A., & Tsao, D. Y. (2010). Functional compartmentalization and viewpoint generalization within the macaque face-processing system. *Science, 330*, 845–851.

Fujita, I., Tanaka, K., Ito, M., & Cheng, K. (1992). Columns for visual features of objects in monkey inferotemporal cortex. *Nature, 360*, 343–346.

Grill-Spector, K., Henson, R., & Martin, A. (2006). Repetition and the brain: Neural models of stimulus-specific effects. *Trends Cogn Sci, 10*, 14–23.

Grill-Spector, K., Knouf, N., & Kanwisher, N. (2004). The fusiform face area subserves face perception, not generic within-category identification. *Nat Neurosci, 7*, 555–562.

Grill-Spector, K., Kushnir, T., Edelman, S., Avidan, G., Itzchak, Y., & Malach, R. (1999). Differential processing of objects under various viewing conditions in the human lateral occipital complex. *Neuron, 24,* 187–203.

Grill-Spector, K., Kushnir, T., Edelman, S., Itzchak, Y., & Malach, R. (1998). Cue-invariant activation in object-related areas of the human occipital lobe. *Neuron, 21,* 191–202.

Grill-Spector, K., Kushnir, T., Hendler, T., Edelman, S., Itzchak, Y., & Malach, R. (1998). A sequence of object-processing stages revealed by fMRI in the human occipital lobe. *Hum Brain Mapp, 6,* 316–328.

Grill-Spector, K., Kushnir, T., Hendler, T., & Malach, R. (2000). The dynamics of object-selective activation correlate with recognition performance in humans. *Nat Neurosci, 3,* 837–843.

Hasson, U., Hendler, T., Ben Bashat, D., & Malach, R. (2001). Vase or face? A neural correlate of shape-selective grouping processes in the human brain. *J Cogn Neurosci, 13,* 744–753.

Haxby, J. V., Gobbini, M. I., Furey, M. L., Ishai, A., Schouten, J. L., & Pietrini, P. (2001). Distributed and overlapping representations of faces and objects in ventral temporal cortex. *Science, 293,* 2425–2430.

Haxby, J. V., Guntupalli, J. S., Connolly, A. C., Halchenko, Y. O., Conroy, B. R., Gobbini, M. I., ... Ramadge, P. J. (2011). A common, high-dimensional model of the representational space in human ventral temporal cortex. *Neuron, 72,* 404–416.

Hemond, C. C., Kanwisher, N. G., & Op de Beeck, H. P. (2007). A preference for contralateral stimuli in human object- and face-selective cortex. *PLoS One, 2,* e574.

Hubel, D. H., & Wiesel, T. N. (1962). Receptive fields, binocular interaction and functional architecture in the cat's visual cortex. *J Physiol, 160,* 106–154.

Huth, A. G., Nishimoto, S., Vu, A. T., & Gallant, J. L. (2012). A continuous semantic space describes the representation of thousands of object and action categories across the human brain. *Neuron, 76,* 1210–1224.

Issa, E. B., & DiCarlo, J. J. (2012). Precedence of the eye region in neural processing of faces. *J Neurosci, 32,* 16666–16682.

Jacques, C., Witthoft, N., Weiner, K. S., Foster, B. L., Miller, K. J., Hermes, D., ... Grill-Spector, K. (2013). Electrocorticography of category-selectivity in human ventral temporal cortex: Spatial organization, responses to single images, and coupling with fMRI. *J Vis, 13*(9), 495.

Jiang, X., Bradley, E., Rini, R. A., Zeffiro, T., Vanmeter, J., & Riesenhuber, M. (2007). Categorization training results in shape- and category-selective human neural plasticity. *Neuron, 53,* 891–903.

Kanwisher, N. (2010). Functional specificity in the human brain: A window into the functional architecture of the mind. *Proc Natl Acad Sci USA, 107,* 11163–11170.

Kanwisher, N., McDermott, J., & Chun, M. M. (1997). The fusiform face area: A module in human extrastriate cortex specialized for face perception. *J Neurosci, 17,* 4302–4311.

Kietzmann, T. C., Swisher, J. D., Konig, P., & Tong, F. (2012). Prevalence of selectivity for mirror-symmetric views of faces in the ventral and dorsal visual pathways. *J Neurosci, 32,* 11763–11772.

Konkle, T., & Oliva, A. (2012). A real-world size organization of object responses in occipitotemporal cortex. *Neuron, 74,* 1114–1124.

Kourtzi, Z., & Kanwisher, N. (2001). Representation of perceived object shape by the human lateral occipital complex. *Science, 293,* 1506–1509.

Kreiman, G., Koch, C., & Fried, I. (2000). Category-specific visual responses of single neurons in the human medial temporal lobe. *Nat Neurosci, 3,* 946–953.

Kriegeskorte, N., Mur, M., Ruff, D. A., Kiani, R., Bodurka, J., Esteky, H., ... Bandettini, P. A. (2008). Matching categorical object representations in inferior temporal cortex of man and monkey. *Neuron, 60,* 1126–1141.

Levy, I., Hasson, U., Avidan, G., Hendler, T., & Malach, R. (2001). Center-periphery organization of human object areas. *Nat Neurosci, 4,* 533–539.

Liu, H., Agam, Y., Madsen, J. R., & Kreiman, G. (2009). Timing, timing, timing: Fast decoding of object information from intracranial field potentials in human visual cortex. *Neuron, 62,* 281–290.

Malach, R., Levy, I., & Hasson, U. (2002). The topography of high-order human object areas. *Trends Cogn Sci, 6,* 176–184.

Malach, R., Reppas, J. B., Benson, R. R., Kwong, K. K., Jiang, H., Kennedy, W. A., ... Tootell, R. B. (1995). Object-related activity revealed by functional magnetic resonance imaging in human occipital cortex. *Proc Natl Acad Sci USA, 92,* 8135–8139.

Marr, D. (1982). *Vision: A computational approach.* San Francisco, CA: Freeman.

Martin, A. (2007). The representation of object concepts in the brain. *Annu Rev Psychol, 58,* 25–45.

McGugin, R. W., Gatenby, J. C., Gore, J. C., & Gauthier, I. (2012). High-resolution imaging of expertise reveals reliable object selectivity in the fusiform face area related to perceptual performance. *Proc Natl Acad Sci USA, 109,* 17063–17068.

Mendola, J. D., Dale, A. M., Fischl, B., Liu, A. K., & Tootell, R. B. (1999). The representation of illusory and real contours in human cortical visual areas revealed by functional magnetic resonance imaging. *J Neurosci, 19,* 8560–8572.

Moutoussis, K., & Zeki, S. (2002). The relationship between cortical activation and perception investigated with invisible stimuli. *Proc Natl Acad Sci USA, 99,* 9527–9532.

Nasr, S., Liu, N., Devaney, K. J., Yue, X., Rajimehr, R., Ungerleider, L. G., & Tootell, R. B. (2011). Scene-selective cortical regions in human and nonhuman primates. *J Neurosci, 31,* 13771–13785.

Orlov, T., Makin, T. R., & Zohary, E. (2010). Topographic representation of the human body in the occipitotemporal cortex. *Neuron, 68,* 586–600.

Parvizi, J., Jacques, C., Foster, B. L., Witthoft, N., Rangarajan, V., Weiner, K. S., & Grill-Spector, K. (2012). Electrical stimulation of human fusiform face-selective regions distorts face perception. *J Neurosci, 32,* 14915–14920.

Peelen, M. V., & Downing, P. E. (2007). The neural basis of visual body perception. *Nat Rev Neurosci, 8,* 636–648.

Rauschecker, A. M., Bowen, R. F., Parvizi, J., & Wandell, B. A. (2012). Position sensitivity in the visual word form area. *Proc Natl Acad Sci USA, 109,* E1568–E1577.

Riesenhuber, M., & Poggio, T. (1999). Hierarchical models of object recognition in cortex. *Nat Neurosci, 2,* 1019–1025.

Saygin, Z. M., Osher, D. E., Koldewyn, K., Reynolds, G., Gabrieli, J. D., & Saxe, R. R. (2012). Anatomical connectivity patterns predict face selectivity in the fusiform gyrus. *Nat Neurosci, 15,* 321–327.

Sayres, R., & Grill-Spector, K. (2008). Relating retinotopic and object-selective responses in human lateral occipital cortex. *J Neurophysiol, 100*(1), 249–267.

Serre, T., Oliva, A., & Poggio, T. (2007). A feedforward architecture accounts for rapid categorization. *Proc Natl Acad Sci USA, 104,* 6424–6429.

Tanaka, K. (1996). Inferotemporal cortex and object vision. *Annu Rev Neurosci, 19,* 109–139.

Tanaka, K., Saito, H., Fukada, Y., & Moriya, M. (1991). Coding visual images of objects in the inferotemporal cortex of the macaque monkey. *J Neurophysiol, 66,* 170–189.

Thorpe, S., Fize, D., & Marlot, C. (1996). Speed of processing in the human visual system. *Nature, 381,* 520–522.

Tong, F., Nakayama, K., Vaughan, J. T., & Kanwisher, N. (1998). Binocular rivalry and visual awareness in human extrastriate cortex. *Neuron, 21,* 753–759.

Tsao, D. Y., Freiwald, W. A., Tootell, R. B., & Livingstone, M. S. (2006). A cortical region consisting entirely of face-selective cells. *Science, 311,* 670–674.

Ullman, S., Vidal-Naquet, M., & Sali, E. (2002). Visual features of intermediate complexity and their use in classification. *Nat Neurosci, 5,* 682–687.

Ungerleider, L. G., & Haxby, J. V. (1994). "What" and "where" in the human brain. *Curr Opin Neurobiol, 4,* 157–165.

Van Essen, D. C., Anderson, C. H., & Felleman, D. J. (1992). Information processing in the primate visual system: An integrated systems perspective. *Science, 255,* 419–423.

Vinberg, J., & Grill-Spector, K. (2008). Representation of shapes, edges, and surfaces across multiple cues in the human visual cortex. *J Neurophysiol, 99*(3), 1380–1393.

Vuilleumier, P., Henson, R. N., Driver, J., & Dolan, R. J. (2002). Multiple levels of visual object constancy revealed by event-related fMRI of repetition priming. *Nat Neurosci, 5,* 491–499.

Walther, D. B., Caddigan, E., Fei-Fei, L., & Beck, D. M. (2009). Natural scene categories revealed in distributed patterns of activity in the human brain. *J Neurosci, 29,* 10573–10581.

Wandell, B. A., & Winawer, J. (2011). Imaging retinotopic maps in the human brain. *Vis Res, 51*(7), 718–737.

Weiner, K. S., & Grill-Spector, K. (2010). Sparsely-distributed organization of face and limb activations in human ventral temporal cortex. *NeuroImage, 52,* 1559–1573.

Weiner, K. S., & Grill-Spector, K. (2012). Improbable simplicity of the fusiform face area. *Trends Cogn Sci, 16,* 251–254.

Weiner, K. S., & Grill-Spector, K. (2013). Neural representations of faces and limbs neighbor in human high-level visual cortex: Evidence for a new organization principle. *Psychol Res, 77,* 74–97.

Weiner, K. S., Golarai, G., Caspers, J., Mohlberg, H., Zilles, K., Amunts, K., & Grill-Spector, K. (2014). The mid-fusiform sulcus: A landmark identifying both cytoarchitectonic and functional divisions of human ventral temporal cortex. *NeuroImage, 84,* 453–465.

Weiner, K. S., Sayres, R., Vinberg, J., & Grill-Spector, K. (2010). MRI-adaptation and category selectivity in human ventral temporal cortex: Regional differences across time scales. *J Neurophysiol, 103,* 3349–3365.

Winawer, J., Horiguchi, H., Sayres, R. A., Amano, K., & Wandell, B. A. (2010). Mapping hV4 and ventral occipital cortex: The venous eclipse. *J Vis, 10*(5), 1.

Witthoft, N., Nguyen, M. L., Golarai, G., LaRocque, K., Liberman, A., & Grill-Spector, K. (2013). Where is human V4? Predicting the location of hV4 and VO1 from cortical folding. *Cereb Cortex.* doi: 10.1093/cercor/bht092.

Zeki, S., & Shipp, S. (1988). The functional logic of cortical connections. *Nature, 335,* 311–317.

V

MOTOR SYSTEMS AND ACTION

Introduction

SCOTT T. GRAFTON AND PETER L. STRICK

In this edition, we emphasize that any meaningful investigation of motor systems must go far beyond the science of understanding how physical movement is generated and controlled. Whether we are seeking a reward, finding a mate, or building a shelter, humans along with many other complex organisms move to accomplish goal-directed actions. Many of the fundamental ideas in movement science, including optimality principles, learning mechanisms, and trade-offs in the evolutionary design of the physical plant, can only be interpreted when they are evaluated within this larger domain of goal-directed action. With this in mind, the chapters in this section provide a rich survey of some of the fundamental issues in understanding complex motor behavior. The chapters range from detailed summaries of motor system anatomy and physiology to the choice as well as the formation of specific motor behaviors.

The section begins with an "update" on the anatomic organization of motor circuits involving the basal ganglia and cerebellum by Richard Dum, Andreea Bostan, and Peter Strick. While the presence of cortical-subcortical loops through either the basal ganglia or cerebellum is well established, these circuits have historically been assumed to be independent, nonoverlapping pathways. Recent studies using viral tracing methods in nonhuman primates challenge this assumption. There are significant interconnections between the two subcortical nuclei. These interconnections are intriguing from a number of functional perspectives. The basal ganglia and cerebellum may provide different mechanisms to support supervised learning, and the interconnections between the structures may enable the algorithms of one structure to improve the outcomes of the other. On the other hand, in patients with Parkinson's, many neuroimaging studies have shown augmented cerebellar activity that normalizes with dopamine therapy, pallidotomy, and deep brain stimulation. Thus, it is possible that the heightened cerebellar activity contributes to the motor dysfunction associated with the disease.

The detailed analysis of motor behavior in terms of kinematics, coupled with neuronal recordings, provides a robust platform for understanding what role the basal ganglia play in controlling or shaping motor behavior. In their chapter on the basal ganglia, Rob Turner and Benjamin Pasquereau provide a detailed critique of some of the classic basal ganglia functions, including skill storage and movement selection. Based on neuronal recordings as well as inferences that can be drawn from patients, they shift the focus to other functions, including skill acquisition through supervised learning and the control of movement "vigor."

Continuing with the theme of subcortical circuits, Joern Diedrichsen and Amy Bastian offer an invaluable primer on cerebellar anatomy and functionality. They draw from a rich history of clinical observations in patients with cerebellar lesions to set a framework for understanding some of the basic ways in which the cerebellum facilitates ongoing movement. They discuss evidence that the cerebellum is involved in the formation of an internal model for motor behavior. They also consider the role of the cerebellum in adaptation for a broad range of motor behaviors, including movement through force fields or walking on split treadmills.

When faced with unexpected perturbations, such as a bump to the arm during a reach, humans are capable of extremely rapid movement corrections to stay on target. In their chapter, Frédéric Crevecoeur, Tyler Cluff, and Steve Scott link these rapid corrections to long loop reflexes generated in part by the primary motor cortex. They extend the classic single joint studies of Evarts to show how corrections are distributed over multiple joints. Perhaps more important, they summarize the basic principles of optimal control theory and use this framework to interpret the computations performed by the primary motor cortex during error correction.

Since the last edition of this book, there has been a rapid development of brain-machine interfaces that can generate movement through robotic systems or the electrical stimulation to peripheral muscles. These systems hold enormous promise for enhancing motor function with spinal cord or brainstem damage, motor neuron disease, and other forms of paralysis. Many of these systems rely on grid electrode arrays capable of multi-unit recording. In his chapter, Aaron Batista reviews the different ways these rich data sets can be analyzed. Emphasis is placed on the degree to which different neurons can generate coordinated activity patterns. These correlation methods and their analogs in functional imaging will be increasingly important throughout cognitive neuroscience.

Ultimately, motor cortex is involved in the generation of descending commands to the spinal cord to generate a repertoire of complex actions. In his chapter, Roger Lemon considers how the cortex might represent action vocabularies. Using object grasping and manipulation as an experimental framework, he evaluates current theories about how the motor cortex actually *causes* movement. The versatility of grasping is exceptional, and cannot be explained by the simple recombination of a few muscle synergies. Lemon summarizes how the diversity of motor cortex activity for grasping is shaped within a larger visuomotor grasping circuit involving premotor and parietal systems.

Actions are a result of a sophisticated decision-making process. Organisms must be able to effectively navigate among action choices to survive. In their chapter, Rushworth and colleagues consider the neural systems that allow for the estimation of value among competing action choices, risks, and rewards. They review evidence from lesion studies, single neuronal recording, and functional MRI to show how cortical areas performing different computations result in the selection of particular behaviors.

The section closes with a chapter by Scott Frey and Scott Grafton that restages the notion of affordance in a modern conceptual framework. They introduce the theory of reciprocal affordance. This proposes that the way an animal perceives the world is closely intertwined with the way in which its motor system can form actions in the environment. Having an internal model of what is graspable provides a neuronal mechanism for automatic, preparatory motor planning from available stored action vocabularies. This framework is particularly useful for understanding how the nervous system adapts perception through physical experience, how the body schema can be reorganized in the setting of amputation, and how the design of brain-machine interfaces might be improved.

36 Basal Ganglia and Cerebellar Circuits with the Cerebral Cortex

RICHARD P. DUM, ANDREEA C. BOSTAN, AND PETER L. STRICK

ABSTRACT The neural connections of the basal ganglia and cerebellum provide important insights into their function. The outputs of the basal ganglia and cerebellum target motor, premotor, prefrontal, posterior parietal, and inferotemporal areas of the cerebral cortex. In addition, the basal ganglia and cerebellum are interconnected. These basal ganglia and cerebellar circuits form the anatomical substrates for influencing not only the control of movement, but also cognitive behaviors like planning, working memory, reward-based learning, sequential behavior, visuospatial perception, and attention. Similarly, abnormal activity in specific basal ganglia and cerebellar circuits may contribute to a variety of neuropsychiatric disorders, including autism, addiction, attention-deficit/hyperactivity disorder, obsessive-compulsive disorder, and Tourette syndrome. Thus, the anatomical substrate exists for the basal ganglia and cerebellum to be involved in a broad range of motor and nonmotor functions.

What are the functions of the basal ganglia and cerebellum? Numerous reports describe the motor deficits associated with damage to these subcortical structures. As a consequence, concepts about basal ganglia and cerebellar function have focused primarily on their contributions to the generation and control of movement. Examination of the macro-organization of basal ganglia and cerebellar connections with the cerebral cortex has led to a different perspective. In this chapter, we focus on two critical questions. First, which cortical areas are the target of the outputs from the basal ganglia and the cerebellum? Second, do the basal ganglia and cerebellum interact at sites other than the cerebral cortex? The answers to these questions lead to some novel and important insights about basal ganglia and cerebellar function.

Classically, the macro-organization of basal ganglia and cerebellar circuitry is described using a relatively simple hierarchical model. The "input layer" of basal ganglia processing is represented by the striatum (caudate, putamen, and ventral striatum). The analogous level in cerebellar circuits is represented by specific pontine nuclei that send mossy fiber inputs to cerebellar cortex. A major source of afferents to the input layers of both circuits originates from widespread regions of the cerebral cortex, including motor, sensory, posterior parietal, prefrontal, cingulate, orbitofrontal, and temporal cortical areas. The "output layer" of basal ganglia processing is represented by the internal segment of the globus pallidus (GPi), the pars reticulata of the substantia nigra (SNpr), and the ventral pallidum. The comparable structures for cerebellar processing are the deep cerebellar nuclei: dentate, interpositus, and fastigial. Neurons in the output layers of both circuits send their axons to the thalamus and, by this route, project back upon the cortex. Thus, a major structural feature of basal ganglia and cerebellar circuits is that they form loops with the cerebral cortex (e.g., Allen & Tsukahara, 1974; Brooks & Thach, 1981; Kemp & Powell, 1971). These loops were thought to function largely in the domain of motor control by funneling information from diverse cortical areas to influence motor output at the level of the primary motor cortex (M1). This view has been supported by the obvious motor symptoms that can result from basal ganglia and cerebellar dysfunction (for reviews, see Bhatia & Marsden, 1994; Brooks & Thach, 1981; DeLong & Georgopoulos, 1981).

Over the past 30 years, an accumulation of information about basal ganglia and cerebellar anatomy has led a number of investigators to challenge this view (e.g., Alexander, DeLong, & Strick, 1986; Goldman-Rakic & Selemon, 1990; Schell & Strick, 1984). It is now clear that basal ganglia and cerebellar efferents terminate in different subdivisions of the ventrolateral thalamus (for a review, see Percheron, François, Talbi, Yelnik, & Fénelon, 1996) which, in turn, project to a myriad of cortical areas. Thus, the outputs from the basal ganglia and cerebellum influence more widespread regions of the cerebral cortex than previously recognized.

Based on these and other anatomical results, Alexander et al. (1986) proposed that the basal ganglia participate in at least five separate loops with the cerebral cortex. These loops were based in part on their cortical targets of the output layer of processing and were designated the skeletomotor, oculomotor, dorsolateral prefrontal, lateral orbitofrontal, and anterior cingulate circuits. According to this scheme, the output of the basal ganglia has the potential to influence not only the control of movement, but also higher-order cognitive and limbic functions subserved by prefrontal, orbitofrontal, and anterior cingulate cortex.

Similarly, Leiner, Leiner, and Dow (1986, 1991, 1993) suggested that cerebellar output is directed to prefrontal as well as to motor areas of the cerebral cortex. They noted that in the course of hominid evolution, the lateral output nucleus of the cerebellum—the dentate—undergoes a marked expansion that parallels the expansion of cerebral cortex in the frontal lobe. They argued that the increase in the size of the dentate is accompanied by an increase in the extent of the cortical areas in the frontal lobe that are influenced by dentate output. As a consequence, Leiner et al. (1986, 1991, 1993) proposed that cerebellar function in humans has expanded to include involvement in certain language and cognitive tasks.

The use of neurotropic viruses (herpes simplex virus type 1 [HSV1] and rabies virus) as transneuronal tracers has been essential for testing these proposals (for review, see Kelly & Strick, 2000, 2003; Strick & Card, 1992; Strick, Dum, & Fiez, 2009). This tracing method can effectively label a chain of up to three synaptically linked neurons in a single experiment (Bostan, Dum, & Strick, 2010; Kelly & Strick, 2003, 2004). In this chapter, we review some of the new observations that result from using viruses to map the basal ganglia and cerebellar projections to the cerebral cortex. These observations have led to important insights about the cortical targets of these circuits and the functional domains they influence.

Primary motor cortex

Retrograde transneuronal transport of HSV1 was used to examine the organization of basal ganglia and cerebellar outputs to M1 (Hoover & Strick, 1993, 1999). In separate animals, virus was injected into M1 regions where face, arm, or leg movements were evoked by intracortical stimulation. The survival time was set to allow retrograde transport of the virus to "first-order" neurons in the thalamus and then retrograde transneuronal transport of virus from these first-order neurons to "second-order" neurons that are the origin of basal ganglia– and cerebello-thalamocortical inputs to M1 (Kelly & Strick, 2000; Strick & Card, 1992).

These experiments produced three major results. First, M1 is richly innervated by the output nuclei of the basal ganglia and cerebellum. The densest projections originate from GPi (figures 36.1 and 36.2) and the dentate (figures 36.3 and 36.4). Less dense projections originate from portions of SNpr and interpositus. Second, both GPi and the dentate are somatotopically organized with separate face, arm, and leg areas that project via the thalamus to the face, arm, and leg areas of M1. Third, and perhaps most surprising, the projections to M1 originate from only 15% of the volume of GPi and about 30% of the volume of the dentate. Furthermore, the output to M1 originates from restricted portions of each subcortical nucleus. These results imply that the majority of the output from the basal ganglia and cerebellum is directed to other cortical areas.

Premotor areas

Basal ganglia and cerebellar projections to the arm representations of premotor areas in the frontal lobe were examined following injections of virus into the ventral premotor area (PMv) or the supplementary motor area (SMA; Akkal, Dum, & Strick, 2007; Hoover & Strick, 1993). These injections consistently labeled neurons in the middle of GPi rostrocaudally. Within

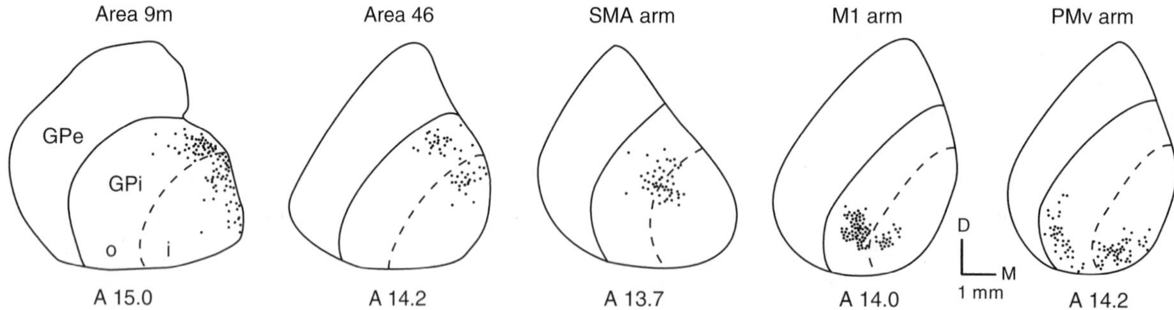

FIGURE 36.1 Origin of pallidal projections to M1, PMv, SMA, area 46, and area 9. Labeled neurons (dots) from several adjacent sections are displayed on coronal sections through the GPi of animals that received virus injections into different cortical areas. The anterior-posterior location of each section is indicated. Abbreviations: GPe, external segment of globus pallidus; GPi, internal segment of the globus pallidus; o, outer portion of the internal segment of globus pallidus; I, inner portion of the internal segment of globus pallidus; M1 arm, arm area of the primary motor cortex; PMv arm, arm area of the ventral premotor area; SMA arm, arm area of the supplementary motor area; area 46 and 9m, cytoarchitectonic regions in the prefrontal cortex. (Adapted from Middleton & Strick, 2000, with permission from Elsevier.)

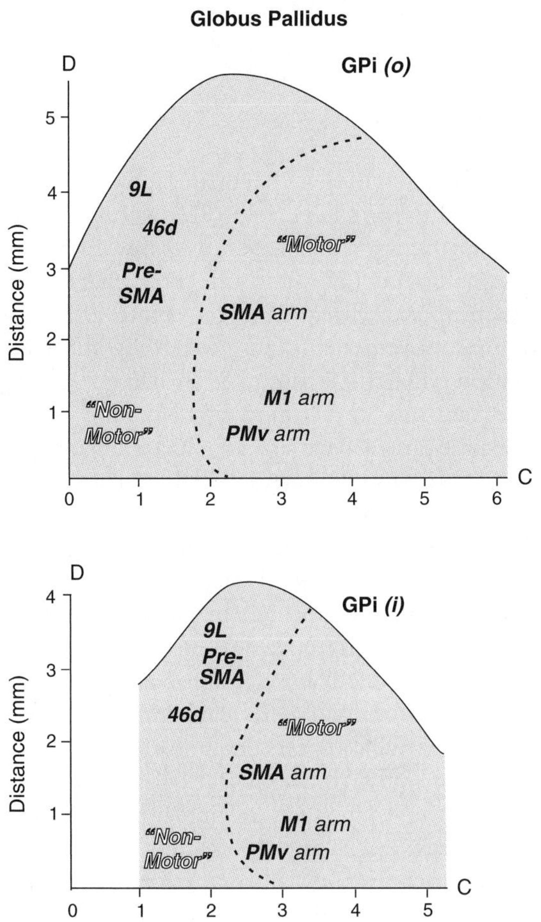

FIGURE 36.2 Summary map of basal ganglia output channels. The outer and inner segments of the GPi are shown as separate unfolded maps (for details of unfolding, see Akkal et al., 2007). In this planar view, the cortical target of each output channel is placed at the site of its densest labeling following retrograde transneuronal transport of virus from that cortical area. The GPi can be divided into "motor" and "nonmotor" domains based on the grouping of output channels that target functionally similar cortical areas. Abbreviations: D, dorsal; C, caudal; see also figure 36.1. (Adapted from Akkal et al., 2007, with permission.)

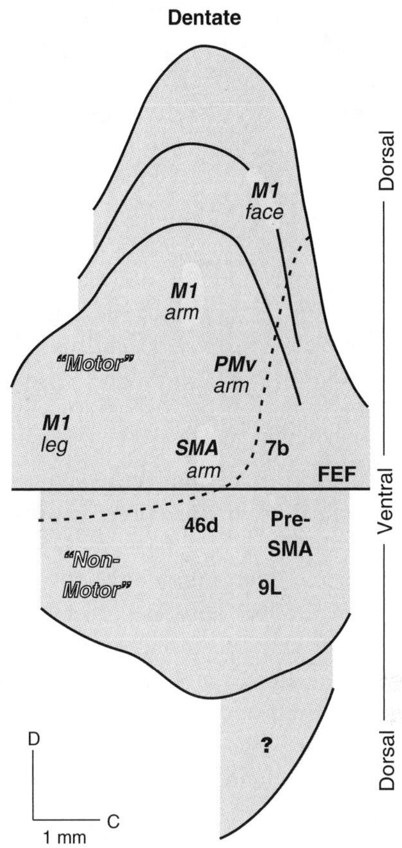

FIGURE 36.4 Summary map of dentate output channels. The dentate is displayed as an unfolded map (for details of unfolding, see Dum & Strick, 2003). The cortical target of each output channel is placed at the site of its peak labeling following retrograde transneuronal transport of virus from that cortical area. The dentate can be divided into "motor" and "nonmotor" domains based on the grouping of output channels that target functionally similar cortical areas. Abbreviations as in figures 36.2 and 36.3. (Adapted from Dum & Strick, 2003, and Akkal et al., 2007, with permission.)

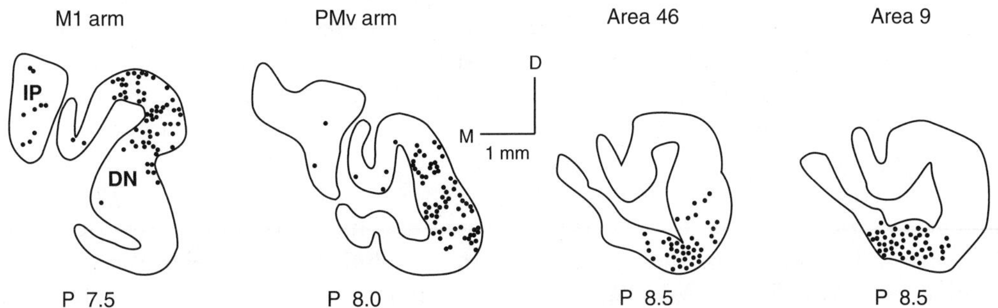

FIGURE 36.3 Origin of cerebellar projections to M1, PMv, area 46, and area 9. Representative coronal sections through the dentate and interpositus nuclei of animals that received virus injections into different cortical areas. Conventions are according to figure 36.2. Abbreviations: DN, dentate nucleus; IP, interpositus nucleus; M, medial. (Adapted from Middleton & Strick, 1998, with permission from Elsevier.)

this region, neurons labeled after injections into the SMA, M1, or PMv formed separate clusters in a dorsal to ventral arrangement (figures 36.1 and 36.2). These observations indicate that pallidal output is not confined to M1, but projects via the thalamus to multiple premotor areas in the frontal lobe (see also Inase & Tanji, 1995; Jinnai, Nambu, Tanibuch, & Yoshida, 1993; Saga et al., 2011; Sakai, Inase, & Tanji, 1999). Furthermore, the arm representation of each motor area receives input from a topographically distinct set of GPi neurons. We have proposed that this arrangement creates distinct "output channels" in the sensorimotor portion of GPi (Akkal et al., 2007; Hoover & Strick, 1993).

The output neurons of the dentate have a similar topographic organization. Injections of virus into the arm representations of M1, PMv, and SMA labeled clusters of neurons in the middle of the dentate rostrocaudally (figures 36.3 and 36.4; Akkal et al., 2007; Middleton & Strick, 1998; see also Hashimoto et al., 2010; Lu, Miyachi, Ito, Nambu, & Takada, 2007). The "hot spot" of each cluster appeared to be centered in a slightly different region of the dentate. The hot spots for the different motor areas are shown on an unfolded map of the dentate (figure 36.4). This diagram emphasizes two important observations. First, the dentate, like GPi, contains distinct output channels that innervate the arm representations of the different cortical motor areas. Second, the output channels to these different arm representations are clustered together. This observation suggests that these output channels are in register within the nucleus and raises the possibility that they form a single map for each body part.

Results from single-neuron recording experiments in awake trained monkeys provide physiological support for the existence of distinct output channels in GPi and dentate (Mushiake & Strick, 1993, 1995; Strick, Dum, & Picard, 1995). These studies suggest that individual output channels are involved in different aspects of motor behavior. Specifically, some output channels appear to be especially concerned with movements that are internally generated, whereas others appear to be devoted to movements guided by exteroceptive cues. Taken together, these observations indicate that the basal ganglia and cerebellum have the capacity to influence a broad range of motor behavior using output channels that project to the premotor areas in the frontal lobe, as well as to M1. Thus, the skeletomotor circuit of Alexander et al. (1986) is more accurately viewed as multiple discrete channels to each of the cortical motor areas (figure 36.5; Middleton & Strick, 2001b). A similar arrangement of output channels characterizes skeletomotor output from the dentate.

Frontal eye field

The results of studies with conventional tracers suggested that the frontal eye field (FEF) receives input via the thalamus from three subcortical nuclei: SNpr, the superior colliculus (SC), and the cerebellar nuclei. To test this proposal, virus was injected into FEF regions, where eye movements were evoked by intracortical stimulation (Lynch, Hoover, & Strick, 1994). Virus-infected neurons were found in SNpr, the optic and intermediate gray layers of the SC, and the dentate nucleus. Within the dentate, labeled neurons were confined to its posterior pole, where some neurons exhibit activity correlated with saccadic eye movements (van Kan, Houk, & Gibson, 1993). Within the basal ganglia, labeled neurons were located in posterior and lateral portions of SNpr (figure 36.6, FEF) where neurons display changes in activity related to saccadic eye movements (Hikosaka & Wurtz, 1983a, 1983b). The regions of the basal ganglia and cerebellum that were labeled after injections of virus into the FEF were strikingly different from those labeled after injections into the skeletomotor areas of the frontal lobe. Thus, the oculomotor output channels in the basal ganglia and cerebellum are distinct from those concerned with skeletomotor function.

Prefrontal cortex

Areas 9, 12, and 46 of the prefrontal cortex appear to be involved in cognitive operations such as the guidance of behavior based on transiently stored information rather than immediate external cues ("working memory"; for review, see Fuster, 1997; Goldman-Rakic, 1996; Passingham, 1993). These regions of prefrontal cortex are known to project to the input stage of basal ganglia and cerebellar processing (e.g., Glickstein, May, & Mercier, 1985; Haber, Kim, Mailly, & Calzavara, 2006; Kemp & Powell, 1971; Schmahmann & Pandya, 1997). We tested whether regions of prefrontal cortex not only project to the basal ganglia and cerebellum, but also are the target of basal ganglia and cerebellar outputs.

Virus injections into areas 9, 12, or 46 labeled many neurons in the output nuclei of the basal ganglia (figures 36.1, 36.2, and 36.6; Middleton & Strick, 1994, 2002). Injections into area 12 labeled neurons in a localized portion of SNpr. In contrast, injections into area 46 labeled neurons largely in GPi. Area 9 injections labeled neurons in both SNpr and GPi. The topographic nature of basal ganglia projections to

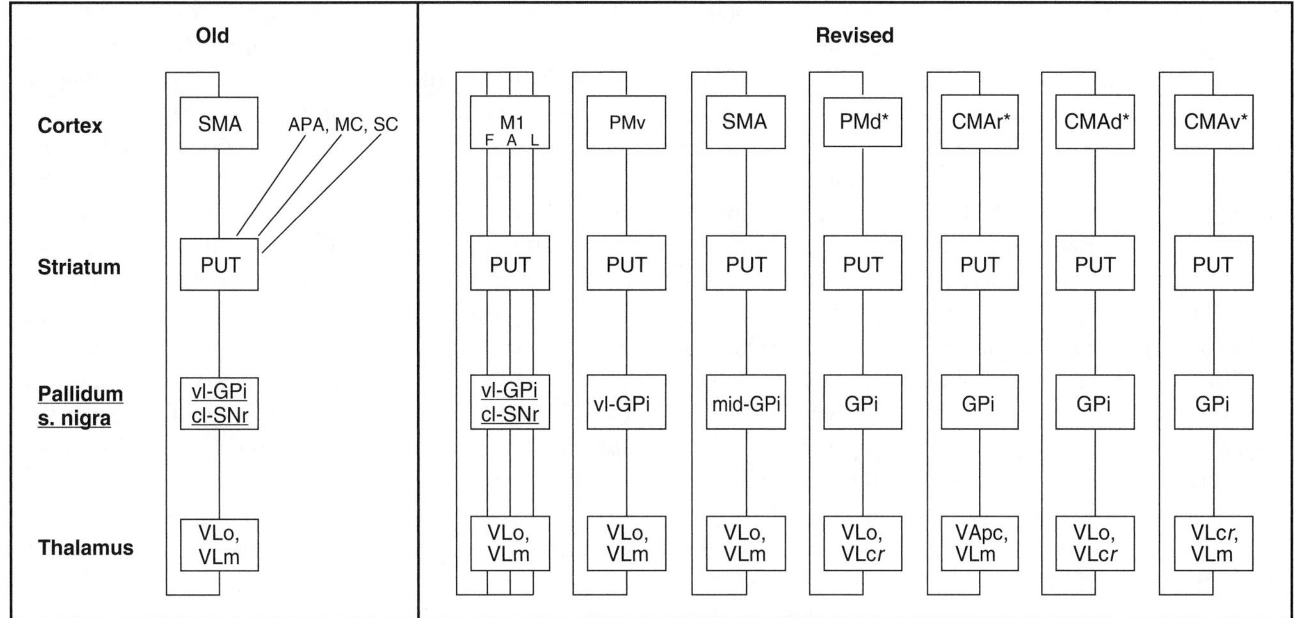

FIGURE 36.5 The original skeletomotor circuit proposed by Alexander, DeLong, and Strick (1986) and our revised scheme. Asterisks indicate loops whose existence is predicted but not specifically tested using virus transport. Cortical abbreviations: CMAd, dorsal cingulate motor area; CMAr, rostral cingulate motor area; CMAv, ventral cingulate motor area; M1, primary motor cortex; PMd, dorsal premotor area; PMv, ventral premotor area; SMA, supplementary motor area. Basal ganglia abbreviations; GPi, internal segment of globus pallidus; PUT, putamen; SNr, substantia nigra pars reticulata; cl, caudolateral; mid, middle; vl, ventrolateral. Thalamic abbreviations: VApc, nucleus ventralis anterior, parvocellular division; VLcc, nucleus ventralis lateralis pars caudalis, caudal division; VLcr, nucleus ventralis lateralis pars caudalis, rostral division; VLm, nucleus ventralis lateralis pars medialis; VLo, nucleus ventralis lateralis pars oralis. (From Middleton & Strick, 2001b, with permission from Elsevier.)

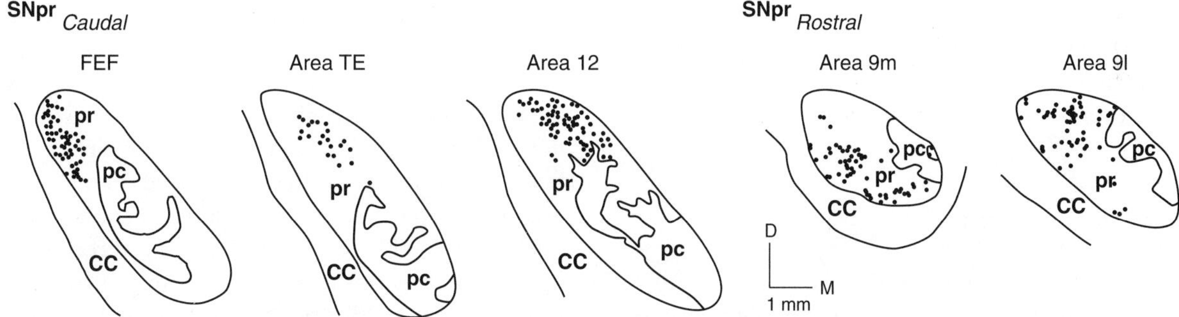

FIGURE 36.6 Origin of nigral projections to the FEF, area TE, area 12, area 9m, and area 9l. Coronal sections indicating the location of labeled neurons in the caudal and rostral regions of the SNpr following virus injections into the different cortical areas. Abbreviations: CC, crus cerebri; pc, pars compacta; pr, pars reticulata. (Adapted from Middleton & Strick, 2000, with permission from Elsevier.)

prefrontal cortex is further emphasized by the finding that different regions within rostral SNpr project to medial and lateral portions of area 9 (figure 36.6, areas 9m and 9l). In all cases, the locations of the neurons labeled in GPi and SNpr after injections into prefrontal areas of cortex were different from the locations of neurons labeled after injections into motor areas of cortex.

Virus injections into areas 9 and 46 (but not area 12) labeled neurons in ventral regions of the dentate nucleus (figure 36.3; Middleton & Strick, 1998, 2001a). The neurons labeled after area 9 injections were found largely medial and caudal to those labeled by area 46 injections. The ventral regions of the dentate that project to these nonmotor areas in the frontal lobe clearly differ from the more dorsal regions of this nucleus that innervate motor areas of the cortex (figures 36.3 and 36.4). Thus, both the basal ganglia and the cerebellum project via the thalamus to multiple areas of prefrontal cortex. Moreover, the output channels in the basal ganglia and cerebellum that influence prefrontal areas of cortex are separate from those that influence motor areas of cortex. This observation suggests that GPi and the dentate can be divided into distinct motor and nonmotor domains (figures 36.2 and 36.4; Akkal et al., 2007; Dum & Strick, 2003).

The presupplementary motor area (pre-SMA) has traditionally been included with the motor areas of the frontal lobe. However, a number of recent observations emphasize the nonmotor nature of this cortical area (Picard & Strick, 2001). For example, unlike the cortical motor areas, the pre-SMA does not project directly to M1 or to the spinal cord. Instead, the pre-SMA is densely interconnected with regions of prefrontal cortex. Virus tracing was used to test whether basal ganglia and cerebellar projections to the pre-SMA originate from the motor or the nonmotor domains of GPi and the dentate (Akkal et al., 2007). Virus injected into the pre-SMA labeled neurons dorsally in the rostral portion of GPi (figure 36.2) and in a ventral portion of the dentate (figure 36.4; Akkal et al., 2007). Thus, the output channels to the pre-SMA from GPi and dentate are grouped with output channels that project to regions of prefrontal cortex rather than output channels to the cortical motor areas (figures 36.2 and 36.4). These observations provide further support for the proposal that the pre-SMA is more similar to regions of prefrontal cortex than it is to the cortical motor areas (Akkal et al., 2007; Picard & Strick, 2001).

Posterior parietal cortex

Areas 5 and 7 in posterior parietal cortex also are known to project to the input stage of basal ganglia and cerebellar processing (e.g., Cavada & Goldman-Rakic, 1991; Glickstein et al., 1985; Kemp & Powell, 1971; Schmahmann & Pandya, 1997; Yeterian & Pandya, 1993). This raised the possibility that these regions of cortex also are the target of basal ganglia and cerebellar outputs. This possibility has been tested in several experiments (Clower, Dum, & Strick, 2005; Clower, West, Lynch, & Strick, 2001; Prevosto, Graf, & Ugolini, 2010). Virus tracing demonstrated that both portions of area 7b in the intraparietal sulcus (IPS) and on the cortical surface are the target of output from the dentate nucleus, whereas only the portion of area 7b on the cortical surface is the target of output from SNpr (Clower et al., 2001, 2005). In addition, area MIP in the anterior bank of the IPS and area LIP in the posterior bank of the IPS receive output from the cerebellum (Prevosto et al., 2010). These results clearly indicate that the sphere of influence of basal ganglia and cerebellar output extends to include portions of the posterior parietal cortex. Although the implications of basal ganglia and cerebellar projections to posterior parietal cortex cannot be fully addressed here, we will highlight two specific proposals about these circuits. The cerebellar projection to posterior parietal cortex may provide signals that contribute to sensory recalibration during some adaptation paradigms (Clower et al., 2001; Prevosto et al., 2010). In addition, we have suggested (Clower et al., 2005) that abnormal signals in the basal ganglia projection to the posterior parietal cortex may contribute to the visuospatial deficits observed in some patients with basal ganglia disorders (Karnath, Himmelbach, & Rorden, 2002).

Inferotemporal cortex

In general, each cortical area that projects to the basal ganglia or cerebellum is known to receive projections back from these subcortical nuclei. Area TE, a region of inferotemporal cortex, is known to project to the input stage of basal ganglia processing (i.e., the tail of the caudate and ventral portions of the putamen; Saint-Cyr, Ungerleider, & Desimone, 1990), but not to the input stage of cerebellar processing (Glickstein et al., 1985; Schmahmann & Pandya, 1997). Consistent with these observations, virus injections into area TE did not result in any labeled neurons in the cerebellar nuclei, but did label a distinct cluster of neurons in SNpr (Middleton & Strick, 1996; figure 36.6, area TE). Most of these neurons were located in a dorsal region of caudal SNpr that appears to be separate from the regions that influence the FEF or subdivisions of prefrontal cortex. Thus, area TE is both a source of input to, and a target of output from, a distinct portion of the basal ganglia. In contrast, TE neither projects to nor receives input from the cerebellum.

TE is known to play a critical role in the visual recognition and discrimination of objects (e.g., Gross, 1972; Miyashita, 1993; Tanaka, Saito, Fukuda, & Moriya, 1991). Physiological studies have shown that the region of SNpr that influences TE contains some neurons that are responsive to the presentation of visual stimuli (e.g., Hikosaka & Wurtz, 1983a). These observations, together with our anatomical results, provide evidence that basal ganglia output is involved in higher-order aspects of visual processing, as well as in motor and cognitive function.

Macro-architecture of subcortical loops with the cerebral cortex

Cortical areas that receive output from the basal ganglia and cerebellum also project to the input stage of these subcortical structures. This observation suggests that closed-loop circuits represent a fundamental architectural feature of basal ganglia and cerebellar connections with the cerebral cortex. This proposal was tested by utilizing two complementary virus-tracing approaches. First, retrograde transneuronal transport of rabies virus was used to define the regions of cerebellar cortex that *project to* the arm area of M1 or to area 46 in the prefrontal cortex (Kelly & Strick, 2003). Briefly, this approach showed that the arm area of M1 receives input from Purkinje cells located mainly in lobules IV-VI of cerebellar cortex (figure 36.7, left). In contrast, area 46 receives input from Purkinje cells located mainly in crus II of the ansiform lobule (figure 36.7, right). Thus, M1 and area 46 are the target of output from separate regions of the cerebellar cortex.

Subsequently, anterograde transneuronal transport of the H129 strain of HSV1 was used to define the regions of the cerebellar cortex that *receive input from* M1 or area 46 (Kelly & Strick, 2003). This approach demonstrated that granule cells in lobules IV-VI of cerebellar cortex are the target of input from the arm area of M1. Lobules IV-VI are the same region of the cerebellar cortex that projects to M1 (figure 36.7, left). Similarly, granule cells in crus II of the cerebellar cortex are the target of input from area 46. Crus II is the same region of the cerebellar cortex that projects to area 46 (figure 36.7, right). These observations provide strong support for our proposal that multiple closed-loop circuits represent a fundamental architectural feature of cerebrocerebellar interactions. Similar closed-loop circuits are also likely to be a fundamental feature of cerebro–basal ganglia interactions (Kelly & Strick, 2004).

Considerable support for the separation of cerebellar cortex into motor and nonmotor domains in humans comes from recent imaging studies (for a review, see Manni & Petrosini, 2004). These studies used functional connectivity MRI (fcMRI) and functional MRI (fMRI) to reveal the presence of two somatotopic representations of the body in the cerebellar cortex of humans (Buckner et al., 2011; Grodd, Hülsmann, Lotze, Wildgruber, & Erb, 2001; Wiestler, McGonigle, & Diedrichsen, 2011). The orientation of the body map is inverted in the anterior lobe and upright in lobule VIII of the posterior lobe.

Studies using fcMRI have provided complementary evidence that the nonmotor domain in the human cerebellar cortex is functionally coupled to association areas of the cerebral cortex. Activity in large portions of the cerebellar cortex, including hemispheric lobule VII, is correlated with activity in frontoparietal association areas, including the cognitive control and default networks (Buckner et al., 2011; Habas et al., 2009; Krienen & Buckner, 2009; O'Reilly, Beckmann, Tomassini, Ramnani, & Johansen-Berg, 2009). Interestingly, some association cortical regions that have limited or no known projections to the cerebellum in the monkey (such as ventral area 46 and regions in orbitofrontal cortex) show functional coupling to cerebellar cortex in the human. Thus, an expanded array of neocortical areas may interact with the cerebellum in humans.

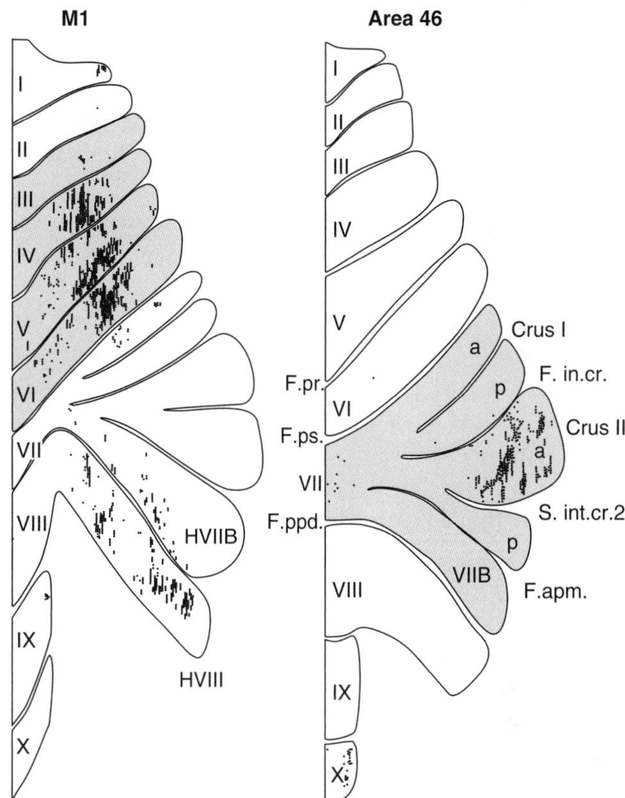

FIGURE 36.7 Origin of cerebellar cortical projections to M1 and area 46. The *dots* on the flattened surface maps of the cerebellar cortex represent Purkinje cells that were labeled by retrograde transneuronal transport of rabies virus from the arm area of M1 (*left panel*) or from area 46 (*right panel*). Note that the Purkinje cells that project to M1 are located in separate lobules from those that project to area 46. Nomenclature and abbreviations are according to Larsell (1970). (Adapted from Kelly & Strick, 2003, with permission.)

There is abundant evidence for activation of cerebellar cortex in a wide variety of nonmotor processes, including executive function, working memory, language, timing, music, and emotion. Since several recent reviews discuss results from fMRI of the human cerebellum and describe the functional topography in cerebellar cortex (e.g., E, Chen, Ho, & Desmond, 2012; Ito, 2008, 2011; Ramnani, 2006; Stoodley, 2012; Stoodley & Schmahmann, 2009; Stoodley, Valera, & Schmahmann, 2012; Strick et al., 2009), this subject will not be presented in detail here. Nevertheless, it is clear that the sites of activation in cognitive and emotion tasks are quite separate from those observed in motor tasks, and the results of these studies are fully consistent with the view that cerebellar cortex has distinct motor and nonmotor domains.

The cerebellum is interconnected with the basal ganglia

The loops that link the cerebellum with the cerebral cortex have traditionally been considered to be anatomically and functionally distinct from those that link the basal ganglia with the cerebral cortex (Doya, 2000; Graybiel, 2005). The outputs from the cerebellum and basal ganglia to the cerebral cortex are relayed through separate thalamic nuclei (Percheron et al., 1996; Sakai, Inase, & Tanji, 1996). Any interactions between cerebrocerebellar and cerebro–basal ganglia loops were thought to occur primarily at the neocortical level. Results from recent anatomical experiments challenge this perspective and provide evidence for disynaptic pathways that directly link the cerebellum with the basal ganglia.

Transneuronal transport of virus demonstrated that the dentate nucleus projects disynaptically to the striatum (caudate and putamen; figure 36.8; Hoshi, Tremblay, Féger, Carras, & Strick, 2005). Projections to the striatum originate from motor and nonmotor domains in the dentate and terminate in regions of putamen and caudate known to be within the "sensorimotor" and "associative" territories of these nuclei (see Parent & Hazrati, 1995). These findings indicate that the disynaptic pathway from the dentate to the striatum enables the cerebellum to influence both nonmotor and motor function within the basal ganglia.

In comparable experiments, virus transport demonstrated that the subthalamic nucleus (STN) projects disynaptically to cerebellar cortex (figures 36.8 and 36.9; Bostan et al., 2010). Projections to the cerebellar cortex originate from motor and nonmotor domains within the STN (see Parent & Hazrati, 1995). Furthermore, the projections terminate in motor and nonmotor regions of the cerebellar cortex. These findings indicate that the disynaptic pathway from the STN to the cerebellar cortex enables the basal ganglia to influence both nonmotor and motor function within the cerebellum. Taken together, these studies indicate that the cerebellum and basal ganglia are components of an interconnected network concerned with motor and nonmotor aspects of behavior.

Functional implications

Clearly, the outputs from the basal ganglia and cerebellum gain access to more widespread and diverse areas of cortex than previously imagined. To date, our studies have shown that the output nuclei of the basal ganglia and cerebellum project (via the thalamus) to skeletomotor, oculomotor, prefrontal, and posterior parietal areas of cortex. In addition, a portion of SNpr projects to inferotemporal cortex. Thus, the anatomical substrate exists for the basal ganglia and cerebellum to influence higher-order aspects of cognition like planning, working memory, sequential behavior, visuospatial perception, and attention as well as skeletomotor and oculomotor function. As a consequence, a sizeable component of basal ganglia and cerebellar output operates outside of the domain of motor control.

The new insights gained from virus tracing have important implications for hypotheses about basal ganglia and cerebellar contributions to normal and abnormal behavior. Detailed discussions of this issue have been presented in our recent reviews (Bostan, Dum, & Strick, 2013; Middleton & Strick, 1998; Strick et al., 2009), and therefore only some examples will be presented here. Abnormal activity in basal ganglia and cerebellar loops with cortical motor areas has been shown to contribute to the symptoms of certain motor disorders, particularly Parkinson's disease and dystonia (for reviews, see Filip, Lungu, & Bareš, 2013; Sadnicka, Hoffland, Bhatia, van de Warrenburg, & Edwards, 2012; Wu & Hallett, 2013). In Parkinson's disease, the loss of dopaminergic neurons of the substantia nigra pars compacta results in tremor, rigidity, bradykinesia, and akinesia (Wichmann, DeLong, Guridi, & Obeso, 2011). However, cerebellar activity is also abnormal in Parkinson's patients (Catalan, Ishii, Honda, Samii, & Hallett, 1999; Ghaemi et al, 2002; Rascol et al., 1997). In parkinsonian patients (Lenz et al., 1988; Ohye, Saito, Fukamachi, & Narabayashi, 1974) and in monkey models of the disease (Guehl et al., 2003), oscillatory activity at tremor frequencies has been recorded in regions of the thalamus that receive cerebellar, not basal ganglia, efferents. Furthermore, the cerebellar receiving thalamus is one of the most effective surgical sites for treating parkinsonian tremor (Narabayashi, Maeda, & Yokochi, 1987). These results suggest that abnormal activity in cerebellar circuits may account for parkinsonian tremor. Furthermore, deep-brain stimulation of the STN is not only highly effective in reducing the

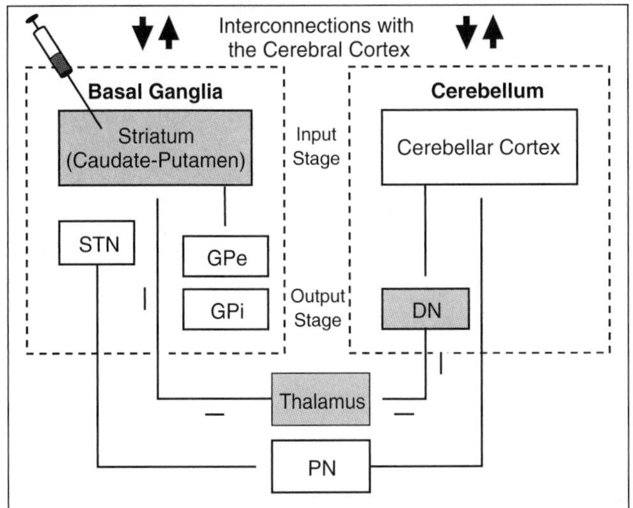

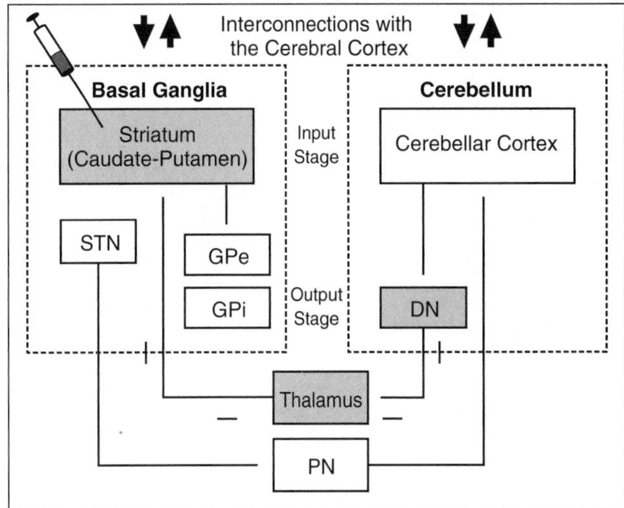

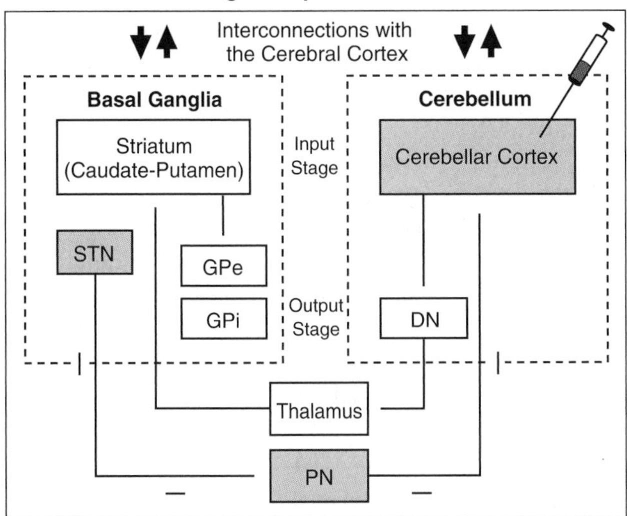

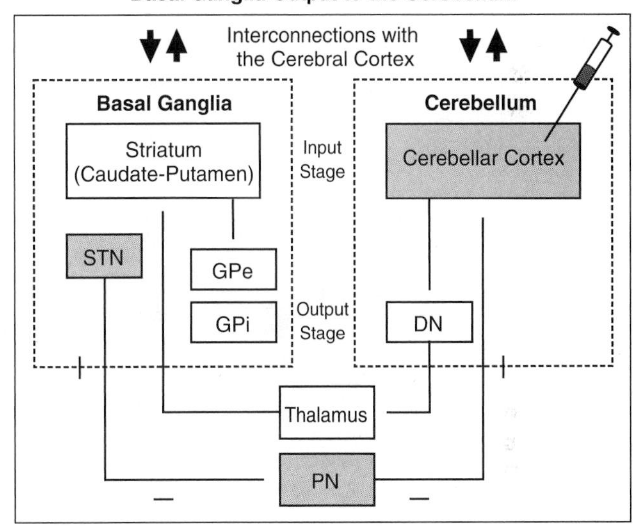

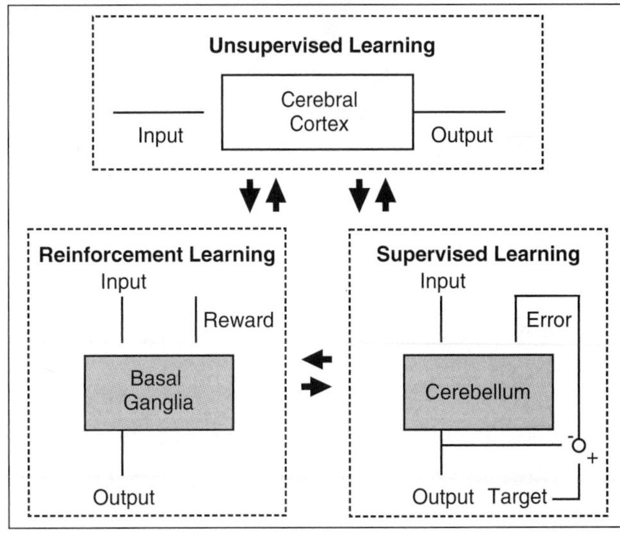

FIGURE 36.8 Circuits interconnecting the basal ganglia and the cerebellum. The top panel depicts the cerebellar (gray shading) projection to the basal ganglia. Rabies virus injected into the striatum went through two stages of transport: retrograde transport to first-order neurons in the thalamus that innervate the injection site, and then retrograde transneuronal transport to second-order neurons in the dentate nucleus (DN) that innervate the first-order neurons. The bottom panel depicts the basal ganglia (gray shading) projection to the cerebellum. Rabies virus injected into the cerebellar cortex was retrogradely transported to first-order neurons in the pontine nuclei (PN), and then transported transneuronally to second-order neurons in the subthalamic nucleus (STN). These interconnections enable two-way communication between the basal ganglia and the cerebellum. The small arrows indicate the direction of virus transport. Abbreviations as in figure 36.1. (Adapted from Bostan et al., 2013, with permission from Elsevier.)

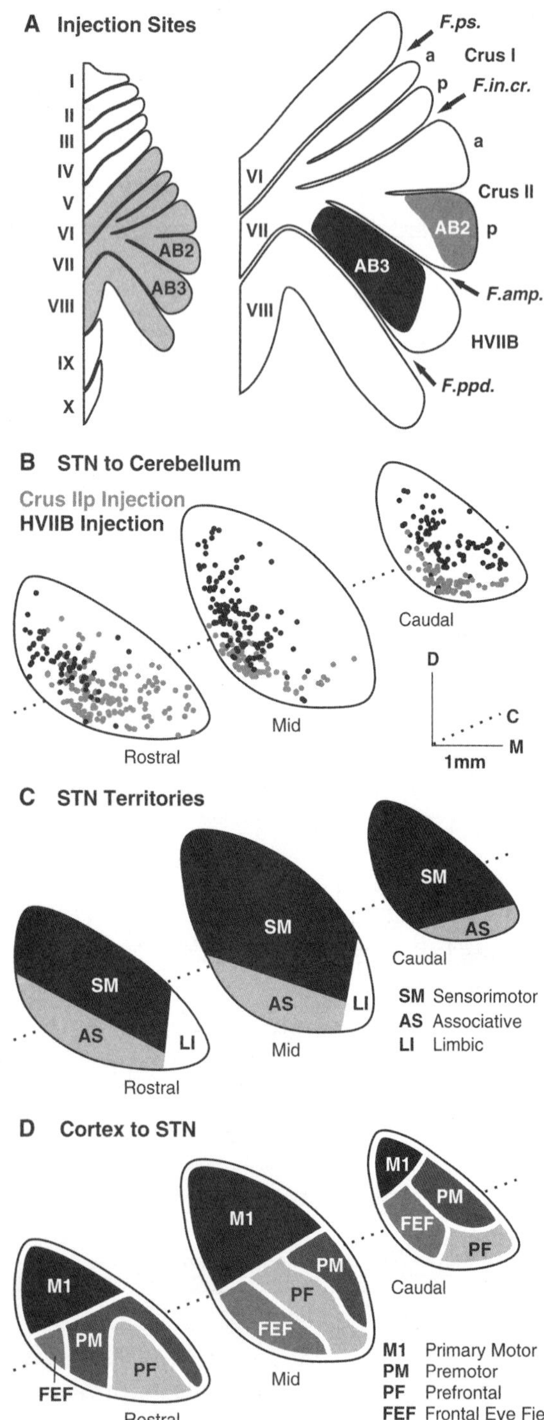

Figure 36.9 The subthalamic nucleus (STN) projects to the cerebellar hemisphere. (A) Rabies virus injection sites in the cerebellar cortex. *Left*: Flattened map of the cerebellar cortex in a Cebus monkey. *Right*: Shaded region on the left map is expanded to show an injection site in Crus IIp (AB2, gray) and another in HVIIB (AB3, black). (B) Combined charts of STN neurons labeled by injections of rabies virus into crus IIp (gray dots) or HVIIB (black dots) illustrate the topographic distribution of STN projections to different cerebellar cortical lobules. (C) Organization of the STN according to the functional subdivisions of the basal ganglia. (D) Schematic summary of projections from the cerebral cortex to the STN. C: caudal; D: dorsal; M: medial. (Adapted from Bostan et al., 2013, with permission from Elsevier.)

motor symptoms in Parkinson's disease (Krack, Fraix, Mendes, Benabid, & Pollak, 2002), but also normalizes cerebellar activity and function (Geday, Østergaard, Johnsen, & Gjedde, 2009; Grafton et al., 2006; Hilker et al., 2004; Payoux et al., 2004; Trost et al., 2006). The disynaptic connection from the STN to the cerebellum may be the anatomical substrate that mediates this effect of STN stimulation (figure 36.8).

Dystonia, another motor disorder often attributed to the basal ganglia (Neychev, Gross, Lehéricy, Hess, & Jinnah, 2011), is characterized by involuntary muscle contractions, twisting movements and abnormal postures (Bhatia & Marsden, 1994). However, dystonia can also arise from cerebellar dysfunction and may be better described as a network disorder involving the basal ganglia and cerebellum (LeDoux, 2011; Neychev et al., 2011). Human carriers of genetic mutations associated with dystonia exhibit abnormalities in both the basal ganglia and the cerebellum (Argyelan et al., 2009; Carbon et al., 2008; Carbon, Argyelan, & Eidelberg, 2010; Carbon & Eidelberg, 2009; Eidelberg, 1998; Ghilardi et al., 2003; Trost et al., 2002). In addition, abnormal cerebellar activity drives dystonic movements both in a pharmacological model of dystonia in normal mice and in mutant tottering mice (Campbell & Hess, 1998; Campbell, North, & Hess, 1999; Chen et al., 2009; Pizoli, Jinnah, Billingsley, & Hess, 2002; Neychev, Fan, Mitev, Hess, & Jinnah, 2008). Overall, these findings support the importance of functional interactions between the cerebellum and the basal ganglia in the manifestation of motor disorders typically associated with the basal ganglia.

Several lines of evidence also implicate the interconnections between the basal ganglia and cerebellum in nonmotor processes, such as associative reward-related learning, as well as in neuropsychiatric dysfunction. The cerebellum and basal ganglia are typically viewed as segregated modules that participate in different aspects of learning. The cerebellum is thought to be involved in adaptive modification of behavior and error-based learning, whereas the basal ganglia is thought to be involved in reward-prediction and reward-based learning (see Doya, 2000; Houk, 2005). Accounts of reward-related learning have emphasized the role of the basal ganglia based on the hypothesis that dopamine neurons reflect reward-prediction error and facilitate reinforcement learning in striatal target neurons (Schultz, Dayan, & Montague, 1997). Human fMRI studies have shown that activity in the striatum is correlated with reward-prediction error in Pavlovian reward-association tasks (O'Doherty, Dayan, Friston, Critchley, & Dolan, 2003; O'Doherty et al., 2004). Strikingly, reward-prediction error in these imaging studies is also strongly correlated with cerebellar signals (O'Doherty et al., 2003). Moreover, there is substantial evidence for cerebellar contributions to associative learning (see Swain, Kerr, & Thompson, 2011; Thompson, Swain, Clark, & Shinkman, 2000). The cerebellum is both necessary and sufficient for the establishment of classical conditioning with aversive stimuli (Brogden & Gnatt, 1942) and is activated in neuroimaging studies of aversive conditioning in humans, along with the striatum (Pohlack, Nees, Ruttorf, Schad & Flor, 2012; Seymour et al., 2004). The co-activations of the cerebellum and the basal ganglia in these tasks (O'Doherty et al., 2003; Pohlack et al., 2012; Seymour et al., 2004; Tanaka et al., 2004) suggest that they may interact in support of processes involving reward-related learning (for review, see Swain et al., 2011; Liljeholm & O'Doherty, 2012).

Cerebellar and basal ganglia interactions in reward-related learning may explain, in part, why lesions in both regions impair reward-based reversal learning (Bellebaum, Koch, Schwarz, & Daum, 2008; Thoma, Bellebaum, Koch, Schwarz, & Daum, 2008) and may help interpret findings that implicate the cerebellum in addiction. Although dopaminergic function and reinforcement learning implemented in the basal ganglia are considered key elements in the process of addiction (for review, see Koob & Volkow, 2010), the cerebellum may also be involved in this disorder (for review, see Miquel, Toledo, García, Coria-Avila, & Manzo, 2009). For example, an imaging study reported that cognitive deficits in addicted individuals were associated with abnormal cerebellar activity (Hester & Garavan, 2004). Furthermore, imaging studies consistently report co-activation of the cerebellum and basal ganglia when addicts are exposed to drug-related cues that increase craving (e.g., David et al., 2005; McClernon, Kozink, Lutz, & Rose, 2009; Olbrich et al., 2006; Yalachkov, Kaiser, & Naumer, 2009). The cerebellar activation occurs during responses to smoking cues (David et al., 2005), alcohol cues (Schneider et al., 2001), heroin cues (Yang et al., 2009), and cocaine cues (Bonson et al., 2002; Grant et al., 1996). Two main explanations have been proposed for cerebellar activations in cue-reactivity paradigms. First, the cerebellum through its prefrontal loop may be active as part of a distributed memory network that subserves emotional and cognitive links between environmental cues and drug craving (Grant et al., 1996). Second, the cerebellum through its motor loops may be active as part of a distributed sensorimotor network that subserves automatized behavioral reactions toward drug-related stimuli (Yalachkov et al., 2009; Yalachkov, Kaiser, & Naumer, 2010).

There is considerable evidence that cerebellar damage can lead to deficits in the performance of cognitive tasks that require rule-based learning, judgment of temporal intervals, visuospatial analysis, and shifting attention between sensory modalities, as well as working memory and planning (see reviews by Akshoomoff & Courchesne, 1992; Botez, Botez, Elie, & Attig, 1989; Fiez, Petersen, Cheney, & Raichle, 1992; Grafman et al., 1992; Ivry & Keele, 1989; Leiner et al., 1986, 1991, 1993; Schmahmann, 1991, 1997; Schmahmann & Sherman, 1998). Many of these deficits reflect functions normally thought to be subserved by areas of prefrontal cortex. Likewise, abnormal activity in basal ganglia and cerebellar loops with nonmotor areas of the cerebral cortex could lead to a broad range of psychiatric and neurological symptoms, such as those associated with autism, addiction, obsessive-compulsive disorder, and Tourette syndrome (for a review, see Lichter & Cummings, 2000). For example, Courchesne and colleagues (Courchesne et al., 1988, 1994; Courchesne, 1997) have suggested that alterations in the cerebellum and its projections to posterior parietal cortex may underlie some of the deficits seen in autistic patients. Rapoport and Wise (1988) have proposed that dysfunction in basal ganglia circuits with anterior cingulate and orbital frontal cortex may explain some of the features of obsessive-compulsive disorder. In Tourette syndrome, the cerebellum and the basal ganglia are likely to be concurrently involved in tic generation (O'Halloran, Kinsella, & Storey, 2012). Tourette syndrome patients can be differentiated from control subjects by an abnormal metabolic pattern that includes increased cerebellar and decreased basal ganglia metabolism (Lerner et al., 2007). Thus, both cerebellar and basal ganglia dysfunction and interactions between the two subcortical nuclei may be important contributors to neuropsychiatric disorders.

In summary, virus tracing has revealed that the output of the basal ganglia and cerebellum targets motor, premotor, prefrontal, posterior parietal, and inferotemporal areas of cortex. These connections provide the basal ganglia and cerebellum with the anatomical substrate to influence not only the control of movement, but also many aspects of cognitive behavior like planning, working memory, sequential behavior, visuospatial perception, and attention. Similarly, there is growing evidence that disorders like autism, addiction, attention-deficit disorder, obsessive-compulsive disorder, and Tourette syndrome are associated with alterations in basal ganglia or cerebellar function. Thus, it is possible that abnormal activity in specific basal ganglia and cerebellar loops with the cerebral cortex results in identifiable sets of neuropsychiatric symptoms. Moreover, we have provided evidence that the cerebellum and basal ganglia are interconnected. These interconnections may allow the two major subcortical nuclei to cooperate in the generation of a broad range of normal motor and nonmotor functions; the interconnections may also enable dysfunction in one circuit to recruit abnormal activity in its partner and, thus, provoke a diverse set of motor and neuropsychiatric disorders.

ACKNOWLEDGMENTS This work was supported in part by funds from the Office of Research and Development, Medical Research Service, and Department of Veterans Affairs, and by National Institutes of Health Grants R01 NS24328 (PLS), R01 MH56661 (PLS), P40 OD010996 (PLS), and P30 NS076405 (PLS).

REFERENCES

Akkal, D., Dum, R. P., & Strick, P. L. (2007). Supplementary motor area and presupplementary motor area: Targets of basal ganglia and cerebellar output. *J Neurosci, 27,* 10659–10673.

Akshoomoff, N. A., & Courchesne, E. (1992). A new role for the cerebellum in cognitive function. *Behav Neurosci, 106,* 731–738.

Alexander, G. E., DeLong, M. R., & Strick, P. L. (1986). Parallel organization of functionally segregated circuits linking basal ganglia and cortex. *Annu Rev Neurosci, 9,* 357–381.

Allen, G. I., & Tsukahara, N. (1974). Cerebrocerebellar communication systems. *Physiol Rev, 54,* 957–1006.

Argyelan, M., Carbon, M., Niethammer, M., Ulug, A. M., Voss, H. U., Bressman, S. B., ... Eidelberg, D. (2009). Cerebellothalamocortical connectivity regulates penetrance in dystonia. *J Neurosci, 29,* 9740–9747.

Bellebaum, C., Koch, B., Schwarz, M., & Daum, I. (2008). Focal basal ganglia lesions are associated with impairments in reward-based reversal learning. *Brain, 131,* 829–841.

Bhatia, K. P., & Marsden, C. D. (1994). The behavioural and motor consequences of focal lesions of the basal ganglia in man. *Brain, 117,* 859–876.

Bonson, K. R., Grant, S. J., Contoreggi, C. S., Links, J. M., Metcalfe, J., Weyl, H. L., ... London, E. D. (2002). Neural systems and cue-induced cocaine craving. *Neuropsychopharmacology, 26,* 376–386.

Bostan, A. C., Dum, R. P., & Strick, P. L. (2010). The basal ganglia communicates with the cerebellum. *Proc Natl Acad Sci USA, 107,* 8452–8456.

Bostan, A. C., Dum, R. P., & Strick, P. L. (2013). Cerebellar networks with the cerebral cortex and basal ganglia. *Trends Cogn Sci, 17,* 241–254.

Bostan, A. C., & Strick, P. L. (2010). The cerebellum and basal ganglia are interconnected. *Neuropsychol Rev, 30,* 261–270.

Botez, M. I., Botez, T., Elie, R., & Attig, E. (1989). Role of the cerebellum in complex human behavior. *Ital J Neurol Sci, 10,* 291–300.

Brogden, W. J., & Gnatt, W. H. (1942). Intraneural conditioning: Cerebellar conditioned reflexes. *Arch Neurol Psychol, 48,* 18.

Brooks, V. B., & Thach, W. T. (1981). Cerebellar control of posture and movement. In V. B. Brooks (Ed.), *Handbook of physiology, Section 1: The nervous system,* Vol. 2, *Motor control, Part II* (pp. 877–946). Bethesda, MD: American Physiological Society.

Buckner, R. L., Krienen, F. M., Castellanos, A., Diaz, J. C., & Yeo, B. T. (2011). The organization of the human cerebellum estimated by intrinsic functional connectivity. *J Neurophysiol, 106,* 2322–2345.

Campbell, D. B., & Hess, E. J. (1998). Cerebellar circuitry is activated during convulsive episodes in the tottering (tg/tg) mutant mouse. *Neuroscience, 85,* 773–783.

Campbell, D. B., North, J. B., & Hess, E. J. (1999). Tottering mouse motor dysfunction is abolished on the Purkinje cell degeneration (pcd) mutant background. *Exp Neurol, 160,* 268–278.

Carbon, M., & Eidelberg, D. (2009). Abnormal structure-function relationships in hereditary dystonia. *Neuroscience, 164,* 220–229.

Carbon, M., Ghilardi, M. F., Argyelan, M., Dhawan, V., Bressman, S. B., & Eidelberg, D. (2008). Increased cerebellar activation during sequence learning in DYT1 carriers: An equiperformance study. *Brain, 131,* 146–154.

Carbon, M., Argyelan, M., & Eidelberg, D. (2010). Functional imaging in hereditary dystonia. *Eur J Neurosci, 17,* 58–64.

Catalan, M. J., Ishii, K., Honda, M., Samii, A., & Hallett, M. (1999). A PET study of sequential finger movements of varying length in patients with Parkinson's disease. *Brain, 122,* 483–495.

Cavada, C., & Goldman-Rakic, P. S. (1991). Topographic segregation of corticostriatal projections from posterior parietal subdivisions in the macaque monkey. *Neuroscience, 42,* 683–696.

Chen, G., Popa, L. S., Wang, X., Gao, W., Barnes, J., Hendrix, C. M., ... Ebner, T. J. (2009). Low-frequency oscillations in the cerebellar cortex of the tottering mouse. *J Neurophysiol, 101,* 234–245.

Clower, D. M., West, R. A., Lynch, J. C., & Strick, P. L. (2001). The inferior parietal lobule is the target of output from the superior colliculus, hippocampus and cerebellum. *J Neurosci, 21,* 6283–6291.

Clower, D. M., Dum, R. P., & Strick, P. L. (2005). Basal ganglia and cerebellar inputs to "AIP." *Cereb Cortex, 15,* 913–920.

Courchesne, E. (1997). Brainstem, cerebellar and limbic neuroanatomical abnormalities in autism. *Curr Opin Neurobiol, 7,* 269–278.

Courchesne, E., Townsend, J., Akshoomoff, N., Saitoh, O., Yeung-Courchesne, R., Lincoln, A., ... Lau, L. (1994). Impairment in shifting attention in autistic and cerebellar patients. *Behav Neurosci, 108,* 848–865.

Courchesne, E., Yeung-Courchesne, R., Press, G. A., Hesselink, J. R., & Jernigan, T. L. (1988). Hypoplasia of cerebellar vermal lobules VI and VII in autism. *N Engl J Med, 318*, 1349–1354.

David, S. P., Munafò, M. R., Johansen-Berg, H., Smith, S. M., Rogers, R. D., Matthews, P. M., & Walton, R. T. (2005). Ventral striatum/nucleus accumbens activation to smoking-related pictorial cues in smokers and nonsmokers: A functional magnetic resonance imaging study. *Biol Psychiatry, 58*, 488–494.

DeLong, M. R., & Georgopoulos, A. P. (1981). Motor functions of the basal ganglia. In V. B. Brooks (Ed.), *Handbook of physiology, Section I: The nervous system*, Vol. 2, *Motor control* (pp. 1017–1061). Bethesda, MD: American Physiological Society.

Doya, K. (2000). Complementary roles of basal ganglia and cerebellum in learning and motor control. *Curr Opin Neurobiol, 10*, 732–739.

Dum, R. P., & Strick, P. L. (2003). An unfolded map of the cerebellar dentate nucleus and its projections to the cerebral cortex. *J Neurophysiol, 89*, 634–639.

E, K.-H., Chen, S.-H., Ho, M. H., & Desmond, J. E. (2012). A meta-analysis of cerebellar contributions to higher cognition from PET and fMRI studies. *Hum Brain Mapp, 35*, 593–615.

Eidelberg, D. (1998). Functional brain networks in movement disorders. *Curr Opin Neurol, 11*, 319–326.

Fiez, J. A., Petersen, S. E., Cheney, M. K., & Raichle, M. E. (1992). Impaired non-motor learning and error detection associated with cerebellar damage. *Brain, 115*, 155–178.

Filip, P., Lungu, O. V., & Bareš, M. (2013). Dystonia and the cerebellum: A new field of interest in movement disorders? *Clin Neurophysiol, 124*(7), 1269–1276.

Fuster, J. M. (1997). *The prefrontal cortex*. New York, NY: Raven Press.

Ghaemi, M., Raethjen, J., Hilker, R., Rudolf, J., Sobesky, J., Deuschl, G., & Heiss, W. D. (2002). Monosymptomatic resting tremor and Parkinson's disease: A multitracer positron emission tomographic study. *Mov Disord, 17*, 782–788.

Ghilardi, M.-F., Carbon, M., Silvestri, G., Dhawan, V., Tagliati, M., Bressman, S., ... Eidelberg, D. (2003). Impaired sequence learning in carriers of the DYT1 dystonia mutation. *Ann Neurol, 54*, 102–109.

Glickstein, M., May, J. G., & Mercier, B. E. (1985). Corticopontine projection in the macaque: The distribution of labelled cortical cells after large injections of horseradish peroxidase in the pontine nuclei. *J Comp Neurol, 235*, 343–359.

Goldman-Rakic, P. S. (1996). The prefrontal landscape: Implications of functional architecture for understanding human mentation and the central executive. *Philos Trans R Soc Lond B Biol Sci, 351*, 1445–1453.

Goldman-Rakic, P. S., & Selemon, L. D. (1990). New frontiers in basal ganglia research. Introduction. *Trends Neurosci, 13*, 241–244.

Grafman, J., Litvan, I., Massaquoi, S., Stewart, M., Sirigu, A., & Hallett, M. (1992). Cognitive planning deficit in patients with cerebellar atrophy. *Neurology, 42*, 1493–1496.

Grafton, S. T., Turner, R. S., Desmurget, M., Bakay, R., Delong, M., Vitek, J., & Crutcher, M. (2006). Normalizing motor-related brain activity: Subthalamic nucleus stimulation in Parkinson disease. *Neurology, 66*, 1192–1199.

Grant, S., London, E. D., Newlin, D. B., Villemagne, V. L., Liu, X., Contoreggi, C., ... Margolin, A. (1996). Activation of memory circuits during cue-elicited cocaine craving. *Proc Natl Acad Sci USA, 93*, 12040–12045.

Graybiel, A. M. (2005). The basal ganglia: Learning new tricks and loving it. *Curr Opin Neurobiol, 15*, 638–644.

Geday, J., Østergaard, K., Johnsen, E., & Gjedde, A. (2009). STN-stimulation in Parkinson's disease restores striatal inhibition of thalamocortical projection. *Hum Brain Mapp, 30*, 112–121.

Grodd, W., Hülsmann, E., Lotze, M., Wildgruber, D., & Erb, M. (2001). Sensorimotor mapping of the human cerebellum: FMRI evidence of somatotopic organization. *Hum Brain Mapp, 13*, 55–73.

Gross, C. G. (1972). Visual functions of inferotemporal cortex. In R. Jung (Ed.), *Handbook of sensory physiology* (pp. 451–482). Berlin: Springer-Verlag.

Guehl, D., Pessiglione, M., François, C., Yelnik, J., Hirsch, E. C., Féger, J., & Tremblay, L. (2003). Tremor-related activity of neurons in the "motor" thalamus: Changes in firing rate and pattern in the MPTP vervet model of parkinsonism. *Eur J Neurosci, 17*, 2388–2400.

Habas, C., Kamdar, N., Nguyen, D., Prater, K., Beckmann, C. F., Menon, V., & Greicius, M. D. (2009). Distinct cerebellar contributions to intrinsic connectivity networks. *J Neurosci, 29*, 8586–8594.

Haber, S. N., Kim, K. S., Mailly, P., & Calzavara, R. (2006). Reward-related cortical inputs define a large striatal region in primates that interface with associative cortical connections, providing a substrate for incentive-based learning. *J Neurosci, 26*, 8368–8376.

Hashimoto, M., Takahara, D., Hirata, Y., Inoue, K., Miyachi, S., Nambu, A., ... Hoshi, E. (2010). Motor and non-motor projections from the cerebellum to rostrocaudally distinct sectors of the dorsal premotor cortex in macaques. *Eur J Neurosci, 31*, 1402–1413.

Hester, R., & Garavan, H. (2004). Executive dysfunction in cocaine addiction: Evidence for discordant frontal, cingulate, and cerebellar activity. *J Neurosci, 24*, 11017–11022.

Hikosaka, O., & Wurtz, R. H. (1983a). Visual and oculomotor functions of monkey substantia nigra pars reticulata. I. Relation of visual and auditory responses to saccades. *J Neurophysiol, 49*, 1230–1253.

Hikosaka, O., & Wurtz, R. H. (1983b). Visual and oculomotor functions of monkey substantia nigra pars reticulata. III. Memory-contingent visual and saccade responses. *J Neurophysiol, 49*, 1268–1284.

Hilker, R., Voges, J., Weisenbach, S., Kalbe, E., Burghaus, L., Ghaemi, M., ... Heiss, W. D. (2004). Subthalamic nucleus stimulation restores glucose metabolism in associative and limbic cortices and in cerebellum: Evidence from a FDG-PET study in advanced Parkinson's disease. *J Cereb Blood Flow Metab, 24*, 7–16.

Hoover, J. E., & Strick, P. L. (1993). Multiple output channels in the basal ganglia. *Science, 259*, 819–821.

Hoover, J. E., & Strick, P. L. (1999). The organization of cerebello- and pallido-thalamic projections to primary motor cortex: An investigation employing retrograde transneuronal transport of herpes simplex virus type 1. *J Neurosci, 19*, 1446–1463.

Hoshi, E., Tremblay, L., Féger, J., Carras, P. L., & Strick, P. L. (2005). The cerebellum communicates with the basal ganglia. *Nat Neurosci, 8*, 1491–1493.

Houk, J. C. (2005). Agents of the mind. *Biol Cybern, 92,* 427–437.

Inase, M., & Tanji, J. (1995). Thalamic distribution of projection neurons to the primary motor cortex relative to afferent terminal fields from the globus pallidus in the macaque monkey. *J Comp Neurol, 353,* 415–426.

Ito, M. (2008). Control of mental activities by internal models in the cerebellum. *Nat Rev Neurosci, 9,* 304–313.

Ito, M. (2011). *The cerebellum: Brain for an implicit self.* Upper Saddle River, NJ: FT Press.

Ivry, R. B., & Keele, S. W. (1989). Timing functions of the cerebellum. *J Cog Neurosci, 1,* 136–152.

Jinnai, K., Nambu, A., Tanibuch, I., & Yoshida, S. (1993). Cerebello- and pallido-thalamic pathways to areas 6 and 4 in the monkey. *Stereotact Funct Neurosurg, 60,* 70–79.

Karnath, H. O., Himmelbach, M., & Rorden, C. (2002). The subcortical anatomy of human spatial neglect: Putamen, caudate nucleus and pulvinar. *Brain, 125,* 350–360.

Kelly, R. M., & Strick, P. L. (2000). Rabies as a transneuronal tracer of circuits in the central nervous system. *J Neurosci Methods, 103,* 63–71.

Kelly, R. M., & Strick, P. L. (2003). Cerebellar loops with motor cortex and prefrontal cortex of a nonhuman primate. *J Neurosci, 12,* 8432–8444.

Kelly, R. M., & Strick, P. L. (2004). Macro-architecture of basal ganglia loops with the cerebral cortex: Use of rabies virus to reveal multisynaptic circuits. *Prog Brain Res, 143,* 449–459.

Kemp, J. M., & Powell, T. P. S. (1971). The connexions of the striatum and globus pallidus: Synthesis and speculation. *Philos Trans R Soc Lond B Biol Sci, 262,* 441–457.

Koob, G. F., & Volkow, N. D. (2010). Neurocircuitry of addiction. *Neuropsychopharmacology, 35,* 217–238.

Krack, P., Fraix, V., Mendes, A., Benabid, A. L., & Pollak, P. (2002). Postoperative management of subthalamic nucleus stimulation for Parkinson's disease. *Mov Disord, 17*(Suppl. 3), S188–197.

Krienen, F. M., & Buckner, R. L. (2009). Segregated fronto-cerebellar circuits revealed by intrinsic functional connectivity. *Cereb Cortex, 19,* 2485–2497.

Larsell, O. (1970). *The comparative anatomy and histology of the cerebellum from monotremes through apes.* Minneapolis: University of Minnesota.

LeDoux, M. S. (2011). Animal models of dystonia: Lessons from a mutant rat. *Neurobiol Dis, 42,* 152–161.

Leiner, H. C., Leiner, A. L., & Dow, R. S. (1986). Does the cerebellum contribute to mental skills? *Behav Neurosci, 100,* 443–454.

Leiner, H. C., Leiner, A. L., & Dow, R. S. (1991). The human cerebro-cerebellar system: Its computing, cognitive, and language skills. *Behav Brain Res, 44,* 113–128.

Leiner, H. C., Leiner, A. L., & Dow, R. S. (1993). Cognitive and language functions of the human cerebellum. *Trends Neurosci, 16,* 444–447.

Lenz, F. A., Tasker, R. R., Kwan, H. C., Schnider, S., Kwong, R., Murayama, Y., … Murphy, J. T. (1988). Single unit analysis of the human ventral thalamic nuclear group: Correlation of thalamic "tremor cells" with the 3–6 Hz component of Parkinsonian tremor. *J Neurosci, 8,* 754–764.

Lerner, A., Bagic, A., Boudreau, E. A., Hanakawa, T., Pagan, F., Mari, Z., … Hallett, M. (2007). Neuroimaging of neuronal circuits involved in tic generation in patients with Tourette syndrome. *Neurology, 68*(23), 1979–1987.

Lichter, D. G., & Cummings, J. L. (2000). *Frontal-subcortical circuits in psychiatry and neurology.* New York, NY: Guilford.

Liljeholm, M., & O'Doherty, J. P. (2012). Contributions of the striatum to learning, motivation, and performance: An associative account. *Trends Cogn Sci, 16,* 467–475.

Lu, X., Miyachi, S., Ito, Y., Nambu, A., & Takada, M. (2007). Topographic distribution of output neurons in cerebellar nuclei and cortex to somatotopic map of primary motor cortex. *Eur J Neurosci, 25,* 2374–2382.

Lynch, J. C., Hoover, J. E., & Strick, P. L. (1994). Input to the primate frontal eye field from the substantia nigra, superior colliculus, and dentate nucleus demonstrated by transneuronal transport. *Exp Brain Res, 100,* 181–186.

McClernon, F. J., Kozink, R. V., Lutz, A. M., & Rose, J. E. (2009). 24-h smoking abstinence potentiates fMRI-BOLD activation to smoking cues in cerebral cortex and dorsal striatum. *Psychopharmacology (Berl), 204*(1), 25–35.

Manni, E., & Petrosini, L. (2004). A century of cerebellar somatotopy: A debated representation. *Nat Rev Neurosci, 5,* 241–249.

Middleton, F. A., & Strick, P. L. (1994). Anatomical evidence for cerebellar and basal ganglia involvement in higher cognitive function. *Science, 266,* 458–461.

Middleton, F. A., & Strick, P. L. (1996). The temporal lobe is a target of output from the basal ganglia. *Proc Natl Acad Sci USA, 93,* 8683–8687.

Middleton, F. A., & Strick, P. L. (1998). Cerebellar output: Motor and cognitive channels. *Trends Cogn Sci, 2,* 348–354.

Middleton, F. A., & Strick, P. L. (2000). Basal ganglia output and cognition: Evidence from anatomical, behavioral, and clinical studies. *Brain Cogn, 42,* 183–200.

Middleton, F. A., & Strick, P. L. (2001a). Cerebellar projections to the prefrontal cortex of the primate. *J Neurosci, 21,* 700–712.

Middleton, F. A., & Strick, P. L. (2001b). A revised neuroanatomy of frontal subcortical circuits. In D. G. Lichter & J. L. Cummings (Eds.), *Frontal-subcortical circuits in psychiatry and neurology* (pp. 44–58). New York, NY: Guilford.

Middleton, F. A., & Strick, P. L. (2002). Basal ganglia "projections" to the prefrontal cortex. *Cereb Cortex, 12,* 926–935.

Miquel, M., Toledo, R., García, L. I., Coria-Avila, G. A., & Manzo, J. (2009). Why should we keep the cerebellum in mind when thinking about addiction? *Curr Drug Abuse Rev, 2*(1), 26–40.

Miyashita, Y. (1993). Inferior temporal cortex: Where visual perception meets memory. *Annu Rev Neurosci, 16,* 245–263.

Mushiake, H., & Strick, P. L. (1993). Preferential activity of dentate neurons during limb movements. *J Neurophysiol, 70,* 2660–2664.

Mushiake, H., & Strick, P. L. (1995). Pallidal neuron activity during sequential arm movements. *J Neurophysiol, 74,* 2754–2758.

Narabayashi, H., Maeda, T., & Yokochi, F. (1987). Long-term follow-up study of nucleus ventralis intermedius and ventrolateralis thalamotomy using a microelectrode technique in parkinsonism. *Appl Neurophysiol, 50,* 330–337.

Neychev, V. K., Fan, X., Mitev, V. I., Hess, E. J., & Jinnah, H. A. (2008). The basal ganglia and cerebellum interact in the expression of dystonic movement. *Brain, 131,* 2499–2509.

Neychev, V. K., Gross, R. E., Lehéricy, S., Hess, E. J., & Jinnah, H. A. (2011). The functional neuroanatomy of dystonia. *Neurobiol Dis, 42,* 185–201.

O'Doherty, J. P., Dayan, P., Friston, K., Critchley, H., & Dolan, R. J. (2003). Temporal difference models and reward-related learning in the human brain. *Neuron, 38,* 329–337.

O'Doherty, J., Dayan, P., Schultz, J., Deichmann, R., Friston, K., & Dolan, R. J. (2004). Dissociable roles of ventral and dorsal striatum in instrumental conditioning. *Science, 304,* 452–454.

Ohye, C., Saito, U., Fukamachi, A., & Narabayashi, H. (1974). An analysis of the spontaneous rhythmic and non-rhythmic burst discharges in the human thalamus. *J Neurol Sci, 22,* 245–259.

Olbrich, H. M., Valerius, G., Paris, C., Hagenbuch, F., Ebert, D., & Juengling, F. D. (2006). Brain activation during craving for alcohol measured by positron emission tomography. *Aust N Z J Psychiatry, 40,* 171–178.

O'Halloran, C. J., Kinsella, G. J., & Storey, E. (2012). The cerebellum and neuropsychological functioning: A critical review. *J Clin Exp Neuropsychol, 34,* 35–56.

O'Reilly, J. X., Beckmann, C. F., Tomassini, V., Ramnani, N., & Johansen-Berg, H. (2009). Distinct and overlapping functional zones in the cerebellum defined by resting state functional connectivity. *Cereb Cortex, 20,* 953–965.

Parent, A., & Hazrati, L. N. (1995). Functional anatomy of the basal ganglia. I. The cortico-basal ganglia-thalamo-cortical loop. *Brain Res Brain Res Rev, 20,* 91–127.

Passingham, R. (1993). *The frontal lobes and voluntary action.* Oxford, UK: Oxford University Press.

Payoux, P., Remy, P., Damier, P., Miloudi, M., Loubinoux, I., Pidoux, B., ... Agid, Y. (2004). Subthalamic nucleus stimulation reduces abnormal motor cortical overactivity in Parkinson disease. *Arch Neurol, 61,* 1307–1313.

Percheron, G., François, C., Talbi, B., Yelnik, J., & Fénelon, G. (1996). The primate motor thalamus. *Brain Res Brain Res Rev, 22,* 93–181.

Picard, N., & Strick, P. L. (2001). Imaging the premotor areas. *Curr Opin Neurobiol, 11,* 663–672.

Pizoli, C. E., Jinnah, H. A., Billingsley, M. L., & Hess, E. J. (2002). Abnormal cerebellar signaling induces dystonia in mice. *J Neurosci, 22,* 7825–7833.

Pohlack, S. T., Nees, F., Ruttorf, M., Schad, L. R., & Flor, H. (2012). Activation of the ventral striatum during aversive contextual conditioning in humans. *Biol Psychol, 91,* 74–80.

Prevosto, V., Graf, W., & Ugolini, G. (2010). Cerebellar inputs to intraparietal cortex areas LIP and MIP: Functional frameworks for adaptive control of eye movements, reaching, and arm/eye/head movement coordination. *Cereb Cortex, 20,* 214–228.

Ramnani, N. (2006). The primate cortico-cerebellar system: Anatomy and function. *Nat Rev Neurosci, 7,* 511–522.

Rapoport, J. L., & Wise, S. P. (1988). Obsessive-compulsive disorder: Evidence for basal ganglia dysfunction. *Psychopharm Bull, 24,* 380–384.

Rascol, O., Sabatini, U., Fabre, N., Brefel, C., Loubinoux, I., Celsis, P., ... Chollet, F. (1997). The ipsilateral cerebellar hemisphere is overactive during hand movements in akinetic parkinsonian patients. *Brain, 120,* 103–110.

Sadnicka, A., Hoffland, B. S., Bhatia, K. P., van de Warrenburg, B. P., & Edwards, M. J. (2012). The cerebellum in dystonia—help or hindrance? *Clin Neurophysiol, 123,* 65–70.

Saint-Cyr, J. A., Ungerleider, L. G., & Desimone, R. (1990). Organization of visual cortical inputs to the striatum and subsequent outputs to the pallido-nigral complex in the monkey. *J Comp Neurol, 298,* 129–156.

Saga, Y., Hirata, Y., Takahara, D., Inoue, K., Miyachi, S., Nambu, A., ... Hoshi, E. (2011). Origins of multisynaptic projections from the basal ganglia to rostrocaudally distinct sectors of the dorsal premotor area in macaques. *Eur J Neurosci, 33,* 285–297.

Sakai, S. T., Inase, M., & Tanji, J. (1996). Comparison of cerebellothalamic and pallidothalamic projections in the monkey (*Macaca fuscata*): A double anterograde labeling study. *J Comp Neurol, 368,* 215–228.

Sakai, S. T., Inase, M., & Tanji, J. (1999). Pallidal and cerebellar inputs to thalamocortical neurons projecting to the supplementary motor area in *Macaca fuscata*: A triple-labeling light microscopic study. *Anat Embryol (Berl), 199,* 9–19.

Schell, G. R., & Strick, P. L. (1984). The origin of thalamic inputs to the arcuate premotor and supplementary motor areas. *J Neurosci, 4,* 539–560.

Schmahmann, J. D. (1991). An emerging concept. The cerebellar contribution to higher function. *Arch Neurol, 48,* 1178–1187.

Schmahmann, J. D. (1997). Rediscovery of an early concept. *Int Rev Neurobiol, 41,* 3–27.

Schmahmann, J. D., & Pandya, D. N. (1997). The cerebrocerebellar system. *Int Rev Neurobiol, 41,* 31–60.

Schmahmann, J. D., & Sherman, J. C. (1998). The cerebellar cognitive affective syndrome. *Brain, 121,* 561–579.

Schneider, F., Habel, U., Wagner, M., Franke, P., Salloum, J. B., Shah, N. J., ... Zilles, K. (2001). Subcortical correlates of craving in recently abstinent alcoholic patients. *Am J Psychiatry, 158,* 1075–1083.

Schultz, W., Dayan, P., & Montague, P. R. (1997). A neural substrate of prediction and reward. *Science, 275,* 1593–1599.

Seymour, B., O'Doherty, J. P., Dayan, P., Koltzenburg, M., Jones, A. K., Dolan, R. J., ... Frackowiak, R. S. (2004). Temporal difference models describe higher-order learning in humans. *Nature, 429,* 664–667.

Stoodley, C. J. (2012). The cerebellum and cognition: Evidence from functional imaging studies. *Cerebellum, 11,* 352–365.

Stoodley, C. J., & Schmahmann, J. D. (2009). Functional topography in the human cerebellum: A meta-analysis of neuroimaging studies. *NeuroImage, 44,* 489–501.

Stoodley, C. J., Valera, E. M., & Schmahmann, J. D. (2012). Functional topography of the cerebellum for motor and cognitive tasks: An fMRI study. *NeuroImage, 59,* 1560–1570.

Strick, P. L., & Card, J. P. (1992). Transneuronal mapping of neural circuits with alpha herpesviruses. In J. P. Bolam (Ed.), *Experimental neuroanatomy: A practical approach* (pp. 81–101). Oxford, UK: Oxford University Press.

Strick, P. L., Dum, R. P., & Fiez, J. A. (2009). Cerebellum and non-motor function. *Annu Rev Neurosci, 32,* 413–434.

Strick, P. L., Dum, R. P., & Picard, N. (1995). Macro-organization of the circuits connecting the basal ganglia with the cortical motor areas. In J. C. Houk, J. L. Davis, &

D. G. Beiser (Eds.), *Models of information processing in the basal ganglia* (pp. 117–130). Cambridge, MA: MIT Press.
Swain, R. A., Kerr, A. L., & Thompson, R. F. (2011). The cerebellum: A neural system for the study of reinforcement learning. *Front Behav Neurosci, 5*, 8.
Tanaka, K., Saito, H.-A., Fukuda, Y., & Moriya, M. (1991). Coding visual images of objects in the inferotemporal cortex of the macaque monkey. *J Neurophysiol, 66*, 170–189.
Tanaka, S. C., Doya, K., Okada, G., Ueda, K., Okamoto, Y., & Yamawaki, S. (2004). Prediction of immediate and future rewards differentially recruits cortico-basal ganglia loops. *Nat Neurosci, 7*, 887–893.
Thoma, P., Bellebaum, C., Koch, B., Schwarz, M., & Daum, I. (2008). The cerebellum is involved in reward-based reversal learning. *Cerebellum, 7*, 433–443.
Thompson, R. F., Swain, R., Clark, R., & Shinkman, P. (2000). Intracerebellar conditioning—Brogden and Gantt revisited. *Behav Brain Res, 110*, 3–11.
Trost, M., Carbon, M., Edwards, C., Ma, Y., Raymond, D., Mentis, M. J., ... Eidelberg, D. (2002). Primary dystonia: Is abnormal functional brain architecture linked to genotype? *Ann Neurol, 52*, 853–856.
Trost, M., Su, S., Su, P., Yen, R. F., Tseng, H. M., Barnes, A., ... Eidelberg, D. (2006). Network modulation by the subthalamic nucleus in the treatment of Parkinson's disease. *NeuroImage, 31*, 301–307.
van Kan, P. L. E., Houk, J. C., & Gibson, A. R. (1993). Output organization of intermediate cerebellum of the monkey. *J Neurophysiol, 69*, 57–73.
Wichmann, T., DeLong, M. R., Guridi, J., & Obeso, J. A. (2011). Milestones in research on the pathophysiology of Parkinson's disease. *Mov Disord, 26*, 1032–1041.
Wiestler, T., McGonigle, D. J., & Diedrichsen, J. (2011). Integration of sensory and motor representations of single fingers in the human cerebellum. *J Neurophysiol, 105*, 3042–3053.
Wu, T., & Hallett, M. (2013). The cerebellum in Parkinson's disease. *Brain, 136*, 696–709.
Yalachkov, Y., Kaiser, J., & Naumer, M. J. (2009). Brain regions related to tool use and action knowledge reflect nicotine dependence. *J Neurosci, 29*, 4922–4929.
Yalachkov, Y., Kaiser, J., & Naumer, M. J. (2010). Sensory and motor aspects of addiction. *Behav Brain Res, 207*, 215–222.
Yang, Z., Xie, J., Shao, Y. C., Xie, C. M., Fu, L. P., Li, D. J., ... Li, S. J. (2009). Dynamic neural responses to cue-reactivity paradigms in heroin-dependent users: An fMRI study. *Hum Brain Mapp, 30*, 766–775.
Yeterian, E. H., & Pandya, D. N. (1993). Striatal connections of the parietal association cortices in rhesus monkeys. *J Comp Neurol, 332*, 175–197.

37 Basal Ganglia Function

ROBERT S. TURNER AND BENJAMIN PASQUEREAU

ABSTRACT The basal ganglia (BG) is composed of several heavily interconnected nuclei at the base of the cerebrum. A series of anatomically distinct parallel circuits through the BG receive afferent projections from and project back to cortical regions that mediate skeletomotor, oculomotor, frontal associative, and limbic functions. Because of this parallel circuit organization, it is fair to say that the BG contributes to a panoply of brain functions ranging from motor control to cognition and motivation. The specific contribution of the BG to these brain functions (i.e., the algorithm or information transform that it performs) remains a topic of active research. All circuits share a common intrinsic organization captured largely by the heuristic model of direct, indirect, and hyperdirect pathways that link BG input stations to outputs. Activation of the direct pathway may facilitate movement, whereas the indirect and hyperdirect pathways may suppress movement. Neuromodulators such as dopamine have different effects on activity and synaptic plasticity in the three pathways. Clinical disorders that involve the BG are often associated with movement disorders. Although the actual contributions of the BG to brain function are still actively debated, growing evidence suggests that it acts as a reinforcement-driven tutor for the learning of automatic behavioral routines.

The basal ganglia (BG) is composed of a group of nuclei situated at the central base of the forebrain (figure 37.1A). Heavily interconnected with the cerebral cortex, this structural network provides a site at which segregated functional information from diverse cortical areas can interact and be influenced by multiple neuromodulators such as dopamine. Outputs from the BG are directed primarily back to the cortex through the thalamus (especially toward the frontal lobe), and to brainstem structures. Inputs that originate from functionally distinct cortical areas are processed in different parallel loop circuits through the BG (Alexander, Crutcher, & DeLong, 1990). Consistent with this anatomical organization, the activity of BG output neurons in these circuits carries signals related to a variety of functions, including voluntary movement of the limbs, eye movement, decision making, and motivational control.

How is this information processed, combined, and modulated in the BG? How do BG disorders disrupt stable posture and smooth efficient movement? What does this tell us about the role of the BG in the normal control of behavior?

Basal ganglia: A group of interconnected subcortical nuclei

Four major nuclei make up the BG: the striatum, the globus pallidus, the subthalamic nucleus (STN), and the substantia nigra. The approximate location of these nuclei in the primate brain is illustrated in figure 37.1A. The striatum, the largest component of the BG, includes the caudate nucleus and the putamen. The globus pallidus is divided into external (GPe) and internal (GPi) segments. The substantia nigra, located in the midbrain, gets its name from its compact dorsal part (SNc), which is heavily pigmented in primates due to the presence of brown neuromelanin in dopamine neurons. Its ventral part, with a more reticulated appearance, is the pars reticulata (SNr).

Because the connections between the nuclei of the BG are complex, it is most interesting to consider these circuits from the perspective of the information that enters and leaves the BG. Where does it come from, how is it directed within the BG, what happens to it within the BG, where does it go when it leaves the BG, and what influence does it exert on behavior?

Information reaching the BG comes from two major sources: the cerebral cortex and the thalamus. Most information enters the BG at the level of the striatum, although some key inputs enter at the STN. Signals leave the BG from two structures: the GPi and the SNr. Because there are multiple routes from inputs to outputs, information flow through the BG may be complex and involve multiple steps of processing (figure 37.1).

Anatomically and functionally distinct parallel circuits

The BG is organized as a series of anatomically segregated loop circuits (Alexander, DeLong, & Strick, 1986; Middleton & Strick, 2000). This topic is reviewed in depth elsewhere in this volume (see chapter 36 by Dum, Bostan, & Strick) and is discussed briefly here. Different areas of cortex terminate within specific territories of the BG that are connected to similarly specific portions of the thalamus, which in turn project back to innervate the areas of cortex from which the circuit originates.

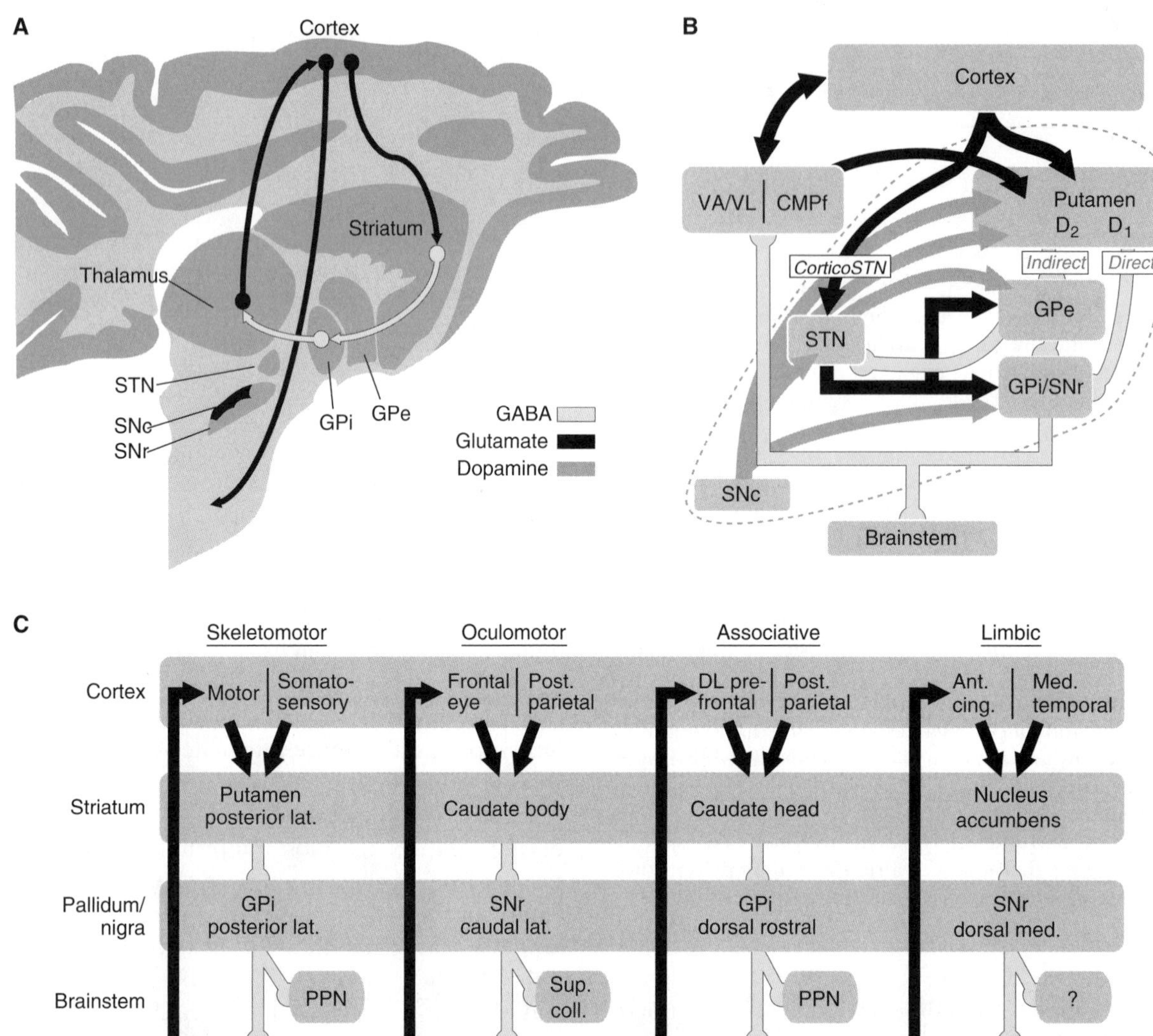

FIGURE 37.1 Circuit diagrams of the basal ganglia (BG) and associated input-output connections. (A) The positions of key BG structures involved in skeletomotor control and their basic input-output connectivity are indicated in a parasagittal section through the macaque brain. The basic loop circuit includes an excitatory glutamatergic (Glu) projection from the neocortex to the striatum (caudate nucleus and putamen) and then inhibitory (γ-amino butyric acid-containing; GABAergic) striatal projection (the "direct pathway") to the internal globus pallidus (GPi). GABAergic neurons in GPi project to targets in the thalamus and brainstem. The main thalamic target of this basic circuit (VA/VL, ventro-anterior/ventrolateral nuclei of the thalamus) projects to the frontal cortex, including parts of the premotor and primary motor cortex. (B) The internal connectivity of the BG (inside the dotted line). Direct and indirect pathways start in striatal projection neurons that express D1- and D2-type dopamine receptors, respectively. D2-type neurons project to the external globus pallidus (GPe). GPe projects to the subthalamic nucleus (STN) and GPi. STN also receives monosynaptic Glu input from the motor cortex and projects to GPi and GPe. GPi sends GABAergic projections to VA/VL and the center median–parafascicular intralaminar complex (CMPf) of the thalamus. CMPf closes another loop by projecting back to the striatum. GPi also projects to brainstem regions such as the pedunculopontine nucleus. Dopaminergic (DA) neurons of the substantia nigra pars compacta (SNc) innervate the striatum and, less densely, the GP and STN. (C) Anatomically segregated parallel circuits through the BG subserve skeletomotor, oculomotor, associative, and limbic functions. All circuits obey a common internal organization. Distant yet functionally related regions of cortex send converging projections to a subregion of the striatum. Medium spiny neurons in each striatal subregion project to distinct subregions of the globus pallidus, which in turn project to different regions of thalamus. (Only direct pathway structures are shown.) BG-receiving subregions of thalamus project back to a subset of the cortical areas that project into each circuit. Descending projections from the GPi terminate in the pedunculopontine nucleus (PPN) or superior colliculus (sup. coll). The target of descending projections from the limbic circuit is not established. Abbreviations: DL, dorsolateral; lat., lateral; med., medial.

Figure 37.1A illustrates an example of this re-entrant loop organization for the circuit that subserves the primary motor cortex. Although the striatum receives afferents from nearly all cortical areas, any one striatal site receives inputs from a small set of cortical areas that have related functions (Flaherty & Graybiel, 1991, 1993). For example, inputs from primary motor and primary sensory cortex—areas involved in skeletomotor function—converge into one specific territory located in the posterolateral putamen. Information from each striatal region is conveyed through separate territories of the BG and back to the cortical areas that originally provide input to its circuit (figure 37.1C). At least four major circuits have been distinguished, including loops associated with skeletomotor, oculomotor, associative (cognitive), and limbic (emotion) functions (Alexander et al., 1986; Hoover & Strick, 1993; Parent & Hazrati, 1995a). This list is by no means exhaustive, given the evidence for loop circuits that include temporal and parietal cortical areas (Clower, Dum, & Strick, 2005; Middleton & Strick, 1996). Several thalamic nuclei also project into the BG with an arrangement that follows the general rule of anatomically segregated parallel functional circuits (McFarland & Haber, 2000; Smith, Raju, Pare, & Sidibe, 2004).

Research continues on whether the anatomical segregation of functional circuits is absolute and the BG-thalamo-cortical circuits are strictly closed loops (Haber & Calzavara, 2009; Joel & Weiner, 1994). For example, growing evidence suggests that limbic regions of the striatum "project" (via a multisynaptic pathway) to non-limbic frontal cortical areas (Kelly & Strick, 2004; Miyachi et al., 2006; Saga et al., 2011). These developments are significant because they provide a pathway for relatively direct communication between functionally disparate areas of cortex (e.g., between limbic and motor cortices). Another important proviso for the segregated loop model is that a significant degree of convergence and "funneling" does occur within each circuit (Percheron & Filion, 1991). The strongest evidence for this notion is the simple fact that the BG output nuclei (GPi and SNr) contain far fewer neurons than the principal input nucleus (the striatum; Harman & Carpenter, 1950; Kemp & Powell, 1971). This funneling and compressing of information may represent one aspect of the core operations performed by this network (Bar-Gad & Bergman, 2001).

Behavioral and electrophysiological experiments also support the concept of functionally distinct BG circuits. Focal lesions and pharmacologic manipulations in the BG produce behavioral abnormalities that differ depending on the region manipulated. For instance, excitation of neuronal activity in the skeletomotor region of the striatum produces abnormal limb movements, whereas disinhibition in associative and limbic regions produce disorders that resemble attention deficit and obsessive-compulsive disorders, respectively (Worbe et al., 2009; Worbe, Epinat, Feger, & Tremblay, 2011). Similarly, neurons that show changes in activity related to arm movements and eye movements are concentrated in different subregions of the BG output nuclei (Anderson & Horak, 1985; DeLong, 1971; Hikosaka & Wurtz, 1983). In all regions, however, many of the neurons are additively influenced by cognitive and motivation-related signals (Hikosaka, Nakamura, & Nakahara, 2006; Kimura, Aosaki, Hu, Ishida, & Watanabe, 1992; Pasquereau et al., 2007; Turner & Anderson, 2005). These observations suggest that many types of information are routed through separate BG channels, but that some types of information may be shared across more than one channel. The crucial question is: what operations does a BG circuit perform on the information it receives?

Organization and physiology of the motor circuit

The BG motor circuit has been studied intensively with respect to its anatomy, physiology, and involvement in behavior. Furthermore, disorders of movement are important components of most BG-associated clinical disorders. Although this section focuses on the motor circuit, it is important to keep in mind that knowledge gained by analysis of the motor circuit is likely to provide insights into the operations of the other parallel BG circuits as well, because of the similar internal connectivity and physiology of all BG circuits.

Figure 37.1B illustrates the basic connectivity of the BG motor circuit. Afferent projections into the BG from skeletomotor-related areas of cortex terminate exclusively in the dorsolateral region of both striatum (putamen) and STN (Parent & Hazrati, 1995a). The putamen also receives glutamatergic projections from the thalamus (specifically, the center median-parafascicular complex, "CMPf"). Efferent projections out of the BG to motor-related regions of the thalamus (VA/VL) originate from the posterolateral GPi. Much of the connectivity between input (striatum/STN) and output (GPi) stages is captured by a simple heuristic model of three principal pathways: (1) the "direct" pathway; (2) the "indirect" pathway; and (3) the "cortico-STN" pathway (figure 37.1B; Albin, Young, & Penney, 1989; DeLong, 1990; Nambu, 2004).

The direct and indirect pathways arise from neurons in the striatum. About 95% of the neurons in the striatum are projection neurons that release the inhibitory neurotransmitter γ-amino-butyric acid (GABA).

These neurons are called medium spiny neurons because of their size and the profusion of spiny processes on their dendrites (Wilson & Groves, 1980). Glutamatergic afferents from the cortex synapse primarily on the dendritic spines (Smith & Bolam, 1990). Thalamic afferents tend to synapse instead on dendritic shafts (Sadikot, Parent, Smith, & Bolam, 1992; Smith & Bolam, 1990). Medium spiny neurons are divided into two distinct types based on their projection patterns (Feger & Crossman, 1984; Smith, Bevan, Shink, & Bolam, 1998), morphology (Gertler, Chan, & Surmeier, 2008), neurochemistry (Gerfen et al., 1990; Shen, Flajolet, Greengard, & Surmeier, 2008), and electrophysiology (Day, Wokosin, Plotkin, Tian, & Surmeier, 2008). One class expresses D1-type dopamine receptors and the neuropeptides dynorphin and substance P. These medium spiny cells project directly to GPi, where they synapse on the GABAergic projection neurons of that nucleus, thereby forming the "direct" pathway. The second type expresses D2-type dopamine receptors and the neuropeptide enkephalin. Those neurons do not project directly to GPi, but instead innervate GABAergic neurons in GPe. From GPe, multiple routes lead to BG output, including a monosynaptic GPe→GPi projection and a bisynaptic GPe→STN→GPi pathway. In this way, D1- and D2-containing medium spiny neurons constitute the starting points for the direct and indirect pathways to the GPi.

The third pathway links cortex to BG output via the glutamatergic STN (Nambu, Tokuno, & Takada, 2002). The STN is an important node in the indirect path because it receives input from both GPe and cortex, and its axons provide the major excitatory input to both pallidal segments (Parent & Hazrati, 1995b). Thus, the STN provides a site at which cortical inputs interact with information processed in the indirect pathway. The internal connectivity of the BG is actually more complicated than stated in the heuristic model outlined here. However, the direct/indirect pathway model captures central features of BG connectivity and accounts for a wide range of physiological and clinical data (Kravitz et al., 2010; Nambu et al., 2002; Shen et al., 2008).

Many of the operations of this network can be understood in terms of changes in firing rates in neurons that belong to the direct and indirect pathways. GPi output neurons generate action potentials spontaneously at very high rates of 50–100 spikes per second (DeLong, 1971; see figure 37.2A for example). Activation of D1-type (direct pathway) medium spiny neurons will inhibit the tonic firing of GPi neurons, which in turn will facilitate thalamocortical activity through the process of disinhibition. In contrast, activation of D2-containing (indirect pathway) medium spiny neurons will inhibit the tonic firing of GPe neurons and, via the multiple indirect pathways, increase GPi firing rates and suppress thalamocortical activity. Activation of STN neurons by direct cortical excitation will also increase GPi firing rates and suppress thalamocortical activity. Because of this arrangement, activation of direct pathway neurons is thought to facilitate thalamocortical activity and thereby movement execution. Activation of indirect and cortico-STN pathways inhibits thalamocortical activity and, consequently, movement (Kravitz et al., 2010; Nambu et al., 2002).

How is neuronal activity in the motor circuit related to movement? Probably the most direct answer to this question comes from considering the spiking activity of the circuit's output neurons. The activity of GPi neurons represents the final product of motor control–related processing in the BG; it is the signal that is sent from the BG to influence other skeletomotor control regions. In the GPi, most neurons show changes in discharge that begin immediately before movements (Anderson & Horak, 1985; DeLong, 1971; Mink & Thach, 1991; Turner & Anderson, 1997). Many of these neurons also respond to proprioceptive stimulation (e.g., tendon

FIGURE 37.2 Abnormal discharge of neurons in GPi is a critical step in the network imbalances associated with Parkinson's disease (B) and Huntington's disease (C). In a normally functioning BG (A), balanced activation of direct and indirect pathways (starting in striatal D1 and D2 neurons, left) leads to intermediate GPi firing rates and random discharge patterns (right). (B) In Parkinson's disease, the dopaminergic innervation of the striatum, pallidum, and STN degenerates (dotted outline, left). The best understood consequence of this is increased activation of D2-type indirect pathway neurons and decreased activation of D1-type direct pathway neurons. The effects of dopamine denervation on pallidum and STN activity are less well understood. The cascading effects result in abnormal GPi activity marked by increased firing rates and rhythmic bursty firing patterns (right). C. In early-stage Huntington's disease, D2-type indirect pathway neurons degenerate selectively (dotted outline, left). The net effect on GPi activity is decreased firing rates and rhythmic bursty firing patterns (right). Diagrams on the left indicate the level of tonic activity in a pathway by the thickness of lines. Traces on the right show brief (~2 second) epochs of action potential discharges (i.e., "spike trains") recorded from three actual GPi neurons under each clinical condition. The spike train in A was recorded from the GPi of a normal macaque monkey. Single-unit spike trains in B and C were recorded from human subjects undergoing surgery for deep brain stimulation. At the time scale used, individual action potentials appear as thin vertical lines of similar height above the level of background noise (thick horizontal bands).

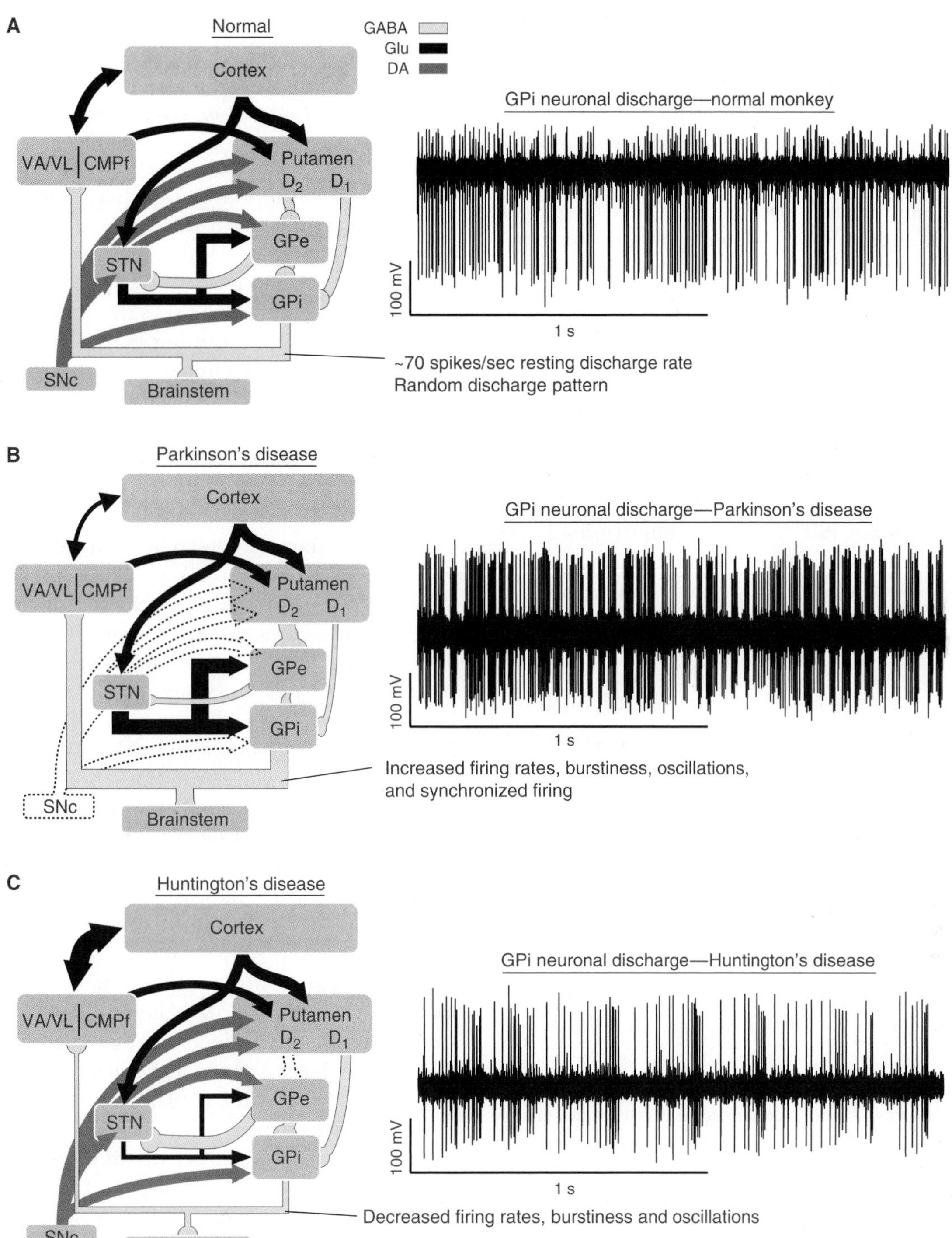
A
Normal
GABA
Glu
DA
Cortex
VA/VL
CMPf
Putamen
D2
D1
GPe
STN
GPi
SNc
Brainstem
GPi neuronal discharge—normal monkey
100 mV
1 s
~70 spikes/sec resting discharge rate
Random discharge pattern
B
Parkinson's disease
Cortex
VA/VL
CMPf
Putamen
D2
D1
GPe
STN
GPi
SNc
Brainstem
GPi neuronal discharge—Parkinson's disease
100 mV
1 s
Increased firing rates, burstiness, oscillations, and synchronized firing
C
Huntington's disease
Cortex
VA/VL
CMPf
Putamen
D2
D1
GPe
STN
GPi
SNc
Brainstem
GPi neuronal discharge—Huntington's disease
100 mV
1 s
Decreased firing rates, burstiness and oscillations

taps; DeLong, Crutcher, & Georgopoulos, 1985). Movement-related changes in firing are almost always influenced by specific characteristics of a movement such as its direction, amplitude, and speed (Turner & Anderson, 1997). Three key observations constrain the role that GPi may play in motor control:

1. Movement-related discharge in the GPi seldom begins early enough to contribute to movement initiation (Mink & Thach, 1991; Turner & Anderson, 1997). Onset latencies are around the time of earliest muscle activation (50–80 milliseconds before movement onset) and, notably, after the activation of primary motor cortex (~120 milliseconds before movement). This observation makes it unlikely that signals processed by the BG contribute to the initiation of movement. Consequently, the common view that the BG plays a central role in the selection of actions (Mink, 1996; Redgrave, Prescott, & Gurney, 1999), a mechanism that obviously must precede movement execution, appears to be untenable. Based on timing, GPi activity may be involved in monitoring movement, updating ongoing motor execution, or both.
2. Movement-related changes in discharge consist of an increase in firing in 60–80% of GPi neurons (DeLong, 1971; Mink & Thach, 1991; Turner & Anderson, 1997). Given that increases in GPi firing are considered to be movement-suppressive, this observation suggests that one of the important functions of output from the BG motor circuit is to suppress or inhibit thalamocortical activity that would otherwise be inappropriate or in conflict with the movement being performed (Mink, 1996).
3. Movement-related activity in the GPi is often influenced by the context of the behavioral task being performed. Movement-related activity can differ depending on the memory requirements of a task (i.e., for movements to remembered target locations versus visible target locations; Turner & Anderson, 2005); whether the movement is discrete or part of a movement sequence (Mushiake & Strick, 1995); or the reward contingencies of the task (Pasquereau et al., 2007). Several studies have confirmed that these cognitive and motivation-related effects cannot be attributed to differences in movement kinematics. Consequently, the BG motor circuit is not involved in the control of movement per se, but rather in integrating cognitive and motivation-related signals with information related to motor commands (Pasquereau et al., 2007; Turner & Anderson, 2005).

The actual role of the BG motor circuit in motor control remains a topic of intense study. Before addressing this topic in more detail, it is helpful to discuss the neuromodulatory mechanisms that affect BG function and the pathophysiology of BG-related disorders.

Modulation of circuit activity by dopamine

Probably the best known and strongest neuromodulatory influence in the striatum is from the nigrostriatal dopaminergic projection. Axons of the dopaminergic cells of the SNc arborize widely and densely in the striatum (gray arrows in figure 37.2), where they terminate most often on the narrow necks that connect spines to the dendritic shafts of medium spiny neurons (Freund, Powell, & Smith, 1984; Groves, Linder, & Young, 1994). This arrangement may provide a mechanism by which dopamine release modulates the influence of cortical inputs to a medium spiny neuron (Nieoullon & Kerkerian-Le Goff, 1992; West, Floresco, Charara, Rosenkranz, & Grace, 2003). Dopaminergic neurons have been shown by Wolfram Schultz to be particularly responsive to unexpected rewards or signals that predict reward (Schultz, 1998). What might be the consequence of this reward-associated signal? Dopamine affects the activity of striatal cells in several ways, and those effects differ for D1- and D2-type medium spiny neurons. First, dopamine has immediate effects on the responsiveness of medium spiny neurons. An increase in dopamine makes D1-type neurons more responsive to corticostriatal excitation and D2-type neurons less responsive (Gerfen & Surmeier, 2011). Second, dopamine has a strong influence on the plasticity of corticostriatal synapses (Gerfen & Surmeier, 2011; Reynolds & Wickens, 2000). Dopaminergic stimulation promotes long-term potentiation (LTP) in D1-type medium spiny neurons and long-term depression (LTD) of excitability in D2-type medium spiny neurons (Kreitzer & Malenka, 2008; Shen et al., 2008). This dopamine-dependent LTP and LTD may be the basis for reward-dependent learning and habit formation in the striatum (Aosaki, Graybiel, & Kimura, 1994). If corticostriatal activation of a D1-type (direct pathway) neuron coincides with an increase in striatal dopamine (e.g., due to delivery of an unexpected reward), then that corticostriatal synapse will be potentiated. Remember that activation of D1-type direct pathway neurons is thought to be movement facilitatory, so it would be appropriate to potentiate movement-facilitating pathways whose activation resulted in delivery of an unexpected reward (Nakamura & Hikosaka, 2006).

Clinical conditions associated with basal ganglia dysfunction

Nearly all of the clinical conditions that involve the BG can be traced to some form of dysfunction in the striatum. For example, loss of the dopaminergic innervation of the striatum results in the motor signs of Parkinson's disease (DeLong & Wichmann, 2009; Franco & Turner, 2012); lesions in the striatum often result in secondary dystonia (Bhatia & Marsden, 1994; Krystkowiak et al., 1998); the D2-type medium spiny neurons of the striatum degenerate selectively in early stages of Huntington's disease (Augood, Faull, & Emson, 1997; Reiner et al., 1988); a specific subpopulation of striatal interneurons is lost in Tourette syndrome (Kalanithi et al., 2005); and striatal neurons appear to have abnormal metabolic activity in obsessive-compulsive disorder (Graybiel & Rauch, 2000). The specific type of behavioral disorder induced by these disparate pathologies is related to the region of the striatum most affected. Dysfunction localized to the posterolateral region of the striatum leads to disorders of movement, whereas dysfunctions of dorsomedial and ventromedial striatal regions lead to disorders in executive control (e.g., attention-deficit/hyperactivity disorder) and disorders in motivation (e.g., obsessive-compulsive disorders), respectively (Worbe et al., 2009). This clinical-anatomical correlation is consistent with the segregated loop circuit organization described previously (Alexander et al., 1986).

Lesions in BG output nuclei, in contrast, have relatively subtle effects on behavior. In fact, as will be discussed in greater depth below, surgical ablation of the motor territory of the GPi (pallidotomy) is an effective treatment for striatal-associated disorders such as Parkinson's disease and dystonia (Baron et al., 1996; Laitinen & Hariz, 1990; Lozano et al., 1997). Together, these observations lead to an apparent paradox—BG-associated disorders arise primarily from dysfunction in the principal input nucleus, the striatum, and can be alleviated by lesions of BG output nuclei. The seeming contradiction can be explained by the concept that it is better to block BG output completely than allow faulty signals from the BG to pervert the normal operations of motor areas that receive BG output (DeLong & Wichmann, 2007). Abnormalities in striatal function, including the clinical disorders listed above, induce grossly abnormal "pathologic" patterns of neuronal activity in the inhibitory output neurons of the BG. These abnormal firing patterns are thought to disrupt the normal operations of BG-recipient brain regions, although the actual mechanisms mediating that disruption remain to be determined. In summary, the GPi constitutes a critical bottleneck in the pathophysiology of BG disorders, and abnormal "pathologic" activity in these neurons is an important step in the expression of BG-associated motor signs.

Parkinson's Disease Patients with Parkinson's disease exhibit akinesia (a paucity of spontaneous movements and slowed movement initiation), bradykinesia (slowed movement execution, including postural responses), rigidity (excessive muscle activity at rest that is enhanced when the muscle is stretched), and a 3- to 8-Hz tremor that diminishes when the individual makes voluntary movements (Paulson & Stern, 1997). Individuals with Parkinson's disease often "freeze," and they have problems combining sequential or simultaneous movements into a smooth motor act.

The changes in BG activity that follow degeneration of the nigrostriatal dopamine system have been well studied in a nonhuman primate model of Parkinson's disease induced by the neurotoxin MPTP (1-methyl-4-phenyl-1,2,3,6-tetrahydropyridine; Bankiewicz et al., 1986; Langston, 1987). Seminal experiments using MPTP found that the activity of neurons in the GPi is altered dramatically following the intoxication (Filion & Tremblay, 1991; Miller & DeLong, 1988); GPi neurons discharged at higher rates than normal, action potentials tended to occur rhythmically in clusters ("bursts"), and the spiking activity of neighboring pallidal neurons was highly synchronized (Raz, Vaadia, & Bergman, 2000). Subsequent neuronal-recording studies in Parkinson's disease patients undergoing surgical treatment confirmed that similar abnormalities in GPi activity are present in the human disease (Levy et al., 2001; Magnin, Morel, & Jeanmonod, 2000; Sterio et al., 1994). (Figure 37.2B shows an example of the abnormal discharge observed in a single neuron recorded from the GPi of a Parkinson's disease patient.) Observations in MPTP-treated monkeys combined with many other lines of evidence have yielded a simple but useful "rate" model to explain how loss of dopamine from the striatum results in abnormal activity in GPi (DeLong & Wichmann, 2007). (See diagram in figure 37.2B.) Because of the opposing effects of dopamine on the two classes of striatal projection neurons, loss of striatal dopamine causes D1-type direct pathway neurons to become less active and D2-type indirect pathway neurons to become more active (Shen et al., 2008). The resulting imbalance favoring activation of the indirect pathway leads to overactivation of GPi neurons and excessive inhibition of BG-receiving regions of the thalamus and cortex involved in motor control. The mechanisms that lead to abnormal firing patterns are less well understood.

Several surgical interventions have been developed to directly interrupt the abnormal excessive GPi activity. These include electrolytic destruction of neurons in the skeletomotor portion of the GPi (pallidotomy; see citations above) or interruption of the abnormal activity by high-frequency electrical stimulation ("deep brain stimulation," DBS) in ventrolateral GPi or in STN (Benabid, 2003; Wichmann & Delong, 2006). How DBS yields clinical benefit remains unclear, although recent evidence suggests that DBS works by blocking the transmission of abnormal pathologic firing patterns in GPi efferent projections (Rubin, McIntyre, Turner, & Wichmann, 2012).

Hyperkinetic Disorders In contrast to the hypokinetic symptoms of Parkinson's disease, involuntary or compulsive movements that cannot be suppressed are found in a variety of other BG disorders, which include Huntington's disease, dystonia, obsessive-compulsive disorder, and the tardive dyskinesia induced by drugs used to treat neuropsychiatric disorders. The abnormal movements in these disorders range from the facial grimaces or jerky flicks of the hands (choreiform movements) characteristic of early Huntington's disease to repetition of more complex acts, such as washing of the hands, that plagues individuals with an obsessive-compulsive disorder. The dyskinesias induced by prolonged L-DOPA therapy for Parkinson's disease also resemble the choreiform movements of Huntington's disease. Many of these "hyperkinetic" disorders are known to be accompanied by a GPi firing rate that is much lower than found in Parkinson's disease, along with an abnormally bursty firing pattern (Papa, Desimone, Fiorani, & Oldfield, 1999; Starr, Kang, Heath, Shimamoto, & Turner, 2008; Starr et al., 2005; Vitek et al., 1999; e.g., see figure 37.2C.)

The rate model can account for many aspects of the hyperkinetic disorders. For example, the abnormally low GPi firing rates in Huntington's disease can be explained as an imbalance that is roughly the converse of the situation in Parkinson's disease: D2-type indirect pathway neurons degenerate selectively in Huntington's, leading to reduced inhibition in the striatum→ GPe pathway and increased firing rate in GPe. This leads, in turn, to increased inhibition of GPi by the GPe→GPi pathway and reduced excitation by the GPe→STN→GPi pathway (figure 37.2C). The net downstream effect of abnormal reduced GPi activity in Huntington's disease appears to be an overfacilitation of the thalamocortical motor circuit, leading to abnormal excessive movements.

Impulse Control Disorders and the Corticosubthalamic Nucleus "Hyperdirect" Pathway The cortico-STN pathway provides a short-latency route to transmit cortical excitatory inputs to the BG output neurons of the GPi and SNr. Because of the directness of this pathway and the fast conduction velocity of its axons, this pathway has also been labeled the "hyperdirect" pathway (Nambu et al., 2002). Keeping in mind that increases in firing in the GPi are thought to suppress movement, the hyperdirect pathway may provide a way to rapidly suppress or control potentially interfering or inappropriate actions. Indeed, a growing list of studies associate this pathway with a subject's ability to suppress or control movements. Most famously, lesions or stroke in the region of the STN can lead to hemiballism, a striking hyperkinetic movement disorder marked by violent flailing movements of one or more limb contralateral to the site of damage (Guridi & Obeso, 2001). Like other hyperkinetic disorders, hemiballism is associated with abnormally low firing rates in the GPi (Hamada & DeLong, 1992; Vitek et al., 1999). In recent years, the cortico-STN pathway has also been implicated in behavioral disorders related to an impaired ability to suppress or delay imminent actions. For example, high-frequency stimulation of the STN, as applied in DBS for PD, increases a subject's impulsivity in a variety of tasks (Cavanagh et al., 2011; Frank, Samanta, Moustafa, & Sherman, 2007; Thobois et al., 2007). The concept that the hyperdirect pathway plays a role in the rapid suppression or delay of imminent actions is also supported by functional imaging (Aron & Poldrack, 2006) and neuronal recording studies (Isoda & Hikosaka, 2008; R. Schmidt, Leventhal, Mallet, Chen, & Berke, 2013). These results all support the idea that abnormal function of the cortico-STN-GPi pathway may play a key role in societally important impulse control disorders, such as those associated with compulsive gambling and substance abuse (Coxon, van Impe, Wenderoth, & Swinnen, 2012; Lhommee et al., 2012; Pelloux & Baunez, 2013).

Problems with the Classical Rate Model of Basal Ganglia Pathophysiology One glaring and unsolved difficulty for the simple rate model is the fact that many hyperkinetic disorders improve dramatically in response to pallidotomy (Cersosimo et al., 2008; Guridi & Obeso, 2001; Lozano et al., 1997). If reduced GPi firing rate alone could induce hyperkinetic disorders, then pallidotomy, which is an ablation of the motor GPi, should exacerbate hyperkinetic disorders rather than improve them. Because of this disparity, many investigators now believe that abnormalities in firing pattern (burstiness, oscillations, or synchronized firing) are just as important in the causation of parkinsonism and hyperkinetic disorders as altered mean firing rate (DeLong & Wichmann, 2009; Rubin et al., 2012). Which firing-pattern abnormality is the critical one, and why, are current topics of research.

It is common to try to deduce the normal functions of the BG by studying the behavioral impairments that accompany BG disorders. This reasoning, that the symptoms of BG disorders represent a kind of "negative image" of normal BG functions, is based on the premise that the primary problem in these disorders is a loss of normal BG function (i.e., loss or disruption of the normal task-related signals transmitted). The success of BG-directed surgical therapies, however, brings that reasoning into question. For example, pallidotomy blocks the transmission of both pathologic activity (the abnormal firing rate and pattern, discussed above) and, importantly, normal task-related signals from the ablated region of the GPi. The fact that transmission-blocking manipulation of the GPi is an effective therapy for both parkinsonism and hyperkinetic disorders suggests that much of the problem in these diseases is not a loss of normal BG signaling, but rather, a disruption of function in other BG-receiving brain regions caused by pathologic output from the BG. Surgical therapies release cortical motor areas from the disrupting influence of pathologic output from the BG and thereby allow those motor control circuits to function more normally. Results from functional imaging studies have supported this idea by showing that pallidotomy and DBS for Parkinson's disease produce a normalization of neuronal activity patterns in non-BG brain regions (Grafton et al., 2006).

Auto-Activation Deficit The final clinical condition to be discussed is an unusual behavioral syndrome termed "athymhormia," or auto-activation deficit, that is associated with focal damage (e.g., ischemic stroke) to the globus pallidus (Bhatia & Marsden, 1994; Habib, 2004; Schmidt et al., 2008). Athymhormia is a disorder of motivation marked by loss or reduction of desire and interest toward activities that were previously highly motivating. Patients with bilateral lesions of the pallidum can perform motor and cognitive tasks normally when prompted by external sensory cues or commands, but they engage in few activities and make few actions spontaneously. Although such cases are rare, they may provide important insights to the contributions of the BG to normal behavior and motor control. Indeed, lesions that include the output nuclei of the BG may disconnect motor-control centers of the brain from the motivation-related signals that normally energize and drive movement (Turner & Desmurget, 2010).

Functions of the basal ganglia using the motor circuit as a model

A remarkably wide range of functions have been suggested for the BG as a whole, ranging from the control of specific aspects of movement to cognitive functions such as working memory, motivation, learning, and procedural memory. From one perspective, such a diversity of function should not come as a surprise given what we know about the macrocircuit organization of the BG (described above; see "Anatomically and functionally distinct parallel circuits"). According to this perspective, multiple BG functions are performed in parallel, with the function of each BG circuit determined by the information that is conveyed into the circuit via cortical and thalamic inputs.

A more probing approach to the analysis of BG functions is to ask: What algorithm (or computational transform) does the canonical BG circuit perform? All of the parallel BG circuits share the same fundamental internal organization, with similarly connected pathways (direct, indirect, and cortico-STN) and similar physiology at each processing step. Thus, it is quite likely that all circuits apply a similar transform to the information that they receive. (This is similar to the perspective used to analyze the parallel circuit organization of the cerebellum. See Diedrichsen & Bastian, chapter 38, this volume.) From this point of view, analysis of the operations and functions of the motor circuit will shed light on how oculomotor, associative, and limbic circuits of the BG work as well. Recent studies suggest three general potential roles for the BG in the control of normal movement: (1) the suppression or facilitation of movements (Mink, 1996; Redgrave et al., 1999); (2) reinforcement-driven learning, especially of habits or skills (Bar-Gad, Morris, & Bergman, 2003; Graybiel, 2005); and (3) the regulation of movement vigor (Turner & Desmurget, 2010). As will be seen, each of these hypotheses is supported, at least partially, by current evidence.

The loop organization of BG-thalamocortical circuits makes it difficult to disentangle the relative roles of the different stages of a circuit. One productive approach to this problem is to investigate how a circuit changes or "transforms" the information it receives from cortical inputs. Using this approach, researchers have compared task-related activity in a certain region of the BG with activity in the cortical area that specifically projects to that circuit. A consistent finding is that task-related activity in the striatum and globus pallidus begins later than activity in connected regions of cortex (Antzoulatos & Miller, 2011; Crutcher & Alexander, 1990; Fujii & Graybiel, 2005; Seo, Lee, & Averbeck, 2012). Because of the late onset times, it is unlikely that the responses of neurons in the BG motor circuit contribute to the selection or initiation of movement under these conditions. A few researchers have studied cortical and striatal activity while animals learn a new skill, such as

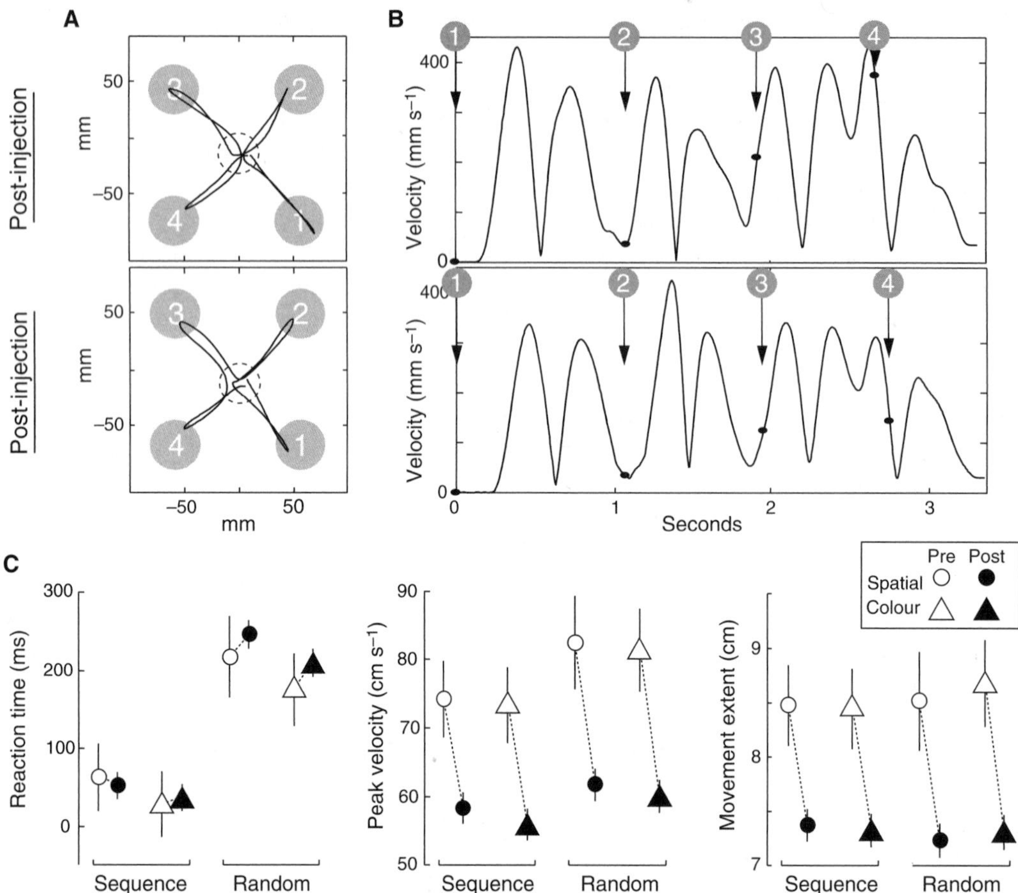

FIGURE 37.3 Inactivation of the BG skeletomotor circuit does not impair movement initiation or sequencing in a neurologically normal animal, but has a selective effect on movement speed and extent. (A) Single behavioral trials illustrate performance of a sequential movement task before and after an injection of muscimol (a long-acting GABAergic inhibitory agent) into the GPi. Animals moved a joystick through a series of four out-and-back component movements. Targets were presented in fixed order (1→4). Spatial trajectories (A) and tangential velocities (B) of the joystick are plotted. Outward movements to capture a peripheral target often began before the instruction cue was presented (↓). GPi inactivation did not impair the smooth uninterrupted execution of the sequence, even though movement extent and velocity were reduced ("post-injection"). (C) Inactivation in the skeletomotor region of GPi had a negligible effect on reaction times (RTs); left, compare pre-injection (open symbols) versus post-injection means (filled symbols). This was true irrespective of whether animals performed overlearned fixed sequences ("Sequence") or random sequences ("Random"), or whether cues provided information by their spatial location (circles) or color (triangles). In contrast, muscimol injections consistently reduced movement velocity (middle) and extent (right) under all conditions. Error bars plus or minus standard error of mean.

choosing the correct response when presented with a novel visual stimulus. In this situation, activity that reflects the newly learned association appears earlier in the striatum than in cortical neurons across the course of learning (Brasted & Wise, 2004; Pasupathy & Miller, 2005; Williams & Eskandar, 2006). Interestingly, task-related activation of the BG appears to become less common as a skill becomes well learned (Tang et al., 2009). These results suggest that the BG contributes selectively to early stages of skill learning (Ashby, Turner, & Horvitz, 2010; Jueptner, Frith, Brooks, Frackowiak, & Passingham,1997).

Another fruitful approach to disentangling BG functions is to determine what aspects of motor behavior are impaired following lesions or inactivations of different circuits. A handful of studies have interrupted GPi activity temporarily by injection of a pharmacological inhibitory agent (i.e., muscimol, $GABA_A$ agonist) into that structure (Desmurget & Turner, 2008; Inase, Buford, & Anderson, 1996; Kato & Kimura, 1992). Inactivation of the skeletomotor region induces significant impairments in the speed and extent of arm movement accompanied by modest postural instability. Remarkably, acute disconnections of the motor circuit do not affect (a) reaction time (i.e., the time required to select and initiate movement); (b) the control of movement direction; (c) feedback mechanisms that mediate on-the-fly movement correction; (d) the fluid performance of familiar sequences of movements; or (e) the expression of familiar stimulus-response habits. Figure 37.3A illustrates examples of single behavioral trials of a familiar sequence of arm movements performed by a monkey before (top) and during (bottom) inactivation of the contralateral GPi by microinjection of muscimol. Before the inactivation, the monkey performed the sequence as a fluid, uninterrupted series of gestures. Individual movements often began before presentation of the visual instruction cue (vertical arrows). During inactivation in the GPi (bottom), sequences were still performed with the fluid anticipatory responding that is characteristic of a well-learned motor skill. In a series of experiments (figure 37.3B), GPi inactivation did not impair movement initiation (i.e., reaction time), but did slow movements substantially ("peak velocity") and reduced their extent (Desmurget & Turner, 2008, 2010).

Given results from inactivation studies like this one, along with clinical observations (e.g., the therapeutic efficacy of pallidotomy) and neuronal recording results, it appears unlikely that the BG motor circuit contributes to movement selection or initiation. It also appears unlikely that the motor circuit is involved in retaining or executing well-learned motor skills like motor sequences or stimulus-response associations. Is there a tenable alternative hypothesis? The observation of circumscribed impairments in regulating movement speed and extent is consistent with the idea that the motor circuit regulates a form of implicit "motor motivation" that influences how vigorously individual actions are performed (Baraduc, Thobois, Gan, Broussolle, & Desmurget, 2013; Mazzoni, Hristova, & Krakauer, 2007; Turner & Desmurget, 2010). The motor vigor hypothesis is also supported by the peculiar symptoms observed in clinical cases of damage to the GP (see "Auto-activation deficit," above), and by the consistent observation in BG structures of neuronal activity that scales with the vigor of movement (i.e., with movement speed, amplitude, or force; Anzak et al., 2012; Brucke et al., 2012; Georgopoulos, DeLong, & Crutcher, 1983; Turner, Desmurget, Grethe, Crutcher, & Grafton, 2003). It remains to be determined if the vigor idea applies equally to all functional circuits of the BG (i.e., that each BG functional circuit links motivation to its own form of action) or if it is rather a specific function of the BG skeletomotor circuit.

Many studies have associated the BG with particular types of learning. As mentioned earlier, patients with lesions involving the BG often have difficulty learning new sequences or skills (Brown et al., 2003; Exner, Koschack, & Irle, 2002; Obeso et al., 2009; Sage et al., 2003). Imaging data from humans indicate that different parts of the BG may be activated as a sequence is learned explicitly versus performed repeatedly with greater speed and accuracy. The associative circuit of the BG is activated during early stages of skill learning, whereas the motor circuit becomes active later, as further practice of a skill leads to habit-like performance (Lehericy et al., 2005). Importantly, learning-related activity appears earlier in the striatum than in connected cortical regions (Pasupathy & Miller, 2005). Combined with the observation that the performance of well-learned skills is not affected by GPi lesions or inactivations, these observations suggest that the BG functions as a kind of tutor, being important during procedural learning but not necessary for storage or recall of already learned information. Very similar ideas have been proposed by Ashby and colleagues with respect to the involvement of associative BG circuits in the learning of visual categories (Ashby et al., 2010) and by the community of scientists who study the roles of the avian BG in song learning (Andalman & Fee, 2009; Brainard & Doupe, 2000). This concept also fits well with the prominent idea that the responses of nigrostriatal dopamine neurons mediate fast reinforcement-driven plasticity in the BG (Bromberg-Martin, Matsumoto, & Hikosaka, 2010; Schultz, 2007).

Thus, multiple lines of evidence indicate that the BG promotes new skill learning, but that other parts of the brain (cortex in particular) take over the storage and production of well-practiced skills. The unique neuromodulatory milieu of the striatum provides an ideal substrate for rapid reinforcement-driven plasticity, but cortex is better suited for long-term retention and execution.

Conclusion

The BG has rich access to information from multiple cortical areas. Anatomically, this input is divided into functionally distinct pathways, but also combined (i.e., funneled) and regulated by multiple neuromodulatory inputs. Dopamine is a neuromodulator of particular interest, because its phasic release is perfectly timed to serve as a learning signal and it has contrasting effects on the synaptic plasticity of direct and indirect pathways through the BG. The final output of the BG is an inhibition of target neurons that is sculpted in association with specific tasks. In large part, this information is directed back to motor and associative areas of the frontal lobes. Over the course of many repetitions, we combine sensory, motivational, and cognitive information to develop behaviors that can be carried out without conscious attention. The BG appears to be a site at which reinforcement acts to modify synaptic action so that these automatic behaviors can be developed.

ACKNOWLEDGMENTS This work was supported in part by the National Institute of Neurological Disorders and Stroke at the National Institutes of Health (grant number NS044393 and NS070865 to RST and the Center for Neuroscience Research in Non-Human Primates, 1P30NS076405).

REFERENCES

Albin, R. L., Young, A. B., & Penney, J. B. (1989). The functional anatomy of basal ganglia disorders. *Trends Neurosci, 12,* 366–375.

Alexander, G. E., Crutcher, M. D., & DeLong, M. R. (1990). Basal ganglia thalamo-cortical circuits: Parallel substrates for motor, oculomotor, "prefrontal" and "limbic" functions. *Prog Brain Res, 85,* 119–146.

Alexander, G. E., DeLong, M. R., & Strick, P. L. (1986). Parallel organization of functionally segregated circuits linking basal ganglia and cortex. *Ann Rev Neurosci, 9,* 357–381.

Andalman, A. S., & Fee, M. S. (2009). A basal ganglia-forebrain circuit in the songbird biases motor output to avoid vocal errors. *Proc Natl Acad Sci USA, 106*(30), 12518–12523.

Anderson, M. E., & Horak, F. B. (1985). Influence of the globus pallidus on arm movements in monkeys. III. Timing of movement-related information. *J Neurophysiol, 54*(2), 433–448.

Antzoulatos, E. G., & Miller, E. K. (2011). Differences between neural activity in prefrontal cortex and striatum during learning of novel abstract categories. *Neuron, 71*(2), 243–249.

Anzak, A., Tan, H., Pogosyan, A., Foltynie, T., Limousin, P., Zrinzo, L., … Brown, P. (2012). Subthalamic nucleus activity optimizes maximal effort motor responses in Parkinson's disease. *Brain, 135*(Pt 9), 2766–2778.

Aosaki, T., Graybiel, A. M., & Kimura, M. (1994). Effect of the nigrostriatal dopamine system on acquired neural responses in the striatum of behaving monkeys. *Science, 265,* 412–415.

Aron, A. R., & Poldrack, R. A. (2006). Cortical and subcortical contributions to Stop signal response inhibition: Role of the subthalamic nucleus. *J Neurosci, 26*(9), 2424–2433.

Ashby, F. G., Turner, B. O., & Horvitz, J. C. (2010). Cortical and basal ganglia contributions to habit learning and automaticity. *Trends Cogn Sci, 14*(5), 208–215.

Augood, S. J., Faull, R. L., & Emson, P. C. (1997). Dopamine D1 and D2 receptor gene expression in the striatum in Huntington's disease. *Ann Neurol, 42*(2), 215–221.

Bankiewicz, K. S., Oldfield, E. H., Chiueh, C. C., Doppman, J. L., Jacobowitz, D. M., & Kopin, I. J. (1986). Hemiparkinsonism in monkeys after unilateral internal carotid artery infusion of 1-methyl-4-phenyl-1,2,3,6-tetrahydropyridine (MPTP). *Life Sci, 39,* 7–16.

Baraduc, P., Thobois, S., Gan, J., Broussolle, E., & Desmurget, M. (2013). A common optimization principle for motor execution in healthy subjects and parkinsonian patients. *J Neurosci, 33*(2), 665–677.

Bar-Gad, I., & Bergman, H. (2001). Stepping out of the box: Information processing in the neural networks of the basal ganglia. *Curr Opin Neurobiol, 11*(6), 689–695.

Bar-Gad, I., Morris, G., & Bergman, H. (2003). Information processing, dimensionality reduction and reinforcement learning in the basal ganglia. *Prog Neurobiol, 71*(6), 439–473.

Baron, M. S., Vitek, J. L., Bakay, R. A. E., Green, J., Kaneoke, Y., Hashimoto, T., … DeLong, M. R. (1996). Treatment of advanced Parkinson's disease by posterior GPi pallidotomy: 1-year results of a pilot study. *Ann Neurol, 40,* 355–366.

Benabid, A. L. (2003). Deep brain stimulation for Parkinson's disease. *Curr Opin Neurobiol, 13*(6), 696–706.

Bhatia, K. P., & Marsden, C. D. (1994). The behavioural and motor consequences of focal lesions of the basal ganglia in man. *Brain, 117*(Pt 4), 859–876.

Brainard, M. S., & Doupe, A. J. (2000). Interruption of a basal ganglia-forebrain circuit prevents plasticity of learned vocalizations. *Nature, 404*(6779), 762–766.

Brasted, P. J., & Wise, S. P. (2004). Comparison of learning-related neuronal activity in the dorsal premotor cortex and striatum. *Eur J Neurosci, 19*(3), 721–740.

Bromberg-Martin, E. S., Matsumoto, M., & Hikosaka, O. (2010). Dopamine in motivational control: Rewarding, aversive, and alerting. *Neuron, 68*(5), 815–834.

Brown, R. G., Jahanshahi, M., Limousin-Dowsey, P., Thomas, D., Quinn, N. P., & Rothwell, J. C. (2003). Pallidotomy and incidental sequence learning in Parkinson's disease. *Neuroreport, 14*(1), 21–24.

Brucke, C., Huebl, J., Schonecker, T., Neumann, W. J., Yarrow, K., Kupsch, A., … Kuhn, A. A. (2012). Scaling of movement is related to pallidal gamma oscillations in patients with dystonia. *J Neurosci, 32*(3), 1008–1019.

Cavanagh, J. F., Wiecki, T. V., Cohen, M. X., Figueroa, C. M., Samanta, J., Sherman, S. J., & Frank, M. J. (2011). Subthalamic nucleus stimulation reverses mediofrontal influence over decision threshold. *Nat Neurosci, 14*(11), 1462–1467.

Cersosimo, M. G., Raina, G. B., Piedimonte, F., Antico, J., Graff, P., & Micheli, F. E. (2008). Pallidal surgery for the treatment of primary generalized dystonia: Long-term follow-up. *Clin Neurol Neurosurg, 110*(2), 145–150.

Clower, D. M., Dum, R. P., & Strick, P. L. (2005). Basal ganglia and cerebellar inputs to "AIP." *Cereb Cortex, 15*(7), 913–920.

Coxon, J. P., van Impe, A., Wenderoth, N., & Swinnen, S. P. (2012). Aging and inhibitory control of action: Cortico-subthalamic connection strength predicts stopping performance. *J Neurosci, 32*(24), 8401–8412.

Crutcher, M. D., & Alexander, G. E. (1990). Movement-related neuronal activity selectively coding either direction or muscle pattern in three motor areas of the monkey. *J Neurophysiol, 64,* 151–163.

Day, M., Wokosin, D., Plotkin, J. L., Tian, X., & Surmeier, D. J. (2008). Differential excitability and modulation of striatal medium spiny neuron dendrites. *J Neurosci, 28*(45), 11603–11614.

DeLong, M. R. (1971). Activity of pallidal neurons during movement. *J Neurophysiol, 34,* 414–427.

DeLong, M. R. (1990). Primate models of movement disorders of basal ganglia origin. *Trends Neurosci, 13,* 281–285.

DeLong, M. R., Crutcher, M. D., & Georgopoulos, A. P. (1985). Primate globus pallidus and subthalamic nucleus: Functional organization. *J Neurophysiol, 53,* 530–543.

DeLong, M. R., & Wichmann, T. (2007). Circuits and circuit disorders of the basal ganglia. *Arch Neurol, 64*(1), 20–24.

DeLong, M. R., & Wichmann, T. (2009). Update on models of basal ganglia function and dysfunction. *Parkinsonism Relat Disord, 15*(Suppl. 3), S237–240.

Desmurget, M., & Turner, R. S. (2008). Testing basal ganglia motor functions through reversible inactivations in the posterior internal globus pallidus. *J Neurophysiol, 99*(3), 1057–1076.

Desmurget, M., & Turner, R. S. (2010). Motor sequences and the basal ganglia: Kinematics, not habits. *J Neurosci, 30*(22), 7685–7690.

Exner, C., Koschack, J., & Irle, E. (2002). The differential role of premotor frontal cortex and basal ganglia in motor sequence learning: Evidence from focal basal ganglia lesions. *Learn Mem, 9*(6), 376–386.

Feger, J., & Crossman, A. R. (1984). Identification of different subpopulations of neostriatal neurons projecting to globus pallidus or substantia nigra in the monkey: A retrograde fluorescence double-labelling study. *Neurosci Lett, 49,* 7–12.

Filion, M., & Tremblay, L. (1991). Abnormal spontaneous activity of globus pallidus neurons in monkeys with MPTP-induced parkinsonism. *Brain Res, 547*(1), 142–151.

Flaherty, A. W., & Graybiel, A. M. (1991). Corticostriatal transformations in the primate somatosensory system. Projections from physiologically mapped body-part representations. *J Neurophysiol, 66,* 1249–1263.

Flaherty, A. W., & Graybiel, A. M. (1993). Two input systems for body representations in the primate striatal matrix: Experimental evidence in the squirrel monkey. *J Neurosci, 13,* 1120–1137.

Franco, V., & Turner, R. S. (2012). Testing the contributions of striatal dopamine loss to the genesis of parkinsonian signs. *Neurobiol Dis, 47*(1), 114–125.

Frank, M. J., Samanta, J., Moustafa, A. A., & Sherman, S. J. (2007). Hold your horses: Impulsivity, deep brain stimulation, and medication in parkinsonism. *Science, 318*(5854), 1309–1312.

Freund, T. F., Powell, J. F., & Smith, J. D. (1984). Tyrosine hydroxylase-immunoreactive boutons in synaptic contact with identified striatonigral neurons, with particular reference to dendritic spines. *Neurosci Lett, 13*(4), 1189–1215.

Fujii, N., & Graybiel, A. M. (2005). Time-varying covariance of neural activities recorded in striatum and frontal cortex as monkeys perform sequential-saccade tasks. *Proc Natl Acad Sci USA, 102*(25), 9032–9037.

Georgopoulos, A., DeLong, M., & Crutcher, M. (1983). Relations between parameters of step-tracking movements and single cell discharge in the globus pallidus and subthalamic nucleus of the behaving monkey. *J Neurosci, 3,* 1586–1598.

Gerfen, C. R., Engber, T. M., Mahan, L. C., Susel, Z., Chase, T. N., Monsma Jr., F. J., & Sibley, D. R. (1990). D1 and D2 dopamine receptor-regulated gene expression of striatonigral and striatopallidal neurons. *Science, 250,* 1429–1432.

Gerfen, C. R., & Surmeier, D. J. (2011). Modulation of striatal projection systems by dopamine. *Annu Rev Neurosci, 34,* 441–466.

Gertler, T. S., Chan, C. S., & Surmeier, D. J. (2008). Dichotomous anatomical properties of adult striatal medium spiny neurons. *J Neurosci, 28*(43), 10814–10824.

Grafton, S. T., Turner, R. S., Desmurget, M., Bakay, R., DeLong, M., Vitek, J., & Crutcher, M. (2006). Normalizing motor-related brain activity: Subthalamic nucleus stimulation in Parkinson disease. *Neurology, 66*(8), 1192–1199.

Graybiel, A. M. (2005). The basal ganglia: Learning new tricks and loving it. *Curr Opin Neurobiol, 15*(6), 638–644.

Graybiel, A. M., & Rauch, S. L. (2000). Toward a neurobiology of obsessive-compulsive disorder. *Neuron, 28*(2), 343–347.

Groves, P. M., Linder, J. C., & Young, S. J. (1994). 5-hydroxydopamine-labeled dopaminergic axons: Three-dimensional reconstructions of axons, synapses and postsynaptic targets in rat neostriatum. *Neuroscience, 58*(3), 593–604.

Guridi, J., & Obeso, J. A. (2001). The subthalamic nucleus, hemiballismus and Parkinson's disease: Reappraisal of a neurosurgical dogma. *Brain, 124*(Pt 1), 5–19.

Haber, S. N., & Calzavara, R. (2009). The cortico-basal ganglia integrative network: The role of the thalamus. *Brain Res Bull, 78*(2–3), 69–74.

Habib, M. (2004). Athymhormia and disorders of motivation in Basal Ganglia disease. *J Neuropsychiatry Clin Neurosci, 16*(4), 509–524.

Hamada, I., & DeLong, M. R. (1992). Excitotoxic acid lesions of the primate subthalamic nucleus result in reduced pallidal neuronal activity during active holding. *J Neurophysiol, 68,* 1859–1866.

Harman, P. J., & Carpenter, M. B. (1950). Volumetric comparisons of the basal ganglia of various primates including man. *J Comp Neurol, 93*(1), 125–137.

Hikosaka, O., Nakamura, K., & Nakahara, H. (2006). Basal ganglia orient eyes to reward. *J Neurophysiol, 95*(2), 567–584.

Hikosaka, O., & Wurtz, R. H. (1983). Visual and oculomotor functions of monkey substantia nigra pars reticulata. I. Relation of visual and auditory responses to saccades. *J Neurophysiol, 49,* 1230–1253.

Hoover, J. E., & Strick, P. L. (1993). Multiple output channels in the basal ganglia. *Science, 259,* 819–821.

Inase, M., Buford, J. A., & Anderson, M. E. (1996). Changes in the control of arm position, movement, and thalamic discharge during local inactivation in the globus pallidus of the monkey. *J Neurophysiol, 75*(3), 1087–1104.

Isoda, M., & Hikosaka, O. (2008). Role for subthalamic nucleus neurons in switching from automatic to controlled eye movement. *J Neurosci, 28*(28), 7209–7218.

Joel, D., & Weiner, I. (1994). The organization of the basal ganglia-thalamocortical circuits: Open interconnected rather than closed segregated. *Neuroscience, 63*(2), 363–379.

Jueptner, M., Frith, C. D., Brooks, D. J., Frackowiak, R. S., & Passingham, R. E. (1997). Anatomy of motor learning. II. Subcortical structures and learning by trial and error. *J Neurophysiol, 77*(3), 1325–1337.

Kalanithi, P. S., Zheng, W., Kataoka, Y., DiFiglia, M., Grantz, H., Saper, C. B., ... Vaccarino, F. M. (2005).

Altered parvalbumin-positive neuron distribution in basal ganglia of individuals with Tourette syndrome. *Proc Natl Acad Sci USA, 102*(37), 13307–13312.

Kato, M., & Kimura, M. (1992). Effects of reversible blockade of basal ganglia on a voluntary arm movement. *J Neurophysiol, 68*, 1516–1534.

Kelly, R. M., & Strick, P. L. (2004). Macro-architecture of basal ganglia loops with the cerebral cortex: Use of rabies virus to reveal multisynaptic circuits. *Prog Brain Res, 143*, 449–459.

Kemp, J. M., & Powell, T. P. S. (1971). The connections of the striatum and globus pallidus: Synthesis and speculation. *Philos Trans R Soc Lond B Biol Sci, 262*, 441–457.

Kimura, M., Aosaki, T., Hu, Y., Ishida, A., & Watanabe, K. (1992). Activity of primate putamen neurons is selective to the mode of voluntary movement: Visually guided, self-initiated or memory-guided. *Exp Brain Res, 89*(3), 473–477.

Kravitz, A. V., Freeze, B. S., Parker, P. R., Kay, K., Thwin, M. T., Deisseroth, K., & Kreitzer, A. C. (2010). Regulation of parkinsonian motor behaviours by optogenetic control of basal ganglia circuitry. *Nature, 466*(7306), 622–626.

Kreitzer, A. C., & Malenka, R. C. (2008). Striatal plasticity and basal ganglia circuit function. *Neuron, 60*(4), 543–554.

Krystkowiak, P., Martinat, P., Defebvre, L., Pruvo, J. P., Leys, D., & Destee, A. (1998). Dystonia after striatopallidal and thalamic stroke: Clinicoradiological correlations and pathophysiological mechanisms. *J Neurol Neurosurg Psychiatry, 65*(5), 703–708.

Laitinen, L. V., & Hariz, M. I. (1990). Pallidal surgery abolishes all parkinsonian symptoms. *Mov Disord, 5*(1), 82.

Langston, J. W. (1987). MPTP: The promise of a new neurotoxin. In C. D. Marsden & S. Fahn (Eds.), *Movement disorders 2* (pp. 73–90). London: Butterworths.

Lehericy, S., Benali, H., van de Moortele, P. F., Pelegrini-Issac, M., Waechter, T., Ugurbil, K., & Doyon, J. (2005). Distinct basal ganglia territories are engaged in early and advanced motor sequence learning. *Proc Natl Acad Sci USA, 102*(35), 12566–12571.

Levy, R., Dostrovsky, J. O., Lang, A. E., Sime, E., Hutchison, W. D., & Lozano, A. M. (2001). Effects of apomorphine on subthalamic nucleus and globus pallidus internus neurons in patients with Parkinson's disease. *J Neurophysiol, 86*(1), 249–260.

Lhommee, E., Klinger, H., Thobois, S., Schmitt, E., Ardouin, C., Bichon, A., ... Krack, P. (2012). Subthalamic stimulation in Parkinson's disease: Restoring the balance of motivated behaviours. *Brain, 135*(Pt 5), 1463–1477.

Lozano, A. M., Kumar, R., Gross, R. E., Giladi, N., Hutchison, W. D., Dostrovsky, J. O., & Lang, A. E. (1997). Globus pallidus internus pallidotomy for generalized dystonia. *Mov Disord, 12*(6), 865–870.

Magnin, M., Morel, A., & Jeanmonod, D. (2000). Single-unit analysis of the pallidum, thalamus and subthalamic nucleus in parkinsonian patients. *Neuroscience, 96*(3), 549–564.

Mazzoni, P., Hristova, A., & Krakauer, J. W. (2007). Why don't we move faster? Parkinson's disease, movement vigor, and implicit motivation. *J Neurosci, 27*(27), 7105–7116.

McFarland, N. R., & Haber, S. N. (2000). Convergent inputs from thalamic motor nuclei and frontal cortical areas to the dorsal striatum in the primate. *J Neurosci, 20*(10), 3798–3813.

Middleton, F. A., & Strick, P. L. (1996). The temporal lobe is a target of output from the basal ganglia. *Proc Natl Acad Sci USA, 93*(16), 8683–8687.

Middleton, F. A., & Strick, P. L. (2000). Basal ganglia and cerebellar loops: Motor and cognitive circuits. *Brain Res Brain Res Rev, 31*(2–3), 236–250.

Miller, W. C., & DeLong, M. R. (1988). Parkinsonian symptomatology: An anatomical and physiological analysis. *Ann N Y Acad Sci, 515*, 287–302.

Mink, J. (1996). The basal ganglia: Focused selection and inhibition of competing motor programs. *Prog Neurobiol, 50*, 381–425.

Mink, J., & Thach, W. (1991). Basal ganglia motor control. II. Late pallidal timing relative to movement onset and inconsistent pallidal coding of movement parameters. *J Neurophysiol, 65*, 301–329.

Miyachi, S., Lu, X., Imanishi, M., Sawada, K., Nambu, A., & Takada, M. (2006). Somatotopically arranged inputs from putamen and subthalamic nucleus to primary motor cortex. *Neurosci Res, 56*(3), 300–308.

Mushiake, H., & Strick, P. L. (1995). Pallidal neuron activity during sequential arm movements. *J Neurophysiol, 74*, 2754–2758.

Nakamura, K., & Hikosaka, O. (2006). Role of dopamine in the primate caudate nucleus in reward modulation of saccades. *J Neurosci, 26*(20), 5360–5369.

Nambu, A. (2004). A new dynamic model of the cortico-basal ganglia loop. *Prog Brain Res, 143*, 461–466.

Nambu, A., Tokuno, H., & Takada, M. (2002). Functional significance of the cortico-subthalamo-pallidal "hyperdirect" pathway. *Neurosci Res, 43*(2), 111–117.

Nieoullon, A., & Kerkerian-Le Goff, L. (1992). Cellular interactions in the striatum involving neuronal systems using "classical" neurotransmitters: Possible functional implications. *Mov Disord, 7*(4), 311–325.

Obeso, J. A., Jahanshahi, M., Alvarez, L., Macias, R., Pedroso, I., Wilkinson, L., ... Rothwell, J. C. (2009). What can man do without basal ganglia motor output? The effect of combined unilateral subthalamotomy and pallidotomy in a patient with Parkinson's disease. *Exp Neurol, 220*(2), 283–292.

Papa, S. M., Desimone, R., Fiorani, M., & Oldfield, E. H. (1999). Internal globus pallidus discharge is nearly suppressed during levodopa-induced dyskinesias. *Ann Neurol, 46*(5), 732–738.

Parent, A., & Hazrati, L. N. (1995a). Functional anatomy of the basal ganglia. I. The cortico-basal ganglia-thalamo-cortical loop. *Brain Res Brain Res Rev, 20*, 91–127.

Parent, A., & Hazrati, L. N. (1995b). Functional anatomy of the basal ganglia. II. The place of subthalamic nucleus and external pallidum in basal ganglia circuitry. *Brain Res Brain Res Rev, 20*(1), 128–154.

Pasquereau, B., Nadjar, A., Arkadir, D., Bezard, E., Goillandeau, M., Bioulac, B., ... Boraud, T. (2007). Shaping of motor responses by incentive values through the basal ganglia. *J Neurosci, 27*(5), 1176–1183.

Pasupathy, A., & Miller, E. K. (2005). Different time courses of learning-related activity in the prefrontal cortex and striatum. *Nature, 433*(7028), 873–876.

Paulson, H. L., & Stern, M. B. (1997). Clinical manifestations of Parkinson's disease. In R. L. Watts & W. C. Koller

(Eds.), *Movement disorders: Neurologic principles and practice* (pp. 183–199). New York, NY: McGraw-Hill.

Pelloux, Y., & Baunez, C. (2013). Deep brain stimulation for addiction: Why the subthalamic nucleus should be favored. *Curr Opin Neurobiol, 23,* 713–720.

Percheron, G., & Filion, M. (1991). Parallel processing in the basal ganglia: Up to a point. *Trends Neurosci, 14,* 55–56.

Raz, A., Vaadia, E., & Bergman, H. (2000). Firing patterns and correlations of spontaneous discharge of pallidal neurons in the normal and the tremulous 1-methyl-4-phenyl-1,2,3,6-tetrahydropyridine vervet model of parkinsonism. *J Neurosci, 20*(22), 8559–8571.

Redgrave, P., Prescott, T. J., & Gurney, K. (1999). The basal ganglia: A vertebrate solution to the selection problem? *Neuroscience, 89*(4), 1009–1023.

Reiner, A., Albin, R. L., Anderson, K. D., D'Amato, C. J., Penney, J. B., & Young, A. B. (1988). Differential loss of striatal projection neurons in Huntington disease. *Proc Natl Acad Sci USA, 85,* 5733–5737.

Reynolds, J. N., & Wickens, J. R. (2000). Substantia nigra dopamine regulates synaptic plasticity and membrane potential fluctuations in the rat neostriatum, in vivo. *Neuroscience, 99*(2), 199–203.

Rubin, J. E., McIntyre, C. C., Turner, R. S., & Wichmann, T. (2012). Basal ganglia activity patterns in parkinsonism and computational modeling of their downstream effects. *Eur J Neurosci, 36*(2), 2213–2228.

Sadikot, A. F., Parent, A., Smith, Y., & Bolam, J. P. (1992). Efferent connections of the centromedian and parafascicular thalamic nuclei in the squirrel monkey: A light and electron microscopic study of the thalamostriatal projection in relation to striatal heterogeneity. *J Comp Neurol, 320,* 228–242.

Saga, Y., Hirata, Y., Takahara, D., Inoue, K., Miyachi, S., Nambu, A., ... Hoshi, E. (2011). Origins of multisynaptic projections from the basal ganglia to rostrocaudally distinct sectors of the dorsal premotor area in macaques. *Eur J Neurosci, 33*(2), 285–297.

Sage, J. R., Anagnostaras, S. G., Mitchell, S., Bronstein, J. M., de Salles, A., Masterman, D., & Knowlton, B. J. (2003). Analysis of probabilistic classification learning in patients with Parkinson's disease before and after pallidotomy surgery. *Learn Mem, 10*(3), 226–236.

Schmidt, L., D'Arc, B. F., Lafargue, G., Galanaud, D., Czernecki, V., Grabli, D., ... Pessiglione, M. (2008). Disconnecting force from money: Effects of basal ganglia damage on incentive motivation. *Brain, 131*(Pt 5), 1303–1310.

Schmidt, R., Leventhal, D. K., Mallet, N., Chen, F., & Berke, J. D. (2013). Canceling actions involves a race between basal ganglia pathways. *Nat Neurosci, 16*(8), 1118–1124.

Schultz, W. (1998). Predictive reward signal of dopamine neurons. *J Neurophysiol, 80*(1), 1–27.

Schultz, W. (2007). Behavioral dopamine signals. *Trends Neurosci, 30*(5), 203–210.

Seo, M., Lee, E., & Averbeck, B. B. (2012). Action selection and action value in frontal-striatal circuits. *Neuron, 74*(5), 947–960.

Shen, W., Flajolet, M., Greengard, P., & Surmeier, D. J. (2008). Dichotomous dopaminergic control of striatal synaptic plasticity. *Science, 321*(5890), 848–851.

Smith, A. D., & Bolam, J. P. (1990). The neural network of the basal ganglia as revealed by the study of synaptic connections of identified neurones. *Trends Neurosci, 13,* 259–265.

Smith, Y., Bevan, M. D., Shink, E., & Bolam, J. P. (1998). Microcircuitry of the direct and indirect pathways of the basal ganglia. *Neuroscience, 86*(2), 353–387.

Smith, Y., Raju, D. V., Pare, J. F., & Sidibe, M. (2004). The thalamostriatal system: A highly specific network of the basal ganglia circuitry. *Trends Neurosci, 27*(9), 520–527.

Starr, P. A., Kang, G. A., Heath, S., Shimamoto, S., & Turner, R. S. (2008). Pallidal neuronal discharge in Huntington's disease: Support for selective loss of striatal cells originating the indirect pathway. *Exp Neurol, 211*(1), 227–233.

Starr, P. A., Rau, G. M., Davis, V., Marks, W. J., Jr., Ostrem, J. L., Simmons, D., ... Turner, R. S. (2005). Spontaneous pallidal neuronal activity in human dystonia: Comparison with Parkinson's disease and normal macaque. *J Neurophysiol, 93*(6), 3165–3176.

Sterio, D., Beric, A., Dogali, M., Fazzini, E., Alfaro, G., & Devinsky, O. (1994). Neurophysiological properties of pallidal neurons in Parkinson's disease. *Ann Neurol, 35*(5), 586–591.

Tang, C. C., Root, D. H., Duke, D. C., Zhu, Y., Teixeria, K., Ma, S., ... West, M. O. (2009). Decreased firing of striatal neurons related to licking during acquisition and overtraining of a licking task. *J Neurosci, 29*(44), 13952–13961.

Thobois, S., Hotton, G. R., Pinto, S., Wilkinson, L., Limousin-Dowsey, P., Brooks, D. J., & Jahanshahi, M. (2007). STN stimulation alters pallidal-frontal coupling during response selection under competition. *J Cereb Blood Flow Metab, 27*(6), 1173–1184.

Turner, R. S., & Anderson, M. E. (1997). Pallidal discharge related to the kinematics of reaching movements in two dimensions. *J Neurophysiol, 77,* 1051–1074.

Turner, R. S., & Anderson, M. E. (2005). Context-dependent modulation of movement-related discharge in the primate globus pallidus. *J Neurosci, 25*(11), 2965–2976.

Turner, R. S., & Desmurget, M. (2010). Basal ganglia contributions to motor control: A vigorous tutor. *Curr Opin Neurobiol, 20*(6), 704–716.

Turner, R. S., Desmurget, M., Grethe, J., Crutcher, M. D., & Grafton, S. T. (2003). Motor subcircuits mediating the control of movement extent and speed. *J Neurophysiol, 90*(6), 3958–3966.

Vitek, J. L., Chockkan, V., Zhang, J. Y., Kaneoke, Y., Evatt, M., DeLong, M. R., ... Bakay, R. A. (1999). Neuronal activity in the basal ganglia in patients with generalized dystonia and hemiballismus. *Ann Neurol, 46*(1), 22–35.

West, A. R., Floresco, S. B., Charara, A., Rosenkranz, J. A., & Grace, A. A. (2003). Electrophysiological interactions between striatal glutamatergic and dopaminergic systems. *Ann N Y Acad Sci, 1003,* 53–74.

Wichmann, T., & DeLong, M. R. (2006). Deep brain stimulation for neurologic and neuropsychiatric disorders. *Neuron, 52*(1), 197–204.

Williams, Z. M., & Eskandar, E. N. (2006). Selective enhancement of associative learning by microstimulation of the anterior caudate. *Nat Neurosci, 9*(4), 562–568.

Wilson, C. J., & Groves, P. M. (1980). Fine structure and synaptic connections of the common spiny neuron of the rat neostriatum: A study employing intracellular injection of horseradish peroxidase. *J Comp Neurol, 194*, 599–615.

Worbe, Y., Baup, N., Grabli, D., Chaigneau, M., Mounayar, S., McCairn, K., ... Tremblay, L. (2009). Behavioral and movement disorders induced by local inhibitory dysfunction in primate striatum. *Cereb Cortex, 19*(8), 1844–1856.

Worbe, Y., Epinat, J., Feger, J., & Tremblay, L. (2011). Discontinuous long-train stimulation in the anterior striatum in monkeys induces abnormal behavioral states. *Cereb Cortex, 21*(12), 2733–2741.

38 Cerebellar Function

JÖRN DIEDRICHSEN AND AMY BASTIAN

ABSTRACT The cerebellum is a large subcortical brain structure that influences movement, sensation, and cognitive behaviors through interactions with the cerebral cortex and brainstem. Cerebellar damage does not abolish these behaviors, but instead reduces their accuracy and flexibility. It has been hypothesized that the cerebellum contributes to its various functions by providing predictions about future sensory and motor states, and by implementing a fast error-driven learning process, which calibrates well-trained behaviors to environmental changes.

Cell types and physiology

The cerebellum consists of a highly folded cortical sheet surrounding white matter and deep cerebellar nuclei. A remarkable feature of the cerebellum is its three-layered cortex, which is homogeneous in both cell type and arrangement (figure 38.1).

The main source of input to the cerebellum is provided by mossy fibers, which transmit information from the neocortex via the pontine nuclei, from brainstem structures including the vestibular and reticular nuclei, or directly from the spinal cord via the spinocerebellar tracts. Mossy fibers synapse on granule cells (figure 38.1), which account for over half of the neurons in the human brain (Azevedo et al., 2009). Granule cells send axons into the molecular layer, where they branch into parallel fibers and form excitatory synapses onto Purkinje cells. Purkinje cells have flat but elaborate dendritic trees, branching orthogonally to the direction of the parallel fibers. In this way, each Purkinje cell can receive input from ~174,000 granule cells. At rest, Purkinje cells fire action potentials, known as simple spikes, at a rate of 40–70 Hz. Climbing fibers provide more sparse input to the cerebellum. Each Purkinje cell receives input from a single climbing fiber, while each climbing fiber branches to innervate ~10 Purkinje cells. Climbing fibers cause Purkinje cells to fire powerful action potentials with complex waveforms, termed complex spikes, at the low rate of 1–4 Hz.

Purkinje cells inhibit cells in the deep cerebellar nuclei, which provide the cerebellum's only output. Pauses in the firing of specific sets of Purkinje cells release inhibition of the deep cerebellar nuclei cells, which in turn emit an excitatory burst of action potentials. The largest of these nuclei, the dentate nucleus, receives input from the lateral hemispheres of the cerebellar cortex and projects via the thalamus back to contralateral premotor, prefrontal, and parietal neocortical regions. More medially, the interposed (emboliform and globose) nuclei receive input from intermediate cerebellar cortex (paravermis) and project to the spinal cord via the red nucleus and to primary motor cortex via the thalamus. The most medial nucleus, the fastigial nucleus, receives input from the midline areas of cerebellar cortex (vermis) and projects to neocortex and brainstem targets, influencing descending vestibulospinal and reticulospinal pathways.

Functional anatomy

The cerebral cortex can be parcellated into different regions based on cytoarchitectonic organization. In contrast, the neural circuit of the cerebellar cortex is remarkably invariant, even though the human cerebellum contains many distinct functional modules. In the medial-lateral direction, it consists of a set of parasaggital compartments, which can be revealed by labeling the expression of aldolase C or zebrin (Sugihara & Shinoda, 2004). In the anterior-posterior direction, a series of horizontally running fissures divide the cerebellum into a set of lobules (figure 38.2), labeled I-X based on Olof Larsell's careful comparative work (Larsell & Jansen, 1972).

Lobules I–V form the anterior cerebellum and are mainly concerned with motor control. Lobules IV and V have reciprocal connections with primary motor cortex (Kelly & Strick, 2003) and a reliable somatotopic organization (Grodd, Hulsmann, Lotze, Wildgruber, & Erb, 2001). The feet are represented in superior and medial aspects of lobule IV, and the arm and hand in the hemispheric part of lobule V. The latter also contains representations of single finger movements (Wiestler, McGonigle, & Diedrichsen, 2011). Upper-limb representations are strongly lateralized. Hand movements activate the ipsilateral anterior lobe almost exclusively. The representation of orofacial movements extends from lobule V into the neighboring lobule VI.

Complex limb movements preferentially activate lobule VI (Schlerf, Verstynen, Ivry, & Spencer, 2010). In contrast to lobule V, lobule VI generally shows

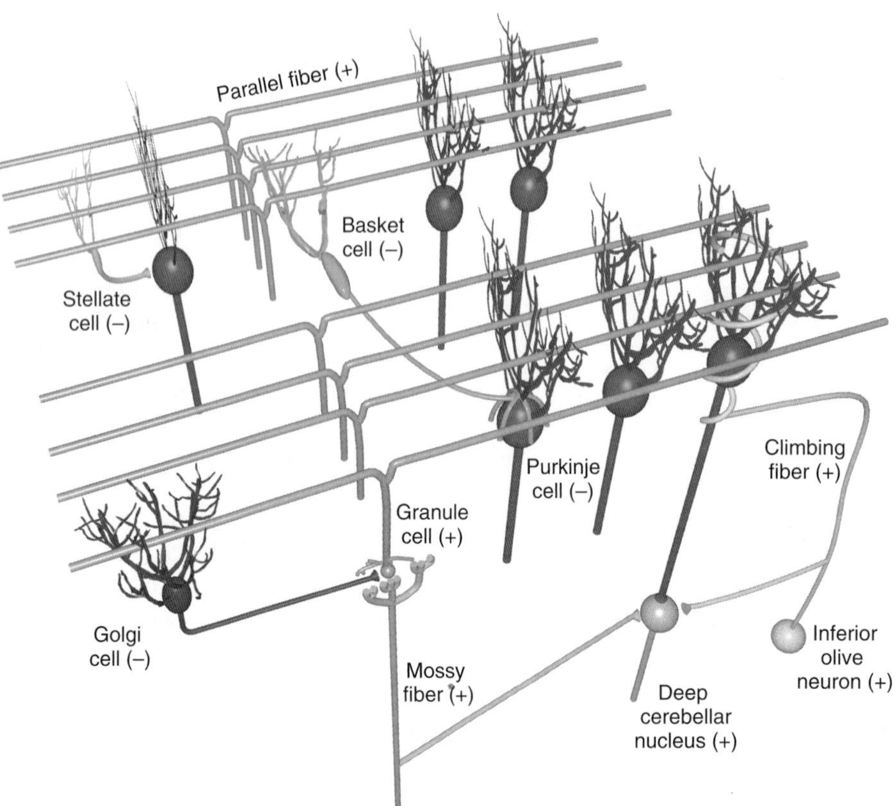

FIGURE 38.1 Main anatomical layout of the cerebellar cortex displaying major cell types and connections. The sign indicates whether the cell gives rise to excitatory (+) or inhibitory (–) connections.

bilateral activation for unilateral hand movements. This suggests that it acts in concert with cortical secondary motor areas such as premotor and supplementary motor area. Indeed, functional connectivity studies show high correlations between activity in lobule VI and activity in the contralateral premotor cortex (Buckner, Krienen, Castellanos, Diaz, & Yeo, 2011). Language tasks, including verb generation, also activate right lobule VI (and crus I of lobule VII), even when motor output-related activity is controlled for. Left lobule VI may be associated with spatial tasks (Stoodley & Schmahmann, 2009). The vermis of lobules VI and VII controls eye movements and is known as the oculomotor vermis (Prsa & Thier, 2011).

Lobule VII is the largest lobule of the human cerebellum, accounting for roughly half the cerebellar gray matter volume (Diedrichsen, Balsters, Flavell, Cussans, & Ramnani, 2009). It is subdivided into lobule VIIa (consisting of crus I and II) and lobule VIIb. In the monkey, the hemispheres of crus II have reciprocal connections with Brodman area 46 (Kelly & Strick, 2003), and human functional connectivity data indicate that lobule VII participates in at least three subnetworks including dorsolateral prefrontal, inferior parietal, and lateral temporal areas (Buckner et al., 2011). Functionally, various language, working memory, and executive function tasks activate lobule VII (Stoodley & Schmahmann, 2009).

Lobules VIIIa and VIIIb are part of a second motor cortical–cerebellar loop (Kelly & Strick, 2003). As with the anterior motor representation, these lobules have a convergent representation of movement and sensory information from the whole body. While there is discernable somatotopy in these regions, it is weaker than that in the anterior lobe.

The hemispheres of lobule IX connects to the anterior cingulate and precuneus (Buckner et al., 2011). The vermal aspect of lobule IX appears to be involved in the regulation of the autonomic nervous system, for example, in cardiovascular control. Finally, the relatively small lobule X (flocculus + nodulus) is involved in vestibular functions and eye movement control.

Together, the cerebellar cortex comprises a patchwork of areas, each with its own pattern of connectivity and functional specialization. The information-processing role that each cerebellar region fulfills within its respective loop remains unclear. However, given the homogenous functional architecture of the cerebellar cortex, many researchers agree that a single computational function should characterize the role of the

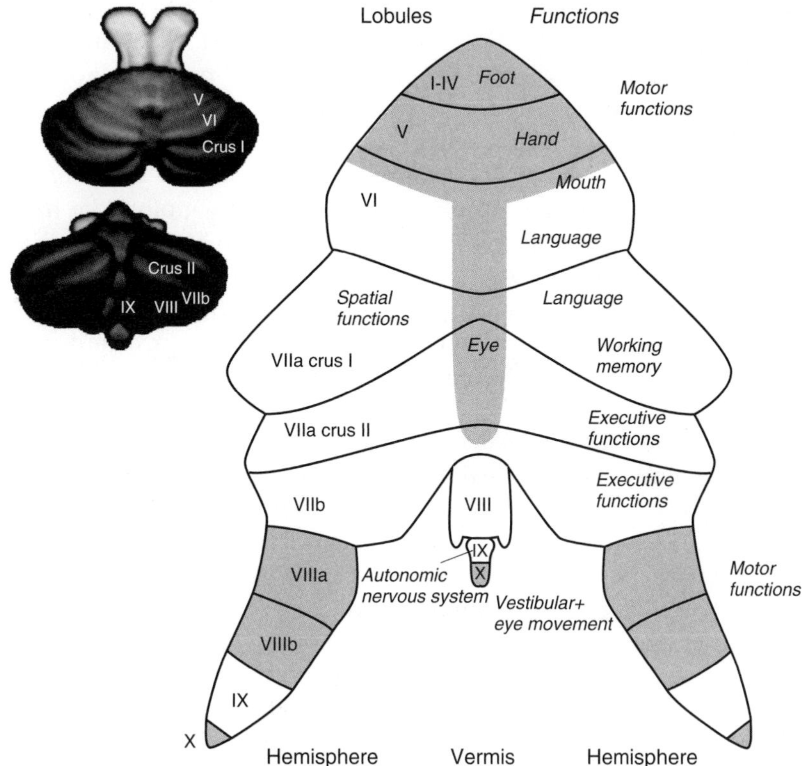

FIGURE 38.2 Functional anatomy of the human cerebellum shown on a flattened representation of the cerebellar cortex (based on Van Essen, 2002). Lobule surface area is roughly proportional to real size. The upper inset shows dorsal and ventral views of a cerebellar 3D reconstruction. Roman numerals indicate lobules following Larsell's notation; italics indicate likely involvement in behavioral functions. Gray shading indicates sensory-motor involvement and connectivity to motor regions.

cerebellum across motor and nonmotor domains (Ivry, Spencer, Zelaznik, & Diedrichsen, 2002; Schmahmann, 2004). However, what constitutes this "universal cerebellar transform" is a much-debated matter.

Clinical aspects

Neurological diseases, tumors, malformations, and stroke can all damage the cerebellum. Loss of movement coordination or "ataxia" is the most common and obvious deficit. Because the cerebellum projects to contralateral neocortical areas, right cerebellar damage affects right-sided movements. Recovery of motor function often occurs, but is generally worse if deep cerebellar nuclei are affected. In a recent study, children and adults recovered better from surgical tumor removal when deep cerebellar nuclei were spared (Konczak, Schoch, Dimitrova, Gizewski, & Timmann, 2005).

Although ataxia is often used to describe specific deficits (e.g., arm ataxia during reaching, or gait ataxia during walking), it is a global term that refers to specific features of dyscoordination. For example, during ataxic reaching movements, hand paths are highly curved and tend to be quite variable from reach to reach (figure 38.3A). An underlying inability to account for complex limb mechanics may cause this abnormal curvature (Bastian, Martin, Keating, & Thach, 1996). Consistent with this idea, patients show greater deficits when they reach quickly and move many joints or body parts simultaneously. Patients compensate by moving more slowly and breaking movements down into simpler components, a phenomenon originally described by Gordon Holmes as decomposition of movement (Holmes, 1939).

One common element of ataxia is dysmetria, poor control over the extent of movement. People with cerebellar damage often overshoot (hypermetria) or undershoot (hypometria) targets. As a result, their movements show oscillations when approaching the target, a phenomenon known as intention tremor. Intention tremor likely represents a series of corrective movements, and may stem from the use of time-delayed sensory feedback (see below). Consistent with this idea, intention tremor is reduced or absent when patients point to a target with their eyes closed, reducing visually guided corrections (Day, Thompson, Harding, & Marsden, 1998).

Walking ataxia has been described as an irregular, somewhat drunken-looking gait pattern involving widened stance, irregular steps, and oscillations in trunk control. Interestingly, it is not necessarily due to leg control deficits—patients often can make normal isolated leg motions yet still show profound walking ataxia (Morton & Bastian, 2003). This suggests that cerebellar influence over walking is distinct from that over voluntary control of isolated leg movements, with the former involving midline, and the latter intermediate and lateral structures.

There are clear oculomotor deficits associated with cerebellar damage, depending upon the region involved. Dysmetria of saccadic eye movements occurs with damage to the oculomotor vermis or the fastigial nucleus. Nystagmus, an involuntary eye movement consisting of slow drift and fast resetting phases, often occurs after cerebellar damage. Abnormalities in the gain of the vestibule-ocular reflex also occur following cerebellar damage, particularly when lobule X is damaged.

Recent work suggests that cerebellar damage may impair somatosensory function, but only in the context of active movement. Cerebellar patients show normal proprioceptive acuity when a limb is passively moved (Maschke, Gomez, Tuite, & Konczak, 2003), but exhibit clear deficits in force perception under active conditions (Bhanpuri, Okamura, & Bastian, 2012).

In contrast to the pronounced disturbances of sensory-motor function, the consequences of cerebellar damage on cognitive processes are much less obvious. This is puzzling, as much of the cerebellum projects to and receives input from frontal, parietal, and temporal regions of neocortex, and because cognitive tasks often activate the cerebellum in functional MRI experiments. Cognitive deficits have been reported in tests of language, executive function, emotion, and attention after cerebellar damage (Ivry et al., 2002; Schmahmann, 1997). Yet there is considerable variation between studies, with patients performing normally in one study but not another. A review by Timmann and Daum (2010) suggests that cognitive deficits are most pronounced in people with acute lesions. Thus, variability may be due to longer-term neocortical compensations for cerebellar loss. However, cognitive deficits may be more robust and long-lasting when cerebellar damage is acquired in childhood. Thus, cognitive deficits arising from cerebellar damage are difficult to characterize. What is clear is that cerebellar loss affects motor function more robustly than cognitive function—due perhaps to the fact that neocortical regions for movement control compensate less for cerebellar loss than do those involved in cognitive processes.

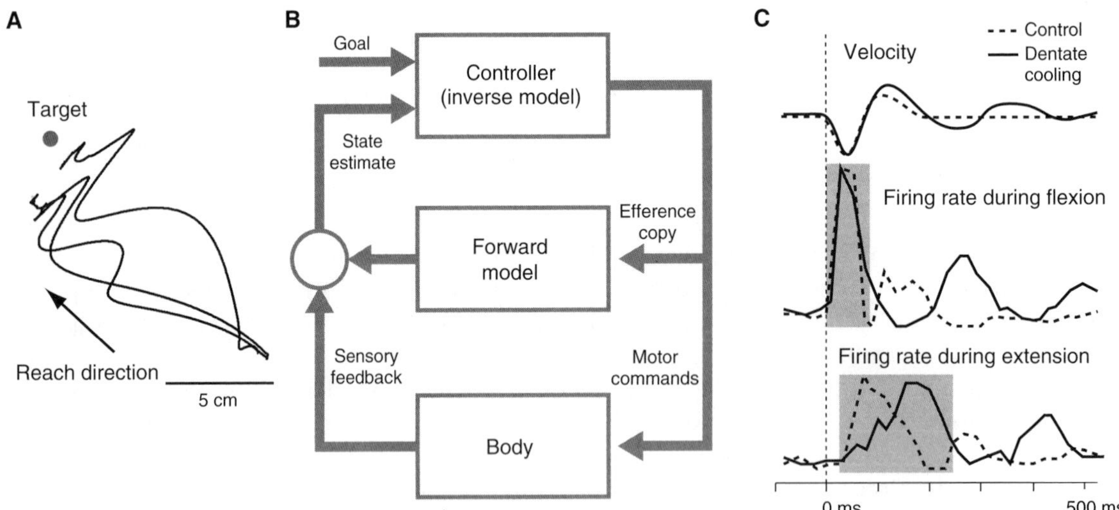

Figure 38.3 The cerebellum as a predictive forward model. (A) Hand paths made by a cerebellar patient reaching toward a target. Note features of ataxia, including dysmetria and endpoint oscillations. (B) A controller generates a motor command based on a behavioral goal and an estimate of the body's state. It sends an efference copy of the command to the forward model (cerebellum), which predicts the command's influence on body state, thereby avoiding problems associated with delayed sensory feedback. (C) Velocity and firing rate of a motor cortical neuron during typical movement (control; dashed line) and when cerebellar input is blocked through dentate cooling (solid line). (Data redrawn from Hore and Flament, 1988.)

Cerebellum as a predictive device for motor control

A central theory about the cerebellum is that it serves as a predictive forward model for motor control (Miall, Weir, Wolpert, & Stein, 1993; Wolpert, Miall, & Kawato, 1998). Forward models are used in engineering to establish stable control in systems with sensory and motor delays. Consider the diagram in figure 38.3B. A controller (i.e., motor cortex) computes motor commands based on the goal and an estimate of the body's current state (i.e., position, velocity). Motor commands are sent to the muscles, causing them to contract and move the body. Sensory feedback about these changes is then used to update the state estimate, closing the feedback loop. However, there is one important problem with such control architecture. Given both motor (~30 ms) and sensory (~30–70 ms) delays, a feedback controller may become unstable during fast movements: the controller may command a reaching movement to stop when sensory feedback indicates that the hand approaches its target. However, by that time the hand may have overshot the target, and antagonist muscle activity for decelerating will arrive too late. The same problem applies to movement corrections. Thus, closed-loop control of a device with considerable sensory delay can cause dysmetria and end-point oscillations, similar to the ataxia exhibited in cerebellar disease. One solution many patients choose to correct this problem is to move more slowly, meaning that sensory feedback has time to catch up with the actual state of the limb.

Fast movement execution requires a different approach. The engineering solution is to employ a predictive device, which uses a copy of the motor command (efference copy) to predict the future state of a limb. This prediction is then integrated (accounting for sensory delays) with actual sensory feedback (Miall et al., 1993). Replacing real sensory feedback with a prediction about limb state allows for smooth and stable real-time control.

The similarity between symptoms of ataxia and a control system using delayed feedback is compelling, and argues that the cerebellum may serve as a forward model in motor control. Early evidence for this idea comes from experiments in which a monkey must hold a handle at a specific position and counteract unexpected perturbations (Hore & Flament, 1988). Normally when perturbed (figure 38.3C, 0 ms), the monkey quickly returns the handle to the specified position. When cerebellar input to motor cortex is disrupted by cooling the dentate nucleus, the arm initially overshoots and then oscillates around the hold position (solid line, upper panel). This is not due to poor control of the agonist muscle that initially corrects the movement. Even during dentate cooling, the initial activity in primary motor cortex neurons appeared normal when a given muscle (triceps) acted as the agonist (figure 38.3C, gray box, middle panel). Rather, oscillations were caused by delayed timing of the antagonist (breaking) muscle that *terminates the corrective response*: when the triceps acted as an antagonist, dentate cooling delayed the predictive response (figure 38.3C, gray box, bottom panel). This suggests that the intact cerebellum plays a role in advancing responses of primary motor cortex, changing it from reactive to predictive control.

The role of prediction in motor control is especially evident in tasks requiring coordination between multiple joints or limbs, as in the fast, isolated elbow movements used when whipping eggs. Although the task does not require shoulder movement, the monoarticular muscles around the shoulder joint activate to counteract the torques induced to the shoulder by the elbow joint's acceleration. Cerebellar damage impairs active compensation for interaction torques during voluntary elbow movements, causing considerable shoulder instability (Bastian et al., 1996). Interestingly, this predictive control even occurs during fast feedback responses, in which the motor system modulates shoulder-muscle responding, based on induced motion to the elbow joint (Kurtzer, Pruszynski, & Scott, 2008). Such mechanisms can be understood as the consequence of a forward model, in which motor commands to correct elbow joint position after perturbation allow predictive corrections of shoulder perturbations. Indeed, long-latency reflexes are modulated by intersegmental information 20–30 ms after the onset of an initial response (Pruszynski et al., 2011). After cerebellar damage, coordinative feedback responses are still present, but are additionally delayed by 10 ms and reduced in size (Kurtzer et al., 2013). These changes may account for the observation that movement deficits in cerebellar patients become exaggerated during multi- versus single-joint movements (Bastian et al., 1996).

Another example of prediction in motor control is the so-called waiter task, in which someone supports a loaded tray with one hand, and then removes the load using the other hand. Healthy individuals reduce the activity of the load-bearing muscles 50 ms before any load change occurs (Hugon, Massion, & Wiesendanger, 1982). This predictive behavior effectively minimizes possible destabilization effects that might occur when the tray is removed. Indeed, when the load is lifted by an external agent (even a predictable agent), no anticipatory response occurs, and the tray shows a small upward displacement, requiring feedback mechanisms to restabilize the system. Cerebellar damage causes this anticipatory response to be poorly calibrated in both size and onset (Diedrichsen, Verstynen, Lehman, & Ivry, 2005).

While these findings provide evidence that the cerebellum acts as a predictive device in motor control, they also show that the cerebellum is not the sole source of predictions. In cerebellar damage, predictive responses are often ill-timed and inappropriately scaled, but nonetheless clearly present, indicating that other brain structures can also generate well-learned predictions. Thus, rather than being the exclusive site of a predictive forward model, the cerebellum may control the exact timing (Ivry et al., 2002) of forward models stored elsewhere, or adapt these to changes in task dynamics. Consistent with latter idea, cerebellar patients do not adjust their anticipatory responses in the waiter task after a catch trial (Diedrichsen et al., 2005).

Cerebellum and error-based learning

The problems experienced by patients with cerebellar pathology in using predictive mechanisms points to a second major domain of cerebellar function: adaptation or error-based learning. Error-based learning in this context refers to the modification of behaviors based on performance errors. This definition excludes learning based on reward prediction errors; that is, sensory feedback signaling only the relative success or failure of a movement (Schultz, Dayan, & Montague, 1997). Instead, error-based learning is driven by discrepancies between actual and predicted movement outcomes (Tseng, Diedrichsen, Krakauer, Shadmehr, & Bastian, 2007). Importantly, prediction errors signal not only *that* a movement has failed, but exactly *how* it has failed. One defining feature of error-based learning is that it utilizes a single prediction error to change behavior in the desired direction on the next attempt.

Take the example of throwing darts at a dartboard: in a typical context, there is a match between where a person is looking and where they aim the movement. However, if you wear laterally displacing prism glasses that shift the visual world, the dart will miss the target (figure 38.4A). Thus, the initial error is equal to the perturbation. Error-based learning will partially correct the next throw in the opposite direction to the shift induced by the prisms. For this to happen, an error-based learning system needs to translate the perceived error direction into a desired motor command change.

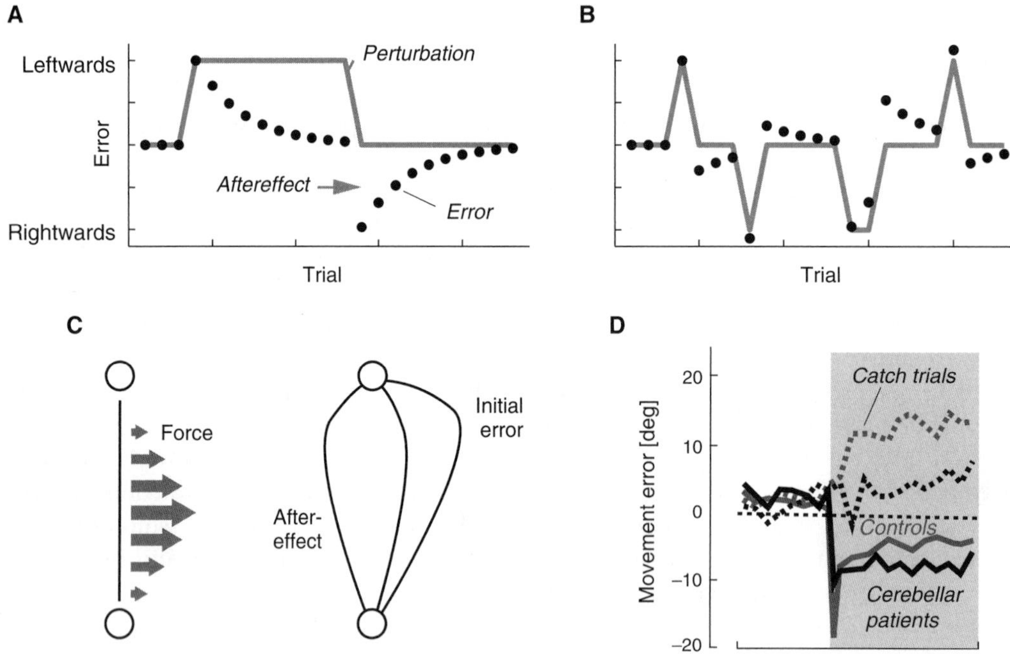

FIGURE 38.4 The role of cerebellum in error-based learning. (A) Traditional adaptation experiment in which a perturbation suddenly occurs, leading to an exponential learning curve. Upon perturbation removal, an aftereffect is observed. (B) A key feature of error-based learning is that it occurs even during random perturbations, which induce small but measurable changes in behavior. (C) During a force-field reaching task, a robotic device exerts a force perpendicular to the direction of movement. After an initial error, the movement adapts, resuming a nearly straight trajectory (gray line). When the force is suddenly removed, an aftereffect in the opposite direction occurs. (D) Deficit of cerebellar patients in adapting to force fields. The force field is switched on (gray box), causing a large initial error. Healthy individuals show increasing aftereffects on trials with no force field, but patients with cerebellar degeneration do not (Smith & Shadmehr, 2005).

To avoid instabilities due to overcorrection, the motor system usually corrects for much less than 100% of the error, meaning that it takes repeated throws to fully adapt to prism glasses. After adaptation, removal of the glasses reveals an aftereffect: the throws show the opposite error and then readapt to the original calibration. Such aftereffects are unavoidable, arising even when participants know that the perturbation has been removed. These aftereffects are considered to signal genuine error-based learning as opposed to conscious or strategic action re-planning (Mazzoni & Krakauer, 2006).

Error-based learning is even observed when perturbations are inconsistent and random across trials (figure 38.4B). Each induced error leads to a measurable small adjustment on the next trial (Thoroughman & Shadmehr, 2000), demonstrating that the error-based mechanism adjusts for single performance errors, even when participants know that all perturbations are random (Diedrichsen, Hashambhoy, Rane, & Shadmehr, 2005). Thus, error-based learning is continuously engaged to correct for small environment changes or performance fluctuations, keeping well-trained motor behaviors calibrated.

Over the last 15 years, substantial evidence has accumulated pointing to a critical role for the cerebellum in error-based learning. An early demonstration comes from prism adaptation during dart throwing (Martin, Keating, Goodkin, Bastian, & Thach, 1996). Patients with focal cerebellar lesions did not adjust throwing movements to displacement prisms, despite repeated unsuccessful throws. Importantly, after removing the glasses, the patients did not show the aftereffects that are found in healthy participants. Thus, cerebellar damage appeared to disrupt error-based learning at a basic level.

Since this initial report, deficits across a range of adaptation tasks have been reported in cerebellar patients. In force-field adaptation tasks, a robotic device is used to introduce perturbations when people make point-to-point reaching movements (figure 38.4C). Error-based learning leads to adjustments in the forces generated by the arm, attenuating the effects of the perturbing forces. To probe this form of anticipatory control, the robotic device is switched off on "catch trials," revealing large and robust aftereffects. Cerebellar patients are highly impaired in this learning task, as evidenced by increased errors and reduced aftereffects (Maschke, Gomez, Ebner, & Konczak, 2004; Smith & Shadmehr, 2005). Similar deficits in adaptation have been reported during split-belt treadmill walking (Morton & Bastian, 2006) and reaching under novel visuomotor transformations (Tseng et al., 2007). The latter paper also shows that observing unexpected movement outcomes, rather than correcting unsuccessful movements, is the likely source of the error signal. Finally, cerebellar damage impairs adaptation of outward saccadic eye movements, although adaptation of inward saccadic eye movements appears to be intact (Golla et al., 2008). However, this is likely due to uncompensated central fatigue, rather than error-driven adaptive processes (Prsa & Thier, 2011).

Cellular mechanisms of cerebellar learning

The idea that the cerebellum implements error-based learning is especially attractive, because we have a relatively detailed picture of how behavioral learning phenomena may relate to the underlying cellular mechanisms.

Marr (1969) and Albus (1971) proposed an influential computational model of error-based learning within the cerebellar neuronal circuit (figure 38.1). They suggested that the mossy fiber-parallel fiber system carries information about the motor command (an efference copy) and the current state of the system to Purkinje cells. The climbing fibers, which emanate from the inferior olive, transmit error signals that induce learning via complex spike activity (CS). Consistent with this hypothesis, climbing fibers discharge during unexpected sensory events, such as the slip of a retinal image or a misreach (for a review, see de Zeeuw et al., 1998). The resulting CS in the Purkinje cell triggers an influx of Ca++, inducing long-term depression. This process is specific to the parallel fiber–Purkinje synapses that were active 100–200 ms before climbing fiber input (Wang, Denk, & Hausser, 2000). Consequently, the arrival of a similar pattern of parallel fiber input causes a reduced simple spike (SS) firing rate in Purkinje cells, releasing deep cerebellar nuclei from inhibition.

This model has been supported in studies of arm control (Gilbert & Thach, 1977), eye-blink conditioning (Thompson, 2005) and adaptation of the vestibulo-ocular reflex (Lisberger, 1988). Recent work by Medina and Lisberger (2008) provides a particularly elegant demonstration. In their task, a monkey was rewarded for visually tracking a moving target. At 0 ms, a target appeared and moved downward; at 250 ms, it deviated from the path, moving slightly to the right (figure 38.5A). Purkinje cells in flocculus (the hemispheric part of lobule X) showed SS activity related to eye-movement pursuit direction. Figure 38.5B shows a cell that increases its activity for downward movement, but decreases activity for rightward movements. The sudden deviation of the target from the downward trajectory induces a slip of the retinal image and a CS (gray triangle). On the next trial, the cell showed a lower SS rate. Importantly,

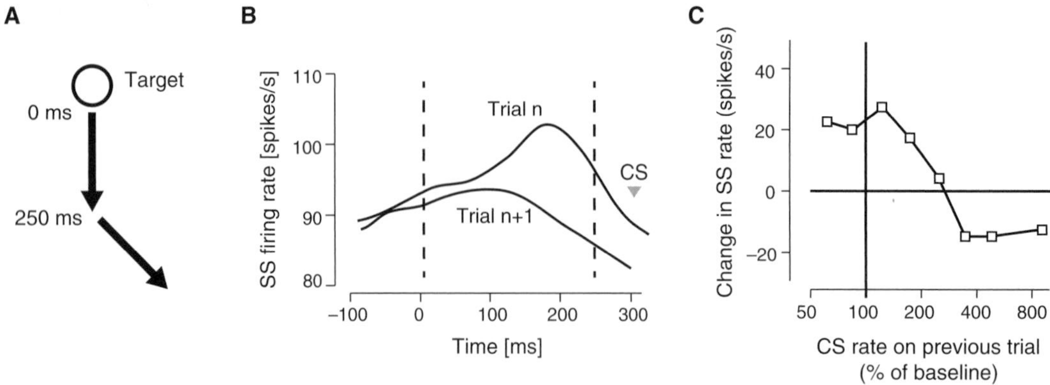

FIGURE 38.5 Cellular mechanisms underlying cerebellar learning. (A) In the task, a monkey fixated a moving target that changed direction 250 ms after movement onset. (B) Trial-to-trial change in the SS rate of a Purkinje cell. The gray triangle shows the occurrence of a complex spike (CS) on trial n. (C) Relationship between CS probability and change in SS rate. (Based on Medina and Lisberger, 2008.)

this SS decrease preceded the CS and hence the moment of directional change. Thus, the Purkinje cell is predicting the error. The decrease in firing rate related to an anticipatory rightward movement of the eye and increased tracking performance accuracy. To prevent SS firing rates from decreasing indefinitely, this system also requires a counteracting (i.e., potentiating) learning mechanism. This can be seen in the potentiating effect of trials in which CS activity does not change from baseline (figure 38.5C). These results provide strong evidence in support of the Marr and Albus hypothesis that the climbing fiber provides an error signal. However, the results also point to the operation of other learning mechanisms within the cerebellar circuitry.

Concluding remarks and open questions

Many open questions about cerebellar function remain. Although the Marr-Albus model of cerebellar learning provides an elegant account of many experimental findings, there is also substantial evidence opposing this theory. For example, the low-frequency CS activity (1–5 Hz) may not be high enough to provide the temporal resolution necessary for error-based adaptation of finely controlled motor behaviors (Horn, Pong, & Gibson, 2004; but see Kitazawa, Kimura, & Yin, 1998). Furthermore, during saccade adaptation the pattern of CS activity increases with learning, consistent with a role in adaptive output production, rather than decreasing as would be predicted if it encoded an error signal (Catz, Dicke, & Thier, 2005; but see Soetedjo, Kojima, & Fuchs, 2008).

On a systems level, it remains unclear whether the cerebellum implements predictive forward models or whether it merely adapts forward models stored elsewhere. Clearly, some well-learned predictive motor mechanisms are preserved in cerebellar patients. Furthermore, recent evidence using transcranial direct current stimulation (Galea, Vazquez, Pasricha, Orban de Xivry, & Celnik, 2011) suggests that the cerebellum may produce short-term modifications of forward models, whereas the neocortex may store longer-lasting motor memories.

At the whole-brain level, it has become clear that the cerebellum is not simply a motor control device, but has functional involvement in a wide array of mental abilities. How the cerebellum contributes to these nonmotor functions remains unclear. We hope that the next generation will fulfill the dream of a universal theory that explains cerebellar function from cellular to behavioral levels, as envisioned by David Marr more than 40 years ago.

ACKNOWLEDGMENTS The authors thank Dr. Erin Heerey and Prof. Richard Ivry for comments on an earlier draft. Thanks to Prof. Daniel Wolpert for assistance with figure 38.1.

REFERENCES

ALBUS, J. S. (1971). A theory of cerebellar function. *Math Biosci, 10*, 25–61.

AZEVEDO, F. A., CARVALHO, L. R., GRINBERG, L. T., FARFEL, J. M., FERRETTI, R. E., LEITE, R. E., … HERCULANO-HOUZEL, S. (2009). Equal numbers of neuronal and nonneuronal cells make the human brain an isometrically scaled-up primate brain. *J Comp Neurol, 513*(5), 532–541.

BASTIAN, A. J., MARTIN, T. A., KEATING, J. G., & THACH, W. T. (1996). Cerebellar ataxia: Abnormal control of interaction torques across multiple joints. *J Neurophysiol, 76*(1), 492–509.

BHANPURI, N. H., OKAMURA, A. M., & BASTIAN, A. J. (2012). Active force perception depends on cerebellar function. *J Neurophysiol, 107*(6), 1612–1620.

Buckner, R. L., Krienen, F. M., Castellanos, A., Diaz, J. C., & Yeo, B. T. (2011). The organization of the human cerebellum estimated by intrinsic functional connectivity. *J Neurophysiol, 106*(5), 2322–2345.

Catz, N., Dicke, P. W., & Thier, P. (2005). Cerebellar complex spike firing is suitable to induce as well as to stabilize motor learning. *Curr Biol, 15*(24), 2179–2189.

Day, B. L., Thompson, P. D., Harding, A. E., & Marsden, C. D. (1998). Influence of vision on upper limb reaching movements in patients with cerebellar ataxia. *Brain, 121*(Pt 2), 357–372.

De Zeeuw, C. I., Simpson, J. I., Hoogenraad, C. C., Galjart, N., Koekkoek, S. K., & Ruigrok, T. J. (1998). Microcircuitry and function of the inferior olive. *Trends Neurosci, 21*(9), 391–400.

Diedrichsen, J., Balsters, J. H., Flavell, J., Cussans, E., & Ramnani, N. (2009). A probabilistic MR atlas of the human cerebellum. *NeuroImage, 46*(1), 39–46.

Diedrichsen, J., Hashambhoy, Y. L., Rane, T., & Shadmehr, R. (2005). Neural correlates of reach errors. *J Neurosci, 25*(43), 9919–9931.

Diedrichsen, J., Verstynen, T., Lehman, S. L., & Ivry, R. B. (2005). Cerebellar involvement in anticipating the consequences of self-produced actions during bimanual movements. *J Neurophysiol, 93*(2), 801–812.

Galea, J. M., Vazquez, A., Pasricha, N., Orban de Xivry, J. J., & Celnik, P. (2011). Dissociating the roles of the cerebellum and motor cortex during adaptive learning: The motor cortex retains what the cerebellum learns. *Cereb Cortex, 21*(8), 1761–1770.

Gilbert, P. F., & Thach, W. T. (1977). Purkinje cell activity during motor learning. *Brain Res, 128*(2), 309–328.

Golla, H., Tziridis, K., Haarmeier, T., Catz, N., Barash, S., & Thier, P. (2008). Reduced saccadic resilience and impaired saccadic adaptation due to cerebellar disease. *Eur J Neurosci, 27*(1), 132–144.

Grodd, W., Hulsmann, E., Lotze, M., Wildgruber, D., & Erb, M. (2001). Sensorimotor mapping of the human cerebellum: FMRI evidence of somatotopic organization. *Hum Brain Mapp, 13*(2), 55–73.

Holmes, G. (1939). The cerebellum of man. *Brain, 62,* 1–30.

Hore, J., & Flament, D. (1988). Changes in motor cortex neural discharge associated with the development of cerebellar limb ataxia. *J Neurophysiol, 60*(4), 1285–1302.

Horn, K. M., Pong, M., & Gibson, A. R. (2004). Discharge of inferior olive cells during reaching errors and perturbations. *Brain Res, 996*(2), 148–158.

Hugon, M., Massion, J., & Wiesendanger, M. (1982). Anticipatory postural changes induced by active unloading and comparison with passive unloading in man. *Pflugers Arch, 393*(4), 292–296.

Ivry, R. B., Spencer, R. M., Zelaznik, H. N., & Diedrichsen, J. (2002). The cerebellum and event timing. *Ann N Y Acad Sci, 978,* 302–317.

Kelly, R. M., & Strick, P. L. (2003). Cerebellar loops with motor cortex and prefrontal cortex of a nonhuman primate. *J Neurosci, 23*(23), 8432–8444.

Kitazawa, S., Kimura, T., & Yin, P. B. (1998). Cerebellar complex spikes encode both destinations and errors in arm movements. *Nature, 392*(6675), 494–497.

Konczak, J., Schoch, B., Dimitrova, A., Gizewski, E., & Timmann, D. (2005). Functional recovery of children and adolescents after cerebellar tumour resection. *Brain, 128*(Pt. 6), 1428–1441.

Kurtzer, I., Pruszynski, J. A., & Scott, S. H. (2008). Long-latency reflexes of the human arm reflect an internal model of limb dynamics. *Curr Biol, 18*(6), 449–453.

Kurtzer, I., Trautman, P., Rasquinha, R. J., Bhanpuri, N. H., Scott, S. H., & Bastian, A. J. (2013). Cerebellar damage diminishes long-latency responses to multijoint perturbations. *J Neurophysiol, 109*(8), 2228–2241.

Larsell, O., & Jansen, J. (1972). *The comparative anatomy and histology of the cerebellum: The human cerebellum, cerebellar connections, and cerebellar cortex.* Minneapolis: University of Minnesota Press.

Lisberger, S. G. (1988). The neural basis for learning of simple motor skills. *Science, 242*(4879), 728–735.

Marr, D. (1969). A theory of cerebellar cortex. *J Physiol, 202,* 437–470.

Martin, T. A., Keating, J. G., Goodkin, H. P., Bastian, A. J., & Thach, W. T. (1996). Throwing while looking through prisms: I. Focal olivocerebellar lesions impair adaptation. *Brain, 119*(4), 1183–1198.

Maschke, M., Gomez, C. M., Ebner, T. J., & Konczak, J. (2004). Hereditary cerebellar ataxia progressively impairs force adaptation during goal-directed arm movements. *J Neurophysiol, 91*(1), 230–238.

Maschke, M., Gomez, C. M., Tuite, P. J., & Konczak, J. (2003). Dysfunction of the basal ganglia, but not the cerebellum, impairs kinaesthesia. *Brain, 126*(Pt. 10), 2312–2322.

Mazzoni, P., & Krakauer, J. W. (2006). An implicit plan overrides an explicit strategy during visuomotor adaptation. *J Neurosci, 26*(14), 3642–3645.

Medina, J. F., & Lisberger, S. G. (2008). Links from complex spikes to local plasticity and motor learning in the cerebellum of awake-behaving monkeys. *Nat Neurosci, 11*(10), 1185–1192.

Miall, R. C., Weir, D. J., Wolpert, D. M., & Stein, J. F. (1993). Is the cerebellum a Smith predictor? *J Mot Behav, 25*(3), 203–216.

Morton, S. M., & Bastian, A. J. (2003). Relative contributions of balance and voluntary leg-coordination deficits to cerebellar gait ataxia. *J Neurophysiol, 89*(4), 1844–1856.

Morton, S. M., & Bastian, A. J. (2006). Cerebellar contributions to locomotor adaptations during splitbelt treadmill walking. *J Neurosci, 26*(36), 9107–9116.

Prsa, M., & Thier, P. (2011). The role of the cerebellum in saccadic adaptation as a window into neural mechanisms of motor learning. *Eur J Neurosci, 33*(11), 2114–2128.

Pruszynski, J. A., Kurtzer, I., Nashed, J. Y., Omrani, M., Brouwer, B., & Scott, S. H. (2011). Primary motor cortex underlies multi-joint integration for fast feedback control. *Nature, 478*(7369), 387–390.

Schlerf, J. E., Verstynen, T. D., Ivry, R. B., & Spencer, R. M. (2010). Evidence of a novel somatopic map in the human neocerebellum during complex actions. *J Neurophysiol, 103*(6), 3330–3336.

Schmahmann, J. D. (1997). *The cerebellum and cognition.* San Diego, CA: Academic Press.

Schmahmann, J. D. (2004). Disorders of the cerebellum: Ataxia, dysmetria of thought, and the cerebellar cognitive affective syndrome. *J Neuropsychiatry Clin Neurosci, 16*(3), 367–378.

Schultz, W., Dayan, P., & Montague, P. R. (1997). A neural substrate of prediction and reward. *Science, 275*(5306), 1593–1599.

Smith, M. A., & Shadmehr, R. (2005). Intact ability to learn internal models of arm dynamics in Huntington's disease but not cerebellar degeneration. *J Neurophysiol, 93*(5), 2809–2821.

Soetedjo, R., Kojima, Y., & Fuchs, A. (2008). Complex spike activity signals the direction and size of dysmetric saccade errors. *Prog Brain Res, 171*, 153–159.

Stoodley, C. J., & Schmahmann, J. D. (2009). Functional topography in the human cerebellum: A meta-analysis of neuroimaging studies. *NeuroImage, 44*(2), 489–501.

Sugihara, I., & Shinoda, Y. (2004). Molecular, topographic, and functional organization of the cerebellar cortex: A study with combined aldolase C and olivocerebellar labeling. *J Neurosci, 24*(40), 8771–8785.

Thompson, R. F. (2005). In search of memory traces. *Annu Rev Psychol, 56*, 1–23.

Thoroughman, K. A., & Shadmehr, R. (2000). Learning of action through adaptive combination of motor primitives. *Nature, 407*(6805), 742–747.

Timmann, D., & Daum, I. (2010). How consistent are cognitive impairments in patients with cerebellar disorders? *Behav Neurol, 23*(1–2), 81–100.

Tseng, Y. W., Diedrichsen, J., Krakauer, J. W., Shadmehr, R., & Bastian, A. J. (2007). Sensory prediction errors drive cerebellum-dependent adaptation of reaching. *J Neurophysiol, 98*(1), 54–62.

Van Essen, D. C. (2002). Surface-based atlases of cerebellar cortex in the human, macaque, and mouse. *Ann N Y Acad Sci, 978*, 468–479.

Wang, S. S., Denk, W., & Hausser, M. (2000). Coincidence detection in single dendritic spines mediated by calcium release. *Nat Neurosci, 3*(12), 1266–1273.

Wiestler, T., McGonigle, D. J., & Diedrichsen, J. (2011). Integration of sensory and motor representations of single fingers in the human cerebellum. *J Neurophysiol, 105*(6), 3042–3053.

Wolpert, D. M., Miall, R. C., & Kawato, M. (1998). Internal models in the cerebellum. *Trends Cogn Sci, 2*(9), 313–321.

39 Computational Approaches for Goal-Directed Movement Planning and Execution

FRÉDÉRIC CREVECOEUR,* TYLER CLUFF,* AND STEPHEN H. SCOTT

ABSTRACT The apparent ease of skilled motor behavior masks the complex neural processes involved in the control of movement. To handle the complexity of most motor tasks, optimal feedback control suggests that the brain continuously processes sensory feedback and selectively compensates for noise disturbances as well as external perturbations. Motivated by this powerful prediction, recent studies have used perturbation paradigms to investigate the neural control of movement. After introducing the basic mathematical concepts of optimal feedback control, we review evidence that the brain generates flexible feedback strategies following perturbations perceived from visual feedback (e.g., a sudden change in the target location), as well as mechanical perturbations applied to the limb. Importantly, we highlight evidence that the motor system can generate goal-directed responses in as little as 50–60 ms following a mechanical perturbation. A transcortical feedback pathway through primary motor cortex appears to play an important role in these rapid corrective responses.

Elite athletes push the limits of the sensorimotor system, integrating sensory information to generate rapid yet remarkably precise movements. A good example is a hockey player breaking away from his defenders. In a split second, the player has to read the goaltender's movement and decide whether to shoot the puck or fake out the goaltender and wait for an opening. Even the simplest movements that we perform in daily life, such as reaching for a cup of coffee, also involve complex sensorimotor coordination. The ability to use sensory information to flexibly guide, correct, or modify our actions is the hallmark of skilled biological control.

Recently, optimal feedback control (OFC) has been used as a model of how the brain processes sensory information to control movement. A powerful feature of this model is that it describes how the motor system should handle performance errors caused by neural variability or environmental disturbances. This feature has renewed interest in feedback response strategies, in particular because they may provide a window into the voluntary control of movement. In this chapter, we introduce the basic notions underlying optimal control theory and discuss how this approach may help us identify problems that the brain must solve to perform even the simplest movements. Then, we review recent findings that emphasize the similarity between biological control and the sophistication of optimal control models. In particular, we focus on perturbation paradigms showing that many aspects of optimal control models are observed in corrective responses generated by humans: (1) continuous processing of sensory feedback underlies the control of movement, and (2) feedback control is tailored to the constraints imposed by the task at hand. We finish our chapter by briefly discussing how sophisticated motor behavior may be linked to processing in distributed brain circuits, highlighting recent evidence that neural processing in primary motor cortex (M1) may possess some of the attributes required for flexible feedback control.

Optimal control: Definitions and applications in neuroscience

Flash and Hogan (1985) introduced optimal control principles to movement neuroscience almost 30 years ago. Intuitively, this approach is based on the assumption that the brain selects motor commands that maximize or minimize a performance criterion. In the context of reaching movements, Flash and Hogan suggested that the brain selects motor plans that minimize the derivative of the hand's acceleration (or the jerk). This hypothesis was justified by the fact that point-to-point movements tend to be smooth with a bell-shaped velocity profile, which is reproduced nicely by minimizing the derivative of the hand's acceleration during reaching movements. Following Flash and Hogan's work, a wealth of studies have proposed biological cost functions that incorporate kinetic parameters (e.g.,

*These authors contributed equally to this work.

minimum torque change; Nakano et al., 1999; Uno, Kawato, & Suzuki, 1989) and energy consumption (Biess, Liebermann, & Flash, 2007; Berret, Chiovetto, Nori, & Pozzo, 2011) to identify movement variables that the motor system may control and optimize.

This approach is grounded in the theory of optimal control, a formalism that applies optimization principles to describe what the motor system should do according to how our limbs move, as well as the performance criteria (e.g., energy consumption) and constraints (e.g., time constraint) associated with movement. The following section outlines the basic notions of control theory and defines (optimal) control problems that are often encountered in movement neuroscience.

Dynamical Systems and Control Problems At the basis of control engineering is the notion of a *dynamical system*: a set of variables evolving as a function of time. For instance, the angular motion of a body segment can be seen as a dynamical system described by *state variables*, such as the joint angle and velocity. In general, the evolving state of a dynamical system can be described by a differential equation of the form:

$$\dot{x} = f(x,u), \tag{1}$$

where x is the vector of state variables and the dot expresses its time derivative. This derivative is a function of the state vector itself and an additional variable, u, called the control vector. It is assumed that the *controller* influences the state of the system by changing the value of the control vector. For example, varying muscle activity changes the torque acting on a joint and alters segmental motion through the dynamical properties captured in the function f. With these definitions, we can define a control problem as follows: find a time-varying control vector that steers the state variables to a desired location. Assuming that the initial state x_0 is known, the problem is to find a control function $u(t)$, $t_0 <= t <= t_f$, such that the solution of Eq. 1,

$$x(t) = x_0 + \int_{t_0}^{t} f(x(s), u(s))ds, \tag{2}$$

meets the constraints of the control problem. A reaching movement, for example, may be described as a control problem where the brain must steer the hand to the spatial location of a goal target. Expressing a simple task such as reaching as a control problem is a powerful approach to identify the challenges that motor control presents for the brain. As we will see, even the simplest movements impose complex sensorimotor transformations.

Deterministic Optimal Control In general, the solution of a particular control problem is not unique. This also applies to biological motor control, where reaching movements may follow distinct paths to the same target or even have different velocities along the same movement path. Given that each of these movements satisfies the goal of reaching the target (i.e., motor equivalence), extensive research has been conducted to identify how the brain selects one control solution among infinitely many alternatives. One way to reduce the set of possible movement solutions is to constrain the problem using a cost function. For instance, we may be interested in applying a sequence of joint torques that allows us to reach a target while minimizing the intensity of muscle activity to avoid fatigue. In this example, the cost is directly related to the motor command, and the problem is to find the reaching path that minimizes muscle energy expenditure. This approach uses optimization principles to determine the best way to reach the target among all possible movement solutions.

A typical cost function contains a final cost, $g(x)$, and a running cost, $L(x,u)$, that accumulates along the trajectory followed by the state variables:

$$J(x,u) = g(x(t_f)) + \int_{t_0}^{t_f} L(x(t), u(t))dt. \tag{3}$$

With these definitions, an optimal control problem can be defined as follows: find a control function that minimizes $J(x,u)$ (Eq. 3), subject to the initial condition $x(t_0) = x_0$, and to the system dynamics (Eq. 1). It can be shown that, under the optimal solution, $J(x,u)$ satisfies the Hamilton-Jacobi-Bellman equation (U represents the set of admissible control actions):

$$-\frac{\partial J(x,u)}{\partial t} = \min_{u \in U}\left\{L(x,u) + \frac{\partial J(x,u)}{\partial x} f(x,u)\right\}. \tag{4}$$

The function $J(x,u)$ is the cost to accumulate from the present time until the end of the problem horizon, t_f. This quantity is called the cost-to-go and plays a central role in the derivation of numerical solutions (Todorov, 2006). An intuitive interpretation of Eq. 4 is that the control vector can vary the orientation of the instantaneous direction of the state trajectory ($f(x,u)$), which should ideally follow the direction that is opposite to the gradient of the cost-to-go ($\partial J/\partial x$). However, the optimization also takes into account the instantaneous cost ($L(x,u)$), and as a result, Eq. 4 achieves the best compromise between the gradient of the cost-to-go and the instantaneous cost ($L(x,u)$). This compromise determines the instantaneous variation of the cost-to-go ($\partial J/\partial t$).

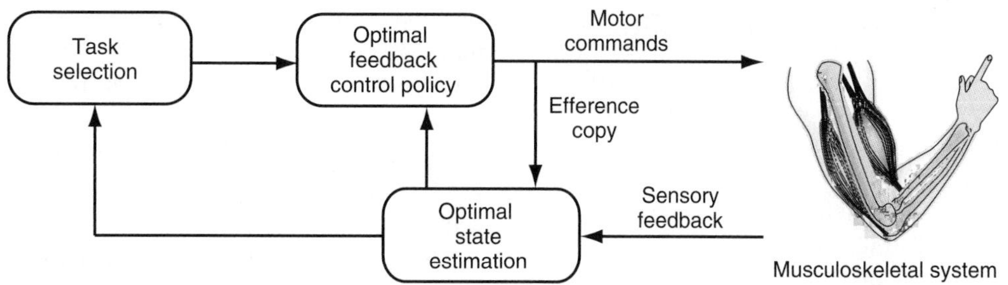

FIGURE 39.1 Illustration of basic processes expected under the OFC framework. The selection of the behavioral task determines the feedback control policy (task selection). The purpose of the feedback control policy is to continuously process and convert sensory data into motor commands that best satisfy the task demand (optimal feedback control policy). Once the sensory feedback is processed, motor commands are sent to the peripheral motor system (biomechanical plant or musculoskeletal system). An efference copy of the descending motor command is used internally to predict the consequences of motor actions. These internal predictions are combined with feedback from sensory receptors to compute the posterior estimate of the state of the body (optimal state estimator). This state estimate is used to adjust the motor commands during the ongoing motor action.

The many studies that have used this approach have agreed on the general conclusion that healthy motor systems favor smooth and efficient movements. The shortcoming of this approach, however, is that because the laws of physics relate all movement parameters, virtually every meaningful cost function in the form of Eq. 3 partially captures the smoothness and efficiency of biological motor control. Also, this approach does not systematically account for the continuous update of motor commands that is necessary to correct for motor errors.

STOCHASTIC OPTIMAL CONTROL In general, factors that can induce motor errors and trial-to-trial variability fall into two broad categories. First, the variable activation of neural circuits induces variable motor behavior. Neural noise can be found in sensory systems, movement preparatory activity, and the activation of muscles in the motor periphery (Churchland, Afshar, & Shenoy, 2006; Faisal, Selen, & Wolpert, 2008; Osborne, Lisberger, & Bialek, 2005; Scott & Loeb, 1994; van Beers, Haggard, & Wolpert, 2004). Additionally, motor errors can be produced by external disturbances resulting from our interaction with the environment. Both neural variability and external disturbances require that the brain continuously update motor commands based on the available sensory data to produce successful behavior.

Harris and Wolpert (1998) considered the influence of neural variability on movement planning and suggested that motor commands are selected to minimize the variance of movement end points, assuming the intensity of motor noise scales with the size of the motor command. While this model explicitly considers the effect of motor noise, it does not address the online adjustment of motor commands required when disturbances alter performance. The control of stochastic processes was introduced to address this limitation. In this framework, feedback is essential to update motor commands and compensate for neural variability. Online monitoring and control are often described in terms of a state estimator combined with a controller (figure 39.1). The state estimator typically combines sensory and motor signals to compute the present state of the body (figure 39.1; optimal state estimator), and the controller uses the estimated state of the body to select control actions that best reflect the goal and constraints of the task (figure 39.1; optimal feedback control policy). This section presents the basic formalism of the problems of estimation and control of stochastic processes.

Because random (Brownian) motion does not have finite instantaneous variations, the control problem is formulated in discrete rather than continuous time (Arnold, 1974). To begin, the analog of Eq. 1 becomes

$$dX = F(X,u)dt + G(X,u)dW. \quad (5)$$

In Eq. 5, the capital X signifies that the state vector is now a stochastic variable. Eq. 5 expresses that small changes in the state (dX) follow a deterministic law described by the function F that captures the system dynamics as in Eq. 1, and a stochastic term that captures random disturbances in the process (dW). We now convert this equation to discrete time by considering changes in the process over a time step of δt. From the definition of Brownian motion (Arnold, 1974), the accumulation of random noise over δt follows a Gaussian distribution with zero mean and variance equal to δt. We use $\xi(t)$ to designate these Gaussian disturbances at each time step. The formulation in discrete time becomes

$$X(t+\delta t)=M(X(t),u(t))+N(X(t),u(t))\xi(t), \quad (6)$$

where $M(.)$ and $N(.)$ are the discrete versions of the functions F and G introduced in Eq. 5. The analog of Eq. 4 also becomes a discrete equation. The optimization can no longer be computed over the true state variables, because independent simulations of the same process will lead to distinct sample paths. Instead, for stochastic control problems, the controller should optimize the expected outcome, which ensures that performance will be optimal in the most likely scenario. The analogue of Eq. 4 becomes

$$J_t(X_t,u_t)=\min_u\{L(X_t,u_t)+E[J_{t+1}(X_{t+1},u_{t+1})\mid X_t,u_t]\}, \quad (7)$$

where $E[.]$ denotes the expected value of the argument and the explicit dependency on time was replaced by the subscript t. Eq. 7 means that the best control action minimizes the present running cost, as well as the expected value of the future cost-to-go given the selected control action (figure 39.1; optimal feedback control policy).

A clear difficulty encountered with stochastic processes is that the state of the system (X_t) may not be known exactly, because sensory feedback is also corrupted by noise. Filtering techniques can be used to estimate the state of a stochastic process (figure 39.1; optimal state estimator). These techniques combine imperfect sensory data with internal assumptions about the state of the system (or internal priors). Prior assumptions are often formulated as the output of a forward model, making a prediction of the present state of the body given the available information, including the motor command sent to the muscles (Wolpert & Flanagan, 2001). Let Y_t denote the information available at time t. We assume that the conditional distribution of Y_t given the state variable X_t is known. First, we can predict the distribution of X_t given the distribution of X_{t-1} by using Eq. 6 (forward model). This prediction step gives a prior belief about the system state. Second, the feedback data, Y_t, can be used to correct the prior distribution by applying Bayes's theorem, which gives the posterior distribution of the state (X_t) given the sensory feedback (Y_t). In other words, the best estimate of the system state is obtained by performing Bayesian integration of sensory signals and forward predictions at each time step. The resulting estimate is optimal in the sense that the variance of the posterior distribution is minimized. In the particular case of linear dynamics affected by additive Gaussian disturbances, this procedure is known as a Kalman filter (Kalman, 1960).

Linear-Quadratic-Gaussian Regulator Let us examine the simplest case where an analytic solution of the optimal control and filtering problems is available. The main assumptions are that the system follows linear dynamics, coupled with a quadratic cost function and subject to additive Gaussian noise (linear quadratic Gaussian, or LQG). The function $M(.)$ introduced in Eq. 6 becomes a linear function of the state and control vectors (represented by the state-space matrices A and B), and the noise disturbance is captured by additive Gaussian noise (ξ_t). With these definitions, the system dynamics become

$$X_{t+1}=AX_t+Bu_t+\xi_t. \quad (8)$$

The sensory feedback available at time t is of the form

$$Y_t=HX_t+\omega_t, \quad (9)$$

where ω_t is Gaussian random variable and H is the mapping between the system state and sensory data. This matrix may express that information about some state variables is not available, such as an external perturbation that can only be indirectly measured through its effect on the motion of the body. The matrix H may also express that some sensory signals provide information about a combination of state variables, as for instance muscle spindles provide sensory feedback that is a mixture of the muscle's length, velocity, and higher derivatives. The cost function is a quadratic form in the state and control variables, defined by weight matrices Q_t and R_t, respectively:

$$L(X_t,u_t)=X_t^TQ_tX_t+u_t^TR_tu_t. \quad (10)$$

The optimal control problem consists in finding a sequence of control variables that minimize the total expected cost, that is,

$$J=E\left[\sum_{t=1}^{N}L(X_t,u_t)\right]. \quad (11)$$

For this class of control problems, the optimal control policy turns out to be a linear function of the estimated state, denoted $\hat{x}_t$:

$$u_t=C_t\hat{x}_t. \quad (12)$$

The sequence of optimal feedback gains, C_t, is determined by the cost matrices (Q_t and R_t in Eq. 10) and by the system dynamics (A and B in Eq. 8). The result is a time-varying feedback gain that tells us how the motor system should transform the estimated state of the system into motor commands.

The optimal estimate of the state is calculated in two steps (Kalman filter). First, the prediction of the next system state is obtained by simulating the dynamics over one time step given the current motor command, u_t, and taking the expected value of the outcome:

$$\hat{x}_t^p=A\hat{x}_{t-1}+Bu_{t-1}. \quad (13)$$

This prediction (or prior belief) is then corrected with the difference between actual and expected feedback, weighted by the Kalman gain:

$$\hat{x}_t = \hat{x}_t^p + K_t\left(Y_t - H\hat{x}_t^p\right). \quad (14)$$

The Kalman gains are determined by the system dynamics (Eq. 8), feedback (Eq. 9), and the covariance matrices of the motor and feedback noise. The Kalman gain (K_t) specifies how much the prior estimate (Eq. 13) should be changed according to the available sensory data. Recent studies have extended the LQG framework to take properties of biological motor control into account, such as the scaling of noise variability with the intensity of the neural signal (Crevecoeur, Sepulchre, Thonnard, & Lefèvre, 2011; Qian, Jiang, Jiang, & Mazzoni, 2013; Todorov, 2005), as well as the nonlinearity of the musculoskeletal system (Li & Todorov, 2007). A complete derivation of the optimal feedback gains and Kalman gains is beyond the scope of the present review, but can be found elsewhere (Åström, 1970; Brown, 1983; Bryson & Ho, 1975).

Figure 39.1 recapitulates the different components of the full control algorithm (estimation and control) and illustrates how they may describe some aspects of motor control. In the framework of optimal control, task selection specifies the movement goal and constraints and can be seen as the definition of the cost function (Eqs. 3 or 10). Indeed, the cost function determines which state variables will be constrained, and how much motor commands will be penalized to attain that goal. Once the cost function is defined, an optimal control policy can be derived as the solution of a well-defined problem. In this respect, the approach based on optimal control makes a theoretical link between motor behavior and the control of a biomechanical plant, which are two fundamental aspects of movement neuroscience (Scott, 2004). Optimal control provides a normative tool to describe how the biomechanical plant should be controlled according to a given behavioral objective. The online control of movement is then realized by applying the feedback control policy (figure 39.1 and Eq. 12) to the estimated state of the body (figure 39.1 and Eqs. 13 and 14).

Optimal Control and Internal Models The notion of an *internal model* is an important conceptual framework in motor neuroscience that is often defined as a group of neurons or circuits that mimic the input-output relationship of the peripheral motor system and environment. A traditional view was that *inverse models* transform a desired movement into a sequence of motor commands, and *forward models* predict the consequences of these motor commands (for a review, see Kawato, 1999). Pairing inverse and forward models according to the intended movement was proposed as a model for sensorimotor coordination (Wolpert & Kawato, 1998).

Optimal feedback control generalizes the neural computations identified within the framework of internal models and bridges the gap between movement planning and execution by providing a goal-related feedback control policy (Eq. 12 for linear systems). In the optimal control framework, the controller must transform the movement goal expressed by the cost function (Eqs. 3 or 11) into a control policy. This operation is similar to the one performed by inverse models, in the sense that they both use knowledge of the body dynamics to map intended movements into motor commands. The notion of a forward model is also present in the optimal control framework. A common perspective is that forward models predict the consequences of motor actions, thereby allowing the brain to compensate for their effect ahead of sensory information. The computation of a prior belief about the state of the body can be seen as a forward prediction of the future state of the body (Eq. 13), based on the current motor commands and internal knowledge of the system dynamics (represented by the matrices A and B; figure 39.1, musculoskeletal system). Motor prediction from forward models is therefore a critical component of optimal feedback control models.

Aside from the problems related to the derivation of the control policy, an important challenge for the motor system is the computation of the state estimate. Indeed, we have only partially addressed the problem of state estimation with the Kalman filter, dealing with the variability of prediction and sensory signals. Additionally, the brain must cope with time delays resulting from the transmission of neural signals along the nerves. Given the presence of sensory delays, the hypothesis that the brain uses state estimation requires converting delayed sensory feedback into present estimates of the state of the body, which involve prediction based on sensory signals that is independent from the motor prediction (Ariff, Donchin, Nanayakkara, & Shadmehr, 2002; Mehta & Schaal, 2002). Compatible with this prediction, we recently showed that rapid sensory predictions are performed following a perturbation (Crevecoeur & Scott, 2013).

Perspective on Alternative Control Approaches: Trading Efficiency for Robustness A central assumption of optimal (feedback) control approaches in movement neuroscience is that the brain knows the system's dynamics exactly. In other words, OFC models give us the best solution, assuming that a correct internal representation of the body and environmental dynamics is available, while uncertainty is handled by considering the presence of random noise in the system. However, uncertainty in the internal model of dynamics can also alter the control of a movement. For instance, the inertia of the limb differs slightly according to the weight of a watch, clothing, or hand-held objects. In addition, muscle dynamics may vary depending on the level of background activity, biochemical factors, or fatigue (Zahalak, 1981). This class of model disturbances does not fall under those disturbances modeled by random noise, as they potentially introduce systematic biases during movement.

In general, researchers have approached this problem with learning or adaptation studies (Shadmehr, Smith, & Krakauer, 2010; Wolpert, Diedrichsen, & Flanagan, 2011). In this framework, changes in motor commands reflect adaptive adjustments of how the brain represents environmental dynamics. While tremendous progress has been made with this approach, a clear shortcoming is that the body and environment can change more rapidly than the adaptation processes typically investigated in motor learning studies (e.g., learning curves varying over tens to hundreds of trials). For instance, muscle dynamics rapidly change during effort without giving us the chance to practice tens of trials to adapt to those changes.

Engineers have developed an approach based on the concept of robustness to deal with these internal model uncertainties. The idea is to make the control design as insensitive to model errors as possible (Bhattacharyya, Chapellat, & Keel, 1995; Doyle, Francis, & Tannenbaum, 1992). This approach typically focuses on properties of the controller rather than on the actual system trajectories emphasized in the classical optimal control approach. An important theoretical result is that the controllers that are the most robust against model errors do not always correspond to the controllers that are the most efficient (Boulet & Duan, 2007; Michiels & Niculescu, 2007). In other words, improving the robustness of control may degrade performance, whereas optimizing a performance criterion can make the control design more fragile to model errors. Compatible with these principles, previous studies have suggested that motor performance is altered to maintain performance or preserve movement smoothness in conditions of higher uncertainty (Crevecoeur, McIntyre, Thonnard, & Lefèvre, 2010; Ronsse, Thonnard, Lefèvre, & Sepulchre, 2008). However, to our knowledge, the trade-off between the efficiency and robustness of biological motor control and its influence on motor planning and execution has not been thoroughly investigated. Given that internal models of body and environmental dynamics can never be known exactly, robustness may be an important consideration in motor neuroscience.

Application to biological control: Flexible sensorimotor control strategies

The motor system has a remarkable ability to perform successfully while never reproducing exactly the same movement. This consistent success in the presence of variability suggests the central nervous system is well aware of the constraints of the task at hand and is less concerned about errors that do not affect performance. This tendency is often referred to as the *minimum intervention principle* (Todorov & Jordan, 2002) and is captured by the ability to ignore limb deviations that do not interfere with task completion. The same idea is reflected in the notion of *structured variability* or an *uncontrolled manifold*, where limb and whole-body motion are more variable along dimensions that are irrelevant for the task (Balasubramaniam, Riley, & Turvey, 2000; Cluff et al., 2011; Scholz & Schoner, 1999; Valero-Cuevas, Venkadesan, & Todorov, 2009).

Optimal feedback control provides a framework for us to understand task-related error corrections. Because motor commands have a cost, there is no need to control movement errors that do not interfere with the intended goal. This trade-off between behavioral performance and motor costs, expressed in a straightforward way by the quadratic cost function in Eq. 10, is only possible if the brain continuously processes sensory data to select control actions that are appropriate for the goal and constraints of the task. In agreement with this principle, we review several studies emphasizing that this type of flexible, task-dependent feedback control underlies both voluntary motor behavior and responses to external perturbations.

Visuomotor Feedback Responses Liu and Todorov (2007) provided compelling evidence for flexible biological control strategies by demonstrating that feedback responses depend on the hand's position when a goal target changes location during reaching. In this experiment, visual target perturbations were introduced at the start (early) or near the end of a reaching movement (late; figure 39.1A) in a task where subjects were instructed to stop at a peripheral target. When the target location was perturbed early in the movement, the participants corrected their hand path smoothly (figure 39.1B). In contrast, hand path corrections were incomplete when the same perturbation was introduced late in the movement (figure 39.1B). Liu and Todorov suggested that the dependency of the correction upon the timing of the target jump was caused by the inherent trade-off between end point accuracy and motor costs. In order to stop near the target, the controller became more sensitive to movement velocity than end point accuracy, leading to consistent under-compensation for target errors introduced at the end of the reaching movement. This systematic under-compensation was reduced when the task instructions were to hit rather than stop at the target (figure 39.1B). Hitting the target removes the constraint on final hand velocity and allows participants to fully correct for positional errors introduced near the target.

Several other attributes of visuomotor control have been addressed in the context of target or cursor (i.e., hand feedback) jumps. In agreement with the principles of stochastic optimal control, several studies have shown that visual perturbation responses are modulated by the reliability of the cursor or target location (Izawa & Shadmehr, 2008; Körding & Wolpert, 2004), by the shape of the goal target (Knill, Bondada, & Chhabra, 2011), and by the relevance of the visual cursor jump relative to the reaching target (Franklin & Wolpert, 2008). As we mentioned earlier, the aim of OFC models is not to eliminate all variability, but rather, allow it to accumulate in dimensions that do not interfere with performance (Todorov, 2004) while minimizing it in dimensions that are relevant for task completion. Knill and colleagues (2011) outlined the same type of selective motor corrections during visuomotor control. Indeed, Knill and colleagues showed that motor corrections following lateral cursor jumps were nearly twice as large for rectangular targets oriented parallel to the movement path compared to when the target was perpendicular to the reach.

Another clear example of flexible visuomotor control is when subjects reach in the presence of visual perturbations that may or may not affect reaching performance (Franklin & Wolpert, 2008). Franklin and Wolpert characterized these selective corrections by unexpectedly shifting hand feedback during point-to-point reaching movements. The hand cursor disturbance either persisted until the end of movement (relevant) or returned to the veridical hand location before the end of the movement (irrelevant, figure 39.2D). By shifting the visual cursor location, it was shown that the motor system selectively corrects hand feedback perturbations that affect the outcome of the task (figure 39.2E), while ignoring perturbations that do not affect performance (figure 39.2F).

In summary, visuomotor perturbation studies have shown that the brain produces distinct feedback responses when the same perturbation is encountered in different behavioral contexts. Consistent with optimal control models, these results reveal that the motor system selectively corrects for visual perturbations that jeopardize task performance while taking into consideration the constraints of the task. The latency of visuomotor corrections was consistently observed in 150–230 ms (Franklin & Wolpert, 2008; Knill et al., 2011), which, after removing delays associated with signal transmission, suggests the brain can rapidly implement flexible feedback responses after a perturbation. In the following section, we present results from mechanical perturbation studies suggesting that flexible feedback responses can be implemented in as little as ~50 ms when they are mediated by rapid changes in limb afferent feedback.

Mechanical Perturbations Investigating how quickly the motor system implements flexible control strategies can provide important insight into the neural pathways involved in feedback control. Mechanical perturbations offer a powerful means to address this problem, because muscle afferent feedback evokes motor responses on a time scale of tens of milliseconds. In fact, when the limb is displaced by a mechanical perturbation, the motor system produces a stereotyped sequence of muscle activity, beginning with the short-latency stretch reflex (response epoch called R1: 20–50 ms post-perturbation) and ending with a voluntary response (>100 ms). The short-latency stretch reflex is the earliest of these responses, and due to its timing can be attributed to spinal processing. These spinal stretch responses are sensitive to joint motion but show little of the functional complexity expressed during voluntary behavior (Pruszynski, Kurtzer, & Scott, 2008).

Between the short-latency and voluntary responses is the long-latency stretch response (R2/R3: 50–105 ms post-perturbation), which includes responses generated from multiple neural substrates (Pruszynski, Kurtzer, & Scott, 2011), including spinal (Ghez & Shinoda, 1978; Matthews, 1984; Schuurmans et al., 2009) and supraspinal pathways (Evarts, 1973; Phillips, 1969). A robust observation in motor physiology studies is that feedback corrections in the long-latency time window exhibit remarkable flexibility and can be modified by the subject's voluntary intent (Crago, Houk, & Hasan, 1976; Hammond, 1956; Rothwell, Traub, & Marsden, 1980) or the demands of an ongoing motor action (see Hasan, 2005; Matthews, 1991; Pruszynski & Scott, 2012, for a comprehensive review).

As we mentioned earlier, OFC models suggest the ability to alter feedback responses to a mechanical perturbation is a direct consequence of task-dependent sensorimotor processing (Scott, 2004). Nashed and

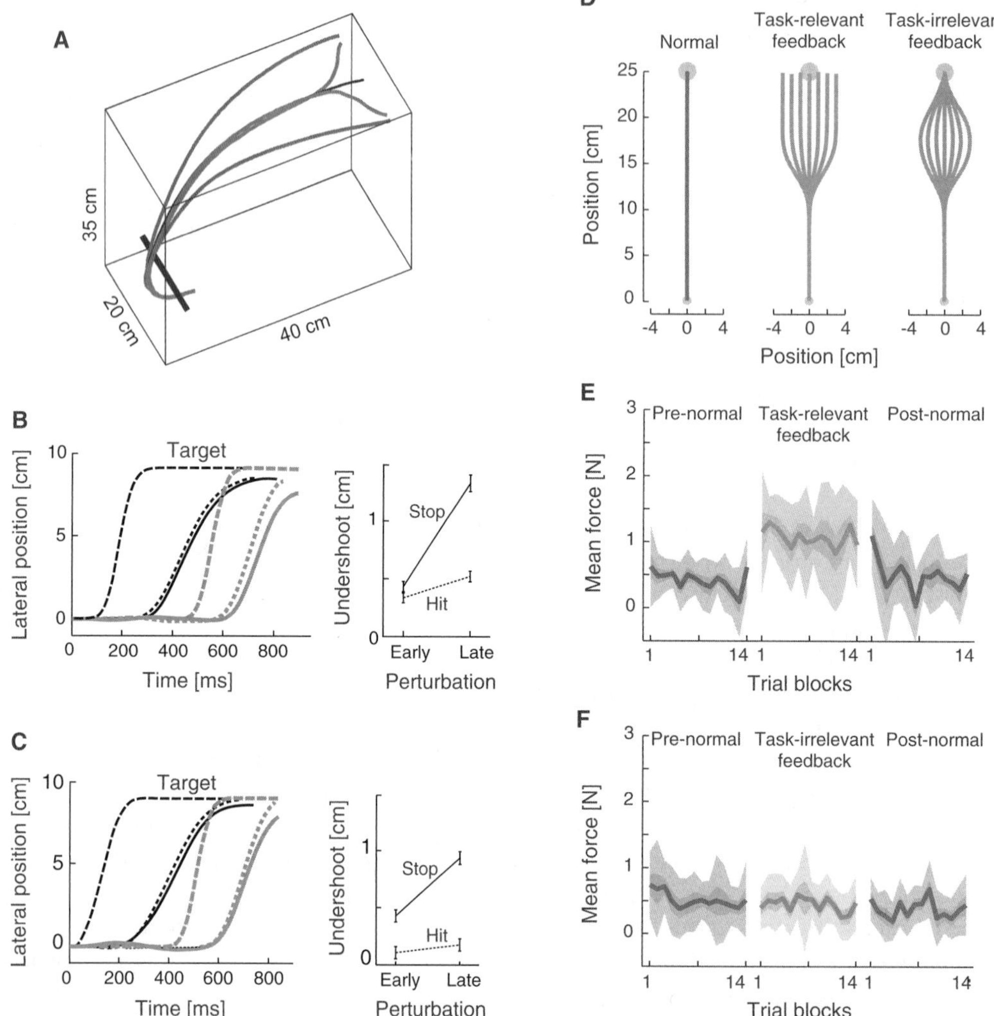

FIGURE 39.2 Visuomotor responses account for the goal of the ongoing task. (A) Subjects were instructed to reach and stop at a target ("stop" condition) that either stayed in the central location or jumped in the lateral direction after the start of the movement. Data are the population average trajectories when participants were instructed to stop at the target. Color code: black, baseline; red, early perturbation; blue, late perturbation. (B) Hand path deviation in the direction of the displaced target. Note that subjects were unable to compensate when the target jumped late in the movement, demonstrating that feedback responses depend on the task constraints. Color scheme is the same as in A. Dashed lines, "hit" condition; solid lines, "stop" condition. (C) Results from target intercept experiment. In this experiment, the target jumped laterally and then moved downward at a fixed rate. The subjects were instructed to "stop" or "hit" the target before it stopped moving. (Adapted from Liu & Todorov, 2007.) (D) Visuomotor responses only correct for errors that are relevant to the ongoing task. In the normal condition, the hand feedback cursor reproduced the hand trajectory. In the task-relevant feedback condition (orange traces), the visual cursor moved away from the hand trajectory and remained at this point for the rest of the movement. In contrast, in the task-irrelevant feedback condition (blue traces), the hand feedback cursor moved away from the hand trajectory but returned to the true hand position by the end of the movement. (E) Time course of adaptation to task-relevant feedback perturbations. Data are the mean (solid line) and standard deviation (shaded region) force difference across subjects between right and left visual perturbations (180–230 ms after cursor jump). Note that subjects produced larger corrections when the cursor displacements were relevant to the ongoing task. (F) Time course of adaptation to task-irrelevant feedback perturbations. Note that subjects did not adapt when the perturbations were irrelevant to the ongoing task. (Adapted from Franklin & Wolpert, 2008.) (See color plate 35.)

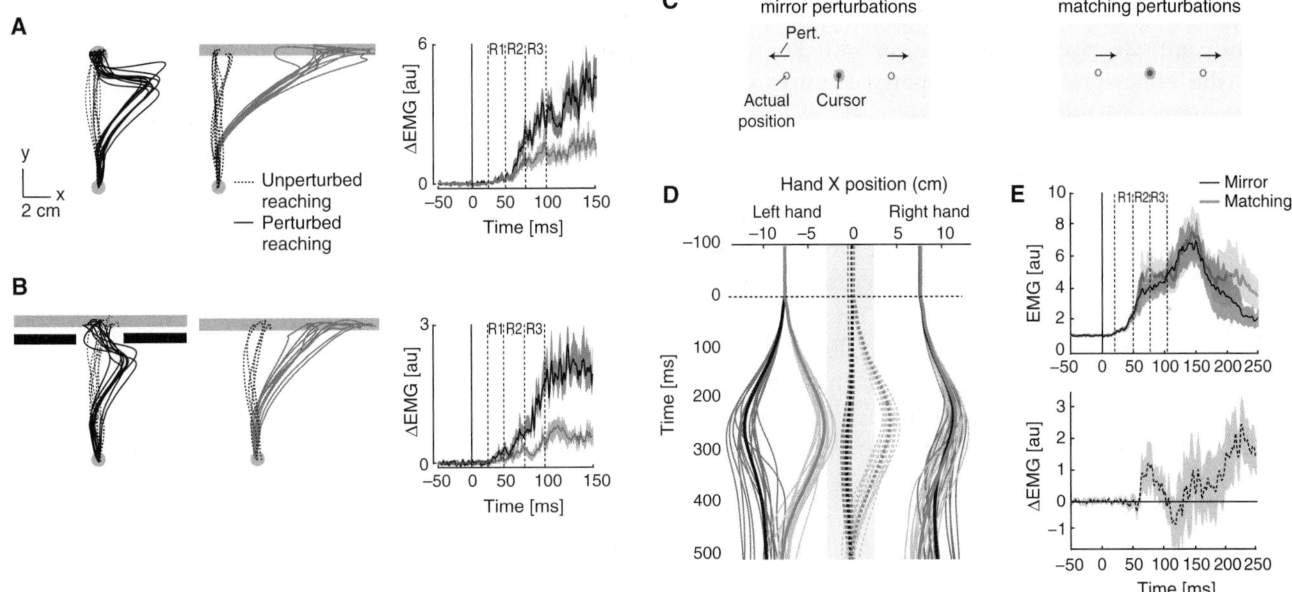

FIGURE 39.3 Corrective responses to mechanical perturbations applied during reaching and postural control. (A) Differences in corrective responses while reaching to a circle target (left panel) or rectangular bar (middle panel). Black dotted lines are unperturbed reaching movements to each target shape. Extensor perturbations were applied at the shoulder and elbow on random trials to elicit elbow motion just after movement onset. Corrective responses were rapidly directed back to the circular target (black solid lines), but in contrast, were directed to new locations on the rectangular bar (gray lines). Note that differences between corrections can be observed in brachioradialis stretch responses (population data, elbow flexor) within ~60 ms of the perturbation (R2 and R3; right panel). Vertical lines denote perturbation onset, and dashed lines denote separation of the different phases of the stretch response. (B) Differences in corrective responses while reaching to a rectangular bar occluded by environmental objects (left panel) and an unoccluded rectangular bar (middle panel). Corrective responses directed the hand between the objects to the target (black solid lines), but directed the hand to new locations on the unoccluded rectangular bar (gray lines). Note that the stretch response begins to reflect the target shape ~60 ms after the perturbation. (Adapted from Nashed et al., 2012.) (C) Corrective responses in a bimanual postural control task. In the mirror perturbation condition (left panel), the hands are perturbed in opposite directions (black arrows). In the matching perturbations condition (right panel), the hands are perturbed in the same direction (gray arrows), causing displacement of the feedback cursor that is displayed at the spatial average position of the two hands. (D) Lateral hand path kinematics. Thin lines correspond to individual subject data; thick lines denote the population average for each condition. Color scheme is the same as in C. Shaded box is the acceptable target region. (E) Pectoralis major stretch responses in the mirror (black lines) and matching conditions (gray lines). Data are the population average response, and shaded region corresponds to the SEM. Bottom panel is the difference between stretch responses in the matching and mirror conditions (Matching – Mirror). Dotted line denotes average difference across participants, shaded region is the SEM. Note that differences in muscle stretch responses emerge ~75 ms after the perturbation, demonstrating that right limb responses depend on feedback about the left arm's motion. (Adapted from Omrani et al., 2013.)

colleagues (2012) recently investigated whether long-latency responses selectively correct for task-relevant errors while subjects made reaching movements to a circular target or rectangular bar oriented perpendicular to the reach (figure 39.3A). On certain trials, a mechanical perturbation was applied to displace the hand in the lateral direction. When the perturbation pushed the hand away from the circular target, the participants performed rapid corrective responses to direct their hand back to the target. In contrast, when the same perturbation was applied while subjects reached to the rectangular bar, the participants redirected their hand to new locations on the bar. Similar context-dependent responses were evoked when obstacles in the environment required that the participants navigate to the target through a narrow channel (figure 39.3B). These behavioral results were reproduced by an optimal feedback control model with differing sensitivity to lateral hand errors. Further, the model predicted that the shape of the goal target should influence the feedback response as early as sensory feedback about the perturbation became available. In agreement with these model predictions, differences in muscle responses between tasks were observed in as little as 70 ms following the perturbation, establishing that rapid motor corrections (i.e., long-latency responses) integrate muscle stretch information with knowledge about the behavioral goal and spatial features of the environment (Nashed et al., 2012).

FLEXIBLE BIMANUAL FEEDBACK CONTROL In the context of sudden mechanical perturbations, Marsden and colleagues (Marsden, Merton, & Morton, 1981) first outlined the task dependency of interlimb responses, noting that long-latency muscle stretch responses in the right arm after left arm perturbations reflected the task the right arm was performing. If the right arm held a table for support, the extensors were activated to stabilize the participant following the left arm perturbation. Remarkably, subjects even reversed their corrective responses and activated the flexor muscles of the right arm if they had to stabilize a cup of tea. Perhaps the most powerful example of how the task modulates interlimb feedback responses is that muscle responses were absent if the subject grasped a loose handle and benefitted little from right arm responses. The coordinated responses observed in the muscles of the unperturbed arm clearly emphasize that online feedback control is not hard-wired but can be engaged at will, depending on the context and on the intended behavior.

Bimanual control therefore provides a remarkable tool to address how sophisticated feedback control can be distributed across different body parts. Motivated by the capacity of OFC models to exploit many different ways to attain the same movement goal (i.e., task redundancy), recent studies have examined feedback corrections in bimanual tasks by comparing motor responses when the two arms act independently or are coupled by the task demand. A compelling example of this flexibility is when one hand is gradually perturbed in a task that requires bimanual reaching movements (Diedrichsen, 2007). If each hand controls its own cursor while reaching to separate targets (two-cursor task), only the perturbed hand shows a corrective response. However, when the two arms control a single cursor (one-cursor task), displayed as the spatial average position of the two hands, perturbations applied to one hand elicit bilateral responses to correct the cursor's trajectory (Diedrichsen, 2007). These flexible responses were reproduced by expressing the task constraints (Q_t in Eq. 7) for each hand independently, or as the average of the two hands. The flexibility of bilateral corrections is not restricted to changes in the size of the response, as even the direction of these coordinated responses can be reversed if required by the task (Diedrichsen & Gush, 2009).

Studies have since confirmed that sensory information from one limb in both reaching (Mutha & Sainburg, 2009) and posture (Dimitriou, Franklin, & Wolpert, 2012; Omrani, Diedrichsen, & Scott, 2013) rapidly modifies corrective responses in the other limb during the long-latency time window. When the two hands control independent cursors and are perturbed in opposite directions, robust long-latency responses are observed in both arms to counter the perturbation (Omrani et al., 2013). In contrast, long-latency stretch responses are substantially smaller when the two hands are perturbed in opposite directions while controlling a single cursor displayed at the spatial average position of the two hands. In this context, corrective responses are unnecessary because the position of the single feedback cursor is not disturbed by the perturbation. When the direction of the left arm perturbation was not predictable, stretch responses in the right arm depended on proprioceptive input from the left arm in the one-cursor condition, and were larger if both arms were perturbed in the same direction (figure 39.3C, D, and E). These differential feedback responses were not observed when each hand controlled its own independent cursor. Why should the brain distribute corrective responses when several effectors are involved in the task? Optimal feedback control predicts this behavioral pattern, since dividing the response across effectors reduces the effort and variability of motor corrections (Diedrichsen & Dowling, 2009).

INTERNAL MODELS OF MULTIJOINT DYNAMICS We have so far focused on evidence that the motor system continuously processes sensory data to generate task-dependent feedback responses. It is important to recognize, however, that task-dependent feedback responses can only be achieved if the motor system has knowledge of how the body should move in response to external forces that arise from our environmental interactions or forces generated by muscles. An important feature of body dynamics is the presence of interaction torques between joints that require coordinated multijoint responses to control the motion at each joint. Extensive evidence has shown that the voluntary motor system compensates for these interaction torques to produce straight reaching movements (Gribble & Ostry, 1999; Hollerbach & Flash, 1982), but an interesting question is whether corrective responses also reflect knowledge of limb mechanics.

Previous work emphasized that mechanical perturbations evoked rapid responses in muscles that are not directly stretched by the perturbation, suggesting that coordinated motor responses occur in the long-latency time window (Gielen, Ramaekers, & van Zuylen, 1988;

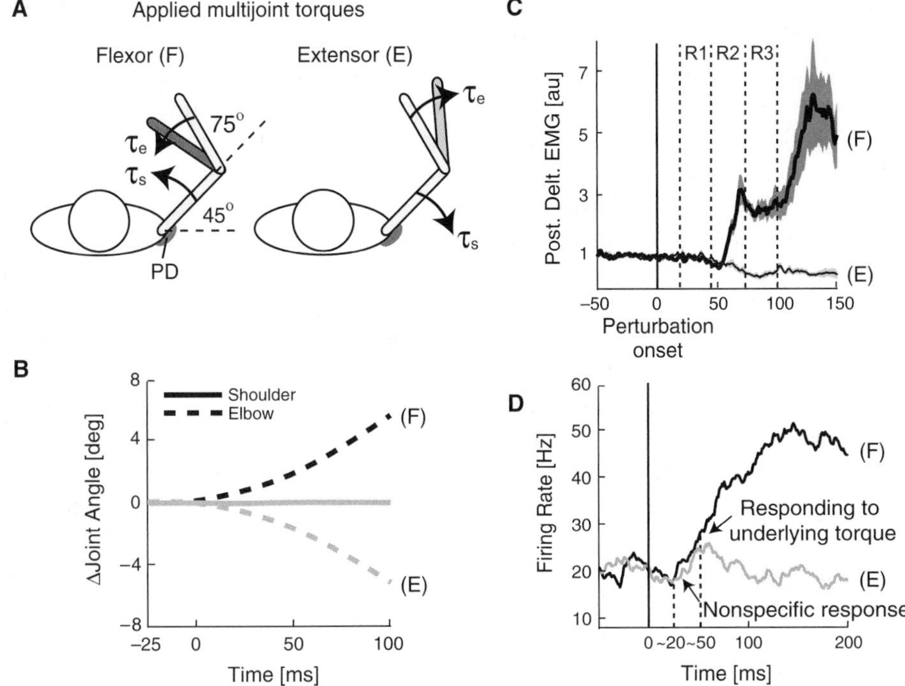

FIGURE 39.4 Long-latency responses express knowledge of multijoint limb mechanics. (A) Subjects were instructed to maintain their hand at a central target, and step-torque perturbations were applied to the shoulder and elbow joints (flexor torques at both joints, dark gray shading and denoted by (F), extensor torques at both joints, light gray shading and denoted by (E)). Stretch responses were recorded from the posterior deltoid muscle (PD) (B) Limb configuration and applied multijoint torques were selected to cause substantial elbow motion (dashed lines) but minimal shoulder motion (solid lines). (C) Posterior deltoid (shoulder extensor) muscle activity aligned on perturbation onset. Note that although there was no change in the length of the posterior deltoid muscle, there is still a robust long-latency response (excitatory and inhibitory). Data are the population-level muscle response, and shaded region corresponds to SEM. Same color scheme as in A and B. (Adapted from Kurtzer et al., 2008.) (D) Population-level response of shoulder-like M1 neurons. Shoulder-like neurons respond to the underlying shoulder torque even though local information from the shoulder is ambiguous about the underlying torque. Note that the response to the underlying torque begins about 50 ms after the onset of the shoulder and elbow perturbations. (Adapted from Pruszynski et al., 2012.)

Soechting & Lacquaniti, 1988). The question of whether these responses relate to limb dynamics was recently addressed by applying different combinations of perturbations to the shoulder and elbow. In one experiment, multijoint loads applied to the shoulder and elbow did not produce motion at the shoulder but led to either flexion or extension motion at the elbow (Kurtzer, Pruszynski, & Scott, 2008, 2009; figure 39.4A, B, and C). The authors found that there was no short-latency muscle stretch response in the posterior deltoid, a shoulder extensor, highlighting that this spinal reflex is not elicited without overt motion at the joint (figure 39.4B, C). In contrast, the long-latency response (50–105 ms post-perturbation) integrated motion information from both joints and generated a large response in the posterior deltoid to counter the shoulder flexor torque, even though there was no motion at the joint (figure 39.4C). These results clearly showed that rapid motor corrections map the sensed motion of the shoulder and elbow joints onto a response that is appropriate for the actual underlying torque rather than the observed motion pattern.

It is important to emphasize that OFC as a theory of motor behavior can only tell us what the optimal motor solution should look like. The studies presented above highlight feedback responses that possess an impressive degree of flexibility and can be modified to suit the needs of many behavioral tasks. It has been demonstrated that long-latency stretch responses are modified in different dynamic environments (Ahmadi-Pajouh, Towhidkhah, & Shadmehr, 2012; Kimura & Gomi, 2009; Krutky, Ravichandran, Trumbower, & Perreault, 2010), and a direct prediction is that corrective responses should be modulated by the novel dynamical context. Changes in corrective responses to a gradual perturbation have been observed over longer time scales (Wagner & Smith, 2008). A recent study shows that long-latency responses also express knowledge of internal models acquired during motor learning (Cluff & Scott, 2013). This result emphasizes that adaptive changes to novel dynamics alter voluntary behavior and rapid feedback responses.

In summary, the studies outlined above emphasized the following principles: (1) upper-limb postural control and reaching involve continuous sensory processing; (2) feedback control processes selectively compensate for errors that interfere with the ongoing task, producing motor strategies appropriate for the movement goal; (3) these principles describe voluntary motor behavior as well as responses to visual and mechanical perturbations; and (4) sophisticated feedback responses emerge ~50–60 ms following a mechanical perturbation, coinciding with the long-latency stretch response.

M1 as part of a flexible feedback controller

A robust observation across the studies outlined above is that short-latency responses (~20–50 ms) are predominantly sensitive to muscle stretch (Pruszynski & Scott, 2012), whereas task-dependent responses consistently emerge in the long-latency time window (~50–100 ms). Long-latency responses coincide with the contribution of long-loop mechanisms, including a transcortical pathway through M1 (Cheney & Fetz, 1984; Desmedt, 1978; Matthews, 1991). Indeed, single-unit recordings in monkeys indicate that M1 receives somatosensory feedback (Evarts & Fromm, 1977; Scott & Kalaska, 1997) or mechanical perturbations (Herter, Korbel, & Scott, 2009; Picard & Smith, 1992). Moreover, human studies using transcranial magnetic stimulation emphasize a causal link between M1 processing and long-latency stretch responses, since motor cortex stimulation disrupts the task-dependent features of long-latency responses (Capaday, Forget, Fraser, & Lamarre, 1991; Day, Riescher, Struppler, Rothwell, & Marsden, 1991; Kimura, Haggard, & Gomi, 2006, but see also Shemmell, An, & Perreault, 2009). Given the involvement of M1 in the generation of voluntary behavior (Porter & Lemon, 1993) and rapid feedback pathways, this brain region is a clear candidate to implement flexible feedback control strategies (Scott, 2004).

A number of neurophysiological studies in nonhuman primates have highlighted that a transcortical feedback through M1 provides important task-dependent processing following mechanical perturbations. In a seminal study, Evarts and Tanji (1976) found that within 40 ms of an upper limb perturbation, the responses of pyramidal tract neurons differ depending on whether the monkey was instructed to push or pull a handle. The authors suggested that these task-dependent neural responses contribute to volitional control of the limb during the long-latency time window. This idea was recently tested in a study examining whether the transcortical feedback pathway through M1 exhibits knowledge of the limb's biomechanical properties (Pruszynski et al., 2011) using the paradigm developed by Kurtzer et al. (2008). The authors applied different combinations of shoulder and elbow torques evoking flexor or extensor motion at the elbow and no motion at the shoulder. As a result of these multijoint perturbation loads, shoulder motion (none, in this case) is ambiguous about the applied perturbation, and the motor system can only appropriately counter the underlying torque by taking elbow motion into consideration. The responses of shoulder-related neurons in MI were found to appropriately respond to the applied load at ~50 ms, about 15 ms before long-latency responses were recorded in shoulder muscles (figure 39.4D).

How this transcortical feedback pathway resolves this multijoint integration problem is unclear. One clue is that the earliest activity in MI, from 20 to 50 ms after a perturbation, does not reflect specific features of the perturbation (figure 39.4D). That is, regardless of perturbation direction, all neurons sensitive to shoulder or elbow motion display similar responses until 50 ms after the perturbation is applied. This may suggest that M1 requires ~30 ms to identify the appropriate response, or that knowledge of limb mechanics is computed elsewhere in sensorimotor circuits. Several brain regions that receive sensory feedback from the limb project to M1, including primary somatosensory cortex, parietal area 5, and cerebellum (Fromm & Evarts, 1982; Martin, Cooper, Hacking, & Ghez, 2000; Mason, Miller, Baker, & Houk, 1998). There was considerable interest in examining how brain regions including M1 responded to sensory feedback in the 1970s (for a review, see Desmedt, 1978), but since then the role of sensory feedback in M1 processing has received little attention from the scientific community. Given the tight link between voluntary control and sensory feedback processing, the use of OFC as a framework to understand voluntary motor control has led to a renewed interest in how different cortical and subcortical circuits participate in feedback processing for motor control (Scott, 2012).

Conclusion

We have argued that OFC is a powerful tool for understanding biological motor control and that it has shed light on many of the complexities the brain must consider to move successfully in the presence of neural variability or environmental disturbances. Perhaps the most important contribution of OFC in movement neuroscience has been to unify motor planning and feedback responses to perturbations in a common framework. These two aspects of motor control have been almost relegated to distinct fields of investigation (Scott, 2008). The major conceptual advance of OFC is the idea that movement planning and execution are two sides of the same story: a goal-directed feedback control policy. We expect that future research will address how the computations underlying flexible feedback control are distributed across different brain regions.

REFERENCES

Ahmadi-Pajouh, M. A., Towhidkhah, F., & Shadmehr, R. (2012). Preparing to reach: Selecting an adaptive long-latency feedback controller. *J Neurosci, 32*(28), 9537–9545.

Ariff, G., Donchin, O., Nanayakkara, T., & Shadmehr, R. (2002). A real-time state predictor in motor control: Study of saccadic eye movements during unseen reaching movements. *J Neurosci, 22*(17), 7721–7729.

Arnold, L. (1974). *Stochastic differential equations: Theory and applications.* New York, NY: Wiley.

Åström, K. (1970). *Introduction to stochastic control theory.* New York, NY: Academic Press.

Balasubramaniam, R., Riley, M. A., & Turvey, M. (2000). Specificity of postural sway to the demands of a precision task. *Gait Posture, 11*(1), 12–24.

Berret, B., Chiovetto, E., Nori, F., & Pozzo, T. (2011). Evidence for composite cost functions in arm movement planning: An inverse optimal control approach. *PLoS Comput Biol, 7*(10), e1002183.

Bhattacharyya, S. P., Chapellat, H., & Keel, L. H. (1995). *Robust control: The parametric approach.* Englewood Cliffs, NJ: Prentice Hall.

Biess, A., Liebermann, D. G., & Flash, T. (2007). A computational model for redundant human three-dimensional pointing movements: Integration of independent spatial and temporal motor plans simplifies movement dynamics. *J Neurosci, 27*(48), 13045–13064.

Boulet, B., & Duan, X. (2007). The fundamental tradeoff between performance and robustness: A new perspective on loop shaping. *IEEE Contr Syst Mag, 27*(3), 30–44.

Brown, R. (1983). *Introduction to random signal analysis and Kalman filtering.* New York, NY: Wiley.

Bryson, A., & Ho, Y.-C. (1975). *Applied optimal control: Optimization, estimation, and control.* New York, NY: Hemisphere Publishing.

Capaday, C., Forget, R., Fraser, R., & Lamarre, Y. (1991). Evidence for a contribution of the motor cortex to the long-latency stretch reflex of the human thumb. *J Physiol, 440*(1), 243–255.

Cheney, P. D., & Fetz, E. E. (1984). Corticomotoneuronal cells contribute to long-latency stretch reflexes in the rhesus monkey. *J Physiol, 349,* 249–272.

Churchland, M. M., Afshar, A., & Shenoy, K. V. (2006). A central source of movement variability. *Neuron, 52*(6), 1085–1096.

Cluff, T., Aspasia, M., Lee, T. D., & Balasubramaniam, R. (2011). Multijoint error compensation mediates unstable object control. *J Neurophysiol, 108,* 1167–1175.

Cluff, T., & Scott, S. H. (2013). Rapid feedback responses correlate with reach adaptation and properties of novel upper limb loads. *J Neurosci, 33*(40), 15903–15914.

Crago, P. E., Houk, J. C., & Hasan, Z. (1976). Regulatory actions of human stretch reflex. *J Neurophysiol, 39*(5), 925–935.

Crevecoeur, F., McIntyre, J., Thonnard, J.-L., & Lefèvre, P. (2010). Movement stability under uncertain internal models of dynamics. *J Neurophysiol, 104,* 1301–1313.

Crevecoeur, F., & Scott, S. H. (2013). Priors engaged in long-latency responses to mechanical perturbations suggest a rapid update in state estimation. *PLoS Comput Biol, 9*(8), e1003177.

Crevecoeur, F., Sepulchre, R. J., Thonnard, J.-L., & Lefèvre, P. (2011). Improving the state estimation for optimal control of stochastic processes subject to multiplicative noise. *Automatica, 47*(3), 591–596.

Day, B. L., Riescher, H., Struppler, A., Rothwell, J. C., & Marsden, C. D. (1991). Changes in the response to magnetic and electrical stimulation of the motor cortex following muscle stretch in man. *J Physiol, 433*(1), 41–57.

Desmedt, J. (1978). *Cerebral motor control in man: Long loop mechanisms.* New York, NY: Karger.

Diedrichsen, J. (2007). Optimal task-dependent changes of bimanual feedback control and adaptation. *Curr Biol, 17*(19), 1675–1679.

Diedrichsen, J., & Dowling, N. (2009). Bimanual coordination as task-dependent linear control policies. *Hum Movement Sci, 28*(3), 334–347.

Diedrichsen, J., & Gush, S. (2009). Reversal of bimanual feedback responses with changes in task goal. *J Neurophysiol, 101*(1), 283–288.

Dimitriou, M., Franklin, D. W., & Wolpert, D. M. (2012). Task-dependent coordination of rapid bimanual motor responses. *J Neurophysiol, 107*(3), 890–901.

Doyle, J. C., Francis, B. A., & Tannenbaum, A. R. (1992). *Feedback control theory.* New York, NY: Macmillan.

Evarts, E. V. (1973). Motor cortex reflexes associated with learned movement. *Science, 179*(4072), 501–503.

Evarts, E. V., & Fromm, C. (1977). Sensory responses in motor cortex neurons during precise motor control. *Neurosci Lett, 5*(5), 267–272.

Evarts, E. V., & Tanji, J. (1976). Reflex and intended responses in motor cortex pyramidal tract neurons of monkey. *J Neurophysiol, 39,* 1069–1080.

Faisal, A. A., Selen, L. P. J., & Wolpert, D. M. (2008). Noise in the nervous system. *Nat Rev Neurosci, 9*(4), 292–303.

Flash, T., & Hogan, N. (1985). The coordination of arm movements: An experimentally confirmed mathematical model. *J Neurosci, 5*(7), 1688–1703.

Franklin, D. W., & Wolpert, D. M. (2008). Specificity of reflex adaptation for task-relevant variability. *J Neurosci, 28*(52), 14165–14175.

Fromm, C., & Evarts, E. V. (1982). Pyramidal tract neurons in somatosensory cortex: Central and peripheral inputs during voluntary movement. *Brain Res, 238*(1), 186–191.

Ghez, C., & Shinoda, Y. (1978). Spinal mechanisms of the functional stretch reflex. *Exp Brain Res, 32,* 55–68.

Gielen, C. C., Ramaekers, L., & van Zuylen, E. J. (1988). Long-latency stretch reflexes as co-ordinated functional responses in man. *J Physiol, 407*(1), 275–292.

Gribble, P. L., & Ostry, D. J. (1999). Compensation for interaction torques during single- and multijoint limb movement. *J Neurophysiol, 82*(5), 2310–2326.

Hammond, P. H. (1956). The influence of prior instruction to the subject on an apparently involuntary neuro-muscular response. *J Physiol, 132*(1), 17–18P.

Harris, C. M., & Wolpert, D. M. (1998). Signal-dependent noise determines motor planning. *Nature, 394*(6695), 780–784.

Hasan, Z. (2005). The human motor control system's response to mechanical perturbation: Should it, can it and does it ensure stability? *J Motor Behav, 37*(6), 484–493.

Herter, T. M., Korbel, T., & Scott, S. H. (2009). Comparison of neural responses in primary motor cortex to transient and continuous loads during posture. *J Neurophysiol, 101*(1), 150–163.

Hollerbach, J., & Flash, T. (1982). Dynamic interactions between limb segments during planar arm movement. *Biol Cybern, 44,* 67–77.

Izawa, J., & Shadmehr, R. (2008). On-line processing of uncertain information in visuomotor control. *J Neurosci, 28*(44), 11360–11368.

Kalman, R. (1960). A new approach to linear filtering and prediction problems. *J Basic Eng-T ASME, 82,* 35–45.

Kawato, M. (1999). Internal models for motor control and trajectory planning. *Curr Opin Neurobiol, 9*(6), 718–727.

Kimura, T., & Gomi, H. (2009). Temporal development of anticipatory reflex modulation to dynamical interactions during arm movement. *J Neurophysiol, 102*(4), 2220–2231.

Kimura, T., Haggard, P., & Gomi, H. (2006). Transcranial magnetic stimulation over sensorimotor cortex disrupts anticipatory reflex gain modulation for skilled action. *J Neurosci, 26*(36), 9272–9281.

Knill, D. C., Bondada, A., & Chhabra, M. (2011). Flexible, task-dependent use of sensory feedback to control hand movements. *J Neurosci, 31*(4), 1219–1237.

Körding, K. P., & Wolpert, D. M. (2004). Bayesian integration in sensorimotor learning. *Nature, 427*(6971), 244–247.

Krutky, M. A., Ravichandran, V. J., Trumbower, R. D., & Perreault, E. J. (2010). Interactions between limb and environmental mechanics influence stretch reflex sensitivity in the human arm. *J Neurophysiol, 103*(1), 429–440.

Kurtzer, I. L., Pruszynski, J. A., & Scott, S. H. (2008). Long-latency reflexes of the human arm reflect an internal model of limb dynamics. *Curr Biol, 18*(6), 449–453.

Kurtzer, I., Pruszynski, J. A., & Scott, S. H. (2009). Long-latency responses during reaching account for the mechanical interaction between the shoulder and elbow joints. *J Neurophysiol, 102*(5), 3004–3015.

Li, W., & Todorov, E. (2007). Iterative linearization methods for approximately optimal control and estimation of non-linear stochastic system. *Int J Control, 80*(9), 1439–1453.

Liu, D., & Todorov, E. (2007). Evidence for the flexible sensorimotor strategies predicted by optimal feedback control. *J Neurosci, 27*(35), 9354–9368.

Marsden, C. D., Merton, P. A., & Morton, H. B. (1981). Human postural responses. *Brain, 104*(3), 513–534.

Martin, J. H., Cooper, S. E., Hacking, A., & Ghez, C. (2000). Differential effects of deep cerebellar nuclei inactivation on reaching and adaptive control. *J Neurophysiol, 83*(4), 1886–1899.

Mason, C. R., Miller, L. E., Baker, J. F., & Houk, J. C. (1998). Organization of reaching and grasping movements in the primate cerebellar nuclei as revealed by focal muscimol inactivations. *J Neurophysiol, 79*(2), 537–554.

Matthews, P. B. (1984). Evidence from the use of vibration that the human long-latency stretch reflex depends upon spindle secondary afferents. *J Physiol, 348*(1), 383–415.

Matthews, P. B. C. (1991). The human stretch reflex and the motor cortex. *Trends Neurosci, 14*(3), 87–91.

Mehta, B., & Schaal, S. (2002). Forward models in visuomotor control. *J Neurophysiol, 88*(2), 942–953.

Michiels, W., & Niculescu, S.-I. (2007). Stability and stabilization of time-delay systems. An eigenvalue-based approach. In *Advances in design and control, 12.* Philadelphia, PA: Society for Industrial and Applied Mathematics (SIAM) Publications.

Mutha, P. K., & Sainburg, R. L. (2009). Shared bimanual tasks elicit bimanual reflexes during movement. *J Neurophysiol, 102*(6), 3142–3155.

Nakano, E., Imamizu, H., Osu, R., Uno, Y., Gomi, H., Yoshioka, T., & Kawato, M. (1999). Quantitative examinations of internal representations for arm trajectory planning: Minimum commanded torque change model. *J Neurophysiol, 81*(5), 2140–2155.

Nashed, J. Y., Crevecoeur, F., & Scott, S. H. (2012). Influence of the behavioral goal and environmental obstacles on rapid feedback responses. *J Neurophysiol, 108*(4), 999–1009.

Omrani, M., Diedrichsen, J., & Scott, S. H. (2013). Rapid feedback corrections during a bimanual postural task. *J Neurophysiol, 109*(1), 147–161.

Osborne, L. C., Lisberger, S. G., & Bialek, W. (2005). A sensory source for motor variation. *Nature, 437*(7057), 412–416.

Phillips, C. G. (1969). The 1968 Ferrier lecture: Motor apparatus of the baboon's hand. *Proc R Soc Lond B Biol Sci, 173*(1031), 141–174.

Picard, N., & Smith, A. M. (1992). Primary motor cortical responses to perturbations of prehension in the monkey. *J Neurophysiol, 68*(5), 1882–1894

Porter, R., & Lemon, R. N. (1993). *Corticospinal function and voluntary movement.* New York, NY: Clarendon Press.

Pruszynski, J. A., Kurtzer, I., Nashed, J. Y., Omrani, M., Brouwer, B., & Scott, S. H. (2011). Primary motor cortex underlies multi-joint integration for fast feedback control. *Nature, 478*(7369), 387–390.

Pruszynski, J. A., Kurtzer, I., & Scott, S. H. (2008). Rapid motor responses are appropriately tuned to the metrics of a visuospatial task. *J Neurophysiol, 100*(1), 224–238.

Pruszynski, J. A., Kurtzer, I., & Scott, S. H. (2011). The long-latency reflex is composed of at least two functionally independent processes. *J Neurophysiol, 106*(1), 449–459.

Pruszynski, J. A., & Scott, S. H. (2012). Optimal feedback control and the long-latency stretch response. *Exp Brain Res, 218,* 341–359.

Qian, N., Jiang, Y., Jiang, J.-P., & Mazzoni, P. (2013). Movement duration, Fitts's law, and an infinite-horizon optimal feedback control model for biological motor systems. *Neural Comput, 25,* 697–724.

Ronsse, R., Thonnard, J.-L., Lefèvre, P., & Sepulchre, R. (2008). Control of bimanual rhythmic movements: Trading efficiency for robustness depending on the context. *Exp Brain Res, 187,* 193–205.

ROTHWELL, J. C., TRAUB, M. M., & MARSDEN, C. D. (1980). Influence of voluntary intent on the human long-latency stretch reflex. *Nature, 286*(5772), 496–498.

SCHOLZ, J., & SCHONER, G. (1999). The uncontrolled manifold concept: Identifying control variables for a functional task. *Exp Brain Res, 126,* 289–306.

SCHUURMANS, J., DE VLUGT, E., SCHOUTEN, A., MESKERS, C., DE GROOT, J., & VAN DER HELM, F. (2009). The monosynaptic Ia afferent pathway can largely explain the stretch duration effect of the long latency M2 response. *Exp Brain Res, 193,* 491–500.

SCOTT, S. H. (2012). The computational and neural basis of voluntary motor control and planning. *Trends Cogn Sci, 16*(11), 541–549.

SCOTT, S. H. (2004). Optimal feedback control and the neural basis of volitional motor control. *Nat Rev Neurosci, 5*(7), 532–546.

SCOTT, S. H. (2008). Inconvenient truths about neural processing in primary motor cortex. *J Physiol, 586*(5), 1217–1224.

SCOTT, S. H., & KALASKA, J. F. (1997). Reaching movements with similar hand paths but different arm orientation. I. Activity of individual cells in motor cortex. *J Neurophysiol, 77,* 826–853.

SCOTT, S. H., & LOEB, G. E. (1994). The computation of position sense from spindles in mono- and multiarticular muscles. *J Neurosci, 14*(12), 7529–7540.

SHADMEHR, R., SMITH, M. A., & KRAKAUER, J. W. (2010). Error correction, sensory prediction, and adaptation in motor control. *Annu Rev Neurosci, 33,* 89–108.

SHEMMELL, J., AN, J. H., & PERREAULT, E. J. (2009). The differential role of motor cortex in stretch reflex modulation induced by changes in environmental mechanics and verbal instruction. *J Neurosci, 29*(42), 13255–13263.

SOECHTING, J. F., & LACQUANITI, F. (1988). Quantitative evaluation of the electromyographic responses to multidirectional load perturbations of the human arm. *J Neurophysiol, 59*(4), 1296–1313.

TODOROV, E. (2004). Optimality principles in sensorimotor control. *Nat Neurosci, 7*(9), 907–915.

TODOROV, E. (2005). Stochastic optimal control and estimation methods adapted to the noise characteristics of the sensorimotor system. *Neural Comput, 17*(5), 1084–1108.

TODOROV, E. (2006). Optimal control theory. In K. Doya (Ed.), *Bayesian brain: Probabilistic approaches to neural coding* (pp. 269–298). Cambridge, MA: MIT Press.

TODOROV, E., & JORDAN, M. I. (2002). Optimal feedback control as a theory of motor coordination. *Nat Neurosci, 5*(11), 1226–1235.

UNO, Y., KAWATO, M., & SUZUKI, R. (1989). Formation and control of optimal trajectory in human multijoint arm movement. *Biol Cybern, 61,* 89–101.

VALERO-CUEVAS, F. J., VENKADESAN, M., & TODOROV, E. (2009). Structured variability of muscle activations supports the minimal intervention principle of motor control. *J Neurophysiol, 102*(1), 59–68.

VAN BEERS, R. J., HAGGARD, P., & WOLPERT, D. M. (2004). The role of execution noise in movement variability. *J Neurophysiol, 91*(2), 1050–1063.

WAGNER, M. J., & SMITH, M. A. (2008). Shared internal models for feedforward and feedback control. *J Neurosci, 28*(42), 10663–10673.

WOLPERT, D. M., & FLANAGAN, J. R. (2001). Motor prediction. *Curr Biol, 11*(18), R729–R732.

WOLPERT, D. M., & KAWATO, M. (1998). Multiple paired forward and inverse models for motor control. *Neural Netw, 11*(7–8), 1317–1329.

WOLPERT, D. M., DIEDRICHSEN, J., & FLANAGAN, J. R. (2011). Principles of sensorimotor learning. *Nat Rev Neurosci, 12,* 739–751.

ZAHALAK, G. I. (1981). A distribution-moment approximation for kinetic theories of muscular contraction. *Math Biosci, 55,* 89–114.

40 Multineuronal Views of Information Processing

AARON BATISTA

ABSTRACT Cognitive neuroscientists have only recently gained the ability to record from many neurons at once while animals behave. The multineuronal view of information processing provides new insights into the collective action of neurons. Some of the principles emerging from these studies are (1) there is flexibility in the exchange of information between brain areas; (2) populations of neurons together embody underlying computational structures; (3) neurons are active in response to specific patterns of synaptic input, and these input patterns can provide a more accurate description of neural "tuning" than do external covariates. A fuller understanding of how neural networks transform and manipulate information is emerging from the perspectives provided by multineuronal recording techniques and the computational tools being developed to interpret multineuronal data.

Neuroscience has made tremendous strides in discovering how the sensory environment and ensuing actions are represented by individual neurons. The question of how information is transformed as it flows through the brain is just as important, but we know far less about it. Understanding how neural information is altered, elaborated, and refined—essentially, neural computation—requires observing more than one neuron at a time.

This chapter surveys the history of multineuronal investigations of the brain. We begin with paired recordings and the work of George Gerstein and his colleagues in the 1960s. They introduced the first tools to probe information flow between pairs of neurons. Next we survey other paired-recording techniques and what they have contributed to our understanding of the neural mechanisms of sensory, motor, and cognitive processes. Then we move beyond pairwise interactions to consider recently introduced multineuronal recording and analysis techniques, and the new perspectives on neuronal information processing that they are enabling. The chapter concludes with a perspective on the types of questions whose answers might be near at hand using multielectrode techniques.

Paired recording techniques

CROSS-CORRELATION Perkel, Gerstein, and Moore (1967) introduced a statistical framework to analyze the interactions between pairs of neurons. Their technique, *cross-correlation* (now often called *synchrony*), involved measuring the probability of neuron 2 firing an action potential with respect to the time when neuron 1 fires (figure 40.1). Connectivity schemes such as direct excitation, direct inhibition, shared excitation, and shared inhibition could be inferred from the shape of the cross-correlation histogram. The central operation is

$$C(\tau) = \sum_t S_A(t)S_B(t+\tau), \tag{1}$$

where C is the cross-correlation histogram (or *correlogram*) and S is the time of action potentials from one neuron. The time lag between action potentials from the two cells is τ. Note that this equation is nearly identical to a convolution (differing only by a sign). The cross-correlation measures how similar two signals are over a range of temporal offsets. C is constructed for each individual trial, then averaged over trials and corrected for chance to yield correlograms like those in figure 40.1. The appearance of a peak in a correlogram can indicate an interaction between two neurons, but Carlos Brody (1999) pointed out that such plots must be interpreted with caution. Peaks can appear spuriously, for example, if two neurons that are not actually connected respond to the same stimulus, or are oscillating in unison. Statistical techniques and a dose of good judgment can help resolve whether a peak in a correlogram actually indicates neural connectivity.

SOME APPLICATIONS OF CROSS-CORRELATION Reid and Alonso (1995) used the cross-correlation technique to examine connectivity between the lateral geniculate nucleus and the primary visual cortex. They discovered that the connections between these areas were highly specific: they were tightly constrained to overlapping receptive field locations and matching light-dark polarity. Even when a correlation was present, it was fairly weak: the probability of a V1 spike appearing in response to the spike of a particular lateral geniculate nucleus cell was less than 8%. It is reasonable to expect the lateral geniculate nucleus–to-V1 projection to be one of

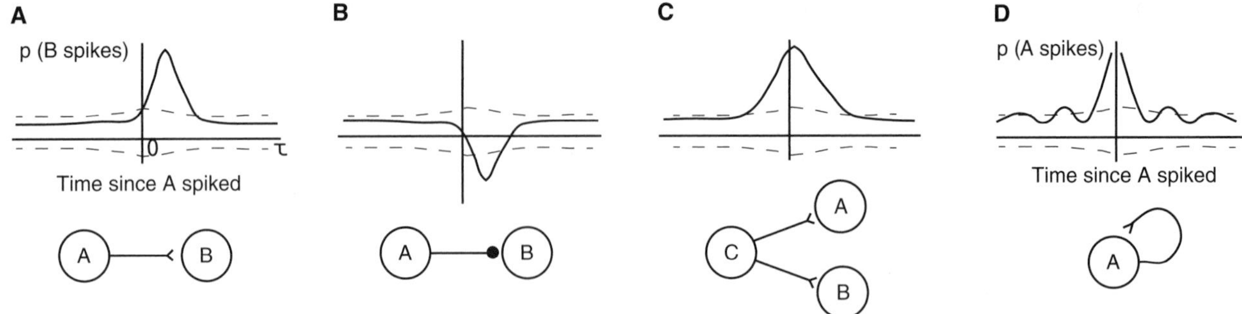

FIGURE 40.1 Covariogram shapes and the connectivity they imply. (A) Direct excitation; (B) direct inhibition; (C) shared excitation from an unobserved third neuron; (D) an autocorrelation reveals oscillations. The large peak at 0 is suppressed. The autapse in the schematic symbolizes any influence (direct or indirect) that A has on itself. The dashed lines show the null hypothesis of independence between the cells, constructed by shuffling data across trials.

the stronger feedforward pathways in the nervous system. Although the specificity of connectivity between the areas is perhaps not surprising, the weakness of the connection strength between particular neurons is a surprise.

At the opposite pole of the nervous system—motor output—a technique conceptually related to cross-correlation has been used to illustrate connectivity in the motor system. In the *spike-triggered average* technique, electromyographic (EMG) activity is recorded along with the activity of a neuron in primary motor cortex (M1). When EMG signals are aligned on the time of spiking of an M1 cell, a distinct peak sometimes emerges after activity has been averaged over thousands of spikes. M1 neurons that can be shown in this manner to cause EMG activity are termed *corticospinal* (CS) cells. They project to the spinal cord, and as such they provide more direct control of muscles and movements than do other cortical neurons. A special subclass of CS cells consists of the *corticomotoneuronal* (CM) cells. These project directly to the motor neurons in the spinal cord. They are best identified anatomically, but physiological evidence for them comes from short-latency peaks in the spike-triggered averaged EMG signal. CM cells may be essential for the dextrous control of the fingers (Bortoff & Strick, 1993).

An important open question is whether correlations among neurons carry information beyond what is conveyed by the activity of the neurons taken individually. To this end, investigators have related the presence of enhanced correlation to the neural encoding of sensory (Panzeri, Schultz, Treves, & Rolls, 1999) and motor events. For example, Nicho Hatsopoulos and colleagues (Hatsopoulos, Ojakangas, Paninski, & Donoghue, 1998) applied cross-correlation to neurons recorded in M1. They observed synchrony between pairs of neurons that carried information about the timing of movement, beyond what was evident in individual cells. This confirmed and extended earlier observations (Riehle, Grun, Diesmann, & Aertsen, 1997), using a related technique called unitary event analysis, of coincident spiking in M1 that conveyed information about the cognitive aspects of movement.

For many years, *synchronized oscillations* have been proposed to serve a special role in cortical function. The demonstration of synchronized oscillations between a pair of cells relies on the cross-correlation technique. *Synchrony* is the tendency for two cells to fire together—that is, to bear a significant correlation at a time lag near zero. An *oscillation*, of course, is the tendency for a single neuron to fire at a particular frequency. If the *autocorrelation* of a cell is plotted (figure 40.1D), where S_B in Eq. 1 is replaced with S_A, oscillations appear as periodic side bands. When the cross-correlation between two neurons reveals periodic side bands, it indicates a synchronized oscillation between those cells. Charlie Gray and colleagues (Gray, König, Engel, & Singer, 1989) found that visual cortex neurons exhibit synchronized oscillations that might function in the integration of separate visual features into a cohesive image. The electroencephalographic (EEG) signal is a consequence of large-scale synchronized oscillations among populations of neurons.

COHERENCE METHODS Cross-correlation studies generally reveal fairly weak interactions between pairs of neurons, even when they are near each other. Typically very few neuron pairs exhibit synchronous spiking, whether they are examined in sensory (Smith et al., 2013) or motor (Jackson, Gee, Baker, & Lemon, 2003) cortex. The scarcity of synchronized neurons is surely exacerbated when the two neurons are not in the same brain area. *Coherence analysis* improves on the resolving power of cross-correlation by examining the correlation within specific frequency bands separately. For rhythmic neural phenomena, such as the *local field potential* (LFP), this is manifestly a good idea. LFP is an aggregate measure of neural activity recorded with a sharp electrode. LFP probably reflects synaptic potentials and local membrane dynamics as well as action potentials of cells near as well as distant from the electrode tip (Buzsaki, Anastassiou, & Koch, 2012). The physiological basis of the LFP is probably close to the mechanisms that generate EEG signals. Like EEG, LFP signals exhibit power at specific frequency bands, which probably reflect local excitatory-inhibitory circuit resonances and long-range reverberant circuits. Sensory, motor set, motor execution, and cognitive functions modulate LFP power in specific frequency bands (Kilavik, Zaepffel, Brovelli, MacKay, & Riehle, 2013; Rickert, 2005), which are similar between LFP and EEG measurements, further suggesting a connection between these processes.

In a *coherogram*, the coherency between two signals (that is, the degree to which they are phase-aligned) is plotted as a function of frequency. Coherency can evolve over time, according to task events, thus coherograms have two independent axes (time and frequency), with the amplitude of coherency on the dependent axis (figure 40.2A).

Some Applications of Coherence Methods Coherency can reveal communication between disparate cortical areas. That communication can depend on the animal's cognitive set. Evidence for this includes a study by Pesaran, Nelson, and Andersen (2008) in which they measured the coherency between spiking activity in the premotor cortex and the local field potential recorded in parietal cortex. They compared the coherency in a task where a monkey was free to search among several possible targets for a reward, and a task where the animal had to touch the targets in a particular sequence to receive its reward. Coherency was greater under free search (figure 40.2A, left panel). The authors posit that this increase in coherency is the signature of a more tightly orchestrated exchange of information between the areas during the more difficult decision-making task.

This result is consistent with a theory of neural communication through coherence proposed by Pascal Fries (2005). Fries argued that neuronal coherence may underlie the flexible exchange of information between cortical areas in a variety of contexts. Coherence may serve to open and close "windows" of information transfer, with different frequency bands providing for communication between different brain areas, or for different cognitive processes; for example, attention (Fries, Reynolds, Rorie, & Desimone, 2001). Fries argues against the classic view of neural communication which holds that information is passed through the firing rates of neurons. He proposes instead that the "communication through coherence" scheme is the main mode of information transfer in the nervous system.

How are coherent oscillations orchestrated? Saalmann and colleagues (Saalmann, Pinsk, Wang, Li, & Kastner, 2012) provided strong evidence that the thalamus provides that orchestration. Using coherence methods, they found that attention modulates synchronized oscillations between cortical areas in the alpha band (around 10 Hz), which could in turn entrain activity in the gamma range (around 40 Hz), classically affiliated with visual attention. That synchrony between cortical areas was mediated at least in part by a particular region of the thalamus, the pulvinar. The authors identified the portion of the pulvinar that was anatomically connected with cortical areas using diffusion tensor imaging, a noninvasive magnetic resonance imaging–based technique that can reveal anatomical connectivity in a living brain. Neurons in this region were activated by attention, and they also exhibited coherent oscillations with cortical areas. Thus, cortical synchronization during attention appears to be orchestrated by the pulvinar.

Identified Neurons In his second career as a neuroscientist, the great Francis Crick was known to interrupt seminar speakers with one innocent question: "The neurons whose properties you just described for us—can you tell us where they project?" The question was intended as a rallying cry: physiology is more meaningful within the context of anatomy. If we find a signal in the brain, or a range of signals within one cortical area, how well does that finding serve our understanding of brain function unless we also know where those neurons project? That question has a complement: to understand how response properties emerge in a neuron, we must know something about the neurons that drive it. Structure/function relationships were key to understanding DNA, as they will be to understanding sensory, motor, and cognitive functions, perhaps even consciousness itself (Crick & Koch, 2005).

Often, a diversity of neural signals coexist in a cortical area. Primary visual cortex contains neurons responsive to different aspects of the visual scene (Van Essen, Anderson, & Felleman, 1992). Neurons in the lateral bank of the intraparietal sulcus (area LIP), a portion of the parietal ("association") cortex, reflect visual attention (Bisley & Goldberg, 2003; Liu, Yttri, & Snyder, 2010) as well as motor planning (Liu, Yttri, & Snyder 2010; Snyder, Batista, & Andersen, 1997). Primary motor cortex contains neurons that encode muscle activity (Evarts, 1968; Kakei, Hoffman, & Strick, 1999) and movements (Georgopoulos, Schwartz, & Kettner, 1986; Kakei et al., 1999). It is tempting to speculate that neurons with different functional properties make distinct projections. Also, we can reason that those different properties might arise from different computations on the same inputs. Testing these ideas requires a way to identify neurons that project from one brain area to another.

If we wish to know whether a neuron ("B" in figure 40.2B) under study receives input from another part of the brain (area "1" in figure 40.2B), we can employ the *orthodromic test*. We insert an electrode in area 1 and deliver a single pulse of microstimulation through it. If neuron B reliably fires an action potential at a brief and consistent lag, that indicates that B receives synaptic input from some neuron near the electrode tip in area 1. Note that using the cross-correlation and coherence techniques described above we can only infer the existence of a connection—other network topologies are consistent with a significant correlation. The orthodromic test is a causal test, and it can decisively demonstrate a connection, although it is quite difficult to perform.

If we are recording from a neuron ("A" in figure 40.2B), and we wish to know whether it projects to area 2, we would use the *antidromic test*. By stimulating in area "2," we can record an action potential in neuron A. This is because axons can be driven bidirectionally. In normal nervous system function, neurons exhibit polarity: action potentials are initiated at the soma and they flow down the axon to the synapses. But, under microstimulation, axons can sustain action potential movement in the retrograde direction: depolarization at the synapse generates an action potential, which flows up the axon to the soma.

The antidromic test is evidence, but not proof, that neuron A projects to area 2. It might be that A receives input from neurons in area 2. To distinguish between those possibilities, we can use the *collision test*. It is the same as the antidromic test, except microstimulation on the electrode in area 2 is triggered by an action potential in neuron A. When this happens, no action potential is recorded from neuron A in response to the microstimulation. I encourage you to pause here to consider why this might be. The explanation is that following the endogenous action potential, the axon is in its refractory period due to the temporary inactivation of sodium channels. This prevents the retrograde passage of the stimulation-triggered action potential. Essentially, the two action potentials collide.

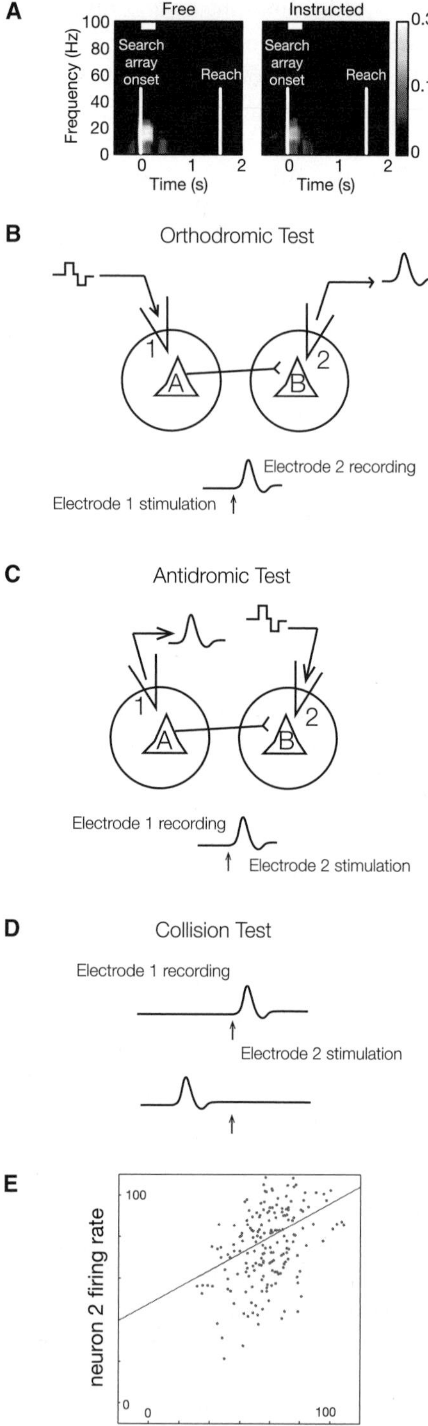

FIGURE 40.2 Other pairwise measurements. (A) Coherograms plot the amplitude of the coherency between spiking activity in the premotor cortex and LFP in the parietal cortex. Coherency is greater in the 15 Hz band when the monkey freely selects targets in a foraging task than when his choices are instructed (Reprinted by permission from Macmillan Publishers Ltd: *Nature*, from Pesaran et al., 2008). Identifying projection neurons: (B) the orthodromic test; (C) the antidromic test; (D) the collision test. Stimulation on electrode 2 causes a spike in neuron A only if A has not spiked recently. (E) Noise correlation. (See color plate 36.)

SOME APPLICATIONS OF IDENTIFIED NEURON TECHNIQUES The tests described above allow investigators to identify neurons that project to another area, or receive input from an area. In a heroic study, Marc Sommer and Bob Wurtz (2006) used both the orthodromic test and the antidromic test to identify neurons in the thalamus that relay information from the superior colliculus (the midbrain region responsible for commanding saccadic eye movements) to the frontal eye fields in the cerebral cortex. They hypothesized that these "relay neurons" inform the cortex about impending eye movements, perhaps to aid in perceptual stability across eye movements. To test this, they reversibly inactivated the thalamus in the vicinity of those relay neurons by injecting muscimol, which mimics the inhibitory neurotransmitter GABA. Indeed, animals were not impaired in making saccades (presumably because the pathway from the frontal eye fields to the superior colliculus was still intact), but they were impaired in keeping track of where their eyes ended up after the saccade: a second saccade after the first was inaccurate. The inaccuracy was consistently in a direction that would have been correct if the first saccade had not occurred.

In the first section above, we discussed how the cross-correlation technique was used to show that projections from the thalamus to the visual cortex were very precise. Tony Movshon and Bill Newsome (1996) used the antidromic technique to show that projections are also precise in one of the next stages in visual processing: the projection from primary visual cortex to the motion-sensitive extrastriate area MT. Among 745 V1 neurons they tested, twelve were identified as MT projection neurons using the collision test. Those projection neurons were themselves sensitive to motion, but responded only to the motion of the component features of the image, rather than to the motion of the overall pattern. This indicates that area MT does not compute motion information *de novo* from its inputs, but rather computes richer motion information (i.e., more closely resembling our motion percepts) from its inputs: V1 cells which are themselves sensitive to motion.

Only about one million cortical neurons project to the spinal cord. These special cells, the *pyramidal tract neurons*, are identified using the collision test. As part of his classic studies that pioneered the technique of recording neural activity in awake, behaving primates, Edward Evarts showed that pyramidal tract neurons correlate with muscle force. This led to the discovery, years later, that different populations of neurons in primary motor cortex coexist that carry information about the activity of muscles, and also about the direction of movements (Kakei et al., 1999).

Using the collision test to identify neurons that receive input from or project to another area requires patience and dedication. These studies tend to have low yields. However, they provide detailed information about how information is transformed as it flows between brain areas that cannot yet be attained using any other technique.

Noise Correlation Of course, functional networks of neurons are large and contain many neurons that do not bear a synaptic connection between them. If we wish to broaden our scope to investigate interactions between pairs of neurons that are not necessarily directly connected, we can use the *noise correlation* (or *spike count correlation*) technique (reviewed in Cohen & Kohn, 2011; Zohary, Shadlen, & Newsome, 1994). This technique can detect interactions on broader time scales than those typically revealed by the cross-correlation technique. It relies on the noise that is inherent in neural responses: when a neuron is presented with the same stimulus repeatedly, its responses are not identical. Trial-by-trial deviations from the average response can be sizable—neural spiking is fairly well described as a Poisson process, where the variance equals the mean. If two neurons are recorded simultaneously, their responses may cofluctuate relative to their means: when one neuron fires higher than its mean, the other neuron might tend to fire higher (or lower) than its own mean. This departure from independence is the noise correlation. It is taken to indicate the cells are part of the same functional network. The noise correlation is mathematically equal to the normalized area under the covariogram (Bair, Zohary, & Newsome, 2001), so the techniques are related, but they emphasize different time scales of interaction. A significant noise correlation can occur even when no short-time scale interaction is evident between the neurons; in this sense, it provides a more inclusive test for pairwise interactions.

The noise correlation $r_{A,B}$ is computed as the covariance between the neurons, normalized by their individual deviations:

$$r_{A,B} = \frac{\mathrm{cov}(A, B)}{\mathrm{sd}(A)\mathrm{sd}(B)} \tag{2}$$

Some Applications of Noise Correlation One of the most exciting applications of the noise correlation technique has been in revealing changes in functional circuitry due to context. It must be true that the nervous system can rapidly reconfigure the flow of information, to flexibly enable responses suitable to the circumstances at hand. As a simple example, the sight of a yellow traffic light might lead a driver to brake, or to accelerate, depending on factors having nothing to do with the visual stimulus. Marlene Cohen and Bill

Newsome (2008) showed that pairs of neurons in area MT change their functional coupling (as assessed through context-dependent changes in their noise correlation) based on the task instructions given to the animal. The visual stimulus did not change, but the manner in which the animal had to act upon the information in the stimulus was altered, and that was sufficient to change the network organization within area MT.

Adam Kohn and Matt Smith (2005) conducted a clever study that showed that brief–time scale interactions between neurons (assessed using cross-correlation) and longer–time scale interactions (assessed with noise correlation) can be decoupled, and thus they are distinct neural phenomena. While recording pairs of neurons in V1, Kohn and Smith were able to change the short–time scale correlation by changing the orientation of the stimulus. This is consistent with the findings described above of Reid and Alonso, and of Movshon and Newsome, which showed that cross-correlation is only present between neurons with precisely aligned tuning properties. As the visual stimulus was moved away from the region of common drive to both neurons, the cross-correlation decreased. The noise correlation was unaffected by changing orientation, but it was influenced by the contrast of the visual stimulus: noise correlation increased when visual contrast decreased. The authors reason that the increase in noise correlation with decreasing stimulus contrast indicates the V1 network shifts from a stimulus-driven mode to a mode in which intrinsic network dynamics dominate its activity.

A reflection before we leave the topic of pairwise techniques. The noise correlation technique hinges on the notion that the correlation between neurons is not driven by a sensory stimulus, but rather by some underlying (unobserved) process that affects both cells. This process need not be "noise" (that is, random and

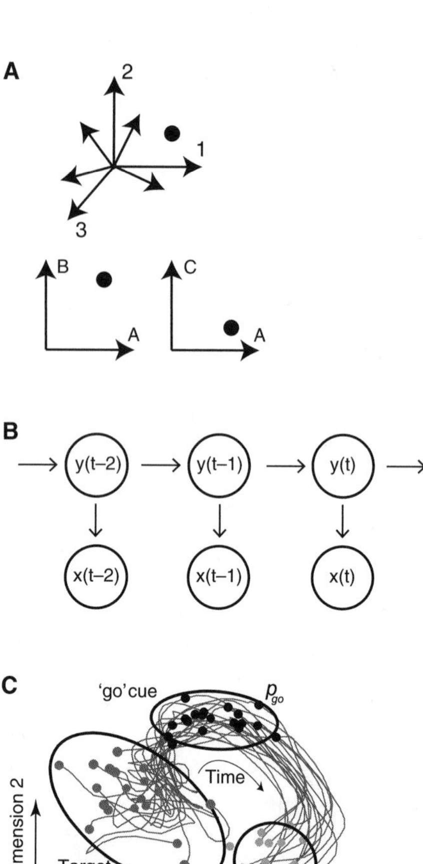

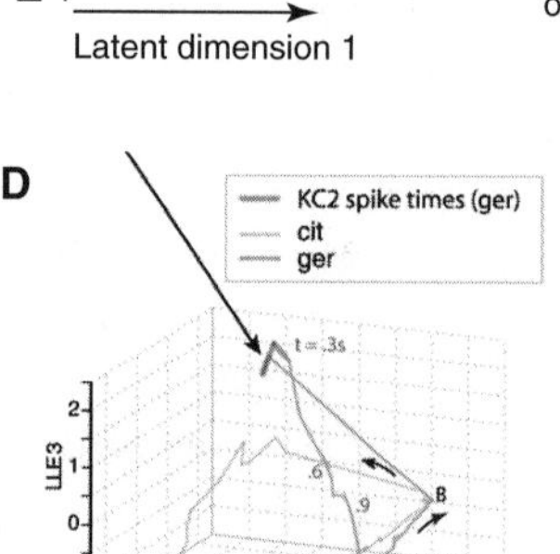

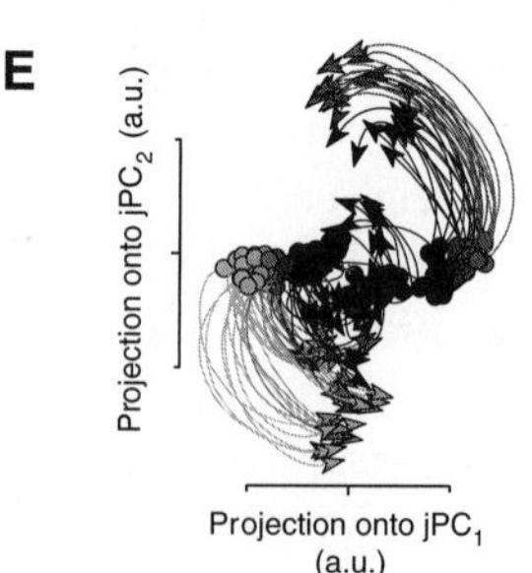

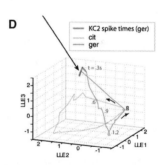

FIGURE 40.3 Multineuronal interactions. (A) Visualizing high-dimensional neural data. A point in seven-dimensional space, and two 2D projections. 1, 2, 3, etc., represent the activity of individual neurons; A, B, and C are linear combinations of the seven original axes. (B) Depiction of a simple latent variable model. (C) "Neural trajectories" through state space. Each line is an individual trial. (Adapted from Afshar et al., 2011, with permission from Elsevier.) (D) Tuning in state space. (Adapted from Broome et al., 2006, with permission from Elsevier.) The arrow indicates the position in state space for which the Kenyon cell is tuned. (E) Intrinsic dynamics in motor cortex. (Adapted by permission from Macmillan Publishers Ltd: *Nature*, from Churchland et al., 2012.) (See color plate 37.)

uninteresting)—it could in fact be a meaningful signal; it is just one that is not under experimental control. By the time an external stimulus can affect a cortical neuron, it has passed through several stages of processing. The neural activity patterns originally generated by the sensory stimulus have been combined, branched, and transformed by the network action of neurons. Under these considerations, the extent to which measurable external covariates can explain the activity of a cortical neuron must be small. The meaningful driver of neural activity is the activity of other neurons. Pairwise techniques suggest this is the case. Direct evidence for this perspective is presented below.

Multineuronal interactions

So far we have considered interactions among pairs of neurons. Modern multielectrode techniques allow the simultaneous recording of several hundred neural units (Collinger et al., 2013; Ifft et al., 2013), and that number is growing steadily. What types of relationships might we observe among dozens or hundreds of neurons? There are not many hypotheses yet about how they might interact. Exploratory data visualization can help us build intuitions about multineuronal function, which can inspire novel hypotheses.

Multineuronal Visualization Techniques To visualize the activity of a population of neurons of arbitrary size, we can use *state space* tools (figure 40.3A). Imagine the firing rate of a neuron is represented as a value along an axis. For two neurons, their firing rates viewed together is a point on a plane (e.g., figure 40.2E). The joint firing rate of N neurons is a point in an N-dimensional space. When $N = 3$, we can plot the data in a cube, and visualize it by rotating the cube using a computer with a fast graphics processor.

Beyond $N = 3$, we must project the data into a lower-dimensional space to visualize it. We could simply consider all pairwise and three-way combinations of neurons in our population and view them one after another, but surely there are more interesting projections of the data. Suppose that two neurons are tightly correlated. In that case it is not informative to view them both; we would rather view a slice through the N-dimensional space where one axis is aligned with the comodulation of those neurons. If we extend this logic to correlations among many neurons, we can ask which projections of multineuronal data are the most interesting and informative. Dimensionality reduction techniques suit this objective—they can provide the maximally informative views of the original data. A well-known technique is *principal components analysis* (PCA). It finds the set of axes that best account for the data. The first principal component (PC) captures the greatest variability in a data set. The second PC captures the greatest variability that remains in the data after the variability accounted for by the first PC has been removed, and so on. For N-dimensional data, N PCs exist, which together yield a complete accounting for the original data. They are just a rotation of the axes of the space in which the data were collected. The beauty of PCA is that you can take just the first few PCs, and have a fairly accurate reconstruction of the original data, using far fewer parameters. You can "dial in" the fidelity with which you wish to reconstruct the original data—you can determine how many PCs it would take to account for (say) 95% of the original data. It is natural to think that the first few principal components that account for the vast majority of the data are capturing the essence of the data set, and that further dimensions are mostly just fitting noise.

PCA may be the best-known dimensionality-reduction technique, but others exist, and may be more suited to the properties of the data. For example, locally linear embedding (Roweis & Saul, 2000) and Isomap (Tenenbaum, de Silva, & Langford, 2000) provide better reconstructions of data sets that have strongly nonlinear shape to them, for example, if the data lie along a curved manifold. Factor analysis (Santhanam et al., 2009; Yu et al., 2009) allows each neuron to have its own noise properties, which given the Poisson-like nature of neural firing (Dayan & Abbott, 2005) is a justifiable and powerful provision. When factor analysis is combined with temporal smoothing (by assuming the data evolve following a *Gaussian process*), a richly informative low-dimensional projection of the time course of the data emerges. Ben Cowley, Byron Yu, and their colleagues (2013) recently developed a software visualization suite that allows users to interact with high-dimensional data and rapidly hone in on interesting projections of those data. Their code is freely available, and it comes with example data sets so that you can familiarize yourself with these powerful new tools.

Latent Variable Models In the sections that follow, we discuss some of the principles of neural function that are emerging from multineuronal visualization techniques. But first an important theoretical tool is needed: the concept of a *latent variable*. Our ability to reduce the dimensionality of neural data relies on the fact that the neurons under study are correlated. Are correlations good or bad? Some theoretical studies (e.g., Zohary et al., 1994) indicate that correlations reduce the information-carrying capacity of the brain (but see Averbeck & Lee, 2004, and Averbeck, Latham,

& Pouget, 2006, for a fuller treatment of the issue). Perhaps correlations are an undesirable but inevitable consequence of brain connectivity. The latent variable perspective offers an alternative view: correlations may indicate that individual neurons together comprise larger information-processing assemblies. Those assemblies, termed *latent variables*, may have properties of their own, which we can now begin to study. By this perspective, a set of latent variables govern information processing in the brain, and individual neurons provide only a noisy, narrow window into the properties of those latent variables.

We are already comfortable with the concept of latent variables in other contexts. For example, imagine the pattern of pixel intensities on a television screen. They may not make much sense taken individually, but when viewed together, it becomes evident that the movements of the actors and objects on the screen drive the activity of individual pixels. The scene makes more sense when we abstract away from the individual pixels and instead focus on how the actors are moving. Another feature of this analogy applies to multineuronal recordings: notice that it requires being able to image the individual pixels well in order to be able to see how they comprise an actor. If you watch television with your glasses off, it only makes it harder to resolve the individual actors. It is only through the collective action of the well-resolved components that the deeper principles can be apprehended. Another intuitive example of a latent variable model is trying to understand the behavior of insects. The flight path of an individual bee might not make much sense on its own, but if the individuals comprising a swarm can be observed, their objectives are clearer.

Figure 40.3B depicts the notion of a latent variable model, using a simplified example, a hidden Markov model. There is some internal process, y, that evolves according to rules of its own. Periodically, y generates phenomena we can observe, x. The challenge then becomes, given a history of observations of x, can we infer the rules that govern y? Formally,

$$\vec{y}(t) = f[\vec{y}(t-1), \vec{y}(t-2), \ldots, \vec{y}(0)]$$
$$\vec{x}(t) = g(\vec{y}(t)), \qquad (3)$$

where y evolves according to its own history and internal dynamics, given by f. In most applications, the simplifying assumption is made that f depends only on the previous time step, but in general y at time t can depend on its history; x are observations of y governed by g. In general x is of higher dimensionality than y (that is, there are more neurons than latent variables), and g injects some noise, so it is not a simple task to invert g to get y from x. The objective of latent variable modeling is to peer through the observations, x, to infer the latent states, y, and ultimately the traits of f that govern how the latent states evolve.

In his influential 1982 book, *Vision*, David Marr proposed that information processing can be described at three levels: The *computational* level expresses the goal of the process. The *algorithmic* level describes the steps and rules by which the computation is performed. The *implementation* level describes the physical operations that instantiate the algorithm. Applying this hierarchy, we can posit that individual neurons implement the algorithms of neural information processing, and those algorithms are best understood at the level of latent variables. By this view, the dimensionality reduction techniques described above are not just visualization tools—they actually reveal deeper organization of neural information processing. The sections that follow will provide evidence for this view. In each example, a multidimensional view is used to provide insight into neural function that is not as readily apparent by observing the activity of individual neurons.

A final note on latent variable models: they are nothing new in neuroscience. Although the formalization is fairly recent (Sahani, 1999), the concept has been around for a long time. In 1986, Georgopoulos, Schwartz, and Kettner introduced the *population vector algorithm* (PVA). A collection of noisy, broadly tuned neurons can be combined using the PVA to reconstruct the precise direction in which an animal reaches. The intended reach direction is the latent variable, and the activity of individual neurons each provides a noisy, narrow view of that process. Recently, Chase, Schwartz, and Kass (2010) showed that when an animal's intended movement direction is taken as a latent variable, it provides a better fit to the motor cortex firing-rate data than does the actual movement. This is a direct validation of the latent variable perspective. In 1985, Moran and Desimone reported that visual attention alters the response properties of extrastriate neurons. Attention is a latent variable—it is a process that can be observed only indirectly, through its effects on behavior and neural activity. Through careful studies of neural activity and behavior, researchers have been able to describe the rules by which attention functions (Petersen & Posner, 2012).

Applications of Multineuronal Techniques

Single-trial analysis The vast majority of behavioral neuroscience studies average over many repetitions of quasi-identical trial conditions. In this way, subtle signals can be discerned, and main trends emerge. However, trial-to-trial variability is often large, and some important information can be subsumed by the mean response. In 2011, Afshar and colleagues examined multineuronal activity during repeats of the same arm movement without averaging them together. They observed that the multineuronal population neural activity followed a fairly stereotyped trajectory through neural state space (figure 40.3C). The animal's reaction time in the task could be predicted based on the location in state space at the time the instruction was provided. When population neural activity happened to be near a particular subregion of the state space, the behavioral response was rapid. When neural activity was further from that optimal subregion, responses were slower, as it took additional time for the neural activity to reach the optimal subregion. Reaction times were not predicted as well when neurons were considered individually. Thus, the multineuronal view of neural activity yielded a stronger prediction of the reaction time on individual trials than did conventional single-neuron analyses.

Neurons are tuned to their inputs What causes a neuron to fire? Clearly, it is the activity of other neurons—the collection of 1,000 or so cells that synapse onto it. However, most behavioral neuroscience studies provide an account for the activity of neurons only in terms of external covariates: the visual image, or movement kinematics. In the absence of being able to observe the neurons that impinge on a given neuron, our options are limited when it comes to providing an account of neural responses. Now, with multineuronal recordings, we can build more accurate predictions of neural responses.

Broome, Jayaraman, and Laurent (2006) provided a description of the tuning properties of neurons in the insect olfactory system. In locusts, odorant molecules picked up by the antennae trigger activity in a distributed population of "projection neurons" (PNs). The PNs transmit information to a set of higher-order neurons, the Kenyon cells (KCs). Although each KC receives input from a vast number of PNs, these cells have a low baseline firing rate and exhibit highly specific and reliable responses to combinations of odorants. This two-stage circuit implements a transition from a broadly distributed neural coding scheme to a sparse coding scheme. How is the selectivity of the KCs achieved? The researchers were able to predict when a KC would fire based on the location in state space of the population response of the PN cells (figure 40.3D). This description yielded a more robust explanation of KC activity than did tuning in the space of the odorants themselves. It appears that the KCs are each "tuned" to specific configurations of firing rates across the population of their input PNs.

To formalize the intuitions conveyed by these experiments, we can use a *generalized linear model* (GLM) framework. In a GLM, the activity of an individual neuron is predicted based on three things: (1) external covariates (e.g., the stimulus), (2) the activity of other neurons, and (3) the neuron's own history. These models have found widespread use lately in explaining responses in the retina (Pillow et al., 2008), visual cortex (Kelly, Kass, Smith, & Lee, 2010), premotor cortex (Zhao et al., 2011), and motor and sensory cortices (Stevenson, 2008; Stevenson et al., 2012).

Two overarching observations are emerging from GLM approaches. First, the responses of cortical neurons are explained predominantly by the activity of other cortical neurons; external covariates play far less of a role. Second, neurons are not all that noisy. When more explanatory signals are available, the amount of noise (that is, unexplained variability) exhibited by individual neurons is dramatically reduced. This observation is reminiscent of influential earlier studies on single neurons that showed when the inputs to a neuron are precisely controlled and repeated, the cell's response is very consistent (Bair & Koch, 1996; Bialek, Rieke, de Ruyter van Steveninck, & Warland, 1991; Mainen & Sejnowski, 1995).

Rich information can be extracted from populations of neurons The previous section explained how seemingly complex tuning properties can be made simpler by expressing them in the proper space—that of the impinging neurons, rather than external stimuli. The converse also appears to be true: seemingly complex responses of individual neurons can provide a basis for extracting simple, useful information. Two groups working independently—Rigotti and colleagues (2013) and Mante and colleagues (2013)—observed that neurons in the prefrontal cortex exhibit complex and idiosyncratic tuning properties when examined individually. However, when machine learning classification techniques are applied to a population of PFC neurons, simple and important parameters about the task can be extracted. These studies argue that a population of neurons each exhibiting complex and idiosyncratic tuning is required for the extraction of a diverse range of simple information. The information that is extracted can vary depending on the needs of the current circumstances (Mante et al., 2013). Neuroscientists tend to seek simple explanations for what causes a cell to fire; the view is emerging that it is actually the neurons whose tuning properties elude simple explanations that convey the most information. Certainly, more evidence will be needed before this perspective becomes widespread.

Neural networks have intrinsic dynamics Any physical system is bound to have its own intrinsic dynamics. A notorious example of intrinsic dynamics is a resonant frequency—the rate at which a system is prone to oscillate, even if the driving input is very small. In manufactured systems ranging from headphones to bridges, good engineering design is required to predict and control (and occasionally, to exploit) the resonances. In the nervous system, uncontrolled resonances can lead to seizures. Even under normal conditions, the nervous system will oscillate at particular frequencies. These resonances result from phenomena across multiple scales, from the ion channel composition in the membranes of individual cells, to the inhibitory-excitatory topology of local networks, to the long recurrent loops that the cerebral cortex makes with subcortical areas (Buzsaki et al., 2012).

Even if they are an unavoidable artifact of physics and network topology, oscillations in the brain accompany useful functions, and may be integral to them. Oscillations around 40 Hz ("gamma") have been linked extensively to attention and cognition (Fries et al., 2001; Jensen, Kaiser, & Lachaux, 2007). Oscillations in central pattern-generator circuits generate useful behaviors such as locomotion and feeding (Marder & Calabrese, 1996). Mark Churchland and colleagues (2012) argue that evolution has harnessed rhythmic oscillations to implement nonrhythmic functions. They performed multielectrode recordings in the motor and premotor cortex of reaching monkeys. Low-dimensional projections of the neural activity patterns revealed oscillatory structure (figure 40.3E). Those oscillations appeared to stem from intrinsic network properties of the cortex itself; neither the muscle activity nor the resulting behavior were overtly rhythmic, although the study demonstrated through simulations that these nonrhythmic signals could be generated from the rhythmic cortical activity they observed.

An intriguing additional observation made by Churchland and colleagues is that the motor preparatory activity preceding the onset of the reach appears to initialize the dynamical system, allowing it to unfold along different neural paths to execute reaches in different directions. Extrapolating from this view, not only our behavior, but also our perceptual experience and cognitive flexibility, may arise from the interplay between intrinsic neural dynamics and external drive.

Primary visual cortex may also exhibit an interaction between intrinsic dynamics and external drive. Tal Kenet and colleagues (2003) identified default network states in primary visual cortex. The network would spontaneously switch between different patterns, and these patterns appeared to correspond to activity patterns that would be driven by visual input. The following year, Fiser, Chiu and Weliky (2004) verified these intrinsic dynamics in awake animals, and extended the findings to show that visual inputs act mostly by changing the intrinsic dynamics of the visual cortex network. An observation mentioned above made by Kohn and Smith (2005) supports this perspective. They saw that noise correlation among V1 neurons increased as the stimulus contrast was reduced. They interpreted this as V1 moving back into a state where its activity is governed by its own intrinsic dynamics.

It appears that activity in the brain is governed in part by its own intrinsic dynamics, and when a stimulus arrives, it interacts with those dynamics. In a large comparative study involving data collected in 13 cortical areas, Mark Churchland and colleagues (2010) found in all cortical areas they examined that the appearance of a stimulus reduces the variability in neural activity.

These observations of intrinsic dynamics interacting with driven activity are reminiscent of the important discovery of a default mode network in the human brain (Raichle et al., 2001). A widespread network of intrinsic dynamics can be observed in positron emission tomography and functional MRI while subjects are at rest. This is a robust phenomenon, which has been confirmed in many studies. Cognitive tasks and sensory inputs reduce the activity in this rest state network gradually as the cognitive demands rise (Greicius, Krasnow, Reiss, & Menon, 2003). The intrinsic dynamics of the default mode network have a fascinating psychological correlate: they appear to be related to the experiential state of mind wandering (Mason et al., 2007).

Brain-computer interfaces A brain-computer interface (BCI) is a technology designed to restore motor function to paralyzed individuals. A variety of neural signals have successfully provided BCI control, including EEG recordings (Wolpaw, Birbaumer, McFarland, Pfurtscheller, & Vaughan, 2002) and multielectrode array recordings from humans (Collinger et al., 2013; Hochberg et al., 2006) and monkeys (Gilja et al., 2012). Users can then control an external effector, such as a robotic arm or a computer cursor. A BCI embodies the latent variable standpoint—here the latent variable is the animal's movement intention. The observed variables are the firing rates of dozens of neurons. Our objective is to infer the movement intention from the firing rates, so we can render their intentions back to the subject as the movement of the effector. Then, the subject can modify their intentions as needed to steer the cursor to the target.

This mapping from neural activity to intended movements is performed by a *decode algorithm*. An effective decode algorithm is provided by the Kalman filter. Originally introduced to track satellites (Kalman, 1960), the Kalman filter embodies the latent variable framework explicitly. It has a *state model*, which predicts the next position (or velocity) of the effector based on its own history (the state is the latent variable), and an *observation model*, whereby the effector's position can be estimated from sensor readings (in a BCI, the observations are neural activity). Both sources of information provide an estimate of the system's location; the estimates are combined in a weighting that depends on the reliability of the source of information. If the sensor data are noisy, the kinematic predictions are made mostly using the state model. In the context of BCI control, the state model essentially smooths the estimation of cursor position given by the neural activity. That smoothing is done in a manner that captures the physics of how limbs tend to move.

Some future directions

Multineuronal recording approaches are beginning to introduce new concepts about how neurons function together to convey and transform information. How far might this approach take us toward understanding the relationship between neural activity and human behavior, with all of its flexibility, spontaneity, and richness? In the near future the number of individual neurons whose activity we can monitor will continue to increase dramatically, and new visualization approaches (e.g., Chung & Deisseroth, 2013) will allow us to observe anatomical connections among those neurons. New questions, testing new concepts and hypotheses about the links between brain and behavior, are bound to be inspired by these techniques. We conclude with a thought experiment that motivates the types of questions that might soon become tractable.

If we could stand on a neuron's axon hillock and look out at the patterns of incoming activity that cause it to spike, we would probably see strikingly different patterns of activity preceding each spike. If we face in the other direction and watch the effect of spikes emanating from a single neuron, we are likely to see that its influence on downstream cells differs markedly depending on the patterns of spiking flowing in from other neurons, and perhaps even on the cell's own recent history of spiking. Not all spikes are the same: they do not all have the same antecedent causes, and they do not have the same downstream effects. Neural information processing must be understood in terms of networks of neurons, and new experimental and analysis tools are allowing those rules to be discovered.

Neural information is transformed and manipulated as it flows through networks. Those networks must also impose constraints on how information can move. Can we observe those "rules of the road" with multineuronal recordings? Those constraints may manifest themselves in observable behaviors, facilitating some, while making others more difficult to perform. In this manner, the space of possibilities that define our individual memories, skills, and personality may be determined by the idiosyncratic shapes of our particular networks of neurons.

ACKNOWLEDGMENTS I thank Matt Smith, Steve Chase, Byron Yu, Richard Dum, and Scott Grafton for comments on the manuscript.

REFERENCES

Afshar, A., Santhanam, G., Yu, B. M., Ryu, S. I., Sahani, M., & Shenoy, K. V. (2011). Single-trial neural correlates of arm movement preparation. *Neuron, 71*(3), 555–564.

Averbeck, B. B., Latham, P. E., & Pouget, A. (2006). Neural correlations, population coding and computation. *Nat Rev Neurosci, 7*(5), 358–366.

Averbeck, B. B., & Lee, D. (2004). Coding and transmission of information by neural ensembles. *Trends Neurosci, 27*(4), 225–230.

Bair, W., & Koch, C. (1996). Temporal precision of spike trains in extrastriate cortex of the behaving macaque monkey. *Neural Comput, 8*(6), 1185–1202.

Bair, W., Zohary, E., & Newsome, W. (2001). Correlated firing in macaque visual area MT: Time scales and relationship to behavior. *J Neurosci, 21*(5), 1676–1697.

Bialek, W., Rieke, F., de Ruyter van Steveninck, R. R., & Warland, D. (1991). Reading a neural code. *Science, 252*(5014), 1854–1857.

Bisley, J., & Goldberg, M. (2003). Neuronal activity in the lateral intraparietal area and spatial attention. *Science, 299*(5603), 81–86.

Bortoff, G. A., & Strick, P. L. (1993). Corticospinal terminations in two new-world primates: Further evidence that corticomotoneuronal connections provide part of the neural substrate for manual dexterity. *J Neurosci, 13*(12), 5105–5118.

Brody, C. (1999). Correlations without synchrony. *Neural Comput, 11*(7), 1537–1551.

Broome, B. M., Jayaraman, V., & Laurent, G. (2006). Encoding and decoding of overlapping odor sequences. *Neuron, 51*(4), 467–482.

Buzsaki, G., Anastassiou, C. A., & Koch, C. (2012). The origin of extracellular fields and currents—EEG, ECoG, LFP and spikes. *Nat Rev Neurosci, 13*(6), 407–420.

Chase, S. M., Schwartz, A. B., & Kass, R. E. (2010). Latent inputs improve estimates of neural encoding in motor cortex. *J Neurosci, 30*(41), 13873–13882.

Chung, K., & Deisseroth, K. (2013). CLARITY for mapping the nervous system. *Nat Methods, 10*(6), 508–513.

Churchland, M. M., Cunningham, J. P., Kaufman, M. T., Foster, J. D., Nuyujukian, P., Ryu, S. I., & Shenoy, K. V.

(2012). Neural population dynamics during reaching. *Nature*, 1–8.

Churchland, M. M., Yu, B. M., Cunningham, J. P., Sugrue, L. P., Cohen, M. R., Corrado, G. S., et al. (2010). Stimulus onset quenches neural variability: A widespread cortical phenomenon. *Nat Neurosci*, *13*(3), 369–378.

Cohen, M. R., & Kohn, A. (2011). Measuring and interpreting neuronal correlations. *Nat Neurosci*, *14*(7), 811–819.

Cohen, M. R., & Newsome, W. T. (2008). Context-dependent changes in functional circuitry in visual area MT. *Neuron*, *60*(1), 162–173.

Collinger, J. L., Wodlinger, B., Downey, J. E., Wang, W., Tyler-Kabara, E. C., Weber, D. J., et al. (2013). High-performance neuroprosthetic control by an individual with tetraplegia. *Lancet*, *381*(9866), 557–564.

Cowley, B. R., Kaufman, M. T., Butler, Z. S., Churchland, M. M., Ryu, S. I., Shenoy, K. V., & Yu, B. M. (2013). DataHigh: Graphical user interface for visualizing and interacting with high-dimensional neural activity. *J Neural Eng*, *10*(6), 1–19.

Crick, F. C., & Koch, C. (2005). What is the function of the claustrum? *Philos Trans R Soc Lond B Biol Sci*, *360*(1458), 1271–1279.

Dayan, P., & Abbott, L. (2005). *Theoretical neuroscience: Computational and mathematical modeling of neural systems*. Cambridge, MA: MIT Press.

Evarts, E. (1968). Relation of pyramidal tract activity to force exerted during voluntary movement. *J Neurophysiol*, *31*(1), 14–27.

Fiser, J., Chiu, C., & Weliky, M. (2004). Small modulation of ongoing cortical dynamics by sensory input during natural vision. *Nature*, *431*(7008), 573–578.

Fries, P. (2005). A mechanism for cognitive dynamics: Neuronal communication through neuronal coherence. *Trends Cogn Sci*, *9*(10), 474–480.

Fries, P., Reynolds, J. H., Rorie, A. E., & Desimone, R. (2001). Modulation of oscillatory neuronal synchronization by selective visual attention. *Science*, *291*(5508), 1560–1563.

Georgopoulos, A. P., Schwartz, A. B., & Kettner, R. E. (1986). Neuronal population coding of movement direction. *Science*, *233*(4771), 1416–1419.

Gilja, V., Nuyujukian, P., Chestek, C. A., Cunningham, J. P., Yu, B. M., Fan, J. M., ... Shenoy, K. V. (2012). A high-performance neural prosthesis enabled by control algorithm design. *Nat Neurosci*, *15*(12), 1752–1757.

Gray, C. M., König, P., Engel, A. K., & Singer, W. (1989). Oscillatory responses in cat visual cortex exhibit inter-columnar synchronization which reflects global stimulus properties. *Nature*, *338*(6213), 334–337.

Greicius, M. D., Krasnow, B., Reiss, A. L., & Menon, V. (2003). Functional connectivity in the resting brain: A network analysis of the default mode hypothesis. *Proc Natl Acad Sci USA*, *100*(1), 253–258.

Hatsopoulos, N., Ojakangas, C., Paninski, L., & Donoghue, J. (1998). Information about movement direction obtained from synchronous activity of motor cortical neurons. *Proc Natl Acad Sci USA*, *95*(26), 15706–15711.

Hochberg, L., Serruya, M., Friehs, G., Mukand, J., Saleh, M., Caplan, A., et al. (2006). Neuronal ensemble control of prosthetic devices by a human with tetraplegia. *Nature*, *442*(7099), 164–171.

Ifft, P. J., Shokur, S., Li, Z., Lebedev, M. A., & Nicolelis, M. A. L. (2013). A brain-machine interface enables bimanual arm movements in monkeys. *Sci Trans Med*, *5*(210), 1–13.

Jackson, A., Gee, V., Baker, S., & Lemon, R. (2003). Synchrony between neurons with similar muscle fields in monkey motor cortex. *Neuron*, *38*(1), 115–125.

Jensen, O., Kaiser, J., & Lachaux, J.-P. (2007). Human gamma-frequency oscillations associated with attention and memory. *Trends Neurosci*, *30*(7), 317–324.

Kakei, S., Hoffman, D. S., & Strick, P. L. (1999). Muscle and movement representations in the primary motor cortex. *Science*, *285*(5436), 2136–2139.

Kalman, R. E. (1960). A new approach to linear filtering and prediction problems. *J Basic Eng-T ASME*, *82*, 35–45.

Kelly, R. C., Kass, R. E., Smith, M. A., & Lee, T. S. (2010). Accounting for network effects in neuronal responses using L1 regularized point process models. *Adv Neural Inf Process Syst*, *23*(2), 1099–1107.

Kenet, T., Bibitchkov, D., Tsodyks, M., Grinvald, A., & Arieli, A. (2003). Spontaneously emerging cortical representations of visual attributes. *Nature*, *425*(6961), 954–956.

Kilavik, B. E., Zaepffel, M., Brovelli, A., MacKay, W. A., & Riehle, A. (2013). The ups and downs of β oscillations in sensorimotor cortex. *Exp Neurol*, *245*, 15–26.

Kohn, A., & Smith, M. (2005). Stimulus dependence of neuronal correlation in primary visual cortex of the macaque. *J Neurosci*, *25*(14), 3661–3673.

Liu, Y., Yttri, E. A., & Snyder, L. H. (2010). Intention and attention: Different functional roles for LIPd and LIPv. *Nat Neurosci*, *13*(4), 495–500.

Mainen, Z. F., & Sejnowski, T. J. (1995). Reliability of spike timing in neocortical neurons. *Science*, *268*(5216), 1503–1506.

Mante, V., Sussillo, D., Shenoy, K. V., & Newsome, W. T. (2013). Context-dependent computation by recurrent dynamics in prefrontal cortex. *Nature*, *503*(7474), 78–84.

Marder, E., & Calabrese, R. L. (1996). Principles of rhythmic motor pattern generation. *Physiol Rev*, *76*(3), 687–717.

Marr, D. (1982). *Vision*. Cambridge, MA: MIT Press.

Mason, M. F., Norton, M. I., Van Horn, J. D., Wegner, D. M., Grafton, S. T., & Macrae, C. N. (2007). Wandering minds: The default network and stimulus-independent thought. *Science*, *315*(5810), 393–395.

Moran, J., & Desimone, R. (1985). Selective attention gates visual processing in the extrastriate cortex. *Science*, *229*(4715), 782–784.

Movshon, J. A., & Newsome, W. T. (1996). Visual response properties of striate cortical neurons projecting to area MT in macaque monkeys. *J Neurosci*, *16*(23), 7733–7741.

Panzeri, S., Schultz, S. R., Treves, A., & Rolls, E. T. (1999). Correlations and the encoding of information in the nervous system. *Proc R Soc Lond B Biol Sci*, *266*(1423), 1001–1012.

Perkel, D. H., Gerstein, G. L., & Moore, G. P. (1967). Neuronal spike trains and stochastic point processes. II. Simultaneous spike trains. *Biophys J*, *7*(4), 419–440.

Pesaran, B., Nelson, M. J., & Andersen, R. A. (2008). Free choice activates a decision circuit between frontal and parietal cortex. *Nature*, *453*(7193), 406–409.

Petersen, S. E., & Posner, M. I. (2012). The attention system of the human brain: 20 years after. *Annu Rev Neurosci*, *35*, 73–89.

Pillow, J. W., Shlens, J., Paninski, L., Sher, A., Litke, A. M., Chichilnisky, E. J., & Simoncelli, E. P. (2008). Spatio-temporal correlations and visual signalling in a complete neuronal population. *Nature, 454*(7207), 995–999.

Raichle, M. E., MacLeod, A. M., Snyder, A. Z., Powers, W. J., Gusnard, D. A., & Shulman, G. L. (2001). A default mode of brain function. *Proc Natl Acad Sci USA, 98*(2), 676–682.

Reid, R. C., & Alonso, J. M. (1995). Specificity of monosynaptic connections from thalamus to visual cortex. *Nature, 378*(6554), 281–284.

Rickert, J. (2005). Encoding of movement direction in different frequency ranges of motor cortical local field potentials. *J Neurosci, 25*(39), 8815–8824.

Riehle, A., Grun, S., Diesmann, M., & Aertsen, A. (1997). Spike synchronization and rate modulation differentially involved in motor cortical function. *Science, 278*(5345), 1950–1953.

Rigotti, M., Barak, O., Warden, M. R., Wang, X.-J., Daw, N. D., Miller, E. K., & Fusi, S. (2013). The importance of mixed selectivity in complex cognitive tasks. *Nature, 497*(7451), 585–590.

Roweis, S., & Saul, L. (2000). Nonlinear dimensionality reduction by locally linear embedding. *Science, 290*(5500), 2323–2326.

Saalmann, Y. B., Pinsk, M. A., Wang, L., Li, X., & Kastner, S. (2012). The pulvinar regulates information transmission between cortical areas based on attention demands. *Science, 337*(6095), 753–756.

Sahani, M. (1999). Latent variable models for neural data analysis. Computation and neural systems program (PhD doctoral dissertation). California Institute of Technology, Pasadena.

Santhanam, G., Yu, B. M., Gilja, V., Ryu, S. I., Afshar, A., Sahani, M., & Shenoy, K. V. (2009). Factor-analysis methods for higher-performance neural prostheses. *J Neurophysiol, 102*(2), 1315–1330.

Smith, M., Jia, X., Zandvakili, A., & Kohn, A. (2013). Laminar dependence of neuronal correlations in visual cortex. *J Neurophysiol, 109*, 940–947.

Snyder, L., Batista, A., & Andersen, R. (1997). Coding of intention in the posterior parietal cortex. *Nature, 386*(6621), 167–170.

Sommer, M., & Wurtz, R. (2006). Influence of the thalamus on spatial visual processing in frontal cortex. *Nature, 444*(7117), 374–377.

Stevenson, I. H., London, B. M., Oby, E. R., Sachs, N. A., Reimer, J., Englitz, B., et al. (2012). Functional connectivity and tuning curves in populations of simultaneously recorded neurons. *PLoS Comput Biol, 8*(11), e1002775.

Stevenson, I. K. K. (2008). Inferring functional connections between neurons. *Curr Opin Neurobiol, 18*(6), 582–588.

Tenenbaum, J., de Silva, V., & Langford, J. (2000). A global geometric framework for nonlinear dimensionality reduction. *Science, 290*(5500), 2319–2323.

Van Essen, D., Anderson, C., & Felleman, D. (1992). Information processing in the primate visual system: An integrated systems perspective. *Science, 255*(5043), 419–423.

Wolpaw, J. R., Birbaumer, N., McFarland, D. J., Pfurtscheller, G., & Vaughan, T. M. (2002). Brain-computer interfaces for communication and control. *Clin Neurophysiol, 113*(6), 767–791.

Yu, B. M., Cunningham, J. P., Santhanam, G., Ryu, S. I., Shenoy, K. V., & Sahani, M. (2009). Gaussian-process factor analysis for low-dimensional single-trial analysis of neural population activity. *J Neurophysiol, 102*(1), 614–635.

Zhao, M., Batista, A., Cunningham, J. P., Chestek, C., Rivera-Alvidrez, Z., Kalmar, R., et al. (2011). An L 1-regularized logistic model for detecting short-term neuronal interactions. *J Comput Neurosci.* doi:10.1007/s10827-011-0365-5

Zohary, E., Shadlen, M., & Newsome, W. (1994). Correlated neuronal discharge rate and its implications for psychophysical performance. *Nature, 370*(6485), 140–143.

41 Action Vocabularies and Motor Commands

ROGER N. LEMON

"Voltaire said of Sir Isaac Newton that with all his science he knew not how his hand moved."
—John Napier, 1971

ABSTRACT This chapter focuses on action vocabularies and motor commands with reference to object grasp and manipulation. First, I highlight the language used in the motor control literature related to motor "commands," which emphasizes the concept of supraspinal structures generating movement by passing a command to the spinal cord. Ideas about commands for movement have been strongly influenced by the fact that electrical stimulation of the motor cortex actually *causes* movement. I highlight critical differences between evoked and natural movements, and especially differences in timing and patterning of neural activity emanating from the primary motor cortex. I discuss the rich versatility of skilled grasp and suggest that this versatility cannot be explained by combination of just a few muscle synergies. I emphasize that the visuomotor grasping circuit primes the motor network for grasp of objects; other (prefrontal) circuits are involved in how the object should be used. I describe some of the work on grasp selectivity in both ventral premotor and primary motor cortex, and on the interactions between them. I discuss the evidence for motor cortex pyramidal tract neurons acting as potential "command neurons." These pyramidal neurons exhibit some features that fit well with such a role, including some neurons intercalated in "smart" transcortical reflexes and with fast, direct access to motoneurons. These features are discussed in the light of theories of motor control, such as optimal feedback control and active inference. Activation of pyramidal tract neurons is not restricted to their role as command neurons: their discharge can also show mirror properties, being modulated during observation of others' actions, without any sign of concomitant electromyographic activity.

Much of the literature on motor control theory makes free use of the term "motor command." By definition, a command must be both delivered and received. Although neural activity in a number of different structures has been identified as having the properties of such a command signal, far less is known about how these signals are received by spinal interneurons and motoneurons and transformed into muscle activity and movement. A further point is that commands for movements should not only initiate the movement, but also continue to guide and update it until the action is completed and the goal achieved. The command signal should also contain high-level information appropriate for the context of the movement, which may vary from trial to trial.

Skilled grasp is a distinctive component of human motor control. Visual guidance of skilled prehension and manipulation of objects is essential for human creative art, sculpture, and music, as well as for technological development, tool use, and manufacture. Therefore, in this chapter, I focus on the status of research on commands for grasp control. I begin by looking at evidence for activity in motor and premotor cortex that is selective for particular types of grasp carried out under visuomotor control. I then consider how activity in the visuomotor grasping circuit results in activity in corticospinal neurons with projections to the cervical spinal cord, and targeting motor nuclei innervating the upper limb muscles recruited during grasp.

Although some features of this corticospinal output from primary motor cortex (M1) fit well with a role as command neurons, others certainly do not. In addition, we now know that some of the same neurons recruited when movements are generated are also recruited during other states such as action observation, so we have to understand what precisely it is that distinguishes the level and pattern of activity that is specifically associated with movement generation. These approaches may help define exactly what we mean by motor commands for grasp.

The language of motor commands: The impact of stimulation studies

It is generally assumed that the brain's motor system generates motor commands that cause movements to occur (Franklin & Wolpert, 2011; Frey et al., 2011). The term "command" is closely related to that of agency in the motor system, since by implication the motor system, by issuing command signals for movement, reinforces our sense that these commands cause things to happen.

She brought her forefinger closer to her face, urging it to move. It remained still because she was pretending ... because willing it to move or being about to move it was not the same as actually moving it. And when she did crook it finally, the action seemed to start in the finger itself, not in some part of her mind. When did it know to move, when did she know to move it? (McEwan, 2001, pp. 35–36)

This quotation nicely sums up the differences between thinking about a movement and actually making one, and also underlines fundamental questions about volition and agency (Haggard, 2008; Tallis, 2004). Hughlings Jackson suggested that movements could be categorized according to the level of volitional control exerted over them. He classified movements involving fine control of the hand as "least automatic"—that is, requiring conscious control. Jackson was struck by the vulnerability to disease of the most advanced functions of the human motor system, including speech and skilled manipulation, reflecting relatively recent evolutionary development. The degradation or loss of these functions was a process of "dissolution" versus evolution (Hughlings Jackson, 1884). This historical account points us firmly in the direction of skilled grasp as a useful target for understanding how these "least automatic" movements are generated and controlled.

While the least automatic movements involved the cortex, the most automatic or reflex movements were considered to originate within the spinal cord (Phillips & Porter, 1977, p. 3). The links between the different levels of the motor system are provided by the "descending" (implying commanding) motor pathways from brain to spinal cord. An early account of the linkage provided by the "motor bundle" or pyramidal tract was given by François Frank in 1887: "The motor bundle gathers up the voluntary motor commands from the surface of the brain and transmits them to the effector apparatus of the brainstem and spinal cord" (quoted in Phillips & Porter, 1977, p. 5).

This concept is long enshrined in the terms "upper" and "lower motoneuron," widely used (Phillips & Landau, 1990) and still very useful in clinical neurology: the pyramidal output neurons of the motor cortex are identified as controlling activity in spinal alpha motoneurons supplying limb muscles. However, there is plentiful evidence that the activity of cortical output neurons (upper motoneurons) and muscles innervated by lower motoneurons can be completely dissociated (Fetz & Finnochio, 1971; Schieber, 2011; see below), so these terms are no longer appropriate for describing cortico-muscular interactions during voluntary movements.

The concept of motor commands resonates strongly with the motor effects of electrical stimulation of the motor cortex, newly discovered during the time that Hughlings Jackson was publishing his seminal essays. The pioneering observations by Fritsch and Hitzig in 1870 on movements evoked from the dog's motor cortex showed that stimulation of the cortex *caused* movement to occur, a discovery that heralded a long sequence of investigations into the "motor" cortex of both humans and experimental animals (see Lemon, 2008a). The introduction of transcranial magnetic stimulation in the 1980s completed a full century of motor cortical stimulation studies. These helped to define how electrical stimulation acts upon the human cortex, and particularly the preferential activation of axons rather than cell bodies (di Lazzero, Ziemann, & Lemon, 2008; Maier, Kirkwood, Brochier, & Lemon, 2013).

Questions soon arose: Were these evoked movements, or the central neural activity that precedes them, the same as occurred during "natural" movement? Certainly some features are characteristic of natural actions, for example, the order of recruitment of motor units (Bawa & Lemon, 1993; Gandevia & Rothwell, 1987). However, we should be cautious. Leyton and Sherrington (1917) had been the first to note the "fractional quality" of the evoked movements, such that they typically would have formed only part of a more complex action. Movements can appear purposeful or goal-related if the cortex is subjected to long (500 ms) and intense (>100 μA) periods of stimulation (Graziano, Taylor, & Moore, 2002). But Phillips and Porter (1977) warned: "Thus faradic stimulation is disqualified, equally with 'galvanic,' as a tool for evoking natural function," but it can be used as "a tool for mapping the outputs that are available for selection by the intracortical activities that it cannot itself evoke" (p. 37).

Thus, although electrical stimulation continues to provide a useful probe for exploring connections within the motor system and for probing the processes within it, we know that the patterns of activity generated by electrical stimulation are quite different from those observed during natural movements. Electrical stimulation of the motor cortex causes intense transynaptic bombardment of pyramidal output neurons. This results in a characteristic repetitive discharge of "I" waves in the corticospinal tract at high frequencies (~600 Hz; Edgley, Eyre, Lemon, & Miller, 1997; Patton & Amassian, 1954). Even a single intracortical stimulus can recruit indirect, transsynaptic responses in a high proportion of corticospinal outputs (Maier et al., 2013). However, discharges at these high frequencies have not been observed in recordings from these same pyramidal neurons during natural movements (di Lazzero et al., 2008).

There are also striking differences in the central delays involved for stimulus evoked versus naturally generated movements. For example, the intense, synchronized output generated by electrical stimulation of the macaque cortex evokes responses in hand and digit muscles with brief onset latencies of about 10 ms. These latencies are much shorter than the 60–100 ms between the onset of M1 activity and subsequent muscle activity during voluntarily generated movements (Cheney & Fetz, 1980; Porter & Lemon, 1993).

Thus, the generation of motor responses by electrical stimulation of the cortex reinforced the idea that the brain commands movement. However, the evidence is that natural voluntary movements are the result of a different and much longer process that far exceeds physiological conduction times. This difference may reflect a slow buildup to some kind of threshold that must be reached before movements can begin (Porter & Lemon, 1993), or it may represent a shift in neural dynamics from one neural state space to another (Shenoy, Sahani, & Churchland, 2013). Many investigators have used perturbations of ongoing movements to illuminate the process of motor control (Evarts & Tanji, 1976; Johansson, Lemon, & Westling, 1994; Scott, 2012). However, even in these cases, the delays involved in generating outputs that update the ongoing movement are still considerably longer than physiological conduction times (Johansson et al., 1994; Scott, 2012).

Action vocabularies for the hand: Reach and grasp

The control of grasp presents particularly complex questions for theories of motor control. The enormous versatility of the hand depends upon the many degrees of freedom that it provides. The position and orientation of any object we wish to grasp can be defined by six variables (three positions and three angles). In contrast, there are already seven degrees of freedom in the arm (three at the shoulder, two at the elbow, and two at the wrist), while the hand (carpals, metacarpals, and phalanges) contribute a further 20 degrees of freedom. It is a remarkable achievement of the grasp-control system to apply a controlled level of force, with a particular vector, to a pen or an instrument held between the finger tips, since this force has to be developed at the very end of a long, articulated bony chain. The transition from moving a digit (such as the index finger) toward an object and then exerting a controlled force vector upon it is particularly demanding. Such an action, rather than being explained by the combination of a small number of basic muscle "synergies," requires the contribution of an extended number of principle components in the activity of muscles acting on the index finger (Valero-Cuevas, Venkadesan, & Todorov, 2009; Venkadesan & Valero-Cuevas, 2008).

Premotor and motor cortex activity that is selective for grasp

Because of the complexity of movements involved in reaching toward an object, grasping, and manipulating it, how can one hope to detect grasp-specific command signals? Early investigations did this by comparing neuronal activity for different grasps (Muir & Lemon, 1983). Hideo Sakata and his colleagues documented neuronal activity while monkeys were successively presented with a wide range of different objects, each evoking a different type of grasp. This approach has been used to explore activity in a number of cortical regions (Murata et al., 1997; Raos, Umilta, Murata, Fogassi, & Gallese, 2006; Taira, Mine, Georgopoulos, Murata, & Sakata, 1990; Umilta, Brochier, Spinks, & Lemon, 2007).

In a landmark paper published in 1995, Jeannerod, Arbib, Rizzolatti, and Sakata suggested that

> the transformation of an object's intrinsic properties into specific grips takes place in a circuit that is formed by the inferior parietal lobule and the inferior premotor area (area F5). Neurons in both these areas code size, shape and orientation of objects, and specific types of grip that are necessary to grasp them. (Jeannerod et al., 1995)

Visual control of grasp depended on a visuomotor grasping circuit that received information via the dorsal visual stream about the shape, size, and location of graspable objects. It was proposed that this circuit transformed visual information in the anterior intraparietal area into vocabularies for grasp in F5, which were in turn executed via M1 descending projections to the hand. All three nodes within this circuit have been demonstrated to contain neurons with grasp-specific selectivity in their discharge.

Umilta et al. (2007) compared activity of neurons recorded simultaneously from area F5 and from M1 during a delayed visuomotor reach-to-grasp task which involved six different grasps. Normalized population activity of F5 and M1 neurons was significantly different across the six grasps. For F5 neurons, activity selective for the different grasps emerged early, soon after the object to be grasped became visible to the monkey, but before the cue to grasp was given. In contrast, M1 selectivity did not appear until shortly before the monkey began the reach to grasp action. There was a high degree of variation in preparatory activity from one neuron to the next (cf. Shenoy et al., 2013).

Significant differences in neuronal discharge across the six grasps continued throughout the trial—including reach, grasp, displacement, and holding of the object. For F5 neurons, activity across different phases of the trial was highly correlated, suggesting that F5 activity reflects the same goal throughout. For M1, correlations were lower, probably because M1 activity is closely associated with the different movements making up each part of the total action.

"Canonical" neurons in F5 (Rizzolatti, Fogassi, & Gallese, 2002) respond with very short latency (~100 ms) to the presentation of graspable objects, and this is considered to reflect the low-level characteristic priming of the visuomotor circuit in preparation for grasp of visible objects. The term "affordance" refers to the grasp employed to manipulate a particular object in a specific manner. Thus, the initial priming provided by the visuomotor grasping circuit can be overridden by a presumably higher-level action selection mechanism (e.g., to grasp a knife not by the handle, but by the blade, in order to pass it to another). The ventral premotor cortex is a possible node for the interaction between these low-level and high-level inputs, since it receives a significant input from prefrontal cortex (area 10) (Dum & Strick, 2005; Gerbella, Belmalih, Borra, Rozzi, & Luppino, 2011).

Premotor–motor cortex interactions

When area F5 is inactivated, monkeys can no longer preshape the hand appropriately for grasp of a particular object (Fogassi et al., 2001). Although area F5 could, in principle, exert its influence over grasp movements via its direct corticospinal projection (Borra, Belmalih, Gerbella, Rozzi, & Luppino, 2010; Dum & Strick, 1991), the major pathway appears to be through its cortico-cortical projections to M1. There are three subdivisions of area F5: F5a, F5p, and F5c (Gerbella et al., 2011). The posterior region (area F5p) is a hand-related field in which the canonical visuomotor neurons responsive to observation and grasp of 3D objects are particularly prominent (Raos et al., 2006; Umilta et al., 2007). This subdivision of F5 has strong reciprocal connections to the anterior intraparietal area in the posterior parietal cortex (Gharbawie, Stepniewska, Qi, & Kaas, 2011), and provides the main connections with the M1 hand area (Godschalk, Lemon, Kuypers, & Ronday, 1984), which are also reciprocal. Unlike the more anterior region of F5 (F5a), it has only weak connections to prefrontal areas. All three subdivisions (including F5c, in which mirror neurons are primarily located; see below) are densely interconnected.

The physiological effects of activating F5 projections to M1 have been the subject of a series of studies (see Lemon, 2012). In summary, single-pulse intracortical stimulation of F5 (mainly the F5p subdivision) exerts strong excitatory and inhibitory effects on single neurons in the M1 hand area (Kraskov, Prabhu, Quallo, Lemon, & Brochier, 2011). Most responses occur at short latency (1.8–3 ms), suggesting that they are mediated by a relatively direct, cortico-cortical route. F5 stimulation exerts substantial facilitation of corticospinal outputs from M1 (Shimazu, Maier, Cerri, Kirkwood, & Lemon, 2004), which results in boosting of responses in motoneurons supplying distal hand muscles (Shimazu et al., 2004) and electromyographic (EMG) responses in these muscles (Cerri, Shimazu, Maier, & Lemon, 2003). In the awake monkey and in human volunteers, premotor–motor cortex interactions are grasp-specific (Davare, Lemon, & Olivier, 2008; Prabhu et al., 2009); in some cases, F5 stimulation can cause significant suppression of the test M1 response.

M1 corticospinal neurons as command neurons

From the early studies of Evarts (1964, 1968), it seemed clear that the large corticospinal neurons in primary motor cortex (antidromically identified from the pyramidal tract and referred to as pyramidal tract neurons, or PTNs) were possible candidates for command neurons. Evarts showed that when monkeys performed fast wrist movements, these M1 PTNs discharged in advance of movement, and their discharge frequency could be correlated with parameters of the upcoming movement, such as its direction and force. These PTNs are also among the fastest-conducting axons in the central nervous system, and this may be related to the need to reduce conduction delays in transmitting activity from cortex to spinal cord. However, fast-conducting axons make up only a few percent of the corticospinal tract.

Other eligible features of PTNs as command neurons include their collateralization to important subcortical motor structures, such as the pontine nuclei (Ugolini & Kuypers, 1986), thereby providing an efference copy of commands to the cerebellum. But PTNs appear to form a quite separate population from cortical neurons projecting to the striatum (Turner & Delong, 2000).

A further key feature of M1 PTNs is that some make direct corticomotoneuronal (CM) connections to alpha motoneurons (Porter & Lemon, 1993; Rathelot & Strick, 2006; Zinger, Harel, Gabler, Israel, & Prut, 2013). CM synapses on motoneurons are not subject to presynaptic inhibition (Jackson, Baker, & Fetz, 2006), suggesting that other systems (e.g., peripheral afferent inputs from the moving limb) do not use this mechanism to modulate or cancel out these CM inputs to motoneurons, allowing fidelity of central commands during centrally generated movements.

Spike-triggered averaging of EMG demonstrated that the natural activity of PTNs could exert a direct

CM action on the target muscle (Fetz & Cheney, 1980; Lemon, Mantel, & Muir, 1986) and allowed insights into how spinal motoneurons processed CM command signals (Lemon, 2008b; Zinger et al., 2013). These studies concentrated on motor cortex as a causal generator of movement, rather than being a central representation of motor parameters (Fetz, 1992; Shenoy et al., 2013). In macaque hand muscles, post-spike facilitation of EMG activity by CM cells begins around 11 ms, in keeping with conduction time estimates from stimulation studies (Lemon et al., 1986; Porter & Lemon, 1993). While this demonstrates unequivocally that pathways revealed by electrical stimulation are utilized during natural movements, it is also the case that at the start of movement, even identified CM cells show much longer delays between discharge onset and EMG onset than predicted on the basis of these conduction times (Cheney & Fetz, 1980; see above).

A recent study showed that the greatest number of corticospinal terminations in the cervical spinal gray matter, from the M1 hand area, was found among the intermediate layers, in which most of the segmental interneurons are located. The second-highest count was for terminations among the motor nuclei of the ventral horn (Morecraft, Ge, Stilwell-Morecraft, McNeal, Pizzimenti, & Darling, 2013). So the CM system does not work in isolation from other descending (Baker, 2011; Lemon, 2008b), segmental (Takei & Seki, 2010), and propriospinal systems (Isa, Ohki, Alstermark, Petterson, & Sasaki, 2007). However, the CM system can provide a significant proportion of the drive needed to maintain motoneuron discharge in steady-state conditions (Cheney, Fetz, & Mewes, 1991).

CM cells have been shown to be active for a whole range of different limb movements, including during tool use by macaques; CM connections are particularly well developed in primates with a high level of dexterity and who use tools (Quallo, Kraskov, & Lemon, 2012).

M1 command neurons and neural implementation of motor control theories

Optimal feedback control is one of a number of current theories of movement planning and control (Franklin & Wolpert, 2011; Todorov & Jordan, 2002). This theory stresses the role of sensory inputs in motor control and suggests that the brain makes an optimal state estimation by integrating efference copy signals with delayed sensory feedback: "The derivation of the optimal control policy (i.e., feedback gains) uses knowledge of the system dynamics, such as properties of the musculoskeletal system, to achieve an optimal balance between behavioural performance and associated motor costs" (Scott, 2012, p. 541).

The motor system is richly supplied with information from a variety of peripheral receptors that provide a detailed account of the current sensory state of a limb and any movement of it. Multiple ascending sensory pathways make this information available to all supraspinal motor centers. The motor cortex itself receives fast somatosensory feedback from the limb: for the M1 hand area, over 80% of neurons receive such inputs (Porter & Lemon, 1993).

Evidence for the operation of optimal feedback control has been gained by studying the effects of perturbing voluntary movements (Scott, 2012). In arm muscles, the response to perturbations of the intended movement consists of a series of reflexes. The shortest latency component arises through a spinal reflex, but its contribution to the response is small and does not show context-related adaptation. In contrast, the slightly later contribution of a transcortical component shows all the features of a "smart" reflex (Evarts & Tanji, 1976). CM cells respond to peripheral inputs (Porter & Lemon, 1993) and have been shown to be recruited during long-latency transcortical reflexes (Cheney & Fetz, 1984).

There are strong transcortical reflexes to perturbation of grasp. When lifting an object between the finger tips, application of sheer forces to the object results in a fast slip response, with rapid upgrade of the grip force to prevent the object slipping from the grasp (Johansson et al., 1994). This response is triggered by slip-related signals in cutaneous afferents and results in force upgrades after only 60 ms. This response can be strongly boosted by transcranial magnetic stimulation over the contralateral motor cortex (Johansson et al., 1994). Again, the spinal component is small and ineffectual.

Active inference theory suggests that activity in descending pathways, such as the corticospinal tract, does not represent commands at all, but instead reflects a prior-based prediction of the proprioceptive input associated with the upcoming movement (Adams, Shipp, & Friston, 2013; Friston, 2010). In this model, spinal reflex mechanisms take on special importance in terms of generating the movement: movement arises as a result of spinal reflex activity signaling an error between the predicted and the actual proprioceptive input. This prediction error is generated in segmental interneurons, which then relay the error signal to the appropriate motoneurons and muscles.

The existence of the CM system poses a considerable challenge for the active inference model. Because CM cells by definition produce monosynaptic input to motoneurons, these inputs must be integrated with many other segmental inputs to the same motoneurons, and therefore they are implemented at the same level of the motor system: the motoneuron cannot then distinguish between a prediction (from a corticospinal input) and a prediction error (from a local interneuron). In addition, it is likely that CM neurons also terminate on last-order interneurons (Porter & Lemon, 1993; Morecraft et al., 2013), and the same signal cannot represent both a prediction and a prediction error (Adams et al., 2013).

Adams et al. (2013) suggest a way around this problem. They propose that the CM system acts to predict the precision or gain of the sensory input by modulating the postsynaptic gain in motoneurons to the prediction error input. There seem to be two fundamental problems with this idea: First, these neuromodulatory signals require activation of relatively slow NMDA channels by CM inputs, whereas the existing evidence is that they act through fast AMPA channels. Second, as pointed out above, spinal reflexes to perturbations are generally weak (particularly in hand muscles) and lack context-dependent properties (Scott, 2012).

Despite these criticisms, the active interference model does have attractive features: not least, it embodies the concept that much of the descending activity is concerned with controlling and interrogating somatosensory input, rather than generating movement. Descending control of sensory input may have evolved much earlier than movement generation (Lemon, 2008b).

Corticospinal mirror neurons and motor commands

Further insights into the possible role of the corticospinal system in movement initiation has come about as a result of recent work that demonstrates that the discharge of PTNs in both premotor and motor cortex can be modulated by simply observing the actions of others, without any sign of concomitant EMG activity (Kraskov, Dancause, Quallo, Shepherd, & Lemon, 2009; Vigneswaran, Philipp, Lemon, & Kraskov, 2013). Neurons that discharge during both self-initiated movement and during action observation were originally termed "mirror neurons" by Rizzolatti and his collaborators (Gallese, Fadiga, Fogassi, & Rizzolatti, 1996).

Some of these PTNs showed contrasting properties for execution versus observation of grasp: they fired vigorously during the monkey's own grasp, but discharge was either partly or completely suppressed during action observation. These suppression mirror neurons (Kraskov et al., 2009) may be involved in inhibiting the monkey's own movement as it observes the experimenter's actions. In F5, execution- and observation-related firing rates are rather similar (Gallese et al., 1996), whereas in M1, PTNs are much more active during execution than observation. As a result, there is a marked disfacilitation of PTN output to the spinal cord during action observation (Vigneswaran et al., 2013). This withdrawal of corticospinal input to spinal centers may assist in suppressing unwanted movement during observation.

While the wider functions of mirror neurons are still being debated, these observations add to many others suggesting that the motor cortex can be active in a number of different states, all of which are quite distinct from movement itself (see Schieber, 2011). These include preparation for movement (Shenoy et al., 2013), mental rehearsal, and imagination (Cisek & Kalaska, 2004; Dushanova & Donoghue, 2010; Macuga & Frey, 2012). Further evidence comes from operant conditioning of M1 outputs (Fetz & Finnochio, 1971) and the use of M1 activity to control a brain-machine interface: a situation in which the monkey is rewarded for activating an ensemble of cortical neurons, but where no muscle activity or movement is required (Schwartz, 2007). Unfortunately, in nearly all of these studies, the neurons have not been identified (see Vigneswaran, Kraskov, & Lemon, 2011). We do know that PTNs are embedded in the M1 cortical microcircuit (Jackson, Spinks, Freeman, Wolpert, & Lemon, 2002). If these circuits play a part in processes such as action observation and mental rehearsal, it is not surprising that M1 PTNs are also involved.

Conclusion: Do M1 PTNs qualify as command neurons?

The M1 PTN is well placed to sum central and peripheral drives and produce an appropriate output to the spinal cord. However, since PTNs receive powerful proprioceptive and kinesthetic inputs from the periphery, as well as from a number of cortical and subcortical centers, it is clear that M1 itself does not generate commands for movement. Rare intracellular recordings in the awake monkey by Matsumura (1979) showed that the phasic bursts of M1 pyramidal neurons that occur before movement onset are preceded by a slow buildup of subthreshold excitation. Even when the motor output from M1 is damaged, it can be shown that processing of the relevant planning and selection of movement is intact (Johnson, Sprehn, & Saykin, 2002). Thus, movement preparation and motor execution involve a widespread cortical and subcortical network that argues strongly against a hierarchical command structure. Within M1 itself, the activity associated with movement is heterogeneous in nature (Churchland, Afshar, & Shenoy, 2006), but unless neurons are identified, this may be difficult to explain. The corticospinal tract has multiple functions in addition to that of a descending motor pathway (Lemon, 2008b), so some heterogeneity is to be expected.

Two key observations should now lead future investigations. First, we need to know more about the processing that occurs within the network between the first signs of discharge and movement onset: this delay may well signal a change in the neural state (Shenoy et al., 2013), but we have yet to understand what the impact of these changes is on the motor apparatus: does this represent a slow buildup of excitability in interneurons and motoneurons, or are other central processes involved? Second, it is important to focus on the differences in discharge frequency and temporal pattern for movement versus other states, such as action observation. This might lead to important clues to both how we "command" movements, and also how we withhold them.

ACKNOWLEDGMENTS Work reported in this chapter was supported by the Wellcome Trust. I fully acknowledge the contributions of a great many of my colleagues and collaborators in the studies cited here.

REFERENCES

Adams, R. A., Shipp, S., & Friston, K. J. (2013). Predictions not commands: Active inference in the motor system. *Brain Struct Funct, 218*, 611–643.

Baker, S. N. (2011). The primate reticulospinal tract, hand function and functional recovery. *J Physiol, 589*, 5603–5612.

Bawa, P., & Lemon, R. N. (1993). Recruitment of motor units in response to transcranial magnetic stimulation in man. *J Physiol, 471*, 445–464.

Borra, E., Belmalih, A., Gerbella, M., Rozzi, S., & Luppino, G. (2010). Projections of the hand field of the macaque ventral premotor area F5 to the brainstem and spinal cord. *J Comp Neurol, 518*, 2570–2591.

Cerri, G., Shimazu, H., Maier, M. A., & Lemon, R. N. (2003). Facilitation from ventral premotor cortex of primary motor cortex outputs to macaque hand muscles. *J Neurophysiol, 90*, 832–842.

Cheney, P. D., & Fetz, E. E. (1980). Functional classes of primate corticomotoneuronal cells and their relation to active force. *J Neurophysiol, 44*, 773–791.

Cheney, P. D., & Fetz, E. E. (1984). Corticomotoneuronal cells contribute to long-latency stretch reflexes in the rhesus monkey. *J Physiol, 349*, 249–272.

Cheney, P. D., Fetz, E. E., & Mewes, K. (1991). Neural mechanisms underlying corticospinal and rubrospinal control of limb movements. *Prog Brain Res, 87*, 213–252.

Churchland, M. M., Afshar, A., & Shenoy, K. V. (2006). A central source of movement variability. *Neuron, 52*, 1085–1096.

Cisek, P., & Kalaska, J. F. (2004). Neural correlates of mental rehearsal in dorsal premotor cortex. *Nature, 431*, 993–996.

Davare, M., Lemon, R., & Olivier, E. (2008). Selective modulation of interactions between ventral premotor cortex and primary motor cortex during precision grasping in humans. *J Physiol, 586*, 2735–2742.

di Lazzero, V., Ziemann, U., & Lemon, R. N. (2008). State of the art: Physiology of transcranial motor cortex stimulation. *Brain Stimul, 1*(4), 345–362.

Dum, R. P., & Strick, P. L. (1991). The origin of corticospinal projections from the premotor areas in the frontal lobe. *J Neurosci, 11*, 667–689.

Dum, R. P., & Strick, P. L. (2005). Frontal lobe inputs to the digit representations of the motor areas on the lateral surface of the hemisphere. *J Neurosci, 25*, 1375–1386.

Dushanova, J., & Donoghue, J. (2010). Neurons in primary motor cortex engaged during action observation. *Eur J Neurosci, 31*, 386–398.

Edgley, S. A., Eyre, J. A., Lemon, R. N., & Miller, S. (1997). Comparison of activation of corticospinal neurones and spinal motoneurones by magnetic and electrical stimulation in the monkey. *Brain, 120*, 839–853.

Evarts, E. V. (1964). Temporal patterns of discharge of pyramidal tract neurons during sleep and waking in the monkey. *J Neurophysiol, 27*, 152–171.

Evarts, E. V. (1968). Relation of pyramidal tract activity to force exerted during voluntary movement. *J Neurophysiol, 31*, 14–27.

Evarts, E. V., & Tanji, J. (1976). Reflex and intended responses in motor cortex pyramidal tract neurons of monkey. *J Neurophysiol, 39*, 1069–1080.

Fetz, E. E. (1992). Are movement parameters recognizably coded in activity of single neurons? *Behav Brain Sci, 15*, 679–690.

Fetz, E. E., & Cheney, P. D. (1980). Postspike facilitation of forelimb muscle activity by primate corticomotoneuronal cells. *J Neurophysiol, 44*, 751–772.

Fetz, E. E., & Finocchio, D. V. (1971). Operant conditioning of specific patterns of neural and muscular activity. *Science, 174*, 431–435.

Fogassi, L., Gallese, V., Buccino, G., Craighero, L., Fadiga, L., & Rizzolatti, G. (2001). Cortical mechanism for the visual guidance of hand grasping movements in the monkey: A reversible inactivation study. *Brain, 124*, 571–586.

Franklin, D. W., & Wolpert, D. M. (2011). Computational mechanisms of sensorimotor control. *Neuron, 72*, 425–442.

Frey, S. H., Fogassi, L., Grafton, S., Picard, N., Rothwell, J. C., Schweighofer, N., ... Fitzpatrick, S. M. (2011). Neurological principles and rehabilitation of action disorders: Computation, anatomy, and physiology (CAP) model. *Neurorehabil Neural Repair, 25*(Suppl. 5), 6S–20S.

Friston, K. (2010). The free-energy principle: A unified brain theory? *Nat Rev Neurosci, 11*, 127–138.

Gallese, V., Fadiga, L., Fogassi, L., & Rizzolatti, G. (1996). Action recognition in the premotor cortex. *Brain, 119*, 593–609.

Gandevia, S. C., & Rothwell, J. C. (1987). Knowledge of motor commands and the recruitment of human motoneurons. *Brain, 110*, 1117–1130.

Gerbella, M., Belmalih, A., Borra, E., Rozzi, S., & Luppino, G. (2011). Cortical connections of the anterior (F5a) subdivision of the macaque ventral premotor area F5. *Brain Struct Funct, 216*, 43–65.

Gharbawie, O. A., Stepniewska, I., Qi, H., & Kaas, J. H. (2011). Multiple parietal-frontal pathways mediate grasping in macaque monkeys. *J Neurosci, 31*, 11660–11677.

Godschalk, M., Lemon, R. N., Kuypers, H. G. J. M., & Ronday, H. K. (1984). Cortical afferents and efferents of monkey postarcuate area: An anatomical and electrophysiological study. *Exp Brain Res, 56*, 410–424.

Graziano, M. S. A., Taylor, C. S. R., & Moore, T. (2002). Complex movements evoked by microstimulation of precentral cortex. *Neuron, 34*, 841–851.

Haggard, P. (2008). Human volition: Towards a neuroscience of will. *Nat Rev Neurosci, 9*, 934–946.

Hughlings Jackson, J. (1884). Croonian lectures on the evolution and dissolution of the nervous system. *Br Med J, 1*(1215), 703–707.

Isa, T., Ohki, Y., Alstermark, B., Petterson, L. G., & Sasaki, S. (2007). Direct and indirect cortico-motoneuronal pathways and control of hand/arm movements. *Physiology (Bethesda), 22*, 145–152.

Jackson, A., Baker, S. N., & Fetz, E. E. (2006). Tests for presynaptic modulation of corticospinal terminals from peripheral afferents and pyramidal tract in the macaque. *J Physiol, 573*, 107–120.

Jackson, A., Spinks, R. L., Freeman, T. C. B., Wolpert, D. M., & Lemon, R. N. (2002). Rhythm generation in monkey motor cortex explored using pyramidal tract stimulation. *J Physiol, 541*, 685–699.

Jeannerod, M., Arbib, M. A., Rizzolatti, G., & Sakata, H. (1995). Grasping objects: The cortical mechanisms of visuomotor transformation. *Trends Neurosci, 18*, 314–320.

Johansson, R. S., Lemon, R. N., & Westling, G. (1994). Time-varying enhancement of human cortical excitability mediated by cutaneous inputs during precision grip. *J Physiol, 481*, 761–775.

Johnson, S. H., Sprehn, G., & Saykin, A. J. (2002). Intact motor imagery in chronic upper limb hemiplegics: Evidence for activity-independent action representations. *J Cogn Neurosci, 14*, 841–852.

Kraskov, A., Dancause, N., Quallo, M. M., Shepherd, S., & Lemon, R. N. (2009). Corticospinal neurons in macaque ventral premotor cortex with mirror properties: A potential mechanism for action suppression? *Neuron, 64*, 922–930.

Kraskov, A., Prabhu, G., Quallo, M. M., Lemon, R. N., & Brochier, T. (2011). Ventral premotor-motor cortex interactions in the macaque monkey during grasp: Response of single neurons to intracortical microstimulation. *J Neurosci, 31*, 8812–8821.

Lemon, R. N. (2008a). An enduring map of the motor cortex. *Exp Physiol, 93*, 798–802.

Lemon, R. N. (2008b). Descending pathways in motor control. *Annu Rev Neurosci, 31*, 195–218.

Lemon, R. N. (2012). Interactions between premotor and motor cortices in non-human primates. In R. R. J. Chen (Ed.), *Cortical connectivity: Brain stimulation for assessing and modulating cortical connectibility and function* (pp. 23–46). Heidelberg: Springer.

Lemon, R. N., Mantel, G. W. H., & Muir, R. B. (1986). Corticospinal facilitation of hand muscles during voluntary movement in the conscious monkey. *J Physiol, 381*, 497–527.

Leyton, S. S. F., & Sherrington, C. S. (1917). Observations on the excitable cortex of the chimpanzee, orangutan and gorilla. *Q J Exp Physiol, 11*, 135–222.

Macuga, K. L., & Frey, S. H. (2012). Neural representations involved in observed, imagined, and imitated actions are dissociable and hierarchically organized. *NeuroImage, 59*, 2798–2807.

Maier, M. A., Kirkwood, P. A., Brochier, T. G., & Lemon, R. N. (2013). Responses of single corticospinal neurons to intracortical stimulation of primary motor and premotor cortex in the anesthetized macaque monkey. *J Neurophysiol, 109*(12), 2982–2998.

Matsumura, M. (1979). Intracellular synaptic potentials of primate motor cortex neurons during voluntary movement. *Brain Res, 163*, 33–48.

McEwan, I. (2001). *Atonement.* New York, NY: Vintage.

Morecraft, R. J., Ge, J., Stilwell-Morecraft, K. S., McNeal, D. W., Pizzimenti, M. A., & Darling, W. G. (2013). Terminal distribution of the corticospinal projection from the hand/arm region of the primary motor cortex to the cervical enlargement in rhesus monkey. *J Comp Neurol, 521*(18), 4205–4235.

Muir, R. B., & Lemon, R. N. (1983). Corticospinal neurons with a special role in precision grip. *Brain Res, 261*, 312–316.

Murata, A., Fadiga, L., Fogassi, L., Gallese, V., Raos, V., & Rizzolatti, G. (1997). Object representation in the ventral premotor cortex (area F5) of the monkey. *J Neurophysiol, 78*, 2226–2230.

Napier, J. (1971). *The roots of mankind.* London: George Allen and Unwin.

Patton, H. D., & Amassian, V. E. (1954). Single- and multiple-unit analysis of cortical stage of pyramidal tract activation. *J Neurophysiol, 17*, 345–363.

Phillips, C. G., & Landau, W. M. (1990). Clinical neuromythology. VIII. Upper and lower motor neuron: The little old synecdoche that works. *Neurology, 40*, 884–886.

Phillips, C. G., & Porter, R. (1977). *Corticospinal neurons: Their role in movement.* New York, NY: Academic Press.

Porter, R., & Lemon, R. N. (1993). Corticospinal function and voluntary movement. *Monogr Physiol Soc, 45*, 247–273.

Prabhu, G., Shimazu, H., Cerri, G., Brochier, T., Spinks, R. L., Maier, M. A., & Lemon, R. N. (2009). Modulation of primary motor cortex outputs from ventral premotor cortex during visually guided grasp in the macaque monkey. *J Physiol, 587*, 1057–1069.

Quallo, M. M., Kraskov, A., & Lemon, R. N. (2012). The activity of primary motor cortex corticospinal neurons during tool use by macaque monkeys. *J Neurosci, 32*, 17351–17364.

Raos, V., Umilta, M. A., Murata, A., Fogassi, L., & Gallese, V. (2006). Functional properties of grasping-related neurons in the ventral premotor area F5 of the macaque monkey. *J Neurophysiol, 95*, 709–729.

Rathelot, J. A., & Strick, P. L. (2006). Muscle representation in the macaque motor cortex: An anatomical perspective. *Proc Natl Acad Sci USA, 103*, 8257–8262.

Rizzolatti, G., Fogassi, L., & Gallese, V. (2002). Motor and cognitive functions of the ventral premotor cortex. *Curr Opin Neurobiol, 12*, 149–154.

SCHIEBER, M. H. (2011). Dissociating motor cortex from the motor. *J Physiol, 589,* 5613–5624.

SCOTT, S. H. (2012). The computational and neural basis of voluntary motor control and planning. *Trends Cogn Sci, 16,* 541–549.

SCHWARTZ, A. B. (2007). Useful signals from motor cortex. *J Physiol, 579,* 581–601.

SHENOY, K. V., SAHANI, M., & CHURCHLAND, M. M. (2013). Cortical control of arm movements: A dynamical systems perspective. *Annu Rev Neurosci, 36,* 337–359.

SHIMAZU, H., MAIER, M. A., CERRI, G., KIRKWOOD, P. A., & LEMON, R. N. (2004). Macaque ventral premotor cortex exerts powerful facilitation of motor cortex outputs to upper limb motoneurones. *J Neurosci, 24,* 1200–1211.

TAIRA, M., MINE, S., GEORGOPOULOS, A. P., MURATA, A., & SAKATA, H. (1990). Parietal cortex neurons of the monkey related to the visual guidance of hand movement. *Exp Brain Res, 83,* 29–36.

TAKEI, T., & SEKI, K. (2010). Spinal interneurons facilitate coactivation of hand muscles during a precision grip task in monkeys. *J Neurosci, 30,* 17041–17050.

TALLIS, R. (2004). *The hand: A philosophical enquiry into human being.* Edinburgh, UK: Edinburgh University Press.

TODOROV, E., & JORDAN, M. I. (2002). Optimal feedback control as a theory of motor coordination. *Nat Neurosci, 5,* 1226–1235.

TURNER, R. S., & DELONG, M. R. (2000). Corticostriatal activity in primary motor cortex of the macaque. *J Neurosci, 20,* 7096–7108.

UGOLINI, G., & KUYPERS, H. G. (1986). Collaterals of corticospinal and pyramidal fibres to the pontine grey demonstrated by a new application of the fluorescent fibre labelling technique. *Brain Res, 365,* 211–227.

UMILTA, M. A., BROCHIER, T. G., SPINKS, R. L., & LEMON, R. N. (2007). Simultaneous recording of macaque premotor and primary motor cortex neuronal populations reveals different functional contributions to visuomotor grasp. *J Neurophysiol, 98,* 488–501.

VALERO-CUEVAS, F. J., VENKADESAN, M., & TODOROV, E. (2009). Structured variability of muscle activations supports the minimal intervention principle of motor control. *J Neurophysiol, 102,* 59–68.

VENKADESAN, M., & VALERO-CUEVAS, F. J. (2008). Neural control of motion-to-force transitions with the fingertip. *J Neurosci, 28,* 1366–1373.

VIGNESWARAN, G., KRASKOV, A., & LEMON, R. N. (2011). Large identified pyramidal cells in macaque motor and premotor cortex exhibit "thin spikes": Implications for cell type classification. *J Neurosci, 31,* 14235–14242.

VIGNESWARAN, G., PHILIPP, R., LEMON, R. N., & KRASKOV, A. (2013). M1 corticospinal mirror neurons and their role in movement suppression during action observation. *Curr Biol, 23,* 236–243.

ZINGER, N., HAREL, R., GABLER, S., ISRAEL, Z., & PRUT, Y. (2013). Functional organization of information flow in the corticospinal pathway. *J Neurosci, 33,* 1190–1197.

42 Choice Values: The Frontal Cortex and Decision Making

MATTHEW F. S. RUSHWORTH, B. K. H. CHAU, U. SCHÜFFELGEN, F.-X. NEUBERT, AND N. KOLLING

ABSTRACT Actions are chosen on the basis of expectations about the benefits they will yield. Areas in the brain's frontal lobes are important for making decisions about which choice it is best to take. In functional magnetic resonance imaging experiments, the ventromedial prefrontal cortex exhibits signals related to the values of potential choices, and the activity pattern is consistent with the emergence of a decision. Ventromedial prefrontal cortex lesions disrupt value-guided decision making. Another frontal lobe area, the anterior cingulate cortex, is also important for value-guided decision making. There is some evidence suggesting that anterior cingulate cortex is especially involved in the selection of an action in the light of the rewards that will follow and the effort that will be invested. More fundamentally, it may be concerned with the computation and comparison of values needed for decision making during foraging. During foraging, the key decision is whether to engage and pursue the default choice, or whether the foraging opportunities available elsewhere in the environment mean it is better to switch away from the current behavior and pursue other alternatives.

When we make a movement, we need to make sure that it is executed correctly. We have to ensure that muscles and joints move in such a way that the limb is directed to the desired target location and the hand is configured correctly. Rather than looking at the neural mechanisms that determine *how* an action is executed, this chapter instead focuses on the cortical mechanisms that underlie *why* one action is chosen instead of another. It focuses on the cortical mechanisms (figure 42.1) that underlie selection of a choice on the basis of the value of its consequences.

Value signals in prefrontal cortex for guiding decisions

The ventromedial prefrontal cortex (vmPFC) and adjacent medial orbitofrontal cortex (mOFC) play a central role in choice selection. For some time, it has been clear that activity in the vmPFC/mOFC region is correlated with the values of stimuli and choices (Kable & Glimcher, 2007; Lebreton, Jorge, Michel, Thirion, & Pessiglione, 2009) and it now seems likely that the vmPFC/mOFC is involved in the actual making of the decision itself. When people make a decision between two potential choices, the vmPFC/mOFC exhibits a signal that suggests it is comparing the values of the choices. For example, Boorman, Behrens, Woolrich, and Rushworth (2009) used functional magnetic resonance imaging (fMRI) to measure blood oxygen level–dependent (BOLD) indices of brain activity in human subjects while they made a series of decisions between two different choices associated with different numbers of reward points. The points won by the subjects were translated into a monetary payment at the end of the experiment. The decisions were difficult in the sense that the reward associations contained two elements: a certain number of points (*reward magnitude*), which changed randomly from trial to trial, and a certain probability of the points being delivered (*reward probability*), which slowly changed over the course of a series of decisions. Therefore, in order to choose effectively, subjects had to consider both of the elements that composed each option.

Boorman and colleagues found vmPFC/mOFC encoded the difference in value between the option that was chosen and the option that was rejected in each decision (figure 42.2). They referred to this as a relative chosen value signal. At the time that a choice is made, the vmPFC/mOFC BOLD signal becomes positively correlated with the value of the option that is chosen and negatively correlated with the value of the option that is being rejected (figure 42.2). One interpretation of this pattern is that vmPFC/mOFC activity increasingly reflects the value of the option that is to be chosen, and at the same time it is less and less activated by the value of the option that is to be rejected. When this process is complete, the decision is made.

Several other fMRI studies have confirmed the existence of relative chosen value signals in vmPFC/mOFC (de Martino, Fleming, Garrett, & Dolan, 2013; FitzGerald, Seymour, & Dolan, 2009; Philiastides, Biele, & Heekeren, 2010; Wunderlich, Dayan, & Dolan, 2012).

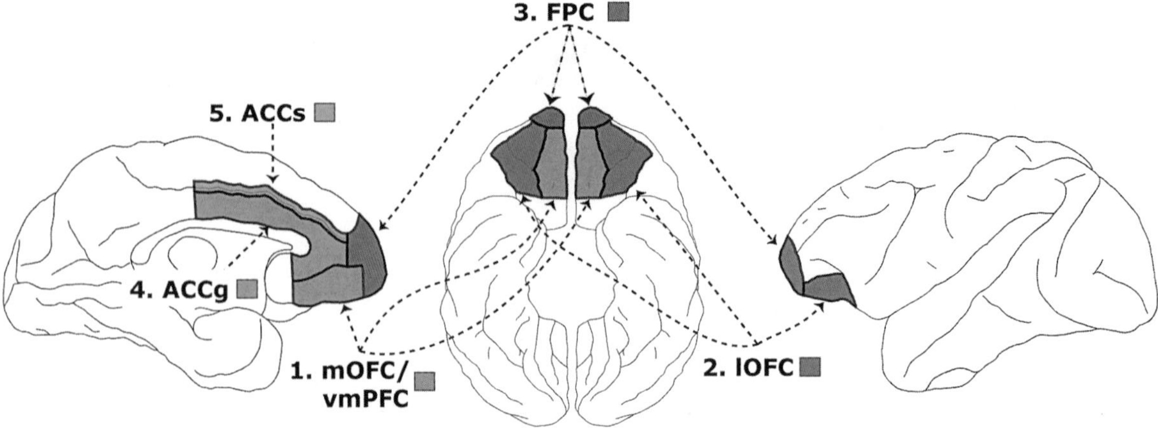

FIGURE 42.1 Frontal brain regions, in the macaque, involved in reward-guided learning and decision making. Abbreviations: vmPFC/mOFC, ventromedial prefrontal cortex/medial orbitofrontal cortex; lOFC, lateral orbitofrontal cortex; ACCs, anterior cingulate cortex sulcus; ACCg, anterior cingulate cortex gyrus. (Adapted from Rushworth et al., 2011.)

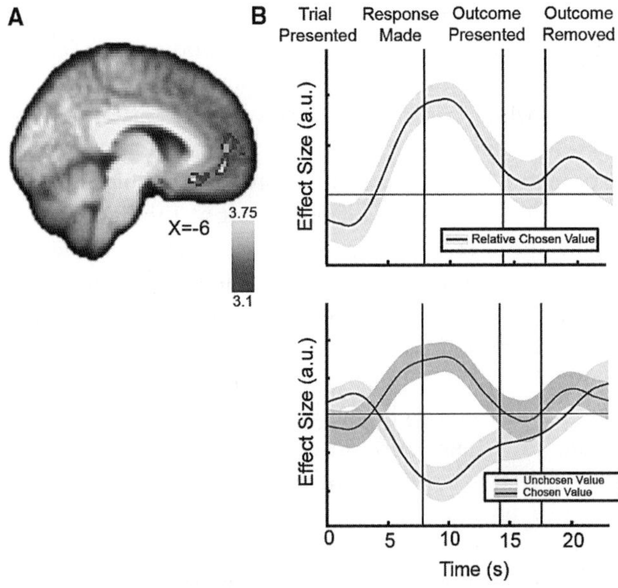

FIGURE 42.2 (A) Sagittal slices through z-statistic maps relating to the relative chosen value (chosen–unchosen expected value) of two options during decision making. (B) *Top panel*: Time course for the effect size of the relative chosen value in the vmPFC is shown throughout the duration of the trial. *Bottom panel*: The same time course is shown with the signal decomposed into chosen and unchosen action values. There is a positive correlation with chosen value and a negative correlation with unchosen value during the decision-making phase. (Adapted from Boorman et al., 2009.)

Some studies, however, have reported slightly different results. For example, Hare, Schultz, Camerer, O'Doherty, and Rangel (2011) reported that vmPFC/mOFC BOLD activity reflected the sum of the values of the choices subjects were offered (rather than the relative difference in their values) while Daw, O'Doherty, Dayan, Seymour, and Dolan (2006) reported vmPFC/mOFC activity reflected the value of the choice that was taken (rather than the difference between it and the rejected option's value).

The time course and emergence of the value difference signal

One way to reconcile these apparently diverging results is to think about them almost as snapshots taken at different time points during the making of the choice. First, the vmPFC/mOFC may represent the value of both potential choices (as in the study by Hare et al., 2011), then it may begin to compare those values, and in the course of doing so a relative value signal will emerge (as in some other studies: Boorman et al., 2009; de Martino et al., 2013; FitzGerald et al., 2009; Philiastides et al., 2010; Wunderlich et al., 2012), and finally, once the comparison process is complete, it is left just carrying a signal representing the value of just the option that has been chosen (as in the study by Daw et al., 2006). This account suggests that if we had a more time-resolved measure of vmPFC/mOFC activity, then we would be able to see the transition from one type of signal to the next.

Hunt and colleagues (2012) tried to take just such a measurement from human subjects during decision making by using magnetoencephalography (MEG) to record brain activity rather than fMRI. First of all, however, Hunt and colleagues attempted to make quantitative predictions about exactly how a neural network ought to make decisions and what signals a network would generate if only the average activity of a brain area, rather than the activity of individual neurons, could be measured. The models they used to predict the activity were first proposed by Wang (2002), and they employed neuron elements with biophysically plausible properties. Wang's model was originally proposed to explain how neurons in the macaque might mediate choices between one of two eye-movement responses on the basis of graded visual evidence, but it can also be adapted to look at choices made on the basis of different values. At the heart of the model is the idea that the neurons can be assigned to two different pools, each encoding one choice (figure 42.3). Each pool of neurons becomes active in proportion to an input it receives that is, in turn, proportional to the value of the choice it represents. Within a pool of neurons there is recurrent excitatory activity, but the interactions between pools of neurons are mediated by inhibitory interneurons. In essence, this means that each pool becomes active in proportion to the value of the choice it encodes and exerts an inhibitory influence on the other pool that is also proportional to the value of the choice it encodes. Ultimately, the network ends up in one of two possible attractor states in which one pool of neurons is active and the other is not, and this activity pattern constitutes the decision. In general, the pool representing the more valuable of the two options remains active.

When Hunt and colleagues looked at the mean field activity in the model (the sum of the postsynaptic activity throughout the model), they found that it was initially dominated by the sum of the values of the two choice options. Subsequently, the mean field activity reflected the value difference between the choices. They then looked throughout the brains of their human subjects for areas with MEG signals that resembled those they had seen in the model—signals that rose and fell in proportion to, first, the sum of the values of the choices and then the difference in value between choices. They identified two areas with such

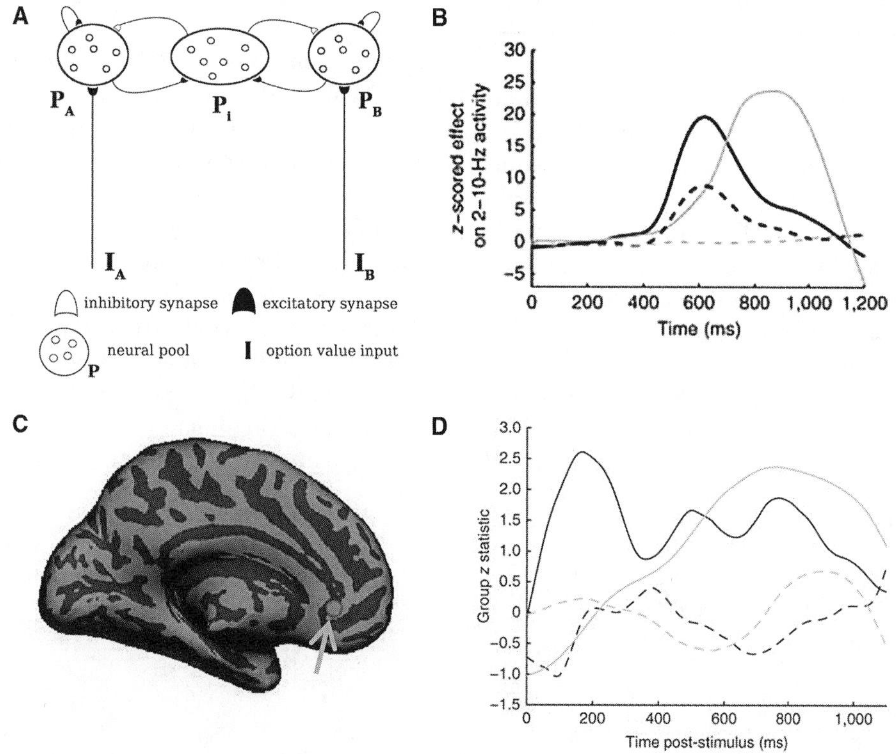

FIGURE 42.3 (A) The biophysical model Hunt and colleagues (2012) used to predict a transition from activity that is proportional to the sum of choice values to the value of the chosen value. It contains two pools of excitatory pyramidal neurons (P_A, P_B) corresponding to choices of either option A or option B. There is recurrent excitation between neurons within pools, but inhibitory interneurons (P_i) mediate competition between pools. Activity in each pool is initially affected by an input proportional (I_A, I_B) to the value of each option, but inhibition between pools leaves only a single pool in a high-firing attractor state. The corresponding option is then chosen. (B) Z-scored effect of overall sum of choice values (on frequency range 3–9 Hz; black lines) and choice value difference (on frequency range 2–4.5 Hz; gray lines) on biophysical model activity; solid lines are correct trials, dashed lines, incorrect trials. (C) vmPFC activity shows several value-related hallmarks of the biophysical network model. (D) Effect of overall value (3–9 Hz; black) and value difference (2–4.5 Hz; gray) on correct/error trials (solid/dashed lines, respectively) during first half of experiment in vmPFC. The analysis used here was performed on human vmPFC but was equivalent to that performed in panel 42.3B on the biophysical model. (Adapted from Hunt et al., 2012.)

signals—one was in the parietal cortex, but the other was in vmPFC/mOFC. In summary, the results are consistent with the idea that a value-comparison process takes place in vmPFC/mOFC, and that during the course of the comparison vmPFC/mOFC is dominated by different types of value signal.

Disrupting vmPFC/mOFC impairs value-guided decision making

If it is true that vmPFC/mOFC discriminates between choices on the basis of their value, then we might expect that vmPFC/mOFC lesions would impair value-guided decision making. We might expect that choice discrimination would become more and more impaired as the difference in choice values decreased. This prediction is based on what happens when mechanisms for color discrimination are impaired by brain lesions in visual association cortex; discrimination performance becomes worse and worse as differences in the color of two stimuli are made smaller and smaller (Buckley, Gaffan, & Murray, 1997).

The prediction was borne out when lesions were made in vmPFC/mOFC in macaques (Noonan et al., 2010). Noonan and colleagues trained their macaques to choose between three stimuli with different probabilistic associations with reward. Each day, the macaques learned about three new stimuli, but on each day one of the stimuli was a high-value stimulus because it was associated with a high probability of reward (0.6), and one stimulus was a low-value stimulus because it had no association with reward (0 probability of reward). The third stimulus was important because on some testing days it was just as poor in value as the worst stimulus (0 probability of reward), but on other days its reward probability began to approach that of the best stimulus (either 0.2 or 0.375 probability of reward). Discriminating the best option from the second-best option therefore became more and more difficult as the second-best option's value approached that of the best option. In the control state macaques were able to make the decision in all three cases, but they learned the identity of the best option more slowly when they were performing testing sessions in which the best option and second-best option were close in value (figure 42.4). Macaques with vmPFC/mOFC lesions were significantly worse when the second-best option was close in value to the best option. In other words, vmPFC/mOFC lesions made macaques worse at taking the most difficult value-guided decisions. Moreover, the disruptive impact on difficult decisions was specific to vmPFC/mOFC lesions and did not occur after lesions in other frontal brain areas that contain neurons with reward-related activity patterns, such as anterior cingulate cortex (ACC) or lateral orbital frontal cortex (lOFC) (Noonan et al., 2010; Rudebeck et al., 2008; Walton, Behrens, Buckley, Rudebeck, & Rushworth, 2010).

Another way to assess value-guided decision making is to examine the consistency and transitivity of choices. Normally, if a person or a monkey prefers option X over Y and Y over Z, then their preferences are said to be transitive if they also prefer X over Z. It is just such transititivity that is lost after vmPFC/mOFC lesions (Rudebeck & Murray, 2011). Once again, the effect is specific to vmPFC/mOFC lesions and is not seen after lOFC lesions. Similar patterns of impairment in value-guided decision making have also been reported in human patients with vmPFC/mOFC lesions (Camille,

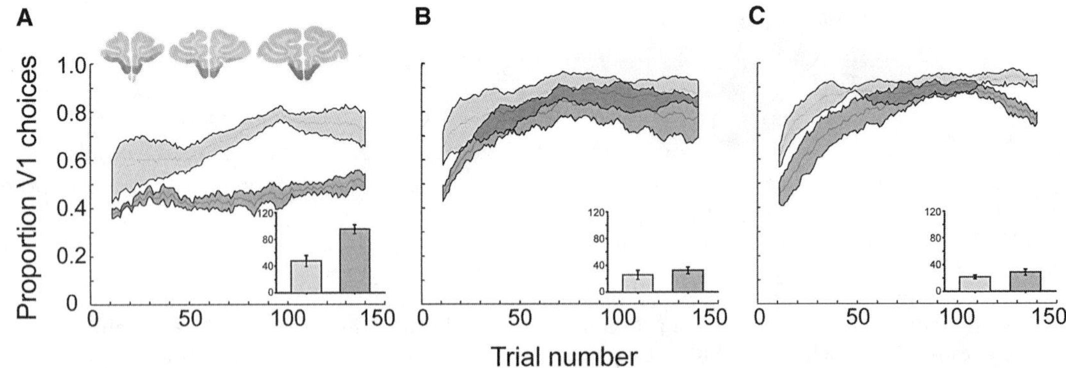

Figure 42.4 Effect of choice option value proximity for macaques with mOFC lesions. Proportion of choices which were of the best value option when the difference in value between the best and second-best option was small (A, 0.2), medium (B, 0.4) and large (C, 0.6). Control pre-lesion (light gray), post-mOFC lesion (dark gray) performance. Insets show number of trials to reach 70% V1 choices. Lesions of mOFCs caused impairments when the best and second-best value differences were small (A). The mOFC lesion locations are represented on an unoperated control brain, with darkness indicating lesion overlap (overlap in one to four animals). (Adapted from Noonan et al., 2010.)

Griffiths, Vo, Fellows, & Kable, 2011; Fellows, 2011; Henri-Bhargava, Simioni, & Fellows, 2012).

The activity of neurons in vmPFC/mOFC

Despite the abundance of neuroimaging studies of vmPFC/mOFC, we have relatively little knowledge of this region's activity at the level of individual neurons. A handful of very recent studies have reported that the activity of neurons in the vmPFC/mOFC of the macaque is modulated when there is a possibility that a reward rather than a punishment might be received (Monosov & Hikosaka, 2012). Firing rate is modulated as a function of the reward size (Bouret & Richmond, 2010; Kaping, Vinck, Hutchison, Everling, & Womelsdorf, 2011) and reward probability (Monosov & Hikosaka, 2012) that is associated with a stimulus that is chosen.

What is less clear is quite how vmPFC/mOFC activity evolves during the course of a decision. Padoa-Schioppa and colleauges (Padoa-Schioppa, 2009; Padoa-Schioppa & Assad, 2006) have presented a detailed account of single-neuron activity patterns when macaques choose between different amounts and types of rewards. In a number of ways, the activity is reminescent of the MEG signals reported by Hunt and colleagues (2012) because it is, initially, related to the value of the potential choices that might be taken. This signal is then rapidly followed by another pattern of activity that is related to the value of the choice that is actually being taken. However, the region in which recordings were made is probably just lateral to the one that has been the focus of the neuroimaging and lesion studies of reward-guided decision making and probably lies in lOFC. Despite the important differences between vmPFC/mOFC and lOFC and their anatomical separation, there are interconnections between the two regions (Carmichael & Price, 1996).

Translating values into actions: Integration across frontal lobe systems

When vmPFC/mOFC makes a decision, it does not do so in the sense of choosing a specific action, but instead the type of decision that it makes is the selection of a reward goal that becomes the focus of behavior. This appears to be especially the case when there are multiple competing alternative choices and attention must be directed while critical comparisons are made between the various options (Noonan et al., 2010). Once the reward goal is selected, then a second type of decision has to be made about which action should be made to obtain the reward goal.

There was little evidence in OFC of encoding of the actions that monkeys made when they were choosing between different types and amounts of reward (Cai & Padoa-Schioppa, 2012; Padoa-Schioppa & Assad, 2006). By contrast, when ACC neurons are recorded in the same paradigm, they do not encode the values of the potential choices, but they do encode the value of the choice taken, and do so with a longer latency than OFC neurons (Cai & Padoa-Schioppa, 2012). ACC neurons did, however, encode the direction of the action used to affect the choice. In other words, it seems that, at least in the paradigm studied by Padoa-Schioppa and colleagues, ACC neurons are only encoding the output of a decision process rather than the decision itself, but that they are then encoding features of the action that will be used to make the choice. Some human neuroimaging studies might be interpreted in a similar manner (Noonan, Mars, & Rushworth, 2011).

The ACC region that carried both chosen value and action direction signals is well placed to influence the motor system. It lies on the dorsal bank of the cingulate sulcus and it is adjacent and interconnected with a region called the rostral cingulate motor area (van Hoesen, Morecraft, & Vogt, 1993). The rostral cingulate motor area in turn sends projections to the spinal cord that probably allow it to influence movement in a relatively direct manner (Dum & Strick, 1991; He, Dum, & Strick, 1995).

In summary, one way in which different frontal lobe areas might interact during decision making is for vmPFC/mOFC and possibly adjacent OFC areas to make a choice between different possible reward goals and for ACC to decide between different possible actions. It is, however, important to emphasize that in the experiments of Padoa-Schioppa and colleagues the only meaningful associations with reward are with stimuli. A given stimulus appears at different locations on different trials, so that the actions that are made to select the stimulus change from trial to trial and actions made to particular locations do not have a consistent relationship with a reward if the stimulus at that location changes. It is possible that another role for the ACC might be revealed if the animals are given the opportunity to learn direct associations between actions and rewards that do not involve any mediation by visual stimuli. Rudebeck and colleagues (2008) and Luk and Wallis (2013) taught macaques tasks in which associations had to be learned between actions and rewards in the absence of any mediating stimuli, and compared them with the more commonly studied type of task in which macaques learned associations between stimuli and rewards. More ACC neurons encoded the choices animals made when they were performing the action-reward association task than the stimulus-reward association task, whereas the opposite was true in lOFC

(Luk & Wallis, 2013). ACC lesions impaired action-reward association learning but not stimulus-reward association learning, while the opposite was true after lOFC lesions (Rudebeck et al., 2008). A similar dissociation between the effects of ACC and OFC lesion deficits has been reported in human patients (Camille, Tsuchida, & Fellows, 2011).

Such patterns of neural activity and lesion effects suggest the existence of different mechanisms for decision making in ACC versus vmPFC/mOFC and lOFC. It is possible that these different mechanisms work in a completely parallel manner, with one set of brain regions learning associations between actions and rewards and another set learning associations between stimuli and rewards. However, it seems unlikely that this will be the case. While values may adhere to particular stimuli and objects in the world in a relatively constant fashion, it is not clear that this is always likely to be true of actions. For example, the value of an action directed to the left or the right might, in many real-world scenarios, change depending on which way a person is facing. What might, however, be pertinent for a system that can bring together information about reward and actions is to decide whether it is worth persisting with a given action or whether it might be better to try an alternative one. These are the types of decisions that animals often have to make when there are no specific stimulus-reward associations available to guide their behavior. In addition, it might be useful for such a mechanism to compute whether the effort entailed by an action outweighs the benefit that it bestows. The next section considers such possibilities in the context of the foraging choices that are important for many animals.

Distinct mechanisms for decision making in ACC and vmPFC/mOFC

The last section discussed how vmPFC/mOFC might be a mechanism for making decisions between potential choices. Although such decisions are commonly studied by psychologists, cognitive neuroscientists, and economists, they may be surprisingly rare outside of the laboratory. This is simply because in many natural situations, when an animal is foraging, potential food items are encountered sequentially rather simultaneously (Freidin & Kacelnik, 2011). It might be only on some occasions that a macaque foraging in the wild is given the opportunity to make the type of choice that the vmPFC/mOFC appears to make—for example, a choice between an apple and an orange. Instead, what is more likely to happen is that the macaque might first see an opportunity to pursue a course of action that might lead to one piece of fruit and only later perceive the opportunity to pursue another piece of fruit by taking another course of action. The critical choice for the foraging animal is, therefore, whether to *engage* with a potential food option or whether to carry on *searching* for better opportunities elsewhere.

In contrast to cognitive neuroscientists, behavioral ecologists have long been interested in such choices. Whether or not an animal engages with the potential option it encounters is determined by the value of the encountered option, the effort that will be expended in engaging with it, whether the environment is, on average, sufficiently rich that better options are likely to be encountered frequently if the animal carries on searching, and how effortful it will be to continue searching (Charnov, 1976; Stephens & Krebs, 1986). In summary, foraging decisions should be governed by the average value of continuing to search (*search value*), the effort entailed by searching (*search costs*), and the value of each option encountered (*encounter value*). Despite the ecological importance of such choices, surprisingly little has been known about their neural mechanisms.

Kolling, Behrens, Mars, and Rushworth (2012) attempted to compare the neural processes underlying engage/search foraging choices with the neural mechanisms underlying binary decision making. On each trial, human subjects could alternate between two different styles of decision making while fMRI data were collected. Before the subjects entered the scanner, they had learned about a set of visual stimuli in which each stimulus was associated with a different number of points that were translated into a monetary payment at the end of the trial. At the beginning of each trial in the fMRI experiment, subjects saw two of these stimuli (figure 42.5). The options encountered in this way were referred to as the *encounter* options. They were intended to simulate the opportunity that a foraging animal might engage with. The investigators refered to their values as *encounter values*. A box at the top of the screen included a number of alternative stimuli. These were intended to represent the richness of the subject's current foraging environment and to indicate the *search value*; if the subject did not engage with the encountered option, then he or she opted to search for better options that would be drawn at random from the stimuli in the box and become the encounter options on the next trial. The first decision that subjects made on each trial was therefore a foraging-style choice to either engage with an encountered option or to search for better alternatives. To simulate search cost, subjects lost points if they chose to search (the cost on each trial was indicated by the color of the box surrounding the search options). Subjects could keep opting to search

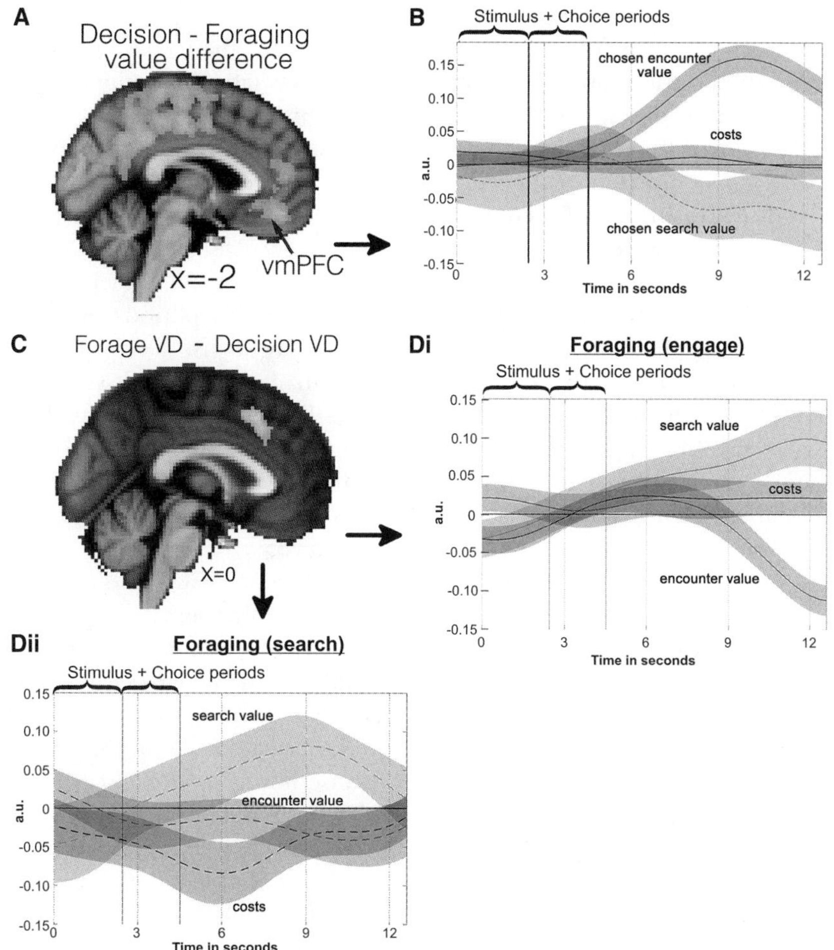

FIGURE 42.5 While vmPFC/mOFC is more active during decision making than foraging (A, B), an ACC region is more active during foraging than decision making (C). The ACC BOLD is positively correlated with the search value and negatively correlated with the encounter value regardless of whether subjects choose to stick with the option encountered (Di) or to search for potential alternatives when the search cost is also represented (Dii). By contrast, vmPFC/mOFC BOLD is positively correlated with the encounter value, when it is chosen, but there is no representation of search values or costs regardless of the choice ultimately made (B). (Adapted from Kolling et al., 2012.)

for better alternatives as many times as they wanted. However, as soon as they opted to engage, they were then able to make the second type of decision—a binary comparitive decision between the two component stimuli that constituted the encounter option. This second decision was therefore of the same type as those that are known to be related to the vmPFC/mOFC.

Several aspects of the results suggested binary choices and foraging choices were being mediated by the vmPFC/mOFC and the ACC, respectively. First, several brain areas, including ACC, were more active when subjects made foraging-style engage/search choices, while other brain regions, including vmPFC/mOFC, were more active when subjects made binary comparison decisions (figure 42.5). Perhaps even more importantly, ACC activity reflected all three key factors that should determine foraging: encounter values, search values, and search costs. By contrast, vmPFC/mOFC only carried an encounter value signal but no representations of search value or search costs. Other fMRI experiments have also reported integrated reward/effort signals in ACC as opposed to vmPFC (Croxson, Walton, O'Reilly, Behrens, & Rushworth, 2009).

Another reason for thinking the two regions, vmPFC/mOFC and ACC, were mediating binary comparative decisions and foraging choices, respectively, could be found in examination of individual differences in activity patterns in the two regions. Individual differences in the strength of ACC signals, but not vmPFC signals, were correlated with individual differences in foraging behavior. Individual differences in the strength of vmPFC/mOFC signals, but not ACC signals, were correlated with individual differences in the way binary comparative choices were made.

The way in which ACC encoded values for foraging choices was quite distinct from the manner in which vmPFC/mOFC encoded values during binary decision making. As earlier sections have described, vmPFC/mOFC encodes a relative value difference signal, with the BOLD signal increasing as a function of the value of the chosen option and decreasing as a function of the value of the rejected option. By contrast, rather than encoding decision values in a framework that is relative to the choice that is made, the ACC encodes foraging choice values in a fixed manner; ACC activity always increases as a function of the search value and decreases as a function of the encounter value. The search value signal, however, ramps up more quickly when subjects are going to make the search choice rather than the engage choice (figure 42.5).

In the foraging task it was noticeable that by default subjects often seemed simply to engage with the option placed in front of them; subjects made more of these choices and did not move away from them unless the value of searching considerably outweighed the value of engaging. One way of thinking about the ACC's choice signals is that they are configured in such a way as to indicate the value of changing away from the current default behavior and of searching for better alternatives.

Reinterpreting the role of ACC

The hypothesis that ACC encodes value signals that are needed for foraging-style choices can explain some of the other signals that have been reported in ACC in the past. For example, it has been noted that during binary choices the ACC often carries a signal that is inversely proportional to the difference in value between the choices available, rather than proportional to the differences in value between the choices as in vmPFC/mOFC (Hare et al., 2011). According to the foraging hypothesis, the inverse value signal recorded during a binary comparison task can be interpreted as a signal reporting how beneficial it would be to not take the choice that is about to be taken but, instead, to switch to the other choice.

Another important strand of research has linked ACC to detecting when there is conflict between possible responses that might be made (Botvinick, 2007). Response conflict occurs when there is a similar amount of evidence favoring more than one response, and once again the ACC may be signaling the value of taking a course of action that is alternative to the one that is being taken. It is not clear, however, that a theory that focuses just on conflict monitoring could explain some of the results found in the foraging task.

If ACC has a mechanism for valuation and promotion of behavioral change and search, then this may suggest a reinterpretation of some other findings made during earlier investigations of ACC. The action-reward learning tasks that are impaired by ACC lesions (Camille, Tsuchida et al., 2011; Kennerley, Walton, Behrens, Buckley, & Rushworth, 2006; Rudebeck et al., 2008) typically involve alternation between actions but no informative stimuli. Repetitive selection of an action interleaved with periods of exploration of alternative actions may be just the sort of behavior normally under the control of a foraging system. It may be that the ACC lesion did not disrupt a mechanism for linking specific actions to outcomes but that, instead, it disrupted a mechanism with which animals decide whether to persist with one choice versus switching to an alternative.

The human ACC region that carries search/engage signals for foraging is well placed to influence whether or not a successful action will be repeated or whether the value of an alternative action will be explored. It lies just adjacent and anterior to the human rostral cingulate motor area (Amiez & Petrides, 2014; Beckmann, Johansen-Berg, & Rushworth, 2009; Picard & Strick, 1996), and as we have already seen a similar region is interconnected with the rostral cingulate motor area (van Hoesen et al., 1993), which has connections to a number of motor regions.

ACC neurons during searching and foraging

A longstanding problem that has dogged attempts to understand ACC has been the difficulty of finding single-neuron activity patterns that are consistent with some of the most influential theories of ACC function. For example, it has proven difficult to identify ACC neurons that detect conflict (Nakamura, Roesch, & Olson, 2005). By contrast, there are several reports of ACC neuron activity that suggest that it is concerned with searching the environment for the best choices to take and with foraging-style search/engage decisions.

Quilodran, Rothé, and Procyk (2008) taught macaques a task in which they had to work out which one of four options was associated with reward. The macaques solved the task by searching through each of the four possible choices. When they identified the choice that was associated with reward, they were allowed to repeat it again for several trials and received rewards on each occasion. After several trials, however, the reward was reassigned to another option and the cycle of searching for the best option, and then repeatedly choosing that option, began again. The animals, therefore, alternated between periods of searching for good choices to take and, once a good choice was identified, of exploiting that choice by taking it repeatedly. The firing rate of many ACC neurons was modulated when the outcomes of the choices were revealed to the monkeys in the search phase of the task, and this was apparent both when the choice led to nonreward as well as when it led, for the first time, to a reward (the first reward, of course, defined the end of the search period). As the monkey shifted into the repetition phase of the task, however, ACC neuron outcome-related activity diminished and instead activity became more prominent at the moment that choices were reinitiated (figure 42.6).

Other features of ACC activity also attest to the importance of the transition between searching and repetitive behavior. Both nonreward and reward outcomes in the search period were followed by an increase in high gamma (60–140 Hz) power in the ACC, which was then followed by a similar increase in lateral prefrontal cortex (Rothé, Quilodran, Sallet, & Procyk, 2011). By contrast, in the repetition phase, post-choice outcome-related activity in ACC and lateral prefrontal cortex was correlated in the beta (10–20 Hz) band.

Other experiments have made even more explicit attempts to simulate the type of foraging decision that animals have to make in the wild. Hayden, Pearson, and Platt (2011) designed a laboratory task for macaques that simulated the types of foraging choices that are considered by Charnov's (1976) marginal value theorem. The marginal value theorem was proposed to explain how animals should spend their time in an environment composed of patchily distributed food sources. As an animal forages in its patch, it gradually depletes the resources it contains. For example, a monkey that continued to forage in the same tree all the time would gradually deplete the fruit it contained. In a patchy environment, an animal should continue foraging within its patch for longer when that patch is richer in resources than the average patch or if considerable effort is going to have to be invested in traveling a large distance to the next patch. Hayden and colleagues simulated such a situation by giving their macaques a target to which they could saccade in order to indicate they wanted to remain foraging in the same patch. Each time they made this decision they received a reward, but the reward gradually declined in size just as the returns from a food patch in the wild might decline if it is continually exploited. The macaques could indicate that they wanted to leave the patch by saccading to another stimulus. There was then a delay until they had their next chance of reward (the delay simulates travel time to the next patch), but the reward level in the new patch was high (this simulates how the new patch has not yet been depleted).

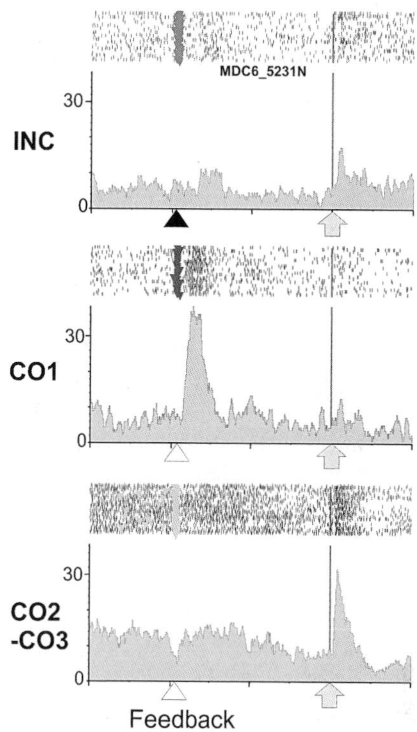

FIGURE 42.6 Macaque monkeys were taught to identify which of four possible directions of response was the correct one. The correct direction remained the same for a few trials but then changed. In this example, there is an increase in activity after feedback (indicated by arrow head) informs the monkey that the action just made is now correct and reward is delivered, but this occurred only on the first occasion that the action was established as the correct one (CO1, second row) and not on subsequent trials (bottom row). This neuron also indicated the direction or sign of the change in action value; it was active when the action was rewarded for the first time, but not when an exploratory choice established that the action was not rewarded (INC, top row). Other neurons exhibited the opposite pattern of activity. Once the action is established as rewarding, the ACC activity occurs in expectation of reward at the time that the lever is touched (vertical line), rather than at the time of feedback (CO2-CO3, bottom row). In other words, ACC activity contains information both about action value expectations and errors in predictions when actions were better or worse than expected. Each row of the figure shows both raster diagrams of the action potentials recorded on individual trials at the top and the mean frequency of action potentials at the bottom. (Adapted from Quilodran et al., 2008.)

The pattern of results suggested that ACC neurons were monitoring the reward intake rate in the current patch and contributing to the monkeys' decisions to leave the patches (figure 42.7). ACC neurons were active in response to the delivery of rewards received each time the monkey carried on foraging in its patch,

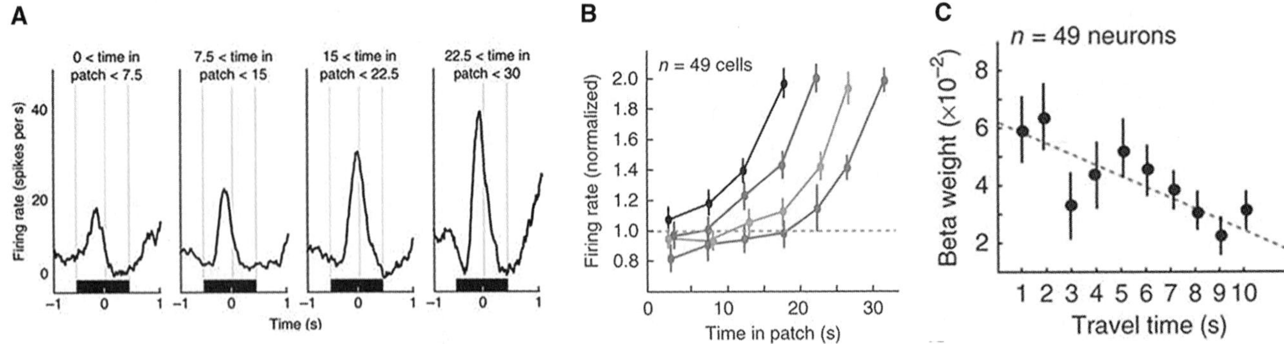

FIGURE 42.7 The activity of neurons in ACC in monkeys during a foraging task. Outcome-related activity increases with each outcome that is received for foraging in a given patch in an example neuron (A) and, on average, the gain of the outcome-related response with time in the patch co-varied with the speed of departure from the patch (B). The rate of gain was a function of the search costs that were to be paid; here, rate of gain is indicated by a regression slope (beta weight) relating time spent foraging in a patch with firing rate, and it can be seen that it decreases with the search cost that will have to be paid in order to travel to a new patch. (Adapted from Hayden et al., 2011.)

but the reward-related firing rates gradually increased as the time approached for the patch-leaving decision. After the reward-related firing rate reached a certain level, the monkeys tended to switch to a new patch on the next trial. The steepest rises in firing rates were seen on those trials where the animal left the patch soonest and when the travel time to the next patch was shortest.

A more direct demonstration that ACC neurons might integrate information, not just about reward expectations, but also about the effort that is going to be invested in pursuing a course of action comes from a multidimensional choice task that involved macaques making choices between different pictures that were each associated with different outcomes (Kennerley, Dahmubed, Lara, & Wallis, 2009). The three critical dimensions that defined each outcome were the size of reward payoff (volume of juice delivered), effort (number of lever presses necessary to earn the reward), and the probability (probability that reward would be delivered). Kennerley and colleagues found that neurons in all three areas they investigated, ACC, lOFC, and lateral prefrontal cortex, encoded the three different factors that determined choice values. However, ACC neurons more frequently encoded each of the three value-determining factors than did lOFC or lateral prefrontal neurons. Individual ACC neurons were also especially likely to multiplex information across all three value dimensions so that information about, for example, both the reward benefits and the effort costs of a choice was integrated within the firing rate of single ACC neurons. Such activity patterns might underlie the integration of reward/effort expectations in ACC activity reported in fMRI studies (Croxson et al., 2009) and would, of course, enable ACC to make the type of reward benefit/effort cost decisions that have been associated with ACC activity (Kolling et al., 2012).

Lesion studies also suggest ACC plays a critical role in integrating information about reward and effort expectations during decision making. Rats can be trained to make choices between two arms of a T-maze that are associated with different reward benefits (for example, two versus four reward pellets) and different effort costs (such as climbing over a barrier versus no barrier). Normally rats are prepared to pay some cost in order to receive a bigger reward benefit, but this changes after lesions are made in the Cg1/Cg2 fields of the ACC (Rudebeck, Walton, Smyth, Bannerman, & Rushworth, 2006; Schweimer & Hauber, 2005; Schweimer, Saft, & Hauber, 2005; Walton, Bannerman, Alterescu, & Rushworth, 2003; Walton, Bannerman, & Rushworth, 2002; Walton et al., 2009). By contrast, lesions of orbitofrontal cortex or prelimbic cortex, in the rat, do not have the same effect.

Conclusion

We have emphasized the different functional contributions of frontal lobe regions to value-guided decision making. We have emphasized a flexible vmPFC/mOFC system that makes comparisons between choice values by focusing on the most relevant aspects of value (for example, reward probability or reward magnitude). In some cases such a system may work in series with brain areas such as the ACC, which might select an action compatible with the reward goal that is the focus of attention. In some situations, however, the ACC might operate in a quite independent manner to take simple stay-switch decisions on the basis of different types of value information such as average search values and effort costs. Although such decisions have received comparatively little attention from cognitive neuroscientists, they may be just the sort of decisions that many animals make when they are foraging, and they may be the types of decisions that many brain regions have evolved to take.

ACKNOWLEDGMENTS Funded by the Medical Research Council and the Wellcome Trust.

REFERENCES

AMIEZ, C., & PETRIDES, M. (2014). Neuroimaging evidence of the anatomo-functional organization of the human cingulate motor areas. *Cereb Cortex, 24*(3), 563–578.

BECKMANN, M., JOHANSEN-BERG, H., & RUSHWORTH, M. F. (2009). Connectivity-based parcellation of human cingulate cortex and its relation to functional specialization. *J Neurosci, 29*(4), 1175–1190.

BOORMAN, E. D., BEHRENS, T. E., WOOLRICH, M. W., & RUSHWORTH, M. F. (2009). How green is the grass on the other side? Frontopolar cortex and the evidence in favor of alternative courses of action. *Neuron, 62*(5), 733–743.

BOTVINICK, M. M. (2007). Conflict monitoring and decision making: Reconciling two perspectives on anterior cingulate function. *Cogn Affect Behav Neurosci, 7*(4), 356–366.

BOURET, S., & RICHMOND, B. J. (2010). Ventromedial and orbital prefrontal neurons differentially encode internally and externally driven motivational values in monkeys. *J Neurosci, 30*(25), 8591–8601.

BUCKLEY, M. J., GAFFAN, D., & MURRAY, E. A. (1997). Functional double dissociation between two inferior temporal cortical areas: Perirhinal cortex versus middle temporal gyrus. *J Neurophysiol, 77,* 587–598.

CAI, X., & PADOA-SCHIOPPA, C. (2012). Neuronal encoding of subjective value in dorsal and ventral anterior cingulate cortex. *J Neurosci, 32*(11), 3791–3808.

CAMILLE, N., GRIFFITHS, C. A., VO, K., FELLOWS, L. K., & KABLE, J. W. (2011). Ventromedial frontal lobe damage disrupts value maximization in humans. *J Neurosci, 31*(20), 7527–7532.

CAMILLE, N., TSUCHIDA, A., & FELLOWS, L. K. (2011). Double dissociation of stimulus-value and action-value learning in humans with orbitofrontal or anterior cingulate cortex damage. *J Neurosci, 31*(42), 15048–15052.

CARMICHAEL, S. T., & PRICE, J. L. (1996). Connectional networks within the orbital and medial prefrontal cortex of macaque monkeys. *J Comp Neurol, 371,* 179–207.

CHARNOV, E. (1976). Optimal foraging: The marginal value theorem. *Theor Popul Biol, 9,* 129–136.

CROXSON, P. L., WALTON, M. E., O'REILLY, J. X., BEHRENS, T. E., & RUSHWORTH, M. F. (2009). Effort-based cost-benefit valuation and the human brain. *J Neurosci, 29*(14), 4531–4541.

DAW, N. D., O'DOHERTY, J. P., DAYAN, P., SEYMOUR, B., & DOLAN, R. J. (2006). Cortical substrates for exploratory decisions in humans. *Nature, 441*(7095), 876–879.

DE MARTINO, B., FLEMING, S. M., GARRETT, N., & DOLAN, R. J. (2013). Confidence in value-based choice. *Nat Neurosci, 16*(1), 105–110.

DUM, R. P., & STRICK, P. L. (1991). The origin of corticospinal projections from the premotor areas in the frontal lobe. *J Neurosci, 11,* 667–689.

FELLOWS, L. K. (2011). Orbitofrontal contributions to value-based decision making: Evidence from humans with frontal lobe damage. *Ann N Y Acad Sci, 1239,* 51–58.

FITZGERALD, T. H., SEYMOUR, B., & DOLAN, R. J. (2009). The role of human orbitofrontal cortex in value comparison for incommensurable objects. *J Neurosci, 29*(26), 8388–8395.

FREIDIN, E., & KACELNIK, A. (2011). Rational choice, context dependence, and the value of information in European starlings *(Sturnus vulgaris). Science, 334,* 1000–1002.

HARE, T. A., SCHULTZ, W., CAMERER, C. F., O'DOHERTY, J. P., & RANGEL, A. (2011). Transformation of stimulus value signals into motor commands during simple choice. *Proc Natl Acad Sci USA, 108*(44), 18120–18125.

HAYDEN, B. Y., PEARSON, J. M., & PLATT, M. L. (2011). Neuronal basis of sequential foraging decisions in a patchy environment. *Nat Neurosci, 14*(7), 933–939.

HE, S.-Q., DUM, R. P., & STRICK, P. L. (1995). Topographic organization of corticospinal projections from the frontal lobe: Motor areas on the medial surface of the hemisphere. *J Neurosci, 15,* 3284–3306.

HENRI-BHARGAVA, A., SIMIONI, A., & FELLOWS, L. K. (2012). Ventromedial frontal lobe damage disrupts the accuracy, but not the speed, of value-based preference judgments. *Neuropsychologia, 50*(7), 1536–1542.

HUNT, L. T., KOLLING, N., SOLTANI, A., WOOLRICH, M. W., RUSHWORTH, M. F., & BEHRENS, T. E. (2012). Mechanisms underlying cortical activity during value-guided choice. *Nat Neurosci, 15*(3), 470–476.

KABLE, J. W., & GLIMCHER, P. W. (2007). The neural correlates of subjective value during intertemporal choice. *Nat Neurosci, 10*(12), 1625–1633.

KAPING, D., VINCK, M., HUTCHISON, R. M., EVERLING, S., & WOMELSDORF, T. (2011). Specific contributions of ventromedial, anterior cingulate, and lateral prefrontal cortex for attentional selection and stimulus valuation. *PLoS Biol, 9*(12), e1001224.

KENNERLEY, S. W., DAHMUBED, A. F., LARA, A. H., & WALLIS, J. D. (2009). Neurons in the frontal lobe encode the value of multiple decision variables. *J Cogn Neurosci, 21*(6), 1162–1178.

KENNERLEY, S. W., WALTON, M. E., BEHRENS, T. E., BUCKLEY, M. J., & RUSHWORTH, M. F. (2006). Optimal decision making and the anterior cingulate cortex. *Nat Neurosci, 9,* 940–947.

KOLLING, N., BEHRENS, T. E., MARS, R. B., & RUSHWORTH, M. F. (2012). Neural mechanisms of foraging. *Science, 336*(6077), 95–98.

LEBRETON, M., JORGE, S., MICHEL, V., THIRION, B., & PESSIGLIONE, M. (2009). An automatic valuation system in the human brain: Evidence from functional neuroimaging. *Neuron, 64*(3), 431–439.

LUK, C. H., & WALLIS, J. D. (2013). Choice coding in frontal cortex during stimulus-guided or action-guided decision-making. *J Neurosci, 33*(5), 1864–1871.

MONOSOV, I. E., & HIKOSAKA, O. (2012). Regionally distinct processing of rewards and punishments by the primate ventromedial prefrontal cortex. *J Neurosci, 32*(30), 10318–10330.

NAKAMURA, K., ROESCH, M. R., & OLSON, C. R. (2005). Neuronal activity in macaque SEF and ACC during performance of tasks involving conflict. *J Neurophysiol, 93*(2), 884–908.

Noonan, M. P., Mars, R. B., & Rushworth, M. F. (2011). Distinct roles of three frontal cortical areas in reward-guided behavior. *J Neurosci, 31*(40), 14399–14412.

Noonan, M. P., Walton, M. E., Behrens, T. E., Sallet, J., Buckley, M. J., & Rushworth, M. F. (2010). Separate value comparison and learning mechanisms in macaque medial and lateral orbitofrontal cortex. *Proc Natl Acad Sci USA, 107*(47), 20547–20552.

Padoa-Schioppa, C. (2009). Range-adapting representation of economic value in the orbitofrontal cortex. *J Neurosci, 29*(44), 14004–14014.

Padoa-Schioppa, C., & Assad, J. A. (2006). Neurons in the orbitofrontal cortex encode economic value. *Nature, 441*(7090), 223–226.

Philiastides, M. G., Biele, G., & Heekeren, H. R. (2010). A mechanistic account of value computation in the human brain. *Proc Natl Acad Sci USA, 107*(20), 9430–9435.

Picard, N., & Strick, P. L. (1996). Motor areas of the medial wall: A review of their location and functional activation. *Cereb Cortex, 6*, 342–353.

Quilodran, R., Rothé, M., & Procyk, E. (2008). Behavioral shifts and action valuation in the anterior cingulate cortex. *Neuron, 57*(2), 314–325.

Rothé, M., Quilodran, R., Sallet, J., & Procyk, E. (2011). Coordination of high gamma activity in anterior cingulate and lateral prefrontal cortical areas during adaptation. *J Neurosci, 31*(31), 11110–11117.

Rudebeck, P. H., Behrens, T. E., Kennerley, S. W., Baxter, M. G., Buckley, M. J., Walton, M. E., & Rushworth, M. F. (2008). Frontal cortex subregions play distinct roles in choices between actions and stimuli. *J Neurosci, 28*(51), 13775–13785.

Rudebeck, P. H., & Murray, E. A. (2011). Dissociable effects of subtotal lesions within the macaque orbital prefrontal cortex on reward-guided behavior. *J Neurosci, 31*(29), 10569–10578.

Rudebeck, P. H., Walton, M. E., Smyth, A. N., Bannerman, D. M., & Rushworth, M. F. (2006). Separate neural pathways process different decision costs. *Nat Neurosci, 9*(9), 1161–1168.

Rushworth, M. F., Noonan, M. P., Boorman, E. D., Walton, M. E., & Behrens, T. E. (2011). Frontal cortex and reward-guided learning and decision-making. *Neuron, 70*(6), 1054–1069.

Schweimer, J., & Hauber, W. (2005). Involvement of the rat anterior cingulate cortex in control of instrumental responses guided by reward expectancy. *Learn Mem, 12*(3), 334–342.

Schweimer, J., Saft, S., & Hauber, W. (2005). Involvement of catecholamine neurotransmission in the rat anterior cingulate in effort-related decision making. *Behav Neurosci, 119*(6), 1687–1692.

Stephens, D. W., & Krebs, J. R. (1986). *Foraging theory*. Princeton, NJ: Princeton University Press.

van Hoesen, G. W., Morecraft, R. J., & Vogt, B. A. (1993). Connections of the monkey cingulate cortex. In B. A. Vogt & M. Gabriel (Eds.), *Neurobiology of cingulate cortex and limbic thalamus*. Boston, MA: Birkhauser.

Walton, M. E., Bannerman, D. M., Alterescu, K., & Rushworth, M. F. S. (2003). Functional specialization within medial frontal cortex of the anterior cingulate for evaluating effort-related decisions. *J Neurosci, 23*, 6475–6479.

Walton, M. E., Bannerman, D. M., & Rushworth, M. F. S. (2002). The role of rat medial frontal cortex in effort-based decision making. *J Neurosci, 22*(24), 10996–11003.

Walton, M. E., Behrens, T. E., Buckley, M. J., Rudebeck, P. H., & Rushworth, M. F. (2010). Separable learning systems in the macaque brain and the role of orbitofrontal cortex in contingent learning. *Neuron, 65*(6), 927–939.

Walton, M. E., Groves, J., Jennings, K. A., Croxson, P. L., Sharp, T., Rushworth, M. F., & Bannerman, D. M. (2009). Comparing the role of the anterior cingulate cortex and 6-hydroxydopamine nucleus accumbens lesions on operant effort-based decision making. *Eur J Neurosci, 29*(8), 1678–1691.

Wang, X. J. (2002). Probabilistic decision making by slow reverberation in cortical circuits. *Neuron, 36*(5), 955–968.

Wunderlich, K., Dayan, P., & Dolan, R. J. (2012). Mapping value based planning and extensively trained choice in the human brain. *Nat Neurosci, 15*(5), 786–791.

43 Finding the Actor in Reciprocal Affordance

SCOTT H. FREY AND SCOTT T. GRAFTON

ABSTRACT Actions are distinguished from movements on the basis of their being goal-directed. At any given moment, the possibilities available for goal-directed action are defined by a complex dynamic between perceived attributes of the environment and the actor's unique goals and behavioral capacities. While much attention has been given to the perceptual side, cognitive neuroscience has all but overlooked the essential role of the actor. For instance, a keyboard may enable an individual to type a letter, but not if he or she is wearing mittens, has no experience with electronics, or suffers from paralysis. Theories of action must accommodate the contextual nature of the situation. Here, we introduce the concept of *reciprocal affordance*, which explicitly acknowledges that actors actively shape, and are shaped by, their environments. This offers a way to integrate a diversity of factors that influence our actions, including development, disability, and the use of technology.

The characteristic that distinguishes action from movement is its goal-directedness, and how we represent and select actions is the fundamental problem in understanding goal-directed behavior. There is consensus that solving this puzzle involves understanding perceptual processing of objects and surfaces, on the one hand, and volitional motor planning and control on the other. However, this dichotomy, while experimentally convenient, is suspect. Evidence is accumulating in both the behavioral sciences and neurosciences that perceptual, motor, and, indeed, a broad set of action-related cognitive functions are intertwined to the point of being largely inseparable. As a consequence, action science must grapple with challenging issues that span all three of these research domains.

Several decades of action research have yielded a deep understanding of perceptual processes that mediate motor behavior. Very little attention, however, has been given to the fact that actions are determined by the *relationship between perceptual attributes of the environment and the unique cognitive and motor capabilities of the actor*. The fact that actors, like their environments, are highly dynamic has been almost entirely ignored. This dynamism is due to rapid changes in goal or motivational states and access to tools or other technologies, as well as to more gradual alterations in behavioral capacity (e.g., those related to growth, development, skill acquisition, senescence, and injury or illness). To address this shortcoming, we introduce the concept of *reciprocal affordance*, which places the strengths of J. J. Gibson's insights (Gibson, 1979) within a modern cognitive neuroscience framework. Our starting point will be a brief introduction to several influential contemporary theories of action, each of which has moved the field forward significantly while also failing to address the dynamic interplay between environment and actor adequately. We then review evidence supporting reciprocal affordance as a new and important conceptual framework, and identify a number of areas that may prove especially amenable to this perspective.

Contemporary theories of action

The emergence of action as an area of inquiry, distinct from perception and motor control, has yielded a number of significant advances. Here, we briefly consider treatment of the actor in shaping goal-directed behavior by several of the more influential theoretical perspectives.

For more than two decades, Goodale and Milner's (1991) dual systems theory has been the dominant theoretical framework. The key idea is that two functionally and anatomically separable pathways are involved in the representation of visual information for perceptual judgments (occipital-temporal ventral stream) versus goal-directed actions (occipital-parietal dorsal stream). There is extensive evidence for these two pathways and their distinct functional properties. Most notable is patient D. F., whose remarkable preservation of visually guided action in spite of severe apperceptive agnosia was the impetus behind the insight that perception for action is supported by mechanisms distinct from those involved in perceptual judgments. Despite its many virtues, this approach is heavily focused on perception of the environment and gives short shrift to the role of the actor in shaping actions. Little attention is devoted, for instance, to the role of the actor's goals and motivations and how unique cognitive and physical capabilities shape their actions.

The *theory of event coding* (TEC) is perhaps the first contemporary framework to emphasize the role of the actor in goal-directed action (Hommel, Musseler, Aschersleben, & Prinz, 2001). This theory makes the important point that studying perception and action separately is an artifice, with perceptual information and action plans represented in a common system of event codes. This significant step acknowledges the need for representational processes that capture not only properties of the environment but also of the actors and their intended actions in the world. There is, however, no clear account in the theory of whether the stimulus determines the behavior (a reframing of behaviorism, where the event code is the stimulus-response association), or the desired behavior sets up conditions for the event code to occur. Real action involves active sensing to maximize physical performance. Furthermore, it is unclear how action choices are made in this framework. Action is built from an appreciation of one's capabilities, desires, motivational states, and experience.

Relatedly, Cisek's affordance competition hypothesis (ACH) is based on extensive recording studies in nonhuman primates that are trained to make reaching movements to different targets (Cisek, 2007). A key observation from this work is that, following the onset of visual targets, there is a buildup of object-selective premovement activity within parietal neurons. These responses are viewed as representations of potential goal-directed actions and escalate over the subsequent delay interval. Following presentation of a cue indicating the action target, there is a selective decrease in activity associated with the uncued target. In Cisek's model, action selection (the perceptually based choice) and specification (how an action is to be physically executed) are intertwined representations, and multiple action possibilities are represented and then resolved through a competitive process.

Cisek's hypothesis goes on to provide an elaborate, neurally based framework that describes how an action results from a multitude of influences arising in widely distributed brain regions. Conspicuously absent from the factors that can influence action choice in ACH are those pertaining to characteristics of the actor. How might the system behave, for instance, if the visual targets were located beyond the monkey's reach, or if the animal's hands were restrained, or paralyzed? In both of these examples, the targets would no longer signal possible actions. To the extent that the premovement neurophysiological responses indicate affordances, one would therefore expect them to be eliminated by these manipulations. In the natural world, the physical state and internal motivation of the actor are dynamic. At present, ACH does not address this key challenge.

Furthermore, the available evidence in support of ACH exclusively involves reaching, which can be characterized entirely by direct sensorimotor matching of the superficial location of the target and the associated limb trajectory. It is unclear how this selection mechanism would need to be enhanced to account for more complex goal-directed behaviors such as grasping, where the choice of a grasp can be affected not only by the perceptual attributes of objects, but also by the intended functional outcome of the action. For example, our choice of grasping a hammer depends on whether we intend to pound or pull a nail, or pass it to another worker, as well as its perceived size, orientation, form, and weight (Johnson & Grafton, 2003).

The dynamic actor

The fundamental challenge for theories of perception is explaining how it is possible to construct stable representations of objects and surfaces in the environment (distal stimuli) from information received by the senses (proximal stimuli) that is highly dynamic due to changes in a variety of factors including lighting, distance, and viewpoint. Retinal images of objects are dramatically affected by all of these variables, and yet we somehow perceive a world of objects with constant properties of color, size, and form (Johnson, 2001; Rock, di Vita, & Barbeito, 1981). Though having received considerably less attention, theories of action must grapple with an additional challenge: the possibilities for goal-directed actions aimed at these stable objects are themselves *dynamic*. The reason is that possibilities for action depend on the capabilities of the actor, which change over several different time scales due to factors including growth and development, senescence, injury or disease, the acquisition of skills, and even the availability of tools and other technologies. The button on the corridor wall presents an opportunity to summon the elevator when pressed, but not to the small child or wheelchair-bound adult who cannot reach it. Likewise, for the adult from a remote culture who has never experienced this technology. Allowing the child to stand on a stool, giving the person in the wheelchair a cane to extend their reach, or demonstrating the procedure to the foreign visitor can rapidly alter the action possibilities presented by that very same button.

In short, the action possibilities in an environment cannot be understood solely by focusing on the perceptual attributes of the environment, because *these opportunities are only defined relative to the actor.* This critical point was understood by Gibson and captured in his original notion of affordances. As discussed below, this concept inspired a substantial body of behavioral research into actors' perceptions of the opportunities for action provided by the environment. We contend that a cognitive neuroscience of action has much to gain by building on this construct in three ways: extending the role of the actor, allowing for representations to mediate affordances, and allowing for deep (inferred) properties of objects and task environments to serve as potential affordances.

Representation of affordances

Since Gibson first introduced the term, much has been written about the meaning(s) of "affordances" in the philosophical and psychological literatures (Chemero, 2003), and it has been become part of the jargon in robotics, human factors engineering, and psychology. Between, and even within, these disciplines, the term has been used inconsistently. Far from being merely an issue of semantics, this definition has practical consequences for how action scientists design experiments and interpret data.

The concept of affordance has begun to appear in the cognitive neuroscience literature as well, where it is often used synonymously with "action-relevant object property" (Barde, Buxbaum, & Moll, 2007; Borghi, Flumini, Natraj, & Wheaton, 2012). While Gibson did appear to use the term in several ways over the years, this differs fundamentally from his basic notion. For Gibson, affordances are *objectively measurable, latent action possibilities* in the environment's objects and surfaces defined *in relation to the perceiver's physical capabilities,* but *not dependent on the perceiver's awareness* (Gibson, 1979). Properties of the world exist independent of actors, while affordances exist only if there is an actor capable of exploiting them.

Gibson's definition of affordances has a number of significant implications for cognitive neuroscientists who are interested in understanding mechanisms of goal-directed action, and we advocate it as a powerful starting position. He viewed perception as a *reciprocal* process, with actors who engage in gathering information from the environment through active use of their senses, and whose possibilities for action are in turn constrained by that environment. He explicitly recognized that organisms are equipped with a variety of what we might best refer to as *sensory effectors.* They move parts of the body that are considered effectors and/or sensors (e.g., eyes, ears, tongue, hands) to investigate the environment. These facts effectively blur any line that might be drawn between what is motor and what is sensory.

If one accepts this assertion, then there are important implications for the way that we design and interpret our experiments. Perhaps the most obvious is that how actors are allowed to respond in our research paradigms matters; relying too heavily on button presses, for example, may limit the relevance of our results to other actions. Because affordances are seen as relative to the actor, the approach also forces scientists to confront the possibility of interspecies differences. The same environment holds different affordances depending on a species' capabilities for action, which are products of evolutionary history. This point is important to consider when generalizing results from animal models to humans.

To remain accurate, perceptions of affordances must adapt as the capabilities of an actor change. Evidence suggests that this is the case even in response to acute manipulations, such as artificially increasing leg length by strapping blocks to the feet (Mark, 1987), extending reach with a tool (Witt, Proffitt, & Epstein, 2005), or altering the range of motion and dynamics of manual grasping thorough use of a handheld device (Jacobs, Danielmeier, & Frey, 2010; Martin, Jacobs, & Frey, 2011). In an upcoming section, we argue that the same should also be true for more gradual changes brought about by growth and development, senescence, and skill acquisition, as well as the onset of injury or disease.

A final implication is that the physical properties of the effectors themselves constrain the affordances in an environment. Of relevance here is recent work showing that problems of action specification are simplified by the physical constraints of the effectors (Overduin, d'Avella, Carmena, & Bizzi, 2012; Valero-Cuevas, Yi, Brown, McNamara, Paul, & Lipson, 2007).

Reciprocal Affordance: Empirical Studies For Gibson, the environment was rich with information and possessed invariant properties that could be used to solve the relationship between distal and proximal stimuli *directly,* without the need for elaborate constructive processes, including internal representations. The first question to answer was, what information is available to the organism in an environment? There have been important successes in determining invariant sources of information (Cutting, Proffitt, & Kozlowski, 1978; Cutting, Springer, Braren, & Johnson, 1992; Turvey, Solomon, & Burton, 1989; Warren, 1984), and investigations of affordance perception continue to be fruitful in the experimental psychology literature (Adolph, Eppler, & Gibson, 1993; Gibson, Owsley, Walker, & Megaw-Nyce, 1979). Without moving a muscle, healthy adults are able to perceive accurately whether a stool is low enough to sit on, a stair is too high to step (Mark, 1987; Warren, 1984), an object is within reach (Fischer, 2000; Rochat & Wraga, 1997), or an opening is of sufficient size to accommodate their hand (Ishak, Adolph, & Lin, 2008). Nevertheless, the direct perception approach in pure form has proven limited. Due largely to its rejection of the need for internal representation, it fell out of the mainstream during the "cognitive revolution," which laid the foundation upon which cognitive neuroscience was erected. Unfortunately, this dismissal needlessly sacrificed a number of valuable insights. As recognized by Neisser and others, many strengths of the Gibsonian approach are compatible with those of information processing, and this amalgamation has led to notable successes in the domain of cognition (Neisser, 1976). We advocate for the utility of a similar hybrid approach to action.

We use the terminology "reciprocal affordance" to underscore the interactive relationship between the dynamic actor and the world in defining opportunities for action. We place a very strong emphasis on the fact that the actor is an active participant in defining these opportunities, which are represented in the brain. The actor supplies novel opportunities for action through their goals, motives, and active manipulation of the environment (including the creation of new objects and technologies), and by virtue of their own capacity to change due to maturation and experience.

A growing body of behavioral evidence is consistent with the hypothesis that actors represent reciprocal affordances, and a useful analogy to consider is the relationship of locks (the environment) and keys (the actor). The actor, interacting with objects and surfaces, has to accommodate the environmental and task requirements, just as a key must fit the lock. However, in this analogy, both the key and the lock are changeable; that is, what is possible with an object can change, and what an actor can do is not fixed.

The state of the key can change the apparent lock Across many natural circumstances, both the state and the behavioral capacity of the actor can have powerful influences on how affordances are perceived. For example, in sports, balls that are easier to stop look like they are moving more slowly than balls that are more difficult to stop (Witt & Sugovic, 2012). The apparent size of an object is also positively related to the grasping ability of the observer (Linkenauger, Witt, & Proffitt, 2011) and using an object will increase perception of related objects (Witt & Brockmole, 2012). As demonstrated in infant locomotion many years earlier (Adolph et al., 1993), physical motion itself alters perception, while in a virtual world, a person's judgment about whether she can squeeze through a closing gap will change if she is walking, independent of perceptual information induced by visual flow (Fajen & Matthis, 2011). The use of a tool to elongate a limb can also alter fundamental perceptual properties, such as parallax and object shape (Witt, 2011). The neural mechanisms underlying these effects remain to be determined. It would seem, however, that the motor command and efference copy, which are used to estimate the physical position of the actor in space in real time (Desmurget & Grafton, 2000; Mulliken, Musallam, & Andersen, 2008), must influence the structure of sensory feedback used in this state estimation.

The key interacts with unseen functional attributes of a lock The actor's functional knowledge plays a role in defining affordances. For instance, the way that actors choose to grasp a familiar tool is affected by their knowledge of the tools' learned functions rather than just the tool's perceptual attributes (orientation, size, form). This functional knowledge is reflected in hand shaping, which is far more elegant and supple than a simple opposition space formed by the thumb and index finger. There is enormous flexibility in how the grasp aperture is actually formed from all of the fingers to match the requirements of a given affordance (Mon-Williams & Bingham, 2011). The kinematics of hand shaping are closely tied to the functional attributes of the object, not just the superficial object properties (Hughes et al., 2012). Tools requiring a specialized grip, such as scissors, are a good example.

Evidence for functional knowledge can also be obtained from analyses of reaction-time data reflecting grasp planning prior to movement onset. For example, the fact that planning times are strongly influenced by the compatibility between the orientation of the object's functionally relevant axis (e.g., handle) and the hand indicates that the grasp is planned by retrieving stored representations of functionally relevant information (Tucker & Ellis, 1998). In interpreting reaction-time studies, it is noteworthy that hand shape for a given object can be generated entirely from prior knowledge, without any online or haptic feedback (Bingham, Coats, & Mon-Williams, 2007). Many recent studies have used priming to establish the specificity of grasp planning for different functional attributes. In a clever study, regular and reverse pliers were used in a paradigm that switched between different tools and task goals from trial to trial. Reaction times demonstrated priming effects related to the action (clamp/release) independent of the specific tool (i.e., hand movement) used to generate that property (Beisert, Massen, & Prinz, 2010). In another study, actors were asked to grasp a beer mug on its side and reorient it vertically using a biomechanically comfortable action. The starting position of the mug was manipulated, resulting in an upright or inverted mug that could hold beer or not (Masson, Bub, & Breuer, 2011). Priming effects only occur if the starting position and handle are in an orientation that results in the more functionally useful end-state: an upright mug. Both of these studies show dramatically that action selection is not based solely on the perceptual attributes of the object, but includes representations of the object's functions, as well as the anticipated demands associated with completing the actors' goal (Rosenbaum, Halloran, & Cohen, 2006).

Experience reshapes the key, leading to new locks Reciprocal affordance allows for increases or decreases in experience to change the actor, resulting in new affordances. This has been observed in many cross-sectional studies of expert athletes. For example, successful archers perceive the target as larger than nonexperts (Lee, Lee, Carello, & Turvey, 2012). Expert rock climbers perceive climbing handholds differently than novices (Handy, Tipper, Borg, Grafton, & Gazzaniga, 2006) and can visualize more routes up a climbing wall than nonexperts (Pezzulo, Barca, Bocconi, & Borghi, 2010). More direct evidence can be found in longitudinal training studies that avoid the pitfalls of cross-sectional comparisons. Once people learn to tie a specific set of knots, their capacity to visually discriminate knots (both familiar and new) is enhanced (Cross et al., 2012). There also appears to be a social aspect to this kind of change. Having learned how to use a tool will strongly bias expectations of how other people will use the same tool (Jacquet, Chambon, Borghi, & Tessari, 2012). These experience-dependent behavioral effects are also reflected in brain activity. Once knots are physically learned, there is greater activity in grasp-related areas of the parietal cortex (Cross et al., 2012). In addition, activity in parietal cortex shows greater specificity for familiar versus learned tools (Valyear, Gallivan, McLean, & Culham, 2012). An important implication of these studies is the localization of reciprocal affordance for objects to areas at the end of the dorsal visual stream, where haptics, motor commands, and vision merge to plan grasping actions.

Experience with technology can change or add new keys to the keychain The relationship between tool use and reciprocal affordance is complex, because the tool both changes and augments the actor. A dramatic example of this was described in tetraplegic patients who learn to use wheelchairs for mobility. Learning this new tool leads to renewed capacity as well as a shift in reciprocal affordance. They are better at perceiving if they can fit through barriers with the wheelchair than naive ambulatory subjects sitting in a wheelchair (Higuchi, Hatano, Soma, & Imanaka, 2009).

The tool can change the actor by creating a new spatial geometry of interactions. When a stick is used to extend reach, what is perceived as being reachable is fundamentally altered (Witt, Proffitt, & Epstein, 2005). One framework to explain this is the "distalization of the end-effector," where the internal body schema is elongated with the tool (Arbib, Bonaiuto, Jacobs, & Frey, 2009; Cardinali et al., 2009). In addition, peripersonal space is defined by the body schema, which in turn is modified with tools (Brozzoli, Pavani, Urquizar, Cardinali, & Farnè, 2009).

Tools can do much more than extend reach; for example, a screw only provides an affordance if one has the appropriate driver. In addition to providing unusually shaped effectors (screw drivers, sewing needles) that provide access to otherwise inaccessible affordances, tools by virtue of their unique mechanical properties can dramatically alter the range of motion for grasping movements (Jacobs, Danielmeier, & Frey, 2010; Martin, Jacobs, & Frey, 2011). When people learn to use a new grasping device, there is striking overlap in brain areas involved in natural hand grasping and novel tool grasping, suggesting that intrinsic mechanisms for planning natural hand grasps are readily adapted for planning new types of grasps.

Old or broken keys may no longer release the same locks When an individual's capacity for action is diminished due to senescence, injury, or disease, there could be a change in reciprocal affordances. An important question is whether changes of physical capacity are reflected in patients' perceptions of action possibilities. In normal aging, some, but not all, aspects of reciprocal affordance adapt to the declining physical strength. As we grow old, our estimation of the climbability of a flight of stairs is appropriately reduced (Konczak, Meeuwsen, & Cress, 1992). However, in other situations, older adults may fail to update adequately their representations of reciprocal affordances to accommodate their diminishing physical capacity. This can lead them to overestimate their ability to stand on an incline or to clear an obstacle (Lafargue, Noel, & Luyat, 2013). Relatedly, the question of how brain or bodily injury might influence perception of affordances remains largely unexplored, but there are some intriguing leads.

Although physically impossible, many patients with dense acute (Johnson, 2000a) or chronic (Johnson, Sprehn, & Saykin, 2002) post-stroke hemiparesis retain the ability to select the least awkward option for grasping an object; that is, to evaluate reciprocal affordances (Johnson, 1998; Johnson, 2000b). Preservation of this ability in hemiparetics might be attributed to continued visual or somatosensory feedback, or both, from the affected limb were it not for the fact that chronic upper-limb amputation also demonstrates preserved grip-selection judgments (Philip & Frey, 2011). Why are these populations not updating their representations of reciprocal affordances in light of their disabilities? A plausible interpretation is that once formed, the integrity of reciprocal-affordance representations does not require continued motor output or sensory feedback to be maintained, but that such activity may be required for updating (Jenkinson, Edelstyn, & Ellis, 2009). If so, then this could have important implications for rehabilitation and assistive technologies. Similar to the overestimation of affordances by the elderly noted earlier (Lafargue, Noel, & Luyat, 2013), more work is needed to address whether this might contribute to incidents such as increased risk of falling in certain patient populations.

In contrast to patients with unilateral hemiparesis or amputation, those with classic ideomotor apraxia do show deficits in representing reciprocal affordances, as tested with a version of the grip-selection task (Buxbaum, Johnson-Frey, & Bartlett-Williams, 2005). Lesions in this case are typically localized to left inferior parietal or posterior inferior prefrontal cortex in a network that is widely acknowledged to play a central role in action cognition, particularly in relationship to the organization of complex goal-oriented behavior, object manipulation, and tool use (Haaland, Harrington, & Knight, 2000; Johnson-Frey, Newman-Norlund, & Grafton, 2005). Note that these regions overlap to some extent with those implicated by functional MRI as being involved in representing reciprocal affordances in the grip-selection task; specifically, the anterior intraparietal sulcus and left ventral premotor cortex (Johnson et al., 2002; Marangon, Jacobs, & Frey, 2011), areas commonly implicated in visually guided grasping (Grafton, 2010). Activity in these areas increases if the subjects must choose the most natural grip for rotating the handle to place a cued end downward, imposing the additional complexity of the calculation of end-state comfort. This suggests that even when subjects must plan multiple steps in advance to achieve end-state comfort, a single neural system is utilized in computing the affordance (Marangon, Jacobs, & Frey, 2011).

Another approach for investigating reciprocal affordance in patients is to determine if their reaching to objects is influenced by the compatibility or incompatibility between object orientation and the hand (e.g., a pot with a handle pointing toward or away from the responding hand; Tucker & Ellis, 1998). Assessing this in a patient is complicated by the potential co-occurrence of unilateral spatial neglect that could lead to a rightward bias irrespective of the functional relevance of the oriented object. To sort this out, the relationship between spatial compatibility of the object location and the responding hand (the Simon effect) and, separately, the orientation of the functional object (the affordance effect), must be disentangled. In healthy subjects, spatial incompatibility between functionally relevant object orientations and the response hand are reflected in prolongation of reaction times, when subjects are selecting or planning the response or retrieving semantically related functional information (Iani, Baroni, Pellicano, & Nicoletti, 2011). Most evidence in healthy subjects shows that spatial compatibility effects and functional affordance effects operate independently, although there are exceptions (Iani et al., 2011). If the Simon effect is considered as a proxy for the spatial attention processes that are impacted in neglect, then results from stroke patients support the argument that these two effects represent fundamentally independent planning operations. In patients with anosagnosia and profound neglect, there is striking preservation of the affordance effect, suggesting that functional knowledge can overcome the neglect that would otherwise cause the patient to ignore the object (Humphreys & Riddoch, 2001; Humphreys, Wulff, Yoon, & Riddoch, 2010). The critical point from these patient studies is the demonstration that reciprocal affordance is a representational process that is relatively independent of spatial attention.

Conclusion

The characteristic that distinguishes action from movement is its goal-directedness. As a result, action scientists face the sizeable challenge of understanding how our actions are determined by the attributes of the objects and surfaces in the environment and by our unique goals and behavioral capacities. The key lies in developing an appreciation of the complex dynamic that exists between actor and environment in defining possibilities for action. We offer the concept of reciprocal affordance as an important step in this direction. By virtue of its explicit acknowledgment that actors shape and are shaped by their environments, it offers a way to integrate a diversity of factors that influence our actions, including development, disability, and the use of technology. The reciprocal-affordance approach will become increasingly important as goal specification is translated to the clinic. The rapid evolution of brain computer systems introduces fundamental challenges in matching the action goals of the actor to the requirements of nonhuman actuators. Any robotic solution will require a dynamic estimation of the state of the actor as a goal-directed behavior unfolds. In the area of stroke recovery, increasing emphasis is being given to functional, rather than motor recovery.

ACKNOWLEDGMENTS Supported by grants to S.T.G. from the PHS (NS44393), the U.S. Army Research Office (contract W911NF-09-0001), and the James S. McDonnell Foundation; and by grants to S.H.F from the PHS grant (NS083377-01), USAMRAA (W81XWH-10-1-1020), and James S. McDonnell Foundation (220020190).

REFERENCES

Adolph, K. E., Eppler, M. A., & Gibson, E. J. (1993). Crawling versus walking infants' perception of affordances for locomotion over sloping surfaces. *Child Dev, 64,* 1158–1174.

Arbib, M. A., Bonaiuto, J. B., Jacobs, S., & Frey, S. H. (2009). Tool use and the distalization of the end-effector. *Psychol Res, 73,* 441–462.

Barde, L. H., Buxbaum, L. J., & Moll, A. D. (2007). Abnormal reliance on object structure in apraxics' learning of novel object-related actions. *J Int Neuropsychol Soc, 13,* 997–1008.

Beisert, M., Massen, C., & Prinz, W. (2010). Embodied rules in tool use: A tool-switching study. *J Exp Psychol Hum Percept Perform, 36,* 359–372.

Bingham, G., Coats, R., & Mon-Williams, M. (2007). Natural prehension in trials without haptic feedback but only when calibration is allowed. *Neuropsychologia, 45,* 288–294.

Borghi, A. M., Flumini, A., Natraj, N., & Wheaton, L. A. (2012). One hand, two objects: Emergence of affordance in contexts. *Brain Cogn, 80,* 64–73.

Brozzoli, C., Pavani, F., Urquizar, C., Cardinali, L., & Farnè, A. (2009). Grasping actions remap peripersonal space. *NeuroReport, 20*(10), 913–917.

Buxbaum, L. J., Johnson-Frey, S. H., & Bartlett-Williams, M. (2005). Deficient internal models for planning hand-object interactions in apraxia. *Neuropsychologia, 43,* 917–929.

Cardinali, L., Frassinetti, F., Brozzoli, C., Urquizar, C., Roy, A. C., & Farnè, A. (2009). Tool-use induces morphological updating of the body schema. *Curr Biol, 19*(12), R478–479.

Chemero, A. (2003). An outline of a theory of affordances. *Ecol Psychol, 15,* 181–195.

Cisek, P. (2007). Cortical mechanisms of action selection: The affordance competition hypothesis. *Philos Trans R Soc Lond B Biol Sci, 362,* 1585–1599.

Cross, E. S., Cohen, N. R., Hamilton, A. F. D. C., Ramsey, R., Wolford, G., & Grafton, S. T. (2012). Physical experience leads to enhanced object perception in parietal cortex: Insights from knot tying. *Neuropsychologia, 50,* 3207–3217.

Cutting, J. E., Proffitt, D. R., & Kozlowski, L. T. (1978). A biomechanical invariant for gait perception. *J Exp Psychol Hum Percept Perform, 4,* 357–372.

Cutting, J. E., Springer, K., Braren, P. A., & Johnson, S. H. (1992). Wayfinding on foot from information in retinal, not optical, flow. *J Exp Psychol Gen, 121,* 41–72.

Desmurget, M., & Grafton, S. (2000). Forward modeling allows feedback control for fast reaching movements. *Trends Cogn Sci, 4,* 423–431.

Fajen, B. R., & Matthis, J. S. (2011). Direct perception of action-scaled affordances: The shrinking gap problem. *J Exp Psychol Hum Percept Perform, 37,* 1442–1457.

Fischer, M. H. (2000). Estimating reachability: Whole-body engagement or postural stability? *Hum Mov Sci, 19,* 297–318.

Gibson, E. J., Owsley, C. J., Walker, A., & Megaw-Nyce, J. (1979). Development of the perception of invariants: Substance and shape. *Perception, 8,* 609–619.

Gibson, J. J. (1979). *The ecological approach to visual perception.* Hillsdale, NJ: Erlbaum.

Goodale, M. A., Milner, A. D., Jakobson, L. S., & Carey, D. P. (1991). A neurological dissociation between perceiving objects and grasping them. *Nature, 349,* 154–156.

Grafton, S. T. (2010). The cognitive neuroscience of prehension: Recent developments. *Exp Brain Res, 204,* 475–491.

Haaland, K. Y., Harrington, D. L., & Knight, R. T. (2000). Neural representations of skilled movement. *Brain, 123*(Pt. 11), 2306–2313.

Handy, T., Tipper, C. M., Borg, J. S., Grafton, S., & Gazzaniga, M. (2006). Motor experience with graspable objects reduces their implicit analysis in visual- and motor-related cortex. *Brain Res, 1097,* 156–166.

Higuchi, T., Hatano, N., Soma, K., & Imanaka, K. (2009). Perception of spatial requirements for wheelchair locomotion in experienced users with tetraplegia. *J Physiol Anthropol, 28,* 15–21.

Hommel, B., Musseler, J., Aschersleben, G., & Prinz, W. (2001). The theory of event coding (TEC): A framework for perception and action planning. *Behav Brain Sci, 24*(5), 849–878; discussion 878–937.

Hughes, C. M. L., Seegelke, C., Spiegel, M. A., Oehmichen, C., Hammes, J., & Schack, T. (2012). Corrections in grasp posture in response to modifications of action goals. *PLoS ONE, 7,* e43015.

Humphreys, G. W., & Riddoch, M. J. (2001). Detection by action: Neuropsychological evidence for action-defined templates in search. *Nat Neurosci, 4,* 84–88.

Humphreys, G. W., Wulff, M., Yoon, E. Y., & Riddoch, M. J. (2010). Neuropsychological evidence for visual- and motor-based affordance: Effects of reference frame and object-hand congruence. *J Exp Psychol Learn Mem Cogn, 36,* 659–670.

Iani, C., Baroni, G., Pellicano, A., & Nicoletti, R. (2011). On the relationship between affordance and Simon effects: Are the effects really independent? *J Cogn Psychol, 23,* 121–131.

Ishak, S., Adolph, K. E., & Lin, G. C. (2008). Perceiving affordances for fitting through apertures. *J Exp Psychol Hum Percept Perform, 34,* 1501–1514.

Jacobs, S., Danielmeier, C., & Frey, S. H. (2010). Human anterior intraparietal and ventral premotor cortices support representations of grasping with the hand or a novel tool. *J Cogn Neurosci, 22,* 2594–2608.

Jacquet, P. O., Chambon, V., Borghi, A. M., & Tessari, A. (2012). Object affordances tune observers' prior expectations about tool-use behaviors. *PLoS ONE, 7,* e39629.

Jenkinson, P. M., Edelstyn, N. M., & Ellis, S. J. (2009). Imagining the impossible: Motor representations in anosognosia for hemiplegia. *Neuropsychologia, 47*, 481–488.

Johnson, S. H. (1998). Cerebral organization of motor imagery: Contralateral control of grip selection in mentally represented prehension. *Psychol Sci, 9*, 219–222.

Johnson, S. H. (2000a). Imagining the impossible: Intact motor representations in hemiplegics. *Neuroreport, 11*, 729–732.

Johnson, S. H. (2000b). Thinking ahead: The case for motor imagery in prospective judgments of prehension. *Cognition, 74*, 33–70.

Johnson, S. H. (2001). Seeing two sides at once: Effects of viewpoint and object structure on recognizing three-dimensional objects. *J Exp Psychol Human, 27*, 1468–1484.

Johnson, S. H., & Grafton, S. T. (2003). From "acting on" to "acting with": The functional anatomy of object-oriented action schemata. *Prog Brain Res, 142*, 127–139.

Johnson, S. H., Sprehn, G., & Saykin, A. J. (2002). Intact motor imagery in chronic upper limb hemiplegics: Evidence for activity-independent action representations. *J Cogn Neurosci, 14*, 841–852.

Johnson, S., Rotte, M., Grafton, S., Hinrichs, H., Gazzaniga, M., & Heinze, H. (2002). Selective activation of a parietofrontal circuit during implicitly imagined prehension. *NeuroImage, 17*, 1693–1704.

Johnson-Frey, S. H., Newman-Norlund, R., & Grafton, S. T. (2005). A distributed left hemisphere network active during planning of everyday tool use skills. *Cereb Cortex, 15*, 681–695.

Konczak, J., Meeuwsen, H. J., & Cress, M. E. (1992). Changing affordances in stair climbing: The perception of maximum climbability in young and older adults. *J Exp Psychol Hum Percept Perform, 18*, 691–697.

Lafargue, G., Noel, M., & Luyat, M. (2013). In the elderly, failure to update internal models leads to over-optimistic predictions about upcoming actions. *PLoS One, 8*, e51218.

Lee, Y., Lee, S., Carello, C., & Turvey, M. T. (2012). An archer's perceived form scales the "hitableness" of archery targets. *J Exp Psychol Hum Percept Perform, 38*, 1125–1131.

Linkenauger, S. A., Witt, J. K., & Proffitt, D. R. (2011). Taking a hands-on approach: Apparent grasping ability scales the perception of object size. *J Exp Psychol Hum Percept Perform, 37*, 1432–1441.

Marangon, M., Jacobs, S., & Frey, S. H. (2011). Evidence for context sensitivity of grasp representations in human parietal and premotor cortices. *J Neurophysiol, 105*, 2536–2546.

Mark, L. S. (1987). Eyeheight-scaled information about affordances: A study of sitting and stair climbing. *J Exp Psychol Hum Percept Perform, 13*, 361–370.

Martin, K., Jacobs, S., & Frey, S. H. (2011). Handedness-dependent and -independent cerebral asymmetries in the anterior intraparietal sulcus and ventral premotor cortex during grasp planning. *NeuroImage, 57*, 502–512.

Masson, M. E. J., Bub, D. N., & Breuer, A. T. (2011). Priming of reach and grasp actions by handled objects. *J Exp Psychol Hum Percept Perform, 37*, 1470–1484.

Mon-Williams, M., & Bingham, G. P. (2011). Discovering affordances that determine the spatial structure of reach-to-grasp movements. *Exp Brain Res, 211*, 145–160.

Mulliken, G. H., Musallam, S., & Andersen, R. A. (2008). Forward estimation of movement state in posterior parietal cortex. *Proc Natl Acad Sci USA, 105*, 8170–8177.

Neisser, U. (1976). *Cognition and reality*. New York, NY: W. H. Freeman.

Overduin, S. A., d'Avella, A., Carmena, J. M., & Bizzi, E. (2012). Microstimulation activates a handful of muscle synergies. *Neuron, 76*, 1071–1077.

Pezzulo, G., Barca, L., Bocconi, A. L., & Borghi, A. M. (2010). When affordances climb into your mind: Advantages of motor simulation in a memory task performed by novice and expert rock climbers. *Brain Cogn, 73*, 68–73.

Philip, B. A., & Frey, S. H. (2011). Preserved grip selection planning in chronic unilateral upper extremity amputees. *Exp Brain Res, 214*, 437–452.

Rochat, P., & Wraga, M. (1997). An account of the systematic error in judging what is reachable. *J Exp Psychol Hum Percept Perform, 23*, 199–212.

Rock, L., di Vita, J., Barbeito, R. (1981). The effect on form perception of change of orientation in the third dimension. *J Exp Psychol Hum Percept Perform, 7*, 719–732.

Rosenbaum, D. A., Halloran, E. S., & Cohen, R. G. (2006). Grasping movement plans. *Psychon Bull Rev, 13*, 918–922.

Tucker, M., & Ellis, R. (1998). On the relations between seen objects and components of potential actions. *J Exp Psychol Hum Percept Perform, 24*, 830–846.

Turvey, M. T., Solomon, H. Y., & Burton, G. (1989). An ecological analysis of knowing by wielding. *J Exp Anal Behav, 52*, 387–407.

Valero-Cuevas, F., Yi, J., Brown, D., McNamara, R., Paul, C., & Lipson, H. (2007). The tendon network of the fingers performs anatomical computation at a macroscopic scale. *IEEE Trans Biomed Eng, 54*, 1161–1166.

Valyear, K. F., Gallivan, J. P., Mclean, D. A., & Culham, J. C. (2012). MRI repetition suppression for familiar but not arbitrary actions with tools. *J Neurosci, 32*, 4247–4259.

Warren, W. H., Jr. (1984). Perceiving affordances: Visual guidance of stair climbing. *J Exp Psychol Hum Percept Perform, 10*, 683–703.

Witt, J. K. (2011). Tool use influences perceived shape and perceived parallelism, which serve as indirect measures of perceived distance. *J Exp Psychol Hum Percept Perform, 37*, 1148–1156.

Witt, J. K., & Brockmole, J. R. (2012). Action alters object identification: Wielding a gun increases the bias to see guns. *J Exp Psychol Hum Percept Perform, 38*, 1159–1167.

Witt, J. K., & Sugovic, M. (2012). Does ease to block a ball affect perceived ball speed? Examination of alternative hypotheses. *J Exp Psychol Hum Percept Perform, 38*, 1202–1214.

Witt, J. K., Proffitt, D. R., & Epstein, W. (2005). Tool use affects perceived distance, but only when you intend to use it. *J Exp Psychol Hum Percept Perform, 31*, 880–888.

VI
MEMORY

Introduction

ANTHONY D. WAGNER

MEMORY INFLUENCES all aspects of cognitive function, as learning through past experience gives rise to predictions (memories) that shape current perception, thinking, and action. Approximately 20 years ago, Endel Tulving introduced this section of the first edition of *The Cognitive Neurosciences* by highlighting two themes central to memory theory and research. First, to understand memory, one must characterize the processes governing memory encoding, storage, and retrieval. Second, memory is a nonunitary entity, as it consists of multiple forms of knowledge expression that depend on distinct neural substrates and computations. In the intervening years, technological advances in the neurosciences have enabled increased anatomical and computational precision in addressing these central topics, while simultaneously sparking novel themes and challenges to some earlier ideas.

This section of this volume documents the field's progress toward increasingly precise models of memory, such as specifying the distinct computations and forms of remembering supported by regions in the medial temporal lobe, delineating the multiple mechanisms that underlie memory performance over short time scales, and illuminating how "context" is neurally represented and serves to influence memory behavior. This section also documents some of the newly emerging themes that have shaped the field in recent years, such as the integration of memory and decision theory, the unexpected role of parietal cortex in memory, and the fundamental link between memory and prediction. The remarkable strides summarized in these chapters evidence past progress and predict a future rich with continued discovery.

Theories of memory have long emphasized that memories operate at different time scales. A common distinction is between short-term forms of memory that enable representations to persist across brief temporal gaps, and long-term forms of memory that enable knowledge acquired even in the distant past to influence current cognition. Canonically, performance after brief delays is thought to depend on short-term (or working) memory processes, wherein sustained attention to internal representations of just-encountered stimuli keeps the representations active in mind, allowing knowledge from the recent past to shape ongoing thinking and action. Recently, this view has been challenged by findings suggesting that performance after brief delays can be supported by multiple mechanisms, some that may correspond to short-term memory processes and others to long-term memory processes. In the first chapter in this section, Ranganath, Hasselmo, and Stern discuss candidate neural mechanisms, brain systems, and cognitive processes that can support performance on short time scales (on the order of tens of seconds). They review how performance after brief delays may depend on active maintenance of neural activity, as well as on short-term changes in synaptic efficacy or neural excitability. At the neural systems level, they consider how active maintenance may emerge from interactions between prefrontal and posterior cortical areas, and they discuss the controversial idea that the medial temporal lobe may also contribute to performance at short delays partly through maintenance processes. Ranganath and colleagues conclude by considering a number of influential cognitive theories of short-term memory.

Beginning with the observation of Henry Molaison's (H.M.) amnesia following surgical resection of his medial temporal lobes some 60 years ago, extensive research has established the critical role of these structures in long-term (declarative) memory for events (episodes) and facts (semantic knowledge). Models of medial temporal lobe function have become increasingly precise in recent years, with considerable work focused on understanding the forms of memory and underlying computations supported by distinct subregions of the medial temporal lobe (which consists of the hippocampal formation and surrounding perirhinal, parahippocampal, and entorhinal cortical areas). Davachi and Preston review the field's progress in specifying the neural processes that encode and reactivate (retrieve) episodic memories, including evidence that the medial temporal cortex (specifically, perirhinal and parahippocampal cortex) and hippocampus subserve distinct computations. In so doing, they illustrate how the advent of event-related functional MRI (fMRI) and multivariate data analyses have enabled quantitative measurement of the neural responses at encoding and retrieval that relate to memory performance, and they discuss extant data supporting the view that event recollection partially depends on hippocampally mediated reinstatement of the distributed patterns of cortical activation that were present at the moment of memory encoding. Davachi and Preston further consider how measures of neural reinstatement are being leveraged to understand the processes that support memory-based inference, generalization, and the integration of knowledge across distinct life events. Leutgeb and Leutgeb review parallel efforts to specify how separable medial temporal cortical processing streams and distinct intra-hippocampal computations support memory. Drawing on an increasingly rich literature in rodents, and to a lesser extent primates, they consider exciting developments in understanding entorhinal cortical function, including the discovery and elucidation of the functional role of medial entorhinal grid cells, and they discuss evidence suggesting that the subregions of the hippocampal formation—dentate gyrus, cornu ammonis (CA) subfields of hippocampus, and subiculum—perform specialized computations during the encoding, storage, and retrieval of memories. The complementary discoveries highlighted in these two chapters illustrate the bidirectional influences of rodent and human research on medial temporal lobe function.

A critical tenet in cognitive models of episodic memory is that event memories entail the binding of event features with representations of the "context" in which the event unfolds. The explanatory power of context for understanding memory behavior was insightfully demonstrated in seminal work by William Estes in the 1950s, work that has inspired the inclusion of context representations in subsequent computational theories of memory. While a dominant theoretical construct, precise specification of what constitutes context and how context is neurally represented have remained elusive. Manning, Kahana, and Norman consider how contextual effects in memory emerge from the influence of representations that persist and slowly drift with time. From their view, context does not consist of a particular type of event content (e.g., representations of the spatial environment in which an event has occurred), but rather consists of information that emerges from our experiences and that persists with different time scales across time. Manning and colleagues review how context can explain important aspects of memory behavior, and they discuss factors that drive contextual drift and updating. At the neural level, they consider possible mechanisms that may give rise to context and review recent neurophysiological and functional imaging evidence for the role of prefrontal cortex and medial temporal lobe structures in representing context.

Given the established dependence of episodic memory on medial temporal lobe binding and reactivation computations, one of the more surprising developments in recent years is the observation that fMRI activity in multiple regions of lateral parietal cortex consistently varies with episodic retrieval behavior. It is perhaps unsurprising that the medial temporal lobe is not the only region to exhibit activity that covaries with retrieval processing, given that binding and reactivation, in isolation, are insufficient to support memory-guided behavior. Indeed, to remember and to act on the products of remembering requires multiple additional processes, from attending to the cues that trigger memory retrieval, to maintaining reactivated information that may serve as the basis for memory-guided decisions, to integrating reactivated evidence with decision criteria, and ultimately to selecting an appropriate action. Two chapters in this section consider the role of parietal cortex in episodic retrieval through the broader lens of remembering as decision making. Uncapher, Gordon, and Wagner introduce a number of prominent hypotheses recently advanced to account for lateral parietal contributions to episodic retrieval and briefly review the literature in relation to these hypotheses. These authors argue that any model of lateral parietal contributions to retrieval must take into account the rich, fine-grained functional heterogeneity evident in lateral parietal cortex, and they offer a working model of lateral parietal function that posits that multiple parietal processes—including processes of attention, multifeatural binding and maintenance, decision making, and action intention—as well as interactions between these processes ultimately support memory-guided behavior. Miller and Dobbins similarly argue that memory inherently entails decision making, and as such many of the neural correlates of episodic retrieval evidenced by functional imaging likely reflect various decision and control processes, rather than memory reactivation processes per se. These authors review an emerging literature that suggests that lateral prefrontal and parietal cortex may contribute to memory-guided behavior, in part through control processes that tune and bias memory judgments. Miller and Dobbins posit a biasing and orienting framework wherein frontoparietal processes bias memory-guided judgments for more efficient processing when biases are confirmed and, alternatively, support attentional orienting and more thorough exploration of memory contents when biases are disconfirmed. Both chapters illustrate how efforts to understand the cognitive neuroscience of memory are increasingly considering memory as decision making, wherein perception ultimately guides action through decision processes that weigh retrieved mnemonic evidence in relation to decision criteria.

From the multiple memory systems perspective, different forms of memory are supported by distinct computations and neural substrates. One central distinction is between declarative memory, which consists of episodic and semantic memory, and procedural memory, which underlies habits and skilled behavior. Further illustrating the intersection between theorizing about memory and decision making, Shohamy and Daw consider the evidence for these distinct memory systems in parallel with the literature on decision making and learning from reward. These authors highlight two exciting theoretical developments. First, they discuss how decision research on reward learning has specified a neural mechanism for learning about actions, and they consider how this mechanism may partly underlie the habits and skills (procedural memory) that have been the focus of much of memory research. Second, they discuss how important aspects of reward-mediated decision behavior are not well accounted for by this "habit-like" mechanism, sparking interest in understanding the mechanisms governing goal representation and goal-directed action learning. Highlighting the reciprocal influences between decision and memory research, Shohamy and Daw consider how declarative memory mechanisms may support goal-directed action learning. The explanatory power of viewing memory and decision behavior as linked promises continued integration of these fields, and the emergence of models in which core mechanisms of memory and decision making are isomorphic.

Memory as decision making directly stems from the demands of life, in which people and other organisms constantly face the challenge of representing the current state of the world and determining how to act to achieve goals and optimize outcomes. Learning enables organisms to draw on predictive models of how past states of the world mapped to outcomes, such that memory-based predictions shape current interpretations of the world and favor selection of appropriate actions. While the conceptualization of memory as prediction has a long history, this view is increasingly influential in theorizing about memory. Not only is it central to theories of procedural memory (habits and skills), priming, and other nondeclarative forms of memory, the idea has also emerged as central to theories of episodic memory, wherein the recollection of details of a past event may be viewed as an episodic-based prediction about what the present may hold. Strikingly, a recent theme in memory research extends this idea to understand how episodic memory enables individuals to imagine possible experiences that may unfold well in the future. Mullally and Maguire discuss the relationship between prediction, imagination, and recollection, and review a wealth of evidence in support of this integrative perspective. In doing so, they consider leading hypotheses for how the medial temporal lobe supports memory, scene construction, and prediction, and they make the powerful point that our conscious realities can be temporally stretched by medial temporal lobe computations that not only project our thoughts (and brain states) back in time (recollecting events past) but also forward into imagined futures (constructing memory-based simulations).

In the fourth edition of *The Cognitive Neurosciences*, Daniel Schacter noted: "We cannot know with any certainty what path memory research will follow in the upcoming years, but we can be confident that it will be exciting to find out." The chapters in this section reveal some of the paths taken and the remarkable discoveries that followed. There is reason to remain optimistic about the exciting science that lies ahead.

44 Short-Term Memory: Neural Mechanisms, Brain Systems, and Cognitive Processes

CHARAN RANGANATH, MICHAEL E. HASSELMO, AND CHANTAL E. STERN

ABSTRACT The idea that memory across delays of a few seconds, or short-term memory (STM), can be distinguished from memory across longer delays, or long-term memory (LTM), has a long history but remains highly controversial. The conflict has come about, in part, because of the challenges in integrating findings from studies of STM at different levels of analysis, using different task paradigms, and in different species. A review of this literature reveals that there is no single neural mechanism, brain system, or cognitive process that supports performance on STM tasks. Furthermore, at the neural level, STM and LTM may be intricately linked.

"Memory is a convenient chapter heading designating certain kinds of problems that scientists study. Methods of science have been brought to bear on the problems of memory for over a hundred years, in many different organisms, and at many different levels of analysis, extending from molecular mechanisms to the phenomena of conscious awareness."
—Endel Tulving, introduction to "Memory" section of the first *Cognitive Neurosciences* (1995)

The study of short-term memory serves as a beautiful example of both the stunning progress in cognitive neuroscience over the past 50 years, and the number of open questions that remain. Researchers have made great strides by relating work from elegant behavioral paradigms in animal models, studies on humans with brain damage, and neuroimaging studies of activity in the healthy human brain. These findings have yielded a series of answers to very specific questions posed at different levels of analysis. Our current challenge is to bridge these different levels of analysis.

Here, we use the terms "short-term memory" (STM) and "long-term memory" (LTM) as operational terms to describe memory performance at various retention intervals, with the understanding that multiple processes and mechanisms (to be considered in this chapter) could differentially support performance on STM and LTM measures. We use STM to refer to time scales up to tens of seconds, and LTM to refer to memory spanning minutes or longer. It is important to note that different literatures use "short term" and "long term" to refer to different concepts and time scales. For example, in the synaptic plasticity literature, the terms short term and long term are used to refer to longer time scales.

It is probably fair to say that most cognitive neuroscientists have been exposed to the general idea that LTM (i.e., the ability to retain information across longer delays on the order of minutes or longer) is supported by lasting, experience-dependent changes in synaptic strength in the medial temporal lobe (MTL) memory system, and that this system is not necessary for retention across very short delays. Instead, the early view held by some was that STM (defined here, operationally, as the ability to retain information across delays of several seconds) was supported by the prefrontal cortex (PFC). At present, few researchers who study STM and LTM would endorse this view, as it presents an overly simplistic, and largely incorrect, mapping between neural mechanisms (persistent activity for STM vs. synaptic plasticity for LTM[1]), brain systems (PFC for STM vs. MTL for LTM), cognitive processes (active maintenance vs. episodic memory encoding and retrieval), and conscious experiences (the attention-demanding process of having something remain present in consciousness vs. the experience of becoming aware of information that is associated with a sense of a different time and place).

In this chapter, we will summarize available evidence regarding the neural mechanisms and brain systems that support STM. We start off by reviewing the basic

[1]Note that we are not using the term "working memory" (WM) in this chapter, though many researchers equate the term WM with STM. This is because we are describing memory mechanisms on short time scales independent of their use for cognitive processes. In contrast, we use WM in a manner that is consistent with its use by Baddeley and others—that is, to describe various cognitive processes that might support STM performance. The definitions of STM and WM are reviewed in more detail in Aben, Stapert, and Blokland (2012).

neural mechanisms that could conceivably support memory at short time scales. Next, we will summarize the available evidence regarding the involvement of areas across the brain in STM, based on evidence from single-unit recording, lesion, and neuroimaging studies. We will then review concepts from different models of STM and describe how these models might help to build links between the mechanisms described above.

Candidate neural mechanisms for short-term memory

There are a number of neural mechanisms that have been proposed to contribute to STM. We will first consider potential mechanisms for the active maintenance of neural activity involving either network level interactions or intrinsic cellular mechanisms. We will then consider other mechanisms involving short-term changes in synaptic efficacy or neuronal excitability.

Most theories of neural mechanisms for memory assume a distributed representation across neurons, in which the items and relationships held in memory are represented by the pattern of neural activity across a population of neurons, with some neurons being more active and other neurons being less active or silent. Hebb (1949) proposed the concept of the cell assembly, defined as a population of active neurons that maintain an active memory trace that remains for a period of time after some stimulus has been perceived. Hebb conceived of the cell assembly as arising from a reverberatory interaction in which activity spreads between neurons in the brain, and this basic idea describes many theories of active maintenance (Amit, 1988; Cohen & Grossberg, 1983; Hopfield, 1984; Tegner, Compte, & Wang, 2002; Wilson & Cowan, 1972). This neural mechanism for active maintenance involves feedback excitation, in which previous spiking activity induces further spiking activity that persists for an extended period. The potentially explosive growth of spiking activity must be countered by feedback inhibition, resulting in fixed point or oscillatory attractor dynamics (Amit & Treves, 1989; Hasselmo, Schnell, & Barkai, 1995; Wilson & Cowan, 1972). Models of spiking activity have commonly used excitatory recurrent connections to model the persistent activity of prefrontal neurons during the retention-delay period of STM tasks (Amit & Brunel, 1997; Zipser, Kehoe, Littlewort, & Fuster, 1993). Notably, this mechanism depends upon a preexisting structure of synaptic connectivity, established through previous learning or through neural connections established during development. Critically, such networks are not specialized for STM, in the sense that they can support transient, input-driven activation of a cell assembly (in the service of LTM) and persistent activation of the cell assembly over short durations in the service of STM tasks.

To the extent that a novel stimulus does not closely correspond to preexisting representations, active maintenance dependent on preexisting synaptic connectivity may not provide a mechanism for short-term retention (Hasselmo & Stern, 2006). As an alternative, active maintenance could be implemented through intrinsic cellular mechanisms within single neurons that do not require synaptic interactions. One theory of the maintenance of novel stimuli (Hasselmo & Stern, 2006) was initially motivated by intracellular single-cell recordings of potential memory mechanisms (Egorov, Hamam, Fransén, Hasselmo, & Alonso, 2002; Fransén, Tahvildari, Egorov, Alonso, & Hasselmo, 2006; Klink & Alonso, 1997; Shalinsky, Magistretti, Ma, & Alonso, 2002). Recordings in the laboratory of Angel Alonso demonstrated that single neurons in slice preparations of the entorhinal cortex, isolated from other neurons by pharmacological blockade of excitatory and inhibitory synaptic transmission, can maintain memory for prior input in the form of persistent spiking activity. This maintenance of persistent spiking depends on the application of acetylcholine or other drugs that activate muscarinic acetylcholine receptors (Egorov et al., 2002; Shalinsky et al., 2002) and is blocked by muscarinic antagonists such as atropine (Egorov et al., 2002; Klink & Alonso, 1997). In the absence of acetylcholine, a depolarizing intracellular current injection causes the neuron to fire a train of spikes during the depolarization, but when the current injection is stopped, the membrane potential of the cell falls back to resting potential and the firing stops. However, in the presence of acetylcholine, a current injection of only a few hundred milliseconds in layer II causes the membrane potential to remain depolarized after the current injection is stopped, and the cell continues to generate spikes for an extended period of many seconds or even minutes without current injection or synaptic input.

The phenomenon underlying self-sustained spiking is sometimes referred to as an afterdepolarization or plateau potential. This phenomenon continues even when both excitatory and inhibitory synaptic transmission are blocked, indicating its dependence on intrinsic mechanisms in a single neuron, rather than on excitatory recurrent connectivity between neurons (Klink & Alonso, 1997). The persistent spiking can return after brief periods of hyperpolarizing current injection, indicating resistance to transient distractors. In addition to the entorhinal cortex, this effect has also been described in other structures of the medial temporal lobe, including the presubiculum (Yoshida & Hasselmo, 2009), the perirhinal cortex (Navaroli, Zhao, Boguszewski, & Brown, 2012), and hippocampus (Knauer, Jochems, Valero-Aracama, & Yoshida, 2013). A similar plateau-potential phenomenon has also been described in layer V of prefrontal cortical slices (Haj-Dahmane & Andrade, 1998) and in the cingulate cortex (Zhang & Seguela, 2010).

In the entorhinal cortex, plateau potentials appear to arise from a calcium-sensitive nonspecific cation current (CAN current) activated by muscarinic acetylcholine receptors (Egorov et al., 2002; Fransén, Alonso, & Hasselmo, 2002; Fransén et al., 2006; Klink & Alonso, 1997) that strongly depolarizes neurons. In the presence of acetylcholine, the generation of spiking by intracellular current injection or synaptic stimulation causes the neuron to enter an internal regenerative cycle of sustained spiking. Each new spike activates voltage-sensitive calcium channels, and the new influx of calcium activates the CAN current (Fransén et al., 2002). This current causes additional depolarization, leading to another spike, which again activates voltage-sensitive calcium channels that further perpetuate the cycle. The cellular processes in the entorhinal and perirhinal cortex provide a mechanism of persistent spiking suitable for retaining representations of novel stimuli in an active state (Hasselmo & Stern, 2006), which could also underlie some of the data from functional MRI (fMRI) and unit-recording studies in parahippocampal structures described below.

The cellular mechanisms for generation of persistent activity could not only support the temporary retention of novel information, but also drive synaptic changes that can support retention over long delays (Hasselmo & Stern, 2006; Jensen, Idiart, & Lisman, 1996; Jensen & Lisman, 2005). Specifically, persistent spiking could generate strengthening of synaptic connections based on presynaptic and postsynaptic activity (Jansen et al., 1996; Jensen & Lisman, 2005), thereby forming patterns of synaptic connectivity that could allow a similar presynaptic cue to evoke the associated postsynaptic activity. The strengthening of long-term synaptic connections can mediate LTM performance by allowing later activity to cue the retrieval of an associated item (Hasselmo et al., 1995; Hasselmo & Wyble, 1997; Jensen et al., 1996; McNaughton & Morris, 1987; Treves & Rolls, 1992). Once retrieval of a long-term memory has been cued, the pattern of recurrent excitation could allow the retrieved memory to be held in active maintenance for an arbitrary period of time (Hasselmo & Wyble, 1997; Zilli & Hasselmo, 2008).

As an alternative to the persistent activity mechanisms described above, STM could also be supported by transient changes in synaptic efficacy that decay over a period of seconds. This form of synaptic change is called "short-term potentiation" or "post-tetanic potentiation" and can be induced by calcium influx in presynaptic terminals (Bliss & Lomo, 1973; McNaughton, Douglas, & Goddard, 1978). Even if the rapidly decaying synaptic modification does not depend upon both presynaptic and postsynaptic activity, it can be used to store memory (Mongillo, Barak, & Tsodyks, 2008). Alternately, any short-term change in excitability could allow selective reactivation of the most recently presented stimulus, including the CAN current described above that can decay over a period of seconds when it does not cause persistent spiking (Klink & Alonso, 1997; Yoshida & Hasselmo, 2009).

Neuronal oscillations have been proposed to contribute to another potential STM mechanism. For instance, Lisman and colleagues proposed a model in which individual items could be transiently maintained by activation of specific ensembles of neurons within a single cycle of a gamma (30–80 Hz) oscillation (Jensen & Lisman, 2005; Lisman & Idiart, 1995; Lisman & Jensen, 2013). In the model, activation of multiple item representations is regulated by low-frequency theta oscillations (4–8 Hz in the human), such that each item representation is sequentially activated during a different theta phase (Lisman & Idiart, 1995). Afterdepolarization due to mechanisms such as the CAN current allow the reactivation of the representation on the next theta cycle, with first-in first-out replacement of items that can model the precession of place cell spiking activity relative to theta (Jensen & Lisman, 1996; Koene & Hasselmo, 2007; Lisman & Idiart, 1995; Navratilova, Giocomo, Fellous, Hasselmo, & McNaughton, 2012). In this framework, the afterdepolarization corresponds to a mechanism of short-term memory for maintaining representations for spatial location (Navratilova et al., 2012; de Almeida, Idiart, Villavicencio, & Lisman, 2012).

Jensen and Lisman (2005) argued that the sequential activation of neocortical item representations on each theta cycle could support both temporary maintenance of multiple items and also the long-term retention of temporal sequence information. The temporal distance between activation of successive item representations in their model is substantially compressed to ~20 ms, which corresponds roughly to the duration of one cycle of gamma oscillation and is within the time window required for long-term potentiation (LTP; Bi & Poo, 1998; Markram, Lubke, Frotscher, & Sakmann, 1997). Thus, the sequential activation of items in a theta cycle could allow for the item representations to become directly linked through Hebbian plasticity mechanisms and, as a result, support long-term retention of multi-item memories. The idea that single-unit activity and gamma oscillations are modulated by the phase of low-frequency oscillations has received strong support from studies reporting direct recordings of single-unit and local field potential activity (Bragin, Jando, Nadasdy, Hetke, Wise, & Buzsaki, 1995; Colgin et al., 2009). Additionally, work by Rainer and colleagues (Liebe, Hoerzer, Logothetis, & Rainer, 2012) has shown that stimulus-selective activity of V4 neurons during the retention period of an STM task was phase-locked to the ongoing theta oscillation. These findings support the idea that theta oscillations might regulate the activation of relevant stimulus information in the service of STM task performance.

Involvement of brain systems in short-term memory

Prefrontal Cortex and Posterior Cortical Areas As noted above, at the level of brain systems, the PFC has been the predominant area of focus in studies of STM. This was largely due to two major findings from studies of the delayed-response task in monkeys. In the delayed-response task, a monkey sees food being put into one of two wells, but then the monkey must wait until after a delay before reaching for the food. Lesion studies, beginning notably with the work of Jacobsen (1938), demonstrated that PFC lesions severely impaired retention of the food location when the delay extended over several seconds. Subsequent single-unit recording studies of the task by Niki, Fuster, and others (e.g., Funahashi, Bruce, & Goldman-Rakic, 1991; Fuster, 1973; Fuster & Alexander, 1971; Kubota & Niki, 1971) revealed that prefrontal neurons showed persistent spiking activity selective to the information that the monkey was to maintain across the delay. Intuitively, these findings would seem to indicate that the PFC is the site of short-term storage, but this interpretation does not explain why the delayed-response deficit in PFC-lesioned monkeys is eliminated if the lights are turned off during the retention interval (Malmo, 1942). Furthermore, PFC lesions in monkeys do not appear to affect retention of object information across short delays (Kowalska, Bachevalier, & Mishkin, 1991), but rather the impairment seems specific to tasks that require a motor response. How does one explain these discrepancies?

The contemporary explanation seems to be that PFC lesions do not directly impact the temporary storage of item information, but rather the active maintenance of the task that is currently relevant. That is, the prefrontal cortex may represent information about "the rules of the game"—the context-dependent mappings between stimuli and actions that are associated with a particular goal. This explanation fits well with single-unit recording data showing that PFC neurons carry a great deal of information about the task rules that are currently relevant (Fuster, 1995; Miller & Cohen, 2001; Stokes et al., 2013; Wallis, Anderson, & Miller, 2001; Wallis & Miller, 2003).

According to this explanation, prefrontal dysfunction impairs the use of higher-order goals to direct behavior, such that behavior is now driven by simple stimulus-response associations. Thus, PFC damage would not be expected to disrupt STM performance under all circumstances, but more specifically under conditions in which one must actively maintain an internal task set in the face of distracting stimuli. Indeed, as noted above, the effects of PFC lesions on STM can be minimized by reducing distractions (Malmo, 1942). Additionally, humans with PFC lesions do not show reductions in the capacity of information that can be held in STM, but they do show deficits on STM tasks that require retention of information in the face of distraction or tasks that require manipulation of recently presented information (D'Esposito & Postle, 1999).

The evidence described above therefore indicates that the PFC is not the site of all short-term storage, but rather that it represents a particular type of information that allows one to retain a task set across short delays. Indeed, it is now clear from single-unit recording and fMRI studies that there is no particular site of short-term storage. Instead, percepts, concepts, and actions are represented at various levels of analysis in different cortical areas (e.g., Felleman & Van Essen, 1991), and any of these representations may be actively retained in the service of task goals (Fuster, 1995; Jonides et al., 2008; Postle, 2006; Ranganath & Blumenfeld, 2005). For instance, single-unit recording data have shown that neurons in visual, auditory, and somatosensory areas show stimulus-selective persistent activity when their preferred stimuli are being actively maintained across short delays (Fuster, 1973, 1995; Miller, Brody, Romo, & Wang, 2003; Pasternak & Greenlee, 2005; Romo, Hernandez, Zainos, Lemus, & Brody, 2002), and fMRI studies have shown that category-selective regions show persistent activity increases when stimuli from the preferred category are actively maintained (Postle, 2006; Ranganath, 2006; Ranganath & D'Esposito, 2005; Schon, Hasselmo, Lopresti, Tricarico, & Stern, 2004). Additional work with multivoxel pattern analysis techniques has shown that even primary sensory cortical areas carry information about stimuli that are actively maintained (e.g., Harrison & Tong, 2009).

Medial Temporal Lobes Perhaps the most controversial idea to emerge recently in the STM literature is that areas in the MTL (like other cortical areas) play a role in STM processes. This idea was controversial because initial studies (Corkin, 1984; Scoville & Milner, 1957) showed that patients with MTL damage exhibit a normal capacity for short-term retention of simple stimuli or recall of sequences of simple digits or spatial locations (i.e., "immediate serial recall"). These studies, along with more recent findings, suggest that patients with MTL damage can show severe impairments on LTM measures but still retain simple information, such as the locations of colored squares, across short delays (e.g., Cave & Squire, 1992). Given the findings described above, however, this is not surprising because simple, familiar, and highly discriminable stimuli could be maintained by persistent activity of representations in cortical areas outside of the MTL. The important question, then, is whether there are any conditions under which MTL regions *do* contribute to STM.

It is, in fact, quite clear that neocortical areas in the MTL—that is, the perirhinal, entorhinal, and parahippocampal cortex—play a critical role in STM when the relevant information is being actively maintained. Early evidence for this idea came from studies of neurons in "inferior temporal" areas (e.g., Fuster & Jervey, 1982), including perirhinal cortex (E. K. Miller, Erickson, & Desimone, 1996; Naya, Yoshida, Takeda, Fujimichi, & Miyashita, 2003), that showed persistent, object-selective activity during short retention delays. Similar effects have been reported in entorhinal cortical neurons (Suzuki, Miller, & Desimone, 1997). Additionally, damage to the perirhinal cortex has been associated with severe impairments in the retention of object information, even across very short delays of a few seconds (Baxter & Murray, 2001), and humans with extensive perirhinal damage show comparable impairments in STM for complex objects (Buffalo, Reber, & Squire, 1998). These findings make sense, as most of the available evidence suggests that perirhinal cortex represents detailed information about the perceptual and semantic characteristics of objects (Burke et al., 2012; Murray, Bussey, & Saksida, 2007; Ranganath & Ritchey, 2012; Tyler et al., 2013). Although less is known about the parahippocampal cortex, this area shows persistent activity when visual scene information is actively maintained (Schon et al., 2004; Schon et al., 2005; Stern, Sherman, Kirchhoff, & Hasselmo, 2001), and damage to this area has been associated with spatial STM impairments (e.g., Hartley et al., 2007; Pierrot-Deseilligny, Muri, Rivaud-Pechoux, Gaymard, & Ploner, 2002; Ploner et al., 2000).

The role of the hippocampus proper in STM processes is more controversial, but several studies have shown that the hippocampus is critical for STM under certain circumstances. Several fMRI studies, for instance, have shown a role for the hippocampus in the active maintenance of novel, complex stimuli. The first of these studies, by Stern and colleagues (2001), investigated activity while participants performed a "two-back" task with novel, complex scene stimuli. On each trial, participants were required to decide whether each scene was the same as one presented two trials previously. The hippocampus showed increased activity during blocks with novel scenes, as compared to blocks with scenes that were highly familiar. Stern et al. also examined activation in the hippocampus during performance of a target detection task that placed minimal demands on STM maintenance. Hippocampal activation did not differentiate between blocks of target detection with novel stimuli and blocks with highly familiar stimuli. This finding suggests that hippocampal activation during the STM task was not driven by passive processing of novel stimuli, but rather by the demand to actively attend to and actively maintain these stimuli.

In a separate set of experiments, Ranganath & D'Esposito (2001) used event-related fMRI to investigate the neural correlates of STM for novel faces. In the first experiment, participants performed a delayed-recognition task, in which a novel sample face was to be maintained across a 7-sec delay in anticipation of a recognition probe. Critically, as in the Stern et al., study, the stimuli presented on each trial were novel and not repeated on subsequent trials. This study found that hippocampal activation was increased during the memory delay, consistent with a role in maintaining the face stimulus. Control conditions demonstrated that the hippocampal activation seen during the STM task was not driven simply by the demand to encode or perform recognition decisions on faces. In the second experiment, the authors replicated the findings of hippocampal activation during maintenance of novel faces in a new sample of participants, and extended it by demonstrating that delay-period activation in the same region was increased during maintenance of novel faces, as compared with familiar faces.

The findings from the studies by Stern et al. (2001) and Ranganath & D'Esposito (2001) prompted several groups to reexamine the question of whether STM is intact following MTL damage. For example, one group replicated the fMRI findings of Ranganath and D'Esposito (2001) and also demonstrated that amnesic participants with medial temporal lobe damage due to anoxia or encephalitis were impaired at this task, even though it only required retention across a 7-sec delay (Nichols, Kao, Verfaillie, & Gabrieli, 2006). A similar study conducted by Olson and colleagues also found that amnesic patients with medial temporal damage were impaired at retaining novel faces across a 4-sec delay (Olson, Moore, Stark, & Chatterjee, 2006; Shrager, Levy, Hopkins, & Squire, 2008[2]).

Although the above lesion studies do not precisely implicate the hippocampus, as it is essentially impossible to rule out damage to other areas in human patients, the convergence with the fMRI results suggests that, at least under some circumstances, the hippocampus plays a role in STM. It is not clear exactly when and how the hippocampus contributes, but the findings described above suggest that novelty and complexity play a role. As described above, most models of the neural mechanisms for active maintenance of familiar stimuli require learning of a previous pattern of synaptic connections. In contrast, novel stimuli may not correspond to previous patterns of synaptic connectivity for representing the configuration of item features or the relationship between items. The active maintenance of novel stimuli might require intrinsic single-cell mechanisms for persistent spiking that have been demonstrated in MTL regions, including the entorhinal cortex (Egorov et al., 2002; Fransén et al., 2006; Klink & Alonso, 1997), the perirhinal cortex (Navaroli et al., 2012), and hippocampus (Knauer et al., 2013). Consistent with this potential role in memory for novel stimuli, MTL structures appear to selectively gate sensory responses based on novelty, showing much stronger responses to novel stimuli than to familiar, repeated stimuli in unit recording (Brown & Xiang, 1998; Riches, Wilson, & Brown, 1991; Xiang & Brown, 2004) and fMRI studies (Stern et al., 2001).

[2]Although Shrager et al. (2008) found STM impairments in their patients, they made the counterintuitive conclusion that this effect was really based on impaired LTM. Their conclusion was based on their finding that healthy individuals were not affected on the same STM task when interfering stimuli were presented during the retention delay. They concluded that because the task was not sensitive to distraction, it was really a measure of LTM (at short delays). Their conclusion, however, is contingent on the assumption that there is a dedicated system for STM that is sensitive to interference, and that there is a separate LTM system that is not sensitive to interference. To our knowledge, this assumption is unsubstantiated, and no theory of memory makes this counterintuitive assumption. Numerous studies have shown that LTM processes are sensitive to interference, and in fact, dual store models tend to make the opposite assumption (i.e., that loss of information in STM is based on decay, rather than interference). Thus, the results from the behavioral study of Shrager et al. (2008) do not suggest a clear explanation, and the results of the patient study are consistent with results from other studies showing that MTL damage impairs retention of novel face information even across short delays.

In addition to maintaining novel, complex stimuli, available evidence suggests a particularly essential role for the hippocampus in short-term retention of spatial and relational information. The basis for this idea comes from models suggesting that the hippocampus is critical for encoding representations of arbitrary relationships between specific aspects of an event (e.g., remembering where you put your keys, the name associated with a familiar face, etc.) in a manner that can support performance on episodic LTM tests (Cohen & Eichenbaum, 1993; Diana, Yonelinas, & Ranganath, 2007; Eichenbaum, Yonelinas, & Ranganath, 2007). If the hippocampus rapidly encodes representations of arbitrary relationships, these representations could support short-term retention of these relationships (Hannula, Tranel, & Cohen, 2006; Olson, Page, Moore, Chatterjee, & Verfaillie, 2006; Ranganath & D'Esposito, 2001).

Evidence from recent imaging studies suggests that the hippocampus may be critical for temporary retention of relational information. For example, at least three imaging studies have reported persistent hippocampal activation during the delay periods of STM tasks that required temporary retention of object-location associations (Mitchell, Johnson, Raye, & D'Esposito, 2000; Piekema, Kessels, Mars, Petersson, & Fernandez, 2006). Another study demonstrated that transient hippocampal activation during encoding and recognition phases was related to successful retention of object-location bindings across a 10-sec delay (Hannula & Ranganath, 2008), consistent with the transient synaptic plasticity mechanism described above. Additional work using multivoxel pattern analysis demonstrated that the perirhinal and parahippocampal cortex maintained representations of object and spatial location information, respectively, and that hippocampal voxel patterns carried information about the specific configuration of object-location conjunctions to be maintained across the delay (Libby, Hannula, & Ranganath, 2013). The retention of object-location associations in such tasks may involve spatial firing patterns of neurons, including place cells in the hippocampus (O'Keefe & Burgess, 2005), grid cells in medial entorhinal cortex (Hafting, Fyhn, Molden, Moser, & Moser, 2005), and object-sensitive cells in lateral entorhinal cortex (Deshmukh & Knierim, 2011; Tsao, Moser, & Moser, 2013).

Studies of amnesic patients generally converge with the imaging results in demonstrating a role for the MTL in short-term maintenance of relational information (Hannula et al., 2006; Hartley et al., 2007; Olson, Page, et al., 2006). For example, Hannula et al. found that amnesics with damage limited to the hippocampus showed impaired *immediate* memory for the locations of objects within a complex scene (Hannula et al., 2006). In a follow-up study, these researchers demonstrated that the effect was not limited to spatial relations by showing that patients were also impaired when tested on immediate memory for arbitrary associations between faces and scenes. Another study investigated memory for object-location associations (Olson, Page, et al., 2006) by using a paradigm adapted from an imaging study (Mitchell et al., 2000), in which subjects remembered the locations of objects in a two-dimensional 3 × 3 grid. Relative to healthy controls, amnesic patients with MTL damage were significantly impaired at retaining the object-location associations across an 8-sec delay. Similar results were found in other studies of object-location binding (Braun et al., 2011; Finke et al., 2011) or scene memory (Hartley et al., 2007; King, Burgess, Hartley, Vargha-Khadem, & O'Keefe, 2002). Collectively, these studies suggest that short-term retention of relational information may take place in the same areas implicated in long-term retention.

Additional evidence (Axmacher et al., 2009; Schon, Quiroz, Hasselmo, & Stern, 2009) suggests that MTL regions may be involved in STM tasks that use complex stimuli and manipulate the number of items to be retained ("memory load"). Using intracranial recordings in humans during a WM task with face stimuli, Axmacher and colleagues (2007) demonstrated that hippocampal and MTL cortical gamma power increased with increasing STM load, and they subsequently reported modulation of hippocampal gamma power by theta phase when multiple items were maintained (Axmacher et al., 2010). Several fMRI studies have also shown increases in hippocampal activity during memory delays with increasing memory load (e.g., Axmacher et al., 2007; Schon et al., 2009; Schon et al., 2013). A functional connectivity study using a Sternberg task with unfamiliar faces reported that the correlation of activity between the inferior frontal gyrus (IFG) and the hippocampus increases with increasing STM load during a delay period (Rissman, Gazzaley, & D'Esposito, 2008). Collectively, the results of these studies suggest encoding and retention of higher loads across a delay results in activity within the MTL, including roles for the hippocampus, entorhinal, and perirhinal cortices.

Although it is clear that sequences of verbalizable items can be maintained independently of the hippocampus (as in the digit span task), it is possible that the hippocampus may nonetheless play a special role in maintenance of temporal sequence information. Recent data show that neurons in the hippocampus may code different temporal intervals within behavioral tasks (Kraus, Robinson, White, Eichenbaum, & Hasselmo, 2013; MacDonald, Lepage, Eden, & Eichenbaum, 2011; Pastalkova, Itskov, Amarasingham, & Buzsaki, 2008). The selective firing response to temporal intervals could allow learning of items or events that occur at specific time points (Hasselmo, 2012). This could mediate the role of hippocampus in memory for temporal order (Hunsaker & Kesner, 2008). Computational modeling shows how these responses could arise from persistent spiking activity of neurons in the entorhinal cortex (Hasselmo, 2012).

Cognitive processes that may support short-term memory

In the 1950s and 1960s, several researchers developed models proposing that information that is active could be held in a capacity-limited short-term store, and that the act of processing this information would result in the development of a memory trace that could be accessed even after long delays. Interestingly, these models did not necessarily assume that short- and long-term stores were supported by different brain regions. For instance, Atkinson & Shiffrin (1971) noted: "One might consider the short-term store simply as being a temporary activation of some portion of the long-term store" (p. 278).

Some of the evidence for dual-store models came from analyses of serial position effects in verbal learning studies. Specifically, the likelihood of recalling a word from a previously studied list is increased for words at the beginning (primacy) and end (recency) of the list. Whereas some manipulations disproportionately impact the primacy effect and recall of middle-list items (e.g., presentation rate) others disproportionately affect the recency effect (e.g., lag between end of list and recall test). Intuitively, it would seem sensible to assume that primacy and middle-item memory is an index of LTM, whereas recency is additionally influenced by processes that support STM (e.g., phonological rehearsal). In fact, more recent studies have shown that the magnitude of the primacy effect is actually influenced by rehearsal (Tan & Ward, 2000), and that robust recency effects can be observed even when phonological rehearsal is not feasible (see Howard & Kahana, 1999, for review). Indeed, the factors influencing primacy and recency effects remain controversial. Some suggest that these effects can largely be accounted for by a single store (Sederberg, Howard, & Kahana, 2008), whereas others suggest that a temporary activation buffer (akin to the idea proposed by Hebb, 1949) additionally contributes to recency (Davelaar, Goshen-Gottstein, Ashkenazi, Haarmann, & Usher, 2005).

Baddeley and Hitch (1974) proposed a model that further elaborated on the processes that support STM performance. Their working memory (WM) model was innovative in two ways. First, the WM model proposed a distinction between the short-term representation of phonological and visual information. A second innovation of the WM model was that it proposed a separation between short-term storage (or "maintenance") and the manipulation of information in the service of task goals. That is, in the original model, two slave systems (the phonological loop and visuospatial sketchpad) were proposed to mediate maintenance, whereas a different component, the central executive, was proposed to mediate the selection, inhibition, and manipulation of information in WM.

One shortcoming of the original Baddeley and Hitch (1974) model is that it did not account for the short-term retention of materials extending beyond the phonological and visuospatial domains (Postle, 2006). To deal with the temporary retention of other materials and the problem of integration of information across modalities, Baddeley (2000) added a new component to the model, termed the "episodic buffer." At present, this component is less specified than other components of the model.

An alternative theoretical approach advocated by many researchers is that mechanisms for attentional selection play a role in the transient activation of stimulus representations, and consequently the limits of attentional focus constrain the number of items that can be actively retained in the service of STM task performance (Cowan, 1997; Ruchkin, Grafman, Cameron, & Berndt, 2003). Consistent with this idea, fMRI studies suggest that activity in the posterior parietal cortex—a critical region for goal-directed attentional selection—is correlated with STM capacity limits (Todd & Marois, 2004). In many respects, this view is compatible with many of the important ideas originally proposed by Baddeley, but is also more general because it proposes that working memory reflects the temporary activation of conceptual, perceptual, and action representations in the service of task goals. Put another way, there may not be a dedicated system for the processes described in the WM model. Instead, WM processes may "arise through the coordinated recruitment, via attention, of brain systems that have evolved to accomplish sensory- , representation- , and action-related functions" (Postle, 2006, p. 23).

Conclusion

As demonstrated here, recent research in neuroscience has revealed the complexity of processes that support STM. Our review highlights the point that there is no single neural mechanism, brain system, or cognitive process that supports performance on STM tasks. Furthermore, the processes that support STM performance appear to involve interactions in multiple anatomical systems, including even MTL regions. Finally, this review highlights the possibility that, at the neural level, STM and LTM are intricately linked. Neurocognitive processes that may specifically support short-term retention of information may also enhance LTM for that information. Likewise, performance on STM tasks may rely on sustained attention to established stimulus representations that were acquired recently, remotely, or early in development. Full understanding of the neural mechanisms of memory function will benefit from continued integration across these levels of analysis, to better understand the complex dynamics of neural circuits mediating memory behavior across different time scales.

ACKNOWLEDGMENTS Research supported by R01 MH068721, R01 MH083734, R01 MH60013, R01 MH61492, Silvio O. Conte Center P50 MH094263, and the Office of Naval Research MURI grant N00014-10-1-0936.

REFERENCES

Aben, B., Stapert, S., & Blokland, A. (2012). About the distinction between working memory and short-term memory. *Front Psychol, 3*, 301–303.

Amit, D. J. (1988). *Modeling brain function: The world of attractor neural networks.* Cambridge, UK: Cambridge University Press.

Amit, D. J., & Brunel, N. (1997). Model of global spontaneous activity and local structured activity during delay periods in the cerebral cortex. *Cereb Cortex, 7*(3), 237–252.

Amit, D. J., & Treves, A. (1989). Associative memory neural networks with low temporal spiking rates. *Proc Natl Acad Sci USA, 86*, 7671–7673.

Atkinson, R. C., & Shiffrin, R. M. (1971). The control of short-term memory. *Sci Am, 225*(2), 82–90.

Axmacher, N., Haupt, S., Cohen, M. X., Elger, C. E., & Fell, J. (2009). Interference of working memory load with long-term memory formation. *Eur J Neurosci, 29*, 1501–1513.

Axmacher, N., Henseler, M. M., Jensen, O., Weinreich, I., Elger, C. E., & Fell, J. (2010). Cross-frequency coupling supports multi-item working memory in the human hippocampus. *Proc Natl Acad Sci USA, 107*(7), 3228–3233.

Axmacher, N., Mormann, F., Fernandez, G., Cohen, M. X., Elger, C. E., & Fell, J. (2007). Sustained neural activity patterns during working memory in the human medial temporal lobe. *J Neurosci, 27*, 7807–7816.

Baddeley, A. (2000). The episodic buffer: A new component of working memory? *Trends Cogn Sci, 4*(11), 417–423.

Baddeley, A., & Hitch, G. J. (1974). Working memory. In G. Bower (Ed.), *Recent advances in learning and motivation* (Vol. VIII, pp. 47–90). New York, NY: Academic Press.

Baxter, M. G., & Murray, E. A. (2001). Opposite relationship of hippocampal and rhinal cortex damage to delayed nonmatching-to-sample deficits in monkeys. *Hippocampus, 11*(1), 61–71.

Bi, G., & Poo, M. (1998). Synaptic modification in cultured hippocampal neurons: Dependence on spike timing, synaptic strength and postsynaptic cell type. *J Neurosci, 18*(24), 10464–10472.

Bliss, T. V., & Lomo, T. (1973). Long-lasting potentiation of synaptic transmission in the dentate area of the anaesthetized rabbit following stimulation of the perforant path. *J Physiol, 232*(2), 331–356.

Bragin, A., Jando, G., Nadasdy, Z., Hetke, J., Wise, K., & Buzsaki, G. (1995). Gamma (40–100 Hz) oscillation in the hippocampus of the behaving rat. *J Neurosci, 15,* 47–60.

Braun, M., Weinrich, C., Finke, C., Ostendorfm F., Lehmann, T. N., & Ploner, C. J. (2011). Lesions affecting the right hippocampal formation differentially impair short-term memory of spatial and nonspatial associations. *Hippocampus, 21*(3), 309–318.

Brown, M. W., & Xiang, J. Z. (1998). Recognition memory: Neuronal substrates of the judgement of prior occurrence. *Prog Neurobiol, 55*(2), 149–189.

Buffalo, E. A., Reber, P. J., & Squire, L. R. (1998). The human perirhinal cortex and recognition memory. *Hippocampus, 8*(4), 330–339.

Burke, S. N., Maurer, A. P., Hartzell, A. L., Nematollahi, S., Uprety, A., Wallace, J. L., & Barnes, C. A. (2012). Representation of three-dimensional objects by the rat perirhinal cortex. *Hippocampus, 22*(10), 2032–2044.

Cave, C. B., & Squire, L. R. (1992). Intact verbal and nonverbal short-term memory following damage to the human hippocampus. *Hippocampus, 2*(2), 151–163.

Cohen, M. A., & Grossberg, S. (1983). Absolute stability of global pattern formation and parallel memory storage by competitive neural networks. *IEEE Trans Syst Man Cybern, 13,* 815–826.

Cohen, N. J., & Eichenbaum, H. (1993). *Memory, amnesia, and the hippocampal system.* Cambridge, MA: MIT Press.

Colgin, L. L., Denninger, T., Fyhn, M., Hafting, T., Bonnevie, T., Jensen, O., … Moser, E. I. (2009). Frequency of gamma oscillations routes flow of information in the hippocampus. *Nature, 462,* 353–357.

Corkin, S. (1984). Lasting consequences of bilateral medial temporal lobectomy: Clinical course and experimental findings in H. M. *Semin Neurol, 4,* 249–259.

Cowan, N. (1997). *Oxford psychology series, Vol. 26: Attention and memory: An integrated framework.* New York, NY: Oxford University Press.

D'Esposito, M., & Postle, B. R. (1999). The dependence of span and delayed-response performance on prefrontal cortex. *Neuropsychologia, 37*(11), 1303–1315.

Davelaar, E. J., Goshen-Gottstein, Y., Ashkenazi, A., Haarmann, H. J., & Usher, M. (2005). The demise of short-term memory revisited: Empirical and computational investigations of recency effects. *Psychol Rev, 112*(1), 3–42.

De Almeida, L., Idiart, M., Villavicencio, A., & Lisman, J. (2012). Alternating predictive and short-term memory modes of entorhinal grid cells. *Hippocampus, 22*(8), 1647–1651.

Deshmukh, S. S., & Knierim, J. J. (2011). Representation of non-spatial and spatial information in the lateral entorhinal cortex. *Front Behav Neurosci, 5,* 69.

Diana, R. A., Yonelinas, A. P., & Ranganath, C. (2007). Imaging recollection and familiarity in the medial temporal lobe: A three-component model. *Trends Cogn Sci, 11*(9), 379–386.

Egorov, A. V., Hamam, B. N., Fransén, E., Hasselmo, M. E., & Alonso, A. A. (2002). Graded persistent activity in entorhinal cortex neurons. *Nature, 420*(6912), 173–178.

Eichenbaum, H., Yonelinas, A. R., & Ranganath, C. (2007). The medial temporal lobe and recognition memory. *Annu Rev Neurosci, 30,* 123–152.

Felleman, D. J., & Van Essen, D. C. (1991). Distributed hierarchical processing in the primate cerebral cortex. *Cereb Cortex, 1*(1), 1–47.

Finke, C., Ostendorf, F., Braun, M., & Ploner, C. J. (2011). Impaired representation of geometric relationships in humans with damage to the hippocampal formation. *PLoS One, 6*(5), e19507.

Fransén, E., Alonso, A. A., & Hasselmo, M. E. (2002). Simulations of the role of the muscarinic-activated calcium-sensitive nonspecific cation current I_{NCM} in entorhinal neuronal activity during delayed matching tasks. *J Neurosci, 22*(3), 1081–1097.

Fransén, E., Tahvildari, B., Egorov, A. V., Hasselmo, M. E., & Alonso, A. A. (2006). Mechanisms of graded persistent cellular activity of entorhinal cortex layer V neurons. *Neuron, 49*(5), 735–746.

Funahashi, S., Bruce, C. J., & Goldman-Rakic, P. S. (1991). Neuronal activity related to saccadic eye movements in the monkey's dorsolateral prefrontal cortex. *J Neurophysiol, 65*(6), 1464–1483.

Fuster, J. M. (1973). Unit activity in prefrontal cortex during delayed-response performance: Neuronal correlates of transient memory. *J Neurophysiol, 36,* 61–78.

Fuster, J. M. (1995). *Memory in the cerebral cortex.* Cambridge, MA: MIT Press.

Fuster, J. M., & Alexander, G. E. (1971). Neuron activity related to short-term memory. *Science, 173,* 652–654.

Fuster, J. M., & Jervey, J. P. (1982). Neuronal firing in the inferotemporal cortex of the monkey in a visual memory task. *J Neurosci, 2,* 361–375.

Hafting, T., Fyhn, M., Molden, S., Moser, M. B., & Moser, E. I. (2005). Microstructure of a spatial map in the entorhinal cortex. *Nature, 436*(7052), 801–806.

Haj-Dahmane, S., & Andrade, R. (1998). Ionic mechanism of the slow afterdepolarization induced by muscarinic receptor activation in rat prefrontal cortex. *J Neurophysiol, 80*(3), 1197–1210.

Hannula, D. E., Libby, L. A., Yonelinas, A. P., & Ranganath, C. (2013). Medial temporal lobe contributions to cued retrieval of items and contexts. *Neuropsychologia, 51*(12), 2322–2332.

Hannula, D. E., & Ranganath, C. (2008). Medial temporal lobe activity predicts successful relational memory binding. *J Neurosci, 28*(1), 116–124.

Hannula, D. E., Tranel, D., & Cohen, N. J. (2006). The long and the short of it: Relational memory impairments in amnesia, even at short lags. *J Neurosci, 26*(32), 8352–8359.

Harrison, S. A., & Tong, F. (2009). Decoding reveals the contents of visual working memory in early visual areas. *Nature, 458*(7238), 632–635.

Hartley, T., Bird, C. M., Chan, D., Cipolotti, L., Husain, M., Vargha-Khadem, F., & Burgess, N. (2007). The

hippocampus is required for short-term topographical memory in humans. *Hippocampus, 17*(1), 34–48.
HASSELMO, M. E. (2012). *How we remember: Brain mechanisms of episodic memory.* Cambridge, MA: MIT Press.
HASSELMO, M. E., SCHNELL, E., & BARKAI, E. (1995). Dynamics of learning and recall at excitatory recurrent synapses and cholinergic modulation in rat hippocampal region CA3. *J Neurosci, 15*(7), 5249–5262.
HASSELMO, M. E., & STERN, C. E. (2006). Mechanisms underlying working memory for novel information. *Trends Cogn Sci, 10*(11), 487–493.
HASSELMO, M. E., & WYBLE, B. P. (1997). Free recall and recognition in a network model of the hippocampus: Simulating effects of scopolamine on human memory function. *Behav Brain Res, 89*(1–2), 1–34.
HEBB, D. O. (1949). *The organization of behavior.* New York, NY: Wiley.
HOPFIELD, J. J. (1984). Neurons with graded response have collective computational properties like those of two-state neurons. *Proc Natl Acad Sci USA, 81*(10), 3088–3092.
HOWARD, M. W., & KAHANA, M. J. (1999). Contextual variability and serial position effects in free recall. *J Exp Psychol Learn Mem Cogn, 25*(4), 923–941.
HUNSAKER, M. R., & KESNER, R. P. (2008). Evaluating the differential roles of the dorsal dentate gyrus, dorsal CA3, and dorsal CA1 during a temporal ordering for spatial locations task. *Hippocampus, 18,* 955–964.
JACOBSEN, C. F. (1938). Studies of cerebral function in primates. *Comp Psychol Monogr, 13,* 1–68.
JENSEN, O., IDIART, M. A., & LISMAN, J. E. (1996). Physiologically realistic formation of autoassociative memory in networks with theta/gamma oscillations: Role of fast NMDA channels. *Learn Mem, 3,* 243–256.
JENSEN, O., & LISMAN, J. E. (1996). Hippocampal CA3 region predicts memory sequences: Accounting for the phase precession of place cells. *Learn Mem, 3,* 279–287.
JENSEN, O., & LISMAN, J. E. (2005). Hippocampal sequence-encoding driven by a cortical multi-item working memory buffer. *Trends Neurosci, 28*(2), 67–72.
JONIDES, J., LEWIS, R. L., NEE, D. E., LUSTIG, C. A., BERMAN, M. G., & MOORE, K. S. (2008). The mind and brain of short-term memory. *Annu Rev Psychol, 59,* 193–224.
KING, J. A., BURGESS, N., HARTLEY, T., VARGHA-KHADEM, F., & O'KEEFE, J. (2002). Human hippocampus and viewpoint dependence in spatial memory. *Hippocampus, 12*(6), 811–820.
KLINK, R., & ALONSO, A. (1997). Muscarinic modulation of the oscillatory and repetitive firing properties of entorhinal cortex layer II neurons. *J Neurophysiol, 77*(4), 1813–1828.
KNAUER, B., JOCHEMS, A., VALERO-ARACAMA, M. J., & YOSHIDA, M. (2013). Long-lasting intrinsic persistent firing in rat CA1 pyramidal cells: A possible mechanism for active maintenance of memory. *Hippocampus, 23*(9), 820–831.
KOENE, R. A., & HASSELMO, M. E. (2007). First-in first-out item replacement in a model of short-term memory based on persistent spiking. *Cereb Cortex, 17,* 1766–1781.
KOWALSKA, D. M., BACHEVALIER, J., & MISHKIN, M. (1991). The role of the inferior prefrontal convexity in performance of delayed nonmatching-to-sample. *Neuropsychologia, 29*(6), 583–600.
KRAUS, B. J., ROBINSON, R. J. 2nd, WHITE, J. A., EICHENBAUM, H., & HASSELMO, M. E. (2013). Hippocampal "time cells": Time versus path integration. *Neuron, 78,* 1090–1101.
KUBOTA, K., & NIKI, H. (1971). Prefrontal cortical unit activity and delayed alternation performance in monkeys. *J Neurophysiol, 34*(3), 337–347.
LIBBY, L. A., HANNULA, D. E., & RANGANATH, C. (Submitted). Medial temporal lobe coding of item and spatial information during relational binding in working memory.
LIEBE, S., HOERZER, G. M., LOGOTHETIS, N. K., & RAINER, G. (2012). Theta coupling between V4 and prefrontal cortex predicts visual short-term memory performance. *Nat Neurosci, 15*(3), 456–462.
LISMAN, J. E., & IDIART, M. A. (1995). Storage of 7 +/− 2 short-term memories in oscillatory subcycles. *Science, 267,* 1512–1515.
LISMAN, J. E., & JENSEN, O. (2013). The θ-γ neural code. *Neuron, 77*(6), 1002–1016.
MACDONALD, C. J., LEPAGE, K. Q., EDEN, U. T., & EICHENBAUM, H. (2011). Hippocampal "time cells" bridge the gap in memory for discontiguous events. *Neuron, 71*(4), 737–749.
MALMO, R. B. (1942). Interference factors in delayed response in monkey after removal of the frontal lobes. *J Neurophysiol, 4,* 295–308.
MARKRAM, H., LUBKE, J., FROTSCHER, M., & SAKMANN, B. (1997). Regulation of synaptic efficacy by coincidence of postsynaptic APs and EPSPs. *Science, 275,* 213–215.
MCNAUGHTON, B. L., DOUGLAS, R. M., & GODDARD, G. V. (1978). Synaptic enhancement in fascia dentata: Cooperativity among coactive afferents. *Brain Res, 157,* 277–293.
MCNAUGHTON, B. L., & MORRIS, R. G. M. (1987). Hippocampal synaptic enhancement and information storage within a distributed memory system. *Trends Neurosci, 10,* 408–415.
MILLER, E. K., & COHEN, J. D. (2001). An integrative theory of prefrontal cortex function. *Annu Rev Neurosci, 24,* 167–202.
MILLER, E. K., ERICKSON, C. A., & DESIMONE, R. (1996). Neural mechanisms of visual working memory in prefrontal cortex of the macaque. *J Neurosci, 16*(16), 5154–5167.
MILLER, P., BRODY, C. D., ROMO, R., & WANG, X. J. (2003). A recurrent network model of somatosensory parametric working memory in the prefrontal cortex. *Cereb Cortex, 13*(11), 1208–1218.
MITCHELL, K. J., JOHNSON, M. K., RAYE, C. L., & D'ESPOSITO, M. (2000). MRI evidence of age-related hippocampal dysfunction in feature binding in working memory. *Brain Res Cogn Brain Res, 10*(1–2), 197–206.
MONGILLO, G., BARAK, O., & TSODYKS, M. (2008). Synaptic theory of working memory. *Science, 319*(5869), 1543–1546.
MURRAY, E. A., BUSSEY, T. J., & SAKSIDA, L. M. (2007). Visual perception and memory: A new view of medial temporal lobe function in primates and rodents. *Annu Rev Neurosci, 30,* 99–122.
NAVAROLI, V. L., ZHAO, Y., BOGUSZEWSKI, P., & BROWN, T. H. (2012). Muscarinic receptor activation enables persistent firing in pyramidal neurons from superficial layers of dorsal perirhinal cortex. *Hippocampus, 22*(6), 1392–1404.
NAVRATILOVA, Z., GIOCOMO, L. M., FELLOUS, J. M., HASSELMO, M. E., & MCNAUGHTON, B. L. (2012). Phase precession and variable spatial scaling in a periodic attractor map model of medial entorhinal grid cells with realistic after-spike dynamics. *Hippocampus, 22*(4), 772–789.
NAYA, Y., YOSHIDA, M., TAKEDA, M., FUJIMICHI, R., & MIYASHITA, Y. (2003). Delay-period activities in two

subdivisions of monkey inferotemporal cortex during pair association memory task. *Eur J Neurosci, 18*(10), 2915–2918.

Nichols, E. A., Kao, Y. C., Verfaellie, M., & Gabrieli, J. D. (2006). Working memory and long-term memory for faces: Evidence from fMRI and global amnesia for involvement of the medial temporal lobes. *Hippocampus, 16*(7), 604–616.

O'Keefe, J., & Burgess, N. (2005). Dual phase and rate coding in hippocampal place cells: Theoretical significance and relationship to entorhinal grid cells. *Hippocampus, 15*(7), 853–866.

Olson, I. R., Moore, K. S., Stark, M., & Chatterjee, A. (2006). Visual working memory is impaired when the medial temporal lobe is damaged. *J Cogn Neurosci, 18*(7), 1087–1097.

Olson, I. R., Page, K., Moore, K. S., Chatterjee, A., & Verfaellie, M. (2006). Working memory for conjunctions relies on the medial temporal lobe. *J Neurosci, 26*(17), 4596–4601.

Pastalkova, E., Itskov, V., Amarasingham, A., & Buzsaki, G. (2008). Internally generated cell assembly sequences in the rat hippocampus. *Science, 321*(5894), 1322–1327.

Pasternak, T., & Greenlee, M. W. (2005). Working memory in primate sensory systems. *Nat Rev Neurosci, 6*(2), 97–107.

Piekema, C., Kessels, R. P., Mars, R. B., Petersson, K. M., & Fernandez, G. (2006). The right hippocampus participates in short-term memory maintenance of object-location associations. *NeuroImage, 33*(1), 374–382.

Pierrot-Deseilligny, C., Muri, R. M., Rivaud-Pechoux, S., Gaymard, B., & Ploner, C. J. (2002). Cortical control of spatial memory in humans: The visuooculomotor model. *Ann Neurol, 52*(1), 10–19.

Ploner, C. J., Gaymard, B. M., Rivaud-Péchoux, S., Baulac, M., Clémenceau, S., Samson, S., & Pierrot-Deseilligny, C. (2000). Lesions affecting the parahippocampal cortex yield spatial memory deficits in humans. *Cereb Cortex, 10*(12), 1211–1216.

Postle, B. R. (2006). Working memory as an emergent property of the mind and brain. *Neuroscience, 139*(1), 23–38.

Ranganath, C. (2006). Working memory for visual objects: Complementary roles of inferior temporal, medial temporal, and prefrontal cortex. *Neuroscience, 139*(1), 277–289.

Ranganath, C., & Blumenfeld, R. S. (2005). Doubts about double dissociations between short- and long-term memory. *Trends Cogn Sci, 9*(8), 374–380.

Ranganath, C., & D'Esposito, M. (2001). Medial temporal lobe activity associated with active maintenance of novel information. *Neuron, 31*, 865–873.

Ranganath, C., & D'Esposito, M. (2005). Directing the mind's eye: Prefrontal, inferior and medial temporal mechanisms for visual working memory. *Curr Opin Neurobiol, 15*(2), 175–182.

Ranganath, C., & Ritchey, M. (2012). Two cortical systems for memory-guided behaviour. *Nat Rev Neurosci, 13*(10), 713–726.

Riches, I. P., Wilson, F. A., & Brown, M. W. (1991). The effects of visual stimulation and memory on neurons of the hippocampal formation and the neighboring parahippocampal gyrus and inferior temporal cortex of the primate. *J Neurosci, 11*(6), 1763–1779.

Rissman, J., Gazzaley, A., & D'Esposito, M. (2008). Dynamic adjustments in prefrontal, hippocampal, and inferior temporal interactions with increasing visual working memory load. *Cereb Cortex, 18*(7), 1618–1629.

Romo, R., Hernandez, A., Zainos, A., Lemus, L., & Brody, C. D. (2002). Neuronal correlates of decision-making in secondary somatosensory cortex. *Nat Neurosci, 5*(11), 1217–1225.

Ruchkin, D. S., Grafman, J., Cameron, K., & Berndt, R. S. (2003). Working memory retention systems: A state of activated long-term memory. *Behav Brain Sci, 26*(6), 709–728; discussion 728–777.

Schon, K., Atri, A., Hasselmo, M. E., Tricarico, M. D., LoPresti, M. L., & Stern, C. E. (2005). Scopolamine reduces persistent activity related to long-term encoding in the parahippocampal gyrus during delayed matching in humans. *J Neurosci, 25*(40), 9112–9123.

Schon, K., Hasselmo, M. E., LoPresti, M. L., Tricarico, M. D., & Stern, C. E. (2004). Persistence of parahippocampal representation in the absence of stimulus input enhances long-term encoding: A functional magnetic resonance imaging study of subsequent memory after a delayed match-to-sample task. *J Neurosci, 24*(49), 11088–11097.

Schon, K., Quiroz, Y. T., Hasselmo, M. E., & Stern, C. E. (2009). Greater working memory load results in greater medial temporal activity at retrieval. *Cereb Cortex, 19*, 2561–2571.

Schon, K., Ross, R. S., Hasselmo, M. E., & Stern, C. E. (2013). Complementary roles of medial temporal lobes and mid-dorsolateral prefrontal cortex for working memory for novel and familiar trial-unique visual stimuli. *Eur J Neurosci, 37*(4), 668–678.

Scoville, W. B., & Milner, B. (1957). Loss of recent memory after bilateral hippocampal lesions. *J Neurol Neurosurg Psych, 20*, 11–21.

Sederberg, P. B., Howard, M. W., & Kahana, M. J. (2008). A context-based theory of recency and contiguity in free recall. *Psychol Rev, 115*(4), 893–912.

Shalinsky, M. H., Magistretti, J., Ma, L., & Alonso, A. A. (2002). Muscarinic activation of a cation current and associated current noise in entorhinal-cortex layer-II neurons. *J Neurophysiol, 88*(3), 1197–1211.

Shrager, Y., Levy, D. A., Hopkins, R. O., & Squire, L. R. (2008). Working memory and the organization of brain systems. *J Neurosci, 28*(18), 4818–4822.

Stern, C. E., Sherman, S. J., Kirchhoff, B. A., & Hasselmo, M. E. (2001). Medial temporal and prefrontal contributions to working memory tasks with novel and familiar stimuli. *Hippocampus, 11*(4), 337–346.

Stokes, M. G., Kusunoki, M., Sigala, N., Nili, H., Gaffan, D., & Duncan, J. (2013). Dynamic coding for cognitive control in prefrontal cortex. *Neuron, 78*(2), 364–375.

Suzuki, W. A., Miller, E. K., & Desimone, R. (1997). Object and place memory in the macaque entorhinal cortex. *J Neurophysiol, 78*(2), 1062–1081.

Tan, L., & Ward, G. (2000). A recency-based account of the primacy effect in free recall. *J Exp Psychol Learn Mem Cogn, 26*(6), 1589–1625.

Tegner, J., Compte, A., & Wang, X. J. (2002). The dynamical stability of reverberatory neural circuits. *Biol Cybern, 87*(5–6), 471–481.

Todd, J. J., & Marois, R. (2004). Capacity limit of visual short-term memory in human posterior parietal cortex. *Nature, 428*(6984), 751–754.

Treves, A., & Rolls, E. T. (1992). Computational constraints suggest the need for two distinct input systems to the hippocampal CA3 network. *Hippocampus, 2*(2), 189–199.

Tsao, A., Moser, M. B., & Moser, E. I. (2013). Traces of experience in the lateral entorhinal cortex. *Curr Biol, 23*(5), 399–405.

Tyler, L. K., Chiu, S., Zhuang, J., Randall, B., Devereux, B. J., Wright, P., ... Taylor, K. I. (2013). Objects and categories: Feature statistics and object processing in the ventral stream. *J Cogn Neurosci, 25*(10), 1723–1735.

Wallis, J. D., & Miller, E. K. (2003). From rule to response: Neuronal processes in the premotor and prefrontal cortex. *J Neurophysiol, 90*(3), 1790–1806.

Wallis, J. D., Anderson, K. C., & Miller, E. K. (2001). Single neurons in prefrontal cortex encode abstract rules. *Nature, 411*, 953–956.

Wilson, H. R., & Cowan, J. D. (1972). Excitatory and inhibitory interactions in localized populations of model neurons. *Biophys J, 12*(1), 1–24.

Xiang, J. Z., & Brown, M. W. (2004). Neuronal responses related to long-term recognition memory processes in prefrontal cortex. *Neuron, 42*(5), 817–829.

Yoshida, M., & Hasselmo, M. E. (2009). Persistent firing supported by an intrinsic cellular mechanism in a component of the head direction system. *J Neurosci, 29*(15), 4945–4952.

Zhang, Z., & Seguela, P. (2010). Metabotropic induction of persistent activity in layers II/III of anterior cingulate cortex. *Cereb Cortex, 20*(12), 2948–2957.

Zilli, E. A., & Hasselmo, M. E. (2008). Modeling the role of working memory and episodic memory in behavioral tasks. *Hippocampus, 18*(2), 193–209.

Zipser, D., Kehoe, B., Littlewort, G., & Fuster, J. (1993). A spiking network model of short-term active memory. *J Neurosci, 13*(8), 3406–3420.

45 The Medial Temporal Lobe and Memory

LILA DAVACHI AND ALISON PRESTON

ABSTRACT Memory is our access to the past. Recent work in cognitive neuroscience has provided evidence that our access to past events is guided by the reinstatement of the same neural patterns of activity that were present during original experiences. In essence, when we remember, the brain returns to a prior brain state, allowing us to reexperience our past. In this chapter, we highlight existing evidence for the role of the medial temporal lobe in forming new episodic memories and reinstating those memories during remembering.

What did she say? Where did you go? What did you eat? Answering questions like these requires that you access previous experiences, or episodes, of your life. Episodic memory is memory for past experiences that occurred at a particular time and place. By contrast, other forms of memory do not require that you access specifics about a prior encounter, such as knowing what a giraffe is (semantic memory) or knowing how to ride a bike (procedural memory). Thus, episodic memory is a record of our personal experiences that make up the narrative of our lives.

The first link between the medial temporal lobe (MTL) and episodic memory occurred through the study of a patient, Henry Molaison, famously known as H.M. At the age of 27, H.M.'s MTL was surgically removed to alleviate epileptic seizures; that surgery, while successful in treating his epilepsy, also left him with dense amnesia. He was no longer able to form new episodic memories, a condition known as *anterograde amnesia*. The specificity of the deficit was remarkable. His intelligence was intact, and he appeared to have normal working memory and procedural memory, but he was confined to living "in the moment" because the present disappeared into the past without a trace. With H.M., it became clear that the MTL is critical for episodic memory. This major discovery set the stage for neuroscientists to determine exactly how the MTL contributes to episodic memory formation and retrieval.

Critically, the MTL is not a single structure but rather is made up of different regions. These include the hippocampus as well as the entorhinal (ERc), perirhinal (PRc), and parahippocampal (PHc) cortices. In this chapter, we review our current understanding of the neural underpinnings of episodic memory, focusing on processes taking place during the experience itself, or *encoding*, and those involved in reactivating or *retrieving* memories at a later time point. We present a model of memory formation and retrieval that has received the most support to date. This model has motivated most empirical investigations into episodic memory processing in the brain.

Memory as reinstatement: The MAR model

The memory as reinstatement (MAR) model (Davachi & Danker, 2013) represents a combination of current theory and knowledge regarding how episodic memories are formed and subsequently accessed. It consists of elements drawn from many influential models of episodic memory and, thus, should not be considered a new model, but rather a summary of the common elements of many existing models (e.g., Alvarez & Squire, 1994; McClelland, McNaughton, & O'Reilly, 1995; Moscovitch et al., 2005; Norman & O'Reilly, 2003).

During an experience, the MAR model proposes that the ongoing, dynamic representation of that experience is represented in distributed cortical and subcortical patterns of neural activation. These activation patterns are driven by sensory-perceptual (visual, auditory, somatosensory) experience, actions, internal thoughts, and emotions, to name a few. Thus, neural activation patterns at any one time point can be thought of as representing the current episode and state of the organism. Critically, it is thought that the distributed pattern of cortical and subcortical firing filters into the MTL cortices and converges on the hippocampus, where a "microrepresentation" of the current episode is created. The connections between hippocampal neurons that make up the memory are thought to undergo subsequent strengthening via the process of long-term potentiation. Specifically, connections between concurrently active hippocampal neurons are more likely to become strengthened compared with those that are not (Hebb, 1949). What results from a successfully encoded experience is a hippocampal neural pattern (HNP) *and* a corresponding cortical neural pattern (CNP). Importantly, the HNP is thought to contain the critical connections between representations that allow the CNP to be accessed later and attributed to a particular event.

Importantly, episodic memory retrieval is thought to involve cue processing that, if successful, will lead to a reinstatement of the HNP. Retrieval cues (e.g., an external stimulus or internal thought) are thought to serve as "keys" that unlock the HNP associated with a prior experience, a process referred to as *hippocampal pattern completion*. Pattern completion refers to the idea that a complete pattern (a memory) can be reconstructed from only a subset of the elements making up that pattern. Thus, successful retrieval is thought to involve reinstatement of all or some of the HNP established during encoding. Finally, reinstatement of the HNP is then thought to be instrumental in reinstating the corresponding CNP, resulting in the concurrent reactivation of disparate cortical regions that were initially active during the experience. Importantly, it is thought that this final stage of cortical reinstatement underlies the subjective experience of recollection and drives mnemonic decision making.

Functional neuroimaging of episodic encoding

The processes through which individual experiences get transformed into long-lasting memory traces have collectively been referred to as *encoding mechanisms* (Davachi, 2006). The predominant neuroimaging approach linking brain activation and episodic encoding mechanisms has been the comparison of trial-by-trial estimates of blood oxygen level–dependent, or BOLD, activation during experiences that are later remembered relative to activation during events that are not remembered. This approach has been referred to as the *difference in memory* (DM) or subsequent memory paradigm (Paller & Wagner, 2002). Using this approach, measures of brain activity during encoding can be related to a variety of memory outcomes, measured by different retrieval tests. For example, one can determine whether each presented item was or was not remembered as well as whether contextual details surrounding that were also recovered.

In two initial groundbreaking studies using the DM approach, encoding activation in the parahippocampal gyrus was greater for words (Wagner et al., 1998) and scenes (Brewer, Zhao, Desmond, Glover, & Gabrieli, 1998) that were successfully remembered relative to stimuli that were forgotten. These two studies thus showed that MTL activation correlates with successful episodic memory formation. These initial findings paved the way for future work further refining the use of the DM paradigm to ask more specific questions about how distinct MTL subregions contribute to memory formation. This question is hotly debated both in the animal (Eichenbaum, Sauvage, Fortin, Komorowski, & Lipton, 2012) and human literatures (Davachi, 2006; Diana, Yonelinas, & Ranganath, 2007; Eichenbaum, Yonelinas, & Ranganath, 2007; Wixted & Squire, 2011) and will be discussed in further detail in the next section.

Hippocampal activity during encoding predicts later associative memory

The hippocampus receives direct input from the MTL cortical regions (ERc, PRc, and PHc), each of which receives a distinct pattern of inputs from other neocortical and subcortical regions. Most researchers now agree that the function of the underlying MTL cortex is distinct from the core function of the hippocampus proper. While the precise nature of this division remains unclear, several studies over the past 10 years have provided a broadly consistent picture of the different functions of the hippocampus and MTL cortex in episodic memory. These studies were motivated by an influential computational model of MTL function that posits that item and associative encoding are supported by distinct, yet complementary, learning systems implemented within the hippocampus and PRc (Marr, 1971; McClelland et al., 1995; Norman & O'Reilly, 2003; O'Reilly & Rudy, 2000).

These studies have employed the DM paradigm to differentiate patterns of brain activation during encoding that relate to later successful item-recognition memory from those related to later recovery of associated items, context, or source. Many of these studies have shown that the magnitude of hippocampal-encoding activation relates to participants' later memory for the contextual details associated with individual events (Awipi & Davachi, 2008; Davachi, Mitchell, & Wagner, 2003; Hannula & Ranganath, 2008; Kirwan & Stark, 2004; Ranganath et al., 2004; Staresina & Davachi, 2008, 2009; Staresina, Duncan, & Davachi, 2011; Uncapher, Otten, & Rugg, 2006; Yu, Johnson, & Rugg, 2012). Furthermore, in many such studies, PRc activation during encoding related to whether items were later recognized, regardless of whether additional contextual details were also available at the time of retrieval (Davachi et al., 2003; Haskins, Yonelinas, Quamme, & Ranganath, 2008; Kirwan & Stark, 2004; Ranganath et al., 2004; Staresina & Davachi, 2008, 2009). These highly consistent results across different memory paradigms provide strong evidence for a clear division of labor across MTL regions in their respective contributions to item and associative memory formation.

Interestingly, these distinctions between encoding mechanisms in human PRc and hippocampus correspond with similar distinctions from single-cell recordings in animals (Brown & Aggleton, 2001; Eichenbaum, Fortin, Sauvage, Robitsek, & Farovik, 2010; Komorowski, Manns, & Eichenbaum, 2009; Sauvage, Fortin, Owens, Yonelinas, & Eichenbaum, 2008). Additionally, there is notable evidence from human patient work that damage to the hippocampus disproportionately impairs recollection, compared with item recognition based on familiarity (Giovanello, Verfaellie, & Keane, 2003; Vann et al., 2009; Yonelinas et al., 2002; but see Wixted & Squire, 2004). In contrast, one recent seminal report showed that selective damage to left PRc resulted in a higher than average propensity to recollect events with little evidence of familiarity-based memory (Bowles et al., 2007). This finding is critical because it is consistent with the growing body of literature linking the PRc with item-encoding mechanisms that allow one to later know that item has previously occurred, even in the absence of remembering the specific episodic context.

In addition to dissociations between item and associative processes across MTL regions, there is growing appreciation that PRc and PHc show preferential responses to different kinds of event content and, thus, may contribute to episodic encoding in a *domain-specific* manner, whereas the hippocampus may be important in *domain-general* binding of the various event elements (Davachi, 2006). A variety of evidence supports this framework. First, it has been demonstrated that the human PRc responds more to objects and faces than scenes, and that PHc shows the opposite response pattern: greater activation to scenes than objects and faces (Litman, Awipi, & Davachi, 2009; Liang, Wagner, & Preston, 2013). Second, when study items were scenes and the associated "context" was devised to be one of six repeating objects, PRc encoding activation now predicted later recollection, whereas successful scene memory was supported by PHc (Awipi & Davachi, 2008). Third, both hippocampal and PRc encoding activation are related to whether object details will be later recalled, whereas only hippocampal activation additionally predicts the recovery of other contextual details (Staresina & Davachi, 2008). Finally, in a tightly controlled study where study items were always words but participants' task was to use each cue word to either imagine an object or a scene, it was shown that PRc activation predicted later source memory for the object-imagery trials, and that PHc activation predicted later source memory for the scene-imagery trials (Staresina et al., 2011). Taken together, it is evident that involvement of MTL cortex in encoding is largely dependent on the *content* of the episode and on what aspects of the episode are attended. By contrast, hippocampal activation is selectively related to whether associated details are later recovered, irrespective of the content of those details. These results also highlight that the role of PRc and PHc in item versus associative encoding will vary depending on the nature of the stimuli being treated as the "item" and the "context" (Staresina et al., 2011).

Thus, taken together, there is strong support for the idea that hippocampal activation during an event is correlated with the later recovery of the details associated with that event. More recent work has turned toward examining multivoxel activation patterns in hippocampus and cortex to allow for greater precision in measuring the HNP and CNP associated with encoding to ask whether they are, indeed, related to later memory. Whereas standard functional MRI (fMRI) analyses compare the mean response of a group of contiguous voxels within a region across experimental conditions, multivoxel pattern analysis uses a computer algorithm, known as a classifier, to determine whether or not the pattern of activity across a set of voxels differs between two conditions (see Lewis-Peacock & Norman, chapter 77 in this volume). Using this approach, it has been recently shown that, to the extent to which prefrontal and temporal lobe cortical activation patterns during encoding are similar to category level information, those items are better remembered on a subsequent memory test (Kuhl, Rissman, & Wagner, 2012). A related experiment further showed that the similarity of an item's pattern of cortical activation across repeated study trials with the item predicted better memory (Xue et al., 2010).

Much less is known about the relationship between MTL-encoding patterns and memory formation. However, a recent study showed that cortical similarity (in PRc and PHc) and hippocampal similarity had distinct relationships to later memory. Specifically, similarity between the PRc response to a particular stimulus and other stimuli from the same perceptual category was related to enhanced memory; the reverse relationship was observed in the hippocampus (LaRocque et al., 2013). Finally, recent work also suggests that similarity in the representation of events may be diagnostic for successful encoding even when the stimuli presented are different. Specifically, hippocampal similarity between items in a sequence during encoding is higher for pairs of items later rated as having occurred close together in time (Ezzyat & Davachi, 2014). This finding suggests that similarity in hippocampal response patterns may also be related to binding items with their temporal context (see Manning, Norman, & Kahana, chapter 47 in this volume).

Hippocampal activation relates to successful retrieval

The MAR model proposes that episodic retrieval results from hippocampal pattern-completion processes in response to partial retrieval cues that drive reinstatement of CNP. Accordingly, successful remembering should be associated with enhanced engagement of the hippocampus during episodic retrieval. Consistent with this prediction, human neuroimaging studies have shown increased hippocampal activation accompanies the conscious recollection of studied episodes (Eldridge, Knowlton, Furmanski, Bookheimer, & Engel, 2000; Wheeler & Buckner, 2004) and recollection of contextual details surrounding an item's prior encounter (Cansino, Maquet, Dolan, & Rugg, 2002; Dobbins, Rice, Wagner, & Schacter, 2003).

Hippocampal activation is also related to successful recognition of arbitrary item associations (Giovanello, Schnyer, & Verfaellie, 2004; Kirwan & Stark, 2004). In one such study, hippocampal activation during presentation of a partial retrieval cue predicted participants' memory accuracy on an associative memory probe moments later (Chen, Olsen, Preston, Glover, & Wagner, 2011). The finding that increased hippocampal activation prior to a memory decision was associated with improved performance likely reflects successful pattern completion to the encoded associate from the partial cue. Converging evidence from rodents also demonstrates the critical link between the hippocampus and episodic remembering, as lesions to the hippocampus selectively impair recollection-like memory while leaving intact familiarity processes that enable simple judgments of whether or not a stimulus is old or new (Fortin, Wright, & Eichenbaum, 2004).

Two recent multivoxel pattern analysis (MVPA) studies have further shown that the distributed pattern of hippocampal activation evoked during vivid recall distinguishes between individual memories for real-world actions performed by people in different contexts (Chadwick, Hassabis, Weiskopf, & Maguire, 2010), even when those events share overlapping features (Chadwick, Hassabis, & Maguire, 2011). These findings are consistent with the idea that the hippocampus magnifies the distinctions between highly similar events—a process termed *pattern separation* (for expanded discussion, see Leutgeb & Leutgeb, chapter 46 in this volume)—resulting in separable memory traces that reduce susceptibility to mnemonic interference. This empirical evidence thus suggests that recollecting a particular event requires reactivating the specific hippocampal representation associated with that event.

Hippocampal and cortical reinstatement during retrieval

A key prediction of the MAR model is that successful episodic remembering is the result of pattern-completion processes that lead to reactivation of the same hippocampal and cortical neurons and patterns of neural activity that were active during the initial experience. While the research reviewed above indicates that MTL responses at retrieval relate to the subjective experience of memory, such studies do not directly compare encoding and retrieval activity patterns, which would be the key test of the reinstatement hypothesis.

In the last decade, several neuroimaging studies have supported the idea that partial cues lead to the reactivation of CNP during retrieval. For instance, in a seminal study, participants learned associations between words (e.g., "dog") and either corresponding sounds ("Woof!") or corresponding pictures (a picture of a dog). During retrieval, words were presented as cues, and participants indicated whether the cue word was presented with a sound or picture (Wheeler, Petersen, & Buckner, 2000). Consistent with the reinstatement hypothesis, retrieval cue words that were paired with sounds elicited activation in the same auditory cortical regions that were active during encoding, while the cue words paired with pictures elicited activation of visual cortical regions preferentially activated by pictures during encoding. More recent work using MVPA has also shown that patterns of cortical activation during category-specific retrieval resemble patterns of activation during encoding (Polyn, Natu, Cohen, & Norman, 2005; Ritchey, Wing, Labar, & Cabeza, 2012). Moreover, Polyn et al. (2005) found that reactivation of category-specific neural encoding patterns preceded participants' verbal recall responses, providing an important link between cortical reinstatement and behavior.

While these findings provide clear evidence that the CNP during encoding is reinstated during successful memory retrieval, they do not speak to the hypothesis that the hippocampus mediates such reinstatement. Using a multivariate analysis approach known as *representational similarity* (Kriegeskorte, Mur, & Bandettini, 2008), Ritchey et al. (2012) found that remembered scenes were associated with greater similarity between the encoding and retrieval patterns evoked for individual scenes. One caveat, however, was that the same scenes were presented both at encoding and retrieval, raising the question of how much of the similarity measures were driven by stimulus processing versus memory-related reinstatement. Nonetheless, they found a significant correlation between encoding-retrieval pattern similarity in cortex and hippocampal activation during retrieval, with hippocampal engagement mediating the link between cortical encoding-retrieval similarity and memory success. These results are consistent with the idea that the hippocampus facilitates reinstatement of cortical memory traces during successful remembering.

However, these findings do not speak to a core aspect of the reinstatement hypothesis—that hippocampal and MTL cortical-encoding patterns themselves are reinstated during memory retrieval. Leading memory models posit that retrieval should not simply activate the MTL, but would specifically activate the same MTL neurons that were active during the original encoding experience. A key demonstration of this principle comes from a landmark study using intracranial single-cell recordings in epileptic patients (Gelbard-Sagiv, Mukamel, Harel, Malach, & Fried, 2008). Hippocampal and MTL cortical neurons were recorded while patients viewed and later recalled a series of short video clips (e.g., a scene from *The Simpsons*). They found that a subset of MTL neurons showed selective firing during initial viewing of the video clips (e.g., one hippocampal neuron fired specifically during *The Simpsons* clip). Critically, they found that these selective MTL neurons were also active during verbal recall of the same videos (i.e., the "*Simpsons*" neuron fired again when the participant recalled *The Simpsons* video clip). Consistent with this electrophysiological work in humans, physiological studies in rodents have also demonstrated reactivation of hippocampal memory traces during retrieval (for a review, see Carr, Jadhav, & Frank, 2011). For instance, hippocampal place cells representing a spatial trajectory through a well-learned environment are replayed in sequence at remote time points as the animals experience new environments (Karlsson & Frank, 2009)—a phenomenon referred to as *hippocampal replay*. Moreover, recent work has shown that interrupting hippocampal replay impairs rodents' ability to correctly navigate to rewards in a well-learned environment (Jadhav, Kemere, German, & Frank, 2012), providing further evidence for the link between hippocampal replay and memory-guided decision making. Collectively, these findings provide strong evidence that the reinstatement of hippocampal memory traces plays an important role in guiding behavior and choice.

Until the recent advent of multivariate fMRI analysis methods, documenting reactivation of MTL-encoding patterns during retrieval in the human brain had proved challenging. One recent study achieved this goal by showing that activation patterns in human MTL cortex for unique word-scene associations during encoding were reinstated during retrieval (Staresina, Henson, Kriegeskorte, & Alink, 2012). Participants were presented with word cues during retrieval and asked to recall the corresponding scene; reinstatement of MTL cortical-encoding patterns was observed only when subjects successfully recalled the scene, providing the first evidence for MTL reinstatement in the human brain. Notably, while this study did not measure hippocampal reinstatement, the degree of MTL cortical reinstatement was correlated with hippocampal retrieval activation, consistent with a role for the hippocampus in coordinating CNP reinstatement.

Memory reinstatement during new encoding

As highlighted in the previous section, memories for past events are often reinstated during new experiences. These findings emphasize that memory encoding and retrieval are not performed in isolation, but rather are interactive processes. Recent human neuroimaging research indicates that new learning influences, and is influenced by, reactivation of existing memories. In one such study, participants learned overlapping (e.g., watch-sink and later watch-pipe) and nonoverlapping (e.g., peanut—moose) pairs of pictures during fMRI scanning (Kuhl, Shah, DuBrow, & Wagner, 2010). The findings revealed that the degree to which prior memories (e.g., memory for the watch-sink event) were reinstated during encoding of new overlapping experiences (watch-pipe) was associated with greater retention of the originally learned information when compared to memory for nonoverlapping information. Moreover, hippocampal engagement during encoding of the overlapping pairs was related to both cortical memory reinstatement and improved memory retention. These findings indicate that memory reinstatement during new encoding helps reduce forgetting of past events. One possibility is that reduced forgetting was not just the result of strengthening of reactivated memories, but may also have resulted from a hippocampal-mediated integrative encoding mechanism, whereby newly encountered information is integrated with existing memories at the time of learning (Shohamy & Wagner, 2008; for a review, see Zeithamova, Schlichting, & Preston, 2012).

According to this mechanism, the fundamental role of the hippocampus in memory is not only to form relationships among elements within an individual experience, but also to construct memory representations that link memory elements across discrete experiences. This constructive hypothesis about hippocampal function dates back to Tolman's concept of a "cognitive map" (Tolman, 1948) and is further exemplified in more modern models of hippocampal function (Cohen & Eichenbaum, 1993; Kumaran & McClelland, 2012; O'Keefe & Nadel, 1978). Integrative encoding proposes that when a new event shares a common feature, or features, with an existing memory trace, the common features elicit memory reinstatement through hippocampal pattern completion. New experiences would then not only be encoded in the context of presently available information in the external world, but would also be bound to reactivated representations of prior related memories. Furthermore, this mechanism makes the fundamental prediction that by combining information across discrete events, hippocampal memory representations would include information that goes beyond direct experience. Accordingly, integrative encoding would not only support strengthening of existing memories (e.g., Kuhl et al., 2010), but would also support novel inferences about the relationships between memory elements that were experienced at different times and generalization of knowledge to entirely new situations, a hallmark of episodic memory.

In recent years, several human neuroimaging studies have set out to test the hypothesis that the hippocampus supports integrative encoding through memory reinstatement. These studies typically employ similar paradigms wherein participants learn a set of overlapping associations (e.g., A goes with B, B goes with C) during fMRI scanning. The existence of an integrated memory organization is demonstrated by the ability of participants to express knowledge about the inferential relationships among elements of overlapping pairs that were not explicitly studied together (i.e., A is related to C). Using this paradigm, two neuroimaging studies have shown that hippocampal activation during initial encoding of overlapping pairs is uniquely related to participants' ability to make subsequent inference judgments (Shohamy & Wagner, 2008; Zeithamova & Preston, 2010).

A more recent study using MVPA further demonstrated that participants reinstate memories for prior associations in cortex (e.g., the A stimulus from an AB pairing) while encoding new related experiences (BC pairs; Zeithamova, Dominick, & Preston, 2012). Moreover, the degree to which participants reinstated prior events during encoding of new associations was related to their ability to infer relationships between distinct events that shared content, with hippocampal encoding activation further tracking the successful use of integrated memories. Other work has shown that memory reinstatement and hippocampal-mediated integration takes place even when participants do not explicitly recall overlapping memories during new event encoding (Wimmer & Shohamy, 2012). Thus, converging evidence from these studies provides evidence for a hippocampal-mediated encoding mechanism whereby overlapping experiences are integrated into a network of related memories as they are learned. Together, these findings provide a deeper understanding of how remembering the past influences how we learn about the present.

Conclusion

In summary, there is strong evidence for hippocampal and cortical contributions to encoding and retrieval and growing evidence for the MAR model supporting the widely held view that memory is supported by the reinstatement of brain patterns that characterize a prior encoding experience. Furthermore, recent work using multivariate fMRI approaches suggests that reinstatement not only strengthens existing memories but also contributes to establishing links across related experiences. There is still much to learn, however, as understanding what aspects of a prior encounter are reinstated is still unknown. Furthermore, very little is known about the temporal relationship between hippocampal and cortical reinstatement.

REFERENCES

Alvarez, P., & Squire, L. R. (1994). Memory consolidation and the medial temporal lobe: A simple network model. *Proc Natl Acad Sci USA, 91*(15), 7041–7045.

Awipi, T., & Davachi, L. (2008). Content-specific source encoding in the human medial temporal lobe. *J Exp Psychol Learn Mem Cogn, 34*(4), 769–779.

Bowles, B., Crupi, C., Mirsattari, S. M., Pigott, S. E., Parrent, A. G., Pruessner, J. C., ... Köhler, S. (2007). Impaired familiarity with preserved recollection after anterior temporal-lobe resection that spares the hippocampus. *Proc Natl Acad Sci USA, 104*(41), 16382–16387.

Brewer, J. B., Zhao, Z., Desmond, J. E., Glover, G. H., & Gabrieli, J. D. (1998). Making memories: Brain activity that predicts how well visual experience will be remembered. *Science, 281*(5380), 1185–1187.

Brown, M. W., & Aggleton, J. P. (2001). Recognition memory: What are the roles of the perirhinal cortex and hippocampus? *Nat Rev Neurosci, 2*(1), 51–61.

Cansino, S., Maquet, P., Dolan, R. J., & Rugg, M. D. (2002). Brain activity underlying encoding and retrieval of source memory. *Cereb Cortex, 12*(10), 1048–1056.

Carr, M. F., Jadhav, S. P., & Frank, L. M. (2011). Hippocampal replay in the awake state: A potential substrate for memory consolidation and retrieval. *Nat Neurosci, 14*(2), 147–153.

Chadwick, M. J., Hassabis, D., & Maguire, E. A. (2011). Decoding overlapping memories in the medial temporal lobes using high-resolution fMRI. *Learn Mem, 18*(12), 742–746.

Chadwick, M. J., Hassabis, D., Weiskopf, N., & Maguire, E. A. (2010). Decoding individual episodic memory traces in the human hippocampus. *Curr Biol, 20*(6), 544–547.

Chen, J., Olsen, R. K., Preston, A. R., Glover, G. H., & Wagner, A. D. (2011). Associative retrieval processes in the human medial temporal lobe: Hippocampal retrieval success and CA1 mismatch detection. *Learn Mem, 18*(8), 523–528.

Cohen, N. J., & Eichenbaum, H. E. (1993). *Memory, amnesia, and the hippocampal system.* Cambridge, MA: MIT Press.

Davachi, L. (2006). Item, context and relational episodic encoding in humans. *Curr Opin Neurobiol, 16*(6), 693–700.

Davachi, L., & Danker, J. F. (2013). The cognitive neuroscience of episodic memory. In K. N. Ochsner & S. M. Kosslyn (Eds.), *The Oxford handbook of cognitive neuroscience* (Vol. I, pp. 375–388). New York, NY: Oxford University Pres.

Davachi, L., Mitchell, J. P., & Wagner, A. D. (2003). Multiple routes to memory: Distinct medial temporal lobe processes build item and source memories. *Proc Natl Acad Sci USA, 100*(4), 2157–2162.

Diana, R. A., Yonelinas, A. P., & Ranganath, C. (2007). Imaging recollection and familiarity in the medial temporal lobe: A three-component model. *Trends Cogn Sci, 11*(9), 379–386.

Dobbins, I. G., Rice, H. J., Wagner, A. D., & Schacter, D. L. (2003). Memory orientation and success: Separable neurocognitive components underlying episodic recognition. *Neuropsychologia, 41*(3), 318–333.

Eichenbaum, H., Fortin, N., Sauvage, M., Robitsek, R. J., & Farovik, A. (2010). An animal model of amnesia that uses Receiver Operating Characteristics (ROC) analysis to distinguish recollection from familiarity deficits in recognition memory. *Neuropsychologia, 48*(8), 2281–2289.

Eichenbaum, H., Sauvage, M., Fortin, N., Komorowski, R., & Lipton, P. (2012). Towards a functional organization of episodic memory in the medial temporal lobe. *Neurosci Biobehav Rev, 36*(7), 1597–1608.

Eichenbaum, H., Yonelinas, A. P., & Ranganath, C. (2007). The medial temporal lobe and recognition memory. *Annu Rev Neurosci, 30,* 123–152.

Eldridge, L. L., Knowlton, B. J., Furmanski, C. S., Bookheimer, S. Y., & Engel, S. A. (2000). Remembering episodes: A selective role for the hippocampus during retrieval. *Nat Neurosci, 3,* 1149–1152.

Ezzyat, Y., & Davachi, L. (2014). Similarity breeds proximity: Pattern similarity within and across contexts is related to later mnemonic judgments of temporal proximity. *Neuron, 81*(5), 1179–1189.

Fortin, N. J., Wright, S. P., & Eichenbaum, H. (2004). Recollection-like memory retrieval in rats is dependent on the hippocampus. *Nature, 431*(7005), 188–191.

Gelbard-Sagiv, H., Mukamel, R., Harel, M., Malach, R., & Fried, I. (2008). Internally generated reactivation of single neurons in human hippocampus during free recall. *Science, 322*(5898), 96–101.

Giovanello, K. S., Schnyer, D. M., & Verfaellie, M. (2004). A critical role for the anterior hippocampus in relational memory: Evidence from an fMRI study comparing associative and item recognition. *Hippocampus, 14*(1), 5–8.

Giovanello, K. S., Verfaellie, M., & Keane, M. M. (2003). Disproportionate deficit in associative recognition relative to item recognition in global amnesia. *Cogn Affect Behav Neurosci, 3*(3), 186–194.

Hannula, D. E., & Ranganath, C. (2008). Medial temporal lobe activity predicts successful relational memory binding. *J Neurosci, 28*(1), 116–124.

Haskins, A. L., Yonelinas, A. P., Quamme, J. R., & Ranganath, C. (2008). Perirhinal cortex supports encoding and familiarity-based recognition of novel associations. *Neuron, 59*(4), 554–560.

Hebb, D. O. (1949). Temperament in chimpanzees; method of analysis. *J Comp Physiol Psychol, 42*(3), 192–206.

Jadhav, S. P., Kemere, C., German, P. W., & Frank, L. M. (2012). Awake hippocampal sharp-wave ripples support spatial memory. *Science, 336*(6087), 1454–1458.

Karlsson, M. P., & Frank, L. M. (2009). Awake replay of remote experiences in the hippocampus. *Nat Neurosci, 12*(7), 913–918.

Kirwan, C. B., & Stark, C. E. (2004). Medial temporal lobe activation during encoding and retrieval of novel face-name pairs. *Hippocampus, 14*(7), 919–930.

Komorowski, R. W., Manns, J. R., & Eichenbaum, H. (2009). Robust conjunctive item-place coding by hippocampal neurons parallels learning what happens where. *J Neurosci, 29*(31), 9918–9929.

Kriegeskorte, N., Mur, M., & Bandettini, P. (2008). Representational similarity analysis: Connecting the branches of systems neuroscience. *Front Syst Neurosci, 2*(4), 1–28.

Kuhl, B. A., Rissman, J., & Wagner, A. D. (2012). Multi-voxel patterns of visual category representation during episodic encoding are predictive of subsequent memory. *Neuropsychologia, 50*(4), 458–469.

Kuhl, B. A., Shah, A. T., DuBrow, S., & Wagner, A. D. (2010). Resistance to forgetting associated with hippocampus-mediated reactivation during new learning. *Nat Neurosci, 13*(4), 501–506.

Kumaran, D., & McClelland, J. L. (2012). Generalization through the recurrent interaction of episodic memories: A model of the hippocampal system. *Psychol Rev, 119*(3), 573–616.

LaRocque, K. F., Smith, M. E., Carr, V. A., Witthoft, N., Grill-Spector, K., & Wagner, A. D. (2013). Global similarity and pattern separation in the human medial temporal lobe predict subsequent memory. *J Neurosci, 33*(13), 5466–5474.

Liang, J. C., Wagner, A. D., & Preston, A. R. (2013). Content representation in the human medial temporal lobe. *Cereb Cortex, 23*(1), 80–96.

Litman, L., Awipi, T., & Davachi, L. (2009). Category-specificity in the human medial temporal lobe cortex. *Hippocampus, 19*(3), 308–319.

Marr, D. (1971). Simple memory: A theory for archicortex. *Philos Trans R Soc Lond B Biol Sci, 262*(841), 23–81.

McClelland, J. L., McNaughton, B. L., & O'Reilly, R. C. (1995). Why there are complementary learning systems in the hippocampus and neocortex: Insights from the successes and failures of connectionist models of learning and memory. *Psychol Rev, 102,* 419–457.

Moscovitch, M., Rosenbaum, R. S., Gilboa, A., Addis, D. R., Westmacott, R., Grady, C., & Nadel, L. (2005). Functional neuroanatomy of remote episodic, semantic and spatial memory: A unified account based on multiple trace theory. *J Anat, 207*(1), 35–66.

Norman, K. A., & O'Reilly, R. C. (2003). Modeling hippocampal and neocortical contributions to recognition memory: A complementary-learning-systems approach. *Psychol Rev, 110*(4), 611–646.

O'Keefe, J., & Nadel, L. (1978). *The hippocampus as a cognitive map*. London: Clarendon.

O'Reilly, R. C., & Rudy, J. W. (2000). Computational principles of learning in the neocortex and hippocampus. *Hippocampus, 10*(4), 389–397.

Paller, K. A., & Wagner, A. D. (2002). Observing the transformation of experience into memory. *Trends Cogn Sci, 6*(2), 93–102.

Polyn, S. M., Natu, V. S., Cohen, J. D., & Norman, K. A. (2005). Category-specific cortical activity precedes retrieval during memory search. *Science, 310*(5756), 1963–1966.

Ranganath, C., Yonelinas, A. P., Cohen, M. X., Dy, C. J., Tom, S. M., & D'Esposito, M. (2004). Dissociable correlates of recollection and familiarity within the medial temporal lobes. *Neuropsychologia, 42*(1), 2–13.

Ritchey, M., Wing, E. A., Labar, K. S., & Cabeza, R. (2012). Neural similarity between encoding and retrieval is related to memory via hippocampal interactions. *Cereb Cortex, 23*(12), 2818–2828.

Sauvage, M. M., Fortin, N. J., Owens, C. B., Yonelinas, A. P., & Eichenbaum, H. (2008). Recognition memory: Opposite effects of hippocampal damage on recollection and familiarity. *Nat Neurosci, 11*(1), 16–18.

Shohamy, D., & Wagner, A. D. (2008). Integrating memories in the human brain: Hippocampal-midbrain encoding of overlapping events. *Neuron, 60*(2), 378–389.

Staresina, B. P., & Davachi, L. (2008). Selective and shared contributions of the hippocampus and perirhinal cortex to episodic item and associative encoding. *J Cogn Neurosci, 20*(8), 1478–1489.

Staresina, B. P., & Davachi, L. (2009). Mind the gap: Binding experiences across space and time in the human hippocampus. *Neuron, 63*(2), 267–276.

Staresina, B. P., Duncan, K. D., & Davachi, L. (2011). Perirhinal and parahippocampal cortices differentially contribute to later recollection of object- and scene-related event details. *J Neurosci, 31*(24), 8739–8747.

Staresina, B. P., Henson, R. N., Kriegeskorte, N., & Alink, A. (2012). Episodic reinstatement in the medial temporal lobe. *J Neurosci, 32*(50), 18150–18156.

Tolman, E. C. (1948). Cognitive maps in rats and men. *Psychol Rev, 55*(4), 189–208.

Uncapher, M. R., Otten, L. J., & Rugg, M. D. (2006). Episodic encoding is more than the sum of its parts: An fMRI investigation of multifeatural contextual encoding. *Neuron, 52*(3), 547–556.

Vann, S. D., Tsivilis, D., Denby, C. E., Quamme, J. R., Yonelinas, A. P., Aggleton, J. P., ... Mayes, A. R. (2009). Impaired recollection but spared familiarity in patients with extended hippocampal system damage revealed by 3 convergent methods. *Proc Natl Acad Sci USA, 106*(13), 5442–5447.

Wagner, A. D., Schacter, D. L., Rotte, M., Koutstaal, W., Maril, A., Dale, A. M., ... Buckner, R. L. (1998). Building memories: Remembering and forgetting of verbal experiences as predicted by brain activity. *Science, 281*(5380), 1188–1191.

Wheeler, M. E., & Buckner, R. L. (2004). Functional-anatomic correlates of remembering and knowing. *NeuroImage, 21*(4), 1337–1349.

Wheeler, M. E., Petersen, S. E., & Buckner, R. L. (2000). Memory's echo: Vivid remembering reactivates sensory-specific cortex. *Proc Natl Acad Sci USA, 97*(20), 11125–11129.

Wimmer, G. E., & Shohamy, D. (2012). Preference by association: How memory mechanisms in the hippocampus bias decisions. *Science, 338*(6104), 270–273.

Wixted, J. T., & Squire, L. R. (2004). Recall and recognition are equally impaired in patients with selective hippocampal damage. *Cogn Affect Behav Neurosci, 4*(1), 58–66.

Wixted, J. T., & Squire, L. R. (2011). The medial temporal lobe and the attributes of memory. *Trends Cogn Sci, 15*(5), 210–217.

Xue, G., Dong, Q., Chen, C., Lu, Z., Mumford, J. A., & Poldrack, R. A. (2010). Greater neural pattern similarity across repetitions is associated with better memory. *Science, 330*(6000), 97–101.

Yonelinas, A. P., Kroll, N. E., Quamme, J. R., Lazzara, M. M., Sauve, M. J., Widaman, K. F., & Knight, R. T. (2002). Effects of extensive temporal lobe damage or mild hypoxia on recollection and familiarity. *Nat Neurosci, 5*(11), 1236–1241.

Yu, S. S., Johnson, J. D., & Rugg, M. D. (2012). Hippocampal activity during recognition memory co-varies with the accuracy and confidence of source memory judgments. *Hippocampus, 22*(6), 1429–1437.

Zeithamova, D., Dominick, A. L., & Preston, A. R. (2012). Hippocampal and ventral medial prefrontal activation during retrieval-mediated learning supports novel inference. *Neuron, 75*(1), 168–179.

Zeithamova, D., & Preston, A. R. (2010). Flexible memories: Differential roles for medial temporal lobe and prefrontal cortex in cross-episode binding. *J Neurosci, 30*(44), 14676–14684.

Zeithamova, D., Schlichting, M. L., & Preston, A. R. (2012). The hippocampus and inferential reasoning: Building memories to navigate future decisions. *Front Hum Neurosci, 6*, 70.

46 The Contribution of Hippocampal Subregions to Memory Coding

JILL K. LEUTGEB AND STEFAN LEUTGEB

ABSTRACT Episodic memories contain information about what happened, where, when, and in which sequence. The formation of episodic memories requires the medial temporal lobe, which consists of subregions with unique connection patterns and specialized cell types. Each subregion can perform specialized computations to support memory processing. Particular computations may selectively support the acquisition, storage, or retrieval phase, or may be flexibly used across the different phases of memory coding. This chapter describes processing streams in the medial temporal lobe, specialized circuits in each subregion of the hippocampal formation, and how they act in concert to support the various aspects of episodic memory.

Hierarchical processing across subregions in the medial temporal lobe

The medial temporal lobe is essential for memories of facts and events. Patient H.M. and similar patients with extensive lesions in the medial temporal lobe have revealed that the loss of these brain regions results in a profound impairment in forming new semantic and episodic memories and in remembering events that occurred over the previous months and years before the brain damage (Scoville & Milner, 1957; Squire & Alvarez, 1995). However, the extent of brain damage in patients may go beyond the areas that are essential for memory or only include a subset of the areas that can support memory. The subset of structures in the medial temporal lobe required for memory has thus been defined with better precision in animal models of amnesia in which the damage can be targeted to selected subregions (Zola-Morgan & Squire, 1993). The core circuit for event memories includes, in primates, the perirhinal cortex, the parahippocampal cortex, the entorhinal cortex, the dentate gyrus, the CA regions of the hippocampus proper (cornu ammonis, or Ammon's horn), and the subiculum. These brain regions are phylogenetically conserved divisions of the cortex and can thus be readily identified across all mammalian lineages. A consistent nomenclature is generally used in different mammalian species, with the notable exception that the parahippocampal cortex in primates and the postrhinal cortex in rodents are corresponding (figure 46.1A, B). Throughout this chapter we will use the rodent nomenclature, focus on rodent circuits, and mention notable differences to primates.

The regions of the medial temporal lobe that comprise the core circuit for memory are arranged in three tiers that are reciprocally connected (figure 46.1C). The first tier integrates information from various cortical association areas and consists of the perirhinal and postrhinal cortex. The second tier consists of the entorhinal cortex, which can be subdivided into lateral and medial divisions (anterior and posterior divisions in nonhuman primates). The lateral and medial divisions are preferentially connected with the perirhinal and postrhinal cortex, respectively. The third tier includes the dentate gyrus, the hippocampal CA regions, and the subiculum. This tier is commonly referred to as the hippocampal formation. In addition to the reciprocal connections between the first and the second and between the second and the third tier, subregions within a tier have uniquely arranged local circuits and consist of distinct anatomical and functional cell types (van Strien, Cappaert, & Witter, 2009). Each subregion is thus thought to act as a specialized processing module that contributes in a unique way to semantic and, in particular, episodic memories.

Episodic memories are composed of three fundamental elements, namely, "where," "when," and "what" (Tulving, 1983). The "where" and "what" elements are initially computed in separate processing streams from sensory cortices to association cortices and are eventually bound together into a unified representation in the hippocampus (Eichenbaum, Sauvage, Fortin, Komorowski, & Lipton, 2011). The "when" element may be directly generated by intrahippocampal computations. This chapter describes how the separate processing streams converge and how the information is further processed in the hippocampus. These computations are known to give rise to place cells in the hippocampus of rodents (O'Keefe & Nadel, 1978) and to related cell types in the hippocampus of primates (Ekstrom et al., 2003; Rolls, 1999). Place cells are

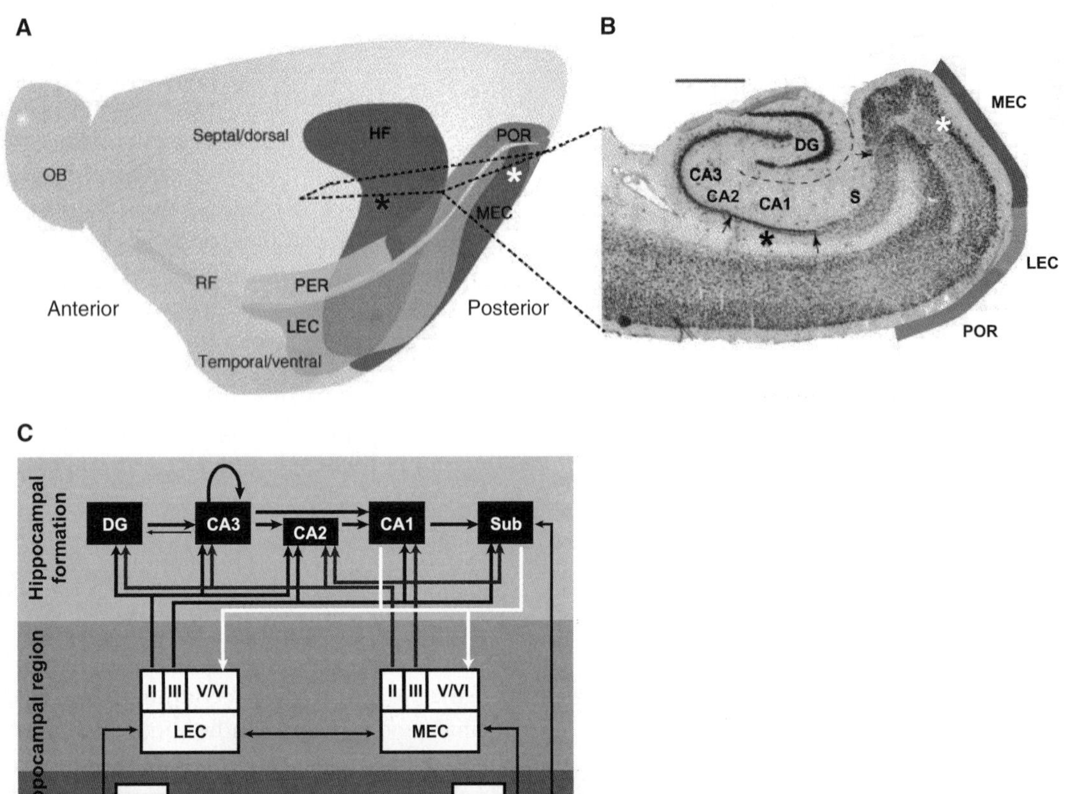

FIGURE 46.1 Anatomy and connectivity of the rodent medial temporal lobe. (A) Schematic of a rat brain, lateral view of the left hemisphere. Areas important for memory and spatial navigation are highlighted and labeled. The dashed box indicates the position of a horizontal section taken through the hippocampal formation (HF) and adjacent cortical structures as shown in (B). (B) Stain of neurons in the HF: dentate gyrus (DG), CA3, CA2, CA1, and subiculum (S); and in the surrounding input and output structures: medial and lateral entorhinal cortex (MEC and LEC), perirhinal cortex (PER), and postrhinal cortex (POR). Asterisks in black and white indicate corresponding positions across the two images. Scale bar, 100 μm. (Adapted from Witter et al., 2006, with permission from Elsevier.) (C) Circuit diagram of brain regions for memory processing. Different box shades depict three tiers in the cortical hierarchy. (Adapted by permission from Macmillan Publishers Ltd: *Nature Reviews Neuroscience*, from van Strien et al., 2009.)

principal neurons that reliably fire action potentials in a particular spatial location (figure 46.2A). However, consistent with the convergence of multiple pathways in the hippocampus, place cells can also integrate other types of information or even switch between representing nonspatial and spatial information (Wood, Dudchenko, & Eichenbaum, 1999). For example, modulation of firing rates at the preferred firing location of a place cell can code for differences in context or in internal state (Kennedy & Shapiro, 2009; S. Leutgeb et al., 2005b). Importantly, the hippocampal code is distributed and combinatorial, such that some cells may be more exclusively spatially tuned while others may be particularly responsive to differences in visual stimuli, odors, elapsed time, or task demands.

Hippocampal cell populations do not only receive information to conjointly represent different types of information, but can also perform a number of unique network computations, such as making input patterns more distinct (i.e., pattern separation) or processing sequence and temporal information. The outcomes of these computations can then be relayed back to cortical areas to generate enriched representations that include object, spatial, and temporal information as well as information about the sequences of current and past events.

Separate processing streams throughout cortical modules

The notion that different types of information are initially processed separately is based on anatomical and physiological evidence. In primates, cortical processing is characterized by parallel pathways (e.g., dorsal and ventral visual stream, and polymodal and unimodal inputs to parahippocampal and perirhinal cortices;

Suzuki & Amaral, 1994; Ungerleider & Mishkin, 1982). In rodents, posterior association areas process visuospatial information and project predominantly to postrhinal cortex, whereas anterior association areas process unimodal sensory information and project predominantly to perirhinal cortex (van Strien, Cappaert, & Witter, 2009). The segregated anatomical pathways result in neural representations for "what" information primarily in the perirhinal cortex and for "where" information primarily in the postrhinal cortex. This segregation is largely retained in the next processing stage. The medial entorhinal cortex (MEC) receives most of its input from the postrhinal cortex. The lateral entorhinal cortex (LEC) receives most of its input from the perirhinal cortex. Accordingly, MEC or postrhinal lesions selectively affect navigation and pathfinding, whereas LEC or perirhinal lesions primarily affect nonspatial information processing and, more generally, object-context associations (Eichenbaum et al., 2011; Suzuki & Amaral, 2004). The idea for a segregation of function at the level of the entorhinal cortex is also supported by the selective presence of specialized cell types in each subregion, as described in the following sections.

Medial Entorhinal Cortex In rodents, MEC can be distinguished from LEC by differences in cytoarchitecture and in the input and output pathways. In addition, the principal neurons of these two entorhinal subregions have distinct firing patterns (Witter & Moser, 2006). The most remarkable cell type in MEC is the grid cell. Grid cells are principal neurons that fire action potentials at multiple spatial locations within an environment. The spatial receptive fields of these neurons are arranged in a highly regular triangular grid (figure 46.2B). Grid cells were first described when performing extracellular recordings from the MEC in awake, behaving rats that explored two-dimensional environments (Hafting, Fyhn, Molden, Moser, & Moser, 2005). Corresponding cells are also found in the primate entorhinal cortex (Killian, Jutras, & Buffalo, 2012). In primates, the grid pattern emerges while a spatial scene is explored by eye movements. During eye movements, the firing peaks for where on the screen the eyes fixate are arranged in a triangular grid pattern (figure 46.2C). Although grid cells are a prominent cell type in the most dorsal pole of MEC in rodents and a common cell type in the posterior entorhinal cortex of primates, there are also numerous other functional cell types in entorhinal cortex. Some of these cell types have obvious spatial firing patterns, such as firing along the borders of a box in rodents or along the border of a screen in primates (Killian et al., 2012; Solstad, Boccara, Kropff, Moser, & Moser, 2008). However, there are also cells

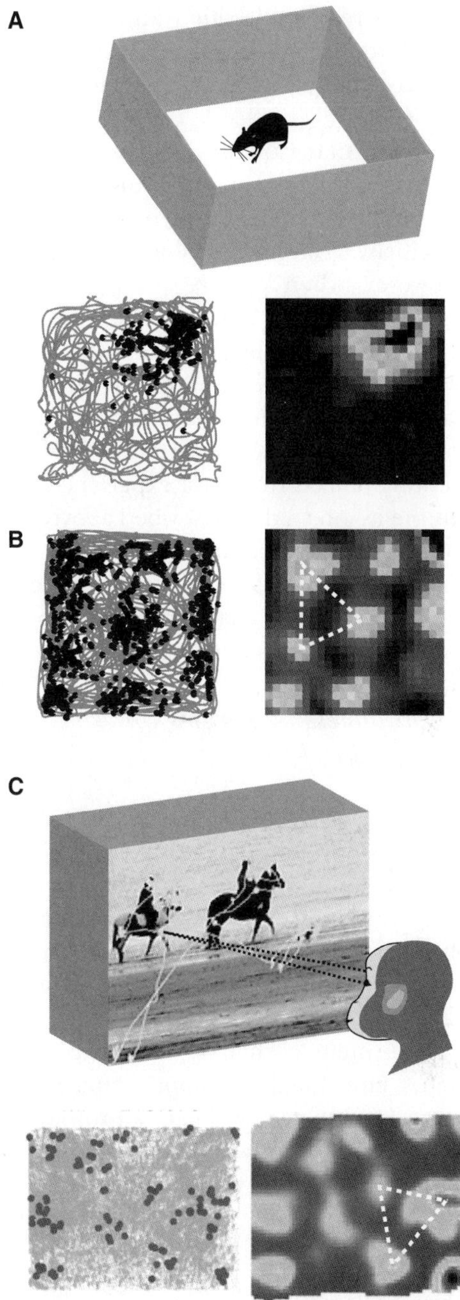

Figure 46.2 Place cells and grid cells. (A) Spatially tuned principal neurons are found in all subregions of the rodent hippocampal formation. A gray line indicates the path of the animal while exploring a 1 m × 1 m open field. Each action potential of the CA1 neuron is shown as a red dot superimposed on the position of the animal when the spike occurred (left). The firing rate of the same CA1 neuron is represented by a heat map with high firing rates in warmer colors (right). Hippocampal neurons are place cells with a single place field, whereas (B) grid cells are found in all layers of medial entorhinal cortex (MEC). The multiple firing peaks of grid cells are arranged as equilateral triangles (white dashed lines) that form a grid-like representation (right). (C) Grid cells in the primate MEC during the visual exploration of spatial scenes. (Reprinted by permission from Macmillan Publishers Ltd: *Nature*, Killian et al., 2012.) (See color plate 38.)

that cannot be readily classified and do not exhibit spatial firing properties, and those are more abundant in primates compared to rodents (Killian et al., 2012).

Lateral Entorhinal Cortex Unlike MEC, in which individual neurons exhibit spatial firing properties, the firing patterns of LEC principal neurons show little spatial specificity (Hargreaves, Rao, Lee, & Knierim, 2005). However, when objects are placed within a recording environment, LEC cells fire selectively at locations relative to the object (Deshmukh & Knierim, 2013) or in specific places where objects were previously located (Tsao, Moser, & Moser, 2013). The firing of LEC neurons in the vicinity of discrete objects and in representing objects from past experiences suggests, along with the lesion studies described above, that LEC cells support object-place memory or item-context memory, hence serving as the only subdivision of the entorhinal cortex that to a larger degree supports the "what" component of episodic memory.

The different processing streams converge in hippocampal subregions

The entorhinal cortex is the gateway to the dentate gyrus and to the hippocampal CA fields. All major excitatory inputs to hippocampus originate from neurons in the entorhinal cortex, and the entorhinal cortex receives back-projections from CA1 and the subiculum. Because the entorhino-hippocampal loops are at the core of memory processing and because disorders with memory impairment each have different patterns of cell loss across entorhinal and hippocampal subregions, it can be expected that we will gain a profound understanding of memory by unraveling the information flow within the circuit. Although questions about memory processing will eventually need to be addressed in complex memory tasks across different species, substantial progress has been made from examining how spatial firing patterns emerge and are processed throughout the entorhino-hippocampal circuit in rodents. This section therefore focuses on our current understanding of the processing of spatial information within the circuit along with key insights from studies on memory processing.

Early Processing Stages of the Hippocampal Formation

Connections of the dentate gyrus and the CA3 subregion The dentate gyrus (DG) is classically considered the first processing stage of the hippocampal formation. The inputs of entorhinal cortex layer II neurons to the principal neurons of the DG, the dentate granule neurons, comprise the first synapse of the "tri-synaptic pathway." The dentate granule neurons send projections, the mossy fibers, which terminate on proximal CA3 dendrites. This is the second synapse. The third synapse is from CA3 to CA1. While the first and the third synapses in the pathway are similar to typical cortical synapses, the synapses between dentate granule cells and CA3 pyramidal neurons are rather exceptional. They are extremely large, and transmitter release from a single bouton is sufficiently strong to reliably result in an action potential in the target CA3 cell (Henze, Wittner, & Buzsaki, 2002). However, each granule cell only makes contact with approximately 50 CA3 cells.

The hippocampal CA3 region receives, in addition to inputs from dentate granule cells, direct inputs from layer II of LEC and MEC. Both LEC and MEC terminate onto the distal portion of CA3 dendrites. The inputs that CA3 cells receive from entorhinal cortex are often from the same en passant axons that also form synapses onto dentate granules cells. Principal neurons in DG and in CA3 thus receive convergent inputs from LEC and MEC (Witter, 2007).

Processing within dentate gyrus Predictions about a separate role of DG in memory processing are based on elements of its architecture that are unique for the DG subregion as well as on the sparse firing of dentate neurons. The firing in DG is sparse in that the mean firing rates of dentate granule neurons are low and in that the proportion of active neurons at any given time is also low (Piatti, Ewell, & Leutgeb, 2013). The sparse activity in DG is the result of a rich inhibitory network. Such strong inhibition shuts off a large number of neurons and makes the DG a competitive network in which only a few neurons fire action potentials (Rolls, 2010). Even though dentate activity is sparse, it is coupled to CA3 through the powerful unidirectional mossy fiber projections. Computational models thus predict that the few active neurons can "detonate" the CA3 network, thereby generating new patterns of neural activity and strengthening the connections between activated cells such that new memories are encoded (O'Reilly & McClelland, 1994; Rolls, 2010).

During the storage of new memories, it is not only important that new activity patterns are imposed onto the CA3 network, but also that the new patterns are distinct from patterns that were stored on previous occasions. A critical step in the encoding of a new episodic memory is therefore the amplification of the differences between the representations that are encoded compared to those that already exist in the network. The powerful effect of the dentate in activating CA3 is therefore paired with a computation that is referred to as "pattern separation." Partially overlapping experiences are made more distinct by amplifying sensory differences before a representation is encoded (figure 46.3A). In behavioral studies in rodents in which neurons of the DG were selectively lesioned or manipulated, it was found that rats were unable to discriminate between adjacent spatial locations while they could easily discern between spatial locations that were separated by greater distances (Gilbert, Kesner, & Lee, 2001). The notion that the DG is necessary for distinguishing between similar spatial locations is also supported by studies of place cells in the DG. Pronounced changes in dentate firing patterns are observed for small differences in sensory input. Very separate neuronal representations can therefore be generated in conditions of high input similarity to the network. Pattern separation can also be observed in the CA3 network, but CA3 requires a higher degree of input difference before it responds in this way. The dentate network can thus perform pattern separation in conditions in which the neuronal activity in CA3 is not yet distinct (figure 46.3B; Guzowski, Knierim, & Moser, 2004; Leutgeb, Leutgeb, Moser, & Moser, 2007).

Pattern separation in the dentate/CA3 region can also be observed with high-resolution functional MRI (fMRI) when human subjects are presented with images of objects that were not seen before (novel), were a repetition of a previously viewed object (repeat), or were similar but not identical to a previously viewed object (lure; figure 46.3C). Results revealed that activity in the dentate/CA3 region increased during the viewing of lure images to the same extent as during the viewing of novel images (Bakker, Kirwan, Miller, & Stark, 2008). This increase in neural activity was not observed in other hippocampal subregions (figure 46.3D). The processing of lures as novel events in the dentate/CA3 region supports the notion that this region is biased toward pattern-separation processes during the passive encoding of information. Although fMRI imaging in humans currently does not have the resolution to separate the DG from CA3 (Carr, Rissman, & Wagner, 2010), the convergent evidence from the data in humans and animals points toward a central role of the competitive DG network for pattern separation.

Processing within CA3 CA3 receives direct inputs from the DG and from the entorhinal cortex and sends forward projections to the hippocampal CA1 area and back-projections to mossy cells in the dentate hilus. In addition, the most exceptional feature of the CA3 subregion is the major intrinsic CA3 to CA3 projection. Because the number of synapses that a neuron can receive is limited by the size of the dendritic tree, a high connection probability between neurons within a network can only be achieved if the number of neurons in the network is not too high. In the rat, the total number of CA3 cells is about 300,000 per hemisphere (Rolls & Treves, 1998). In comparisons between primates and rodents, it is interesting to note that the increased size of the primate hippocampus is associated with a larger increase in the number of dentate and CA1 neurons rather than of CA3 neurons (West, 1990), such that the high intrinsic connectivity within the CA3 subregion can be maintained.

The high connection probability within the network of excitatory neurons gives rise to a recurrent or autoassociative architecture. Recurrent networks can perform several classes of computations (Amit, 1989). First, they can hold information online when a subgroup of CA3 cells activates other subgroups of CA3 cells. Neuronal activity patterns can thus reverberate within the circuit to sustain a short-term memory. Second, the ongoing reverberation can be used to associate incoming patterns with patterns that are intrinsically retained. Third, the network can also perform a long-term memory function. Long-term memory relies on first storing a configuration of synaptic strengths during ongoing activity within the circuit. When the recurrent network is later presented with incomplete inputs or with inputs that somewhat differ from previously learned patterns, the subset of neurons that is first activated can iteratively recover the neuronal activity pattern that fully corresponds to the pattern that was stored at the time of learning (McNaughton & Morris, 1987). The recovery of complete neuronal firing patterns from either partial inputs or from partially distinct inputs supports memory retrieval and is a neural computation that is referred to as "pattern completion" (figure 46.3A). Because pattern completion moves network activity toward a stored firing pattern, it must also shift neural activity away from other stored patterns. Pattern completion and pattern separation thus occur in concert, and the transitions between the modes can be sudden (figure 46.3B). However, the firing patterns of CA3 cells can also respond gradually for sensory configurations that are intermediate to learned configurations (J. K. Leutgeb et al., 2005a). Importantly, gradual changes in activity patterns not only code for intermediate sensory inputs, but can also code task parameters that are relevant for memory performance (Allen, Rawlins, Bannerman, & Csicsvari, 2012).

The Output Areas of the Hippocampus

The CA2 subregion Based upon anatomical analysis of hippocampal neurons, several CA subregions have

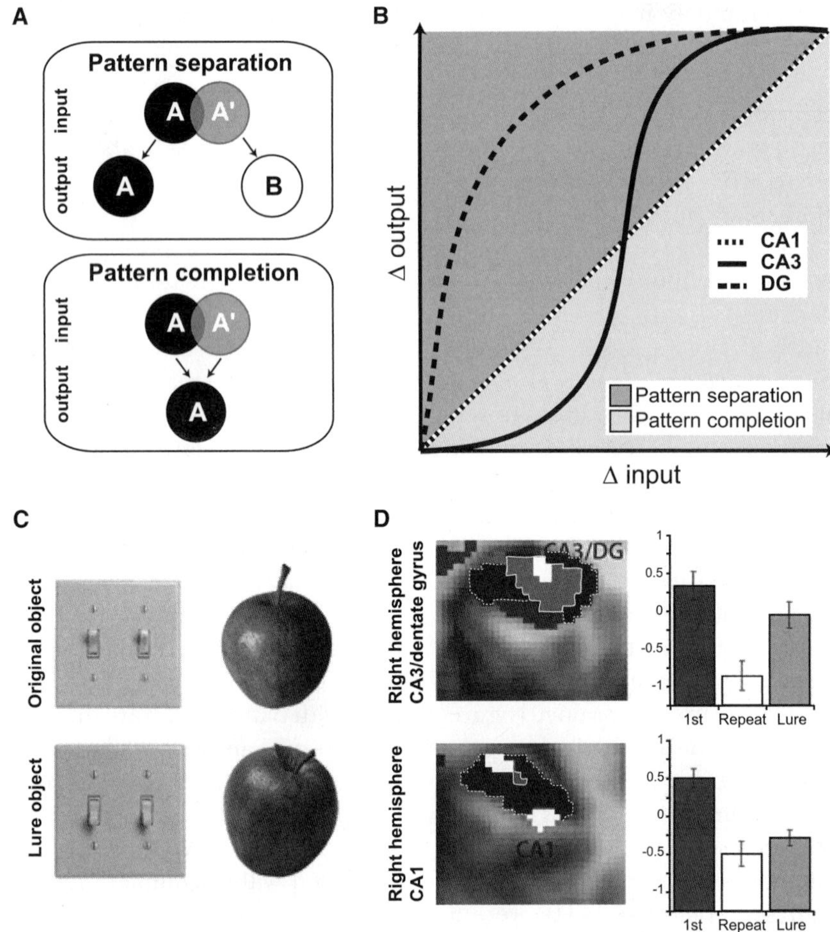

FIGURE 46.3 Distinct computations in hippocampal subfields. (A) Pattern separation mechanisms transform similar input patterns into more orthogonal output patterns (top). Pattern completion processes achieve the opposite and recover the original input pattern despite subtle changes in the input pattern (bottom). (Adapted from Yassa & Stark, 2011, with permission from Elsevier.) (B) Output of a neural network while gradually changing the input pattern. Different computations are observed in each hippocampal subregion. The CA1 network (stippled line) reflects the gradual change in input. In DG (dashed line), subtle input differences are amplified and become orthogonalized. CA3 (solid black line) performs both pattern completion and pattern separation depending on the degree of change in the incoming input patterns. (C) Examples of sample stimuli (original and lure versions) viewed by patients during fMRI imaging. (D) *Left:* Regions of activity in the medial temporal lobe are overlaid in white in the DG/CA3 region (gray surrounded by solid white line) and in the CA1 region (black surrounded by dashed white line). *Right:* Mean activity in each region of interest for each trial condition (First presentation, Repeat, and Lure). In the DG/CA3 region, lures elicit similar levels of activity as the first presentation, which indicates pattern separation. In CA1, lures evoke the same level of activity as repeated trials, which indicates pattern completion. (Adapted from Bakker et al., 2008. Reprinted with permission from AAAS.)

been described (Lorente de No, 1934; Ramon y Cajal, 1893). Although frequently left out of standard depictions of the hippocampal circuit, CA2 is a distinct anatomical region based on protein- and gene-expression profiles (Lein, Callaway, Albright, & Gage, 2005) and on its anatomical connectivity. The hippocampal CA2 subregion is unique among the CA subregions in that it receives inputs from both layer II and III of entorhinal cortex. In addition, unlike CA3 and CA1, CA2 neurons are more strongly excited by entorhinal cortex inputs to their distal dendrites than by CA3 inputs to their more proximal dendrites (Chevaleyre & Siegelbaum, 2010). The unique convergence of strong excitatory drive from entorhinal cortex to CA2 may result in distinct contributions to episodic memory. Behavioral studies in mice, in which the function of the CA2 region was selectively silenced, have revealed a selective deficit in storing the "when" component of memories (DeVito et al., 2009; Wersinger, Ginns, O'Carroll, Lolait, & Young, 2002). However, additional experiments and computational models are needed to describe how such a contribution to memory could be performed and how temporal information in CA2 may influence the neural network mechanisms that represent time in CA1.

The CA1 subregion CA1 is essential for routing information from earlier processing stages in the hippocampal loop back out to the neocortex. Recurrent excitatory connections within the CA1 network are weak such that the processing occurs in a feedforward manner that transforms an input pattern (from CA3, CA2, and layer III of entorhinal cortex) into an output pattern (to subiculum and the deep layers of entorhinal cortex). The three main excitatory inputs to hippocampal CA1 cells are from CA3 cells, from CA2 cells, and directly from layer III cells in the entorhinal cortex. Why would CA1 receive input from entorhinal cortex directly, but also indirectly through longer loops in which input from layer II of entorhinal cortex is routed through the DG, CA3, and CA2 to CA1? A major difference in the organization of entorhinal inputs is that dentate, CA3, and CA2 cells receive convergent inputs from layer II of LEC and MEC, while layer III projects to two separate zones within the CA1 area. The direct input from entorhinal cortex layer III to CA1 therefore more closely represents incoming cortical processing streams, while the indirect pathway from CA3 may add information from long-term or short-term memory. The differential processing along the indirect pathway may result in some degree of mismatch compared to the direct inputs, which can be detected in CA1 and can be flexibly used in updating its representations and in generating arbitrary associations (Lisman & Grace, 2005; McNaughton & Morris, 1987; Mizumori, Smith, & Puryear, 2007).

The different input streams can also be used to interleave incoming firing patterns and to generate the sequential activation of CA1 cells at time scales over which ongoing neuronal activity can be associated within the hippocampal circuit (Eichenbaum, 2013; O'Keefe & Recce, 1993). However, there is also a fundamentally different mechanism for temporal coding, which does not depend on continuously sustained hippocampal firing. It has been proposed that a stable neural code can be compared to a neural code that, when reinstated, has changed over time (Howard & Kahana, 2002). Although this mechanism could be used over any time period, the neuronal activity patterns in CA1 show such drift over periods of hours to weeks. Importantly, neuronal activity in CA3 remains stable over the corresponding time intervals. The retention of a stable reference representation in CA3 enables a comparison in which the similarity between the stable and the time-varying representation can serve as an indication of how long ago an event occurred. Importantly, the neuronal population code in CA1 changes over time in a way such that temporal information can be gained without a major loss of spatial and contextual information (Mankin et al., 2012). The "where," "what," and "when" aspects of episodic memories can thus be simultaneously represented in the activity patterns of CA1 neurons. Behavioral studies in which the CA1 region was selectively lesioned also support a role for CA1 in binding temporal associations to other aspects of memory (Kesner, Hunsaker, & Ziegler, 2010).

Subiculum The subiculum is the last processing stage in the long loop through hippocampal subregions back to the entorhinal cortex. The subiculum receives inputs from CA1 pyramidal neurons, and similar to CA1 receives inputs from layer III of LEC and MEC to separate regions of the cell layer. The subiculum in turn sends projections to the deep layers of the entorhinal cortex (Witter, 2006). The contribution of the subiculum to memory processing is unclear, although recent human fMRI imaging studies suggest that the subiculum and CA1 are selectively activated during periods of memory retrieval rather than periods of memory encoding (Carr et al., 2010). Neural mechanisms in the subiculum that may support memory retrieval are unknown. However, the subiculum differs from CA1 in the degree to which it supports spatial memory. Lesions of the hippocampus result in deficits in spatial memory, while lesions of the subiculum do not lead to similar deficits (O'Mara, 2005). Such a dissociation is consistent with a circuit diagram in which all hippocampal information flows through CA1, but in which CA1 can support circuit function by either directly projecting to entorhinal cortex or by routing information through the subiculum.

Integrated function across medial temporal lobe subregions

By combining the unique computations of neural networks within each subregion, the medial temporal lobe generates neuronal processing streams that enable the acquisition and initial retention of fact and event memories. The position of a subregion within the circuit, the unique intrinsic neuronal architecture of a subregion, and the specialized neuronal firing patterns within each subregion determine which specialized computations are performed. A particular computation may selectively support the acquisition, storage, or retrieval of memories, or a computation may be flexible such that it can support memory processing across different phases.

For example, memory acquisition is thought to activate input stages from entorhinal cortex to DG and from DG to CA3. In this loop, the dentate activity is thought to be particularly important for first generating a distinct code and for then strongly activating a population of CA3 cells such that the information can be stored as a separate pattern in a recurrent network that includes back-projections from CA3 to dentate and intrinsic projections within CA3. In contrast, retrieval may more directly activate CA1 as well as CA3, and CA3 might have an important role in retrieving events from incomplete inputs. Such retrieval can either result in reactivating previously learned information or in a mismatch between the information that is retrieved from memory and the direct information from entorhinal cortex. Such a mismatch can participate in generating a novelty signal, which may, in turn, activate the earlier processing stages to initiate storage of new information. Although this model is consistent with a specialized role of hippocampal subregions, it also illustrates that there is not a one-to-one mapping of memory phases or of representing particular aspects of memories, such as "where," "what," and "when," to a particular anatomical subregion in the hippocampus. Each subcircuit is, rather, specialized for a particular computation that can be used flexibly to support different aspects of memory.

During memory processing, the flow of information is thought to be from entorhinal cortex to hippocampus and back to entorhinal cortex. However, our knowledge about how neuronal activity in the different subregions influences each other is rudimentary. Most of the current information is based on following the transformation of spatial firing patterns through the circuit. Because grid firing in entorhinal cortex is abundant and spatially precise, it has been widely assumed that entorhinal grid cells are required for generating hippocampal place cells. However, abolishing grid firing does not substantially interfere with spatial firing of hippocampal place cells (Koenig, Linder, Leutgeb, & Leutgeb, 2011). Conversely, abolishing the firing of hippocampal place cells results in the loss of grid cell firing (Bonnevie et al., 2013).

These findings illustrate that information processing is not just feedforward, even though the neuronal connectivity in the loop is predominantly unidirectional. Instead, the circuit can be conceptualized as a loop in which information iteratively emerges, which can account for the finding that most recorded cells in entorhinal cortex and hippocampus respond to some aspect of memory tasks during which they are recorded. The matching responses of cells across subregions do not necessarily imply that different subregions are redundant with respect to each other, but can rather be interpreted as evidence of a series of specialized computations that each contribute to the emergence of a common set of neuronal responses throughout the loop.

ACKNOWLEDGMENTS This work was supported by an Ellison Medical Foundation New Scholar Award in Aging, the Walter F. Heiligenberg Professorship, and the Ray Thomas Edwards Foundation.

REFERENCES

ALLEN, K., RAWLINS, J. N., BANNERMAN, D. M., & CSICSVARI, J. (2012). Hippocampal place cells can encode multiple trial-dependent features through rate remapping. *J Neurosci, 32,* 14752–14766.

AMIT, D. J. (1989). *Modeling brain function: The world of attractor neural networks.* Cambridge, UK: Cambridge University Press.

BAKKER, A., KIRWAN, C. B., MILLER, M., & STARK, C. E. (2008). Pattern separation in the human hippocampal CA3 and dentate gyrus. *Science, 319,* 1640–1642.

BONNEVIE, T., DUNN, B., FYHN, M., HAFTING, T., DERDIKMAN, D., KUBIE, J. L., ... MOSER, M. B. (2013). Grid cells require excitatory drive from the hippocampus. *Nat Neurosci, 16,* 309–317.

CARR, V. A., RISSMAN, J., & WAGNER, A. D. (2010). Imaging the human medial temporal lobe with high-resolution fMRI. *Neuron, 65,* 298–308.

CHEVALEYRE, V., & SIEGELBAUM, S. A. (2010). Strong CA2 pyramidal neuron synapses define a powerful disynaptic cortico-hippocampal loop. *Neuron, 66,* 560–572.

DESHMUKH, S. S., & KNIERIM, J. J. (2013). Influence of local objects on hippocampal representations: Landmark vectors and memory. *Hippocampus, 23*(4), 253–267.

DEVITO, L. M., KONIGSBERG, R., LYKKEN, C., SAUVAGE, M., YOUNG, W. S. 3RD, & EICHENBAUM, H. (2009). Vasopressin 1b receptor knock-out impairs memory for temporal order. *J Neurosci, 29,* 2676–2683.

EICHENBAUM, H. (2013). Memory on time. *Trends Cogn Sci, 17,* 81–88.

EICHENBAUM, H., SAUVAGE, M., FORTIN, N., KOMOROWSKI, R., & LIPTON, P. (2011). Towards a functional organization of episodic memory in the medial temporal lobe. *Neurosci Biobehav Rev, 36*(7), 1597–1608.

EKSTROM, A. D., KAHANA, M. J., CAPLAN, J. B., FIELDS, T. A., ISHAM, E. A., NEWMAN, E. L., & FRIED, I. (2003). Cellular networks underlying human spatial navigation. *Nature, 425,* 184–188.

GILBERT, P. E., KESNER, R. P., & LEE, I. (2001). Dissociating hippocampal subregions: Double dissociation between dentate gyrus and CA1. *Hippocampus, 11,* 626–636.

GUZOWSKI, J. F., KNIERIM, J. J., & MOSER, E. I. (2004). Ensemble dynamics of hippocampal regions CA3 and CA1. *Neuron, 44,* 581–584.

HAFTING, T., FYHN, M., MOLDEN, S., MOSER, M. B., & MOSER, E. I. (2005). Microstructure of a spatial map in the entorhinal cortex. *Nature, 436,* 801–806.

HARGREAVES, E. L., RAO, G., LEE, I., & KNIERIM, J. J. (2005). Major dissociation between medial and lateral entorhinal input to dorsal hippocampus. *Science, 308,* 1792–1794.

HENZE, D. A., WITTNER, L., & BUZSAKI, G. (2002). Single granule cells reliably discharge targets in the hippocampal CA3 network in vivo. *Nat Neurosci, 5,* 790–795.

Howard, M. W., & Kahana, M. J. (2002). A distributed representation of temporal context. *J Math Psychol, 46,* 269–299.

Kennedy, P. J., & Shapiro, M. L. (2009). Motivational states activate distinct hippocampal representations to guide goal-directed behaviors. *Proc Natl Acad Sci USA, 106,* 10805–10810.

Kesner, R. P., Hunsaker, M. R., & Ziegler, W. (2010). The role of the dorsal CA1 and ventral CA1 in memory for the temporal order of a sequence of odors. *Neurobiol Learn Mem, 93,* 111–116.

Killian, N. J., Jutras, M. J., & Buffalo, E. A. (2012). A map of visual space in the primate entorhinal cortex. *Nature, 491,* 761–764.

Koenig, J., Linder, A. N., Leutgeb, J. K., & Leutgeb, S. (2011). The spatial periodicity of grid cells is not sustained during reduced theta oscillations. *Science, 332,* 592–595.

Lein, E. S., Callaway, E. M., Albright, T. D., & Gage, F. H. (2005). Redefining the boundaries of the hippocampal CA2 subfield in the mouse using gene expression and 3-dimensional reconstruction. *J Comp Neurol, 485,* 1–10.

Leutgeb, J. K., Leutgeb, S., Moser, M. B., & Moser, E. I. (2007). Pattern separation in the dentate gyrus and CA3 of the hippocampus. *Science, 315,* 961–966.

Leutgeb, J. K., Leutgeb, S., Treves, A., Meyer, R., Barnes, C. A., McNaughton, B. L., ... Moser, E. I. (2005a). Progressive transformation of hippocampal neuronal representations in "morphed" environments. *Neuron, 48,* 345–358.

Leutgeb, S., Leutgeb, J. K., Barnes, C. A., Moser, E. I., McNaughton, B. L., & Moser, M. B. (2005b). Independent codes for spatial and episodic memory in hippocampal neuronal ensembles. *Science, 309,* 619–623.

Lisman, J. E., & Grace, A. A. (2005). The hippocampal-VTA loop: Controlling the entry of information into long-term memory. *Neuron, 46,* 703–713.

Lorente de No, R. (1934). Studies on the structure of the cerebral cortex. II. Continuation of the study of the ammonic system. *J Psychol Neurol, 46,* 113–117.

Mankin, E. A., Sparks, F. T., Slayyeh, B., Sutherland, R. J., Leutgeb, S., & Leutgeb, J. K. (2012). Neuronal code for extended time in the hippocampus. *Proc Natl Acad Sci USA, 109,* 19462–19467.

McNaughton, B. L., & Morris, R. G. M. (1987). Hippocampal synaptic enhancement and information-storage within a distributed memory system. *Trends Neurosci, 10,* 408–415.

Mizumori, S. J., Smith, D. M., & Puryear, C. B. (2007). Hippocampal and neocortical interactions during context discrimination: Electrophysiological evidence from the rat. *Hippocampus, 17,* 851–862.

O'Keefe, J., & Nadel, L. (1978). *The hippocampus as a cognitive map.* Oxford, UK: Oxford University Press.

O'Keefe, J., & Recce, M. L. (1993). Phase relationship between hippocampal place units and the EEG theta rhythm. *Hippocampus, 3,* 317–330.

O'Mara, S. (2005). The subiculum: What it does, what it might do, and what neuroanatomy has yet to tell us. *J Anat, 207,* 271–282.

O'Reilly, R. C., & McClelland, J. L. (1994). Hippocampal conjunctive encoding, storage, and recall: Avoiding a trade-off. *Hippocampus, 4,* 661–682.

Piatti, V. C., Ewell, L. A., & Leutgeb, J. K. (2013). Neurogenesis in the dentate gyrus: Carrying the message or dictating the tone. *Front Neurosci, 7,* 50.

Ramon y Cajal, S. (1893). Estructura del asta de Ammon y fascia dentata. *Ann Soc Esp Hist Nat, 22.*

Rolls, E. T. (1999). Spatial view cells and the representation of place in the primate hippocampus. *Hippocampus, 9,* 467–480.

Rolls, E. T. (2010). A computational theory of episodic memory formation in the hippocampus. *Behav Brain Res, 215,* 180–196.

Rolls, E. T., & Treves, A. (1998). *Neural networks and brain function.* New York, NY: Oxford University Press.

Scoville, W. B., & Milner, B. (1957). Loss of recent memory after bilateral hippocampal lesions. *J Neurol Neurosurg Psychiatry, 20,* 11–21.

Solstad, T., Boccara, C. N., Kropff, E., Moser, M. B., & Moser, E. I. (2008). Representation of geometric borders in the entorhinal cortex. *Science, 322,* 1865–1868.

Squire, L. R., & Alvarez, P. (1995). Retrograde amnesia and memory consolidation: A neurobiological perspective. *Curr Opin Neurobiol, 5,* 169–177.

Suzuki, W. A., & Amaral, D. G. (1994). Perirhinal and parahippocampal cortices of the macaque monkey: Cortical afferents. *J Comp Neurol, 350,* 497–533.

Suzuki, W. A., & Amaral, D. G. (2004). Functional neuroanatomy of the medial temporal lobe memory system. *Cortex, 40,* 220–222.

Tsao, A., Moser, M. B., & Moser, E. I. (2013). Traces of experience in the lateral entorhinal cortex. *Curr Biol, 23,* 399–405.

Tulving, E. (1983). *Elements of episodic memory.* Oxford, UK: Clarendon Press.

Ungerleider, L. G., & Mishkin, M. (1982). Two cortical visual systems. In D. J. Ingle, M. A. Goodale, & R. J. W. Mansfield (Eds.), *Analysis of visual behavior* (pp. 549–586). Cambridge, MA: MIT Press.

van Strien, N. M., Cappaert, N. L., & Witter, M. P. (2009). The anatomy of memory: An interactive overview of the parahippocampal-hippocampal network. *Nat Rev Neurosci, 10,* 272–282.

Wersinger, S. R., Ginns, E. I., O'Carroll, A. M., Lolait, S. J., & Young, W. S., 3rd (2002). Vasopressin V1b receptor knockout reduces aggressive behavior in male mice. *Mol Psychiatry, 7,* 975–984.

West, M. J. (1990). Stereological studies of the hippocampus: A comparison of the hippocampal subdivisions of diverse species including hedgehogs, laboratory rodents, wild mice and men. *Prog Brain Res, 83,* 13–36.

Witter, M. P. (2006). Connections of the subiculum of the rat: Topography in relation to columnar and laminar organization. *Behav Brain Res, 174,* 251–264.

Witter, M. P. (2007). The perforant path: Projections from the entorhinal cortex to the dentate gyrus. *Prog Brain Res, 163,* 43–61.

Witter, M. P., & Moser, E. I. (2006). Spatial representation and the architecture of the entorhinal cortex. *Trends Neurosci, 29,* 671–678.

Wood, E. R., Dudchenko, P. A., & Eichenbaum, H. (1999). The global record of memory in hippocampal neuronal activity. *Nature, 397,* 613–616.

Yassa, M. A., & Stark, C. E. (2011). Pattern separation in the hippocampus. *Trends Neurosci, 34,* 515–525.

Zola-Morgan, S., & Squire, L. R. (1993). Neuroanatomy of memory. *Annu Rev Neurosci, 16,* 547–563.

47 The Role of Context in Episodic Memory

JEREMY R. MANNING, MICHAEL J. KAHANA, AND KENNETH A. NORMAN

ABSTRACT In this chapter we discuss the role of context in organizing episodic memories. We define context as *slowly drifting* information (i.e., information that persists over a relatively long time scale in a person's brain; for example, a representation of a person's location). First, we use the temporal context model of memory search (Howard & Kahana, 2002) to illustrate how binding slowly drifting contextual information to more transient representations (e.g., of studied words) serves to organize our memories of the more transient information. We next present electrophysiological studies examining the role of slowly drifting representations in organizing episodic memories, and we provide an overview of the brain systems involved in representing slowly drifting information. Finally, we discuss sources of variability in the rate of contextual drift.

Context is simultaneously one of the most fundamental and elusive concepts in memory research. Memory researchers often define context by exclusion: in a memory experiment, there is a set of *items* that the participant is being asked to memorize (e.g., a list of words), and then there is *context*, which reflects everything else that is represented in the person's brain during the experiment. Context might include, for example, information about the external environment, mood, thoughts about recently encountered items, plans concerning the future, and incidental features of the stimuli such as the color and the spatial location of a word on the screen (for a review, see Smith & Vela, 2001). This list makes it clear that defining context in terms of the type of information being represented is futile. Under this definition, anything can be context—for example, if you test memory for which colors were seen, rather than word identity, then the background colors become the items and word identities become part of the context.

Instead of defining context in terms of specific types of information (and the roles they play in particular experiments), we focus here on the time scale of information representation. Figure 47.1 illustrates, in timeline form, what might be going through a person's head as he or she learns a list of words in a typical memory experiment. The figure shows how information is represented at different time scales. At the bottom, there are sensory representations of words that persist only while the word is being presented; above, there are representations of thoughts or mental states that persist over longer time scales.

The central idea of this chapter is that slowly drifting information (i.e., information that persists over relatively long time scales) can be used to *time-stamp* and organize more quickly drifting information. This time-stamping is accomplished by means of the hippocampus binding co-active representations together (O'Reilly & Rudy, 2001), including representations that are drifting at different rates. In the scenario shown in figure 47.1, these bindings allow words to cue retrieval of co-active *contextual threads* and vice versa. For example, if the participant recalls the word **shark**, he or she might also recall being hungry when that word was presented, which in turn might trigger retrieval of the fact that **leaf** was also presented while the person was hungry. When **leaf** is retrieved, this might trigger retrieval of having been worried about exams, and then the participant might recall also hearing the word **skull** when worrying about exams, and so on. The set of contextual threads that were active during a previously experienced event (e.g., studying the word **skull**) constitutes a unique time-stamp for that event, and we will refer to the process of reactivating these contextual threads as *contextual reinstatement.* Effectively, this contextual reinstatement process allows the participant to mentally "jump back in time" to when the contextual threads were initially active. If the participant succeeds in reinstating a large number of contextual threads that were linked to an item at study, they will all convergently cue the associated event, thereby boosting the probability of retrieving the event.

To understand the role of context in episodic memory, we need to understand the factors that give rise to slow drift in the brain, and the role that slow drift plays in time-stamping memories. Speaking generally, there are two different (and non-mutually exclusive) ways to get slow drift. The first possibility is that slow drift can arise from the brain representing slowly drifting features of the world. For example, because we cannot teleport

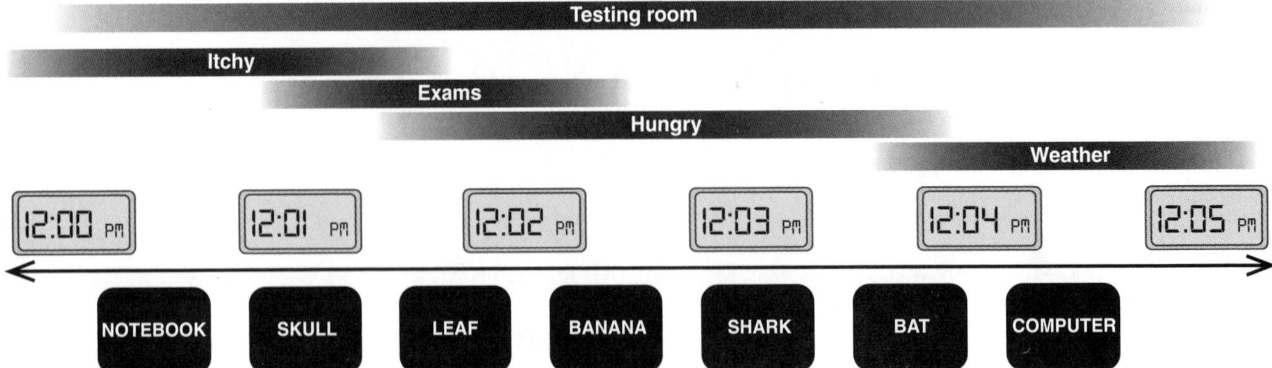

FIGURE 47.1 Illustration of contextual drift. A hypothetical participant studies the seven words in the bottom row over the course of five minutes (times are indicated by the digital clock faces). In addition to thoughts related to the studied words, the participant experiences thoughts related to other external stimuli (e.g., the testing room) as well as internal states (feeling itchy, worrying about upcoming exams, etc.). These thoughts persist for different amounts of time and can overlap.

between locations, our location in one moment will be similar to our location in the next moment. Therefore brain areas that represent current location information will show the requisite slow drift property. The second possibility is *intrinsic maintenance*. Even if features of the world disappear quickly, the brain has the ability to sustain patterns of neural firing corresponding to both external features of the world and internal thoughts (see "Brain Systems Involved in Representing Context" below and also chapter 44, this volume).

The rest of this chapter is divided into three main parts. In the next section, we describe a computational model of memory search to illustrate how the aforementioned psychological principles (slow drift and contextual reinstatement) can account for detailed patterns of memory data. We also discuss electrophysiological evidence for slow drift and contextual reinstatement. We then provide an overview of the brain systems involved in representing slowly drifting information. In the last section, we discuss sources of variability in contextual drift.

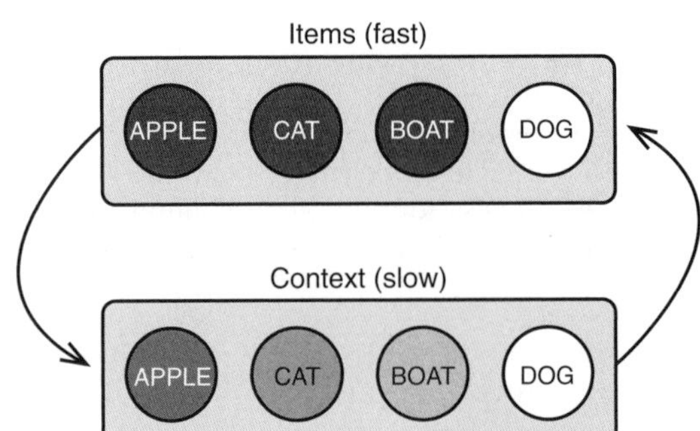

FIGURE 47.2 The temporal context model. Nodes (represented by circles) in the item layer correspond to the sensory representations of the words on the study list. When an item (in this case, **dog**) is studied, its node in the item layer is activated. The shadings of the nodes reflect their activations, where brighter shading reflects greater activation. Nodes in the context layer represent a running average of item-layer activations. Here the activations in the context layer reflect the state of the model as the participant studies the fourth word on a list, **dog**, after studying **apple**, **cat**, and **boat**. The arrows denote interactions between the item and context layers.

Explaining behavioral memory data using the temporal context model

The worth of a psychological theory may be measured by how well it explains detailed patterns of behavioral and neural data. Toward this end, researchers have built computational models that instantiate the principles outlined above (slow drift and contextual reinstatement) and fit the models to detailed patterns of memory data. Here we focus on modeling data from the *free-recall* paradigm, followed by a brief survey of other relevant paradigms. In the free-recall paradigm, participants study lists of items (typically words) and then attempt to recall the items in any order. Free recall is a useful test bed for models of context and memory because it provides data regarding both accuracy (i.e., which memories were recalled) and the order in which items were recalled.

To provide an intuition for how slow drift and contextual reinstatement can account for behavioral memory data, we will focus our discussion on one specific model, the temporal context model (TCM; Howard & Kahana, 2002), schematized in figure 47.2. TCM is a member of a large family of models that incorporate slow drift and contextual reinstatement (e.g., Davelaar, Goshen-Gottstein, Ashkenazi, Haarmann, & Usher,

2005; Dennis & Humphreys, 2001; Polyn, Norman, & Kahana, 2009; Sederberg, Howard, & Kahana, 2008; Shankar & Howard, 2012). The purpose of this section is not to differentiate between models within this family, but rather to illustrate these models' shared predictions.

In TCM, there are two interconnected layers of nodes (or computing elements). These two layers represent the same information but at different time scales. The *item layer* represents the item that is currently being studied on the list, and the *context layer* represents a running average of recently studied items. As the participant studies each item on a list (e.g., the words **apple**, **cat**, **boat**, **dog**), the corresponding node in the item layer is activated. In the figure, the participant is currently viewing the word **dog**, so the **dog** node is active (denoted by bright shading) and the **apple**, **cat**, and **boat** nodes are inactive (denoted by darker shading). As each node of the item layer is activated, the item becomes associated with the current state of the context layer. The context layer is continually updated by averaging together the current item representation with the previous state of context. Because the context layer computes a running average, the state of context evolves gradually as the participant studies the list.

Recall is simulated in TCM by using the current state of context as a retrieval cue. Each item's node in the item layer is activated according to how similar that item's associated state of context is to the current state of context. Items stochastically compete to be recalled, with more active items serving as stronger competitors. When an item wins the recall competition, two things happen. First, the recalled item is incorporated into the current state of context (e.g., if you recall **dog**, that is blended into the context vector). Second, the current state of context is also updated with *retrieved mental context* (i.e., the state of mental context that was linked to the just-retrieved item at study). The updated state of context is then used to probe memory for other items. As discussed earlier in this chapter, reinstating an item's study-phase context can be construed as mentally jumping back to the moment when the just-recalled item was studied. Cuing with reinstated contextual information increases the chances that the next item recalled will be one associated with a similar context.

Modeling Key Regularities in Free-Recall Behavior TCM and similar models have enjoyed extensive success at explaining many stereotyped behaviors observed during free recall and other episodic memory-related tasks. One fundamental regularity is termed the *recency effect*, which refers to participants' ability to more easily retrieve information pertaining to recent experiences (e.g., items from the end of a just-studied list) than information pertaining to long-ago experiences (Murdock, 1962). The *contiguity effect* is another fundamental regularity in free recall, and refers to participants' tendency to successively recall items that occupied nearby positions in the study lists (Kahana, 1996). For example, if a list contained the sub-sequence **apple cat boat** and the participant recalled the word **cat**, it is far more likely that the next response would be either **boat** or **apple** than some other list item.

Both the recency effect and the contiguity effect exhibit *time-scale invariance*, meaning that these same patterns are observed at short and long time scales. For example, recency effects are observed in immediate free-recall experiments (where participants are given a list of words and are tested immediately afterwards) and also in situations where the items on the list are spaced out across multiple days (Glenberg, Bradley, Kraus, & Renzaglia, 1983) or even longer (Moreton & Ward, 2010). Likewise, contiguity effects are observed in immediate free recall and also in *continual distractor free recall*, where participants perform a distracting task after studying each item (Bjork & Whitten, 1974; Howard & Kahana, 1999). Contiguity effects have even been observed across word lists (i.e., given that a participant has just recalled an item from the fourth list, he or she is more likely to recall the next item from the third or fifth list than from the second or sixth lists; Howard, Youker, & Venkatadass, 2008).

The recency and contiguity effects (along with some degree of time-scale invariance) emerge naturally out of retrieved-context models of episodic memory like TCM. According to these models, recency effects arise because the context present at test is *relatively* more similar to the context associated with the end-of-list items than the context associated with earlier items. This fact is true regardless of whether the items were presented within a few seconds of each other or days apart, which explains why the model predicts recency effects at both short and long time scales.

Retrieved-context models like TCM explain the contiguity effect in terms of slow drift at study and contextual reinstatement at test. When participants retrieve an item, they retrieve contextual features associated with that item, and then they use the retrieved contextual features to probe their memories for more items. Due to slow drift at study, items studied nearer in time will have been associated with (relatively) more similar states of context than far-apart items. Just as with the recency effect, this *relative match* property holds regardless of the time delay between successive item presentations, explaining why retrieved-context models predict contiguity effects at a variety of time scales.

The fact that context models can explain recency and contiguity does not mean that contextual drift is always responsible for these effects. For example, it may be possible to explain recency effects in immediate free recall in terms of participants actively maintaining end-of-list items in working memory, and then directly recalling these items from working memory (e.g., Atkinson & Shiffrin, 1968). Likewise, it is possible to explain contiguity effects in immediate free recall in terms of participants directly forming links between adjacently studied items. However, these alternative accounts cannot explain why recency and contiguity effects persist across long time scales.

Importantly, many retrieved-context models (including TCM) are not fully time-scale invariant. For example, TCM has a characteristic contextual drift rate which determines how rapidly the state of context evolves to incorporate new information. Events that occur at time scales slower than this drift rate will not show substantial recency or contiguity effects. A recently developed theory from Shankar and Howard (2012) extends TCM to account for precise timing and full time-scale invariance by positing that drift occurs across a spectrum of time scales (also see Howard & Eichenbaum, 2013).

Other Findings That Can Be Addressed Using This Framework While modelers interested in how context shapes episodic memory have focused primarily on free recall, it is important to note that contextual tagging and contextual reinstatement contribute to some degree to virtually every episodic memory paradigm. For example, in tests of recognition memory, presenting information about a past event (e.g., displaying a previously studied word) can reinstate the context in which the word was studied, such that words studied in similar contexts are subsequently remembered more easily (Schwartz, Howard, Jing, & Kahana, 2005). Contextual reinstatement can also lead to memory errors. For example, suppose that a participant studies two lists of items, *A* and *B*. If the participant is reminded of studying list *A* prior to studying list *B*, the contextual threads associated with list *A* may be reinstated and bound to the list *B* items. This binding can lead participants to *misattribute* memories of list *B* items to list *A* (Gershman, Schapiro, Hupbach, & Norman, 2013; Hupbach, Gomez, Hardt, & Nadel, 2007; Sederberg, Gershman, Polyn, & Norman, 2011). Just as contextual reinstatement can facilitate access to memories linked to those contextual threads, contextual shifts can inhibit access to memories associated with out-of-date contextual threads. Evidence from the *list-method directed-forgetting paradigm* suggests that when participants are asked to forget list *A* items prior to studying list *B*, they respond to this *forget cue* instruction by deliberately shifting their state of context, thereby making the list *A* items less accessible (for a review of the relevant evidence see Sahakyan, Delaney, Foster, & Abushanab, 2013).

Contextual matching may also be used to explain how people judge the order in which a series of events occurred. Specifically, if our current context is more similar to the context associated with event *y* than with event *x*, we may judge event *y* to have occurred more recently than event *x* (see discussion of Manns, Howard, & Eichenbaum, 2007, below, for neural data relating recency judgments to contextual drift).

Electrophysiological Evidence for Slow Drift and Contextual Reinstatement Recent advances in neural recording and analysis methods have allowed researchers to test whether neural patterns during episodic memory experiments are consistent with retrieved-context models. As described above, retrieved-context models predict that recalling a studied item should lead to the reinstatement of a gradually evolving contextual representation. To test this prediction, Manning, Polyn, Baltuch, Litt, and Kahana (2011) recorded electrical signals from electrodes implanted throughout the brains of human neurosurgical patients as they participated in a delayed free-recall experiment (figure 47.3A, B). The researchers first sought to isolate slowly drifting neural patterns that might be involved in representing context. Next, the researchers examined whether those (putative) context representations were reinstated as the participants recalled the studied words.

The researchers isolated candidate context representations by identifying *temporally autocorrelated* neural patterns (i.e., patterns that were more similar during the study of nearby words than temporally distant words) as the participants studied the words. After the patients had studied and recalled words from many lists, the researchers computed the similarity between the neural patterns recorded just prior to recalling a word and the neural patterns recorded at study.

If the neural patterns recorded as a participant recalls an item reflect only item information, then they should match the neural patterns recorded when the item was studied and also possibly afterwards (to the extent that the item representations persisted in the participant's brain). In contrast, if the neural pattern at retrieval reflects reinstated, slowly drifting context from the study phase, then it should match the patterns recorded *both* before and after that item was studied, with similarity falling off gradually in both directions. Figure 47.3C

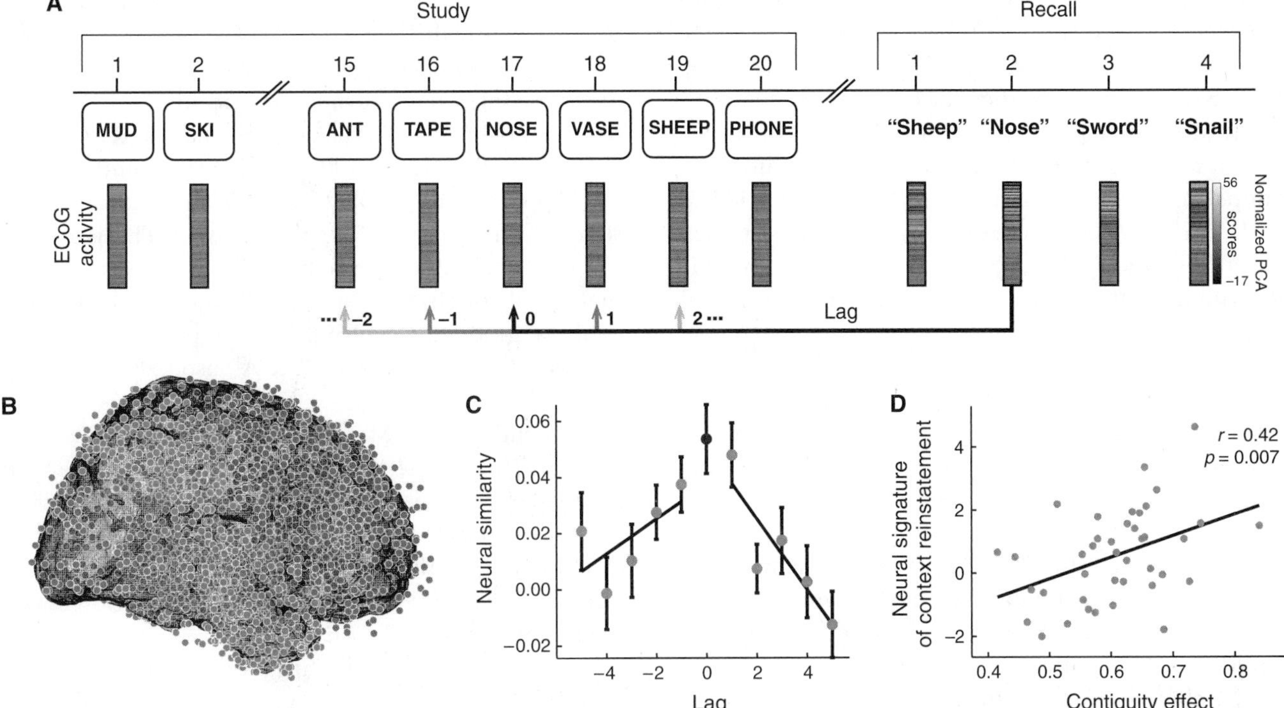

FIGURE 47.3 Neural evidence for contextual reinstatement in humans. (A) After studying a list of 15 or 20 words and performing a brief distraction task, participants recalled as many words as they could remember, in any order. ECoG activity was recorded during each study and recall event. The similarity between the recorded patterns was computed as a function of lag. (B) Each dot marks the location of a single electrode implanted in the brain of a neurosurgical patient. (C) Similarity between the activity recorded during recall of a word from serial position i and study of a word from serial position i + lag (the black dot denotes the study and recall of the same word; i.e., lag = 0). (D) Participants who exhibited stronger neural signatures of context reinstatement also exhibited more pronounced contiguity effects.

shows that the data matched this latter pattern, thereby supporting retrieved-context models. Furthermore, the degree to which individual patients exhibited this neural signature of contextual reinstatement was correlated with the behavioral contiguity effect (i.e., participants' tendency to successively recall neighboring list items; figure 47.3D). In a related study, Howard, Viskontas, Shankar, and Fried (2012) collected extracellular recordings from various medial temporal lobe regions in humans during a recognition memory test, and found a similar neural signature of contextual reinstatement.

Brain systems involved in representing context

PREFRONTAL CORTEX As reviewed by Polyn and Kahana (2008), prefrontal cortex (PFC) has several properties that make it an especially good candidate for representing contextual information. In particular, PFC can actively maintain patterns of neural firing in the face of distraction (e.g., Miller, Erickson, & Desimone, 1996). This capacity for active maintenance should cause neural patterns in PFC to change more slowly than they would otherwise (see Ranganath, Hasselmo, & Stern, chapter 44 in this volume). In keeping with this idea, a number of neural-recording studies have found direct evidence that neural patterns in PFC drift slowly. For example, Hyman, Ma, Balaguer-Ballester, Durstewitz, and Seamans (2012) recorded from dozens of neurons in the rodent medial prefrontal cortex (mPFC) as rats navigated in two environments. They found that the firing rates of mPFC neurons were temporally autocorrelated. Furthermore, Jenkins and Ranganath (2010) found that patterns of functional MRI (fMRI) activity in the right lateral PFC drifted slowly while participants studied lists of pictures; the rate of neural drift predicted how accurately participants could remember when particular pictures were presented over the time course of the experiment. Converging evidence for the PFC's role in representing contextual information comes from studies of patients with frontal lobe damage. These studies have found that damage to frontal regions impairs memory performance on strongly contextually mediated tasks like free recall, whereas performance is relatively spared on tasks where context plays a lesser role (e.g., Shimamura, 1994; Wheeler, Stuss, & Tulving, 1995).

MTL Structures Studies have found evidence for slowly drifting patterns of neural activity in several medial temporal lobe structures, including parahippocampal cortex (PHC) and hippocampus (e.g., Howard et al., 2012). We elaborate on the roles of each of these structures below (for additional discussion of MTL contributions to episodic memory, see chapter 45, this volume).

Parahippocampal cortex and the posterior medial system Several recent papers have argued that PHC represents a person's inference about the *situation* they are currently in (e.g., reading a book, listening to music, cooking dinner, etc.). For example, Bar and Aminoff (2003) found that the PHC shows greater activation in response to objects that are strongly diagnostic of situational context (e.g., a roulette wheel or a beach chair) than to objects without a strong associated context (e.g., a cherry or a fly), although see Epstein and Ward (2010) for an alternative interpretation. Insofar as a person's representation of the situation they are in changes gradually over time (except at event boundaries; see "What Drives Contextual Drift?" below), brain regions that represent situational information should exhibit gradually changing neural patterns. In keeping with this view, the slow drift property predicted by context models has been demonstrated in PHC using both electrophysiology (Howard et al., 2012) and fMRI (Turk-Browne, Simon, & Sederberg, 2012).

Although PHC plays an important role in representing situational context, it is not the only such region. As reviewed by Ranganath and Ritchey (2012), PHC is part of a densely interconnected network of regions called the *posterior medial system* that includes retrosplenial cortex, the mamillary bodies, anterior thalamic nuclei, presubiculum, parasubiculum, posterior cingulate, precuneus, angular gyrus, and ventromedial PFC. Ranganath and Ritchey (2012) propose that structures in the posterior medial system work together to match incoming cues about the current context to internal *situation models* that specify the spatial, temporal, and causal relationships that define specific situations. For example, the situation model for going to a movie might describe the properties of movie theaters, the typical time sequence of events during a movie-going outing, movie theater etiquette, and so on (Zacks, Speer, Swallow, Braver, & Reynolds, 2007; Zwaan & Radvansky, 1998).

The hippocampus The hippocampus is the key structure responsible for binding quickly drifting information (e.g., sensory representations of words on a study list) to more slowly drifting information, thereby making it possible for studied items to cue reinstatement of slowly drifting contextual information, and vice versa (Cohen & Eichenbaum, 1993; Diana, Yonelinas, & Ranganath, 2007; O'Reilly & Rudy, 2001). To play this binding role, the hippocampus needs to receive inputs from areas representing slowly drifting information. This is accomplished via connections from areas like PFC and PHC that go through the entorhinal cortex (which itself exhibits gradually evolving neural patterns; Egorov, Hamam, Fransen, Hasselmo, & Alonso, 2002) into the hippocampus.

A recent study by Manns et al. (2007) provides clear evidence that patterns of hippocampal firing evolve gradually over time, and that these gradually evolving patterns are behaviorally relevant. In their study, Manns et al. (2007) simultaneously recorded activity from multiple neurons in the CA1 hippocampal subregion as rats sequentially sampled a "list" of odors. After sampling the sequence of odors, the rat had to choose which of two odors in the sequence had been presented more recently. The authors found that patterns of neural firing in CA1 changed gradually as animals sampled the odors, and that the degree of neural drift over the course of the list predicted behavioral accuracy on a recency-discrimination test. The authors interpret this finding in terms of the idea that greater neural drift indicates greater contextual separation, which (in turn) makes it easier to temporally discriminate between items on the recency test. Notably, in addition to showing a within-list neural drift effect, Manns et al. (2007) also observed slow drift across lists of odors (for discussion, see Howard & Eichenbaum, 2013); this fits with the idea (mentioned above in the section "Explaining Behavioral Memory Data Using the Temporal Context Model") that context drifts at multiple time scales (Shankar & Howard, 2012). For a related finding showing slow drift in CA1, see Mankin et al. (2012).

Temporal Receptive Windows In addition to the aforementioned regions, how can we discover other areas involved in representing context? Naïvely, one could just look for regions exhibiting gradually evolving neural patterns using fMRI. The problem with this approach is that the fMRI signal is constrained to drift slowly due to the sluggishness of the blood flow response that it measures, regardless of the drift rates of the underlying neurocognitive processes. Therefore the mere presence of slow drift in the fMRI signal is not diagnostic of slow drift in the person's thoughts.

To address this problem, Hasson, Yang, Vallines, Heeger, and Rubin (2008) devised a new technique for measuring a brain region's sensitivity to information at different time scales. Instead of directly measuring the drift rate of neural patterns, they measured the *history-dependence* of neural activity in a region. Specifically, they manipulated what came before a particular stimulus (by rearranging scenes in a movie), and asked whether the response of a region to a particular scene was altered by changing the scenes that came before it. For example, if a region's response to a scene is altered by changing what happened 5 minutes previously (but not 10 minutes ago), this indicates that the region retains information from 5 minutes ago (but not 10 minutes ago). Hasson et al. (2008) define the *temporal receptive window* (or TRW) for a particular region as the length of time, prior to the stimulus presentation, during which the presentation of other information may affect the neural response to the stimulus.

Using a variant of this approach, Lerner, Honey, Silbert, and Hasson (2011) found that the temporal parietal junction (including the angular gyrus and supramarginal gyrus) and precuneus were sensitive to the previous sentence in an auditory story, and medial PFC showed an even longer TRW (extending to the previous paragraph and possibly further). Hasson et al. (2008) identified a similar set of "long TRW" regions using a movie stimulus, including precuneus and the temporal parietal junction. Importantly, there is strong overlap between the set of long TRW regions and the posterior medial network regions identified by Ranganath and Ritchey (2012). We should emphasize that regions can show long TRWs for a variety of reasons—for example, a region might have a long TRW because it has intrinsic integrator properties (e.g., Arnsten, Wang, & Paspalas, 2012), or because it can actively maintain specific patterns of activity (as in PFC), or because it is receiving information from other regions involved in memory storage (e.g., the hippocampus).

What drives contextual drift?

One of the main goals of theories of context and memory is understanding *variability* in contextual drift by explaining the circumstances that result in mental context changing more or less quickly. A key implication of the situation model view described above in the section "Brain Systems Involved in Representing Context" is that mental context will change sharply when a person's (inferred) situation changes.

This view is supported by data showing that *event boundaries* (moments when participants infer a change in their situation; e.g., shifting from eating dinner to washing dishes) can cause forgetting. In the event-processing literature, several behavioral studies have found that (controlling for elapsed time) participants are impaired at recalling details from the previous event compared to the current event (e.g., Radvansky & Copeland, 2006; Swallow, Zacks, & Abrams, 2009). Using a long-term memory paradigm, Ezzyat and Davachi (2011) found that participants had difficulty recalling associations between adjacent sentences that spanned an event boundary, compared to sentences that were part of the same event. These behavioral findings are consistent with the idea that event boundaries induce a sharp discontinuity in context, resulting in decreased accessibility of details from the previous event and also decreased contiguity effects in long-term memory tests (for relevant neural evidence, see Swallow et al., 2011).

Explaining these event-boundary effects poses a major challenge for computational models of contextual drift. Computational models that update context via a simple integration process (i.e., by computing a running average of recently encountered stimuli) posit that, when the situation shifts, information about previously encountered stimuli will gradually fade out of context rather than exhibiting a rapid shift. As discussed by Polyn et al. (2009), this gradual fade is not enough to explain the sharp drop in recall observed at event boundaries. Polyn et al. (2009) created situational shifts at study by having participants switch (multiple times) between encoding tasks as they studied lists of words, and then had participants freely recall the studied items. To model the effects of these task switches on free recall, Polyn et al. (2009) had to incorporate an extra context-disruption mechanism that was triggered whenever participants switched between encoding tasks.

While this context-disruption mechanism helps to fit the data, it does not provide a clear mechanistic account of why context is disrupted at event boundaries (it just posits that it happens). Modeling work by Shankar, Jagadisan, and Howard (2009) may provide some insight into this issue. Their model (which they call pTCM, for predictive TCM) modifies TCM such that, instead of updating context with item information, context is updated with a prediction of which items will be presented next. Insofar as event boundaries are marked by sharp changes in predictions (i.e., what you predict at the end of one event is very different from what you predict at the beginning of another), this model may be able to simulate the findings described above (also see Reynolds, Zacks, & Braver, 2007).

Conclusion

Context has long been the "dark matter" of memory theories. Researchers have found it necessary to posit a gradually evolving context representation in order to explain patterns of memory data from free recall and other tasks. This gradually evolving representation (and the idea that it can be reinstated during retrieval) is the glue that holds together most modern theories of memory retrieval. However, until recently, no one had been able to observe contextual drift or reinstatement directly. Instead, the role of context during memory encoding and retrieval had been something indirectly inferred through its effects on behavioral memory performance.

In this chapter, we have reviewed recent progress in the cognitive neuroscience of memory that has allowed us to start bringing the dark matter of context into the light. Neurophysiological and fMRI studies have given us a much better idea of which regions are most strongly involved in representing contextual information and, more importantly, they have given us the ability to track how neural activity drifts within those regions. In the coming years, the ability to track this drift and relate it to memory behavior will allow us to develop even more powerful models of how our brains time-stamp our memories and how these time-stamps allow us to retrieve information concerning the past.

REFERENCES

Arnsten, A. F. T., Wang, M. J., & Paspalas, C. D. (2012). Neuromodulation of thought: Flexibilities and vulnerabilities in prefrontal cortical network synapses. *Neuron, 76,* 223–239.

Atkinson, R. C., & Shiffrin, R. M. (1968). Human memory: A proposed system and its control processes. In K. W. Spence & J. T. Spence (Eds.), *The psychology of learning and motivation* (Vol. 2, pp. 89–105). New York, NY: Academic Press.

Bar, M., & Aminoff, E. (2003). Cortical analysis of visual context. *Neuron, 38,* 347–358.

Bjork, R. A., & Whitten, W. B. (1974). Recency-sensitive retrieval processes in long-term free recall. *Cognit Psychol, 6,* 173–189.

Cohen, N. J., & Eichenbaum, H. (1993). *Memory, amnesia, and the hippocampal system.* Cambridge, MA: MIT Press.

Davelaar, E. J., Goshen-Gottstein, Y., Ashkenazi, A., Haarmann, H. J., & Usher, M. (2005). The demise of short-term memory revisited: Empirical and computational investigations of recency effects. *Psychol Rev, 112,* 3–42.

Dennis, S., & Humphreys, M. S. (2001). A context noise model of episodic word recognition. *Psychol Rev, 108,* 452–478.

Diana, R. A., Yonelinas, A. P., & Ranganath, C. (2007). Imaging recollection and familiarity in the medial temporal lobe: A three-component model. *Trends Cogn Sci, 11*(9), 379–386.

Egorov, A., Hamam, B., Fransen, E., Hasselmo, M., & Alonso, A. (2002). Graded persistent activity in entorhinal cortex neurons. *Nature, 420*(6912), 173–178.

Epstein, R., & Ward, E. (2010). How reliable are visual context effects in the parahippocampal place area? *Cereb Cortex, 20*(2), 294.

Ezzyat, Y., & Davachi, L. (2011). What constitutes an episode in episodic memory? *Psychol Sci, 22*(2), 243–252.

Gershman, S. J., Schapiro, A. C., Hupbach, A., & Norman, K. A. (2013). Neural context reinstatement predicts memory misattribution. *J Neurosci, 33*(20), 8590–8595.

Glenberg, A. M., Bradley, M. M., Kraus, T. A., & Renzaglia, G. J. (1983). Studies of the long-term recency effect: Support for a contextually guided retrieval theory. *J Exp Psychol Learn Mem Cogn, 12,* 413–418.

Hasson, U., Yang, E., Vallines, I., Heeger, D. J., & Rubin, N. (2008). A hierarchy of temporal receptive windows in human cortex. *J Neurosci, 28*(10), 2539–2550.

Howard, M. W., & Eichenbaum, H. (2013). The hippocampus, time, and memory across scales. *J Exp Psychol Gen, 142*(4), 1211–1230.

Howard, M. W., & Kahana, M. J. (1999). Contextual variability and serial position effects in free recall. *J Exp Psychol Learn Mem Cogn, 25,* 923–941.

Howard, M. W., & Kahana, M. J. (2002). A distributed representation of temporal context. *J Math Psychol, 46,* 269–299.

Howard, M. W., Viskontas, I. V., Shankar, K. H., & Fried, I. (2012). Ensembles of human MTL neurons "jump back in time" in response to a repeated stimulus. *Hippocampus, 22,* 1833–1847.

Howard, M. W., Youker, T. E., & Venkatadass, V. (2008). The persistence of memory: Contiguity effects across hundreds of seconds. *Psychon Bull Rev, 15,* 58–63.

Hupbach, A., Gomez, R., Hardt, O., & Nadel, L. (2007). Reconsolidation of episodic memories: A subtle reminder triggers integration of new information. *Learn Mem, 14,* 47–53.

Hyman, J., Ma, L., Balaguer-Ballester, E., Durstewitz, D., & Seamans, J. (2012). Contextual encoding by ensembles of medial prefrontal cortex neurons. *Proc Natl Acad Sci USA, 109*(13), 5086–5091.

Jenkins, L. J., & Ranganath, C. (2010). Prefrontal and medial temporal lobe activity at encoding predicts temporal context memory. *J Neurosci, 30*(46), 15558–15565.

Kahana, M. J. (1996). Associative retrieval processes in free recall. *Mem Cognit, 24,* 103–109.

Lerner, Y., Honey, C. J., Silbert, L. J., & Hasson, U. (2011). Topographic mapping of a hierarchy of temporal receptive windows using a narrated story. *J Neurosci, 31*(8), 2906–2915.

Mankin, E. A., Sparks, F. T., Slayyeh, B., Sutherland, R. J., Leutgeb, S., & Leutgeb, J. K. (2012). Neuronal code for extended time in the hippocampus. *Proc Natl Acad Sci USA, 109*(47), 19462–19467.

Manning, J. R., Polyn, S. M., Baltuch, G., Litt, B., & Kahana, M. J. (2011). Oscillatory patterns in temporal lobe reveal context reinstatement during memory search. *Proc Natl Acad Sci USA, 108*(31), 12893–12897.

Manns, J. R., Howard, M. W., & Eichenbaum, H. (2007). Gradual changes in hippocampal activity support remembering the order of events. *Neuron, 56*(3), 530–540.

Miller, E. K., Erickson, C. A., & Desimone, R. (1996). Neural mechanisms of visual working memory in prefrontal cortex of the macaque. *J Neurosci, 16,* 5154.

MORETON, B. J., & WARD, G. (2010). Time scale similarity and long-term memory for autobiographical events. *Psychon Bull Rev, 17*(4), 510–515.

MURDOCK, B. B. (1962). The serial position effect of free recall. *J Exp Psychol, 64*, 482–488.

O'REILLY, R. C., & RUDY, J. W. (2001). Conjunctive representations in learning and memory: Principles of cortical and hippocampal function. *Psychol Rev, 108*(2), 311–345.

POLYN, S. M., & KAHANA, M. J. (2008). Memory search and the neural representation of context. *Trends Cogn Sci, 12*, 24–30.

POLYN, S. M., NORMAN, K. A., & KAHANA, M. J. (2009). A context maintenance and retrieval model of organizational processes in free recall. *Psychol Rev, 116*(1), 129–156.

RADVANSKY, G. A., & COPELAND, D. E. (2006). Walking through doorways causes forgetting: Situation models and experienced space. *Mem Cognit, 34*(5), 1150–1156.

RANGANATH, C., & RITCHEY, M. (2012). Two cortical systems for memory-guided behavior. *Nat Rev Neurosci, 13*, 713–726.

REYNOLDS, J. R., ZACKS, J. M., & BRAVER, T. S. (2007). A computational model of event segmentation from perceptual prediction. *Cognitive Sci, 31*, 613–643.

SAHAKYAN, L., DELANEY, P. F., FOSTER, N. L., & ABUSHANAB, B. (2013). List-method directed forgetting in cognitive and clinical research: A theoretical and methodological review. *Psychol Learn Motiv, 59*, 131–189.

SCHWARTZ, G., HOWARD, M. W., JING, B., & KAHANA, M. J. (2005). Shadows of the past: Temporal retrieval effects in recognition memory. *Psychol Sci, 16*, 898–904.

SEDERBERG, P. B., GERSHMAN, S. J., POLYN, S. M., & NORMAN, K. A. (2011). Human memory reconsolidation can be explained using the temporal context model. *Psychon Bull Rev, 18*(3), 455–468.

SEDERBERG, P. B., HOWARD, M. W., & KAHANA, M. J. (2008). A context-based theory of recency and contiguity in free recall. *Psychol Rev, 115*(4), 893–912.

SHANKAR, K. H., & HOWARD, M. W. (2012). A scale-invariant internal representation of time. *Neural Comp, 24*, 134–193.

SHANKAR, K. H., JAGADISAN, U. K. K., & HOWARD, M. W. (2009). Sequential learning using temporal context. *J Math Psychol, 53*, 474–485.

SHIMAMURA, A. P. (1994). Memory and frontal lobe function. In M. S. Gazzaniga (Ed.), *The cognitive neurosciences* (pp. 803–815). Cambridge, MA: MIT Press.

SMITH, S. M., & VELA, E. (2001). Environmental context-dependent memory: A review and meta-analysis. *Psychon Bull Rev, 8*(2), 203–220.

SWALLOW, K. M., BARCH, D. M., HEAD, D., MALEY, C. J., HOLDER, D., & ZACKS, J. M. (2011). Changes in events alter how people remember recent information. *J Cogn Neurosci, 23*(5), 1052–1064.

SWALLOW, K. M., ZACKS, J. M., & ABRAMS, R. A. (2009). Event boundaries in perception affect memory encoding and updating. *J Exp Psychol Gen, 138*(2), 236–257.

TURK-BROWNE, N. B., SIMON, M. G., & SEDERBERG, P. B. (2012). Scene representations in parahippocampal cortex depend on temporal context. *J Neurosci, 32*(21), 7202–7207.

WHEELER, M. A., STUSS, D. T., & TULVING, E. (1995). Frontal lobe damage produces episodic memory impairment. *J Int Neuropsych Soc, 1*, 525–536.

ZACKS, J. M., SPEER, N. K., SWALLOW, K. M., BRAVER, T. S., & REYNOLDS, J. R. (2007). Event perception: A mind-brain perspective. *Psychol Bull, 133*, 273–293.

ZWAAN, R. A., & RADVANSKY, G. A. (1998). Situation models in language comprehension and memory. *Psychol Bull, 123*(2), 162–185.

48 Parietal Lobe Mechanisms Subserving Episodic Memory Retrieval

MELINA R. UNCAPHER, ALAN M. GORDON, AND ANTHONY D. WAGNER

ABSTRACT Episodic memory enables conscious remembrance of events past and the recognition of previously encountered stimuli. Functional neuroimaging investigations of episodic retrieval indicate that retrieval-related activity extends well beyond the medial temporal lobe and prefrontal cortex, and consistently includes multiple, functionally distinct retrieval effects in lateral posterior parietal cortex (PPC). Here we review what is known about PPC activity at retrieval and how this activity relates to the broad range of cognitive operations subserved by PPC (e.g., attention, multifeatural binding, decision making, and action intention). We conclude by introducing a working model of PPC contributions to episodic remembering.

Episodic memory enables humans to discriminate novel from previously encountered stimuli and to retrieve details of past events. Given its importance for everyday functioning and how it is disrupted in disease, extensive neuroimaging research has sought to delineate the neural mechanisms that subserve episodic remembering in healthy humans. Extant functional MRI (fMRI) data implicate several large-scale brain regions and their network interactions in episodic retrieval, including the medial temporal lobe (MTL), prefrontal cortex (PFC), and posterior parietal cortex (PPC; figure 48.1). While rich neuropsychological literatures demonstrate that MTL and PFC damage can lead to noticeable memory impairments (Shimamura, 1995; Squire, 1992), lateral PPC damage is associated with subtle memory changes (Berryhill, 2012). This puzzling result has sparked debate about the function of lateral PPC (hereafter PPC) during retrieval.

Interest in PPC contributions to episodic retrieval emerged from event-related fMRI observations of "old/new" or "retrieval success" effects, wherein PPC activity (predominantly left-lateralized) is greater during recognition of previously encountered stimuli as "old" (hits) relative to classification of novel stimuli as "new" (correct rejections). Initial studies suggested a dorsal/ventral axis of PPC functional organization (Cabeza, Ciaramelli, Olson, & Moscovitch, 2008; Vilberg & Rugg, 2008; Wagner, Shannon, Kahn, & Buckner, 2005). In general, fMRI blood oxygen level–dependent (BOLD) activity in dorsal PPC (dPPC) tracks differences in item-memory strength or *familiarity*—that is, the sense of prior encounter that is unaccompanied by remembrance of contextual detail. By contrast, ventral PPC (vPPC) is engaged during *recollection*—that is, the retrieval of contextual details of a prior event. While this coarse dorsal/ventral distinction in PPC old/new effects has prompted multiple hypotheses and motivated novel experimentation, recent meta-analyses and empirical studies indicate a more fine-grained parcellation of PPC retrieval-related effects (figure 48.2; e.g., Hutchinson, Uncapher, & Wagner, 2009; Hutchinson et al., 2014; Sestieri, Shulman, & Corbetta, 2010), calling into question whether two-process (dorsal/ventral) models are sufficient to account for PPC computations during retrieval.

This chapter summarizes fMRI research on PPC function during episodic retrieval, and considers how multiple PPC mechanisms may influence memory-guided action. We first introduce prominent hypotheses regarding PPC contributions to retrieval and briefly review the current state of the field. Critically, we argue that any model of PPC contributions to retrieval must take into account the fine-grained functional heterogeneity in PPC. We then offer a working model of PPC function that incorporates seemingly conflicting hypotheses. Broadly, we suggest that multiple PPC functions—including mechanisms of attention, multifeatural binding, decision making, and action intention—may each explain a particular subset of PPC retrieval effects and interact to support retrieval-guided behavior.

Attention

ATTENTION TO MEMORY From one perspective, PPC old/new effects reflect the differential engagement of attention depending on retrieval outcomes. A particularly influential attention-based hypothesis—termed *attention to memory* (AtoM; Cabeza et al., 2008; Ciaramelli, Grady, & Moscovitch, 2008)—builds on a model of perceptual attention (Corbetta, Patel, & Shulman, 2008) in which dPPC supports goal-directed ("top-down") orienting of attention, while vPPC supports

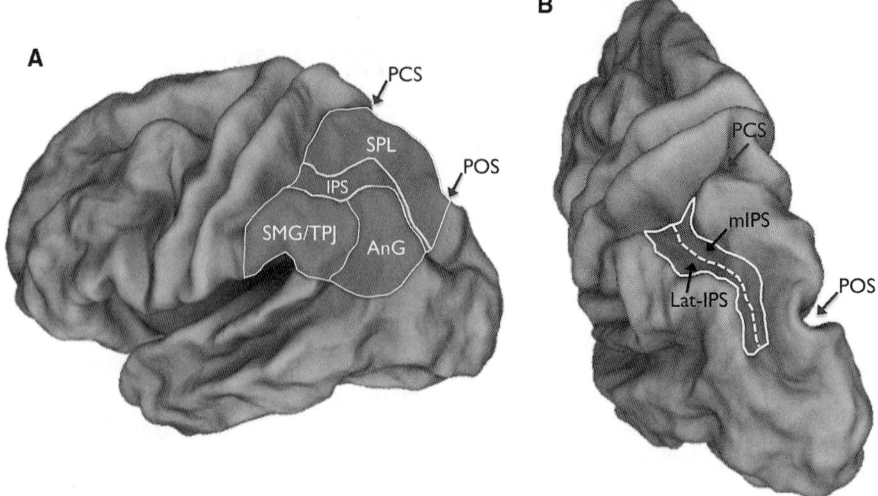

FIGURE 48.1 PPC anatomy. (A) Lateral view, with macroanatomical regions labeled: SPL, superior parietal lobule; IPS, intraparietal sulcus; SMG, supramarginal gyrus; TPJ, temporoparietal junction; AnG, angular gyrus; PCS, postcentral sulcus; POS, parietal-occipital sulcus. (B) Dorsal view, with medial IPS (mIPS) and lateral IPS (lat-IPS) labeled (dotted line; fundus of the IPS).

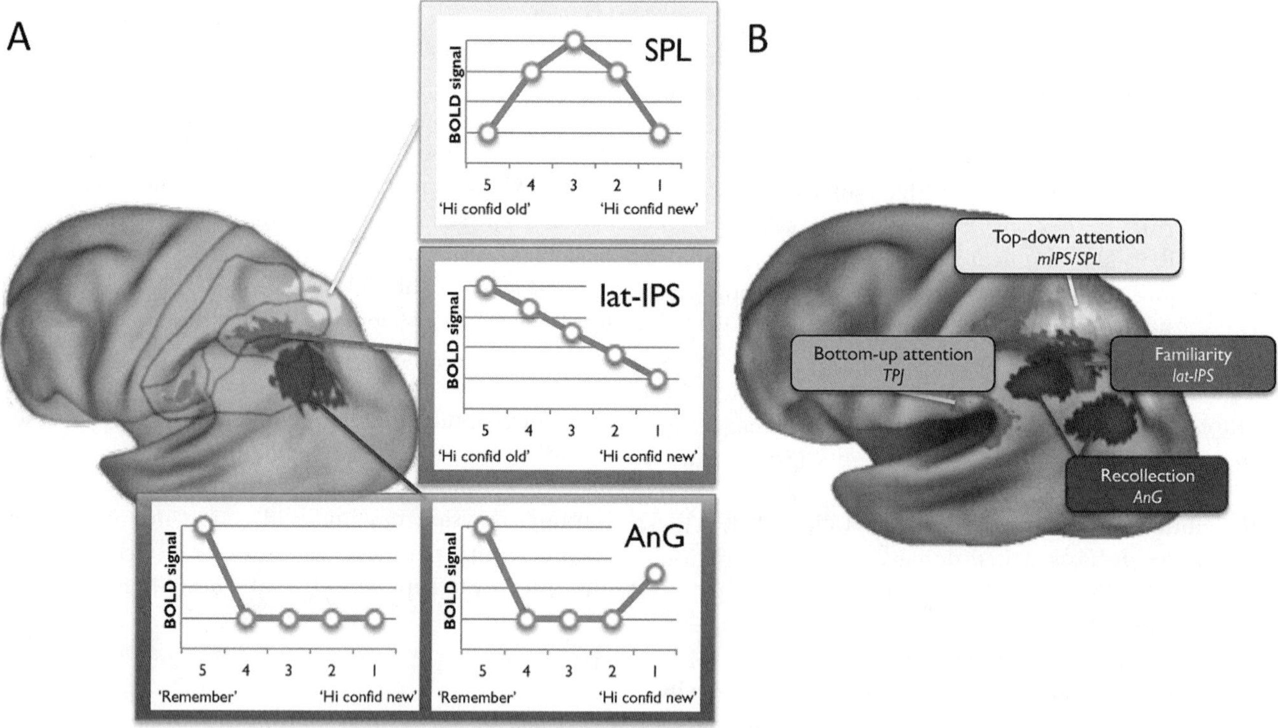

FIGURE 48.2 PPC functional dissociations. (A) Hutchinson et al. (2014) revealed a quadruple dissociation of activity patterns during retrieval: lat-IPS (red) activity showed a monotonic relationship with item-memory strength; AnG (blue) activity tracked recollection of event details; mIPS/SPL (yellow) demonstrated a decision uncertainty effect, with TPJ (green) qualitatively showing the reverse. Black outlines on the brain indicate parcellation identified by Nelson et al. (2010) using graph-theoretic analyses of resting-state and task fMRI data. Line graphs depict schematics of representative activity patterns. Note: AnG activity tracks recollection (left graph) and sometimes also shows a novelty effect (right graph). "Remember," Remember responses; "hi confid," high confidence. (B) Meta-analyses of top-down and bottom-up attention effects, and recollection and familiarity-based retrieval effects reveal largely nonoverlapping parietal regions supporting attention and memory (adapted from Hutchinson et al., 2014). (See color plate 39.)

stimulus-driven ("bottom-up") reorienting of attention. Extending this attention model from perception to memory, AtoM posits that dPPC mechanisms are recruited to allocate attention to the goal of memory retrieval (e.g., when effortful pre- and/or post-retrieval processing is required to make a memory decision), and that relevant memory cues or recollected memories result in vPPC bottom-up attentional capture.

The dorsal component of AtoM garners support from studies in which demands on top-down attention at retrieval are thought to differ across conditions. First, relative to high-confidence decisions, low-confidence decisions, which should require greater top-down attention, elicit greater dPPC activity (Cabeza et al., 2008; Vilberg & Rugg, 2008). Second, recognition decisions accompanied by familiarity may be associated with greater dPPC activity relative to recollective decisions largely because, in the framework of many studies, familiarity-based decisions are more effortful and thus more demanding of top-down attention (Cabeza et al., 2008). When a recollective task was designed to be more effortful than an item-memory task, greater dPPC activity was observed in the recollection condition (Ciaramelli, Grady, Levine, Ween, & Moscovitch, 2010). Third, attentional orienting during a visual search task was shown to engage dPPC regions that are similar to those engaged during a memory search task (Cabeza et al., 2011).

The ventral component of AtoM also garners support from multiple lines of evidence. First, the detection of recollected or high-strength mnemonic information is proposed to be analogous to the detection of target information in perception (Ciaramelli et al., 2008); such retrieval events are thought to be salient and thus should elicit bottom-up attentional capture (putatively associated with vPPC activity). Consistent with this account, greater vPPC activity is observed (1) when recognition is accompanied by the subjective report of "*remembering*" the past event—putatively indicating retrieval of some detail(s) of the original study episode—relative to when recognition is accompanied by the subjective sense of familiarity ("*know*" and high-confidence "*familiar*"/"*old*" responses), and (2) when participants objectively recollect details associated with a test probe's past encounter, including retrieval of a specific contextual (source) detail, retrieval of an associate of a cue, and retrieval of more rather than fewer event details (Cabeza et al., 2008; Rugg, Johnson, & Uncapher, in press; Vilberg & Rugg, 2008; Wagner et al., 2005). Second, consistent with bottom-up attention being engaged when expectations are violated (e.g., oddball effects and invalid vs. valid trials during attention cueing; Corbetta, Patel, & Shulman, 2008), *memory-based* expectation violations elicit activity that overlaps old/new effects in vPPC (O'Connor, Han, & Dobbins, 2010). Similarly, old/new effects overlap with effects posited to be a proxy for bottom-up attention (Ciaramelli et al., 2010): greater vPPC activity was observed when participants were presented with recombined word pairs that may elicit a memory-based expectation violation relative to intact word pairs (which presumably confirm memory-based expectations). Third, in a study that investigated (bottom-up) target detection in a memory task and a perceptual task, activity during mnemonic and perceptual target detection overlapped in vPPC (Cabeza et al., 2011). While none of these studies directly relate PPC activity during recollection or high-confidence memory decisions to that during conditions known to demand bottom-up attention, they can be interpreted as suggesting that vPPC-mediated attentional processes are engaged under various retrieval conditions.

Beyond Dual-Attention Accounts While the preceding findings (and others) lend support to AtoM, a growing literature demonstrates that the PPC regions engaged during attention tasks are not the same as those typically showing old/new effects. First, Sestieri and colleagues (2010) demonstrated that top-down perceptual and mnemonic search tasks elicit activity in adjacent but nonoverlapping regions of dPPC, with the perceptual task recruiting medial intraparietal sulcus (IPS) and the memory task recruiting lateral IPS. Second, Hutchinson and colleagues (2014) reported complementary findings (figure 48.2A): while BOLD activity in angular gyrus (AnG) tracked recollection, and activity in lateral IPS tracked item-memory strength, medial IPS/superior parietal lobule (SPL) activity tracked (1) top-down visuospatial attention (as evidenced by overlap with "attendotopic maps"—that is, topographic maps indicating where in visual space top-down attention is allocated) and (2) retrieval decision uncertainty (i.e., low-confidence > high-confidence recognition decisions). Moreover, activity in a fourth region—temporoparietal junction (TPJ)—demonstrated a pattern that qualitatively resembled the inverse of that in medial IPS/SPL. In a third line of evidence, Uncapher and colleagues (Uncapher, DuBrow, Hutchinson, & Wagner, 2011) manipulated when attention and memory operations were likely to occur during a retrieval task, and demonstrated parallel dissociations as reported in Hutchinson et al. (2014): AnG activity tracked recollection success, SPL activity tracked top-down attentional orienting, and TPJ activity exhibited bottom-up attentional reorienting effects.

Collectively, these findings (and others) point to a quadruple dissociation in PPC (figure 48.2A). In dPPC, medial IPS/SPL regions appear to support top-down attention and are engaged during uncertain retrieval decisions, whereas lateral IPS supports a mechanism (or mechanisms) that positively varies with item-memory strength. In vPPC, TPJ appears to track bottom-up attention, which may be disengaged or suppressed during uncertain retrieval decisions, whereas AnG supports a mechanism (or mechanisms) that varies with event recollection. The partitioning of PPC into four functional regions is further supported by meta-analyses of the attention and retrieval literatures (figure 48.2B). Specifically, Hutchinson and colleagues (2009, 2014) observed a quadruple dissociation within these broader literatures, with the four retrieval-identified PPC foci (figure 48.2A) anatomically overlapping the four regions observed in the meta-analyses (figure 48.2B). As such, there is now considerable evidence for at least four functionally separable PPC regions in which activity varies during episodic retrieval, with lateral IPS and AnG demonstrating old/new effects distinct from medial IPS/SPL and TPJ regions demonstrating attention effects.

Posterior Parietal Cortex Anatomical and Functional Heterogeneity A broader literature demonstrates that PPC comprises multiple subregions, each with unique receptor composition and structural connectivity (Nelson et al., 2013). For instance, cytoarchitectonic parcellation of PPC reveals at least 11 distinct subregions: seven in vPPC (Caspers et al., 2006), and at least four in dPPC (Scheperjans et al., 2008). Such structural partitioning relates to functional partitioning, as regions exhibiting similar receptor architectonics have been shown to belong to the same functional network (e.g., Zilles & Amunts, 2009). Diffusion tensor imaging (DTI) studies also reveal fine-grained parcellation of PPC, with multiple subregions of vPPC showing macroscopically distinct structural connectivity profiles (Caspers et al., 2011; Uddin et al., 2010).

Heterogeneity within PPC is also reflected in functional profiles. First, at least seven dPPC regions (IPS0–IPS5 and SPL1) contain "attendotopic maps" (Silver & Kastner, 2009); these regions appear separable from the lateral IPS region showing item-memory strength effects, and partially overlap with the medial IPS/SPL region demonstrating a retrieval decision uncertainty effect (Hutchinson et al., 2014). Second, "resting-state functional connectivity" analyses indicate that distinct aspects of PPC functionally connect with distinct large-scale networks (Nelson et al., 2013). For example, four to eight large-scale networks include PPC nodes (Yeo et al., 2011). Within vPPC, (1) SMG and AnG show differing, if not opposite, connectivity with the MTL, and only AnG appears to belong to a functional network that includes parahippocampal gyrus and hippocampus; (2) boundary detection and graph theoretic analyses reveal an abrupt transition between SMG and AnG in their global functional connectivity profiles; and (3) these boundaries are functionally meaningful, with AnG—but not SMG—seeds being sensitive to retrieval outcomes (for visualization of boundaries in relation to the retrieval effects discussed above, see figure 48.2A). As such, extensive evidence indicates that considerable structural and functional heterogeneity exists in PPC, making it unlikely that a coarse dorsal/ventral account can explain the full pattern of PPC activity during retrieval.

Proponents of AtoM argue that extensive PPC functional heterogeneity is more apparent than real, particularly in vPPC, since they interpret the literature as revealing "largely overlapping [effects] with some differences around the edges" (Cabeza, Ciaramelli, & Moscovitch, 2012). This position seems difficult to reconcile with the considerable evidence detailed above, including the retrieval-related dissociations between AnG and TPJ (figure 48.2). Moreover, Nelson, McDermott, and Petersen (2012) point to the sharp connectivity boundary between AnG and TPJ/SMG as strong evidence against such a view of vPPC functional organization. Following criteria for defining a distinct cortical area—a region possessing unique function, architectonics, connections, and topography (Felleman & Van Essen, 1991)—the findings reviewed above suggest that PPC contains at least four, but more likely six or more, major subdivisions. Given extant data, the AtoM hypothesis may explain retrieval-related effects in medial IPS/SPL (top-down attention to retrieval cues and/or mnemonic evidence) and TPJ (bottom-up attention that is suppressed during top-down attention allocation required for retrieval; figure 48.3). However, AtoM leaves unspecified the functional significance of the old/new effects in AnG and lateral IPS (which appear sensitive to recollection and item-memory strength, respectively). Indeed, while proponents of AtoM posit that the anatomical separability of memory and visual attention effects in dPPC and in vPPC do not pose a challenge to AtoM, as they may reflect common attentional computations oriented either to external inputs (medial IPS/SPL and TPJ) or internal representations (lateral IPS and AnG; Cabeza et al., 2012), it is unclear how this can account for the dissociable activity profiles that are observed during retrieval (figure 48.2A)—specifically, the monotonic function in lateral IPS together with the nonmonotonic function in medial IPS/SPL;

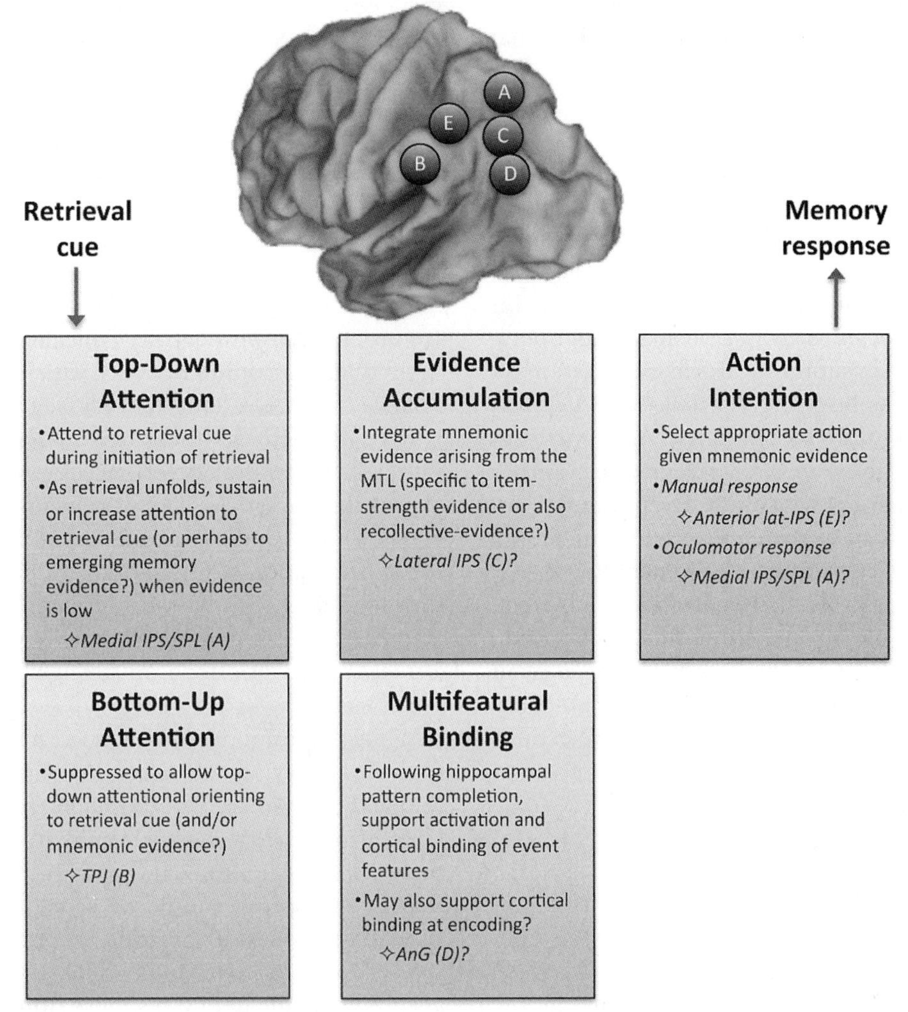

FIGURE 48.3 Working model of PPC operations during episodic retrieval (see text for details).

the same holds for the distinct functional profiles in TPJ and AnG. We next discuss other hypotheses advanced to understand retrieval effects in AnG and lateral IPS.

Buffer/binding accounts of angular gyrus function

As reviewed above, extensive evidence indicates that AnG activity tracks recollection at retrieval. Two closely related hypotheses—the *episodic buffer* (Vilberg & Rugg, 2008) and *cortical binding of relational activity* (CoBRA; Shimamura, 2011) accounts—posit that AnG operates on the products of hippocampally mediated pattern completion (see chapters 45 and 46 in this volume). AnG is thought to be well positioned to represent and operate over the multifeatural information reinstated by the MTL, given that AnG (1) receives strong disynaptic input from the hippocampus by way of parahippocampal gyrus and retrosplenial cortex, and (2) functionally couples with these MTL structures at rest and during tasks.

The episodic buffer hypothesis proposes that AnG acts as an interface between episodic memory operations and executive processes engaged in service of memory-guided decisions. By this account, AnG mechanisms serve to *maintain* recollected memory content, so that the retrieved mnemonic information can be interrogated in service of retrieval goals (Vilberg & Rugg, 2008). Consistent with this account, AnG demonstrates a "recollection load effect" (Hutchinson et al., 2014; Rugg et al., in press; Vilberg & Rugg, 2008)—that is, activity is greater when more versus less information (or more specific vs. more general information) is recollected, and thus presumably when more information must be maintained. Moreover, Vilberg and Rugg (2012) demonstrated that, whereas hippocampus is only transiently engaged during recollection, AnG activity persists for variable durations depending on how long recollected information must be held to meet task demands, consistent with the notion of a "buffer" that maintains retrieved representations.

CoBRA draws from the connectivity of AnG, wherein AnG serves as a cortical convergence zone of multimodal information, and thus may be engaged during retrieval to facilitate reactivation of the disparate details of a recollected event (Shimamura, 2011). As such, CoBRA is closely related to the episodic buffer account in that AnG mechanisms are proposed to operate on the output of hippocampal pattern completion operations, in this case to support cortical reactivation of event features. It has been argued that CoBRA differs from the episodic buffer account in that AnG processes operate on cortically stored representations, rather than acting as a separate store of these representations. In this manner, it can be considered an activation model of the episodic buffer hypothesis (Shimamura, 2011). While both accounts predict a recollection load effect, given the putative role of AnG in multimodal or multifeatural integration, the binding account may explain why AnG activity is sometimes greater for high-confidence relative to low-confidence correct rejections (perhaps reflecting the encoding/binding of novel associations during retrieval; figure 48.2A).

Mnemonic accumulator accounts of lateral intraparietal sulcus and angular gyrus

Many acts of memory retrieval can be construed as a decision process, in which retrieved mnemonic evidence is computed in relation to decision criteria (Ratcliff, 1988). The mechanisms governing memory-based decisions, which depend on internally generated evidence, may parallel those supporting perceptual decisions, which depend on external inputs. Theoretical models, which posit that evidence is accumulated toward one or more bounds over the course of a decision (Ratcliff, 1988; Usher & McClelland, 2001), can successfully account for a range of behavioral phenomena from mnemonic and perceptual decision tasks, including speed-accuracy trade-offs, the positive skew in reaction time (RT) distributions, and slower RTs for incorrect versus correct decisions.

In two-choice perceptual decision tasks, neurons in lateral intraparietal cortex (LIP, the putative nonhuman primate homolog to human IPS; e.g., Van Essen et al., 2001) gradually increase their firing rates until a decision is made (Shadlen & Newsome, 2001), with greater perceptual evidence eliciting steeper slopes in LIP firing rates. These data suggest that LIP neurons, along with neurons in the frontal eye fields (FEF) and principal sulcus of PFC, code for a cumulative decision variable during perceptual decisions. In humans, BOLD activity in IPS and FEF is greater (e.g., Heekeren, Marrett, Bandettini, & Ungerleider, 2004; Kayser, Buchsbaum, Erickson, & D'Esposito, 2010) and more temporally extended (Ho, Brown, & Serences, 2009; see also Ploran et al., 2007) during low- versus high-evidence perceptual decisions, suggesting a role for these regions in an accumulative decision process.

As reviewed above, BOLD activity in lateral IPS tracks item-memory strength during recognition decisions (figure 48.2A), with activity typically being greater for high- than low-confidence hits, which in turn is greater relative to misses and correct rejections. Moreover, lateral IPS activity tracks *perceived* memory strength rather than veridical experience, being greater for false alarms than misses or correct rejections (Wagner et al., 2005). One hypothesis is that lateral IPS contributes to the accumulation of mnemonic evidence toward a decision bound. By this account, greater memory strength leads to greater accumulative activity, perhaps reflecting a steeper response slope on stronger memory strength trials.

While the accumulator framework may provide a way to link activity in LIP neurons with BOLD data in human IPS, there are several challenges for this account. First, it is unclear how accumulative activity in LIP neurons translates to BOLD activity. Some have argued that *lower* evidence should be accompanied by greater BOLD responses in regions where decision signals are accumulated (Ho et al., 2009; Kayser et al., 2010), assuming (1) the neural response slope is shallower and thus takes longer to reach the bound, and (2) accumulative processes terminate once the bound is reached. The BOLD pattern predicted by this view would appear to better fit that seen in medial IPS/SPL, which demonstrates greater activity for lower-confidence memory decisions (a decision uncertainty effect). Alternatively, it is possible that decision evidence may continue to accumulate even after the decision bound is reached, which would give rise to greater activity in situations of greater mnemonic evidence, and is the pattern seen in lateral IPS. Second, retrieval-related activity in multiple PPC regions varies with decision bias, including when bias is shifted (see chapter 49, this volume); it is presently unknown how shifts in decision bounds impact BOLD activity reflecting accumulative processes. Third, from one perspective, activity in a mnemonic accumulator region might be expected to track greater strength of evidence, regardless of whether the evidence favors an "old" or a "new" response. That is, a test probe may elicit stronger evidence that it is old (high- vs. low-confidence hits) or that it is novel (high- vs. low-confidence correct rejections). If evidence for novelty is accumulated (rather than correct rejections being based on the absence of strong evidence that the probe is old), then a U-shaped activity function might be predicted (rather than a monotonic item-memory strength function). Such a pattern is sometimes observed in AnG, rather than lateral IPS, during recognition decisions (figure 48.2B).

Others have posited that AnG implements an accumulative integration-to-bound mechanism during retrieval (Criss, Wheeler, & McClelland, 2013). Criss and colleagues observed that AnG BOLD activity peaked more quickly when participants rejected foils in a block that contained strongly encoded old items relative to a block that contained weakly encoded old items, which offers tentative support for the view that the level of evidence (as indexed by perceived novelty) influences the rise rate of AnG activity (N.B.: this result is also consistent with binding/buffer accounts). Other data potentially challenge an accumulator account of AnG function. Guerin and Miller (2011) had participants encode faces that occurred with low (once or twice) or high frequency (five or six times), and then make a frequency discrimination between pairs of faces ("which face appeared more often?"). Test pairs consisted of easy versus difficult discriminations (low-high pairs vs. low-low or high-high pairs). Importantly, AnG activity was greater when frequency discriminations involved high-frequency faces (i.e., high-high > low-high > low-low), rather than when discriminations were easy versus difficult, suggesting that AnG activity may scale with the amount of retrieved information independent of decision certainty. This pattern appears to favor binding/buffer accounts of AnG function during retrieval.

Action intention

PPC computations support the planning of movements, as evidenced by extensive data on the role of PPC in eye, arm, and hand movements (Andersen & Cui, 2009). The vast majority of retrieval studies required participants to indicate their memory decisions by manual button press, raising the possibility that aspects of PPC activity during retrieval reflect action-intention processes (which may interact with processes of attention, binding, and evidence accumulation). From this perspective, some PPC old/new effects may reflect the preparation of memory-guided actions, with the strength of memory evidence serving to modulate the strength or duration of action-intention processes. This hypothesis predicts that (1) existing PPC old/new effects will localize to regions that exhibit preference for hand movements, and (2) the localization of old/new effects will shift depending on the memory-guided action to be performed (e.g., manual vs. oculomotor responses). To date, no study has directly tested either prediction. Here, we review evidence from the action-intention literature that suggests a pattern of effector-specificity compatible with this hypothesis.

In nonhuman primates, effector-specificity is observed in PPC during preparation of eye, arm, and hand movements (Andersen & Cui, 2009). Neurons in the "parietal reach region" of PPC exhibit preferential activity during preparation of arm movements, anterior intraparietal cortical neurons show preference for grasping movements, and a "parietal eye field" in PPC shows strong preference for eye movements. Although monkey-human homologies are underspecified, it is thought the human homolog of the parietal eye field is medial IPS, anterior intraparietal area is thought to be homologous to anterior IPS, and the parietal reach region may dissociate into two regions in humans, medial IPS and superior parietal-occipital sulcus (Vesia & Crawford, 2012).

There is growing fMRI evidence for effector-specificity in human PPC (Vesia & Crawford, 2012), with a saccade bias in more posterior/medial areas and a hand/arm bias in more anterior/lateral areas. For instance, Beurze, de Lange, Toni, and Medendorp (2009) demonstrated that anterior IPS exhibits hand-specific activity, the superior parietal-occipital sulcus exhibits eye-specific activity, and medial IPS is sensitive to both effectors. Interestingly, recent data suggest that this posterior-anterior gradient may not reflect effector-specificity *per se*, but instead may relate to the different functional goals that different effectors enact (i.e., saccade goals operate on eye-centered reference frames, whereas limb goals interact with both eye- and body-reference frames; Vesia & Crawford, 2012).

Given the central role of dPPC regions in action intention, a largely unexplored hypothesis is that PPC activity during retrieval—particularly in IPS—reflects the transformation of mnemonic evidence to action intention. As noted, during manually signaled memory decisions, medial IPS/SPL demonstrates a decision uncertainty effect, whereas lateral IPS (often spanning mid- to anterior IPS) monotonically tracks perceived item-memory strength, raising an intriguing possibility that retrieval-related dissociations may partially reflect the role of action computations during recognition decisions. In the only study (to our knowledge) to investigate action intention during retrieval (Shannon & Buckner, 2004), participants made button presses either to old items or to new items, withholding responses to the other class of items (manipulated between experiments). While PPC old/new effects did not vary according to response contingency, which seems to argue against an action-intention account of retrieval-related PPC activity, it remains possible that participants prepared (but did not execute) a response on every trial. Given the apparent co-localization of some dPPC retrieval-related effects with dPPC regions implicated in action intention, further studies are needed to directly investigate the degree to which retrieval-related activity varies with the specific effector used to make memory-guided actions.

Working model

Retrieval-guided behavior emerges from multiple neurocognitive processes—including processes mediated by PPC—that are triggered by retrieval cues and end in action. PPC subregions influence retrieval through their participation in large-scale functional networks (Nelson et al., 2010), and retrieval-guided behavior likely depends on dynamic interactions between these networks. Here, we briefly introduce a working model that posits putative roles for specific PPC regions during retrieval (figure 48.3), and we highlight fundamental open questions.

Retrieval is initiated by cues that are encountered in the external environment or are internally generated. Top-down attention is likely engaged during retrieval initiation, supporting the allocation of attention to retrieval cues; subsequently, top-down attentional allocation likely unfolds dynamically over the course of the retrieval act, depending on the nature of the emerging memory signals. In particular, top-down attention to retrieval cues, and perhaps also to the products of retrieval, is likely to be maintained or even increase over the course of uncertain or difficult retrieval trials (giving rise to the decision uncertainty effect). Extant evidence suggests that medial IPS/SPL supports top-down attention, and that engagement of these regions is accompanied by suppression of TPJ-mediated bottom-up attention during retrieval attempts; however, more evidence is needed to fully assess both hypotheses.

Attended retrieval cues can elicit two types of mnemonic signals (see chapter 45, this volume): (1) MTL cortex-dependent item-memory strength evidence that supports familiarity-based retrieval decisions, and (2) hippocampal-dependent pattern completion that drives reinstatement of event features and supports recollection-based retrieval decisions. Extant data suggest that pattern completion gives rise to MTL interactions with AnG, which may contribute to the reactivation and maintenance of multifeatural event details in cortex. A role for AnG in multifeatural binding may also be present during the encoding of novel events. At present, it is unclear whether pattern-completed evidence is combined with item-memory evidence during memory-based decisions, or whether recollective and item-memory signals are independently accumulated.

The role of parietal cortex in memory decisions is also complicated by questions related to the transfer function that maps neuronal firing rate to BOLD activity, and to the anatomical overlap between parietal regions that support perceptual and mnemonic decisions. Specifically, it is unclear how accumulative decision processes observed at the neuronal level should manifest in BOLD data. It also remains to be seen whether the posterior IPS region implicated in the accumulation of perceptual evidence is dissociable from the lateral IPS region that has been shown to track item-memory strength.

Another important question concerns the extent to which mnemonic evidence accumulation and action-intention processes are separable. During manual actions, which may be supported by anterior IPS in humans, it is possible that processes that select such actions are at least partially distinct from the item-memory strength patterns seen in lateral IPS (which likely only partially overlap with anterior IPS). Although no data are presently available, we hypothesize that memory-guided oculomotor actions will result in medial IPS/SPL activity that will be distinct from the observed lateral IPS item-memory strength effects, establishing the partial functional independence of putative evidence accumulation and action-selection processes.

Conclusion

The present review and working model may serve as a starting point for considering how the fine-grained functional heterogeneity of PPC influences memory-guided behavior. Continued progress in understanding PPC contributions to retrieval is fundamental to building cognitive neuroscience models of memory, and will have implications for understanding parietal contributions to attention, decision making, and action. Further progress likely will come through a next wave of experimentation that leverages methods with higher spatial and temporal specificity. First, marked across-participant anatomical and functional variability in PPC may necessitate a shift toward more subject-level analyses, employing within-subject comparisons of regions supporting perceptual decision making, manual versus oculomotor action selection, and top-down and bottom-up attention. Second, many of the hypothesized roles of PPC in retrieval entail dynamically unfolding computations that play out with distinct temporal profiles, and could thus be further characterized using methods with greater temporal resolution. For example, a better understanding of PPC mnemonic accumulative processes could be obtained with measurements of the slope of neural responses during retrieval. We are optimistic that such investigations will continue to reveal the unexpected, multifaceted nature of PPC contributions to memory.

REFERENCES

Andersen, R. A., & Cui, H. (2009). Intention, action planning, and decision making in parietal-frontal circuits. *Neuron, 63*(5), 568–583.

Berryhill, M. E. (2012). Insights from neuropsychology: Pinpointing the role of the posterior parietal cortex in episodic and working memory. *Front Integr Neurosci, 6,* 31.

Beurze, S. M., de Lange, F. P., Toni, I., & Medendorp, W. P. (2009). Spatial and effector processing in the human parietofrontal network for reaches and saccades. *J Neurophysiol, 101*(6), 3053–3062.

Cabeza, R., Ciaramelli, E., & Moscovitch, M. (2012). Cognitive contributions of the ventral parietal cortex: An integrative theoretical account. *Trends Cogn Sci, 16*(6), 338–352.

Cabeza, R., Ciaramelli, E., Olson, I. R., & Moscovitch, M. (2008). The parietal cortex and episodic memory: An attentional account. *Nat Rev Neurosci, 9*(8), 613–625.

Cabeza, R., Mazuz, Y. S., Stokes, J., Kragel, J. E., Woldorff, M. G., Ciaramelli, E., ... Moscovitch, M. (2011). Overlapping parietal activity in memory and perception: Evidence for the attention to memory model. *J Cogn Neurosci, 23*(11), 3209–3217.

Caspers, S., Eickhoff, S. B., Rick, T., von Kapri, A., Kuhlen, T., Huang, R., ... Zilles, K. (2011). Probabilistic fibre tract analysis of cytoarchitectonically defined human inferior parietal lobule areas reveals similarities to macaques. *NeuroImage, 58*(2), 362–380.

Caspers, S., Geyer, S., Schleicher, A., Mohlberg, H., Amunts, K., & Zilles, K. (2006). The human inferior parietal cortex: Cytoarchitectonic parcellation and interindividual variability. *NeuroImage, 33*(2), 430–448.

Ciaramelli, E., Grady, C., Levine, B., Ween, J., & Moscovitch, M. (2010). Top-down and bottom-up attention to memory are dissociated in posterior parietal cortex: Neuroimaging and neuropsychological evidence. *J Neurosci, 30*(14), 4943–4956.

Ciaramelli, E., Grady, C. L., & Moscovitch, M. (2008). Top-down and bottom-up attention to memory: A hypothesis (AtoM) on the role of the posterior parietal cortex in memory retrieval. *Neuropsychologia, 46*(7), 1828–1851.

Corbetta, M., Patel, G., & Shulman, G. L. (2008). The reorienting system of the human brain: From environment to theory of mind. *Neuron, 58*(3), 306–324.

Criss, A. H., Wheeler, M. E., & McClelland, J. L. (2013). A differentiation account of recognition memory: Evidence from fMRI. *J Cogn Neurosci, 25*(3), 421–435.

Felleman, D. J., & Van Essen, D. C. (1991). Distributed hierarchical processing in the primate cerebral cortex. *Cereb Cortex, 1*(1), 1–47.

Guerin, S. A., & Miller, M. B. (2011). Parietal cortex tracks the amount of information retrieved even when it is not the basis of a memory decision. *NeuroImage, 55*(2), 801–807.

Heekeren, H. R., Marrett, S., Bandettini, P. A., & Ungerleider, L. G. (2004). A general mechanism for perceptual decision-making in the human brain. *Nature, 431*(7010), 859–862.

Ho, T. C., Brown, S., & Serences, J. T. (2009). Domain general mechanisms of perceptual decision making in human cortex. *J Neurosci, 29*(27), 8675–8687.

Hutchinson, J. B., Uncapher, M. R., & Wagner, A. D. (2009). Posterior parietal cortex and episodic retrieval: Convergent and divergent effects of attention and memory. *Learn Memory, 16*(6), 343–356.

Hutchinson, J. B., Uncapher, M. R., Weiner, K. S., Bressler, D. W., Silver, M. A., Preston, A. R., & Wagner, A. D. (2014). Functional heterogeneity in posterior parietal cortex across attention and episodic memory retrieval. *Cereb Cortex, 24*(1), 49–66.

Kayser, A. S., Buchsbaum, B. R., Erickson, D. T., & D'Esposito, M. (2010). The functional anatomy of a perceptual decision in the human brain. *J Neurophysiol, 103*(3), 1179–1194.

Nelson, S. M., Cohen, A. L., Power, J. D., Wig, G. S., Miezin, F. M., Wheeler, M. E., ... Petersen, S. E. (2010). A parcellation scheme for human left lateral parietal cortex. *Neuron, 67*(1), 156–170.

Nelson, S. M., McDermott, K. B., & Petersen, S. E. (2012). In favor of a fractionation view of ventral parietal cortex: Comment on Cabeza et al. *Trends Cogn Sci, 16*(8), 399–400.

Nelson, S. M., McDermott, K. B., Wig, G. S., Schlaggar, B. L., & Petersen, S. E. (2013). The critical roles of localization and physiology for understanding parietal contributions to memory retrieval. *Neuroscientist, 19*(6), 578–591

O'Connor, A. R., Han, S., & Dobbins, I. G. (2010). The inferior parietal lobule and recognition memory: Expectancy violation or successful retrieval? *J Neurosci, 30*(8), 2924–2934.

Ploran, E. J., Nelson, S. M., Velanova, K., Donaldson, D. I., Petersen, S. E., & Wheeler, M. E. (2007). Evidence accumulation and the moment of recognition: Dissociating perceptual recognition processes using fMRI. *J Neurosci, 27*(44), 11912–11924.

Ratcliff, R. (1988). Continuous versus discrete information processing modeling accumulation of partial information. *Psychol Rev, 95*(2), 238–255.

Rugg, M. D., Johnson, J. D., & Uncapher, M. R. (in press). Encoding and retrieval in episodic memory: Insights from fMRI. In A. Duarte, M. Berense, & D. R. Addis (Eds.), *Cognitive neuroscience of memory.* Hoboken, NJ: Wiley-Blackwell.

Scheperjans, F., Eickhoff, S. B., Hömke, L., Mohlberg, H., Hermann, K., Amunts, K., & Zilles, K. (2008). Probabilistic maps, morphometry, and variability of cytoarchitectonic areas in the human superior parietal cortex. *Cereb Cortex, 18*(9), 2141–2157.

Sestieri, C., Shulman, G. L., & Corbetta, M. (2010). Attention to memory and the environment: Functional specialization and dynamic competition in human posterior parietal cortex. *J Neurosci, 30*(25), 8445–8456.

Shadlen, M. N., & Newsome, W. T. (2001). Neural basis of a perceptual decision in the parietal cortex (area LIP) of the rhesus monkey. *J Neurophysiol, 86*(4), 1916–1936.

Shannon, B. J., & Buckner, R. L. (2004). Functional-anatomic correlates of memory retrieval that suggest nontraditional processing roles for multiple distinct regions within posterior parietal cortex. *J Neurosci, 24*(45), 10084–10092.

Shimamura, A. P. (1995). Memory and frontal lobe function. In M. S. Gazzaniga (Ed.), *The cognitive neurosciences* (pp. 803–814). Cambridge, MA: MIT Press.

Shimamura, A. P. (2011). Episodic retrieval and the cortical binding of relational activity. *Cogn Aff Behav Neurosci, 11*(3), 277–291.

Silver, M. A., & Kastner, S. (2009). Topographic maps in human frontal and parietal cortex. *Trends Cogn Sci, 13*(11), 488–495.

Squire, L. R. (1992). Memory and the hippocampus: A synthesis from findings with rats, monkeys, and humans. *Psychol Rev, 99*(2), 195–231.

Uddin, L. Q., Supekar, K., Amin, H., Rykhlevskaia, E., Nguyen, D. A., Greicius, M. D., & Menon, V. (2010). Dissociable connectivity within human angular gyrus and intraparietal sulcus: Evidence from functional and structural connectivity. *Cereb Cortex, 20*(11), 2636–2646.

Uncapher, M. R., DuBrow, S., Hutchinson, J. B., & Wagner, A. D. (2011). Temporal interplay of attention and memory during associative retrieval. Paper presented at the 41st Annual Meeting of the Society for Neuroscience, Washington, DC.

Usher, M., & McClelland, J. L. (2001). The time course of perceptual choice: The leaky, competing accumulator model. *Psychol Rev, 108*(3), 550–592.

Van Essen, D. C., Lewis, J. W., Drury, H. A., Hadjikhani, N., Tootell, R. B., Bakircioglu, M., & Miller, M. I. (2001). Mapping visual cortex in monkeys and humans using surface-based atlases. *Vis Res, 41*(10–11), 1359–1378.

Vesia, M., & Crawford, J. D. (2012). Specialization of reach function in human posterior parietal cortex. *Exp Brain Res, 221*(1), 1–18.

Vilberg, K. L., & Rugg, M. D. (2008). Memory retrieval and the parietal cortex: A review of evidence from a dual-process perspective. *Neuropsychologia, 46*(7), 1787–1799.

Vilberg, K. L., & Rugg, M. D. (2012). The neural correlates of recollection: Transient versus sustained fMRI effects. *J Neurosci, 32*(45), 15679–15687.

Wagner, A. D., Shannon, B. J., Kahn, I., & Buckner, R. L. (2005). Parietal lobe contributions to episodic memory retrieval. *Trends Cogn Sci, 9*(9), 445–453.

Yeo, B. T., Krienen, F. M., Sepulcre, J., Sabuncu, M. R., Lashkari, D., Hollinshead, M., ... Buckner, R. L. (2011). The organization of the human cerebral cortex estimated by functional connectivity. *J Neurophysiol, 106*(3), 1125–1165.

Zilles, K., & Amunts, K. (2009). Receptor mapping: Architecture of the human cerebral cortex. *Curr Opin Neurol, 22*(4), 331–339.

49 Memory as Decision Making

MICHAEL B. MILLER AND IAN G. DOBBINS

ABSTRACT Memory researchers investigating the brain regions involved in successfully retrieving a past event have revealed a robust and ubiquitous pattern of activity across lateral regions of the left prefrontal and parietal cortex. While some of these regions may play a direct causal role in episodic memory retrieval, such as the accumulation of mnemonic evidence or the buffering of retrieved representations, we present evidence that much of the activity can be accounted for by decision and control processes and not memory retrieval per se. Two general lines of research demonstrate that brain activity greater for hits than correct rejections can reflect decision biases and/or reorientation. We present a biasing and orienting model of parietal contribution that converges with studies of patients with parietal lobe damage and syntheses with research on visual attention and valuation.

Episodic or event memory pervades our everyday lives, but its study is challenging because inferences about memory functioning are fairly indirect. These inferences rely on establishing the links between the original events that yield the potential for subsequent memory expression and a retrieval demand that may take place minutes, days, or years following the events. The researcher cannot directly manipulate raw memory signals in the same manner he or she might manipulate frequency in, say, an audition experiment. Instead, the properties of memory representations must be inferred through far more indirect methods. Hence, the advent of functional brain imaging using positron emission tomography (PET) and then functional magnetic resonance imaging (fMRI) provided the opportunity to more directly observe the process of episodic memory retrieval (a.k.a. ecphory) that guides our behavior in a wide variety of contexts. The overarching message of this chapter is that while these techniques have greatly informed our understanding of episodic recognition memory, in doing so, they suggest that much of the observed brain activation reflects various decision and control processes and not memory-retrieval processes per se. Thus the "simple" act of recognition appears to tap a host of complex control processes geared toward tuning and biasing memory judgments.

In everyday life, episodic memories are retrieved in various ways, including spontaneous encounters with cues that trigger the retrieval of a memory, as well as through purposeful searches of past events. We also make decisions throughout the course of the day that depend on episodic memory evidence; for example, did I park the car in this lot? have I read that book before? did my wife tell me to pick up the kids? Strategic decision processes are involved in each of these mnemonically guided decisions, although sometimes in subtle ways. For example, judging whether one read a particular book (a recognition task) may be influenced by factors other than the memory "signal" itself, such as an assessment of one's general familiarity with the author, knowledge of the release date of the book, or other factors, which may bias one toward a positive or negative final conclusion (Johnson, Hashtroudi, & Lindsay, 1993; Mandler, 1980; Schacter, 1996; Tulving, 1983). Recognition tests have been used to probe the contents of memory as early as 1913, when Hollingworth stated in his report in *The American Journal of Psychology* that "the value of a single presentation is greater in recognition than in recall, and the difference between the values of repetitions becomes still greater the more meaning the material possesses" (Hollingworth, 1913, p. 543). As George Mandler put it, to "recognize is the act of perceiving something as previously known ... the recognition of the prior occurrence of an event" (1980, p. 252). He postulated that recognition consists of two separate and additive processes: (1) the *recognition of familiarity* which is a continuous value retrieved quickly and automatically based largely on the perceptual characteristics of the previous exposure, and (2) the *identification as a result of a retrieval process* that is thought to be initiated once the familiarity judgment fails to provide an unequivocal decision. These two processes are now commonly known as familiarity and recollection (Jacoby, 1991; Yonelinas, 1997), although current models generally assume that recollection is sought during every recognition trial in standard recognition tests, and not merely those trials that yield ambiguous sensations of familiarity, and it is thought that familiarity is often based on conceptual characteristics of the previous exposure.

Mandler's well-known analogy of these two separable forms of memory content invites you to imagine walking onto a bus and seeing a man whose unexpected familiarity convinces you that you have seen him before, but you cannot immediately recall where or when. Following a deliberate search of memory where various candidate possibilities are considered, you realize that the man is a butcher in your neighborhood. In order to understand the neural representations of these various forms of memory content, cognitive neuroscientists typically contrast the recognition of a previously encountered item (hit) with the correct judgment that a memory probe is newly encountered (correct rejection). The natural inclination is to assume that regions demonstrating greater activation for the former than the latter are directly involved in the retrieval or representation of memory evidence. However, further inspection reveals that many original candidate regions support processes aside from episodic retrieval itself. It is important to emphasize that functional imaging researchers, like animal researchers, have relied heavily on recognition tasks for the study of episodic memory for the simple reason that they are amenable to available methodologies. Not only are recognition tasks relatively easy to train nonhuman animals on, but they also enable controlled timing and trial intermixing that is essential for event-related fMRI approaches. Thus, we know considerably less about other types of memory demands, such as free recall and the recall of other types of complex materials (cf. Long, Öztekin, & Badre, 2010).

The elusive "successful retrieval" map

One of the goals of the early neuroimaging work on memory was to distinguish between regions signaling successful episodic retrieval from those supporting the ability to deliberately engage in a retrieval attempt (Buckner, Koustall, Schacter, Dale et al., 1998; Kapur et al., 1995; Rugg, Fletcher, Frith, Frackowiak, & Dolan, 1996; Schacter, Alpert, Savage, Rauch, & Albert, 1996; Tulving et al., 1994). Brain activity associated with new items on a recognition test would necessarily contain little to no information regarding the study session, and therefore the attempt to retrieve episodic content from new items would contain more information about the attempt itself than the product of that attempt. On the other hand, brain activity associated with old items would contain information both about the retrieval attempt as well as the product of that retrieval attempt. These early neuroimaging studies necessarily relied on contrasts between blocks of trials. Therefore, the typical design included blocks of mostly old items compared with blocks of mostly new items, often "hidden" within the context of periods in which old and new items were equally frequent so as to avoid triggering nonmemorial response strategies. Comparison of mostly old and mostly new blocks revealed a pattern of activity that was commonly referred to as the "old/new" effect or the "successful retrieval" effect. While many of these early studies produced differential activity within the prefrontal cortex (Buckner, Koutstaal, Schacter, Dale et al., 1998; Rugg et al., 1996; Schacter, Alpert et al., 1996), Kapur and colleagues (1995) did find greater activity for mostly "old" blocks versus mostly "new" blocks in parietal regions, including the medial precuneus and the left lateral parietal cortex. Since both conditions were thought to include equal amounts of retrieval effort, the extra activity associated with mostly "old" blocks was interpreted as activity associated with ecphory, that is, the successful retrieval of episodic information in response to the recognition memory probe.

An obvious limitation of the block design is that the signal must be averaged across trial types other than the trial type of interest. For example, a mostly "old" block must include some new items in order to keep the subject honest, and it must also include incorrect responses to old items (i.e., misses). Another less obvious problem is that the design may force unintended psychological effects. For example, blocking trials according to the probability of a target may affect the subjects' strategies (Buckner, Koutstaal, Schacter, Wagner et al., 1998) or the saliency of the targets (Herron, Henson, & Rugg, 2004) depending on the block. The advent of event-related designs in fMRI allowed for a more precise comparison of old and new items by removing unsuccessful trials from the successful retrieval effect. Trials could now be selected based on the response of the subject, allowing a direct comparison of hits (correctly recognizing an old item) to correct rejections (correctly rejecting a new item). No longer did errors need to muddy the interpretations. The consequence of this advancement was that studies produced a more robust pattern of activation associated with successful retrieval (Buckner Koutstaal, Schacter, Wagner et al., 1998; Herron et al., 2004; Konishi, Wheeler, Donaldson, & Buckner, 2000; McDermott, Jones, Petersen, Lageman, & Roediger, 2000; Nolde, Johnson, & D'Esposito, 1998). Recent meta-analyses reveal that the typical pattern of activations for this effect includes regions of the anterior prefrontal cortex (PFC), anterior insula, thalamus, anterior cingulate, dorsolateral prefrontal cortex, medial prefrontal cortex, medial parietal cortex, and lateral posterior parietal cortex (e.g., Spaniol et al., 2009; see figure 49.1).

This widespread activation pattern associated with successful retrieval was quite surprising, because many of the regions that were found to exhibit successful retrieval activity were not thought to play a role in simple recognition judgment. For example, neuropsychological studies had shown that although the PFC was

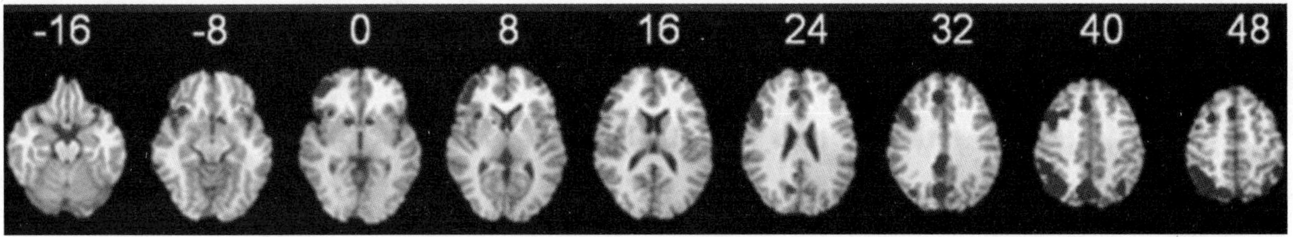

FIGURE 49.1 Activation likelihood map of "retrieval success" effect during recognition. (Taken from Spaniol et al. 2009.)

important for source-memory attribution and other complex recall tasks, patients with fairly extensive PFC damage were thought to be quite normal on simple verbal recognition tasks (Janowsky, Shimamura, & Squire, 1989; Milner, Corsi, & Leonard, 1991; Shimamura, Janowsky, & Squire, 1990; though see Wheeler, Stuss, & Tulving, 1997). These neuropsychological studies suggested that memory tasks that made demands on executive functioning or inhibitory control, such as free recall, source memory, and temporal ordering, were impaired by PFC damage because these secondary processes were damaged; however, basic recognition ability remained intact (Shimamura, 1995). The prefrontal cortex has reciprocal connections throughout the cortex, and it appears to filter and control much of the flow of information between sensory inputs and motor outputs (Miller & Cohen, 2001). However, these connections vary greatly across PFC regions, indicating considerable functional heterogeneity in this large region.

Functional specialization in prefrontal cortex during retrieval attempts

Although considerable uncertainty remains, a better understanding of the functional contribution of some of the PFC regions to memory is beginning to develop. Here we briefly consider two regions and outline putative functions supported by functional imaging findings.

LEFT VENTROLATERAL PREFRONTAL CORTEX (~BA 47) This region, located along the inferior frontal gyrus, is not only revealed by the retrieval success contrast, but it is also implicated in studies examining source memory and semantic memory. In source-memory studies, subjects encode information that originated from a particular context within the study session. Then, at test, on some trials subjects are asked to determine the particular context through which the item was encoded, whereas on other trials, they are asked to make an item memory judgment and are not required to recall the context. Brain activity associated with source-memory judgments is then compared to activity associated with item memory judgments alone. Even when stimulus materials are completely matched across source and item memory conditions, source judgments are associated with greater activity in left ventrolateral PFC (Dobbins, Foley, Schacter, & Wagner, 2002). However, the level of activation is quite similar regardless of whether or not the source attribution is correct (Dobbins, Rice, Wagner, & Schacter, 2003). In addition, the magnitude of activity in this region during the source task appears to be linked to the degree that the probe's semantic as opposed to perceptual characteristics are potentially relevant for the source judgment (Dobbins & Wagner, 2005). Consistent with this finding, damage to this region (Thompson-Schill et al., 1998) and disruption via transcranial magnetic stimulation (Gough, Nobre, & Devlin, 2005) have impaired semantic processing. Collectively, these findings suggest that the left ventrolateral PFC region supports the controlled or strategic semantic processing of probes during source-memory attempts and presumably during memory attributions in general. This process, known as semantic elaboration, can improve retrieval outcomes to the extent that the semantic features evident during the initial encounter match the features attended to at test (Roediger & Geraci, 1990; Tulving & Thomson, 1973).

LEFT DORSOLATERAL PREFRONTAL CORTEX (~BA 6/8) This region is often activated in close conjunction with dorsolateral parietal cortex, with which it is directly anatomically connected. In controlled judgment domains outside of memory, these two regions are often described as members of a frontoparietal control network (Dosenbach et al., 2007). The left DLPFC also demonstrates greater activation during source- versus item-memory decisions for matched verbal probes; however, unlike the left ventrolateral region, it appears to be engaged even prior to the arrival of the probes and thus demonstrates a greater activation when the retrieval question dictates a source requirement instead of an item requirement (Dobbins & Han, 2006). The region then increases activation during both types of tasks when the probes are presented. As with left ventrolateral PFC, activation is not modulated by the success of the source-memory judgment (Dobbins et al., 2003). This pattern is consistent with the long-standing role of the region during verbal working-memory demands (Nystrom et al., 2000; Rypma, Prabhakaran, Desmond, Glover, & Gabrieli, 1999) and led to the hypothesis that the region supports the online maintenance of descriptions of sought-after source information along with the maintenance of candidate probes if more than one is available during verbal episodic memory demands. In the cognitive literature, "retrieval descriptions" are quite important and are hypothesized to bias retrieval by foregrounding the general characteristics that the observer believes should be present in to-be-recovered content (Norman & Bobrow, 1979). These characteristics can be gleaned from the retrieval query, but may also be informed by experience with the task and other preexperimental beliefs. Computationally, this may be thought of as a coarse way of incorporating statistical priors into the judgment of recovered memory content. Additionally, recent work using transcranial magnetic stimulation more directly demonstrated that this general area plays a role in the foregrounding or biasing of representations in posterior cortex (Feredoes et al., 2011), a conclusion also suggested by reversible cooling studies in nonhuman primates (Chafee & Goldman-Rakic, 2000).

Functional specialization in parietal cortex during retrieval attempts

Given the neuropsychological link between the PFC and source- and working-memory processes, it was not surprising that PFC activations were observed with the retrieval success contrast. However, this is not the case with the prominent parietal lobe activations that were also observed with this contrast. This region has never been associated with episodic memory in the long-standing neuropsychological literature, since extensive damage to the region leaves basic recognition abilities intact. Given the robust successful retrieval activation in parietal cortex (perhaps the most reliable of any regions exhibiting this effect), the race to functionally explain the role of the parietal lobe has become something of a quest in the last 10 years (Wagner, Shannon, Kahn, & Buckner, 2005). Wagner and colleagues conducted an influential meta-analysis in 2005 that noted that posterior parietal activity (PPC) was typically modulated by (1) the subjective perception that an item was old—activity for false alarms was greater than for misses; (2) the retrieval orientation of the subject—the goals of the task-modulated PPC activity regardless of mnemonic history; and (3) recollection-based versus familiarity-based recognition. These general observations led to the development of three prominent hypotheses that are still currently under debate (see chapter 48 in this volume). These hypotheses include the attention to internal representations hypothesis, which states that PPC could be involved in shifting attention away from external stimuli to internal representations of memory that were presumably arising from other regions, such as the medial temporal lobe. This hypothesis would certainly be supported by the neuroimaging work showing that the PPC is sensitive to changes in the retrieval goals and orientation (Dobbins et al., 2003; Dobbins & Wagner, 2005). For instance, in a study that separated activations to retrieval cue type (source or item memory) from activation to the actual memory probes, the left lateral parietal response (unlike the lateral premotor PFC) was insensitive to differences in the cues, but then demonstrated a prominent response to probes during source- but not item-memory judgments (Dobbins & Han, 2006). One putative interpretation of this pattern of results is that the response signaled the shift of attention toward recovered recollective mnemonic content that is critical for making source-memory attributions, but less important for endorsing items merely based on familiarity.

Similar ideas were later incorporated into a parietal memory framework termed the attention to memory (AtoM) model, which builds on the visual attention work of Corbetta and Shulman (2002) and posits both top-down and bottom-up attention mechanisms supporting memory in parietal cortex (Cabeza, Ciaramelli, & Moscovitch, 2012). Top-down mechanisms are held to take place in the superior parietal lobule (SPL) and reflect directed attention toward weak memory signals under difficult retrieval circumstances. In contrast, bottom-up processes, held to take place in supramarginal gyrus regions, are thought to reflect the capture of attention by recollective content arising from memory systems, presumably in the medial temporal lobes.

An alternative to the assumption that inferior lateral parietal responses reflect the capture of attention during memory is the idea that these responses reflect the accumulation of recovered memory evidence toward a decision bound. This hypothesis draws its inspiration from studies using single-cell recordings of monkeys in monkey area LIP while making a simple choice that is based on the integration of sensory signals until a decision is reached (Shadlen & Newsome, 2001), and behavioral research supports the idea that recognition evidence accumulates during the course of trials as well (e.g., Van Zandt & Maldonado-Molina, 2004). Support for the accumulation hypothesis of parietal activation has been gained by demonstrating that activation tracks the number of original contextual details that are recovered by the participant (Vilberg & Rugg, 2009a). However, a recent fMRI study on frequency judgments found that parietal activity tracked with the absolute amount of information even when that information was not the basis of the decision (Guerin & Miller, 2011). Finally, a related hypothesis suggests that parietal responses reflect the operation of an episodic-memory output buffer, in which the recovered contents of episodic memory are temporarily stored in a buffer that makes them rapidly accessible to decision making. Such a buffer had been proposed by Baddeley (2000) as the missing component to his working memory model, and it would work similarly to visual and verbal working memory buffers.

Criteria for fashioning a parietal lobe functional model

Despite a host of candidates, there remains little consensus on the functional roles of the parietal lobe during recognition memory. Building on the discussion of Wagner, Shannon, Kahn, and Buckner (2005), we propose a series of criteria that a successful model must achieve.

1. The characterization must be compatible with extant neuropsychological findings and be able to account for the historical absence of any link between parietal damage and recognition-memory impairment.

2. The characterization must clearly distinguish between causal and noncausal functional models with respect to memory-retrieval ability. A causal interpretation reflects a characterization that would be essential for successful retrieval behavior to occur. For example, the episodic buffer account is causal because damage or removal of the buffer would yield behavioral amnesia, since the contents of episodic memory would be unavailable. In contrast, a model that assumed the response reflected the implementation of a decision bias would be noncausal, in that damage would yield an inability to flexibly bias memory decisions but would not prevent basic memory functioning.

3. The characterization should incorporate both condition-level blood oxygen level–dependent, or BOLD, effects (e.g., hits greater than correct rejection) and it should anticipate individual differences. For example, the episodic buffer model would predict that individual differences in accuracy should track with increases in blood oxygen level–dependent activity.

A biasing and orienting model of parietal contributions

We propose a model that arises out of two general lines of research: one in which observers are required to shift memory decision biases adaptively during blocks of trials, and one in which they are required to incorporate biases into their recognition judgments on a trial-by-trial basis. Both lines of research suggest that parietal regions are important for the implementation and adjustment of decision biases during recognition. Before briefly describing these paradigms and their findings, it is important to note why bias during decision making is critical, particularly in the case of memory decisions.

The phrase "successful retrieval effect" implies that the effect directly supports retrieval processes; however, a more appropriate phrase might be the "recognition judgment effect," which does not presuppose a direct link between the activations and the availability of memory content. Critically, optimal decision making in recognition (and all discrimination tasks) depends not only upon the current stimulus evidence, but also upon the context in which that evidence is encountered. This is formalized under Bayesian reasoning through the incorporation of prior probabilities with current observed evidence in order to arrive at a posterior probability. Equivalently, under the signal-detection framework, the observer uses a decision bias that can be adjusted in order to maximize outcomes. Thus the use of informative priors in Bayesian reasoning and flexible biases (a.k.a. criteria) in signal-detection theory serve an identical function, namely, to modulate decisions using information other than that which is directly perceived or remembered. For example, even if recognition evidence were quite strong, one would nonetheless want to be highly cautious in judging an individual as recognized in a context in which there was little prior probability that familiar individuals would be encountered (e.g., a foreign airport terminal). In such situations, under the signal-detection framework, ideal observers would use what is termed a "strict" or "conservative" decision bias, requiring extremely high levels of evidence to judge an individual as recognized. In contrast, in more familiar environments they would use a more "lax" or "liberal" bias. The adopted bias can also be informed by the relative benefits and costs of correct and incorrect judgments as well as the prior probabilities (Gold & Shadlen, 2001; Green & Swets, 1966; Macmillan & Creelman, 2005). Thus, unlike the pejorative lay use of the term "decision bias," ideal responding actually requires biasing one's judgments; the same

TABLE 49.1

Sample subjects from two studies: Aminoff et al. (2012), using a target probability manipulation, and Kantner, Vettel, and Miller (submitted), using a payoff manipulation. In the initial condition in each study, the subjects have equal discrimination ability (d′) and they are both relatively equivalent in criterion (c). But in the contrasting condition only one of the two subjects appropriately adapts his or her criteria, which benefits that subject's proportion correct (PC) relative to the other subject. (H = hit rate; CR = correct rejection rate.)

	High target probability					Low target probability					When adapting to the conservative condition ...
Subject	H	CR	d′	c	PC	H	CR	d′	c	PC	
111	.81	.41	.64	−.56	.68	.48	.79	.77	.43	.70	now avoids false alarms.
051	.90	.27	.64	−.93	.70	.85	.39	.76	−.65	.53	still avoids misses.
	Conservative payoff					Liberal payoff					When adapting to the liberal condition ...
Subject	H	CR	d′	c	$	H	CR	d′	c	$	
161	.43	.71	.39	.37	4.00	.86	.20	.23	−.95	4.40	now avoids misses.
59	.34	.77	.34	.57	4.60	.29	.77	.18	.65	−2.60	still avoids false alarms.

recognition evidence should *not* lead to the same response on all occasions.

Recent behavioral work from our labs demonstrates that there are considerable individual differences in observers' ability to appropriately bias their recognition judgments using blockwise manipulations of target probabilities, blockwise manipulations of payoffs, or trialwise manipulations of target probability (Aminoff et al., 2012; Selmeczy & Dobbins, 2012). An example of this variability and its consequences is shown in table 49.1.

THE LINK BETWEEN PARIETAL CORTEX AND THE BIASING OF DECISIONS In the visual attention literature, the link between biasing judgments and parietal cortex is well established. When observers use predictive cues that reliably anticipate the spatial location of subsequent perceptual probes, activation is increased in the PPC, specifically in the intraparietal sulcus (IPS), compared to situations in which environmental transients occur at the same location as an upcoming target (e.g., Kincade, Abrams, Astafiev, Shulman, & Corbetta, 2005). The former is typically referred to as an endogenous shift of spatial attention; however, it is equally appropriate to refer to it as a spatial judgment bias. In contrast, when probes are encountered at unexpected locations, compared to validly endogenously cued locations, there is increased activation in ventral parietal regions surrounding the temporoparietal junction. As noted by Hutchinson, Uncapher, and Wagner (2009) and Nelson, McDermott, and Petersen (2012), these regions do not coincide with those implicated during recognition memory. However, as suggested by Cabeza et al. (2012), they may serve similar roles at the algorithmic level.

An early fMRI study suggested a role for recognition decision bias in the parietal cortex (Miller, Handy, Cutler, Inati, & Wolford, 2001). Using a block design, Miller and colleagues crossed manipulations of criteria with manipulations of sensitivity. A group-level analysis revealed that more activation occurred in dorsolateral prefrontal and dorsolateral parietal cortex during blocks in which the decision criterion shifted on a trial-by-trial basis compared to blocks in which the decision criterion remained stable. However, a similar analysis with blocks of high d' (sensitivity) compared to blocks of low d' revealed only medial activations, with no activations in parietal cortex whatsoever. A more recent event-related fMRI study of 95 subjects found that similar regions were activated for trials in which the criterion shifted compared to trials in which the criterion remained the same (Aminoff et al., submitted; see figure 49.2). Further, Aminoff and colleagues found that the regions sensitive to criterion shifting significantly overlapped with regions that exhibited greater activity for hits compared to correct rejections. Even the regions of successful retrieval activity that did not overlap with regions of criterion-shift activity using the strict thresholds were significantly correlated with individual differences in the conservativeness of the criterion (see below).

In a recent fMRI study, Vilberg and Rugg (2009b) manipulated the test-wide prevalence of targets (25% or 75%) during a combined source- and item-recognition judgment task. At test, subjects indicated that they recognized the items and remembered one of two source contexts, merely recognized the items but couldn't recollect the source context, or believed the items to be new. Critically, they found that regions in the lateral post-central gyrus and dorsal superior parietal lobule that demonstrated "retrieval success" effects were modulated by the target base rates. However, they also found that more lateral and inferior parietal regions, along with the middle portion of the IPS, demonstrated a "retrieval success" effect that was insensitive to the listwide target probabilities, with the authors suggesting this pattern was directly reflective of retrieval.

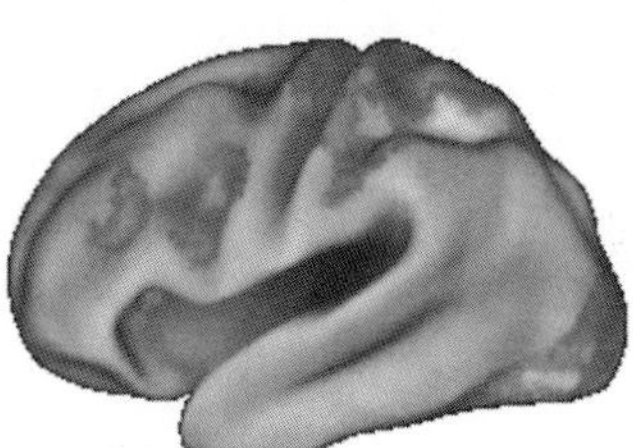

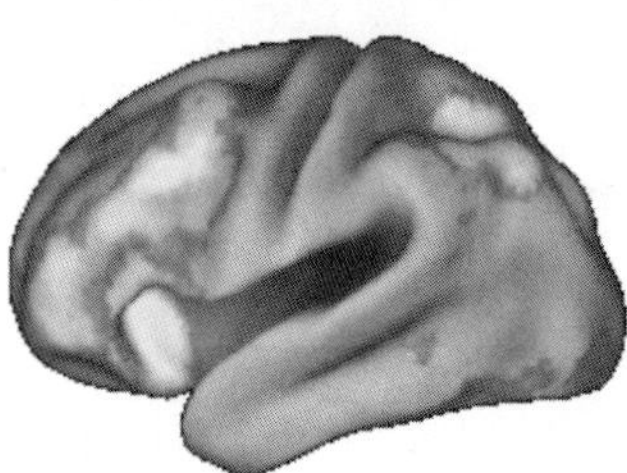

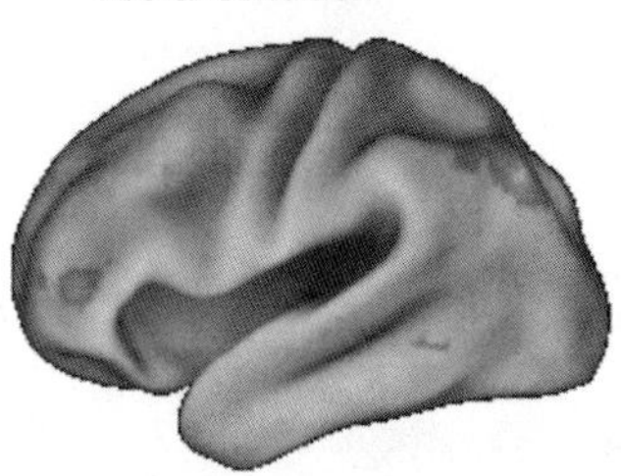

FIGURE 49.2 A comparison of activation maps (n = 95 subjects; p > .05, false-discovery rate corrected) for the shifting criterion contrast and the "retrieval success" contrast in the conservative and liberal conditions (see Aminoff et al., submitted).

But this interpretation was hampered by the fact that the subjects were explicitly instructed to ignore any changes in the base rate of the targets, and the changes in the target probabilities were not cued in any way. Several behavioral studies have demonstrated that criterion shifts only occur if the subject is aware of the change in the conditions (Estes & Maddox, 1995; Rhodes & Jacoby, 2007; Wixted & Stretch, 2000). Indeed, the listwide density manipulation in the Vilberg and Rugg (2009b) study failed to induce any behavioral decision biases. As we discuss below, in paradigms yielding clear behavioral evidence of biases, whether due to the shifting or the conservativeness of the bias, much of the lateral parietal response is affected.

For example, Aminoff et al. (submitted) used a blockwise biasing paradigm in which target probability (70% or 30%) was indicated by a color cue for small blocks of six to nine trials for both face and word stimuli that were tested in separate runs. Participants were informed of this contingency and displayed prominent behavioral biases induced by these cues. In terms of the behavioral results, Aminoff and colleagues (2012) recently reported clear individual differences in the degree that biases were induced across participants. These were stable domain-general differences that were observed across materials. As for the brain activation, there were three critical effects that emerged from the study (Aminoff et al., submitted). First, as noted above, when observers were required to shift the bias, there was a transient response at the beginning of the block in the SPL that extended along the IPS, regardless of the direction of shift or the accuracy of the response. Second, although an apparent "retrieval success" effect was observed in dorsolateral parietal cortex during blocks inducing a conservative bias (i.e., strong expectation of new materials), the effect was virtually eliminated under blocks that induced a liberal bias (i.e., strong expectation of old materials; see figure 49.2). Finally, activity associated with "retrieval success" in both the conservative and liberal conditions was significantly related to individual differences in criterion but not to individual differences in accuracy. Overall, these data suggest that the parietal retrieval response must take into account the adopted decision biases.

In order to examine the interaction between expectations and memory evidence, O'Connor, Han, and Dobbins (2010) used a trial-wise cuing paradigm in which each recognition memory probe was preceded by a verbal cue (Likely Old or Likely New) that indicated its probable memory status. During critical runs, these cues were valid on 80% of trials. Consistent with Aminoff et al. (submitted), these cues yielded prominent decision biases. With respect to brain activity, the critical comparison of invalidly versus validly cued trials (figure 49.3) resulted in prominent prefrontal and parietal activations. The results also demonstrated that invalid cuing activations occurred in supramarginal and angular gyrus regions for both old and new materials that were correctly identified. This invalid cued response partially overlaps with the response demonstrated when observers must shift the criterion applied to memory judgments (figure 49.2); however, it appears to extend more laterally. Because correctly identified new materials are unlikely to have any episodic content, these researchers concluded that the functional role of these regions cannot reflect episodic retrieval or the accumulation or buffering of episodic content. Furthermore, in an independent fMRI data set of basic uncued recognition, activation in these parietal regions was shown to correlate with individual differences in adopted decision bias, and not retrieval accuracy. Specifically, this analysis demonstrated that for individuals who were increasingly conservative, there was an increase in the hits>correct rejections signal difference. This corresponds to an increase during identification of the memory class least expected by the participant, and the lateral parietal region was held to signal disconfirmations of decision biases that were either cue-induced or

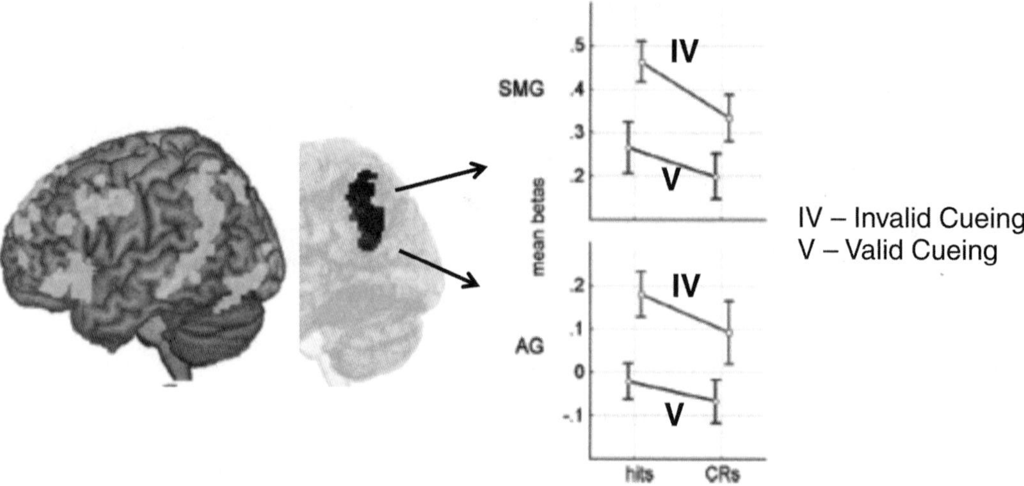

FIGURE 49.3 The violation of cued recognition memory expectations. Left panel shows activations that result when recognition memoranda violate expectations cued on a trialwise basis (invalid > validly cued recognition memory probes; voxelwise threshold $p < .001$ with greater than 5 continuous voxel extent). This invalid memory cueing effect occurs in both supramarginal and angular gyri (middle panel) and results in an increased response for both correctly identified old materials (hits) and new materials (correct rejections). Right panel: taken from O'Connor et al. (2010).

general characteristics of the participants (see also Herron & Rugg, 2003).

A final recent fMRI study looking at trialwise decision biases afforded sufficient power to separately examine the two types of violations possible, namely, unexpected familiarity following the Likely New cue and unexpected novelty following the Likely Old cue (Jaeger, Konkel, & Dobbins, 2013). Figure 49.4 shows three regions demonstrating three different patterns, including sensitivity to unexpected familiarity, sensitivity to unexpected novelty, and sensitivity to both unexpected familiarity and unexpected novelty. Postcentral gyrus and medial SPL areas demonstrated greater activation for new than old materials following the Likely Old cue, consistent with orienting toward unexpected novelty. This interpretation is further supported by the failure to detect any discernible signal difference in these regions when observers instead expected new materials following the Likely New cue. The opposite pattern of results occurred in anterior angular gyrus. Here differential activation occurred following the Likely New cues, in which there was a stronger response for old versus new materials, which is consistent with orienting toward unexpected familiarity. There was no signal difference between the materials when the observers instead expected familiarity following the Likely Old cues. Finally, the intraparietal sulcus demonstrated increased activation for whichever class of memoranda was unexpected. This pattern of activation led to the conclusion that this region supported the orienting of attention toward unexpected memory content, and hence may be critical for overriding adopted decision biases. From this overall perspective, the saliency of memory signals is governed by the bias induced by the cues. This notion is further supported by the results of an individual differences analysis of the anterior angular gyrus response. Within this region, the unexpected familiarity response (hits vs. correct rejections following the Likely New cue) was strongly associated with individual differences in accuracy. This might have been mistakenly construed as evidence that the region supports retrieval in a causal manner. However, the activations under the Likely Old and uncued conditions bore no relationship to individual differences in accuracy within those conditions. This is precisely the marker of saliency anticipated under an orienting model, in that those individuals who discriminate well showed the most marked differential response to unexpectedly familiar stimuli that conflicted with the Likely New bias versus expected novel stimuli that confirmed the Likely New bias.

Critically, the findings in Jaeger et al. (2013) point to considerable functional heterogeneity in parietal cortex during recognition judgment. Neither the unexpected familiarity nor unexpected novelty responses have been previously isolated, but their close proximity to the more general mid-IPS response may mean that these various responses may have been collapsed in prior designs and discussions. Furthermore, the three patterns of response in figure 49.4 corresponded well with recent parcellations of parietal cortex suggested by the analysis of resting connectivity data via graph theory (Nelson et al., 2010). Moving forward, then, it may be important to distinguish between general control

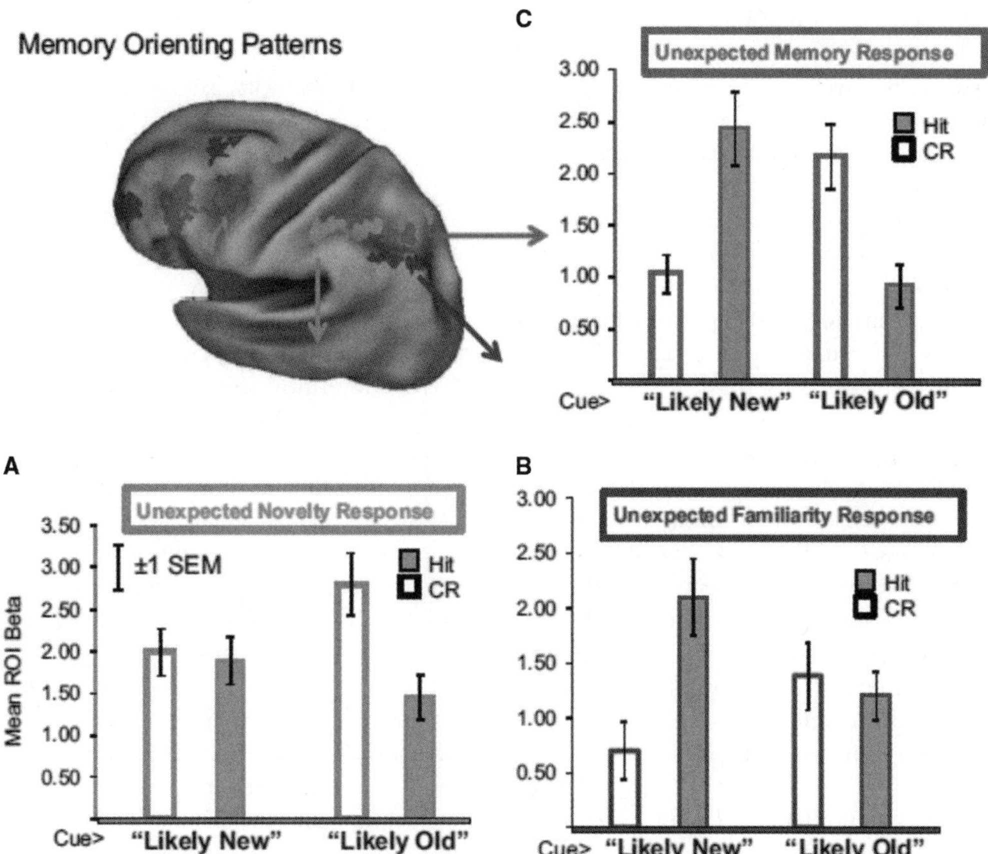

FIGURE 49.4 Selective responses to particular types of violations of memory expectations. Targeted contrasts and masking procedures isolated three different patterns of response in left lateral parietal cortex. The left anterior IPS/PoCG region in green illustrated an unexpected novelty response pattern, with bar plot (A) illustrating the pattern across cue conditions and item types. The posterior anterior angular gyrus region in blue demonstrated an unexpected familiarity response, with bar plot (B), illustrating the pattern of response across cue conditions and item types. The mid-IPS region in red demonstrated a general unexpected memory effect, with bar plot (C) illustrating the pattern of response across cue conditions and item types. (Taken from Jaeger et al., 2013.) (See color plate 40.)

mechanisms that underlie criterion shifts and violations of adopted criteria (that presumably engender shifts) from material specific effects, such as unexpected novelty, and unexpected familiarity responses that may be linked to mobilization of different investigatory responses depending on the nature of the violation. For example, figure 49.2 illustrates that the most prominent response to facing a requirement to shift the recognition criterion occurs along the IPS (figure 49.2, left panel), regardless of the direction of shift. Somewhat analogously, the mid-IPS response in figure 49.4 occurred whenever a memorandum violated the expectation held under the current bias/criterion. Additionally, both contrasts also implicated similar lateral PFC regions, and so both may capture a process or processes that are important when an observer must update or reevaluate the adequacy of the current decision criterion, regardless of whether it is liberal or conservative.

In summary, recent fMRI findings demonstrate that adopting a liberal decision bias eliminates previously observed successful retrieval activation differences in intraparietal sulcus and supramarginal gyrus regions. Furthermore, the requirement to shift a recognition decision bias yields prominent activations in these regions as well. There also appears to be functional specificity in the parietal cortex when adopted biases are violated versus confirmed by memory materials, both in the case of novelty (in regions not typically associated with "retrieval success" effects) and familiarity (in regions typically associated with "retrieval success" effects). Finally, individual difference analyses using ROIs in or near the anterior angular gyrus do not support a direct role in retrieval. Instead they suggest that activation for old materials is governed by the salience of familiarity or perhaps recollection signals, which is a function of both the cue condition and the observer's basic discrimination ability. Thus fMRI data are beginning to converge on the idea that orienting or reorienting toward novelty or familiarity may critically rely upon parietal regions, and that this operation may be critically linked to the initial biasing of memory judgments. Conversely, the data do not suggest a strong role for the regions identified in the actual retrieval or buffering of episodic content. Critically, however, one region typically linked to this function, the posterior/ventral angular gyrus, has not been implicated in studies of recognition bias and remains a viable candidate for a more direct role in retrieval itself.

Convergence with neuropsychology

Additional evidence against a direct causal role for PPC and recognition memory retrieval can be seen in recent neuropsychological investigations on patients with parietal lesions. Since previous studies on patients with parietal damage had not focused on memory impairments, several recent studies targeted memory tests similar to the ones used in neuroimaging studies. Yet they have generally found little to no effect on basic recognition memory accuracy (Ally, Simons, McKeever, Peers, & Budson, 2008; Ciaramelli, Grady, Levine, Ween, & Moscovitch, 2010; Dobbins, Jaeger, Studer, & Simons, 2012; Simons, Peers, Mazuz, Berryhill, & Olson, 2010). While parietal patients may freely report fewer details on long-term autobiographical memory tests, they appear to perform normally when given specific probes (Berryhill, 2012; Berryhill, Picasso, Arnolds, Drowos, & Olson, 2010). This lack of an effect due to damage to the parietal lobe is difficult to reconcile with parietal models of memory that ascribe a function that depends on this region to accumulate or buffer the contents of episodic retrieval. Furthermore, Dobbins et al. (2012) examined parietal patients using the trial-wise biasing manipulation and found that unlike frontal patients and controls, the one deficit in their behavior appeared to be the inability to use Likely New cues to appropriately bias their recognition decisions, suggesting a deficit in the ability to integrate externally cued biases into the assessment of recognition evidence. Interestingly, it is the Likely New cue condition that led to the unexpected familiarity response in anterior angular gyrus in Jaeger et al. (2013), and this region was likely heavily compromised in the parietal patients of Dobbins et al. (2012).

Synthesis with visual attention and valuation research

As noted above and suggested by Cabeza, Ciaramelli, and colleagues (2008), different regions of parietal cortex may perform functions that are algorithmically similar, albeit in vastly different domains. In the case of recognition memory, there remains controversy over the degree to which memory-linked activations overlap with those involved in visuospatial attention. In a direct within-subjects comparison of the parietal regions involved in memory and visuospatial attention, Hutchinson and colleagues (2014) used a field-mapping approach to demonstrate that the medial IPS and SPL response was likely common across visual and memory domains, perhaps reflecting the sustained deployment of visual attention to external probes in proportion to reaction times. In contrast, lateral IPS and angular gyrus memory-linked activations were disjoint with spatial attention regions. Although these researchers cast the lateral IPS and angular gyrus responses in terms of familiarity versus recollection-based retrieval products, respectively, as discussed above, these regions are also heavily modulated by the requirement to shift memory decision biases, and by memoranda that conflict with these adopted decision biases. In addition, these regions demonstrate a pattern of brain-behavior correlation at the subject level that tracks tonic observer biases in uncued recognition, not observer accuracy. In conjunction with the failure to observe even minor recognition accuracy deficits following large parietal lesions, and the initial demonstration that such lesions may impair the ability to flexibly adopt cued recognition biases, the data would suggest that these regions may be important for the differential weighting of novelty, familiarity, and perhaps recollective information depending upon contextually cued biases and not memory retrieval per se. Under this interpretation, memory evidence is necessary, but not sufficient, to drive activation. In addition, the map that results during recognition reflects the relative salience of particular types of memory information in light of the adopted processing biases and other factors discussed below.

The concept of memory orienting is not novel in the cognitive literature (e.g., Mandler, 1980) or in the context of functional brain imaging (Herron & Rugg, 2003). However, it is relatively unexplored. Critically, if memory information, like visual information, is multidimensional or multifeatural, then the relative salience of these different features of recovered memory evidence (e.g., familiarity, novelty, and recollection) is governed by a host of factors, including (1) the adopted bias of the observer (i.e., which features are expected given general task emphasis, biasing cues, task experience/practice, and subject traits), (2) individual differences in core retrieval abilities which govern the basic availability of types of memory evidence, (3) encoding manipulations that determine the availability of episodic content, and (4) the motivational significance of particular features, which is known to drive parietal activation during saccade decisions in nonhuman primates (Leathers & Olson, 2012; e.g., cells driven by the motivational salience of the cue but not the action value). Thinking about parietal activation at retrieval as a function of the relative salience of different types of memory information is a fairly unexplored framework. However, it is clear that externally cued decision biases, which govern expectations about the types of memory signals that should be encountered and also heavily influence behavioral response patterns, concomitantly have very large effects on the distribution of and nature of parietal responses during recognition. Additionally, the salience of expectations will be modulated by the discrimination ability of the subjects. In other words, those who discriminate well should also show greater response to violations because the conflicting signals will be strong (Jaeger, Konkel, & Dobbins, 2013). Nonetheless, a biasing and orienting framework assumes the functional significance of these activations lies not in directly supporting retrieval success, but instead in biasing judgments for more efficient processing when those biases are confirmed and, alternatively, orienting and exploring memory contents more thoroughly when those biases are disconfirmed.

ACKNOWLEDGMENTS We wish to thank Danielle King for helpful suggestions. The research cited in this chapter was supported by the Institute for Collaborative Biotechnologies through contract no. W911NF-09-D-0001 from the U.S. Army Research Office to MBM and by the National Institute of Mental Health R01-MH073982 to IGD.

REFERENCES

Ally, B. A., Simons, J. S., McKeever, J. D., Peers, P. V., & Budson, A. E. (2008). Parietal contributions to recollection: Electrophysiological evidence from aging and patients with parietal lesions. *Neuropsychologia, 46*(7), 1800–1812.

Aminoff, E. M., Clewett, D., Freeman, S., Frithsen, A., Tipper, C., Johnson, A., … Miller, M. B. (2012). Individual differences in shifting decision criterion: A recognition memory study. *Mem Cognit, 40,* 1016–1030.

Aminoff, E. M., Freeman, S., Clewett, D., Tipper, C., Frithsen, A., Johnson, A., … Miller, M. B. (submitted). Adapting to a cautious state of mind during a recognition test: A large-scale fMRI study.

Baddeley, A. (2000). The episodic buffer: A new component of working memory? *Trends Cogn Sci, 4*(11), 417–423.

Berryhill, M. E. (2012). Insights from neuropsychology: Pinpointing the role of the posterior parietal cortex in episodic and working memory. *Front Integr Neurosci, 6,* 31.

Berryhill, M. E., Picasso, L., Arnolds, R. A., Drowos, D. B., & Olson, I. R. (2010). Similarities and differences between parietal and frontal patients in autobiographical and constructed experience tasks. *Neuropsychologia, 48,* 1385–1393.

Buckner, R. L., Koutstaal, W., Schacter, D. L., Dale, A. M., Rotte, M. R., & Rosen, B. R. (1998). Functional-anatomic study of episodic retrieval. II. Selective averaging of event-related fMRI trials to test the retrieval success hypothesis. *NeuroImage, 7,* 163–175.

Buckner, R. L., Koutstaal, W., Schacter, D. L., Wagner, A. D., & Rosen, B. R. (1998). Functional-anatomic study of episodic retrieval using fMRI. I. Retrieval effort versus retrieval success. *NeuroImage, 7,* 151–162.

Cabeza, R., Ciaramelli, E., & Moscovitch, M. (2012). Cognitive contributions of the ventral parietal cortex: An integrative theoretical account. *Trends Cogn Sci, 16*(6), 338–352.

Cabeza, R., Ciaramelli, E., Olson, I. R., & Moscovitch, M. (2008). The parietal cortex and episodic memory: An attentional account. *Nat Rev Neurosci, 9,* 613–625.

Chafee, M. V., & Goldman-Rakic, P. S. (2000). Inactivation of parietal and prefrontal cortex reveals interdependence of neural activity during memory-guided saccades. *J Neurophysiol, 83*(3), 1550–1566.

Ciaramelli, E., Grady, C., Levine, B., Ween, J., & Moscovitch, M. (2010). Top-down and bottom-up attention to memory are dissociated in posterior parietal cortex: Neuroimaging and neuropsychological evidence. *J Neurosci, 30*(14), 4943–4956.

Corbetta, M., & Shulman, G. L. (2002). Control of goal-directed and stimulus-driven attention in the brain. *Nat Rev Neurosci, 3,* 201–215.

Dobbins, I. G., Foley, H., Schacter, D. L., & Wagner, A. D. (2002). Executive control during episodic retrieval: Multiple prefrontal processes subserve source memory. *Neuron, 35*(5), 989–996.

Dobbins, I. G., & Han, S. (2006). Cue- versus probe-dependent prefrontal cortex activity during contextual remembering. *J Cogn Neurosci, 18*(9), 1439–1452.

Dobbins, I. G., Jaeger, A., Studer, B., & Simons, J. S. (2012). Use of explicit memory cues following parietal lobe lesions. *Neuropsychologia, 50,* 2992–3003.

Dobbins, I. G., Rice, H. J., Wagner, A. D., & Schacter, D. L. (2003). Memory orientation and success: Separable neurocognitive components underlying episodic recognition. *Neuropsychologia, 41,* 318–333.

Dobbins, I. G., & Wagner, A. D. (2005). Domain-general and domain-sensitive prefrontal mechanisms for recollecting events and detecting novelty. *Cereb Cortex, 15*(11), 1768–1778.

Dosenbach, N. U. F., Fair, D. A., Miezin, F. M., Cohen, A. L., Wenger, K. K., Dosenbach, R. A. … Petersen, S. E. (2007). Distinct brain networks for adaptive and stable task control in humans. *Proc Natl Acad Sci USA, 104*(26), 11073–11078.

Estes, W. K., & Maddox, W. T. (1995). Interactions of stimulus attributes, base-rate and feedback in recognition. *J Exp Psychol Learn Mem Cogn, 21,* 1075–1095.

Feredoes, E., Heinen, K., Weiskopf, N., Ruff, C., & Driver, J. (2011). Causal evidence for frontal involvement in memory target maintenance by posterior brain areas during distracter interference of visual working memory. *Proc Natl Acad Sci USA, 108*(42), 17510–17515.

Gold, J. I., & Shadlen, M. N. (2001). Neural computations that underlie decisions about sensory stimuli. *Trends Cogn Sci, 5*(1), 10–16.

Gough, P. M., Nobre, A. C., & Devlin, J. T. (2005). Dissociating linguistic processes in the left inferior frontal cortex with transcranial magnetic stimulation. *J Neurosci, 25*(35), 8010–8016.

Green, D. M., & Swets, J. A. (1966). *Signal detection theory and psychophysics.* New York, NY: Wiley.

Guerin, S. A., & Miller, M. B. (2011). Parietal cortex tracks the amount of information retrieved even when it is not the basis of a memory decision. *NeuroImage, 55*(2), 801–807.

Herron, J. E., Henson, R. N., & Rugg, M. D. (2004). Probability effects on the neural correlates of retrieval success: An fMRI study. *NeuroImage, 21*(1), 302–310.

Herron, J. E., & Rugg, M. D. (2003). Retrieval orientation and the control of recollection. *J Cogn Neurosci, 15*(6), 843–854.

Hollingworth, H. L. (1913). Characteristic differences between recall and recognition. *Am J Psychol, 24*(4), 532–544.

Hutchinson, J. B., Uncapher, M. R., & Wagner, A. D. (2009). Posterior parietal cortex and episodic retrieval: Convergent and divergent effects of attention and memory. *Learn Mem, 16*(6), 343–356.

Hutchinson, J. B., Uncapher, M. R., Weiner, K. S., Bressler, D. W., Silver, M. A., Preston, A. R., & Wagner, A. D. (2014). Functional heterogeneity in posterior parietal cortex across attention and episodic memory retrieval. *Cereb Cortex, 24*(1), 49–66.

Jacoby, L. L. (1991). A process dissociation framework: Separating automatic from intentional uses of memory. *J Mem Lang, 30,* 513–541.

Jaeger, A., Konkel, A., & Dobbins, I. G. (2013). Unexpected novelty and familiarity orienting responses in lateral parietal cortex during recognition judgment. *Neuropsychologia, 51*(6), 1061–1076.

Janowsky, J. S., Shimamura, A. P., & Squire, L. R. (1989). Source memory impairment in patients with frontal lobe lesions. *Neuropsychologia, 27,* 1043–1056.

Johnson, M. K., Hashtroudi, S., & Lindsay, D. S. (1993). Source monitoring. *Psychol Bull, 114*(1), 3.

Kantner, J., Vettel, J. M., & Miller, M. B. (submitted). Dubious decision evidence and criterion flexibility in recognition memory.

Kapur, S., Craik, F. I. M., Jones, C., Brown, G. M., Houle, S., & Tulving, E. (1995). Functional role of the prefrontal cortex in retrieval of memories: A PET study. *NeuroReport, 6,* 1880–1884.

Kincade, J. M., Abrams, R. A., Astafiev, S. V., Shulman, G. L., & Corbetta, M. (2005). An event-related functional magnetic resonance imaging study of voluntary and stimulus-driven orienting of attention. *J Neurosci, 25*(18), 4593–4604.

Konishi, S., Wheeler, M. E., Donaldson, D. I., & Buckner, R. L. (2000). Neural correlates of episodic retrieval success. *NeuroImage, 12,* 276–286.

Leathers, M. L., & Olson, C. R. (2012). In monkeys making value-based decisions, LIP neurons encode cue salience and not action value. *Science, 338,* 132–135.

Long, N. M., Öztekin, I., & Badre, D. (2010). Separable prefrontal cortex contributions to free recall. *J Neurosci, 30*(33), 10967–10976.

Macmillan, N. A., & Creelman, C. D. (2005). *Detection theory: A user's guide* (2nd ed.). Mahwah, NJ: Erlbaum.

Mandler, G. (1980). Recognizing: The judgment of previous occurrence. *Psychol Rev, 87*(3), 252–271.

McDermott, K. B., Jones, T. C., Petersen, S. E., Lageman, S. K., & Roediger, H. L. (2000). Retrieval success is accompanied by enhanced activation in anterior prefrontal cortex during recognition memory: An event-related fMRI study. *J Cogn Neurosci, 12,* 965–976.

Miller, E. K., & Cohen, J. D. (2001). An integrative theory of prefrontal cortex function. *Annu Rev Neurosci, 21,* 167–202

Miller, M. B., Handy, T. C., Cutler, J., Inati, S., & Wolford, G. L. (2001). Brain activations associated with shifts in response criterion on a recognition test. *Can J Exp Psychol, 55*(2), 164–175.

Milner, B., Corsi, P., & Leonard, G. (1991). Frontal-lobe contribution to recency judgments. *Neuropsychologia, 29*(6), 601–618.

Nelson, S. M., Cohen, A. L., Power, J. D., Wig, G. S., Miezin, F. M., Wheeler, M. E., ... Petersen, S. E. (2010). A parcellation scheme for human left lateral parietal cortex. *Neuron, 67*(1), 156–170.

Nelson, S. M., McDermott, K. B., & Petersen, S. E. (2012). In favor of a "fractionation" view of ventral parietal cortex: Comment on Cabeza et al. *Trends Cogn Sci, 16*(8), 399–400.

Nolde, S. F., Johnson, M. K., & D'Esposito, M. (1998). Left prefrontal activation during episodic remembering: An event-related fMRI study. *NeuroReport, 9,* 3509–3514.

Norman, D. A., & Bobrow, D. G. (1979). Descriptions: An intermediate stage in memory retrieval. *Cognit Psychol, 11*(1), 107–123.

Nystrom, L. E., Braver, T. S., Sabb, F. W., Delgado, M. R., Noll, D. C., & Cohen, J. D. (2000). Working memory for letters, shapes, and locations: FMRI evidence against stimulus-based regional organization in human prefrontal cortex. *NeuroImage, 11,* 424–446.

O'Connor, A. R., Han, S., & Dobbins, I. G. (2010). The inferior parietal lobule and recognition memory: Expectancy violation or successful retrieval? *J Neurosci, 30*(8), 2924–2934.

Rhodes, M. G., & Jacoby, L. L. (2007). On the dynamic nature of response criterion in recognition memory: Effects of base rate, awareness, and feedback. *J Exp Psychol Learn Mem Cogn, 33*(2), 305–320.

Roediger, H. L., & Geraci, L. (1990). Implicit memory. *Am Psychol, 45*(9), 1043–1056.

Rugg, M. D., Fletcher, P. C., Frith, C. D., Frackowiak, R. S. J., & Dolan, R. J. (1996). Differential activation of the prefrontal cortex in successful and unsuccessful memory retrieval. *Brain, 119,* 2073–2083.

Rypma, B., Prabhakaran, V., Desmond, J. E., Glover, G. H., & Gabrieli, J. D. E. (1999). Load-dependent roles of prefrontal cortical regions in the maintenance of working memory. *NeuroImage, 9,* 216–226.

Schacter, D. L. (1996). Illusory memories: A cognitive neuroscience analysis. *Proc Natl Acad Sci USA, 93*(24), 13527–13533.

Schacter, D. L., Alpert, N. M., Savage, C. R., Rauch, S. L., & Albert, M. S. (1996). Conscious recollection and the human hippocampal formation: Evidence from positron emission tomography. *Proc Natl Acad Sci USA, 93,* 321–325.

Selmeczy, D., & Dobbins, I. G. (2012). Metacognitive awareness and adaptive recognition biases. *J Exp Psychol Learn Mem Cogn, 39*(3), 678–690.

Shadlen, M. N., & Newsome, W. T. (2001). Neural basis of a perceptual decision in the parietal cortex (area LIP) of the rhesus monkey. *J Neurophysiol, 86,* 1916–1936.

Shimamura, A. P. (1995). Memory and frontal lobe function. In M. S. Gazzaniga (Ed.), *The cognitive neurosciences* (pp. 803–813). Cambridge, MA: MIT Press.

Shimamura, A. P., Janowsky, J. S., & Squire, L. R. (1990). Memory for the temporal order of events in patients with frontal lobe lesions and amnesic patients. *Neuropsychologia, 28,* 803–813.

Simons, J. S., Peers, P. V., Mazuz, Y. S., Berryhill, M. E., & Olson, I. R. (2010). Dissociation between memory accuracy and memory confidence following bilateral parietal lesions. *Cereb Cortex, 20,* 479–485.

Spaniol, J., Davidson, P. S. R., Kim, A. S. N., Han, H., Moscovitch, M., & Grady, C. L. (2009). Event-related fMRI studies of episodic encoding and retrieval: Meta-analyses using activation likelihood estimation. *Neuropsychologia, 47,* 1765–1779.

Thompson-Schill, S. L., Swick, D., Farah, M. J., D'Esposito, M., Kan, I. P., & Knight, R. T. (1998). Verb generation in patients with focal frontal lesions: A neuropsychological test of neuroimaging findings. *Proc Natl Acad Sci USA, 95,* 15855–15860.

Tulving, E. (1983). *Elements of episodic memory.* Oxford, UK: Clarendon.

Tulving, E., Kapur, S., Markowitsch, H. J., Craik, F. I. M., Habib, R., & Houle, S. (1994). Neuroanatomical correlates of retrieval in episodic memory: Auditory sentence recognition. *Proc Natl Acad Sci USA, 91,* 2012–2015.

Tulving, E., & Thomson, D. M. (1973). Encoding specificity and retrieval processes in episodic memory. *Psychol Rev, 80*(5), 352–373.

Van Zandt, T., & Maldonado-Molina, M. M. (2004). Response reversals in recognition memory. *Cognition, 30*(6), 1147–1166.

Vilberg, K. L., & Rugg, M. D. (2009a). Functional significance of retrieval-related activity in lateral parietal cortex: Evidence from fMRI and ERPs. *Hum Brain Mapp, 30*(5), 1490–1501.

Vilberg, K. L., & Rugg, M. D. (2009b). An investigation of the effects of relative probability of old and new test items on the neural correlates of successful and unsuccessful source memory. *NeuroImage, 45,* 562–571.

Wagner, A. D., Shannon, B. J., Kahn, I., & Buckner, R. L. (2005). Parietal lobe contributions to episodic memory retrieval. *Trends Cogn Sci, 9,* 445–453.

Wheeler, M. E., Stuss, D. T., & Tulving, E. (1997). Toward a theory of episodic memory: The frontal lobes and autonoetic consciousness. *Psychol Bull, 121*(3), 331–354.

Wixted, J. T., & Stretch, V. (2000). The case against a criterion-shift account of false memory. *Psychol Rev, 107*(2), 368–376.

Yonelinas, A. P. (1997). Recognition memory ROCs for item and associative information: The contribution of recollection and familiarity. *Mem Cognit, 25*(6), 747–763.

50 Habits and Reinforcement Learning

DAPHNA SHOHAMY AND NATHANIEL D. DAW

ABSTRACT A central challenge in memory research has been to characterize the cognitive and neural mechanisms by which habits are formed. Here we review recent advances in understanding how habits are learned and how they interact with other forms of learning to guide decisions and actions. We address these questions by reviewing research on multiple memory systems alongside a complementary body of work on decision making and learning from reward, highlighting converging evidence from animal physiology, computational models, human functional MRI, and patient research. Together, this work has demonstrated that reward prediction models of midbrain dopamine neurons successfully account for a wide—but also limited—range of motivated behaviors; specifically, those that underlie habitual, stimulus-response learning. We further review emerging research that has begun to answer new questions, such as how habit learning interacts with goal-directed behavior, how other forms of memory contribute to decision making, and how behavior transitions from goal-directed to habitual control.

How does memory support behavior? Research on the neurobiology of learning and memory has led to the proposal that different forms of memory are supported by different brain structures. A focus of research within this framework is to determine which structures contribute to memory for facts and events, termed *declarative memory*, and which to memory for habits or skills, often termed *procedural memory*.

Procedural memory, by its very definition, is more closely linked to action than declarative memory. Yet in studies of memory, declarative memory has traditionally been more fully investigated, whereas many questions remain about the neural, psychological, and computational mechanisms supporting procedural memories. At the same time, declarative memories clearly inform decisions and other day-to-day behaviors, but the links to behavior from the more abstract cognitive capacity as studied in the laboratory remain largely underexplored.

In this chapter, we address these questions by reviewing research on multiple memory systems alongside a complementary body of work on decision making and learning from reward. Much research in the latter area has detailed a neural mechanism for learning about actions, which appears to correspond, at least in part, to procedural learning as envisioned by memory researchers. Importantly, however, there has been increasing consciousness among decision researchers that these habit-like action mechanisms are insufficient to explain many sorts of reward-directed decision behavior, leading to a recent focus on dissociating them from other, more deliberative or explicit influences on decisions. Though not yet fully understood, these sorts of behaviors—known as goal-directed or model-based actions—may serve as a bridge from declarative memory to behavior.

The convergence between memory and decision making thus serves to fill in gaps in either field. Here we survey these developments and discuss how understanding the convergence between these areas provides a framework for understanding the mechanisms by which habits are learned, how different forms of memory guide decisions, and the neural mechanisms supporting these behaviors. Viewing these systems together, as jointly supporting decision behavior, has begun to lead to an enriched view of both, and particularly of their interactions.

Multiple memory systems supporting learning and memory

A prominent idea in cognitive neuroscience is that there are different kinds of memory, each supported by different neural systems (Cohen & Eichenbaum, 1993; Gabrieli, 1998; Squire & Zola, 1996). Decades of research into *declarative* memory originated from studies with patients with memory deficits, such as patient H.M. As the case of H.M. demonstrated, damage to the medial temporal lobe (MTL), including the hippocampus and surrounding cortical regions, leads to severe, and selective, *declarative memory* impairments—an inability to learn new facts and to form new memories for events (Scoville & Milner, 1957). This initial discovery set the stage for decades of detailed investigation into the role of the hippocampus and surrounding MTL cortices in memory. Although the precise role is still debated, substantial progress has been made both in anatomical precision and in functional characterization (as described in chapters 45 and 46, this volume).

The selectivity of memory impairments in patient H.M. and others highlighted another important fact about memory organization: damage to the MTL did not impair other forms of learning: the ability to learn new procedures and skills, often referred to as *procedural* or *habit learning*, remained intact, leaving open questions about the neural and cognitive mechanisms that support such forms of learning.

Although the research presented in the remainder of this chapter elaborates and somewhat complicates the traditional view, habit learning is classically characterized in opposition to declarative memory: as gradual, implicit, and outside conscious awareness. Converging evidence suggests that habit learning depends on the basal ganglia (Yin & Knowlton, 2006). Much of this evidence comes from studies of patients with Parkinson's disease, a progressive neurological disorder that involves a loss of dopaminergic input to the striatum due to a reduction in dopamine neurons in the substantia nigra pars compacta (Kish, Shannak, & Hornykiewicz, 1988; Peran et al., 2010). Patients with Parkinson's disease, who famously suffer from motor impairments, also have cognitive and mnemonic impairments, even in the earliest stages of the disease when loss of dopaminergic input is relatively selective to the striatum.

Thus, individuals with Parkinson's disease provide a useful model of basal ganglia dysfunction in humans and can be compared with individuals with MTL dysfunction to ask questions about the unique contribution of each brain region to different kinds of memory. Such comparisons, detailed below, have revealed that the pattern of memory impairments found in patients with Parkinson's disease is quite different from that found in patients with MTL damage. In particular, patients with Parkinson's disease are impaired on tasks that involve gradual learning of stimulus-response or action-outcome associations, but are spared on tests of declarative memory.

THE BASAL GANGLIA, PROBABILISTIC LEARNING, AND PARKINSON'S DISEASE Probabilistic category-learning paradigms have been central to understanding the contributions of the basal ganglia to learning and memory. One widely used measure, the "weather-prediction" task (see figure 50.1), requires that participants use trial-and-error feedback to learn to predict categorical outcomes (sun or rain) based on four different visual cues (simple shapes). On each trial between one and three of the four are presented, yielding 14 possible stimuli (all possible combinations of the cues without displaying all four or none at all). The relationship between cues and outcomes is probabilistic, such that across all trials each cue predicts an outcome only some of the time. The complex cue structure and probabilistic nature of the associations, with no consistent one-to-one mapping between stimuli and outcomes, was originally thought to hamper attempts at explicitly encoding cue-outcome associations, making improved performance dependent on gradual, implicit learning (Gluck, Shohamy, & Myers, 2002; Knowlton, Mangels, & Squire, 1996).

Indeed, early studies demonstrated that amnesics were capable of displaying some incremental learning

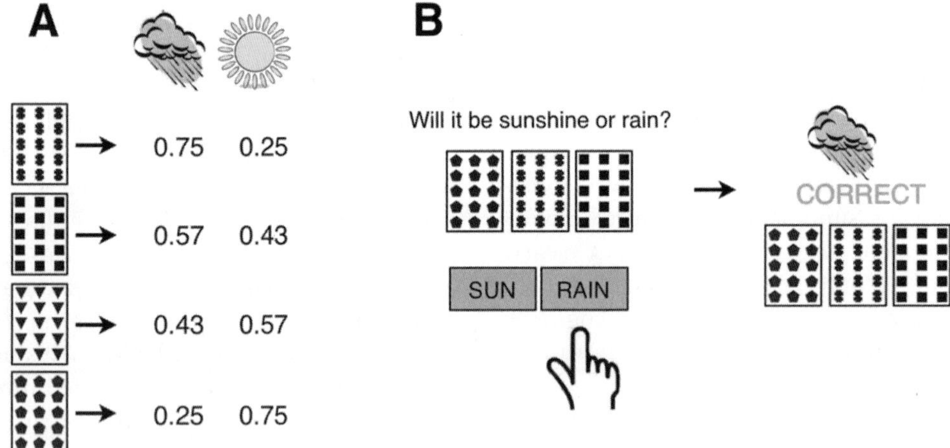

FIGURE 50.1 Structure and task events from a probabilistic classification task, often referred to as the "weather-prediction" task. (A) Each of four visual cues—cards with shapes—is independently and probabilistically associated with either "rain" or "sun." (B) On each trial, a combination of one to three cards is shown. Subjects respond based on their prediction of the weather for that trial, and receive response-contingent feedback. Learning this task is thought to involve implicit habit-learning mechanisms. Patients with disrupted striatal function due to Parkinson's disease are slow to learn the probabilistic associations—and healthy participants show increased activation in the striatum—when learning is driven by immediate, response-contingent feedback. Performing this task when there is a distracting secondary task at the same time—a behavioral hallmark of habits—is also related to increased activation in the striatum.

(measured as increased accuracy in performance) despite having no explicit memory for the testing episode (revealed by multiple-choice questions about details of the testing events). By contrast, Parkinson's disease patients, especially those with severe symptoms, were impaired at incremental learning of the task, but had intact explicit memory for the task events (Knowlton et al., 1996; Knowlton, Squire, & Gluck, 1994).

This double dissociation was consistent with other similar double dissociations obtained in animal research between different forms of learning, such as place versus response learning (e.g., Packard, 1999). Together, such findings were central in advancing the notion that the basal ganglia and MTL support two dissociable memory systems. In particular, the findings provided evidence that the basal ganglia are necessary for nonmotor, incremental learning of stimulus-response associations. These findings in humans converge with a large body of animal lesion work also demonstrating a key role for the basal ganglia in incremental, stimulus-response learning (e.g., Foerde & Shohamy, 2011; Mishkin, Malamut, & Bachevalier, 1984; Packard, 1999; Yin & Knowlton, 2006).

These findings highlighted the basal ganglia as a potential substrate of habit learning, raising many questions about the mechanisms by which such learning takes place. First, while Parkinson's disease patients are impaired on some types of implicit learning, they are not impaired at others, indicating the need for a careful characterization of what exactly a habit is, and how habit learning differs from other sorts of nondeclarative memories. Second, questions were raised about whether participants completing the "weather-prediction" task were necessarily learning the contingencies implicitly; subsequent studies demonstrated that healthy subjects could use simple, explicit strategies to support learning. Furthermore, studies with functional imaging in humans found that the hippocampus is activated during learning of the "weather-prediction task," in addition to the basal ganglia (Poldrack et al., 2001; Shohamy, Myers, Onlaor, & Gluck, 2004). Indeed, as subsequent studies demonstrated, both healthy people and Parkinson's disease patients might in fact rely on the hippocampus when learning in this task (Foerde, Knowlton, & Poldrack, 2006; Moody, Bookheimer, Vanek, & Knowlton, 2004). Thus, even in what is considered to be a prototypical implicit learning task, multiple memory systems may work together to contribute to learning.

As outlined below, many similar questions—and some answers—have come up in a different line of research demonstrating a role for midbrain dopamine neurons in learning.

Dopamine and habits

A different line of research suggests a mechanism that might underlie a dopaminergic role in habit formation. The dopaminergic input to striatum (and indeed to other parts of the forebrain, including hippocampus) arises from neurons whose cell bodies are located in two small midbrain nuclei, the substantia nigra pars compacta and the ventral tegmental area. These are the neurons whose degeneration is primarily implicated in Parkinson's disease.

Recordings from these dopaminergic neurons in behaving monkeys suggest that they carry a very suggestive signal related to reward. These neurons tend to behave similarly to one another; they have a low background firing rate punctuated by brief, phasic excitations and inhibitions. Following a series of reports describing various circumstances under which these phasic firing modulations occur (see Schultz, 1992, for an early review), seminal computational modeling work pointed out that many of these responses could collectively be understood as a signal known as a *reward prediction error* (Houk, Adams, & Barto, 1995; Montague, Dayan, & Sejnowski, 1996; Schultz, Dayan, & Montague, 1997). A reward prediction error is the difference between the reward you expected and the reward you got; in computer science and engineering, such signals are commonly used for learning to make such predictions, since they carry information about how to adjust errant predictions in light of experience (Sutton & Barto, 1998).

Figure 50.2A shows a typical response in a dopamine neuron (Fiorillo, Tobler, & Schultz, 2003). In this case, a monkey was trained that five different pictures would be followed by a drop of juice, or not, with five different probabilities ranging in steps from 0 to 100 percent. The figure shows the neural response when a drop of juice is received following each of these pictures. If the monkey has learned the contingencies (which his behavior demonstrates he has), then he should expect reward to different degrees, ranging from not at all to certainty. When a reward is actually received, then, the prediction error should range from completely unpredicted to no error at all, with various degrees of surprise in between. Accordingly, across the conditions the neuron shows a phasic excitation to reward proportional to this error. The figure also shows that when a reward is expected but omitted, the neuron shows an inhibitory response consistent with the (negative) prediction error. However, due to the low background firing rate of the neurons, it has been harder to detect a quantitative relationship between the size of the negative error and the magnitude of the inhibition

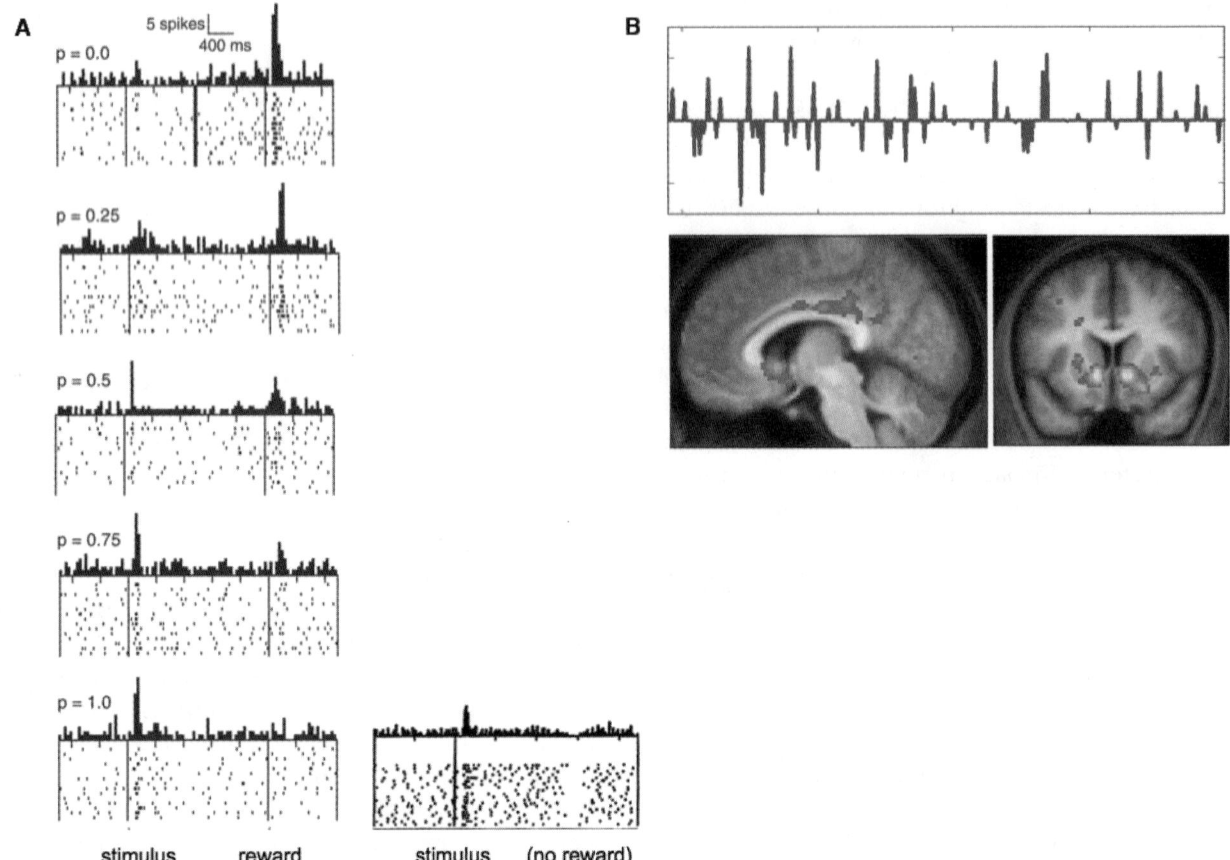

FIGURE 50.2 (A) Responses of dopamine neurons recorded from the primate midbrain, reprinted from Fiorillo et al. (2003) and Schultz et al. (1997) with permission. Five stimuli predict juice reward with different probabilities ($p = 0$ to 1, top to bottom). The neuron responds to each stimulus proportional to its associated reward probability, but when reward is delivered, the response is larger the less expected was the reward. When an expected reward is omitted, dopamine neurons are inhibited (right). These responses are consistent with a reward prediction error. (B) BOLD activity in areas of the human brain, notably ventral striatum, correlates with a reward prediction error extracted from a trial-by-trial computational model of each subject's learning (top trace). Displayed at $p < .001$ uncorrected; data replotted from Daw et al. (2011). (See color plate 41.)

(Bayer & Glimcher, 2005; Bayer, Lau, & Glimcher, 2007; Fiorillo et al., 2003).

Figure 50.2A also shows that in addition to the response to received or omitted primary rewards, the neuron also responds to cues (here, the pictures) that are predictive of reward, and more so for those most predictive of reward. This can also be understood as a reward prediction error, in a particular class of models called temporal-difference learning models (Sutton, 1988). Here the goal is to predict not just immediate rewards but cumulative future rewards, and reward prediction errors occur whenever there is an unexpected change in your future prospects: here, when an unexpected cue comes on signaling the availability of future reward. In computer science, these sorts of predictive responses are particularly useful for learning tasks requiring multiple behaviors in sequence, such as mazes or chess: they help chain together behaviors by providing feedback for actions that set the stage for further successful ones.

These basic findings have now been replicated and extended in many studies in monkeys and rodents. Human functional imaging in similar tasks also tells a similar story: the blood oxygen level–dependent (BOLD) signal at dopaminergic targets (principally ventral striatum; see figure 50.2B) shows positive and negative excursions related to positive and negative reward prediction errors (McClure, Berns, & Montague, 2003; O'Doherty, Dayan, Friston, Critchley, & Dolan, 2003), and indeed this activity looks very much like the dopaminergic responses of figure 50.2A, but smeared out by a slow hemodynamic response. Of course, the metabolic activity detected by functional MRI (fMRI) is not specifically diagnostic of a particular underlying neural cause such as dopamine. However, there are a number of indications, for example, from pharmacological and Parkinson's disease studies (Pessiglione, Seymour, Flandin, Dolan, & Frith, 2006; Schonberg et al., 2010) that the prediction-error related portion of the striatal BOLD signal is affected by dopamine.

Dopamine and Movement How does all this relate to behavior, and more specifically to habit learning? A basic abstract concept is that reward prediction errors are useful for learning to predict rewards, and learning to predict rewards is useful for learning how to choose the most rewarding actions. But how does this actually work?

One clue is that the medium spiny neurons of striatum (MSNs; the projection neurons of the structure, which are targeted by dopaminergic inputs) are strongly implicated in movement, for instance, due to the prominent motor symptoms of Parkinson's disease. A classic understanding of their functional anatomy (Alexander & Crutcher, 1990; DeLong, 1990) holds that they receive inputs from frontal cortex (notably, motor-related areas), and ultimately reciprocate them indirectly, via a series of connections through additional basal ganglia and thalamic structures. Two such pathways from striatum back to cortex have been classically described (although there are likely more), called the direct and indirect pathways. Due to the number of inhibitory synapses along the route, it is believed that the direct pathway has a net excitatory and the indirect a net inhibitory effect on cortex. The concept is that these two pathways act a bit like a gas pedal and brakes for behavior, exciting or inhibiting cortical neurons that control movement.

Thus, dopamine is well positioned to affect movement behavior via impinging upon striatal MSNs in the direct and indirect pathways. Indeed, one effect of dopamine is thought to be directly exciting MSNs projecting into the direct pathway and inhibiting those projecting into the indirect pathway, so as to have an overall invigorating effect on movement. The removal of this input (and the consequent imbalance favoring the indirect pathway) is a classic explanation for many of the movement difficulties in Parkinson's disease (DeLong, 1990).

However, dopamine does more than directly excite or inhibit MSNs; it also modulates the plasticity of their inputs from cortex (Reynolds & Wickens, 2002). This is a candidate site for learning driven by the dopaminergic prediction error described in the previous section. In short, MSNs appear to be involved in producing or withholding movements, and plasticity driven by phasic dopamine activity might, by strengthening or weakening synapses onto these neurons, therefore strengthen or weaken the chance of repeating those movements in the future. Such a rule—strengthening actions that lead to positive reward prediction error and weakening those that lead to negative prediction error—is used in computational models of action learning, known as the actor-critic. In psychology, learning rules of this sort have a long history going back to Thorndike's "law of effect" (Thorndike, 1911).

Dopamine, Learning, and the Basal Ganglia A dopaminergic role in learning from reward prediction errors, then, is a candidate mechanism for habit learning and some of the cognitive symptoms in Parkinson's disease. But so far, we have presented mostly correlational (unit recording) evidence supporting this idea. Do manipulations of dopamine actually affect learning in a manner predicted by these theories? A very suggestive first clue is that dopamine in striatum appears to be a common link across drugs of abuse: many, like cocaine and amphetamines, mimic or release dopamine directly, whereas many others (such as nicotine and alcohol) appear to do so indirectly. These effects have been modeled in terms of the reward prediction error learning theories described above (Redish, 2004); the idea is simply that drugs amplify or mimic positive reward prediction errors and promote further drug-seeking.

In humans, research in Parkinson's patients, building on the early work with the weather-prediction task mentioned above, has sought cognitive deficits in reward learning more directly related to the emerging picture of dopaminergic function. For instance, Shohamy, Myers, Onlaor et al. (2004), inspired by the temporal-difference model of predictive dopaminergic responses such as the cue-related responses in figure 50.2, tested Parkinson's patients on a multistep chaining task in which they had to learn a sequence of behaviors by trial and error. Patients were specifically impaired at such learning over multiple steps relative to a simple single-step association. Second, based on the idea that phasic dopamine promotes repeating successful actions (and the absence of dopamine favors avoiding unsuccessful ones), Frank, Seeberger, and O'Reilly (2004) combined a probabilistic learning task with a transfer phase that tested to what extent the resulting behavior was supported by learning to repeat successful versus learning to avoid unsuccessful actions. Parkinson's patients tested off their dopamine-replacement medication tended to learn via avoidance, while medication produced more learning from success. Similar asymmetries between learning from positive and negative feedback have now been observed in several other learning tasks (Bodi et al., 2009; Cools, Altamirano, & D'Esposito, 2006).

There is also a long line of evidence from animal experiments supporting a dopaminergic role in reinforcement (see Wise, 2004, for a review). Consistent with their role as drugs of abuse in humans, drugs that agonize dopamine are potent rewards in animals, supporting lever-pressing and conditioned place preferences. Conversely, low doses of drugs that antagonize dopamine can have extinction-like effects, causing animals to gradually "unlearn" appetitive behaviors. There is also a long-standing literature on how animals will learn to work to activate certain neurons electrically. One of the most effective targets for such "brain-stimulation reward" is the medial forebrain bundle, which contains (among other fibers) the ascending axons from the dopaminergic neurons.

The recent advent of molecular tools has enabled more targeted experiments selectively zeroing in on the dopaminergic neurons. Animals learn to nose-poke, lever-press, or go to a location to obtain stimulation of their dopamine neurons (Adamantidis et al., 2011; Kim et al., 2012; Tsai et al., 2009), and, conversely, to avoid locations where dopamine neurons are inhibited (Tan et al., 2012). Similar reinforcing and punishing effects can be produced by simulating the (dopamine-recipient) medium spiny neurons in the direct and indirect pathways, respectively (Kravitz, Tye, & Kreitzer, 2012). Together, these data seem to provide causal support for the hypothesis of dopaminergic habit learning.

What is a habit?

There are thus both correlational and causal data suggesting that dopamine is involved in a particular mechanism for procedural learning. One way to sharpen our understanding of this hypothesized mechanism is to consider what it is *not*. If—as the dopamine theories suggest—the essence of this sort of learning is strengthening a tendency to repeat actions that have been rewarded in the past, then the behaviors it produces should have a telltale inflexibility.

For instance, habits learned in this way can take you back to a restaurant you've enjoyed in the past, but they can't navigate to a new restaurant you've just heard about, even if it's in a familiar neighborhood. Similarly, if you enjoy sushi every day for lunch but suddenly develop an allergy to fish, the learning mechanism we have described would be expected to rigidly take you back to your usual restaurant the following day at lunchtime (it having always been rewarded in the past) rather than choosing a different restaurant appropriate to your new circumstances. This is because all that these mechanisms learn is how well actions have done previously; they cannot reason prospectively so as to base choices on any other information such as knowledge about menus and maps or your allergy diagnoses or appetites for particular foods.

Accordingly, a long tradition in psychology has aimed to identify behaviors that are habit-like in this sense and to distinguish them from others that might be more informed or deliberative (Dickinson, 1985; Dickinson & Balleine, 2002; Tolman, 1948). The latter are called "goal-directed" actions, because they are based on knowledge of the particular, desirable goal (such as sushi) and that the action will produce it. They are also known as "model-based" decisions, after a family of algorithms in computer science that learn such knowledge (an "internal model" of the task or environment) and use it to evaluate options and guide decisions (Daw, Niv, & Dayan, 2005). One key insight of this research is that many behaviors can be ambiguous. A rat pressing a lever, and receiving food, might in principle be doing so because that action has been reinforced in the past, or alternatively because he knows that the lever produces food and that he is hungry for it.

These two possibilities can be distinguished using a number of experimental manipulations, such as reward devaluation. Consider a hungry rat trained to lever-press for food. If the rat is then fed to satiety, or the particular outcome food devalued (by pairing it with drug-induced illness to induce an aversion), the question arises whether the rat will—on his next opportunity with the lever—still work for the food that he no longer wants. Habit mechanisms such as the temporal-difference learning theory associated with dopamine predict that the animal should, somewhat paradoxically, lever-press in this circumstance, at least until he reexperiences the action producing the undesired outcome. (Only at this point can such learning strategies determine that the action is no longer rewarding.) In contrast, if an animal adjusts his choices away from the devalued action *without* such experience, this demonstrates that the decision to lever-press was based on information not just about the lever's previous reward history (as habit mechanisms predict) but about the particular rewarding consequences (the food) expected for the action.

The outcome of many such experiments is that rodents, and also humans, can at different times display both sorts of behaviors (see Balleine & O'Doherty, 2010, for review). Which one is observed depends on the circumstances. For instance, extensively trained behaviors tend to be devaluation-insensitive—consistent with the sort of habits predicted by the procedural learning mechanism discussed here—whereas more moderately trained behaviors are devaluation *sensitive*, that is, goal-directed in a manner inconsistent with such theories. These two modes appear to be learned in parallel, but to trade off in controlling behavior, as though in competition. For instance, lesion studies in animals and functional imaging in humans suggest that goal-directed and habitual behaviors arise from dissociable neural pathways, involving distinct loops connecting regions of frontal cortex to associated parts of the dorsal striatum. Animals with lesions to areas involved with goal-directed action, such as the dorsomedial striatum, can learn to lever-press for food, but even early in training (when goal-directed action would normally dominate), devaluation probes demonstrate this behavior is habitual. Conversely, lesions to habit areas, such as the dorsolateral striatum, render even overtrained lever-pressing perpetually goal-directed: unlike in intact animals, it does not transition to devaluation-insensitive (see Yin & Knowlton, 2006, for review). Results like these suggest that procedural learning does not strictly depend on extensive training—rather, it is learned even early in training but normally dominates only later.

All this brings us back to dopamine. The temporal-difference learning hypothesis predicts that dopamine should affect behavior by producing habits. This is the sort of learning for which the signals recorded from dopamine neurons (described above) appear to be appropriate. The experiments described in this section verify that habits of this sort do indeed exist in humans and animals, although they appear to be accompanied by an additional, goal-directed decision mechanism. But can they be tied to dopamine? For the most part, the experiments that tie dopamine to decisions in a causal sense, from the weather-prediction task to recent optogenetic studies, demonstrate that dopamine is involved in guiding choices without using a manipulation such as reward devaluation to examine whether these choices are specifically of the habitual sort. There are as yet only a couple of studies suggesting that interfering with dopamine specifically blocks the formation of habits, while leaving goal-directed behavior intact (Faure, Haberland, Conde, & El Massioui, 2005; Wang et al., 2011).

Relating memory to action

How does all this relate, then, to the declarative versus procedural distinction—and the classic work with the weather-prediction task, with which we began?

There has been much additional work using the weather-prediction task, testing and confirming many predictions that derive from our understanding of how dopamine, prediction errors, and the striatum support learning, as well as testing more extensively the habitual nature of the representations that are learned.

First, the reinforcement-driven nature of the learning appears to be an essential element for determining whether or not the striatum is involved. When healthy participants learn the same task by observing the correct outcome, instead of making a response and receiving feedback, learning in healthy people depends more on the hippocampus and less on the striatum. Removing the feedback-based component of learning in this task also ameliorates the learning deficits in patients with Parkinson's disease (Poldrack et al., 2001; Shohamy, Myers, Grossman et al., 2004). In fact, the same is true even when the feedback on each trial is simply delayed by several seconds, as shown in figure 50.3 (Foerde, Race, Verfaellie, & Shohamy, 2013; Foerde & Shohamy, 2011). In addition, and as predicted by the reward prediction errors recorded in dopamine neurons, the learning impairments in Parkinson's disease differ based on whether learning is driven by rewards or by punishments, and each is differentially impacted by dopaminergic medication. Notably, under most conditions people are poor at explicitly expressing what they have learned. Thus, the striatum appears to be specialized for learning from immediate, response-contingent feedback, regardless of the implicit versus explicit nature of the learning.

Second, manipulating the circumstances and the neural systems driving probabilistic learning can change the representation of what was learned from flexible to inflexible in systematic ways. One hallmark feature of habits is that they are more likely to guide behavior when people are distracted. Indeed, learning in the weather-prediction task engages the striatum more, and results in more habitual (less flexible) knowledge, when learned under dual-task conditions compared with learning under single-task conditions, which elicits more hippocampal activation and more flexibility (Foerde et al., 2006). Learning under dual-task conditions has also been shown to shift learning to be less goal-directed and more habitual in other learning tasks that bear a closer logical resemblance to the classic devaluation work in rodents (Otto, Gershman, Markman, & Daw, 2013).

Together, these findings suggest that which systems support learning depends on the circumstances under which learning takes place, rather than the structure of the knowledge to be learned, and that which system is engaged during learning will have important

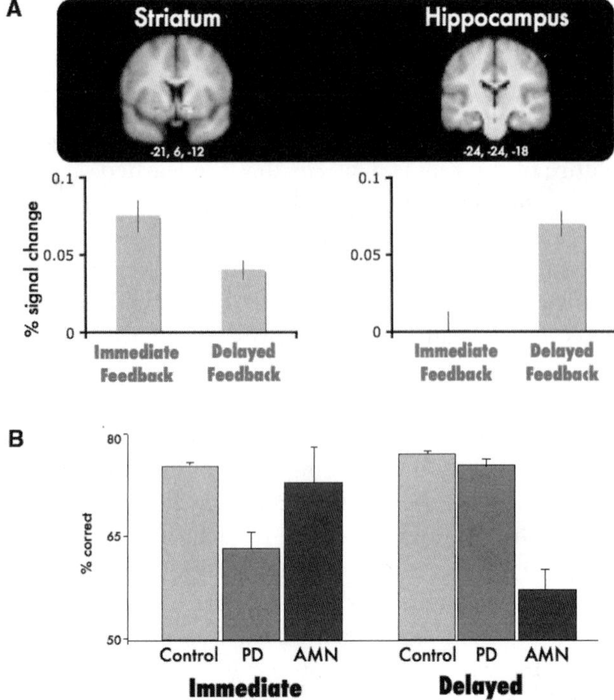

FIGURE 50.3 Probabilistic learning shifts from the striatum to the hippocampus when response-contingent feedback is delayed, as demonstrated by converging fMRI and patient studies. (A) FMRI reveals prediction-error related signals in humans during learning a probabilistic task. The striatum and the hippocampus display differential magnitudes of feedback-related signals for conditions wherein learning is driven by feedback that is immediate (within 1 second) versus delayed (by 7 seconds). (Data from Foerde & Shohamy, 2011.) (B) Data from patients support the selective role for the striatum in learning from immediate feedback. Patients with disrupted striatal function due to Parkinson's disease ("PD") are impaired at learning from immediate feedback but intact when learning from delayed feedback. Patients with amnesia due to MTL damage ("AMN") show the opposite pattern: intact learning from immediate feedback but impaired learning from delayed feedback. (Data from Foerde et al., 2013.) (See color plate 42.)

implications for the representation of what was learned. An important question is whether such effects are due to differences in neural systems engaged during learning, or whether they primarily exert an effect by changing the expression of what was learned.

FROM GOAL-DIRECTED AND HABITUAL TO DECLARATIVE AND PROCEDURAL We have defined habitual behavior in contrast to a second category of so-called goal-directed behavior. If habits as defined here correspond to procedural learning, do goal-directed behaviors draw on declarative memory and on related neural substrates such as the MTL memory system? This seems like a plausible hypothesis (Dickinson, 1985) since what defines goal-directed choices is that they depend on a map or model of the task contingencies, such as the specific outcome expected for the action: in effect, a set of declarative facts.

The jury is still out on this, but there are certainly some suggestions to the affirmative. First, the earliest demonstrations that habitual learning could not fully explain decision behavior were in spatial navigation tasks, where Tolman demonstrated that animals could draw on knowledge about the layout of the environment (a "cognitive map") to plan novel routes, rather than being strictly bound to repeat previously rewarded ones. It has long been suggested that the hippocampus subserves this "cognitive map" for spatial navigation (O'Keefe & Nadel, 1978). More recently, recordings of place cells in the hippocampus have demonstrated representations of locations "running ahead" of animals' actual locations at choice points, suggesting a direct neural mechanism for prospective evaluation of candidate routes in spatial navigation as in model-based learning (Johnson & Redish, 2007; Pfeiffer & Foster, 2013).

Accordingly, in spatial navigation tasks also, there can be transitions from a more deliberative and flexible strategy to a more habitual one with overtraining, mirroring that for rat lever-pressing. For instance, rats trained to turn right in repeated runs on a plus maze can be tested in probe trials where they start from the opposite end of the maze (for review see Packard, 1999). Early in training, they turn left—approaching the same goal location, as though guided by a cognitive map—but following overtraining, they instead emit the same response, turning right to visit a different location. Lesions suggest these more location-and response-based behaviors are dependent on hippocampus and striatum, respectively. These spatial response habits appear to be mechanistically similar to overtrained lever-pressing: for instance, spatial responses trained on a t-maze have been shown to be insensitive to reward devaluation in the same way as overtrained lever-pressing, and to be sensitive to inactivation of a prefrontal area (infralimbic cortex) that also controls habitual lever-pressing (Smith, Virkud, Deisseroth, & Graybiel, 2012).

But outside the spatial domain, is hippocampus (and flexible, relational, or declarative memory) involved in goal-directed decisions? The record is mixed. Hippocampal lesions have had little effect on devaluation sensitivity in rodent lever-pressing (Corbit & Balleine, 2000; Corbit, Ostlund, & Balleine, 2002). On the other hand, two related effects—transitive inference and acquired equivalence—both also measure the ability to flexibly transfer learning from one context to another, and both are closely tied to hippocampus in rodents and humans (Dusek & Eichenbaum, 1997; Shohamy & Wagner, 2008).

We earlier mentioned that the traditional view of procedural memory holds that it differs from declarative memory in being inflexible, implicit, and gradual. The experiments and theories discussed above formalize the sense in which habitual behaviors are more inflexible than goal-directed ones. What about the other characteristics?

In memory research, traditionally a great deal of weight has been put on consciousness as a defining feature of different memory systems. Although the essential role of consciousness in memory is now questioned (e.g., Henke, 2010), a common assumption going back to the multiple memory systems theory is that declarative memories are consciously aware, while procedural learning is not. In contrast, this feature has not, historically, been considered a core characteristic of goal-directed versus habitual behaviors, which were instead defined more operationally in terms of the sorts of associations or representations that drive the actions (Dickinson, 1985). Interestingly, recent work in memory has raised questions about whether consciousness awareness is actually a fundamental and necessary feature of the distinction between memory systems. Recent reports demonstrate that flexible, hippocampally guided behavior often occurs in the absence of awareness (Henke, 2010; Schapiro, Kustner, & Turk-Browne, 2012; Wimmer & Shohamy, 2012). This suggests that a more computational or process-based perspective on the nature of the representations may be more useful, and leaves open the question of how, when, and if consciousness plays into learning and memory.

Similarly, the speed of learning (one shot vs. incremental over many experiences) is another feature that has been traditionally associated with multiple memory systems, but again does not play a defining role in the decision work. This, too, most likely does not align in so simple a way with different neural memory systems. For instance, the aforementioned acquired equivalence and transitive inference effects are both characteristic examples of flexible, MTL-mediated learning, but at least some forms of them are not one-shot; instead, the relationships that support flexible transfer on these tasks must be built up slowly over multiple episodes. Meanwhile, the lesion studies discussed above suggest that habits do not arise only after extensive, incremental overtraining, but can support the initial acquisition of lever-pressing as well, albeit perhaps not one-shot learning. Instead here too a critical factor is likely the representation—what is learned—rather than the speed with which it is learned.

Interactions Finally, viewing the declarative versus procedural distinction in the broader context of how these systems might ultimately give rise to action also helps to showcase different possibilities for how they must interact in the service of behavior. From this view, the simple picture of two independent systems underlying memory for different sorts of information seems unlikely to explain the rich interactions between these mechanisms, both psychologically and neurally. Understanding these interactions is a major topic of current research in both the memory and decision areas.

First, the circuitry associated with these two systems is deeply interconnected. Hippocampus provides a major input to ventral striatum. Dopamine, famous for its role in driving learning in the striatum, also innervates the hippocampus (Gasbarri, Packard, Campana, & Pacitti, 1994; Groenewegen, Vermeulen-Van der Zee, te Kortschot, & Witter, 1987; Samson, Wu, Friedman, & Davis, 1990). It has been suggested that this innervation is so important that dopamine is necessary for long-term plastic changes in hippocampal neurons (e.g., Huang & Kandel, 1995; for review, see Bethus, Tse, & Morris, 2010; Lisman, Grace, & Duzel, 2011). Understanding the role of dopamine in hippocampal learning—and how this role interacts with and complements its role in driving learning in the basal ganglia—is still an active area of investigation, and many open questions remain. Nonetheless, there is evidence to suggest that learning in the hippocampus, like in the basal ganglia, is sensitive to motivation and to outcomes, effects that are likely mediated by dopaminergic inputs to both of these systems (for review, see Shohamy & Adcock, 2010).

Similarly, research on the brain's action systems has now clearly rejected the traditional view of a subcortical (basal ganglia) habit system alongside a cortical (prefrontal or hippocampal) goal-directed action system. Such a view was always suspicious given the "loop" structure by which frontal cortex and striatum are tightly interconnected (Alexander & Crutcher, 1990). Accordingly, the functional anatomy from lesion and imaging studies now makes clear that both sorts of action each rely on different parts of both basal ganglia and cortex, and in particular that there are parts of striatum associated with goal-directed action and parts of frontal cortex involved in habits (Yin & Knowlton, 2006).

Anatomical features like these challenge the traditional view of the hippocampus and the basal ganglia as independent and distinct. Consistent with this, there is much evidence to suggest that the hippocampus and the basal ganglia interact in complex ways to learn and to guide behavior, with reports of both competitive and cooperative interactions between them.

Initial evidence for oppositional interactions between striatal and MTL systems came from lesion studies in animals (Chang & Gold, 2003a, 2003b; Eichenbaum, Fagan, Mathews, & Cohen, 1988; McDonald & White, 1995; Mitchell & Hall, 1988; Packard, Hirsh, & White, 1989; Rabe & Haddad, 1969; Schroeder, Wingard, & Packard, 2002). These studies showed that damage to one system could lead to improved performance by the other system. Further indirect evidence for competition between systems came from fMRI studies reporting evidence for "neural competition" between the hippocampus and the striatum during learning of the "weather-prediction" task. Specifically, during learning, there was a negative correlation between BOLD activity in the striatum (caudate) and the hippocampus (Poldrack et al., 2001).

One limitation with such results is that this sort of "neural competition" does not reveal whether interactions reflect interference with learning per se, or only with the ability to *express* what is being learned (that is, the behavioral output of learning). Additionally, these negative interactions found in imaging studies have not been linked to behavioral evidence for competition in humans as they have in the animal work. Moreover, evidence for this type of competition has been scant in neuropsychological populations (but see Cavaco, Anderson, Allen, Castro-Caldas, & Damasio, 2004). Thus, in humans, these results are still in need of further explication.

FMRI has also provided evidence for synergistic relationships between the hippocampus and the striatum. In particular, functional connectivity between the hippocampus and the striatum was found to support the transfer of reward value across related items. Here, the association of a single item with reward occurred at the same time as the spread of this reward value to other items associated in memory, supporting later flexible decisions among items that had never been chosen between before—a hallmark feature of hippocampal relational flexibility displayed in the context of value-guided decisions (Wimmer & Shohamy, 2012).

Additionally, cooperative interactions have been reported between the dopaminergic midbrain and MTL, as correlations between these regions predict how well individuals perform behaviorally on tests of memory (Adcock, Thangavel, Whitfield-Gabrieli, Knutson, & Gabrieli, 2006; Duzel et al., 2009; Shohamy & Wagner, 2008; Wittmann et al., 2005). Consistent with these results, a number of theories and findings have proposed that such midbrain-hippocampus interactions are critical for long-term episodic memory (Lisman & Grace, 2005; Redondo & Morris, 2011; Shohamy & Adcock, 2010).

Finally, in addition to their interactions with one another, a point of major current interest is how these two systems interact with other networks in the brain subserving other cognitive capacities. In both the memory and decision domains, for instance, the role of cognitive control and related prefrontal structures is beginning to be appreciated. Indeed, it has been suggested that, in some cases, the prefrontal cortex may play a role in mediating between multiple systems (Poldrack & Rodriguez, 2004). Thus, a current challenge is to understand learning of different kinds in terms of broad networks of interacting systems, rather than independent modules.

Conclusion

The past several years have led to substantial progress in understanding the neural and cognitive mechanisms of habit learning. This progress was a direct result of converging data from two traditionally distinct domains—memory systems and decision making—and from multiple levels of analysis, including electrophysiology, computational modeling, human brain imaging, and neuropsychology. Together, this work has helped elucidate the basic mechanisms by which reinforcement can drive learning, guide decisions, and build habits.

This work has demonstrated that reward prediction models of midbrain dopamine neurons successfully account for a wide—but also limited—range of motivated behaviors; specifically, those that underlie habitual, stimulus-response learning. This progress opens the door to many new questions. Having understood how separate systems work, we can now turn to asking questions about how they work together to guide behavior. How does habit learning interact with goal-directed behavior? How does episodic memory contribute to other forms of decision making? What are the mechanisms that control the transition of behavior from goal-directed to habitual, and can such mechanisms be leveraged to promote adaptive behavior? Answering such integrative questions represents the next challenge for research in both memory and decision making.

REFERENCES

Adamantidis, A. R., Tsai, H. C., Boutrel, B., Zhang, F., Stuber, G. D., Budygin, E. A., ... de Lecea, L. (2011). Optogenetic interrogation of dopaminergic modulation of the multiple phases of reward-seeking behavior. *J Neurosci, 31*(30), 10829–10835.

Adcock, R. A., Thangavel, A., Whitfield-Gabrieli, S., Knutson, B., & Gabrieli, J. D. (2006). Reward-motivated learning: Mesolimbic activation precedes memory formation. *Neuron, 50*(3), 507–517.

Alexander, G. E., & Crutcher, M. D. (1990). Functional architecture of basal ganglia circuits: Neural substrates of parallel processing. *Trends Neurosci, 13*(7), 266–271.

Balleine, B. W., & O'Doherty, J. P. (2010). Human and rodent homologies in action control: Corticostriatal determinants of goal-directed and habitual action. *Neuropsychopharmacology, 35*(1), 48–69.

Bayer, H. M., & Glimcher, P. W. (2005). Midbrain dopamine neurons encode a quantitative reward prediction error signal. *Neuron, 47*(1), 129–141.

Bayer, H. M., Lau, B., & Glimcher, P. W. (2007). Statistics of midbrain dopamine neuron spike trains in the awake primate. *J Neurophysiol, 98*(3), 1428–1439.

Bethus, I., Tse, D., & Morris, R. G. (2010). Dopamine and memory: Modulation of the persistence of memory for novel hippocampal NMDA receptor-dependent paired associates. *J Neurosci, 30*(5), 1610–1618.

Bodi, N., Keri, S., Nagy, H., Moustafa, A., Myers, C. E., Daw, N., … Gluck, M. A. (2009). Reward-learning and the novelty-seeking personality: A between- and within-subjects study of the effects of dopamine agonists on young Parkinson's patients. *Brain, 132*(Pt. 9), 2385–2395.

Cavaco, S., Anderson, S. W., Allen, J. S., Castro-Caldas, A., & Damasio, H. (2004). The scope of preserved procedural memory in amnesia. *Brain, 127*(Pt. 8), 1853–1867.

Chang, Q., & Gold, P. E. (2003a). Intra-hippocampal lidocaine injections impair acquisition of a place task and facilitate acquisition of a response task in rats. *Behav Brain Res, 144*(1–2), 19–24.

Chang, Q., & Gold, P. E. (2003b). Switching memory systems during learning: Changes in patterns of brain acetylcholine release in the hippocampus and striatum in rats. *J Neurosci, 23*(7), 3001–3005.

Cohen, N. J., & Eichenbaum, H. (1993). *Memory, amnesia, and the hippocampal system.* Cambridge, MA: MIT Press.

Cools, R., Altamirano, L., & D'Esposito, M. (2006). Reversal learning in Parkinson's disease depends on medication status and outcome valence. *Neuropsychologia, 44*(10), 1663–1673.

Corbit, L. H., & Balleine, B. W. (2000). The role of the hippocampus in instrumental conditioning. *J Neurosci, 20*(11), 4233–4239.

Corbit, L. H., Ostlund, S. B., & Balleine, B. W. (2002). Sensitivity to instrumental contingency degradation is mediated by the entorhinal cortex and its efferents via the dorsal hippocampus. *J Neurosci, 22*(24), 10976–10984.

Daw, N. D., Gershman, S. J., Seymour, B., Dayan, P., & Dolan, R. J. (2011). Model-based influences on humans' choices and striatal prediction errors. *Neuron, 69*(6), 1204–1215.

Daw, N. D., Niv, Y., & Dayan, P. (2005). Uncertainty-based competition between prefrontal and dorsolateral striatal systems for behavioral control. *Nat Neurosci, 8*(12), 1704–1711.

DeLong, M. R. (1990). Primate models of movement disorders of basal ganglia origin. *Trends Neurosci, 13*(7), 281–285.

Dickinson, A. (1985). Actions and habits: The development of behavioural autonomy. *Philos Trans R Soc Lond B Biol Sci, 308*(1135), 67–78.

Dickinson, A., & Balleine, B. (2002). The role of learning in the operation of motivational systems. In C. R. Gallistel (Ed.), *Stevens' handbook of experimental psychology,* Vol. 3: *Learning, motivation and emotion* (3rd ed., pp. 497–534). New York, NY: Wiley.

Dusek, J. A., & Eichenbaum, H. (1997). The hippocampus and memory for orderly stimulus relations. *Proc Natl Acad Sci USA, 94*(13), 7109–7114.

Duzel, E., Bunzeck, N., Guitart-Masip, M., Wittmann, B., Schott, B. H., & Tobler, P. N. (2009). Functional imaging of the human dopaminergic midbrain. *Trends Neurosci, 32*(6), 321–328.

Eichenbaum, H., Fagan, A., Mathews, P., & Cohen, N. J. (1988). Hippocampal system dysfunction and odor discrimination learning in rats: Impairment or facilitation depending on representational demands. *Behav Neurosci, 102*(3), 331–339.

Faure, A., Haberland, U., Conde, F., & El Massioui, N. (2005). Lesion to the nigrostriatal dopamine system disrupts stimulus-response habit formation. *J Neurosci, 25*(11), 2771–2780.

Fiorillo, C. D., Tobler, P. N., & Schultz, W. (2003). Discrete coding of reward probability and uncertainty by dopamine neurons. *Science, 299*(5614), 1898–1902.

Foerde, K., Knowlton, B. J., & Poldrack, R. A. (2006). Modulation of competing memory systems by distraction. *Proc Natl Acad Sci USA, 103*(31), 11778–11783.

Foerde, K., Race, E., Verfaellie, M., & Shohamy, D. (2013). A role for the medial temporal lobe in feedback-driven learning: Evidence from amnesia. *J Neurosci, 33*(13), 5698–5704.

Foerde, K., & Shohamy, D. (2011). Feedback timing modulates brain systems for learning in humans. *J Neurosci, 31*(37), 13157–13167.

Frank, M. J., Seeberger, L. C., & O'Reilly, R. C. (2004). By carrot or by stick: Cognitive reinforcement learning in parkinsonism. *Science, 306*(5703), 1940–1943.

Gabrieli, J. D. (1998). Cognitive neuroscience of human memory. *Annu Rev Psychol, 49,* 87–115.

Gasbarri, A., Packard, M. G., Campana, E., & Pacitti, C. (1994). Anterograde and retrograde tracing of projections from the ventral tegmental area to the hippocampal formation in the rat. *Brain Res Bull, 33*(4), 445–452.

Gluck, M. A., Shohamy, D., & Myers, C. (2002). How do people solve the "weather prediction" task? Individual variability in strategies for probabilistic category learning. *Learn Mem, 9*(6), 408–418.

Groenewegen, H. J., Vermeulen-Van der Zee, E., te Kortschot, A., & Witter, M. P. (1987). Organization of the projections from the subiculum to the ventral striatum in the rat. A study using anterograde transport of *Phaseolus vulgaris* leucoagglutinin. *Neuroscience, 23*(1), 103–120.

Henke, K. (2010). A model for memory systems based on processing modes rather than consciousness. *Nat Rev Neurosci, 11*(7), 523–532.

Houk, J. C., Adams, J. L., & Barto, A. G. (1995). A model of how the basal ganglia generate and use neural signals that predict reinforcement. In J. C. Houk, J. L. Davis, & D. G. Beiser (Eds.), *Models of information processing in the basal ganglia* (pp. 249–270). Cambridge, MA: MIT Press.

Huang, Y. Y., & Kandel, E. R. (1995). D1/D5 receptor agonists induce a protein synthesis-dependent late potentiation in the CA1 region of the hippocampus. *Proc Natl Acad Sci USA, 92*(7), 2446–2450.

Johnson, A., & Redish, A. D. (2007). Neural ensembles in CA3 transiently encode paths forward of the animal at a decision point. *J Neurosci, 27*(45), 12176–12189.

Kim, K. M., Baratta, M. V., Yang, A., Lee, D., Boyden, E. S., & Fiorillo, C. D. (2012). Optogenetic mimicry of the transient activation of dopamine neurons by natural reward is sufficient for operant reinforcement. *PLoS One, 7*(4), e33612.

Kish, S. J., Shannak, K., & Hornykiewicz, O. (1988). Uneven pattern of dopamine loss in the striatum of patients with idiopathic Parkinson's disease. Pathophysiologic and clinical implications. *N Engl J Med, 318*(14), 876–880.

Knowlton, B. J., Mangels, J. A., & Squire, L. R. (1996). A neostriatal habit learning system in humans. *Science, 273*(5280), 1399–1402.

Knowlton, B. J., Squire, L. R., & Gluck, M. A. (1994). Probabilistic classification learning in amnesia. *Learn Mem, 1*(2), 106–120.

Kravitz, A. V., Tye, L. D., & Kreitzer, A. C. (2012). Distinct roles for direct and indirect pathway striatal neurons in reinforcement. *Nat Neurosci, 15*(6), 816–818.

Lisman, J. E., & Grace, A. A. (2005). The hippocampal-VTA loop: Controlling the entry of information into long-term memory. *Neuron, 46*(5), 703–713.

Lisman, J., Grace, A. A., & Duzel, E. (2011). A neoHebbian framework for episodic memory; role of dopamine-dependent late LTP. *Trends Neurosci, 34*(10), 536–547.

McClure, S. M., Berns, G. S., & Montague, P. R. (2003). Temporal prediction errors in a passive learning task activate human striatum. *Neuron, 38*(2), 339–346.

McDonald, R. J., & White, N. M. (1995). Hippocampal and nonhippocampal contributions to place learning in rats. *Behav Neurosci, 109*(4), 579–593.

Mishkin, M., Malamut, B., & Bachevalier, J. (1984). Memories and habits: Two neural systems. In G. Lynch, J. L. McGaugh, & N. M. Weinberger (Eds.), *The neurobiology of learning and memory* (pp. 65–88). New York, NY: Guilford.

Mitchell, J. A., & Hall, G. (1988). Caudate-putamen lesions in the rat may impair or potentiate maze learning depending upon availability of stimulus cues and relevance of response cues. *Q J Exp Psychol B, 40*(3), 243–258.

Montague, P. R., Dayan, P., & Sejnowski, T. J. (1996). A framework for mesencephalic dopamine systems based on predictive Hebbian learning. *J Neurosci, 16*(5), 1936–1947.

Moody, T. D., Bookheimer, S. Y., Vanek, Z., & Knowlton, B. J. (2004). An implicit learning task activates medial temporal lobe in patients with Parkinson's disease. *Behav Neurosci, 118*(2), 438–442.

O'Doherty, J. P., Dayan, P., Friston, K., Critchley, H., & Dolan, R. J. (2003). Temporal difference models and reward-related learning in the human brain. *Neuron, 38*(2), 329–337.

O'Keefe, J., & Nadel, L. (1978). *The hippocampus as a cognitive map*, Vol. 3. Oxford, UK: Clarendon.

Otto, A. R., Gershman, S. J., Markman, A. B., & Daw, N. D. (2013). The curse of planning: Dissecting multiple reinforcement-learning systems by taxing the central executive. *Psychol Sci, 24*(5), 751–761.

Packard, M. G. (1999). Glutamate infused posttraining into the hippocampus or caudate-putamen differentially strengthens place and response learning. *Proc Natl Acad Sci USA, 96*(22), 12881–12886.

Packard, M. G., Hirsh, R., & White, N. M. (1989). Differential effects of fornix and caudate nucleus lesions on two radial maze tasks: Evidence for multiple memory systems. *J Neurosci, 9*(5), 1465–1472.

Peran, P., Cherubini, A., Assogna, F., Piras, F., Quattrocchi, C., Peppe, A., ... Sabatini, U. (2010). Magnetic resonance imaging markers of Parkinson's disease nigrostriatal signature. *Brain, 133*(11), 3423–3433.

Pessiglione, M., Seymour, B., Flandin, G., Dolan, R. J., & Frith, C. D. (2006). Dopamine-dependent prediction errors underpin reward-seeking behaviour in humans. *Nature, 442*(7106), 1042–1045.

Pfeiffer, B. E., & Foster, D. J. (2013). Hippocampal place-cell sequences depict future paths to remembered goals. *Nature, 497*(7447), 74–79.

Poldrack, R. A., Clark, J., Paré-Blagoev, E. J., Shohamy, D., Creso Moyano, J., Myers, C., & Gluck, M. A. (2001). Interactive memory systems in the human brain. *Nature, 414*(6863), 546–550.

Poldrack, R. A., & Rodriguez, P. (2004). How do memory systems interact? Evidence from human classification learning. *Neurobiol Learn Mem, 82*(3), 324–332.

Rabe, A., & Haddad, R. K. (1969). Acquisition of two-way shuttle-box avoidance after selective hippocampal lesions. *Physiol Behav, 4*(3), 319–323.

Redish, A. D. (2004). Addiction as a computational process gone awry. *Science, 306*(5703), 1944–1947.

Redondo, R. L., & Morris, R. G. (2011). Making memories last: The synaptic tagging and capture hypothesis. *Nat Rev Neurosci, 12*(1), 17–30.

Reynolds, J. N., & Wickens, J. R. (2002). Dopamine-dependent plasticity of corticostriatal synapses. *Neural Netw, 15*(4–6), 507–521.

Samson, Y., Wu, J. J., Friedman, A. H., & Davis, J. N. (1990). Catecholaminergic innervation of the hippocampus in the cynomolgus monkey. *J Comp Neurol, 298*(2), 250–263.

Schapiro, A. C., Kustner, L. V., & Turk-Browne, N. B. (2012). Shaping of object representations in the human medial temporal lobe based on temporal regularities. *Curr Biol, 22*(17), 1622–1627.

Schonberg, T., O'Doherty, J. P., Joel, D., Inzelberg, R., Segev, Y., & Daw, N. D. (2010). Selective impairment of prediction error signaling in human dorsolateral but not ventral striatum in Parkinson's disease patients: Evidence from a model-based fMRI study. *NeuroImage, 49*(1), 772–781.

Schroeder, J. P., Wingard, J. C., & Packard, M. G. (2002). Post-training reversible inactivation of hippocampus reveals interference between memory systems. *Hippocampus, 12*(2), 280–284.

Schultz, W., Dayan, P., & Montague, P. R. (1997). A neural substrate of prediction and reward. *Science, 275*(5306), 1593–1599.

Schultz, W. (1992). Activity of dopamine neurons in the behaving primate. *Semin Neurosci, 4*(2), 129–138.

Scoville, W. B., & Milner, B. (1957). Loss of recent memory after bilateral hippocampal lesions. *J Neurol Neurosurg Psychiatry, 20*(1), 11–21.

Shohamy, D., & Adcock, R. A. (2010). Dopamine and adaptive memory. *Trends Cogn Sci, 14*(10), 464–472.

Shohamy, D., Myers, C. E., Grossman, S., Sage, J., Gluck, M. A., & Poldrack, R. A. (2004). Cortico-striatal contributions to feedback-based learning: Converging data from neuroimaging and neuropsychology. *Brain, 127*(Pt. 4), 851–859.

Shohamy, D., Myers, C. E., Onlaor, S., & Gluck, M. A. (2004). Role of the basal ganglia in category learning: How do patients with Parkinson's disease learn? *Behav Neurosci, 118*(4), 676–686.

Shohamy, D., & Wagner, A. D. (2008). Integrating memories in the human brain: Hippocampal-midbrain encoding of overlapping events. *Neuron, 60*(2), 378–389.

Smith, K. S., Virkud, A., Deisseroth, K., & Graybiel, A. M. (2012). Reversible online control of habitual behavior by optogenetic perturbation of medial prefrontal cortex. *Proc Natl Acad Sci USA, 109*(46), 18932–18937.

Squire, L. R., & Zola, S. M. (1996). Structure and function of declarative and nondeclarative memory systems. *Proc Natl Acad Sci USA, 93*(24), 13515–13522.

Sutton, R. S. (1988). Learning to predict by the methods of temporal differences. *Mach Learn, 3*(1), 9–44.

Sutton, R. S., & Barto, A. G. (1998). *Reinforcement learning: An introduction.* Cambridge, MA: MIT Press.

Tan, K. R., Yvon, C., Turiault, M., Mirzabekov, J. J., Doehner, J., Labouebe, G., ... Luscher, C. (2012). GABA neurons of the VTA drive conditioned place aversion. *Neuron, 73*(6), 1173–1183.

Thorndike, E. L. (1911). *Animal intelligence: Experimental studies.* New York, NY: Macmillan.

Tolman, E. C. (1948). Cognitive maps in rats and men. *Psychol Rev, 55*(4), 189–208.

Tsai, H. C., Zhang, F., Adamantidis, A., Stuber, G. D., Bonci, A., de Lecea, L., & Deisseroth, K. (2009). Phasic firing in dopaminergic neurons is sufficient for behavioral conditioning. *Science, 324*(5930), 1080–1084.

Wang, L. P., Li, F., Wang, D., Xie, K., Wang, D., Shen, X., & Tsien, J. Z. (2011). NMDA receptors in dopaminergic neurons are crucial for habit learning. *Neuron, 72*(6), 1055–1066.

Wimmer, G. E., & Shohamy, D. (2012). Preference by association: How memory mechanisms in the hippocampus bias decisions. *Science, 338*(6104), 270–273.

Wise, R. A. (2004). Dopamine, learning and motivation. *Nat Rev Neurosci, 5*(6), 483–494.

Wittmann, B. C., Schott, B. H., Guderian, S., Frey, J. U., Heinze, H. J., & Duzel, E. (2005). Reward-related FMRI activation of dopaminergic midbrain is associated with enhanced hippocampus-dependent long-term memory formation. *Neuron, 45*(3), 459–467.

Yin, H. H., & Knowlton, B. J. (2006). The role of the basal ganglia in habit formation. *Nat Rev Neurosci, 7*(6), 464–476.

51 Prediction, Imagination, and Memory

SINÉAD L. MULLALLY AND ELEANOR A. MAGUIRE

ABSTRACT On the face of it, prediction, imagination, and memory seem to be distinct cognitive functions. However, metacognitive, cognitive, neuropsychological, and neuroimaging evidence is emerging that they are not, suggesting intimate links in their underlying processes. Here we explore these empirical findings and the evolving theoretical frameworks that seek to explain how a common neural system supports our recollection of times past, imagination, and our attempts to predict the future.

"We predominantly stand in the present facing the future rather than looking back to the past."
—Suddendorf and Corballis (2007)

Although not immediately intuitive, the idea that memory, imagination, and predicting what might happen in the future are intimately linked is not new. Throughout the centuries, this notion has consistently reemerged within philosophical, psychological, and contemporary work, along with the belief that the role of recollection is to serve imagination and prediction of the future. For instance, in 1798 Immanuel Kant noted that "Recalling the past (remembering) occurs only with the intention of making it possible to foresee the future" (Kant, 2006, p. 79). In 1871, the White Queen in Lewis Carroll's *Through the Looking Glass* astutely observed, "It's a poor sort of memory that only works backwards" (Carroll, 1994, ch. 5), while in 2006 Suddendorf argued, "It is accurate prediction of the future, more so than accurate memory of the past per se, that conveys adaptive advantage" (Suddendorf, 2006, p. 1007).

There is behavioral evidence supporting the connection between memory and imagination of the future. For instance, D'Argembeau and van der Linden (2004) asked participants to mentally "reexperience" personal past events (episodic memory) or to "preexperience" (episodic future thinking; Atance & O'Neill, 2001) possible future events that had or would occur in the close or distant past or future. For the past and future, temporally close events were associated with more sensorial and contextual details and evoked stronger feelings of reexperiencing (or preexperiencing) than the temporally distant equivalents. Similarly, D'Argembeau and van der Linden (2006) showed that individual differences, such as capacity for visual imagery, affected the phenomenological experience of episodic memory and episodic future thinking. Notably, specific errors made when recollecting the past were also evident when people engaged in predicting the future (for review, see Gilbert & Wilson, 2007).

If memory and imagination are intimately linked, it is natural to ask whether they are supported by the same neural structures. This has been examined in two ways: one with a focus on the hippocampus, and the other with an eye to an extended set of brain areas—including the medial and lateral prefrontal cortices, posterior cingulate, retrosplenial and lateral temporal cortices, and the medial temporal lobe (MTL)—often called the "core network" for episodic memory and imagination (Buckner & Carroll, 2007; Spreng, Mar, & Kim, 2009). Considering first the hippocampus, since the seminal work of Scoville and Milner (1957), the MTL, and particularly the hippocampus, have been recognized as playing a pivotal role in our ability to recollect past experiences. This paper described the case of H.M., who underwent bilateral temporal lobectomy for the relief of intractable epilepsy, rendering him amnesic, or unable to acquire new episodic memories. They noted "after the operation this young man could no longer recognize the hospital staff nor find his way to the bathroom, and he seemed to recall nothing of the day-to-day events of his hospital life" (Scoville & Milner, 1957, p. 14). The case of H.M. precipitated 50 years of subsequent work examining the role of the hippocampus in memory (for reviews, see Corkin: 2002; Squire, 2004).

Episodic memory, however, is not the only function that has been ascribed to the hippocampus. In the 1970s, O'Keefe and Dostrovsky (1971) discovered cells in the rat hippocampus that displayed location-specific firing (so-called "place cells"), while damage to the hippocampus severely disrupted spatial navigation ability (Morris, Garrud, Rawlins, & O'Keefe, 1982). This evidence prompted O'Keefe and Nadel (1978) to suggest the hippocampus may provide the spatial scaffold for episodic memories. While this idea has been debated (Cohen & Eichenbaum, 1991), the onus remains on theoretical accounts of hippocampal function to explain the mnemonic (Spiers, Maguire, & Burgess, 2001) and navigation (Maguire, Nannery, & Spiers, 2006) deficits observed in patients following bilateral hippocampal damage (Burgess, Maguire, & O'Keefe, 2002). But it seems that even explaining memory and navigation is not sufficient; as the links between episodic memory, imagination, and thinking about the future have crystallized, evidence has started to accrue implicating the hippocampus and core network in these latter functions also. In fact, there has been an explosion of interest in this domain, with Klein (2013) noting a 10-fold increase in investigative activity in the last five years. So what is the evidence that a common neural system, which includes the hippocampus, underpins episodic memory, imagination, and prediction of the future?

Neuropsychological evidence

Patients with Bilateral Hippocampal Damage and Amnesia Initial interest in the neural substrates of imagining future scenarios can be traced to early neuropsychological observations, which tentatively suggested that patients with severe amnesia also had difficulties imagining and planning their personal future (e.g., Banks & Karam, 1996; Talland, 1965). This paved the way for more detailed investigations. For instance, patient K.C., who became profoundly amnesic after suffering widespread brain damage (including to the MTL) appeared unable to imagine his personal future (Tulving, 1985; see also Rosenbaum et al., 2005). Similarly, patient D.B., with central and peripheral brain atrophy, displayed episodic memory impairments equal in severity to K.C.'s and was also unable to project himself into the future (Klein, Loftus, & Kihlstrom, 2002). As these patients suffered widespread neurological damage, it was not possible to localize the ability to think about the future to specific brain regions.

No formal study was published concerning H.M.'s ability to imagine fictitious events, but anecdotal evidence suggests his ability to predict his personal future was impaired. When, in 1992, H.M. was asked what he believed he would do tomorrow, he replied "whatever is beneficial" and appeared to have "no database to consult when asked what he would do the next day, week, or in years to come" (S. Corkin, personal communication; cited in de Vito & Della Sala, 2011, p. 1019). Similarly, when H.M. was asked to make a prediction about his personal future, he would respond with a happening from the distant past or he did not respond at all (S. Steinvorth and S. Corkin, personal communication; cited in Buckner, 2010). This suggests that the MTL, including the hippocampus, supports the recollection of the past and imagination of the future.

It was not until 2007, however, that the first systematic study of imagination ability in patients with selective bilateral hippocampal damage was published (Hassabis, Kumaran, Vann, & Maguire, 2007). These profoundly amnesic patients were unable to construct atemporal fictitious scenes (i.e., scenes with no past or future temporal connotations) in the mind's eye or to imagine future events involving themselves. For example, when asked to imagine simple fictional scenes such as "imagine lying on a white sandy beach in a beautiful tropical bay" they, unlike controls, struggled to construct a coherent response (see figure 51.1). When formally measured, the patients' scenes were significantly less vivid and more spatially fragmented than those of controls, effects that have since been replicated in a new cohort of seven focal hippocampal-damaged amnesic patients (Mullally, Intraub, & Maguire, 2012). Interestingly, this deficit did not appear to be a generalized imagination deficit, since the patients were able to vividly imagine single acontextual objects. Nor did it appear to be solely attributable to their memory deficits, because when they were presented with all of the individual scene components required to construct scenes, the patients remained unable to use these elements to form cohesive scene representations (Hassabis, Kumaran, Vann, et al., 2007). Hassabis et al. proposed that there was something specific about the imagination of spatially coherent fictitious or future scenes that requires the hippocampus.

Other groups have since reported deficits in imagining scenes or events in patients with bilateral hippocampal damage (Andelman, Hoofien, Goldberg, Aizenstein, & Neufeld, 2010; Race, Keane, & Verfaellie, 2011). For example, Race et al. (2011) found that amnesic patients with MTL damage were impaired at recollecting their remote and recent past and at imagining their near and distant futures. Critically, these patients were able to generate appropriate and detailed story-based narratives when presented with drawings of scenes, indicating that a generalized narrative problem is unlikely to account for the deficits observed. Of note, several studies involving developmental amnesic patients, whose hippocampal damage occurred early in life, have documented an apparently preserved ability to construct scenes (Cooper, Vargha-Khadem, Gadian, & Maguire, 2011; Hurley, Maguire, & Vargha-Khadem, 2011; Maguire, Vargha-Khadem, & Hassabis, 2010). However, it appears that this is based on their intact semantic memory and world knowledge (Klein, 2013),

Cue: *"Imagine you are lying on a white sandy beach in a beautiful tropical bay"*

P03: As for seeing I can't really, apart from just sky. I can hear the sound of seagulls and of the sea... um... I can feel the grains of sand between my fingers... um... I can hear one of those ship's hooters [laughter]... um... that's about it. *Are you're actually seeing this in your mind's eye?* No, the only thing I can see is blue. *So if you look around what can you see?* Really all I can see is the colour of the blue sky and the white sand, the rest of it, the sounds and things, obviously I'm just hearing. *Can you see anything else?* No, it's like I'm kind of floating...

Control: It's very hot and the sun is beating down on me. The sand underneath me is almost unbearably hot. I can hear the sounds of small wavelets lapping on the beach. The sea is a gorgeous aquamarine colour. Behind me is a row of palm trees and I can hear rustling every so often in the slight breeze. To my left the beach curves round and becomes a point. And on the point there are a couple of buildings, wooden buildings, maybe someone's hut or a bar of some sort. The other end of the beach, looking the other way, ends in big brown rocks. There's no one else around. Out to sea is a fishing boat. It's quite an old creaking looking boat, chugging past on its small engine. It has a cabin in the middle and pile of nets in the back of the boat. There's a guy in the front and I wave at him and he waves back...[continues]...

FIGURE 51.1 Example of an imagined scenario from Hassabis, Kumaran, Vann, et al. (2007). The cue is shown at the top, below which is an excerpt from P03, a patient with bilateral hippocampal damage, followed by that of a control participant who was age-, education-, and IQ-matched to P03. Interviewer's probing comments are in italics. Relevant background information is noted in square brackets.

and does not involve true visualization of an imagined scene, or engagement of the hippocampus (Mullally, Vargha-Khadem, & Maguire, 2014). In one study that failed to observe imagination-based scene deficits in patients with hippocampal damage acquired in adulthood (Squire et al., 2010), the patients were in fact not amnesic, showing instead only mild and nonsignificant memory deficits when compared with a control group (Maguire & Hassabis, 2011; Mullally, Hassabis, & Maguire, 2012). This suggests that an impaired scene imagination profile may only occur in the context of adult-acquired hippocampal damage and concomitant, severe memory disturbance.

SCENE IMAGINATION DEFICITS IN OTHER POPULATIONS Scene/event imagination deficits have also been observed in older adults and in other patient groups; populations in which hippocampal function and other neural components of the core network are known to be compromised (for review, see Schacter, Addis, & Buckner, 2008). In one study, Addis, Wong, and Schacter (2008) noted deficits in future simulations in older adults. In a subsequent study, Addis, Musicaro, Pan, and Schacter (2010) sought to investigate whether the impairment could simply be an expression of an episodic memory deficit whereby elderly adults simply "recast" entire remembered events into the future. Using a recombination paradigm (see also Addis, Pan, Vu, Laiser, & Schacter, 2009) participants first provided a set of autobiographical memories and were later asked to imagine novel events containing a combination of specific details taken from these episodic memories. The older adults continued to generate fewer episodic details for imagined events, suggesting that recollection difficulties alone cannot explain their impaired scene/event imagination (see also Addis, Roberts, & Schacter, 2011; Gaesser, Sacchetti, Addis, & Schacter, 2011; Romero & Moscovitch, 2012).

Co-existing episodic memory and scene/event imagination impairments have also been noted in a number of other populations. Williams et al. (1996) reported that suicidal, depressed patients' recollection of the past and simulation of the future lacked specific details and appeared "over-general" relative to controls, while D'Argembeau, Raffard, & van der Linden (2008) found that patients with schizophrenia generated significantly fewer episodic details for both past and future events. Interestingly, hippocampal atrophy has been documented in depression (Bremner et al., 2000), schizophrenia (Herold et al., 2012), and in the aging brain (Driscoll & Sutherland, 2005), suggesting that damage to this brain region may be a critical factor underlying these disparate cognitive impairments.

Looking beyond the hippocampus, there have been few neuropsychological studies focussed on the role of brain areas outside the MTL in imagination of fictitious or future scenes. In one such study, Berryhill, Picasso, Arnold, Drowos, and Olson (2010) tested patients with parietal and prefrontal cortex lesions. Both patient types were impaired, although the precise reasons for this were unclear. The authors suggest, for example, that the frontal patients may have had difficulty with accessing and/or selecting elements for inclusion in the imagined scenes. Nevertheless, this study illustrates that the hippocampus does not act alone in scene imagination and predicting the future.

Neuroimaging evidence

A wider network beyond the hippocampus has been highlighted in particular by neuroimaging studies. Episodic memory has been consistently linked with activation of a core network (reviewed in Maguire, 2001; Svoboda, McKinnon, & Levine, 2006). Attention then turned to probing whether imagination of fictitious and future events are also supported by a wider set of brain areas beyond the hippocampus, as suggested by the neuropsychological findings, and if these are the same regions that support episodic memory (Okuda et al., 2003). Addis, Wong, and Schacter (2007) showed extensive overlap in activity of this network (the medial and lateral prefrontal cortices, posterior cingulate cortex, retrosplenial cortex, lateral temporal cortices, and the MTL) when participants recollected or imagined detailed events. This suggests that the episodic system is not solely involved in memory but may also support event imagination–based processes. This was later extended to include scene construction processes when Hassabis, Kumaran, and Maguire (2007) reported activation of the core network when participants recollected episodic memories, recollected previously imagined fictitious scenes, or constructed entirely novel fictitious scenes (see also Addis et al., 2009; Botzung, Denkova, & Manning, 2008; Spreng et al., 2009; Szpunar, Watson, & McDermott, 2007). In summary, these data, coupled with the neuropsychological evidence, strongly indicate that a core network of brain regions that includes the hippocampus supports both mnemonic and scene imagination–based processes. But what mechanisms might be involved?

Theoretical accounts

Mental Time Travel into the Future A number of theories have been proposed that attempt to explain the overlap between episodic memory and scene/event imagination. One such account is based on Tulving's proposal that recollection of episodic memories involves a mental journey into the past (i.e., mental time travel) in which one has a subjective sense of self over time ("autonoesis"; Tulving, 2002). This concept of mental time travel or self-projection, believed to depend upon the integrity of the hippocampus, can also be applied to imagining the future and possibly spatial navigation (Buckner & Carroll, 2007; Suddendorf & Corballis, 2007; Szpunar et al., 2007). This account, however, struggles to explain why hippocampal-damaged patients are unable to imagine atemporal fictitious scenes (Hassabis, Kumaran, Vann, et al., 2007; Mullally, Intraub, & Maguire, 2012). Indeed, a recent functional MRI study found that frontal and parietal cortices, but not the hippocampus, supported mental time travel (Nyberg, Kim, Habib, Levine, & Tulving, 2010), and Andrews-Hanna, Reidler, Sepulcre, Poulin, & Buckner (2010) showed that imagining scenes best accounted for activity in the hippocampus and MTL, while other regions were concerned with the self and with time. Thus, the idea of mental time travel has undoubted heuristic value and may account for the contributions of some areas in the core network to imagining the future, but not the hippocampus.

Constructive Episodic Simulation Hypothesis The constructive episodic simulation hypothesis (reviewed in Schacter et al., 2012) proposes that episodic memory and thinking about the future are supported by a similar neural network because they are both constructive in nature. In this way, episodic memory is conceptualized as a constructive process (Bartlett, 1932), whose critical function is to make available the information required for the simulation/construction of future events. Thus, information is processed in a manner that enables the relevant details to be flexibly recombined to form a novel event. Moreover, this hypothesized recombination process dovetails with existing theoretical accounts of hippocampal function that emphasize the role of the hippocampus in binding arbitrary or accidentally occurring relations among individual elements within an experience (the relational theory; Cohen & Eichenbaum, 1993; Konkel & Cohen, 2009) and/or to the specific scene context (the binding of items and contexts model; Ranganath, 2010). In this way the constructive episodic simulation hypothesis elegantly explains the co-activation of the core network during self-relevant mnemonic and future simulation processes. However, this framework has not explicitly sought to account for the striking navigation deficits observed in patients with hippocampal damage and amnesia (Burgess et al., 2002). Moreover, recent evidence that amnesic patients are able to flexibly recombine elements within an experience (Mullally & Maguire, submitted) and bind items with each other and with a context (Mullally, Intraub, & Maguire, 2012) are also challenges for this view.

Scene Construction Theory A third account is scene construction theory (SCT; Hassabis & Maguire, 2007, 2009; Maguire & Mullally, 2013), which was originally proposed following the observation that patients with hippocampal damage and amnesia are unable to imagine scenes. In contrast to the constructive episodic simulation hypothesis, which had a broader focus on the core network, SCT attempts to account specifically for the role of the hippocampus in scene imagination, episodic memory, and spatial navigation. In essence, SCT proposes that the hippocampus primarily acts to facilitate the construction of atemporal scenes, and in doing so allows the event details of episodic memories and imagined future experiences a foundation upon which to reside. In this way, hippocampal-dependent scene construction processes are held to underpin and support episodic memory, predicting the future, spatial navigation, and perhaps even dreaming and mind-wandering, with the addition of self-related and temporal processing implemented via the recruitment of other regions within the core network.

SCT thus places scenes at the center of hippocampal information processing, underpinning (but not wholly responsible for) critical functions such as episodic memory, prospection, and spatial navigation. This has intuitive appeal—for most people, recalling the past, thinking about the future, and planning how to get somewhere typically involve imagining scenes. This also resonates with the patients' experiences of trying to imagine scenes: "There is no scene in front of me here. It's frustrating because I feel like there should be. I feel like I'm listening to the radio instead of watching it on the TV. I'm trying to imagine different things happening but there's no visual scene opening out in front of me"; "It's hard trying to get the space, it keeps getting squashed" (Mullally, Intraub, & Maguire, 2012). The appeal of the SCT is that it offers a unified account of why such a wide range of seemingly disparate functions are impaired following hippocampal damage.

Further recent evidence appears to place scene construction at the heart of hippocampal processing. Boundary extension (BE; Intraub & Richardson, 1989) is a ubiquitous cognitive phenomenon whereby we erroneously remember seeing more of a scene than was present in the sensory input, and occurs because when we view a scene, we implicitly extrapolate beyond the borders to form an extended representation of that scene. In the absence of the original visual input, this extended scene is misremembered instead of the original input, causing a memory error. Of note, BE only occurs in relation to scenes and not to single, isolated objects (Gottesman & Intraub, 2002), a dissociation that mirrors the imagination dichotomy observed in amnesic patients (Hassabis, Kumaran, Vann, et al., 2007). Critically, however, BE depends upon an intact ability to construct scenes. Thus, patients with bilateral hippocampal damage and amnesia, who are unable to construct scenes, should be unable to form this extended representation and therefore fail to commit the BE memory error. This would result in a situation where amnesic patients display superior performance relative to healthy controls, and is exactly what Mullally, Intraub, and Maguire (2012) recently demonstrated on a variety of BE paradigms (see figure 51.2).

These results enabled Mullally, Intraub, and Maguire (2012) to conclude that the patients' attenuated BE could not be attributed to memory impairment, since they paradoxically outperformed the controls on a range of memory tasks. Therefore, in this context, impaired memory did not lead to impaired scene construction. Instead, impaired scene construction actually led to better memory, thus successfully separating these two processes.

These data also suggest that a function of the hippocampus is the implicit and continuous prediction of the upcoming environment; that is, the hippocampus is continually constructing scenes, extrapolating beyond the boundaries of our current field of view. A recent neuroimaging study (Chadwick, Mullally, & Maguire, 2013), which investigated the role of the hippocampus in BE in control participants, supports this hypothesis. Specifically, they found robust activity in the hippocampus during the presentation of the original scene stimulus. Significantly, this activity was observed only on trials where participants later committed the BE error. This suggests that the hippocampus is involved early at the initial stage of the BE effect where the predictive scene extension process (attenuated in the hippocampal patients) is hypothesized to occur. The BE data (Chadwick et al., 2013; Mullally, Intraub, & Maguire, 2012), in addition to the original scene construction findings (Hassabis, Kumaran, Vann, et al., 2007; Hassabis, Kumaran, & Maguire, 2007; Mullally, Intraub, & Maguire, 2012), support the idea that the primary function of the hippocampus may not be mnemonic (Maguire & Mullally, 2013; see also Graham, Barense, & Lee, 2010) but may instead be to predict the nature of the world beyond the immediate sensorium (Bar, 2011).

In conclusion, we believe that metacognitive, cognitive, neuropsychological, and neuroimaging evidence leaves in no doubt the close ties between episodic memory, scene imagination, and predicting the future. An understanding of this relationship is still in its

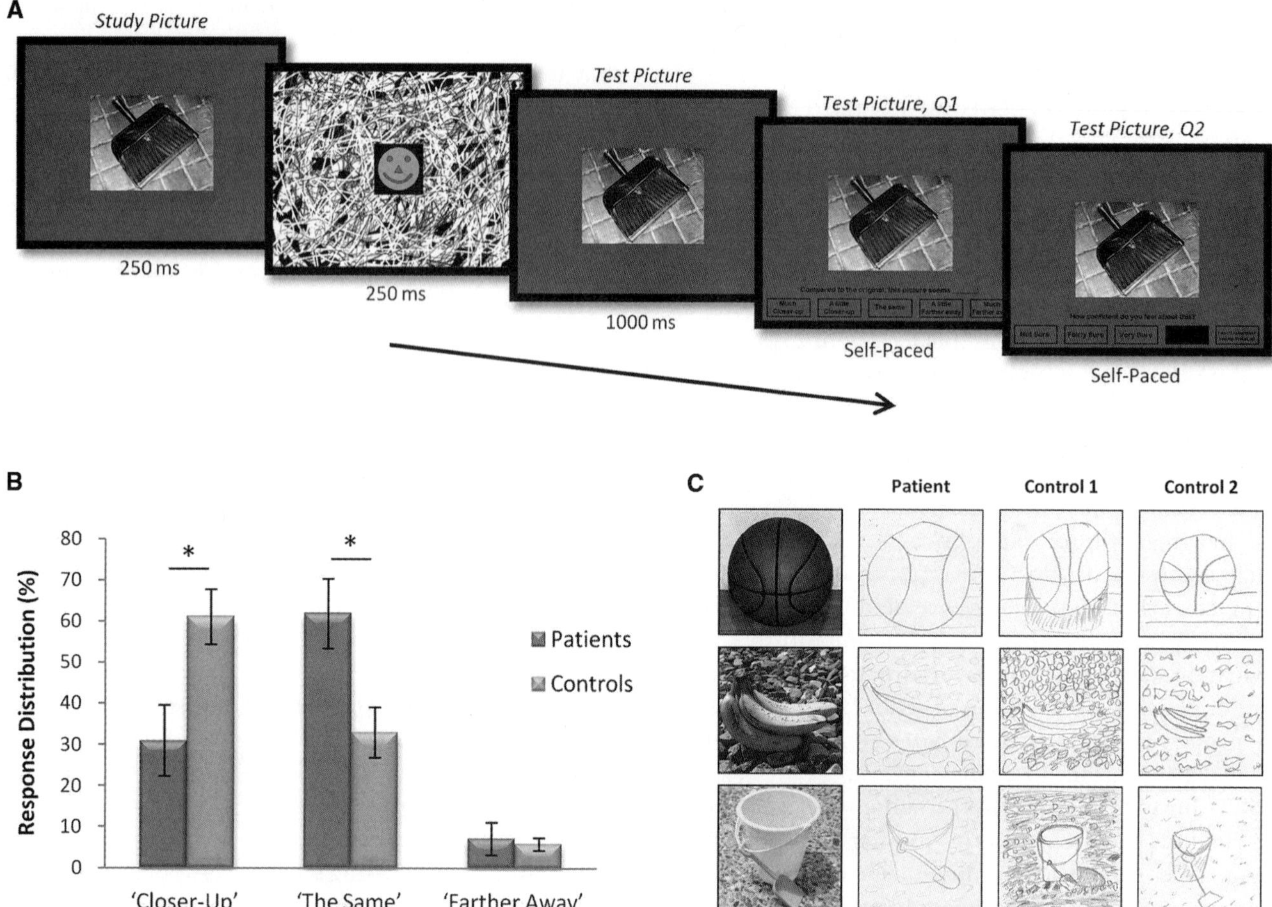

FIGURE 51.2 Examples of boundary extension (BE) tasks (Mullally, Intraub, & Maguire, 2012). (A) Timeline of an example trial from a rapid serial visual presentation BE task. The initial photograph of a simple scene was presented briefly followed by a dynamically changing mask. The second (test) picture (which unbeknownst to the participants was always identical to the original picture) immediately followed the mask. The task was to rate the second picture relative to the first. There were five options ranging from "much closer-up" to "much farther away," including the correct response "the same." (B) BE is revealed by disproportionally larger number of "closer-up" responses. Overall, control participants made significantly more of these erroneous responses, while patients with bilateral hippocampal amnesia made significantly more accurate ("the same") responses, and thus showed significantly reduced BE relative to controls. Means (± standard error of mean); $*P < 0.05$. (C) In a drawing task, three scene photographs (left panel) were studied for 15 seconds and immediately drawn from memory. Drawings by an example hippocampal-damaged amnesic patient (middle left panel) and two matched control participants (middle right and right panels) are displayed. As is evident, this patient more accurately depicted the proportional size of the object relative to the background while the control participants' drawings expose how they extrapolated beyond the given view. (See color plate 43.)

infancy. Interestingly, electrophysiological studies documenting preplay (e.g., Johnson & Redish, 2007) add to this evolving picture by hinting at animal parallels in predicting what might occur in the future. Further studies in humans will need to move beyond describing this relationship to focus on the mechanisms involved, the precise functions of each area within the core network, and the connectivity between them. In this way, we are confident that the next five years will hasten important new insights into this question that has intrigued through the ages.

ACKNOWLEDGMENT E.A.M. is supported by the Wellcome Trust. Note that this chapter is an edited version of an article that appeared in *The Neuroscientist*: Mullally, S. L., and Maguire, E.A. (2014). Memory, imagination and predicting the future: A common brain mechanism? *The Neuroscientist, 20,* 220–234.

REFERENCES

ADDIS, D. R., MUSICARO, R., PAN, L., & SCHACTER, D. L. (2010). Episodic simulation of past and future events in older adults: Evidence from an experimental recombination task. *Psychol Aging, 25*(2), 369–376.

Addis, D. R., Pan, L., Vu, M. A., Laiser, N., & Schacter, D. L. (2009). Constructive episodic simulation of the future and the past: Distinct subsystems of a core brain network mediate imagining and remembering. *Neuropsychologia, 47*(11), 2222–2238.

Addis, D. R., Roberts, R. P., & Schacter, D. L. (2011). Age-related neural changes in autobiographical remembering and imagining. *Neuropsychologia, 49*(13), 3656–3669.

Addis, D. R., Wong, A. T., & Schacter, D. L. (2007). Remembering the past and imagining the future: Common and distinct neural substrates during event construction and elaboration. *Neuropsychologia, 45*(7), 1363–1377.

Addis, D. R., Wong, A. T., & Schacter, D. L. (2008). Age-related changes in the episodic simulation of future events. *Psychol Sci, 19*(1), 33–41.

Andelman, F., Hoofien, D., Goldberg, I., Aizenstein, O., & Neufeld, M. Y. (2010). Bilateral hippocampal lesion and a selective impairment of the ability for mental time travel. *Neurocase, 16*(5), 426–435.

Andrews-Hanna, J. R., Reidler, J. S., Sepulcre, J., Poulin, R., & Buckner, R. L. (2010). Functional-anatomic fractionation of the brain's default network. *Neuron, 65*(4), 550–562.

Atance, C. M., & O'Neill, D. K. (2001). Episodic future thinking. *Trends Cogn Sci, 5*(12), 533–539.

Banks, W. P., & Karam, S. J. (1996). Medico-psychological study of a memory disorder. *Conscious Cogn, 5*(1–2), 2–21.

Bar, M. (Ed.). (2011). *Prediction in the brain: Using our past to generate a future.* Oxford, UK: Oxford University Press.

Bartlett, F. C. (1932). *Remembering: A study in experimental and social psychology.* Cambridge, UK: Cambridge University Press.

Berryhill, M. E., Picasso, L., Arnold, R., Drowos, D., & Olson, I. R. (2010). Similarities and differences between parietal and frontal patients in autobiographical and constructed experience tasks. *Neuropsychologia, 48*(5), 1385–1393.

Botzung, A., Denkova, E., & Manning, L. (2008). Experiencing past and future personal events: Functional neuroimaging evidence on the neural bases of mental time travel. *Brain Cogn, 66*(2), 202–212.

Bremner, J. D., Narayan, M., Anderson, E. R., Staib, L. H., Miller, H. L., & Charney, D. S. (2000). Hippocampal volume reduction in major depression. *Am J Psychiatry, 157*(1), 115–118.

Buckner, R. L. (2010). The role of the hippocampus in prediction and imagination. *Annu Rev Psychol, 61*, 27–48.

Buckner, R. L., & Carroll, D. C. (2007). Self-projection and the brain. *Trends Cogn Sci, 11*(2), 49–57.

Burgess, N., Maguire, E. A., & O'Keefe, J. (2002). The human hippocampus and spatial and episodic memory. *Neuron, 35*(4), 625–641.

Carroll, L. (1994). *Through the looking glass.* London: Penguin. Originally published 1871.

Chadwick, M. J., Mullally, S. L., & Maguire, E. A. (2013). The hippocampus extrapolates beyond the view in scenes: An fMRI study of boundary extension. *Cortex, 49*(8), 2067–2079.

Cohen, N. J., & Eichenbaum, H. (1991). The theory that wouldn't die: A critical look at the spatial mapping theory of hippocampal function. *Hippocampus, 1*(3), 265–268.

Cohen, N. J., & Eichenbaum, H. (1993). *Memory, amnesia, and the hippocampal system.* Cambridge, MA: MIT Press.

Cooper, J. M., Vargha-Khadem, F., Gadian, D. G., & Maguire, E. A. (2011). The effect of hippocampal damage in children on recalling the past and imagining new experiences. *Neuropsychologia, 49*(7), 1843–1850.

Corkin, S. (2002). What's new with the amnesic patient HM? *Nat Rev Neurosci, 3*, 153–160.

D'Argembeau, A., Raffard, S., & van der Linden, M. (2008). Remembering the past and imagining the future in schizophrenia. *J Abnorm Psychol, 117*(1), 247–251.

D'Argembeau, A., & van der Linden, M. (2004). Phenomenal characteristics associated with projecting oneself back into the past and forward into the future: Influence of valence and temporal distance. *Conscious Cogn, 13*(4), 844–858.

D'Argembeau, A., & van der Linden, M. (2006). Individual differences in the phenomenology of mental time travel: The effect of vivid visual imagery and emotion regulation strategies. *Conscious Cogn, 15*(2), 342–350.

De Vito, S., & Della Sala, S. (2011). Predicting the future. *Cortex, 47*(8), 1018–1022.

Driscoll, I., & Sutherland, R. J. (2005). The aging hippocampus: Navigating between rat and human experiments. *Rev Neurosci, 16*(2), 87–121.

Gaesser, B., Sacchetti, D. C., Addis, D. R., & Schacter, D. L. (2011). Characterizing age-related changes in remembering the past and imagining the future. *Psychol Aging, 26*(1), 80–84.

Gilbert, D. T., & Wilson, T. D. (2007). Prospection: Experiencing the future. *Science, 317*(5843), 1351–1354.

Gottesman, C. V., & Intraub, H. (2002). Surface construal and the mental representation of scenes. *J Exp Psychol Human, 28*(3), 589–599.

Graham, K. S., Barense, M. D., & Lee, A. C. (2010). Going beyond LTM in the MTL: A synthesis of neuropsychological and neuroimaging findings on the role of the medial temporal lobe in memory and perception. *Neuropsychologia, 48*(4), 831–853.

Hassabis, D., Kumaran, D., & Maguire, E. A. (2007). Using imagination to understand the neural basis of episodic memory. *J Neurosci, 27*(52), 14365–14374.

Hassabis, D., Kumaran, D., Vann, S. D., & Maguire, E. A. (2007). Patients with hippocampal amnesia cannot imagine new experiences. *Proc Natl Acad Sci USA, 104*(5), 1726–1731.

Hassabis, D., & Maguire, E. A. (2007). Deconstructing episodic memory with construction. *Trends Cogn Sci, 11*(7), 299–306.

Hassabis, D., & Maguire, E. A. (2009). The construction system of the brain. *Philos Trans R Soc Lond B Biol Sci, 364*(1521), 1263–1271.

Herold, C. J., Lässer, M. M., Schmid, L. A., Seidl, U., Kong, L., Fellhauer, I., … Schröder, J. (2012). Hippocampal volume reduction and autobiographical memory deficits in chronic schizophrenia. *Psychiatry Res, 211*(3), 189–194.

Hurley, N. C., Maguire, E. A., & Vargha-Khadem, F. (2011). Patient HC with developmental amnesia can construct future scenarios. *Neuropsychologia, 49*(13), 3620–3628.

Intraub, H., & Richardson, M. (1989). Wide-angle memories of close-up scenes. *J Exp Psychol Learn Mem Cogn, 15*(2), 179–187.

Johnson, A., & Redish, A. D. (2007). Neural ensembles in CA3 transiently encode paths forward of the animal at a decision point. *J Neurosci, 27*(45), 12176–12189.

Kant, I. (2006). *Anthropology from a pragmatic point of view.* Originally published in 1798. Trans. Robert B. Louden. Cambridge Texts in the History of Philosophy. Cambridge: Cambridge University Press.

Klein, S. B. (2013). The complex act of projecting oneself into the future. *WIREs Cogn Sci, 4,* 63–79.

Klein, S. B., Loftus, J., & Kihlstrom, J. J. (2002). Memory and temporal experience: The effects of episodic memory loss on an amnesic patient's ability to remember the past and imagine the future. *Soc Cogn, 20,* 353–379.

Konkel, A., & Cohen, N. J. (2009). Relational memory and the hippocampus: Representations and methods. *Front Neurosci, 3*(2), 166–174.

Maguire, E. A. (2001). Neuroimaging studies of autobiographical event memory. *Philos Trans R Soc Lond B Biol Sci, 356*(1413), 1441–1451.

Maguire, E. A., & Hassabis, D. (2011). Role of the hippocampus in imagination and future thinking. *Proc Natl Acad Sci USA, 108*(11), E39.

Maguire, E. A., & Mullally, S. L. (2013). The hippocampus: A manifesto for change. *J Exp Psychol Gen, 142*(4), 1180–1189.

Maguire, E. A., Nannery, R., & Spiers, H. J. (2006). Navigation around London by a taxi driver with bilateral hippocampal lesions. *Brain, 129*(11), 2894–2907.

Maguire, E. A., Vargha-Khadem, F., & Hassabis, D. (2010). Imagining fictitious and future experiences: Evidence from developmental amnesia. *Neuropsychologia, 48*(11), 3187–3192.

Morris, R. G., Garrud, P., Rawlins, J. N., & O'Keefe, J. (1982). Place navigation impaired in rats with hippocampal lesions. *Nature, 297*(5868), 681–683.

Mullally, S. L., Hassabis, D., & Maguire, E. A. (2012). Scene construction in amnesia: An fMRI study. *J Neurosci, 32*(16), 5646–5653.

Mullally, S. L., Intraub, H., & Maguire, E. A. (2012). Attenuated boundary extension produces a paradoxical memory advantage in amnesic patients. *Curr Biol, 22*(4), 261–268.

Mullally, S. L., & Maguire, E. A. (submitted). Counterfactual thinking in patients with amnesia.

Mullally, S. L., Vargha-Khadem, F., & Maguire, E. A. (2014). Scene construction in developmental amnesia: An fMRI study. *Neuropsychologia, 52,* 1–10.

Nyberg, L., Kim, A. S., Habib, R., Levine, B., & Tulving, E. (2010). Consciousness of subjective time in the brain. *Proc Natl Acad Sci USA, 107*(51), 22356–22359.

O'Keefe, J., & Dostrovsky, J. (1971). The hippocampus as a spatial map. Preliminary evidence from unit activity in the freely-moving rat. *Brain Res, 34*(1), 171–175.

O'Keefe, J., & Nadel, L. (1978). *The hippocampus as a cognitive map.* Oxford, UK: Clarendon.

Okuda, J., Fujii, T., Ohtake, H., Tsukiura, T., Tanji, K., Suzuki, K., ... Yamadori, A. (2003). Thinking of the future and past: The roles of the frontal pole and the medial temporal lobes. *NeuroImage, 19*(4), 1369–1380.

Race, E., Keane, M. M., & Verfaellie, M. (2011). Medial temporal lobe damage causes deficits in episodic memory and episodic future thinking not attributable to deficits in narrative construction. *J Neurosci, 31,* 10262–10269.

Ranganath, C. (2010). A unified framework for the functional organization of the medial temporal lobes and the phenomenology of episodic memory. *Hippocampus, 20*(11), 1263–1290.

Romero, K., & Moscovitch, M. (2012). Episodic memory and event construction in aging and amnesia. *J Mem Lang, 67*(2), 270–284.

Rosenbaum, R. S., Köhler, S., Schacter, D. L., Moscovitch, M., Westmacott, R., Black, S. E., ... Tulving, E. (2005). The case of K. C.: Contributions of a memory-impaired person to memory theory. *Neuropsychologia, 43*(7), 989–1021.

Schacter, D. L., Addis, D. R., & Buckner, R. L. (2008). Episodic simulation of future events: Concepts, data, and applications. *Ann NY Acad Sci, 1124,* 39–60.

Schacter, D. L., Addis, D. R., Hassabis, D., Martin, V. C., Spreng, R. N., & Szpunar, K. K. (2012). The future of memory: Remembering, imagining, and the brain. *Neuron, 76*(4), 677–694.

Scoville, W. B., & Milner, B. (1957). Loss of recent memory after bilateral hippocampal lesions. *J Neurol Neurosurg Psychiatry, 20*(1), 11–21.

Spiers, H. J., Maguire, E. A., & Burgess, N. (2001). Hippocampal amnesia. *Neurocase, 7*(5), 357–382.

Spreng, R. N., Mar, R. A., & Kim, A. S. (2009). The common neural basis of autobiographical memory, prospection, navigation, theory of mind, and the default mode: A quantitative meta-analysis. *J Cogn Neurosci, 21*(3), 489–510.

Squire, L. R. (2004). Memory systems of the brain: A brief history and current perspective. *Neurobiol Learn Mem, 82*(3), 171–177.

Squire, L. R., van der Horst, A. S., McDuff, S. G., Frascino, J. C., Hopkins, R. O., & Mauldin, K. N. (2010). Role of the hippocampus in remembering the past and imagining the future. *Proc Natl Acad Sci USA, 107*(44), 19044–19048.

Suddendorf, T. (2006). Foresight and evolution of the human mind. *Science, 312*(5776), 1006–1007.

Suddendorf, T., & Corballis, M. C. (2007). The evolution of foresight: What is mental time travel, and is it unique to humans? *Behav Brain Sci, 30*(3), 299–351.

Svoboda, E., McKinnon, M. C., & Levine, B. (2006). The functional neuroanatomy of autobiographical memory: A meta-analysis. *Neuropsychologia, 44*(12), 2189–2208.

Szpunar, K. K., Watson, J. M., & McDermott, K. B. (2007). Neural substrates of envisioning the future. *Proc Natl Acad Sci USA, 104*(2), 642–647.

Talland, G. A. (1965). *Deranged memory: A psychonomic study of the amnesic syndrome.* New York, NY: Academic Press.

Tulving, E. (1985). Memory and consciousness. *Can Psychol, 26,* 1–12.

Tulving, E. (2002). Episodic memory: From mind to brain. *Annu Rev Psychol, 53,* 1–25.

Williams, J. M., Ellis, N. C., Tyers, C., Healy, H., Rose, G., & MacLeod, A. K. (1996). The specificity of autobiographical memory and imageability of the future. *Mem Cognit, 24*(1), 116–125.

VII
LANGUAGE AND ABSTRACT THOUGHT

Introduction

PETER HAGOORT

An adequate neurobiological model of our uniquely human language faculty has to meet the following two requirements: (1) it should decompose complex language skills such as speaking and listening into the contributing types of knowledge and processing steps (the cognitive architecture); (2) it should specify how these are instantiated in, and supported by, the organization of the human brain (the neural architecture; Hagoort, 2013). Until not too long ago, the neurobiological model that dominated the field was the Wernicke-Lichtheim-Geschwind (WLG) model (Levelt, 2013). In this model, the human language faculty was situated in the left perisylvian cortex, with a strict division of labor between the frontal and temporal regions. Wernicke's area in left temporal cortex was assumed to subserve the comprehension of speech, whereas Broca's area in left inferior frontal cortex was claimed to subserve language production. The arcuate fasciculus connected these two areas. Although Broca's area, Wernicke's area, and adjacent cortex are still considered to be key nodes in the language network, the distribution of labor between these regions is different than was claimed in the WLG model. Lesions in Broca's region are long since known to impair not only language production but also language comprehension, whereas lesions in Wernicke's region also affect language production.

More recently, neuroimaging studies provided further evidence that the classical view on the role of these regions is no longer tenable. For example, central aspects of language production and comprehension are subserved by shared neural circuitry (Segaert, Menenti, Weber, Petersson, & Hagoort, 2012). Moreover, the classical model focused on single-word processing, whereas a neurobiological account of language processing in its full extent should also take into account what goes on beyond production and comprehension of single words. As a consequence of the mounting evidence against the classical WLG model, in recent years alternative neurobiological models for language have been proposed. In addition, due to new innovative ways of measuring the living human brain in healthy participants, such as diffusion tractography, novel ideas about the architecture of the language cortex arose (Catani et al., 2005; Catani et al., 2013).

This section of the current volume summarizes some exciting developments that characterize the progress in the field since the previous version of this handbook.

One important development relates to the neuroanatomical characteristics of the language-ready brain. Language-relevant cortex is mapped out in much more detail than before. Relevant areas such as Broca's area and Wernicke's area are further parcellated in terms of their cytoarchitectonic and receptorarchitectonic properties (Amunts et al., 2010). Furthermore, diffusion tractography studies have provided a much more extended view on the major connections between language-relevant areas, above and beyond the arcuate fasciculus (see chapter 52 by Amunts & Catani). This infrastructure supports a system that is characterized by an amazing pace.

One of the central aspects of understanding language is the speed at which it occurs. We can produce and easily recognize three to five words per second. Mainly based on electrophysiological evidence (ERPs), there is nowadays firm and quite detailed evidence that fast comprehension is, in part, achieved by predictions that are made continuously during the updating of the input representation. Quite detailed evidence is available that these predictions happen at different levels, all the way from the event schemas and syntactic characteristics of lexical items such as their grammatical gender to the actual words that might follow (see chapter 55 by Kutas, Urbach, & Federmeier).

The role of prediction and most other features of language processing is based on studies of only a handful of—mostly Indo-European—languages. Classically, the neurobiology of language has focused on core features of language such as words and syntax, with an eye on those properties that are universally shared among the many languages in the world. However, a central feature of human languages is the remarkable variation between them. Our capacity for language is deeply rooted in our biological makeup. We all share the capacity to acquire language within the first few years of life, without any formalized teaching program. Despite its complexity, we master our native language well before we can lace our shoes or perform simple calculations. This is all based on the universal availability of a language-ready brain. At the same time, there seems to be no other cognitive system in humans that shows as much variability as language. Language comes in very different forms, at all levels of organization. The more than 6,000 languages still in existence today vary widely in their sound repertoires, their grammatical structures, or the meaning that the lexical items code for. For instance, some languages have a sound repertoire of only a dozen phonemes, whereas others have more than a hundred; some languages require every sentence to be tensed, others require every referent to be specified for visibility/invisibility. Further, sign languages are expressed by movements of hands and face, whereas spoken languages are expressed by movements of the vocal tract. How the specific characteristic of any particular language modulates the organization and recruitment of the underlying neurobiological infrastructure is still a largely open question. It is clear that cross-language comparisons are needed. One example is provided in chapter 54 by Marslen-Wilson, Bozic, and Tyler. They provide evidence for the domain of word formation, where language families vary substantially in inflectional and derivational morphology. Another domain is sign language. The language of the deaf is expressed in a very different format than speech (see chapter 56 by Emmorey and Özyürek). These differences are shown to create variations on a common theme in the neurobiology of language.

The common theme is codetermined by the intrinsic constraints that the brain brings to bear for any given cognitive task. In effect, the brain is a Kantian machine (Dehaene & Brannon, 2010). The consequence of this idea is that reading and speaking might have developed in a way that adapts their structural features optimally to intrinsic brain organization (Christiansen & Chater, 2008; Christiansen & Müller, chapter 58, this volume; Dehaene & Cohen, 2007). One novel and exciting idea is that speech might have adapted itself to the intrinsic rhythms of the brain. This sheds new light on language evolution and on how central structural characteristics of speech (phonemes, syllables, prosodic contours) might be shaped by the characteristics of intrinsic brain rhythms (delta, theta, gamma; see chapter 53 by Ghazanfar and Poeppel). The lesson here is that key features of human speech might have adapted themselves to the organization of the brain, rather than, or in addition to, the reverse scenario.

So far, what has been discussed is the machinery to encode and decode verbal utterances. However, in communicative settings, the coded meaning of these utterances is very often not the same as the intended message.

For instance, the statement "It is hot here" produced in the right context will often be interpreted as a request for action; for example, to open the window or turn on the air conditioner. In communication, the listener tries to infer the intention behind the utterance. What is understood is often unsaid but requires making inferences. This step from coded meaning to what is often called speaker meaning involves the theory of mind, or ToM, network (see chapter 57 by Hagoort and Levinson). To understand the use of language in its full extent, we need to understand what neurobiological systems, in addition to those representing a lexicon and syntax, are needed to determine the communicative value of language. This has become part of the research agenda in recent years. It shows clearly that any account that claims language comprehension can be fully explained by a mirror neuron system will fail.

Language has often been claimed to enable thinking at a level of abstraction unmatched by conceptual processes in other species. However, just as in the case of language, an adequate understanding of thinking requires decomposition in its core constituents. Our thinking, abstract as it may be, as well as the choice of our linguistic expressions are deeply connected to neural mechanisms of voluntary choice. The basic outlines of the human and monkey systems for making decisions begin to emerge (see chapter 59 by Glimcher).

Language and thinking are among the most complex functions the human mind commands. Its neurobiological infrastructure is still understood only very partially. It is beyond the scope of this section to do full justice to the progress that nevertheless has been made. In recent years new theoretical orientations have been developed and novel topics have appeared on the research agenda, all indicating that we are faced with a field on the move.

REFERENCES

Amunts, K., Lenzen, M., Friederici, A. D., Schleicher, A., Morosan, P., Palomero-Gallagher, N., & Zilles, K. (2010). Broca's region: Novel organizational principles and multiple receptor mapping. *PLoS Biology, 8*(9), e1000489.

Catani, M., Jones, D. K., & Ffytche, D. H. (2005). Perisylvian language networks of the human brain. *Ann Neurol, 57,* 8–16.

Catani, M., Mesulam, M. M., Jakobsen, E., Malik, F., Martersteck, A., Wieneke, C., ... Rogalski, E. (2013). A novel frontal pathway underlies verbal fluency in primary progressive aphasia. *Brain, 37,* 1724–1737.

Christiansen, M. H., & Chater, N. (2008). Language as shaped by the brain. *Behav Brain Sci, 3,* 489–558.

Dehaene, S., & Brannon, E. M. (2010). Space, time, and number: A Kantian research program. *Trends Cogn Sci, 14,* 517–519.

Dehaene, S., & Cohen, L. (2007). Cultural recycling of cortical maps. *Neuron, 56,* 384–398.

Hagoort, P. (2013). MUC (memory, unification, control) and beyond. *Front Psychol, 4*(416).

Levelt, W. J. M. (2013). *A history of psycholinguistics: The pre-Chomskyan era.* Oxford, UK: Oxford University Press.

Segaert, K., Menenti, L., Weber, K., Petersson, K. M., & Hagoort, P. (2012). Shared syntax in language production and comprehension—an fMRI study. *Cereb Cortex, 22,* 1662–1670.

52 Cytoarchitectonics, Receptorarchitectonics, and Network Topology of Language

KATRIN AMUNTS AND MARCO CATANI

ABSTRACT Anatomical models of language include a core left perisylvian network between Wernicke's region in the temporal lobe and Broca's region in the frontal lobe. In addition, a number of other cortical areas and subcortical nuclei contribute to language functions. This extended language network comprises the inferior parietal cortex (i.e., Geschwind's region), the anterior temporal lobe, the anterior insula, the presupplementary motor area (preSMA), and the lateral occipitotemporal cortex (i.e., visual word form area, VWFA). At the subcortical level, the thalamus, basal ganglia, and cerebellum are reciprocally connected to cortical language regions and modulate their activity. In this chapter, we describe the anatomy of the language cortical regions as defined by their neuronal composition and layering (i.e., cytoarchitectonics) and the receptor distribution for the main neurotransmitters of the central nervous system (i.e., receptorarchitectonics). The principal association, projection, and commissural language pathways described by recent diffusion tractography studies will also be reviewed. A particular focus of this chapter is to characterize the anatomical interindividual variability and its relevance to function in the normal brain and in language disorders.

Language development in humans occurred as a direct consequence of the functional specialization of cortical areas located in the temporal (i.e., Wernicke's region) and frontal lobes (i.e., Broca's region) of the left hemisphere. These areas are primarily dedicated to language, and for this reason are identified as core language regions. Other more bilaterally distributed cortical areas connect to these left-lateralized core regions and form an extended language network. In this chapter, we aim to introduce the reader to current approaches used to map the main language areas as defined by cytoarchitectonics, receptorarchitectonics, and their networks as defined by *in vivo* diffusion tractography. Although our approach is fundamentally anatomical, we refer to possible functional correlates and associated symptoms whenever possible.

Core language regions

Methods and theoretical approaches to anatomically based definitions of Broca's region have changed throughout the history of its study, resulting in the proposal of many different parcellation schemes (for a review, see Amunts & Zilles, 2012). On the surface, Broca's region corresponds to the pars opercularis and triangularis of the inferior frontal gyrus. Cytoarchitectonically, the pars opercularis and triangularis are largely occupied by areas 44 and 45, respectively. Their most distinct feature is the presence of very large pyramidal cells in deep layer III. These two areas differ from each other for the thickness of layer IV, well developed in area 45, but thinner and invaded by pyramidal cells from neighboring layers III and V in area 44. Layer IV is known to process, among others, input from the thalamus.

Probabilistic cytoarchitectonic maps of areas 44 and 45 have been generated using quantitative and statistical criteria for delineating their borders (figure 52.1A; Amunts et al., 1999). These maps reveal important anatomical features of Broca's region. Intersubject variability of the overall surface and extent of these two areas is evident in the healthy human brain (Amunts et al., 2004). In addition, a large portion of these two areas is buried in the depths of the sulci, which prevents any possibility of predicting their exact extent and location solely on the basis of the surface anatomy of the inferior frontal gyrus (figure 52.1A), which is in itself highly variable among people (Ono, Kubik, & Abernathey, 1990).

One advantage of probabilistic maps is that they offer a quantitative approach to cortical localization in patients with brain lesion and in functional activation studies (see www.jubrain.fz-juelich.de; Amunts et al., 2004; Fischl et al., 2007). For example, it has been shown that the maximum probability map for area 44 is more likely to overlap with activations generated by

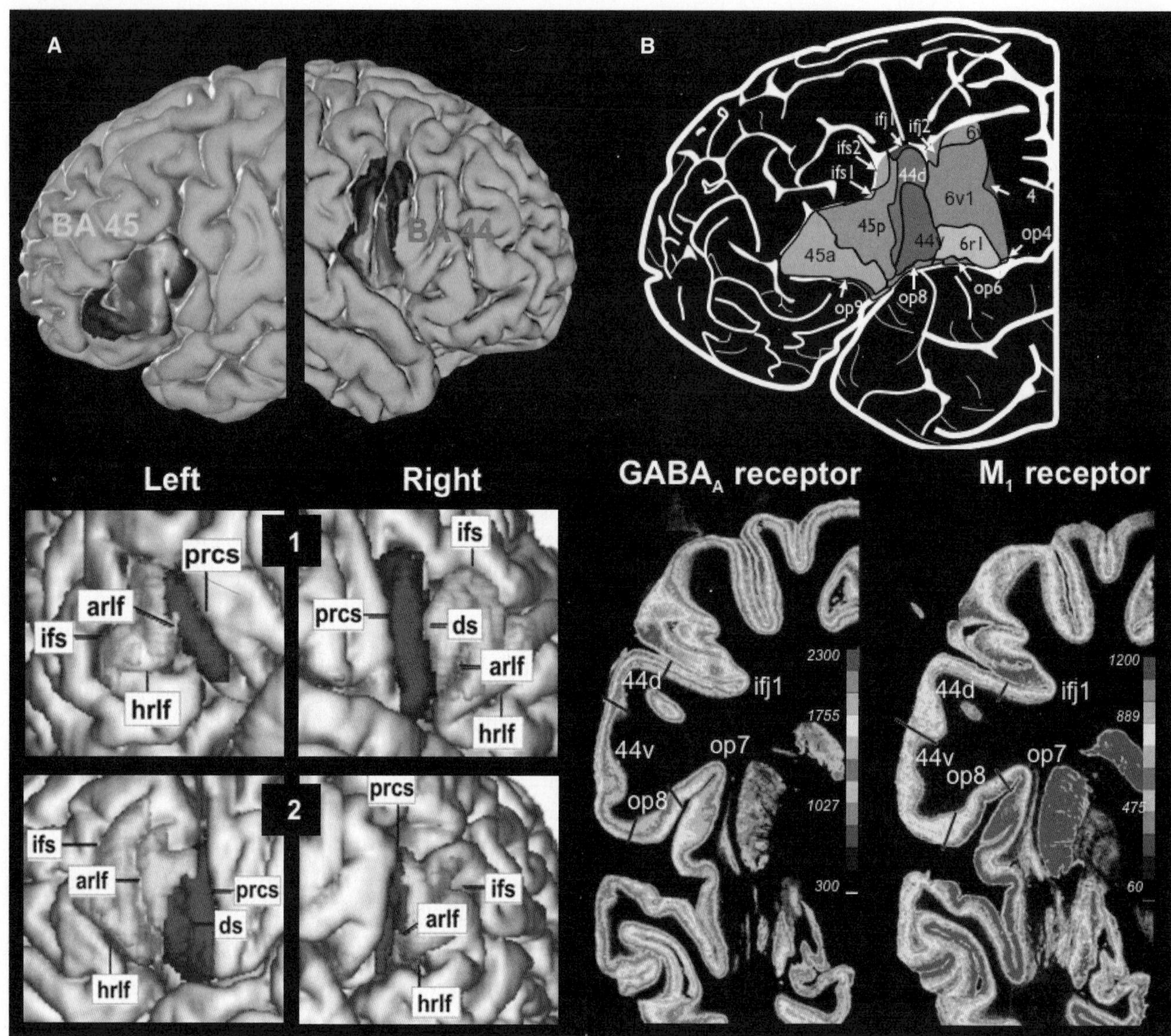

FIGURE 52.1 Cortical mapping of Broca's region based on post-mortem cytoarchitectonic and receptorarchitectonic analysis. (A) Top, surface view of the probabilistic maps for Brodmann's area 44 (BA44) and 45 (BA45) derived from quantitative cytoarchitectonic measurements of 10 individual brains. The probabilistic maps are intended to quantify the interindividual variability of the areas that compose Broca's region, where "hot" tones correspond to a high anatomical overlap, and "cold" tones to higher variability. This interindividual variability can be appreciated in the bottom panel, where the left and right maps for BA44 (red) and BA45 (yellow) are reported for two representative brains. Note, for example, the difference in the left-right asymmetry of BA 44 and 45. In addition, the borders of the two areas are not always well identified by the main sulci of this region: inferior frontal sulcus (ifs); horizontal ramus of the lateral fissure (hrlf); ascending ramus of the lateral fissure (arlf); precentral sulcus (prcs); diagonal sulcus (ds), (data from Amunts et al., 2004). (B) Further segregation of Broca's region based on neurotransmitter receptors distribution. This analysis can reveal unique characteristics of subregions that may share similar cytoarchitectonic morphology. Area 44v and op8, for example, have a similar density of GABA$_A$ or alternately GABA(A) receptors but different density of cholinergic muscarine receptors M1. (See color plate 44.)

tasks involving syntactic processing (Friederici & Kotz, 2003) and lexical decision (Heim, Eickhoff, Ischebeck, & Amunts, 2007), whereas area 45 is more involved in semantic tasks (Amunts et al., 2004). The cytoarchitectonic maps can also be used as seed regions to study structural and functional connectivity of language networks (Heim, Eickhoff, & Amunts, 2011). These are only a few applications of many. An important aspect to consider is that functional activations often do not correspond to the cytoarchitectonic subdivision of Broca's

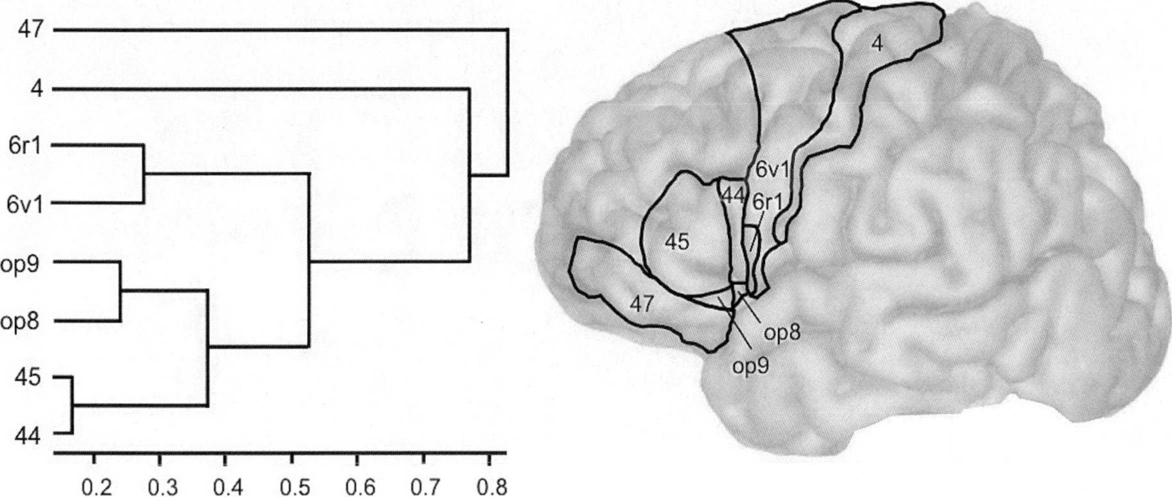

Figure 52.2 The cluster tree illustrates similarities between frontal areas based on their receptor distribution profile. (Modified from Amunts et al., 2010.)

region into two areas when language-related activations are superimposed on probabilistic maps. Indeed, often activations overlap only in part with these areas and extend to other neighboring areas (Indefrey et al., 2001). Several reasons may explain the lack of an exact correspondence between functional activations and cytoarchitectonic maps, including methodological shortcomings of both techniques, which may limit our ability to capture the complex segregation of Broca's region.

The application of novel receptorarchitectonic analysis to cortical areas (Zilles & Amunts, 2009) shows that Broca's region is indeed more complex than previously assumed. Amunts et al. (2010) analyzed the receptor distribution of different neurotransmitter systems (glutamate, GABA, serotonin, acetylcholine) in serial sections of the posterior frontal cortex. The receptor analysis showed that not only do areas 44 and 45 differ in their receptor distribution profile, but each area can be further subdivided into multiple areas (figure 52.1B). These subareas show a similar receptorarchitectonic pattern to areas 44 and 45, which arguably suggests their involvement in language-related processes. Interestingly, ventral area 6 has a pattern that is more similar to areas 44 and 45 than to area 47, which suggests a more distant structural and possibly functional relatedness between areas of the same gyrus (i.e., areas 44, 45, and 47 are parts of the inferior frontal gyrus) than between areas lying in different gyri (areas 44 and 45 in the inferior frontal gyrus and area 6 in the precentral gyrus; figure 52.2).

The anatomical correlates of Wernicke's region are more ambiguously defined than those of Broca's region. In its modern definition, Wernicke's region occupies the posterior aspect of the superior temporal gyrus and part of the middle temporal gyrus and is involved in multiple aspects of language processing (Binder et al., 2000; Hickok & Poeppel, 2007). Cytoarchitectonically, Wernicke's region includes areas 22 and 42 and part of area 37 (Aboitiz & Garcia, 1997), with further segmentations identified using probabilistic cytoarchitectonic and receptorarchitectonic mapping. The superior temporal gyrus, for example, is occupied by three distinct areas (figure 52.3). Area Te1, the primary auditory cortex, occupies a large portion of the Heschl gyrus (Morosan et al., 2001). Area Te2 is partially located on the most lateral aspect of the Heschl gyrus and on the opercular surface of the superior temporal gyrus, whereas area Te3 occupies approximately the posterior two-thirds of the free surface of the superior temporal gyrus (Morosan et al., 2005) and differs from its neighboring areas for the prominent size and high density of pyramidal neurons in layer IIIc and a high cellular density in layer V. With respect to receptor architecture, Te3 is characterized by a low density of muscarinic M2 receptors (Morosan et al., 2005). Te3 comprises further subdivisions (Rivier & Clarke, 1997; Wallace, Johnston, & Palmer, 2002) that have some correspondence with findings from functional activation studies (Friederici & Kotz, 2003).

Mesgarani and colleagues (2014) have recently shown that the superior temporal gyrus responds selectivity to distinct phonetic features of spoken language and contains the acoustic-phonetic representation of speech. In their comment to this paper, Grodzinsky and Nelken (2014) emphasized that speech representation in the experiment was dominated by abstract and linguistically defined features (Grodzinsky & Nelken, 2014).

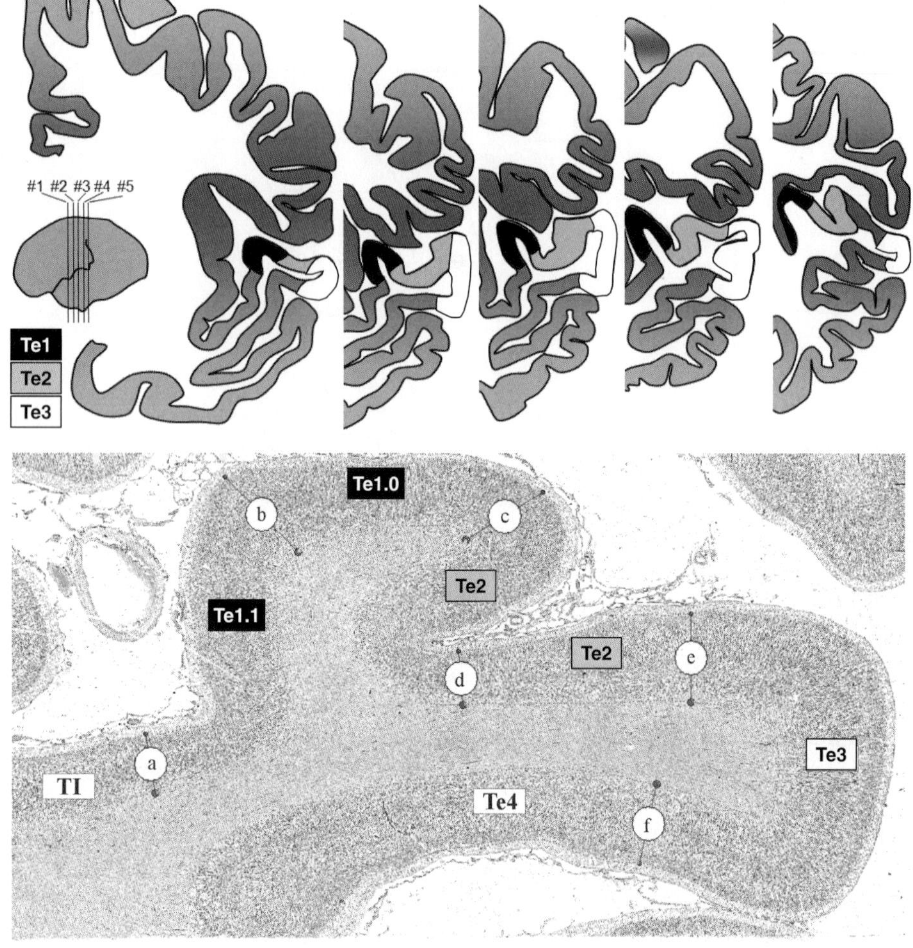

FIGURE 52.3 Cytoarchitectonic of the language areas in the superior temporal region (Morosan et al., 2005).

Other cortical regions

In addition to Broca's and Wernicke's regions, other cortical areas are relevant to speech and language. Geschwind's region in the inferior parietal lobule involves area 40 in the supramarginal gyrus and area 39 in the angular gyrus (Catani et al., 2005). A more recent parcellation of this region into seven cytoarchitectonic areas has been proposed, with five subdivisions covering area 40 and two subdivisions covering area 39 (Caspers et al., 2006). Of the five anterior subdivisions, the three most anterior ones show a receptor architecture similar to that of area 44 in the frontal lobe (Caspers et al., 2013). Geschwind's region is part of an extended network that links core language regions to areas involved in memory, semantic knowledge, and social cognition (Binder et al., 2009; Jacquemot & Scott, 2006; Meyer, Obleser, Anwander, & Friederici, 2012; Vilberg & Rugg, 2008).

In the mesial aspect of the superior frontal gyrus, area 6 and, in part, posterior areas 8 and 32 form an important language region that participates in planning, initiation, and monitoring of speech (Paulesu, Frith, & Frackowiak, 1993; Penfield & Roberts, 1959). This region, which corresponds to the anterior cingulate cortex and preSMA, has also been characterized using modern cytoarchitectonic or receptorarchitectonic mapping in the human and macaque brains (Geyer et al., 1998; Palomero-Gallagher, Mohlberg, Zilles, & Vogt, 2008). The preSMA is involved in action monitoring, distinguishing self from others' actions, and low-level aspects of mentalizing (Lombardo et al., 2010; Yoshida et al., 2011). The anterior insula is highly connected to the frontal operculum and Broca's region (Catani et al., 2012; Cerliani et al., 2012) and is often activated in language-production tasks (Dronkers, 1996). In the ventral occipitotemporal region, the visual word form area is specialized for processing written strings (Cohen et al., 2000).

Finally, the anterior temporal lobe, which includes area 38 and anterior portions of areas 20, 21, 22, and 36, has been recognized as an important language hub involved in semantic processing (Catani et al., 2013; Mesulam et al., 2013).

Subcortical structures

Cortical language areas form complex networks with the thalamus, basal ganglia, and cerebellum (Cappa & Vallar, 1992; Schmahmann & Pandya, 2008). Damage to these subcortical structures can manifest with either isolated aphasia or language deficits that are part of more complex neurological syndromes. Anatomical and clinical studies suggest that these cortico-subcortical networks are segregated (figure 52.4). Broca's region and preSMA, for example, receive their main afferents from the anterior thalamus (i.e., ventral anterior nucleus). Stuttering and reduced spontaneous speech have been described in association with damage of the ventral anterior nucleus or its connections to language frontal areas (Watkins, Smith, Davis, & Howell, 2008). Wernicke's and Geschwind's regions receive projections primarily from the pulvinar and the lateral posterior nuclei, respectively. Reduced comprehension is more commonly associated with lesions to these posterior thalamic nuclei (Cappa & Vallar, 1992).

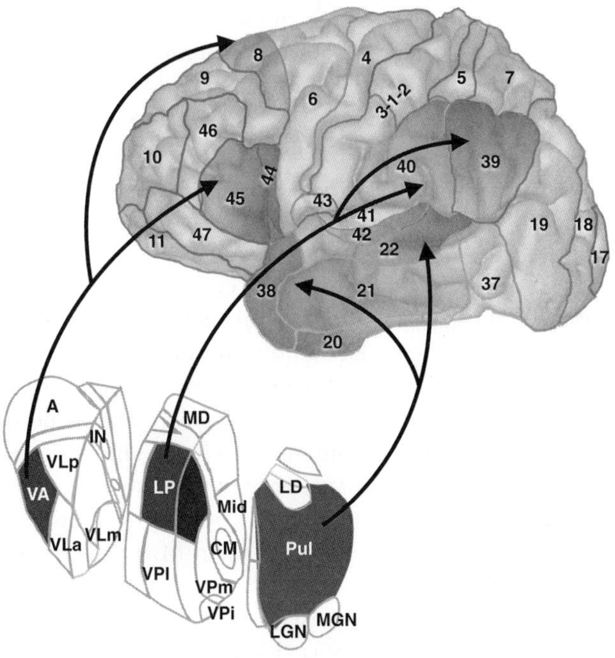

FIGURE 52.4 Thalamocortical connections of the language areas. A: anterior group; LGN: lateral geniculate nucleus; LP: lateroposterior; MD: mediodorsal; MGN: medial geniculate nucleus; Mid: midline group; Pul: pulvinar; VA: ventral anterior; VLa: ventral lateral anterior; VPi: ventral posterior inferior; VPl: ventral posterior lateral; VPm: ventral posterior medial; VLp: ventral lateral posterior. (Modified from Catani & Thiebaut de Schotten, 2012.)

Pure lesions of the thalamus do not conform with classical neurological taxonomy of aphasia syndromes, and a precise clinical-anatomical correlation is often difficult to perform in these patients since lesions are often large and involve multiple thalamic nuclei and surrounding white matter (Cappa & Vallar, 1992). In general, repetition deficits are rare and comprehension deficits of mild intensity. Severe writing difficulties and semantic paraphasias are relatively more common.

The striatum composed of the caudate and putamen is another important relay station of the cortico-subcortical loops. Stimulation of the striatum is associated with involuntary language production, while lesions more frequently cause nonfluent speech (Crosson, 1992). In bilinguals the left caudate has an important role in language switch (Crinion et al., 2006).

The cerebellum receives cortical afferents from language areas through the pontine nuclei and projects to the same areas via the thalamus. In patients with cerebellar lesions, verbal fluency can be impaired to the point of telegraphic speech or mutism, but true aphasic disorders are rare. Anomia, agrammatic speech, and abnormal prosody (i.e., high-pitched, hypophonic whining, etc.) have been also reported in these patients (Schmahmann & Pandya, 2008).

White matter pathways

Language areas are reciprocally connected through a complex system of association and commissural tracts whose anatomy has been studied in humans with postmortem blunt dissections and more extensively with diffusion imaging tractography (figure 52.5A).

The arcuate fasciculus is a dorsal perisylvian association tract connecting Wernicke's, Geschwind's, and Broca's regions. Within the arcuate fasciculus, two parallel pathways have been distinguished: the medial, direct pathway connecting Wernicke's with Broca's region (i.e., the arcuate fasciculus *sensu strictu* or long segment); and the indirect pathway, consisting of an anterior segment that links Broca's to Geschwind's region and a posterior segment between Geschwind's and Wernicke's region (Catani et al., 2005). López-Barroso et al. (2013) showed that performance in word learning correlates with microstructural properties and strength of functional connectivity of the direct connections between Broca's and Wernicke's regions. This study demonstrates that our ability to learn new words relies on an efficient and fast communication between

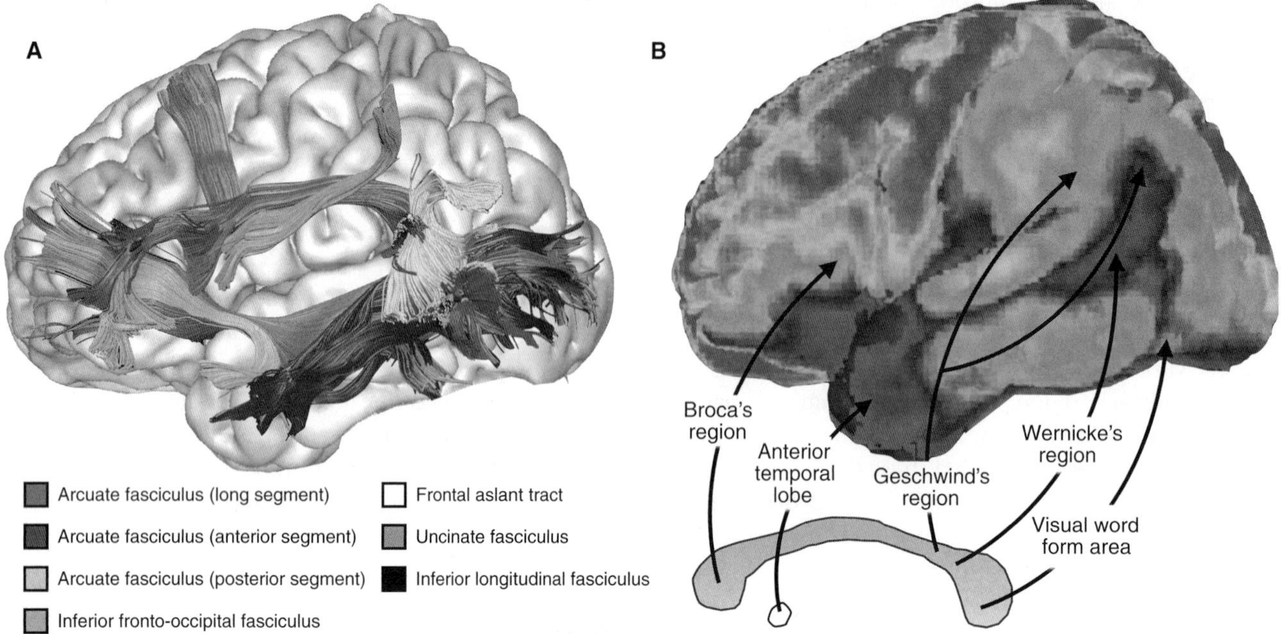

FIGURE 52.5 Language networks visualized with diffusion tractography. (A) Association pathways of left hemisphere connecting the main language regions. (B) Density of callosal cortical projections based on diffusion imaging. "Hot" tones indicate higher degree of transcallosal connections. Note the reduced interhemispheric connectivity for most of the language regions. (See color plate 45.)

auditory temporal and motor frontal regions. Schulze, Vargha-Khadem, and Mishkin (2012) suggested that the absence of these connections in nonhuman primates might explain our unique ability to learn new words. The long segment is also important for syntactic processing (Tyler et al., 2011; Wilson et al., 2011) and words repetition (Parker Jones et al., 2014). Within the indirect pathway, the anterior and posterior segments have different roles in reading (Thiebaut de Schotten et al., 2014), phonological and semantic processing (Binder et al., 2009; Newhart et al., 2012), verbal working memory (Jacquemot & Scott, 2006), and pragmatic interpretation (Hagoort, 2013).

Language areas of the temporal and frontal lobes are also interconnected by a set of ventral longitudinal tracts (Catani & Mesulam, 2008). The inferior longitudinal fasciculus carries visual information from occipital and posterior temporal areas to the anterior temporal lobe (Catani, Jones, Donato, & Ffytche, 2003) and plays an important role in visual object recognition, reading, and linking object representations to their lexical labels. The uncinate fasciculus connects the anterior temporal lobe to the orbitofrontal region and part of the inferior frontal gyrus and may play an important role in lexical retrieval, semantic associations, and naming (Catani et al., 2013). The uncinate fasciculus is severely damaged in those patients with the semantic variant of primary progressive aphasia (Catani et al., 2013). The inferior fronto-occipital fasciculus provides direct connections between occipital (and perhaps posterior temporal cortex) and frontal cortex in the human brain (Forkel et al., 2012). The relevance of this fasciculus to reading, writing, and other aspects of language remains to be established (Duffau et al., 2005; Forkel et al., 2012).

The frontal aslant tract is a newly described pathway connecting Broca's region with medial frontal areas including preSMA and cingulate cortex (Catani et al., 2012; Lawes et al., 2008). Medial regions of the frontal lobe facilitate speech initiation through direct connection to the pars opercularis and triangularis of the inferior frontal gyrus. Patients with lesions to these areas present with various degrees of speech impairment from a total inability to initiate speech (i.e., mutism) to mild altered fluency (Catani et al., 2012; Naeser et al., 1989). The frontal aslant tract is damaged in patients with the nonfluent/agrammatic form of primary progressive aphasia (Catani et al., 2013).

The frontal operculum is connected to the insula through a system of short U-shaped fronto-insular tracts (Catani et al., 2012; Cerliani et al., 2012). Direct insular inputs to Broca's region from the insula provide visceral and emotional information for speech-output modulation according to internal states. Lesions to these insular connections may result in motor aprosodia (e.g., flat intonation) and apraxia of speech (Dronkers, 1996).

In addition to the above tracts, data from axonal tracing studies in the monkey have suggested the existence of additional pathways, including the middle longitudinal fasciculus (Seltzer & Pandya, 1984) and the extreme capsule tract (Schmahmann & Pandya, 2006). The exact role of these tracts in humans remains to be demonstrated.

Interhemispheric asymmetry and gender differences

Lateralization of language functions and underlying structural asymmetry are characteristic features of human brain organization. Several hypotheses have been proposed in the past to explain language lateralization. One of these proposals suggests that evolutionary pressures guiding brain size and lateralization of specialized language areas are accompanied by reduced interhemispheric connectivity and greater intrahemispheric connectivity (Aboitiz, Scheibel, Fisher, & Zaidel, 1992). Thus, by being connected through less transcallosal fibers or fibers of smaller and slower-conducting diameter, specialized areas reduce "unnecessary crosstalk" and maintain separate parallel processing systems through association tracts (i.e., arcuate fasciculus; Doron & Gazzaniga, 2008). Indeed, diffusion tractography in humans reveals that language areas show reduced transcallosal connections compared to nonlanguage regions (figure 52.5B).

Cytoarchitectonic asymmetry, although already present in one-year-old infants, becomes progressively more evident throughout childhood (Amunts, Schleicher, Ditterich, & Zilles, 2003). In adults, leftwards asymmetry has been reported for Broca's region using cytoarchitectonic analysis (Brodmann area 44 in the pars opercularis) (Amunts et al., 1999; Uylings et al., 2006). Significant leftward asymmetry in the volume of the pars opercularis has also been reported based on in vivo MRI-based measurements (Keller et al., 2007). Gray matter concentration differences in the posterior part of the inferior frontal gyrus (pars opercularis) have been found to correlate with language dominance assessed by the sodium amytal procedure (Dorsaint-Pierre et al., 2006). Asymmetry has also been observed for receptor architecture. A larger concentration of cholinergic M2 receptors has been found in left areas 44v and 44d compared to the right counterparts. Other receptors show no interhemispheric differences (Amunts et al., 2010). Structural asymmetry at the microstructural level has been reported for other regions of the temporal lobe, including areas 42 and 22 (Buxhoeveden, Switala, Litaker, Roy, & Casanova, 2001; Galaburda, Sanides, & Geschwind, 1978; Hutsler & Galuske, 2003; Jacobs, Schall, & Scheibel, 1993).

Left-right asymmetries have been investigated with respect to the underlying white matter (Catani et al., 2007; Catani et al., 2012; López-Barroso et al., 2013; Nijhuis, van Cappellen van Walsum, & Norris, 2013). An extreme degree of leftward lateralization of the long segment of the arcuate fasciculus has been reported in approximately 60% of the normal population (figure 52.6A; Catani et al., 2007). The remaining 40% of the population show either a mild leftward lateralization (20%) or a bilateral, symmetrical pattern (20%). Overall, females are more likely to have a bilateral pattern compared to males (figure 52.6B). The degree of lateralization of the long segment has important clinical implications. Forkel et al. (2014) showed that patients with aphasia are more likely to recover normal language at six months if at the time of the stroke they have a larger volume of the long segment in the right unaffected hemisphere (figure 52.6C). The frontal aslant tract, although left-lateralized in most people, shows a more bilateral pattern compared to the arcuate fasciculus, which could explain the prompt recovery of speech functions in those patients with unilateral damage of this tract (Catani et al., 2012).

Conclusion

In this chapter, we suggest an anatomical language model encompassing a core and an extended language network. The core network includes the arcuate fasciculus connecting Broca's and Wernicke's regions. This network is left-lateralized in most human brains. In addition, an extended network provides access to social, emotional, and attentional inputs necessary for successful language performances. One unique feature of language as cognitive process is the extraordinary ability to access, process, and bind together information derived from memory and sensorial perception to create and convey messages. Perhaps this ability derives from the development of a unique pattern of connections that typifies human brains. A precise correspondence between contemporary neurocognitive models of language (e.g., Hagoort, 2013; Jeon & Friederici, 2013; Hickok & Poeppel, 2007), functional activation, and anatomically defined characteristics of cortical areas and connecting pathways remains a challenging task in contemporary brain mapping.

ACKNOWLEDGMENTS We would like to thank Valentina Bambini, Stephanie Forkel, and other members of the Neuroanatomy and Tractography Laboratory (www.natbrainlab.com), the Institute of Neuroscience and Medicine of the Research Centre Juelich, and C. and O. Vogt Institute for Brain Research of the Heinrich Heine University Duesseldorf for their helpful

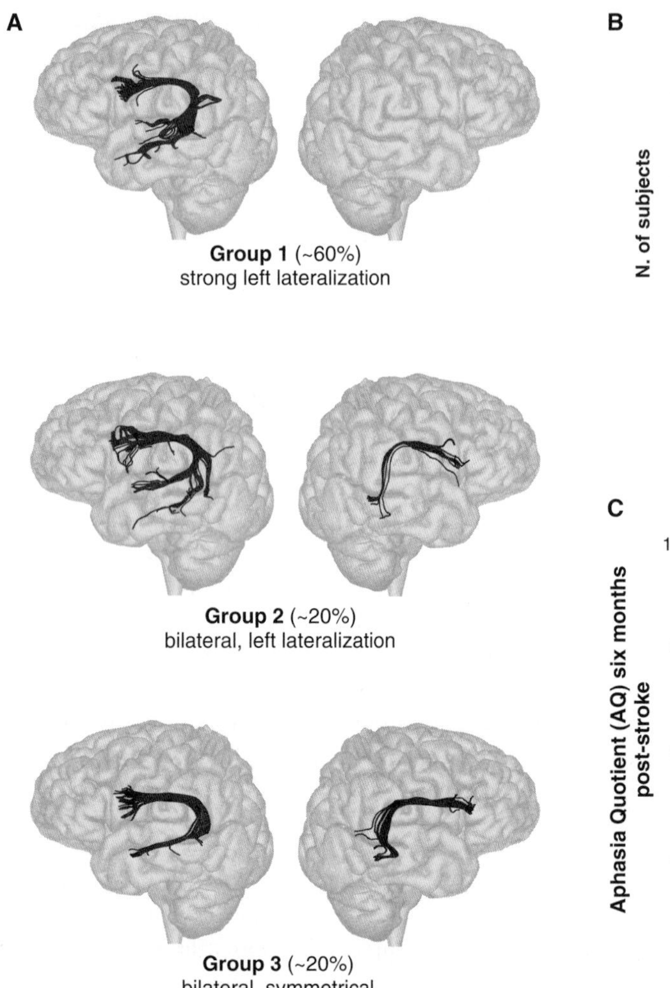

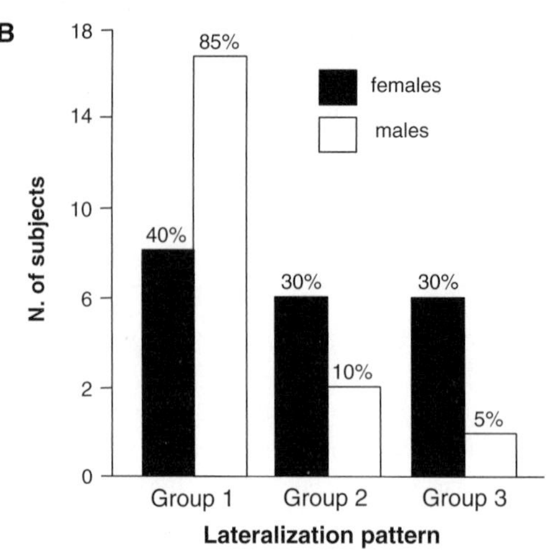

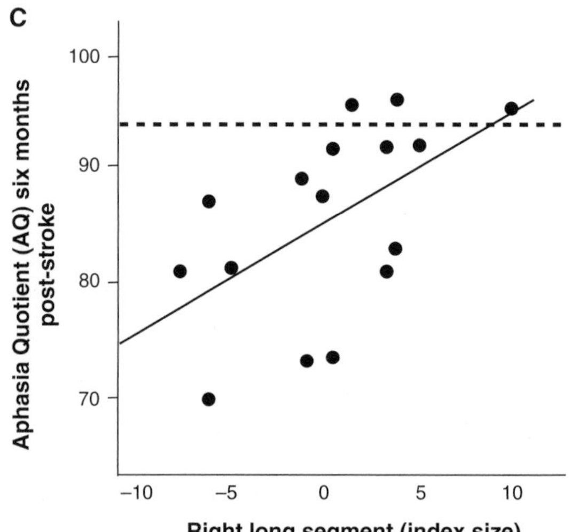

FIGURE 52.6 Distribution of the pattern of lateralization of the long segment in (A) three groups and (B) differences in lateralization between genders (modified from Catani et al., 2007). (C) Recovery of language six months after stroke is correlated with the volume of the right long segment (Forkel et al., 2014).

comments on the manuscript. Michel Thiebaut de Schotten kindly provided the data set for figure 52.5.

REFERENCES

ABOITIZ, F., & GARCIA, G. L. (1997). The evolutionary origin of language areas in the human brain. A neuroanatomical perspective. *Brain Res Rev, 25,* 381–396.

ABOITIZ, F., SCHEIBEL, A. B., FISHER, R. S., & ZAIDEL, E. (1992). Fiber composition of the human corpus callosum. *Brain Res, 598,* 143–153.

AMUNTS, K., LENZEN, M., FRIEDERICI, A. D., SCHLEICHER, A., MOROSAN, P., PALOMERO-GALLAGHER, N., & ZILLES, K. (2010). Broca's region: Novel organizational principles and multiple receptor mapping. *PLoS Biol, 8,* e1000489.

AMUNTS, K., SCHLEICHER, A., BÜRGEL, U., MOHLBERG, H., UYLINGS, H. B. M., & ZILLES, K. (1999). Broca's region revisited: Cytoarchitecture and intersubject variability. *J Comp Neurol, 412,* 319–341.

AMUNTS, K., SCHLEICHER, A., DITTERICH, A., & ZILLES, K. (2003). Broca's region: Cytoarchitectonic asymmetry and developmental changes. *J Comp Neurol, 465,* 72–89.

AMUNTS, K., WEISS, P. H., MOHLBERG, H., PIEPERHOFF, P., EICKHOFF, S., GURD, J., SHAH, J. N., MARSHALL, J. C., FINK, G. R., & ZILLES, K. (2004). Analysis of the neural mechanisms underlying verbal fluency in cytoarchitectonically defined stereotaxic space—the role of Brodmann's areas 44 and 45. *NeuroImage, 22,* 42–56.

AMUNTS, K., & ZILLES, K. (2012). Architecture and organizational principles of Broca's region. *Trends Cogn Sci, 16,* 418–426.

BINDER, J. R., DESAI, R. H., GRAVES, W. W., & CONANT, L. L. (2009). Where is the semantic system? A critical review and meta-analysis of 120 functional neuroimaging studies. *Cereb Cortex, 19,* 2767–2796.

BINDER, J. R., FROST, J. A., HAMMEKE, T. A., BELLGOWAN, P. S. F., SPRINGER, J. A., KAUFMAN, J. N., & POSSING, E. T. (2000). Human temporal lobe activation by speech and nonspeech sounds. *Cereb Cortex, 10,* 512–528.

Buxhoeveden, D. P., Switala, A. E., Litaker, M., Roy, E., & Casanova, M. F. (2001). Lateralization of minicolumns in human planum temporale is absent in nonhuman primate cortex. *Brain Behav Evolut, 57*(6), 349–358.

Cappa, S. F., & Vallar, G. (1992). Neuropsychological disorders after subcortical lesions: Implications for neural models of language and spatial attention. In G. Valla, S. F. Cappa, & Claus-W. Wallesch (Eds.), *Neuropsychological disorders associated with subcortical lesions* (pp. 7–41). New York, NY: Oxford University Press.

Caspers, S., Geyer, S., Schleicher, A., Mohlberg, H., Amunts, K., & Zilles, K. (2006). The human inferior parietal cortex: Cytoarchitectonic parcellation and interindividual variability. *NeuroImage, 33*, 430–448.

Caspers, S., Schleicher, A., Bacha-Trams, M., Palomero-Gallagher, N., Amunts, K., & Zilles, K. (2013). Organization of the human inferior parietal lobule based on receptor architectonics. *Cereb Cortex, 23*, 615–628.

Catani, M., Allin, M. P., Husain, M., Pugliese, L., Mesulam, M. M., Murray, R. M., & Jones, D. K. (2007). Symmetries in human brain language pathways correlate with verbal recall. *Proc Natl Acad Sci USA, 104*, 17163–17168.

Catani, M., Dell'acqua, F., Vergani, F., Malik, F., Hodge, H., Roy, P., … Thiebaut de Schotten, M. (2012). Short frontal lobe connections of the human brain. *Cortex, 48*, 273–291.

Catani, M., Jones, D. K., Donato, R., & Ffytche, D. H. (2003). Occipito-temporal connections in the human brain. *Brain, 126*, 2093–2107.

Catani, M., Jones, D. K., & Ffytche, D. H. (2005). Perisylvian language networks of the human brain. *Ann Neurol, 57*, 8–16.

Catani, M., & Mesulam, M. (2008). The arcuate fasciculus and the disconnection theme in language and aphasia: History and current state. *Cortex, 44*, 953–961.

Catani, M., Mesulam, M. M., Jakobsen, E., Malik, F., Martersteck, A., Wieneke, C., … Rogalski, E. (2013). A novel frontal pathway underlies verbal fluency in primary progressive aphasia. *Brain, 37*, 1724–1737.

Catani, M., & Thiebaut de Schotten, M. (2012). *Atlas of human brain connections.* Oxford, UK: Oxford University Press.

Cerliani, L., Thomas, R. M., Jbabdi, S., Siero, J. C., Nanetti, L., Crippa, A., … & Keysers, C. (2012). Probabilistic tractography recovers a rostrocaudal trajectory of connectivity variability in the human insular cortex. *Hum Brain Mapp, 33*, 2005–2034.

Cohen, L., Dehaene, S., Naccache, L., Lehéricy, S., Dehaene-Lambertz, G., Hénaff, M. A., & Michel, F. (2000). The visual word form area: Spatial and temporal characterization of an initial stage of reading in normal subjects and posterior split-brain patients. *Brain, 123*, 291–307.

Crinion, J., Turner, R., Grogan, A., Hanakawa, T., Noppeney, U., Devlin, J. T., … Price, C. J. (2006). Language control in the bilingual brain. *Science, 9*, 1537–1540.

Crosson, B. (1992). Subcortical functions in language: A working model. *Brain Lang, 25*, 257–292.

Doron, K. W., & Gazzaniga, M. S. (2008). Neuroimaging techniques offer new perspectives on callosal transfer and interhemispheric communication. *Cortex, 44*, 1023–1029.

Dorsaint-Pierre, R., Penhune, V. B., Watkins, K. E., Neelin, P., Lerch, J. P., Bouffard, M., & Zatorre, R. J. (2006). Asymmetries of the planum temporale and Heschl's gyrus: Relationship to language lateralization. *Brain, 129* (Pt. 5), 1164–1176.

Dronkers, N. F. (1996). A new brain region for coordinating speech articulation. *Nature, 384*, 159–161.

Duffau, H., Gatignol, P., Mandonnet, E., Peruzzi, P., Tzourio-Mazoyer, N., & Capelle, L. (2005). New insights into the anatomo-functional connectivity of the semantic system: A study using cortico-subcortical electrostimulations. *Brain, 128*, 797–810.

Fischl, B., Rajendran, N., Busa, E., Augustinack, J., Hinds, O., Yeo, B. T. T., … Zilles, K. (2007). Cortical folding patterns and predicting cytoarchitecture. *Cereb Cortex, 18*, 1973–1980.

Forkel, S. J., Thiebaut de Schotten, M., Dell'Acqua, F., Kalra, L., Murphy, D., Williams, S., & Catani, M. (2014). Right arcuate fasciculus volume predicts recovery from left stroke aphasia. *Brain* (in press).

Forkel, S. J., Thiebaut de Schotten, M., Kawadler, J. M., Dell'acqua, F., Danek, A., & Catani, M. (2012). The anatomy of fronto-occipital connections from early blunt dissections to contemporary tractography. *Cortex.* doi:10.1016/j.cortex.2012.09.005

Friederici, A. D., & Kotz, S. A. (2003). The brain basis of syntactic processes: Functional imaging and lesions studies. *NeuroImage, 20*(Suppl. 1), S8–S17.

Galaburda, A. M., Sanides, F., & Geschwind, N. (1978). Human brain. Cytoarchitectonic left-right asymmetries in the temporal speech region. *Arch Neurol, 35*, 812–817.

Geyer, S., Matelli, M., Luppino, G., Schleicher, A., Jansen, Y., Palomero-Gallagher, N., & Zilles, K. (1998). Receptor autoradiographic mapping of the mesial motor and premotor cortex of the macaque monkey. *J Comp Neurol, 397*, 231–250.

Grodzinsky, Y., & Nelken, I. (2014). Neuroscience: The neural code that makes us human. *Science, 343*, 978–979.

Hagoort, P. (2013). MUC (memory, unification, control) and beyond. *Front Psychol, 416*, 1–13.

Heim, S., Eickhoff, S. B., & Amunts, K. (2011). Different roles of cytoarchitectonic BA 44 and BA 45 in phonological and semantic verbal fluency as revealed by dynamic causal modelling. *NeuroImage, 48*, 616–624.

Heim, S., Eickhoff, S. B., Ischebeck, A. K., & Amunts, K. (2007). Modality-independent involvement of the left BA 44 during lexical decision making. *Brain Struct Funct, 212*, 95–106.

Hickok, G., & Poeppel, D. (2007). The cortical organization of speech processing. *Nat Rev Neurosci, 8*, 393–402.

Hutsler, J., & Galuske, R. A. W. (2003). Hemispheric asymmetries in cerebral cortical networks. *Trends Neurosci, 26*(8), 429–435.

Indefrey, P., Brown, C. M., Hellwig, F., Amunts, K., Herzog, H., Seitz, R. J., & Hagoort, P. (2001). A neural correlate of syntactic encoding during speech production. *Proc Natl Acad Sci USA, 98*, 5933–5936.

Jacobs, B., Schall, M., & Scheibel, A. B. (1993). A quantitative dendritic analysis of Wernicke's area in humans. II. Gender, hemispheric, and environmental factors. *J Comp Neurol, 327*, 97–111.

Jacquemot, C., & Scott, S. K. (2006). What is the relationship between phonological short-term memory and speech processing? *Trends Cogn Sci, 10,* 480–486.

Jeon, H. A., & Friederici, A. D. (2013). Two principles of organization in the prefrontal cortex are cognitive hierarchy and degree of automaticity. *Nat Commun, 4,* 2041.

Keller, S. S., Highley, J. R., Garcia Finana, M., Sluming, V., Rezaie, R., & Roberts, N. (2007). Sulcal variability, stereological measurement and asymmetry of Broca's area on MR images. *J Anat, 211,* 534–555.

Lawes, I. N., Barrick, T. R., Murugam, V., Spierings, N., Evans, D. R., Song, M., & Clark, C. A. (2008). Atlas-based segmentation of white matter tracts of the human brain using diffusion tensor tractography and comparison with classical dissection. *NeuroImage, 39,* 62–79.

Lombardo, M. V., Chakrabarti, B., Bullmore, E. T., Wheelright, S. J., Sadek, S. A., Suckling, J., & Baron-Cohen, S. (2010). Shared neural circuits for mentalizing about the self and others. *J Cogn Neurosci, 22,* 1623–1635.

López-Barroso, D., Catani, M., Ripollésa, P., Dell'Acqua, F., Rodríguez-Fornells, A., & de Diego-Balaguera, R. (2013). Word learning is mediated by the left arcuate fasciculus. *Proc Natl Acad Sci USA, 110,* 13168–13173.

Mesgarani, N., Cheung, G., Johnson, K., & Chang, E. F. (2014). Phonetic feature encoding in human superior temporal gyrus. *Science, 343,* 1006–1010.

Mesulam, M. M., Wieneke, C., Hurley, R., Rademaker, A., Thompson, C. K., Weintraub, S., & Rogalski, E. J. (2013). Words and objects at the tip of the left temporal lobe in primary progressive aphasia. *Brain, 136,* 601–618.

Meyer, L., Obleser, J., Anwander, A., & Friederici, A. D. (2012). Linking ordering in Broca's area to storage in left temporo-parietal regions: The case of sentence processing. *NeuroImage, 62,* 1987–1998.

Morosan, P., Rademacher, J., Schleicher, A., Amunts, K., Schormann, T., & Zilles, K. (2001). Human primary auditory cortex: Cytoarchitectonic subdivision and mapping into a spatial reference system. *NeuroImage, 13,* 684–701.

Morosan, P., Schleicher, A., Amunts, K., & Zilles, K. (2005). Multimodal architectonic mapping of human superior temporal gyrus. *Anat Embryol, 210,* 401–406.

Naeser, M. A., Palumbo, C. L., Helm-Estabrooks, N., Stiassny-Eder, D., & Albert, M. (1989). Severe nonfluency in aphasia. Role of the medial subcallosal fasciculus and other white matter pathways in recovery of spontaneous speech. *Brain, 112,* 1–38.

Newhart, M., Trupe, L. A., Gomez, Y., Cloutman, J., Molitoris, J. J., Davis, C., ... Hillis, A. E. (2012). Asyntactic comprehension, working memory, and acute ischemia in Broca's area versus angular gyrus. *Cortex, 48,* 1288–1297.

Nijhuis, E. H., van Cappellen van Walsum, A. M., & Norris, D. G. (2013). Topographic hub maps of the human structural neocortical network. *PLoS One, 8,* e65511.

Ono, M., Kubik, S., & Abernathey, C. D. (1990). *Atlas of the cerebral sulci.* Stuttgart: Thieme.

Palomero-Gallagher, N., Mohlberg, H., Zilles, K., & Vogt, B. J. (2008). Cytology and receptor architecture of human anterior cingulate cortex. *J Comp Neurol, 508,* 906–926.

Parker Jones, O., Prejawa, S., Hope, T. M., Oberhuber, M., Seghier, M. L., Leff, A. P., Green, D. W., & Price, C. J. (2014). Sensory-to-motor integration during auditory repetition: A combined fMRI and lesion study. *Front Hum Neurosci, 8,* 24.

Paulesu, E., Frith, C. D., & Frackowiak, R. S. J. (1993). The neural correlates of the verbal component of working memory. *Nature, 362,* 342–345.

Penfield, W., & Roberts, L. (1959). *Speech and brain mechanisms.* Princeton, NJ: Princeton University Press.

Rivier, F., & Clarke, S. (1997). Cytochrome oxidase, acetylcholinesterase, and NADPH-diaphorase staining in human supratemporal and insular cortex: Evidence for multiple auditory areas. *NeuroImage, 6,* 288–304.

Schmahmann, J., & Pandya, D. N. (2006). *Fiber pathways of the brain.* New York, NY: Oxford University Press.

Schmahmann, J. D., & Pandya, D. N. (2008). Disconnection syndromes of basal ganglia, thalamus, and cerebrocerebellar systems. *Cortex, 44,* 1037–1066.

Schulze, K., Vargha-Khadem, F., & Mishkin, M. (2012). Test of a motor theory of long-term auditory memory. *Proc Natl Acad Sci USA, 109,* 7121–7125.

Seltzer, B., & Pandya, D. N. (1984). Further observations on parieto-temporal connections in the rhesus monkey. *Exp Brain Res, 55,* 301–312.

Thiebaut de Schotten, M., Cohen, L., Amemiya, E., Braga, L. W., & Dehaene, S. (2014). Learning to read improves the structure of the arcuate fasciculus. *Cereb Cortex, 24,* 989–995.

Tyler, L. K., Marslen-Wilson, W. D., Randall, B., Wright, P., Devereux, B. J., Zhuang, J., Papoutsi, M., & Stamatakis, E. A. (2011). Left inferior frontal cortex and syntax: Function, structure and behaviour in patients with left hemisphere damage. *Brain, 134,* 415–431.

Uylings, H. B. M., Jacobsen, A. M., Zilles, K., & Amunts, K. (2006). Left-right asymmetry in volume and number of neurons in adult Broca's area. *Cortex, 42,* 652–658.

Vilberg, K. L., & Rugg, M. D. (2008). Memory retrieval and the parietal cortex: A review of evidence from a dual-process perspective. *Neuropsychologia, 46,* 1787–1799.

Wallace, M. N., Johnston, P. W., & Palmer, A. R. (2002). Histochemical identification of cortical areas in the auditory region of the human brain. *Exp Brain Res, 143,* 499–508.

Watkins, K. E., Smith, S. M., Davis, S., & Howell, P. (2008). Structural and functional abnormalities of the motor system in developmental stuttering. *Brain, 131,* 50–59.

Wilson, S. M., Galantucci, S., Tartaglia, M. C., Rising, K., Patterson, D. K., Henry, M. L., ... Gorno-Tempini, M. L. (2011). Syntactic processing depends on dorsal language tracts. *Neuron, 72,* 397–403.

Yoshida, K., Saito, N., Iriki, A., & Isoda, M. (2011). Representation of others' action by neurons in monkey medial frontal cortex. *Curr Biol, 21,* 249–253.

Zilles, K., & Amunts, K. (2009). Receptor mapping: Architecture of the human cerebral cortex. *Curr Opin Neurol, 22,* 331–339.

53 The Neurophysiology and Evolution of the Speech Rhythm

ASIF A. GHAZANFAR AND DAVID POEPPEL

ABSTRACT Speech research has typically focused on processing single speech events, that is, vowels, syllables, or words. A different approach focuses on connected speech. This work points to new mechanisms for understanding human speech perception and production in the context of neurophysiology as well as in the context of comparative and evolutionary studies. The concept at the center of this research program concerns rhythms—in the acoustic signal, the neural activity, and the motoric output systems. There is a remarkable correspondence between average durations of speech or speech-like expressions and the frequency ranges of cortical oscillations. Here we summarize a multitime-resolution hypothesis about the perception of connected speech, capitalizing on the role that neuronal oscillations play in processing speech. We then examine the notion of the speech rhythm from a comparative perspective, demonstrating that lip-smacking behavior in macaques shares crucial features with speech. The human and nonhuman primate data jointly make a strong case for the coordination of neural and vocal rhythms in communication.

Across languages, speech typically exhibits a 3–8 Hz rhythm (Chandrasekaran, Trubanova, Stillittano, Caplier, & Ghazanfar, 2009; Crystal & House, 1982; Elliot & Theunissen, 2009; Greenberg, Carvey, Hitchcock, & Chang, 2003; Malecot, Johonson, & Kizziar, 1972), reflected in the envelope of the speech waveform. This slow energy modulation corresponds roughly to the syllabic sequence (or syllabic "chunking") of speech. Syllabic structure as reflected by the speech envelope is perceptually critical because it signals speaking rate, carries stress and tonal contrasts, and can be viewed as the carrier of the linguistic (question, statement, etc.) or affective (happy, sad, etc.) prosody of an utterance (Rosen, 1992). Consequently, high sensitivity to envelope structure and dynamics is critical for successful perception. Disrupting this rhythm significantly reduces intelligibility (Drullman, Festen, & Plomp, 1994; Elliot & Theunissen, 2009; Ghitza & Greenberg, 2009; Saberi & Perrott, 1999; Shannon, Zeng, Kamath, Wygonski, & Ekelid, 1995; Smith, Delgutte, & Oxenham, 2002), as does disrupting the visual component arising from mouth and facial movements (Vitkovitch & Barber, 1996). Thus, the speech rhythm parses the signal into basic units from which information on a finer (faster) temporal scale can be extracted (Ghitza, 2011).

Based on linguistic, psychophysical, and physiological considerations, it has been proposed that speech is analyzed in parallel at multiple time scales (Boemio, Fromm, Braun, & Poeppel, 2005; Poeppel, 2003; Poeppel, Idsardi, & van Wassenhove, 2008). Both local-to-global and global-to-local analyses are carried out concurrently (multitime resolution processing). The principal motivations for this hypothesis are twofold. First, one single, short temporal integration window that underpins hierarchical processing—that is, increasingly larger analysis units as one ascends the processing hierarchy—fails to account for the spectral and temporal sensitivity of the auditory system and is hard to reconcile with behavioral performance. Second, the computational strategy of analyzing information on multiple scales is widely used in engineering and biological systems, and the neuronal infrastructure exists to support multiscale computation (Buzsaki & Draguhn, 2004). We conjecture that speech is chunked into segments of roughly featural or phonemic length. In parallel, there is a "global" analysis that yields coarse, syllabic-scale inferences about speech and that subsequently refines segmental analysis.

The speech signal contains events of different durations: short energy bursts and formant transitions occur over 10–80 ms, whereas syllabically carried information occurs over 150–300 ms. The processing of both types of events is compatible with a hierarchical model (smaller acoustic units are concatenated into larger units like syllables), a parallel model (both temporal units are extracted independently and then combined), or a reverse hierarchical model (coarser temporal analysis precedes/guides finer analysis). Although there is no decisive evidence, at the behavioral level, the data favor the parallel or reverse hierarchy models. There is some independence in the processing of long (slow modulation) and short (fast modulation) units. For instance, speech is intelligible when it is first segmented into units up to 60 ms, and these local units are temporally reversed (Saberi & Perrott, 1999). This observation rules out the idea that speech processing relies solely on hierarchical processing of shorter before larger units. That is, the correct extraction of short units is not a prerequisite for comprehension. Overall, there exists a grouping of psychophysical phenomena such that some cluster at thresholds of ~50 ms and below and others cluster at ~200 ms and above (a similar clustering is observed for temporal properties in vision; Holcombe, 2009).

Gamma and theta rhythms in auditory cortex

One mechanistic hypothesis about chunking speech and other sounds is that cortical oscillations could be instruments of sampling the continuous speech signal (Giraud & Poeppel, 2012; Poeppel, 2003). Neural oscillations reflect synchronous activity of neuronal assemblies that are either intrinsically coupled or coupled by a common input (Wang, 2010). Oscillations can be observed in the absence of stimulation and are modulated by sensory stimulation. In this way, cortical oscillations are thought to shape the output of neurons in the form of spike-timing and to generate phases of high and low neuronal excitability (Kayser, Logothetis, & Panzeri, 2010; Schroeder & Lakatos, 2009). Segmental and suprasegmental analyses are carried out simultaneously due to neuronal oscillations at different rates (figure 53.1A). Considering a mean phoneme length of 25–80 ms and a mean syllabic length of 150–300 ms, dual-scale segmentation will involve two sampling mechanisms, one in the low gamma range and one in the theta range.

With a period of ~25 ms, gamma oscillations provide a 10–15 ms window for integrating spectrotemporal information (low spiking rate) followed by a 10–15 ms window for propagating the output (high spiking rate). However, a 10–15 ms integration window is too short to characterize a ~50 ms phoneme. How many gamma cycles are required to correctly encode phonemes? This question has been addressed by computational modeling. Using a PING (pyramidal interneuron gamma) model of gamma oscillations that modulate activity in a coding neuronal population, Shamir, Ghitza, Epstein, and Kopell (2009) show that a sawtooth input signal designed to have the typical duration and amplitude modulation of a diphone (~50 ms; typically a consonant-vowel or vowel-consonant transition) can be represented by three gamma cycles that act as a three-bit code. This code has the required capacity to distinguish different shapes of the stimulus and is therefore a plausible means to distinguish phonemes. That 50 ms diphones are correctly discriminated with three gamma cycles suggests that phonemes could be sampled with one/two gamma cycles. Consequently, the frequency of neural oscillations in auditory cortex might constitute a biophysical determinant with respect to the size of the minimal acoustic unit that can be manipulated for linguistic purposes.

Gamma and theta rhythms work together, and the phase of theta oscillations determines the power (and possibly phase) of gamma oscillations (Schroeder, Lakatos, Kajikawa, Partan, & Puce, 2008). This relationship is referred to as "nesting." Theta oscillations can be phase-reset by several means, including through multimodal cortico-cortical pathways (Arnal, Morillon, Kell, & Giraud, 2009), but most probably by stimulus onsets. The largest cortical auditory evoked response measured with electroencephalography and magnetoencephalography (MEG), about 100 ms after stimulus onset, corresponds to the phase reset of theta activity (Arnal, Wyart, & Giraud, 2011). This phase reset aligns the speech signal and the cortical theta rhythm, the proposed instrument of speech segmentation, into syllabic units. As speech is amplitude modulated precisely at the theta rate, this results in aligning neuronal excitability with those parts of the speech signal that are most informative in terms of energy and spectrotemporal content.

Recent psychophysical research emphasizes the importance of aligning the acoustic speech signal with the brain's oscillatory/quasi-rhythmic activity. Ghitza and Greenberg (2009) demonstrated that intelligibility is restored by inserting periods of silence in a speech signal that was made unintelligible by time-compressing it by a factor of 3. Adding silent periods to speech to restore an optimal temporal rate, equivalent to restoring "syllabicity," improves performance even though the speech segments that remained acoustically available are not more intelligible. The implicated time constants suggested a phenomenological model involving three nested rhythms in the theta (5 Hz), beta or low gamma (20–40 Hz), and gamma (80 Hz) domains (Ghitza, 2011).

Multitime resolution processing: Asymmetric sampling in time

Poeppel (2003) attempted to integrate four lines of evidence concerning speech and oscillations. First, speech contains information on at least two critical time scales, correlating with segmental and syllabic information. Second, many auditory psychophysical phenomena also fall in two groups, with integration constants of ~25–50 ms and 200–300 ms. Third, both patient and imaging data reveal cortical asymmetries: both hemispheres participate in auditory analysis but are

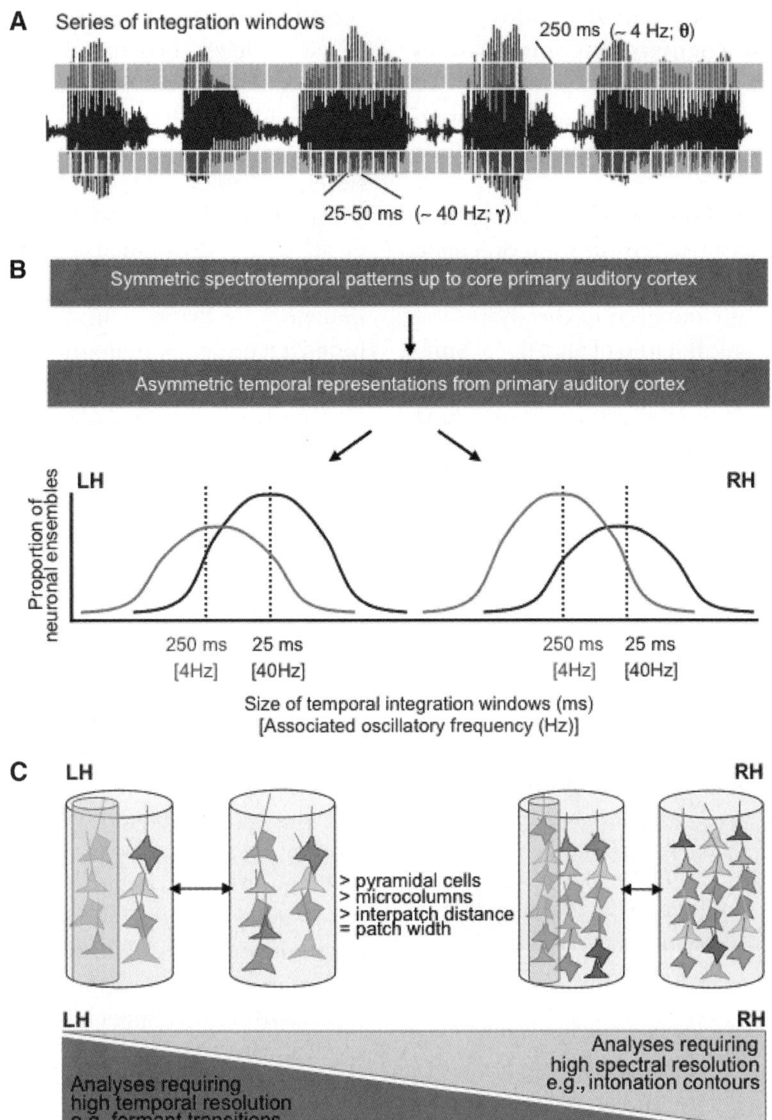

FIGURE 53.1 (A) Temporal relationship between the speech waveform and the two proposed integration time scales (in ms) and associated brain rhythms (in Hz). (B) Proposed mechanisms for asymmetric speech parsing: the left auditory cortex (LH) contains a larger proportion of neurons able to oscillate at gamma frequency than the right one (RH). (C) Differences in cytoarchitectonic organization between right and left auditory cortices. Left auditory cortex contains larger pyramidal cells in superficial cortical layers and exhibits bigger micro columns, a larger patch width, and interpatch distance.

optimized for different types of processing. Finally, neuronal oscillations might relate in a principled way to temporal integration constants of different sizes. The proposal holds that there exist hemispherically asymmetric distributions of neuronal ensembles with preferred shorter versus longer integration constants; these cell groups "sample" the input on different time scales: left auditory cortex has a relatively higher proportion of short (gamma) integrating ensembles, right auditory cortex has more long (theta) integrating neuronal ensembles (figure 53.1B). As a consequence, left-hemisphere auditory cortex is better equipped for parsing speech at the segmental scale, and right auditory cortex for parsing speech at the syllabic time scale.

The AST hypothesis accounts for various psychophysical and neuroimaging results that show the left temporal cortex responds better to many aspects of rapidly modulated auditory content, while right temporal cortex responds better to slowly modulated signals, including music, voices, and other sounds (Zatorre, Belin, & Penhune, 2002). Systematic differences in integration windows between left and right auditory cortices explain speech functional asymmetry by a better sensitivity of left auditory cortex to information carried in fast temporal modulations that convey brief cues. A specialization of the right to slower modulations grants it better sensitivity to slower cues such as harmonicity and periodicity (Rosen, 1992)—cues important to identify vowels, syllables, and speaker identity (figure 53.1C).

Consistent with AST, temporally extended stimuli built from short segments of different durations elicit rightward asymmetry when longer time segments were used (e.g., 300 ms) as compared to the short-time structure signals (e.g., 25 ms; Boemio et al., 2005). Similarly, there is significant rightward lateralization for stimuli with increasing length of spectrotemporal time windows (Overath, Kumar, von Kriegstein, & Griffiths, 2008). In contrast, rapidly modulated speech and non-speech signals generate leftward lateralization (Jamison, Watkins, Bishop, & Matthews, 2006; Zaehle, Wuestenberg, Meyer, & Jaencke, 2004). While still controversial, there is emerging consensus that temporal parameters of the sort discussed here play a central role in decoding auditory signals in neocortex. Figure 53.2 illustrates

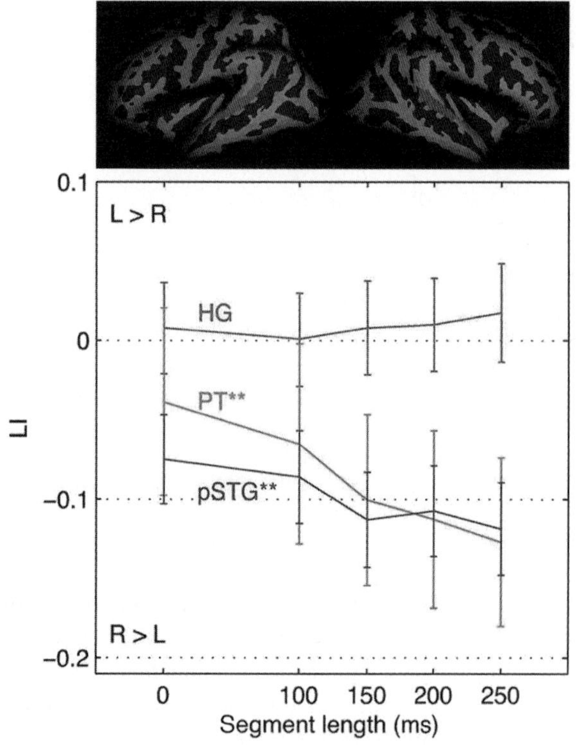

FIGURE 53.2 The functional lateralization index (LI) in anatomically defined regions of interest is displayed; positive values indicate L > R lateralization. HG: Heschl's gyrus (red); PT: planum temporale (green); pSTG: posterior superior temporal gyrus (blue). $P < 0.001$. Error bars indicate 1 standard error of the mean. (Figure reprinted with permission from Liem et al., 2013. Copyright © 2013 Wiley Periodicals, Inc.) (See color plate 46.)

recent imaging data that show the rightward asymmetry of more slowly modulated speech information (Liem, Hurschler, Jaencke, & Meyer, 2013), a data pattern that is readily replicated.

Are asymmetric sampling properties architectural system features or, instead, driven into the system by stimulus properties? Giraud and colleagues (Giraud et al., 2007; Morillon, Liegeois-Chauvel, Arnal, Benar, & Giraud, 2012) have measured the lateralization of neuronal oscillations in subjects at rest. Using combined electroencephalography and functional MRI as well as electrocorticography, stronger expression of gamma rhythm in left auditory cortex and a stronger expression in theta rhythm in right auditory cortex have been observed.

Visual influences on the speech rhythm

Facial motion enhances speech perception, in terms of intelligibility (Sumby & Pollack, 1954) and response times (Chandrasekaran, Lemus, Trubanova, Gondan, & Ghazanfar, 2011; van Wassenhove, Grant, & Poeppel, 2005). Facial motion helps because of robust temporal correspondences between mouth movements and the speech envelope (Chandrasekaran et al., 2009): mouth motion, like the speech envelope, is temporally modulated in the theta frequency range. Moreover, the timing of mouth movements relative to the onset of the voice is consistently between 100 and 300 ms. Thus, facial dynamics predict upcoming speech details. The statistics of audio-*visual* speech in the context of the auditory cortical rhythms described above, therefore, suggest that speech communication is a reciprocally coupled, multisensory event wherein the outputs of the signaler are matched to the neural processes of the receiver (Chandrasekaran et al., 2009).

MEG recordings from participants viewing audiovisual movies show that the phase of auditory theta activity carries robust information for parsing the temporal structure of stimulus dynamics in both sensory modalities concurrently (Luo, Liu, & Poeppel, 2010). Auditory cortex can track visual dynamics just as it does the auditory speech envelope. Recent data demonstrate how such rhythmic visual inputs enhance both perception and cortical signal processing. Zion Golumbic, Cogan, Schroeder, and Poeppel (2013) investigated whether congruent visual input of an attended speaker enhances cortical selectivity in auditory cortex, leading to diminished representation of ignored stimuli. They recorded MEG signals from participants attending to segments of natural continuous speech. Viewing a speaker's face enhances the capacity of auditory cortex to track the temporal speech envelope of that speaker. Moreover, this mechanism was most effective in a "cocktail party" scenario, emphasizing preferential tracking of the attended speaker. Since visual cues in speech precede the associated auditory signals (Chandrasekaran et al., 2009), they likely serve a predictive role in auditory speech processing.

How did the speech rhythm evolve?

Given the importance of rhythms in audiovisual speech, understanding its evolution requires investigating the origins of its rhythmic structure. Moreover, the gamma and theta oscillations of the human auditory cortex are also present in the auditory cortex of monkeys (and probably all mammals). Other mammals do not have vocal signals that rival the human speech signal in complexity; it is thus of significant interest to understand how speech evolved to exploit those preexisting brain rhythms. There are many similarities in multisensory vocal communication between monkeys and humans (and, in some cases, apes; Ghazanfar, 2013). Like humans, monkeys match individual identity and expression types across modalities (Ghazanfar & Logothetis, 2003; Sliwa, Duhamel, Pascalis, & Wirth, 2011), segregate competing voices in noisy conditions using vision (Jordan, Brannon, Logothetis, & Ghazanfar, 2005), use formant frequencies to estimate the body size of conspecifics (Ghazanfar et al., 2007), and use facial motion to speed up their reaction times to vocalizations (Chandrasekaran et al., 2011). However, there are also important differences in how humans produce speech (Ghazanfar & Rendall, 2008), and these differences enhance multisensory communication above and beyond what monkeys can do. One universal feature of speech—lacking in monkey vocalizations—is its multisensory (auditory and visual) rhythm. That is, when humans speak both the acoustic output and the movements of the mouth are highly rhythmic and tightly correlated with each other (Chandrasekaran et al., 2009).

One theory posits that the rhythm of speech evolved through the modification of rhythmic facial movements in ancestral primates (MacNeilage, 1998). Such facial movements are extremely common as visual communicative gestures in primates. Lip-smacking, for example, is an affiliative signal observed in many genera (Hinde & Rowell, 1962; Redican, 1975; van Hooff, 1962), including chimpanzees (Parr, Cohen, & de Waal, 2005). It is characterized by regular cycles of vertical jaw movement, often involving a parting of the lips. While lip-smacking by both monkeys and chimpanzees is often produced during grooming interactions, monkeys also exchange lip-smacking bouts during face-to-face interactions (Ferrari et al., 2009; van Hooff, 1962). Lip-smacks are among the first facial expressions produced by infant monkeys (Ferrari et al., 2006; de Marco & Visalberghi, 2007) and used during mother-infant interactions (Ferrari, Paukner, Ionica, & Suomi, 2009). According to MacNeilage (1998), during the course of speech evolution, such nonvocal rhythmic facial expressions were coupled to vocalizations to produce the audiovisual components of babbling-like (i.e., consonant-vowel-like) speech expressions.

While direct tests of such evolutionary hypotheses are difficult, there are many lines of evidence that demonstrate that the production of lip-smacking in macaque monkeys is similar to the orofacial rhythms produced during speech. Both speech and lip-smacking are distinct from chewing, another rhythmic orofacial motion that uses the same effectors. Importantly, in contrast to chewing movements (which are slower), lip-smacking exhibits a speech-like rhythm in the 3–8 Hz frequency range (Ghazanfar, Chandrasekaran, & Morrill, 2010).

Developmental Parallels If the mechanisms underlying rhythm in monkey lip-smacks and human speech are homologous, their developmental trajectories should be similar (Gottlieb, 1992; Schneirla, 1949). Moreover, this common trajectory should be distinct from the developmental trajectory of other rhythmic mouth movements.

In humans, the earliest form of rhythmic vocal behavior occurs sometime after 6 months of age, when vocal babbling abruptly emerges (Locke, 1993; Preuschoff, Quartz, & Bossaerts, 2008; Smith & Zelaznik, 2004). Babbling is characterized by the production of canonical syllables that have acoustic characteristics similar to adult speech. Their production involves rhythmic sequences of a mouth close-open alternation (Davis & MacNeilage, 1995; Lindblom, Krull, & Stark, 1996; Oller, 2000). This alternation results in a consonant-vowel syllable—representing the only syllable type present in all the world's languages (Bell & Hooper, 1978). However, babbling does not emerge with the same rhythmic structure as adult speech, but rather there is a sequence of structural changes in the rhythm. There are at least two changes: frequency and variability. In adults, the speech rhythm is ~5 Hz (Chandrasekaran et al., 2009; Crystal & House, 1982; Dolata, Davis, & MacNeilage, 2008; Greenberg et al., 2003; Malecot et al., 1972) in infant babbling, the rhythm is considerably slower. Between 2 and 12 months, infants produce speech at a slower rate of roughly 2.8 to 3.4 Hz (Dolata et al., 2008; Levitt & Wang, 1991; Lynch, Oller, Steffens, & Buder, 1995; Nathani, Oller, & Cobo-Lewis, 2003). This developmental trajectory from babbling to speech is distinct from another cyclical mouth movement, *chewing*. The frequency of chewing movements in humans is highly stereotyped: it is slow in frequency and remains virtually unchanged from early infancy into adulthood (Green et al., 1997; Kiliaridis, Karlsson, & Kjellberge, 1991).

Measuring the rhythmic frequency and variability of lip-smacking in nonhuman primates across neonatal, juvenile, and adult age groups is one way to uncover any similarities with the developmental trajectory of the human speech rhythm (Dolata et al., 2008; Morrill, Paukner, Ferrari, & Ghazanfar, 2012). Macaques are the most accessible model system because it is very difficult (if not impossible) to acquire developmental data from great apes. It turns out that the developmental trajectory of monkey lip-smacking parallels speech development (Locke, 2008; Morrill et al., 2012). Measurements of the rhythmic frequency and variability of lip-smacking across individuals in the three different age groups revealed that young individuals produce slower, more variable mouth movements; as they get older, these movements become faster and less variable (Dolata et al., 2008; Morrill et al., 2012). This is exactly as speech develops, from babbling to adult consonant-vowel production (Dolata et al., 2008; Morrill et al., 2012). Furthermore, as in human speech development (Smith & Zelaznik, 2004), the variability and frequency changes in lip-smacking are independent in that juveniles have the same rhythmic lip-smacking frequency as adult monkeys, but the lip-smacking is more variable. Importantly, the developmental trajectory for lip-smacking was different from that of chewing (Morrill et al., 2012). The developmental differences between lip-smacking and chewing are identical to those reported in humans for speech and chewing (Moore & Ruark, 1996; Steeve, 2010; Steeve, Moore, Green, Reilly, & McMurtrey, 2008).

THE COORDINATION OF EFFECTORS Given human speech and monkey lip-smacking parallels, commonalities in the effectors' use are expected. Specifically, homologous structures participating both in speech and lip-smacking are predicted to be similarly coordinated. Evidence for this comes from motor control. During speech, the functional coordination between key vocal tract anatomical structures (jaw/lips, tongue, and hyoid) is more loosely coupled than during chewing (Hiiemae et al., 2002; Hiiemae & Palmer, 2003; Matsuo & Palmer, 2010; Moore, Smith, & Ringel, 1988; Ostry & Munhall, 1994). X-ray cineradiography visualizing the dynamics of the macaque vocal tract during lip-smacking and chewing revealed that lips, tongue, and hyoid move during lip-smacks (as in speech) and do so with a speech-like 3–8 Hz rhythm (Ghazanfar, Takahashi, Mathur, & Fitch, 2012). Relative to lip-smacking, movements during chewing were significantly slower for each of these structures. Most importantly, the temporal coordination of these structures was distinct for each behavior. Although the hyoid moves continuously during lip-smacking, there is no coupling of the hyoid with lips and tongue movements. During chewing there is more coordination between the three structures, consistent with what is observed in human speech and chewing (Hiiemae et al., 2002; Hiiemae & Palmer, 2003).

Facial electromyographic studies of muscle coordination during lip-smacking and chewing also revealed distinct activity patterns (Shepherd, Lanzilotto, & Ghazanfar, 2012). Lip-smacking showed rhythmic modulation in the lower facial muscles, whereas chewing behavior exhibited less structured muscle activity, probably reflecting the variable manipulation of the food. This supports the view that lip-smacking and speech, but not chewing, are similar at the level of the rhythmicity and functional coordination of effectors.

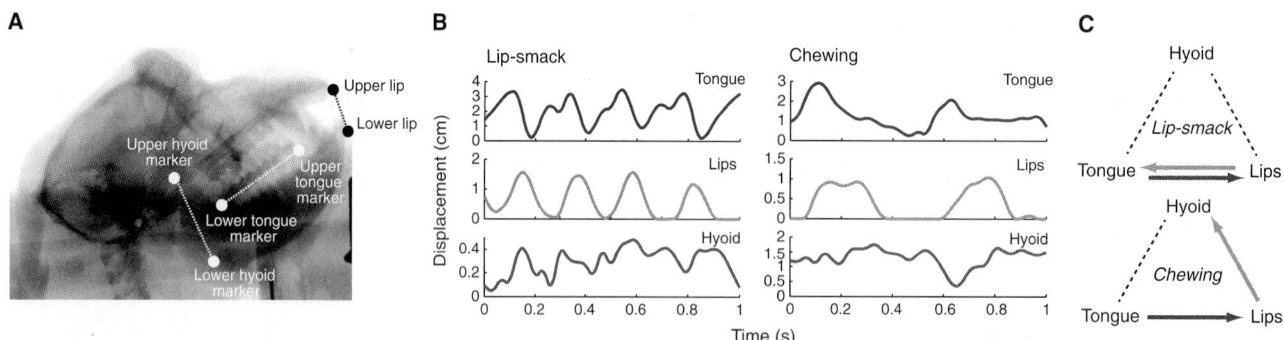

FIGURE 53.3 (A) The anatomy of the macaque monkey vocal tract as imaged with cineradiography. The key vocal tract structures are labeled: the lips, tongue, and hyoid. (B) Time-displacement plot of the tongue, inter-lip distance, and hyoid for one exemplar each of lip-smacking and chewing. (C) Arrow schematics show the direction of significant influence from each structure onto the other two as measured by the partial directed coherence analysis of signals such as those in B. (Figure reprinted with permission from Ghazanfar et al., 2012, with permission from Elsevier.)

Perceptual Tuning The 3–8 Hz rhythm is critical to speech perception: disrupting the auditory component of this rhythm significantly reduces intelligibility (Drullman et al., 1994; Elliot & Theunissen, 2009; Saberi & Perrott, 1999; Shannon et al., 1995; Smith et al., 2002), as does disrupting the visual dynamics arising from mouth and facial movements (Vitkovitch & Barber, 1996). A preferential-looking procedure was used to test whether monkeys were differentially sensitive to lip-smacking produced with a rhythmic frequency in the species-typical range (mean 4–6 Hz; Ghazanfar, Morrill, & Kayser, 2013). Computer-generated monkey avatars produced varying lip-smacking frequencies but had otherwise identical features (Chandrasekaran et al., 2011; Steckenfinger & Ghazanfar, 2009). The use of avatar faces allowed for the control of additional factors that could influence looking times, such as head and eye movements and lighting conditions for face and background. Avatar faces were generated to produce three different lip-smacking rhythms: 3, 6, and 10 Hz. In every test session, one avatar was lip-smacking at 6 Hz while the other was lip-smacking at either 3 or 10 Hz. Preferential looking was measured by calculating the looking times to one or the other avatar while these were presented on a wide screen in front of the monkey. Monkeys showed an overall preference for the natural rhythm when compared to the perturbed rhythms. This lends behavioral support for the hypothesis that perceptual processes are similarly tuned to the natural frequencies of communication signals across primate species.

Differences Between Lip-Smacking and Speech Production Two core features of speech production—its rhythmical structure and temporal coordination of vocal elements—are shared with lip-smacking. There are also striking differences. Most obviously, lip-smacking lacks a vocal component (a quiet, consonant-like bilabial plosive, or /p/, sound is incidentally produced during lip-smacking). Thus, the capacity to produce *intentional vocalizations* during rhythmic vocal tract movements seems to be a human feature. How can lip-smacking be related to speech if there is no vocal component? In human and nonhuman primates, the basic mechanisms of voice production are broadly similar and consist of two distinct components: the laryngeal source and the vocal tract filter (Fant, 1970; Fitch & Hauser, 1995; Ghazanfar & Rendall, 2008). Voice production involves (1) a sound generated by air pushed by the lungs through the vibrating vocal folds within the *larynx* (the source) and (2) the modification through filtering of this sound by the *vocal tract* resonators above the larynx (the filter). The filter consists of the nasal and oral cavities whose shapes can be changed by movements of the jaw, tongue, hyoid, and lips. These two basic components of the vocal apparatus behave and interact in complex ways to generate a wide range of sounds. The lip-smacking data only address the evolution of vocal tract movements (the filter component) involved in speech production.

Other differences between lip-smacking and speech are the range of hyoid movements and the coupling of the lips with the tongue. For the latter, the coupling of

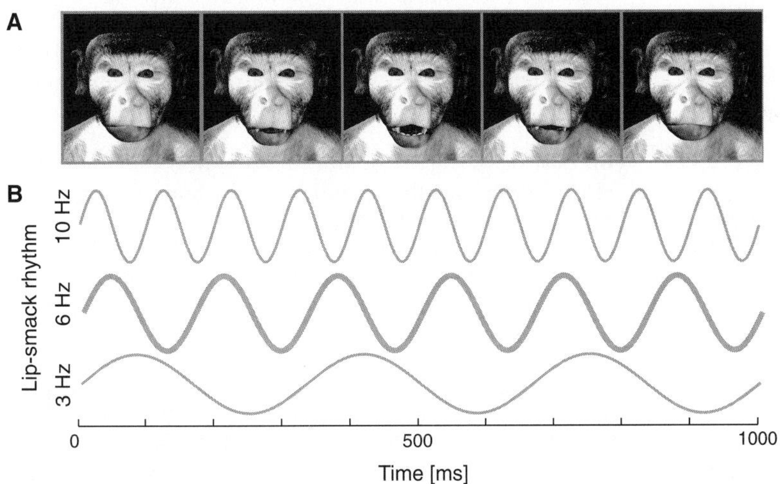

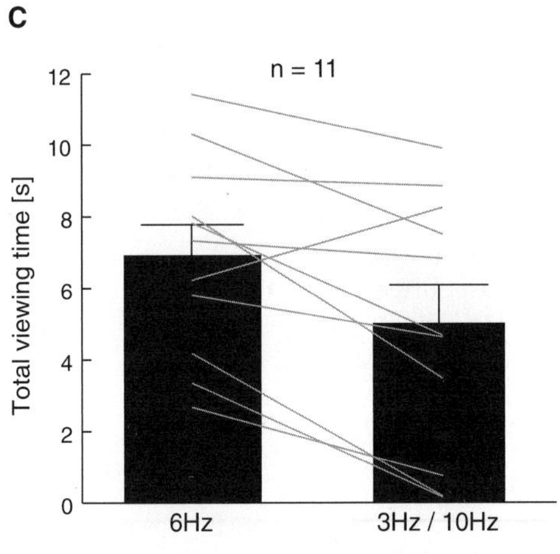

Figure 53.4 (A) Example frame sequence from a video clip showing an avatar face producing a lip-smacking gesture. Lip-smacking is characterized by regular cycles of vertical jaw movement, often involving a parting of the lips. (B) Synthetic lip-smack rates were faster (10 Hz) or slower (3 Hz) than the natural rate (6 Hz). (C) Total viewing times in seconds for individual subjects (lines) and grand total (mean and standard error). All but one subject showed a preference for the avatar with biological lip-smack rate. (Figure reprinted with permission from Ghazanfar et al., 2013.)

the lips and tongue during lip-smacking is unlikely to be the case for human speech where their independence allows for the production of a wide range of sounds. For the former, the data show that the hyoid occupies the same active space during lip-smacking and chewing (Ghazanfar et al., 2012). In contrast, cineradiography studies of human speech versus chewing show a dichotomy in hyoid movement patterns (Hiiemae et al., 2002). These movement-range differences of the hyoid in humans versus macaques could be due to functional differences in suprahyoid muscle length, the degree of neural control over this muscle group, and/or species differences in hyoid position. During human development, the position of the hyoid relative to the mandible and tongue shifts (Lieberman, McCarthy, Hiiemae, & Palmer, 2001). This increases the range of tongue movements, and possibly hyoid movements, relative to nonhuman primates. Movements of either or both effectors could influence the active space of the hyoid.

Conclusion

This chapter summarized a research program that builds on a fortuitous alignment of ideas across a range of disciplines: acoustics, linguistics, perception, neurophysiology, evolution, and comparative anatomy and physiology. The concept at the center—rhythms, as observed acoustically, behaviorally, and neurophysiologically—has allowed for the formulation of mechanistic hypotheses about how communication sounds are perceived and produced. The temporally corresponding rhythms can be used to systematically investigate both human and nonhuman vocalizations, from connected speech to lip-smacking, and in multisensory contexts. This suggests an important evolutionary role for how rhythmic neural activity undergirds quasi-rhythmic vocal behavior typical of speech.

ACKNOWLEDGMENTS The preparation of this manuscript was supported by NIH 2R01NS05489 to AAG and NIH 2R01DC05660 to DP.

REFERENCES

Arnal, L. H., Morillon, B., Kell, C. A., & Giraud, A. L. (2009). Dual neural routing of visual facilitation in speech processing. *J Neurosci, 29*, 13445–13453.

Arnal, L. H., Wyart, V., & Giraud, A. L. (2011). Transitions in neural oscillations reflect prediction errors generated in audiovisual speech. *Nat Neurosci, 14*, 797–801.

Bell, A., & Hooper, J. B. (1978). *Syllables and segments*. Amsterdam: North-Holland.

Boemio, A., Fromm, S., Braun, A., & Poeppel, D. (2005). Hierarchical and asymmetric temporal sensitivity in human auditory cortices. *Nat Neurosci, 8*, 389–395.

Buzsaki, G., & Draguhn, A. (2004). Neuronal oscillations in cortical networks. *Science, 304*, 1926–1929.

Chandrasekaran, C., Lemus, L., Trubanova, A., Gondan, M., & Ghazanfar, A. A. (2011). Monkeys and humans share a common computation for face/voice integration. *PLoS Comput Biol, 7*, e1002165.

Chandrasekaran, C., Trubanova, A., Stillittano, S., Caplier, A., & Ghazanfar, A. A. (2009). The natural statistics of audiovisual speech. *PLoS Comput Biol, 5*, e1000436.

Crystal, T., & House, A. (1982). Segmental durations in connected speech signals: Preliminary results. *J Acoust Soc Am, 72*, 705–716.

Davis, B. L., & MacNeilage, P. F. (1995). The articulatory basis of babbling. *J Speech Hear Res, 38*, 1199–1211.

De Marco, A., & Visalberghi, E. (2007). Facial displays in young tufted capuchin monkeys (*Cebus apella*): Appearance, meaning, context and target. *Folia Primatol (Basel), 78*, 118–137.

Dolata, J. K., Davis, B. L., & MacNeilage, P. F. (2008). Characteristics of the rhythmic organization of vocal babbling: Implications for an amodal linguistic rhythm. *Infant Behav Dev, 31*, 422–431.

Drullman, R., Festen, J. M., & Plomp, R. (1994). Effect of reducing slow temporal modulations on speech reception. *J Acoust Soc Am, 95*, 2670–2680.

Elliot, T. M., & Theunissen, F. E. (2009). The modulation transfer function for speech intelligibility. *PLoS Comput Biol, 5*, e1000302.

Fant, G. (1970). *Acoustic theory of speech production* (2nd ed.). Paris: Mouton.

Ferrari, P., Visalberghi, E., Paukner, A., Fogassi, L., Ruggiero, A., & Suomi, S. (2006). Neonatal imitation in rhesus macaques. *PLoS Biol, 4*, 1501.

Ferrari, P. F., Paukner, A., Ionica, C., & Suomi, S. (2009). Reciprical face-to-face communication between rhesus macaque mothers and their newborn infants. *Curr Biol, 19*, 1768–1772.

Fitch, W. T., & Hauser, M. D. (1995). Vocal production in nonhuman-primates: Acoustics, physiology, and functional constraints on "honest" advertisement. *Am J Primatol, 37*, 191–219.

Ghazanfar, A. A. (2013). Multisensory vocal communication in primates and the evolution of rhythmic speech. *Behav Ecol Sociobiol, 67*, 830–839.

Ghazanfar, A. A., Chandrasekaran, C., & Morrill, R. J. (2010). Dynamic, rhythmic facial expressions and the superior temporal sulcus of macaque monkeys: Implications for the evolution of audiovisual speech. *Eur J Neurosci, 31*, 1807–1817.

Ghazanfar, A. A., & Logothetis, N. K. (2003). Facial expressions linked to monkey calls. *Nature, 423*, 937–938.

Ghazanfar, A. A., Morrill, R. J., & Kayser, C. (2013). Monkeys are perceptually tuned to facial expressions that exhibit a theta-like speech rhythm. *Proc Natl Acad Sci USA, 110*, 1959–1963.

Ghazanfar, A. A., & Rendall, D. (2008). Evolution of human vocal production. *Curr Biol, 18*, R457–R460.

Ghazanfar, A. A., Takahashi, D. Y., Mathur, N., & Fitch, W. T. (2012). Cineradiography of monkey lipsmacking reveals the putative origins of speech dynamics. *Curr Biol, 22,* 1176–1182.

Ghazanfar, A. A., Turesson, H. K., Maier, J. X., van Dinther, R., Patterson, R. D., & Logothetis, N. K. (2007). Vocal tract resonances as indexical cues in rhesus monkeys. *Curr Biol, 17,* 425–430.

Ghitza, O. (2011). Linking speech perception and neurophysiology: Speech decoding guided by cascaded oscillators locked to the input rhythm. *Front Psychol, 2,* 130.

Ghitza, O., & Greenberg, S. (2009). On the possible role of brain rhythms in speech perception: Intelligibility of time-compressed speech with periodic and aperiodic insertions of silence. *Phonetica, 66,* 113–126.

Giraud, A. L., Kleinschmidt, A., Poeppel, D., Lund, T. E., Frackowiak, R. S. J., & Laufs, H. (2007). Endogenous cortical rhythms determine cerebral specialization for speech perception and production. *Neuron, 56,* 1127–1134.

Giraud, A. L., & Poeppel, D. (2012). Cortical oscillations and speech processing: Emerging computational principles and operations. *Nat Neurosci, 15,* 511–517.

Gottlieb, G. (1992). *Individual development and evolution: The genesis of novel behavior.* New York, NY: Oxford University Press.

Green, J. R., Moore, C. A., Ruark, J. L., Rodda, P. R., Morvee, W. T., & van Witzenberg, M. J. (1997). Development of chewing in children from 12 to 48 months: Longitudinal study of EMG patterns. *J Neurophysiol, 77,* 2704–2727.

Greenberg, S., Carvey, H., Hitchcock, L., & Chang, S. (2003). Temporal properties of spontaneous speech—a syllable-centric perspective. *J Phonetics, 31,* 465–485.

Hiiemae, K. M., & Palmer, J. B. (2003). Tongue movements in feeding and speech. *Crit Rev Oral Biol M, 14,* 413–429.

Hiiemae, K. M., Palmer, J. B., Medicis, S. W., Hegener, J., Jackson, B. S., & Lieberman, D. E. (2002). Hyoid and tongue surface movements in speaking and eating. *Arch Oral Biol, 47,* 11–27.

Hinde, R. A., & Rowell, T. E. (1962). Communication by posture and facial expressions in the rhesus monkey (*Macaca mulatta*). *Proc Zool Soc Lond, 138,* 1–21.

Holcombe, A. O. (2009). Seeing slow and seeing fast: Two limits on perception. *Trends Cogn Sci, 13,* 216–221.

Jamison, H. L., Watkins, K. E., Bishop, D. V., & Matthews, P. M. (2006). Hemispheric specialization for processing auditory nonspeech stimuli. *Cereb Cortex, 16,* 1266–1275.

Jordan, K. E., Brannon, E. M., Logothetis, N. K., & Ghazanfar, A. A. (2005). Monkeys match the number of voices they hear with the number of faces they see. *Curr Biol, 15,* 1034–1038.

Kayser, C., Logothetis, N. K., & Panzeri, S. (2010). Millisecond encoding precision of auditory cortex neurons. *Proc Natl Acad Sci USA, 107,* 16876–16881.

Kiliaridis, S., Karlsson, S., & Kjellberge, H. (1991). Characteristics of masticatory mandibular movements and velocity in growing individuals and young adults. *J Dent Res, 70,* 1367–1370.

Levitt, A., & Wang, Q. (1991). Evidence for language-specific rhythmic influences in the reduplicative babbling of French- and English-learning infants. *Lang Speech, 34,* 235–239.

Lieberman, D. E., McCarthy, R. C., Hiiemae, K. M., & Palmer, J. B. (2001). Ontogeny of postnatal hyoid and larynx descent in humans. *Arch Oral Biol, 46,* 117–128.

Liem, F., Hurschler, M. A., Jaencke, L., & Meyer, M. (2013). On the planum temporale lateralization in suprasegmental speech perception: Evidence from a study investigating behavior, structure, and function. *Hum Brain Mapp.* Advance online publication. doi:10.1002/hbm.22291

Lindblom, B., Krull, D., & Stark, J. (1996). Phonetic systems and phonological development. In B. de Boysson-Bardies, S. de Schonen, P. Jusczyk, P. F. MacNeilage, & J. Morton (Eds.), *Developmental neurocognition: Speech and face processing in the first year of life* (pp. 399–410). Dordrecht, NL: Kluwer.

Locke, J. L. (1993). *The child's path to spoken language.* Cambridge, MA: Harvard University Press.

Locke, J. L. (2008). Lipsmacking and babbling: Syllables, sociality, and survival. In B. L. Davis & K. Zajdo (Eds.), *The syllable in speech production* (pp. 111–129). New York, NY: Erlbaum.

Luo, H., Liu, Z., & Poeppel, D. (2010). Auditory cortex tracks both auditory and visual stimulus dynamics using low-frequency neuronal phase modulation. *PLoS Biology, 8,* e1000445.

Lynch, M., Oller, D. K., Steffens, M., & Buder, E. (1995). Phrasing in prelinguistic vocalizations. *Dev Psychobiol, 23,* 3–25.

MacNeilage, P. F. (1998). The frame/content theory of evolution of speech production. *Behav Brain Sci, 21,* 499–546.

Malecot, A., Johonson, R., & Kizziar, P.-A. (1972). Syllable rate and utterance length in French. *Phonetica, 26,* 235–251.

Matsuo, K., & Palmer, J. B. (2010). Kinematic linkage of the tongue, jaw, and hyoid during eating and speech. *Arch Oral Biol, 55,* 325–331.

Moore, C. A., & Ruark, J. L. (1996). Does speech emerge from earlier appearing motor behaviors? *J Speech Hear Res, 39,* 1034–1047.

Moore, C. A., Smith, A., & Ringel, R. L. (1988). Task specific organization of activity in human jaw muscles. *J Speech Hear Res, 31,* 670–680.

Morillon, B., Liegeois-Chauvel, C., Arnal, L. H., Benar, C. G., & Giraud, A. L. (2012). Asymmetric function of theta and gamma activity in syllable processing: An intra-cortical study. *Front Psychol, 3,* 248.

Morrill, R. J., Paukner, A., Ferrari, P. F., & Ghazanfar, A. A. (2012). Monkey lip-smacking develops like the human speech rhythm. *Dev Sci, 15,* 557–568.

Nathani, S., Oller, D. K., & Cobo-Lewis, A. (2003). Final syllable lengthening (FSL) in infant vocalizations. *J Child Lang, 30,* 3–25.

Oller, D. K. (2000). *The emergence of the speech capacity.* Mahwah, NJ: Erlbaum.

Ostry, D. J., & Munhall, K. G. (1994). Control of jaw orientation and position in mastication and speech. *J Neurophysiol, 71,* 1528–1545.

Overath, T., Kumar, S., von Kriegstein, K., & Griffiths, T. D. (2008). Encoding of spectral correlation over time in auditory cortex. *J Neurosci, 28,* 13268–13273.

Parr, L. A., Cohen, M., & de Waal, F. (2005). Influence of social context on the use of blended and graded facial displays in chimpanzees. *Intl J Primatol, 26,* 73–103.

Poeppel, D. (2003). The analysis of speech in different temporal integration windows: Cerebral lateralization as "asymmetric sampling in time." *Speech Commun, 41,* 245–255.

Poeppel, D., Idsardi, W. J., & van Wassenhove, V. (2008). Speech perception at the interface of neurobiology and linguistics. *Philos Trans R Soc Lond B Biol Sci, 363,* 1071–1086.

Preuschoff, K., Quartz, S. R., & Bossaerts, P. (2008). Human insula activation reflects risk prediction errors as well as risk. *J Neurosci, 28,* 2745–2752.

Redican, W. K. (1975). Facial expressions in nonhuman primates. In L. A Rosenblum (Ed.), *Primate behavior: Developments in field and laboratory research* (pp. 103–194). New York, NY: Academic Press.

Rosen, S. (1992). Temporal information in speech: Acoustic, auditory and linguistic aspects. *Philos Trans R Soc Lond B Biol Sci, 336,* 367–373.

Saberi, K., & Perrott, D. R. (1999). Cognitive restoration of reversed speech. *Nature, 398,* 760.

Schneirla, T. C. (1949). Levels in the psychological capacities of animals. In R. W. Sellars, V. J. McGill, & M. Farber (Eds.), *Philosophy for the future* (pp. 243–286). New York, NY: Macmillan.

Schroeder, C. E., & Lakatos, P. (2009). Low-frequency neural oscillations as instruments of sensory selection. *Trends Neurosci, 32,* 9–18.

Schroeder, C. E., Lakatos, P., Kajikawa, Y., Partan, S., & Puce, A. (2008). Neuronal oscillations and visual amplification of speech. *Trends Cogn Sci, 12,* 106–113.

Shamir, M., Ghitza, O., Epstein, S., & Kopell, N. (2009). Representation of time-varying stimuli by a network exhibiting oscillations on a faster time scale. *PLoS Comput Biol, 5,* e1000370.

Shannon, R. V., Zeng, F.-G., Kamath, V., Wygonski, J., & Ekelid, M. (1995). Speech recognition with primarily temporal cues. *Science, 270,* 303–304.

Shepherd, S. V., Lanzilotto, M., & Ghazanfar, A. A. (2012). Facial muscle coordination during rhythmic facial expression and ingestive movement. *J Neurosci, 32,* 6105–6116.

Sliwa, J., Duhamel, J. R., Pascalis, O., & Wirth, S. (2011). Spontaneous voice-face identity matching by rhesus monkeys for familiar conspecifics and humans. *Proc Natl Acad Sci USA, 108,* 1735–1740.

Smith, A., & Zelaznik, H. N. (2004). Development of functional synergies for speech motor coordination in childhood and adolescence. *Dev Psychobiol, 45,* 22–33.

Smith, Z. M., Delgutte, B., & Oxenham, A. J. (2002). Chimaeric sounds reveal dichotomies in auditory perception. *Nature, 416,* 87–90.

Steckenfinger, S. A., & Ghazanfar, A. A. (2009). Monkey visual behavior falls into the uncanny valley. *Proc Natl Acad Sci USA, 106,* 18362–18466.

Steeve, R. W. (2010). Babbling and chewing: Jaw kinematics from 8 to 22 months. *J Phonetics, 38,* 445–458.

Steeve, R. W., Moore, C. A., Green, J. R., Reilly, K. J., & McMurtrey, J. R. (2008). Babbling, chewing, and sucking: Oromandibular coordination at 9 months. *J Speech Lang Hear Res, 51,* 1390–1404.

Sumby, W. H., & Pollack, I. (1954). Visual contribution to speech intelligibility in noise. *J Acoust Soc Am, 26,* 212–215.

van Hooff, J. A. R. A. M. (1962). Facial expressions of higher primates. *Symp Zool Soc Lond, 8,* 97–125.

van Wassenhove, V., Grant, K. W., & Poeppel, D. (2005). Visual speech speeds up the neural processing of auditory speech. *Proc Natl Acad Sci USA, 102,* 1181–1186.

Vitkovitch, M., & Barber, P. (1996). Visible speech as a function of image quality: Effects of display parameters on lipreading ability. *Appl Cognitive Psych, 10,* 121–140.

Wang, X.-J. (2010). Neurophysiological and computational principles of cortical rhythms. *Physiol Rev, 90,* 1195–1268.

Zaehle, T., Wuestenberg, T., Meyer, M., & Jaencke, L. (2004). Evidence for rapid auditory perception as the foundation of speech processing: A sparse temporal sampling fMRI study. *Eur J Neurosci, 20,* 2447–2456.

Zatorre, R. J., Belin, P., & Penhune, V. B. (2002). Structure and function of auditory cortex: Music and speech. *Trends Cogn Sci, 6,* 37–46.

Zion Golumbic, E., Cogan, G. B., Schroeder, C. E., & Poeppel, D. (2013). Visual input enhances selective speech envelope tracking in auditory cortex at a "cocktail party." *J Neurosci, 33,* 1417–1426.

54 Morphological Systems in Their Neurobiological Contexts

WILLIAM D. MARSLEN-WILSON, MIRJANA BOZIC, AND LORRAINE K. TYLER

ABSTRACT The neural framework for human language and communication is analyzed in terms of two intersecting but evolutionarily distinguishable neurobiological systems: a left-hemisphere frontotemporal system required for core syntactic functions, integrated with a bihemispheric system shared with our primate relatives that supports sound to meaning mapping and semantic and pragmatic interpretation. This view of language as a dynamic coalition of interacting brain systems provides the framework for a cross-linguistic examination (in English, Polish, and Arabic) of the way in which basic generative systems for inflectional and derivational word formation are distributed across these two systems and whether (and how) different languages vary in these respects.

The ability to communicate using language is fundamental to the distinctive and remarkable success of the modern human. It is this capacity that separates us most decisively from our primate cousins, despite all that we have in common across species as intelligent social primates. A major scientific challenge for the cognitive neurosciences is to understand exactly this relationship. What is the neurobiological context in which human language and communication have emerged, and what are the special human properties that make language itself possible?

Scientific (and popular) thinking about this relationship has been dominated by our strong phenomenological experience, as speakers and listeners, that we command and inhabit a unified social communication system built around language. In neuroscientific terms, this has long been expressed in the form of a single, central language system built around a left-hemisphere cortical network, linking critical areas in the left posterior temporal lobe ("Wernicke's area" in figure 54.1) and in the left inferior frontal lobe ("Broca's area" in figure 54.1). This classic concept, established through the pioneering research of the nineteenth-century neurologists Paul Broca and Carl Wernicke, reflects the selective effects of left-hemisphere brain damage on basic grammatical functions involved in language comprehension and language production.

Recent research in our laboratories (and others), focusing on the neural systems supporting different aspects of syntactic, semantic, and morphological processing, suggests major limitations to this classic approach to language and the brain. The Broca-Wernicke diagram captures (imperfectly) one important aspect of the neural substrate for language function—the key role of the left-hemisphere network—but it obscures another, equally important one. This is the role of bihemispheric systems and processes, where both left and right hemispheres work together to provide the fundamental underpinnings for human communicative processes. A more fruitful approach to human language and communication requires an extended neurobiological framework, where these capacities are supported by intersecting but evolutionarily and functionally distinguishable neural systems. This leads to a view of language as a dynamic coalition of interacting brain systems, where we ask how core linguistic and communicative functions are distributed across these systems and whether this distribution of functions varies between languages.

In this chapter we will develop this framework as a basis for examining how a core property of human language, dealing with words and their internal structure, is neurally instantiated and how this varies across a contrastive sample of different languages. This aspect of language, known as morphology, is built around the *morpheme*—the minimal meaning-bearing linguistic unit—and the ways in which different morphemes—typically a stem morpheme (e.g., *sad, jump*) and a bound grammatical morpheme (e.g., *–ness, -ed*)—combine to create new surface forms (*sadness, jumped*). A basic distinction is made between *inflectional* and *derivational* processes, where inflectional morphology subserves primarily grammatical functions, while derivational morphology is associated with the creation and representation of new words in the language. As we will see below, these different aspects of morphological function can be linked to broader, neurobiologically based distinctions between left-hemisphere systems essential for combinatorial morphosyntax, and bilateral systems underpinning lexical representation and semantic interpretation.

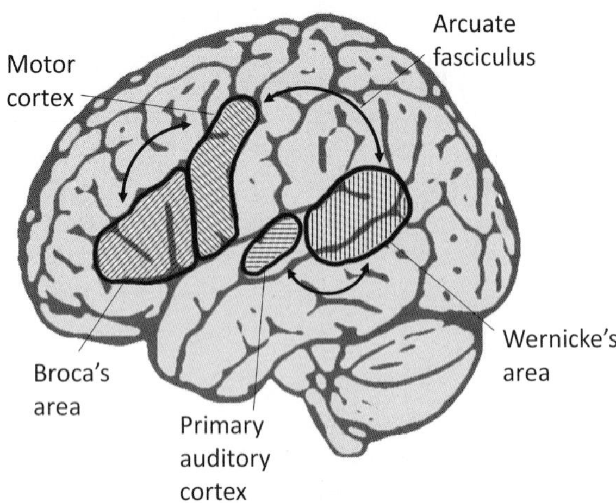

FIGURE 54.1 The classic Broca-Wernicke model of the neural basis for human language.

Neurobiological substrates for language and communication

From an extended neurobiological perspective, strong evolutionary continuity is provided by a distributed, bihemispheric set of capacities, shared with our primate relatives, that support the dynamic interpretation of multimodal sensory inputs, most relevantly in the context of social communication between members of the same species. We propose that this basic architecture underpins related communicative functions in the human.

Research in the neurosciences assumes a direct continuity between the structural and functional architecture of the major neural systems seen in nonhuman primates and those that underpin the corresponding systems in the modern human. In the domain of visual perception, for example, the concept of dorsal and ventral processing pathways is taken as fundamental to a neuroscientific account of the human perceptual system, just as it is in the macaque. Similarly, the architecture of cortical auditory processing in the macaque (Kaas & Hackett, 2000) forms the foundation for current understanding of human auditory function. These parallels can also be traced for systems that support communicative functions more generally.

An influential example of this is the extension of the dorsal/ventral distinction into the auditory perceptual domain (Rauschecker & Tian, 2000). This distinction dominates current thinking about human processing of complex auditory inputs, and of speech and language in particular (Friederici, 2011; Hickok & Poeppel, 2007). Another important domain relates to the intricate neural systems that support the integration of auditory and visual modalities in macaque communicative signals (e.g., Ghazanfar, Chandrasekaran, & Logothetis, 2008). Still other research suggests bilateral frontotemporal interactions in the macaque brain, elicited by the presence of conspecific calls, that serve to underpin the interpretation of these calls in their broader behavioral context (e.g., Gil-da-Costa et al., 2006).

Our hypothesis here is that these (and other) systems, already well established in the macaque to support their ecological niche as intelligent primates living in dynamically structured social groupings, are largely present in the modern human. In the context of human language comprehension, these bihemispheric systems not only support the ability to identify words in the speech input—integrating visual and auditory cues in face-to-face interaction—but also provide the basis for our ability to make sense of these word-meanings in the general context of the listener's knowledge of the world and of the specific context of speaking.

At the same time, the bihemispheric systems inherited from our primate ancestors are not sufficient to support modern human language. Primate communication systems are not remotely comparable to human language in their expressive capacities (Tomasello, 2008). A critical divergence lies in the domain of grammatical (or syntactic) function. Human language is much more than a set of signs that stand for things. It constitutes a powerful and flexible set of grammatical devices for organizing the flow of linguistic information and its interpretation, allowing us to represent and combine abstract linguistic elements, where these elements (words and morphemes) convey not only meaning but also the structural cues that indicate how these elements are linked together.

The left-hemisphere (LH) frontotemporal system provides the key neurobiological substrate for these core grammatical capacities. This system depends, furthermore, on neuroanatomical developments that are significantly more extensive in the human than in even our closest primate relatives. There are substantial increases in the size and complexity of the frontal and temporal areas critical for language function, coupled with the emergence of major fiber tracts connecting these areas (Rilling et al., 2008). While the functional consequences of these changes are not fully understood, it is clear that their presence is essential. When the LH system is damaged, the parallel right-hemisphere (RH) regions cannot take over their functions, even when damage is sustained early in childhood.

A central claim in this extended neurobiological framework is that the LH system neither replaces nor displaces the bihemispheric system for social communication and action, so that we are dealing in the modern human with functionally distinct neurobiological systems, however well integrated they may be in ordinary language use. The strongest evidence for this comes from their functional separability. Using behavioral and neuroimaging techniques, we see this both in unimpaired young adults (Bozic, Tyler, Ives, Randall, & Marslen-Wilson, 2010) and in patients with left-hemisphere brain damage. Neuropsychological evidence, dissociating a syntactically focused left-lateralized system from a bilateral lexically and semantically oriented system, comes from research that focuses on the patterns of preservation and loss of these aspects of language function in patients with LH lesions and varying residual language abilities (Griffiths et al., 2012; Tyler et al., 2011; Wright, Stamatakis, & Tyler, 2012).

Performance on tests of syntactic comprehension correlates with tissue integrity in left inferior frontal gyrus (LIFG), in left posterior superior and middle temporal gyri and in the major white matter tracts connecting them, as well as with neural activity in these regions during the processing of syntactically complex materials. The combination of structural and functional effects shows that syntax depends on a LH frontotemporal network with both dorsal and ventral components (Friederici, 2011; Griffiths et al., 2012; Rolheiser, Stamatakis, & Tyler, 2011) and where, notably, recruitment of RH activity is unable to support recovery of syntax after damage to the left (Tyler et al., 2011).

Critically, however, the amount of LH damage, and the extent to which it involved key LH frontotemporal circuits, did not affect the patients' ability to identify the words being spoken or to understand the messages being communicated—so long as syntactic cues are not required to do so (Wright et al., 2012). This is consistent with the observation that nonfluent LH patients often have preserved semantic comprehension (Crinion & Price, 2005; Hagoort, Wassenaar, & Brown, 2003; Tyler & Marslen-Wilson, 2008) and with evidence for RH involvement in the semantic and pragmatic interpretation of spoken language (Indefrey & Cutler, 2004). Overall, these findings demonstrate the functional dissociability of contrasting left-lateralized and bilateral neural systems supporting core linguistic and general communicative processes.

Dissociating morphological systems within and between languages

If human language is indeed supported by an underlying coalition of dissociable neural systems, then what are the implications of this for the set of computational functions attributed to human language? How far are they concentrated within a single central system, as opposed to being distributed across different systems? Perhaps the most salient of these properties—certainly in the Chomskyan tradition—is the *generative capacity* of human language, whereby an infinite range of communicative meanings can be expressed using the combination and recombination of a finite set of linguistic elements. This is not just a property of syntactic combination across phrases and clauses, but also of combinatorial processes realized at lexical and morphological levels.

Both types of morphological process under examination here—derivation and inflection—exemplify the core generative function of *recursion* (Hauser, Chomsky, & Fitch, 2002), where two or more basic linguistic elements (in this case morphemes) combine to form larger units that themselves enter into further combinatorial operations. We ask here whether the neural realization of this generative process in inflectional and derivational morphology is indeed reserved, cross-linguistically, to the LH frontotemporal system—as claimed, for example, by Berwick, Friederici, Chomsky, and Bolhuis (2012).

For inflectional morphology, a particular focus has been the English past tense morpheme {-ed}, as in forms like *jumped*, analyzable as the stem {*jump*} + {-ed}. In our earlier research on this topic, we focused on the patterns of relative impairment and preservation of function seen in nonfluent aphasic patients with frontotemporal LH damage. These patients exhibit a selective impairment for regular inflectional morphology but complete sparing of stem-based access (Longworth, Marslen-Wilson, Randall, & Tyler, 2005; Marslen-Wilson & Tyler, 1998). They show, for example, normal semantic priming for simple verb stem primes in pairs like *hope/wish*, and also for irregular past tense primes like *shook/tremble*, where these forms are also accessed and represented as whole forms. In contrast, we see no priming for regular past tense primes in pairs like *accused/blame*, where lexical access requires morphophonological parsing that breaks the prime words down into their component morphemes ({*accuse*} + {*-ed*}). Other patient studies, using correlational techniques to relate variations in neural integrity to performance with regular past tense primes (Tyler, Marslen-Wilson, & Stamatakis, 2005), confirm these dissociations and pinpoint LIFG and superior temporal gyrus as critical for intact performance in the comprehension of regularly inflected forms. As with syntax more generally, there is no sign that the RH can take over the critical functions that support regular inflectional morphology.

In further studies, using neuroimaging measures with unimpaired participants, we confirmed both the specificity of the link between regular inflectional morphology and left frontotemporal cortex, and the dissociability of these left-lateralized processes from a bilateral set of processes engaged by the perceptual processing of morphologically simple stems (Bozic et al., 2010). Listeners heard words that were either linguistically complex regularly inflected words like *played* or linguistically simple words like *ramp* that are nonetheless perceptually complex by virtue of their onset-embedded competitor (here the word *ram*). Using a factor analysis technique that identified the separable dimensions of the brain's neural response to these sets of words, a specifically left-lateralized frontotemporal component was elicited by the linguistically complex words. In contrast, words that were perceptually but not linguistically complex (like *ramp*) activated a bihemispheric set of regions, partially overlapping in left Brodmann area 45 with the linguistic component. This dissociation, seen in the unimpaired brain, complements the dissociations described earlier in patients with left frontotemporal damage.

The neurocognitive properties of regular inflectional morphology (as seen in English) are consistent with the defining properties of inflectional morphology as linguistically defined (Bickel & Nichols, 2007; Marslen-Wilson, 2007). First, across languages, inflectional morphology does not create new words with new meanings that require new lexical entries. Inflectional variants like *cat* and *cats* or *walk* and *walked* are not listed as separate headwords in standard dictionaries, and their meaning is fully predictable from the meaning of the stem combined with the grammatical properties of the inflection. Secondly, inflections are responsive, in a regular and predictable way, to the properties of the syntactic environment in which they occur (e.g., Bickel & Nichols, 2007). The information carried by the inflection is not just about the stem itself, but about the phrasal and sentential environment to which that stem relates.

The compositionality and contextual dependence of regularly inflected forms is consistent with a neurocognitive account where inflectional morphemes are stripped from their stems early in processing, and where the grammatical information they carry engages the combinatorial linguistic mechanisms supported by LH perisylvian brain regions (Marslen-Wilson & Tyler, 2007). It is unclear, however, how far these left-lateralized effects reflect the specifically linguistic demands associated with grammatical morphemes—for example, their implications for the structural organization of an incoming utterance—and how far they reflect control processes involved in the online parsing of complex input strings into stems and affixes (Bozic et al., 2010; Tyler, Cheung, Devereux, & Clarke, 2013).

Derivationally complex forms, though also constructed by combining a stem with a grammatical morpheme, have quite different properties. Whereas inflectional morphemes result in different forms of the same word, responsive to its current syntactic environment, derivational processes produce new words—in linguistic terms, new lexemes with new lexical entries (Matthews, 1991)—whose meaning and syntactic function is much more context-independent. The meaning of these words is less predictable from the simple combination of the meaning of their constituent morphemes. This lack of compositionality—and the range of variation from opaque forms like *department* to transparent forms like *bravely*—has prompted long-standing psycholinguistic controversies as to whether derived words are represented and processed on a decompositional morphemic basis, or as whole forms with no underlying representation of their morphological structure (e.g., Clahsen, Sonnenstuhl, & Blevins, 2003; Marslen-Wilson, Tyler, Waksler, & Older, 1994). Previous neuropsychological and neuroimaging research has not provided a consistent view of derivational processes, with studies varying in outcomes, languages, tasks, and methods.

In the broader dual-systems context, a recent functional MRI (fMRI) study (Bozic, Tyler, Su, Wingfield, & Marslen-Wilson, 2013) specifically addressed the question of whether derivationally complex words in English selectively engage the LH system in the same way as inflected forms—implying decompositional access processes—or whether they engage the distributed bihemispheric system, consistent with whole-form, noncompositional accounts. In the latter case, lexical access for derived stems would be handled by bilateral temporal mechanisms for stem access. The study varied the semantic transparency and affix productivity of a set of English-derived and pseudoderived words, forming a gradient from nondecomposable opaque forms like *breadth* and *archer* to potentially decomposable transparent forms (varying in affix productivity) like *warmth* and *farmer*. There was no evidence for selective activation of the LH frontotemporal system, even for the potentially most decomposable *farmer* forms. Instead, we saw increased activity at bilateral frontotemporal locations, both for semantically opaque derived words and for transparent words with unproductive affixes (like *warmth*). This activation reflects cohort-based perceptual competition between the whole forms and their embedded stems or pseudostems. Transparent forms with productive affixes, arguably retaining some representation of their morphological constituency (Bozic, Tyler, Su et al., 2013; Clahsen et al., 2003; Marslen-Wilson, 2007), do not

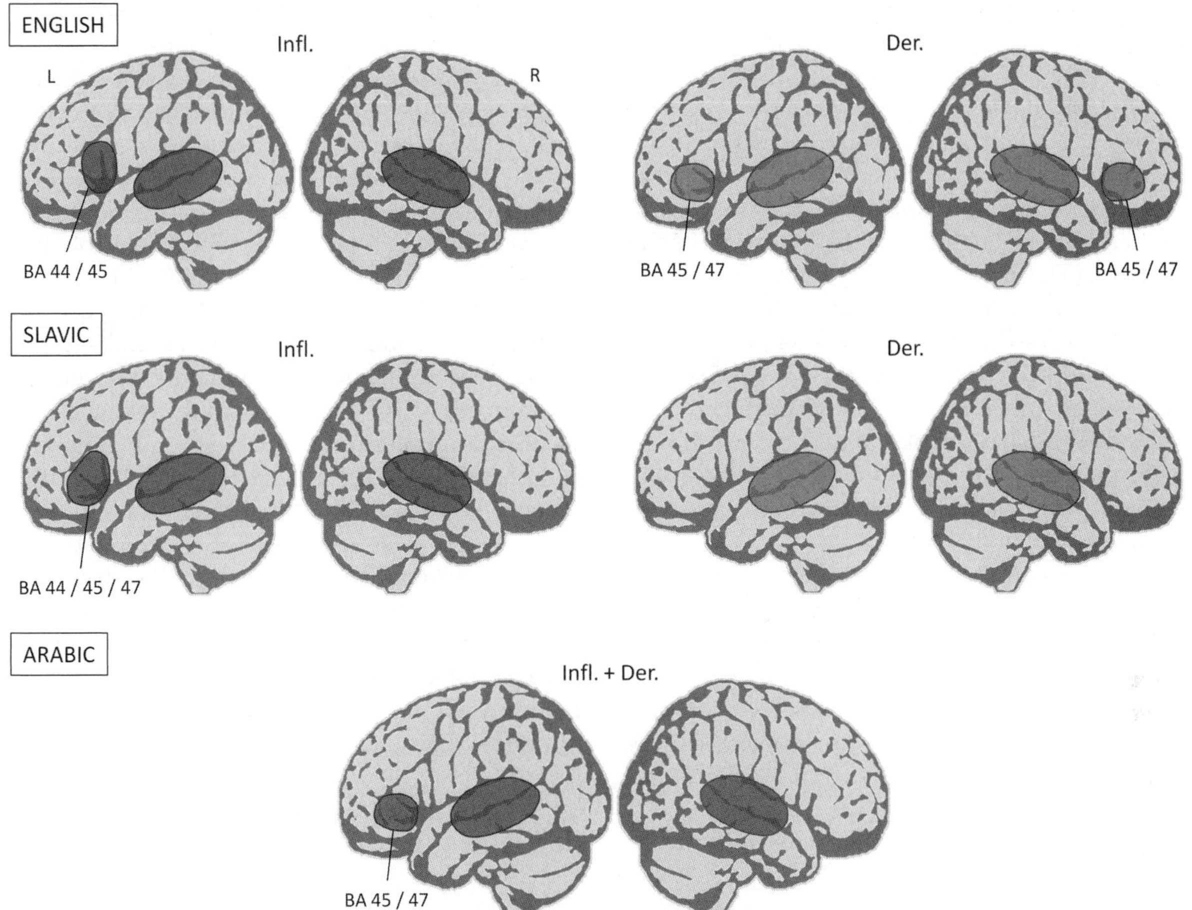

FIGURE 54.2 Schematic overview of inflectional (Infl.) and derivational (Der.) results for parallel fMRI experiments in English, Slavic (Polish and Russian), and Arabic. Presence or absence of left-hemisphere and right-hemisphere inferior frontal activation is shown (including left inferior frontal gyrus Brodmann area (BA) subcomponents), together with indicative left and right temporal (superior temporal gyrus and middle temporal gyrus) activations. (See color plate 47.)

generate comparable levels of competition, but nonetheless show no signs of selective LH activation (see figure 54.2).

This pattern of results for English suggests a strong distinction between inflectional morphology and derivational morphology, with inflectional morphemes selectively triggering decompositional, primarily syntactic linguistic processes, while derivational morphemes apparently play their role as part of the whole-word semantic representations created for each derived lexeme, preferentially engaging bilateral temporal systems. The generality and the cross-linguistic applicability of such an analysis needs to be carefully examined, however, given the particular properties of English morphology.

Inflectional morphology in English is very reduced in scope, and the majority of words in the language are produced as bare stems. This leads to a strong distributional contrast between words that are overtly morphologically complex (*jumped, yards,* etc.) and those that are not (*jump, yard*). Furthermore, the complex forms themselves are distributionally extreme, with typically just one dominant affix. The regular {*-ed*} past tense morpheme applies across the board to more than 10,000 English verbs, with a subset of about 160 idiosyncratic irregular forms, while the noun plural {*-s*} applies to every English noun with just a handful of exceptions (*geese, oxen,* etc.). The language learning environment induced by these distributional properties may lead to sharper or different distinctions between inflected and uninflected forms, compared to languages with more inflection classes and with no clear distinction between regular and irregular forms. Where derivationally complex forms are concerned, these occur in English in a much less decompositional processing environment than in languages where almost every surface form is morphologically complex (as in the Slavic and Semitic languages discussed below), and which will also vary in the complexity and prevalence of the parsing operations that these forms require.

Morphological systems in Slavic and Semitic languages

Two sets of studies have examined the scope of the English derivational-inflectional distinction from a dual-systems perspective, in two language families (Slavic and Semitic) that contrast strongly both with each other and with English. Polish, a member of the Slavic language family, is typologically similar to English in the distinction it makes between inflectional and derivational processes, and in its use of *concatenative* word-formation processes (where stems and affixes are separate phonological forms that are strung together in a sequence to construct complex surface forms). It contrasts with English both in the prevalence of morphological complexity and in the richness and diversity of its inflectional system.

All content words in Polish are morphologically complex, realized within rich inflectional paradigms. Every noun is inflectionally marked for case, number, and gender, and similarly for adjectives, numerals, and pronouns. The neuter nominal form *badanie*, "inspection," for example, from *badać*, "to inspect," breaks down into the derived stem *bad-ani*, made up of the verbal root {*bad-*} and the derivational suffix {*-ani-*}, which in turn combines with the inflectional morpheme {*-e*} that marks the nominative case for a neuter singular noun. Polish verbal inflection is complex and heterogeneous, marking three tenses, with six person/number categories, together with affixes that express mood, voice, tense, and aspect. Verb stems fall into 11 major inflection classes, most divided into subclasses, with no dominant "regular" inflectional class.

Despite these major distributional and linguistic differences between Polish and English, two fMRI experiments suggest close similarities in how the two languages interface with a dual processing framework. Szlachta, Bozic, Jelowicka, and Marslen-Wilson (2012) followed the Bozic et al. (2010) experiment in contrasting inflectional complexity with the presence or absence of competition-based perceptual complexity. Every word in the experiment was inflectionally complex and selectively activated LIFG and not RIFG. Words with onset-embedded competing stems—for example, the noun *kotlet* "cutlet," where *kot* "cat" is the competitor stem—showed a bilateral pattern of activation. Perceptual competition effects were strongest at bilateral temporal sites (superior and middle temporal gyri), with no sign of selective left frontotemporal activation.

Parallel outcomes were seen in a second study (Bozic, Szlachta, & Marslen-Wilson, 2013) focusing on derivational complexity. There was no evidence for selective left-lateralized activation, even for transparent and productive forms like *żabka*, "little frog" (from *żaba* "frog"), or *czytanie*, "reading" (from *czytać*, "to read"). Activation associated with the presence of derivational complexity was bilateral, primarily middle temporal, and paralleled the English results in terms of the distribution of competition effects. As in English, the transparent productive forms did not generate cohort competition, again indicating a potential role for morphological structure but without engaging LIFG.

The pattern of results for Polish suggests that the neural distribution of morpholexical functions seen for English is not attributable simply to the idiosyncratic statistics of English inflectional morphology. These conclusions are reinforced for Slavic more generally by a recent fMRI study in Russian (Klimovich-Smith, Bozic, Marslen-Wilson, 2013), which found strong selective LIFG activation for simple and complex inflected forms, but only bilateral temporal effects for derived words (and, as in Polish, no bilateral frontal effects).

Across both English and Slavic (see figure 54.2), regular inflectional morphemes, carrying structurally relevant grammatical information and requiring morphophonological parsing to separate these morphemes from their stems, selectively engage a LH circuit that critically includes the LIFG. Derivationally complex forms, in contrast, do not selectively engage the LIFG and are processed primarily by bilateral temporal structures, already identified as central to the mapping of stems onto meanings. Parallel research in Finnish—a typologically different "agglutinative" language with a rich affixing morphology—also suggests a similar division of labor, with inflectional complexity engaging left frontotemporal regions, and derivational complexity eliciting stronger bilateral and RH effects (e.g., Leminen, Leminen, Kujala, & Shtyrov, 2013).

However, before we conclude that these cross-linguistic results reflect potential universals in the mapping of different linguistic functions onto different neurobiological systems, we need to consider recent fMRI results from Arabic. As a Semitic language, Arabic (like Hebrew) presents a more radical cross-linguistic contrast, diverging in two fundamental ways from Indo-European languages like English and Polish. The first is a basic difference in Arabic word-formation, which is *nonconcatenative* in nature. An Arabic surface word—for example, the verb *katab*, "write"—is not constructed by sequentially combining a surface stem with a grammatical affix, but by interleaving two abstract underlying morphemic elements—the consonantal root {ktb}, with the semantic field of "writing," and the word pattern {-a-a-} with the grammatical meaning "verbal active perfective." The root {ktb} combines (as do most Arabic roots) with several different verbal and nominal word patterns, with each resulting surface form being a different word. Depending on the word pattern, the root {ktb} can surface as a noun, such as *kitaab*, "book," or *kaatib*, "writer," or as a verb, such as *kataba*, "write," or *kutiba*, "it was written," and so forth.

This is a much more synchronically active system of lexical representation than we see in Indo-European languages. Psycholinguistic evidence from Arabic (and Hebrew) shows that both types of morpheme are actively parsed and identified during the recognition of spoken and written forms (e.g., Boudelaa & Marslen-Wilson, 2005, 2011; Frost, Forster, & Deutsch, 1997). Dynamic morphological decomposition seems to be obligatory for all eligible surface forms in Arabic, and (unlike English or Polish) is independent of semantic factors, with opaque forms like *katiibatun*, "squadron" (where the {ktb} root does not have a meaning related to "writing"), being just as effective in root or word pattern priming tasks as transparent forms like *kitaab*, "book."

The second major difference is that the linguistic functions supported by inflectional and derivational processes in languages like English seem to be allocated quite differently in Semitic languages like Arabic. In particular, the word pattern morpheme—in addition to its phonological role in providing the prosodic structure of the surface form—supports both derivational and inflectional functions as standardly defined. As illustrated above, the word pattern not only serves to generate new lexemes (noun or verb stems), but also to convey information about grammatical properties of the resulting forms—such as tense, mood, and aspect—that are generally regarded as being inflectional in nature.

At the same time, Arabic also has concatenative morphological processes, purely inflectional in character, which attach affixes to nonconcatenatively derived stems to mark properties such as person, number, and possession. The word *katam* "hide," for example, is generated nonconcatenatively by interleaving the root {ktm} with the word pattern {-a-a-}. An inflectional person/number suffix can then be attached to this stem, generating the form *katamuu*, "they hid." This secondary affixing process parallels the inflectional affixing processes examined in English and Slavic.

Given this complex mix of properties, how do they map onto the dual neurobiological systems under investigation here, and how do they relate to the patterns we see in English and Slavic? A recent fMRI experiment, paralleling these earlier studies, probed the activation patterns elicited by purely nonconcatenatively complex stems (e.g., *tarak* "leave" or *kalb* "dog") and by stems with concatenated inflectional affixes—e.g., *katamuu*, "they hid" (from *katam* "hide" + {*uu*} "they") or *qalbuhu*, "his heart" (from *qalb* "heart" + {*hu*} "his"). These were compared with simple forms that could not be decompositionally analyzed, either concatenatively or nonconcatenatively, and that elicited temporal lobe activation bilaterally, with no significant inferior frontal gyrus involvement (Carota, Boudelaa, Bozic, Su, & Marslen-Wilson, 2014; see also Boudelaa, Pulvermüller, Hauk, Shtyrov, & Marslen-Wilson, 2010).

The results for the two complex sets were very similar. Both the purely nonconcatenative forms and the forms with added concatenative suffixes showed selective activation of LIFG, combined with strong bilateral temporal effects, but with no sign of RIFG activation (combined results are shown in figure 54.2). Most strikingly, this means that the nonconcatenative processes active in the perception of Arabic spoken words—involving morphemes with both derivational and inflectional properties—call selectively on the same LIFG regions that are required cross-linguistically for the analysis and interpretation of concatenatively suffixed inflectional forms.

This outcome suggests, in turn, that a critical variable, cross-linguistically, is not whether a given morphological process is linguistically labeled as derivational or inflectional, but whether the neural realization of this process is synchronically decompositional. Access to lexical representations in Arabic—perhaps because these combine inflectional as well as derivational information—is actively decompositional, with a complex time-sensitive dependence between the extraction of cues to the identity of the consonantal root and the identification of the accompanying word-pattern (Boudelaa & Marslen-Wilson, 2005, 2011). Inflectional morphology in English and Slavic is similarly decompositional, with cues to the presence of an inflectional suffix analyzed in the context of cues to the identity of the stem.

In each case we see selective activation of LIFG and no activation in RIFG. It is not clear, however, whether this is for specifically linguistic reasons—for example, because these cases invoke recursive combinatorial operations—or whether it reflects more domain-general functions. The parsing processes required to separate inflectional suffixes from stems or word-patterns from roots will generate temporary ambiguities and competing analysis paths that make strong demands on dynamic processes of cognitive control. It is doubtful, however, that data from fMRI will be sufficient to resolve these questions.

Using MEG combined with multivariate analysis techniques (cf. Su, Fonteneau, Marslen-Wilson, & Kriegeskorte, 2012), Tyler et al. (2013) were able to separate LIFG roles in cognitive control from middle temporal roles in lexical access during the analysis and reanalysis of syntactically ambiguous phrases. The use of similarly time-sensitive measures may be necessary to resolve the respective roles, cross-linguistically, of domain-specific and domain-general factors in the perceptual processing of morphologically complex forms.

Conclusion

The research outlined here makes the case that an explanatory neuroscientific account of human language and communication must be founded in the neurobiological properties of the relevant brain systems, themselves rooted in their primate evolutionary contexts, and that the resulting research program must be systematically cross-linguistic. It is only by studying the neural realization of different linguistic functions in the context of contrasting language systems—as evidenced here by our preliminary studies in English, Slavic, and Arabic—that we can begin to understand the special role of the left hemisphere in human language and how this relates to language functions distributed bilaterally across the brain.

ACKNOWLEDGMENTS This research was supported by grants to W. M. W. and to L. K. T. from the UK Medical Research Council, and by ERC Advanced Investigator Grants 230570 to W. M. W. and 249640 to L. K. T. We thank our research groups for their many contributions to the work presented here, and Peter Hagoort and an anonymous reviewer for their comments on this chapter.

REFERENCES

Berwick, R. C., Friederici, A. D., Chomsky, N., & Bolhuis, J. (2012). Evolution, brain, and the nature of language. *Trends Cogn Sci, 17*(2), 89–98.

Bickel, B., & Nichols, J. (2007). Inflectional morphology. In T. Shopen (Ed.), *Language typology and syntactic description,* Volume III: *Grammatical categories and the lexicon* (pp. 169–240). Cambridge, UK: Cambridge University Press.

Boudelaa, S., & Marslen-Wilson, W. D. (2005). Discontinuous morphology in time: Incremental masked priming in Arabic. *Lang Cogn Proc, 20,* 207–260.

Boudelaa, S., & Marslen-Wilson, W. D. (2011). Productivity and priming: Morphemic decomposition in Arabic. *Lang Cogn Proc, 26*(4), 624–652.

Boudelaa, S., Pulvermüller, P., Hauk, O., Shtyrov, Y., & Marslen-Wilson, W. D. (2010). Arabic morphology in the neural language system: A mismatch negativity study. *J Cogn Neurosci, 22,* 998–1010.

Bozic, M., Szlachta, Z., & Marslen-Wilson, W. D. (2013). Cross-linguistic parallels in processing derivational morphology. *Brain Lang, 127*(3), 533–538.

Bozic, M., Tyler, L. K., Ives, D. T., Randall, B., & Marslen-Wilson, W. (2010). Bihemispheric foundations for human speech comprehension. *Proc Natl Acad Sci USA, 107*(40), 17439–17444.

Bozic, M., Tyler, L. K., Su, L., Wingfield, C., & Marslen-Wilson, W. D. (2013). Neurobiological systems for lexical representation and analysis in English. *J Cogn Neurosci, 25*(10), 1678–1691.

Carota, F., Boudelaa, S., Bozic, M., Su, L., & Marslen-Wilson, W. D. (2014). Neurocognitive properties of concatenative and nonconcatenative morphology in Arabic: Evidence from multivariate fMRI analyses. Paper presented at the 21st Annual Meeting of the Cognitive Neuroscience Society, Boston, MA.

Clahsen, H., Sonnenstuhl, I., & Blevins, J. P. (2003). Derivational morphology in the German mental lexicon: A dual mechanism account. In R. H. Baayen & R. Schreuder (Eds.), *Morphological structure in language processing* (pp. 125–155). Berlin: Mouton de Gruyter.

Crinion, J., & Price, C. J. (2005). Right anterior superior temporal activation predicts auditory sentence comprehension following aphasic stroke. *Brain, 128,* 2858–2871.

Friederici, A. D. (2011). The brain basis of language processing: From structure to function. *Physiol Rev, 91,* 1357–1392

Frost, R., Forster, K. I., & Deutsch, A. (1997). What can we learn from the morphology of Hebrew: A masked priming investigation of morphological representation. *J Exp Psychol Learn Mem Cogn, 23,* 829–856.

Ghazanfar, A. A., Chandrasekaran, C., & Logothetis, N. K. (2008). Interactions between the superior temporal sulcus and auditory cortex mediate dynamic face/voice integration in rhesus monkeys. *J Neurosci, 28,* 4457–4469.

Gil-da-Costa, R., Martin, A., Lopes, M. A., Munoz, M., Fritz, J. B., & Braun, A. R. (2006). Species-specific calls activate homologs of Broca's and Wernicke's areas in the macaque. *Nat Neurosci, 9,* 1064–1070.

Griffiths, J. D., Marslen-Wilson, W. D., Stamatakis, E., & Tyler, L. K. (2012). Functional organization of the neural language system: Dorsal and ventral pathways are critical for syntax. *Cereb Cortex, 23*(1), 139–147.

Hagoort, P., Wassenaar, M., & Brown, C. (2003). Real-time semantic compensation in patients with agrammatic comprehension: Electrophysiological evidence for multiple-route plasticity. *Proc Natl Acad Sci USA, 100,* 4340–4345.

Hauser, M. D., Chomsky, N., & Fitch, W. T. (2002). The faculty of language: What it is, who has it, and how did it evolve? *Science, 298,* 1569–1579.

Hickok, G., & Poeppel, D. (2007). The cortical organization of speech processing. *Nat Rev Neurosci, 8,* 393–402.

Indefrey, P., & Cutler, A. (2004). Pre-lexical and lexical processing in listening. In M. S. Gazzaniga (Ed.), *The cognitive neurosciences* (3rd ed., pp 759–774). Cambridge, MA: MIT Press.

Kaas, J., & Hackett, T. A. (2000). Subdivisions of auditory cortex and processing streams in primates. *Proc Natl Acad Sci USA, 97,* 11793–11799.

Klimovich-Smith, A., Bozic, M., & Marslen-Wilson, W. D. (2013). Neural interfaces between morphology and syntax: Evidence from Russian. Paper presented at the Fifth Annual Meeting of the Society for the Neurobiology of Language, San Diego, CA.

Leminen, A., Leminen, M., Kujala, T., & Shtyrov, Y. (2013). Neural dynamics of inflectional and derivational morphology processing in the human brain. *Cortex, 49,* 2758–2771.

Longworth, C. E., Marslen-Wilson, W. D., Randall, B., & Tyler, L. K. (2005). Getting to the meaning of the regular past tense: Evidence from neuropsychology. *J Cogn Neurosci, 17*(7), 1087–1097.

Marslen-Wilson, W. D. (2007). Morphological processes in language comprehension. In G. Gaskell (Ed.), *Oxford handbook of psycholinguistics.* Oxford, UK: Oxford University Press.

Marslen-Wilson, W. D., & Tyler, L. K. (1998). Rules, representations, and the English past tense. *Trends Cogn Sci, 2,* 428–435.

Marslen-Wilson, W. D., & Tyler, L. K. (2007). Morphology, language and the brain: The decompositional substrate for language comprehension. *Philos Trans R Soc B Biol Sci, 362,* 823–836.

Marslen-Wilson, W. D., Tyler, L. K., Waksler, R., & Older, L. (1994). Morphology and meaning in the English mental lexicon. *Psychol Rev, 101,* 3–33.

Matthews, P. H. (1991). *Morphology* (2nd ed.). Cambridge, UK: Cambridge University Press.

Rauschecker, J. P., & Tian, B. (2000). Mechanisms and streams for processing of "what" and "where" in auditory cortex. *Proc Natl Acad Sci USA, 97,* 11800–11806.

Rilling, J. K., Glasser, M. F., Preuss, T. M., Ma, X., Zhao, T., Hu, X., & Behrens, T. E. J. (2008). The evolution of the arcuate fasciculus revealed with comparative DTI. *Nat Neurosci, 11,* 426–428.

Rolheiser, T., Stamatakis, E. A., & Tyler, L. K. (2011). Dynamic processing in the human language system: Synergy between the arcuate fascicle and extreme capsule. *J Neurosci, 31,* 16949–16957.

Su, L., Fonteneau, E., Marslen-Wilson, W. D., & Kriegeskorte, N. (2012). Spatiotemporal searchlight representational similarity analysis in EMEG source space. *Proc IEEE International Workshop on Pattern Recognition in NeuroImaging* (pp. 97–100).

Szlachta, Z., Bozic, M., Jelowicka, A., & Marslen-Wilson, W. D. (2012). Neurocognitive dimensions of lexical complexity in Polish. *Brain Lang, 121*(3), 219–225.

Tomasello, M. (2008). *Origins of human communication.* Cambridge, MA: MIT Press.

Tyler, L. K., Cheung, T. P., Devereux, B. J., & Clarke, A. (2013). Syntactic computations in the language network: Characterizing dynamic network properties using representational similarity analysis. *Front Psychol, 4,* 271.

Tyler, L. K., & Marslen-Wilson, W. D. (2008). Frontotemporal brain systems supporting spoken language comprehension. *Philos Trans R Soc Lond B Biol Sci, 363,* 1037–1054.

Tyler, L.K., Marslen-Wilson, W.D., & Stamatakis, E. (2005). Differentiating lexical form, meaning, and structure in the neural language system. *Proc Natl Acad Sci USA, 102*(23), 8375–8380.

Tyler, L. K., Marslen-Wilson, W. D., Randall, B., Wright, P., Devereux, B. J., Zhuang, J., ... Stamatakis, E. A. (2011). Left inferior frontal cortex and syntax: Function, structure and behaviour in left-hemisphere damaged patients. *Brain, 134*(2), 415–431.

Wright, P., Stamatakis, E., & Tyler, L. K. (2012). Differentiating hemispheric contributions to syntax and semantics in patients with left-hemisphere lesions. *J Neurosci, 32,* 8149–8157.

55 The "Negatives" and "Positives" of Prediction in Language

MARTA KUTAS, KARA D. FEDERMEIER, AND THOMAS P. URBACH

ABSTRACT We review event-related brain potential (ERP) evidence for the rapid deployment of information from semantic memory and current context during language comprehension. We summarize studies showing how linguistic expectancies can emerge at various levels as sentential context accumulates. We detail various qualitatively different processes (as indexed by the N400, frontal and parietal late positive components, and a frontal negativity) engaged when predictions are or are not made—and are or are not met—as people make sense of sentences.

In making sense of language, people typically draw on diverse verbal and nonverbal information from past experience and the current environment—that is, context. Although effects of contextual variables may be clear, the causal mechanisms of context effects are not. A long-standing and influential idea is that biological information-processing systems routinely utilize contextual information to prepare in advance for what might come (see Bar, 2011). The hypothesis that expectancies (prediction, anticipation, prospection) play a causal role in perception and cognition more generally is widespread in the cognitive neurosciences—for example, the role of lateral prefrontal cortex in the "expectation of, and preparation for, anticipated events" (Fuster, 2001, p. 325), "perceptual predictions" (Zacks, Speer, Swallow, Braver, & Reynolds, 2007, p. 273), and the "prospective brain" (Schacter, Addis, & Buckner, 2007). However, within the domain of language comprehension, the role of prediction has been more controversial.

Contextual variables impact real-time language processing, and sentential contexts in particular have been shown to modulate, for example, signal-to-noise thresholds for speech recognition in background noise (Miller, Heise, & Lichten, 1951), the duration required for visual word identification (Tulving & Gold, 1963), lexical decision times (Fischler & Bloom, 1979), eye movements in reading (Ehrlich & Rayner, 1981), and event-related brain potential (ERP) amplitudes in word-by-word reading (Kutas & Hillyard, 1980; Lau, Stroud, Plesch, & Phillips, 2006). At issue is whether these effects arise due to the post hoc match between information derived from a current stimulus and that previously gleaned from the context (e.g., "integration"; see Brown & Hagoort, 1993) or are instead mediated by real-time predictive processing (Federmeier, 2007).

For present purposes, prediction involves the activation of or information about likely upcoming stimuli, prior to their receipt, that plays a causal role in stimulus processing. Linguistic predictions need not be conscious and typically are not. Their representational content may be fine-grained and specific, for example, for a particular expected word or sound, or for semantic features or gist or other types of stimulus attributes such as grammatical structure. At a given point in time, there may be multiple (even incompatible) predictions with varying strengths, all subject to dynamic revision.

Without unequivocal evidence of representational activation prior to the actual occurrence of some "predicted" input, facilitated processing can be attributed to nonpredictive factors such as increased ease of integration of the input with the message-level representation. Thus, determining when and how predictions, as defined above, play a role in real-time language comprehension is nontrivial. ERPs have proven an especially useful tool for disentangling the contributions of different language-processing mechanisms, including prediction, to context effects. In particular, ERPs provide a multidimensional window into the consequences of lexicosentential (in)congruity, which have been measured relatively early in time, as amplitude modulations of the N400 (see also P2 amplitude effects) at and prior to the apprehension of critical stimuli, and in post-N400 positivities and negativities; we discuss each in turn.

The N400

The N400 refers to a relative negativity between 250 and 600 ms in the ERP to a potentially meaningful item (e.g., spoken or written word, acronym, pseudoword, picture, face, gesture, etc.). The N400 reflects the summed postsynaptic activity of a highly distributed brain network that includes higher-level perceptual areas and multimodal processing and storage areas in the left superior/middle temporal gyri and medial and anterior temporal lobes of both hemispheres (Lau, Phillips, & Poeppel, 2008; van Petten & Luka, 2006). The N400 is especially large to nouns that do not fit with the meaning of their preceding context compared to the contextually most predictable noun (e.g., "*dog*" in "*I take my coffee with cream and dog*" versus "*sugar*"; Kutas & Hillyard, 1980). N400s are also present to all but the most highly predictable nouns, even when they are semantically congruent. All other factors (word frequency, concreteness, orthographic neighborhood size, word class) held constant, N400 amplitude is inversely correlated ($r > .8$) with an item's cloze probability in an offline paper-and-pencil completion task. For some researchers the N400 sensitivity to cloze probability reflects the activity of a predictive language system, whereas for others it reflects variable resource needs for a contextual integration operation. Of course, the N400 could reflect both, and from attempts to address this issue have emerged novel ERP paradigms that have informed the functional organization of semantic memory and incremental language processing beyond semantic memory access.

A case in point is the "related anomaly" paradigm (Kutas & Hillyard, 1984; Kutas, Lindamood, & Hillyard, 1984). While all low cloze words that complete a highly constraining context elicit large N400s relative to the high cloze best completions (BC), those for words related to the meaning of the BC are reliably smaller. For example, the N400 to "*weights*" is smaller than that to "*collars*," both anomalous endings for "*The barbells the strongman lifted were very* ____," presumably because of the relationship between "*weights*" and the BC (*heavy*), although the relationship was not controlled in early studies.

Federmeier and Kutas (1999a, 1999b) investigated categorical relations. They used sentence pairs to establish an expectation for a particular member of a category, which was violated in a third of the sentences by a member of a different category (between-category violation) and in a third by a different category member (within-category violation). Although the two category-violation types were equally implausible on average, the within-category violation enjoyed greater perceptuo-semantic-feature overlap with the expected category exemplar (e.g., *He caught the pass and scored another touchdown. There was nothing he enjoyed more than a good game of football/ baseball/ monopoly.*). This pattern was observed for written words, spoken words, and line drawings (Federmeier & Kutas, 2001; Federmeier, McLennan, de Ochoa, & Kutas, 2002). Critically, the N400 reduction increased as a function of the cloze probability of—and thus likely strength of the prediction for—the unpresented BC. Moreover, in a visual half-field version, in which the category member was briefly (200 ms) flashed in the right or left visual field, this related anomaly N400 effect was limited to the right visual field (which biases processing to the left hemisphere). These results underscore the impact of memory structure (in this case, by category-based featural similarity) on word processing in sentences and suggest the use of predictive processing mechanisms during normal comprehension, mediated by the left hemisphere (Federmeier, 2007).

Metusalem et al. (2012) adapted this paradigm to test the hypothesis that language comprehension is similarly influenced by event knowledge–based organization of semantic memory. Participants read short passages—for example, *A huge blizzard ripped through town last night. My kids ended up getting the day off from school. They spent the whole day outside building a big* ...—continued by the BC (*snowman*) or by one of two types of linguistically unlicensed, less probable words either related to the general event (*playing in the snow*) activated by the passage (*jacket*) or not (*towel*). Both types of improbable words elicited larger N400s than the BC, but those to the linguistically unlicensed but event-related continuations were reliably smaller (figure 55.1A). These results point to the immediacy and incrementality of event knowledge use during language comprehension and to events as an organizing principle of semantic memory.

Laszlo and Federmeier (2009) similarly employed the related anomaly paradigm to show that the information brought online by sentence context affects processing of inputs based on their orthography. Participants read sentences such as "*The genie was ready to grant his third and final*" that were completed either by the BC (*wish*) or by an unexpected word, pseudoword, or illegal string that was either an orthographic neighbor or not (figure 55.1B, words only). Relative to nonneighbors, all neighbors—regardless of their lexical status—showed facilitation, that is, smaller N400s. This result is difficult to explain via mere integrative processes (for another example, see Rommers, Meyer, Praamstra, & Huettig, 2013). Overall, these results show that context acts through the structure of event knowledge, semantic categories, and orthography, and further suggests that the system can anticipate features of upcoming words along these dimensions.

Arguably, stronger electrophysiological evidence for prediction comes from ERP effects present *before* the predicted item of interest. There are a handful of such studies with the same general design—a lead-in context that sets up an expectation for a particular noun,

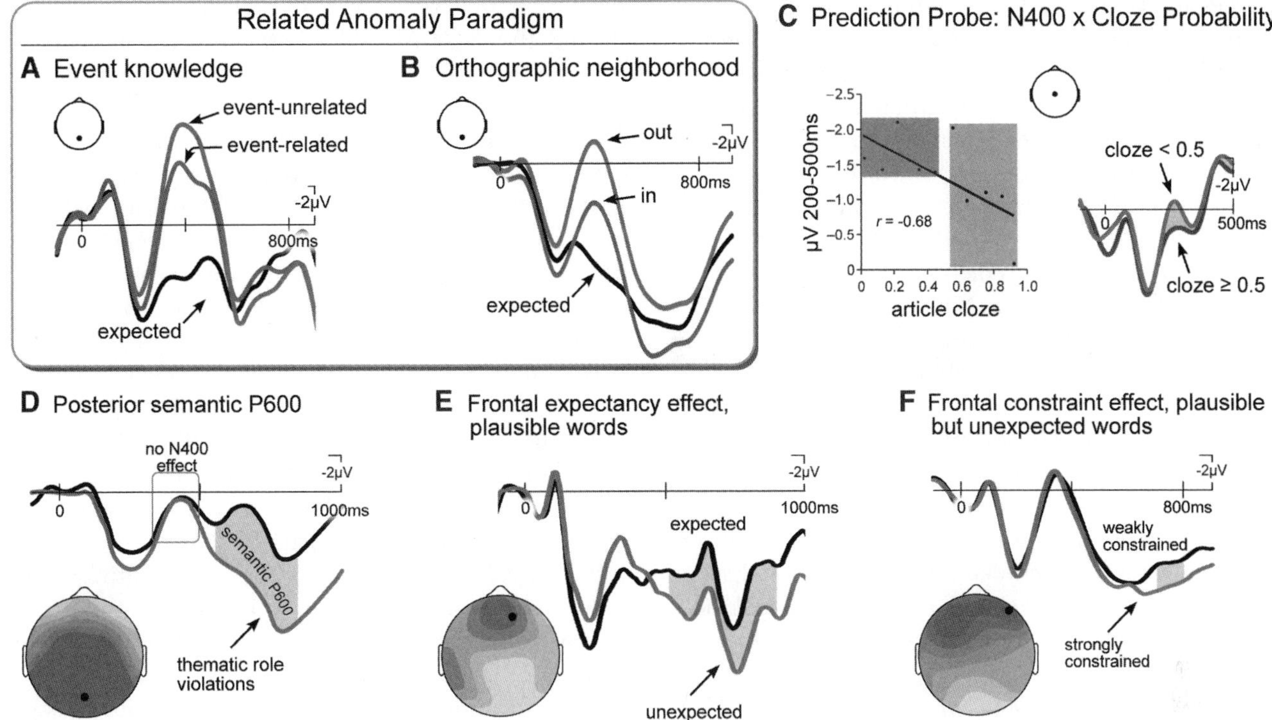

FIGURE 55.1 Grand average ERPs. (A) Expected, linguistically unlicensed event-related and event-unrelated continuations in discourse context. (B) Expected, unexpected in and out of orthographic neighborhood. (C) Correlations and ERPs to articles based on cloze-probability median split. (D) Semantic P600 for thematic role violations. (E) Frontal positivity to unexpected vs. expected plausible words (DeLong & Kutas, unpublished data). (F) Frontal positivity to unexpected plausible words as a function of contextual constraint. (See color plate 48.)

preceded by a word (e.g., article, adjective) that either matches or mismatches the upcoming noun (e.g., in grammatical gender, morphology, or phonology). Critically, either class of prenominal words could reasonably be followed by a congruent matching noun, and is not a violation *per se*. As such, these should not elicit differing ERPs, unless the comprehender expects a noun that is only consistent with one of the alternative prenominal words. Several such investigations report differing ERPs to the noun-matching versus mismatching prenominal alternatives.

In a series of experiments, sentential contexts (spoken or written) in Spanish were designed to create an expectation for a noun of feminine or masculine gender (verbal or depicted), occasionally "violated" by an article of the other grammatical gender. The results indicate that the language-processing system was anticipating a word of a particular gender, as a differential ERP pattern was found for articles of the expected versus unexpected gender (Wicha, Moreno, & Kutas, 2003; Wicha, Moreno, & Kutas, 2004; Wicha, Bates, Moreno, & Kutas, 2003). Using a similar logic, van Berkum, Brown, Zwitserlood, Kooijman, & Hagoort (2005) inferred syntactic prediction from differential ERPs at the gender-marked adjectives in Dutch (also see Otten, Nieuwland, & van Berkum, 2007; Otten & van Berkum, 2008, 2009).

DeLong, Urbach, and Kutas (2005) argued for prediction of phonological word forms based on differential ERPs to the indefinite articles *a* and *an*, which have the same meaning but are predictive of the initial phoneme class of the subsequent word (words beginning with a consonant versus vowel sound, respectively). Participants read sentences ranging in constraint from low to high and continuing with a range of more or less expected indefinite article-noun pairings; for example, "*Because it frequently rains in London, Nigel always carries* ..." for which *an umbrella* is highly expected and *a newspaper*, while plausible, is less so. There was a statistically reliable inverse correlation between N400 amplitudes and not only the target nouns but also the articles (figure 55.1C). These data provide clear evidence of graded linguistic prediction at a phonological level. Yet another variant of this design indicates prediction for semantically defined classes of words in Polish (Szewczyk & Schriefers, 2013).

Taken together, these studies point to the availability of a predictive language processer that activates word features and forms in advance of their input. They demonstrate that linguistic expectancies at many levels, from semantic and syntactic to orthographic and phonological, can emerge as sentential context accumulates. The fact that the system is capable of making predictions, however, does not mean that it always can or will do so (e.g., there are hemispheric and age-related differences in the tendency to predict; Federmeier, 2007). Nor will predictions, when made, always be correct. Indeed, whereas N400 amplitude varies with many lexicosemantic variables, its latency is relatively stable, even though the availability of information and processes necessary for prediction and incremental comprehension are considerably more variable across time, as a function of the input, task demands, and individual differences, among other factors. This, together with the need to refine and revise initial interpretations when predictions are incorrect, would seem to imply comprehension and prediction-related processes post-N400. Next we discuss evidence for these.

Posterior late positivities: The late positive complex and semantic P600

Processing differences between expected and semantically anomalous words are sometimes seen post-N400 (500–900 ms), with anomalous words eliciting increased centroparietal positivity. This is often referred to as an effect on the late positive complex (LPC). Semantic anomaly effects on the LPC seem to be more dependent on attention and less obligatory than N400 anomaly effects. Whereas N400 effects can be observed even during the attentional blink or under masking, these obliterate LPC effects (Luck, Vogel, & Shapiro, 1996; Misra & Holcomb, 2003). Diminished LPC effects have been described in children (e.g., Juottonen, Revonsuo, & Lang, 1996) as well as in adults with schizophrenia (e.g., Koyama et al., 1994), whose N400 effects in the same conditions are similar to adult controls. There is notable variability in the tendency to elicit LPC effects to semantic anomalies, which does not seem to be straightforwardly linked to stimulus factors such as contextual constraint (Kos, van den Brink, & Hagoort, 2012; Van Petten & Luka, 2012) or to individual differences in working memory capacity (Kos et al., 2012). Even with similar stimulus and task conditions, only about a third of published studies show an LPC post-N400 effect (Van Petten & Luka, 2012), perhaps because, in a given sample, only about half of the people seem to be eliciting such effects (Kos et al., 2012).

The functional significance of LPC anomaly effects remains underspecified and says little about prediction *per se*. Typically, these effects are interpreted as indicating that an anomaly—detected by the system during the N400 time window—is being (explicitly) noted by the participants, who are "reviewing the prior context to determine what went wrong and if the problem might be repaired" (Van Petten & Luka, 2012). This view resembles accounts proffered for posterior late positive effects (P600s) coincident with syntactic processing difficulties, including agreement errors, phrase structure violations, and garden path constructions (e.g., Gouvea, Phillips, Kazanina, & Poeppel, 2009).

In some cases, semantic anomalies elicit posterior LPC effects *without* accompanying N400 effects. (i.e., indistinguishable N400s for expected vs. anomalous words). For example, using sentences like "For breakfast, the boys/eggs would only eat ... ." Kuperberg, Sitnikova, Caplan, and Holcomb (2003) observed no N400 differences but a larger posterior positivity to *eat* following *eggs* than *boys* (figure 55.1D; see also Hoeks, Stowe, & Doedens, 2004; Kim & Osterhout, 2005; van Herten, Kolk, & Chwilla, 2005, for similar reports). A number of different accounts have been put forward for these so-called semantic P600 effects. Many interpret the N400 facilitation (amplitude reduction) to the anomalies as indicating that the comprehension system was subject to a temporary semantic illusion—that is, in contrast to cases wherein LPC and N400 effects co-occur, that the system initially erroneously interprets the anomalous word as plausible. Several researchers have postulated that the N400 reflects a processing stage insensitive to certain types of (for example, combinatorial) processes essential for appreciating the semantic implausibility of a given word in a particular sentence position (e.g., Kuperberg, 2007). The P600 in this case is taken to reflect processes involved in syntactic revision or in aligning the conflicting interpretations arising in different processing streams. Others (Brouwer, Harmut, & Hoeks, 2012) argue convincingly against such "multistream" accounts. If the N400 is an index of lexicosemantic retrieval (subject to preactivation) and *not* a direct reflection of contextual integration (as elaborated in Kutas & Federmeier, 2011), then the lack of N400 effect in these cases could mean that semantic access for the anomalous words was facilitated (for example, because of the semantic relationship between *breakfast, eggs,* and *eating* in the above example); critically, it does *not* (necessarily) mean that the comprehension system was ever "fooled." On this view, the lack of N400 may reflect preactivation, and the (semantic) P600 effect indicates that the system has detected the implausibility and that it is, parsimoniously, functionally equivalent to LPCs following N400 anomaly effects.

Thus, some types of unexpected words, including semantic anomalies, as well as syntactic violations, spelling mistakes (Munte, Schiltz, & Kutas, 1998), and unexpected switches between different languages (Moreno, Federmeier, & Kutas, 2002), are associated with posterior late positivities that have gone by different names: LPC, (semantic) P600. An anomaly (or misparse or error), however, is not a necessary condition for these effects. For example, Kandhadai & Federmeier (2010a, 2010b) linked LPC effects to the flexible "reordering" of pairs of words (e.g., fish-hook, hook-fish), so that participants could best appreciate the semantic relationship between them. Repetition also modulates the LPC to words in lists and texts (Rugg, 1990; Van Petten, Kutas, Kluender, Mitchiner, & McIsaac, 1991), and posterior late positivities more generally have been associated with memory-retrieval processes (e.g., review by Wilding & Ranganath, 2012).

Although the degree of neural and functional overlap between these positivities remains controversial, they bear a "family resemblance" to one another and to the well-characterized P300 component of the ERP. The P300 (a.k.a. P3b) is a domain-general potential observed for the processing of motivationally significant, task-relevant stimuli. Like the positivities seen in language, the P300 has been linked to the processing of anomalous or unexpected stimuli ("oddballs") because its amplitude is inversely correlated with stimulus probability, where probability is multiply determined and subjectively assessed (see Coulson, King, & Kutas, 1998, showing sensitivity of syntactic P600 to probability). P300 amplitude is also affected by attention and task demands (although the P600 might not be; see Swaab, Ledoux, Camblin, & Boudewyn, 2011). P300 latency reflects an upper limit on the time necessary to evaluate a stimulus in the context of the task (Kutas, McCarthy, & Donchin, 1977); thus, it makes sense that P300s to, for example, oddball tones would occur earlier than late positivities for stimuli unexpected by virtue of their semantic or syntactic properties.

Given the morphological and functional similarity of the P300 to language-related posterior late positivities, research into the neurobiological bases of the P300 may help to inform the neurobiology of language. Bringing together a large body of research across multiple disciplines, Nieuwenhuis, Aston-Jones, and Cohen (2005; following Pineda, Foote, & Neville, 1989) theorized that the P300 reflects phasic activity of the locus-coeruleus (LC) norepinephrine (NE) system. On their account, detection of a motivationally salient stimulus elicits a release of NE, which acts as a gain control mechanism, potentiating responses to that stimulus. The NE system has been hypothesized to signal state changes ("unexpected uncertainty") within a behavioral context by increasing the influence of bottom-up signals to processing and facilitating the development of a new attentional set (Avery, Nitz, Chiba, & Krichmar, 2012; Yu & Dayan, 2005). An appealing aspect of this theory is that it can explain the electrophysiological and functional similarities in various P300-like potentials, while still allowing for differences between them, given the broad connectivity of the NE system. Thus, for example, it has been found that the syntactic P600 is disrupted by basal ganglia lesions, whereas the P300 to sensory feature-based oddballs is unaffected (Frisch, Kotz, von Cramon, & Friederici, 2003). On the LC-NE theory, this would suggest that the basal ganglia may be important for determining the significance of syntactic (but not simple perceptual) features in the input, as a necessary precursor to the elicitation of a phasic NE response.

Frontal late positivity and negativity

Wholly implausible words (both not predicted and also difficult to integrate) often elicit posterior post-N400 positivity effects. A different pattern emerges from comparisons between expected words and those that are unexpected but plausible. In this case, unexpected words elicit an increased positivity over frontal rather than posterior recording sites (reviewed in Van Petten & Luka, 2012; see figure 55.1E). For example, in a category-verification task (e.g., the cue *A type of bird,* followed by high-typicality targets like *robin,* low-typicality targets like *chicken,* and anomalous targets like *potato*), Federmeier, Kutas, and Schul (2010) observed (for young adults) frontal positivity to low typicality—but *not* wholly anomalous—completions.

This difference between expected and unexpected words seems likely to be made up of at least two, separable effects: a positive-going deflection for unexpected words that violate predictions, as well as a negative-going deflection to some expected words encountered under moderately high constraint. Wlotko and Federmeier (2012a, 2012b) describe a frontal negativity that is larger to words completing sentences where there are only a few less probable alternative completions than when there are many. They link this negativity to adjustments in the interpretation of the prior context information—a "frame shift," as in joke comprehension (e.g., Coulson & Kutas, 2001)—and/or possibly to working memory use (Ruchkin, Johnson, Grafman, Canoune, & Ritter, 1992) and/or semantic selection (Lee & Federmeier, 2009). Critically, this frontal negativity does not seem to require prediction (Wlotko & Federmeier, 2012a; Wlotko, Federmeier, & Kutas, 2012).

The frontal positivity to prediction violations, instead, can be characterized most cleanly when equally unexpected words (in the ideal case, the same lexical items; Federmeier, Wlotko, De Ochoa-Dewald, & Kutas, 2007) are compared as a function of contextual constraint. For example, the word *log* has matched, low-cloze probability in both contexts:

(1) He was cold most of the night and finally got up to get another …

(2) He fell on the floor after tripping on the …

However, context (1) leads readers to have strong predictions for the alternative ending, *blanket*, whereas context (2) does not lead to consistent expectations for an alternative word. Relative to (2) the ERP to (1) is characterized by increased prefrontal positivity between 500 and 900 ms (figure 55.1F).

The link between this frontal positivity and prediction-related processes is strengthened by its absence in most older adults (Federmeier, Kutas, & Schul, 2010; Wlotko et al., 2012), who similarly tend not to show other prediction-related ERP patterns (Federmeier et al., 2002). These effects can, however, be observed in a subset of older adults with high verbal fluency (DeLong, Groppe, Urbach, & Kutas, 2012; Federmeier et al., 2010), who also may show young-like N400 prediction effects (Federmeier et al., 2002). Federmeier and colleagues have linked the frontal positivity to processes involved in message-level meaning construction in the face of failed predictions. We speculate that this response may arise from activity related to the basal forebrain cholinergic system, which has been shown to be sensitive to the degree of "expected uncertainty" in the environment, and which controls the balance of top-down versus bottom-up influence on processing, without changing the attentional set (Avery, Nitz, Chiba, & Krichmar, 2012; Yu & Dayan, 2005). Indeed, recent work has shown that the use of top-down predictions in sentence processing, even within young adults who seem likely to be predicting, is importantly modulated by the utility of those predictions (Wlotko & Federmeier, 2011). For example "fooling" the predictive processing system by presenting unexpected synonyms of the best completion—that is, words whose semantics but not lexical form are expected—eradicates the frontal positivity when participants are reading for comprehension (and thus care most about semantics) but not when they are making lexical decisions (and do care about word form).

Progress and future directions

With particular attention to the thorny issue of predictive processing, the experiments reviewed herein demonstrate how the latency, polarity, and scalp distribution of ERP measures such as the N400 and late positivities, in suitable experimental designs, can shed light on the mechanisms by which context variables impact the time course of interpretation. Some provide strong evidence for contextually supported and seemingly graded predictive and incremental processes.

These studies are existence proofs that prediction does occur in language processing. The project now is to characterize principles governing predictive processes and the biological mechanisms at work. Of all the contextually available information (verbal and nonverbal), which does the system use to generate predictions? How fast do they come online, and does this vary with any trait and/or state variables and environmental factors? Is prediction automatic and obligatory, or does it come online under only some conditions? When predictions are made, what kinds of information are predicted, with what specificity, and what degree of strength? Do contextual variables constrain the space of possible predictions to winner take all, with unsuccessful competitors squashed or even inhibited? If predictions turn out to be incorrect, what is the cost (c.f., Smith & Levy, 2011)? Finally, what is the relation if any, of prediction to the speed and depth of the interpretations ultimately constructed? Clearly, much remains to be discovered, and it seems likely that ERPs will continue to play an important explanatory part.

REFERENCES

Avery, M. C., Nitz, D. A., Chiba, A. A., & Krichmar, J. L. (2012). Simulation of cholinergic and noradrenergic modulation of behavior in uncertain environments. *Front Comput Neurosci, 6*, 5.

Bar, M. (2011). *Predictions in the brain: Using our past to generate a future.* New York, NY: Oxford University Press.

Brouwer, H., Harmut, F., & Hoeks, J. (2012). Getting real about semantic illusions: Rethinking the functional role of the P600 in language comprehension. *Brain Res, 1446*, 127–143.

Brown, C., & Hagoort, P. (1993). The processing nature of the N400: Evidence from masked priming. *J Cogn Neurosci, 5*, 34–44.

Coulson, S., King, J. W., & Kutas, M. (1998). ERPs and domain specificity: Beating a straw horse. *Lang Cogn Proc, 13*, 653–672.

Coulson, S., & Kutas, M. (2001). Getting it: Human event-related brain response to jokes in good and poor comprehenders. *Neurosci Lett, 316*, 71–74.

DeLong, K. A., Groppe, D. M., Urbach, T. P., & Kutas, M. (2012). Thinking ahead or not? Natural aging and anticipation during reading. *Brain Lang, 121*, 226–239.

DeLong, K. A., Urbach, T. P., & Kutas, M. (2005). Probabilistic word pre-activation during language comprehension inferred from electrical brain activity. *Nat Neurosci, 8,* 1117–1121.

Ehrlich, S. F., & Rayner, K. (1981). Contextual effects on word perception and eye-movements during reading. *J Verb Learn Verb Behav, 20,* 641–655.

Federmeier, K. D. (2007). Thinking ahead: The role and roots of prediction in language comprehension. *Psychophysiology, 44,* 491–505.

Federmeier, K. D., & Kutas, M. (1999a). A rose by any other name: Long-term memory structure and sentence processing. *J Mem Lang, 41,* 469–495.

Federmeier, K. D., & Kutas, M. (1999b). Right words and left words: Electrophysiological evidence for hemispheric differences in meaning processing. *Cognit Brain Res, 8,* 373–392.

Federmeier, K. D., & Kutas, M. (2001). Meaning and modality: Influences of context, semantic memory organization, and perceptual predictability on picture processing. *J Exp Psychol Learn Mem Cogn, 27,* 202–224.

Federmeier, K. D., Kutas, M., & Schul, R. (2010). Age-related and individual differences in the use of prediction during language comprehension. *Brain Lang, 115,* 149–161.

Federmeier, K. D., McLennan, D. B., De Ochoa, E., & Kutas, M. (2002). The impact of semantic memory organization and sentence context information on spoken language processing by younger and older adults: An ERP study. *Psychophysiology, 39,* 133–146.

Federmeier, K. D., Wlotko, E. W., De Ochoa-Dewald, E., & Kutas, M. (2007). Multiple effects of sentential constraint on word processing. *Brain Res, 1146,* 75–84.

Fischler, I., & Bloom, P. A. (1979). Automatic and attentional processes in the effects of sentence contexts on word recognition. *J Verb Learn Verb Behav, 18,* 1–20.

Frisch, S., Kotz, S. A., von Cramon, D. Y., & Friederici, A. D. (2003). Why the P600 is not just a P300: The role of the basal ganglia. *Clin Neurophysiol, 114,* 336–340.

Fuster, J. M. (2001). The prefrontal cortex—an update: Time is of the essence. *Neuron, 30,* 319–333.

Gouvea, A. C., Phillips, C., Kazanina, N., & Poeppel, D. (2009). The linguistic processes underlying the P600. *Lang Cogn Proc, 25,* 149–188.

Hoeks, J. C. J., Stowe, L. A., & Doedens, G. (2004). Seeing words in context: The interaction of lexical and sentence level information during reading. *Cogn Brain Res, 19,* 59–73.

Juottonen, K., Revonsuo, A., & Lang, H. (1996). Dissimilar age influences on two ERP waveforms (LPC and N400) reflecting semantic context effect. *Brain Res Cogn Brain Res, 4,* 99–107.

Kandhadai, P., & Federmeier, K. D. (2010a). Automatic and controlled aspects of lexical associative processing in the two cerebral hemispheres. *Psychophysiology, 47,* 774–785.

Kandhadai, P., & Federmeier, K. D. (2010b). Hemispheric differences in the recruitment of semantic processing mechanisms. *Neuropsychologia, 48,* 3772–3781.

Kim, A., & Osterhout, L. (2005). The independence of combinatory semantic processing: Evidence from event-related potentials. *J Mem Lang, 52,* 205–225.

Kos, M., van den Brink, D., & Hagoort, P. (2012). Individual variation in the late positive complex to semantic anomalies. *Front Psychol, 3,* 318.

Koyama, S., Hokama, H., Miyatani, M., Ogura, C., Nageishi, Y., & Shimokochi, M. (1994). ERPs in schizophrenic patients during word recognition task and reaction times. *Electroen Clin Neurophysiol, 92,* 546–554.

Kuperberg, G. R. (2007). Neural mechanisms of language comprehension: Challenges to syntax. *Brain Res, 1146,* 23–49.

Kuperberg, G. R., Sitnikova, T., Caplan, D., & Holcomb, P. J. (2003). Electrophysiological distinctions in processing conceptual relationships within simple sentences. *Brain Res Cogn Brain Res, 17,* 117–129.

Kutas, M., & Federmeier, K. D. (2011). Thirty years and counting: Finding meaning in the N400 component of the event-related brain potential (ERP). *Annu Rev Psychol, 62,* 621–647.

Kutas, M., & Hillyard, S. A. (1980). Reading sensless sentences: Brain potentials reflect semantic incongruity. *Science, 207,* 203–205.

Kutas, M., & Hillyard, S. A. (1984). Brain potentials during reading reflect word expectancy and semantic association. *Nature, 307,* 161–163.

Kutas, M., Lindamood, T. E., & Hillyard, S. A. (1984). Word expectancy and event-related brain potentials during sentence processing. In S. Kornblum & J. Requin (Eds.), *Preparatory states and processes* (pp. 217–237). Hillsdale, NJ: Erlbaum.

Kutas, M., McCarthy, G., & Donchin, E. (1977). Augmenting mental chronometry: The P300 as a measure of stimulus evaluation time. *Science, 197,* 792–795.

Laszlo, S., & Federmeier, K. D. (2009). A beautiful day in the neighborhood: An event-related potential study of lexical relationships and prediction in context. *J Mem Lang, 61,* 326–338.

Lau, E. F., Phillips, C., & Poeppel, D. (2008). A cortical network for semantics: (De)constructing the N400. *Nat Rev Neurosci, 9,* 920–933.

Lau, E. F., Stroud, C., Plesch, S., & Phillips, C. (2006). The role of structural prediction in rapid syntactic analysis. *Brain Lang, 98,* 74–88.

Lee, C. L., & Federmeier, K. D. (2009). Wave-ering: An ERP study of syntactic and semantic context effects on ambiguity resolution for noun/verb homographs. *J Mem Lang, 61,* 538–555.

Luck, S. J., Vogel, E. K., & Shapiro, K. L. (1996). Word meanings can be accessed but not reported during the attentional blink. *Nature, 383,* 616–618.

Metusalem, R., Kutas, M., Urbach, T. P., Hare, M., McRae, K., & Elman, J. L. (2012). Generalized event knowledge activation during online sentence comprehension. *J Mem Lang, 66,* 545–567.

Miller, G. A., Heise, G. A., & Lichten, W. (1951). The intelligibility of speech as a function of the context of the test materials. *J Exp Psychol, 41,* 329–335.

Misra, M., & Holcomb, P. J. (2003). Event-related potential indices of masked repetition priming. *Psychophysiology, 40,* 115–130.

Moreno, E. M., Federmeier, K. D., & Kutas, M. (2002). Switching languages, switching palabras (words): An electrophysiological study of code switching. *Brain Lang, 80,* 188–207.

Munte, T. F., Schiltz, K., & Kutas, M. (1998). When temporal terms belie conceptual order. *Nature, 395,* 71–73.

Nieuwenhuis, S., Aston-Jones, G., & Cohen, J. D. (2005). Decision making, the P3, and the locus coeruleus-norepinephrine system. *Psychol Bull, 131,* 510–532.

Otten, M., Nieuwland, M. S., & van Berkum, J. J. A. (2007). Great expectations: Specific lexical anticipation influences the processing of spoken language. *BMC Neurosci, 8*, 89.

Otten, M., & van Berkum, J. (2008). Discourse-based word anticipation during language processing: Prediction or priming? *Discourse Process, 45*, 464–496.

Otten, M., & van Berkum, J. J. A. (2009). Does working memory capacity affect the ability to predict upcoming words in discourse? *Brain Res, 1291*, 92–101.

Pineda, J. A., Foote, S. L., & Neville, H. J. (1989). Effects of locus coeruleus lesions on auditory, long-latency, event-related potentials in monkey. *J Neurosci, 9*, 81–93.

Rommers, J., Meyer, A. S., Praamstra, P., & Huettig, F. (2013). The contents of predictions in sentence comprehension: Activation of the shape of objects before they are referred to. *Neuropsychologia, 51*, 437–447.

Ruchkin, D. S., Johnson, R., Jr., Grafman, J., Canoune, H., & Ritter, W. (1992). Distinctions and similarities among working memory processes: An event-related potential study. *Brain Res Cogn Brain Res, 1*, 53–66.

Rugg, M. D. (1990). Event-related brain potentials dissociate repetition effects of high- and low-frequency words. *Mem Cognit, 18*, 367–379.

Schacter, D. L., Addis, D. R., & Buckner, R. L. (2007). Remembering the past to imagine the future: The prospective brain. *Nat Rev Neurosci, 8*, 657–661.

Smith, N. J., & Levy, R. (2011). Cloze but no cigar: The complex relationship between cloze, corpus, and subjective probabilities in language processing. Paper presented at the Proceedings of the 33rd Annual Meeting of the Cognitive Sciences Conference, Austin, TX.

Swaab, T. Y., Ledoux, K., Camblin, C. C., & Boudewyn, M. A. (2011). Language-related ERP components. In S. J. Luck & E. S. Kappenman (Eds.), *The Oxford handbook of event-related potential components* (pp. 397–440). New York, NY: Oxford University Press.

Szewczyk, J. M., & Schriefers, H. (2013). Prediction in language comprehension beyond specific words: An ERP study on sentence comprehension in Polish. *J Mem Lang, 68*, 297–314.

Tulving, E., & Gold, C. (1963). Stimulus information and contextual information as determinants of tachistoscopic recognition of words. *J Exp Psychol, 66*, 319.

van Berkum, J. J. A., Brown, C. M., Zwitserlood, P., Kooijman, V., & Hagoort, P. (2005). Anticipating upcoming words in discourse: Evidence from ERPs and reading times. *J Exp Psychol Learn Mem Cogn, 31*, 443–467.

van Herten, M., Kolk, H. H., & Chwilla, D. J. (2005). An ERP study of P600 effects elicited by semantic anomalies. *Brain Res Cogn Brain Res, 22*, 241–255.

Van Petten, C., & Luka, B. J. (2006). Neural localization of semantic context effects in electromagnetic and hemodynamic studies. *Brain Lang, 97*, 279–293.

Van Petten, C., & Luka, B. J. (2012). Prediction during language comprehension: Benefits, costs, and ERP components. *Int J Psychophysiol, 83*, 176–190.

Van Petten, C., Kutas, M., Kluender, R., Mitchiner, M., & McIsaac, H. (1991). Fractionating the word repetition effect with event-related potentials. *J Cogn Neurosci, 3*, 131–150.

Wicha, N. Y., Bates, E. A., Moreno, E. M., & Kutas, M. (2003). Potato not pope: Human brain potentials to gender expectation and agreement in Spanish spoken sentences. *Neuro Let, 346*(3), 165–168.

Wicha, N. Y., Moreno, E. M., & Kutas, M. (2003). Expecting gender: An event-related brain potential study on the role of grammatical gender in comprehending a line drawing within a written sentence in Spanish. *Cortex, 39*(3), 483–508.

Wicha, N. Y., Moreno, E. M., & Kutas, M. (2004). Anticipating words and their gender: An event-related brain potential study of semantic integration, gender expectancy, and gender agreement in Spanish sentence reading. *J Cogn Neurosci, 16*(7), 1272–1288.

Wilding, E. L., & Ranganath, C. (2012). Electrophysiological correlates of episodic memory processes. In S. Luck & E. Kappenman (Eds.), *The Oxford handbook of event-related potential components* (pp. 373–396). New York, NY: Oxford University Press.

Wlotko, E., & Federmeier, K. D. (2011). Flexible implementation of anticipatory language comprehension mechanisms. *J Cogn Neurosci*, Suppl. 1, 233.

Wlotko, E. W., & Federmeier, K. D. (2012a). Age-related changes in the impact of contextual strength on multiple aspects of sentence comprehension. *Psychophysiology, 49*, 770–785.

Wlotko, E. W., & Federmeier, K. D. (2012b). So that's what you meant! Event-related potentials reveal multiple aspects of context use during construction of message-level meaning. *NeuroImage, 62*, 356–366.

Wlotko, E. W., Federmeier, K. D., & Kutas, M. (2012). To predict or not to predict: Age-related differences in the use of sentential context. *Psychol Aging, 27*, 975–988.

Yu, A. J., & Dayan, P. (2005). Uncertainty, neuromodulation, and attention. *Neuron, 46*, 681–692.

Zacks, J. M., Speer, N. K., Swallow, K. M., Braver, T. S., & Reynolds, J. R. (2007). Event perception: A mind-brain perspective. *Psychol Bull, 133*, 273–293.

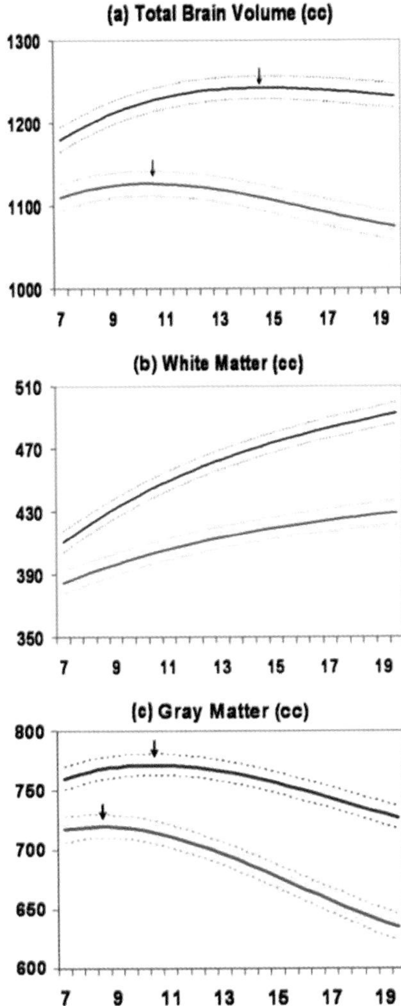

PLATE 1 Mean volume (with 95% confidence intervals) by age in years for males (blue) and females (red) for (a) total brain volume, (b) white matter volume and (c) gray matter volume. (Reproduced from Lenroot et al., 2007.) (See figure 2.2.)

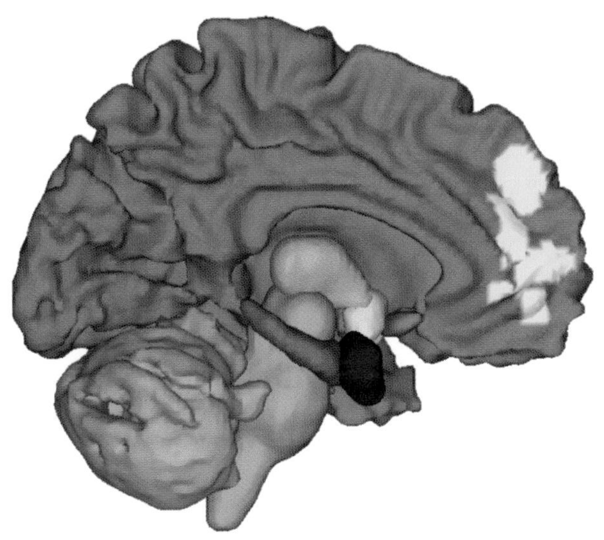

PLATE 2 A qualitative meta-analysis of the region of dmPFC that consistently shows decreased activity during mentalizing tasks between late childhood and adulthood. This meta-analysis shows voxels in mPFC (in yellow) that are within 10 mm of the peak voxel found to have a significant negative relationship with age in three or more of eight published developmental fMRI studies of social cognition (Blakemore et al., 2007; Burnett et al., 2009; Gunther Moor et al., 2012; Güroğlu et al., 2009; Pfeifer et al., 2007, 2009; Sebastian et al., 2011; van den Bos et al., 2011). Meta-analysis was performed using Neurosynth software (www.neurosynth.org). (See figure 4.2.)

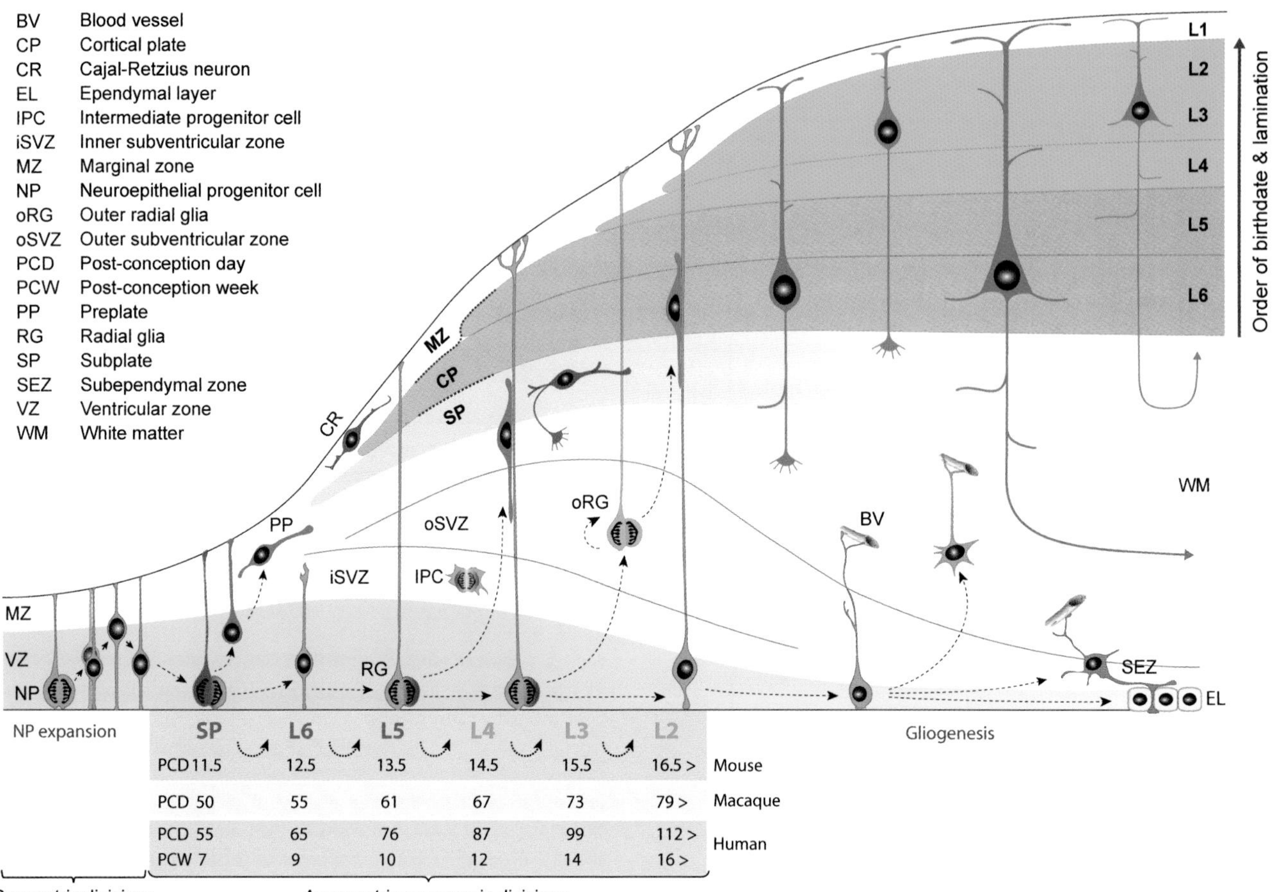

PLATE 3 Schematic of generation and migration of projection neurons and glia in the neocortex. Projection neurons are generated by progenitor cells in the ventricular zone (VZ) and subventricular zone (SVZ). Their generation and migration into the cortical plate (CP) occurs in an inside-first, outside-last manner. At the end of neurogenesis, radial glial (RG) cells lose their polarity and generate glia. (Adapted from Kwan et al., 2012.) (See figure 8.2.)

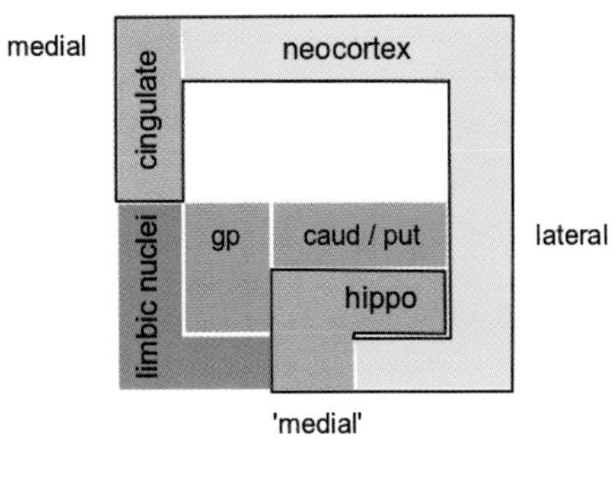

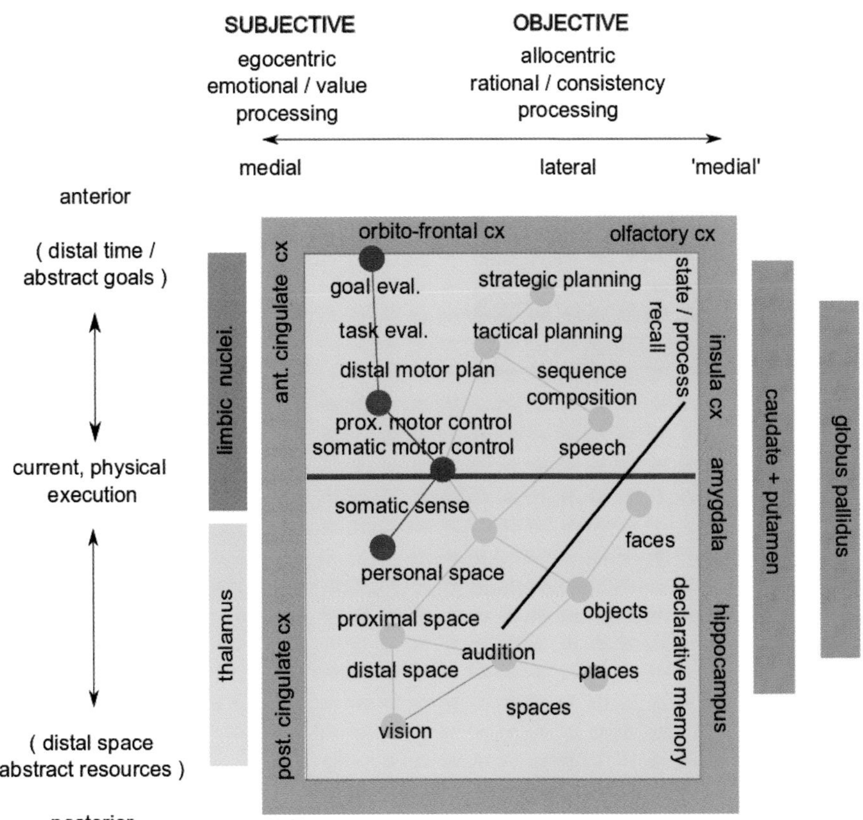

PLATE 4 Organization of the cortical sheet: schematic showing functional organization of cortical sheet and some of its related structures in the right hemisphere. *Above*: Transverse section shows how limbic nuclei, basal ganglia (gp, globus pallidus; caud/put, caudate and putamen), and hippocampus (hippo) are folded and come to occupy more "medial" positions. *Below*: Plan view of the cortical sheet (cyan) with central sulcus indicated by thick red horizontal line. The sheet is surrounded by limbic cortical regions (brown; e.g., cingulate cortex, insula) and associated limbic nuclei (red; e.g., septal nuclei, amygdala). Dynamically evolving behaviors are represented schematically as "nodes," representing regions of active processing, and "edges," which represent the axonal communication channels between active nodes. The channels act directly through cortico-cortical connections, or indirectly via thalamus and basal ganglia. Multiple behaviors may evolve simultaneously (green graph; see text), while the red graph represents the various functional relations of the behavior currently being executed. (See figure 9.3.)

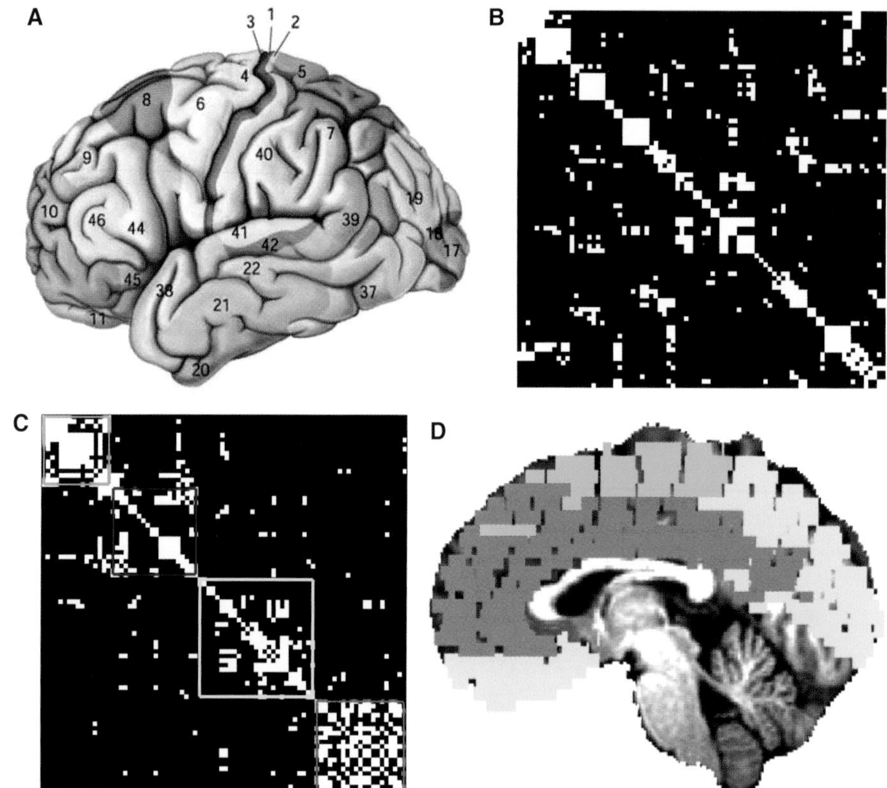

Plate 5 Clustering of brain neural activity into modules and calculation of modularity. (A) The Brodmann's areas, onto which the measured fMRI neural activity data are projected (Dubuc). (B) Correlation matrix of the neural activity between the different Brodmann's areas, with only the largest elements retained (white). (C) Reordered correlation matrix, showing the grouping of the areas into four modules. The Brodmann's areas are grouped into modules by permuting the rows and columns of the correlation matrix. The modularity calculated from Eq. 1 is 0.6441, and the contributions to M are 0.1585, 0.1310, 0.1592, and 0.1954 from the green, red, yellow, and orange modules, respectively. (D) The Brodmann's areas grouped into the four modules. In this slice, only two voxels of the second module are visible, with the majority of the voxels in a different region of the brain not visible in this slice. From Chen & Deem (2014). (See figure 12.1.)

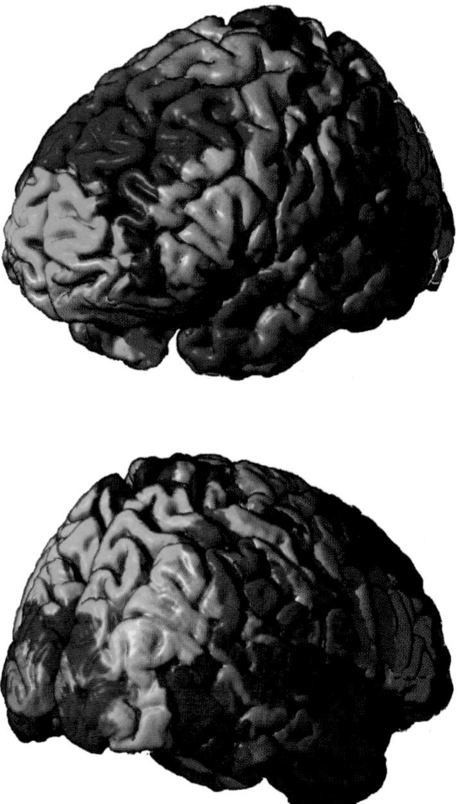

Plate 6 The Brodmann areas of the human cerebral cortex, onto which fMRI data are projected for further analysis. (From Dow.) Each Brodmann area is shaded a different color. (See figure 12.7.)

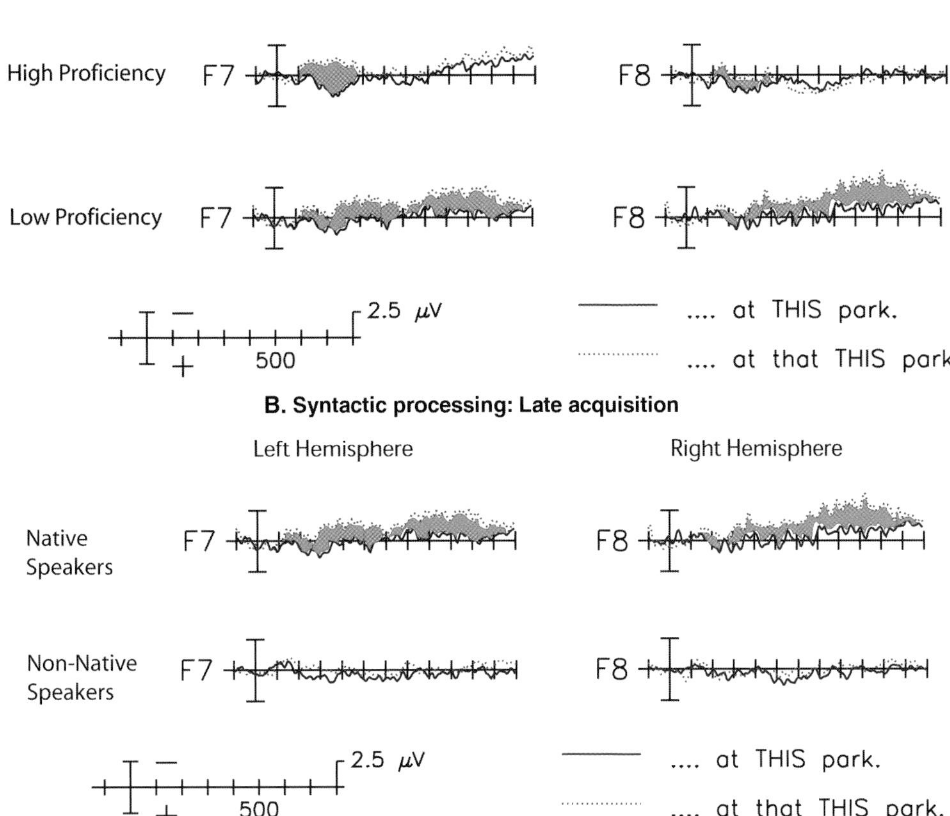

PLATE 7 Event-related brain potential (ERP) to syntactic violations in spoken sentences showing independent effects of age of acquisition and proficiency on neural processing for language. (A) Comparison of high-proficiency second-language learners to native speakers of equivalent proficiency. (B) Comparison of native speakers with higher versus lower levels of English language proficiency. (Data from Pakulak & Neville, 2010, and Pakulak & Neville, 2011.) (See figure 13.2.)

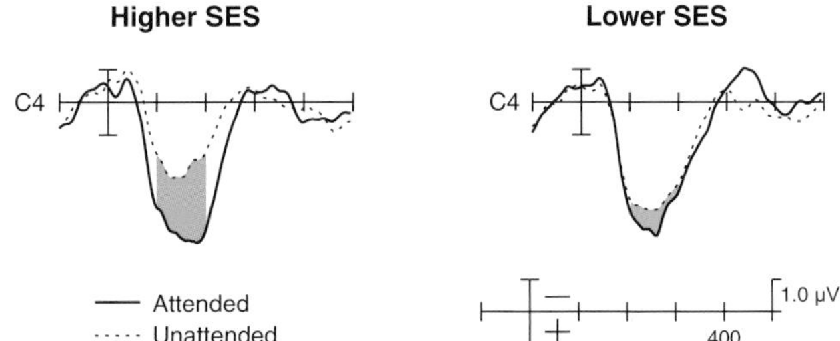

PLATE 8 Effects of selective attention on neural processing in children aged 3–8 years from higher versus lower socioeconomic status (SES) backgrounds. Children from higher SES backgrounds had significantly greater effects of selective attention on neural processing than children from lower SES backgrounds. (Data from Stevens, Lauinger, & Neville, 2009.) (See figure 13.3.)

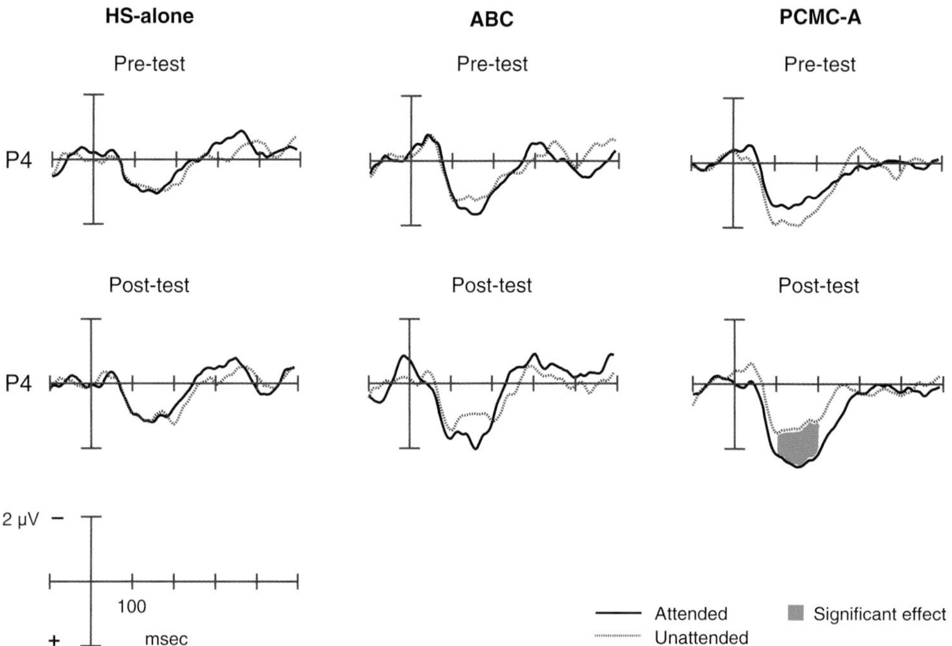

PLATE 9 Effects of selective attention on neural processing in children aged 3–5 years from lower socioeconomic status backgrounds, both before and after an eight-week training period. Only children in the family-based program with a greater emphasis on parent training ("PCMC-A") showed an increase in the effects of attention on neural processing. Children in Head Start alone ("HS-alone") or in a comparison training program ("ABC") did not show significant changes in the effects of attention on neural processing from pre- to post-training. Children in the PCMC-A program also made significantly greater gains than either comparison group on standardized measures of receptive language and nonverbal IQ, as well as parent reports of child behavior. See main text for details. (Data from Neville et al., 2013.) (See figure 13.4.)

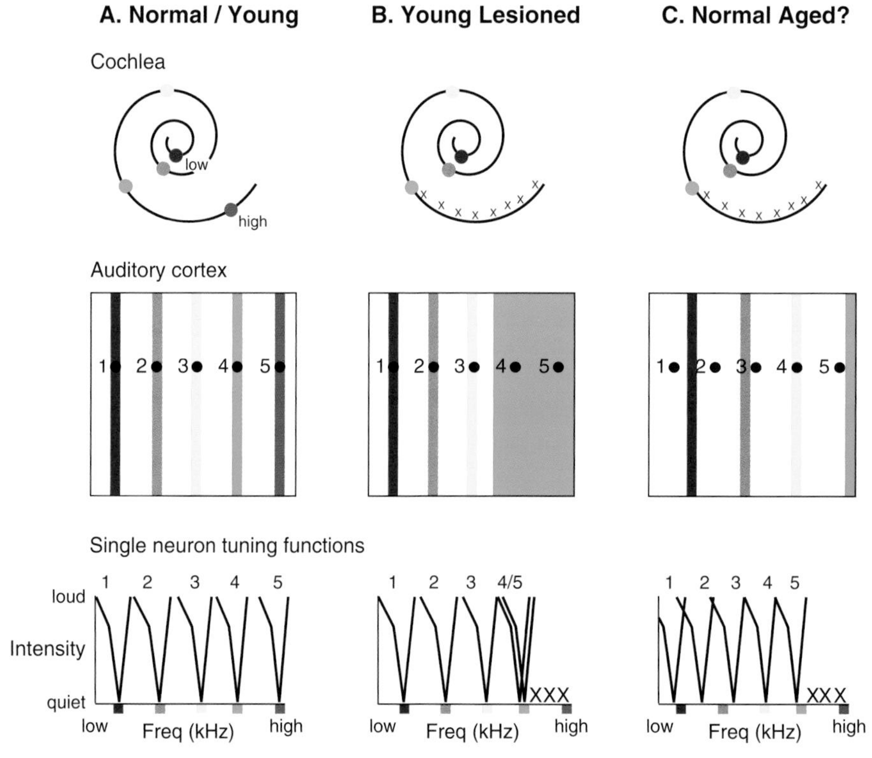

PLATE 10 Cortical map changes after cochlear damage. (A) Schematic representation of the cochlea (top) with low frequencies represented at the apex (red) and progressively higher frequencies represented toward the basal end (blue). The middle panel shows a schematic of the primary auditory cortex and the isofrequency bands where neurons have similar frequency tuning corresponding to the different cochlear locations. Black dots (numbered 1–5) denote recording sites that correspond to the tuning functions of individual neurons shown in the bottom row. In the normal animal, there is an orderly and progressive shift of best frequency across A1. (B) Results from lesions of the basal turn of the cochlea (high frequencies, denoted by X) and the subsequent expansion of the representation of the frequencies at the edge of the lesion, with little change in the representation of the other frequencies. (C) One possible outcome of high-frequency hearing loss as a consequence of natural aging, where the representation of many frequencies are altered to regain an orderly, topographic representation in A1. (See figure 14.1.)

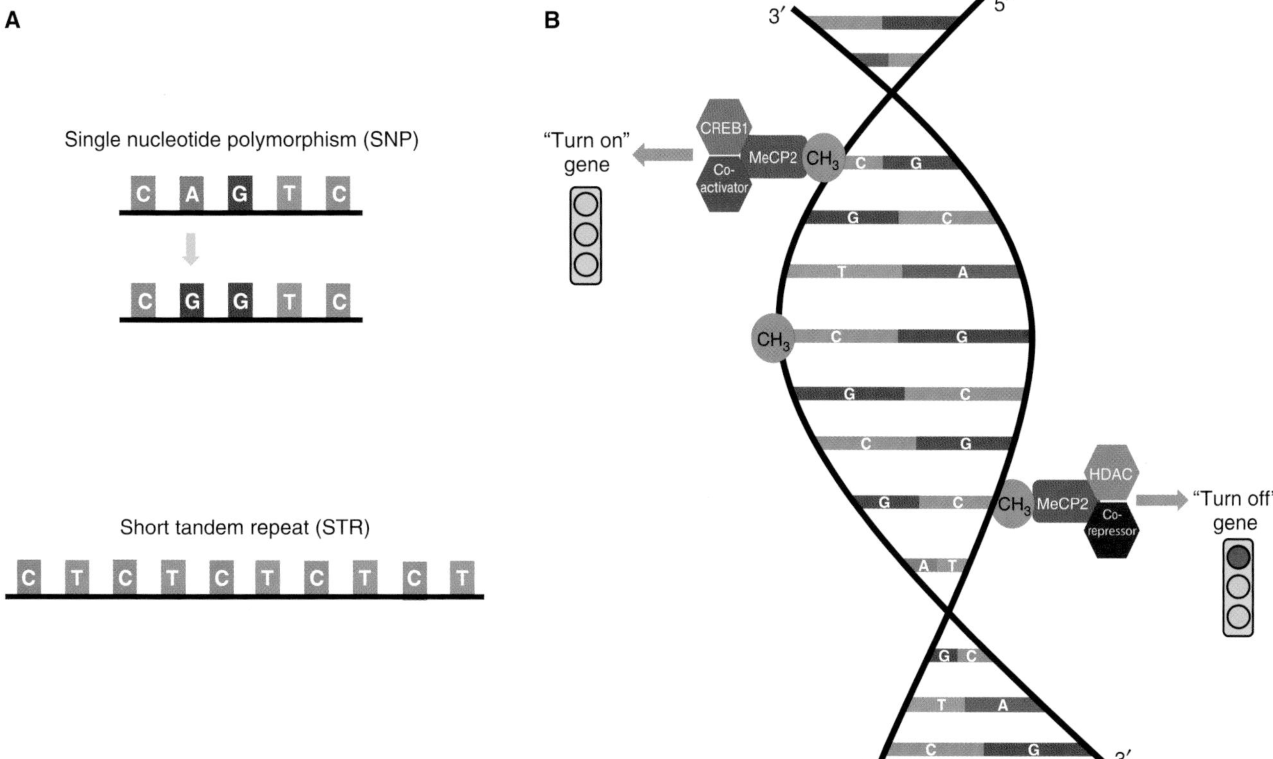

Plate 11 Schematic of common gene polymorphisms and DNA methylation. (A) Single nucleotide polymorphisms (SNPs) consist of a single nucleotide variation between individuals. Short tandem repeats (STRs) contain multiple repeats of a certain short sequence of nucleotides. (B) DNA methylation refers to the addition of methyl groups to specific cytosine residues. MeCP2 binds to methylated DNA and can either recruit HDACs and other co-repressors to suppress gene transcription or can recruit cAMP response element-binding (CREB) protein and other transcriptional co-activators to enhance gene expression. (See figure 15.1.)

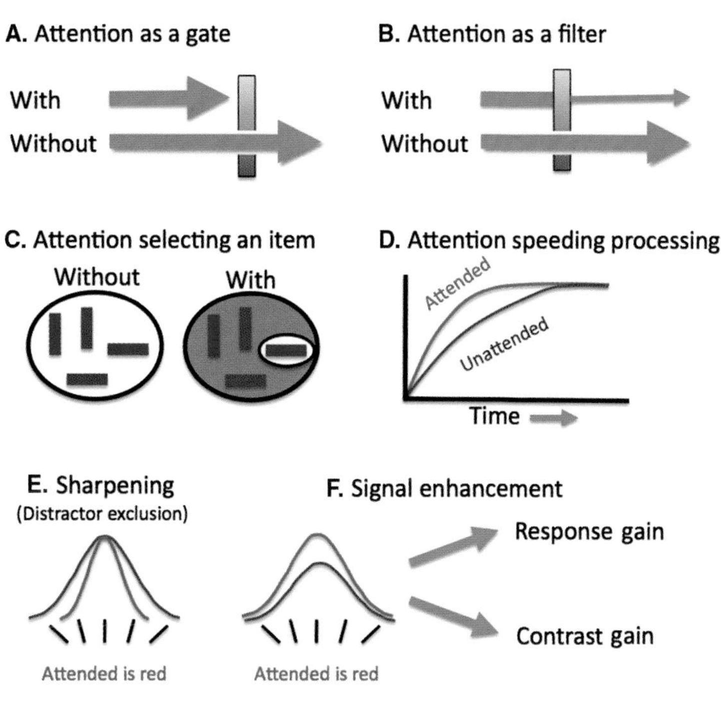

Plate 12 A sampling of the ways that attention could alter a signal. (See figure 16.4.)

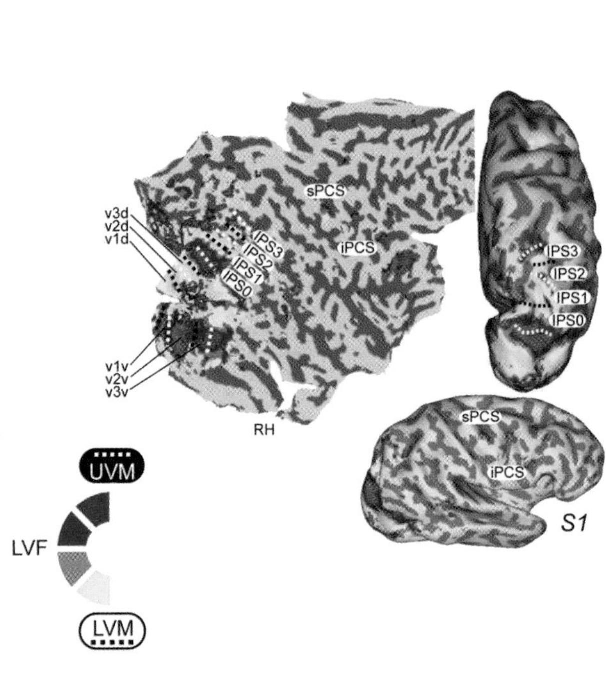

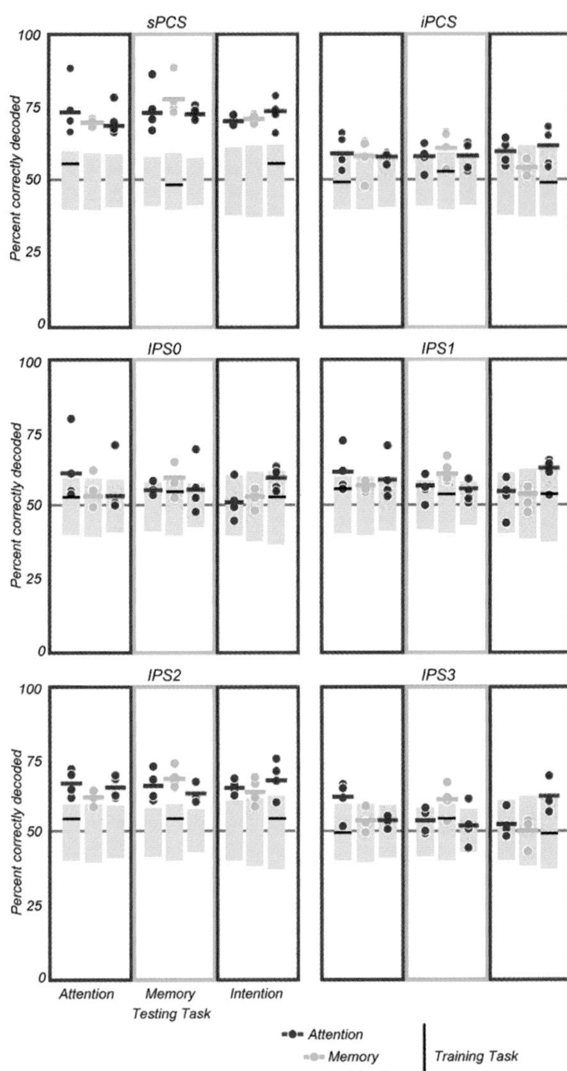

Plate 13 Task general representation of attentional priority in parietal and frontal cortex. *Left:* Topographic maps in the right hemisphere of a single subject. As indicated in the central color wheel key, warm colors reflect the position of the stimulus in the contralateral left visual field (LVF). The borders of topographic areas in occipital and parietal cortex are demarked by dotted black lines reflecting the lower visual meridian (LVM) and dotted white lines reflecting the upper visual meridian (UVM). In this subject, anterior and dorsal to early visual areas V1–V3, four topographic areas are found along the caudal-rostral intraparietal sulcus (IPS0–IPS3). In the prefrontal cortex (PFC), two topographic areas are found along the dorsal—ventral precentral sulcus (sPCS and iPCS; LH, left hemisphere; RH, right hemisphere). Note that the sPCS and the iPCS in these figures correspond to the superior and inferior precentral cortex (PCC) in the main text, respectively. *Right:* Classifier results for decoding the visual field that was currently prioritized with attention. For each topographic area, the percentage of correctly decoded trials is plotted. Each dot is an individual subject, and each horizontal line is the mean performance across subjects. The color of the dots keys the task used to train the classifier. The color of the boxes keys the task used to test the classifier. Within-task classification, dot and box colors match; across-task classification, dot and box colors do not match. The gray boxes represent the 2.5th and 97.5th percentile of the null distribution generated by random permutation analysis. Dots and bars beyond these cutoffs are significantly different from chance. The multivoxel pattern of delay period activity only in sPCS (superior PCC) and IPS2 predicts the prioritized hemifield both within and across the three spatial cognitive tasks. The black horizontal bars are the mean performance of the control analyses, in which the mean signal difference of all voxels in the left and right hemisphere topographic areas was used to predict the prioritized hemifield. (All data adapted and reprinted with permission from Jerde et al., 2012.) (See figure 17.2.)

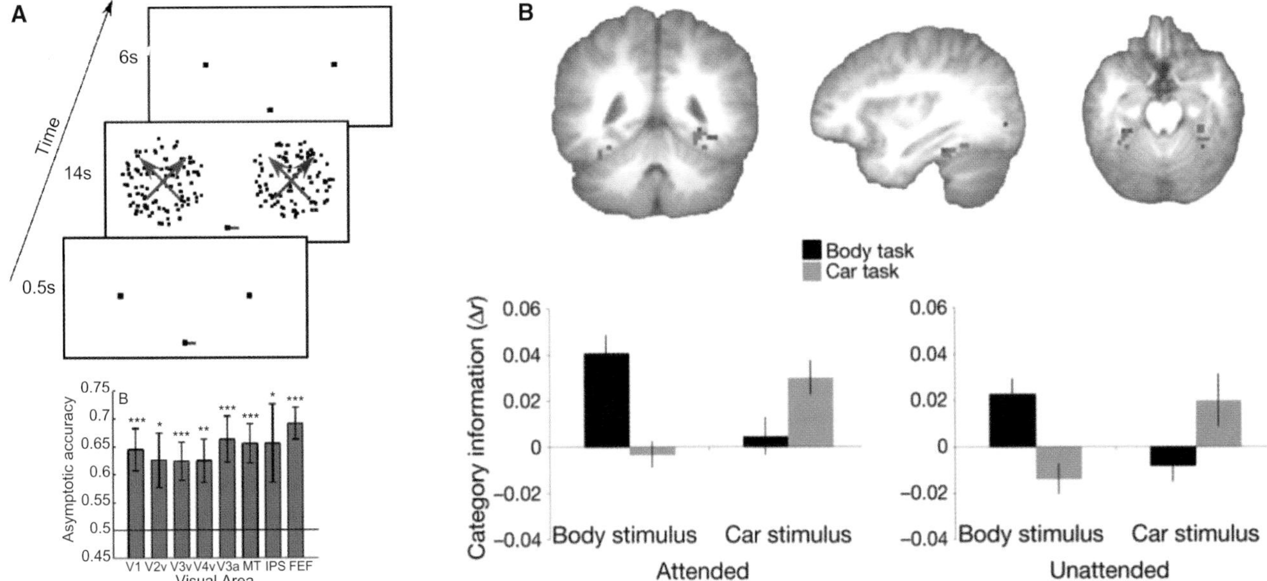

Plate 14 Feature- and object category–based attention effects. (A) *Top panel*: Sequence of events on a trial where the observer was attending to 45° motion in the right stimulus aperture. One-half of the dots in each stimulus aperture moved at 45°, and the other half moved at 135°, for the duration of the 14 sec presentation period. Targets were defined as a brief slowing of the dots at the attended location that moved in the currently attended direction (45° in this figure); distractors were defined as a brief slowing of the dots at the attended location that moved in the unattended direction (135° in this figure). *Bottom-left panel*: Asymptotic multi-voxel classification accuracy for the attended direction of motion based on fMRI activation patterns in each visual region (error bars, ±SEM across observers). (Data adapted and reprinted with permission from Serences and Boynton, 2007.) (B) *Top panel*: Clusters of object-selective activations in ventral temporal cortex (as determined by contrasting activations evoked by viewing intact vs. scrambled objects) in a group-average analysis at $P < 0.005$ (Talairach coordinates of peak: $x = 35$, $y = -41$, $z = -18$; $N = 10$). The lower panel shows category information as a function of category, task, and attention in individually defined object-selective cortex. Category information was calculated by taking the difference between within-category comparisons and between-category comparisons, and reflects the amount of category information in multi-voxel patterns of activation. Significant category information depended solely on task instruction and not on spatially attended location. Task-relevant information was processed to the categorical level, even when unattended, whereas task-irrelevant information was not represented at that level, even when attended. Error bars indicate ± SEM. (Data adapted and reprinted with permission from Peelen et al., 2009.) (See figure 17.3.)

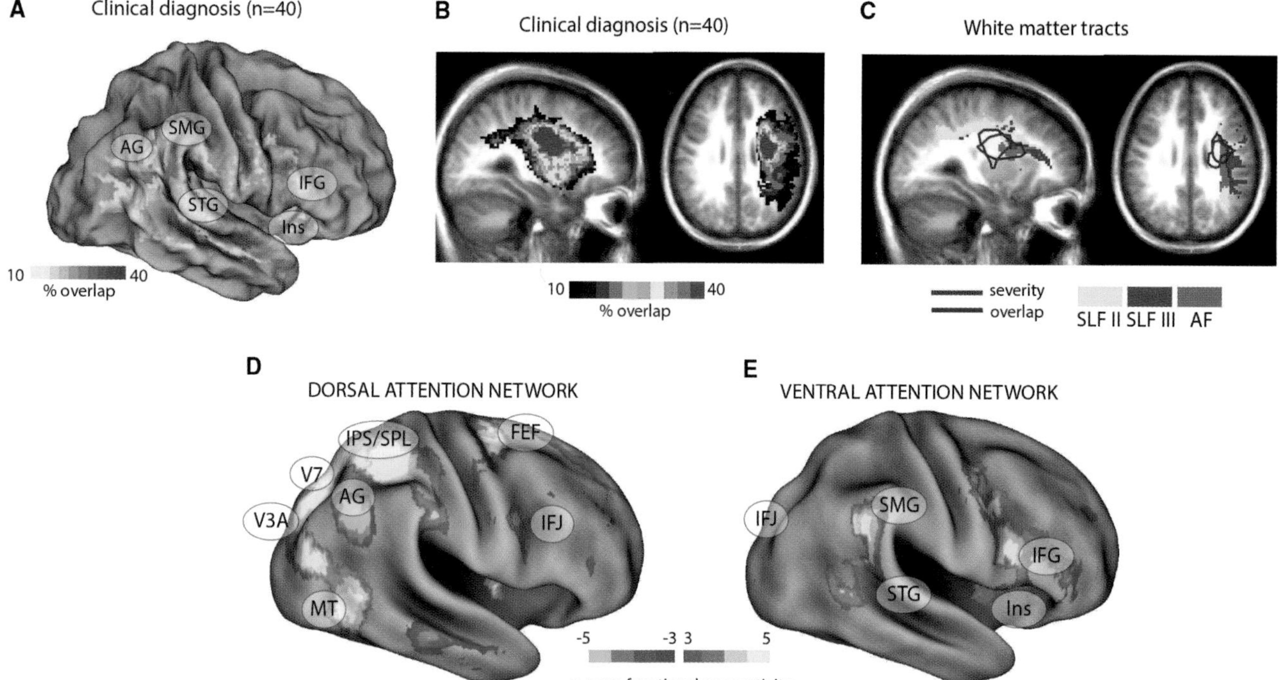

PLATE 15 (A) Anatomy of neglect defined based on clinical diagnosis by treating therapist ($N = 40$ patients). Note ventral distribution of lesions with involvement especially of superior temporal gyrus (STG), supramarginal gyrus (SMG), angular gyrus (AG), inferior frontal gyrus (IFG), and insula (Ins). (B) Lesions extend in the periventricular white matter both ventrally and dorsally. White matter damage is more common in severe neglect. (C) The region of maximal damage in the white matter overlaps with common white matter pathways: superior longitudinal fasciculus (SLF) and arcuate (AF), which connect frontal and parietal regions, and ventral frontal to dorsal parietal cortex. (D) Functional connectivity maps of dorsal attention network with core regions identified. (E) Functional connectivity maps of ventral attention network with core regions identified. (See figure 19.3.)

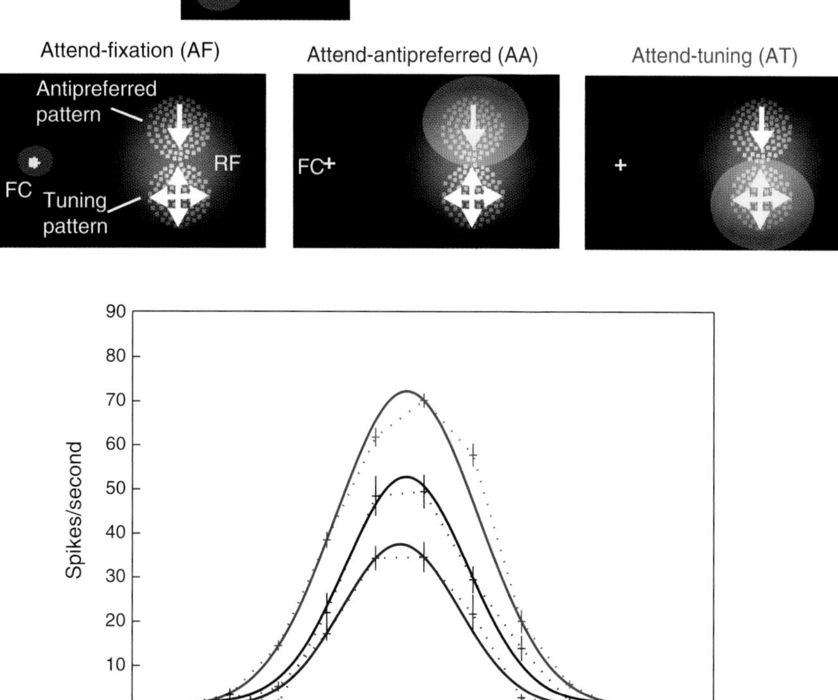

PLATE 16 Switching attention between two moving stimuli inside the RF of MT neurons. The top panels illustrate the experimental design of Treue and Martinez-Trujillo (1999). Two RDPs were presented inside the RF of an MT neuron (white area) while the animal fixated a dot (small square). One RDP always moved in the neurons antipreferred direction (AP pattern) and the other could change direction from trial to trial (tuning pattern). The lower panel illustrates tuning curves (responses as a function of the tuning pattern direction) obtained when the animal attended to the fixation point (black), to the tuning pattern (red) and to the AP pattern (blue). (See figure 20.1.)

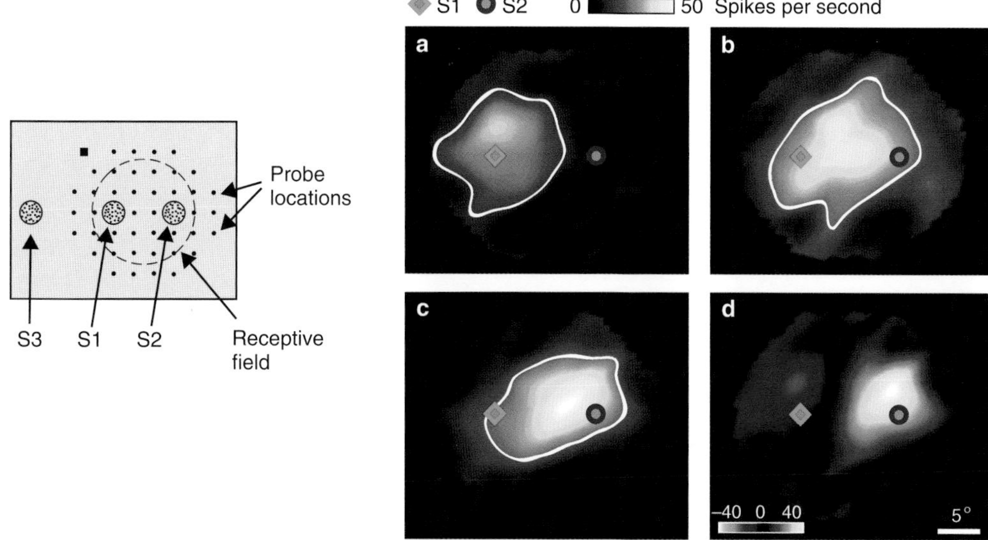

PLATE 17a *Left:* Sketch of the layout in the study of Womelsdorf et al. (2006), depicting an example of the placement of the three moving random dot patterns (shown here as textured circles) that were present in every trial as well as the grid of locations at which a series of small probes could briefly appear within a trial. The single dot represents the fixation point where the animal has to maintain his gaze throughout every trial. *Right:* RF profiles of an example neuron, when attention was directed inside the RF, to stimulus S1 (panel a) or S2 (panel c), or when attention was directed outside the RF, to S3 (panel b). The surface color at each point in the plots indicates the increase in the neuron's response elicited by the presentation of a probe stimulus at that position, over the response observed in the absence of a probe (that is, when only S1 and S2 were present). Panel d depicts a difference map computed by subtracting the RF when attention was on S1 from the RF when attention was on S2. The map illustrates that shifting attention from S1 to S2 enhances responsiveness around S2 and reduces it near S1. (See figure 20.2A.)

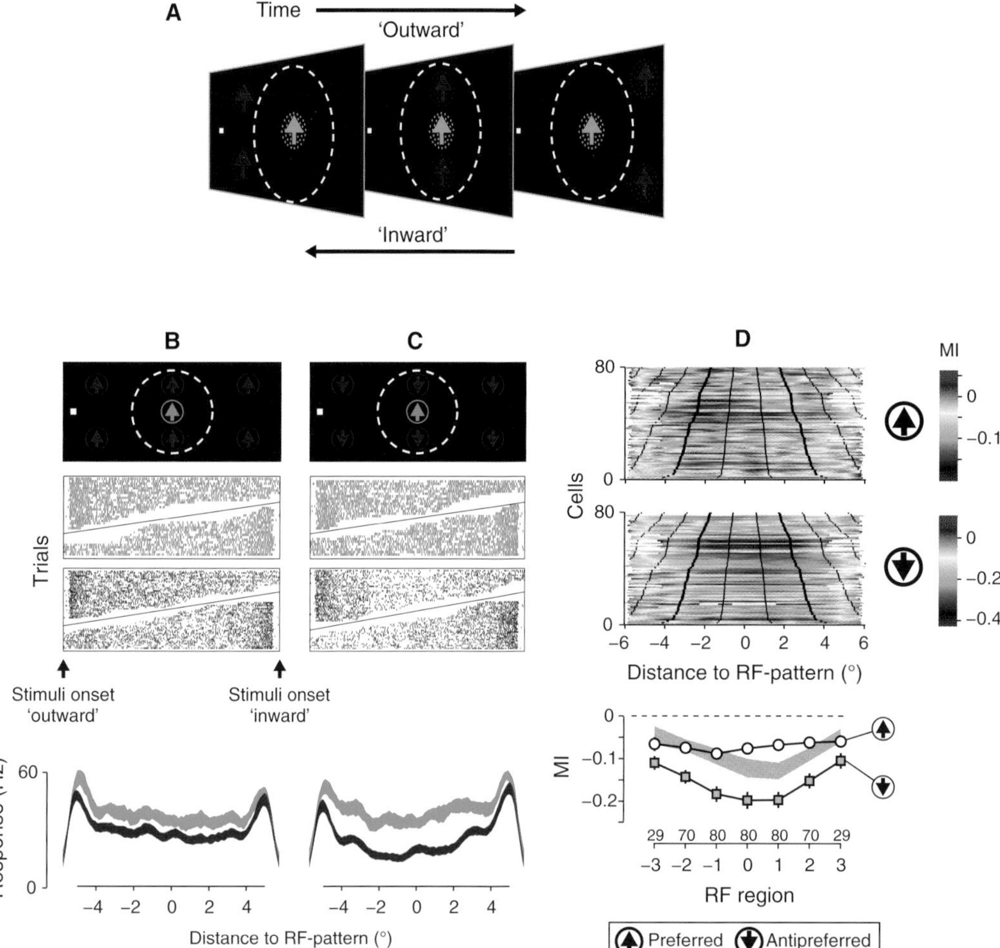

PLATE 17b Splitting receptive fields with attention. (A) Sketch of the experimental paradigm of Niebergall et al. (2011b). The experiment was identical to the one described in figure 20.4, except that a third RDP (with local dot motion in the neuron's preferred direction) was placed and stayed in the center of the RF, and the animal was instructed to attend either to the translating RDPs (*tracking*) or to the RF stimulus (*attend RF*). (B) Responses for *attend RF (green rasters)* and *tracking (red rasters)* trials. Data from an example cell for the stimulus configuration with the translating RDPs dots locally moving in the preferred direction. The bottom plot shows average responses (± standard error) as a function of the translating RDP position during *attend RF* (green) and *tracking* (red) trials. (C) Responses when the translating pattern dots moved in the anti-preferred direction. (D) Response modulation for all neurons. The cells have been sorted according to their RF size, and the strength of the modulation appears in color. The arrows indicate the direction of the translating patterns (up: preferred, down: anti-preferred). The bottom graph plots the mean modulation for both directions (two lines) as well as the difference in modulation between them (gray band, representing the average difference of the two line plots ± 95% confidence interval). (See figure 20.2B.)

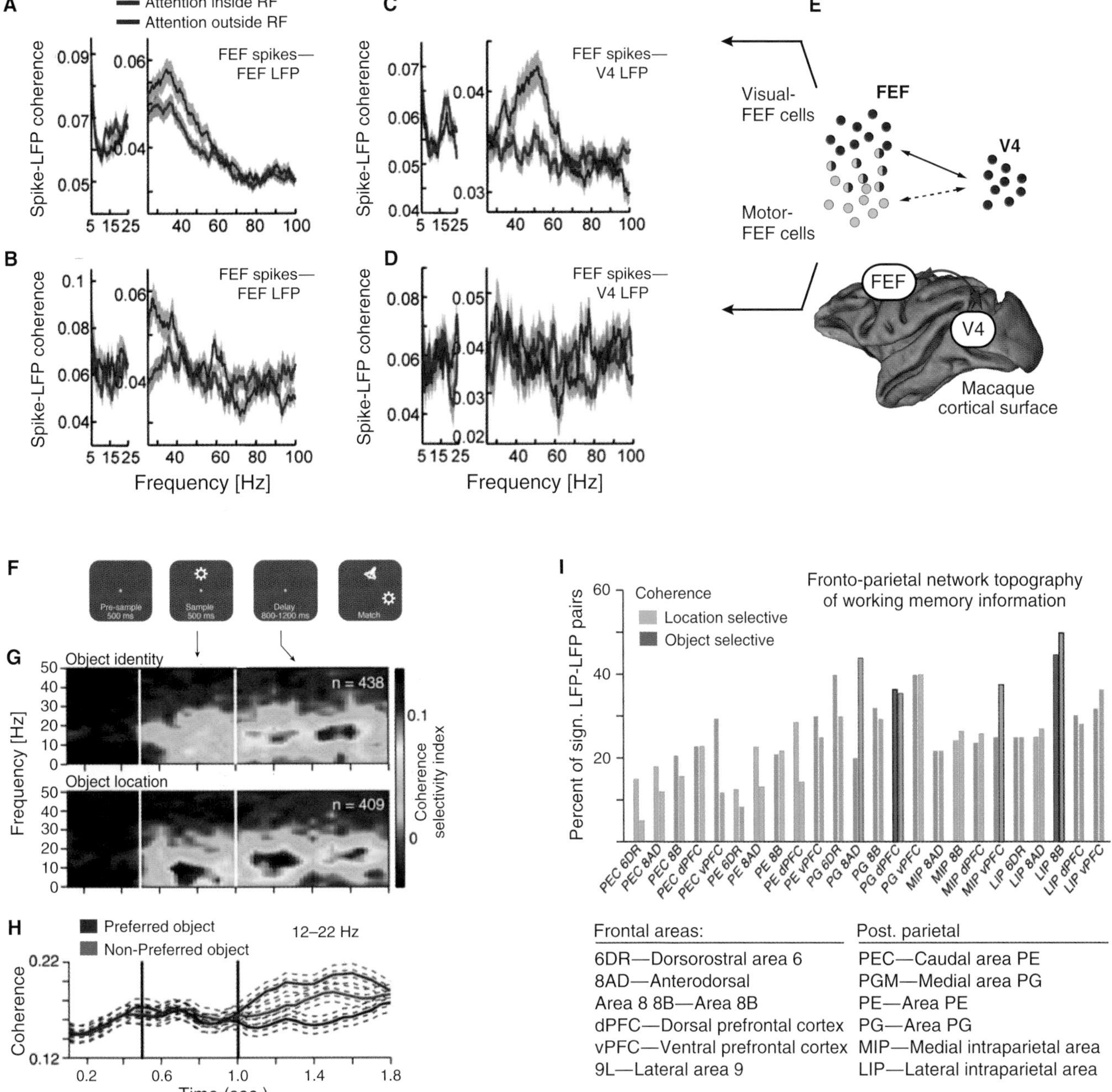

Plate 18 (A) Average coherence of spike-trains of visually tuned neurons in the frontal eye field (FEF) to the local field potential (LFP) when the monkeys attended inside (red) or outside (blue) the visual receptive field of the FEF cell. (B) Same as in (A), but for "motor" cells in the FEF that are tuned to saccadic target directions and are located predominantly in deeper cortical layers in the FEF. (C) Long-range coherence of visually tuned FEF cells with the LFP recorded in visual area V4 FEF. (D) Long-range coherence of motor-tuned FEF cells with the LFP recorded in visual area V4. (E) Illustration of cells in the FEF that are primarily responsive to visual stimuli and located in supra-granular layers (red), visuomotor neurons (half-red, half-green), and motor-tuned neurons (green) that are predominantly located in deeper layers. The solid/dashed arrows to V4 cells indicate selective coupling of visual FEF to visual V4 cells, but a lack of coupling from motor-related FEF cells. (F) The working memory task required encoding of a sample object or sample object location. Following a 0.8–1.2 sec delay, the macaque monkeys had to identify which of two objects or object locations matched the sample. (G) Average selectivity of coherence for object identity (*upper panel*) and location (*bottom panel*) during sample and delay epochs (separated by white vertical lines). The LFP coherence was computed between pairs of frontal and posterior parietal sites. (H) Average coherence, for each of three objects, rank-ordered for each recorded pair according to their object preference. The time course shows differences in coherence with the start of the delay period. (I) The proportion of frontoparietal LFP-LFP pairs that showed individually significant selectivity of beta band coherence, split according to their anatomical locations. (Panels (A)–(D) adapted from Gregoriou et al., 2012; panels (F)–(I) adapted from Salazar et al., 2012. Reprinted with permission from AAAS.) (See figure 21.3.)

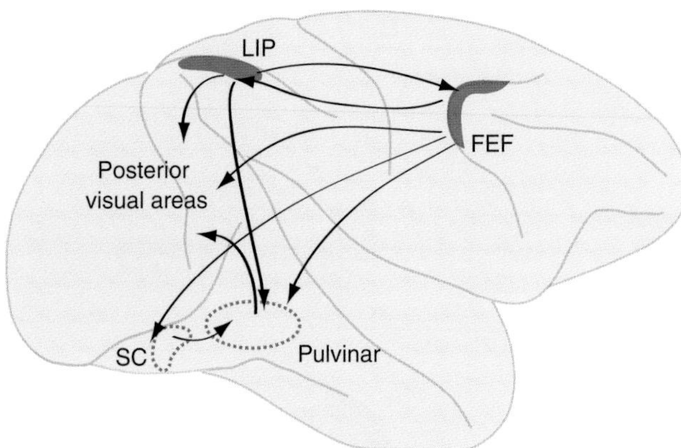

Plate 19 Macaque brain highlighting structures implicated in the control of visual selective attention and their connections, namely, the lateral intraparietal area (LIP) in posterior parietal cortex, the frontal eye field (FEF) in prefrontal cortex, the superior colliculus (SC) in the midbrain, and the pulvinar nuclei in the thalamus. Dotted outlines indicate subcortical structures. (See figure 22.1.)

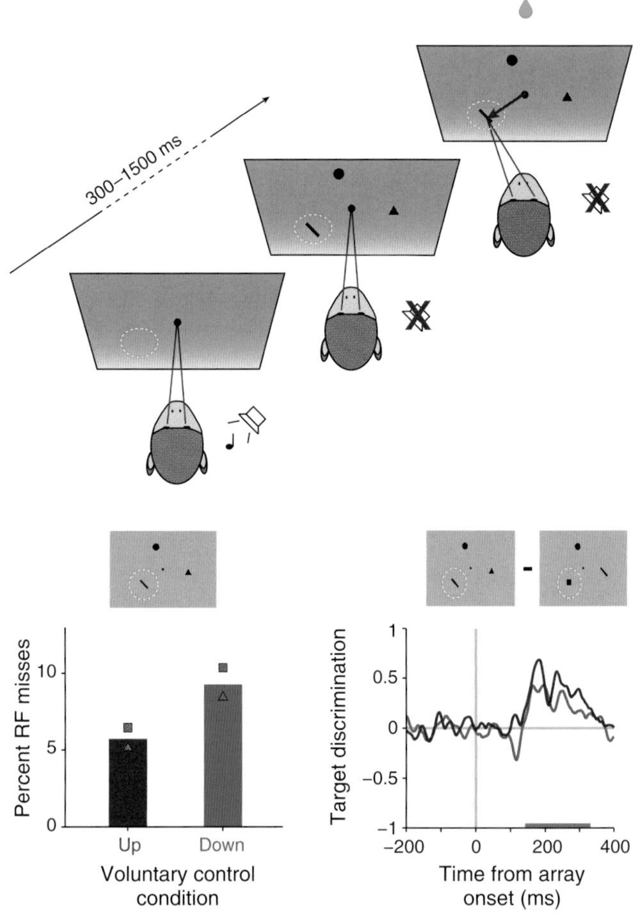

Plate 20 Operant control of FEF neurons and its effects on selective attention measured behaviorally and neurophysiologically. *Top*: The operant control task in which the monkey fixated a central spot on an otherwise blank video display and was rewarded for increasing or decreasing the firing rate of FEF neurons. Dotted circle shows the FEF RF. Speaker icon and musical note depict auditory feedback of FEF neuronal activity (spike train). Subsequent panels depict the visual-search probe trials in which a search array appeared, the auditory feedback ceased ("x" on speaker icon) and the monkey was rewarded (blue droplet) for directing a saccade toward an oriented bar target. *Bottom left*: The frequency of trials in which the monkey failed to respond to the target ("misses") opposite the RF was increased during DOWN operant control of FEF activity in both monkeys (square and triangle symbols). *Bottom right*: Target discrimination by FEF neurons was increased during upward (red) operant control relative to downward (blue). (Adapted from Schafer and Moore, 2011.) (See figure 22.2.)

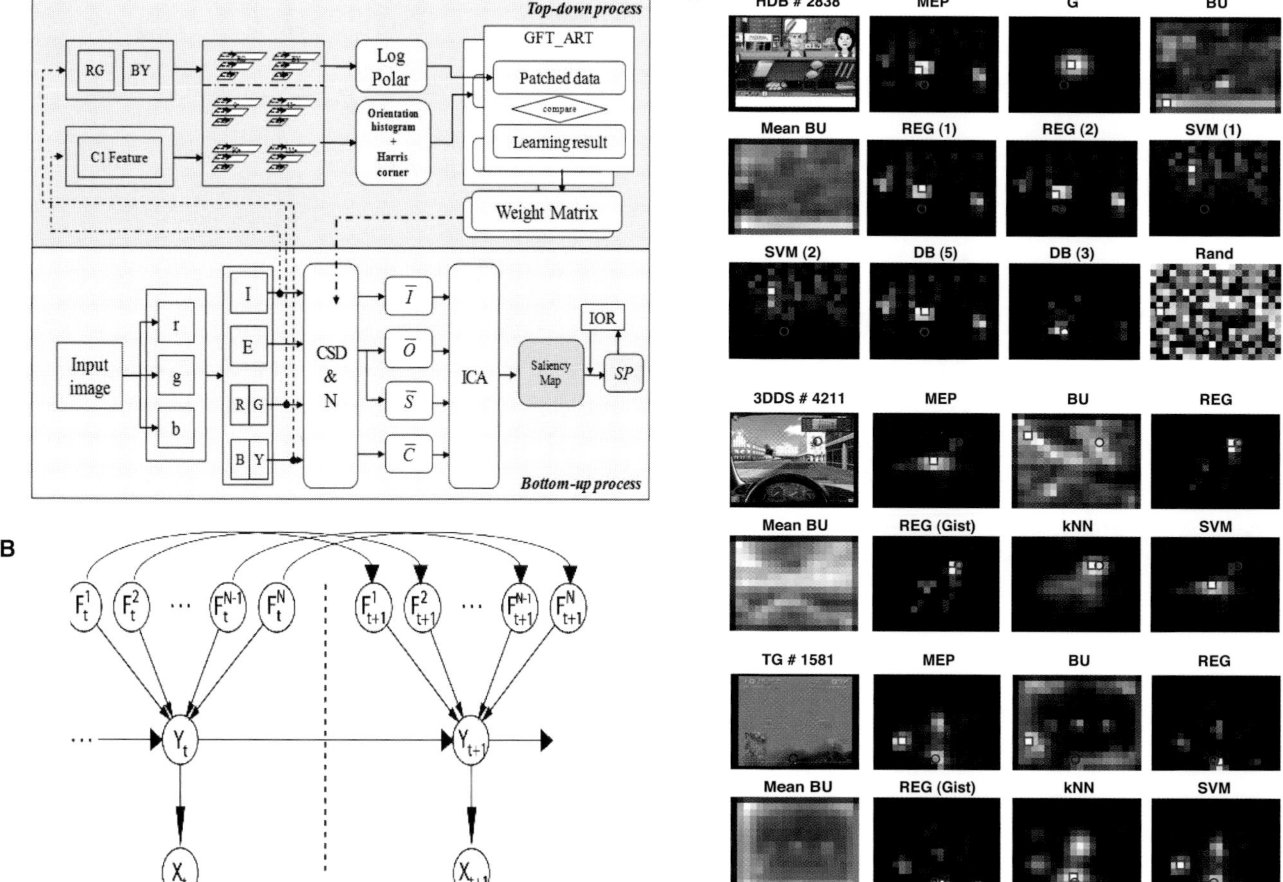

PLATE 21 Examples of recent top-down models. (A) Model of Ban et al. (2010), which integrates bottom-up and top-down components; r, g, b: red, green, and blue color channels. I: intensity feature. E: edges. R, G: red-green color. B, Y: blue-yellow color. CSD&N: center-surround differences and normalization. ICA: independent component analysis. GFT_ART: growing fuzzy topology adaptive resonance theory. SP: saliency point. (B) Graphical representation of the dynamic Bayesian network (DBN) approach of Borji et al. (2012b), unrolled over two time-slices. X_t is the current saccade position, Y_t is the currently attended object, and F_t^i is the function that describes object i at the current scene. All variables are discrete. It also shows a time-series plot of probability of objects being attended and a sample frame with tagged objects and eye fixation overlaid. (C) Sample predicted saccade maps of the DBN model (shown in B) on three video games and tasks: running a hot-dog stand (HDB; top three rows), driving (3DDS; middle two rows), and flight combat (TG, bottom two rows). Each red circle indicates the observer's eye position superimposed with each map's peak location (blue squares). Smaller distance indicates better prediction. Models compared are as follows. MEP: mean eye position over all frames during the game play (control model). G: trivial Gaussian map at the image center. BU: bottom-up saliency map of the Itti model. Mean BU: average saliency maps over all video frames. REG(1): regression model that maps the previous attended object to the current attended object and fixation location. REG(2): similar to REG(1), but the input vector consists of the available objects at the scene augmented with the previously attended object. SVM(1) and SVM(2) correspond to REG(1) and REG(2) but using a support vector machine (SVM) classifier. Similarly, DB(5) and DB(3) correspond to REG(1) and REG(2), meaning that in DB(5) the network considers just one previously attended object, while in DB(3) each network slice consists of the previously attended object as well as information about the previous objects in the scene. REG(Gist): regression based only on the gist of the scene. kNN: k-nearest-neighbors classifier. Rand: white noise random map (control). Overall, DB(3) performed best at predicting where the player would look next (Borji et al., 2012b). (See figure 23.3.)

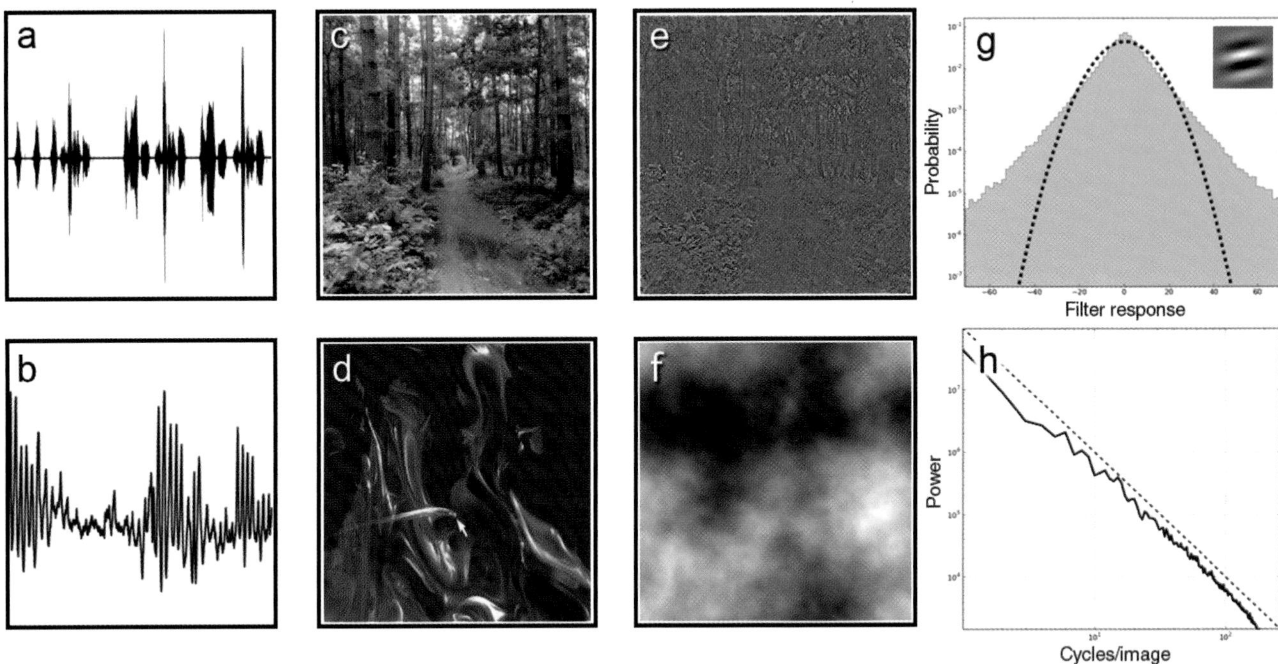

PLATE 22 Examples and properties of natural stimuli. (A) A natural auditory stimulus, the sound pressure wave of a zebra finch's song (M. Fee). (B) The position of a rat's vibrissa during natural whisking (data courtesy of D. Kleinfeld). (C) A natural image (photograph by Ruben Holthuijsen). (D) An underwater olfactory plume passing a lobster antenna (photograph by M. Koehl). (E) The image in (C), whitened. (F) A Gaussian random image with spectral characteristics matching those of natural scenes. (G) The probability density, gray, of the output of a Gabor filter (inset) acting on the image in (C); differs strongly from Gaussian (dotted). (H) Natural images have a close to $1/f^2$ power spectrum. (See figure 26.1.)

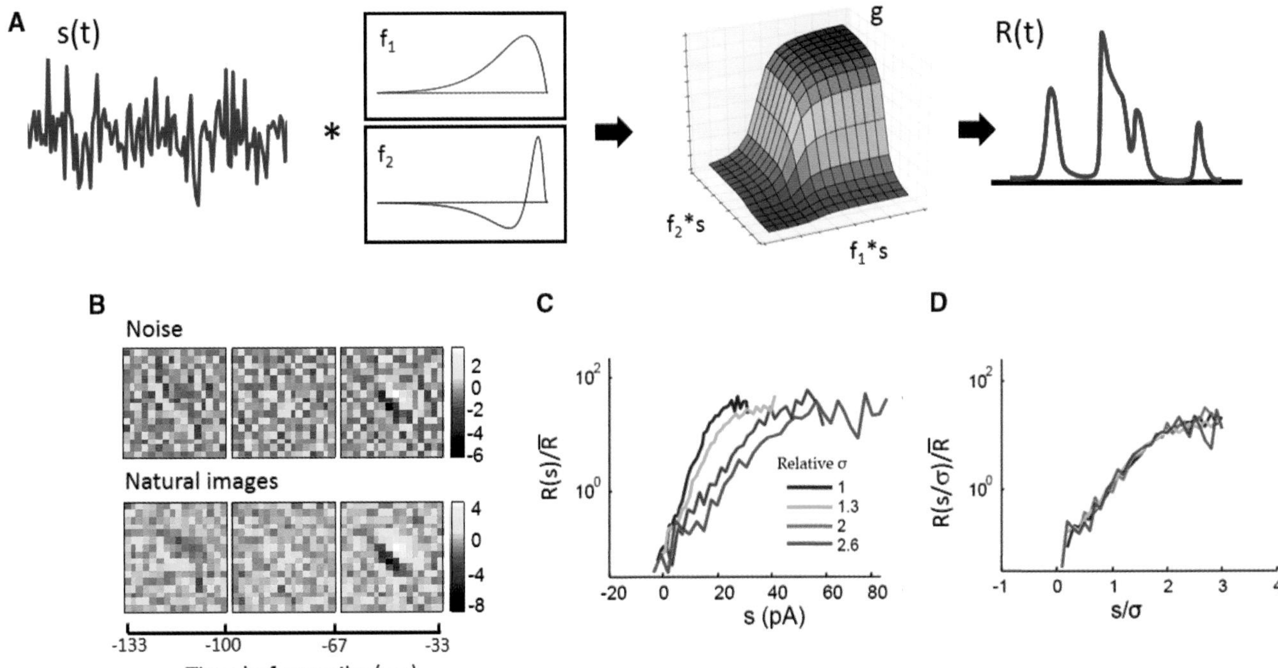

PLATE 23 Linear-nonlinear models and adaptation. (A) Multidimensional linear-nonlinear cascade model: the stimulus $s(t)$ is filtered through one or more filters f_i and passed through a nonlinear function $g(f_1*s, f_2*s, \ldots)$. Here, f_1 is a leaky integrating filter while f_2 is closer to a derivative. The filters f_i and g can both change with adaptation to stimulus context. (B) Example of filter adaptation: spatiotemporal filters from V1 computed using the maximally informative dimension method (Sharpee et al., 2004) when driven either by a flickering noise input or a natural image ensemble. The filters show changes in low-frequency content (adapted from Sharpee et al., 2006). (C) The nonlinearity g can also adapt. Input-output curves from a cortical neuron driven by a filtered white noise current input (R. Mease). The stimulus s is the current filtered through the spike-triggered average. (D) The same input-output curves plotted against the stimulus normalized by the standard deviation of the current. (See figure 26.2.)

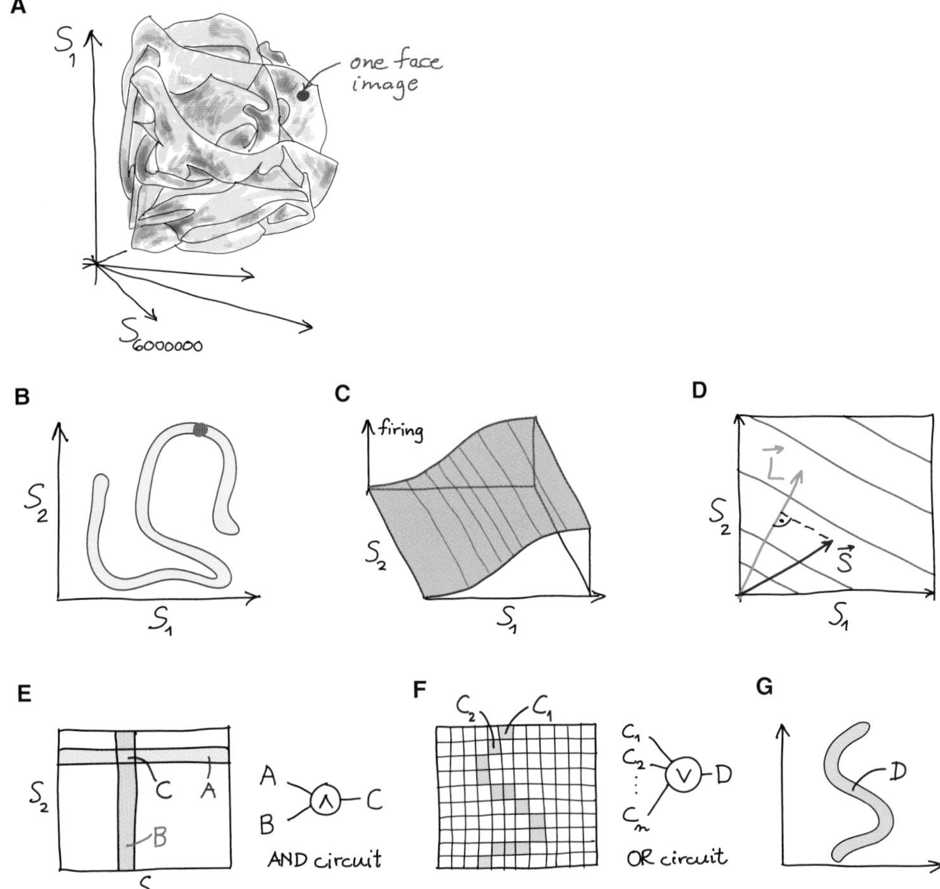

PLATE 24 A geometric view of sensory computations. (A) Each stimulus is a vector in a high-dimensional space, whose axes are defined by the excitation of all the sensory receptors, for example, the intensity on the 6 million cones in a human eye. One retinal image of a face is a point in that space. The same face seen under different views and illuminations defines a high-dimensional surface (DiCarlo & Cox, 2007). (B) Two-dimensional stimulus space. A "face cell" responds only when the stimulus is in a thin, convoluted, shoelace-shaped region. (C) The response of a receptor as a function of the stimulus variables. Note the contour lines are straight and orthogonal to the receptor's axis of sensitivity (here the S_1 axis). (D) A receptive field is represented by a vector **L** in stimulus space. Integration of the stimulus **S** by this receptive field is the projection of **S** onto **L**. The response of a linear-nonlinear (LN) neuron depends only on this projection: the contour lines of its response function are straight and orthogonal to **L**. (E–G) How to create a shoelace-shaped response region in three simple steps. See text for details. (See figure 28.1.)

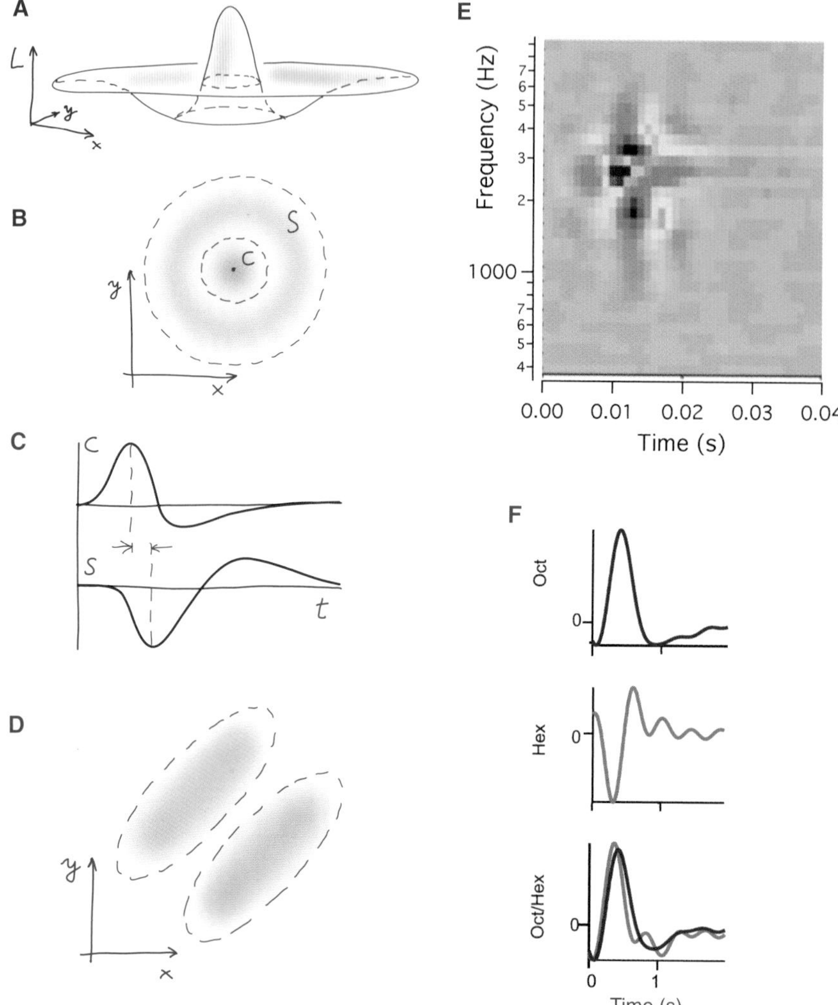

PLATE 25 Receptive fields in various sensory systems. (A–C) Spatiotemporal receptive fields in the early visual system. Here the components of the stimulus vector **S** are the light intensity at different locations and time points. Any given stimulus vector consists of a short "movie" $S(x,y,t)$. The receptive field $L(x,y,t)$ can similarly be viewed as a movie. For displays on paper, one often shows a single frame $L(x,y,t_0)$ of this movie (panels A–B) along with the time course $L(x_0,y_0,t)$ at a particular location (panel C). Note that L has both positive and negative components, meaning that light is excitatory (red) at some locations and times, and inhibitory (blue) at others. For retinal bipolar cells, ganglion cells, and thalamic relay cells, the receptive fields look very similar except for some scaling along the spatial and temporal axes. The spatial profile (B) contains a center region (C) and a surround region (S) in which light has opposite action. When displayed in three dimensions (A), this profile has the appearance of a "Mexican hat." The time course of the receptive field (C) is biphasic, both in the center and the surround. Generally, the surround response is slightly delayed relative to the center (C). (D) For a simple cell in primary visual cortex, the receptive field profile typically shows elongated side-by-side regions in which light has opposite action. The time course again is biphasic, as in panel C. (E) Spectrotemporal receptive field of a neuron in the auditory area field L of the zebra finch (Nagel & Doupe, 2008). Here the stimulus $S(f,t)$ is represented as a spectrogram, plotting sound power as a function of time and frequency. The receptive field of this cell shows excitatory (red) and inhibitory (blue) regions. Along the spectral dimension one finds a "Mexican hat" profile, with a central region of excitatory frequencies and adjacent inhibitory regions. The action of the surround is delayed relative to the center. In the time domain, the receptive field is strongly biphasic or even triphasic. This is only one of many receptive field shapes encountered in this brain region. (F) Osmotemporal receptive field of a neuron in the olfactory pathway of the locust (Geffen et al., 2009). Here the stimulus $S(o,t)$ is the time-varying concentration of two discrete odors, octanone or hexanal. The receptive field $L(o,t)$ specifies the contribution to the response of each odor at different times. *Top and middle*: Receptive field derived from experiments in which only one odor was presented at a time. *Bottom*: Receptive field measured when both odors were varied simultaneously. Note the neuron's response to hexanal changes polarity under this condition. (See figure 28.2.)

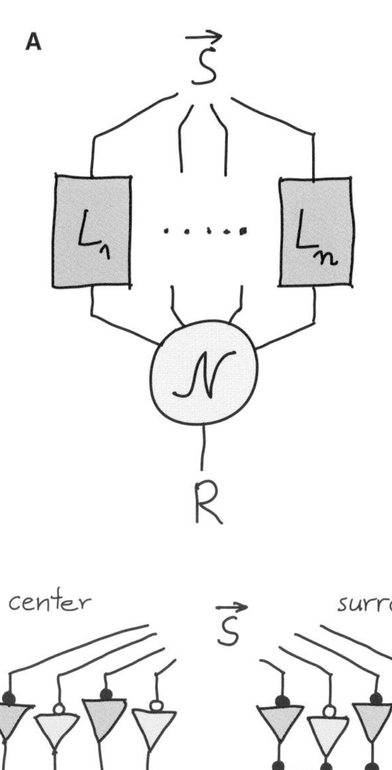

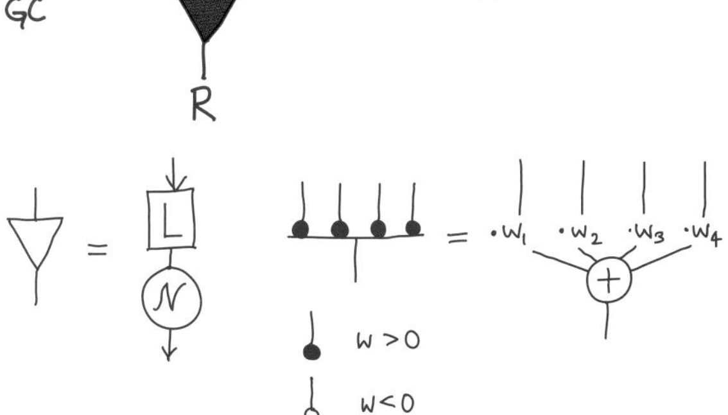

PLATE 26 Response models beyond the receptive field. (A) Multi-LN model. Here the stimulus, **S**, is processed linearly through multiple parallel filters. Their outputs are then combined in a single static nonlinearity, yielding the response, *R*. (B) Circuit model with a cascade of LN stages that may be arranged in series or parallel, including feed-forward and feedback pathways. This example takes inspiration from retinal circuitry, and each participating neuron type (On and Off bipolar cells, BC; amacrine cells, AC; ganglion cell**s**, GC) is represented by a simple LN model: a weighted summation of inputs (weights w_i) followed with a temporal filter and a nonlinear response function. With appropriate choice of those parameters, the circuit makes quantitatively accurate predictions for the response of the so-called object motion sensitive ganglion cells that sense differential motion between foreground and background (Ölveczky et al., 2003). (See figure 28.3.)

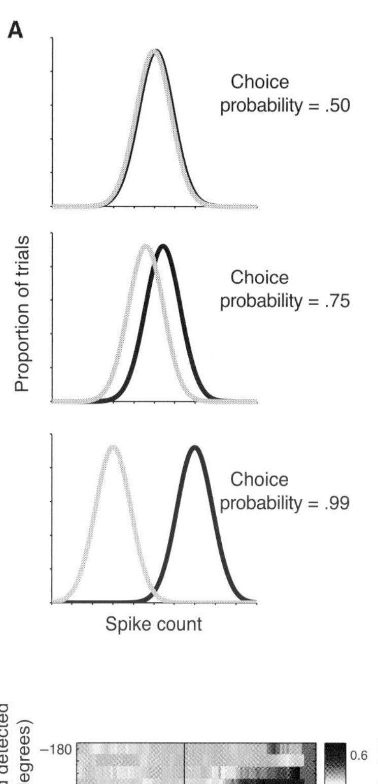

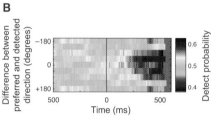

PLATE 27 (A) Choice probability is calculated by comparing the distributions of neural responses on trials when the subject made each of two choices. This schematic shows pairs of distributions with different choice probabilities. (B) Choice probabilities (referred to as "detect probabilities" in this detection task) decrease for neurons whose preferred direction does not match the direction of motion being detected. (Adapted from Bosking & Maunsell, 2011.) (See figure 31.3.)

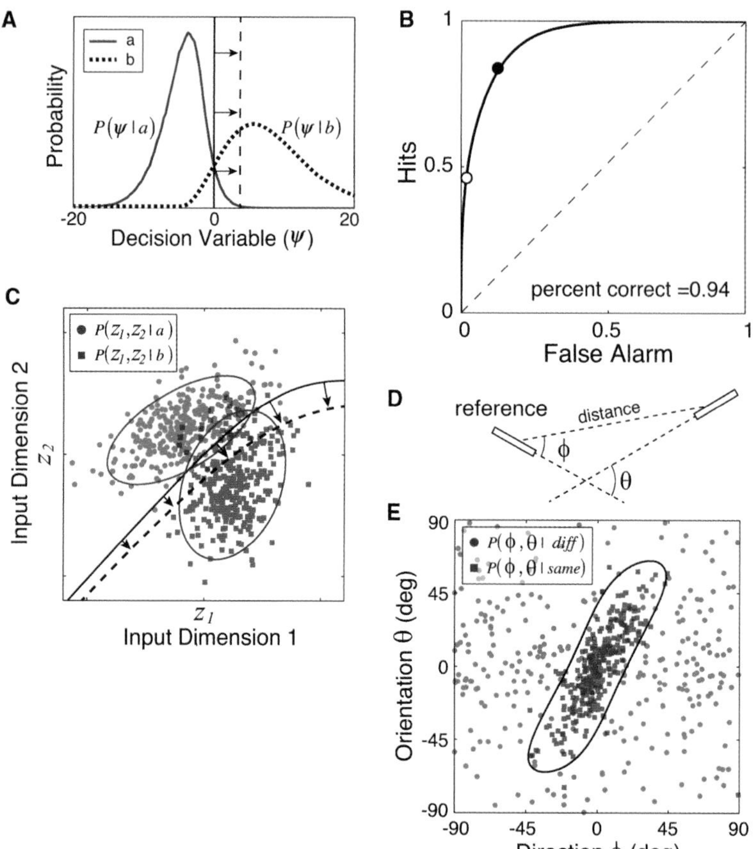

PLATE 28 Detection and discrimination. (A) Distribution of decision variable values conditioned on two different states of the world (a and b). (B) The receiver operating characteristic for the distributions in A. (C) Samples from hypothetical Gaussian likelihood distributions for the two stimulus categories in a yes-no detection task. The log of the ratio of these two likelihood distributions at any point is the decision variable in A; that is, $\Psi = \log[p(z_1,z_2|b)/p(z_1,z_2|a)]$. The solid and dashed contours correspond to the two criteria in A. (D) Three parameters describing the possible geometrical relationships between two contour elements. (E) Samples of geometrical relationships for pairs of contour elements belonging to the same physical contour (blue symbols) and belonging to different physical contours (red symbols), for a particular distance between the elements. The black curve shows the optimal decision bound given equal prior probabilities. (Data from Geisler & Perry, 2009.) (See figure 32.1.)

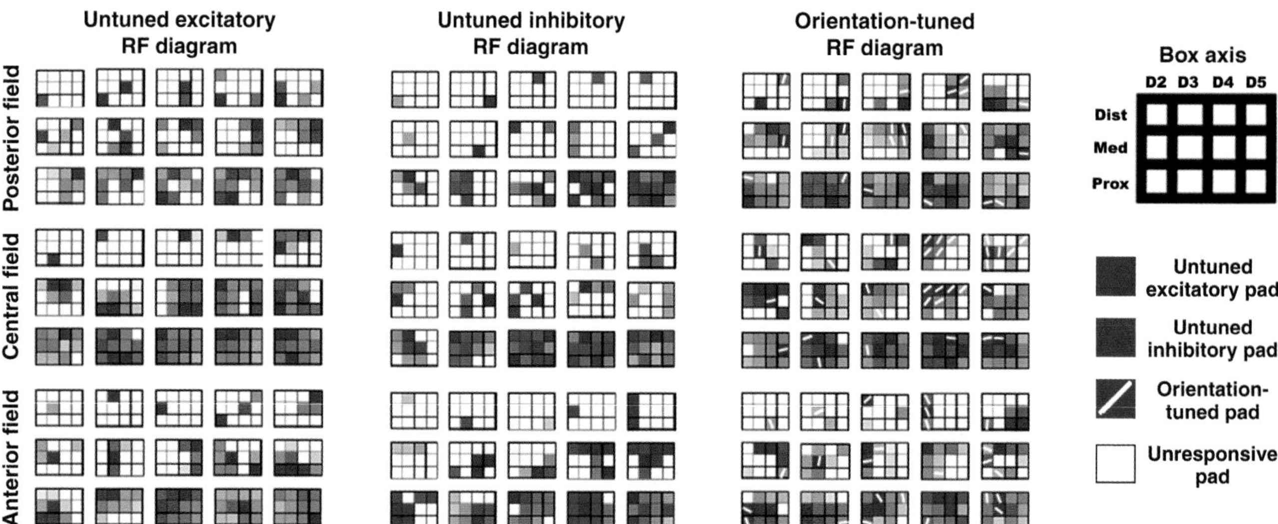

PLATE 29 Receptive fields of SII neurons. Each box of twelve squares represents the RF of a single neuron, with each square representing a finger pad. Red: excitatory; Blue: inhibitory. (Adapted with permission from Fitzgerald et al., 2006.) (See figure 33.5.)

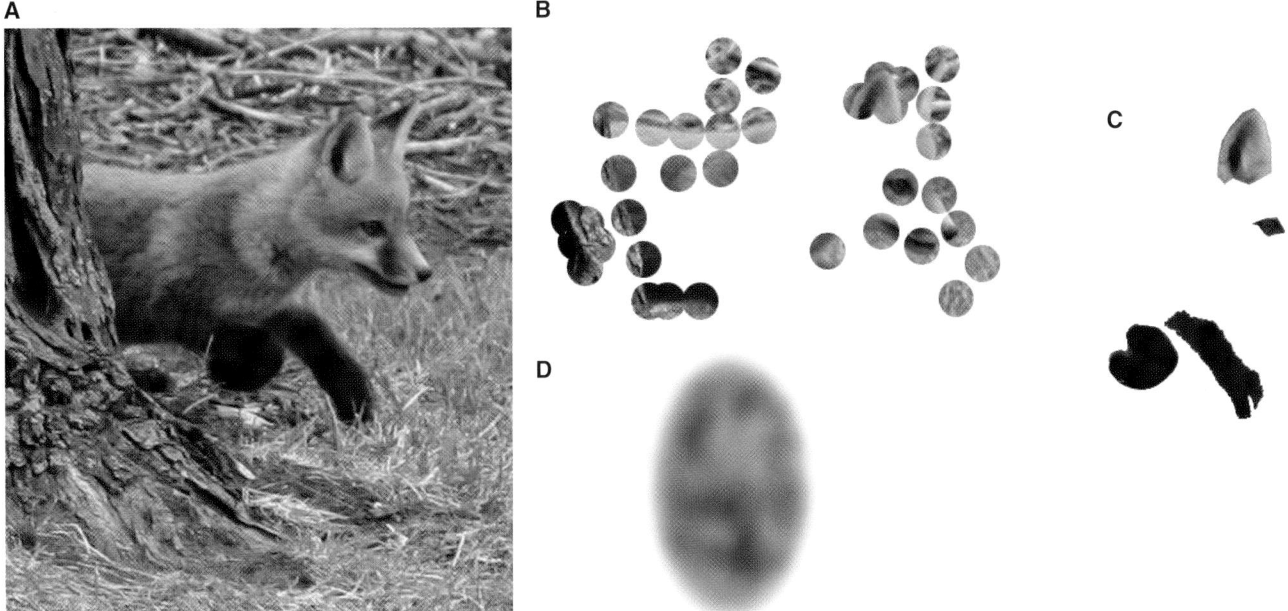

Plate 30 (A) This figure illustrates two problems. (1) How can local measurements made from small patches (B), using neurons with small receptive fields, be integrated to recognize objects and patterns (e.g., fox, tree trunk, grass)? (2) How does the visual system support a limitless number of descriptions of a single scene? Answers need to account for flexible access to information of various types over a range of spatial scales, such as the various edge and textural properties of local regions B, the shape of parts C, and intermediate- and higher-level concepts such as "head" D, respectively. There is a bootstrapping problem, in that the accurate interpretation of any local patch is ambiguous without knowledge of the rest. (See figure 34.1.)

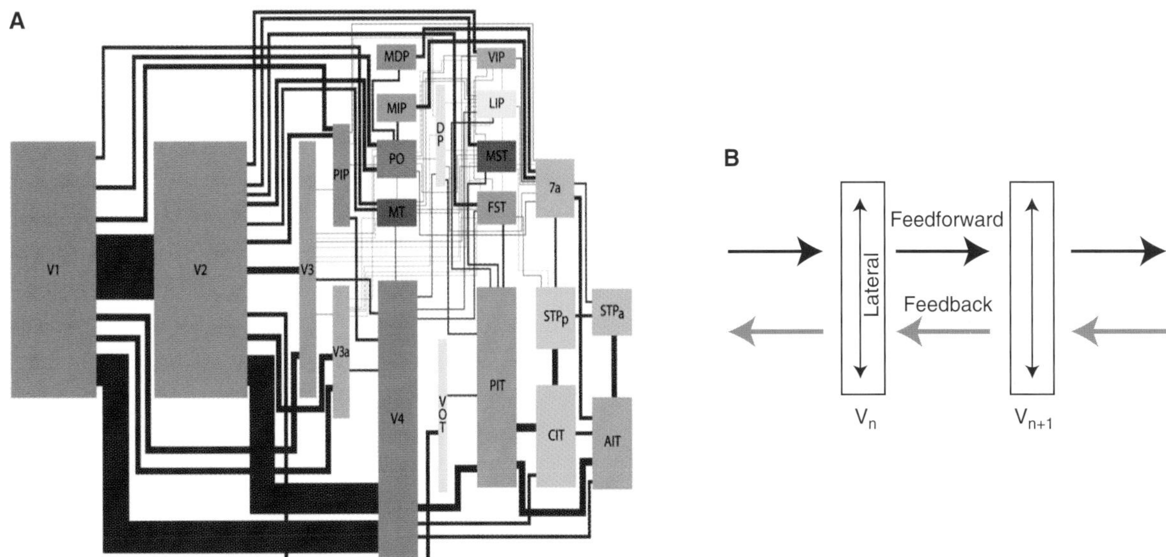

Plate 31 (A) Schematic of macaque monkey visual cortex. (Reprinted from Wallisch & Movshon, 2008, with permission from Elsevier; see also Lennie, 1998). The colored rectangles represent visual areas (see Felleman & van Essen, 1991). The gray lines show the connections between areas, with the thickness proportional to estimates of the number of feedforward fibers. Areas in warm and cool tones belong to the dorsal and ventral streams, respectively. (B) Feedforward and feedback connections represent transmission of feedforward and feedback signals between visual areas. Lateral (also called "horizontal") organization within areas, representing features of similar types and level of abstraction. (See figure 34.2.)

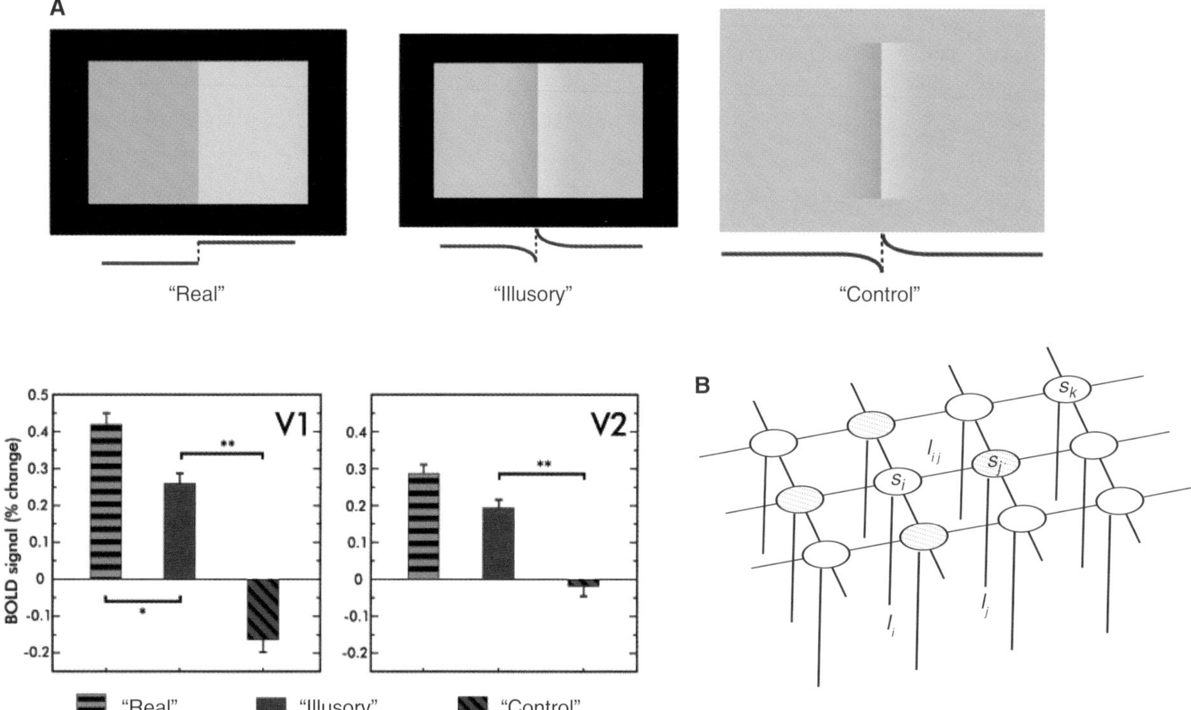

Plate 32 (A) The upper-middle panel shows a classic illusion known as the Craik-O'Brien effect. Away from the vertical border, the left and right rectangles have the same luminance, as indicated by the red line, which shows how light intensity varies from left to right. The interesting perceptual observation is that the left rectangle looks darker than the right. In fact, there is little difference between the appearance of a real intensity difference (upper left), and the illusory one. The lower graphs show that voxels in both V1 and V2 respond to apparent changes in lightness almost as strongly as real changes, as compared with a control. (Reprinted from Boyaci et al., 2007, with permission from Elsevier.) (B) An undirected graph (Markov random field) can be used to formulate prior probabilities representing lateral, spatial statistical dependencies for contours and surface properties such as reflectance (cf. Marroquin et al., 1987; Kersten, 1991). (See figure 34.4.)

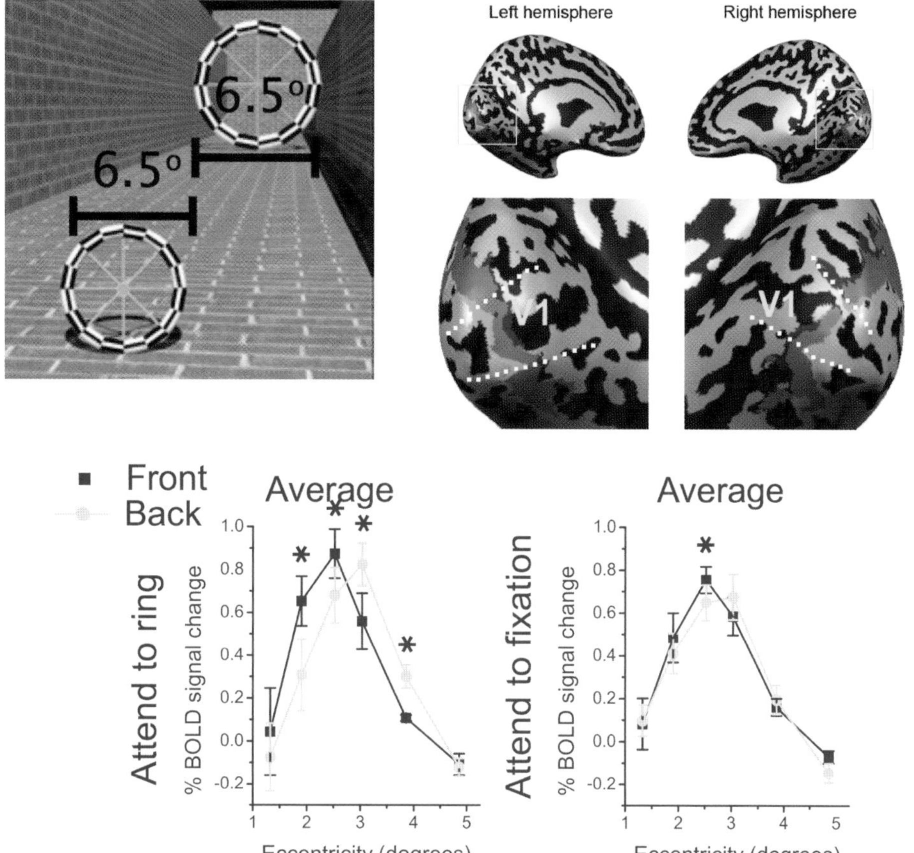

PLATE 33 This figure illustrates how global, contextual information for 3D depth can shift the spatial extent of activity in human V1. Upper left panel: both rings are the same size in the image, but the ring in the back of the hallway is perceived to be bigger than the one in front because it appears to be further away. Upper right panel: the spatial extent of activity in human cortical area V1 is larger for the back ring (red) than the front ring (green), similar to what would be found if the image of the back ring, in the absence of the 3D depth context, actually was bigger. Lower panel: when a person attends to the ring, the increase in spatial extent as indicated by a shift in the ring response with eccentricity, is more robust (bottom left) than when attending to the fixation mark (bottom right). (Reprinted from Fang et al., 2008, with permission from Elsevier.) (See figure 34.7.)

A Lateral-medial gradient of large-scale maps in VTC

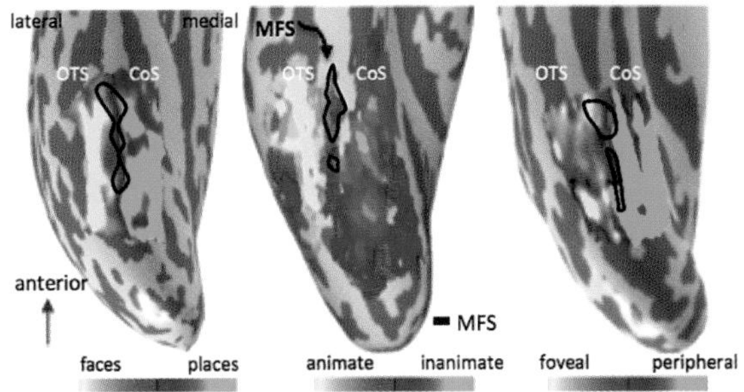

B Fine-scale organization of VTC

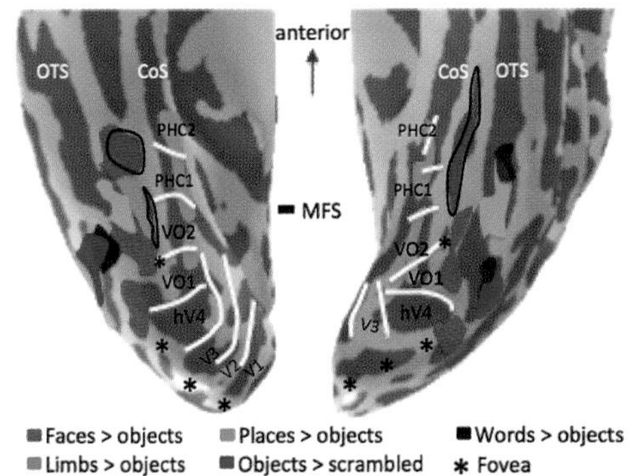

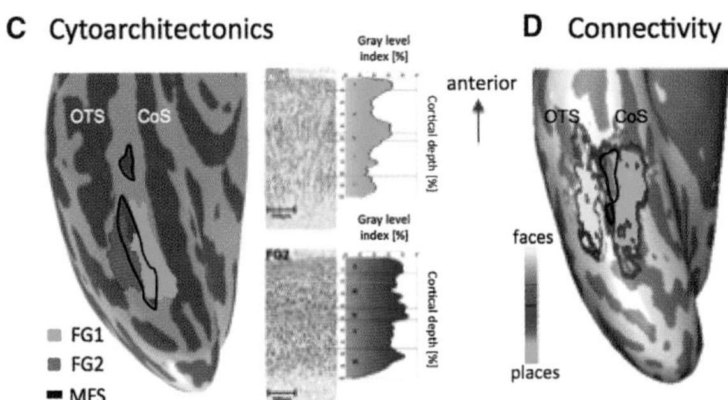

Plate 34 Functional architecture of VTC. (A) Several large-scale functional maps tile the cortical sheet with the same lateral-medial gradient. From left: domain specificity, animacy (the second principal component [PC2] from Haxby et al., 2011), and eccentricity bias on an example right hemisphere. (B) Fine-scale organization in a representative subject measured with 1.8 mm voxels. Functional regions are located in predictable cortical positions relative to macroanatomical landmarks and relative to one another. *White:* borders of retinotopic regions. (C) *Left*: inflated cortical surface with cytoarchitectonic regions of VTC in one representative postmortem subject showing the cytoarchitectonic transition between FG1/FG2 occurring within the MFS (black). *Right*: cytoarchitectonic characteristics of FG1 and FG2 (left) with profiles of gray level indices (right). FG1 displays a columnar arrangement of small pyramidal cells, while FG2 does not, and FG2 displays a higher neuronal density than FG1. (Adapted from Caspers et al., 2013, and Weiner et al., 2014.) (D) Correspondence between structural connectivity and functional division of VTC in a representative subject: thresholded boundaries of inferotemporal (dark red) and lingual (dark blue) connectivity are nearly identical to the faces vs. places functional map. (Adapted from Saygin et al., 2012.) (See figure 35.4.)

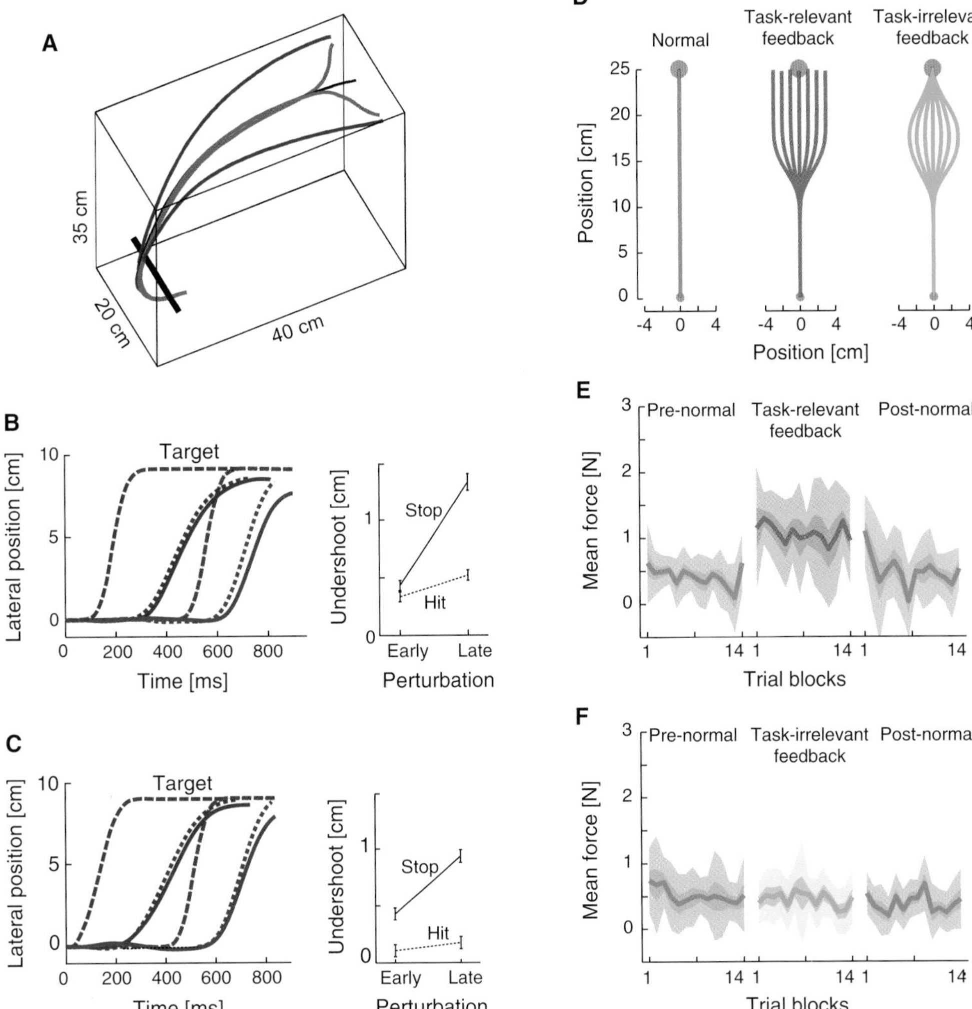

Plate 35 Visuomotor responses account for the goal of the ongoing task. (A) Subjects were instructed to reach and stop at a target ("stop" condition) that either stayed in the central location or jumped in the lateral direction after the start of the movement. Data are the population average trajectories when participants were instructed to stop at the target. Color code: black, baseline; red, early perturbation; blue, late perturbation. (B) Hand path deviation in the direction of the displaced target. Note that subjects were unable to compensate when the target jumped late in the movement, demonstrating that feedback responses depend on the task constraints. Color scheme is the same as in A. Dashed lines, "hit" condition; solid lines, "stop" condition. (C) Results from target intercept experiment. In this experiment, the target jumped laterally and then moved downward at a fixed rate. The subjects were instructed to "stop" or "hit" the target before it stopped moving. (Adapted from Liu & Todorov, 2007.) (D) Visuomotor responses only correct for errors that are relevant to the ongoing task. In the normal condition, the hand feedback cursor reproduced the hand trajectory. In the task-relevant feedback condition (orange traces), the visual cursor moved away from the hand trajectory and remained at this point for the rest of the movement. In contrast, in the task-irrelevant feedback condition (blue traces), the hand feedback cursor moved away from the hand trajectory but returned to the true hand position by the end of the movement. (E) Time course of adaptation to task-relevant feedback perturbations. Data are the mean (solid line) and standard deviation (shaded region) force difference across subjects between right and left visual perturbations (180–230 ms after cursor jump). Note that subjects produced larger corrections when the cursor displacements were relevant to the ongoing task. (F) Time course of adaptation to task-irrelevant feedback perturbations. Note that subjects did not adapt when the perturbations were irrelevant to the ongoing task. (Adapted from Franklin & Wolpert, 2008.) (See figure 39.2.)

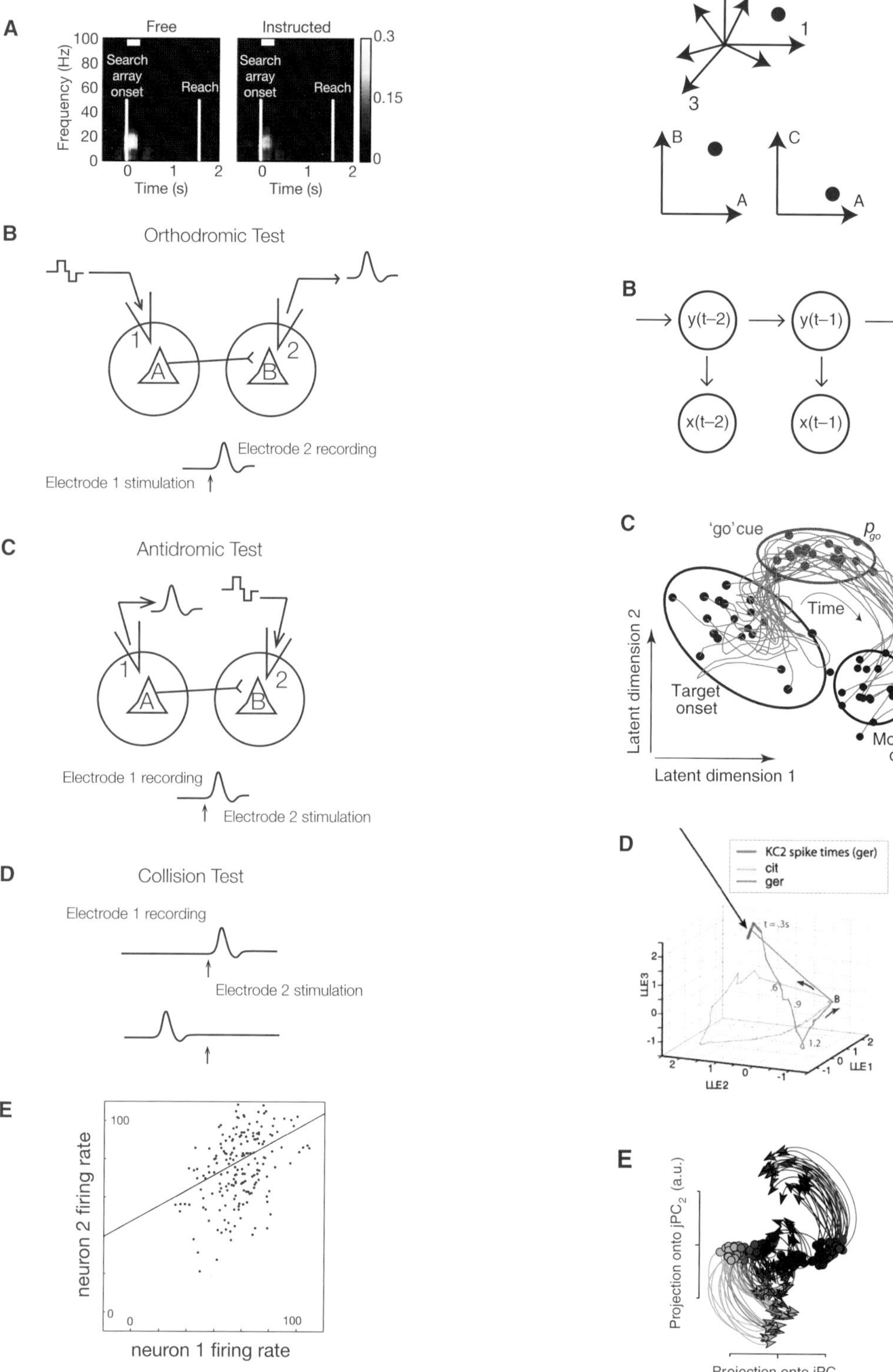

Plate 36 Other pairwise measurements. (A) Coherograms plot the amplitude of the coherency between spiking activity in the premotor cortex and LFP in the parietal cortex. Coherency is greater in the 15 Hz band when the monkey freely selects targets in a foraging task than when his choices are instructed (Reprinted by permission from Macmillan Publishers Ltd: *Nature*, Pesaran et al., 2008). Identifying projection neurons: (B) the orthodromic test; (C) the antidromic test; (D) the collision test. Stimulation on electrode 2 causes a spike in neuron A only if A has not spiked recently. (E) Noise correlation. (See figure 40.2.)

Plate 37 Multineuronal interactions. (A) Visualizing high-dimensional neural data. A point in seven-dimensional space, and two 2D projections. 1, 2, 3, etc., represent the activity of individual neurons; A, B, and C are linear combinations of the seven original axes. (B) Depiction of a simple latent variable model. (C) "Neural trajectories" through state space. Each line is an individual trial. (Adapted from Afshar et al., 2011, with permission from Elsevier.) (D) Tuning in state space. (Adapted from Broome et al., 2006, with permission from Elsevier). The arrow indicates the position in state space for which the Kenyon cell is tuned. (E) Intrinsic dynamics in motor cortex. (Adapted by permission from Macmillan Publishers Ltd: *Nature*, from Churchland et al., 2012.) (See figure 40.3.)

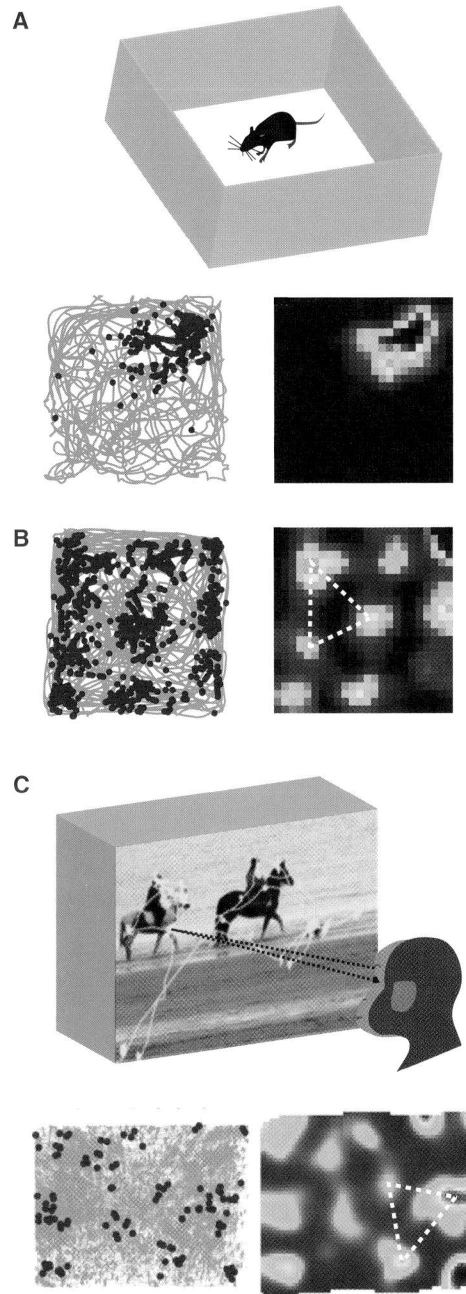

Plate 38 Place cells and grid cells. (A) Spatially tuned principal neurons are found in all subregions of the rodent hippocampal formation. A gray line indicates the path of the animal while exploring a 1 m × 1 m open field. Each action potential of the CA1 neuron is shown as a red dot superimposed on the position of the animal when the spike occurred (left). The firing rate of the same CA1 neuron is represented by a heat map with high firing rates in warmer colors (right). Hippocampal neurons are place cells with a single place field, whereas (B) grid cells are found in all layers of medial entorhinal cortex (MEC). The multiple firing peaks of grid cells are arranged as equilateral triangles (white dashed lines) that form a grid-like representation (right). (C) Grid cells in the primate MEC during the visual exploration of spatial scenes. (Reprinted by permission from Macmillan Publishers Ltd.: *Nature*, Killian et al., 2012.) (See figure 46.2.)

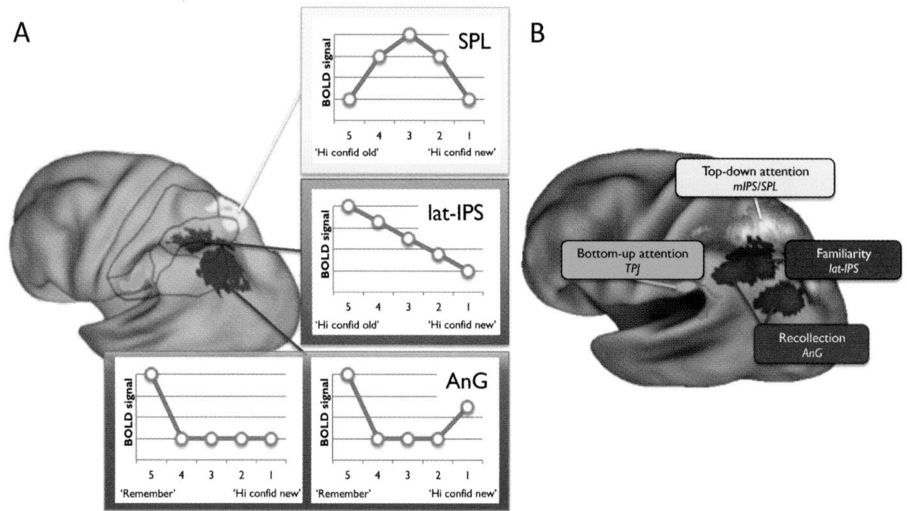

PLATE 39 PPC functional dissociations. (A) Hutchinson et al. (2014) revealed a quadruple dissociation of activity patterns during retrieval: lat-IPS (red) activity showed a monotonic relationship with item-memory strength; AnG (blue) activity tracked recollection of event details; mIPS/SPL (yellow) demonstrated a decision uncertainty effect, with TPJ (green) qualitatively showing the reverse. Black outlines on the brain indicate parcellation identified by Nelson et al. (2010) using graph-theoretic analyses of resting-state and task fMRI data. Line graphs depict schematics of representative activity patterns. Note: AnG activity tracks recollection (left graph) and sometimes also shows a novelty effect (right graph). "Remember," Remember responses; "hi confid," high confidence. (B) Meta-analyses of top-down and bottom-up attention effects, and recollection and familiarity-based retrieval effects reveal largely nonoverlapping parietal regions supporting attention and memory (adapted from Hutchinson et al., 2014). (See figure 48.2.)

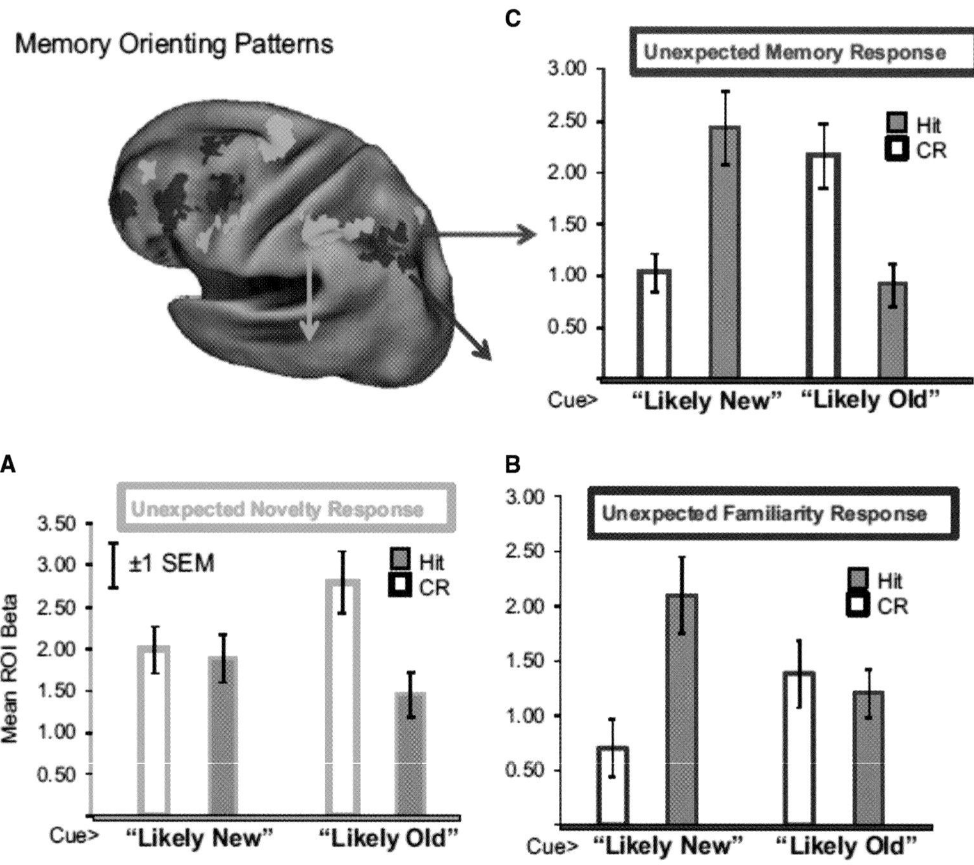

PLATE 40 Selective responses to particular types of violations of memory expectations. Targeted contrasts and masking procedures isolated three different patterns of response in left lateral parietal cortex. The left anterior IPS/PoCG region in green illustrated an unexpected novelty response pattern, with bar plot (A) illustrating the pattern across cue conditions and item types. The posterior anterior angular gyrus region in blue demonstrated an unexpected familiarity response, with bar plot (B), illustrating the pattern of response across cue conditions and item types. The mid-IPS region in red demonstrated a general unexpected memory effect, with bar plot (C) illustrating the pattern of response across cue conditions and item types. (See figure 49.4.)

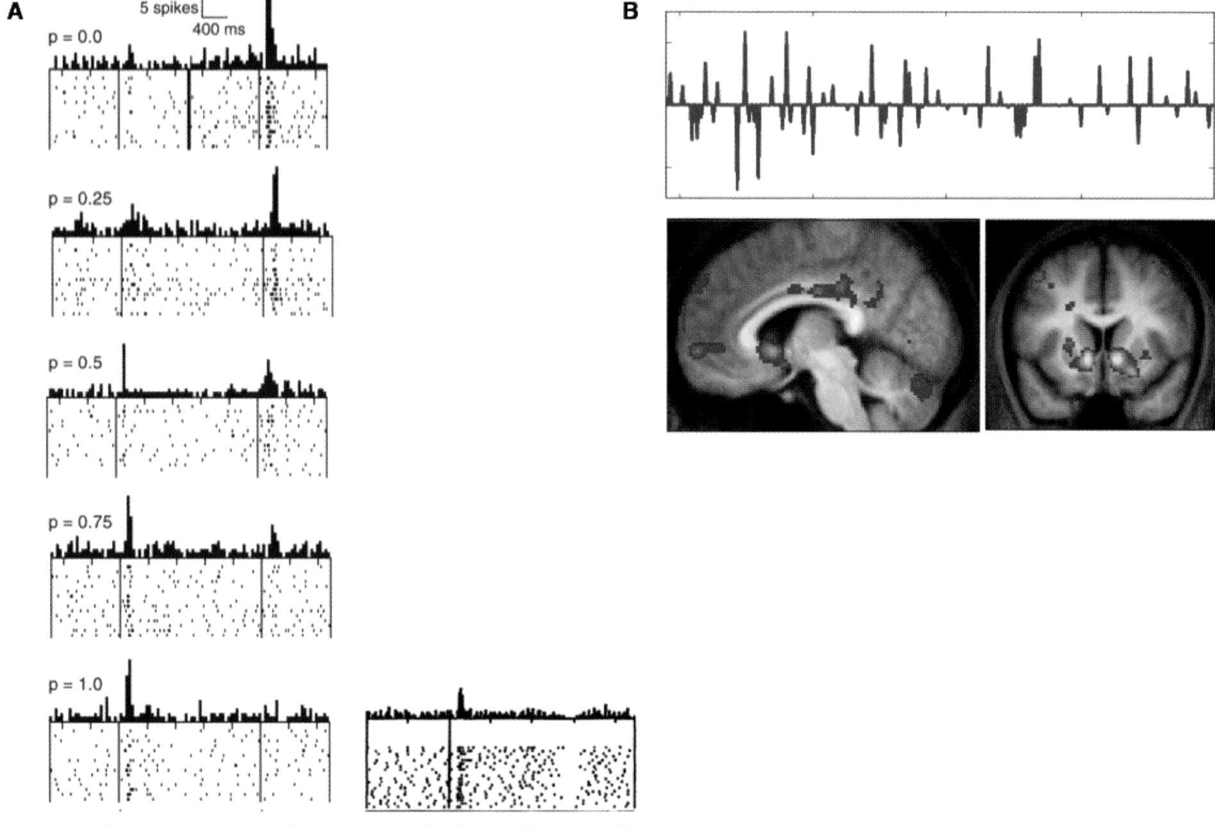

PLATE 41 (A) Responses of dopamine neurons recorded from the primate midbrain, reprinted from Fiorillo et al. (2003) and Schultz et al. (1997) with permission. Five stimuli predict juice reward with different probabilities ($p = 0$ to 1, top to bottom). The neuron responds to each stimulus proportional to its associated reward probability, but when reward is delivered, the response is larger the less expected was the reward. When an expected reward is omitted, dopamine neurons are inhibited (right). These responses are consistent with a reward prediction error. (B) BOLD activity in areas of the human brain, notably ventral striatum, correlates with a reward prediction error extracted from a trial-by-trial computational model of each subject's learning (top trace). Displayed at $p < .001$ uncorrected; data replotted from Daw et al. (2011). (See figure 50.2.)

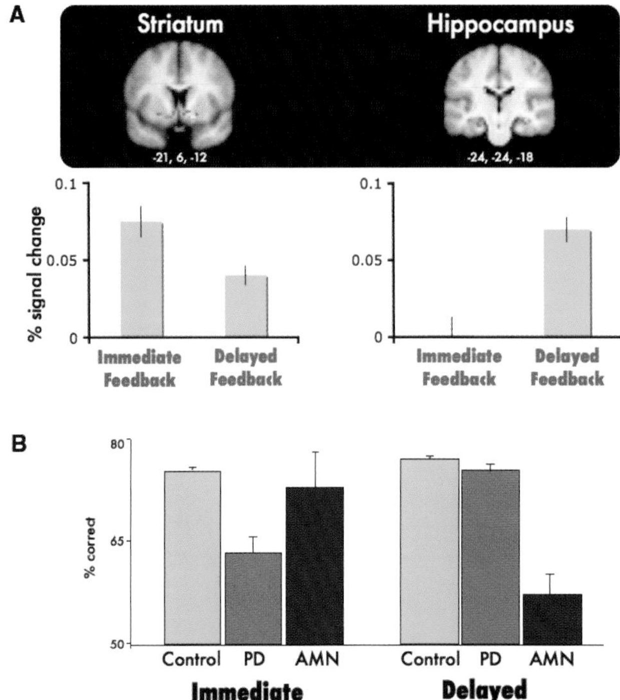

Plate 42 Probabilistic learning shifts from the striatum to the hippocampus when response-contingent feedback is delayed, as demonstrated by converging fMRI and patient studies. (A) FMRI reveals prediction-error related signals in humans during learning a probabilistic task. The striatum and the hippocampus display differential magnitudes of feedback-related signals for conditions wherein learning is driven by feedback that is immediate (within 1 second) versus delayed (by 7 seconds). (Data from Foerde & Shohamy, 2011.) (B) Data from patients support the selective role for the striatum in learning from immediate feedback. Patients with disrupted striatal function due to Parkinson's disease ("PD") are impaired at learning from immediate feedback but intact when learning from delayed feedback. Patients with amnesia due to MTL damage ("AMN") show the opposite pattern: intact learning from immediate feedback but impaired learning from delayed feedback. (Data from Foerde et al., 2013.) (See figure 50.3.)

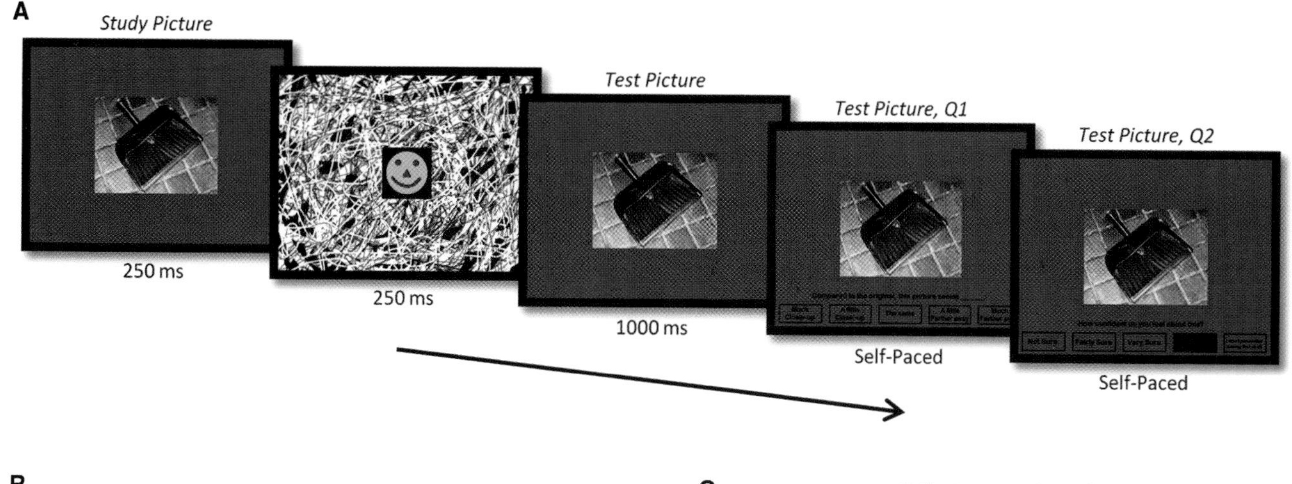

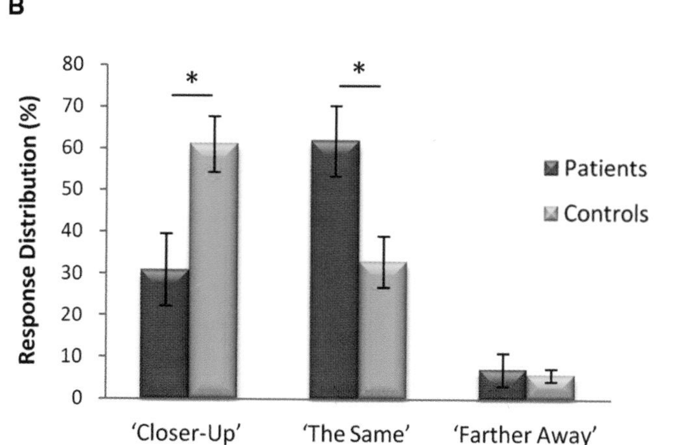

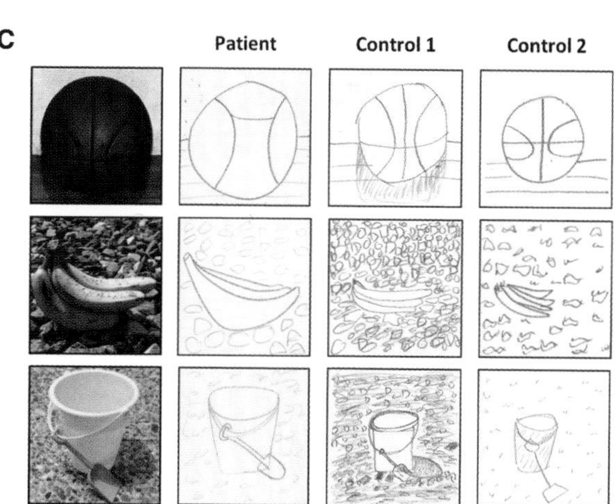

PLATE 43 Examples of boundary extension (BE) tasks (Mullally et al., 2012a). (A) Timeline of an example trial from a rapid serial visual presentation BE task. The initial photograph of a simple scene was presented briefly followed by a dynamically changing mask. The second (test) picture (which unbeknownst to the participants was always identical to the original picture) immediately followed the mask. The task was to rate the second picture relative to the first. There were five options ranging from "much closer-up" to "much farther away," including the correct response "the same." (B) BE is revealed by disproportionally larger number of "closer-up" responses. Overall, control participants made significantly more of these erroneous responses, while patients with bilateral hippocampal amnesia made significantly more accurate ("the same") responses, and thus showed significantly reduced BE relative to controls. Means (± SEM); $^*P < 0.05$. (C) In a drawing task, three scene photographs (left panel) were studied for 15 seconds and immediately drawn from memory. Drawings by an example hippocampal-damaged amnesic patient (middle left panel) and two matched control participants (middle right and right panels) are displayed. As is evident, this patient more accurately depicted the proportional size of the object relative to the background whilst the control participants' drawings expose how they extrapolated beyond the given view. (See figure 51.2.)

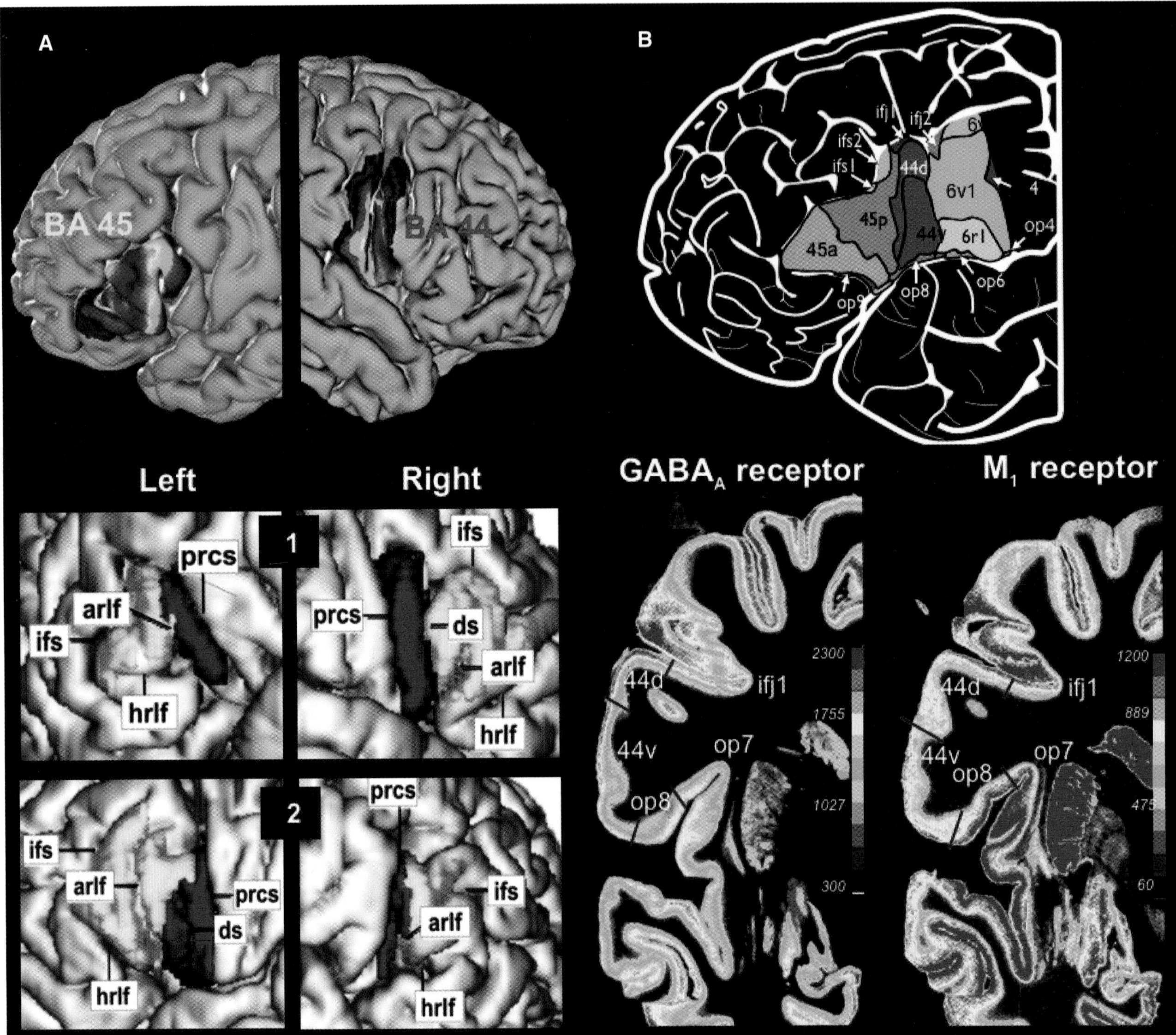

Plate 44 Cortical mapping of Broca's region based on postmortem cytoarchitectonic and receptorarchitectonic analysis. (A) Top, surface view of the probabilistic maps for Brodmann's area 44 (BA44) and 45 (BA45) derived from quantitative cytoarchitectonic measurements of 10 individual brains. The probabilistic maps are intended to quantify the interindividual variability of the areas that compose Broca's region, where "hot" tones correspond to a high anatomical overlap, and "cold" tones to higher variability. This interindividual variability can be appreciated in the bottom panel, where the left and right maps for BA44 (red) and BA45 (yellow) are reported for two representative brains. Note, for example, the difference in the left-right asymmetry of BA 44 and 45. In addition, the borders of the two areas are not always well identified by the main sulci of this region: inferior frontal sulcus (ifs); horizontal ramus of the lateral fissure (hrlf); ascending ramus of the lateral fissure (arlf); precentral sulcus (prcs); diagonal sulcus (ds), (data from Amunts et al., 2004). (B) Further segregation of Broca's region based on neurotransmitter receptors distribution. This analysis can reveal unique characteristics of subregions that may share similar cytoarchitectonic morphology. Area 44v and op8, for example, have a similar density of GABAA receptors but different density of cholinergic muscarine receptors M1.B) Further segregation of Broca's region based on neurotransmitter receptors distribution. This analysis can reveal unique characteristics of subregions that may share similar cytoarchitectonic morphology. Area 44 v and op8, for example, have a similar density of $GABA_A$ or alternately GABA(A) receptors but different density of cholinergic muscarine receptors M1. (See figure 52.1.)

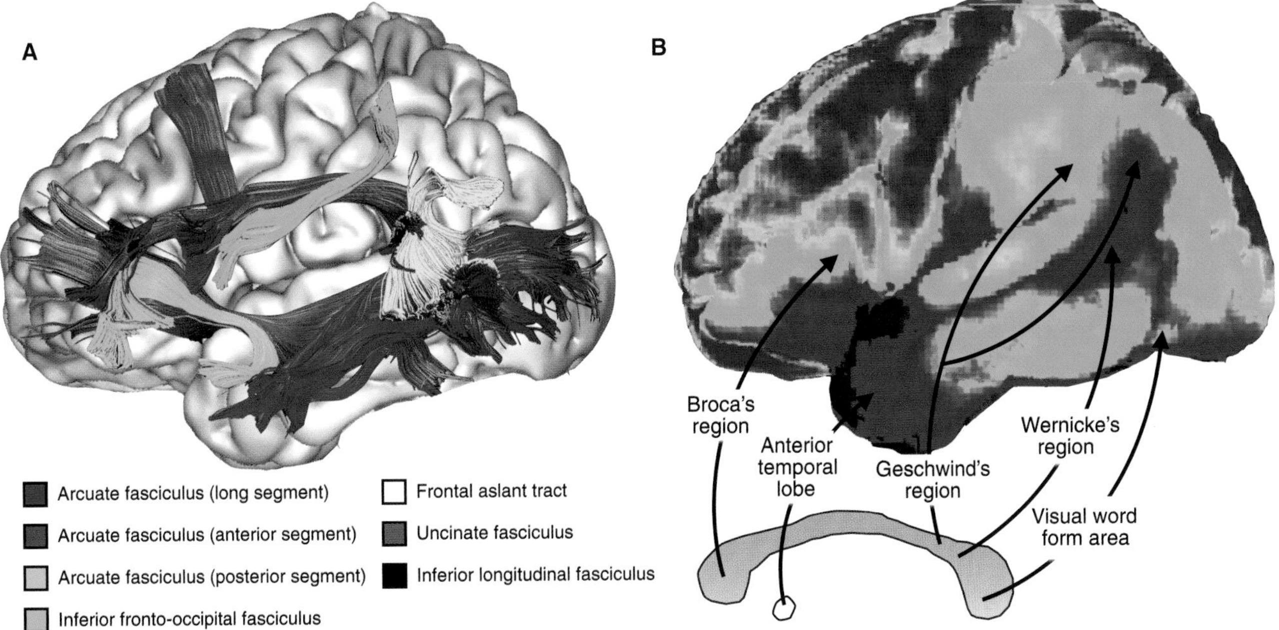

Plate 45 Language networks visualized with diffusion tractography. (A) Association pathways of left hemisphere connecting the main language regions. (B) Density of callosal cortical projections based on diffusion imaging. "Hot" tones indicate higher degree of transcallosal connections. Note the reduced interhemispheric connectivity for most of the language regions. (See figure 52.5.)

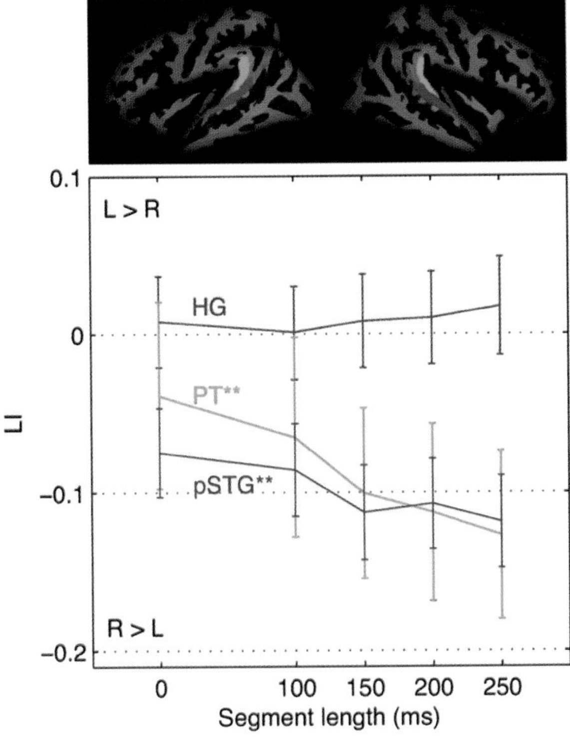

Plate 46 The functional lateralization index (LI) in anatomically defined regions of interest is displayed; positive values indicate L > R lateralization. HG: Heschl's gyrus (red); PT: planum temporale (green); pSTG: posterior superior temporal gyrus (blue). $P < 0.001$. Error bars indicate 1 standard error of the mean. (Figure reprinted with permission from Liem et al., 2013. Copyright © 2013 Wiley Periodicals, Inc.) (See figure 53.2.)

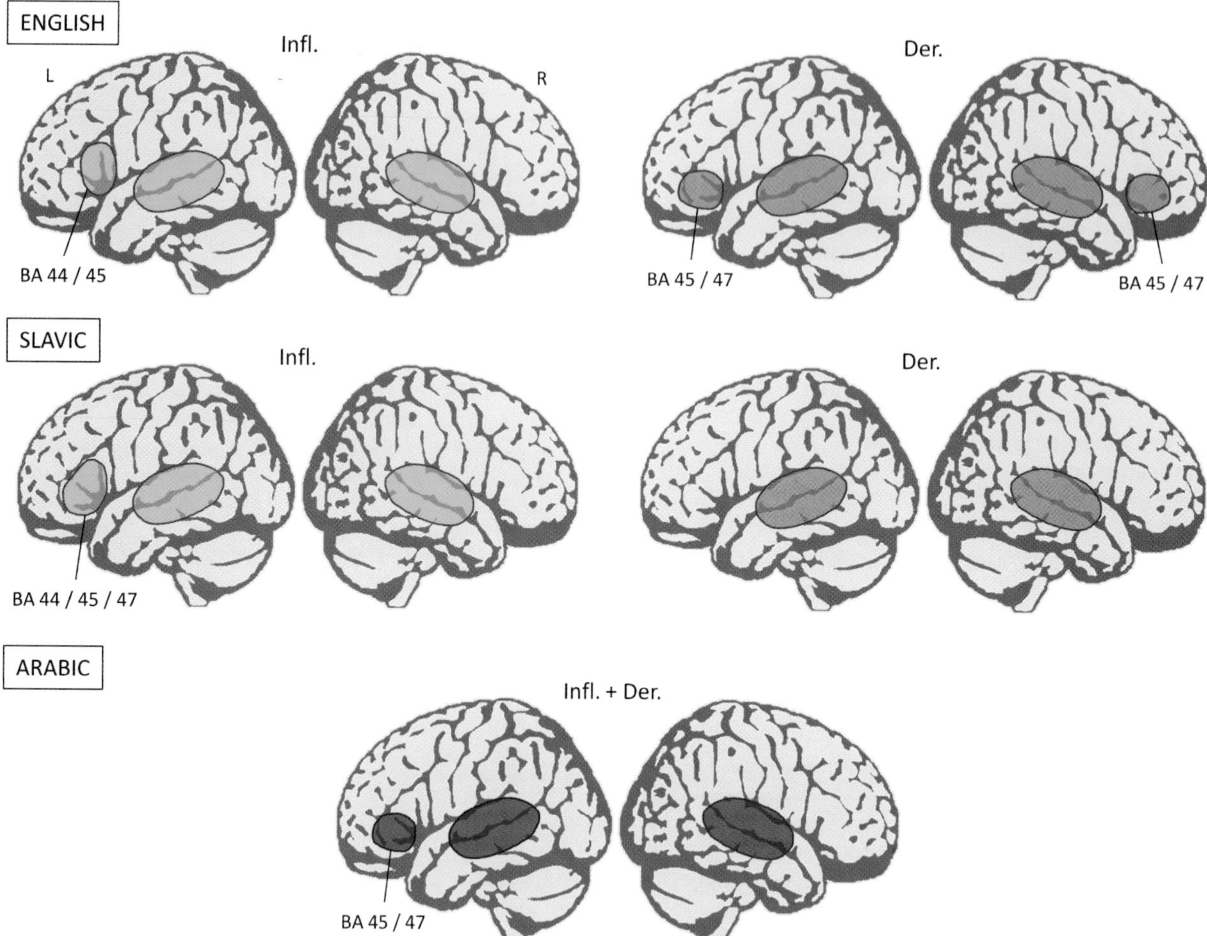

PLATE 47 Schematic overview of inflectional (Infl.) and derivational (Der.) results for parallel fMRI experiments in English, Slavic (Polish and Russian), and Arabic. Presence or absence of left-hemisphere and right-hemisphere inferior frontal activation is shown (including left inferior frontal gyrus Brodmann area (BA) subcomponents), together with indicative left and right temporal (superior temporal gyrus and middle temporal gyrus) activations. (See figure 54.2.)

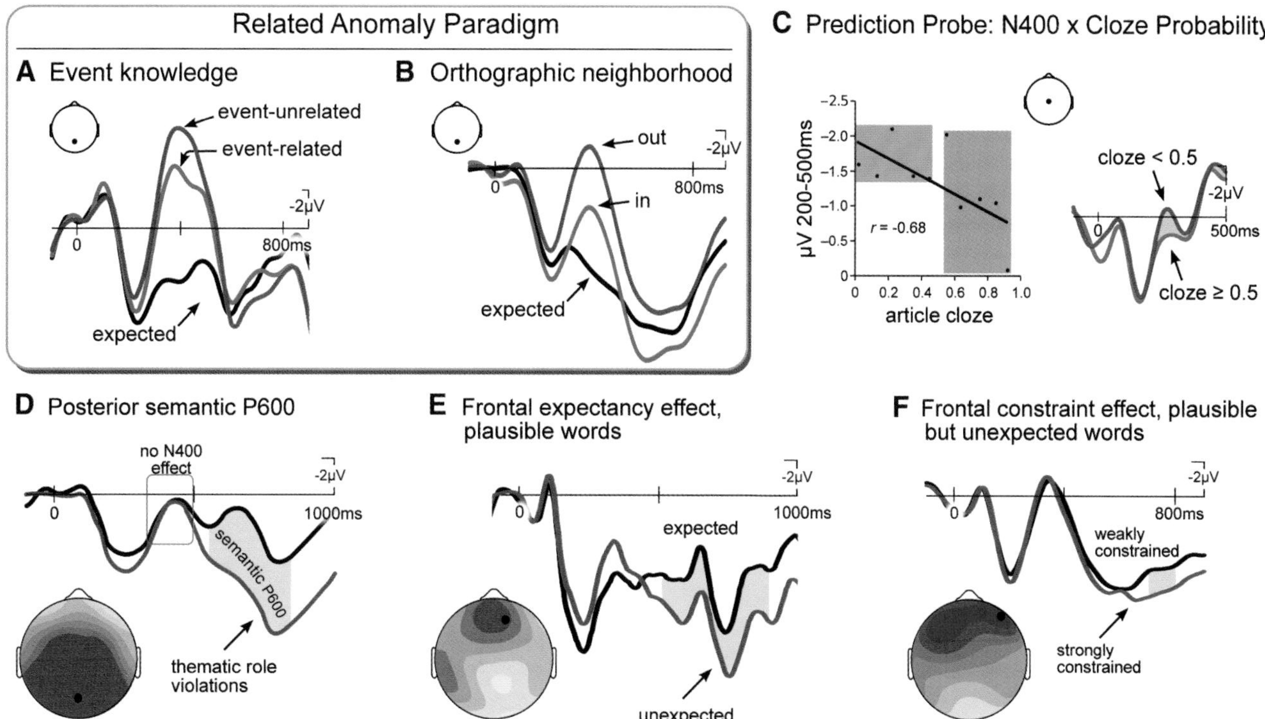

PLATE 48 Grand average ERPs. (A) Expected, linguistically unlicensed event-related and event-unrelated continuations in discourse context. (B) Expected, unexpected in and out of orthographic neighborhood. (C) Correlations and ERPs to articles based on cloze probability median split. (D) Semantic P600 for thematic role violations. (E) Frontal positivity to unexpected *vs.* expected plausible words (DeLong & Kutas, unpublished data). (F) Frontal positivity to unexpected plausible words as a function of contextual constraint. (See figure 55.1.)

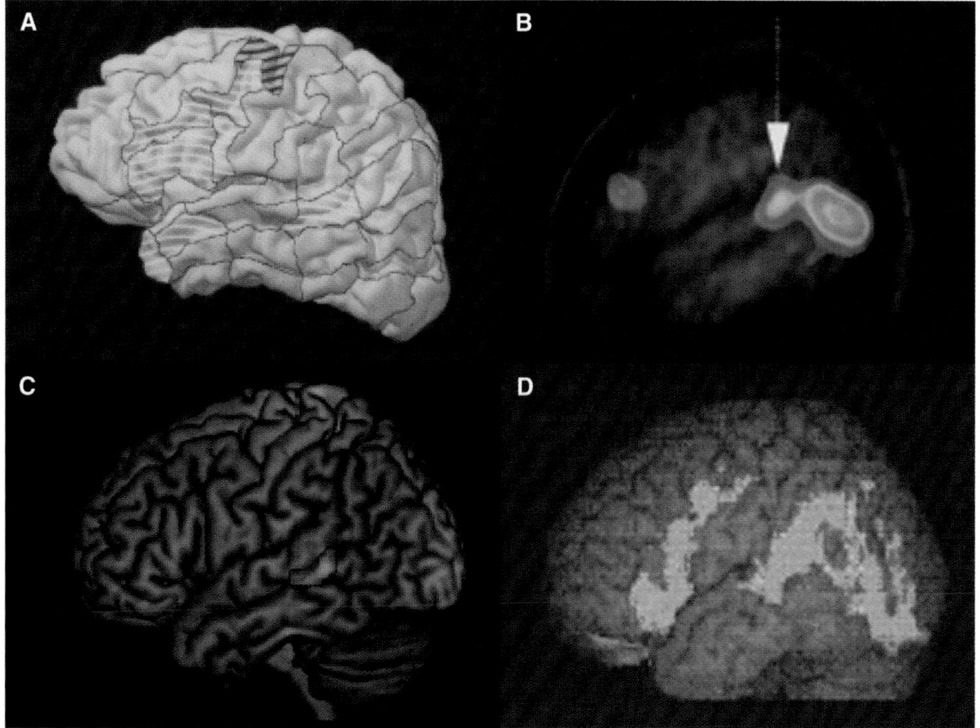

PLATE 49 Left-hemisphere activations when deaf signers comprehend sign language. (A) Signed sentences vs. nonsense signs from Neville et al. (1998). (B) Signs and pseudosigns vs. fixation baseline from Petitto et al. (2000). (C) Greater activation for deaf signers than for hearing nonsigners for pseudosigns from Emmorey, Xu, & Braun (2011). (D) Signed sentences vs. a still image of the signer from MacSweeney et al. (2002). (See figure 56.1.)

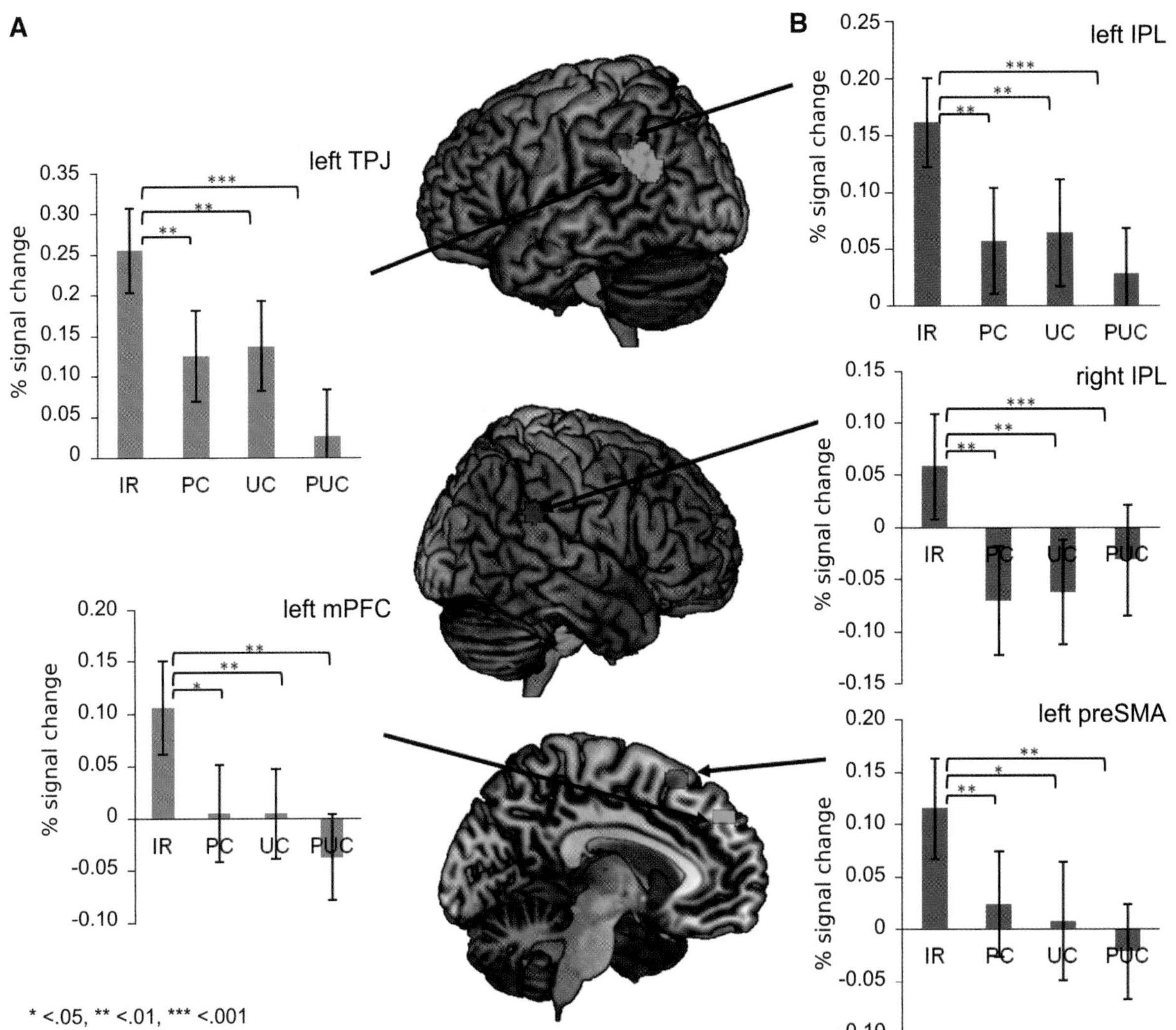

Plate 50 Regions of interest (ROIs) were interrogated for activations to an indirect request (IR) for action (as in "It is hot here" in the presence of a picture of a window) relative to three control conditions: PC (picture control), UC (utterance control), and BC (baseline control). The image shows all ROIs superimposed on a brain template. The bar diagrams illustrate mean percent signal change for each condition. The error bars depict the standard error. (A) Green ROIs show regions from the theory of mind (ToM) localizer (medial prefrontal cortex, mPFC; and temporoparietal junction, TPJ). (B) Red ROIs refer to regions that were activated during action execution (pre-supplementary motor area, preSMA; bilateral inferior parietal lobule, IPL; van Ackeren et al., 2012; reprinted with permission). (See figure 57.1.)

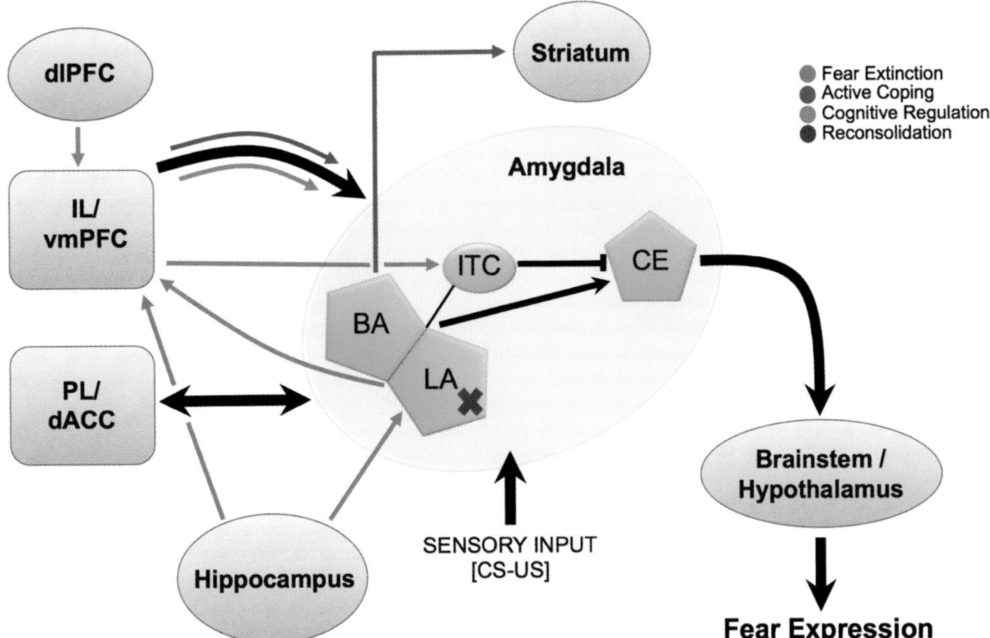

PLATE 51 Neurocircuitry of fear learning and regulation: the lateral nucleus (LA) of the amygdala receives afferent sensory information regarding the conditioned stimulus (CS) and unconditioned stimulus (US) relationship and is the site of storage of the fear memory. The LA and basal nuclei (BA) are interconnected and both project to the central nucleus (CE), which has outputs to brainstem and hypothalamic regions that control the expression of the conditioned response (CR). Following conditioning, the prelimbic (PL) region of the ventromedial prefrontal cortex (vmPFC) is activated following CS presentations and drives the expression of conditioned fear. During fear-extinction learning and consolidation, connections are established between the infralimbic (IL) subregion of the vmPFC and the inhibitory intercalated (ITC) cell masses, which inhibit activity in the CE (circuit denoted in orange). During extinction recall, these connections are activated, inhibiting fear expression. Contextual modulation of extinction expression is mediated by projections from the hippocampus to the vmPFC and/or amygdala. During active coping (circuit denoted in blue), information from the LA is routed not to the CE, which drives fear expression, but to the BA, which in turn projects to the striatum. The IL/vmPFC supports the learning active coping responses, and mediates the subsequent inhibition of fear expression outside of the instrumental learning context. During cognitive regulation, the dorsolateral prefrontal cortex (dlPFC) regulates fear expression through projections to the vmPFC, which in turn inhibits amygdala activity (circuit denoted in green). Reconsolidation (denoted by red X) diminishes conditioned fear expression through alteration of the original CS-US association stored in the LA. (See figure 60.1.)

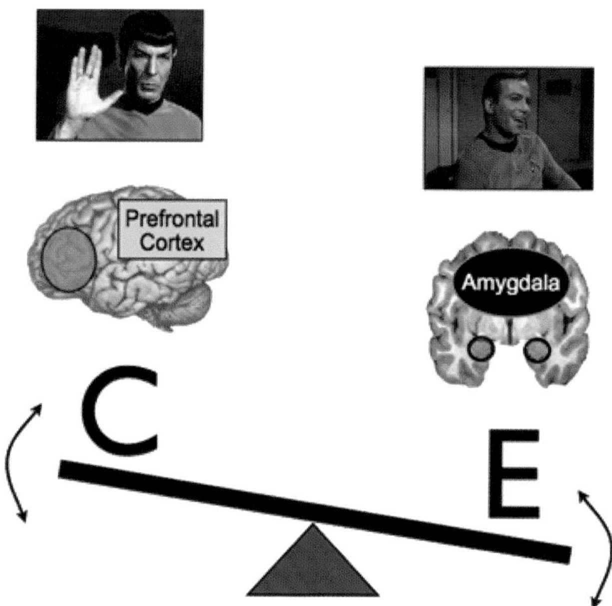

Plate 52 A schematic representation of the seesaw relationship between cognitive control and emotion postulated by many lay and scientific theories. Such theories posit that cognitive control and emotion are inherently antagonistic. Such an antagonistic relationship is exemplified by the opposing points of view represented by popular television characters like Mr. Spock (who embodies the value of control) and Captain Kirk (who embodies the value of impulsive emotion) from the 1960s television series *Star Trek*. In contemporary theories, specific brain systems are often assisted with control (prefrontal cortex) and emotion (amygdala), and scientific questions revolve around the situations in which the activity of prefrontal cortex and the amygdala seesaw with one another. (See figure 62.1.)

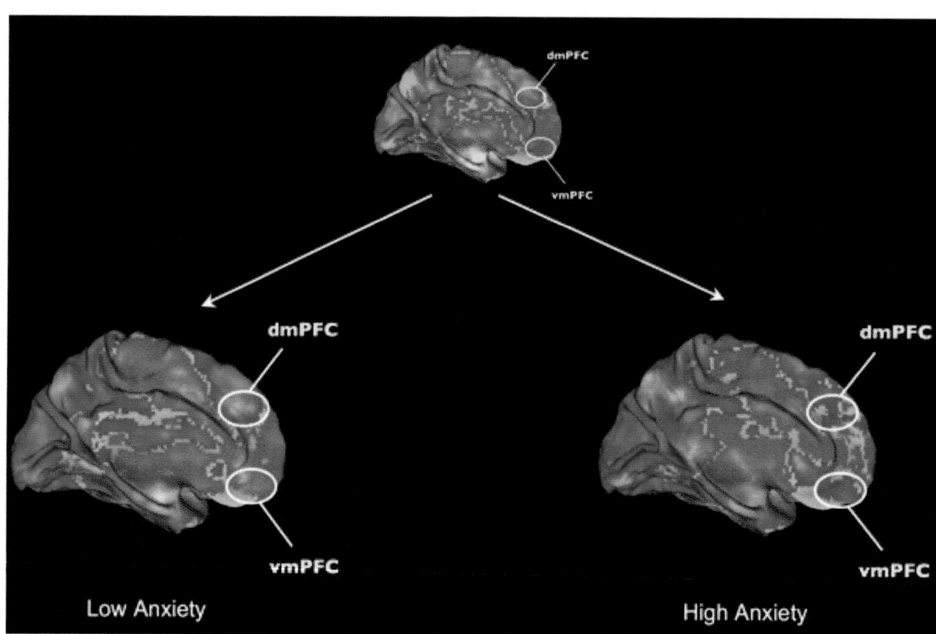

Plate 53 Amygdala-prefrontal functional connectivity at rest. While all subjects show that amygdala activity is positively correlated with vmPFC activity and negatively correlated with dorsomedial PFC activity at rest, highly anxious subjects show a reverse pattern in the vmPFC compared to low anxious subjects. (Adapted from Kim et al., 2011.) (See figure 64.1.)

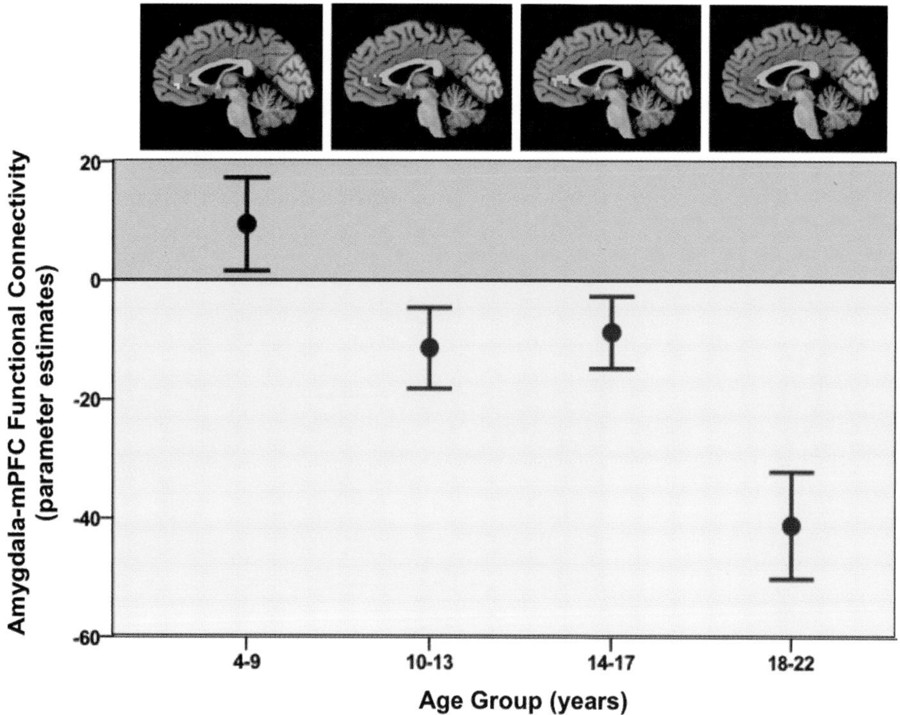

PLATE 54 Development of amygdala-prefrontal functional connectivity. A developmental switch from positive to negative functional connectivity between the amygdala and mPFC was observed during the transition from childhood to adolescence. (Adapted from Gee et al., 2013.) (See figure 64.2.)

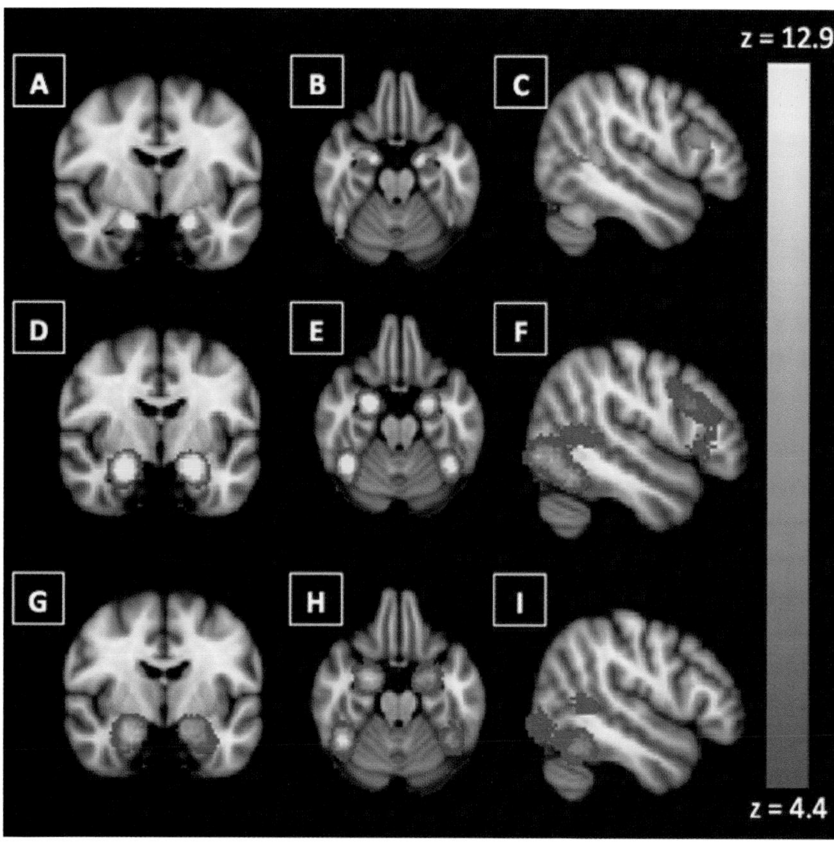

PLATE 55 Group analysis of functional localizer data from 215 participants, with NeuroSynth validation. Panels A through C depict brain regions showing greater BOLD responses to face vs. control stimuli ($p < .05$, corrected), including bilateral amygdala. We used NeuroSynth—a meta-analytic tool capable of synthesizing results from a database of neuroimaging studies—to externally validate these data. Panels D through F represent the results of a forward inference analysis, showing activation across the NeuroSynth database for the term "face." These maps indicate the consistency of activation for the term. Panels G through I show the results of a reverse inference analysis. These maps indicate the relative selectivity of activation for the term (see Yarkoni et al., 2011, for details). (This figure originally appeared in Mende-Siedlecki, Verosky, Turk-Browne, & Todorov, 2013.) (See figure 66.1.)

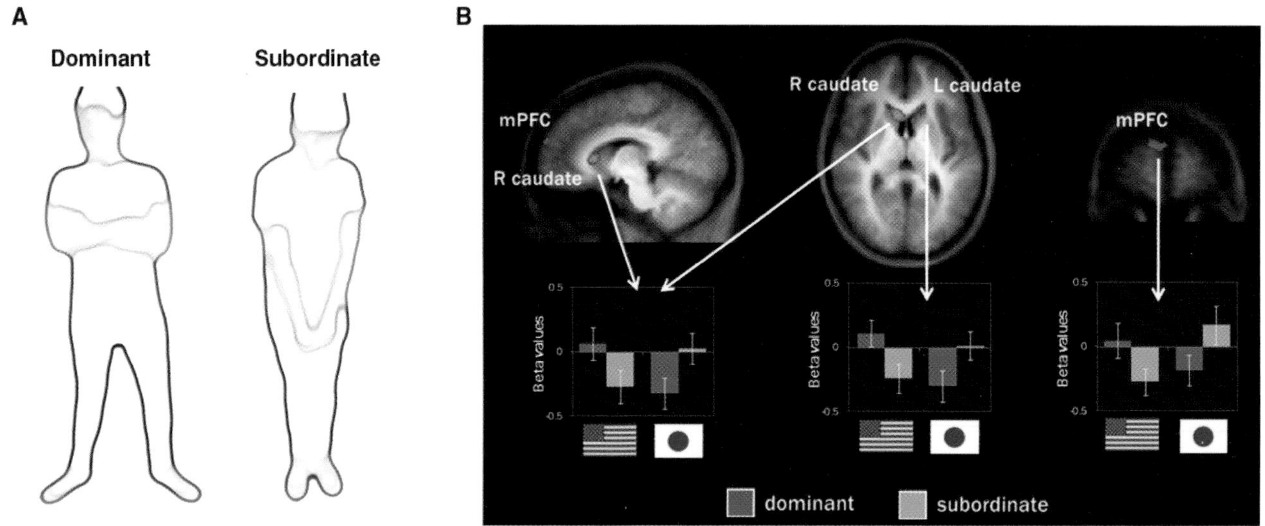

PLATE 56 Cultural specificity to (A) dominant and subordinate figural outlines in (B) caudate and dorsal medial prefrontal cortex (mPFC) response. (Adapted from Freeman et al., 2009.) (See figure 67.1.)

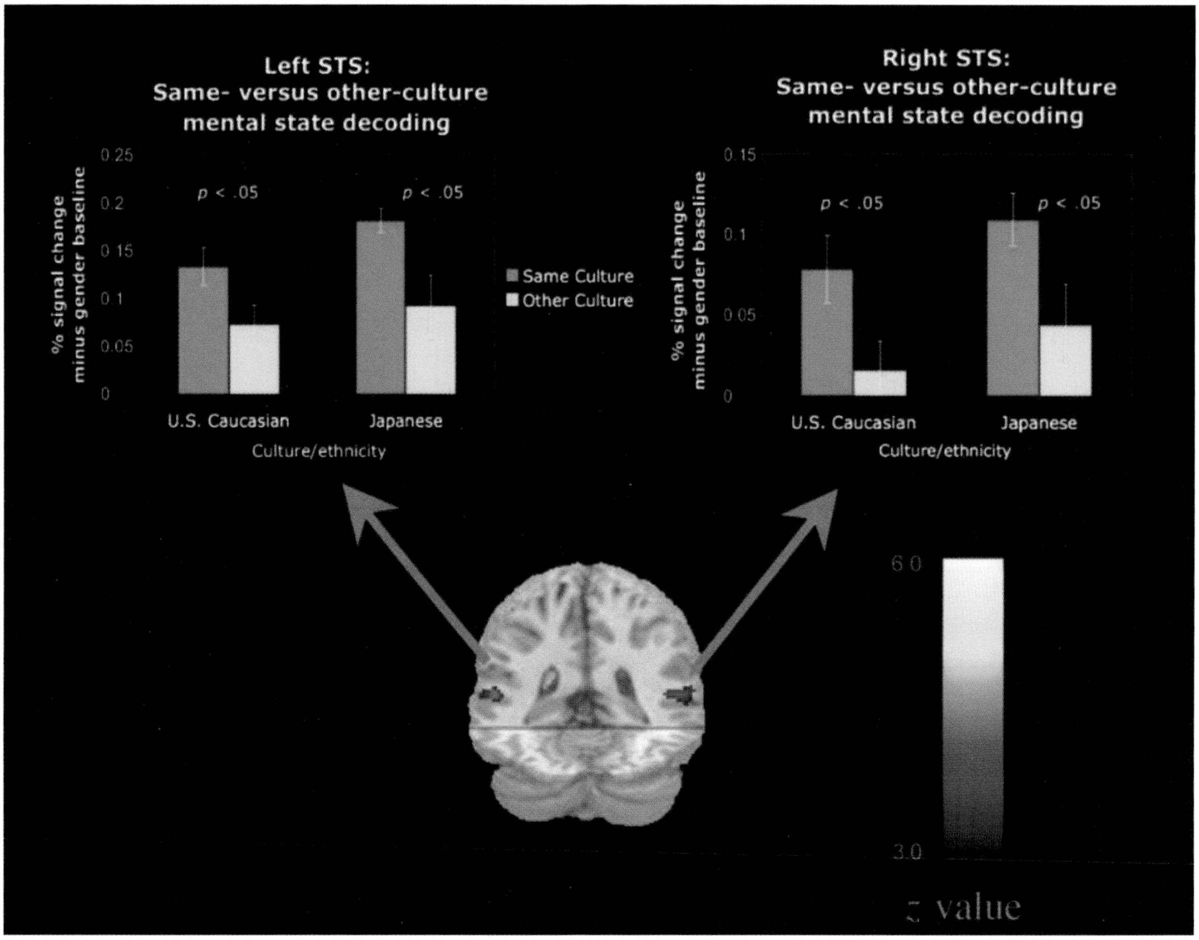

PLATE 57 Cultural specificity in (A) left and (B) right posterior superior temporal sulcus (STS) response to the reading the mind in the eyes (RME) task. (Adapted from Adams et al., 2009.) (See figure 67.2.)

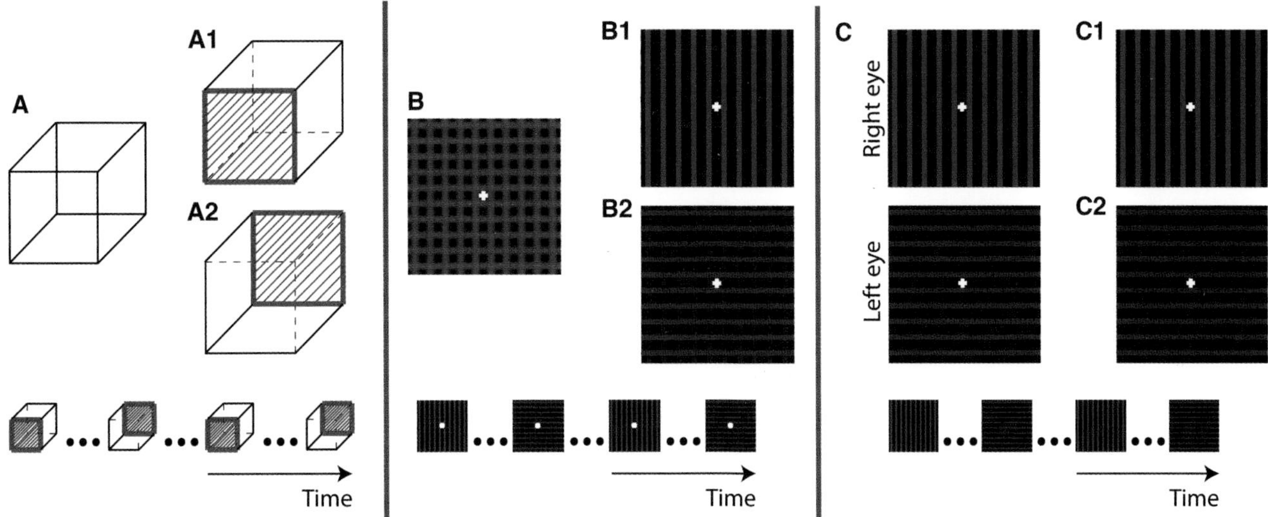

PLATE 58 Example of bistable percepts. (A) The Necker cube can be interpreted in two possible configurations (A1 and A2). Upon viewing the Necker cube, subjects can voluntarily switch from one configuration to the other (bottom). (B) Monocular rivalry. Upon fixating on the grid of horizontal red lines and vertical blue lines, subjects report a percept that alternates between predominantly blue lines (B1) and predominantly red lines (B2). (C) Binocular rivalry. When presenting different stimuli to each eye, subjects report a percept that alternates between the two inputs (C1, C2) in a seemingly random fashion. During binocular rivalry, it is difficult to voluntarily switch between the two percepts. (See figure 68.1.)

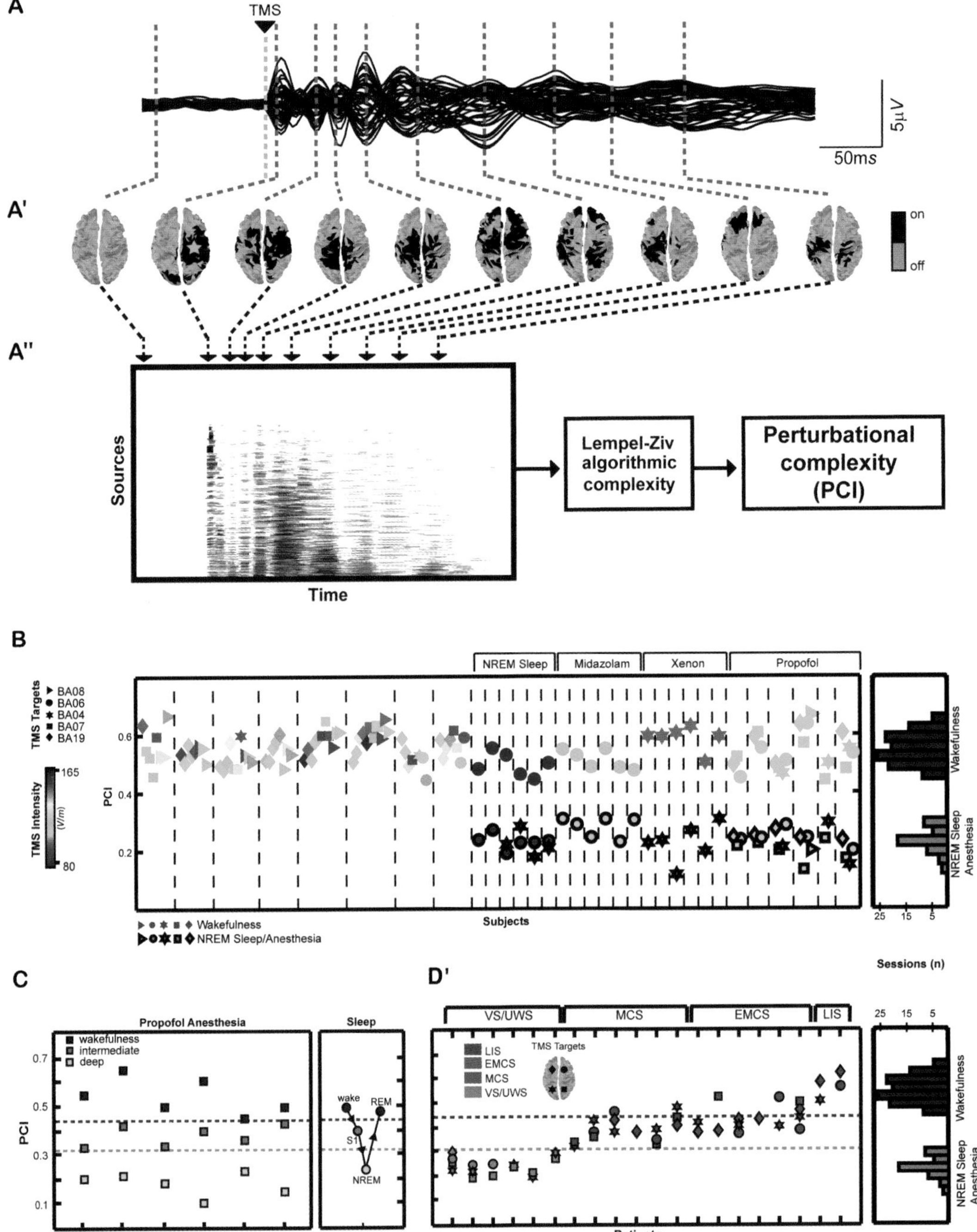

Plate 59 Calculating and testing the perturbational complexity index (PCI). Operationally, PCI is defined as the normalized Lempel-Ziv complexity of the overall spatiotemporal pattern of significant cortical activation triggered by a direct perturbation. Its calculation starts from scalp TMS-evoked potentials (A) and requires performing source-modeling and nonparametric statistics to detect the time course of significant cortical activations triggered by TMS (A′). This procedure results in a binary spatiotemporal matrix of cortical activation (A″), which is compressed using the same algorithm (Lempel & Ziv, 1976) that is commonly employed to zip digital files. Complex patterns of activation that are, at once, distributed and differentiated cannot be efficiently compressed and result in high values of perturbational complexity. (B) PCI values are shown for 152 TMS sessions collected from 32 healthy subjects. The histograms on the right display the distributions of PCI across subjects during alert wakefulness (dark gray bars) and loss of consciousness in NREM sleep and different forms of anesthesia (light gray bars). PCI does not depend on stimulation site or TMS intensity but is solely sensitive to changes in the level of consciousness. PCI calculated during wakefulness (110 sessions) ranges between 0.44 and 0.67 (mean: 0.55 ± 0.05), whereas PCI calculated after loss of consciousness (42 sessions) ranges between 0.12 and 0.31 (mean: 0.23 ± 0.04), giving rise to two separated distributions. (C) PCI yields intermediate values during intermediate levels of propofol sedation and during sleep stage 1. (D) PCI values are shown for 48 TMS sessions collected from 20 brain-injured patients. PCI followed the level of consciousness (as clinically assessed with coma recovery scale, or CRS-R) progressively increasing from vegetative (VS) through a minimally conscious state (MCS) to recovery of functional communication (emergence from the minimally conscious state; EMCS) and attaining levels of healthy awake subjects in locked-in syndrome (LIS). Patient results are within the frame of reference obtained in awake, sleeping, and anesthetized control subjects. (Modified from Casali et al., 2013.) (See figure 69.2.)

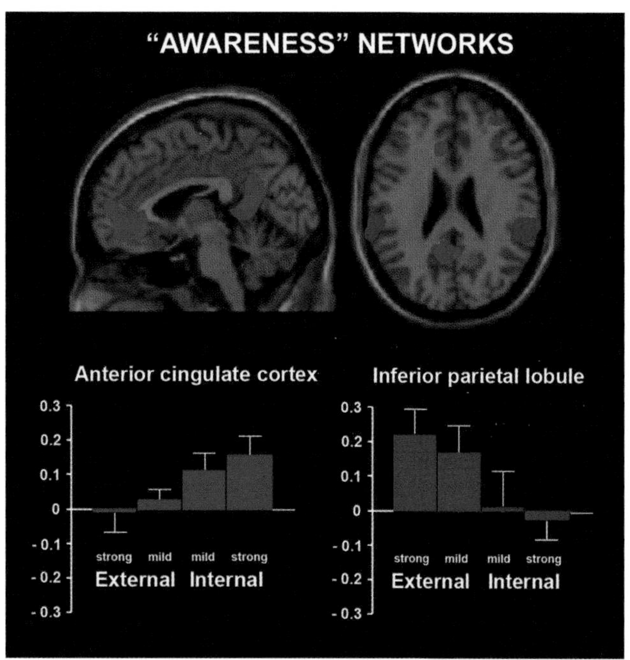

Plate 60 Subjective ratings about awareness are coupled to the activity of the resting brain. Behavioral reports of increased internal awareness (blue areas) correlate with connectivity in cerebral regions of the so-called default mode network, encompassing anterior and posterior cingulate cortices. Inversely, when subjects report being more externally oriented toward their environment, areas of a frontoparietal attentional network (red areas) are recruited (adapted from Vanhaudenhuyse et al., 2011). These data suggest that resting-state fMRI acquisitions could be an alternative paradigm to study residual brain function in human populations where communication is hindered (e.g., due to motor paralysis, language comprehension, or lack of motivation), such as in patients with disorders of consciousness. (See figure 70.1.)

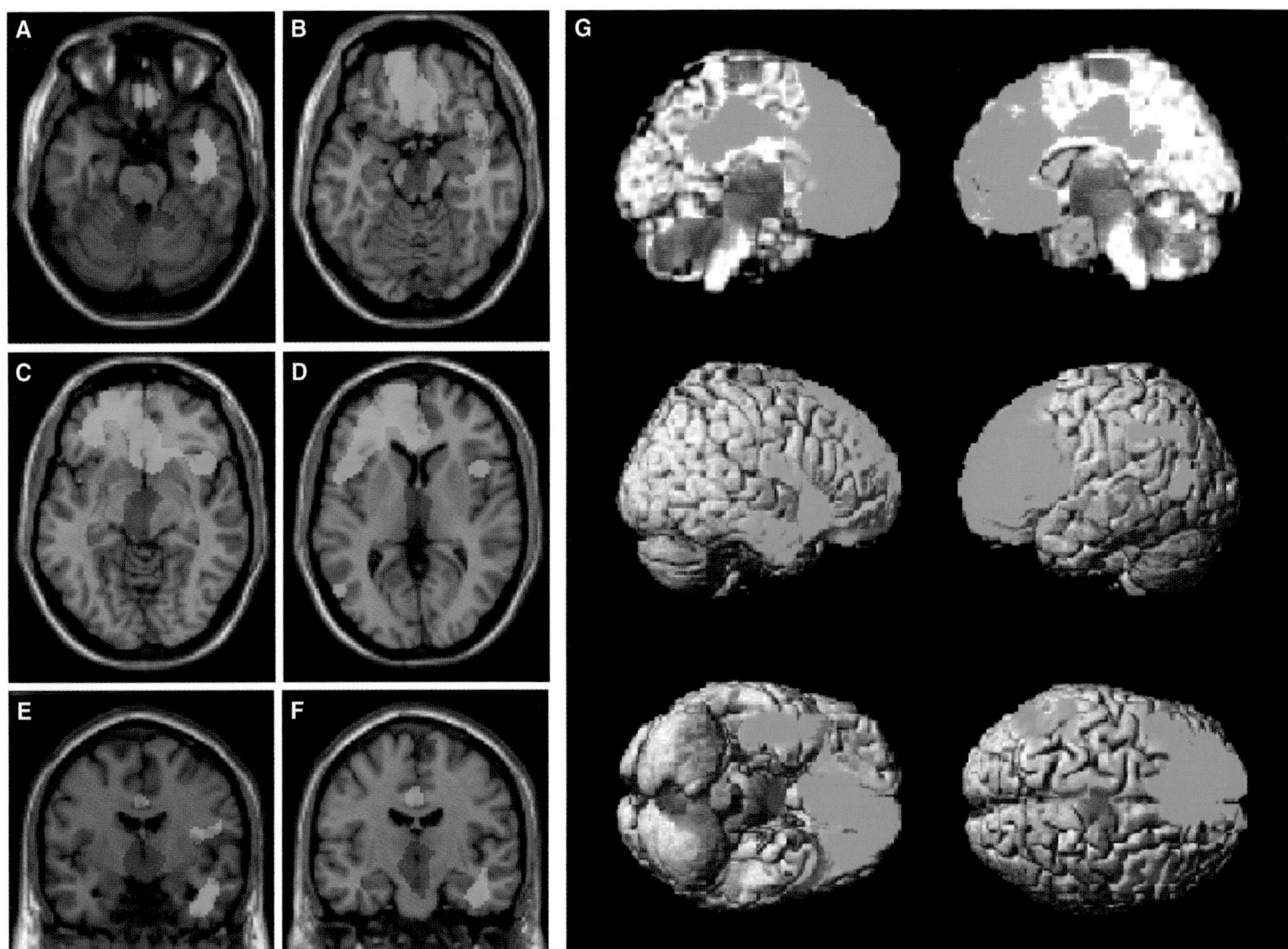

Plate 61 Cerebral blood flow imaging in temporal lobe seizures with impaired consciousness. Complex partial seizures arising from the temporal lobe are associated with significant cerebral blood flow increases and decreases in widespread brain regions. Statistical parametric maps depict SPECT increases in red and decreases in green. Changes ipsilateral to seizure onset are shown on the left side of the brain, and contralateral changes on the right side of the brain (combining patients with left and right onset seizures; $n = 10$). Data are from >90 sec after seizure onset, when consciousness was markedly impaired. Note that at earlier times there were SPECT increases in the ipsilateral mesial temporal lobe (not shown). (A–F) Horizontal sections progressing from inferior to superior (A–D), and coronal sections (E, F) progressing from anterior to posterior showing blood flow increases in the bilateral midbrain, hypothalamus, medial thalamus, and midbrain. Decreases are seen in the bilateral association cortex. (G) Three-dimensional surface renderings show increases mainly in the bilateral medial diencephalon, upper brainstem, and medial cerebellum, while decreases occur in the ipsilateral contralateral frontal and parietal association cortex (same data as A–F). Extent threshold, $k = 125$ voxels (voxel size = $2 \times 2 \times 2$ mm). Height threshold, $P = 0.01$. (Reproduced from Blumenfeld, McNally, et al., 2004, by permission of Oxford University Press.) (See figure 71.4.)

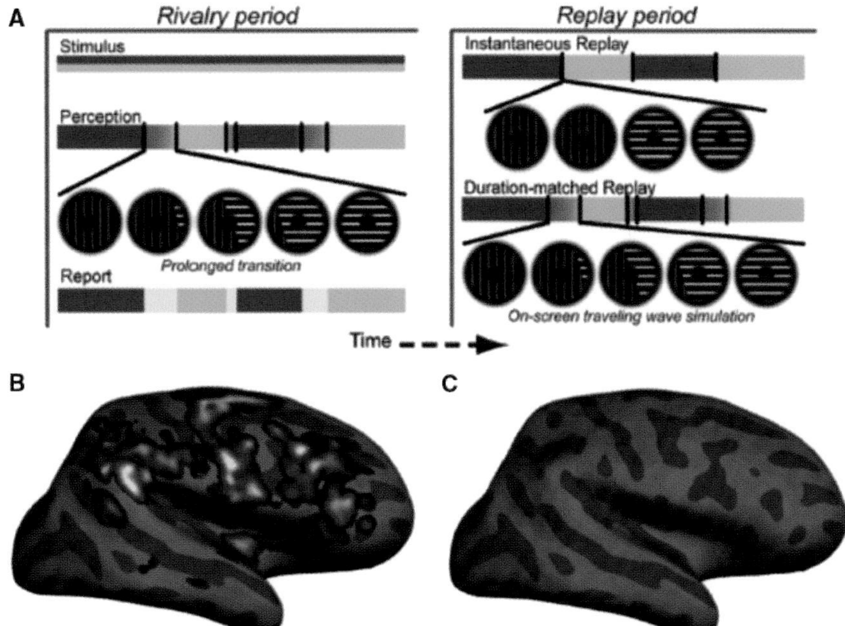

Plate 62 Confounding activation in frontoparietal areas can be introduced by a less faithful replay of what subjects really experience. (A) An exemplar time course of stimulus and perception during genuine binocular rivalry. One eye is stimulated by a vertical red and the other eye by a horizontal green grating (top). The subjective experience is highly complex, involving various durations of intermediate percepts (rows 2 and 3). As a control condition, a perceptually matched replay movie can be presented to both eyes without interocular conflict. However, if the replay is reproduced from binary responses, the variable nature of each perceptual transition cannot be captured (e.g., instantaneous replay; top right of the panel A). With a third response option of accurately reporting the transitional states, a more faithful replay can be reproduced (e.g., Duration-matched replay; bottom right of panel A). (B) Brain activity contrasted at the perceptual transition between genuine rivalry and instantaneous replay (without faithfully reproduced transitions), indicating a widespread right-lateralized frontoparietal activity, replicating previous studies. (C) No difference in the brain activity between genuine rivalry and duration-matched replay. (Modified from Knapen, et al., 2011.) (See figure 72.4.)

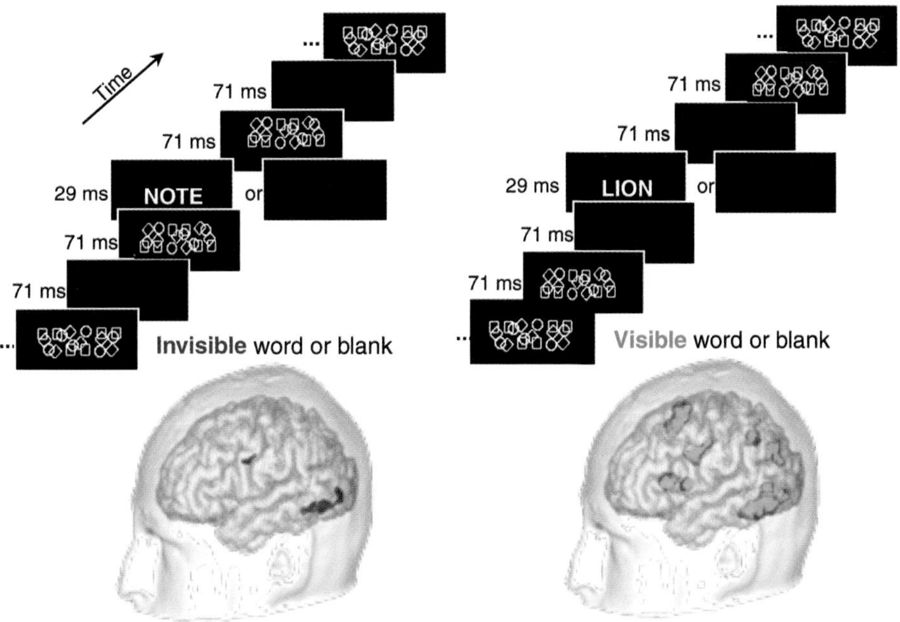

Plate 63 Contrast of unconscious and conscious processing in fMRI (Dehaene et al., 2001). While the neural activation induced by invisible words is primarily restricted to occipitotemporal regions (left panel), conscious perception is associated with the involvement of a parietofrontal network (right panel). (See figure 73.1.)

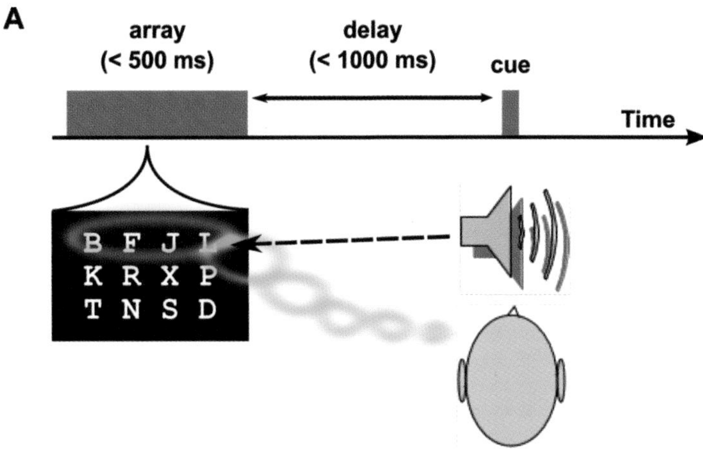

B

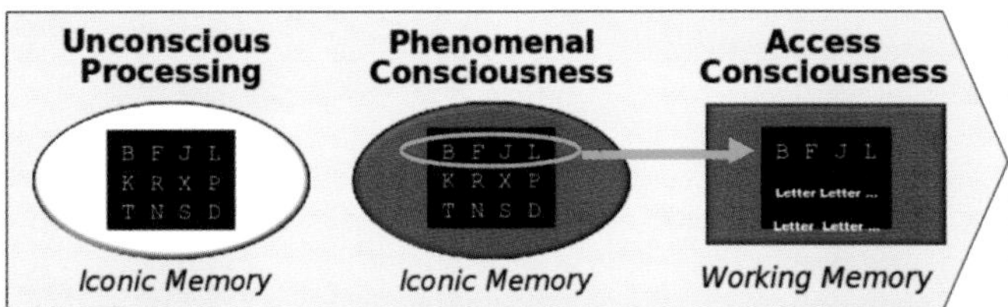

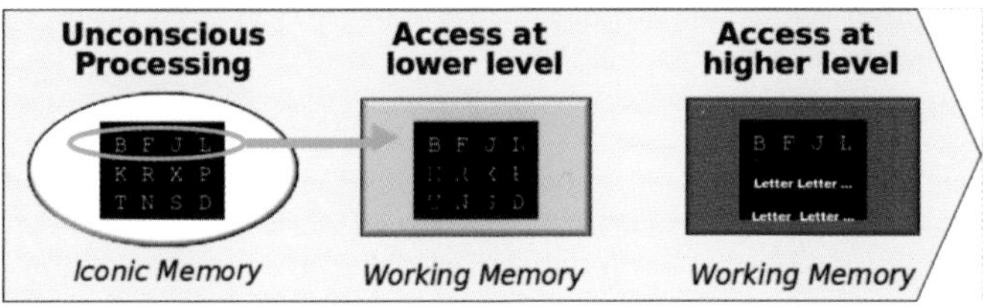

Plate 64 The Sperling paradigm (Sperling, 1960) and its interpretations. (A) Experimental procedure for the cued report. A brief array of letters is shown, followed by a random tone cue (high tone in this example). The pitch of the cue (low, medium, high) instructs subjects to report one of the three rows (lower, middle, or higher row, respectively). When participants are not cued and have to report all letters in the array, performance is restricted to about 4 out of 12 items. However, when using the post-stimulus cue to report a specific row, performance increased to 3 out of 4 items. This suggests that a large amount of information is available but decays by the time of reporting. (B) Two interpretations of the results. Interpretation 1 assumes that subjects are phenomenally conscious of the whole content in iconic memory demonstrated by the high-level performances at short delays. Interpretation 2 hypothesizes that subjects access both high- and low-level information from iconic memory. Low-level information is reconstructed at higher levels. (See figure 73.2.)

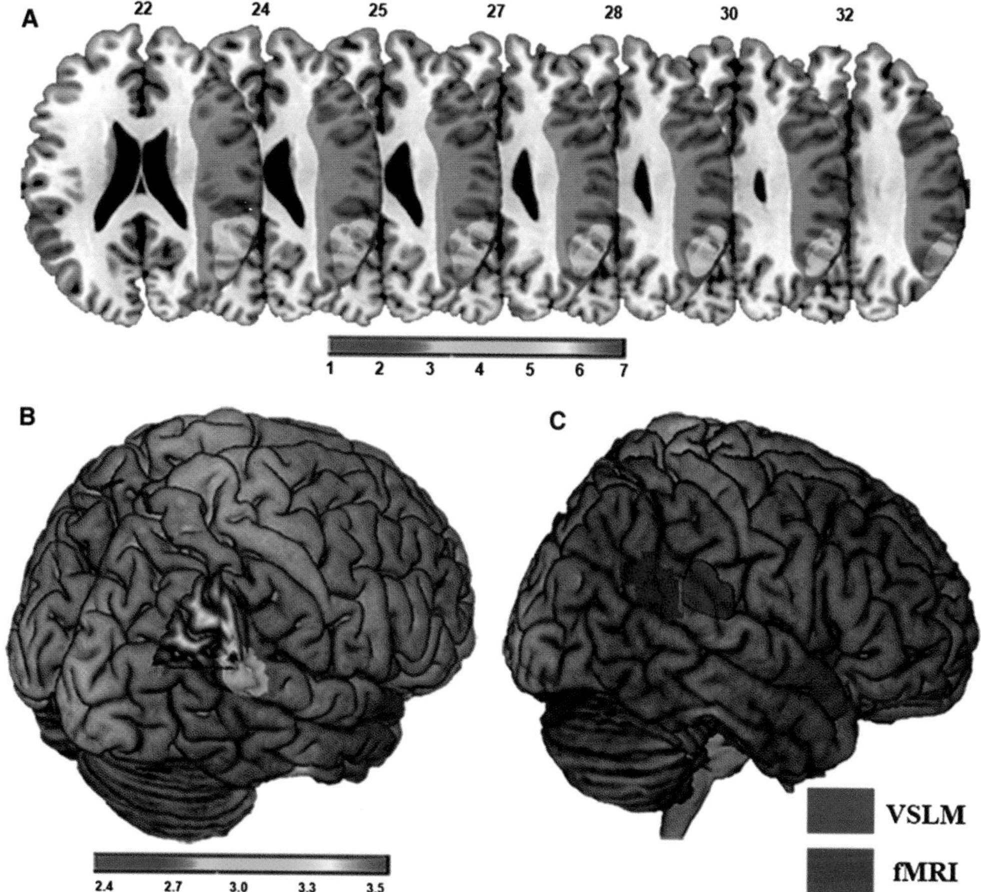

PLATE 65 Brain damage leading to changes in bodily self-consciousness (out-of-body experiences, or OBE). Brain damage and results of lesion overlap analysis (top) in nine patients with OBEs due to focal brain damage are shown. Maximal lesion overlap centers at the right temporoparietal junction (TPJ) at the angular gyrus (red). Overlap color code ranges from violet (one patient) to red (seven patients). Voxel-based lesion symptom mapping (VLSM; bottom left) of focal brain damage leading to OBEs. The violet-to-red cluster shows the region that VLSM analysis associated statistically with OBEs as compared to control patients. The color code indicates significant Z scores of the respective voxels, showing maximal involvement of the right TPJ. (Modified from Ionta et al., 2011.) (See figure 74.1.)

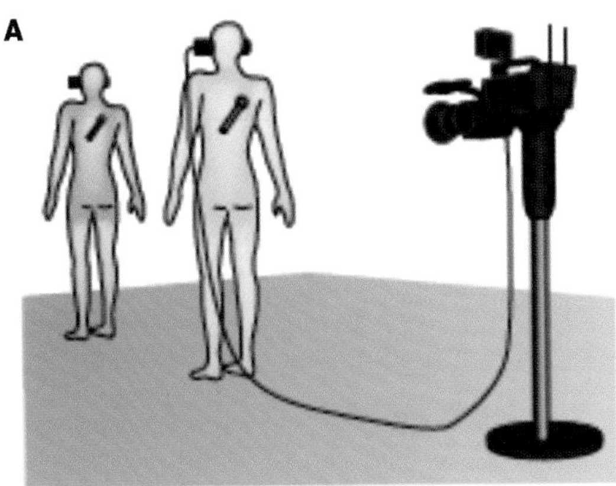

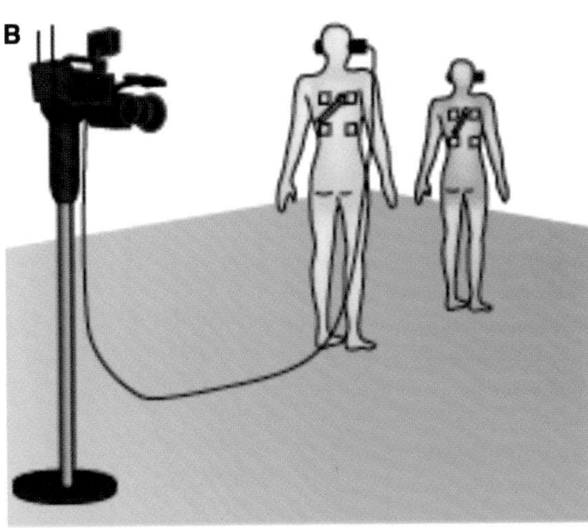

Plate 66 (A) Experimental setup during the full-body illusion. A participant (light color) sees his own back through video goggles, as if a virtual body (dark color) were located a few meters in front. An experimenter administers tactile stroking to the participant's back, which the participant sees on the video goggles as visual stroking on the virtual body. Synchronous but not asynchronous visuotactile stimulation results in illusory self-identification with and self-location toward the virtual body. (Modified from Blanke, 2012.) (B) Full-body illusion and visuotactile integration. The experimental setup of the full-body illusion was adapted to acquire repeated behavioral measurements related to visuotactile perception (i.e., the cross-modal congruency effect; CCE). In addition to the visuotactile stroking (as in A), participants wore vibrotactile devices and saw visual stimuli (light-emitting diodes) on their backs while viewing their bodies through video goggles. The CCE is a behavioral measure that indicates whether or not a visual and a touch stimulus are perceived to be at identical spatial locations. Participants were asked to indicate where they perceived a single touch stimulus (i.e., short vibration) applied either just below the shoulder or on the lower back. Distracting visual stimuli (i.e., short light flashes) were also presented on the back, either at the same or at a different position (and were filmed by the camera). Under these conditions, participants are faster to detect a touch stimulus if the visual distractor was presented at the same location (i.e., congruent trial) compared to touches co-presented with a more distanced visual distractor (i.e., incongruent trial). CCE measurements were carried out while illusory self-identification was modulated by visuotactile stroking as described in A. The effect of congruency on reaction times was larger during synchronous than asynchronous visuotactile stroking, indicating greater interference of irrelevant visual stimuli during illusory self-identification with the virtual body. (Modified from Blanke, 2012.) (See figure 74.2.)

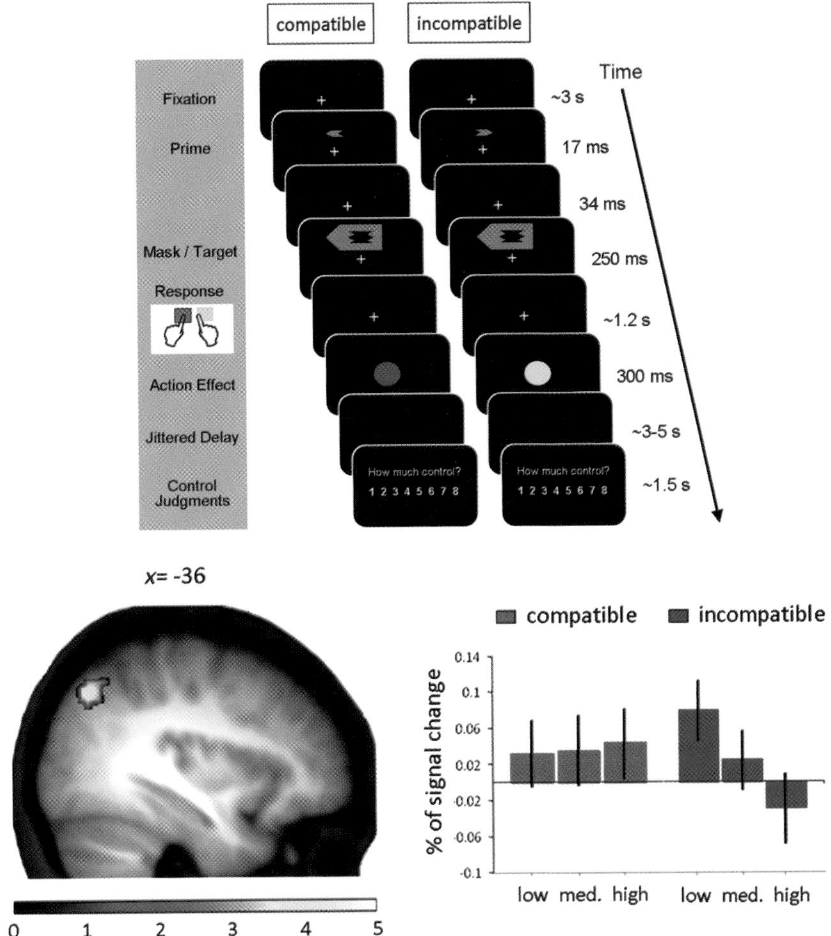

Plate 67 Subliminal priming of actions influences sense of agency *prospectively*. *Upper panel*: subliminal primes facilitate or impair actions in response to target arrow stimuli. Responses are followed by appearance of a color patch, and participants judge how much control they have over the color patch. *Lower panel*: activation of the angular gyrus associated with the target-response event varies negatively with perceived level of control, but only following incompatible priming. (Reproduced from Chambon et al., 2013, by permission of Oxford University Press.) (See figure 75.3.)

Structural model

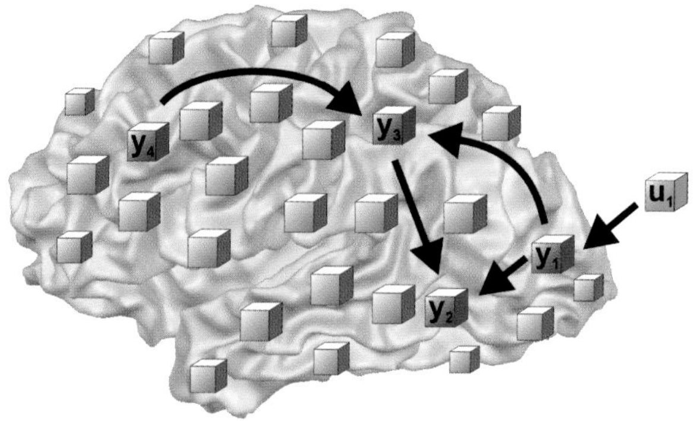

Dynamical model

$$y_1 = f_1(u_1)$$
$$y_2 = f_2(y_1, y_3)$$
$$y_3 = f_3(y_1, y_4)$$

u_1

y_1

y_2

y_3

y_4

Plate 68 A partitioning of functional MRI (fMRI) connectivity models. Models to estimate causal connectivity from fMRI time series can be partitioned into a structural model and a dynamical model. The structural model contains a selection of the structures in the brain that are assumed to be of importance in the cognitive process or task under investigation. Specifically, it specifies which regions of interest (ROIs) in the spatially rich high-dimensional fMRI data set will be considered for further analysis, as illustrated by the selection of the red boxes $y_1 \ldots y_4$. The structural model can also define the possible interactions between the ROIs in the form of one or more directed graph models that might be compared in a later model comparison step. Finally, the structural model also defines where exogenous inputs (that may be under control of the experimenter) can exert effects on the network. The dynamical model embeds the structural model assumptions into parameterized equations that relate the selected measurements and inputs to each other. Connectivity modeling involves the estimation of the parameters in the dynamical model from actual measurements y_j and, possibly, inputs u_k. (Adapted from Roebroeck, Formisano et al., 2011.) (See figure 76.1.)

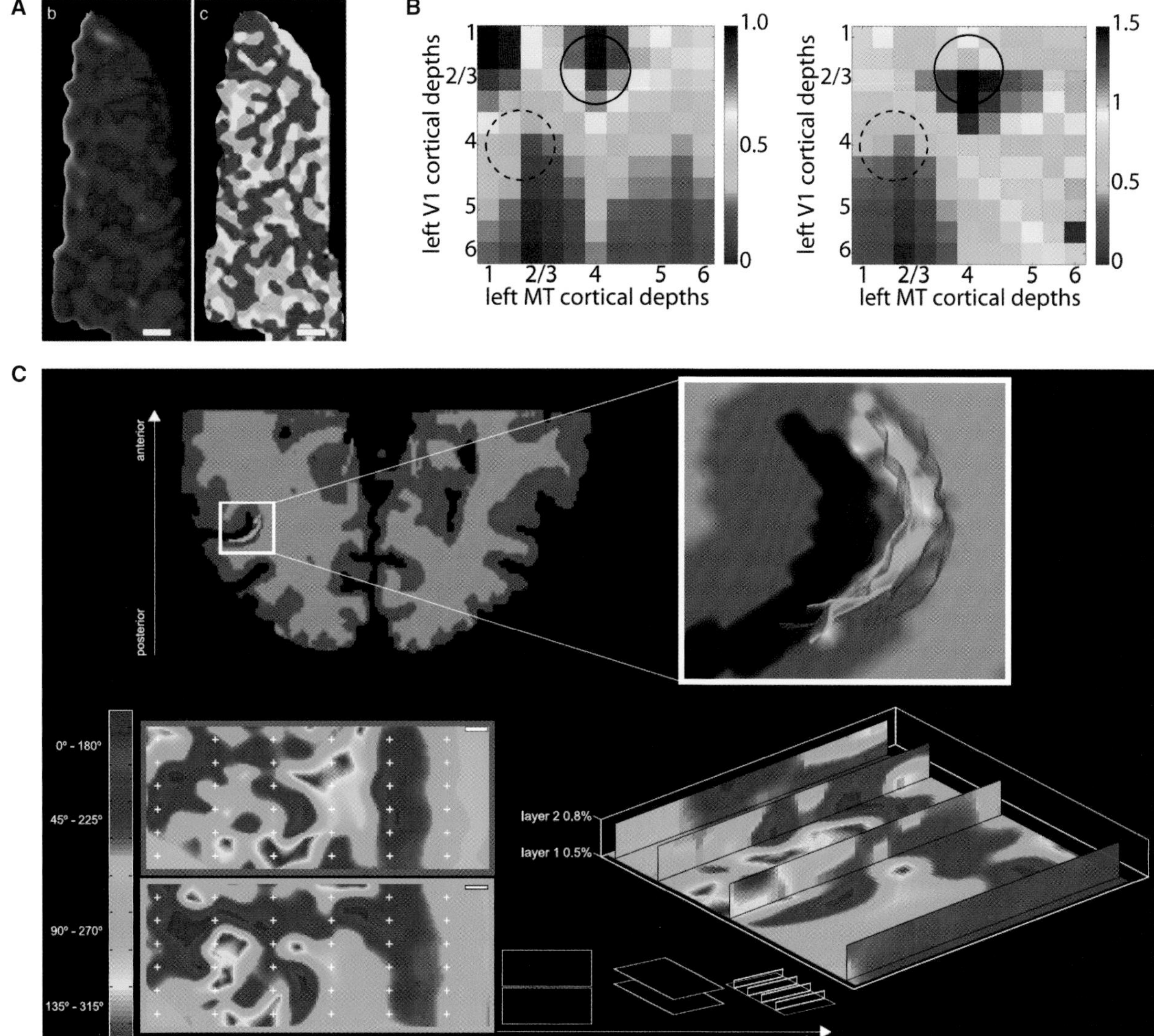

Plate 69 High spatial resolution 7T fMRI in the human visual system with cortical-column and layer-level specificity. (A) Region of interest selection in V1 on a T1-weighted anatomical image. *Left*: Ocular dominance map in human V1. *Right*: Orientation preference map (from Yacoub et al., 2008). (B) Cortical depth resolved resting-state fMRI correlation matrices between human V1 and middle-temporal (MT) areas, showing increased V1 layer IV to MT layers II–III correlations. *Left*: V1 to MT correlations normalized to 1. *Right*: V1 to MT correlations normalized to diagonal elements (from Polimeni et al., 2010). (C) Axis-of-motion preference map in human MT showing columnar-resolution motion selectivity and its modulation across cortical depth (Zimmermann et al., 2011). (See figure 76.3.)

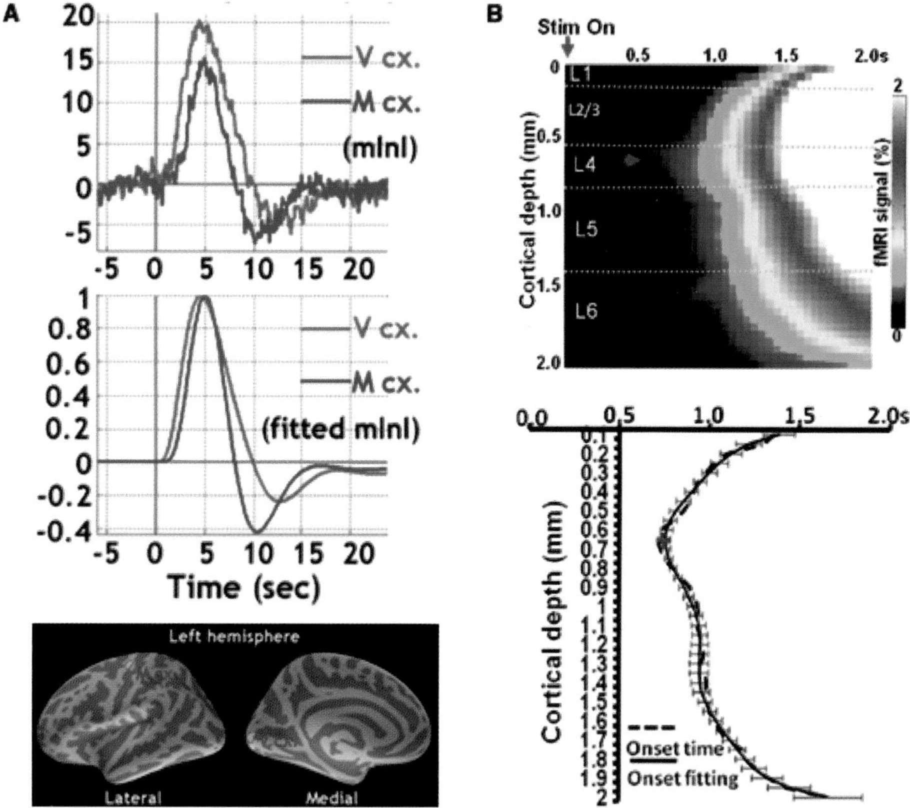

Plate 70 High temporal resolution fMRI showing relative timing between and within cortical areas. (A) High temporal resolution evolution of hemodynamic responses in human visual and motor cortex in a visuomotor task sampled at 10 Hz. *Top*: Regions of interest, visual cortex (V cx) and motor cortex (M cx), on reconstructed cortical surfaces. *Middle*: Measured responses. *Bottom*: Model-fitted responses showing visual cortex signal increases before motor cortex in the early stages of the hemodynamic response function, reflecting the task-induced order of neuronal events, despite different hemodynamic response function shapes (from Tsai et al., 2012). (B) High spatial and temporal evolution of hemodynamic responses in rat barrel cortex. *Top*: fMRI signal change for different cortical layers (L1–L6) against post-stimulus time. *Bottom*: Recorded and fitted onset time of the hemodynamic response against post-stimulus time (Yu et al., 2013). (See figure 76.4.)

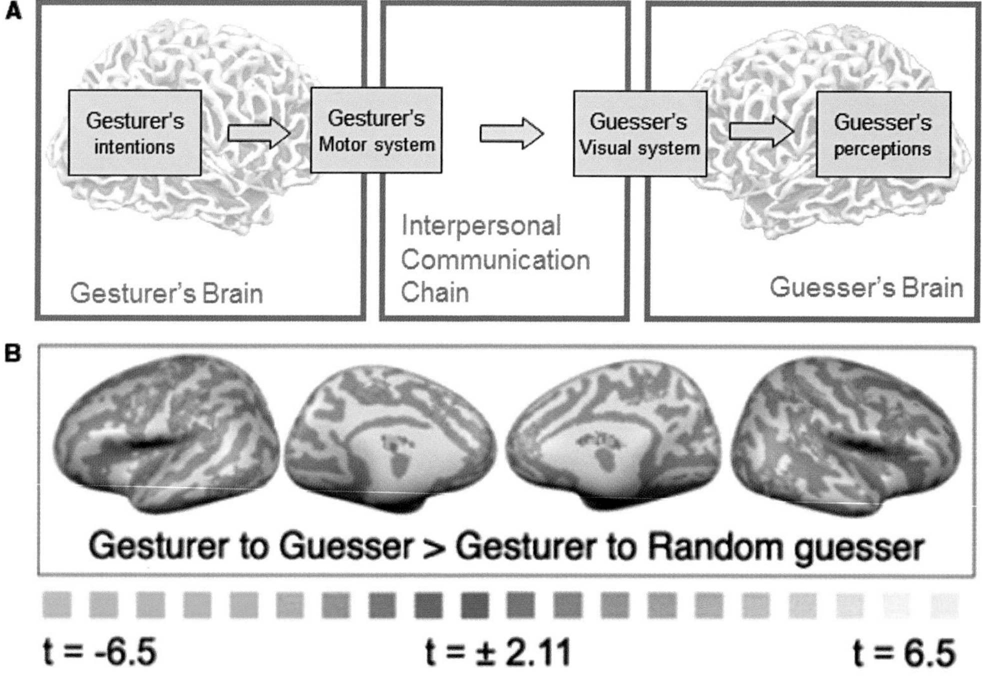

Plate 71 Between-brain causality in a social communication study, illustrating causality inference when a detailed generative model is missing. (A) The causal chain of information that goes from the sender of information (the gesturer in charades) to the receiver of information (the guesser) can span seconds. (B) The time-lagged flow of information between communicating brains is resolved with G-causality analysis of fMRI. (Adapted from Schippers et al., 2010.) (See figure 76.6.)

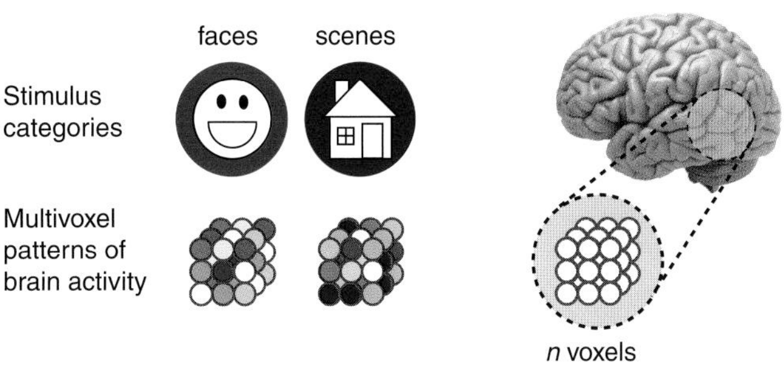

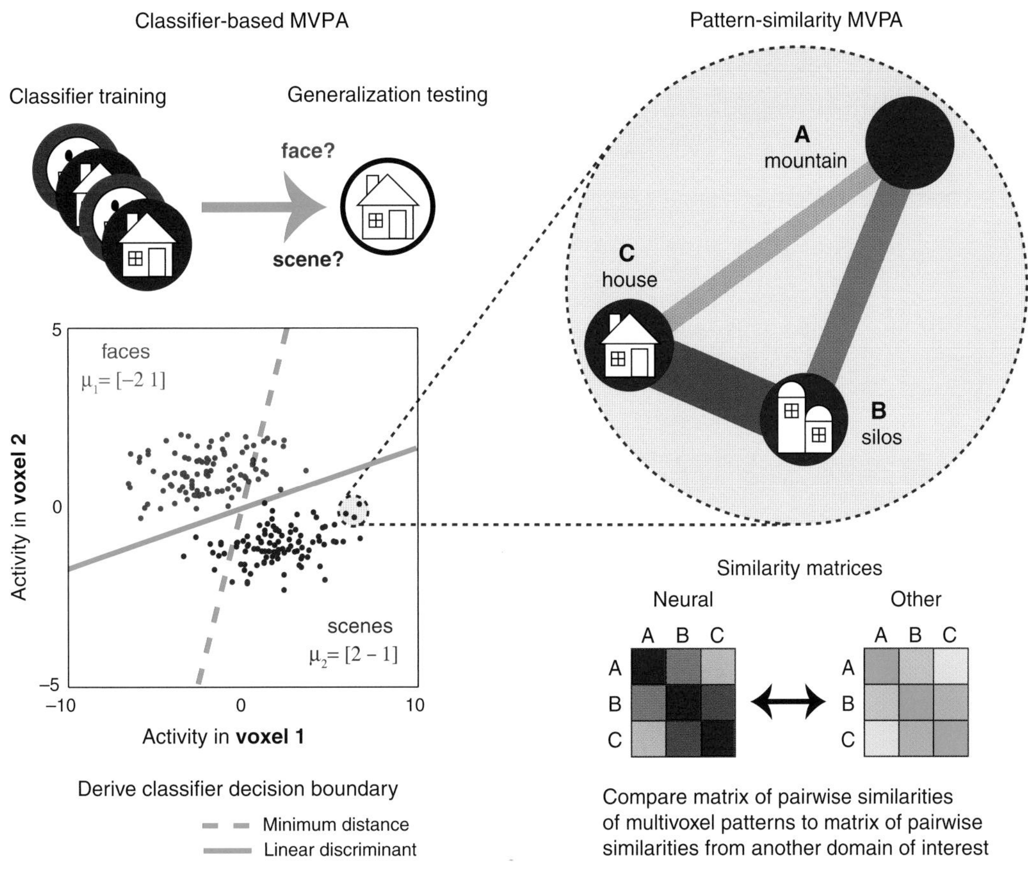

Plate 72 The two main types of multivoxel pattern analysis of functional MRI data. (See figure 77.2.)

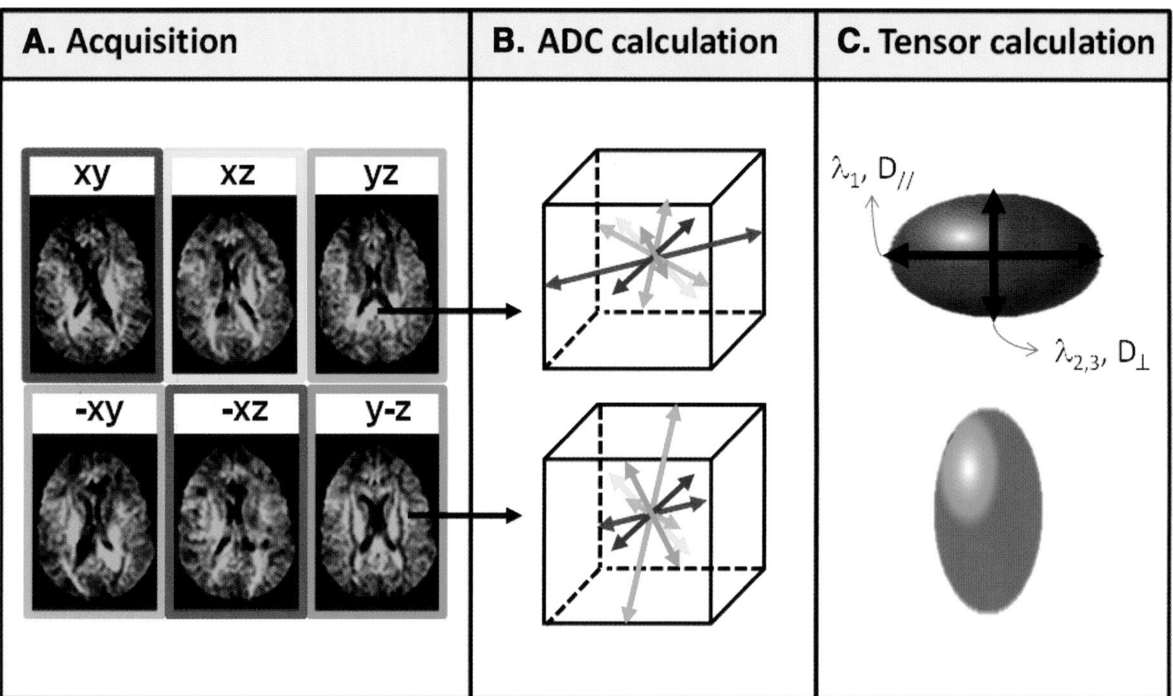

PLATE 73 Diffusion tensor imaging (DTI) pipeline including (A) acquisition at multiple gradient directions homogenously covering space (the example in the figure includes the basic six-direction scheme). (B) Apparent diffusion coefficient (ADC) calculation for each direction. The two graphs represent two voxels—the top graph refers to a voxel in the corpus callosum and the bottom graph for a voxel is superior longitudinal fascicules. (C) Tensor analysis of the data in panel B resulting in ellipsoids colored according to their orientation (left-right in red, anterior-posterior in green) representing the fiber orientation in the two selected voxels. (See figure 78.7.)

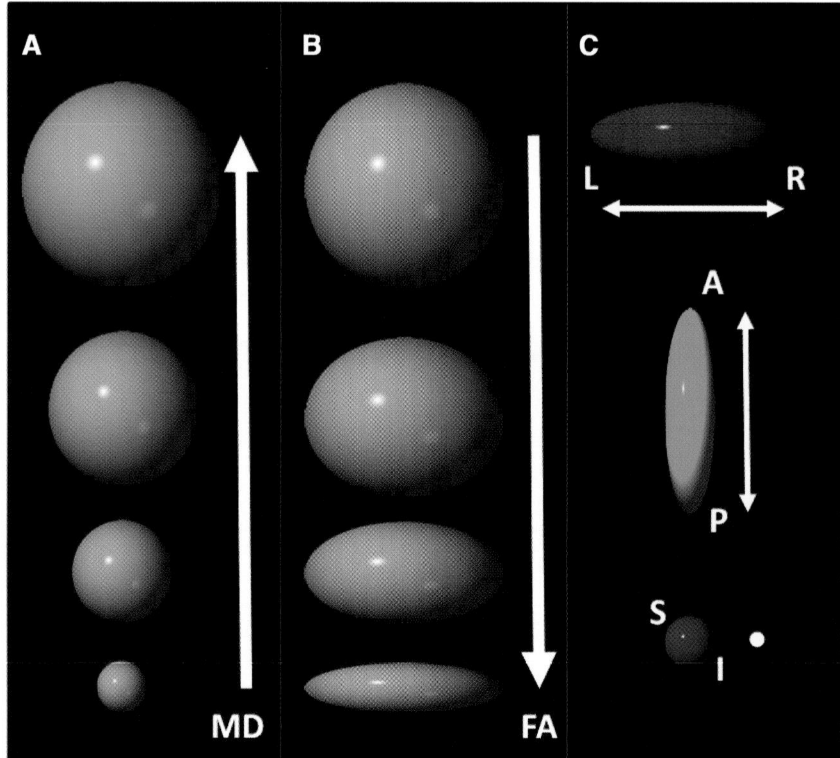

PLATE 74 The meaning of mean diffusivity (MD) and fractional anisotropy (FA), and fiber orientation on the ellipsoid extracted from DTI analysis. (A) Changes in MD will affect the size of the ellipsoid; increase in MD will cause an increase the size of the ellipsoid. (B) Changes in FA will affect the shape of the ellipsoid; low FA (as in cerebrospinal fluid) will appear as a pure sphere, whereas highly ordered white matter bundles will appear as a very elongated cigar-shape ellipsoid. (C) The orientation of the ellipsoid is indicated by its color; ellipsoids that lie in the left-right direction are colored red, ellipsoids in the anterior-posterior position are colored green, and ellipsoids in the inferior-superior position are colored blue. (See figure 78.8.)

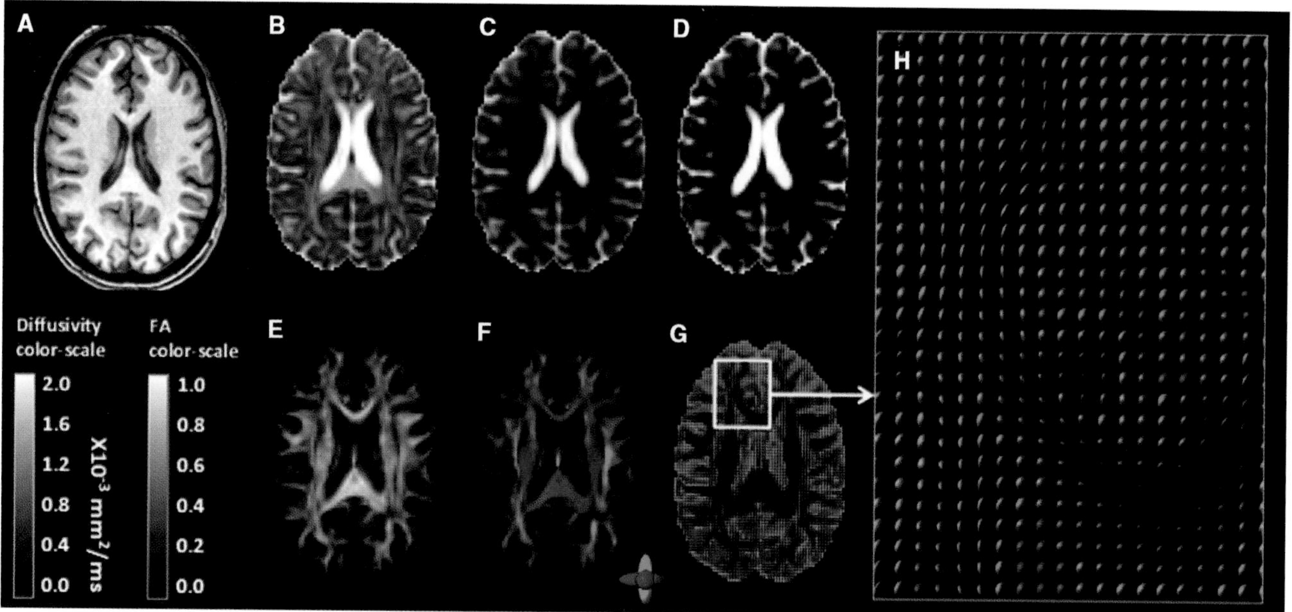

PLATE 75 Summary of quantitative indices extracted from DTI. (A) Reference T_1-weighted scan; (B) Axial diffusivity; (C) Radial diffusivity; (D) mean diffusivity; (E) fractional anisotropy; (F) color-coded FA map; (G) ellipsoid glyph map with section at the genu of the corpus callosum enlarge in (H). Color scale for (B–D) is given in diffusivity color scale (bottom left). Color scale for FA map is given in the FA color scale. (See figure 78.9.)

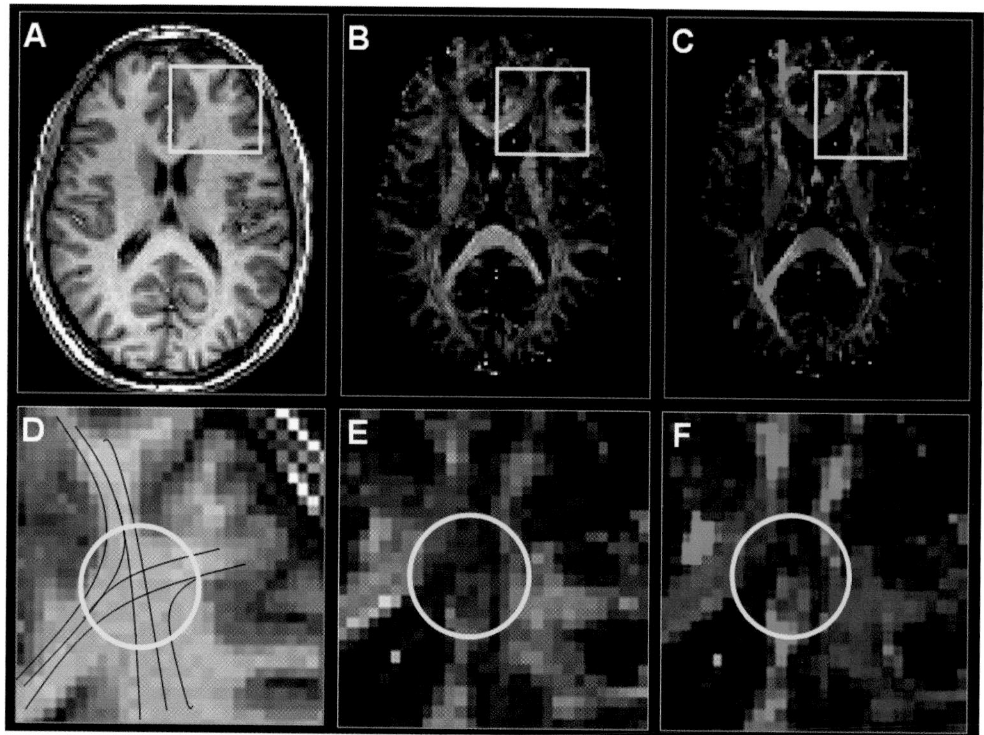

PLATE 76 Example for partial volume artifact in DTI. (A) T_1-weighted anatomical image with the right frontal region enlarged at (D), including line drawings of the different fibers system the passes in that region: one arriving from the genu of the corpus callosum and one from the thalamic radiation of the internal capsule, and both projecting into the frontal lobe. (B) FA maps enlarged at (E) showing low anisotropy area (marked by yellow circle) in the area of the crossing fiber system. This is also shown in the color-coded FA maps, ((C) enlarged at (F)) where one can follow the direction of the fiber (red for the corpus callosum and green for the thalamic radiation). In the crossing fiber region, the observed color is a mix of the two-fiber system. Since two crossing fibers reside within the same pixel in the region, the diffusion is hindered in all measured directions, which artifactually can be characterized by isotropic diffusion and thus reduced FA. (See figure 78.10.)

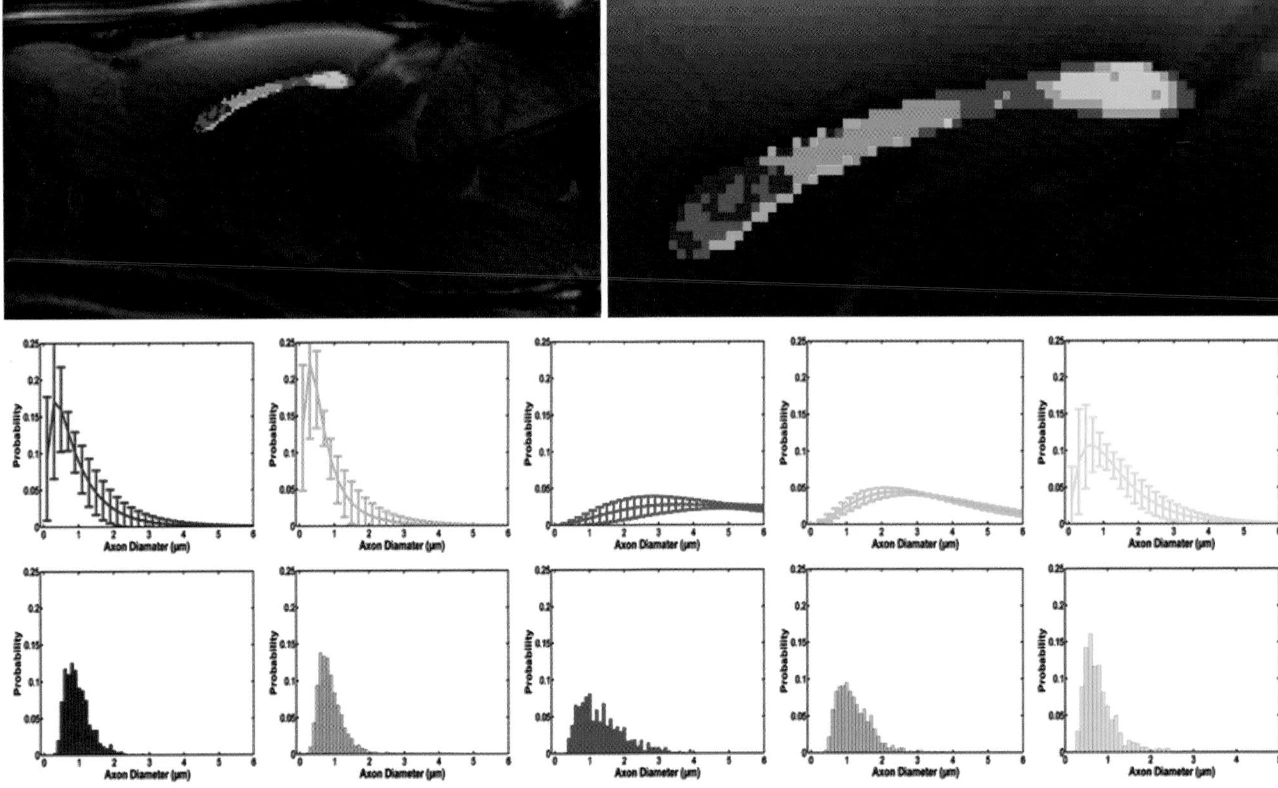

PLATE 77 Cluster analysis of the axon diameter distribution along the corpus callosum: (top row, left). A midsaggital T_2-weighted MRI with the AxCaliber clusters superimposed, enlarged at right. The AxCaliber averaged ADDs for the different clusters are shown in the middle row; note that the colors of the graphs match the clusters' colors. In the bottom row are the histological ADDs of the same clusters. (See figure 78.11.)

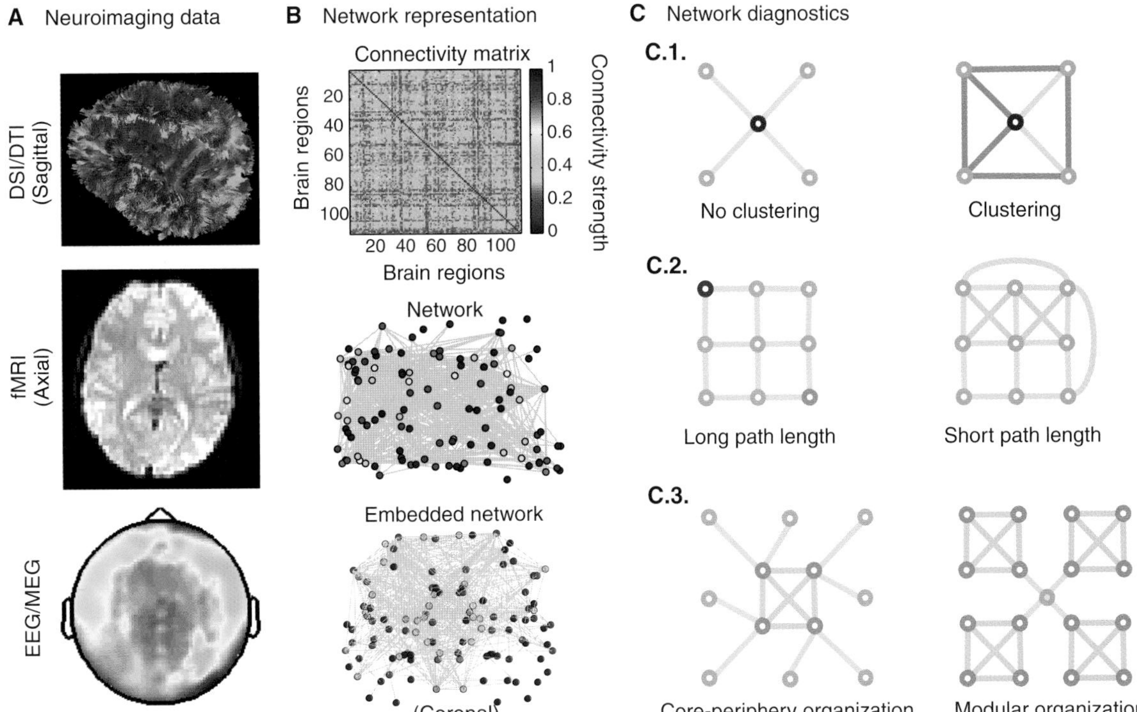

Plate 78 From data to diagnostics: The stages of a network study. (A) Data acquisition. Neuroimaging data can capture structural connectivity (e.g., diffusion spectrum imaging, DSI; diffusion tensor imaging, DTI) or functional connectivity (e.g., functional magnetic resonance imaging, fMRI; electroencephalography, EEG; or magnetoencephalography, MEG). (B) Representations of a network. *Top:* A connectivity matrix in which matrix elements (or pixels in the grid) represent one connection between two brain regions, and the color indicates the strength of that connection. *Center:* A topographical network visualization in which brain regions that are strongly (weakly) connected to one another lie close to (far from) each other in the plane. *Bottom:* An embedded network visualization in which nodes are placed in anatomically accurate locations. (C) Network diagnostics. (C.1) The clustering coefficient is a diagnostic of local network structure. The left panel contains a network with zero connected triangles and therefore no clustering, while the right panel contains a network in which additional edges have been added to close the connected triples (i.e., three nodes connected by two edges; green) to form triangles (i.e., three nodes connected by three edges; brown), thereby leading to higher clustering. (C.2) The average shortest path length is a diagnostic of global network structure. The left panel contains a network with a relatively long average path length. For example, to move from the purple node (*top left*) to the orange node (*bottom right*) requires one to traverse at least four edges. The right panel contains a network in which addition edges have been added (green) to form triangles or to link distant nodes (peach), thereby leading to a shorter average path length by comparison. (C.3) Mesoscale network structure can take many forms. The left panel contains a network with a core of densely connected nodes (red circles; green edges) and a periphery of sparsely connected nodes (blue circles; gray edges). The right panel contains a network with four densely connected modules (red circles; green edges) and a connector hub (blue circle; gray edges) that links these modules to one another. (See figure 79.1.)

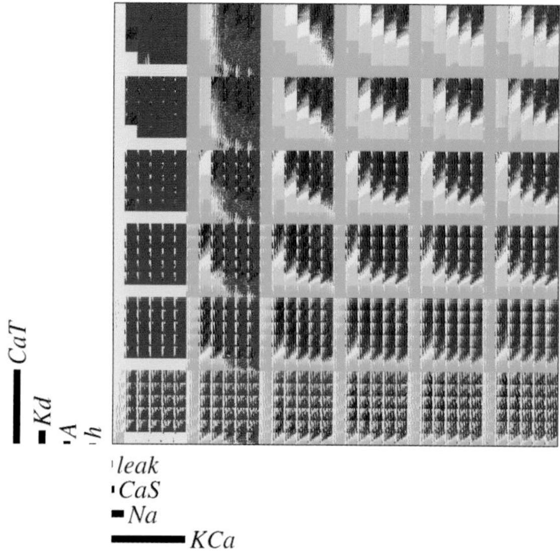

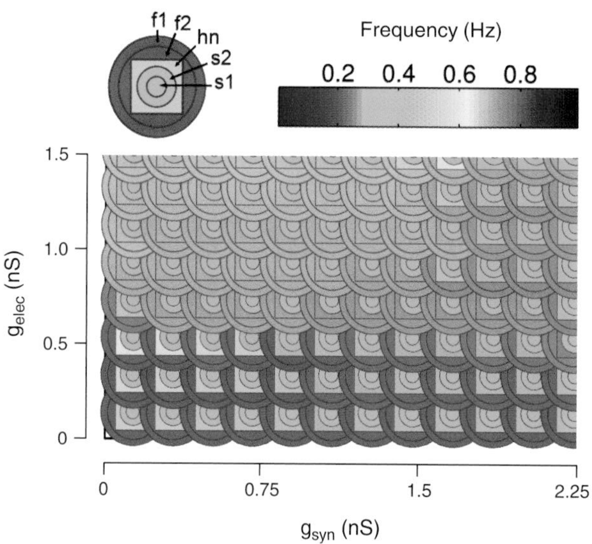

Plate 79 Methods for visualizing high-dimensional data. *Top:* Dimensional stacking nests axes within one another to achieve a 2D representation of higher-dimensional data. Each point in the image represents a single parameter combination, and the color represents a property of that model (for example, bursting vs. tonic spiking). The outer, largest edges of the square represent increments of the transient calcium (CaT) (*y*-axis) and calcium-dependent potassium (KCa) conductances (*x*-axis). Nested within each increment of these two parameters are all increments of delayed-rectifier (Kd) (*y*-axis) and sodium (Na) conductances (*x*-axis), which repeat within each interval of the outer axis. This nesting continues until all parameter combinations have been represented. The order of nesting is important for the picture that emerges. In this case, the nesting order is automatically determined to maximize contiguous areas of model behavior (represented by color). *Bottom:* An example "parameterscape" plots five measurements from each model in a population (frequencies of five oscillators: f1,2, fast oscillator 1,2; s1,2, slow oscillator 1,2; hn: "hub neuron" oscillator) relative to two parameters of the model: g_{syn}, a synaptic conductance (nS), and g_{elec}, a gap junction conductance (nS). Color of each shape represents frequency. The juxtaposition of the shapes in this manner gives a quick view of which oscillators are firing at similar frequencies and whether other models nearby in the parameter space exhibit similar frequency behavior. (See figure 80.5.)

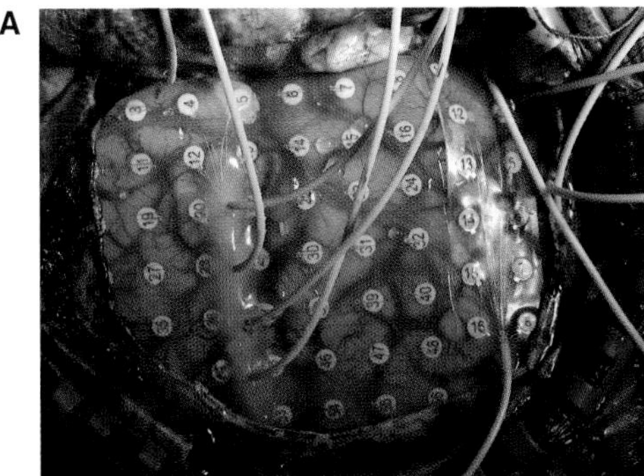

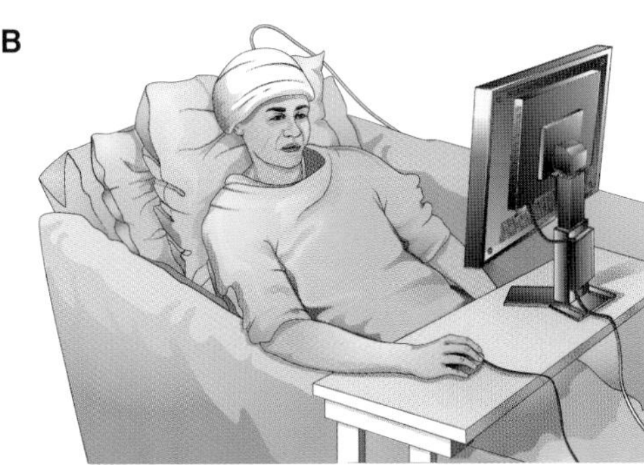

Plate 80 (A) An example of a subdural grid-electrode array implanted on the cortical surface of a patient with medically intractable epilepsy. After implantation, the craniotomy is closed with electrode tails (insulated white wires) tunneled outside the scalp to be connected to a recording system in an epilepsy monitoring unit (EMU). (B) Research studies can be conducted in awake and cooperative patients with implanted intracranial electrodes who are undergoing clinical testing in the EMU. While resting comfortably in their hospital bed, subjects can make behavioral responses with a computer mouse. (See figure 81.1.)

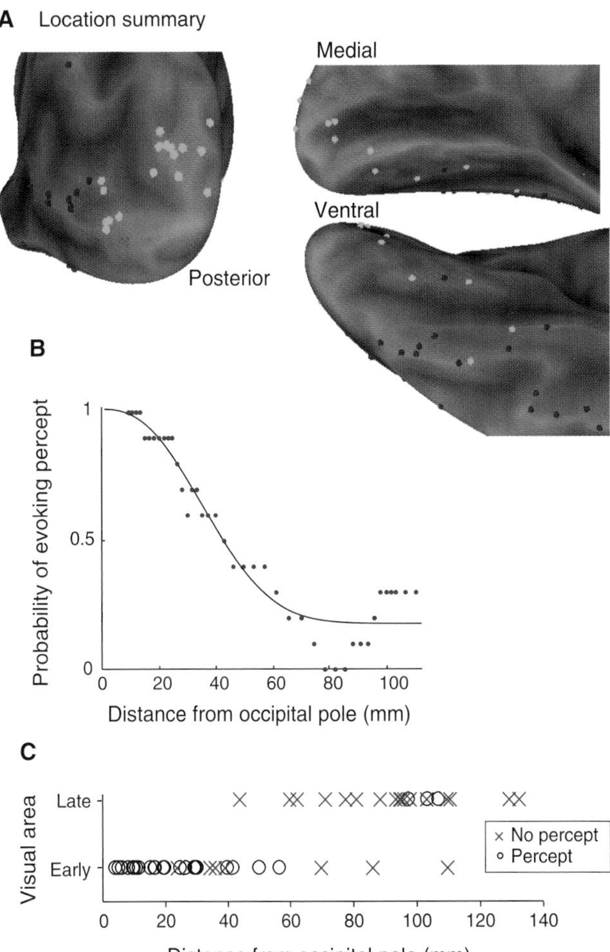

Plate 81 (A) The locations of 50 electrodes across 10 subjects are plotted as spheres on a single, inflated left hemisphere. The left hemisphere is shown from posterior, medial, and ventral views. Green color indicates that electrical stimulation of the electrode produced a percept. Red color indicates that it did not. (B) In each subject, the distance between the occipital pole and the electrode along the cortical surface was measured. For each distance, the probability of evoking a percept was computed. Each blue point shows the average cortical surface distance and probability for 10 electrodes, calculated with a moving-window average. The black curve shows the best-fit Weibull function. (C) Electrodes were also classified depending on their position in the visual hierarchy as either early (electrodes located in areas V1, V2, V3, V3a, and V4) or late (all other areas). Electrodes that produced a percept are shown as a black O; electrodes that did not are shown as a red X. Most early electrodes (symbols at the bottom) produced a percept, most late electrodes (symbols at the top) did not. There was a rough correspondence between early and late classification and distance from the occipital pole (shown on the *x*-axis). (Reprinted with permission from Murphey, Maunsell, Beauchamp, & Yoshor, 2009.) (See figure 81.2.)

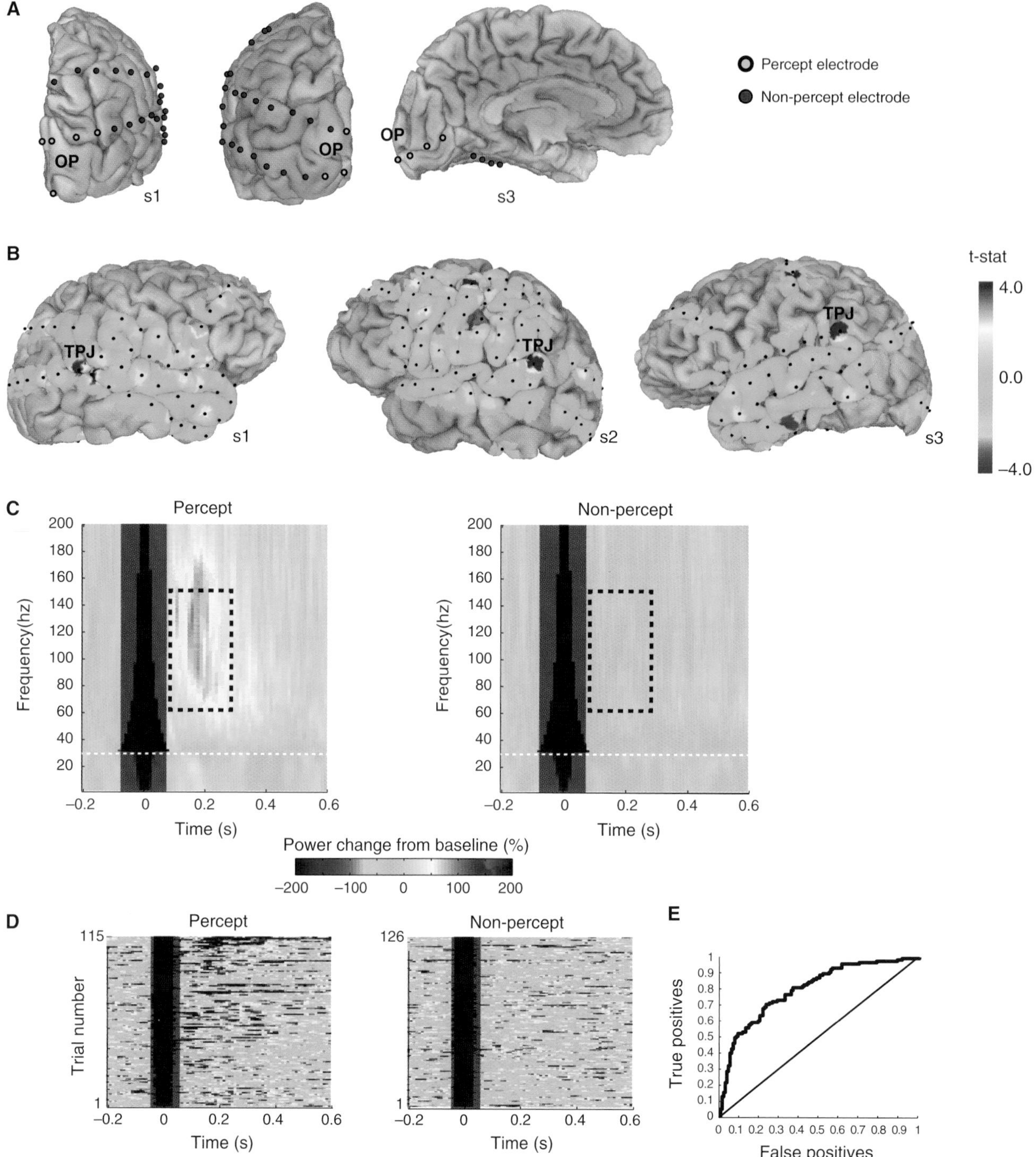

Plate 82 (A) Percept electrodes (green) that produced a phosphene upon electrical stimulation and nonpercept electrodes (red) that did not in three subjects. Subject 1 (s1) shows a posterior view of the right hemisphere; s2, posterior view of left hemisphere; s3, medial view of left hemisphere. Electrodes were implanted only in a single hemisphere for each subject (right for s1, left for s2 and s3). OP: occipital pole. (B) Maps showing the difference in gamma power between percept electrode stimulation and nonpercept electrode stimulation in three subjects (same subjects as in [A]). For each subject, one percept electrode and the nearest nonpercept electrode in the implanted hemisphere was repeatedly stimulated, and the significance of post-stimulation difference in gamma power at each electrode (except for the stimulation electrodes) was calculated and mapped to the cortical surface. Black spheres show electrode locations. TPJ: temporoparietal junction. (C) TPJ response during electrical stimulation of occipital electrodes that did (left) or did not (right) produce a phosphene, averaged across subjects. Color scale indicates power at each frequency. Dark gray bar centered at $t = 0$ indicates stimulation artifact. Dashed white line at $f = 30$ Hz indicates the boundary between two different frequency-estimate techniques. Dashed black line indicates gamma band f-t window used to estimate power for single-trial analysis. (D) TPJ gamma responses for every trial during stimulation of a percept (left) or nonpercept (right) electrode. Each horizontal line (raster) shows the power in a single trial over time, collapsed across 60–150 Hz. Same color scale as (C). (E) Receiver-operating curve analysis of single trial data. (Reprinted with permission from Beauchamp et al., 2012.) (See figure 81.4.)

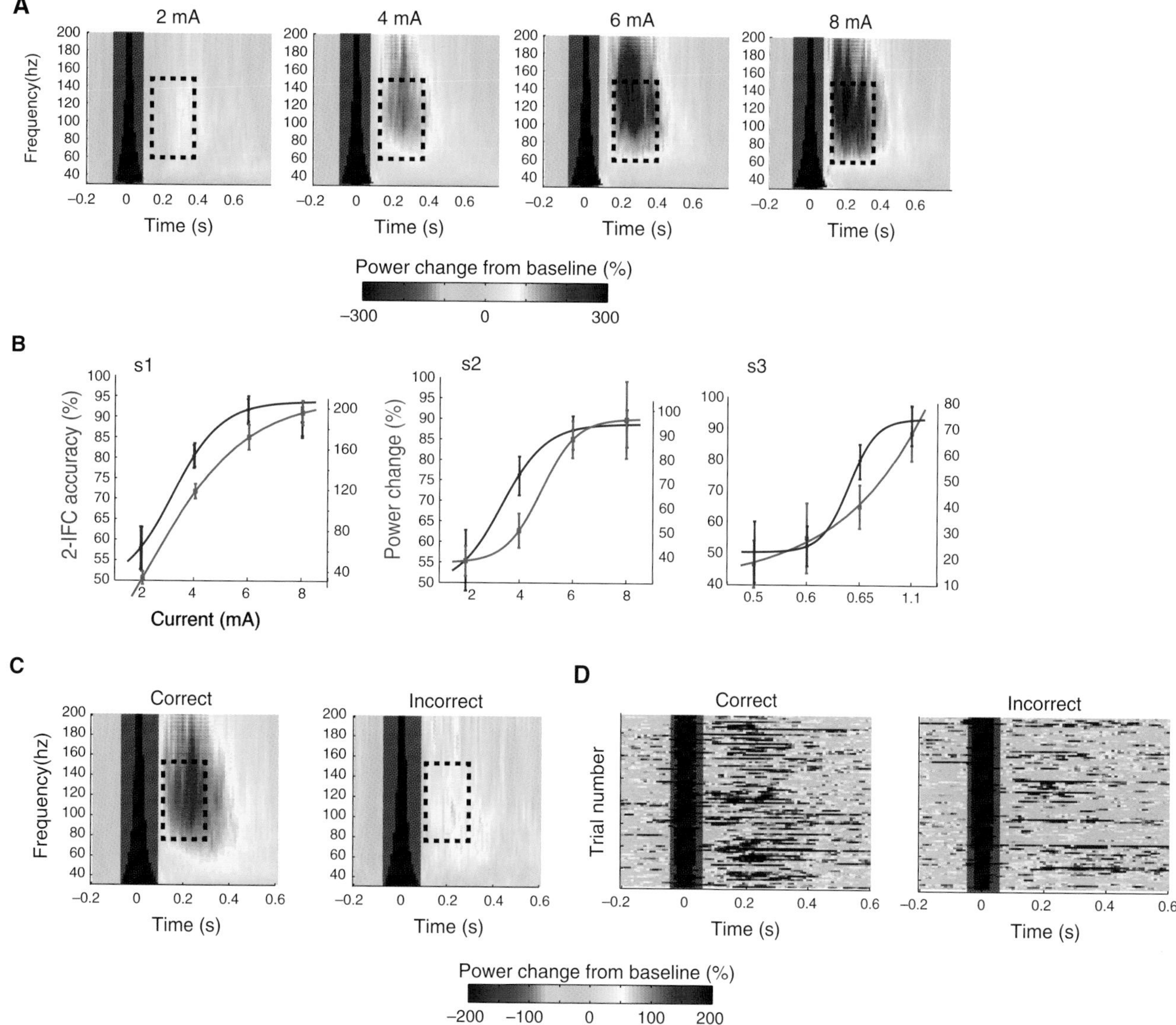

PLATE 83 (A) The average TPJ response during electrical stimulation of three percept electrodes in the occipital lobe in s1 at varying stimulation currents (2–8 mA). (B) Psychometric (blue) and neurometric (red) functions for subjects s1, s2, and s3. The psychometric curve (left *y*-axis) shows the behavioral performance during the two-interval forced-choice (2-IFC) task at different stimulation currents; performance was near chance (50%) at low currents and near ceiling (100%) at high currents (error bars show 75% confidence interval from the binomial distribution). The neurometric curve (right *y*-axis) shows the TPJ gamma power at the same currents (error bars show standard error of the mean). (C) TPJ response during percept electrode stimulation with near-threshold currents, averaged across subjects. Data averaged from trials in which subjects correctly (left) or incorrectly (right) discriminated which of two intervals contained electrical stimulation. Stimulation current was the same for correct and incorrect trials (4 mA for s1 and s2, 0.65 mA for s3). (D) TPJ response in the gamma band for single correct (left) and incorrect (right) trials of electrical stimulation at the same current. Each horizontal line (raster) shows the power in a single trial over time, collapsed across the gamma band. (Reprinted with permission from Beauchamp et al., 2012.) (See figure 81.5.)

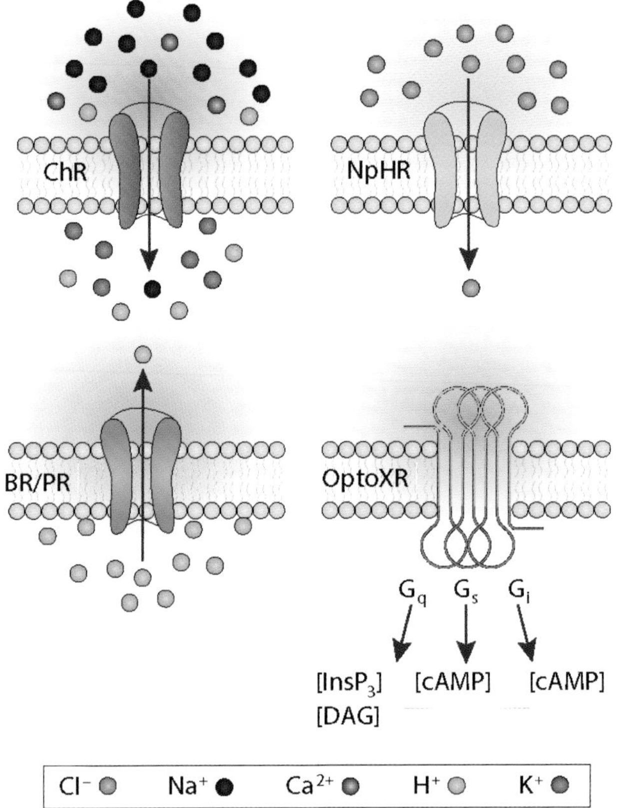

PLATE 84 Single-component optogenetic tool categories. Four major classes of opsin commonly used in optogenetics experiments, each encompassing light sensation and effector function within a single gene, include (1) channelrhodopsins (ChR), which are light-activated cation channels that give rise to inward (excitatory) currents under physiological conditions; (2) halorhodopsins (NpHR, shown), which are inhibitory (outward-current) chloride pumps; (3) bacteriorhodopsins and proteorhodopsins (BR/PR), proton pumps that tend to be inhibitory and include archaerhodopsins; and (4) optoXRs, which modulate secondary messenger signaling pathways. (Adapted with permission from Tye & Deisseroth, 2012.) (See figure 82.1.)

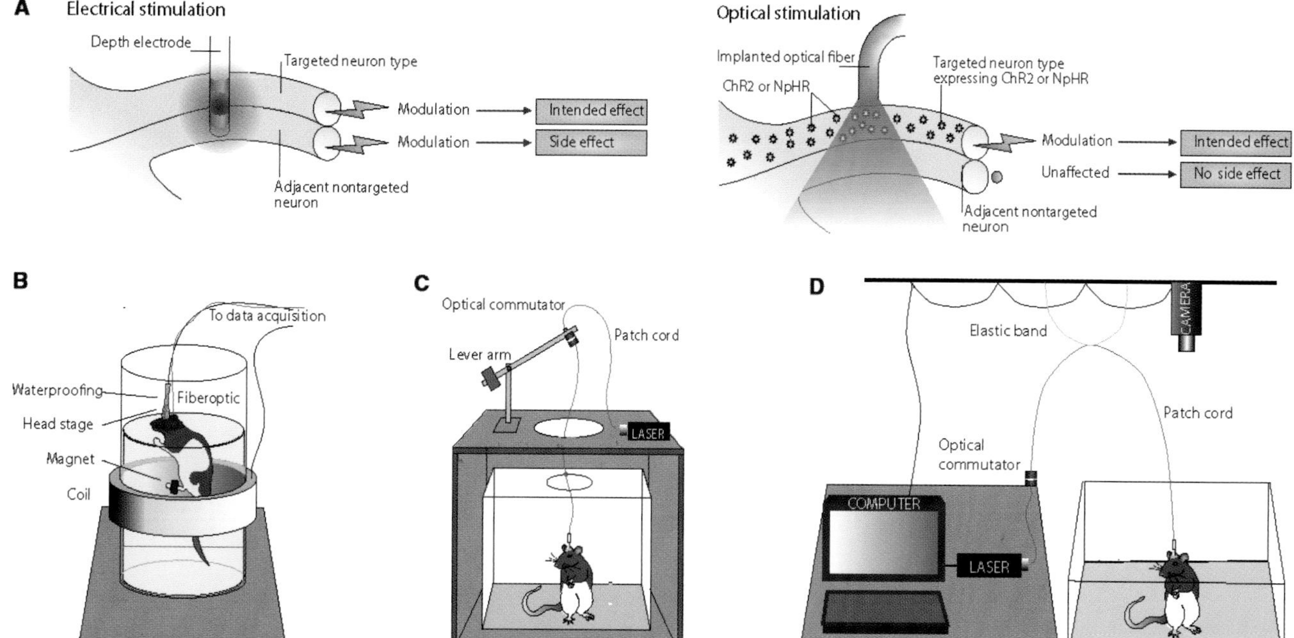

Plate 85 Integrating optogenetics with behavior. (A) Genetic targeting of channelrhodopsin (ChR2) or halorhodopsin (NpHR) into defined classes of neurons allows cell-specific neuromodulation and avoids inadvertent stimulation of irrelevant circuit elements, as occurs with electrical stimulation. (Adapted with permission from Zhang et al., 2007.) (B) Diverse behavioral rigs can be outfitted for mammalian optogenetic experimentation. The forced swim test has been automated with magnetic induction-based detection of kicks combined with optogenetic stimulation and electrical recording. (C) Operant behavior can also be combined with optogenetics. The chamber itself is modified to accommodate the entry of fiberoptics and recording wires, and the stimulation/recording assembly is kept out of the reach of the rat with a counterweighted lever arm. (D) Optogenetic manipulation can also be combined with behavior in open fields or large mazes. In these experiments, an elastic band, rather than a lever arm, is used to support stimulation/recording equipment. Video recording combined with custom or commercially available software can be used to synchronize optical stimulation with behavior. (Adapted with permission from Zalocusky & Deisseroth, 2013.) (See figure 82.2.)

A

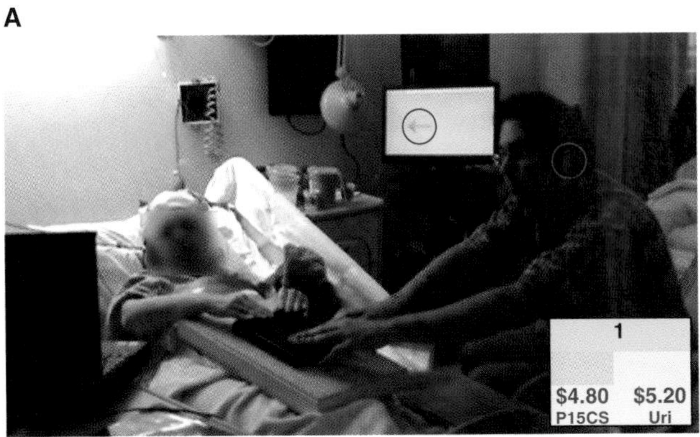

B

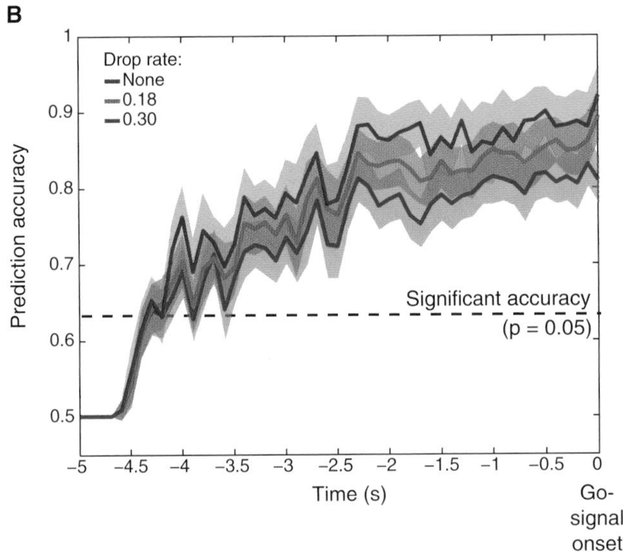

PLATE 86 (A) The experimental setup in the clinic. The patient and experimenter are watching the game screen (inset on bottom right) on a computer (bottom left) displaying the countdown to the go-signal and still pressing down the buttons of the response box. The real-time system already computed a prediction, and so displays an arrow on the screen behind the patient and plays a tone in the experimenter's ear ipsilateral to the hand it predicted he should raise to beat the patient. (B) Across-subjects average of the prediction accuracy (mean ± standard error of mean shaded) versus time before go-signal onset. Values above the dashed horizontal line are significant at $p = 0.05$. (Adapted with permission from Maoz et al., 2012.) (See figure 87.2.)

56 Language in Our Hands: Neural Underpinnings of Sign Language and Co-Speech Gesture

KAREN EMMOREY AND ASLI ÖZYÜREK

ABSTRACT The human ability to communicate information is not exclusive to speech—meaning can also be conveyed by the hands in sign languages and in co-speech gestures. Traditionally, the two have been studied separately, partially because each has very different properties (e.g., co-speech gesture is integrated with the speech channel; sign language is an independent, hierarchical system). This chapter brings together for the first time findings on the neural underpinnings of signs and co-speech gestures. Sign language and co-speech gestures both recruit the left hemisphere for production and involve the right hemisphere for representation of spatial events. Sign, co-speech gesture, and speech also implicate similar neural substrates for comprehension: left inferior frontal cortex, posterior superior temporal cortex, and bilateral middle temporal gyri, as well as similar electrophysiological responses to semantic violations (N400) and syntactic violations (P600). This review underscores the fact that the brain's general infrastructure for language and communication treats meaningful information coming from all channels and formats in highly similar ways.

Meaning can be conveyed by the hands in a variety of ways: through signing (used by deaf communities), pantomimed actions and emblematic gestures (often used in the absence of speech), or through gestures produced while speaking (i.e., gesticulation). This chapter focuses primarily on the first and the last phenomena, which form the endpoints of the gesture continuum (McNeill, 1992). As one moves from gesticulation toward sign language, idiosyncratic gestures used with speech are replaced by conventionalized/pantomimic gestures, linguistic properties increase, and the obligatory use of speech declines. Unlike co-speech gesture, sign languages are complete linguistic systems, exhibiting structural and language-specific constraints at the phonological, lexical, syntactic, and discourse levels.

Traditionally, the world's languages have been grouped into two classes, signed and spoken languages, based on the modality through which communicative messages are transmitted: visual-manual versus auditory-vocal. However, in the last decade it has become clear that this simple modality distinction does not capture the fundamental multimodal complexity of the human language faculty. Spoken languages also exploit the visual-gestural modality for expression and use gestures accompanying speech, with the hands, face, and body as articulators. For example, speakers can move the fingers of an inverted V-hand in a wiggling manner while saying "he walked across" or use bodily demonstrations of reported actions as they tell narratives. Gestures contribute semantic, syntactic, and pragmatic information to the verbal part of an utterance. To be able to understand the neural infrastructure of our language faculty as a whole, we need to take into account the composite multimodal utterances of spoken languages—that is, both speech and co-speech gestures. Similarly, mapping the neural architecture for manual gestures that form a hierarchical linguistic system in sign languages is also necessary for a general account of the neural bases of the human language faculty.

Because language on the hands has very different properties for sign languages and for co-speech gesture, we discuss their neural underpinnings in separate sections. When appropriate, we highlight relevant parallels and differences in each section and situate their neural processing in the context of what we know for spoken language.

The neurobiology of sign language

Sign languages exploit visual-spatial contrasts at the phonological level (e.g., locations on the body constitute contrastive units of form), at the morphological level (e.g., temporal aspect is marked by distinct patterns of movement), at the syntactic level (e.g., grammatical roles are distinguished through the use of locations in signing space), and at the discourse level (e.g., topics can be associated with spatial locations). Despite this dependence on visual-spatial processing, the left, rather than the right, hemisphere is critical for producing and comprehending sign language. Damage to the left hemisphere produces sign language aphasias that are parallel to spoken language aphasias, while damage to the right hemisphere does not. Signers with damage to left frontal cortex often exhibit nonfluent aphasia, characterized by effortful signing with reduced grammatical morphology. Lesions to more posterior regions within the left hemisphere produce fluent aphasias, characterized by fluent signing with persistent paraphasias (phonological or semantic errors). Signers with left temporal lobe damage also have much poorer sign comprehension compared to those who suffer damage to the right hemisphere or to other left-hemisphere regions.

In this chapter, we focus on a few key left-hemisphere brain circuits that have been identified as supporting language functions, and consider whether they are specific to spoken language and whether adaptations within these neural circuits are observed for sign language. Although there is clear overlap between the neural substrate for both language comprehension and production, these functions are discussed separately in order to explore the particular effects of distinct motor systems on language production and different perceptual systems on language comprehension.

Sign Language Production The left inferior frontal gyrus (IFG), including Broca's area (BA 44 and 45), has long been known to be involved in speech production. Several types of data indicate that this region is equally important for sign production. Using positron emission tomography combined with probabilistic cytoarchitectonic mapping, Horwitz et al. (2004) examined the roles of BA 44 and 45 in the production of spoken and signed narratives by hearing bilinguals who were native users of American Sign Language (ASL) and English. Horwitz et al. (2004) found involvement of BA 45, but little or no involvement of BA 44, during language production (either signing or speaking) compared to a baseline condition in which participants produced complex nonlinguistic oral or manual articulations. Similarly, extensive activation in BA 44, but not in BA 45, was observed for the nonlinguistic oral or manual control tasks compared to rest. This pattern of results suggests that BA 44, rather than BA 45, is engaged during the production of complex movements of the oral or manual articulators and that BA 45 is more likely engaged in articulator-independent aspects of language production.

Consistent with this hypothesis, Emmorey, Mehta, and Grabowski (2007) found that left IFG (primarily BA 45) was equally engaged when deaf signers and hearing speakers overtly named pictures (contrasted with a standard baseline task that required overt motor responses without lexical access). Further, Corina, San Jose-Robertson, Guillemin, High, and Brown (2003) reported activation in left inferior frontal cortex (primarily BA 45, extending into BA 47) when deaf right-handed signers generated verb signs with either their right or left hand, compared to repeating one-handed noun signs. These results suggest that more anterior regions of left inferior frontal cortex (BA 45 and 47) are engaged in modality-independent lexical search and retrieval processes and that this functional specialization is unaffected by the left-handed production of signs.

One striking difference between signed and spoken language production is the involvement of the left superior parietal lobule (SPL) during signing but not speaking (e.g., Emmorey et al., 2007). The precise role of left SPL in sign production is unclear, but this region is known to be involved in the online control and programming of reach movements to target locations, and for sign production, the hand must move to specific locations on the torso, face, or nondominant hand. In addition, SPL plays an important role in the assessment and monitoring of self-generated hand and arm movements. Because signing is not visually guided (i.e., signers do not look at their hands), sign language production must involve somatosensory monitoring of hand and arm movements and of hand postures and configurations.

Finally, the neural systems that support the production of pantomimed gestures and signs are nonidentical, even when signs are indistinguishable from pantomimes (e.g., the signs BRUSH-HAIR and HAMMER resemble the actions of hairbrushing and hammering). Emmorey, McCullough, et al. (2011) asked deaf signers to generate either pantomimed actions or iconic ASL verbs in response to pictures of manipulable objects. Verb generation engaged left IFG, as found in previous studies of both signed and spoken word generation, whereas pantomime generation engaged bilateral superior parietal cortex. These findings are consistent with data from lesion studies in which aphasic signers were reported to be impaired in sign production (e.g., making phonological and semantic errors), but their production of pantomimes was spared (Corina et al., 1992; Marshall, Atkinson, Smulovitch, Thacker, & Woll, 2004).

In sum, very similar left-hemisphere regions are engaged for signed and spoken language production. Left inferior frontal cortex is critically involved in the production of both signs and words, with the more posterior region (BA 44) engaged in phonetic and phonological encoding and more anterior regions (BA 45 and 47) engaged in lexical-semantic processes. Left superior parietal cortex is more involved in signing than speaking, most likely because this region is functionally specialized for the control and monitoring of arm and hand movements with respect to a body-centered framework. Finally, pantomimed actions and sign production rely on partially segregated neural systems, with sign production relying differentially on left frontotemporal cortices involved in lexical retrieval.

Sign Language Comprehension Not surprisingly, speech comprehension engages primary and secondary auditory cortices within superior temporal cortex bilaterally, and recent research indicates that phonological processing for spoken language is associated with the posterior superior temporal sulcus (STS) bilaterally (see Hickok & Poeppel, 2007). Given that sign languages are perceived visually rather than auditorally, is there any evidence that phonological processing of signs engages superior temporal cortex? The answer is *yes.*

First, many neuroimaging studies report bilateral activation in posterior superior temporal cortex during sign language comprehension (see figure 56.1). Second, evidence that posterior STS is engaged in phonological processing comes from studies that examined linguistically structured pseudosigns. Petitto et al. (2000) found that viewing both pseudosigns and real signs engaged superior temporal cortex bilaterally for deaf signers, but no activation in this region was observed for hearing individuals who had not acquired a sign-based phonological system (figure 56.1B). Emmorey, Xu, and Braun (2011) reported that pseudosigns activated left posterior STS to a greater extent for deaf ASL signers than for hearing nonsigners (figure 56.1C). Increased left STS activation for deaf signers was hypothesized to reflect heightened sensitivity to body movements that conform to the phonological structure of ASL. Left STS may be significantly more active for deaf signers than for hearing nonsigners because neurons in this region become particularly receptive to body movements that are linguistically structured and constrained.

The location and size of the neural response to signs within superior temporal cortex is modulated by hearing status and linguistic knowledge. In a functional MRI (fMRI) study investigating the comprehension of British Sign Language (BSL) sentences, MacSweeney et al. (2002) found that deaf signers exhibited greater activation in a more anterior region of left superior temporal gyrus (STG) compared to hearing signers (figure 56.1D).

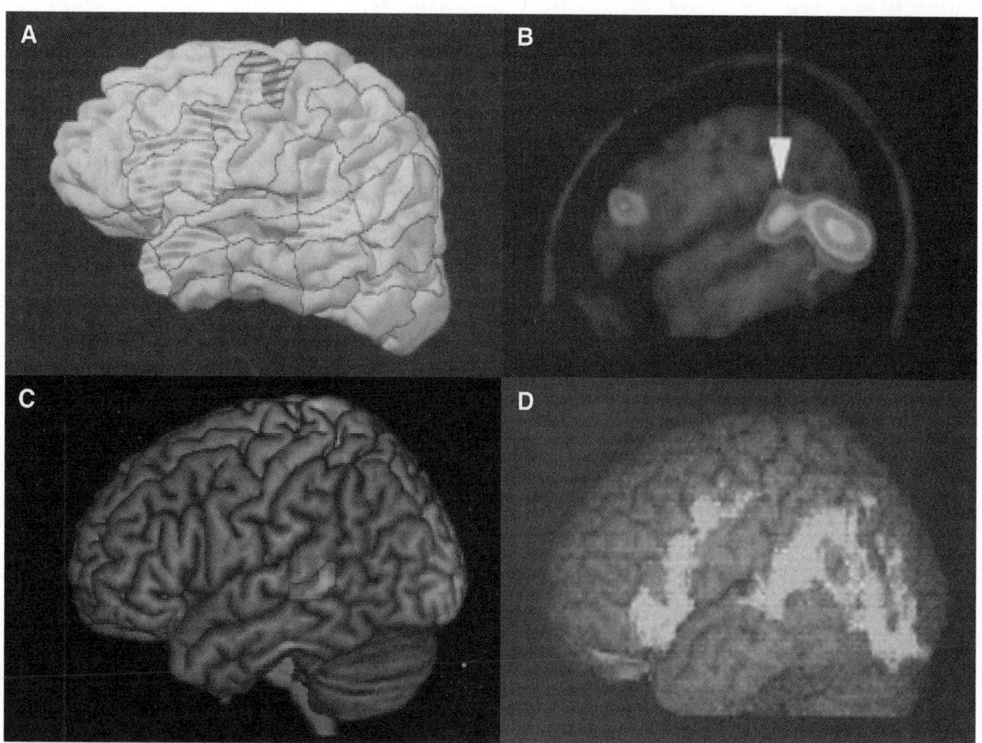

Figure 56.1 Illustration of left-hemisphere activations when deaf signers comprehend sign language. (A) Signed sentences vs. nonsense signs from Neville et al. (1998). (B) Signs and pseudosigns vs. fixation baseline from Petitto et al. (2000). (C) Greater activation for deaf signers than for hearing nonsigners for pseudosigns from Emmorey, Xu, & Braun (2011). (D) Signed sentences vs. a still image of the signer from MacSweeney et al. (2002). (See color plate 49.)

These authors suggested that for hearing signers anterior left STG may be privileged for processing heard speech and therefore is not engaged during sign language processing. In the absence of auditory input, deaf signers recruit auditory regions for processing sign language, as well as other nonlinguistic visual and somatosensory stimuli (e.g., Karns, Dow, & Neville, 2012). For deaf individuals, activation extends more anteriorly along STG when comprehending sign language compared to hearing signers. This more anterior region may not be as engaged for hearing sign-speech bilinguals because this area is adjacent to primary auditory cortex, which may preferentially respond to auditory speech over visual sign language input.

In addition, a recent MEG study by Leonard et al. (2012) revealed that both signs and words activated superior temporal cortex during a relatively late time window associated with lexical-semantic processing (300–500 ms after stimulus onset), but only speech for hearing individuals activated these regions during early sensory processing (80–120 ms). Leonard et al. (2012) concluded that activation in superior temporal cortex is associated with lexical processing of signs and does not result from a rewiring of visual sensory input to auditory cortices due to congenital deafness. Further, Cardin et al. (2013) found that when viewing signs, increased activation in left STS was driven by linguistic knowledge rather than by auditory deprivation. Deaf individuals who were not signers ("oral deaf") did not show increased activation in left STS compared to hearing nonsigners when viewing signed input. Thus, left-hemisphere STG/STS activation when comprehending sign language appears to arise from linguistic processing rather than from early (low-level) visual processing.

Several event-related potential (ERP) studies have found that semantic violations in sign language elicit an N400 response that is parallel to what has been reported for spoken and written language. The N400 is hypothesized to index semantic processing and integration. In addition, Capek et al. (2009) reported that syntactic violations in ASL that involved reversed verb agreement elicited a left anterior negativity (LAN) followed by a later, more broadly distributed P600 response. For spoken language processing, the LAN component is hypothesized to index early automatic syntactic processes, while the P600 is hypothesized to reflect syntactic reanalysis and error correction. These results indicate that syntactic and semantic processes are supported by distinct brain systems for both signed and spoken languages.

Finally, the comprehension of pantomimed gestures can be preserved in the face of sign language comprehension deficits (Corina et al., 1992; Marshall et al., 2004), suggesting that neural systems that support symbolic gesture and sign language processing are nonidentical. However, there are no reports indicating a double dissociation in which pantomime comprehension is impaired but sign language comprehension is spared. Thus, it is likely that at some level of processing, pantomimed gestures and sign language share a common neural substrate. Emmorey, Xu, Gannon, Goldin-Meadow, and Braun (2010) found very similar patterns of activation within bilateral posterior temporal cortex when deaf signers passively viewed pantomimed actions and ASL signs, but with evidence for greater activation in left IFG when viewing ASL signs. Xu, Gannon, Emmorey, Smith, and Braun (2009) found that comprehending symbolic gestures (pantomimes and social emblems, such as the "be quiet" gesture) and their spoken language equivalents both engaged left IFG and left posterior middle temporal gyrus (MTG). The authors suggested that these areas are part of a domain-general semantic network for human communication. Recently, Rong, Xu, Emmorey, Braun, and Hickok (2012) reanalyzed the Xu et al. (2009) data using a region of interest (ROI)–based multivariate pattern classification method and reported modality-specific responses within the posterior MTG ROI. More medial and posterior voxels responded preferentially to symbolic gesture and more anterior and lateral voxels responded preferentially to speech. This suggests that there are subregions in posterior MTG that are sensitive to the modality of input.

In sum, signed and spoken languages rely on very similar neural substrates for comprehension, including left inferior frontal cortex (see figure 56.1A and 56.1D), superior temporal cortex, and bilateral middle temporal cortex. Furthermore, the same electrophysiological responses to semantic violations (N400) and syntactic violations (LAN and P600) are observed for both language types. Lastly, sign language comprehension can be impaired in the face of successful pantomime comprehension, but it appears that understanding both signs and pantomimic gestures engages left posterior MTG.

The neurobiology of co-speech gesture

Investigations of the neurobiology of co-speech gesture have focused primarily on how information conveyed by a manual gesture is integrated with information conveyed by the accompanying speech. Co-speech gestures can display semiotic complexity of different types (e.g., points, demonstrations of objects and events), have different communicative functions (e.g., emphasis, disambiguation, speech acts), and vary in their relation to speech (e.g., conveying redundant or complementary information). Neuroimaging research has focused mostly on "iconic" gestures that represent objects and events by bearing partial formal resemblance to them (e.g., a chopping gesture while describing the steps of a recipe). While such gestures may resemble pantomimic actions, the meaning of co-speech gestures is often ambiguous and depends on speech for interpretation (e.g., Habets, Kita, Shao, Özyürek, and Hagoort, 2011).

Co-Speech Gesture Production Unlike sign languages, little is currently known about the neural regions engaged for gestures produced while speaking because they are difficult to study using neuroimaging techniques due to movement artifacts. Lesion studies (e.g., aphasia) often involve only a few case studies (e.g., Rose, 2006), and aphasic patients often use gestures in the absence of speech, making it difficult to infer the localization of co-speech gesture production.

Nonetheless, data from healthy participants suggests that hand preference for co-speech gestures is determined by the lateralization of language. Kimura (1973a, 1973b) found that right-handers with an (assumed) left-hemisphere language dominance produced more right-hand gestures than left-hand gestures, and vice-versa. However, a study with split-brain patients (Kita & Lausberg, 2008) showed that some gestures that involve spatial imagery can be generated by the right hemisphere (specifically, iconic gestures that situate referents in gesture space, such as an inverted V-shaped hand with wiggling fingers to depict someone walking). Interestingly, such gestures resemble classifier predicates used in spatial language in sign languages, which also recruit the right hemisphere in both production and comprehension (e.g., Emmorey et al., 2013; Hickok, Pickell, Klima, & Bellugi, 2009).

Co-speech gesture comprehension: Event-related potential studies

ERP studies of co-speech gestures have mostly focused on the N400 component to investigate whether co-speech gestures evoke semantic processing. The earliest studies isolated gestures from the accompanying speech and examined how they modulated the upcoming words or how they were processed following images. Kelly, Kravitz, and Hopkins (2004) found that ERPs to spoken words (targets) were modulated when these words were preceded by gestures (primes) that contained information about the size and shape of objects that the target words referred to (e.g., tall, wide, etc.). Compared to matching target words, mismatching words evoked an early P1/N2 effect, followed by an N400 effect, suggesting an influence of gesture on spoken words first at the level of "sensory/phonological" processing and later at the level of semantic processing. Wu and Coulson (2007) found that semantically incongruous gestures presented after cartoon images elicited a negative-going ERP effect around 450 msec, in comparison to gestures congruent with the cartoon image.

Gestures presented in a speech context produced similar results. Holle and Gunter (2007) asked subjects to listen to a sentence in which an ambiguous noun was accompanied by a gesture that disambiguated the word (e.g., "ball" with a playing ball gesture vs. a dancing gesture). An N400 effect was observed to a relevant word later in the sentence if its meaning did not match the meaning indicated by the gesture earlier in the sentence. In an ERP study, Özyürek, Willems, Kita, and Hagoort (2007) investigated the integration of co-speech gestures and spoken words to a previous sentence context. Sentences had critical words accompanied by gestures. Either the word or the gesture could be semantically anomalous or congruent with respect to the context set up by the sentence. Semantically anomalous gestures and anomalous words elicited identical N400 effects. The time course of integration of meaningful information derived from gestures did not differ from that conveyed through spoken words.

Finally, Habets et al. (2011) showed that the closer speech and gesture were temporally to each other, the more likely they were to be integrated with each other (i.e., when speech and gesture were simultaneous or when speech was delayed by 160 or 360 msec in relation to the gesture). ERPs time-locked to the speech onset showed a significant difference between semantically congruent versus incongruent gesture-speech combinations for the N400 component with SOAs of 0 and 160 msec, but not for the 360 msec SOA.

Few studies have investigated the neural infrastructure of noniconic beat gestures, which are short, rhythmic hand movements that co-occur with an emphasized segment of the speech. Holle et al. (2012) showed that such gestures facilitate the syntactic processing of speech. German-speaking participants were shown sentences with either the canonical (dominant) subject-object-verb (SOV) German word order or sentences with the less dominant object-subject-verb (OSV) word order. The sentences were presented either with or without beat gestures that co-occurred with the grammatical subject. The OSV sentences without an accompanying gesture elicited a P600 effect at the verb compared to the canonical SOV sentences. However, the P600 effect disappeared when the beat gesture emphasized the grammatical subject in the OSV sentences. Beat gestures made the grammatical roles clear early on in the sentence, and thus a syntactic reanalysis was not required.

The neural correlates of semantic processing for iconic co-speech gestures seems quite similar to that of spoken words and manual signs, as indexed by the N400 ERP component. In addition, the temporal overlap of gestures with speech modulates their semantic processing, thus corroborating the dependence of co-speech gestures on the speech channel (in contrast to signs that are not dependent on another system). Furthermore, gestures (beats) can influence syntactic processing of an utterance, as indexed by modulations of the P600 ERP component.

Co-speech gesture comprehension: Functional MRI studies

Functional MRI studies have attempted to locate the brain areas involved in the perception of iconic gestures in relation to speech. In general, these studies found left frontal and left posterior temporal cortices to be implicated in interpreting gestures and/or integrating gestures with speech, more specifically the left IFG, posterior superior temporal sulcus (STSp), and bilateral MTG (e.g., Dick, Mok, Raja Beharelle, Goldin-Meadow, & Small, 2012; Straube, Weis, Green, & Kircher, 2012; Willems, Özyürek, & Hagoort, 2007, 2009). Interestingly, studies examining spoken language comprehension alone have also found that increased semantic processing results in increased activity in these regions, especially left IFG and MTG (e.g., Snijders et al., 2009).

The contribution of left IFG to semantic integration of speech and gesture was first reported by Willems et al. (2007). Participants heard sentences in which a critical word was accompanied by a gesture (same stimuli from Özyürek et al., 2007). Either the word or the gesture could be semantically anomalous (i.e., incongruent) with respect to the context set up by the sentence. The incongruent word or gesture (demanding more semantic processing) elicited greater activity than congruent conditions in left IFG.

Left IFG has also been found to respond more strongly to metaphoric gestures, that is, gestures with abstract meaning (e.g., a "high" gesture accompanying speech like "the level of presentation was high"), compared to iconic gestures accompanying the same speech (Kircher et al., 2009). Dick et al. (2012) also found left IFG to be more active for complementary gestures (speech: "work"; gesture: typing) than for redundant gestures (speech: "typing"; gesture: typing). Complementary gestures, like metaphoric gestures, add information and require more semantic processing than redundant gestures. Finally, Skipper, Goldin-Meadow, Nusbaum, and Small (2009) showed that when iconic gestures were related to the accompanying speech, they exhibited a weaker influence on other motor and language-relevant cortical areas (including left IFG) compared to when the hand movements were meaningless (i.e., grooming gestures) or when there were no hand movements.

Thus, left IFG is responsive to iconic gestures, especially to those with an increased semantic processing load, that is, when gestures are difficult to integrate into the previous or overlapping co-speech context (i.e., incongruent, metaphoric, or complementary iconic gestures).

Posterior temporal regions are also involved in the semantic integration of gesture and speech. While MTG is more frequently found to be involved in semantic integration of speech and gesture, the role of STS has been more controversial. Holle et al. (2008) suggested that activity in STSp reflects sensitivity to the semantic integration of gesture and speech. In that study, STSp (but not left IFG) was more active for speech (dominant or subordinate homonyms) accompanied by meaningful iconic gestures than to speech accompanied by nonmeaningful grooming movements. However, Dick et al. (2012) and Willems et al. (2007, 2009) did not find activation in this area. Dick et al. (2012) argued that STSp may be involved in connecting information from the visual and auditory modalities in general, but not in semantic integration *per se.*

A stronger consensus has been achieved with regard to activation of left and/or right posterior MTG (MRTp). Green et al. (2009) found that left MTGp responded more strongly to sentences accompanied by unrelated gestures (hard to make sense of in relation to speech) than to those accompanied by related gestures. Dick et al. (2012) also found this area to be sensitive to complementary gestures, in comparison to redundant gestures. Willems et al. (2009) found that the left and right MTGp responded more to speech accompanied by incongruent pantomimes (conventionalized actions with objects such as ironing or twisting—the meaning of which would be clear without speech) than to the same speech accompanied by congruent pantomimes. However, MTGp was not activated for incongruent pairs of speech and co-speech gestures (these gestures would be ambiguous without speech) compared to congruent pairs. Incongruent speech-gesture pairs activated only left IFG and not MTGp. The authors argued that bilateral MTG is more likely to be involved in matching two input streams for which there is a relatively stable common object representation (i.e., "twist" in speech with a twisting gesture), parallel to the notion that both the sight of a dog and the sound of its barking form part of a representation of our knowledge about dogs (Hein et al., 2007). However, when integration of gesture and speech requires a new representation of the input streams, the increased semantic processing of iconic gestures results in increased activation of left IFG. At this point, these characterizations should be seen more as tendencies rather than exclusive functions of left IFG and MTG in speech and gesture integration.

Straube et al. (2012) attempted to isolate brain activation for iconic gestures (regardless of their involvement in speech integration). They compared activation for meaningful spoken sentences (S+) to sentences from an unknown language (S–), and they also compared activation for co-speech gestures presented without their accompanying speech (G+) with meaningless gestures (G–). Meaningful iconic gestures activated left IFG, bilateral parietal cortex, and bilateral temporal areas, but the overlap of activations for meaningful speech and meaningful gestures occurred in left IFG and bilateral MTG. These findings are consistent with the hypothesis by Xu et al. (2009) that left IFG and MTGp are involved in meaning extraction for communicative gestures (with or without speech) as well as for speech alone.

Functional MRI studies on semantic comprehension of co-speech iconic gestures and their integration indicate a similar neural signature to that for the semantic comprehension of words in context: a critical role for left IFG (sensitive to the increase in the semantic load required to process iconic gestures) and MTG (activated when similar information is conveyed in the two input streams). STG and STS may be engaged in the integration of gesture and speech at the audio-visual level, in addition to playing a possible role in meaning integration.

Conclusion

This review has revealed a surprising degree of overlap between the cortical regions and processes that support both signed and spoken language and co-speech gesture processing, in spite of the differences in their properties at many levels. Even though less is known for co-speech gesture production than for sign, the existing evidence suggests left-hemisphere dominance for the production of sign, gesture, and speech. However, both sign and gesture seem to involve the right hemisphere for some types of representations. Furthermore, for both speech and sign, left IFG is involved in lexical production, despite differences in the articulators. Left superior parietal cortex seems to be critically involved in sign language production, possibly due to the articulatory demands of manual phonology and somatosensory output monitoring.

Regarding comprehension, evidence from ERP studies indicates that the N400 response indexes semantic processing for both language modalities. The integration of meaningful information appears to follow the same time course for signs and words within a sentence context and for gestures in a speech context. In addition, the P600 component is sensitive to syntactic violations in signed and spoken languages, and this response can be mitigated by the presence of a co-speech gesture that reduces the need for syntactic reanalysis. Evidence from fMRI indicates a role for posterior superior temporal cortex in comprehending co-speech gesture and sign language, although the precise function of this region is likely to be different for the two. For sign language, STG/STS may be more engaged in phonological and lexical decoding of manual signs, whereas for co-speech gesture, STG/STS may be involved in audio-visual integration. Finally, both left IFG and MTG are engaged in semantic processing for language (both spoken and signed) and for iconic co-speech gestures.

The commonalities in the brain's processing of communicative information from co-speech gesture and sign (and in many ways speech) are striking, in spite of the differences in modalities (e.g., gesture is dependent on speech, but sign language is unimodal) and differences in the categorical versus gradient nature of representations. These commonalities underscore the fact that the brain's general infrastructure for language and communication can be recruited by different formats through which communicative information is transmitted.

ACKNOWLEDGMENTS Preparation of this chapter was supported in part by a grant from the National Institute on Deafness and other Communication Disorders (R01 DC010997) to K. E. and S. D. S. U. and by a European Research Council (ERC) Starting Grant to A. O.

REFERENCES

Capek, C. M., Grossi, G., Newman, A. J., McBurney, S. L., Corina, D., & Roeder, B. (2009). Brain systems mediating semantic and syntactic processing in deaf native signers: Biological invariance and modality specificity. *Proc Natl Acad Sci USA, 106*(21), 8784–8789.

Cardin, V., Orfanidou, E., Rönnberg, J., Capek, C. M., Rudner, M., & Woll, B. (2013). Dissociating cognitive and sensory neural plasticity in human superior temporal cortex. *Nat Commun, 4,* 1–5.

Corina, D. P., Poizner, H., Bellugi, U., Feinberg, T., Dowd, D., & O'Grady-Batch, L. (1992). Dissociation between linguistic and non-linguistic gestural systems: A case for compositionality. *Brain Lang, 43,* 414–447.

Corina, D. P., San Jose-Robertson, L., Guillemin, A., High, J., & Braun, A. R. (2003). Language lateralization in a bimanual language. *J Cogn Neurosci, 15*(5), 718–730.

Dick, A. S., Mok, E., Raja Beharelle, A., Goldin-Meadow, S., & Small, S. L. (2012). Frontal and temporal contributions to understanding the iconic co-speech gestures that accompany speech. *Hum Brain Mapp,* doi:10.1002/hbm.22222

Emmorey, K., McCullough, S., Mehta, S., Ponto, L. B., & Grabowski, T. (2011). Sign language and pantomime production differentially engage frontal and parietal cortices. *Lang Cogn Proc, 26*(7), 878–901.

Emmorey, K., McCullough, S., Mehta, S. H., Ponto, L. B., & Grabowski, T. J. (2013). The biology of linguistic expression impacts neural correlates for spatial language. *J Cogn Neurosci, 25*(4), 517–533.

Emmorey, K., Mehta, S., & Grabowski, T. J. (2007). The neural correlates of sign and word production. *NeuroImage, 36,* 202–208.

Emmorey, K., Xu, J., & Braun, A. (2011). Neural responses to meaningless pseudosigns: Evidence for sign-based phonetic processing in superior temporal cortex. *Brain Lang, 117,* 34–38.

Emmorey, K., Xu, J., Gannon, P., Goldin-Meadow, S., & Braun, A. (2010). CNS activation and regional connectivity during pantomime observation: No engagement of the mirror neuron system for deaf signers. *NeuroImage, 49,* 994–1005.

Green, A., Straube, B., Weis, S., Jansen, A., Willmes, K., Konrad, K., & Kircher, T. (2009). Neural integration of iconic and unrelated coverbal gestures: A functional MRI study. *Hum Brain Mapp, 30,* 3309–3324.

Habets, B., Kita, S., Shao, Z., Özyürek, A., & Hagoort, P. (2011). The role of synchrony and ambiguity in speech-gesture integration during comprehension. *J Cogn Neurosci, 23,* 1845–1854.

Hickok, G., & Poeppel, D. (2007). The cortical organization of speech processing. *Nat Rev Neurosci, 8,* 393–402.

Hein, G., Doehrmann, O., Müller, N. G., Kaiser, J., Muckli, L., & Naumer, M. J. (2007). Object familiarity and semantic congruency modulate responses in cortical audiovisual integration areas. *J Neurosci, 27*(30), 7881–7887.

Hickok, G., Pickell, H., Klima, E., & Bellugi, U. (2009). Neural dissociation in the production of lexical versus classifier signs in ASL: Distinct patterns of hemispheric asymmetry. *Neuropsychologia, 47,* 382–387.

Holle, H., & Gunter, T. C. (2007). The role of iconic gestures in speech disambiguation: ERP evidence. *J Cogn Neurosci, 19,* 1175–1192.

Holle, H., Gunter, T. C., Rüschemeyer, S. A., Hennenlotter, A., & Iacoboni, M. (2008). Neural correlates of the processing of co-speech gestures. *NeuroImage, 39,* 2010–2024.

Holle, H., Obermeier, C., Schmidt-Kassow, M., Friederici, A. D., Ward, J., & Gunter, T. C. (2012). Gesture facilitates the syntactic analysis of speech. *Front Psychol, 3,* 74.

Horwitz, B., Amunts, K., Bhattacharyya, R., Patkin, D., Jeffries, K., Zilles, K., & Braun, A. R. (2004). Activation of Broca's area during the production of spoken and signed language: A combined cytoarchitectonic mapping and PET analysis. *Neuropsychologia, 41,* 1868–1876.

Karns, C. M., Dow, M. W., & Neville, H. J. (2012). Altered cross-modal processing in the primary auditory cortex of congenitally deaf adults: A visual-somatosensory fMRI study with a double-flash illusion. *J Neurosci, 32,* 9626–9638.

Kelly, S. D., Kravitz, C., & Hopkins, M. (2004). Neural correlates of bimodal speech and gesture comprehension. *Brain Lang, 89,* 253–260.

Kimura, D. (1973a). Manual activity during speaking—I. Right-handers. *Neuropsychologia, 11,* 45–50.

Kimura, D. (1973b). Manual activity during speaking—II. Left-handers. *Neuropsychologia, 11,* 51–55.

Kircher, T., Straube, B., Leube, D., Weis, S., Sachs, O., & Willmes, K., et al. (2009). Neural interaction of speech and gesture: Differential activations of metaphoric co-verbal gestures. *Neuropsychologia, 47,* 169–179.

Kita, S., & Lausberg, H. (2008). Generation of co-speech gestures on spatial imagery from the right hemisphere: Evidence from split-brain patients. *Cortex, 44,* 131–139.

Leonard, M. K., Ferjan Ramirez, N., Torres, C., Travis, K. E., Hatrak, M., Mayberry, R. I., & Halgren, E. (2012). Signed words in the congenitally deaf evoke typical late lexicosemantic responses with no early visual responses in left superior temporal cortex. *J Neurosci, 32,* 9700–9705.

MacSweeney, M., Woll, B., Campbell, R., McGuire, P. K., David, A. S., Williams, S. C., ... Brammer, M. J. (2002). Neural systems underlying British Sign Language and audio-visual English processing in native users. *Brain, 125*(Pt. 7), 1583–1593.

Marshall, J., Atkinson, J., Smulovitch, E., Thacker, A., & Woll, B. (2004). Aphasia in a user of British Sign Language: Dissociation between sign and gesture. *Cogn Neuropsychol, 21*(5), 537–554.

McNeill, D. (1992). *Hand and mind: What gestures reveal about thoughts.* Chicago, IL: University of Chicago Press.

Neville, H., Bavelier, D., Corina, D., Rauschecker, J., Karni, A., Lalwani, A., ... Turner, R. (1998). Cerebral organization for language in deaf and hearing subjects: Biological constraints and effects of experience. *Proc Natl Acad Sci USA, 95,* 922–929.

Özyürek, A., Willems, R. M., Kita, S., & Hagoort, P. (2007). On-line integration of semantic information from speech and gesture: Insights from event-related brain potentials. *J Cogn Neurosci, 19,* 605–616.

Petitto, L. A., Zatorre, R. J., Gauna, K., Nikelski, E. J., Dostie, D., & Evans, A. C. (2000). Speech-like cerebral activity in profoundly deaf people processing signed languages: Implications for the neural basis of human language. *Proc Natl Acad Sci USA, 97*(25), 13961–13966.

Rong, F., Xu, J., Emmorey, K., Braun, A., & Hickok, G. (2012). Modality-specificity is evident in the micro-organization of "amodal" conceptual-access areas. Poster presented at the Neurobiology of Language Conference, San Sebastian, Spain.

Rose, M. L. (2006). The utility of arm and hand gestures in the treatment of aphasia. *Adv Speech Lang Pathol, 8*(2), 92–109.

Skipper, J. I., Goldin-Meadow, S., Nusbaum, H. C., & Small, S. L. (2009). Gestures orchestrate brain networks for language understanding. *Curr Biol, 19,* 661–667.

Snijders, T., Vosse, T., Kempen, G., van Berkum, J., Petersson, K. M., & Hagoort, P. (2009). Retrieval and unification of syntactic structure in sentence comprehension: An fMRI study using word-category ambiguity. *Cereb Cortex, 19,* 1493–1503.

Straube, B., Green, A., Weis, S., & Kircher, T. (2012). A supramodal neural network for speech and gesture semantics: An fMRI study. *PLoS ONE, 7,* e51207.

Willems, R., Özyürek, A., & Hagoort, P. (2007). When language meets action: The neural integration of gesture and speech. *Cereb Cortex, 17,* 2322–2333.

Willems, R., Özyürek, A., & Hagoort, P. (2009). Differential roles for left inferior frontal and superior temporal cortex in multimodal integration of action and language. *NeuroImage, 47,* 1992–2004.

Wu, Y., & Coulson, S. (2007). How iconic gestures enhance communication: An ERP study. *Brain Lang, 101,* 234–245.

Xu, J., Gannon, P., Emmorey, K., Smith, J. F., & Braun, A. R. (2009). Symbolic gestures and spoken language are processed by a common neural system. *Proc Natl Acad Sci USA, 106*(49), 20664–20669.

57 Neuropragmatics

PETER HAGOORT AND STEPHEN C. LEVINSON

ABSTRACT Linguistic expressions are often underdetermined with respect to the meaning that they convey. That is why context is needed to establish the message that is intended to be communicated by the linguistic expressions that are used. Sometimes context information is explicitly provided, but often it relies on implicit background knowledge shared between speaker and listener. The context of communication needs to be invoked in understanding and producing linguistic codes. What is required, in addition to the linguistic code itself, to determine the meaning of an expression is usually referred to as pragmatics. The neurobiology of pragmatics is an area that has come under investigation only recently. Here we discuss central aspects of pragmatics. On the basis of recent functional MRI (fMRI) studies, it has become clear that the theory of mind network is involved in making the step from coded meaning to what is often called "speaker meaning," that is the intended message for the listener. Event-related potential studies have investigated whether pragmatic processing is superimposed on and follows in time the computation of coded meaning or, alternatively, whether context information is immediately integrated with lexical, syntactic, and semantic sources of information. On the whole, the evidence supports the immediacy assumption, suggesting that linguistic and nonlinguistic context information is immediately used for utterance interpretation. However, quite some individual variation is found in different aspects of pragmatic processing.

Neurobiological models of language not only need to address the circuitry that is crucial for encoding and decoding the content of an utterance, but they also need to specify the neural infrastructure for inferring what the speaker intended to communicate by uttering a sentence (i.e., speaker meaning; Grice, 1989; Noveck & Reboul, 2008). This is what we broadly refer to as neuropragmatics (Bambini, 2010; Bara, 2010). We will first discuss what is covered by the term *pragmatics* and why it is crucial in an account of human communication and language. This is followed by an overview of recent studies on the neural basis of pragmatics.

Pragmatics

Pragmatics is the study of the way in which context, including the discourse, beliefs, and inferences of participants, contributes to the meanings of utterances (see, e.g., Levinson, 1983; Sperber & Wilson, 1987). The mechanisms involved are diverse, but they give rise to substantial inferences that are not actually coded in what is said. Consider the following interchange:

A: Hey, I wonder who ate some of the chocolates I was going to give to Anne?

B: Oh, I heard the kids in the kitchen earlier.

From A's utterance, one may infer that (the parentheticals indicate some of the different categories of pragmatic inference referred to later):

(1) A wants to know who is responsible for eating some of the chocolates (speech act conditions).

(2) A thinks B might know the answer to his implied question, knows about the existence of the chocolates, and so on (audience design: utterances are formulated and adapted for specific listeners).

(3) A thinks that not all of the chocolates were eaten (scalar implicature from the use of *some*).

Likewise, from B's utterance in the context of A's question much further information can be inferred:

(4) B doesn't know for sure who ate some of the chocolates, but B is nevertheless trying to provide some kind of answer (adjacency pair constraint: answers should follow questions; Grice's maxim of relevance).

(5) B knows or guesses that A thinks the chocolates were in the kitchen (bridging inference, audience design).

(6) B is suggesting that perhaps the kids ate some of the chocolates (indirect speech act, conversational implicature), via additional premises like "kids tend to do naughty things" (common ground: implicit background knowledge shared between speakers and listeners).

None of these inferences are logical deductions from what is said. They can only be derived by making additional assumptions about the purposes and intentions with which we engage in verbal interaction and about common understandings about how we should use language. For example, other things being equal, we assume that people will recognize our speech acts (questions, requests, assertions, accusations, etc.) even though they are often not coded directly—here A's *I wonder who* is not syntactically a question, but rather specifies a precondition for asking one (the basis for inference 1 above), which A thinks should be enough for B to recognize the intention (inference 2 above). A doesn't seem to think all the chocolates have been eaten (3 above), because otherwise he wouldn't have said *some of the chocolates*—*some* means literally "at least one" so is compatible with "all" (cf. *some in fact all of the professors are stupid*), but since *all* wasn't said, the contrast suggests "not all."

Even more elaborate are the inferences we attribute to B's utterance, largely because of its position immediately after A's. If A's utterance is an indirect question, given that questions should be followed by answers, we try and interpret B's response as an indirect answer (inference 4), suggesting specifically that the kids may be to blame (inference 6). That in turn requires lots of further assumptions, like the chocolates being in the kitchen (inference 5), while drawing on common presumptions like the possible moral turpitude of younger members of the species, and so forth (6).

What is clear from any such little interchange is that what is coded in words and linguistic constructions far underdetermines what is obviously meant or intended: meaning in a broad sense is an iceberg, with the little linguistically coded part riding on a large submersed mountain of mutual inference.

Not all the things we infer from utterances come loaded with this intentional baggage. For example, we may detect from an accent that a speaker is not a native English speaker, or on the telephone from the acoustics that the speaker is a woman—these are usually not part of what the speaker is trying to communicate. The philosopher Grice (1957) made an important distinction between the kind of meaning that is intended to be communicated ("nonnatural meaning," or meaning-nn) and the kind of meaning that is simply conveyed unintentionally ("natural meaning"). He defined "nonnatural meaning" as the effect a communicator intends to cause in a recipient just by getting the recipient to recognize that intention: for example, I can yawn (itself a potential natural symptom of tiredness) in such a way that you can recognize that I am finding you boring. The intention-laden yawn is for Grice (and the pragmaticist in general) the curious phenomenon at the heart of human communication; when we use words we use them in the same kind of heavily intentional environment, where the choice of expressions are scrutinized for their underlying purpose and intentions. In the language domain, meaning-nn amounts to a theory of *speaker meaning*—what the speaker intends to achieve by the use of his or her utterance in a specific context, which can contrast with *sentence meaning*, what is abstractly coded and independent of context.

Grice's theory of meaning sought to characterize the whole domain of intentional communication, but it provides no mechanisms for recognizing communicative intentions. Here the levers seem to be provided by a wide range of background conventions or understandings. For example, we advance our communicative purposes in conversation by producing utterances in turn, where each utterance can be attributed with at least one main point or speech act (Searle, 1969)—so that, for example, one may recognize in A's turn above a question, which will require an answer in response.

Yet another kind of lever for making inferences in conversation is a general presumption of rational cooperation. Grice's (1975) second great contribution was the theory of conversational implicature, which spelt out some of the parameters of this presumptive cooperation in terms of four main "maxims of conversation." B's utterance in the interchange above (*Oh, I heard the kids in the kitchen earlier*) has no obvious connection to A's. But we make the assumption that A and B are trying to aid their mutual conversational undertaking, and that allows a range of detailed inferences. Grice posits a maxim of relevance, which requires a timely and pertinent response (hence the need to figure out how B's utterance ties to A's). He also posits a maxim of quantity, specifying that information provided should be not too much nor too little for the purposes in hand: a *Who did it?* question can be succinctly answered by a name like *Bill*, for example. Here there seem to be quite specific rules of thumb that can guide pragmatic inference. For example, the maxim of quantity presupposes some brevity metric, and linguistic systems provide sets of alternates, which will typically be of roughly equal brevity, for example, {*white, red, yellow, blue, …*}. By the maxim of quantity, saying *He's waving a white flag* can thus be said to suggest—or *conversationally implicate*, as Grice would have it—that the flag is purely white; if it had been white, blue, and red, by the maxim of quantity requiring sufficient information, you should have said so (see Levinson, 2000). Further, such sets of alternates are often ordered by informativity, or forming ordered scales; for example <*all, most, many, several, some*>, <*must, may*>, <*and, or*>, etc. Here informationally stronger items to the left entail items further to the right: *the kids ate all the chocolates* entails "the kids ate (at least) some of the chocolates." We can now state one rule for generating general pragmatic inferences, so-called *generalized conversational implicatures* (or more specifically here, *scalar implicatures*), namely, that asserting a weaker item on a scale conversationally implicates a stronger item doesn't obtain (otherwise, by the maxim of quantity, you should have said the stronger one). That's the reason that we can make inferences like (3) from the little interchange between A and B above.

Another lever is a background assumption of audience design. When we refer to persons or things, we use the terms we think the recipient will be able to use to recognize the referent, choosing between *Mike, the author of "The Ethical Brain," the head of the Sage Centre,* etc. In our snippet above, the definite articles in *the chocolates, the kitchen,* and *the kids* implicate the mutual recognizability of the referents (inference 2 above). Normally, we introduce new inanimate referents with indefinite articles, and thereafter refer to them with definite ones. But we may also introduce assumptions without ever making overt reference to them, as in *We're so sorry we're late. Our hire car broke down; the steering wheel came off,* where the definite phrase *Our hire car* introduces the fact that we hired a car, and *the steering wheel* has to be understood as part of the same car (Clark, 1996). We trade all the time on presumed "common ground," assuming that we'll both keep track of references previously introduced, and that we can use stereotypes (like the naughtiness of kids) as premises in unarticulated inferences.

This brief introduction to the scope of pragmatic inference already makes clear, first, how fundamental the role of pragmatic inference is in language understanding, and second, the extent to which it depends massively on theory of mind, and specifically on reasoning about the other's purposes and intentions, together with constant updating of what we think our recipients presume or know (or do not know), and what we think they can infer from what we implicate but do not say. Pragmatics is, in other words, the science of the understood but unsaid.

The neural infrastructure for pragmatics in language

Despite the central contribution of pragmatic inferences to communication, most research on the neurobiology of language has focused on either single-word processing or on the syntactic and semantic operations involved in decoding the content of an utterance. Perisylvian cortex, with a left-hemisphere dominance, is known to be crucial for decoding propositional content (Hagoort, 2005, 2013; Hagoort & Poeppel, 2013). However, as we have seen above, communication involves much more than this. Knowledge about context and speaker needs to be invoked to infer the intended message (speaker meaning) and to engage in successful communication (Bambini, 2010; Levinson, 2000). Relatively few studies have investigated the neural infrastructure for effective communication beyond the core linguistic machinery for word retrieval, syntax, and semantics. Many aspects of pragmatics that we discussed above have not yet been investigated at the level of brain organization. Here we will review and summarize our limited current knowledge in the domain of neuropragmatics.

Recent studies investigated different aspects of communicative intentions, such as conversational implicatures and indirect requests. Bašnáková et al. (2013) contrasted direct and indirect replies—two classes of utterances whose speaker meanings are more and less similar to their coded meaning. In their study, participants listened to natural spoken dialogue in which the final and critical utterance—for example, "It is hard to give a good presentation"—had different meanings depending on the dialogue context and the immediately preceding question. This critical utterance either served as a direct reply (to the question "How hard is it to give a good presentation?") or an indirect reply (to "Did you like my presentation?"). One of the major motivations for speakers to reply indirectly in conversations is to mutually protect one another's public self (e.g., Brown & Levinson, 1987; Goffman, 1967; Holtgraves, 1999). Half of the indirect utterances represented such emotionally charged face-saving situations, such as attempts not to offend the person asking the question. The other half of the indirect replies represented more neutral situations, in which the speaker's motivation for indirectness was simply to provide more information than just a simple "no." In the indirectness effect, there were activations in the medial prefrontal cortex (mPFC) extending into the right anterior part of the supplementary motor area, and in the right temporoparietal junction (TPJ), a pattern typical for tasks that involve mentalizing based on a theory of mind (ToM; Amodio & Frith, 2006; Mitchell, Macrae, & Banaji, 2006; Saxe, Moran, Scholz, & Gabrieli, 2006). Although the exact role of all the individual ToM regions is not yet clearly established, both mPFC and right TPJ constitute core regions in ToM research (Carrington & Bailey, 2009). The most specific hypothesis about the role of the posterior part of (right) TPJ (Mars et al., 2012) in the mentalizing network is that it is implicated in mental state reasoning, that is, thinking about other people's beliefs, emotions, and desires (Saxe, 2010). Activation in the right TPJ also correlates with severity of autism in a self-other mental state reasoning task (Lombardo, Chakrabarti, Bullmore, & Baron-Cohen, 2011).

The mPFC cortex is a large cortical region with a variety of roles characteristic of social cognition (Amodio & Frith, 2006; Saxe & Powell, 2006). The peaks of the activation in the Bašnáková et al. study (2013) fall in the anterior and posterior rostral divisions, which are associated with complex sociocognitive processes such as mentalizing and thinking about the intentions of others or about oneself (Amodio & Frith, 2006). Interestingly, the involvement of these regions is also consistently observed in discourse comprehension (e.g., Mar, 2011; Mason & Just, 2009). This might come as no surprise, since it is likely that the motivations, goals, and desires of fictional characters are accessed in a similar manner as with real-life protagonists (Mar & Oatley, 2008). In fact, an influential model from the discourse-processing literature (Mason & Just, 2009) ascribes the dorsomedial part of the frontal cortex and the right TPJ a functional role as a *protagonist perspective network*, which generates expectations about how the protagonists of stories will act based on understanding their intentions.

A recent fMRI study on the processing of indirect requests (van Ackeren, Casasanto, Bekkering, Hagoort, & Ruschemeyer, 2012) confirmed the role of the ToM network in inferring speaker meaning. Participants were presented with sentences in the presence of a picture. In one condition, the sentence in combination with the picture could be interpreted as an indirect request for action. For example, the utterance "It is hot here" combined with a picture of a door is likely to be interpreted as a request to open the door. However, the same utterance combined with the picture of a desert will be interpreted as a statement. Van Ackeren et al. found that sentences in the indirect-request condition activated the ToM network much more strongly than the very same sentences in the control conditions with the same pictures and sentences but without the possibility to interpret the sentence in the context of the picture as an indirect request. The recognition of a speech act induced by an utterance in combination with its context requires the inferential machinery instantiated in the ToM network. Interestingly, van Ackeren et al. (2012) also found action-related regions more strongly activated in the indirect request condition. The indirect request for action seems to induce action preparation automatically, even in sentences that do not contain any action words. For a summary of the results, see figure 57.1.

In their fMRI study on conversational implicatures, Jang et al. (2013) manipulated the level of explicitness in question-answer pairs, from very explicit (A: "Is Dr. Smith in his office now?" B: "Dr. Smith is in his office now") to highly implicit (A: "Is Dr. Smith in his office now?" B: "The black car is parked outside the building"). The implicit answers generated stronger activations in mPFC and posterior cingulate cortex. The involvement of these areas can be attributed to the mentalizing operations subserved by the ToM network, although in this study TPJ was not found to be activated in relation to the conversational implicatures. In addition, the angular gyrus and the anterior temporal lobes were more strongly activated, presumably due to the top-down influence on processing of the different concepts in the answers (Binder & Desai, 2011; Patterson, Nestor, & Rogers, 2007).

Another type of conversational implicature is irony. Understanding irony requires inferring the speaker's attitude towards the linguistic expression (e.g., "what a wonderful talk" said ironically to convey that the speaker found the talk quite awful). A handful of studies on irony (for a review, see Spotorno, Koun, Prado, Van der Henst, & Noveck, 2012) reported mPFC involvement. So far, only one study (Spotorno et al., 2012) reported activation in TPJ. Spotorno et al. (2012) had their participants read stories in which the critical sentence was either a literal or an ironic statement. They found stronger activation for the target sentences in the ironic stories compared to the literal stories in all four regions of the ToM network: the right TPJ, the left TPJ, the mPFC, and the precuneus. Interestingly, the authors also report an increased functional connectivity between the ventral mPFC and the left inferior frontal gyrus for the ironic target sentences, suggesting some form of interaction between areas for mentalizing and semantic unification (Hagoort, 2013).

As we discussed above, meaning broadly construed is an iceberg, with the linguistically coded part riding on a large submersed mountain of mutual inference. This mutual inference is also at stake in situations where the linguistic code is absent. Recent studies investigated how human communication develops in a novel communicative action for which the constraints from coded meanings are lacking (Blokpoel et al., 2012; de Ruiter et al., 2010; van Rooij et al., 2011). The results of these studies indicate that human communication relies on inferential mechanisms shared across interlocutors (Stolk et al., 2013), rather than on sensorimotor brain-to-brain couplings (Rizzolatti & Graighero, 2007). It was found that the creation of novel shared symbols upregulates activity in the right temporal lobe across pairs of communicators, and over temporal scales unrelated to transient sensorimotor events (Stolk et al., 2013). These results suggest that this part of the brain supports the updating of common ground during a communicative interaction.

Although the number of studies on the neuropragmatics of language is still limited, there is a remarkable consistency in the finding that understanding the communicative intent of an utterance requires mentalizing. Since the linguistic code underdetermines speaker meaning, the ToM network needs to be invoked to get from coded meaning to speaker meaning. Despite the great popularity of the view that the mirror neuron system is sufficient for action understanding (Rizzolatti & Sinigaglia, 2010), this system does not provide the crucial neural infrastructure for inferring speaker

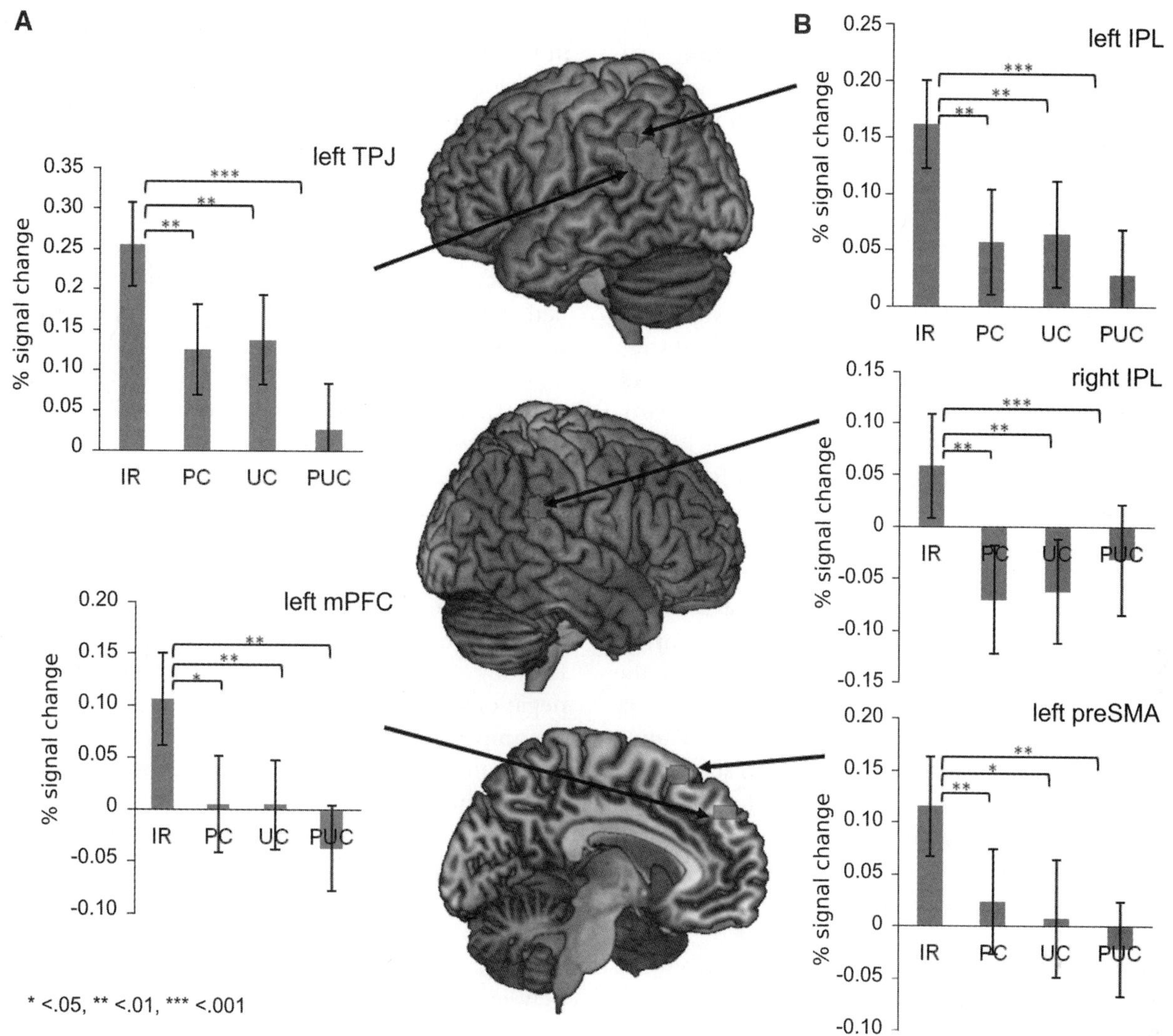

FIGURE 57.1 Regions of interest (ROIs) were interrogated for activations to an indirect request (IR) for action (as in "It is hot here" in the presence of a picture of a window) relative to three control conditions: PC (picture control), UC (utterance control), and BC (baseline control). The image shows all ROIs superimposed on a brain template. The bar diagrams illustrate mean percent signal change for each condition. The error bars depict the standard error. (A) Green ROIs show regions from the theory of mind (ToM) localizer (medial prefrontal cortex, mPFC; and temporoparietal junction, TPJ). (B) Red ROIs refer to regions that were activated during action execution (pre-supplementary motor area, preSMA; bilateral inferior parietal lobule, IPL; van Ackeren et al., 2012; reprinted with permission). (See color plate 50.)

meaning. Next to core areas for retrieving lexical information from memory and unification of the lexical building blocks in producing and understanding multiword utterances, other brain networks are needed to realize language-driven communication to its full extent.

The immediacy of pragmatic processing

A central issue of debate is to what extent pragmatic inferences are generated automatically and by default (Levinson, 2000) or, alternatively, are linked with processing effort and increased processing time, as is claimed to follow from the relevance theory of Sperber and Wilson (1995; cf. Noveck & Reboul, 2008). The area in which the alternative accounts are tested most explicitly is that of scalar implicatures. For example, the quantifier *some*, although logically equivalent to *all*, is pragmatically often interpreted as *some but not all* (see above). The speaker is assumed to use *some* for a reason, which is to convey information that would not be provided by *all* (Grice's maxim of quantity). The question is whether scalar implicatures are computed by default, or only if licensed by context. In the first case, a term such as *some* will automatically be interpreted pragmatically, that is, as *some but not all*. In the latter case, this will need to be induced by context. So far, most studies on the processing of scalar implicatures were behavioral. Only a few studies have exploited electrophysiological brain responses to investigate the processing of scalar implicatures (Hartshorne, Snedeker, & Kim, 2013; Nieuwland, Ditman, & Kuperberg, 2010; Noveck & Posada, 2003). Nieuwland et al. (2010) investigated how quickly pragmatic knowledge is recruited during informative and underinformative usage of quantifiers. In their study, they compared the ERPs to critical words in sentences with an underinformative usage of *some* (e.g., *"Some people have lungs, ..."*) to those with an informative usage (e.g., *"Some people have pets, ..."*). The results showed quite some individual variation. Participants who scored high on scales for pragmatic competence showed an N400 effect to the target words (*lungs/pets*) in the underinformative compared to the informative statements (i.e., to *lungs* vs. *pets*). Based on the pragmatic interpretation of *some* (*some but not all*), they showed an immediate N400 response when this reading was violated (as in the case of "lungs"). No such effect was observed in participants who scored low on pragmatic ability. This result suggests that scalar implicatures can be immediately incorporated during sentence comprehension, albeit that this is done with quite a bit of individual variation.

Hartshorne, Snedeker, and Kim (2013) introduced an additional control, using "only." In "*Only some politicians are corrupt,*" the semantic reading that in fact all politicians are corrupt is excluded. This is different for *"Some politicians are corrupt,"* which could be extended as follows, *"Some politicians, in fact all politicians, are corrupt."* With the addition of this control condition, they compared declarative sentences such as (1) "Addison ate some of the cookies before breakfast this morning, and *the rest* are on the counter," with a conditional version as in (2) "If Addison ate some of the cookies before breakfast this morning, then *the rest* are on the counter." The noun phrase *the rest* in (1) is only felicitous if Addison has not eaten all of the cookies, which is what the scalar implicature entails. The assumption is that, instead, conditional sentences suppress the implicature (Noveck, Chiercia, Chevaux, Guelminger, & Sylvestre, 2002). The crucial results in this study are the interactions between the "declarative/conditional" and the "some/only some" conditions. The ERP results showed the absence of an interaction to *some*. However, at *the rest* an interaction was observed. The authors conclude that although context did not affect the processing of the quantifier that triggered the scalar implicature (*some*), it did affect the processing of subsequent words in the sentence.

Overall, the results on scalar implicatures indicate that their processing costs are relatively minor, although some contextual modulation can be obtained with quite some individual variation. However, the number of studies is limited, and more definitive answers need to be based on further investigations.

Although scalar implicatures have been used as a test bed for alternative theoretical claims, many other studies investigated the influence of nonlinguistic contexts on sentence comprehension. These included information about the speaker (van Berkum, Van Den Brink, Tesink, Kos, & Hagoort, 2008), co-speech gestures (Özyürek, Willems, Kita, & Hagoort, 2007), and world knowledge (Hagoort, Bastiaansen, Hald, & Petersson, 2004). These studies found the same effects for nonlinguistic compared to linguistic information on the amplitude and latency of ERPs such as the N400. In addition, discourse information (Hald, Steenbeek-Planting, & Hagoort, 2007) and pragmatically licensed negations (Nieuwland & Kuperberg, 2008) influence the amplitude of the N400 at exactly the same latencies as semantic anomalies do. This suggests that both linguistic and nonlinguistic contexts have an immediate impact on the interpretation of an utterance (Hagoort & van Berkum, 2007).

In summary, the empirical evidence is still inconclusive with respect to the speed and automaticity of the inferential steps that are needed to close the gap between coded meaning and speaker meaning (Noveck & Reboul, 2008). Very likely there are developmental and individual differences in the degree to which pragmatic inferences are automatically integrated with lexical, syntactic, and semantic processing operations. At the same time, it is clear that a full understanding of the neurobiological infrastructure for language requires a specification of the neural circuitry that establishes the common ground between speaker and listener (Clark, 1996) and that gets us from coded meaning to speaker meaning.

Conclusion

The neurobiology of language has focused mainly on lexical and sublexical processes, and beyond the single-word level, especially on syntactic operations. Pragmatic aspects of language have become part of the neurobiological agenda only recently (cf. van Berkum, 2009). As we have shown, this is a necessary step if we want to understand how linguistic codes get their communicative value. Moreover, integration of neuropragmatics into the overall picture is helpful for avoiding theoretical pitfalls, such as the idea that sensorimotor simulation and the mirror neuron system are even close to being sufficient for language understanding. Finally, in addition to core areas for language in mainly left perisylvian cortex, additional neuronal circuits are involved to get from coded meaning to speaker meaning.

REFERENCES

Amodio, D. M., & Frith, C. D. (2006). Meeting of minds: The medial frontal cortex and social cognition. *Nat Rev Neurosci, 7*(4), 268–277.

Bambini, V. (2010). Neuropragmatics: A foreword. *Ital J Linguist, 22*(1), 1–20.

Bara, B. (2010). *Cognitive pragmatics.* Cambridge, MA: MIT Press.

Bašnáková, J., Weber, K., Petersson, K. M., van Berkum, J., & Hagoort, P. (2013). Beyond the language given: The neural correlates of inferring speaker meaning. *Cereb Cortex.* doi:10.1093/cercor/bht112

Binder, J. R., & Desai, R. H. (2011). The neurobiology of semantic memory. *Trends Cogn Sci, 15*(11), 527–536.

Blokpoel, M., van Kesteren, M., Stolk, A., Haselager, P., Toni, I., & van Rooij, I. (2012). Recipient design in human communication: Simple heuristics or perspective taking? *Front Hum Neurosci, 6*, 253.

Brown, P., & Levinson, S. C. (1987). *Politeness: Some universals in language usage.* Cambridge, UK: Cambridge University Press.

Carrington, S. J., & Bailey, A. J. (2009). Are there theory of mind regions in the brain? A review of the neuroimaging literature. *Hum Brain Mapp, 30*(8), 2313–2335.

Clark, H. H. (1996). *Using language.* Cambridge, UK: Cambridge University Press.

De Ruiter, J. P., Noordzij, M. L., Newman-Norlund, S., Newman-Norlund, R., Hagoort, P., Levinson, S. C., & Toni, I. (2010). Exploring the cognitive infrastructure of communication. *Interact Stud, 11*(1), 51–77.

Goffman, E. (1967). *Interaction ritual: Essays on face-to-face interaction.* New York, NY: Pantheon.

Grice, H. P. (1957). Meaning. *The philosophical review, 66*(3), 377–388.

Grice, H. P. (1975). Logic and conversation. In P. Cole & J. L. Morgan (Eds.), *Syntax and semantics: Speech acts* (Vol. 3, pp. 41–58). New York, NY: Academic Press.

Grice, H. P. (1989). *Studies in the way of words.* Cambridge, MA: Harvard University Press.

Hagoort, P. (2005). On Broca, brain, and binding: A new framework. *Trends Cogn Sci, 9*, 416–423.

Hagoort, P. (2013). MUC (Memory, Unification, Control) and beyond. *Front Psychol, 4*, 416.

Hagoort, P., Hald, L., Bastiaansen, M., & Petersson, K. M. (2004). Integration of word meaning and world knowledge in language comprehension. *Science, 304*, 438–441.

Hagoort, P., & Poeppel, D. (2013). The infrastructure of the language-ready brain. In M. A. Arbib (Ed.), *Language, music, and the brain: A mysterious relationship* (pp. 233–255). Cambridge, MA: MIT Press.

Hagoort, P., & van Berkum, J. (2007). Beyond the sentence given. *Philos Trans R Soc Lond B Biol Sci, 362*(1481), 801–811.

Hald, L. A., Steenbeek-Planting, E. G., & Hagoort, P. (2007). The interaction of discourse context and world knowledge in online sentence comprehension. Evidence from the N400. *Brain Res, 1146*, 210–218.

Hartshorne, J. K., Snedeker, J., & Kim, A. (2013). The neural computation of scalar implicature. Paper presented at the Annual Meeting of the Cognitive Science Society, Berlin, Germany.

Holtgraves, T. (1999). Comprehending indirect replies: When and how are their conveyed meanings activated? *J Mem Lang, 41*, 519–540.

Jang, G., Yoon, S.-A., Lee, S.-E., Park, H., Kim, J., Ko, J. H., & Park, H.-J. (2013). Everyday conversation requires cognitive inference: Neural bases of comprehending implicated meanings in conversations. *NeuroImage, 81*, 61–72.

Levinson, S. C. (1983). *Pragmatics.* Cambridge, UK: Cambridge University Press.

Levinson, S. C. (2000). *Presumptive meanings.* Cambridge, MA: MIT Press.

Lombardo, M. V., Chakrabarti, B., Bullmore, E. T., & Baron-Cohen, S. (2011). Specialization of right temporoparietal junction for mentalizing and its relation to social impairments in autism. *NeuroImage, 56*(3), 1832–1838.

Mar, R. A. (2011). The neural bases of social cognition and story comprehension. *Annu Rev Psychol, 62*, 103–134.

Mar, R. A., & Oatley, K. (2008). The function of fiction is the abstraction and simulation of social experience. *Perspect Psychol Sci, 3*(3), 173–192.

Mars, R. B., Sallet, J., Schuffelgen, U., Jbabdi, S., Toni, I., & Rushworth, M. F. S. (2012). Connectivity-based subdivisions of the human right "temporoparietal junction area": Evidence for different areas participating in different cortical networks. *Cereb Cortex, 22*(8), 1894–1903.

Mason, R. A., & Just, M. A. (2009). The role of the theory-of-mind cortical network in the comprehension of narratives. *Lang Linguist Compass, 3*(1), 157–174.

Mitchell, J. P., Macrae, C. N., & Banaji, M. R. (2006). Dissociable medial prefrontal contributions to judgments of similar and dissimilar others. *Neuron, 50*(4), 655–663.

Nieuwland, M. S., Ditman, T., & Kuperberg, G. R. (2010). On the incrementality of pragmatic processing: An ERP investigation of informativeness and pragmatic abilities. *J Mem Lang, 63*(3), 324–346.

Nieuwland, M. S., & Kuperberg, G. R. (2008). When truth is not too hard to handle. *Psychol Sci, 19*(12), 1213–1218.

Noveck, I. A., & Posada, A. (2003). Characterizing the time course of an implicature: An evoked potentials study. *Brain Lang, 85*(2), 203–210.

Noveck, I. A., & Reboul, A. (2008). Experimental pragmatics: A Gricean turn in the study of language. *Trends Cogn Sci, 12*(11), 425–431.

Noveck, I. A., Chiercia, G., Chevaux, F., Guelminger, R., & Sylvestre, E. (2002). Linguistic-pragmatic factors in interpreting disjunctions. *Think Reasoning, 8*, 297–326.

Özyürek, A., Willems, R. M., Kita, S., & Hagoort, P. (2007). On-line integration of semantic information from speech and gesture: Insights from event-related brain potentials. *J Cogn Neurosci, 4*, 605–616.

Patterson, K., Nestor, P. J., & Rogers, T. T. (2007). Where do you know what you know? The representation of semantic knowledge in the human brain. *Nat Rev Neurosci, 8*(12), 976–987.

Rizzolatti, G., & Craighero, L. (2007). Language and mirror neurons. In M. G. Gaskell (Ed.), *Oxford handbook of psycholinguistics* (pp. 771–785). Oxford, UK: Oxford University Press.

Rizzolatti, G., & Sinigaglia, C. (2010). The functional role of the parieto-frontal mirror circuit: Interpretations and misinterpretations. *Nat Rev Neurosci, 11,* 264–274.

Saxe, R. (2010). The right-temporo-parietal junction: A specific brain region for thinking about thoughts. In A. Leslie & T. German (Eds.), *Handbook of theory of mind.* Philadelphia, PA: Psychology Press.

Saxe, R., Moran, J. M., Scholz, J., & Gabrieli, J. (2006). Overlapping and non-overlapping brain regions for theory of mind and self reflection in individual subjects. *Soc Cogn Aff Neurosci, 1*(3), 229–234.

Saxe, R., & Powell, L. J. (2006). It's the thought that counts: Specific brain regions for one component of theory of mind. *Psychol Sci, 17*(8), 692–699.

Searle, J. R. (1969). *Speech acts: An essay in the philosophy of language.* Cambridge, UK: Cambridge University Press.

Sperber, D., & Wilson, D. (1987). Précis of *Relevance: Communication and cognition. Behav Brain Sci, 10,* 697–710.

Sperber, D., & Wilson, D. (1995). *Relevance: Communication and cognition* (2nd ed.). Oxford, UK: Blackwell.

Spotorno, N., Koun, E., Prado, J., Van Der Henst, J. B., & Noveck, I. A. (2012). Neural evidence that utterance-processing entails mentalizing: The case of irony. *NeuroImage, 63,* 25–39.

Stolk, A., Noordzij, M. L., Verhagen, L., Volman, I., Schoffelen, J.-M., Oostenveld, O., … Toni, I. (2013). Cerebral coherence between communicators marks the emergence of meaning. Paper presented at the 43rd Annual Meeting of the Society for Neuroscience, San Diego, CA.

van Ackeren, M. J., Casasanto, D., Bekkering, H., Hagoort, P., & Rueschemeyer, S.-A. (2012). Pragmatics in action: Indirect requests engage theory of mind areas and the cortical motor network. *J Cogn Neurosci, 24*(11), 2237–2247.

van Berkum, J. J. A. (2009). The neuropragmatics of "simple" utterance comprehension: An ERP review. In K. Yatsushiro & U. Sauerland (Eds.), *Semantics and pragmatics: From experiment to theory.* Basingstroke, U.K.: Palgrave Macmillan.

van Berkum, J. J. A., Van Den Brink, D., Tesink, C., Kos, M., & Hagoort, P. (2008). The neural integration of speaker and message. *J Cogn Neurosci, 20,* 580–591.

van Rooij, I., Kwisthout, J., Blokpoel, M., Szymanik, J., Wareham, T., & Toni, I. (2011). Intentional communication: Computationally easy or difficult? *Front Hum Neurosci, 5,* 1–18.

58 Cultural Recycling of Neural Substrates During Language Evolution and Development

MORTEN H. CHRISTIANSEN AND RALPH-AXEL MÜLLER

ABSTRACT Cultural evolution has emerged as a key source of explanation for the emergence of complex linguistic structure in the human lineage. In this chapter, we argue that the cultural evolution of language has been shaped by nonlinguistic constraints deriving from the human brain. By analogy to reading, novel cortical networks for acquiring and using language are suggested to have emerged through the cultural recycling of preexisting neural substrates. These language networks inherited the structural properties and limitations of their component cortical circuits. In support for this perspective on the neurobiology of language, we discuss evidence regarding the multifunction nature of Broca's area—often considered to be a canonical language region—and the distributed nature of lexicosemantic representations. We conclude by noting that more research is needed to explore how the cultural evolution perspective may provide new insights into the neurobiology of language.

Research on language evolution aims to answer some of the most fundamental questions about the nature of our linguistic abilities: Why is language the way it is, and how did it come to be that way? Fueled by theoretical constraints derived from recent advances in the brain and cognitive sciences, the past couple of decades have seen an explosion of research on language evolution. This research was initially prompted by Pinker and Bloom's (1990) groundbreaking article arguing for the natural selection of biological structures dedicated to language. The new millennium, however, has seen a shift toward explaining language evolution in terms of cultural evolution rather than biological adaptation. Nonetheless, although the cultural evolution of language has had a substantial impact on the cognitive sciences, it has received relatively little attention within cognitive neuroscience (though see, e.g., Arbib, 2010; Deacon, 1997, for exceptions).

In this chapter, we outline how the cultural evolution of language may be consistent with recent thinking about the cognitive neuroscience of language. First, we discuss the logical problem of language evolution faced by theories proposing biological adaptations for arbitrary features of language. As an alternative, we argue that the cultural evolution of language provides a solution to this problem, indicating how we can explain the close fit between the structure of language and the mechanisms employed for acquiring and using language. We then review recent proposals about neuronal recycling and how they may provide a neural foundation for the cultural evolution of language by analogy with a human skill that we know is the product of cultural evolution: reading. Finally, we discuss some of the implications for the neurobiology of language, by highlighting specific nonlinguistic neural substrates that provide the bases upon which language networks emerge during development.

A solution to the logical problem of language evolution: Language shaped by the brain

The acquisition of language is subject to a number of biological, species-specific constraints. After all, only humans have language; no other animal communication system comes close to the complexity and diversity of forms we see in human language (e.g., Evans & Levinson, 2009). A key question is, however, whether these biological constraints necessarily have to be specific to language or whether they may be broader in nature, deriving from constraints on nonlinguistic neural mechanisms that have been pressed into use in language.

A long-standing influential approach is to assume that language acquisition is constrained by a Universal Grammar (UG): a genetic language-specific neural system analogous to the visual system (e.g., Maynard-Smith & Szathmáry, 1997; Pinker, 1997). As such, UG provides a possible explanation for the close fit between the structure of language and how it is acquired and used. But the idea of linguistically driven biological adaptations as the origin of a genetically specified UG faces a *logical problem of language evolution* (Christiansen & Chater, 2008). UG is meant to characterize a set of universal grammatical principles that hold across all languages (e.g., Chomsky, 1981). It is a central assumption that these principles are *arbitrary*, and not determined by functional considerations, such as constraints on learning, memory, cognitive abilities, or communicative effectiveness. This creates an evolutionary problem because any combination of arbitrary principles will be equally adaptive. A possible solution is to construe the principles as constituting a communicative protocol by analogy with inter-computer communication: it does not matter what specific settings (principles) are adopted as long as everyone adopts the same set of settings (Pinker & Bloom, 1990). However, this solution faces three fundamental difficulties relating to the dispersion of human populations, language change, and the question of what is genetically encoded (Christiansen & Chater, 2008).

First, the problem of divergent populations of language users arises across a range of different scenarios concerning language evolution and human migration. In all cases, it would seem that the evolution of UG would require a process of gradual adaptation prior to the dispersion of human populations and an abrupt cessation of such adaptation afterwards to avoid genetic assimilation to diverging local linguistic environments (Baronchelli, Chater, Pastor-Satorras, & Christiansen, 2012). Second, the adaptationist account of UG faces the problem that within a single population, linguistic conventions change much more rapidly than genes, thus creating a "moving target" for natural selection. Computational simulations have shown that under conditions of relatively slow linguistic change, arbitrary principles do not become genetically fixed—even when the genetic makeup of the learners is allowed to affect the direction of linguistic change (Chater, Reali, & Christiansen, 2009). Third, natural selection produces adaptations designed to fit the specific environment in which selection occurs. It is thus puzzling that an adaptation for UG would have resulted in the genetic underpinnings of a system capturing the abstract features of all possible human linguistic environments, rather than fixing the superficial properties of the immediate linguistic environment in which the first language originated.

It remains possible, though, that language did have a substantial impact on human genetic evolution. The above arguments only preclude biological adaptations for arbitrary features of language, whereas there might be features that are universally stable across linguistic environments (such as the need for enhanced memory capacity, or complex pragmatic inferences; Givón & Malle, 2002) that might lead to biological adaptation (Christiansen, Reali, & Chater, 2011). However, these language features are likely to be functional, to facilitate language *use*—and thus would typically not be considered part of UG.

But without UG, how can we explain the apparent close fit between the structure of language and the mechanisms by which it is acquired and used? Instead of asking how the brain may have been adapted for language, we suggest that we may get more insight into language evolution by asking the opposite question: How has language been adapted to the brain? This question highlights the fact that language cannot exist independently of human brains. Without our brains, there would be no language. Thus, there is a stronger selective pressure on language to adapt to the human brain than the other way around. Processes of cultural evolution involving repeated cycles of learning and use are hypothesized to have shaped language into what we can observe today. The solution to the logical problem of language evolution is, then, that cultural evolution has shaped language to fit the human brain (Christiansen & Chater, 2008).

The last decade has seen a growing body of work suggesting that language may have evolved primarily by way of cultural evolution rather than biological adaptation. Evidence in support of this perspective on language evolution comes from computational modeling, behavioral experimentation, linguistic analyses, and many other lines of scientific inquiry (see Dediu et al., 2013, for a review). A key hypothesis emerging from this work is that the cultural evolution of language primarily has been shaped by nonlinguistic constraints deriving from neural mechanisms existing prior to the emergence of language (see Christiansen & Chater, 2008, for a review of the historical pedigree of this perspective). Language is viewed as an evolving complex system in its own right; features that make language easier to learn and use, or are more communicatively efficient, will tend to proliferate, whereas features that hinder communication will tend to disappear (or not come into existence in the first place).

Christiansen and Chater (2008) describe four different types of constraints that act together to shape the cultural evolution of language. One source of constraints derives from the perceptual and motor machinery that supports language. For example, the serial nature of vocal (and sign) production forces a sequential construction of messages with a strong bias toward local information due to the limited capacity of perceptual memory. The nature of our cognitive architecture provides a second type of constraints on the cultural evolution of language through limitations on learning, memory, and processing. Limitations on working memory, for instance, will constrain the number and length of dependencies between nonadjacent elements in a sentence. The structure of our mental representations and reasoning abilities constitutes a third kind of constraints on language evolution. For example, human basic categorization abilities appear to be reflected in the structure of lexical representations. Finally, sociopragmatic considerations provide yet another source of constraints on how language can evolve. As an example, consider how a shared pragmatic context may lighten the informational load on a particular sentence (i.e., it does not have to carry the full meaning by itself). Importantly, these four types of constraints do not act independently of one another; rather, specific linguistic patterns arise from a combination of several of these constraints acting in unison. Individual languages emerge through a gradual historical process of tinkering, recruiting different constellations of constraints, and thus giving rise to the diversity of languages.[1]

The idea of language as shaped by cultural evolution to fit preexisting constraints from the human brain also promises to simplify the problem of language acquisition. When children acquire their native language(s), their biases will be the right biases because language has been optimized by past generations of learners to fit those very biases (Chater & Christiansen, 2010; Zuidema, 2003). This does not, however, trivialize the problem of language acquisition but instead suggests that children tend to make the right guesses about how their language works—not because of an innate UG—but because language has been shaped by cultural evolution to fit the nonlinguistic constraints that they bring to bear on language acquisition. A key remaining question, though, to which we turn next, is whether it is possible to provide a more detailed account of the neural bases supporting the development and cultural evolution of language.

Cultural recycling of neural substrates during development

Over the past decade, a new perspective on the functional architecture of the brain has emerged (see Anderson, 2010, for a review). Instead of viewing various brain regions as being dedicated to broad cognitive domains such as language, vision, memory, or reasoning, it is proposed that low-level neural circuits that have evolved for one specific purpose are redeployed as part of another neuronal network to accommodate a new function. This general perspective has been developed independently in a number of different theoretical proposals, including the "neural exploitation" theory (Gallese, 2008), the "shared circuits model" (Hurley, 2008), the "neuronal recycling" hypothesis (Dehaene & Cohen, 2007), and the "massive redeployment hypothesis" (Anderson, 2010). The basic premise is that reusing existing neural circuits to accomplish a new function is more likely from an evolutionary perspective than evolving a completely new circuit *de novo* (cf. Jacob, 1977)

If this hypothesis is correct, we should expect most brain areas to participate in multiple, potentially diverse behavioral functions. Supporting this prediction, Anderson (2010) reviews results from 1,469 subtraction-based fMRI studies involving eleven different task domains, ranging from action execution, vision, and attention to memory, reasoning, and language, finding that any given cortical region is typically active for most of these task domains. That is, a specific neural circuit that is active in a particular cognitive task, such as language, is generally also active for multiple other tasks.

The cultural recycling hypothesis further predicts that cognitive functions that have emerged more recently in human evolution should be more widely distributed across the cerebral cortex than older ones. This is because these more recent traits will be able to rely on a wider variety of cortical circuits with different, potentially useful properties in order to produce the optimal network for this novel function, and there is no *a priori* reason that these neural circuits should be placed next to one another (Anderson, 2010). Thus, if the neural mechanisms involved in language are primarily the product of recycling of older neural substrates, as proposed by cultural evolution theorists, then we would expect to find the brain areas involved in language to be widely distributed across the brain. Analyzing the co-activation of Brodmann areas for eight different task domains in 472 fMRI experiments, Anderson (2008) found that language was the task domain for which co-activation patterns were the most widely scattered across the brain. Following language in terms of the degree of distribution of neural co-activation patterns came reasoning, memory, emotion, mental imagery, visual perception, action and, lastly, attention. Indeed, language was significantly more widely distributed than the latter three task domains: visual perception, action, and attention.

Importantly, as existing neural circuits take on new roles by participating in new networks to accommodate

[1]A possible reason for why extant non-human primates do not have language may be that humans have gone through a number of biological adaptations, most of which are not specific to language, but which provided the right kind of perceptuo-motor, cognitive, conceptual, and socio-pragmatic foundations for language to "take off" by way of cultural evolution.

novel functions, they retain their original function (though, the latter may in some cases be affected by properties of the new function through developmental processes[2]). The limitations and computational constraints of the original workings of those circuits will therefore be inherited by the new function, creating a "neuronal niche" (Dehaene & Cohen, 2007) for cultural evolution. In other words, the emerging new function will be shaped by constraints deriving from the recycled neural circuits as it evolves culturally. Thus, this is the sense in which we argue that language has been shaped by the brain through the cultural recycling of preexisting neural substrates.

Reading as a Product of Cultural Recycling Writing systems are only about 7000 years old and for most of this time the ability to read and write was confined to a small group of individuals. Thus, reading is a culturally evolved ability for which humans would be unlikely to have any specialized biological adaptations. This makes reading a prime candidate for a cognitive skill that is the product of cultural recycling of prior neural substrates.

Dehaene and Cohen (2007) argue that skills resulting from culturally mediated neuronal recycling, such as reading, should have certain characteristics. First, variability in the neural representations of the skill should be limited across individuals and cultures. With regard to reading, the visual word form area, which is located in the left occipito-temporal sulcus, has been consistently associated with word processing across different individuals and writing systems. Second, there should be considerable similarity across cultures in the manifestation of the skill itself. Consistent with this prediction, Dehaene and Cohen (2007) note that individual characters in writings systems across the world consist of an average of three strokes, and the intersection contours of the parts of these characters follow the same frequency distribution (e.g., T, Y, Z, Δ). Third, there should be some continuity in terms of both neural biases and abilities for learning in nonhuman primates. That reading might build (at least in part) on the recruitment of evolutionary older mechanism for object recognition is supported by recent results from a study of orthographic processing in baboons (Grainger et al., 2012) indicating that they were able to distinguish English words from nonsense words.

The available data regarding the neural representation of reading, combined with analyses of writing systems and experiments with nonhuman primates, suggest that writing systems have been shaped by a neuronal niche that includes the left ventral occipito-temporal cortex. Next, we extend this argument to include language more generally, outlining the neuronal niche within which language has evolved by cultural evolution.

Nonlinguistic neural constraints on language

Evidence predominantly drawn from functional neuroimaging in adults supports the hypothesis of language having adapted to the brain. We discuss two important sources of evidence, one related to the functional diversity of Broca's area, the other to the distributed brain organization for lexicosemantic representations.

Broca's area (comprising Brodmann areas [BAs] 44 and 45 in the left inferior frontal gyrus [LIFG]) is considered crucial among the "language regions" of the human brain (Price, 2010). Some proposals even assign an exclusive linguistic or syntactic role to LIFG (e.g., Grodzinsky, 2000). As described above, however, exclusive language specialization would be unexpected from an evolutionary perspective. Such proposals also overlook extensive neuroimaging evidence (reviewed in Müller, 2009). Here, we focus on one example, the role of LIFG in motor-related processing and action perception.

Outside neurolinguistics, Broca's area is often considered a premotor (rather than a language) region (e.g., Curtis & D'Esposito, 2003). Indeed, postmortem cellular evidence shows that BA 44 is "dysgranular" cortex (containing few cells with sensory afferents in cortical layer IV), a feature shared with primary motor cortex (Amunts et al., 1999). Unsurprisingly, the role of LIFG in language has been related to its motor specialization (Rizzolatti, Fogassi, & Gallese, 2002). Relevant evidence originates from monkey studies (Rizzolatti & Gentilucci, 1988). Specifically, neurons in monkey area F5, which may correspond to human BA 44 (Rizzolatti & Arbib, 1998), respond to object-directed action when presented visually or auditorily, *in the absence* of any motor response (Kohler et al., 2002). Since these *mirror neurons* are not directly involved in motor execution, they are considered important for imitation as well as for detecting and recognizing the actions of others (Rizzolatti & Craighero, 2004). These functions are supported by connectivity between ventral premotor and inferior parietal cortex (Fabbri-Destro & Rizzolatti, 2008), with possible additional participation of the superior temporal sulcus (Rizzolatti & Craighero, 2004).

[2]For example, exposure to the specific patterns of occurrence of center-embedded clauses in German appears to affect sequential learning of nonadjacent dependencies more generally (de Vries et al., 2012).

Ample imaging evidence suggests that the mirror neuron system (MNS) exists in the human brain in regions corresponding to those identified in monkey studies, that is, bilateral IFG and inferior parietal cortex (Molenberghs, Cunnington, & Mattingley, 2012), possibly suggesting some role in the emergence of language (Arbib, 2010; Rizzolatti & Craighero, 2004). From this perspective, the existence of mirror neurons in Broca's area is not a coincidence, but indicates a crucial role for imitation and action recognition as building blocks of language (Nishitani, Schurmann, Amunts, & Hari, 2005). In the framework of the cultural recycling model, this implies that LIFG is not a "language area" that happens to also play a role in other, apparently nonlinguistic functions. Rather, LIFG had initially developed action-related functions, which secondarily made it suited for its role in language emergence. Indeed, it has been suggested that LIFG's action-related functions develop early in infancy through associative learning (Cook et al., 2014) prior to the development of language. This view of sequence and causality can be applied both to child development, where action recognition and imitation may be considered building blocks of language acquisition (Glenberg & Gallese, 2012), and to evolution (Corballis, 2010), where the existence of the MNS in "preverbal" nonhuman primates is known (as described above).[3] However, while the MNS and the imitative abilities and action recognition it affords may be necessary phylo- and ontogenetic conditions for the emergence of language, they cannot be sufficient: Macaque monkeys (as studied by Rizzolatti and colleagues) possess an MNS but never developed language. Language evolution must therefore rely on other building blocks with distributed brain organization other than imitation and action recognition, highlighting the role of connectivity.

The relevance of connectivity for functional specialization in cortical regions, such as LIFG, is fundamental. As proposed by Passingham, Stephan, and Kotter (2002), the functional role of each brain region may be largely determined by its afferent and efferent connectivity patterns, that is, by which other brain regions it "hears from" (receives synaptic input from) and "talks to" (sends axons with synaptic terminals to). The connectivity-based principle implies that local specializations in cortex are not arbitrary. This is well understood for sensorimotor regions (e.g., primary visual cortex has visual functions because it receives input from the retina via the thalamus), but specializations of association cortices are often not understood in analogous ways. For example, why is a major "language region" located in LIFG?

As mentioned, a neural action recognition system is not a sufficient condition for language. Correspondingly, the connectivity of IFG is far more complex than described above in the context of action recognition (Anwander, Tittgemeyer, von Cramon, Friederici, & Knosche, 2007). While the arcuate fasciculus indeed connects IFG with inferior parietal and lateral temporal regions in posterior perisylvian cortex (Catani, Jones, & Ffytche, 2005), the functional relevance of these connections goes beyond those ascribed to the MNS, relating, for example, to spatial processing and attention (Sack, 2009) and auditory processing, auditory-visual integration, and face processing (Hein & Knight, 2008). IFG furthermore connects with both the dorsal stream, crucial for visuospatial processing and visuomotor coordination (Goodale & Westwood, 2004), and the ventral stream (Saur et al., 2008), which provides meaningful interpretation of visual and auditory stimuli (Grill-Spector & Malach, 2004). Although not all of these functions may appear immediately relevant to language emergence, the status of Broca's area as a convergence zone (Mesulam, 1998; Meyer & Damasio, 2009), where connections with numerous other sensorimotor and association cortices come together, provides a promising neuroscientific account of its crucial role in language.

Related to connectivity-based functional specialization, a second example of "language adapting to the brain" concerns lexicosemantic organization. Beyond LIFG and its crucial role in lexicosemantic representations (Binder, Desai, Graves, & Conant, 2009), the distributed organization of semantic representations is reflected in the principle of category-specificity, which was first observed in patients with semantic deficits that differentially affect specific classes of objects. Some patients, for example, show dissociations between impaired animate and retained inanimate items (Mahon & Caramazza, 2009). Warrington and McCarthy (1987) point out that sensory features are important for distinguishing between living items, while action semantics are more important for inanimate items, like tools; loss of sensory or action knowledge could therefore differentially disrupt the semantic representations of living and nonliving items, respectively. Crucially, category-specificity does not reflect impaired processing of sensory input or motor output, but impairment at the conceptual level (Mahon & Caramazza, 2009).

The clinical evidence of category-specificity is supported by imaging findings in healthy adults. For

[3] Importantly, though, we see this perspective as being agnostic with regard to the question of whether language originated in the gestural or vocal modality.

example, the processing of semantic representations related to action and function is associated with activation in left (pre)motor cortex (Chao, Weisberg, & Martin, 2002; Goldberg, Perfetti, & Schneider, 2006; Lubrano, Filleron, Demonet, & Roux, 2014) and left posterior middle temporal cortex (Chao, Haxby, & Martin, 1999; Hwang, Palmer, Basho, Zadra, & Müller, 2009)—the latter being important for nonbiological object motion perception (Beauchamp, Lee, Haxby, & Martin, 2003). Conversely, animate categories are linked to activations in visual cortices, such as the fusiform gyrus (Chao et al., 2002) and the superior temporal sulcus (Chao et al., 1999; Tyler et al., 2003), an area involved in the perception of biological motion (Pelphrey, Morris, Michelich, Allison, & McCarthy, 2005).

A meta-analysis (Chouinard & Goodale, 2010) corroborated activation differences for naming of animals (temporooccipital regions) vs. naming of tools (left prefrontal, premotor, and somatosensory regions), highlighting the role of sensorimotor cortices in lexicosemantic representations. Examples are premotor and primary motor cortex (Chao & Martin, 2000; Hauk, Johnsrude, & Pulvermüller, 2004), visual cortices in fusiform gyrus and occipital lobe (Goldberg et al., 2006; Pulvermüller & Hauk, 2006), and orbitofrontal olfactory regions (Goldberg et al., 2006; Gonzalez et al., 2006). Action words related to different body parts (face, arms, legs) are associated with patterns of activation corresponding to the somatotopic organization in primary motor and somatosensory cortices (Carota, Moseley, & Pulvermüller, 2012), further supporting the sensorimotor bases of lexicosemantic representations.

Several theoretical models have been developed to account for the lesion and imaging evidence. According to sensory/functional (Warrington & McCarthy, 1987) and sensorimotor models (Martin, 2007), semantic representations are distributed throughout the brain as a reflection of sensorimotor processes crucial to their *acquisition*, with weighted participation based on the relative importance of each sensorimotor region. Alternatively, the domain-specific hypothesis (Caramazza & Mahon, 2003) proposes that differential brain organization is based on evolutionary importance—for example, animals forming a separate category because they may be predators or prey and are thus crucial for survival. However, both of these approaches, as well as the related theory of "grounded cognition" (Barsalou, 2008), are in agreement that lexicosemantic representations and the underlying object knowledge do not constitute separate brain systems but are intimately tied to sensorimotor systems. This implies a hierarchical principle, but one without strict division between sensory and conceptual realms.

Semantic representations (possibly with the exception of abstract words; see Shallice & Cooper, 2013) can thus be considered highly complex sets of sensorimotor-based representations, whose complexity is reflected in distributed brain organization. This implies that the lexicosemantic system adapts to preexisting brain systems that support sensorimotor and other nonverbal functions. It also illustrates biological and evolutionary economy. Thus, the emergence of language in hominid evolution did not require the launch of an entirely novel set of brain "modules," as suggested by Chomsky (e.g., 1972) and followers (Fodor, 1983), but made use of existing neural machinery, that is, brain systems for sensation, perception, and motor functions.

Conclusion

Much recent work on the evolution of language has focused on the role of cultural transmission across language learners and users in the emergence of complex linguistic structure. This work has suggested that much of language may have been shaped by neural mechanisms predating the origin of language (e.g., Christiansen & Chater, 2008). In this chapter, we have sought to understand the evolution of language in terms of the cultural recycling of neural substrates, suggesting that language may have "recruited" preexisting networks in development to support the evolution of various language functions. Just as our reading ability relies on a network of cortical circuits that existed before the invention of writing systems, so—we argue—has language largely come to rely on networks involving brain mechanisms not dedicated to language. However, much work still needs to be done and we hope that the present chapter might serve as a starting point for future cognitive neuroscience research on the cultural evolution of language.

REFERENCES

Amunts, K., Schleicher, A., Burgel, U., Mohlberg, H., Uylings, H. B., & Zilles, K. (1999). Broca's region revisited: Cytoarchitecture and intersubject variability. *J Comp Neurol, 412*(2), 319–341.

Anderson, M. L. (2008). Circuit sharing and the implementation of intelligent systems. *Connect Sci, 20*, 239–251.

Anderson, M. L. (2010). Neural reuse: A fundamental organizational principle of the brain. *Behav Brain Sci, 33*, 245–313.

Anwander, A., Tittgemeyer, M., von Cramon, D. Y., Friederici, A. D., & Knosche, T. R. (2007). Connectivity-based parcellation of Broca's area. *Cereb Cortex, 17*(4), 816–825.

Arbib, M. A. (2010). Mirror system activity for action and language is embedded in the integration of dorsal and ventral pathways. *Brain Lang, 112*(1), 12–24.

Baronchelli, A., Chater, N., Pastor-Satorras, R., & Christiansen, M. H. (2012). The biological origin of linguistic diversity. *PLoS ONE, 7*(10), e48029.

Barsalou, L. W. (2008). Grounded cognition. *Annu Rev Psychol, 59,* 617–645.

Beauchamp, M. S., Lee, K. E., Haxby, J. V., & Martin, A. (2003). FMRI responses to video and point-light displays of moving humans and manipulable objects. *J Cogn Neurosci, 15*(7), 991–1001.

Binder, J. R., Desai, R. H., Graves, W. W., & Conant, L. L. (2009). Where is the semantic system? A critical review and meta-analysis of 120 functional neuroimaging studies. *Cereb Cortex, 19*(12), 2767–2796.

Caramazza, A., & Mahon, B. Z. (2003). The organization of conceptual knowledge: The evidence from category-specific semantic deficits. *Trends Cogn Sci, 7*(8), 354–361.

Carota, F., Moseley, R., & Pulvermüller, F. (2012). Body-part-specific representations of semantic noun categories. *J Cogn Neurosci, 24*(6), 1492–1509.

Catani, M., Jones, D. K., & Ffytche, D. H. (2005). Perisylvian language networks of the human brain. *Ann Neurol, 57*(1), 8–16.

Chao, L. L., & Martin, A. (2000). Representation of manipulable man-made objects in the dorsal stream. *NeuroImage, 12*(4), 478–484.

Chao, L. L., Haxby, J. V., & Martin, A. (1999). Attribute-based neural substrates in temporal cortex for perceiving and knowing about objects. *Nat Neurosci, 2*(10), 913–919.

Chao, L. L., Weisberg, J., & Martin, A. (2002). Experience-dependent modulation of category-related cortical activity. *Cereb Cortex, 12*(5), 545–551.

Chater, N., & Christiansen, M. H. (2010). Language acquisition meets language evolution. *Cognitive Sci, 34,* 1131–1157.

Chater, N., Reali, F., & Christiansen, M. H. (2009). Restrictions on biological adaptation in language evolution. *Proc Natl Acad Sci USA, 106,* 1015–1020.

Chomsky, N. (1972). *Language and mind.* New York, NY: Harcourt.

Chomsky, N. (1981). *Lectures on government and binding.* New York, NY: Foris.

Chouinard, P. A., & Goodale, M. A. (2010). Category-specific neural processing for naming pictures of animals and naming pictures of tools: An ALE meta-analysis. *Neuropsychologia, 48*(2), 409–418.

Christiansen, M. H., & Chater, N. (2008). Language as shaped by the brain. *Behav Brain Sci, 31,* 489–558.

Christiansen, M. H., Reali, F., & Chater, N. (2011). Biological adaptations for functional features of language in the face of cultural evolution. *Hum Biol, 83,* 247–259.

Cook, R., Bird, G., Catmur, C., Press, C., & Heyes, C. (2014). Mirror neurons: From origin to function. *Behav Brain Sci, 37,* 177–241.

Corballis, M. C. (2010). Mirror neurons and the evolution of language. *Brain Lang, 112*(1), 25–35.

Curtis, C. E., & D'Esposito, M. (2003). Persistent activity in the prefrontal cortex during working memory. *Trends Cogn Sci, 7*(9), 415–423.

De Vries, M. H., Geukes, S., Zwitserlood, P., Petersson, K. M., & Christiansen, M. H. (2012). Processing multiple non-adjacent dependencies: Evidence from sequence learning. *Philos Trans R Soc Lond B Biol Sci, 367,* 2065–2076.

Deacon, T. W. (1997). *The symbolic species: The co-evolution of language and the brain.* New York, NY: Norton.

Dediu, D., Cysouw, M., Levinson, S. C., Baronchelli, A., Christiansen, M. C., Croft, W., ... Lieven, E. (2013). Cultural evolution of language. In P. J. Richerson & M. H. Christiansen (Eds.), *Cultural evolution: Society, technology, language and religion* (pp. 303–332). Cambridge, MA: MIT Press.

Dehaene, S., & Cohen, L. (2007). Cultural recycling of cortical maps. *Neuron, 56,* 384–398.

Evans, N., & Levinson, S. (2009). The myth of language universals: Language diversity and its importance for cognitive science. *Behav Brain Sci, 32,* 429–492.

Fabbri-Destro, M., & Rizzolatti, G. (2008). Mirror neurons and mirror systems in monkeys and humans. *Physiology (Bethesda), 23,* 171–179.

Fodor, J. A. (1983). *The modularity of mind.* Cambridge, MA: MIT Press.

Gallese, V. (2008). Mirror neurons and the social nature of language: The neural exploitation hypothesis. *Soc Neurosci, 3,* 317–333.

Givón, T., & Malle, B. F. (Eds.). (2002). *The evolution of language out of pre-language.* Amsterdam: Benjamins.

Glenberg, A. M., & Gallese, V. (2012). Action-based language: A theory of language acquisition, comprehension, and production. *Cortex, 48*(7), 905–922.

Goldberg, R. F., Perfetti, C. A., & Schneider, W. (2006). Perceptual knowledge retrieval activates sensory brain regions. *J Neurosci, 26*(18), 4917–4921.

Gonzalez, J., Barros-Loscertales, A., Pulvermüller, F., Meseguer, V., Sanjuan, A., Belloch, V., & Avila, C. (2006). Reading cinnamon activates olfactory brain regions. *NeuroImage, 32*(2), 906–912.

Goodale, M. A., & Westwood, D. A. (2004). An evolving view of duplex vision: Separate but interacting cortical pathways for perception and action. *Curr Opin Neurobiol, 14*(2), 203–211.

Grainger, J., Dufau, S., Montant, M., Ziegler, J. C., & Fagot, J. (2012). Orthographic processing in baboons (*Papio papio*). *Science, 336,* 245–248.

Grill-Spector, K., & Malach, R. (2004). The human visual cortex. *Annu Rev Neurosci, 27,* 649–677.

Grodzinsky, Y. (2000). The neurology of syntax: Language use without Broca's area. *Behav Brain Sci, 23*(1), 1–71.

Hauk, O., Johnsrude, I., & Pulvermüller, F. (2004). Somatotopic representation of action words in human motor and premotor cortex. *Neuron, 41*(2), 301–307.

Hein, G., & Knight, R. T. (2008). Superior temporal sulcus—It's my area: Or is it? *J Cogn Neurosci, 20*(12), 2125–2136.

Hurley, S. L. (2008). The shared circuits model (SCM): How control, mirroring, and simulation can enable imitation, deliberation, and mindreading. *Behav Brain Sci, 31,* 1–58.

Hwang, K., Palmer, E. D., Basho, S., Zadra, J. R., & Müller, R.-A. (2009). Category-specific activations during word generation reflect experiential sensorimotor modalities. *NeuroImage, 48,* 717–725.

Jacob, F. (1977). Evolution and tinkering. *Science, 196,* 1161–1166.

Kohler, E., Keysers, C., Umilta, M. A., Fogassi, L., Gallese, V., & Rizzolatti, G. (2002). Hearing sounds, understanding actions: Action representation in mirror neurons. *Science, 297*(5582), 846–848.

Lubrano, V., Filleron, T., Demonet, J. F., & Roux, F. E. (2014). Anatomical correlates for category-specific naming of objects and actions: A brain stimulation mapping study. *Hum Brain Mapp, 35*(2), 429–443.

Mahon, B. Z., & Caramazza, A. (2009). Concepts and categories: A cognitive neuropsychological perspective. *Annu Rev Psychol, 60,* 27–51.

Martin, A. (2007). The representation of object concepts in the brain. *Annu Rev Psychol, 58,* 25–45.

Maynard-Smith, J., & Szathmáry, E. (1997). *Major transitions in evolution.* New York, NY: Oxford University Press.

Mesulam, M.-M. (1998). From sensation to cognition. *Brain, 121,* 1013–1052.

Meyer, K., & Damasio, A. (2009). Convergence and divergence in a neural architecture for recognition and memory. *Trends Neurosci, 32*(7), 376–382.

Molenberghs, P., Cunnington, R., & Mattingley, J. B. (2012). Brain regions with mirror properties: A meta-analysis of 125 human fMRI studies. *Neurosci Biobehav Rev, 36*(1), 341–349.

Müller, R.-A. (2009). Language universals in the brain: How linguistic are they? In M. H. Christiansen, C. Collins, & S. Edelman (Eds.), *Language universals* (pp. 224–252). Oxford, UK: Oxford University Press.

Nishitani, N., Schurmann, M., Amunts, K., & Hari, R. (2005). Broca's region: From action to language. *Physiology (Bethesda), 20,* 60–69.

Passingham, R. E., Stephan, K. E., & Kotter, R. (2002). The anatomical basis of functional localization in the cortex. *Nat Rev Neurosci, 3*(8), 606–616.

Pelphrey, K. A., Morris, J. P., Michelich, C. R., Allison, T., & McCarthy, G. (2005). Functional anatomy of biological motion perception in posterior temporal cortex: An fMRI study of eye, mouth and hand movements. *Cereb Cortex, 15*(12), 1866–1876.

Pinker, S. (1997). *How the mind works.* New York, NY: Norton.

Pinker, S., & Bloom, P. (1990). Natural language and natural selection. *Brain Behav Sci, 13,* 707–727.

Price, C. J. (2010). The anatomy of language: A review of 100 fMRI studies published in 2009. *Ann NY Acad Sci, 1191,* 62–88.

Pulvermüller, F., & Hauk, O. (2006). Category-specific conceptual processing of color and form in left frontotemporal cortex. *Cereb Cortex, 16*(8), 1193–1201.

Rizzolatti, G., & Arbib, M. A. (1998). Language within our grasp. *Trends Neurosci, 21*(5), 188–194.

Rizzolatti, G., & Craighero, L. (2004). The mirror-neuron system. *Annu Rev Neurosci, 27,* 169–192.

Rizzolatti, G., Fogassi, L., & Gallese, V. (2002). Motor and cognitive functions of the ventral premotor cortex. *Curr Opin Neurobiol, 12*(2), 149–154.

Rizzolatti, G., & Gentilucci, M. (1988). Motor and visual-motor functions of the premotor cortex. In P. Rakic & W. Singer (Eds.), *Neurobiology of neocortex* (pp. 269–284). New York, NY: Wiley.

Sack, A. T. (2009). Parietal cortex and spatial cognition. *Behav Brain Res, 202*(2), 153–161.

Saur, D., Kreher, B. W., Schnell, S., Kummerer, D., Kellmeyer, P., Vry, M. S., ... Weiller, C. (2008). Ventral and dorsal pathways for language. *Proc Natl Acad Sci USA, 105*(46), 18035–18040.

Shallice, T., & Cooper, R. P. (2013). Is there a semantic system for abstract words? *Front Hum Neurosci, 7,* 175.

Tyler, L. K., Bright, P., Dick, E., Tavares, P., Pilgrim, L., Fletcher, P., ... Moss, H. (2003). Do semantic categories activate distinct cortical regions? Evidence for a distributed neural semantic system. *Cogn Neuropsychol, 20*(3), 541–559.

Warrington, E. K., & McCarthy, R. A. (1987). Categories of knowledge. Further fractionations and an attempted integration. *Brain, 110*(Pt. 5), 1273–1296.

Zuidema, W. (2003). How the poverty of the stimulus solves the poverty of the stimulus. In S. Becker, S. Thrun, & K. Obermayer (Eds.), *Advances in neural information processing systems 15*(pp. 51–58). Cambridge, MA: MIT Press.

59 The Emerging Standard Model of the Human Decision-Making Apparatus

PAUL GLIMCHER

ABSTRACT Just over a decade ago, neurobiologists knew almost nothing about the neural mechanisms of voluntary choice. Today, the basic outlines of the human and monkey systems for making decisions are now clearly beginning to emerge. Speaking broadly, it is now known that a set of frontal cortical and basal ganglia store and represent the idiosyncratic values we place on all kinds of things. This value system, a complex of many areas, synthesizes a single common representation of the values of all of the many choice options that we face at any given moment. These so-called subjective values are then sent to a set of frontal and parietal areas that actually perform the choice process. This chapter provides an outline of the human and monkey value and choice systems.

Over the course of the past decade, enormous progress has been made toward understanding the basic mechanism by which the human brain makes choices. This review focuses on what is known about the mechanisms we employ when choosing between goods or options that have different intrinsic values to us. This kind of decision making is known within neuroscientific circles as *value-based* decision making. Contemporary studies suggest that we can think of value-based decision making as reflecting two sequentially arranged neural mechanisms: a *valuation mechanism* that learns, stores, and retrieves the values of goods or actions under consideration, and a *choice mechanism* that takes the output of the valuation circuit as its starting point and generates a choice from among those options. This segregation of the decision-making system into two neat components is part pedagogy and part reality (Padoa-Schioppa, 2011), but it serves as a starting point for understanding how value-based choice arises in the brain.

Value-based studies of decision making in neuroscience emerged originally from studies of movement control. Scholars of movement control had long been interested in how animals choose what movement to make: if I can look at (or reach toward) any point in extra-personal space, how do I select one movement from the set of all of those possible movements? By the early 1990s, movement scholars had begun to turn toward older economic theories like expected utility theory (von Neumann & Morgenstern, 1944) in an effort to understand this process. Unlike those studying perceptual decision making who approached choice from the realm of sensation (Newsome, Britten, & Movshon, 1989; Shadlen, Britten, Newsome, & Movshon, 1996), these scholars found themselves trying to understand the idiosyncratic "preferences" that guide human decision making from a cognitive neuroscientific point of view. This led them to search for hidden internal representations that might guide decision-making processes within the central nervous system.

Economists had a long history of developing theories that specified the minimally complex internal representation of preferences that could, in principle, account for a given set of observed decisions (Houthakker, 1950; McFadden, 2005), and this group of cognitive neuroscientists took this as a starting point (Platt & Glimcher, 1999). They hypothesized that economic theories might define the internal variables used by the nervous system to guide choice. What resulted from that hypothesis was a new discipline at the borders of neuroscience, psychology, and economics: neuroeconomics. This chapter reviews one of the central threads in that discipline, which is often called the *standard model.*

An overview of the standard model for value-based decision making

THE CHOICE CIRCUIT We begin our overview by examining the neural circuits that actually make choices, once an internal representation of the values of the options under consideration has already been expressed with the nervous system. For an economist, it is natural to assume that decision makers hold some "internal" representation of the subjective values of the goods or actions under consideration. This internal representation, if it is to support logically consistent comparisons, must by definition lie on a single common scale (Glimcher, 2011; Sugrue, Corrado, & Newsome, 2005), and we begin by making the assumption that such a representation exists. (We turn to the details of that representation, and the assumptions it reflects, in the next section.)

For a neoclassical economist, choosers "choose" simply by executing the mathematical operation that identifies the option in the current choice set that has the highest value on that common scale—the *argmax* operation. For a biologist, however, this argmax process is anything but trivial, and serves as our current focus. How does the algorithmic machine of the human brain encode the values of items in the current choice set? How does it deliberate so as to identify the most highly valued option within that representation? And how does the well-studied stochasticity of the mammalian brain (Churchland et al., 2010; Glimcher, 2005; Tolhurst, Movshon, & Dean, 1983) interact with this process in a way that accounts for the stochasticity observed and modeled behaviorally by economists like the Nobel laureate Daniel McFadden (2005; see also Loomes, Graham, & Sugden, 1995)?

Answering these three questions—how are subjective values (the neural correlates of the economic objects called *utilities*) encoded in the choice circuit? how is the argmax operation performed? and how does stochasticity arise in the nervous system?—has been a principle goal of a number of laboratories over the course of the last two decades, and so we turn to each of these questions in turn. It is, however, worth noting that most work on these particular questions has been conducted in the brains of nonhuman primates because of current technological limitations. As a result, it is through studies of the monkey that most of our understanding of the choice process (as opposed to the valuation process, which has been extensively studied in both humans and other animals) has been developed.

How are the subjective values that encode our preferences encoded in the choice circuit?

A tremendous amount is known about how many things are represented in the mammalian brain, a fact to which this book is a testament. We know that most classes of information represented in the cerebral cortex are topographically encoded on anatomically two-dimensional "maps." The cortex is made up of dozens of these small topographic maps. The two-dimensional structure of the primary visual cortex, for example, provides a topographic map of the world as seen by the retina, as shown in figure 59.1 (Hubel & Wiesel, 1974).

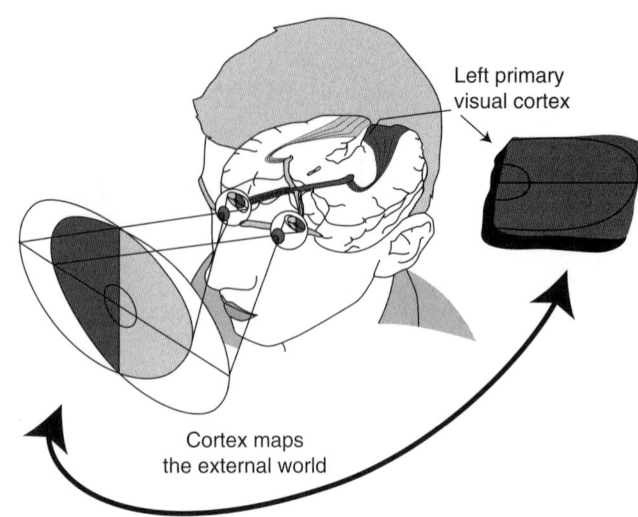

FIGURE 59.1 The primary visual cortex forms a topographic map of the visual world as seen through the retina.

At each point on that map, the firing rates of neurons tell us something about what the retina sees at that location in the visual world. What is hugely important is that this basic organizational structure has turned out to be almost entirely universal in the mammalian brain. And, interestingly, this seems also to be the way in which subjective values are encoded in the choice circuit. Consider the prototypical example of this, the lateral intraparietal area, or *area LIP* (Gnadt & Andersen, 1988). Area LIP organizes information about the actions a monkey can choose to take, and organizes information about what orienting eye movement to make next (as well as encoding information about what is important at each of those locations). Firing rates on this map encode (among other things; see, for example, Gottlieb, 2007) the subjective values of each eye movement in the set of all possible eye movements (Glimcher, 2003). Just as the visual cortex is a topographic map of the properties of the important visual world, LIP appears to include a topographic map of one of the most important properties of each movement—how valuable it is to the organism at any moment in time (Sereno, Pitzalis, & Martinez, 2001; Schluppeck, Glimcher, & Heeger, 2005). To make this clearer, consider figure 59.2. Here, each point on the three-dimensional (3D) graph represents neural activity at one point on the LIP map as it would be seen if we could observe the entire map at one time. The inset cartoon shows the location of two visual targets. If the monkey looks at one of these targets he gets a reward, and the reward magnitudes vary by target. If he looks anywhere else, he gets no reward. What we can see in this reconstruction is that activity on the map encodes the value, to the monkey, of every possible orienting eye movement.

FIGURE 59.2 A simulation of the map of eye movements in the lateral intraparietal area (area LIP). Each vertex represents a neuron on the LIP map, and brightness and height indicate the firing rate of that neuron. The inset shows the two targets presented to a monkey subject. The right target, if looked at, yields twice as much reward as the left target. (Courtesy of Ryan Webb.)

One interesting feature of the map is that the representation of each of the three valuable movements is quite broadly distributed. This reflects the fact that adjacent neurons are strongly connected to one another and that the strength of that connection falls off as a function of distance (Gilbert & Wiesel, 1979). A feature not immediately obvious in this reconstruction is that each neuron in the map is also connected in an *inhibitory* way to other neurons. The exact pattern of these inhibitory connections varies from area to area, but for our purposes let us consider these inhibitory connections to be universal; each neuron is connected with a fixed level of inhibition to every other neuron in the topography (Haider, Hausser, & Carandini, 2013; Lee, Helms, Augustine, & Hall, 1997). The peaks of activity, thus, in some sense, "fight" with each other for control of the map.

How is the argmax operation performed?

It should be obvious that choosing, for a network like this, amounts to identifying the peak that is highest on maps like the one portrayed in figure 59.2, and then passing that information on to the circuits that generate movements. Studies of brain slices (Ozen, Helms, & Hall, 2003) suggest that when the monkey chooses, the strength of *all* of the short-range excitatory connections in the topographic map increases. As short-range excitation grows over all of the map, the largest peak becomes stronger and thus more effective at suppressing its neighbors. As those neighbors are suppressed, they become less effective at suppressing the largest peak. The result is a self-reinforcing growth of the largest peak, sometimes called a *winner-take-all* computation (Edelman & Keller, 1996; Maass, 2000; van Gisbergen, van Opstal, & Tax, 1987).

How then does the growing, largest peak trigger the selected movement? Some of these topographic maps include a biophysical threshold that makes that possible. In the superior colliculus, which receives topographically mapped connections from LIP, when the firing rates of neurons at any one location on the map experience firing rates over about 100 Hz, these neurons change state and burst at about 1,000 Hz for a roughly fixed period of time (Hall & Moschovakis, 2004; van Opstal & van Gisbergen, 1989). This very high rate completely suppresses all other movement-triggering activity on that map, and then passes on to influence the movement-control circuits of the brainstem.

It is important to understand, however, that it is not one small brain map that controls all eye movement–related decision making. A cascade of reciprocally connected maps, as shown in figure 59.3, performs this function. The specializations of these maps are a subject of intense current inquiry and uncertainty, but we do know about a few key specializations. We know, for example, that the last of the maps in the eye-movement cascade is the superior colliculus, and we know that it is only this map on which the biophysical threshold mechanism operates (Hall & Moschovakis, 2004).

How does behavioral stochasticity arise?

We know that neuronal firing rates are in fact quite stochastic. If we were to direct two inputs to the LIP

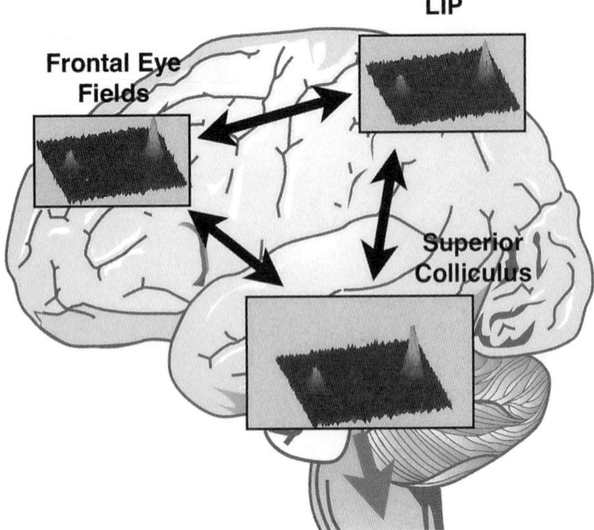

FIGURE 59.3 An example of the interconnected cascade of topographic maps that appear to represent the values of current options and execute the choice process, at least for simple orienting eye-movement decision making. The actual cascade of eye movement networks is much larger, including areas like the supplementary eye fields and a number of other areas. (Courtesy of Ryan Webb.)

map indicating that a particular movement was worth 0.5ml of fruit juice, the neurons in that brain area would show a time-varying firing rate. Intrinsic randomness in the map representation causes these peaks to bounce around a bit from moment to moment, as shown in figure 59.4. Thus, whenever the values of two options get close to one another, neuronal "noise" has an effect on the winner-take-all process. Thus variability emerges as a key feature of the operation of this network (Dorris & Glimcher, 2004), exactly the kind described by the random utility theory models (e.g., McFadden, 2005) of economics.

What about other kinds of movement-related decisions?

Is what we know about eye movements true in other systems? We do know that a similar network of areas operates in a similar way to control decisions about movements of the hand (Cisek & Kalaska, 2010). And we have some data suggesting a similar architecture for many other classes of movements (Colby & Duhamel, 1991). But what about more abstract decisions about goods, rather than actions? The truth is that we simply do not know the answer to that question (Padoa-Schioppa, 2011). But what we do know comes from the study of more anterior (frontal) parts of the brain thought to be intimately involved in storing and representing the values of goods and actions. Ultimately, it must be these areas that provide the subjective value signals on which the neurons in areas like LIP operate. So with that, we turn to the valuation circuits of the primate brain.

THE VALUATION CIRCUIT Studies of the valuation circuits of the brain emerged, in large part, from studies of the reinforcement learning systems of the brain during the 1990s. Just as studies of the choice circuit were gathering steam, it began to be generally accepted that the dopamine neurons of the midbrain played a critical role in learning the values of actions (Montague, Dayan, & Sejnowsky, 1996). That suggested that one way to begin to understand valuation circuits was through the study of dopamine. At the same time that these studies of dopamine were being undertaken, a second line of inquiry with regard to value also got underway: the search for functional MRI (fMRI) signals that correlated with the subjective values of goods, actions, and events expressed in various ways by human subjects (Breiter, Aharon, Kahneman, Dale, & Shizgal, 2001; Delgado, Nystrom, Fissell, Noll, & Fiez, 2000; Elliott, Friston, & Dolan, 2000; Knutson, Westdorp, Kaiser, & Hommer, 2000; McCabe, Houser, Ryan, Smith, & Trouard, 2001). Both of these approaches converged on two particular brain areas: the striatum and a portion of the medial prefrontal cortex. Activity in these two areas was found to consistently predict people's preferences—and preferences of literally *all* kinds. If someone was a *steep temporal discounter*, she preferred small immediate rewards to waiting for larger rewards, and activity in these areas in response to delayed rewards was steeply "discounted" (Kable & Glimcher, 2007). If someone placed a higher value on food rewards than on monetary rewards, then so did these areas in that person (Chib, Rangel, Shimojo, & O'Doherty, 2009; Levy & Glimcher, 2011). If someone was inclined to co-operate in a trust game, then activity in these areas was higher for offers that required cooperation (Sanfey, Rilling, Aronson, Nystrom, & Cohen, 2003).

The utility theories of economics and their descendants—for example, the *prospect theory* of Kahneman and Tversky (1979)—propose that decision makers trade off different kinds of rewards just as if they were comparing those rewards in a single common currency. The observation that activity in the medial prefrontal cortex and the striatum seem to encode human preferences for rewards of all types in a single common neural currency immediately suggested that activity in these brain areas plays a role in behavior similar to that played by utility-like objects in economic theory. But it is important to note that these utility-like signals, now typically called *subjective value signals*, differ from utilities in

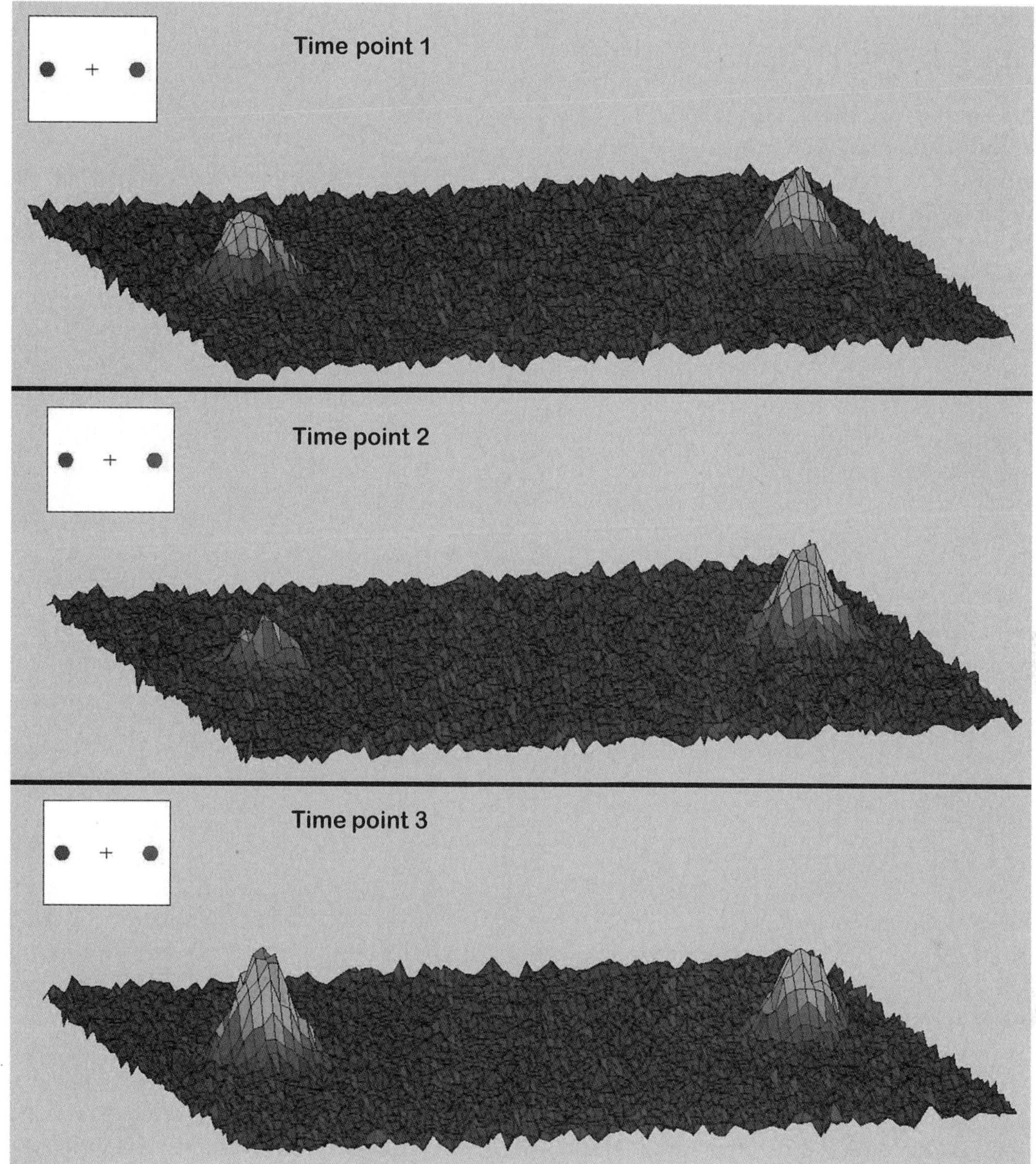

FIGURE 59.4 Three snapshots of LIP activity engendered to left and right options that have equal true values. Neuronal variability causes the heights of the peaks to vary stochastically from moment to moment. (Courtesy of Ryan Webb.)

several important ways. The most important of these ways is that subjective value signals always correlate with and predict choice behavior—even when that choice behavior is incompatible with a utility-based theory; these neural signals appear to be much more closely related to behavior than they are to traditional economic theories.

The medial prefrontal cortex and the striatum

While some important debate about this point remains, nearly all models of choice thus now posit the existence of a utility-like signal in the medial prefrontal cortex and the ventral striatum that combines the outputs of many antecedent brain areas onto a single common scale appropriate for direct comparison and choice (Fehr & Rangel, 2011; Glimcher, 2011; Kable & Glimcher, 2009; and see Wallis & Rushworth, 2013, for an important extension of this approach). Brain activity in the medial prefrontal cortex and striatum clearly reveals the idiosyncratic values people place on goods, actions, or rewards. One can predict how people will trade off delays to future rewards, different kinds of rewards, social rewards, even co-operation from measurements of activity in these two areas (Levy & Glimcher, 2012). So it is natural to hypothesize that the choice circuits we encountered in the preceding section receive inputs from these areas—although those connections have not yet been unambiguously identified.

The most common hypothesis is thus that somewhere in the brain the values of all different kinds of rewards must be stored—perhaps in very abstract and incomplete ways that are highly context-dependent and that likely involve the middle temporal lobe of the brain. When a subject is offered a choice between two or more goods or actions, activity in the medial prefrontal cortex and the striatum comes to represent the values of these rewards on a single common scale for comparison. It cannot be understated how much data supports this basic conclusion. Meta-studies of the available literature robustly show that activity in these areas predicts preferences in nearly every paper that has ever been published on this subject (Levy & Glimcher, 2012). The medial prefrontal cortex and the striatum are hot spots in choice and valuation that almost certainly serve as a critical input to the choice circuitry. The bulk of the evidence available today suggests that these two areas serve as a *final common path for valuation* in the human brain. (Although it may also be the case that some kinds of costs are not reflected in this representation; Kolling, Behrens, Mars, & Rushworth, 2012.)

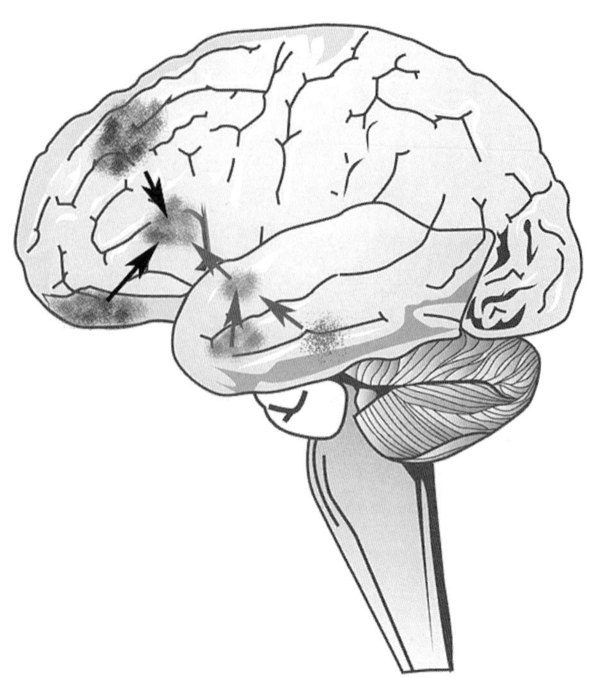

FIGURE 59.5 A cartoon of the valuation cascade in frontal cortex and the basal ganglia. Dark gray cortical areas include the medial prefrontal cortex, the dorsolateral prefrontal cortex, and the orbitofrontal cortex. Light gray subcortical areas include the striatum, the amygdala, and the hypothalamus.

Inputs to common value areas

So from where do valuation signals observed in the medial prefrontal cortex and the ventral striatum come? Both fMRI and single-neuron recordings suggest that a large number of brain areas contribute to these common value signal representations. Studies of risk aversion over food rewards (Levy & Glimcher, 2011) point at the hypothalamus as a critical input to the medial prefrontal cortex for valuing these kinds of rewards. Studies of cooperative behavior (Fehr, 2009) and self-control (Hare, Camerer, & Rangel, 2009) point at the dorsolateral prefrontal cortex as a critical input to this area for valuing social cooperation and goods that require or invoke self-control processes. Studies of the orbitofrontal cortex (O'Doherty, Kringelbach, Rolls, Hornak, & Andrews, 2001; Padoa-Schioppa, 2011; Tremblay & Schultz, 1999) point to this area as critical to the valuation of many consumable rewards. Studies of the amygdala (Phelps & LeDoux, 2005; Salzman & Fusi, 2010) point to this area as playing a role in the emotional regulation of reward values. What emerges is a fairly complex network of brain areas, schematized in figure 59.5, that construct in medial prefrontal cortex and in the striatum a subjective value signal that guides choice.

It is important to stress, however, that models of the valuation circuit are much less well developed than models of the choice circuit. Much of the work on valuation signals has been conducted in humans using fMRI. The result is that we know much more about where valuation signals arise in the nervous system than about how they are encoded. For this reason, we also know much less about how the choice and valuation networks interact. We do not know, for example, how people choose among goods when no action is required in the choice process. Does a winner-take-all process like that observed in LIP occur within one of the valuation circuit maps? That seems likely (Padoa-Schioppa, 2011), but at this point we simply do not know.

What does seem clear is a basic outline. When a subject is offered a choice between two or more options, a network of frontal and striatal areas constructs subjective value signals, on a common scale, for those goods, actions, or rewards in the medial prefrontal cortex and in some parts of the striatum. Next, we know that if we ask subjects to select from among the rewards being valued in these areas by making movements, the choice circuits come to represent these subjective values and to converge as a choice is made. We do not know how the goods and rewards represented in areas like the medial prefrontal cortex are mapped into "action space" (the movement-based topographic representation that has been so well studied in monkeys), nor do we know how all classes of "costs" are incorporated into these representations. It may be that the striatum plays a role in this process. It may be that actions are represented in the medial prefrontal cortex. But we know that a mapping of this type must take place; activity in the parietal choice maps closely mirrors mPFC subjective value signals and guides choice to a precision that simply cannot reflect a lucky coincidence.

Putting it all together: Relating value-based decision to perceptual decision

The chapter in this volume by Shadlen and colleagues describes one of the most important lines of research in neurobiological studies of decision making, the random-dot experiments of Bill Newsome, Michael Shadlen, and their colleagues (Newsome, Britten, Salzman, & Movshon, 1990). Their studies suggest that perceptual signals originating in extrastriate visual cortex's area MT are passed to area LIP, the frontal eye fields, and the superior colliculus for decision making. As we have seen in this chapter, these three areas are known as choice areas in value-based decision-making models. So what does this mean? Do these choice areas do two different jobs, one value-related and one perception-related? Or is an integrated view of value-based and perceptual-based decision making possible?

Over the past decade, there has been growing evidence that an integrated view is not only possible, but now almost required. Both the value-based and perceptual decision-making traditions have identified the frontal and parietal choice areas as a key stage in many kinds of decision making. The value-based studies described in this chapter have identified frontal cortical and basal ganglia inputs to those choice circuits. The perceptual studies have identified the inputs of perceptual circuits to those same choice circuits. What this suggests is a tremendous degree of unity. It suggests that the so-called choice circuits can operate on a broad range of inputs to produce coherent behavior under many circumstances. But just how completely can these two views be integrated?

Value-based studies of decision making have tended to be about temporally slow decisions. Subjects, whether humans or animals, are presented with two or more options, and then after a delay asked to choose between them. The data we have today suggests that subjective value signals are injected onto the choice maps during the first part of that process. Then, when the subject is asked to choose, the global excitatory and inhibitory balance of one or more of the choice networks is suddenly changed. Studies of perceptual decision making, in contrast, have tended to examine reaction-time tasks in which subjects integrate sensory data as fast as they can and then make choices as soon as possible. Can the same neural systems, using the same computational properties, produce both behaviors?

Soltani and colleagues (Soltani & Wang, 2008) were the first scholars to demonstrate that the answer to that question is an unambiguous yes. The same circuits really can perform both of these functions. What Soltani and colleagues demonstrated is that what differs between these two classes of models comes down to the inhibitory-excitatory balance of the network as the choice process progresses. In temporally discrete value-based decision making, the network begins in a very noncompetitive mode. Peaks of activity on the maps like those shown in figure 59.2 coexist. When a choice is made, changes in the excitatory-inhibitory balance of the network cause the peaks of activity to compete, implementing the winner-take-all process. In dynamic perceptual decision making, in contrast, the inhibitory-excitatory balance of the network can be thought of as preset, so that some degree of competition is always present. This preset balance means that as soon as some critical level of evidence is achieved, the network automatically converges to a single choice. In this kind of decision making, it is thus the preset ratio of excitation to inhibition that serves to implement the decision threshold used in models like *drift diffusion* (Gold & Shadlen, 2007; Ratcliff, 1978). While the details of that synthesis are beyond the scope of this chapter (see also Louie & Glimcher, 2012, for more on this), what is important is that these two classes of models are now really beginning to fit together in our understanding of human decision making.

REFERENCES

BREITER, H. C., AHARON, I., KAHNEMAN, D., DALE, A., & SHIZGAL, P. (2001). Functional imaging of neural responses to expectancy and experience of monetary gains and losses. *Neuron, 30*(2), 619–639.

CHIB, V. S., RANGEL, A., SHIMOJO, S., & O'DOHERTY, J. P. (2009). Evidence for a common representation of decision values for dissimilar goods in human ventromedial prefrontal cortex. *J Neurosci, 29*(39), 12315–12320.

CHURCHLAND, M. M., YU, B. M., CUNNINGHAM, J. P., SUGRUE, L. P., COHEN, M. R., CORRADO, G. S., ... SHENOY, K. V. (2010). Stimulus onset quenches neural variability: A widespread cortical phenomenon. *Nat Neurosci, 13*(3), 369–378.

CISEK, P., & KALASKA, J. F. (2010). Neural mechanisms for interacting with a world full of action choices. *Annu Rev Neurosci, 33,* 269–298.

COLBY, C. L., & DUHAMEL, J. R. (1991). Heterogeneity of extrastriate visual areas and multiple parietal areas in the macaque monkey. *Neuropsychologia, 29*(6), 517–537.

Delgado, M. R., Nystrom, L. E., Fissell, C., Noll, D. C., & Fiez, J. A. (2000). Tracking the hemodynamic responses to reward and punishment in the striatum. *J Neurophysiol, 84*(6), 3072–3077.

Dorris, M. C., & Glimcher, P. W. (2004). Activity in posterior parietal cortex is correlated with the relative subjective desirability of action. *Neuron, 44*(2), 365–378.

Edelman, J. A., & Keller, E. L. (1996). Activity of visuomotor burst neurons in the superior colliculus accompanying express saccades. *J Neurophysiol, 76*(2), 908–926.

Elliott, R., Friston, K. J., & Dolan, R. J. (2000). Dissociable neural responses in human reward systems. *J Neurosci, 20*(16), 6159–6165.

Fehr, E. (2009). On the economics and biology of trust. *J Eur Econ Assoc, 7*(2–3), 235–266.

Fehr, E., & Rangel, A. (2011). Neuroeconomic foundations of economic choice—recent advances. *J Econ Perspect, 25*(4), 3–30.

Gilbert, C. D., & Wiesel, T. N. (1979). Morphology and intracortical projections of functionally characterized neurons in the cat visual cortex. *Nature, 280,* 120–125.

Glimcher, P. W. (2003). The neurobiology of visual-saccadic decision making. *Annu Rev Neurosci, 26,* 133–179.

Glimcher, P. W. (2005). Indeterminacy in brain and behavior. *Annu Rev Psychol, 56,* 25–56.

Glimcher, P. W. (2011). *Foundations of neuroeconomic analysis.* New York, NY: Oxford University Press.

Gnadt, J. W., & Andersen, R. A. (1988). Memory related motor planning activity in posterior parietal cortex of macaque. *Exp Brain Res, 70*(1), 216–220.

Gold, J. I., & Shadlen, M. N. (2007). The neural basis of decision making. *Annu Rev Neurosci, 30,* 535–574.

Gottlieb, J. (2007). From thought to action: The parietal cortex as a bridge between perception, action, and cognition. *Neuron, 53*(1), 9–16.

Haider, B., Hausser, M., & Carandini, M. (2013). Inhibition dominates sensory responses in the awake cortex. *Nature, 493*(7430), 97–100.

Hall, W. C., & Moschovakis, A. (2004). *The superior colliculus: New approaches for studying sensorimotor integration.* Boca Raton, FL: CRC Press.

Hare, T. A., Camerer, C. F., & Rangel, A. (2009). Self-control in decision-making involves modulation of the vmPFC valuation system. *Science, 324*(5927), 646–648.

Houthakker, H. S. (1950). Revealed preference and the utility function. *Economica (New Series), 17*(66), 159–174.

Hubel, D. H., & Wiesel, T. N. (1974). Sequence regularity and geometry of orientation columns in the monkey striate cortex. *J Comp Neurol, 158*(3), 267–293.

Kable, J. W., & Glimcher, P. W. (2007). The neural correlates of subjective value during intertemporal choice. *Nat Neurosci, 10*(12), 1625–1633.

Kable, J. W., & Glimcher, P. W. (2009). The neurobiology of decision: Consensus and controversy. *Neuron, 63*(6), 733–745.

Kahneman, D., & Tversky, A. (1979). Prospect theory: An analysis of decision under risk. *Econometrica, 47*(2), 263–291.

Knutson, B., Westdorp, A., Kaiser, E., & Hommer, D. (2000). FMRI visualization of brain activity during a monetary incentive delay task. *NeuroImage, 12*(1), 20–27.

Kolling, N., Behrens, T. E., Mars, R. B., & Rushworth, M. F. (2012). Neural mechanisms of foraging. *Science, 336*(6077), 95–98.

Lee, P. H., Helms, M. C., Augustine, G. J., & Hall, W. C. (1997). Role of intrinsic synaptic circuitry in collicular sensorimotor integration. *Proc Natl Acad Sci USA, 94*(24), 13299–13304.

Levy, D. J., & Glimcher, P. W. (2011). Comparing apples and oranges: Using reward-specific and reward-general subjective value representation in the brain. *J Neurosci, 31*(41), 14693–14707.

Levy, D. J., & Glimcher, P. W. (2012). The root of all value: A neural common currency for choice. *Curr Opin Neurobiol, 22*(6), 1027–1038.

Loomes, G., & Sugden, R. (1995). Incorporating a stochastic element into decision theories. *Eur Econ Rev, 39*(3–4), 641–648.

Louie, K., & Glimcher, P. W. (2012). Efficient coding and the neural representation of value. *Ann NY Acad Sci, 1251,* 13–32.

Maass, W. (2000). On the computational power of winner-take-all. *Neural Comput, 12*(11), 2519–2535.

McCabe, K., Houser, D., Ryan, L., Smith, V., & Trouard, T. (2001). A functional imaging study of cooperation in two-person reciprocal exchange. *Proc Natl Acad Sci USA, 98*(20), 11832–11835.

McFadden, D. L. (2005). Revealed stochastic preference: A synthesis. *Econ Theor, 26*(2), 245–264.

Montague, P. R., Dayan, P., & Sejnowsky, T. J. (1996). A framework for mesencephalic dopamine systems based on predictive Hebbian learning. *J Neurosci, 16*(5), 1936–1947.

Newsome, W. T., Britten, K. H., & Movshon, J. A. (1989). Neuronal correlates of a perceptual decision. *Nature, 341*(6237), 52–54.

Newsome, W. T., Britten, K. H., Salzman, C. D., & Movshon, J. A. (1990). Neuronal mechanisms of motion perception. *Cold Spring Harb Symp Quant Biol, 55,* 697–705.

O'Doherty, J., Kringelbach, M. L., Rolls, E. T., Hornak, J., & Andrews, C. (2001). Abstract reward and punishment representations in the human orbitofrontal cortex. *Nat Neurosci, 4*(1), 95–102.

Ozen, G., Helms, M. C., & Hall, W. C. (2003). The intracollicular neuronal network. In W. C. Hall & A. Moschovakis (Eds.), *The superior colliculus: New approaches for studying sensorimotor integration* (pp. 147–158). Boca Raton, FL: CRC Press.

Padoa-Schioppa, C. (2011). Neurobiology of economic choice: A good-based model. *Annu Rev Neurosci, 34,* 333–359.

Phelps, E. A., & LeDoux, J. E. (2005). Contributions of the amygdala to emotion processing: From animal models to human behavior. *Neuron, 48*(2), 175–187.

Platt, M. L., & Glimcher, P. W. (1999). Neural correlates of decision variables in parietal cortex. *Nature, 400*(6741), 233–238.

Ratcliff, R. (1978). A theory of memory retrieval. *Psychol Rev, 85,* 59–108.

Salzman, C. D., & Fusi, S. (2010). Emotion, cognition, and mental state representation in amygdala and prefrontal cortex. *Annu Rev Neurosci, 33,* 173–202.

Sanfey, A. G., Rilling, J. K., Aronson, J. A., Nystrom, L. E., & Cohen, J. D. (2003). The neural basis of economic decision-making in the ultimatum game. *Science, 300*(5626), 1755–1758.

Schluppeck, D., Glimcher, P., & Heeger, D. J. (2005). Topographic organization for delayed saccades in human posterior parietal cortex. *J Neurophysiol, 94*(2), 1372–1384.

Sereno, M. I., Pitzalis, S., & Martinez, A. (2001). Mapping of contralateral space in retinotopic coordinates by a parietal cortical area in humans. *Science, 294*(5545), 1350–1354.

Shadlen, M. N., Britten, K. H., Newsome, W. T., & Movshon, J. A. (1996). A computational analysis of the relationship between neuronal and behavioral responses to visual motion. *J Neurosci, 16*(4), 1486–1510.

Soltani, A., & Wang, X. J. (2008). From biophysics to cognition: Reward-dependent adaptive choice behavior. *Curr Opin Neurobiol, 18*(2), 209–216.

Sugrue, L. P., Corrado, G. S., & Newsome, W. T. (2005). Choosing the greater of two goods: Neural currencies for valuation and decision making. *Nat Rev Neurosci, 6*(5), 363–375.

Tolhurst, D. J., Movshon, J. A., & Dean, A. F. (1983). The statistical reliability of signals in single neurons in cat and monkey visual cortex. *Vision Res, 23*(8), 775–785.

Tremblay, L., & Schultz, W. (1999). Relative reward preference in primate orbitofrontal cortex. *Nature, 398*(6729), 704–708.

van Gisbergen, J. A., van Opstal, A. J., & Tax, A. M. M. (1987). Collicular ensemble coding of saccades based on vector summation. *Neuroscience, 21*(2), 541–555.

van Opstal, A. J., & van Gisbergen, J. A. (1989). A nonlinear model for collicular spatial interactions underlying the metrical properties of electrically elicited saccades. *Biol Cybern, 60*(3), 171–183.

von Neumann, J., & Morgenstern, O. (1944). *Theory of games and economic behavior.* Princeton, NJ: Princeton University Press.

Wallis, J. D., & Rushworth, M. F. S. (2013). Integrating benefits and costs in decision-making. In P. W. Glimcher & E. Fehr (Eds.), *Neuroeconomics: Decision-making and the brain* (2nd ed., pp. 399–421). Waltham, MA: Academic Press.

VIII
SOCIAL NEUROSCIENCE AND EMOTION

Introduction

ELIZABETH A. PHELPS AND TODD F. HEATHERTON

SOCIAL AND affective neuroscience have become core topics within the broader context of cognitive neuroscience. As we noted in the prior edition, the social and emotional aspects of the brain are inexorably linked, the adaptive significance of emotions being closely linked to their social value, and nearly all social interaction produces affective responses. The chapters in this section converge to show that social and emotional processes are important for understanding many of the most basic aspects of cognition and learning.

Previous editions of *The Cognitive Neurosciences* have highlighted the important role of the amygdala across species as a critical structure in the neuroscience of emotion, and this edition is no exception. However, as the chapters in this edition indicate, there is an increasing interest and emphasis on how the amygdala interacts with other brain circuits and systems to promote the adaptive expression of emotion. Hartley, Moscarello, Quirk, and Phelps describe a range of techniques that can be used to control fear across species and outline the overlapping and unique circuitry underlying each technique, discussing research from rodents to humans. A key component of the control of fear is the interaction of amygdala and prefrontal cortex. This circuitry is also highlighted in the model of self-regulation described by Wagner and Heatherton. In their model, there is a dynamic tension between domain-general frontal control mechanisms and context-specific subcortical emotion and reward circuits. Self-regulation fails when there is a loss of connectivity or balance between these regions. Ochsner discusses the limitations of models, such as that proposed by Wagner and Heatherton, in explaining several important aspects of emotional experience. He proposes a parallel constraint satisfaction model for the interaction of cognition and emotion.

These chapters describe research based on studies of the adult brain, but we know that these brain circuits vary over the life span. One of the hallmarks of developmental neuroscience is that developmental trajectory is not uniform across brain regions. Somerville and Casey explore how brain development alters emotional reactivity and regulation. Specifically, they highlight adolescence as a time period where there is both asymmetry in the development of the brain systems of emotion expression and its control, and also tremendous change. Gee and Whalen explore how developmental changes interact with the amygdala's primary role in responding to uncertainty and learning to predict biologically relevant outcomes. They demonstrate how exploring blood oxygen level–dependent, or BOLD, responses to the biologically relevant stimuli of facial expressions can inform our understanding of neural changes linked to psychopathology.

How we respond to face stimuli is also explored by Mende-Siedlecki and Todorov, who discuss evidence that the amygdala possesses face-selective properties that can be observed in the absence of emotional expressions. As such, they argue that the amygdala plays a core role in face perception. This link between perception and emotion is the core theme of the chapter by Koscik, White, Chapman, and Anderson. They discuss the bidirectional nature of the relationship between early sensory processes and higher-order knowledge regarding social and emotional information. The final chapter in this section by Ambady and Freeman examines the interplay of cultural and genetic factors on the neural bases of human perception. They describe evidence to suggest that cultural experiences influence both basic perceptual and attentional mechanisms as well as higher-order social domains, such as emotional expression recognition and theory of mind.

Although the chapters in this section take us from the rodent in the laboratory trying to avoid a shock to the influence of culture experiences on the human brain, there is significant overlap in many of the brain systems that are highlighted, but also some important differences. Across these diverse approaches, it is becoming increasingly clear that we need to consider the interaction of networks of activity rather than the specific activations in discrete brain regions. This trend is likely to continue as neural models of social and affective processes gain the necessary specificity and complexity that characterize social and emotional processes.

60 The Cognitive Neuroscience of Fear and Its Control: From Animal Models to Human Experience

CATHERINE A. HARTLEY, JUSTIN M. MOSCARELLO, GREGORY J. QUIRK, AND ELIZABETH A. PHELPS

ABSTRACT The complex social and emotional cognition characteristic of human experience relies upon more basic forms of learning that enable us to assign affective value to stimuli. Our neuroscientific understanding of these associative learning processes has grown out of decades of detailed research in animal models using Pavlovian conditioning paradigms. In this chapter, we begin by reviewing this foundational research, describing the basic learning mechanisms through which we assign negative value to environmental stimuli. We then examine various means by which we control the expression of these learned fear associations. We focus specifically on the functional architecture underlying four distinct types of regulatory processes: extinction, active coping, cognitive emotion regulation, and reconsolidation. In each section, we review what is known about the neurocircuitry of the regulatory method from the nonhuman animal literature, as well as from studies in humans.

The growth of cognitive neuroscience over the last few decades has brought a dramatic increase in the study of affective and social cognition. The foundation of much of this work lies in research in animal models, which has yielded a detailed neuroscientific account of the associative mechanisms through which we evaluate and anticipate salient environmental events. Through recent work examining more complex social and emotional cognition, it has become increasingly clear that these higher-order processes recruit a common underlying neural circuitry as these basic forms of evaluative learning. On one hand, this convergence is intuitive. Associative learning imbues our world with emotional significance, rendering objectively neutral stimuli ominous or alluring, evoking thoughts and feelings, and motivating our actions and decisions. Much of this emotional experience stems from social interaction. Conversely, social cognition is intrinsically emotional, as our thoughts about other people are laden with learned evaluations. On the other hand, these overlapping substrates of both basic and complex learning and behavior present a vexing puzzle, challenging scientists to understand how these shared neural circuits give rise to the profound complexity of human social and emotional experience. In this chapter, we will attempt to lay a foundation for this inquiry, describing the basic learning mechanisms through which we assign negative value to environmental stimuli, the means by which we control the expression of these learned associations, and the neurocircuitry underlying these processes.

The survival of any organism depends on its ability to detect and evade environmental threat. Cues and contexts that signal danger must be learned rapidly so that serious threats can be anticipated with few actual encounters. These memories should also be long-lasting. A perilous situation might be infrequent, but an organism should not need to learn about a previous threat anew. Across a range of species, the neurocircuitry governing fear learning represents an evolutionarily conserved solution to these shared environmental demands, enabling the rapid formation of persistent fear memories. The expression of these learned fears diverts the organism's cognitive, behavioral, and physiological resources toward the detection and response to threat. While this recruitment of resources confers robust adaptive benefit in the face of danger, it also means that fear expression in the absence of imminent threat is costly. In humans, excessive fear that persists even in safe contexts is a cardinal feature of anxiety disorders. Thus, the failure to regulate the expression of fear in accordance with the true presence of environmental threat may be a critical factor underlying vulnerability to psychopathology.

In our daily lives, we flexibly control the expression of fear in multiple ways. We may recognize that a stimulus that previously predicted threat is perfectly safe in a different setting (e.g., hearing the growl of a bear while in the forest versus in the zoo). We may take actions to make a potentially unpleasant situation less daunting (e.g., bringing a friend to a party where we will know few others). Sometimes simply changing the way we think about a dreaded future event is sufficient to disarm our fear (e.g., the vaccine shot will hurt, but it's quick and necessary to stay well). Finally, recent research suggests that experiences that violate our negative expectations may, under certain conditions, directly overwrite this invalidated information.

In this chapter, we examine the functional architecture underlying the acquisition and control of fear. We begin with a brief discussion of the neurocircuitry of fear acquisition and then explore four types of regulatory processes that can be utilized to control fear: extinction, active coping, cognitive regulation, and reconsolidation. An overview of the basic circuitry discussed is depicted in figure 60.1.

Fear acquisition

Humans and animals alike use environmental cues to determine the presence or absence of danger. These cues can acquire their emotional valence through a process of Pavlovian learning, which occurs when a previously neutral conditioned stimulus (CS) comes to predict an aversive unconditioned stimulus (US). In order for the CS to evoke a conditioned fear response, it must temporally coincide with the US, and it must carry information (i.e., reduce uncertainty) about the timing and likelihood of US delivery (Gallistel, 2003; Pavlov, 1927; Rescorla, 1988). Through this learning, a CS can evoke an array of behavioral and physiological reactions that comprise a preparatory response to predicted threat. In rodents, commonly studied conditioned responses (CRs) include freezing as well as changes in heart rate, respiration, and potentiated startle responses. In humans, CRs include changes in skin conductance, potentiated startle, and pupil dilation.

Across species, the amygdala is a key locus for the acquisition, consolidation, and expression of Pavlovian fear memory. In rodents, these mnemonic processes are typically studied using auditory fear conditioning, in which a tone CS comes to predict an electrical shock

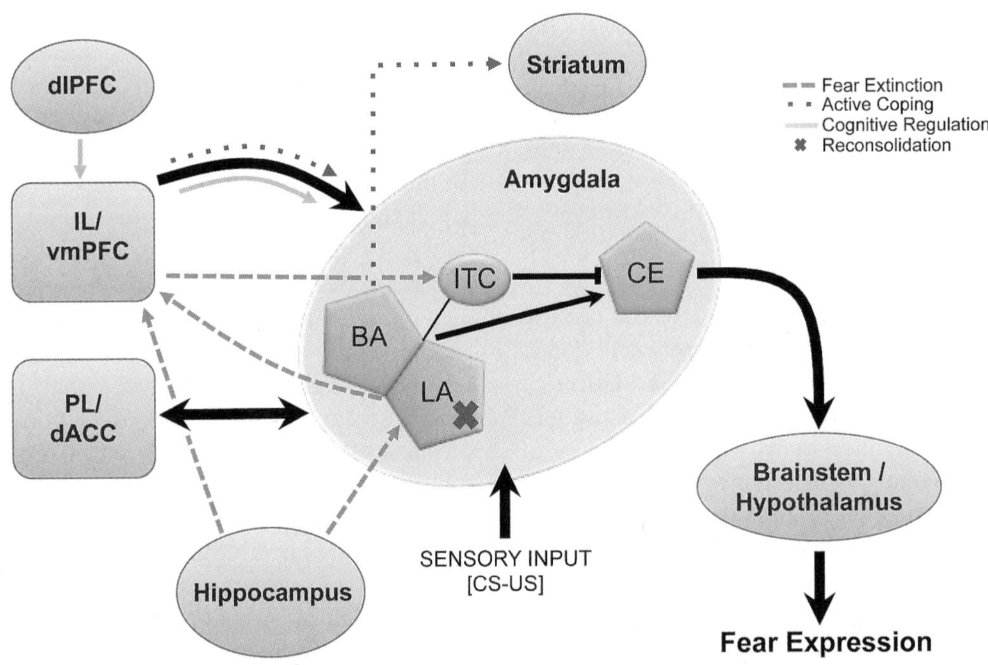

FIGURE 60.1 Neurocircuitry of fear learning and regulation: the lateral nucleus (LA) of the amygdala receives afferent sensory information regarding the conditioned stimulus (CS) and unconditioned stimulus (US) relationship and is the site of storage of the fear memory. The LA and basal nuclei (BA) are interconnected and both project to the central nucleus (CE), which has outputs to brainstem and hypothalamic regions that control the expression of the conditioned response (CR). Following conditioning, the prelimbic (PL) region of the ventromedial prefrontal cortex (vmPFC) is activated following CS presentations and drives the expression of conditioned fear. During fear-extinction learning and consolidation, connections are established between the infralimbic (IL) subregion of the vmPFC and the inhibitory intercalated (ITC) cell masses, which inhibit activity in the CE (circuit denoted in orange). During extinction recall, these connections are activated, inhibiting fear expression. Contextual modulation of extinction expression is mediated by projections from the hippocampus to the vmPFC and/or amygdala. During active coping (circuit denoted in blue), information from the LA is routed not to the CE, which drives fear expression, but to the BA, which in turn projects to the striatum. The IL/vmPFC supports the learning active coping responses and mediates the subsequent inhibition of fear expression outside of the instrumental learning context. During cognitive regulation, the dorsolateral prefrontal cortex (dlPFC) regulates fear expression through projections to the vmPFC, which in turn inhibits amygdala activity (circuit denoted in green). Reconsolidation (denoted by red X) diminishes conditioned fear expression through alteration of the original CS-US association stored in the LA. (See color plate 51.)

US. Lesion and pharmacological studies demonstrate the role of particular amygdalar subregions, focusing on lateral (LA) and central nuclei (CE) as key substrates (Goosens & Maren, 2001; Nader, Majidishad, Amorapanth, & LeDoux, 2001; Wilensky, Schaffe, Kristense, & LeDoux, 2006). The LA contains multimodal neurons that respond to both tones and shock (Romanski, Clugnet, Bordi, & LeDoux, 1993) and appears to be the site of both acquisition and storage of the fear memory (Schafe, Doyère, & LeDoux, 2005). CS-evoked LA activity and conditioned freezing develop along the same time course (Repa et al., 2001). Fear conditioning strengthens LA synaptic inputs to CE neurons (Li et al., 2013), which drive the expression of conditioned fear responses via descending projections to brainstem and hypothalamic nuclei.

Research in humans corroborates the central role of the amygdala in the acquisition, storage, and expression of conditioned fear (Phelps & LeDoux, 2005). Human functional imaging studies reveal increases in the blood oxygen level dependent (BOLD) signal in the amygdala in response to a conditioned threat versus neutral stimulus (Büchel, Morris, Dolan, & Friston, 1998; Cheng, Knight, Smith, & Helmstetter, 2006; LaBar et al., 1998). Patients with unilateral (LaBar et al., 1995) or bilateral (Bechara et al., 1995) amygdala lesions exhibit impaired physiological expression fear conditioning, despite intact declarative or episodic knowledge of the CS-US contingency, which has been shown to depend on the hippocampus (Bechara et al., 1995; LaBar, Phelps, et al., 2005).

Although the focus of this chapter is the basic mechanisms of Pavlovian fear acquisition and its control, it is important to acknowledge that in humans fears are often learned indirectly, without direct aversive experience, through cognitive and social means, such as through verbal communication or social observation. While the verbal communication of fears is unique to humans, fears acquired through social observation have been reported in some other species. Importantly, these socially acquired fears have been shown to rely on circuitry overlapping with Pavlovian fear conditioning and appear to be dependent on the amygdala (see Olsson & Phelps, 2007, for a review). In other words, social fear learning, which may be predominant in everyday human life, takes advantage of phylogenetically older mechanisms of fear conditioning.

Extinction

Of all the techniques that can be used to control fear, extinction has been most thoroughly investigated both neurobiologically and behaviorally. Extinction is a process by which presentation of a CS without the US leads to the gradual suppression of previously acquired CRs. This decrease in fear does not reflect the erasure of the original Pavlovian memory, but rather a distinct form of inhibitory safety learning that counteracts the behavioral output of conditioned fear. The persistence of the original memory is evidenced by experimental manipulations that cause the reemergence of fear CRs following extinction, such as contextual shifts (renewal), unsignaled US presentation (reinstatement), or the mere passage of time (spontaneous recovery; Herry et al., 2010; Myers & Davis, 2002; Sotres-Bayon & Quirk, 2010). In a typical extinction experiment, CS-US pairings are followed by extinction training (i.e., unreinforced CSs). A subsequent extinction recall test consisting of one or more unreinforced CSs enables assessment of whether extinction learning is retained.

The study of extinction in rodents has revealed that fear extinction recruits a network of cortical and subcortical structures. Initial extinction learning depends on the amygdala, with substrates in both the lateral and basal (BA) nuclei. The acquisition of extinction attenuates many of the CS-evoked LA neuronal responses. However, a subset of LA neurons maintains robust responses to the CS even after extinction (Repa et al., 2001), resulting in a persistent fear memory trace that allows for the rapid return of extinguished CRs in circumstances such as renewal, reinstatement, or spontaneous recovery. BA is crucial for the expression of both conditioned fear and extinction (Anglaga-Figueroa & Quirk, 2005; Herry et al., 2008; Sierra-Mercado, Padilla-Coreano, & Quirk, 2011). These contrasting functions are reflected in the activity of two discrete cell populations within BA (Herry et al., 2008). One population, described as fear neurons, shows decreasing levels of CS-evoked activity across extinction training, while the other, described as extinction neurons, increases its firing rate across the same period (Herry et al., 2008).

Extinction neurons in BA are reciprocally connected with the infralimbic division (IL) of the medial prefrontal cortex (mPFC; Herry et al., 2008), which plays a key role in extinction learning, consolidation, and retrieval (Burgos-Robles, Vidal-Gonzalez, Santini, & Quirk, 2007; Morgan, Romanski, & LeDoux, 1993; Morgan & LeDoux, 1995; Quirk, Russo, Barron, & Lebron, 2000). Following extinction training, IL neurons respond to the CS during extinction recall (Milad & Quirk, 2002). This cortical activity is thought to feed back onto the amygdala via the intercalated cell masses (ITC). The ITC is an inhibitory cell population interposed between the BA/LA and CE. By inhibiting LA signals to the CE, the IL indirectly inhibits the CE, and thus the expression of conditioned fear responses (Berretta,

Pantazopoulos, Caldera, Pantazopoulos, & Paré, 2005; Likhtik et al., 2008; Royer, Martina, & Paré, 1999). In contrast to the IL, the neighboring prelimbic subregion (PL) of mPFC opposes extinction, driving the expression of fear CRs (Corcoran & Quirk, 2007; Sierra-Mercado et al., 2011; Sotres-Bayon, Sierra-Mercado, Pardilla-Delgado, & Quirk, 2012). During extinction training, CS-evoked responses in PL correlate with the subsequent failure of extinction recall (Burgos-Robles, Vidal-Gonzalez, & Quirk, 2009), suggesting that the interplay between IL and PL cortical regions determines the success or failure of extinction learning.

While cued fear extinction depends on the interplay of the mPFC and amygdala, information about whether the current context signals threat or safety is encoded by the hippocampus, which can also mediate the competition between functional fear inhibition and excitation circuits to ensure that fear extinction is adaptively expressed (Maren, Phan, & Liberzon, 2013). The ventral hippocampus projects directly to interneurons in the PL, allowing inhibition of the principle cells that promote conditioned fear expression (Sotres-Bayon et al., 2012). This region also directly activates BA fear neurons (Herry et al., 2008). Thus the connectivity of the hippocampus enables the facilitation of either extinction learning or fear expression depending on contextual information. The influence of context on fear expression is mediated by the dorsal hippocampus. Inactivation of the dorsal hippocampus prevents the contextual renewal of extinguished fears in a novel context (Corcoran & Maren, 2004). Additionally, when fear conditioning, extinction, and recall occur in the same place, dorsal hippocampal inactivation prior to recall prevents the fear recovery (Corcoran & Maren, 2004). Together, these data suggest that the dorsal hippocampus is sensitive to the relationship between the current context and the predicted US, driving the expression of learned fear in ambiguous environments.

Functional neuroimaging studies in humans suggest that the circuitry governing fear extinction in rodents is largely conserved across species. The BOLD signal in a subgenual region of the ventromedial prefrontal cortex (vmPFC), a putative homologue of the rodent infralimbic region, increases across the course of extinction learning, whereas activation in the amygdala shows a corresponding decrease (Phelps, Delgado, Nearing, & LeDoux, 2004). This inverse pattern of activation in the vmPFC and amygdala is also observed when extinction memories are retrieved (Kalisch et al., 2006; Milad, Wright, et al., 2007; Phelps et al., 2004), and both the magnitude of the vmPFC BOLD signal change (Milad, Wright et al., 2007) as well as the thickness of the cortical gray matter in this region (Hartley, Fischl, & Phelps, 2011; Milad et al., 2005) have been shown to correlate with fear extinction recall. BOLD activation in the dorsal anterior cingulate cortex in fear-conditioning studies commonly increases in response to CS presentation. Cortical thickness in this region and CS-evoked BOLD activation correlate with fear expression (Milad, Quirk, et al., 2007), motivating the proposal that the dorsal anterior cingulate cortex may represent a human homologue of the rodent prelimbic region. As in the rodent, the human hippocampus appears to play a critical role in the context-dependent expression of fear and extinction learning. Experimental paradigms in which fear and extinction learning are associated with visually distinct contextual stimuli have observed hippocampal BOLD activation during context-specific extinction retrieval (Kalisch et al., 2006; Milad, Wright, et al., 2007), and the magnitude of this signal change correlated positively with activation in the vmPFC (Milad, Wright, et al., 2007). The absence of context-dependent fear reinstatement in patients with hippocampal lesions (LaBar & Phelps, 2005) comports with parallel findings in experimental lesion studies in rodents (Wilson, Brooks, & Bouton, 1995).

INDIVIDUAL VARIATION AND CLINICAL IMPLICATIONS OF EXTINCTION LEARNING Extinction learning and retention are impaired in a number of anxiety-related disorders (Graham & Milad, 2011; Lissek, Pine, & Grillon, 2006; Milad et al., 2009), suggesting that attenuated fear extinction may confer psychiatric vulnerability. However, the factors governing an individual's ability to form and retain fear extinction memories are not presently well understood. Extinction learning and retention appear to be relatively stable individual characteristics (Bush, Sotres-Bayon, & LeDoux, 2007; Fredrikson, Annas, Georgiades, Hursti, & Tersman, 1993; Zeidan et al., 2012), and recent research has begun to clarify how individual variation in genetic background, sex, and developmental stage modulate fear extinction.

Fear conditioning is substantially heritable (Hettema, Annas, Neale, Kendler, & Fredrikson, 2003), likely reflecting the interactive contribution of multiple genes that influence fear learning and retrieval processes. Convergent evidence in both humans and genetically modified mice suggests that genetically mediated reduction in the expression of brain-derived neurotrophic factor impairs fear extinction learning (Soliman et al., 2010), whereas reduction in the expression of the serotonin transporter selectively impairs extinction recall (Hartley et al., 2012; Wellman et al., 2007). Genetic variation in the dopaminergic (Lonsdorf et al., 2009;

Raczka et al., 2011) and endocannabinoid system (Gunduz-Cinar et al., 2012) also appears to modulate extinction learning and retention, consistent with pharmacological evidence of the importance of these neuromodulatory systems in fear extinction (Haaker et al., 2013; Rabinak et al., 2013).

The prevalence of anxiety and mood disorders is twice as high in women as in in men (Breslau et al., 1998; Pigott, 2003). Recent studies in both rodents and humans examining sex differences in fear extinction across the menstrual cycle indicate that cyclic phases of low estrogen are associated with diminished extinction retention (Milad et al., 2010; Zeidan et al., 2012). This work suggests that normal hormonal fluctuations in women are accompanied by changes in fear memory consolidation and retrieval that might contribute to the sex disparity in the vulnerability to anxiety.

The neurocircuitry supporting fear regulation undergoes substantial changes across development, paralleled by marked qualitative changes in fear extinction processes (Casey, Pattwell, Glatt, & Lee, 2012). Extinction learning in preweanling rats does not recruit the vmPFC (Kim, Hamlin, & Richardson, 2009), and these animals do not exhibit typical fear reemergence phenomena following extinction (Kim & Richardson, 2007; Yap & Richardson, 2007). These data suggest that extinction in juveniles may effectively overwrite or erase the amygdala-based representation of a learned fear, instead of establishing a competing inhibitory memory. Data from recent studies in both rodents and humans suggest that during adolescence, both the acquisition (Pattwell et al., 2012) and retention (McCallum, Kim, & Richardson, 2010) of extinction learning are compromised, suggesting that adolescence may be a period of particular vulnerability to persistent fear. Collectively, these data suggest that developmental changes in extinction processes may represent windows of opportunity or resistance for the treatment of fears acquired early in life.

Extinction-based exposure therapies, in which patients confront a fear-eliciting memory or stimulus in a safe clinical context (Foa, 2011), are a common behavioral approach to the treatment of anxiety disorders. While exposure therapy has demonstrated efficacy in a range of anxiety disorders, not all patients are responsive. Individual variation in the ability to acquire and retain extinction learning, influenced by the factors described in the previous section, might modulate treatment efficacy. Recent research has identified several pharmacological agents that improve extinction learning and retention and could potentially be useful adjuncts to behavioral therapy. Diverse classes of drugs have been shown to facilitate extinction learning and/or retention, including the NMDA agonist D-cycloserine (DCS; Mao, Hsiao, & Gean, 2006; Walker, Ressler, Lu, & Davis, 2002), cannabinoids (Lafenetre, Chaouloff, & Marsicano, 2007; Rabinak et al., 2013), adrenergic antagonists (Morris & Bouton, 2007), estradiol (Milad et al., 2009), corticosterone (Gourley, Kedves, Olausson, & Taylor, 2009), brain-derived neurotrophic factor agonists (Andero et al., 2011), growth factors (Graham & Richardson, 2011), and dopamine precursor L-DOPA (Haaker et al., 2013). Multiple studies have reported increased efficacy of behavioral therapy when performed in conjunction with DCS administration (Hofmann et al., 2006; Otto et al., 2010; Ressler et al., 2004). Interestingly, D-cycloserine has no effect on fear extinction learning in healthy human subjects, suggesting that this drug is most effective in circumstances of pathological fear (Grillon, 2009; Guastella, Lovibond, Dadds, Mitchell, & Richardson, 2007). These data suggest that the treatment of anxiety can benefit by combining extinction-based therapies with substances that ameliorate the underlying associative learning processes upon which these techniques depend.

Active coping

In the field of behavioral ecology, the term *coping style* is used to describe the way an organism typically responds to an aversive stimulus or situation (Koolhaas et al., 1999). Coping styles are thought to exist along a continuum from proactive to reactive, where proactive coping involves responses designed to actively manipulate an individual's environment and reactive styles are dominated by behavioral inhibition and withdrawal. Relative to a proactive style, reactive coping is associated with an enhanced likelihood of deleterious health outcomes (Koolhaas et al., 1999). Although passively freezing in response to a CS is a deeply ingrained reactive behavior in the rodent stemming from innate defenses against predation (Fanselow & Lester, 1988), animals can also learn to respond to anticipated threat in a proactive manner. Coping actively with a fear requires a sequence of distinct learning processes. First, one must learn that a stimulus poses a threat (i.e., fear conditioning). Next, one must learn an instrumental action that can be taken to avoid, escape, or otherwise exert control over the feared stimulus. Recent research suggests the exercise of instrumental control over a stressor can yield persistent neural changes that alter subsequent fear expression. Through these processes, active coping may foster control over fear responses in both present and future aversive situations.

In instrumental conditioning paradigms such as escape from fear (EFF), animals learn to replace a reactive behavior with a proactive one that allows them to limit their contact with an aversive CS. EFF is divided into three phases: an initial CS-US fear-conditioning phase, an instrumental phase, in which the animal can inactivate the CS by performing a specific behavior (e.g., rearing), and a test phase, in which a retrieval CS that cannot be instrumentally controlled is presented. Animals that successfully learn to terminate the aversive CS through instrumental action show far less freezing and more active behavior during the test compared to yoked controls, for whom the unreinforced CSs experienced during the instrumental phase are equivalent to an extinction session (Cain & LeDoux, 2007). Lesions of LA and BA, but not CE, disrupt the acquisition of EFF (Amorapanth et al., 2000). Instrumental learning is thought to depend upon direct BA projections to the striatum, a known substrate of goal-directed action (Cain & LeDoux, 2008). This is intriguing, because while lesion or inactivation of LA and CE disrupt fear conditioning (LeDoux, Cicchetti, Xagoraris, & Romanski, 1990; LeDoux, Iwata, Cicchetti, & Reis, 1988; Wilensky et al., 2006), pretraining lesions of BA have no impact on Pavlovian learning (Anglaga-Figueroa & Quirk, 2005). This double dissociation between the CE and BA suggests that instrumental and Pavlovian learning are processed through distinct amygdalar pathways emanating out from LA. Inactivation of the CE promotes active avoidance behavior in animals that previously showed no evidence of instrumental learning (Lazaro-Munoz, LeDoux, & Cain, 2010), suggesting that these nuclei may mediate a direct competition between passive and active responses to threat.

Actively responding to an aversive CS also recruits the infralimbic mPFC (IL) pathway, which is implicated in suppressing learned fear during extinction. Experiments using a signaled active-avoidance paradigm, in which an action performed during the CS prevents US delivery, suggest that inactivation of the IL slows avoidance learning and increases the mutually incompatible conditioned freezing response. Crucially, active avoidance causes an IL-dependent attenuation of conditioned freezing both in and out of the training environment, suggesting that instrumental memory yields a generalizable inhibition of conditioned fear (Moscarello & LeDoux, 2013). This lasting effect of instrumental control is consistent with evidence from alternative paradigms suggesting that active control over aversive situations yields plastic changes in the IL that diminish subsequent conditioned fear expression (Baratta et al., 2007; Baratta et al., 2009) and buffer the subsequent response to even uncontrollable stressors at later time (Amat et al., 2005, 2006). Because both active coping and extinction learning recruit the IL to suppress conditioned fear, this structure appears to be a final common pathway for the inhibition of innate physiological and behavioral reactions that constitute fear expression.

Few studies in humans have probed the neural mechanisms underlying active coping. One recent functional MRI (fMRI) study in which subjects could avoid a cued shock by pressing a key observed heightened correlation of activity within the amygdala and the ventral striatum in the avoidance condition (Delgado et al., 2009), consistent with the role of this pathway in the shift from passive to active responses to fear (Amorapanth, LeDoux, & Nader, 2000). In another study in which subjects had to avoid a virtual predator that would ultimately deliver shock (Mobbs et al., 2009), activity in the vmPFC increased during avoidance phase, consistent with a role for the IL in supporting active avoidance in rodents (Moscarello & LeDoux, 2013). A recent behavioral study examining the effects of avoidance learning on subsequent fear conditioning indicates that, as in rodents (Baratta et al., 2007), experiences of instrumental control can diminish the subsequent expression of conditioned fear (Hartley et al., 2013), suggesting a mechanism by which active coping may promote psychological resilience (Maier & Watkins, 2010). Collectively, this work suggests that in environments that afford the ability to escape or disarm a present threat through action, inhibiting passive fear expression in favor of proactive coping responses can have lasting beneficial consequences.

Cognitive regulation

Studies of active coping investigate how overt, instrumental behaviors can diminish fear expression. However, another means by which humans can actively control fear is to change their thoughts. Theories of emotion highlight how the interpretation or appraisal of an event can influence the emotional response (Scherer, 2005). Although a few cognitive techniques have been shown to diminish fear, most studies examine strategies that emphasize reinterpreting the emotional significance of the event (see Ochsner & Gross, 2008, for a review and chapter 62 in this volume). By actively manipulating this appraisal process, fear or negative affect can be diminished.

In a typical study of cognitive emotion regulation, participants are presented negative emotional scenes and asked to reinterpret the events depicted in the scene in such a way as to reduce their negative affective response, such as imagining that a scene depicting a bloody wound is fake. This type of active reappraisal has been shown to be effective at reducing negative affect using both self-report and physiological measures of emotion (Ochsner & Gross, 2008). BOLD studies of reappraisal consistently report decreased amygdala activation and increased activation of the dorsolateral prefrontal cortex (dlPFC) and/or ventrolateral prefrontal cortex, along with some involvement of medial PFC (mPFC) regions. From these studies, a general model of the cognitive regulation of fear or negative affect has emerged, in which the dlPFC is involved in the effortful manipulation or interpretation of the stimulus and the changes in the amygdala are the result of the top-down modulation of the emotional meaning of the stimulus (see chapters 61 and 62 in this volume for more detailed reviews). One important aspect of this model is that the dlPFC does not project directly to the amygdala (Barbas, 2000; McDonald, Mascagni, & Guo, 1996). Instead, its influence is likely mediated by mPFC regions that have stronger connections with the amygdala (Urry et al., 2006).

An important aspect of the studies of cognitive regulation of fear is that PFC inhibition of the amygdala is critical to the control of fear. In an effort to directly compare the function of the PFC in the inhibition of the amygdala across extinction and cognitive regulation, Delgado, Nearing, LeDoux, and Phelps (2008) examined the regulation of conditioned fear. In this study, the cognitive regulation technique resulted in decreased fear expression accompanied by increased BOLD activation of the dlPFC, decreased activation of the amygdala, and increased activation of a region of the vmPFC overlapping with that observed in fear extinction (Delgado et al., 2008; Phelps et al., 2004). In a direct comparison with data from an extinction study, similar patterns of activation were observed in the amygdala and vmPFC when CRs were diminished through either extinction or cognitive regulation. However, only the cognitive regulation paradigm resulted in increased BOLD activation of the dlPFC, consistent with a function for this region in the online manipulation or reinterpretation of the meaning of the CS. When comparing responses across these regions during cognitive regulation, it was found that BOLD responses in the vmPFC were correlated with both the dlPFC and the amygdala. These results suggest a model by which the dlPFC inhibition of the amygdala during cognitive regulation is mediated through the same vmPFC region thought to mediate the inhibition of fear with extinction. It is possible that, much like the generation of fear through cognitive and social means relies on the amygdala for expression, the inhibition of fear through cognitive regulation relies on a phylogenetically shared vmPFC-amygdala circuitry. Although most cognitive regulation techniques are unique to humans, by linking components of the neural circuitry of extinction with regulation, we gain some insight into additional potential details of the neural mechanisms underlying the cognitive control of fear.

Laboratory studies of cognitive emotion regulation techniques are proposed to mirror the processes used during cognitive therapy in the clinic. A primary goal of cognitive therapy is to enable the patient to form more realistic evidence-based appraisals of a situation, thereby regulating the associated emotional responses (Allen, McHugh, & Barlow, 2008). This training likely uses the neural pathways engaged during cognitive regulation. Consistent with this suggestion, a recent study reported that fMRI activation in response to fearful faces in the amygdala and vmPFC predicts success of cognitive-behavioral therapy treatment in PTSD patients (Bryant et al., 2007). This suggests that the efficacy of such treatment may rely on the functional integrity of the earlier discussed neural circuitry and the success with which individuals are able to engage these regulatory mechanisms.

Reconsolidation

The fear-regulation techniques reviewed thus far all require the inhibition of the fear memory via the PFC. Because the original fear memory is not significantly altered, the fear may return under a variety of circumstances. For example, the neurohormonal changes that occur with nonspecific stress may lead to functional impairment of the PFC (Arnsten, 2009) and recent studies have shown that the success of both extinction retrieval (Raio, Brignoni-Perez, Goldman, & Phelps, 2014) and cognitive regulation (Raio et al., 2013) diminishes under stress. A potentially more persistent technique to control fear is to target the reconsolidation of the original fear memory. After initial acquisition, memories go through a consolidation process that transitions the labile, short-term representation to a stable form for long-term storage. Retrieval of a previously consolidated memory can make it labile again, and a process of reconsolidation must occur in order for it to regain stability. If the reconsolidation process is disrupted or modified, the original memory will not be accessible for more than a brief period beyond retrieval (for review, see Johansen, Cain, Ostroff, & LeDoux, 2011; Nader & Einarsson, 2010; Nader, Schafe, & LeDoux, 2000).

The consolidation and reconsolidation of Pavlovian fear memory require the synthesis of new proteins in LA. The injection of a protein-synthesis inhibitor into that structure disrupts both processes, destroying evidence of the memory at long-term time points. While all long-term memories go through a protein synthesis–dependent consolidation period, not all instances of retrieval trigger reconsolidation. In rodent studies of Pavlovian fear learning, retrieval involves the presentation of a reminder stimulus that reactivates the original memory. Reconsolidation only occurs if this stimulus somehow violates expectations, suggesting that this process functions to introduce new information into the existing memory. If the retrieval stimulus in a CS-US pairing is identical to those presented during training, protein-synthesis inhibitors injected into LA have no effect on the later expression of fear CRs (Díaz-Mataix, Ruiz Martinez, Schafe, LeDoux, & Doyère, 2013; Wang, de Oliveira Alvares, & Nader, 2009)—suggesting that no reconsolidation occurred. However, if retrieval is prompted by the CS or the US presented alone (Díaz-Mataix, Debiec, LeDoux, & Doyère, 2011), or by a US presented at a novel time during the CS (Debiec, Díaz-Mataix, Bush, Doyère, & Ledoux, 2010; Díaz-Mataix et al., 2013; Wang et al., 2009), then protein-synthesis inhibitors injected into LA strongly diminish fear CRs during a subsequent test.

Reconsolidation may be triggered by a mismatch between the reminder and the original training, but to change the fear memory an additional manipulation of reconsolidation is necessary. As outlined above, reconsolidation can be disrupted with a pharmacological agent, but it also possible to change the memory by introducing new information into the existing memory during reconsolidation. Research on reconsolidation-extinction boundaries demonstrates that the effectiveness of extinction is strongly increased if it occurs during the reconsolidation window, when the memory is in a temporarily labile state. Following a single retrieval CS with an extinction session 10 or 60 minutes later leads to a more lasting reduction of conditioned fear than extinction sessions 6 or 24 hours after the retrieval CS, when the reconsolidation window is presumably closed (Monfils, Cowansage, Klann, & LeDoux, 2009). Crucially, this manipulation is relatively insensitive to renewal, reinstatement, and spontaneous recovery (Monfils et al., 2009). While reconsolidation-extinction boundaries are an area of open inquiry (for review, see Auber, Tedesco, Jones, Monfils, & Chiamulera, 2013), these data suggest that reconsolidation can be used to reduce fear without the use of potentially toxic drugs, and underscore the idea that old memories can be altered with a well-timed intervention.

In humans, the ability to persistently target fear memories without toxic pharmacological manipulations is crucial for its potential translation to clinical disorders. Although there is some evidence that a relatively safe drug, propranolol, can influence fear reconsolidation in humans, its effects do not extend to multiple measures of conditioned fear (Kindt, Soeter, & Vervliet, 2009), suggesting it may target the expression of specific fear-output measures as opposed to the fear memory itself (see Schiller & Phelps, 2011, and Lonergan, Olivera-Gigueroa, Pitman, & Brunet, 2013, for reviews). However, introducing extinction training after fear-memory retrieval in humans has been shown to persistently diminish the expression of fear for up to a year (Schiller et al., 2009) and reduce subsequent amygdala BOLD responses (Agren et al., 2012). It is proposed that, in contrast to standard extinction that inhibits the fear memory via the vmPFC, extinction training during the reconsolidation window serves to update and modify the original fear memory. A recent fMRI study compared BOLD responses during standard extinction training and extinction during reconsolidation. Although the extinction procedure was identical, extinction during reconsolidation failed to engage the vmPFC and reduced amygdala-vmPFC connectivity, relative to standard extinction (Schiller et al., 2013). These results are consistent with rodent studies suggesting that timing extinction training to coincide with reconsolidation induces plasticity in the LA that may alter the original fear memory (Monfils et al., 2009; Clem & Huganir, 2010), thus reducing the need to inhibit it.

It is important to note that the strength of initial training can impose a temporary boundary condition on reconsolidation. With strong training in rodents, presentation of the CS alone cannot trigger reconsolidation in the first week after learning, but can do so one month later. In comparison, when initial training is weaker, the CS alone triggers reconsolidation just 24 hours after conditioning (Monfils et al., 2009; Wang et al., 2009). Intriguingly, reconsolidation can be triggered 24 hours after strong training if the US occurs at an earlier than expected time during the CS (Díaz-Mataix et al., 2013). In order to be able to translate this potentially exciting technique to clinical interventions, future research will need to more fully illuminate the potential constraints of fear memory reconsolidation—including the relationship between strength of training, the age of memory, and the precise mismatch between the reminder and learning.

Conclusion

In spite of the tremendous advances in human neuroscience techniques in the last 20 years, our ability to investigate the specificity of neural circuits in humans is still relatively limited. It is because of these limitations that exploring the similarities of neural circuits across species is especially important. However, it is equally important to acknowledge that everyday human fears are much more complex and nuanced than fears induced in a laboratory setting. By using animal models as a basis for understanding the neural systems of human emotion and social processing, one can begin to explore when and how these models can be extended to human experience, and when they cannot. In this chapter, we have outlined detailed animal models of the neurobiology of associative fear learning and its control and attempted to show how they can inform our understanding of the human cognitive neuroscience of emotion. From this basis, we can begin to explore the involvement of these processes and neural circuits in more complex and, one might argue, more interesting human experiences.

REFERENCES

Agren, T., Engman, J., Frick, A., Björkstrand, J., Larsson, E. M., Furmark, T., & Fredrikson, M. (2012). Disruption of reconsolidation erases a fear memory trace in the human amygdala. *Science, 337*(6101), 1550–1552.

Allen, L. B., McHugh, R. K., & Barlow, D. H. (2008). Emotional disorders: A unified protocol. In D. H. Barlow (Ed.), *Clinical handbook of psychological disorders* (4th ed., pp. 216–249). New York, NY: Guilford Press.

Amat, J., Baratta, M. V., Paul, E., Bland, S. T., Watkins, L. R., & Maier, S. F. (2005). Medial prefrontal cortex determines how stressor controllability affects behavior and dorsal raphe nucleus. *Nat Neurosci, 8*(3), 365–371.

Amat, J., Paul, E., Zarza, C., Watkins, L., & Maier, S. (2006). Previous experience with behavioral control over stress blocks the behavioral and dorsal raphe nucleus activating effects of later uncontrollable stress: Role of the ventral medial prefrontal cortex. *J Neurosci, 26*(51), 13264–13272.

Amorapanth, P., LeDoux, J. E., & Nader, K. (2000). Different lateral amygdala outputs mediate reactions and actions elicited by a fear-arousing stimulus. *Nat Neurosci, 3*(1), 74–79.

Andero, R., Heldt, S. A., Ye, K., Liu, X., Armario, A., & Ressler, K. J. (2011). Effect of 7, 8-dihydroxyflavone, a small-molecule TrkB agonist, on emotional learning. *Am J Psychiatry, 168*, 163–172.

Anglaga-Figueroa, D., & Quirk, G. J. (2005). Lesions of the basal amygdala block expression of conditioned fear but not extinction. *J Neurosci, 25*(42), 9680–9685.

Arnsten, A. F. (2009). Stress signalling pathways that impair prefrontal cortex structure and function. *Nat Rev Neurosci, 10*(6), 410–422.

Auber, A., Tedesco, V., Jones, C. E., Monfils, M. H., & Chiamulera, C. (2013). Post-retrieval extinction as reconsolidation interference: Methodological issues or boundary conditions? *Psychopharmacology (Berl), 226*(4), 631–647.

Baratta, M. V., Zarza, C. M., Gomez, D. M., Campeau, S., Watkins, L. R., & Maier, S. F. (2009). Selective activation of dorsal raphe nucleus-projecting neurons in the ventral medial prefrontal cortex by controllable stress. *Eur J Neurosci, 30*(6), 1111–1116.

Baratta, M., Christianson, J., Gomez, D., Zarza, C., Amat, J., Masini, C., ... Maier, S. (2007). Controllable versus uncontrollable stressors bi-directionally modulate conditioned but not innate fear. *Neuroscience, 146*(4), 1495–1503.

Barbas, H. (2000). Connections underlying the synthesis of cognition, memory, and emotion in primate prefrontal cortices. *Brain Res Bull, 52*, 319–330.

Bechara, A., Tranel, D., Damasio, H., Adolphs, R., Rockland, C., & Damasio, A. R. (1995). Double dissociation of conditioning and declarative knowledge relative to the amygdala and hippocampus in humans. *Science, 269*, 1115–1118.

Berretta, S., Pantazopoulos, H., Caldera, M., Pantazopoulos, P., & Paré, D. (2005). Infralimbic cortex activation increases c-Fos expression in intercalated neurons of the amygdala. *Neuroscience, 132*(4), 943–953.

Breslau, N., Kessler, R. C., Chilcoat, H. D., Schultz, L. R., Davis, G. C., & Andreski, P. (1998). Trauma and posttraumatic stress disorder in the community: The 1996 Detroit Area Survey of Trauma. *Arch Gen Psychiatry, 55*, 626–632.

Bryant, R. A., Felmingham, K., Kemp, A., Das, P., Hughes, G., Peduto, A., & Williams, L. (2007). Amygdala and ventral anterior cingulate activation predicts treatment response to cognitive behaviour therapy for post-traumatic stress disorder. *Psychol Med, 38*, 556–561.

Büchel, C., Morris, J., Dolan, R. J., & Friston, K. J. (1998). Brain systems mediating aversive conditioning: An event-related fMRI study. *Neuron, 20*(5), 947–957.

Burgos-Robles, A., Vidal-Gonzalez, I., & Quirk, G. J. (2009). Sustained conditioned responses in prelimbic prefrontal neurons are correlated with fear expression and extinction failure. *J Neurosci, 29*(26), 8474–8482.

Burgos-Robles, A., Vidal-Gonzalez, I., Santini, E., & Quirk, G. J. (2007). Consolidation of fear extinction requires NMDA receptor-dependent bursting in the ventromedial prefrontal cortex. *Neuron, 53*(6), 871–880.

Bush, D. E., Sotres-Bayon, F., & LeDoux, J. E. (2007). Individual differences in fear: Isolating fear reactivity and fear recovery phenotypes. *J Trauma Stress, 20*(4), 413–422.

Cain, C. K., & LeDoux, J. E. (2007). Escape from fear: A detailed behavioral analysis of two atypical responses reinforced by CS termination. *J Exp Psychol Anim Behav Process, 33*(4), 451–463.

Cain, C. K., & LeDoux, J. E. (2008). Brain mechanisms of Pavlovian and instrumental aversive conditioning. In R. J. Blanchard, D. C. Blanchard, G. Griebel, & D. J. Nutt (Eds.), *Handbook of anxiety and fear* (pp. 103–124). New York, NY: Elsevier.

Casey, B. J., Pattwell, S. S., Glatt, C. E., & Lee, F. S. (2012). Treating the developing brain: implications from human imaging and mouse genetics. *Annu Rev Med, 64*, 427–439.

Cheng, D. T., Knight, D. C., Smith, C. N., & Helmstetter, F. J. (2006). Human amygdala activity during the expression of fear responses. *Behav Neurosci, 120*(6), 1187.

Clem, R. L., & Huganir, R. L. (2010). Calcium-permeable AMPA receptor dynamics mediate fear memory erasure. *Science, 330*(6007), 1108–1112.

Corcoran, K. A., & Maren, S. (2004). Factors regulating the effects of hippocampal inactivation on renewal of conditional fear after extinction. *Learn Mem, 11*(5), 598–603.

Corcoran, K. A., & Quirk, G. J. (2007). Activity in prelimbic cortex is necessary for the expression of learned, but not innate, fears. *J Neurosci, 27*(4), 840–844.

Debiec, J., Díaz-Mataix, L., Bush, D. E., Doyère, V., & LeDoux, J. E. (2010). The amygdala encodes specific sensory features of an aversive reinforcer. *Nat Neurosci, 13*(5), 536–537.

Delgado, M. R. (2009). Avoiding negative outcomes: Tracking the mechanisms of avoidance learning in humans during fear conditioning. *Front Behav Neurosci, 3,* 1–9.

Delgado, M. R., Nearing, K. I., LeDoux, J. E., & Phelps, E. A. (2008). Neural circuitry underlying the regulation of conditioned fear and its relation to extinction. *Neuron, 59,* 829–838.

Díaz-Mataix, L., Debiec, J., LeDoux, J. E., & Doyère, V. (2011). Sensory-specific associations stored in the lateral amygdala allow for selective alteration of fear memories. *J Neurosci, 31*(26), 9538–9543.

Díaz-Mataix, L., Ruiz Martinez, R. C., Schafe, G. E., LeDoux, J. E., & Doyère, V. (2013). Detection of a temporal error triggers reconsolidation of amygdala-dependent memories. *Curr Biol, 23*(6), 467–472.

Fanselow, M. S., & Lester, L. S. (1988). A functional behavioristic approach to aversively motivated behavior: Predatory imminence as a determinant of the topography of defensive behavior. In R. C. Bolles & M. D. Beecher (Eds.), *Evolution and learning* (pp. 185–212). Hillsdale, NJ: Erlbaum.

Foa, E. B. (2011). Prolonged exposure therapy: Past, present, and future. *Depress Anxiety, 28*(12), 1043–1047.

Fredrikson, M., Annas, P., Georgiades, A., Hursti, T., & Tersman, Z. (1993). Internal consistency and temporal stability of classically conditioned skin conductance responses. *Biol Psychol, 35*(2), 153–163.

Gallistel, C. R. (2003). Conditioning from an information processing perspective. *Behav Process, 62*(1–3), 89–101.

Goosens, K. A., & Maren, S. (2001). Contextual and auditory fear conditioning are mediated by the lateral, basal, and central amygdaloid nuclei in rats. *Learn Mem, 8*(3), 148–155.

Gourley, S. L., Kedves, A. T., Olausson, P., & Taylor, J. R. (2009). A history of corticosterone exposure regulates fear extinction and cortical NR2B, GluR2/3, and BDNF. *Neuropsychopharmacology, 34,* 7077–7160.

Graham, B. M., & Milad, M. R. (2011). The study of fear extinction: Implications for anxiety disorders. *Am J Psychiatry, 168*(12), 1255–1265.

Graham, B. M., & Richardson, R. (2011). Fibroblast growth factor-2 alters the nature of extinction. *Learn Mem, 18*(2), 80–84.

Grillon, C. (2009). D-cycloserine facilitation of fear extinction and exposure-based therapy might rely on lower-level, automatic mechanisms. *Biol Psychiatry, 66*(7), 636–641.

Guastella, A. J., Lovibond, P. F., Dadds, M. R., Mitchell, P., & Richardson, R. (2007). A randomized controlled trial of the effect of D-cycloserine on extinction and fear conditioning in humans. *Behav Res Ther, 45*(4), 663–672.

Gunduz-Cinar, O., Macpherson, K. P., Cinar, R., Gamble-George, J., Sugden, K., Williams, B., & Holmes, A. (2012). Convergent translational evidence of a role for anandamide in amygdala-mediated fear extinction, threat processing and stress-reactivity. *Mol Psychiatry, 18*(7), 813–823.

Haaker, J., Gaburro, S., Sah, A., Gartmann, N., Lonsdorf, T. B., Meier, K., ... Kalisch, R. (2013). Single dose of L-dopa makes extinction memories context-independent and prevents the return of fear. *Proc Natl Acad Sci USA, 110*(26), 2428–2436.

Hartley, C. A., Gorun, A., Reddan, M. C., & Phelps, E. A. (2013). Stressor controllability modulates fear extinction in humans. *Neurobiol Learn Mem.* Advance online publication. doi:10.1016/j.nlm.2013.12.003

Hartley, C. A., Fischl, B., & Phelps, E. A. (2011). Brain structure correlates of individual differences in the acquisition and inhibition of conditioned fear. *Cereb Cortex, 21*(9), 1954–1962.

Hartley, C. A., McKenna, M. C., Salman, R., Holmes, A., Casey, B., Phelps, E. A., & Glatt, C. E. (2012). Serotonin transporter polyadenylation polymorphism modulates the retention of fear extinction memory. *Proc Natl Acad Sci USA, 109*(14), 5493–5498.

Herry, C., Ciocchi, S., Senn, V., Demmou, L., Müller, C., & Lüthi, A. (2008). Switching on and off fear by distinct neuronal circuits. *Nature, 454*(7204), 600–606.

Herry, C., Ferraguti, F., Singewald, N., Letzkus, J. J., Ehrlich, I., & Lüthi, A. (2010). Neuronal circuits of fear extinction. *Eur J Neurosci, 31*(4), 599–612.

Hettema, J. M., Annas, P., Neale, M. C., Kendler, K. S., & Fredrikson, M. (2003). A twin study of the genetics of fear conditioning. *Arch Gen Psychiatry, 60*(7), 702.

Hofmann, S. G., Schulz, S. M., Meuret, A. E., Moscovitch, D. A., & Suvak, M. (2006). Sudden gains during therapy of social phobia. *J Consult Clin Psychol, 74*(4), 687–697.

Johansen, J. P., Cain, C. K., Ostroff, L. E., & LeDoux, J. E. (2011). Molecular mechanisms of fear learning and memory. *Cell, 147*(3), 509–524.

Kalisch, R., Korenfeld, E., Klaas, S., Weiskopf, N., Seymour, B., & Dolan, R. (2006). Context-dependent human extinction memory is mediated by a ventromedial prefrontal and hippocampal network. *J Neurosci, 26*(37), 9503–9511.

Kim, J. H., & Richardson, R. (2007). A developmental dissociation of context and GABA effects on extinguished fear in rats. *Behav Neurosci, 121*(1), 131–139.

Kim, J. H., Hamlin, A. S., & Richardson, R. (2009). Fear extinction across development: The involvement of the medial prefrontal cortex as assessed by temporary inactivation and immunohistochemistry. *J Neurosci, 29*(35), 10802–10808.

Kindt, M., Soeter, M., & Vervliet, B. (2009). Beyond extinction: Erasing human fear responses and preventing the return of fear. *Nat Neurosci, 12*(3), 256–258.

Koolhaas, J. M., Korte, S. M., de Boer, S. F., van der Vegt, B. J., van Reenen, C. G., Hopster, H., ... Blokhuis, H. J. (1999). Coping styles in animals: Current status in behavior and stress-physiology. *Neurosci Biobehav Rev, 23,* 925–935.

LaBar, K. S., Gatenby, C., Gore, J. C., LeDoux, J. E., & Phelps, E. A. (1998). Human amygdala activation during conditioned fear acquisition and extinction: A mixed trial fMRI study. *Neuron, 20,* 937–945.

LaBar, K. S., LeDoux, J. E., Spencer, D. D., & Phelps, E. A. (1995). Impaired fear conditioning following unilateral temporal lobectomy in humans. *J Neurosci, 15*(10), 6846–6855.

LaBar, K. S., & Phelps, E. A. (2005). Reinstatement of conditioned fear in humans is context-dependent and impaired in amnesia. *Behav Neurosci, 119*, 677–686.

Lafenetre, P., Chaouloff, F., & Marsicano, G. (2007). The endocannabinoid system in the processing of anxiety and fear and how CB1 receptors may modulate fear extinction. *Pharmacol Res, 56*, 367–381.

Lazaro-Munoz, G., LeDoux, J. E., & Cain, C. K. (2010). Sidman instrumental avoidance initially depends on lateral and basal amygdala and is constrained by central amygdala-mediated Pavlovian processes. *Biol Psychiatry, 15*, 1120–1127.

LeDoux, J. E., Cicchetti, P., Xagoraris, A., & Romanski, L. M. (1990). The lateral amygdaloid nucleus: Sensory interface of the amygdala in fear conditioning. *J Neurosci, 10*(4), 1062–1069.

LeDoux, J. E., Iwata, J., Cicchetti, P., & Reis, D. J. (1988). Different projections of the central amygdaloid nucleus mediate autonomic and behavioral correlates of conditioned fear. *J Neurosci, 8*(7), 2517–2529.

Li, H., Penzo, M. A., Taniguchi, H., Kopec, C. D., Huang, Z. J., & Li, B. (2013). Experience-dependent modification of a central amygdala fear circuit. *Nat Neurosci, 16*(3), 332–339.

Likhtik, E., Popa, D., Apergis-Schoute, J., Fidacaro, G. A., & Paré, D. (2008). Amygdala intercalated neurons are required for expression of fear extinction. *Nature, 454*(7204), 642–645.

Lissek, S., Pine, D. S., & Grillon, C. (2006). The strong situation: A potential impediment to studying the psychobiology and pharmacology of anxiety disorders. *Biol Psychol, 72*, 265.

Lonergan, M. H., Olivera-Figueroa, L. A., Pitman, R. K., & Brunet, A. (2013). Propranolol's effects on the consolidation and reconsolidation of long-term emotional memory in healthy participants: A meta-analysis. *J Psychiatry Neurosci, 38*, 222–231.

Lonsdorf, T. B., Weike, A. I., Nikamo, P., Schalling, M., Hamm, A. O., & Öhman, A. (2009). Genetic gating of human fear learning and extinction: Possible implications for gene-environment interaction in anxiety disorder. *Psychol Sci, 20*(2), 198–206.

Maier, S. F., & Watkins, L. R. (2010). Role of the medial prefrontal cortex in coping and resilience. *Brain Res, 1355*, 52–60.

Mao, S. C., Hsiao, Y. H., & Gean, P. W. (2006). Extinction training in conjunction with a partial agonist of the glycine site on the NMDA receptor erases memory trace. *J Neurosci, 26*(35), 8892–8899.

Maren, S., Phan, K. L., & Liberzon, I. (2013). The contextual brain: Implications for fear conditioning, extinction and psychopathology. *Nat Rev Neurosci, 14*(6), 417–428.

McCallum, J., Kim, J. H., & Richardson, R. (2010). Impaired extinction retention in adolescent rats: Effects of D-cycloserine. *Neuropsychopharmacology, 35*(10), 2134–2142.

McDonald, A. J., Mascagni, F., & Guo, L. (1996). Projections of the medial and lateral prefrontal cortices to the amygdala: A *Phaseolus vulgaris* leucoagglutinin study in the rat. *Neuroscience, 71*, 55–75.

Milad, M. R., Pitman, R. K., Ellis, C. B., Gold, A. L., Shin, L. M., Lasko, N. B., ... Rauch, S. L. (2009). Neurobiological basis of failure to recall extinction memory in posttraumatic stress disorder. *Biol Psychiatry, 66*, 1075–1082.

Milad, M., Quinn, B., Pitman, R., Orr, S., & Fischl, B. (2005). Thickness of ventromedial prefrontal cortex in humans is correlated with extinction memory. *Proc Natl Acad Sci USA, 102*(30), 10706–10711.

Milad, M. R., & Quirk, G. J. (2002). Neurons in medial prefrontal cortex signal memory for fear extinction. *Nature, 420*(6911), 70–74.

Milad, M. R., Quirk, G. J., Pitman, R. K., Orr, S. P., Fischl, B., & Rauch, S. L. (2007). A role for the human dorsal anterior cingulate cortex in fear expression. *Biol Psychiatry, 62*(10), 1191–1194.

Milad, M. R., Wright, C. I., Orr, S. P., Pitman, R. K., Quirk, G. J., & Rauch, S. L. (2007). Recall of fear extinction in humans activates the ventromedial prefrontal cortex and hippocampus in concert. *Biol Psychiatry, 62*, 446–454.

Milad, M. R., Zeidan, M. A., Contero, A., Pitman, R. K., Klibanski, A., Rauch, S. L., & Goldstein, J. M. (2010). The influence of gonadal hormones on conditioned fear extinction in healthy humans. *Neuroscience, 168*(3), 652–658.

Mobbs, D., Marchant, J. L., Hassabis, D., Seymour, B., Tan, G., Gray, M., ... Frith, C. D. (2009). From threat to fear: The neural organization of defensive fear systems in humans. *J Neurosci, 29*(39), 12236–12243.

Monfils, M. H., Cowansage, K. K., Klann, E., & LeDoux, J. E. (2009). Extinction-reconsolidation boundaries: Key to persistent attenuation of fear memories. *Science, 324*(5929), 951–955.

Morgan, M. A., & LeDoux, J. E. (1995). Differential contribution of dorsal and ventral medial prefrontal cortex to the acquisition and extinction of conditioned fear in rats. *Behav Neurosci, 109*(4), 681–688.

Morgan, M. A., Romanski, L. M., & LeDoux, J. E. (1993). Extinction of emotional learning: Contribution of medial prefrontal cortex. *Neurosci Lett, 163*,109–113.

Morris, R. W., & Bouton, M. E. (2007). The effect of yohimbine on the extinction of conditioned fear: A role for context. *Behav Neurosci, 121*(3), 501.

Moscarello, J. M., & LeDoux, J. E. (2013). Active avoidance learning requires prefrontal suppression of amygdala-mediated defensive reactions. *J Neurosci, 33*(9), 3815–3823.

Myers, K. M., & Davis, M. (2002). Behavioral and neural analysis of extinction. *Neuron, 36*(4), 567–584.

Nader, K., & Einarsson, E. Ö. (2010). Memory reconsolidation: An update. *Ann NY Acad Sci, 1191*(1), 27–41.

Nader, K., Majidishad, P., Amorapanth, P., & LeDoux, J. E. (2001). Damage to the lateral and central, but not other, amygdaloid nuclei prevents the acquisition of auditory fear conditioning. *Learn Mem, 8*(3), 156–163.

Nader, K., Schafe, G. E., & LeDoux, J. E. (2000). The labile nature of consolidation theory. *Nat Rev Neurosci, 1*(3), 216–219.

Ochsner, K. N., & Gross, J. J. (2008). Cognitive emotion regulation insights from social cognitive and affective neuroscience. *Curr Dir Psychol Sci, 17*(2), 153–158.

Olsson, A., & Phelps, E. A. (2007). Social learning of fear. *Nat Neurosci, 10*(9), 1095–1102.

Otto, M. W., Tolin, D. F., Simon, N. M., Pearlson, G. D., Basden, S., Meunier, S. A., ... Pollack, M. H. (2010). Efficacy of D-cycloserine for enhancing response to cognitive-behavior therapy for panic disorder. *Biol Psychiatry, 67*(4), 365–370.

Pattwell, S. S., Duhoux, S., Hartley, C. A., Johnson, D. C., Jing, D., Elliott, M. D., … Lee, F. S. (2012). Altered fear learning across development in both mouse and human. *Proc Natl Acad Sci USA, 109*(40), 16318–16323.

Pavlov, I. P. (1927). *Conditioned reflexes.* London: Routledge & Kegan Paul.

Phelps, E. A., Delgado, M. R., Nearing, K. I., & LeDoux, J. E. (2004). Extinction learning in humans: Role of the amygdala and vmPFC. *Neuron, 43*(6), 897–905.

Phelps, E. A., & LeDoux, J. E. (2005). Contributions of the amygdala to emotion processing: From animal models to human behavior. *Neuron, 48*(2), 175–187.

Pigott, T. A. (2003). Anxiety disorders in women. *Psychiatr Clin North Am, 26*(3), 621–672.

Quirk, G. J., Russo, G. K., Barron, J. L., & Lebron, K. (2000). The role of ventromedial prefrontal cortex in the recovery of extinguished fear. *J Neurosci, 20,* 6225–6231.

Rabinak, C. A., Angstadt, M., Sripada, C. S., Abelson, J. L., Liberzon, I., Milad, M. R., & Phan, K. L. (2013). Cannabinoid facilitation of fear extinction memory recall in humans. *Neuropharmacology, 64,* 396–402.

Raczka, K. A., Mechias, M. L., Gartmann, N., Reif, A., Deckert, J., Pessiglione, M., & Kalisch, R. (2011). Empirical support for an involvement of the mesostriatal dopamine system in human fear extinction. *Transl Psychiatry, 1*(6), e12.

Raio, C. M., Brignoni-Perez, E., Goldman, R., & Phelps, E. A. (2014). Acute stress impairs the retrieval of extinction memory in humans. *Neurobiol Learn Mem.* Advance online publication. doi:10.1016/j.nlm.2014.01.015

Raio, C. M., Orederu, T. A., Palazzolo, L., Shurick, A. A., & Phelps, E. A. (2013). Cognitive emotion regulation fails the stress test. *Proc Natl Acad Sci USA, 110*(37), 15139–15144.

Repa, J. C., Muller, J., Apergis, J., Desrochers, T. M., Zhou, Y., & LeDoux, J. E. (2001). Two different lateral amygdala cell populations contribute to the initiation and storage of memory. *Nat Neurosci, 4*(7), 724–731.

Rescorla, R. A. (1988). Behavioral studies of Pavlovian conditioning. *Annu Rev Neurosci, 11,* 329–352.

Ressler, K. J., Rothbaum, B. O., Tannenbaum, L., Anderson, P., Graap, K., Zimand, E., … Davis, M. (2004). Cognitive enhancers as adjuncts to psychotherapy: Use of D-cycloserine in phobic individuals to facilitate extinction of fear. *Arch Gen Psychiatry, 61*(11), 1136–1144.

Romanski, L. M., Clugnet, M. C., Bordi, F., & LeDoux, J. E. (1993). Somatosensory and auditory convergence in the lateral nucleus of the amygdala. *Behav Neurosci, 107*(3), 444–450.

Royer, S., Martina, M., & Paré, D. (1999). An inhibitory interface gates impulse traffic between the input and output stations of the amygdala. *J Neurosci, 19,* 10575–10583.

Schafe, G. E., Doyère, V., & LeDoux, J. E. (2005). Tracking the fear engram: The lateral amygdala is an essential locus of fear memory storage. *J Neurosci, 25*(43), 10010–10014.

Scherer, K. R. (2005). What are emotions? And how can they be measured? *Soc Sci Inform, 44*(4), 695–729.

Schiller, D., Kanen, J. W., LeDoux, J. E., Monfils, M. H., & Phelps, E. A. (2013). Extinction during reconsolidation of threat memory diminishes prefrontal cortex involvement. *Proc Natl Acad Sci USA, 110*(50), 20040–20045.

Schiller, D., Monfils, M. H., Raio, C. M., Johnson, D. C., LeDoux, J. E., & Phelps, E. A. (2009). Preventing the return of fear in humans using reconsolidation update mechanisms. *Nature, 463*(7277), 49–53.

Schiller, D., & Phelps, E. A. (2011). Does reconsolidation occur in humans? *Front Behav Neuro, 5,* 24.

Sierra-Mercado, D., Padilla-Coreano, N., & Quirk, G. J. (2011). Dissociable roles of prelimbic and infralimbic cortices ventral hippocampus and basolateral amygdala in the expression and extinction of conditioned fear. *Neuropsychopharmacology, 36*(2), 529–538.

Soliman, F., Glatt, C. E., Bath, K. G., Levita, L., Jones, R. M., Pattwell, S. S., … Casey, B. J. (2010). A genetic variant BDNF polymorphism alters extinction learning in both mouse and human. *Science, 327*(5967), 863–866.

Sotres-Bayon, F., & Quirk, G. J. (2010). Prefrontal control of fear: More than just extinction. *Curr Opin Neurobiol, 20*(2), 231–235.

Sotres-Bayon, F., Sierra-Mercado, D., Pardilla-Delgado, E., & Quirk, G. J. (2012). Gating of fear in prelimbic cortex by hippocampal and amygdala inputs. *Neuron, 76*(4), 804–812.

Urry, H. L., van Reekum, C. M., Johnstone, T., Kalin, N. H., Thurow, M. E., Schaefer, H. S., … Davidson, R. J. (2006). Amygdala and ventromedial prefrontal cortex are inversely coupled during regulation of negative affect and predict the diurnal pattern of cortisol secretion among older adults. *J Neurosci, 26*(16), 4415–4425.

Walker, D. L., Ressler, K. J., Lu, K. T., & Davis, M. (2002). Facilitation of conditioned fear extinction by systemic administration or intra-amygdala infusions of D-cycloserine as assessed with fear-potentiated startle in rats. *J Neurosci, 22*(6), 2343–2351.

Wang, S. H., de Oliveira Alvares, L., & Nader, K. (2009). Cellular and systems mechanisms of memory strength as a constraint on auditory fear reconsolidation. *Nat Neurosci, 12*(7), 905–912.

Wellman, C. L., Izquierdo, A., Garrett, J. E., Martin, K. P., Carroll, J., Millstein, R., … Holmes, A. (2007). Impaired stress-coping and fear extinction and abnormal corticolimbic morphology in serotonin transporter knockout mice. *J Neurosci, 27*(3), 684–691.

Wilensky, A. E., Schaffe, G. E., Kristense, M. P., & LeDoux, J. E. (2006). Rethinking the fear circuit: The central nucleus of the amygdala is required for the acquisition, consolidation and expression of Pavlovian fear conditioning. *J Neurosci, 26,* 12387–12396.

Wilson, A., Brooks, D. C., & Bouton, M. E. (1995). The role of the rat hippocampal system in several effects of context in extinction. *Behav Neurosci, 109*(5), 828–836.

Yap, C. S., & Richardson, R. (2007). Extinction in the developing rat: An examination of renewal effects. *Dev Psychobiol, 49*(6), 565–575.

Zeidan, M. A., Igoe, S. A., Linnman, C., Vitalo, A., Levine, J. B., Klibanski, A., … Milad, M. R. (2011). Estradiol modulates medial prefrontal cortex and amygdala activity during fear extinction in women and female rats. *Biol Psychiatry, 70*(10), 920–927.

Zeidan, M. A., Lebron-Milad, K., Thompson-Hollands, J., Im, J. J., Dougherty, D. D., Holt, D. J., & Milad, M. R. (2012). Test–retest reliability during fear acquisition and fear extinction in humans. *CNS Neurosci Ther, 18*(4), 313–317.

61 Self-Regulation and Its Failures

DYLAN D. WAGNER AND TODD F. HEATHERTON

ABSTRACT The ability to flexibly alter behavior in the service of future goals is one of the key evolutionary adaptations that has enabled humankind to flourish. Self-regulation refers to a set of mental processes for overriding impulses, selectively attending to goal-relevant information, and monitoring thoughts and behavior for signs of failure. Although self-regulation is of fundamental importance to an individual's success, failures of self-regulation are common. In this chapter we focus on the role of exposure to tempting cues, negative emotions, and limited cognitive resources in bringing about self-regulation failure. We review evidence from neuropsychology and functional neuroimaging that successful self-regulation depends on the interaction between brain structures in the prefrontal cortex involved in representing goals and directing attention away from goal-irrelevant stimuli and cortical and subcortical structures involved in representing the value of rewards during decision making. When the balance between these countervailing systems is disrupted as a result of overwhelming impulses, negative affect, or deficient top-down control, self-regulation failure ensues.

More than any other species, humans are especially talented at controlling their own behavior in order to follow their goals and abide by rules and laws. Outside of certain psychiatric and neurological conditions, even the most impulsive human is still leaps and bounds more capable of not blowing off work, of not eating everything in the pantry, or of not stealing their spouse's food than are the most precocious of nonhuman primates. Despite humankind's enormous advantage in this domain, successful self-regulation remains difficult and failures are common (Baumeister & Heatherton, 1996; Wagner & Heatherton, in press). How is it, then, that humans are capable of inhibiting urges and pushing aside temptations in order to pursue their goals?

One commonly held view is that humans have evolved specific mental faculties that allow for superior planning and behavioral flexibility. These adaptations underlie humans' apparent superiority at self-regulation, and it has been hypothesized that they arose from a disproportionate amount of cortical expansion of the prefrontal cortex (PFC) over the course of human evolution (Rilling, 2006). Given the known role of the PFC in self-control, it was reasonable to assume that the brain would show specific structural changes to support these putative cognitive adaptations. However, recent work suggests that relatively larger size of the PFC in humans compared to nonhuman primates may have been overstated (Semendeferi, Lu, Schenker, & Damasio, 2002). Instead, it has been suggested that, rather than overall size, the human PFC demonstrated increased white matter connectivity (i.e., Schoenemann, Sheehan, & Glotzer, 2005; although see Barton & Venditti, 2013). Regardless of whether the human PFC shows evidence of specialized morphological enlargement, that humans possess a unique capacity for planning and self-regulation compared to all other animals appears irrefutable.

Or does it? Just how unique is the human capacity for self-regulation? Comparative psychological research on nonhuman primates and other animals has generally found that, with the exception of some domesticated animals (e.g., Miller, Pattison, DeWall, Rayburn-Reeves, & Zentall, 2010), most nonhuman animals display remarkably poor ability to inhibit prepotent responses in order to obtain later, larger rewards (i.e., delay of gratification). For example, most nonhuman animals, including many primates, will tolerate delays of only a few seconds before consuming a desired food item (Green, Myerson, Holt, Slevin, & Estle, 2004; Ramseyer, Pelé, Dufour, Chauvin, & Thierry, 2006). There are, however, some rare exceptions. For instance, great apes and some species of birds will tolerate delays as long as a few minutes provided the expected reward greatly exceeds the value of the currently available item (e.g., Beran, 2002; Dufour, Wascher, Braun, Miller, & Bugnyar, 2012). Given the studies discussed above, it appears that outside of certain primates and species of birds, evidence of self-regulation-like behavior is sparse among nonhuman animals. Why is it that humans have evolved this complex capacity to self-regulate?

The importance of self-regulation for human social groups

Unlike many animals, humans display a prolonged period of development and are unable to care for themselves for the first decade of life. Being a member of a social group brings with it tremendous advantages, from sharing the burden of child rearing to cooperative hunting and food sharing (Buss & Kenrick, 1998). As human safety and survival has long depended on living in groups, it has been suggested that humans have a fundamental need to belong (Baumeister & Leary, 1995) that motivates them to avoid behaviors that could lead to their expulsion from the group (e.g., theft of common resources) as this would greatly lessen their chances for survival (Goodall, 1986; Heatherton, 2011). Among humans and possibly some great apes (see Jensen, Call, & Tomasello, 2007), cheaters who fail to share with other members of the group could face expulsion from the group (Kurzban & Leary, 2001). In fact, it is precisely this punishment of cheating behavior that enables cooperation to be an evolutionarily sound strategy for some social animals (Boyd, Gintis, & Bowles, 2010). Thus, by allowing individuals to inhibit impulses and bring their behavior in line with group standards, self-regulation has permitted humans to benefit from all the perks of living in social groups, such as cooperation and food sharing but also the transmission of knowledge and culture.

What is self-regulation?

At its core, self-regulation is concerned with starting, stopping, or modifying thoughts, emotions, or behavior in order to pursue goals or stay in line with societal norms. Self-regulation encompasses both internal modes of control, such as when people regulate their thoughts or attempt to change their emotional states, and also external ones, such as when people initiate or stop a behavior (e.g., starting work or stopping oneself from overeating). Although many models of self-regulation exist, most share a similar framework characterized by the capacity to set goals, regulate thoughts, behaviors, or emotions, and monitor for signs of failure (Bandura, 1991; Baumeister & Heatherton, 1996; Carver & Scheier, 1981; Heatherton, 2011; Metcalfe & Mischel, 1999). As will be apparent, self-regulation shares many similarities with components of executive function in cognitive psychology and neuropsychology (e.g., Norman & Shallice, 1986) such as working memory and attention control. Indeed, aspects of executive function have since become incorporated into theories of self-regulation (Hofmann, Schmeichel, & Baddeley, 2012).

Cognitive neuroscience of self-regulation failure

Self-regulation failures typically occur upon exposure to a desired stimulus or following some precipitating event, such as emotional distress, alcohol consumption, or exhaustion of self-regulatory resources. Successful self-regulation relies on a delicate balance between the strength of urges and impulses on the one hand and the capacity to keep them in check on the other. Self-regulation failure then can occur due to a particularly strong impulse or when the capacity to engage in self-control is impaired or the motivation is absent. Below, we review three common threats to this balance: exposure to tempting cues (e.g., food, drugs), emotional and social distress, and depletion of self-regulatory resources.

Cue Exposure and Impulse Inhibition Modern-day humans are surrounded by a wealth of indulgences. At no other time in history has such a plethora of pleasures and vices been so readily available. It is perhaps little wonder, then, that the dominant form of self-regulation in daily life is impulse control (see Hofmann, Baumeister, Förster, & Vohs, 2011). In the psychological literature, an impulse typically refers to an urge or desire to consume a particular item or engage in a pleasurable behavior. With some exceptions, impulses are inherently rewarding behaviors that invade people's attention and require effort to inhibit (Metcalfe & Mischel, 1999). One of the most common ways an impulse can arise is from viewing an activating stimulus, such as food advertisements or the sight and smell of a cigarette. Studies show that physiological measures such as heart rate and salivary responses are increased following exposure to food cues in dieters (Brunstrom, Yates, & Witcomb, 2004) and cigarette cues in smokers (Drobes & Tiffany, 1997). Perhaps not surprisingly, then, exposure to desired items also increases craving for and consumption of the substance (e.g., Carter & Tiffany, 1999; Federoff, Polivy, & Herman, 1997; Sayette, Martin, Wertz, Shiffman, & Perrott, 2001). Perhaps more perniciously, being exposed to tempting substances can also have effects that many people are not consciously aware of. For example, tempting cues can capture people's attention, even when presented incidentally as part of a film (Lochbuehler, Voogd, Scholte, & Engels, 2011) and may activate motor schemas for using the substance (e.g., the action of holding and smoking a cigarette; see Tiffany, 1990). This last point is especially interesting in light of recent functional neuroimaging work suggesting that among smokers, viewing scenes of other people smoking activates brain regions associated with representing goal-directed actions (Wagner, Dal Cin, Sargent, Kelley, & Heatherton, 2011).

Nonhuman animal neurophysiology studies show that consuming rewards (foods, drugs) or engaging in rewarding activities (e.g., sex) is associated with activation of the mesolimbic dopamine system (i.e., the ventral tegmental area and nucleus accumbens/ventral striatum) and the orbitofrontal cortex (Damsma, Pfaus, Wenkstern, Phillips, & Fibiger, 1992; Kringelbach, 2005; Schilström, Svensson, Svensson, & Nomikos, 1998).

In humans, functional neuroimaging research has similarly shown that activity in the ventral striatum and orbitofrontal cortex increases when consuming (Gottfried, O'Doherty, & Dolan, 2003; Kringelbach, O'Doherty, Rolls, & Andrews, 2003) or viewing cues associated with appetitive rewards such as food or attractive faces, as well as abstract rewards such as money (Cloutier, Heatherton, Whalen, & Kelley, 2008; Knutson, Taylor, Kaufman, Peterson, & Glover, 2005; van der Laan, de Ridder, Viergever, & Smeets, 2011). Given the role of the reward system in motivating behavior, there has been considerable effort aimed at investigating whether people who are at risk for obesity or substance-abuse problems show any abnormalities in reward processing. So far the evidence seems to indicate that increased striatal responses to food (Demos, Heatherton, & Kelley, 2012) or drug cues (Janes et al., 2010; McClernon, Kozink, & Rose, 2008) are predictive of real-world behavior. For instance, people who are rated as being particularly sensitive to rewards show heightened food cue–related activity in the ventral striatum and orbitofrontal cortex (Beaver et al., 2006). A similar finding was demonstrated by Demos and colleagues (2012), in which individual differences in ventral striatal responses to food cues and erotic scenes predicted subsequent weight gain and degree of sexual activity in a six-month follow-up. Taken together, these findings suggest that individual differences in cue reactivity may reflect a stable sensitivity to rewards, the excess of which makes it more difficult for some individuals to maintain self-control. An extreme example of this comes from a study by Casey and colleagues (2011), which showed that people who had difficulty delaying gratification as children exhibited heightened activity in the ventral striatum when viewing appetitive stimuli over 40 years later.

Thus far we have discussed findings in which passive viewing of food or drug cues led to heightened reward-related neural responses in the striatum and orbitofrontal cortex. However, an important question for understanding self-regulation failure is what happens when people are explicitly engaging in self-control in order to reduce their craving for appetitive stimuli. In general, it has been found that explicit regulation of cravings and desires involves the lateral PFC and the anterior cingulate cortex. For instance, the lateral PFC and ACC show increased activity when attempting to down-regulate responses to cigarette paraphernalia among smokers (Brody et al., 2007; Kober et al., 2010), to food cues among dieters (Siep et al., 2012), drug cues among substance abusers (Volkow et al., 2010), and to monetary rewards for most people (Delgado, Gillis, & Phelps, 2008). Importantly, across all these studies, engaging in explicit self-regulation, whether through cognitive reappraisal or other means, resulted in decreased activity in the striatum and orbitofrontal cortex.

Social and Emotional Distress A frequently reported cause of self-regulation failure is the experience of emotional and social distress. For example, negative affect often precedes binge eating and binge drinking episodes (Haedt-Matt & Keel, 2011; Witkiewitz & Villarroel, 2009). Laboratory inductions of negative mood or social distress (e.g., social rejection) similarly show that experiencing negative affect leads to disinhibited behavior (e.g., Twenge, Baumeister, Tice, & Stucke, 2001). For instance, negative mood inductions lead dieters to subsequently overeat (Heatherton, Herman, & Polivy, 1991; Heatherton, Striepe, & Wittenberg, 1998) and smokers to crave smoking (Willner & Jones, 1996). Similarly, inducing social rejection has been shown to increase consumption of unhealthy foods, reduce task persistence, and interfere with the ability to sustain attention (Baumeister, DeWall, Ciarocco, & Twenge, 2005).

A number of mechanisms have been proposed to explain how emotional and social distress influence self-regulation (for a review, see Wagner & Heatherton, 2013b). Common among them is the notion that people are motivated to repair their mood and change their negative emotional state. This often takes the form of increased efforts to regulate emotions, which can come at a cost to self-regulation since the increased cognitive load incurred is thought to impair monitoring of ongoing behavior (e.g., Johns, Inzlicht, & Schmader, 2008) and reduce cognitive resources that could otherwise be used in the service of self-regulation.

Another mechanism whereby negative affect can lead to self-regulation failure comes from research suggesting that negative affect increases the perceived reward value of temptations, rendering them more difficult to inhibit. For instance, negative affect has been shown to reduce people's ability to delay gratification, biasing them toward accepting immediate monetary rewards over waiting for larger delayed payments (e.g., Lerner, Li, & Weber, 2013; Mischel, Ebbesen, & Zeiss, 1973; Twenge et al., 2001). In addition, experiencing negative affect is associated with increased cravings for carbohydrate-rich foods (Christensen & Pettijohn, 2001) and, among smokers, increases both the intensity of smoking (McKee et al., 2011) and the amount of pleasure people report from smoking a cigarette (Zinser, Baker, Sherman, & Cannon, 1992). Together, these studies suggest that negative affect may serve to ramp up the gain on temptations and pleasurable activities.

Indeed, research in nonhuman animals suggests that emotional distress (usually via social isolation) can elicit reward-seeking behavior owing to a stress-induced sensitization of brain regions involved in reward processing (e.g., Peciña, Schulkin, & Berridge, 2006; Piazza & Le Moal, 1996; Ramsey & van Ree, 1993). Research in humans provides converging evidence in the form of affect-related modulation of activity in the ventral striatum and orbitofrontal cortex (OFC) to appetitive cues (e.g., Killgore & Yurgelun-Todd, 2006; Wagner, Boswell, Kelley, & Heatherton, 2012). For instance, the experimental induction of negative mood has been shown to increase food-cue related activity in the OFC (Wagner et al., 2012).

Research on social rejection provides another means for examining the role of emotional distress on self-regulation. As mentioned above, a number of studies have shown that social rejection can bring about self-regulation failure (Baumeister et al., 2005). Although there is considerable research on the neural correlates of experiencing social rejection (for a review, see Eisenberger, 2012), there are far fewer cognitive neuroscience studies examining the link between social exclusion and self-regulation failure. Of the few extant studies, results show that social rejection leads to reduced activity in the PFC during executive function tasks. For example, social rejection was found to reduce activity in the lateral PFC as well as impair accuracy when completing complex math problems (Campbell et al., 2006). Similarly, Peake and colleagues (2013) demonstrated that social exclusion increased risk-taking behavior (operationalized as amount of crashes in a driving simulator) and reduced lateral PFC activity during crash trials.

Taken together, the results of these studies suggest two mechanisms whereby the experience of emotional and social distress impairs self-regulation. First, the experience of negative affect may serve to sensitize the ventral striatum and OFC to the incentive value of appetitive rewards (e.g., Peciña et al., 2006; Piazza & Le Moal, 1996). And second, negative affect may also act directly on self-control, reducing the capacity to engage in sustained attention or inhibit prepotent responses (e.g., Campbell et al., 2006; Peake, Dishion, Stormshak, Moore, & Pfeifer, 2013).

Depletion of Limited Self-Regulatory Resources

In the preceding section, we briefly touched upon the notion that one of the means by which negative affect can sabotage self-control is through imposing a cognitive load as individuals juggle regulating affect with regulating their behavior in other domains (e.g., food or drug consumption). One consequence of this sustained mental effort is that it can temporarily deplete cognitive resources required for self-regulation, thereby leaving people vulnerable to temptations. This view of self-regulation as being resource-limited and subject to fatigue is the central tenant of the limited resource, or *strength*, model of self-regulation (e.g., Baumeister & Heatherton, 1996). Since its formulation, the strength model has received support from a large number of studies within the laboratory (for a meta-analysis, see Hagger, Wood, Stiff, & Chatzisarantis, 2010) as well as more naturalistic experiments using experience-sampling methods (Hofmann et al., 2011). For example, engaging in a prior effortful self-control task can subsequently make people more vulnerable to temptations such as food and alcohol (Muraven, Collins, & Nienhaus, 2002; Vohs & Heatherton, 2000), less able to control their emotions (Schmeichel, 2007), and more likely to violate social norms (DeBono, Shmueli, & Muraven, 2011; Vohs, Baumeister, & Ciarocco, 2005).

In recent years, there have been several efforts aimed at defining the mechanisms underlying self-regulatory depletion effects, with researchers suggesting that depletion reflects temporary decreases in circulating blood glucose availability (Gailliot et al., 2007), lay beliefs in self-control (Job, Dweck, & Walton, 2010), or shifts in motivation and attention away from effortful control and toward more rewarding activities (Beedie & Lane, 2012; Inzlicht & Schmeichel, 2012). More recently, it has been suggested that the strength model may have overemphasized the role of self-regulatory depletion impairing top-down control, neglecting the possibility that depletion may also increase the strength of temptations and impulses, thereby making them more difficult to resist. Evidence for this last conjecture comes from research demonstrating that following self-regulatory depletion, people rate emotions as more extreme, pain as more intense, and desires as more strongly felt than nondepleted individuals (Vohs, Baumeister, Mead, Ramanathan, & Schmeichel, manuscript submitted for publication).

Research on the neural basis of self-regulatory depletion is still in its infancy, with only a handful of studies investigating the effects of depletion on the neural systems involved in self-control. Within the cognitive domain, three studies have examined the effects of self-regulatory depletion on the subsequent recruitment of prefrontal brain regions during self-control tasks. The first of these studies measured the error-related negativity, an index of conflict monitoring thought to originate in the ACC, and found that relative to a control group, depleted participants showed a reduced error-related negativity during a subsequent task requiring self-control and that this effect mediated behavioral performance impairments (Inzlicht & Gutsell, 2007). Two other experiments examined the aftereffect of depletion using functional neuroimaging. In both cases it was found that the right lateral PFC showed reduced activity in participants who first completed an effortful self-control task, compared to control subjects (Friese, Binder, Luechinger, Boesiger, & Rasch, 2013; Hedgcock, Vohs, & Rao, 2012).

As mentioned above, recent behavioral work suggests that another possible aftereffect of self-regulatory depletion is the intensification of desires and emotions (e.g., Schmeichel, 2007; Vohs et al., 2013). Consistent with this theory are results from two functional neuroimaging studies, one in the emotional domain and the other using tempting food cues. In the first of these, participants were depleted using a difficult attention-control task and subsequently exposed to a series of emotional scenes. Compared to nondepleted individuals, depleted subjects viewing negatively valenced emotional scenes showed an exaggerated response in the amygdala, a region involved in the detection of threat (Wagner & Heatherton, 2013a). Moreover, this was accompanied by reduced functional connectivity between the amygdala and the ventromedial prefrontal cortex, a region commonly involved in emotion regulation, when viewing negatively valenced emotional scenes (Wagner & Heatherton, 2013a). Within the appetitive domain, another study found that dieters who completed a prior self-regulation task subsequently showed an exaggerated reward response in the OFC when viewing highly appetizing food cues. Moreover, compared to nondepleted dieters, depleted dieters showed reduced functional connectivity between the OFC and the lateral PFC during appetizing food trials (Wagner, Altman, Boswell, Kelley, & Heatherton, 2013). Together, the results of these two studies suggest that self-regulatory depletion may serve to impair the functional connectivity between regions important for self-regulation, such as the lateral PFC for appetitive stimuli (Delgado et al., 2008; Kober et al., 2010; Somerville, Hare, & Casey, 2011) and the ventromedial prefrontal cortex for emotional stimuli (e.g., Johnstone, van Reekum, Urry, Kalin, & Davidson, 2007; Somerville et al., 2012). This breakdown in functional connectivity leaves responses in reward- and emotion-related brain regions unchecked, thereby leading to an exaggerated response to appetitive (Wagner et al., 2013) and emotionally charged stimuli (Wagner & Heatherton, 2013a).

Conclusion

Based on the research reviewed above, successful self-regulation can be conceptualized as involving a balance between impulse strength and the capacity to override or otherwise control them. This "balance" model (Heatherton & Wagner, 2011) suggests that when people experience strong urges brought on by exposure to temptations, to the degree that these urges outweigh the capacity to control them, self-regulation failure becomes more likely. Conversely, when the capacity to engage in self-regulation is impaired either by negative affect or prior expenditure of self-regulatory resources, then temptations can exert a greater sway over behavior, again leading to self-regulatory failure. Evidence from cognitive neuroscience suggests that this tug-of-war involves brain regions important for representing the reward value of temptations and the intensity of emotions as well as lateral and medial regions of the PFC important for implementing goal-directed behavior and inhibiting cravings and desires. This relatively simple model of self-regulation failure shares much in common with developmental models of self-control that emphasize differential maturation rates between the prefrontal cortex and regions involved in representing rewards and emotions (e.g., Somerville, Jones, & Casey, 2010). For instance, in this line of research it has been found that compared to adults adolescents show exaggerated responses to appetitive and rewarding stimuli in regions such as the ventral striatum (May et al., 2004; Somerville et al., 2011), and that this is largely due to a failure to appropriately recruit regions of the lateral PFC.

Further evidence for a causal role of the lateral PFC in response inhibition comes from research using transcranial magnetic stimulation as well as transcranial direct current stimulation, which has shown that inactivation of the right lateral PFC increases impulsive behavior and risky decision making (Chambers et al., 2006; Knoch, Pascual-Leone, Meyer, Treyer, & Fehr, 2006) whereas stimulating the function of this region with transcranial direct current stimulation pushes people to become risk-averse and better able to inhibit impulses (Fecteau et al., 2007; Jacobson, Javitt, & Lavidor, 2011). In line with the imaging work reviewed above demonstrating a role for the lateral PFC in regulating cravings and desires (e.g., Delgado et al., 2008; Kober et al., 2010; Wagner et al., 2013), these studies suggest any event that impairs lateral PFC function may precipitate self-regulation failure, particularly when an individual is faced with strong temptations or desires. Moreover, the work reviewed in this chapter suggests that absent the inhibitory influence of the lateral PFC, brain regions involved in representing reward value, emotional valence, or both may become sensitized to environmental triggers, which in turn bias individuals toward seeking out rewards, acting on desires, and ultimately abandoning their long-term goals.

Finally, it is important to note that, although we have taken the view that cue exposure, mood, and self-regulatory depletion all interfere with the ability to initiate top-down control over impulses and desires, the data from cognitive neuroscience are largely consistent with alternative interpretations such as motivational explanations, which state that the failure to engage in self-control may, in some cases, reflect a conscious choice on the part of individuals to abandon long-term goals in favor of instant gratification. Arbitrating between these competing accounts of how cues, negative affect, and resource depletion bring about self-regulatory collapse presents an important next step for theories of self-regulation failure—one in which, we believe, cognitive neuroscience promises to play a pivotal role.

ACKNOWLEDGMENTS Preparation of this chapter was supported by grants from the National Institute on Drug Abuse (R01DA22582) and the National Heart, Lung, and Blood Institute (R21HL114092).

REFERENCES

Bandura, A. (1991). Social cognitive theory of self-regulation. *Organ Behav Hum Dec, 50*(2), 248–287.

Barton, R. A., & Venditti, C. (2013). Human frontal lobes are not relatively large. *Proc Natl Acad Sci USA, 110*(22), 9001–9006.

Baumeister, R. F., DeWall, C. N., Ciarocco, N. J., & Twenge, J. M. (2005). Social exclusion impairs self-regulation. *J Pers Soc Psychol, 88*(4), 589–604.

Baumeister, R. F., & Heatherton, T. (1996). Self-regulation failure: An overview. *Psychol Inq, 7*(1), 1–15.

Baumeister, R. F., & Leary, M. R. (1995). The need to belong: Desire for interpersonal attachments as a fundamental human motivation. *Psychol Bull, 117*(3), 497–529.

Beaver, J. D., Lawrence, A. D., Ditzhuijzen, J. V., Davis, M. H., Woods, A., & Calder, A. J., (2006). Individual differences in reward drive predict neural responses to images of food. *J Neurosci, 26*, 5160–5166.

Beedie, C. J., & Lane, A. M. (2012). The role of glucose in self-control: Another look at the evidence and an alternative conceptualization. *Pers Soc Psychol Rev, 16*(2), 143–153.

Beran, M. J. (2002). Maintenance of self-imposed delay of gratification by four chimpanzees (*Pan troglodytes*) and an orangutan (*Pongo pygmaeus*). *J Gen Psychol, 129*(1), 49–66.

Boyd, R., Gintis, H., & Bowles, S. (2010). Coordinated punishment of defectors sustains cooperation and can proliferate when rare. *Science, 328*(5978), 617–620.

Brody, A. L., Mandelkern, M. A., Olmstead, R. E., Jou, J., Tiongson, E., Allen, V., ... Cohen, M. S. (2007). Neural substrates of resisting craving during cigarette cue exposure. *Biol Psychiatry, 62*(6), 642–651.

Brunstrom, J. M., Yates, H. M., & Witcomb, G. L. (2004). Dietary restraint and heightened reactivity to food. *Physiol Behav, 81*(1), 85–90.

Buss, D. M., & Kenrick, D. T. (1998). Evolutionary social psychology. In D. T. Gilbert, S. T. Fiske, & G. Lindzey (Eds.), *The handbook of social psychology*, Vol. 2 (4th ed., pp. 982–1026). New York, NY: McGraw-Hill.

Campbell, W. K., Krusemark, E. A., Dyckman, K. A., Brunell, A. B., McDowell, J. E., Twenge, J. M., & Clementz, B. A. (2006). A magnetoencephalography investigation of neural correlates for social exclusion and self-control. *Soc Neurosci, 1*(2), 124–134.

Carter, B. L., & Tiffany, S. T. (1999). Meta-analysis of cue-reactivity in addiction research. *Addiction, 94*(3), 327–340.

Carver, C. S., & Scheier, M. (1981). *Attention and self-regulation: A control-theory approach to human behavior.* New York, NY: Springer-Verlag.

Casey, B. J., Somerville, L. H., Gotlib, I., Ayduk, O., Franklin, N., Askren, M. K., ... Shoda, Y. (2011) Behavioral and neural correlates of delay of gratification 40 years later. *Proc Natl Acad Sci USA, 108*(36), 14998–15003.

Chambers, C. D., Bellgrove, M. A., Stokes, M. G., Henderson, T. R., Garavan, H., Robertson, I. H., ... Mattingley, J. B. (2006). Executive "brake failure" following deactivation of human frontal lobe. *J Cogn Neurosci, 18*(3), 444–455.

Christensen, L., & Pettijohn, L. (2001). Mood and carbohydrate cravings. *Appetite, 36*(2), 137–145.

Cloutier, J., Heatherton, T. F., Whalen, P. J., & Kelley, W. M. (2008). Are attractive people rewarding? Sex differences in the neural substrates of facial attractiveness. *J Cogn Neurosci, 20*(6), 941–951.

Damsma, G., Pfaus, J. G., Wenkstern, D., Phillips, A. G., & Fibiger, H. C. (1992). Sexual behavior increases dopamine transmission in the nucleus accumbens and striatum of male rats: Comparison with novelty and locomotion. *Behav Neurosci, 106*(1), 181–191.

DeBono, A., Shmueli, D., & Muraven, M. (2011). Rude and inappropriate: The role of self-control in following social norms. *Pers Soc Psychol Bull, 37*(1), 136–146.

Delgado, M. R., Gillis, M. M., & Phelps, E. A. (2008). Regulating the expectation of reward via cognitive strategies. *Nat Neurosci, 11*(8), 880–881.

Demos, K. E., Heatherton, T. F., & Kelley, W. M. (2012). Individual differences in nucleus accumbens activity to food and sexual images predict weight gain and sexual behavior. *J Neurosci, 32*(16), 5549–5552.

Drobes, D. J., & Tiffany, S. T. (1997). Induction of smoking urge through imaginal and in vivo procedures: Physiological and self-report manifestations. *J Abnorm Psychol, 106*(1), 15–25.

Dufour, V., Wascher, C. A. F., Braun, A., Miller, R., & Bugnyar, T. (2012). Corvids can decide if a future exchange is worth waiting for. *Biol Lett, 8*(2), 201–204.

Eisenberger, N. I. (2012). Broken hearts and broken bones: A neural perspective on the similarities between social and physical pain. *Curr Dir Psychol Sci, 21*(1), 42–47.

Fecteau, S., Pascual-Leone, A., Zald, D. H., Liguori, P., Théoret, H., Boggio, P. S., & Fregni, F. (2007). Activation of prefrontal cortex by transcranial direct current stimulation reduces appetite for risk during ambiguous decision making. *J Neurosci, 27*(23), 6212–6218.

Federoff, I. D. C., Polivy, J., & Herman, C. P. (1997). The effect of pre-exposure to food cues on the eating behavior of restrained and unrestrained eaters. *Appetite, 28*(1), 33–47.

FRIESE, M., BINDER, J., LUECHINGER, R., BOESIGER, P., & RASCH, B. (2013). Suppressing emotions impairs subsequent stroop performance and reduces prefrontal brain activation. *PLoS ONE, 8*(4), e60385.

GAILLIOT, M. T., BAUMEISTER, R. F., DEWALL, C. N., MANER, J. K., PLANT, E. A., TICE, D. M., ... SCHMEICHEL, B. J. (2007). Self-control relies on glucose as a limited energy source: Willpower is more than a metaphor. *J Pers Soc Psychol, 92*(2), 325–336.

GOODALL, J. (1986). *The chimpanzees of Gombe: Patterns of behavior.* Cambridge, MA: Belknap Press of Harvard University Press.

GOTTFRIED, J. A., O'DOHERTY, J., & DOLAN, R. J. (2003). Encoding predictive reward value in human amygdala and orbitofrontal cortex. *Science, 301*(5636), 1104–1107.

GREEN, L., MYERSON, J., HOLT, D. D., SLEVIN, J. R., & ESTLE, S. J. (2004). Discounting of delayed food rewards in pigeons and rats: Is there a magnitude effect? *J Exp Anal Behav, 81*(1), 39–50.

HAEDT-MATT, A. A., & KEEL, P. K. (2011). Revisiting the affect regulation model of binge eating: A meta-analysis of studies using ecological momentary assessment. *Psychol Bull, 137*(4), 660–681.

HAGGER, M. S., WOOD, C., STIFF, C., & CHATZISARANTIS, N. L. D. (2010). Ego depletion and the strength model of self-control: A meta-analysis. *Psychol Bull, 136*(4), 495–525.

HEATHERTON, T. F. (2011). Neuroscience of self and self-regulation. *Annu Rev Psychol, 62*(1), 363–390.

HEATHERTON, T. F., HERMAN, C. P., & POLIVY, J. (1991). Effects of physical threat and ego threat on eating behavior. *J Pers Soc Psychol, 60*(1), 138–143.

HEATHERTON, T. F., STRIEPE, M., & WITTENBERG, L. (1998). Emotional distress and disinhibited eating: The role of self. *Pers Soc Psychol Bull, 24*(3), 301–313.

HEATHERTON, T. F., & WAGNER, D. D. (2011). Cognitive neuroscience of self-regulation failure. *Trends Cogn Sci, 15*(3), 132–139.

HEDGCOCK, W. M., VOHS, K. D., & RAO, A. R. (2012). Reducing self-control depletion effects through enhanced sensitivity to implementation: Evidence from fMRI and behavioral studies. *J Consum Psychol, 22*(4), 486–495.

HOFMANN, W., BAUMEISTER, R. F., FÖRSTER, G., & VOHS, K. D. (2011). Everyday temptations: An experience sampling study of desire, conflict, and self-control. *J Pers Soc Psychol, 102*(6), 1318–1335.

HOFMANN, W., SCHMEICHEL, B. J., & BADDELEY, A. D. (2012). Executive functions and self-regulation. *Trends Cogn Sci, 16*(3), 174–180.

INZLICHT, M., & GUTSELL, J. N. (2007). Running on empty: Neural signals for self-control failure. *Psychol Sci, 18*(11), 933–937.

INZLICHT, M., & SCHMEICHEL, B. J. (2012). What is ego depletion? Toward a mechanistic revision of the resource model of self-control. *Perspect Psychol Sci, 7*(5), 450–463.

JACOBSON, L., JAVITT, D. C., & LAVIDOR, M. (2011). Activation of inhibition: Diminishing impulsive behavior by direct current stimulation over the inferior frontal gyrus. *J Cogn Neurosci, 23*(11), 3380–3387.

JANES, A. C., PIZZAGALLI, D. A., RICHARDT, S., DEB FREDERICK, B. CHUZI, S., PACHAS, G., ... KAUFMAN, M. J. (2010). Brain reactivity to smoking cues prior to smoking cessation predicts ability to maintain tobacco abstinence. *Biol Psychiatry, 67*(8), 722–729.

JENSEN, K., CALL, J., & TOMASELLO, M. (2007). Chimpanzees are rational maximizers in an ultimatum game. *Science, 318*(5847), 107–109.

JOB, V., DWECK, C. S., & WALTON, G. M. (2010). Ego depletion—is it all in your head? Implicit theories about willpower affect self-regulation. *Psychol Sci, 21*(11), 1686–1693.

JOHNS, M., INZLICHT, M., & SCHMADER, T. (2008). Stereotype threat and executive resource depletion: Examining the influence of emotion regulation. *J Exp Psychol Gen, 137*(4), 691–705.

JOHNSTONE, T., VAN REEKUM, C. M., URRY, H. L., KALIN, N. H., & DAVIDSON, R. J. (2007). Failure to regulate: Counterproductive recruitment of top-down prefrontal-subcortical circuitry in major depression. *J Neurosci, 27*(33), 8877–8884.

KILLGORE, W. D. S., & YURGELUN-TODD, D. A. (2006). Affect modulates appetite-related brain activity to images of food. *Int J Eat Disorder, 39*(5), 357–363.

KNOCH, D., PASCUAL-LEONE, A., MEYER, K., TREYER, V., & FEHR, E. (2006). Diminishing reciprocal fairness by disrupting the right prefrontal cortex. *Science, 314*(5800), 829–832.

KNUTSON, B., TAYLOR, J., KAUFMAN, M., PETERSON, R., & GLOVER, G. (2005). Distributed neural representation of expected value. *J Neurosci, 25*(19), 4806–4812.

KOBER, H., MENDE-SIEDLECKI, P., KROSS, E. F., WEBER, J., MISCHEL, W., HART, C. L., & OCHSNER, K. N. (2010). Prefrontal-striatal pathway underlies cognitive regulation of craving. *Proc Natl Acad Sci USA, 107*(33), 14811–14816.

KRINGELBACH, M. L. (2005). The human orbitofrontal cortex: Linking reward to hedonic experience. *Nat Rev Neurosci, 6*(9), 691–702.

KRINGELBACH, M. L., O'DOHERTY, J., ROLLS, E. T., & ANDREWS, C. (2003). Activation of the human orbitofrontal cortex to a liquid food stimulus is correlated with its subjective pleasantness. *Cereb Cortex, 13*(10), 1064–1071.

KURZBAN, R., & LEARY, M. R. (2001). Evolutionary origins of stigmatization: The functions of social exclusion. *Psychol Bull, 127*(2), 187–208.

LERNER, J. S., LI, Y., & WEBER, E. U. (2013). The financial costs of sadness. *Psychol Sci, 24*(1), 72–79.

LOCHBUEHLER, K., VOOGD, H., SCHOLTE, R. H. J., & ENGELS, R. C. M. E. (2011). Attentional bias in smokers: Exposure to dynamic smoking cues in contemporary movies. *J Psychopharmacol, 25*(4), 514–519.

MAY, J. C., DELGADO, M. R., DAHL, R. E., STENGER, V. A., RYAN, N. D., FIEZ, J. A., & CARTER, C. S. (2004). Event-related functional magnetic resonance imaging of reward-related brain circuitry in children and adolescents. *Biol Psychiatry, 55*(4), 359–366.

MCCLERNON, F. J., KOZINK, R. V., & ROSE, J. E. (2008). Individual differences in nicotine dependence, withdrawal symptoms, and sex predict transient fMRI-BOLD responses to smoking cues. *Neuropsychopharmacology, 33*(9), 2148–2157.

MCKEE, S. A., SINHA, R., WEINBERGER, A. H., SOFUOGLU, M., HARRISON, E. L., LAVERY, M., & WANZER, J. (2011). Stress decreases the ability to resist smoking and potentiates smoking intensity and reward. *J Psychopharmacol, 25*(4), 490–502.

METCALFE, J., & MISCHEL, W. (1999). A hot/cool system analysis of delay of gratification: Dynamics of willpower. *Psychol Rev, 106*(1), 3–19.

Miller, H. C., Pattison, K. F., DeWall, C. N., Rayburn-Reeves, R., & Zentall, T. R. (2010). Self-control without a "self"? Common self-control processes in humans and dogs. *Psychol Sci, 21*(4), 534–538.

Mischel, W., Ebbesen, E. B., & Zeiss, A. R. (1973). Selective attention to the self: Situational and dispositional determinants. *J Pers Soc Psychol, 27*(1), 129–142.

Muraven, M., Collins, R. L., & Nienhaus, K. (2002). Self-control and alcohol restraint: An initial application of the self-control strength model. *Psychol Addict Behav, 16*(2), 113–120.

Norman, D. A., & Shallice, T. (1986). Attention to action: Willed and automatic control of behavior. In R. Davidson, R. Schwartz, & D. Shapiro (Eds.), *Consciousness and self-regulation: Advances in research and theory*, Vol. 4 (pp. 1–18). New York, NY: Plenum Press.

Peake, S. J., Dishion, T. J., Stormshak, E. A., Moore, W. E., & Pfeifer, J. H. (2013). Risk-taking and social exclusion in adolescence: Neural mechanisms underlying peer influences on decision-making. *NeuroImage, 82*, 23–34.

Peciña, S., Schulkin, J., & Berridge, K. C. (2006). Nucleus accumbens corticotropin-releasing factor increases cue-triggered motivation for sucrose reward: Paradoxical positive incentive effects in stress? *BMC Biol, 4*, 8.

Piazza, P. V., & Le Moal, M. L. (1996). Pathophysiological basis of vulnerability to drug abuse: Role of an interaction between stress, glucocorticoids, and dopaminergic neurons. *Annu Rev Pharmacol, 36*, 359–378.

Ramsey, N. F., & van Ree, J. M. (1993). Emotional but not physical stress enhances intravenous cocaine self-administration in drug-naive rats. *Brain Res, 608*(2), 216–222.

Ramseyer, A., Pelé, M., Dufour, V., Chauvin, C., & Thierry, B. (2006). Accepting loss: The temporal limits of reciprocity in brown capuchin monkeys. *Philos Trans R Soc Lond B Biol Sci, 273*(1583), 179–184.

Rilling, J. K. (2006). Human and nonhuman primate brains: Are they allometrically scaled versions of the same design? *Evol Anthropol, 15*(2), 65–77.

Sayette, M. A., Martin, C. S., Wertz, J. M., Shiffman, S., & Perrott, M. A. (2001). A multi-dimensional analysis of cue-elicited craving in heavy smokers and tobacco chippers. *Addiction, 96*(10), 1419–1432.

Schilström, B., Svensson, H. M., Svensson, T. H., & Nomikos, G. G. (1998). Nicotine and food induced dopamine release in the nucleus accumbens of the rat putative role of α7 nicotinic receptors in the ventral tegmental area. *Neuroscience, 85*(4), 1005–1009.

Schmeichel, B. J. (2007). Attention control, memory updating, and emotion regulation temporarily reduce the capacity for executive control. *J Exp Psychol Gen, 136*(2), 241–255.

Schoenemann, P. T., Sheehan, M. J., & Glotzer, L. D. (2005). Prefrontal white matter volume is disproportionately larger in humans than in other primates. *Nat Neurosci, 8*(2), 242–252.

Semendeferi, K., Lu, A., Schenker, N., & Damasio, H. (2002). Humans and great apes share a large frontal cortex. *Nat Neurosci, 5*(3), 272–276.

Siep, N., Roefs, A., Roebroeck, A., Havermans, R., Bonte, M., & Jansen, A. (2012). Fighting food temptations: The modulating effects of short-term cognitive reappraisal, suppression and up-regulation on mesocorticolimbic activity related to appetitive motivation. *NeuroImage, 60*(1), 213–220.

Somerville, L. H., Hare, T., & Casey, B. J. (2011). Frontostriatal maturation predicts cognitive control failure to appetitive cues in adolescents. *J Cogn Neurosci, 23*(9), 2123–2134.

Somerville, L. H., Jones, R. M., & Casey, B. J. (2010). A time of change: Behavioral and neural correlates of adolescent sensitivity to appetitive and aversive environmental cues. *Brain Cognit, 72*(1), 124–133.

Somerville, L. H., Wagner, D. D., Wig, G. S., Moran, J. M., Whalen, P. J., & Kelley, W. M. (2012). Interactions between transient and sustained neural signals support the generation and regulation of anxious emotion. *Cereb Cortex, 23*(1), 49–60.

Tiffany, S. T. (1990). A cognitive model of drug urges and drug-use behavior: Role of automatic and nonautomatic processes. *Psychol Rev, 97*(2), 147–168.

Twenge, J. M., Baumeister, R. F., Tice, D. M., & Stucke, T. S. (2001). If you can't join them, beat them: Effects of social exclusion on aggressive behavior. *J Pers Soc Psychol, 81*(6), 1058–1069.

Van der Laan, L. N., de Ridder, D. T. D., Viergever, M. A., & Smeets, P. A. M. (2011). The first taste is always with the eyes: A meta-analysis on the neural correlates of processing visual food cues. *NeuroImage, 55*(1), 296–303.

Vohs, K. D., Baumeister, R. F., & Ciarocco, N. J. (2005). Self-regulation and self-presentation: Regulatory resource depletion impairs impression management and effortful self-presentation depletes regulatory resources. *J Pers Soc Psychol, 88*(4), 632–657.

Vohs, K. D., Baumeister, R. F., Mead, N. L., Hofmann, W., Ramanathan, S., & Schmeichel, B. J. (2013). Engaging in self-control heightens urges and feelings. Manuscript submitted for publication.

Vohs, K. D., & Heatherton, T. F. (2000). Self-regulatory failure: A resource-depletion approach. *Psychol Sci, 11*(3), 249 –254.

Volkow, N. D., Fowler, J. S., Wang, G.-J., Telang, F., Logan, J., Jayne, M., ... Swanson, J. M. (2010). Cognitive control of drug craving inhibits brain reward regions in cocaine abusers. *NeuroImage, 49*(3), 2536–2543.

Wagner, D. D., Altman, M., Boswell, R. G., Kelley, W. M., & Heatherton, T. F. (2013). Self-regulatory depletion enhances neural responses to rewards and impairs top-down control. *Psychol Sci, 24*(11), 2262–2271.

Wagner, D. D., Boswell, R. G., Kelley, W. M., & Heatherton, T. F. (2012). Inducing negative affect increases the reward value of appetizing foods in dieters. *J Cogn Neurosci, 24*(7), 1625–1633.

Wagner, D. D., Dal Cin, S., Sargent, J. D., Kelley, W. M., & Heatherton, T. F. (2011). Spontaneous action representation in smokers when watching movie characters smoke. *J Neurosci, 31*(3), 894–898.

Wagner, D. D., & Heatherton, T. F. (2013a). Self-regulatory depletion increases emotional reactivity in the amygdala. *Soc Cogn Aff Neurosci, 8*(4), 410–417.

Wagner, D. D., & Heatherton, T. F. (2013b). Emotion and self-regulation failure. In J. J. Gross (Ed.), *Handbook of emotion regulation* (2nd ed., pp. 613–628). New York, NY: Guilford.

Wagner, D. D., & Heatherton, T. F. (in press). Self-regulation and its failure: Seven deadly threats to self-regulation. In

E. Borgida & J. Bargh (Eds.), *APA Handbook of personality and social psychology,* Vol. 1: *Attitudes and social cognition.* Washington, DC: American Psychological Association.

Willner, P., & Jones, C. (1996). Effects of mood manipulation on subjective and behavioural measures of cigarette craving. *Behav Pharmacol,* 7(4), 355–363.

Witkiewitz, K., & Villarroel, N. A. (2009). Dynamic association between negative affect and alcohol lapses following alcohol treatment. *J Consult Clin Psychol,* 77(4), 633–644.

Zinser, M. C., Baker, T. B., Sherman, J. E., & Cannon, D. S. (1992). Relation between self-reported affect and drug urges and cravings in continuing and withdrawing smokers. *J Abnorm Psychol,* *101*(4), 617–629.

62 What Is the Role of Control in Emotional Life?

KEVIN N. OCHSNER

ABSTRACT The question of how cognition—and cognitive control in particular—interacts with emotion has been of long-standing interest in popular culture and scientific inquiry. This chapter presents a fresh take on their interaction, offered as a counterpoint to popular theories suggesting that emotional processes and higher-level cognitive processes, such as those supporting self-control, are antagonists that compete for control over behavior, with the rise of one leading to the fall of the other. While such "seesaw" models can be very useful for explaining specific classes of phenomena, they may not be able to account for the full range of interactions between affective and cognitive processes of which humans are capable—and on which they depend on a daily basis. The alternative account of their relationship is motivated by a mix of old-school appraisal theory and contemporary cognitive neuroscience data, whereby top-down cognitive control mechanisms play an integral role in at least four aspects of our emotional lives: generating emotions, reporting on and understanding our own emotions, understanding the emotions of others, and regulating our emotions. This account suggests that cognitive and affective processes are intimately intertwined and provides a foundation for understanding core functional relationships among multiple kinds of behavior and for asking new kinds of questions about them.

If there is one thing humans are fond of it is dichotomies. Up versus down. Left versus right. Good versus bad. Such dichotomies are simple and easy to understand and apply. They typically have traction because they capture something important about the world. One dichotomy critically important for psychology is the distinction between cognition and emotion.

Many lay and scientific accounts of their relationship focus on one form of cognition—cognitive control, or the ability to engage in deliberate and reasoned thinking—and the way in which it stifles emotional responses. This relationship is epitomized by the relationship between the two central characters of my favorite 1960s television series: *Star Trek*. On one hand there is Spock, science officer on the Starship Enterprise. Spock is a Vulcan, and Vulcans have spent centuries eliminating base emotional impulses in favor of higher-order cognitive control and reasoning. On the other hand, there is James T. Kirk, the all-too-human starship captain. Kirk is the embodiment of impulsive emotion, driven as he is by a pulsating libido and an overweening machismo.

Many *Star Trek* episodes depict the struggle between cognitively controlled Spock and emotionally impulsive Kirk as they argue, cajole, and sometimes outright battle each other in a seesawing war to prove which human faculty—cognitive control or emotion—is supreme. This model of the cognition-emotion relationship embodies a key assumption common to both lay and scientific theories: namely, that cognitive control is inherently antagonistic to emotion (see figure 62.1).

Is this seesaw view of the cognitive control–emotion relationship sufficient to explain all possible ways in which control and emotion interact? If it is not, what other kinds of models of their relationship might we consider?

The goal of this chapter is to begin addressing these questions. Toward that end, it is divided into three parts. The first reviews a handful of salient exemplars of the seesaw model, using them as foils for considering what other kinds of cognition-emotion models might be useful. The second presents research from my lab and others that suggests a particular alternative view of the control-emotion relationship. Here, we will consider examples of emotion generation, self-reports, perception, and regulation that either are not spoken to by seesaw models or are inconsistent with (at least strong versions of) them. In the third and last section, we will consider one potential model that fits these data and its implications for theoretical approaches to understanding cognition, emotion, and future research on their relationship.

Seesaw models of cognitive control and emotion

Historically, the idea that cognition and emotion seesaw back-and-forth with one another dates back at least to Descartes's ideas about the dichotomy between passion and reason (Damasio, 1994) and carry forward in contemporary ideas about development (Somerville & Casey, 2010), moral judgment (Greene et al., 2001), and a range of other phenomena. For present purposes, it is germane to review a few examples that

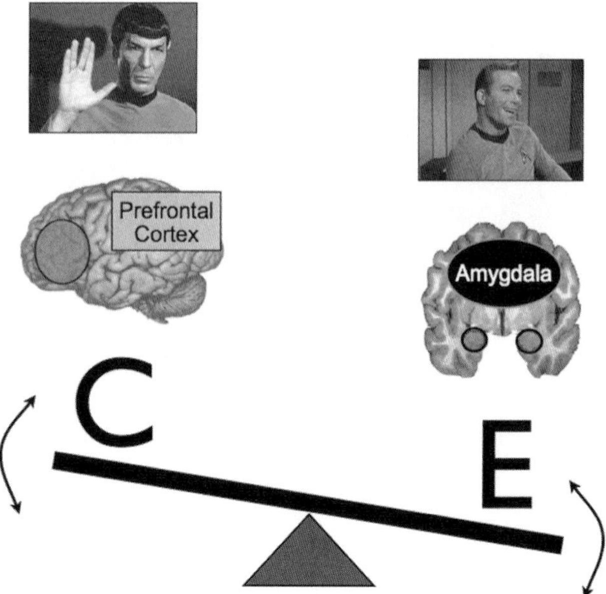

FIGURE 62.1 A schematic representation of the seesaw relationship between cognitive control and emotion postulated by many lay and scientific theories. Such theories posit that cognitive control and emotion are inherently antagonistic. Such an antagonistic relationship is exemplified by the opposing points of view represented by popular television characters like Mr. Spock (who embodies the value of control) and Captain Kirk (who embodies the value of impulsive emotion) from the 1960s television series *Star Trek*. In contemporary theories, specific brain systems are often assisted with control (prefrontal cortex) and emotion (amygdala), and scientific questions revolve around the situations in which the activity of prefrontal cortex and the amygdala seesaw with one another. (See color plate 52.)

illustrate the core ideas. Here we briefly review three seesaw type models that have a neuroscience foundation.

WHAT THE MODELS PROPOSE Both human and animal research have reliably associated cognitive control and emotional responding with distinct but overlapping sets of brain systems. While it is beyond the scope of this chapter to review these systems here, for present purposes what's important are the regions highlighted in the context of seesaw models.

On one hand, cognitive control has been shown to depend upon regions of lateral prefrontal cortex thought to be critical for a host of subprocesses essential for deliberate and effortful control over a thought, behavior, or emotion (Beer, Knight, & D'Esposito, 2006; Buhle et al., 2013; Hartley & Phelps, 2012; Miller & Cohen, 2001; Ochsner, Silvers, & Buhle, 2012). These subprocesses depend upon prefrontal regions essential for keeping goal-relevant information in mind, inhibiting prepotent responses, and looking information up in memory. As such, prefrontal activity is often taken as a marker for the operation of cognitive control processes.

On the other hand, emotional responding has been shown to involve a number of cortical and subcortical structures, and perhaps foremost among them is the amygdala. The amygdala receives multimodal sensory inputs and associates them with body states that reflect the relevance of stimuli to one's current or chronic goals, wants, and needs (Cunningham & Brosch, 2012; Cunningham & Zelazo, 2007; Kim et al., 2011). These associations may trigger components of affective responses ranging from heightened attention to freezing and escape behaviors. In so doing, the amygdala has a natural bias toward detecting, encoding, and triggering appropriate responses to stimuli that might be potentially threatening (Anderson, Christoff, Panitz, De Rosa, & Gabrieli, 2003). As such, amygdala activity is often taken as a marker for threat-related emotional processes.

This essential distinction between cognitive control and emotion, and their prefrontal versus amygdalar bases, has played itself out in at least three influential scientific theories.

The first example comes from the 1990s, when Drevets and Raichle proposed a "balance model" of the relationship between cognition and emotion (Drevets & Raichle, 1998). This model was based on early positron emission tomography studies of higher-order cognitive abilities, such as working memory, reasoning, and language. They noted that whenever participants were asked to engage in an effortful higher-level cognitive task, heightened prefrontal activity was observed, along with relative decreases in amygdala activity (among other subcortical structures). These data prompted the hypothesis that in healthy adults, the engagement of higher cognitive processes naturally "turned off" emotional processes. In clinical populations, however, this naturally antagonistic relationship between cognition and emotion was out of balance. Instead, certain populations, like those with major depression, showed relatively decreased prefrontal activity accompanied by increased amygdala activity when not asked to do anything at all (i.e., when at "rest").

A second example comes from the 2000s in the form of Lieberman's affect-labeling theory (e.g., Lieberman et al., 2007; Lieberman, Inagaki, Tabibnia, & Crockett, 2011; Payer, Baicy, Lieberman, & London, 2012; Torrisi, Lieberman, Bookheimer, & Altshuler, 2013). In a typical affect-labeling study, participants might be asked to decide which of two exemplar stimuli best matches a target photograph of an emotionally expressive face (e.g., a fear expression). In one case, the exemplars are themselves emotional expressions (e.g., fear and disgust faces) and participants must match them to the target on the basis of their perceptual features. In another case, the exemplars are words that could label the target expression (e.g., the words "fear" and "disgust") and participants must match them to the target on the basis of their meaning. In a finding replicated many times across variants of the procedure, the perceptual case elicits greater amygdala and lesser prefrontal activity, whereas the verbal case elicits greater prefrontal and lesser amygdala activity. The explanation is that affective processes are disrupted by the cognitive processes involved in selecting a label for an affective stimulus.

A third example from the 2010s was presented in a review paper by Heatherton and Wagner on the brain systems underlying self-regulation failure (Heatherton & Wagner, 2011). This paper proposes a model to explain why various factors (e.g., negative mood) reliably lead to failures in the ability to engage in goal-directed, adaptive self-regulation (e.g., staying on a diet). The idea is that self-regulation failure becomes more likely to the extent that a given factor either disrupts prefrontal activity, thereby reducing the capacity for cognitive control, or enhances the response of subcortical structures involved in emotion or motivation, thereby making emotional or motivated responses less controllable. In a key figure, this model is visualized as a seesaw with various factors pushing the cognitive or emotional ends of the seesaw up or down.

Limitations of Scope Although it is beyond the purview of this review to weigh all of the evidence cited in the support of each model, for present purposes suffice to say that each model accounts very well for a particular kind of interaction between cognitive control and emotion that can be usefully modeled in terms of a seesaw. Critically, such interactions typically involve cases where cognitive control is engaged for the purpose of stopping, inhibiting, or blocking an emotional response. This can be done either in the service of an explicit regulatory goal (e.g., to diet to lose weight) or in the service of an implicit, or incidental, regulatory goal (e.g., as is posited in affect-labeling theory).

There is an additional core feature of these and other related models, however. To differing degrees, they implicitly or explicitly assume that in the context of emotional responses, cognitive control always limits their expression. Or, put another way, any time cognitive control is engaged, emotional responding is blocked, disrupted, or halted.

Is the antagonistic, either/or assumption about the relationship between cognitive control and emotion warranted? Or are there varieties of interaction between cognitive control and emotion not explained by seesaw theories and that may even be inconsistent with them? The section that follows makes the case that there are at least three other kinds of core emotional phenomena—and even some emotion regulatory phenomena—that fit this bill. In the end, we will see that seesaw theories have their clear domains of usefulness but may be limited in scope to a set of phenomena that involve the use of cognitive control to down-regulate emotion.

Before moving on, it should be acknowledged that this chapter glosses over important nuances for each of the seesaw models. For example, current theories don't simply posit a seesawing between prefrontal and subcortical regions, but rather more complex context-dependent changes in connectivity between them (e.g., Torrisi et al., 2013; Townsend et al., 2012; Wagner, Altman, Boswell, Kelley, & Heatherton, 2013) that may underlie examples of self-regulation success or failure. Furthermore, most of these models—and the others cited at the outset of this section—were not formulated with the intention of accounting for the range of control-emotion interactions under consideration here. For example, none of them were intended to account for the role of control in emotion generation.

With these caveats in mind, this chapter takes the liberty of being glossy for a couple of reasons. First, the glossy version of seesaw models fits key aspects of the *lay* conception of control-emotion interactions—as epitomized by the Kirk-Spock dynamic, for example—that this chapter seeks to complexify, if not outright counter. And second, it allows us to present seesaw models as a bit of a "straw man" hypothesis that begs the pointed question of whether seesaw models are by themselves complete. With seesaw models as the foil, let us now consider data that suggest an alternative account of control-emotion interactions.

Can cognitive control and emotion interact in ways not accounted for by seesaw models?

To begin building an alternative account of the interactions between cognitive control and emotion, this section reviews functional imaging research on a wider range of emotional phenomena than is typically taken into account by seesaw models.

Imagine that the circle in figure 62.2 represents the full set of phenomena that involve interactions between cognitive control and emotion. The top left quadrant represents emotion generation, the top right represents emotion self-reports, the lower left emotion regulation, and the lower right emotion perception. Let's now

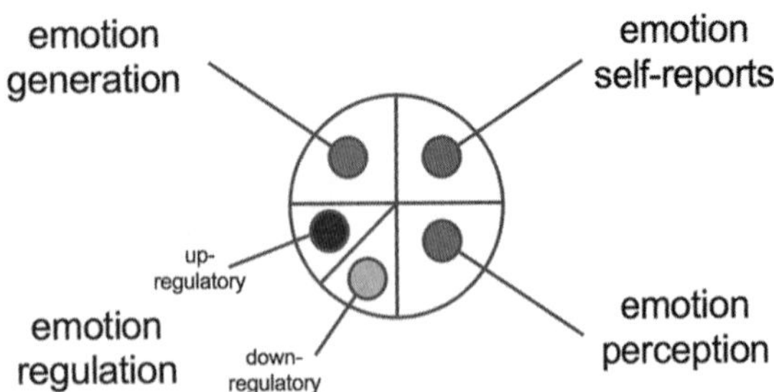

FIGURE 62.2 A schematic representation the range of phenomena that could involve interactions between cognitive control and emotion. As discussed in the chapter, control and emotion may interact in the context of emotion generation, emotion self-reports, emotion perception, and emotion regulation. Within the domain of emotion regulation, control may have both up- and down-regulatory effects. The chapter argues that seesaw models best account for down-regulatory interactions between control and emotion but that up-regulatory interactions also are possible, as are other types of interactions in the context of the full range of phenomena that involve controlling emotion.

review examples of each type of phenomena that exemplify cognitive control—emotion interactions that may not be well accounted for by seesaw models. After doing so, we'll consider what implications these data have for theory and research.

EMOTION GENERATION A fundamental question in psychology has been how emotions are generated. So important is this question that the field was preoccupied with it during the 1980s in what became popularly known as the cognition-emotion debate. On one side was the Berkeley psychologist Richard Lazarus, who argued that emotions resulted from the operation of cognitive appraisal processes that interpreted the meaning of events (Lazarus, 1982, 1991). You felt good or bad about an event depending upon the cognitive meaning ascribed to it. On the other side was the Michigan psychologist Robert Zajonc, who famously argued that preferences need no inferences (Zajonc, 1980, 1984). Put another way, he argued that affective responses are primal and basic, veritable involuntary reflexes that did not require any intervening cognitive processes. For some time, Lazarus and Zajonc argued about whether cognition was necessary for emotions to be generated. Over time, the field came to realize that the answer to this question hinged upon one's definition of cognition. If you believe that cognitive processes can be either automatic or controlled, then the terms of the debate change: the question is no longer whether cognition is necessary for emotion, but whether the emotion-generation process itself is automatic or could involve controlled processes as well.

This is where the debate about emotion generation becomes relevant to our present concerns. If emotions can be generated both by relatively bottom-up, amygdala-dependent automatic processes and by prefrontally dependent, top-down controlled processes, then this would be inconsistent with strong versions of the seesaw model—versions positing that cognitive control always disrupts affective processes.

In a functional imaging experiment, we asked what types of brain systems are involved in the controlled top-down generation of emotion as compared to the bottom-up, stimulus-driven generation of emotion (Ochsner et al., 2009). To accomplish this, we asked participants to view negative or neutral photographic images on three types of trials. Two of these trials involved participants looking at images and letting themselves respond naturally. By directly comparing brain activation when participants were responding naturally to a negative as compared to a neutral image, we sought to isolate brain regions involved in the bottom-up generation of emotion. The third trial type presented neutral images but asked participants to construct a negative interpretation of what they saw happening. For example, if they saw an image of an unmade bed, they might use cognitive control to construct a back story whereby a nominally happy couple got out of that bed in the morning and was tragically struck by a car and killed. By comparing brain activation when participants viewed neutral images and used top-down processes to construct negative interpretations, as opposed to letting themselves respond naturally to neutral images, we sought to isolate the brain regions involved in the controlled top-down generation of emotion.

Three key findings emerged. First, we found that the top-down as compared to the bottom-up generation of emotion activated a set of prefrontal systems typically involved in cognitive control, including ventrolateral regions associated with selection and/or inhibition (Aron & Poldrack, 2005; Badre & Wagner, 2007; Thompson-Schill, 2003) and dorsal regions typically involved in selective attention or working memory (Wager, Jonides, & Reading, 2004; Wager & Smith, 2003). In addition, there was activation of dorsal medial prefrontal regions implicated in making attributions about mental states (Amodio & Frith, 2006; Olsson & Ochsner, 2008; Zaki & Ochsner, 2012) as well as dorsal cingulate regions thought to monitor conflicts between desired and actual responses (Botvinick, Cohen, & Carter, 2004). We hypothesized that these systems are involved in selecting the elements of, and keeping in mind, negative interpretations that generate an affective response from the top down.

Second, we found that bottom-up emotion generation strongly activated the amygdala bilaterally. Intriguingly, top-down emotion generation also activated an overlapping subregion of the left amygdala. This suggested that bottom-up emotion generation depended more strongly on the amygdala, but that top-down emotion generation could also trigger activity here simply by interpreting a neutral event in aversive terms.

Third, we found that activity in the amygdala correlated with the strength of affective responses when emotions were generated from the bottom up, but that activity in medial prefrontal cortex correlated with the strength of affective responses when emotions were generated from the top down.

Together, these data suggested that bottom-up and top-down emotion generation had in common their ability to trigger responses in the amygdala, but differed in their dependence on the amygdala as a bottom-up triggering system for emotion. Whereas bottom-up responses depended quite strongly on the amygdala, top-down responses depended more strongly on prefrontal regions implicated in higher cognitive processes, including those involved in cognitive control.

It is not clear how seesaw models might account for such findings. In this experiment, we saw engagement of prefrontal regions, including ventral prefrontal regions most strongly associated with seesaw-type theories, when affective responses were increasing both behaviorally and neurally (i.e., in terms of amygdala activity). Instead of inhibiting or disrupting amygdala response, this experiment highlights a way in which prefrontal activity, by elaborating the affective meaning of a stimulus, can actually engender amygdala activity and heighten experienced negative affect.

Emotion Self-Reports The ability to introspectively access and verbally describe our emotions is typically termed a "self-report of emotion" within the psychological literature. As such, it is a primary dependent measure in many studies of emotion. Emotion self-reports are important in everyday life as well. In circumstances ranging from telling romantic partners how we feel, to managing conflicts with coworkers, to meditating on how our day is going, to talking with our therapists, we take for granted the ability to self-report on our emotional states.

The very ubiquity of emotion self-reports may lead them to be overlooked, however, as a focal topic of study. Recently, we went to the experimental neuroscience literature to ask whether anyone had unpacked the mechanisms involved in self-reports of emotion. While meta-analyses had documented a myriad of regions active during emotional experience in general, few studies had attempted to describe the mechanisms involved in self-reporting on emotion, per se (Wager et al., 2008).

To begin addressing this issue, we first drew on constructivist theories of emotion (Gendron & Barrett, 2009; Lindquist & Barrett, 2008) to formulate a hypothesis about the psychological processes and brain systems that enable us to construct self-reports. In brief, we thought that there might be three kinds of brain systems involved (Satpute, Shu, Weber, Roy, & Ochsner, 2013). First, there might be brain systems, like the amygdala, that generate one's initial bottom-up, stimulus-driven, emotional response. Second, there might be brain systems, like the medial prefrontal cortex, involved in attending to and conceptualizing mental states that could play a key role in generating a set of possible descriptors for one's emotional response. Third, regions of lateral prefrontal cortex important for the controlled selection of context-appropriate responses might play a key role in selecting among potential emotion descriptors the one that best encapsulates, or categorizes, one's current emotional state.

To test this account, we presented participants with photographic images that varied in their emotional intensity, from neutral to moderate negativity and extending up to highly negative images (Satpute et al., 2013). By identifying brain regions whose activity varied as a function of image intensity, we sought to identify regions involved in triggering the initial affective response, like the amygdala. While viewing these images, participants were asked to attend to either the *perceptual* aspects of these images—judging whether the images had relatively more curvy or straight lines—or to their *emotional response* to the image—judging whether they felt good, bad, or somewhere in between. By comparing brain activation when participants were attending to the *perceptual* features as compared to their *emotional response*, we sought to identify regions involved in being aware of and conceptualizing one's emotional response, such as medial prefrontal cortex. Finally, participants made these judgments in one of two modes. On *continuous* trials, participants were given a scale that graded continuously from one end of a continuum to another (either from bad to good or from curvy to straight). On *categorical* trials, participants were given one of three discrete response categories to choose from (bad, neutral; good, curvy; equal, straight). By comparing brain activation on *categorical* versus *continuous* trials, we sought to identify brain regions involved in the controlled selection of context-appropriate descriptions. Here, it might be noted that both *continuous* and *categorical* judgment modes require that experience be translated into conceptual terms. The key is that *categorical* trials require selecting between a set of discrete response options, which taxes these processes to a greater extent than simply clicking somewhere on a continuous scale that doesn't require one to sharply distinguish among qualitatively different kinds of experience.

What we found supported our initial hypotheses. Amygdala responses co-varied with the intensity of the images. A region of dorsal medial prefrontal cortex typically involved in making attributions about mental states was more engaged when participants attended to their emotional responses as compared to the perceptual features of images. And right ventral lateral prefrontal cortex was more active when judgments were made categorically as compared to using a continuous scale.

These results are problematic for seesaw models, insofar as they show engagement of regions of prefrontal cortex involved in cognitive control and higher-level cognition in a situation that doesn't involve down-regulation of affective responses—either behaviorally or neurally. Neither self-reports of negative emotion nor amygdala responses were lower on categorical trials as compared to continuous trials, even though prefrontal engagement was greater in that same comparison. Instead, this experiment shows that prefrontal regions can help identify and understand the nature of emotional responses by helping conceptualize what one's response might be, then selecting the most appropriate conceptualization for self-report.

EMOTION PERCEPTION Most research on emotion perception focuses on the correspondence between specific facial features or movements and specific underlying emotional responses. For example, turning up the corners of your mouth and crinkling the corners of your eyes are known to be key components of the facial expression of happiness. But do such facial displays always connote that one is happy? Consider, for example, the case of a used car salesman. When he approaches you on a used car lot with a broad smile on his face, is he genuinely happy to see you? Or is he using that smile as a deliberate means to make you like and trust him and therefore purchase the cars he has to offer? We encounter such apparent conflicts between the nonverbal cues to emotion that people express and the information we have about the emotional meaning of the context in which those cues are encountered. Facial expressions may look neutral at a funeral not because people lack feeling for the events transpiring but because they're masking their feelings of grief. People smile and nod in conversation not because they're necessarily happy, but because we have social norms in the West that dictate this is a polite thing to do. And so on.

We have termed these conflicts between different types of cues to emotion *social cognitive conflicts* because they require the engagement of fundamentally social cognitive processes to draw inferences about why and what emotion people are experiencing (Zaki, Hennigan, Weber, & Ochsner, 2010).

What brain systems are involved in resolving these social cognitive conflicts (Zaki et al., 2010)? To begin addressing this question, we adapted a methodology we've used previously to study empathy (Zaki, Bolger, & Ochsner, 2009; Zaki, Weber, Bolger, & Ochsner, 2009). In our prior empathy work, we collected short video clips of real people talking about real-life experiences that elicited varying degrees of positive or negative emotion. We also had them make continuous ratings of how they felt as they were talking about these experiences. This allowed us to select 10-second segments of these videos where participants had self-reported strong positive or negative emotions. In our experiment on social cognitive conflict, we then presented these 10-second clips without sound. On *consistent* trials, these silent video clips were presented along with phrases that described same valenced events. For example, if the video clip had come from a positive experience, and the nonverbal cues present in the video therefore connoted positive emotion, it would be paired with a positive event, like having lunch with a friend. On *conflict* trials, the silent video clips were presented along with phrases that described opposite valenced events. For example, a positive video clip might be paired with a negative event description, like one's dog dying. The key idea is that the video clips presented nonverbal cues to positive or negative emotion that could then be perceived in the context of knowledge about events that were themselves either positive or negative. The job of a participant was to view each video of a target person and its accompanying event caption and decide how the target was feeling, taking both types of cues into account. This required weighing the relative contributions of each cue to one's overall sense of the target's emotional state, which required resolving social cognitive conflicts on conflict trials.

Using this method, we observed both behavioral and neural signatures that social cognitive conflict had successfully been elicited. The time it took to make judgments about a target person's emotions was longer on conflict than on consistent trials, and was accompanied by activation of right ventral lateral prefrontal cortex and dorsal anterior cingulate cortex—two regions typically activated in studies of response conflict and attempts to resolve them using cognitive control (Botvinick, Braver, Barch, Carter, & Cohen, 2001; Botvinick et al., 2004).

To understand how participants resolve these conflicts, we calculated a trial-by-trial index of the extent to which a participant's judgment of a target's emotions reflected relative reliance on the nonverbal or contextual cues. Here, we took advantage of the fact that we'd had a separate group of participants make normative ratings of each video taken alone and each event description taken alone. For any given video-and-event combination, we could average these ratings and set that average value as the benchmark for emotion judgments that reflected equal weighting of nonverbal and contextual cues. To the extent that a participant gave a rating of target emotions that was more similar to the rating given to the video or context description taken alone, we could calculate the difference between that rating and the average to quantify the extent to which a given judgment was relying more strongly on the video or context cues. In this way, for each trial we calculated a single number that represented the extent to which participants equally or differentially weighted the video and context cues when making their judgments of the target's emotions, and then entered these values into a parametric brain-imaging analysis. We found that as participants relied more on the nonverbal cues, they showed greater activity in parietal and premotor regions typically described as part of the mirror neuron system (Gallese, Keysers, & Rizzolatti, 2004). These regions likely supported representations of the motor intentions underlying nonverbal cues to emotion. As participants relied more on the contextual cues, they showed greater activity in medial prefrontal and temporal-parietal regions implicated in making attributions about mental states and representing semantic knowledge more generally (Saxe, 2006).

A final analysis related the regions implicated in resolving conflict to those implicated in the relative reliance on video or contextual cues. We asked whether functional connectivity (essentially, a correlation between activation in two regions across time) was greater between regions involved in resolving cognitive conflict and the mirror neuron system whenever participants relied on that system more to guide their judgments and, conversely, whether connectivity was greater between cognitive conflict regions and the prefrontal and parietal regions representing the meaning of event contexts as participants relied more on those systems to guide their judgments. That is exactly what we found.

We interpreted these data in light of biased competition models of prefrontal control (Miller & Cohen, 2001). These models suggest that prefrontal regions control behavior, thought, and emotion by biasing ongoing processing in favor of goal-relevant representations in posterior and inferior cortical and subcortical regions. When judging the emotions of another person on the basis of nonverbal and contextual cues that provide conflicting information about that person's emotions, prefrontal regions may bias competition between regions that represent nonverbal and contextual information as each makes differential contributions to your perception and judgment.

This kind of prefrontal involvement in emotion perception falls outside the range of phenomena explained by seesaw models of emotion-control interactions. As such, it provides evidence that cognitive control can play an important role in perceiving and understanding the emotions expressed by others.

Emotion Regulation We began this chapter by briefly describing seesaw models that appear suited for describing situations where cognitive control has down-regulatory effects on affective responses. Research on emotion regulation is often cited as providing additional support for these models. For example, studies of the use of cognitive reappraisal to reframe the meaning of affectively charged stimuli in ways that lessen their emotional impact typically show activation of prefrontal control regions accompanied by diminished activity in affective response regions (Heatherton & Wagner, 2011; Payer et al., 2012).

A problem for seesaw models arises, however, with the finding that reappraisal can be used to up-regulate emotion as well. Using reappraisal to increase negative emotion, for example, by cognitively elaborating the painful experiences and possible negative outcomes of people depicted in aversive photos, activates prefrontal control regions while at the same time increasing activity and subcortical regions like the amygdala (Buhle et al., 2013; Ochsner et al., 2004; Urry et al., 2006). These data parallel those in the section on emotion generation above. When elaborating negative meanings for events, heightened prefrontal activity can be accompanied by heightened activity and affect-triggering regions like the amygdala (Ochsner et al., 2009).

If strong versions of seesaw models were correct, then engagement of prefrontal cortex and emotional context should blunt amygdala response. But the opposite was observed. Taking these data together with the other data reviewed above, we suggest that an expanded view of possible interactions between cognitive control and affect/emotion might be warranted.

Beyond the seesaw: An alternative account of the relationship between cognitive control and emotion, and its implications for theory and research

This chapter began with the compelling view that emotion and cognitive control are often in opposition. As a shorthand, we referred to these theoretical accounts as seesaw models that posited the engagement of cognitive control in the pursuit of explicit or implicit regulatory goals as the effect of diminishing neural and behavioral signatures of affective response.

Using the glossy, popular version of these models as a foil, we then reviewed four kinds of evidence that cognitive control in affective processes can interact in ways not well accounted for by seesaw models (summarized in figure 62.3). First, we saw that the use of cognitive control to elaborate the affective meaning of an event can generate emotion via the recruitment of prefrontal control systems and the triggering of activation in subcortical affect systems like the amygdala. Second, we saw that control regions could help select stimulus-appropriate words for describing and reporting on one's current emotional state. Third, in regard to perceiving other people's emotions, we saw that control regions may help resolve social cognitive conflicts between interpretations of different kinds of cues that provide different kinds of information about others' emotions. Fourth and last, we saw that prefrontal control regions can be used to up-regulate emotion, and in so doing, up-regulate responses in subcortical affect systems like the amygdala.

If these data either are inconsistent with or fall outside the intended explanatory realm of seesaw models, then how should we conceive of cognitive control–emotion interactions?

	Control system activation	Emotion system activation	Kind of interaction
Generation			control generates and elaborates affective meanings that elicit affective responses
Regulation		or	control generates and maintains alternative meanings that up- or down-regulate affect
Self-reports			control generates & selects among possible descriptors for one's own
Perception			control biases attention towards cues to other's emotions

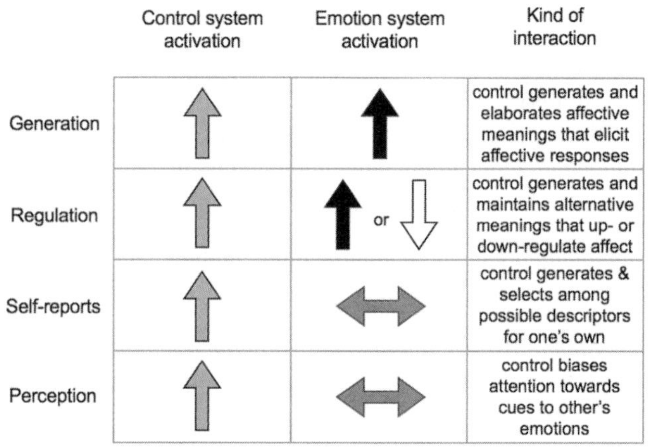

FIGURE 62.3 A tabular summary of the ways in which control systems in emotion systems interact during the four kinds of emotional phenomenon discussed in this chapter. As illustrated in the table, in every case control systems are activated, but only in a subset of emotion-regulatory phenomena is activation in prefrontal cortex associated with diminished activity in the amygdala or other emotion-related systems. For emotion generation, self-reports, and perception, control is associated with heightened activation in emotion systems, or it does not clearly up- or down-regulate it—either because control systems may make use of, but not directly influence, representations in emotion systems (as when self-reporting emotion), or because control shifts attention between neural representations of emotion-relevant information (as when perceiving another person's emotions). See text for details. *Note*: Upward-pointing gray arrows indicate greater activation and control systems; upward-pointing black or downward-pointing white arrows indicate greater or lesser activation of emotion systems; double-headed sideways arrows represent activation and emotion systems that are not simply up- or down- regulated by control systems.

CONSTRAINT SATISFACTION AND THE MAKING OF MEANING In their classic book, *The Measurement of Meaning*, Osgood, Suci, and Tannenbaum showed that approximately two-thirds of the meaning we ascribe to things in the world arises from our valenced evaluation of those things as good or bad, positive or negative, pleasant or unpleasant (Osgood, Suci, & Tannenbaum, 1957). Fundamentally, this highlights that our affective responses form a core element of what events, people, places, things, and so on mean to us. Indeed, the most profound and important experiences of our daily lives wouldn't mean as much if they were bereft of their emotional potency.

That said, how are these meanings formed and represented in the brain? That really is the fundamental question at hand in this chapter. Whether we are generating emotion, reporting on our own emotion, perceiving the emotions of others, or regulating our emotions, the particular emotional response we have reflects the meaning we ascribe to the stimulus or stimuli we are encountering (Scherer, Schorr, & Johnstone, 2001).

Viewed this way, seesaw models imply that the meaning of people, places, and things is derived from the relative balance of affective and cognitive control processes. Although theoretically control and emotion could be in perfect equilibrium, a greater contribution from one source leads necessarily to a lesser contribution from the other. As noted earlier in this chapter, seesaw models may best account for the class of emotional phenomena where cognitive control is being used in service of implicit or explicit goals to down-regulate affective responses.

Given evidence that other types of interactions are possible, we offer an alternative means of modeling control-emotion interactions (see figure 62.4). Since the 1970s and 1980s, a particular class of computational

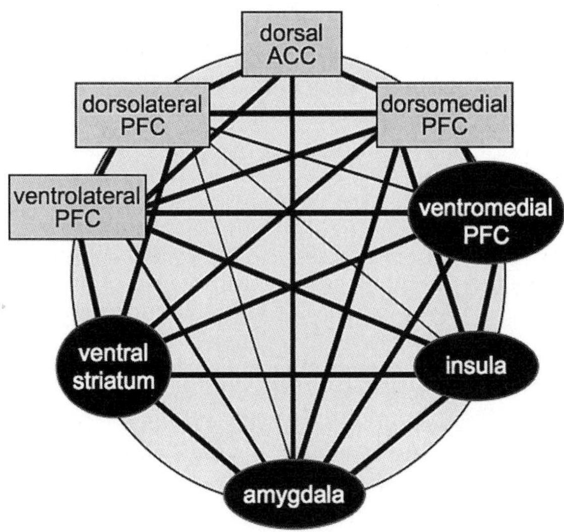

Figure 62.4 A visual representation of a hypothetical constraint satisfaction model of emotion. In this model, emotional state is represented by the profile of activation across the sphere and by interactions between a network of systems that could be described as supporting cognitive control processes (black ovals) or emotional responding (gray rectangles). As described in the text, each kind of system can make a greater or lesser contribution to one's current emotional response. To the extent that the highlighted systems are making greater contributions, we might describe emotion as being influenced by cognitive control. Depending on the specific meaning control systems ascribe to a given stimulus or stimuli, activity in emotion systems will increase, decrease, or be changed in some other way.

model has proved useful for modeling complex behaviors that arise from the contributions of multiple underlying processing systems (Maia & Cleeremans, 2005; Read, Vanman, & Miller, 1997; Shoda, LeeTiernan, & Mischel, 2002; Shultz & Lepper, 1996). These models rest on a principle known as parallel constraint satisfaction, whereby behavior results from an overall pattern of activation produced by a network that represents the possible relations and interactions between constituent processing elements. In the case of cognitive control–emotion interactions, each can be broken down into subcomponent processing systems that each comprise a key element in a constraint satisfaction network. In this chapter thus far we've only talked about two of these kinds of regions, but the list of potentially important regions can be expanded a bit here to illustrate the argument.

On the cognitive control side, key processing elements might include ventral lateral prefrontal regions important for selecting context-appropriate and inhibiting context-inappropriate responses (Aron, Robbins, & Poldrack, 2004; Payer et al., 2012; Wager et al., 2005), dorsal lateral prefrontal and parietal regions important for holding goals and task-relevant information in mind (Wager et al., 2004; Wager & Smith, 2003), dorsal cingulate regions important for monitoring conflicts between desired behaviors and actual behaviors (Botvinick et al., 2004). and dorsal medial prefrontal regions implicated in representing higher-level thoughts about mental states (Zaki & Ochsner, 2012).

On the emotion side, we might list the amygdala, which is important for detecting, encoding into memory, and triggering appropriate responses to goal-relevant and, especially, threatening stimuli (Davis & Whalen, 2001; Hariri & Whalen, 2011; Neta & Whalen, 2011; Whalen, 1998); striatal regions important for learning what stimuli predict rewarding or reinforcing outcomes in promoting behaviors that attain these outcomes, including emotional states (Kober et al., 2008; O'Doherty, 2004); the insula, which may integrate information about affect-relevant body states and represent that information in awareness (Craig, 2009; Zaki, Davis, & Ochsner, 2012); and ventral medial prefrontal regions implicated in integrating and updating these multiple sources of information about one's current affective state (Hare, O'Doherty, Camerer, Schultz, & Rangel, 2008; Rangel, Camerer, & Montague, 2008; Schoenbaum, Saddoris, & Stalnaker, 2007; Schoenbaum, Takahashi, Liu, & McDannald, 2011).

On this view, one's emotional state—and the emotional meaning one derives from a situation—reflects a profile of activation across all of these systems (and likely many more; Barrett, Mesquita, Ochsner, & Gross, 2007; Freeman & Ambady, 2011; Ochsner & Barrett, 2001; Ochsner & Gross, 2013). More strongly activated representations have a greater impact on one's current emotional experience, interoceptive assessment, judgment, behavior, and so on. In this way, apparently different kinds of emotional phenomena can arise from different kinds of interactions among the same network of underlying brain systems. Data consistent with this idea were presented in the prior section: highly overlapping combinations of prefrontal control and subcortical affect systems may be engaged, and presumably interact differently, during emotion generation, self-report, perception, and regulation. What's important is the *way* in which the systems interact, the kind of consistent—or conflicting—meanings represented within and across each system, and how strongly activated each of these multiple meaning representations are.

To make this concrete, imagine that you live in Manhattan and late one night you find yourself walking down the street to your apartment. As you get close to your building, you see two young men walking toward you. One reaches into his coat pocket when you are only a few feet away. For the amygdala, this might trigger an alarm call that you could be in danger and should be afraid. Medial prefrontal systems might bring to mind thoughts about these guys looking untrustworthy, the thought that you might be afraid, and thoughts about whether they look aggressive. If prefrontal control systems are engaged, however, they might generate and represent a potential alternative meaning for the situation. Ventral lateral prefrontal regions support retrieval from memory of the facts that this is November, that you live near an undergraduate dorm on the Columbia campus, and therefore these are probably just students returning home from an evening out, just like you. To the extent that these prefrontal representations are the most strongly activated and exert a greater influence on your overall emotional response than do the amygdala-based representations, you may feel less fearful and more confident that you won't be mugged as you walk up the steps to your apartment door and the two young men pass harmlessly behind you.

This example is, of course, overly simplistic, but it captures some of the essence of constraint satisfaction models. And these models can apply to all the other kinds of emotional phenomena described in the prior section. During emotion generation, either subcortical or cortical systems can represent meaningful interpretations of stimuli that have the effect of engendering an emotional response. When you are trying to understand and report on your emotional state, subcortical and prefrontal control systems each make a contribution to those reports. And when you're perceiving another person's emotion, control systems can help bias activation to favor brain systems representing one or another type of clue to their emotional state.

WHAT'S NEXT? While this model relies as much on metaphor as a seesaw model, it may provide a framework for accounting for a broader range of interactions between cognitive control and emotion. As such, it suggests two directions for future research.

First, we might expand our experimental purview to study not just cases where cognitive control appears to down-regulate some type of affective (or more generally, maladaptive) response, but to study the myriad other situations where we generate, introspect on, perceive, and regulate emotion in ourselves and others in ways that enhance, transform, integrate, and arbitrate between potential affective responses, interpretations, judgments, and so on. Second, it highlights just how far future research needs to go in providing increasingly specific descriptions of underlying neural mechanisms. As a field, we are increasingly moving beyond descriptions of neural bases that specify only relative levels of activation in different systems, à la seesaw models. Instead, more complicated path, mediation, and multivoxel pattern-based models are appearing that specify complex combinations of systems and the way in which they interact. As we study a wider range of phenomena, we need a more complex set of models to account for them. In this regard, we can follow the lead of research testing certain aspects of the glossy seesaw model, which have begun to use exactly these kinds of methods to understand the neural underpinnings of self-regulatory success and failure (e.g., Torrisi et al., 2013; Townsend et al., 2012; Wagner et al., 2013).

In the end, this way of thinking prompts a return to the *Star Trek* example that kicked off the chapter. To be sure, the Spock versus Kirk dynamic illustrated something fundamental about the way in which different aspects of our nature interact. But like the seesaw models this dynamic exemplifies, it is only part of the *Star Trek* worldview. Indeed, Kirk and Spock represented but two viewpoints within a polyglot crew who each had their own viewpoints. And that crew represented but one node in a larger Starfleet of perspectives. Ultimately, actions taken at the level of Starfleet represented some weighted combination of all of the opinions of its members. If we take the analogy between our brain and Starfleet seriously, our future task is to boldly go where no researchers have gone before, exploring new ways in which cognitive control and emotion can interact.

ACKNOWLEDGMENTS Completion of this chapter was supported by grant R01HD069178-03 from NICHD and grant R01AG043463 from NIA.

REFERENCES

AMODIO, D. M., & FRITH, C. D. (2006). Meeting of minds: The medial frontal cortex and social cognition. *Nat Rev Neurosci, 7*(4), 268–277.

ANDERSON, A. K., CHRISTOFF, K., PANITZ, D., DE ROSA, E., & GABRIELI, J. D. (2003). Neural correlates of the automatic processing of threat facial signals. *J Neurosci, 23*(13), 5627–5633.

ARON, A. R., & POLDRACK, R. A. (2005). The cognitive neuroscience of response inhibition: Relevance for genetic research in attention-deficit/hyperactivity disorder. *Biol Psychiatry, 57*(11), 1285–1292.

ARON, A. R., ROBBINS, T. W., & POLDRACK, R. A. (2004). Inhibition and the right inferior frontal cortex. *Trends Cogn Sci, 8*(4), 170–177.

BADRE, D., & WAGNER, A. D. (2007). Left ventrolateral prefrontal cortex and the cognitive control of memory. *Neuropsychologia, 45*(13), 2883–2901.

BARRETT, L. F., MESQUITA, B., OCHSNER, K. N., & GROSS, J. J. (2007). The experience of emotion. *Annu Rev Psychol, 58*, 373–403.

Beer, J. S., Knight, R. T., & D'Esposito, M. (2006). Controlling the integration of emotion and cognition: The role of frontal cortex in distinguishing helpful from hurtful emotional information. *Psychol Sci, 17*(5), 448–453.

Botvinick, M. M., Braver, T. S., Barch, D. M., Carter, C. S., & Cohen, J. D. (2001). Conflict monitoring and cognitive control. *Psychol Rev, 108*(3), 624–652.

Botvinick, M. M., Cohen, J. D., & Carter, C. S. (2004). Conflict monitoring and anterior cingulate cortex: An update. *Trends Cogn Sci, 8*(12), 539–546.

Buhle, J. T., Silvers, J. A., Wager, T. D., Lopez, R., Onyemekwu, C., Kober, H., ... Ochsner, K. N. (2013). Cognitive reappraisal of emotion: A meta-analysis of human neuroimaging studies. *Cereb Cortex.* Advance online publication. doi:10.1093/cercor/bht154

Craig, A. D. (2009). How do you feel—now? The anterior insula and human awareness. *Nat Rev Neurosci, 10*(1), 59–70.

Cunningham, W. A., & Brosch, T. (2012). Motivational salience: Amygdala tuning from traits, needs, values, and goals. *Curr Dir Psychol Sci, 21*(1), 54–59.

Cunningham, W. A., & Zelazo, P. D. (2007). Attitudes and evaluations: A social cognitive neuroscience perspective. *Trends Cogn Sci, 11*(3), 97–104.

Damasio, A. (1994). *Descartes' error: Emotion, reason, and the human brain.* New York, NY: Putnam.

Davis, M., & Whalen, P. J. (2001). The amygdala: Vigilance and emotion. *Mol Psychiatry, 6*(1), 13–34.

Drevets, W. C., & Raichle, M. E. (1998). Reciprocal suppression of regional cerebral blood flow during emotional versus higher cognitive processes: Implications for interactions between emotion and cognition. *Cogn Emot, 12*(3), 353–385.

Freeman, J. B., & Ambady, N. (2011). A dynamic interactive theory of person construal. *Psychol Rev, 118*(2), 247–279.

Gallese, V., Keysers, C., & Rizzolatti, G. (2004). A unifying view of the basis of social cognition. *Trends Cogn Sci, 8*(9), 396–403.

Gendron, M., & Barrett, L. F. (2009). Reconstructing the past: A century of ideas about emotion in psychology. *Emot Rev, 1*(4), 316–339.

Greene, J. D., Sommerville, R. B., Nystrom, L. E., Darley, J. M., & Cohen, J. D. (2001). An fMRI investigation of emotional engagement in moral judgment. *Science, 293*(5537), 2105–2108.

Hare, T. A., O'Doherty, J., Camerer, C. F., Schultz, W., & Rangel, A. (2008). Dissociating the role of the orbitofrontal cortex and the striatum in the computation of goal values and prediction errors. *J Neurosci, 28*(22), 5623–5630.

Hariri, A. R., & Whalen, P. J. (2011). The amygdala: Inside and out. *F1000 Biol Rep, 3*, 2.

Hartley, C. A., & Phelps, E. A. (2012). Anxiety and decision-making. *Biol Psychiatry, 72*(2), 113–118.

Heatherton, T. F., & Wagner, D. D. (2011). Cognitive neuroscience of self-regulation failure. *Trends Cogn Sci, 15*(3), 132–139.

Kim, M. J., Loucks, R. A., Palmer, A. L., Brown, A. C., Solomon, K. M., Marchante, A. N., & Whalen, P. J. (2011). The structural and functional connectivity of the amygdala: From normal emotion to pathological anxiety. *Behav Brain Res, 223*(2), 403–410.

Kober, H., Barrett, L. F., Joseph, J., Bliss-Moreau, E., Lindquist, K., & Wager, T. D. (2008). Functional grouping and cortical-subcortical interactions in emotion: A meta-analysis of neuroimaging studies. *NeuroImage, 42*(2), 998–1031.

Lazarus, R. S. (1982). Thoughts on the relations between emotion and cognition. *Am Psychol, 37*(9), 1019–1024.

Lazarus, R. S. (1991). Progress on a cognitive-motivational-relational theory of emotion. *Am Psychol, 46*(8), 819–834.

Lieberman, M. D., Eisenberger, N. I., Crockett, M. J., Tom, S. M., Pfeifer, J. H., & Way, B. M. (2007). Putting feelings into words: Affect labeling disrupts amygdala activity in response to affective stimuli. *Psychol Sci, 18*(5), 421–428.

Lieberman, M. D., Inagaki, T. K., Tabibnia, G., & Crockett, M. J. (2011). Subjective responses to emotional stimuli during labeling, reappraisal, and distraction. *Emotion, 11*(3), 468–480.

Lindquist, K. A., & Barrett, L. F. (2008). Constructing emotion: The experience of fear as a conceptual act. *Psychol Sci, 19*(9), 898–903.

Maia, T. V., & Cleeremans, A. (2005). Consciousness: Converging insights from connectionist modeling and neuroscience. *Trends Cogn Sci, 9*(8), 397–404.

Miller, E. K., & Cohen, J. D. (2001). An integrative theory of prefrontal cortex function. *Annu Rev Neurosci, 24*, 167–202.

Neta, M., & Whalen, P. J. (2011). The primacy of negative interpretations when resolving the valence of ambiguous facial expressions. *Psychol Sci, 21*(7), 901–907.

Ochsner, K. N., & Barrett, L. F. (2001). A multiprocess perspective on the neuroscience of emotion. In T. J. Mayne & G. A. Bonanno (Eds.), *Emotions: Current issues and future directions* (pp. 38–81). New York, NY: Guilford.

Ochsner, K. N., & Gross, J. J. (2013). The neural bases of emotion and emotion regulation: A valuation perspective. In J. J. Gross & R. H. Thompson (Eds.), *The handbook of emotion regulation* (pp. 23–42). New York, NY: Guilford.

Ochsner, K. N., Ray, R. D., Cooper, J. C., Robertson, E. R., Chopra, S., Gabrieli, J. D. E., & Gross, J. J. (2004). For better or for worse: Neural systems supporting the cognitive down- and up-regulation of negative emotion. *NeuroImage, 23*(2), 483–499.

Ochsner, K. N., Ray, R. D., Hughes, B., McRae, K., Cooper, J., Weber, J., ... Gross, J. J. (2009). Bottom-up and top-down processes in emotion generation: Common and distinct neural mechanisms. *Psychol Sci, 20*(11), 1322–1331.

Ochsner, K. N., Silvers, J. A., & Buhle, J. T. (2012). Functional imaging studies of emotion regulation: A synthetic review and evolving model of the cognitive control of emotion. *Ann NY Acad Sci, 1251*, E1–24.

O'Doherty, J. P. (2004). Reward representations and reward-related learning in the human brain: Insights from neuroimaging. *Curr Opin Neurobiol, 14*(6), 769–776.

Olsson, A., & Ochsner, K. N. (2008). The role of social cognition in emotion. *Trends Cogn Sci, 12*(2), 65–71.

Osgood, C. E., Suci, G. J., & Tannenbaum, P. H. (1957). *The measurement of meaning.* Urbana: University of Illinois Press.

Payer, D. E., Baicy, K., Lieberman, M. D., & London, E. D. (2012). Overlapping neural substrates between intentional and incidental down-regulation of negative emotions. *Emotion, 12*(2), 229–235.

Rangel, A., Camerer, C., & Montague, P. R. (2008). A framework for studying the neurobiology of value-based decision making. *Nat Rev Neurosci, 9*(7), 545–556.

Read, S. J., Vanman, E. J., & Miller, L. C. (1997). Connectionism, parallel constraint satisfaction processes, and Gestalt principles: (Re)introducing cognitive dynamics to social psychology. *Pers Soc Psychol Rev, 1*(1), 26–53.

Satpute, A. B., Shu, J., Weber, J., Roy, M., & Ochsner, K. N. (2013). The functional neural architecture of self-reports of affective experience. *Biol Psychiatry, 73*(7), 631–638.

Saxe, R. (2006). Uniquely human social cognition. *Curr Opin Neurobiol, 16*(2), 235–239.

Scherer, K. R., Schorr, A., & Johnstone, T. (Eds.). (2001). *Appraisal processes in emotion: Theory, methods, research.* New York, NY: Oxford University Press.

Schoenbaum, G., Saddoris, M. P., & Stalnaker, T. A. (2007). Reconciling the roles of orbitofrontal cortex in reversal learning and the encoding of outcome expectancies. *Ann NY Acad Sci, 1121,* 320–335.

Schoenbaum, G., Takahashi, Y., Liu, T. L., & McDannald, M. A. (2011). Does the orbitofrontal cortex signal value? *Ann NY Acad Sci, 1239,* 87–99.

Shoda, Y., LeeTiernan, S., & Mischel, W. (2002). Personality as a dynamical system: Emergency of stability and distinctiveness from intra- and interpersonal interactions. *Pers Soc Psychol Rev, 6*(4), 316–325.

Shultz, T. R., & Lepper, M. R. (1996). Cognitive dissonance reduction as constraint satisfaction. *Psychol Rev, 103*(2), 219–240.

Somerville, L. H., & Casey, B. J. (2010). Developmental neurobiology of cognitive control and motivational systems. *Curr Opin Neurobiol, 20*(2), 236–241.

Thompson-Schill, S. L. (2003). Neuroimaging studies of semantic memory: Inferring "how" from "where." *Neuropsychologia, 41*(3), 280–292.

Torrisi, S. J., Lieberman, M. D., Bookheimer, S. Y., & Altshuler, L. L. (2013). Advancing understanding of affect labeling with dynamic causal modeling. *NeuroImage, 82,* 481–488.

Townsend, J. D., Torrisi, S. J., Lieberman, M. D., Sugar, C. A., Bookheimer, S. Y., & Altshuler, L. L. (2012). Frontal-amygdala connectivity alterations during emotion down-regulation in bipolar I disorder. *Biol Psychiatry, 73*(2), 127–135.

Urry, H. L., van Reekum, C. M., Johnstone, T., Kalin, N. H., Thurow, M. E., Schaefer, H. S., ... Davidson, R. J. (2006). Amygdala and ventromedial prefrontal cortex are inversely coupled during regulation of negative affect and predict the diurnal pattern of cortisol secretion among older adults. *J Neurosci, 26*(16), 4415–4425.

Wager, T. D., Barrett, L. F., Bliss-Moreau, E., Lindquist, K., Duncan, S., Kober, H., ... Mize, J. (2008). The neuroimaging of emotion. In M. Lewis, J. M. Haviland-Jones, & L. F. Barrett (Eds.), *The handbook of emotion* (3rd ed., pp. 249–271). New York, NY: Guilford.

Wager, T. D., Jonides, J., & Reading, S. (2004). Neuroimaging studies of shifting attention: A meta-analysis. *NeuroImage, 22*(4), 1679–1693.

Wager, T. D., & Smith, E. E. (2003). Neuroimaging studies of working memory: A meta-analysis. *Cogn Aff Behav Neurosci, 3*(4), 255–274.

Wager, T. D., Sylvester, C. Y., Lacey, S. C., Nee, D. E., Franklin, M., & Jonides, J. (2005). Common and unique components of response inhibition revealed by fMRI. *NeuroImage, 27*(2), 323–340.

Wagner, D. D., Altman, M., Boswell, R. G., Kelley, W. M., & Heatherton, T. F. (2013). Self-regulatory depletion enhances neural responses to rewards and impairs top-down control. *Psych Sci, 11,* 2262–2267.

Wagner, D. D., & Heatherton, T. F. (2013). Self-regulatory depletion increases emotional reactivity in the amygdala. *Soc Cogn Aff Neuro, 8,* 410–417.

Whalen, P. J. (1998). Fear, vigilance, and ambiguity: Initial neuroimaging studies of the human amygdala. *Curr Dir Psychol Sci, 7*(6), 177–188.

Zajonc, R. B. (1980). Feeling and thinking: Preferences need no inferences. *Am Psychol, 35*(2), 151–175.

Zajonc, R. B. (1984). On the primacy of affect. *Am Psychol, 39*(2), 117–123.

Zaki, J., Bolger, N., & Ochsner, K. (2009). Unpacking the informational bases of empathic accuracy. *Emotion, 9*(4), 478–487.

Zaki, J., Davis, J. I., & Ochsner, K. N. (2012). Overlapping activity in anterior insula during interoception and emotional experience. *NeuroImage, 62*(1), 493–499.

Zaki, J., Hennigan, K., Weber, J., & Ochsner, K. N. (2010). Social cognitive conflict resolution: Contributions of domain-general and domain-specific neural systems. *J Neurosci, 30*(25), 8481–8488.

Zaki, J., & Ochsner, K. (2012). The neuroscience of empathy: Progress, pitfalls and promise. *Nat Neurosci, 15*(5), 675–680.

Zaki, J., Weber, J., Bolger, N., & Ochsner, K. (2009). The neural bases of empathic accuracy. *Proc Natl Acad Sci USA, 106*(27), 11382–11387.

63 Emotional Reactivity and Regulation Across Development

LEAH H. SOMERVILLE AND BJ CASEY

ABSTRACT Fears of the dark, sadness following peer rejection, and disappointment from losing a job are examples of emotions that we experience throughout our lives. The intensity of these emotions can vary by age and across individuals. In this chapter, we examine the development of two aspects of emotional behavior: emotional reactivity and emotional regulation. Emotional *reactivity* can be defined as the mental processes that support the detection and assignment of salience to environmental inputs. Emotion *regulation* processes are those that would support a modification of an emotional response most commonly assessed in terms of the reduction of an initial emotional response. Although these emotional processes shift in important ways from infancy to adulthood, we focus on changes that take place during transitions into and out of adolescence.

A 2-year-old's temper tantrum and an adolescent's oscillating mood swings illustrate how emotional processes may change in important ways with development. Through human development, we develop the ability to filter the constant stream of environmental inputs and determine which are important and which hold emotional significance. Emotional responses to this information affect our actions and how attention is deployed to them. In this chapter, we present evidence for developmental changes in emotional reactivity and emotional regulation. We provide examples of how positive and negative emotional cues engage cognitive processes that support the detection, appraisal, and regulation of emotions.

Two terms used throughout this chapter are emotional reactivity and emotion regulation. Emotion *reactivity* may be defined as the mental processes that support the detection and assignment of salience to negative or positive information and our reaction to that information. Emotion *regulation* processes are those that support a modification of an emotional response (Gross & Thompson, 2009), most commonly assessed in terms of the reduction of an initial emotional response. Although these emotional processes shift in important ways from infancy to adulthood, we focus on behavioral and brain changes that take place during adolescence, given the high prevalence of anxiety and mood disorders during this developmental window (Beesdo, Knappe, & Pine, 2009).

The emotional landscape of adolescence

Emotional reactivity and regulation are influenced by a number of factors—some transient and developmentally regulated, others stable within an individual (Somerville, Jones, & Casey, 2010). A large body of psychological research has focused on whether differences in behavioral and emotional reactivity early in life are indicative of stable, persistent, life-long emotional reactivity patterns. Classic work by Ainsworth and Bowlby on attachment (Ainsworth & Bowlby, 1991), Kagan on behavioral inhibition (Kagan, Snidman, Arcus, & Reznick, 1994), Mischel on delay of gratification (Mischel, Shoda, & Peake, 1988) and Rothbart (Rothbart, 2007) on temperament suggests that emotion reactivity and regulation can reflect trait-like tendencies early in life that statistically predict long-term mental and physical health and illness.

These emotional tendencies vary by age and environmental factors. For example, increasing environmental demands (e.g., stress) placed on an individual can exacerbate emotional reactivity, while the development of regulatory abilities can help to quiet emotions (Mischel & Baker, 1975; Posner & Rothbart, 2007). Environmental challenges and stressors are thought to reach a normative peak during adolescence, when the individual transitions from dependence on the parent to relative independence while regulatory abilities are still developing—a period of life Hall described as ridden with emotional upheaval, negative mood, and chronic stress (Arnett, 1999; Hall, 1904).

The period of adolescence begins around the onset of puberty and ends ambiguously when an individual achieves adult-like levels of independence. The adolescent faces unique challenges with more independence than ever before, but also more responsibilities at home, school, and in social interactions (Larson, 2001; Spear, 2000). The social landscape takes on a heightened importance, while individuals strive to identify with increasingly complex and salient social groups (Pfeifer & Peake, 2012; Somerville, 2013). Simultaneously, physical development mediated by hormonal changes of puberty results in widespread physiological shifts that can be unpredictable, intense, and stressful (Peper & Dahl, 2013; Sisk & Zehr, 2005). These changes are supported by self-reports of more frequent, intense, and rapid fluctuations in negative and positive emotions by adolescents relative to children or adults (Larson, Moneta, Richards, & Wilson, 2002; Rutter, Graham, Chadwick, & Yule, 1976).

The challenges faced by the adolescent are not specific to humans but observed across species, as evidenced by similar increases in interactions with peers, novelty seeking, and reward sensitivity during this developmental phase. These behaviors are thought to have evolved to serve adaptive functions (Spear & Varlinskaya, 2010) related to successful mating and attainment of resources necessary for survival. We highlight significant changes in brain maturation and function that parallel these behavioral changes.

Emotional reactivity and regulatory circuitry

Emotional behavior is thought to emerge from interactions between brain regions that make up frontolimbic circuitry. One essential region of this circuitry is the *amygdala*, located in the medial temporal lobe. The amygdala is important for detecting salient information, assigning it emotional importance, and modulating neural circuits and physiological reactivity through dense feedback projections. As such, this region prioritizes information processing of positive or negative cues and mobilizes behavior in service of reacting to and learning about salient information (Davis & Whalen, 2001; Phelps & LeDoux, 2005). One way in which this region influences behavior is by forming and storing learned associations between environmental cues and positive and negative outcomes (e.g., conditioning; LeDoux, 2000). Its response can be regulated through direct and indirect connections with cortical regions such as the ventromedial and ventrolateral prefrontal cortices, which can up- or down-regulate amygdala responses based on the contextual appropriateness of the emotional response, additional learned associations, or explicit attempts to regulate an emotional reaction (LaBar, Gatenby, Gore, LeDoux, & Phelps, 1998; Milad & Quirk, 2002; Ochsner & Gross, 2005).

Signaling the salience of environmental inputs also draws on the processing resources of the basal ganglia, a collection of subcortical nuclei including the caudate, putamen, globus pallidus, and nucleus accumbens (also termed *ventral striatum* in humans). These regions play a critical role in reinforcement learning, integrating motivational signals, and facilitating instrumental behavior (Casey et al., 2000; Delgado, 2007; Pasupathy & Miller, 2005; Schultz, Tremblay, & Hollerman, 1998).

The amygdala and ventral striatum are often characterized as "negative" and "positive" emotion centers, respectively. However, the role of these structures in emotional processing is not straightforwardly valence-specific (Levita et al., 2009; Lindquist, Wager, Kober, Bliss-Moreau, & Barrett, 2012; Pagnoni, Zink, Montague, & Berns, 2002). Rather, the amygdala and ventral striatum are thought to play complementary roles in detecting and learning about salience based on associative properties and prediction error, respectively (Li, Schiller, Schoenbaum, Phelps, & Daw, 2011). Dense, hierarchical striatal-prefrontal cortical projections ("loops") comprise reciprocal circuits that integrate motivational signals with context-specific and regulatory demands to ultimately guide motivated behavior (Haber, Kim, Mailly, & Calzavara, 2006; Haber & Knutson, 2010).

Development of emotional reactivity and regulatory brain circuitry

Emotions can exert potentially powerful influences over our thoughts, actions, and decisions. We and others have put forth a hypothesis that normative trajectories of brain development might lead to unique emotion-cognition interactions during adolescence (Casey & Caudle, 2013; Casey, Getz, & Galvan, 2008; Casey, Jones, & Hare, 2008; Ernst, Pine, & Hardin, 2006; Geier & Luna, 2009; Somerville et al., 2010; Steinberg, 2008). This view builds on existing evidence from behavioral, structural, and functional magnetic resonance imaging studies of humans, and incorporates findings from nonhuman primate and rodent models. The central tenet is that normative trajectories of brain development lead subcortical and cortical systems of the brain to interact differentially during childhood, adolescence, and adulthood. On one hand, subcortical systems such as the amygdala and striatum are temporarily heightened in their response strength during adolescence. This heightened response could be due to multiple factors, including relatively early emerging subcortical gray matter development and connectivity paired with temporarily heightened neuromodulatory dopamine signaling (Laviola, Pascucci, & Pieretti, 2001; Teicher, Andersen, & Hostetter, 1995). On the other hand, the prefrontal cortex that can implement cognitive and emotional regulation (and interact extensively with subcortical regions as part of reciprocal circuitry) continues to undergo substantial structural and functional development during adolescence.

As a result, subcortical-cortical circuitry interactions are thought to manifest differently during childhood, adolescence, and adulthood. During childhood, dopaminergic signals are not yet sensitized, and cortical and subcortical structures and connectivity are continuing to refine. In adulthood, prefrontal cortical signaling is more structurally and functionally mature as a result of continued development and experience, leading to a stronger capacity for prefrontal control properties to facilitate behavioral regulation. During adolescence, subcortical circuitry generates strong signals that are countered by still-maturing regulatory capacity, leading to a strong influence of emotional and motivational signals over choices and actions. The day-to-day challenges and stress during adolescence may further enhance this shift in balance (Arnsten, Wang, & Paspalas, 2012). With age and experience, the connectivity between these regions is strengthened and provides a mechanism for top-down modulation of motivated and emotional behavior (Somerville, Fani, & McClure-Tone, 2011; van den Bos et al., 2011).

Emotion reactivity and regulation in adolescence

As described above, adolescence is a period of the life span marked by intensified daily emotional experiences. Strong-signaling subcortical circuitry, paired with still-maturing cortical circuitry, could partially account for adolescents' emotional behavior. In the next section, we review evidence relating cognitive and emotional functioning, and their interactions, with circuit-level brain function as it changes from childhood to adulthood.

Amplified Emotional Signaling in Adolescence

One component of the neurodevelopmental hypothesis is that emotional information is processed robustly, perhaps even in an exaggerated fashion, by the adolescent brain. A growing body of work has probed for unique patterns of functional recruitment in adolescents to salient positive and negative cues. One experimental approach presents static images of facial expressions that vary in emotional qualities for participants to passively view or make a simple judgment on. These studies show exaggerated recruitment of the amygdala while viewing facial expressions of emotion in adolescents relative to children (Hare et al., 2008) and adults (Guyer et al., 2008; Hare et al., 2008; Monk et al., 2003; Williams et al., 2006; see figure 63.1A, and see also Somerville, Fani, & McClure-Tone, 2011).

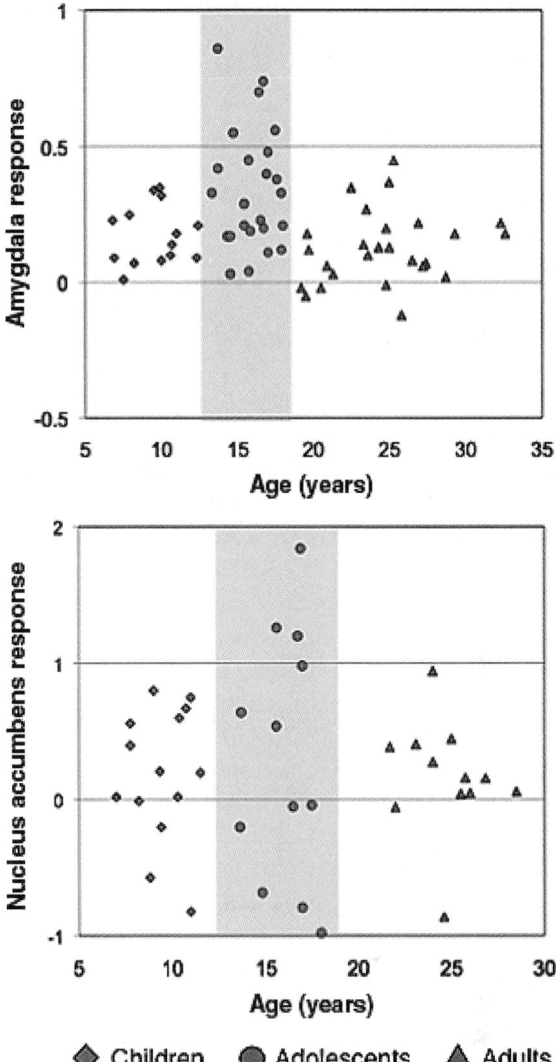

Figure 63.1 *Top:* Amygdala response to facial expressions of emotion was significantly greater in adolescents than in children or adults. *Bottom:* Ventral striatal response to receiving a large monetary reward was significantly greater in adolescents than in children or adults. (Reproduced from Somerville et al., 2010, with permission from Elsevier.)

It should be noted that these effects are not wholly consistent across studies (see McClure et al., 2004; Pine et al., 2001; and Spear, 2011, for a differential viewpoint on aversive processes in adolescence), possibly attributable to different age groupings of comparison or task demands and documented low test-retest reliability in adolescent participants' amygdala reactivity (van den Bulk et al., 2013). Further, these effects are influenced by pubertal development. Longitudinal investigations have demonstrated that the amygdala and ventrolateral prefrontal cortex response while viewing sad facial expressions (but not other expressions) is significantly greater after pubertal onset than just prior (Moore et al., 2012), and that the ventral striatum is significantly more reactive to a wide range of emotional facial displays after puberty's onset relative to before (Pfeifer et al., 2011). Together, these studies suggest that emotional facial expressions are assigned higher salience in adolescents than in younger and older individuals, perhaps reflecting their social communication value during a time of life characterized by social reorientation (Crone & Dahl, 2012; Nelson, Leibenluft, McClure, & Pine, 2005).

Numerous studies have begun to examine emotion reactivity in adolescents to the receipt or loss of monetary incentive or prizes. These studies show that the magnitude (Ernst et al., 2005; Galvan et al., 2006; van Leijenhorst et al., 2010) and temporal sustainment (Fareri, Martin, & Delgado, 2008) of responses in the ventral striatum are greater in adolescents than in children and adults (figure 63.1B). Importantly, this pattern has been recently extended in human imaging to contexts of primary rewards such as juice (Galvan & McGlennen, 2013), indicating that even fundamental rewards may elicit heightened value signaling during adolescence. However, like the amygdala findings, results do not always support an account of adolescent striatal hypersensitivity (see Bjork et al., 2004, and Galvan, 2010).

A complementary approach examines the extent to which participants' neural activity reflects computational learning signals that are thought to reflect dopaminergic neuronal activity (Schultz, Dayan, & Montague, 1997). These studies examine how the expectations of receiving a reward based on prior outcomes and its value (e.g., low or high magnitude) impact future choices. Cohen and colleagues (2010) showed that compared to children and adults, adolescents respond more quickly to stimuli that have been previously associated with a high (relative to low) reward value and show greater ventral striatal activity to unexpected reward. This heightened positive prediction error and striatal activity has been suggested as a possible mechanism for increased risky decisions during adolescence. Alternatively, van den Bos, Crone, and colleagues (van den Bos et al., 2011) have shown neural representation of prediction errors to be similar across age. According to their work, the functional connectivity between the ventral striatum and prefrontal cortex is what is changing as a function of age. As such, developmental changes in behavior during adolescence may reflect how learning signals guide behavior and expectations (van den Bos et al., 2009, 2011).

Emotion Regulation in Adolescence In our daily lives, we rarely allow ourselves to experience emotions with complete abandonment. This control of emotions is an important societal function, since emotions can disrupt other goal-directed activity through wide-ranging mechanisms—from fundamental ways in which emotion captures our attention (Vuilleumier, 2005), to the disruptive effect of high physiological arousal on cognitive performance (Yerkes & Dodson, 1908), to the sociocultural display rules that can motivate us to suppress the expression of our true feelings (Gross & Levenson, 1993). As a result, human emotion is ubiquitously subject to regulation attempts that vary from implicit and automatic to intentional and effortful.

Emotion regulation is accomplished through interactions between emotion circuitry and regions of the brain that are capable of exerting regulation or control over emotional responses. Given the continuing development of cognitive control capabilities (de Luca et al., 2003; Huizinga, Dolan, & van der Molen, 2006) throughout youth, one could hypothesize that regulatory influences operate with reduced efficiency during childhood and adolescence compared to adulthood.

Here we focus on the developmental trajectories of three types of emotion regulation: forms of regulation whereby participants must disengage from emotional material in order to accomplish a secondary task, processes whereby emotional information is reappraised to reduce its affective impact, and emotion regulation that is the result of associative learning.

Disengagement from Emotional Material The need to disengage from emotional aspects of stimuli in order to complete a secondary task is a form of emotion regulation that draws on cognitive and attention control resources. One task that has been used to elucidate developmental emotion-cognition interactions is the emotional go/no-go task, where participants must follow instructions to respond to (or withhold a response to) stimuli that vary in appetitive or aversive salience. Adolescents have more difficulty than children and adults withholding a response toward an emotionally arousing happy face relative to emotionally neutral faces, suggesting that the appetitive salience associated with happy faces selectively disrupts their impulse control (Somerville, Hare, & Casey, 2011). Hare and colleagues demonstrated that adolescents are slower to respond to fearful faces (Hare et al., 2008), suggesting that negative emotional content is more disruptive to their cognitive performance during adolescence as well. The incidental presence of emotionally negative images reduces young adolescents' performance during a go/no-go task more than in children or older individuals (Cohen-Gilbert & Thomas, 2013), likely by capturing attentional and cognitive resources (Grose-Fifer,

Rodrigues, Hoover, & Zottoli, 2013). However, other studies have found that adolescent regulation capacity is intermediate relative to children and adults when faced with emotional cues, though adolescents respond more quickly in the presence of emotional cues to which they ought not respond (Tottenham, Hare, & Casey, 2011). Generally, this pattern suggests that when adolescents should engage in cognitive control over their behavior, the presence of emotional cues more strongly interrupts that ability compared to younger and older individuals.

These findings may be explained by strong subcortical responding while viewing emotional images in adolescents (Hare et al., 2008; Somerville, Hare, et al., 2011) during a time when recruitment of and connectivity with cognitive control circuitry, including ventromedial and lateral prefrontal cortices, has not yet reached adult-level maturity (Asato, Terwilliger, Woo, & Luna, 2010; Luna, Garver, Urban, Lazar, & Sweeney, 2004; Luna, Padmanabhan, & O'Hearn, 2010; Rubia et al., 2006). Thus, in contexts in which emotional cues conflict with cognitive control demands, regulatory success is reduced in adolescents. As such, the disruption of cognition by emotion during adolescence may be the result of robust subcortical responding with a regulatory capacity that is constrained by still-developing prefrontal systems.

Cognitive Reappraisal Reappraisal is a form of emotion regulation whereby an individual attempts to alter the meaning of an emotional cue through cognitive reinterpretation (see also Ochsner, chapter 62 in this volume). Greater reductions in negative affect are observed after reappraisal than other emotion regulation strategies (Gross, 2002), indicating that reappraisal is a particularly effective form of emotion regulation. It is striking that despite the efficacy of reappraisal and the strength of adolescent emotional reactions, adolescents tend not to utilize reappraisal strategies as much as adults do in their daily lives (Garnefski, Legerstee, Kraaij, van den Kommer, & Teerds, 2002). Reappraisal relies on numerous, complex cognitive processes, including working memory, abstract reasoning, and selection (Ochsner & Gross, 2005). This has led scientists to speculate that reappraisal ability might be a late-developing process whose efficacy is constrained by neurodevelopmental maturation.

A recent study (Silvers et al., 2012) asked participants ranging in age from 10 to 22 to view emotionally negative images and reappraise their reaction to them so as to reduce their negative impact. Results showed robust age differences in the degree to which reappraisal reduced negative affect, with greater age predicting greater regulatory success. When images contained social content, younger adolescents showed even less regulatory efficiency, which offers the initial suggestion that social content is a relevant dimension that might particularly challenge adolescents' regulatory abilities. A brain imaging study with similar methodology replicated the continual improvement of reappraisal success with greater age, and was paralleled by greater recruitment in the lateral prefrontal cortex with development (McRae et al., 2012). Thus, late-developing trajectories of the prefrontal cortex might result in a lessened capacity for adolescents to capitalize on reappraisal in reducing their affective responses.

Emotional Learning and Regulation A third form of emotion regulation critically depends on fundamental forms of associative learning. Classical conditioning is a form of emotional learning whereby an individual observes associative pairings between an otherwise neutral stimulus and an inherently appetitive or aversive outcome. This associative linkage is critically dependent on synaptic plasticity that stores the learned association at the level of individual synapses within the amygdala and hippocampus (Phillips & LeDoux, 1992; Schafe, Nader, Blair, & LeDoux, 2001). Importantly, expression of these learned associations is dynamically updated by changing environmental contingencies. For example, if a cue that had previously predicted an electric shock does not any longer, the organism generates a new memory that signifies the new learned association with safety. This process is called extinction. The ventromedial prefrontal cortex is necessary for the storage and expression of extinction memories, in that its signal of safety propagates to the amygdala via direct signaling that blocks the expression of fear (Maren & Quirk, 2004). In this way, the expression of extinction learning can regulate emotional responses in a contextually appropriate manner.

Research in this area is suitable for translational approaches that allow for cross-species comparison that critically links levels of analysis that range from molecular physiology to human behavior. Given that extinction learning is mediated by the ventromedial prefrontal cortex (Milad & Quirk, 2002; Phelps, Delgado, Nearing, & LeDoux, 2004), one could predict that the capacity to regulate emotional responses via extinction might vary across the developmental course. Recent examinations of fear acquisition and extinction in rodent and human developmental samples (from childhood to adulthood) demonstrate that even young individuals can acquire fear memories through classical conditioning. The expression of this association (freezing in rodents; galvanic skin response in humans) is

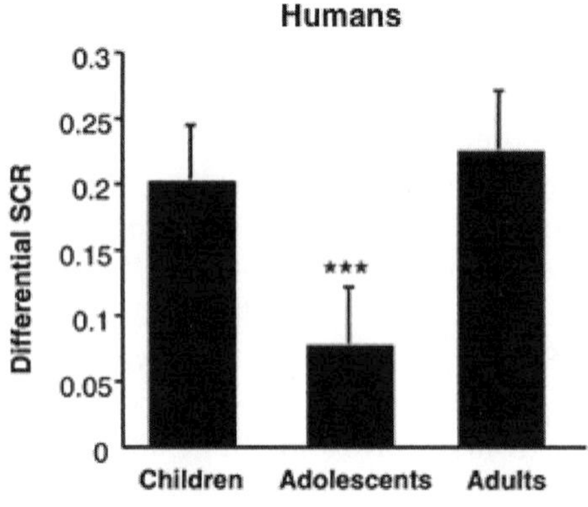

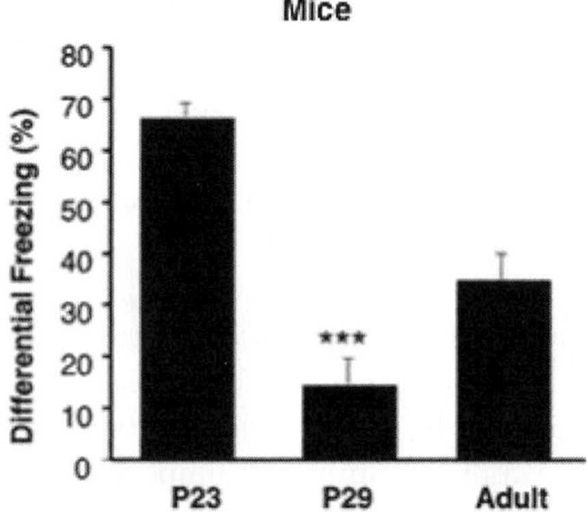

FIGURE 63.2 A lack of extinction learning and retention of extinction memory is observed in adolescent humans (*top*) and mice (*bottom*). P23, "child"; P29, "periadolescent." (Reproduced with permission from Pattwell et al., 2012.)

comparable in magnitude across ages, suggesting that conditioned learning is evident very early in life. However, the extent to which humans and rodents express extinction learning varies nonlinearly across development. Specifically, adolescents show a reduction in the expression of extinction learning relative to both children and adults (figure 63.2; Pattwell et al., 2012). In this way, adolescents do not seem to be able to capitalize on learned extinction cues. Interestingly, this effect has a nonlinear developmental course: younger and older individuals show stronger extinction than adolescents.

Conclusion

The work highlighted in this chapter illustrates the complexity of emotional processes and their development. Increasing daily challenges associated with striving for independence, paired with heightened signaling of subcortical regions during adolescence, can exacerbate emotions, while the development of regulatory abilities supported by refinement of cortical connections can regulate emotions. Collectively, the studies presented in this chapter contribute to our understanding of neural responses to positive and negative emotional information and how they may influence adolescent behavior differentially from that of children and adults. These findings are significant for informing the developmental trajectory of psychopathologies like anxiety and mood disorders that emerge during this developmental period.

During adolescence, we believe that unique patterns of subcortical-cortical interactions impact emotional processes at the psychological level. This developmental neurobehavioral pattern underscores the conceptualization of development as not only linear in nature, but representing complex trajectories as well. Behaviorally, emotional tendencies seem to shift during adolescence in a way that sensitizes them to emotional information differently than during younger or older ages. At the neural level, nonlinear recruitment, structural changes, and connectivity challenge assumptions that developmental processes "fill in" slowly and steadily. This pattern reflects the multitude of changes that take place in adolescence, ranging from hormonal, physical, and psychosocial that adolescents must navigate with less "regulatory input" from authority figures. Tuning of brain circuitry underlying emotion regulation reflects normative, experience-dependent maturation that results from adolescents interacting with their environments. Thus, the differences reported here should not be pathologized or interpreted as adolescent deficiencies, but rather as normal developmental progressions, much like a child learning to walk.

There is a growing recognition that adolescence is a time of life associated with numerous preventable health risks that result from dysregulation of emotional responses, including the deleterious outcomes of risk taking, experimentation with drugs and alcohol, relational aggression, the emergence of mood anxiety disorders, and suicide, to name a few. Despite the fundamental importance of this developmental phase, the majority of neurobehavioral findings reported here have emerged within the past decade. The degree to which adolescent neurodevelopment research has recently achieved recognition as an important, yet underspecified, topic of study within the emotion and emotion regulation research disciplines is truly exciting. With a greater volume of comprehensive and careful empirical research, this area will continue to advance toward the ultimate goal of translating neurodevelopmental findings in ways that improve adolescent mental and physical health.

ACKNOWLEDGMENTS Supported by NIMH grants R00 MH087813 (LHS) and P50 MH 079513 (BJC).

REFERENCES

Ainsworth, M. S., & Bowlby, J. (1991). An ethological approach to personality development. *Am Psychol, 46*(4), 333.

Arnett, J. (1999). Adolescent storm and stress, reconsidered. *Am Psychol, 54,* 317–326.

Arnsten, A. F., Wang, M. J., & Paspalas, C. D. (2012). Neuromodulation of thought: Flexibilities and vulnerabilities in prefrontal cortical network synapses. *Neuron, 76*(1), 223–239.

Asato, M. R., Terwilliger, R., Woo, J., & Luna, B. (2010). White matter development in adolescence: A DTI study. *Cereb Cortex, 20*(9), 2122–2131.

Beesdo, K., Knappe, S., & Pine, D. S. (2009). Anxiety and anxiety disorders in children and adolescents: Developmental issues and implications for DSM-V. *Psych Clin North Am, 32*(3), 483–524.

Bjork, J. M., Knutson, B., Fong, G. W., Caggiano, D. M., Bennett, S. M., & Hommer, D. W. (2004). Incentive-elicited brain activation in adolescents: Similarities and differences from young adults. *J Neurosci, 24*(8), 1793–1802.

Casey, B. J., & Caudle, K. (2013). The teenage brain: Self control. *Curr Dir Psychol Sci, 22*(2), 82–87.

Casey, B. J., Getz, S., & Galvan, A. (2008). The adolescent brain. *Dev Rev, 28*(1), 62–77.

Casey, B. J., Jones, R. M., & Hare, T. (2008). The adolescent brain. *Ann NY Acad Sci, 1124,* 111–126.

Casey, B. J., Thomas, K. M., Welsh, T. F., Badgaiyan, R., Eccard, C., Jennings, J. R., & Crone, E. A. (2000). Dissociation of response conflict, attentional control, and expectancy with functional magnetic resonance imaging (fMRI). *Proc Natl Acad Sci USA, 97,* 8727–8733.

Cohen, J. R., Asarnow, R. F., Sabb, F. W., Bilder, R. M., Bookheimer, S. Y., Knowlton, B. J., & Poldrack, R. A. (2010). A unique adolescent response to reward prediction errors. *Nat Neurosci, 13*(6), 669–671.

Cohen-Gilbert, J. E., & Thomas, K. M. (2013). Inhibitory control during emotional distraction across adolescence and early adulthood. *Child Dev, 84*(6), 1954–1966.

Crone, E. A., & Dahl, R. E. (2012). Understanding adolescence as a period of social-affective engagement and goal flexibility. *Nat Rev Neurosci, 13*(9), 636–650.

Davis, M., & Whalen, P. J. (2001). The amygdala: Vigilance and emotion. *Mol Psychiatry, 6*(1), 13–34.

De Luca, C. R., Wood, S. J., Anderson, V., Buchanan, J.-A., Proffitt, T. M., Mahony, K., & Pantelis, C. (2003). Normative data from the CANTAB. I: Development of executive function over the life span. *J Clin Exp Neuropsychol, 25*(2), 242–254.

Delgado, M. R. (2007). Reward-related responses in the human striatum. *Ann NY Acad Sci, 1104,* 70–88.

Ernst, M., Nelson, E. E., Jazbec, S., McClure, E. B., Monk, C. S., Leibenluft, E., ... Pine, D. S. (2005). Amygdala and nucleus accumbens in responses to receipt and omission of gains in adults and adolescents. *NeuroImage, 25*(4), 1279–1291.

Ernst, M., Pine, D. S., & Hardin, M. (2006). Triadic model of the neurobiology of motivated behavior in adolescence. *Psychol Med, 36*(3), 299–312.

Fareri, D. S., Martin, L. N., & Delgado, M. R. (2008). Reward-related processing in the human brain: Developmental considerations. *Dev Psychopathol, 20,* 1191–1211.

Galvan, A. (2010). Adolescent development of the reward system. *Front Hum Neurosci, 4,* 6.

Galvan, A., Hare, T. A., Parra, C. E., Penn, J., Voss, H., Glover, G., & Casey, B. J. (2006). Earlier development of the accumbens relative to orbitofrontal cortex might underlie risk-taking behavior in adolescents. *J Neurosci, 26*(25), 6885–6892.

Galvan, A., & McGlennen, K. M. (2013). Enhanced striatal sensitivity to aversive reinforcement in adolescents versus adults. *J Cogn Neurosci, 25*(2), 284–296.

Garnefski, N., Legerstee, J., Kraaij, V., van den Kommer, T., & Teerds, J. A. N. (2002). Cognitive coping strategies and symptoms of depression and anxiety: A comparison between adolescents and adults. *J Adolesc, 25,* 603–611.

Geier, C., & Luna, B. (2009). The maturation of incentive processing and cognitive control. *Pharmacol Biochem Behav, 93*(3), 212–221.

Grose-Fifer, J., Rodrigues, A., Hoover, S., & Zottoli, T. (2013). Attentional capture by emotional faces in adolescence. *Adv Cogn Psychol, 9*(2), 81–91.

Gross, J. J. (2002). Emotion regulation: Affective, cognitive, and social consequences. *Psychophysiology, 39,* 281–291.

Gross, J. J., & Levenson, R. W. (1993). Emotional suppression: Physiology, self-report, and expressive behavior. *J Pers Soc Psychol, 64,* 970–970.

Gross, J. J., & Thompson, R. A. (2009). Emotion regulation: Conceptual and empirical foundations. In J. J. Gross (Ed.), *Handbook of emotion regulation* (pp. 3–22). New York, NY: Guilford.

Guyer, A. E., Monk, C. S., McClure-Tone, E. B., Nelson, E. E., Roberson-Nay, R., Adler, A. D., ... Ernst, M. (2008). A developmental examination of amygdala response to facial expressions. *J Cogn Neurosci, 20*(9), 1565–1582.

Haber, S. N., Kim, K. S., Mailly, P., & Calzavara, R. (2006). Reward-related cortical inputs define a large striatal region in primates that interface with associative cortical connections, providing a substrate for incentive-based learning. *J Neurosci, 26*(32), 8368–8376.

Haber, S. N., & Knutson, B. (2010). The reward circuit: Linking primate anatomy and human imaging. *Neuropsychopharmacology, 1,* 1–23.

Hall, G. S. (1904). *Adolescence: Its psychology and its relation to physiology, anthropology, sociology, sex, crime, religion, and education,* Vols. I & II. Englewood Cliffs, NJ: Prentice-Hall.

Hare, T. A., Tottenham, N., Galvan, A., Voss, H. U., Glover, G. H., & Casey, B. J. (2008). Biological substrates of emotional reactivity and regulation in adolescence during an emotional go-nogo task. *Biol Psychiatry, 63*(10), 927–934.

Huizinga, M., Dolan, C. V., & van der Molen, M. W. (2006). Age-related change in executive function: Developmental trends and a latent variable analysis. *Neuropsychologia, 44,* 2017–2036.

Kagan, J., Snidman, N., Arcus, D., & Reznick, J. S. (1994). *Galen's prophecy: Temperament in human nature.* New York, NY: Basic Books.

LaBar, K. S., Gatenby, J. C., Gore, J. C., LeDoux, J. E., & Phelps, E. A. (1998). Human amygdala activation during conditioned fear acquisition and extinction: A mixed-trial fMRI study. *Neuron, 20*(5), 937–945.

Larson, R. W. (2001). How U.S. children and adolescents spend time: What it does (and doesn't) tell us about their development. *Curr Dir Psychol Sci, 10*(4), 160–164.

Larson, R. W., Moneta, G., Richards, M. H., & Wilson, S. (2002). Continuity, stability, and change in daily emotional experience across adolescence. *Child Dev, 73*(4), 1151–1165.

Laviola, G., Pascucci, T., & Pieretti, S. (2001). Striatal dopamine sensitization to D-amphetamine in periadolescent but not adult rats. *Pharmacol Biochem Behav, 68,* 115–124.

LeDoux, J. E. (2000). Emotion circuits in the brain. *Annu Rev Neurosci, 23,* 155–184.

Levita, L., Hare, T. A., Voss, H. U., Glover, G., Ballon, D. J., & Casey, B. J. (2009). The bivalent side of the nucleus accumbens. *NeuroImage, 44*(3), 1178–1187.

Li, J., Schiller, D., Schoenbaum, G., Phelps, E. A., & Daw, N. D. (2011). Differential roles of human striatum and amygdala in associative learning. *Nat Neurosci, 14*(10), 1250–1252.

Lindquist, K. A., Wager, T. D., Kober, H., Bliss-Moreau, E., & Barrett, L. F. (2012). The brain basis of emotion: A meta-analytic review. *Behav Brain Sci, 35*(3), 121–143.

Luna, B., Garver, K. E., Urban, T. A., Lazar, N. A., & Sweeney, J. A. (2004). Maturation of cognitive processes from late childhood to adulthood. *Child Dev, 75*(5), 1357–1372.

Luna, B., Padmanabhan, A., & O'Hearn, K. (2010). What has fMRI told us about the development of cognitive control through adolescence? *Brain Cogn, 72*(1), 101–113.

Maren, S., & Quirk, G. J. (2004). Neuronal signalling of fear memory. *Nat Rev Neurosci, 5*(11), 844–852.

McClure, E. B., Monk, C. S., Nelson, E. E., Zarahn, E., Leibenluft, E., Bilder, R. M., ... Pine, D. S. (2004). A developmental examination of gender differences in brain engagement during evaluation of threat. *Biol Psychiatry, 55*(11), 1047–1055.

McRae, K., Gross, J. J., Weber, J., Robertson, E. R., Sokol-Hessner, P., Ray, R. D., ... Ochsner, K. N. (2012). The development of emotion regulation: An fMRI study of cognitive reappraisal in children, adolescents and young adults. *Soc Cogn Affect Neurosci, 7*(1), 11–22.

Milad, M. R., & Quirk, G. J. (2002). Neurons in medial prefrontal cortex signal memory for fear extinction. *Nature, 420*(6911), 70–74.

Mischel, W., & Baker, N. (1975). Cognitive transformations of reward objects through instructions. *J Pers Soc Psychol, 31*(2), 254–261.

Mischel, W., Shoda, Y., & Peake, P. K. (1988). The nature of adolescent competencies predicted by preschool delay of gratification. *J Pers Soc Psychol, 54*(4), 687.

Monk, C. S., McClure, E. B., Nelson, E. E., Zarahn, E., Bilder, R. M., Leibenluft, E., & Pine, D. S. (2003). Adolescent immaturity in attention-related brain engagement to emotional facial expressions. *NeuroImage, 20,* 420–428.

Moore, W. E., Pfeifer, J. H., Masten, C. L., Mazziotta, J. C., Iacoboni, M., & Dapretto, M. (2012). Facing puberty: Associations between pubertal development and neural responses to affective facial displays. *Soc Cogn Affect Neurosci, 7*(1), 35–43.

Nelson, E. E., Leibenluft, E., McClure, E. B., & Pine, D. S. (2005). The social re-orientation of adolescence: A neuroscience perspective on the process and its relation to psychopathology. *Psychol Med, 35,* 163–174.

Ochsner, K. N., & Gross, J. J. (2005). The cognitive control of emotion. *Trends Cogn Sci, 9*(5), 242–249.

Pagnoni, G., Zink, C. F., Montague, P. R., & Berns, G. S. (2002). Activity in human ventral striatum locked to errors of reward prediction. *Nat Neurosci, 5*(2), 97–98.

Pasupathy, A., & Miller, E. K. (2005). Different time courses of learning-related activity in the prefrontal cortex and striatum. *Nature, 433,* 873–876.

Pattwell, S. S., Duhoux, S., Hartley, C. A., Johnson, D. C., Jing, D., Elliott, M. D., ... Yang, R. R. (2012). Altered fear learning across development in both mouse and human. *Proc Natl Acad Sci USA, 109*(40), 16318–16323.

Peper, J. S., & Dahl, R. E. (2013). The teenage brain: Surging hormones—brain-behavior interactions during puberty. *Curr Dir Psychol Sci, 22*(2), 134–139.

Pfeifer, J. H., Masten, C. L., Moore, W. E., Oswald, T. M., Mazziotta, J. C., Iacoboni, M., & Dapretto, M. (2011). Entering adolescence: Resistance to peer influence, risky behavior, and neural changes in emotion reactivity. *Neuron, 69*(5), 1029–1036.

Pfeifer, J. H., & Peake, S. J. (2012). Self-development: Integrating cognitive, socioemotional, and neuroimaging perspectives. *Dev Cogn Neurosci, 2*(1), 55–69.

Phelps, E. A., Delgado, M. R., Nearing, K. I., & LeDoux, J. E. (2004). Extinction learning in humans: Role of the amygdala and vmPFC. *Neuron, 43*(6), 897–905.

Phelps, E. A., & LeDoux, J. E. (2005). Contributions of the amygdala to emotion processing: From animal models to human behavior. *Neuron, 48*(2), 175–187.

Phillips, R. G., & LeDoux, J. E. (1992). Differential contribution of the amygdala and hippocampus to cued and contextual fear conditioning. *Behav Neurosci, 106,* 274–285.

Pine, D. S., Grun, J., Zarahn, E., Fyer, A., Koda, V., Li, W., ... Bilder, R. M. (2001). Cortical brain regions engaged by masked emotional faces in adolescents and adults: An fMRI study. *Emotion, 1*(2), 137–147.

Posner, M. I., & Rothbart, M. K. (2007). Research on attention networks as a model for the integration of psychological science. *Annu Rev Psychol, 58,* 1–23.

Rothbart, M. K. (2007). Temperament, development, and personality. *Curr Dir Psychol Sci, 16*(4), 207–212.

Rubia, K., Smith, A. B., Woolley, J., Nosarti, C., Heyman, I., Taylor, E., & Brammer, M. J. (2006). Progressive increase of frontostriatal brain activation from childhood to adulthood during event-related tasks of cognitive control. *Hum Brain Mapp, 27,* 973–993.

Rutter, M., Graham, P., Chadwick, O. F. D., & Yule, W. (1976). Adolescent turmoil: Fact or fiction? *J Child Psychol Psychiatry, 17,* 35–56.

Schafe, G. E., Nader, K., Blair, H. T., & LeDoux, J. E. (2001). Memory consolidation of Pavlovian fear conditioning: A cellular and molecular perspective. *Trends Neurosci, 24*(9), 540–546.

Schultz, W., Dayan, P., & Montague, P. R. (1997). A neural substrate of prediction and reward. *Science, 275*(5306), 1593–1599.

Schultz, W., Tremblay, L., & Hollerman, J. R. (1998). Reward prediction in primate basal ganglia and frontal cortex. *Neuropharmacology, 37*(4–5), 421–429.

Silvers, J. A., McRae, K., Gabrieli, J. D. E., Gross, J. J., Remy, K. A., & Ochsner, K. N. (2012). Age-related differences in emotional reactivity, regulation, and rejection sensitivity in adolescence. *Emotion, 12*(6), 1235–1247.

Sisk, C. L., & Zehr, J. L. (2005). Pubertal hormones organize the adolescent brain and behavior. *Front Neuroendocrin, 26*(3–4), 163–174.

Somerville, L. H. (2013). The teenage brain: Sensitivity to social evaluation. *Curr Dir Psychol Sci, 22*(2), 129–135.

Somerville, L. H., Fani, N., & McClure-Tone, E. B. (2011). Behavioral and neural representations of emotional facial expressions across the life span. *Dev Neuropsychol, 36*(4), 408–428.

Somerville, L. H., Hare, T., & Casey, B. J. (2011). Frontostriatal maturation predicts cognitive control failure to appetitive cues in adolescents. *J Cogn Neurosci, 23*(9), 2123–2134.

Somerville, L. H., Jones, R. M., & Casey, B. J. (2010). A time of change: Behavioral and neural correlates of adolescent sensitivity to appetitive and aversive environmental cues. *Brain Cogn, 72*(124–133).

Spear, L. P. (2000). The adolescent brain and age-related behavioral manifestations. *Neurosci Biobehav Rev, 24*(4), 417–463.

Spear, L. P. (2011). Rewards, aversions and affect in adolescence: Emerging convergences across laboratory animal and human data. *Dev Cogn Neurosci, 1*(4), 390–403.

Spear, L. P., & Varlinskaya, E. I. (2010). Sensitivity to ethanol and other hedonic stimuli in an animal model of adolescence: Implications for prevention science? *Dev Psychobiol, 52*(3), 236–243.

Steinberg, L. (2008). A social neuroscience perspective on adolescent risk-taking. *Dev Rev, 28,* 78–106.

Teicher, M. H., Andersen, S. L., & Hostetter, J. C., Jr. (1995). Evidence for dopamine receptor pruning between adolescence and adulthood in striatum but not nucleus accumbens. *Dev Brain Res, 89*(2), 167–172.

Tottenham, N., Hare, T. A., & Casey, B. J. (2011). Behavioral assessment of emotion discrimination, emotion regulation, and cognitive control in childhood, adolescence, and adulthood. *Front Psychol, 2,* 39.

van den Bos, W., Cohen, M. X., Kahnt, T., & Crone, E. A. (2011). Striatum–medial prefrontal cortex connectivity predicts developmental changes in reinforcement learning. *Cereb Cortex, 22*(6), 1247-1255.

van den Bos, W., Güroğlu, B., van den Bulk, B. G., Rombouts, S. A., & Crone, E. A. (2009). Better than expected or as bad as you thought? The neurocognitive development of probabilistic feedback processing. *Front Hum Neurosci, 3,* 52.

van den Bulk, B. G., Koolschijn, P., Meens, P. H., van Lang, N. D., van der Wee, N. J., Rombouts, S. A., … Crone, E. A. (2013). How stable is activation in the amygdala and prefrontal cortex in adolescence? A study of emotional face processing across three measurements. *Dev Cogn Neurosci, 4,* 65–76.

Van Leijenhorst, L., Gunther Moor, B., op de Macks, Z., Rombouts, S. A., Westenberg, P. M., & Crone, E. A. (2010). Adolescent risky decision-making: Neurocognitive development of reward and control regions. *NeuroImage, 51,* 345–355.

Vuilleumier, P. (2005). How brains beware: Neural mechanisms of emotional attention. *Trends Cogn Sci, 9*(12), 585.

Williams, L. M., Brown, K. J., Palmer, D., Liddell, B. J., Kemp, A. H., Olivieri, G., … Gordon, E. (2006). The mellow years? Neural basis of improving emotional stability with age. *J Neurosci, 26*(24), 6422–6430.

Yerkes, R. M., & Dodson, J. D. (1908). The relation of strength of stimulus to rapidity of habit-formation. *J Comp Neurol Psychol, 18*(5), 459–482.

64 The Amygdala: Relations to Biologically Relevant Learning and Development

DYLAN G. GEE AND PAUL J. WHALEN

ABSTRACT The amygdala plays a critical role in learning about cues in the environment that predict biologically relevant outcomes and informing our behavioral responses in anticipation of these outcomes. Consistent with classic principles of learning theory, the amygdala is particularly responsive to uncertainty, which enhances learning. Much of the extant research on the human amygdala has employed facial expressions to elucidate its contributions to emotional learning. In addition to their role in nonverbal communication, facial expressions can be considered conditioned stimuli based on their reinforcement history in prior social situations. Through interactions between the amygdala and prefrontal cortex (PFC), bottom-up and top-down processing shape this social learning. Stronger amygdala-prefrontal connectivity begets better behavioral outcomes, and disrupted cross-talk between these regions underlies emotion dysregulation in healthy and clinical populations. Moreover, the amygdala and its connections with PFC undergo dynamic changes throughout development, which likely contribute to developmental changes in emotional behavior. Given the neurodevelopmental nature of many disorders and the widespread implication of amygdala-prefrontal circuitry in psychopathology, understanding how the amygdala changes in typical development and possible disruptions in biologically relevant learning are both fundamental to treating psychopathology. The following chapter will detail how the amygdala contributes to biologically relevant learning and gives rise to individual differences in emotional behavior, as well as how these processes change with development and risk for psychopathology.

The amygdala and biologically relevant learning

The amygdala plays a central role in learning about biologically relevant events, which can signal the relative safety or danger of a given environment. In the framework of classical conditioning, the amygdala responds to biologically relevant events that require no previous learning (i.e., unconditioned stimuli, or USs, that are inherently emotionally salient, such as a shock), as well as the events that predict these biologically relevant events (i.e., conditioned stimuli, or CSs, which begin as neutral but take on emotional salience through conditioning). This form of learning depends on the amygdala in both humans and nonhuman animals (see LeDoux, 1996). While nonhuman animal studies still comprise the majority of our knowledge about amygdala contributions to biologically relevant learning during classical conditioning (see Davis & Whalen, 2001), neuroimaging studies have documented a largely similar role for the human amygdala. For example, functional neuroimaging studies have demonstrated contributions of the amygdala during classical conditioning paradigms with a variety of CS-US contingencies, including colored squares predicting electric shock (LaBar, Gatenby, Gore, LeDoux, & Phelps, 1998).

These findings demonstrate in a clear manner that the human amygdala is involved in learning about environmental cues that predict biologically relevant outcomes as well as informing behavioral responses in anticipation of these outcomes (LeDoux & Schiller, 2009). Research in humans has extended this initial work to show that the amygdala might function in a more general manner, facilitating all kinds of biologically relevant learning. For example, the amygdala is responsive to both aversive and appetitive CSs (Paton, Belova, Morrison, & Salzman, 2006). Data such as these fit well with attentional hypotheses (Kapp, Whalen, Supple, & Pascoe, 1992) and seminal work demonstrating a critical role for the amygdala in associative changes observed in appetitive reward paradigms in the rat (Gallagher & Holland, 1994). That is, amygdala activation to a tone that predicts shock is but a part of an affective information-processing system that directs our attention to important events in the environment. The amygdala functions as an orienting subsystem for the rest of the brain, which allows us to respond appropriately to those events (Davis & Whalen, 2001; Gallagher & Holland, 1994). In alerting other systems at times when it would be advantageous to gather information (i.e., learn), the amygdala supports adaptive functions that cross the categorical boundaries of constructs such as motivation, emotion, vigilance, attention, and cognition.

THE ROLE OF UNCERTAINTY IN BIOLOGICALLY RELEVANT LEARNING Derived from principles of learning theory, the idea that there is more to learn in the face of uncertainty is fundamental to how the amygdala responds and orients the rest of the brain to the environment. The Rescorla-Wagner model is based on the notion that an increase in the associative value of a CS will be greatest (i.e., learning is most likely) on trials when the individual finds the occurrence of the US to be surprising (Rescorla & Wagner, 1972). Together with the magnitude of the US, the predictive value of the CS dictates the degree of surprise created by the US. The predictive value of the CS increases quickly in early conditioning trials, when the organism makes new associations, but decreases during late trials (see Bouton, 2007). A related but separate model of classical conditioning, the Pearce-Hall model, focuses on the amount of attention to the CS that is recruited by the degree of surprise created by the US on the preceding trial (Pearce & Hall, 1980). Interestingly, this model has helped to guide research on the influence of amygdala activity on attention. Particularly relevant is work establishing a role for the central nucleus of the amygdala in enhancing CS associability when expectations are violated (Holland & Gallagher, 1999) and in nonspecific attention or arousal observed in the service of learning (Kapp et al., 1992). Recent work has extended these principles to the human amygdala by testing a hybrid model positing that the degree of surprise to the US (i.e., prediction errors) drives learning, but that associability of the CS affects learning rates. Specifically, a reversal-learning task revealed a functional dissociation of amygdala subregions during associative learning, such that a more dorsal portion of the human amygdala (where the central nucleus resides) was more sensitive to immediate surprise at the time of the US, whereas the ventral portion was more sensitive to associability at the time of the CS (Boll, Gamer, Gluth, Finsterbusch, & Büchel, 2013). Taken together, these studies highlight the contributions of the amygdala in processing uncertainty that is fundamental to learning.

Facial Expressions as Conditioned Stimuli Humans have a great deal of experience with facial expressions of emotion (Somerville & Whalen, 2006) and have learned that they predict important outcomes. Facial expressions play a crucial role in nonverbal communication. Not only do they communicate the internal state of the expressor, but they also convey important information about the immediate environment. Thus, information gleaned from facial expressions of emotion can be used to guide future behavior. When conceptualized in this manner, facial expressions are conditioned stimuli. That is, they are environmental cues that have been associated with important outcomes in the past. When they are encountered, we use their past reinforcement history to predict what will happen next.

Indeed, human neuroimaging research shows that presentations of static pictures of facial expressions engage neural systems similar to those engaged during associative learning tasks, including the amygdala. Like associative learning, responses to facial expressions can be implicit (Whalen et al., 1998). In humans, patients with selective amygdala lesions displayed deficits in processing the facial expression of fear (Adolphs, Tranel, Damasio, & Damasio, 1995), leading to numerous neuroimaging studies using presentations of fearful faces to probe amygdala function. These studies have shown that the amygdala is particularly responsive to fearful faces compared to other expressions, including angry, happy, and neutral faces (e.g., Breiter et al., 1996; Morris et al., 1996). Consistent with the importance of uncertainty in learning, since the amygdala responds more to fearful faces than angry faces, which embody a direct threat, one function of the amygdala may be to augment cortical function to assist in the resolution of predictive uncertainty (Whalen et al., 2001). That is, the inherent ambiguity of fearful faces, in that they predict the increased probability of threat without providing information about its nature or location, leads to greater activation of the amygdala.

Given that the amygdala plays a major role in the resolution of predictive uncertainty associated with fearful faces, surprised faces provide a particularly important comparison expression. Indeed, surprise may be the second-most compromised expression in patients with selective amygdala damage, following fear (Adolphs, Tranel, Damasio, & Damasio, 1994). Fearful and surprised faces have common features (e.g., eye-widening), and both expressions indicate the detection of a significant but *unknown* eliciting event. Consistent with this logic, the amygdala shows robust activation to surprised faces (Kim et al., 2004). Further, neural responses to surprised expressions show a greater resistance to extinction, presumably due to their inconsistent reinforcement history. Likening facial expressions of emotion to conditioned stimuli allows us to draw parallels between two seemingly distinct avenues of research: those that characterize the neural processes associated with learning about stimuli that predict biologically relevant outcomes (i.e., Pavlovian conditioning) and those that characterize reactions to stimuli that have acquired similar predictive value through our experiences in the social world (i.e., facial expressions of emotion) (Davis, Johnstone, Mazzulla, Oler, & Whalen, 2010).

Amygdala-prefrontal connectivity

Bottom-up and top-down connections between the amygdala and medial PFC (mPFC) are fundamental to emotional behavior. While attention to salient stimuli can be biased through bottom-up processes driven by the amygdala, this reactivity is thought to be modulated through top-down cognitive control by the mPFC (Ochsner, Bunge, Gross, & Gabrieli, 2002). Competition between these bottom-up and top-down processes is highlighted in behavioral phenomena such as emotion regulation, fear conditioning, and extinction (Bishop, 2007; Quirk & Beer, 2006). While numerous studies have assessed the separate contributions of the amygdala and mPFC to top-down and bottom-up interactions in emotion, respectively, more recent studies suggest that the structural and functional connectivity between these two regions is a better predictor of behavioral outcomes than the activity of either region alone (Banks, Eddy, Angstadt, Nathan, & Phan, 2007; Kim, Gee, Loucks, Davis, & Whalen, 2011; Pezawas et al., 2005). Stronger coupling between the amygdala and the mPFC begets better behavioral outcomes in terms of effective emotion regulation and lower anxiety.

Amygdala-Prefrontal Structural Connectivity Data from nonhuman primate brains have revealed strong bidirectional anatomical connections between the amygdala and mPFC. A heavy projection of afferent fibers to the amygdala originates in the mPFC (including orbitofrontal cortex and ventral anterior cingulate cortex; Aggleton, Burton, & Passingham, 1980; Leichnetz & Astruc, 1976). There is also a heavy projection from a posterior but dorsal portion of mPFC to the amygdala (Pandya, van Hoesen, & Mesulam, 1981). Reciprocally, the amygdala sends efferent projections to these same ventral and dorsal mPFC regions (Amaral & Price, 1984; Barbas & de Olmos, 1990; Carmichael & Price, 1995; Ghashghaei & Barbas, 2002), and to a slightly more rostral portion of the mPFC (Ghashghaei, Hilgetag, & Barbas, 2007). The mPFC regulates activity of the amygdala through input to the basolateral nuclei of the amygdala and the intercalated cells, which regulate inputs from the basolateral nuclei to the central nucleus and thus inhibit amygdala output (Harris & Westbrook, 1998; Milad & Quirk, 2002).

The majority of our knowledge about structural connections in this circuitry is based on the literature of nonhuman animal studies due to the invasive nature of lesion and tracing studies, which are difficult to employ in humans. However, a number of studies using noninvasive methods such as diffusion tensor imaging have identified an amygdala-prefrontal axonal pathway in the human, with a specific focus on amygdala connectivity with the dorsal and ventral regions of the mPFC (Croxson et al., 2005; Johansen-Berg et al., 2008; Kim & Whalen, 2009). These regions may be analogous to the dorsal and ventral prefrontal regions identified in the previous studies of connectivity in nonhuman primates.

Amygdala-Prefrontal Functional Connectivity at Rest While structural connectivity characterizes the neuroanatomical architecture of amygdala-prefrontal circuitry, functional connectivity measures provide information about the moment-to-moment interactions between the amygdala and mPFC in response to particular stimuli or at rest. Based on the extensive anatomical connections between the amygdala and mPFC shown in human and nonhuman primates, a number of investigations have used functional connectivity to assess the strength of amygdala-mPFC coupling and its relationship with behavioral outcomes. Resting-state functional MRI provides unique information about the intrinsic connections between brain regions in the absence of explicit task instructions and has been shown to index the functional integrity and maintenance of network connections (Biswal, Yetkin, Haughton, & Hyde, 1995). Resting-state maps have revealed distinct patterns of connectivity for each of the human amygdala subdivisions (Roy et al., 2009). At rest, the amygdala was positively coupled with ventral regions and negatively coupled with dorsal regions. Amygdala coupling at rest can be used to predict individual differences. For example, the amygdala was positively coupled with ventromedial prefrontal cortex (vmPFC) in participants reporting lower levels of anxiety, but not in subjects reporting higher levels of anxiety (Kim et al., 2011; figure 64.1). In future examinations, resting-state connectivity may be used to predict amygdala-prefrontal activation or how well someone regulates their emotional responses when challenged with a particular task.

Amygdala-Prefrontal Functional Connectivity Predicting Adaptive Behavior The majority of examinations of amygdala functional connectivity have been conducted during explicit tasks, with findings demonstrating the central role of amygdala-prefrontal interactions for behaviors such as emotion regulation, extinction, and resolving ambiguity. Emotion regulation is a classic example of how top-down and bottom-up processes compete and interact to produce optimal (or counterproductive) behavioral outcomes. For example, one's instinctive reaction to a frightening scene in a horror movie may include an urge to scream or run out of the room. Normally, this bottom-up reaction is controlled by a top-down intervention (e.g., reminding oneself that this is only a movie). Recent findings suggest that the degree of efficient crosstalk

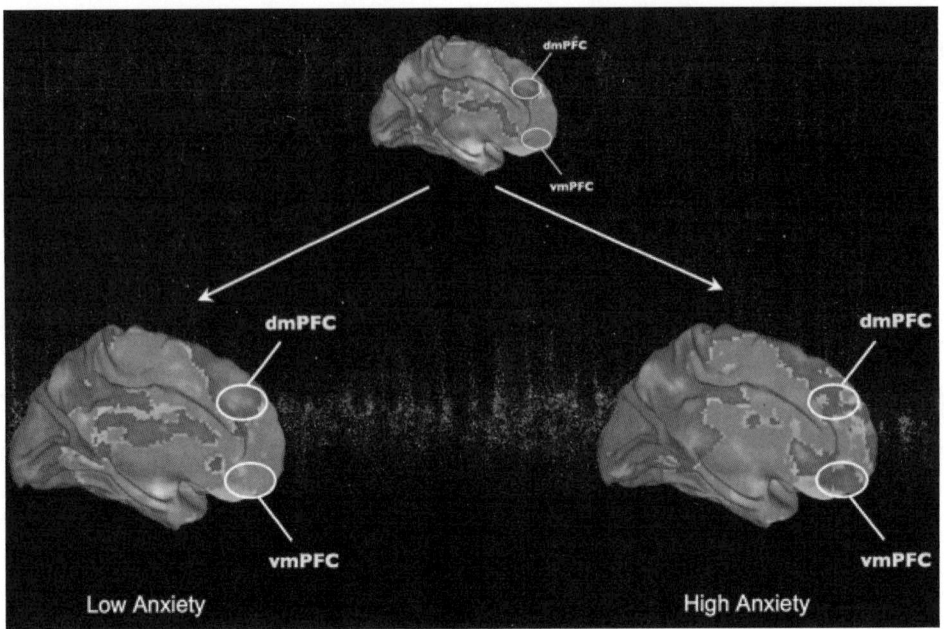

FIGURE 64.1 Amygdala-prefrontal functional connectivity at rest. While all subjects show that amygdala activity is positively correlated with vmPFC activity and negatively correlated with dorsomedial PFC activity at rest, highly anxious subjects show a reverse pattern in the vmPFC compared to low anxious subjects. (Adapted from Kim et al., 2011.) (See color plate 53.)

between the amygdala and the PFC corresponds to one's ability to regulate one's emotions in this way.

In a study of emotion regulation, functional connectivity between the amygdala and mPFC increased during reappraisal, and the strength of this connectivity was associated with participants' self-report of how effective they were at regulating their emotions (Banks et al., 2007). A selective increase in the functional coupling of the amygdala with the vmPFC and dorsolateral PFC during emotion regulation has also been reported (Erk et al., 2010). Similarly, numerous studies have demonstrated increased prefrontal activity and concomitant decreased amygdala activity during successful emotion regulation (e.g., Ochsner et al., 2002; Wager, Davidson, Hughes, Lindquist, & Ochsner, 2008), highlighting the importance of amygdala-prefrontal interactions in successful top-down control of emotion. Supporting the idea that stronger amygdala-prefrontal connectivity begets better emotion regulation, stronger amygdala-mPFC functional coupling has also been associated with greater amygdala habituation (Hare et al., 2008) and lower anxiety (e.g., Gee et al., 2013; Pezawas et al., 2005). Although correlational measures cannot inform the directional nature of regional influences, negative (inverse) amygdala-prefrontal coupling has been theorized to reflect top-down regulation.

Functional interactions between the amygdala and mPFC also support the extinction of fear conditioning, which involves the suppression of a previously learned pairing between a CS and US (Quirk, 2002; Rescorla, 2001). The inhibition of CS-US associations depends critically on top-down regulatory input from the mPFC to the basolateral amygdala (Milad & Quirk, 2002). For example, in rodents, electric stimulation of the mPFC resulted in the inhibition of conditioned responses. In humans, increased vmPFC activation and structural volume have been associated with successful extinction (Hartley, Fischl, & Phelps, 2011; Phelps, Delgado, Nearing, & LeDoux, 2004). These results suggest that emotion regulation and extinction rely on overlapping neural mechanisms (Delgado, Nearing, LeDoux, & Phelps, 2008), consistent with the fact that both processes involve reevaluating biologically relevant stimuli (Quirk & Beer, 2006). These structural and functional findings highlight the importance of amygdala-mPFC interactions for the regulation and inhibition necessary for extinction learning and memory.

Amygdala-prefrontal interactions further support the resolution of ambiguity in our environment. Unlike fearful faces, surprised faces do not predict the valence of an unknown eliciting event such that they can be subjectively interpreted as either positive or negative (Neta, Norris, & Whalen, 2009). Individual differences in valence judgments of surprised faces correspond to distinct patterns of brain activity involving the amygdala and vmPFC (Kim, Somerville, Johnstone, Alexander, &

Whalen, 2003). Specifically, lower amygdala and greater vmPFC activity were observed during positive interpretations, with the opposite pattern to negative interpretations. Here, the vmPFC plays a theorized role in resolving the emotional ambiguity of surprised faces, similar to its role in top-down regulatory input to the amygdala during fear extinction or emotion regulation (Quirk & Beer, 2006). Indeed, greater vmPFC activity predicts both (1) more positive ratings of surprise (Kim et al., 2003) and (2) more positive interpretations of an extinguished tone (i.e., tone now predicts no shock; Oler, Quirk, & Whalen, 2009). Amygdala reactivity to surprised faces can also be modulated by context. When surprised faces were paired with positive or negative sentences that provided a clear resolution to the ambiguity of the face, greater amygdala, weaker vmPFC, and greater ventrolateral PFC activity were observed in response to faces in negative contexts (Kim et al., 2004). Taken together, these studies suggest that emotionally ambiguous stimuli incite competition between top-down and bottom-up processes, with the balance of activity in this circuitry reflecting the resolution of ambiguity.

Though various forms of top-down control depend critically on the amygdala and PFC, the extent to which different emotional processes rely on the same regions of PFC is less clear. While extinction specifically involves the ventromedial region of PFC, emotion regulation paradigms highlight the role of ventral and dorsal *lateral* PFC, in addition to vmPFC, in regulating amygdala reactivity (Erk et al., 2010; Wager et al., 2008). Specifically, mPFC may mediate the top-down effects of lateral PFC on the amygdala (Delgado et al., 2008; Lieberman et al., 2007). When comparing specific types of emotion regulation, suppression and reappraisal strategies were both characterized by decreased activity of the amygdala and increased activity of the PFC—usually including both medial and lateral PFC (Ochsner & Gross, 2005). These same regions also support more automatic, incidental forms of emotion regulation such as affect labeling (Hariri, Bookheimer, & Mazziotta, 2000; Lieberman et al., 2007). These findings provide functional and structural evidence for shared neural mechanisms during different types of top-down control of amygdala reactivity.

Amygdala-Prefrontal Connectivity and Psychopathology An emphasis on amygdala-prefrontal interactions is particularly relevant to clinical studies, as impaired interaction between bottom-up and top-down processes is believed to be a hallmark of many psychiatric disorders (Davis & Whalen, 2001). As detailed earlier in this chapter, individual differences in normative anxiety relate to the strength of amygdala activity and its connections with mPFC (e.g., Hare et al., 2008; Pezawas et al., 2005). Not surprisingly, pathological anxiety in psychiatric disorders is marked by disruption in this circuitry, particularly an imbalance between hyperactivity of the amygdala and hypoactivity of the PFC (e.g., Shin et al., 2005). While atypical amygdala-prefrontal circuitry is a classic neurobiological model for anxiety disorders, disturbances in amygdala-prefrontal function have now been implicated in a range of psychopathology including depression, bipolar disorder, schizophrenia, borderline personality disorder, psychopathy, and attention-deficit/hyperactivity disorder (e.g., Blair, 2008; Foland et al., 2008; Taylor et al., 2012). Future research is needed to understand how abnormal amygdala-prefrontal interactions contribute to psychopathology. For example, it may be the case that a failure of top-down regulatory control allows bottom-up responses to disrupt typical functioning. Alternatively, initial bottom-up reactions might be so exaggerated that they override the normally functioning top-down control system (LeDoux, 1996). Given the prominent role of amygdala-prefrontal circuitry in typical and atypical behavior, future research in this domain will be critical for informing the etiology of psychiatric disorders, understanding the nature and timing of their onset, and developing novel treatments.

Developmental trajectories of amygdala circuitry

Typical development is marked by dramatic changes in emotional behavior and regulation (e.g., Bunge, Dudukovic, Thomason, Vaidya, & Gabrieli, 2002; Tottenham, Hare, & Casey, 2011). Moreover, given the neurodevelopmental nature of many psychiatric disorders and the increase in risk for psychopathology in adolescence (Pine, Cohen, Gurley, Brook, & Ma, 1998), research on the healthy maturation of amygdala circuitry can elucidate how neurodevelopmental processes may go awry in patients who develop disorders characterized by amygdala abnormalities (Gee et al., 2012).

Amygdala Development in Childhood and Adolescence Consistent with behavioral changes in emotional learning across development, relevant networks in the brain change significantly across the course of typical development (reviewed in Somerville, Fani, & McClure-Tone, 2011). Structurally, the amygdala is a rapidly developing region (reviewed in Tottenham & Sheridan, 2009). The fastest rate of structural volume growth in the amygdala occurs within the first two postnatal weeks and stabilizes by eight months old in nonhuman primates (Payne, Machado, Bliwise, &

Bachevalier, 2010). In humans, the basic neuroanatomical architecture of the amygdala is present at birth (Humphrey, 1968; Ulfig, Setzer, & Bohl, 2003). Longitudinal examination in humans has demonstrated continued structural development of the amygdala through two years of age, and evidence suggests that structural growth is complete by age four in girls but shows a modest but significant linear increase beyond childhood in boys (Giedd et al., 1996; Gilmore et al., 2012). Taken together, these studies suggest that the amygdala undergoes early structural development.

However, functionally, the amygdala displays protracted changes throughout development. The amygdala shows functionality early in life, since it responds to emotional stimuli in childhood (Baird et al., 1999; Thomas et al., 2001). Though children show reliable amygdala signal to facial expressions of emotion, evidence suggests that patterns of amygdala activation to distinct expressions may differ between children and adults. For example, whereas adults show greater amygdala activation to fearful than neutral faces (Whalen et al., 2001), distinctive patterns of amygdala activation have been observed in youth, with some evidence that children display greater activation to neutral than fearful faces (Thomas et al., 2001; Tottenham, Hare, Millner, et al., 2011). However, other studies have found similar patterns of differential amygdala response to sad and disgusted faces in children, as seen in adults (e.g., Lobaugh, Gibson, & Taylor, 2006). Thus, the extent to which the amygdala responds to facial expressions uniquely in children remains unclear.

Evidence suggests that significant changes in amygdala function occur from early childhood throughout adolescence; however, findings on the direction of change in amygdala activation across development differ depending on task design and age groups studied. To date, studies comparing amygdala activation to facial expressions of emotion across children, adolescents, and adults have demonstrated important functional changes across the life span (Gee et al., 2013; Hare et al., 2008). During the presentation of fearful faces, a linear decrease in amygdala activation was observed from 4 through 22 years of age, such that amygdala reactivity was highest in childhood and decreased with age (Gee et al., 2013). When activation was collapsed across facial expressions (fearful, happy, and calm), amygdala reactivity was higher in adolescence compared with children and adults (Hare et al., 2008). Though fewer investigations have examined amygdala function during childhood, studies directly comparing adolescents with adults have shown decreases in amygdala activation from adolescence to adulthood (Gee et al., 2013; Guyer et al., 2008; Hare et al., 2008; Monk et al., 2003). Recent research has also shown changes in amygdala function related to puberty. Consistent with a decrease in amygdala activation with development, amygdala activation was higher in prepubertal and early pubertal adolescents compared with mid- and late-pubertal adolescents (Forbes, Phillips, Silk, Ryan, & Dahl, 2011). In addition, amygdala function related more strongly to pubertal stage at age 13 than at age 10, though there were no significant differences in activation at these ages (Moore et al., 2012). Future research will aid in characterizing the trajectory of amygdala function across a broader developmental period, as well as clarifying how the observation of linear or nonlinear developmental trajectories may depend on factors such as task.

Development of Amygdala-Prefrontal Connectivity

Given the fundamental role of amygdala-prefrontal interactions for effective emotion regulation, fear conditioning, and extinction learning, it is critical to investigate how connections between the amygdala and mPFC emerge. Rodent models indicate key changes in amygdala-prefrontal connectivity during adolescence (Cunningham, Bhattacharyya, & Benes, 2002; Kim & Richardson, 2009). While less is known about the development of amygdala connectivity in humans, prior work has demonstrated that the strength of amygdala-mPFC functional connectivity was associated with amygdala habituation in adolescents, suggesting a functional role of connectivity in adolescence similar to that in adulthood (Hare et al., 2008). In addition, prior research has shown that the strength of amygdala-prefrontal effective connectivity increases with age (Perlman & Pelphrey, 2011). To characterize changes in connectivity during typical development, a recent study examined amygdala-mPFC functional connectivity to fearful faces from early childhood through early adulthood. Findings revealed a developmental switch in the valence of connectivity, such that the amygdala and mPFC were positively coupled in early childhood and became negatively coupled across development (Gee et al., 2013). Specifically, amygdala-mPFC coupling switched from an immature phenotype (positive coupling) to a mature phenotype (negative coupling) during the transition to adolescence (figure 64.2).

Moreover, connectivity became more strongly negative with age, replicating studies of amygdala-mPFC coupling in adults (e.g., Hare et al., 2008; Kim et al., 2003). In this study, the developmental switch in amygdala-mPFC functional connectivity mediated the relationship between age and a developmentally normative decrease in separation anxiety. Given the role of mPFC in regulating amygdala reactivity, evidence of stronger

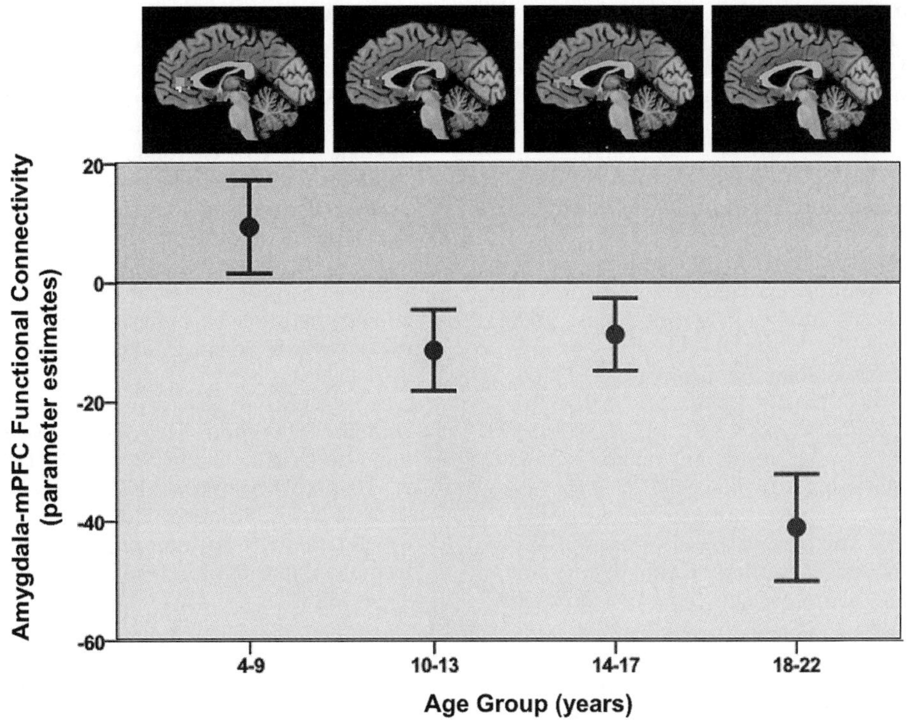

FIGURE 64.2 Development of amygdala-prefrontal functional connectivity. A developmental switch from positive to negative functional connectivity between the amygdala and mPFC was observed during the transition from childhood to adolescence. (Adapted from Gee et al., 2013.) (See color plate 54.)

negative coupling and reduced amygdala reactivity with age may provide a neurobiological basis for developmental improvements in emotion regulation.

Conclusion

Emotion, whether typical or in the context of psychopathology, can be conceived as a constant interplay between an organism's reactions to biologically relevant stimuli (bottom-up processing) and its attempts to modulate these responses (top-down processing). A wealth of animal and human studies demonstrate the central role of the amygdala and its connections with PFC in these processes. Across the life span, amygdala function and connectivity undergo dynamic changes, which likely support changes in learning and social behavior that are fundamental to development. For example, learning about the relative safety or danger of a given environment plays a critical role in key developmental transitions, such as when the organism leaves the nest or enters adolescence. Moreover, it is through amygdala contributions in development that faces take on predictive value through experiences in the social world, building the reinforcement history that gives rise to individual differences in neural and behavioral responses. While these neurodevelopmental changes facilitate adaptive behavior such as stronger emotion regulation, changes in amygdala-prefrontal circuitry may also render individuals vulnerable in certain stages of development or with risk for psychopathology. The coming decades will be an exciting time for research on affective neuroscience, as we better understand how the amygdala interacts with a broader brain circuitry to give rise to what we currently call emotion. This line of research will play a critical role in detailing how amygdala circuitry facilitates adaptive behavior, contributes to developmental changes, and influences risk for psychopathology.

REFERENCES

ADOLPHS, R., TRANEL, D., DAMASIO, H., & DAMASIO, A. (1994). Impaired recognition of emotion in facial expressions following bilateral damage to the human amygdala. *Nature, 372*(6507), 669–672.

ADOLPHS, R., TRANEL, D., DAMASIO, H., & DAMASIO, A. R. (1995). Fear and the human amygdala. *J Neurosci, 15*(9), 5879–5891.

AGGLETON, J. P., BURTON, M. J., & PASSINGHAM, R. E. (1980). Cortical and subcortical afferents to the amygdala of the rhesus monkey (*Macaca mulatta*). *Brain Res, 190*(2), 347–368.

AMARAL, D. G., & PRICE, J. L. (1984). Amygdalo-cortical projections in the monkey (*Macaca fascicularis*). *J Comp Neurol, 230*(4), 465–496.

Baird, A. A., Gruber, S. A., Fein, D. A., Maas, L. C., Steingard, R. J., Renshaw, P. F., ... Yurgelun-Todd, D. A. (1999). Functional magnetic resonance imaging of facial affect recognition in children and adolescents. *J Am Acad Child Psychol, 38*(2), 195–199.

Banks, S. J., Eddy, K. T., Angstadt, M., Nathan, P. J., & Phan, K. L. (2007). Amygdala–frontal connectivity during emotion regulation. *Soc Cogn Affect Neurosci, 2*(4), 303–312.

Barbas, H., & de Olmos, J. (1990). Projections from the amygdala to basoventral and mediodorsal prefrontal regions in the rhesus monkey. *J Comp Neurol, 300*(4), 549–571.

Bishop, S. J. (2007). Neurocognitive mechanisms of anxiety: An integrative account. *Trends Cogn Sci, 11*(7), 307–316.

Biswal, B., Yetkin, F. Z., Haughton, V. M., & Hyde, J. S. (1995). Functional connectivity in the motor cortex of resting human brain using echo-planar MRI. *Magnet Reson Med, 34*(4), 537–541.

Blair, R. J. R. (2008). The amygdala and ventromedial prefrontal cortex: Functional contributions and dysfunction in psychopathy. *Philos Trans R Soc Lond B Biol Sci, 363*(1503), 2557–2565.

Boll, S., Gamer, M., Gluth, S., Finsterbusch, J., & Büchel, C. (2013). Separate amygdala subregions signal surprise and predictiveness during associative fear learning in humans. *Eur J Neurosci, 37*(5), 758–767.

Bouton, M. E. (2007). Theories of conditioning. In M. E. Bouton (Ed.), *Learning and behavior: A contemporary synthesis* (pp. 102–142). Sunderland, MA: Sinauer.

Breiter, H. C., Etcoff, N. L., Whalen, P. J., Kennedy, W. A., Rauch, S. L., Buckner, R. L., ... Rosen, B. R. (1996). Response and habituation of the human amygdala during visual processing of facial expression. *Neuron, 17*(5), 875–887.

Bunge, S. A., Dudukovic, N. M., Thomason, M. E., Vaidya, C. J., & Gabrieli, J. D. E. (2002). Immature frontal lobe contributions to cognitive control in children: Evidence from fMRI. *Neuron, 33*(2), 301–311.

Carmichael, S. T., & Price, J. L. (1995). Limbic connections of the orbital and medial prefrontal cortex in macaque monkeys. *J Comp Neurol, 363*(4), 615–641.

Croxson, P. L., Johansen-Berg, H., Behrens, T. E. J., Robson, M. D., Pinsk, M. A., Gross, C. G., ... Rushworth, M. F. S. (2005). Quantitative investigation of connections of the prefrontal cortex in the human and macaque using probabilistic diffusion tractography. *J Neurosci, 25*(39), 8854–8866.

Cunningham, M. G., Bhattacharyya, S., & Benes, F. M. (2002). Amygdalo-cortical sprouting continues into early adulthood: Implications for the development of normal and abnormal function during adolescence. *J Comp Neurol, 453*(2), 116–130.

Davis, F. C., Johnstone, T., Mazzulla, E. C., Oler, J. A., & Whalen, P. J. (2010). Regional response differences across the human amygdaloid complex during social conditioning. *Cereb Cortex, 20*(3), 612–621.

Davis, M., & Whalen, P. J. (2001). The amygdala: Vigilance and emotion. *Mol Psychiatry, 6*(1), 13–34.

Delgado, M. R., Nearing, K. I., LeDoux, J. E., & Phelps, E. A. (2008). Neural circuitry underlying the regulation of conditioned fear and its relation to extinction. *Neuron, 59*(5), 829–838.

Erk, S., Mikschl, A., Stier, S., Ciaramidaro, A., Gapp, V., Weber, B., & Walter, H. (2010). Acute and sustained effects of cognitive emotion regulation in major depression. *J Neurosci, 30*(47), 15726–15734.

Foland, L. C., Altshuler, L. L., Bookheimer, S. Y., Eisenberger, N., Townsend, J., & Thompson, P. M. (2008). Evidence for deficient modulation of amygdala response by prefrontal cortex in bipolar mania. *Psychiat Res Neuroim, 162*(1), 27–37.

Forbes, E. E., Phillips, M. L., Silk, J. S., Ryan, N. D., & Dahl, R. E. (2011). Neural systems of threat processing in adolescents: Role of pubertal maturation and relation to measures of negative affect. *Dev Neuropsychol, 36*(4), 429–452.

Gallagher, M., & Holland, P. C. (1994). The amygdala complex: Multiple roles in associative learning and attention. *Proc Natl Acad Sci USA, 91*(25), 11771–11776.

Gee, D. G., Humphreys, K. L., Flannery, J., Goff, B., Telzer, E. H., Shapiro, M., ... Tottenham, N. (2013). A developmental shift from positive to negative connectivity in human amygdala-prefrontal circuitry. *J Neurosci, 33*(10), 4584–4593.

Gee, D. G., Karlsgodt, K. H., van Erp, T. G. M., Bearden, C. E., Lieberman, M. D., Belger, A., ... Cannon, T. D. (2012). Altered age-related trajectories of amygdala-prefrontal circuitry in adolescents at clinical high risk for psychosis: A preliminary study. *Schizophr Res, 134*(1), 1–9.

Ghashghaei, H. T., & Barbas, H. (2002). Pathways for emotion: Interactions of prefrontal and anterior temporal pathways in the amygdala of the rhesus monkey. *Neuroscience, 115*(4), 1261–1279.

Ghashghaei, H. T., Hilgetag, C. C., & Barbas, H. (2007). Sequence of information processing for emotions based on the anatomic dialogue between prefrontal cortex and amygdala. *NeuroImage, 34*(3), 905–923.

Giedd, J. N., Vaituzis, A. C., Hamburger, S. D., Lange, N., Rajapakse, J. C., Kaysen, D., ... Rapoport, J. L. (1996). Quantitative MRI of the temporal lobe, amygdala, and hippocampus in normal human development: Ages 4–18 years. *J Comp Neurol, 366*(2), 223–230.

Gilmore, J. H., Shi, F., Woolson, S. L., Knickmeyer, R. C., Short, S. J., Lin, W., ... Shen, D. (2012). Longitudinal development of cortical and subcortical gray matter from birth to 2 years. *Cereb Cortex, 22*(11), 2478–2485.

Guyer, A. E., Monk, C. S., McClure-Tone, E. B., Nelson, E. E., Roberson-Nay, R., Adler, A. D., ... Ernst, M. (2008). A developmental examination of amygdala response to facial expressions. *J Cogn Neurosci, 20*(9), 1565–1582.

Hare, T. A., Tottenham, N., Galvan, A., Voss, H. U., Glover, G. H., & Casey, B. J. (2008). Biological substrates of emotional reactivity and regulation in adolescence during an emotional go-nogo task. *Biol Psychiatry, 63*(10), 927–934.

Hariri, A. R., Bookheimer, S. Y., & Mazziotta, J. C. (2000). Modulating emotional responses: Effects of a neocortical network on the limbic system. *NeuroReport, 11*(1), 43–48.

Harris, J. A., & Westbrook, R. F. (1998). Evidence that GABA transmission mediates context-specific extinction of learned fear. *Psychopharmacology, 140*(1), 105–115.

Hartley, C. A., Fischl, B., & Phelps, E. A. (2011). Brain structure correlates of individual differences in the acquisition and inhibition of conditioned fear. *Cereb Cortex, 21*(9), 1954–1962.

Holland, P. C., & Gallagher, M. (1999). Amygdala circuitry in attentional and representational processes. *Trends Cogn Sci, 3*(2), 65–73.

Humphrey, T. (1968). The development of the human amygdala during early embryonic life. *J Comp Neurol, 132*(1), 135–165.

Johansen-Berg, H., Gutman, D. A., Behrens, T. E. J., Matthews, P. M., Rushworth, M. F. S., Katz, E., ... Mayberg, H. S. (2008). Anatomical connectivity of the subgenual cingulate region targeted with deep brain stimulation for treatment-resistant depression. *Cereb Cortex, 18*(6), 1374–1383.

Kapp, B. S., Whalen, P. J., Supple, W. F., & Pascoe, J. P. (1992). Amygdaloid contributions to conditioned arousal and sensory information processing. In J. P. Aggleton (Ed.), *The amygdala: Neurobiological aspects of emotion, memory, and mental dysfunction* (pp. 229–254). New York, NY: Wiley-Liss.

Kim, H., Somerville, L. H., Johnstone, T., Alexander, A. L., & Whalen, P. J. (2003). Inverse amygdala and medial prefrontal cortex responses to surprised faces. *NeuroReport, 14*(18), 2317–2322.

Kim, H., Somerville, L. H., Johnstone, T., Polis, S., Alexander, A. L., Shin, L. M., & Whalen, P. J. (2004). Contextual modulation of amygdala responsivity to surprised faces. *J Cogn Neurosci, 16*(10), 1730–1745.

Kim, J. H., & Richardson, R. (2009). The effect of the mu-opioid receptor antagonist naloxone on extinction of conditioned fear in the developing rat. *Learn Mem, 16*(3), 161–166.

Kim, M. J., & Whalen, P. J. (2009). The structural integrity of an amygdala–prefrontal pathway predicts trait anxiety. *J Neurosci, 29*(37), 11614–11618.

Kim, M. J., Gee, D. G., Loucks, R. A., Davis, F. C., & Whalen, P. J. (2011). Anxiety dissociates dorsal and ventral medial prefrontal cortex functional connectivity with the amygdala at rest. *Cereb Cortex, 21*(7), 1667–1673.

LaBar, K. S., Gatenby, J. C., Gore, J. C., LeDoux, J. E., & Phelps, E. A. (1998). Human amygdala activation during conditioned fear acquisition and extinction: A mixed-trial fMRI study. *Neuron, 20*(5), 937–945.

LeDoux, J. (1996). *The emotional brain: The mysterious underpinnings of emotional life.* New York, NY: Simon & Schuster.

LeDoux, J. E., & Schiller, D. (2009). The human amygdala: Insights from other animals. In P. J. Whalen & E.A. Phelps (Eds.), *The human amygdala* (pp. 43–60). New York, NY: Guilford.

Leichnetz, G. R., & Astruc, J. (1976). The efferent projections of the medial prefrontal cortex in the squirrel monkey (*Saimiri sciureus*). *Brain Res, 109*(3), 455–472.

Lieberman, M. D., Eisenberger, N. I., Crockett, M. J., Tom, S. M., Pfeifer, J. H., & Way, B. M. (2007). Putting feelings into words: Affect labeling disrupts amygdala activity in response to affective stimuli. *Psychol Sci, 18*(5), 421–428.

Lobaugh, N. J., Gibson, E., & Taylor, M. J. (2006). Children recruit distinct neural systems for implicit emotional face processing. *NeuroReport, 17*(2), 215–219.

Milad, M. R., & Quirk, G. J. (2002). Neurons in medial prefrontal cortex signal memory for fear extinction. *Nature, 420*(6911), 70–74.

Monk, C. S., McClure, E. B., Nelson, E. E., Zarahn, E., Bilder, R. M., Leibenluft, E., ... Pine, D. S. (2003). Adolescent immaturity in attention-related brain engagement to emotional facial expressions. *NeuroImage, 20*(1), 420–428.

Moore, W. E., 3rd, Pfeifer, J. H., Masten, C. L., Mazziotta, J. C., Iacoboni, M., & Dapretto, M. (2012). Facing puberty: Associations between pubertal development and neural responses to affective facial displays. *Soc Cogn Affect Neurosci, 7*(1), 35–43.

Morris, J. S., Frith, C. D., Perrett, D. I., Rowland, D., Young, A. W., Calder, A. J., & Dolan, R. J. (1996). A differential neural response in the human amygdala to fearful and happy facial expressions. *Nature, 383*(6603), 812–815.

Neta, M., Norris, C. J., & Whalen, P. J. (2009). Corrugator muscle responses are associated with individual differences in positivity-negativity bias. *Emotion, 9*(5), 640–648.

Ochsner, K. N., Bunge, S. A., Gross, J. J., & Gabrieli, J. D. E. (2002). Rethinking feelings: An fMRI study of the cognitive regulation of emotion. *J Cogn Neurosci, 14*(8), 1215–1229.

Ochsner, K. N., & Gross, J. J. (2005). The cognitive control of emotion. *Trends Cogn Sci, 9*(5), 242–249.

Oler, J. A., Quirk, G. J., & Whalen, P. J. (2009). Cinguloamygdala interactions in surprise and extinction: Interpreting associative ambiguity. In B. A. Vogt (Ed.), *Cingulate neurobiology and disease* (pp. 207–218). New York, NY: Oxford University Press.

Pandya, D. N., van Hoesen, G. W., & Mesulam, M. M. (1981). Efferent connections of the cingulate gyrus in the rhesus monkey. *Exp Brain Res, 42*(3–4), 319–330.

Paton, J. J., Belova, M. A., Morrison, S. E., & Salzman, C. D. (2006). The primate amygdala represents the positive and negative value of visual stimuli during learning. *Nature, 439*(7078), 865–870.

Payne, C., Machado, C. J., Bliwise, N. G., & Bachevalier, J. (2010). Maturation of the hippocampal formation and amygdala in *Macaca mulatta*: A volumetric magnetic resonance imaging study. *Hippocampus, 20*(8), 922–935.

Pearce, J. M., & Hall, G. (1980). A model for Pavlovian learning: Variations in the effectiveness of conditioned but not of unconditioned stimuli. *Psychol Rev, 87*(6), 532–552.

Perlman, S. B., & Pelphrey, K. A. (2011). Developing connections for affective regulation: Age-related changes in emotional brain connectivity. *J Exp Child Psychol, 108*(3), 607–620.

Pezawas, L., Meyer-Lindenberg, A., Drabant, E. M., Verchinski, B. A., Munoz, K. E., Kolachana, B. S., ... Weinberger, D. R. (2005). 5-HTTLPR polymorphism impacts human cingulate-amygdala interactions: A genetic susceptibility mechanism for depression. *Nat Neurosci, 8*(6), 828–834.

Phelps, E. A., Delgado, M. R., Nearing, K. I., & LeDoux, J. E. (2004). Extinction learning in humans: Role of the amygdala and vmPFC. *Neuron, 43*(6), 897–905.

Pine, D. S., Cohen, P., Gurley, D., Brook, J., & Ma, Y. (1998). The risk for early-adulthood anxiety and depressive disorders in adolescents with anxiety and depressive disorders. *Arch Gen Psychiatry, 55*(1), 56–64.

Quirk, G. J. (2002). Memory for extinction of conditioned fear is long-lasting and persists following spontaneous recovery. *Learn Mem, 9*(6), 402–407.

Quirk, G. J., & Beer, J. S. (2006). Prefrontal involvement in the regulation of emotion: Convergence of rat and human studies. *Curr Opin Neurobiol, 16*(6), 723–727.

Rescorla, R. A. (2001). Retraining of extinguished Pavlovian stimuli. *J Exp Psychol Anim Behav Proc, 27*(2), 115–124.

Rescorla, R. A., & Wagner, A. D. (1972). A theory of Pavlovian conditioning: Variations in the effectiveness of reinforcement and nonreinforcement. In A. H. Black & W. F. Prokasy (Eds.), *Classical conditioning II.* New York, NY: Appleton-Century-Crofts.

Roy, A. K., Shehzad, Z., Margulies, D. S., Kelly, A. M. C., Uddin, L. Q., Gotimer, K., … Milham, M. P. (2009). Functional connectivity of the human amygdala using resting state fMRI. *NeuroImage, 45*(2), 614–626.

Shin, L. M., Wright, C. I., Cannistraro, P. A., Wedig, M. M., McMullin, K., Martis, B., … Rauch, S. L. (2005). A functional magnetic resonance imaging study of amygdala and medial prefrontal cortex responses to overtly presented fearful faces in posttraumatic stress disorder. *Arch Gen Psychiatry, 62*(3), 273–281.

Somerville, L. H., & Whalen, P. J. (2006). Prior experience as a stimulus category confound: An example using facial expressions of emotion. *Soc Cogn Affect Neurosci, 1*(3), 271–274.

Somerville, L. H., Fani, N., & McClure-Tone, E. B. (2011). Behavioral and neural representation of emotional facial expressions across the lifespan. *Dev Neuropsychol, 36*(4), 408–428.

Taylor, S. F., Kang, J., Brege, I. S., Tso, I. F., Hosanagar, A., & Johnson, T. D. (2012). Meta-analysis of functional neuroimaging studies of emotion perception and experience in schizophrenia. *Biol Psychiatry, 71*(2), 136–145.

Thomas, K. M., Drevets, W. C., Whalen, P. J., Eccard, C. H., Dahl, R. E., Ryan, N. D., & Casey, B. J. (2001). Amygdala response to facial expressions in children and adults. *Biol Psychiatry, 49*(4), 309–316.

Tottenham, N., Hare, T. A., & Casey, B. J. (2011). Behavioral assessment of emotion discrimination, emotion regulation, and cognitive control in childhood, adolescence, and adulthood. *Front Dev Psychol, 2*, 39.

Tottenham, N., Hare, T. A., Millner, A., Gilhooly, T., Zevin, J. D., & Casey, B. J. (2011). Elevated amygdala response to faces following early deprivation. *Dev Sci, 14*(2), 190–204.

Tottenham, N., & Sheridan, M. (2009). A review of adversity, the amygdala and the hippocampus: A consideration of developmental timing. *Front Hum Neurosci, 3*, 68.

Ulfig, N., Setzer, M., & Bohl, J. (2003). Ontogeny of the human amygdala. *Ann NY Acad Sci, 985*, 22–33.

Wager, T. D., Davidson, M. L., Hughes, B. L., Lindquist, M. A., & Ochsner, K. N. (2008). Prefrontal-subcortical pathways mediating successful emotion regulation. *Neuron, 59*(6), 1037–1050.

Whalen, P. J., Rauch, S. L., Etcoff, N. L., McInerney, S., Lee, M. B., & Jenike, M. A. (1998). Masked presentations of emotional facial expressions modulate amygdala activity without explicit knowledge. *J Neurosci, 18*(1), 411–418.

Whalen, P. J., Shin, L. M., McInerney, S. C., Fischer, H., Wright, C. I., & Rauch, S. L. (2001). A functional MRI study of human amygdala responses to facial expressions of fear versus anger. *Emotion, 1*(1), 70–83.

65 Sensory Foundations of Socioemotional Perception

TIMOTHY R. KOSCIK, NICOLE WHITE, HANAH A. CHAPMAN, AND ADAM K. ANDERSON

ABSTRACT Accumulating evidence suggests that social cognitive processing and emotion are integrally intertwined. This link encompasses early sensory processes and high-order, conceptual processing, and likely stems from the adaptive, functional role that emotions fulfill in regulating behavior toward biologically relevant goals. In this chapter, we discuss the link between sensory and socioemotional processes from converging perspectives, including shared neural substrates emphasizing critical roles of the amygdala, ventromedial prefrontal cortex, and insular cortex; commonalities in connection with chemosensory, visual, and motor systems; and bidirectional interactions of perceptual and socioemotional processes. We discuss the evolutionary origins of socioemotional systems, and emphasize their function in biasing behavior, suggesting that complex emotional and social cognition neurobehavioral programs originate and continue to share a deep functional integration with simpler basic sensory and perceptual processes.

Traditional models of cognition view higher processes such as reasoning and decision making as separate from the workings of dedicated affective neural systems, but growing evidence supports the notion of an integral link between affective and cognitive processes (e.g., for reviews, see Pessoa, 2009; Phelps, 2006). Of particular relevance in the present chapter is the relationship between emotion and social cognition. Social processes and behavior appear to be heavily influenced by emotional states and abilities, and emotional experiences may serve an important function as social signals (Adolphs, 2002, 2003). In this chapter, we review the neuroanatomy of socioemotional systems, and in particular discuss the relationship between early sensory processes and higher-order knowledge regarding social and emotional information. We emphasize the importance of sensory and motor processes in understanding information in socioemotional contexts and argue that these lower-level processes are critical in shaping our understanding of emotional and social events in the world. Rather than acting as mere way stations for the relaying of sensory inputs and outputs from higher-level neural processing, we argue that perception and action are integral to socioemotional processing.

Theories of emotion: Form and function

Darwin (1872) was the first scientist to speculate about the importance of emotional expression in humans. He suggested that emotional expression serves a functional role in regulating behavior in response to salient environmental events and in signaling important social information to others. More recently, Shariff and Tracy (2011) suggest that early in our evolutionary lineage, emotions served as *cues* (e.g., for action) but were later exploited as *signals* (e.g., communicating one's emotional state to others). Theoretical classifications of the range of human emotion impact and shape our understanding of the functional role of emotional processes in the brain. Many models of emotion focus on the form or content of emotion experiences and emphasize the importance of classifying distinct subjective emotion experiences (e.g., Ekman, 1999).

Other researchers argue that our best chance at understanding emotion experience will be borne of attempts to understand their adaptive functions. LeDoux (2012) argues for conceptualizing the range of phenomena typically under the umbrella of emotions based on how the underlying physiological processes serve the pursuit of motivational drives for optimizing survival. He writes: "By focusing on the subjective state, emotion theories tend to gloss over the underlying details of emotional processing for the sake of converging on a single word that symbolizes diverse underlying states mediated by different kinds of circuits" (LeDoux, 2012, p. 655). For instance, aggression may be motivated by an array of underlying motivational drives, including personal survival and drives to reproduce. In each case, the same emotional experience might be induced by different, adaptively functional motivations. By adopting this framework, LeDoux abstracts away from the notion of basic emotions in favor of examining the function of sensory and motor processes in regulating behavior to optimize survival.

Farb, Chapman, and Anderson (2013) also argue that focusing on the function of emotions will be more informative for understanding the neural processes related to emotional experience, though they do not go so far as to ignore the basic emotions altogether. They suggest that the adaptive function of emotions is to shape perception, cognition, and behavior to aid the organism in fulfilling relevant goals, in line with the survival-circuit framework presented by LeDoux (2012). Both frameworks emphasize function over form and make conceptualization of emotional experience more flexible. If the function of physiological experiences serving emotions is to adaptively regulate behavior to fulfill goals and desires, then we can extend the function of emotions into any cognitive domain requiring an organism to regulate its behavior to negotiate challenges and opportunities to achieve desired goals. There seems to be a considerable overlap of such emotional processes for the regulation of social behavior.

Social cognition and emotion: Overlapping systems?

Assuming that the function of emotions is to allow organisms to behave adaptively, one particularly relevant function of emotions is *social signaling*. The evolutionary exploitation of emotional expressions as communicative signals may have resulted in the subsequent evolution of "self-conscious" emotions, which serve exclusively social functions (Shariff & Tracy, 2011). For instance, Tracy and Robins (2008) suggest that pride, which is recognized equivalently across cultures, may have evolved to indicate social status. Adolphs (2002, 2003) argues that emotions not only serve important functions in social communication but also regulate social behavior. Deficits in processing emotion-related information are often associated with deficits in the social domain (Adolphs, Baron-Cohen, & Tranel, 2002; Fine & Blair, 2000; Milders, Fuchs, & Crawford, 2003). Arguably, recognition of the perceptually or semantically salient aspects of an expression is an inherently social task. On this view, any act of emotion recognition that is not self-oriented is a social process, and so it is not surprising that there is extensive overlap between social and emotional brain regions. This is not to say that social cognition can be reduced to emotion-related processes; rather, in the vast majority of social interactions, some component of neural processing is likely related to decoding the emotional state of another individual.

Neuroanatomical substrates of socioemotional processes

The link between social and emotional processes extends deep into the evolutionary heritage of the vertebrate brain. The anatomy of social-emotional brain regions has roots in primitive sensory systems, particularly the chemosensory systems. The brain of the last common ancestor of vertebrates contained telencephalic components that were most likely entirely olfactory with an additional striatal region (Aboitiz, Montiel, Morales, & Concha, 2002; Aboitiz, Morales, & Montiel, 2003; Butler & Hodos, 2005). Aboitiz and colleagues (2003) have suggested that early mammals possessed largely olfactory-based brains and navigated their environments (spatially and socially) via odor-labeled routes, places, and objects. Over time, olfactory-based representations became elaborated to include associative networks of other modalities, eventually leading to the unique mammalian neocortex (Aboitiz et al., 2002). The evolutionary expansion of neocortex in mammals may be due to a series of adaptations to nocturnal life that favored the elaboration of olfactory systems, thereby expanding the telencephalon, which may then have been exapted to receive thalamic inputs from other senses (Aboitiz, 1992).

Functions of the Chemosensory Systems Chemosensory systems provide vital information about the physical and chemical makeup of places, objects, and others. Chemical sampling provides direct and difficult-to-feign information, which helps guide diverse behavioral repertoires. Indeed, the ability to detect chemical compounds, particularly in the bodily secretions of conspecifics, allows mammals to communicate biologically relevant information, including sex, reproductive status, social status, immunological status, and so on, and is involved in sexual behavior, parental behavior, aggression, territorial marking, and individual discrimination (for reviews, see Brennan & Keverne, 2004; Halpern & Martınez-Marcos, 2003; Sanchez-Andrade & Kendrick, 2009).

Chemosensory evaluations, particularly those in the olfactory domain (Haddad, Lapid, Harel, & Sobel, 2008), prioritize the analysis of stimulus hedonic value (i.e., pleasantness or unpleasantness). Thus it is reasonable to expect that other hedonic processes, particularly emotion, should show considerable overlap in neural structures with chemosensory systems.

Overlap of Chemosensory and Socioemotional Neural Systems The interaction between chemosensory and socioemotional neural systems is demonstrated in humans by the interaction of emotion and olfaction. Unsurprisingly, pleasant odors can elicit positive affect (Seubert, Rea, Loughead, & Habel, 2009), reduce anxiety, and improve mood (Lehrner, Eckersberger, Walla, Pötsch, & Deecke, 2000; Lehrner, Marwinski,

Lehr, Johren, & Deecke, 2005), whereas unpleasant odors can reduce feelings of calmness and alertness and negatively impact mood (Weber & Heuberger, 2008). Conversely, viewing unpleasant emotional images reduces olfactory sensitivity (Pollatos et al., 2007); perceived emotions impact subjective pleasantness of odors in the same direction as emotional valence (Pollatos et al., 2007); and video-based emotion induction alters the subjective intensity of odors in men (Chen & Dalton, 2005). Social cognition and olfaction also interact. Empathy is positively correlated with odor identification for smells (Spinella, 2002). Moreover, face recognition can be enhanced by certain odors or disrupted by others (Seubert et al., 2010; Walla, Mayer, Deecke, & Lang, 2005). Likewise, recognition of facial expressions of disgust is enhanced by the presentation of an odor regardless of the valence of the odor (Seubert et al., 2010).

Three neural regions exemplify the overlap between socioemotional and chemosensory systems, namely, the amygdalae, ventromedial prefrontal cortex (including orbitofrontal and medial prefrontal regions), and insular cortex. Each of these regions plays important roles in chemosensory processing as well as having well-known roles in social and emotional processing.

Amygdala

Neuroanatomy of the amygdalae The amygdalae are located in the anterior medial temporal lobe and are made up of more than 10 nuclei differentiable in terms of cytoarchitecture and connection patterns. These nuclei include a basolateral nucleus and cortical-like, centromedial, and other nuclei, including the anterior amygdala area, the amygdalohippocampal area, and the intercalated nuclei (for a review, see Sah, Faber, Lopez de Armentia, & Power, 2003). Connections to the amygdalae are diverse, coming from virtually all neural systems. The amygdalae receive visual, auditory, olfactory, gustatory, somatosensory, and visceral input from higher-order sensory cortices (Höistad & Barbas, 2008; McDonald, 1998). Although there may be less evidence for direct projections from primary auditory or visual regions to the amygdalae, the amygdalae do receive early-stage olfactory and gustatory information (McDonald, 1998). In addition to unimodal sensory inputs, the amygdala receives polymodal and highly processed input from the hippocampus and perirhinal regions as well as prefrontal cortex, particularly from ventromedial prefrontal regions (Aggleton, Burton, & Passingham, 1980; Ghashghaei & Barbas, 2002; Ghashghaei, Hilgetag, & Barbas, 2007). Subcortical afferents to the amygdalae include the hypothalamus, substantia innominata, diagonal band, thalamus, periaqueductal gray, and peripeduncular nucleus (Aggleton et al., 1980). Generally, connections with the amygdalae are reciprocal, positioning the amygdalae as important nodes for influencing broad swaths of cognitive, emotional, and peripheral bodily processes. Notably, the amygdalae have strong projections to hypothalamic and brainstem centers, allowing them to influence behavior relatively directly (for a review, see Höistad & Barbas, 2008).

Chemosensory roles of the amygdalae The amygdalae, along with other structures located near the uncus in the medial temporal lobe, receive direct input from the olfactory bulb (Hadley, Orlandi, & Fong, 2004; Scalia & Winans, 1975; Scott, 1986). In most mammals, excluding humans and closely related nonhuman primates, the accessory olfactory bulb (of the vomeronasal system) connects directly to the amygdalae (Halpern & Martınez-Marcos, 2003; Meredith, 1991; Scalia & Winans, 1975). The amygdalae serve as an important site where information from the main olfactory and vomeronasal systems interact in most mammals, particularly the posteromedial cortical amygdala (Halpern & Martınez-Marcos, 2003).

In terms of chemosensory function, the human amygdalae have been shown to be involved in processing the intensity, more than the valence, of pleasant and unpleasant odors (Anderson et al., 2003; Gottfried, Deichmann, Winston, & Dolan, 2002). Intensity representations in the amygdalae may be for valenced odors only, suggesting a complex combination of intensity and valence (Winston, Gottfried, Kilner, & Dolan, 2005). Perhaps because aversive odors are hedonically intense, the amygdalae are activated by highly aversive odors, and activations have been differentially related to subjective ratings of aversiveness (Gottfried et al., 2002; Royet, Plailly, Delon-Martin, Kareken, & Segebarth, 2003; Zald & Pardo, 1997). In addition, the amygdalae are involved in encoding and retrieving olfactory information, independent of the hedonic properties of the odors (Jung et al., 2006). Temporal lobe resections, which often include the amygdala (but probably other olfactory-related regions as well), result in deficits in odor discrimination, but not detection (Jones-Gotman & Zatorre, 1988).

In addition to amygdalar involvement in olfaction, the amygdalae are involved in gustatory processing. The amygdalae receive direct input from primary gustatory insular cortex as well as subcortical gustatory nuclei (McDonald, 1998). Similar to olfaction, the amygdalae respond to the intensity of taste valence (Small et al., 2003). Amygdalar response to taste may also signify some sort of nutritional and thus motivational relevance, given that the amygdalae are activated more strongly by high-fat foods than low-fat ones (Grabenhorst, Rolls, Parris, & D'Souza, 2010).

Emotional roles of the amygdalae Although the amygdalae at one time were thought to most strongly respond to fear-related stimuli and negative affect, they clearly respond to other emotions, including positive affect. For example, happy and sad faces activate the amygdalae, though to a lesser degree than fearful faces, while angry and disgusted faces do not (for a review, see Fusar-Poli et al., 2009). Equally salient pleasant and unpleasant odors result in similar magnitude of amygdala response (Anderson et al., 2003). It has been suggested that the human amygdalae may be involved in the appraisal of biologically relevant stimuli more broadly, including social stimuli (Sander, Grafman, & Zalla, 2003). For example, Cunningham and colleagues (Cunningham, van Bavel, & Johnsen, 2008) demonstrate that amygdalae activity is modulated by evaluative goals and motivational salience, in addition to a bias toward negative information. That the amygdalae are associated with emotionally but not perceptually salient events (Anderson & Phelps, 2001; Todd, Talmi, Schmitz, Susskind, & Anderson, 2012) and specifically emotional influences on learning and memory (McGaugh, 2004; Phelps & LeDoux, 2005) is consistent with their role not as a stimulus-driven processor of specific stimuli, but a flexible emotional modulator of information processing.

Social roles of the amygdalae The emerging view of the role of the amygdalae in social cognition and behavior is that this region is critical for evaluating salience or relevance. Insofar as social behaviors are extremely relevant for humans, the realm of social cognition is highly salient and thus recruits the amygdalae (for reviews, see Adolphs, 2010; Amaral, 2003; Amaral et al., 2003; Sander et al., 2003). There are numerous examples of amygdalar involvement in social cognitive processes. Bilateral amygdala damage results in deficits in recognizing social emotions (Adolphs, 2002), in a tendency to prefer abnormally reduced interpersonal distance (Kennedy, Gläscher, Tyszka, & Adolphs, 2009), reduced eye contact (Kennedy & Adolphs, 2010), and reduced social network size (Becker et al., 2012). The amygdalae play an extensive role in the evaluation of faces (Mende-Siedlecki, Said, & Todorov, 2012; Todorov, 2012); single-cell recordings indicate that neurons in the amygdalae represent parts of faces as well as whole faces (Rutishauser et al., 2011). The amygdalae are also involved in implicit social judgments, including subliminal racial biases (Cunningham et al., 2004; Lieberman et al., 2005). Lastly, the amygdalae are also involved in cooperation and interpersonal trust (Engell, Haxby, & Todorov, 2007; Koscik & Tranel, 2011; Singer et al., 2004; van Honk, Eisenegger, Terburg, Stein, & Morgan, 2013).

Ventromedial Prefrontal Cortex

Neuroanatomy of the ventromedial prefrontal cortex The ventromedial prefrontal cortex (vmPFC) is a large region of neocortex that comprises medial portions of the orbitofrontal cortex (OFC) as well as portions of the medial walls of the prefrontal cortex (PFC), including portions of the anterior and subgenual cingulate cortex. Interestingly, the vmPFC receives direct amygdalar and amygdalothalmic input, whereas dorsolateral portions of the PFC lack amygdalar input (Porrino, Crane, & Goldman-Rakic, 1981). The vmPFC connections to the amygdalae are most dense with posterior orbitofrontal and posterior medial prefrontal regions (Ghashghaei et al., 2007). Efferent vmPFC neurons converge on the thalamus, which projects to the brainstem and autonomic centers; these hypothalamic centers also connect to the amygdalae. This descending pathway provides a rapid means for the vmPFC to influence the autonomic system and is likely involved in the experience and expression of emotion (Barbas, Saha, Rempel-Clower, & Ghashghaei, 2003). As a portion of the PFC, and unlike other cortical regions, vmPFC has widespread connections with the reticular nucleus of the thalamus. Moreover, the mediodorsal thalamus, the principal thalamic nucleus for the PFC, has similar extensive connections with the reticular nucleus. This suggests that the PFC, including the vmPFC, is intimately involved in controlling and selecting the flow of information important for selective attention (Zikopoulos & Barbas, 2007). On a larger scale, within the vmPFC there is evidence for a trend where monitoring of reward value involves medial OFC, and evaluating punishment value involves lateral OFC. Additionally, representations of reinforcers become more complex from posterior to anterior regions (Kringelbach & Rolls, 2004).

Chemosensory roles of the ventromedial prefrontal cortex The vmPFC plays an important role in both olfaction and gustation, and it receives relatively direct input from these systems. Regarding olfaction, branches of the olfactory tract from the olfactory bulb send projections directly to a region of vmPFC located ventral to the genu and rostrum of the corpus callosum (Carmichael, Clugnet, & Price, 1994; Ongür, Ferry, & Price, 2003;

Tanabe et al., 1975; Yarita et al., 1980). As well, primary olfactory regions, in piriform cortex, project to secondary olfactory regions in the OFC (Carmichael et al., 1994). Data from macaques suggest that orbitofrontal regions are critical for odor discrimination; however, the homologous region in humans may be significantly more anterior and in a different cytoarchitectonic region (Gottfried & Zald, 2005).

In humans, the OFC is involved in higher-order olfactory processing (Gottfried et al., 2002). OFC is activated more by the valence of olfactory stimuli than by their intensity (Anderson et al., 2003). Furthermore, the OFC is involved in emotional olfactory processing (Anderson et al., 2003; Royet et al., 2003; Zald & Pardo, 1997) and odor discrimination and recognition, but not odor detection (Jones-Gotman & Zatorre, 1988; Potter & Butters, 1980; Zatorre & Jones-Gotman, 1991). Pleasant odors have been found to activate different OFC regions compared to unpleasant odors: medial OFC is activated for pleasant but not unpleasant odors, and lateral OFC regions are activated for unpleasant odors (Anderson et al., 2003; Rolls, Kringelbach, & de Araujo, 2003).

The other major chemosensory system present in humans, the gustatory system, is represented in caudolateral OFC alongside the olfactory representations mentioned above (for reviews, see Rolls, 2004; Rolls & Grabenhorst, 2008). The OFC responds to the valence of taste stimuli, with the right caudolateral OFC more responsive to pleasant than unpleasant tastes (Small et al., 2003). Mid-OFC and anterior cingulate cortex activity correlates with the subjective pleasantness of fat texture and flavors; the pregenual cingulate cortex in particular represents highly pleasant food stimuli (Grabenhorst et al., 2010). Overall, the OFC has two roles in gustatory processing. Since it is involved in olfactory encoding, including retronasal olfaction present during feeding, it is likely that the OFC binds sensory inputs to produce a flavor percept. In addition, given the OFC's role in reward, this region likely combines flavor and reward value to guide feeding behavior (Small et al., 2007).

Emotional roles of the ventromedial prefrontal cortex Converging evidence from humans, macaques, and rats suggests that the OFC plays a role in simple emotional responses through its representation of reward information, including primary reinforcers and abstract reinforcers (e.g., money and social reinforcers; Rolls, 2004; Rudebeck, Bannerman, & Rushworth, 2008). OFC activation relates to the subjective experience of emotion, and OFC damage impairs emotional behaviors, feeling, and learning (for a review, see Rolls & Grabenhorst, 2008). Medial PFC regions, especially anterior cingulate cortex, appear to be involved in more complex emotions (Rudebeck et al., 2008). Damage to the vmPFC has been shown to reduce interpersonal disgust (Ciaramelli, Sperotto, Mattioli, & di Pellegrino, 2013), which is thought to arise from the primitive distaste response (Chapman et al., 2009).

In addition to simple emotional responses, the vmPFC plays an important role in emotional regulation of behavior, particularly at the intersection of emotion and decision making. The somatic marker hypothesis suggests that somatic states, through which emotions are expressed, influence and bias decision making (Damasio, Tranel, & Damasio, 1990; Damasio, 1994, 1996). Damage to the vmPFC, including medial OFC, results in both blunted emotional responses, demonstrated by reduced psychophysiological responses to emotional stimuli (Damasio et al., 1990), and abnormal decision making, demonstrated by abnormally disadvantageous choices in the Iowa gambling task (Bechara, Tranel, Damasio, & Damasio, 1996).

Social roles of the ventromedial prefrontal cortex Perhaps the best-known evidence for the role of the vmPFC in social cognition is the case of Phineas Gage (Harlow, 1868), who exhibited profound personality changes, including changes in social behavior, following a traumatic injury that likely included portions of his vmPFC (Damasio, Grabowski, Frank, Galaburda, & Damasio, 1994). More recently, patient E.V.R. demonstrates that damage to the vmPFC results in profound changes in personality and social behavior despite intact intellectual capabilities (Eslinger & Damasio, 1985). Damage to the vmPFC results in abnormally high rejection of unfair offers in the ultimatum game (Koenigs & Tranel, 2007) and a lack of distinction between human and computerized opponents on a similar task (Moretti, Dragone, & di Pellegrino, 2009); increased moral permissibility of attempted harm (Young et al., 2010); and a decrease in normal correspondence bias (Koscik & Tranel, 2013). Regions within the OFC respond to facial attractiveness (Winston, O'Doherty, Kilner, Perrett, & Dolan, 2007). For an extensive review of the role of the vmPFC in social cognition and behavior, see Forbes and Grafman (2010).

It has recently been proposed that the role of the vmPFC in social evaluation has been important for the evolution of the human brain. In most mammals, chemosensory evaluation of conspecifics was one of the most important roles for the brain regions that are homologous to the human vmPFC. Given that humans are highly social creatures faced with similar needs for conspecific evaluation despite a conspicuous (though not total) lack of chemosensory communication, Koscik and Tranel (2012) have proposed that the human vmPFC retains its role in social evaluation. However, instead of relying on chemosensory information, the multimodal, integrative nature of the vmPFC has been adapted for inferring social value. This shift from chemosensory processing to inferential computation increased the processing power necessary to make social evaluations, which in turn provided a driving force for the enlargement of the human brain (Koscik & Tranel, 2012).

Insula

Neuroanatomy of the insula The insular cortex is tucked deep in the sylvian fissure beneath the frontal, temporal, and parietal opercula. The insula receives multimodal information, which likely acts to coordinate internal and external sources of information through subjective emotional awareness. Lesions to the insula can cause distinct deficits, including gustatory, olfactory, auditory, somatosensory, body awareness, affective, volitional, and linguistic impairments, as well as addiction (for reviews, see Ibañez, Gleichgerrcht, & Manes, 2010; Jones, Ward, & Critchley, 2010).

There is active debate over the best way to parcellate the insula and the scale on which to do so (for a review, see Kelly et al., 2012). In terms of gross anatomy, the insula appears divisible into anterior and posterior segments based on the central insular sulcus, which divides the insula roughly into dorsoanterior and ventroposterior portions. Based on studies of functional connectivity, perhaps the most popular view is that the insula can be divided into three subdivisions: posterior, dorsal anterior, and ventral anterior regions (Chang, Yarkoni, Khaw, & Sanfey, 2013; Deen, Pitskel, & Pelphrey, 2011). The dorsal anterior region is associated with cognitive functions and is connected to the dorsal anterior cingulate cortex. The ventral anterior subdivision is associated with affective and chemosensory processes and is primarily connected to pregenual anterior cingulate cortex, while the posterior region is associated with and functionally connected to primary and secondary sensorimotor cortices (Chang et al., 2013; Deen et al., 2011). For an alternative four-subregion parcellation scheme, see Kurth and colleagues (2010).

Chemosensory roles of the insula Primary gustatory cortex is located in the anterior insula and adjoining frontal operculum (for a review, see Rolls, 2008). The human gustatory cortex is located more caudal than would be predicted from the location of homologous regions in primates, perhaps due to the addition of a human-unique area in the anterior insula rostral to this region (Small, 2010). The anterior insula and frontal operculum respond to the valence of taste stimuli (Small et al., 2003) and whether taste stimuli are passively or actively evaluated (Bender, Veldhuizen, Meltzer, Gitelman, & Small, 2009). However, it may be better to conceptualize insular taste cortex as a multimodal region for integrating oral sensations rather than unimodal cortex (Small, 2010). Ventral anterior insula responds to oral stimulation regardless of modality or edibility; however, connectivity to feeding-related brain regions, including the hypothalamus, is greater when tasting potentially nutritive stimuli (Rudenga, Green, Nachtigal, & Small, 2010). In addition to taste stimuli, the right anterior insula and ventral striatal area increase activity during exposure to disgusting odors, whereas the left anterior insula responds to all odors regardless of valence (Heining et al., 2006).

Emotional roles of the insula Consistent with its relation to gustation, meta-analysis reveals that the insula responds strongly to disgust faces (Fusar-Poli et al., 2009), and direct recordings in the ventral anterior insula in epilepsy patients reveal a crucial role for this region in categorizing facial expressions of disgust (Krolak-Salmon et al., 2003). Both dorsal and ventral anterior insula respond to other categories of disgusting images, although posterior insula does not (Deen et al., 2011; Wright et al., 2004). Disgust sensitivity correlates with ventral anterior insular responses to images of disgusting foods (Calder et al., 2007). Despite this evidence for the involvement of the insula in disgust processing, the insula may not be a disgust center per se (Chapman & Anderson, 2012). For example, the insula is also activated by anger (Damasio et al., 2000), fear (Schienle et al., 2002), anxiety (Critchley, Wiens, Rotshtein, Ohman, & Dolan, 2004), and pain (Peyron, Laurent, & García-Larrea, 2000).Thus, although disgust may be a particularly strong stimulus for the insula, insular activity is clearly not exclusive to disgust.

Social roles of the insula Several lines of evidence suggest that the insula plays a role in social cognition. Given its role in disgust, the insula may be involved in social forms of disgust. For example, observing disgust in others activates the anterior insula (Wicker, Keysers, Plailly, & Royet, 2003), and the insular response to disgusted facial expressions is correlated with scores on an empathy scale (Hein & Singer, 2008; Jabbi, Swart, & Keysers, 2007; Singer, 2006). Disgust appears to be closely connected to morality; similar facial activity is evoked in response to both physically disgusting stimuli (e.g., cockroaches, feces) and moral transgressions

(Cannon, Schnall, & White, 2010; Chapman, Kim, Susskind, & Anderson, 2009), and experimentally inducing disgust increases moral wrongness judgments (Schnall, Benton, & Harvey, 2008; Wheatley & Haidt, 2005). The insula may support the connection between the primitive distaste response, disgust, and morality (for reviews, see Chapman & Anderson, 2012, 2013). These data provide critical support for the intersection of emotion, social cognition, and moral judgment, suggesting the moral regulation of interpersonal behavior may originate in primitive chemosensory precursors related to affective evaluations of what may be good and bad to eat (Chapman & Anderson, 2013).

The insula's role in social cognition also extends beyond disgust. Observing pain in other people consistently activates the insula (Singer, Seymour et al., 2004), and the insula displays increased activation for faces of cooperators in the Prisoner's Dilemma compared to neutral faces (Singer, Kiebel, et al., 2004). In addition, activity in the right insula increases when forming bad impressions of others' personalities; moreover, greater interaction between the insula and hippocampus relates to better memory for faces associated with negative personality impressions (Tsukiura, Shigemune, Nouchi, Kambara, & Kawashima, 2012).

Effects of emotion on sensory processes

Given the neuroanatomical relationships between sensory and socioemotional systems, we may ask: To what extent does socioemotional processing affect sensory process? There is considerable evidence indicating that emotion can modulate selective attention, and there may be an important role for emotions in early sensory gating.

PERCEIVING EMOTION IN OTHERS AFFECTS SENSORY PROCESSES Emotions, particularly emotional expressions, can communicate important information about the environment to other individuals. For example, when others are expressing fear, it would be wise to increase one's own vigilance so as to prepare for danger. Indeed, presenting fear expressions before a discrimination task enhances contrast sensitivity (Phelps, Ling, & Carrasco, 2006). In addition, observing the eye-widening associated with fearful facial expressions, compared to neutral or disgust expressions, enhances gaze-direction discrimination and facilitates peripheral-target detection (Lee, Susskind, & Anderson, 2013). Thus, visual processing is enhanced by the presence of a fear facial expression, which may indicate the presence of a threat that needs to be detected (Whalen, 1998).

The effects of perceived emotion on sensory processes are evident in activity in sensory cortices. Exposure to fearful faces enhances early responses in visual cortex (Keil, Stolarova, Moratti, & Ray, 2007; Pourtois, Grandjean, Sander, & Vuilleumier, 2004). Task-irrelevant affective information, including fearful expressions and shock-pairing, enhances activity in early visual cortex, including V1, and alters the functional connectivity of early visual cortex (Damaraju, Huang, Barrett, & Pessoa, 2009). Lastly, fearful faces presented unilaterally increase activity in the contralateral visual system, which may indicate temporary reduction in activation thresholds for retinotopic locations previously containing fearful faces (Carlson, Reinke, LaMontagne, & Habib, 2011).

EXPRESSING EMOTION AFFECTS SENSORY PROCESSES Emotional impacts on perception are not limited to utilizing the facial expressions of others as communicative signals that bias one's own perception. Expressing emotion oneself also serves to bias one's own perception to meet behavioral goals (Darwin, 1872; Lee et al., 2013; Susskind et al., 2008). If emotions are thought of as preparatory action tendencies, then facial expressions might be thought of as part of the mechanism that prepares an individual or organism for appropriate behaviors.

For example, fear and disgust have distinct—and opposing—effects on sensory perception (Susskind et al., 2008). Making a fear expression results in a significant increase in the upper visual field size relative to neutral, consistent with increased opening of the eyes and raising of the eyebrows (Lee et al., 2013; Susskind et al., 2008). By contrast, making disgust expressions result in narrowing of the upper and lower visual field relative to neutral, consistent with lowering of the brow and raising of the cheeks. Fear expressions also result in faster saccades and enhanced stimulus detection at further visual field eccentricities, whereas disgust expressions result in slower saccades and reduced stimulus detection. In addition to alterations of the visual apparatus when posing fear and disgust expressions, fear expressions resulted in dilation of the nostrils, while disgust expressions resulted in restriction of the nostrils. In summary, Susskind and colleagues (2008) suggest that emotional expressions alter properties of the facial sensory apparatus that fundamentally affect the ability to collect environmental information. In the case of fear, the effect is to open up the sensory apparatus to enhance detection and vigilance; while in the case of disgust, the effect is closure of the sensory apparatus to mitigate exposure to contaminants. Via facial expressions, emotions can gate sensory processing and thus influence perception at its earliest stage, limiting how photons reach the retina and molecules the olfactory mucosa. Although co-opted for the purposes of social communication, facial expressions may have originated as a sensory adaptation for the sender, suggesting that the more recent socioemotional roles of facial expression arise from an original sensory function.

Experiencing Emotion Oneself Affects Sensory Processes In addition to perceiving or making emotional facial expressions, subjectively experiencing emotion can also affect perception. Todd and colleagues (2012) have shown that emotionally arousing images are indeed perceived to be more perceptually vivid than neutral images. When identical visual noise was superimposed on arousing and neutral images, observers perceived the arousing images to be less noisy than the neutral images. This enhanced perceptual vividness was mediated by amygdalar influences on extrastriate cortex and occurred relatively early on in perceptual processing (~200 msec following stimulus onset; Todd et al., 2012). Similarly, auditory stimuli conditioned with aversive experiences are perceived to be louder (Asutay & Västfjäll, 2012). Emotional enhancement of visual processing appears to come at a cost, however. In particular, processing of spatially or temporally peripheral information is reduced, such as in emotion-induced blindness (Ciesielski, Armstrong, Zald, & Olatunji, 2010; Kennedy & Most, 2012; Riggs et al., 2011).

While Todd and colleagues found similar effects of positive and negative emotional arousal on perceptual vividness, other evidence suggests that positive and negative affect may bias perceptual gating in opposing ways. Schmitz, de Rosa, and Anderson (2009) induced positive, negative, and neutral affect using International Affective Picture System images. Following mood induction, participants were presented with compound face-house stimuli, with a relatively small face overlaid on a relatively larger house and asked to attend to the sex of the faces. Negative affect resulted in decreased activity in the parahippocampal place area, suggesting decreased encoding of peripheral information (i.e., information about the house). In contrast, positive affect resulted in increased activity in the parahippocampal place area, suggesting an increase in encoding of peripheral information (Schmitz et al., 2009). Further, positive and negative emotions were shown to have opposing effects on functional connectivity between the parahippocampal place areas with early visual cortex, suggesting that emotions differentially gate information flow from early perceptual cortices. This work demonstrates that emotional states, rather than emotional stimuli, differently affect perceptual encoding.

Perceptual contributions to socioemotional processes

In addition to the effects of emotional information on sensory (via facial expressions) and perceptual (via experienced emotion) processes, there is evidence of the inverse relationship. At the most basic level, sensory systems are required for perceiving others. Beyond this basic function, however, perceptual systems may support higher-order socioemotional processes. Recognizing actions and emotions from multimodal perceptual information (i.e., visual, auditory, and olfactory processes involved in social perception) forms the basis for understanding the intentions and feelings of others and allows us to successfully navigate our social world.

Visual Contributions Visual processes play a critical role in the ability to recognize other agents from visual cues, including form and motion, and multiple regions of extrastriate cortex are specialized for processing various visual aspects of social stimuli. Moreover, activity in these visual processing regions may be instrumental in supporting higher socioemotional abilities. The fusiform face area is preferentially involved in processing face information (e.g., Grill-Spector, Knouf, & Kanwisher, 2004; Kanwisher, McDermott, & Chun, 1997), while the extrastriate body area and superior temporal sulcus are involved in processing human forms from static and dynamic information, respectively (see Peelen & Downing, 2007, for a review of body perception). Much of our social experience makes use of the visual system, and some researchers suggest that information in these perceptual regions forms part of our understanding of emotional and social information (e.g., Scholl & Tremoulet, 2000; Schultz, 2005).

The role of the visual system in social processing is at least in part bottom-up. A variety of studies examining processing of animate motion in nonhuman animals, human infants, and clinical populations support this view. Newly hatched chicks show innate biases toward perceiving the motion of animate entities that are likely not attributable to elaborate top-down mechanisms. For example, chicks presented with animations of geometric shapes preferentially approach shapes that appear to be alive, that is shapes that demonstrate self-propelled motion, a hallmark of animacy (Mascalzoni, Regolin, & Vallortigara, 2010; Tremoulet & Feldman, 2000). Chicks can also discriminate between complex instances of biological form from motion relative to non-biological, rigid, or rotating shapes (Vallortigara,

Regolin, & Marconato, 2005). Such biases toward animate entities may have critical social consequences for survival, since this mechanism is likely involved in imprinting (e.g., Jaynes, 1957; Ramsay & Hess, 1954). Human infants as young as two days old also exhibit spontaneous preferences for biological over nonbiological motion, as indexed by longer looking times (Simion, Regolin, & Bulf, 2008). These cross-species data provide support for the notion of innate sensitivity of the visual system to animate or biological motion. Since newborn animals do not possess the depth of higher-level associations of animate motion with social context that might influence perceptual processing, innate preference for biological motion may form a bottom-up foundation for social cognition, in that early visual processing of socially relevant stimuli supports the development of more complex socioemotional perception.

Further support for the bottom-up role of visual processes in socioemotional understanding comes from studies of visual information processing in autism spectrum disorder (ASD). ASD is largely characterized by deficits in social intelligence, but some authors suggest that the observed socioemotional impairments in ASD may reflect (at least in part) a more general, lower-level impairment in processing socially relevant perceptual information. For instance, individuals with ASD commonly show deficits in processing both faces and facial expressions of emotion (for a review, see Harms, Martin, & Wallace, 2010) that may be related to a perceptual bias in ASD toward local rather than global stimulus features (Behrmann, Thomas, & Humphreys, 2006; Brosnan, Scott, Fox, & Pye, 2004; Frith & Happé, 1994). Whereas faces are typically processed holistically, those with ASD may be biased to attend to individual features, impairing recognition of faces as gestalts. Eye-tracking of individuals with ASD indicates differential face scanning compared to healthy individuals (Pelphrey et al., 2002). Further research on local processing bias in ASD shows that performance on face-processing tasks can be improved by cueing individuals with ASD to attend to appropriate facial features (López, Donnelly, Hadwin, & Leekam, 2004). Local processing bias may also account for deficits in animate motion perception observed in ASD. In animate displays depicting multiple dynamic geometric shapes, healthy individuals tend to describe motion in social terms (e.g., chasing), while those with ASD show impairments in perceiving the socially interactive elements of animate motion (Castelli, Frith, Happé, & Frith, 2002). If those with ASD are biased toward local stimulus features, this could explain why they are less able to describe interactivity of animate agents, the recognition of which is contingent on attention to the global correlated motion of multiple individual elements.

Motor Contributions There is also evidence for a role of the motor system in facilitating socioemotional processes. Generally, these findings fall under the umbrella of the *simulation theory of social cognition* (Gallese, Keysers, & Rizzolatti, 2004; Gallese, 2007) and its extension to emotional experiences (Nielsen, 2002). The basic tenet of these theories is that others' emotions and mental states are understood, at least in part, through covert motor simulations of their expressions and actions (Rizzolatti & Sinigaglia, 2010). In other words, we may recognize the emotional expressions and actions of other individuals by engaging our own motor systems in covert simulation of the activity we are observing. In this way, internal feedback signals from the motor system aid us in resolving the meaning of observed motor activity in others. This kind of embodied simulation of observed actions may play a critical role in our ability to understand the feelings and intentions of others.

In support of simulation theory, Neal and Chartrand (2011) demonstrate that internal feedback signals resulting from mimicking facial expressions influence the accuracy of identifying those expressions in others. In particular, individuals who received Botox injections, which render the muscles around the eyes less effective at contracting, are significantly worse at identifying facial expressions than controls. In a second experiment, the authors applied a face mask that created a tight film on the surface of the skin, increasing the resistance of the skin to facial muscle activity and thereby amplifying muscle signals (Neal & Chartrand, 2011). Stronger facial feedback signals elicited in this condition were associated with significant improvement in recognizing expressions. Thus, inhibition of facial musculature impaired recognition, while amplification of musculature feedback improved recognition, providing support for a direct role of facial mimicry and feedback signals in recognition of emotional expressions.

The motor system is also involved in perception of human forms from motion information. Point-light biological motion stimuli (Johansson, 1973) depict human figures using only about a dozen points of light, each representing a major joint on the human form. These stimuli are often used in action-perception research because they convey such compelling information about the dynamic human form without introducing overly complex visual features (e.g., faces, clothing). Further, by randomizing the starting position of each point-light, scrambled control stimuli can be created that completely disrupt perception of human forms but contain identical motion information. Using functional MRI, human biological motion (relative to the scrambled control stimulus) significantly engaged regions in the premotor cortex (Saygin, Wilson, Hagler, Bates, & Sereno, 2004). Likewise, lesions to premotor regions are associated with significant impairment on biological motion recognition relative to healthy individuals, indicating that this region is necessary for accurate perception of human actions from biological motion (Saygin, 2007).

Chemosensory Contributions In addition to the effects of visual and motor systems on social-emotional perception, growing evidence supports the involvement of the chemosensory system in social processing. For example, Eskine, Kacinik, and Prinz (2011) found that drinking an unpleasant liquid made moral judgments more severe. Conversely, participants who made judgments about moral transgressions rated a beverage as being more disgusting than did participants who made judgments about neutral actions (Eskine, Kacinik, & Webster, 2012). Similarly, participants who recalled a time when they were treated unfairly rated a mildly unpleasant taste as being stronger than participants who recalled a time they were treated fairly (Skarlicki, Hoegg, Aquino, & Nadisic, 2013). These results extended to unfairness directed toward others: participants who viewed a video depicting interpersonal unfairness rated the taste as being stronger than did participants who viewed a control video. These sources of evidence suggest an overlap and bidirectional influence between basic chemosensory perception and sociomoral cognition, consistent with the origins of complex social and emotional judgment in basic perceptual processes related to distaste (Chapman et al., 2009; Chapman & Anderson, 2013).

Conclusion

In this chapter, we have reviewed evidence from multiple domains of research indicating the involvement of lower-level perceptual mechanisms in what are traditionally thought of as higher-level socioemotional capacities. Both social and emotional processes seem to involve largely overlapping activity in neural regions commonly activated in perceptual tasks, whether proximal chemosensory systems such as smell and taste, or distal systems such as vision. We have also reviewed evidence for a bidirectional relationship between socioemotional and perceptual processes: facial expressions of emotion alter sensory gating, and affective states can bias early perceptual regions to influence the course of information processing, while conversely, perceptual processes may also shape the course of socioemotional responses. Given that socioemotional neural systems once evolved from simpler sensory systems, but also continue to the present day to demonstrate overlapping behavioral and neural substrates, we suggest that socioemotional and sensory systems have maintained a deep functional integration.

REFERENCES

Aboitiz, F. (1992). The origin of the mammalian brain as a case of evolutionary irreversibility. *Med Hypotheses, 38*, 301–304.

Aboitiz, F., Montiel, J., Morales, D., & Concha, M. (2002). Evolutionary divergence of the reptilian and the mammalian brains: Considerations on connectivity and development. *Brain Res Brain Res Rev, 39*(2–3), 141–153.

Aboitiz, F., Morales, D., & Montiel, J. (2003). The evolutionary origin of the mammalian isocortex: Towards an integrated developmental and functional approach. *Behav Brain Sci, 26*(5), 535–586.

Adolphs, R. (2002). Neural systems for recognizing emotion. *Curr Opin Neurobiol, 12*(2), 169–177.

Adolphs, R. (2003). Cognitive neuroscience of human social behaviour. *Nat Rev Neurosci, 4*(3), 165–178.

Adolphs, R. (2010). What does the amygdala contribute to social cognition? *Ann NY Acad Sci, 1191*, 42–61.

Adolphs, R., Baron-Cohen, S., & Tranel, D. (2002). Impaired recognition of social emotions following amygdala damage. *J Cogn Neurosci, 14*(8), 1264–1274.

Aggleton, J., Burton, M., & Passingham, R. (1980). Cortical and subcortical afferents to the amygdala of the rhesus monkey (*Macaca mulatta*). *Brain Res, 190*(2), 347–368.

Amaral, D. (2003). The amygdala, social behavior, and danger detection. *Ann NY Acad Sci, 1000*(1), 337–347.

Amaral, D., Capitanio, J., Jourdain, M., Mason, W., Mendoza, S., & Prather, M. (2003). The amygdala: Is it an essential component of the neural network for social cognition? *Neuropsychologia, 41*(2), 235–240.

Anderson, A., & Phelps, E. (2001). Lesions of the human amygdala impair enhanced perception of emotionally salient events. *Nature, 411*(6835), 305–309.

Anderson, A. K., Christoff, K., Stappen, I., Panitz, D., Ghahremani, D. G., Glover, G., ... Sobel, N. (2003). Dissociated neural representations of intensity and valence in human olfaction. *Nat Neurosci, 6*(2), 196–202.

Asutay, E., & Västfjäll, D. (2012). Perception of loudness is influenced by emotion. *PLoS ONE, 7*(6), e38660.

Barbas, H., Saha, S., Rempel-Clower, N., & Ghashghaei, T. (2003). Serial pathways from primate prefrontal cortex to autonomic areas may influence emotional expression. *BMC Neurosci, 4*, 25.

Bechara, A., Tranel, D., Damasio, H., & Damasio, A. (1996). Failure to respond autonomically to anticipated future outcomes following damage to prefrontal cortex. *Cereb Cortex, 6*(2), 215–225.

Becker, B., Mihov, Y., Scheele, D., Kendrick, K. M., Feinstein, J. S., Matusch, A., ... Hurlemann, R. (2012). Fear processing and social networking in the absence of a functional amygdala. *Biol Psychiatry, 72*(1), 70–77.

Behrmann, M., Thomas, C., & Humphreys, K. (2006). Seeing it differently: Visual processing in autism. *Trends Cogn Sci, 10*(6), 258–264.

Bender, G., Veldhuizen, M., Meltzer, J., Gitelman, D., & Small, D. (2009). Neural correlates of evaluative compared with passive tasting. *Eur J Neurosci, 30*(2), 327–338.

Brennan, P., & Keverne, E. (2004). Something in the air? New insights into mammalian pheromones. *Curr Biol, 14*(2), R81–R89.

Brosnan, M., Scott, F., Fox, S., & Pye, J. (2004). Gestalt processing in autism: Failure to process perceptual relationships and the implications for contextual understanding. *J Child Psychol Psychiatry, 45*(3), 459–469.

Butler, A., & Hodos, W. (2005). *Comparative vertebrate neuroanatomy: Evolution and adaptation.* New York, NY: Wiley-Liss.

Calder, A., Beaver, J., Davis, M., van Ditzhuijzen, J., Keane, J., & Lawrence, A. (2007). Disgust sensitivity predicts the insula and pallidal response to pictures of disgusting foods. *Eur J Neurosci, 25*(11), 3422–3428.

Cannon, P., Schnall, S., & White, M. (2010). Transgressions and expressions: Affective facial muscle activity predicts moral judgments. *Soc Psychol Pers Sci, 2*(3), 325–331.

Carlson, J., Reinke, K., LaMontagne, P., & Habib, R. (2011). Backward masked fearful faces enhance contralateral occipital cortical activity for visual targets within the spotlight of attention. *Soc Cogn Affect Neurosci, 6*(5), 639–645.

Carmichael, S., Clugnet, M., & Price, J. (1994). Central olfactory connections in the macaque monkey. *J Comp Neurol, 346*(3), 403–434.

Castelli, F., Frith, C., Happé, F., & Frith, U. (2002). Autism, Asperger syndrome and brain mechanisms for the attribution of mental states to animated shapes. *Brain, 125*(8), 1839–1849.

Chang, L., Yarkoni, T., Khaw, M., & Sanfey, A. (2013). Decoding the role of the insula in human cognition: Functional parcellation and large-scale reverse inference. *Cereb Cortex, 23*(3), 739–749.

Chapman, H., & Anderson, A. (2012). Understanding disgust. *Ann NY Acad Sci, 1251*, 62–76.

Chapman, H., & Anderson, A. (2013). Things rank and gross in nature: A review and synthesis of moral disgust. *Psychol Bull, 139*(2), 300–327.

Chapman, H., Kim, D., Susskind, J., & Anderson, A. (2009). In bad taste: Evidence for the oral origins of moral disgust. *Science, 323*(5918), 1222–1226.

Chen, D., & Dalton, P. (2005). The effect of emotion and personality on olfactory perception. *Chem Senses, 30*(4), 345–351.

Ciaramelli, E., Sperotto, R., Mattioli, F., & di Pellegrino, G. (2013). Damage to the ventromedial prefrontal cortex reduces interpersonal disgust. *Soc Cogn Affect Neurosci, 8*(2), 171–180.

Ciesielski, B., Armstrong, T., Zald, D., & Olatunji, B. (2010). Emotion modulation of visual attention: Categorical and temporal characteristics. *PLoS ONE, 5*(11), e13860.

Critchley, H., Wiens, S., Rotshtein, P., Ohman, A., & Dolan, R. (2004). Neural systems supporting interoceptive awareness. *Nat Neurosci, 7*(2), 189–195.

Cunningham, W., Johnson, M., Raye, C., Gatenby, J., Gore, J., & Banaji, M. (2004). Separable neural components in the processing of black and white faces. *Psychol Sci, 15*(12), 806–813.

Cunningham, W., van Bavel, J., & Johnsen, I. (2008). Affective flexibility: Evaluative processing goals shape amygdala activity. *Psychol Sci, 19*(2), 152–160.

Damaraju, E., Huang, Y.-M., Barrett, L., & Pessoa, L. (2009). Affective learning enhances activity and functional connectivity in early visual cortex. *Neuropsychologia, 47*(12), 2480–2487.

Damasio, A. (1994). *Descartes' error: Emotion, reason, and the human brain.* New York, NY: Putnam.

Damasio, A. (1996). The somatic marker hypothesis and the possible functions of the prefrontal cortex. *Philos Trans R Soc Lond B Biol Sci, 351*(1346), 1413–1420.

Damasio, A., Grabowski, T., Bechara, A., Damasio, H., Ponto, L., Parvizi, J., & Hichwa, R. (2000). Subcortical and cortical brain activity during the feeling of self-generated emotions. *Nat Neurosci, 3*(10), 1049–1056.

Damasio, A., Tranel, D., & Damasio, H. (1990). Individuals with sociopathic behavior caused by frontal damage fail to respond autonomically to social stimuli. *Behav Brain Res, 41*(2), 81–94.

Damasio, H., Grabowski, T., Frank, R., Galaburda, A., & Damasio, A. (1994). The return of Phineas Gage: Clues about the brain from the skull of a famous patient. *Science, 264*(5162), 1102–1105.

Darwin, C. (1872). *The expression of the emotions in man and animals.* London: John Murray.

Deen, B., Pitskel, N., & Pelphrey, K. (2011). Three systems of insular functional connectivity identified with cluster analysis. *Cereb Cortex, 21*(7), 1498–1506.

Ekman, P. (1999). Basic emotions. In T. Dalgleish & M. Power (Eds.), *Handbook of cognition and emotion* (pp. 45–60). New York, NY: Wiley.

Engell, A. D., Haxby, J. V., & Todorov, A. (2007). Implicit trustworthiness decisions: Automatic coding of face properties in the human amygdala. *J Cogn Neurosci, 19*(9), 1508–1519.

Eskine, K. J., Kacinik, N. A., & Prinz, J. J. (2011). A bad taste in the mouth: Gustatory disgust influences moral judgment. *Psychol Sci, 22*(3), 295–299.

Eskine, K., Kacinik, N., & Webster, G. (2012). The bitter truth about morality: Virtue, not vice, makes a bland beverage taste nice. *PLoS ONE, 7*(7), e41159.

Eslinger, P., & Damasio, A. (1985). Severe disturbance of higher cognition after bilateral frontal lobe ablation: Patient EVR. *Neurology, 35*(12), 1731–1731.

Farb, N., Chapman, H., & Anderson, A. (2013). Emotions: Form follows function. *Curr Opin Neurobiol*, 1–6.

Fine, C., & Blair, R. (2000). The cognitive and emotional effects of amygdala damage. *Neurocase, 6*(6), 435–450.

Forbes, C., & Grafman, J. (2010). The role of the human prefrontal cortex in social cognition and moral judgment. *Annu Rev Neurosci, 33*, 299–324.

Frith, U., & Happé, F. (1994). Autism: Beyond "theory of mind." *Cognition, 50*(1–3), 115–132.

Fusar-Poli, P., Placentino, A., Carletti, F., Landi, P., Allen, P., Surguladze, S., ... Politi, P. (2009). Functional atlas of emotional faces processing: A voxel-based meta-analysis of 105 functional magnetic resonance imaging studies. *J Psychiatry Neurosci, 34*(6), 418–432.

Gallese, V. (2007). Before and below "theory of mind": Embodied simulation and the neural correlates of social cognition. *Philos Trans R Soc Lond B Biol Sci, 362*(1480), 659–669.

Gallese, V., Keysers, C., & Rizzolatti, G. (2004). A unifying view of the basis of social cognition. *Trends Cogn Sci, 8*(9), 396–403.

Ghashghaei, H., & Barbas, H. (2002). Pathways for emotion: Interactions of prefrontal and anterior temporal pathways in the amygdala of the rhesus monkey. *Neuroscience, 115*(4), 1261–1279.

Ghashghaei, H., Hilgetag, C., & Barbas, H. (2007). Sequence of information processing for emotions based on the anatomic dialogue between prefrontal cortex and amygdala. *NeuroImage, 34*(3), 905–923.

Gottfried, J., Deichmann, R., Winston, J., & Dolan, R. (2002). Functional heterogeneity in human olfactory cortex: An event-related functional magnetic resonance imaging study. *J Neurosci, 22*(24), 10819–10828.

Gottfried, J., & Zald, D. (2005). On the scent of human olfactory orbitofrontal cortex: Meta-analysis and comparison to non-human primates. *Brain Res Brain Res Rev, 50*(2), 287–304.

Grabenhorst, F., Rolls, E., Parris, B., & D'Souza, A. (2010). How the brain represents the reward value of fat in the mouth. *Cereb Cortex, 20*(5), 1082–1091.

Grill-Spector, K., Knouf, N., & Kanwisher, N. (2004). The fusiform face area subserves face perception, not generic within-category identification. *Nat Neurosci, 7*(5), 555–562.

Haddad, R., Lapid, H., Harel, D., & Sobel, N. (2008). Measuring smells. *Curr Opin Neurobiol, 18*(4), 438–444.

Hadley, K., Orlandi, R., & Fong, K. (2004). Basic anatomy and physiology of olfaction and taste. *Otolaryng Clin N Am, 37*(6), 1115–1126.

Halpern, M., & Martinez-Marcos, A. (2003). Structure and function of the vomeronasal system: An update. *Prog Neurobiol, 70*(3), 245–318.

Harlow, J. (1868). Recovery from the passage of an iron rod through the head. *Publ Mass Med Soc, 2*, 327–347.

Harms, M., Martin, A., & Wallace, G. (2010). Facial emotion recognition in autism spectrum disorders: A review of behavioral and neuroimaging studies. *Neuropsychol Rev, 20*(3), 290–322.

Hein, G., & Singer, T. (2008). I feel how you feel but not always: The empathic brain and its modulation. *Curr Opin Neurobiol, 18*(2), 153–158.

Heining, M., Young, A., Ioannou, G., Andrew, C., Brammer, M., Gray, J., & Phillips, M. (2006). Disgusting smells activate human anterior insula and ventral striatum. *Ann NY Acad Sci, 1000*(1), 380–384.

Höistad, M., & Barbas, H. (2008). Sequence of information processing for emotions through pathways linking temporal and insular cortices with the amygdala. *NeuroImage, 40*(3), 1016–1033.

Ibañez, A., Gleichgerrcht, E., & Manes, F. (2010). Clinical effects of insular damage in humans. *Brain Struct Funct, 214*(5–6), 397–410.

Jabbi, M., Swart, M., & Keysers, C. (2007). Empathy for positive and negative emotions in the gustatory cortex. *NeuroImage, 34*(4), 1744–1753.

Jaynes, J. (1957). Imprinting: The interaction of learned and innate behavior: II. The critical period. *J Comp Physiol Psychol, 50*(1), 6–10.

Johansson, G. (1973). Visual perception of biological motion and a model for its analysis. *Percept Psychophys, 14*(2), 201–211.

Jones, C., Ward, J., & Critchley, H. (2010). The neuropsychological impact of insular cortex lesions. *J Neurol Neurosurg Psychiatry, 81*(6), 611–618.

Jones-Gotman, M., & Zatorre, R. J. (1988). Olfactory identification deficits in patients with focal cerebral excision. *Neuropsychologia, 26*(3), 387–400.

Jung, J., Hudry, J., Ryvlin, P., Royet, J.-P., Bertrand, O., & Lachaux, J.-P. (2006). Functional significance of olfactory-induced oscillations in the human amygdala. *Cereb Cortex, 16*(1), 1–8.

Kanwisher, N., McDermott, J., & Chun, M. (1997). The fusiform face area: A module in human extrastriate cortex specialized for face perception. *J Neurosci, 17*(11), 4302–4311.

Keil, A., Stolarova, M., Moratti, S., & Ray, W. (2007). Adaptation in human visual cortex as a mechanism for rapid discrimination of aversive stimuli. *NeuroImage, 36*(2), 472–479.

Kelly, C., Toro, R., di Martino, A., Cox, C., Bellec, P., Castellanos, F., & Milham, M. (2012). A convergent functional architecture of the insula emerges across imaging modalities. *NeuroImage, 61*(4), 1129–1142.

Kennedy, B., & Most, S. (2012). Perceptual, not memorial, disruption underlies emotion-induced blindness. *Emotion, 12*(2), 199–202.

Kennedy, D., & Adolphs, R. (2010). Impaired fixation to eyes following amygdala damage arises from abnormal bottom-up attention. *Neuropsychologia, 48*(12), 3392–3398.

Kennedy, D., Gläscher, J., Tyszka, J., & Adolphs, R. (2009). Personal space regulation by the human amygdala. *Nat Neurosci, 12*(10), 1226–1227.

Koenigs, M., & Tranel, D. (2007). Irrational economic decision-making after ventromedial prefrontal damage: Evidence from the Ultimatum Game. *J Neurosci, 27*(4), 951–956.

Koscik, T., & Tranel, D. (2011). The human amygdala is necessary for developing and expressing normal interpersonal trust. *Neuropsychologia, 49*(4), 602–611.

Koscik, T., & Tranel, D. (2012). Brain evolution and human neuropsychology: The inferential brain hypothesis. *J Int Neuropsychol Soc, 18*(3), 394–401.

Koscik, T., & Tranel, D. (2013). Abnormal causal attribution leads to abnormal (but advantageous) economic decision-making following ventromedial prefrontal cortex damage. *J Cogn Neurosci, 25*(8), 1372–1382.

Kringelbach, M., & Rolls, E. (2004). The functional neuroanatomy of the human orbitofrontal cortex: Evidence from neuroimaging and neuropsychology. *Prog Neurobiol, 72*(5), 341–372.

Krolak-Salmon, P., He, M.-A., Isnard, J., Tallon-Baudry, C., Gue, M., Vighetto, A., & Bertrand, O. (2003). An attention modulated response to disgust in human ventral anterior insula. *Ann Neurol, 53*, 446–453.

Kurth, F., Zilles, K., Fox, P., Laird, A., & Eickhoff, S. (2010). A link between the systems: Functional differentiation and integration within the human insula revealed by meta-analysis. *Brain Struct Funct, 214*(5–6), 519–534.

LeDoux, J. (2012). Rethinking the emotional brain. *Neuron, 73*(4), 653–676.

Lee, D., Susskind, J., & Anderson, A. (2013). Social transmission of the sensory benefits of fear eye-widening. *Psychol Sci, 24*(6), 957–965.

Lehrner, J., Eckersberger, C., Walla, P., Pötsch, G., & Deecke, L. (2000). Ambient odor of orange in a dental office reduces anxiety and improves mood in female patients. *Physiol Behav, 71*(1–2), 83–86.

Lehrner, J., Marwinski, G., Lehr, S., Johren, P., & Deecke, L. (2005). Ambient odors of orange and lavender reduce anxiety and improve mood in a dental office. *Physiol Behav, 86*(1–2), 92–95.

Lieberman, M., Hariri, A., Jarcho, J., Eisenberger, N., & Bookheimer, S. (2005). An fMRI investigation of race-related amygdala activity in African-American and Caucasian-American individuals. *Nat Neurosci, 8*(6), 720–722.

López, B., Donnelly, N., Hadwin, J., & Leekam, S. (2004). Face processing in high-functioning adolescents with autism: Evidence for weak central coherence. *Vis Cognit, 11*(6), 673–688.

Mascalzoni, E., Regolin, L., & Vallortigara, G. (2010). Innate sensitivity for self-propelled causal agency in newly hatched chicks. *Proc Natl Acad Sci USA, 107*(9), 4483–4485.

McDonald, A. (1998). Cortical pathways to the mammalian amygdala. *Prog Neurobiol, 55*(3), 257–332.

McGaugh, J. L. (2004). The amygdala modulates the consolidation of memories of emotionally arousing experiences. *Annu Rev Neurosci, 27*, 1–28.

Mende-Siedlecki, P., Said, C. P., & Todorov, A. (2012). The social evaluation of faces: A meta-analysis of functional neuroimaging studies. *Soc Cogn Affect Neurosci, 8*(3), 285–299.

Meredith, M. (1991). Sensory processing in the main and accessory olfactory systems: Comparisons and contrasts. *J Steroid Biochem, 39*(4), 601–614.

Milders, M., Fuchs, S., & Crawford, J. (2003). Neuropsychological impairments and changes in emotional and social behaviour following severe traumatic brain injury. *J Clin Exp Neuropsychol, 25*(2), 157–172.

Moretti, L., Dragone, D., & di Pellegrino, G. (2009). Reward and social valuation deficits following ventromedial prefrontal damage. *J Cogn Neurosci, 21*(1), 128–140.

Neal, D., & Chartrand, T. (2011). Embodied emotion perception: Amplifying and dampening facial feedback modulates emotion perception accuracy. *Soc Psychol Pers Sci, 2*(6), 673–678.

Nielsen, L. (2002). The simulation of emotion experience: On the emotional foundations of theory of mind. *Phenomenol Cogn Sci, 1*, 255–286.

Ongür, D., Ferry, A., & Price, J. (2003). Architectonic subdivision of the human orbital and medial prefrontal cortex. *J Comp Neurol, 460*(3), 425–449.

Peelen, M. V., & Downing, P. E. (2007). The neural basis of visual body perception. *Nat Rev Neurosci, 8*(8), 636–648.

Pelphrey, K., Sasson, N., Reznick, J., Paul, G., Goldman, B., & Piven, J. (2002). Visual scanning of faces in autism. *J Autism Dev Disord, 32*(4), 249–261.

Pessoa, L. (2009). How do emotion and motivation direct executive control? *Trends Cogn Sci, 13*(4), 160–166.

Peyron, R., Laurent, B., & García-Larrea, L. (2000). Functional imaging of brain responses to pain. A review and meta-analysis. *Neurophysiol Clin, 30*(5), 263–288.

Phelps, E. (2006). Emotion and cognition: Insights from studies of the human amygdala. *Annu Rev Psychol, 57*, 27–53.

Phelps, E., & LeDoux, J. (2005). Contributions of the amygdala to emotion processing: From animal models to human behavior. *Neuron, 48*(2), 175–187.

Phelps, E., Ling, S., & Carrasco, M. (2006). Emotion facilitates perception and potentiates the perceptual benefits of attention. *Psychol Sci, 17*(4), 292–299.

Pollatos, O., Kopietz, R., Linn, J., Albrecht, J., Sakar, V., Anzinger, A., ... Wiesman, M. (2007). Emotional stimulation alters olfactory sensitivity and odor judgment. *Chem Senses, 32*(6), 583–589.

Porrino, L., Crane, A., & Goldman-Rakic, P. (1981). Direct and indirect pathways from the amygdala to the frontal lobe in rhesus monkeys. *J Comp Neurol, 136*, 121–136.

Potter, H., & Butters, N. (1980). An assessment of olfactory deficits in patients with damage to prefrontal cortex. *Neuropsychologia, 18*(6), 621–628.

Pourtois, G., Grandjean, D., Sander, D., & Vuilleumier, P. (2004). Electrophysiological correlates of rapid spatial orienting towards fearful faces. *Cereb Cortex, 14*(6), 619–633.

Ramsay, A., & Hess, E. (1954). A laboratory approach to the study of imprinting. *Wilson Bull, 66*(3), 196–206.

Riggs, L., McQuiggan, D., Farb, N., Anderson, A., & Ryan, J. (2011). The role of overt attention in emotion-modulated memory. *Emotion, 11*(4), 776–785.

Rizzolatti, G., & Sinigaglia, C. (2010). The functional role of the parieto-frontal mirror circuit: Interpretations and misinterpretations. *Nat Rev Neurosci, 11*(4), 264–274.

Rolls, E. (2004). The functions of the orbitofrontal cortex. *Brain Cogn, 55*(1), 11–29.

Rolls, E. (2008). Functions of the orbitofrontal and pregenual cingulate cortex in taste, olfaction, appetite and emotion. *Acta Physiol Hung, 95*(2), 131–164.

Rolls, E., & Grabenhorst, F. (2008). The orbitofrontal cortex and beyond: From affect to decision-making. *Prog Neurobiol, 86*(3), 216–244.

Rolls, E., Kringelbach, M., & de Araujo, I. (2003). Different representations of pleasant and unpleasant odours in the human brain. *Eur J Neurosci, 18*(3), 695–703.

Royet, J.-P., Plailly, J., Delon-Martin, C., Kareken, D., & Segebarth, C. (2003). FMRI of emotional responses to odors: Influence of hedonic valence and judgment, handedness, and gender. *NeuroImage, 20*(2), 713–728.

Rudebeck, P. H., Bannerman, D. M., & Rushworth, M. F. S. (2008). The contribution of distinct subregions of the ventromedial frontal cortex to emotion, social behavior, and decision making. *Cogn Aff Behav Neurosci, 8*(4), 485–497.

Rudenga, K., Green, B., Nachtigal, D., & Small, D. (2010). Evidence for an integrated oral sensory module in the human anterior ventral insula. *Chem Senses, 35*(8), 693–703.

Rutishauser, U., Tudusciuc, O., Neumann, D., Mamelak, A., Heller, A., Ross, I., ... Adolphs, R. (2011). Single-unit responses selective for whole faces in the human amygdala. *Curr Biol, 21*(19), 1654–1660.

Sah, P., Faber, E., Lopez de Armentia, M., & Power, J. (2003). The amygdaloid complex: Anatomy and physiology. *Physiol Rev, 83*(3), 803–834.

Sanchez-Andrade, G., & Kendrick, K. (2009). The main olfactory system and social learning in mammals. *Behav Brain Res, 200*(2), 323–335.

Sander, D., Grafman, J., & Zalla, T. (2003). The human amygdala: An evolved system for relevance detection. *Rev Neurosci, 14*(4), 303–316.

Saygin, A. (2007). Superior temporal and premotor brain areas necessary for biological motion perception. *Brain, 130*(Pt 9), 2452–2461.

Saygin, A., Wilson, S., Hagler, D., Bates, E., & Sereno, M. (2004). Point-light biological motion perception activates human premotor cortex. *J Neurosci, 24*(27), 6181–6188.

Scalia, F., & Winans, S. (1975). The differential projections of the olfactory bulb and accessory olfactory bulb in mammals. *J Comp Neurol, 161*(1), 31–55.

Schienle, A., Stark, C., Walter, B., Blecker, C., Ott, U., Kirsch, P., ... Vaitl, D. (2002). The insula is not specifcally involved in disgust processing: An fMRI study. *NeuroReport, 13*(16), 2023–2026.

Schmitz, T., de Rosa, E., & Anderson, A. (2009). Opposing influences of affective state valence on visual cortical encoding. *J Neurosci, 29*(22), 7199–7207.

Schnall, S., Benton, J., & Harvey, S. (2008). With a clean conscience: Cleanliness reduces the severity of moral judgments. *Psychol Sci, 19*(12), 1219–1222.

Scholl, B., & Tremoulet, P. (2000). Perceptual causality and animacy. *Trends Cogn Sci, 4*(8), 299–309.

Schultz, R. T. (2005). Developmental deficits in social perception in autism: The role of the amygdala and fusiform face area. *Int J Dev Neurosci, 23*(2–3), 125–141.

Scott, J. W. (1986). The olfactory bulb and central pathways. *Experientia, 42*(3), 223–232.

Seubert, J., Kellermann, T., Loughead, J., Boers, F., Brensinger, C., Schneider, F., & Habel, U. (2010). Processing of disgusted faces is facilitated by odor primes: A functional MRI study. *NeuroImage, 53*(2), 746–756.

Seubert, J., Rea, A., Loughead, J., & Habel, U. (2009). Mood induction with olfactory stimuli reveals differential affective responses in males and females. *Chem Senses, 34*(1), 77–84.

Shariff, A., & Tracy, J. (2011). What are emotion expressions for? *Curr Dir Psychol Sci, 20*(6), 395–399.

Simion, F., Regolin, L., & Bulf, H. (2008). A predisposition for biological motion in the newborn baby. *Proc Natl Acad Sci USA, 105*(2), 809–813.

Singer, T. (2006). The neuronal basis and ontogeny of empathy and mind reading: Review of literature and implications for future research. *Neurosci Biobehav Rev, 30*(6), 855–863.

Singer, T., Kiebel, S., Winston, J., Dolan, R., & Frith, C. (2004). Brain responses to the acquired moral status of faces. *Neuron, 41*, 653–662.

Singer, T., Seymour, B., O'Doherty, J., Kaube, H., Dolan, R., & Frith, C. (2004). Empathy for pain involves the affective but not sensory components of pain. *Science, 303*(5661), 1157–1162.

Skarlicki, D., Hoegg, J., Aquino, K., & Nadisic, T. (2013). The bitter truth about morality: Virtue, not vice, makes a bland beverage taste nice. *J Exp Soc Psychol.*

Small, D. (2010). Taste representation in the human insula. *Brain Struct Funct, 214*(5–6), 551–561.

Small, D., Bender, G., Veldhuizen, M., Rudenga, K., Nachtigal, D., & Felsted, J. (2007). The role of the human orbitofrontal cortex in taste and flavor processing. *Ann NY Acad Sci, 1121*, 136–151.

Small, D., Gregory, M., Mak, Y., Gitelman, D., Mesulam, M., & Parrish, T. (2003). Dissociation of neural representation of intensity and affective valuation in human gustation. *Neuron, 39*, 701–711.

Spinella, M. (2002). A relationship between smell identification and empathy. *Int J Neurosci, 112*(6), 605–612.

Susskind, J., Lee, D., Cusi, A., Feiman, R., Grabski, W., & Anderson, A. (2008). Expressing fear enhances sensory acquisition. *Nat Neurosci, 11*(7), 843–850.

Tanabe, T., Yarita, H., Iino, M., Ooshima, Y., & Takagi, S. F. (1975). An olfactory cortex projection area in orbitofrontal of the monkey. *J Neurophysiol, 38*(5), 1269–1283.

Todd, R., Talmi, D., Schmitz, T., Susskind, J., & Anderson, A. (2012). Psychophysical and neural evidence for emotion-enhanced perceptual vividness. *J Neurosci, 32*(33), 11201–12.

Todorov, A. (2012). The role of the amygdala in face perception and evaluation. *Motiv Emotion, 36*(1), 16–26.

Tracy, J., & Robins, R. (2008). The nonverbal expression of pride: Evidence for cross-cultural recognition. *J Pers Soc Psychol, 94*(3), 516–530.

Tremoulet, P., & Feldman, J. (2000). Perception of animacy from the motion of a single object. *Perception, 29*(8), 943–951.

Tsukiura, T., Shigemune, Y., Nouchi, R., Kambara, T., & Kawashima, R. (2012). Insular and hippocampal contributions to remembering people with an impression of bad personality. *Soc Cogn Affect Neurosci.* doi:10.1093/scan/nss025

Vallortigara, G., Regolin, L., & Marconato, F. (2005). Visually inexperienced chicks exhibit spontaneous preference for biological motion patterns. *PLoS Biol, 3*(7), e208.

van Honk, J., Eisenegger, C., Terburg, D., Stein, D. J., & Morgan, B. (2013). Generous economic investments after basolateral amygdala damage. *Proc Natl Acad Sci USA, 110*(7), 2506–2510.

Walla, P., Mayer, D., Deecke, L., & Lang, W. (2005). How chemical information processing interferes with face processing: A magnetoencephalographic study. *NeuroImage, 24*(1), 111–117.

Weber, S., & Heuberger, E. (2008). The impact of natural odors on affective states in humans. *Chem Senses, 33*(5), 441–447.

Whalen, P. J. (1998). Fear, vigilance, and ambiguity: Initial neuroimaging studies of the human amygdala. *Curr Dir Psychol Sci, 7*, 177–188.

Wheatley, T., & Haidt, J. (2005). Transgressions and expressions: Affective facial muscle activity predicts moral judgments. *Psychol Sci, 16*(10), 780–784.

Wicker, B., Keysers, C., Plailly, J., & Royet, J.-P. (2003). Both of us disgusted in my insula: The common neural basis of seeing and feeling disgust. *Neuron, 40*, 655–664.

Winston, J., Gottfried, J., Kilner, J., & Dolan, R. (2005). Integrated neural representations of odor intensity and affective valence in human amygdala. *J Neurosci, 25*(39), 8903–8907.

Winston, J., O'Doherty, J., Kilner, J., Perrett, D., & Dolan, R. (2007). Brain systems for assessing facial attractiveness. *Neuropsychologia, 45*(1), 195–206.

Wright, P., He, G., Shapira, N., Goodman, W., & Liu, Y. (2004). Disgust and the insula: FMRI responses to pictures of mutilation and contamination. *NeuroReport, 15*(15), 2347–2351.

Yarita, H., Iino, M., Tanabe, T., Kogure, S., & Takagi, S. F. (1980). A transthalamic olfactory pathway to orbitofrontal cortex in the monkey. *J Neurophysiol, 43*(1), 69–85.

Young, L., Bechara, A., Tranel, D., Damasio, H., Hauser, M., & Damasio, A. (2010). Damage to ventromedial prefrontal cortex impairs judgment of harmful intent. *Neuron, 65*(6), 845–851.

Zald, D., & Pardo, J. (1997). Emotion, olfaction, and the human amygdala: Amygdala activation during aversive olfactory stimulation. *Proc Natl Acad Sci USA, 94*(8), 4119–4124.

Zatorre, R., & Jones-Gotman, M. (1991). Human olfactory discrimination after unilateral frontal or temporal lobectomy. *Brain, 114*(Pt. 1), 71–84.

Zikopoulos, B., & Barbas, H. (2007). Circuits for multisensory integration and attentional modulation through the prefrontal cortex and the thalamic reticular nucleus in primates. *Rev Neurosci, 18*(6), 417–438.

66 The Role of the Amygdala in Face Processing

PETER MENDE-SIEDLECKI AND ALEXANDER TODOROV

ABSTRACT Historically, the amygdala's role in face processing has been relegated to the domain of emotion, while a network of posterior cortical regions is typically considered the core network involved in face processing. In this chapter, we examine several decades of neurophysiological and neuroimaging work suggesting that the amygdala possesses face-selective properties, which can be observed in the absence of emotional expressions. We also review a recent large-scale group analysis of functional MRI data from 215 participants in face-localizer tasks observing robust responses to faces in the amygdala, which were reliable over time and identifiable in a majority of participants. Taken together, these data suggest that the amygdala deserves consideration as a key node in the core face-processing network.

Successful navigation of our social universe requires an understanding of other social agents. Harmonious interaction depends on accurate inferences regarding a rich compendium of social information—identity, race, gender, age, emotions, mental states, and personality, just to name a few. Much of this information can be inferred from the human face. Behavioral research shows that we are able to rapidly extract information regarding identity (Grill-Spector & Kanwisher, 2005; Yip & Sinha, 2002), race and gender (Cloutier, Mason, & Macrae, 2005; Martin & Macrae, 2007), emotional expression (Esteves & Öhman, 1993; Whalen et al., 1998), and attractiveness (Locher, Unger, Sociedade, & Wahl, 1993; Olson & Marshuetz, 2005), and even make inferences about personality such as aggressiveness (Bar, Neta, & Linz, 2006) and trustworthiness (Willis & Todorov, 2006; Todorov, Pakrashi, & Oosterhof, 2009) based upon minimal exposure to faces. Moreover, these inferences and evaluations have behavioral consequences (Todorov, Mende-Siedlecki, & Dotsch, 2013). Given the importance of faces, it is not surprising that they have been the subject of an extensive study in psychology and cognitive neuroscience (Calder, Rhodes, Johnson, & Haxby, 2011).

The initial groundwork for research on the neural bases of face perception was laid in the 1970s and early 1980s by researchers mapping the visual properties of inferior temporal (IT) cortex (e.g., Gross, Rocha-Miranda, & Bender, 1972; see also Gross, 1994, 2008). Single-unit recordings in nonhuman primates indicate that populations of neurons in IT cortex and superior temporal sulcus (STS) respond selectively to faces, as compared to comparably complex object categories (Bruce, Desimone, & Gross, 1981; Desimone, 1991; Desimone, Albright, Gross, & Bruce, 1984; Perrett, Rolls, & Caan, 1982; Perrett et al., 1984). These results have been widely and robustly replicated (Eifuku, de Souza, Tamura, Nishijo, & Ono, 2004; Hasselmo, Rolls, & Baylis, 1989; Kiani, Esteky, & Tanaka, 2005; Tsao, Freiwald, Knutsen, Mandeville, & Tootell, 2003), generalized to human subjects (Ojemann, Ojemann, & Lettich, 1992), and confirmed in combination with functional magnetic resonance imaging (fMRI; Tsao, Freiwald, Tootell, & Livingston, 2006). More recent work has refined and revised theories on the neural architecture of face perception, suggesting that there may be multiple face-selective patches (Freiwald & Tsao, 2012; Rajimehr, Young, & Tootell, 2009; Weiner & Grill-Spector, 2010, 2012). However, contemporaneous research using identical methodology suggests that the neural bases of face processing are not limited to posterior cortical areas.

Single-unit recordings of face-selective amygdala responses

In 1979, Sanghera and colleagues set out to examine whether neurons in the amygdala respond to visual input, citing previous anatomical work describing projections from IT cortex to the amygdala (Herzog & Hoesen, 1976; Jones & Powell, 1970). Indeed, of 1,754 neurons analyzed, 113 units in the lateral aspect of the amygdala were deemed responsive to visual stimuli. Moreover, in the last paragraph of the paper, the authors wrote, "It should also perhaps be noted that nine of the amygdaloid neurons with visual responses were found to respond primarily to faces or photographs of faces" (Sanghera, Rolls, & Roper-Hall, 1979, p. 624). Over the next 30 years, research confirming these initial findings has steadily accumulated, leading to the conclusion that there are neurons within the amygdala that display robust selectivity for faces.

To begin with, a number of additional single-unit recording studies of nonhuman primates have replicated Sanghera and colleagues' findings (Leonard, Rolls, Wilson, & Baylis, 1985; Nakamura, Mikami, & Kubota, 1992; Rolls, 1984), with later studies examining the effects of novelty (Wilson & Rolls, 1993) as well as identity and expression (Gothard, Battaglia, Erickson, Spitler, & Amaral, 2007) on face-selective responses in the amygdala. Subsequently, these findings were corroborated by high-resolution fMRI work with nonhuman primates (Hoffman, Gothard, Schmid, & Logothetis, 2007; Logothetis, Guggenberger, Peled, & Pauls, 1999), and generalized to humans in neurophysiological work with patients undergoing treatment for epilepsy (Fried, MacDonald, & Wilson, 1997; Fried, Cameron, Yashar, Fong, & Morrow, 2002; Kreiman, Koch, & Fried, 2000; Mormann et al., 2008; Quiroga, Reddy, Kreiman, Koch, & Fried, 2005; Rutishauser et al., 2011; Seeck et al., 1995; Steinmetz, 2008; Viskontas, Quiroga, & Fried, 2009; see also Sato et al., 2012, for similar findings using intracranial field potentials).

Across the majority of this work, several patterns have emerged. First, based on single-unit recording work in both humans and nonhuman primates, it is clear that only a subset of neurons within the amygdala is truly selective for faces. A conservative estimate of the number of face-selective amygdala neurons among the visually responsive neurons would likely lie somewhere between 5% and 10% (see Todorov, 2012), though some authors have suggested that as many as 20% might display selectivity for whole faces (Rutishauser et al., 2011). Second, face-selective responses in the amygdala are qualitatively different from those in posterior cortical regions on at least a few metrics. Direct comparison between recordings in amygdala and STS suggest that face-selective responses in the amygdala have longer response latencies than those in STS and, moreover, generalize somewhat less across faces than those in STS (Leonard et al., 1985). Third, face-selective neurons in the amygdala are less clustered, as compared to those in IT cortex and STS. Finally, it seems that no single facial characteristic or quality (e.g., species specificity, emotional expression, identity, familiarity, etc.) can completely account for the amygdala's responses to faces. For example, Leonard and colleagues observed that only four face-selective neurons in the amygdala (out of 19 tested) responded preferentially to faces with emotional expressions, while Gothard and colleagues observed that 64% of face-selective amygdala neurons responded to both emotional expression and identity.

Neuroimaging evidence of the amygdala's face-selective properties

There is a lack of complete convergence between the single-unit recording research and human neuroimaging work, with respect to which brain areas are ultimately classified as face-selective. Functional MRI studies of human face perception have classically relied upon paradigms employing functional localizers (Kanwisher, McDermott, & Chun, 1997; McCarthy, Puce, Gore, & Allison, 1997), which allow researchers to compare between blood oxygen level–dependent, or BOLD, responses to faces and to comparatively complex objects (e.g., houses). In this way, localizer studies and single-unit recording studies share the same fundamental logic—subjects are presented with stimuli from various categories (e.g., faces, body parts, objects, scrambled faces, etc.), and subsequently, researchers test for units (areas of the brain or neurons, respectively) that respond preferentially to a given category.

Despite relatively similar approaches in terms of tasks and stimuli, neuroimaging studies employing functional localizers consistently identify only regions in posterior cortical areas, including the fusiform face area (FFA; Kanwisher et al., 1997; McCarthy et al., 1997; Tong, Nakayama, Moscovitch, Weinrib, & Kanwisher, 2000), occipital face area (OFA; Gauthier, Skudlarski, Gore, & Anderson, 2000; Puce, Allison, Asgari, Gore, & McCarthy, 1996), and posterior superior temporal sulcus (pSTS; Allison, Puce, & McCarthy, 2000; Puce et al., 1996). Taken together, the FFA, OFA, and pSTS are considered to comprise a network of core regions involved in the perception of faces (Haxby & Gobbini, 2012; Haxby, Hoffman, & Gobbini, 2000; Said, Haxby, & Todorov, 2011).

However, face-localizer studies do not typically classify the amygdala as a face-selective region. When considered in the context of regions involved in face perception, the amygdala is occasionally referenced as a supporting node in the extended face network (Haxby et al., 2000; Haxby & Gobbini, 2012), tasked with processing emotional expressions. However, a number of human neuroimaging experiments have reported increased amygdala response to neutral faces (for example, Dubois et al., 1999; Fitzgerald, Angstadt, Jelsone, Nathan, & Phan, 2006; Johnstone et al., 2005; Kesler-West et al., 2001; van der Gaag, Minderaa, & Keysers, 2007; Wright & Liu, 2006). While some researchers have argued that the "extended" aspects of the face-processing network deserve more attention and prominence (Ishai, 2008), this perspective has faced its share of criticism (Wiggett & Downing, 2008).

Three related issues likely contribute to this divergence between the neuroimaging and the neurophysiological literatures with respect to the amygdala's face-selectivity: (1) measurement limitations of fMRI, (2) low statistical power, and (3) theoretical biases (see Todorov, 2012). First, the amygdala is subject to a decreased signal-to-noise ratio relative to cortical regions due to its location and small size, and as a result is comparatively difficult to image (LaBar, Gitelman, Mesulam, & Parrish, 2001; Zald, 2003). Second, traditional practices associated with the functional localizer approach, namely, relying on individual brain analysis, result in decreased statistical power, which may hamper researchers' ability to isolate face-selective responses in the amygdala. Given single-unit recording work suggesting that only about 10% of neurons in the amygdala display face-selective properties (Fried et al., 1997; Kreiman et al., 2000; Rutishauser et al., 2011), it is possible that these responses would not be detectable on the individual subject level with fMRI, especially when stringent statistical thresholds are imposed. The tendency not to perform group analyses of localizer data (an effort aimed at avoiding cortical misalignment) only further reduces statistical power. Finally, it is possible that overarching theoretical biases have contributed to the problem. Vision scientists and emotion researchers' interest in posterior cortical regions and the amygdala, respectively, while grounded in previous research, may have resulted in a narrow view of the amygdala's function.

In an effort to overcome these challenges, we performed a large-scale, whole-brain group analysis of functional localizer data (Mende-Siedlecki, Verosky, Turk-Browne, & Todorov, 2013). A total of 215 participants were recruited to 10 separate studies employing face localizers, which, in turn, used a variety of stimuli and behavioral tasks. We reasoned that this large sample size would afford us adequate statistical power necessary to counteract the limitations of fMRI.[1]

To begin, we conducted a simple contrast between responses to faces and control stimuli. We found robust face-selective responses in the amygdala, as well as in the posterior network of cortical regions observed in prior studies—including FFA, right pSTS, and OFA (figure 66.1A, B, C). This analysis also revealed face-selective responses in right dorsolateral prefrontal cortex (dlPFC) and superior colliculus (SC). We performed an additional conjunction analysis across the set of 10 studies that indicated that face-selective responses in bilateral amygdala, bilateral FFA, and right pSTS were robust with respect to task differences.

Our next step was to validate our initial findings using the NeuroSynth platform (http://neurosynth.org; Yarkoni, Poldrack, Nichols, Van Essen, & Wager, 2011), which draws upon a database of more than 4,000 extant neuroimaging studies to simulate interactive brain maps for specific search terms (i.e., "face"). NeuroSynth offers both forward inference and reverse-inverse capabilities. While the former technique tests the likelihood of activation in a region given the term—that is, "face"—the latter tests the likelihood that a term is used in a study given the presence of activation. We replicated the majority of regions observed in our initial group analysis: bilateral amygdala, bilateral FFA, and right pSTS, using both forward inference (figure 66.1D, E, F) and reverse inference techniques (figure 66.1G, H, I).

Subsequently, we tested whether face-selective responses in the amygdala are detectable in individual participants. Indeed, we were able to observe face-selective regions in the amygdala in a majority of participants (86.5% in right amygdala, 84.2% in left amygdala, at $p < .05$), but these responses were especially sensitive to increases in the strictness of statistical thresholding, much more so than in FFA. This difference may be a direct consequence of decreased signal-to-fluctuating-noise ratios in the amygdala compared to posterior cortical regions. To ascertain that the face-selective regions in the amygdala were not an artifact of liberal statistical thresholding, we compared individual face-selective responses in the amygdala to those extracted from a control region defined anatomically in A1. We confirmed that face-selective responses in the amygdala were more extensive and stronger than those in the A1 control region.

Complimenting these standard univariate approaches, we conducted a separate set of multivariate analyses in order to test the reliability of responses in face-selective regions over time, as well as the covariance of these reliability scores across face-selective regions. Since a little more than half of our data came from participants in multirun versions of the face-localizer task ($N = 119$), we focused just on these participants and tested whether the multivoxel patterns of face-selective responses were reliably correlated across runs. We first conducted a group analysis of the first-run data for all multirun participants, which yielded a similar set of face-selective regions of interest (ROIs) as observed in the full group analysis, including bilateral FFA,

[1]Here, we present data specific to face-selective responses in the amygdala. A more detailed overview of the methods, analyses, and results of this investigation—including those related to other regions deemed to be face-selective by these analyses—can be found in Mende-Siedlecki, Verosky, Turk-Browne, and Todorov (2013).

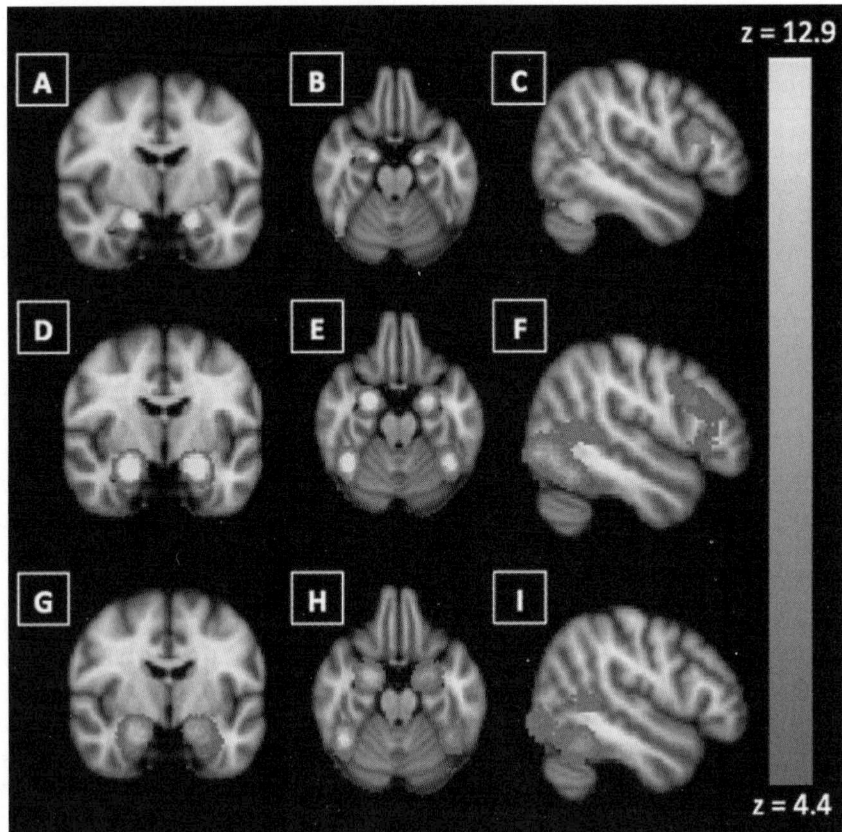

FIGURE 66.1 Group analysis of functional localizer data from 215 participants, with NeuroSynth validation. Panels A through C depict brain regions showing greater BOLD responses to face vs. control stimuli ($p < .05$, corrected), including bilateral amygdala. We used NeuroSynth—a meta-analytic tool capable of synthesizing results from a database of neuroimaging studies—to externally validate these data. Panels D through F represent the results of a forward inference analysis, showing activation across the NeuroSynth database for the term "face." These maps indicate the consistency of activation for the term. Panels G through I show the results of a reverse inference analysis. These maps indicate the relative selectivity of activation for the term (see Yarkoni et al., 2011, for details). (This figure originally appeared in Mende-Siedlecki, Verosky, Turk-Browne, & Todorov, 2013.) (See color plate 55.)

bilateral amygdala, and right pSTS. Next, for each ROI and for each participant, we extracted voxel-by-voxel parameter estimates from the Faces > Controls contrast, for both runs of the localizer task, and computed the Pearson correlation between the two runs. On average, across the 119 multirun participants, face-selective responses in the amygdala were reliable over time, and this reliability was significantly greater than zero; the average cross-run reliability was also computed for an anatomically defined control region in A1.

Moving forward, we sought to explore systematic covariances of these multivoxel reliability correlations across the network of face-selective ROIs. We observed that the right amygdala and right FFA were most highly intercorrelated with those of other ROIs, based upon correlations of cross-run reliability scores. Specifically, reliability in right amygdala correlated with left amygdala, right FFA, left FFA, right pSTS, and SC. A principle components analysis identified two distinct sources of variance in the reliability of face-selective responses. On one hand, cortical regions (right and left FFA, right dlPFC) loaded strongly onto the first principal component, while on the other hand, subcortical regions (right and left amygdala, SC) loaded strongly onto the second principal component. (We note that the right pSTS loaded equally onto both principal components, with relatively weak strength.) These first two components accounted for approximately 50% of the total variance.

Finally, we performed a series of psychophysical interaction (PPI) analyses aimed at isolating face-specific patterns of functional connectivity between our face-selective ROIs and other brain areas. We placed seeds in right and left amygdala and observed enhanced connectivity with bilateral fusiform, primary visual cortex, and right dlPFC during the presentation of faces, as compared to control stimuli.

Conclusion

The data described above provide evidence for robust face-selective responses in the human amygdala, which are reliable over time, consistent over multiple tasks and stimuli, and identifiable in individual participants. Taken together, this set of analyses indicates that there are regions in the amygdala that are face-selective. Critically, we arrived at this conclusion based on the same criteria that are used to classify the posterior cortical regions (FFA, OFA, and pSTS) as being face-selective. While we do not claim that the amygdala's primary function is the processing of faces, we argue that the robustness and reliability of the face-selective responses detected in the amygdala merit consideration of the amygdala as a core node of the face-processing network.

These results are highly consistent with recent neuroimaging studies implicating the amygdala in general face processing (Engell & McCarthy, 2013; Rossion, Hanseeuw, & Dricot, 2012), beyond the perception of emotional expressions. Rossion and colleagues (2012) sought to identify regions involved in face processing by performing a large-scale factorial functional localizer ($N = 40$), where participants performed an N-back task on images of faces, cars, scrambled faces, and scrambled cars. A conservatively thresholded conjunction between a Faces > Cars contrast and a Faces > Scrambled Faces contrast yielded robust activity in right and left amygdala, as well as FFA, OFA, and pSTS. In a related vein, Engell and McCarthy (2013) set out to examine overlap between regions involved in the processing of faces and biological motion. Large-scale analyses of localizer tasks in which participants passively viewed faces and houses ($N = 124$), as well as faces and scenes ($N = 79$) both identified the right and left amygdala as responding preferentially to face stimuli.

Despite this strong convergence, these results stop short of defining the amygdala's specific role in face processing. Moreover, these data are unlikely to determine conclusively whether face-selective responses observed in the amygdala are reflective of the visual properties of faces, attentional changes associated with faces, the inherent social value of faces, or some combination of these possibilities. That said, the amygdala's connectivity—with both cortical and subcortical structures—offers important context regarding its contributions to face processing.

While correlational in nature, the results of the principle components analysis described above suggest an intriguing dissociation between cortical and subcortical structures. This implication is consistent with prior literature proposing two distinct streams of face processing (de Gelder, van Honk, & Tamietto, 2011; Garrido, Barnes, Sahani, & Dolan, 2012; Johnson, 2005). By and large, this previous work discusses the amygdala's role in face processing in the context of early detection of environmental threats—specifically, fearful faces (Cecere, Bertini, & Làdavas, 2013; Jiang & He, 2006; Lerner et al., 2012; Pasley, Mayes, & Schultz, 2004; Pessoa, 2005; Williams et al., 2006). In this framework, information traveling along the subcortical route is transmitted from the retina to the SC, and then via the pulvinar to the amygdala.

However, recent research suggests a broader role, beyond fear, threat, or negative affect, whereby the amygdala is acting to facilitate rapid assessment of behaviorally relevant sensory input (Garrido et al., 2012). For example, Santos and colleagues (Santos, Mier, Kirsch, & Meyer-Lindenberg, 2011) demonstrated that faces that are salient elicit amygdala activity, regardless of emotional expression or threat. In this study, participants viewed 3×3 arrays of faces and were directed to detect target faces (happy expression, angry expression, blue mouth and eyebrows, or red mouth and eyebrows), which appeared on 50% of the trials. Contrasting trials containing targets and trials with no targets yielded robust activity in right and left amygdala. At the same time, the amygdala activity did not differentiate between the various target conditions. Moreover, Santos and colleagues found that amygdala activity predicted target detection. These results are consistent with an ongoing shift away from initial characterizations of the amygdala as purely a "fear module" and toward a broader role of detecting situationally or motivationally salient stimuli (Adolphs, 2010; Cunningham & Brosch, 2012; Cunningham, van Bavel, & Johnsen, 2008; Sander, Grafman, & Zalla, 2003; Todorov, 2012)

Work on the amygdala's connectivity—both anatomical and functional—offers additional insight into the amygdala's influence on face perception. In our PPI analysis, we observed that both right and left amygdala seeds showed a pattern of face-specific connectivity with bilateral fusiform gyri, primary visual cortex, and right dlPFC. This result is highly consistent with anatomical work in nonhuman primates, which observes both reciprocal, long-range connections between amygdala and IT cortex, along with back-projections to striate and extrastriate cortex (Amaral, Behniea, & Kelly, 2003; Amaral, Price, Pitkänen, & Carmichael, 1992). Furthermore, our PPI analysis is in line with diffusion tensor imaging work in humans that also demonstrates strong connectivity between the amygdala and primary visual areas (Avidan, Hadj-Bouziane, Liu, Ungerleider, &

Behrmann, 2013; Catani, Jones, Donato, & Ffytche, 2003; Gschwind, Pourtois, Schwartz, van de Ville, & Vuilleumier, 2012; Pugliese et al., 2009). To exert its excitatory effects on sensory regions in the cortex, the amygdala transmits through the nucleus basalis of Meynert, which maintains direct cholinergic projections to the cortex (Kapp, Whalen, Supple, & Pascoe, 1992; Whalen, 1998). Taken together, this profile of connectivity suggests that the amygdala is uniquely suited to rapidly process faces that are in some way behaviorally relevant and subsequently decrease the sensory threshold in cortical regions (in this case, primary visual areas), accelerating more fine-grained aspects of face processing.

In identifying the amygdala's precise contributions to face processing, future work must inevitably confront additional open questions—with likely interrelated answers. For example, standard fMRI's spatial resolution makes it difficult to determine which amygdala nuclei are responsible for the face-selective responses we observed. The amygdala comprises many individual subnuclei with varying functionality and connectivity (Aggleton, 2000; Freese & Amaral, 2009), and with dissociable response patterns in the context of face processing (Hoffman et al., 2007; Kim, Somerville, Johnstone, Alexander, & Whalen, 2003; Whalen et al., 2001). Diffusion tensor imaging may potentially provide a more exact specification of the source of face-selective responses in the amygdala (Bach, Behrens, Garrido, Weiskop, & Dolan, 2011; Saygin, Osher, Augustinack, Fischl, & Gabrieli, 2011). Moreover, a precise account of which amygdala subnuclei contribute to face processing may help to reconcile the present data with work on the amygdala's role in processing emotional expressions (Morris et al., 1996; Whalen et al., 1998; Winston, O'Doherty, & Dolan, 2003; Yang et al., 2002) and social evaluations of faces (Said, Baron, & Todorov, 2009; Said, Dotsch, & Todorov, 2010; Todorov & Engell, 2008; Todorov, Said, Oosterhof, & Engell, 2011; Winston, O'Doherty, Kilner, Perrett, & Dolan, 2007).

Furthermore, future work must assess the degree to which face-selective responses in the amygdala differ from those in posterior cortical regions like FFA and pSTS. While our data categorize the amygdala as face-selective based on the same metrics and criteria by which we would classify the FFA or pSTS as face-selective, the patterns of neuronal selectivity underlying these responses may differ considerably. The primate literature provides evidence for this divergence. While face-selective neurons in the amygdala are likely to be found intermixed with other neurons that do not display face-selective properties and, moreover, may not even be selective for any class of visual stimuli (Leonard et al., 1985; Nakamura et al., 1992; Sanghera et al., 1979), the face-selective regions observed in posterior cortical areas seem to be much more homogenously composed of face-selective neurons (Tsao et al., 2006). A better understanding of the fundamental differences between cortical and subcortical aspects is critical for building a comprehensive neural model of face processing.

Convergent findings across multiple methods, paradigms, and species populations have robustly identified responses in the amygdala that are preferential for faces. Although additional investigations are needed to characterize the exact nature and purpose of the amygdala's contributions to face processing, there is ample evidence to suggest that the amygdala deserves consideration as a core face-selective region. The populations of face-selective neurons found within the amygdala are tasked with detecting and processing one of the most important stimuli we social creatures can perceive—the face.

REFERENCES

Adolphs, R. (2010). What does the amygdala contribute to social cognition? *Ann NY Acad Sci, 1191*, 42–61.

Aggleton, J. P. (2000). *The amygdala: A functional analysis.* Oxford, UK: Oxford University Press.

Allison, T., Puce, A., & McCarthy, G. (2000). Social perception from visual cues: Role of the STS region. *Trends Cogn Sci, 4*, 267–278.

Amaral, D. G., Behniea, H., & Kelly, J. L. (2003). Topographic organization of projections from the amygdala to the visual cortex in the macaque monkey. *J Neurosci, 118*, 1099–1120.

Amaral, D. G., Price, J. L., Pitkänen, A., & Carmichael, S. T. (1992). Anatomical organization of the primate amygdaloid complex. In J. P. Aggleton (Ed.), *The amygdala: Neurobiological aspects of emotion, memory, and mental dysfunction* (pp. 1–66). New York, NY: Wiley-Liss.

Avidan, G., Hadj-Bouziane, F., Liu, N., Ungerleider, L., & Behrmann, M. (2013). Selective dissociation between core and extended regions in the face processing network in congenital prosopagnosia. *Cereb Cortex.* Advance online publication. doi:10.1093/cercor/bht007

Bach, D. R., Behrens, T. E., Garrido, L., Weiskop, N., & Dolan, R. J. (2011). Deep and superficial amygdala nuclei projections revealed in vivo by probabilistic tractography. *J Neurosci, 31*, 618–623.

Bar, M., Neta, M., & Linz, H. (2006). Very first impressions. *Emotion, 6*, 269–278.

Bruce, C., Desimone, R., & Gross, C. G. (1981). Visual properties of neurons in a polysensory area in superior temporal sulcus of the macaque. *J Neurophysiol, 46*, 369–384.

Calder, A. J., Rhodes, G., Johnson, M. H., & Haxby, J. V. (2011). *The Oxford handbook of face perception.* Oxford, UK: Oxford University Press.

Catani, M., Jones, D. K., Donato, R., & Ffytche, D. H. (2003). Occipito-temporal connections in the human brain. *Brain, 126*, 2093–2107.

Cecere, R., Bertini, C., & Làdavas, E. (2013). Differential contribution of cortical and subcortical visual pathways to the implicit processing of emotional faces: A tDCS study. *J Neurosci, 33,* 6469–6475.

Cloutier, J., Mason, M. F., & Macrae, C. N. (2005). The perceptual determinants of person construal: Reopening the cognitive toolbox. *J Pers Soc Psychol, 88,* 885–894.

Cunningham, W. A., & Brosch, T. (2012). Motivational salience: Amygdala tuning from traits, needs, values, and goals. *Curr Dir Psychol Sci, 21,* 54–59.

Cunningham, W. A., van Bavel, J. J., & Johnsen, I. R. (2008). Affective flexibility: Evaluative processing goals shape amygdala activity. *Psychol Sci, 19,* 152–160.

De Gelder, B., van Honk, J., & Tamietto, M. (2011). Emotion in the brain: Of low roads, high roads and roads less travelled. *Nat Rev Neurosci, 15,* 425.

Desimone, R. (1991). Face-selective cells in the temporal cortex of monkeys. *J Cogn Neurosci, 3,* 1–8.

Desimone, R., Albright, T. D., Gross, C. G., & Bruce, C. (1984). Stimulus-selective properties of inferior temporal neurons in the macaque. *J Neurosci, 4,* 2051–2062.

Dubois, S., Rossion, B., Schiltz, C., Bodart, J. M., Michel, C., Bruyer, R., & Crommelinck, M. (1999). Effect of familiarity on the processing of human faces. *NeuroImage, 9,* 278–289.

Eifuku, S., de Souza, W. C., Tamura, R., Nishijo, H., & Ono, T. (2004). Neuronal correlates of face identification in the monkey anterior temporal cortical areas. *J Neurophysiol, 91,* 358–371.

Engell, A. D., & McCarthy, G. (2013). Probabilistic atlases for face and biological motion perception: An analysis of their reliability and overlap. *NeuroImage, 74,* 140–151.

Esteves, F., & Öhman, A. (1993). Masking the face: Recognition of emotional facial expressions as a function of the parameters of backward masking. *Scand J Psychol, 34,* 1–18.

Fitzgerald, D. A., Angstadt, M., Jelsone, L. M., Nathan, P. J., & Phan, K. L. (2006). Beyond threat: Amygdala reactivity across multiple expressions of facial affect. *NeuroImage, 30,* 1441–1448.

Freese, J. L., & Amaral, D. G. (2009). Neuroanatomy of the primate amygdala. In P. J. Whalen & E. A. Phelps (Eds.), *The human amygdala* (pp. 3–42). New York, NY: Guilford.

Freiwald, W., & Tsao, D. (2012). Taking apart the neural machinery of face processing. In A. Calder, G. Rhodes, M. Johnson, J. Haxby, & J. Keane (Eds.), *Handbook of face perception* (pp. 707–719). Oxford, UK: Oxford University Press.

Fried, I., Cameron, K. A., Yashar, S., Fong, R., & Morrow, J. W. (2002). Inhibitory and excitatory responses of single neurons in the human medial temporal lobe during recognition of faces and objects. *Cereb Cortex, 12,* 575–584.

Fried, I., MacDonald, K. A., & Wilson, C. (1997). Single neuron activity in human hippocampus and amygdala during recognition of faces and objects. *Neuron, 18,* 753–765.

Garrido, M. I., Barnes, G. R., Sahani, M., & Dolan, R. J. (2012). Functional evidence for a dual route to the amygdala. *Curr Biol, 22,* 129–134.

Gauthier, I., Skudlarski, P., Gore, J. C., & Anderson, A. W. (2000). Expertise for cars and birds recruits brain areas involved in face recognition. *Nat Neurosci, 3,* 191–197.

Gothard, K. M., Battaglia, F. P., Erickson, C. A., Spitler, K. M., & Amaral, D. G. (2007). Neural responses to facial expression and face identity in the monkey amygdala. *J Neurophysiol, 97,* 1671–1683.

Grill-Spector, K., & Kanwisher, N. (2005). Visual recognition: As soon as you know it is there, you know what it is. *Psychol Sci, 16,* 152–160.

Gross, C. G. (1994). How inferior temporal cortex became a visual area. *Cereb Cortex, 5,* 455–469.

Gross, C. G. (2008). Single neuron studies of inferior temporal cortex. *Neuropsychologia, 46*(3), 841–852.

Gross, C. G., Rocha-Miranda, C. E., & Bender, D. B. (1972). Visual properties of neurons in inferotemporal cortex of the macaque. *J Neurophysiol, 35,* 96–111.

Gschwind, M., Pourtois, G., Schwartz, S., van de Ville, D., & Vuilleumier, P. (2012). White-matter connectivity between face-responsive regions in the human brain. *Cereb Cortex, 22,* 1564–1576.

Hasselmo, M. E., Rolls, E. T., & Baylis, G. C. (1989). The role of expression and identity in the face-selective responses of neurons in the temporal visual cortex of the monkey. *Behav Brain Res, 32,* 203–218.

Haxby, J. V., & Gobbini, M. I. (2012). Distributed neural systems for face perception. In A. Calder, G. Rhodes, M. Johnson, J. Haxby, & J. Keane (Eds.), *Handbook of face perception* (pp. 93–111). Oxford, UK: Oxford University Press.

Haxby, J. V., Hoffman, E. A., & Gobbini, M. I. (2000). The distributed human neural system for face perception. *Trends Cogn Sci, 4,* 223–233.

Herzog, A. G., & van Hoesen, G. W. (1976). Temporal neocortical afferent connections to the amygdala in the rhesus monkey. *Brain Res, 115,* 57–69.

Hoffman, K. L., Gothard, K. M., Schmid, M. C., & Logothetis, N. K. (2007). Facial-expression and gaze-selective responses in the monkey amygdala. *Curr Biol, 17,* 766–772.

Ishai, A. (2008). Let's face it: It's a cortical network. *NeuroImage, 40*(2), 415–419.

Jiang, Y., & He, S. (2006). Cortical responses to invisible faces: Dissociating subsystems for facial-information processing. *Curr Biol, 16*(20), 2023–2029.

Johnson, M. H. (2005). Subcortical face processing. *Nat Rev Neurosci, 6,* 766–774.

Johnstone, T., Somerville, L. H., Alexander, A. L., Oakes, T. R., Davidson, R. J., Kalin, N. H., & Whalen, P. J. (2005). Stability of amygdala BOLD response to fearful faces over multiple scan sessions. *NeuroImage, 25,* 1112–1123.

Jones, E. G., & Powell, T. P. S. (1970). An anatomical study of converging sensory pathways within the cerebral cortex of monkey. *Brain Res, 93,* 793–820.

Kanwisher, N., McDermott, J., & Chun, M. M. (1997). The fusiform face area: A module in human extrastriate cortex specialized for face perception. *J Neurosci, 17,* 4302–4311.

Kapp, B. S., Whalen, P. J., Supple, W. F., & Pascoe, J. P. (1992). Amygdaloid contributions to conditioned arousal and sensory information processing. In J. P. Aggleton (Ed.), *The amygdala: Neurobiological aspects of emotion, memory, and mental dysfunction* (pp. 229–254). New York, NY: Wiley-Liss.

Kesler-West, M. L., Andersen, A. H., Smith, C. D., Avison, M. J., Davis, C. E., Kryscio, R. J., & Blonder, L. X. (2001). Neural substrates of facial emotion processing using fMRI. *Cogn Brain Res, 11,* 213–226.

Kiani, R., Esteky, H., & Tanaka, K. (2005). Differences in onset latency of macaque inferotemporal neural responses to primate and non-primate faces. *J Neurophysiol, 94,* 1587–1596.

Kim, H., Somerville, L. H., Johnstone, T., Alexander, A., & Whalen, P. J. (2003). Inverse amygdala and medial prefrontal cortex responses to surprised faces. *NeuroReport, 14,* 2317–2322.

Kreiman, G., Koch, C., & Fried, I. (2000). Category-specific visual responses of single neurons in the human medial temporal lobe. *Nat Neurosci, 3,* 946–953.

LaBar, K. S., Gitelman, D. R., Mesulam, M. M., & Parrish, T. B. (2001). Impact of signal-to-noise on functional MRI of the human amygdala. *NeuroReport, 12,* 3461–3464.

Leonard, C. M., Rolls, E. T., Wilson, F. A. W., & Baylis, G. C. (1985). Neurons in the amygdala of the monkey with responses selective for faces. *Behav Brain Res, 15,* 159–176.

Lerner, Y., Singer, N., Gonen, T., Weintraub, Y., Cohen, O., Rubin, N., & Ungerleider, L. G. (2012). Feeling without seeing? Engagement of ventral, but not dorsal, amygdala during unaware exposure to emotional faces. *J Cogn Neurosci, 24,* 531–542.

Locher, P., Unger, R., Sociedade, P., & Wahl, J. (1993). At first glance: Accessibility of the physical attractiveness stereotype. *Sex Roles, 28,* 729–743.

Logothetis, N. K., Guggenberger, H., Peled, S., & Pauls, J. (1999). Functional imaging of the monkey brain. *Nat Neurosci, 2,* 555–562.

Martin, D., & Macrae, C. N. (2007). A face with a cue: Exploring the inevitability of person categorization. *Eur J Soc Psychol, 37,* 806–816.

McCarthy, G., Puce, A., Gore, J. C., & Allison, T. (1997). Face-specific processing in the human fusiform gyrus. *J Cogn Neurosci, 9,* 605–610.

Mende-Siedlecki, P., Verosky, S. C., Turk-Browne, N. B., & Todorov, A. (2013). Robust selectivity for faces in the human amygdala in the absence of expressions. *J Cogn Neurosci, 25*(12), 2086–2106.

Mormann, F., Kornblith, S., Quiroga, R. Q., Kraskov, A., Cerf, M., Fried, I., & Koch, C. (2008). Latency and selectivity of single neurons indicate hierarchical processing in the human medial temporal lobe. *J Neurosci, 28,* 8865–8872.

Morris, J. S., Frith, C. D., Perrett, D. I., Rowland, D., Young, A. W., Calder, A. J., & Dolan, R. J. (1996). A differential neural response in the human amygdala to fearful and happy facial expressions. *Nature, 383,* 812–815.

Nakamura, K., Mikami, A., & Kubota, K. (1992). Activity of single neurons in the monkey amygdala during performance of a visual discrimination task. *J Neurophysiol, 67,* 1447–1463.

Ojemann, J. G., Ojemann, G. A., & Lettich, E. (1992). Neuronal activity related to faces and matching in human right nondominant temporal cortex. *Brain, 115,* 1–13.

Olson, I. R., & Marshuetz, C. (2005). Facial attractiveness is appraised in a glance. *Emotion, 5,* 498–502.

Pasley, B. N., Mayes, L. C., & Schultz, R. T. (2004). Subcortical discrimination of unperceived objects during binocular rivalry. *Neuron, 42,* 163–172.

Perrett, D. I., Rolls, E. T., & Caan, W. (1982). Visual neurons responsive to faces in the monkey temporal cortex. *Exp Brain Res, 47,* 329–342.

Perrett, D. I., Smith, P. A. J., Potter, D. D., Mistlin, A. J., Head, A. S., Milner, A. D., & Jeeves, M. A. (1984). Neurons responsive to faces in the temporal cortex: Studies of functional organization, sensitivity to identity and relation to perception. *Hum Neurobiol, 3,* 197–208.

Pessoa, L. (2005). To what extent are emotional stimuli processed without attention and awareness? *Curr Opin Neurobiol, 15,* 188–196.

Puce, A., Allison, T., Asgari, M., Gore, J. C., & McCarthy, G. (1996). Differential sensitivity of human visual cortex to faces, letterstrings, and textures: A functional magnetic resonance imaging study. *J Neurosci, 16,* 5205–5215.

Pugliese, L., Catani, M., Ameis, S., Dell'Acqua, F., Thiebaut de Schotten, M., Murphy, C., ... Murphy, D. G. (2009). The anatomy of extended limbic pathways in Asperger syndrome: A preliminary diffusion tensor imaging tractography study. *NeuroImage, 47,* 427–434.

Quiroga, R. Q., Reddy, L., Kreiman, G., Koch, C., & Fried, I. (2005). Invariant visual representation by single neurons in the human brain. *Nature, 435,* 1102–1107.

Rajimehr, R., Young, J. C., & Tootell, R. B. H. (2009). An anterior temporal face patch in human cortex, predicted by macaque maps. *Proc Natl Acad Sci USA, 106,* 1995–2000.

Rolls, E. T. (1984). Neurons in the cortex of the temporal lobe and in the amygdala of the monkey with responses selective for faces. *Hum Neurobiol, 3,* 209–222.

Rossion, B., Hanseeuw, B., & Dricot, L. (2012). Defining face perception areas in the human brain: A large-scale factorial fMRI face localizer analysis. *Brain Cogn, 79,* 138–157.

Rutishauser, U., Tudusciuc, O., Neumann, D., Mamelak, A. N., Heller, A. C., Ross, I. B., ... Adolphs, R. (2011). Single-unit responses selective for whole faces in the human amygdala. *Curr Biol, 21,* 1654–1660.

Said, C. P., Baron, S., & Todorov, A. (2009). Nonlinear amygdala response to face trustworthiness: Contributions of high and low spatial frequency information. *J Cogn Neurosci, 21,* 519–528.

Said, C. P., Dotsch, R., & Todorov, A. (2010). The amygdala and FFA track both social and non-social face dimensions. *Neuropsychologia, 48,* 3596–3605.

Said, C. P., Haxby, J. V., & Todorov, A. (2011). Brain systems for assessing the affective value of faces. *Philos Trans R Soc Lond B Biol Sci, 336,* 1660–1670.

Sander, D., Grafman, J., & Zalla, T. (2003). The human amygdala: An evolved system for relevance detection. *Rev Neurosci, 14,* 303–316.

Sanghera, M. F., Rolls, E. T., & Roper-Hall, A. (1979). Visual response of neurons in the dorsolateral amygdala of the alert monkey. *Exp Neurol, 63,* 61–62.

Santos, A., Mier, D., Kirsch, P., & Meyer-Lindenberg, A. (2011). Evidence for a general face salience signal in human amygdala. *NeuroImage, 14,* 3111–3116.

Sato, W., Kochiyama, T., Uono, S., Matsuda, K., Usui, K., Inoue, Y., & Toichi, M. (2012). Temporal profile of amygdala gamma oscillations in response to faces. *J Cogn Neurosci, 24,* 1420–1433.

Saygin, Z. M., Osher, D. E., Augustinack, J., Fischl, B., & Gabrieli, J. D. (2011). Connectivity-based segmentation of human amygdala nuclei using probabilistic tractography. *NeuroImage, 56,* 1353–1361.

Seeck, M., Schomer, D., Mainwaring, N., Ives, J., Dubuisson, D., Blume, H., ... Mesulam, M. M. (1995). Selectively distributed processing of visual object recognition in the temporal and frontal lobes of the human brain. *Ann Neurol, 37,* 538–545.

STEINMETZ, P. N. (2008). Alternate task inhibits single-neuron category-selective responses in the human hippocampus while preserving selectivity in the amygdala. *J Cogn Neurosci, 21,* 347–358.

TODOROV, A. (2012). The role of the amygdala in face perception and evaluation. *Motiv Emotion, 36,* 16–26.

TODOROV, A., & ENGELL, A. (2008). The role of the amygdala in implicit evaluation of emotionally neutral faces. *Soc Cogn Affect Neurosci, 3,* 303–312.

TODOROV, A., MENDE-SIEDLECKI, P., & DOTSCH, R. (2013). Social judgments from faces. *Curr Opin Neurobiol, 23*(3), 373–380.

TODOROV, A., PAKRASHI, M., & OOSTERHOF, N. N. (2009). Evaluating faces on trustworthiness after minimal time exposure. *Soc Cognition, 27,* 813–833.

TODOROV, A., SAID, C. P., OOSTERHOF, N. N., & ENGELL, A. D. (2011). Task-invariant brain responses to the social value of faces. *J Cogn Neurosci, 23,* 2766–2781.

TONG, F., NAKAYAMA, K., MOSCOVITCH, M., WEINRIB, O., & KANWISHER, N. (2000). Response properties of human fusiform face area. *Cogn Neuropsychol, 12,* 257–279.

TSAO, D. Y., & FREIWALD, W. A. (2006). What's so special about the average face? *Trends Cogn Sci, 10,* 391–393.

TSAO, D. Y., FREIWALD, W. A., KNUTSEN, T. A., MANDEVILLE, J. B., & TOOTELL, R. B. (2003). Faces and objects in macaque cerebral cortex. *Nat Neurosci, 6,* 989–995.

TSAO, D. Y., FREIWALD, W. A., TOOTELL, R. B. H., & LIVINGSTONE, M. S. (2006). A cortical region consisting entirely of face-selective cells. *Science, 311,* 670–674.

VAN DER GAAG, C., MINDERAA, R. B., & KEYSERS, C. (2007). The BOLD signal in the amygdala does not differentiate between dynamic facial expressions. *Soc Cogn Affect Neurosci, 2,* 93–103.

VISKONTAS, I. V., QUIROGA, R. Q., & FRIED, I. (2009). Human medial temporal lobe neurons respond preferentially to personally relevant images. *Proc Natl Acad Sci USA, 106,* 21329–21334.

WEINER, K. S., & GRILL-SPECTOR, K. (2010). Sparsely-distributed organization of face and limb activations in human ventral temporal cortex. *NeuroImage, 52,* 1559–1573.

WEINER, K. S., & GRILL-SPECTOR, K. (2012). The improbable simplicity of the fusiform face area. *Trends Cogn Sci, 16,* 251–254.

WHALEN, P. J. (1998). Fear, vigilance, and ambiguity: Initial neuroimaging studies of the human amygdala. *Curr Dir Psychol Sci, 7,* 177–188.

WHALEN, P. J., RAUCH, S. L., ETCOFF, N. L., MCINERNEY, S. C., LEE, M., & JENIKE, M. A. (1998). Masked presentations of emotional facial expressions modulate amygdala activity without explicit knowledge. *J Neurosci, 18,* 411–418.

WHALEN, P. J., SHIN, L. M., MCINERNEY, S. C., FISCHER, H., WRIGHT, C. I., & RAUCH, S. L. (2001). A functional MRI study of human amygdala responses to facial expressions of fear vs. anger. *Emotion, 1,* 70–83.

WIGGETT, A. J., & DOWNING, P. E. (2008). The face network: Overextended? (Comment on: "Let's face it: It's a cortical network" by Alumit Ishai). *NeuroImage, 40,* 420–422.

WILLIAMS, L. M., DAS, P., LIDDELL, B. J., KEMP, A. H., RENNIE, C. J., & GORDON, E. (2006). Mode of functional connectivity in amygdala pathways dissociates level of awareness for signals of fear. *J Neurosci, 6,* 9264–9271.

WILLIS, J., & TODOROV, A. (2006). First impressions: Making up your mind after 100 ms exposure to a face. *Psychol Sci, 17,* 592–598.

WILSON, F. A. W., & ROLLS, E. T. (1993). The effects of novelty and familiarity on neuronal activity recorded in the amygdala of monkeys performing recognition memory tasks. *Exp Brain Res, 93,* 367–382.

WINSTON, J., O'DOHERTY, J., & DOLAN, R. J. (2003). Common and distinct neural responses during direct and incidental processing of multiple facial emotions. *NeuroImage, 20,* 84–97.

WINSTON, J. S., O'DOHERTY, J., KILNER, J. M., PERRETT, D. I., & DOLAN, R. J. (2007). Brain systems for assessing facial attractiveness, *Neuropsychologia, 7,* 195–206.

WRIGHT, P., & LIU, Y. (2006). Neutral faces activate the amygdala during identity matching. *NeuroImage, 29,* 628–636.

YANG, T. T., MENON, V., ELIEZ, S., BLASEY, C., WHITE, C. D., REID, A. J., ... REISS, A. L. (2002). Amygdalar activation associated with positive and negative facial expressions. *NeuroReport, 13,* 1737–1741.

YARKONI, T., POLDRACK, R. A., NICHOLS, T. E., VAN ESSEN, D. C., & WAGER, T. D. (2011). Large-scale automated synthesis of human functional neuroimaging data. *Nat Methods, 8,* 665–670.

YIP, A., & SINHA, P. (2002). Role of color in face recognition. *Perception, 31,* 995–1003.

ZALD, D. H. (2003). The human amygdala and the emotional evaluation of sensory stimuli. *Brain Res Rev, 41,* 88–123.

67 The Cultural Neuroscience of Human Perception

NALINI AMBADY AND JONATHAN B. FREEMAN

ABSTRACT Culture and the brain were once thought of as mutually exclusive views on behavioral variation—an idea that is changing with the emerging field of cultural neuroscience. In this chapter, we discuss recent research examining the interplay of cultural and genetic factors on the neural bases of human perception. We conclude that cultural experience readily impacts basic mechanisms underlying perception, ranging from lower-level nonsocial domains (e.g., attentional deployment and object perception) to higher-order social domains (e.g., emotion recognition and theory of mind). More broadly, we discuss the promise and pitfalls of a cultural neuroscience approach to psychological processes and explain how this multilevel approach can contribute to both cognitive neuroscience as well as to social and cultural psychology.

The cultural neuroscience of human perception

Humans are biological systems embedded in larger social systems. We see, think, and act in the context of others. We exist in cultural environments in which specific meanings, practices, and institutions organize our perceptual, cognitive, and behavioral tendencies. In spite of this fact, culture and the brain were once thought of as mutually exclusive views on variation in behavior and were subjects of study divided by relatively strict divisions across the natural and social sciences. Although it is clear now that culture and social experience may readily shape brain function, this was not always the case. Even with respect to more general, nonsocial experience, the broader field of cognitive neuroscience early on was surprised by such influences on the brain. In the early 1990s, Posner (1993, p. 674) remarked in *Science*: "If the neural systems used for a given task can change with 15 minutes of practice … how can we any longer separate organic structures from their experience in the organism's history?"

The answer, we now know, is simply that we cannot. Recent research examining the neural mechanisms underlying the effects of cultural experience on the brain and, conversely, examining how the brain gives rise to cultural experience has led to the emergence of a new field: cultural neuroscience (Ambady & Bharucha, 2009; Chiao & Ambady, 2007; Han & Northoff, 2008; Han et al., 2013; Kitayama & Park, 2010). Accumulating evidence for neuroplasticity and the coextension of culture and the brain has made clear that cultural and neural processes are not inherently distinct objects of inquiry; instead, they may richly interact. A complete and accurate understanding of behavioral variation requires the comprehensive study of these tandem processes.

Arguably, the human brain evolved to permit elaborate social behavior, particularly so that such behavior may be adapted to the social structures and patterns in which individuals find themselves. Cultural neuroscience is centered around the assumption of a bidirectional relationship between culture and the brain: the brain adapts to cultural processes, and cultural processes adapt to neural constraints (Ambady & Bharucha, 2009). As such, this burgeoning field aims to provide a fuller understanding of psychological phenomena at multiple levels of analysis.

Neuroplasticity

Progress in the emerging field of cultural neuroscience has been bolstered by accumulating evidence for neuroplasticity. The human brain has long been known to be intrinsically malleable, with environmental and experiential factors determining both its function and structure. Occipital regions centrally involved in vision, for example, can be recruited to process sounds in blind individuals (Gougoux et al., 2009), and primary auditory cortex can be co-opted to process visual stimuli in deaf individuals (Finney et al., 2001). Thus, in the face of impairments, the brain is able to flexibly reorganize itself. Even beyond recovering from impairments, recent evidence has documented the astounding flexibility of neural mechanisms to high-level social and cultural experience.

Consider a study examining the influences of juggling experience on neural activity. All participants lacked juggling experience; half were asked to teach themselves to be able to juggle for at least one minute continuously, and were given three months to practice. Following the three months, voxel-based morphometry revealed that those who had learned to juggle exhibited increased gray matter in two areas associated with visual and motor activity—the mid-temporal area and the posterior intraparietal sulcus. After an additional three months, during which the participants had stopped juggling, these regions' gray matter returned to original size (Draganski et al., 2004). In another study, expert male dancers underwent functional MRI (fMRI) while watching videos of various dance moves. When viewing dance moves that the experts had been trained to perform relative to moves they had not, the experts showed greater activation in regions associated with the mirror neuron system involved in the action simulation, including the premotor cortex, posterior intraparietal sulcus, and posterior superior temporal sulcus (STS; Calvo-Merino, Glaser, Grezes, Passingham, & Haggard, 2005).

Similar effects for experience and exposure on neural activation were found in a study on taxi drivers in London (Maguire et al., 2000). London taxi drivers undergo extensive training over a two- to four-year period that involves learning, among other things, the layout of 25,000 streets in the city in order to obtain an operating license. Using voxel-based morphometry, Maguire and colleagues found that the gray matter volume of the posterior hippocampus was enlarged and anterior hippocampus gray matter volume was reduced in London taxi cab drivers, when compared with an age-matched control group. Further, hippocampal volume was positively correlated with taxi-driving experience. The longer the taxi drivers had driven in London, the greater the posterior hippocampal gray matter volume and the more decreased the anterior gray matter volume. This suggested that environmental demands and, specifically, spatial navigation experience due to one's chosen occupation, dynamically shapes the structure of the hippocampus. Together, all these studies show that the brain continually adapts to the cultural environment and is plastic with respect to both its function and structure. Such work has been critical in providing a foundation for a neuroscience of culture.

Culture mapping and source analysis

There are two overarching goals of the emerging field of cultural neuroscience (Ambady & Bharucha, 2009). The first goal is for researchers to map out the differences and similarities across cultures at the behavioral and neural levels—that is, the goal of *cultural mapping*. The majority of past research in cultural neuroscience fits within the purview of this goal, seeking to understand the tuning of neural processes to the cultural environment. On the one hand, cultural mapping can show how the environments or cues of different cultures are processed differently by individuals from a given culture. Alternatively, culture mapping can show how the same environment or same cues are processed differently by individuals from different cultures. Culture mapping can also reveal how identical cues are processed differently across cultures.

Once such multilevel cultural convergences and divergences are mapped out, the second, perhaps more challenging, goal of cultural neuroscience is to determine the actual sources or causes of these differences and similarities—that is, the goal of *source analysis*. By and large, cultural neuroscience studies typically compare two different cultural groups (e.g., Americans and Chinese), with the assumption that behavioral and neural differences that arise are due to the wealth of cultural learning, experiences, and contexts that pervade the lives of those two groups. However, in some cases that assumption may be premature, as the cultural factors may covary with genetic factors.

Although the vast majority of genetic variation exists within populations (~93–95% of genetic variation; Rosenberg et al., 2002), there is variation between populations of different ancestral origins (~3–5% of genetic variation). Thus, it is possible that cognitive and neural processes associated with specific genes that vary in allelic frequency may show variable functioning across populations, independent of cultural influences. Indeed, more than 70% of genes are expressed at the neural level (Hariri, Drabant, & Weinberger, 2006). Thus, one major question in cultural neuroscience is the extent to which behavioral and neural differences across cultures arise due to cultural versus genetic factors, or their unique interaction (Chiao & Ambady, 2007; Way & Lieberman, 2010).

There a number of genes important to brain and behavior that show variation in allelic frequency across geographical regions, including the serotonin transporter (5-HTTLPR) and dopamine D4 receptor (DRD4) exon III polymorphisms. For example, evidence indicates that the S allele of 5-HTTLPR is associated with increased negative emotion and heightened anxiety, and the S allele is more prevalent in East Asian populations (e.g., 70–80% S carriers) relative to other nations (e.g., 50% or less S carriers). Recently, Chiao and Blizinsky (2010) examined the prevalence of 5-HTTLPR in different countries and correlated the relative proportion of short and long alleles in each population with a measure of individualism and collectivism for each country. Overall, Western societies tend to be characterized by independence and individualism, emphasizing individuals' goals and achievements. East Asian societies, on the other hand, tend to be more interdependent and collectivist, emphasizing relationships, and roles. Chiao and Blizinsky found that 5-HTTLPR S carriers were considerably more prevalent in collectivistic relative to individualistic populations. Counterintuitively, however, collectivist populations contain considerably fewer depressed individuals, while having an increased frequency of the 5-HTTLPR short allele linked to heightened anxiety and depression. Chiao and Blizinsky argued that this seeming paradox may be explained by cultural ideology and norms of collectivism that act as a buffer and provide social support in those populations. This interpretation is tentative but suggests that the unique interplay of cultural and genetic factors may have a profound impact on brain and behavior.

Human perception

Cultural psychologists have long documented that culture can exert a deep influence on the way individuals think and behave. Culture influences how we perceive ourselves and others (Markus & Kitayama, 1991) as well as how we perceive and interpret more basic, lower-level visual and perceptual cues. Some of the biggest strides in cultural neuroscience have come from the domain of human visual perception.

Two cultures whose social structure and practices differ considerably in such a way as to influence visual processing are Western culture and East Asian culture. As mentioned, Western societies tend to be more individualist and independent, whereas East Asian societies tend to be more collectivist and interdependent. These two different sociocultural systems are known to give rise to dissimilar patterns of cognition (Nisbett, Peng, Choi, & Norenzayan, 2001). Recent work has shown that these systems are also likely to influence visual attention to aspects of the environment (e.g., Kitayama, Duffy, Kawamura, & Larsen, 2003; Masuda & Nisbett, 2001). Specifically, practices and ideas in Western societies tend to require separating objects from their contexts and interpreting independent and absolute aspects of environmental stimuli (i.e., analytic thinking). Practices and ideas in East Asian societies, however, tend to require interpreting objects in conjunction with their context and understanding the relatedness among environmental stimuli (i.e., holistic thinking). In contrast, Western societies (emphasizing independence) place more value on salient objects and one's own relationship to those objects. This should lead to Westerners directing more attention to these, without as much concern for context. Indeed, East Asians are more likely to perceive objects and scenes as wholes and to pay attention to contextual information (referred to as holistic perception), whereas Westerners are more likely to perceive objects and scenes according to their distinct parts and ignore contextual information (referred to as analytic perception; Nisbett et al., 2001; Nisbett & Miyamoto, 2005).

Object Perception Overall, Americans engage in more analytic perception and Japanese engage in more holistic perception. The framed-line test (Kitayama et al., 2003) has been especially useful in demonstrating how these two cultures shape divergent patterns of visual perception and attentional deployment. In the framed-line test, participants are shown a square figure with a vertical line hanging from its top edge (but not spanning the entire height of the square), located in the horizontal center. After briefly inspecting this arrangement, participants are shown a new square figure of a different size. In the absolute condition, participants are asked to draw a line in this new square that is identical in absolute length to the vertical line previously seen. In the relative condition, however, they are asked to draw a line that has identical proportion to the context (i.e., the surrounding square frame) as that of the vertical line previously seen. Thus, performance in the absolute task depends on analytic processing of a salient stimulus and characteristics that are independent of context. Performance in the relative task, however, depends on holistic processing that includes the surrounding square frame, and the relationship between the salient stimulus and its context. Consistently, Americans perform better in the absolute task than in the relative task, whereas Japanese show the reverse pattern, performing better in the relative task than in the absolute task (Kitayama et al., 2003).

Building on these behavioral findings, Hedden and colleagues (2008) examined neural activity during this task using fMRI. Participants were asked to judge the size of a vertical line either incorporating (relative condition) or ignoring (absolute) contextual information, the surrounding square frame. Results revealed cultural variation in neural responses to the extent that distinct brain regions were recruited to perform the relative and absolute line judgment tasks in relation to the perceiver's culture. Participants recruited frontal and parietal regions associated with attentional control to a greater extent when engaged in a task that was incongruent with their cultural patterns. Thus, activity in frontoparietal regions increased when people of East Asian descent ignored contextual information and people of European descent incorporated contextual information during line size judgments. Moreover, the degree of activation during the incongruent relative to the congruent judgment task was negatively correlated with degree of individualism in people of European descent and the degree of acculturation in people of East Asian descent. Thus, the brain's attentional network was more strongly engaged by culturally nonpreferred perceptual judgments. This line of work also illustrates how cultural experience modulates the function of neural mechanisms in a sensitive, graded fashion.

Beyond somewhat simplistic line and frame stimuli, Gutchess, Welsh, Boduroglu, and Park (2006) converged on similar results in higher-level object perception. East Asian Americans and non-Asian Americans performed a task involving the recognition of complex pictures showing an object against a background while their neural responses were measured using fMRI. East Asian Americans and non-Asian Americans performed equally well but recruited distinct brain regions during the task. Non-Asian Americans showed more activation in the object-processing areas in the ventral visual cortex than did the East Asians, who showed more activation in the left occipital and fusiform areas associated with perceptual analysis. Goh and colleagues (2007) also found that elderly East Asian, Singaporean participants showed less of an fMRI adaptation response in object-processing areas compared to older Western adults. Westerners who were presented with images of an object showed reduced neural activation to the object with subsequent presentations, indicating that they had adapted to seeing the object. In contrast, East Asians continued to show an equally strong neural response during subsequent presentations of the same object, with all iterations showing a response as if they were seeing the object for the first time. These findings suggest that Westerners allocate greater attention to objects than do East Asians, whose attention may be directed elsewhere (such as to the background).

Thus, a perceiver's cultural background shapes the neural mechanisms underlying the perception of objects and their surrounding context in a manner adaptive for the values of one's culture. Culture equips individuals with different perceptual strategies that are evident at the behavioral and neural levels.

Emotion Perception Successfully reading others' emotions is important because they avail the perceiver with information about upcoming behaviors or environmental conditions. As others' facial expressions warn and ready perceivers for impending action, and because such actions are most likely to happen within one's culture, the emotions that are most ecologically relevant are those that are expressed by members of one's own culture (Weisbuch & Ambady, 2008). Indeed, it has been proposed for over two decades that one's cultural background may influence the recognition of others' emotions (Lutz & White, 1986). Thus, one question of interest to cultural neuroscientists is whether members of a given culture exhibit a selective ability to recognize the emotions of members of one's own culture. It is possible that acculturation leads to the unique tuning of the perceptual system to emotional expressions of other members of that same culture. Elfenbein and Ambady (2002b) conducted a meta-analysis of studies involving face emotion-recognition tasks across multiple cultures. Indeed, analysis of the results from these studies led to the conclusion that individuals are better at recognizing own-culture expressions relative to other-culture expressions, pointing to a robust cultural specificity in emotion recognition.

An early cultural neuroscience study examined the neural basis of this cultural specificity. American and Japanese participants were presented with American and Japanese faces exhibiting angry, fearful, happy, or neutral expressions. Both American and Japanese participants showed a significantly greater bilateral amygdala response to the perception of fear faces when posed by same-culture, ingroup members as compared to fear expressions posed by other-culture, outgroup members. Thus, American participants showed a stronger amygdala response to fearful American faces, and Japanese participants showed a stronger amygdala response to fearful Japanese faces. No significant differences were observed for anger, happy, or neutral expressions (Chiao et al., 2008). One possible explanation for these results is that a fear expression may have particular communicative value for ingroup members because it provides information about dangers in the environment that might be especially relevant to them. In addition, expressions of fear may be particularly valuable for eliciting the help of others (Marsh, Kozak, & Ambady, 2007). Recognizing fear signals from ingroup members might motivate helping behavior and contribute to the success and survival of the group. Thus, the amygdala exhibits selective responses to own-culture fear displays, providing a neural correlate of the cultural specificity in emotion recognition.

A more recent study found that culture influences neural responses to fear faces depending on the direction of gaze. In Western cultures, eye gaze is generally seen as a sign of respect (Argyle, 1976). A failure to make eye contact may therefore be seen as disingenuous. In East Asian cultures, however, direct eye contact can be perceived as impolite and inappropriate. In these cultures, averted eye gaze, especially downward shifts in gaze, is generally perceived as respectful (Knapp & Hall, 2002). Eye gaze also affects the perception of emotions. For instance, Adams and Kleck (2003, 2005) found that direct relative to averted gaze affected perceptions of approach-oriented emotional facial expressions (e.g., anger and joy) by facilitating speed of processing and increasing recognition accuracy. Averted relative to direct gaze, on the other hand, exerted a similar influence on the perception of avoidance-oriented emotions (e.g., fear and sadness). Within the same culture, a differential neural response in the amygdala has also been found to direct-gaze as compared to averted-gaze emotional faces (Adams, Gordon, Baird, Ambady, & Kleck, 2003).

In one recent fMRI study, both Japanese and US participants showed stronger neural activation to same-culture fear faces with averted gaze and other-culture fear faces with direct gaze in several areas of the brain associated with face, gaze, and emotion processing such as the bilateral fusiform gyri, the left caudate, and the right insula (Adams, Franklin, et al., 2010). Interestingly, however, all participants showed greater activation in face-processing areas, such as the bilateral fusiform gyri, to direct relative to averted gaze when displayed on Japanese faces, whereas the opposite was true for US Caucasian faces. This finding suggests that both Japanese and US Caucasian participants may share a common understanding of the distinct cultural meanings associated with gaze behavior. In this case, participants from both cultural groups showed greater activation to incongruous eye gaze behaviors, based on what is generally considered most culturally appropriate. This finding indicates that social expectations and cultural norms may be transmitted and processed consistently across different cultures. Further, underlying an overall behavioral difference across two cultures, one set of neural mechanisms may exhibit an analogous difference in responding, whereas other mechanisms may exhibit a cultural convergence.

Such simultaneous convergence and divergence was also observed in a study by Freeman, Rule, Adams, and Ambady (2009). American and Japanese participants were presented with figural outlines of dominant and subordinate bodily displays (figure 67.1A). In response to the same bodily cues, reward-related mesolimbic regions including the head of the caudate nucleus, bilaterally, and a dorsal aspect of the medial prefrontal cortex showed a mirror-image pattern of responding across the two cultures. Specifically, the caudate and medial prefrontal cortex showed stronger responses to dominant displays in Americans, whereas these same regions showed stronger responses to subordinate displays in Japanese individuals (figure 67.1B). Further, the magnitude of activation in these regions was related to self-reported dominant or subordinate behavioral tendencies. These mesolimbic regions are typically activated by rewarding and motivationally significant stimuli, and in responding to such stimuli they help to coordinate learning and behavior. This finding suggests that perceiving other individuals' nonverbal power displays triggers reward-related responses that may help contribute to high-level social behaviors. Moreover, the results demonstrate additional neural correlates of the individualism characteristic of Americans, emphasizing dominance and elevation of the self, versus the collectivism characteristic of Japanese individuals, emphasizing deference and obligation to others, elevation of one's social group, and subordination.

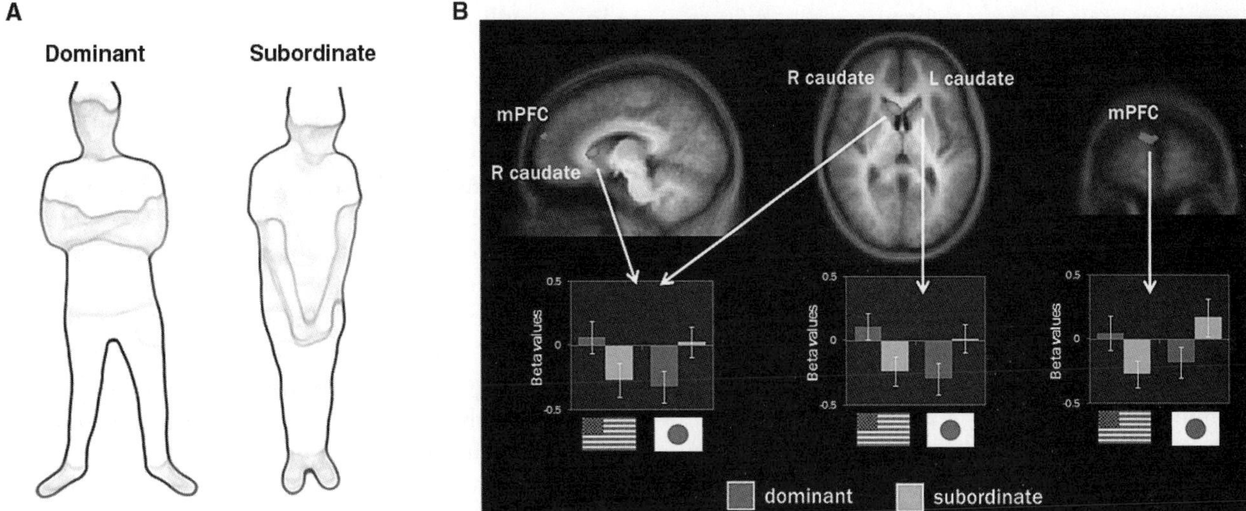

Figure 67.1 Cultural specificity to (A) dominant and subordinate figural outlines in (B) caudate and dorsal medial prefrontal cortex (mPFC) response. (Adapted from Freeman et al., 2009.) (See color plate 56.)

Mind Perception Beyond recognizing expressions of the face or body, the ability to understand others' thoughts is one of the most defining attributes of human behavior (e.g., Saxe & Baron-Cohen, 2006). It is often referred to as "theory of mind," since it requires theorizing that others have minds like one's own and that one may therefore be able to use one's own mind to understand what is occurring in others' minds (see Gallagher & Frith, 2003, for a review). Naturally, the ability to infer the internal states of others' minds is susceptible to broader cultural influences.

Kobayashi, Glover, and Temple (2007) examined theory of mind by using false-belief and cartoon tasks with 8- to 12-year-old American and Japanese children while measuring their brain responses using fMRI. In a typical false-belief task, someone places an object into a cupboard in the presence of an observer. The observer then leaves and the object is moved from the cupboard to another location. The test, then, is to see whether the child will understand that the observer still thinks the object is in the original location—the cupboard (since this is where the observer saw it placed) or if the child will mistakenly apply his or her own knowledge about the object's true, current location. A child with a theory of mind should be able to take the perspective of the observer and assume the observer will look for the object in the cupboard. The study found activation in the ventromedial prefrontal cortex and precuneus across both groups of children, suggesting these areas are important for universal understanding of intentionality. But there was also evidence of cultural specificity in other brain areas. Japanese children showed stronger activation in the inferior frontal gyrus, similar to Japanese adults in a previous study by the same group. Thus, the neural substrates underlying the understanding of others' intentions do seem to be affected by culture. A more recent study was able to clarify these findings.

Behavioral data suggests that people are better at reading the minds of others from their own cultures as compared to those from other cultures (Adams, Rule, et al., 2010). The "reading the mind in the eyes" test involves presenting participants with cropped photographs of faces so that only the eyes are visible and was developed by Baron-Cohen and his colleagues (Baron-Cohen, Wheelwright, Hill, Raste, & Plump 2001). Participants have to choose from adjectives that best describe the target's mental state. This task is believed to require the perceiver to take the perspective of the target in order to infer his or her state of mind. Individuals who lack mental inference abilities (such as patients with neurological damage) show severe impairment in choosing which adjectives best describe the targets' mental states (Adolphs, Baron-Cohen, & Tranel, 2002). Adams et al. (2010) found that own-culture "reading the mind in the eyes" judgments more strongly engaged the STS, a region important to social inferences and theory of mind. Specifically, American participants showed stronger bilateral STS activity when inferring the mental states of American targets, and Japanese participants showed stronger bilateral STS activity when inferring the mental states of Japanese targets (figure 67.2). Thus, culture equips its perceivers with a culturally tuned ability to infer others' mental states, reflected in activity of the STS. In this case, the eyes of different cultures are differentially processing in the STS, with preference for one's own culture.

Conclusion

The emergence of cultural neuroscience thus far has allowed for a deeper understanding of the processes underlying human perception. This new field has shed light on the nature of cross-cultural similarities and differences previously observed in behavioral work, and extended our understanding of the phenomena to the neural level. We reviewed evidence showing that culture shapes basic perceptual processes at varying levels of complexity and across both nonsocial and social domains. We highlighted how aspects of the sociocultural environment shape perceptual processing and give rise to culturally specific behavioral and neural responses. Much of this research involved identifying the neural correlates of established cross-cultural differences in perception. As discussed earlier, however, there is an increasing focus on the interplay of genetic and cultural factors; we expect this can only continue, as the role of culture-gene interactions becomes easier to examine with methodological advances and increases in the ability to measure genetic polymorphisms.

In addition, a cultural neuroscience approach to human perception holds great potential for an increased understanding of how culture influences behavior, the mind, and the brain more broadly. For instance, one way to examine plasticity is to examine how the brain changes with exposure to a new culture. Examining the unique effects of acculturation in immigrants and how cultural influences manifest at the neural level in bicultural and multicultural individuals will be of great importance. Interestingly, there might be critical stages—stages of development during which the brain is more likely to change than other stages—that future research could address (Han et al., 2013). Cultural neuroscience shows promise in helping not only to understand the neural basis of psychological processes, but also in helping to uncover the plasticity of the brain's response to its culture and environment.

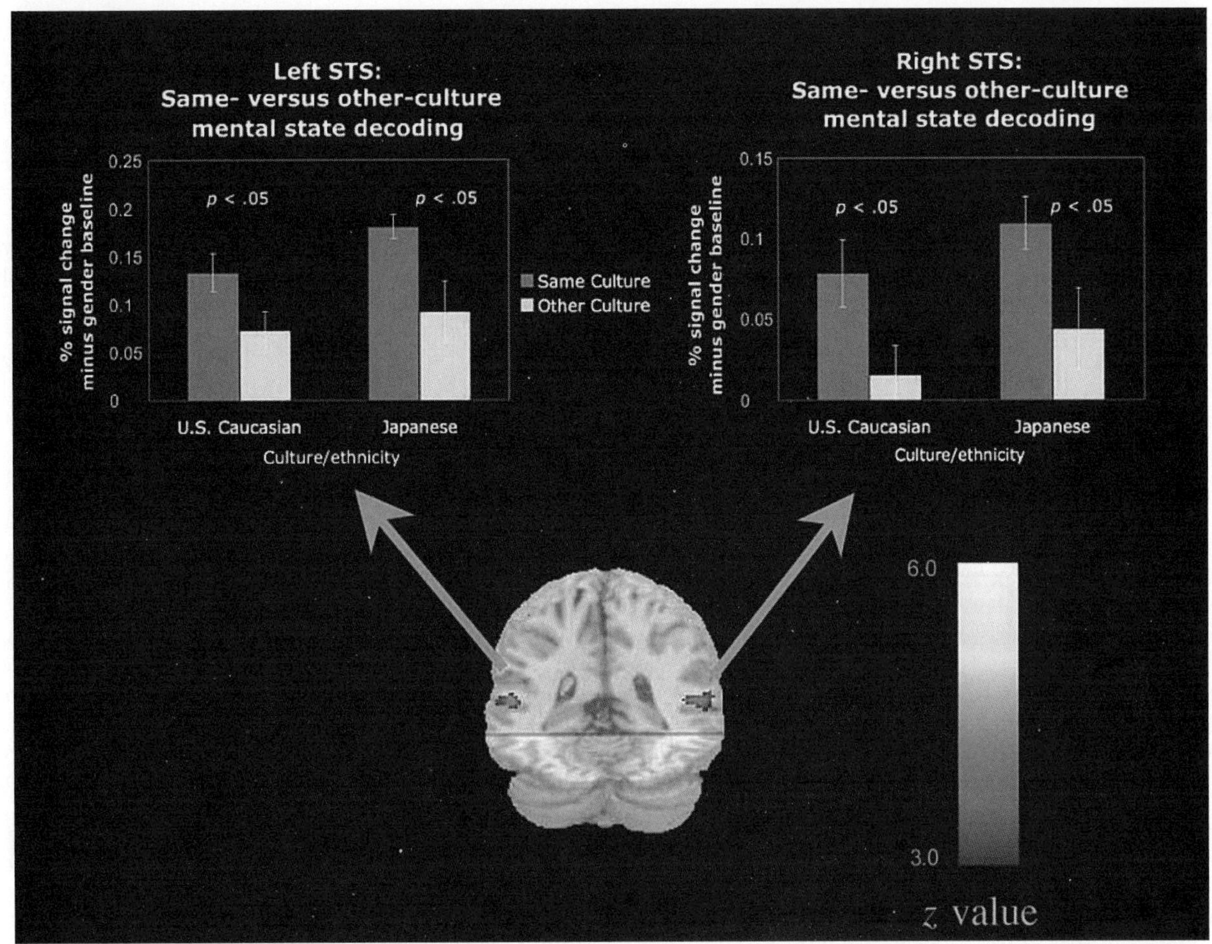

FIGURE 67.2 Cultural specificity in (A) left and (B) right posterior superior temporal sulcus (STS) response to the reading the mind in the eyes (RME) task. (Adapted from Adams et al., 2009.) (See color plate 57.)

That being said, there are a number of logistical challenges in conducting research in the area of cultural neuroscience. Attention must be paid to ensure proper control is implemented, given differing scanner sites. Further, almost all the recent discoveries in cognitive neuroscience have come from richer industrialized nations. For example, to our knowledge, there have been no studies in cultural neuroscience in Africa, South America, or the Middle East. Within the field of psychology, 95% of psychological samples come from countries with only 12% of the world's population (Arnett, 2008), typically Western industrialized nations (Henrich et al., 2010). Within the field of human neuroimaging alone, 90% of peer-reviewed neuroimaging studies come from Western countries (Chiao & Ambady, 2007). Nevertheless, the cultural neuroscience framework represents an unprecedented opportunity to link human diversity across multiple levels of analysis, from genes and brain to mind and behavior. As outlined throughout this chapter, it is revealing novel insights into the influences of culture on a broad range of perceptual phenomena at multiple levels of analysis, from lower-level attention to higher-order social processes. Advances in concepts and methodology in cognitive neuroscience provide a solid foundation for examining the mutual interplay of cultural, neural, and genetic forces throughout the life span.

Historically, psychology has vacillated in its focus; at some times, popular approaches focus on influences of learning, experience, and culture, and at other times they focus on innateness and fixity. Like most dichotomies, rarely does either side provide a full and complete characterization of the phenomenon. Theoretical and empirical work identifying the various factors underlying what is learned and what is innate and how they interact will be critical, and the cultural neuroscience approach would seem highly promising in this regard. The cultural neuroscience approach provides the exciting opportunity to examine the mutual interplay of culture and biology across multiple levels of analysis, from genes and brain to mind and behavior, and ultimately to move toward a more complete and accurate understanding of human perception.

ACKNOWLEDGMENTS This work was supported, in part, by a National Institutes of Health R01 MH070833-01A1 grant. Correspondence should be addressed to Nalini Ambady, Department of Psychology, Stanford University, Jordan Hall, Bldg. 420, Stanford, CA 94305. Email: nambady@stanford.edu.

REFERENCES

Adams, R. B., Jr., Franklin, R., Rule, N., Freeman, J., Yoshikawa, S., Kveraga, K., … Ambady, N. (2010). Culture, gaze, and the neural processing of fear expressions: An fMRI investigation. *Soc Cogn Affect Neurosci, 5*, 340–348.

Adams, R. B., Jr., Gordon, H. L., Baird, A. A., Ambady, N., & Kleck, R. E. (2003). Effects of gaze on amygdala sensitivity to anger and fear faces. *Science, 300*(5625), 1536.

Adams, R. B., Jr., & Kleck, R. E. (2003). Perceived gaze direction and the processing of facial displays of emotion. *Psychol Sci, 14*(6), 644–647.

Adams, R. B., Jr., & Kleck, R. E. (2005). Effects of direct and averted gaze on the perception of facially communicated emotion. *Emotion, 5*(1), 3–11.

Adams, R. B., Jr., Rule, N. O., Franklin, R. G., Jr., Wang, E., Stevenson, M. T., Yoshikawa, S., … Ambady, N. (2010). Cross-cultural reading the mind in the eyes: An fMRI investigation. *J Cogn Neurosci, 22*(1), 97–108.

Adolphs, R., Baron-Cohen, S., & Tranel, D. (2002). Impaired recognition of social emotions following amygdala damage. *J Cogn Neurosci, 14*, 1264–1274.

Ambady, N., & Bharucha, J. (2009). Culture and the brain. *Curr Dir Psychol Sci, 18*, 342–345.

Argyle, M. (1976). Social interactions. *Science, 194*(17840301), 1046–1047.

Arnett, J. J. (2008). The neglected 95%: Why American psychology needs to become less American. *Amer Psychol, 63*(7), 602–614.

Baron-Cohen, S., Wheelwright, S., Hill, J., Raste, Y., & Plump, I. (2001). The "reading the mind in the eyes" test, revised version: A study with normal adults, and adults with Asperger syndrome or high-functioning autism. *J Child Psychol Psychiatry, 42*, 241–251.

Calvo-Merino, B., Glaser, D. E., Grezes, J., Passingham, R. E., & Haggard, P. (2005). Action observation and acquired motor skills: An fMRI study with expert dancers. *Cereb Cortex, 15*(8), 1243–1249.

Chiao, J. Y., & Ambady, N. (2007). Cultural neuroscience: Parsing universality and diversity across levels of analysis. In S. Kitayama & D. Cohen (Eds.), *Handbook of cultural psychology* (pp. 237–254). New York, NY: Guilford.

Chiao, J. Y., & Blizinsky, K. D. (2010). Culture–gene coevolution of individualism–collectivism and the serotonin transporter gene. *Proc Biol Sci, 277*(1681), 529–537.

Chiao, J. Y., Iidaka, T., Gordon, H. L., Nogawa, J., Bar, M., Aminoff, E., & Sadato, N. (2008). Cultural specificity in amygdala response to fear faces. *J Cogn Neurosci, 20*, 2167–2174.

Draganski, B., Gaser, C., Busch, V., Schuierer, G., Bogdahn, U., & May, A. (2004). Neuroplasticity: Changes in grey matter induced by training. *Nature, 427*(6972), 311–312.

Elfenbein, H. A., & Ambady, N. (2002b). Is there an in-group advantage in emotion recognition? *Psychol Bull, 128*(11931518), 243–249.

Finney, E. M., Fine, I., & Dobkins, K. R. (2001). Visual stimuli activate auditory cortex in the deaf. *Nat Neurosci, 4*(12), 1171–1173.

Freeman, J. B., Rule, N. O., Adams, R. B., Jr., & Ambady, N. (2009). Culture shapes a mesolimbic response to signals of dominance and subordination that associates with behavior. *NeuroImage, 47*(1), 353–359.

Gallagher, H. L., & Frith, C. D. (2003). Functional imaging of "theory of mind." *Trends Cogn Sci, 7*, 77–83.

Goh, J., Chee, M., Tan, J., Venkatraman, V., Hebrank, A., & Leshikar, E. (2007). Age and culture modulate object processing and object-scene binding in the ventral visual area. *Cogn Aff Behav Neurosci, 7*(1), 44–52.

Gougoux, F., Belin, P., Voss, P., Lepore, F., Lassonde, M., & Zatorre, R. J. (2009). Voice perception in blind persons: A functional magnetic resonance imaging study. *Neuropsychologia, 47*(13), 2967–2974.

Gutchess, A., Welsh, R., Boduroglu, A., & Park, D. C. (2006). Cultural differences in neural function associated with object processing. *Cogn Aff Behav Neurosci, 6*(2), 102–109.

Han, S., & Northoff, G. (2008). Culture-sensitive neural substrates of human cognition: A transcultural neuroimaging approach. *Nat Rev Neurosci, 9*(8), 646–654.

Han, S., Northoff, G., Vogeley, K., Wexler, B. E., Kitayama, S., & Varnum, M. E. (2013). A cultural neuroscience approach to the biosocial nature of the human brain. *Annu Rev Psychol, 64*, 335–359.

Hariri, A., Drabant, E., & Weinberger, D. (2006). Imaging genetics: Perspectives from studies of genetically driven variation in serotonin function and corticolimbic affective processing. *Biol Psychiatry, 59*, 888–897.

Hedden, T., Ketay, S., Aron, A., Markus, H. R., & Gabrieli, J. D. E. (2008). Cultural influences on neural substrates of attentional control. *Psychol Sci, 19*(1), 12–17.

Henrich, J., Heine, S. J., & Norenzayan, A. (2010). Most people are not WEIRD. *Nature, 466*(7302), 29–29.

Kitayama, S., Duffy, S., Kawamura, T., & Larsen, J. T. (2003). Perceiving an object and its context in different cultures: A cultural look at new look. *Psychol Sci, 14*, 201–206.

Kitayama, S., & Park, J. (2010). Cultural neuroscience of the self: Understanding the social grounding of the brain. *Soc Cogn Affect Neurosci, 5*(2–3), 111–129.

Knapp, M. L., & Hall, J. (2002). *Nonverbal behavior in human interaction*. New York, NY: Wadsworth.

Kobayashi, C., Glover, G. H., & Temple, E. (2007). Cultural and linguistic influence on neural bases of "theory of mind": An fMRI study with Japanese bilinguals. *Brain Lang, 98*, 210–220.

Lutz, C., & White, G. M. (1986). The anthropology of emotions. *Annu Rev Anthropol, 15*, 405–436.

Maguire, E. A., Gadian, D. G., Johnsrude, I. S., Good, C. D., Ashburner, J., & Frackowiak, R. S. (2000). Navigation-related structural change in the hippocampi of taxi drivers. *Proc Natl Acad Sci USA, 97*(8), 4398–4403.

Markus, H. R., & Kitayama, S. (1991). Culture and the self: Implications for cognition, emotion, and motivation. *Psychol Rev, 98,* 224–253.

Marsh, A. A., Kozak, M. N., & Ambady, N. (2007). Accurate identification of fear facial expressions predicts prosocial behavior. *Emotion, 7*(2), 239–251.

Masuda, T., & Nisbett, R. E. (2001). Attending holistically versus analytically: Comparing the context sensitivity of Japanese and Americans. *J Pers Soc Psychol, 81,* 922–934.

Nisbett, R. E., & Miyamoto, Y. (2005). The influence of culture: Holistic versus analytic perception. *Trends Cogn Sci, 9*(10), 467–473.

Nisbett, R. E., Peng, K., Choi, I., & Norenzayan, A. (2001). Culture and systems of thought: Holistic versus analytic cognition. *Psychol Rev, 108,* 291–310.

Posner, M. I. (1993). Seeing the mind. *Science, 262,* 673–674.

Rosenberg, N., Pritchard, J., Weber, J., Cann, H., Kidd, K., Zhivotovsky, L., & Feldman, M. (2002). Genetic structure of human populations. *Science, 298,* 2381–2385.

Saxe, R., & Baron-Cohen, S. (2006). Editorial: The neuroscience of theory of mind. *Soc Neurosci, 1,* 1–9.

Way, B., & Lieberman, M. (2010). Is there a genetic contribution to cultural differences? Collectivism, individualism, and genetic markers of social sensitivity. *Soc Cogn Aff Neurosci, 5*(2–3), 203–211.

Weisbuch, M., & Ambady, N. (2008). Affective divergence: Automatic responses to others' emotions dependent on group membership. *J Pers Soc Psychol, 95,* 1063–1079.

IX

CONSCIOUSNESS

Introduction

GIULIO TONONI

SCIENTISTS ARE finally approaching the relationship between the brain and consciousness in several complementary ways. Over the past 20 years, many have attempted to identify the neural correlates of consciousness (NCC)—the minimal neuronal mechanisms jointly sufficient for any one specific conscious percept. For example, one can examine how brain activity changes when a sensory stimulus is experienced or not, everything else being held as similar as possible. In his chapter, Kreiman illustrates what can be learned through such an approach using binocular rivalry as a case study. He then moves on to consider what we know about the NCC of voluntary actions, which have now been investigated using electroencephalography (EEG), functional MRI (fMRI), and intracranial recordings of neuronal activity.

A complementary strategy is to consider conditions in which consciousness is globally diminished and ask what has changed in the brain. The chapter by Massimini considers the loss of consciousness that occurs in dreamless sleep, in general anesthesia, and after massive brain lesions. Recent work using transcranial magnetic stimulation and high-density EEG shows that level of consciousness can be measured objectively, without requiring behavioral reports, by considering to what extent brain activity is both integrated and differentiated across these diverse conditions and in individual subjects. Demertzi and Laureys examine in detail how a combination of careful clinical tests, passive paradigms that use EEG, positron emission tomography, fMRI, and a combination of transcranial magnetic stimulation and EEG to evaluate brain responses to sensory stimuli, and active paradigms that examine the neural responses to commands in behaviorally unresponsive patients, can shed light on level of consciousness. Blumenfeld summarizes what we have learned about how and when consciousness is lost in epileptic seizures such as absences, generalized tonic-clonic seizures, and complex partial seizures. This is an area of study that was nearly nonexistent even 10 years ago, but has by now revealed a remarkable diversity of individual differences in the way consciousness changes during a seizure and in underlying mechanisms.

A further strategy is to try to distinguish consciousness and its neural correlates from other aspects of cognition, both conceptually and experimentally. Tsuchiya and Koch make a strong case for a double dissociation between consciousness and attention. They also point out that some putative NCC may actually be related to stimulus processing or response preparation rather than to consciousness itself. This is a conceptual distinction that will certainly promote new experimental paradigms and lead to much-needed refinements. Kouider and Sackur address another classic distinction in consciousness research—that between phenomenal and access consciousness. They argue instead for a unified framework in which consciousness is pragmatically identified with access to cognitive content, but access itself can be more or less partial, and discuss the underlying neural mechanisms. In Blanke's survey, one sees clearly how a combination of neuroimaging with clinical and behavioral studies has already refined our understanding of what awareness of the body contributes to consciousness itself, and of how one can distinguish the awareness of one's body (self-identification), of where it is in space (self-location), and from where one perceives the world (first-person perspective). The chapter by Haggard examines instead the way awareness of agency—the intentional initiation and execution of actions—relates to the underlying brain mechanisms, both in experimental settings and in pathological conditions. He also evaluates what our growing understanding of the control of intention, decision making, action, and movement says about free will and social responsibility.

A final strategy that should complement experimental and clinical studies is the development of a theoretical framework that clarifies what consciousness is, how it can be generated by a physical system, and how it can be measured. This brings us back to the chapter by Massimini, which proposes a theoretically motivated measure of level of consciousness and tests theoretical predictions against data from clinical and experimental neuroscience. Altogether, the section on consciousness shows clearly how much we have learned since the days when research on consciousness was nearly barred from psychology and neuroscience or banished to the afterthoughts of a textbook, pointing to a future in which consciousness may instead regain its central seat in the cognitive neurosciences.

68 Neural Correlates of Consciousness: Perception and Volition

GABRIEL KREIMAN

ABSTRACT Consciousness is the result of interactions among neuronal networks in our brains. Although our mechanistic understanding of consciousness remains tentative, the last two decades have seen intense and increasing efforts aimed toward elucidating the neural circuits and spatiotemporal dynamics underlying certain aspects of conscious experience. Experimental paradigms such as those involving bistable perception (wherein a constant stimulus can give rise to different percepts), combined with neurophysiological recordings and computational theories, have provided hints at neuronal signals that correlate with subjective perception. It is also revealing to determine which firing patterns do not correlate with consciousness or which ones correlate with preconscious sensations. Both nonconscious and preconscious signals have been described in perceptual studies as well as in studies of volitional decisions. Discriminating neuronal activity linked to internal perceptual changes in the absence of external changes provides empirical constraints and initial glimpses into how signaling cascades in the brain can give rise to consciousness.

The phenomenological feeling of *consciousness* is central to our moment-to-moment experiences. What we see, hear, feel, or reminisce forms the content of conscious sensations. Most scientists would agree that consciousness is ultimately encoded and orchestrated by the activity of neurons in the brain, but the elucidation of where, when, how, and why neural ensemble dynamics lead to consciousness remains a deep and fundamental mystery.

Our brains are physical entities. In many ways, the chemical components of brains are similar to those encountered in plants and trees and distinct from those in chairs and tables. An important distinguishing feature of the organization of the chemical components in our brains is the existence of interconnected neurons—yet interconnected neural circuits are also present in *Caenorhabditis elegans* and related worms. Whether the neuronal circuits in worms lead to conscious sensations somewhat akin to the ones experienced by humans is not clear. Out of the bewildering complexity (Tononi, 2004; Tononi & Edelman, 1998) that results from vast numbers of interconnected neurons arises the feelings of love and pain and the capacity to plan our future and prove new mathematical theorems. An elegant theoretical framework to define when, how, and why a given circuit of neurons can give rise to consciousness whereas an apparently similar circuit or ensemble firing pattern does not has been proposed by Tononi. The theory proposes that "the level of consciousness of a physical system is related to the repertoire of causal states (*information*) available to the system as a whole (*integration*)" (2004, p. 253, emphasis in original). This theoretical framework can help distinguish why some physical systems may experience consciousness while others do not, which species can experience consciousness, why different neuronal circuits show distinct contributions to consciousness, and even why different activity patterns in the same neuronal circuit may show different correlations with conscious experience.

To formulate a principled strategy to begin to investigate the problem of which aspects of brain function correlate with specific contents of conscious sensations and which do not, Crick and Koch (1990) defined the neural correlates of consciousness (NCC). The NCC represents a "minimal set of neuronal events and mechanisms sufficient for a specific conscious percept" (Koch, 2005). Over the last two decades, growing enthusiasm and blossoming efforts have capitalized on tools to interrogate brain activity at the circuit level in order to take initial steps toward investigating the NCC.

There are multiple fascinating aspects of conscious experience that need to be explained. We focus the discussion here on two of them, sensory perception and volition. With some degree of approximation, we can think of these as representative of brain input and output, respectively. We aim to provide an overview of the advances, difficulties, experiments, and theories that have shaped the discussions around the relationship between neural activity and conscious perception in a few specific instances, which we cover more thoroughly at the expense of other themes in the field. In doing so, we are not doing justice to a large body of heroic efforts. We refer the reader to several reviews (see Baars, 1989; Blanke, 2012; Blumenfeld, 2011; Cotterill, 2001; Crick, Koch, Kreiman, & Fried, 2004;

Dehaene & Changeux, 2011; Jackendoff, 1987; Kim & Blake, 2005; Koch, 2005; Kouider & Dehaene, 2007; Logothetis, 1998; Metzinger, 2000; Rees, Kreiman, & Koch, 2002; Posner, 1994; Searle, 1998, 2005; Singer, 1998; Tononi, 2005; Tononi & Koch, 2011).

Articulating a set of requisites to guide the search for neural correlates of consciousness

It is helpful to ponder what it would take to *understand* the neural circuits that orchestrate a given aspect of consciousness. For the sake of argument, let us consider a given putative NCC for a certain percept *P*. The putative NCC could take the form of activity from a particular subset of neurons, in a specific brain area, at a certain time point and with a structured firing pattern (Koch, 2005). We discuss below what we do and do not know about those neurons, areas, times, and patterns; for the moment, we refer to these as putative NCC for short. We are also not specifying what the percept *P* is. As an example and to frame the discussion, consider the perception of a given face. What sort of evidence would reinforce our belief in a certain activity pattern to be part of the NCC? To a reasonable first-order approximation, we can borrow from the articulation of requisites to correlate neurophysiological responses with behavior (Parker & Newsome, 1998):

(1) Activity in the putative NCC should correlate with the percept *P*. The putative NCC should be elicited by any situation that leads to percept *P* (e.g., presentation of stimulus *P*, visual imagery of *P*, dreaming about *P*, perception of *P* during a bistable perception or a masking experiment, etc.).

(2) Conversely, the putative NCC should *not* be elicited by situations that do *not* lead to percept *P* regardless of the presence or absence of *P* (e.g., presentation of a different percept *Q*, perception of *Q* during a bistable perception test even when the stimulus that would otherwise lead to *P* may be present on the screen, etc.). Note that the same neurons can still be involved in representing *P* and *not P* as long as the NCC as a whole is correlated only with *P*.

(3) No NCC, no consciousness. If we somehow abolish the activity related to the NCC, we expect to observe a concomitant impairment in consciousness. Abolishing activity could take the form of a deliberate lesion in animal models, a certain neurological condition in patients, pharmacological interventions, optogenetic interventions, and so on.

(4) Activation of the NCC should lead to the percept *P*. This activation could be caused by presentation of an external stimulus, electrical stimulation, optogenetic techniques, or any other means. According to this proposal, how the NCC is activated is not relevant; detection of the NCC implies eliciting percept *P*.

This set of requirements implies that we should be able to "read out" percepts *P* by interrogating neural activity patterns. We should only be able to detect NCC whenever the subject experiences *P*. Detecting the NCC in the absence of percept *P* or vice versa would violate (1–4). Violations of (1–4) suggest that we have not fully captured the NCC. Admittedly, this is a tall order.

Consciousness and perception

We focus here on the quest to characterize the neuronal circuits involved in eliciting visual percepts because there has been more work in this domain. Several aspects of the discussion here will also be applicable to other sensory modalities, and perhaps even to other aspects of conscious experience. Following on the definition outlined above for the NCC, we seek to describe the correlates of a specific visual percept (say, seeing a particular face) in terms of specific brain areas, neuronal types, ensemble activity patterns, and timing.

Visual inputs are rapidly transformed into conscious perception. Visual signals impinging on the retinae go through a cascade of processes that lead to perception in a fraction of a second (Blumberg & Kreiman, 2010; Hung, Kreiman, Poggio, & DiCarlo, 2005; Lamme & Roelfsema, 2000; Liu, Agam, Madsen, & Kreiman, 2009; Logothetis & Sheinberg, 1996; Schmolesky et al., 1998; Thorpe, Fize, & Marlot, 1996; VanRullen & Thorpe, 2002). When exactly perception arises along this rapid and approximately hierarchical sequence of steps remains unclear (Baars, 1989; Fisch et al., 2009; Koch, 2005). A rough outline of the brain areas and pathways involved in this cascade has been documented through anatomical and physiological studies, particularly in monkeys (Bullier, 2001; Felleman & Van Essen, 1991). Significant efforts are currently being directed toward describing these cascades in rodents (Wang & Burkhalter, 2007). Due to the difficulties inherent in anatomical mapping and neurophysiological recordings, much less is known about the architecture of the human visual system.

Information from the retina is needed to trigger a visual percept, yet several arguments suggest that we are not *directly* aware of the activity of neurons in our retinae. Investigators have argued that one of the requirements for the NCC is that of an *explicit* representation. A set of neurons is said to explicitly represent a certain aspect of information *y* if an accurate estimate of *y* can be obtained by a one-layer network acting on the output of those neurons (Koch, 2005). Information about local contrast changes is explicitly encoded in the retina. Information about the presence or absence of a face is *not* explicitly encoded in the retina (even though the information can be decoded from retinal activity through a cascade of multiple operations). Several aspects of the properties of retinal photoreceptors argue against their explicit representation of conscious information: (1) retinal photoreceptors can follow rapid spatiotemporal changes that we are not aware of (e.g., a monitor's refresh rate); (2) there are no photoreceptors at the blind spot, but we do not see a hole in the corresponding location in an image; (3) there is no feedback to the retina from other parts of the brain; (4) visual percepts can be elicited through imagery or dreaming in the absence of retinal activation; (5) the type of visual information that we are typically conscious of cannot be decoded in a single step from the retina; (6) retinal photoreceptors show major activity changes in response to multiple small eye movements that we constantly make and that we are completely unaware of. Similar comments can be made about the output cells in the retina (retinal ganglion cells) as well as their target cells in the thalamus (within the lateral geniculate nucleus, or LGN). Several, but not all, of these arguments are also pertinent in the case of primary visual cortex (V1; Crick & Koch, 1995; Leopold, 2012).

Experimental Paradigms As we ascend through the visual hierarchy, neuronal responses become more sophisticated, and the neurons' firing preferences gradually begin to acquire some of the properties that we associate with our subjective visual percepts (Connor, Brincat, & Pasupathy, 2007; Logothetis & Sheinberg, 1996; Tanaka, 1996). The arguments used above to rule out a role for the retina or LGN in conscious perception lose their weight, and there is a need to use more sophisticated experimental paradigms in an attempt to dissociate perception and sensory inputs. Common to many of these empirical approaches is to consider two conditions in which the external inputs are identical, where other internal variables are as similar as possible and the percepts are different (Kim & Blake, 2005). Consider the famous Necker cube (figure 68.1), which can be perceived in two different configurations. Let us assume that fixation, attention, arousal, and other variables are identical in two trials in which the viewer perceives the two different possible configurations (figure 68.1B, C). By assumption (identical fixation), retinal activity should be similar in both trials (except that neurons are capricious and may not show the exact same spiking pattern in two seemingly identical trials, not even in the retina; see, e.g., Van Steveninck, Lewen, Strong, Koberle, & Bialek, 1997). Yet, for the percept to be distinct in the two conditions, something *must* be different in the brain.

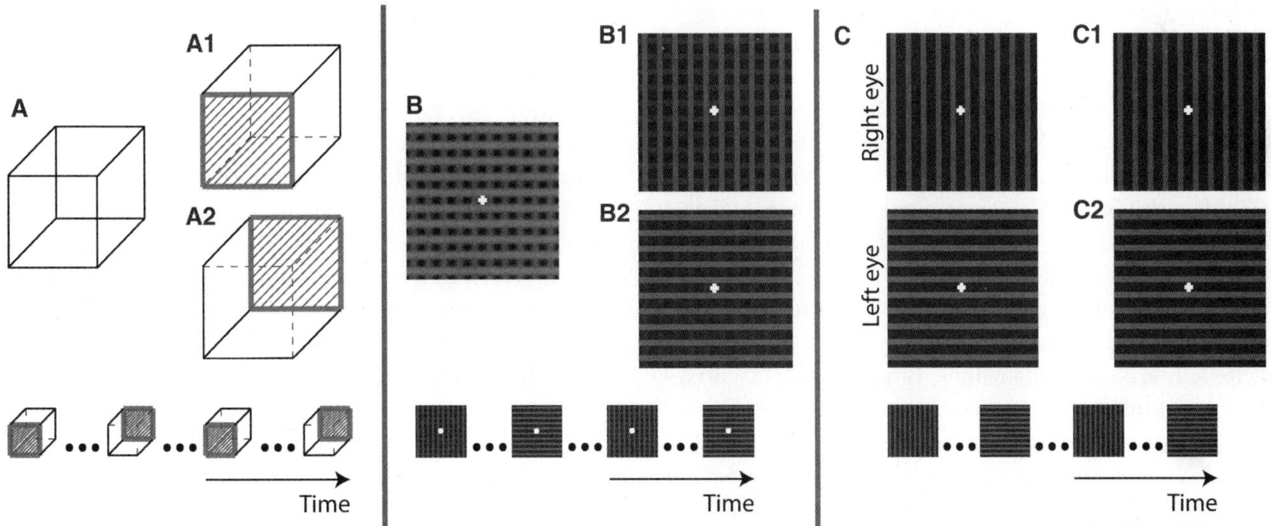

Figure 68.1 Example of bistable percepts. (A) The Necker cube can be interpreted in two possible configurations (A1 and A2). Upon viewing the Necker cube, subjects can voluntarily switch from one configuration to the other (bottom). (B) Monocular rivalry. Upon fixating on the grid of horizontal red lines and vertical blue lines, subjects report a percept that alternates between predominantly blue lines (B1) and predominantly red lines (B2). (C) Binocular rivalry. When presenting different stimuli to each eye, subjects report a percept that alternates between the two inputs (C1, C2) in a seemingly random fashion. During binocular rivalry, it is difficult to voluntarily switch between the two percepts. (See color plate 58.)

We know almost nothing about what changes in cortex when perception switches while viewing the Necker cube. A few other paradigms have proven more fruitful for investigation. We focus below on discussing bistable percepts, including binocular rivalry (Blake & Logothetis, 2002), flash suppression (Wilke, Logothetis, & Leopold, 2003; Wolfe, 1984), and structure from motion (Bradley, Chang, & Andersen, 1998). Several other experimental paradigms have been examined, including near-threshold or degraded stimulation, visual crowding, visual masking (Kouider & Dehaene, 2007; Macknik, 2006), and different forms of transiently induced "blindness" such as motion-induced blindness (Bonneh, Cooperman, & Sagi, 2001), inattentional blindness (Rees, Russell, Frith, & Driver, 1999), change blindness (O'Regan, Rensink, & Clark, 1999; Simons & Rensink, 2005), and the attentional blink (Raymond, Shapiro, & Arnell, 1992; Sergent, Baillet, & Dehaene, 2005).

Neurophysiological Changes Correlated with Bistable Percepts Several experimental paradigms exploit the observation that our brains impose a single interpretation on the outside world at any given time (figure 68.1). In cases where the visual input is ambiguous and is compatible with two (or more) different possible interpretations, our brains settle on one of them. Under many circumstances, this decision is only transient, and our brains may toggle between one interpretation and the other. The Necker cube discussed above constitutes one such case. In ambiguous depth from motion stimuli, a set of dots is set to rotate in such a way that the image can be interpreted to rotate in either of two possible directions (Bradley et al., 1998; Siegel & Andersen, 1988). Another example is the phenomenon of monocular rivalry (Campbell & Howell, 1972; figure 68.1B). Consider a grid of horizontal red bars and vertical blue bars; perception mostly alternates between horizontal red bars and vertical blue bars at the expense of an interpretation containing both horizontal and vertical bars. Rivalry is much stronger when elicited in a binocular fashion. Binocular rivalry arises when a stimulus R is shown to the right eye and a different stimulus L is shown to the left eye (figure 68.1C). Typically, perception alternates between R and L in a seemingly random fashion. The rate of alternation depends on the characteristics of R and L. In a trivial case, if L is blank, R completely dominates perception (e.g., if you cover the left eye). If R and L are very large (say more than 5–6 degrees of visual angle), in addition to perceptual alterations, parts of the image are perceived as a mixture of R and L (known as piecemeal rivalry). Stimuli with high contrast typically dominate perception. The psychophysics of binocular rivalry and the variables that govern dominance and alterations have been carefully studied (Alais & Blake, 2005).

Which neuronal changes accompany the perceptual changes evident upon viewing bistable stimuli? Access to neuronal responses in the human brain is rather limited (Engel, Moll, Fried, & Ojemann, 2005; Kreiman, 2007; Mukamel & Fried, 2012); most neurophysiological studies of binocular rivalry to date have focused on examining responses in macaque monkeys. Before delving into the neurophysiological responses in macaque monkeys, it is important to ask whether monkeys perceive bistable stimuli in the same way that humans do. Because it is difficult to access subjective experiences in monkeys, investigators have used ingenious controls and behavioral measurements to evaluate perceptual alternations while monkeys view bistable stimuli. For example, embedded within a binocular rivalry experiment are catch trials consisting of monocular stimulus presentation where there is no ambiguity about what the behavioral responses should be. Throughout these studies, the answer is consistently that monkeys seem to perceive changes in the interpretation of bistable stimuli in the same way that humans do (Libedinsky, Savage, & Livingstone, 2009; Macknik & Livingstone, 1998; Myerson, Miezin, & Allman, 1981; Sheinberg & Logothetis, 1997; Siegel & Andersen, 1988).

In binocular rivalry experiments, ascending through the visual hierarchy, there is a progressive increase in the proportion of neurons that show a correlation with perception (Leopold & Logothetis, 1999). While binocularly presenting orthogonal drifting gratings to fixating monkeys, none of the neurons in the LGN showed any evidence of alternations indicative of binocular rivalry (but it should be noted that monkeys were passively fixating and not reporting their perceptual alterations; Lehky & Maunsell, 1996). Similar conclusions were reached in a motion-induced perceptual suppression experiment wherein the activity of LGN neurons did not correlate with subjective perception, but pulvinar neurons did (Wilke, Mueller, & Leopold, 2009).

Logothetis and colleagues conducted an elegant series of experiments in monkeys that were trained to report their alternating percepts during binocular rivalry by pressing and holding levers. In primary visual cortex, only about 18% of the neurons showed perceptual modulation (Leopold, 2012; Leopold & Logothetis, 1996). This small percentage of neurons showing perceptual modulation in primary visual cortex is consistent with the idea that perceptual alternations arise from competitive mechanisms beyond V1 (Leopold & Logothetis, 1996, 1999; Logothetis, Leopold, &

Sheinberg, 1996; Wilke, Leopold, & Logothetis, 2002; see, however, Blake, 1989; Blake & Logothetis, 2002). In areas V4 and MT, 38% and 43% of the neurons showed perceptual modulation (Leopold & Logothetis, 1996; Logothetis & Schall, 1989). Intriguingly, in area MT, half of the neurons increased their firing rate when their preferred stimulus was perceived, while the other half showed the reverse pattern; that is, they increased their activity when their preferred stimulus was suppressed. The correlation between subjective perception and the activity of some, but not all, MT neurons was also apparent in a structure-from-motion task during conditions in which the input was ambiguous (Bradley et al., 1998). The bewildering variety of neurons that show different degrees of correlation with perception is perhaps a reflection of the intricacy of connectivity patterns in neural circuits. The advent of techniques based on optogenetics may enable the possibility of interrogating (and manipulating) different layers and different types of neurons. Analysis of circuitry at the level of "brain areas" may prove insufficient to uncover the NCC. The neurons in V1, V4, and MT that show stronger correlation with subjective perception may have distinct properties (in terms of their location, inputs, and outputs, and how they interact with other neurons) from their neighboring neurons whose firing is oblivious or anticorrelated with the percepts.

In the highest echelons of the ventral visual stream, 90% of the neurons in the inferior temporal cortex and the superior temporal sulcus showed a correlation between their firing rate and the subjective report of the neuron's preferred stimulus (Sheinberg & Logothetis, 1997). In a variant of binocular rivalry–denominated flash suppression, a stimulus is flashed monocularly followed by presentation of another stimulus to the other eye (Wolfe, 1984). The flashed stimulus dominates perception, even though the initial stimulus remains present. In parallel to the binocular rivalry results, neurons in the macaque inferior temporal cortex and superior temporal sulcus show a strong correlation with the percept (Sheinberg & Logothetis, 1997). Similarly, most neurons in the human medial temporal lobe also show a correlation with subjective perception during flash suppression (Kreiman, Fried, & Koch, 2002).

Attention, Imaging, and Stimulus Configuration Dependencies In contrast to neurophysiological recordings in macaque V1, several human functional neuroimaging studies have suggested that activity in primary visual cortex also correlates with subjective perception (Haynes, Deichmann, & Rees, 2005; Polonsky, Blake, Braun, & Heeger, 2000; Tong & Engel, 2001; see also similar claims in the LGN: Wunderlich, Schneider, & Kastner, 2005). These discrepancies between blood flow measurements and neurophysiological signals have also been observed in other experiments (Logothetis, 2002; Posner & Gilbert, 1999; Sirotin & Das, 2009; Watanabe et al., 2011). An elegant study by Maier and colleagues examined blood oxygen level–dependent (BOLD) functional MRI signals, local field potentials (LFPs), and spiking responses in primary visual cortex during binocular rivalry (Maier et al., 2008). During physical removal of a stimulus, BOLD signals, LFPs, and spiking responses agreed with each other. However, during perceptual suppression, there were small changes in BOLD signals and LFP responses that were not observed at the spiking level. These observations highlight the notion that blood flow and spiking signals measure different aspects of brain function, and the nature of the relationship between these signals may prove to be important to gain further mechanistic insights into the circuitry involved in conscious perception (Leopold, 2012). A potential explanation of these results is that blood flow signals in the LGN and in primary visual cortex reflect feedback modulation from higher visual areas after perceptual rivalry has been resolved, consistent with the notion that V1 does not have a direct role in visual awareness (Crick & Koch, 1995; Leopold, 2012).

Strong modulation in blood flow signals in early visual areas in the absence of concomitant strong modulation at the spike level has also been observed in studies of attentional modulation (Kastner & Ungerleider, 2000; Reynolds & Chelazzi, 2004). It has been argued that modulatory changes observed during binocular rivalry experiments could reflect attentional fluctuations (Macknik & Martinez-Conde, 2009). The extent to which attention and awareness can be dissociated has been a matter of significant debate (Dehaene & Changeux, 2011; Koch & Tsuchiya, 2012; Posner, 1994). Under most everyday circumstances, attention and awareness go hand in hand, yet psychophysical experiments suggest that it is possible to attend to a stimulus even in the absence of awareness (e.g., Koch & Tsuchiya, 2012; van Boxtel, Tsuchiya, & Koch, 2010). Furthermore, a functional imaging study showed that blood flow signals in V1 are modulated by attention but not by changes in awareness (Watanabe et al., 2011). These results are consistent with recent neurophysiological recordings that lend further support to distinct signals giving rise to attentional modulation and awareness (Maier, 2012).

Other factors beyond attention can also influence the relationship between neuronal activity and awareness. The discussion in the previous section described differences in the correlations between subjective perception and neuronal responses across areas. In a particularly intriguing study, Maier and colleagues asked whether, for a given individual neuron, this correlation depended on the details of the stimulus configuration (Maier, Logothetis, & Leopold, 2007). The authors recorded neurons in area MT in the macaque monkey during binocular rivalry flash suppression. For a given stimulus, the results were consistent with earlier recordings (Logothetis & Schall, 1989). However, when the authors changed the stimulus configuration (e.g., different motion directions or drifting gratings instead of random dots), the extent to which the neuron signaled subjective percepts was significantly altered. For example, a neuron may show changes that correlate with perception when the right eye sees a left moving grating and the left eye sees a downward moving grating, but not when the stimulus in the left eye is an upward-moving grating. These idiosyncratic correlations force us to revisit the notion that the NCC for a particular percept may invoke a fixed set of neurons. These puzzling observations further suggest the urgent need to relate theoretical ideas of how consciousness arises to neurophysiological recordings (Tononi, 2004).

Consciousness and volition

Consciousness is clearly not restricted to awareness of sensory events. At the other end of the sensory/motor spectrum, volitional actions carry a strong sensation of ownership (e.g., "I want to raise my hand"). Before discussing the neuronal manifestations that correlate with our awareness of intention to act, we need to discuss the controversial notion that our volition and intentions are dictated by neurons.

Volition and Free Will Studying the neural signatures that correlate with consciousness about volitional actions may have important implications in settling the age-old questions about "free will." Intuitively defined, the word "free" associated with will implies that, for most of our actions, we experience the strong subjective feeling that we could have opted to act otherwise (see also Haggard, chapter 75 in this volume). If we are asked to pick between a blue pen and a black pen and we pick the black one, it seems that we might as well have picked the blue one.

The extent to which free will is truly free or merely an illusion has been a matter of debate for millennia, with strong advocates on both sides (for an overview, see Haggard, 2008; Heisenberg, 2009). At least two main cautionary notes should be discussed here. The first involves the distinction between "determinism" and "chaos." A system is said to be deterministic if its future state is entirely defined by the initial conditions (and any external forces). A system is said to be chaotic if it displays extreme sensitivity to initial conditions (minuscule differences in the initial state can lead to widely different future states, as in the famous parable of the butterfly effect; Devaney, 2003). These two words are *not* antagonistic. A system can be both deterministic and chaotic. Consider the act of flipping a coin: obtaining heads or tails can depend on a lot of factors (exact initial angle, torque, speed, wind, properties of the surface where the coin lands, etc.) but the physics underlying the problem are well defined and purely deterministic. Nobody would claim that the coin "wanted" to land on heads. The other cautionary note has to do with *computability*. There are many reasons why certain functions may be difficult (or impossible) to compute: there are problems that are not computable (Garey & Johnson, 1979), and there are computable problems that require unrealistic computational resources or data that we do not have access to. Chaos and questions about computational resources may make it very difficult and perhaps impractical to make predictions in certain systems but neither speaks against determinism or in favor of free will.

Neurophysiological Correlates of Volitional Decisions The majority of studies about decision making have focused on situations in which a cue indicates the target behavior. The cue may be noisy, interpreting the cue may require training and memory, and the relationship between the cue and reward could be a probabilistic one. Yet common to many of these experiments are a temporal trigger and an incentive to choose one action versus another. In contrast, the study of volitional decisions requires situations where different actions are equally likely and attractive.

A few studies in macaque monkeys have examined neurophysiological responses in the parietal and frontal cortex while monkeys performed volitional decisions (Maimon & Assad, 2006a, 2006b; Okano & Tanji, 1987; Romo & Schultz, 1992). Single neurons in the lateral intraparietal area, cortical area 5, the basal ganglia, and frontomotor areas exhibited gradual increases in firing rate during execution of volitional arm movements. Some of the neurons showed activation during both visually triggered movements and proactive movements. However, the slow ramp in firing rates was characteristic of internally generated movements only.

This slow increase in activity is reminiscent of gradual changes in scalp electroencephalographic signals in the human brain during execution of volitional movements (Brass & Haggard, 2008; Deecke et al., 1987; Haggard, 2008). In a variant of this type of experiment,

Libet asked subjects to tap their index finger at will and also report the time of their intention to act based on an analog clock present on the screen during the experiment (Libet, 1985; Libet, Gleason, Wright, & Pearl, 1983). These experiments revealed that the averaged scalp electroencephalographic signals preceded the conscious intention of the urge to move by several hundreds of milliseconds. The interpretation of these experiments has been the subject of much debate in the field (e.g., see Libet, 1985, 2002, and discussions in the same issue).

In a recent study, Fried and colleagues took advantage of a rare clinical opportunity to record from >1,000 neurons in the human frontal and temporal lobes when subjects performed willed action and reported the time of volition onset, as in the Libet experiment. Consistent with earlier studies using noninvasive methods, they found evidence at the single neuron level in humans for an anatomically localized early frontal cortex signal that preceded conscious will. Over a time period of ~1,500 msec prior to the awareness of will, an increasing number of neurons in two specific brain regions, the supplementary motor area and anterior cingulate cortex, were progressively recruited. The subjectively reported onset of volition could be accurately predicted on a single trial basis based on neural activity in the supplementary motor area well before the subject's awareness. Based on these findings, the authors proposed a computational model and a biophysically plausible mechanism for the emergence of conscious will in humans based on progressive recruitment of neuronal ensembles in frontal cortex until a threshold is crossed. The model is consistent with the notion that the all-or-none nature of consciousness is the result of gradual accumulation reaching a threshold (Crick & Koch, 2003). Furthermore, in another study, Fried showed that electrical stimulation in the human supplementary motor area triggered an "urge" to perform motor actions (Fried et al., 1991).

Outlook

The fundamental problem of understanding how neuronal circuits give rise to conscious sensations has risen from nebulous beginnings and debates to become a major effort in cognitive neuroscience. Progress in the field should be interpreted with cautious optimism. Caution is important because this is undoubtedly a difficult problem, and current theories are as diverse as they could be. And yet there is optimism and steady progress. The last two decades have seen the blossoming of a young generation of energetic and heroic investigators who have dared to ask difficult questions and approach them with a new arsenal of tools that is making rapid strides in elucidating other aspects of cognition, including multielectrode arrays, computational modeling, microstimulation, optogenetics, and so on.

Advances and controversies in trying to correlate neural signals and conscious perception have led to a theoretical framework that provides quantitative definitions of how neuronal interactions could lead to consciousness (Tononi, 2004). Additionally, initial but significant steps have been made toward better defining questions about conscious processing (e.g., dissociating pure attentional effects from conscious perception; Koch and Tsuchiya, 2012), toward sharpening experimental tools (e.g., noticing that blood flow signals may not reveal underlying spiking; Maier, 2008), and toward a richness of experimental paradigms and approaches (Dehaene & Changeux, 2011).

There is currently significant excitement in cognitive neuroscience with the advent of tools that enable the manipulation of circuits at unprecedented resolution (e.g., Han et al., 2009). These tools open the doors to exciting and promising opportunities to attempt to transiently inactivate and also directly stimulate local circuits, and thus bias subjective decisions in ways that have not been possible before. While it is anyone's guess whether the quest for the NCC will be resolved in the near future, there is no question that we should expect fascinating surprises and novel insights in the community's efforts to elucidate how physical systems lead to consciousness.

REFERENCES

Alais, D., & Blake, R. (2005). *Binocular rivalry.* Cambridge, MA: MIT Press.

Baars, B. (1989). *A cognitive theory of consciousness.* Cambridge, UK: Cambridge University Press.

Blake, R. (1989). A neural theory of binocular rivalry. *Psychol Rev, 96,* 145–167.

Blake, R., & Logothetis, N. (2002). Visual competition. *Nat Rev Neurosci, 3,* 13–21.

Blanke, O. (2012). Multisensory brain mechanisms of bodily self-consciousness. *Nat Rev Neurosci, 13,* 556–571.

Blumberg, J., & Kreiman, G. (2010). How cortical neurons help us see: Visual recognition in the human brain. *J Clin Invest, 120,* 3054–3063.

Blumenfeld, H. (2011). Epilepsy and the consciousness system: Transient vegetative state? *Neurol Clin, 29,* 801–823.

Bonneh, Y., Cooperman, A., & Sagi, D. (2001). Motion-induced blindness in normal observers. *Nature, 411,* 798–801.

Bradley, D. C., Chang, G. C., & Andersen, R. A. (1998). Encoding of 3D structure from motion by primate area MT neurons. *Nature, 392,* 714–717.

Brass, M., & Haggard, P. (2008). The what, when, whether model of intentional action. *Neuroscientist, 14,* 319–325.

Bullier, J. (2001). Integrated model of visual processing. *Brain Res Brain Res Rev, 36,* 96–107.

Campbell, F. W., & Howell, E. R. (1972). Monocular alternation: A method for the investigation of pattern vision. *J Physiol, 225,* 19–21.

Connor, C. E., Brincat, S. L., & Pasupathy, A. (2007). Transformation of shape information in the ventral pathway. *Curr Opin Neurobiol, 17,* 140–147.

Cotterill, R. (2001). Evolution, cognition and consciousness. *J Consciousness Stud, 8,* 3–17.

Crick, F., & Koch, C. (1990). Some reflections on visual awareness. *Cold Spring Harbor Symposia on Quantitative Biology, 55,* 953–962.

Crick, F., & Koch, C. (1995). Are we aware of neural activity in primary visual cortex? *Nature, 375,* 121–123.

Crick, F., & Koch, C. (2003). A framework for consciousness. *Nat Neurosci, 6,* 119–126.

Crick, F., Koch, C., Kreiman, G., & Fried, I. (2004). Consciousness and neurosurgery. *Neurosurgery, 55,* 273–282.

Deecke, I., Lang, W., Heller, H., Hufnagl, M., & Kornhuber, H. (1987). Bereitschaftspotential in patients with unilateral lesions of the supplementary motor area. *J Neurol, 50,* 4.

Dehaene, S., & Changeux, J. P. (2011). Experimental and theoretical approaches to conscious processing. *Neuron, 70,* 200–227.

Devaney, R. L. (2003). *An introduction to chaotic dynamical systems.* Boulder, CO: Westview.

Engel, A. K., Moll, C. K., Fried, I., & Ojemann, G. A. (2005). Invasive recordings from the human brain: Clinical insights and beyond. *Nat Rev Neurosci, 6,* 35–47.

Felleman, D. J., & Van Essen, D. C. (1991). Distributed hierarchical processing in the primate cerebral cortex. *Cereb Cortex, 1,* 1–47.

Fisch, L., Privman, E., Ramot, M., Harel, M., Nir, Y., Kipervasser, S., ... Malach, R. (2009). Neuronal "ignition": Non-linear activation at the threshold of perceptual awareness in human ventral stream visual cortex. *Neuron, 64*(4), 562–574.

Fried, I., Katz, A., McCarthy, G., Sass, K. J., Williamson, P., Spencer, S. S., & Spencer, D. D. (1991). Functional organization of human supplementary motor cortex studied by electrical stimulation. *J Neurosci, 11,* 3656–3666.

Garey, M., & Johnson, D. (1979). *Computers and intractability: A guide to the theory of NP-completeness.* New York, NY: Freeman.

Haggard, P. (2008). Human volition: Towards a neuroscience of will. *Nat Rev Neurosci, 9,* 934–946.

Han, X., Qian, X., Bernstein, J. G., Zhou, H. H., Franzesi, G. T., Stern, P., ... Boyden, E. S. (2009). Millisecond-timescale optical control of neural dynamics in the nonhuman primate brain. *Neuron, 62,* 191–198.

Haynes, J. D., Deichmann, R., & Rees, G. (2005). Eye-specific effects of binocular rivalry in the human lateral geniculate nucleus. *Nature, 438,* 496–499.

Heisenberg, M. (2009). Is free will an illusion? *Nature, 459,* 2.

Hung, C. P., Kreiman, G., Poggio, T., & DiCarlo, J. J. (2005). Fast read-out of object identity from macaque inferior temporal cortex. *Science, 310,* 863–866.

Jackendoff, R. (1987). *Consciousness and the computational mind.* Cambridge, MA: MIT Press.

Kastner, S., & Ungerleider, L. G. (2000). Mechanisms of visual attention in the human cortex. *Annu Rev Neurosci, 23,* 315–341.

Kim, C. Y., & Blake, R. (2005). Psychophysical magic: Rendering the visible "invisible." *Trends Cogn Sci, 9,* 381–388.

Koch, C. (2005). *The quest for consciousness.* Los Angeles, CA: Roberts.

Koch, C., & Tsuchiya, N. (2012). Attention and consciousness: Related yet different. *Trends Cogn Sci, 16,* 103–105.

Kouider, S., & Dehaene, S. (2007). Levels of processing during non-conscious perception: A critical review of visual masking. *Philos Trans R Soc Lond B Biol Sci, 362,* 857–875.

Kreiman, G. (2007). Single neuron approaches to human vision and memories. *Curr Opin Neurobiol, 17,* 471–475.

Kreiman, G., Fried, I., & Koch, C. (2002). Single neuron correlates of subjective vision in the human medial temporal lobe. *Proc Natl Acad Sci USA, 99,* 8378–8383.

Lamme, V. A., & Roelfsema, P. R. (2000). The distinct modes of vision offered by feedforward and recurrent processing. *Trends Neurosci, 23,* 571–579.

Lehky, S. R., & Maunsell, J. H. R. (1996). No binocular rivalry in the LGN of alert monkeys. *Vis Res, 36,* 1225–1234.

Leopold, D. A. (2012). Primary visual cortex: Awareness and blindsight. *Annu Rev Neurosci, 35,* 91–109.

Leopold, D. A., & Logothetis, N. K. (1996). Activity changes in early visual cortex reflect monkeys' percepts during binocular rivalry. *Nature, 379,* 549–553.

Leopold, D. A., & Logothetis, N. K. (1999). Multistable phenomena: Changing views in perception. *Trends Cogn Sci, 3,* 254–264.

Libedinsky, C., Savage, T., & Livingstone, M. (2009). Perceptual and physiological evidence for a role for early visual areas in motion-induced blindness. *J Vis, 9*(1), 1–10.

Libet, B. (1985). Unconscious cerebral initiative and the role of conscious will in voluntary action. *Behav Brain Sci, 8,* 529–566.

Libet, B. (2002). The timing of mental events: Libet's experimental findings and their implications. *Conscious Cogn, 11,* 291–299.

Libet, B., Gleason, C., Wright, E., & Pearl, D. (1983). Time of conscious intention to act in relation to onset of cerebral activity (readiness-potential). *Brain, 106,* 623–642.

Liu, H., Agam, Y., Madsen, J. R., & Kreiman, G. (2009). Timing, timing, timing: Fast decoding of object information from intracranial field potentials in human visual cortex. *Neuron, 62,* 281–290.

Logothetis, N. K. (1998). Single units and conscious vision. *Philos Trans R Soc Lond B Biol Sci, 353,* 1801–1818.

Logothetis, N. K. (2002). The neural basis of the blood-oxygen-level-dependent functional magnetic resonance imaging signal. *Philos Trans R Soc Lond B Biol Sci, 357,* 1003–1037.

Logothetis, N. K., Leopold, D. A., & Sheinberg, D. L. (1996). What is rivaling during binocular rivalry? *Nature, 380,* 621–624.

Logothetis, N. K., & Schall, J. D. (1989). Neuronal correlates of subjective visual perception. *Science, 245,* 761–763.

Logothetis, N. K., & Sheinberg, D. L. (1996). Visual object recognition. *Annu Rev Neurosci, 19,* 577–621.

Macknik, S. (2006). Visual masking approaches to visual awareness. *Prog Brain Res, 155,* 177–215.

Macknik, S. L., & Livingstone, M. S. (1998). Neuronal correlates of visibility and invisibility in the primate visual system. *Nat Neurosci, 1,* 144–149.

Macknik, S. L., & Martinez-Conde, S. (2009). The role of feedback in visual attention and awareness. In M. Gazzaniga, *The cognitive neurosciences* (Cambridge, MA: MIT Press, 2009), 1165–1179.

Maier, A. (2012). The cortical microcircuitry of conscious perception and selective attention. Paper presented the Annual Meeting of the Society for Neuroscience, New Orleans, LA.

Maier, A., Logothetis, N., & Leopold, D. (2007). Context-dependent perceptual modulation of single neurons in primate visual cortex. *Proc Natl Acad Sci USA, 104,* 5620–5625.

Maier, A., Wilke, M., Aura, C., Zhu, C., Ye, F. Q., & Leopold, D. A. (2008). Divergence of fMRI and neural signals in V1 during perceptual suppression in the awake monkey. *Nat Neurosci, 11,* 1193–1200.

Maimon, G., & Assad, J. A. (2006a). Parietal area 5 and the initiation of self-timed movements versus simple reactions. *J Neurosci, 26,* 2487–2498.

Maimon, G., & Assad, J. A. (2006b). A cognitive signal for the proactive timing of action in macaque LIP. *Nat Neurosci, 9,* 948–955.

Metzinger, T. (2000). *Neural correlates of consciousness: Empirical and conceptual questions.* Cambridge, MA: MIT Press.

Mukamel, R., & Fried, I. (2012). Human intracranial recordings and cognitive neuroscience. *Annu Rev Psychol, 63,* 511–537.

Myerson, J., Miezin, F., & Allman, J. (1981). Binocular rivalry in macaque monkeys and humans: A comparative study in perception. *Behav Anal Lett, 1,* 149–159.

Okano, K., & Tanji, J. (1987). Neuronal activities in the primate motor fields of the agranular frontal cortex preceding visually triggered and self-paced movement. *Exp Brain Res, 66,* 155–166.

O'Regan, J., Rensink, R., & Clark, J. (1999). Change-blindness as a result of mudsplashes. *Nature, 398,* 34.

Parker, A. J., & Newsome, W. T. (1998). Sense and the single neuron: Probing the physiology of perception. *Annu Rev Neurosci, 21,* 227–277.

Polonsky, A., Blake, R., Braun, J., & Heeger, D. (2000). Neuronal activity in human primary visual cortex correlates with perception during binocular rivalry. *Nat Neurosci, 3,* 1153–1159.

Posner, M. I. (1994). Attention: The mechanisms of consciousness. *Proc Natl Acad Sci USA, 91,* 7398–7403.

Posner, M. I., & Gilbert, C. D. (1999). Attention and primary visual cortex. *Proc Natl Acad Sci USA, 96,* 2585–2587.

Raymond, J. E., Shapiro, K. L., & Arnell, K. M. (1992). Temporary suppression of visual processing in an RSVP task: An attentional blink? *J Exp Psychol Hum Percept Perform, 18,* 849–860.

Rees, G., Kreiman, G., & Koch, C. (2002). Neural correlates of consciousness in humans. *Nat Rev Neurosci, 3,* 261–270.

Rees, G., Russell, C., Frith, C. D., & Driver, J. (1999). Inattentional blindness versus inattentional amnesia for fixated but ignored words. *Science, 286,* 2504–2507.

Reynolds, J. H., & Chelazzi, L. (2004). Attentional modulation of visual processing. *Annu Rev Neurosci, 27,* 611–647.

Romo, R., & Schultz, W. (1992). Role of primate basal ganglia and frontal cortex in the internal generation of movements. III. Neuronal activity in the supplementary motor area. *Exp Brain Res, 91*(3), 396–407.

Schmolesky, M., Wang, Y., Hanes, D., Thompson, K., Leutgeb, S., Schall, J., & Leventhal, A. (1998). Signal timing across the macaque visual system. *J Neurophysiol, 79,* 3272–3278.

Searle, J. (1998). How to study consciousness scientifically. *Philos Trans R Soc Lond B Biol Sci, 353,* 1935–1942.

Searle, J. (2005). Consciousness: What we still don't know [Review of the book *The Quest for Consciousness* by C. Koch]. *The New York Review of Books.* Available from http://www.nybooks.com/articles/archives/2005/jan/13/consciousness-what-we-still-dont-know/.

Sergent, C., Baillet, S., & Dehaene, S. (2005). Timing of the brain events underlying access to consciousness during the attentional blink. *Nat Neurosci, 8,* 1391–1400.

Sheinberg, D. L., & Logothetis, N. K. (1997). The role of temporal areas in perceptual organization. *Proc Natl Acad Sci USA, 94,* 3408–3413.

Siegel, R. M., & Andersen, R. A. (1988). Perception of three-dimensional structure from motion in monkey and man. *Nature, 331,* 259–261.

Simons, D. J., & Rensink, R. A. (2005). Change blindness: Past, present, and future. *Trends Cogn Sci, 9,* 16–20.

Singer, W. (1998). Consciousness and the structure of neuronal representations. *Philos Trans R Soc Lond B Biol Sci, 353,* 1829–1840.

Sirotin, Y. B., & Das, A. (2009). Anticipatory haemodynamic signals in sensory cortex not predicted by local neuronal activity. *Nature, 457,* 475–479.

Tanaka, K. (1996). Inferotemporal cortex and object vision. *Annu Rev Neurosci, 19,* 109–139.

Thorpe, S., Fize, D., & Marlot, C. (1996). Speed of processing in the human visual system. *Nature, 381,* 520–522.

Tong, F., & Engel, S. (2001). Interocular rivalry revealed in the human cortical blind-spot representation. *Nature, 411,* 195–199.

Tononi, G. (2004). An information integration theory of consciousness. *BMC Neurosci, 5,* 42.

Tononi, G. (2005). Consciousness, information integration, and the brain. *Prog Brain Res, 150,* 109–126.

Tononi, G., & Edelman, G. (1998). Consciousness and complexity. *Science, 282*(5395), 1846–1851.

Tononi, G., & Koch, C. (2011). The neural correlates of consciousness: An update. *Ann NY Acad Sci, 1124,* 239–261.

Van Boxtel, J. J., Tsuchiya, N., & Koch, C. (2010). Opposing effects of attention and consciousness on afterimages. *Proc Natl Acad Sci USA, 107,* 8883–8888.

Van Steveninck, R., Lewen, G. D., Strong, S. P., Koberle, R., & Bialek, W. (1997). Reproducibility and variability in neural spike trains. *Science, 275,* 1805–1808.

VanRullen, R., & Thorpe, S. (2002). Surfing a spike wave down the ventral stream. *Vis Res, 42,* 2593–2615.

Wang, Q., & Burkhalter, A. (2007). Area map of mouse visual cortex. *J Comp Neurol, 502,* 339–357.

Watanabe, M., Cheng, K., Murayama, Y., Ueno, K., Asamizuya, T., Tanaka, K., & Logothesis, N. (2011). Attention but not awareness modulates the BOLD signal in the human V1 during binocular suppression. *Science, 334*(6057), 829–831.

Wilke, M., Leopold, D., & Logothetis, N. (2002). Flash suppression without interocular conflict. *Soc Neurosci Abstracts, 161.15.*

Wilke, M., Logothetis, N. K., & Leopold, D. A. (2003). Generalized flash suppression of salient visual targets. *Neuron, 39,* 1043–1052.

Wilke, M., Mueller, K. M., & Leopold, D. A. (2009). Neural activity in the visual thalamus reflects perceptual suppression. *Proc Natl Acad Sci USA, 106,* 9465–9470.

Wolfe, J. (1984). Reversing ocular dominance and suppression in a single flash. *Vis Res, 24,* 471–478.

Wunderlich, K., Schneider, K. A., & Kastner, S. (2005). Neural correlates of binocular rivalry in the human lateral geniculate nucleus. *Nat Neurosci, 8,* 1595–1602.

69 Toward an Objective Index of Consciousness

MARCELLO MASSIMINI

ABSTRACT We usually assess another individual's level of consciousness based on her/his ability to connect to the surrounding environment and produce appropriate responses. However, we know that consciousness can be entirely generated within the brain, even in the absence of any interaction with the external world; this happens almost every night, while we dream, and may occur during certain forms of anesthesia as well as in brain-injured patients who emerge from coma and remain unresponsive. Yet, to this day, we still lack a way to assess the level of consciousness that is independent of processing sensory inputs and producing appropriate motor outputs. The aim of this chapter is to suggest that establishing an objective index of the level of consciousness that does not rely on a subject's capacity to access or respond to the surrounding environment is a challenging undertaking, but not an impossible one. One way to accomplish this task involves three fundamental steps. First, start from theoretical principles that suggest which intrinsic properties are fundamental for a physical system to give rise to conscious experience. Second, devise a practical means to gauge these properties in human brains. Third, test the candidate metric in different controlled conditions—such as wakefulness, sleep, anesthesia, and brain injury—in which consciousness is known to be present, diminished, or lost. To the extent that this procedure yields a reliable, graded measurement scale along the unconsciousness-consciousness spectrum, an objective frame of reference may be available to assess subjects who are completely disconnected from the external environment.

Everyone knows what consciousness is: it is what vanishes when we fall into dreamless sleep or general anesthesia and reappears when we wake up or when we dream—in other words, it is synonymous with experience. Though we are very familiar, via a first-person perspective, with the transition from consciousness to unconsciousness and back, we still lack a scientifically well-grounded method to assess the level of consciousness of other individuals. How do we judge if somebody is conscious—experiencing things such as sights, sounds, and maybe pains? Usually, if we observe purposeful behavior and appropriate responses to sensory stimuli or commands, we decide that the person is conscious. If in doubt, as when someone is resting with eyes closed, we can ask: if she answers that she was thinking or daydreaming, we infer she was conscious.

But sometimes matters are less clear: someone fast asleep shows no purposeful activity and will not respond to questions. If awakened, at times she may say she was experiencing nothing; at other times that she was dreaming, and recall a vivid experience. During anesthesia, some people may regain consciousness yet be unable to signal it. Similarly, some patients with brain damage may be behaviorally unresponsive and thus judged clinically unconscious, yet they may be able to generate brain signals indicating they understood a question or a command. In general, the problem is that, while we assess level of consciousness based on an individual's ability to access and respond to the external environment, these features are not necessary.

Dissociations between consciousness and responsiveness

SLEEP Behaviorally, the most striking consequence of falling asleep is a progressive disconnection from the external environment. Reaction times to auditory tones become longer prior to falling asleep, and responses are absent coincident with the transition to nonrapid eye movement (NREM) sleep (Ogilvie & Wilkinson, 1984). With the deepening of NREM sleep, responses can be obtained only with progressively louder tones; this feature is known as "high arousal threshold," and it persists during REM sleep (Rechtschaffen, Hauri, & Zeitlin, 1966).

Conscious experience during sleep can be assessed by studying subjective reports obtained after awakenings from different stages or at different times of night. Generally, during NREM sleep, there are longer reports later in the night and short reports early in the night. Notably, during the first NREM episode, when slow waves are prevalent in the electroencephalogram (EEG), a substantial number of awakenings yield no report whatsoever (McNamara et al., 2010). Thus, NREM sleep early in the night is the only phase of adult life during which healthy human subjects may deny that they were experiencing anything at all. In this case, responsiveness and consciousness are concurrently reduced.

Indeed, the most remarkable dissociation between responsiveness and conscious experience in physiological conditions occurs during REM sleep. Awakening from REM sleep yields reports of conscious experience 80–90% of the time and, especially in the morning hours, the percentage is close to 100%, which is of course the report rate of wakefulness. Most REM reports have the characteristic of typical dreams: complex, temporally unfolding hallucinatory episodes that can be as vivid as waking experience—yet the subject remains unresponsive, and sensory stimuli are ignored to the point that they are rarely incorporated in dreams (Koulack, 1969).

The neural mechanisms of unresponsiveness to sensory events during sleep are still unclear. During NREM sleep, thalamocortical neurons become hyperpolarized and generate intrinsic oscillatory activity that may decouple inputs from outputs. Thus, it has been suggested that the "thalamic gate" to the cerebral cortex is partially closed (Steriade, 2000). This putative mechanism of sensory disconnection, however, cannot be invoked during REM sleep, when thalamocortical neurons return to be steadily depolarized, as in quiet wakefulness. Thus, it has been suggested that the gating of sensory inputs may occur somewhere at the cortical level (Nir & Tononi, 2010). In addition to sensory disconnection, muscle paralysis brought about by descending inhibition from brainstem centers contributes to unresponsiveness, especially during REM sleep (Hobson, Pace-Schott, & Stickgold, 2000).

Anesthesia As during physiological sleep, during anesthesia responsiveness and consciousness may also decouple. Some dissociative anesthetic agents, such as ketamine at high doses, are known to induce a dreamlike hallucinatory state associated with sensory disconnection and complete unresponsiveness. In addition, during general anesthesia subjects may recover consciousness and regain sensory connectedness to the environment but still remain unable to respond. Once every 1,000 to 2,000 operations, a patient may wake up and report having experienced events of surgery (Sandin, Enlund, Samuelsson, & Lennmarken, 2000). This condition, where wakefulness-like consciousness cannot be communicated to the outside world, is termed "anesthesia awareness" and may have long-lasting psychological effects. Lack of responsiveness in these cases can be largely ascribed to muscle-paralyzing agents that are usually administered to achieve gross patient immobility during surgical procedures. However, studies employing the isolated forearm technique, in which an inflated cuff prevents the paralyzing agent from reaching the hand muscles, have shown that direct pharmacological muscle blockage is not the only mechanism responsible for impaired responsiveness in subjects who become aware during anesthesia; though patients undergoing the isolated forearm protocol may use their nonparalyzed hand to respond when prompted, they rarely show spontaneous responses and voluntary activity (Sanders, Tononi, Laureys, & Sleigh, 2012). This striking reduction of motor initiative points to a central mechanism for decreased responsiveness. One possibility is that since most anesthetics suppress neuronal activity in the putamen and the amygdala (Mhuircheartaigh et al., 2010), they may produce unresponsiveness by impairment of motivation or decision-making action selection.

Brain Injury In patients who survive severe brain injury, consciousness is assessed clinically based on the subject's ability to connect and respond to the environment. Patients who remain unresponsive even though their eyes may be open are considered unconscious (vegetative state). The appearance of nonreflexive behaviors, such as visual tracking or responding to simple commands, are sufficient clinical criteria for the minimally conscious state (MCS), while functional communication marks the emergence from the minimally conscious state (EMCS; Giacino et al., 2002). Because of concurrent lesion of motor systems and pathways, however, it may happen that brain-injured patients recover consciousness but are unable to signal it behaviorally. For this reason, neuroimaging protocols (Owen et al., 2006) have been developed to probe for signs of awareness even in patients who are completely unable to move. In these protocols, subjects are instructed verbally to enter and sustain specific mental states (such as imagining playing tennis) while their brain activity is recorded; in this way, some vegetative patients may signal that they are aware by producing specific neural responses and, in exceptional cases, they can establish a basic form of communication (Monti et al., 2010). However, willful brain responses to commands, as well as cognitive potentials triggered by sensory stimuli, are often absent in conscious brain-injured patients, resulting in a significant rate of false negatives (Bardin et al., 2011; Fischer, Luaute, & Morlet, 2010; Höller et al., 2011). In fact, there may be many cases, such as in aphasia, akinetic mutism, catatonic depression, or diffuse dopaminergic lesions, where a patient, although aware, is not able to understand or willing to respond. More fundamentally, a brain-injured subject may not respond to verbal commands or sensory stimuli simply because a peripheral or central lesion prevents sensory inputs from being transmitted and processed effectively.

The empirical search for an objective index of consciousness: Advances and problems

The physiological, pharmacological, and pathological examples outlined above demonstrate that conscious experience can be present in subjects who are unable to access sensory events and/or unable to produce appropriate (behavioral or neural) responses. Ideally, since sensory and motor functions are not necessary, one should try to identify and measure directly in the brain the minimal set of neuronal mechanisms that are necessary and sufficient for consciousness to emerge, the so-called neural correlates of consciousness (NCC). Practically, these aspects of neuronal activity should always be detectable when consciousness is present and always absent when consciousness is lost.

Over the past two decades, great progress has been made in the search for the NCC. This approach, spearheaded by Francis Crick and Christof Koch (Crick & Koch, 2003) is often implemented by comparing brain responses to sensory stimuli that are or are not perceived, as indicated by the subject's verbal report. Stimuli can be embedded in noise so they are at threshold for visibility, they can be masked by strong stimuli presented briefly afterwards, or they can be made visible/invisible through manipulations such as binocular rivalry (see Dehaene & Changeux, 2011, for review). In this way, it has been possible to ask whether the visibility of a stimulus is correlated with changes in activity in primary versus higher-order visual areas, whether visibility is accompanied by late EEG potentials (e.g., P300), or by changes in power and coherence in various frequency bands. While this work is essential in advancing the scientific study of consciousness, it should be remembered that in these experimental paradigms subjects remain fully conscious throughout—it is only the content of consciousness that changes. Thus, the level of consciousness is not the variable under study. Crucially, these experimental approaches depend on a subject's capacity to decide whether she saw or heard something and her willingness to communicate her decision: in other words, in these paradigms it is difficult to dissociate consciousness from sensory processing and executive function. Yet, as described in the previous section, these functions may be impaired or absent independently of consciousness, especially in pathological conditions.

Undoubtedly, studying the neural correlates of perception has added new insights, suggesting several possibilities concerning the neural mechanisms that may underlie conscious experience. Thus, it has been proposed that consciousness may require the ignition of strong, widespread brain activations. For example, a crucial role has been proposed for high-frequency neural activity and the dynamic formation of neuronal coalitions, including prefrontal areas (Crick & Koch, 2003); the occurrence of reentrant interactions between the front and the back of the brain (Edelman, 2001); a high level of cortical depolarization with a background of high-frequency activity (Llinás, Ribary, Contreras, & Pedroarena, 1998); the involvement of a "global workspace" (Baars, 2005) encompassing frontoparietal areas (Dehaene & Changeux, 2011); the activation of higher-order association cortices (Laureys, 2004); the long-range, high-frequency synchronization of brain activity (Singer, 2001); and related ideas.

While these proposals have heuristic value, the empirical evidence does not provide criteria for necessity and sufficiency. For example, prefrontal areas are certainly important for evaluating and reflecting and deciding upon an experience, but why would they be necessary for the experience itself? Indeed, consciousness appears to survive large prefrontal lesions (Markowitsch & Kessler, 2000). Reentry can favor cooperative interactions between distant areas, but there is abundant opportunity for reentry in structures that do not seem to participate directly in generating consciousness, such as the hippocampal formation (Crick & Koch, 1998). High-frequency firing may signal strong activation, but it can occur in the absence of consciousness, and vice versa. For example, strong, high-frequency activity, and even reactivity to sensory stimuli, can occur during both deep sleep and anesthesia (Kakigi et al., 2003; Kroeger & Amzica, 2007; Steriade, 2000), and both humans and rats undergoing inhalation anesthesia show loss of consciousness (in rats, loss of righting reflex) despite increased gamma power in the EEG (Imas et al., 2005). Late event-related potentials, such as the mismatch negativity and P300, reflect widespread brain activations that occur when subjects report detecting a stimulus, thus representing a neural correlate of conscious access to sensory events (Dehaene & Changeux, 2011). However, more often than not, these components are absent in conscious brain-injured patients (Fischer et al., 2010; Höller et al., 2011; King et al., 2013). Synchronization usually reflects the occurrence of distributed interactions, which is presumably important for consciousness, but there are well-known situations, such as generalized seizures, in which frontoparietal areas and high-order associative networks are highly active and massively synchronized, yet consciousness is lost (Arthuis et al., 2009). Similarly, coherence and Granger causality can actually be increased when consciousness is reduced in NREM sleep (Duckrow & Zaveri, 2005) and during loss of consciousness induced by general anesthesia (Barret et al., 2012; Supp, Siegel, Hipp, & Engel, 2011).

In parallel with measures that index the strength, extent, and synchronization of brain activity, alternative empirical markers of level of consciousness have been proposed that quantify the information or spectral content of brain signals. In anesthesiology, the most commonly used monitor is the bispectral index (BIS). The BIS evaluates the frontal EEG based on proprietary algorithms that compound indices of EEG beta activity, fast and slow synchronization, and burst-suppression patterns, yielding a value between 0 (inactive EEG) and 100 (fully alert subject). Another algorithm—the spectral entropy algorithm—is based on evaluating a frontal EEG channel as well as muscle activity (E-Entropy). Although on average such indices have lower values during deep anesthesia than when the subject is alert (Avidan et al., 2008), they cannot reliably assess consciousness in individual subjects due to wide variation across subjects (Kaskinoro et al., 2011). Equally, they are not reliable in discriminating at the individual level between conscious and unconscious brain-injured patients (Schnakers, Majerus, & Laureys, 2005).

In conclusion, based on empirical evidence, several brain-derived measures have been considered as candidate markers of level of consciousness. All these measures evaluate either the spatial extent of neuronal activations or their spectral content/entropy. None of them, however, discriminates reliably between consciousness and unconsciousness on a subject-by-subject basis in different conditions.

A theory-driven index of consciousness

Theoretical Principles Given the difficulty of identifying a reliable marker of consciousness based on empirical data, a theoretical approach grounded in first principles may represent a useful complement (Boly, Massimini, & Tononi, 2009). Such an approach should start from self-evident axioms in order to establish what physical properties are fundamental for consciousness and how they can be measured.

Naturally, in the case of consciousness, evidence can only be gathered from phenomenology, the first-person observation of subjective experience itself. Phenomenologically, each conscious experience is both differentiated—it has many specific features that distinguish it from a large repertoire of other experiences—and integrated—it cannot be divided into independent components (Tononi, 2004). Mechanistically, these fundamental properties of subjective experience are thought to rely on the ability of multiple, functionally specialized cortical areas to interact rapidly and effectively to form an integrated whole. Hence, an emerging idea in theoretical neuroscience is that consciousness relies on an optimal balance between functional integration and functional differentiation in thalamocortical networks—otherwise defined as brain complexity (Seth, Dienes, Cleeremans, Overgaard, & Pessoa, 2008; Tononi, 2004, 2008; Tononi & Edelman, 1998). This notion implies that brain complexity should be high when consciousness is present and low whenever consciousness is lost in sleep, anesthesia, or coma.

Based on this general principle, theoretical measures have been designed to assess the joint presence of differentiation and integration in neural systems. For example, neural complexity (C_N; Tononi, Sporns, & Edelman, 1994) is high when small subsets of elements tend to show independence (differentiation) but large subsets show increasing dependence (integration). A related metric, called causal density (C_d; Seth, Barrett, & Barnett, 2011), is based on Granger causality and is high if a system's elements are both globally integrated (they predict each other's activity) and differentiated (they contribute to these predictions in different ways). Finally, Φ, a measure that is directly derived from the information integration theory of consciousness (IITC; Tononi, 2008), is based on perturbing a system in all possible ways in order to count the number of different states (differentiation) that can be discriminated through causal interactions within the system as a whole (integration).

These theoretical measures differ in some respect, but they all share the insight that level of consciousness depends on the extent to which neural elements can engage in complex activity patterns that are, at once, distributed within a system of causally interacting cortical areas (integrated) and differentiated in space and time (information-rich). So far, however, the proposed theoretical metrics can only be applied to simple systems of simulated elements or under highly restrictive assumptions.

A Practical Tool to Approximate Theoretical Measures Clearly, testing the general hypothesis that changes in level of consciousness are invariably linked to changes in brain complexity entails the development of practical approximations. As recently proposed (Massimini, Boly, Casali, Rosanova, & Tononi, 2009), a rather straightforward way to gauge the conjoint presence of integration and information in real brains involves directly probing the cerebral cortex (in order to avoid possible subcortical filtering and gating) by employing a perturbational approach (thus testing causal interactions rather than temporal correlations) and examining to what extent cortical regions can interact as a whole (integration) to produce differentiated responses (information). According to this proposal, a signature of consciousness is that the thalamocortical system should respond to perturbations with complex, rapidly changing activity patterns (information) that affect a distributed set of cortical areas (integration). On the other hand, it can be predicted that during loss of consciousness, whether this is caused by sleep, anesthesia, or coma, the brain should react to perturbations with a response that is local (loss of integration) and/or stereotypical (loss of information).

Practically, these predictions can be tested in human brains by employing a combination of transcranial magnetic stimulation (TMS) and high-density EEG, a technique that allows stimulating directly a subset of cortical neurons and measuring, with good spatial-temporal resolution, the effects produced by this perturbation on the rest of the thalamocortical system (Ilmoniemi et al., 1997). As shown in figure 69.1, in healthy, awake subjects TMS triggers a complex EEG response involving different cortical areas at different times (Massimini et al., 2005; Rosanova et al., 2009).

Conversely, when subjects lose consciousness during NREM sleep, TMS pulses invariably produce a simple wave of activation that remains localized to the site of stimulation, indicating a breakdown of communication and a loss of integration within thalamocortical networks (Massimini et al., 2005). The disappearance of a long-range, differentiated pattern of cortical activation is not simply due to a reduction of responsiveness of hyperpolarized cortical neurons. In fact, increasing TMS intensity only results in a larger, simple positive-negative wave, closely resembling a spontaneous sleep slow wave (Massimini et al., 2007). In this case, the response to TMS is stereotypical and spreads like an oil spot to vast regions of the cortex, revealing a loss of differentiation. Similar local and/or stereotypical responses are invariably also found during general anesthesia (Ferrarelli et al., 2010) as well as in brain-injured patients with an unambiguous clinical diagnosis of a vegetative state (Rosanova et al., 2012). Crucially, wakefulness-like, complex responses always recover during REM sleep (Massimini et al., 2010) in minimally conscious patients, who show signs of nonreflexive

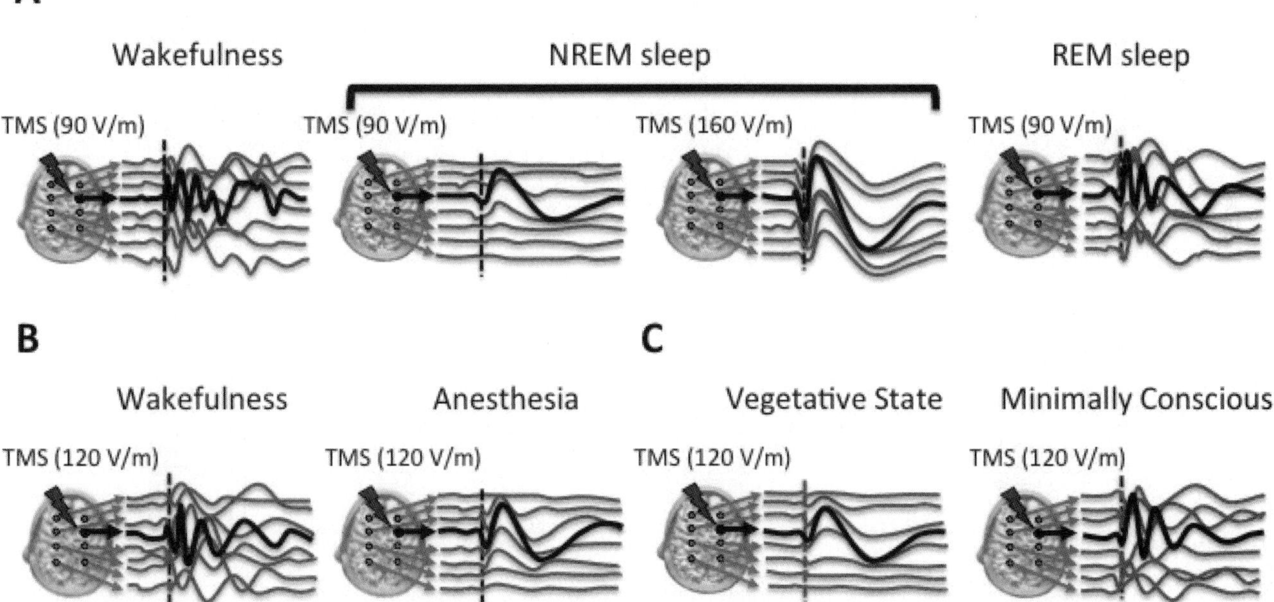

Figure 69.1 Examples of transcranial magnetic stimulation (TMS)-evoked cortical activations in sleep (A), anesthesia (B), and brain-injured patients (C). The gray arrows indicate the cortical sites of TMS. The values in the parentheses indicate the intensity of the electric field (volt/meter) induced by TMS on the cortical surface. The traces show the TMS-evoked currents recorded from eight cortical sources in both hemispheres (gray circles); the thick trace highlights the activation recorded from the cortical areas located under the stimulator. During wakefulness, TMS triggers a sustained response that engages distributed cortical sources in spatially and temporally differentiated patterns of activation. During nonrapid eye movement (NREM) sleep, anesthesia, and the vegetative state, the thalamocortical system, despite being active and reactive, loses its ability to engage in distributed, complex activity patterns; it either breaks down in casually independent modules (loss of integration) or, when TMS is delivered at high intensity, bursts in a large and stereotypical response (loss of differentiation). During rapid eye movement (REM) sleep and in the minimally conscious state, the TMS response recovers its spatial spread and differentiation.

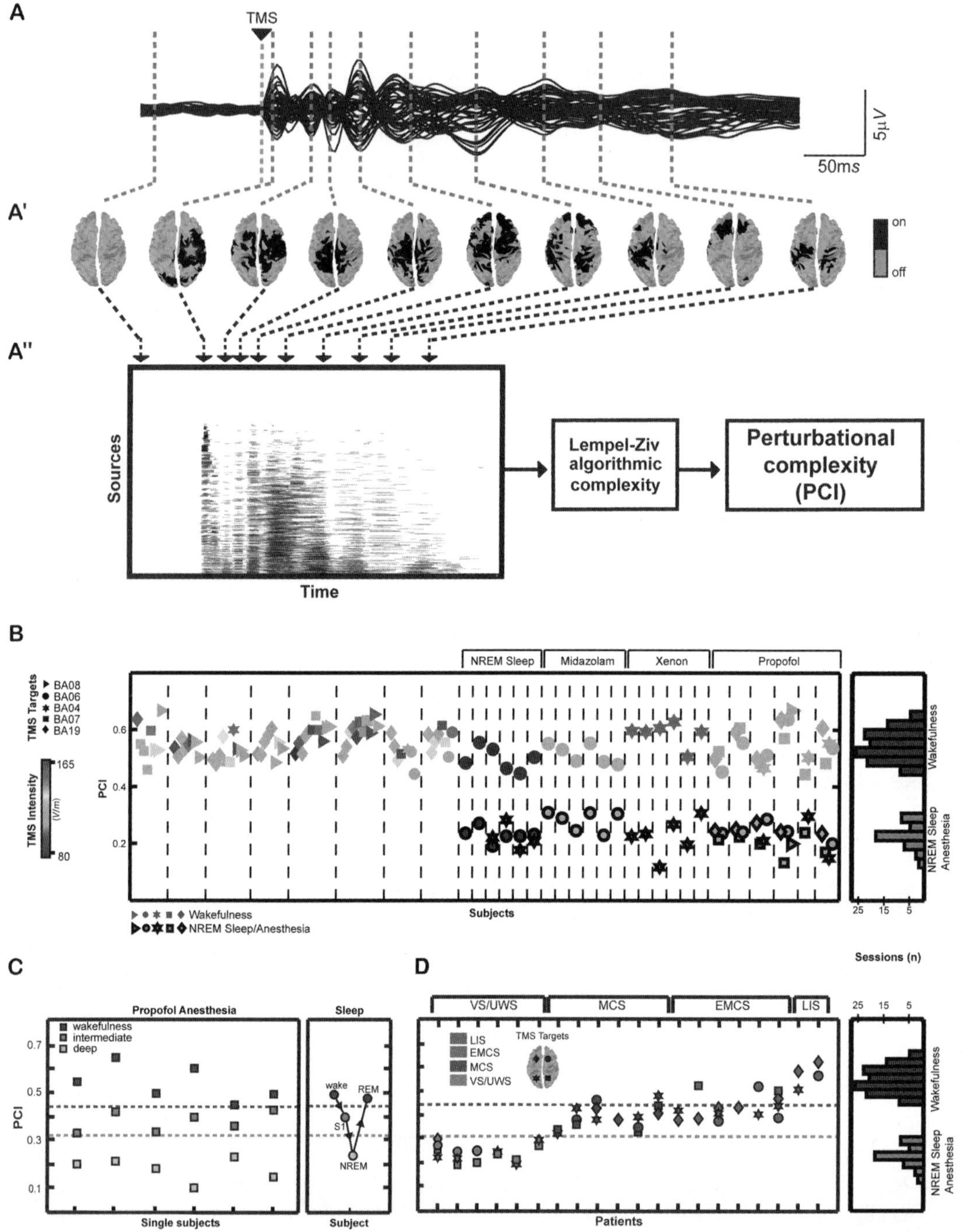

FIGURE 69.2 Calculating and testing the perturbational complexity index (PCI). Operationally, PCI is defined as the normalized Lempel-Ziv complexity of the overall spatiotemporal pattern of significant cortical activation triggered by a direct perturbation. Its calculation starts from scalp TMS-evoked potentials (A) and requires performing source-modeling and nonparametric statistics to detect the time course of significant cortical activations triggered by TMS (A′). This procedure results in a binary spatiotemporal matrix of cortical activation (A″), which is compressed using the same algorithm (Lempel & Ziv, 1976) that is commonly employed to zip digital files. Complex patterns of activation that are, at once, distributed and differentiated cannot be efficiently compressed and result in high values of perturbational complexity. (B) PCI values are shown for 152 TMS sessions collected from 32 healthy subjects. The histograms on the right display the distributions of PCI across subjects during alert wakefulness (dark gray bars) and loss of consciousness in NREM sleep and different forms of anesthesia (light gray bars). PCI does not depend on stimulation site or TMS intensity but is solely sensitive to changes in the level of consciousness. PCI calculated during wakefulness (110 sessions) ranges between 0.44 and 0.67 (mean: 0.55 ± 0.05), whereas PCI calculated after loss of consciousness (42 sessions) ranges between 0.12 and 0.31 (mean: 0.23 ± 0.04), giving rise to two separated distributions. (C) PCI yields intermediate values

activity, and in locked-in syndrome (LIS) subjects, who are totally paralyzed except for vertical eye movements through which they signal that they are aware (Rosanova et al., 2012).

ESTABLISHING A MEASURING SCALE Besides providing qualitative support to basic theoretical predictions, TMS and high-density EEG measurements open the possibility of developing a quantitative index of the level of consciousness, a necessary step to construct a measuring scale. To this end, a novel empirical measure called the perturbational complexity index (PCI) was recently introduced (Casali et al., 2013). Calculating PCI involves two fundamental steps: (1) perturbing the cortex with TMS to engage distributed interactions in the brain (*integration*) and (2) "zipping" (i.e., compressing) the resulting electrocortical responses to measure their algorithmic complexity (*information*). The underlying idea is that PCI should be low if causal interaction among cortical areas is reduced (loss of integration), because the matrix of activation engaged by TMS is spatially restricted; PCI is also expected to be low if many interacting areas react to the perturbation but they do so in a stereotypical way (loss of differentiation) because, in this case, the resulting matrix is large but redundant and can be effectively compressed. In fact, PCI should reach high values only if the initial perturbation is transmitted to a large set of integrated areas that react in a differentiated way, giving rise to a spatiotemporal pattern of deterministic activation that cannot be easily reduced. In these terms, PCI provides a rough estimation of the theoretical measure of Φ, which can be defined as the amount of irreducible information that a system generates above its parts (Tononi, 2008).

In recent work, PCI was tested on a large data set of TMS-evoked potentials recorded in healthy subjects during wakefulness, dreaming, NREM sleep, and different levels of sedation induced by various anesthetic agents (midazolam, xenon, and propofol), as well as in brain-injured patients who emerged from coma and entered different clinical states (vegetative, MCS, EMCS, LIS). As shown in figure 69.2, PCI is reproducible within and across subjects and depends in a graded fashion on the level of consciousness. Crucially, in healthy (awake, sleeping, and anesthetized) subjects—whose level of consciousness can be known based on subjective reports upon awakening—PCI provides a reproducible and reliable scale along the unconsciousness-consciousness spectrum. This scale can then be used as an independent frame of reference to assess more challenging cases. For example, brain-injured patients (MCS, EMCS, and LIS) who show minimal signs of consciousness attained values of brain complexity that were invariably above the maximum value obtained in unconscious, anesthetized, or sleeping, healthy subjects.

To the extent that this measurement scale is further validated by empirical data, it may be then applicable to disconnected/unresponsive individuals whose level of consciousness is unknown. Already in the context of the present results, finding a PCI value above the sleep-anesthesia distribution in a patient who is otherwise completely disconnected from the external environment would suggest that she/he is conscious to some extent.

Crucially, PCI is measured by evaluating the compressibility of the deterministic brain response to TMS, a perturbation that engages large portions of the thalamocortical system directly without requiring the subjects to perform any sensory, motor, or cognitive task. In this way, the brain's capacity for consciousness can be assessed based on the complexity of cortical interactions, independent of the subject's capacity to access and react to external stimuli.

Conclusion

Compelling evidence suggests that subjective experience can be generated within a brain that is disconnected from the external world on the input and output side. In these cases, behavioral and neurophysiological measures of consciousness that rely on the brain's capacity to access or respond to the external environment are not applicable by definition. In the present chapter, we suggest that resorting to a theoretical approach, combined with systematic empirical testing in controlled conditions, may help overcome such circularity. Theoretical principles suggest that in order to obtain a reliable index of level of consciousness, measuring either the spread or the entropy of neuronal activations is not enough; instead, one should assess the

during intermediate levels of propofol sedation and during sleep stage 1. (D) PCI values are shown for 48 TMS sessions collected from 20 brain-injured patients. PCI followed the level of consciousness (as clinically assessed with coma recovery scale, or CRS-R) progressively increasing from vegetative (VS) through a minimally conscious state (MCS) to recovery of functional communication (emergence from the minimally conscious state; EMCS) and attaining levels of healthy awake subjects in locked-in syndrome (LIS). Patient results are within the frame of reference obtained in awake, sleeping, and anesthetized control subjects (modified from Casali et al., 2013). (See color plate 59.)

intrinsic brain's capacity for information integration, a fundamental mechanism of consciousness as suggested by phenomenology. A first attempt in this direction is represented by PCI, an empirical measure that gauges, albeit coarsely, both the information content and the integration of the overall output of the thalamocortical system in response to a direct perturbation. In parallel, reliable indices that gauge the brain capacity to integrate information based on spontaneous EEG activity alone (Barret & Seth, 2011; King et al., 2013; Marinazzo et al., 2014) may be developed. At the moment, perturbing the brain to measure PCI overcomes a fundamental limitation of current brain-based metrics of consciousness since it allows establishing, for the first time, an objective scale that is reliable across subjects and conditions (sleep, anesthesia, and coma). This is a crucial step; yet the validity of these theory-driven measures may be questioned by further testing. As an example, future experiments may demonstrate that brain complexity is low in subjects anesthetized with ketamine, who are unresponsive but report vivid dreams upon awakening. Similarly, it will be important to demonstrate that the brain's capacity for information integration is high in vegetative patients who show willful neuronal activation in the MRI scanner. In all cases, it is likely that a joint development of theories and novel measurement tools, where precise hypotheses are systematically validated—or rejected—by means of extensive testing in different conditions will benefit the search for a scientifically grounded marker of consciousness.

ACKNOWLEDGMENTS Research funded by PRIN 2010 from Italian Ministry of Education, University and Research; EU grant FP7-ICT-2011-9 (n. 600806, "Corticonics"); and the James S. McDonnell Foundation Scholar Award 2013.

REFERENCES

ARTHUIS, M., VALTON, L., RÉGIS, J., CHAUVEL, P., WENDLING, F., NACCACHE, L., ... BARTOLOMEI, F. (2009). Impaired consciousness during temporal lobe seizures is related to increased long-distance cortical-subcortical synchronization. *Brain, 132*(8), 2091–2101.

AVIDAN, M. S., ZHANG, L., BURNSIDE, B. A., FINKEL, K. J., SEARLEMAN, A. C., SELVIDGE, J. A., ... EVERS, A. S. (2008). Anesthesia awareness and the bispectral index. *N Engl J Med, 358*(11), 1097–1108.

BAARS, B. J. (2005). Global workspace theory of consciousness: Toward a cognitive neuroscience of human experience. *Prog Brain Res, 150*, 45–53.

BARDIN, J. C., FINS, J. J., KATZ, D. I., HERSH, J., HEIER, L. A., TABELOW, K., ... VOSS, H. U. (2011). Dissociations between behavioural and functional magnetic resonance imaging-based evaluations of cognitive function after brain injury. *Brain, 134*(3), 769–782.

BARRETT, A. B., MURPHY, M., BRUNO, M.-A., NOIRHOMME, Q., BOLY, M., LAUREYS, S., & SETH, A. K. (2012). Granger causality analysis of steady-state electroencephalographic signals during propofol-induced anesthesia. *PLoS ONE, 7*(1), e29072.

BARRET, A. B., & SETH, A. K. (2011). Practical measures of integrated information on time-series data. *PLoS Comput Biol, 7*(1), e1001052

BOLY, M., MASSIMINI, M., & TONONI, G. (2009). Theoretical approaches to the diagnosis of altered states of consciousness. *Prog Brain Res, 177*, 383–398.

CASALI, A., GOSSERIES, O., ROSANOVA, M., BOLY, M., SARASSO, S., CASALI, K., ... MASSIMINI, M. (2013). A theoretically based index of consciousness independent of sensory processing and behavior. *Sci Transl Med, 14; 5*(198), 198ra105.

CRICK, F., & KOCH, C. (1998). Consciousness and neuroscience. *Cereb Cortex, 8*(2), 97–107.

CRICK, F., & KOCH, C. (2003). A framework for consciousness. *Nat Neurosci, 6*(2), 119–126.

DEHAENE, S., & CHANGEUX, J.-P. (2011). Experimental and theoretical approaches to conscious processing. *Neuron, 70*(2), 200–227.

DUCKROW, R. B., & ZAVERI, H. P. (2005). Coherence of the electroencephalogram during the first sleep cycle. *Clin Neurophysiol, 116*(5), 1088–1095.

EDELMAN, G. (2001). Consciousness: The remembered present. *Ann NY Acad Sci, 929*, 111–122.

FERRARELLI, F., MASSIMINI, M., SARASSO, S., CASALI, A., RIEDNER, B. A., ANGELINI, G., ... PEARCE, R. A. (2010). Breakdown in cortical effective connectivity during midazolam-induced loss of consciousness. *Proc Natl Acad Sci USA, 107*(6), 2681–2686.

FISCHER, C., LUAUTE, J., & MORLET, D. (2010). Event-related potentials (MMN and novelty P3) in permanent vegetative or minimally conscious states. *Clin Neurophysiol, 121*(7), 1032–1042.

GIACINO, J. T., ASHWAL, S., CHILDS, N., CRANFORD, R., JENNETT, B., KATZ, D. I., ... ZASLER, N. D. (2002). The minimally conscious state: Definition and diagnostic criteria. *Neurology, 58*(3), 349–353.

HOBSON, J. A., PACE-SCHOTT, E. F., & STICKGOLD, R. (2000). Dreaming and the brain: Toward a cognitive neuroscience of conscious states. *Behav Brain Sci, 23*(6), 793–842; discussion 904–1121.

HÖLLER, Y., BERGMANN, J., KRONBICHLER, M., CRONE, J. S., SCHMID, E. V., GOLASZEWSKI, S., & LADURNER, G. (2011). Preserved oscillatory response but lack of mismatch negativity in patients with disorders of consciousness. *Clin Neurophysiol, 122*(9), 1744–1754.

ILMONIEMI, R. J., VIRTANEN, J., RUOHONEN, J., KARHU, J., ARONEN, H. J., NÄÄTÄNEN, R., & KATILA, T. (1997). Neuronal responses to magnetic stimulation reveal cortical reactivity and connectivity. *NeuroReport, 8*(16), 3537–3540.

IMAS, O. A., ROPELLA, K. M., WARD, B. D., WOOD, J. D., & HUDETZ, A. G. (2005). Volatile anesthetics enhance flash-induced gamma oscillations in rat visual cortex. *Anesthesiology, 102*(5), 937–947.

KAKIGI, R., NAKA, D., OKUSA, T., WANG, X., INUI, K., QIU, Y., ... HOSHIYAMA, M. (2003). Sensory perception during sleep in humans: A magnetoencephalograhic study. *Sleep Med, 4*(6), 493–507.

KASKINORO, K., MAKSIMOW, A., LÅNGSJÖ, J., AANTAA, R., JÄÄSKELÄINEN, S., KAISTI, K., ... SCHEININ, H. (2011). Wide inter-individual variability of bispectral index and spectral entropy at loss of consciousness during increasing concentrations of dexmedetomidine, propofol, and sevoflurane. *Br J Anaesth, 107*(4), 573–580.

King, J. R., Faugeras, F., Gramfort, A., Schurger, A., El Karoui, I., Sitt, J. D., ... Dehaene, S. (2013). Single-trial decoding of auditory novelty responses facilitates the detection of residual consciousness, *NeuroImage, 83,* 726–738.

King, J. R., Sitt, J. D., Faugeras, F., Rohaut, B. A., El Karoui, I., Cohen, L., ... Dehaene, S. (2013). Information sharing in the brain indexes consciousness in noncommunicative patients, *Curr Biol, 23*(19), 1914–1919.

Koulack, D. (1969). Effects of somatosensory stimulation on dream content. *Arch Gen Psychiatry, 20*(6), 718–725.

Kroeger, D., & Amzica, F. (2007). Hypersensitivity of the anesthesia-induced comatose brain. *J Neurosci, 27*(39), 10597–10607.

Laureys, S. (2004). Functional neuroimaging in the vegetative state. *NeuroRehabilitation, 19*(4), 335–341.

Lempel, A., & Ziv, J. (1976). On the complexity of finite sequences. *IEEE Trans Inform Theory, 22,* 75–81.

Llinás, R., Ribary, U., Contreras, D., & Pedroarena, C. (1998). The neuronal basis for consciousness. *Philos Trans R Soc Lond B Biol Sci, 353*(1377), 1841–1849.

Markowitsch, H. J., & Kessler, J. (2000). Massive impairment in executive functions with partial preservation of other cognitive functions: The case of a young patient with severe degeneration of the prefrontal cortex. *Exp Brain Res, 133*(1), 94–102.

Marinazzo, D., Gosseries, O., Boly, M., Ledoux, D., ... Laureys, S. (2014). Directed information transfer in scalp electroencephalographic recordings: Insights on disorders of consciousness. *Clin EEG Neurosci, 45,* 33–39.

Massimini, M., Boly, M., Casali, A., Rosanova, M., & Tononi, G. (2009). A perturbational approach for evaluating the brain's capacity for consciousness. *Prog Brain Res, 177,* 201–214.

Massimini, M., Ferrarelli, F., Esser, S. K., Riedner, B. A., Huber, R., Murphy, M., ... Tononi, G. (2007). Triggering sleep slow waves by transcranial magnetic stimulation. *Proc Natl Acad Sci USA, 104*(20), 8496–8501.

Massimini, M., Ferrarelli, F., Huber, R., Esser, S. K., Singh, H., & Tononi, G. (2005). Breakdown of cortical effective connectivity during sleep. *Science, 309*(5744), 2228–2232.

Massimini, M., Ferrarelli, F., Murphy, M., Huber, R., Riedner, B., Casarotto, S., & Tononi, G. (2010). Cortical reactivity and effective connectivity during REM sleep in humans. *Cogn Neurosci, 1*(3), 176–183.

McNamara, P., Johnson, P., McLaren, D., Harris, E., Beauharnais, C., & Auerbach, S. (2010). REM and NREM sleep mentation. *Int Rev Neurobiol, 92,* 69–86.

Mhuircheartaigh, R. N., Rosenorn-Lanng, D., Wise, R., Jbabdi, S., Rogers, R., & Tracey, I. (2010). Cortical and subcortical connectivity changes during decreasing levels of consciousness in humans: A functional magnetic resonance imaging study using propofol. *J Neurosci, 30*(27), 9095–9102.

Monti, M. M., Vanhaudenhuyse, A., Coleman, M. R., Boly, M., Pickard, J. D., Tshibanda, L., ... Laureys, S. (2010). Willful modulation of brain activity in disorders of consciousness. *N Engl J Med, 362*(7), 579–589.

Nir, Y., & Tononi, G. (2010). Dreaming and the brain: From phenomenology to neurophysiology. *Trends Cogn Sci, 14*(2), 88–100.

Ogilvie, R. D., & Wilkinson, R. T. (1984). The detection of sleep onset: Behavioral and physiological convergence. *Psychophysiology, 21*(5), 510–520.

Owen, A. M., Coleman, M. R., Boly, M., Davis, M. H., Laureys, S., & Pickard, J. D. (2006). Detecting awareness in the vegetative state. *Science, 313*(5792), 1402.

Rechtschaffen, A., Hauri, P., & Zeitlin, M. (1966). Auditory awakening thresholds in REM and NREM sleep stages. *Percept Mot Skills, 22*(3), 927–942.

Rosanova, M., Casali, A., Bellina, V., Resta, F., Mariotti, M., & Massimini, M. (2009). *J Neurosci, 29*(24), 7679–7685.

Rosanova, M., Gosseries, O., Casarotto, S., Boly, M., Casali, A. G., Bruno, M.-A., ... Massimini, M. (2012). Recovery of cortical effective connectivity and recovery of consciousness in vegetative patients. *Brain, 135*(Pt 4), 1308–1320.

Sanders, R. D., Tononi, G., Laureys, S., & Sleigh, J. W. (2012). Unresponsiveness ≠ unconsciousness. *Anesthesiology, 116*(4), 946–959.

Sandin, R. H., Enlund, G., Samuelsson, P., & Lennmarken, C. (2000). Awareness during anesthesia: A prospective case study. *Lancet, 355*(9205), 707–711.

Schnakers, C., Majerus, S., & Laureys, S. (2005). Bispectral analysis of electroencephalogram signals during recovery from coma: Preliminary findings. *Neuropsychol Rehabil, 15*(3–4), 381–388.

Seth, A. K., Barrett, A. B., & Barnett, L. (2011). Causal density and integrated information as measures of conscious level. *Philos Transact A Math Phys Eng Sci, 369,* 3748.

Seth, A. K., Dienes, Z., Cleeremans, A., Overgaard, M., & Pessoa, L. (2008). Measuring consciousness: Relating behavioural and neurophysiological approaches. *Trends Cogn Sci, 12,* 314.

Singer, W. (2001). Consciousness and the binding problem. *Ann NY Acad Sci, 929,* 123–146.

Steriade, M. (2000). Corticothalamic resonance, states of vigilance and mentation. *Neuroscience, 101*(2), 243–276.

Supp, G. G., Siegel, M., Hipp, J. F., & Engel, A. K. (2011). Cortical hypersynchrony predicts breakdown of sensory processing during loss of consciousness. *Curr Biol, 21*(23), 1988–1993.

Tononi, G. (2004). An information integration theory of consciousness. *BMC Neurosci, 5,* 42.

Tononi, G. (2008). Consciousness as integrated information: A provisional manifesto. *Biol Bull, 215*(3), 216–242.

Tononi, G., & Edelman, G. M. (1998). Consciousness and complexity. *Science, 282*(5395), 1846–1851.

Tononi, G., Sporns, O., & Edelman, G. M. (1994). A measure for brain complexity: Relating functional segregation and integration in the nervous system. *Proc Natl Acad Sci USA, 91,* 5033.

70 Consciousness Alterations After Severe Brain Injury

ATHENA DEMERTZI AND STEVEN LAUREYS

ABSTRACT The past fifteen years have provided an unprecedented collection of discoveries that bear upon our scientific understanding of consciousness in the human brain following severe brain damage. Highlighted among these discoveries are unique demonstrations that patients with little or no behavioral evidence of conscious awareness may retain critical cognitive capacities. These first scientific demonstrations support the possibility that some severely brain-injured patients in long-standing conditions of limited behavioral responsiveness may nonetheless retain latent capacities for awareness. Such capacities include the human functions of language and higher-level cognition that, either spontaneously or by thought-directed interventions, may reemerge even at long time intervals or can remain unrecognized. Functional neuroimaging, such as positron emission tomography and functional magnetic resonance imaging, as well as electroencephalography and evoked potential studies, have offered the possibility to objectively approach covert cognitive processes in patients who are otherwise incapable of intelligible or sustained behavioral expression. Such studies have used experimental protocols to assess brain function during resting-state conditions and after external stimulation. These technologies have further permitted the detection of nonverbal command-following and even established muscle-independent means of communication with some behaviorally unresponsive patients. Such advances are expected to shed light on the gray zones between the clinical entities of consciousness and help resolve medical and ethical controversies around the management of such challenging situations.

Defining consciousness from a clinical perspective

Consciousness is a multifaceted term for which there is no universal definition (Zeman, 2001). Clinical practice dealing with patients with disorders of consciousness teaches that we can define consciousness by reducing it to two components: wakefulness and awareness (Posner, Saper, Schiff, & Plum, 2007). Clinically, the level of wakefulness can be gauged based on eye opening, which may be absent, stimulus-induced, or spontaneous and sustained. Awareness is more difficult to define and more challenging to assess behaviorally. At the clinical level, we can only infer awareness by asking patients to follow simple commands or by observing nonreflex behaviors such as visual pursuit. The relationship between wakefulness and awareness can be described in a linear manner. For instance, every night when falling asleep, we experience a decrease of the level of wakefulness up to the point where we lose awareness of our environment. Clinical conditions resulting from severe brain injury, where the relationship between wakefulness and awareness is violated, challenge our understanding of how consciousness works in healthy situations.

Consciousness alterations after severe brain injury

BRAIN DEATH Classically, brain death is caused by a massive brain lesion, such as trauma, intracranial hemorrhage, or anoxia. The diagnostic guidelines for brain death are (Quality Standards Subcommittee of the American Academy of Neurology, 1995):

1. demonstration of coma
2. evidence for the cause of coma
3. absence of confounding factors (hypothermia, drugs, electrolyte, endocrine disturbances)
4. absence of brainstem reflexes
5. absent motor responses
6. positive apnea testing
7. a repeat evaluation in six hours (but the time period is considered arbitrary)
8. confirmatory laboratory tests (only when specific components of the clinical testing cannot be reliably evaluated)

No recovery from brain death has ever been reported over the last 50 years in a patient fulfilling the abovementioned clinical criteria (Laureys, 2005).

COMA Coma can result from bihemispheric diffuse cortical or white matter damage or bilateral brainstem lesions, affecting the subcortical reticular arousing systems. Coma is a time-limited condition leading to death, recovery of consciousness, or transition to a vegetative state (Laureys, 2007). Many factors such as etiology, the patient's general medical condition, age, clinical signs, and complimentary examinations influence the management and prognosis of coma. In terms of clinical signs, after three days of observation, a negative outcome is heralded by absence of pupillary or corneal reflexes, stereotyped or absent motor response to noxious stimulation, absent bilateral cortical responses of somatosensory evoked potentials, and, for anoxic coma, biochemical markers (i.e., high levels of serum neuron-specific enolase) (Wijdicks, Hijdra, Young, Bassetti, & Wiebe, 2006).

Vegetative State, or Unresponsive Wakefulness Syndrome The vegetative state (VS) is usually caused by diffuse lesions to gray and white matter. It can be a transition to further recovery, or it may be permanent. This is the case for VS that lasts more than one year after traumatic injury or three months after nontraumatic injury. At present, there are no validated paraclinical prognostic markers for individual patients except that the chances of recovery depend on a patient's age, etiology, and time spent in this condition (Multi-Society Task Force on PVS, 1994b).

The VS is usually described as a "state of arousal without awareness." The criteria for the diagnosis of VS are (Multi-Society Task Force on PVS, 1994a):

1. no evidence of awareness of self or environment and an inability to interact with others
2. no evidence of sustained, reproducible, purposeful, or voluntary behavioral responses to visual, auditory, tactile, or noxious stimuli
3. no evidence of language comprehension or expression
4. intermittent wakefulness manifested by the presence of sleep-wake cycles
5. sufficiently preserved hypothalamic and brainstem autonomic functions to permit survival with medical and nursing care
6. bowel and bladder incontinence
7. variably preserved cranial-nerve and spinal reflexes

More recently, it has been recognized that some in the health care community, media, and lay public feel uncomfortable using the unintentionally denigrating "vegetable-like" connotation, seemingly intrinsic to the term VS. Hence, the European Task Force on Disorders of Consciousness proposed the alternative name "unresponsive wakefulness syndrome" (UWS), a more neutral and descriptive term pertaining to patients showing a number of clinical signs of unresponsiveness (i.e., lacking response to commands or oriented voluntary movements) in the presence of wakefulness (Laureys et al., 2010).

Minimally Conscious State The minimally conscious state (MCS) was defined as a disorder of consciousness in 2002 by the Aspen Workgroup to differentiate it from VS/UWS. According to the defining criteria (Giacino et al., 2002), patients in MCS manifest at least one of the following:

1. purposeful behavior (including movements or affective behavior) contingent to relevant environment stimuli and not due to reflexive activity, such as visual pursuit or sustained fixation occurring in direct response to moving or salient stimuli, smiling or crying in response to verbal or visual emotional but not neutral stimuli, reaching for objects demonstrating a relationship between object location and direction of reach, touching or holding objects in a manner that accommodates the size and shape of the object, and vocalizations or gestures occurring in direct response to the linguistic content of questions
2. following simple commands
3. gestural or verbal yes/no response, regardless of accuracy
4. intelligible verbalization

Like the VS/UWS, MCS may be chronic and sometimes permanent. Emergence from MCS is defined by the ability to exhibit functional interactive communication or functional use of objects.

Locked-In Syndrome Locked-in syndrome (LIS) can result from a bilateral ventral pontine lesion (Posner, Saper, Schiff, & Plum, 2007b), but mesencephalic lesions have also been reported (for a review, see Laureys et al., 2005). In LIS there is no dissociation between arousal and awareness, but it is mentioned here since it can be misdiagnosed as a disorder of consciousness. According to the American Congress of Rehabilitation Medicine criteria (1995), LIS patients demonstrate:

1. sustained eye opening (bilateral ptosis should be ruled out as a complicating factor)
2. quadriplegia or quadriparesis
3. aphonia or hypophonia
4. a primary mode of communication via vertical or lateral eye movements or blinking of the upper eyelid to signal yes/no responses.
5. preserved cognitive abilities

Based on motor capacities, LIS can be divided into three categories (Bauer, Gerstenbrand, & Rumpl, 1979):

a. classic LIS, which is characterized by quadriplegia and anarthria with eye-coded communication

b. incomplete LIS, which is characterized by remnants of voluntary responsiveness other than eye movements

c. total LIS, which is characterized by complete immobility, including all eye movements, combined with preserved consciousness

Once a LIS patient becomes medically stable and is given appropriate medical care, life expectancy is estimated to be up to several decades (Laureys et al., 2005). Even if the chances of good motor recovery are very limited, existing eye-controlled computer-based communication technologies currently allow these patients to control their environment (Chatelle et al., 2012; Stoll et al., 2013).

Detecting consciousness in severely brain-injured patients

Behavior The existing behavioral scales mainly focus on deducing awareness of the environment. For example, the widely used Glasgow Coma Scale scores eye, verbal, and motor responses to external stimuli (Teasdale & Jennett, 1974). The Full Outline of Unresponsiveness (FOUR) scale recognizes the difficulty to assess verbal responses, especially in cases of intubated patients or patients with tracheotomy, and requires patients to show nonverbal conscious behaviors such as eye blinking or hand signing to command (Wijdicks et al., 2005). In that case, the FOUR can detect patients with LIS but cannot always differentiate VS/UWS from MCS patients. The Revised Coma Recovery Scale (CRS-R) is the most sensitive scale to differentiate MCS from unresponsive patients because it assesses auditory, visual, motor, oromotor, and communication abilities next to arousal assessment, covering the diagnostic criteria for MCS (Giacino, Kalmar, & Whyte, 2004). Among the existing scales, the CRS-R has been recommended as the most appropriate tool to evaluate patients with disorders of consciousness (Seel et al., 2010).

To date, bedside evaluation is the gold standard for diagnosing patients with consciousness impairments. Nevertheless, incorrect diagnosis is not rare (Schnakers et al., 2009). In order to minimize misdiagnosis rate, neuroimaging and electrophysiology procedures have begun to assume an adjunctive role in the diagnostic assessment of patients with disorders of consciousness.

Assisting Technologies Functional neuroimaging, such as positron emission tomography and functional magnetic resonance imaging (fMRI) as well as electroencephalography (EEG) and evoked potential studies, have offered the possibility to objectively approach covert cognitive processes in patients who are otherwise incapable of intelligible or sustained behavioral expression. Such studies have used experimental protocols to assess brain function during resting-state conditions and after external sensory stimulation.

Neuroimaging In resting conditions, when we do not perform any task and receive no external stimulation, our brains are engaged in some kind of typical cognitive activity when the mind is unconstrained (e.g., Mason et al., 2007; Raichle & Snyder, 2007). We have recently proposed to reduce the phenomenological complexity of such cognitive process into two components: *external* awareness, namely, everything we perceive through our senses, and *internal* awareness, or stimulus-independent thoughts (Demertzi, Soddu, & Laureys, 2013). Internal and external awareness have been shown to negatively correlate both behaviorally and at the brain level (Vanhaudenhuyse et al., 2011). More particularly, when healthy subjects were asked to independently self-rate their external and internal awareness as these were before an auditory prompt (and hence considered to reflect a resting condition), they reported an anticorrelated pattern between these two states. Interestingly, the alternation between the external and internal milieu not only was found to characterize overt behavioral reports but also had a cerebral correlate: behavioral reports of internal awareness were linked to fMRI activity of midline anterior cingulate/mesiofrontal areas as well as posterior cingulate/precuneal cortices classically coined as the default mode network (DMN). Inversely, subjective ratings for external awareness correlated with the activity of lateral fronto-parieto-temporal regions often linked to attentional processing (figure 70.1). These data suggest a coupling between brain and behavior during the resting state and underscore that certain brain areas seem to play an important role in sustaining consciousness (Heine et al., 2012). Indeed, in brain death, no DMN functional connectivity could be identified (Boly et al., 2009; Soddu et al., 2011). But alongside the spectrum of consciousness impairment, ranging from healthy controls and patients with LIS toward MCS, VS/UWS, and coma, functional connectivity is reduced in DMN (Vanhaudenhuyse et al., 2010) and other congitive-related networks (Demertzi et al., 2014). Such data suggest that fMRI resting-state acquisitions could be used as an assisting means to gain insight in patients' diagnostic picture.

Using positron emission tomography, when patients in VS/UWS were compared to healthy subjects in a resting state, they were characterized by reduced levels of global metabolism; nevertheless, recovery from VS/UWS did not coincide with resumption of global metabolic activity (Laureys, Owen, & Schiff, 2004). Rather, patients in VS/UWS showed impaired metabolism in a widespread network encompassing midline and lateral

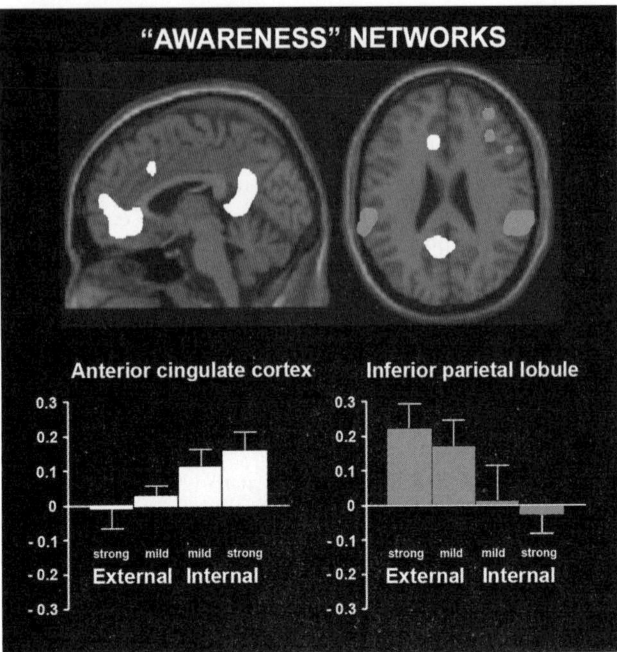

Figure 70.1 Subjective ratings about awareness are coupled to the activity of the resting brain. Behavioral reports of increased internal awareness (blue areas) correlate with connectivity in cerebral regions of the so-called default mode network, encompassing anterior and posterior cingulate cortices. Inversely, when subjects report being more externally oriented toward their environment, areas of a frontoparietal attentional network (red areas) are recruited (adapted from Vanhaudenhuyse et al., 2011). These data suggest that resting-state fMRI acquisitions could be an alternative paradigm to study residual brain function in human populations where communication is hindered (e.g., due to motor paralysis, language comprehension, or lack of motivation), such as in patients with disorders of consciousness. (See color plate 60.)

associative cortices compared to healthy controls (Laureys et al., 1999). Importantly, functional connections of these areas with the thalami were restored after recovery from VS/UWS (Laureys et al., 2000). It was recently shown that, compared to healthy controls, patients in VS/UWS exhibit metabolic dysfunction in both external and internal awareness networks as well as in the thalami. In contrast, patients in MCS compared to healthy controls showed dysfunction mostly in internal awareness network and thalami, which could reflect an altered self-awareness in these patients that is difficult to quantify at the bedside (Thibaut et al., 2012).

Brain responses to external stimuli provide valuable information not only about the preserved functional (and to some degree anatomical) connectivity among distinct brain regions but also about the nature of detected responses. Pain is among the most robust stimulations allowing for the investigation of residual brain function in noncommunicating clinical conditions. Preserved pain perception in MCS patients is suggested by findings of cerebral correlates of pain processing in a network similar to that in healthy controls (Boly et al., 2008). The activation pattern observed in patients in MCS was also much more widespread than in unresponsive patients, suggesting a difference in capacity for pain perception. The type of administered stimuli also seems to make a difference to the observed neural responses and therefore further assist in the inference of awareness. Stimuli with emotional valence, for example, such as infant cries and the patient's own name, induced a much more widespread activation in patients in MCS than did meaningless noise (Laureys, Perrin, et al., 2004). The activation pattern was comparable with that previously obtained in healthy controls. Patients also showed higher fMRI activity in the anterior cingulate cortex after listening to their own name as compared to listening to a familiar name, and this activity correlated with the behaviorally assessed level of consciousness of the patient (Qin et al., 2010). Such results imply that self-referential stimuli, like one's own name, are attention-grabbing and therefore can be used in the assessment of residual brain function of these patients (Demertzi, Vanhaudenhuyse et al., 2013).

Electrophysiology Various EEG paradigms have made an effort to differentiate between the clinical entities of disorders of consciousness. Fifteen-minute EEG resting-state acquisitions showed that patients in VS/UWS had significantly higher correlated oscillations than patients in MCS in the delta-frequency band (Lehembre et al., 2012). Such delta-frequency activity is represented as a high-amplitude brain wave with an oscillation between 0 and 4 Hz usually associated with the deepest stages of sleep, also known as slow-wave sleep. As previously shown, power in the delta band also increases with severity of disorders of consciousness (Leon-Carrion, Martin-Rodriguez, Damas-Lopez, Barroso y Martin, & Dominguez-Morales, 2008). Similarly, the bispectral index, a measure of the depth of anesthesia, was shown to discriminate between unresponsive and patients in MCS (Schnakers, Majerus, & Laureys, 2005). The bispectral index was also positively correlated with behavioral scores of awareness at the time of testing and associated with outcome results at one year post-trauma. Additionally, an EEG entropy score of 52 (value ranging from 0 to 91, with higher scores indicating higher consciousness level) was shown to be able to differentiate acutely unconscious from patients in MCS with 89% sensitivity and 90% specificity (Gosseries et al., 2011). Other efforts have also been made to use EEG signal patterns as a prognostic tool for these patients. For example, it has been observed that patients in VS/UWS who made a behavioral recovery, at three-month follow-up they showed higher occipital source power in the alpha band of resting EEG when compared to those who did not (Babiloni et al., 2009). Normally, high power of prestimulus cortical alpha rhythms (about 8–12 Hz) underlies conscious perception in healthy subjects. As such, cortical sources of resting alpha rhythms might predict recovery in VS/UWS patients.

EEG studies measuring effective connectivity also seem to be able to differentiate between unresponsive patients and those in MCS. Effective connectivity is a measure of the causal relationship between brain areas. One study using a mismatch negativity paradigm and applying dynamic causal modeling found that the only significant difference in functional connectivity between VS/UWS and MCS was an impairment of backward connectivity from frontal to temporal cortices (Boly et al., 2011). In addition, measurement of EEG effective connectivity after the application of transcranial magnetic stimulation (TMS) revealed that unresponsive patients showed a simple, local response after the TMS pulses. In contrast, patients in MCS showed more complex activations after the TMS pulses, which involved distant cortical areas ipsilateral and contralateral to the site of stimulation (Rosanova et al., 2012). Importantly, an EEG-TMS objective index has recently been proposed allowing for accurate patient classification at the single-subject level (Casali et al., 2013).

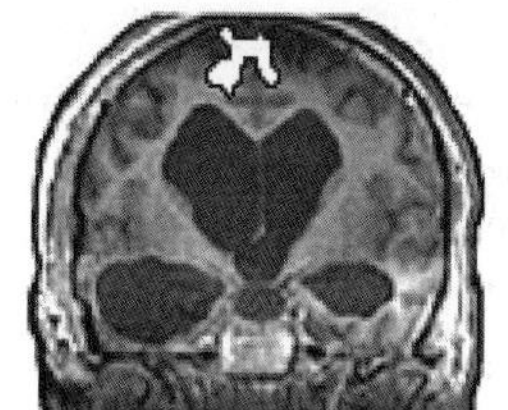
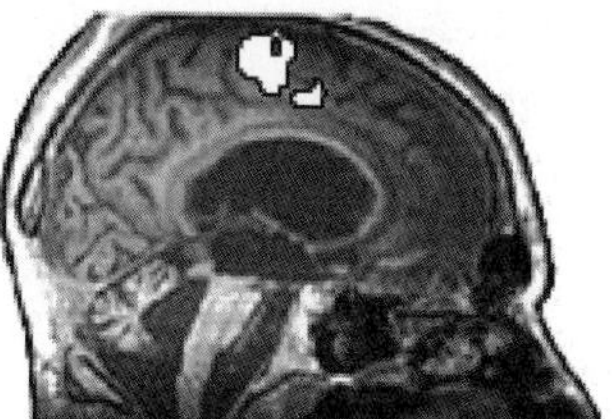
Is your father's name Alexander?

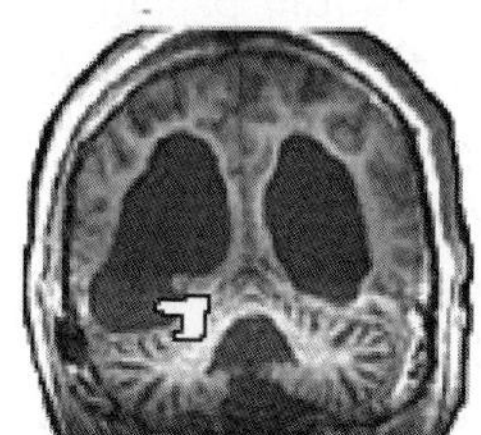
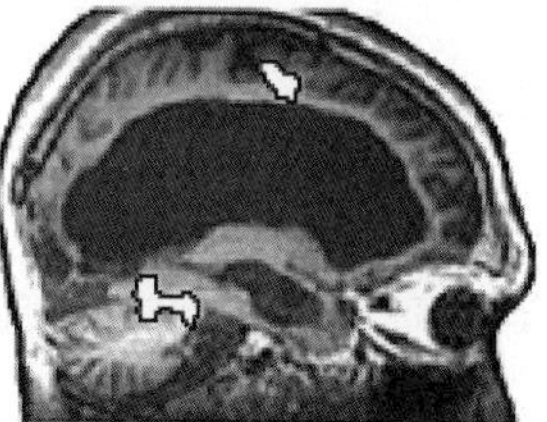
Is your father's name Thomas?

FIGURE 70.2 Muscle-independent fMRI communication with a patient clinically diagnosed as "vegetative"/unresponsive. Following a certain communication protocol, where the instruction to imagine playing tennis was used as a coded "yes" response (encompassing supplementary motor area; upper panel in the illustration) and the command to imagine navigating in one's house corresponded to a "no" answer (parahippocampal area; lower panel in the illustration), this patient was able to willfully modulate the brain activity and provide responses to a series of autobiographical questions (adapted from Monti et al., 2010).

FUNCTIONAL NEUROIMAGING AND ELECTROPHYSIOLOGY AS TOOLS FOR COMMAND-FOLLOWING AND COMMUNICATION A willful modulation of brain activity to a certain command can be an alternative way to evidence awareness in the absence of motor output. Using fMRI, it was demonstrated that a posttraumatic patient behaviorally diagnosed as in VS/UWS could repeatedly and for a prolonged time follow commands when assessed using a novel fMRI paradigm of mental imagery (Owen et al., 2006). When asked to imagine playing tennis and walking through her house, the patient activated the supplementary motor area and parahippocampal areas, respectively. These specific activation patterns were not different from those previously observed in a cohort of healthy volunteers (Boly et al., 2007). Such a command-following paradigm has been further developed to be implemented as a communication system. In such a protocol, differentiation between brain responses was used as a proxy for behavioral responses. Using the previously described "tennis paradigm," it was demonstrated that of 54 patients, 5 were able to follow the commands to imagine playing tennis and navigating around their house, and one patient was able to use the modulation of brain activity to answer reliably "yes" or "no" to simple questions, even though no signs of communication had ever been found through bedside examination (Monti et al., 2010). The automated user-independent analysis of the acquired fMRI data classified the brain's responses as a "yes" or "no" answer to a series of simple questions (figure 70.2).

Similarly, active paradigms have been attempted with cheaper and portable EEG-based technologies. In the case of a young comatose woman who failed to show any motor signs of conscious awareness, only EEG-evoked potential based on command-following allowed the diagnosis of total LIS at the intensive care unit (Schnakers et al., 2008). The task was to count a target name or her own name in a list of other names. A previous study using this task demonstrated that while most patients in MCS exhibited increased amplitude of the P300 when instructed to count, no task-related P300 changes were observed in unresponsive patients. A number of studies have demonstrated that EEG power spectral analysis can be also used as a flexible bedside tool to demonstrate awareness in brain-injured patients who are otherwise unable to communicate (Chatelle et al., 2012; Cruse et al., 2011). What remains to be shown is whether such technologies can be used as evidence of the expressed will of a competent patient (Gantner, Bodart, Laureys, & Demertzi, 2012). For example, how can a negative response of an "unresponsive" patient to the question of whether he or she wants to continue to live be considered a reliable response to be respected? Similarly, should pain treatment in a patient in MCS change once he or she has communicated that he or she is suffering? Should proving consciousness in these patients be considered a piece of evidence to be celebrated, or can it work against patients' and families' best interests (Jox, Bernat, Laureys, & Racine, 2012)? These aforementioned questions require answers that future establishment of ethical and legal provisions can provide.

Medico-ethical implications of dealing with patients with disorders of consciousness

Ever since disorders of consciousness appeared in the clinical setting, clinicians, scholars, theologians, and ethicists have wondered what it is like to be in a state of profoundly disturbed consciousness (e.g., Thompson, 1969). Such controversies mainly stem from how different people regard indefinite survival in disorders of consciousness (Demertzi, Laureys, & Bruno, 2011; Jennett, 2002). Despite the general view that quality of life is diminished in disease as a result of limited capacities to functionally engage in everyday living, one needs to consider that these attitudes are formulated from a third-person perspective. Consequently, only rough estimations about what it is like to be in such a situation can be made with the risk of imminent bias. Such bias could be attributed to the fact that patients' quality of life evaluations are made from the perspective of healthy individuals who tend to underestimate patients' subjective well-being (Demertzi, Gosseries, Ledoux, Laureys, & Bruno, 2013a; Nizzi et al., 2012). Indeed, it was recently showed that patients in LIS expressed a positive subjective quality of life, contrary to what could be expected in this condition (Bruno et al., 2011). In this self-reporting survey, it was shown that the majority of patients in chronic LIS, despite mentioning severe restrictions in community reintegration, professed good subjective well-being. Self-reported happiness status was associated with longer duration in this condition, the ability to produce speech via assisting technologies, and lower rates of anxiety. These findings suggest that healthy persons who are not in direct contact with this patient population can have distorted pictures about what life is like in these severely constrained situations. But what about the opinions of health care workers, those who are more likely to interact with patients?

When clinicians were asked to express their opinions on possible pain perception in VS/UWS, a significant number of medical doctors ascribed pain perception in VS/UWS (56%) despite formal guidelines suggesting the opposite (e.g., The Multi-Society Task Force on PVS, 1994b). For MCS, there was no discrepancy in opinions, and the majority (97%) of respondents thought that MCS patients feel pain (Demertzi, Racine et al., 2013; Demertzi et al., 2009), in line with neuroimaging data strongly suggesting preserved pain perception in MCS (Boly et al., 2008). The issue of pain management in unresponsive patients becomes more challenging when withdrawal from life-supporting treatments, such as artificial nutrition and hydration, has been agreed upon. In a wide survey around Europe, we showed that health care workers' opinions on end of life in disorders of consciousness differed depending on the diagnosis (i.e., respondents supported treatment withdrawal more often for patients in VS/UWS than in MCS), professional background (i.e., when physicians imagined being in MCS, they preferred more often to be kept alive compared to paramedical professionals), region of origin (i.e., Northern Europeans agreed with treatment withdrawal more often compared to Central and Southern European respondents), and religious beliefs (i.e., religious respondents agreed less with treatment limitation in both VS/UWS and MCS compared to nonreligious respondents; Demertzi, Ledoux, et al., 2011). These data show that personal opinions about ethical issues in disorders of consciousness differ, and hence different clinical practice can be expected. For example, in the European survey, the majority of participants approved of stopping treatment in VS/UWS (66%), much more than in MCS (28%). In this case, patients in VS/UWS may run the risk of being left without administration of opioids or other analgesic drugs during their dying process (Fins, 2006; Laureys, 2005) on the grounds that they are unable to experience suffering due to hunger or thirst.

From such studies on clinicians' attitudes and attitudes of patients' families (e.g., Kuehlmeyer, Borasio, & Jox, 2012), it becomes evident that medical and ethical controversies continue to exist for patients with disorders of consciousness. In order to resolve them, at least to a certain degree, we need to improve our current understanding of how these patients function. The use of objective biomarkers may help us to better determine the differences in underlying pathophysiology characterizing the clinical entities of consciousness. Consequently, clinicians should learn about patients' values and preferences and focus attention on changes in patient status, keeping the patients' best interests in mind (Jox et al., 2012).

ACKNOWLEDGMENTS This work was supported by the Belgian National Funds for Scientific Research (FNRS), the European Commission, the James McDonnell Foundation, the European Space Agency, Mind Science Foundation, the French-Speaking Community Concerted Research Action, the Public Utility Foundation, Université Européenne du Travail, Fondazione Europea di Ricerca Biomedica, and the University and University Hospital of Liège.

REFERENCES

AMERICAN CONGRESS OF REHABILITATION MEDICINE. (1995). Recommendations for use of uniform nomenclature pertinent to patients with severe alterations of consciousness. *Arch Phys Med Rehabil, 76*(2), 205–209.

BABILONI, C., SARA, M., VECCHIO, F., PISTOIA, F., SEBASTIANO, F., ONORATI, P., ... ROSSINI, P. M. (2009). Cortical sources of resting-state alpha rhythms are abnormal in persistent vegetative state patients. *Clin Neurophysiol, 120*(4), 719–729.

BAUER, G., GERSTENBRAND, F., & RUMPL, E. (1979). Varieties of the locked-in syndrome. *J Neurol, 221*(2), 77–91.

BOLY, M., COLEMAN, M. R., DAVIS, M. H., HAMPSHIRE, A., BOR, D., MOONEN, G., ... OWEN, A. M. (2007). When thoughts become action: An fMRI paradigm to study volitional brain activity in non-communicative brain injured patients. *NeuroImage, 36*(3), 979–992.

BOLY, M., FAYMONVILLE, M.-E., SCHNAKERS, C., PEIGNEUX, P., LAMBERMONT, B., PHILLIPS, C., ... LAUREYS, S. (2008). Perception of pain in the minimally conscious state with PET activation: An observational study. *Lancet Neurol, 7*(11), 1013–1020.

BOLY, M., GARRIDO, M. I., GOSSERIES, O., BRUNO, M.-A., BOVEROUX, P., SCHNAKERS, C., ... FRISTON, K. (2011). Preserved feedforward but impaired top-down processes in the vegetative state. *Science, 332*(6031), 858–862.

BOLY, M., TSHIBANDA, L., VANHAUDENHUYSE, A., NOIRHOMME, Q., SCHNAKERS, C., LEDOUX, D., ... LAUREYS, S. (2009). Functional connectivity in the default network during resting state is preserved in a vegetative but not in a brain dead patient. *Hum Brain Mapp, 30*, 2393–2400.

BRUNO, M.-A., BERNHEIM, J., LEDOUX, D., PELLAS, F., DEMERTZI, A., & LAUREYS, S. (2011). A survey on self-assessed well-being in a cohort of chronic locked-in syndrome patients: Happy majority, miserable minority. *BMJ Open, 1*(1), e000039.

CASALI, A. G., GOSSERIES, O., ROSANOVA, M., BOLY, M., SARASSO, S., CASALI, K. R., ... MASSIMINI, M. (2013). A theoretically based index of consciousness independent of sensory processing and behavior. *Sci Transl Med, 5*(198), 198ra105.

CHATELLE, C., CHENNU, S., NOIRHOMME, Q., CRUSE, D., OWEN, A. M., & LAUREYS, S. (2012). Brain-computer interfacing in disorders of consciousness. *Brain Inj, 26*(12), 1510–1522.

CRUSE, D., CHENNU, S., CHATELLE, C., BEKINSCHTEIN, T. A., FERNANDEZ-ESPEJO, D., PICKARD, J. D., ... OWEN, A. M. (2011). Bedside detection of awareness in the vegetative state: A cohort study. *Lancet, 378*(9809), 2088–2094.

DEMERTZI, A., GÓMEZ, F., CRONE, J. S., VANHAUDENHUYSE, A., TSHIBANDA, L., & NOIRHOMME, Q. (2014). Multiple fMRI system-level baseline connectivity is disrupted in patients with consciousness alterations. *Cortex, 52*, 35–46.

DEMERTZI, A., GOSSERIES, O., LEDOUX, D., LAUREYS, S., & BRUNO, M.-A. (2013). Quality of life and end-of-life decisions after brain injury. In N. Warren & L. Manderson (Eds.), *Reframing disability and quality of life: A global perspective* (pp. 95–110). Dordrecht: Springer.

DEMERTZI, A., LAUREYS, S., & BRUNO, M.-A. (2011). The ethics in disorders of consciousness. In J. L. Vincent (Ed.), *Annual update in intensive care and emergency medicine* (pp. 675–682). Berlin: Springer-Verlag.

DEMERTZI, A., LEDOUX, D., BRUNO, M.-A., VANHAUDENHUYSE, A., GOSSERIES, O., SODDU, A., ... LAUREYS, S. (2011). Attitudes towards end-of-life issues in disorders of consciousness: A European survey. *J Neurol, 258*(6), 1058–1065.

DEMERTZI, A., RACINE, E., BRUNO, M. A., LEDOUX, D., GOSSERIES, O., VANHAUDENHUYSE, A., ... LAUREYS, S. (2013). Pain perception in disorders of consciousness: Neuroscience, clinical care, and ethics in dialogue. *Neuroethics, 6*(1), 37–50.

DEMERTZI, A., SCHNAKERS, C., LEDOUX, D., CHATELLE, C., BRUNO, M.-A., VANHAUDENHUYSE, A., ... LAUREYS, S. (2009). Different beliefs about pain perception in the vegetative and minimally conscious states: A European survey of medical and paramedical professionals. *Prog Brain Res, 177*, 329–338.

DEMERTZI, A., SODDU, A., & LAUREYS, S. (2013). Consciousness supporting networks. *Curr Opin Neurobiol, 23*(2), 239–244.

DEMERTZI, A., VANHAUDENHUYSE, A., BREDART, S., HEINE, L., DI PERRI, C., & LAUREYS, S. (2013). Looking for the self in pathological unconsciousness. *Front Hum Neurosci, 7*, 538.

FINS, J. J. (2006). Affirming the right to care, preserving the right to die: Disorders of consciousness and neuroethics after Schiavo. *Palliat Support Care, 4*(2), 169–178.

GANTNER, I. S., BODART, O., LAUREYS, S., & DEMERTZI, A. (2012). Our rapidly changing understanding of acute and chronic disorders of consciousness: Challenges for neurologists. *Fut Neurol, 8*(1), 43–54.

GIACINO, J. T., ASHWAL, S., CHILDS, N., CRANFORD, R., JENNETT, B., KATZ, D. I., ... ZASLER, N. D. (2002). The minimally conscious state: Definition and diagnostic criteria. *Neurology, 58*(3), 349–353.

GIACINO, J. T., KALMAR, K., & WHYTE, J. (2004). The JFK coma recovery scale-revised: Measurement characteristics and diagnostic utility. *Arch Phys Med Rehab, 85*(12), 2020–2029.

GOSSERIES, O., SCHNAKERS, C., LEDOUX, D., VANHAUDENHUYSE, A., BRUNO, M.-A., DEMERTZI, A., ... LAUREYS, S. (2011). Automated EEG entropy measurements in coma, vegetative state/unresponsive wakefulness syndrome and minimally conscious state. *Funct Neurol, 36*(1), 25–30.

HEINE, L., SODDU, A., GOMEZ, F., VANHAUDENHUYSE, A., TSHIBANDA, L., THONNARD, M., ... DEMERTZI, A. (2012). Resting state networks and consciousness: Alterations of multiple resting state network connectivity in physiological, pharmacological and pathological consciousness states. *Front Psychol, 3*, 1–12.

JENNETT, B. (2002). Attitudes to the permanent vegetative state. In B. Jennett (Ed.), *The vegetative state: Medical facts, ethical and legal dilemmas* (pp. 97–125). Cambridge, UK: Cambridge University Press.

Jox, R. J., Bernat, J. L., Laureys, S., & Racine, E. (2012). Disorders of consciousness: Responding to requests for novel diagnostic and therapeutic interventions. *Lancet Neurol, 11*(8), 732–738.

Kuehlmeyer, K., Borasio, G. D., & Jox, R. J. (2012). How family caregivers' medical and moral assumptions influence decision making for patients in the vegetative state: A qualitative interview study. *J Med Ethics, 38*(6), 332–337.

Laureys, S. (2005). Science and society: Death, unconsciousness and the brain. *Nat Rev Neurosci, 6*(11), 899–909.

Laureys, S. (2007). Eyes open, brain shut. *Sci Am, 296*(5), 84–89.

Laureys, S., Celesia, G., Cohadon, F., Lavrijsen, J., Leon-Carrion, J., Sannita, W. G., ... European Task Force on Disorders of Consciousness. (2010). Unresponsive wakefulness syndrome: A new name for the vegetative state or apallic syndrome. *BMC Med, 8*(1), 68.

Laureys, S., Faymonville, M.-E., Luxen, A., Lamy, M., Franck, G., & Maquet, P. (2000). Restoration of thalamocortical connectivity after recovery from persistent vegetative state. *Lancet, 355*(9217), 1790–1791.

Laureys, S., Goldman, S., Phillips, C., van Bogaert, P., Aerts, J., Luxen, A., ... Maquet, P. (1999). Impaired effective cortical connectivity in vegetative state: Preliminary investigation using PET. *NeuroImage, 9*(4), 377–382.

Laureys, S., Owen, A. M., & Schiff, N. D. (2004). Brain function in coma, vegetative state, and related disorders. *Lancet Neurol, 3*(9), 537–546.

Laureys, S., Pellas, F., van Eeckhout, P., Ghorbel, S., Schnakers, C., Perrin, F., ... Goldman, S. (2005). The locked-in syndrome : What is it like to be conscious but paralyzed and voiceless? *Prog Brain Res, 150*, 495–511.

Laureys, S., Perrin, F., Faymonville, M.-E., Schnakers, C., Boly, M., Bartsch, V., ... Maquet, P. (2004). Cerebral processing in the minimally conscious state. *Neurology, 63*(5), 916–918.

Lehembre, R., Bruno, M.-A., Vanhaudenhuyse, A., Chatelle, C., Cologan, V., Leclercq, Y., ... Noirhomme, Q. (2012). Resting-state EEG study of comatose patients: A connectivity and frequency analysis to find differences between vegetative and minimally conscious states. *Funct Neurol, 27*(1), 41–47.

Leon-Carrion, J., Martin-Rodriguez, J. F., Damas-Lopez, J., Barroso y Martin, J. M., & Dominguez-Morales, M. R. (2008). Brain function in the minimally conscious state: A quantitative neurophysiological study. *Clin Neurophysiol, 119*(7), 1506–1514.

Mason, M. F., Norton, M. I., van Horn, J. D., Wegner, D. M., Grafton, S. T., & Macrae, C. N. (2007). Wandering minds: The default network and stimulus-independent thought. *Science, 315*(5810), 393–395.

Monti, M. M., Vanhaudenhuyse, A., Coleman, M. R., Boly, M., Pickard, J. D., Tshibanda, L., ... Laureys, S. (2010). Willful modulation of brain activity in disorders of consciousness. *N Engl J Med, 362*(7), 579–589.

Multi-Society Task Force on PVS. (1994a). Medical aspects of the persistent vegetative state (1). *N Engl J Med, 330*(21), 1499–1508.

Multi-Society Task Force on PVS. (1994b). Medical aspects of the persistent vegetative state (2). *N Engl J Med, 330*(22), 1572–1579.

Nizzi, M. C., Demertzi, A., Gosseries, O., Bruno, M.-A., Jouen, F., & Laureys, S. (2012). From armchair to wheelchair: How patients with a locked-in syndrome integrate bodily changes in experienced identity. *Conscious Cogn, 21*(1), 431–437.

Owen, A. M., Coleman, M. R., Boly, M., Davis, M. H., Laureys, S., & Pickard, J. D. (2006). Detecting awareness in the vegetative state. *Science, 313*(5792), 1402.

Posner, J., Saper, C., Schiff, N. D., & Plum, F. (Eds). (2007). *Plum and Posner's diagnosis of stupor and coma.* New York, NY: Oxford University Press.

Qin, P., Di, H., Liu, Y., Yu, S., Gong, Q., Duncan, N., ... Northoff, G. (2010). Anterior cingulate activity and the self in disorders of consciousness. *Hum Brain Mapp, 31*(12), 1993–2002.

Quality Standards Subcommittee of the American Academy of Neurology. (1995). Practice parameters for determining brain death in adults (summary statement). *Neurology, 45*(5), 1012–1014.

Raichle, M. E., & Snyder, A. Z. (2007). A default mode of brain function: A brief history of an evolving idea. *NeuroImage, 37*(4), 1083–1090; discussion 1097–1099.

Rosanova, M., Gosseries, O., Casarotto, S., Boly, M., Casali, A. G., Bruno, M.-A., ... Massimini, M. (2012). Recovery of cortical effective connectivity and recovery of consciousness in vegetative patients. *Brain, 135*(Pt 4), 1308–1320.

Schnakers, C., Majerus, S., & Laureys, S. (2005). Bispectral analysis of electroencephalogram signals during recovery from coma. *Neuropsychol Rehabil, 15*(3–4), 381–388.

Schnakers, C., Perrin, F., Schabus, M., Majerus, S., Ledoux, D., Damas, P., ... Laureys, S. (2008). Voluntary brain processing in disorders of consciousness. *Neurology, 71*(20), 1614–1620.

Schnakers, C., Vanhaudenhuyse, A., Giacino, J. T., Ventura, M., Boly, M., Majerus, S., ... Laureys, S. (2009). Diagnostic accuracy of the vegetative and minimally conscious state: Clinical consensus versus standardized neurobehavioral assessment. *BMC Neurol, 9*, 35.

Seel, R. T., Sherer, M., Whyte, J., Katz, D. I., Giacino, J. T., Rosenbaum, A. M., ... Zasler, N. (2010). Assessment scales for disorders of consciousness: Evidence-based recommendations for clinical practice and research. *Arch Phys Med Rehabil, 91*(12), 1795–1813.

Soddu, A., Vanhaudenhuyse, A., Demertzi, A., Bruno, M.-A., Tshibanda, L., Di, H., ... Noirhomme, Q. (2011). Resting state activity in patients with disorders of consciousness. *Funct Neurol, 26*(1), 37–43.

Stoll, J., Chatelle, C., Carter, O., Koch, C., Laureys, S., & Einhauser, W. (2013). Pupil responses allow communication in locked-in syndrome patients. *Curr Biol, 23*(15), R647–648.

Teasdale, G., & Jennett, B. (1974). Assessment of coma and impaired consciousness. A practical scale. *Lancet, 2*(7872), 81–84.

Thibaut, A., Bruno, M.-A., Chatelle, C., Gosseries, O., Vanhaudenhuyse, A., Demertzi, A., ... Laureys, S. (2012). Metabolic activity in external and internal awareness networks in severely brain-damaged patients. *J Rehabil Med, 44*(5), 487–494.

Thompson, G. T. (1969). An appeal to doctors. *Lancet, 2*, 1353.

Vanhaudenhuyse, A., Demertzi, A., Schabus, M., Noirhomme, Q., Bredart, S., Boly, M., ... Laureys, S.

(2011). Two distinct neuronal networks mediate the awareness of environment and of self. *J Cogn Neurosci, 23*(3), 570–578.

Vanhaudenhuyse, A., Noirhomme, Q., Tshibanda, L. J., Bruno, M.-A., Boveroux, P., Schnakers, C., … Boly, M. (2010). Default network connectivity reflects the level of consciousness in non-communicative brain-damaged patients. *Brain, 133*(Pt. 1), 161–171.

Wijdicks, E. F. M., Bamlet, W. R., Maramattom, B. V., Manno, E. M., & McClelland, R. L. (2005). Validation of a new coma scale: The FOUR score. *Ann Neurol, 58*(4), 585–593.

Wijdicks, E. F. M., Hijdra, A., Young, G. B., Bassetti, C. L., & Wiebe, S. (2006). Practice parameter: Prediction of outcome in comatose survivors after cardiopulmonary resuscitation (an evidence-based review): Report of the Quality Standards Subcommittee of the American Academy of Neurology. *Neurology, 67*(2), 203–210.

Zeman, A. (2001). Consciousness. *Brain, 124*(7), 1263–1289.

71 Consciousness and Seizures

HAL BLUMENFELD

ABSTRACT Why do seizures temporarily interrupt consciousness? To answer this question, recent studies have combined behavioral, electrophysiological, and neuroimaging measurements to explore brain networks crucial for both normal consciousness and its disruption by seizures. The most common seizure types leading to impaired consciousness are absence seizures, generalized tonic-clonic seizures, and temporal lobe complex partial seizures. Although these seizure types differ in many ways, they share a final common set of anatomical structures leading to impaired consciousness. These structures, which can be referred to as the "consciousness system," comprise regions well known to play a role in controlling the overall level of conscious arousal, including the upper brainstem activating systems, thalamus, basal forebrain, and higher-order frontoparietal association cortex. Interestingly, different seizures affect these same structures in different ways. Absence seizures lead to abnormal increased activity in the thalamus, with a mixture of increased and decreased activity in the frontoparietal cortex. Generalized tonic-clonic seizures are accompanied by increases in activity in many brain regions, but following seizures while consciousness is still impaired there is decreased activity in frontoparietal cortex associated with persistently increased activity in the cerebellum. Temporal lobe complex partial seizures exhibit an interesting pattern of increased activity in the temporal lobe and thalamus, along with decreased activity in frontoparietal cortex resembling slow-wave sleep. Human studies and animal models support a model in which temporal lobe seizures inhibit the subcortical arousal systems provoking a transition to sleep-like cortical activity. Better understanding of impaired consciousness in seizures may shed light on both normal consciousness mechanisms and on potential new treatment approaches for people living with epilepsy.

Transient impaired consciousness in epileptic seizures provides a window into mechanisms of normal consciousness. Brain circuits operating in optimal fashion generate consciousness, but are disrupted by specific anatomical and physiological alterations during epileptic seizures. Understanding these mechanisms has enormous importance for patients with epilepsy, since transient loss of consciousness has a negative impact on safety, productivity, emotional health, and quality of life (Sperling, 2004; Vickrey et al., 2000). New insights gained from understanding impaired consciousness in epilepsy may also provide novel treatment avenues for other disorders of consciousness.

Several types of seizures can cause impaired consciousness, including absence, generalized tonic-clonic, and temporal lobe complex partial seizures. Although these seizures differ from each other markedly in their physiology and behavioral features, they converge on a common set of neuroanatomical structures when consciousness is impaired (table 71.1). These include the frontoparietal association cortex and subcortical arousal networks, often affected in other disorders of consciousness as well (Laureys & Schiff, 2009; Laureys & Tononi, 2008). These key structures, which can be referred to as the "consciousness system" (Blumenfeld, 2009, 2010), are affected through different mechanisms in different epileptic seizures, which may reflect important aspects of normal brain network function.

This chapter will first introduce the consciousness system, its main components, and how it is involved in disorders of consciousness, including epilepsy. The discussion will next focus on recent advances in behavioral, electrophysiological, and neuroimaging techniques that have shed new light on the mechanisms of impaired consciousness in epileptic seizures (table 71.1). Hopefully, with further work, we will soon be able to prevent this devastating interruption in the lives of people with epilepsy and provide them the full benefits of normal consciousness.

The consciousness system

Although philosophical discussions of consciousness can be enlightening (Baars, Ramsoy, & Laureys, 2003; Chalmers, 1996; Dennett, 1991; Nagel, 1974; Searle, 1997), neurologists tend to think of consciousness and its disruption as a problem in functional neuroanatomical localization. In studying coma and related disorders, Plum and Posner introduced a classic distinction between brain systems that control the *level of consciousness* and those that generate the *content of consciousness* (Plum & Posner, 1972, 1982). The content of consciousness includes all of the hierarchically organized sensory and motor systems, memory, and emotions and drives. Individual brain networks with specialized functions each contribute to the content of consciousness. The level of consciousness depends on other specialized brain systems that control whether we are alert, attentive, and aware (mnemonic: AAA; Blumenfeld, 2010).

TABLE 71.1

Seizures and impaired consciousness: Summary of behavior, electrophysiology, and neuroimaging findings

Seizure Type	Behavior	Electrophysiology	Neuroimaging
Absence seizures	Behavioral arrest typically 3–10 sec with minor eyelid or hand movements and rapid return to baseline. Simple repetitive tasks can often continue during seizures.	Widespread bilateral 3–4 Hz spike-wave discharges with maximum amplitude in midline anterior frontal region and possibly precuneus. Animal models suggest focal bilateral onset with sparing of some regions.	Functional MRI shows increases in thalamus, but complicated early increases in some areas (e.g., medial frontal cortex, precuneus) preceding EEG onset by several seconds, and later widespread frontoparietal association cortex decreases lasting long after EEG end.
Generalized tonic-clonic seizures	Rigid tonic extension and clonic jerking of limbs usually lasting 1–2 minutes with profound unresponsiveness continuing into the post-ictal period.	High-frequency polyspike discharge in tonic phase, rhythmic polyspike and wave in clonic phase, generalized suppression post-ictally. Human intra-cranial EEG shows some regions spared in "generalized" seizures.	Focal CBF increases in frontoparietal association cortex and thalamus. Post-ictal CBF increases in cerebellum correlated with thalamic increases and frontoparietal decreases. Animal model supports relatively focal bilateral cortical increases based on fMRI.
Complex partial (temporal lobe[a]) seizures	Behavioral arrest lasting 1–2 minutes commonly with oral and manual automatisms, and confusion in the post-ictal period.	High-frequency discharge in medial temporal lobe, and sleep-like delta slow waves in frontoparietal cortex. Slow waves continue post-ictally. Animal models suggest depressed subcortical arousal impacts cortex.	CBF increases in temporal lobe and medial diencephalon-upper brainstem with CBF decreases in frontoparietal association cortex. Animal model shows fMRI increases in lateral septum and anterior hypothalamus; these regions may inhibit subcortical arousal.

[a]We focus here on complex partial seizures of temporal lobe origin, since less is known about the pathophysiology of impaired consciousness in complex partial seizures initiated from other cortical regions.

Reproduced from Blumenfeld, 2012, with permission from Elsevier.

Our level of consciousness affects each of the many contents of consciousness.

The specialized brain networks that regulate the level of consciousness have been studied extensively over the past century. Early work on human brain disorders (Penfield, 1950; Von Economo, 1930) and experimental animal models (Bremer, 1955; Moruzzi & Magoun, 1949) has been complemented by more recent studies demonstrating the importance of both cortical and subcortical structures in controlling level of consciousness (Steriade & McCarley, 2010). In analogy to sensory, motor, and other cortical-subcortical brain systems, the specialized structures regulating level of consciousness can be called the "consciousness system" (Blumenfeld, 2009, 2010; figure 71.1).

The consciousness system includes cortical components such as the medial frontal, anterior cingulate, posterior cingulate, and medial parietal (precuneus, retrosplenial) cortex on the medial surface (figure 71.1A), as well as the lateral frontal, orbital frontal, and lateral temporal-parietal association cortex on the lateral surface (figure 71.1B). Portions of the insula (not shown) also likely participate. Subcortical components include the basal forebrain, hypothalamus, thalamus, and upper brainstem activating systems (figure 71.1A), as well as portions of the basal ganglia, cerebellum, and amygdala (not shown). The anatomy of the consciousness system was discussed in greater detail in a recent review (Blumenfeld, 2012). Of note, the cortical components of the consciousness system include both the recently described default mode network, which is important for internally directed processing (Raichle et al., 2001), as well as other cortical regions important for externally oriented attention (Asplund, Todd, Snyder, & Marois, 2010; Buschman & Miller, 2007; Dosenbach et al., 2007; Vanhaudenhuyse et al., 2011).

Impaired consciousness in epilepsy and other disorders occurs when the systems providing either the content or level of consciousness are disturbed. Selective loss of individual systems serving the content of

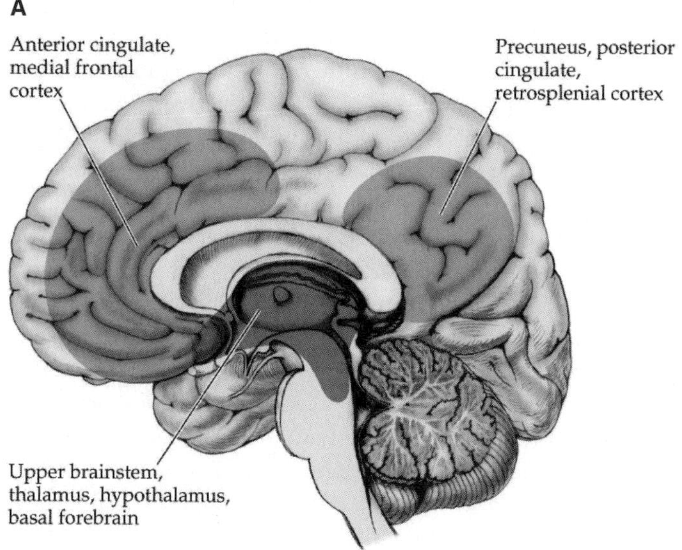

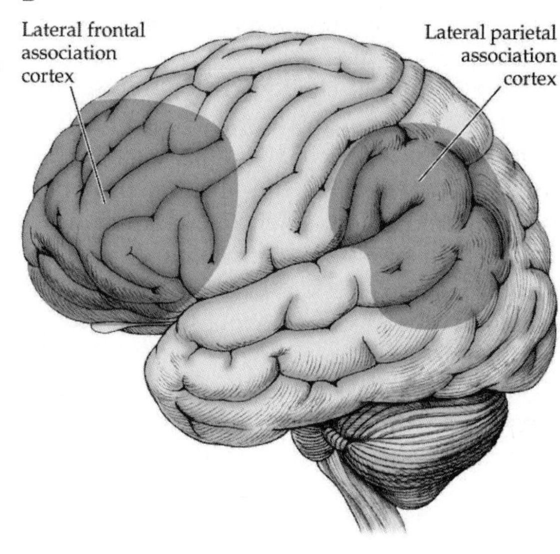

FIGURE 71.1 The consciousness system. Anatomical structures known to regulate the level of consciousness. (A) Medial view. (B) Lateral view. Cortical components of the consciousness system (*shown in blue*) include the medial and lateral frontoparietal association cortex, anterior and posterior cingulate, precuneus, and retrosplenial cortex. Subcortical components (*shown in red*) include the basal forebrain, hypothalamus, thalamus, and upper brainstem activating systems. Note that other circuits, such as the basal ganglia and cerebellum, may also participate in attention and other aspects of consciousness. (Reproduced with permission from Blumenfeld, 2010.)

consciousness, such as a lesion in the visual cortex or in the language areas, causes specific deficits in that aspect of consciousness but is not usually considered a disorder of consciousness *per se*. On the other hand, if many or nearly all contents of consciousness are impaired together, this represents a disorder of consciousness. Most commonly, this occurs when there is dysfunction in the consciousness system (figure 71.1). Disorders of the consciousness system cause an impaired level of consciousness, which leads to widespread dysfunction in most contents of consciousness. Examples include coma, vegetative state, minimally conscious state (Laureys, Owen, & Schiff, 2004; Laureys & Schiff, 2009; Laureys & Tononi, 2008), and certain types of epileptic seizures discussed in the next section.

Seizures associated with impaired consciousness

Seizures are classified into those that are partial (focal)—involving more localized areas of the brain—and those that are generalized—affecting widespread regions bilaterally (Berg et al., 2010; International League Against Epilepsy, 1981, 1989). Not all seizures cause impaired consciousness. Focal seizures may cause localized symptoms, such as hand twitching or an odd sensation in the epigastric area, without alterations in overall alertness, attention, or awareness. Partial seizures without impaired consciousness have traditionally been called simple partial seizures, while those with impaired consciousness are called complex partial seizures. Generalized seizures can also have variable manifestations depending on the intensity, duration, and physiological pattern of the epileptic discharges. The most common generalized seizure types are generalized tonic-clonic seizures and absence seizures, both of which usually cause marked impairment of consciousness. Because absence, generalized tonic-clonic, and complex partial seizures are the major seizure types associated with impaired consciousness, the remainder of our discussion will focus on these three types of seizures (table 71.1). Interestingly, despite the differences between these three seizure types, all converge on a final common set of anatomical regions, leading to impaired consciousness (table 71.1).

Absence (*petit mal*) seizures have been recognized in the medical literature since the 1700s (Temkin, 1971) and consist of brief episodes of staring and unresponsiveness. Most commonly seen in childhood absence epilepsy, absence seizures can also occur in adolescents and adults. The electroencephalogram (EEG) shows a characteristic widespread bilateral 3–4 Hz spike-wave discharge, which begins and ends abruptly along with the absence behavioral event. Although classified as a form of generalized epilepsy, recent work suggests that absence seizures involve selective bilateral cortical and subcortical networks while sparing others (Blumenfeld,

2005a; Meeren, van Luijtelaar, Lopes da Silva, & Coenen, 2005), which may help explain specific deficits in consciousness.

Generalized tonic-clonic (*grand mal*) seizures are well-recognized, whole-body convulsive episodes in which consciousness is deeply impaired, both during seizures and for a variable time period afterward. Again, despite being classified as generalized seizures, recent work suggests intense focal involvement of the consciousness system (figure 71.1), which may be related to the deep state of unconsciousness seen in generalized tonic-clonic seizures.

Complex partial seizures are episodes of staring and unresponsiveness, but unlike absence seizures, they involve focal brain discharges on EEG, usually last longer (typically 1–2 minutes), and involve focal "automatisms" such as repetitive chewing or hand wiping. Complex partial seizures are seen most commonly in temporal lobe epilepsy, but can occur in focal seizures arising from other brain regions as well. One important puzzle has been why focal seizures should cause impaired consciousness, but as we will discuss, at least in the case of temporal lobe seizures it has been shown that seizures disrupt the consciousness system, leading to bilateral cortical dysfunction (Blumenfeld, McNally et al., 2004; Englot et al., 2008, 2009, 2010).

Much work has been done in recent years to further advance our understanding of the pathophysiology of impaired consciousness in these three major seizure types. The sections that follow will review the main behavioral, electrophysiological, and neuroimaging findings shedding new light on impaired consciousness in epilepsy (table 71.1).

Behavior

The behavioral deficits in epilepsy are similar to other disorders of consciousness such as coma and vegetative and minimally conscious states, except that the changes are more transient in epilepsy (Blumenfeld, 2011). For example, behavioral arousal in generalized tonic-clonic seizures transiently resembles coma, because in both conditions patients are deeply unresponsive to external stimuli. One interesting difference is that in coma the eyes are closed (Plum & Posner, 1982), whereas in generalized tonic-clonic seizures the eyes are usually open. Absence seizures are very transiently similar to vegetative or minimally conscious states, with either no response or occasional simple responses to external stimuli (Blumenfeld, 2005b). In both absence and complex partial seizures the eyes are usually open, and there may be automatism, or orienting responses toward stimuli as in the vegetative state (Escueta, Bacsal, & Treiman, 1982; Sadleir, Scheffer, Smith, Connolly, & Farrell, 2009) and simple motor responses (McPherson et al., 2012) resembling the minimally conscious state.

Behavioral evaluation of consciousness during seizures has employed a variety of testing procedures. The most extensive testing has been accomplished in absence seizures, ranging from simple reaction time or motor tasks to tests of verbal responsiveness and memory (reviewed in Blumenfeld, 2005b). For other seizure types, retrospective questionnaires have been used (Ali et al., 2010; Cavanna et al., 2008) as well as scales based on review of video during seizures (Arthuis et al., 2009; Blumenfeld, McNally et al., 2004; Blumenfeld et al., 2009; Englot et al., 2010; Lambert, Arthuis, McGonigal, Wendling, & Bartolomei, 2012; Lee et al., 2002). Recently our group has investigated prospective methods for testing patients during inpatient video/EEG monitoring of seizures, including video games meant to simulate real-world situations such as driving (Yang et al., 2010) and standardized prospective behavioral testing batteries, including both verbal and nonverbal items (Bauerschmidt et al., 2013; McPherson et al., 2012; Yang et al., 2012).

It has been pointed out that lack of external responses does not prove loss of consciousness during seizures (Gloor, 1986). Patients might be conscious during seizures but unable to respond due to motor or language deficits, and may not report their experiences later due to amnesia. Similar considerations apply to other states of impaired consciousness in which internal awareness may persist despite external unresponsiveness (Sanders, Tononi, Laureys, & Sleigh, 2012). Recent advances in functional neuroimaging may offer new avenues for testing internal states of awareness in apparently unresponsive patients (Monti et al., 2010; Owen et al., 2006).

Absence seizures have the largest body of literature in which detailed behavioral testing has been done, most likely because these brief events can be captured in outpatient EEGs and absence seizures in some children will occur multiple times in one recording session (Blumenfeld, 2005b; Kostopoulos, 2001). Deficits in behavior during absence seizures are not absolute (Blumenfeld, 2005b; Chipaux, Vercueil, Kaminska, Mahon, & Charpier, 2013), and although a complete understanding of the mechanisms has not been reached, there are several factors that influence the variable severity of impairment during absence seizures. One important factor is the difficulty of the behavioral task. Impairment during absence seizures is more severe for tasks requiring verbal responses or complex decision making, whereas tasks involving simple repetitive actions can sometimes continue right through seizures (figure 71.2A). This is similar to other disorders of consciousness, such as global encephalopathy, which classically impair higher-order functions more severely than simple tasks (Mesulam, 2000).

Another factor that influences behavioral impairment during absence seizures is the precise timing of tasks relative to seizure onset and end. There is often some recovery toward the end of seizures, and depending on the task there may also be some initial sparing just after seizure onset, resulting in a U-shaped time course of deficits (figure 71.2A). It has been claimed that spike-wave discharges lasting less than three seconds do not cause deficits, and although this may be true for clinically obvious absence seizures, with careful behavioral testing transient deficits can be detected for even brief episodes lasting less than one second (Berman et al., 2010; Browne, Penry, Porter, & Dreifuss, 1974). In addition to deficits during seizures, it has been shown that patients with childhood absence epilepsy often have significant attention deficits in the interictal period even when no spike-wave discharges are occurring (Killory et al., 2011; Levav et al., 2002; Mirsky & van Buren, 1965; Vega et al., 2010).

Generalized tonic-clonic seizures usually last for about two minutes, with profound impairment of consciousness during seizures and for a variable time in the post-ictal period. Onset may be focal or bilateral, and behavior progresses through a series of stages, including bilateral clonic, tonic, vibratory, and clonic activity, followed by post-ictal lethargy (Blumenfeld et al., 2009; Jobst, 2001; Theodore et al., 1994; Varghese et al., 2009). During and following generalized tonic-clonic seizures, patients are deeply unresponsive to even basic tasks such as ball grasp, visual tracking, or blink to visual threat (McPherson et al., 2012). Amnesia commonly occurs for events around the time of seizures. Interestingly, in rare cases patients remain conscious during generalized tonic-clonic seizures and can reliably describe their experiences afterwards (Bell, Walczak, Shin, & Radtke, 1997; Botez, Serbanescu, & Stoica, 1966; Weinberger & Lusins, 1973). The mechanisms for this spared function should be investigated further, but it has been speculated that such seizures may involve bilateral frontal regions while sparing other areas necessary for consciousness.

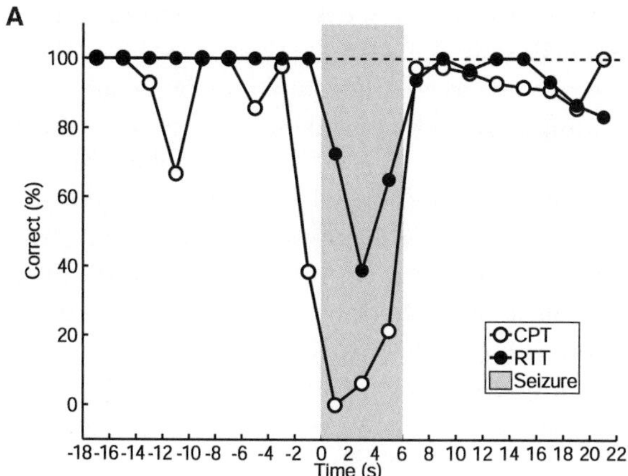

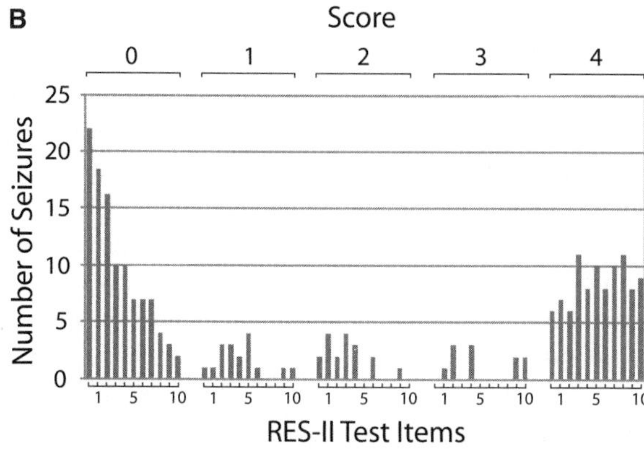

Figure 71.2 Behavioral changes in seizures. (A) Variable and transient behavioral impairment during childhood absence seizures. Percent correct responses are shown over time (2 sec time bins) before, during, and after seizures (shaded region). Performance on the more difficult continuous performance task (CPT) declined rapidly for letters presented just before seizure onset and recovered quickly after seizure end. Impaired performance on the simpler repetitive tapping task (RTT) was more transient than on CPT, did not begin until after seizure onset, and was less severely impaired during seizures than the CPT task ($F = 15.3$, $P = 0.017$; ANOVA). Results are based on a total of 53 seizures in 8 patients. (B) Bimodal distribution of impaired consciousness in partial seizures. Impairment on behavioral tasks is bimodally distributed, suggesting that most partial seizures can readily be separated into those with overall impairment (left cluster in the histogram) versus those without overall impairment (right cluster). Multiple standardized behavioral tasks were administered prospectively during partial seizures using the revised Responsiveness in Epilepsy Scale (RES-II), including both verbal and nonverbal items (items 1–10) and responses were scored based on video/EEG review. Scores show a bimodal distribution, with the large majority receiving a score of either "0" (no response whatsoever) or "4" (normal, unimpaired response). Data are from 33 partial seizures in 11 patients, 4 with temporal lobe epilepsy and 7 with neocortical or unlocalized partial epilepsy. A bimodal pattern of behavioral test scores was also seen in 35 partial seizures from 14 patients using an earlier prospective testing battery (Yang et al., 2012) as well as in a recent study with larger sample size (Cunningham et al., 2014). A bimodal pattern has not been observed with similar testing items in other disorders of consciousness (Giacino, Kalmar, & Whyte, 2004). (Reproduced with permission from Bai et al., 2010, panel A; and Bauerschmidt et al., 2013, panel B, with permission from Elsevier.)

Complex partial seizures, most commonly arising from the temporal lobe, consist of staring and unresponsiveness lasting for about one to two minutes. Often "automatisms" are seen, such as lip-smacking or repetitive semi-purposeful hand or leg movements (Escueta, Kunze, Waddell, Boxley, & Nadel, 1977; Hoffmann, Elger, & Kleefuss-Lie, 2008; Penfield, 1950). Consciousness is commonly spared for the initial part of the seizure, is maximally impaired in the later ictal period, and then remains impaired into the post-ictal period. The deficits in complex partial seizures are often less severe than during generalized tonic-clonic seizures. For example, simple responses such a grasping a ball or visual tracking are preserved in over half of complex partial seizures (McPherson et al., 2012). However, tasks that require more meaningful responses such as command-following, decision making, or speaking are severely impaired (Cunningham et al., 2014).

One important and somewhat controversial question is whether complex partial seizures disrupt the content or the level of consciousness. In other words, are the deficits in complex partial seizures caused by selective problems in one or several specific cognitive systems (e.g., language dysfunction, memory dysfunction, etc.), or is the overall level of behavioral arousal impaired because of depressed function of the consciousness system (figure 71.1)? Most likely, both mechanisms contribute to impaired consciousness in complex partial seizures. Sometimes, however, partial seizures are difficult to classify with regard to consciousness when selective deficits in cognitive function occur *without* impairment in overall level of behavioral arousal—this can occur, for example, when focal seizures cause déjà vu, amnesia, aphasia, hallucinations, "forced thinking," or altered self-perception without deficits in overall level of consciousness (Ali, Rickards, & Cavanna, 2012; Cavanna, Rickards, & Ali, 2011; Heydrich, Dieguez, Grunwald, Seeck, & Blanke, 2010; Picard & Craig, 2009). For these reasons, the latest report on classification of seizures recommends eliminating consciousness as a major classifying feature of partial (focal) seizures (Berg et al., 2010). However, recent behavioral studies have shown that the vast majority of partial seizures show either deficits in multiple verbal and nonverbal cognitive functions during seizures, or relative sparing of most functions (figure 71.2B; Bauerschmidt et al., 2013; Cunningham et al., 2014; Yang et al., 2012). This bimodal distribution of deficits (figure 71.2B) suggests that cognitive impairment in partial seizures is usually a broad phenomenon affecting virtually all functions, as would be expected in a disorder of overall level of consciousness.

Electrophysiology

Since first introduced for human use in the late 1920s (Berger, 1929), EEG has provided tremendous insight into the pathophysiology of epileptic seizures. Low-density scalp EEG recordings have recently been supplemented by high-density >200 channel scalp or intracranial recordings with higher sampling frequencies, yielding unprecedented spatial and temporal resolution. Animal models provide invasive recordings of single neurons or neuronal ensembles. The full potential of these techniques has not yet been realized, but has begun to be applied to understanding impaired consciousness in epilepsy.

Absence seizures are accompanied by a characteristic large amplitude 3–4 Hz spike-wave discharge on EEG typically lasting less than ~10 sec with relatively sudden onset and end (Ebersole & Pedley, 2003). Although the discharges are bilateral and widely distributed, they are not truly "generalized" in the sense that the voltage is maximum in the anterior midline contacts. Localized maximum amplitude of spike wave has been confirmed using conventional EEG (Rodin & Ancheta, 1987; Weir, 1965), high-density EEG (Holmes, Brown, & Tucker, 2004), and magnetoencephalography (Westmijse, Ossenblok, Gunning, & van Luijtelaar, 2009). Electrophysiological studies in animal models of absence epilepsy also demonstrate focal bilateral spike wave with relative sparing of more posterior regions (Meeren, Pijn, van Luijtelaar, Coenen, & Lopes da Silva, 2002; Nersesyan, Herman, Erdogan, Hyder, & Blumenfeld, 2004; Vergnes, Marescaux, & Depaulis, 1990). These findings suggest that deficits in consciousness during absence seizures may be related to intense involvement of particular brain regions, even in so-called generalized seizures (Kostopoulos, 2001; Pavone & Niedermeyer, 2000). Several early studies reported that impaired consciousness in absence seizures is associated with EEG features, including spike-wave amplitude, duration, rhythmicity, frontocentral distribution, and generalization (Browne et al., 1974; Jus & Jus, 1960; Mirsky & van Buren, 1965). However, since other investigators claimed there is no relation between EEG and absence behavioral severity (Boudin, Barbizet, & Masson, 1958; Davidoff & Johnson, 1964; Gastaut, 1954), this may be a topic worthy of further investigation.

In generalized tonic-clonic seizures, the EEG during the tonic phase shows widespread polyspike discharges or low-voltage fast activity, which gives way to polyspike-and-wave during the clonic phase and then generalized suppression post-ictally. There is some evidence based on EEG that generalized tonic-clonic seizures, like absence seizures, may affect some brain regions more strongly than others. For example, secondarily generalized tonic-clonic seizures show some EEG asymmetry even during the generalized phase (Kriss, Halliday, Halliday, & Pratt, 1978; McNally & Blumenfeld, 2004), and intracranial EEG has demonstrated that generalized tonic-clonic seizures can spare some brain regions (Schindler, Leung, Lehnertz, & Elger, 2007).

Temporal lobe complex partial seizures exhibit focal 5–7 Hz theta frequency discharges over the temporal lobe on scalp EEG, often accompanied by more widespread slower delta- and theta-frequency activity. Intracranial EEG shows periodic spikes or low-voltage fast activity initially in one medial temporal lobe, followed by theta-frequency polyspike-and-wave activity extending to the lateral temporal cortex unilaterally or bilaterally (figure 71.3A, B, C). Loss of consciousness is more common with bilateral involvement (figure 71.3E, G) but can also occur with unilateral temporal lobe seizure activity (Englot et al., 2010; Gloor, Olivier, & Ives, 1980; Lux et al., 2002). Seizures with left-sided onset are more commonly associated with impaired consciousness; however, this effect may be caused at least in part by a bias favoring verbal testing methods; impaired consciousness can certainly be seen in temporal lobe seizures with right-sided onset as well (Englot et al., 2010).

We recently observed an interesting phenomenon in which sleep-like slow-wave activity occurs in the frontoparietal association cortex during temporal lobe seizures with impaired consciousness (Blumenfeld, Rivera, et al., 2004). This slow-wave activity had been observed previously but interpreted as seizure propagation (Lieb, Dasheiff, & Engel, 1991). However, the frontoparietal slow-wave activity during temporal lobe seizures more closely resembles the EEG of slow-wave sleep, coma, or encephalopathy than seizures. Fundamental studies in an animal model demonstrated very similar phenomena and confirmed that this ictal neocortical slow-wave activity has reduced neuronal firing, blood flow, and metabolism, similar to deep anesthesia or sleep, whereas seizure activity shows increases in all of these parameters (Englot et al., 2008, 2009). We concluded that focal seizures in the temporal lobe can induce a unique state of depressed cortical function, which may be crucial for impaired consciousness in this disorder. In support of this, we found that bilateral delta-frequency slow-wave activity on intracranial EEG recordings from the frontoparietal cortex is strongly associated with impaired consciousness in temporal lobe epilepsy (figure 71.3A–F). Temporal lobe seizures without neocortical slow-wave activity (figure 71.3G) are not associated with impaired consciousness.

Depth electrode recordings from patients with temporal lobe epilepsy have revealed abnormally enhanced thalamocortical synchrony, which may contribute to these physiological changes (Arthuis et al., 2009; Bartolomei, 2012; Guye et al., 2006; Rosenberg et al., 2006). Although less is known about impaired consciousness in complex partial seizures originating outside the temporal lobe, a recent study showed similar findings using intracranial EEG in parietal lobe epilepsy (Lambert et al., 2012).

Neuroimaging

Functional neuroimaging has provided important new insights into the mechanisms of impaired consciousness in epilepsy. Earlier work based on lower-resolution methods has given way to functional MRI (fMRI) and other techniques. Single-photon emission computed tomography (SPECT) does not have the resolution of fMRI, but allows imaging of cerebral blood flow despite patient movement during seizures. SPECT is based on injection of a radiotracer during seizures, providing a "snapshot" of cerebral blood flow at the time of injection. The tracer is rapidly taken up by the brain and remains relatively stable, allowing neuroimaging to be done 1–2 hours later when the patient is no longer moving (Devous, Leroy, & Homan, 1990; Kim, Zubal, & Blumenfeld, 2009; McNally et al., 2005). Caution must be exercised in interpreting functional neuroimaging studies, since they are only indirectly related to neural activity through neurometabolic and neurovascular coupling. Direct measurements of neuronal activity in animal models have revealed unexpected relationships and shed important light on the interpretation of neuroimaging signals in epilepsy and under normal conditions (Englot et al., 2008; Hyder et al., 2010; Mishra et al., 2011; Schridde et al., 2008; Suh, Ma, Zhao, Sharif, & Schwartz, 2006).

Neuroimaging in patients with absence epilepsy has recently used simultaneous EEG-fMRI to demonstrate changes in all parts of the consciousness system during seizures. Spike-wave discharges in these patients produce fMRI increases in the thalamus as well as the primary visual, sensorimotor, and auditory cortex bilaterally, while predominantly fMRI decreases are seen in default-mode cortical regions such as the medial frontoparietal cortex and lateral parietal cortex (Archer, Abbott, Waites, & Jackson, 2003; Berman et al., 2010; Carney et al., 2010; Gotman et al., 2005; Moeller, LeVan, et al., 2010; Moeller, Siebner, Wolff, Muhle, Granert, et al., 2008; Salek-Haddadi et al., 2003). The lateral frontal cortex shows variable changes (Carney, Masterton, Flanagan, Berkovic, & Jackson, 2012). It was recently recognized that the standard "canonical" hemodynamic response function does not adequately

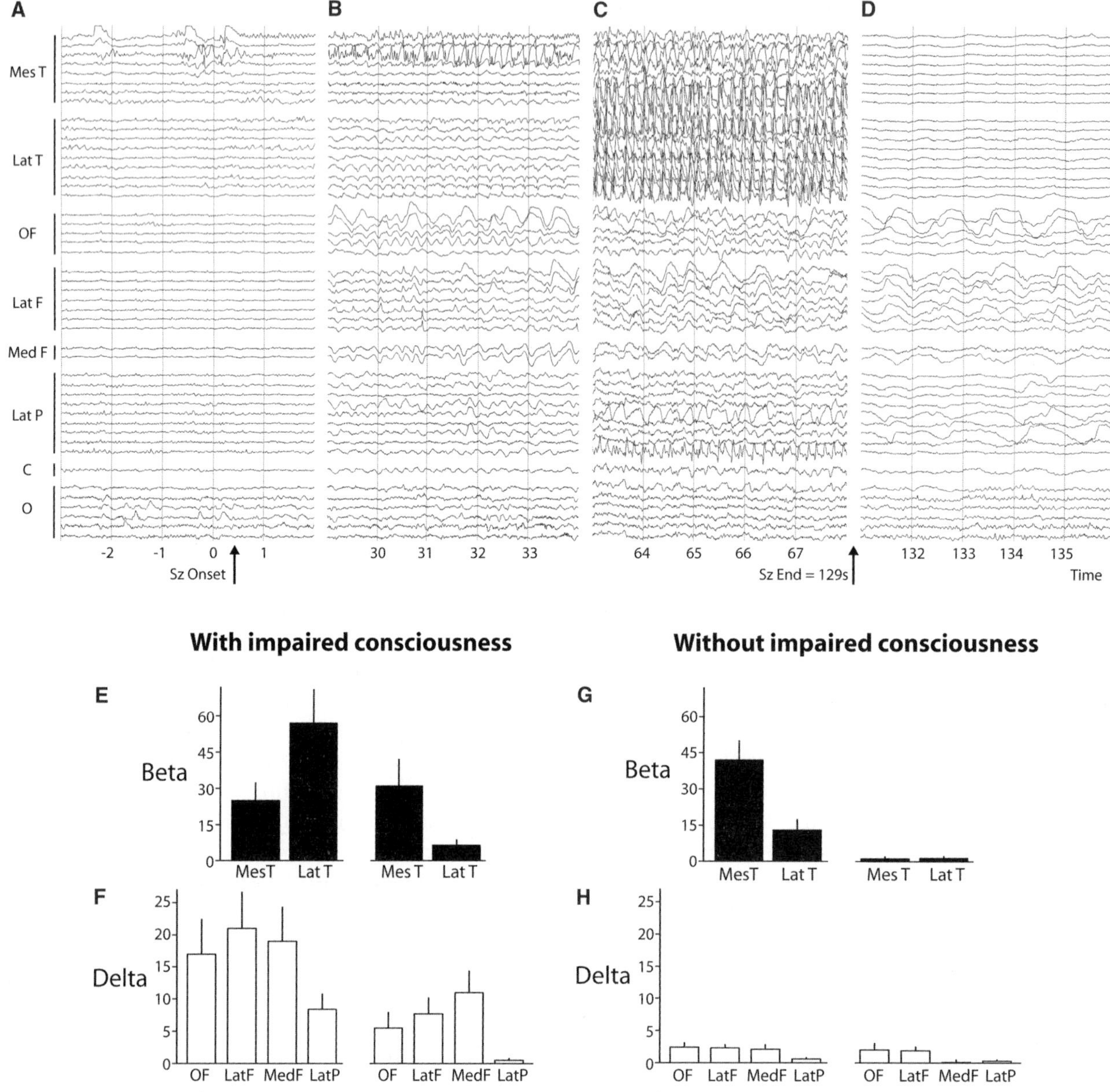

FIGURE 71.3 EEG in temporal lobe seizures with impaired consciousness shows cortical slow-wave activity. (A–D) Time course of intracranial EEG changes during typical temporal lobe seizure with impaired consciousness. Only ipsilateral contacts are shown. Bars along left margin indicate electrode contacts from different strips, rows, or depth electrodes in the indicated brain regions. A subset of representative electrodes are shown of the 128 studied in this patient. Calibration bar on right is 3 mV. Montage is referential to mastoid. (A) Seizure onset with low-voltage fast activity emerging from periodic spiking in the mesial temporal contacts. (B) Sample of EEG from early seizure. Rhythmic polyspike and sharp wave activity develops in the mesial temporal lobe, while the frontal and parietal contacts show large-amplitude irregular slow activity. (C) Sample of EEG from mid-seizure. Polyspike and wave activity is present in the mesial and lateral temporal lobe contacts, with ongoing slow waves in the association cortex. Paracentral Rolandic and occipital contacts are relatively spared. (D) Post-ictal suppression is seen in temporal lobe contacts, with continued irregular slowing in the frontoparietal neocortex. (E–H) Group data. Focal temporal lobe seizures with impaired consciousness have bilateral increases in temporal beta frequency and frontoparietal delta frequency activity, while seizures without impaired consciousness show mainly increases in ipsilateral temporal lobe beta. Mean fractional changes (± standard error of mean) in intracranial EEG power compared to 60 sec preseizure baseline. (E) Temporal lobe beta in seizures with impaired consciousness. (F) Neocortical delta in seizures with impaired consciousness. (G) Temporal lobe beta in seizures without impaired consciousness. (H) Neocortical delta in seizures without impaired consciousness. Bilateral temporal lobe beta activity and frontoparietal delta activity were significantly higher in seizures with vs. without impaired consciousness ($P < 0.05$, Mann Whitney U test; $n = 38$ seizures with impaired consciousness and 25 seizures without impaired consciousness in 26 patients). Mes T, mesial temporal; Lat T, lateral temporal; OF, orbital frontal; LatF, lateral frontal; MedF, medial frontal; LatP, lateral parietal; C, perirolandic pre- and post-central gyri; O, occipital. (Reproduced from Englot et al., 2010, by permission of Oxford University Press.)

fit fMRI changes in absence seizures. Thus, fMRI increases may begin well before the EEG onset of seizures, and a complex sequence of fMRI increases and decreases continue in different brain regions long after EEG seizure offset (Bai et al., 2010; Carney et al., 2010; Moeller, Siebner, Wolff, Muhle, Boor, et al., 2008). To fully relate the fMRI changes in absence seizures to impaired consciousness, it will be necessary to more carefully analyze these complicated signals. So far, only a few studies have attempted to relate fMRI and behavioral impairment in absence seizures (Berman et al., 2010; Li et al., 2009; Moeller, Muhle, et al., 2010), showing simply that more extensive fMRI changes occur when behavior is impaired. Studies employing larger sample sizes and data-driven approaches to fMRI signal analysis will hopefully soon yield more information about network impairment in absence seizures (Guo et al., 2011).

Generalized tonic-clonic seizures have been imaged mainly with ictal SPECT, which is lower resolution than fMRI but alleviates problems with patient safety and movement artifact during seizures. Neuroimaging of generalized tonic-clonic seizures has shown dramatic changes in the consciousness system (figure 71.1) both during partial seizures with secondary generalization (Blumenfeld et al., 2009; Blumenfeld, Westerveld, et al., 2003; Lee et al., 1987; Rowe et al., 1989; Shin, Hong, Tae, & Kim, 2002; Varghese et al., 2009), as well as in tonic-clonic seizures induced by electroconvulsive therapy (Bajc et al., 1989; Blumenfeld, McNally, Ostroff, & Zubal, 2003; Blumenfeld, Westerveld, et al., 2003; Enev et al., 2007; McNally & Blumenfeld, 2004; Takano et al., 2007; Vollmer-Haase, Folkerts, Haase, Deppe, & Ringelstein, 1998). Specifically, during generalized tonic-clonic seizures, increases are observed in the bilateral lateral frontal and parietal cortex, medial parietal cortex, thalamus, and upper brainstem, while decreases are seen in the medial frontal and cingulate cortex. Post-ictally, cerebral blood flow decreases are seen in medial and lateral frontoparietal association cortex (Blumenfeld et al., 2009; Enev et al., 2007). Again, these changes overlap but are not limited to the default-mode network. The cerebellum shows an interesting progressive increase in cerebral blood flow in the late ictal and early post-ictal periods, beginning in the midline and progressing into the lateral cerebellar hemispheres (Blumenfeld et al., 2009). Prior work from a cat model of generalized tonic-clonic seizures showed a similar progressive increase in cerebellar activity (Salgado-Benitez, Briones, & Fernandez-Guardiola, 1982), which was proposed to contribute to seizure termination and post-ictal suppression of forebrain activity. Cerebellar Purkinje cells have strong inhibitory outputs via the deep cerebellar nuclei and thalamus. To investigate this mechanism further, we performed a correlation analysis of SPECT signals in the cerebellum versus the rest of the brain during and following generalized tonic-clonic seizures across patients. This showed a strong correlation between cerebellar increases and *decreased* activity in the frontoparietal consciousness system structures (Blumenfeld et al., 2009). The neural mechanisms for these changes should be investigated further; however, the findings suggest that cerebellar inhibitory outputs may be important for depressed forebrain function and impaired consciousness in generalized tonic-clonic seizures.

Complex partial seizures have also been imaged mainly with ictal SPECT. These studies have shown ictal increases in cerebral blood flow in the upper brainstem, medial thalamus, and hypothalamus (figure 71.4; Blumenfeld, McNally, et al., 2004; Hogan, Kaiboriboon, Bertrand, Rao, & Acharya, 2006; Lee et al., 2002; Mayanagi, Watanabe, & Kaneko, 1996; Tae et al., 2005). These subcortical changes were shown to be associated with impaired consciousness (Blumenfeld, McNally, et al., 2004; Lee et al., 2002). In addition, cortical SPECT *decreases* were found in the bilateral frontoparietal association cortex in temporal lobe complex partial seizures (figure 71.4; Blumenfeld, McNally, et al., 2004; Chassagnon et al., 2009; van Paesschen, Dupont, van Driel, van Billoen, & Maes, 2003), in the same regions in which slow-wave activity was reported on intracranial EEG (Englot et al., 2010). The cortical and subcortical changes in complex partial seizures again involve the main components of the consciousness system, including the bilateral upper brainstem/medial diencephalon and frontoparietal association cortex. In contrast, simple partial temporal lobe seizures do not show these network effects, but instead show localized changes in one medial temporal lobe (Blumenfeld, McNally, et al., 2004). The relationship between cortical and subcortical changes in temporal lobe complex partial seizures was further strengthened by correlation analysis of SPECT data across patients, showing a strong relationship between increases in the medial thalamus and *decreases* in the bilateral frontoparietal association cortex (Blumenfeld, McNally, et al., 2004).

Based on these findings, we proposed a "network inhibition hypothesis" to explain the puzzle of why focal seizures in the temporal lobe often cause loss of consciousness (figure 71.5; Blumenfeld, 2009, 2012; Blumenfeld & Taylor, 2003; Englot & Blumenfeld, 2009; Englot et al., 2008, 2009, 2010; Norden & Blumenfeld, 2002; Yu & Blumenfeld, 2009). Under normal conditions, consciousness is maintained by bidirectional interactions between the cortical and subcortical

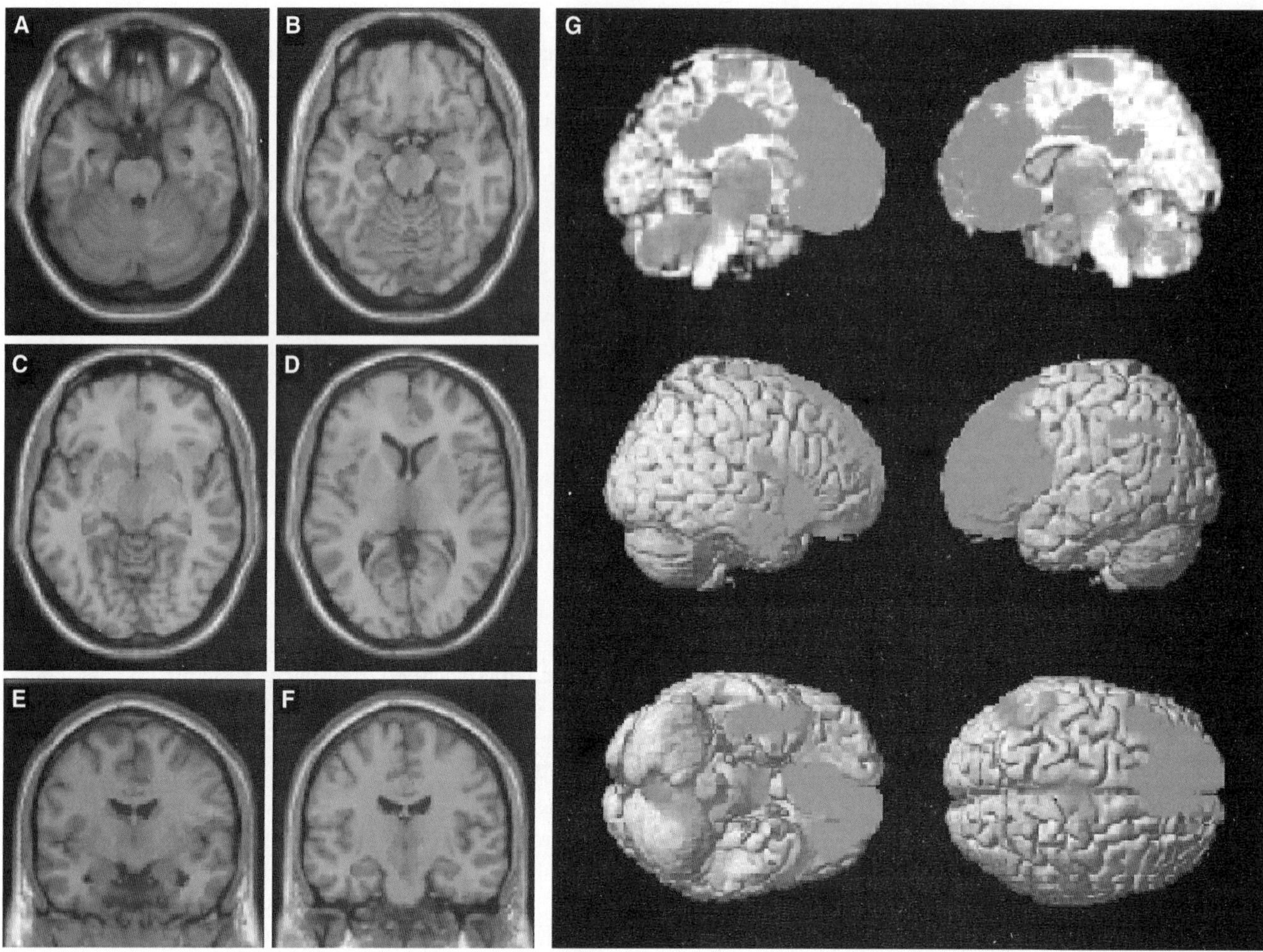

FIGURE 71.4 Cerebral blood flow imaging in temporal lobe seizures with impaired consciousness. Complex partial seizures arising from the temporal lobe are associated with significant cerebral blood flow increases and decreases in widespread brain regions. Statistical parametric maps depict SPECT increases in red and decreases in green. Changes ipsilateral to seizure onset are shown on the left side of the brain, and contralateral changes on the right side of the brain (combining patients with left and right onset seizures; $n = 10$). Data are from >90 sec after seizure onset, when consciousness was markedly impaired. Note that at earlier times there were SPECT increases in the ipsilateral mesial temporal lobe (not shown). (A–F) Horizontal sections progressing from inferior to superior (A–D), and coronal sections (E, F) progressing from anterior to posterior showing blood flow increases in the bilateral midbrain, hypothalamus, medial thalamus, and midbrain. Decreases are seen in the bilateral association cortex. (G) Three-dimensional surface renderings show increases mainly in the bilateral medial diencephalon, upper brainstem, and medial cerebellum, while decreases occur in the ipsilateral contralateral frontal and parietal association cortex (same data as A–F). Extent threshold, $k = 125$ voxels (voxel size = $2 \times 2 \times 2$ mm). Height threshold, $P = 0.01$. (Reproduced from Blumenfeld, McNally, et al., 2004, by permission of Oxford University Press.) (See color plate 61.)

components of the consciousness system (figure 71.5A). Focal temporal lobe seizures produce abnormal polyspike discharges in the medial temporal lobe (figure 71.5B). Seizures propagate along known anatomical pathways to subcortical structures, particularly to GABAergic inhibitory neurons in the lateral septal nuclei, anterior hypothalamic ventrolateral preoptic area, thalamic reticular nucleus, habenula, substantia nigra pars reticulata, ventral pallidum, and cerebellar cortex (figure 71.5C). These regions may powerfully inhibit subcortical arousal systems in the upper brainstem, thalamus, hypothalamus, and basal forebrain (figure 71.5D). Removal of subcortical arousal leads to widespread cortical slow-wave activity (figure 71.5D) resembling coma or deep sleep and produces impaired consciousness.

In addition to the human data, neuroimaging and other measurements in animal models have provided valuable insights into the mechanisms of impaired consciousness in complex partial temporal lobe seizures. Rat hippocampal seizures show behavioral arrest and neocortical slow-wave activity similar to humans (Englot

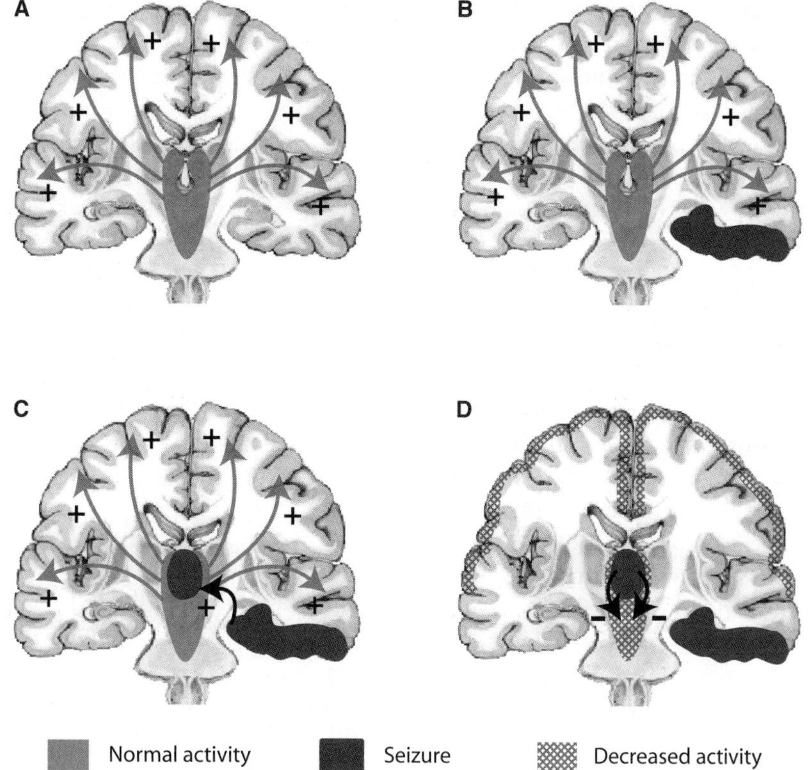

FIGURE 71.5 Network inhibition hypothesis for impaired consciousness in temporal lobe complex partial seizures. (A) Under normal conditions, the upper brainstem-diencephalic activating systems interact with the cerebral cortex to maintain normal consciousness. (B) Focal seizure involving the mesial temporal lobe. If the seizure remains confined, then a simple partial seizure will occur without impairment of consciousness. (C) Spread of seizure activity from the temporal lobe to midline subcortical structures. Propagation often occurs to the contralateral mesial temporal as well (not shown). (D) Inhibition of subcortical activating systems leads to depressed activity in bilateral frontoparietal association cortex and to loss of consciousness. (Modified with permission from Blumenfeld & Taylor, 2003.)

et al., 2008). Multiunit recordings revealed up and down states of neuronal firing, closely resembling coma, deep sleep, or encephalopathy (Englot et al., 2008; Haider, Duque, Hasenstaub, & McCormick, 2006; Steriade, Contreras, Curro Dossi, & Nunez, 1993). Rat fMRI and electrophysiology confirmed that ictal neocortical slow activity is a unique state of depressed cortical function, associated with a mean *decrease* in neuronal firing, cerebral blood flow, cerebral blood volume, and cerebral metabolic rate of oxygen consumption (Englot et al., 2008). In contrast, hippocampal or neocortical seizure activity elicits *increases* in all of these variables. Further experiments demonstrated that cutting the fornix prevented subcortical spread of seizures, neocortical slow activity, and behavioral arrest (Englot et al., 2009). In addition, stimulation of inhibitory subcortical structures such as the lateral septum replicated slow activity and behavioral arrest (Englot et al., 2009) and single-unit recordings revealed a shutdown of neurons involved in subcortical arousal, such as cholinergic neurons in the penunculopontine tegmental nucleus (Motelow et al., 2012). The combination of human data and work from the animal model suggests that therapeutic neurostimulation of subcortical arousal systems (Schiff et al., 2007) may be a reasonable future treatment option, with a goal of improving consciousness in medically and surgically refractory complex partial seizures.

Summary and future directions

We have seen that absence, generalized tonic-clonic, and complex partial seizures all converge on the same set of anatomical structures when consciousness is impaired. These seizures differ in behavior, physiology, and neuroimaging, yet all converge on the consciousness system (figure 71.1) through different mechanisms (table 71.1). Seizures can disrupt the consciousness system either through direct propagation or through indirect network effects, leading to impaired level of consciousness. Mechanisms involving changes in information integration or the global workspace have also been proposed (Bartolomei, 2012; Tononi, 2005; Tononi & Koch, 2008). In absence seizures, the thalamus shows abnormal increased activity, but the complex time course of fMRI signals has made the exact nature of cortical changes less certain. Although the data so far suggest that the consciousness system is involved, additional studies are needed that take into account the complex fMRI time course in relation to behavior. In addition, improved animal models may add to the fundamental understanding of cortical and subcortical changes in absence seizures. Neuroimaging and electrophysiology measurements in generalized tonic-clonic seizures suggest that focal bilateral regions of the consciousness system are involved most intensely. The cerebellar changes seen late in seizures and continuing into the post-ictal period may play an important role in seizure termination as well as in impaired consciousness, and should be investigated further. Complex partial seizures have been a puzzle, since it was unclear why focal seizures produce this sleepwalk-like state, with impairment in multiple cognitive functions. The network inhibition hypothesis explains why the level of consciousness is decreased in complex partial seizures. Evidence from human as well as animal model neuroimaging and electrophysiology support a mechanism in which focal seizures depress subcortical arousal systems, producing sleep-like slow-wave activity in the association cortex and impaired consciousness. Additional work is needed to fully understand the neurobiology of suppressed cortical arousal during seizures. Furthermore, the mechanisms of impaired consciousness in nontemporal lobe complex partial seizures (with frontal, parietal, or occipital onset) are another important topic of future study.

For people living with epilepsy, the impact of impaired consciousness on quality of life is substantial (Charidimou & Selai, 2011). Unpredictable loss of consciousness during seizures causes increased risk of driving hazards, other accidents, and injuries such as burns, falls, and drowning, impaired school and work performance, and social stigmatization (de Boer, Mula, & Sander, 2008; Nei & Bagla, 2007; Yang et al., 2010). Depressed arousal during and following seizures may contribute to respiratory compromise and sudden unexplained death in epilepsy (Richerson & Buchanan, 2011). There is also evidence that impaired consciousness in the peri-ictal time period is a major factor leading to unreliable reporting of seizure occurrence by patients to their physicians (Blum, Eskola, Bortz, & Fisher, 1996; Ezeani et al., 2012; Hoppe, Poepel, & Elger, 2007).

It is hoped that further work will lead to a greater understanding of the behavior and mechanisms in epileptic unconsciousness, leading the way to improved treatments. While stopping seizures remains the goal, approximately one-fifth of patients with epilepsy are refractory to all medical and surgical treatments and suffer from disabling seizures with impaired consciousness. Treatments aimed at preserving the level of consciousness during seizures could greatly improve the quality of life for these patients, and will hopefully be achieved in the near future.

ACKNOWLEDGMENTS William Chen and Robert Kim helped prepare the figures. This work was supported by NIH R01NS055829, R01NS066974, R01MH67528, R01HL059619, P30NS052519, U01NS045911, a Donaghue Foundation Investigator Award, and the Betsy and Jonathan Blattmachr Family.

REFERENCES

Ali, F., Rickards, H., Bagary, M., Greenhill, L., McCorry, D., & Cavanna, A. E. (2010). Ictal consciousness in epilepsy and nonepileptic attack disorder. *Epilepsy Behav, 19*, 522–525.

Ali, F., Rickards, H., & Cavanna, A. E. (2012). The assessment of consciousness during partial seizures. *Epilepsy Behav, 23*, 98–102.

Archer, J. S., Abbott, D. F., Waites, A. B., & Jackson, G. D. (2003). FMRI "deactivation" of the posterior cingulate during generalized spike and wave. *NeuroImage, 20*, 1915–1922.

Arthuis, M., Valton, L., Regis, J., Chauvel, P., Wendling, F., Naccache, L., … Bartolomei, F. (2009). Impaired consciousness during temporal lobe seizures is related to increased long-distance cortical-subcortical synchronization. *Brain, 132*, 2091–2101.

Asplund, C. L., Todd, J. J., Snyder, A. P., & Marois, R. (2010). A central role for the lateral prefrontal cortex in goal-directed and stimulus-driven attention. *Nat Neurosci, 13*, 507–512.

Baars, B. J., Ramsoy, T. Z., & Laureys, S. (2003). Brain, conscious experience and the observing self. *Trends Neurosci, 26*, 671–675.

Bai, X., Vestal, M., Berman, R., Negishi, M., Spann, M., Vega, C., … Blumenfeld, H. (2010). Dynamic time course of typical childhood absence seizures: EEG, behavior, and functional magnetic resonance imaging. *J Neurosci, 30*, 5884–5893.

Bajc, M., Medved, V., Basic, M., Topuzovic, N., Babic, D., & Ivancevic, D. (1989). Acute effect of electroconvulsive therapy on brain perfusion assessed by Tc99m-hexamethylpropyleneamineoxim and single photon emission computed tomography. *Acta Psychiatr Scand, 80*, 421–426.

Bartolomei, F. (2012). Coherent neural activity and brain synchronization during seizure-induced loss of consciousness. *Arch Ital Biol, 150*, 164–171.

Bauerschmidt, A., Koshkelashvili, N., Ezeani, C. C., Yoo, J. Y., Zhang, Y., Manganas, L. N., … Blumenfeld, H. (2013). Prospective assessment of ictal behavior using the revised Responsiveness in Epilepsy Scale (RES-II). *Epilepsy Behav, 26*(1), 25–28.

Bell, W. L., Walczak, T. S., Shin, C., & Radtke, R. A. (1997). Painful generalised clonic and tonic-clonic seizures with retained consciousness. *J Neurol Neurosurg Psychiatry, 63*, 792–795.

Berg, A. T., Berkovic, S. F., Brodie, M. J., Buchhalter, J., Cross, J. H., van Emde Boas, W., ... Scheffer, I. E. (2010). Revised terminology and concepts for organization of seizures and epilepsies: Report of the ILAE Commission on Classification and Terminology, 2005–2009. *Epilepsia, 51*, 676–685.

Berger, H. (1929). Ueber das elektrenkephalogramm des menschen. *Arch Psychiatr Nervenkr, 87*, 527.

Berman, R., Negishi, M., Vestal, M., Spann, M., Chung, M., Bai, X., ... Blumenfeld, H. (2010). Simultaneous EEG, fMRI, and behavioral testing in typical childhood absence seizures. *Epilepsia, 51*(10), 2011–2022.

Blum, D. E., Eskola, J., Bortz, J. J., & Fisher, R. S. (1996). Patient awareness of seizures. *Neurology, 47*, 260–264.

Blumenfeld, H. (2005a). Cellular and network mechanisms of spike-wave seizures. *Epilepsia, 46*(Suppl. 9), 21–33.

Blumenfeld, H. (2005b). Consciousness and epilepsy: Why are patients with absence seizures absent? *Prog Brain Res, 150*, 271–286.

Blumenfeld, H. (2009). Epilepsy and consciousness. In S. Laureys & G. Tononi (Eds.), *The neurology of consciousness: Cognitive neuroscience and neuropathology* (pp. 15–30). New York, NY: Academic.

Blumenfeld, H. (2010). *Neuroanatomy through clinical cases* (2nd ed.). Sunderland, MA: Sinauer.

Blumenfeld, H. (2011). Epilepsy and the consciousness system: Transient vegetative state? *Neurol Clin, 29*, 801–823.

Blumenfeld, H. (2012). Impaired consciousness in epilepsy. *Lancet Neurol, 11*, 814–826.

Blumenfeld, H., McNally, K. A., Ostroff, R. B., & Zubal, I. G. (2003). Targeted prefrontal cortical activation with bifrontal ECT. *Psychiatry Res, 123*, 165–170.

Blumenfeld, H., McNally, K. A., Vanderhill, S. D., Paige, A. L., Chung, R., Davis, K., ... Spencer, S. S. (2004). Positive and negative network correlations in temporal lobe epilepsy. *Cereb Cortex, 14*, 892–902.

Blumenfeld, H., Rivera, M., McNally, K. A., Davis, K., Spencer, D. D., & Spencer, S. S. (2004). Ictal neocortical slowing in temporal lobe epilepsy. *Neurology, 63*, 1015–1021.

Blumenfeld, H., & Taylor, J. (2003). Why do seizures cause loss of consciousness? *Neuroscientist, 9*, 301–310.

Blumenfeld, H., Varghese, G., Purcaro, M. J., Motelow, J. E., Enev, M., McNally, K. A., ... Paige, A. L. (2009). Cortical and subcortical networks in human secondarily generalized tonic-clonic seizures. *Brain, 132*, 999–1012.

Blumenfeld, H., Westerveld, M., Ostroff, R. B., Vanderhill, S. D., Freeman, J., Necochea, A., ... Zubal, I. G. (2003). Selective frontal, parietal and temporal networks in generalized seizures. *NeuroImage, 19*, 1556–1566.

Botez, M. I., Serbanescu, T., & Stoica, I. (1966). The problem of focal epileptic seizures on both parts of the body without loss of consciousness. *Psychiatr Neurol Neurochir, 69*, 431–437.

Boudin, G., Barbizet, J., & Masson, S. (1958). Etude de la dissolution de la conscience dans 3 cas de petit mal avec crises prolongées. *Rev Neurol, 99*, 483–487.

Bremer, F. (1955). Interrelationships between cortex and subcortical structures; introductory remarks. *Electroen Clin Neurophysiol, 1*(Suppl. 4), 145–148.

Browne, T. R., Penry, J. K., Porter, R. J., & Dreifuss, F. E. (1974). Responsiveness before, during and after spike-wave paroxysms. *Neurology, 24*, 659–665.

Buschman, T. J., & Miller, E. K. (2007). Top-down versus bottom-up control of attention in the prefrontal and posterior parietal cortices. *Science, 315*, 1860–1862.

Carney, P. W., Masterton, R. A., Flanagan, D., Berkovic, S. F., & Jackson, G. D. (2012). The frontal lobe in absence epilepsy: EEG-fMRI findings. *Neurology, 78*, 1157–1165.

Carney, P. W., Masterton, R. A., Harvey, A. S., Scheffer, I. E., Berkovic, S. F., & Jackson, G. D. (2010). The core network in absence epilepsy. Differences in cortical and thalamic BOLD response. *Neurology, 75*, 904–911.

Cavanna, A. E., Mula, M., Servo, S., Strigaro, G., Tota, G., Barbagli, D., ... Monaco, F. (2008). Measuring the level and content of consciousness during epileptic seizures: The Ictal Consciousness Inventory. *Epilepsy Behav, 13*, 184–188.

Cavanna, A. E., Rickards, H., & Ali, F. (2011). What makes a simple partial seizure complex? *Epilepsy Behav, 22*, 651–658.

Chalmers, D. J. (1996). *The conscious mind: In search of a fundamental theory.* Oxford, UK: Oxford University Press.

Charidimou, A., & Selai, C. (2011). The effect of alterations in consciousness on quality of life (QoL) in epilepsy: Searching for evidence. *Behav Neurol, 24*, 83–93.

Chassagnon, S., Namer, I. J., Armspach, J. P., Nehlig, A., Kahane, P., Kehrli, P., ... Hirsch, E. (2009). SPM analysis of ictal-interictal SPECT in mesial temporal lobe epilepsy: Relationships between ictal semiology and perfusion changes. *Epilepsy Res, 85*, 252–260.

Chipaux, M., Vercueil, L., Kaminska, A., Mahon, S., & Charpier, S. (2013). Persistence of cortical sensory processing during absence seizures in human and an animal model: Evidence from EEG and intracellular recordings. *PLoS ONE, 8*(3), e58180.

Cunningham, C., Chen, W. C., Shorten, A., McClurkin, M., Choezom, T., Schmidt, C.P., ... Blumenfeld, H. (2014). Impaired consciousness in partial seizures is bimodally distributed. *Neurology, 82*(19), 1736–1744.

Davidoff, R. A., & Johnson, L. C. (1964). Paroxysmal EEG activity and cognitive-motor performance. *Electroen Clin Neurophysiol, 16*, 343–354.

De Boer, H. M., Mula, M., & Sander, J. W. (2008). The global burden and stigma of epilepsy. *Epilepsy Behav, 12*, 540–546.

Dennett, D. (1991). *Consciousness explained.* New York, NY: Little, Brown.

Devous, M. D., Sr., Leroy, R. F., & Homan, R. W. (1990). Single photon emission computed tomography in epilepsy. *Semin Nucl Med, 20*, 325–341.

Dosenbach, N. U., Fair, D. A., Miezin, F. M., Cohen, A. L., Wenger, K. K., Dosenbach, R. A., ... Petersen, S. E. (2007). Distinct brain networks for adaptive and stable task control in humans. *Proc Natl Acad Sci USA, 104*, 11073–11078.

Ebersole, J. S., & Pedley, T. A. (2003). *Current practice of clinical electroencephalography* (3rd ed.). Philadelphia, PA: Lippincott Williams & Wilkins.

Enev, M., McNally, K. A., Varghese, G., Zubal, I. G., Ostroff, R. B., & Blumenfeld, H. (2007). Imaging onset and propagation of ECT-induced seizures. *Epilepsia, 48*, 238–244.

Englot, D. J., & Blumenfeld, H. (2009). Consciousness and epilepsy: Why are complex-partial seizures complex? *Prog Brain Res, 177,* 147–170.

Englot, D. J., Mishra, A. M., Mansuripur, P. K., Herman, P., Hyder, F., & Blumenfeld, H. (2008). Remote effects of focal hippocampal seizures on the rat neocortex. *J Neurosci, 28*(36), 9066–9081.

Englot, D. J., Modi, B., Mishra, A. M., DeSalvo, M., Hyder, F., & Blumenfeld, H. (2009). Cortical deactivation induced by subcortical network dysfunction in limbic seizures. *J Neurosci, 29*(41), 13006–13018.

Englot, D. J., Yang, L., Hamid, H., Danielson, N., Bai, X., Marfeo, A., ... Blumenfeld, H. (2010). Impaired consciousness in temporal lobe seizures: Role of cortical slow activity. *Brain, 133*(12), 3764–3777.

Escueta, A. V., Bacsal, F. E., & Treiman, D. M. (1982). Complex partial seizures on closed-circuit television and EEG: A study of 691 attacks in 79 patients. *Ann Neurol, 11,* 292–300.

Escueta, A. V., Kunze, U., Waddell, G., Boxley, J., & Nadel, A. (1977). Lapse of consciousness and automatisms in temporal lobe epilepsy: A videotape analysis. *Neurology, 27,* 144–155.

Ezeani, C., Detyniecki, K., Bauerschmidt, A., Winstanley, S., Duckrow, R., Hirsch, L., & Blumenfeld, H. (2012). Accuracy of patients' seizure reporting during video EEG monitoring. *AES Abstracts.* Retreived from https://www.aesnet.org/meetings_events/annual_meeting_abstracts

Gastaut, H. (1954). The brain stem and cerebral electrogenesis in relation to consciousness. In J. F. Delafresnaye (Ed.), *Brain mechanisms and consciousness* (pp. 249–283). Springfield, IL: Thomas.

Giacino, J. T., Kalmar, K., & Whyte, J. (2004). The JFK Coma Recovery Scale-Revised: Measurement characteristics and diagnostic utility. *Arch Phys Med Rehab, 85,* 2020–2029.

Gloor, P. (1986). Consciousness as a neurological concept in epileptology: A critical review. *Epilepsia, 27*(Suppl. 2), S14-S26.

Gloor, P., Olivier, A., & Ives, J. (1980). Loss of consciousness in temporal lobe epilepsy: Observations obtained with stereotaxic depth electrode recordings and stimulations. In R. Canger, F. Angeleri, & J. K. Penry (Eds.), *Advances in epileptology: The XIth Epilepsy International Symposium* (pp. 349–353). New York, NY: Raven.

Gotman, J., Grova, C., Bagshaw, A., Kobayashi, E., Aghakhani, Y., & Dubeau, F. (2005). Generalized epileptic discharges show thalamocortical activation and suspension of the default state of the brain. *Proc Natl Acad Sci USA, 102,* 15236–15240.

Guo, J. N., Gonzalez, J. L., Bai, X., Negishi, M., Danielson, N., Han, X., ... Blumenfeld, H. (2011). EEG and fMRI correlates of variable performance during typical childhood absence seizures. *Soc Neurosci Abstracts, 400.*15.

Guye, M., Regis, J., Tamura, M., Wendling, F., McGonigal, A., Chauvel, P., & Bartolomei, F. (2006). The role of corticothalamic coupling in human temporal lobe epilepsy. *Brain, 129*(7), 1917–1928.

Haider, B., Duque, A., Hasenstaub, A. R., & McCormick, D. A. (2006). Neocortical network activity in vivo is generated through a dynamic balance of excitation and inhibition. *J Neurosci, 26,* 4535–4545.

Heydrich, L., Dieguez, S., Grunwald, T., Seeck, M., & Blanke, O. (2010). Illusory own body perceptions: Case reports and relevance for bodily self-consciousness. *Conscious Cogn, 19,* 702–710.

Hoffmann, J. M., Elger, C. E., & Kleefuss-Lie, A. A. (2008). Lateralizing value of behavioral arrest in patients with temporal lobe epilepsy. *Epilepsy Behav, 13,* 634–636.

Hogan, R. E., Kaiboriboon, K., Bertrand, M. E., Rao, V., & Acharya, J. (2006). Composite SISCOM perfusion patterns in right and left temporal seizures. *Arch Neurol, 63,* 1419–1426.

Holmes, M. D., Brown, M., & Tucker, D. M. (2004). Are "generalized" seizures truly generalized? Evidence of localized mesial frontal and frontopolar discharges in absence. *Epilepsia, 45,* 1568–1579.

Hoppe, C., Poepel, A., & Elger, C. E. (2007). Epilepsy: Accuracy of patient seizure counts. *Arch Neurol, 64,* 1595–1599.

Hyder, F., Sanganahalli, B. G., Herman, P., Coman, D., Maandag, N. J., Behar, K. L., ... Rothman, D. L. (2010). Neurovascular and neurometabolic couplings in dynamic calibrated fMRI: Transient oxidative neuroenergetics for block-design and event-related paradigms. *Front Neuroenergetics, 2,* 18.

International League Against Epilepsy. (1981). Proposal for revised clinical and electroencephalographic classification of epileptic seizures. From the Commission on Classification and Terminology of the International League Against Epilepsy. *Epilepsia, 22,* 489–501.

International League Against Epilepsy. (1989). Proposal for revised classification of epilepsies and epileptic syndromes. From the Commission on Classification and Terminology of the International League Against Epilepsy. *Epilepsia, 30,* 389–399.

Jobst, B. C., Williamson, P. D., Neuschwander, T. B., Darcey, T. M., Thadani, V. M., & Roberts, D. W. (2001). Secondarily generalized seizures in mesial temporal epilepsy: Clinical characteristics, lateralizing signs, and association with sleep-wake cycle. *Epilepsia, 42,* 1279–1287.

Jus, A., & Jus, C. (1960). Etude électro-clinique des altérations de conscience dans le petit mal. *Stud Cercet Neurol, 5,* 243–254.

Killory, B. D., Bai, X., Negishi, M., Vega, C., Spann, M. N., Vestal, M., ... Blumenfeld, H. (2011). Impaired attention and network connectivity in childhood absence epilepsy. *NeuroImage, 56*(4), 2209–2217.

Kim, S. H., Zubal, I. G., & Blumenfeld, H. (2009). Epilepsy localization by ictal and interictal SPECT. In R. L. van Heertum, M. Ichise, & R. S. Tikofsky (Eds.), *Functional cerebral SPECT and PET imaging* (pp. 131–148). Philadelphia, PA: Lippincott Williams & Wilkins.

Kostopoulos, G. K. (2001). Involvement of the thalamocortical system in epileptic loss of consciousness. *Epilepsia, 42*(Suppl. 3), 13–19.

Kriss, A., Halliday, A. M., Halliday, E., & Pratt, R. T. (1978). EEG immediately after unilateral ECT. *Acta Psychiatr Scand, 58,* 231–244.

Lambert, I., Arthuis, M., McGonigal, A., Wendling, F., & Bartolomei, F. (2012). Alteration of global workspace during loss of consciousness: A study of parietal seizures. *Epilepsia, 53*(12), 2104–2110.

Laureys, S., Owen, A. M., & Schiff, N. D. (2004). Brain function in coma, vegetative state, and related disorders. *Lancet Neurol, 3,* 537–546.

Laureys, S., & Schiff, N. D. (Eds.). (2009). *Disorders of consciousness.* Annals of the New York Academy of Sciences, Vol. 1157. New York, NY: Wiley-Blackwell.

Laureys, S., & Tononi, G. (2008). *The neurology of consciousness: Cognitive neuroscience and neuropathology.* Waltham, MA: Academic Press.

Lee, B. I., Markand, O. N., Wellman, H. N., Siddiqui, A. R., Mock, B., Krepshaw, J., & Kung, H. (1987). HIPDM single photon emission computed tomography brain imaging in partial onset secondarily generalized tonic-clonic seizures. *Epilepsia, 28,* 305–311.

Lee, K. H., Meador, K. J., Park, Y. D., King, D. W., Murro, A. M., Pillai, J. J., & Kaminski, R. J. (2002). Pathophysiology of altered consciousness during seizures: Subtraction SPECT study. *Neurology, 59,* 841–846.

Levav, M., Mirsky, A. F., Herault, J., Xiong, L., Amir, N., & Andermann, E. (2002). Familial association of neuropsychological traits in patients with generalized and partial seizure disorders. *J Clin Exp Neuropsychol, 24,* 311–326.

Li, Q., Luo, C., Yang, T., Yao, Z., He, L., Liu, L., ... Zhou, D. (2009). EEG-fMRI study on the interictal and ictal generalized spike-wave discharges in patients with childhood absence epilepsy. *Epilepsy Res, 87,* 160–168.

Lieb, J. P., Dasheiff, R. B., & Engel, J., Jr. (1991). Role of the frontal lobes in the propagation of mesial temporal lobe seizures. *Epilepsia, 32,* 822–837.

Lux, S., Kurthen, M., Helmstaedter, C., Hartje, W., Reuber, M., & Elger, C. E. (2002). The localizing value of ictal consciousness and its constituent functions: A video-EEG study in patients with focal epilepsy. *Brain, 125,* 2691–2698.

Mayanagi, Y., Watanabe, E., & Kaneko, Y. (1996). Mesial temporal lobe epilepsy: Clinical features and seizure mechanism. *Epilepsia, 37*(Suppl. 3), 57–60.

McNally, K. A., & Blumenfeld, H. (2004). Focal network involvement in generalized seizures: New insights from electroconvulsive therapy. *Epilepsy Behav, 5,* 3–12.

McNally, K. A., Paige, A. L., Varghese, G., Zhang, H., Novotny, E. J., Spencer, S. S., ... Blumenfeld, H. (2005). Localizing value of ictal-interictal SPECT analyzed by SPM (ISAS). *Epilepsia, 46,* 1450–1464.

McPherson, A., Rojas, L., Bauerschmidt, A., Ezeani, C. C., Yang, L., Motelow, J. E., ... Blumenfeld, H. (2012). Testing for minimal consciousness in complex partial and generalized tonic-clonic seizures. *Epilepsia, 53*(10), e180–183.

Meeren, H. K., Pijn, J. P., van Luijtelaar, E. L., Coenen, A. M., & Lopes da Silva, F. H. (2002). Cortical focus drives widespread corticothalamic networks during spontaneous absence seizures in rats. *J Neurosci, 22,* 1480–1495.

Meeren, H., van Luijtelaar, G., Lopes da Silva, F., & Coenen, A. (2005). Evolving concepts on the pathophysiology of absence seizures: The cortical focus theory. *Arch Neurol, 62,* 371–376.

Mesulam, M. M. (2000). *Principles of behavioral and cognitive neurology.* New York, NY: Oxford University Press.

Mirsky, A. F., & van Buren, J. M. (1965). On the nature of the "absence" in centrencephalic epilepsy: A study of some behavioral, electroencephalographic, and autonomic factors. *Electroen Clin Neurophysiol, 18,* 334–348.

Mishra, A. M., Ellens, D. J., Schridde, U., Motelow, J. E., Purcaro, M. J., DeSalvo, M. N., ... Blumenfeld, H. (2011). Where fMRI and electrophysiology agree to disagree: Corticothalamic and striatal activity patterns in the WAG/Rij rat. *J Neurosci, 31,* 15053–15064.

Moeller, F., LeVan, P., Muhle, H., Stephani, U., Dubeau, F., Siniatchkin, M., & Gotman, J. (2010). Absence seizures: Individual patterns revealed by EEG-fMRI. *Epilepsia, 51,* 2000–2010.

Moeller, F., Muhle, H., Wiegand, G., Wolff, S., Stephani, U., & Siniatchkin, M. (2010). EEG-fMRI study of generalized spike and wave discharges without transitory cognitive impairment. *Epilepsy Behav, 18*(3), 313–316.

Moeller, F., Siebner, H. R., Wolff, S., Muhle, H., Boor, R., Granert, O., ... Siniatchkin, M. (2008). Changes in activity of striato-thalamo-cortical network precede generalized spike wave discharges. *NeuroImage, 39*(4), 1839–1849.

Moeller, F., Siebner, H. R., Wolff, S., Muhle, H., Granert, O., Jansen, O., ... Siniatchkin, M. (2008). Simultaneous EEG-fMRI in drug-naive children with newly diagnosed absence epilepsy. *Epilepsia, 49*(9), 1510–1519.

Monti, M. M., Vanhaudenhuyse, A., Coleman, M. R., Boly, M., Pickard, J. D., Tshibanda, L., ... Laureys, S. (2010). Willful modulation of brain activity in disorders of consciousness. *N Engl J Med, 362,* 579–589.

Moruzzi, G., & Magoun, H. W. (1949). Brain stem reticular formation and activation of the EEG. *Electroen Clin Neurophysiol, 1,* 455–473.

Motelow, J. E., Gummadavelli, A., Zayyad, Z., Mishra, A. M., Sachdev, R. N. S., Sanganahalli, B. G., ... Blumenfeld, H. (2012). Brainstem cholinergic and thalamic dysfunction during limbic seizures: Possible mechanism for cortical slow oscillations and impaired consciousness. *Soc Neurosci Abstracts, 487.*25.

Nagel, T. (1974). What is it like to be a bat? *Philos Rev, 82,* 435–456.

Nei, M., & Bagla, R. (2007). Seizure-related injury and death. *Curr Neurol Neurosci Rep, 7,* 335–341.

Nersesyan, H., Herman, P., Erdogan, E., Hyder, F., & Blumenfeld, H. (2004). Relative changes in cerebral blood flow and neuronal activity in local microdomains during generalized seizures. *J Cereb Blood Flow Metab, 24,* 1057–1068.

Norden, A. D., & Blumenfeld, H. (2002). The role of subcortical structures in human epilepsy. *Epilepsy Behav, 3,* 219–231.

Owen, A. M., Coleman, M. R., Boly, M., Davis, M. H., Laureys, S., & Pickard, J. D. (2006). Detecting awareness in the vegetative state. *Science, 313,* 1402.

Pavone, A., & Niedermeyer, E. (2000). Absence seizures and the frontal lobe. *Clin Electroencephalogr, 31,* 153–156.

Penfield, W. (1950). Epileptic automatism and the centrencephalic integrating system. *Res Publ Assoc Res Nerv Ment Dis, 30,* 513–528.

Picard, F., & Craig, A. D. (2009). Ecstatic epileptic seizures: A potential window on the neural basis for human self-awareness. *Epilepsy Behav, 16,* 539–546.

Plum, F., & Posner, J. B. (1972). The diagnosis of stupor and coma. *Contemp Neurol Ser, 10,* 1–286.

Plum, F., & Posner, J. B. (1982). *The diagnosis of stupor and coma.* Philadelphia, PA: Davis.

Raichle, M. E., MacLeod, A. M., Snyder, A. Z., Powers, W. J., Gusnard, D. A., & Shulman, G. L. (2001). A default mode of brain function. *Proc Natl Acad Sci USA, 98,* 676–682.

Richerson, G. B., & Buchanan, G. F. (2011). The serotonin axis: Shared mechanisms in seizures, depression, and SUDEP. *Epilepsia, 52*(Suppl. 1), 28–38.

Rodin, E., & Ancheta, O. (1987). Cerebral electrical fields during petit mal absences. *Electroen Clin Neurophysiol, 66,* 457–466.

Rosenberg, D. S., Mauguiere, F., Demarquay, G., Ryvlin, P., Isnard, J., Fischer, C., ... Magnin, M. (2006). Involvement of medial pulvinar thalamic nucleus in human temporal lobe seizures. *Epilepsia, 47,* 98–107.

Rowe, C. C., Berkovic, S. F., Sia, S. T., Austin, M., McKay, W. J., Kalnins, R. M., & Bladin, P. F. (1989). Localization of epileptic foci with postictal single photon emission computed tomography. *Ann Neurol, 26,* 660–668.

Sadleir, L. G., Scheffer, I. E., Smith, S., Connolly, M. B., & Farrell, K. (2009). Automatisms in absence seizures in children with idiopathic generalized epilepsy. *Arch Neurol, 66,* 729–734.

Salek-Haddadi, A., Lemieux, L., Merschhemke, M., Friston, K. J., Duncan, J. S., & Fish, D. R. (2003). Functional magnetic resonance imaging of human absence seizures. *Ann Neurol, 53,* 663–667.

Salgado-Benitez, A., Briones, R., & Fernandez-Guardiola, A. (1982). Purkinje cell responses to a cerebral penicillin-induced epileptogenic focus in the cat. *Epilepsia, 23,* 597–606.

Sanders, R. D., Tononi, G., Laureys, S., & Sleigh, J. W. (2012). Unresponsiveness not equal unconsciousness. *Anesthesiology, 116,* 946–959.

Schiff, N. D., Giacino, J. T., Kalmar, K., Victor, J. D., Baker, K., Gerber, M., ... Rezai, A. R. (2007). Behavioural improvements with thalamic stimulation after severe traumatic brain injury. *Nature, 448,* 600–603.

Schindler, K., Leung, H., Lehnertz, K., & Elger, C. E. (2007). How generalised are secondarily "generalised" tonic clonic seizures? *J Neurol Neurosurg Psychiatry, 78,* 993–996.

Schridde, U., Khubchandani, M., Motelow, J., Sanganahalli, B. G., Hyder, F., & Blumenfeld, H. (2008). Negative BOLD with large increases in neuronal activity. *Cereb Cortex, 18,* 1814–1827.

Searle, J. R. (1997). Reductionism and the irreducibility of consciousness. In N. Block, O. J. Flanagan, & G. Guzeldere (Eds.), *The nature of consciousness* (pp. 451–460). Cambridge, MA: MIT Press.

Shin, W. C., Hong, S. B., Tae, W. S., & Kim, S. E. (2002). Ictal hyperperfusion patterns according to the progression of temporal lobe seizures. *Neurology, 58,* 373–380.

Sperling, M. R. (2004). The consequences of uncontrolled epilepsy. *CNS Spectr, 9,* 98–101.

Steriade, M., Contreras, D., Curro Dossi, R., & Nunez, A. (1993). The slow (<1 Hz) oscillation in reticular thalamic and thalamocortical neurons: Scenario of sleep rhythm generation in interacting thalamic and neocortical networks. *J Neurosci, 13,* 3284–3299.

Steriade, M. M., & McCarley, R. W. (2010). *Brain control of wakefulness and sleep* (2nded.). New York, NY: Springer.

Suh, M., Ma, H., Zhao, M., Sharif, S., & Schwartz, T. H. (2006). Neurovascular coupling and oximetry during epileptic events. *Mol Neurobiol, 33,* 181–197.

Tae, W. S., Joo, E. Y., Kim, J. H., Han, S. J., Suh, Y.-L., Kim, B. T., ... Hong, S. B. (2005). Cerebral perfusion changes in mesial temporal lobe epilepsy: SPM analysis of ictal and interictal SPECT. *NeuroImage, 24,* 101–110.

Takano, H., Motohashi, N., Uema, T., Ogawa, K., Ohnishi, T., Nishikawa, M., ... Matsuda, H. (2007). Changes in regional cerebral blood flow during acute electroconvulsive therapy in patients with depression: Positron emission tomographic study. *Br J Psychiatry, 190,* 63–68.

Temkin, O. (1971). *The falling sickness: A history of epilepsy from the Greeks to the beginnings of modern neurology* (2nd ed.). Baltimore, MD: Johns Hopkins.

Theodore, W. H., Porter, R. J., Albert, P., Kelley, K., Bromfield, E., Devinsky, O., & Sato, S. (1994). The secondarily generalized tonic-clonic seizure: A videotape analysis. *Neurology, 44,* 1403–1407.

Tononi, G. (2005). Consciousness, information integration, and the brain. *Prog Brain Res, 150,* 109–126.

Tononi, G., & Koch, C. (2008). The neural correlates of consciousness: An update. *Ann NY Acad Sci, 1124,* 239–261.

van Paesschen, W., Dupont, P., van Driel, G., van Billoen, H., & Maes, A. (2003). SPECT perfusion changes during complex partial seizures in patients with hippocampal sclerosis. *Brain, 126,* 1103–1111.

Vanhaudenhuyse, A., Demertzi, A., Schabus, M., Noirhomme, Q., Bredart, S., Boly, M., ... Laureys, S. (2011). Two distinct neuronal networks mediate the awareness of environment and of self. *J Cogn Neurosci, 23,* 570–578.

Varghese, G., Purcaro, M. J., Motelow, J. E., Enev, M., McNally, K. A., Levin, A. R., ... Blumenfeld, H. (2009). Clinical use of ictal SPECT in secondarily generalized tonic-clonic seizures. *Brain, 132*(8), 2102–2113.

Vega, C., Vestal, M., DeSalvo, M., Berman, R., Chung, M., Blumenfeld, H., & Spann, M. N. (2010). Differentiation of attention-related problems in childhood absence epilepsy. *Epilepsy Behav, 19,* 82–85.

Vergnes, M., Marescaux, C., & Depaulis, A. (1990). Mapping of spontaneous spike and wave discharges in Wistar rats with genetic generalized non-convulsive epilepsy. *Brain Res, 523,* 87–91.

Vickrey, B. G., Berg, A. T., Sperling, M. R., Shinnar, S., Langfitt, J. T., Bazil, C. W., ... Spencer, S. S. (2000). Relationships between seizure severity and health-related quality of life in refractory localization-related epilepsy. *Epilepsia, 41,* 760–764.

Vollmer-Haase, J., Folkerts, H. W., Haase, C. G., Deppe, M., & Ringelstein, E. B. (1998). Cerebral hemodynamics during electrically induced seizures. *NeuroReport, 9,* 407–410.

Von Economo, C. (1930). Sleep as a problem of localization. *J Nerv Ment Dis, 71,* 249–259.

Weinberger, J., & Lusins, J. (1973). Simultaneous bilateral focal seizures without loss of consciousness. *Mt Sinai J Med, 40,* 693–696.

Weir, B. (1965). The morphology of the spike-wave complex. *Electroen Clin Neurophysiol, 19,* 284–290.

Westmijse, I., Ossenblok, P., Gunning, B., & van Luijtelaar, G. (2009). Onset and propagation of spike and slow wave discharges in human absence epilepsy: A MEG study. *Epilepsia, 50*(12), 2538–2548.

Yang, L., Morland, T. B., Schmits, K., Rawson, E., Narasimhan, P., Motelow, J. E., ... Blumenfeld, H.

(2010). A prospective study of loss of consciousness in epilepsy using virtual reality driving simulation and other video games. *Epilepsy Behav, 18*, 238–246.

Yang, L., Shklyar, I., Lee, H. W., Ezeani, C. C., Anaya, J., Balakirsky, S., … Blumenfeld, H. (2012). Impaired consciousness in epilepsy investigated by a prospective responsiveness in epilepsy scale (RES). *Epilepsia, 53*, 437–447.

Yu, L., & Blumenfeld, H. (2009). Theories of impaired consciousness in epilepsy. *Ann NY Acad Sci, 1157*, 48–60.

72 On the Relationship Between Consciousness and Attention

NAOTSUGU TSUCHIYA AND CHRISTOF KOCH

ABSTRACT Over the last 20 years, our understanding of the neuronal basis of perceptual consciousness and selective attention has greatly progressed. This advancement was facilitated by research using visual illusions and task designs that keep sensory input constant yet vary internal factors such as top-down attention and subjective visibility. To isolate the neuronal mechanisms of consciousness and attention, however, it has become increasingly clear that keeping the sensory input constant is not enough. Unless manipulated independently, consciousness and attention usually covary. Recent studies that independently vary both consciousness and attention have found that the behavioral and neuronal effects of consciousness and attention can be dissociated, implying that their neuronal mechanisms may be largely independent. Yet, even if independent neuronal mechanisms underlie consciousness and attention, there remains a conceptual dispute over the exact relationship between these processes. It is now generally accepted that subjects can selectively attend to attributes of events or objects without becoming aware of them. Whether the converse is also true is much more contentious. That is, is attentional amplification of the neural representation of an event or object always necessary to experience it (i.e., is consciousness without attention possible?). We argue that attentional amplification is necessary to experience an object only when it needs to be "selected" among other objects that compete with it in space and time. In a situation without any competition (e.g., an isolated object or a uniform texture), selective attention may not play any significant role. Accordingly, we argue that the neuronal mechanisms that give rise to consciousness need to be carefully disentangled from the neuronal mechanisms that resolve competition. Using an isolated-object paradigm, future studies can test the possibility of consciousness with no top-down attentional amplification in mice or monkeys by inactivating synaptic inputs from frontoparietal attentional areas back to visual areas using the rapidly advancing technology of optogenetics.

Conceptual issues of consciousness and attention

DEFINITION OF CONSCIOUSNESS AND ATTENTION Both "consciousness" and "attention" are words used in everyday conversation with different meanings. As in previous papers (Koch & Tsuchiya, 2007; van Boxtel, Tsuchiya, & Koch, 2010a), we restrict our discussion to "consciousness" in the sense of "contents of consciousness," to be distinguished from the meaning inherent in "levels of consciousness," as in sleep or arousal (Laureys, 2005). Though the content of consciousness includes notions related to the self, we focus here on the vastly better-studied aspects of visual consciousness (for other modalities, see several articles featured in Tsuchiya & van Boxtel, 2013). Furthermore, we focus on top-down selective attention, which is an endogenous, goal-oriented, and volitionally controlled component of attention, as opposed to bottom-up selective attention, which is an exogenous, automatic, stimulus-driven component of attention (Itti & Koch, 2001).

Top-down attention and consciousness are both mental processes that can be manipulated without changes in sensory inputs. As such, they are amenable to contrastive analysis (Baars, 1997) without stimulus confound. Contrastive methods have been extensively used in the search for the neuronal correlates of consciousness (NCC). The NCC is defined as the minimal set of neuronal mechanisms jointly sufficient for any one conscious percept (Koch, 2004). To dissociate the NCC from neural activity that merely correlates with the physical characteristics of the stimulus, an array of visual illusions has been extensively used (Kim & Blake, 2005).

Researchers have investigated behavioral and neuronal effects of top-down attention under identical sensory inputs while instructing subjects to direct their attention in one way or the other. This contrasts with studies of bottom-up attention, where some aspects of the stimulus array (e.g., an abrupt cue) are manipulated, making it likely that some of the observed attentional effects are directly caused by external factors.

Ultimately, the function of selective attention is to reduce the onslaught of sensory information streaming in from sensory arrays to a trickle that can be processed in real time by the organism. At the neuronal level (Kastner & Ungerleider, 2000), both top-down and bottom-up attention act in at least three ways: (1) to increase the baseline neuronal activity (but also see Otazu, Tai, Yang, & Zador, 2009); (2) to amplify the neuronal response to the selected location, feature, and/or object; this may occur by reducing noise correlations among neighboring neurons (Cohen & Maunsell, 2009; Mitchell, Sundberg, & Reynolds, 2009); and

(3) to suppress the neuronal response to locations, features, objects, or events that are not selected. In the discussion on the relationship between consciousness and attention, the second aspect of attention, attentional amplification, is most relevant. The third aspect is also critical when attention resolves competition among visual objects, as we emphasize in this chapter.

There are two core questions with regard to the relationship between consciousness and attention: Can top-down attention select and amplify the neuronal activity representing an object, even when that object is not consciously perceived? Is attentional amplification necessary to perceive an object consciously? Before reviewing recent studies on these issues, we elaborate what we mean by the contents of consciousness.

Qualia in a Broad and Narrow Sense The fundamental question at the heart of the mind-body problem is how electrochemical activity in the brain generates subjective conscious experience. Consciousness is always experienced, at any given moment, as a unified whole. This single moment of conscious experience is sometimes called "a quale in a broad sense," as defined in Balduzzi and Tononi (2009; a quale is the singular form for the Latin word *qualia*). In a narrow sense, a quale refers to a particular aspect within a quale in the broad sense. Examples of the narrow-sense qualia are the redness of a Rothko painting or the dull pain of a toothache. A quale in the narrow sense can be considered an elemental unit of conscious experience and corresponds to the "one conscious percept" that the NCC is minimally sufficient for. Note that narrow-sense qualia are the targets of most psychophysical studies, since broad-sense qualia are more difficult to study. In a typical psychophysical experiment, the redness of one object is compared with that of another object. Although two objects presented at different positions or time points produce different broad-sense qualia, they can be comparable in the narrow sense (Kanai & Tsuchiya, 2012).

Does top-down attention alter qualia in either the broad or the narrow sense? In this chapter, we argue that the role of top-down attention depends on the nature of the sensory input. If the input is cluttered with many competing objects, top-down attention resolves the competition among the stimuli for access to neuronal resources and alters qualia (Liu, Abrams, & Carrasco, 2009; see Carrasco, 2011, for a review). It is unlikely, however, for top-down attention to have any major effects on qualia if an object is presented in isolation.

Phenomenal Consciousness and Cognitive Access A closely related concept to qualia is "phenomenal consciousness," which is distinguished from "access consciousness" (Block, 1996, 2007). Phenomenal consciousness, defined as "what it is like to have any one specific experience," is closely related to both broad- and narrow-sense qualia. A representation is access-conscious if it is posed for direct control of reasoning, reporting, and action (Block, 1996). Cognitively accessed contents of consciousness are stored for working memory and flexibly guide present and future behaviors. Such reportable conscious contents are sometimes called "awareness" and are sometimes distinguished from phenomenal consciousness (Chalmers, 1995). To avoid confusion, we will not use the term "awareness" in this chapter.

Here, we will discuss the neuronal mechanisms that are directly responsible for narrow-sense qualia rather than the neuronal mechanisms that enable cognitive access. Cognitive access mechanisms may involve selective attention, especially when sensory inputs involve competition among objects. Thus, phenomenal consciousness without cognitive access is closely related to qualia without top-down attention (Block, 2007; Campana & Tallon-Baudry, 2013; Koch & Tsuchiya, 2008; Lamme, 2010).

Possible relationships between consciousness and attention

Consciousness Is Not Necessary for Attention What is the relationship between consciousness, as in narrow- or broad-sense qualia, and attention? While some philosophers argue that attention without consciousness is not possible, largely on a theoretical basis (de Brigard & Prinz, 2010; Marchetti, 2012; Mole, 2008), accumulating evidence strongly suggests that attention can amplify neuronal activity that remains, however, inaccessible to consciousness. By last count, at least 37 papers report such an effect (for reviews, see Cohen, Cavanagh, Chun, & Nakayama, 2012; Dehaene, Changeax, Naccache, Sackur, & Sargent, 2006; Koch & Tsuchiya, 2007; Tallon-Baudry, 2012; van Boxtel et al., 2010a). In particular, bottom-up spatial attention can be attracted by invisible stimuli (Hsieh, Colas, & Kanwisher, 2011; also see Zhaoping, 2008, 2012), top-down spatial and feature-based attention can be directed to invisible stimuli (Faivre & Kouider, 2011; Kaunitz, Fracasso, & Melcher, 2011), invisible objects can elicit object-based attention (Norman, Heywood, & Kentridge, 2013; but also see Tapia, Breitmeyer, Jacob, & Broyles, 2012), and temporal attention improves priming for invisible objects (Naccache, Blandin, & Dehaene, 2002). Given this mounting evidence, we conclude that visual consciousness is not necessary for selective visual attention to operate. Put differently, subjects can attend to a feature or to an entire object without consciously seeing that feature or object.

Is Attention Necessary for Consciousness? While attention without consciousness is largely agreed upon among visual psychologists, the issue of "consciousness without attention" has been more controversial (e.g., Cohen et al., 2011, 2012; Tsuchiya, Block, & Koch, 2012; Tallon-Baudry, 2012; Tsuchiya & van Boxtel, 2013).

A popular conception of attention is to view it as a "gateway" to consciousness. In this view, a subject can consciously experience some aspect of the sensory input only if he or she attends to it. It postulates that some amount of attention is always needed to amplify neuronal activity in order for any aspects of the input to be consciously experienced at all.

This view has been challenged by evidence that suggests some classes of percepts and behaviors (e.g., iconic memory, partial reportability, gist, and animal and gender detection in dual tasks) can give rise to consciousness with little, or perhaps no, top-down attentional amplification (Block, 2011; Hardcastle, 1997; Iwasaki, 1993; Koch & Tsuchiya, 2007; Lamme, 2010).

Supportive evidence comes, for example, from a series of dual-task experiments (Braun & Julesz, 1998), where subjects perform a demanding task at the fixation point while simultaneously carrying out another task at the periphery. The dual-task paradigm has revealed that subjects carry out certain peripheral discrimination tasks, such as form (triangular vs. circular; Braun, 1994) and conjunction of orientation and color (Braun & Julesz, 1998), at the same level of performance when these tasks are performed alone, (that is, as single tasks) and when they are performed together with the demanding central task (the performance level is adjusted to be ~75% correct for each task when performed alone).

The features that can be discriminated with little or no top-down attention include not only a number of low- and mid-level features (Braun, 1994; Lee, Koch, & Braun, 1997) but also some high-level object categories such as faces (Reddy, Reddy, & Koch, 2006; Reddy, Wilken, & Koch, 2004) and animals (Li, VanRullen, Koch, & Perona, 2002). Surprisingly, other classes of seemingly simpler discrimination (e.g., color-bisected disks, rotated letter "L" vs. "T") cannot be performed together with the attention-demanding task (VanRullen, Reddy, & Koch, 2004). When detection or discrimination thresholds in visual tasks are measured under single- or dual-task conditions, the effects of attention are nearly absent for simple detection or coarse discrimination (Lee, Itti, Koch, & Braun, 1999; Morrone, Denti, & Spinelli, 2002; Tsuchiya & Braun, 2007), while the attentional effects increase as the discrimination task involves more intense competition among stimuli or fine discrimination of an object that requires high spatial resolution (e.g., Landolt-square and vernier resolution, Yeshurun & Carrasco, 1999). In other words, these studies suggest that certain kinds of conscious discriminations can be performed with little or perhaps no top-down attentional amplification, while others cannot. Unfortunately, however, these results do not prove "consciousness without attention," because the dual-task paradigm involves reporting when top-down attention is drained away; it cannot be ruled out that some minimal attentional resources have not been allocated to these stimuli.

Why do certain types of discrimination substantially benefit from top-down attention, while others do not? Some studies suggest that natural stimuli are processed so efficiently that they can proceed with little or no top-down amplification (Fei-Fei, VanRullen, Koch, & Perona, 2005; VanRullen, Reddy, & Koch, 2004). A feedforward computational model of bottom-up object recognition supports this idea, showing that natural scenes can be categorized with little or no top-down attention (Serre, Oliva, & Poggio, 2007). Recent studies on the nature of peripheral vision suggest that summary statistics accessible at the periphery may be sufficient for certain types of discrimination, possibly accounting for why these seemingly complex stimuli can be discriminated at the periphery without much attentional amplification (Freeman & Simoncelli, 2011; Rosenholtz, Huang, & Ehinger, 2012; figure 72.1).

Recently, Cohen and colleagues (2012) argued for the lack of any direct evidence supporting "consciousness without attention." As to the dual-task results, they point out that even discrimination of natural scenes suffers when top-down attention is sufficiently withdrawn via more demanding tasks (Cohen, Alvarez, & Nakayama, 2011; Walker, Stafford, & Davis, 2008).

The scope of the attentional gateway hypothesis, however, seems limited to situations where intense competition among objects is present. Because attention selectively enhances neural responses for a target among possible distractors, it is not surprising that any visual discrimination becomes more dependent on top-down attention when the target is exposed to sufficiently strong competition. The conclusion obtained in highly competitive situations is unlikely to generalize to situations where only a lone object is presented in isolation or where a uniform texture is perceived. In these cases, the object or texture is experienced, while top-down attention has no possibility to select a target among competing visual stimuli (for there are none). We will return to this point later in this chapter.

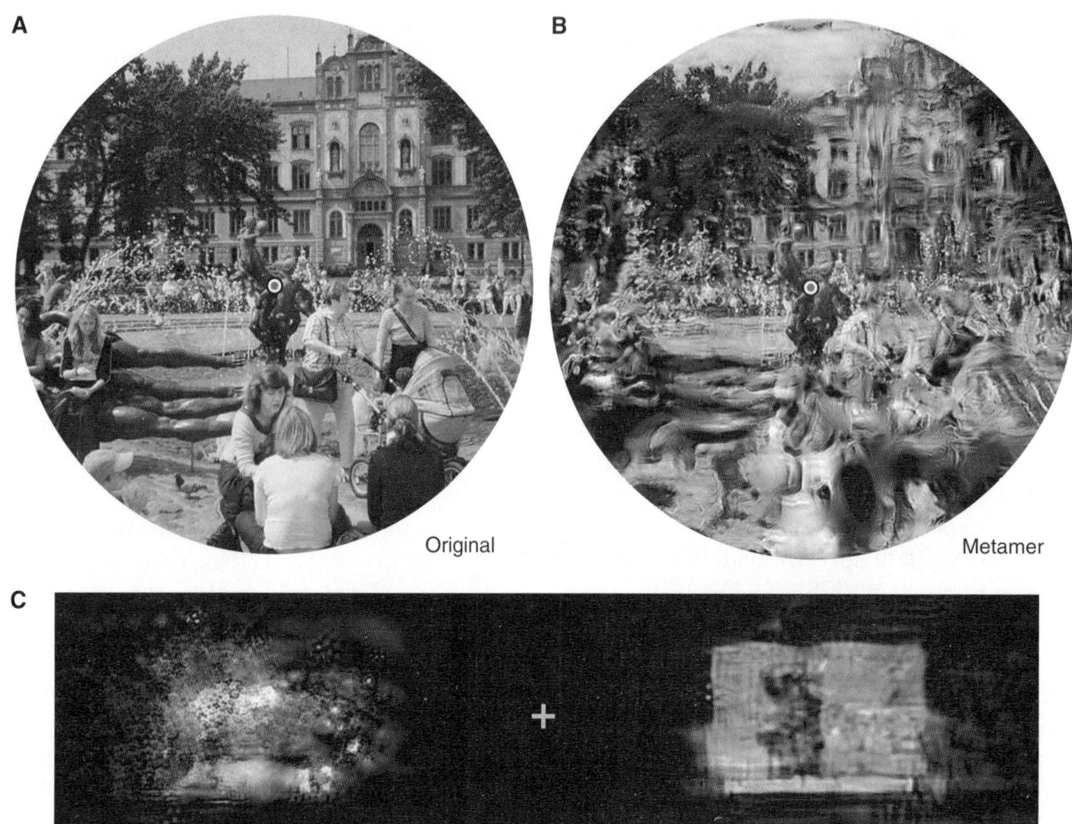

FIGURE 72.1 Why can we perceive natural scenes at the periphery with little or no top-down attention? (A, B) Perceptual metamers (Freeman & Simoncelli, 2011), and (C) mongrels (Rosenholtz et al., 2012). The nature of peripheral vision has been studied using psychophysics and computational modeling. (A, B) One such computational model of the ventral visual stream responds identical to (B) as it does to the original photograph (A). Indeed, when fixated at the center, both look subjectively the same; that is, they produce similar qualia in the broad sense. (C) Impoverished representation at the periphery can still be sufficient to discriminate natural scenes. Thus, degraded representation at the periphery is rich enough to generate vivid but coarse conscious phenomenology of a gist of the scene without top-down attentional amplification (Campana & Tallon-Baudry, 2013).

EXPECTATION, ATTENTION, AND CONSCIOUSNESS Proponents of "the gateway theory" (i.e., that attention is necessary for consciousness) emphasize visual phenomena in which a salient stimulus is not noticed unless top-down attention is directed to the stimulus (Cohen et al., 2012; O'Regan & Noe, 2001).

In *change blindness* (Simons & Rensink, 2005), a large change between two otherwise identical images tends to be missed until attention is directed to the changed location. In *inattentional blindness* (Mack & Rock, 1998), a stimulus that appears at an unexpected location, away from attention, goes unnoticed. In *attentional blink* (Chun & Potter, 1995; Raymond, Shapiro, & Arnell, 1992), two targets are embedded in a rapid temporal sequence of distractors. When the first target is detected, the second target tends to be missed if it occurs ~200–500 msec after the first target.

There have been suggestions that these phenomena may occur not because of the failure of attention, but due to a failure of other mental processes, such as expectation (Braun, 2001; also see Cartwright-Finch & Lavie, 2007; Mack, 2001) and immediate memory (Landman, Spekreijse, & Lamme, 2003; Sligte, Scholte, & Lamme, 2008; Wolfe, 1999). Change/inattentional blindness occur especially when the change is not expected and is less familiar (Werner & Thies, 2000). While expectation and attention are both top-down mental processes, they are likely to be supported by distinct neuronal mechanisms, serving different biological functions (Summerfield & Egner, 2009). A recent functional MRI (fMRI) study (Kok et al., 2012), which independently manipulated attention and expectation, found that fMRI signals in V1 are enhanced or reduced by attention only when the target is expected (figure 72.2), exposing the interdependency between attention and expectation. In the larger perspective, it suggests that the question of the relation between consciousness and attention will need

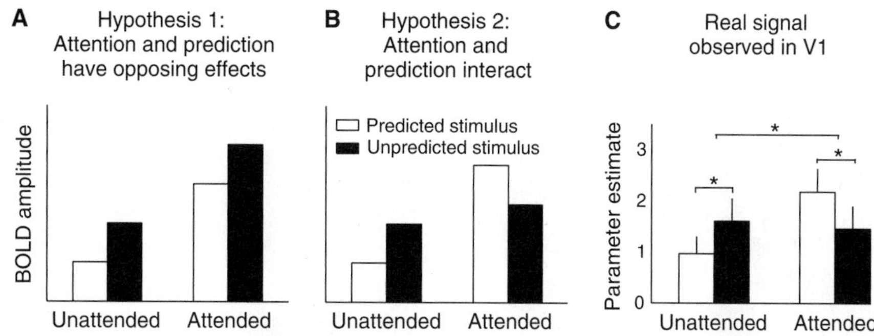

FIGURE 72.2 Dissociating expectation from attention using functional brain imaging (Kok et al., 2012). (A) Hypothesis 1 states that attention amplifies neuronal activation while expectation (or prediction) leads to reduction of the response in a predictive-coding framework (Friston, 2005; Rao & Ballard, 1999). (B) Hypothesis 2 states that the effects of expectation (or prediction) reverse depending on whether a stimulus is attended (Friston, 2009; Rao, 2005). Attention amplifies responses to a predicted stimulus compared to an unpredicted stimulus, while it has little effect on an unpredicted stimulus. (C) Functional MRI response in V1 is consistent with hypothesis 2, dissociating the effects of attention and expectation. (Modified from Kok et al., 2012.) Future studies need to investigate the relationship between attention and consciousness, while controlling for expectation of the stimuli.

to be reformulated in broader terms to include expectation. The next step in the investigation of the neural correlates of consciousness is to determine how attention, expectation, and consciousness relate to each other.

The attentional blink paradigm is claimed to be a sensitive assay for the necessity of attention for consciousness (Cohen et al., 2012). However, it is a rather complicated phenomenon, combining forms of backward and forward masking, and relies on working memory (Akyurek, Hommel, & Jolicoeur, 2007; Colzato, Spape, Pannebakker, & Hommel, 2007; for a review, see Dux & Marois, 2009). It is unclear whether the attentional blink paradigm purely manipulates top-down attention without other confounds.

AN OPTIMAL PSYCHOPHYSICAL TEST FOR CONSCIOUSNESS WITHOUT ATTENTION? What is the ideal test that could determine whether or not attentional amplification is necessary for consciousness? As argued above, change/inattentional blindness as well as attentional blink may induce perceptual blindness because of a failure of interaction between top-down attention and other processes, such as expectation and working memory. On the other hand, the dual-task paradigm has fewer complications, being a purer manipulation of top-down attention. However, as long as subjects report on a poorly attended (rather than a completely unattended) stimulus, some amount of attention might be directed to the stimulus, making the dual-task paradigm unsuited to attain a "no attention" condition.

Indeed, achieving zero top-down attention without any confound may be nearly impossible within a standard psychophysical paradigm in humans (Koch & Tsuchiya, 2007; van Boxtel, Tsuchiya, & Koch, 2010b; but see also Block, 2013). In the future, however, it may become possible to test consciousness with no attention using animal models, which we will discuss later in this chapter.

Given the practical difficulty of establishing consciousness with no attention, we turn to recent studies that try to test if consciousness and attention are supported by different neuronal mechanisms. By independently manipulating consciousness and attention, these studies measure the behavioral and neuronal consequences of such manipulations. The results are mostly consistent with the proposition that consciousness and attention are supported by distinctive neuronal mechanisms (Tallon-Baudry, 2012).

Independent manipulations of consciousness and attention

Recently, researchers have started manipulating consciousness and attention independently using the same stimuli within a single task structure (Kanai, Tsuchiya, & Verstraten, 2006; van Boxtel et al., 2010b; van den Bussche, Hughes, Humbeeck, & Reynvoet, 2010; Watanabe et al., 2011; Wyart & Tallon-Baudry, 2008; figure 72.3A). These studies have typically manipulated the conscious visibility of a target stimulus via backward masking or continuous flash suppression (Tsuchiya & Koch, 2005), treating visibility as a proxy for conscious awareness; that is, if the stimulus is not seen and not reported on, it is not consciously perceived. Then, they simultaneously manipulated bottom-up attention by

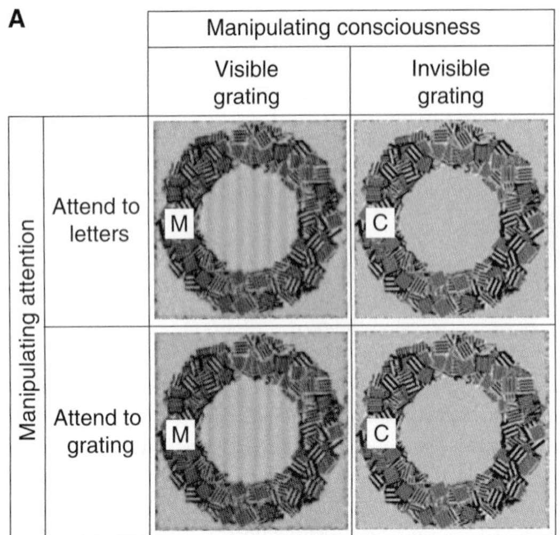

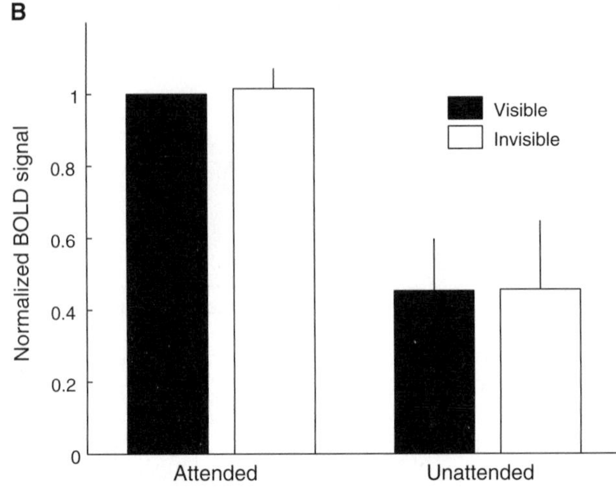

FIGURE 72.3 (A) 2-by-2 factorial design for independent manipulation of top-down attention and conscious visibility of the stimulus (Watanabe et al., 2011). Subjects are asked to carry out one of two attention tasks while viewing either a visible or an invisible target stimulus. At the same time, dependent variables, such as hemodynamic responses in the brain, are measured. Here, we illustrate roughly what participants perceived in each condition (not the physical stimulus) in the study by Watanabe and colleagues. Subjects either had to report the presence of a target letter at fixation or the visibility of a target grating inside the continuous flash suppression, arranged in a ring configuration, presented in the periphery. (B) Functional MRI responses in V1 are strongly modulated by top-down attention but not by conscious visibility of the grating. Modified based on figures 2 and S2 (7 subjects in total) in Watanabe et al. (2011). The data was provided by the original author. The mean BOLD signal over 7–18 sec from the block onset is normalized to the attended and visible condition. The error bar represents a 95% confidence interval.

cueing, or they modulated the spread of top-down attention using a dual-task design.

When manipulated independently, consciousness and attention affect the duration of afterimages in opposing ways (van Boxtel et al., 2010b). A number of previous experiments showed, paradoxically, that attending to visual stimuli reduces the duration of the induced afterimages (Baijal & Srinivasan, 2009; Lak, 2008; Lou, 2001; Suzuki & Grabowecky, 2003; Wede & Francis, 2007) and, independently, that perceptual invisibility reduces the duration of the afterimages (Gilroy & Blake, 2005; Tsuchiya & Koch, 2005). Because the effects of consciousness depend on the details of the afterimage-inducing stimuli (e.g., spatial frequency; Brascamp, van Boxtel, Knapen, & Blake, 2010), it is critical to use the same stimuli to study the effects of consciousness and attention.

Watanabe and colleagues (2011) likewise independently manipulated consciousness and attention while recording fMRI signals from the primary visual cortex (V1) in humans (figure 72.3A, B). They found that the V1 hemodynamic response is strongly modulated by attention but not by the visibility of a grating. Similar effects have been reported for neuronal activity recorded via microelectrodes from monkey V1 (Maier, Cox, Reavis, Adams, & Leopold, 2011). These two experiments challenge many previous fMRI studies (Haynes, Deichmann, & Rees, 2005; Polonsky, Blake, Braun, & Heeger, 2000; Wunderlich, Schneider, & Kastner, 2005) that used binocular rivalry and located the neuronal correlates of consciousness to V1 and even the lateral geniculate nucleus of the thalamus (also see Maier et al., 2008, on the difference between single neuron activity and hemodynamic response in monkey V1). This is a prime example of the necessity of separating consciousness from attention (Koch & Tsuchiya, 2011; Tse, Martinez-Conde, Schlegel, & Macknik, 2005); unless explicitly manipulated independently, the neuronal correlates of consciousness can, and usually will, covary with the neuronal correlates of attention.

Taken together, the behavioral and neuronal effects of consciousness and attention do not always vary together, especially when consciousness and attention are independently manipulated, in contrast with what one might expect from the gateway theory. This is easiest to explain assuming that consciousness and attention are independent biological processes.

Untangling the neuronal correlates of consciousness from those of cognitive access

As we have seen in the previous section, the neuronal mechanisms that support consciousness and attention may well be distinct. Yet, at the level of behaviors and perception, the effects of consciousness and attention are often closely related. In particular, when we attend to a feature, object, or event, we usually become conscious of it. Furthermore, neuroimaging studies suggest that similar brain regions, such as frontoparietal areas, are commonly activated when probing for consciousness or attention (Bor & Seth, 2012; Rees & Lavie, 2001). Without explicit and independent manipulation, however, neuronal activity in these areas might reflect the neuronal correlates of cognitive access rather than those of narrow-sense qualia (Frässle, Sommer, Jansen, Naber, & Einhäuser, 2014). Next, we consider potential reasons for this surface similarity and discuss the role of the task designs that have been extensively utilized for studying consciousness and attention.

THE ROLE OF COMPETITION IN THE STUDY OF THE NEURONAL CORRELATES OF ATTENTION AND CONSCIOUSNESS Studying top-down attention inherently necessitates a stage for competition among stimuli, where attention selects one or a few target items and suppresses distracting stimuli, as is the case for the dual-task paradigm, change/inattentional blindness, and attentional blink. In all these cases, a target competes against other distractors that are presented within spatiotemporal proximity.

Is competition also necessary to study the NCC? Do we consciously experience an item only when it competes against other objects? Obviously not; we can see perfectly well an isolated stimulus or perceive a uniform texture when there are no other distracting stimuli to be suppressed by attention. However, most NCC studies use experimental paradigms that introduce competition among objects (Kim & Blake, 2005). Bistable figures induce competition between two possible interpretations. Binocular rivalry and continuous flash suppression render an object invisible via interocular competition. While binocular rivalry is less controllable by top-down attention than bistable figures (Meng & Tong, 2004), a recent study suggests that binocular rivalry requires visual attention (Zhang, Jamison, Engel, He, & He, 2011). Crowding renders a target unrecognizable by surrounding objects in close proximity at the periphery (Faivre & Kouider, 2011; He, Cavanagh, & Intriligator, 1996). Motion-induced blindness (Bonneh, Cooperman, & Sagi, 2001) results from competition between a moving surface and an object. Therefore, when these visual illusions are used as the main tools for the search for the NCC, the putative NCC can involve the neuronal mechanisms that resolve competition, in addition to those that are directly responsible for generating qualia. Therefore, it is desirable to have experimental designs in the NCC literature that do not involve competition and thus do not engage attentional mechanisms.

THE NEURONAL CORRELATES OF ACCESS AND CONSCIOUSNESS The research program for finding the NCC assumes that for a given narrow-sense quale X, there exists a minimal set of neurons whose activity pattern is jointly sufficient to generate the quale X (or, more generally, brain mechanisms sufficient to generate the quale X; Crick, 1994; Metzinger, 2000; Tononi & Koch, 2008). Activating this NCC—by TMS, electrical stimulation, or optogenetics—would induce the associated quale while inactivating the NCC would eliminate the quale. It is also assumed that an object that elicits a quale X is always experienced (as a part of a broad-sense quale) when the NCC sustains a particular pattern of activity. This is assumed to be true whether or not there are competing stimuli for the object for the quale X.

Now we consider those neuronal mechanisms that enable the subject to cognitively access a quale X, a process we call the neuronal correlates of access. It involves those mechanisms that hold the representation of the object in working memory and enable the subject to report on it using some response modality.

In this framework, the NCC that have been identified in most previous studies are an aggregate of the mechanisms underlying phenomenal consciousness, such as the coalition of neurons that maximizes the integrated information in Tononi's integrated information theory of consciousness (Oizumi, Albantakis, & Tononi, 2014; Tononi, 2012) or activity in the global neuronal workspace of Dehaene and Changeux (Dehaene & Changeux, 2011), together with the additional mechanisms that mediate attention as well as access (Aru, Bachmann, Singer, & Melloni, 2012; de Graaf, Hsieh, & Sack, 2012). Untangling all three will not be easy.

The proponents of the gateway theory, according to which every conscious percept must first pass through the crucible of attention, claim that a quale X without cognitive access cannot be scientifically studied, or that it may not even exist (Cohen et al., 2012; Cohen & Dennett, 2011; Dehaene et al., 2006; O'Regan & Noe, 2001). We argue, however, that cognitive access is only necessary when the object for the quale X needs to be selectively amplified to win the competition among other objects. The exact amount of attentional amplification needed would depend on a category of the object for the quale X (e.g., less for faces and natural scenes, more for rotated letters) as well as the intensity of competition.

This raises the question of whether access is needed at all for phenomenal consciousness to occur (Block,

2007, 2011; Campana & Tallon-Baudry, 2013; Lamme, 2010). Conscious access and reports might enhance a quale X to the extent that it can be held in working memory and reported accurately (Kouider, de Gardelle, Sackur, & Dupoux, 2010; Lau & Rosenthal, 2011). In fact, from a third-person perspective, it seems impossible to assess a conscious experience without any report from the subject. However, so-called mind-reading techniques are being developed that allow perceptual content to be read out from the hemodynamic response in many regions of the brain, including volunteers watching movies (Nishimoto et al., 2011), patients with grave neurological damage and in a minimally conscious state (Monti et al., 2010; Owen et al., 2006), or visual imagery experienced during sleep onset (Horikawa, Tamaki, Miyawaki, & Kamitani, 2013). Furthermore, from a first-person perspective, on a day-to-day basis we rarely report the contents of consciousness outside the lab. That is, most of us don't go through our daily life and keep up a constant verbal report of what we momentarily experience. Keeping this argument in mind, next, we review recent studies, which suggest frontoparietal areas as the core of the NCC.

The Putative Neuronal Correlates of Consciousness in Frontoparietal Areas Based on the meta-analysis of the NCC studies, frontoparietal areas have often been suggested as a common and central component for conscious perception (Bor & Seth, 2012; Rees & Lavie, 2001). However, this claim needs to be reevaluated for several reasons.

First, given a large number of distinct cytoarchitectonically defined regions within frontoparietal cortex, these claims are too unspecific. Even narrowing them to within dorsolateral prefrontal cortex and posterior parietal cortex, they are still rather coarse. Without identifying particular subareas within these areas for each individual subject, it is unclear if the same regions are engaged in both consciousness- and attention-related tasks for a given subject.

Second, the studies that point to activities in frontoparietal areas as the NCC typically utilize visual illusions that involve competition among objects, as we saw in the last section. Studies that use illusions that minimize stimulus competition tend not to find frontoparietal activation as the NCC (Tse et al., 2005).

Lastly, as discussed below, the act of reporting by button press on one's own perception gives rise to activity in frontal regions that are not present under conditions of unreported perception (Frässle et al., 2014).

Another recent fMRI study also points to the role of frontoparietal areas in solving competition between stimuli. In previous studies, frontoparietal cortex has been activated at the time of perceptual switches while viewing ambiguous stimuli and binocular rivalry (Kleinschmidt, Buchel, Zeki, & Frackowiak, 1998; Lumer, Friston, & Rees, 1998; Lumer & Rees, 1999). However, a recent replication of this effect by Knapen and colleagues (2011) discovered that the effects critically depend on the exact nature of the control stimulus. When they created the control stimulus, whose dynamics during transitions is comparable to that during genuine binocular rivalry (see figure 72.4), the difficulty in deciding which stimulus was perceptually dominant became comparable between real rivalry and the control. As a result, the difference in frontoparietal activity between the rivalry and control conditions disappeared. This study implies that activity in frontoparietal areas can be caused by, rather than cause, difficult perceptual decisions that are required during perceptual transitions.

On the Importance of a No-Report Paradigm While subjective reports are the gold standard in consciousness research, the development of mind-reading techniques allows researchers to infer the contents of consciousness in the absence of explicit reports from subjects. Certain conscious visual content can be decoded in healthy participants (Brown & Norcia, 1997; Garcia, Srinivasan, & Serences, 2013; Haxby et al., 2001; Haynes & Rees, 2005; Horikawa et al., 2013; Kamitani & Tong, 2005; Nishimoto et al., 2011) and in some nonresponsive and noncommunicable patients (Monti et al., 2010; Owen et al., 2006).

Consider binocular rivalry for moving gratings (Frässle et al., 2014). Here, both optokinetic nystagmus (objective) or verbal (subjective) reports can be used to infer the dominant direction of perceived motion. When both objective and subjective reports are used, the usual occipital, parietal, and frontal regions are identified as fMRI correlates of perceptual alternations. However, when subjects passively experience rivalry without verbally reporting their percept and the dominant direction of motion is inferred via the objective optokinetic nystagmus, differential activity in frontal areas disappeared and only activity in occipital and parietal regions remained.

The contents of consciousness can also be inferred without overt reports in nonhuman primates, especially using robust visual illusions. Wilke and colleagues presented a salient red circle that could be masked using generalized flash suppression (GFS; Wilke, Logothetis, & Leopold, 2003; Wilke, Mueller, & Leopold, 2009; figure 72.5A) to monkeys who were trained to report their percept with a lever. By titrating the stimulus parameters for GFS, they rendered the illusion

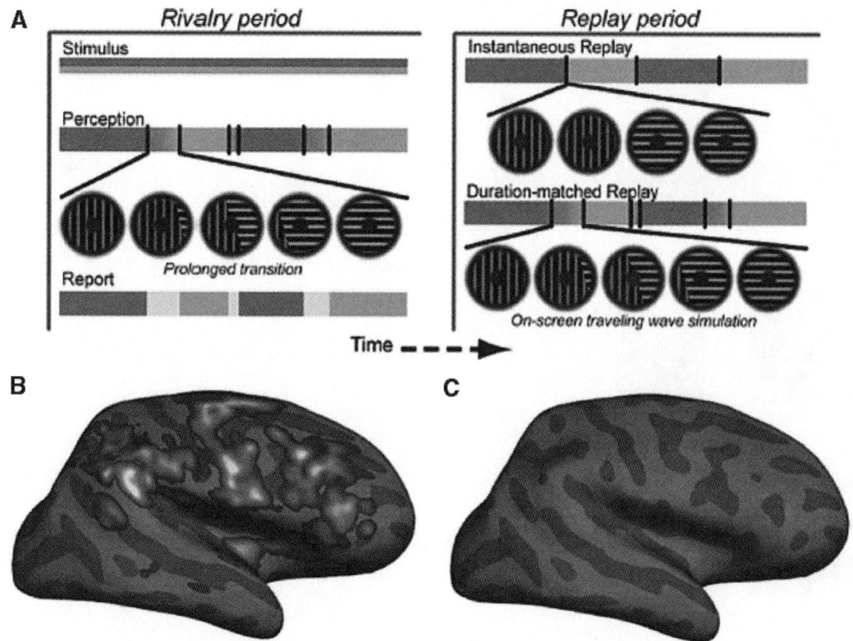

FIGURE 72.4 Confounding activation in frontoparietal areas can be introduced by a less faithful replay of what subjects really experience. (A) An exemplar time course of stimulus and perception during genuine binocular rivalry. One eye is stimulated by a vertical red and the other eye by a horizontal green grating (top). The subjective experience is highly complex, involving various durations of intermediate percepts (rows 2 and 3). As a control condition, a perceptually matched replay movie can be presented to both eyes without interocular conflict. However, if the replay is reproduced from binary responses, the variable nature of each perceptual transition cannot be captured (e.g., instantaneous replay; top right of panel A). With a third response option of accurately reporting the transitional states, a more faithful replay can be reproduced (e.g., duration-matched replay; bottom right of panel A). (B) Brain activity contrasted at the perceptual transition between genuine rivalry and instantaneous replay (without faithfully reproduced transitions), indicating a widespread right-lateralized frontoparietal activity, replicating previous studies. (C) No difference in the brain activity between genuine rivalry and duration-matched replay. (Modified from Knapen et al., 2011.) (See color plate 62.)

ambiguous, so that monkeys sometimes saw a target and sometimes did not. When they recorded single-unit activity and local field potentials (LFP) from the dorsal and ventral pulvinar, a visual thalamic nucleus, they found that, on average, both single-neuron activity and low-frequency power (9–30Hz) of the LFP were correlated with fluctuating perceptual reports on a trial-by-trial basis (figures 72.5B, D).

They then carefully tailored the stimulus parameters to control the visibility of the circular target without requiring perceptual reports from the monkeys. In the absence of such report, the single neuron spikes in the pulvinar continued to correlate with the visibility of the target (figure 72.5C), while the low-frequency power of LFP did not (figure 72.5E). This is a striking dissociation of multimodal neuronal responses within a single brain area, with low-frequency power coding for a neuronal correlate of access while spikes encoded the neuronal correlates of qualia. The former disappeared without explicit reports, but the latter remained the same regardless of the presence or absence of the reports (Wilke et al., 2009).

An alternative way to dissociate the neuronal correlates of qualia from that of access is to ask subjects to report their percept with different response modalities (e.g., Imamoglu, Kahnt, Koch, & Haynes, 2012). The report-modality specific activity should be considered as a part of the neuronal basis of access but not of qualia.

In sum, these experiments support the claim that what we experience every waking minute of our lives—consciousness for objects, thoughts, events, or memories—exists whether or not we report on them. Studies that employ a no-report paradigm can distinguish access from qualia per se, while NCC studies exploiting visual phenomena that rely on competition among stimuli, objects, or eyes are likely to be contaminated by access mechanisms (Macknik & Martinez-Conde, 2009; Tse et al., 2005). Another alternative way to isolate the neuronal mechanisms of qualia is to focus on a single, isolated object without any competition.

CONSCIOUS PERCEPTION OF AN ISOLATED OBJECT WITHOUT TOP-DOWN ATTENTION In our opinion, the

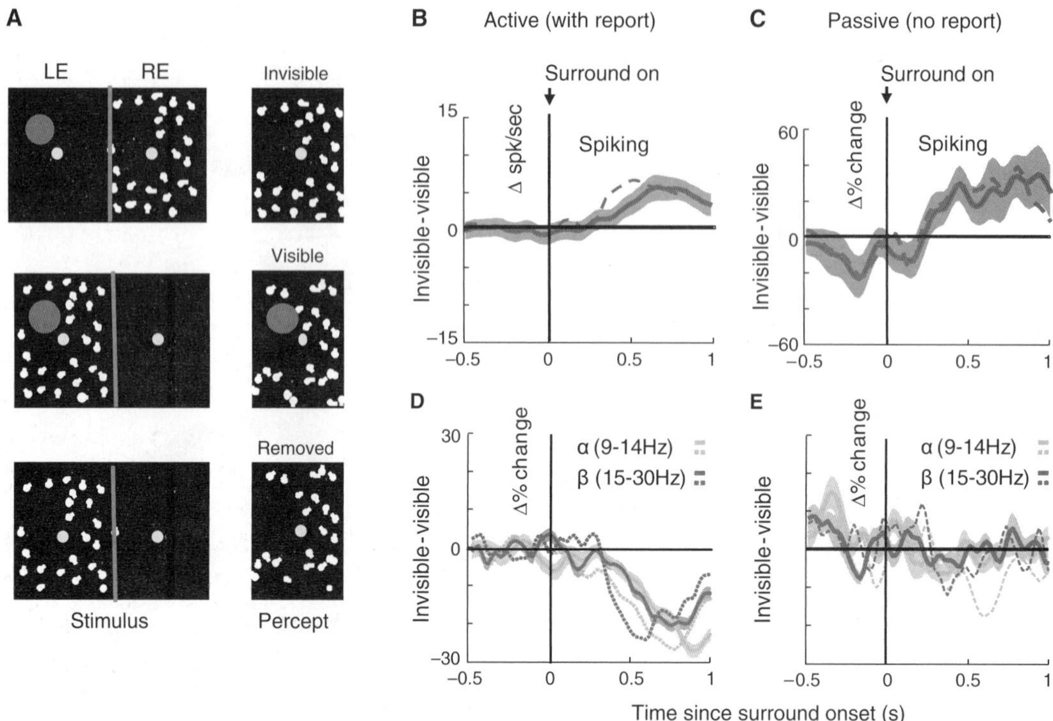

FIGURE 72.5 Neural correlates of access and qualia. (A) Wilke and colleagues (2009) used a binocular masking technique to manipulate the visibility of a disk. ("LE" = left eye; "RE" = right eye.) The neuronal activity is compared with the control condition where the stimulus was physically removed. (B, C) While recording spiking activity in the pulvinar, some neurons signaled the visibility of the stimulus independent of the presence (B) or absence (C) of report, suggesting that these spiking activities are a correlate of subjective experience. Solid lines indicate the difference in firing rate between visible vs. (perceptually) invisible trials. Dotted lines indicate the difference in firing rate between visible vs. physical removal trials. (D, E) The low-frequency (9–30 Hz) power of the local field potentials (LFP) in the pulvinar distinguished the visibility only when monkeys reported the visibility (D), but not when they did not report the visibility of the stimulus (E), making them neuronal correlates of access but not of qualia.

detection or coarse discrimination of a single object is unlikely to suffer even if top-down attentional mechanisms were completely eliminated. However, any psychophysical manipulation may not be complete and cannot assure total removal of attentional resources. How could one prove that no top-down attention is required to detect a target?

One way to do so is to transiently knock out the axons that project to early visual cortices from frontal and parietal areas as well as higher-order visual cortices. In the limit, one would obtain a purely feedforward set of cortical regions (with local feedback, of course) but no modulatory input from higher cortical regions reaching down into superficial layers in earlier cortical regions.

Recent development of optogenetic techniques promises an opportunity to test such an idea of "consciousness with no top-down attention" in animals, especially in mice (Deisseroth, 2011; Yizhar, Fenno, Davidson, Mogri, & Deisseroth,, 2011) and to a lesser extent in monkeys (Diester et al., 2011; Gerits et al., 2012; Han et al., 2009). The neuronal source of top-down attentional amplification is likely to be some parts of frontoparietal cortex (Bressler et al., 2008; Noudoost, Chang, Steinmetz, & Moore, 2010). The extraordinarily rapid development of ever more refined transgenic mice as well as viral techniques, both of which target specific, molecular, and projectional defined neuronal populations that can be labeled and turned on or off with millisecond precision from anywhere from milliseconds to hours, has given systems neuroscience an amazing ability to delicately, reversibly, and transiently intervene and to observe the phenotype at the behavioral and the circuit levels (Huang & Zeng, 2013). In particular, it is possible to either locally or globally turn off (and back on) feedback pathways originating in frontal or parietal regions and projecting back into one of the 10 or more visual regions in the common laboratory mouse (Cruikshank, Urabe, Nurmikko, & Connors, 2010; Oh et al., 2014; Wang, Gao, & Burkhalter, 2011). The mouse can be trained in a visual foraging task, and its top-down attention can be selectively manipulated using optogenetic and pharmacogenetic tools.

If top-down attention were to be transiently inactivated by these interventionist optogenetic techniques, we predict that the affected animal would not be able to perform a visual task if the target is presented together with distractors, whereas it would be able to perform the task if the target is presented in isolation. Without inactivation, the same animal would perform the task as well as before the inactivation, with or without any distractors.

To make sure these tasks are performed consciously, the mice or monkeys can be trained to report their confidence via post-decision wagering (Kepecs, Uchida, Zariwala, & Mainen, 2008; Kiani & Shadlen, 2009; Persaud, McLeod, & Cowey, 2007). As the confidence or metacognitive judgment is likely to be mediated by medial orbitofrontal areas (Fleming, Weil, Nagy, Dolan, & Rees, 2010; but also see Komura, Nikkuni, Hirashima, Uetake, & Miyamoto, 2013), knocking out top-down attention in dorsal prefrontal or parietal areas might not affect confidence judgments.

Conclusion

Studies of visual attention and consciousness have greatly benefited from task designs that keep input stimuli constant while the contents of consciousness or the focus of top-down attention are shifted. This strategy has been particularly successful, and continues to be popular in the search for the NCC (Koch, 2004; Logothetis, 1998). Now we know that such an approach is, by itself, not enough. Competing stimuli in visual illusions invite involvement of attentional mechanisms, a part of the neuronal correlates of access, which is distinct from the neuronal correlates of qualia. In this chapter, we reviewed several ways to disentangle access and attention from consciousness, including (1) independent manipulation of consciousness and attention, (2) removing reports or modifying report modalities, and (3) using an isolated stimulus to diminish the role of selection.

The refinement of experimental procedures is coupled to the refinement of the conceptual framework for consciousness, attention, and access (see also Aru et al., 2012; De Graaf, 2012; Hohwy, 2009). The concepts of consciousness and attention will be further refined, as has been the case for the cognitive neuroscience of memory. The concept of memory has faced many revisions, subdivided into iconic memory (Sperling, 1960), fragile memory (Sligte et al., 2008), working memory, and declarative and non-declarative long-term memory (Milner, Squire, & Kandel, 1998).

Even such a seemingly purely metaphysical question as "does consciousness exist with no attention" can be addressed with elegant experimentation, possibly utilizing the development of optogenetics in transgenic animals. Consciousness research has successfully turned many philosophical questions into empirical questions, and it will continue to do so in the future.

ACKNOWLEDGMENTS NT is supported by the JST PRESTO fellowship (Japan), the ARC Future Fellowship (Australia), and the ARC Discovery Project (Australia), and CK is supported by the G. Harold & Leila Y. Mathers Charitable Foundation (US). The authors thank Jakob Hohwy, Jeroen van Boxtel, Lisandro Kaunitz, Fabiano Baroni, and Ned Block for comments on an earlier version of the manuscript.

REFERENCES

Akyurek, E. G., Hommel, B., & Jolicoeur, P. (2007). Direct evidence for a role of working memory in the attentional blink. *Mem Cognit, 35*(4), 621–627.

Aru, J., Bachmann, T., Singer, W., & Melloni, L. (2012). Distilling the neural correlates of consciousness. *Neurosci Biobehav Rev, 36*(2), 737–746.

Baars, B. J. (1997). Some essential differences between consciousness and attention, perception, and working memory. *Conscious Cogn, 6*(2–3), 363–371.

Baijal, S., & Srinivasan, N. (2009). Types of attention matter for awareness: A study with color afterimages. *Conscious Cogn, 18*(4), 1039–1048.

Balduzzi, D., & Tononi, G. (2009). Qualia: The geometry of integrated information. *PLoS Comput Biol, 5*(8), e1000462.

Block, N. (1996). How can we find the neural correlate of consciousness? *Trends Neurosci, 19*(11), 456–459.

Block, N. (2007). Consciousness, accessibility, and the mesh between psychology and neuroscience. *Behav Brain Sci, 30*(5–6), 481–499; discussion 499–548.

Block, N. (2011). Perceptual consciousness overflows cognitive access. *Trends Cogn Sci, 15*(12), 567–575.

Block, N. (2013). The grain of vision and the grain of attention. *Thought, 1*, 170–184.

Bonneh, Y. S., Cooperman, A., & Sagi, D. (2001). Motion-induced blindness in normal observers. *Nature, 411*(6839), 798–801.

Bor, D., & Seth, A. K. (2012). Consciousness and the prefrontal parietal network: Insights from attention, working memory, and chunking. *Front Psychol, 3*, 63.

Brascamp, J. W., van Boxtel, J. J., Knapen, T., & Blake, R. (2010). A dissociation of attention and awareness in phase-sensitive but not phase-insensitive visual channels. *J Cogn Neurosci, 22*(10), 2326–2344.

Braun, J. (1994). Visual search among items of different salience: Removal of visual attention mimics a lesion in extrastriate area V4. *J Neurosci, 14*(2), 554–567.

Braun, J. (2001). It's great but not necessarily about attention. *Psyche, 7*(6).

Braun, J., & Julesz, B. (1998). Withdrawing attention at little or no cost: Detection and discrimination tasks. *Percept Psychophys, 60*(1), 1–23.

Bressler, S. L., Tang, W., Sylvester, C. M., Shulman, G. L., & Corbetta, M. (2008). Top-down control of human visual cortex by frontal and parietal cortex in anticipatory visual spatial attention. *J Neurosci, 28*(40), 10056–10061.

Brown, R. J., & Norcia, A. M. (1997). A method for investigating binocular rivalry in real-time with the steady-state VEP. *Vision Res, 37*(17), 2401–2408.

Campana, F., & Tallon-Baudry, C. (2013). Anchoring visual subjective experience in a neural model: The coarse vividness hypothesis. *Neuropsychologia, 51*(6), 1050–1060.

Carrasco, M. (2011). Visual attention: The past 25 years. *Vis Res, 51*, 1484–1525.

Cartwright-Finch, U., & Lavie, N. (2007). The role of perceptual load in inattentional blindness. *Cognition, 102*(3), 321–340.

Chalmers, D. J. (1995). Facing up to the problem of consciousness. *J Consciousness Stud, 2*(3), 200–219.

Chun, M. M., & Potter, M. C. (1995). A two-stage model for multiple target detection in rapid serial visual presentation. *J Exp Psychol Hum Percept Perform, 21*(1), 109–127.

Cohen, M. A., Alvarez, G. A., & Nakayama, K. (2011). Natural-scene perception requires attention. *Psychol Sci, 22*(9), 1165–1172.

Cohen, M. A., Cavanagh, P., Chun, M. M., & Nakayama, K. (2012). The attentional requirements of consciousness. *Trends Cogn Sci, 16*(8), 411–417.

Cohen, M. A., & Dennett, D. C. (2011). Consciousness cannot be separated from function. *Trends Cogn Sci, 15*(8), 358–364.

Cohen, M. R. & Maunsell, J. H. (2009). *Nat Neurosci, 12*, 1594–1600.

Colzato, L. S., Spape, M., Pannebakker, M. M., & Hommel, B. (2007). Working memory and the attentional blink: Blink size is predicted by individual differences in operation span. *Psychon Bull Rev, 14*(6), 1051–1057.

Crick, F. (1994). *Astonishing hypothesis: The scientific search for the soul.* New York, NY: Scribner.

Cruikshank, S. J., Urabe, H., Nurmikko, A. V., & Connors, B. W. (2010). Pathway-specific feedforward circuits between thalamus and neocortex revealed by selective optical stimulation of axons. *Neuron, 65*, 230–245.

De Brigard, F., & Prinz, J. (2010). Attention and consciousness. *Wiley Interdiscip Rev Cog Sci, 1*, 51–59.

De Graaf, T. A., Hsieh, P. J., & Sack, A. T. (2012). The "correlates" in neural correlates of consciousness. *Neurosci Biobehav Rev, 36*(1), 191–197.

Dehaene, S., & Changeux, J. P. (2011). Experimental and theoretical approaches to conscious processing. *Neuron, 70*(2), 200–227.

Dehaene, S., Changeux, J. P., Naccache, L., Sackur, J., & Sergent, C. (2006). Conscious, preconscious, and subliminal processing: A testable taxonomy. *Trends Cogn Sci, 10*(5), 204–211.

Deisseroth, K. (2011). Optogenetics. *Nat Methods, 8*(1), 26–29.

Diester, I., Kaufman, M. T., Mogri, M., Pashaie, R., Goo, W., Yizhar, O., ... Shenoy, K. V. (2011). An optogenetic toolbox designed for primates. *Nat Neurosci, 14*(3), 387–397.

Dux, P. E., & Marois, R. (2009). The attentional blink: A review of data and theory. *Atten Percept Psychophys, 71*(8), 1683–1700.

Faivre, N., & Kouider, S. (2011). Multi-feature objects elicit nonconscious priming despite crowding. *J Vis, 11*(3), 2.

Fei-Fei, L., VanRullen, R., Koch, C., & Perona, P. (2005). Why does natural scene categorization require little attention? Exploring attentional requirements for natural and synthetic stimuli. *Vis Cognit, 12*(6), 893–924.

Fleming, S. M., Weil, R. S., Nagy, Z., Dolan, R. J., & Rees, G. (2010). Relating introspective accuracy to individual differences in brain structure. *Science, 329*(5998), 1541–1543.

Frässle, S., Sommer, J., Jansen, A., Naber, M., & Einhäuser, W. (2014). Binocular rivalry: Frontal activity relates to introspection and action but not to perception. *J Neurosci, 34*, 1738–1747.

Freeman, J., & Simoncelli, E. P. (2011). Metamers of the ventral stream. *Nat Neurosci, 14*(9), 1195–1201.

Friston, K. (2005). A theory of cortical responses. *Philos Trans R Soc Lond B Biol Sci, 360*(1456), 815–836.

Friston, K. (2009). The free-energy principle: A rough guide to the brain? *Trends Cogn Sci, 13*(7), 293–301.

Garcia, J. O., Srinivasan, R., & Serences, J. T. (2013). Near-real-time feature-selective modulations in human cortex. *Curr Biol, 23*(6), 515–522.

Gerits, A., Farivar, R., Rosen, B. R., Wald, L. L., Boyden, E. S., & Vanduffel, W. (2012). Optogenetically induced behavioral and functional network changes in primates. *Curr Biol, 22*(18), 1722–1726.

Gilroy, L. A., & Blake, R. (2005). The interaction between binocular rivalry and negative afterimages. *Curr Biol, 15*(19), 1740–1744.

Han, X., Qian, X., Bernstein, J. G., Zhou, H. H., Franzesi, G. T., Stern, P., ... Boyden, E. S. (2009). Millisecond-timescale optical control of neural dynamics in the nonhuman primate brain. *Neuron, 62*(2), 191–198.

Hardcastle, V. G. (1997). Attention versus consciousness: A distinction with a difference. *Cogn Stud, 4*, 56–66.

Haxby, J. V., Gobbini, M. I., Furey, M. L., Ishai, A., Schouten, J. L., & Pietrini, P. (2001). Distributed and overlapping representations of faces and objects in ventral temporal cortex. *Science, 293*(5539), 2425–2430.

Haynes, J. D., Deichmann, R., & Rees, G. (2005). Eye-specific effects of binocular rivalry in the human lateral geniculate nucleus. *Nature, 438*(7067), 496–499.

Haynes, J. D., & Rees, G. (2005). Predicting the orientation of invisible stimuli from activity in human primary visual cortex. *Nat Neurosci, 8*(5), 686–691.

He, S., Cavanagh, P., & Intriligator, J. (1996). Attentional resolution and the locus of visual awareness. *Nature, 383*(6598), 334–337.

Horikawa, T., Tamaki, M., Miyawaki, Y., & Kamitani, Y. (2013). Neural decoding of visual imagery during sleep. *Science, 340*(6132), 639–642.

Hohwy, J. (2009). The neural correlates of consciousness: New experimental approaches needed? *Consciousness Cogn, 18*, 428–438.

Hsieh, P. J., Colas, J. T., & Kanwisher, N. (2011). Pop-out without awareness: Unseen feature singletons capture attention only when top-down attention is available. *Psychol Sci, 22*(9), 1220–1226.

Huang, Z. J., & Zeng, H. (2013). Genetic approaches to neural circuits in the mouse. *Annu Rev Neurosci, 36*, 183–215.

Imamoglu, F., Kahnt, T., Koch, C., & Haynes, J. D. (2012). Changes in functional connectivity support conscious object recognition. *NeuroImage, 63*(4), 1909–1917.

Itti, L., & Koch, C. (2001). Computational modelling of visual attention. *Nat Rev Neurosci, 2*(3), 194–203.

IWASAKI, S. (1993). Spatial attention and two modes of visual consciousness. *Cognition, 49*(3), 211–233.

KAMITANI, Y., & TONG, F. (2005). Decoding the visual and subjective contents of the human brain. *Nat Neurosci, 8*(5), 679–685.

KANAI, R., & TSUCHIYA, N. (2012). Qualia. *Curr Biol, 22*(10), R392–396.

KANAI, R., TSUCHIYA, N., & VERSTRATEN, F. A. (2006). The scope and limits of top-down attention in unconscious visual processing. *Curr Biol, 16*(23), 2332–2336.

KASTNER, S., & UNGERLEIDER, L. G. (2000). Mechanisms of visual attention in the human cortex. *Annu Rev Neurosci, 23,* 315–341.

KAUNITZ, L., FRACASSO, A., & MELCHER, D. (2011). Unseen complex motion is modulated by attention and generates a visible aftereffect. *J Vis, 11*(13), 10.

KEPECS, A., UCHIDA, N., ZARIWALA, H. A., & MAINEN, Z. F. (2008). Neural correlates, computation and behavioural impact of decision confidence. *Nature, 455*(7210), 227–231.

KIANI, R., & SHADLEN, M. N. (2009). Representation of confidence associated with a decision by neurons in the parietal cortex. *Science, 324*(5928), 759–764.

KIM, C. Y., & BLAKE, R. (2005). Psychophysical magic: Rendering the visible "invisible." *Trends Cogn Sci, 9*(8), 381–388.

KLEINSCHMIDT, A., BUCHEL, C., ZEKI, S., & FRACKOWIAK, R. S. (1998). Human brain activity during spontaneously reversing perception of ambiguous figures. *Philos Trans R Soc Lond B Biol Sci, 265*(1413), 2427–2433.

KNAPEN, T., BRASCAMP, J., PEARSON, J., VAN EE, R., & BLAKE, R. (2011). The role of frontal and parietal brain areas in bistable perception. *J Neurosci, 31*(28), 10293–10301.

KOCH, C. (2004). *The quest for consciousness: A neurobiological approach.* Denver, CO: Roberts.

KOCH, C., & TSUCHIYA, N. (2007). Attention and consciousness: Two distinct brain processes. *Trends Cogn Sci, 11*(1), 16–22.

KOCH, C., & TSUCHIYA, N. (2008). Phenomenology without conscious access is a form of consciousness without top-down attention. *Behav Brain Sci, 30*(5/6), 509–510.

KOCH, C., & TSUCHIYA, N. (2011). Attention and consciousness: Related yet different. *Trends Cogn Sci, 16*(2), 103–105.

KOK, P., RAHNEV, D., JEHEE, J. F., LAU, H. C., & DE LANGE, F. P. (2012). Attention reverses the effect of prediction in silencing sensory signals. *Cereb Cortex, 22*(9), 2197–2206.

KOMURA, Y., NIKKUNI, A., HIRASHIMA, N., UETAKE, T., & MIYAMOTO, A. (2013). Responses of pulvinar neurons reflect a subject's confidence in visual categorization. *Nat Neurosci, 16*(6), 749–755.

KOUIDER, S., DE GARDELLE, V., SACKUR, J., & DUPOUX, E. (2010). How rich is consciousness? The partial awareness hypothesis. *Trends Cogn Sci, 14*(7), 301–307.

LAK, A. (2008). Attention during adaptation weakens negative afterimages of perceptually colour-spread surfaces. *Can J Exp Psychol, 62*(2), 101–109.

LAMME, V. A. (2010). How neuroscience will change our view on consciousness. *Cogn Neurosci, 1*(3), 204–220.

LANDMAN, R., SPEKREIJSE, H., & LAMME, V. A. (2003). Large capacity storage of integrated objects before change blindness. *Vision Res, 43*(2), 149–164.

LAU, H., & ROSENTHAL, D. (2011). Empirical support for higher-order theories of conscious awareness. *Trends Cogn Sci, 15*(8), 365–373.

LAUREYS, S. (2005). The neural correlate of (un)awareness: Lessons from the vegetative state. *Trends Cogn Sci, 9*(12), 556–559.

LEE, D. K., ITTI, L., KOCH, C., & BRAUN, J. (1999). Attention activates winner-take-all competition among visual filters. *Nat Neurosci, 2*(4), 375–381.

LEE, D. K., KOCH, C., & BRAUN, J. (1997). Spatial vision thresholds in the near absence of attention. *Vision Res, 37*(17), 2409–2418.

LI, F. F., VANRULLEN, R., KOCH, C., & PERONA, P. (2002). Rapid natural scene categorization in the near absence of attention. *Proc Natl Acad Sci USA, 99*(14), 9596–9601.

LIU, T., ABRAMS, J., & CARRASCO, M. (2009). Voluntary attention enhances contrast appearance. *Psychol Sci, 20*(3), 354–362.

LOGOTHETIS, N. K. (1998). Single units and conscious vision. *Philos Trans R Soc Lond B Biol Sci, 353*(1377), 1801–1818.

LOU, L. (2001). Effects of voluntary attention on structured afterimages. *Perception, 30*(12), 1439–1448.

LUMER, E. D., FRISTON, K. J., & REES, G. (1998). Neural correlates of perceptual rivalry in the human brain. *Science, 280*(5371), 1930–1934.

LUMER, E. D., & REES, G. (1999). Covariation of activity in visual and prefrontal cortex associated with subjective visual perception. *Proc Natl Acad Sci USA, 96*(4), 1669–1673.

MACK, A. (2001). Inattentional blindness: Reply to commentaries. *Psyche, 7*(16).

MACK, A., & ROCK, I. (1998). *Inattentional blindness.* Cambridge, MA.: MIT Press.

MACKNIK, S. L., & MARTINEZ-CONDE, S. (2009). The role of feedback in visual attention and awareness. In M. S. Gazzaniga (Ed.), *The cognitive neurosciences* (4th ed., pp. 1165–1175). Cambridge, MA: MIT Press.

MAIER, A. V., COX, M. A., REAVIS, E. A., ADAMS, G. K., & LEOPOLD, D. A. (2011). Perceptual awareness and selective attention differentially modulate neuronal responses in primary visual cortex. Paper presented at the Society for Neuroscience, Washington, DC.

MAIER, A., WILKE, M., AURA, C., ZHU, C., YE, F. Q., & LEOPOLD, D. A. (2008). Divergence of fMRI and neural signals in V1 during perceptual suppression in the awake monkey. *Nat Neurosci, 11*(10), 1193–1200.

MARCHETTI, G. (2012). Against the view that consciousness and attention are fully dissociable. *Front Psychol, 3,* 36.

MENG, M., & TONG, F. (2004). Can attention selectively bias bistable perception? Differences between binocular rivalry and ambiguous figures. *J Vis, 4*(7), 539–551.

METZINGER, T. (2000). *Neural correlates of consciousness: Empirical and conceptual questions.* Cambridge, MA: MIT Press.

MILNER, B., SQUIRE, L. R., & KANDEL, E. R. (1998). Cognitive neuroscience and the study of memory. *Neuron, 20*(3), 445–468.

MITCHELL, J. F., SUNDBERG, K. A., & REYNOLDS, J. H. (2009). *Neuron, 63,* 879–888.

MOLE, C. (2008). Attention in the absence of consciousness? *Trends Cogn Sci, 12*(2), 44; author reply 44–45.

MONTI, M. M., VANHAUDENHUYSE, A., COLEMAN, M. R., BOLY, M., PICKARD, J. D., TSHIBANDA, L., ... LAUREYS, S. (2010). Willful modulation of brain activity in disorders of consciousness. *N Engl J Med, 362*(7), 579–589.

Morrone, M. C., Denti, V., & Spinelli, D. (2002). Color and luminance contrasts attract independent attention. *Curr Biol, 12*(13), 1134–1137.

Naccache, L., Blandin, E., & Dehaene, S. (2002). Unconscious masked priming depends on temporal attention. *Psychol Sci, 13*(5), 416–424.

Nishimoto, S., Vu, A. T., Naselaris, T., Benjamini, Y., Yu, B., & Gallant, J. L. (2011). Reconstructing visual experiences from brain activity evoked by natural movies. *Curr Biol, 21*(19), 1641–1646.

Norman, L. J., Heywood, C. A., & Kentridge, R. W. (2013). Object-based attention without awareness. *Psychol Sci, 24*(6), 836–843.

Noudoost, B., Chang, M. H., Steinmetz, N. A., & Moore, T. (2010). Top-down control of visual attention. *Curr Opin Neurobiol, 20*(2), 183–190.

Oh, S., Harris, J. A., Ng, L., Winslow, B., Cain, N., … Zeng, H. (2014). A mesoscale connectome of the mouse brain. *Nature, 508*, 207–214.

Oizumi, M., Albantakis, L., & Tononi, G. (2014). From the phenomenology to the mechanisms of consciousness: Integrated Information Theory 3.0. *PLoS Comput Biol, 10*(5): doi: e1003588.

O'Regan, J. K., & Noe, A. (2001). A sensorimotor account of vision and visual consciousness. *Behav Brain Sci, 24*(5), 939–973; discussion 973–1031.

Otazu, G. H., Tai, L. H., Yang, Y., & Zador, A. M. (2009). Engaging in an auditory task suppresses responses in auditory cortex. *Nat Neurosci, 12*(5), 646–654.

Owen, A. M., Coleman, M. R., Boly, M., Davis, M. H., Laureys, S., & Pickard, J. D. (2006). Detecting awareness in the vegetative state. *Science, 313*(5792), 1402.

Persaud, N., McLeod, P., & Cowey, A. (2007). Post-decision wagering objectively measures awareness. *Nat Neurosci, 10*(2), 257–261.

Polonsky, A., Blake, R., Braun, J., & Heeger, D. J. (2000). Neuronal activity in human primary visual cortex correlates with perception during binocular rivalry. *Nat Neurosci, 3*(11), 1153–1159.

Rao, R. P. (2005). Bayesian inference and attentional modulation in the visual cortex. *NeuroReport, 16*(16), 1843–1848.

Rao, R. P., & Ballard, D. H. (1999). Predictive coding in the visual cortex: A functional interpretation of some extra-classical receptive-field effects. *Nat Neurosci, 2*(1), 79–87.

Raymond, J. E., Shapiro, K. L., & Arnell, K. M. (1992). Temporary suppression of visual processing in an RSVP task: An attentional blink? *J Exp Psychol Hum Percept Perform, 18*(3), 849–860.

Reddy, L., Reddy, L., & Koch, C. (2006). Face identification in the near-absence of focal attention. *Vision Res, 46*(15), 2336–2343.

Reddy, L., Wilken, P., & Koch, C. (2004). Face-gender discrimination is possible in the near-absence of attention. *J Vis, 4*(2), 106–117.

Rees, G., & Lavie, N. (2001). What can functional imaging reveal about the role of attention in visual awareness? *Neuropsychologia, 39*(12), 1343–1353.

Rosenholtz, R., Huang, J., & Ehinger, K. A. (2012). Rethinking the role of top-down attention in vision: Effects attributable to a lossy representation in peripheral vision. *Front Psychol, 3*, 13.

Serre, T., Oliva, A., & Poggio, T. (2007). A feedforward architecture accounts for rapid categorization. *Proc Natl Acad Sci USA, 104*(15), 6424–6429.

Simons, D. J., & Rensink, R. A. (2005). Change blindness: Past, present, and future. *Trends Cogn Sci, 9*(1), 16–20.

Sligte, I. G., Scholte, H. S., & Lamme, V. A. (2008). Are there multiple visual short-term memory stores? *PLoS ONE, 3*(2), e1699.

Sperling, G. (1960). The information available in brief visual presentations. *Psychol Monogr, 74*, 1–29.

Summerfield, C., & Egner, T. (2009). Expectation (and attention) in visual cognition. *Trends Cogn Sci, 13*(9), 403–409.

Suzuki, S., & Grabowecky, M. (2003). Attention during adaptation weakens negative afterimages. *J Exp Psychol Hum Percept Perform, 29*(4), 793–807.

Tallon-Baudry, C. (2012). On the neural mechanisms subserving consciousness and attention. *Front Psychol, 2*, 397.

Tapia, E., Breitmeyer, B. G., Jacob, J., & Broyles, E. C. (2012). Spatial attention effects during conscious and nonconscious processing of visual features and objects. *J Exp Psychol Hum Percept Perform, 39*(3), 745–756.

Tononi, G. (2012). Integrated information theory of consciousness: An updated account. *Arch Ital Biol, 150*(2–3), 56–90.

Tononi, G., & Koch, C. (2008). The neural correlates of consciousness: An update. *Ann NY Acad Sci, 1124*, 239–261.

Tse, P. U., Martinez-Conde, S., Schlegel, A. A., & Macknik, S. L. (2005). Visibility, visual awareness, and visual masking of simple unattended targets are confined to areas in the occipital cortex beyond human V1/V2. *Proc Natl Acad Sci USA, 102*(47), 17178–17183.

Tsuchiya, N., Block, N., & Koch, C. (2012). Top-down attention and consciousness: Comment on Cohen et al. *Trends Cogn Sci, 16*(11), 527.

Tsuchiya, N., & Braun, J. (2007). Contrast thresholds for component motion with full and poor attention. *J Vis, 7*(3), 1.

Tsuchiya, N., & Koch, C. (2005). Continuous flash suppression reduces negative afterimages. *Nat Neurosci, 8*(8), 1096–1101.

Tsuchiya, N., & van Boxtel, J. J. (2013). Introduction to research topic: Attention and consciousness in different senses. *Front Psychol, 4*, 249.

Van Boxtel, J. J., Tsuchiya, N., & Koch, C. (2010a). Consciousness and attention: On sufficiency and necessity. *Front Psychol, 1*, 217.

Van Boxtel, J. J., Tsuchiya, N., & Koch, C. (2010b). Opposing effects of attention and consciousness on afterimages. *Proc Natl Acad Sci USA, 107*(19), 8883–8888.

Van den Bussche, E., Hughes, G., Humbeeck, N. V., & Reynvoet, B. (2010). The relation between consciousness and attention: An empirical study using the priming paradigm. *Conscious Cogn, 19*(1), 86–97.

VanRullen, R., Reddy, L., & Koch, C. (2004). Visual search and dual tasks reveal two distinct attentional resources. *J Cogn Neurosci, 16*(1), 4–14.

Walker, S., Stafford, P., & Davis, G. (2008). Ultra-rapid categorization requires visual attention: Scenes with multiple foreground objects. *J Vis, 8*(4), 1–12.

Wang, Q., Gao, E., & Burkhalter, A. (2011). Gateways of ventral and dorsal streams in mouse visual cortex. *J Neurosci, 31*(5), 1905–1918.

Watanabe, M., Cheng, K., Murayama, Y., Ueno, K., Asamizuya, T., Tanaka, K., & Logothetis, N. (2011). Attention but not awareness modulates the BOLD signal in the human V1 during binocular suppression. *Science, 334*(6057), 829–831.

Wede, J., & Francis, G. (2007). Attentional effects on afterimages: Theory and data. *Vision Res, 47*(17), 2249–2258.

Werner, S., & Thies, B. (2000). Is "change blindness" attenuated by domain-specific expertise? An expert–novices comparison of change detection in football images. *Vis Cognit, 7*(1–3), 163–173.

Wilke, M., Logothetis, N. K., & Leopold, D. A. (2003). Generalized flash suppression of salient visual targets. *Neuron, 39*(6), 1043–1052.

Wilke, M., Mueller, K. M., & Leopold, D. A. (2009). Neural activity in the visual thalamus reflects perceptual suppression. *Proc Natl Acad Sci USA, 106*(23), 9465–9470.

Wolfe, J. M. (1999). Inattentional amnesia. In V. Coltheart (Ed.), *Fleeting memories* (pp. 71–94). Cambridge, MA: MIT Press.

Wunderlich, K., Schneider, K. A., & Kastner, S. (2005). Neural correlates of binocular rivalry in the human lateral geniculate nucleus. *Nat Neurosci, 8*(11), 1595–1602.

Wyart, V., & Tallon-Baudry, C. (2008). Neural dissociation between visual awareness and spatial attention. *J Neurosci, 28*(10), 2667–2679.

Yeshurun, Y., & Carrasco, M. (1999). Spatial attention improves performance in spatial resolution tasks. *Vision Res, 39*(2), 293–306.

Yizhar, O., Fenno, L. E., Davidson, T. J., Mogri, M., & Deisseroth, K. (2011). Optogenetics in neural systems. *Neuron, 71*(1), 9–34.

Zhang, P., Jamison, K., Engel, S., He, B., & He, S. (2011). Binocular rivalry requires visual attention. *Neuron, 71*(2), 362–369.

Zhaoping, L. (2008). Attention capture by eye of origin singletons even without awareness—a hallmark of a bottom-up saliency map in the primary visual cortex. *J Vis, 8*(5), 1–18.

Zhaoping, L. (2012). Gaze capture by eye-of-origin singletons: Interdependence with awareness. *J Vis, 12*(2), 17.

73 Consciousness and Its Access Mechanisms

SID KOUIDER AND JÉRÔME SACKUR

ABSTRACT Consciousness is a fundamental dimension of our mental life that involves both cognitive functions (attention, verbalization, working memory, and so on), and subjective, experiential aspects. During the past two decades, thanks to conceptual and methodological progress, a cognitive neuroscience of consciousness has emerged and gained full respectability. However, this science remains challenged regarding whether the subjective dimension of experience can be fully accounted for by the neuronal and cognitive mechanisms underlying conscious access. In this chapter, we first review the progress and challenges of the cognitive neuroscience of consciousness. We then discuss recent proposals that vindicate specific approaches to the subjective, phenomenal dimension of consciousness while denying the importance of access mechanisms. In contrast to these proposals, we argue for a unified approach to consciousness, whereby experiential and cognitive dimensions of consciousness rely on the same set of core neural mechanisms.

Despite all the progress made recently in the scientific study of consciousness, there are still intense controversies regarding what a theory of consciousness should be. In particular, the lack of consensus concerning the psychological definition of consciousness has rendered the study of its neural basis somewhat inconclusive. In this chapter, we review the progress made so far in uncovering the neurocognitive mechanisms of conscious access, and we emphasize the dependence of such progress on precise, operational definitions of consciousness. We outline some of the main issues that research on this topic faces today, in particular the issue of whether consciousness can be envisioned independently of its access mechanisms.

Moving on without definition

Nowadays, the vast majority of scientists reject dualist interpretations of consciousness that imply a separation between mind (consciousness, thoughts) and matter (the brain, neurons). Consciousness is amenable to a materialistic approach, with a biological perspective: we will understand how and why we are conscious by studying the cerebral and neuronal features of the brain. Yet the lack of a consensus definition of consciousness and the reduction of mental states to neuronal structures are daunting challenges that consciousness researchers face no less than in other fields of cognitive neurosciences. How then, could any progress be achieved in the field? Two strategies have eased these issues and led to considerable progress over the last two decades, and both critically depend on a psychological operationalization of "consciousness." The *contrastive approach* put forward by Bernard Baars (1989) allows moving on without formal definition, while the search for a *neural correlate of consciousness* (hereafter NCC) put forward by Francis Crick and Christof Koch (1995) allows moving on without focusing too much, at least for now, on the necessity of a reduction to elementary brain structures and processes. The combination of these two approaches has constituted the core of most recent successes in the scientific study of consciousness.

The contrastive approach

The contrastive approach is based on the idea that even if we don't know *what* consciousness is, not to mention *why* we are conscious in the first place, we at least know *when* it happens. Consciousness is thus considered as an outcome variable (absent/present), allowing us to compare situations where it occurs to close situations where, all other things being equal, it doesn't. This approach enables researchers not only to delineate the conditions for a stimulus to access consciousness, but also to specify the extents of unconscious processes.

This strategy has been most successfully applied to perceptual consciousness, that is, consciousness about an external event. Many experiments use very brief visual stimuli that are sometimes visible and sometimes invisible. Subjects have to report whether they saw a stimulus, which is taken as an index of whether they were conscious of it or not. By comparing the two situations, one can tell apart which cognitive mechanisms are shared, and which are specific to conscious processing. Using this methodology, it was shown that some high-level processes are triggered in the absence of perceptual consciousness (for instance, extracting the semantic information of a word or digit; Marcel, 1983a,

1983b; Dehaene et al., 1998), while some others seem to require that participants be conscious of the stimulus (e.g., applying a rule; Sackur & Dehaene, 2009; but see Sklar et al., 2012). Notice that, critically, the use of this methodology implies that one can trust participants in their reports of whether they are conscious or not.

The neural correlates of consciousness

According to Crick and Koch (1995), the best starting strategy for a neurobiological science of consciousness is to search for the NCC. These are defined as "the minimal set of neuronal mechanisms or events jointly sufficient for a specific conscious percept or experience" (Koch, 2004, p. 16). A candidate for the NCC would therefore be a structure involved only during conscious experience and would never be active outside conscious experience. According to this approach, neuroscientists should actually leave aside, at least for the moment, the problem of reducing conscious mental events to their associated physiological structure and processes. Rather, they should focus first on "correlating" them and finding out about their relations, which will ultimately lead to a better understanding of the whole issue.

In experimental practice, this strategy implies a contrastive approach, aimed at characterizing the neural, rather than cognitive, features that are specifically involved during conscious as opposed to unconscious processing. Here also, the search for NCCs depends on the admission of subjective reports as valid experimental data: in order to test whether any brain structure is an NCC, one has to trust subjects regarding the classification of their own mental states as "conscious" or "not conscious."

The prefrontal cortex

Early uses of the contrastive method in the search for NCCs relied heavily on the progress made in neuroimaging methods in the 1990s. Earlier studies used binocular rivalry and pointed to the ventral stream of the brain as critical for visual awareness. Binocular rivalry consists of presenting a stimulus (e.g., a face) to one eye and another stimulus (e.g., a house) to the other eye. Perceptually, the two objects do not merge but give rise to alternating percepts: we see the house for a few seconds, then the face for a few seconds, then the house returns, etc. Thus, while the input stimulation is constant, the content of consciousness varies, and one can directly estimate which regions are associated with conscious percepts, everything else being equal. This technique revealed that the primary visual cortex is activated by objects regardless of whether they are perceived consciously or not (Leopold & Logothetis, 1996). These posterior regions thus cannot be considered as NCCs. Instead, regions located more anteriorly and more ventrally in the inferotemporal cortex were involved specifically during conscious perception of a face as opposed to an alternative percept (Leopold & Logothetis, 1996; Tong, Nakayama, Vaughan, & Kanwisher, 1998). The visual ventral stream could thus be considered a potential NCC.

But in the early 2000s, Stanislas Dehaene and his colleagues showed that this was not always the case (Dehaene et al., 2001). They used the method of visual masking: a word was presented very briefly (about 30 msec) and temporally surrounded by abstract and meaningless shapes (the masks) that render the word invisible (subjects see a flicker of shapes, but report seeing no word). By removing the temporally surrounding masks, one can render the word visible—thus creating a contrastive situation (see figure 73.1). Dehaene and colleagues showed that the inferotemporal cortex, which is part of the ventral visual stream, is activated by unconscious masked words. Although the strength of this activation was much lower than for conscious perception, these findings ruled out the ventral stream as an NCC. By contrast, in this study, the parietal and prefrontal cortices were activated exclusively in the conscious situation. Since then, numerous studies have shown the particular importance of the prefrontal cortex for consciousness (see Dehaene & Changeux, 2011, for an extensive review), making this region a candidate NCC.

The use by the same group of alternative imaging methods with better temporal resolutions, such as magnetoencephalography and electroencephalography, has more recently led to a better understanding of the temporal dynamics giving rise to conscious experience. They reveal that perceptual consciousness is a relatively late phenomenon that is preceded by a cascade of neural events operating in an unconscious manner. Indeed, they found evidence for a two-stage mechanism for visual awareness (Del Cul, Baillet, & Dehaene, 2007; Sergent, Baillet, & Dehaene, 2005): in a first stage, lasting for about 200–300 msec, visual stimulations induce activations in the visual areas of the brain, more specifically in the occipitotemporal cortex. Activity in these sensory areas tends to increase with the strength of the visual stimulus (duration, contrast, energy, etc.), irrespective of whether it is consciously perceived.

Conscious representations arise only afterward, when neuronal activity exceeds a certain threshold. Activity induced by the perceived object suddenly spreads to the prefrontal cortex and is dispatched to other cortices. In

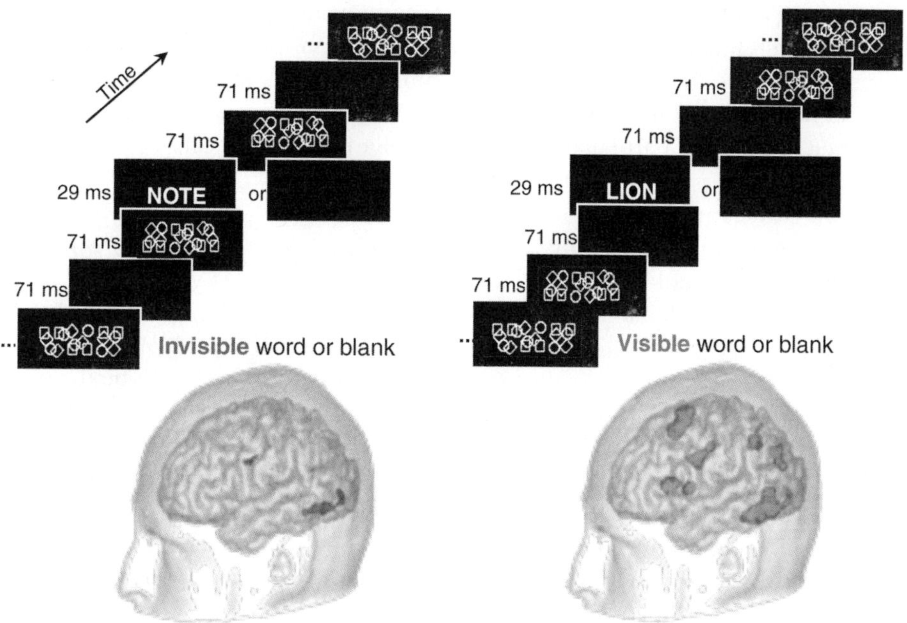

FIGURE 73.1 Contrast of unconscious and conscious processing in functional MRI (Dehaene et al., 2001). While the neural activation induced by invisible words is primarily restricted to occipitotemporal regions (left panel), conscious perception is associated with the involvement of a parietofrontal network (right panel). (See color plate 63.)

other words, the neural mechanisms that are specifically implicated in consciousness are only involved at the final stage of a long chain of unconscious events. Further, these studies pointed to the prefrontal cortex as an area where neural information converges, creating global brain activity and allowing sensory areas to interact with other, task-relevant regions.

The global neuronal workspace

Such findings have been integrated in a theoretical framework by Dehaene and his colleagues, Jean-Pierre Changeux and Lionel Naccache (Dehaene & Changeux, 2011; Dehaene et al., 2006; Dehaene & Naccache, 2001). Called the global neuronal workspace, it is a neurobiological extension of the cognitive global workspace theory originally proposed by the psychologist Bernard Baars (1989).

According to this theory, the cerebral architecture is composed of two qualitatively distinct types of elements. The first type is represented by a large network of domain-specific processors, in both cortical and subcortical regions, that are each attuned to the processing of a particular type of information. For instance, the occipitotemporal cortex is constituted of many such domain-specific processors, or "cerebral modules" (movement processing in MT/V5, face processing in the fusiform face area, etc.). Although these neural processors can differ widely in complexity and domain specificity, they share several common properties: they are triggered automatically, they are encapsulated (their internal computations are not available to other processors) and, importantly, they operate unconsciously.

Consciousness involves a second type of element, namely, the cortical "workspace" neurons that are particularly dense in prefrontal, cingulate, and parietal regions. These neurons send and receive projections to many distant areas through long-range excitatory axons, breaking the modularity of the nervous system and allowing the domain-specific processors to exchange information. The global workspace provides a common communication protocol by allowing the broadcasting of information to multiple neural targets. A mental state is conscious if two conditions are met. First, the content of the mental state must be represented as an explicit neuronal firing pattern that can reach workspace neurons. Second, top-down amplification mechanisms mobilizing the long-distance workspace connections must render the representation accessed, sharpened, and maintained. A mental state, even if it respects the first condition (explicit firing pattern available to workspace neurons), will remain unconscious until its neural signal is amplified. This amplification is the neural counterpart of top-down attention, which, in this framework, is a necessary condition for consciousness. Whether consciousness requires top-down attention is a highly debated issue (Cohen, Cavanagh, Chun, & Nakayama, 2012; Dehaene et al., 2006; Koch

& Tsuchiya, 2007; Lamme, 2006). This framework enriches the traditional search for the NCC: according to the global neuronal workspace theory of consciousness, no single area is viewed as necessary and sufficient for consciousness. Rather, it stresses a particular type of neural interaction between a set of interconnected areas. The necessary and sufficient condition for consciousness of a mental representation is that the information that implements this representation should be distributed and shared among a global network of densely connected areas.

The hard problem of consciousness

The strategy of looking for neural correlates of consciousness has, up to now, been fruitful. We now have a better view of which neural mechanisms are important for consciousness, and scientific theories provide functional descriptions and testable predictions regarding conscious processing. Yet many have criticized the very foundations of this approach, arguing that functional explanations come at the price of sacrificing the "phenomenal" aspects of consciousness: functional explanations are restricted to the cognitive mechanisms (i.e., attention, working memory, etc.) underlying access to conscious contents, ignoring the problem of how these contents arise in the first place. Indeed, some philosophers (Chalmers, 1996) have concluded that there is not one single problem, but actually two problems of consciousness: they distinguish between the "easy problem" and "hard problem." In a nutshell, the easy problem consists in explaining the functional properties of conscious representations. They are intrinsically accessible: one can verbalize to some extent any conscious content, reencode its information in any format available, store it in memory, integrate it in reasoning, focus attention on it, and so on. These properties can be studied by means of the usual objective methods of experimental cognitive neuroscience.

By contrast, the hard problem consists in explaining the subjective, qualitative side of conscious representation—using the phrase of Thomas Nagel (1974), the sense of "what it is like" to be conscious. It is argued that even if all the functional cognitive properties of conscious representations were unfolded, there would still be a subjective remainder. With the help of cognitive neuroscience, we can hope to understand how we put to work the representation of a red signpost on the side of the road: why we notice it, how we associate it with specific behaviors, and so on; still, the specific subjective feeling that *this red* elicits in the observer would, according to this perspective, stand as something of a mystery. The functional aspects of consciousness are considered "easy" from an epistemological standpoint (although they may be immensely intricate and complex empirically) because they constitute information-processing challenges; the problem of qualia is "hard" because it involves crossing the objective/subjective, public/private divides.

Dissociative approaches to consciousness

With respect to the epistemic distinction between an easy and a hard problem, Ned Block has proposed that consciousness should be dissociated into two components, namely, access and phenomenal consciousness (Block, 1995, 2007). Phenomenal consciousness is related to the private, first-person experience. Access consciousness corresponds to the fact that some representations are "poised for direct control of thought and action" (Block, 1995); it designates the functional cognitive properties of conscious contents, which can be explained in terms of computational mechanisms and are linked to global broadcasting (Block, 2005) in agreement with workspace theories of consciousness.

Several neuroscientists have adopted Block's dissociation and explicitly distinguish between two neural correlates of consciousness. For instance, the duplex vision theory of Milner and Goodale (1995) has recently been updated to associate sustained ventral stream activity with phenomenal consciousness, while only the involvement of more anterior (e.g., prefrontal) regions supports conscious access (Goodale, 2007). Similarly, Semir Zeki (2007) has recently linked micro- and macro-consciousness in his original theory (Zeki & Bartels, 1999) with phenomenal consciousness of specific attributes (colors, contrasts, etc.) and bound objects, respectively, while unified consciousness is somewhat analogous to access consciousness. In the local recurrence theory of Victor Lamme (2006), phenomenal experience is explicitly associated with any recurrent neuronal activity (i.e., local or global loops), while conscious access occurs only with global recurrence. Although all these theories diverge in many respects, they all link phenomenal consciousness with posterior (i.e., occipitotemporal) regions, while anterior (i.e., prefrontal, workspace) areas are linked to conscious access (see Kouider, 2009, for a review). They are also motivated by the possibility of probing consciousness in the absence of subjective reports (Lamme, 2006) and are thus committed to the hypothesis that there exists a form of phenomenal consciousness that might be irreductible to access mechanisms. We now turn to the empirical and epistemological consequences of this commitment.

Neural purity and the overflow argument

Two main empirical arguments, which we termed the *overflow argument* and the *neural purity argument* (see Kouider, de Gardelle, Sackur, & Dupoux, 2010), have been offered by proponents of the access/phenomenal consciousness dissociation. The overflow argument is rooted in the intuition that we are conscious of much more than we can describe and manipulate. This intuition was operationalized by Sperling over half a century ago (Sperling, 1960), who used letter arrays to quantify the amount of information available at a given time after presentation of a complex visual scene (see figure 73.2). Using short presentation times and a pioneering cued report method, Sperling showed that the information available for a short period of time after stimulus presentation vastly exceeded the information subjects could spontaneously report. This has been taken as an indication that phenomenal consciousness does indeed overflow access (Block, 2007). Yet, as we discuss below, it remains controversial whether the large amount of available information in cued reports reflects phenomenally conscious representations or unconscious processing that becomes reportable by virtue of the cues (Block, 2007; de Gardelle, Sackur, & Kouider, 2009; Dehaene et al., 2006; Sergent et al., 2012).

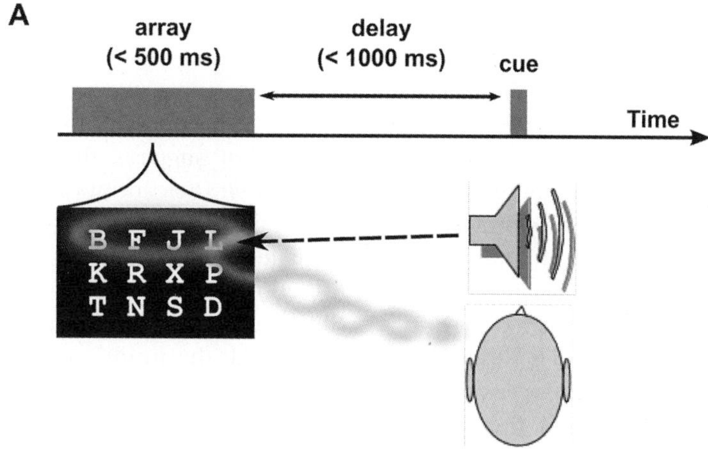

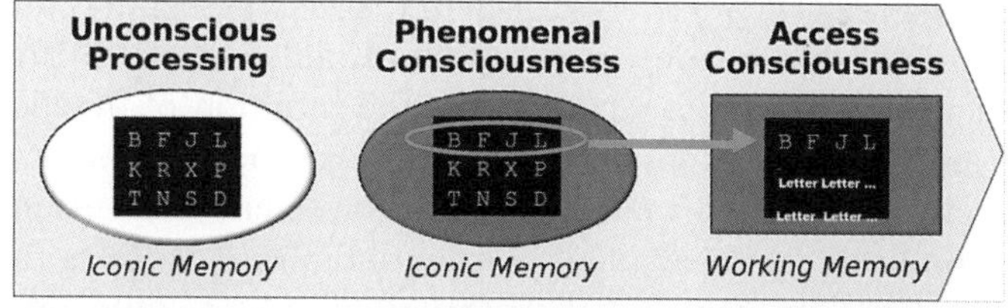

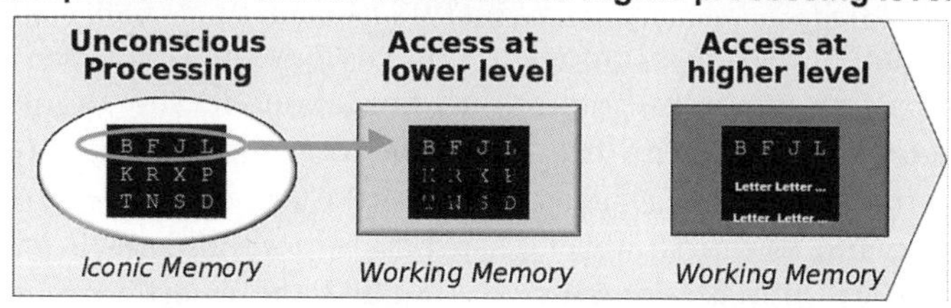

FIGURE 73.2 The Sperling paradigm (Sperling, 1960) and its interpretations. (A) Experimental procedure for the cued report. A brief array of letters is shown, followed by a random tone cue (high tone in this example). The pitch of the cue (low, medium, high) instructs subjects to report one of the three rows (lower, middle, or higher row, respectively). When participants are not cued and have to report all letters in the array, performance is restricted to about 4 out of 12 items. However, when using the post-stimulus cue to report a specific row, performance increased to 3 out of 4 items. This suggests that a large amount of information is available but decays by the time of reporting. (B) Two interpretations of the results. Interpretation 1 assumes that subjects are phenomenally conscious of the whole content in iconic memory demonstrated by the high-level performances at short delays. Interpretation 2 hypothesizes that subjects access both high- and low-level information from iconic memory. Low-level information is reconstructed at higher levels. (See color plate 64.)

The neural purity argument follows from the assumption that there exist specific neural mechanisms for phenomenal experience (e.g., local neural recurrence). Such mechanisms allegedly constitute pure indices of consciousness, more reliable than subjective reports, which are limited by verbal, memory, and attentional abilities. For instance, Block and Lamme argue that in paradigms where subjects cannot report the presence of a stimulus due to inattention (e.g., change blindness, inattentional blindness, attentional blink), they might still be phenomenally conscious of the stimulus as long as it induces local recurrence in perceptual brain regions (Block, 2007; Lamme, 2006).

The interplay between the neural purity and overflow arguments is complex with respect to whether one should trust subjective reports. On the one hand, the overflow argument depends on the intuition that there is more to a given conscious experience than we can report. It thus depends on a negative statement: "there is something in my conscious experience that I cannot report." If this statement is to be meaningful, it is a second-order report (a meta-report) of consciousness, because it states the incompleteness of some access-consciousness report. Subjects should be trusted regarding this intuition. On the other hand, the neural purity argument implies that by studying brain activations, we know more precisely than subjects themselves whether they are conscious or not. In other words, their reports should not be trusted.

The limits of dissociative approaches

Arguments for a dissociative approach to consciousness suffer from serious flaws. We have put forward the fact that the phenomenal overflow argument is confounded with situations of partial awareness, while the neural purity argument reflects the confusion between phenomenal consciousness and unconscious perceptual processes (Kouider et al., 2010, 2012).

We start with the overflow argument. First, it is important to stress that limits on (verbal) reportability should not be equated with limits on access. Perception involves nonconceptual contents that are difficult to verbalize, such as shades of colors, smells, and so on. However, the relative poverty of verbal reports in these domains should not be equated with poverty in access. Indeed, the hallmark of psychophysics is precisely to uncover the rich, graded, and multidimensional aspects of domains such as color or smell perception using indirect measures like similarity judgments (Gescheider, 1997; Sackur, 2013). Furthermore, as verbal reports take time and are performed in a sequential manner, accessible information may have disappeared prior to verbalization. Nonetheless, subjects' performance on nonverbal tasks such as detection or discrimination shows that information can be accessed before it fades away. In other words, the overflow argument might only show that access overflows verbal report. Further, the demarcation between expressible and ineffable contents may not be clear-cut: it is well known, for instance, that experienced wine tasters acquire a vocabulary and develop descriptive skills to finely capture nuances of sensory experiences that seem elusive at first. Similarly, early introspective psychologists of the Külpe and Titchener schools developed impressive fine-grained skills in order to describe visual impressions created by stimuli very similar to those later used by Sperling (see, for instance, Dallenbach, 1920). These examples indicate that descriptive powers can be improved, to the point that there may not be any fixed limit to what aspects of conscious experience are reportable versus those that are not. This does not logically rebut the overflow argument, but suggests that whether subjects are to be trusted on their intuition about overflow is itself something that should be put under experimental scrutiny. This leads to the second line of argument against the overflow argument, namely, that its apparent compelling force might be illusory.

Indeed, the intuition of a rich phenomenal experience on which the overflow argument is built might be overstated. Observers might overestimate both the quantity and accuracy of the information they experience at one given moment, lured either by a nonspecific "cognitive illusion of seeing" (O'Regan & Noe, 2001), or by perceptual illusions (de Gardelle et al., 2009; Kouider et al., 2010). In addition, if we admit that the intuition of overflow is a meta-report of consciousness, the possibility of consciousness without the involvement of access mechanisms is methodologically dubious: if subjects do not have access to their experience, how could one determine that they are conscious of it? Actually, someone experiencing phenomenology without access should not only be unable to talk about it, she should not even *know* anything about it! In other words, reporting a "rich but unaccessed visual experience" demonstrates that we have access to *some* kind of information.

Finally, the assertion that phenomenal experiences can arise in the absence of access leads to an epistemological impasse: in order to prove that a particular content is phenomenal, one has to ask the subject about it. But if the subject is attempting to report about her experience, it also means that she is attempting to access it. Hence, one faces an *observer effect*: any observation of the internal states of a system changes the state of the system (Kouider, 2009). As such, any attempts to observe internal states prior to access will necessarily be contaminated by access mechanisms themselves.

A potential escape from the problems outlined above might be to accept the neural purity argument, according to which phenomenological consciousness can be probed regardless of reportability, through neural indices. However, this strategy is circular, since validating the neural index in the first place necessarily requires reliance on access mechanisms. Indeed, demonstrating that a specific neural mechanism (e.g., local recurrence) is sufficient for consciousness initially requires the assessment of neural events while probing whether the subject is conscious. As the sole uncontroversial way to prove consciousness relies on access mechanisms, it appears impossible to map neural and phenomenal states without depending on access. This is not to say that we cannot, in some situation, infer conscious contents from brain states. As we gain more insights into the nature of the brain mechanisms associated with conscious experience, we can reapply this knowledge in cases where reports are impossible, for instance, in cases of patients with locked-in syndrome and in vegetative states (Laureys et al., 2005; Owen et al., 2006), as well as in the case of preverbal infants (Kouider et al., 2013). But clearly, this extrapolation beyond the domain of reportability is justified, because we had first relied on conscious reports, and thus on access mechanisms.

Finally, the neural purity argument largely reflects a theoretical confusion: it merely shows that the brain processes information without consciousness, but not that there is phenomenal experience associated with these processes. A supposed neural index of phenomenal consciousness in the absence of access may thus simply reflect unconscious processes (Dehaene et al., 2006; Kouider, Dehaene, Jobert, & Le Bihan, 2007). Yet, because one cannot demonstrate whether phenomenal experience is involved or not, the neural purity argument becomes unfalsifiable: if, say, local recurrence is observed in the absence of conscious access, stipulating alternative forms of consciousness, instead of unconscious processing, cannot be verified and simply becomes a matter of faith.

Partial awareness and the illusion of phenomenal richness

Nevertheless, while phenomenal consciousness seems dubious both from methodological and epistemological standpoints, phenomenality in itself is a reality. Our conscious mental content does seem to exceed all possible reports, and it has a qualitative and subjective "feel" that is private. Here, we explain how, by means of the notions of partial awareness, confidence evaluation, and expectations, access mechanisms can mechanistically account for *phenomenality* without reliance on specific and dedicated mechanisms for *phenomenal consciousness.*

The notion of "levels of representation" is one of the most venerable notions in cognitive psychology: for instance, a written word might be encoded at the level of nonspecific geometrical features, letter fragments, specific letter shapes, or abstract letters, and then at lexical, phonological, and semantic levels (Vinckier et al., 2007). We know from numerous psycholinguistic tasks that these levels of representation are somewhat independent, in the sense that some tasks can require access to one specific level. This kind of representational hierarchy is implicit and basic in most areas of cognitive psychology, but has been largely ignored for consciousness. Recently, we proposed that different levels of representation of one and the same stimulus might be separately consciously accessed and lead to global broadcasting independently from one another (Kouider et al., 2010). For instance, because of some degradation, a visual stimulus may only be accessed at some lower levels, making it only partially conscious. Thus a word might be accessed at the level of letter features, while remaining unaccessed at higher levels having to do with the whole word form (which does not preclude unconscious processing at these higher levels). But conscious contents are not simply stimulus driven: the cognitive system has some a priori knowledge about the world, with some confidence level about the likelihood of sensory signals. Hence, access to partial information is combined with prior knowledge of what should be perceived: if participants expect to be shown letters and are partially conscious of letter fragments, they illusorily *see* letters (de Gardelle et al., 2009). The intuition of a rich, elusive phenomenality comes from real-life situations, where stimuli are complex and span a large portion of the visual field. Thus, at each moment, various parts of the scene are accessed at different levels, with restricted levels for eccentric and crowded stimuli. Since the pioneering work of McConkie and Rayner (1975), who used eye-tracking methodology to blur a text beyond a window centered at fixation, it has been known that we do not need rich and detailed information over the entire visual field to produce a visual consciousness with the impression of richness. More recently, Freeman and Simoncelli (2011), using more controlled methods, constructed stimuli that looked exactly alike in spite of systematic distortions at the periphery. Again, this suggests that our visual system accesses only low-level geometrical information in the periphery of the visual field, and creates on this basis a conscious representation that is illusorily detailed. Our visual experience is always a mixture of detailed and coarse information: information at fixation is accessed at the highest possible level, while information in the periphery is only accessed at the level of coarse features. However, the visual system does not assume that the world in the periphery is blurred. Rather, our confidence that there is potentially detailed information in the periphery is high. The integration of low-level conscious access and of high confidence about what is potentially discriminable mechanistically yields an impression of ineffable richness, which is precisely the characteristic of phenomenality. The interaction between neuronal processes dealing with sensory signals, prior expectations, and confidence evaluation may thus constitute core mechanisms of conscious phenomenality.

Conclusion

As we have seen in the previous sections, a core issue in cognitive neuroscience is whether consciousness should be extended beyond its access mechanisms. We explained how the idea that one should dissociate access and phenomenal consciousness on the basis of separated neuronal and functional properties was both epistemologically and empirically dubious. With a few simple assumptions involving hierarchized representational levels, prior expectations, and confidence evaluation, one can reframe the issue of dissociable forms of consciousness into dissociable levels of conscious access.

Yet it would be presumptuous to assume that we now fully understand how conscious contents arise from this kind of neurocognitive architecture. Even if a neurocognitive description could account for and predict the occurrence and content of a specific conscious experience, some would certainly still not be convinced that this explains how one goes from the neural level to the experiential one. This has recently led some of the most recognized scientists in the field, such as Christof Koch, who originally proposed the NCC approach, and Giulio Tononi, to abandon this perspective, considering the whole reductionist approach as being intrinsically limited in addressing this issue of how consciousness arises in the first place (Koch, 2012; Tononi & Koch, 2008). Instead, consciousness should be envisioned in terms of complex systems having more to do with information theory than specific properties of the brain. In contrast to this radical shift from the neurobiological approach, we advocate an empirical stance toward the hard problem of consciousness and phenomenality: we believe that within the traditional perspective of cognitive neuroscience, finer-grained distinctions of levels of access and more complex (e.g., Bayesian) mechanisms of integration with priors and expectations may provide a progressive bridging of the gap between functional mechanisms and subjective experience.

ACKNOWLEDGMENTS This work was supported by funding from the Agence National de la Recherche and from the European Research Council ("DynaMind" project). We thank Vincent de Gardelle and Emmanuel Dupoux for fruitful discussions on the topic of this chapter.

REFERENCES

Baars, B. J. (1989). *A cognitive theory of consciousness.* Cambridge, UK: Cambridge University Press.

Block, N. (1995). On a confusion about a function of consciousness. *Behav Brain Sci, 18*(2), 227–287.

Block, N. (2005). Two neural correlates of consciousness. *Trends Cogn Sci, 9*(2), 46–52.

Block, N. (2007). Consciousness, accessibility, and the mesh between psychology and neuroscience. *Behav Brain Sci, 30*(5–6), 481–499; discussion 499–548.

Chalmers, D. (1996). *The conscious mind.* New York, NY: Oxford University Press.

Cohen, M. A., Cavanagh, P., Chun, M. M., & Nakayama, K. (2012). The attentional requirements of consciousness. *Trends Cogn Sci, 16*(8), 411–417.

Crick, F., & Koch, C. (1995). Are we aware of neural activity in primary visual cortex? *Nature, 375*, 121–123.

Dallenbach, K. M. (1920). Introspection and general methods. *Psychol Bull, 17*(10), 313–321.

De Gardelle, V., Sackur, J., & Kouider, S. (2009). Perceptual illusions in brief visual presentations. *Conscious Cogn, 18*(3), 569–577.

Dehaene, S., & Changeux, J. P. (2011). Experimental and theoretical approaches to conscious processing. *Neuron, 70*(2), 200–227.

Dehaene, S., Changeux, J. P., Naccache, L., Sackur, J., & Sergent, C. (2006). Conscious, preconscious, and subliminal processing: A testable taxonomy. *Trends Cogn Sci, 10*(5), 204–211.

Dehaene, S., & Naccache, L. (2001). Towards a cognitive neuroscience of consciousness: Basic evidence and a workspace framework. *Cognition, 79*(1–2), 1–37.

Dehaene, S., Naccache, L., Cohen, L., Bihan, D. L., Mangin, J. F., Poline, J. B., & Riviere, D. (2001). Cerebral mechanisms of word masking and unconscious repetition priming. *Nat Neurosci, 4*(7), 752–758.

Dehaene, S., Naccache, L., Le Clec, H. G., Koechlin, E., Mueller, M., Dehaene-Lambertz, G., Van de Moortele, P. F., & Le Bihan, D. (1998). Imaging unconscious semantic priming. *Nature, 395*, 597–600.

Del Cul, A., Baillet, S., & Dehaene, S. (2007). Brain dynamics underlying the nonlinear threshold for access to consciousness. *PLoS Biol, 5*(10), e260.

Freeman, J., & Simoncelli, E. P. (2011). Metamers of the ventral stream. *Nat Neurosci, 14*(9), 1195–1201.

Gescheider, G. (1997). *Psychophysics: The fundamentals.* Mahwah, NJ: Erlbaum.

Goodale, M. (2007). Duplex vision: Separate cortical pathways for conscious perception and the control of action. In

M. Velmans & S. Schneider (Eds.), *The Blackwell companion to consciousness* (pp. 616–627). Oxford, UK: Blackwell.

Koch, C. (2004). *The quest for consciousness: A neurobiological approach.* Denver, CO: Roberts.

Koch, C. (2012). *Consciousness: Confessions of a romantic reductionist.* Cambridge, MA: MIT Press.

Koch, C., & Tsuchiya, N. (2007). Attention and consciousness: Two distinct brain processes. *Trends Cogn Sci, 11*(1), 16–22.

Kouider, S. (2009). Neurobiological theories of consciousness. In W. Banks (Ed.), *Encyclopedia of consciousness,* Vol. 2 (pp. 87–100). Oxford, UK: Elsevier.

Kouider, S., Dehaene, S., Jobert, A., & Le Bihan, D. (2007). Cerebral bases of subliminal and supraliminal priming during reading. *Cereb Cortex, 17*(9), 2019–2029.

Kouider, S., de Gardelle, V., Sackur, J., & Dupoux, E. (2010). How rich is consciousness? The partial awareness hypothesis. *Trends Cogn Sci, 14*(7), 301–307.

Kouider, S., Sackur, J., & de Gardelle, V. (2012). Do we still need phenomenal consciousness? *Trends Cogn Sci, 16*(3), 140–141.

Kouider, S., Stahlhut, C., Gelskov, S., Barbosa, L., de Gardelle, V., Dutat, M., ... Dehaene-Lambertz, G. (2013). A neural marker of perceptual consciousness in infants. *Science, 340*(6130), 376–380.

Lamme, V. A. (2006). Towards a true neural stance on consciousness. *Trends Cogn Sci, 10*(11), 494–501.

Laureys, S., Pellas, F., van Eeckhout, P., Ghorbel, S., Schnakers, C., Perrin, F., ... Goldman, S. (2005). The locked-in syndrome: What is it like to be conscious but paralyzed and voiceless? *Prog Brain Res, 150,* 495–511.

Leopold, D. A., & Logothetis, N. K. (1996). Activity changes in early visual cortex reflect monkeys' percepts during binocular rivalry. *Nature, 379*(6565), 549–553.

McConkie, G. W., & Rayner, K. (1975). The span of the effective stimulus during a fixation in reading. *Percept Psychophys, 17,* 578–586.

Milner, A. D., & Goodale, M. A. (1995). *The visual brain in action.* New York, NY: Oxford University Press.

Nagel, T. (1974). What is it like to be a bat? *Philos Rev, 83*(4), 435–450.

O'Regan, J. K., & Noe, A. (2001). A sensorimotor account of vision and visual consciousness. *Behav Brain Sci, 24*(5), 939–973; discussion 973–1031.

Owen, A. M., Coleman, M. R., Boly, M., Davis, M. H., Laureys, S., & Pickard, J. D. (2006). Detecting awareness in the vegetative state. *Science, 313*(5792), 1402.

Sackur, J. (2013). Two dimensions of visibility revealed by multidimensional scaling of metacontrast. *Cognition, 126*(2), 173–180.

Sackur, J., & Dehaene, S. (2009). The cognitive architecture for chaining of two mental operations. *Cognition, 111,* 187–211.

Sergent, C., Baillet, S., & Dehaene, S. (2005). Timing of the brain events underlying access to consciousness during the attentional blink. *Nat Neurosci, 8*(10), 1391–1400.

Sergent, C., Wyart, V., Babo-Rebelo, M., Cohen, L., Naccache, L., & Tallon-Baudry, C. (2012). Cueing attention after the stimulus is gone can retrospectively trigger conscious perception. *Curr Biol, 23,* 150–155.

Sklar, A. Y., Levy, N., Goldstein, A., Mandel, R., Maril, A., & Hassin, R. R. (2012). Reading and doing arithmetic nonconsciously. *Proc Natl Acad Sci USA, 109,* 19614–19619.

Sperling, G. (1960). The information available in brief visual presentation. *Psychol Monogr, 74,* 1–29.

Tong, F., Nakayama, K., Vaughan, J. T., & Kanwisher, N. (1998). Binocular rivalry and visual awareness in human extrastriate cortex. *Neuron, 21*(4), 753–759.

Tononi, G., & Koch, C. (2008). The neural correlates of consciousness: An update. *Ann NY Acad Sci, 1124,* 239–261

Vinckier, F., Dehaene, S., Jobert, A., Dubus, J. P., Sigman, M., & Cohen, L. (2007). Hierarchical coding of letter strings in the ventral stream: Dissecting the inner organization of the visual word-form system. *Neuron, 55,* 143–156.

Zeki, S. (2007). A theory of micro-consciousness. In M. Velmans & S. Schneider (Eds.), *The Blackwell companion to consciousness* (pp. 580–588). Oxford, UK: Blackwell.

Zeki, S., & Bartels, A. (1999). Toward a theory of visual consciousness. *Conscious Cogn, 8*(2), 225–259.

74 Bodily Self-Consciousness

OLAF BLANKE

ABSTRACT Recent data have linked self-consciousness to the processing of multisensory bodily signals in temporoparietal and premotor cortex. Studies in which subjects receive ambiguous multisensory information about the location and appearance of their own body have shown that activity in these brain areas reflects the conscious experience of identifying with the body (self-identification), with the experience of where "I" am in space (self-location), and with the experience of the perspective from where "I" perceive the world (first-person perspective). I argue that these data may form the basis for a neurobiological model of self-consciousness, grounding higher-order notions of self-consciousness and personhood in multisensory brain mechanisms.

Humans experience a "real me" that "resides" in "my" body and is the subject or "I" of experience and thought. This is self-consciousness, the feeling that conscious experiences are bound to the self. Thus, experiences happen for someone, the subject or the self, in an immediate way. They are generally felt as "*my*" experiences and seem to be present whenever "I" am perceiving a color, have a thought, or feel pain. Experiences are felt as belonging to "somebody," and it is this unitary entity, the "I," that is often considered to be one of the most astonishing features of the human mind. So far self-consciousness has mainly been approached by philosophical inquiry, and has led to an impressive body of work. However, there is an overabundance of diverging theories about the nature of the self and self-consciousness that are not data-driven (i.e., Aikins, 2005; Bermudez, 1998; Gallagher & Shear, 1999).

A powerful approach to investigate self-consciousness experimentally has been to target brain mechanisms that process bodily signals (i.e., bodily self-consciousness; Blanke & Metzinger, 2009; Christoff, Cosmelli, Legrand, & Thompson, 2011; Damasio & Meyer, 2009; Jeannerod, 2003; Knoblich, 2002; Legrand, 2007; de Vignemont, 2011). The study of such bodily signals is complex since bodily input is continuously present and updated and conveyed by many senses, including tactile, proprioceptive, and visual signals about the body. Although prominent accounts of self-consciousness have also highlighted the importance of motor and interoceptive/homeostatic signals (Craig, 2009; Damasio, 1999; Frith, 2005; Jeannerod, 2003; Seth, Suzuki, & Critchley, 2012), the present review will focus on multisensory mechanisms and highlight recent research that has used video, virtual reality, and robotics technologies to study bodily self-consciousness. These findings in healthy subjects will be compared with evidence from abnormal states of bodily self-consciousness in two neurological conditions, out-of-body experiences and heautoscopy (Brugger, 2002; Devinsky, Feldmann, Burrowes, & Bromfield, 1989). Recent data about three aspects of bodily self-consciousness will be discussed, and in particular how these three aspects relate to the processing of bodily signals and which functional and neural mechanisms they may share. These are self-identification with the body (i.e., the experience of owning a body), self-location (i.e., the experience of where I am in space), and the first-person perspective (i.e., the experience from where I perceive the world).

Starting with the breakdown of bodily self-consciousness, the results of detailed observations in neurological patients who have had out-of-body experiences (characterized by abnormal self-identification, self-location, and first-person perspective) will be described. Next, some of the major experimental paradigms, behavioral results, and neuroimaging findings will be reviewed. In a final section, I will highlight how the principles of bodily self-consciousness extend to interoceptive signals, describe interactions with visual consciousness, and sketch some limitations of the present approach.

When the self leaves the body: The out-of-body experience

If you ever while lying in bed suddenly had the distinct impression of floating up near the ceiling and looking back down at your body on the bed, then it is likely that you had an out-of-body experience (OBE). OBEs are striking phenomena because they challenge our everyday experience of the spatial unity of self and body: they challenge our experience of a "real me" that "resides" in my body and is the subject, or "I," of experience and thought. As Sylvan Muldoon, one of the first modern authors to describe an OBE, wrote: "I was floating in the very air, rigidly horizontal, a few feet above the bed. ... I was moving toward the ceiling, horizontal and powerless. ... I managed to turn around and there ... was another 'me' lying quietly upon the bed" (Muldoon & Carrington, 1929).

Although OBEs may be considered a bizarre departure from normal human experience, they are more than a mere curiosity, have allowed study of the bodily foundations of self-consciousness, and have impacted cognitive neuroscience research in healthy subjects. OBEs have been reported since time immemorial and have been estimated to occur in about 5% of the general population (Blanke, Landis, Spinelli, & Seeck, 2004). During an OBE, the subject is awake and experiences the "self," or center of awareness, as being located outside the physical body at a somewhat elevated level. Thus self-location is abnormal, and this change is described by most individuals undergoing an OBE as the most astonishing OBE aspect. It is from this elevated extrapersonal location that the subject's body and the world are perceived (i.e., abnormal first-person perspective) (Blanke et al., 2004; Brugger, 2002; Devinsky et al., 1989). Most subjects experience seeing their own body as lying on the ground or in bed, and the experience tends to be described as vivid and realistic, differing from the degree of realism described in dreams or dream-like states. Thus, self-identification with a body—that is, the sensation of owning a body—is experienced at the elevated, disembodied location and not at the location of the physical body (i.e., abnormal self-identification). What causes this disunity between self and body and changes in self-identification, self-location, and our everyday body-centered, first-person perspective?

The neurology of bodily self-consciousness

OBEs of neurological origin have been reported in patients suffering from many different neurological diseases (Blanke et al., 2004; Brugger, 2002; Devinsky et al., 1989; Maillard, Vignal, Anxionnat, Taillandier, & Vespignani, 2004). Most common are migraine and epilepsy, but OBEs have also been described after focal electrical cortical stimulation (Blanke, Ortigue, Landis, & Seeck, 2002; de Ridder, van Laere, Dupont, Menovsky, & van de Heyning, 2007), general anesthesia, typhoid fever, and spinal cord damage. Anatomically, OBEs have been associated with damage to the right and left temporoparietal junction (TPJ; Blanke et al., 2004), in particular the posterior superior temporal gyrus, angular gyrus, and the supramarginal gyrus. Outside the TPJ, several cases with damage in the precuneus (de Ridder et al., 2007) and frontotemporal cortex have also been described (Devinsky et al., 1989). Although these studies have implicated a wide variety of brain regions, a recent lesion-analysis study using voxel-based, lesion-symptom mapping—in the largest sample of patients with OBEs due to focal brain damage to date—implicated a well-centered region in the right hemisphere at the junction of the right angular gyrus with the posterior superior temporal gyrus (figure 74.1; Ionta et al., 2011).

Based on the frequent association of OBEs with visuo-somatosensory illusions, abnormal vestibular sensations (Lopez, Halje, & Blanke, 2008), and the known role of the TPJ in multisensory integration (i.e., Bremmer et al., 2001; Calvert, Campbell, & Brammer, 2000), it has further been proposed that OBEs (and abnormal self-identification, self-location, and first-person perspective) occur due to disturbed multisensory integration of bodily signals in peripersonal space (somatosensory, visual, and proprioceptive signals) and extrapersonal space (visual and vestibular signals; Blanke et al., 2004; Blanke, 2012). These clinical data have informed the design of experimental approaches in healthy subjects, and it has been shown that multisensory conflicts of bodily signals can alter bodily self-consciousness by inducing altered states of self-identification, self-location, and the first-person perspective.

Experimental approaches to self-identification and self-location

Using video, virtual reality, and/or robotic devices, researchers have studied bodily self-consciousness by inducing experimentally controlled changes in self-location, self-identification, and first-person perspective (Ehrsson, 2007; Lenggenhager, Tadi, Metzinger, & Blanke, 2007). The experimental protocols are based on paradigms from multisensory perception that had previously been tested mostly in studies on upper-limb perception (for review see Blanke, 2012; Makin, Holmes, & Ehrsson, 2008; Serino et al., 2011; Tsakiris, 2010) and exploit visuotactile and visuovestibular conflicts to induce full-body illusions or out-of-body illusions.

In most full-body illusion paradigms, a tactile stroking stimulus is repeatedly applied to the back or chest of a participant who is being filmed and simultaneously views (through a head-mounted display) the stroking of a human body in a real-time film or virtual-reality animation. A video camera is placed 2 meters behind the person, filming the participant's back from behind (figure 74.2A). In an experiment based on OBEs of neurological origin, participants view a video image of their body (the "virtual body") from an "outside," third-person perspective (Lenggenhager et al., 2007) while an experimenter stroked their backs with a stick. The stroking in these paradigms was felt by the participants on their backs and also seen on the backs of the virtual bodies. The head-mounted display showed the stroking of the virtual bodies either in real time or not (using

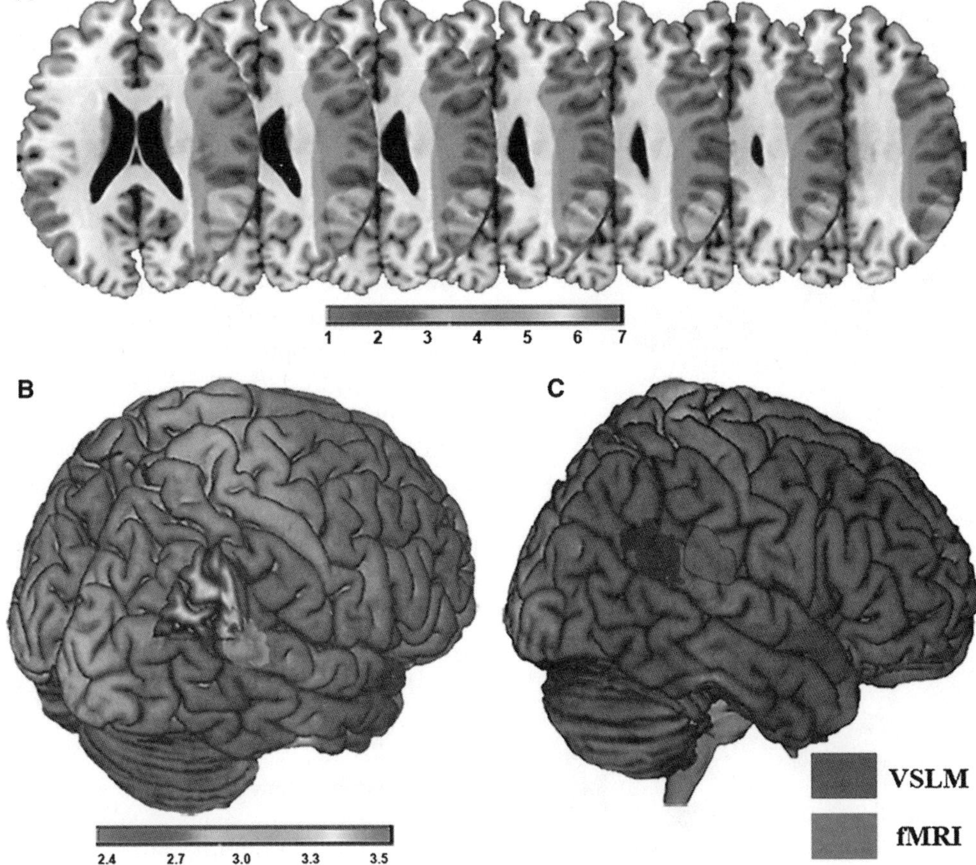

FIGURE 74.1 Brain damage leading to changes in bodily self-consciousness (out-of-body experiences, or OBE). Brain damage and results of lesion overlap analysis (top) in nine patients with OBEs due to focal brain damage are shown. Maximal lesion overlap centers at the right temporoparietal junction (TPJ) at the angular gyrus (red). Overlap color code ranges from violet (one patient) to red (seven patients). Voxel-based lesion symptom mapping (VLSM; bottom left) of focal brain damage leading to OBEs. The violet-to-red cluster shows the region that VLSM analysis associated statistically with OBEs as compared to control patients. The color code indicates significant Z scores of the respective voxels, showing maximal involvement of the right TPJ. (Modified from Ionta et al., 2011.) (See color plate 65.)

an online video-delay or offline prerecorded data), generating synchronous and asynchronous visuotactile stimulation, respectively.

Under these conditions, subjects self-identified with the seen virtual body, and such illusory self-identification with the virtual body was stronger during synchronous than during asynchronous stroking conditions (Lenggenhager et al., 2007; for a similar approach, see Ehrsson, 2007). Concerning self-location, a forward drift (as predicted) was found. Thus, self-location was experienced at a position that was closer to the virtual body, as if subjects were located "in front" of the position where they had been standing during the experiment, compatible with a breakdown of the experienced spatial unity between self and body. Later work confirmed that self-location toward and self-identification with the virtual body are systematically influenced by different visuotactile conflicts, and that these changes can also be achieved in the supine position (Ionta et al., 2011; Lenggenhager, Mouthon, & Blanke, 2009).

Changes in bodily self-consciousness are also associated with an alteration in how visual stimuli interfere with the perception of tactile stimuli that are applied while stroking is applied to induce full-body and out-of-body illusions. Such visuotactile interference is a behavioral index of whether visual and tactile stimuli are functionally perceived to be in the same spatial location (i.e., visuotactile cross-modal congruency effect; Igarashi, Kimura, Spence, & Ichihara, 2008; Pavani, Spence, & Driver, 2000; Shore, Barnes, & Spence, 2006; Spence, Pavani, & Driver, 2004). When testing such cross-modal congruency effects during the full-body illusion, it was found that visual stimuli seen at a position 2 meters in front of the subject's back and tactile stimuli (applied

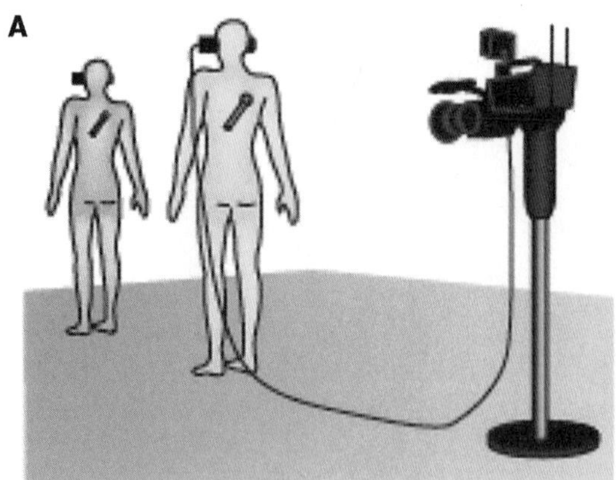

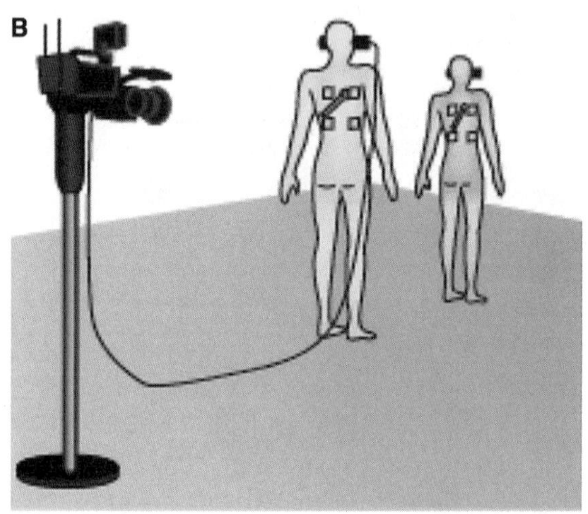

FIGURE 74.2 (A) Experimental setup during the full-body illusion. A participant (light color) sees his own back through video goggles, as if a virtual body (dark color) were located a few meters in front. An experimenter administers tactile stroking to the participant's back, which the participant sees on the video goggles as visual stroking on the virtual body. Synchronous but not asynchronous visuotactile stimulation results in illusory self-identification with and self-location toward the virtual body. (Modified from Blanke, 2012.) (B) Full-body illusion and visuotactile integration. The experimental setup of the full-body illusion was adapted to acquire repeated behavioral measurements related to visuotactile perception (i.e., the cross-modal congruency effect; CCE). In addition to the visuotactile stroking (as in A), participants wore vibrotactile devices and saw visual stimuli (light-emitting diodes) on their backs while viewing their bodies through video goggles. The CCE is a behavioral measure that indicates whether or not a visual and a touch stimulus are perceived to be at identical spatial locations. Participants were asked to indicate where they perceived a single touch stimulus (i.e., short vibration) applied either just below the shoulder or on the lower back. Distracting visual stimuli (i.e., short light flashes) were also presented on the back, either at the same or at a different position (and were filmed by the camera). Under these conditions, participants are faster to detect a touch stimulus if the visual distractor was presented at the same location (i.e., congruent trial) compared to touches co-presented with a more distanced visual distractor (i.e., incongruent trial). CCE measurements were carried out while illusory self-identification was modulated by visuotactile stroking as described in A. The effect of congruency on reaction times was larger during synchronous than asynchronous visuotactile stroking, indicating greater interference of irrelevant visual stimuli during illusory self-identification with the virtual body. (Modified from Blanke, 2012.) (See color plate 66.)

on the subject's back) were functionally perceived to be in the same spatial location (figure 74.2B). Moreover, this interference was modulated in states of illusory self-identification and self-location (Aspell, Lenggenhager, & Blanke, 2009; see also Aspell, Lavanchy, Lenggenhager, & Blanke, 2010). These data provide online perceptual evidence that self-identification and self-location with a virtual body alter how the brain perceives stimuli applied to the subject's own physical body. Research has shown that alterations of bodily self-consciousness are also associated with physiological changes (i.e., skin conductance response to a threat directed toward the virtual body; Ehrsson, 2007; Petkova & Ehrsson, 2008) and nociceptive changes (Hänsel et al., 2011; figure 74.2B). Thus, not only is tactile perception altered, but it was also found that pain thresholds increase (Hänsel et al., 2011) and body temperature (Salomon, Lim, Herbelin, Hesselmann, & Blanke, 2013) decreases during the full-body illusion.

Multisensory brain mechanisms of self-identification, self-location, and first-person perspective

Neuroimaging data on bodily self-consciousness are unfortunately still sparse, and have linked self-identification and self-location to several brain regions using different paradigms and techniques. Thus, a functional MRI study (Petkova et al., 2011) reported that self-identification is associated with activity in bilateral ventral premotor cortex, left posterior parietal cortex, and the left putamen. Both cortical regions, premotor and parietal cortex, have also been linked to illusory ownership of an arm as manipulated in the rubber hand illusion using visuotactile stimulation (Ehrsson, Spence, & Passingham, 2004; Makin et al., 2008; Blanke, 2012). In an EEG study, self-identification and self-location with a virtual body were associated with differential suppression of alpha-band power (8–13 Hz) oscillations in bilateral medial sensorimotor regions and medial premotor cortex (Lenggenhager, Halje, & Blanke, 2011). Alpha-band oscillations over central areas (that is, the mu rhythm) have previously been linked to sensorimotor processing (Pineda, 2005), and mu rhythm suppression is thought to reflect increased cortical activation in sensorimotor and/or premotor cortices (Oakes et al., 2004). Another functional MRI study (Ionta et al., 2011) found that self-identification with a virtual body is associated with activation in the right middle-inferior temporal cortex (partially overlapping with the extrastriate body area), a region that is like the premotor cortex involved in the multisensory processing of human bodies (Astafiev, Stanley, Shulman, & Corbetta, 2004; Downing et al., 2001; Grossman and Blake, 2002). Future work is needed to distinguish regions associated with ownership for the entire body from those related to ownership for body parts such as an arm or the face (for review see Blanke, 2012; Makin et al., 2008).

Recently changes in the experienced direction of the first-person perspective have also been achieved using robotic stimulation while participants were in a supine position and viewed a virtual body that was filmed from an elevated position (Ionta et al., 2011; Pfeiffer et al., 2013). Despite identical visuotactile stimulation, half of the participants experienced looking upward toward the virtual body (Up group), and half experienced looking down on the virtual body (Down group), and these subjective perspectival changes were associated with consistent changes in self-location (Ionta et al., 2011; Pfeiffer et al., 2013). In addition, subjective reports of elevated self-location and sensations of flying, floating, rising, and lightness (which are common in out-of-body experiences) were frequent in the Down group and rare in the Up group. These data show that self-location depends on the experienced direction of the first-person perspective. It also shows that self-location versus self-identification are associated with different multisensory mechanisms, as self-identification was not found to depend on the first-person perspective.

Changes in self-location and the first-person perspective were reflected in bilateral TPJ activity (Ionta et al., 2011). TPJ activity peaked in the posterior superior temporal gyri, differed between synchronous and asynchronous stroking conditions, and depended on the experienced direction of the first-person perspective. More recent work was able to extend these findings and show that these subjective changes in the first-person perspective are associated with interindividual differences of visuovestibular integration (Pfeiffer et al., 2013) such as visual verticality judgments using the rod-and-frame test (Isableu, Ohlmann, Cremieux, & Amblard, 1997; Lopez, Lacour, Magnan, & Borel, 2006; Young, Oman, Watt, Money, & Lichtenberg, 1984). Based on these data, it can be concluded that participants from the Up group rely more strongly on vestibular cues from the physical body (indicating downward gravity directed toward the physical body) than on visual gravitational cues from the virtual body (indicating downward gravity directed away from the physical body), whereas participants from the Down group show the opposite pattern (relying more strongly on visual gravitational cues from the virtual body than on vestibular cues from the physical). The role of vestibular processing in the first-person perspective and bodily self-consciousness is supported by the proximity of the vestibular cortex to TPJ (Lopez & Blanke, 2011), related changes in first-person perspective during the so-called inversion illusion (Lackner, 1992; Lackner & DiZio, 2002), and frequent vestibular disturbances during OBEs. Accordingly, it has been argued that changes in the experienced direction of the first-person perspective are due to abnormal signal integration of otolithic vestibular cues and visual cues (Blanke, 2012; for a more extensive review of visuovestibular-somatosensory interactions, see Blanke, 2012; Lackner & DiZio, 2002).

Bodily self-consciousness

It has been argued elsewhere (Blanke & Metzinger, 2009) that the three aspects of bodily self-consciousness —self-identification, self-location, and first-person perspective—are the necessary constituents of the simplest form of self-consciousness, which arises when the brain encodes the origin of the first-person perspective from within a spatial frame of reference (i.e., self-location) that is associated with self-identification. The reviewed data reveal that self-identification depends on somatosensory and visual signals and likely involves bimodal visuotactile neurons that have also been described for arm-related manipulations (i.e., Ehrsson et al., 2004; Graziano, Cooke, & Taylor, 2000; Iriki, Tanaka, & Iwamura, 1996; Maravita & Iriki, 2004), whereas self-location and the first-person perspective depend on the additional integration of these bodily signals with vestibular cues, probably in trimodal visuotactile-vestibular neurons in posterior parietal and temporoparietal cortex (i.e., Bremmer, Klam, Duhamel, Ben Hamed, & Graf, 2002; Duhamel, Colby, & Goldberg, 1998; Grüsser, Pause, & Schreiter, 1990). These differences between self-identification versus self-location and first-person perspective are corroborated by neuroimaging and neurological data, showing that self-identification recruits primarily bilateral premotor and parietal regions, whereas self-location and the first-person perspective recruit posterior parietal–TPJ regions (including the insula; Ionta et al., 2014) with a right hemispheric predominance.

Although it is obviously not necessary to master concepts (like the first-person pronoun "I") or symbolic language to enjoy the form of self-consciousness described here, many recent authors have thought brain mechanisms of agency are necessary for bodily self-consciousness (Frith, 2005; Jeannerod, 2003; Pacherie, 2008). While the experience of being the author of one's actions and being able to selectively control a body are important for bodily self-consciousness, the reviewed data show how the passive, multisensory, and globalized experiences of self-identification, self-location, and first-person perspective are sufficient for a minimally conscious selfhood (Blanke & Metzinger, 2009): global motor control and agency (Kannape & Blanke, 2013; Kannape, Schwabe, Tadi, & Blanke, 2010) or cognitive self-reference based on language or memory (see below) are not necessary conditions.

Bodily self-consciousness and interoception

Are interoceptive bodily signals relevant for bodily self-consciousness? The large majority of reviewed studies on bodily self-consciousness focused on the manipulation of exteroceptive sources of information about the body (i.e., vision and touch). Yet recent behavioral, imaging, and neurological data also link the insular cortex and interoceptive processing to bodily self-consciousness, and evidence has been put forward that the brain's representations of internal bodily states (Critchley, Wiens, Rotshtein, Ohman, & Dolan, 2004) are equally or even more important for the self (Craig, 2009; Damasio, 1999). Although it is not yet known whether exteroceptive and interoceptive systems form a common or distinct system for bodily self-consciousness, recent evidence suggests this is likely to be the case (Aspell et al., 2013; Suzuki, Garfinkel, Critchley, & Seth, 2013; Tsakiris et al., 2011). Interoceptive and homeostatic processing (e.g., heart, lungs, intestinal tract, blood pressure) have been associated with the insular cortex (Craig, 2009) or the posterior medial parietal cortex (Damasio, 1999; Damasio & Meyer, 2009), and several recent studies have revealed that level of interoceptive awareness (as measured in a heartbeat-awareness task) modulates measures of bodily self-consciousness (Tajadura-Jimenez et al., 2013; Tsakiris et al., 2011). Stronger evidence for a common system for bodily self-consciousness has been provided by research based on experiments that investigated whether a conflict between an interoceptive signal (the heartbeat) and an exteroceptive (visual) signal modulates bodily self-consciousness (in a comparable fashion to the previously described visuo-tactile conflicts in the full-body illusion). Presenting cardio-visual illumination of the virtual body in the full-body illusion (Aspell et al., 2013) so that a flashing silhouette was either temporally synchronous or asynchronous with respect to the participant's heartbeats (figure 74.3A), changes in illusory self-identification and self-location were induced that were of comparable magnitude as in those using purely exteroceptive (visuo-tactile) conflicts (figure 74.3b; for related work, see Suzuki et al., 2013). It was further found that such cardio-visual conflict modulated tactile perception (in the cross-modal congruency task), and that this was reflected by somatosensory activity in parietal cortex (as measured by somatosensory evoked potentials).

These findings are compatible with proposals that both exteroceptive and interoceptive signals are important for bodily self-consciousness (see also Seth et al., 2012). Given that the reviewed data show exteroceptive and interoceptive signals are combined and that they are potent modulators of bodily self-consciousness, it may be argued that signals from inside and outside the human body form an integrated cortical system for bodily self-consciousness that includes the insula. Such insula involvement in bodily self-consciousness has been confirmed using functional connectivity analysis (Ionta et al., 2014) and neurological data in patients suffering from heautoscopy (Brugger, 2002; Heydrich & Blanke, 2013). Patients with heautoscopy have damage centering in the left posterior insula (Heydrich & Blanke, 2013) and suffer from abnormal self-identification and self-location. Affective and interoceptive symptoms are frequently reported by such patients, and it has been argued that abnormal self-identification and self-location in patients with heautoscopy is caused by a breakdown of self-other discrimination regarding affective somatosensory experience due to a disintegration of visuosomatosensory signals with interoceptive (and/or emotional) bodily signals. These findings highlight the role of interoceptive and emotional brain mechanisms for self-identification.

Bodily self-consciousness and visual consciousness

How does self-consciousness relate to the larger field of consciousness studies? Empirical research into brain mechanisms of consciousness has mainly focused on behavioral and neural differences between conscious and unconscious processing of stimuli, mostly in the visual system (i.e., Crick & Koch, 1990; Dehaene & Changeux, 2011). Can such consciousness arise without self-consciousness? This is an interesting topic for future research, and would require integrating studies of bodily self-consciousness with earlier research of consciousness that investigated conscious and unconscious

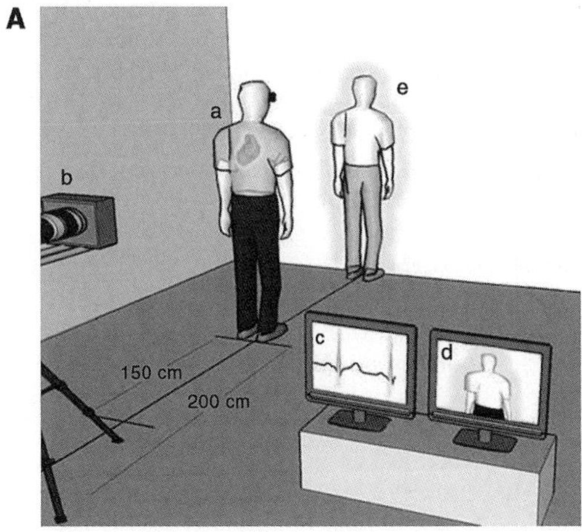

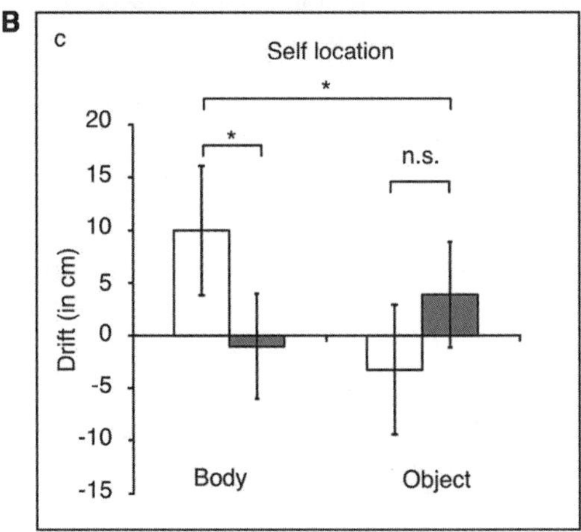

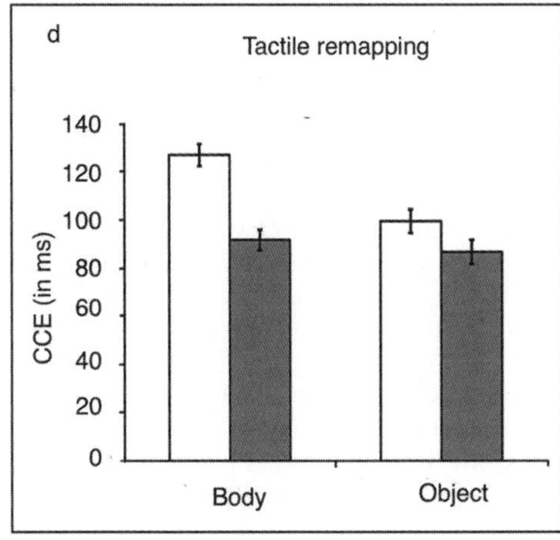

FIGURE 74.3 (A) Cardiovisual full-body illusion setup. Participants (a) stood with their backs facing a video camera (b) placed 200 cm behind them. An electrocardiogram was recorded and R-peaks were detected in real-time (c), triggering a flashing silhouette (d) outlining the participant's body (virtual body) (e). The video showing the virtual body was projected in real time onto a head-mounted display. It appeared visually that the virtual body was standing 200 cm in front of the participant. (Modified from Aspell et al., 2013.) (B) Self-location and visuotactile integration. Self-location was modulated by cardiovisual synchrony only in the body conditions (shown on the left; "Body"). A greater change of self-location toward the virtual body was found in the synchronous versus asynchronous body condition. No significant difference was found between the synchronous and asynchronous control conditions (shown on the right; "Object"). The cross-modal congruency effect (CCE, see text) was larger during cardiovisual synchrony and had greater magnitude when the vibration and the visual distractor were on the same side, reflecting associated changes in self-location (B, right).

target detection, binocular rivalry, bistable perception, continuous flash suppression, and other perceptions in well-defined psychophysical settings (i.e., Tsuchiya & Koch, 2005). Roy Salomon has taken first steps in this direction and was able to show that proprioceptive and vestibular signals as well as changes in self-identification and self-location (in the full-body illusion) modulate visual consciousness as quantified in a continuous flash suppression task (Salomon et al., 2013). These last experiments suggest that bodily self-consciousness may turn out to be an important component for consciousness generally (Damasio, 1999). As Gerald Edelman stated, "It is not enough to say that the mind (and consciousness) is embodied. You also have to say how" (1992). Bodily self-consciousness may provide this link, bringing the "I" of conscious perception with consciousness studies.

Conclusion

The "I" of conscious experience is one of the most astonishing features of the human mind. The reviewed investigations of self-identification, self-location, and first-person perspective have described some of the multisensory brain processes that give rise to bodily self-consciousness. As argued elsewhere (Blanke & Metzinger, 2009), these three aspects are the necessary constituents of the simplest form of self-consciousness, which arises when the brain encodes the origin of the first-person perspective from within a spatial frame of reference (i.e., self-location) associated with self-identification. Cognitive psychologists and neuroscientists have studied many different aspects of self-related processes (e.g., Arzy, Thut, Mohr, Michel, & Blanke, 2006; Christoff et al., 2011; Esslen, Metzler, Pascual-Marqui, & Jancke, 2008; Gillihan & Farah, 2005; Heatherton et al., 2006; Jeannerod, 2003; Knoblich, 2002; Legrand & Ruby, 2009; Macrae, Moran, Heatherton, Banfield, & Kelley, 2004; Northoff et al., 2006; Perrin et al., 2005; Platek et al., 2006; Vogeley & Fink, 2003). Mechanisms of bodily self-consciousness overlap with several of these self-related processes, and it will be an exciting endeavor to better understand how the brain mechanisms of bodily self-consciousness are linked to memory and language (Dennett, 1991; Gazzaniga, LeDoux, & Wilson, 1977) as well as other important higher-order ("narrative" and "extended") aspects of self-consciousness and personhood.

REFERENCES

Aikins, K. (Ed.). (2005). *Self and subjectivity*. Hoboken, NJ: Wiley-Blackwell.

Arzy, S., Thut, G., Mohr, C., Michel, C. M., & Blanke, O. (2006). Neural basis of embodiment: Distinct contributions of temporoparietal junction and extrastriate body area. *J Neurosci, 26,* 8074–8081.

Aspell, J. E., Heydrich, L., Herbelin, B., & Blanke, O. (2013). Turning body and self inside out. Cardio-visual illumination modulates bodily self consciousness and tactile perception. *Psychol Sci, 24,* 2445–2453.

Aspell, J. E., Lavanchy, T., Lenggenhager, B., & Blanke, O. (2010). Seeing the body modulates audiotactile integration. *Eur J Neurosci, 31,* 1868–1873.

Aspell, J. E., Lenggenhager, B., & Blanke, O. (2009). Keeping in touch with one's self: Multisensory mechanisms of self-consciousness. *PLoS ONE, 4,* e6488.

Astafiev, S. V., Stanley, C. M., Shulman, G. L., & Corbetta, M. (2004). Extrastriate body area in human occipital cortex responds to the performance of motor actions. *Nat Neurosci, 7,* 542–548.

Bermudez, J. L. (1998). *The paradox of self-consciousness*. Cambridge, MA: MIT Press.

Blanke, O. (2012). Multisensory brain mechanisms of bodily self-consciousness. *Nat Rev Neurosci, 13,* 556–571.

Blanke, O., Landis, T., Spinelli, L., & Seeck, M. (2004). Out-of-body experience and autoscopy of neurological origin. *Brain, 127,* 243–258.

Blanke, O., & Metzinger, T. (2009). Full-body illusions and minimal phenomenal selfhood. *Trends Cogn Sci, 13,* 7–13.

Blanke, O., Ortigue, S., Landis, T., & Seeck, M. (2002). Stimulating illusory own-body perceptions. *Nature, 419,* 269–270.

Bremmer, F., Klam, F., Duhamel, J. R., Ben Hamed, S., & Graf, W. (2002). Visual-vestibular interactive responses in the macaque ventral intraparietal area (VIP). *Eur J Neurosci, 16,* 1569–1586.

Bremmer, F., Schlack, A., Shah, N. J., Zafiris, O., Kubischik, M., Hoffmann, K.-P., … Fink, G. R. (2001). Polymodal motion processing in posterior parietal and premotor cortex: A human fMRI study strongly implies equivalencies between humans and monkeys. *Neuron, 29,* 287–296.

Brugger, P. (2002). Reflective mirrors: Perspective-taking in autoscopic phenomena. *Cogn Neuropsychiatry, 7,* 179–194.

Calvert, G. A., Campbell, R., & Brammer, M. J. (2000). Evidence from functional magnetic resonance imaging of crossmodal binding in the human heteromodal cortex. *Curr Biol, 10,* 649–657.

Christoff, K., Cosmelli, D., Legrand, D., & Thompson, E. (2011). Specifying the self for cognitive neuroscience. *Trends Cogn Sci, 15,* 104–112.

Craig, A. D. (2009). How do you feel—now? The anterior insula and human awareness. *Nat Rev Neurosci, 10,* 59–70.

Crick, F., & Koch, C. (1990). Some reflections on visual awareness. *Cold Spring Harb Symp Quant Biol, 55,* 953–962.

Critchley, H. D., Wiens, S., Rotshtein, P., Ohman, A., & Dolan, R. J. (2004). Neural systems supporting interoceptive awareness. *Nat Neurosci, 7,* 189–195.

Damasio, A. R. (1999). *The feeling of what happens: Body and emotion in the making of consciousness.* San Diego, CA: Harcourt Brace.

Damasio, A., & Meyer, D. E. (2009). Consciousness: An overview of the phenomenon and of its possible neural basis. In S. Laureys & G. Tononi (Eds.), *The neurology of consciousness* (pp. 3–14). London: Elsevier.

de Ridder, D., van Laere, K., Dupont, P., Menovsky, T., & van de Heyning, P. (2007). Visualizing out-of-body experience in the brain. *N Engl J Med, 357,* 1829–1833.

de Vignemont, F. (2011). Embodiment, ownership and disownership. *Conscious Cogn, 20,* 82–93.

Dehaene, S., & Changeux, J. P. (2011). Experimental and theoretical approaches to conscious processing. *Neuron, 70,* 200–227.

Dennett, D. C. (1991). *Consciousness explained.* New York, NY: Penguin.

Devinsky, O., Feldmann, E., Burrowes, K., & Bromfield, E. (1989). Autoscopic phenomena with seizures. *Arch Neurol, 46,* 1080–1088.

Downing, P. E., Jiang, Y., Shuman, M., & Kanwisher, N. (2001). A cortical area selective for visual processing of the human body. *Science, 293,* 2470–2473.

Duhamel, J. R., Colby, C. L., & Goldberg, M. E. (1998). Ventral intraparietal area of the macaque: Congruent visual and somatic response properties. *J Neurophysiol, 79,* 126–136.

Edelman, G. (1992). *Bright air, brilliant fire.* New York, NY: BasicBooks.

Ehrsson, H. H. (2007). The experimental induction of out-of-body experiences. *Science, 317,* 1048.

Ehrsson, H. H., Spence, C., & Passingham, R. E. (2004). That's my hand! Activity in premotor cortex reflects feeling of ownership of a limb. *Science, 305,* 875–877.

Esslen, M., Metzler, S., Pascual-Marqui, R., & Jancke, L. (2008). Pre-reflective and reflective self-reference: A spatiotemporal EEG analysis. *NeuroImage, 42,* 437–449.

Frith, C. (2005). The self in action: Lessons from delusions of control. *Conscious Cogn, 14,* 752–770.

Gallagher, S., & Shear, J. (1999). *Models of the self.* Thorverton, UK: Imprint Academic.

Gazzaniga, M. S., LeDoux, J. E., & Wilson, D. H. (1977). Language, praxis, and the right hemisphere: Clues to some mechanisms of consciousness. *Neurology, 27,* 1144–1147.

Gillihan, S. J., & Farah, M. J. (2005). Is self special? A critical review of evidence from experimental psychology and cognitive neuroscience. *Psychol Bull, 131,* 76–97.

Graziano, M. S., Cooke, D. F., & Taylor, C. S. (2000). Coding the location of the arm by sight. *Science, 290,* 1782–1786.

Grossman, E. D., & Blake, R. (2002). Brain areas active during visual perception of biological motion. *Neuron, 35,* 1167–1175.

Grüsser, O. J., Pause, M., & Schreiter, U. (1990). Vestibular neurones in the parieto-insular cortex of monkeys (*Macaca fascicularis*): Visual and neck receptor responses. *J Physiol, 430,* 559–583.

Hänsel, A., Lenggenhager, B., von Kanel, R., Curatolo, M., & Blanke, O. (2011). Seeing and identifying with a virtual body decreases pain perception. *Eur J Pain, 15,* 874–879.

Heatherton, T. F., Wyland, C. L., Macrae, C. N., Demos, K. E., Denny, B. T., & Kelley, W. M. (2006). Medial prefrontal activity differentiates self from close others. *Soc Cogn Affect Neurosci, 1,* 18–25.

Heydrich, L., & Blanke, O. (2013). Distinct illusory own-body perceptions caused by damage to posterior insula and extrastriate cortex. *Brain, 136,* 790–803.

Igarashi, Y., Kimura, Y., Spence, C., & Ichihara, S. (2008). The selective effect of the image of a hand on visuotactile interactions as assessed by performance on the crossmodal congruency task. *Exp Brain Res, 184,* 31–38.

Ionta, S., Heydrich, L., Lenggenhager, B., Mouthon, M., Fornari, E., Chapuis, D., ... Blanke, O. (2011). Multisensory mechanisms in temporo-parietal cortex support self-location and first-person perspective. *Neuron, 70,* 363–374.

Ionta, S., Martuzzi, R., Salomon R., & Blanke, O. (2014). The brain network reflecting bodily self-consciousness: A functional connectivity study. *Soc Cogn Affect Neurosci.* Advance online publication. doi:10.1093/scan/nst185

Iriki, A., Tanaka, M., & Iwamura, Y. (1996). Coding of modified body schema during tool use by macaque postcentral neurones. *NeuroReport, 7,* 2325–2330.

Isableu, B., Ohlmann, T., Cremieux, J., & Amblard, B. (1997). Selection of spatial frame of reference and postural control variability. *Exp Brain Res, 114,* 584–589

Jeannerod, M. (2003). The mechanism of self-recognition in humans. *Behav Brain Res, 142,* 1–15.

Kannape, O. A., & Blanke, O. (2013). The self in motion: Sensorimotor and cognitive mechanisms in gait agency. *J Neurophysiol, 110,* 1837–1847.

Kannape, O. A., Schwabe, L., Tadi, T., & Blanke, O. (2010). The limits of agency in walking humans. *Neuropsychologia, 48,* 1628–1636.

Knoblich, G. (2002). Self-recognition: Body and action. *Trends Cogn Sci, 6,* 447–449.

Lackner, J. R. (1992). Spatial orientation in weightless environments. *Perception, 21,* 803–812.

Lackner, J. R., & DiZio, P. (2002). Somatosensory and proprioceptive contributions to body orientation, sensory localization, and self-calibration. *Adv Exp Med Biol, 508,* 69–78.

Legrand, D. (2007). Pre-reflective self-as-subject from experiential and empirical perspectives. *Conscious Cogn, 16,* 583–599.

Legrand, D., & Ruby, P. (2009). What is self-specific? Theoretical investigation and critical review of neuroimaging results. *Psychol Rev, 116,* 252–282.

Lenggenhager, B., Halje, P., & Blanke, O. (2011). Alpha band oscillations correlate with illusory self-location induced by virtual reality. *Eur J Neurosci, 33,* 1935–1943.

Lenggenhager, B., Mouthon, M., & Blanke, O. (2009). Spatial aspects of bodily self-consciousness. *Conscious Cogn, 18,* 110–117.

Lenggenhager, B., Tadi, T., Metzinger, T., & Blanke, O. (2007). Video ergo sum: Manipulating bodily self-consciousness. *Science, 317,* 1096–1099.

Lopez, C., & Blanke, O. (2011). The thalamocortical vestibular system in animals and humans. *Brain Res Rev, 67,* 119–146.

Lopez, C., Halje, P., & Blanke, O. (2008). Body ownership and embodiment: Vestibular and multisensory mechanisms. *Neurophysiol Clin, 38,* 149–161.

Lopez, C., Lacour, M., Magnan, J., & Borel, L. (2006). Visual field dependence-independence before and after unilateral vestibular loss. *NeuroReport, 17,* 797–803.

Macrae, C. N., Moran, J. M., Heatherton, T. F., Banfield, J. F., & Kelley, W. M. (2004). Medial prefrontal activity predicts memory for self. *Cereb Cortex, 14,* 647–654.

Maillard, L., Vignal, J. P., Anxionnat, R., Taillandier, L., & Vespignani, H. (2004). Semiologic value of ictal autoscopy. *Epilepsia, 45,* 391–394.

Makin, T. R., Holmes, N. P., & Ehrsson, H. H. (2008). On the other hand: Dummy hands and peripersonal space. *Behav Brain Res, 191,* 1–10.

Maravita, A., & Iriki, A. (2004). Tools for the body (schema). *Trends Cogn Sci, 8,* 79–86.

Muldoon, S. J., & Carrington, H. (1929). *The projection of the astral body.* London: Rider.

Northoff, G., Heinzel, A., de Greck, M., Bermpohl, F., Dobrowolny, H., & Panksepp, J. (2006). Self-referential processing in our brain—a meta-analysis of imaging studies on the self. *NeuroImage, 31,* 440–457.

Oakes, T. R., Pizzagalli, D. A., Hendrick, A. M., Horras, K. A., Larson, C. L., Abercrombie, H. C., ... Davidson, R. J. (2004). Functional coupling of simultaneous electrical and metabolic activity in the human brain. *Hum Brain Mapp, 21,* 257–270.

Pacherie, E. (2008). The phenomenology of action: A conceptual framework. *Cognition, 107,* 179–217.

Pavani, F., Spence, C., & Driver, J. (2000). Visual capture of touch: Out-of-the-body experiences with rubber gloves. *Psychol Sci, 11,* 353–359.

Perrin, F., Maquet, P., Peigneux, P., Ruby, P., Degueldre, C., Balteau, E., ... Laureys, S. (2005). Neural mechanisms involved in the detection of our first name: A combined ERPs and PET study. *Neuropsychologia, 43,* 12–19.

Petkova, V. I., Björnsdotter, M., Gentile, G., Jonsson, T., Li, T. Q., & Ehrsson, H. H. (2011). From part- to whole-body ownership in the multisensory brain. *Curr Biol, 21,* 1118–1122.

Petkova, V. I., & Ehrsson, H. H. (2008). If I were you: Perceptual illusion of body swapping. *PLoS ONE, 3,* e3832.

Pfeiffer, C., Lopez, C., Schmutz, V., Duenas, J. A., Martuzzi, R., & Blanke, O. (2013). Multisensory origin of the subjective first-person perspective: Visual, tactile, and vestibular mechanisms. *PLoS ONE, 8,* e61751.

Pineda, J. A. (2005). The functional significance of mu rhythms: Translating "seeing" and "hearing" into "doing." *Brain Res Brain Res Rev, 50,* 57–68.

Platek, S. M., Loughead, J. W., Gur, R. C., Busch, S., Ruparel, K., Phend, N., ... Langleben, D. D. (2006). Neural substrates for functionally discriminating self-face from personally familiar faces. *Hum Brain Mapp, 27,* 91–98.

Salomon, R., Lim, M., Herbelin, B., Hesselmann, G., & Blanke, O. (2013). Posing for awareness: Proprioception modulates access to visual consciousness in a continuous flash suppression task. *J Vis, 13,* 2.

Serino, A., Canzoneroni, E., & Avenanti, A. (2011). Fronto-parietal areas necessary for a multisensory representation of peripersonal space in humans: An rTMS study. *J Cogn Neurosci, 23,* 2956–2967.

Seth, A. K., Suzuki, K., & Critchley, H. D. (2012). An interoceptive predictive coding model of conscious presence. *Front Psychol, 2,* 395.

Shore, D. I., Barnes, M. E., & Spence, C. (2006). Temporal aspects of the visuotactile congruency effect. *Neurosci Lett, 392,* 96–100.

Spence, C., Pavani, F., & Driver, J. (2004). Spatial constraints on visual-tactile cross-modal distractor congruency effects. *Cogn Aff Behav Neurosci, 4,* 148–169.

Suzuki, K., Garfinkel, S. G., Critchley, H. D., & Seth, A. K. (2013). Multisensory integration across interoceptive and exteroceptive domains modulates self-experience in the rubber-hand illusion. *Neuropsychologia, 51*(13), 2909–2917.

Tajadura-Jimenez, A., & Tsakiris, M. (2013). Balancing the "inner" and the "outer" self: Interoceptive sensitivity modulates self-other boundaries. *J Exp Psychol Gen.* Advance online publication.

Tsakiris, M. (2010). My body in the brain: A neurocognitive model of body-ownership. *Neuropsychologia, 48,* 703–712.

Tsakiris, M., Tajadura-Jimenez, A., & Costantini, M. (2011). Just a heartbeat away from one's body: Interoceptive sensitivity predicts malleability of body-representations. *Proc Biol Sci, 278,* 2470–2476.

Tsuchiya, N., & Koch, C. (2005). Continuous flash suppression reduces negative afterimages. *Nat Neurosci, 8,* 1096–1101.

Vogeley, K., & Fink, G. R. (2003). Neural correlates of the first-person-perspective. *Trends Cogn Sci, 7,* 38–42.

Young, L. R., Oman, C. M., Watt, D. G., Money, K. E., & Lichtenberg, B. K. (1984). Spatial orientation in weightlessness and readaptation to earth's gravity. *Science, 225,* 205–208.

75 Intention and Agency

PATRICK HAGGARD

ABSTRACT Cognitive neuroscience makes an important distinction between actions that are triggered by a specific stimulus in the external world (i.e., reactions) and actions that are relatively independent of any individual stimulus. These internally-generated actions instead result in a combination of motivation to act and an internal decision about what action to make and when to make it. Some internally-generated actions are accompanied by a characteristic conscious experience of "willing" or intending to move, which is clearly linked to the ability to initiate the action. This chapter focuses on the brain mechanisms and computations underlying intentional action. The medial frontal cortex plays a key role in generating intentional actions and, in conjunction with the parietal cortex, in the subjective experience of intention. The second part of the chapter focuses on the sense of agency—the experience that our actions aim at external outcomes and are the causes of those outcomes. The same frontal brain circuits that initiate intentional action are also responsible for linking actions to outcomes. Some evidence suggests the action-outcome link is a prospective prediction of outcomes within action-related circuits, rather than a retrospective causal inference within outcome-related circuits. The chapter ends with a brief discussion of the implications of cognitive neuroscience for social and moral responsibility.

Intentions are mental states that represent one's future actions. However, not all actions are intentional. The key tasks of the cognitive neuroscience of human intentional action are to identify the specific features of intentional actions as opposed to other kinds of actions, to investigate the specific mechanisms of intentional action in the brain, and to consider the implications of these mechanisms for human cognition in general.

The cognitive neurosciences' interest in intentional action emerges from two sources: the classic philosophical interest in free will and the increasing mechanistic understanding of the motor circuits in the brain. Healthy human adults generally have the experience that they can choose their actions for themselves on at least some occasions. This makes them responsible for an action and its outcomes. However, the neurosciences view actions as events in the motor system, caused by neural mechanisms in cortical and subcortical cognitive and motor areas. The first view is person-centered and subjective, while the second is system-centered and mechanistic. Reconciling these two very different views of human action represents an important frontier in our scientific understanding of ourselves. Moreover, scientific investigation of intention and agency may be relevant for the diagnosis and treatment of the many neuropsychiatric disorders that affect capacity for voluntary action. Third, and perhaps most importantly, neuroscientific findings may have important social implications, because human society depends on a concept of individual autonomy and responsibility for action.

What is intentional action?

Ludwig Wittgenstein famously asked, "What is left over if I subtract the fact that my arm goes up from the fact that I raise my arm?"(Wittgenstein, 1953, §621). We all know that intentionally lifting one's arm and having one's arm lifted feel quite different, but cognitive psychology has nevertheless struggled to provide a good definition of intentional actions, where there is a direct and immediate correspondence between stimulus and action. Intentional actions are not driven by any obvious or immediate external stimulus. For example, my leg may kick forward because of a sharp tap on the tendon just below the kneecap (a reflex), or because I just decide to kick my leg forward. In essence, this approach defines intentional actions as internally-generated, rather than externally-triggered (Passingham, Bengtsson, & Lau, 2010). This definition has dominated recent thinking about intentional action, and has both advantages and disadvantages. The main disadvantage is that the definition says what intentional actions are not, but does not say what they are in a positive sense. In particular, the causes of intentional action remain largely undefined. On the other hand, this definition gives a clear operational definition of intentional action that can be used in experimental designs. Indeed, many studies have compared the brain mechanisms for internally-generated versus externally-triggered actions. These studies are reviewed in the next section.

Several studies have compared the brain processes for internally-generated versus externally-triggered actions. The design of these studies aims to make the physical movement, and any physical stimulation, balanced across experimental conditions. Differences between conditions can then be linked to the different causes of action. For example, an early positron emission tomography study (Deiber et al., 1991) asked participants to move a joystick at regular intervals. The direction in which to move the joystick was either instructed by an external signal or freely chosen by the participants. The results showed increased neural activity in the supplementary motor cortex when actions were selected by internal cues, compared to external cues. A subsequent study (Deiber, Ibañez, Sadato, & Hallett, 1996) replicated this difference even when the external cues allowed the same of degree of motor preparation as free selection of action, and showed that the centers specifically involved with internal generation of action were located more anteriorly within the supplementary motor cortex, in the region subsequently labelled the presupplementary motor area (preSMA).

More recent functional MRI studies have contrasted free selection with external guidance for several different dimensions of action decision. In the studies of Krieghoff, Brass, Prinz, and Waszak (2009) and Mueller, Brass, Waszak, and Prinz (2007), participants selected which of two actions to make on the basis of external cues or their own free choice. In another condition, the action itself was fixed, but participants could decide when to make it, again based on either external cues or free choice. Freely selecting which action to make, as opposed to being instructed which action to make, activated the rostral cingulate zone (Mueller et al., 2007). A subsequent study replicated this finding, and also showed that freely selecting when to act activated a slightly more posterior area on the surface of the superior frontal gyrus (Krieghoff et al., 2009).

The above findings fit with a general account of a fundamental gradient in the specialization of frontal cortex, between internal guidance of action on the medial aspect and external triggering on the lateral aspect. Medial and lateral frontal regions have different connectivity: the lateral frontal cortex receives from the inferior parietal regions that form the extension of the dorsal visual stream, while the medial frontal regions receive from limbic and superior parietal regions (Averbeck, Battaglia-Mayer, Guglielmo, & Caminiti, 2009).

Perhaps the strongest evidence for a concept of internally guided action comes from lesion studies. Monkeys learned to perform an arbitrary movement (raising the arm) to receive food. When the medial part of the premotor cortex was surgically ablated, the frequency of this action was sharply reduced, although the animals could make the action quite normally in response to a tone (Thaler, Chen, Nixon, Stern, & Passingham, 1995). Recordings from single units in the medial and lateral premotor cortex confirmed this preference. Monkeys learned a sequence of three successive movements. They then performed these movements either on the basis of internally-stored information or in a condition where an external visual cue specified each movement in turn. The majority of cells in the medial frontal areas, notably the supplementary motor area (SMA), showed a preference for internally-generated movement, while the majority of cells located more laterally in the premotor cortex showed a preference for externally-cued movement (Halsband, Matsuzaka, & Tanji, 1994).

Intentional action as internal generation of motor information

Internally-guided and externally-triggered actions are opposite ends of a cognitive continuum. Most everyday actions probably lie somewhere in between, and involve a mixture of one's own decisions and responses to the external environment. Many readers of this book will have had the experience of participating in a psychology experiment. You typically decide for yourself to participate (signing the consent form being the external marker of this decision), but then you may respond to the stimuli the experimenter shows you. It is hard to say exactly which parts of the activity are intentional, as opposed to stimulus-triggered, and what specific factor makes them intentional. In fact, intentional actions are characterized by the presence of several different factors, none of which may be necessary, but which may be jointly sufficient. We describe some of these below.

First, intentional actions result from internal *decisions* about what to do, when to do it, and even whether to do it at all. Mike Shadlen has proposed the term "freedom from immediacy" for this core property of human and some nonhuman action selection (Shadlen & Gold, 2004). Clearly, actions can be more or less constrained by the environment, so the degree of internally-generated decision making must be graded.

Second, intentional actions normally have a distinctive *motivation*: they are goal-directed or aim at producing a particular outcome. This indicates a tight linkage between intentional action and neural reward systems. Humans and animals internally generate actions that produce appropriate reinforcement, even when there is no external stimulus cuing the action. Somatic states such as hunger and thirst may form the basis of intentional action (Damasio, 2000). However, the human experimental literature has struggled to include this motivational aspect in experimental designs. Given a transparent choice between more rewarding and less rewarding actions, or between more and less moral outcomes, people generally choose the most desirable action (Moretto, Walsh, & Haggard, 2011). For example, if I offer someone a $10 bill or a $20 bill, they will probably reach for the $20 bill—but this seems more like a direct response to the $20 stimulus than an internally-generated action. Because of this ambiguity, many human experiments used arbitrary, valueless action decisions. Such experiments clearly cannot capture how everyday intentional actions follow from goals and reasons, but they may be able to study the process of generating intentional action.

A third critical element of intention is the *timing of initiation* of action itself. The combination of a motivation to act and a decision about which action to make is not always sufficient to trigger action. For example, action may be physically prevented by external constraint, or not appropriate given current circumstances, competing actions, or competing goals. For these reasons, the brain needs a mechanism to maintain intentions and release them only at the appropriate time: another form of "freedom from immediacy." The temporal aspect of intention has received more experimental attention than any other, and forms the topic of the next section.

The combination of decision, motivation, and timing of initiation gives intentional actions a level of flexibility and complexity that stimulus-driven actions lack. The traditional definition of intentional action can sometimes be misinterpreted as implying an action with no apparent cause—something that is clearly nonsensical. Perhaps it is better to think of internally generated actions as responses where a combination of several factors jointly cause action (Schüür & Haggard, 2011).

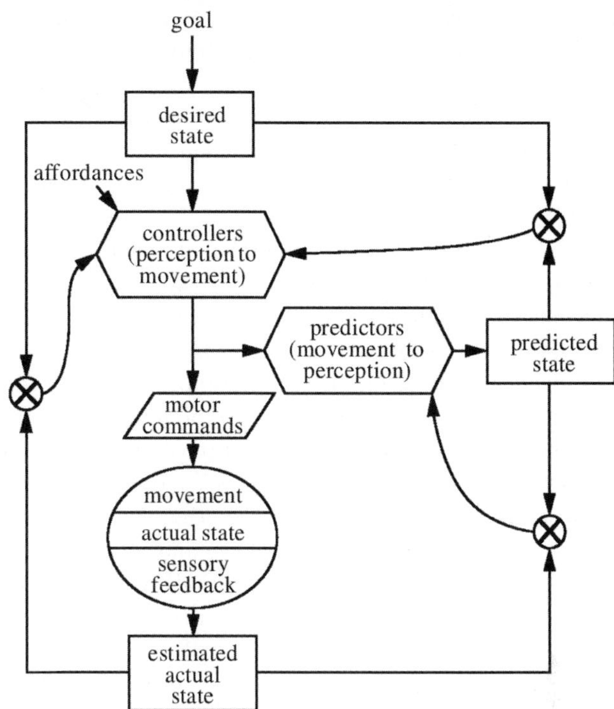

FIGURE 75.1 Computational motor control model for goal-directed action. (Reprinted with permission from Frith et al., 2000. Copyright © 2000, The Royal Society.)

Mental chronometry of intentions

Mental chronometry has been a rich line of investigation in many areas of psychology, and no more so than in intentional action. Recent computational models (see figure 75.1) provide a useful framework for considering intentions and time.

In these models (see chapter 39), actions begin with specification of a goal. A "planner," or inverse model, converts the goal into a set of motor commands to achieve the goal. These commands are sent from the brain to the spinal cord and the muscles, causing physical movement and finally bringing about the goal state. A copy of the motor command is sent to an internal forward model, which predicts the effects of the motor command. Therefore, the system potentially has access to several different kinds of information about a current action, each relating to a different time: a description of the future or goal state, an estimate of the current state, and a delayed description based on feedback from the body and the external world.

This model of intentional action begins with goals, but the concept of action goal has been viewed in two quite different ways within cognitive neuroscience. In cognitive psychology, intentional action goals are often considered as prospective memories to perform a desired action at a later, appropriate time. In sensorimotor control, the goal is a state of the body, such as getting the hand to a particular target position in space. The main difference between the two approaches is how close the person is to the time of action. Prospective memory involves long-term or prior intentions: other events and other actions may occur before the intended action. Sensorimotor control models, in contrast, begin with the immediate current intention or goal and are assumed to operate over the short term. They therefore deal with intentions-in-action rather than long-range intentions (Searle, 1983). An external triggering event, or an internal representation of elapsed time, is required to retrieve a relevant prior intention from long-term memory and transform it to an intention-in-action: this process marks the trigger or decision to make the action *now*. This progression toward triggering action is associated with a gradient across the prefrontal cortex: more anterior regions deal with more abstract, longer-range intentions, while more posterior regions deal with specific action details (Koechlin & Summerfield, 2007; Pacherie, 2008).

Neuroimaging studies identify the lateral part of Brodmann's area (BA) 10 in the prefrontal cortex as a key area for *maintaining* intentions (Burgess, Gonen-Yaacovi, & Volle, 2011). However, fewer studies have investigated how a stored intention is transformed into an action. One recent study used multivariate pattern analysis of fMRI data to identify the regions of the brain from which it was possible to decode which of two tasks a person would perform, and also when they would switch to the new task (Momennejad & Haynes, 2012). Two distinct regions of the medial prefrontal cortex carried information relevant to timing of intention. The lateral prefrontal cortex in both hemispheres *maintained* information about the forthcoming switch. At the time of retrieval, and at switching to the new task, information about timing was present more medially in BA10 as well as bilaterally in the SMA. While this study cannot causally identify the trigger signal that means "act now," it confirms a lateral-to-medial as well as an anterior-posterior gradient in prefrontal cortex in transforming long-range intentions into current actions.

Conscious intention and brain activity

A key question in cognitive neuroscience of intentional action has been the relation between conscious intention and the brain processes that trigger action. The processes of action selection, motivation, and temporal flexibility that characterize intentional action are all associated with specific conscious experiences, of choosing, of urge, and of commitment to act. I have the experience that "I" am the cause of my own actions, while the conscious experiences associated with reflex actions and with habitual actions are quite different. In reflex action, we are generally aware of the triggering stimulus and are sometimes surprised by our own motor responses. In habitual or routine actions such as walking or typing, we experience a background buzz of ongoing activity, without clear awareness of each individual motor command.

In the last decades, several neuroscientists have investigated the neural basis of the characteristic conscious experiences accompanying intentional action. Sometimes this work has been used to consider metaphysical questions, such as whether humans have free will or not. Here we limit ourselves to a "natural history" approach to intention, asking what conscious experiences accompany intentional action and what mechanisms produce those experiences. Whether this data constitutes evidence of "free will" is a philosophical rather than a neuroscientific point, and the subject of many excellent books (e.g., Kane, 2005).

Inquiry in this field began with the famous "Libet experiment" (Libet, Gleason, Wright, & Pearl, 1983). Libet asked participants to make a simple voluntary action (a movement of the wrist) at a time of their own choosing. EEG was recorded from the scalp throughout the experiment. In addition, participants watched a spot that rotated continuously on a clock face. After each action, participants indicated where the spot had been when they first "felt the urge to move." Libet's finding was simple, but striking (figure 75.2). People experienced the urge to move on average 206 msec before the physical onset of muscle contraction. However, the preparatory activity in the brain, or "readiness potential" (Kornhuber & Deecke, 1965) was present as much as 1 sec prior to this point. This temporal sequence, in which brain activity comes first and conscious awareness of impending movement comes later, seems to rule out the Cartesian dualist concept of conscious free will, according to which the conscious controls movements of the body via the intermediate mechanisms of the brain. Libet concluded that voluntary acts are initiated by unconscious brain mechanisms, rather than by conscious intention.

This result has attracted considerable attention in philosophy, and in the public imagination. The experimental method suffers from several drawbacks. For example, simple voluntary actions often do not produce a single, discrete, vivid experience of "urge," or intention, or decision. Estimating the time of conscious intention is more difficult than estimating the time of an auditory onset or visual onset. Some have questioned whether such judgments make sense at all. Nevertheless, the basic result seems to replicate: people can report an experience a few hundred milliseconds prior to voluntary action. It is less clear what gives rise to this experience, and what the content of the experience is.

From a neuroscientific point of view, the Libet experiment can seem almost trivial: what could conscious intention be other than a product of brain activity?

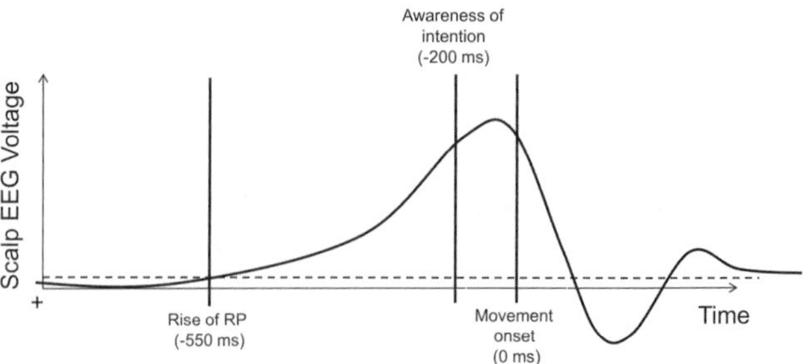

Figure 75.2 The Libet experiment. The buildup of a "readiness potential" preceding voluntary action occurs several hundred milliseconds before action onset. In contrast, the subjective experience of intention is reported to occur around 200 msec before action onset. Libet argued from the temporal order of these events that conscious intention could not be the trigger for action.

However, since numerous brain processes contribute to preparation and control of actions, an important scientific question is to identify those that underlie the experience of intention. Fried and colleagues (Fried, Mukamel, & Kreiman, 2011) recorded from 1,019 individual neurons in the brains of patients undergoing preoperative monitoring before epilepsy surgery, while the patients performed the Libet experiment. They found several neurons in the medial frontal cortex that progressively modulated their firing rate before the patients' average reported time of conscious intention. Interestingly, these neurons could either increase or decrease their firing rates as the action onset approached, suggesting that medial frontal areas may initiate action by a balance of excitatory and inhibitory drives onto motor-execution areas. We shall return to this point later. They also showed that they could accurately predict the time of the conscious intention to move from a population of a few hundred such medial frontal neurons. Neurons in other brain areas, such as the temporal lobe, showed much less association with conscious intention.

"Negative volition" and inhibition of action

An interesting recent debate has focused on decisions to inhibit intentional actions. Behavioral sciences have often struggled to study inhibition of action, largely because it produces no measurable behavioral output. As a result, the contribution of inhibitory processes to intentional action has generally been underestimated. Nevertheless, classical neuropsychological studies showed that medial frontal lobe damage could cause a form of excessive, compulsive action, in which the patient's actions would be triggered automatically by objects that happened to be present in the environment, even when no particular reason or desire to perform the action could be detected. These were termed utilization behaviors (Lhermitte, 1983). This finding suggests that one normal function of the healthy medial frontal cortex may be to inhibit those actions that we currently do not wish to make, and not simply to initiate those actions we do wish to make. A dramatic example comes from the neurological syndrome of anarchic hand. This unilateral form of utilization behavior sometimes follows a medial frontal and callosal lesion. Della Sala and colleagues (Della Sala, Marchetti, & Spinnler, 1991) report the case of a patient whose anarchic right hand reached out involuntarily to grab a hot beverage immediately after the patient announced that she had decided to wait for the drink to cool before taking it. This compelling example suggests that the maintenance, and then eventual release, of inhibition may be part of the normal regulatory process that allows us to make an action only when it is appropriate, and not at other times when it would be inappropriate. Recent work in healthy humans (van den Wildenberg et al., 2010) and monkeys (Wardak, 2011) confirms that deciding *whether* to act and deciding *when* to act are closely linked. Even in a simple reaction task, cognitive motor areas maintain a tonic inhibitory influence on the primary motor cortex until just before the time of the expected "go" signal (Duque et al., 2010).

Intentional inhibition has sometimes been viewed dualistically, as a form of "conscious veto" underlying self-control (Libet, 2005). However, the neuropsychological cases described above show that a specific frontal brain mechanism is responsible for inhibiting actions that might, in principle, be executed. Neuroimaging studies of inhibition broadly agree with this view. The classic experimental paradigms for studying inhibition generally involve an external *stop* or *no go* signal—yet in everyday life, healthy adults are expected to inhibit inappropriate actions endogenously, without any explicit instruction (Aron, 2011; Filevich, Kühn, & Haggard, 2012). Therefore, intentional paradigms such as the Libet experiment have also been used to investigate inhibition (Brass & Haggard, 2007). Participants were asked to prepare voluntary key-press actions, but then to cancel their action at the last possible moment on a freely chosen subset of trials. These intentional inhibition trials present an unusual experimental challenge, because the input is not directly controlled by the experimenter's instructions, nor is there any behavioral output to measure! Crucially, however, participants indicated the time at which they experienced willing the action using a rotating clock hand, even if the action was subsequently canceled. This subjective marker allowed the authors to investigate brain activity time-locked to intending actions and then inhibiting them. A region of BA 9 in the dorsomedial prefrontal cortex was found to be activated in inhibition trials, but deactivated in action trials. A later study, using a different experimental paradigm, additionally showed that this area had a strong effective connectivity with the preSMA during inhibition trials, relative to action trials (Kühn, Haggard, & Brass, 2009).

Based on this evidence, it was suggested that this dorsomedial frontal area may express the intention to inhibit actions, and may do so by exerting control over action-preparation circuits elsewhere in the frontal cortex. Interestingly, a similar prefrontal region, with similar connectivity to preSMA, was activated when participants intentionally decided to resist feeling the emotion suggested by unpleasant visual stimuli (Kühn, Haggard, & Brass, 2013). This region may therefore form an important part of a general brain circuit underlying inhibitory self-control. Consistent with this view, hypoactivity of the medial prefrontal cortex during inhibition tasks has regularly been reported in studies of ADHD (Rubia et al., 1999; Smith et al., 2006).

Sense of agency

The sense of agency refers to the feeling that one controls one's own actions and, through them, events in the external world. Sense of agency thus refers to the experience and mental representation of the relation between one's own intentional actions and their external sensory consequences. It is difficult to imagine a more fundamental distinctive feature of human mental life: all human endeavors, technologies, and transformations of our environment are ultimately based on being aware of the consequences of our actions. Moreover, sense of agency plays a crucial social role: agents can only be held responsible for the consequences of their actions if there is general agreement on who performed the action. For example, many legal systems allow a defense based on reduced or absent sense of agency.

Animals can learn and perform instrumental or goal-directed actions very readily. A rat that presses a lever for food perhaps experiences a kind of agency. However, the range and sophistication of human instrumental action is clearly much wider. Agency is best treated as a psychophysical problem: how do people represent and perceive the relation between intentional actions and outcomes? Like all psychophysical investigations, we begin with physical reality: these are the *facts* of agency, whether the agent performed the action or not. Next, we can consider the agent's awareness deriving from these facts. The term "sense of agency" refers to the normal awareness of initiating one's voluntary actions, and thus controlling their immediate outcomes. For many habitual actions, the sense of agency is thin, but it remains present as a background "buzz," accompanying normal mental life (Synofzik, Vosgerau, & Newen, 2008). Sense of agency moves to the foreground of consciousness during significant action decisions or when action outcomes are important.

Most scientific studies have not tackled the conscious experience of agency directly, but have asked participants to make binary *judgments* about whether they did or did not cause a particular action outcome. These studies have identified ambiguous situations in which judgments of agency may be incorrect, and have also identified the physical clues used for making such judgments. Farrer and colleagues (2008) instructed participants to move a set of wooden pegs into holes while in an fMRI scanner. Participants viewed a video image of their own action with a variable delay. In a first condition, participants simply detected whether the image was delayed or not, while in a second condition, participants judged whether the video they saw showed their own action or not—that is, they made judgments of agency. The behavioral results from the first condition showed that delays in visual feedback became detectable only if they lasted 200–300 msec or more. The results from the second condition showed that this time window produced an ambiguous sense of agency, with participants sometimes judging the observed action as their own and sometimes not. Comparing positive and negative judgments of agency for these intermediate delays did not identify any brain area coding for positive judgments of agency. However, the angular gyrus showed a stronger activation for trials where participants denied agency, compared to trials where they accepted it. Interestingly, neuroimaging results from the first, delay-detection condition also showed that angular gyrus activation increased with the video delay. This study confirms and extends previous reports of angular gyrus involvement in agency jugement (Sirigu, Daprati, Pradat-Diehl, Franck, & Jeannerod, 1999). It further suggests that the sense of being in control of one's actions is a default mode of the brain, not specifically coded as an individual state. Second, timing provides an important cue for the function of this circuit.

The computational motor control model (figure 75.1) has been used to explain the sense of agency. If the comparison between the state of the limb estimated by an internal model matches that reported by sensory signals, the current sensory input is attributed to one's own action. This causes a sense of agency over the current sensory input. In the model, as in the human neuroimaging literature, agency is not defined by any positive signal but only by an absence of errors. Moreover, the model insists that sense of agency is necessarily retrospective: the brain waits for delayed sensory feedback before the computations that lead to sense of agency can begin. This link between feedback delay and agency recalls the pattern of results of the neuroimaging results described above.

The same model has been used to explain why the consequences of self-produced action are often not perceived: conscious awareness is associated not with actual sensory feedback but with the element of sensory input that is not predicted by one's own motor command (Blakemore, Wolpert, & Frith, 2000; Gentsch & Schütz-Bosbach, 2011). However, it may seem paradoxical to use the same model both to account for the experience of controlling external events and to account for the relative imperception of external events caused by one's own agency.

A recent study resolved some of these uncertainties with a novel interpretation of the role of the angular gyrus in sense of agency. Chambon and colleagues (Chambon, Wenke, Fleming, Prinz, & Haggard, 2013; see figure 75.3) used subliminal arrow stimuli to prime participants' actions, so that selecting the appropriate action in response to a left or right target arrow was either facilitated or inhibited. After the participants responded, one of several color patches appeared on the screen after a short delay. Participants reported how much control they felt they had over the color appearing on the screen. The results showed a higher sense of control over action outcomes when participants' responses had been facilitated by subliminal priming, relative to when they had been impaired. Interestingly, the angular gyrus showed increasing activation as the participants' feeling of control reduced, but only for incompatibly primed trials. Crucially, the primes in this experiment did not predict the color patches that appeared after each action. Therefore, the influence of priming on sense of agency was interpreted as a prospective fluency effect: people feel a stronger sense of control when it is easy to *select* which action to take, irrespective of the statistical relation between their action and outcome. People mistake the ease of choosing what to do for actually achieving something. Interestingly, the angular gyrus showed increasing levels of activation as the subjective sense of control over the color path decreased, but only for incompatible trials. This result confirms the negative coding for nonagency in the angular gyrus. However, this code may be prospectively based on what we think we may achieve, and not just retrospectively based on what we have actually achieved.

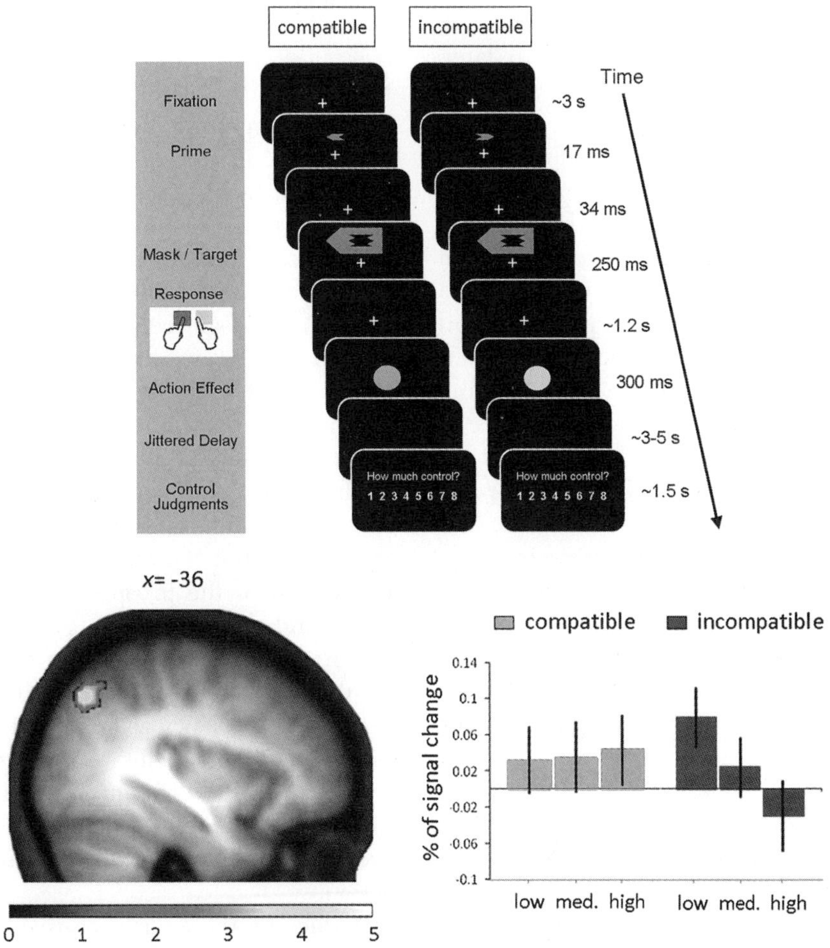

FIGURE 75.3 Subliminal priming of actions influences sense of agency *prospectively*. *Upper panel*: subliminal primes facilitate or impair actions in response to target arrow stimuli. Responses are followed by appearance of a color patch, and participants judge how much control they have over the color patch. *Lower panel*: activation of the angular gyrus associated with the target-response event varies negatively with perceived level of control, but only following incompatible priming. (Reproduced from Chambon et al., 2013, by permission of Oxford University Press.) (See color plate 67.)

Low-level agency perception

Several studies have shown that human judgments of agency may be biased. In social situations, people tend to think they caused events that were in fact caused by the actions of others (Wegner, 2003; Wegner & Wheatley, 1999). The bias to overestimate one's own agency in the case of simple sensorimotor events may form an instance of a more general tendency to believe in one's own self-efficacy (Bandura, 2001). However, it remains unclear whether these biases actually alter low-level perception of agency, or whether they are merely response biases. Investigating this question requires an implicit measure of agency. One candidate is the "intentional binding" effect. In the original demonstration (Haggard, Clark, & Kalogeras, 2002), participants used the clock method developed by Libet and colleagues to indicate the time at which they made a voluntary key-press action. In one block of trials, each action was followed after a short delay by an auditory tone, while in another block of trials no tone occurred. The perceived time of the actions was shifted later, toward the subsequent tone, in the block where tones occurred. In the same fashion, the perceived time of the tone was compared between a baseline block where participants simply heard a tone occurring at random intervals, without making any action, and an experimental block in which participants caused the tone through their own voluntary action. The perceived time of the tone was shifted earlier, toward the action that caused it, compared to the baseline block. Taken together, the action binding and the tone binding implied a perceptual compression of the time interval between action and outcome. Crucially, neither of these effects occurred when the voluntary key-press action was replaced by an involuntary twitch of the same muscles induced by transcranial magnetic stimulation (Haggard, Clark, & Kalogeras, 2002; see figure 75.4), or by a second tone. On this basis, it was suggested that intentional binding may reflect the temporal association between intentional actions and their outcomes. It may therefore provide an implicit marker of sense of agency.

The intentional binding effect involves independent measures of action binding and tone binding. Therefore, intentional binding allows sense of agency to be broken down into distinct components, in a way that judgment of agency attribution does not. For example, by varying the probability of a tone following an action, Moore and Haggard (2008) showed that action binding involved two distinct processes, one prospective and one retrospective. In a block where only 50% of actions were followed by tones, action binding was stronger on trials where the tone occurred than on trials where it did not, suggesting a change in the perceived time of the action triggered retrospectively by the ensuing tone. In a block where most actions (75%) were followed by tones, action binding was again observed on trials where the tone did occur, but also on trials without any tone. This pattern of results suggests that the action was shifted prospectively toward the tone, in anticipation of its likely occurrence. Voss and colleagues used a similar design to compare the predictive and retrospective processes of action binding between a group of psychotic patients and age-matched healthy controls (Voss et al., 2010). They found similar levels of binding in both groups, but for very different reasons. In the healthy controls, action binding was dominated by the prospective component, suggesting it depended on predicting the likely outcomes of actions. In the patients, action binding was dominated by the retrospective component. Disorganization of thought and agency are cardinal symptoms of schizophrenic psychosis. This result suggests that even low-level experience of simple sensorimotor events in psychosis depends on opportunistic conjunctions of events, rather than learned regularities of instrumental action.

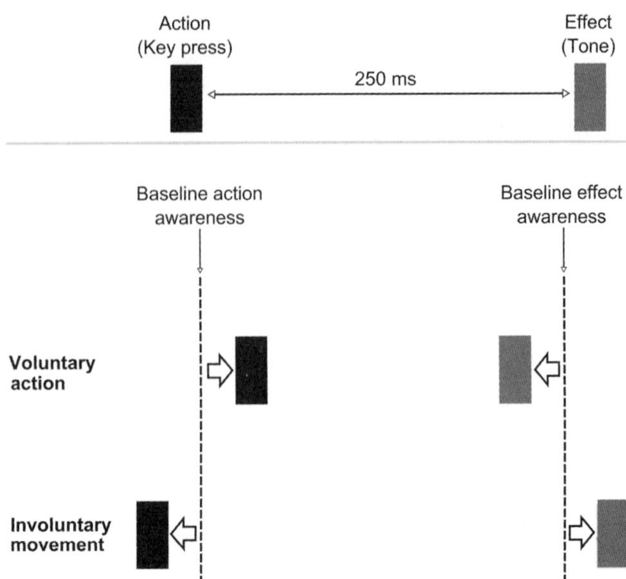

FIGURE 75.4 The intentional binding effect. (Reproduced from Moore et al., 2013, with permission from Elsevier.)

Studies of the neural basis of intentional binding have suggested a key role for several frontal lobe circuits. Importantly, these show considerable overlap with the areas involved in generating intentional action. They are also quite distinct from the parietal areas generally identified with explicit judgments of agency.

This evidence can only be reviewed briefly here. First, intentional binding effects were reduced after inactivation of the preSMA caused by repetitive transcranial magnetic stimulation (Moore, Ruge, Wenke, Rothwell, & Haggard, 2010). Second, Parkinsonian patients showed significantly stronger intentional binding after taking their dopaminergic medication than immediately before, suggesting that dopaminergic drive in the basal-ganglia-thalamocortical loop strongly contributes to sense of agency (Moore, Schneider, et al., 2010). Finally, an fMRI study showed that activation of the caudal SMA correlated more strongly with the perceived interval between voluntary actions and tones than between passive movements and identical tones (Kühn, Brass, & Haggard, 2012).

Context and belief in the conscious experience of intentional action

Taken together, these results suggest that the processes of preparing and initiating voluntary actions in the frontal lobes also involve a prediction of the anticipated consequences of action, much as the computational motor control model suggests. Importantly, the experience of agency, or causing the effects of one's action, appears to derive from the same circuits that develop action itself. Thus, the sense of agency may not simply be an inference or confabulation of authorship, but a measurable internal signal within the motor system. There is currently a lively controversy regarding the division of labor between the frontal and parietal lobes in both intention and agency. The conscious urge to move that was previously identified with medial frontal cortex was also recently reported after direct electrical stimulation of parietal cortex (Desmurget et al., 2009), while patients with parietal lesions had a delayed experience of intention to move (Sirigu et al., 2004). The involvement of parietal cortex in agency is also clear, since the angular gyrus is routinely activated in judgments of (non)agency, as we have seen. However, frontal activations did correlate with implicit markers of sense of agency, whereas parietal activations did not (Kühn et al., 2012). Therefore, one interim hypothesis suggests that the frontal lobe houses the core machinery for voluntary action, including the prospective construction of experience of goal-directed action and the low-level experience of agency. The parietal lobes, in contrast, would monitor the motor instructions generated frontally. By comparing these instructions with later sensory inputs, a parietal comparator could detect nonagency and contribute to explicit judgments of agency.

Responsibility for action

All human societies appear to hold individuals responsible for their actions. In many systems of law, theories of responsibility are explicitly dualist. The agents' capacity for rational thought, their conscious choice over their actions, and their freedom to do otherwise—all make them responsible for what they do. In dramatic contrast, neuroscience ultimately views intentional actions as consequences of specific, deterministic electrical and chemical brain processes. Modern cognitive neuroscience rejects dualist notions of brain-independent consciousness, and instead views an individual's choices, actions, and even his or her character as mechanistic functions of his or her brain. For the neuroscientist, the conscious, rational thought is simply an additional brain process, which is presumably just as determined as all others. These opposing views form the basis of a lively debate between law and neuroscience. For example, if psychopathy is a neuropsychiatric disorder, are psychopaths truly responsible for their actions? Can neuroscience help to guide such individuals toward more prosocial behaviors (Anderson & Kiehl, 2013)? Greene and Cohen argue that if neurobiological determinism is true, then retributive justice and punishment are misguided. To the extent that a criminal's brain "made them do it," then there is little point in punishing criminals for their actions (Greene & Cohen, 2004). Of course, the same determinist neuroscientific theories of action offer ready support for a rehabilitative view of justice. Rehabilitation of offenders could, in principle, provide learning experiences that change the brain, and thus behavior.

Humans are social animals, and our brain mechanisms have coevolved with our increasingly large and diverse societies. To participate in such a society requires a brain that is able to follow, or learn to follow, the social and moral codes by which societies work. Equally, societies will generally try to protect themselves against possible harm from individuals who do not follow these codes. The methods of protection are costly, since they involve either costly care for those who are incapable of following such codes or costly punishment for those who are judged capable, but who transgress. For a strict neurobiological determinist, individuals are punished not so much for the action that they performed, but for having the kind of brain they have—since it was their brain that caused their action. The consequences of this view are highly unpalatable and morally unacceptable! However, even a determinist may agree that responsibility and punishment are more acceptable in a society where all individuals have an equal opportunity to learn to follow legal and moral codes of behavior than in a society where some individuals have little or no such opportunity to learn. Education and culture are society's ways of influencing, or constructing, individual brains that produce appropriate behaviors.

The debate on responsibility often ignores this crucial role of learning: responsibility for action presupposes a brain that is capable of learning codes of behavior, and also a brain that has been trained for action by the appropriate learning experiences. If equal access to learning experiences is absent, socially organized responsibility and punishment could simply be seen as social tools for individuals with one kind of brain to control individuals with "less desirable" brains. Perhaps socially organized responsibility and punishment only make sense in societies that ensure equal access to appropriate learning experiences for all.

ACKNOWLEDGMENTS Preparation of this chapter was supported by an ESRC Professorial Fellowship (ES/J023140/1) and by an ERC Advanced Grant HUMVOL (323943).

REFERENCES

Anderson, N. E., & Kiehl, K. A. (2013). Psychopathy: Developmental perspectives and their implications for treatment. *Restor Neurol Neurosci, 32*(1), 103–117.

Aron, A. R. (2011). From reactive to proactive and selective control: Developing a richer model for stopping inappropriate responses. *Biol Psychiatry, 69*(12), e55–68.

Averbeck, B. B., Battaglia-Mayer, A., Guglielmo, C., & Caminiti, R. (2009). Statistical analysis of parieto-frontal cognitive-motor networks. *J Neurophysiol, 102*(3), 1911–1920.

Bandura, A. (2001). Social cognitive theory: An agentic perspective. *Annu Rev Psychol, 52*, 1–26.

Blakemore, S. J., Wolpert, D., & Frith, C. (2000). Why can't you tickle yourself? *NeuroReport, 11*(11), R11–16.

Brass, M., & Haggard, P. (2007). To do or not to do: The neural signature of self-control. *J Neurosci, 27*(34), 9141–9145.

Burgess, P. W., Gonen-Yaacovi, G., & Volle, E. (2011). Functional neuroimaging studies of prospective memory: What have we learnt so far? *Neuropsychologia, 49*(8), 2246–2257.

Chambon, V., Wenke, D., Fleming, S. M., Prinz, W., & Haggard, P. (2013). An online neural substrate for sense of agency. *Cereb Cortex, 23*(5), 1031–1037.

Damasio, A. (2000). *The feeling of what happens: Body, emotion and the making of consciousness.* New York, NY: Vintage.

Deiber, M. P., Ibañez, V., Sadato, N., & Hallett, M. (1996). Cerebral structures participating in motor preparation in humans: A positron emission tomography study. *J Neurophysiol, 75*(1), 233–247.

Deiber, M. P., Passingham, R. E., Colebatch, J. G., Friston, K. J., Nixon, P. D., & Frackowiak, R. S. (1991). Cortical areas and the selection of movement: A study with positron emission tomography. *Exp Brain Res, 84*(2), 393–402.

Della Sala, S., Marchetti, C., & Spinnler, H. (1991). Right-sided anarchic (alien) hand: A longitudinal study. *Neuropsychologia, 29*(11), 1113–1127.

Desmurget, M., Reilly, K. T., Richard, N., Szathmari, A., Mottolese, C., & Sirigu, A. (2009). Movement intention after parietal cortex stimulation in humans. *Science, 324*(5928), 811–813.

Duque, J., Lew, D., Mazzocchio, R., Olivier, E., & Ivry, R. B. (2010). Evidence for two concurrent inhibitory mechanisms during response preparation. *J Neurosci, 30*(10), 3793–3802.

Farrer, C., Frey, S. H., van Horn, J. D., Tunik, E., Turk, D., Inati, S., & Grafton, S. T. (2008). The angular gyrus computes action awareness representations. *Cereb Cortex, 18*(2), 254–261.

Filevich, E., Kühn, S., & Haggard, P. (2012). Intentional inhibition in human action: The power of "no." *Neurosci Biobehav Rev, 36*(4), 1107–1118.

Fried, I., Mukamel, R., & Kreiman, G. (2011). Internally generated preactivation of single neurons in human medial frontal cortex predicts volition. *Neuron, 69*(3), 548–562.

Frith, C. D., Blakemore, S-J., & Wolpert, D. M. (2000). Explaining the symptoms of schizophrenia: Abnormalities in the awareness of action. *Brain Res Rev, 31*(2-3), 357–363.

Gentsch, A., & Schütz-Bosbach, S. (2011). I did it: Unconscious expectation of sensory consequences modulates the experience of self-agency and its functional signature. *J Cogn Neurosci, 23*, 3817–3828.

Greene, J., & Cohen, J. (2004). For the law, neuroscience changes nothing and everything. *Philos Trans R Soc Lond B Biol Sci, 359*(1451), 1775–1785.

Haggard, P., Clark, S., & Kalogeras, J. (2002). Voluntary action and conscious awareness. *Nat Neurosci, 5*(4), 382–385.

Halsband, U., Matsuzaka, Y., & Tanji, J. (1994). Neuronal activity in the primate supplementary, pre-supplementary and premotor cortex during externally and internally instructed sequential movements. *Neurosci Res, 20*(2), 149–155.

Kane, R. (2005). *A contemporary introduction to free will.* New York, NY: Oxford University Press.

Koechlin, E., & Summerfield, C. (2007). An information theoretical approach to prefrontal executive function. *Trends Cogn Sci, 11*(6), 229–235.

Kornhuber, H. H., & Deecke, L. (1965). Changes in the brain potential in voluntary movements and passive movements in man: Readiness potential and reafferent potentials. *Pflugers Arch Gesamte Physiol Menschen Tiere, 284*, 1–17.

Krieghoff, V., Brass, M., Prinz, W., & Waszak, F. (2009). Dissociating what and when of intentional actions. *Front Hum Neurosci, 3*, 3.

Kühn, S., Brass, M., & Haggard, P. (2012). Feeling in control: Neural correlates of experience of agency. *Cortex, 49*(7), 1935–1942.

Kühn, S., Haggard, P., & Brass, M. (2009). Intentional inhibition: How the "veto-area" exerts control. *Hum Brain Mapp, 30*(9), 2834–2843.

Kühn, S., Haggard, P., & Brass, M. (2013). Differences between endogenous and exogenous emotion inhibition in the human brain. *Brain Struct Funct* (E-pub ahead of print).

Lhermitte, F. (1983). "Utilization behaviour" and its relation to lesions of the frontal lobes. *Brain, 106*(Pt. 2), 237–255.

Libet, B. (2005). *Mind time: The temporal factor in consciousness.* Cambridge, MA: Harvard University Press.

Libet, B., Gleason, C. A., Wright, E. W., & Pearl, D. K. (1983). Time of conscious intention to act in relation to onset of cerebral activity (readiness-potential). The unconscious initiation of a freely voluntary act. *Brain, 106*(Pt. 3), 623–642.

Momennejad, I., & Haynes, J.-D. (2012). Human anterior prefrontal cortex encodes the "what" and "when" of future intentions. *NeuroImage, 61*(1), 139–148.

Moore, J. W., Cambridge, V. C., Morgan, H., Giorlando, F., Adapa, R., & Fletcher, P. C. (2013). Time, action and psychosis: Using subjective time to investigate the effects of ketamine on sense of agency. *Neuropsychologia, 51*(2), 377–384.

Moore, J. W., Ruge, D., Wenke, D., Rothwell, J., & Haggard, P. (2010). Disrupting the experience of control in the human brain: Pre-supplementary motor area contributes to the sense of agency. *Philos Trans R Soc Lond B Biol Sci, 277*(1693), 2503–2509.

Moore, J. W., Schneider, S. A., Schwingenschuh, P., Moretto, G., Bhatia, K. P., & Haggard, P. (2010). Dopaminergic medication boosts action-effect binding in Parkinson's disease. *Neuropsychologia, 48*(4), 1125–1132.

Moore, J., & Haggard, P. (2008). Awareness of action: Inference and prediction. *Conscious Cogn, 17*(1), 136–144.

Moretto, G., Walsh, E., & Haggard, P. (2011). Experience of agency and sense of responsibility. *Conscious Cogn, 20*(4), 1847–1854.

Mueller, V. A., Brass, M., Waszak, F., & Prinz, W. (2007). The role of the preSMA and the rostral cingulate zone in internally selected actions. *NeuroImage, 37*(4), 1354–1361.

Pacherie, E. (2008). The phenomenology of action: A conceptual framework. *Cognition, 107*(1), 179–217.

Passingham, R. E., Bengtsson, S. L., & Lau, H. C. (2010). Medial frontal cortex: From self-generated action to reflection on one's own performance. *Trends Cogn Sci, 14*(1), 16–21.

Rubia, K., Overmeyer, S., Taylor, E., Brammer, M., Williams, S. C., Simmons, A., & Bullmore, E. T. (1999). Hypofrontality in attention deficit hyperactivity disorder during higher-order motor control: A study with functional MRI. *Am J Psychiatry, 156*(6), 891–896.

Schüür, F., & Haggard, P. (2011). What are self-generated actions? *Conscious Cogn, 20*(4), 1697–1704.

Searle, J. R. (1983). *Intentionality.* Cambridge, UK: Cambridge University Press.

Shadlen, M. N., & Gold, J. I. (2004). The neurophysiology of decision-making as a window on cognition. In M. S. Gazzaniga (Ed.), *The cognitive neurosciences* (3rd ed., pp. 1229–1241). Cambridge, MA: MIT Press.

Sirigu, A., Daprati, E., Ciancia, S., Giraux, P., Nighoghossian, N., Posada, A., & Haggard, P. (2004). Altered awareness of voluntary action after damage to the parietal cortex. *Nat Neurosci, 7*(1), 80–84.

Sirigu, A., Daprati, E., Pradat-Diehl, P., Franck, N., & Jeannerod, M. (1999). Perception of self-generated movement following left parietal lesion. *Brain, 122*(Pt. 10), 1867–1874.

Smith, A. B., Taylor, E., Brammer, M., Toone, B., & Rubia, K. (2006). Task-specific hypoactivation in prefrontal and temporoparietal brain regions during motor inhibition and task switching in medication-naive children and adolescents with attention deficit hyperactivity disorder. *Am J Psychiatry, 163*(6), 1044–1051.

Synofzik, M., Vosgerau, G., & Newen, A. (2008). Beyond the comparator model: A multifactorial two-step account of agency. *Conscious Cogn, 17*(1), 219–239.

Thaler, D., Chen, Y. C., Nixon, P. D., Stern, C. E., & Passingham, R. E. (1995). The functions of the medial premotor cortex. I. Simple learned movements. *Exp Brain Res, 102*(3), 445–460.

van den Wildenberg, W. P. M., Burle, B., Vidal, F., van der Molen, M. W., Ridderinkhof, K. R., & Hasbroucq, T. (2010). Mechanisms and dynamics of cortical motor inhibition in the stop-signal paradigm: A TMS study. *J Cogn Neurosci, 22*(2), 225–239.

Voss, M., Moore, J., Hauser, M., Gallinat, J., Heinz, A., & Haggard, P. (2010). Altered awareness of action in schizophrenia: A specific deficit in predicting action consequences. *Brain, 133*(10), 3104–3112.

Wardak, C. (2011). The role of the supplementary motor area in inhibitory control in monkeys and humans. *J Neurosci, 31*(14), 5181–5183.

Wegner, D. M. (2003). *The illusion of conscious will.* Cambridge, MA: MIT Press.

Wegner, D. M., & Wheatley, T. (1999). Apparent mental causation. Sources of the experience of will. *Am Psychol, 54*(7), 480–492.

Wittgenstein, L. (1953). *Philosophical investigations.* Hoboken, NJ: Blackwell.

X

ADVANCES IN METHODOLOGY

Introduction

B. A. WANDELL

There was a time, not so long ago, when the experimental and theoretical tools available to a cognitive neuroscientist were relatively limited. Data presentation in a talk or paper might comprise some colored spots, suggestive of neural activity, superimposed on anatomical images of brain slices. Or, there might be an X-ray scan or post-mortem image showing the general site of a brain lesion. Such data would be accompanied by a diagram comprised of a series of boxes with arrows coming in and out. These diagrams were overviews of potential information-processing algorithms.

Those days are over. The methodological advances in the whole of neuroscience are quite breathtaking; new technologies for measuring and analyzing the living human brain are surely among the most remarkable. Today's young cognitive neuroscientist is expected to understand an enormous range of technologies and computational methods. Moreover, the next generation of cognitive neuroscientists must understand not only the methods that are specific to measuring human brain and behavior, but also the methods in neighboring fields of neuroscience working on animal models. In recognition of the striking advances in methodology, this fifth edition of *The Cognitive Neurosciences* includes a new section to provide readers with an introduction to some of the exciting new methods for measuring and interpreting brain structure and function.

The existence of many different methods opens many opportunities but also poses many important challenges for cognitive neuroscience. Each method assesses brain function by its own unique dependent measure, with its own units. The methods probe the brain at a wide range of length scales and time scales. Some of the methods are noninvasive and applicable to healthy volunteers, whereas other methods are only appropriate for animal models or only made as part of clinical procedures. One of the great challenges going forward will be to find ways to integrate the data from different measurements into a single theoretical understanding of brain structures and functions. To meet this task, we must understand the different technologies and their associated theoretical tools.

The chapters in this section provide an overview of the current state of several technologies. Functional MRI (fMRI), which is only 20 years old, is now a common and widely accepted technology. In recent years, it has been augmented by a set of analytical tools that were once hard to envision. The measurements themselves have advanced, so that it is now possible to obtain an fMRI signal within submillimeter voxels containing a few tens of thousands of neurons. These measurements can be obtained at a sampling rate of less than a second. A strength of fMRI is that it has a large field of view of activity in the living human brain. Several chapters introduce modern theoretical tools that take advantage of the field of view. Signals from different parts of the brain influence one another, and in some cases there will be a causal relationship—one region is the principal driver of another. Roebroeck and Goebel explain the theory and methods for assessing causality. These methods are applicable to several neuroscience methods, including fMRI and scalp measurements (electroencephalography and magnetoencephalography). Lewis-Peacock and Norman describe theoretical methods for using information from multiple voxels at once, a set of techniques called multivoxel pattern analysis. These methods have improved sensitivity, but they also suggest that there is meaning in the pattern of activation that cannot be found in the activation within a single localized region. Thus, the large field of view in MRI has enabled cognitive neuroscientists to extend their thinking from the response of a single cell or a single voxel to the pattern of activity in the whole brain.

The last 10 years have seen a substantial advance in ways to measure using MRI. One of the most important is diffusion-weighted imaging. This measurement method, along with computational methods called tractography, has been used to characterize the massive number of connections in the human brain. Assaf describes the measurement methods, including diffusion tensor imaging and diffusion spectrum imaging. He also describes other quantitative MRI techniques for assessing human brain tissue. In a rather surprising turn of events, the act of learning something new changes the diffusivity in the living human brain. These new methods provide insight about both connections and function.

There are many long-range connections in the human brain; the significance and function of these connections must depend on the absolute size of the brain. The human brain is about 2,500 times the volume of the mouse brain, and the ratio of human to mouse white matter is even larger, about 50,000 to 1. Hence, the challenge of understanding white matter cannot be understood simply as a long list of connections; understanding it will require theory. Bassett and Lynall describe network theory, a branch of mathematics that has provided a set of tools for characterizing the network of connections in the human brain. Network theory can be applied to several types of measurements, including white matter connections or causal links derived from functional measurements.

Simple model systems provide us with an opportunity to undertake fundamental measurements that cannot be ethically or practically undertaken in humans. Kispersky, O'Leary, and Marder use a simple model system to address a fundamental question about key biological components—the membrane channels that are essential for neural communication. They show that neural network behavior can be precisely replicated despite considerable variability in the number of channels. Channels are essential for signaling. But the absolute number or distribution is not. A network of cells can perform its function despite considerable variation at the channel level.

The methods reviewed to this point measure system properties. Experimental work benefits greatly from the ability to perturb the system one is measuring. Yoshor and Beauchamp describe experiments in the living human brain that involve both measuring and exciting neural tissue in visual cortex. Using intracranial electrodes in the human brain—now called electrocorticography—they measure local mean fields, and they also perturb these fields. As they explain, these technologies can be used in very controlled experimental designs that inform us about visual perception and at the same time provide valuable information for the design of prosthetics. Zalocusky and Deisseroth describe the relatively new technique of optogenetics. This technique is now being widely applied in the mouse brain, and a few investigators have implemented the method in nonhuman primates. Optogenetic methods provide the scientist with the ability to initiate signals in specific classes of cells with high temporal precision, much higher than classical pharmacology. The signals initiated by optogenetics spread through the neural circuits and influence behavior in a variety of ways.

The growth of cognitive neuroscience technologies has also led to an enormous growth in the amount of data. The days of a simple lab notebook, and a methods section comprising a few paragraphs, have been replaced by terabytes of data and sophisticated algorithms of high complexity. Science is based on transparency and replication. Future generations of scientists must develop tools for data and computational sharing. Marcus describes the problems of data stewardship and sharing and discusses an approach that he is developing. The field of neuroinformatics will be an essential part of cognitive neuroscience as we go forward together.

The chapters in this section describe recent cutting-edge tools; they are a subset of the large number of methods now being used to understand the brain. Other sections of this volume include methods that are also important, including multiunit electrode recordings, calcium imaging, and behavior measured locally or at scale on the Internet. No doubt, by the time the next edition of the book is produced, yet more techniques will be added to our arsenal. To work together as a field, we must find ways to speak across these different methodologies. The proliferation of these methods provides new opportunities and exciting times for young cognitive neuroscientists.

76 Computational Causal Modeling of High-Resolution Functional MRI Data

ALARD ROEBROECK AND RAINER GOEBEL

ABSTRACT Functional MRI is increasingly used to study the connectivity in large-scale brain networks that supports perception, action and cognition. This chapter focuses on the modeling of causality (directed influence or effective connectivity) and how this can be applied to fMRI data obtained in well-designed cognitive, perceptual, or motor tasks. This kind of modeling entails the estimation of multivariate mathematical models from the data record which embed assumptions that allow us to move beyond mere correlations between activity signals. We discuss the main challenges of causal fMRI data modeling: the missing region problem, the missing time problem, and the missing model problem. We then suggest how greatly increased spatial and temporal resolution of acquisition can assist in addressing these challenges and, potentially, move computational causal fMRI modeling into the realm of cortical microcircuit dynamics.

Understanding how interactions between brain structures support the performance of specific cognitive tasks or perceptual and motor processes is a prominent goal in cognitive neuroscience. Neuroimaging methods, such as electroencephalography (EEG), magnetoencephalography (MEG), and functional magnetic resonance imaging (fMRI), are ever more employed to address questions of functional connectivity, inter-region coupling, and networked computation that go beyond the "where" and "when" of task-related activity (Friston, 1994; Horwitz, Friston, & Taylor, 2000; McIntosh, 2004; Salmelin & Kujala, 2006; Valdes-Sosa, Kotter, & Friston, 2005). A network perspective onto the parallel and distributed processing in the brain—even on the scale accessible by neuroimaging methods—is a promising approach to enlarge our understanding of perceptual, cognitive, and motor functions. Functional MRI, in particular, is increasingly used, not only to localize structures involved in cognitive and perceptual processes, but also to study the connectivity in large-scale brain networks that support these functions.

Generally, a distinction is made between three types of brain connectivity (Sporns, 2010). *Structural connectivity* (or anatomical connectivity) refers to the physical presence of an axonal projection from one brain area to another. Identification of large axon bundles connecting remote regions in the brain has recently become possible noninvasively in vivo by diffusion MRI (dMRI) and computational tractography analysis (Johansen-Berg & Behrens, 2009; Johansen-Berg & Jbabdi, 2011; Jones, 2010). *Functional connectivity* refers to the correlation structure (or, more generally, any order of statistical dependency) in the data such that brain areas can be grouped into interacting networks. Finally, *effective connectivity* modeling moves beyond statistical dependency to measures of directed influence and causality within networks (Friston, 1994).

This chapter focuses on the latter: modeling of directed influence, effective connectivity, or causality and how this can be applied to fMRI data obtained in well-designed cognitive, perceptual, or motor tasks. Causal modeling of this kind entails the estimation of multivariate mathematical models from the data record. These models embed assumptions (implicitly or explicitly) that allow us to move beyond correlations to directions of information flow. Statistical inference on parameters estimated in the context of such models then provides quantification of directed connectivity, and inference can take place on individual parameters or on a group of parameters together; the latter is often referred to as model comparison. We can partition models of causal influence into two parts, each necessitating choices and assumptions: the structural model and the dynamical model (figure 76.1).

The *structural model* contains (1) a selection of the regions of interest (ROIs) in the brain that are assumed to be of importance in the cognitive process or task under investigation, (2) the possible interactions between those structures, and (3) the possible effects of exogenous inputs onto the network. Note that although ideally the structural model within a causal modeling framework is guided by measurement or knowledge of direct structural connectivity (as defined above), it is not identical to the concept of structural connectivity. For instance, a structural model can contain connections between ROIs that are hypothesized or thought to exist only indirectly via several relay stations. The exogenous inputs may be under control of the experimenter and often have the form of a simple

Structural model

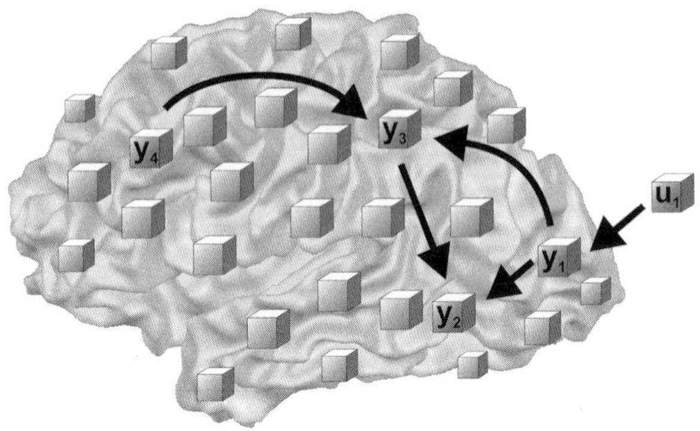

Dynamical model

$$y_1 = f_1(u_1)$$
$$y_2 = f_2(y_1, y_3)$$
$$y_3 = f_3(y_1, y_4)$$

u_1

y_1

y_2

y_3

y_4

FIGURE 76.1 A partitioning of functional MRI (fMRI) connectivity models. Models to estimate causal connectivity from fMRI time series can be partitioned into a structural model and a dynamical model. The structural model contains a selection of the structures in the brain that are assumed to be of importance in the cognitive process or task under investigation. Specifically, it specifies which regions of interest (ROIs) in the spatially rich high-dimensional fMRI data set will be considered for further analysis, as illustrated by the selection of the red boxes $y_1 \dots y_4$. The structural model can also define the possible interactions between the ROIs in the form of one or more directed graph models that might be compared in a later model comparison step. Finally, the structural model also defines where exogenous inputs (that may be under control of the experimenter) can exert effects on the network. The dynamical model embeds the structural model assumptions into parameterized equations that relate the selected measurements and inputs to each other. Connectivity modeling involves the estimation of the parameters in the dynamical model from actual measurements y_j and, possibly, inputs u_k. (Adapted from Roebroeck, Formisano et al., 2011.) (See color plate 68.)

indicator function that can represent, for instance, the presence or absence of a visual stimulus. Thus, they represent the primary way in which experimental task design enters connectivity analysis. The *dynamical model* consists of parameterized equations that relate the signals of the selected structures and exogenous inputs to each other. The functional form of these equations can embed assumptions on signal dynamics, temporal precedence, or physiological processes from which signals originate. Connectivity modeling involves the estimation of (and inference on) the parameters in the dynamical model from actual measurements and, possibly, exogenous inputs. The number of parameters to be estimated (i.e., the total model complexity) is directly dependent on the complexity of the structural model (i.e., how many ROIs and connections are included) and the complexity of the dynamical model.

In this chapter, we discuss causal modeling of fMRI data, its challenges, and how greatly increased spatial and temporal resolution of acquisition can assist in addressing some of these. We first briefly review some of the spatial and temporal properties of the fMRI signal and the modeling of its mechanisms. We also focus on a few recent techniques that have greatly increased acquisition resolution. We then discuss two popular methods for causal modeling of fMRI time series: G-causality and dynamic causal modeling. This leads to the challenges one faces when using these techniques as tools to answer cognitive neuroscience questions, formulated as three problems: the missing region problem, the missing time problem, and the missing model problem. We conclude with summary and discussion of future directions.

Spatial and temporal properties of the blood oxygen level–dependent functional MRI signal

The fMRI signal noninvasively reflects the activity within neuronal populations with excellent spatial resolution (millimeters down to hundreds of micrometers at high field strength), good temporal resolution (seconds down to hundreds of milliseconds), and whole-brain coverage of the human or animal brain (Logothetis, 2008). Although fMRI is possible with a few different techniques, the blood oxygen level–dependent (BOLD) contrast mechanism is employed in the great majority of cases. In short, the BOLD fMRI signal is sensitive to changes in blood oxygenation, blood flow, and blood volume that result from oxidative glucose metabolism, which, in turn, is needed to fuel local neuronal activity (Buxton, Uludag, Dubowitz, & Liu, 2004). This is why fMRI is usually classified as a "metabolic" or "hemodynamic" neuroimaging modality. Its superior spatial resolution, in particular, distinguishes it from other functional brain imaging modalities used in humans, such as EEG, MEG, and positron emission tomography. A strength of fMRI as a neuroimaging technique is that an adjustable trade-off is available to the user between spatial resolution, spatial coverage, temporal resolution, and signal-to-noise ratio (SNR) of the acquired data. For instance, although fMRI can achieve excellent spatial resolution at good SNR and reasonable temporal resolution, one can choose to sacrifice some spatial resolution to gain a better temporal resolution for any given study. This trade-off is important in causal modeling that tries to use high spatial or temporal resolution to improve sensitivity, and makes clear that using *both* very high spatial and temporal resolution is difficult to achieve. However, the continuous improvements in MRI hardware and pulse sequences regularly increase the breadth of the achievable spatiotemporal resolution for a given contrast-to-noise level and field of view. Note that this primarily concerns the nominal resolution and SNR of the data *acquisition.* It is important to realize that the physiology underlying the BOLD fMRI signal can put fundamental limitations on the effective resolution (spatial and temporal) and SNR that is achieved in relation to the neuronal processes of interest.

The BOLD fMRI signal arises from a complex chain of processes that we can classify into neuronal, physiological, and physical processes (Uludag, Dubowitz, & Buxton, 2005), each of which contain some crucial parameters and variables and which have been modeled in various ways (see figure 76.2). Each set of processes has a profound influence on the effective spatial and temporal resolution of BOLD fMRI. Moreover, modeling these processes correctly and parsimoniously (in the dynamical model) greatly affects the capacity for correct causal inference. Since the fMRI signal is sensitive to the local oxidative metabolism in the brain, it mainly reflects the most energy-consuming of the neuronal processes. In primates, postsynaptic processes account for the great majority (about 75%) of the metabolic costs of neuronal signaling events (Attwell & Iadecola, 2002). Indeed, BOLD fMRI has been shown in various experimental paradigms to be more sensitive to postsynaptic activity, rather than axon generation and propagation or "spiking" (Logothetis, Pauls, Augath, Trinath, & Oeltermann, 2001; Rauch, Rainer, & Logothetis, 2008; Thomsen, Offenhauser, & Lauritzen, 2004). On the physiological level, the main variables that mediate the BOLD contrast in fMRI are cerebral blood flow, cerebral blood volume, and the cerebral metabolic rate of oxygen. These physiological variables change the oxygen saturation of the blood, which is the basis of the BOLD signal contrast. BOLD fMRI is directly sensitive to the relative amount of oxyhemoglobin and deoxyhemoglobin and to the fraction of cerebral tissue that is occupied by blood (the cerebral blood volume), which are controlled by local neurovascular coupling processes. Neurovascular processes, in turn, are tightly coupled to neurometabolic processes controlling the rate of oxidative glucose

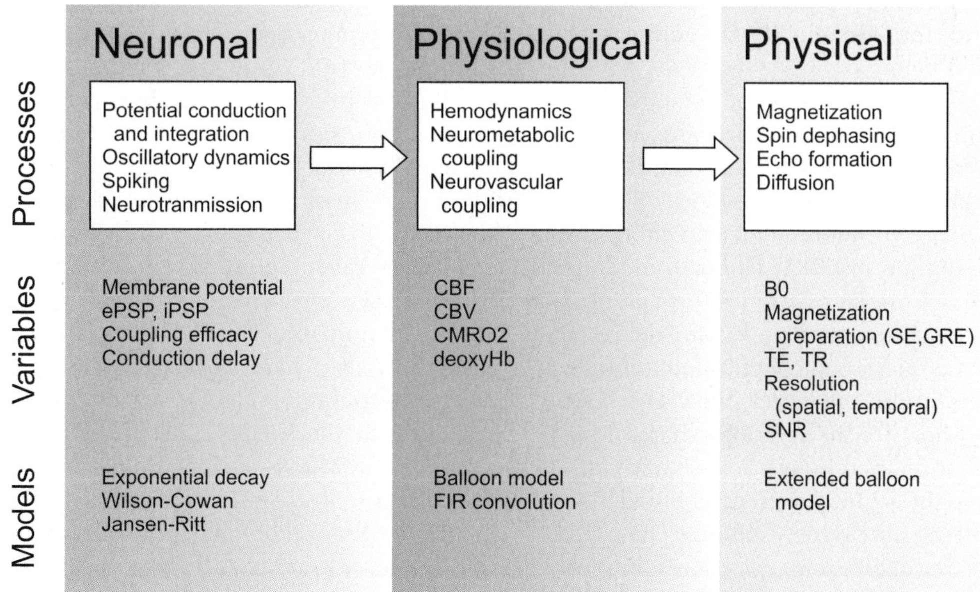

FIGURE 76.2 The neuronal, physiological, and physical processes (top row) and variables and parameters involved (middle row) in the complex causal chain of events that leads to the formation of the fMRI signal (ePSP, excitatory postsynaptic potential; iPSP, inhibitory postsynaptic potential; FIR, finite impulse response; B_0, main magnetic field, CBF: cerebral blood flow, CBV: cerebral blood volume, $CMRO_2$: cerebral metabolic rate of oxygen, TE: time to echo, TR: time to repeat, SE: spin echo, GRE: gradient echo, SNR: signal-to-noise ratio). The bottom row lists some mathematical models of the subprocesses that play a role in the analysis and causal modeling of fMRI signals. (Adapted from Roebroeck, Seth, et al., 2011.)

metabolism (the cerebral metabolic rate of oxygen) that is needed to fuel neural activity.

The canonical *temporal* dynamics of blood flow and blood volume, the hemodynamics, show a robust BOLD signal increase 1–2 seconds after neuronal activity, which peaks at 4–6 seconds if activity ceases quickly. The initial neuronal activity is quickly followed by a large cerebral blood flow increase, an overcompensating response that supplies much more oxyhemoglobin to the local blood system than has been metabolized. As a consequence, the oxygenation of the blood increases and the magnetic resonance signal increases. The increased flow also induces a "ballooning" of the blood vessels, increasing cerebral blood volume, the proportion of volume taken up by blood, further increasing the signal. A mathematical characterization of the hemodynamic processes in BOLD fMRI at 1.5–3 T has been given in the biophysical balloon model (Buxton, Wong, & Frank, 1998; Buxton et al., 2004). A simplification of the full balloon model has become important in causal models of brain connectivity (Friston, Mechelli, Turner, & Price, 2000), as discussed below. It is the temporal specificity of the hemodynamics that mostly determines the ultimate effective temporal resolution of BOLD fMRI. The hemodynamic response to a brief neural activity event is sluggish and delayed. This means that the signal is a delayed and low-pass filtered version of underlying neuronal activity. Moreover, differences in hemodynamics in different parts of the brain form a potential confound for dynamic brain connectivity models, as discussed below.

It is an interesting *spatial* property of hemodynamic processes that, although they are characterized by a large overcompensating reaction to neuronal activity, their effects are highly local. The level of locality of the hemodynamic response to neuronal activity limits the effective spatial resolution of fMRI. The path of blood inflow in the brain is from large arteries through arterioles into capillaries, where exchange with neuronal tissue takes place at a microscopic level. Blood outflow takes place via venules into the larger veins. The main regulators of blood flow are the arterioles that are surrounded by smooth muscle, although arteries and capillaries are also thought to be involved in blood flow regulation (Attwell et al., 2010). Different hemodynamic parameters have different spatial resolutions. While cerebral blood volume and cerebral blood flow change in all compartments but mostly in venules, oxygenation changes occur mostly in venules and veins. Thus, the achievable spatial resolution with fMRI is limited by its specificity to the smaller arterioles and venules and microscopic capillaries supplying the tissue, rather than the larger supplying arteries and draining veins. The larger vessels have a larger domain of supply or extraction and, as a consequence, their signal is blurred and mislocalized with respect to neuronally active tissue. Here, physiology and physics interact in an important way. It can be shown theoretically (by the effects of thermal motion of spin diffusion over time and the distance of the spins to deoxyhemoglobin) that the origin of the BOLD signal in spin-echo (SE) pulse sequences at high main field strengths (larger than 3T) is much more specific to the microscopic vasculature than to the larger arteries and veins (Uludag, Muller-Bierl, & Ugurbil, 2009). This does not hold for gradient-echo sequences or for SE sequences at lower field strengths. The cost of this greater specificity and higher effective spatial resolution is that SE-BOLD has a lower intrinsic SNR than gradient-echo–BOLD.

High spatial and temporal resolution functional MRI

Ultra-high field (UHF) scanners of 7T and above are increasingly used for high-resolution fMRI studies because there is a large increase in the SNR available for fMRI compared to 1.5T or 3T systems. Combined with advances in parallel imaging (De Zwart, van Gelderen, Golay, Ikonomidou, & Duyn, 2006; Heidemann et al., 2006; Pruessmann, 2004; Wiesinger et al., 2006) and the necessary radio frequency coil technology, UHF fMRI at 7T and beyond has increased the level of spatial and temporal detail accessible with fMRI by at least an order of magnitude over the last decade. This also allows us to exploit the higher spatial specificity to neuronal events provided by SE pulse sequences.

A few 7T fMRI studies in the human visual system with enough spatial resolution to achieve specificity at a level close to cortical columns and layers illustrate the possibilities. Yacoub and colleagues (Yacoub, Harel, & Ugurbil, 2008) were able to characterize ocular dominance maps and orientation preference maps in human V1 (figure 76.3A). The achieved resolution at the level of cortical columns is an important finding, since it indicates that the small point-spread function of the employed SE fMRI is specific enough to differentiate fine-grained cortical representations within cortical topographic maps. Polimeni and colleagues (Polimeni, Fishl, Greve, & Wald, 2010) found resting-state BOLD correlations between output layers II–III of V1 and input layer IV of MT, higher than between any other combinations of granular, infragranular, or supragranular layers between these areas (figure 76.1B). This is in correspondence with the known layer IV to layer II–III forward connectivity between V1 and MT (Felleman & Van Essen, 1991). This result presents a clear indication

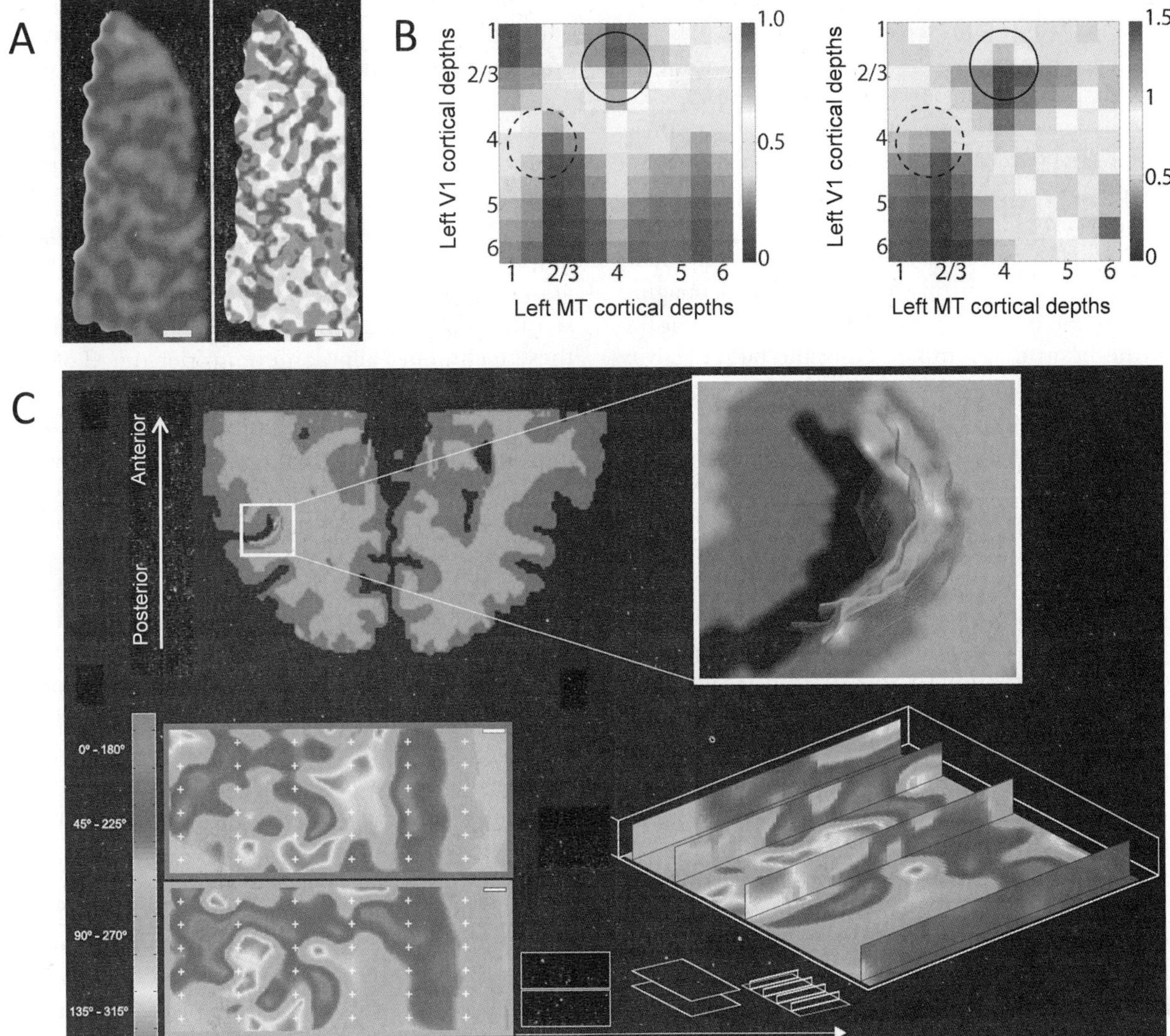

FIGURE 76.3 High spatial resolution 7T fMRI in the human visual system with cortical-column and layer-level specificity. (A) Region of interest selection in V1 on a T1-weighted anatomical image. *Left*: Ocular dominance map in human V1. *Right*: Orientation preference map (from Yacoub et al., 2008). (B) Cortical depth resolved resting-state fMRI correlation matrices between human V1 and middle-temporal (MT) areas, showing increased V1 layer IV to MT layers II–III correlations. *Left*: V1 to MT correlations normalized to 1. *Right*: V1 to MT correlations normalized to diagonal elements (from Polimeni et al., 2010). (C) Axis-of-motion preference map in human MT showing columnar-resolution motion selectivity and its modulation across cortical depth (Zimmermann et al., 2011). (See color plate 69.)

that BOLD fMRI at UHF has enough spatial specificity to discriminate (small contiguous sets of) layers and sample information about their connectivity and functional roles in cortical activity dynamics. Zimmermann and colleagues (2011) show an example of a human visual system study with high spatial resolution UHF fMRI that illustrates the possibility of combining the cortical tangential and radial specificity. In their study, these authors distinguished column-level and layer-level axis-of-motion selectivity in MT simultaneously (figure 76.1C).

Complementary to tangential, and especially radial, specificity afforded by very high spatial resolution fMRI, there are other ways of deriving directionality from space that rely on sophistication of analysis. A case in point is the study from Heinzle, Kahnt, and Haynes (2011), who use a support vector machine–based regression to explain extrastriate human response from those in V1. The support vector machine helps reduce the dimensionality of the large number of candidate voxels in V1 by selecting a sparse set of best-predicting voxels. They find a clear, topographically organized connectivity structure between visual areas, in what they term "cortico-cortical receptive fields" (CCRF), even between spontaneous activity during the resting state in complete darkness. The CCRF of a voxel is given by the weight distribution of the optimal linear combination of voxels in V1 that best predicts the signal fluctuations of a voxel in a downstream area. For the same goal, Haak et al. (2012) developed the method of connective field modeling. The spatial connective field of a population of neurons in a target fMRI imaging voxel is estimated as the contiguous set of voxels in the rest of the brain that jointly predict the target activity best. Concretely, the center and width of a connective field are estimated as the peak and width of a two-dimensional (2D) Gaussian on the cortical surface that maximizes the fit to the signal of the target when used as a local spatial weighting function. This complements and extends the population receptive field estimation method (Dumoulin & Wandell, 2008) in which activity is explained as a function of stimulus-space information. Interestingly, the connectivity field generates a concept of causality-from-space based on cortical tangential information, complementary to the concept from radial depth discussed above. This is derived from the convergence or divergence of (functional) connectivity between areas. For instance, as the authors point out, if the connective field size for V1 to V2 is larger than for V2 to V1, this would indicate that visual information converges from V1 to V2.

Recent advances in enhancing the temporal acquisition resolution of fMRI have come largely from using the spatial encoding capacity of radio frequency–phased arrays with many receive coils for three-dimensional (3D) encoding. When a larger part of the sample is simultaneously excited (either as a contiguous slab or as a set of noncontiguous 2D slices), the combination with 3D encoding can speed up whole-brain acquisition by an order of magnitude. Fast fMRI techniques that apply subsets of these principles include CAIPIRINHA (Breuer et al., 2006); simultaneous multislice and multiplexed 2D echo planar imaging (EPI; Feinberg et al., 2010; Moeller et al., 2010; Setsompop et al., 2012); 3D EPI or echo volumar imaging (Posse et al., 2012); inverse imaging (InI; Lin et al., 2012); and magnetic resonance encephalography (MR-EG; Hennig, Zhong, & Speck, 2007). Multiplexed 2D and accelerated 3D fMRI techniques typically solve *overdetermined* inverse problems (from radio frequency–coil data to voxel signals) because gradient encoding and coil sensitivities completely determine the spatial solution and together fully sample the voxel grid. In contrast, MR-EG and InI image reconstruction solve intrinsically *underdetermined* inverse problems akin to those solved in EEG/MEG distributed source analysis. MR-EG and InI can achieve very high temporal resolution, because a lot of the time-consuming gradient encoding steps are omitted. This means regularization and prior constraints are needed in the computerized image reconstruction to come to a unique spatial problem solution. This solution will have intrinsically lower spatial resolution, which, depending on the kind and degree of remaining gradient encoding, can be inhomogeneous (typically higher at the periphery of the brain, i.e., the cortex, where coil sensitivity information is largest). Thus in a real sense these techniques allow an even wider range of trading in spatial resolution for much higher temporal resolution. In the limit of no gradient encoding at all, one would be imaging by the so-called one-voxel-one-coil principle (Hennig et al., 2007), with a coarse spatial resolution entirely determined by coil sensitivities and an almost unlimited temporal resolution.

The greatly expanded breadth over which SNR, spatial coverage and temporal and spatial resolution can be traded off have mostly been used to the advantage of higher temporal resolution. This has allowed the relative timing of BOLD signal variations to be evaluated directly at the level of hundreds and even tens of milliseconds. For instance, Tsai et al. (2012) used a multiprojection version of the InI technique to achieve 4 mm isotropic nominal spatial resolution across the whole human cortex with a temporal resolution of 10 Hz (a time-to-repetition, or TR, of 100 msec). The authors apply this acquisition to a simple sensorimotor task in which lateralized visual stimuli have to be responded to as fast as possible with an ipsilateral button-press. They observe a temporal offset of several hundreds of milliseconds with V1 activating before M1, reflecting the task-induced order of neuronal events despite differing hemodynamic response function shapes (figure 76.4A). In Chang et al. (2013), the inverse imaging reconstruction principle is combined with an echo-shifting technique that allows a TR shorter than the time-to-echo. This is used to achieve a whole-brain coverage at a staggering temporal sampling rate of 40 Hz (TR = 25 msec), about an 80-fold speed-up compared to conventional EPI with less than a 2-fold SNR loss. The authors report that this fMRI acquisition allows detection of visual stimulus timing offsets from hemodynamic response function shape analysis of 400 msec at the individual level and as low as 50 msec at the group level. Finally, in a rat study, Yu, Qian, Chen, Dodd, and Koretsky (2013) show the possible ultimate combined spatial and temporal specificity of BOLD contrast fMRI (figure 76.4B). By employing a line-scanning technique (that does not have tangential cortical dissolving power), a very high radial cortical depth resolution is achieved at a 20 Hz sampling rate (TR = 50 msec).

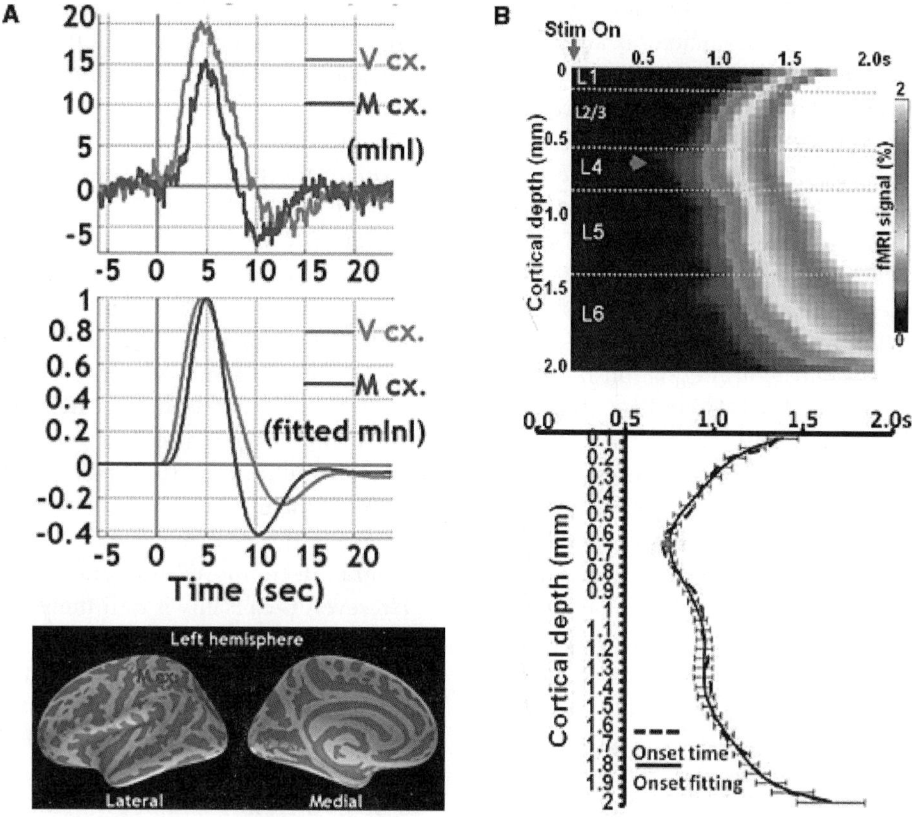

FIGURE 76.4 High temporal resolution fMRI showing relative timing between and within cortical areas. (A) High temporal resolution evolution of hemodynamic responses in human visual and motor cortex in a visuomotor task sampled at 10 Hz. *Top*: Regions of interest, visual cortex (V cx) and motor cortex (M cx), on reconstructed cortical surfaces. *Middle*: Measured responses. *Bottom*: Model-fitted responses showing visual cortex signal increases before motor cortex in the early stages of the hemodynamic response function, reflecting the task-induced order of neuronal events, despite different hemodynamic response function shapes (from Tsai et al., 2012). (B) High spatial and temporal evolution of hemodynamic responses in rat barrel cortex. *Top*: fMRI signal change for different cortical layers (L1–L6) against post-stimulus time. *Bottom*: Recorded and fitted onset time of the hemodynamic response against post-stimulus time (Yu et al., 2013). (See color plate 70.)

BOLD fMRI responses after whisker stimulation show an earlier response onset at layer IV in the barrel cortex. In contrast, BOLD responses in motor cortex through cortico-cortical somatomotor connections had a simultaneous early onset in layer II–III and layer V. These results show strong agreement of laminar fMRI onset (i.e., both spatially and temporally) with input projections into the cortex.

Causal time-series modeling of fMRI

Recently, causal modeling techniques that make use of the temporal dynamics in the fMRI signal and employ time-series analysis and systems identification theory have become popular. This modeling of temporal fMRI signal structure, and its use for causal direction estimation, sets these *dynamic* techniques apart from earlier methods that do not use temporal structure (and that we call *static* techniques). Within the class of dynamic techniques, two separate developments have been most used: Granger causality analysis (GCA; Goebel, Roebroeck, Kim, & Formisano, 2003; Roebroeck, Formisano, & Goebel, 2005; Valdes-Sosa, 2004) and dynamic causal modeling (DCM; Daunizeau, David, & Stephan, 2009; Friston, Harrison, & Penny, 2003). Despite the common goal, there are differences between the two methods. Whereas GCA explicitly models temporal precedence and uses the concept of Granger causality (or G-causality), usually formulated in a discrete time-series analysis framework, DCM employs a biophysically motivated generative model formulated in a continuous-time, dynamic-system framework. And although these approaches have recently started developing in an integrated single direction (Valdes-Sosa, Roebroeck, Daunizeau, & Friston, 2011), initially each was focused on separate issues that pose challenges for the estimation of causal influence from fMRI data. Whereas DCM is formulated as an explicit state space model to account for the temporal convolution of neuronal events by sluggish hemodynamics in a small number of areas, GCA has mostly been aimed at solving the problem of region selection in the enormous spatial dimensionality of fMRI data. Although these two approaches to statistical analysis of causal influence have the focus in this chapter, they are complemented by many other methods (most of them static), such as psychophysiological interactions (Friston et al., 1997), covariance structural equation modeling (McIntosh & Gonzalez-Lima, 1994), and causal graphical models (Ramsey et al., 2009).

Granger causality, or G-causality, was proposed by the economist Clive Granger (Granger, 1969, 1980) and partially based upon earlier ideas of Norbert Wiener (1956). Almost simultaneously with Granger's work, Akaike (1968), and Schweder (1970) introduced similar concepts of influence, prompting Valdes-Sosa and colleagues (Valdes-Sosa et al., 2011) to coin the term "WAGS influence" (for Wiener-Akaike-Granger-Schweder). This is a generalization of a proposal by Aalen (Aalen, 1987; Aalen & Frigessi, 2007), who was among the first to point out the connections between Granger's and Schweder's influence concepts. The general concept of WAGS influence gives an operational definition of what "causality" or "influence" could mean for observations, structured in time, for multiple variables of interest. In economics, the variables of interest might be interest rates, employment numbers, and the federal budget deficit. In neuroscience, the variables could be invasive electrode recordings, intracranial EEG, noninvasive EEG, MEG, or fMRI time series from different parts of the brain. The idea of WAGS influence or G-causality is that a variable *A* G-causes another variable *B* if the prediction of *B*'s values improves when we use past values of *A*, given that all other relevant information is taken into account. Two more things need to be specified when we want to apply this idea to our data: (1) which model we use to make predictions and (2) what "all other relevant information" is. The second point is dealt with in the structural model selection process, which entails the selection of a reasonable set of relevant variables (e.g., voxels, channels, or ROIs). Structural model selection is rarely an easy modeling step, as we will discuss below in the context of the "missing region problem." The most common answer to the first point is the linear autoregressive (AR) model for discretely sampled data. The AR model is a simple model that can flexibly represent a wide range of signal dynamics, auto- and cross-correlation patterns, and spectral characteristics, and is easy to estimate from data records. The initial developments in AR modeling of fMRI data led to a number of applications studying human mental states and cognitive processes, such as task switching (Roebroeck et al., 2005), gestural communication (Schippers, Roebroeck, Renken, Nanetti, & Keysers, 2010), top-down control of visual spatial attention (Bressler, Tang, Sylvester, Shulman, & Corbetta, 2008), switching between executive control and default-mode networks (Sridharan, Levitin, & Menon, 2008), fatigue (Deshpande, LaConte, James, Peltier, & Hu, 2009), and the resting state (Uddin, Kelly, Biswal, Xavier Castellanos, & Milham, 2009). Additional variants of AR modeling applied to fMRI include time-varying influence (Havlicek, Jan, Brazdil, & Calhoun, 2010), blockwise (or "cluster-wise") influence from one group of variables to another (Barrett, Barnett, & Seth, 2010; Sato et al., 2010), and frequency-decomposed influence (Sato et al., 2009).

However, G-causality is definitely not tied exclusively to the standard linear AR model, especially when viewed in the wider context of WAGS influence. It can be equally well instantiated in nonlinear models (Freiwald et al., 1999) and time-varying models for nonstationary data (Hesse, Moller, Arnold, & Schack, 2003), and it can be framed in terms of nonparametric spectral factorization (Dhamala, Rangarajan, & Ding, 2008). In addition, WAGS influence has been instantiated in Markov processes and more general stochastic processes, based on Martingale theory (Aalen & Frigessi, 2007) and in continuous-time signal models (Florens & Fougere, 1996). Nonetheless, it will be informative to compare the class of linear stochastic models (LSMs), of which the AR model is a special case, with the signal model in the DCM approach to discuss their subtle distinctions.

Both LSMs and DCMs can be given a state-space formulation (figure 76.5). In a state-space representation, the relations between measured variables y_j (e.g., fMRI data) and exogenous input variables u_k (e.g., stimulus functions) are modeled through *unobservable* state variables z_i. State-space representations generally consist of two sets of equations. The *transition equations*, or *state equations*, describe the evolution of the dynamic system over time, capturing relations among the hidden state variables z_i themselves and the influence of exogenous inputs u_k. The *observation equations*, or *measurement equations*, describe how the measurement variables y_j are obtained from the hidden state variables z_i and the inputs u_k. The LSM accommodates equivalent representation of the general class of autoregressive moving average models with exogenous inputs (ARMAX models; Reinsel, 1997). Connectivity modeling of neuroimaging data involves the estimation of the elements

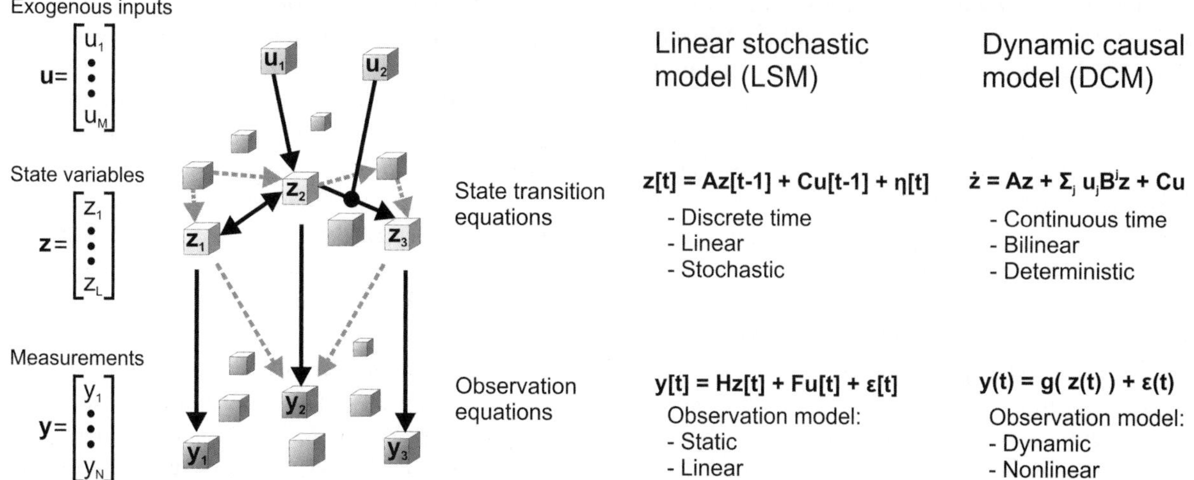

FIGURE 76.5 State-space representations of dynamic connectivity models. The state-space representations for a linear stochastic model (often employed in Granger causality analysis) and a dynamic causal model are shown and compared with respect to their mathematical properties. In a state-space representation, the relations between measured variables $\mathbf{y} = (y_1, \ldots, y_N)$ and, possibly, exogenous input variables $\mathbf{u} = (u_1, \ldots, u_M)$ are modeled through unobservable state variables $\mathbf{z} = (z_1, \ldots, z_L)$. State-space equations generally consist of two sets of equations. The transition equations or state equations describe the evolution of the dynamic system over time, capturing relations among state variables $\mathbf{z}$ themselves and the influence of exogenous inputs $\mathbf{u}$. The observation equations or measurement equations relate the measurement variables $\mathbf{y}$ to the state variables $\mathbf{z}$ and inputs $\mathbf{u}$. Connectivity modeling of neuroimaging data involves the estimation of the elements in the coefficient matrices $\mathbf{A}$, $\mathbf{B}_j$, and $\mathbf{C}$ from measurements $\mathbf{y}[t]$ and, possibly, inputs $\mathbf{u}[t]$. Whereas a linear stochastic model employs linear stochastic transition equations, those in dynamic causal modeling are bilinear and deterministic. (Adapted from Roebroeck, Formisano, et al., 2011.)

in the coefficient matrices (A, B^j, and C in figure 76.5) from measurements $\mathbf{y}[t]$ and, possibly, the inputs $\mathbf{u}[t]$.

The first important difference in modeling signal dynamics is that LSMs employ linear stochastic transition equations, whereas those in the most-used forms of DCM are bilinear and deterministic. The stochastic term in the LSM transition equation allows for variation in the state variables that cannot be explained by the inputs $\mathbf{u}[t]$. In fact, in the case of a purely autoregressive model, exogenous inputs are absent and all signal variation is modeled as driven by uncorrelated stochastic processes (called "innovations"). This forces all dynamic and spectral complexity in the observed signals to be represented in the model parameters. It is exactly this property of comprehensive and flexible representation of signal dynamics and spectral properties that has made autoregressive models a popular tool in analyzing complex biophysical signals (Bernasconi & Konig, 1999; Brovelli et al., 2004; Ding, Bressler, Yang, & Liang, 2000; Harrison, Penny, & Friston, 2003; Kaminski, Ding, Truccolo, & Bressler, 2001). In contrast, the transition equation in DCM for fMRI, in its most used form, does not have a stochastic term. As a consequence, any and all signal dynamics that it can capture are limited to the signal subspace spanned by the assumed inputs. In other words, it assumes that all neural population dynamics can be captured without error from the chosen inputs and the transformation of that input in its "flow" through the DCM network. The exogenous inputs mostly have a very simple form, such as a stimulus function that represents the presence or absence of a visual stimulus or level of experimental manipulation, such as attention left versus right. The inability of DCM to model signal variations beyond those implied by the exogenous inputs makes its connectivity estimation highly dependent on the exact number and form of the assumed inputs and the form of the structural model. Stochastic extensions to DCM have been developed (Daunizeau, Friston, & Kiebel, 2009; Friston, Stephan, Li, & Daunizeau, 2010). These developments clearly have the potential to eliminate one of the differentiating aspects of LSMs and deterministic DCMs and bring the models even closer together. Interestingly, the inclusion of noise in the state equations makes inference on stochastic DCMs usefully interpretable in the stochastic framework of WAGS influence (Valdes-Sosa et al., 2011).

The second important difference in modeling signal dynamics is that, in DCM, the state variables are given a definite physical interpretation within a generative model of the data. For every selected region, a single state variable represents the neuronal or synaptic activity of a local population of neurons, and (in DCM for BOLD fMRI) four or five more (Stephan, Weiskopf, Drysdale, Robinson, & Friston, 2007) represent hemodynamic quantities such as capillary blood volume, blood flow, and deoxyhemoglobin content. All state variables (and the equations governing their dynamics) that serve the mapping of neuronal activity to the fMRI measurements $\mathbf{y}[t]$ (including the observation equation) can be called the *observation model*. Most of the physiologically motivated generative model in DCM for fMRI is therefore concerned with an observation model encapsulating *hemodynamics*. The remaining state variables and transition equations can then be said to model *neurodynamics*. In contrast, in LSM and GCA, the state variables may or may not have a definite physical interpretation, depending on the particular representation chosen. However, in the most straightforward representations, LSM state variables are very simple functions of the measurements $\mathbf{y}[t]$. In its standard formulation, LSM/GCA does not use a biophysical model of hemodynamics and, as a consequence, models neurodynamics and hemodynamics with the same simple dynamical model. In short, the LSM observation model amounts to a linear combination of the state variables at the same moment in time (and hence, is static), whereas the observation model in DCM is nonlinear and dynamic.

The observation model in DCM for fMRI is a biophysical model of hemodynamic coupling largely based on the balloon (Buxton et al., 1998) and Windkessel models (Mandeville et al., 1999). The parameters in this model, such as transit time and autoregulation, are estimated conjointly with the parameters quantifying neuronal connectivity. Thus, the forward biophysical model of hemodynamics is "inverted" in the estimation procedure to achieve a deconvolution of fMRI time series and obtain estimates of the underlying neuronal states. It is important to note that the specific biophysical model for the interactions between neuronal states (neurodynamics) on one hand, and the model for the hemodynamics on the other hand, largely dictate which of these models will absorb given aspects of the observed data. For instance, if there are delayed coherent variations between variables in the observed data and the hemodynamic model has much more affordance for delays than the neurodynamic model (as is the case in DCM for fMRI), then the delay will be put into the hemodynamics in the fitting of the model.

Challenges

Causal modeling aspires to use high-resolution spatial and temporal fMRI signals and good experimental design to resolve neuronal population interactions. However, the accompanying challenges are mainly caused by the enormous dimensionality of the data that contains hundreds of thousands of channels (voxels) and the temporal convolution of neuronal events by sluggish hemodynamics that can differ between remote parts of the brain. We formulate the ensuing challenges here as the "missing region problem," the "missing time problem," and the "missing model problem" and discuss how each can be ameliorated or avoided.

Missing Region Problem A distinction can be made between exploratory and confirmatory *structural* model selection procedures for brain connectivity, that is, procedures that choose the set of applicable structural models. Exploratory techniques use information in the data to investigate the relative applicability of many models. As such, they have the potential to detect "missing" regions in structural models. Confirmatory approaches test hypotheses about connectivity within a small set of models assumed to be applicable. DCM is mostly used as a confirmatory technique in structural model selection. This means a small number of models are compared, where each model specifies the nodes and edges in a directed (possibly cyclic) structural graph model. Applications of DCM often use very simple structural models (typically employing 3–6 ROIs) in combination with its complex, parameter-rich dynamical model. The clear danger with overly simple structural models is that of spurious influence, or the *missing region problem*: an erroneous influence found between two selected regions that in reality is due to interactions with additional regions that have been ignored (Roebroeck, Formisano, & Goebel, 2011). Prototypical examples of the missing region problem of relevance in brain connectivity are those between unconnected structures A and B that receive common input from, or are intervened by, an unmodeled region C.

Early applications of G-causality to fMRI data were aimed at counteracting the problems with overly restrictive anatomical models by employing more permissive structural models in combination with a simple dynamical model (Goebel et al., 2003; Roebroeck et al., 2005; Valdes-Sosa, 2004). These applications reflect the observation that estimation of mathematical models from time-series data generally has two important aspects: model selection and model identification (Ljung, 1999). In the *model selection* stage, a class of models is chosen by the researcher that is deemed suitable for the problem at hand. In the *model identification* stage, the parameters in the chosen model class are estimated from the observed data record. In practice, model selection and identification often occur in a somewhat interactive fashion where, for instance, model selection can be informed by the fit of different models to the data achieved in an identification step. The important point is that model selection involves a mixture of choices and assumptions on the part of the researcher and the information gained from the data record itself.

The technique of Granger causality mapping was developed to explore all regions in the brain that interact with a single selected reference region using autoregressive modeling of fMRI time series (Roebroeck et al., 2005). By employing a simple bivariate model containing the reference region and, in turn, every other voxel in the brain, the sources and targets of influence for the reference region can be mapped. It was shown that such an exploratory mapping approach can form an important tool in structural model selection. Although a bivariate model does not discern direct from indirect influences, the mapping approach locates potential sources of common input and areas that could act as intervening network nodes. Other applications of autoregressive modeling to fMRI data have considered full multivariate models on large sets of selected brain regions, illustrating the possibility of estimating high-dimensional dynamical models. For instance, Valdés-Sosa et al. (Valdes-Sosa, 2004; Valdes-Sosa, Sanchez-Bornot, et al., 2005) applied these models to parcellations of the entire cortex in conjunction with sparse regression approaches that enforce an implicit structural model selection within the set of parcels. In another example (Deshpande, Hu, Stilla, & Sathian, 2008), a full multivariate model was estimated over 25 ROIs (that were found to be activated in the investigated task) together with an explicit reduction procedure to prune regions from the full model as a structural model-selection procedure. Exploratory structural model selection is not principally tied to GCA. Indeed, methodological developments for DCM have arisen that address the comparison of a large number of structural models, such as family wise inference (Penny et al., 2010) and post hoc inference (Friston & Penny, 2011; Rosa, Friston, & Penny, 2012).

Missing Time Problem Causal computational modeling of the BOLD fMRI signal that aims to use the temporal structure of the signal faces a set of challenges that we together refer to as the missing time problem (cf. Valdes-Sosa et al., 2011). The hemodynamic response to a brief neural activity event is sluggish and delayed, entailing that the fMRI BOLD signal is a delayed and low-pass filtered version of underlying neuronal activity. More than the distorting effects of hemodynamic processes on the temporal structure of fMRI signals per se, it is the *difference* in hemodynamics in different parts of the brain that forms a severe confound for dynamic brain-connectivity models. Particularly, the delay imposed upon fMRI signals with respect to the underlying neural activity is known to vary between subjects and, to a lesser extent, between different brain regions of the same subject (Aguirre, Zarahn, & D'Esposito, 1998; Saad, Ropella, Cox, & DeYoe, 2001). Finally, the sluggish and delayed BOLD signal is relatively sparsely temporally sampled such that, in a real sense, time is missing in our data.

David et al. (2008) aimed at direct comparison of GCA and DCM for fMRI time series and explicitly pointed at deconvolution of variable hemodynamics for causal inference. The authors created a controlled animal experiment where gold-standard validation of neuronal connectivity estimation was provided by intracranial EEG measurements. As discussed extensively in Friston (2009) and Roebroeck, Formisano, et al. (2011), such a validation experiment can provide important information on best practices in fMRI-based brain connectivity modeling that, however, need to be carefully discussed and weighed. In David et al.'s study, simultaneous fMRI, EEG, and intracranial EEG were measured in 6 rats during epileptic episodes in which spike-and-wave discharges spread through the brain. Functional MRI was used to map the hemodynamic response throughout the brain to seizure activity, where ictal and interictal states were quantified by the simultaneously recorded EEG. Three structures were selected by the authors as the crucial nodes in the network that generates and sustains seizure activity and further analyzed with (1) DCM, (2) simple AR modeling of the fMRI signal, and (3) AR modeling applied to neuronal-state variable estimates obtained with a hemodynamic deconvolution step. By applying G-causality analysis to deconvolved fMRI time series, the stochastic dynamics of the linear state-space model are augmented with the complex, biophysically motivated observation model in DCM. The results showed both AR analysis after deconvolution and DCM analysis to be in accordance with the gold-standard intracranial EEG analyses, identifying the most pertinent influence relations undisturbed by variations in hemodynamic response function latencies. In contrast, the final result of simple AR modeling of the fMRI signal showed less correspondence with the gold standard, due to the confounding effects of different hemodynamic latencies in different regions that are not accounted for in the model.

The lack of causal models to account for the varying hemodynamics convolving the signals of interest and aggregation of dynamics between time samples has prompted a set of validation studies evaluating the conditions under which discrete AR models can provide reliable connectivity estimates. In Roebroeck et al. (2005), simulations were performed to validate the use of bivariate AR models in the face of hemodynamic convolution and sampling. They showed that under these conditions (even without variability in hemodynamics), AR estimates for a unidirectional influence are biased toward inferring bidirectional causality, a well-known problem when dealing with aggregated time series (Wei, 1990). They then went on to show that unbiased nonparametric inference for bivariate AR models can be based on a difference of influence term (X->Y—Y->X). In addition, they posited that inference on such influence estimates should always include experimental modulation of influence, in order to rule out hemodynamic variation as an underlying reason for spurious causality. In Deshpande, Sathian, and Hu (2010), the authors simulated fMRI data by manipulating the causal influence and neuronal delays between local field potentials acquired from the macaque cortex and varying the hemodynamic delays of a convolving hemodynamic response function, the SNR and the sampling period of the final simulated fMRI data. They found that in multivariate simulations with four regions and hemodynamic and neuronal delays drawn from a uniform random distribution, correct network detection from fMRI was well above chance and was up to 90% under conditions of fast sampling and low measurement noise. Other studies confirmed the observation that techniques with intermediate temporal resolution, such as fMRI, can yield good estimates of the causal connections based on AR models (Stevenson & Kording, 2010), even in the face of variable hemodynamics (Ryali, Supekar, Chen, & Menon, 2010). In contrast with these results, another recent simulation study investigating a host of connectivity methods concluded low detection performance (sensitivity) of directed influence by AR models (Smith et al., 2010) compared to static models that use the instantaneous dependence structure in the data. This may be explained by the mechanism used to *generate* the gold-standard connectivity in this study, which, in contrast to the earlier studies, did not use an explicit mechanism to create delayed influences at the neuronal level with specified latencies.

An insightful comprehensive analysis of the combined issues of hemodynamic variability and coarse temporal sampling is provided in Seth, Chorley, and Barnett (2013). It is shown, both with theoretical arguments and detailed simulations, that G-causality based on discrete linear AR models is invariant to the convoluting effects of hemodynamics, even when these are different between brain areas. The theoretical argument stems from the fact that a general measure of influence remains unchanged if channels are each premultiplied with different invertible lag operators (Geweke, 1982). In Seth et al. (2013), it is shown by simulations that this invariance property generalizes to the relatively mild nonlinearities of the balloon model equations. This means that, when sampling fast enough, G-causality at the neuronal level is accurately detected in the BOLD signal even though hemodynamic convolution reverses the order of signal peaks in the analyzed BOLD time series. Although no definitive statements could be made, this seemed to require sampling rates on the order of 10 Hz or higher, that is, TRs of 100 msec or less. The authors make the case that it is a low sampling rate, rather than hemodynamic convolution, that is the main potential confounder in AR analysis of fMRI data. This is especially relevant given the advances in fast fMRI pulse sequences that were discussed above. It is important to emphasize again that this analysis is specific to linear discrete AR models. More general discrete vector autoregressive moving average models have further invariances that can account for combinations of convolution, sampling, and additive noise situations (Amendola, Niglio, & Vitale, 2010; Roebroeck, Seth, & Valdes-Sosa, 2011; Solo, 2006).

Missing Model Problem Foregoing discussions about missing region and missing time problems might give an impression that increasing model complexity is the way forward for causal connectivity models for fMRI—either structural model complexity by including more regions, or dynamical model complexity by formulating realistic parameter-rich generative models, or both. There certainly is a lot to gain in that direction, especially when supported by the advances in spatiotemporal fMRI resolution. However, in resorting to model complexity-increasing solutions to missing region and time problems, one faces what we will call the missing model problem. Part one of this problem is very straightforward: sometimes models do not exist. Although almost pedantically simple and obvious to state, this has important repercussions in cognitive neuroscience investigations of causal brain connectivity. It begins with the realization that the absence of complex models does not necessarily mean that no useful investigation can be performed. A particularly powerful illustration has recently been reported in Schippers et al. (2010), where between-brain (rather than within-brain) influence is investigated in the context of human social communication by gestures in the game of charades (see figure 76.6). Here, the causal chain of information that goes from the sender of information (the gesturer in charades) to the receiver of information (the guesser) is investigated with conventional fMRI time resolution.

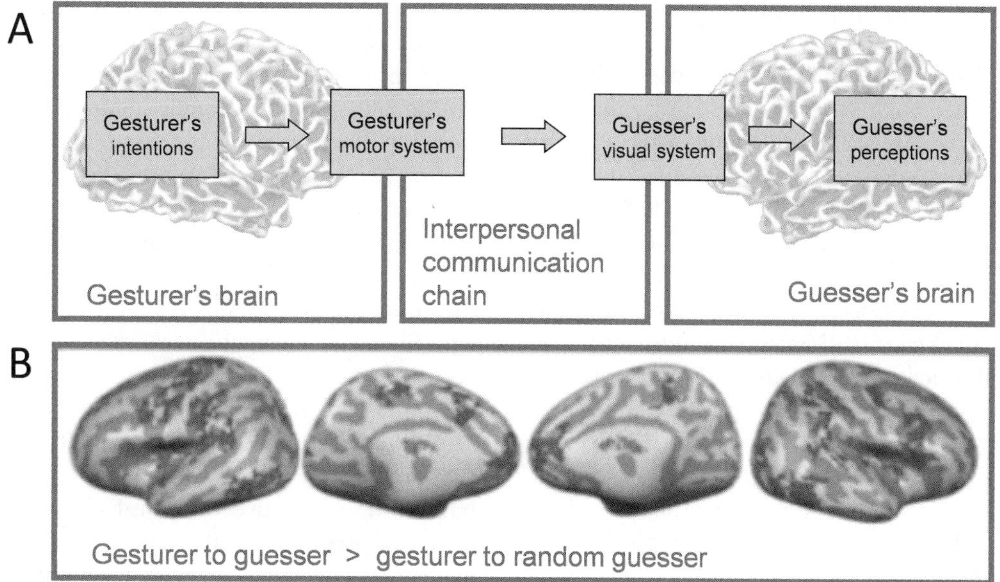

FIGURE 76.6 Between-brain causality in a social communication study, illustrating causality inference when a detailed generative model is missing. (A) The causal chain of information that goes from the sender of information (the gesturer in charades) to the receiver of information (the guesser) can span seconds. (B) The time-lagged flow of information between communicating brains is resolved with G-causality analysis of fMRI. (Adapted from Schippers et al., 2010.) (See color plate 71.)

Since the total chain of temporal causality in social communication can span intervals of seconds (figure 76.6A), it can be very informative to investigate time-delayed information between the brains of the communicators. The time-lagged flow of information was resolved with G-causality analysis and, as expected, the between-brain G-causality from gesturer (information sender) to guesser (information receiver) was found to be significant, in contrast to that in the anticausal direction (figure 76.6B). More importantly as a finding in social neuroscience research, the brain areas found to be communicating were found to be part of the putative human mirror system in both brains. This illustrates the use of the nongeneratively modeled G-causality concept to answer interesting neuroscience questions that (1) are not easily formulated by a generative model, for instance, incorporating the entire interpersonal communication chain, and (2) are concerned with the identification of the unknown areas that interact to perform a perceptual, cognitive, or motor task.

Part two of the missing model problem is more subtle, and states that many existing realistic models contain too many interrelated parameters to estimate from the available data records. This stipulates the interaction between models and data in model identifiability. A model is identifiable from a given data record if there is a unique setting of its parameter values that fits the data best. If there is a number (or even an infinity) of possible parameter settings that all fit the data equally well, we call the model unidentifiable. If we are faced with an unidentifiable model, we have only a few options: give up (cf. part one of the missing model problem), simplify the model (and inevitably make it less realistic), try to acquire data that supports its identification better, or increase the number and weights of assumptions or Bayesian priors that force parameters to a priori values. It could be supposed that part two of the missing model problem is largely solved in model comparison frameworks that use a mixture of model fit and model complexity penalty as a criterion, such as the Bayesian information criterion (BIC) or free energy. Indeed, the use of such model comparison criteria is to be preferred over model fit alone in any model selection or comparison step. However, the weights of complexity and model fit in the balance are dictated by the criterion chosen by the researcher. What is the "best" model (among those considered) according to one criterion could be ranked lower by another criterion that, for example, weights model fit more and complexity less. Moreover, it is common to perform model comparison between dynamical model parameters within a confined structural model (cf. the missing region problem). This means complexity is balanced with model fit in a situation where the above choices have already been made, since the data have already been recorded and structural and dynamical models have been preselected. The missing model problem is about general choices to be made before this phase.

Discussion and future directions

We discussed causal modeling of fMRI data, focusing on two popular dynamic methods (G-causality and dynamic causal modeling) and their distinguishing characteristics, as has been discussed extensively in the literature (e.g., Roebroeck, Formisano, et al., 2011, and its comments and references). This led us to the formulation of the most important challenges for causal time-series modeling of fMRI as the missing region problem, missing time problem, and missing model problem. The main differences between methods can be viewed as the relative prominence and the preferred methodological solutions they give to these three problems. Interestingly, the most recent developments in both methods (see, e.g., Roebroeck, Seth et al., 2011; Valdes-Sosa et al., 2011) have largely adopted typical strong points of the other, such as structural model exploration (Friston & Penny, 2011; Rosa et al., 2012) or generative modeling of hemodynamics (Ryali et al., 2010; Smith, Pillai, Chen, & Horwitz, 2009). This means computational causal modeling of fMRI data as a tool for the cognitive neuroscientist is now even less a choice between one concrete method and the other. Rather, it is about choosing appropriate solutions that provide a balanced answer to the three problems. The balanced answer will be different each time for different types of experimental questions. Performing structural model exploration and considering larger structural models addresses the missing region problem. Inverting generative models of hemodynamics is a methodological answer to the missing time problem. The missing model problem forces one to consider that such model-complexity-increasing solutions cannot always be combined and fully exploited in each situation. Of course, DCM and GCA are by no means the only causal modeling methods for fMRI. Among other popular methods are graphical causal models (Ramsey et al., 2009), psychophysiological interactions (Friston et al., 1997), and covariance structural equation modeling (McIntosh & Gonzalez-Lima, 1994). We do not mean to discourage the usage of these methods. In fact, the challenges of causal fMRI modeling and much of the discussion above apply just as much to these methods.

At the beginning of this chapter, we posed the question of how increased spatial and temporal resolution of acquisition can assist in causal fMRI modeling. We can now revisit this question with a future prospect that can be formulated most easily for causal time-series models such as GCA and DCM.

Part of the excitement about high spatial resolution fMRI is that its capability of resolving cortical laminar structure provides an experimental link to cortical microcircuits. The canonical cortical microcircuit thesis postulates that neocortex is organized as a sheet of many microcircuits (Douglas & Martin, 1991, 2004, 2007a, 2007b). Each radially organized circuit stretches over the cortical layers and organizes feedforward and feedback outputs and inputs in discrete layers (figure 76.7). As such, high spatial resolution fMRI has the potential to obtain causal direction from intracortical space (see figure 76.3B), completely complementary to the concept of causal direction from time in G-causality. The theory of cortical microcircuits is well established and can be used to give dynamical computational accounts of perceptual or cognitive mechanisms, such as predictive coding (Bastos et al., 2012). As the animal study in figure 76.4B illustrates, BOLD fMRI potentially has enough effective temporal resolution to capture some of the temporal precedence in layer activity dynamics stemming from the structural microcircuit organization. The prospect is that increased spatial and temporal resolution of fMRI acquisition can support the identification of microcircuit models of a reasonable level of sophistication. This will require causal modeling approaches that take both intracortical space and temporal signal structure into account, possibly along with aspects such as cortico-cortical connective field signatures. In turn, this creates the potential to tie causal analysis of fMRI data to computational simulation models of cognition and perception (Peters, Jans, van de Ven, De Weerd, & Goebel, 2010) and add physiological realism to neurodynamics models. Although models of hemodynamics for causal fMRI analysis have reached a reasonable level of complexity, the models of neuronal dynamics used to date have remained simple, comprising one or two state variables for an entire cortical region or subcortical structure. Realistic microcircuit models of neuronal activity have a long history and have reached a high level of sophistication (Deco, Jirsa, Robinson, Breakspear, & Friston, 2008; Markram, 2006). Furthermore, microcircuit models have been used in combination with structural connectivity data from dMRI to capture the slow temporal dynamics observed with fMRI in humans in the resting state (Deco, Jirsa, & McIntosh, 2011; Honey et al., 2009; Honey, Thivierge, & Sporns, 2010).

The use of dMRI to guide model selection and comparison, and possibly aid in avoiding the missing region problem, has also begun to enter causal modeling of cognitive neuroscience investigations with a task design (Stephan & Tittgemeyer, 2009). Extending the multimodal information possibilities, several studies have aimed at model-driven fusion of simultaneously recorded fMRI and EEG data. This can be achieved by inverting a separate observation model for each

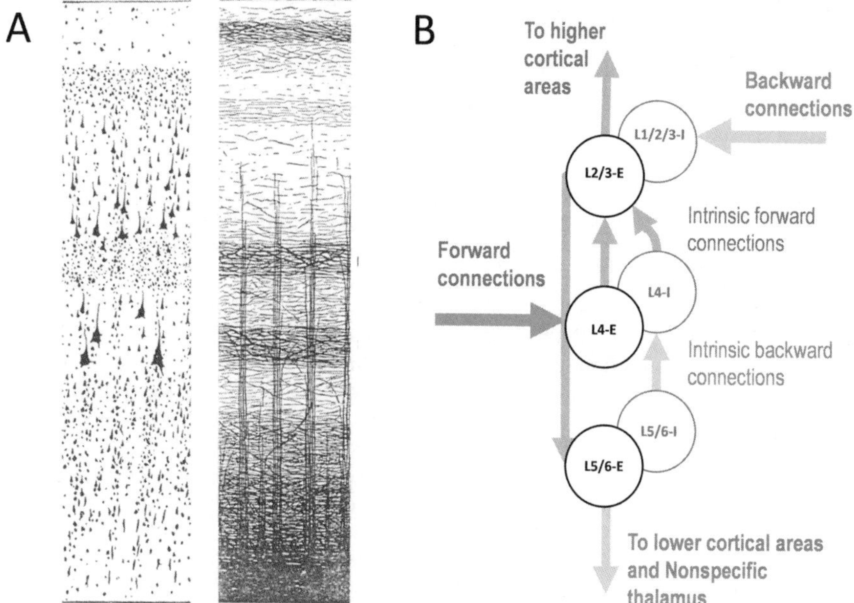

FIGURE 76.7 Cortical microcircuits and the laminar specificity of causal information flow. (A) Typical density variation over cortical layers of cell bodies (left, cytoarchitecture) and myelin (right, myeloarchitecture) (B) A canonical cortical microcircuit model that supports predictive coding. (Adapted from Bastos et al., 2012.)

modality while using the same underlying neuronal model (Deneux & Faugeras, 2010; Riera, Jimenez, Wan, Kawashima, & Ozaki, 2007; Valdes-Sosa et al., 2009). This approach holds great potential to fruitfully combine the superior spatial resolution of fMRI with the superior temporal resolution of EEG. In Valdes-Sosa et al. (2009), structural connectivity information obtained from dMRI is also incorporated. We look in anticipation to what high spatial and temporal resolution fMRI, perhaps in conjunction with other neuroimaging modalities, can bring for causal computational modeling. If the missing model problem is sufficiently considered in this endeavor, we expect human cognitive neuroscience stands to gain very important research tools to answer new questions that are currently out of reach.

REFERENCES

AALEN, O. O. (1987). Dynamic modeling and causality. *Scand Actuar J, 1987*(3–4), 177–190.

AALEN, O. O., & FRIGESSI, A. (2007). What can statistics contribute to a causal understanding? *Scand J Stat, 34,* 155–168.

AGUIRRE, G. K., ZARAHN, E., & D'ESPOSITO, M. (1998). The variability of human, BOLD hemodynamic responses. *NeuroImage, 8,* 360–369.

AKAIKE, H. (1968). On the use of a linear model for the identification of feedback systems. *Ann Inst Stat Math, 20,* 425–439.

AMENDOLA, A., NIGLIO, M., & VITALE, C. (2010). Temporal aggregation and closure of VARMA models: Some new results. In F. Palumbo, C. N. Lauro, & M. J. Greenacre (Eds.), *Studies in classification, data analysis, and knowledge organization: Data analysis and classification* (pp. 435–443). Berlin: Springer.

ATTWELL, D., BUCHAN, A. M., CHARPAK, S., LAURITZEN, M., MACVICAR, B. A., & NEWMAN, E. A. (2010). Glial and neuronal control of brain blood flow. *Nature, 468,* 232–243.

ATTWELL, D., & IADECOLA, C. (2002). The neural basis of functional brain imaging signals. *Trends Neurosci, 25,* 621–625.

BARRETT, A. B., BARNETT, L., & SETH, A. K. (2010). Multivariate Granger causality and generalized variance. *Phys Rev E Stat Nonlin Soft Matter Phys, 81,* 041907.

BASTOS, A. M., USREY, W. M., ADAMS, R. A., MANGUN, G. R., FRIES, P., & FRISTON, K. J. (2012). Canonical microcircuits for predictive coding. *Neuron, 76,* 695–711.

BERNASCONI, C., & KONIG, P. (1999). On the directionality of cortical interactions studied by structural analysis of electrophysiological recordings. *Biol Cybern, 81,* 199–210.

BRESSLER, S. L., TANG, W., SYLVESTER, C. M., SHULMAN, G. L., & CORBETTA, M. (2008). Top-down control of human visual cortex by frontal and parietal cortex in anticipatory visual spatial attention. *J Neurosci, 28,* 10056–10061.

BREUER, F. A., BLAIMER, M., MUELLER, M. F., SEIBERLICH, N., HEIDEMANN, R. M., GRISWOLD, M. A., & JAKOB, P. M. (2006). Controlled aliasing in volumetric parallel imaging (2D CAIPIRINHA). *Magn Reson Med, 55,* 549–556.

BROVELLI, A., DING, M., LEDBERG, A., CHEN, Y., NAKAMURA, R., & BRESSLER, S. L. (2004). Beta oscillations in a large-scale sensorimotor cortical network: Directional influences revealed by Granger causality. *Proc Natl Acad Sci USA, 101,* 9849–9854.

BUXTON, R. B., ULUDAG, K., DUBOWITZ, D. J., & LIU, T. T. (2004). Modeling the hemodynamic response to brain activation. *NeuroImage, 23*(Suppl. 1), S220–S233.

Buxton, R. B., Wong, E. C., & Frank, L. R. (1998). Dynamics of blood flow and oxygenation changes during brain activation: The balloon model. *Magn Reson Med, 39,* 855–864.

Callaway, E. M. (1998). Local circuits in primary visual cortex of the macaque monkey. *Annu Rev Neurosci, 21,* 47–74.

Chang, W. T., Nummenmaa, A., Witzel, T., Ahveninen, J., Huang, S., Tsai, K. W., ... Lin, F. H. (2013). Whole-head rapid fMRI acquisition using echo-shifted magnetic resonance inverse imaging. *NeuroImage, 78,* 325–338.

Daunizeau, J., David, O., & Stephan, K. E. (2009). Dynamic causal modelling: A critical review of the biophysical and statistical foundations. *NeuroImage, 58*(2), 312–322.

Daunizeau, J., Friston, K. J., & Kiebel, S. J. (2009). Variational Bayesian identification and prediction of stochastic nonlinear dynamic causal models. *Physica D, 238,* 2089–2118.

David, O., Guillemain, I., Saillet, S., Reyt, S., Deransart, C., Segebarth, C., & Depaulis, A. (2008). Identifying neural drivers with functional MRI: An electrophysiological validation. *PLoS Biol, 6,* 2683–2697.

De Zwart, J. A., van Gelderen, P., Golay, X., Ikonomidou, V. N., & Duyn, J. H. (2006). Accelerated parallel imaging for functional imaging of the human brain. *NMR Biomed, 19,* 342–351.

Deco, G., Jirsa, V. K., & McIntosh, A. R. (2011). Emerging concepts for the dynamical organization of resting-state activity in the brain. *Nat Rev Neurosci, 12,* 43–56.

Deco, G., Jirsa, V. K., Robinson, P. A., Breakspear, M., & Friston, K. (2008). The dynamic brain: From spiking neurons to neural masses and cortical fields. *PLoS Comput Biol, 4,* e1000092.

Deneux, T., & Faugeras, O. (2010). EEG-fMRI fusion of paradigm-free activity using Kalman filtering. *Neural Comput, 22,* 906–948.

Deshpande, G., Hu, X., Stilla, R., & Sathian, K. (2008). Effective connectivity during haptic perception: A study using Granger causality analysis of functional magnetic resonance imaging data. *NeuroImage, 40,* 1807–1814.

Deshpande, G., LaConte, S., James, G. A., Peltier, S., & Hu, X. (2009). Multivariate Granger causality analysis of fMRI data. *Hum Brain Mapp, 30,* 1361–1373.

Deshpande, G., Sathian, K., & Hu, X. (2010). Effect of hemodynamic variability on Granger causality analysis of fMRI. *NeuroImage, 52,* 884–896.

Dhamala, M., Rangarajan, G., & Ding, M. (2008). Analyzing information flow in brain networks with nonparametric Granger causality. *NeuroImage, 41,* 354–362.

Ding, M., Bressler, S. L., Yang, W., & Liang, H. (2000). Short-window spectral analysis of cortical event-related potentials by adaptive multivariate autoregressive modeling: Data preprocessing, model validation, and variability assessment. *Biol Cybern, 83,* 35–45.

Douglas, R. J., & Martin, K. A. (1991). A functional microcircuit for cat visual cortex. *J Physiol, 440,* 735–769.

Douglas, R. J., & Martin, K. A. (2004). Neuronal circuits of the neocortex. *Annu Rev Neurosci, 27,* 419–451.

Douglas, R. J., & Martin, K. A. (2007a). Mapping the matrix: The ways of neocortex. *Neuron, 56,* 226–238.

Douglas, R. J., & Martin, K. A. (2007b). Recurrent neuronal circuits in the neocortex. *Curr Biol, 17,* R496–500.

Dumoulin, S. O., & Wandell, B. A. (2008). Population receptive field estimates in human visual cortex. *NeuroImage, 39,* 647–660.

Feinberg, D. A., Moeller, S., Smith, S. M., Auerbach, E., Ramanna, S., Gunther, M., ... Yacoub, E. (2010). Multiplexed echo planar imaging for sub-second whole brain FMRI and fast diffusion imaging. *PLoS ONE, 5,* e15710.

Felleman, D. J., & Van Essen, D. C. (1991). Distributed hierarchical processing in the primate cerebral cortex. *Cereb Cortex, 1,* 1–47.

Florens, J. P., & Fougere, D. (1996). Noncausality in continuous time. *Econometrica, 64,* 1195–1212.

Freiwald, W. A., Valdes, P., Bosch, J., Biscay, R., Jimenez, J. C., Rodriguez, L. M., ... Singer, W. (1999). Testing nonlinearity and directedness of interactions between neural groups in the macaque inferotemporal cortex. *J Neurosci Methods, 94,* 105–119.

Friston, K. (1994). Functional and effective connectivity in neuroimaging: A synthesis. *Hum Brain Mapp, 2,* 56–78.

Friston, K. (2009). Causal modelling and brain connectivity in functional magnetic resonance imaging. *PLoS Biol, 7,* e33.

Friston, K., & Penny, W. (2011). Post hoc Bayesian model selection. *NeuroImage, 56,* 2089–2099.

Friston, K., Stephan, K., Li, B., & Daunizeau, J. (2010). Generalised filtering. *Math Probl Eng, 2010,* 1–35.

Friston, K. J., Buechel, C., Fink, G. R., Morris, J., Rolls, E., & Dolan, R. J. (1997). Psychophysiological and modulatory interactions in neuroimaging. *NeuroImage, 6,* 218–229.

Friston, K. J., Harrison, L., & Penny, W. (2003). Dynamic causal modelling. *NeuroImage, 19,* 1273–1302.

Friston, K. J., Mechelli, A., Turner, R., & Price, C. J. (2000). Nonlinear responses in fMRI: The balloon model, Volterra kernels, and other hemodynamics. *NeuroImage, 12,* 466–477.

Geweke, J. F. (1982). Measurement of linear dependence and feedback between multiple time series. *J Am Stat Assoc, 77,* 304–324.

Goebel, R., Roebroeck, A., Kim, D. S., & Formisano, E. (2003). Investigating directed cortical interactions in time-resolved fMRI data using vector autoregressive modeling and Granger causality mapping. *Magn Reson Imaging, 21,* 1251–1261.

Granger, C. W. J. (1969). Investigating causal relations by econometric models and cross-spectral methods. *Econometrica, 37,* 424–438.

Granger, C. W. J. (1980). Testing for causality: A personal viewpoint. *J Econ Dyn Control, 2,* 329–352.

Haak, K. V., Winawer, J., Harvey, B. M., Renken, R., Dumoulin, S. O., Wandell, B. A., & Cornelissen, F. W. (2012). Connective field modeling. *NeuroImage, 66C,* 376–384.

Harrison, L., Penny, W. D., & Friston, K. (2003). Multivariate autoregressive modeling of fMRI time series. *NeuroImage, 19,* 1477–1491.

Havlicek, M., Jan, J., Brazdil, M., & Calhoun, V. D. (2010). Dynamic Granger causality based on Kalman filter for evaluation of functional network connectivity in fMRI data. *NeuroImage, 53,* 65–77.

Heidemann, R. M., Seiberlich, N., Griswold, M. A., Wohlfarth, K., Krueger, G., & Jakob, P. M. (2006). Perspectives and limitations of parallel MR imaging at high field strengths. *Neuroimaging Clin N Am, 16,* 311–320, xi.

Heinzle, J., Kahnt, T., & Haynes, J. D. (2011). Topographically specific functional connectivity between visual field maps in the human brain. *NeuroImage, 56,* 1426–1436.

Hennig, J., Zhong, K., & Speck, O. (2007). MR-encephalography: Fast multi-channel monitoring of brain physiology with magnetic resonance. *NeuroImage, 34,* 212–219.

Hesse, W., Moller, E., Arnold, M., & Schack, B. (2003). The use of time-variant EEG Granger causality for inspecting directed interdependencies of neural assemblies. *J Neurosci Methods, 124,* 27–44.

Honey, C. J., Sporns, O., Cammoun, L., Gigandet, X., Thiran, J. P., Meuli, R., & Hagmann, P. (2009). Predicting human resting-state functional connectivity from structural connectivity. *Proc Natl Acad Sci USA, 106,* 2035–2040.

Honey, C. J., Thivierge, J. P., & Sporns, O. (2010). Can structure predict function in the human brain? *NeuroImage, 52,* 766–776.

Horwitz, B., Friston, K. J., & Taylor, J. G. (2000). Neural modeling and functional brain imaging: An overview. *Neural Netw, 13,* 829–846.

Johansen-Berg, H., & Behrens, T. E. J. (Eds.). (2009). *Diffusion MRI: From quantitative measurement to in-vivo neuroanatomy.* London: Academic Press.

Johansen-Berg, H., & Jbabdi, S. (2011). Tractography: Where do we go from here? *Brain Connect, 1,* 169–183.

Jones, D. K. (Ed.). (2010). *Diffusion MRI: Theory, methods, and applications.* Oxford, UK: Oxford University Press.

Kaminski, M., Ding, M., Truccolo, W. A., & Bressler, S. L. (2001). Evaluating causal relations in neural systems: Granger causality, directed transfer function and statistical assessment of significance. *Biol Cybern, 85,* 145–157.

Lin, F. H., Tsai, K. W., Chu, Y. H., Witzel, T., Nummenmaa, A., Raij, T., ... Belliveau, J. W. (2012). Ultrafast inverse imaging techniques for fMRI. *NeuroImage, 62,* 699–705.

Ljung, L. (1999). *System identification: Theory for the user.* Upper Saddle River, NJ: Prentice-Hall.

Logothetis, N. K. (2008). What we can do and what we cannot do with fMRI. *Nature, 453,* 869–878.

Logothetis, N. K., Pauls, J., Augath, M., Trinath, T., & Oeltermann, A. (2001). Neurophysiological investigation of the basis of the fMRI signal. *Nature, 412,* 150–157.

Mandeville, J. B., Marota, J. J., Ayata, C., Zaharchuk, G., Moskowitz, M. A., Rosen, B. R., & Weisskoff, R. M. (1999). Evidence of a cerebrovascular postarteriole windkessel with delayed compliance. *J Cereb Blood Flow Metab, 19,* 679–689.

Markram, H. (2006). The blue brain project. *Nat Rev Neurosci, 7,* 153–160.

McIntosh, A. R. (2004). Contexts and catalysts: A resolution of the localization and integration of function in the brain. *Neuroinformatics, 2,* 175–182.

McIntosh, A. R., & Gonzalez-Lima, F. (1994). Network interactions among limbic cortices, basal forebrain, and cerebellum differentiate a tone conditioned as a Pavlovian excitor or inhibitor: Fluorodeoxyglucose mapping and covariance structural modeling. *J Neurophysiol, 72,* 1717–1733.

Moeller, S., Yacoub, E., Olman, C. A., Auerbach, E., Strupp, J., Harel, N., & Uğurbil, K. (2010). Multiband multislice GE-EPI at 7 tesla, with 16-fold acceleration using partial parallel imaging with application to high spatial and temporal whole-brain fMRI. *Magn Reson Med, 63,* 1144–1153.

Penny, W., Stephan, K. E., Daunizeau, J., Rosa, M. J., Friston, K., Schofield, T. M., & Leff, A. P. (2010). Comparing families of dynamic causal models. *PLoS Comput Biol, 6*(3), e1000709.

Peters, J. C., Jans, B., van De Ven, V., De Weerd, P., & Goebel, R. (2010). Dynamic brightness induction in V1: Analyzing simulated and empirically acquired fMRI data in a "common brain space" framework. *NeuroImage, 52,* 973–984.

Polimeni, J. R., Fishl, B., Greve, D. N., & Wald, L. L. (2010). Laminar-specific output- to input-layer connections between cortical areas V1 and MT observed with high-resolution restingstate fMRI. *Proc Intl Soc Mag Reson Med, 18.*

Posse, S., Ackley, E., Mutihac, R., Rick, J., Shane, M., Murray-Krezan, C., ... Speck, O. (2012). Enhancement of temporal resolution and BOLD sensitivity in real-time fMRI using multi-slab echo-volumar imaging. *NeuroImage, 63,* 115–130.

Pruessmann, K. P. (2004). Parallel imaging at high field strength: Synergies and joint potential. *Top Magn Reson Imaging, 15,* 237–244.

Ramsey, J. D., Hanson, S. J., Hanson, C., Halchenko, Y. O., Poldrack, R. A., & Glymour, C. (2009). Six problems for causal inference from fMRI. *NeuroImage, 49,* 1545–1558.

Rauch, A., Rainer, G., & Logothetis, N. K. (2008). The effect of a serotonin-induced dissociation between spiking and perisynaptic activity on BOLD functional MRI. *Proc Natl Acad Sci USA, 105,* 6759–6764.

Reinsel, G. C. (1997). *Elements of multivariate time series analysis.* New York, NY: Springer-Verlag.

Riera, J. J., Jimenez, J. C., Wan, X., Kawashima, R., & Ozaki, T. (2007). Nonlinear local electrovascular coupling. II: From data to neuronal masses. *Hum Brain Mapp, 28,* 335–354.

Roebroeck, A., Formisano, E., & Goebel, R. (2005). Mapping directed influence over the brain using Granger causality and fMRI. *NeuroImage, 25,* 230–242.

Roebroeck, A., Formisano, E., & Goebel, R. (2011). The identification of interacting networks in the brain using fMRI: Model selection, causality and deconvolution. *NeuroImage, 58*(2), 296–302.

Roebroeck, A., Seth, A. K., & Valdes-Sosa, P. (2011). Causal time series analysis of functional magnetic resonance imaging data. *Proc Journal of Machine Learning Research Workshop and Conference 12* (pp. 65–94).

Rosa, M. J., Friston, K., & Penny, W. (2012). Post-hoc selection of dynamic causal models. *J Neurosci Methods, 208,* 66–78.

Ryali, S., Supekar, K., Chen, T., & Menon, V. (2010). Multivariate dynamical systems models for estimating causal interactions in fMRI. *NeuroImage, 54*(2), 807–823.

Saad, Z. S., Ropella, K. M., Cox, R. W., & DeYoe, E. A. (2001). Analysis and use of FMRI response delays. *Hum Brain Mapp, 13,* 74–93.

Salmelin, R., & Kujala, J. (2006). Neural representation of language: Activation versus long-range connectivity. *Trends Cogn Sci, 10,* 519–525.

Sato, J. R., Fujita, A., Cardoso, E. F., Thomaz, C. E., Brammer, M. J., & Amaro, E., Jr. (2010). Analyzing the connectivity between regions of interest: An approach based on cluster Granger causality for fMRI data analysis. *NeuroImage, 52,* 1444–1455.

Sato, J. R., Takahashi, D. Y., Arcuri, S. M., Sameshima, K., Morettin, P. A., & Baccala, L. A. (2009). Frequency domain connectivity identification: An application of partial directed coherence in fMRI. *Hum Brain Mapp, 30,* 452–461.

Schippers, M. B., Roebroeck, A., Renken, R., Nanetti, L., & Keysers, C. (2010). Mapping the information flow from one brain to another during gestural communication. *Proc Natl Acad Sci USA, 107,* 9388–9393.

Schweder, T. (1970). Composable Markov processes. *J Appl Probab, 7,* 400–410.

Seth, A. K., Chorley, P., & Barnett, L. C. (2013). Granger causality analysis of fMRI BOLD signals is invariant to hemodynamic convolution but not downsampling. *NeuroImage, 65,* 540–555.

Setsompop, K., Gagoski, B. A., Polimeni, J. R., Witzel, T., Wedeen, V. J., & Wald, L. L. (2012). Blipped-controlled aliasing in parallel imaging for simultaneous multislice echo planar imaging with reduced g-factor penalty. *Magn Reson Med, 67,* 1210–1224.

Smith, J. F., Pillai, A., Chen, K., & Horwitz, B. (2009). Identification and validation of effective connectivity networks in functional magnetic resonance imaging using switching linear dynamic systems. *NeuroImage, 52,* 1027–1040.

Smith, S. M., Miller, K. L., Salimi-Khorshidi, G., Webster, M., Beckmann, C. F., Nichols, T. E., ... Woolrich, M. W. (2010). Network modelling methods for FMRI. *NeuroImage, 54*(2), 875–891.

Solo, V. (2006). On causality, I: Sampling and noise. *Proc IEEE Conference on Decision and Control* (pp. 3634–3639).

Sporns, O. (2010). Connectome. *Scholarpedia, 5*(2), 5584.

Sridharan, D., Levitin, D. J., & Menon, V. (2008). A critical role for the right fronto-insular cortex in switching between central-executive and default-mode networks. *Proc Natl Acad Sci USA, 105,* 12569–12574.

Stephan, K. E., & Tittgemeyer, M. (2009). Tractography-based priors for dynamic causal models. *NeuroImage, 47,* 1628–1638.

Stephan, K. E., Weiskopf, N., Drysdale, P. M., Robinson, P. A., & Friston, K. J. (2007). Comparing hemodynamic models with DCM. *NeuroImage, 38,* 387–401.

Stevenson, I. H., & Kording, K. P. (2010). On the similarity of functional connectivity between neurons estimated across time scales. *PLoS ONE, 5,* e9206.

Thomsen, K., Offenhauser, N., & Lauritzen, M. (2004). Principal neuron spiking: Neither necessary nor sufficient for cerebral blood flow in rat cerebellum. *J Physiol, 560,* 181–189.

Tsai, K. W., Nummenmaa, A., Witzel, T., Chang, W. T., Kuo, W. J., & Lin, F. H. (2012). Multi-projection magnetic resonance inverse imaging of the human visuomotor system. *NeuroImage, 61,* 304–313.

Uddin, L. Q., Kelly, A. M., Biswal, B. B., Xavier Castellanos, F., & Milham, M. P. (2009). Functional connectivity of default mode network components: Correlation, anticorrelation, and causality. *Hum Brain Mapp, 30,* 625–637.

Uludag, K., Dubowitz, D. J., & Buxton, R. B. (2005). Basic principles of functional MRI. In R. R. Edelman & J. R. Hesselink (Eds.), *Clinical magnetic resonance imaging.* San Diego, CA: Elsevier.

Uludag, K., Muller-Bierl, B., & Ugurbil, K. (2009). An integrative model for neuronal activity-induced signal changes for gradient and spin echo functional imaging. *NeuroImage, 48,* 150–165.

Valdes-Sosa, P. A. (2004). Spatio-temporal autoregressive models defined over brain manifolds. *Neuroinformatics, 2,* 239–250.

Valdes-Sosa, P. A., Kotter, R., & Friston, K. J. (2005). Introduction: Multimodal neuroimaging of brain connectivity. *Philos Trans R Soc Lond B Biol Sci, 360,* 865–867.

Valdes-Sosa, P. A., Roebroeck, A., Daunizeau, J., & Friston, K. (2011). Effective connectivity: Influence, causality and biophysical modeling. *NeuroImage, 58*(2), 339–361.

Valdes-Sosa, P. A., Sanchez-Bornot, J. M., Lage-Castellanos, A., Vega-Hernandez, M., Bosch-Bayard, J., Melie-Garcia, L., & Canales-Rodriguez, E. (2005). Estimating brain functional connectivity with sparse multivariate autoregression. *Philos Trans R Soc Lond B Biol Sci, 360,* 969–981.

Valdes-Sosa, P. A., Sanchez-Bornot, J. M., Sotero, R. C., Iturria-Medina, Y., Aleman-Gomez, Y., Bosch-Bayard, J., ... Ozaki, T. (2009). Model driven EEG/fMRI fusion of brain oscillations. *Hum Brain Mapp, 30,* 2701–2721.

Wei, W. W. S. (1990). *Time series analysis: Univariate and multivariate methods.* Redwood City, CA: Addison-Wesley.

Wiener, N. (1956). The theory of prediction. In E. F. Berkenbach (Ed.), *Modern mathematics for engineers.* New York, NY: McGraw-Hill.

Wiesinger, F., van de Moortele, P. F., Adriany, G., de Zanche, N., Ugurbil, K., & Pruessmann, K. P. (2006). Potential and feasibility of parallel MRI at high field. *NMR Biomed, 19,* 368–378.

Yacoub, E., Harel, N., & Ugurbil, K. (2008). High-field fMRI unveils orientation columns in humans. *Proc Natl Acad Sci USA, 105,* 10607–10612.

Yu, X., Qian, C., Chen, D., Dodd, S., & Koretsky, A. (2013). Laminar specificity of fMRI onset times distinguishes top down from bottom up neural inputs mediating cortical plasiticty. *Proc Intl Soc Mag Reson Med, 21.*

Zimmermann, J., Goebel, R., de Martino, F., van de Moortele, P. F., Feinberg, D., Adriany, G., ... Yacoub, E. (2011). Mapping the organization of axis of motion selective features in human area MT using high-field fMRI. *PLoS ONE, 6,* e28716.

77 Multivoxel Pattern Analysis of Functional MRI Data

JARROD A. LEWIS-PEACOCK AND KENNETH A. NORMAN

ABSTRACT The central goal of cognitive neuroscience is to understand how information is processed in the brain. To accomplish this goal, researchers studying human cognition are increasingly relying on multivoxel pattern analysis (MVPA); this method involves analyzing spatially distributed (multivoxel) patterns of functional MRI activity, with the goal of decoding the information that is represented across the ensemble of voxels. In this chapter, we describe the major subtypes of MVPA, we provide examples of how MVPA has been used to study neural information processing, and we highlight recent technical advances in MVPA.

Cognitive neuroscience theories deal with information processing: what information is represented in different brain structures, how is this information transformed over time, and how is it used to guide behavior? Functional MRI (fMRI) constitutes a powerful tool for addressing these questions: while a subject performs a cognitive task, we can obtain estimates of local blood flow (a proxy for local neural processing) from tens of thousands of distinct neuroanatomical locations (*voxels*, or volumetric pixels) within a matter of seconds.

Traditional univariate fMRI analysis methods have focused on characterizing how cognitive variables modulate the activity of individual brain voxels (or small clusters of voxels; e.g., Gonsalves & Cohen, 2010). The goal of this chapter is to describe a different approach to fMRI analysis that focuses on extracting information about a person's cognitive state (i.e., a snapshot of the person's current internal thought space) from spatially distributed, multivoxel patterns of fMRI activity. This approach is referred to as multivoxel pattern analysis (MVPA; Haxby et al., 2001; Haynes & Rees, 2006; Kamitani & Tong, 2005; Norman, Polyn, Detre, & Haxby, 2006; Pereira, Mitchell, & Botvinick, 2009). Over the past decade, this approach has become ubiquitous in fMRI research, and its adoption has led to novel discoveries about the brain bases of perception, attention, imagery and working memory, episodic memory, semantic knowledge, language processing, and decision making (see Rissman & Wagner, 2012; Tong & Pratte, 2012).

Given the goal of detecting the presence of a particular cognitive state in the brain, the primary advantage of MVPA methods over individual voxel–based methods is increased sensitivity for detecting this information. Conventional fMRI analysis methods try to find voxels that show a statistically significant response to the experimental conditions. To increase sensitivity to a particular condition, these methods spatially average across voxels that respond significantly to that condition. Although this approach reduces noise, it also reduces signal in two important ways: first, voxels with weaker (i.e., nonsignificant) responses to a particular condition might carry some information about the presence or absence of that condition. Second, spatial averaging blurs out fine-grained spatial patterns that might discriminate between experimental conditions.

Like conventional methods, the MVPA approach also seeks to boost sensitivity by looking at the contributions of multiple voxels. However, to avoid the signal-loss issues mentioned above, MVPA does not routinely involve uniform spatial averaging of voxel responses. Instead, the MVPA approach uses a weighted average of responses, treating each voxel as a distinct source of information about the participant's cognitive state. The technique finds ways to optimize these weights, and then aggregates this (possibly weak) information across voxels to derive a more precise sense of what the participant is thinking. The multivoxel response can be thought of as a combinatorial code for representing distinctions between cognitive states (see figure 77.1). Because MVPA analyses focus on high spatial-frequency (and often idiosyncratic) patterns of response, they are typically conducted within individual subjects, although recent advances in data alignment procedures have paved the way for the expansion of classification analyses beyond the individual subject (Haxby et al., 2011; see the section "New developments," below).

Broadly speaking, the term MVPA has come to encompass two distinct methods. The first method involves using *pattern classification* methods imported from machine learning to learn a mapping between multivoxel brain states and cognitive state information. This approach flips standard univariate fMRI analysis on its head: standard voxel-based analysis uses multiple

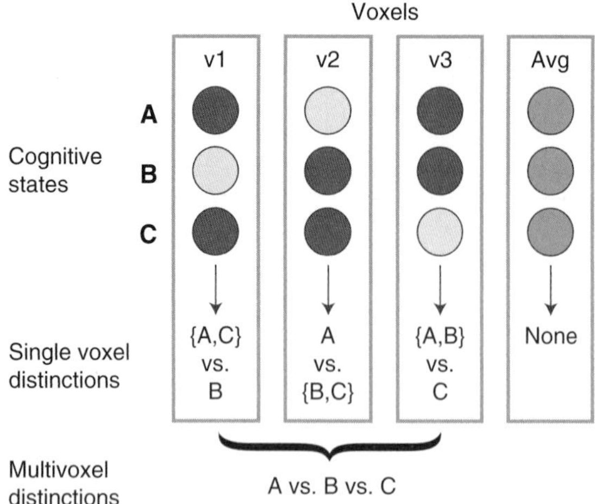

FIGURE 77.1 Observing the multivoxel response pattern allows us to distinguish all three cognitive states, A, B, and C. Considering each voxel in isolation provides only partial discrimination.

regression to predict the activity of individual voxels based on the participant's cognitive state. By contrast, classification-based MVPA uses multiple regression to predict the participant's cognitive state based on the activity of multiple voxels. The second major subtype of MVPA does not use pattern classifiers; rather, it examines the *similarity structure* of multivoxel patterns (i.e., which patterns are similar to one another) and uses this similarity structure information to draw conclusions about what information is reflected in these patterns.

In the section "Mechanics of multivoxel pattern analysis" below, we provide an overview of these two subtypes of MVPA. In the section "Applications of multivoxel pattern analysis," we illustrate how MVPA has been used to study information representation and processing in the brain. Finally, in the section "New developments," we discuss recent advances in MVPA that allow for finer-grained mappings between brain activity and cognitive states within individuals, and also new methods for aligning and combining brain data across individuals.

Importantly, while this chapter is focused on fMRI, we should emphasize that most of the MVPA methods described here can be applied to other imaging modalities as well (for applications to electroencephalography and magnetoencephalography data, see, e.g., Jafarpour, Horner, Fuentemilla, Penny, & Duzel, 2013; for applications to direct neural recording data, see, e.g., Hung, Kreiman, Poggio, & DiCarlo, 2005).

Mechanics of multivoxel pattern analysis

Here, we will review the basic procedures of MVPA. All pattern analyses start with *preprocessing* of the raw fMRI blood oxygen level–dependent, or BOLD, data, including temporal and spatial realignment, noise filtering, and z-scoring of the data (over time, within each voxel) within each run. Next, *feature selection* chooses which voxels will be included in the analysis. All voxels in the brain can be used, but it is often advantageous to limit the analysis to certain voxels. One way to select features is to limit the analysis to specific anatomical regions (e.g., Haxby et al., 2001, focused on ventral temporal cortex in their study of visual object processing). Univariate statistics used in conventional fMRI analysis (e.g., Mitchell et al., 2004) and newer multivariate "wrapper methods" (Guyon & Elisseeff, 2003) can also be used for feature selection (e.g., one can discard the voxels that—taken on their own—do the worst job of discriminating between conditions). Finally, *pattern assembly* involves sorting the data into discrete "brain patterns" corresponding to the pattern of activity across the selected voxels at a particular time in the experiment. Patterns can be assembled using the preprocessed fMRI signal for each trial or, alternatively, by using multiple regression to estimate the unique neural response in each voxel for each trial (Mumford, Turner, Ashby, & Poldrack, 2012). Brain patterns are labeled according to which cognitive state (or experiment condition, stimulus, response, etc.) generated the pattern; this labeling procedure needs to account for the fact that the hemodynamic response measured by the scanner is delayed and smeared out in time, relative to the instigating neural event. Once the patterns have been assembled, MVPA can proceed along two main branches of analysis: *classifier-based MVPA* and *pattern-similarity MVPA*. We will now discuss both methods in turn.

CLASSIFIER-BASED MULTIVOXEL PATTERN ANALYSIS

There are two steps to classifier-based MVPA. The first step, *classifier training*, involves feeding a subset of labeled patterns into a multivariate pattern-classification algorithm. Based on these patterns, the classification algorithm learns a function that maps between voxel activity patterns and cognitive states. As illustrated in figure 77.2, brain patterns can be viewed as points in a multidimensional voxel space; the goal of the classifier is to find a decision boundary in this space that best separates the patterns associated with the to-be-discriminated cognitive states. The second step is generalization testing: given a new pattern of brain activity (not previously presented to the classifier), can the trained classifier correctly determine the cognitive state associated with that pattern?

The most commonly used classifiers are linear classifiers, which derive a linear decision boundary between

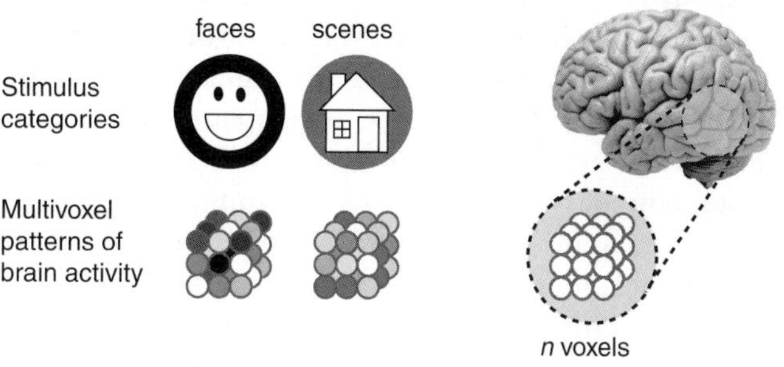

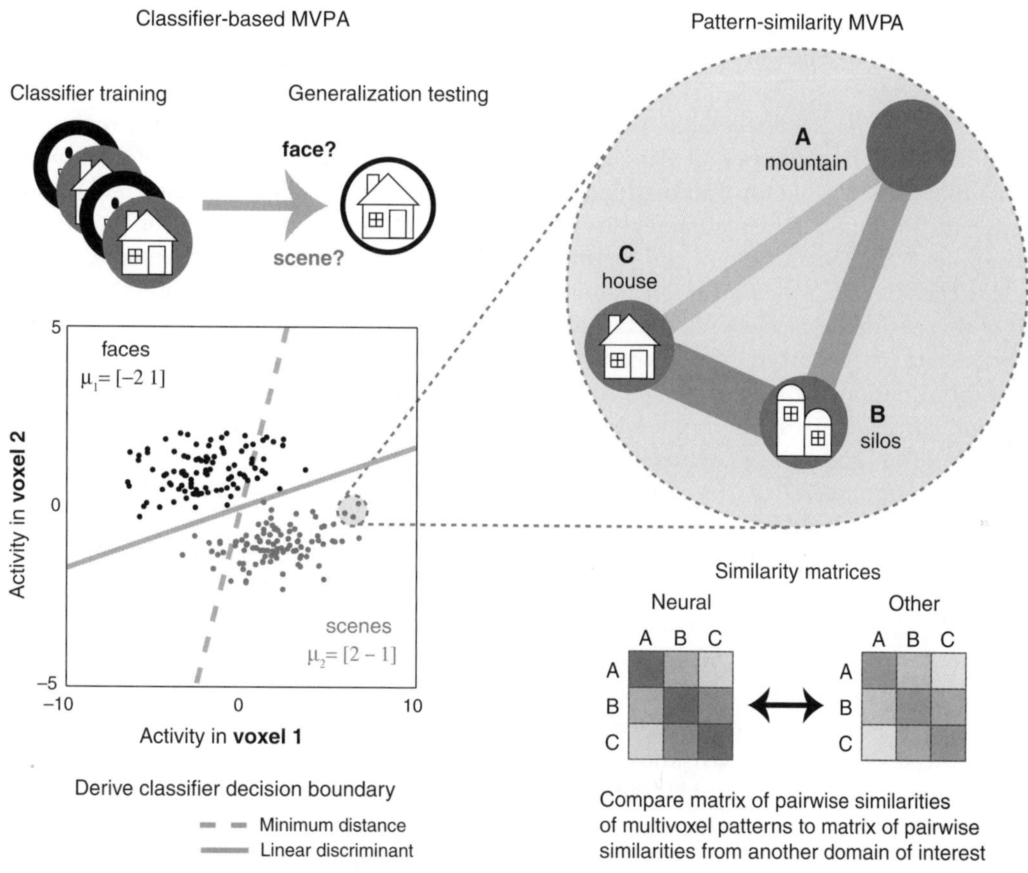

FIGURE 77.2 The two main types of multivoxel pattern analysis of functional MRI data. (See color plate 72.)

classes. At training, the classifier learns a weight for each voxel, plus an intercept term; collectively, these weights determine the equation of the hyperplane (in two dimensions, a line) that forms the decision boundary. At test, the classifier uses these weights (and intercept) to compute a weighted sum of voxel activity values, and it uses this weighted sum to determine whether the test pattern falls on one side or the other of the decision boundary. There is a wide range of linear classification algorithms; the main difference between these algorithms relates to *which features of the data* they use when modeling the data. The multidimensional clouds of data for each class can be characterized in terms of their mean value and also their covariance matrix. This matrix specifies the spread of the cloud along each voxel dimension (i.e., how tall or wide the cloud is) and also the covariance between each pair of dimensions (i.e., the tilt of the cloud). The simplest classifier is the minimum distance classifier (e.g., Haxby et al., 2001), which estimates the mean value for each class based on the training data and then classifies new points based on their proximity to these means. However, figure 77.2 demonstrates how ignoring the covariance matrix can produce nonoptimal decision boundaries. More complex linear classifiers (e.g., Fisher's linear discriminant; Duda, Hart, & Stork, 2012) can converge on the optimal decision boundary by creating a more sophisticated model of the data: in addition to estimating the mean, they also model the class-conditional probability densities (i.e., they estimate the full covariance matrix within and between voxel dimensions for each class). Figure 77.2 shows how the boundary learned by a linear-discriminant classifier factors in the "tilt of the ellipse" for each data cloud. Nonlinear classifiers (e.g., k nearest neighbor, multilayer neural networks) can form even more complex decision boundaries.

Classification of fMRI data is a challenging problem, for several reasons: first, the number of data points (brain patterns) that are available for training tends to be small relative to the number of parameters in the model. For example, Polyn, Natu, Cohen, and Norman (2005) trained a classifier on 450 total brain patterns (150 for each of three stimulus classes) per participant, where each brain pattern consisted of approximately 7,000 voxels. In this situation, the covariance matrix has millions of unique entries (corresponding to all of the voxels, plus all of the unique *pairings* of voxels); each of these entries is a parameter that needs to be estimated based on only a few hundred training patterns. Further adding to the complexity of this problem, the brain patterns are very noisy (i.e., the clouds are highly dispersed). In this kind of situation, where the data are noisy and the number of parameters being estimated by the classifier dwarfs the number of training patterns, classifiers are prone to *overfitting* the noise in the training data: that is, the classifier may learn idiosyncratic features of the training examples rather than the actual distinction between the classes, thereby leading to poor generalization.

Overfitting is the main obstacle to achieving good fMRI classification. One way to combat overfitting is to collect more data, but there are practical limits on collecting more data per participant. In the final section of this chapter, we will discuss new developments in MVPA that allow us to obtain more data by combining across subjects. The other way to combat overfitting is to try to limit the complexity of the classifier. For example, Gaussian naive Bayes classifiers (GNB; Pereira et al., 2009) simplify the modeling of the covariance matrix by treating the n dimensions of the data as independent (such that the off-diagonal elements of the covariance matrix are zero), thus reducing the number of parameters to estimate from n^2 to n. Support vector machines (SVMs; Cox & Savoy, 2003) achieve complexity control by defining the category boundary in terms of a small number of support vectors (i.e., training exemplars close to the decision boundary). Another way to limit the number of free parameters is to limit the number of voxels used for classification (e.g., restricting classification of oriented gratings to low-level visual cortex; Kamitani & Tong, 2005; or restricting classification of faces and scenes to ventral temporal cortex; Kuhl, Rissman, Chun, & Wagner, 2011). This is a useful approach when there is a priori knowledge of strong selectivity for the classes in particular brain regions.

An effective way to reduce the complexity of linear classifiers is to add a *regularization* parameter to the model that punishes undesirable properties of the solution (e.g., large weights on individual voxels). Common forms of regularization are L2 regularization, which penalizes the sum of squares of the voxel weights, and L1 regularization, which penalizes the absolute value of the weights. As the regularization parameter is increased, L2 regularization pulls in extreme voxel weights (resulting in a smoother distribution of weights), whereas L1 regularization causes some weights to be driven to zero (a sparser solution). In both cases, the regularization parameter limits the space of possible solutions, thereby reducing the flexibility of the classifier and reducing overfitting.

Practically speaking, all of the above forms of complexity control (GNBs, SVMs, voxel reduction, and regularization) have been shown to improve generalization performance, relative to linear classifiers that do not incorporate complexity control. The only exception is when the number of voxels is very small or the number of training patterns is very large, at which point it becomes feasible to estimate the full covariance matrix. Importantly, with fMRI data, nonlinear classifiers virtually never outperform linear classifiers on generalization tests—the added flexibility of these classifiers leads to overfitting.

PATTERN-SIMILARITY MULTIVOXEL PATTERN ANALYSIS The second major form of MVPA is pattern-similarity analysis (e.g., Kriegeskorte, Mur, & Bandettini, 2008). Here, brain patterns are viewed as points in high-dimensional voxel space, where the distance between points indicates the similarity of the patterns. Rather than specifying which features of the data to separate with a classifier, pattern-similarity analysis summarizes the space using a matrix that records the distance between each pair of points. This matrix can be viewed as a neural "fingerprint" of the representational space. Although information about the exact positions of the points is lost, the information about similarity structure contained in the pairwise similarity matrix is highly diagnostic of what information is coded in that region (e.g., if items with similar shapes elicit similar neural patterns but items with similar sizes do not, this indicates that the region is more sensitive to shape than size information).

The final step in pattern-similarity MVPA is to compare the neurally derived similarity matrix to some other similarity matrix (e.g., to a matrix holding a cognitive model's predictions about the conceptual similarity between stimuli). The comparison between these matrices is used to evaluate the quality of the model's predictions. A key benefit of the pattern-similarity approach is that—in contrast to the pattern-classification approach outlined above—it is not necessary to explicitly specify (ahead of time) the dimensions of cognitive variance that are of interest. Rather, all of the requisite analyses can be carried out post hoc (e.g., to see if an area represents the size of an object, look at whether objects that are similar in size gave rise to similar neural patterns).

Applications of multivoxel pattern analysis

In this section, we will describe three common uses of MVPA: (1) classifier-based thought tracking, (2) classifier-based information mapping, and (3) information mapping based on pattern similarity. We will discuss the goals of each analysis, and we will review some recent applications of each method.

CLASSIFIER-BASED THOUGHT TRACKING The goal of the classifier-based thought tracking approach is to measure participants' thoughts on a trial-by-trial basis, to characterize the dynamics of these thoughts, and to assess how they relate to behavior. This approach is used when the main concern is tracking a particular latent cognitive state, and there is relatively less concern about how that cognitive state is represented in the brain (although this approach can be applied to specific regions of interest to localize cognitive representations).

Compared to univariate methods, MVPA squeezes more information about the participant's cognitive state out of each snapshot of fMRI data, thereby increasing the effective temporal resolution of fMRI analysis and making it possible to record trajectories of cognitive states over time. However, even with the added sensitivity of MVPA, not all cognitive states are equally "visible" to fMRI. In this situation, researchers often find it useful to take the cognitive state of interest and link it to something that we know is highly visible with fMRI: stimulus *category* information (e.g., faces and scenes). Consider an analogy: when injecting contrast dyes in neuroanatomy, we don't care whether the dye stains cells green or red, so long as the colors are visible under the microscope and so long as the different stains we are using (to measure different cellular properties) have distinct colors. Likewise, when we attach cognitive states to faces or scenes, for example, we don't do this because we care about faces or scenes per se, but rather because thoughts about faces and scenes are highly visible and differentiable with fMRI.

This type of MVPA has been used to study various aspects of memory and cognition. For example, Polyn et al., (2005) used classifiers in a free recall experiment and showed that category-specific patterns of activity emerged about 6 seconds prior to verbal recalls from a given category. In a more recent study, Zeithamova, Dominick, and Preston (2012) used classifier-based thought tracking to explore the process of memory integration. Classifiers tracked the reinstatement of object and scene category information during repeated exposures to AB and BC stimulus pairs (e.g., frog-bucket and bucket-scene). Across subjects, the degree of reactivation of the C item (in this example, the scene) during AB exposures was positively correlated with later performance on a transitive inference memory test for the A-C association; the authors explain this result in terms of participants binding the (reactivated) C item to the A item at encoding. For other recent examples of classifier-based thought tracking, see Lewis-Peacock, Drysdale, Oberauer, and Postle (2012) and Detre, Natarajan, Gershman, and Norman (2013).

Cautionary notes for thought-tracking studies The ideal situation for thought-tracking is to get independent readouts of the relevant cognitive states, but achieving this goal can be difficult. Classifiers are opportunistic: if two categories are anticorrelated in the training set (e.g., all training patterns are either faces or scenes, never both), the classifier will learn this negative correlation, and it will come to treat the lack of scene activity as strong evidence for the presence of faces (see Kuhl et al., 2011, for discussion of this issue). Training on additional categories alleviates this problem by reducing the size of the negative correlation between categories at study (e.g., if there are faces, scenes, and objects, then the absence of faces does not perfectly predict the presence of scenes).

CLASSIFIER-BASED INFORMATION MAPPING A second application of MVPA is less concerned with getting a useful readout of information processing during individual trials, and more concerned with assessing whether a particular fine-grained distinction is represented in a particular brain region (e.g., Pereira & Botvinick, 2011). This analysis is similar in concept to the mass-univariate approach, in that the goal is to determine which brain regions are responsive to a particular cognitive process. However, rather than considering how the activity in each individual voxel is predicted by a person's (presumed) cognitive state, classifier-based information mapping uses information from multiple voxels simultaneously to predict the person's cognitive state.

This analysis can be done using many different a priori regions of interest, or it can be done using the searchlight method (e.g., Kriegeskorte, Goebel, & Bandettini, 2006). This method consists of constructing a "searchlight" of voxels and sliding this searchlight all around the three-dimensional brain volume. For each placement of the searchlight, you consider the multivoxel pattern of activity within that searchlight. A classifier is trained on these patterns and then used to assess how informative these patterns are about the cognitive states of interest.

This approach has been used to discover new insights into cognition and the localization of function in the brain. For example, Soon, Brass, Heinze, and Haynes (2008) used the searchlight technique to discover brain regions whose activity patterns were predictive of future decisions. They found that the outcome of a simple decision (to press a left or right button) could be decoded from prefrontal and parietal cortices up to 10 seconds prior to this decision entering awareness.

Cautionary notes for classification-based information mapping An important caveat for the information-mapping approach is that above-chance decoding—which signals that a brain region contains information about a particular cognitive distinction—does not necessarily imply that this region is involved in guiding behavior based on that distinction (e.g., Williams, Dang, & Kanwisher, 2007). Furthermore, information mapping is opportunistic and may produce false-positive results; see Todd, Nystrom, and Cohen (2013) for discussion of how MVPA can be more susceptible than univariate analysis to experimental confounds (e.g., task difficulty). Finally, multivariate decoding is not necessarily more sensitive than univariate decoding (Jimura & Poldrack, 2012). If the underlying signal has a coarse spatial scale, then univariate approaches using spatial smoothing at this scale will outperform MVPA. In this case, the extra parameters used to model the data in MVPA can lead to overfitting (Kriegeskorte et al., 2006).

Pattern Similarity Analysis The goal of pattern-similarity analysis of fMRI data (e.g., Kriegeskorte, Mur, & Bandettini, 2008) is to make inferences about the similarity of mental concepts based on the similarity of patterns of brain activity elicited by those concepts. The strength and versatility of this approach comes from the many different ways that similarity matrices can be computed and thus permits many types of comparisons. For example, Kriegeskorte, Mur, Ruff, et al. (2008) showed that the similarity structure of neural patterns in human IT cortex (measured using fMRI) resembles the similarity structure of neural patterns in monkey IT cortex (measured using electrophysiology). Furthermore, they showed that the similarity structure of neural patterns in both human and monkey IT (specifically, clustering into animate vs. inanimate objects) could not be explained purely in terms of the low-level visual features of the stimuli.

Pattern-similarity analysis is frequently used to study the representation of item-specific information—the logic here is that regions that differentiate items within a category should show greater pattern similarity between two instances of the same item, compared to two distinct items from the same category. For example, Ritchey, Wing, Labar, and Cabeza (2012) found evidence that encoding-retrieval similarity at the individual item level predicted memory success. Similarity between an item's neural representation and the neural representations of *other* studied items has also been used to predict memory performance. For example, LaRocque et al. (2013) found that greater levels of across-item pattern similarity in perirhinal and parahippocampal cortices were associated with better recognition memory performance.

Cautionary notes for pattern-similarity analysis As described above, classifiers compute weighted combinations of features that discriminate between classes; uninformative or noisy features may be effectively "filtered out" by being assigned small weight values. In contrast, pattern-similarity analyses do not compute weights for each voxel—these analyses treat all voxels as equally important. For this reason, pattern-similarity analyses are more susceptible to contamination from uninformative or noisy features than classifiers. Another concern is that pattern-similarity results can be influenced by univariate effects. For example, imagine that a 33-voxel searchlight contains a 10-voxel subregion that tracks memory strength, such that all 10 of these voxels activate together for remembered (but not forgotten) items. This will increase the average pattern similarity between remembered items. Naively, one might interpret this effect in terms of neural representations "converging" in representational space, when (in fact) it is merely due to a univariate effect being superimposed on the searchlight region.

New developments

In this section, we will discuss recent advances in MVPA that complement and extend existing approaches.

Decoding and Encoding Representational Spaces A major limitation of the classifier studies discussed above is that the classifiers are *specialists*: they can only discriminate between cognitive states that they were trained to identify, and the training process is highly laborious. The classifier needs to be trained on a large number of "snapshots" of these cognitive states (on the order of hundreds or more, depending on how subtle the differences are between the cognitive states) before it can discriminate between them reliably. You can train a classifier to discriminate between brain patterns elicited by lions and camels, but this classifier won't tell you anything about the difference between oranges and grapes. This fact places a strong limitation on the kinds of questions that can be addressed in any particular study.

Recently, several studies have sought to surmount this limitation by reconceptualizing the decoding problem: instead of treating stimulus classes as distinct entities, these studies draw on the psychological literature on representation and conceptualize psychological states as points in a high-dimensional representational space. For example, the meaning of a particular concrete noun can be conceptualized as a point in a high-dimensional "meaning space," where each dimension corresponds to a particular aspect of the noun's meaning—for example, can it be eaten? can it be manipulated? can it be used as shelter? (see Just, Cherkassky, Aryal, & Mitchell, 2010).

Once stimuli have been placed in an *n*-dimensional feature space, classifiers can be trained to decode *each feature dimension* (e.g., what does the brain look like for nouns that describe edible vs. inedible items). These classifiers can then be applied to a novel brain pattern and used to decode the coordinate of that stimulus in the *n*-dimensional feature space. The decoded set of coordinates can then be compared to a "dictionary" of the meaning vectors associated with particular words, and—based on this—the classifier can make a guess about which word the person is thinking about at that moment. Alternatively, some studies have used a complementary *encoding* approach where, instead of predicting feature vectors based on brain patterns, these studies learned to predict brain patterns based on a combination of feature vectors: that is, if a word has a particular meaning vector, what should its fMRI pattern look like? See Naselaris, Kay, Nishimoto, and Gallant (2011) for further discussion of encoding and decoding models.

The power of this "feature space" idea is that it is usually possible to learn the neural correlates of particular feature dimensions based on a limited subset of stimuli; once the brain-to-feature mapping has been learned by a decoding model, the model can be used to decode the feature vector for any stimulus that resides within the representational space, regardless of whether that stimulus appeared at training; likewise, once the feature-to-brain mapping has been learned by an encoding model, it can be used to predict the brain response to any stimulus that resides within the representational space. For example, Mitchell et al. (2008) used this approach to decode which of two novel words (i.e., words not presented during classifier training) the participant was thinking about, with 77% accuracy. Several other studies have used this feature-based decomposition approach to decode the contents of visual stimuli based on brain activity (e.g., Kay, Naselaris, Prenger, & Gallant, 2008).

Importantly, if a particular feature-space model yields above-chance decoding (or above-chance prediction of brain patterns in a particular region), this tells us that the model has *some* relationship to how those stimuli are coded in the brain, but there could be other models that do a better job. Given two competing models of neural coding, one way to discriminate between them is to build encoding models based on the two different "feature spaces," and to see which one of them does a better job of predicting the observed fMRI activity (Serences & Saproo, 2012).

Improving Across-Subject Classification Earlier, we discussed the "data starvation" problem in MVPA analysis: the number of brain snapshots is typically low relative to the number of parameters being estimated by the classifier, resulting in a high danger of overfitting. The easiest way to combat this problem would be to combine data across participants. However, this will only work to our benefit if the brain patterns corresponding to particular cognitive states are reasonably consistent across participants—otherwise, the added within-class variability (resulting from across-participant differences) will offset the beneficial effects of having more data. The key question thus becomes: how can we align data across participants in a manner that minimizes across-participant variability in cognitive representations?

The standard approach to across-subject alignment is to transform each participant's data into a common template space based on anatomical landmarks, and then to combine the transformed data. This procedure has proved to be very useful for standard univariate fMRI analyses, but there have been relatively few reports of anatomical alignment alone leading to good across-subject classification. There are two likely reasons for this: first, the transformations result in spatial blurring, which might erase high-spatial-frequency information in the data that would otherwise be useful for the classifier. In addition, people have different experiential histories that shape how concepts are represented in their brains—no amount of anatomical alignment will correct for such differences.

To address these issues, Haxby and colleagues have developed a new across-subject alignment procedure called *hyperalignment* (Haxby et al., 2011) that aligns brains not based on anatomical landmarks but rather based on the functioning of those brains (i.e., aligning parts of the brain that behave similarly, regardless of "where" exactly in the brain these parts came from). The basic hyperalignment algorithm performs a Procrustes transformation that rotates, scales, and shifts temporal trajectories of voxels to best align data sets from different brains. Haxby and colleagues hyperaligned a data set of 21 participants viewing the movie *Raiders of the Lost Ark* and performed across-subject classification for movie segments, faces and objects, and animal species. They found that hyperalignment produced far superior classification performance compared to alignment based purely on anatomy; hyperalignment even matched the accuracy of within-subject classification. Good classification indicates good alignment of patterns across participants, which suggests that adding more participants to the training set should help generalization even further. Therefore, hyperalignment is an extremely promising approach to minimizing the "data starvation" problem of MVPA.

Conclusion

Multivoxel pattern analysis allows us to detect information in the brain that was not visible using previously developed methods of fMRI analysis. Getting a better handle on the informational contents of a person's brain puts researchers in a better position to test theories of how information processing works in the brain and how cognitive states shape behavior.

MATLAB software for performing a classifier analysis (the Princeton Multi-Voxel Pattern Analysis toolbox) can be found at www.pni.princeton.edu/mvpa. Alternatively, there is a separately developed Python version of the toolbox (PyMVPA) available at http://neuro.debian.net. MATLAB software for performing pattern-similarity MVPA can be found at www.mrc-cbu.cam.ac.uk/methods-and-resources/toolboxes.

Although MVPA offers advantages over other forms of analysis, there are limitations to what it can accomplish (Davis & Poldrack, 2013). Going forward, it will be beneficial to compare MVPA to other measurement and analysis techniques to get a better sense of which aspects of neural processing we can and cannot detect with MVPA. Comparisons of MVPA with univariate analysis (Jimura & Poldrack, 2012) and fMRI adaptation (Epstein & Morgan, 2012) suggest that these methods are interrogating different aspects of the neural code. It will also be useful to compare MVPA to neurophysiology data from human and nonhuman primates (e.g., Kriegeskorte, Mur, Ruff, et al., 2008) to better understand the strengths and limitations of this powerful, but relatively recent, advance in the cognitive neuroscience toolkit.

REFERENCES

Cox, D. D., & Savoy, R. L. (2003). Functional magnetic resonance imaging (fMRI) "brain reading": Detecting and classifying distributed patterns of fMRI activity in human visual cortex. *NeuroImage, 19*(2), 261–270.

Davis, T., & Poldrack, R. A. (2013). Measuring neural representations with fMRI: Practices and pitfalls. *Ann NY Acad Sci, 1296*, 108–134.

Detre, G. J., Natarajan, A., Gershman, S. J., & Norman, K. A. (2013). Moderate levels of activation lead to forgetting in the think/no-think paradigm. *Neuropsychologia, 51*(12), 2371–2388.

Duda, R. O., Hart, P. E., & Stork, D. G. (2012). *Pattern classification*. New York, NY: Wiley.

Epstein, R. A., & Morgan, L. K. (2012). Neural responses to visual scenes reveals inconsistencies between fMRI adaptation and multivoxel pattern analysis. *Neuropsychologia, 50*(4), 530–543.

Gonsalves, B. D., & Cohen, N. J. (2010). Brain imaging, cognitive processes, and brain networks. *Perspect Psychol Sci, 5*(6), 744–752.

Guyon, I., & Elisseeff, A. (2003). An introduction to variable and feature selection. *J Mach Learn Res, 3*, 1157–1182.

Haxby, J. V., Gobbini, M. I., Furey, M. L., Ishai, A., Schouten, J. L., & Pietrini, P. (2001). Distributed and overlapping representations of faces and objects in ventral temporal cortex. *Science, 293*(5539), 2425–2430.

Haxby, J. V., Guntupalli, J. S., Connolly, A. C., Halchenko, Y. O., Conroy, B. R., Gobbini, M. I., & Ramadge, P. J. (2011). A common, high-dimensional model of the representational space in human ventral temporal cortex. *Neuron, 72*(2), 404–416.

Haynes, J. D., & Rees, G. (2006). Decoding mental states from brain activity in humans. *Nat Rev Neurosci, 7*(7), 523–534.

Hung, C. P., Kreiman, G., Poggio, T., & DiCarlo, J. J. (2005). Fast readout of object identity from macaque inferior temporal cortex. *Science, 310*(5749), 863–866.

Jafarpour, A., Horner, A. J., Fuentemilla, L., Penny, W. D., & Duzel, E. (2013). Decoding oscillatory representations and mechanisms in memory. *Neuropsychologia, 51*(4), 772–780.

Jimura, K., & Poldrack, R. A. (2012). Analyses of regional-average activation and multivoxel pattern information tell complementary stories. *Neuropsychologia, 50*(4), 544–552.

Just, M. A., Cherkassky, V. L., Aryal, S., & Mitchell, T. M. (2010). A neurosemantic theory of concrete noun representation based on the underlying brain codes. *PloS ONE, 5*(1), e8622.

Kamitani, Y., & Tong, F. (2005). Decoding the visual and subjective contents of the human brain. *Nat Neurosci, 8*(5), 679–685.

Kay, K. N., Naselaris, T., Prenger, R. J., & Gallant, J. L. (2008). Identifying natural images from human brain activity. *Nature, 452*(7185), 352–355.

Kriegeskorte, N., Goebel, R., & Bandettini, P. (2006). Information-based functional brain mapping. *Proc Natl Acad Sci USA, 103*(10), 3863–3868.

Kriegeskorte, N., Mur, M., & Bandettini, P. (2008). Representational similarity analysis: Connecting the branches of systems neuroscience. *Front Sys Neurosci, 2.*

Kriegeskorte, N., Mur, M., Ruff, D. A., Kiani, R., Bodurka, J., Esteky, H., & Bandettini, P. A. (2008). Matching categorical object representations in inferior temporal cortex of man and monkey. *Neuron, 60*(6), 1126–1141.

Kuhl, B. A., Rissman, J., Chun, M. M., & Wagner, A. D. (2011). Fidelity of neural reactivation reveals competition between memories. *Proc Natl Acad Sci USA, 108*(14), 5903–5908.

LaRocque, K. F., Smith, M. E., Carr, V. A., Witthoft, N., Grill-Spector, K., & Wagner, A. D. (2013). Global similarity and pattern separation in the human medial temporal lobe predict subsequent memory. *J Neurosci, 33*(13), 5466–5474.

Lewis-Peacock, J. A., Drysdale, A. T., Oberauer, K., & Postle, B. R. (2012). Neural evidence for a distinction between short-term memory and the focus of attention. *J Cogn Neurosci, 24*(1), 61–79.

Mitchell, T. M., Hutchinson, R., Niculescu, R. S., Pereira, F., Wang, X., Just, M. A., & Newman, S. D. (2004). Learning to decode cognitive states from brain images. *Mach Learn, 57,* 145–175.

Mitchell, T. M., Shinkareva, S. V., Carlson, A., Chang, K.-M., Malave, V. L., Mason, R. A., & Just, M. A. (2008). Predicting human brain activity associated with the meanings of nouns. *Science, 320*(5880), 1191–1195.

Mumford, J. A., Turner, B. O., Ashby, F. G., & Poldrack, R. A. (2012). Deconvolving BOLD activation in event-related designs for multivoxel pattern classification analyses. *NeuroImage, 59*(3), 2636–2643.

Naselaris, T., Kay, K. N., Nishimoto, S., & Gallant, J. L. (2011). Encoding and decoding in fMRI. *NeuroImage, 56*(2), 400–410.

Norman, K. A., Polyn, S. M., Detre, G. J., & Haxby, J. V. (2006). Beyond mind-reading: Multi-voxel pattern analysis of fMRI data. *Trends Cogn Sci, 10*(9), 424–430.

Pereira, F., & Botvinick, M. (2011). Information mapping with pattern classifiers: A comparative study. *NeuroImage, 56*(2), 476–496.

Pereira, F., Mitchell, T., & Botvinick, M. (2009). Machine learning classifiers and fMRI: A tutorial overview. *NeuroImage, 45*(1), S199–209.

Polyn, S. M., Natu, V. S., Cohen, J. D., & Norman, K. A. (2005). Category-specific cortical activity precedes retrieval during memory search. *Science, 310,* 1963–1966.

Rissman, J., & Wagner, A. D. (2012). Distributed representations in memory: Insights from functional brain imaging. *Annu Rev Psychol, 63,* 101–128.

Ritchey, M., Wing, E. A., LaBar, K. S., & Cabeza, R. (2012). Neural similarity between encoding and retrieval is related to memory via hippocampal interactions. *Cereb Cortex, 23*(12), 2818–2828.

Serences, J. T., & Saproo, S. (2012). Computational advances towards linking BOLD and behavior. *Neuropsychologia, 50*(4), 435–446.

Soon, C. S., Brass, M., Heinze, H.-J., & Haynes, J.-D. (2008). Unconscious determinants of free decisions in the human brain. *Nat Neurosci, 11*(5), 543–545.

Todd, M. T., Nystrom, L. E., & Cohen, J. D. (2013). Confounds in multivariate pattern analysis: Theory and rule representation case study. *NeuroImage, 77,* 157–165.

Tong, F., & Pratte, M. S. (2012). Decoding patterns of human brain activity. *Annu Rev Psychol, 63,* 483–509.

Williams, M. A., Dang, S., & Kanwisher, N. G. (2007). Only some spatial patterns of fMRI response are read out in task performance. *Nat Neurosci, 10*(6), 685–686.

Zeithamova, D., Dominick, A. L., & Preston, A. R. (2012). Hippocampal and ventral medial prefrontal activation during retrieval-mediated learning supports novel inference. *Neuron, 75*(1), 168–179.

78 Quantitative Magnetic Resonance Imaging

YANIV ASSAF

ABSTRACT Quantitative measures of biological tissues such as measuring relaxation times (T_1, T_2, and T_1*), magnetization transfer, diffusion, flow, and exchange have been studied for almost 30 years. But only recently the acquisition of these measurements became fast enough, and the analytical methods robust enough, to enable large population studies. With the great investment in technology development of these methods, parameters such as the myelin water fraction (Deoni, Rutt, Arun, et al., 2008) and the axon diameter (Assaf, Blumenfeld-Katzir, Yovel, & Basser, 2008; Alexander et al., 2010) distribution can now be estimated in vivo and for the whole brain. This is a quantum leap in the ability of MRI to dive into the microstructure of the brain and provide the researcher with invaluable parameters that will provide a more comprehensive characterization of the tissue.

With these new approaches to studying the brain, the link between structure and function can be revisited. A full description of quantitative magnetic resonance imaging (MRI) methods requires a whole book (e.g., Tofts, 2003); in this chapter, we will focus on two methods that are being intensively explored and used, T_1 mapping and diffusion MRI. Both methods provide an unprecedented view of the brain, especially the white matter, allowing exploration of the microstructure, in vivo, at a scale that could have been achieved thus far only by microscopes.

What is quantitative magnetic resonance imaging?

CONTRAST MECHANISMS OF MAGNETIC RESONANCE MRI is a powerful imaging modality for several reasons: it is performed in vivo, noninvasively, and is highly sensitive to tissue composition. MRI is also a very flexible modality, permitting acquisition of a multitude of different contrasts, each sensitive to a different aspect of the tissue (Hashemi, Bradley, & Lisanti, 2004; Stark & Bradley, 1999). Either representing a MRI physical parameter (e.g., relaxation) or a biophysical property of the tissue (e.g., diffusion), MRI has become central in neuroscience, enabling scientists to explore the brain in many ways.

We can divide MRI contrast into three categories:

1. contrast that originates from an MRI physical parameter
2. contrast that represents a biophysical property of the tissue
3. contrast that is induced by external chemicals that are transported into the tissue (contrast agent).

The acquisition of each contrast necessitates the use of different pulse programs, and as a result each MRI experiment should be planned to achieve the suitable contrast to the research. Thus, the researcher faces a difficult problem when performing an MRI experiment: what is the best contrast mechanism to investigate the phenomenon of interest? To answer this question, one should understand the origins of each contrast mechanism.

MRI is not a specific modality in the sense that the vast majority of MRI experiments measure water, which is not specific to a molecule, an organ, a tissue, or any cellular type (Hashemi et al., 2004; Stark & Bradley, 1999). Therefore, increasing the specificity of MRI is usually achieved by manipulating the MRI signal (by changing the pulse sequence) so that it will be more sensitive and specific to certain parts of the tissue (Hashemi et al., 2004).

The most basic MRI sequences provide contrast that is sensitive to the different relaxation times of MRI: T_1, T_2, and T_2*. T_1 contrast provides excellent distinction among the three tissue types in the brain (gray matter, white matter, and cerebrospinal fluid); T_2 is highly sensitive to the water content of the tissue; T_2* is influenced by the magnetic susceptibility of the tissue. More advanced techniques can extract information that is specific to biophysical parameters of the tissue. In this category, information about diffusion, flow, macromolecular content, and chemical exchange can be extracted. Additional approaches combining several contrast mechanisms and mathematical models allow us to relate the MRI signal into specific tissue compartments, such as white matter, myelin, iron concentration, and so on.

Quantitative Versus Qualitative A distinction should be made between quantitative and qualitative MRI data. The initial image reconstructed from the scanner, for any pulse sequence and contrast mechanism, is qualitative, not quantitative. Qualitative images are typically identified by adding the word "weighted" to the description of the contrast mechanism. For example, performing a spin-echo experiment with experimental parameters that enhance T_2 contrast will result in a "T_2-weighted image"; performing an experiment with a pulsed-gradient, spin-echo diffusion sequence will result in a "diffusion-weighted image."

Quantitative information is derived by specific processing of the acquired "weighted" image or a set of such images. Typically, more than one "weighted" image is needed to extract a quantitative image. For example, at least two diffusion-weighted images with different experimental parameters are needed to calculate the apparent diffusion coefficient (Le Bihan, 1995; see below); several inversion recovery or spin-echo images are needed to compute the T_1 or T_2 values (Hashemi et al., 2004; see below).

The quantitative maps are extremely powerful since their accuracy is higher than the weighted image, and their interpretation is more straightforward because they can be compared across subjects or groups of subjects. One drawback to quantitative mapping is that it is still scanner- and pulse sequence–specific. Using the same pulse sequence and methods across scanners (even with the same magnetic field and company brand) may lead to systematic differences in the quantitative information. The main reason for that is that different sources of biased noise affect different scanning sites, and since noise is generally not modeled in most analysis procedures, it will affect the measured parameters (in a different way across scanners). The processing methods must take into account the specific biases and noise characteristics of the magnetic resonance imaging scanner. Different methods to avoid biases and reduce noise can lead to different results. For example, calculation of the T_1 can be achieved by several procedures (e.g., inversion recovery, Hahn, 1949; and DESPOT1, Deoni, Rutt, Arun, et al., 2008); however, with each method different T_1 values will be computed, although the correlation between the methods will be high.

Another limitation of quantitative MRI is that acquisition of several images is typically needed for accurate calculation of the desired parameter. Consequently, quantitative measures may require long acquisition times, which limit the amount of data that can be acquired for a single subject in a single session. Finally, other factors, such as complicated and ill-posed models, limit the applicability of quantitative MRI. Nevertheless, with the advances in MRI technology, more stable gradient systems, more accurate radio-frequency (RF) coils, and user-friendly modeling algorithms, the use of quantitative MRI in brain research increased dramatically in recent years.

T_1/T_2 relaxometry

Relaxometry MRI experimentation entails RF excitation of the hydrogen atoms of water molecules (specifically the magnetic spin of the nucleus). In physics, such excitation is termed perturbation to equilibrium state, following which the system seeks the return to its balanced condition in a process called relaxation (Hashemi et al., 2004; Stark & Bradley, 1999). In many cases, this concept is straightforward because the system can be perturbed in a single way (e.g., nuclear imaging, optical imaging, etc.). The magnetic resonance imaging signal is more complicated than most, because magnetic resonance was blessed with not one but two relaxation mechanisms; each refers to different aspects of the excitation process (Hashemi et al., 2004; Stark & Bradley, 1999).

In MRI, RF excitation causes two parallel physical processes to occur: (1) the spin of the hydrogen atom nuclei is excited to a higher energy state; (2) the oscillation of the entire spin population of the tissue becomes synchronized in coherence (i.e., resonance). The two above-mentioned relaxation mechanisms of MRI refer to these two processes: relaxation time T_1 describes the return of the excited spin magnetization to its equilibrium state, while relaxation time T_2 relates to the loss of spin coherence or resonance following excitation. T_1 and T_2 occur in parallel after the RF excitation. They differ significantly in their time constants and the biophysical processes that affect them.

T_1 is the basic relaxation mechanism in magnetic resonance. Also termed spin-lattice or longitudinal relaxation, T_1 refers to the return of magnetization to its equilibrium state by transferring energy to the environment (lattice). As a consequence, T_1 depends on the interaction of the spins with the environment, and hence T_1 is affected by the external magnetic field level as well as the composition and mobility of the environment (Hahn, 1949). Following that concept, areas with lower mobility of the environment (e.g., fat, rich areas such as white matter) will have different T_1 from fluid-rich areas (e.g, cerebrospinal fluid, or CSF). Thus, T_1 images provide the best contrast between the main tissue types (gray matter, white matter, and CSF)

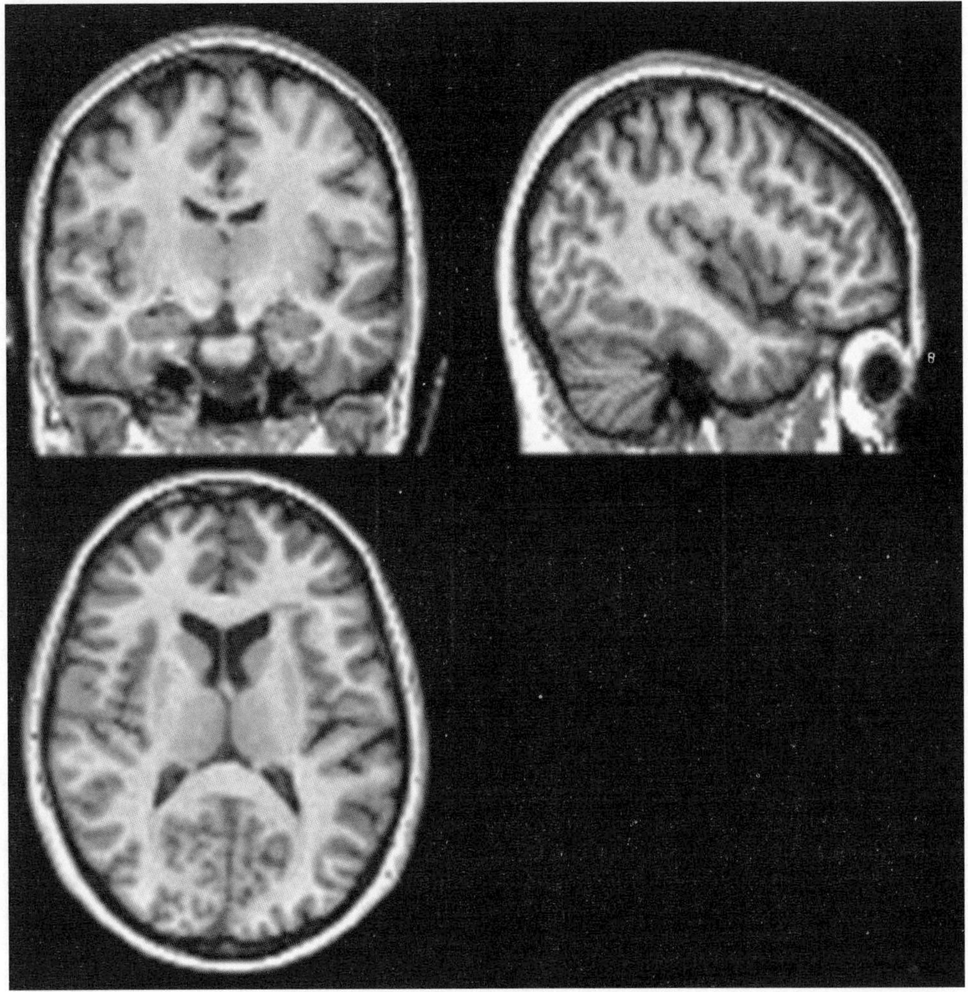

FIGURE 78.1 T_1-weighted MRI of the human brain showing high-contrast among the three main tissue types: cerebro-spinal fluid (CSF), gray matter, and white matter.

and are generally used for segmentation and parcellation of the brain (figure 78.1).

The initial RF excitation causes a coherent precision of the spins. This coherence declines over time, and T_2 refers to the time constant of the coherence loss process (also termed dephasing). As indicated by its additional names, spin-spin or transverse relaxation, T_2 mechanism differs from T_1, and the T_2 value depends on other factors. The loss of spin coherence occurs due to interactions of spins with other spin magnetization (either of water or other molecules). Consequently, T_2 depends on the chemical content and mobility of the spins. Here we differentiate between two sources of interactions: (1) spin coherence loss due to magnetic field inhomogeneities, typically caused by the chemical composition of the tissue (e.g., high concentrations of iron ions); (2) spin coherence loss due to interaction with other hydrogen atom spins (Hashemi et al., 2004; Stark & Bradley, 1999). While both processes lead to similar consequences (i.e., dephasing), the latter can be experimentally separated and is called the true spin-spin interaction relaxation, T_2. This is the basic contrast mechanism used for diagnosis purposes since spin-spin interactions are extremely sensitive for water content, which regularly changes in pathology or following insult to the tissue (Stark & Bradley, 1999).The combination of both spin coherence loss processes is called T_2*. This is the contrast mechanism used to measure the blood oxygenation level that is at the heart of fMRI.

The term *MRI relaxometry* refers to acquisition and analysis methods that quantify the relaxation parameters (mainly T_1 and T_2) for each image pixel in the brain (or any other part of the body).

INVERSION RECOVERY: THE MEASUREMENT OF T_1 After RF excitation, spins exchange energy with the lattice and return to their equilibrium state at an exponential

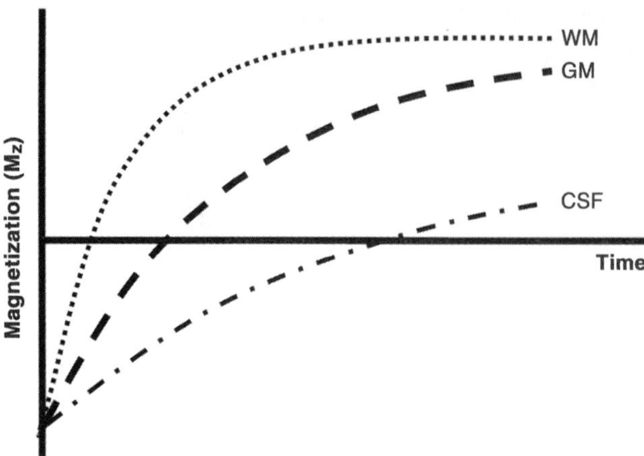

FIGURE 78.2 Scheme of T_1 relaxation curves for different brain tissue types: cerebro-spinal fluid (CSF), gray matter (GM), and white matter (WM).

rate. The exponential time constant, T_1, measures how long spins in a particular neighborhood take to reach the equilibrium of the unperturbed magnetization level (figure 78.2).

Historically, the main challenge in relaxometry was to produce an acquisition protocol that separates T_1 from T_2. This is not straightforward, since the two processes occur simultaneously. Concomitantly, due to the basic physics of the MRI experiment (which are beyond the scope of this chapter), T_1 cannot be measured directly (since it happens on the longitudinal axis of the spin's magnetization, which physically cannot be measured; Hahn, 1949; Hashemi et al., 2004).

This problem was solved by the forefathers of MRI. In 1949, Erwin Hahn developed a pulse sequence, inversion recovery (IR), that measures T_1 and remains a traditional pulse sequence for measuring T_1 (Hahn, 1949). In order to separate T_1 from T_2, IR uses a pulse (180° flip angle) that excites the spins without introducing any coherence. Thus only T_1 relaxation is present without contribution of T_2. Although a 180° pulse produces a very "clean" T_1 process, it cannot be measured (due to the above-mentioned reasons). Therefore, the trick in IR is to "freeze" the process at a certain time point along its development (following a time interval called inversion time, TI) and apply a 90° pulse that introduces a coherence level proportional to the magnitude of longitudinal magnetization at that time.

The application of the IR sequence (180°-TI-90°-Acq.) captures T_1 process at a specific time point (TI) along its evolution. To complete the measurement of T_1, one needs to repeat the experiment several times at different TIs to obtain enough data points to characterize the magnetization growth and compute T_1 according to the following formula:

$$Mz(T_I) = Mz, eq\left(1 - 2\exp\left(-\frac{TI}{T_1}\right)\right),$$

where Mz(TI) is the measured magnetization TI, Mz is the magnetization at equilibrium (which can be measured when TI is set to zero), TI is the inversion time, and T_1 is the time constant of the longitudinal relaxation. This formula becomes complicated if several T_1 components are modeled (i.e., several compartments within the voxel with different T_1s) and even more complicated if there is exchange between the different T_1 pools.

While inversion recovery is a wonderful acquisition and analysis pipeline to measure T_1 relaxation, it has one significant drawback—it takes a lot of time. Accurate T_1 analysis will require roughly 30–60 minutes of acquisition. Yet, once measured, such experiments yielded high differences in the T_1 values of the different brain tissue type; for example in 4 Tesla, the T_1 of white matter is ~940 msec, while that of gray matter is around 1,350 msec (Kim, Hu, & Ugurbil, 1994). Moreover, several papers have shown that T_1 is greatly influenced by myelin content and that myelin is the main contributor for T_1 reduction in white matter and in the deep layers of the cortex (Bydder, Hajnal, & Young, 1998; Duewell et al., 1996; Hajnal et al. 1992; Kim, Hu, & Ugurbil, 1994). In the following sections, additional approaches that allow a more rapid computation of T_1 will be described.

SPIN ECHO: T_2 MEASUREMENT The measurement of T_2 is more straightforward than T_1 since this process occurs at the transverse axis of spin's magnetization, which is detected by MRI receiver coils. The main challenge here is to separate pure spin-spin interactions (T_2) from coherence loss due to magnetic field inhomogeneity (T_2*, see above). In 1950, Erwin Hahn, who introduced inversion recovery, published a new pulse sequence called spin echo that eliminates field inhomogeneity effects and thus permits measurement of T_2 relaxation (Hahn, 1950). Using the pulse sequence 90°-TE/2-180°-TE/2-acq, Hahn showed that using a 90° excitation pulse followed by a 180° pulse generates a signal at a particular point in time, called the echo time (TE), in which the effects of dephasing is absent. While full description of the magnetization processes that occur in spin echo is complicated, the measurement is relatively simple (Hashemi et al., 2004; Tofts, 2003). By repeating the spin-echo experiment at different TE intervals, it is possible to compute the average T_2 of the tissue or, if data quality permits, to estimate different T_2

components. It is well accepted that in brain tissue there are three pools of T_2 components: a long T_2 pool (>100 ms, mainly CSF water), intermediate T_2 pool (~60 ms, cellular T_2) and short T_2 pool (<10 ms, myelin water T_2; Gore, 2001; figure 78.3). It has been proposed that the measurement of the short T_2 pool can serve as an indicator of myelin water. However, the accurate measurement of this short T_2 water pool requires certain implementation of the pulse sequence as well as long acquisition scans, making it less preferable for that purpose.

Advances in Relaxometry Measures Spin echo and inversion recovery are excellent methods that enable relaxation time quantification, albeit with a lengthy acquisition. One of the great challenges in relaxometry was to develop new acquisition and analysis pipelines that enable fast and robust estimation of these relaxation properties. In the last decade, such approaches were indeed developed and are based on the fast imaging sequences, steady-state precession, and spoiled gradient echo.

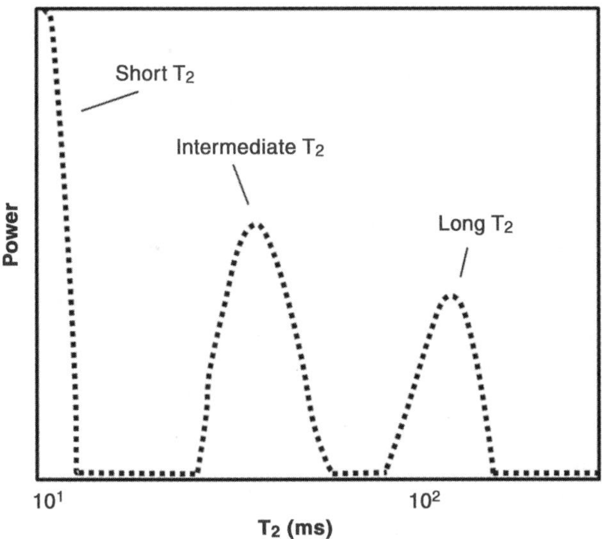

Figure 78.3 Spectrum of the T_2 components in the mouse optic nerve. (Modified from Gore, 2001.)

Steady-state precession refers to sequences in which the acquired magnetization reaches a steady-state level after a series of RF pulses. In spoiled gradient echo, fast acquisition with extremely short TR and small flip angle in combination with a spoiler gradient to remove transverse relaxation achieves excellent T_1 contrast. The acquisition scheme of the fast relaxometry methods is similar to the traditional methods in the sense that in order to calculate the relaxation time (either T_1 or T_2), there is a need for multiple acquisitions of the same slices but with different experimental parameters. The advantage of these methods over the conventional ones is that the acquisition is faster by an order of magnitude.

T_1 maps can be achieved by computing several images of the same slice acquired with spoiled gradient echo sequences with two or more flip angles (a technique that is also called driven-equilibrium single-pulse observation of T_1, or DESPOT1; Deoni, Rutt, Arun, et al., 2008; Deoni, Rutt, & Jones, 2008). DESPOT1 yields T_1 maps with extremely rapid acquisition (a few minutes, depending on resolution; see figure 78.4A), yet it assumes single T_1 population for each voxel. In a similar manner, steady-state precession sequences (e.g., bSSPS) acquired with varying flip angle can be used to produce T_2 maps (figure 78.4B).

Different variants of this method exist with different names (e.g., FIST). The main difference between the methods is not in the acquisition protocol, where the

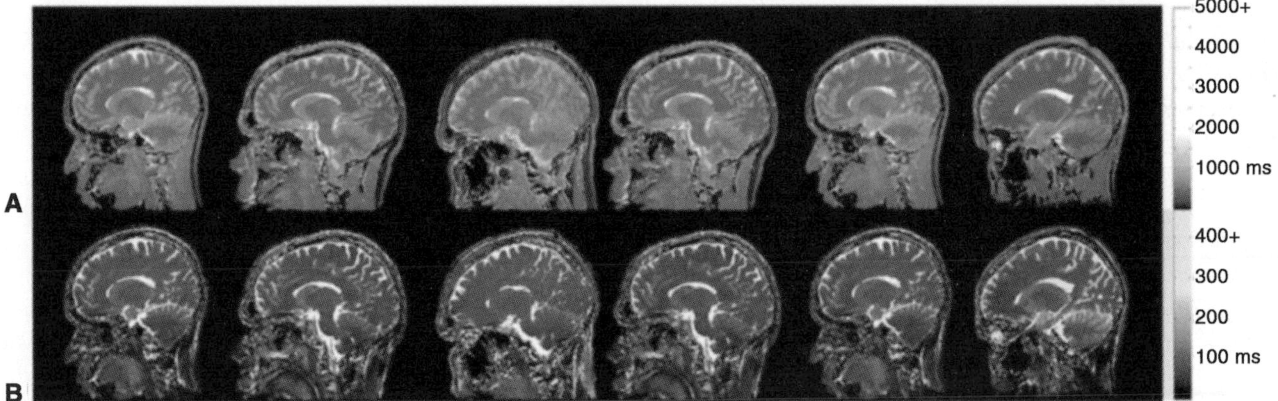

Figure 78.4 Representative sagittal slices (at approximately the same location) through the whole brain. 3D T_1(A) and T_2(B) map volumes for each of six volunteers.

differences are minor, but in the algorithm used for calculation of the relaxation time.

Although the fast acquisition methods are extremely powerful in calculating the relaxation times within a reasonable acquisition time, they do provide a single value of each voxel. Yet it is commonly agreed that even the most homogenous voxel in the brain will contain spectra of different water pools, each with its own relaxation characteristics. For instance, a voxel in white matter will have several water pools that are known to have distinct T_1 values (myelin water exhibiting short T_1, and axonal water and extra-axonal water having longer T_1). It may be that some of these pools exchange with one another, implying that within the time frame of the acquisition a certain water molecule can experience several pools. In order to obtain subvoxel information, more sophisticated acquisition and analysis framework had to be devised. Multicomponent DESPOT (mcDESPOT) is such a model, allowing the calculation of several T_1/T_2 values per water (Deoni, Rutt, Arun et al., 2008). The acquisition in mcDESPOT is similar to DESPOT1/2, although it requires more data points, but the model is completely different and more complicated. Such an approach allows the computation of highly specific contrasts, such as the myelin water fraction map (Deoni, Rutt, Arun, et al., 2008). Other models were recently suggested to improve some of the drawbacks of mcDESPOT and require different acquisition or analysis schemes.

Diffusion imaging

Background Diffusion imaging provides powerful quantitative measures of neural tissue within minutes of acquisition (Johansen-Berg & Behrens, 2009; Jones, 2011; Le Bihan, 1995). Diffusion MRI measures the net displacement of molecules per unit of time. The distinction between diffusion and displacement is important, especially in the understanding and interpretation of the diffusion MRI experiment of neural tissue. In the theoretical and physical description of transport phenomena, diffusion and displacement are linked by Einstein's relation:

$$<x>^2 = 6Dt_D,$$

where $<x>$ is the net displacement, D is the diffusion coefficient, and t_D is the time unit (also called diffusion time). This link suggests that one can easily infer the diffusion coefficient from displacement measurements. However, this relation was developed for a very specific condition where the diffusion is free—that is, there are no barriers to the motion of the molecules. This condition is rarely met in biological systems, thus the distinction between diffusion and displacement will become significant, as will be described in the following sections.

Diffusion imaging is measured using a pulse sequence called pulsed-gradient spin-echo (figure 78.5A; Stejskal & Tanner, 1965). In this pulse sequence, which is based on the basic spin-echo method, two sets of magnetic field gradients are applied, separated briefly in time. Each of the magnetic field gradients tags the spin magnetization according to their location at the time of gradient application. Any change in the location of the spin in the time interval (diffusion time, Δ) between the two gradients will lead to signal loss. The size of the signal loss is proportional to the spin displacement during the diffusion time. Regions containing fast spin diffusion (e.g., CSF) produce low signal; areas with slow spin diffusion (e.g., white matter) will have higher signal (figure 78.5B). In the 1960s, Stejskal and Tanner developed a mathematical relation between the signal loss and the diffusion coefficient (Stejskal & Tanner, 1965):

$$\log(E) = -bD,$$

where E is the signal loss or decay is diffusion experiment, D is the diffusion coefficient, and b (b-value; Le Bihan, 1995) is a parameter that sums various experimental parameters related to the diffusion weighting, including the strength and duration of the diffusion gradient and the diffusion time ($b = \gamma^2 g^2 \delta^2(\Delta - \delta/3)$, where γ is the gyromagnetic ratio, g is the strength of the applied diffusion gradients, δ is the duration of the diffusion gradient and Δ is the diffusion time). The b-value is an important parameter in diffusion MRI, since it controls the diffusion weighting: how much signal will decay due to diffusion. Choosing the b-value in diffusion MRI experiments is critical, since this parameter determines the sensitivity to the diffusion/displacement; for example, if we study molecules that exhibit very slow diffusion or small displacements (e.g., due to many barriers to diffusion), low b-values will lead to little signal loss and thus poor signal-to-noise ratio for estimating the diffusion coefficient. Thus, in experiments where slow diffusion is expected, one should prefer using higher b-values.

Indeed, in experiments performed on post-mortem tissue where the tissue is fixated and the diffusion is slowed, higher b-value is needed to achieve similar signal decays compared to an in vivo situation. Typically the b-value is controlled by the magnitude and duration of the diffusion gradients (see different gray scale levels indicating variable magnitude of the diffusion gradient in figure 78.5; Jones, 2011; Le Bihan, 1995). Examples for the effect of b-values on diffusion imaging contrast are shown in figure 78.5.

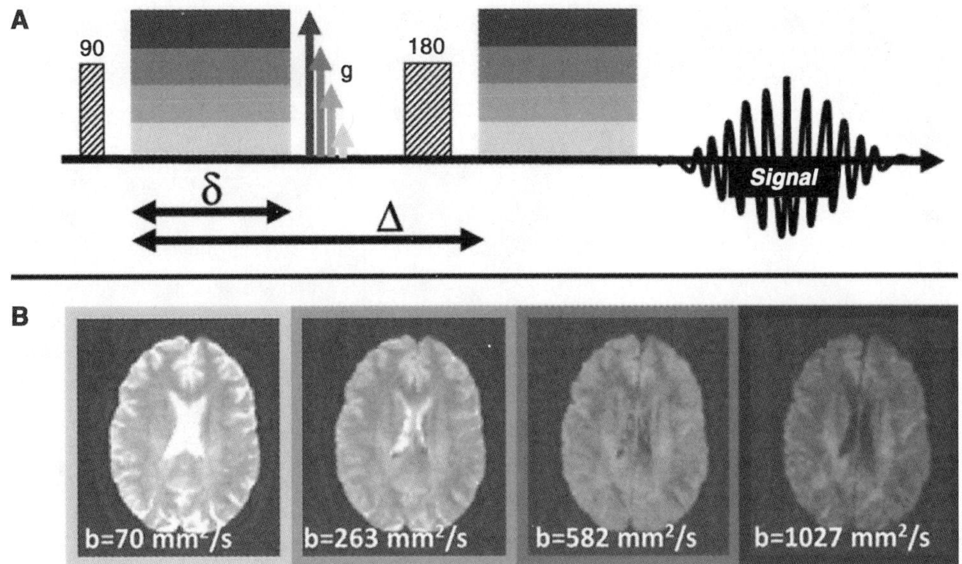

FIGURE 78.5 (A) The pulsed-gradient spin-echo pulse sequence depicting the spin-echo sequence (900 followed by 1,800 pulse) and the pair of pulsed gradients (four versions according to the gray levels). The time intervals (δ and Δ) and gradient amplitude (g) that determine the b-value are outlined. (B) The obtained diffusion-weighted images from the experiment in panel A show the change in contrast and signal intensity with the increase in b-value.

When barriers to the molecule's motion exist (e.g., cell membrane), the displacement will be affected (as the net displacement will be confined to the barrier geometry), while the diffusion coefficient itself remains the same (in theory). Since in diffusion MRI we measure the displacement rather than the diffusion, the exact diffusion coefficient (according to the Stejskal-Tanner relation above) will be miscalculated. Therefore, in a biological system, the computed diffusion coefficient is called the *apparent diffusion coefficient* (ADC) to indicate that the measured value includes the effects of barriers to motion. This is also called hindered diffusion (Le Bihan, 1995).

The diffusion MRI measurement is directional. By specifying the direction of the applied pulsed gradient, the diffusion or displacement will be acquired along that direction. In an aqueous environment, such as the CSF, the signal loss is the same in all directions. But in fibrous and cellular tissue, the signal loss is highly directional. In seminal works from the early 1990s, it was shown that the contrast and signal decay at various brain regions depends dramatically on the orientation of the applied diffusion gradient, especially in white matter (figure 78.6; Moseley et al., 1990). This is the basis for the diffusion tensor imaging (DTI) method.

DIFFUSION TENSOR IMAGING As indicated in figure 78.6, the displacement and ADC in neural tissue, especially in white matter, may differ when measured in different directions. There is greater signal loss in the direction perpendicular to the fibers compared to the parallel direction. The displacement of water molecules in neural tissue is anisotropic—that is, not equal in all directions. When considering the white matter, the interpretation is straightforward—intuitively, water molecules should experience more displacement parallel to the long axis of the neuronal fibers (due to the high packing of the neuronal fibers and their myelin lamellas) than perpendicular to them. Based on this concept, it has been proposed that characterization of white matter can be achieved by measuring the ADC perpendicular and parallel to the fibers' orientation and extracting an anisotropy index reflecting the differences between the two measures (e.g., $D_{\perp}/D_{//}$). Such measurement can be performed on very specific fiber systems, where the orientation of the fibers is such that it allows easy prescription of diffusion gradients in the parallel and perpendicular directions (e.g., the spinal cord, optic nerve, and corpus callosum). However, for most fiber systems in the brain, defining the parallel and perpendicular directions is impossible. Moreover, even if such information were available, due to the heterogeneity in fiber orientations in the brain, each system would have to be acquired separately.

In the mid-1990s, the concept of DTI was proposed (Basser, 1995; Basser, Mattiello, & Le Bihan, 1992, 1994; Basser & Pierpaoli, 1998; Pierpaoli & Basser, 1996; Pierpaoli et al., 1996). In DTI, the whole brain is sampled with diffusion imaging in multiple directions (the same directions for all brain voxels, in at least six directions),

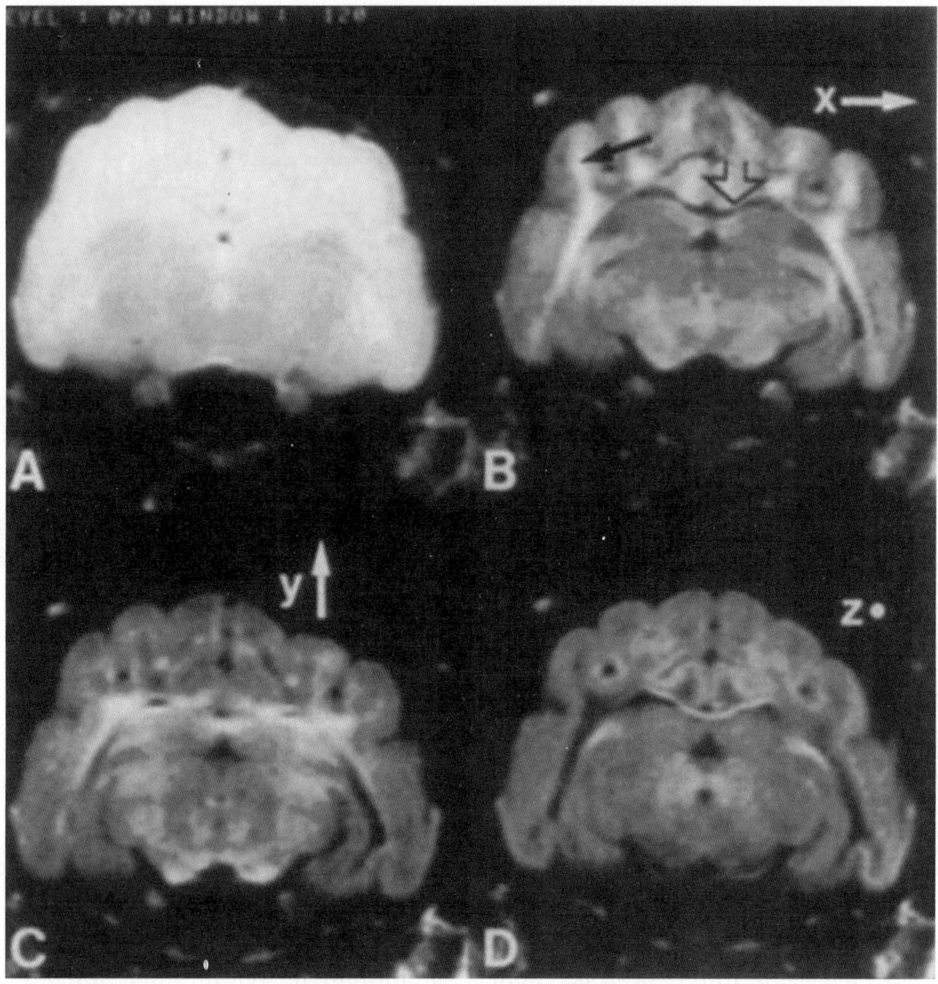

FIGURE 78.6 Pulsed-gradient spin-echo diffusion-weighted images on the cat brain: (A) no diffusion weighting; (B–D) with diffusion weighting when the gradients are applied along the x (B), y (C), and z (D) directions.

and for each direction the ADC is computed. The directional ADC maps are then analyzed using a diffusion tensor model. Mathematically, there are several ways to apply the diffusion tensor model, yet in all of them the result is a three-dimensional ellipsoid that best describes the multiple measures of the ADC in the different directions. The DTI acquisition and analysis pipeline is summarized in figure 78.7.

Once the ellipsoid is calculated in each voxel, it is possible to define the principal directions that will correspond to the diffusivity parallel to the fibers (largest axis of the ellipsoid) and diffusivity perpendicular to the fibers (mean of the two shortest axes of the ellipsoid). The extracted diffusivities are often called axial diffusivity, $D_{//}$ or λ_1 (largest eigenvalue), for the largest axis of the ellipsoid; and radial diffusivity, $D_{\perp}$ or λ_2/λ_3 (smallest eigenvalues), for the mean of the two shortest axes of the ellipsoid (figure 78.7).

The principal diffusivities (axial and radial diffusivities) can be used to compute various summation parameters. The most commonly used are the mean diffusivity (MD) and the fractional anisotropy (FA; Basser & Pierpaoli, 1998; Pierpaoli & Basser, 1996; Pierpaoli et al., 1996). The MD is the average of the three eigen values (three axes of the ellipsoid), while the FA is the normalized variance between them. The formulations for both indices are

$$\mathrm{MD} = (\lambda 1 + \lambda 2 + \lambda 3)/3$$

$$FA = \frac{3}{2} \cdot \frac{\sqrt{(\lambda_1 - \langle D \rangle)^2 + (\lambda_2 - \langle D \rangle)^2 + (\lambda_3 - \langle D \rangle)^2}}{(\lambda_1^2 + \lambda_2^2 + \lambda_3^2)}.$$

The size, shape, and direction of the ellipsoid characterize the diffusion tensor and can be extracted for each voxel individually. The following figure summarizes the different options of the ellipsoid varying in

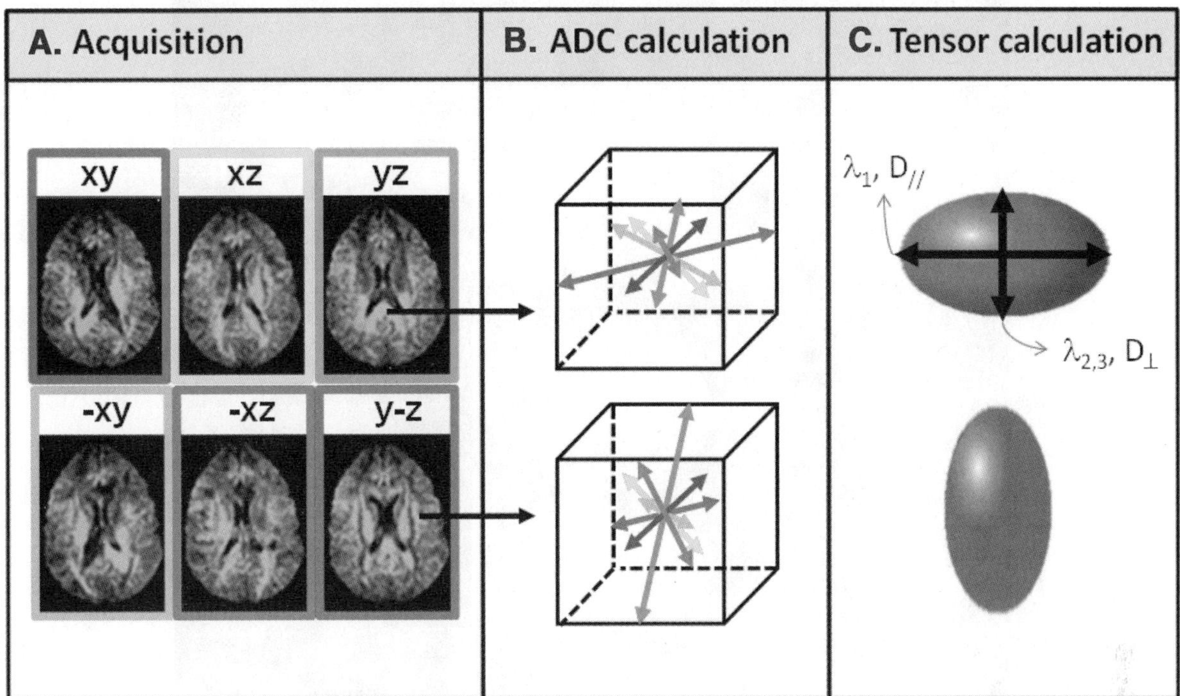

FIGURE 78.7 Diffusion tensor imaging (DTI) pipeline including (A) acquisition at multiple gradient directions homogeneously covering space (the example in the figure includes the basic six-direction scheme). (B) Apparent diffusion coefficient (ADC) calculation for each direction. The two graphs represent two voxels—the top graph refers to a voxel in the corpus callosum and the bottom graph for a voxel is superior longitudinal fascicules. (C) Tensor analysis of the data in panel B resulting in ellipsoids colored according to their orientation (left-right in red, anterior-posterior in green) representing the fiber orientation in the two selected voxels. (See color plate 73.)

size (figure 78.8A) indicating on the MD, varying in shape (figure 78.8B, rounded sphere → elongated ellipsoid) indicating on the FA, and varying in color (figure 78.8C) indicating on the orientation. The latter code, where ellipsoid (fibers) crossing in the left-right direction will be colored in red, ellipsoids that lie in the anterior-posterior direction will be colored in green, and ellipsoids that are orientated from inferior to posterior position will be colored in blue, is extremely widely used and is termed the color-coded FA map (Pajevic & Pierpaoli, 1999). It should be noted that any arbitrary orientation of the ellipsoid can be achieved by combinations of the three colors (red-green-blue). Figure 78.9 summarizes the most widely used parameters extracted from DTI analysis. The information in the color-coded maps is enormous, since it provides three-dimensional information into two-dimensional maps, and the anatomy of the fiber systems becomes apparent.

DTI provided unprecedented information about white matter, its physiology, and architecture. Even so, there are artifacts and error in DTI's acquisition (Jones & Cercignani, 2010). Some of the artifacts in DTI are common to any MRI measurement, and some are specific to this method (Jones & Cercignani, 2010). For example, as in any MRI method, and especially in diffusion MRI when the signal is attenuated, DTI suffers from signal-to-noise limitations. Other acquisition-related issues include gradient stability, eddy current artifacts, gradient sampling schemes, and optimized *b*-value, which have been studied in recent years.

In addition to acquisition artifacts, there are inherent flaws in the processing of DTI images that are manifested in errors in the computed DTI indices. The most common problems arise in voxels that contain more than a single tissue type, such as voxels at the boundary between white matter and ventricles. Another very problematic case arises in white matter regions that contain several intersecting fiber populations. In such areas, the DTI model fits only a single direction, and this may be a compromise direction that fails to represent the underlying architecture (see figure 78.10). This has implication on FA and MD calculation in those areas as well as on tractography (see below).

IMAGING MICROSTRUCTURE DTI's apparent inability, in some cases, to characterize the underlying

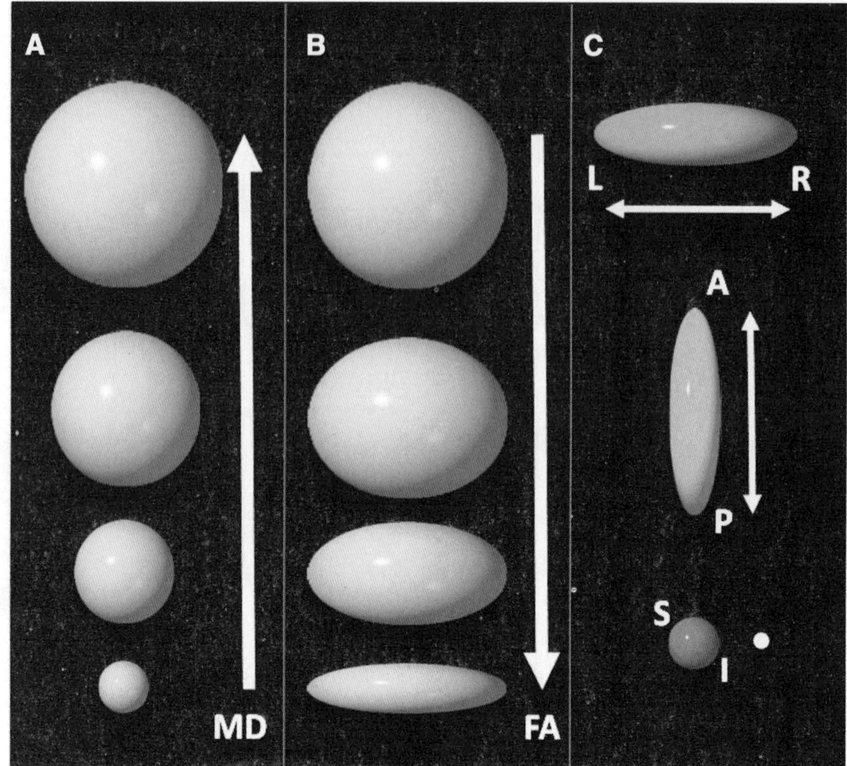

FIGURE 78.8 The meaning of mean diffusivity (MD) and fractional anisotropy (FA), and fiber orientation on the ellipsoid extracted from DTI analysis. (A) Changes in MD will affect the size of the ellipsoid; increase in MD will cause an increase the size of the ellipsoid. (B) Changes in FA will affect the shape of the ellipsoid; low FA (as in cerebrospinal fluid) will appear as a pure sphere, whereas highly ordered white matter bundles will appear as a very elongated cigar-shape ellipsoid. (C) The orientation of the ellipsoid is indicated by its color; ellipsoids that lie in the left-right direction are colored red, ellipsoids in the anterior-posterior position are colored green, and ellipsoids in the inferior-superior position are colored blue. (See color plate 74.)

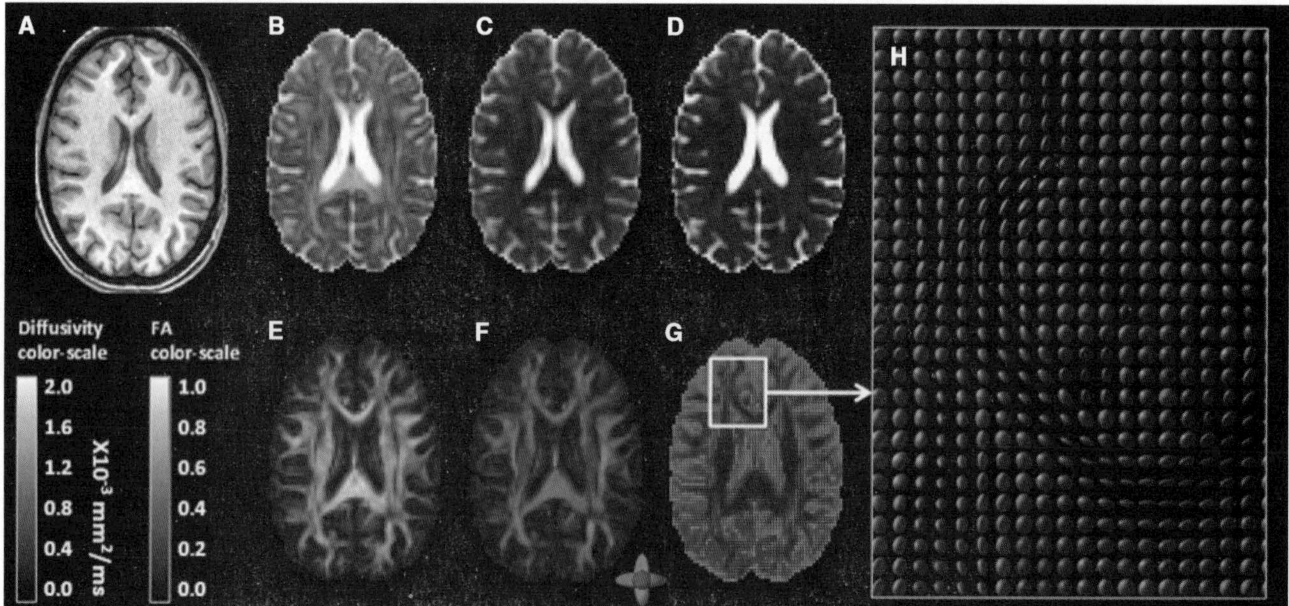

FIGURE 78.9 Summary of quantitative indices extracted from DTI. (A) Reference T_1-weighted scan; (B) Axial diffusivity; (C) Radial diffusivity; (D) mean diffusivity; (E) fractional anisotropy; (F) color-coded FA map; (G) ellipsoid glyph map with section at the genu of the corpus callosum enlarge in (H). Color scale for (B–D) is given in diffusivity color scale (bottom left). Color scale for FA map is given in the FA color scale. (See color plate 75.)

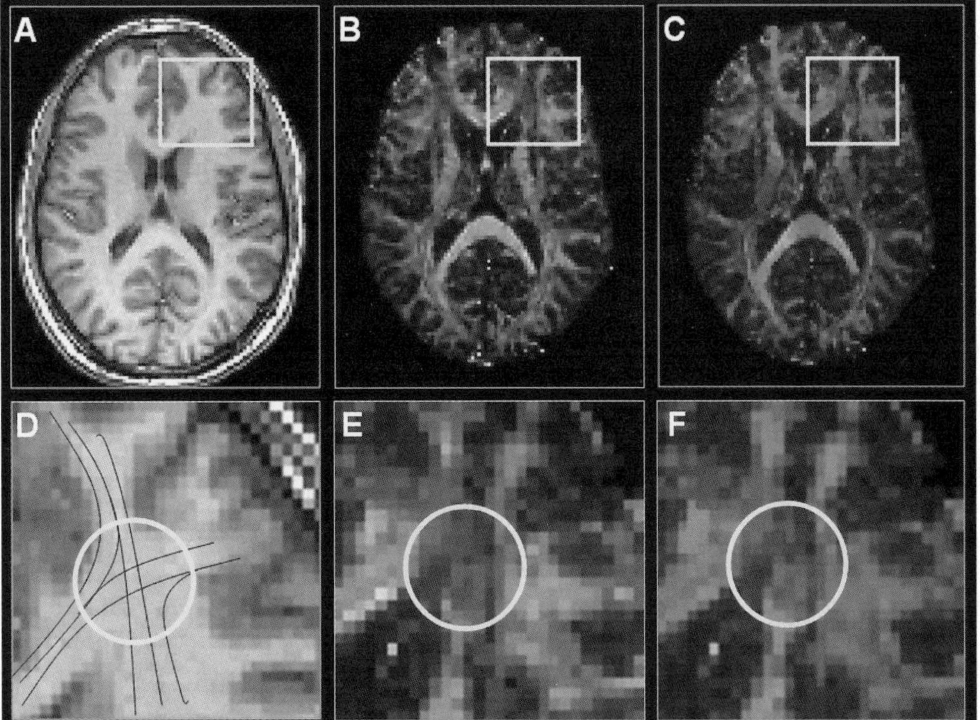

FIGURE 78.10 Example for partial volume artifact in DTI. (A) T_1-weighted anatomical image with the right frontal region enlarged at (D), including line drawings of the different fibers system the passes in that region: one arriving from the genu of the corpus callosum and one from the thalamic radiation of the internal capsule, and both projecting into the frontal lobe. (B) FA maps enlarged at (E) showing low anisotropy area (marked by yellow circle) in the area of the crossing fiber system. This is also shown in the color-coded FA maps, ((C) enlarged at (F)) where one can follow the direction of the fiber (red for the corpus callosum and green for the thalamic radiation). In the crossing fiber region, the observed color is a mix of the two-fiber system. Since two crossing fibers reside within the same pixel in the region, the diffusion is hindered in all measured directions, which artifactually can be characterized by isotropic diffusion and thus reduced FA. (See color plate 76.)

architecture of white matter initiated directed research to overcome this problem. Indeed, technical and computational solutions to some issues (e.g., crossing fibers) were proposed (Assaf & Basser, 2005; Descoteaux, Deriche, Knosche, & Anwander, 2009; Parker & Alexander, 2003; Tournier, Calamante, & Connelly, 2007). However, the same studies indicated that DTI fails to represent the underlying tissue even in homogenous white matter, since DTI's model (Gaussian tensor) is not appropriate for neural tissue (Assaf, Mayk, & Cohen, 2000). The main indication for that was the measurement of diffusion signal decay at a large range of b-values. According to the Stejskal-Tanner equation above, the relation between the signal decay and the diffusion coefficient is exponential, but it was shown that this is not the case for neural tissue, and the dependency of the signal decay on the b-value seems to be composed of multiple exponents whose number cannot be determined easily and thus cannot be assigned to physiological compartments (Assaf & Cohen, 1998, 2000). DTI's Gaussian model is not able to analyze such complex signal dependency, and it is now agreed that DTI is a good first approximation of the diffusion processes in neural tissue, but certainly does not represent or characterize it completely. For more comprehensive characterization of the diffusion MRI signal, alternative models to DTI had to be formulated.

Studying the origins and characteristics of the multiexponential decay in diffusion experiments suggested that diffusion in neural tissue is complicated and that molecules at different subvoxel components (cellular compartments) have different diffusion characteristics. For example, it was found that one can measure two diffusion processes in neural tissue: hindered and restricted (Assaf, Freidlin, Rohde, & Basser, 2004). While restricted diffusion refers to water that is confined within specific environments, hindered diffusion implies water molecules move significantly but encounter an obstacle to their motion every now and then. Further research suggested that the origin of restricted diffusion stems from water molecules that are trapped within axons, and that if one can design an acquisition and analysis pipeline that will enable characterization of restricted diffusion, it will be possible to obtain direct information regarding the axons.

The alternative methods to DTI that analyze diffusion imaging data can be categorized as "model-based" and "model-free" approaches. The "model free" approaches characterize the signal decay by simple mathematical transformations (principal component analysis and Fourier transform, Assaf, Mayk, & Cohen, 2000; kurtosis analysis, Jensen et al., 2005, etc.), yet the interpretation of the output parameters of these methods is sometimes as vague as DTI. However, they are more accurate than DTI simply because these methods can analyze the multiexponential decay, which DTI cannot.

The "model-based" approaches usually differentiate the two types of diffusion in the tissue, hindered and restricted, and try to assign each with a cellular compartment. For example, the composite hindered and restricted model of diffusion (CHARMED; Assaf & Basser, 2005; Assaf et al., 2004) models diffusion signal decay as a composition of two diffusion processes: restricted diffusion in the axonal compartment (modeled by diffusion within impermeable cylinders) and hindered diffusion elsewhere (modeled by a diffusion tensor). The great advantage of such a model is that it is possible to separate the diffusion signal into two components, each with its own characteristics (Dr, diffusion within the axons; Dh, diffusion outside the axons; Fr, axonal density, etc.), and investigate each one of them independently. CHARMED and similar approaches are regarded as a better approximation of tissue microstructure than DTI. Additional frameworks were developed in recent years on the same concepts of CHARMED with some modification and enable modeling different features of neural tissue architecture (Assaf et al., 2013; Zhang, Schneider, Wheeler-Kingshott, & Alexander, 2012).

The specificity of the CHARMED model to the axonal compartment opens a vista of new opportunities in exploring this compartment. The axon diameter distribution is an important factor in brain and white matter physiology since the diameter is linearly correlated with

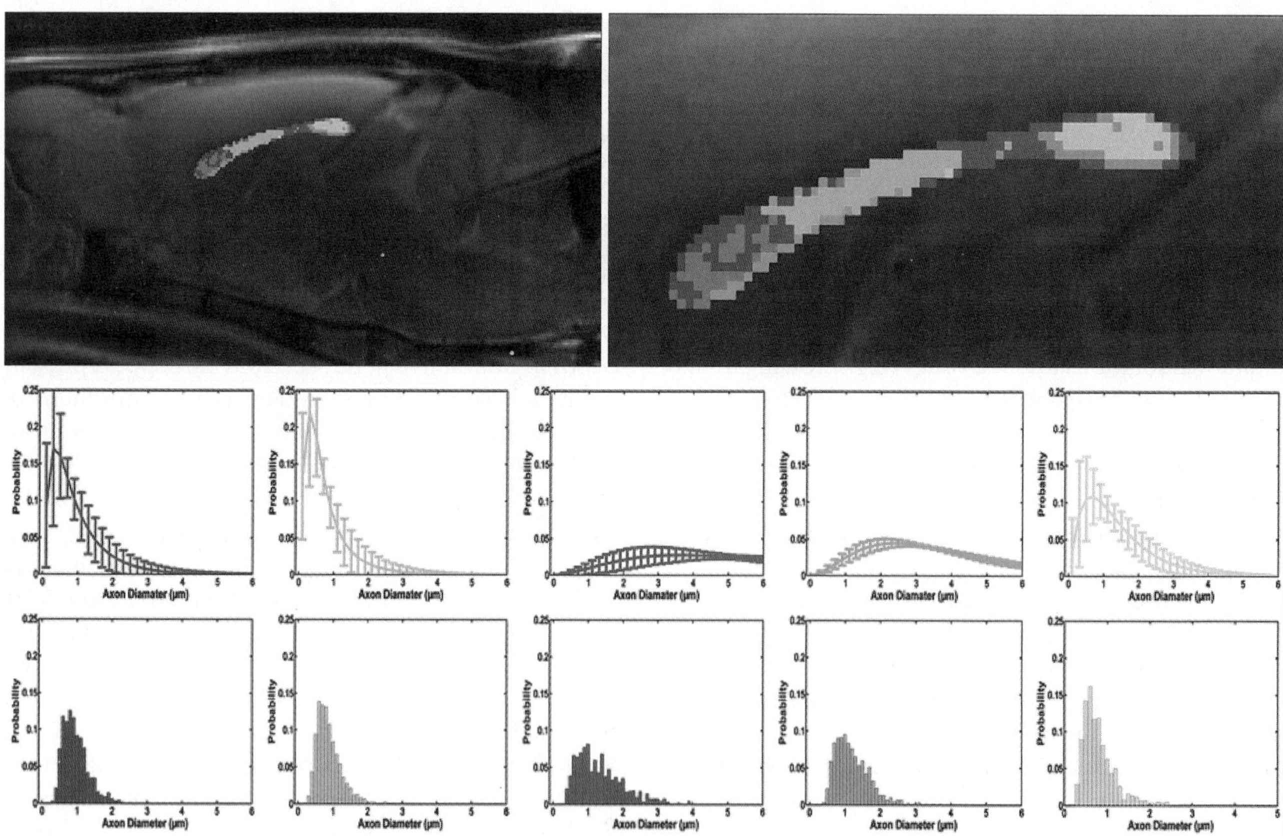

FIGURE 78.11 Cluster analysis of the axon diameter distribution along the corpus callosum: (top row, left). A midsaggital T_2-weighted MRI with the AxCaliber clusters superimposed, enlarged at right. The AxCaliber averaged ADDs for the different clusters are shown in the middle row; note that the colors of the graphs match the clusters' colors. In the bottom row are the histological ADDs of the same clusters. (See color plate 77.)

the conduction velocity (Waxman, 1980). AxCaliber, which is based on CHARMED, enables the estimation of the axon diameter distribution from CHARMED measurement performed at a wide range of diffusion times (Barazany, Basser, & Assaf, 2008, 2009). Changing the diffusion time permits selectively weighting different populations of axons based on their diameter. For example, at short enough diffusion time (~10 msec), only small axons (1–2 microns) will experience restricted diffusion, while larger axons will exhibit only hindrance to diffusion since the water molecules did not have enough time to explore the boundaries of the axons. At longer diffusion times, larger axons will experience restricted diffusion. Analyzing the entire multi-diffusion time data set simultaneously allows the estimation of the axon diameter distribution function (in the original AxCaliber framework this function was modeled by a gamma function). AxCaliber was validated versus electron microscopy on various samples including isolated nerves, spinal cord, and the corpus callosum of the rat (figure 78.11; Barazany et al., 2009). The comparison with histology showed that the axonal size distribution as measured with AxCaliber corresponds to the histology, although there is a factor between the two measures, especially in the in vivo condition. This factor can be explained by two causes: the first is the shrinkage factor in histology due to fixation that is not apparent in the in vivo condition; the second is the inability of AxCaliber to estimate accurately very small diameter axons (<1 m) due to limits in gradient strength. Taken together, these two reasons are probably the causes for the factor between the histological and imaging measures.

One of the main drawbacks of AxCaliber is that the framework was designed for a specific condition when the orientation of the measured fiber system is known. As mentioned above, such limitation does not allow whole-brain analysis. In an alternative approach called ActiveAx, a similar model was suggested that can estimate only the mean axonal size (not the whole distribution) but for any fiber system and the whole brain (Alexander et al., 2010; Zhang, Hubbard, Parker, & Alexander, 2011). ActiveAx was also validated on the human brain as well as the vervet brain, where comparison with histology was also performed, revealing similar observations to AxCaliber.

Conclusion

MRI methods continue to make great progress. The field is not far from reaching the theoretical resolution limits (micron resolution), or exploring many other possible quantitative approaches to provide better microstructural analysis of tissue features. The introduction of quantitative MRI measures will continue to provide new perspectives on brain anatomy and physiology. For example, the relation between structure and function in the brain could be explored and redefined. In addition, the role of tissue microstructure in cognition could be evaluated. The use of quantitative methods in MRI will become more central in neuroscience, since they will allow investigation of the human brain in vivo in great detail, bringing MRI to the verge of virtual microscopy.

REFERENCES

Alexander, D. C., Hubbard, P. L., Hall, M. G., Moore, E. A., Ptito, M., Parker, G. J., & Dyrby, T. B. (2010). Orientationally invariant indices of axon diameter and density from diffusion MRI. *NeuroImage, 52*(4), 1374–1389.

Assaf, Y., Alexander, D. C., Jones, D. K., Bizzi, A., Behrens, T. E., Clark, C. A., … Parker, G. J. (2013). The CONNECT project: Combining macro- and micro-structure. *NeuroImage, 80*, 273–282.

Assaf, Y., & Basser, P. J. (2005). Composite hindered and restricted model of diffusion (CHARMED) MR imaging of the human brain. *NeuroImage, 27*(1), 48–58.

Assaf, Y., Blumenfeld-Katzir, T., Yovel, Y., & Basser, P. J. (2008). AxCaliber: A method for measuring axon diameter distribution from diffusion MRI. *Magn Reson Med, 59*(6), 1347–1354.

Assaf, Y., & Cohen, Y. (1998). Non-mono-exponential attenuation of water and N-acetyl aspartate signals due to diffusion in brain tissue. *J Magn Reson, 131*(1), 69–85.

Assaf, Y., & Cohen, Y. (2000). Assignment of the water slow-diffusing component in the central nervous system using q-space diffusion MRS: Implications for fiber tract imaging. *Magn Reson Med, 43*(2), 191–199.

Assaf, Y., Freidlin, R. Z., Rohde, G. K., & Basser, P. J. (2004). New modeling and experimental framework to characterize hindered and restricted water diffusion in brain white matter. *Magn Reson Med, 52*(5), 965–978.

Assaf, Y., Mayk, A., & Cohen, Y. (2000). Displacement imaging of spinal cord using q-space diffusion-weighted MRI. *Magn Reson Med, 44*(5), 713–722.

Barazany, D., Basser, P. J., & Assaf, Y. (2008). In-vivo measurement of the axon diameter distribution in the rat's corpus callosum. *Proc Intl Soc Magn Reson Med, 16*, 567.

Barazany, D., Basser, P. J., & Assaf, Y. (2009). In vivo measurement of axon diameter distribution in the corpus callosum of rat brain. *Brain, 132*(Pt. 5), 1210–1220.

Basser, P. J. (1995). Inferring microstructural features and the physiological state of tissues from diffusion-weighted images. *NMR Biomed, 8*(7–8), 333–344.

Basser, P. J., Mattiello, J., & Le Bihan, D. (1992). Diagonal and off-diagonal components of the self-diffusion tensor: Their relation to and estimation from the NMR spin-echo signal. *Proc Intl Soc Magn Reson Med, 11*, 1222.

Basser, P. J., Mattiello, J., & Le Bihan, D. (1994). MR diffusion tensor spectroscopy and imaging. *Biophys J, 66*(1), 259–267.

Basser, P. J., & Pierpaoli, C. (1998). A simplified method to measure the diffusion tensor from seven MR images. *Magn Reson Med, 39*(6), 928–934.

Bydder, G. M., Hajnal, J. V., & Young, I. R. (1998). MRI: Use of the inversion recovery pulse sequence. *Clin Radiol, 53*(3), 159–176.

Deoni, S. C., Rutt, B. K., Arun, T., Pierpaoli, C., & Jones, D. K. (2008). Gleaning multicomponent T_1 and T_2 information from steady-state imaging data. *Magn Reson Med, 60*(6), 1372–1387.

Deoni, S. C., Rutt, B. K., & Jones, D. K. (2008). Investigating exchange and multicomponent relaxation in fully-balanced steady-state free precession imaging. *J Magn Reson Imaging, 27*(6), 1421–1429.

Descoteaux, M., Deriche, R., Knosche, T. R., & Anwander, A. (2009). Deterministic and probabilistic tractography based on complex fibre orientation distributions. *IEEE Trans Med Imaging, 28*(2), 269–286.

Duewell, S., Wolff, S. D., Wen, H., Balaban, R. S., & Jezzard, P. (1996). MR imaging contrast in human brain tissue: Assessment and optimization at 4 T. *Radiology, 199*(3), 780–786.

Gore, J. C. (2001). Functional MRI is fundamentally limited by inadequate understanding of the original of fMRI. *Magn Reson Imaging, 19*(3-4), 295–300.

Hahn, E. L. (1949). An accurate nuclear magnetic resonance method for measuring spin-lattice relaxation times. *Phys Rev, 76*(1), 145.

Hahn, E. L. (1950). Spin echoes. *Phys Rev, 80*, 580–594.

Hajnal, J. V., de Coene, B., Lewis, P. D., Baudouin, C. J., Cowan, F. M., Pennock, J. M., ... Bydder, G. M. (1992). High signal regions in normal white matter shown by heavily T_2-weighted CSF nulled IR sequences. *J Comput Assist Tomogr, 16*(4), 506–513.

Hashemi, R. H., Bradley, W. G., & Lisanti, C. J. (2004). *MRI: The basics.* Philadelphia, PA: Lippincott Williams & Wilkins.

Jensen, J. H., Helpern, J. A., Ramani, A., Lu, H., & Kaczynski, K. (2005). Diffusional kurtosis imaging: The quantification of non-gaussian water diffusion by means of magnetic resonance imaging. *Magn Reson Med, 53*(6), 1432–1440.

Johansen-Berg, H., & Behrens, T. E. J. (2009). *Diffusion MRI: From quantitative measurement to in-vivo neuroanatomy.* Boston, MA: Academic Press.

Jones, D. K. (2011). *Diffusion MRI: Theory, methods, and application.* New York, NY: Oxford University Press.

Jones, D. K., & Cercignani, M. (2010). Twenty-five pitfalls in the analysis of diffusion MRI data. *NMR Biomed, 23*(7), 803–820.

Kim, S. G., Hu, X., & Ugurbil, K. (1994). Accurate T_1 determination from inversion recovery images: Application to human brain at 4 Tesla. *Magn Reson Med, 31*(4), 445–449.

Le Bihan, D. (1995). *Diffusion and perfusion magnetic resonance imaging: Applications to functional MRI.* New York, NY: Raven.

Moseley, M. E., Cohen, Y., Kucharczyk, J., Mintorovitch, J., Asgari, H. S., Wendland, M. F., ... Norman, D. (1990). Diffusion-weighted MR imaging of anisotropic water diffusion in cat central nervous system. *Radiology, 176*(2), 439–445.

Pajevic, S., & Pierpaoli, C. (1999). Color schemes to represent the orientation of anisotropic tissues from diffusion tensor data: Application to white matter fiber tract mapping in the human brain. *Magn Reson Med, 42*(3), 526–540.

Parker, G. J., & Alexander, D. C. (2003). Probabilistic Monte Carlo based mapping of cerebral connections utilising whole-brain crossing fibre information. *Inf Process Med Imaging, 18*, 684–695.

Pierpaoli, C., & Basser, P. J. (1996). Toward a quantitative assessment of diffusion anisotropy. *Magn Reson Med, 36*(6), 893–906.

Pierpaoli, C., Jezzard, P., Basser, P. J., Barnett, A., & di Chiro, G. (1996). Diffusion tensor MR imaging of the human brain. *Radiology, 201*(3), 637–648.

Stark, D. D., & Bradley, W. G. (1999). *Magnetic resonance imaging.* St. Louis, MO: Mosby.

Stejskal, E. O., & Tanner, J. E. (1965). Spin diffusion measurements: Spin echoes in presence of a time-dependent field gradient. *J Chem Phys, 42*(1), 288–292.

Tofts, P. (2003). *Quantitative MRI of the brain: Measuring changes caused by disease.* Hoboken, NJ: Wiley.

Tournier, J. D., Calamante, F., & Connelly, A. (2007). Robust determination of the fibre orientation distribution in diffusion MRI: Non-negativity constrained super-resolved spherical deconvolution. *NeuroImage, 35*(4), 1459–1472.

Waxman, S. G. (1980). Determinants of conduction velocity in myelinated nerve fibers. *Muscle Nerve, 3*(2), 141–150.

Zhang, H., Hubbard, P. L., Parker, G. J., & Alexander, D. C. (2011). Axon diameter mapping in the presence of orientation dispersion with diffusion MRI. *NeuroImage, 56*(3), 1301–1315.

Zhang, H., Schneider, T., Wheeler-Kingshott, C. A., & Alexander, D. C. (2012). NODDI: Practical in vivo neurite orientation dispersion and density imaging of the human brain. *NeuroImage, 64*(4), 1000–1016.

79 Network Methods to Characterize Brain Structure and Function

DANIELLE S. BASSETT AND MARY-ELLEN LYNALL

ABSTRACT Network science provides tools that can be used to understand the structure and function of the human brain in novel ways using simple concepts and mathematical representations. Network neuroscience is a rapidly growing field with implications for systems neuroscience, cognitive neuroscience, and clinical medicine. In this chapter, we describe the methodology of network science as applied to neuroimaging data. We cover topics in constructing networks, probing network structure, generating network "diagnostics," and experimental design. We discuss several current frontiers and the associated methodological challenges and considerations. We aim to provide a practical introduction to the field: we supplement the explanations and examples with pointers to resources for students or researchers interested in using these methods to address their own questions in empirical and theoretical neuroscience.

Why network neuroscience?

Each area of the human brain plays a unique role in processing information gleaned from the external world and in driving our responses to that external world via behavior. Mapping these roles has led to enormous insights into the complex and varied contributions of different brain regions to our mental function. However, the brain is far from a set of disconnected building blocks. Instead, at each moment throughout the day, parts of the brain communicate with one another in complex spatiotemporal patterns, like evolving dance partners in a multifarious choreography, which enable the formation of creative thoughts, the acquisition of new skills, and the adaptation of human behavior. Understanding this spatiotemporal complexity requires a paradigmatic shift in our conceptual approaches, empirical goals, and quantitative methods: in short, in the way that we design and interpret our models and experiments.

Systems neuroscience addresses this complexity by seeking to understand the structure and function of large-scale neural circuits and systems. How do individual brain areas interact with one another to enable cognitive function? How is cognition constrained by white matter pathways? How does the brain transition between functions like memory, attention, and movement? How do we control the interactions between different neural circuits in our brains?

To answer these questions, we can use tools from systems neuroscience to perform (1) data analysis to extract characteristic or predictive patterns in the data and (2) forward modeling to build mathematical models of the system from first principles. The distinction here is important: *data analytical* approaches lead to descriptions of an observed process; for example, brain areas A and B tend to be alternately activated during a visual task. In contrast, *modeling* involves the creation of a set of mathematical descriptions (i.e., equations) that describe how components of the system behave, given certain inputs or conditions. For example, the activity in brain area A could be described by an equation that captures its behavior, including its dependency on the activity of brain area B, and vice versa. Crucially, the mathematical descriptions in a model can then be used to *predict* the behavior of the system given a different set of inputs, or in a different context. Naturally, models are much more difficult to create than descriptions.

In this chapter, we focus on newly developed tools for data analysis, which we refer to under the broad term *network neuroscience*, that we envision will dramatically inform the efforts in forward modeling in the coming years. Network neuroscience provides a simple and elegant systems approach to understanding how neural circuits function, how they constrain one another, and how they differ across individuals. A network representation of a biological system (e.g., a genome, proteome, or connectome) treats individual components (e.g., genes, proteins, brain regions) as network nodes and treats interactions between these components as network edges. Network science, an interdisciplinary approach spanning biology, economics, sociology, linguistics, and computer science, provides a battery of quantitative diagnostics that enable us to describe the architecture of this network in a statistically principled manner. One can then study these properties of the network to gain insight into organizational principles and evolutionary drivers of complex cognitive phenomena.

As a basis for discussion throughout the chapter, we use data from a previously published experiment (Bassett, Wymbs, et al., 2011; Bassett, Porter et al., 2013) to illustrate how neuroimaging data can be transformed into a network, how these networks can be studied, and how they can be compared statistically with one another to address a neuroscientific question. It is important to keep in mind during this exposition that there is no generic "correct" way to do a network analysis: to showcase some of the tools available, at each stage of our example analysis, we outline various alternative methodological choices available to the researcher.

A few foundational concepts

In figure 79.1, we illustrate the path from data to diagnostics commonly traversed in a network study. In a first step, data is collected from human subjects using one of the many neuroimaging modalities currently in use: structural MRI, functional MRI, electroencephalography,

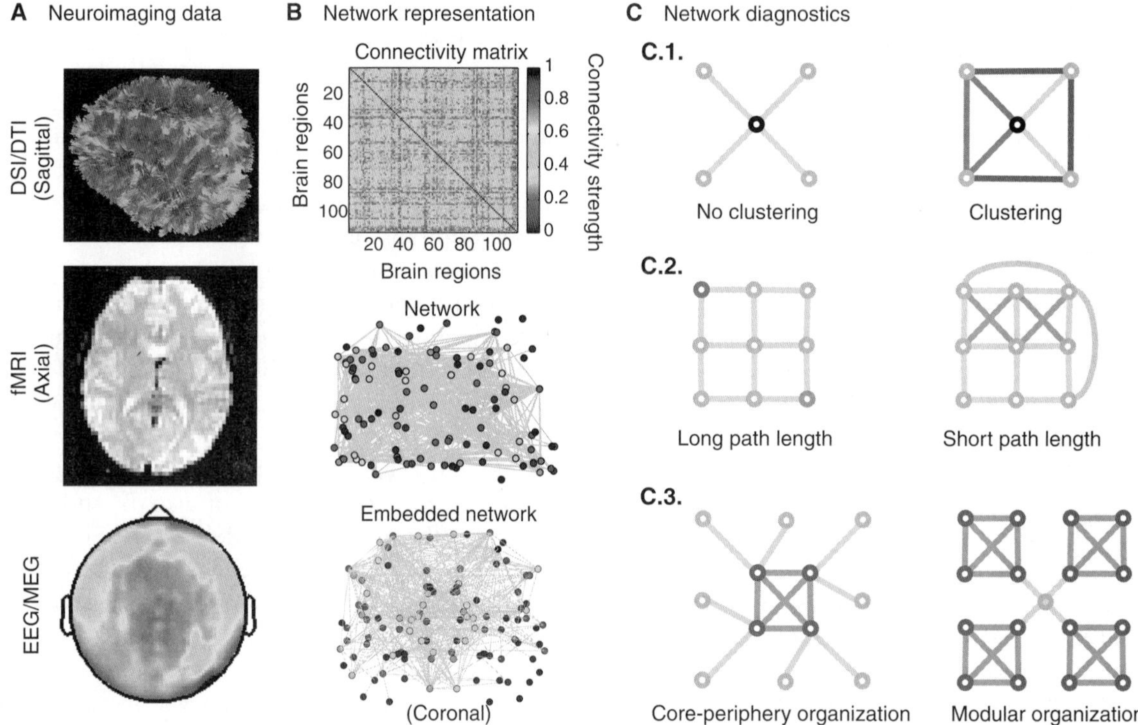

FIGURE 79.1 From data to diagnostics: The stages of a network study. (A) Data acquisition. Neuroimaging data can capture structural connectivity (e.g., diffusion spectrum imaging, DSI; diffusion tensor imaging, DTI) or functional connectivity (e.g., functional magnetic resonance imaging, fMRI; electroencephalography, EEG; or magnetoencephalography, MEG). (B) Representations of a network. *Top:* A connectivity matrix in which matrix elements (or pixels in the grid) represent one connection between two brain regions, and the color indicates the strength of that connection. *Center:* A topographical network visualization in which brain regions that are strongly (weakly) connected to one another lie close to (far from) each other in the plane. *Bottom:* An embedded network visualization in which nodes are placed in anatomically accurate locations. (C) Network diagnostics. (C.1) The clustering coefficient is a diagnostic of local network structure. The left panel contains a network with zero connected triangles and therefore no clustering, while the right panel contains a network in which additional edges have been added to close the connected triples (i.e., three nodes connected by two edges; green) to form triangles (i.e., three nodes connected by three edges; brown), thereby leading to higher clustering. (C.2) The average shortest path length is a diagnostic of global network structure. The left panel contains a network with a relatively long average path length. For example, to move from the purple node (*top left*) to the orange node (*bottom right*) requires one to traverse at least four edges. The right panel contains a network in which addition edges have been added (green) to form triangles or to link distant nodes (peach), thereby leading to a shorter average path length by comparison. (C.3) Mesoscale network structure can take many forms. The left panel contains a network with a core of densely connected nodes (red circles; green edges) and a periphery of sparsely connected nodes (blue circles; gray edges). The right panel contains a network with four densely connected modules (red circles; green edges) and a connector hub (blue circle; gray edges) that links these modules to one another. (See color plate 78.)

magnetoencephalography, or diffusion imaging. Next, the data acquired from each subject are converted to network form. The construction of the network depends upon the researcher's choice in defining network nodes and network edges. Finally, the constructed networks are analyzed statistically to test hypotheses regarding the organization of the networks. Statistical diagnostics come in two forms: (1) previously defined network diagnostics that have proven useful in previous studies, or (2) diagnostics created in an individual study to capture a pattern observed in the data.

In this section, we will describe this path in greater detail, illustrating choices that can be made at each stage of the analysis and the impact these choices can have on the conclusions that can be drawn from the study.

Network Construction Having decided to undertake a network analysis, we most likely have a neuroimaging data set in hand and are faced with the question of how to extract a network from it. To do that, we must define what a network actually is. A network can be defined in mathematical terms as a graph G composed of N nodes, which represent brain regions and E edges between those nodes that represent region-to-region relationships. The use of the term *graph* here differs from the common usage depicting a visual representation of data on axes. Instead, in network science the term *graph* often refers to the join-the-dots pattern of connections (edges) between nodes.

To construct a brain graph, we must choose how to subdivide the brain into network nodes (or brain regions) and how to define the edges (or interactions) between those nodes. The choice of nodes and edges in the extraction of brain networks from neuroimaging data varies widely, and the question of whether a single most appropriate choice exists remains under debate (Bassett & Bullmore, 2009; Bullmore & Bassett, 2011; Bullmore & Sporns, 2009; Bullmore et al., 2009).

Types of brain networks In some cases, the choice of node and edge definition depends upon the type of network under study. In general, there are two types of brain networks. Functional brain networks are constructed from functional neuroimaging data (e.g., functional MRI, electroencephalography, or magnetoencephalography), and network edges represent the functional or effective connectivity patterns between brain areas. Structural brain networks are constructed from diffusion-based neuroimaging data (e.g., diffusion tensor imaging or diffusion spectrum imaging) and network edges represent the "hard-wired" white matter connectivity patterns between brain areas. Functional and structural brain networks each provide different types of information about brain organization and cognitive function. There is no simple relationship between a person's structural and functional networks: for example, areas that have no detectable white-matter connections can be functionally connected. The question of how these two types of networks relate to one another is a source of considerable scientific endeavor (e.g., Hermundstad et al., 2013; Honey et al., 2009).

Node choice: Parcellation To create a network, we subdivide the system that we are studying into components, and we represent these components as network nodes. The components of the brain are often thought of as regions of interest: primary visual cortex, dorsolateral prefrontal cortex, or fusiform gyrus. Each region can then be represented as a node in the brain network.

A map that segregates the many voxels of a neuroimaging data set into regions of interest or network nodes is referred to as a parcellation. There are two basic types of parcellations: (1) those based on neuroanatomy and cytoarchitectonics and (2) those based on data-driven clustering methods. In applying a parcellation to neuroimaging data, our goal is to choose areas of the brain that can be treated as separate units in the brain system (Butts, 2009), where "separate" can be defined in many different ways: functionally, structurally, or anatomically.

Neuroanatomical parcellations define brain regions based on the underlying neuroanatomy. The set of Brodmann areas is an example of a neuroanatomical parcellation, in which each brain region is comprised of tissue with a particular cytoarchitecture—that is, a particular arrangement and appearance of stained neuronal cell bodies, when slices of brain tissue are viewed under a microscope (Brodmann, 1909). A network that uses Brodmann areas as network nodes can be used to probe the relationship between relatively large brain areas such as dorsolateral prefrontal cortex (area 46), primary motor cortex (area 4), and the fusiform gyrus (area 37). Other parcellations that are similar in spirit include the Automated Anatomical Labeling atlas, the Harvard-Oxford atlas, and the LONI Probabilistic Brain Atlas. A key advantage of using neuroanatomical parcellations to define network nodes is that they enable neurobiological interpretation and simplify group-based interpretations and comparisons.

Connectivity-based parcellations define brain regions based on data-driven clustering methods that isolate sets of voxels with similar functional or structural properties. The functional parcellation of Power et al.

(2011) is an example of a connectivity-based parcellation in which each brain region is composed of voxels that show similar neurophysiological activity as measured by fMRI. The resultant network can be used to probe the interactions between functionally distinct areas in a given task, and these areas may or may not adhere to cytoarchitectonic boundaries. Connectivity-based parcellations have also been derived from diffusion imaging scans by clustering white matter tractography data (e.g., Cloutman & Lambon Ralph, 2012; Gorbach et al., 2011; Perrin et al., 2008; Roca, Rivière, Guevara, Poupon, & Mangin, 2009). In comparison to the neuroanatomical parcellations, the connectivity-based parcellations provide a unique window into individual differences in brain structure and function, which can empower the search for biological underpinnings of connectivity while somewhat complicating group comparisons.

After choosing a parcellation, we must choose how to apply that parcellation to the neuroimaging data set at hand. In the context of a functional imaging scan, we detect a time-varying signal (or "time course") from each voxel (or three-dimensional pixel) in the brain. The activity of all voxels within the region can be averaged together to create representative regional time courses (Achard, Salvador, Whitcher, Suckling, & Bullmore, 2006). However, it is possible that by averaging time series together, we lose important information about signal variability within a region. Alternative approaches include calculating the median regional activity or the first principal component of the activity—that is, a single representative signal that accounts for as much as possible of the variability in the signals from all the included voxels. In the context of a structural imaging scan, the white matter tracts terminating in a single region are treated identically, and so the "strength" of a structural connection is simply proportional to the number of detected tracts that connect the two brain regions (Hagmann et al., 2008). An alternative method, which avoids the difficulty and arbitrariness of making either/or decisions about whether tracts near boundaries end in one particular area or another, is to weight tracts according to their spatial placement. To quantify the connections to a given region, tracts located close to the center of mass of tract termination points within that region could be more heavily weighted than tracts located farther away.

What is the effect of node choice on network studies? First, the definition of any particular node affects how one can interpret changes in that node's network properties. If we have defined a node to be the entire primary motor cortex, we must interpret changes in that node differently than if we had defined the node to be only the hand motor cortex. Second, the choice of a parcellation scheme can alter the observed network structure. As a simple example, consider region size. When we use a parcellation scheme with relatively small nodes, we will be probing the network structure of the brain at a higher resolution than if we used a parcellation scheme with relatively large nodes (Bassett, Brown, Deshpande, Carlson, & Grafton, 2011; Bassett et al., 2010; Meunier, Lambiotte, Fornito, Ersche, & Bullmore, 2009; Zalesky et al., 2010). While several studies have demonstrated that qualitative features of network organization are relatively immune to changes in parcellation scheme (Bassett, Brown et al., 2011; de Reus & Heuvel, 2013; Wang et al., 2009), quantitative properties of networks (to be discussed in the next section) and biological interpretations are necessarily altered.

Edge choice: Structural and functional connectivity To characterize the relationships or edges between network nodes, we must define a type of interaction between brain areas. Two basic types of edges are commonly used: (1) those that estimate the "hard-wired" anatomical connectivity between brain regions and (2) those that estimate the functional coherence or real-time interactions between brain regions. The goal of these approaches is to define a single consistent type of interaction between brain regions from which to construct a single type of network (Butts, 2009). Networks with multiple types of links are referred to as multiplex networks (e.g., Bianconi, 2013; Mucha, Richardson, Macon, Porter, & Onnela, 2010), but these are rarely used in neuroimaging studies.

An anatomical edge can be defined by the existence of a white matter tract connecting brain region i to brain region j. This edge could either be given a binary value (e.g., 1 if a tract exists between region i and region j and 0 if a tract does not exist) or a continuous value (e.g., the number of tracts that exist between region i and region j). Alternatively, anatomical edges can be represented by the mean fractional anisotropy or magnetization transfer ratio (a proxy for myelination) along a tract or set of tracts between regions (Hagmann et al., 2008; van den Heuvel, Mandl, Stam, Kahn, & Hulshoff Pol, 2010). Anatomical edges are used to represent the information transmission capabilities between large-scale brain regions (Sporns, 2010).

A functional edge can be defined by coherent oscillatory activity in region i and region j, putatively representing communication (Fries, 2005). This edge could either be given a binary value (e.g., 1 if the coherence between region i and region j is statistically significant and 0 if it is not) or a continuous value (e.g., the magnitude of the coherence between region i and region *j*). Alternatively, functional edges can be defined based on any statistical relationship between regional time series (David, Cosmelli, & Friston, 2004; Dawson, Cha, Lewis, Mendola, & Shmuel, 2013; Gates & Molenaar, 2012; Smith et al., 2011; Watanabe et al., 2013); examples include mutual information, synchronization likelihood, and partial correlation. A common choice in fMRI networks is the Pearson correlation (Zalesky, Fornito, & Bullmore, 2012). If two brain areas have similar activity—however that is defined—network approaches assume that they are in some way functionally linked. These edges are taken to reflect functional interactions such as information transfer, coordination, or shared processing (Bassett & Bullmore, 2009; Bullmore & Bassett, 2011; Bullmore & Sporns, 2009).

PROBING NETWORK STRUCTURE After defining the nodes and edges of a network, we can begin to probe the organization of that network to better understand its structure and, to some extent, its function (Butts, 2009; Proulx, Promislow, & Phillips, 2005). Two key issues inform our next steps: statistical noise in the data and the presence of multiple scales of interest in the network. In the recent literature, several methods have been proposed to address each of these factors, and here we briefly review these approaches.

Statistical noise In empirical measurements of brain activity and anatomy, noise affects our confidence in the estimated strength of edges in both anatomical and functional brain networks. In a functional network, should we treat two regions as connected if their activity profiles are correlated with a Pearson's r of 0.01? Or in an anatomical network, should we treat two regions as being connected if they are linked by a single streamline, as estimated by white matter tractography algorithms? Answers to these questions depend on our understanding of the noise present in the data. Noise in these data sets can stem from biological, measurement, or data-processing sources or a complicated combination of all three.

To maximize the power to detect neurophysiologically relevant connectivity patterns, two main solutions have been proposed. In one approach, we can test the statistical significance of each edge in the network, and then remove statistically nonsignificant edges from the network by setting their value to 0. We then study the organization of the remaining (statistically significant) edges. This approach is often utilized in the context of functional networks. If edges are defined by statistical similarities in regional brain time series, the significance of the elements of the resulting connectivity matrix is affected by the large number of tests that have been performed, which significantly increases the chances of type I (false-positive) errors.

Several multiple-comparison correction methods for these errors, such as Bonferroni and false discovery rate (Benjamini & Hochberg, 1995), have been proffered (e.g., Achard et al., 2006; Bassett, Wymbs, et al., 2011), although some argue that these approaches are too stringent for network-based analyses (Fornito, Zalesky, & Breakspear, 2013). A false-positive correction in which edges are retained if their p values are less than $1/N$, where N is the number of nodes in the network, may be a suitable compromise (e.g., Bassett, Meyer-Lindenberg, Weinberger, Coppola, & Bullmore, 2009; Lynall et al., 2010).

In a second approach, we do not perform any statistical thresholding on the edge weights. Instead, we study the entire network (inclusive of the noise), and then compare that network structure to null models that have been constructed to account for one or more sources of noise (Bassett, Porter, et al., 2013). For example, we can construct a null model for functional brain networks by creating surrogate time series that maintain the mean, variance, and autocorrelation of the original signal. By comparing the real brain network to this null model, we can identify features of the network that cannot simply be accounted for by linear properties of the time series. A critical area of ongoing research is the development of more sophisticated null models that seek to account for more complicated structure in biological networks such as growth and development, temporal dynamics, and physical embedding of the network inside of the skull (Bassett, Porter, et al., 2013; Klimm, Bassett, Carlson, & Mucha, 2013; Sporns, 2006).

Multiscale structure In studying many complex systems, we often try to isolate a single level of the system and study it intently in the hopes of gaining an intuition for how the system works. However, for most systems this is a simplification that often costs us understanding. Complex systems often display multiscale structure, or nontrivial organization, across different scales. The brain is no exception, having intricate architectures that exist and processes that occur across a range of spatial, temporal, and topological network scales (Siebenhuhner & Bassett, 2013).

The presence of multiscale structure in the brain constrains the types of methods that we can use to study brain networks. An important example of multiscale architecture lies in the spatial distribution of edges, which shows different organizational features depending on whether the edges are strong versus weak or short versus long, and on whether they connect regions

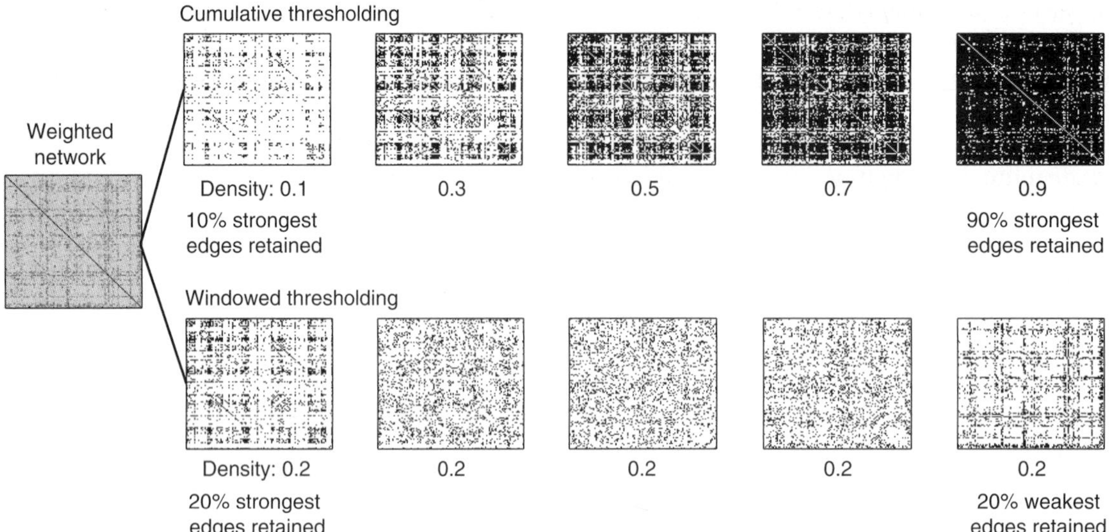

FIGURE 79.2 Probing multiscale architecture in network edge weights. Multiscale structure in a weighted network (left) can be probed using thresholding techniques. Cumulative thresholding (top) is the procedure whereby a connectivity matrix is separated into a set of graphs that contain edges above a certain weight value or threshold. Windowed thresholding (bottom) is the procedure whereby a connectivity matrix is separated into a set of graphs that contain edges whose weight values lie within a given weight range or window (Bassett et al., 2012; Schwarz & McGonigle, 2011).

that are within versus between hemispheres (e.g., Bassett, Nelson, Mueller, Camchong, & Lim, 2012; Hermundstad et al., 2013). Simple binary networks (where edges are treated as either present or absent) and nonembedded networks (where edge locations in the brain are ignored) necessarily neglect this multiscale structure, and in doing so dismiss potentially important biological signatures present in the data.

For simplicity, we will use the remainder of this subsection to discuss the role of edge strength in multiscale brain structure. To study the network topology (the arrangement of the elements in a network) and putative biological utility of edge strength, we can probe weighted networks (where edges maintain estimated strengths) using a variety of thresholding techniques. Cumulative thresholding is the procedure whereby a family of graphs is created from a single connectivity matrix; each graph in the family contains edges above a certain weight value or threshold (see figure 79.2, top). The threshold for each graph in the family is unique, and specifies the density of edges in the graph: small values of the threshold produce dense graphs and large values of the threshold produce sparse graphs. Windowed thresholding is the procedure whereby a different family of graphs is created; each graph in the family contains edges whose weight values lie within a given weight range or window (see figure 79.2, bottom; Bassett et al., 2012; Schwarz & McGonigle, 2011). The width of the weight range specifies the density of each graph in the family: small weight ranges produce sparse graphs and large weight ranges produce dense graphs.

Cumulative and windowed thresholding techniques can provide insight into the multiscale nature of a network's connectivity patterns, embedded in edge weights. The windowed thresholding approach has the advantage of isolating the network structure of strong versus weak edge weights. Strong edges can provide insight into the organization of energetically costly links in anatomical brain networks and of heavily utilized coordination links in functional brain networks. Weak edges, while technically less significant according to some statistical tests (Achard et al., 2006), can imply strongly correlated network states (Schneidman, Berry, Segev, & Bialek, 2006) and can distinguish diseased network structures in schizophrenia (Bassett et al., 2012). Observed weak correlations between regional activity could be driven at least in part by the variability of neuronal activity signals, which plays an important role in cognitive function (Misic, Mills, Taylor, & McIntosh, 2010), development (McIntosh et al., 2010), and recovery from injury (Raja Beharelle, Kovacevic, McIntosh, & Levine, 2012).

GENERATING NETWORK DIAGNOSTICS A network "diagnostic" is simply a named measurement used to describe a network's properties. These properties can then be compared across different networks. For example, a patient's network diagnostics can be compared with those of a healthy participant; or a human brain network can be compared to a worm neuronal network, to see whether the properties captured by that diagnostic are shared by neural systems in different species (Bassett et al., 2010).

Network diagnostics are not quantities like mass, entropy, or replication rate that a priori have a clear physical meaning. Instead, they are mathematical definitions that formalize architectural concepts specific to network science (Newman, 2010). As an outsider, this jargon can at times make the field seem impenetrable. However, upon closer inspection, many diagnostics capture an intuitive property of interconnected systems that can often be easily interpreted in the context of the brain (Rubinov & Sporns, 2009; Sporns, 2010).

Local, global, and mesoscale properties We can use network diagnostics to study the organization of anatomical and functional brain networks across spatial scales, from the neighborhood of the network surrounding a single brain area (captured using *local* statistics) to the architecture of the entire network (captured using *global* statistics). A typical local diagnostic is the clustering coefficient C, which relates to the likelihood that a node's neighbors are connected to one another, forming triangles or loops (see figure 79.1C.1). The clustering coefficient can be calculated for each node separately, and then the values can be averaged across nodes to determine a mean clustering coefficient for the network. A brain-centric interpretation of the clustering coefficient is that it might quantify the amount of local information segregation (Bullmore & Sporns, 2009; Sporns, 2010).

A typical global diagnostic is the average shortest path length L. The shortest path between node i and node j is the smallest number of edges that must be traversed to get from node i to node j. The average shortest path is the mean shortest path over all possible pairs of nodes in the graph (see figure 79.1C.2). A brain-centric interpretation of the average shortest path length is that it might quantify the amount of global information integration (Bullmore & Sporns, 2009; Sporns, 2010). A related concept—the network efficiency (Latora & Marchiori, 2001, 2003)—is also calculated based on shortest paths through a graph. Networks with high efficiency have short path lengths and networks with low efficiency have long path lengths. Network efficiency has been interpreted in relation to the efficiency of information processing in the brain (Achard & Bullmore, 2007; Sporns, 2010).

Mesoscale diagnostics ("meso-" means "middle") capture intermediate-level properties of network organization. Rather than focusing on either the local neighborhood of a node or the global structure of the entire graph, mesoscale diagnostics characterize the organization of *groups* of nodes. For example, core-periphery diagnostics enable us to uncover a core of densely and mutually interconnected nodes and a periphery of sparsely connected nodes (see figure 79.1C.3, left; Borgatti & Everett, 1999; Rombach, Porter, Fowler, & Mucha, 2012). Core-periphery organization might confer robustness to the brain's structural core (van den Heuvel & Sporns, 2011) and enable a balance between stability and adaptivity in brain dynamics (Bassett, Wymbs, et al., 2013). Modularity is another type of mesoscale property in which sets of nodes form densely connected subgroups (see figure 79.1C.3, right; Fortunato, 2010; Porter, Onnela, & Mucha, 2009). Modular organization provides a natural substrate for the combined integration and segregation of information processing arguably required for healthy brain function (Bullmore & Sporns, 2009).

Despite the changing fashions for particular diagnostics (for example, small-worldness, a buzzword of the early 2000s that is now falling somewhat out of favor), no single diagnostic can capture all of the important organizational properties of networks (Newman, 2010). The library of network diagnostics is continually growing as applied mathematicians, physicists, engineers, computer scientists, and others define new mathematical entities to capture previously unexplored patterns in network structure. While each diagnostic has a unique mathematical definition, it is possible for several diagnostics to produce values that are highly correlated with one another across network samples (e.g., across different brains). A current challenge is to determine the families of network diagnostics that provide complementary but not necessarily independent information about functional and anatomical brain organization (Lynall et al., 2010).

Interpretational caveats Network diagnostics can be intuitively interpreted in terms of information processing: high clustering can suggest that information is processed in local domains, while short path length can suggest that information is being transmitted over longer distances within the network. However, the biological meaning of these interpretations requires a conceptual leap from topological to biological terms, which are semantically equivalent, but not necessarily conceptually interchangeable (Rubinov & Bassett, 2011). Biological efficiency, for example, has evolutionary implications that may not apply to network efficiency. More generally, such interpretations of network diagnostics require empirical validation demonstrating the relationship between quantifiable estimates of information processing or biological efficiency and network characteristics. Until then, a cautious interpretation of network diagnostics need not hamper the utility of these approaches in prediction, classification, diagnosis, and monitoring and in the study of system-level dynamics underlying cognitive function.

Using diagnostics to probe brain network structure We have shown how a variety of different networks could be produced from a single brain scan. The network diagnostics generated will vary considerably, depending on whether a weighted or binary network is used, whether we take a cumulative or windowed thresholding approach, and what range of connection densities we choose to consider. For example, we can see in figure 79.3A that the value of the clustering coefficient tends to be small for sparse graphs and large for dense graphs, where neighbors of a node are more likely to also be connected to one another (see figure 79.3A). The value of network efficiency shows the opposite effect: large values characterize sparse graphs and small values characterize dense graphs, where any pair of nodes is likely to be connected.

Given this variation, how do we choose which actual numbers should be used to represent a sample, when it is compared to another? The characteristic curves of diagnostic values as a function of graph density or mean edge weight provide useful signatures of brain networks in different states (e.g., health and disease, or various task states). We can collapse a curve extracted from a single participant into one number by calculating the area under the curve (Ginestet, Nichols, Bullmore, & Simmons, 2011). The advantage of this procedure is that we can then determine group differences in this value by performing a simple *t*-test or permutation test. The disadvantage of this procedure is that we have necessarily lost information about the shape of the curve. Two groups might have identical area-under-the-curve values but quite different curve shapes. An alternative approach is to use functional data analysis, a statistical technique developed for the principled study of curves, to determine whether the shape of the curves is significantly different (Bassett et al., 2012; Siebenhuhner, Weiss, Coppola, Weinberger, & Bassett, 2013; see figure 79.3B).

While network diagnostic values and curves provide insight into the organization of the network, it is the mapping of these values back to brain regions that enables us to make fine-grained biological interpretations of our data. Different network diagnostics can display very different spatial distributions across the surface of the brain (see figure 79.3C). For example, in an early motor skill learning task (Bassett, Porter, et al., 2013; Bassett, Wymbs, et al., 2011), the clustering coefficient is strong in the primary motor cortex and weak in visual association areas, while the network efficiency displays the opposite trend. Such maps can be constructed for individual participants or for groups of participants and enable us to link our results to the large body of neuroimaging literature isolating functions of individual brain regions.

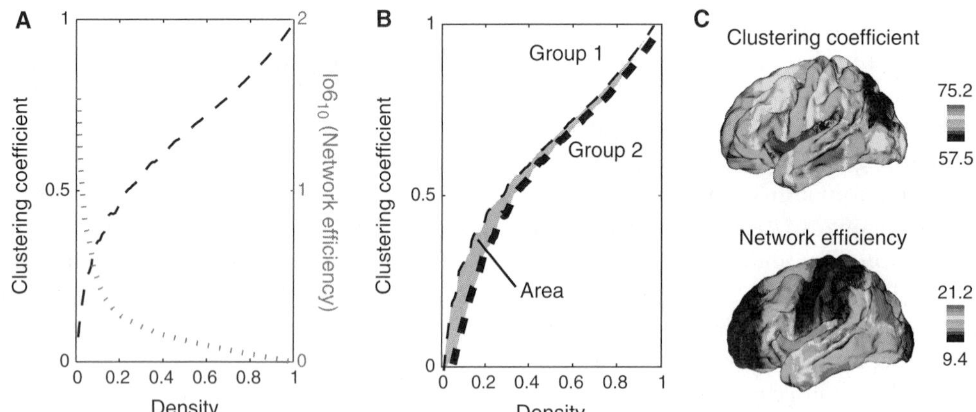

Figure 79.3 Using diagnostics to probe brain network structure. (A) Clustering coefficient (left *y*-axis, black) and network efficiency (right *y*-axis, light gray) as a function of network density as estimated using cumulative thresholding techniques. (B) Schematic of mean clustering coefficient versus density curves for two sets of participants: group 1 (thin, light-gray dashed line) and group 2 (thick, dark-gray dashed line). Functional data analysis is a statistical framework that can be used to determine whether the area between the two curves is significant in comparison to a null model. (C) Area under the clustering coefficient versus density curve (top) and the network efficiency versus density curve (bottom) for 112 regions of the brain, defined according to the Harvard-Oxford atlas (see Bassett, Wymbs, et al., 2011, and Bassett, Porter, et al., 2013, for additional details on this data set).

Experimental Design Since its inception, network neuroscience has predominantly been exploratory in nature. In network-based studies of the brain, researchers are constantly fine-tuning methodological approaches and isolating empirical questions that are amenable to these approaches. Often these studies have included a reanalysis of previously published data, which had initially been examined from a more traditional perspective. The use of previously acquired data is unquestionably justified for many reasons, including but not limited to the use of taxpayer money for research studies, the time spent by subject participants in volunteering for the study, the difficulty in obtaining patient data, and the richness of data acquired by current neuroimaging techniques that cannot be mined completely using a single analytic approach. These studies have provided extensive early validation of network methods and their use in understanding the human brain.

However, the reanalysis of previously acquired data has its disadvantages. The most critical disadvantage is that these studies were often not conceived with network-based hypotheses in mind. New data acquired using experimental paradigms specifically designed to test network hypotheses will open up entirely new fields of inquiry. The development of such paradigms and hypotheses is an important frontier in network neuroscience.

What can network approaches tell us?

Is network neuroscience an enlightening new approach to cognitive science, or is it simply an interesting intellectual exercise for the mathematically inclined? The empirical evidence to date strongly supports the former conclusion.

The first type of evidence comes from the fact that network neuroscience has provided diagnostics that display close relationships with more traditional measurements or known quantities. For example, people with a variety of psychiatric and neurological diseases have differently connected brains than people who are healthy (Bassett & Bullmore, 2009). When people perform different tasks, their brain regions interact differently, leading to alterations in network diagnostics (e.g., (Bassett, Meyer-Lindenberg, Achard, Duke, & Bullmore, 2006). Over both long and short time scales, from years (Meunier, Achard, Morcom, & Bullmore, 2008) to days and minutes (Bassett, Wymbs, et al., 2011), the brain changes in how different regions interact with each other. As your behavior changes, so do your brain networks (Reijmer et al., 2013).

Together, these results provide important validation of the network approach to neuroscience. But not all of these results are surprising or groundbreaking. The results often grab the attention of the media instead because they directly address the perennial problem of Cartesian dualism that pervades both lay and medical thinking: linking the mind to the brain. Indeed, studies demonstrating brain correlates of behaviors or psychiatric disease continually inform, and provoke new developments in, the philosophy of mind. In addition to their philosophical appeal, network methods can also display these results in a quantitative and visually appealing way. But do these results fundamentally advance our understanding of how the human brain works?

Network Neuroscience as Explanation The growing consensus in the community is a resounding "Yes." Network science provides a fundamentally new level of explanation for cognitive function. And what do we mean by an "explanation"? Among other things, an explanation can (1) describe phenomena in terms of more fundamental and general principles, or (2) provide a causal history of the phenomena (Woodward, 2011). Reductionist models of biological phenomena have traditionally provided explanations of the second form (providing causal histories) but in general have difficulty providing explanations of the first form (linking to fundamental principles). Network science, however, provides inherently new information about how the brain works in relation to general mathematical and physical principles, and it is this new information that informs novel hypotheses, interpretations, and empirical studies.

Shifting conceptual paradigms Network science has supported a fundamental paradigm shift in the conceptual framework that we use for neuroscientific inquiry. By placing significant weight on the importance of interactions, network methods stand in contrast to other approaches focused primarily on the localization of cognitive functions to specific brain areas through the study of local brain activity. Instead, the principled investigation of time-dependent communication between brain regions, facilitated by network methods, has enabled the discovery and description of both intrinsic (the default mode network; Snyder & Raichle, 2012) and extrinsic connectivity phenomena.

Revealing organizational principles Network science can be used to uncover organizational principles of complex systems. As an example, consider the use of network science in the identification of evolutionary and metabolic constraints on brain structure (Bassett et al., 2010; Bullmore & Sporns, 2012). By studying network architecture present in natural organisms, from human to worm, we can infer fundamental principles of brain organization, such as cost-efficiency in network organization (Bullmore & Sporns, 2012), and posit their alteration in disease states (Vértes et al., 2012). Extracting these guiding principles is of critical importance in building an expanded theoretical neuroscience, and is a necessary complement to the accrual of increasingly detailed accounts of specific brain areas, pathways, and molecules.

Distilling mechanisms of disease In addition to uncovering organizational principles in the healthy brain, network neuroscience has provided new insights into the mechanisms of disease. In Alzheimer's disease, for example, regions of dense functional connectivity (also known as network hubs) correspond to areas of greatest plaque deposition (Buckner et al., 2009). In contrast to descriptions of plaque density from non-systems approaches, these results from network science suggest a disease mechanism: high metabolic function, information processing, and structural connectivity in brain network hubs might augment the pathological cascade in Alzheimer's disease. Network neuroscience methods have also played a primary role in our growing understanding of schizophrenia as a brain-wide pathology, characterized by extensive dysconnectivity rather than localized abnormal activity (Fornito, Zalesky, Pantelis, & Bullmore, 2012). Even in the context of stroke, network methods have been used to show that communication pathways are altered far from the lesioned site (Crofts et al., 2011), illustrating the wide-ranging possible uses of network methods in the study of both localized injury and distributed disease.

Network Neuroscience as a Young Field Despite these exciting recent advances, network neuroscience is still a very young field, and many challenges remain. Particularly salient frontiers evident in the recent literature include the following:

- Network dynamics. The human brain is a dynamic system (Deco, Jirsa, & McIntosh, 2011) underpinning the complexities of cognitive function. The extension of network methods to characterize the temporal changes in putative communication patterns in the brain is necessary to understand the constantly evolving nature of cognition (Allen et al., 2014; Bassett, Wymbs, et al., 2011; Kramer et al., 2011; Smith et al., 2012).
- Neurophysiological and genetic drivers of network organization. The brain networks that we observe are driven by lower-level physiological processes (e.g., (Bassett et al., 2012; Vaishnavi et al., 2010) and genetic phenomena (e.g., Esslinger et al., 2009; Fornito et al., 2011). Determining the role of molecular and cellular dynamics in large-scale network neuroscience is critical for a mechanistic understanding of brain development and function.
- Network-based prediction. Network approaches provide novel possibilities for classification and prediction. Machine learning, mathematical modeling, and statistical analyses have shown promise in predicting brain state (e.g., Richiardi, Eryilmaz, Schwartz, Vuilleumier, & Van De Ville, 2011), disease progression (e.g., Raj, Kuceyeski, & Weiner, 2012), and potential receptivity to neurorehabilitation efforts (Bassett, Wymbs, et al., 2011, 2013).
- Network approaches to behavior, perception, and evolution. Applications of network methods outside of neuroimaging could provide important insights into cognitive function. For example, the network concept of community structure has been used to capture the organization of human movements (e.g., Wymbs, Bassett, Mucha, Porter, & Grafton, 2012), the temporal relationships between concepts (e.g., Schapiro, Rogers, Cordova, Turk-Browne, & Botvinick, 2013), and the genetic interactions underlying cellular machines impacting on neural function (e.g., Conaco et al., 2012).

Together, these frontiers promise important progress in our understanding of cognitive function, its neurophysiological and genetic underpinnings, and its relationship to behavior.

As with any young field, the scientific excitement of network neuroscience is paired with its growing pains. First, it is not always clear how to translate the findings of network neuroscience to the clinic, be that in the development of antipsychotic drugs or in the rehabilitation of injured patients. Progress in this area requires further advances in clinical systems neuroscience. Second, sources of noise in specific subject populations (e.g., movement in adolescents or in people with schizophrenia) can produce network signatures, which if not adequately corrected for can lead to the inaccurate identification of group differences in network structure. Indeed, the identification and understanding of individual differences in brain connectivity will be an important area of growth in the coming years. Third, some network diagnostics and concepts can appear to be fads: small-worldness and power-law degree distributions were of great interest until it was shown that most real-world networks are small-world and power-law degree distributions do not necessarily imply a specific underlying mechanism (e.g., criticality; Stumpf & Porter, 2012). Finally, simple network representations of complex systems like the brain necessarily abstract away many potentially important biological details: nodes are not all identical, but instead have different structural and functional properties (e.g., Hermundstad et al., 2013). A critical effort in the coming years will be to extend network representations to take into account these inter-regional differences to create more biologically realistic models of this complex data. Despite these growing pains, network science is a vibrant, rapidly growing field that brings with it exceptional promise for both empirical and theoretical neuroscience.

Conclusion

In this chapter, we have discussed why network methods provide an interesting approach to the study of neuroscientific problems in general. The promise, challenges, and controversies of network neuroscience make it an exciting area, ripe for rapid progress, and crying out for new minds. To encourage you to dive in, we have provided a list of reviews and resources to explore in Box 79.1 and a list of helpful toolboxes in Box 79.2. We hope that these tools will be of use to you as you walk the paths of methodological innovation and scientific discovery.

ACKNOWLEDGMENTS We thank Brian Wandell for helpful comments on earlier versions of the manuscript.

Box 79.1: Reviews and Resources

- Several reviews address the general techniques used in applying network science to neuroscience data: Bassett and Bullmore (2006); Reijneveld, Ponten, Berendse, and Stam (2007); Bullmore and Sporns (2009); He and Evans (2010); Bullmore and Bassett (2011); Wig, Schlaggar, and Petersen (2011); Sporns (2011); Kaiser (2011); Bullmore and Sporns (2012); Sporns (2012); Stam and Straaten (2012); Sporns (2012).
- A smaller number of reviews focus on network applications in disease: Bassett and Bullmore (2009); He, Chen, Gong, and Evans (2009); Xie and He (2011); Fornito et al. (2012).
- *Networks: An Introduction* is an excellent network science textbook (Newman, 2010).
- *Networks of the Brain* is a wonderful book about the application of network science to neuroscience (Sporns, 2010).

Box 79.2: Toolboxes and Helpful Code

- The Brain Connectivity Toolbox is a MATLAB toolbox for network characterization (Rubinov & Sporns, 2009).
- The UCLA Multimodal Connectivity Package is a set of Python programs used to calculate connectivity metrics from a variety of neuroimaging modalities (Brown, 2013).
- Netwiki contains shared data and code for network analysis, including methods for dynamic community detection (Mucha, 2013).
- Statnet is a suite of packages for statistical network analysis, focusing especially on exponential random graph models, which test hypotheses about the processes that might have led to the generation of a particular network you have identified (Handcock et al., 2013).
- NetworkX is a Python package with a large user-base focused on the creation, manipulation, and study of complex networks (Hagberg, Schult, & Swart, 2013).
- Pajek is a Windows package that supports the decomposition, analysis, and visualization of large networks (Batagelj & Mrvar, 2013).
- There are multiple options for visualizing static, dynamic (changing over time), and even interactive brain networks. These include Graphviz (or RGraphviz in R; Bilgin et al., 2013), igraph (Csardi & Nepusz, 2013), gephi (Bastian, Heymann, & Jacomy, 2013), d3 (Bostock, 2013), and helpful user contributions on MATLAB's File Exchange (www.mathworks.com/matlabcentral/fileexchange).
- Sometimes there is no substitute for writing your own code.

REFERENCES

Achard, S., & Bullmore, E. (2007). Efficiency and cost of economical brain functional networks. *PLoS Comput Biol, 3*, e17.

Achard, S., Salvador, R., Whitcher, B., Suckling, J., & Bullmore, E. (2006). A resilient, low-frequency, small-world human brain functional network with highly connected association cortical hubs. *J Neurosci, 26*(1), 63–72.

Allen, E. A., Damaraju, E., Plis, S. M., Erhardt, E. B., Eichele, T., & Calhoun, V. D. (2014). Tracking whole-brain connectivity dynamics in the resting state. *Cereb Cortex, 24*(3), 663–676.

Bassett, D. S., Brown, J. A., Deshpande, V., Carlson, J. M., & Grafton, S. T. (2011). Conserved and variable architecture of human white matter connectivity. *NeuroImage, 54*(2), 1262–1279.

Bassett, D. S., & Bullmore, E. T. (2006). Small-world brain networks. *Neuroscientist, 12*, 512–523.

Bassett, D. S., & Bullmore, E. T. (2009). Human brain networks in health and disease. *Curr Opin Neurol, 22*(4), 340–347.

Bassett, D. S., Greenfield, D. L., Meyer-Lindenberg, A., Weinberger, D. R., Moore, S., & Bullmore, E. (2010).

Efficient physical embedding of topologically complex information processing networks in brains and computer circuits. *PLoS Comput Biol, 6*(4), e1000748.

Bassett, D. S., Meyer-Lindenberg, A., Achard, S., Duke, T., & Bullmore, E. (2006). Adaptive reconfiguration of fractal small-world human brain functional networks. *Proc Natl Acad Sci USA, 103,* 19518–19523.

Bassett, D. S., Meyer-Lindenberg, A., Weinberger, D. R., Coppola, R., & Bullmore, E. (2009). Cognitive fitness of cost-efficient brain functional networks. *Proc Natl Acad Sci USA, 106*(28), 11747–11752.

Bassett, D. S., Nelson, B. G., Mueller, B. A., Camchong, J., & Lim, K. O. (2012). Altered resting state complexity in schizophrenia. *NeuroImage, 59*(3), 2196–2207.

Bassett, D. S., Porter, M. A., Wymbs, N. F., Grafton, S. T., Carlson, J. M., & Mucha, P. J. (2013). Robust detection of dynamic community structure in networks. *Chaos, 23,* 1.

Bassett, D. S., Wymbs, N. F., Porter, M. A., Mucha, P. J., Carlson, J. M., & Grafton, S. T. (2011). Dynamic reconfiguration of human brain networks during learning. *Proc Natl Acad Sci USA, 108*(18), 7641–7646.

Bassett, D. S., Wymbs, N. F., Rombach, M. P., Porter, M. A., Mucha, P. J., & Grafton, S. T. (2013). Task-based core-periphery organisation of human brain dynamics. *PLoS Comp Biol, 9*(9), e1003171.

Bastian, M., Heymann, S., & Jacomy, M. (2013). Gephi [Computer software]. Available from https://gephi.org/

Batagelj, V., & Mrvar, A. (2013). Pajek [Computer program]. Available from http://pajek.imfm.si/doku.php

Benjamini, Y., & Hochberg, Y. (1995). Controlling the false discovery rate: A practical and powerful approach to multiple testing. *J R Stat Soc Ser B, 57,* 289–300.

Bianconi, G. (2013). Statistical mechanics of multiplex networks: Entropy and overlap. *Phys Rev E, 87*(6-1), 062806.

Bilgin, A., Caldwell, D., Ellson, J., Gansner, E., Hu, Y., North, S., ... Woodhull, G. (2013). Graphviz [Computer software]. Available from http://www.graphviz.org/

Borgatti, S. P., & Everett, M. G. (1999). Models of core/periphery structures. *Soc Networks, 21,* 375–395.

Bostock, M. (2013) D3 [Computer software]. Available from http://d3js.org/

Brodmann, K. (1909). *Vergleichende lokalisationslehre der grosshirnrinde.* Leipzig: Johann Ambrosius Barth.

Brown, J. A. (2013). UCLA multimodal connectivity package [computer program]. Available from www.ccn.ucla.edu/wiki/index.php/UCLA_Multimodal_Connectivity_Package

Buckner, R. L., Sepulcre, J., Talukdar, T., Krienen, F. M., Liu, H., Hedden, T., ... Johnson, K. A. (2009). Cortical hubs revealed by intrinsic functional connectivity: Mapping, assessment of stability, and relation to Alzheimer's disease. *J Neurosci, 29*(6), 1860–1873.

Bullmore, E. T., & Bassett, D. S. (2011). Brain graphs: Graphical models of the human brain connectome. *Ann Rev Clin Psych, 7,* 113–140.

Bullmore, E., Barnes, A., Bassett, D. S., Fornito, A., Kitzbichler, M., Meunier, D., ... Suckling, J. (2009). Generic aspects of complexity in brain imaging data and other biological systems. *NeuroImage, 47*(3), 1125–1134.

Bullmore, E., & Sporns, O. (2009). Complex brain networks: Graph theoretical analysis of structural and functional systems. *Nat Rev Neurosci, 10*(3), 186–198.

Bullmore, E., & Sporns, O. (2012). The economy of brain network organization. *Nat Rev Neurosci, 13*(5), 336–349.

Butts, C. T. (2009). Revisiting the foundations of network analysis. *Science, 325*(5939), 414–416.

Cloutman, L. L., & Lambon Ralph, M. A. (2012). Connectivity-based structural and functional parcellation of the human cortex using diffusion imaging and tractography. *Front Neuroanat, 6*(34).

Conaco, C., Bassett, D. S., Zhou, H., Arcila, M. L., Degnan, S. M., Degnan, B. M., ... Kosik, K. S. (2012). Functionalization of a protosynaptic gene expression network. *Proc Natl Acad Sci USA, 109*(Suppl. 1), 10612–10618.

Crofts, J. J., Higham, D. J., Bosnell, R., Jbabdi, S., Matthews, P. M., Behrens, T. E., & Johansen-Berg, H. (2011). Network analysis detects changes in the contralesional hemisphere following stroke. *NeuroImage, 54*(1), 161–169.

Csardi, G., & Nepusz, T. (2013). igraph [computer software]. Available from http://igraph.sourceforge.net

David, O., Cosmelli, D., & Friston, K. J. (2004). Evaluation of different measures of functional connectivity using a neural mass model. *NeuroImage, 21,* 659–673.

Dawson, D. A., Cha, K., Lewis, L. B., Mendola, J. D., & Shmuel, A. (2013). Evaluation and calibration of functional network modeling methods based on known anatomical connections. *NeuroImage, 67,* 331–343.

De Reus, M. A., & van den Heuvel, M. P. (2013). The parcellation-based connectome: Limitations and extensions. *NeuroImage, 80,* 397-404.

Deco, G., Jirsa, V. K., & McIntosh, A. R. (2011). Emerging concepts for the dynamical organization of resting-state activity in the brain. *Nat Rev Neurosci, 12*(1), 43–56.

Esslinger, C., Walter, H., Kirsch, P., Erk, S., Schnell, K., Arnold, C., ... Meyer-Lindenberg, A. (2009). Neural mechanisms of a genome-wide supported psychosis variant. *Science, 324*(5927), 605.

Fornito, A., Zalesky, A., Bassett, D., Meunier, D., Yucel, M., Wood, S. J., ... Bullmore, E. (2011). Genetic influences on economical properties of human functional cortical networks. *J Neurosci, 31*(9), 3261–3270.

Fornito, A., Zalesky, A., & Breakspear, M. (2013). Graph analysis of the human connectome: Promise, progress, and pitfalls. *NeuroImage, 80C,* 426–444.

Fornito, A., Zalesky, A., Pantelis, C., & Bullmore, E. T. (2012). Schizophrenia, neuroimaging and connectomics. *NeuroImage, 62*(4), 2296–2314.

Fortunato, S. (2010). Community detection in graphs. *Phys Rep, 486*(3–5), 75–174.

Fries, P. (2005). A mechanism for cognitive dynamics: Neuronal communication through neuronal coherence. *Trends Cogn Sci, 9,* 474–480.

Gates, K. M., & Molenaar, P. C. (2012). Group search algorithm recovers effective connectivity maps for individuals in homogeneous and heterogeneous samples. *NeuroImage, 63*(1), 310–319.

Ginestet, C. E., Nichols, T. E., Bullmore, E. T., & Simmons, A. (2011). Brain network analysis: Separating cost from topology using cost-integration. *PLoS ONE, 6*(7), e21570.

Gorbach, N. S., Schütte, C., Melzer, C., Goldau, M., Sujazow, O., Jitsev, J., ... Tittgemeyer, M. (2011). Hierarchical information-based clustering for connectivity-based cortex parcellation. *Front Neuroinform, 5*(18).

Hagberg, A. A., Schult, D. A., & Swart, P. J. (2013). Networkx [computer software]. Available from http://networkx.github.io/index.html

Hagmann, P., Cammoun, L., Gigandet, X., Meuli, R., Honey, C. J., Wedeen, V. J., & Sporns, O. (2008). Mapping the structural core of human cerebral cortex. *PLoS Biol, 6*(7), e159.

Handcock, M. S., Hunter, D. R., Butts, C. T., Goodreau, S. M., & Morris, M. (2013). Statnet: Software tools for the statistical modeling of network data [computer software]. Available from http://statnetproject.org

He, Y., Chen, Z., Gong, G., & Evans, A. (2009). Neuronal networks in Alzheimer's disease. *Neuroscientist, 15*(4), 333–350.

He, Y., & Evans, A. (2010). Graph theoretical modeling of brain connectivity. *Curr Opin Neurol, 23*(4), 341–350.

Hermundstad, A. M., Bassett, D. S., Brown, K. S., Aminoff, E. M., Clewett, D., Freeman, S., ... Carlson, J. M. (2013). Structural foundations of resting-state and task-based functional connectivity in the human brain. *Proc Natl Acad Sci USA, 110*(15), 6169–6174.

Honey, C. J., Sporns, O., Cammoun, L., Gigandet, X., Thiran, J. P., Meuli, R., & Hagmann, P. (2009). Predicting human resting-state functional connectivity from structural connectivity. *Proc Natl Acad Sci USA, 106*(6), 2035–2040.

Kaiser, M. (2011). A tutorial in connectome analysis: Topological and spatial features of brain networks. *NeuroImage, 57*(3), 892–907.

Klimm, F., Bassett, D. S., Carlson, J. M., & Mucha, P. J. (2013). Resolving structural variability in network models and the brain. arXiv: 1306.2893.

Kramer, M. A., Eden, U. T., Lepage, K. Q., Kolaczyk, E. D., Bianchi, M. T., & Cash, S. S. (2011). Emergence of persistent networks in long-term intracranial EEG recordings. *J Neurosci, 31*(44), 15757–15767.

Latora, V., & Marchiori, M. (2001). Efficient behavior of small-world networks. *Phys Rev Lett, 87*, 198701.

Latora, V., & Marchiori, M. (2003). Economic small-world behavior in weighted networks. *Eur Phys J B, 32*, 249–263.

Lynall, M. E., Bassett, D. S., Kerwin, R., McKenna, P., Muller, U., & Bullmore, E. T. (2010). Functional connectivity and brain networks in schizophrenia. *J Neurosci, 30*(28), 9477–9487.

McIntosh, A. R., Kovacevic, N., Lippe, S., Garrett, D., Grady, C., & Jirsa, V. (2010). The development of a noisy brain. *Arch Ital Biol, 148*(3), 323–337.

Meunier, D., Achard, S., Morcom, A., & Bullmore, E. (2008). Age-related changes in modular organization of human brain functional networks. *NeuroImage, 44*(3), 715–723.

Meunier, D., Lambiotte, R., Fornito, A., Ersche, K. D., & Bullmore, E. T. (2009). Hierarchical modularity in human brain functional networks. *Front Neuroinformatics, 3*, 37.

Misic, B., Mills, T., Taylor, M. J., & McIntosh, A. R. (2010). Brain noise is task dependent and region specific. *J Neurophysiol, 104*(5), 2667–2676.

Mucha, P. J. (2013). Netwiki [Online community database]. Available from http://netwiki.amath.unc.edu/

Mucha, P. J., Richardson, T., Macon, K., Porter, M. A., & Onnela, J.-P. (2010). Community structure in time-dependent, multiscale, and multiplex networks. *Science, 328*(5980), 876–878.

Newman, M. E. J. (2010). *Networks: An introduction.* Oxford, UK: Oxford University Press.

Perrin, M., Cointepas, Y., Cachia, A., Poupon, C., Thirion, B., Rivière, D., ... Mangin, J. F. (2008). Connectivity-based parcellation of the cortical mantle using q-ball diffusion imaging. *Int J Biomed Imaging, 2008*, 368406.

Porter, M. A., Onnela, J.-P., & Mucha, P. J. (2009). Communities in networks. *Not Am Math Soc, 56*(9), 1082–1097, 1164–1166.

Power, J. D., Cohen, A. L., Nelson, S. M., Wig, G. S., Barnes, K. A., Church, J. A., ... Petersen, S. E. (2011). Functional network organization of the human brain. *Neuron, 72*(4), 665–678.

Proulx, S., Promislow, D., & Phillips, P. C. (2005). Network thinking in ecology and evolution. *Trends Ecol Evol, 20*, 345–353.

Raj, A., Kuceyeski, A., & Weiner, M. (2012). A network diffusion model of disease progression in dementia. *Neuron, 73*(6), 1204–1215.

Raja Beharelle, A., Kovacevic, N., McIntosh, A. R., & Levine, B. (2012). Brain signal variability relates to stability of behavior after recovery from diffuse brain injury. *NeuroImage, 60*(2), 1528–1537.

Reijmer, Y. D., Leemans, A., Brundel, M., Jaap Kappelle, L., & Jan Biessels, G. (2013). Disruption of the cerebral white matter network is related to slowing of information processing speed in patients with type 2 diabetes. *Diabetes, 62*(6), 2112–2115.

Reijneveld, J. C., Ponten, S. C., Berendse, H. W., & Stam, C. J. (2007). The application of graph theoretical analysis to complex networks in the brain. *Clin Neurophysiol, 118*(11), 2317–2331.

Richiardi, J., Eryilmaz, H., Schwartz, S., Vuilleumier, P., & Van De Ville, D. (2011). Decoding brain states from fMRI connectivity graphs. *NeuroImage, 56*(2), 616–626.

Roca, P., Rivière, D., Guevara, P., Poupon, C., & Mangin, J. F. (2009). Tractography-based parcellation of the cortex using a spatially-informed dimension reduction of the connectivity matrix. *Med Image Comput Comput Assist Interv, 12*(Pt. 1), 935–942.

Rombach, M. P., Porter, M. A., Fowler, J. H., & Mucha, P. J. (2012). Core-periphery structure in networks. arXiv: 1202.2684.

Rubinov, M., & Bassett, D. S. (2011). Emerging evidence of connectomic abnormalities in schizophrenia. *J Neurosci, 31*(17), 6263–6265.

Rubinov, M., & Sporns, O. (2009). Complex network measures of brain connectivity: Uses and interpretations. *NeuroImage, 52*(3), 1059–1069.

Schapiro, A. C., Rogers, T. T., Cordova, N. I., Turk-Browne, N. B., & Botvinick, M. M. (2013). Neural representations of events arise from temporal community structure. *Nat Neurosci, 16*(4), 486–492.

Schneidman, E., Berry, M. J., Segev, R., & Bialek, W. (2006). Weak pairwise correlations imply strongly correlated network states in a neural population. *Nature, 440*(7087), 1007–1012.

Schwarz, A. J., & McGonigle, J. (2011). Negative edges and soft thresholding in complex network analysis of resting state functional connectivity data. *NeuroImage, 55*(3), 1132–1146.

Siebenhuhner, F., & Bassett, D. S. (2013). Multiscale network organization in the human brain. In M. Pesenson (Ed.), *Multiscale analysis and nonlinear dynamics: From genes to the brain.* New York, NY: Wiley.

Siebenhuhner, F., Weiss, S. A., Coppola, R., Weinberger, D. R., & Bassett, D. S. (2013). Intra- and inter-frequency brain network structure in health and schizophrenia. *PLoS ONE, 8*(8), e72351.

Smith, S. M., Miller, K. L., Moeller, S., Xu, J., Auerbach, E. J., Woolrich, M. W., ... Ugurbil, K. (2012). Temporally independent functional modes of spontaneous brain activity. *Proc Natl Acad Sci USA, 109*(8), 3131–3136.

Smith, S. M., Miller, K. L., Salimi-Khorshidi, G., Webster, M., Beckmann, C. F., Nichols, T. E., ... Woolrich, M. W. (2011). Network modelling methods for FMRI. *NeuroImage, 54*(2), 875–891.

Snyder, A. Z., & Raichle, M. E. (2012). A brief history of the resting state: The Washington University perspective. *NeuroImage, 62*(2), 902–910.

Sporns, O. (2006). Small-world connectivity, motif composition, and complexity of fractal neuronal connections. *Biosystems, 85*(1), 55–64.

Sporns, O. (2010). *Networks of the brain.* Cambridge, MA: MIT Press.

Sporns, O. (2011). The human connectome: A complex network. *Ann NY Acad Sci, 1224*, 109–125.

Sporns, O. (2012). From simple graphs to the connectome: Networks in neuroimaging. *NeuroImage, 62*(2), 881–886.

Stam, C. J., & van Straaten, E. C. (2012). The organization of physiological brain networks. *Clin Neurophysiol, 123*(6), 1067–1087.

Stumpf, M. P. H., & Porter, M. A. (2012). Critical truths about power laws. *Science, 51*, 665–666.

Vértes, P. E., Alexander-Bloch, A. F., Gogtay, N., Giedd, J. N., Rapoport, J. L., & Bullmore, E. T. (2012). Simple models of human brain functional networks. *Proc Natl Acad Sci USA, 109*(15), 5868–5873.

Vaishnavi, S. N., Vlassenko, A. G., Rundle, M. M., Snyder, A. Z., Mintun, M. A., & Raichle, M. E. (2010). Regional aerobic glycolysis in the human brain. *Proc Natl Acad Sci USA, 107*(41), 17757–17762.

van den Heuvel, M. P., Mandl, R. C., Stam, C. J., Kahn, R. S., & Hulshoff Pol, H. E. (2010). Aberrant frontal and temporal complex network structure in schizophrenia: A graph theoretical analysis. *J Neurosci, 30*(47), 15915–15926.

van den Heuvel, M. P., & Sporns, O. (2011). Rich-club organization of the human connectome. *J Neurosci, 31*(44), 15775–15786.

Wang, J., Wang, L., Zang, Y., Yang, H., Tang, H., Gong, Q., ... He, Y. (2009). Parcellation-dependent small-world brain functional networks: A resting-state fMRI study. *Hum Brain Mapp, 30*(5), 1511–1523.

Watanabe, T., Hirose, S., Wada, H., Imai, Y., Machida, T., Shirouzu, I., ... Masuda, N. (2013). A pairwise maximum entropy model accurately describes resting-state human brain networks. *Nat Commun, 4*, 1370.

Wig, G. S., Schlaggar, B. L., & Petersen, S. E. (2011). Concepts and principles in the analysis of brain networks. *Ann NY Acad Sci, 1224*, 126–146.

Woodward, J. (2011). Scientific explanation. In E. N. Zalta (Ed.), *The Stanford encyclopedia of philosophy.* Retrieved from http://plato.stanford.edu/archives/win2011/entries/scientific-explanation

Wymbs, N. F., Bassett, D. S., Mucha, P. J., Porter, M. A., & Grafton, S. T. (2012). Differential recruitment of the sensorimotor putamen and frontoparietal cortex during motor chunking in humans. *Neuron, 74*, 936–946.

Xie, T., & He, Y. (2011). Mapping the Alzheimer's brain with connectomics. *Front Psychiatry, 2*, 77.

Zalesky, A., Fornito, A., & Bullmore, E. (2012). On the use of correlation as a measure of network connectivity. *NeuroImage, 60*(4), 2096–2106.

Zalesky, A., Fornito, A., Harding, I. H., Cocchi, L., Yucel, M., Pantelis, C., & Bullmore, E. T. (2010). Whole-brain anatomical networks: Does the choice of nodes matter? *NeuroImage, 50*(3), 970–983.

80 Using Computational Models to Understand Robustness and Variability in Nervous Systems

TILMAN J. KISPERSKY*, TIMOTHY O'LEARY*, AND EVE MARDER

ABSTRACT Computational models of neurons allow us to understand how cellular components interact to generate neuronal activity. Model activity, and consequently model explanatory power, depends on model parameter values. How do we choose parameters for neuron models in a principled way? Some parameters (such as neuronal resting potential or average spiking rate) can be estimated directly from experimental data; however, many other parameters (such as density of ion channels in the neuronal membrane) cannot be constrained easily unless they are explicitly measured. More importantly, many parameters in biological systems are highly variable, so even with careful measurements it is inaccurate to represent a distribution of measured values with a single model parameter value. Here, we describe several techniques to overcome these problems in the context of conductance-based neuron models. Rather than building one, representative model, we construct a family of models, each with parameters drawn from a distribution of values based on experimental observations. From this population, we select those models that match observed activity and discard the rest. Overall, the techniques we describe result in computer models that explicitly capture the variability observed in neurons and overcome the problem of manually choosing parameters for these models.

To fully understand nervous system function, we need to build quantitative models that incorporate experimentally measured properties of neurons and networks. Models allow us to understand how different components and processes interact to generate emergent behavior and therefore provide a means of articulating and testing theories of how nervous systems work. The most famous example of this is the Hodgkin-Huxley model of action potential generation in the squid giant axon (Hodgkin & Huxley, 1952). This model explains how action potentials, or "spikes," in neurons are generated by an interaction between two different membrane currents, a fast inward (depolarizing) sodium current and a slower outward (hyperpolarizing) potassium current. As is the case with all good models, its power is that the basic principles generalize: given appropriate experimental measurements, this model and its more elaborate cousins that include additional currents can describe the excitable electrical behavior of all types of neurons in any nervous system (Armstrong & Hille, 1998).

Experimental work over many decades has revealed how other components of neurons and networks, such as synapses, dendrites, and other membrane conductances, operate in quantitative terms. With modern imaging and electrophysiological and molecular techniques, it is becoming possible to obtain data in living nervous systems from the level of individual biochemical reactions to the activity patterns in large neural circuits. Similarly, computer power is increasing at a rate that makes detailed simulations of these phenomena possible (Alivisatos et al., 2012; de Schutter, Ekeberg, Kotaleski, Achard, & Lansner, 2005; Markram, 2006). Even with these advances, a major challenge when building neuron models is turning experimental measurements into appropriate model parameters.

Thus, in principle, we can construct biologically realistic models of neuronal networks as a means of understanding nervous system function at any level of detail that can be accessed experimentally. In practice, it is not possible to fully estimate all the parameters of a neural system by experimental observation because even for a single neuron a detailed model can have tens or even hundreds of parameters representing morphology, channel densities, and the kinetics of the resulting membrane conductances. In addition, when attempting to estimate values of model parameters from experimental data, one finds that the measured values are widely variable, even in neurons with virtually identical overall behavior (Marder & Goaillard, 2006). We emphasize that, in this context, these differences are not due to measurement error; they represent genuine biological differences. For example, figure 80.1 shows measurements of membrane currents in two

*These authors contributed equally to this work.

anatomically and functionally equivalent cells taken from the crustacean nervous system (Schulz, Goaillard, & Marder, 2006). In this example, the overall behavior of the two neurons is indistinguishable in spite of large differences in three of the underlying membrane currents.

Observations such as those in figure 80.1 also place us in a practical predicament as modelers. First, neuronal systems have a large number of parameters, making simultaneous measurement of all parameters in a single preparation unfeasible. Second, even when experimental error is taken into account, the measurements of subsets of parameters are highly variable, and we do not typically know how all the parameters covary. There are at least three approaches to dealing with this predicament:

1. adopt a "representative" value for each parameter, for example, the mean of the experimental measurements
2. use the experimental values as "initial guesses" of the model parameters, then search parameter space to find values that are in some sense "optimal." For example, optimal parameters might be sought on the basis of close agreement between the overall behavior of the model and behavior of the system observed experimentally
3. treat the model parameters as a distribution that can vary among different versions of the model. In this approach, we construct a *family of models* rather than a single model, then either systematically vary each parameter over the observed range or randomly sample within that range.

The third approach is the only one that explicitly captures biological variability and is therefore the most relevant for developing a complete understanding of biological systems (Doloc-Mihu & Calabrese, 2011; Goaillard, Taylor, Schulz, & Marder, 2009; Golowasch, Goldman, Abbott, & Marder, 2002; Hudson & Prinz, 2010; O'Leary, Williams, Caplan, & Marder, 2013; Prinz, Billimoria, & Marder, 2003; Prinz, Bucher, & Marder, 2004; Taylor, Goaillard, & Marder, 2009; Williams et al., 2013). For instance, a deep question that is prompted by the experimental data shown above is how nervous systems function in spite of underlying variability in their components and properties (Goldman, Golowasch, Marder, & Abbott, 2001). It is important to construct models that capture biological variability, not only as a means of addressing this question, but as a pragmatic and principled way of deciding whether or not a model accurately describes the underlying population of neurons.

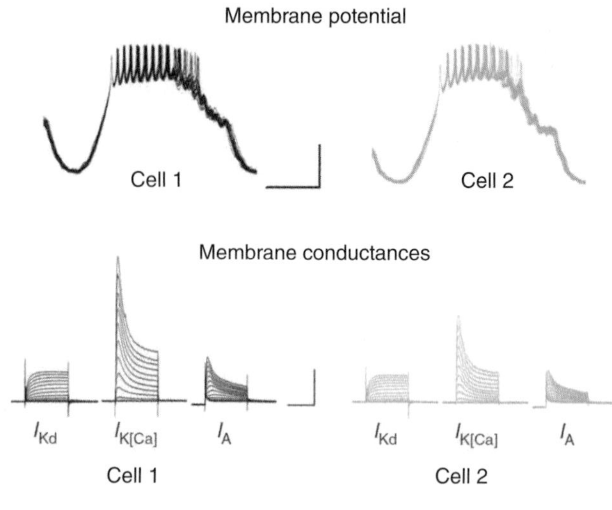

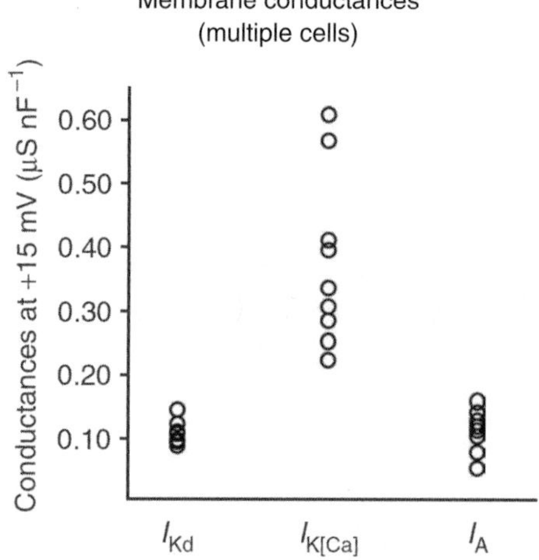

FIGURE 80.1 Recordings from multiple cells reveal biological variability. *Top:* Current-clamp recordings of spontaneous bursting activity from a lateral pyloric (LP) neuron from the *Cancer borealis* stomatogastric ganglion. Each animal has one LP neuron. The LP neuron produces periodic bursts. In both cells shown, several single-burst cycles are overlaid to show the invariance of the burst shape over time. Across the two cells, voltage behavior appears to be qualitatively similar. *Middle:* Voltage-clamp recordings of the neurons shown in (top) to measure individual K^+ conductances. Traces show current through a single population of voltage-dependent ion channels as the membrane voltage is stepped through a range. Each trace shows the current for a single voltage step. Currents recorded are the delayed rectifying K^+ current (IK_d), the Ca^{2+}-dependent K^+ current (IK[Ca]), and the A-current, the transient K^+ current (I_A). The same measurements are done for the second neuron (cell 2, gray). *Bottom:* Across animals, conductances have highly variable values. Each circle represents a single animal's conductance through a subset of channels at +15 mV. (Adapted from Schulz et al., 2006.)

Shortcomings of using unique parameter sets in models

Figure 80.2 (Golowasch et al., 2002) shows the behavior of a computational model of a single cell with several different types of conductances, including a fast sodium (Na^+) conductance and a slower, delayed-rectifier K^+ conductance (K_d). Three different parameter sets are used to generate three different models (figure 80.2A, left panels 1–3). Each parameter set has distinct values of the Na^+ and K_d conductances (figure 80.2A, right panels 1–3). These values were chosen to lie in an experimentally determined range. The behavior of the model is very similar in all three cases, in spite of large variation in these two conductances. Significantly, when a model is constructed from the average values of both conductances, its behavior changes and no longer matches that of all the models used to compute the averages (figure 80.2B).

There are two important lessons to take from this simple example. First, very different underlying parameters can produce very similar model behavior. This means that the task of choosing an "optimal" set of parameters based on overall behavior does not have a well-defined solution (i.e., there are multiple solutions that generate a given behavior). In other words, a consequence of neuronal variability is that many different parameter combinations can produce functioning neurons. Second, the average parameter values can give qualitatively different behavior to three examples chosen within the experimental bounds. This indicates that the average parameter values are not necessarily represented biologically in any single cell.

This prompts caution in adopting strategies (1) and (2) and suggests that model parameters might not be best described by single numerical values. Strategy (3) avoids this problem and allows us to address interesting biological questions, such as how sensitive a biological system is to variation in specific parameters or how the parameters might covary to give a particular global behavior (Marder & Taylor, 2011). However, working with populations of models introduces technical problems, most notably the computing power required to run many instances of a model rather than a single instance. We illustrate these concepts with single-cell examples; however, the methods are quite general and can be used for entire networks.

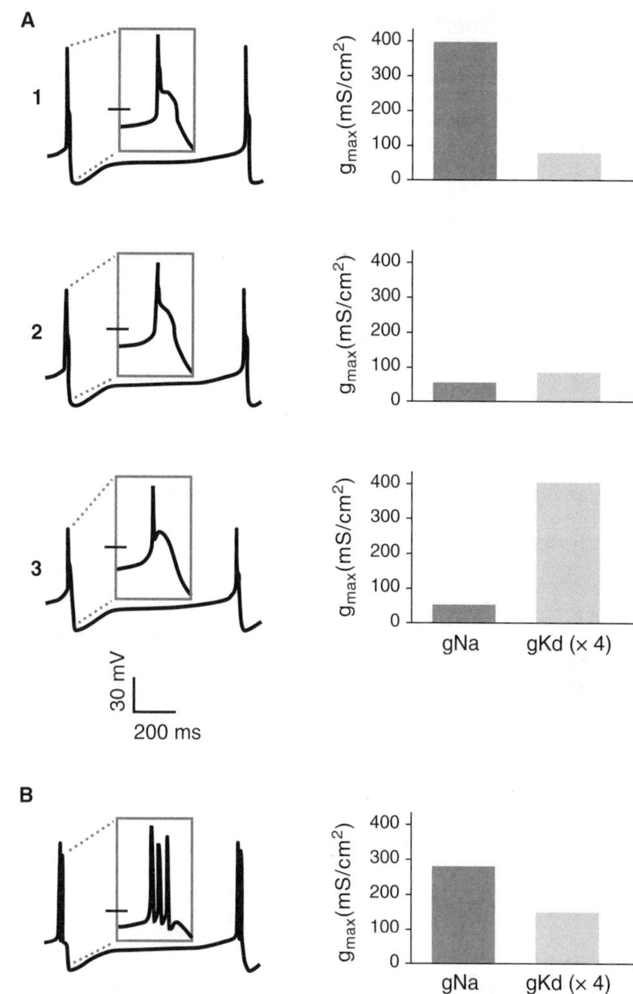

FIGURE 80.2 Different parameter combinations can produce equivalent model behavior. (A) Each row (1–3) shows a different set of parameters for the same model neuron. Sample activity traces (left column) are qualitatively similar for each model (inset: expanded view of a single-spike burst, scale bar: 50 ms). The individual conductance values for the sodium and delayed-rectifier potassium conductances vary substantially over the three different models. (B) A model that uses the average values (right column) of the conductances used in (A). The model does not produce the same qualitative behavior observed in all three models in (A). Specifically, this model has multiple spikes for each burst (left panel, inset: expanded view, scale bar: 50 ms). (Adapted from Golowasch et al., 2002.)

Capturing parameter variability using families of models

Suppose we want to build a model neuron using experimental data. The data consist of intracellular voltage-clamp and current clamp recordings, from which we can estimate a subset of the biophysical parameters, such as the properties of different ionic conductances. Across multiple cells, we would obtain a range of values for each parameter that would provide an estimate of the mean value and the spread of this biological feature. If we have a large number of data points, we can infer something of the shape of the distribution of parameters. For example, we may be able to infer a covariance

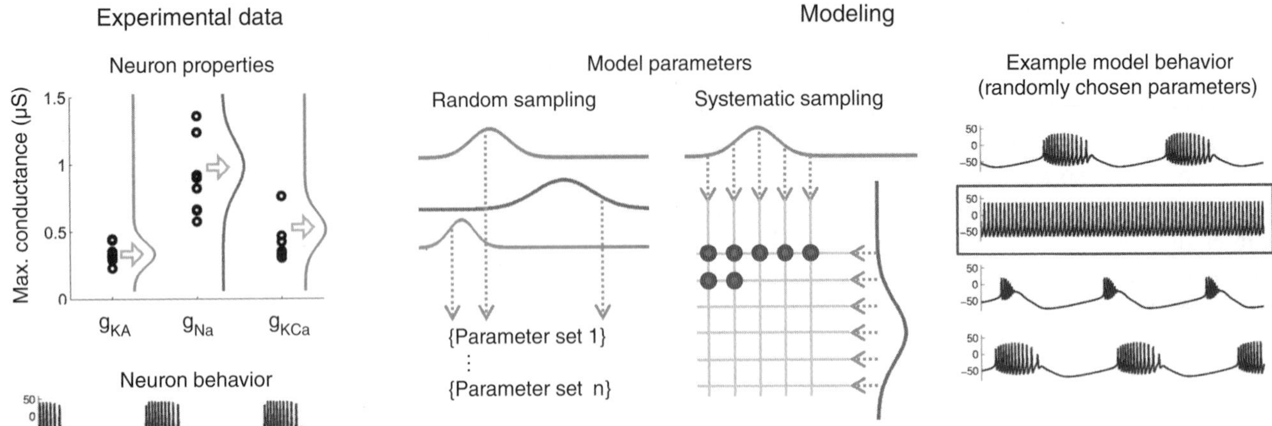

FIGURE 80.3 Procedure for generating populations of model neurons. *Left:* Maximal conductance measurements from hypothetical experiments are taken (black circles). These measurements allow the estimation of the distributions of each property in the population (vertical curves; gray arrows indicate sample means; sample points fit with Gaussians). Below, the output of an example neuron shows periodic bursting behavior, which is typical in this type of neuron (scale bar: 100 ms). *Middle:* Each model in the population is created either by randomly sampling the parameter space (left) or by systematically sampling the space on a grid (right). When randomly sampling, parameter values are drawn according to the probability distributions estimated from the experimental data (dotted lines). When systematically sampling, parameter values are drawn at regular intervals and all possible combinations of the intervals are used. *Right:* Resulting models are simulated. Many resulting neuron models have qualitatively similar bursting behavior to what is observed experimentally (rows 1, 3, 4). However, one of the neurons shows tonic spiking behavior (box), highlighting that different qualitative behaviors are possible from these parameters (scale bar: 100 ms).

structure and then sample from distributions that preserve covariance (such as multidimensional Gaussian distributions with the same covariance).

We can then construct a family of models in one of two ways, as illustrated in figure 80.3. We can randomly sample the measured parameter distributions, choosing values over the experimentally observed range—we refer to this as the *random sampling* method. Alternatively, we can systematically vary each parameter independently over the observed range using prespecified intervals. We call the latter approach *systematic sampling* (or "grid" sampling).

Systematic/grid sampling is attractive because it explicitly traces the dependence of model behavior as each parameter is varied and will generate all combinations of parameter values over the chosen intervals (Prinz et al., 2003). For example, we might be interested in how the firing rate of a neuron changes as Na^+ conductance is increased (for examples, see below) for all possible combinations of other parameters. However, the disadvantage of this approach is that it suffers from the so-called curse of dimensionality. Suppose we have N different parameters and we want to vary each over the experimentally observed range. If we wish to sample each parameter at M distinct values, then the total number of models we need to simulate is M^N. The computational cost of performing this simulation is exponential in the number of parameters. It is easy to see that regardless of how powerful our computing resources are, we only need to add a few parameters before the simulation becomes impractical or unfeasible. For example, suppose it takes 1 second to simulate a population with 10 parameters. Each additional parameter increases the simulation time 10-fold, so simulating this same population with 11 parameters would take 10 seconds, while a simulation with 20 parameters takes 10^{10} seconds, or 317 years.

Even single-neuron models can easily require tens of parameters, so the curse of dimensionality is very real, especially when one considers larger-scale simulations of networks of neurons. Yet it is nonetheless possible to obtain useful information on how underlying parameter variability contributes to model behavior using random sampling in place of systematic sampling (Sobie, 2009; Taylor et al., 2009). While random sampling is never guaranteed to cover specific combinations of parameters, finding how the behavior of the models depends on the distribution of parameters is not subject to the curse of dimensionality. This is because the computational cost of a single simulation is fixed (it can of course, be significant); however, the number of replicates in a family of simulations can be chosen arbitrarily and will only be dependent on the precision required to draw conclusions. For example, suppose we have a single-neuron model with many conductances, and we want to find the proportion of conductance combinations over a fixed range that produces a tonically spiking neuron. An unbiased estimate of this proportion is $p = \frac{t}{N}$, where t is the number of tonic spiking cells and N is the number of "trials" (parameter combinations tested). The standard error in this estimate is $\sqrt{\frac{p(1-p)}{N}}$, which depends only on the number of simulations and the underlying proportion of tonically spiking cells—it is *independent of the number of parameters* in the model.

In situations where the data show interesting structure in the variability, such as systematic correlations between parameters, random (uniform) sampling provides a means to test whether correlations are necessary for the system to function or whether they are a consequence of some additional constraint. For example, correlated variability is found in the expression of ion channels in identified neurons of the crustacean stomatogastric ganglion, but computational studies indicate that these are not necessary for function, implying the existence of a biological constraint in the expression mechanism. We may, however, use experimentally measured correlations in parameters to further constrain populations of models. For example, we can sample from multidimensional Gaussian distributions that have covariance equal to the measured covariance when restricted to the parameters that have been measured simultaneously.

Case studies

Cataloguing Biological Variability in Populations of Models An initial purpose of building variable-neuron populations is purely to catalog the variability observed in experiments. These types of databases are not meant for simulations directly, but rather to provide a repository of individual observations that can serve to constrain specific simulations at a later time. One such database, NEOBASE (Muhammad & Markram, 2005), contains the individual intrinsic conductances measured from many different neurons. In addition, the experimenters collected anatomical reconstructions and gene-expression patterns from each neuron. Often, neurons were recorded simultaneously as pairs, thus yielding information about synaptic coupling between neurons. The database is relational, which means that it explicitly represents relationships between individual entries (i.e., neurons). This means that the specific information about one particular neuron's properties can be correlated with the exact neuron it was coupled with. Similarly, structural information and gene-expression information can be compard with other neurons of the same type as well as synaptically coupled partner neurons. NEOBASE allows other investigators to mine the data in interesting ways by searching for hidden relationships or by identifying statistical correlations. Another similar database, Neuroelectro (www.neuroelectro.org), leverages the work already published by other groups by collecting basic biophysical information from papers archived online. This database captures only intrinsic properties, such as resting membrane potentials and baseline firing rates. However, the information is captured from many different cortical and subcortical cell types. Neuroelectro aims to compare different cell types with one another in terms of their electrical properties. Both of these efforts, and others like them, aim to provide a rich source of data that can be accessed by other scientists for more targeted analyses.

Simulating Model Databases to Assess Features of Parameter Space More commonly, model databases are constructed with model parameters selected automatically by computer, either randomly or by sampling parameter space on a grid (Prinz et al., 2004). Each model is then assessed for its ability to produce biologically realistic behavior. An initial application of database simulation asked the question: How many different combinations of parameters can yield similar circuit behavior? In this example, models were constructed to mimic the pyloric circuit of the crustacean stomatogastric ganglion, a ganglion that produces a triphasic pattern of muscle activation required for proper digestion. This small circuit contains three neuronal types at its core and its connectivity is known. Thus, each model in the database contained the same three model neurons with the same synaptic connections. Models differed in their density of intrinsic conductances and synaptic strengths. The output of the circuit is highly stereotyped and varies minimally between animals. Importantly, this property allowed models to be rejected when their behavior deviated from the physiologically observed behavior. After millions of model circuits had been simulated, models that replicated the behavior observed experimentally were extracted and the parameters analyzed. Remarkably, successful models had a wide variety of parameters and often differed substantially in terms of their intrinsic properties (Prinz et al., 2004). Beyond identifying single successful models, database simulations can be useful for characterizing the shape of multidimensional parameter spaces. Such results can identify entire regions in which successful models reside and identify important properties of these regions, for example, that they are sometimes noncontiguous with one another (Achard & de Schutter, 2006).

A follow-up question that arises from these results: How are individual parameters related to neuron-level properties like spike threshold or spike rate? More generally, this question can be phrased as: How sensitive are models to any given parameter? This can be approached in a population of variable models by analyzing how much one property, like spike rate, changes when removing a single parameter compared with the full model. Measured over many different models and over many different parameters, this approach showed that multiple conductances contribute to neuron-level properties simultaneously in all cases (figure 80.4F). This highlights the ability of neurons to find multiple ways to produce desired output and suggests that nervous systems use variability as a method to ensure robust, functional output (Sobie, 2009; Sobie & Sarkar, 2011; Taylor et al., 2009).

Variable Model Populations Can Show Consistent Response to a Perturbation Model databases can be used to ask what the result of a perturbation would be on a set of realistically variable neurons. In this context, a perturbation refers to a global stimulus like a change in gene expression, a change in ambient temperature or the application of a neuromodulator. In a traditional modeling study, a perturbation would be applied to a single model only. In that case, the response to the perturbation might be a feature of the specific parameter combination of that exact model. To obtain a more general understanding of the effect of a perturbation, it can be applied to a database of models. This process can uncover general effects of the perturbation that are consistent through the entire population and, equally importantly, discover effects that would not be observed when considering only a single model. Further, any effect that occurs in most models is more likely to be "real" in the biological sense. In one study utilizing this approach, the sodium conductance was instantaneously tripled in a set of biophysical neuron models (Kispersky, Caplan, & Marder, 2012). This manipulation can be considered biologically analogous to an increase in gene expression of a single ion channel due to, for example, the application of a neuromodulator. Tripling of the Na^+ conductance in these model neurons was found to have two different effects on excitability depending on the firing rate. Incorporating variability into the model population highlighted that the firing rate–dependent effect of the conductance increase was a general feature of this perturbation and not model-specific.

Using Variable Model Populations to Estimate Parameters Including parameter variability in computer models can be useful when estimating the values of unknown parameters. For example, when constructing a model of a neuron that is not well characterized, assumptions about underlying conductances cannot be avoided. Many neurons are initially studied with current clamp recordings. These recordings provide the response to depolarizing and hyperpolarizing inputs, frequency-current relationships, and the response to noisy stimuli. Such responses are the product of many distinct, underlying conductances that cannot be easily reconstructed from the recordings. In principle, voltage clamp recordings can be used to characterize individual conductances. However, this process depends on being able to pharmacologically isolate each channel population, which is often not possible. Alternatively, a population of models with variability in each conductance parameter can be used to estimate unknown parameters with fewer assumptions. Each model is simulated and evaluated relative to known current clamp behavior. In this manner, a resulting set of models can be discovered that contain parameter combinations that yield realistic behavior. In one case, this method was applied to study olfactory mitral and granule cell ion-channel distribution (Bhalla & Bower, 1993). In these cell types, conductances had been measured at the soma, and geometries of the neurons had been reconstructed. However, the distribution of each ion channel throughout the neuronal arbor was unknown. By synthesizing the known information about somatic conductance and morphology, a model population was generated in which the free parameters were the ion-channel densities. Models that matched known firing behavior were identified. The results suggested that mitral cells concentrate calcium and slow potassium channels in the glomeruli, while sodium and fast potassium channels are concentrated in the soma. A similar approach was used to study how individually variable excitatory postsynaptic potentials (EPSPs) combine to form a larger, composite EPSP observed in cat muscles (Segev, Fleshman, & Burke, 1990). In this system, individual EPSPs have variable amplitudes and time courses and sum nearly linearly. The unknown distribution of single synapses along the muscle contributing to the composite EPSP was investigated in a population of models that combined synapses in many different configurations.

Epilogue: Visualization and analysis

We have seen that modeling neurons and neural systems requires us to grapple with large numbers of parameters.

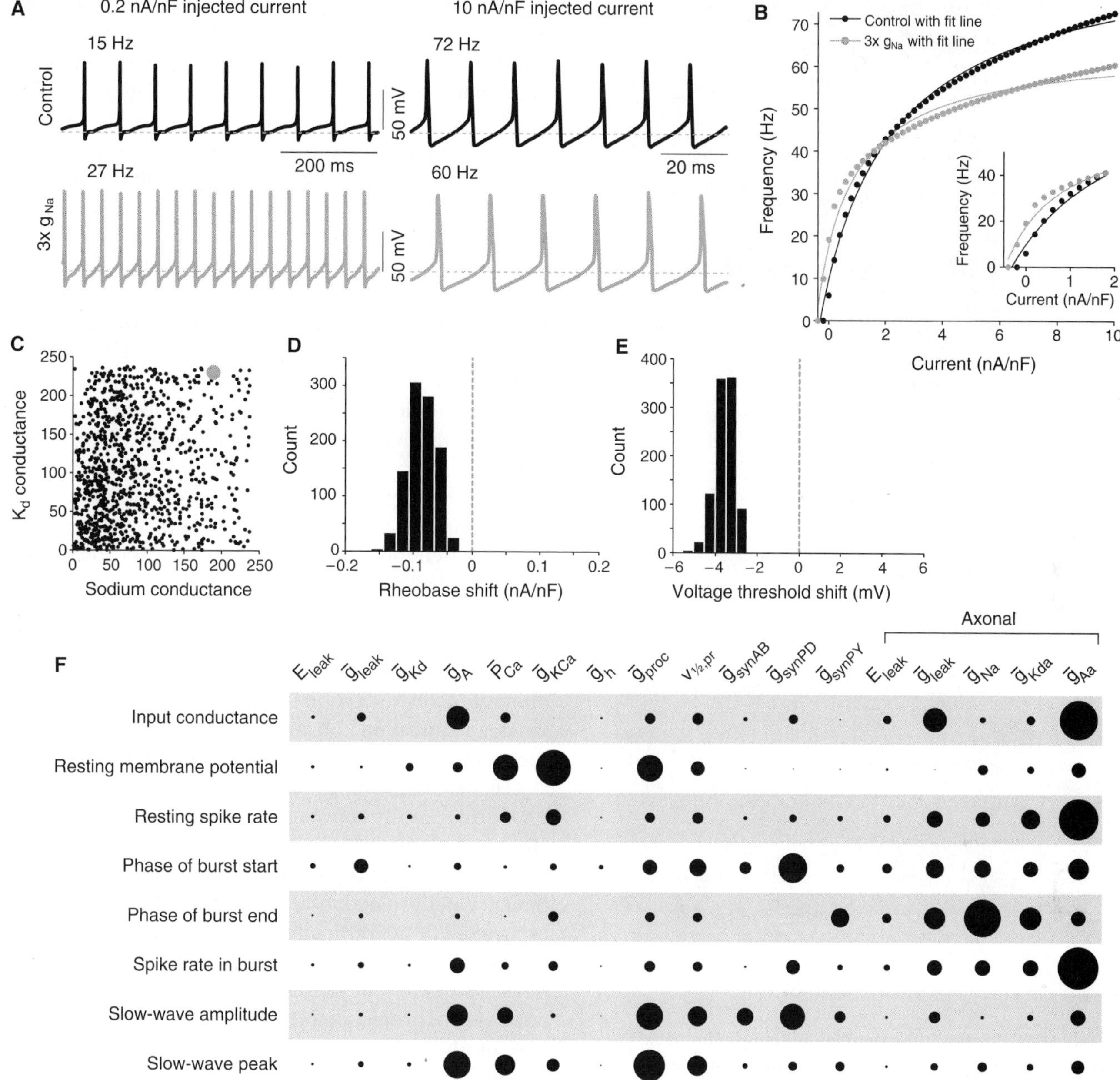

FIGURE 80.4 Population simulations show that each neuron's behavior is sensitive to multiple intrinsic properties. (A) Example model neuron from a population in which the sodium conductance was tripled (black: control; gray: 3x g_{Na}). The left column shows the model neuron at low current injection (firing rate increases with 3x g_{Na}), and the right column shows the model neuron at high current injection (firing rate decreases with 3x g_{Na}). (B) Frequency-current curves for this example neuron with fit lines. (C) A population of 1,000 randomly generated models was considered in this study. Models were only included in the population if they fired tonically at low current injection level. Black points represent the sodium and delayed rectifier potassium conductance of single models in the population. The large gray dot in the upper-right region is the model shown in (A). (D) Over the whole population, tripling the sodium conductance always decreased the rheobase (the minimum current required to fire an action potential). All points in the histogram are below zero (gray line). (E) Similarly, voltage threshold was hyperpolarized for all 1,000 models when the sodium conductance was tripled. (Adapted from Kispersky et al., 2012.) (F) The neuron-level properties (rows) of a population of lateral pyloric neurons are plotted relative to their sensitivities to biophysical parameters (columns). The size of each circle represents how sensitive a given behavior was to that property. In general, each property was sensitive to several parameters at once. (Adapted from Taylor et al., 2009.)

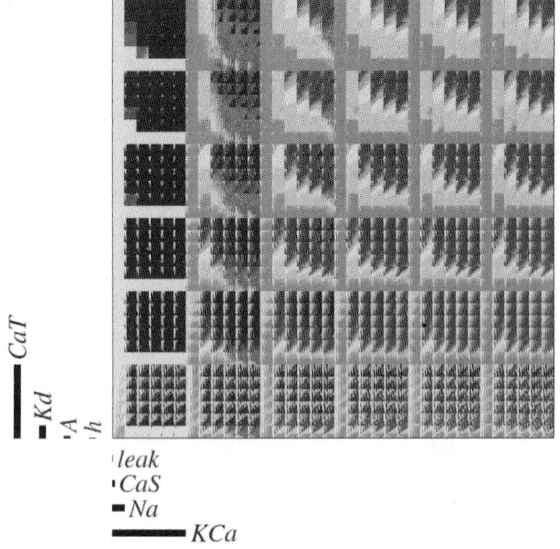

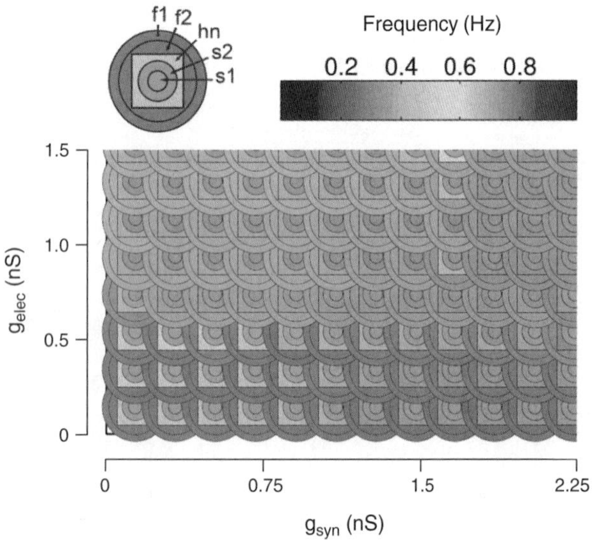

FIGURE 80.5 Methods for visualizing high-dimensional data. *Top:* Dimensional stacking nests axes within one another to achieve a 2D representation of higher-dimensional data. Each point in the image represents a single parameter combination, and the color represents a property of that model (for example, bursting vs. tonic spiking). The outer, largest edges of the square represent increments of the transient calcium (CaT) (*y*-axis) and calcium-dependent potassium (KCa) conductances (*x*-axis). Nested within each increment of these two parameters are all increments of delayed-rectifier (Kd) (*y*-axis) and sodium (Na) conductances (*x*-axis), which repeat within each interval of the outer axis. This nesting continues until all parameter combinations have been represented. The order of nesting is important for the picture that emerges. In this case, the nesting order is automatically determined to maximize contiguous areas of model behavior (represented by color). *Bottom:* An example "parameterscape" plots five measurements from each model in a population (frequencies of five oscillators: f1,2, fast oscillator 1,2; s1,2, slow oscillator 1,2; hn: "hub neuron" oscillator) relative to two parameters of the model: g_{syn}, a synaptic conductance (nS), and g_{elec}, a gap junction conductance (nS). Color of each shape represents frequency. The juxtaposition of the shapes in this manner gives a quick view of which oscillators are firing at similar frequencies and whether other models nearby in the parameter space exhibit similar frequency behavior. (See color plate 79.)

While computational hardware and methods are improving, we are still left with the problem of interpreting and conveying results that necessarily reside in high-dimensional data sets. As with simulations, analyzing data from model populations becomes increasingly difficult when the number of parameters increases. The goal of analyzing data from simulations of high-dimensional model populations is to obtain a qualitative picture of the shape of the underlying parameter space. In general, this requires careful statistical approaches or dimensionality-reduction techniques (Hastie, Tibshirani, & Friedman, 2009). Dimensionality-reduction techniques remove redundancies in high-dimensional data and thus represent structure using fewer dimensions. There are many techniques for achieving this, each with its own set of advantages and drawbacks. For example, principal components analysis performs a high-dimensional rotation of data on a transformed basis, ordered by decreasing variance. This can work well for data with a simple (linear) covariance structure that can be well-approximated by assuming variance components are orthogonal. Both experimental and simulated biological data often have more complex structures that require more sophisticated dimensionality-reduction techniques that preserve local kinks and twists in the distribution. One example of such a technique is locally linear embedding, which fits a piecewise regression model to subsets of variables in a way that locally represents the variance of the data (Roweis & Saul, 2000).

Visualizing high-dimensional data in a two- or three-dimensional projection is another problem that grows exponentially in difficulty with the number of parameters. One method, called dimensional stacking, plots parameters in nested squares, producing a two-dimensional projection of all parameters plotted against one another (Taylor, Hickey, Prinz, & Marder, 2006). The order in which squares are nested changes the resulting image (figure 80.5, left). A selection algorithm is used to automate the order of square nesting such that regions of similar network behavior are co-localized. This produces an overall picture in which regions of model behaviors can be associated with parameters. A second method, called the parameterscape (figure 80.5, right), nests three to six attributes, color-coded for their values, in concentric circles or squares in a two-dimensional plot (Gutierrez, O'Leary, & Marder, 2013). Both of these methods work extremely well for grid searches, and might be adopted for the visualization of random searches by judicious binning of the data. There are other, more general methods of projecting high-dimensional data onto lower-dimensional spaces for visualization purposes, such as a *t*-Distributed Stochastic Neighbor Embedding (van der Maaten & Hinton, 2008), which approximately preserves distance between data points, so that their similarity/dissimilarity can be inferred. As with all methods of analysis, there is no "one-size-fits-all" solution, and the particular features of a given data set and candidate visualization method need to be understood before proceeding.

In general, analyzing and visualizing data from model populations becomes increasingly difficult as the number of parameters increases. Presumably, as time goes on, new methods of visualization of the results from more and more complex sets of models will be developed. That said, the success of both dimensional stacking and the parameterscape methods reveal the importance of trying to overcome the limitations of the human mind in visualizing the results of varying many parameters at the same time. And of course, it is exactly that which is helpful for understanding models in which behavior results from the interaction of multiple nonlinear processes.

REFERENCES

Achard, P., & de Schutter, E. (2006). Complex parameter landscape for a complex neuron model. *PLoS Comput Biol, 2,* e94.

Alivisatos, A. P., Chun, M., Church, G. M., Greenspan, R. J., Roukes, M. L., & Yuste, R. (2012). The brain activity map project and the challenge of functional connectomics. *Neuron, 74,* 970–974.

Armstrong, C. M., & Hille, B. (1998). Voltage-gated ion channels and electrical excitability. *Neuron, 20,* 371–380.

Bhalla, U. S., & Bower, J. M. (1993). Exploring parameter space in detailed single neuron models: Simulations of the mitral and granule cells of the olfactory bulb. *J Neurophysiol, 69,* 1948–1965.

de Schutter, E., Ekeberg, O., Kotaleski, J. H., Achard, P., & Lansner, A. (2005). Biophysically detailed modelling of microcircuits and beyond. *Trends Neurosci, 28,* 562–569.

Doloc-Mihu, A., & Calabrese, R. L. (2011). A database of computational models of a half-center oscillator for analyzing how neuronal parameters influence network activity. *J Biol Phys, 37,* 263–283.

Goaillard, J. M., Taylor, A. L., Schulz, D. J., & Marder, E. (2009). Functional consequences of animal-to-animal variation in circuit parameters. *Nat Neurosci, 12,* 1424–1430.

Goldman, M. S., Golowasch, J., Marder, E., & Abbott, L. F. (2001). Global structure, robustness, and modulation of neuronal models. *J Neurosci, 21,* 5229–5238.

Golowasch, J., Goldman, M. S., Abbott, L. F., & Marder, E. (2002). Failure of averaging in the construction of a conductance-based neuron model. *J Neurophysiol, 87,* 1129–1131.

Gutierrez, G. J., O'Leary, T., & Marder, E. (2013). Multiple mechanisms switch an electrically coupled, synaptically inhibited neuron between competing rhythmic oscillators. *Neuron, 77,* 845–858.

Hastie, T., Tibshirani, R., & Friedman, J. H. (2009). *The elements of statistical learning: Data mining, inference, and prediction* (2nd ed.). New York, NY: Springer.

Hodgkin, A. L., & Huxley, A. F. (1952). A quantitative description of membrane current and its application to conduction and excitation in nerve. *J Physiol, 117,* 500–544.

Hudson, A. E., & Prinz, A. A. (2010). Conductance ratios and cellular identity. *PLoS Comput Biol, 6*(7), e1000838.

Kispersky, T. J., Caplan, J. S., & Marder, E. (2012). Increase in sodium conductance decreases firing rate and gain in model neurons. *J Neurosci, 32,* 10995–11004.

Marder, E., & Goaillard, J. M. (2006). Variability, compensation and homeostasis in neuron and network function. *Nat Rev, 7,* 563–574.

Marder, E., & Taylor, A. L. (2011). Multiple models to capture the variability in biological neurons and networks. *Nat Neurosci, 14,* 133–138.

Markram, H. (2006). The blue brain project. *Nat Rev Neurosci, 7,* 153–160.

Muhammad, A. J., & Markram, H. (2005). NEOBASE: Databasing the neocortical microcircuit. *Stud Health Technol Inform, 112,* 167–177.

O'Leary, T., Williams, A. H., Caplan, J. S., & Marder, E. (2013). Correlations in ion channel expression emerge from homeostatic tuning rules. *Proc Natl Acad Sci USA, 110,* E2645–2654.

Prinz, A. A., Billimoria, C. P., & Marder, E. (2003). Alternative to hand-tuning conductance-based models: Construction and analysis of databases of model neurons. *J Neurophysiol, 90,* 3998–4015.

Prinz, A. A., Bucher, D., & Marder, E. (2004). Similar network activity from disparate circuit parameters. *Nat Neurosci, 7,* 1345–1352.

Roweis, S. T., & Saul, L. K. (2000). Nonlinear dimensionality reduction by locally linear embedding. *Science, 290,* 2323–2326.

Schulz, D. J., Goaillard, J. M., & Marder, E. (2006). Variable channel expression in identified single and electrically coupled neurons in different animals. *Nat Neurosci, 9,* 356–362.

Segev, I., Fleshman, J. W., Jr., & Burke, R. E. (1990). Computer simulation of group Ia EPSPs using morphologically realistic models of cat alpha-motoneurons. *J Neurophysiol, 64,* 648–660.

Sobie, E. A. (2009). Parameter sensitivity analysis in electrophysiological models using multivariable regression. *Biophys J, 96,* 1264–1274.

Sobie, E. A., & Sarkar, A. X. (2011). Regression methods for parameter sensitivity analysis: Applications to cardiac arrhythmia mechanisms. *Proc. IEEE Engineering in Medicine and Biology Society* (pp. 4657–4660).

Taylor, A. L., Goaillard, J. M., & Marder, E. (2009). How multiple conductances determine electrophysiological properties in a multicompartment model. *J Neurosci, 29,* 5573–5586.

Taylor, A. L., Hickey, T. J., Prinz, A. A., & Marder, E. (2006). Structure and visualization of high-dimensional conductance spaces. *J Neurophysiol, 96,* 891–905.

van der Maaten, L., & Hinton, G. (2008). Visualizing data using t-SNE. *J Mach Learn Research, 9,* 2579–2605.

Williams, A. H., Kwiatkowski, M. A., Mortimer, A. L., Marder, E., Zeeman, M. L., & Dickinson, P. S. (2013). Animal-to-animal variability in the phasing of the crustacean cardiac motor pattern: An experimental and computational analysis. *J Neurophysiol, 109,* 2451–2465.

81 Studies of Human Visual Perception Using Cortical Electrical Stimulation

DANIEL YOSHOR AND MICHAEL S. BEAUCHAMP

ABSTRACT One of the best-studied examples of the functional organization of cerebral cortex is the visual system, where distinct cortical areas that respond selectively to visual stimuli have been identified in the primate brain. However, the fact that these areas respond to visual stimuli falls short of proving that they are critical to visual perception. The long-recognized observation that electrical stimulation of human occipital cortex can make a patient experience a visual percept in the absence of an external visual stimulus offers perhaps the strongest evidence that visual cortex has a causal role in visual perception. In this chapter, we review our recent efforts to understand how electrical stimulation of cortex can produce a visual percept and to compare the abilities of different human visual areas to contribute to perception. We also consider implications of this work to understanding of the neural basis of visual perception, and to the future development of a cortical visual prosthetic to restore vision to the blind.

How does activity in the brain lead to visual perception, a term that we use as an umbrella for all of the mysterious processes that result in conscious awareness of an external visual stimulus? The most popular approach to this question is to measure the brain activity that occurs during presentation of natural stimuli, using techniques ranging from microscopic (recordings of individual neurons using penetrating microelectrodes) to macroscopic (measurements of large populations of neurons using blood oxygen level–dependent functional magnetic resonance imaging, or BOLD fMRI). These techniques and others have been used to show that the primate visual system contains over 30 different cortical areas (Orban, Van Essen, & Vanduffel, 2004). However, a correlation between the presentation of a visual stimulus and activity in a specific neuronal population does not necessarily indicate a causal relationship between the measured neural activity and visual perception. To demonstrate such a relationship, it is necessary to activate the relevant neurons and demonstrate that this results in a visual percept (Parker & Newsome, 1998). Electrical stimulation is a technique that enables researchers to artificially activate a specific brain region, affording the opportunity to study how activity in that region influences or even produces a visual percept (Borchers, Himmelbach, Logothetis, & Karnath, 2011; Clark, Armstrong, & Moore, 2011; Histed, Ni, & Maunsell, 2012). When applied to human subjects, electrical stimulation offers a unique opportunity to study the qualitative properties of stimulation-induced visual percepts, because human subjects can readily offer verbal descriptions of percepts. This can offer insights about how activity in visual cortex underlies visual perception, and may advance efforts to restore vision with cortical prosthetics for blind patients.

Early studies of electrical stimulation of human visual cortex

In 1870, Gustav F. Fritsch and Eduard H. Hitzig used a thin metal probe to apply an electrical current to the cortical surface of awake dogs, finding that stimulation of the central part of the hemisphere resulted in contralateral limb movements (Fritsch & Hitzig, 1870). This finding was subsequently corroborated in human patients by the neurosurgeons Fedor K. Krause and Ortfrid F. Foerster (Luders & Luders, 2001). These discoveries helped overturn the previous dogma that cognitive functions are equally distributed across the cerebral cortex (equipotentiality), leading to a new understanding that the brain is divided into functionally specialized regions. In the ensuing years, electrical stimulation to study human brain function was pioneered by Canadian neurosurgeon Wilder Penfield in a series of landmark studies beginning in 1928 (Penfield, 1958; Penfield & Rasmussen, 1950). Penfield used electrical stimulation to study the function of cortical areas in awake and cooperative patients undergoing brain operations under local anesthesia, usually for the treatment of epilepsy. He extended the results of Fritsch and Hitzig to map the organization of motor cortex, the motor homunculus, for which he is perhaps best remembered today. He also examined sensory cortex and was able to elicit reports from patients on perceptual experiences produced with stimulation, including somatosensory, auditory, and visual percepts (Penfield & Perot, 1963; Perot & Penfield, 1960). Penfield and co-workers found that stimulation of the occipital lobe could produce a percept of a simple visual stimulus, a bright, colorless spot of light called a "phosphene."

In the course of carrying out his studies, Penfield noted that stimulation of wide regions of the brain did not produce any discernible response—neither a visible movement nor a report of a percept. In fact, stimulation of many cortical sites was undetected by his human subjects. Collectively, the experience from Penfield's studies in human patients suggests that only stimulation of the brain in or near so-called primary sensory areas will routinely result in the patient reporting a sensory percept. For example, Penfield found that stimulation of the occipital pole (probably at or near the primary visual area, V1) resulted in a percept of a phosphene. But stimulation of other regions in the occipital, temporal, and parietal lobes, regions that include what are now recognized as visual areas that are relatively distant from V1, did not produce reports of a percept (Penfield & Perot, 1963).

It is noteworthy that Penfield's experiments and most other early studies of electrical stimulation of human visual cortex (Brindley & Lewin, 1968; Dobelle & Mladejovsky, 1974) took place in the operating room during brain operations performed under local anesthesia. The time window for carrying out these experiments during an operation was understandably limited by clinical concerns related to prolonging the time of surgery and by patient cooperativity. Acquiring good research data in this time-pressured and stressful environment is challenging, and it is possible that this environment obscured patients from detecting subtle percepts ("false negatives") or led them to spuriously reports percepts that were not actually produced with stimulation ("false positives"). These problems were ameliorated when physiological research in human subjects with intracranial electrodes shifted from the operating room to the epilepsy monitoring unit (EMU) in the 1990s (Engel, Moll, Fried, & Ojemann, 2005; Friedman & Riviello, 2012). After undergoing surgical implantation of an array of intracranial electrodes (figure 81.1A), patients are continuously monitored in an EMU, a hospital unit dedicated for this purpose, over multiple days of video and electrophysiological recordings (from the implanted electrodes) for clinical purposes—in order to capture the onset of a spontaneously occurring seizure and define the epileptic region of the brain. This period of clinical monitoring offers a more hospitable environment for carrying out research studies that require awake and cooperative patients with electrodes directly on the brain; there is less time pressure than in the operating room, and patients are generally more comfortable and cooperative, as well as better able to be tested with multiple trials of stimulation (figure 81.1B). Working in this environment, Lee and colleagues confirmed Penfield's finding that stimulation near the occipital pole commonly resulted in a report of a phosphene percept, and also that many sites in occipital, temporal, and parietal lobes classified as "extrastriate visual cortex" did not produce any perceptual reports (Lee et al., 2000).

A

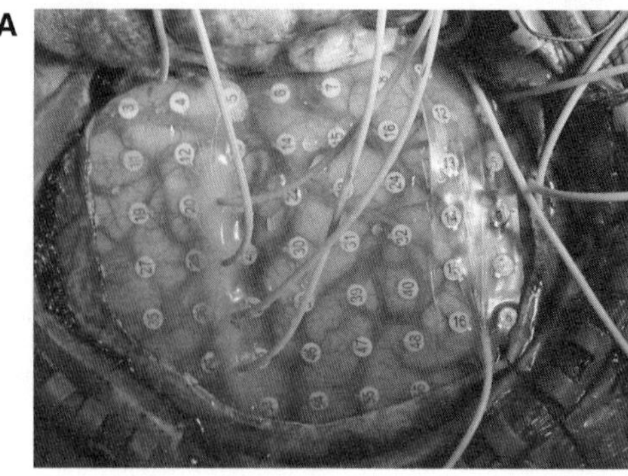

B

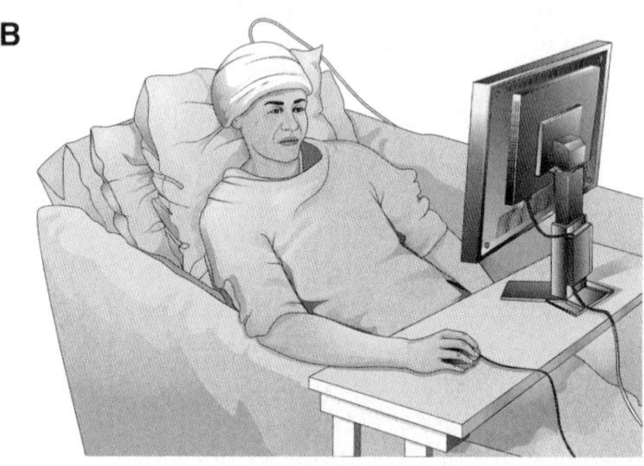

FIGURE 81.1 (A) An example of a subdural grid-electrode array implanted on the cortical surface of a patient with medically intractable epilepsy. After implantation, the craniotomy is closed with electrode tails (insulated white wires) tunneled outside the scalp to be connected to a recording system in an epilepsy monitoring unit (EMU). (B) Research studies can be conducted in awake and cooperative patients with implanted intracranial electrodes who are undergoing clinical testing in the EMU. While resting comfortably in their hospital bed, subjects can make behavioral responses with a computer mouse. (See color plate 80.)

Electrical stimulation of identified human visual areas

While many cortical areas have been identified that respond selectively to visual stimuli ("visual areas"), it remains unclear which of these areas have the ability to support visual perception. Some have proposed that either early visual areas or late visual areas are privileged in supporting perception (Blake & Logothetis, 2002; Tong, 2003), while others have suggested that ventral visual areas support perception and dorsal visual areas do not (Goodale & Milner, 1992). Past studies of cortical electrical stimulation (CES) of human visual cortex (Brindley & Lewin, 1968; Dobelle & Mladejovsky, 1974; Girvin et al., 1979; Lee et al., 2000; Penfield, 1958; Penfield & Rasmussen, 1950) did not define the specific visual areas being stimulated—these studies largely predated functional neuroimaging. Without functional identification of the stimulated visual areas, it remained unclear how the ability to produce a percept varies across different visual areas, and if the character of percepts produced by CES is predicted by the specific visual area that is stimulated.

To better understand the link between CES of specific areas of visual cortex and the production of a visual percept, we performed a systematic study involving stimulation of electrodes located on fMRI-identified visual areas in 10 patients with epilepsy who underwent semichronic implantation of subdural electrode arrays for clinical purposes (Murphey, Maunsell, Beauchamp, & Yoshor, 2009). Only patients with normal visual function on clinical testing (including normal quantitative visual field testing) and anatomically normal visual cortex were enrolled in the study, and electrodes ultimately determined to be located on epileptic cortex were excluded from analysis. The standard clinical electrodes employed in this study were circular platinum alloy discs embedded in sheets of silastic (figure 81.1A); each electrode had an uninsulated center with a diameter of 2.2 mm that contacted the cortical surface directly. BOLD fMRI was used to identify individual visual areas in each subject by using phasic retinotopic stimuli and functional localizers prior to surgery for electrode implantation. Post-implantation computed tomography (CT) scans were merged with the preimplantation fMRI to localize electrodes relative to specific visual areas. Visually evoked responses recorded from the implanted electrodes offered additional confirmation of which electrodes were located on visually responsive areas and agreed well with the characterizations made with fMRI.

All 50 sites of visual cortex that were studied with electrical stimulation were plotted on a single brain in standard space (figure 81.2A). Plotting the probability of percept production against distance of an electrode site from the occipital pole allows for a quantitative measure of location relative to the hierarchy of visual processing; it is well established based on latency and other measures that cortical visual processing begins at the V1 in the occipital pole and follows a roughly posterior to anterior progression (Maunsell, 1995). This plot (figure 81.2B) showed a sharp decline in percept production with increasing distance from the occipital pole. The decreasing probability with distance from V1

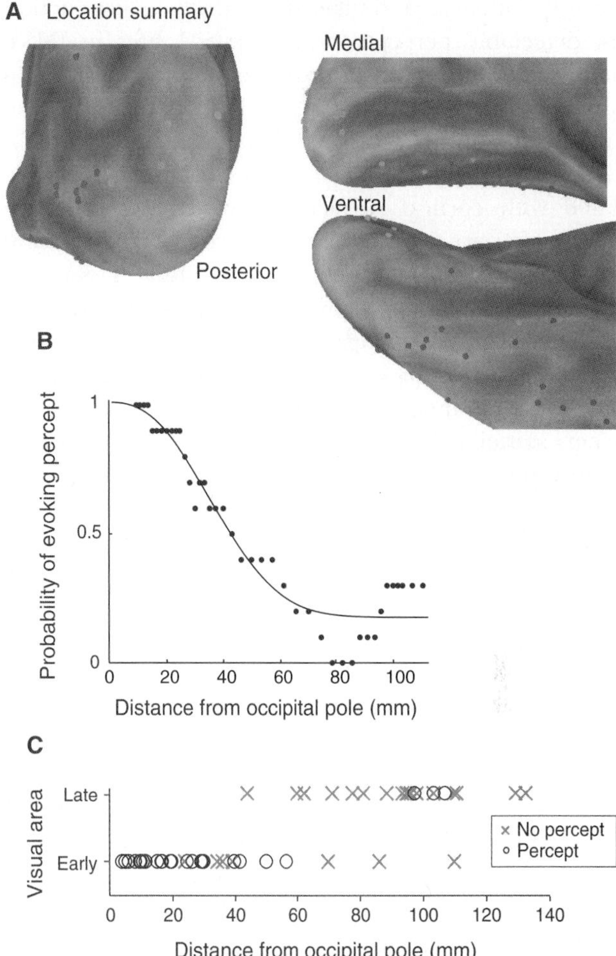

Figure 81.2 (A) The location of 50 electrodes across 10 subjects are plotted as spheres on a single, inflated left hemisphere. The left hemisphere is shown from posterior, medial, and ventral views. Green color indicates that electrical stimulation of the electrode produced a percept. Red color indicates that it did not. (B) In each subject, the distance between the occipital pole and the electrode along the cortical surface was measured. For each distance, the probability of evoking a percept was computed. Each blue point shows the average cortical surface distance and probability for 10 electrodes, calculated with a moving-window average. The black curve shows the best-fit Weibull function. (C) Electrodes were also classified depending on their position in the visual hierarchy as either early (electrodes located in areas V1, V2, V3, V3a, and V4) or late (all other areas). Electrodes that produced a percept are shown as a black O; electrodes that did not are shown as a red X. Most early electrodes (symbols at the bottom) produced a percept, most late electrodes (symbols at the top) did not. There was a rough correspondence between early and late classification and distance from the occipital pole (shown on the *x*-axis). (Reprinted with permission from Murphey, Maunsell, Beauchamp, & Yoshor, 2009.) (See color plate 81.)

corresponded to a similar sharp drop in production of detectable percepts with electrode sites in fMRI-identified early versus late visual areas: 100% of V1 sites produced a percept, while just 11% of fusiform face area (FFA) sites (a late visual area) did (figure 81.2C). These data demonstrate that when electrically stimulated with cortical surface electrodes, early human visual areas are far more likely to evoke a percept than later areas. Multiple sites in later visual areas that showed strong and selective visual responses with both fMRI and LFP recordings failed to produce a detectable percept with stimulation—strong evidence that there are late visual areas that do not produce a visual percept when stimulated (with these methods). However, the ability to produce a percept is not completely restricted to early visual areas, and there is no sharp dichotomy between early and late visual areas in their abilities to support perception.

For the sites that did produce a detectable percept with stimulation, subjects were asked to report on the qualitative properties of the percept. Subjects usually reported a consistent percept with repeated stimulation of a specific individual electrode. Stimulation of many sites in both early and late visual areas elicited descriptions of consistent simple phosphenes (small flashes of white or silver light). Stimulation of some areas elicited reports of more complex percepts, with descriptions of color or more complex shapes. However, no systemic difference was noted between percept complexity for early and late sites. In contrast to previous reports (Blanke, Landis, & Seeck, 2000; Lee et al., 2000; Perot & Penfield, 1960; Puce, Allison, & McCarthy, 1999), no elaborate percepts such as images of faces or objects or memories were described from any site where electrical stimulation produced a detectable percept. This discrepancy could reflect the fact that electrodes that produce complex visual percepts are relatively rare, and our study tested a relatively small number of electrodes. It is also possible that this discrepancy relates to the fact that previous studies did not objectively verify that the subject actually detected stimulation of a particular electrode—perhaps the oft-cited accounts of subjects reporting a complex visual percept would not have held up to this kind of scrutiny (described below).

Unlike previous studies of visual cortex stimulation, we did not wholly rely on the subject's subjective perceptual reports without any psychophysical verification. Instead, an electrode was ultimately defined as being capable of evoking a percept strictly on the basis of data collected from a two-interval forced-choice (2-IFC) task over multiple trials (typically >10 for each current level tested for each electrode). During this task (figure 81.3A, B), subjects had their eyes closed and covered

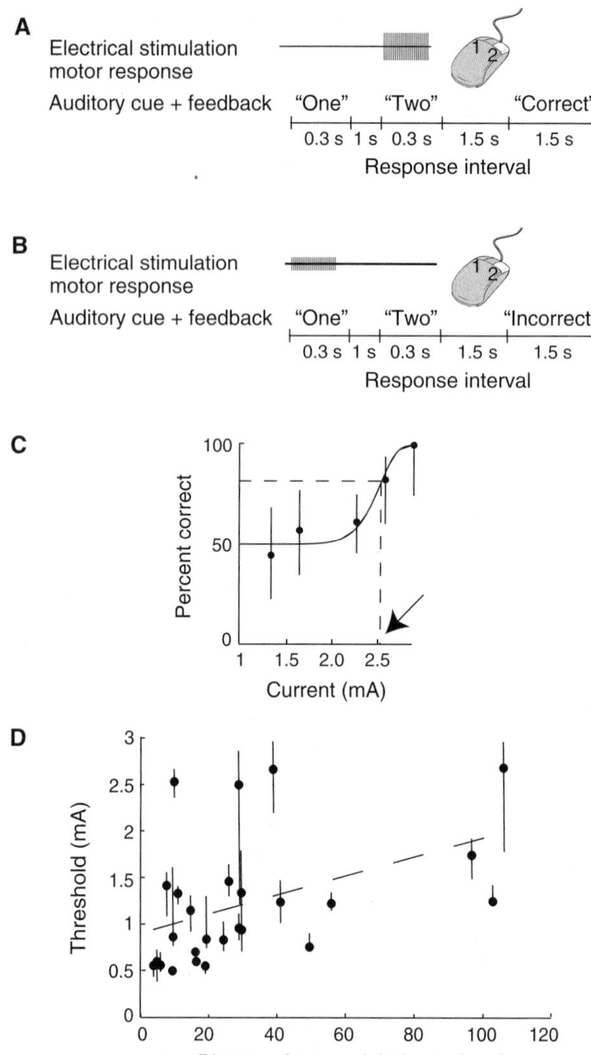

FIGURE 81.3 Determining thresholds for detecting electrical stimulation. (A) Each trial contained two 300-msec epochs, marked by the words "one" and "two." The current amplitude (and the epoch containing the stimulus) varied from trial to trial. The subject's task was to detect the epoch in which the stimulation was delivered. A sample trial is shown in which a high-amplitude current train was delivered in the second epoch, and the subject responded by pressing mouse button two and received positive feedback. (B) A sample trial in which a low-amplitude current was delivered in the first epoch, and the subject responded by pressing mouse button two and received negative feedback. (C) Behavioral performance at a single V2 electrode. Each point shows the performance at different stimulation currents (error bars, 95% confidence intervals, CI). The black curve is the best-fit psychometric function. The dashed line shows the threshold of 2.53 mA (95% CI, 2.36–2.66 μA). (D) Correlation between threshold and distance from the occipital pole across electrodes (error bars, 95% confidence interval for detection threshold). Dashed line shows linear fit ($r = 0.48$, $P = 0.03$). (Reprinted with permission from Murphey et al., 2009.)

with a blindfold. In each trial, 300 msec of a 200 Hz biphasic pulse (200 μs) was randomly delivered in one of two time intervals to an electrode positioned on identified visual cortex. The current intensity in each trial was also varied randomly across a range that spanned the estimated detection threshold determined with preliminary testing for each electrode. Subjects were required to use a button press to indicate during which of the two intervals they detected a percept. The forced-choice design required subjects to make a guess even on trials where no overt stimulus of any kind was detected. If a given level of electrical current at a specific visual electrode produced a detectable phosphene, we expect the subject to correctly identify which of the two intervals contained the electrical stimulation at a performance level significantly better than chance (50%). This provided clear and objective verification that the subject detected a percept linked to electrical stimulation of the brain.

By testing a range of different current amplitudes, we were also able to examine how detection thresholds varied across visual cortex. This is potentially important, because if some areas are functionally more distant from perception and not as able to support it, they might require stronger electrical currents to produce a detectable percept than other areas that lie in closer proximity to brain regions that are most critical for perception. For each electrode, stimulus trains of different strengths were delivered pseudorandomly in one of the two intervals on different trials for the 2-IFC task. The resulting behavioral performance plot was fitted with a sigmoid function to find the threshold for behavioral detection, defined by convention as the current level that resulted in 82% correct detection (figure 81.3C). The mean detection threshold across all electrodes that produced a percept was 1.12 mA (range 0.49–2.65, SD = 0.68). To determine whether there was a relationship between position in the visual cortical hierarchy and detection threshold, the threshold for each electrode was plotted against the cortical surface distance (figure 81.3D). With increasing distance from the occipital pole, there was a slight increase in threshold (0.01 ma/mm, $r = 0.48$, $P = 0.03$) corresponding to a trend toward higher thresholds for late visual areas (1.9 vs. 1.1 mA, $P = 0.08$). This reveals a clear dichotomy in our results. While the probability of evoking a percept varied greatly from early to late areas, the current threshold (if a percept was evoked) varied only slightly.

The decreasing probability of eliciting a percept in early versus late areas may arise from differences in the functional organization of these areas. If perception of a visual stimulus requires activity in a network of brain areas, CES in early areas may more often propagate to this network because of greater extrinsic connectivity in early areas (Tolias et al., 2005). Conversely, stimulation of late areas with fewer extrinsic connections may be less likely to propagate to this network, reducing the likelihood of evoking a percept. Another possibility is that the decreasing probability of eliciting a percept in early versus late areas is a function of electrode size relative to the functional organization visual cortex. Early visual areas that are much more likely to produce a percept are characterized by a highly retinotopic organization; neurons processing information from one part of the visual field are located adjacent to neurons processing information from an adjacent part of the visual field. In contrast, the visual preferences of neurons in later areas such as area IT seem not to be organized in a continuous map—neighboring columns may respond preferentially to quite disparate features (Tanaka, 1996). When a 2.2 mm electrode is used to deliver an electrical current to a site in early visual cortex, it likely activates a spatially contiguous patch of cortex in a manner that is not entirely inconsistent with what might occur with natural vision, and is therefore readily interpreted by the brain without any training and results in a visual percept. In contrast, using the same method to stimulate a discontinuous map in a later visual area may produce a signal that is very different from any pattern of activity produced with natural vision, and is therefore not readily interpretable and does not produce a conscious percept.

It is noteworthy that with experience across thousands of trials (vastly more than used in our human study), nonhuman primates can routinely and reliably detect microstimulation of both early and late visual areas with similar accuracy, in contrast to the limited ability to detect stimulation of late visual areas in human subjects (Murphey & Maunsell, 2007). This apparent discrepancy across species may reflect the much finer scale of electrical stimulation used in the monkey versus the human study—in the monkeys, much smaller penetrating microelectrodes (micron versus millimeter) were used with much lower current levels (microamps versus milliamps). It is also possible that the extreme level of training in the monkey experiments explains the discrepancy. Perhaps after thousands of trials, the monkey learns to detect a highly unnatural pattern of activation of late visual cortex (with or possibly without an accompanying visual percept; Histed et al., 2012).

Stimulation reveals a role for the temporoparietal junction in visual perception

When a percept of a phosphene occurs reliably with stimulation of a specific electrode site, this demonstrates a causal link between the population of neurons that is directly activated by the electrical current and the percept. But this population of directly activated neurons is not solely responsible for the percept of the phosphene. Presumably, the behavioral consequences of CES depend not only on the depolarization of local neuronal cell bodies, but also on depolarization of nearby axons followed by orthodromic and antidromic action potentials, and on indirect neuronal activation that likely extends to brain regions quite distant from the neurons that are directly driven by CES (Clark,et al., 2011; Tehovnik, Tolias, Sultan, Slocum, & Logothetis, 2006). In a second study, we sought to better understand the neuronal activity involved in visual perception by adding electrocorticography recordings to CES. To do this, we compared electrocorticography recordings when stimulation produced a percept to when it did not under several controlled experimental conditions (Beauchamp, Sun, Baum, Tolias, & Yoshor, 2012).

First we screened individual electrodes located on visual cortex (defined with fMRI) to identify electrode sites that produced a percept when stimulated ("percept electrodes") as well as other electrode sites that did not produce a percept ("nonpercept electrodes") by electrically stimulating each electrode with a current level that was well above the expected threshold for detection. As expected, percept electrodes were concentrated over early visual areas near the occipital pole (figure 81.4A). Following screening, one percept electrode and the nearest nonpercept electrode in each subject were selected for the first experiment. Single 5 msec current pulses were repeatedly delivered to both of these electrodes while patients remained seated and alert but did not perform a behavioral task (using very short pulse durations helped minimize stimulation artifacts that would obscure recordings in nearby electrodes). The same suprathreshold stimulation current was used for the percept electrode and the nonpercept electrode in each subject, and this current was sufficient to always produce a phosphene in the percept electrodes with each pulse. Time-locked to the delivery of the stimulation pulses, neurophysiological recordings were collected from all nonstimulated electrodes. Neural oscillations in the gamma range (60–150 Hz) have been found to reflect neuronal spiking and may serve as a general mechanism of information processing (Engel, Fries, & Singer, 2001; Ray & Maunsell, 2011; Whittingstall & Logothetis, 2009). Comparing the gamma activity evoked by percept and nonpercept electrode stimulation revealed a striking difference in gamma activity in the temporoparietal junction (TPJ) for percept versus nonpercept electrode stimulation (figure 81.4B). When percept electrodes were electrically stimulated, a burst of gamma activity was observed in the TPJ beginning at 100 msec after electrical stimulation and continuing for 200 msec (figure 81.4C). Nonpercept electrode stimulation with the same current produced no such TPJ activity. This distribution was reliably present across individual trials (figure 81.4D). A receiver operating curve (ROC) constructed from this data (figure 81.4E) revealed that it was possible to discriminate between a percept and nonpercept electrode based on the TPJ gamma response with a high degree of discriminability (mean d' across subjects, 1.2).

The observation that TPJ gamma activity was present on trials in which percept electrodes were stimulated but not on trials where nonpercept electrodes were stimulated raised the possibility that TPJ activity may be causally related to visual perception. But another possibility is that TPJ gamma power is merely correlated with the location of electrical stimulation (high for stimulation of early visual areas, low for late visual areas) and not truly linked to percept generation. To distinguish these possibilities, we capitalized on the observation that electrical stimulation of percept electrodes

FIGURE 81.4 (A) Percept electrodes (green) that produced a phosphene upon electrical stimulation and nonpercept electrodes (red) that did not in three subjects. Subject 1 (s1) shows a posterior view of the right hemisphere; s2, posterior view of left hemisphere; s3, medial view of left hemisphere. Electrodes were implanted only in a single hemisphere for each subject (right for s1, left for s2 and s3). OP: occipital pole. (B) Maps showing the difference in gamma power between percept electrode stimulation and nonpercept electrode stimulation in three subjects (same subjects as in [A]). For each subject, one percept electrode and the nearest nonpercept electrode in the implanted hemisphere was repeatedly stimulated, and the significance of post-stimulation difference in gamma power at each electrode (except for the stimulation electrodes) was calculated and mapped to the cortical surface. Black spheres show electrode locations. TPJ: temporoparietal junction. (C) TPJ response during electrical stimulation of occipital electrodes that did (left) or did not (right) produce a phosphene, averaged across subjects. Color scale indicates power at each frequency. Dark gray bar centered at $t = 0$ indicates stimulation artifact. Dashed white line at $f = 30$ Hz indicates the boundary between two different frequency-estimate techniques. Dashed black line indicates gamma band f-t window used to estimate power for single-trial analysis. (D) TPJ gamma responses for every trial during stimulation of a percept (left) or nonpercept (right) electrode. Each horizontal line (raster) shows the power in a single trial over time, collapsed across 60–150 Hz. Same color scale as (C). (E) Receiver-operating curve analysis of single trial data. (Reprinted with permission from Beauchamp et al., 2012.) (See color plate 82.)

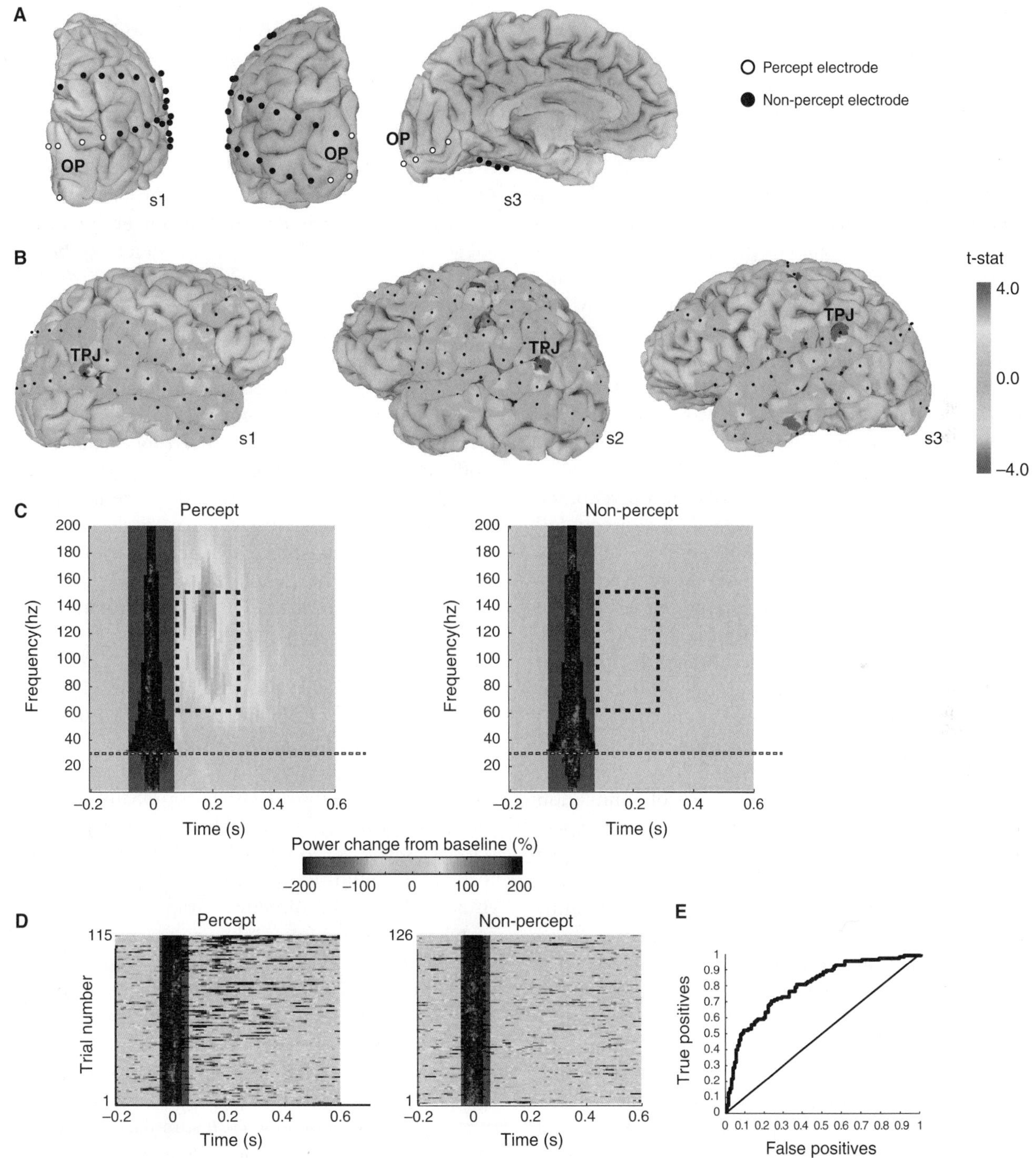

A
Percept electrode
Non-percept electrode
OP
s1
OP
OP
s3
B
TPJ
s1
TPJ
s2
TPJ
s3
t-stat
4.0
0.0
–4.0
C
Percept
Non-percept
Frequency(hz)
Time (s)
Power change from baseline (%)
–200
–100
0
100
200
D
Percept
Non-percept
Trial number
115
126
1
Time (s)
E
True positives
False positives

does not always produce a phosphene: the likelihood of phosphene perception increases with increasing stimulation current (Murphey & Maunsell, 2007; Murphey et al., 2009). In the next experiment, we focused on stimulating the same percept electrode with varying levels of current, using the same 2-IFC task described earlier combined with concurrent recordings from other electrodes, including TPJ electrodes. We found that at low stimulation currents, low levels of TPJ gamma were observed, and as the current was increased the TPJ gamma increased accordingly (figure 81.5A). We then compared the TPJ gamma level with the behavioral performance in detecting a phosphene in the 2-IFC task. At high currents, behavioral performance in the 2-IFC task (a measure of phosphene perception) was nearly perfect, and the corresponding TPJ gamma power was similarly high. At low currents, behavioral performance was no better than chance, and TPJ gamma was also very low. The neurometric and psychometric functions that we plotted based on this data show similar monotonic increases in behavioral performance and in TPJ gammas with increasing stimulation current (figure 81.5B).

The similarity between neurometric and psychometric functions supports the idea of a link between TPJ activity and visual perception. But it still possible that while increasing currents led to both improved discrimination and increased TPJ gamma power, these were independent processes. To test this hypothesis, we examined trials in the 2-IFC task in which the identical near-threshold current was delivered to a percept electrode. As expected, this threshold level of current delivered to the very same electrode produced a mix of correct trials and incorrect trials. If TPJ gamma power was simply correlated with the amount of stimulation current but not related to perception, we would expect no power difference between correct and incorrect trials, because the stimulation current was exactly the same for each trial. Instead we found significantly greater TPJ gamma power in correct trials than in incorrect trials, even though both sets of trials used the exact same electrode and the exact same stimulation parameters (figure 81.5C). An ROC analysis of the individual trial data (figure 81.5D) revealed a significant ability to discriminate correct from incorrect trials based on stimulation-evoked TPJ gamma power (mean d' across subjects, 0.74). Finally, if TPJ gamma power is truly a neural signature of the phosphene percept, an ideal observer should be able to perform the 2-IFC task just by comparing the TPJ gamma power in the two intervals in each individual trial. Indeed, we found that within correct trials there was a very large TPJ power difference between the stimulated and nonstimulated intervals, while this difference was very small in the incorrect trials. An ROC analysis of the individual trials revealed that an ideal observer does very well at using TPJ gamma power to distinguish the two intervals in the correct trials ($d' = 1.1$) but not in the incorrect trials ($d' = 0.3$).

Collectively, these analyses provide compelling evidence that subjects perceived a stimulation-induced phosphene only when high-gamma power was recorded in the TPJ. TPJ activity during phosphene perception was observed both during passive stimulation and while subjects performed a behavioral task, making it difficult to attribute the gamma activity to task performance. Our observation of visual perception–related gamma activity in TPJ is striking because converging evidence suggests that the TPJ is critical for detecting behaviorally relevant stimuli (Corbetta & Shulman, 2002). For example, damage to the ventral region of the parietal lobe, especially the TPJ, is well known to cause spatial neglect (Corbetta & Shulman, 2011; Karnath & Rorden, 2012). This suggests a possible parallel with our results. When electrical stimulation does not produce a phosphene, neural activity is produced locally at the electrode site but it does not propagate through the cortical network to evoke TPJ activity and hence fails to enter conscious awareness, just as may occur with visual stimuli in cases of spatial neglect. In contrast, when neural activity at the stimulation site does propagate to the TPJ, the activity does enter conscious awareness and a phosphene percept is produced.

Future potential of visual cortex stimulation to restore vision to the blind

Electrical stimulation of visual cortex not only offers useful insights into the neural basis of visual perception, but also suggests a potential mechanism for restoring vision to the blind. The possibility of bypassing damaged visual pathways and directly activating the intact visual cortex has long intrigued physicians and scientists. It is already well established that direct electrical stimulation of the occipital lobe can reliably make patients with acquired blindness as well as sighted subjects "see" an individual phosphene (Brindley & Lewin, 1968; Dobelle, Mladejovsky, & Girvin, 1974). A cortical visual prosthetic has been envisioned using a camera and computer to drive precise electrical stimulation of multiple sites in retinotopically organized visual cortex to produce specific spatial patterns of activation, and to thus enable blind subjects to have a meaningful visual experience (Normann et al., 2009; Schiller & Tehovnik, 2008; Tehovnik, Slocum, Smirnakis, & Tolias, 2009; Troyk et al., 2005).

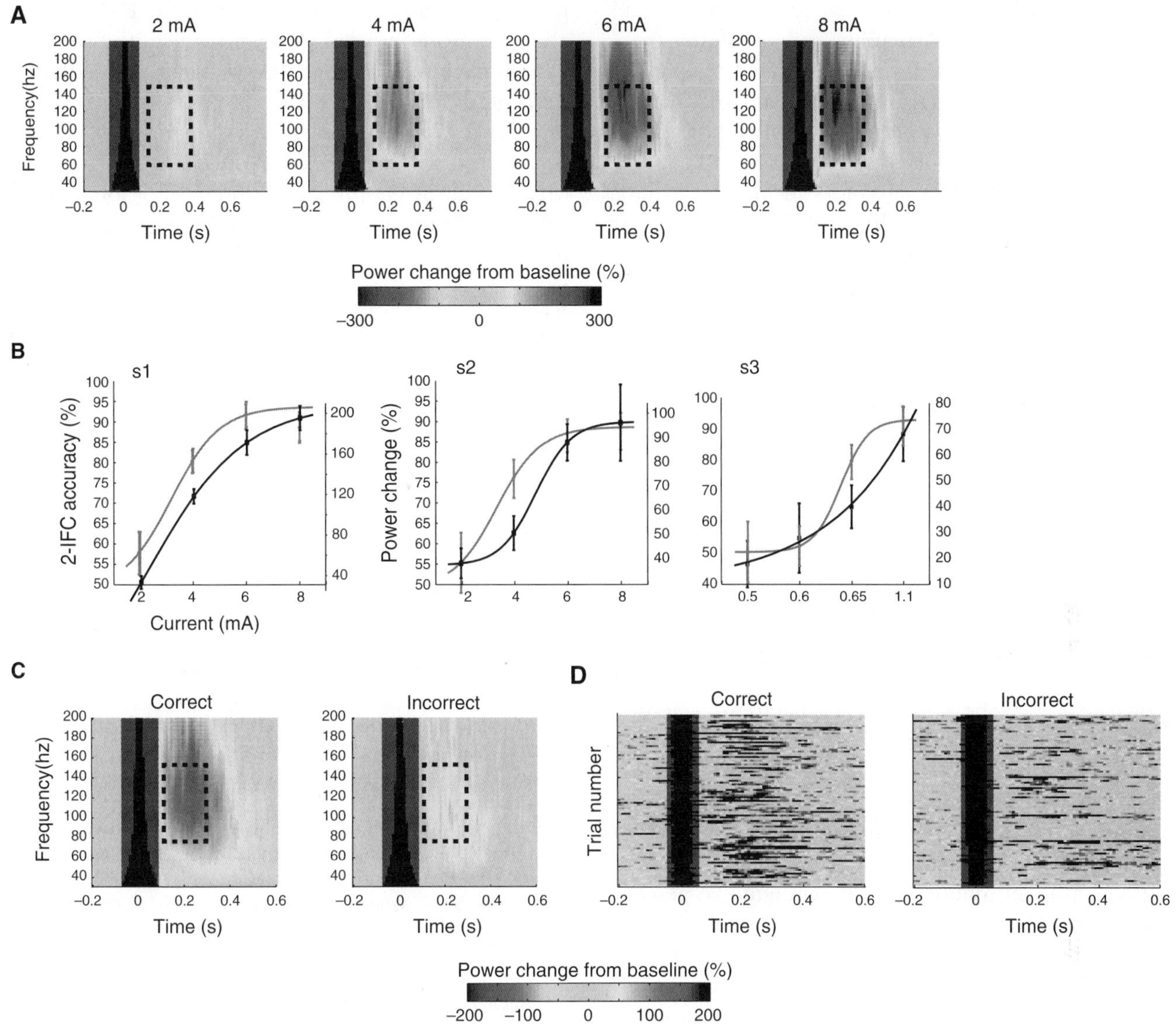

FIGURE 81.5 (A) The average TPJ response during electrical stimulation of three percept electrodes in the occipital lobe in s1 at varying stimulation currents (2–8 mA). (B) Psychometric (blue) and neurometric (red) functions for subjects s1, s2, and s3. The psychometric curve (left *y*-axis) shows the behavioral performance during the two-interval forced-choice (2-IFC) task at different stimulation currents; performance was near chance (50%) at low currents and near ceiling (100%) at high currents (error bars show 75% confidence interval from the binomial distribution). The neurometric curve (right *y*-axis) shows the TPJ gamma power at the same currents (error bars show standard error of the mean). (C) TPJ response during percept electrode stimulation with near-threshold currents, averaged across subjects. Data averaged from trials in which subjects correctly (left) or incorrectly (right) discriminated which of two intervals contained electrical stimulation. Stimulation current was the same for correct and incorrect trials (4 mA for s1 and s2, 0.65 mA for s3). (D) TPJ response in the gamma band for single correct (left) and incorrect (right) trials of electrical stimulation at the same current. Each horizontal line (raster) shows the power in a single trial over time, collapsed across the gamma band. (Reprinted with permission from Beauchamp et al., 2012.) (See color plate 83.)

Several groups have attempted preliminary studies on the feasibility of a cortical visual prosthetic of this type in blind human patients, but it is unclear how much these studies have confirmed the promise of this strategy to restore vision (Dobelle, Quest, Antunes, Roberts, & Girvin, 1979; Schmidt et al., 1996). One problem is that many past studies of visual cortex stimulation in human subjects excessively relied on imprecise and unverified qualitative verbal reports. More recently, there have been attempts to study electrical stimulation of visual cortex in a more systematic and quantitative fashion using nonhuman primates (Bradley et al., 2005;

Davis et al., 2012; Tehovnik et al., 2009; Torab et al., 2011). But these studies in animal subjects suffer from a different drawback, namely, the limited abilities of trained monkeys to report on their perceptual experiences, even with the most elegant experimental designs. In spite of years of important preliminary work, and in spite of all the dramatic advances in brain-computer interface technology over the past decade, there has been relatively little recent progress in the development of a cortical visual prosthetic. While many technical hurdles remain, including engineering and biocompatibility issues, one critical limiting factor is that it is still not clear whether phosphenes can be used as building blocks for producing percepts of complex visual forms, as the basic design of a cortical visual prosthetic assumes. There is good evidence that artificial activation of a site in retinotopic visual cortex will result in perception of a phosphene in a location corresponding to the spatial properties of the stimulated site (Schiller & Tehovnik, 2008), but it is still unknown if an array of multiple phosphenes can be perceived as spatially specific additive elements that can be combined to produce coherent percept of complex forms. It is possible that projections from the retina and thalamus are essential to create anything more than a rudimentary visual sensation, and that phosphene percepts are not readily interpreted by the brain as spatial elements that can be combined into forms. It is also possible that stimulation of multiple sites in visual cortex results in a competitive interaction or inhibition, rather than additive perceptual elements.

The important and ongoing work in monkeys toward the development of a cortical visual prosthetic is challenged in filling this knowledge gap because monkeys cannot directly describe the character of their percepts. While elegant experimental designs have enabled trained monkeys to report on biases in stimulus perception produced with electrical stimulation of visual cortex (Afraz, Kiani, & Esteky, 2006; Salzman, Britten, & Newsome, 1990), it will be very difficult to train a monkey to exhaustively report on the vast array of properties of a complex visual form. Human subjects can directly confirm the existence of an overt visual percept with stimulation and can easily provide a description of the size, shape, color, and other properties of an artificially created percept. Experiments in human subjects that combine quantitative methods of psychophysics with the unique ability of human subjects to use language should be able to most effectively test the potential of a cortical visual prosthetic that uses stimulation of multiple distinct populations of visual neurons concurrently to produce a range of complex percepts useful for guiding behavior. If experiments in human patients with implanted electrodes in the EMU (in parallel with work in monkeys) can confirm the potential of multielectrode stimulation to produce predictable percepts of complex visual forms, there would be strong impetus for the development and testing of a modern prototype of a cortical visual prosthetic in blind patients, with careful psychophysical testing and a longitudinal study of the effects of training on perceptual performance.

ACKNOWLEDGMENTS We thank Ping Sun, Dona Murphey, Sarah Baum, Xiaomei Pei, Inga Schepers, Andreas Tolias, and John Maunsell for their important contributions to this work. We are also grateful to the patients and the clinical staff at St. Luke's Episcopal Hospital. This research was supported in part by NIH R01EY023336 and VA Merit Award 5I01CX000325 (DY), and NIH NS045053 (MSB).

REFERENCES

Afraz, S. R., Kiani, R., & Esteky, H. (2006). Microstimulation of inferotemporal cortex influences face categorization. *Nature, 442*(7103), 692–695.

Beauchamp, M. S., Sun, P., Baum, S. H., Tolias, A. S., & Yoshor, D. (2012). Electrocorticography links human temporoparietal junction to visual perception. *Nat Neurosci, 15*(7), 957–959.

Blake, R., & Logothetis, N. (2002). Visual competition. *Nat Rev Neurosci, 3*(1), 13–21.

Blanke, O., Landis, T., & Seeck, M. (2000). Electrical cortical stimulation of the human prefrontal cortex evokes complex visual hallucinations. *Epilepsy Behav, 1*(5), 356–361.

Borchers, S., Himmelbach, M., Logothetis, N., & Karnath, H. O. (2011). Direct electrical stimulation of human cortex—the gold standard for mapping brain functions? *Nat Rev Neurosci, 13*(1), 63–70.

Bradley, D. C., Troyk, P. R., Berg, J. A., Bak, M., Cogan, S., Erickson, R., ... Xu, H. (2005). Visuotopic mapping through a multichannel stimulating implant in primate V1. *J Neurophysiol, 93*(3), 1659–1670.

Brindley, G. S., & Lewin, W. S. (1968). The sensations produced by electrical stimulation of the visual cortex. *J Physiol, 196*(2), 479–493.

Clark, K. L., Armstrong, K. M., & Moore, T. (2011). Probing neural circuitry and function with electrical microstimulation. *Proc Biol Sci, 278*(1709), 1121–1130.

Corbetta, M., & Shulman, G. L. (2002). Control of goal-directed and stimulus-driven attention in the brain. *Nat Rev Neurosci, 3*(3), 201–215.

Corbetta, M., & Shulman, G. L. (2011). Spatial neglect and attention networks. *Annu Rev Neurosci, 34*, 569–599.

Davis, T. S., Parker, R. A., House, P. A., Bagley, E., Wendelken, S., Normann, R. A., & Greger, B. (2012). Spatial and temporal characteristics of V1 microstimulation during chronic implantation of a microelectrode array in a behaving macaque. *J Neural Eng, 9*(6), 065003.

Dobelle, W. H., & Mladejovsky, M. G. (1974). Phosphenes produced by electrical stimulation of human occipital cortex, and their application to the development of a prosthesis for the blind. *J Physiol, 243*(2), 553–576.

Dobelle, W. H., Mladejovsky, M. G., & Girvin, J. P. (1974). Artifical vision for the blind: Electrical stimulation of visual cortex offers hope for a functional prosthesis. *Science, 183*(4123), 440–444.

Dobelle, W. H., Quest, D. O., Antunes, J. L., Roberts, T. S., & Girvin, J. P. (1979). Artificial vision for the blind by electrical stimulation of the visual cortex. *Neurosurgery, 5*(4), 521-527.

Engel, A. K., Fries, P., & Singer, W. (2001). Dynamic predictions: Oscillations and synchrony in top-down processing. *Nat Rev Neurosci, 2*(10), 704–716.

Engel, A. K., Moll, C. K., Fried, I., & Ojemann, G. A. (2005). Invasive recordings from the human brain: Clinical insights and beyond. *Nat Rev Neurosci, 6*(1), 35–47.

Friedman, D. E., & Riviello, J. J. (2012). Extraoperative brain mapping using chronically implanted subdural electrodes. In D. Yoshor & E. M. Mizrahi (Eds.), *Clinical brain mapping* (pp. 93–103). New York, NY: McGraw-Hill.

Fritsch, G., & Hitzig, E. (1870). Uber die elektrische Errebarkeit des Grosshims. *Arch Anat Physiol Wiss Med, 37*, 300–332.

Girvin, J. P., Evans, J. R., Dobelle, W. H., Mladejovsky, M. G., Henderson, D. C., Abramov, I., ... Turkel, J. (1979). Electrical stimulation of human visual cortex: The effect of stimulus parameters on phosphene threshold. *Sens Processes, 3*(1), 66–81.

Goodale, M. A., & Milner, A. D. (1992). Separate visual pathways for perception and action. *Trends Neurosci, 15*(1), 20–25.

Histed, M. H., Ni, A. M., & Maunsell, J. H. (2012). Insights into cortical mechanisms of behavior from microstimulation experiments. *Prog Neurobiol, 103*, 115–130.

Karnath, H. O., & Rorden, C. (2012). The anatomy of spatial neglect. *Neuropsychologia, 50*(6), 1010–1017.

Lee, H. W., Hong, S. B., Seo, D. W., Tae, W. S., & Hong, S. C. (2000). Mapping of functional organization in human visual cortex: Electrical cortical stimulation. *Neurology, 54*(4), 849–854.

Luders, J., & Luders, H. (2001). Contributions of Fedor Krause and Orrtid Foerster to epilepsy surgery. In H. Luders & Y. G. Comair (Eds.), *Epilepsy surgery* (2nd ed., pp. 23–34). Philadelphia, PA: Lippincott Williams & Wilkins.

Maunsell, J. H. (1995). The brain's visual world: Representation of visual targets in cerebral cortex. *Science, 270*(5237), 764–769.

Murphey, D. K., & Maunsell, J. H. (2007). Behavioral detection of electrical microstimulation in different cortical visual areas. *Curr Biol, 17*(10), 862–867.

Murphey, D. K., Maunsell, J. H., Beauchamp, M. S., & Yoshor, D. (2009). Perceiving electrical stimulation of identified human visual areas. *Proc Natl Acad Sci USA, 106*(13), 5389–5393.

Normann, R. A., Greger, B., House, P., Romero, S. F., Pelayo, F., & Fernandez, E. (2009). Toward the development of a cortically based visual neuroprosthesis. *J Neural Eng, 6*(3), 035001.

Orban, G. A., Van Essen, D., & Vanduffel, W. (2004). Comparative mapping of higher visual areas in monkeys and humans. *Trends Cogn Sci, 8*(7), 315–324.

Parker, A. J., & Newsome, W. T. (1998). Sense and the single neuron: Probing the physiology of perception. *Annu Rev Neurosci, 21*, 227–277.

Penfield, W. (1958). Some mechanisms of consciousness discovered during electrical stimulation of the brain. *Proc Natl Acad Sci USA, 44*(2), 51–66.

Penfield, W., & Perot, P. (1963). The brain's record of auditory and visual experience. A final summary and discussion. *Brain, 86*, 595–696.

Penfield, W., & Rasmussen, T. (1950). *The cerebral cortex in man*. New York, NY: Macmillan.

Perot, P., & Penfield, W. (1960). Hallucinations of past experience and experiential responses to stimulation of temporal cortex. *Trans Am Neurol Assoc, 85*, 80–84.

Puce, A., Allison, T., & McCarthy, G. (1999). Electrophysiological studies of human face perception. III: Effects of top-down processing on face-specific potentials. *Cereb Cortex, 9*(5), 445–458.

Ray, S., & Maunsell, J. H. (2011). Different origins of gamma rhythm and high-gamma activity in macaque visual cortex. *PLoS Biol, 9*(4), e1000610.

Salzman, C. D., Britten, K. H., & Newsome, W. T. (1990). Cortical microstimulation influences perceptual judgements of motion direction. *Nature, 346*(6280), 174–177.

Schiller, P. H., & Tehovnik, E. J. (2008). Visual prosthesis. *Perception, 37*(10), 1529–1559.

Schmidt, E. M., Bak, M. J., Hambrecht, F. T., Kufta, C. V., O'Rourke, D. K., & Vallabhanath, P. (1996). Feasibility of a visual prosthesis for the blind based on intracortical microstimulation of the visual cortex. *Brain, 119*(Pt 2), 507–522.

Tanaka, K. (1996). Inferotemporal cortex and object vision. *Annu Rev Neurosci, 19*, 109–139.

Tehovnik, E. J., Slocum, W. M., Smirnakis, S. M., & Tolias, A. S. (2009). Microstimulation of visual cortex to restore vision. *Prog Brain Res, 175*, 347–375.

Tehovnik, E. J., Tolias, A. S., Sultan, F., Slocum, W. M., & Logothetis, N. K. (2006). Direct and indirect activation of cortical neurons by electrical microstimulation. *J Neurophysiol, 96*(2), 512–521.

Tolias, A. S., Sultan, F., Augath, M., Oeltermann, A., Tehovnik, E. J., Schiller, P. H., & Logothetis, N. K. (2005). Mapping cortical activity elicited with electrical microstimulation using FMRI in the macaque. *Neuron, 48*(6), 901–911.

Tong, F. (2003). Primary visual cortex and visual awareness. *Nat Rev Neurosci, 4*(3), 219–229.

Torab, K., Davis, T. S., Warren, D. J., House, P. A., Normann, R. A., & Greger, B. (2011). Multiple factors may influence the performance of a visual prosthesis based on intracortical microstimulation: Nonhuman primate behavioural experimentation. *J Neural Eng, 8*(3), 035001.

Troyk, P. R., Bradley, D., Bak, M., Cogan, S., Erickson, R., Hu, Z., ... Towle, V. (2005). Intracortical visual prosthesis research—approach and progress. *Proc. IEEE Engineering in Medicine & Biology Society* (pp. 7376–7379).

Whittingstall, K., & Logothetis, N. K. (2009). Frequency-band coupling in surface EEG reflects spiking activity in monkey visual cortex. *Neuron, 64*(2), 281–289.

82 Optogenetics in the Study of Mammalian Behavior

KELLY A. ZALOCUSKY AND KARL DEISSEROTH

ABSTRACT Many of the challenges of modern experimental neuroscience were outlined by Francis Crick. In 1999, Crick described limitations that were present in genetic engineering, electrophysiology, and anatomical tracing, among other technological challenges. In particular, he highlighted one important goal for the field of neurobiology: to be able to turn the firing of one or more types of neuron on and off in the alert animal in a rapid manner, ideally using light. Now, less than 15 years later, the use of light to control neurons in behaving animals is a widespread and indispensable method in neuroscience. The compatibility of this approach with freely moving, complex behaviors in mammals is a crucial feature that was enabled by the development of modern optogenetics. Here we review key elements of optogenetic control as well as major recent cognitive and behavioral findings obtained in freely moving mammals.

Optogenetics is the use of photosensitive proteins (such as those encoded by opsin genes; figure 82.1) to control cells with light and achieve specificity of control that electrodes cannot deliver (figure 82.2; Yizhar, Fenno, Prigge et al., 2011). Currently, typical optogenetic experiments employ one or more of several major types of opsin. These include (1) excitatory cation channels, which are used to depolarize a cell's membrane potential and can be used to elicit action potentials via the native voltage-gated sodium channels, (2) inhibitory ion pumps, which are used to hyperpolarize a cell's membrane potential, inhibiting it from spiking, or (3) photoresponsive G-protein coupled receptors (GPCRs), which are used to manipulate intracellular second-messenger systems (figure 82.1). The direct ion channels and pumps are derived typically from microorganisms such as algae and archaebacteria, and belong to the microbial opsin gene family, which are seven-transmembrane proteins like the GPCRs and the vertebrate rhodopsins but lack primary sequence homology (Zhang et al., 2011).

Excitatory opsins

The channelrhodopsins are light-activated cation channels that can be used to depolarize neurons and thereby drive precisely timed action potentials (Boyden, Zhang, Bamberg, Nagel, & Deisseroth, 2005; Yizhar, Fenno, Davidson, Mogri, & Deisseroth, 2011). These channels have been modified to allow for reliable fast spiking up to 200 Hz (Gunaydin et al., 2010) or, conversely, to generate long–time scale (up to 30 minutes) subthreshold depolarizations from a single millisecond-scale pulse of light (Berndt, Yizhar, Gunaydin, Hegemann, & Deisseroth, 2009; Yizhar, Fenno, Prigge, et al., 2011). These "step-function opsins" have been used to enhance the excitability of a neural population over prolonged experimental sessions without driving spiking directly, and are of particular interest in allowing for complex behavioral experimentation (e.g., social exploration) with no light-delivery hardware in place (Yizhar, Fenno, Prigge, et al., 2011).

Inhibitory opsins

Opsins also can be used to inhibit neurons. Halorhodopsins are light-activated chloride pumps (Lanyi & Oesterhelt, 1982; Matsuno-Yagi & Mukohata, 1977; Zhang et al., 2011). By pumping chloride ions into the cell, they can hyperpolarize and therefore inhibit neurons, preventing spiking. An amber light–activated form derived from the archaebacterium *Natronomonas pharaonis* has been shown to be useful for optogenetic inhibition-mediated control of behavior in mammals (Stuber et al., 2011; Tan et al., 2012; Tye et al., 2011, 2012; Warden et al., 2012; Witten et al., 2010). Another class of hyperpolarizing opsins includes the bacteriorhodopsins (Chow et al., 2010; Gradinaru et al., 2010; Ihara et al., 1999; Mattis et al., 2012; Oesterhelt & Stoeckenius, 1971, 1973; Zhang et al., 2011). Upon illumination, these proteins pump protons out of the cell, thereby hyperpolarizing neurons and inhibiting spiking.

Optically activated G-protein coupled receptors

Finally, light-activated proteins called optoXRs have been engineered to mimic the activity of GPCRs (Airan, Thompson, Fenno, Bernstein, & Deisseroth, 2009; Levitz et al., 2013). Fusions of the extracellular and

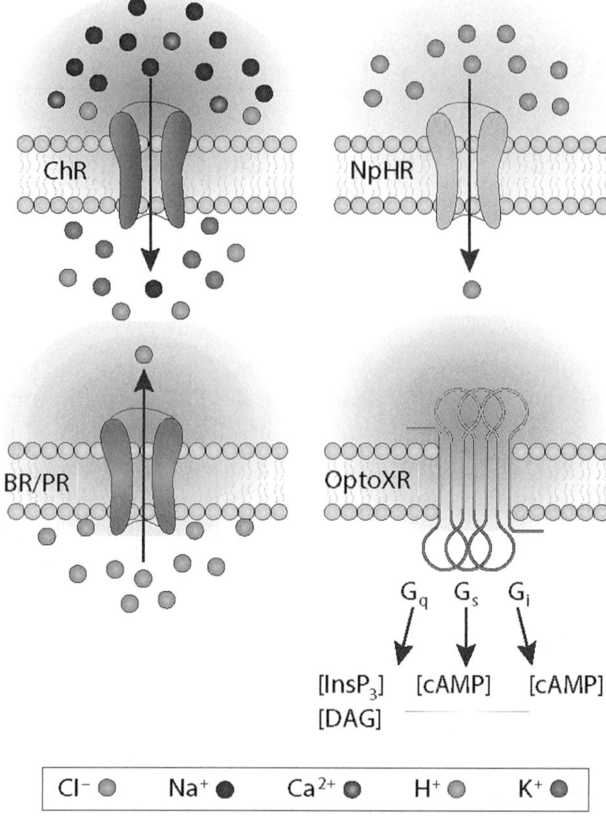

FIGURE 82.1 Single-component optogenetic tool categories. Four major classes of opsin commonly used in optogenetics experiments, each encompassing light sensation and effector function within a single gene, include (1) channelrhodopsins (ChR), which are light-activated cation channels that give rise to inward (excitatory) currents under physiological conditions; (2) halorhodopsins (NpHR, shown), which are inhibitory (outward-current) chloride pumps; (3) bacteriorhodopsins and proteorhodopsins (BR/PR), proton pumps that tend to be inhibitory and include archaerhodopsins; and (4) optoXRs, which modulate secondary messenger signaling pathways. (Adapted with permission from Tye & Deisseroth, 2012.) (See color plate 84.)

transmembrane domains of bovine rhodopsin with intracellular loops of the α_1 (for G_q signaling) or β_2 (for G_s signaling) adrenergic receptors reproduce key specific aspects of the native GPCR second-messenger signaling, in response to light rather than ligands (Airan et al., 2009). Recently, similar strategies have led to creation of a light-activated dopamine D1 receptor (Fenno, Airan, Bernstein, & Deisseroth, 2009) and light-activated recruitment of G_i signaling (Oh, Maejima, Liu, Deneris, & Herlitze, 2010). Organic synthesis methods have led to a separate strategy for generating an optically activated metabotropic glutamate receptor (Levitz et al., 2013). In this instance, a photo-switchable tether was used to connect the ligand and the receptor; when light hits the tether, the tether's conformation changes, allowing the ligand and receptor to interact (Levitz et al., 2013).

By shining specific wavelengths of light onto neurons expressing opsins that respond to those wavelengths (often using an implanted fiberoptic), researchers can excite neurons (driving them to fire action potentials at increased, precisely defined rates), or inhibit neurons (reducing or completely preventing action-potential firing). The key advantages of optogenetics, which we will discuss in detail, are (1) cell-type specificity, the ability to directly target specific populations of neurons based on gene-expression profile, location, and/or projection pattern; (2) temporal specificity, the light-sensitive, membrane-bound proteins utilized in optogenetics respond with millisecond precision; and (3) compatibility with other neuroscience techniques, including patch clamp, extracellular electrophysiology, functional MRI (fMRI), and calcium imaging.

Targeting specificity

Neuroscientists have, for many years, possessed a number of tools to manipulate neural activity. Pharmacological interventions allow for inhibition or modulation of neurons. This manipulation can be systemic, or it can be local, if drugs are infused directly into the brain. Cooling has also been used to inhibit neural activity, and, conversely, electrical microstimulation has been used to enhance the neural signal from the brain region of study (Girard & Bullier, 1989). The spatial extent of these manipulations is poorly defined, however, and pharmacology, in particular, is not temporally precise. Additionally, these techniques are—for the most part—incapable of exciting or inhibiting just one type of neuron in brain regions where many types of neuron are intermingled.

LOCATION SPECIFICITY OF OPTOGENETICS Because opsins are simply proteins, any of the methods generally used for gene delivery can be utilized to express opsins in neurons. The most common technique is virally mediated gene transfer (Yizhar, Fenno, Prigge, et al., 2011). A virus is engineered to carry a genetic payload that includes the opsin gene, expressed under the control of a specific genetic promoter region. In the simplest case, this promoter is chosen because it is active in driving gene expression in all neurons (a panneuronal promoter). The experimenter uses a precisely guided syringe to deliver a small volume of virus (typically less than 1 μL) into the brain at predetermined stereotactic coordinates. In the simplest experimental design, the virus's intended target of infection, or

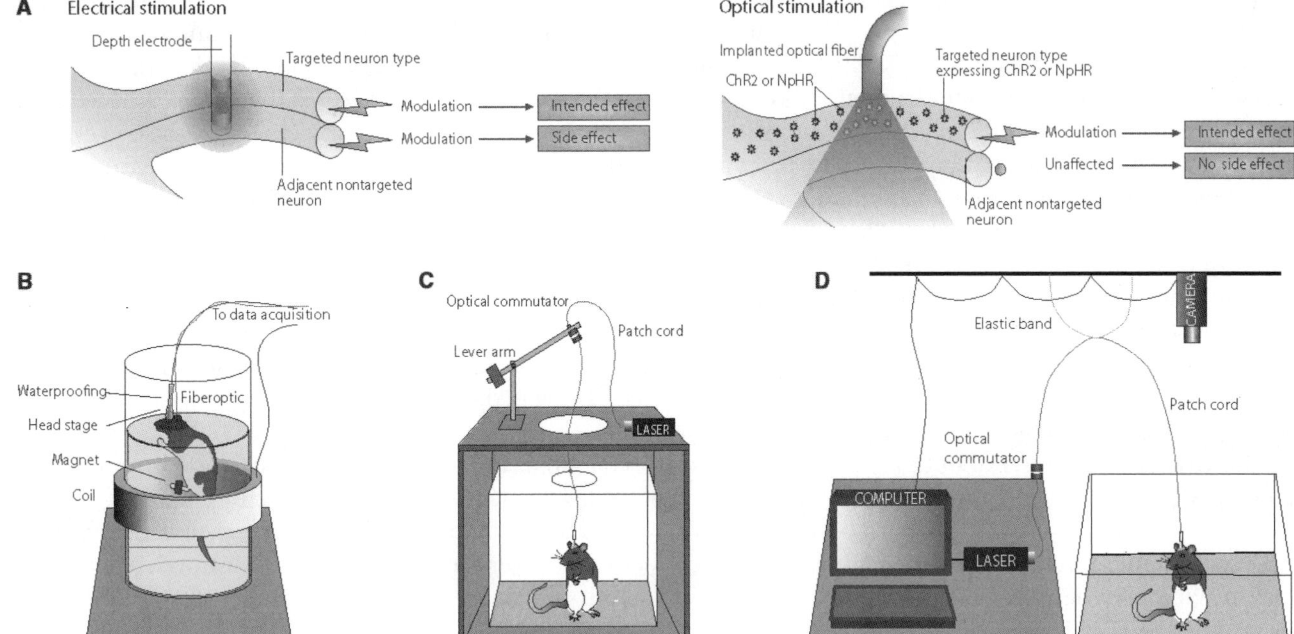

FIGURE 82.2 Integrating optogenetics with behavior. (A) Genetic targeting of channelrhodopsin (ChR2) or halorhodopsin (NpHR) into defined classes of neurons allows cell-specific neuromodulation and avoids inadvertent stimulation of irrelevant circuit elements, as occurs with electrical stimulation. (Adapted with permission from Zhang et al., 2007.) (B) Diverse behavioral rigs can be outfitted for mammalian optogenetic experimentation. The forced swim test has been automated with magnetic induction-based detection of kicks combined with optogenetic stimulation and electrical recording. (C) Operant behavior can also be combined with optogenetics. The chamber itself is modified to accommodate the entry of fiberoptics and recording wires, and the stimulation/recording assembly is kept out of the reach of the rat with a counterweighted lever arm. (D) Optogenetic manipulation can also be combined with behavior in open fields or large mazes. In these experiments, an elastic band, rather than a lever arm, is used to support stimulation/recording equipment. Video recording combined with custom or commercially available software can be used to synchronize optical stimulation with behavior. (Adapted with permission from Zalocusky & Deisseroth, 2013.) (See color plate 85.)

tropism, is all local neuronal cell bodies. Using a virus that infects cell bodies, and a promoter that is active in all neurons, allows the experimenter to control all transduced local neurons—regardless of cell type, genetic profile, or projection pattern.

GENETIC TARGETING Many experiments require specificity of opsin expression based not just on location but also on cellular gene-expression profiles. An experimenter may want to directly control, for example, only those neurons in the brainstem that generate dopamine, or only the excitatory pyramidal cells in a specific region of cortex, in order to determine the specific effects of this population on other cells, physiology, and behavior. This type of specificity can be obtained in certain cases by using cell subtype–specific promoters. Although most such genetic-control regions are too large to package into commonly used viruses, some promoters, including those for serotonin (Benzekhroufa, Liu, Teschemacher, & Kasparov, 2009), CaMKIIα (Lee et al., 2010), and somatostatin (Tan et al., 2008), have been used successfully for this purpose.

A complementary approach (Yizhar, Fenno, Prigge, et al., 2011) involves the use of "recombinase driver" animals (in which a gene encoding a protein such as Cre recombinase is present only in the neuron subtype of interest). In this instance, the researcher would use the same viral-injection approach described above, but would choose a viral payload whose expression is Cre recombinase–dependent—in other words, the virus will infect all local cells, but opsin expression can only occur in cells that also express the Cre recombinase, so that the combination of the transgenic animal and the recombinase-dependent virus allows the researcher to express opsin specifically in those neurons in which Cre recombinase is present. Many lines of transgenic mice (Anikeeva et al., 2012; Carter et al., 2010; Chaudhury et al., 2012; Domingos et al., 2011; Gee et al., 2012; Haubensak et al., 2010; Lammel et al., 2012; Letzkus et al., 2011; Lobo et al., 2010; Olsen, Bortone, Adesnik,

& Scanziani, 2012; Sohal, Zhang, Yizhar, & Deisseroth, 2009; Tan et al., 2012; Threlfell et al., 2012; Tsai et al., 2009; Tye et al., 2012; Witten et al., 2010; Yizhar, Fenno, Prigge, et al., 2011) and a limited number of transgenic rats (Tye et al., 2012; Witten et al., 2011) expressing Cre recombinase under neuron subtype–specific promoters have been used in optogenetics experiments. Because genetic-control regions much larger than those that will fit into a virus can be used when generating a transgenic animal, many more neuronal subtypes can be targeted using this technique.

Projection-Based Targeting If one goal of neuroscience is to understand the form and function of neural circuits, then, in addition to location specificity and cell-type specificity, it is also quite useful to have neural control based on neuronal-projection patterns. An experimenter might be interested, for example, not just in excitatory neurons in primary auditory cortex (A1), but specifically those excitatory neurons in A1 that project to the basal ganglia. This type of specificity is gained primarily by one of two approaches: projection targeting and the use of retrogradely traveling viruses.

Projection targeting (Yizhar, Fenno, Prigge, et al., 2011) is perhaps the most versatile approach to anatomical targeting specificity. Because opsins are trafficked efficiently along neural processes, even distant axons and axon terminals of opsin-expressing neurons become photosensitive. This property enables researchers to transduce neurons with opsin at the site of the cell bodies but illuminate at the site of downstream projection targets—thereby defining the population to be controlled by virtue of (1) cell body location, (2) promoter-expression properties, and (3) axonal-projection pathway. This approach has been widely used in mouse (e.g., Stuber et al., 2011; Threlfell et al., 2012; Tye et al., 2011) and rat (e.g., Gradinaru, Mogri, Thompson, Henderson, & Deisseroth, 2009; Lee et al., 2010; Shabel, Proulx, Trias, Murphy, & Malinow, 2012; Stefanik et al., 2013; Warden et al., 2012; Witten et al., 2011) behavior.

An additional method for targeting a particular neural projection is the use of retrogradely propagating viruses. Take, for example, our hypothetical experimenter, who is interested in basal ganglia–projecting cells in A1. In this instance, she could inject a retrogradely traveling virus carrying the opsin gene into the basal ganglia. The virus would infect the presynaptic terminals there and travel retrogradely to all cells that send projections to that portion of the basal ganglia. By implanting her fiberoptic into A1, and therefore only shining light into A1, the experimenter could directly excite or inhibit only A1 cells that project to the basal ganglia in order to determine the specific effects of this population on circuit dynamics and behavior. Examples of viruses that can transduce axon terminals and give rise to expression in the corresponding cell bodies include herpes simplex virus, rabies virus, and vesicular stomatitis virus (Beier et al., 2011; Osakada et al., 2011; Ugolini, 2010). Unfortunate limitations of most such retrograde strategies include incomplete transduction of local axon terminals and/or toxicity. Further work that will make these approaches more smoothly compatible with optogenetics is needed.

Cortical Layer–Specific Targeting One final targeting strategy is worthy of specific mention, namely, targeting to individual layers of cerebral cortex. Individual layers of cortex have distinct connectivity and functional properties (Hubel & Wiesel, 1974; Szentagothai, 1978) that may help give rise to the hypothesized characteristics of the canonical cortical microcircuit (Douglas, Martin, & Whitteridge, 1989). Optogenetics uniquely enables causal manipulation of these individual layers. There are a small number of transgenic mouse lines that allow for this type of targeting using the virally mediated methods discussed previously (Letzkus et al., 2011; Olsen et al., 2012). A potentially more generalizable method, though experimentally difficult, is in-utero electroporation. In-utero electroporation involves injecting a small amount of DNA (which is inherently negatively charged) into the cerebral ventricles of a developing embryo, then using a small amount of electrical current to move that DNA into the developing cortex (Saito, 2006). Because the cortex develops one layer at a time, the timing of the electroporation can be chosen such that a particular cortical layer is transduced with DNA. This strategy enables anatomical layer–specific opsin expression with no requirement for the neurons' genetic identity at all. Here, again, a cautionary note is that transduction can be so strong that toxicity linked to protein overexpression can be seen, and careful titration of experimental parameters and timing is needed.

Temporal specificity

Optogenetic proteins like the channelrhodopsins are capable of responding to pulses of light with millisecond precision (Yizhar, Fenno, Prigge, et al., 2011). Optogenetics, therefore, uniquely enables temporal specificity in neural excitation or inhibition with respect to a behavioral task or a recorded neural signal. By shining light only during a specific aspect of a behavior—for example, when a rat is in the left half of a maze, or in the half-second after a mouse presses a lever—experimenters can precisely determine the causal role of neural signals during behavior. Optogenetic stimulation triggered on an animal's behavior was first successfully used in 2009 in what has now come to be known as the real-time, or dynamic, conditioned place preference paradigm (Airan et al., 2009). Since then, this type of behavior-triggered stimulation has been used to examine such phenomena as sleep-wake transitions (Carter et al., 2010), intracranial self-stimulation (Witten et al., 2011), and addiction (Witten et al., 2010).

An additional and particularly exciting use of this temporal specificity is optical stimulation triggered on neural events. Electrophysiology has generated many interesting hypotheses about the function of particular neural signals. Detecting and disrupting these signals in real time, in the precise location they are recorded, will allow researchers to test the role of these neural events. In a recent study of focal cortical epilepsy, researchers succeeded in detecting seizure activity, in real time, using electroencephalography, and in terminating those seizures, again in real time, by initiating optogenetic inhibition of thalamocortical neurons whenever a seizure was detected (Paz et al., 2013). In addition to its therapeutic potential, this study defined a role for thalamocortically projecting neurons in this epileptiform activity.

Integration with existing techniques

Functional Magnetic Resonance Imaging Functional MRI has enabled acquisition of useful and interesting data, both regarding localization of function and description of whole-brain networks recruited during complex human behaviors. The fMRI-measured blood oxygenation level–dependent (BOLD) signal, however, is an indirect and temporally complex measure of brain metabolism (Logothetis & Wandell, 2004). Until recently, there had been no way to causally test which circuit elements could give rise to the complex BOLD signal dynamics. In 2010, it was demonstrated that optogenetically eliciting action potentials in principal cell bodies is sufficient to initiate a robust, local, positive BOLD signal that recapitulates key temporal aspects of BOLD responses seen in humans (Lee et al., 2010). In addition to generating complex local BOLD signals, this optical stimulation also elicited BOLD in anatomically connected areas across the brain. These findings have now been borne out by others, who have also highlighted the importance of local action potentials in dictating BOLD signal characteristics (Desai et al., 2011; Kahn et al., 2011, 2013; Lee et al., 2010). As methods for fMRI improve, we anticipate increased collaboration and feedback between cognitive psychology and neuroscience, which will accelerate our ability to generate and causally test hypotheses about behavior in awake and behaving animals, both in normal functioning and in disease models.

Electrophysiology Systems electrophysiology, in which researchers use microelectrodes to record the activity of individual neurons in vivo, has delivered for cognitive neuroscience a great deal of elegant and detailed information about activity of neural-circuit elements that correlate with behavior. In these experiments, researchers typically record from neurons in a particular brain region while the experimental subject performs some repeated behavior. In some cases, by averaging the signals recorded over many repetitions, researchers can determine how the neurons in a particular region typically respond during that behavior. Such work (for example, in studies of rodent spatial navigation, primate visual processing, and primate reward processing; Moser, Kropff, & Moser, 2008; Schultz, 1998; Shadlen & Newsome, 2001) has been conducted in rodents, nonhuman primates, and even a select few human patients.

Limitations of classical electrophysiology include the correlative (noncausal) nature of the data obtained. Moreover, these data generally arise from a small number of individual neurons. To begin to demonstrate causality, many experiments have utilized pharmacological interventions. Again, while useful, such manipulations are by nature imprecise in terms of time, space, and even target. Electrical stimulation is also commonly used, but electrical stimulation and electrophysiological recording are not smoothly compatible, since such stimulation generates massive electrical artifacts in the recordings. Optogenetic stimulation, however, can allow simultaneous in vivo neural manipulation and recording in real time (Gradinaru et al., 2007), allowing researchers to directly observe the electrical dynamics of the circuit during stimulation and behavior.

Another type of limitation associated with classical electrophysiology was also highlighted by Crick: "There is no indication where these different sets of neurons are sending their information, let alone exactly what type of neuron they are ... in the long run it will be essential to know which type of neuron the electrode is recording from. This problem deserves immediate and serious attention" (Crick, 1999, p. 2024). Because the only information gleaned from recorded neurons is electrical activity, it is frequently difficult, if not impossible, to discern corresponding projection or gene-expression patterns. This type of information will be crucial, however, to mapping the function of complete neural networks and describing the types of distributed calculations that likely underlie interesting and complex behaviors.

This class of problems is beginning to be addressed, at least in some circuits, through the use of optogenetics for "optical tagging" (Cohen, Haesler, Vong, Lowell, & Uchida, 2012). Optical tagging is simultaneous electrical recording and optogenetic stimulation to determine the opsin-expression status of a recorded cell. If spikes (exhibiting a characteristic electrical waveform) are directly elicited with a short enough latency after light pulses to exclude an indirect or synaptic excitation mechanism, spikes observed with that characteristic waveform thereafter in the absence of light can be inferred to be derived from an opsin-expressing cell. This approach is not without controversy, and it remains unclear how certain one can be of direct excitation in all circuits. But distinct from electrophysiology, much information on the optogenetically controlled cells can be obtained through cell-filling labels with post hoc subsequent histological or CLARITY-based analysis (Chung et al., 2013) for molecular and anatomical phenotyping of the opsin-expressing cells in that particular organism.

Experimental examples

Optogenetic research has now influenced many fields of cognitive neuroscience. Here we provide brief descriptions of recent insights in two fields, memory and reward, to illustrate the process and behavioral outcomes.

Memory Results from patients have long suggested that recent memories are, in some way, more vulnerable than memories that were laid down in the distant past. Patients with lesions to the hippocampus are often found to have lost recent memories and to be impaired in forming new ones. Interestingly, in many such patients, memories from the distant past, termed "remote memories," are more intact (Ribot, 1881). This clinical evidence, in addition to evidence from experimental lesions in animal models (Anagnostaras, Maren, & Fanselow, 1999; Bontempi, Laurent-Demir, Destrade, & Jaffard, 1999), led to the theory that, over time, memories become distributed in cortex and independent of the hippocampus (the consolidation theory of memory). Yet in other instances, hippocampal lesions have led to complete, temporally flat amnesia, where all memories are lost regardless of whether they were recent or remote (Broadbent, Squire, & Clark, 2006; Riedel et al., 1999). These conflicting data gave rise to distinct theories of memory, including the multiple-trace theory, which posits dynamic interplay between hippocampus and cortex during the retrieval of even very remote memories (Nadel & Moscovitch, 1997).

One potential confound in previous studies of memory, however, was that they were conducted after physical lesions—which can be studied days, months, or even years after they occur—or after pharmacologic lesions, which, while shorter in time scale, still last for hours. During that time, compensatory plasticity could have changed how the memory was stored or retrieved. Goshen and colleagues used the temporal specificity of optogenetics to inhibit principal cells in hippocampus exactly during the memory test (Goshen et al., 2011) in mice. They found that if these hippocampal cells were inhibited only during the 5-minute memory test, ability to recall a remote memory was greatly diminished. If the researchers instead inhibited these same cells for 30 minutes (25 minutes before the test and during the test itself), then the mice were able to recall the remote memory. The authors checked for and found no nonspecific effects of the brief remote inhibition on global brain activity patterns; for example, no effects on nonhippocampal (cued) memories were seen; no changes in baseline or evoked neural activity were observed across the brain in response to this precise hippocampal inhibition as assessed by immediate-early gene expression (c-fos) mapping; and no effects on memory were seen when another major source of synaptic drive to the forebrain, the olfactory bulb, was instead silenced. Together, this evidence suggests the presence of compensatory mechanisms for memory retrieval, as well as a model for memory that is still substantially hippocampus-dependent in the default state, even for older, remote memories.

Reward Early research into the neural substrates of reward included studies of intracranial self-stimulation (ICSS) in rats (Olds & Milner, 1954). This research demonstrated that rats will work for direct electrical stimulation of certain brain regions, and provided the first clues regarding which brain structures mediate behavioral reinforcement. Sites effective for ICSS strongly overlap with major projections of the dopamine (DA) system (Corbett & Wise, 1980), and lesions of DA neurons, or drugs that block DA neurotransmission, diminish ICSS behavior (Fibiger, LePiane, Jakubovic, & Phillips, 1987; German & Bowden, 1974), together suggesting that DA mediates sustained ICSS. Electrical stimulation recruits a spatially complex population of neurons (Histed, Bonin, & Reid, 2009;

Logothetis et al., 2010), however, and not all sites that support ICSS receive dense DA innervation (Wise & Rompre, 1989). Replacing classical electrical stimulation of the ventral tegmental area (VTA) with optogenetic stimulation specific to local dopamine-expressing neurons, a recent study found that phasic optical timulation of VTA-DA neurons was sufficient for the acquisition and maintenance of robust ICSS behavior. Additionally, the authors found that stimulation of the dopaminergic VTA projection to the nucleus accumbens was itself sufficient to support ICSS, in a study integrating rat projection targeting with generation of rat transgenic lines for optogenetics (Witten et al., 2011).

Over the last two decades, it has been widely hypothesized that DA signals could also be used for learning and for reward prediction (Fiorillo, Newsome, & Schultz, 2008; Ljungberg, Apicella, & Schultz, 1992; Mirenowicz & Schultz, 1996; Schultz, Dayan, & Montague, 1997; Schultz, 1998, 2002; Tobler, Fiorillo, & Schultz, 2005), and it has been postulated that the firing of DA neurons represents the reward-prediction error of classical temporal difference learning (Schultz, Dayan, & Montague, 1997; Schultz, 2002). Until recently, however, researchers have been unable to manipulate DA neurons with the temporal specificity needed to address the role of DA in these learning models. Current research utilizing temporally precise optogenetic excitation or inhibition has now succeeded in directly testing and confirming key aspects of these long-held predictions surrounding dopamine signaling, learning, and reward (Steinberg et al., 2013), and in extending reward classes studied to include feeding (Domingos et al., 2011), cocaine (Witten et al., 2010), sexual behavior (discussed below), and other naturalistic appetitive stimuli.

Sex and Violence Multiple experiments across diverse species have implicated a subregion of the hypothalamus, termed the ventromedial hypothalamus (VMH), in both mating and attack behavior. In fact, studies measuring genetic markers of recent neural activity (immediate early genes) have shown increased neural activity in the VMH after both mating and attack (Veening et al., 2005). And remarkably, electrical stimulation in this region is capable of eliciting either attack or mating behavior in laboratory rats (Lammers, Kruk, Meelis, & Van der Poel, 1988; Pfaff & Sakuma, 1979). As previously discussed, however, it is difficult to be certain of the spatial extent or target cell type of electrical stimulation (Histed, Bonin, & Reid, 2009), and the VMH contains multiple types of neurons which may each have distinct behavioral effects.

In a recent study, researchers expressed channelrhodopsin-2, an excitatory opsin, in just one subnucleus of the VMH (the ventrolateral subportion of VMH, or VMHvl). By directly exciting specifically this subpopulation of neurons in the VMH to elicit effects on downstream connected circuits and behavior, the authors were able to elicit intense aggression, not just against male intruder mice, but also against female mice (who typically elicit mating rather than attack behavior) and even against inanimate objects (Lin et al., 2011). Inhibition of this same subpopulation decreased propensity to attack. Additionally, by recording from individual VMHvl neurons during both attack and sexual behavior, the authors showed that neurons that were excited by attack were often progressively inhibited over the course of a sexual encounter. This finding on neural activity was corroborated in behavior; if the researchers optically stimulated the VMHvl *during* a sexual encounter, the previously observed indiscriminate aggressive behaviors were no longer present. In this class of experiment, the location-specificity and the temporal specificity of optogenetics can be integrated to address a long-standing behavioral and neuroanatomical puzzle.

Future directions

Here we have outlined the basic structure and tools of optogenetic experimentation that may be applied to cognitive neuroscience. We expect this field of neuroscience to continue to grow rapidly, and we expect increased collaboration between psychologists studying human behavior and neuroscientists who use optogenetics to causally test hypotheses about neural mechanisms. Together with associated enabling technologies, optogenetics will likely play a crucial role in contributing to our deepening understanding of how diverse classes of neural circuit components interact to give rise to complex behaviors, pathological conditions, and therapeutic responses.

REFERENCES

Airan, R. D., Thompson, K. R., Fenno, L. E., Bernstein, H., & Deisseroth, K. (2009). Temporally precise in vivo control of intracellular signalling. *Nature, 458*(7241), 1025–1029.

Anagnostaras, S. G., Maren, S., & Fanselow, M. S. (1999). Temporally graded retrograde amnesia of contextual fear after hippocampal damage in rats: Within-subjects examination. *J Neurosci, 19*(3), 1106–1114.

Anikeeva, P., Andalman, A. S., Witten, I., Warden, M., Goshen, I., Grosenick, L., ... Deisseroth, K. (2012). Optetrode: A multichannel readout for optogenetic control in freely moving mice. *Nat Neurosci, 15*(1), 163–170.

Beier, K. T., Saunders, A., Oldenburg, I. A., Miyamichi, K., Akhtar, N., Luo, L., … Cepko, C. L. (2011). Anterograde or retrograde transsynaptic labeling of CNS neurons with vesicular stomatitis virus vectors. *Proc Natl Acad Sci USA, 108*(37), 15414–15419.

Benzekhroufa, K., Liu, B.-H., Teschemacher, A. G., & Kasparov, S. (2009). Targeting central serotonergic neurons with lentiviral vectors based on a transcriptional amplification strategy. *Gene Therapy, 16*(5), 681–688.

Berndt, A., Yizhar, O., Gunaydin, L. A., Hegemann, P., & Deisseroth, K. (2009). Bi-stable neural state switches. *Nat Neurosci, 12*(2), 229–234.

Bontempi, B., Laurent-Demir, C., Destrade, C., & Jaffard, R. (1999). Time-dependent reorganization of brain circuitry underlying long-term memory storage. *Nature, 400*(6745), 671–675.

Boyden, E. S., Zhang, F., Bamberg, E., Nagel, G., & Deisseroth, K. (2005). Millisecond-timescale, genetically targeted optical control of neural activity. *Nat Neurosci, 8*(9), 1263–1268.

Broadbent, N. J., Squire, L. R., & Clark, R. E. (2006). Reversible hippocampal lesions disrupt water maze performance during both recent and remote memory tests. *Learn Mem, 13*(2), 187–191.

Carter, M. E., Yizhar, O., Chikahisa, S., Nguyen, H., Adamantidis, A., Nishino, S., … de Lecea, L. (2010). Tuning arousal with optogenetic modulation of locus coeruleus neurons. *Nat Neurosci, 13*(12), 1526–1533.

Chaudhury, D., Walsh, J. J., Friedman, A. K., Juarez, B., Ku, S. M., Koo, J. W., … Han, M.-H. (2012). Rapid regulation of depression-related behaviours by control of midbrain dopamine neurons. *Nature, 493*(7433), 532–536.

Chow, B. Y., Han, X., Dobry, A. S., Qian, X., Chuong, A. S., Li, M., … Boyden, E. S. (2010). High-performance genetically targetable optical neural silencing by light-driven proton pumps. *Nature, 463*(7277), 98–102.

Chung, K., Wallace, J., Kim, S. Y., Kalyanasundaram, S., Andalman, A. S., Davidson, T. J., … Deisseroth, K. (2013). Structural and molecular interrogation of intact biological systems. *Nature, 497*(7449), 332–337.

Cohen, J. Y., Haesler, S., Vong, L., Lowell, B. B., & Uchida, N. (2012). Neuron-type-specific signals for reward and punishment in the ventral tegmental area. *Nature, 482*(7383), 85–88.

Corbett, D., & Wise, R. A. (1980). Intracranial self-stimulation in relation to the ascending dopaminergic systems of the midbrain: A moveable electrode mapping study. *Brain Res, 185*, 1–15.

Crick, F. (1999). The impact of molecular biology on neuroscience. *Philos Trans R Soc Lond B Biol Sci, 354*(1392), 2021–2025.

Desai, M., Kahn, I., Knoblich, U., Bernstein, J., Atallah, H., Yang, A., … Boyden, E. S. (2011). Mapping brain networks in awake mice using combined optical neural control and fMRI. *J Neurophysiol, 105*(3), 1393–1405.

Domingos, A. I., Vaynshteyn, J., Voss, H. U., Ren, X., Gradinaru, V., Zang, F., … Friedman, J. (2011). Leptin regulates the reward value of nutrient. *Nat Neurosci, 14*(12), 1562–1568.

Douglas, R. J., Martin, K. A. C., & Whitteridge, D. (1989). A canonical microcircuit for neocortex. *Neural Comput, 1*(4), 480–488.

Fenno, L. E., Airan, R. D., Bernstein, H., & Deisseroth, K. (2009). Optical control of dopamine receptor signaling. Paper presented at the meeting of the Society for Neuroscience, Chicago, IL.

Fibiger, H. C., LePiane, F. G., Jakubovic, A., & Phillips, A. G. (1987). The role of dopamine in intracranial self-stimulation of the ventral tegmental area. *J Neurosci, 7*(12), 3888–3896.

Fiorillo, C. D., Newsome, W. T., & Schultz, W. (2008). The temporal precision of reward prediction in dopamine neurons. *Nat Neurosci, 11*(8), 966–973.

Gee, S., Ellwood, I., Patel, T., Luongo, F., Deisseroth, K., & Sohal, V. S. (2012). Synaptic activity unmasks dopamine D2 receptor modulation of a specific class of layer V pyramidal neurons in prefrontal cortex. *J Neurosci, 32*(14), 4959–4971.

German, D. C., & Bowden, D. M. (1974). Catecholamine systems as the neural substrate for intracranial self-stimulation: A hypothesis. *Brain Res, 73*(3), 381–419.

Girard, P., & Bullier, J. (1989). Visual activity in area V2 during reversible inactivation of area 17 in the macaque monkey. *J Neurophysiol, 62*(6), 1287–1302.

Goshen, I., Brodsky, M., Prakash, R., Wallace, J., Gradinaru, V., Ramakrishnan, C., & Deisseroth, K. (2011). Dynamics of retrieval strategies for remote memories. *Cell, 147*(3), 678–689.

Gradinaru, V., Mogri, M., Thompson, K. R., Henderson, J. M., & Deisseroth, K. (2009). Optical deconstruction of parkinsonian neural circuitry. *Science, 324*(5925), 354–359.

Gradinaru, V., Thompson, K. R., Zhang, F., Mogri, M., Kay, K., Schneider, M. B., & Deisseroth, K. (2007). Targeting and readout strategies for fast optical neural control in vitro and in vivo. *J Neurosci, 27*(52), 14231–14238.

Gradinaru, V., Zhang, F., Ramakrishnan, C., Mattis, J., Prakash, R., Diester, I., … Deisseroth, K. (2010). Molecular and cellular approaches for diversifying and extending optogenetics. *Cell, 141*(1), 154–165.

Gunaydin, L. A., Yizhar, O., Berndt, A., Sohal, V. S., Deisseroth, K., & Hegemann, P. (2010). Ultrafast optogenetic control. *Nat Neurosci, 13*(3), 387–392.

Haubensak, W., Kunwar, P. S., Cai, H., Ciocchi, S., Wall, N. R., Ponnusamy, R., … Anderson, D. J. (2010). Genetic dissection of an amygdala microcircuit that gates conditioned fear. *Nature, 468*(7321), 270–276.

Histed, M. H., Bonin, V., & Reid, R. C. (2009). Direct activation of sparse, distributed populations of cortical neurons by electrical microstimulation. *Neuron, 63*(4), 508–522.

Hubel, D. H., & Wiesel, T. N. (1974). Uniformity of monkey striate cortex: A parallel relationship between field size, scatter, and magnification factor. *J Comp Neurol, 158*(3), 295–305.

Ihara, K., Umemura, T., Katagiri, I., Kitajima-Ihara, T., Sugiyama, Y., Kimura, Y., & Mukohata, Y. (1999). Evolution of the archaeal rhodopsins: Evolution rate changes by gene duplication and functional differentiation. *J Mol Biol, 285*(1), 163–174.

Kahn, I., Desai, M., Knoblich, U., Bernstein, J., Henninger, M., Graybiel, A. M., … Moore, C. I. (2011). Characterization of the functional MRI response temporal linearity via optical control of neocortical pyramidal neurons. *J Neurosci, 31*(42), 15086–15091.

Kahn, I., Knoblich, U., Desai, M., Bernstein, J., Graybiel, A. M., Boyden, E. S., … Moore, C. I. (2013). Optogenetic drive of neocortical pyramidal neurons generates fMRI signals that are correlated with spiking activity. *Brain Res, 1511*, 33-45.

Lammel, S., Lim, B. K., Ran, C., Huang, K. W., Betley, M. J., Tye, K. M., … Malenka, R. C. (2012). Input-specific control of reward and aversion in the ventral tegmental area. *Nature, 491*(7423), 212–217.

Lammers, J. H. C. M., Kruk, M. R., Meelis, W., & van der Poel, A. M. (1988). Hypothalamic substrates for brain stimulation-induced attack, teeth-chattering and social grooming in the rat. *Brain Res, 449*(1–2), 311–327.

Lanyi, J. K., & Oesterhelt, D. (1982). Identification of the retinal-binding protein in halorhodopsin. *J Biol Chem, 257*(5), 2674–2677.

Lee, J. H., Durand, R., Gradinaru, V., Zhang, F., Goshen, I., Kim, D.-S., … Deisseroth, K. (2010). Global and local fMRI signals driven by neurons defined optogenetically by type and wiring. *Nature, 465*(7299), 788–792.

Letzkus, J. J., Wolff, S. B. E., Meyer, E. M. M., Tovote, P., Courtin, J., Herry, C., & Lüthi, A. (2011). A disinhibitory microcircuit for associative fear learning in the auditory cortex. *Nature, 480*(7377), 331–335.

Levitz, J., Pantoja, C., Gaub, B., Janovjak, H., Reiner, A., Hoagland, A., … Isacoff, E. Y. (2013). Optical control of metabotropic glutamate receptors. *Nat Neurosci, 16*(4), 507–516.

Lin, D., Boyle, M. P., Dollar, P., Lee, H., Lein, E. S., Perona, P., & Anderson, D. J. (2011). Functional identification of an aggression locus in the mouse hypothalamus. *Nature, 470*(7333), 221–226.

Ljungberg, T., Apicella, P., & Schultz, W. (1992). Responses of monkey dopamine neurons during learning of behavioral reactions. *J Neurophysiol, 67*(1), 145–163.

Lobo, M. K., Covington, H. E., Chaudhury, D., Friedman, A. K., Sun, H., Damez-Werno, D., … Nestler, E. J. (2010). Cell type specific loss of BDNF signaling mimics optogenetic control of cocaine reward. *Science, 330*(6002), 385–390.

Logothetis, N. K., Augath, M., Murayama, Y., Rauch, A., Sultan, F., Goense, J., … Merkle, H. (2010). The effects of electrical microstimulation on cortical signal propagation. *Nat Neurosci, 13*(10), 1283–1291.

Logothetis, N. K., & Wandell, B. A. (2004). Interpreting the BOLD Signal. *Annu Rev Physiol, 66*(1), 735–769.

Matsuno-Yagi, A., & Mukohata, Y. (1977). Two possible roles of bacteriorhodopsin; a comparative study of strains of *Halobacterium halobium* differing in pigmentation. *Biochem Biophys Res Comm, 78*(1), 237–243.

Mattis, J., Tye, K. M., Ferenczi, E. A., Ramakrishnan, C., O'Shea, D. J., Prakash, R., … Deisseroth, K. (2012). Principles for applying optogenetic tools derived from direct comparative analysis of microbial opsins. *Nat Methods, 9*(2), 159–172.

Mirenowicz, J., & Schultz, W. (1996). Preferential activation of midbrain dopamine neurons by appetitive rather than aversive stimuli. *Nature, 379*(6564), 449–451.

Moser, E. I., Kropff, E., & Moser, M.-B. (2008). Place cells, grid cells, and the brain's spatial representation system. *Annu Rev Neurosci, 31*(1), 69–89.

Nadel, L., & Moscovitch, M. (1997). Memory consolidation, retrograde amnesia and the hippocampal complex. *Curr Opin Neurobiol, 7*(2), 217–227.

Oesterhelt, D., & Stoeckenius, W. (1971). Rhodopsin-like protein from the purple membrane of *Halobacterium halobium. Nature, 233*(39), 149–152.

Oesterhelt, D., & Stoeckenius, W. (1973). Functions of a new photoreceptor membrane. *Proc Natl Acad Sci USA, 70*(10), 2853–2857.

Oh, E., Maejima, T., Liu, C., Deneris, E., & Herlitze, S. (2010). Substitution of 5-HT1A receptor signaling by a light-activated G protein-coupled receptor. *J Biol Chem, 285*(40), 30825–30836.

Olds, J., & Milner, P. (1954). Positive reinforcement produced by electrical stimulation of septal area and other regions of rat brain. *J Comp Physiol Psychol, 47*(6), 419–427.

Olsen, S. R., Bortone, D. S., Adesnik, H., & Scanziani, M. (2012). Gain control by layer six in cortical circuits of vision. *Nature, 483*(7387), 47–52.

Osakada, F., Mori, T., Cetin, A. H., Marshel, J. H., Virgen, B., & Callaway, E. M. (2011). New rabies virus variants for monitoring and manipulating activity and gene expression in defined neural circuits. *Neuron, 71*(4), 617–631.

Paz, J. T., Davidson, T. J., Frechette, E. S., Delord, B., Parada, I., Peng, K., … Huguenard, J. R. (2013). Closed-loop optogenetic control of thalamus as a tool for interrupting seizures after cortical injury. *Nat Neurosci, 16*(1), 64–70.

Pfaff, D. W., & Sakuma, Y. (1979). Facilitation of the lordosis reflex of female rats from the ventromedial nucleus of the hypothalamus. *J Physiol, 288*, 189–202.

Ribot, T. (1881). *Les maladies de la mémoire*. Paris: L'Harmattan.

Riedel, G., Micheau, J., Lam, A. G., Roloff, E. L., Martin, S. J., Bridge, H., … Morris, R. G. (1999). Reversible neural inactivation reveals hippocampal participation in several memory processes. *Nat Neurosci, 2*(10), 898–905.

Saito, T. (2006). In vivo electroporation in the embryonic mouse central nervous system. *Nat Protoc, 1*(3), 1552–1558.

Schultz, W. (1998). Predictive reward signal of dopamine neurons. *J Neurophysiol, 80*(1), 1–27.

Schultz, W. (2002). Getting formal with dopamine and reward. *Neuron, 36*(2), 241–263.

Schultz, W., Dayan, P., & Montague, P. R. (1997). A neural substrate of prediction and reward. *Science, 275*(5306), 1593–1599.

Shabel, S. J., Proulx, C. D., Trias, A., Murphy, R. T., & Malinow, R. (2012). Input to the lateral habenula from the basal ganglia is excitatory, aversive, and suppressed by serotonin. *Neuron, 74*(3), 475–481.

Shadlen, M. N., & Newsome, W. T. (2001). Neural basis of a perceptual decision in the parietal cortex (area LIP) of the rhesus monkey. *J Neurophysiol, 86*(4), 1916–1936.

Sohal, V. S., Zhang, F., Yizhar, O., & Deisseroth, K. (2009). Parvalbumin neurons and gamma rhythms enhance cortical circuit performance. *Nature, 459*(7247), 698–702.

Stefanik, M. T., Moussawi, K., Kupchik, Y. M., Smith, K. C., Miller, R. L., Huff, M. L., … LaLumiere, R. T. (2013). Optogenetic inhibition of cocaine seeking in rats. *Addict Biol, 18*(1), 50–53.

Steinberg, E. E., Keiflin, R., Boivin, J. R., Witten, I. B., Deisseroth, K., & Janak, P. H. (2013). A causal link between prediction errors, dopamine neurons and learning. *Nat Neurosci, 16*, 966–973.

Stuber, G. D., Sparta, D. R., Stamatakis, A. M., van Leeuwen, W. A., Hardjoprajitno, J. E., Cho, S., … Bonci, A. (2011). Excitatory transmission from the amygdala to nucleus accumbens facilitates reward seeking. *Nature, 475*(7356), 377–380.

Szentagothai, J. (1978). The Ferrier lecture, 1977. The neuron network of the cerebral cortex: A functional interpretation. *Proc R Soc Lond B Biol Sci, 201*(1144), 219–248.

Tan, K. R., Yvon, C., Turiault, M., Mirzabekov, J. J., Doehner, J., Labouèbe, G., ... Lüscher, C. (2012). GABA neurons of the VTA drive conditioned place aversion. *Neuron, 73*(6), 1173–1183.

Tan, W., Janczewski, W. A., Yang, P., Shao, X. M., Callaway, E. M., & Feldman, J. L. (2008). Silencing preBötzinger Complex somatostatin-expressing neurons induces persistent apnea in awake rat. *Nat Neurosci, 11*(5), 538–540.

Threlfell, S., Lalic, T., Platt, N. J., Jennings, K. A., Deisseroth, K., & Cragg, S. J. (2012). Striatal dopamine release is triggered by synchronized activity in cholinergic interneurons. *Neuron, 75*(1), 58–64.

Tobler, P. N., Fiorillo, C. D., & Schultz, W. (2005). Adaptive coding of reward value by dopamine neurons. *Science, 307*, 1642–1645

Tsai, H.-C., Zhang, F., Adamantidis, A., Stuber, G. D., Bonci, A., de Lecea, L., & Deisseroth, K. (2009). Phasic firing in dopaminergic neurons is sufficient for behavioral conditioning. *Science, 324*(5930), 1080–1084.

Tye, K. M., & Deisseroth, K. (2012). Optogenetic investigation of neural circuits underlying brain disease in animal models. *Nat Rev Neurosci, 13*, 251–266.

Tye, K. M., Mirzabekov, J. J., Warden, M. R., Ferenczi, E. A., Tsai, H.-C., Finkelstein, J., ... Deisseroth, K. (2012). Dopamine neurons modulate neural encoding and expression of depression-related behaviour. *Nature, 493*(7433), 537–541.

Tye, K. M., Prakash, R., Kim, S.-Y., Fenno, L. E., Grosenick, L., Zarabi, H., ... Deisseroth, K. (2011). Amygdala circuitry mediating reversible and bidirectional control of anxiety. *Nature, 471*(7338), 358–362.

Ugolini, G. (2010). Advances in viral transneuronal tracing. *J Neurosci Methods, 194*(1), 2–20.

Veening, J. G., Coolen, L. M., de Jong, T. R., Joosten, H. W., de Boer, S. F., Koolhaas, J. M., & Olivier, B. (2005). Do similar neural systems subserve aggressive and sexual behaviour in male rats? Insights from c-Fos and pharmacological studies. *Eur J Pharmacol, 526*(1–3), 226–239.

Warden, M. R., Selimbeyoglu, A., Mirzabekov, J. J., Lo, M., Thompson, K. R., Kim, S.-Y., ... Deisseroth, K. (2012). A prefrontal cortex-brainstem neuronal projection that controls response to behavioural challenge. *Nature, 492*(7429), 428–432.

Wise, R. A., & Rompre, P. P. (1989). Brain dopamine and reward. *Annu Rev Psychol, 40*(1), 191–225.

Witten, I. B., Lin, S.-C., Brodsky, M., Prakash, R., Diester, I., Anikeeva, P., ... Deisseroth, K. (2010). Cholinergic interneurons control local circuit activity and cocaine conditioning. *Science, 330*(6011), 1677–1681.

Witten, I. B., Steinberg, E. E., Lee, S. Y., Davidson, T. J., Zalocusky, K. A., Brodsky, M., ... Deisseroth, K. (2011). Recombinase-driver rat lines: Tools, techniques, and optogenetic application to dopamine-mediated reinforcement. *Neuron, 72*(5), 721–733.

Yizhar, O., Fenno, L. E., Davidson, T. J., Mogri, M., & Deisseroth, K. (2011). Optogenetics in neural systems. *Neuron, 71*(1), 9–34.

Yizhar, O., Fenno, L. E., Prigge, M., Schneider, F., Davidson, T. J., O'Shea, D. J., ... Deisseroth, K. (2011). Neocortical excitation/inhibition balance in information processing and social dysfunction. *Nature, 477*(7363), 171–178.

Zalocusky, K., & Deisseroth, K. (2013). Optogenetics in the behaving rat: Integration of diverse new technologies in a vital animal model. *Optogenetics*, 2013, 1–17.

Zhang F., Aravanis, A. M., Adamantidis, A., de Lecea, L., & Deisseroth, K. (2007). Circuit-breakers: Optical technologies for probing neural signals and systems. *Nat Rev Neurosci, 8*(8), 577–581.

Zhang, F., Vierock, J., Yizhar, O., Fenno, L. E., Tsunoda, S., Kianianmomeni, A., ... Deisseroth, K. (2011). The microbial opsin family of optogenetic tools. *Cell, 147*(7), 1446–1457.

83 Neuroimaging Informatics: Tools to Manage and Share Neuroimaging and Related Data

DANIEL MARCUS

ABSTRACT Neuroimaging informatics is the set of data-centric technologies used to enable the practice of neuroimaging-based science. It includes technologies like data management platforms, databases of stored data and knowledge, data structures for representing images and other data, and a variety of software applications to achieve everything from data capture to high-throughput analysis to data mining. In this chapter, I review the current state of the art in neuroimaging informatics, frequently using the XNAT informatics platform and Human Connectome Project as examples.

Why neuroimaging informatics?

The Human Connectome Project (HCP), a National Institutes of Health–funded project to generate connectivity maps of the human brain using state-of-the-art neuroimaging methods, will study 1,200 individuals over a five-year period, including collecting structural, functional, and diffusion magnetic resonance imaging (MRI), extensive behavioral and cognitive phenotypic data, and next-generation genomics. While the HCP is currently at the cutting edge of neuroimaging capabilities, its methods are quickly being adopted and incorporated into a broad range of other studies. It therefore serves as a useful example of the sort of data challenges that large-scale neuroimaging studies encounter:

1. *Big data.* For each HCP subject, the acquired imaging data is over 10 GB, the preprocessed data is over 12 GB, and the dense connectome data is over 50 GB (Marcus et al., 2013). In total, the HCP data set is expected to exceed 1,000 TB, or the equivalent of over 220,000 DVDs.

2. *Multimodal data.* The HCP imaging protocol includes T1- and T2-weighted structural scans, multiple resting-state functional MRI (fMRI) scans, eight separate task fMRI scans, and diffusion imaging. The scans are acquired in four to five separate imaging sessions (Van Essen et al., 2013). During preprocessing, the modalities are spatially coregistered, and many analytic approaches will synthesize across modalities.

3. *Nonimaging measures.* The HCP behavioral battery includes dozens of assessments and tests, covering cognitive, emotional, and sensory domains (Barch et al., 2013). The battery is acquired on several different computerized testing platforms, each using its own proprietary data format.

4. *Extensive image processing.* The acquired imaging data is processed following a standardized sequence of steps that includes coregistration, distortion correction, denoising, and surface reconstruction (Glasser et al., 2013). On the HCP's compute cluster, each subject's preprocessing requires over 24 hours to execute.

5. *Sensitive subject information.* The high-resolution images acquired by the HCP contain facial characteristics of study participants that may be considered personal health information (PHI) (Chen et al., 2007). In addition, many of the behavioral measures may be considered highly personal by subjects and their families, including drug use, pregnancy status, and psychiatric traits. Human subject protections and federal Health Insurance Portability and Accountability Act (HIPAA) regulations require subject privacy and confidentiality to be maintained when distributing these data.

6. *Collaboration.* The HCP research team consists of 100 personnel from 100 institutions in five countries. In order for this collaborative team to work effectively, the data must be made securely accessible across a geographically dispersed network.

7. *Data sharing.* The HCP grant carries a mandate from the NIH to proactively share its data as openly as possible with the international neuroimaging community.

Given these challenges, not surprisingly, a number of informatics systems have emerged to provide data management, security, sharing, and workflow solutions (Das, Zijdenbos, Harlap, Vins, & Evans, 2011; Ozyurt et al., 2010; Scott et al., 2011). The XNAT imaging informatics platform, developed in my own laboratory, was designed to address many of these issues (Marcus, Olsen, Ramaratnam, & Buckner, 2007). I will use XNAT throughout this chapter to illustrate the sort of informatics capabilities that are important for managing large-scale neuroimaging studies. XNAT is open source and extensible and is widely used in the neuroimaging community, including as the HCP's internal private database and open-access public database.

The XNAT platform is built on a standardized workflow that is intended to mirror the real-world operations of collaborative research studies such as the HCP. The workflow covers a stepwise process from acquisition through data sharing and interweaves the best practices, policy, and sociology of doing neuroimaging science. A quarantine stage, for example, provides a mechanism for reviewing data quality and completeness prior to use by local investigators. Additional stages enable secure data access by specific collaborators and by the broader research community. With each stage, a variety of productivity tools, such as web-based reports, searching tools, and automated processing routines, are provided to facilitate use of the data.

Data organization

Most neuroimaging informatics systems, including XNAT, model the data using a hierarchical organization. The *project level* contains data that are related to one another. Often a project is equivalent to a research study, but it could also be used to hold a subset of data from a study or a superset of data aggregated from multiple studies. Within the HCP, for example, a number of different projects are used to manage various subcomponents of the study, including 3 Tesla (3T) optimization data, 7 Tesla (7T) optimization data, and the primary study. Regardless of the contents of a project, the primary role of the project in the data hierarchy is to provide an element around which data can be conveniently grouped for navigation and for delimiting user access privileges. The *subject level* of the hierarchy represents the individuals on whom measurements are made, usually a human but also nonhuman primates and other model species. The subject level includes demographic and other nonchanging information about the subject and contains the set of experimental data elements obtained from the individual. The *visit level* captures all data obtained from a subject across all methods (e.g., MRI session, positron emission tomography session, neuropsychological evaluation) within a scoped time frame. Visits are often completed in a single day or two but may be open for weeks or months in complex studies. Within the HCP, all subjects undergo a primary visit on the 3T scanner; a subset of subjects will undergo a second 7T visit at a later date. The visit level is especially useful for tracking of data within longitudinal studies that obtain repeated measures over time. The *experiment level* contains the actual experimental data acquired in the study, with specific experiment types defined to capture the specific elements associated with a particular instrument. For example, XNAT includes data experiment types for all Digital Imaging and Communications in Medicine (DICOM) imaging modalities (PET, MRI, CT, etc.) and many common psychological instruments. Custom experiment types can be added to XNAT systems to capture additional instruments.

Much of the functionality of XNAT and similar systems is built around this type of four-level data hierarchy; it enables experimental data to be navigated, grouped, tracked, and validated. For example, XNAT includes a protocol-tracking feature that allows study managers to define the expected timing of visits and the types of data to be collected within each visit. From this definition, XNAT provides user interfaces for entering the data associated with a visit and for navigating existing subject data.

User and programming interfaces

Neuroimaging informatics systems typically provide a web-based user interface. XNAT's web interface enables users to navigate by data type and by project. When navigating by data type, all data of that type (that the user is authorized to view) from across projects are aggregated into a single view. When navigating by project, data of different types within a single project are aggregated together. In the project-based view, users can review the various types of data within the project and administer project settings. From either navigation path, users can select specific subjects and experiments to view in more detail. At each level of navigation, users are presented with a set of context-specific actions for further interacting with the data. From the aggregated data tables, for example, users can download spreadsheets and filter the presented data. From individual experiment reports, they can generate PDF views and email links to the report to colleagues. If a user has sufficient access privileges, they are also presented with options to execute processing pipelines and edit the data.

In addition to the web-based user interface, systems often provide application programming interfaces (APIs) for programmatically interacting with the database. The XNAT API follows a representational state transfer (REST) web-services architecture, which provides distinct and knowable hypertext transfer protocol (HTTP)–based resource locators for all hosted data elements and files (Fielding & Taylor, 2002). The data-access portion of the API closely follows XNAT's data hierarchy, including providing access to custom data types added to an XNAT repository using .xsd file extensions. The API also includes administration, pipeline, and operations components to provide comprehensive interactivity with XNAT functionality. A primary usage of the API is to enable external software to interface with XNAT. The HCP visualization system, Connectome Workbench, for example, uses the XNAT REST API to retrieve images for visualization and to execute dynamic multisubject analyses. Similarly, the API can be used to write scripts that interface with XNAT; scripting libraries have been developed in a number of languages, including Java, R, MATLAB, and Python.

Importing images

The core of a neuroimaging study, of course, is the data acquired at the scanner. In recent years, capturing these data directly from the scanner has been greatly eased by the near-universal implementation of the DICOM standard developed by the scanner manufacturers (National Electrical Manufacturers Association, 2011). The DICOM standard defines both a data format and a network transmission protocol. The data format includes the actual image data as well as a rich set of metadata that details how the images were obtained. Reviewing these metadata is an important component of a rigorous quality control process (see below), and a handful of the fields are essential for subsequent processing and analysis of the images. The DICOM network protocol specifies how image data is sent from the scanner (and other DICOM-based systems) to DICOM-compatible receivers. The most common receivers are the picture archiving and communication systems designed primarily for clinical use. Because they are designed for clinical use, picture archiving and communication systems lack a number of features—longitudinal views, integration, post-processing, data access control, and so on—that are important for organizing and managing neuroimaging research data. A main focus of neuroimaging informatics systems is to implement these functions. XNAT, for example, includes a DICOM receiver and metadata import system that maps incoming DICOM data to study-specific projects, around which security and data access privileges are constructed, and to research subjects and longitudinal study visits. XNAT also provides a mapping to bridge across the varying terminologies used in the clinically focused DICOM standard and by research neuroimagers. In typical practice, a DICOM "study" is equivalent to a research "session," and a "series" is equivalent to a "scan."

In research neuroimaging, .nifti has emerged as the de facto standard file format. However, no scanners produce it, and so it is necessarily a derivative product generated through a file conversion process. The NIfTI metadata model is significantly more limited than DICOM, and much useful data acquisition information is not preserved in the conversion process (e.g., repetition time, flip angle, etc.). As this information is lost, many researchers have fallen into the trap of *assuming* they know the acquisition parameters for their data. Inevitably, with this approach, acquisition protocols will drift over time as scanners are upgraded or they are "tweaked" by colleagues. To avoid this pitfall, it's important to preserve the original DICOM files and to implement quality control and review procedures (see below). Despite this limitation, NIfTI has several important attributes. Most importantly, it implements a coherent coordinate system that accounts for left-right directionality in the data and supports up to 128-bit floating-point binary data. It also includes several convenience features like compact single file storage and an extension mechanism to incorporate additional data or metadata.

While DICOM is the preferred format for archiving data (given the pitfalls described above), neuroinformatics platforms often implement a NIfTI -based data import workflow in addition to or instead of DICOM. Because NIfTI does not define a network-transmission protocol, nor does it include metadata by which study organization can be inferred, data systems must implement data-transfer procedures and mechanisms for inferring data organization. XNAT, for example, includes a NIfTI web-based upload interface that imports the data into projects, sessions, or scans according to directory and file names. Users can customize the naming patterns to match their data. Acquisition metadata for the imported NIfTI data can be supplied using the XNAT programming interface.

Quality control

As neuroimaging studies continue to grow in scale and complexity, the risk of data-collection errors grows as well. Such errors include systematic acquisition with an incorrect protocol or sporadic errors like poor head positioning, subject motion, and susceptibility artifacts. Quality control (QC) procedures built into the informatics workflow can be used to mitigate these risks. QC procedures include image-acquisition validation, manual image review, and automated image analysis. XNAT includes methods to support each of these forms of QC. For acquisition validation, XNAT checks whether the parameters of imported DICOM files match a project-specific protocol and outputs a report detailing which tests pass and fail. The validation protocol, for example, could include a test to verify that an incoming MRI study includes a series type "MPRAGE" and three series of type "BOLD." It could subsequently test that each BOLD series has a repetition time of 2.4 seconds. The acquisition validation service is executed via XNAT's pipeline system (see "Automation" section, below). Manual image review, done soon after acquisition, is the most common and essential QC procedure. Within XNAT, a number of manual review forms have been developed to enable various levels of manual review. Typical review criteria include head positioning, motion artifacts, other artifacts, and image contrast, and may be scored either on a pass/fail or a multilevel (e.g., poor/ok/good/excellent) scale. Automated image analysis provides a quantitative approach to quality review. Using XNAT's pipeline system, specific processing routines can be executed, and quantitative metrics extracted. These quantitative metrics can then be presented in a report and compared against expected ranges or distributions generated from prior acquisitions.

All three of these QC methods are used by the HCP. Acquisition validation runs immediately after the scans are imported and checks over 20 parameters. A research analyst then reviews the structural images and enters scores into the HCP XNAT system. Meanwhile, automated QC pipelines execute to generate quantitative metrics, including temporal signal-to-noise ratio, motion displacement, and DVARS. All of these processes complete within hours of acquisition, and if key scans are deemed to be of insufficient quality, the subject is typically rescanned.

Automation

In order to prepare neuroimaging data for analysis, the acquired scans are typically run through a series of processing steps, including inhomogeneity correction, cross-modal coregistration, and denoising. A handful of widely used software packages and many in-house software libraries are available for accomplishing these tasks. While installing and running these packages is generally straightforward and can be facilitated by preconfigured operating systems like NeuroDebian (Halchenko & Hanke, 2012), the actual execution of such processing often entails running dozens of individual software routines, each with an expansive set of execution options. Keeping track of the exact steps and procedures for a chosen processing strategy can be extremely challenging, yet it's critical to apply a consistent and systematic processing scheme for all data within the study. A number of software frameworks, often referred to as *pipeline systems,* are available for designing and executing repeatable image-processing procedures (Dinov et al., 2010; Gorgolewski et al., 2011; Marcus, Olsen, et al., 2007; Oinn et al., 2004). Pipeline systems, including the pipeline tools built into XNAT, typically entail a number of components. A specification component enables the sequence of processing steps to be defined and documented. An execution component applies a particular pipeline specification to one or more data sets. The execution component is often linked to a clustered computing environment (see below) to enable high-throughput execution. A logging component records error and status messages, including data-provenance information. Together, these components enable investigators to execute repeatable processing over time and across data sets. Pipeline systems can also be used to explore parameterized processing of a data set and to share data-processing schemes between groups. XNAT's pipeline service is fully integrated with XNAT's database, user interface, and web services, allowing users to link specific pipeline configurations to individual projects and to tie pipeline execution to their data in XNAT.

In order to capture the processing history of data generated by a pipeline (or manually, for that matter), *data-provenance* tools record processing details systematically, allowing users to confirm and verify their work and to report their methodologies alongside published data sets. Data-provenance structures typically describe an *entity* that is acted on, the *activity* that did the acting, and the *agent* responsible for an activity taking place. Within XNAT, provenance is recorded using a format developed by the Biomedical Information Research Network (BIRN) that details the input and output files to a processing routine (the entities), the details of the processing routine, including version information and input parameters (the activity), and the user and computing environment responsible for the execution of the routine (the agents; Gadde et al., 2012). For multistep processing streams, such as pipelines, each entity-activity-agent combination is recorded as steps within an overall sequence. Recent efforts within the neuroinformatics community are moving towards the PROV data model (PROV-DM) developed by the W3C standards body as a universal model for documenting provenance of digital and real-world objects (W3C, 2013).

Anyone who has ever attempted to process large amounts of neuroimaging data knows how difficult it can be to monitor multiple ongoing processes and to optimize use of available computing resources. Fortunately, software is available to cluster multiple computers into a coherent resource. With a cluster, users can submit large batches of jobs—indeed, more than can be run simultaneously—to the cluster. The cluster will queue these jobs and manage their execution in an optimized manner across the clustered computing hardware. Open Grid Engine (OGE) and related products are the most widely used tools for organizing and managing computing clusters. While some neuroimaging tools (e.g., FSL) are natively designed to distribute their processing across an OGE, more often users are faced with using OGE's command line tools for posting jobs to a processing queue and tracking their execution. XNAT's pipeline system, along with tools such as LONI Pipeline, natively supports integration with OGE. However, care must be taken in managing the actual computers that are on the cluster. Each of the compute nodes must have the required software (with matching versions) installed. It is also important that the nodes be running the same operating system versions and, ideally, be based on the same hardware platform. Without adequate management of the cluster, computational results may vary depending on the specific node on which a job runs.

Data integration

In addition to the acquired imaging data, neuroimaging studies often include various behavioral, clinical, genetic, and other measures. Many informatics systems provide mechanisms for capturing novel data. These include general-purpose electronic data capture (EDC) systems, such as REDCap (Harris et al., 2009), that provide web-based tools for creating data entry forms. EDCs typically utilize a set of generic key-value database tables that enable an open set of data to be captured. While REDCap and similar tools are powerful data entry systems, they do not support imaging data, so additional steps are needed to integrate nonimaging and imaging outcomes. Some imaging informatics systems, including XNAT, provide EDC-like functionality directly. Extensions to XNAT's data model are implemented using XML Schema, a format for defining XML-based data structures. From these schemas, XNAT automatically generates all of the software components necessary to import and utilize the new data type, including database tables, web-based reports, data entry forms, and programmatic interfaces.

XNAT provides a number of mechanisms for capturing these data into its database. The generated web-based forms can be used to directly enter the data, either in real time by research subjects or a research assistant, or retrospectively by the research assistant transcribing from a paper form or some other external source (e.g., an electronic medical record). Alternatively, electronic data can be imported using XNAT's spreadsheet import service or its programmatic interface. The programmatic interface is particularly useful for implementing controlled extract, transform, and load (ETL) procedures, in which data are extracted from an external database, cleaned, and transformed into an XNAT-compliant format and uploaded to XNAT over the XNAT API. Within the HCP, ETL procedures are used to import data from the NIH Toolbox and University of Pennsylvania behavioral testing systems into XNAT. As an alternative to extending the XNAT database itself, behavioral and other nonimaging data may be stored in an external database and merged with imaging data as a final data set preparation stage prior to analysis. Data federation tools developed by the BIRN and others enable automated cross-system queries and integration (Bug et al., 2008; Zhang et al., 2011).

Security and privacy

A number of factors make security and privacy important components of neuroimaging informatics systems. Neuroimaging data contain anatomic information, including facial features, that may be used to identify the subject. Many studies also collect sensitive information—psychiatric measures, drug and alcohol usage, cognitive performance measures—that subjects expect to be maintained with discretion. These issues are particularly challenging in the context of data sharing. Institutional review boards and the federal HIPAA provide direction and enforcement on how human subject data may be accessed, stored, and distributed. The core requirements include encryption, access control, and de-identification.

Encryption is required for both storage and transport of data. While most recent operating systems natively support encryption of stored data, the feature is often not enabled by default. Encryption of data transport of the network is typically achieved by using network protocols that support secure socket layer protocols, such as HTTPS and secure FTP, which authenticates the data host and encrypts the data while in transit. However, it is important to note that most DICOM devices do *not* send data over encrypted protocols, so it is not advisable to send data from a DICOM device to a DICOM receiver that is located outside of the device's firewalled network. As a secure alternative, one can set up DICOM "relay" software, such as Clinical Trial Processor (RSNA, 2013), inside the firewall to receive data over the DICOM protocol and then forward the data onto a receiver, such as an XNAT server, outside the firewall over HTTPS. Conversely, XNAT Gateway can be used to securely retrieve encrypted data from an XNAT system to DICOM devices like scanners and workstations.

Access control is typically maintained through password-protected user accounts. XNAT maintains an internal registry of users and their assigned roles on projects. Typically, users provide their credentials directly to XNAT, via a login page in XNAT's web application or embedded in the HTTP headers associated with a web service call. XNAT verifies these credentials prior to granting the user access to data. XNAT also supports authenticating users against external identity providers, such as a university personnel directory. Using this alternative mechanism, the user supplies his external credentials to XNAT, which then verifies the credentials against an interface provided by the external identity provider. Once a user is logged in, XNAT limits the user's access to the project to which the user has been explicitly granted access. Users are typically assigned to one of three default roles on a project: owners have read, write, edit, and delete privileges on project data and can alter other users' roles; members have read, write, and edit privileges; and collaborators have read-only privileges. Custom roles can also be created to provide more fine-grained access control. An MRI technician, for example, could be granted member privileges on a project's MRI data, allowing him to upload MRI studies while being denied access to the project's clinical and other data. In a typical configuration, a new project can be created in an XNAT deployment by any authorized user of that system. The user can grant other users access to the project and invite new individuals to become users in the system. At any time, project owners can view who has access to a project's data, their level of access, and a brief history of their access.

De-identification refers to removing information that may be used to identify human subjects, including the 18 HIPAA-defined identifiers—names, dates, phone numbers, social security numbers, and so on (US Department of Health and Human Services, 2013). Similarly, anonymization refers to removal of identifying information as well as any codes that would allow linking back to identified information. Data distributed beyond the immediate research team must be de-identified. Even within a research team, data are often de-identified to minimize the risk of accidental breaches of subject privacy. Data shared openly must be fully anonymized, which can be a difficult task to accomplish with neuroimaging studies. XNAT provides a number of tools to assist in de-identification and anonymization. It includes a DICOM editing language and processor that can be used to remove or replace metadata fields in DICOM that may contain identifying information. All imported DICOM data is automatically processed following site-wide and project-specific DICOM edit scripts. When DICOM files are initially received, either directly over the DICOM receiver or via the HTTP interfaces, a site-wide script is applied. The default site-wide script provided with XNAT replaces the patient ID and patient name fields with XNAT session and subject identifiers. The default scripts can be edited or replaced by system administrators to enforce local privacy rules. As the data are being archived into a specific project, a project-specific script is executed to apply additional rules. If the data are subsequently moved to a different project, the new project's script is applied. Project managers can configure the project-specific scripts to provide little or no way modification, to comply with the DICOM standard for de-identification (DICOM Standards Committee, Working Group 18 Clinical Trials, 2011), or to execute a custom de-identification scheme.

A final aspect of de-identification is the necessity of removing identifying information in the actual image data. Potentially identifying facial features, including ears, must be obscured, while leaving necessary features (brain tissue) fully intact. Common procedures typically first register the scan to a target image, which provides a reference for where identifying features are positioned. An algorithm is then executed to remove or obscure these features, and the resulting image is then transformed back to the original subject space. One widely used approach uses a brain mask or tissue-segmentation algorithm to locate brain, face, and other structures, then blacks out or otherwise obscures areas outside the desired regions (Bischoff-Grethe et al., 2007). These methods are generally effective at rendering anatomic features unrecognizable, while preserving the cranial vault, but often require supervision to ensure that brain tissue is not impacted and may be affected by variation in diagnoses, age groups, MR field inhomogeneity, and other subject- and acquisition-specific variability (Fennema-Notestine et al., 2006). In addition, downstream algorithms may be disrupted by edge artifacts introduced by the defacing. Many projects have therefore moved to approaches that minimally alter the image. The method used by the HCP, for example, identifies the surface containing the face and the ears and then runs an irreversible blurring algorithm on that surface only to reduce anatomic recognizability (Milchenko & Marcus, 2013).

Data sharing

The benefits of open sharing have been well articulated by many advocates of open science (Poline et al., 2012). Inspired by this movement, many investigators are choosing to voluntarily and enthusiastically share their data. Others are mandated to share data by funding agencies that desire to see their investments put to broader use and by journal publishers who believe papers are best understood when paired with the data used to produce them. Within neuroimaging, projects such as the Alzheimer's Disease Neuroimaging Initiative (Jack et al., 2008), Open Access Series of Structural Images (Marcus, Wang, et al., 2007), the HCP (Van Essen et al., 2013), and the 1000 Functional Connectomes (Biswal et al., 2010) have demonstrated that open sharing of high-quality, well-documented data sets is beneficial to both producers and consumers of data. These projects have helped establish best practices for addressing the human-subject regulations and logistical complications that have long been roadblocks to sharing (Mennes, Biswal, Xavier Castellanos, & Milham, 2013; Poline et al., 2012). In concert with these practices, informatics systems provide a technical mechanism for sharing data.

While a dedicated instance of XNAT or other informatics platforms could be deployed to share a data set, a number of organizations have set up public sites for sharing neuroimaging data. These include the International Neuroinformatics Coordinating Facility, the Neuroimaging Informatics Tools Resource Clearinghouse (NITRC), and the Open fMRI project. Using these sites to share data has a number of advantages over operating a dedicated standalone site. For one, these organizations have worked out the technical issues and absorbed the costs of producing and maintaining the required computing, networking, and security infrastructure. Further, as the hosts of multiple data sets, they also have broader visibility within the community. The International Neuroinformatics Coordinating Facility Dataspace, which layers security, access control, and high-speed data, transfers on top of a general file-sharing service. Importantly, the service allows sharing with select collaborators or with the open community. The NITRC Image Repository builds on XNAT to provide a focused neuroimage data-sharing environment that is connected to the NITRC Computational Environment, an Amazon Cloud–based platform for executing processing and analysis routines on shared data. The Open fMRI site, in particular, is a model for open-access data sharing (Poldrack et al., 2013). The operators enforce a careful curation process for ensuring the integrity of the data that includes manual review of defaced images, quality control metrics, and processing output. The site has a clean, ergonomic user interface and clearly documents their standard data file organization and naming conventions and the provenance of hosted data sets. Finally, its data sets are typically distributed under the unrestrictive Public Domain Dedication and License version 1.0, which allows consumers of the data to redistribute and reprocess the data in innovative ways.

REFERENCES

Barch, D. M., Burgess, G. C., Harms, M. P., Petersen, S. E., Schlaggar, B. L., Corbetta, M., ... Van Essen, D. (2013). Function in the human connectome: Task-fMRI and individual differences in behavior. *NeuroImage, 80,* 169–189.

Bischoff-Grethe, A., Ozyurt, I. B., Busa, E., Quinn, B. T., Fennema-Notestine, C., Clark, C. P., ... Fischl, B. (2007). A technique for the deidentification of structural brain MR images. *Hum Brain Mapp, 28*(9), 892–903.

Biswal, B. B., Mennes, M., Zuo, X. N., Gohel, S., Kelly, C., Smith, S. M., ... Milham, M. P. (2010). Toward discovery science of human brain function. *Proc Natl Acad Sci USA, 107*(10), 4734–4739.

Bug, W., Astahkov, V., Boline, J., Fennema-Notestine, C., Grethe, J. S., Gupta, A., ... Martone, M. E. (2008). Data federation in the biomedical informatics research network: Tools for semantic annotation and query of distributed multiscale brain data." *AMIA Annu Symp Proc, 1220.*

Chen, J. J., Siddiqui, K. M., Fort, L., Moffitt, R., Juluru, K., Kim, W., ... Siegel, E. L. (2007). Observer success rates for identification of 3D surface reconstructed facial images and implications for patient privacy and security. *SPIE Proc,* 65161B–65161B–8.

Das, S., Zijdenbos, A. P., Harlap, J., Vins, D., & Evans, A. C. (2011). LORIS: A web-based data management system for multi-center studies. *Front Neuroinform, 5,* 37.

DICOM Standards Committee, Working Group 18 Clinical Trials. (2011). *Digital imaging and communications in medicine,* Suppl. 142, *Clinical trial de-identification profiles.* Retrieved from ftp://medical.nema.org/medical/dicom/final/sup142_ft.pdf

Dinov, I., Lozev, K., Petrosyan, P., Liu, Z., Eggert, P., Pierce, J., ... Toga, A. (2010). Neuroimaging study designs, computational analyses and data provenance using the LONI pipeline. *PloS ONE, 5*(9).

Fennema-Notestine, C., Ozyurt, I. B., Clark, C. P., Morris, S. Bischoff-Grethe, A., Bondi, M. W., ... Brown, G. G. (2006). Quantitative evaluation of automated skull-stripping methods applied to contemporary and legacy images: Effects of diagnosis, bias correction, and slice location. *Hum Brain Mapp, 27*(2), 99–113.

Fielding, R. T., & Taylor, R. N. (2002). Principled design of the modern web architecture. *ACM Trans Internet Technol, 2*(2), 115–150.

Gadde, S., Aucoin, N., Grethe, J. S., Keator, D. B., Marcus, D. S., & Pieper, S. (2012). XCEDE: An extensible schema for biomedical data. *Neuroinformatics, 10*(1), 19–32.

Glasser, M. F., Sotiropoulos, S. N., Wilson, J. A., Coalson, T. S., Fischl, B., Andersson, J. L., ... Van Essen, D. C. (2013). The minimal preprocessing pipelines for the human connectome project. *NeuroImage, 80,* 105–124.

Gorgolewski, K., Burns, C. D., Madison, C., Clark, D., Halchenko, Y. O., Waskom, M. L., & Ghosh, S. S. (2011). Nipype: A flexible, lightweight and extensible neuroimaging data processing framework in Python. *Front Neuroinform, 5*, 13.

Halchenko, Y. O., & Hanke, M. (2012). Open is not enough. Let's take the next step: An integrated, community-driven computing platform for neuroscience. *Front Neuroinform, 6*, 22.

Harris, P. A., Taylor, R., Thielke, R., Payne, J., Gonzalez, N., & Conde, J. G. (2009). Research electronic data capture (REDCap)—a metadata-driven methodology and workflow process for providing translational research informatics support. *J Biomed Inform, 42*(2), 377–381.

Jack, C. R., Jr., Bernstein, M. A., Fox, N. C., Thompson, P., Alexander, G., Harvey, D., … Weiner, M. W. (2008). The Alzheimer's disease neuroimaging initiative (ADNI): MRI methods. *J Magn Reson Imaging, 27*(4), 685–691.

Marcus, D. S., Olsen, T. R., Ramaratnam, M., & Buckner, R. L. (2007). The extensible neuroimaging archive toolkit: An informatics platform for managing, exploring, and sharing neuroimaging data. *Neuroinformatics, 5*(1), 11–34.

Marcus, D. S., Harms, M. P., Snyder, A. Z., Jenkinson, M., Wilson, J. A., Glasser, M. F., … Essen, D. C. (2013). Human connectome project informatics: Quality control, database services, and data visualization. *NeuroImage, 80*, 202–219.

Marcus, D. S., Wang, T. H., Parker, J., Csernansky, J. G., Morris, J. C., & Buckner, R. L. (2007). Open access series of imaging studies (OASIS): Cross-sectional MRI data in young, middle aged, nondemented, and demented older adults. *J Cogn Neurosci, 19*(9), 1498–1507.

Mennes, M., Biswal, B. B., Xavier Castellanos, F., & Milham, M. P. (2013). Making data sharing work: The FCP/INDI experience. *NeuroImage, 82*, 683–691.

Milchenko, M., & Marcus, D. (2013). Obscuring surface anatomy in volumetric imaging data. *Neuroinfomatics, 11*(1), 65–75.

National Electrical Manufacturers Association. (2011). *Digital imaging and communications in medicine.* Retrieved from http://medical.nema.org/standard.html

Oinn, T., Addis, M., Ferris, J., Marvin, D., Senger, M., Greenwood, M., … Li, P. (2004). Taverna: A tool for the composition and enactment of bioinformatics workflows. *Bioinformatics, 20*(17), 3045–3054.

Ozyurt, I. B., Keator, D. B., Wei, D., Fennema-Notestine, C., Pease, K. R., Bockholt, J., & Grethe, J. S. (2010). Federated web-accessible clinical data management within an extensible neuroimaging database. *Neuroinformatics, 8*(4), 231–249.

Poldrack, R. A., Barch, D. M., Mitchell, J. P., Wager, T. D., Wagner, A. D., Devlin, J. T., … Milham, M. P. (2013). Toward open sharing of task-based fMRI data: The OpenfMRI project. *Front Neuroinform, 7*, 12.

Poline, J. B., Breeze, J. L., Ghosh, S., Gorgolewski, K., Halchenko, Y. O., Hanke, M., … Kennedy, D. N. (2012). Data sharing in neuroimaging research. *Front Neuroinform, 6*, 9.

Radiological Society of North America (RSNA). (2013). *Clinical trial processor.* Retrieved from http://mircwiki.rsna.org/index.php?title=CTP_Articles

Scott, A., Courtney, W., Wood, D., de la Garza, R., Lane, S., King, M., … Calhoun, V. D. (2011). COINS: An innovative informatics and neuroimaging tool suite built for large heterogeneous datasets. *Front Neuroinform, 5*, 33.

US Department of Health and Human Services. (2013). The HIPAA Privacy Rule. Retrieved from www.hhs.gov/ocr/privacy/hipaa/administrative/privacyrule/index.html

Van Essen, D. C., Smith, S. M., Barch, D. M., Behrens, T. E. J., Yacoub, E., & Ugurbil, K. (2013). The WU-Minn human connectome project: An overview. *NeuroImage, 80*, 62–79.

World Wide Web Consortium (W3C). (2013). PROV-DM: The PROV Data Model. Retrieved from www.w3.org/TR/prov-dm/

Zhang, J., Haider, S., Baran, J., Cros, A., Guberman, J. M., Hsu, J., … Kasprzyk, A. (2011). BioMart: A data federation framework for large collaborative projects. *Database, 2011.*

XI NEUROSCIENCE AND SOCIETY

Introduction

WALTER SINNOTT-ARMSTRONG AND ADINA ROSKIES

THIS VOLUME provides a comprehensive survey of current work in cognitive neuroscience, but the full impact of this area of research is not realized until we see how its fruits can affect the world beyond the laboratory. Broadly speaking, the social effects of recent developments in cognitive neuroscience are threefold: cognitive neuroscience can *improve our knowledge* of other minds and our understanding of our social milieu, and in so doing *change the way we think* about others and enable us to develop *tools to alter our capacities and behavior*. The work we discuss in this section focuses on societal issues that are affected by our understanding of the brain, and the social and moral questions and implications of that understanding.

Most fundamentally, we relate to other people as people. Each of us has an immediate sense of whether something in our environment is a person rather than an object or nonhuman animal, and our behavior hinges on this perception. But how do we detect persons? The chapter in this section by Lasana Harris, Victoria Lee, and Beatrice Capestany surveys recent findings in neuroscience that illuminate the psychological processes by which we perceive others as people with minds like our own. These authors also reveal some of the ways in which such processes can fail when we dehumanize other people in certain social groups. Our understanding of this kind of dehumanization may someday help us address some of our most serious social problems.

Neuroscience not only illuminates the way in which we perceive mental states, it can also tell us something about the content of the mental states of others. There are many contexts in which we might want to know *which* mental states another person has. Did the witness believe what he said? Did the defendant intend what he did? Does he really love me? What is she thinking? What does he remember? To what extent can we answer these questions with our current techniques? In her chapter, Adina Roskies addresses our ability to use neuroimaging to "read" brains and reconstruct specific mental content. She also considers the ethical questions this ability raises. Roskies concludes that, to date, brain imaging has achieved only very limited success in mind-reading, which should limit the ways in which neuroimaging technologies should be used for practical applications. This also means that, so far, neuroscience does not pose as great a threat to privacy as many opponents fear.

In general, we hold people morally accountable for intentional harms to other people. The moral judgments we make are then crucial to how we relate to each other in society, because they guide our decisions and also affect our attitudes toward other people whose acts we find immoral. Joshua Greene explains how neuroscience is beginning to illuminate the neural bases of moral judgment. If neuroscience can help us understand why we judge some acts and the people who perform them as immoral and others as moral, then it will surely help us understand a central aspect of the way we relate to each other. These judgments also underlie central philosophical and institutional aspects of our social world, which are explored in the subsequent chapters.

The understanding of social relations we gain from neuroscience may also *change* how we view and relate to other members of society. In particular, some people seem to think that if our actions result from electrical and chemical activity in our brains, then people are not ever morally responsible for their actions. In response, the chapter by Uri Maoz and Gideon Yaffe argues that, contrary to some intuitions and extreme claims, neuroscience does not really undermine moral responsibility in general. Rather, it may show that some people in special cases are less responsible than they might initially seem to be. Thus, recent neuroscience can illuminate subtle aspects of responsibility without globally undermining this central social framework.

One such special case is addiction. In their chapter, Mimi Belcher, Nora Volkow, Gerard Moeller, and Sergi Ferré argue that the lessons from neuroscience about addiction do and should affect the way in which we react to addicts and their misbehaviors. An understanding of addiction as a brain disease can make us more sympathetic to the plight of people who struggle with addiction. This understanding and sympathy can then point toward new programs to help addicts deal with their addiction and to make them less destructive to themselves and to society.

Another kind of case is criminal behavior. The chapter by Lyn Gaudet, Nate Anderson, and Kent Kiehl discusses neural and genetic bases of certain types of criminal behavior, but argues there can be no neuroscience of crime or aggression in general, because there are too many differences between the brains of various kinds of criminals. The brains of psychopaths are not like those of gang members or white-collar criminals. Moreover, all genetic and neural differences among criminals interact with the environment in ways that complicate prediction of crime. Neuroscience still can help us a lot in dealing with specific kinds of criminals, and as our understanding grows some treatments may become available, but there is no hope that neuroscience will solve the problem of crime in one fell swoop.

Neuroscience affects society in a third way, when it is used as a *tool*. Perhaps the most dramatic example is neuropsychopharmacology—the use of neuroscience to develop drugs or other treatments to cure brain diseases or mental defects. In his chapter, Anjan Chatterjee discusses these positive uses of neuroscience. However, often these same drugs or treatments can be used off-label, not to cure illnesses but instead to enhance healthy people. For example, many healthy children use Ritalin to increase their focus and performance in school, and healthy adults use modafinil to stay awake and alert longer. Enhancing otherwise normal function raises a host of ethical questions. Chatterjee provides a measured picture of both the promise as well as the perils of the tools that cognitive neuroscience research is making available.

Another prevalent factor that shapes our social lives is economics. In his chapter, Scott Huettel surveys ways in which neuroscientists can work with economists to understand decision making in social contexts. This understanding has relevance to a wide spectrum of social issues. It also has potential to be used to affect our social and economic transactions through neuromarketing. Companies have sprung up in recent years to use neuroscience to control consumer purchases, and many opponents find this use of neuroscience downright scary. Huettel argues that the reality of neuromarketing is not as frightening as its critics claim.

The breadth of issues where cognitive neuroscience meets society is more extensive than could possibly be treated in this short section. For instance, neuroscience research has implications for the ways in which socioeconomic status and social forces can affect brain development and, ultimately, behavior. Understanding aging and perhaps affecting the aging process of the brain through pharmacological and genetic manipulations will undoubtedly also have profound effects on society. These are topics that, for reasons of space, we could not explore here. The topics covered here are only a few illustrations of how new insights and tools from neuroscience hold great promise for the betterment of society. They also raise profound moral quandaries concerning how cognitive neuroscience should be used. Our aim in this section on neuroscience and society is to show that neuroscience can raise the level of debate on these thorny and important issues.

84 The Cognitive Neuroscience of Person Perception

LASANA T. HARRIS, VICTORIA K. LEE, AND BEATRICE H. CAPESTANY

ABSTRACT With the advent of cognitive neuroscience, there has been an increasing use of technologies such as functional MRI and electroencephalography to study person perception—the psychological process by which people perceive other people. But why is it not sufficient to use what we know about how the brain processes objects to study how we process people? Social psychology has taught us that people are not simply objects; human beings, unlike objects, have minds. Minds allow us to form impressions of other people, while other people simultaneously form impressions of us. Though other agents—animals, for instance—may have minds, only people have minds like ours, and perceive us as we perceive them. They may formulate opinions about us and take relevant action toward us, just as we may to them. As a result, knowing what a person has on their mind is relevant. Therefore, person perception is a two-part process: (1) a perception of the physical person (his or her body), and (2) an inference about his or her mental contents (his or her mind). This chapter reviews the cognitive neuroscience of person perception, covering research on biological motion, face perception, mimicry, stereotyping and group processes, status and power, trait inferences, self-perception, empathy and other social emotions, and social decision making.

What makes person perception special?

As you proceed through this chapter, you may wonder what is special about person perception, or how it differs from object perception. How is your perception of the people in your classroom different from your perception of the textbooks and pencils strewn around you? The first difference is that, unlike books and pencils, the people around you have minds—internal states—that can originate their actions. As a result, the way that we perceive people may rely on additional neural processing beyond perceptual category and feature space mapping. Not only do we perceive the shape and form of a person like we would the shape and form of a pencil or book, but we also infer something about the person that we can't see—his or her mind. To do so, we make mental-state inferences about what the person is thinking, feeling, and planning. Therefore, person perception is a two-part process involving physical perception and mental-state inferences. Of course, one may argue for more than two parts, constructing a continuum from physical objects, to objects that are alive, objects with some degree of mental states, and objects with minds like ours (Gray, Gray, & Wegner, 2007). However, for parsimony and because only people have minds like ours, we focus simply on the two extremes when describing person perception.

After describing the parts of the brain involved in person perception, we describe research related to both physical perceptions and mental-state inferences of people. Because people originate their own actions that can help or harm us, we are motivated to predict behavior during social interactions and explain why people behave the way they do. Physical perceptions coupled with mental-state inferences help inform these separate explanation and prediction motives—a unique feature of person perception. Moreover, although we can form an impression about an object (e.g., whether a car can make it through a road trip), people form impressions of us at the same time we are forming impressions of them. Given the basic human need to belong, we try to manage the impressions others form of us. As a result, person perception is complex, since we are often trying to do two things at once—form an impression of others and manage the impression others form about us. This may require complex neural machinery that resides in the hyperdeveloped neocortex of humans, separating us from other primates and making person perception a uniquely human enterprise.

Complex brain machinery for person perception

Cognitive neuroscience has delineated a specific network of brain regions engaged during person perception. When forming impressions of others from personality trait descriptions as opposed to forming impressions of objects using category information (e.g., size, color) there is more engagement of medial prefrontal cortex (MPFC; Mitchell, Macrae, & Banaji, 2005). This part of the brain is central to person perception and integrates complex information coming from brain regions that detect biological motion (superior temporal sulcus, or STS) as well as direct attention (precuneus) and supply affective information (amygdala, insula; Haxby, Gobbini, & Montgomery, 2004). Though other animals possess analogous brain regions of comparable volume, the complexity of the interconnectivity in human brain regions may give rise to uniquely human qualities (Sherwood & Smaers, 2013) like a highly developed sense of self. However, research has shown that there are other integration centers in the person-perception brain network beyond MPFC. Activity in the temporoparietal junction (TPJ) corresponds to the rule matching necessary for the implementation of social norms and moral behavior (Saxe, 2010). Activity in the anterior cingulate cortex (ACC; Eisenberger, Lieberman, & Williams, 2003) and posterior cingulate cortex (PCC; Schiller, Freeman, Mitchell, Uleman, & Phelps, 2009) also corresponds to person-perception tasks, underlying the variety of processes necessary to perceive people. In fact, evolutionary theories of the brain suggest social environment contributed to our ultracooperative and ultracompetitive behavior. Perhaps this context prioritized brain mechanisms underlying person perception because these mechanisms facilitated successful adaptation in such environments. For instance, evolutionary anthropologists looking at cytoarchitectonics across primate species suggest that von Econonomo neurons in the insula and ACC are unique to animals that display complex social cognition abilities—humans, apes, elephants, whales, dolphins, and porpoises—suggesting these distinct neurons in areas implicated in person-perception play a key role in social behavior (Stimpson et al., 2011). Below, we describe nodes in the person-perception brain network as they relate to specific aspects of person perception, including physical perceptions of people and mental-state inferences.

Physical perceptions of people

BIOLOGICAL MOTION The ability to identify biological agents by their movement allows us to direct attention to things that may potentially help or harm us. Imagine hiking in the woods and not recognizing the movement of a snake, or being caught in a physical confrontation without seeing the person charge toward you, fists waving. If the ultimate goal is to predict and explain people's behavior, important cues to the presence of an agent can be obtained by detecting biological motion.

A common method for studying the perception of biological motion is to use point light displays (PLD). To create these stimuli, 10–12 lights (or sensors) are attached to the limbs and joints of an individual before recording his or her movement (e.g., walking, running, dancing, etc.). When presented with only PLD (recordings absent of the person's body), the combination of these lights represents a coherent pattern of motion easily identified by participants as biological motion (as compared to randomly moving lights; Johansson, 1973). This technique, created in the 1970s, continues to be used today to understand how the brain perceives biological motion. Interestingly, we can interpret PLDs of other agents with mental states, such as animals.

Research has identified a network of brain regions specialized for perceiving biological motion. Middle temporal areas (MT) in the dorsal visual pathway detect nonbiological motion, but biological motion activates regions of posterior STS and premotor cortex (Saygin, 2007). Further evidence for separate anatomical structures for biological motion comes from patient studies; while some patients can perceive coherent motion but not biological motion, others can perceive human actions but have trouble detecting motion patterns in nonbiological stimuli (Heberlein, Adolphs, Tranel, & Damasio, 2004). This double dissociation suggests these processes rely on separate regions of the brain.

The roles of premotor cortex and posterior STS in perception of biological motion are further confirmed by a study that employed repetitive transcranial magnetic stimulation to activate these two regions while participants completed a detection task with PLDs (van Kemenade, Muggleton, Walsh, & Saygin, 2012). On each trial, participants viewed either a PLD or a scrambled version of the stimuli and indicated if a person was present. The researchers collected a baseline measure of performance for each participant and then applied repetitive transcranial magnetic stimulation (on separate days) to the premotor cortex, posterior STS, and a control site (vertex) before a retest of the same person-identification task. The measure of interest was the change in performance from baseline after applying repetitive transcranial magnetic stimulation, which over premotor cortex produced significantly more false alarms to biological motion. Similar effects were shown in posterior STS, although to a lesser extent. These results provide causal evidence consistent with neuroimaging results that implicate these two regions in the perception of biological motion.

FACE PERCEPTION Perceiving people involves much more than perceiving their motion. When viewing other people, our attention is primarily focused on the person's face. Faces convey important social information about a person's identity and mental states. But how do we go about recognizing our friend in a crowd? Cognitive neuroscience models of face perception (Haxby, Hoffman, & Gobbini, 2000) have identified different pathways in the brain for different aspects of the phenomenon. Making up part of the "core system" for visual analysis, early perception of facial features occurs in the inferior occipital gyri. From there, a ventral pathway, including the fusiform gyrus and regions of inferior temporal cortex, is responsible for representing features of a face that do not change across situations or expressions, in other words, representing people's identities. On the other hand, a dorsal pathway from inferior occipital gyri processes changeable aspects of the face, such as eye gaze and facial expressions that communicate emotion. This pathway includes more dorsal regions of the temporal lobe, passing through STS and on to an "extended system" for person perception. This extended system includes the intraparietal sulcus (IPS) for spatially directed attention, auditory complex for prelexical speech perception, and amygdala, insula, and limbic system for processing emotion. The anterior temporal lobe is also implicated as a ventral pathway component of this system, representing personal identity, name, and biographical information.

The fusiform face area (FFA) along the ventral pathway has been a source of debate in recent years. Typically, researchers have observed activation in FFA when showing participants pictures of faces versus houses or common objects, suggesting this area is specifically dedicated to processing and individuating faces. However, some studies have suggested that FFA is not specifically dedicated to the processing of faces per se, but stimuli with which people are particularly expert. For example, when bird experts viewed pictures of different kinds of birds, they showed greater activation in FFA than when car experts viewed the same stimuli (Gauthier, Skudlarski, Gore, & Anderson, 2000). This implies that faces represent a category of stimuli that most people are experts at identifying, consistent with social psychological and evolutionary anthropological accounts of social behavior. Similarly, researchers trained participants to become experts in recognizing novel stimuli known as greebles. When tested in the scanner, greeble experts showed a similar pattern of activity in the FFA to greebles and to human faces (Gauthier, Tarr, Anderson, Skudlarski, & Gore, 1999). These studies suggest FFA processing may reflect a more general visual expertise. Stated differently, individual expertise lies in perceiving other people's faces, adding to the unique nature of person perception.

MIMICRY Mimicry may be one of the most pervasive consequences of person perception. It occurs spontaneously when encountering other people, and probably evolved to help humans communicate. Social psychologists have discovered that individuals automatically and spontaneously mimic many facets of social interaction, from gestures and postures to emotions and facial expressions (Chartrand & Bargh, 1999). Although social psychologists have discovered many nuances in mimicry, there has been limited research that explores the brain during mimicry. However, the discovery of mirror neurons in monkeys has led researchers to postulate a similar mechanism in the brains of humans that may account for mimicking behavior (Fabbri-Destro & Rizzolatti, 2008).

Like face-selective neurons, mirror neurons were first discovered in the macaque monkey. These neurons discharge when the monkey performs certain hand movements and when observing another monkey (or even a human) performing similar hand movements. Interestingly, these mirror neurons respond only to goal-directed actions, implying that they are especially important for the representation of general actions (such as goals) rather than simple properties of movements (such as movement direction or muscle activity; Fabbri-Destro & Rizzolatti, 2008). These results led to theorizing about human mirror neurons in the premotor and parietal cortices that appear to respond during the generation of action, as well as during the observation of others' actions. Numerous fMRI studies have obtained results vaguely consistent with the mirror neuron theory, and have concluded that a mirror-neuron system in humans may help us understand and interpret others' actions. However, since fMRI is not yet able to differentiate responses at the neuronal level, the actual existence of human mirror neurons like those observed in monkeys is still debated.

Yet it is interesting to consider the mirror system's role in mimicry during social behaviors. Using electroencephalography Mu wave suppression—decreased sensorimotor rhythms—as an index of mirror-neuron system activation, researchers wanted to investigate the effects of social content on brain activity (Oberman, Pineda, & Ramachandran, 2007). Participants viewed four videos with differing levels of social interactions: a video of white noise, balls being thrown in the air, three individuals tossing a ball to each other, and three individuals tossing a ball to themselves and at the screen toward the viewer. Given the association of Mu wave suppression and sensorimotor neurons, the results suggest the mirror-neuron system is sensitive to the presence of social cues, with the white noise condition showing the least Mu suppression and the condition where the balls were tossed toward the participant showing the most Mu suppression. The graded nature of the Mu suppression, from least social to most social, suggests that a human mirror system, if it exists, may be specialized for processing socially relevant stimuli.

Stereotyping and Group Processes Since the 1950s, social psychologists have been interested in understanding stereotypes (Allport, 1954) because engaging in negative stereotypes can often lead to prejudice and discrimination. Like mimicry, stereotyping is an automatic response when encountering other people. However, stereotypes do not have to be negative. Stereotypes simplify processing by creating ways for the brain to integrate a lot of information about people (Fiske & Neuberg, 1990). Researchers have begun to uncover how the brain perceives and categorizes race and have implicated several brain regions, including the amygdala, ACC, dorsolateral prefrontal cortex, and FFA (Eberhardt, 2005; Kubota, Banaji, & Phelps, 2012).

An influential study on this topic used fMRI to explore the neural substrates involved in the unconscious and implicit evaluation of Black and White social groups (Phelps et al., 2000). White participants in the scanner viewed pictures of unfamiliar White and Black male faces with neutral expressions. During each presentation, participants indicated whether the face was the same or different from the one immediately preceding it. After scanning was complete, participants completed indirect measures of implicit race bias—the race Implicit Association Test (IAT) and an eye-blink startle response task, as well as an explicit measure called the Modern Racism Scale, which measures conscious, self-reported beliefs about Blacks. Results revealed that participants had a significant pro-White bias in the IAT and a trend toward greater potentiated startle eye-blink when viewing Black compared to White faces, indicating that both implicit measures captured pro-White bias. The explicit measure of racial bias, however, revealed the opposite pattern—participants consciously expressed pro-Black attitudes. Imaging results indicate that the majority of the White participants showed greater amygdala activation when viewing unfamiliar Black faces compared to White faces, although there was significant variability between subjects. Interestingly, amygdala activity correlated with IAT scores, as well as the potentiation of eye-blink startle responses. Thus, in both indirect measures of racial bias, a stronger pro-White bias was correlated with stronger amygdala activity. In a follow-up study, the same experimental procedure was carried out, except this time with pictures of famous, positively regarded Black and White males. Imaging data revealed no consistent pattern of amygdala activity, suggesting that differences in brain activity may be due to culturally acquired knowledge.

Similar to stereotyping, group processes have also been investigated with the tools of cognitive neuroscience. Humans are not only quick to categorize individuals into groups, but they are quick to join and prefer an ingroup. Minimal-group paradigms—meaningless, arbitrary groups created in the experiment—have demonstrated that randomly assigning participants to groups elicits strong ingroup preferences (Tajfel, 1970). In one such study, researchers used the minimal-group paradigm with a mixed-race team to explore brain regions associated with processing ingroup and outgroup members, separate from any preexisting stereotypes or attitudes (Van Bavel, Packer, & Cunningham, 2008). White participants were told that they had been randomly assigned to one of two different groups, and in order to continue the study they had to learn their team members by studying pictures of their faces. The teams comprised 12 individuals, always six Black and six White. The learning task consisted of categorizing the pictures as belonging to the participant's group or not. During scanning, participants completed a categorization task, identifying each face according to team membership or race. The imaging results demonstrate that both real-world and minimal ingroup members were associated with greater activity in the fusiform gyri, amygdala, dorsal striatum, and orbitofrontal cortex (OFC). Moreover, participants with greater ingroup bias in OFC activity reported a stronger preference for ingroup over outgroup members. The results of this study provide some evidence that ingroup members are processed differently than outgroup members. Furthermore, this study demonstrates that race is not the only salient category—any arbitrary distinction can create group biases, even to the extent of overcoming preexisting racial bias.

Status When living in groups, one's social status can provide important social information. Imagine going to dinner with a business partner: the events of the evening may be influenced by whether this business partner is your boss (a high-status person) or a person interviewing for a job under your supervision (a low-status person). Interactions with each of these people may be different, such that with your boss you show nonverbal signs of social engagement (e.g., more eye contact, head nods, and laughing) while with the job interviewee you show more signs of social disengagement (e.g., yawning, doodling, wandering eyes; Guinote, 2007). Although humans are quite adept at recognizing differences in social status and power, only in recent years have we begun to understand how the brain tracks information about one's relative standing in a social hierarchy.

Researchers have identified different brain regions sensitive to viewing people of high and low status depending on whether the hierarchy is stable or unstable (Zink et al., 2008). Participants played an initial training game in which they competed with two confederates in a reaction-time task (study 1) and a visual-discrimination task (study 2). Correct responses resulted in monetary rewards. Unknowingly, participants' outcomes were fixed, such that at the end of the training session a social hierarchy was created in which one player was better and one worse than the participant. These ranks were represented visually to the participant in the scanner via a star system (high rank = 3 stars, participant = 2 stars, and low rank = 1 star). While in the scanner, participants played a similar monetary task with the high- and low-status players separately. In study 1, the social hierarchy was fixed such that the participant was always ranked second (two stars) in the hierarchy. In study 2, the social hierarchy was unstable—participants could move up or down in the hierarchy depending on performance during various parts of the game.

Results of study 1 show that when perceiving others who are higher in rank compared to lower in rank relative to the participant, a series of brain regions were active including bilateral occipital/parietal cortex, ventral striatum, parahippocampal cortex, and dorsolateral prefrontal cortex. Results of study 2 replicated that of study 1, but also showed additional brain regions engaged when the social hierarchy was unstable—specifically bilateral thalamus, right amygdala, PCC, MPFC, primary motor cortex, somatosensory cortex, and supplementary motor area, suggesting that when there is a chance to change one's social status, regions of the brain implicated in valuation, planning, and executive control may be activated in order to help accomplish moving up in the hierarchy.

Mental-state inferences about people

PERSONALITY TRAIT INFERENCES Whereas face perception focuses on a physical aspect of a person (his or her face), trait inferences focus on mental representations of other people's minds. Social psychological research demonstrates that human beings integrate statistical information that informs trait inferences (Heider, 1958). Frequency—high consistency across behavior—allows us to infer that a person will always perform that action (e.g., in the past, Doug has almost always donated to charity). Other people's behavior—consensus—allows us to make inferences about a target person (e.g., hardly anyone donates to charity). The indiscriminate nature of the behavior—distinctiveness—allows us to infer something about a person's traits (Doug also helps the elderly, the sick, and the needy). Together, these three bits of statistical information—high consistency, low consensus, and low distinctiveness—interact to drive trait attributions to people (dispositional attributions; Kelley, 1972; McArthur, 1972).

Research demonstrates that dispositional attributions activate regions of the brain implicated in social processing. In one study, participants read sentences describing a person's behavior, such as "John laughs at the comedian," and were given additional information that manipulated consistency, consensus, and distinctiveness information. For example, in some trials they were told that "John always laughs at the comedian, no one else laughs at the comedian, and John laughs at every comedian." This combination of high consistency, low consensus, and low distinctiveness led people to attribute the cause of John's amusement to John—a trait inference that John is easily amused. When making this attribution, MPFC and STS were more active than with any other information combination, suggesting these parts of the brain support trait inferences (Harris, Todorov, & Fiske, 2005). Interestingly, when this paradigm was replicated with objects engaged in action instead of people (e.g., "The pen falls off the table"), the same statistical information was used to attribute internal causation to objects. However, the part of the brain supporting these inferences was the amygdala, not MPFC or STS (Harris & Fiske, 2008). This suggests a dissociation similar to that observed for motion perception: only people can drive specific activity patterns in the social brain, even when using the same statistical information.

EMPATHY AND OTHER SOCIAL EMOTIONS If personality trait inferences are predominantly cognitive, empathy is a predominantly affective concept. In reality, cognition and emotion are not mutually exclusive psychological processes, but such a distinction enables parsimonious explanation. Social psychologists define empathy as the ability to experience what someone else is feeling, accompanied by a motivation to help that other person (Batson, 1991). Empathy is distinct from sympathy, which involves simply feeling sorry for someone. The last decade has seen an explosion of research on empathy and the brain.

Empathy recruits a variety of brain regions. Empathy for pain involves the anterior insula and ACC, while other forms of empathy involve the STS, amygdala, and OFC (Bernhardt & Singer, 2012). Evidence to support an affective link to empathic responding comes primarily from one theory of empathic concern—a motivation to avoid a negative feeling in oneself. An ERP study in physicians nicely illustrates this idea. Physicians and regular control participants were shown pictures of body parts pricked by a needle or a Q-tip (Decety, Yang, & Cheng, 2010). Whereas nonphysicians showed early differentiation in the N110 and later differentiation in the P300, physicians showed no such response differentiation to witnessing pain. This suggests either an early emotion-regulation process in physicians to dissipate the potential negative affect that results from witnessing someone in pain, or a selection effect. In either case, this does not suggest that physicians are not empathic, but regulating empathic feelings may facilitate adequate execution of their job.

Social emotions are not simply complex configurations of basic emotions. Though basic emotions can occur in a social context, social emotions are distinct because they require mental-state inferences to trigger the emotion, and can result in mental-state inferences as a consequence of the experience of the emotion. Social psychology demonstrates that social groups elicit distinct emotional responses beyond simply like and dislike (Fiske, Cuddy, Glick, & Xu, 2002). Social groups perceived as high in warmth (good or ill intentions) and competence (ability to enact intentions) personality traits (e.g., cultural ingroups) elicit pride and admiration, while social groups perceived as high in warmth but low in competence (e.g., the elderly) elicit pity and sympathy. Social groups low in warmth and high in competence (e.g., the rich) elicit envy, while social groups low in both personality traits (e.g., the homeless) elicit disgust. While emotions like pride, envy, pity, and sympathy are social emotions that can only be experienced in the actual, imagined, or implied presence of another person (Harris & Fiske, 2009), both people and objects can elicit disgust. Disgust in a social context triggers negative moral evaluations and can lead to dehumanization. Participants show reduced activity in MPFC and STS when reporting disgust to stereotypic representations of homeless people compared to stereotypic representations of social group members who elicit pride, envy, or even pity (Harris & Fiske, 2006). A separate study demonstrates that when participants are forced to think about the person's preferences (i.e., infer a mental state) before viewing them, they do not show a reduced MPFC and STS response to homeless people (Harris & Fiske, 2007). Together, these studies demonstrate that not all people in all contexts engage parts of the brain implicated in social processing.

Self-Perception Humans engage in self-referential thoughts and processes constantly. William James, the father of psychology, was one of many to question what it meant to think about "me" and the implications of having a self. But the self, if it exists, is merely a byproduct of the brain—a mental-state inference. Some research has suggested that taking a self-perspective engages a distinct neural system from the one required to represent the mental states of others (Kelley et al., 2002; Mitchell, Banaji, & Macrae, 2005). Psychological research demonstrates that adjectives processed in reference to the self are better recalled than items processed for their general meaning only, now known as the self-reference effect (Rogers, Kuiper, & Kirker, 1977). Although the self has not been localized in the brain, neuroscience research is beginning to uncover several key regions that underlie self-processing.

In an early fMRI investigation, researchers wanted to understand why the self-reference effect occurs (Kelley et al., 2002). To answer this question, participants were instructed to judge a trait adjective in one of three ways: either by making a judgment about the self ("Does the adjective describe you?"), other ("Does the adjective describe (then) current U.S. President George Bush?"), or case ("Is the adjective presented in uppercase letters?"). Participants then performed a subsequent memory test, which allowed the researchers to conduct a traditional levels-of-processing contrast, comparing semantic-based processing to surface-based processing. Behavioral results showed that trait adjectives judged in both semantic conditions were later remembered better than adjectives that were judged by their surface features. In the brain during self-versus-other comparisons, results demonstrated greater BOLD activation in MPFC and PCC, while a wider network, including the left inferior frontal cortex and ACC, showed more activity during semantic judgments. This suggests that the self may be the most relevant social agent, serving as an anchor when perceiving other people.

Social Decision Making People make decisions every day in a social context. Whether we are deciding to do a favor for a friend or close a deal with a potential business partner, our decisions have consequences that lead to significant rewards and punishments, such as a better relationship with our friend or a poor business transaction. Therefore, it is important to understand how our decisions are influenced by the presence or absence of other people and how we incorporate social information into our decision-making processes.

Many social decision-making studies use economic games (e.g., the trust, ultimatum, or dictator games, etc.; Rilling & Sanfey, 2011) to study the biological underpinnings of social decision making. These economic games are ideal because they allow researchers to predict what people *should* do (according to rational economic theory) and compare it to what people *actually* do. Participants' behavior in these games often contradicts what economic or game theory suggests people should do, illustrating that social decision making engages processes other than pure reason.

How does the social context change the way decision-making structures like the striatum process information? In one study, participants received biographies suggesting good, bad, and neutral moral character for three hypothetical partners in a trust game. When asked whether they would like to invest with the partner for a chance to earn a larger reward if the partner reciprocated, participants were more likely to invest with the morally good partner than the neutral and bad partners. However, there were no real differences between the partners' reinforcement rates—they all returned a profit only 50% of the time. If social decision making is like any other kind of decision making, participants should learn to treat each of the three partners the same because there are no differences in how likely they are to return a profit; that is, they should update their beliefs based on feedback in the game. However, results show that participants were more likely to invest with the morally good partner than the neutral partner and least often with the morally bad partner, even in later parts of the game when they should have updated beliefs based on feedback. Interestingly, differential neural signals typically associated with positive and negative feedback were only observed for the partner who had a neutral description of his moral character. For the morally good and bad partners, feedback signals in the striatum looked similar, suggesting reduced reliance on feedback signals when prior social information is available (Delgado, Frank, & Phelps, 2005).

Conclusion and societal implications

In closing, the cognitive neuroscience of person perception represents a variety of research using the tools of cognitive neuroscience but informed by theory in social psychology, evolutionary anthropology, behavioral economics, and philosophy. As such, it reflects a truly interdisciplinary approach to a complex phenomenon that has puzzled philosophers for centuries. Getting a peek into the proverbial black box using the tools of cognitive neuroscience illustrates the complexities of the process, holding the promise of facilitating understanding of person perception. This understanding is important because inferring another's mental states ascribes that person full humanity, triggering empathic concern and imbuing that person with moral protections reserved for human beings (Harris & Fiske, 2009). Given the occasional failure of people to infer others' mental states (Harris & Fiske, 2006; 2007; 2009), person perception hints at a possible mechanism to facilitate human atrocities. As the cognitive neuroscience of person-perception advances, the field can better inform policy to combat group-based prejudices and behaviors that stem from a lack of empathic concern.

REFERENCES

Allport, G. W. (1954). *The nature of prejudice.* New York, NY: Addison-Wesley.

Batson, C. D. (1991). *The altruism question: Toward a social-psychological answer.* Hillsdale, NJ: Erlbaum.

Bernhardt, B. C., & Singer, T. (2012). The neural basis of empathy. *Annu Rev Neurosci, 35,* 1–23.

Chartrand, T. L., & Bargh, J. A. (1999). The chameleon effect: The perception-behavior link and social interaction. *J Pers Soc Psychol, 76,* 893–910.

Decety, J., Yang, C. Y., & Cheng, Y. (2010). Physicians down-regulate their pain empathy response: An event-related brain potential study. *NeuroImage, 50,* 873–882.

Delgado, M. R., Frank, R. H., & Phelps, E. A. (2005). Perceptions of moral character modulate the neural systems of reward during the trust game. *Nat Neurosci, 8*(11), 1611–1618.

Eberhardt, J. L. (2005). Imaging race. *Am Psychol, 60,* 181–190.

Eisenberger, N. I., Lieberman, M. D., & Williams, K. D. (2003). Does rejection hurt? An fMRI study of social exclusion. *Science, 302,* 290–292.

Fabbri-Destro, M., & Rizzolatti, G. (2008). Mirror neurons and mirror systems in humans and monkeys. *Physiology, 23,* 171–179.

Fiske, S. T., Cuddy, A. J., Glick, P., & Xu, J. (2002). A model of (often mixed) stereotype content: Competence and warmth respectively follow from perceived status and competition. *J Pers Soc Psychol, 82,* 878–902.

Fiske, S. T., & Neuberg, S. L. (1990). A continuum of impression formation, from category-based to individuating processes: Influences of information and motivation on attention and interpretation. In M. P. Zanna (Ed.), *Advances in experimental social psychology* (Vol. 23, pp. 1–74). New York, NY: Academic Press.

Fiske, S. T., & Taylor, S. E. (2013). *Social cognition: From brains to culture.* London: Sage.

Gauthier, I., Skudlarski, P., Gore, J. C., & Anderson, A. W. (2000). Expertise for cars and birds recruits brain areas involved in face recognition. *Nat Neurosci, 3,* 191–197.

Gauthier, I., Tarr, M. J., Anderson, A. W., Skudlarski, P., & Gore, J. C. (1999). Activation of the middle fusiform "face area" increases with expertise in recognizing novel objects. *Nat Neurosci, 2,* 568–573.

Gray, H. M., Gray, K., & Wegner, D. M. (2007). Dimensions of mind perception. *Science, 315,* 619.

Guinote, A. (2007). Behavioral variability and the situated focus theory of power. *Eur Rev Soc Psychol, 18,* 256–295.

Harris, L. T., & Fiske, S. T. (2006). Dehumanizing the lowest of the low: Neuroimaging responses to extreme outgroups. *Psychol Sci, 17,* 847–853.

Harris, L. T., & Fiske, S. T. (2007). Social groups that elicit disgust are differentially processed in the mPFC. *Soc Cogn Affect Neurosci, 2,* 45–45.

Harris, L. T., & Fiske, S. T. (2008). Brooms in Fantasia: Neural correlates of anthropomorphizing objects. *Soc Cognition, 26,* 209–222.

Harris, L. T., & Fiske, S. T. (2009). Social neuroscience evidence for dehumanised perception. *Eur Rev Soc Psychol, 20,* 192–231.

Harris, L. T., Todorov, A., & Fiske, S. T. (2005). Attributions on the brain: Neuro-imaging dispositional inferences, beyond theory of mind. *NeuroImage, 28*(4), 763–769.

Haxby, J. V., Gobbini, M. I., & Montgomery, K. (2004). Spatial and temporal distribution of face and object representations in the human brain. In M. Gazzaniga (Ed.), *The cognitive neurosciences* (pp. 889–904). Cambridge, MA: MIT Press.

Haxby, J. V., Hoffman, E. A., & Gobbini, M. I. (2000). The distributed human neural system for face perception. *Trends Cogn Sci, 4*(6), 223–233.

Heberlein, A. S., Adolphs, R., Tranel, D., & Damasio, H. (2004). Cortical regions for judgments of emotions and personality traits from point-light walkers. *J Cogn Neurosci, 16,* 1143–1158.

Heider, F. (1958). *The psychology of interpersonal relations.* New York, NY: Wiley.

Johansson, G. (1973). Visual perception of biological motion and a model for its analysis. *Percept Psychophys, 14*(2), 201–211.

Kelley, H. H. (1972). Attribution in social interaction. In: E. E. Jones, D. E. Kanouse, H. H. Kelley, R. E. Nisbett, S. Valins, & B. Weiner (Eds.), *Attribution: Perceiving the cause of behavior* (pp. 1–26). Hillsdale, NJ: Erlbaum.

Kelley, W. M., Macrae, C. N., Wyland, C. L., Caglar, S., Inati, S., & Heatherton, T. F. (2002). Finding the self? An event-related fMRI study. *J Cogn Neurosci, 14*(5), 785–794.

Kubota, J. T., Banaji, M. R., & Phelps, E. A. (2012). The neuroscience of race. *Nat Neurosci, 15,* 940–948.

McArthur, L. Z. (1972). The how and what of why: Some determinants and consequences of causal attribution. *J Pers Soc Psychol, 22,* 171–193.

Mitchell, J. P., Banaji, M. R., & Macrae, C. N. (2005). The link between social cognition and self-referential thought in the medial prefrontal cortex. *J Cogn Neurosci, 18,* 1306–1315.

Mitchell, J. P., Macrae, C. N., & Banaji, M. R. (2005). Forming impressions of people versus inanimate objects: Social-cognitive processing in the medial prefrontal cortex. *NeuroImage, 26,* 251–257.

Oberman, L. M., Pineda, J. A., & Ramachandran, V. S. (2007). The human mirror neuron system: A link between action observation and social skills. *Soc Cogn Aff Neurosci, 2*(1), 62–66.

Phelps, E. A., O'Connor, K. J., Cunningham, W. A., Funayama, E. S., Gatenby, J. C., Gore, J. C., & Banaji, M. R. (2000). Performance on indirect measures of race evaluation predicts amygdala activation. *J Cogn Neurosci, 12*(5), 729–738.

Rilling, J. K., & Sanfey, A. G. (2011). The neuroscience of social decision-making. *Annu Rev Psychol, 62,* 23–48.

Rogers, T. B., Kuiper, N. A., & Kirker, W. S. (1977). Self-reference and the encoding of personal information. *J Pers Soc Psychol, 35,* 677–678.

Saxe, R. (2010). The right temporo-parietal junction: A specific brain region for thinking about thoughts. In A. Leslie & T. German (Eds.), *Handbook of theory of mind.* New York, NY: Psychology Press.

Saygin, A. P. (2007). Superior temporal and premotor brain areas necessary for biological motion perception. *Brain, 130,* 2452–2461.

Schiller, D., Freeman, J. B., Mitchell, J. P., Uleman, J. S., & Phelps, E. A. (2009). A neural mechanism of first impressions. *Nat Neurosci, 12,* 508–514.

Sherwood, C. C., & Smaers, J. B. (2013). What's the fuss over human frontal lobe evolution? *Trends Cogn Sci, 17*(9), 432–433.

Stimpson, C. D., Tetreault, N. A., Allman, J. M., Jacobs, B., Butti, C., Hof, P. R., & Sherwood, C. C. (2011). Biochemical specificity of von Economo neurons in hominoids. *Am J Hum Biol, 23,* 22–28.

Tajfel, H. (1970). Experiments in intergroup discrimination. *Sci Am, 223,* 96–102.

Van Bavel, J. J., Packer, D. J., & Cunningham, W. A. (2008). The neural substrates of in-group bias: A functional magnetic resonance imaging investigation. *Psychol Sci, 19*(11), 1131–1139.

van Kemenade, B. M., Muggleton, N., Walsh, V., & Saygin, A. P. (2012). The effects of TMS over STS and premotor cortex on the perception of biological motion. *J Cogn Neurosci, 24*(4), 896–904.

Zink, C. F., Tong, Y., Chen, Q., Bassett, D. S., Stein, J. L., & Meyer-Lindenberg, A. (2008). Know your place: Neural processing of social hierarchy in humans. *Neuron, 58*(2), 273–283.

85 Mindreading and Privacy

ADINA L. ROSKIES

ABSTRACT Neuroimaging techniques provide unprecedented access to a variety of kinds of information about the brain, including, to some extent, the contents of thoughts. This chapter describes the extent to which fMRI allows us to "read minds," including recent advances in our ability to decode semantic content, to reconstruct visual and auditory perceptions, and to identify memories and tell sincere from deceptive responses. As this chapter chronicles, our current abilities to read minds are more limited than many realize, but even moderate prospects for improvement raise ethical and legal questions about how this information is related to privacy rights and the evidential status of imaging data. These are pressing questions for society that are only now beginning to be explored.

With today's technology, information that was formerly assumed to be private is no longer so. Your mobile phone monitors your physical whereabouts, your employer can read your email, and your supermarket and countless websites track your purchases and preferences. Recently, we learned that the government may secretly monitor your data via various websites (Gallagher, 2013). It might seem that the last secure realms of privacy are within one's own head: the contents of our thoughts generally remain private unless we choose to disclose them to others through speech, gestures, or other voluntary actions. Unlike a lot of other information, the contents of one's thoughts seem to be directly accessible only to the thinker, unless revealed by voluntary disclosure[1] (Gallagher, 2013).

But many worry that the bastion of our mind is or may soon be breached. Over the past few decades, neuroscientific technologies have been developed that now threaten to penetrate even the inner sanctum of our minds. Or so goes the worry. This chapter explores the scope and imminence of this threat, its legal standing, and the ethical challenges it raises.

Brainreading and mindreading

There are a variety of kinds of information that may be accessible via neurotechnologies. For example, brain scans can reveal medically relevant information about a subject's brain, such as early signs of dementia (Teipel et al., 2013), indicators of mental illness (Mueller, Keeser, Reiser, Teipel, & Meindl, 2012; Tang, Wang, Cao, & Tan, 2012; Whalley et al., 2013), or incidental findings such as the unanticipated presence of an aneurysm or tumor (Carré et al., 2013; Paulsen, Carter, Platt, Huettel, & Brannon, 2011; Scott, Murphy, & Illes, 2012). There is evidence that certain kinds of relatively stable character traits, such as risk-aversiveness or anxiety, may be inferred on the basis of neuroimaging data (Carré et al., 2013; Paulsen et al., 2011), and evidence that such traits may be inferred on the basis of data collected from tasks not explicitly probing character traits raises the worry that such information could be gained without the subject's knowledge or consent (Farah, Smith, Gawuga, Lindsell, & Foster, 2008). Along the same lines, there is evidence that neuroimaging data can reveal information about a person's unconscious attitudes and biases (Azevedo et al., 2013; Harlé, Chang, van 't Wout, & Sanfey, 2012; Stanley et al., 2012; Van Bavel, Packer, & Cunningham, 2008), again often by way of passive measures or alternative tasks. Because such information could be obtained without a subject knowing or consenting to the purposes for which it might be used, and because this information could be of interest to third parties (such as insurance companies, potential employers, medical providers, or law enforcement), protections must be in place to ensure that such information is not surreptitiously obtained or misused. There has been significant discussion on this front in the healthcare literature, and many of the provisions designed to protect medical or genetic data provide a potential model for protecting the privacy of this sort of information (Scott, 2000).

In this chapter, I intend to consider a different kind of information that one might glean from neuroimaging: information about the content of thoughts. In contrast to the information mentioned above, which is correlated to diseases or character traits, the information I focus on here is in occurrent mental states with propositional content. It is the information one would obtain if mindreading were possible. These days, many assume that neuroscience technology already has the

[1]It would be a mistake to overlook involuntary signals that provide insight into mental content, such as microfacial expressions: we read minds in a variety of ways. However, this access is rather indirect, and the revealed content is not precise.

capacity to infringe mental privacy, or that it is all but upon us. But is this realistic?

What would it be for a technology to read our minds? "Reading" at a minimum involves a mapping of a physical pattern to meaning. It is useful to consider a spectrum whose extremes are what I'll call "brainreading" on one end and "mindreading," on the other. There is a practical difference between the two related to the fine-grainedness of the content that can be discerned from measurements of brain activity. To the extent that the mapping is effected merely by rough and brute-force empirical correlations between measurements of the physical state and mental functions, we should view it as brainreading. As the techniques progress to allow inferences of fine-grained content on the basis of a generative scheme, they would approach the criteria for mindreading. Practically speaking, the distinction may depend on how systematic and generative the mapping is that we establish between mental states and physical ones.

Brainreading allows one to infer coarse-grained content from brain data largely on the basis of empirical correlations: for example, emotional reactivity or fear can be inferred from amygdala activation, and perception or imagining a face can be inferred from activation in the fusiform face area, or FFA. Thus, although brainreading provides some information regarding mental content, it fails to disambiguate between a large number of semantically different possibilities. Is the subject experiencing fear or anxiety? What is the object of their fear? It is present or imagined? Whose face are they seeing or imagining? Even though brainreading relies upon a nontrivial understanding of the functional anatomy of the brain, it is far from allowing us to precisely fix the content of the mental states of the subject.

On the other hand, mindreading with brain-imaging devices would require us to be able to determine the propositional content of mental states and distinguish approximately the same subtle level of content that language enables us to express. For example, what is the propositional attitude of the subject: belief, disbelief, hope, fear, love? What is the content of their thought? About whom are they thinking? What are the relata and relations that constitute their mental contents? Can we distinguish the thought that "Tom angered Mary" from the thought that "Mary angered Tom"? Can we distinguish "Tom was angry" from "Tom was outraged" from "Tom was disappointed"?[2] Although the distinction between mindreading and brainreading may not be sustainable in principle, determining where one may be on a brainreading/mindreading spectrum may be important in practice, as may what it tells us and how accurately it does so. These issues may be especially relevant when the data applies to realms of human interaction that trade in shades of gray, such as the ethical, social, and legal.

Brainreading is already here: as I relate below, using our knowledge of anatomical/functional correlations, we can in many cases fill in rough information about the kind of cognitive or emotional state a subject is in, and what processes or cognitive functions are active. What are the prospects for mindreading?

To mind-read we would have to be able to reliably correlate propositional content with brain activity patterns. One way to do this would be to generate a giant look-up table, a "dictionary" that provides a profile of functional activation for each word, phrase, or concept that could serve to specify the content of a mental state (and presumably a reverse dictionary so that we could look up the pattern to translate it into content). Depending on the amount of individual variability in functional brain organization, we might have to create a unique look-up table for each subject. A more parsimonious approach to identifying mental contents would be to develop an accurate model of mental representation, such that one could reliably generate accurate patterns of brain activity to novel words, concepts, or propositions. This would be by far the best bet for practical uses of mindreading, since it is impossible to chart empirically the entire space of possible representations. Researchers have made some headway in showing a limited proof of principle for generative models (discussed below), but the degree to which fine-grained content is encoded systematically (i.e., in a way that can be exploited to generalize) rather than fortuitously in the brain is unclear.

But the problem is harder than just this. Dictionaries provide translations for individual words or conceptual elements. But if we are concerned with content, we are concerned not only with the elements of thought, but also with their relations. After all, "The butler did it" and "The butler did not do it" have contrary meanings, as do "George provoked Harry" and "Harry provoked George." Language and thought are infinitely generative. Thus, unless we wanted to build a dictionary that is impossibly large and unwieldy by testing almost all reasonable strings, we would also have to understand how relational structure in thought or inner speech was encoded in the brain. Whether these aspects of content can be decoded from brain signals is an open question. A similar problem exists for the attitudes one takes to

[2]If radical philosophical externalism about mental content is correct, it would be in principle impossible to infer precise mental content purely on the basis of the brain's physical states. However, mind-world regularities might still enable relatively robust inferences about content.

propositions. In addition, whether the way in which concepts combine in terms of their neural signals is compositional in the way that language is compositional is also unknown (see, e.g., Reverberi, Görgen, & Haynes, 2011). The tasks for mindreading are threefold: (1) identifying activity corresponding to individual content elements; (2) identifying activity reflecting conceptual relations; (3) being able to infer content across subjects.

Can we read minds with neuroimaging methods?

In this section, I briefly relate some of the landmark early brainreading studies that point the way to the possibility of mindreading, and then discuss the current status of neuroscientific methods for discerning the contents of our thoughts. Some important questions to bear in mind: How good is the technique for identifying or reconstructing content? How narrow is the hypothesis space of the experiment described? How generalizable are the results? What assumptions do the methods used depend upon, and are those assumptions warranted? What are the technical limitations of the techniques? What spatial and temporal resolution is needed to differentiate between specific mental states? To what extent must the subject cooperate in order for the method to work? Is the information being probed conscious or latent? Thinking about these questions will enable the reader to assess the likelihood that the experiments described will pose a real threat to privacy in real-world contexts.

Early Steps: Brainreading In one of the first brainreading studies, O'Craven and Kanwisher (2000) presented subjects with pictures of faces and places and demonstrated the expected changes in blood oxygen level–dependent (BOLD) signal in the fusiform face area (FFA) and parahippocampal place area (PPA) to perception of faces and places, respectively. They then showed that these same brain regions were active during mental imagery of those same stimulus classes. This turns out to be a common finding: many of the same brain regions involved in processing external stimuli are also active during thoughts about the same type of stimuli (see, e.g., Polyn, Natu, Cohen, & Norman, 2005). Having localized the FFA and PPA in their individual subjects, the researchers then showed they could classify the mental state of the subject on the basis of the brain data. They instructed subjects to imagine looking at faces or places for different epochs of time and found that the variation of the BOLD signal in the FFA and PPA could allow a scientist unaware of the instructions given to the subject to determine the epochs during which the subject was thinking of a face and during which he or she was thinking of a place with ~85% accuracy, significantly higher than chance. Their results demonstrated that specific *classes* of thought content could be determined from brain-activation data.

Around the same time, Tong, Nakayama, Vaughan, and Kanwisher (1998) demonstrated that fMRI had the ability to show differences in conscious perceptual mental states that were not also accompanied by differences in perceptual input. They simultaneously presented subjects with a picture of a face at one eye, and a picture of a building at the other, thereby causing binocular rivalry, a phenomenon in which signals from the two eyes compete for dominance in consciousness. Rather than seeing both stimuli, the subject is aware of only one of the visual stimuli at a time. Subjects indicated which of the two stimuli they were aware of seeing with a button press. Even though the visual input to the eyes did not change, Tong et al. found that activation in the FFA and PPA varied with the conscious experience of a face or place, and the conscious perception could be predicted by the BOLD activation levels, which were as robust as when they were viewing nonrivalrous stimuli.

Importantly, researchers who were "brainreading" in these cases knew that the stimuli fell into one of two broad classes, so they needed only to determine which of two mental state types was more likely than the other. In addition, the studies were done on stimuli for which the brain shows distinct anatomical specificity for processing. The studies reveal nothing about the ability to identify mental states of arbitrary class using fMRI, or about the possibility of distinguishing particulars within these classes. For example, the studies do not address whether it is possible to distinguish thinking about Bill Clinton from thinking about Robin Williams, whether distinctions can be made among arbitrary numbers or kinds of classes for which separate brain areas are not known to mediate specific types of representations, or whether sense can be made of the propositional content of the subjects' mental states.

Multivariate Analysis of Functional MRI Data and Decoding of Semantic Information Neuroimaging underwent a sea change with the advent of multivariate techniques for data analysis (also called multivoxel pattern analysis, or MVPA). MVPA analyzes patterns of brain activity across many voxels, rather than just net change of signal in a localized region. It had long been debated whether various cortical regions were specialized for processing different stimulus classes or whether processing was distributed across many cortical regions. The discovery of the FFA and

PPA supported the modularity thesis, raising the possibility that all kinds of categories had various dedicated cortical-processing modules (whether this is the correct interpretation for certain areas, such as the FFA, is still a matter of dispute). In a seminal study, Haxby, Gobbini, Ishai, Schouten, and Pietrini (2001) presented subjects with pictures of objects from a variety of categories, including faces, shoes, tools, chairs, and cats. Haxby et al. found that activity for all these categories was widespread across the cortex, and, in one of the pioneering uses of multivariate techniques in fMRI analysis, showed that *patterns* of activation differed among brain regions for each stimulus class, even when those regions did not show significant net changes in activity between categories. Moreover, even when one eliminated the information from the brain region responding maximally to a class of stimuli (such as ignoring the information from FFA for face processing), one could still identify the stimulus class to which the item belonged on the basis of activation patterns in other cortical areas. Thus, information encoding the identity of visual categories was widespread throughout cortex. This cast doubt on the highly modular model of visual recognition that is suggested by taking the FFA and PPA to be modules dedicated to processing only information from their preferred categories.

In the early 2000s, researchers began in earnest to develop multivariate techniques for pattern recognition, and it is here that the prospects for mindreading truly began. In a groundbreaking series of studies, a group from Carnegie Mellon University showed that brain signatures related to perceiving individual objects could be recognized, and that a generative model based upon statistical association could to a large degree predict whole-brain fMRI patterns. In an initial study, Shinkareva et al. (2008) showed that they could predict with a mean of 80% accuracy which of a set of 10 line drawings a person was looking at, based up on his fMRI data, and with approximately 10% greater accuracy which of two object categories the drawing was from. They also showed that these patterns were standard enough across subjects that the category of object was predictable when the classifier was trained only on data from other subjects. This study suggests that individual objects have unique and discernable neural signals within individuals, raising the possibility that particular objects of mental states could be decoded if classifiers could be trained on a broad array of data from an individual subject. Perhaps more significantly, it suggests that the overall structure of object encoding and processing is uniform enough across individuals to enable decoding of some mental states based on information obtained from others. In a landmark companion paper, Mitchell et al. (2008) trained a classifier to predict the fMRI signatures of 60 objects drawn from 12 categories. The classifier related the statistics of word associations between the objects and common verbs with the MRI results. Then, when presented with a novel object upon which the classifier had not been trained, the classifier predicted an fMRI activation pattern that was very similar to the actual fMRI pattern observed when the subject saw that object. These results suggest that the way our brains encode object information is systematic enough that a reasonably good model of the semantics of object representation could be developed to generalize to novel stimuli. If so, general pattern-recognition systems could potentially be developed to decode arbitrary kinds of mental content. This study is the first indication of the feasibility in principle of a generative model of object semantics based on brain data.

While these developments are impressive, even startling, it is worth noting the level of proficiency so far attained. For example, a recent paper reports being able to decode modality-independent semantic information from brain data acquired during a categorization task with stimuli presented in various modalities (auditory, visual pictures, written and spoken words; Simanova, Hagoort, Oostenveld, & van Gerven, 2012). The classifier could identify categories across modalities. For instance, it distinguished between recollected tool or animal categories during a free-recall task with accuracies of approximately 70%. Although impressive, it is still far from demonstrating that fine-grained abstract semantic information is decodable from brain data.

Decoding by Reconstruction Most of the work discussed above correlates content with activity in higher cortical areas, which presumably reflect the semantic content of the mental state rather than low-level properties of sensory processes. However, to date the most successful decoding schemes have relied upon the relatively well-understood and regular properties of early stages in cortical sensory processing. The next sections discuss the state of and prospects for various types of sensory and memory reconstruction.

The Current Status of Visual Reconstruction The visual system is perhaps the best understood cortical pathway. Thirion and colleagues (2006) have used insights from the organization of the visual system to reconstruct simple visual stimuli from neuroimages, showing proof of principle that one could infer stimuli from knowledge of the transfer function from visual stimulus to cortex. Drawing on this general approach, Gallant et al. (Kay, Naselaris, Prenger, & Gallant, 2008) developed methods to reconstruct natural visual scenes from brain data. By examining cortical activation profiles to a large set of images, they construct a receptive field model for each voxel in early visual areas (i.e., a model of how various low-level image features at a location of visual space maps to brain activity). Their model described tuning along spatial, orientation, and spatial frequency domains (Kay et al., 2008). Then, they present subjects with a novel image drawn from a large library of images and measure the brain activity. Based on the activation pattern in early visual areas, they can identify the image in the library most likely to have produced that activation pattern. With a library of 120 images, the decoding selected the correct image 92% of the time (chance performance is 8%). With a much larger library (1,000 images), accuracy remained high, falling to 82%. The authors estimate that performance on the entire Google library of images would remain well above chance. The authors also note that decoding was effective with single-trial data. Single-trial data is noisy data, so one would expect that selection accuracy would be lower than that found in the earlier studies. It is, but at approximately 50% (with the library of 120), it remains well above chance. The significance of this finding is that it makes possible the prospect of real-time decoding.

Later work by the same group improved upon the early visual reconstruction paradigm (Naselaris, Prenger, Kay, Oliver, & Gallant, 2009). When there were errors with the earlier method, the images selected by the classifier were structurally similar but semantically quite different than the target image. To address this, Gallant and colleagues combined a visual decoding scheme (like that described above) with a semantic decoder, which relied upon information from anterior brain areas. They also combined this with a Bayesian approach, which used a prior based on the statistics of natural image structure gleaned from an image database to help with image selection. The three approaches combined allowed them to "reconstruct" images that were structurally and semantically similar to the target image (Naselaris et al., 2009). Importantly, even this method does not do pure bottom-up reconstruction: the reconstructions always correspond to an image in the original database. Since for a real-world reconstruction task there are an infinite number of possible images, this method cannot hope to reproduce exactly any arbitrary viewed image. However, with a large enough database, the reconstruction for an arbitrary natural image could still be quite good.

In the most recent work, Gallant and colleagues (Nishimoto et al., 2011) have extended this approach yet further, to enable the reconstruction of dynamic visual scenes (movie clips) from brain data. Again, they relied upon priors obtained from a large library of video clips. When tested on novel clips not included in the library, the algorithm selected the clip in the library most likely to be the stimulus. The authors report a high degree of similarity between the chosen clip and the novel stimulus clip. Work by Hasson and colleagues (Hasson et al., 2008; Hasson, Nir, Levy, Fuhrmann, & Malach, 2004) indicates that human brains share common activity profiles when viewing dynamic natural scenes, which implies that this method will work relatively well across subjects. Nishimoto et al. (2011) suggest that their method for reconstructing dynamic visual stimuli may also be useful for reconstructing dynamic visual imagery from brain data. Of course, the success of this kind of approach will depend upon the library of clips that make up the prior, and the similarity of mental imagery to natural dynamic stimuli.

Prospects for Auditory Reconstruction Just as visual images or scenes can be seen or imagined, so auditory experiences can be heard or imagined. As in vision, the human auditory system follows simple organizational principles in primary cortical areas, with increasing complexity as one ascends the cortical system. This organization has been exploited to enable some aspects of sound to be decoded from fMRI signals. There is evidence, for instance, that different patterns of brain activity encode aspects of the category of acoustic signal (human speech, animal sounds, etc.; Formisano, de Martino, Bonte, & Goebel, 2008). Although no general reconstruction of heard speech from sound has been possible thus far, a recent study may provide proof of principle. In that study (Formisano et al., 2008), subjects were asked to listen to repeated presentations of three vowel sounds, each spoken by three different speakers. Using pattern-recognition methods and training on this data set, experimenters were able to determine with approximately 70% accuracy which of the three sounds was being uttered, and by which speaker, even on trials not in the training set. On the basis of their data Formisano et al. postulate separate distributed regions of cortex for encoding phonemes and speaker identity. They also found that they could train a classifier on vowels from one speaker and correctly classify the vowels spoken by the others with approximately 60% accuracy. This suggests, in addition, that the speech sound cortical representations are acoustically invariant along certain dimensions. This may be the first demonstration of the feasibility of decoding auditory speech information, but it has a number of significant limitations that should make one circumspect about the near-term prospects of decoding speech from brain activity.

For one thing, the classifier only discriminated between three vowels, a highly impoverished set of stimuli relative to the approximately 44 phonemes in English, and many more in some other languages. Secondly, these sounds were presented in isolation, not embedded in a speech stream. Indeed, the temporal order of sounds is a crucial aspect of language—only order disambiguates the phonemic sequences of "super" and "pursue," and grammar is highly dependent on temporal order. Merely distinguishing individual phonemes is a long way from decoding real speech and speech content. Along the same lines, Formisano et al. found that speaker identity was discriminated by the classifier primarily by the fundamental frequency of the auditory signal (Formisano et al., 2008). While this may suffice to discriminate between certain speakers, there are many other aspects of speech sounds that distinguish individual voices, and other markers will have to be found in order to be able in general to recognize the identity of a speaker from a brain signal. Thus, although the study provides reason to anticipate future progress in decoding speech from brain data, it is far from showing that a general decoder would be feasible.

Identifying Memories and Lie Detection One long-standing goal of memory research has been to understand the way in which memories are encoded in the brain. Such knowledge could potentially be leveraged into a method of decoding memory content, or for assessing the veridicality of memory-like signatures (Garoff-Eaton, Kensinger, & Schacter, 2007). However, despite ongoing advances in understanding memory processes, little progress has been made in understanding content-specific aspects of encoding and retrieval.

With regard to aspects of memory neuroimaging relevant to mindreading, most of the work has focused on proof of possibility. For instance, a paper that claims to "decode episodic memory traces" shows that particular memories can be discriminated with MVPA techniques (Chadwick, Hassabis, Weiskopf, & Maguire, 2010). Subjects were repeatedly familiarized with three different short video clips and practiced recalling them by visualization on demand. They then recalled these clips in the scanner, in both a cued and a self-generated order. Pattern classifiers trained on activation data from hippocampus could distinguish which of the three clips was recalled with better than chance accuracy (approximately 45% accuracy, where 33% is chance). This shows that some memory-related information is accessible in temporal lobe structures, but many interpretive questions remain. For instance, it is questionable whether recollection of practiced video clips is paradigmatic of episodic memory. In addition, being able to distinguish among a small set of alternatives is not "decoding": there is no evidence that the full range of information needed for classifying or reconstructing a remembered stimulus is recoverable from imaging data. Claims of abilities to identify intentions with fMRI are subject to the same limitations (see, e.g., Haynes et al., 2007).

Rissman, Greely, and Wagner (2010) explored the extent to which subjective and objective aspects of memory recall can be recovered from imaging data. MVPA was applied to data from a face memory task, in which subjects had to indicate whether a given face stimulus was remembered from a previous study session and to indicate their confidence in its familiarity. Classifiers were trained to distinguish subjective familiarity and objective prior exposure. While these factors are often correlated, they can be disentangled. While the classifiers were fairly good at indicating how subjectively familiar stimuli were (~75% accuracy), they were at chance in classifying the objective status of stimuli when subjective status was controlled for.[3] This study also examined whether decoding ability was affected by the context of retrieval. Subjects were exposed to stimuli in the context of an attractiveness-rating task, and then scanned during an implicit memory (gender-identification) task and later an explicit memory task. Classifiers were at chance at detecting old from new faces in the implicit task. These results suggest that while fMRI might be able to be used to distinguish subjective memory states in forensic contexts, its value will be highly limited in noncooperative contexts or in application to questions of objective veridicality.

Lie-detection techniques have perhaps raised the greatest concerns about privacy in the public sphere. An enormous amount of effort has been directed to adapting neuroimaging techniques to distinguishing lies from sincere responses. While such measures are relatively effective at distinguishing these in the experimental contexts in which they are developed, there are deep problems with external and ecological validity, and little insight into content-related aspects that could elevate them into true mindreading experiments. To the extent that they work, neuroimaging studies of lie detection tend to be better examples of brainreading than of mindreading: they proceed on the basis of correlation between patterns of brain activity and instances

[3]They also had higher accuracy when only high confidence responses were counted.

of lying or truth-telling, but the correlations could have more to do with arousal, oddball effects, and other physiological correlates of lying in such experiments than with deception or mental content itself. The Rissman et al. (2010) study, discussed above, also suggests some limitations for neuroimaging methods of lie detection: they may work better when tasks are explicit rather than implicit, and are better at classifying subjective than objective memory states. It is also doubtful that the methods so far developed are robust in the face of countermeasures or otherwise noncompliant subjects. For a critical review of neuroimaging for lie detection, see Wagner (2010) and Wolpe, Foster, and Langleben (2005).[4] Finally, we still have little ability to discern brain processes that are relevant to conscious experience from those processes that proceed without conscious awareness. Insofar as it may be important to discern occurrent from latent mental content, this inability is severely limiting.

What might neuroscience be able to discern in the future?

It is likely that with improved technology and methods, our ability to reconstruct the contents of mental states will continue to improve. We will be better able to reconstruct perceptions from brain patterns, but the accuracy of these reconstructions will probably be limited by semantic ambiguities and the reliance on assumptions about priors that may not hold for individuals. Individual differences in functional-anatomical organization, whether due to genetic differences or experience-dependent processes, will also pose limits to what can be generically decoded from brain patterns. The extent of such differences is currently not well understood. More will be possible if models can be tailored on the basis of individual data, but this would require extensive construction of individual models on the basis of empirical data derived from those very individuals, which might then place significant limits upon what state actors can extract from brain data for purposes that might be against individual interests.

Undoubtedly, efforts at reconstruction will extend to areas so far largely ignored. For example, although there is little data so far showing that inner speech can be decoded, there is evidence that inner speech leads to activation of brain structures involved in auditory processing (Shergill et al., 2002) and speech production (Marvel & Desmond, 2012), and thus some prospect that at least some aspects of the inner narrative could be decoded. However, no evidence currently exists to suggest that anything like the stream of consciousness of inner speech will be recoverable from brain data.

The experiments described above exhibit both remarkable progress in neuroimaging in discriminating aspects of mental content, and the significant limitations it faces in succeeding as a general mindreading methodology. Technical limitations in spatial and temporal resolution constrain mapping to fine-grained mental content, and individual variability and the need for compliance limit the contexts in which it can be used.

Ethical and legal implications

The Value of Mental Privacy The prospects for reading mental content, even if limited, raise questions about the value of mental privacy. We are a people obsessed with liberty, and privacy ensures a certain kind of freedom: freedom from surveillance and intervention of unwanted parties, including the state. Despite this, there is substantial philosophical controversy about both the nature of and justification for privacy as a right or value, and little written about mental privacy, perhaps because it has for so long been taken for granted.

Legal Protections for Mental Privacy The unsettled nature of the philosophical discourse about privacy is mirrored by the unsettled role for privacy rights in the law. Intimations of the importance of privacy are found in the US Constitution, but nowhere does the Constitution explicitly confer a right to privacy on citizens. The Supreme Court has variously interpreted the First, Fourth, Fifth, Ninth, and Fourteenth Amendments as grounding protections for privacy, most notably in its rulings about substantive due process. The Fifth Amendment prohibits the government from compelling a person to testify against himself in a criminal case, thus providing him a limited right to mental privacy in a particular context. The Fourth Amendment affirms the "right of the people to be secure in their persons, houses, papers, and effects, against unreasonable searches and seizures." Both amendments are suggestive of a right to privacy that extends to the mind, but neither is clear about the scope of the right.

Rulings from a series of Supreme Court cases do little to clarify the scope of mental privacy rights. In *Schmerber v. California* (384 US 757, 1966), a drunk-driving case, the Court affirms a distinction between testimony and physical evidence and holds that the Fifth Amendment protects testimony but not physical evidence.

[4]There are also some impressive examples of using EEG methods for lie detection. See, e.g., Hu and Rosenfeld (2012); Rosenfeld et al. (2008).

Defendants thus can be compelled to produce physical evidence (such as blood, DNA, fingerprints) that could be incriminating, but cannot be compelled to testify (take communicative action) against themselves. The testimonial/physical distinction, while perhaps intuitive, raises difficulties in the application of the law to cases of neuroscience evidence, for novel neuroimaging techniques measure physical brain properties to reveal information that heretofore has only been accessible via communicative acts (Farahany, 2012a). Fourth Amendment cases regarding information for which warrants are or are not required distinguish between information that encompasses content (such as the body of an email) and noncontent information, such as the header and address to which the email is sent (*Smith v. Maryland*, 442 US 735, 1979). Commentators have argued that both these distinctions are untenable given today's technologies (Farahany, 2012b).[5]

To date, the main tests of functional neuroimaging data as legal evidence have been in the context of lie detection. Although in recent cases in which neuroimaging data for lie detection has been introduced as forensic evidence it has been not been admitted, the rationales varied. In the most recent case, the Sixth Circuit Court of Appeals concluded that neuroimaging as evidence for lie detection did not meet standards for scientific evidence (*United States v. Semrau*, 693 F.3d 510, 2012). However, the court explicitly left the door open for future forensic use of fMRI lie detection. In another recent case, from a New York state trial court, the judge maintained it was the purview of juries, not machines, to determine whether a witness was lying, but he also raised questions about the reliability of such evidence. No cases have yet ruled on the question of whether neuroimaging infringes upon legally protected mental privacy.

Conclusion

The foregoing discussion about the scientific prospects for mindreading suggests that real threats of mental invasion from fMRI are limited. It appears that little that neuroimaging can now do rises to the level of a threat to a legally protected right. Moreover, given that neuroimaging requires a compliant subject, there is also little imminent danger of coercive uses of mindreading from the state. This state of affairs notwithstanding, it is clear that current legal doctrine does not provide a coherent theoretical basis for assessing the legal status of mental monitoring via brain imaging. In addition, the way we use social media and other technologies suggests that we may be undergoing a radical cultural shift that involves a devaluation of privacy. Since legal doctrines regarding privacy are based in part on notions of "reasonable expectations," and since which expectations are deemed reasonable is culturally dependent, the emergence of technologies that encourage broad dissemination of personal data threaten the very cultural expectations under which privacy has been enshrined as an inalienable right. The shift in modern culture may do more to threaten mental privacy than will any neurotechnology. We need a renewed inquiry into the value of privacy, and the reasons for which it is valued, in order to combat such devaluation.

[5]The Court is aware of the physical/testimonial ambiguity. In *Schmerber*, it wrote, "a distinction [between physical and testimonial evidence] is not readily drawn. Some tests seemingly directed to obtain 'physical evidence,' for example, lie detector tests measuring changes in body function during interrogation, may actually be directed to eliciting responses which are essentially testimonial" (384 US at 764). It still remains for the Court to satisfactorily clarify the bounds of the "essentially testimonial" (see, e.g., *Crawford v. Washington*, 541 US 36, 2004, and the line of Sixth Amendment cases extending from it).

REFERENCES

Azevedo, R. T., Macaluso, E., Avenanti, A., Santangelo, V., Cazzato, V., & Aglioti, S. M. (2013). Their pain is not our pain: Brain and autonomic correlates of empathic resonance with the pain of same and different race individuals. *Hum Brain Mapp, 34*(12), 3168–3181.

Carré, A., Gierski, F., Lemogne, C., Tran, E., Raucher-Chéné, D., Béra-Potelle, C., ... Limosin, F. (2013). Linear association between social anxiety symptoms and neural activations to angry faces: From subclinical to clinical levels. *Soc Cogn Affect Neurosci.* Advance online publication. doi:10.1093/scan/nst061

Chadwick, M., Hassabis, D., Weiskopf, N., & Maguire, E. A. (2010). Decoding individual episodic memory traces in the human hippocampus. *Curr Biol, 20*, 544–547.

Farah, M. J., Smith, M. E., Gawuga, C., Lindsell, D., & Foster, D. (2008). Brain imaging and brain privacy: A realistic concern? *J Cogn Neurosci, 21*, 119–127.

Farahany, N. (2012a). Incriminating thoughts. *Stanford Law Rev, 64*, 351–408.

Farahany, N. (2012b). Searching secrets. *U Penn Law Rev, 160*, 1239–1308.

Formisano, E., de Martino, F., Bonte, M., & Goebel, R. (2008). "Who" is saying "what"? Brain-based decoding of human voice and speech. *Science, 322*, 970–973.

Gallagher, R. (2013). Fact and fiction in the NSA surveillance scandal. *Slate*, Retrieved from www.slate.com/articles/technology/future_tense/2013/06/edward_snowden_fact_checking_which_surveillance_claims_were_right.html

Garoff-Eaton, R. J., Kensinger, E. A., & Schacter, D. L. (2007). The neural correlates of conceptual and perceptual false recognition. *Learn Mem, 14*, 684–692.

Harlé, K. M., Chang, L. J., van 't Wout, M., & Sanfey, A. G. (2012). The neural mechanisms of affect infusion in social economic decision-making: A mediating role of the anterior insula. *NeuroImage, 61*, 32–40.

Hasson, U., Landesman, O., Knappmeyer, B., Vallines, U., Rubin, N., & Heeger, D. J. (2008). Neurocinematics: The neuroscience of film. *Projections, 2*, 1–26.

Hasson, U., Nir, Y., Levy, I., Fuhrmann, G., & Malach, R. (2004). Intersubject synchronization of cortical activity during natural vision. *Science, 303*, 1634–1640.

Haxby, J. V., Gobbini, M. I., Ishai, A., Schouten, J. L., & Pietrini, P. (2001). Distributed and overlapping representations of faces and objects in visual cortex. *Science, 293*, 2425–2430.

Haynes, J.-D., Sakai, K., Rees, G., Gilbert, S., Frith, C., & Passingham, R. E. (2007). Reading hidden intentions in the human brain. *Curr Biol, 17*, 323–328.

Hu, X., & Rosenfeld, J. P. (2012). Combining the P300-complex trial-based Concealed Information Test and the reaction time-based autobiographical Implicit Association Test in concealed memory detection. *Psychophysiology, 49*, 1090–1100.

Kay, K. N., Naselaris, T., Prenger, R. J., & Gallant, J. L. (2008). Identifying natural images from human brain activity. *Nature, 452*, 352–355.

Marvel, C. L., & Desmond, J. E. (2012). From storage to manipulation: How the neural correlates of verbal working memory reflect varying demands on inner speech. *Brain Lang, 120*, 42–51.

Mitchell, T. M., Shinkareva, S. V., Carlson, A., Chang, K. M., Malave, V. L., Mason, R. A., & Just, M. A. (2008). Predicting human brain activity associated with the meanings of nouns. *Science, 320*, 1191–1195.

Mueller, S., Keeser, D., Reiser, M. F., Teipel, S., & Meindl, T. (2012). Functional and structural MR imaging in neuropsychiatric disorders, Part 2: Application in schizophrenia and autism. *Am J Neuroradiol, 33*, 2033–2037.

Naselaris, T., Prenger, R. J., Kay, K. N., Oliver, M., & Gallant, J. L. (2009). Bayesian reconstruction of natural images from human brain activity. *Neuron, 63*, 902–915.

Nishimoto, S., Vu, A. T., Naselaris, T., Benjamini, Y., Yu, B., & Gallant, J. L. (2011). Reconstructing visual experiences from brain activity evoked by natural movies. *Curr Biol, 21*, 1641–1646.

O'Craven, K. M., & Kanwisher, N. (2000). Mental imagery of faces and places activates corresponding stimulus-specific brain regions. *J Cogn Neurosci, 12*, 1013–1023.

Paulsen, D. J., Carter, R. M., Platt, M. L., Huettel, S. A., & Brannon, E. M. (2011). Neurocognitive development of risk aversion from early childhood to adulthood. *Front Hum Neurosci, 5*, 178.

Polyn, S. M., Natu, V. S., Cohen, J. D., & Norman, K. A. (2005). Category-specific cortical activity precedes retrieval during memory search. *Science, 310*, 1963–1966.

Reverberi, C., Görgen, K., & Haynes, J.-D. (2011). Compositionality of rule representations in human prefrontal cortex. *Cereb Cortex, 22*(6), 1237–1246.

Rissman, J., Greely, H. T., & Wagner, A. D. (2010). Detecting individual memories through the neural decoding of memory states and past experience. *Proc Natl Acad Sci USA, 107*, 9849–9854.

Rosenfeld, J. P., Labkovsky, E., Winograd, M., Lui, M. A., Vandenboom, C., & Chedid, E. (2008). The complex trial protocol (CTP): A new, countermeasure-resistant, accurate, P300-based method for detection of concealed information. *Psychophysiology, 45*, 906–919.

Scott, C. (2000). Is too much privacy bad for your health? An introduction to the law, ethics, and HIPAA rule on medical privacy. *GA St U Law Rev, 17*, 481.

Scott, N. A., Murphy, T. H., & Illes, J. (2012). Incidental findings in neuroimaging research: A framework for anticipating the next frontier. *J Empir Res Hum Res Ethics, 7*, 53–57.

Shergill, S. S., Brammer, M. J., Fukuda, R., Bullmore, E., Amaro, E., Jr., Murray, R. M., & McGuire, P. K. (2002). Modulation of activity in temporal cortex during generation of inner speech. *Hum Brain Mapp, 16*, 219–227.

Shinkareva, S. V., Mason, R. A., Malave, V. L., Wang, W., Mitchell, T. M., & Just, M. A. (2008). Using FMRI brain activation to identify cognitive states associated with perception of tools and dwellings. *PLoS ONE, 3*, e1394.

Simanova, I., Hagoort, P., Oostenveld, R., & van Gerven, M. A. J. (2012). Modality-independent decoding of semantic information from the human brain. *Cereb Cortex, 24*, 426–434.

Stanley, D. A., Sokol-Hessner, P., Fareri, D. S., Perino, M. T., Delgado, M. R., Banaji, M. R., & Phelps, E. A. (2012). Race and reputation: Perceived racial group trustworthiness influences the neural correlates of trust decisions. *Philos Trans R Soc Lond B Biol Sci, 367*, 744–753.

Tang, Y., Wang, L., Cao, F., & Tan, L. (2012). Identify schizophrenia using resting-state functional connectivity: An exploratory research and analysis. *Biomed Eng Online, 11*, 50.

Teipel, S. J., Grothe, M., Lista, S., Toschi, N., Garaci, F. G., & Hampel, H. (2013). Relevance of magnetic resonance imaging for early detection and diagnosis of Alzheimer disease. *Med Clin North Am, 97*, 399–424.

Thirion, B., Duchesnay, E., Hubbard, E., Dubois, J., Poline, J. B., Lebihan, D., & Dehaene, S. (2006). Inverse retinotopy: Inferring the visual content of images from brain activation patterns. *NeuroImage, 33*, 1104–1116.

Tong, F., Nakayama, K., Vaughan, J. T., & Kanwisher, N. (1998). Binocular rivalry and visual awareness in human extrastriate cortex. *Neuron, 21*, 753–759.

Van Bavel, J. J., Packer, D. J., & Cunningham, W. A. (2008). The neural substrates of in-group bias: A functional magnetic resonance imaging investigation. *Psychol Sci, 19*, 1131–1139.

Wagner, A. D. (2010). Can neuroscience identify lies? In A. S. Mansfield (Ed.), *A judge's guide to neuroscience: A concise introduction*, (pp. 13–25). Santa Barbara, CA: University of California, Santa Barbara.

Whalley, H. C., Sussmann, J. E., Romaniuk, L., Stewart, T., Papmeyer, M., Sprooten, E., ... McIntosh, A. M. (2013). Prediction of depression in individuals at high familial risk of mood disorders using functional magnetic resonance imaging. *PLoS ONE, 8*, e57357.

Wolpe, P. R., Foster, K. R., & Langleben, D. D. (2005). Emerging neurotechnologies for lie-detection: Promises and perils. *Am J Bioeth, 5*, 39.

86 The Cognitive Neuroscience of Moral Judgment and Decision Making

JOSHUA D. GREENE

ABSTRACT This article reviews recent advances in the cognitive neuroscience of moral judgment and behavior. This field is conceived, not as the study of a distinct set of neural functions, but as an attempt to understand how the brain's core neural systems coordinate to solve problems that we define, for nonneuroscientific reasons, as "moral." These systems enable the representation of value, cognitive control, the imagination of distal events, and the representation of mental states. Research examines the brains of morally pathological individuals, the responses of healthy brains to prototypically immoral actions, and the brain's responses to more complex moral problems such as philosophical and economic dilemmas.

Cognitive neuroscience aims to understand the mind in physical terms. This endeavor assumes that the mind *can* be understood in physical terms and, insofar as it is successful, validates that assumption. Against this philosophical backdrop, the cognitive neuroscience of moral judgment takes on special significance. Moral judgment is, for many, the quintessential operation of the mind beyond the body, the earthly signature of the soul (Greene, 2011). (In many religious traditions it is, after all, the quality of a soul's moral judgment that determines where it ends up.) Thus, the prospect of understanding moral judgment in physical terms is especially alluring, or unsettling, depending on your point of view. In this brief review I provide a progress report on our attempts to understand how the human brain makes moral judgments and decisions.

The paradox of the "moral brain"

The fundamental problem with the "moral brain" is that it threatens to take over the entire brain, and thus cease to be a meaningful neuroscientific topic. This is not because morality is meaningless, but rather because neuroscience is centrally concerned with physical mechanisms, and it's increasingly clear that morality has few, if any, neural mechanisms of its own (Greene & Haidt, 2002; Parkinson et al., 2011; Young & Dungan, 2012).

By way of analogy, consider the concept of a *vehicle.* Motorcycles and sailboats are vehicles. Lawnmowers and kites are not. But, mechanically speaking, motorcycles have more in common with (gas-powered) lawnmowers than with sailboats, and sailboats operate more like kites than motorcycles. This doesn't mean that the concept of a vehicle is meaningless. Rather, the world's vehicles are united, not by their internal mechanisms, but at a more abstract, functional level. So, too, with morality. More specifically, I (Greene, 2013), like many others (Darwin, 1871/2004; Frank, 1988; Gintis, Bowles, Boyd, & Fehr, 2005; Haidt, 2012) believe that morality is a suite of cognitive mechanisms that enable otherwise selfish individuals to reap the benefits of cooperation. That is, we have psychological features that are straightforwardly moral (such as empathy, righteous indignation, and an aversion to harming innocent people) and others that are not (such as gossip, embarrassment, vengefulness, and ingroup favoritism) because they enable us to achieve goals that we can't achieve through collective selfishness. I won't defend this controversial thesis here. Instead, my point is that *if* this unified theory of morality is correct, it doesn't bode well for a unified theory of moral neuroscience. What's more, as we'll see, the data increasingly bear out this skepticism. In the early days of moral neuroscience, it was thought, perhaps not unreasonably, that one might isolate the distinctive neural mechanisms of moral thought (Moll, Eslinger, & Oliveira-Souza, 2001) and that the human brain might house a dedicated "moral organ" (Hauser, 2006). These views, however, are no longer tenable. It's now clear that the "moral brain" is, more or less, the whole brain, applying its computational powers to problems that we, on nonneuroscientific grounds, identify as "moral."

Understanding this is, itself, a kind of progress, but it leaves the cognitive neuroscience of morality—and the author of a chapter that would summarize it—in an awkward position. To truly understand the neuroscience of morality, we must understand the many neural systems that shape moral thinking, none of which, so far, appears to be specifically moral. These include systems that enable the representation of value and that motivate its pursuit (Knutson, Taylor, Kaufman,

Peterson, & Glover, 2005; Pessoa, 2010; Rangel, Camerer, & Montague, 2008; Schultz, Dayan, & Montague, 1997), systems that orchestrate thought and action in accordance with internal goals (Miller & Cohen, 2001), systems that enable the imagination of complex distal events (Buckner, Andrews-Hanna, & Schacter, 2008; Raichle et al., 2001), and systems that enable the representation of people's hidden mental states (Frith & Frith, 2006; Mitchell, 2009), among others. In short, if you want to understand the neuroscience of morality, you might start by working your way through this weighty volume.

Of course, some neuroscientific topics bear more directly on morality than others, as indicated by my nonrandom list of relevant neural systems. This suggests that the present task isn't hopeless, that we can make some useful generalizations about the cognitive neuroscience of morality, even while acknowledging that the moral brain is not a distinct entity. This field, properly understood, will not isolate and describe the mechanisms essential for morality while the rest of cognitive neuroscience goes about its business. Instead, it provides a set of useful *entry points* into the broader problems of complex cognition and decision making (cf. Buckholtz & Meyer-Lindenberg, 2012, for a parallel view of psychopathology). More specifically, we can study the brains of people who reliably commit basic moral transgressions, the reactions of healthy brains to such transgressions, and the ways in which our brains handle more complex moral problems. Along the way we'll encounter some recurring themes that point the way toward a more encompassing account of moral, and nonmoral, cognition.

Bad brains

In the 1990s, Damasio and colleagues published a series of path-breaking studies of decision making in patients with damage to ventromedial prefrontal cortex (VMPFC), one of the regions damaged in the famous case of Phineas Gage (Damasio, 1994). VMPFC patients were mysterious because their real-life decision making was clearly impaired, but their deficits typically evaded detection using standard neurological measures of executive function (Saver & Damasio, 1991) and moral reasoning (Anderson, Bechara, Damasio, Tranel, & Damasio, 1999). Using a game designed to simulate real-world risky decision making (the Iowa Gambling Task), Bechara, Tranel, Damasio, and Damasio (1996) documented these behavioral deficits and demonstrated, using autonomic measures, that these deficits are emotional. It seems that such patients make poor decisions because they are unable to generate the feelings that guide adaptive decision making in healthy individuals. These early studies, while identifying a key biological substrate for moral choice, also underscore the critical role of learning in moral development. Late-onset VMPFC damage typically results in poor decision making and a deterioration of "moral character" (Damasio, 1994), but children with early-onset VMPFC damage are likely to develop into "sociopathic" adults who, in addition to being reckless and irresponsible, are duplicitous, aggressive, and strikingly lacking in empathy (Anderson et al., 1999; Grattan & Eslinger, 1992).

Studies of psychopaths and other individuals with antisocial personality disorder (APD) underscore the importance of emotion in moral decision making. APD is a catch-all diagnosis for individuals whose behavior is unusually antisocial. Psychopathy, in contrast, is a more specific, somewhat heritable disorder (Viding, Blair, Moffitt, & Plomin, 2005) whereby individuals exhibit a pathological degree of callousness, lack of empathy or emotional depth, and lack of genuine remorse for their antisocial actions (Hare, 1991). Psychopaths tend to engage in instrumental aggression, while other individuals with APD are characterized by reactive aggression (Blair, 2001).

Psychopathy is characterized by profound but selective emotional deficits. Psychopaths exhibit normal electrodermal responses to threat cues (e.g., a picture of shark's open mouth), but reduced responses to distress cues (e.g., a picture of a crying child; Blair, Jones, Clark, & Smith, 1997). In a classic study, Blair (1995) provided evidence that psychopaths fail to distinguish between rules that authorities cannot legitimately change ("moral" rules, e.g., a classroom rule against hitting) from rules that authorities can legitimately change ("conventional" rules, e.g., a rule prohibiting talking out of turn). According to Blair, psychopaths see all rules as *mere* rules because they lack the emotional responses that lead ordinary people to imbue moral rules with genuine, authority-independent moral legitimacy. While this is consistent with what is generally known about psychopathic psychology, a more recent study challenges the original finding that psychopaths do not draw the moral/conventional distinction (Aharoni, Sinnott-Armstrong, & Kiehl, 2012).

Studies of psychopathy and APD implicate a wide range of brain regions including the insula, posterior cingulate cortex, parahippocampal gyrus, and superior temporal gyrus (Kiehl, 2006; Raine & Yang, 2006). However, as emphasized by Blair (2007), two interconnected structures take center stage: the amygdala and the VMPFC. These regions, along with subregions of subgenual anterior cingulate cortex and lateral prefrontal cortex, form a network that is essential for generating and regulating responses to salient stimuli (Pessoa, 2010). Blair (2007) has proposed that psychopathy arises primarily from amygdala dysfunction, which is crucial for stimulus-reinforcement learning (Davis & Whalen, 2001) and thus for normal moral socialization (Oxford, Cavell, & Hughs, 2003). In psychopaths (or individuals with psychopathic traits) the amygdala exhibits weaker responses to fearful faces (Marsh et al., 2008), to emotional words (Kiehl et al., 2001), to pictures indicating moral violations (Harenski, Harenski, Shane, & Kiehl, 2010; Harenski, Kim, & Hamann, 2009), and to dilemmas involving harmful actions (Glenn, Raine, & Schug, 2009). As noted above, the amygdala operates in tight conjunction with the VMPFC, and, consistent with this, psychopathic individuals also exhibit reduced VMPFC responses to morally salient stimuli (Harenski et al., 2010). Beyond the amygdala-VMPFC circuit, psychopaths also exhibit hypoactivity in the default mode network (DMN; Buckner et al., 2008; Raichle et al., 2001) during moral judgment (Pujol et al., 2012), consistent with this network's heightened response to emotionally engaging moral dilemmas in healthy people (Greene et al., 2001). (Note that some of the participants in this study failed to meet standard criteria for psychopathy. See Schaich Borg & Sinnott-Armstrong, 2013.)

Psychopaths, in addition to their weak affective responses to harm, are known for their impulsive behavior (Hare, 1991). The VMPFC serves as part of the frontostriatal pathway, responsible for representing the values of outcomes and actions based on past experience (Knutson et al., 2005; Rangel et al., 2008). Individuals with psychopathic traits (specifically, impulsive antisocial behavior) exhibit heightened responses to reward within this system (Buckholtz et al., 2010) along with increased striatal volume (Glenn, Raine, Yaralian, & Yang, 2010). Finally, their emotional deficits may sometimes cause them to rely more heavily on explicit reasoning, dependent on the frontoparietal control network (Glenn, Raine, Schug, Young & Hauser, 2009; Koenigs, Kruepke, Zeier, & Newman, 2012). Thus, while the origins of psychopathy may lie in one or more discrete neural abnormalities, their influence is felt throughout the brain.

Good brains

Studies of healthy individuals responding to moral transgressions are generally consistent with studies of psychopaths and others with APD. They, too, highlight the importance of the amygdala and VMPFC (Blair, 2007; Decety & Porges, 2011; Heekeren et al., 2005; Moll et al., 2002; Schaich Borg et al., 2006;) and confirm the importance of these structures in moral development (Decety, Michalska, & Kinzler, 2012). For reasons explained below, studies of moral judgment employing text-based narrative stimuli tend to implicate the entire DMN. Several studies highlight the importance of the insula in representing the aversiveness of moral transgressions (Baumgartner, Fischbacher, Feierabend, Lutz, & Fehr, 2009; Decety, Michalska, & Kinzler, 2012; Greene, Nystrom, Engell, Darley, & Cohen, 2004; Schaich Borg, Lieberman, & Kiehl, 2008; Schaich Borg, Sinnott-Armstrong, Calhoun, & Kiehl, 2011). Others indicate that the representation of moral value, like other forms of value, depends on the brain's domain-general valuation mechanisms enabled by the frontostriatal pathway (Decety & Porges, 2011; Moll et al., 2006; Shenhav & Greene, 2010).

One of the most basic distinctions in moral evaluation is between intentional and accidental harm. (As Oliver Wendell Holmes Jr. famously observed, even a dog knows the difference between being tripped over and being kicked.) Young, Saxe, and colleagues have conducted a series of studies examining how the brain represents and applies this distinction in the context of moral judgment. Their work highlights the importance of the temporoparietal junction (TPJ) along with other DMN regions, which are widely implicated in ToM or "mentalizing" (Frith & Frith, 2006; Mitchell, 2009). The TPJ is especially sensitive to attempted harms (Koster-Hale, Saxe, Dungan, & Young, 2013; Young, Cushman, Hauser, & Saxe, 2007), which are wrong because of the agent's mental state, not the action's outcome. Disrupting TPJ activity results in a child-like (Piaget, 1965), "no harm no foul" pattern of judgment in which attempted harms are judged less harshly (Young, Camprodon, Hauser, Pascual-Leone, & Saxe, 2010). We see the same pattern in patients with VMPFC damage (Young, Bechara, et al., 2010) and split-brain patients (Miller et al., 2010), indicating that the use of mental-state information in moral judgment depends, at least in part, on translating this information into an affective signal and on the integration of information across the cerebral hemispheres. Individuals with high-functioning autism exhibit a complementary pattern, "if harm, then foul," judging accidental harms unusually harshly (Moran et al., 2011). Accidental harms appear to set up a tension between outcome-based and intention-based harm. Consistent with this, such harms preferentially engage the frontoparietal control network (Miller & Cohen, 2001).

Puzzled brains

We've considered the two most straightforward entry points into moral neuroscience: the unhealthy brains of people who act badly and the healthy brain's responses to protoypically bad acts. A third approach begins with moral dilemmas. Moral dilemmas are useful, not because they reflect everyday moral experience, but because dilemmas, by their nature, pit competing processes against one another. They are high-contrast stimuli, analogous to the flashing checkerboards of vision scientists, and thus especially useful for revealing cognitive structure (Cushman & Greene, 2012).

The research described above emphasizes the role of emotion in moral judgment (Haidt, 2001), while traditional theories of moral development emphasize the role of controlled cognition (Kohlberg, 1969; Turiel, 2006). I and others have developed a dual-process (Chaiken & Trope, 1999; Kahneman, 2003) theory of moral judgment that synthesizes these perspectives (Greene et al., 2001; Greene, 2007, 2013). According to this theory, both intuitive emotional responses and more controlled cognitive responses play crucial and, in some cases, competing roles. More specifically, this theory associates controlled cognition with utilitarian (or consequentialist) moral judgment aimed at promoting the "greater good" (Mill, 1861/1998) while associating automatic emotional responses with competing deontological judgments that are naturally justified in terms of rights or duties (Kant, 1785/1959).

We developed this theory in response to a long-standing philosophical puzzle known as the Trolley Problem (Foot, 1978; Thomson, 1985). In one version, which I'll call the *switch* case, one can save five people who are mortally threatened by a runaway trolley by hitting a switch. This will turn the trolley onto a side track, where it will run over and kill only one person instead. Here, most people approve of diverting the trolley (Petrinovich, O'Neill, & Jorgensen, 1993), a characteristically utilitarian judgment favoring the greater good. In the contrasting *footbridge* dilemma, a runaway trolley once again threatens five people. The only way to save the five is to push a large person off a footbridge and into the trolley's path, stopping the trolley but killing the person pushed. (Yes, this will work, and, no, you can't stop the trolley yourself.) Here, most people say that it's wrong to trade one life for five, consistent with the deontological perspective favoring the rights of the individual over the greater good. The question: why do people typically say "yes" to hitting the switch, but "no" to pushing?

We hypothesized that this pattern of judgment reflects the outputs of distinct and (in some cases) competing neural systems (Greene et al., 2001). The more "personal"[1] harmful action in the *footbridge* case, pushing the man off the footbridge, triggers a relatively strong negative emotional response, while the relatively impersonal harmful action in the *switch* case does not. This predicts increased activity in emotion-related brain regions in response to "personal" dilemmas, such as the *footbridge* case, as compared to "impersonal" dilemmas, such as the *switch* case.

This emotional response can explain why people say "no" to pushing the man off the footbridge. But why do people say "yes" to hitting the switch? The answer seems obvious enough: hitting the switch saves more lives. We hypothesized that this utilitarian response depends on explicit cost-benefit reasoning enabled by the frontoparietal control network (Miller & Cohen, 2001), including the DLPFC. Thus, we predicted increased DLPFC activity in response to "impersonal" dilemmas, such as the *switch* case, in which this controlled response tends to dominate. Likewise, we predicted increased DLPFC activity when people override a negative emotional response in making a utilitarian judgment, as when people say "yes" to the *footbridge* dilemma.

We first tested this theory using functional MRI (fMRI; Greene et al., 2001), contrasting a (rather heterogeneous) set of "personal" dilemmas with a set of (even more heterogeneous) "impersonal" dilemmas. (More recent studies have been better controlled, focusing on differing responses to "high-conflict" dilemmas such as the *footbridge* case.) We found that the "personal" dilemmas elicited increased activity in what is now known as the DMN (Buckner et al., 2008; Raichle et al., 2001), including large portions of medial prefrontal cortex, medial parietal cortex, and the TPJ, all of which had been previously associated with emotion (e.g., Maddock, 1999). In contrast, the "impersonal" dilemmas elicited relatively greater activity in the frontoparietal control network. Also as predicted, our second fMRI experiment (Greene et al., 2004) found increased DLPFC activity for utilitarian judgment and increased amygdala activity for "personal" dilemmas. These results provided initial support for the dual-process theory, which has been both supported and refined by subsequent research using a broad range of methods.

[1]The personal/impersonal distinction (Greene et al., 2001) has been revised (Greene et al., 2009) since it was originally introduced. For present purposes, one can think of "personal" harms as ones in which the agent actively and intentionally harms the victim using the direct force of his or her muscles.

In retrospect, the DMN's response to "personal" dilemmas is best interpreted as *related* to increased emotional engagement, but not as its proper neural substrate. The DMN is active when people are doing nothing in particular (hence "default") and is most reliably engaged by attention to nonpresent events, as in remembering the past, imaging the future, thinking about contents of other minds, and imaging hypothetical possibilities (Buckner et al., 2008; DeBrigard, Addis, Ford, Schacter, & Giovanello, 2013). Thus, if "personal" dilemmas preferentially engage the DMN, it's probably not because DMN activity reflects emotional engagement per se. Rather, it's because "personal" dilemmas make for especially gripping mental television, which may be both a cause and a consequence of their emotional salience. Consistent with this hypothesis, Amit and Greene (2012) found that individuals with more visual cognitive styles tend to make fewer utilitarian judgments in response to high-conflict personal dilemmas and that disrupting visual imagery while contemplating these dilemmas increases utilitarian judgment.

More direct evidence for the dual-process theory comes from studies of patients with emotion-related deficits. Mendez, Anderson, and Shapira (2005) found that patients with frontotemporal dementia, who are known for their "emotional blunting," were disproportionately likely to approve of the utilitarian action in the *footbridge* dilemma. Likewise, patients with VMPFC lesions make up to five times as many utilitarian judgments in response to standard high-conflict dilemmas (Ciaramelli, Muccioli, Ladavas, & di Pellegrino, 2007; Koenigs et al., 2007) and in response to dilemmas pitting familial duty against the greater good (e.g., your sister vs. five strangers; Thomas, Croft, & Tranel, 2011). VMPFC patients also exhibit correspondingly weak physiological responses when making such judgments (Moretto, Ladàvas, Mattioli, & di Pellegrino, 2010), and healthy people who are more physiologically reactive are less utilitarian (Cushman, Murray, Gordon-Mckeon, Wharton, & Greene, 2012). Low-anxiety psychopaths (Koenigs et al., 2012) and people with high levels of testosterone (Carney & Mason, 2010), which is associated with a higher tolerance for stress, tend to make more utilitarian judgments, as do people with alexithymia (Koven, 2011), a condition that reduces awareness of one's own emotional states. Here, the VMPFC seems to respond specifically to harmful behavior that is active and also intentional, rather than merely foreseen (Schaich Borg et al., 2006).

Other studies highlight the role of the amygdala. As noted above, individuals with psychopathic traits exhibit reduced amygdala responses to personal moral dilemmas (Glenn, Raine, & Schug, 2009). In healthy people, amygdala activity tracks self-reported emotional responses to harmful transgressions and predicts deontological judgments in response to them (Shenhav & Greene, 2014). Studies employing pharmacological interventions paint a consistent picture. Citalopram—a selective serotonin-reuptake inhibitor (SSRI) that, in the short-term, increases emotional reactivity through its influence on the amygdala and VMPFC, among other regions—increases deontological judgment (Crockett, Clark, Hauser, & Robbins, 2010). By contrast, lorazepam, an anti-anxiety drug, has the opposite effect (Perkins et al., 2012). Consistent with the effects of citalopram, variation in the serotonin transporter (*5-HTTLPR*) genotype (S alleles) predicts deontological judgment, but in response dilemmas in which the harm is a foreseen side effect (Marsh et al., 2011).

Most of the evidence linking controlled cognition to utilitarian judgment comes from behavioral studies beyond the scope of this chapter (e.g., Greene et al., 2008; Paxton, Ungar, & Greene, 2012). However, a few neuroscientific studies, in addition to those described above (Greene et al., 2001, 2004), provide further evidence. Sarlo et al. (2012) examined the temporal dynamics of moral judgment using EEG and found a pattern consistent with the results of Greene et al. (2001, 2004). Here, *footbridge*-like dilemmas produced a stronger early neural response (P260) in regions consistent with VMPFC activity, while *switch*-like dilemmas elicited more utilitarian responses and a more pronounced later component consistent with the engagement of the frontoparietal control network. Also consistent with this, activity in the frontoparietal control network is associated with rejecting the deontological distinction between harmful acts and harmful omissions (Cushman et al., 2012). (See also Schaich Borg et al., 2006.) Likewise, VMPFC patients who tend to give more utilitarian responses are thought to do so because their capacity for explicit, cost-benefit reasoning remains intact (Koenigs et al., 2007).

A recent study (Shenhav & Greene, 2014) helps differentiate the functions of the amygdala and VMPFC in moral judgment. As noted above, amygdala signal tracks with self-reports of negative emotional responses to harmful actions and predicts deontological condemnation of those actions. The VMPFC, however, does not. Instead, the VMPFC is most active when people have to make "all things considered" judgments, as compared to simply reporting on emotional reactions or utilitarian considerations. This suggests that the amygdala generates an initial negative response to personally harmful actions (consistent with Glenn, Raine, & Schug, 2009), while the VMPFC weighs that signal against a competing signal reflecting the utilitarian advantages of committing the harmful act. This is consistent with an evolving understanding of the VMPFC as a domain-general integrator of decision weights (Rangel & Hare, 2010). However, this leaves us with a puzzle: if the VMPFC is acting as a neutral broker among competing decision weights, then why does VMPFC damage so reliably increase utilitarian judgment? Our hypothesis is that the frontoparietal control network's explicit utilitarian reasoning can influence behavior independent of the VMPFC, while the amygdala's competing deontological signal requires the VMPFC's integration, at least when competing utilitarian considerations are in play. Thus, if this is correct, VMPFC damage favors utilitarian judgment, not by damaging a region with inherent deontological tendencies, but by damaging a pathway that is necessary for deontological judgment, but not utilitarian judgment, to prevail.

This integrative role for the VMPFC is consistent with its role in integrating other kinds of morally relevant information. Shenhav and Greene (2010) examined people's responses to dilemmas in which failing to save one person can allow one to save a group of others. We varied the size of the group and the probability of saving them. We found that neural sensitivity to the magnitude of the outcome (group size) in the ventral striatum predicts behavioral sensitivity to this variable, and we observed a parallel effect for outcome probability in the insula. The VMPFC, by contrast, responded to the interaction of these two variables, reflecting the probability-discounted magnitude of the moral consequences. In other words, the VMPFC represents "expected moral value," just as it represents expected value in self-interested economic decision making (Knutson et al., 2005). Thus, once again, we see a domain-general system—here, the frontostriatal pathway—operating in the context of moral judgment. This system evolved in mammals to value goods that tend to exhibit diminishing marginal returns. This may explain our puzzling (and highly consequential) tendency to regard the saving of human lives as exhibiting diminishing marginal returns, as if the hundredth life saved is somehow worth less than the first.

In an important theoretical development, Cushman (2013) and Crockett (2013) have proposed that the dissociation between deontological and utilitarian/consequentialist judgment reflects a more general dissociation between model-free and model-based learning systems (Daw & Doya, 2006). Model-free learning mechanisms assign values to actions intrinsically based on past experience, while model-based learning mechanisms attach values to actions based on internal models of causal relations in the world. Thus, an action may seem intrinsically wrong because past experience has associated actions of that type (e.g., pushing people) with negative consequences (e.g., social disapproval), and yet the same action may seem right because it will, according to one's world-model, produce optimal consequences (saving five lives instead of one). Thus, the fundamental tension in normative ethics, reflected in the competing philosophies of Kant and Mill, may find its origins in a competition between distinct, domain-general mechanisms for assigning values to actions.

Cooperative brains

Research on altruism and cooperation does not always fall under the heading of "morality," but it could not be more central to our understanding of the moral brain. The most basic question about the cognitive neuroscience of altruism and cooperation is this: what neural processes enable and motivate people to be "nice"—that is, to pay costs to benefit others?

Consistent with our evolving story, the value of helping others, both in unidirectional altruism and bidirectional cooperation, is represented in the frontostriatal pathway. Activity in this pathway tracks the value of charitable contributions (Moll et al., 2006; Hare et al., 2010), sharing resources with other individuals (Zaki & Mitchell, 2011), cooperation (Rilling et al., 2007), maximizing benefits delivered by a distribution of resource (i.e., "efficiency"), and optimizing the subjective trade-off between efficiency and equality (Hsu, Anen, & Quartz, 2008). Likewise, this pathway tracks the value of punishing individuals who are insufficiently "nice" (de Quervain et al., 2004; Singer et al., 2006). As above, the DMN appears to have a hand in altruism as well. TPJ volume (Morishima, Schunk, Bruhin, Ruff, & Fehr, 2012) and medial PFC activity (Rilling et al., 2007; Waytz, Zaki, & Mitchell, 2012) both predict altruistic behavior.

Thus, the brain uses its endogenous carrots—reward signals—to motivate cooperative behavior. It also uses its sticks—negative affective responses to uncooperative behavior. Activity in the insula, known for its role in the representation of somatic states and the awareness of feelings (Craig, 2009), scales with the magnitude of the unfairness in unfair Ultimatum Game offers (Sanfey, Rilling, Aronson, Nystrom, & Cohen, 2003), predicts aversion to inequality in the distribution of resources (Hsu et al., 2008), and predicts egalitarian behavior and attitudes (Dawes et al., 2012). The insula and the amygdala both respond to the punishment of well-behaved people (Singer, Kiebel, Winston, Dolan, & Frith, 2004).

The dual-process tension between automatic and controlled processes is observed in a range of morally laden economic choices. Accepting unfair Ultimatum Game offers, despite their distastefulness, is associated with increased activity in the frontoparietal control network (Sanfey et al., 2003; Tabibnia, Satpute, & Lieberman, 2008). Perhaps surprisingly, VMPFC damage leads to increased *rejection* of unfair offers (Koenigs & Tranel, 2007). (Consistent with this, psychopaths do the same; Koenigs, Kruepke, & Newman, 2010.) This may be because the VMPFC integrates signals responding both to unfairness and material gain (which compete in the Ultimatum Game) and because, in the absence of such signals, one applies a reciprocity rule. In a study of dishonesty, Greene and Paxton (2009) gave people repeated opportunities to gain money by lying about their accuracy in predicting the outcomes of coin-flips. Consistently honest subjects appeared to be "gracefully" honest, exhibiting no additional engagement of the frontoparietal control network in forgoing dishonest gains. By contrast, subjects who behaved dishonestly (as indicated by improbably high self-reported accuracy) exhibited increased control-related activity, both when lying and when refraining from lying. A follow-up study (Abe & Greene, 2013) traces these behavioral differences to response characteristics of the frontostriatal pathway. Baumgartner et al. (2009) describe a similar dual-process dynamic, in which breaking promises involves increased engagement of the amygdala and the frontoparietal control network. (For a behavioral approach to dual-process cooperation, also see Rand, Greene, & Nowak, 2012.)

Cooperation depends on trust, which in turn requires evaluating individuals (Delgado, Frank, & Phelps, 2005; Singer et al., 2004) and groups (Phelps et al., 2000) as potential cooperation partners. Oxytocin, a neuropeptide known for its role in social attachment and affiliation in mammals (Insel & Young, 2001) appears to be important for both kinds of decisions. Intranasal administration of oxytocin increases investment in a "trust game" (Kosfeld, Heinrichs, Zak, Fischbacher, & Fehr, 2005), but also biases judgment and behavior toward ingroup members and against outgroup members (de Dreu et al., 2010; de Dreu, Greer, Van Kleef, Shalvi, & Handgraaf, 2011). Likewise, genetic variants associated with oxytocin are associated with increased prosocial behavior, particularly when the world is seen as threatening (Poulin, Holman, & Buffone, 2012).

From an evolutionary perspective, the double-edged sword of human morality comes as no surprise. Morality evolved, not as a device for universal cooperation, but as a competitive weapon, as a system for turning Me into Us, which in turn enables Us to outcompete Them. Morality's dark, tribalistic side is powerful, but there's no reason why it must prevail. The flexible thinking enabled by our enlarged prefrontal cortices may enable us to retain the best of our moral impulses while transcending their inherent limitations (Greene, 2013; Pinker, 2011).

Looking back, and ahead

How does the moral brain work? Answer: exactly the way you'd expect it to work if you understand (1) which cognitive functions morality requires and (2) which cognitive functions are performed by the brain's core neural systems. On the one hand, this means that morality has no proprietary neural territory of its own. On the other hand, it means that the cognitive neuroscience of morality, beginning with the entry points described above, can teach important lessons about how the brain's core neural systems interact to solve complex problems.

From its inception, cognitive neuroscience has focused on structure-function relationships. We have a general understanding of what various neural structures do, but when it comes to complex cognition, we're mostly blind to the specific information content shuttled about the brain. We know, for example, that the thought of pushing someone off a footbridge pushes our emotional buttons, but we know almost nothing about how we think such thoughts in the first place. However, with the advent of multivariate analysis methods (Kriegeskorte, Goebel, & Bandettini, 2006; Norman, Polyn, Detre, & Haxby, 2006), we may finally be ready to understand how the brain encodes and manipulates the *contents* of thoughts. When we finally do, we will learn a lot more about morality—and everything else.

ACKNOWLEDGMENTS Many thanks to Joshua Buckholtz, Joe Paxton, Adina Roskies, Walter Sinnott-Armstrong, and Liane Young for helpful comments.

REFERENCES

Abe, N., & Greene, J. D. (2013). Response to anticipated reward in the nucleus accumbens predicts behavior in an independent test of honesty. (Under review.)

Aharoni, E., Sinnott-Armstrong, W., & Kiehl, K. A. (2012). Can psychopathic offenders discern moral wrongs? A new look at the moral/conventional distinction. *J Abnorm Psychol, 121*(2), 484.

Amit, E., & Greene, J. D. (2012). You see, the ends don't justify the means: Visual imagery and moral judgment. *Psychol Sci, 23*(8), 861–868.

Anderson, S. W., Bechara, A., Damasio, H., Tranel, D., & Damasio, A. R. (1999). Impairment of social and moral behavior related to early damage in human prefrontal cortex. *Nat Neurosci, 2,* 1032–1037.

Baumgartner, T., Fischbacher, U., Feierabend, A., Lutz, K., & Fehr, E. (2009). The neural circuitry of a broken promise. *Neuron, 64*(5), 756–770.

Bechara, A., Tranel, D., Damasio, H., & Damasio, A. R. (1996). Failure to respond autonomically to anticipated future outcomes following damage to prefrontal cortex. *Cereb Cortex, 6,* 215–225.

Blair, R. J. (1995). A cognitive developmental approach to mortality: Investigating the psychopath. *Cognition, 57,* 1–29.

Blair, R. J. (2001). Neurocognitive models of aggression, the antisocial personality disorders, and psychopathy. *J Neurol Neurosurg Psychiatry, 71,* 727–731.

Blair, R. J. (2007). The amygdala and ventromedial prefrontal cortex in morality and psychopathy. *Trends Cogn Sci, 11,* 387–392.

Blair, R. J., Jones, L., Clark, F., & Smith, M. (1997). The psychopathic individual: A lack of responsiveness to distress cues? *Psychophysiology, 34,* 192–198.

Buckholtz, J. W., & Meyer-Lindenberg, A. (2012). Psychopathology and the human connectome: Toward a transdiagnostic model of risk for mental illness. *Neuron, 74*(6), 990–1004.

Buckholtz, J. W., Treadway, M. T., Cowan, R. L., Woodward, N. D., Benning, S. D., Li, R., & Zald, D. H. (2010). Mesolimbic dopamine reward system hypersensitivity in individuals with psychopathic traits. *Nat Neurosci, 13*(4), 419–421.

Buckner, R. L., Andrews-Hanna, J. R., & Schacter, D. L. (2008). The brain's default network. *Ann NY Acad Sci, 1124*(1), 1–38.

Carney, D. R., & Mason, M. F. (2010). Decision making and testosterone: When the ends justify the means. *J Exp Soc Psychol, 46*(4), 668–671.

Chaiken, S., & Trope, Y. (Eds.). (1999). *Dual-process theories in social psychology.* New York, NY: Guilford.

Ciaramelli, E., Muccioli, M., Ladavas, E., & di Pellegrino, G. (2007). Selective deficit in personal moral judgment following damage to ventromedial prefrontal cortex. *Soc Cogn Affect Neurosci, 2,* 84–92.

Craig, A. D. (2009). How do you feel—now? The anterior insula and human awareness. *Nat Rev Neurosci, 10*(1), 59–70.

Crockett, M. J. (2013). Models of morality. *Trends Cogn Sci, 17*(8), 363–366.

Crockett, M. J., Clark, L., Hauser, M. D., & Robbins, T. W. (2010). Serotonin selectively influences moral judgment and behavior through effects on harm aversion. *Proc Natl Acad Sci USA, 107*(40), 17433–17438.

Cushman, F. (2013). Action, outcome and value: A dual-system framework for morality. *Pers Soc Psychol Rev, 17*(3), 273–292.

Cushman, F., & Greene, J. D. (2012). Finding faults: How moral dilemmas illuminate cognitive structure. *Soc Neurosci,* 7(3), 269–279.

Cushman, F., Murray, D., Gordon-McKeon, S., Wharton, S., & Greene, J. D. (2012). Judgment before principle: Engagement of the frontoparietal control network in condemning harms of omission. *Soc Cogn Affect Neurosci, 7*(8), 888–895.

Damasio, A. R. (1994). *Descartes' error: Emotion, reason, and the human brain.* New York, NY: G. P. Putnam.

Darwin, C. (1871/2004). *The descent of man.* New York, NY: Penguin.

Davis, M., & Whalen, P. J. (2001). The amygdala: Vigilance and emotion. *Mol Psychiatry, 6,* 13–34.

Daw, N. D., & Doya, K. (2006). The computational neurobiology of learning and reward. *Curr Opin Neurobiol, 16*(2), 199–204.

Dawes, C. T., Loewen, P. J., Schreiber, D., Simmons, A. N., Flagan, T., McElreath, R., & Paulus, M. P. (2012). Neural basis of egalitarian behavior. *Proc Natl Acad Sci USA, 109*(17), 6479–6483.

De Brigard, F., Addis, D. R., Ford, J. H., Schacter, D. L., & Giovanello, K. S. (2013). Remembering what could have happened: Neural correlates of episodic counterfactual thinking. *Neuropsychologia, 51*(12), 2401–2414.

De Dreu, C. K., Greer, L. L., Handgraaf, M. J., Shalvi, S., van Kleef, G. A., Baas, M., & Feith, S. W. (2010). The neuropeptide oxytocin regulates parochial altruism in intergroup conflict among humans. *Science, 328*(5984), 1408–1411.

De Dreu, C. K., Greer, L. L., van Kleef, G. A., Shalvi, S., & Handgraaf, M. J. (2011). Oxytocin promotes human ethnocentrism. *Proc Natl Acad Sci USA, 108*(4), 1262–1266.

De Quervain, D. J., Fischbacher, U., Treyer, V., Schellhammer, M., Schnyder, U., Buck, A., & Fehr, E. (2004). The neural basis of altruistic punishment. *Science, 305,* 1254–1258.

Decety, J., Michalska, K. J., & Kinzler, K. D. (2012). The contribution of emotion and cognition to moral sensitivity: A neurodevelopmental study. *Cereb Cortex, 22*(1), 209–220.

Decety, J., & Porges, E. C. (2011). Imagining being the agent of actions that carry different moral consequences: An fMRI study. *Neuropsychologia, 49*(11), 2994–3001.

Delgado, M. R., Frank, R., & Phelps, E. A. (2005). Perceptions of moral character modulate the neural systems of reward during the trust game. *Nat Neurosci, 8,* 1611–1618.

Foot, P. (1978). The problem of abortion and the doctrine of double effect. In P. Foot, *Virtues and vices, and other essays in moral philosophy* (pp. 19–32). Oxford, UK: Blackwell.

Frank, R. H. (1988). *Passions within reason: The strategic role of the emotions.* New York, NY: Norton.

Frith, C. D., & Frith, U. (2006). The neural basis of mentalizing. *Neuron, 50*(4), 531–534.

Gintis, H. E., Bowles, S. E., Boyd, R. E., & Fehr, E. E. (2005). *Moral sentiments and material interests: The foundations of cooperation in economic life.* Cambridge, MA: MIT Press.

Glenn, A. L., Raine, A., & Schug, R. A. (2009). The neural correlates of moral decision-making in psychopathy. *Mol Psychiatry, 14*(1), 5.

Glenn, A. L., Raine, A., Schug, R. A., Young, L., & Hauser, M. (2009). Increased DLPFC activity during moral decision-making in psychopathy. *Mol Psychiatry, 14*(10), 909–911.

Glenn, A. L., Raine, A., Yaralian, P. S., & Yang, Y. (2010). Increased volume of the striatum in psychopathic individuals. *Biol Psychiatry, 67*(1), 52–58.

Grattan, L. M., & Eslinger, P. J. (1992). Long-term psychological consequences of childhood frontal lobe lesion in patient DT. *Brain Cogn, 20,* 185–195.

Greene, J. (2011). Social neuroscience and the soul's last stand. In A. Todorov, S. Fiske, & D. Prentice (Eds.), *Social neuroscience: Toward understanding the underpinnings of the social mind.* New York, NY: Oxford University Press.

Greene, J. (2013). *Moral tribes: Emotion, reason, and the gap between us and them.* New York, NY: Penguin.

Greene, J., & Haidt, J. (2002). How (and where) does moral judgment work? *Trends Cogn Sci, 6,* 517–523.

Greene, J. D. (2007). The secret joke of Kant's soul. In W. Sinnott-Armstrong (Ed.), *Moral psychology.* Vol. 3: *The neuroscience of morality: Emotion, disease, and development* (pp. 35–79). Cambridge, MA: MIT Press.

Greene, J. D., Cushman, F. A., Stewart, L. E., Lowenberg, K., Nystrom, L. E., & Cohen, J. D. (2009). Pushing moral buttons: The interaction between personal force and intention in moral judgment. *Cognition, 111*(3), 364–371.

Greene, J. D., Morelli, S., Lowenberg, K., Nystrom, L., & Cohen, J. D. (2008). Cognitive load selectively interferes with utilitarian moral judgment. *Cognition, 107,* 1144–1154.

Greene, J. D., Nystrom, L. E., Engell, A. D., Darley, J. M., & Cohen, J. D. (2004). The neural bases of cognitive conflict and control in moral judgment. *Neuron, 44,* 389–400.

Greene, J. D., & Paxton, J. M. (2009). Patterns of neural activity associated with honest and dishonest moral decisions. *Proc Natl Acad Sci USA, 106*(30), 12506–12511.

Greene, J. D., Sommerville, R. B., Nystrom, L. E., Darley, J. M., & Cohen, J. D. (2001). An fMRI investigation of emotional engagement in moral judgment. *Science, 293,* 2105–2108.

Haidt, J. (2001). The emotional dog and its rational tail: A social intuitionist approach to moral judgment. *Psychol Rev, 108,* 814–834.

Haidt, J. (2012). *The righteous mind: Why good people are divided by politics and religion.* New York, NY: Random House.

Hare, R. D. (1991). *The Hare psychopathy checklist.* Revised. Toronto: Multi-Health Systems.

Hare, T. A., Camerer, C. F., Knoepfle, D. T., O'Doherty, J. P., & Rangel, A. (2010). Value computations in ventral medial prefrontal cortex during charitable decision making incorporate input from regions involved in social cognition. *J Neurosci, 30*(2), 583–590.

Harenski, C. L., Harenski, K. A., Shane, M. S., & Kiehl, K. A. (2010). Aberrant neural processing of moral violations in criminal psychopaths. *J Abnorm Psychol, 119*(4), 863.

Harenski, C. L., Kim, S. H., & Hamann, S. (2009). Neuroticism and psychopathy predict brain activation during moral and nonmoral emotion regulation. *Cogn Aff Behav Neurosci, 9*(1), 1–15.

Hauser, M. (2006). The liver and the moral organ. *Soc Cogn Affect Neurosci, 1,* 214–220.

Heekeren, H. R., Wartenburger, I., Schmidt, H., Prehn, K., Schwintowski, H. P., & Villringer, A. (2005). Influence of bodily harm on neural correlates of semantic and moral decision-making. *NeuroImage, 24,* 887–897.

Hsu, M., Anen, C., & Quartz, S. R. (2008). The right and the good: Distributive justice and neural encoding of equity and efficiency. *Science, 320,* 1092–1095.

Insel, T. R., & Young, L. J. (2001). The neurobiology of attachment. *Nat Rev Neurosci, 2,* 129–136.

Kahneman, D. (2003). A perspective on judgment and choice: Mapping bounded rationality. *Am Psychol, 58,* 697–720.

Kant, I. (1785/1959). *Foundation of the metaphysics of morals.* Indianapolis, IN: Bobbs-Merrill.

Kiehl, K. A. (2006). A cognitive neuroscience perspective on psychopathy: Evidence for paralimbic system dysfunction. *Psychiatry Res, 142,* 107–128.

Kiehl, K. A., Smith, A. M., Hare, R. D., Mendrek, A., Forster, B. B., Brink, J., & Liddle, P. F. (2001). Limbic abnormalities in affective processing by criminal psychopaths as revealed by functional magnetic resonance imaging. *Biol Psychiatry, 50,* 677–684.

Koven, N. S. (2011). Specificity of meta-emotion effects on moral decision-making. *Emotion, 11*(5), 1255.

Knutson, B., Taylor, J., Kaufman, M., Peterson, R., & Glover, G. (2005). Distributed neural representation of expected value. *J Neurosci, 25*(19), 4806–4812.

Koenigs, M., Kruepke, M., & Newman, J. P. (2010). Economic decision-making in psychopathy: A comparison with ventromedial prefrontal lesion patients. *Neuropsychologia, 48*(7), 2198–2204.

Koenigs, M., Kruepke, M., Zeier, J., & Newman, J. P. (2012). Utilitarian moral judgment in psychopathy. *Soc Cogn Affect Neurosci, 7*(6), 708–714.

Koenigs, M., & Tranel, D. (2007). Irrational economic decision-making after ventromedial prefrontal damage: Evidence from the Ultimatum Game. *J Neurosci, 27,* 951–956.

Koenigs, M., Young, L., Adolphs, R., Tranel, D., Cushman, F., Hauser, M., & Damasio, A. (2007). Damage to the prefrontal cortex increases utilitarian moral judgements. *Nature, 446,* 908–911.

Kohlberg, L. (1969). Stage and sequence: The cognitive-developmental approach to socialization. In D. A. Goslin (Ed.), *Handbook of socialization theory and research* (pp. 347–480). Chicago, IL: Rand McNally.

Kosfeld, M., Heinrichs, M., Zak, P. J., Fischbacher, U., & Fehr, E. (2005). Oxytocin increases trust in humans. *Nature, 435,* 673–676.

Koster-Hale, J., Saxe, R., Dungan, J., & Young, L. L. (2013). Decoding moral judgments from neural representations of intentions. *Proc Natl Acad Sci USA, 110*(14), 5648–5653.

Kriegeskorte, N., Goebel, R., & Bandettini, P. (2006). Information-based functional brain mapping. *Proc Natl Acad Sci USA, 103*(10), 3863–3868.

Maddock, R. J. (1999). The retrosplenial cortex and emotion: New insights from functional neuroimaging of the human brain. *Trends Neurosci, 22,* 310–316.

Marsh, A. A., Crowe, S. L., Henry, H. Y., Gorodetsky, E. K., Goldman, D., & Blair, R. J. R. (2011). Serotonin transporter genotype (5-HTTLPR) predicts utilitarian moral judgments. *PLoS ONE, 6*(10), e25148.

Marsh, A., Finger, E., Mitchell, D., Reid, M., Sims, C., Kosson, D., & Blair, R. (2008). Reduced amygdala response to fearful expressions in children and adolescents with callous-unemotional traits and disruptive behavior disorders. *Am J Psychiatry, 165*(6), 712–720.

Mendez, M. F., Anderson, E., & Shapira, J. S. (2005). An investigation of moral judgement in frontotemporal dementia. *Cogn Behav Neurol, 18,* 193–197.

MILL, J. S. (1861/1998). *Utilitarianism.* New York, NY: Oxford University Press.

MILLER, E. K., & COHEN, J. D. (2001). An integrative theory of prefrontal cortex function. *Annu Rev Neurosci, 24,* 167–202.

MILLER, M. B., SINNOTT-ARMSTRONG, W., YOUNG, L., KING, D., PAGGI, A., FABRI, M., & GAZZANIGA, M. S. (2010). Abnormal moral reasoning in complete and partial callosotomy patients. *Neuropsychologia, 48*(7), 2215–2220.

MITCHELL, J. P. (2009). Inferences about mental states. *Philos Trans R Soc Lond B Biol Sci, 364*(1521), 1309–1316.

MOLL, J., DE OLIVEIRA-SOUZA, R., ESLINGER, P. J., BRAMATI, I. E., MOURÃO-MIRANDA, J., ANDREIUOLO, P. A., & PESSOA, L. (2002). The neural correlates of moral sensitivity: A functional magnetic resonance imaging investigation of basic and moral emotions. *J Neurosci, 22*(7), 2730–2736.

MOLL, J., ESLINGER, P. J., & OLIVEIRA-SOUZA, R. (2001). Frontopolar and anterior temporal cortex activation in a moral judgment task: Preliminary functional MRI results in normal subjects. *Arq Neuropsiquiatr, 59,* 657–664.

MOLL, J., KRUEGER, F., ZAHN, R., PARDINI, M., DE OLIVEIRA-SOUZA, R., & GRAFMAN, J. (2006). Human fronto-mesolimbic networks guide decisions about charitable donation. *Proc Natl Acad Sci USA, 103,* 15623–15628.

MORAN, J. M., YOUNG, L. L., SAXE, R., LEE, S. M., O'YOUNG, D., MAVROS, P. L., & GABRIELI, J. D. (2011). Impaired theory of mind for moral judgment in high-functioning autism. *Proc Natl Acad Sci USA, 108*(7), 2688–2692.

MORETTO, G., LÀDAVAS, E., MATTIOLI, F., & DI PELLEGRINO, G. (2010). A psychophysiological investigation of moral judgment after ventromedial prefrontal damage. *J Cogn Neurosci, 22*(8), 1888–1899.

MORISHIMA, Y., SCHUNK, D., BRUHIN, A., RUFF, C. C., & FEHR, E. (2012). Linking brain structure and activation in temporoparietal junction to explain the neurobiology of human altruism. *Neuron, 75*(1), 73–79.

NORMAN, K. A., POLYN, S. M., DETRE, G. J., & HAXBY, J. V. (2006). Beyond mind-reading: Multi-voxel pattern analysis of fMRI data. *Trends Cogn Sci, 10*(9), 424–430.

OXFORD, M., CAVELL, T. A., & HUGHES, J. N. (2003). Callous/unemotional traits moderate the relation between ineffective parenting and child externalizing problems: A partial replication and extension. *J Clin Child Adolesc Psychol, 32,* 577–585.

PARKINSON, C., SINNOTT-ARMSTRONG, W., KORALUS, P. E., MENDELOVICI, A., MCGEER, V., & WHEATLEY, T. (2011). Is morality unified? Evidence that distinct neural systems underlie moral judgments of harm, dishonesty, and disgust. *J Cogn Neurosci, 23*(10), 3162–3180.

PAXTON, J. M., UNGAR, L., & GREENE, J. D. (2012). Reflection and reasoning in moral judgment. *Cognitive Sci, 36*(1), 163–177.

PERKINS, A. M., LEONARD, A. M., WEAVER, K., DALTON, J. A., MEHTA, M. A., KUMARI, V., & ETTINGER, U. (2012). A dose of ruthlessness: Interpersonal moral judgment is hardened by the anti-anxiety drug lorazepam. *J Exp Psychol Gen, 142*(3), 612–620.

PESSOA, L. (2010). Emotion and cognition and the amygdala: From "what is it?" to "what's to be done?" *Neuropsychologia, 48*(12), 3416–3429.

PETRINOVICH, L., O'NEILL, P., & JORGENSEN, M. (1993). An empirical study of moral intuitions: Toward an evolutionary ethics. *J Pers Soc Psychol, 64,* 467–478.

PHELPS, E. A., O'CONNOR, K. J., CUNNINGHAM, W. A., FUNAYAMA, E. S., GATENBY, J. C., GORE, J. C., & BANAJI, M. R. (2000). Performance on indirect measures of race evaluation predicts amygdala activation. *J Cogn Neurosci, 12*(5), 729–738.

PINKER, S. (2011). *The better angels of our nature: Why violence has declined.* New York, NY: Viking.

PUJOL, J., BATALLA, I., CONTRERAS-RODRÍGUEZ, O., HARRISON, B. J., PERA, V., HERNÁNDEZ-RIBAS, R., & CARDONER, N. (2012). Breakdown in the brain network subserving moral judgment in criminal psychopathy. *Soc Cogn Affect Neurosci, 7*(8), 917–923.

PIAGET, J. (1965). *The moral judgement of the child.* New York, NY: Free Press.

POULIN, M. J., HOLMAN, E. A., & BUFFONE, A. (2012). The neurogenetics of nice: Receptor genes for oxytocin and vasopressin interact with threat to predict prosocial behavior. *Psychol Sci, 23*(5), 446–452.

RAICHLE, M. E., MACLEOD, A. M., SNYDER, A. Z., POWERS, W. J., GUSNARD, D. A., & SHULMAN, G. L. (2001). A default mode of brain function. *Proc Natl Acad Sci USA, 98*(2), 676–682.

RAND, D. G., GREENE, J. D., & NOWAK, M. A. (2012). Spontaneous giving and calculated greed. *Nature, 489*(7416), 427–430.

RANGEL, A., CAMERER, C., & MONTAGUE, P. R. (2008). A framework for studying the neurobiology of value-based decision making. *Nat Rev Neurosci, 9*(7), 545–556.

RANGEL, A., & HARE, T. (2010). Neural computations associated with goal-directed choice. *Curr Opin Neurobiol, 20*(2), 262–270.

RAINE, A., & YANG, Y. (2006). Neural foundations to moral reasoning and antisocial behavior. *Soc Cogn Affect Neurosci, 1,* 203–213.

RILLING, J., GLENN, A., JAIRAM, M., PAGNONI, G., GOLDSMITH, D., ELFENBEIN, H., & LILIENFELD, S. (2007). Neural correlates of social cooperation and non-cooperation as a function of psychopathy. *Biol Psychiatry, 61,* 1260–1271.

SANFEY, A. G., RILLING, J. K., ARONSON, J. A., NYSTROM, L. E., & COHEN, J. D. (2003). The neural basis of economic decision-making in the Ultimatum Game. *Science, 300,* 1755–1758.

SARLO, M., LOTTO, L., MANFRINATI, A., RUMIATI, R., GALLICCHIO, G., & PALOMBA, D. (2012). Temporal dynamics of cognitive–emotional interplay in moral decision-making. *J Cogn Neurosci, 24*(4), 1018–1029.

SAVER, J., & DAMASIO, A. (1991). Preserved access and processing of social knowledge in a patient with acquired sociopathy due to ventromedial frontal damage. *Neuropsychologia, 29,* 1241–1249.

SCHAICH BORG, J., LIEBERMAN, D., & KIEHL, K. A. (2008). Infection, incest, and iniquity: Investigating the neural correlates of disgust and morality. *J Cogn Neurosci, 20,* 1–19.

SCHAICH BORG, J., HYNES, C., VAN HORN, J., GRAFTON, S., & SINNOTT-ARMSTRONG, W. (2006). Consequences, action, and intention as factors in moral judgments: An fMRI investigation. *J Cogn Neurosci, 18,* 803–817.

SCHAICH BORG, J., & SINNOTT-ARMSTRONG, W. (2013). Do psychopaths make moral judgments? In K. Kiehl & W. Sinnott-Armstrong (Eds.), *Handbook on psychopathy and law* (pp. 107–128). New York, NY: Oxford University Press.

Schaich Borg, J., Sinnott-Armstrong, W., Calhoun, V. D., & Kiehl, K. A. (2011). Neural basis of moral verdict and moral deliberation. *Soc Neurosci, 6*(4), 398–413.

Schultz, W., Dayan, P., & Montague, P. R. (1997). A neural substrate of prediction and reward. *Science, 275*, 1593–1599.

Shenhav, A., & Greene, J. D. (2010). Moral judgments recruit domain-general valuation mechanisms to integrate representations of probability and magnitude. *Neuron, 67*(4), 667–677.

Shenhav, A., & Greene, J. D. (2014). Integrative moral judgment: Dissociating the roles of the amygdala and ventromedial prefrontal cortex. *J Neurosci, 34*(13), 4741–4749.

Singer, T., Kiebel, S., Winston, J., Dolan, R., & Frith, C. (2004). Brain response to the acquired moral status of faces. *Neuron, 41*, 653–662.

Singer, T., Seymour, B., O'Doherty, J. P., Stephan, K. E., Dolan, R. J., & Frith, C. D. (2006). Empathic neural responses are modulated by the perceived fairness of others. *Nature, 439*, 466–469.

Tabibnia, G., Satpute, A. B., & Lieberman, M. D. (2008). The sunny side of fairness: Preference for fairness activates reward circuitry (and disregarding unfairness activates self-control circuitry). *Psychol Sci, 19*, 339–347.

Thomas, B. C., Croft, K. E., & Tranel, D. (2011). Harming kin to save strangers: Further evidence for abnormally utilitarian moral judgments after ventromedial prefrontal damage. *J Cogn Neurosci, 23*(9), 2186–2196.

Thomson, J. (1985). The trolley problem. *Yale Law J, 94*, 1395–1415.

Turiel, E. (2006). Thought, emotions and social interactional processes in moral development. In M. Killen & J. Smetana (Eds.), *Handbook of moral development* (pp. 1–30). Mahwah, NJ: Erlbaum.

Viding, E., Blair, R. J., Moffitt, T. E., & Plomin, R. (2005). Evidence for substantial genetic risk for psychopathy in 7-year-olds. *J Child Psychol Psychiatry, 46*, 592–597.

Waytz, A., Zaki, J., & Mitchell, J. P. (2012). Response of dorsomedial prefrontal cortex predicts altruistic behavior. *J Neurosci, 32*(22), 7646–7650.

Young, L., Bechara, A., Tranel, D., Damasio, H., Hauser, M., & Damasio, A. (2010). Damage to ventromedial prefrontal cortex impairs judgment of harmful intent. *Neuron, 65*(6), 845–851.

Young, L., Camprodon, J. A., Hauser, M., Pascual-Leone, A., & Saxe, R. (2010). Disruption of the right temporoparietal junction with transcranial magnetic stimulation reduces the role of beliefs in moral judgments. *Proc Natl Acad Sci USA, 107*(15), 6753–6758.

Young, L., Cushman, F., Hauser, M., & Saxe, R. (2007). The neural basis of the interaction between theory of mind and moral judgment. *Proc Natl Acad Sci USA, 104*(20), 8235–8240.

Young, L., & Dungan, J. (2012). Where in the brain is morality? Everywhere and maybe nowhere. *Soc Neurosci, 7*(1), 1–10.

Zaki, J., & Mitchell, J. P. (2011). Equitable decision making is associated with neural markers of intrinsic value. *Proc Natl Acad Sci USA, 108*(49), 19761–19766.

87 Cognitive Neuroscience and Criminal Responsibility

URI MAOZ AND GIDEON YAFFE

ABSTRACT A defendant is criminally responsible only if he engaged in a wrongful act, or *actus reus* (e.g., for larceny, voluntarily taking someone else's property without permission), with a guilty mind, or *mens rea* (e.g., knowing he had taken someone else's property without permission, and intending not to return it), and lacks affirmative defenses (e.g., the insanity defense). We therefore first review neuroscientific studies that inform the nature of voluntary action and so could tell us something of importance about the *actus reus* of crimes. Then we look at studies of intention, perception of risk, and other mental states that matter to the *mens rea* of crimes. Lastly, we discuss studies of self-control, which might be relevant to some formulations of the insanity defense. As we show, to date, very little is known about the brain that is of significance for understanding criminal responsibility, although there is potential for work of importance.

Shortly after turning 40, Michael became interested in child pornography. Finding himself attracted to his 12-year-old stepdaughter, he fondled her and was convicted of sexual molestation of a child. The judge gave Michael the chance to avoid jail time by successfully passing a treatment program. He did very poorly in the program; among other things, he came on to staff and other patients. Michael also complained of headaches, and finally a neurologist recommended a structural MRI of Michael's brain, revealing a large orbitofrontal tumor. After recovering from the tumor-removal surgery, Michael no longer had sexual urges directed toward children and passed the treatment program easily. He returned home and seemed to have no problems beyond those stemming from his brain surgery. Months later, however, his sexual desires for children returned. Another MRI revealed that the tumor too had returned. It was again removed, and, again, Michael's urges went away (Burns & Swerdlow, 2003).

While it is at best rare for impulses toward criminal behavior to be traceable to specific brain abnormalities, Michael's case may be the only one to date where the relevant brain abnormality was also correctable through medical intervention. However, Michael's case is also importantly just like *every* case of criminal behavior: criminal behavior is a product of the interaction between the states and functional dispositions of the brain, on the one hand, and the environment, on the other. If we knew enough about the brain, and if our medical technology was sophisticated enough, couldn't we, through medical intervention, always eliminate the impulse that gave rise to a crime?

Consider Michael's case in relationship to a study conducted by Kiehl and colleagues (Aharoni et al., 2013), showing that error-related activity in the anterior cingulate cortex—a limbic region associated with impulse control—during an inhibitory task predicted recidivism in male inmates about to be released from jail. A first and essential step toward appreciating the bearing of neuroscientific results on criminal responsibility is recognizing the sense in which Michael's criminal behavior, and that of the recidivating inmates, is no different from anyone else's: it has its source in the brain and the environment. There are two natural ways to respond: give up on criminal responsibility altogether, or accept that a person can be fully responsible for his bad behavior even though it is explicable, in principle, given enough information about his brain and his environment. Those drawn to the former view may be ready to stop reading this chapter now. If nobody is criminally responsible for his behavior, then neither neuroscience nor any other science can help us to understand criminally responsible behavior better; such behavior does not exist. However, those drawn to the latter view can seek ways in which neuroscience can, or does, inform our understanding of the features of human beings in virtue of which they are criminally responsible for their behavior. This is the approach taken here. We survey recent neuroscientific studies that might be thought to shed light on those facts about people in virtue of which they are, and are held by our legal system to be, criminally responsible for their behavior.

Under what conditions is a person criminally responsible for his behavior? Understanding the answer given by our legal system requires appreciating some legally defined concepts: *actus reus*, *mens rea*, and affirmative defenses. Criminal statutes define crimes. These definitions express a series of conditions that must be met, and must be shown to have been met by the prosecution, to determine guilt. For instance, under typical larceny statutes a defendant is guilty only if he (1) took something (2) that belonged to someone else (3) without permission, and while (i) knowing that he was taking it, (ii) being aware of a substantial risk that it belonged to someone else, (iii) being aware of a substantial risk that he lacked permission to take it, and (iv) intending never to return it. Facts that are not exclusively facts about the defendant's mind, such as (1), (2), and (3), are the crime's "*actus reus.*" The mental facts, such as (i), (ii), (iii), and (iv), are the crime's "*mens rea.*"

However, a defendant who performed the crime's *actus reus* with *mens rea* must also lack an affirmative defense, such as insanity, self-defense, or duress, to be criminally responsible. For instance, someone who is suffering from a severe mental illness that includes deific hallucinations may delusionally believe that he must steal his neighbor's car in order to comply with God's demands. When he does so, he may well meet all of the conditions required for commission of larceny, but he is not criminally responsible for his behavior. A similar point can be made about the other affirmative defenses. Consider someone who steals his neighbor's car under deadly threat from a third party.

Potentially, a neuroscientific result could show that a class of people, or even an individual person, fails to meet, or succeeds in meeting, one of the necessary conditions of criminal responsibility. If it could be shown, for instance, that people with orbitofrontal tumors like Michael's typically meet the law's criteria for insanity, that would support the claim that Michael was not criminally responsible for his behavior. Notice that such a result would not extend to everyone who engages in criminal behavior. Some states of the brain—e.g., low anterior cingulate cortex activity—that might give rise to criminal behavior (coupled with environmental factors) might do so without the agent meeting the law's criterion for insanity. Sanity and insanity are conditions of the brain. And criminal behavior, sane or insane, has its source in the brain. However, if the neural sources of insanity can be identified, that would be of potential use to the legal system.

In addition, neuroscientific studies might illuminate the neural mechanisms that underlie those features of people who are criminally responsible for their behavior, and so help us to understand criminal responsibility better, without thereby supporting an argument for or against holding any person or class of people criminally responsible. Such results about criminal responsibility would add to our knowledge not just of the brain but of one of the most socially important phenomena to which the brain gives rise: crimes for which people are responsible and deserving of punishment. This would be of value even if justice in our legal system would not be increased thereby.

The first section of this chapter discusses neuroscientific studies that bear on the nature of voluntary action, and so could, potentially, tell us something of importance about the *actus reus* of crimes. The second section discusses studies of intention, perception of risk, and other mental states that matter to the *mens rea* of crimes. And the third and last section discusses studies of self-control. While self-control is less relevant to criminal responsibility than one might think, for reasons that are explained, something like self-control is of relevance to some formulations of the insanity defense, and to other issues important to the assessment of criminal responsibility. As we will see in all three sections, the work that has been done so far is just the smallest drop in the bucket. To date, very little is known about the brain that is of significance for understanding criminal responsibility.

Voluntary action

It seems abhorrent to hold people criminally responsible only for their thoughts. And it seems abhorrent to hold people criminally responsible for their status. It would be wrong, that is, to punish people for intending to do things that they take no steps toward doing, and it would be wrong to punish people for, for instance, being poor, or unemployed, or belonging to a particular race or class. These base-level moral intuitions are accommodated in the law in part through a restriction on the *actus reus* of crimes: the *actus reus* must include a voluntary act. There are some exceptions to this—notably, people are also criminally responsible for omitting actions they have a duty to perform—but, as a default, criminal responsibility requires a voluntary act.

The law has employed something like the following definition of a voluntary act: a bodily movement guided by a conscious mental representation of that bodily movement. If a hurricane's wind blows someone through a shop window, shattering it, there is no guilt for destruction of property since the bodily motion that caused property to be destroyed was not guided by the agent's mental activity. The bodily movements during an epileptic seizure, also, do not rise to the legal standard of a voluntary act, since the brain activity that gives rise to them does not guide them in the way, for instance, in which a desire to move one's hand guides one's hand. Consider a more problematic case: in *People v. Newton* (1970), the defendant was shot in the gut during an altercation with police. Moments later, he shot and killed one of the officers before wandering several blocks to a hospital where he collapsed, unconscious. On waking later, he reported no memory of the incident. A doctor testified that it is common for people who have just suffered severely traumatic injury, as Newton had, to engage in complex, seemingly goal-directed bodily movements while unconscious. Eventually, Newton was acquitted on the grounds that the bodily movements in which he engaged when shooting the officer dead were not voluntary acts, since the mental representations that guided them were not conscious.

The nature of voluntary bodily movements is amenable to investigation using neuroscientific tools. And, in fact, a seminal study showed that when subjects were asked to flex their wrist or finger at a time of their choice, a brain potential known to precede voluntary action—named the readiness potential—appeared about 300 msec before the time that subject reported as the onset of the urge to move (Libet, Gleason, Wright, & Pearl, 1983). While independently replicated and expanded, this experiment and its follow-ups were heavily criticized on empirical and conceptual grounds.[1] However, as is detailed elsewhere (Roskies, 2010; Sinnott-Armstrong & Nadel, 2010; Yaffe, 2010), even if taken at face value, it is not warranted to conclude from such work that the law's definition of voluntary action is flawed.

Another set of experiments studied the awareness that people have of initiating, executing, and controlling volitional action—termed "the sense of agency." These experiments relied on patients who underwent invasive brain mapping to minimize the impact of brain tumor removal surgery on their later everyday life. This includes electrical stimulation of various brain regions to determine their functionality (e.g., by eliciting hand movement or arresting speech). It was shown that stimulating right inferior parietal regions triggers an intention for contralateral movement, while left inferior-parietal stimulation provoked an intention to move the lips and talk (Desmurget et al., 2009). At higher stimulation intensities, participants reported having performed these movements, though no movement occurred. In contrast, premotor-region stimulation triggered contralateral limb and mouth movements, while patients firmly denied moving (figure 87.1). This suggests that—at least under rather abnormal circumstances—humans may experience agency over phantom actions and carry out actions with no accompanying sense of agency. Thus, the brain circuitry involved in the production of bodily movement may not be as closely related to the circuitry involved in the sensation of planning, initiating, or even engaging in bodily movement as we might have thought.

Do these experiments show that bodily movements are not guided by conscious mental activity, as required by the law's voluntary act requirement? It is far from clear. A brain-stimulation experiment could presumably be developed in which subjects are shown to be under a certain kind of illusion: they think that they are acting voluntarily when they are not. But the possibility of illusion in one case does not establish its ubiquity in all cases, just like the existence of optical illusions does not render all vision illusory. Further, it is not clear whether guiding one's bodily movements through a conscious mental state necessarily results in a sense of agency, or whether the sense of agency is either necessary or sufficient for consciousness of the kind that the law takes to be involved in voluntary action. These nontrivial questions might be partly empirically answerable. But without the answers, there is no reason to take such experiments to establish the inadequacy of the law's definition of a voluntary act.

Intention and the perception of risk

Imagine you are offered a new iPad on the street at half its retail price. You ask no questions and buy it, only to later be arrested for the crime of receipt of stolen property. "But I didn't *know* it was stolen," you say. This might be true. Perhaps you are very naïve and believed the iPad was not stolen. Or perhaps you were not sure one way or the other, although you recognized there was a good chance it was stolen. Whether you are guilty of the crime depends on what *mens rea* standard is set by the statute defining it. Does the statute require, for guilt, knowledge that the property is stolen? Or will awareness of a large enough risk of that suffice? Or is it enough that a reasonable (not naïve) person in your situation would have known, even if you did not?

Recognizing that people can be in various mental states with respect to a particular fact, and that people vary in their culpability in accordance with these variations in mental state, the law allows for different kinds of *mens rea*. For our purposes here, the most important types are *intent* and *recklessness*. The Model Penal Code,

[1]For a recent discussion of empirical and conceptual problems with this type of experiments see Maoz et al. (2014). In particular, Breitmeyer (1985) and Roskies (2010) discuss why generalizing the Libet results, which focused on random decisions, to deliberate decisions is problematic. Predictability of deliberate decisions is beginning to be studied (Maoz et al., 2012), and preliminary evidence suggests that they may utilize different brain processes than random ones (Mudrik et al., 2013).

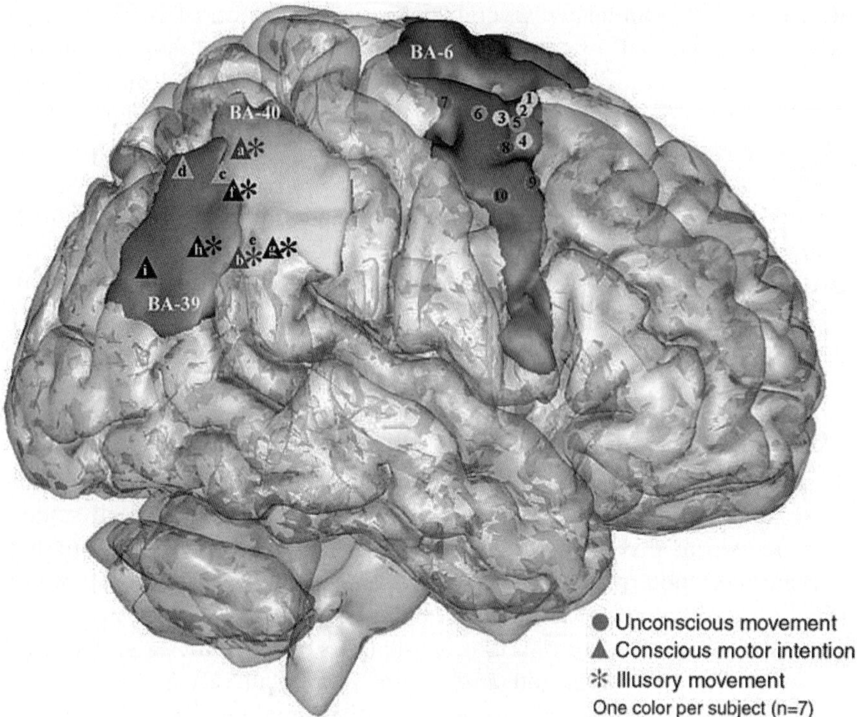

FIGURE 87.1 Premotor and parietal responsive sites shown after registration of individual magnetic resonance images to the Montreal Neurological Institute template. Left-hemisphere stimulations were reported on the right hemisphere. Shaded areas define anatomical boundaries of Brodmann areas 40, 39, and 6. (Adapted from Desmurget et al., 2009. Reprinted with permission from AAAS.)

which has had tremendous influence on the law in the United States,[2] defines "intent" or "purpose" as having a particular act or causal result of that act as one's "conscious object" (ALI, 1962, §2.02(2)(a)(i)). This language is aimed at capturing the idea that those who intend acts or results *aim* at them, or make them their goal. By contrast, the Model Penal Code defines "recklessness" as "awareness of a substantial and unjustifiable risk" that one is acting in a certain way, or that one will bring about a particular result through one's act (ALI, 1962, §2.02(2)(c)). As the receipt of stolen property example suggests, one can be reckless with respect to a particular fact—that the property is stolen—without intending to bring it about. We can assume that you did not intend to receive stolen property; you would have been happy to receive an iPad for half-price, stolen or not. But you might have been aware of a substantial and unjustifiable risk that the property was stolen.

Some neuroscientific work purporting to be investigating neural mechanisms underlying intention is investigating a psychological state importantly different from the one to which the law gives that label (e.g., Shadlen, Kiani, Hanks, & Churchland, 2008, on the neural underpinnings of perceptual discrimination). These may have to do more with perceptual judgments about the optimally rewarded action than with neural substrates of intent, as defined in the Model Penal Code.

Studies in which subjects are induced to have intentions closer to those with legal relevance exist.[3] For

[2]In the United States, there are 51 primary jurisdictions, and so 51 important and distinct bodies of criminal law (the 50 states and the federal law), each of which is both different from and similar to the others in various respects. The Model Penal Code was developed by the American Law Institute in 1962 and has been revised at various times since, with the purpose of assisting legislatures to update and standardize the penal code of the United States. It has been very influential in that respect.

[3]The law is concerned both with so called distal intentions, or intentions to act in the future, and proximate intentions, or intentions to act now. However, distal intentions matter to criminal liability only to the degree to which they are executed through the formation of later proximal intentions. In the study described in the main text, the focus is on the neural activity underlying the formation of the distal intention, but the experimental setup assures that subjects who follow the experimenter's instructions will act on that intention later, presumably through the ultimate formation of a proximate intention.

instance, Haynes and colleagues were able to decode or even arguably predict subjects' intentions about adding or subtracting two numbers (Haynes et al., 2007; Soon, He, Bode, & Haynes, 2013; see chapter 85, this volume). However, in the Haynes study, the subjects had no reason to keep their choices hidden from the experimenter. Yet to what extent could intentions be decoded from neural activity in a competitive situation in which revealing one's intentions is detrimental?

Maoz and colleagues (Maoz, Ye, Ross, Mamelak, & Koch, 2012) decoded the motor intentions of participants in a matching-pennies game, where a subject and opponent had to raise one hand at the go signal. The subject would win \$0.10 from his opponent if the hand he raised was a mirror image of his opponent's. Otherwise he lost \$0.10 to his opponent. Both players started with \$5, and if the subject was the overall winner over 50 trials, he received his final winnings in cash. The subjects—consenting epilepsy patients implanted with intracranial electrodes for clinical purposes—therefore had every incentive to keep their intentions hidden from the experimenter (figure 87.2A), though they were not explicitly informed that their brain activity would be used to decipher their actions. Nevertheless, subjects' intentions could be correctly decoded with 70% accuracy, on average, online and in real time, and used against them in the game. The average prediction accuracy rose to 83% in more rigorous offline analysis, and to 92% when the system was allowed to predict only on the 70% of the trials on which it was most confident (figure 87.2B).

Yet correlation does not entail causation: is the brain activity on the basis of which the subjects' intentions could be decoded the same as that through which the intention is stored, or does it merely accompany it? In addition, there is mounting evidence that the intuitively appealing model of serial decision making guiding some experiments of this kind may not be accurate. Under this intuitive model, decision making involves a three-step process—(1) gathering information (from the senses/memory), (2) deciding and forming an intention to act, and (3) executing the action. Instead, sensory information may be used to continuously specify several potential actions in parallel, and often in the same brain regions that later control the chosen behavior (Cisek & Kalaska, 2010; Freedman & Assad, 2011; Gold & Shadlen, 2007; Kable & Glimcher, 2009; Shadlen, Kiani, Hanks, & Churchland, 2008; Wise, Boussaoud, Johnson, & Caminiti, 1997). These processes continue after movement begins, even when sensory input is stopped, facilitating changes of mind (Resulaj, Kiani, Wolpert, & Shadlen, 2009; Selen, Shadlen, & Wolpert, 2012). Therefore, intention might not be clearly

A

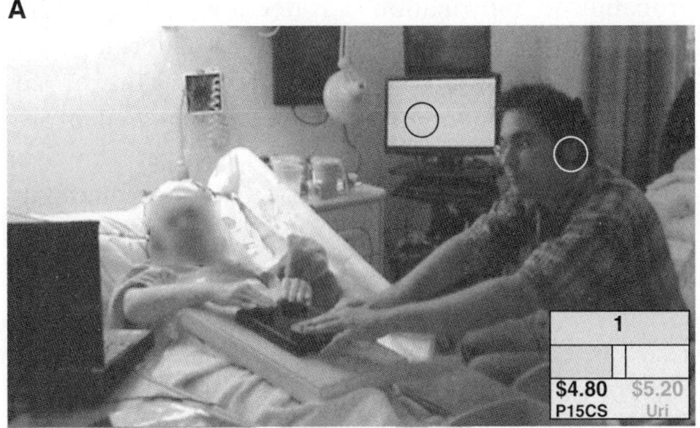

B

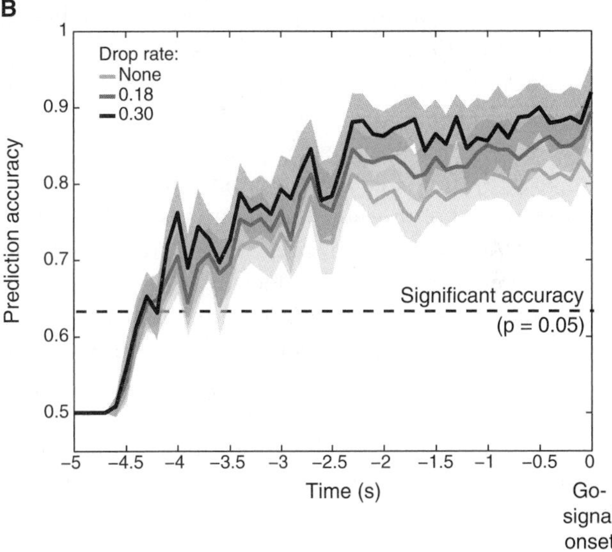

FIGURE 87.2 (A) The experimental setup in the clinic. The patient and experimenter are watching the game screen (inset on bottom right) on a computer (bottom left) displaying the countdown to the go-signal and still pressing down the buttons of the response box. The real-time system already computed a prediction, and so displays an arrow on the screen behind the patient and plays a tone in the experimenter's ear ipsilateral to the hand it predicted he should raise to beat the patient. (B) Across-subjects average of the prediction accuracy (mean ± standard error of mean shaded) versus time before go-signal onset. Values above the dashed horizontal line are significant at $p = 0.05$. (Adapted with permission from Maoz et al., 2012.) (See color plate 86.)

mappable onto a single neural process, and the brain may contain representations of multiple, conflicting action plans. If so, further conceptual work is needed to determine what, exactly, distinguishes an intention, in the sense that matters to criminal responsibility, from other mental states representing action plans.

Turning now to recklessness—awareness of a substantial and unjustifiable risk of a fact of legal importance; for example, the iPad being stolen. Many neuroscientific studies examine how people process probabilistic information for decision making. One famous set used the Iowa Gambling Task, developed by Damasio and colleagues. Normal, healthy control participants and patients with bilateral ventromedial prefrontal cortex (vmPFC) damage picked cards from four decks randomly bearing positive or negative monetary rewards. The positive rewards in decks A and B were double that in decks C and D. But A and B were stacked so that selecting cards from them would result in a loss overall, while selecting from C or D would result in a gain. The players, who were instructed to gain as much money as possible, knew nothing about the setup of the decks or when the game would be stopped. The control participants began to choose advantageously—that is, selecting mainly from C and D—before they could report any knowledge about the advantageous strategy. They also generated anticipatory skin-conductance responses before selecting from the riskier A or B decks, and 70% of them could spell out the decks' setup before the end of the game. The patients neither chose advantageously—generally preferring cards from A and B—nor generated skin-conductance responses, not even the 50% who eventually could explicitly explain the decks' setup (Bechara, Damasio, Damasio, & Anderson, 1994; Bechara, Damasio, Tranel, & Damasio, 1997). It therefore seems that for decisions like those in the Iowa Gambling Task, the vmPFC is required for collecting information about risks for reward-seeking action.

While the representation of probabilistic information is a part of the legal notion of recklessness, it is not exhausted by that idea. First, for recklessness, the probabilities must be "substantial" and the conduct "unjustifiable." But, more importantly, recklessness involves *conscious awareness* of probabilistic information. The legally reckless agent represents information about the risks of harming others while *aware* of the possibility, although not the certainty, that his actions would do harm. Awareness of risks is crucial for the reckless agent to be *criminally culpable.* Where there is awareness of risk, and action that imposes risk, there seems to be a disregard of the importance of the harms being risked. To date, few if any neuroscientific studies have investigated the distinctive nature of *conscious awareness* of risk, distinguishing its neural basis, and role in decision making, from tacit, or unconscious representations of probabilistic information.

One area where the law could particularly use assistance from neuroscience concerns the impact of mental disorders on mental states crucial to criminal responsibility. Broadly speaking, for instance, addiction involves disruption of dopamine signals, and brain areas mediated by dopamine, like the striatum, are crucial for processing and learning from probabilistic information (Balleine, Delgado, & Hikosaka, 2007; Cromwell & Schultz, 2003; chapter 88, this volume). We would therefore expect that addicts are consciously aware of risks in different ways, and in different patterns, from nonaddicts. Exactly *how* addiction modulates conscious awareness of risk is important for the legal system. Many addicts find their way into courtrooms, and often it is crucial whether they were reckless with respect to the harms they caused when they acted. Currently, however, the legal system incorporates no empirical information about how addicts represent and process information regarding risks when judging their criminal responsibility. This is in part because the legal system is slow to incorporate scientific information, and with good reason: the gap between the lab and the world is often too wide to warrant changing the way legal judgments are made. Moreover, scientific studies of decision in the face of risk and their modulation by mental disorders have not been guided by the legal conception of recklessness. Significant progress could be made in this area—progress that might point the way to substantial legal reform—by neuroscientific research guided from the outset by entrenched legal concepts, like that of recklessness. Similar points can be made about the bearing of mental disorders on intention and other mental states that matter to *mens rea.* More work can and should be done in this area.

Self-control

The capacity for self-control seems important to responsibility. Some of those who harmed others were in control; others lost control, but could have held back at the moment they inflicted harm; and still others lost control and could not have maintained it, and consequently could not have refrained from the harmful behavior. These agents seem to differ from a moral standpoint.

Nevertheless, for the most part, the criminal law is insensitive to these moral differences. It is very rare for a difference in treatment under the criminal law—for example, less or no punishment—to turn on whether the defendant was in control, or could have been. There are some exceptions; for instance, a defendant who is very upset when he kills another person will, if various other conditions are met, be guilty not of murder but of the lesser crime of manslaughter. Possibly, the law grants mitigation in such cases through a recognition that at least some who kill while in a state of extremely heightened emotion are less than fully in control of their actions. However, mitigation in such cases is specific to homicide, and thus not available to those who, for instance, commit the crime of destruction of property while very upset, as when a person smashes his girlfriend's windshield after an argument. This indicates the law's stingy attitude toward basing differences in treatment on differences in control. Still, there are other important exceptions. Under one, uncommon formulation of the insanity defense, a defendant who "lacked substantial capacity to conform his conduct to the requirements of law" due to mental illness that gave rise to that conduct would be excused from criminal responsibility (ALI, 1962, §4.01(1)). Under this so-called volitional prong of the insanity defense, those who cannot, or find it extremely difficult to, do what the law requires of them are not subject to criminal punishment. In addition, the Supreme Court has recently ruled that adolescents who have committed very serious crimes, including murder, ought not be punished as severely as otherwise identical adults in part on the grounds that adolescents, as a group, are more impulsive than adults (*Graham v. Florida*, 2010; *Miller v. Alabama*, 2012; *Roper v. Simmons*, 2005). So, the court seems to predicate an important difference in legal treatment of adolescents, although not adults, on a difference in the capacity for self-control. It is also possible that a deeper scientific understanding of control and its limits could inform the law. Perhaps if more were known about the factors that influence and limit self-control, and how to measure it, the legal system would predicate more differences in treatment on differences in control.

Important neuroscientific work on self-control has emerged in recent years, although it remains uncertain how, if at all, it bears on criminal responsibility. For instance, Rangel and colleagues first instructed hungry dieters and nondieters to make choices about either the healthiness or taste of various foods on a five-point scale (figure 87.3A). One item that was rated as neutral on both scales was designated the reference food for each subject, and subjects repeatedly chose between it and different foods. To make the choices concrete, one trial was selected at random and the subjects had to eat the food they had selected. They then divided the subjects into self-controllers (SC) and nonself-controllers (NSC) based on their choices during the experiment (e.g., declining unhealthy, liked items).

Using functional MRI, the researchers found that activity in the vmPFC was correlated with the subjects' choices, regardless of the self-control they exhibited. Also, vmPFC activity was correlated with both taste and health for SC, but only with taste for NSC. In contrast, dorsolateral prefrontal cortex (dlPFC) activity increased during successful self-control trials and then also correlated with vmPFC activity. They therefore suggest that vmPFC reflects short-term goals (taste), which are then modulated by long-term considerations (health) using the dlPFC; and that the extent to which dlPFC can modulate vmPFC activity accounts for much of the difference between successful and unsuccessful self-control (figure 87.3B; Hare, Camerer, & Rangel, 2009).[4]

How work of this kind bears on criminal responsibility is far from clear. Self-control might be variable within subjects, and context-sensitive. Someone who has difficulty resisting tempting food might easily resist

[4]For another vein of work focused on potential strategies for resisting temptation, see, for example, McRae et al. (2010).

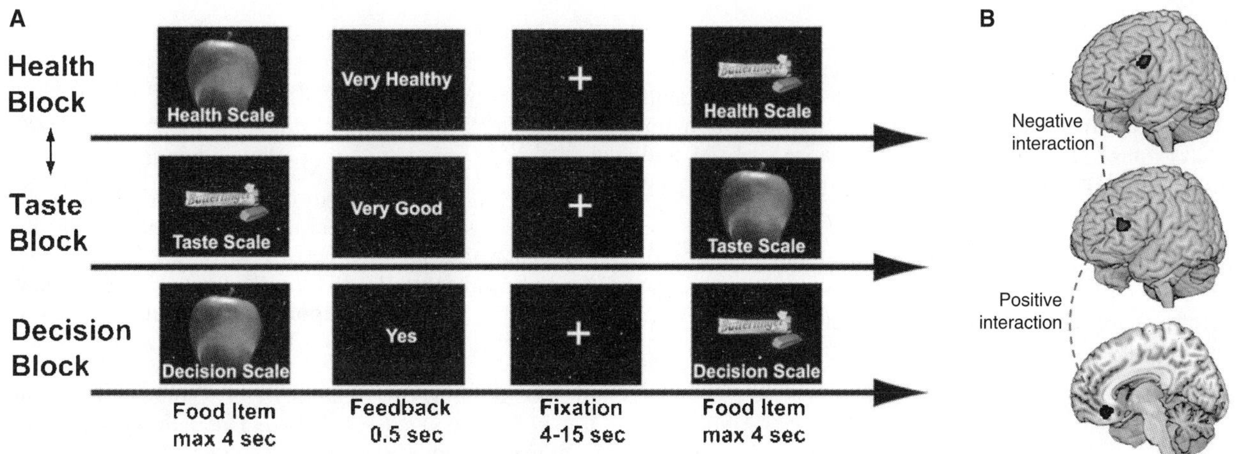

FIGURE 87.3 (A) Experimental progress during the health, taste, and decision blocks (in gray scale). (B) The ventromedial prefrontal cortex (top) may control the dorsolateral prefrontal cortex (bottom) through an intermediate brain region like the inferior frontal gyrus (BA 46; middle). (Adapted from Hare et al., 2009. Reprinted with permission from AAAS.)

temptations to steal. And someone who has trouble resisting the temptation to mildly harm another while playing a game in a lab might not be tempted to harm outside the lab. In addition, how much, and what type, of self-control is required for moral responsibility, much less criminal, is difficult to specify. Hence, very little can be said with confidence about what recent work on the neuroscience of self-control implicates, if anything, about criminal responsibility. Potentially, however, work of this kind could provide the first steps in helping us to determine how particular psychological disorders result in sufficient absence of "substantial capacity to conform one's conduct to law" (ALI, 1962, §4.01(1)) for insanity. Much work, however, would need to be done before such steps could be taken.

Conclusion

Many of the deepest and most difficult questions about criminal responsibility and the brain are neither more nor less tractable in light of recent neuroscientific experiments. Is there something about the dependency of our mental life on the state of a physical organ that is incompatible with criminal responsibility? If so, descriptions of how the mind, and the behavior to which it gives rise, are dependent on that organ will likely tell us nothing about criminal responsibility. Yet on the assumption that criminal responsibility is not just here to stay—as it surely is—but *should be* here to stay, there is room to ask: To what extent can neuroscience illuminate the nature and underlying mechanisms of those features that contribute to and constitute criminal responsibility? The question is very much worth asking, and while some work has already been done, as described in the sections of this chapter, there is much more to do.

ACKNOWLEDGMENTS This work was funded by the Ralph Schlaeger Charitable Foundation and Bial Foundation for U. M., FSU Big Questions in Free Will Initiative for U. M. and G. Y., and the Mellon Foundation New Directions Fellowship for G. Y.

The authors thank Adina Roskies and Walter Sinnott-Armstrong and other participants in a session discussing a draft of this chapter during July, 2013, for helpful comments. They also thank Caitlin Duncan for helpful comments and typesetting.

REFERENCES

Aharoni, E., Vincent, G. M., Harenski, C. L., Calhoun, V. D., Sinnott-Armstrong, W., Gazzaniga, M. S., & Kiehl, K. A. (2013). Neuroprediction of future rearrest. *Proc Natl Acad Sci USA, 110*(15), 6223–6228.

American Law Institute (ALI). (1962). *Model penal code and commentaries.* Philadelphia, PA: American Law Institute.

Balleine, B. W., Delgado, M. R., & Hikosaka, O. (2007). The role of the dorsal striatum in reward and decision-making. *J Neurosci, 27*(31), 8161–8165.

Bechara, A., Damasio, A. R., Damasio, H., & Anderson, S. W. (1994). Insensitivity to future consequences following damage to human prefrontal cortex. *Cognition, 50*(1), 7–15.

Bechara, A., Damasio, H., Tranel, D., & Damasio, A. R. (1997). Deciding advantageously before knowing the advantageous strategy. *Science, 275*(5304), 1293–1295.

Breitmeyer, B. G. (1985). Problems with the psychophysics of intention. *Behav Brain Sci, 8*(4), 539–540.

Burns, J. M., & Swerdlow, R. H. (2003). Right orbitofrontal tumor with pedophilia symptom and constructional apraxia sign. *Arch Neurol, 60*(3), 437–440.

Cisek, P., & Kalaska, J. F. (2010). Neural mechanisms for interacting with a world full of action choices. *Annu Rev Neurosci, 33,* 269–298.

Cromwell, H. C., & Schultz, W. (2003). Effects of expectations for different reward magnitudes on neuronal activity in primate striatum. *J Neurophysiol, 89*(5), 2823–2838.

Desmurget, M., Reilly, K. T., Richard, N., Szathmari, A., Mottolese, C., & Sirigu, A. (2009). Movement intention after parietal cortex stimulation in humans. *Science, 324*(5928), 811–813.

Freedman, D. J., & Assad, J. A. (2011). A proposed common neural mechanism for categorization and perceptual decisions. *Nat Neurosci, 14*(2), 143–146.

Gold, J. I., & Shadlen, M. N. (2007). The neural basis of decision making. *Annu Rev Neurosci, 30,* 535–574.

Hare, T. A., Camerer, C. F., & Rangel, A. (2009). Self-control in decision-making involves modulation of the vmPFC valuation system. *Science, 324*(5927), 646–648.

Haynes, J. D., Sakai, K., Rees, G., Gilbert, S., Frith, C., & Passingham, R. E. (2007). Reading hidden intentions in the human brain. *Curr Biol, 17*(4), 323–328.

Kable, J. W., & Glimcher, P. W. (2009). The neurobiology of decision: Consensus and controversy. *Neuron, 63*(6), 733.

Libet, B., Gleason, C. A., Wright, E. W., & Pearl, D. K. (1983). Time of conscious intention to act in relation to onset of cerebral activity (readiness-potential). The unconscious initiation of a freely voluntary act. *Brain, 106*(3), 623–642.

Maoz, U., Ye, S., Ross, I., Mamelak, A., & Koch, C. (2012). Predicting action content on-line and in real time before action onset—an intracranial human study. *Adv Neural Inf Process Syst, 25.*

Maoz, U., Mudrik, L., Rivlin, R., Ross, I., Mamelak, A., & Yaffe, G. (2014). On reporting the onset of the intention to move. In A. R. Mele (Ed.), *Surrounding free will.* New York, NY: Oxford University Press.

McRae, K., Hughes, B., Chopra, S., Gabrieli, J. D., Gross, J. J., & Ochsner, K. N. (2010). The neural bases of distraction and reappraisal. *J Cogn Neurosci, 22*(2), 248–262.

Mudrik, L., Maoz, U., Xu, D., Duncan, C., Zhang, Q., & Koch, C. (2013). Dissecting different types of decision-making: An ERP study of reasoned vs. unreasoned voluntary decisions. Paper presented at the 43rd Annual Meeting of the Society for Neuroscience, San Diego, CA.

Resulaj, A., Kiani, R., Wolpert, D. M., & Shadlen, M. N. (2009). Changes of mind in decision-making. *Nature, 461*(7261), 263–266.

Roskies, A. L. (2010). How does neuroscience affect our conception of volition? *Annu Rev Neurosci, 33,* 109–130.

Selen, L. P., Shadlen, M. N., & Wolpert, D. M. (2012). Deliberation in the motor system: Reflex gains track evolving evidence leading to a decision. *J Neurosci, 32*(7), 2276–2286.

Shadlen, M. N., Kiani, R., Hanks, T. D., & Churchland, A. K. (2008). Neurobiology of decision making: An intentional framework. In C. Engel & W. Singer (Eds.), *Better than conscious? Decision making, the human mind, and implications for institutions* (pp. 71–102). Cambridge, MA: MIT Press.

Sinnott-Armstrong, W., & Nadel, L. (Eds.). (2010). *Conscious will and responsibility: A tribute to Benjamin Libet.* New York, NY: Oxford University Press. Available from Oxford Scholarship Online. doi:10.1093/acprof:oso/9780195381641.001.0001

Soon, C. S., He, A. H., Bode, S., & Haynes, J. D. (2013). Predicting free choices for abstract intentions. *Proc Natl Acad Sci USA, 110*(15), 6217–6222.

Wise, S. P., Boussaoud, D., Johnson, P. B., & Caminiti, R. (1997). Premotor and parietal cortex: Corticocortical connectivity and combinatorial computations. *Annu Rev Neurosci, 20*(1), 25–42.

Yaffe, G. (2010). Libet and the criminal law's voluntary act requirement. In W. Sinnott-Armstrong & L. Nadel (Eds.), *Conscious will and responsibility* (pp. 189–206). New York, NY: Oxford University Press.

88 Society and Addiction: Bringing Understanding and Appreciation to a Mental Health Disorder

ANNABELLE M. BELCHER, NORA D. VOLKOW, F. GERARD MOELLER, AND SERGI FERRÉ

ABSTRACT Substance abuse–related issues present an enormous global financial burden in terms of health, policy, and treatment. The social stigma associated with addiction defers treatment-seeking, further exacerbating the problem. Abundant scientific data point to addiction as a mental health disorder, with several predispositional, environmental, and drug-induced factors. Yet despite attempts from a small cadre of addiction researchers, public appreciation of addiction as a treatable mental health disorder wanes, and perceptions of voluntariness remain high. We present a summary of existing evidence to substantiate addiction's status as a mental health disorder, and argue that it is incumbent on scientists and clinicians to change their framework to understand addiction as a mental disorder before society's view of addiction as a disease of choice can be changed.

If you have never been addicted, you can have no clear idea what it means to need junk with the addict's special need. You don't decide to be an addict. One morning you wake up sick and you're an addict.
—William S. Burroughs, *Junky*

After several years of chronic cocaine use and relapse, a young man in his mid-twenties finally hit rock bottom. One night, after years of losing several jobs, dropping out of school twice, losing friends, disappointing family, and feeling like a complete failure, he decided to try to change his life, and he went to an Alcoholics Anonymous meeting. The meeting began with personal introductions, and individuals gave their stories and the reasons for seeking a change in their lives. The young man's turn came, and he introduced himself and began to express his desire to quit his destructive habit. He had scarcely mentioned his cocaine dependence, when he was stopped short by one of the men in the group and told that he "ought to find another meeting, one that is more suited for *his type* of addiction." Sensing the unveiled antipathy in the room, the young man took the man's nuanced suggestion, and left abruptly. He didn't go back to meetings for quite some time.

This true story, as expressed by the young man to one of the authors, gives an example of the discrimination felt by many individuals who abuse particular classes of addictive substances. One can only imagine that the public stigmatization from those who *aren't* addicted to substances might be even greater.

Public perception of addiction

According to the latest results from the National Survey on Drug Use and Health, an estimated 18.6% of the adult US population suffered a mental illness in 2012 (SAMHSA, 2013). Three classes of mental health disorders (MHDs) are particularly common in the United States: depressive, anxiety, and substance use disorders (SUD), with lifetime prevalence rates of approximately 17%, 29%, and 35%, respectively (SAMHSA, 2013). Although there is no debate among scientists and those in the medical community that SUD is a MHD, public perception regarding SUD (and MHDs in general) has not been particularly generous. In addition to being a brain disorder of unknown etiology that is characterized by aberrant behaviors, there is a widespread perception that SUD involves a volitional component: in order to be cured, all a person needs to do is to stop using the drug. This notion, unfortunately embraced by some mental health professionals and neuroscientists (Heyman, 2009; Satel & Lillienfeld, 2013), has made it very difficult to change public perception of addiction. Society is less than forgiving about a disorder in which the victim continues to consume the substance that is making him or her ill, particularly when compared to a victim of cancer or amyotrophic lateral sclerosis (Lou Gehrig's disease)—diseases in which there is the perception that the victim's will plays no role whatsoever.

Why do public attitudes about MHDs matter? As Pescosolido and colleagues state, "Public attitudes set the context in which individuals in the community respond to the onset of mental health problems, clinicians respond to individuals who come for treatment, and public policy is crafted" (2010, p. 1324). Additionally, the negative public perception that surrounds an MHD affects whether an afflicted individual seeks treatment at all. Delayed treatment can mean progression of the SUD, which can become even more difficult to treat and have additional nonpsychiatric medical consequences.

Part of the problem with the way in which society views SUD could be in the lack of brain-based (neurobiological) diagnostics available for diagnosing SUD. In contrast, for neurological disorders, a CAT scan reveals a brain tumor, and posthumous plaques and tangles confirm Alzheimer's disease. But there is no biomarker of SUD, and addiction is aptly characterized as a behavioral disorder. The newly released *Diagnostic and Statistical Manual of Mental Disorders*, 5th edition, known as the DSM-V (APA, 2013) submits a definition, and corollary checklist of symptoms, of SUD:

> A maladaptive pattern of substance use, leading to clinically significant impairment or distress, as manifested by at least two of the following, occurring at any time within the same 12-month period:
>
> (1) Failure to fulfill obligations
> (2) Hazardous use
> (3) Social/interpersonal substance-related problems
> (4) Tolerance
> (5) Withdrawal
> (6) Persistent desire/unsuccessful efforts to cut down use
> (7) Using more or for longer than was intended
> (8) Neglect of important activities
> (9) Great deal of time spent in substance-involving activities
> (10) Psychological/physical use-related problems
> (11) Craving

Immediately apparent from the DSM (a handbook that is an essential fixture in every clinician's desktop library) is that the theme of "volition" can be inferred in the description of addiction. Although the checklist in the original text of the manual is a bit lengthier than the pared-down version given above, inherent in the definition is a notion that the addict's behavior is against his or her will (of consuming less drug than intended, or to cease taking drugs); concepts of "effort" and "intention" are pertinent to the diagnosis of someone with SUD.

The search for SUD biomarkers has provided the field with an appreciation of the fact that the brain of the addicted individual is very different from that of someone who is not addicted. Neuroimaging and cognitive evaluations of people with SUD, as well as experimental evidence from animal models of SUD, demonstrate that drugs have a negative impact on brain structure and functioning as well as neurotransmitter systems and their receptors, and at the molecular level, may cause epigenetic changes (Robinson & Nestler, 2011; Volkow et al., 2011). These changes occur following exposure to the drug—a snapshot of what drugs do to the brain. But more recently, the question of predisposition has become a topic of intense investigation, as studies increasingly suggest that a premorbid susceptibility to addiction may set the stage for acquiring SUD. In the following sections, we briefly synthesize the most current research concerning the roles of the drug, the genes, and the environment in the formation of the addicted brain.

Drugs

SUD is not unlike other mental health disorders (e.g., schizophrenia or depression) in the fact that predisposing risk factors may play heavily in the determination of who becomes addicted. But one key aspect that separates SUD from all other MHDs is the fact that the drug itself produces changes that cause one to become addicted. Although this may seem like an all-too obvious (perhaps circular) statement, there is a subtlety in this statement when one considers that the drug acts like a virus in a body. Drugs slowly chip away at the areas of the brain that are responsible for an individual's cognitive control, destroying the ability to exert decision making and the self-control that are both needed in order to abstain from drug use. Concomitantly, drug-induced changes in the brain decrease the salience of all other natural rewards, so that the only pleasure that the individual is truly capable of experiencing is derived from use of the drug. These changes are long-lasting, and set the stage for a cycle of drug use that unfortunately, for many individuals, ends in death.

The attractive property of drugs is determined by their ability to act as strong rewards. As operationally defined by experimental psychology, a reward is an attractive object that incites an individual to approach or to work for it (Wise, 2004), and includes natural rewards such as food or sex. Drugs of abuse are powerful rewards by themselves: the common biochemical effect of all abused drugs lies in their ability to produce a massive and abrupt increase in dopaminergic neurotransmission (Volkow, Wang, Fowler, Tomasi & Telang, 2011; Wise, 2004). The ascending dopaminergic systems are fundamental in the processing of information related to the learning and performance of reward-related behaviors (Wise, 2004). In humans, positron emission tomography studies have shown that acute administration of psychostimulants, nicotine, alcohol, and marijuana increases extracellular dopamine in the striatum, a main target of the ascending dopaminergic system. Critically, these effects are associated with the pleasurable reward-related subjective experience (Volkow et al., 2011).

A very important role of dopamine lies in its ability to establish conditioned rewards: previously neutral stimuli, once paired with the primary reward, also acquire rewarding properties (Wise, 2004). The strong increase in dopaminergic neurotransmission induced by addictive drugs determines the acquisition of multiple and strong conditioned rewards, which in turn act as strong behavioral attractors. A cardinal characteristic of addiction is enhanced motivation to procure drugs. Drug-seeking and drug-taking become the main motivational drives for someone with SUD, overshadowing any interest for other rewards, and the addicted subject is aroused and motivated when seeking drugs but tends to be withdrawn and apathetic when exposed to nondrug activities (Volkow et al., 2011).

So far, our analysis of the effects of addictive drugs can be understood in the frame of their direct effects on the dopaminergic systems and the consequent expected response of the reward-related brain circuits. But positron emission tomography imaging studies have consistently shown long-lasting neurochemical maladaptations in addicted individuals, particularly to psychostimulants. Those include significant reductions in the basal density of dopamine D2 receptors in the striatum and decreased metabolism in the prefrontal cortex, the main cortical area innervated by ascending dopaminergic systems. Moreover, imaging studies in addicted subjects and in subjects at genetic risk for addiction have shown that the reduction in striatal D2 receptors correlates with the reduction in activity in the prefrontal cortex (Volkow et al., 2011). These cortical alterations may be largely responsible for the impairment in executive function seen in individuals with SUD.

Executive functioning is a complex construct and involves several cognitive domains that are necessary for reward-related behaviors, allowing us to anticipate outcomes and adapt to changing situations. It encompasses a set of cognitive abilities that control and regulate other abilities and behaviors, which include the ability to initiate and stop actions, to monitor and change behavior as needed, and to plan future behavior when faced with novel tasks and situations. Decision making is one domain of executive functioning that is particularly impaired in SUD. A hallmark of addiction is the persistent inability to regulate one's behavior; continued drug use, despite full cognizance of negative consequences (health-related, etc.), is but one example of this general impairment in executive function. When tested in the laboratory with experimental tasks of decision-making ability, individuals with SUD consistently make suboptimal choices that, over time, lead to cumulative loss (Bechara, 2005). To use Antoine Bechara's phrase, this "myopia for future consequences" has real-world validity, and reflects the inability of an individual with SUD to give appropriate weight to the negative consequences incurred by using drugs. The orbitofrontal cortex (OFC) is widely accepted to be a key brain region involved in decision making. Individuals with SUD perform remarkably similar to patients with lesions of the OFC on tasks that require sound decision-making skills (Bechara, 2005). The OFC is also the area of the prefrontal cortex that shows the largest basal metabolism decreases in individuals with SUD (Volkow et al., 2011).

Research conducted with SUD individuals does not have the power to distinguish whether deficits in OFC function are induced by the drug, or whether they represent aspects of a premorbid condition. Results obtained with animal models strongly suggest that these neuro-alterations are drug-induced. In experimental animals, repeated psychostimulant exposure is associated with OFC dysfunction (Lucantonio, Stalnaker, Shaham, Niv, & Schoenbaum, 2012), and with significant decreases in striatal D2-receptor density (Everitt et al., 2008; Nader, Czoty, Gould, & Riddick, 2008). The multiple biochemical mechanisms involved in the direct effects of the drug are being elucidated and include epigenetic factors (Robinson & Nestler, 2011). Importantly, these drug-induced brain alterations ultimately translate to drug-induced alterations in behavior, as decision making is compromised, the net result of which promotes drug-seeking.

Genes

A burgeoning literature demonstrates that a certain risk genotype predisposes individuals to SUD. Family and twin epidemiological studies show that genes contribute to the vulnerability to SUD, with estimates of heritability at about 50% (Kendler, Chen, et al., 2012). Some genetic variance seems to be specific to certain classes of drugs (particularly for alcohol and nicotine addiction; Kendler, Chen, et al., 2012); yet most genetic sources of risk for SUD are not highly substance-specific in their effects. Twin studies show that multiple classes of drugs of abuse share a genetic variance. It is the unique environmental experiences that largely determine whether predisposed individuals will use or misuse one class of psychoactive substances rather than another (Kendler, Chen, et al., 2012).

Yet very few genes have been identified as possible risk factors. A broad conclusion is that the genes that predispose to SUD (and other MHDs) might be many, but that the individual genes' contributions are not large. The power of molecular genetics increases, however, when we consider the occurrence of gene-by-gene interactions (also known as epistasis) and gene-by-environment interactions. Gene-by-gene interactions imply that the joint effect of two (or more) genetic variants differ from the sum of their individual, independent effects. Gene-by-environment interactions imply that genotypes must depend, to some degree, on the environmental context in which they are expressed. Findings from these approaches suggest that completely different phenotypes may arise from these interactions. Importantly, although traditional models of psychiatric epidemiology often assume that the relationship between individuals and their environment is unidirectional, from environment to person, twin analyses demonstrate that at least some responses to the environment are heritable (Kendler, Chen, et al., 2012). One recognized example of a clear gene-by-environment interaction in an MHD is the association of stress and a polymorphism (common genetic variant) of the serotonin transporter (5-HTT) gene in the development of depression (discussed below). For the case of SUD, we need to turn to one more concept that increases the power of molecular genetics: the use of personality traits as endophenotypes.

The term "endophenotype" was initially introduced by Gottesman and Shields (1973) as an internal, intermediate phenotype (i.e., not obvious to the unaided eye) that fills the gap in the causal chain between genes and disease. The use of endophenotypes might help to resolve questions about disease etiology, since instead of focusing on observable endpoints of a disorder (which may have several, nonoverlapping etiologies), this more conservative approach may provide simpler clues to genetic underpinnings than the disease syndrome itself. This idea, that clinical phenotypes can be decomposed or deconstructed into smaller subphenotypes, may provide researchers with measurable phenomena that are more closely tied to a genetic basis. In addition to furthering genetic analysis, endophenotypes can clarify classification and diagnosis and foster the development of animal models (Gottesman & Gould, 2003). Endophenotypes should be more tractable to genetic dissection, in part by virtue of their assumed proximity to the genetic antecedents of disease. Endophenotypes of SUD include phenotypes at the cellular level, at the circuit-neural systems level, and at the behavioral level (individual differences in personality).

Personality traits are substantially influenced by genes (Munafo & Flint, 2011), and it is becoming generally accepted that extreme expression of some personality traits plays a key role in the development and expression of mental disorders. If common MHDs, including SUD, are associated with particular personality phenotypes, we should expect MHDs to be associated with abnormal functioning of the brain systems that underlie those phenotypes. If this line of reasoning were correct, then an obvious prediction would be that factors (genetic and environmental) that determine the development of specific personality phenotypes would be the same as those that determine some MHDs. Several personality traits have been repeatedly invoked to be associated with SUD; we are going to focus on two in particular: impulsivity and neuroticism (trait anxiety or negative emotionality; Kotov, Gamez, Schmidt, & Watson, 2010; Moeller, Barratt, Dougherty, Schmitz, & Swann, 2001). Impulsivity is defined as a predisposition toward rapid, unplanned reactions to internal or external stimuli without regard to the negative consequences of these reactions to the individual or others (Moeller et al., 2001). Embedded in this definition is the fact that impulsivity is a complex construct, which includes at least two different components: "impulsive action" and "impulsive choice." Impulsive action can be operationally defined as disinhibition, a failure of volitional motor control or motor inhibition, while impulsive choice implies a tendency to accept small immediate or likely rewards in favor of large delayed or unlikely ones (Dalley, Everitt, & Robbins, 2011).

Disinhibition/constraint is a well-established dimension of personality (Multidimensional Personality Questionnaire; Tellegen, 1982). Recent studies by Ersche, Jones, et al. (2012) strongly suggest that disinhibition constitutes an endophenotype of SUD. As heritable traits, endophenotypes should be able to be measured objectively in both patients and their biological siblings. Importantly, individuals with psychostimulant addiction and their biological siblings without history of drug abuse had very similar deficits in a measure of inhibitory control (stop-signal reaction time), and they also showed similar frontostriatal brain abnormalities, compared to unrelated healthy volunteers (Ersche, Jones, et al., 2012). Those abnormalities included reduced integrity in the fiber tracts adjacent to the right ventrolateral prefrontal cortex (VLPFC), specifically in the right inferior frontal gyrus, a cortical area previously shown to be part of the neural network involved in the stop-signal task (Aron et al., 2007). Specific polymorphisms of the dopamine D4 receptor gene (presence of the D4.7 allele) and the dopamine transporter (DAT; homozygous for the DAT 10 allele) have been shown to be associated with increased disinhibition (Congdon, Lesch, & Canli, 2008; Cornish et al., 2005). Additionally, a significant DAT gene-by-D4 gene interaction exists, as healthy adults expressing the homozygous DAT 10/10 genotype and D4.7 allele show a significant impairment in inhibitory control, as measured with the stop-signal task (Congdon et al., 2008). A long literature suggests that disinhibition is an endophenotype for attention-deficit/hyperactivity disorder (ADHD) and studies draw associations between ADHD and the same D4 and DAT polymorphisms (Congdon et al., 2008; Cornish et al., 2005). Altogether, this suggests that disinhibition can be used as an endophenotype of SUD, which is moderated at least by the D4 and DAT genes through their regulation of specific frontostriatal circuits.

Impairment in decision making overlaps conceptually with the "impulsive choice" dimension of impulsivity (see above). Therefore, the two main dimensions of impulsivity give a frame to characterize two dimensions of the addicted brain: a premorbid personality-related dimension, related to disinhibition, and a second, mostly drug-induced dimension, related to impairment of executive functioning. Several clinical studies support the existence of the two-component impulsivity construct in the pathogenesis of SUD. In cocaine-dependent subjects, there is no correlation in the measurements of decision making and behavioral inhibition (Kjome et al., 2010), underscoring the notion that the two constructs are independent and can be differentially affected in SUD. But it is also important to consider that there might also be a contribution of a premorbid decision-making impairment (Ersche, Turton, et al., 2012).

In addition, striatal D2-receptor density inversely correlates with measures of impulsive action in animals and in healthy and addicted individuals (Everitt et al., 2008; Lee et al., 2009; Ghahremani et al., 2012). Furthermore, D2-receptor density is not only decreased by drug exposure (see above), but is dependent on individual differences that can be modulated by changes in environment (social dominance; Nader et al., 2008). In further support of this notion, D2-receptor availability in the dorsal striatum is significantly negatively correlated with motor inhibition and positively correlated with activation in several prefrontal cortical regions, including the right VLPFC and the dorsal striatum itself (Ghahremani et al., 2012). In summary, the endophenotype of disinhibition appears to be moderated by D2 receptors through both genetic and drug-mediated mechanisms, providing substantial evidence for the role of genetic and environmental factors in the expression of this personality trait.

Neuroticism or negative emotionality is a crucial dimension that surfaces in any study examining trait characteristics of psychopathology. The three most common MHDs, depressive, anxiety, and substance use disorders, are associated with high neuroticism, and there is a high rate of comorbidity between them (Kotov et al., 2010). Individuals with SUD and their nonaddicted biological siblings score higher than controls in neuroticism and in measures of stress sensitivity; applying the use of the endophenotype concept described above, this strongly suggests that neuroticism could constitute an endophenotype of SUD (Ersche, Turton, et al., 2012). It is well accepted that the short allele (S-allele) of a polymorphism in the promoter region of the 5-HTT gene is associated with high neuroticism (reviewed in Caspi, Hariri, Holmes, Uher, & Moffitt, 2010). This polymorphism has become the most investigated genetic variant in neuroscience, because (among other reasons) it has provided a clear example of a gene-by-environment interaction: a significant number of S-allele carriers develop depression after experiencing stressful life events and childhood maltreatment (Caspi et al., 2010). Neuroimaging studies have shown positive correlations between indices of anxiety and amygdala reactivity to threatening stimuli, and S-allele carriers exhibit elevated amygdala reactivity to those stimuli. In addition, the S allele is associated with reduced gray matter in the (right) ventromedial prefrontal cortex (VMPFC) and amygdala, and reduced morphological and functional connectivity between both structures (Caspi et al., 2010). Individual differences in the VMPFC-amygdala pathway correlate with levels of anxiety associated with fearful stimuli, and with reduced involuntary regulation of negative emotions.

Thus, the results of many studies suggest that neuroticism (negative emotionality, reduced involuntary regulation of negative emotions) and disinhibition (impulsive action, reduced voluntary motor control) constitute endophenotypes of SUD. Recent studies also indicate that the same cortical areas involved in volitional motor control are also involved in volitional control of negative emotions, implicating the existence of a functional connection between VLPFC and VMPFC (Tabibnia et al., 2011). There is also experimental evidence that volitional control of negative emotions is impaired in SUD, which correlates with a decreased intensity of gray matter in the right inferior frontal gyrus (Tabibnia et al., 2011). Thus, patients with SUD show impairment of both involuntary and voluntary control of negative emotions.

Environment

Without losing focus of the multiple effects of the drug itself, SUD is influenced both by genetic risk factors (reflecting a specific liability to engage in drug-seeking and drug-taking) and by a range of environmental factors. Psychoactive drugs are almost always provided through social contacts, and social environments can strongly encourage or discourage drug use. In fact, SUD typically arises through a two-stage process. First, during a critical development period (mostly adolescence) individuals are exposed to an environment that encourages substance abuse. Once exposed to the drug, their "vulnerability genes" will facilitate a progression from heavy use to drug dependence. However, exposure to a particular environment also depends on personality traits (we choose our "friends").

One obvious question is why these "vulnerability genes" are so prevalent. It might be expected that a genotype responsible for the production of a disinhibited and anxious organism, if completely disadvantageous, would likely have been eradicated by natural selection. It has been suggested that the high prevalence of those common genetic variants such as the S-allele of the 5-HTT and the D4.7 variant of the D4 receptor imply that the carriers of these genotypes benefit after their interaction with specific environmental experiences (Belsky et al., 2009). Belsky and colleagues argue cogently that they should more properly be called "plasticity genes": apart from being less beneficial upon exposure to adversity (a stressful environment), these genes are beneficial when exposed to enriched environmental conditions. As mentioned above, the carriers of the S-allele show increased vulnerability to depression in the context of environmental stress (Caspi et al., 2010). Yet, S-allele carriers function better than non-S-allele carriers when encountering few or none of the stressors (Belsky et al., 2009). The same for-better-and-for-worse pattern has been observed with the D4.7 variant: children were differentially susceptible to both sensitive and insensitive parenting dependent on the presence of the 7-repeat D4 receptor allele (Bakermans-Kranenburg & van IJzendoorn, 2007). These gene-by-environment interactions can explain why children with a reactive or fearful temperament or a reactive stress-response system appear to suffer most from persistent family conflict or low quality of daycare, but also appear to benefit disproportionately from supportive rearing environments (Bakermans-Kranenburg & van IJzendoorn, 2007; Belsky et al., 2009).

Familial influences do in fact play a role in the development of SUD. A recent large study of adopted children and adoptive relatives showed that a genetic risk index and an environmental risk index are important predicting factors for SUD (Kendler, Sundquist, et al., 2012). The environmental risk index included adoptive parental history of divorce, death, criminal activity, and alcohol problems, as well as an adoptive sibling history of SUD and psychiatric or alcohol problems. Significantly, risk for SUD in adopted children was more strongly predicted by SUD in the adoptive siblings than in the adoptive parents (Kendler, Sundquist, et al., 2012). These results suggest that social influences (e.g., peer deviance and drug availability) shared with adoptive siblings are more potent environmental risks for SUD than an adverse parental influence. This study also revealed a significant positive interaction between both risk indices, constituting a gene-by-environment interaction: adopted children at high genetic risk were more sensitive to the pathogenic effects of the adverse family environments than those at low genetic risk (Kendler, Sundquist, et al., 2012).

Toward public appreciation of a mental health disorder

Billions of dollars are spent annually on the repercussive effects of SUD. We now have substantial evidence that SUD arises from a complex interaction of the direct effects of the drug, the genetic background of the individual, and the environment in which the individual is raised and lives. This mental disorder is similar to other MHDs in that it has clearly associated genetic, environmental, and gene-environment interaction factors. But a major distinction that separates SUD from other MHDs is in the fact that there is a heavy component of the disorder that is attributable to the direct effects of the drug itself. This disorder is not simply an MHD; it is a *drug-induced* MHD. The question is not "why does an individual try drugs?" but "why can't a drug-addicted individual *stop* taking drugs?" SUD bears resemblance to another chronic disorder with strong genetic and environmental influences—that of type II diabetes. As with SUD, individuals with type II diabetes suffer the consequences of poor lifestyle choices; choices which, if avoided, can result in remission of the disease. For the person with type II diabetes, food is the rewarding substance. Importantly, if the diabetic individual is able to control the intake of that rewarding substance, he or she can avoid the multiple negative consequences of the disorder. As is the case with diabetes and excessive food intake, it goes without saying that without the drug, there is no SUD: in addition to its feature as the attractive stimulus, the drug also elicits the establishment of multiple and strong conditioned rewards. But critically, long-term administration of the drug itself induces alterations in the brain that impair decision making (impulsive choice) and inhibitory control (impulsive action). As reviewed above, the drug effectively has the capability to change personality, making the individual more impulsive, which adds to the vulnerability to consume drugs. A compellingly similar scenario holds for type II diabetes: recent evidence suggests that insulin resistance is associated with executive dysfunction, which, like SUD, could exacerbate a vicious cycle of aberrant decision making associated with further consumption of the disease-inducing agent (Schuur et al., 2010). The logical extension is that SUD is not simply a behavioral or lifestyle choice: it is an MHD with several aggravating factors. As such, this disorder presents multiple, nonmutually exclusive therapeutic targets, at the levels of both prevention and treatment. We predict that a deeper understanding by society of the multiple causes of addiction will increase tolerance and sensitivity toward afflicted individuals.

In 1997, Allan Leshner, who at the time was the Director of the National Institute on Drug Abuse, published a seminal work in *Science* entitled "Addiction Is a Brain Disease, and It Matters" (Leshner, 1997). In addition to its importance as a statement of official position from a government representative of science, this now-classic article issued a call for the replacement of the traditional, perhaps moralistic societal response to addiction with scientific study and understanding. Nearly 20 years later, we know substantially more about the neural mechanisms of addiction; yet policy has not kept up with the pace of the bench (lack of addiction-treatment centers and as-yet incompletely resolved insurance coverage represent two salient examples). Although much work remains in the elucidation of the specifics, we have winnowed our searchlight to focus on a specific set of structures in the brain. We now understand that addiction is a complicated, multifactorial brain disorder that results in aberrant behavior and severe distortions of what it means to lead a meaningful, productive life.

We conclude our review with a call to society, and, particularly, to health care providers and scientists to understand addiction as a mental health disorder with predisposing and aggravating factors. Having the "will" to quit using drugs is no more possible for an individual with SUD than the simple act of grasping an object is possible for a person without hands. The brain (specifically, circuits involved in decision making and those that mediate the volitional motor control and automatic and volitional control of negative emotions) of an SUD-affected individual is, in a sense, broken, and drugs reinforce this state of being broken. When society is made to understand this, it will decide to accept scientific knowledge concerning addiction as fact. This release from the stigmatization surrounding individuals with SUD will have profound effects on policy and, ultimately, on the way in which addiction is treated.

ACKNOWLEDGMENTS Supported by the NIDA Intramural Research Program.

REFERENCES

American Psychiatric Association (APA). (2013). *Diagnostic and statistical manual of mental disorders* (5th ed.). Washington, DC: American Psychiatric Publishing.

Aron, A. R., Durston, S., Eagle, D. M., Logan, G. D., Stinear, C. M., & Stuphorn, V. (2007). Converging evidence for a fronto-basal-ganglia network for inhibitory control of action and cognition. *J Neurosci, 27*(44), 11860–11864.

Bakermans-Kranenburg, M. J., & van IJzendoorn, M. H. (2007). Research review: Genetic vulnerability or differential susceptibility in child development: The case of attachment. *J Child Psychol Psychiatry, 48*, 1160–1173.

Bechara, A. (2005). Decision making, impulse control and loss of willpower to resist drugs: A neurocognitive perspective. *Nat Neurosci, 8*, 1458–1463.

Belsky, J., Jonassaint, C., Pluess, M., Stanton, M., Brummett, B., & Williams, R. (2009). Vulnerability genes or plasticity genes? *Mol Psychiatry, 14*, 746–754.

Caspi, A., Hariri, A. R., Holmes, A., Uher, R., & Moffitt, T. E. (2010). Genetic sensitivity to the environment: The case of the serotonin transporter gene and its implications for studying complex diseases and traits. *Am J Psychiatry, 167*, 509–527.

Congdon, E., Lesch, K. P., & Canli, T. (2008). Analysis of DRD4 and DAT polymorphisms and behavioral inhibition in healthy adults: Implications for impulsivity. *Am J Med Genet B Neuropsychiatr Genet, 147B*(1), 27–32.

Cornish, K. M., Manly, T., Savage, R., Swanson, J., Morisano, D., Butler, N., ... Hollis, C. P. (2005). Association of the dopamine transporter (DAT1) 10/10-repeat genotype with ADHD symptoms and response inhibition in a general population sample. *Mol Psychiatry, 10*, 686–698.

Dalley, J. W., Everitt, B. J., & Robbins, T. W. (2011). Impulsivity, compulsivity, and top-down cognitive control. *Neuron, 69*, 680–694.

Ersche, K. D., Jones, P. S., Williams, G. B., Turton, A. J., Robbins, T. W., & Bullmore, E. T. (2012). Abnormal brain structure implicated in stimulant drug addiction. *Science, 335*, 601–604.

Ersche, K. D., Turton, A. J., Chamberlain, S. R., Müller, U., Bullmore, E. T., & Robbins, T. W. (2012). Cognitive dysfunction and anxious-impulsive personality traits are endophenotypes for drug dependence. *Am J Psychiatry, 169*, 926–936.

Everitt, B. J., Belin, D., Economidou, D., Pelloux, Y., Dalley, J. W., & Robbins, T. W. (2008). Neural mechanisms underlying the vulnerability to develop compulsive drug-seeking habits and addiction. *Philos Trans R Soc Lond B Biol Sci, 363*, 3125–3135.

Ghahremani, D. G., Lee, B., Robertson, C. L., Tabibnia, G., Morgan, A. T., de Shetler, N., ... London, E. D. (2012). Striatal dopamine D2/D3 receptors mediate response inhibition and related activity in frontostriatal neural circuitry in humans. *J Neurosci, 32,* 7316–7324.

Gottesman, I. I., & Gould, T. D. (2003). The endophenotype concept in psychiatry: Etymology and strategic intentions. *Am J Psychiatry, 160,* 636–645.

Gottesman, I. I., & Shields, J. (1973). Genetic theorizing and schizophrenia. *Br J Psychiatry, 122,* 15–30.

Heyman, G. M. (2009). *Addiction: A disorder of choice.* Cambridge, MA: Harvard University Press.

Kendler, K. S., Chen, X., Dick, D., Maes, H., Gillespie, N., Neale, M. C., & Riley, B. (2012). Recent advances in the genetic epidemiology and molecular genetics of substance use disorders. *Nat Neurosci, 15,* 181–189.

Kendler, K. S., Sundquist, K., Ohlsson, H., Palmér, K., Maes, H., Winkleby, M. A., & Sundquist, J. (2012). Genetic and familial environmental influences on the risk for drug abuse: A national Swedish adoption study. *Arch Gen Psychiatry, 69,* 690–697.

Kjome, K. L., Lane, S. D., Schmitz, J. M., Green, C., Ma, L., Prasla, I., ... Moeller, F. G. (2010). Relationship between impulsivity and decision making in cocaine dependence. *Psychiatry Res, 178,* 299–304.

Kotov, R., Gamez, W., Schmidt, F., & Watson, D. (2010). Linking "big" personality traits to anxiety, depressive, and substance use disorders: A meta-analysis. *Psychol Bull, 136,* 768–821.

Lee, B., London, E. D., Poldrack, R. A., Farahi, J., Nacca, A., Monterosso, J. R., ... Mandelkern, M. A. (2009). Striatal dopamine D2/D3 receptor availability is reduced in methamphetamine dependence and is linked to impulsivity. *J Neurosci, 29,* 14734–14740.

Leshner, A. I. (1997). Addiction is a brain disease, and it matters. *Science, 278,* 45–47.

Lucantonio, F., Stalnaker, T. A., Shaham, Y., Niv, Y., & Schoenbaum, G. (2012). The impact of orbitofrontal dysfunction on cocaine addiction. *Nat Neurosci, 15,* 358–366.

Moeller, F. G., Barratt, E. S., Dougherty, D. M., Schmitz, J. M., & Swann, A. C. (2001). Psychiatric aspects of impulsivity. *Am J Psychiatry, 158,* 1783–1793.

Munafo, M. R., & Flint, J. (2011). Dissecting the genetic architecture of human personality. *Trends Cogn Sci, 15,* 395–400.

Nader, M. A., Czoty, P. W., Gould, R. W., & Riddick, N. V. (2008). Review. Positron emission tomography imaging studies of dopamine receptors in primate models of addiction. *Philos Trans R Soc Lond B Biol Sci, 363,* 3223–3332.

Pescosolido, B. A., Martin, J. K., Long, J. S., Medina, T. R., Phelan, J. C., & Link, B. G. (2010). "A disease like any other"? A decade of change in public reactions to schizophrenia, depression, and alcohol dependence. *Am J Psychiatry, 167,* 1321–1330.

Robinson, A. J., & Nestler, E. J. (2011). Transcriptional and epigenetic mechanisms of addiction. *Nat Rev Neurosci, 12,* 623–637.

Satel, S., & Lillienfeld, S. O. (2013). *Brainwashed: The seductive appeal of mindless neuroscience.* New York, NY: Basic Books.

Schuur, M., Henneman, P., van Swieten, J. C., Zillikens, M. C., de Koning, I., Janssens, A. C., ... van Duijn, C. M. (2010). Insulin-resistance and metabolic syndrome are related to executive function in women in a large family-based study. *Eur J Epidemiol, 25,* 561–568.

Substance Abuse and Mental Health Services Administration. (2013). *Results from the 2012 National Survey on Drug Use and Health: Mental Health Findings.* NSDUH Series H-47, HHS Publication No. (SMA) 13-4805. Rockville, MD: Substance Abuse and Mental Health Services Administration.

Tabibnia, G., Monterosso, J. R., Baicy, K., Aron, A. R., Poldrack, R. A., Chakrapani, S., ... London, E. D. (2011). Different forms of self-control share a neurocognitive substrate. *J Neurosci, 31,* 4805–4810.

Tellegen, A. (1982). Brief manual for the multidimensional personality questionnaire. Unpublished manuscript, University of Minnesota, MN.

Volkow, N. D., Wang, G. J., Fowler, J. S., Tomasi, D., & Telang, F. (2011). Addiction: Beyond dopamine reward circuitry. *Proc Natl Acad Sci USA, 108,* 15037–15042.

Wise, R. A. (2004). Dopamine, learning and motivation. *Nat Rev Neurosci, 5,* 483–494.

89 Neuroscience of Antisocial Behavior

LYN M. GAUDET, NATHANIEL E. ANDERSON, AND KENT A. KIEHL

ABSTRACT Over half of all incarcerated individuals suffer from a major mental health disorder. Because many mental health issues can contribute to persistent antisocial behavior, there is reason to consider crime both a societal problem and a major public health issue. This chapter describes the relationship between neuroscience and a number of variables—particularly age, psychopathic traits, and acute psychosis—that are associated with an increased risk for engaging in antisocial behavior. This review indicates that these variables are linked to dysfunction in overlapping brain systems that govern variables ranging from simple impulse control to complex moral development. Understanding the basic relationship between these neural systems and behavior is a key to understanding a major relationship between the brain and antisocial behavior and ultimately designing interventions that will promote desistance from crime.

The cost of crime

The annual expenditures for health care in the United States totaled $2.6 trillion in 2010 (Centers for Medicare and Medicaid, 2012). The aggregate cost of crime during the same period was $3.2 trillion—half a trillion *more* than all of the nation's health care (Anderson, 2011). Moreover, this estimate of crime costs does not include the emotional toll crime places on victims. Despite the enormous costs, public discourse rarely devotes much attention to understanding the reasons behind criminal behavior. However, this landscape is changing. Due in part to increasing diffusion of high-profile psychology and neuroscience research in the popular media, there has been an increase in public awareness of mental health issues and their contribution to crime and violence. The U.S. Department of Justice reported that over half of all incarcerated individuals have a major mental health disorder, compared to 11 percent of the general population (James & Glaze, 2006). As an increasing body of research reveals how mental health issues can contribute to persistent antisocial behavior, we now must view crime as both a societal problem and major public health issue.

This goal of this chapter is to describe the current understanding of the relationship between neuroscience and a number of variables that lead to crime. Several mental health issues are known to be associated with higher rates of antisocial behavior. And certain known properties of the brain, impacted by these mental health issues, influence behavior in ways that can increase the likelihood of criminal offending. This review indicates that a variety of psychiatric disorders are linked to dysfunction in overlapping brain systems that govern variables ranging from simple impulse control to complex moral development. Understanding the basic relationship between these neural systems and behavior is a key to understanding a major relationship between the brain and crime.

Crime is not a mental illness

It is important to explain why it is challenging to characterize the relationship between the brain and crime. Crime is a social construct, highly dependent on the laws and mores of a particular place and time. In certain countries, it is acceptable to engage in behaviors that are prohibited in other cultures. Substance use is an excellent example. In the United States, over half of all current federal prison inmates have been convicted for drug-related offenses (Carson & Golinelli, 2013), and 65–85% of incarcerated individuals meet criteria for substance dependence or abuse (Karberg & James, 2005; National Center on Addiction and Substance Abuse, 2010). So, in the United States, where drug-related laws are relatively strict, a strong case can be made for a relationship between illicit substance dependence—a mental health issue—and criminal behavior. But the definition of drug-related crime varies highly from place to place. Under a statute passed in 2009 in Mexico, there is no action taken against those carrying up to a half-gram of cocaine, 40 grams of marijuana, 40 milligrams of methamphetamine, or 50 milligrams of heroin.

The picture is no less complicated if we focus instead on social norms. While there is usually considerable overlap between the categories, socially acceptable behavior can easily differ from morally acceptable behavior, and both of those can differ from what is considered criminal. Incest is an example of behavior that is almost universally considered morally and socially inappropriate, and evokes a strong disgust response at the neural level (Borg, Lieberman, & Kiehl, 2008), but the legality of incest varies widely across the world. Some countries do not have any criminal penalties if the incestual activity is between two consenting adults, while other countries require mandatory imprisonment for any sexual activity between related individuals (e.g., Ireland's Criminal Law [Incest Proceedings] Act of 1995, and Australia's Commonwealth Consolidated Acts).

The intentional killing of another human being is another example. Killing is an act that most would consider a moral violation. However, there are circumstances in which killing another can be morally justified, such as self-defense, or even legally permissible, such as an act of war. Simply put, it is very difficult to define universal wrongs, and criminal behavior is a multifaceted construct. There are many factors, many of which are not pathological, that lead to antisocial behavior. However, a few psychiatric constructs and developmental periods of life are associated with increased risk for engaging in impulsive, antisocial, and even violent behavior. This chapter focuses on a few of these more discretely defined constructs, which are more proximally related to behavioral patterns that can lead to crime.

THE ADOLESCENT BRAIN Before directly addressing specific forms of psychopathology, it will be helpful to discuss normal trajectories of neuropsychological development, which set the stage for understanding of the complex relationship between crime and the brain. Anecdotally, it is not uncommon to perceive adolescents as negligent, irresponsible, and reckless. The Roman philosopher Lucius Annaeus Seneca observed, "It is the failing of youth not to be able to restrain its own violence." It seems now, 2,000 years later, science is confirming that there is more to this view than prejudice, and empirical evidence supports specific physiological causes behind Seneca's insight. These facts have had a conspicuous impact on legal decisions, perhaps most notably as we restrict the imposition of capital punishment on individuals under the age of 18 (*Roper v. Simmons*, 2005).

Being an adolescent is its very own incremental risk factor for delinquent behavior. Indeed, mortality rates for adolescents are disproportionally high due primarily to elevated rates of violence and suicide, reckless driving, risky sexual behaviors, and dangerous levels of drug and alcohol consumption (Institute of Medicine and National Research Council, 2011). Furthermore, evidence suggests that the trajectory of these behaviors is not simply linear; rather, adolescents (compared to both young children and adults) are particularly prone to risky behavior and impulsive decisions (Spear, 2000).

In order to understand these tendencies, it is critical to understand the difference between cognitive *capacity* and cognitive *control*. The cognitive capacity, or logical reasoning abilities, of 15-year-old adolescents are on par with those of adults; they are able to accurately perceive risk and estimate their vulnerability to it (Reyna & Farley, 2006; Steinberg, 2007). Yet knowledge of these risks is not sufficient to change their behavior. Cognitive control regulates behavior, a process that depends on the linkage between the frontal lobe and other areas of the brain that confer information about rewards and punishments. It is this linkage that likely is not yet developed in adolescents. This explains why *knowledge* of risks and adverse consequences does not have an effect on their *behavior*: 90% of American teens have been educated on the risks associated with sex, drugs, and alcohol, yet studies unequivocally show that the education does not result in less risk taking in regard to those behaviors (Steinberg, 2004). Knowing something is wrong or risky and being able to refrain from doing something wrong or risky are two very different things.

Advances in our understanding of brain development have helped to explain why this is true, and these findings set the stage for understanding other instances—that is, psychopathological outcomes—when abnormalities in the brain influence tendencies toward crime. Despite the fact that the brain has very nearly reached its full adult size by age 6, and peaks in size around early adolescence (Giedd et al., 1999), several other important developmental changes occur within the brain throughout adolescence into adulthood. For instance, gray matter in frontal portions of the brain decreases due to the pruning of synapses, making for more efficient neural connections (Lenroot & Giedd, 2006). The white matter tracts, serving as efficient connections between brain regions, increase in density, particularly those connecting the cognitive-control regions in the prefrontal cortex to subcortical, emotion-related regions of the brain (Eluvathingal, Hasan, Kramer, Fletcher, & Ewing-Cobbs, 2007). In addition, cellular and neurochemical changes (dopaminergic activity) that influence reward sensitivity increase in the frontal parts of the brain (Ernst et al., 2005).

Simply put, these changes in the brain influence our behavior in predictable ways. At a time when the neural machinery for cognitive control and abstract appreciation of future consequences have not fully matured, adolescents are also particularly motivated by the potential for reward, and this likely has a strong influence over adolescents' high propensity for risky decision making (Casey, Jones, & Hare, 2008; Galvan et al., 2006; Steinberg, 2007). Despite the apparent consequences of this, evolutionary theorists propose that this developmental trajectory confers particular advantages to adolescents at a critical period of development, facilitating exploration and learning from a wide variety of experiences and encouraging high rates of procreation during a period of high reproductive fitness (e.g., Ellis et al., 2012). At the same time, the adolescent brain provides at least one natural model for how physical properties of the brain can influence decision making with the potential to promote risky, impulsive, or even criminally deviant behavior.

Violence and aggression

Research into the modern "neuroscience of crime" began in the 1990s, when a number of groups began using various brain-imaging methods in an attempt to identify the neural underpinnings of violence and aggression in forensic and psychiatric samples. Between 1995 and 2000, 17 studies collected imaging data on individuals who had committed a violent act but were also diagnosed with a psychiatric illness such as alcoholism (Kuruoglu et al., 1996), dementia (Hirono et al., 2000), and schizophrenia and schizoaffective disorders (Wong et al., 1997), or were referred for psychiatric evaluation by the court (Raine et al., 1994). For the studies that included forensic samples, every one of the participants pled not guilty by reason of insanity, had a major mental or personality disorder (e.g., schizophrenia, paranoid personality disorder), had been found incompetent to stand trial, or had been found guilty and were planning to assert diminished capacity as mitigating evidence in the trial's sentencing phase (Raine et al., 1994; Raine, Buschbaum & LaCasse, 1997; Raine, Meloy, et al., 1998; Raine et al., 1998).

A review of the 17 studies revealed prefrontal, temporal, and (left side) medial-temporal dysfunction as well as an association between the prefrontal cortex and subcortical structures and aggression and violence. One of the functional neuroimaging studies by Raine et al. (1997) used positron emission tomography to compare brains of adults pleading not guilty by reason of insanity who had committed murder to those of sex- and age-matched controls. The authors found prefrontal and parietal glucose metabolism reductions in the psychiatric patients who had committed murder, but no differences in the temporal cortex. Unfortunately, the presence of the comorbid psychiatric issues makes it extremely difficult to determine the extent to which the neurological deficits were specifically related to committing murder, and as the authors note, the findings cannot be generalized from not guilty by reason of insanity murder cases to other types of violent crime. The other early studies in this arena suffer from the same confounds since they all used psychiatric populations. Nevertheless, these early studies were formative and ambitious in applying brain-imaging methodology in the study of criminal behavior. They laid the groundwork for more rigorous examinations of neural mechanisms that might predispose one to violence or other behaviors often associated with crime.

The limitations of studying the neuroscience of violence parallel that of studying the neuroscience of crime in general. Violence can be defined and categorized in a number of ways, and aggressive behavior is not necessarily the result of mental illness. Neuroscientists who study aggression broadly categorize these behaviors into impulsive and premeditated varieties (McEllistrem, 2004), subserved by different brain systems. *Impulsive aggression* is characterized as a reactive, emotional response to some immediate provocation or frustration. At its most basic representation, impulsive aggression is guided by the mammalian *basic threat circuit*, which connects the medial hypothalamus and periaqueductal gray matter (Gregg & Siegel, 2001). Electrical stimulation of this circuit in animals produces an immediate, reflexive defensive-rage response. Many parts of the brain, including the amygdala, cingulate gyrus, and prefrontal cortex, are connected to this circuit and can influence the intensity of such a response. In humans, abnormalities in the ventromedial or orbitofrontal cortex have often been associated with unmoderated reactive aggression, conceivably through decreased inhibitory input to this basic threat circuit (Bufkin & Luttrell, 2005). In contrast, *premeditated aggression* is characterized by predatory actions that serve some instrumental purpose. In animal models it can be exhibited by stimulation of a circuit connecting the central amygdala to ventral periaqueductal gray matter through the lateral hypothalamus, instigating stereotyped predation behavior in mammals (Gregg & Siegel, 2001). For example, a resting cat will rise, stealthily circle an anesthetized rodent (which it was previously ignoring), and deliver a single calculated strike at the prey.

Rather than simply serving a basic instinct to hunt and eat, human predatory aggression may serve purposes as diverse and complicated as revenge, sexual gratification, monetary reward, interpersonal dominance, or even to achieve a political goal. Neither form of aggression, impulsive or premeditated, is necessarily the result of mental illness; however, psychopathology can certainly influence the degree to which we control our impulses to react violently or the degree to which we are willing to eschew moral sensibility for personal gain. In fact, a growing body of research demonstrates that specific developmental deficits in the brain can impact the ways threats are processed and moral decisions are made, and the result is patterns of chronic antisocial behavior and violence observed in individuals called psychopaths.

Neuroscience of psychopathy

Among the various neuropsychological conditions and psychological constructs that are relevant to criminal and violent behavior, psychopathy, or psychopathic personality disorder, is the most relevant. While our understanding of psychopathy has evolved over a very long history, our modern conceptualization of the disorder has its roots in the case studies of Hervey Cleckley (1941), in which he delineated 16 core personality traits that tend to co-occur in those who we call psychopaths. Psychopathy is characterized by stable patterns of impulsivity, irresponsibility, and a lack of empathy, guilt, or remorse across multiple domains of an individual's life. As such, psychopaths show a profound disregard for the rights and well-being of others. At the same time, psychopaths are ordinarily not delusional or psychotic, and they generally perform at normal to high levels of intelligence. This combination of personality traits may be recognized at varying levels among healthy, relatively high-functioning members of society; however, the full clinical manifestation of these traits occurs in less than 1% of the general population (Hare, 2003).

Psychopathy is most commonly assessed using the Hare Psychopathy Checklist (PCL; Hare, 1980), revised in 1991 and 2003 (PCL-R; Hare, 1991, 2003). The PCL-R is administered in two steps: a semistructured interview and collateral file review. A trained administrator of the PCL-R conducts both pieces and scores the individual on the presence and severity of 20 items, each of which is assigned a score of zero, one, or two. A score of two on an individual item means that the item is a character trait or behavior repeatedly demonstrated by the individual in most or all areas of their life; a score of one means the item describes the individual in some areas of their life, and a zero means the item does not apply to the individual. The recommended score for a diagnosis of psychopathy is 30 out of a maximum 40 points.

In contrast to antisocial personality disorder (ASPD), which relies mostly on the occurrence of criminal behavior for its diagnosis, psychopathy is characterized primarily by personality traits, which are more proximal indicators of psychological patterns of thought and motivation and thus more indicative of psychiatric pathology. It is a much-belabored point among researchers and clinicians that psychopathy is not well accounted for by the ASPD diagnostic criteria (Cunningham & Reidy, 1998; Hare, Hart, & Harpur, 1991). The base rates of the two diagnoses highlight that the two disorders are not describing the same population: due to the generality of the symptoms, rates of ASPD in incarcerated populations range up to 80%, whereas the rate of psychopathy is between 15% and 25% (Hare, 2003; Hare et al., 1991).

Why is psychopathy the most relevant disorder to criminal and violent behavior? Psychopathy, as measured by the PCL-R, predicts criminal recidivism (Porter, Birt, & Boer, 2001; Salekin, Rogers, & Sewell, 1996), and is a particularly strong predictor of violent recidivism (Cornell et al., 1996; Harris, Rice, & Cormier, 1991; Porter, Brinke, & Wilson, 2009). Within one year of release from prison, psychopaths are 4–6 times more likely to commit another violent crime than are nonpsychopaths (Hemphill, Hare, & Wong, 1998). Within 10 years after release, over 70% of psychopaths (with a history of violence) commit another violent offense, and 20-year follow-ups indicate that as many as 90% will be rearrested for violent crimes. Recidivism rates for nonpsychopathic violent offenders appear to plateau around 40% (Hare, Clark, Grann, & Thornton, 2000; Harris et al., 1991; Hemphill et al., 1998; Rice & Harris, 1997). This is not to say that psychopaths are necessarily violent; in fact, in Cleckley's studies, he considered violence to be more an exception than the rule. Rather, this evidence more specifically suggests that psychopaths who do behave violently are very likely to continue that pattern of behavior, even after incarceration.

Another reason why psychopathy is among the most important constructs associated with crime is that there is overwhelming scientific evidence that it represents a formal neuropsychiatric disorder, with unique and definitive biological underpinnings (Kiehl, 2006) and strong genetic heritability (Viding, Blair, Moffitt, & Plomin, 2005). A crucial element that has long distinguished psychopathy from more general exhibitions of antisocial behavior is a profound deficit in the ability to utilize emotionally relevant information to guide future behavior. In this case, "emotionally relevant" refers to stimuli that signal threat or reinforcement—information that is relevant to one's safety and survival. This has been evident from early psychophysiological studies demonstrating that psychopaths fail to acquire normal physiological responses to environmental cues that indicate potential threat (Hare, 1968; Hare & Quinn, 1971). Another highly reproduced finding related to emotional processing in psychopaths is their conspicuous deficit in the typical heightening of startle reflexes that accompany anxious or aversive states (Patrick, 1994).

A prominent component of the neural circuitry that governs these physiological responses is the amygdala, which is connected to a host of other brain areas. These areas work together to recognize threat and reward and incorporate emotional information into higher-level cognitive processes (Papez, 1937). This integrated system, which includes the amygdala, ventromedial prefrontal cortex, cingulate gyrus, parahippocampal gyrus, insula, and temporal pole, is referred to as the paralimbic system (Brodmann, 1994). Contemporary neuroscientists use a number of imaging techniques to measure both the structure and function of these and other brain regions noninvasively. Imaging work has found the structure and contextual functioning of these paralimbic brain regions to be abnormal in psychopaths.

Neuroimaging of psychopathy

A review of the neuroimaging literature in psychopathy reveals widespread, context-dependent (on the particular cognitive task) differences in brain activity relative to healthy controls, with the most robust abnormalities commonly found in the limbic, frontal, and temporal regions (Anderson & Kiehl, 2012; Koenigs et al., 2011).

Kiehl and colleagues (2001) were the first to report amygdala dysfunction in criminal psychopaths using functional MRI. Failure to engage the amygdala and orbitofrontal cortex during tasks that require aversive conditioning (i.e., learning to associate a specific behavior with punishment) has also been consistently associated with psychopathy (e.g., Birbaumer et al., 2005; Veit et al., 2002). Various reports have also indicated contextual functional deficits in the cingulate cortex (Birbaumer et al., 2005; Kiehl et al., 2001; Rilling et al., 2007; Veit et al., 2002), insula (Birbaumer et al., 2005; Veit et al., 2002), and extended temporal regions (Müller et al., 2008). Furthermore, these widespread deficits are apparent at a young age; brain-imaging findings in youth with psychopathic traits largely mirror the findings from adults described above (Finger et al., 2008; Finger et al., 2011; Jones, Laurens, Herba, Barker, & Viding, 2009; Marsh et al., 2008).

Investigations of the anatomical features of the psychopathic brain show that structural abnormalities mirror the functional deficits noted above. Many of these are relatively small-scale studies and suffer certain limitations due to restricted sample sizes. Nevertheless, studies have found reduced gray matter volumes in psychopaths' amygdala, orbitofrontal cortex, and cingulate cortex (Boccardi et al., 2011; Yang & Raine, 2009; Yang, Raine, Colletti, Toga, & Narr, 2010). Volume reductions have also been reported in the anterior temporal regions (Yang, Raine, Colletti, Toga, & Narr, 2011) and insula (de Oliveira-Souza et al., 2008). In a large-scale investigation of nearly 300 incarcerated adult participants, psychopathy scores were associated with tissue reductions in the amygdala, orbitofrontal cortex, posterior cingulate, parahippocampal region, and the temporal pole (Ermer, Cope, Nyalakanti, Calhoun, & Kiehl, 2012). These gray matter reductions were replicated in a large sample of incarcerated juveniles ($n = 218$) with psychopathic traits (Ermer, Cope, Nyalakanti, Calhoun, & Kiehl, 2013).

Despite some inevitable variability in empirical findings due to methodological variation and sampling characteristics, what is striking is the overall consensus of findings. The consistent implication of paralimbic and limbic structures in psychopathy across diverse methodologies strengthens the interpretation that these regions are critical for understanding the underlying neural dysfunction in psychopathy and its relationship with persistent criminal behavior. And the structural-imaging literature to date is consistent with the characterization of psychopathy as a profound developmental personality disorder that is present from an early age.

Can genes account for crime?

Some behavioral genetics studies have estimated that around 50% of the variance in antisocial behavior is attributable to genes (Mason & Frick, 1994; Moffitt, 2005). However, genes do not code for behavior, they code for protein, and the links between proteins and patterns of behavior are complicated and circuitous. Genetics, of course, influences physical properties of the brain and nervous system in ways that can bias learning, behavior, and motivation, but only through the complex interaction between genes and environment. The large number of possible genetic variants also creates a kind of microenvironment in which genes depend on other genes to determine a specific phenotype. The complexity of these relationships can be briefly summed up by the concepts of pleiotropy—some genes affect many traits, not just one—and polygenicity—the fact that most traits depend on the influence of many genes, not just one. While the link between genes and behavior can seem hopelessly complex, great progress has been made in understanding the modest influence that a few specific genes have on physiology, and how this can sometimes translate into a variety of criminal outcomes. Prominent among these are genes that code for proteins affecting monoaminergic neurotransmitter systems.

Monoamine neurotransmitters such as serotonin (5-HT), dopamine, and norepinephrine exert wide and varied effects on the brain and behavior, impacting mood states, attention, reinforcement, learning, and general sates of arousal. For example, monoamine oxidase is an enzyme responsible for the breakdown of monoamine neurotransmitters, limiting the strength and duration of their effects. A genetic variant resulting in reduced expression of monoamine oxidase A (MAOA-L) is strongly associated with impulsive, aggressive behavior (Brunner, Nelen, Breakefield, Ropers, & van Oost, 1993; Meyer-Lindenberg et al., 2006).

Meyer-Lindenberg and colleagues (2006) demonstrated that the presence of MAOA-L reliably predicts reduced gray matter volume in paralimbic regions, including the amygdala, orbitofrontal cortex, and cingulate cortex, which was also associated with hyperresponsive amygdala activity and reduced regulatory prefrontal activity during emotional arousal. The authors suggest that this genetic variant very likely promotes reduced effectual control over limbic circuits due to chronically elevated monoamine levels during critical developmental periods. This leads to increased vulnerability to early life emotional stress and amplified threat response and impulsivity in adulthood, highlighting the influence of environmental factors in development of a specific phenotype.

Another prominent genetic variant that can influence development of antisocial behavior codes for a protein called the serotonin transporter, 5-HTT. Many drugs that work as antidepressants act to block the action of this protein, increasing active levels of serotonin in the brain. The serotonin transporter gene is perhaps the most investigated in the monoaminergic system due to serotonin's prominent role in regulating mood and behavior (Caspi, Hariri, Holmes, Uher, & Moffitt, 2010). Individuals carrying a short variant of 5-HTT are more prone to depression and anxiety (Caspi et al., 2003; Kaufman et al., 2004; Kendler, Kuhn, Vittum, Prescott, & Riley, 2005), which again can promote vulnerability to early life stress (Caspi et al., 2010). This again highlights the role of developmental factors, which include environmental influences to determine phenotypes.

Another important monoaminergic neurotransmitter is dopamine, which is critical in the processing of punishment and reward contingencies—ultimately influencing decision making and future planning (Pessiglione, Seymour, Flandin, Dolan, & Frith, 2006). Genes for the dopamine transporter and dopamine receptors (DRD2 and DRD4) in the brain have long-established relationships with externalizing, antisocial behavior. For instance, the 9-repeat/10-repeat variation of the dopamine transporter is associated with hyperresponsivity of reward centers in the brain (Dreher, Kohn, Kolachana, Weinberger, & Berman, 2009), and has often been associated with externalizing problems, such as ADHD and behavior problems in youth (Young et al., 2002). It is also associated with substance abuse in adulthood (Kohnke et al., 2005). The A1 variant of the DRD2 gene reduces the number of viable dopamine binding sites (Ritchie & Noble, 2003), and has the probable consequence of shifting reward sensitivity and motivational salience (Volkow, Fowler, & Wang, 1999). Likewise, the long, 7-repeat variation of the DRD4 genotype (DRD4-7) is linked to high novelty-seeking behavior in adolescents (Becker et al., 2005) and poor response inhibition in cocaine abusers (Congdon, Lesch, & Canli, 2008), as well as externalizing behavior (DeYoung et al., 2006). Schmitt, Fox, and Hamer (2007) propose gene-gene interactions between short 5-HTT and long DRD4 genotypes, suggesting children with this combination are particularly prone to maladaptive social behavior.

These are just a few of the numerous and complicated genetic variants and interactions that have been reported to influence particular traits relevant in our discussion of criminality. Even this brief introduction emphasizes some of the challenges in genetic analysis. While we are aware of specific candidates for gene-behavior linkages, complex relationships between genes and environment and gene-gene interactions make it all but impossible to describe more than weak associations between an individual genetic variant and specific behavioral traits of interest. This is not intended to dismiss this field of inquiry, but rather to emphasize the need for strategic experimental design and appropriate hypothesis testing that acknowledges the limitations of candidate gene studies. The most influential studies will be those that link genes or combinations of genes to a particular physiological mechanism known to be affected by a variation in protein expression. These variations, impacted by environmental influences, may then bias behavior in predictable ways, for instance, by making an individual more impulsive or more sensitive to reward contingencies. It cannot, as the theme of this chapter emphasizes, predetermine criminal behavior.

Neuroscience of psychosis

While aggression can be present in a number of psychiatric conditions, such as post-traumatic stress disorder (Silva et al., 2001), bipolar disorder (Garno, Gunawardane, & Goldberg, 2008), temporal lobe epilepsy (Van Elst et al., 2000), certain types of dementia (Haller, Binder, & McNiel, 1989; Lai et al., 2003), and substance abuse (Boles & Miotto, 2003), this section focuses on violent behavior in psychosis. Psychosis is a break from reality that causes symptoms such as hallucinations, delusions, and disordered thoughts. Psychosis presents itself in disorders like schizophrenia, bipolar disorder, and major depression.

Worldwide scientific studies suggest that psychosis—if untreated—is associated with a fairly significant risk for violence. Prior to beginning antipsychotic medication, 1 in 3 psychosis patients engaged in some type of violent behavior, 1 in 6 engaged in moderately severe violent behavior, and 1 in 100 patients committed a violent act that resulted in serious injury (Large & Nielssen, 2011). In the general population, approximately 1 in 25,000 (Nielssen & Large, 2010) individuals will commit a homicide. Worldwide risk for homicide in first-episode patients with psychosis is 1 in 629 presentations (Nielssen & Large, 2010). The risk drops to 1 in 9,090 presentations if the patient receives treatment. Indeed, there is a 15-fold increase in risk for committing homicide in untreated psychosis patients compared to patients receiving appropriate antipsychotic treatment (Large & Nielssen, 2008). While the relationship between psychosis and risk for violence is extremely complex and mediated by a number of factors (discussed below), the statistics support the argument that early identification and treatment is the best way to reduce the risk for homicide and violence in patients with mental illness. And if 1 in approximately 630 psychosis patients will commit homicide, that represents a 40-fold increased risk compared to an individual in the general population.

While over 20 studies around the world have found a positive relationship between psychosis and violence (Fazel et al., 2009), there has been less research into the mechanisms that mediate this relationship (Witt, van Dorn, & Fazel, 2013). Meta-analyses suggest a number of dynamic factors contribute to the increased risk for violence in psychosis (Fazel et al., 2009; Witt, van Dorn, & Fazel, 2013). And the factors that contribute are not the same as the risk factors for violence in individuals without mental illness (e.g., lower intelligence scores) or risk factors for recidivism (e.g., difficulty with employment; Farrington, Loeber, & Ttofi, 2012), although there is some overlap (e.g., substance abuse). A 2009 meta-analysis of studies reporting risk of interpersonal violence and violent criminality in psychosis patients compared to the general population found most of the elevated risk of violence in psychosis compared to the general population to be mediated by substance abuse comorbidity (Fazel et al., 2009).

Other dynamic factors that increase risk of violence in psychosis include high impulsivity, hostile behavior, poor treatment and medication compliance, and recent alcohol or drug misuse (Witt et al., 2013). Additional incremental risk factors within patients suffering from psychosis are persecutory delusions or ideation, hallucinations with threatening content, and command hallucinations (Bo, Abu-Akel, Kongerslev, Haahr, & Simonsen, 2011). Both delusions and command hallucinations have been associated with reactive and instrumental violence (Felthous, 2008).

One of the challenges in identifying a unifying theory of schizophrenia is accounting for both the positive and negative symptoms. Temporally, the onset of psychosis manifests itself in a prodromal phase marked by negative symptoms and a decline in general functioning before the display of positive symptoms (Fenton & McGlashan, 1994). With this progression of symptoms in mind, one hypothesis is that the neural changes that result in negative symptoms impair the cognitive regulatory control process, which results in additional impairment and cognitive distortions and eventually gives rise to the positive symptoms (Bowins, 2011).

A detailed review of the neural mechanisms underlying schizophrenia is beyond the scope of this chapter. Briefly, studies have found dysregulation in cortical, subcortical, and limbic circuitry, as well as significantly reduced gray matter, especially in the prefrontal and anterior superior temporal cortices (Fornito et al., 2013; Giuliani, Calhoun, Pearlson, Francis, & Buchanan, 2005; Sullivan et al., 1998; Tamminga et al., 1992; Zipursky, Lim, Sullivan, Brown, & Pfefferbaum, 1992). Because the disease impairs multiple aspects of executive function, memory, and attention, it has been hypothesized that the illness results from dysfunction in a number of distributed brain networks (Andreasen et al., 1999). Some of these regions overlap with regions implicated in psychopathy. And while some of these circuits are more profoundly impaired in schizophrenia, there are deficits in areas in common in the two distinct disorders.

Predicting antisocial behavior and improving outcomes

Disinhibited behavior or impulsivity is one of the variables most highly correlated with reoffending. Personality tests, neuropsychological tests, and risk-assessment tools are all designed to assess individuals' cognitive capacities, yet these instruments are merely proxies for measuring the brain's inhibitory and cognitive-control systems (Aharoni et al., 2013). The brain regions associated with impulse control are well documented. Recently, Aharoni et al. (2013) conducted the first prospective neuroprediction study of criminal recidivism. It was hypothesized that deficits in cognitive control, measured by ACC activity during a go/no go task, would help predict whether individuals released from prison would engage in future criminal behavior as measured by re-arrest. ACC activity predicted above and beyond traditional risk-assessment measures: inmates with impaired cognitive control (low ACC activity) were 4.4 times more likely to be re-arrested for a nonviolent crime than inmates with high ACC activity within a four-year period. This study highlights the fact that impaired cognitive control may be a way to identify individuals at risk for life course–persistent antisocial behavior.

If the United States is to make any meaningful reduction in criminal behavior on a large scale, it needs to help break the cycle of recidivism by implementing interventions that promote desistence from crime. In order to help develop programs to rehabilitate offenders, the field of risk assessment, or risk-needs assessment, has sought to understand the variables that promote risk for antisocial behavior. Treatment programs then need to be designed to remediate the risks. For example, alcohol and substance abuse are known predictors of recidivism (Chandler, Fletcher, & Volkow, 2009; Fazel, Bains, & Doll, 2006), and treatment for such problems leads to reductions in antisocial behavior (Butzin, Martin, & Inciardi, 2005). Additionally, release conditions that reward abstinence and punish relapse have shown incredible promise in reducing recidivism (Hawken & Kleiman, 2009).

It is important to recognize that the neural mechanisms that underlie traits such as impulsivity and cognitive control are malleable. Cognitive remediation in schizophrenia has demonstrated significant improvements in cognitive performance, symptoms, and psychosocial functioning (McGurk et al., 2007), supporting the notion that even the most seemingly intractable traits can benefit from treatment.

One treatment program that applies this amenability-to-change philosophy to extremely high risk, antisocial youth is the Mendota Juvenile Treatment Center (MJTC) in Madison, Wisconsin. MJTC has made tremendous strides in improving outcomes and reducing general as well as violent recidivism in adolescents with severe behavioral problems and violent histories (Caldwell, Skeem, Salekin, & van Rybroek, 2006). Given the enormous cost of antisocial behavior that leads to crime, investment in programs like MJTC are likely to show significant benefits for society and the individual.

Conclusion

This review has identified three critical variables associated with an increased risk of antisocial behavior: age, particularly the adolescent years, psychopathic traits, and acute psychosis. One mechanism the three constructs share is weakened or impaired cognitive resources and control. This impairment is likely a major contributor toward the behavioral outcomes associated with these variables.

REFERENCES

Aharoni, E., Vincent, G. M., Harenski, C. L., Calhoun, V. D., Sinnott-Armstrong, W., Gazzaniga, M. S., & Kiehl, K. A. (2013). Neuroprediction of future rearrest. *Proc Natl Acad Sci USA*, *110*(15), 6223–6228.

Anderson, D. A. (2011). The cost of crime. *Found Trends Microecon*, 7(3), 209–265.

Anderson, N. E., & Kiehl, K. A. (2012). The psychopath magnetized: Insights from brain imaging. *Trends Cogn Sci*, *16*(1), 52–60.

Andreasen, N. C., Nopoulos, P., O'Leary, D. S., Miller, D. D., Wassink, T., & Flaum, M. (1999). Defining the phenotype of schizophrenia: Cognitive dysmetria and its neural mechanisms. *Biol Psychiatry*, *46*, 908–920.

Becker, K., Laucht, M., El-Faddagh, M., & Schmidt, M. H. (2005). The dopamine D4 receptor gene exon III polymorphism is associated with novelty seeking in 15-year-old males from a high-risk community sample. *J Neur Trans*, *112*(6), 847–858.

Birbaumer, N., Veit, R., Lotze, M., Erb, M., Hermann, C., Grodd, W., & Flor, H. (2005). Deficient fear conditioning in psychopathy: A functional magnetic resonance imaging study. *Arch Gen Psychiatry*, *62*(7), 799.

Bo, S., Abu-Akel, A., Kongerslev, M., Haahr, U. H., & Simonsen, E. (2011). Risk factors for violence among patients with schizophrenia. *Clin Psychol Rev*, *31*(5), 711–726.

Boccardi, M., Frisoni, G. B., Hare, R. D., Cavedo, E., Najt, P., Pievani, M., ... Repo-Tiihonen, E. (2011). Cortex and amygdala morphology in psychopathy. *Psychiatry Res*, *193*(2), 85–92.

Boles, S. M., & Miotto, K. (2003). Substance abuse and violence: A review of the literature. *Aggr Viol Behav*, *8*(2), 155–174.

Borg, J. S., Lieberman, D., & Kiehl, K. A. (2008). Infection, incest, and iniquity: Investigating the neural correlates of disgust and morality. *J Cogn Neurosci*, *20*(9), 1529–1546.

Bowins, B. (2011). A cognitive regulatory control model of schizophrenia. *Brain Res Bull*, *85*(1), 36–41.

Brodmann, K. (1994). *Localisation in the cerebral cortex.* London: Smith-Gordon.

Brunner, H. G., Nelen, M., Breakefield, X. O., Ropers, H. H., & van Oost, B. A. (1993). Abnormal behavior associated with a point mutation in the structural gene for monoamine oxidase A. *Science*, *262*, 578.

Bufkin, J. L., & Luttrell, V. R. (2005). Neuroimaging studies of aggressive and violent behavior: Current findings and implications for criminology and criminal justice. *Trauma Violence Abuse*, *6*(2), 176–191.

Butzin, C. A., Martin, S. S., & Inciardi, J. A. (2005). Treatment during transition from prison to community and subsequent illicit drug use. *J Subst Abuse Treat*, *28*, 351–358.

Caldwell, M., Skeem, J., Salekin, R., & van Rybroek, G. (2006). Treatment response of adolescent offenders with psychopathy features a 2-year follow-up. *Crim Justice Behav*, *33*(5), 571–596.

Carson, E. A., & Golinelli, D. (2013). Prisoners in 2012: Trends in admissions and releases, 1991–2012. Washington, DC: U.S. Deptartment of Justice, Bureau of Justice Statistics. NCJ243920, Table 5, p. 3, and Appendix Table 10, p. 43.

Casey, B. J., Jones, R. M., & Hare, T. A. (2008). The adolescent brain. *Ann NY Acad Sci, 1124*, 111–126.

Caspi, A., Hariri, A. R., Holmes, A., Uher, R., & Moffitt, T. E. (2010). Genetic sensitivity to the environment: The case of the serotonin transporter gene and its implications for studying complex diseases and traits. *Am J Psychiatry, 167*, 509–527.

Caspi, A., Sugden, K., Moffitt, T. E., Taylor, A., Craig, I. W., Harrington, H., … Poulton, R. (2003). Influence of life stress on depression: Moderation by a polymorphism in the 5-HHT gene. *Science, 301*, 291–293.

Centers for Medicare and Medicaid, Office of the Actuary, National Health Statistics Group. (2012). National health expenditure data.

Chandler, R. K., Fletcher, B. W., & Volkow, N. D. (2009). Treating drug abuse and addiction in the criminal justice system: Improving public health and safety. *JAMA, 301*, 183–190.

Cleckley, H. (1941). *The mask of sanity: An attempt to reinterpret the so-called psychopathic personality.* Oxford, UK: Mosby.

Cornell, D. G., Warren, J., Hawk, G., Stafford, E., Oram, G., & Pine, D. (1996). Psychopathy in instrumental and reactive violent offenders. *J Consult Clin Psychol, 64*(4), 783.

Congdon, E., Lesch, K. P., & Canli, T. (2008). Analysis of DRD4 and DAT polymorphisms and behavioral inhibition in healthy adults: Implications for impulsivity. *Am J Med Genet B Neuropsychiatr Genet, 147B*, 27–32.

Cunningham, M. D., & Reidy, T. J. (1998). Antisocial personality disorder and psychopathy: Diagnostic dilemmas in classifying patterns of antisocial behavior in sentencing evaluations. *Behav Sci Law, 16*(3), 333–351.

De Oliveira-Souza, R., Hare, R. D., Bramati, I. E., Garrido, G. J., Azevedo Ignácio, F., Tovar-Moll, F., & Moll, J. (2008). Psychopathy as a disorder of the moral brain: Fronto-temporo-limbic grey matter reductions demonstrated by voxel-based morphometry. *NeuroImage, 40*(3), 1202–1213.

DeYoung, C. G., Peterson, J. B., Séguin, J. R., Mejia, J. M., Pihl, R. O., Beitchman, J. H., … Palmour, R. M. (2006). The dopamine D4 receptor gene and moderation of the association between externalizing behavior and IQ. *Arch Gen Psychiatry, 63*, 1410–1416.

Dreher, J., Kohn, P., Kolachana, B., Weinberger, D. R., & Berman, K. F. (2009). Variation in dopamine genes influences responsivity of the human reward system. *Proc Natl Acad Sci USA, 106*, 617–622.

Ellis, B. J., DelGiudice, M., Dishion, T. J., Figueredo, A. J., Gray, P., Griskevicius, V., … Wilson, D. S. (2012). The evolutionary basis of risky adolescent behavior: Implications for science, policy, and practice. *Dev Psychol, 48*, 598–623.

Eluvathingal, T., Hasan, K., Kramer, L., Fletcher, J., & Ewing-Cobbs, L. (2007). Quantitative diffusion tensor tractography of association and projection fibers in normally developing children and adolescents. *Cereb Cortex, 17*, 2760–2768.

Ernst, M., Nelson, E. E., Jazbec, S., McClure, E. B., Monk, C. S., … Pine, D. S. (2005). Amygdala and nucleus accumbens in responses to receipt and omission of gains in adults and adolescents. *NeuroImage, 25*, 1279–1291.

Ermer, E., Cope, L. M., Nyalakanti, P. K., Calhoun, V. D., & Kiehl, K. A. (2012). Aberrant paralimbic gray matter in criminal psychopathy. *J Abnorm Psychol, 121*(3), 649.

Ermer, E., Cope, L. M., Nyalakanti, P. K., Calhoun, V. D., & Kiehl, K. A. (2013). Aberrant paralimbic gray matter in incarcerated male adolescents with psychopathic traits. *J Am Acad Child Psychol, 52*(1), 94.

Farrington, D. P., Loeber, R., & Ttofi, M. M. (2012). Risk and protective factors for offending. In B. C. Welsh & D. P. Farrington (Eds.), *The Oxford handbook of crime prevention* (pp. 46–69). New York, NY: Oxford University Press.

Fazel, S., Bains, P., & Doll, H. (2006). Substance abuse and dependence in prisoners: A systematic review. *Addiction, 101*, 181–191.

Fazel, S., Gulati, G., Linsell, L., Geddes, J. R., & Grann, M. (2009). Schizophrenia and violence: Systematic review and meta-analysis. *PLoS Med, 6*(8), e1000120.

Felthous, A. R. (2008). Schizophrenia and impulsive aggression: A heuristic inquiry with forensic and clinical implications. *Behav Sci Law, 26*(6), 735–758.

Fenton, W. S., & McGlashan, T. H. (1994). Antecedent, symptoms progression, and long-term outcome of the deficit syndrome in schizophrenia. *Am J Psychiatry, 151*(3), 351–356.

Finger, E. C., Marsh, A. A., Blair, K. S., Reid, M. E., Sims, C., Ng, P., … Blair, R. J. R. (2011). Disrupted reinforcement signaling in the orbitofrontal cortex and caudate in youths with conduct disorder or oppositional defiant disorder and a high level of psychopathic traits. *Am J Psychiatry, 168*(2), 152–162.

Finger, E. C., Marsh, A. A., Mitchell, D. G., Reid, M. E., Sims, C., Budhani, S., … Leibenluft, E. (2008). Abnormal ventromedial prefrontal cortex function in children with psychopathic traits during reversal learning. *Arch Gen Psychiatry, 65*(5), 586.

Fornito, A., Harrison, B. J., Goodby, E., Dean, A., Ooi, C., Nathan, P. J., & Bullmore, E. T. (2013). Functional dysconnectivity of corticostriatal circuitry as a risk phenotype for psychosis. *JAMA Psychiatry, 70*(11), 1143–1151.

Galvan, A., Hare, T. A., Parra, C. E., Penn, J., Voss, H., Glover, G., & Casey, B. J. (2006). Earlier development of the accumbens relative to orbitofrontal cortex might underlie risk-taking behavior in adolescents. *J Neurosci, 26*, 6885–6892.

Garno, J. L., Gunawardane, N., & Goldberg, J. F. (2008). Predictors of trait aggression in bipolar disorder. *Bipolar Disord, 10*(2), 285–292.

Giedd, J. N., Blumenthal, J., Jeffries, N. O., Castellanos, F. X., Liu, H., Zijdenbos, A., … Rapoport, J. L. (1999). Brain development during childhood and adolescence: A longitudinal MRI study. *Nat Neurosci, 2*, 861–863.

Gregg, T. R., & Siegel, A. (2001). Brain structures and neurotransmitters regulating aggression in cats: Implications for human aggression. *Prog Neuropsychopharmacol Biol Psychiatry, 25*(1), 91–140.

Giuliani, N. R., Calhoun, V. D., Pearlson, G. D., Francis, A., & Buchanan, R. W. (2005). Voxel-based morphometry versus region of interest: A comparison of two methods for analyzing gray matter differences in schizophrenia. *Schizophr Res, 74*(2), 135–147.

Haller, E., Binder, R. L., & McNiel, D. E. (1989). Violence in geriatric patients with dementia. *Bull Am Acad Psychiatry Law, 17*(2), 183–188.

Hare, R. D. (1968). Psychopathy, autonomic functioning, and the orienting response. *J Abnorm Psychol, 73*, 1.

Hare, R. D. (1980). A research scale for the assessment of psychopathy in criminal populations. *Pers Indiv Diff, 1*(2), 111–119.

Hare, R. D. (1991). *The psychopathy checklist-revised.* Toronto: Multi-Health Systems, Inc.

Hare, R. D. (2003). *The Hare psychopathy checklist-revised.* Toronto: Multi-Health Systems, Inc.

Hare, R. D., Clark, D., Grann, M., & Thornton, D. (2000). Psychopathy and the predictive validity of the PCL-R: An international perspective. *Behav Sci Law, 18*(5), 623–645.

Hare, R. D., Hart, S. D., & Harpur, T. J. (1991). Psychopathy and the DSM-IV criteria for antisocial personality disorder. *J Abnorm Psychol, 100*(3), 391.

Hare, R. D., & Quinn, M. J. (1971). Psychopathy and autonomic conditioning. *J Abnorm Psychol, 77*(3), 223.

Harris, G. T., Rice, M. E., & Cormier, C. A. (1991). Psychopathy and violent recidivism. *Law Hum Behav, 15*(6), 625.

Hawken, A., & Kleiman, M. (2009). Managing drug involved probationers with swift and certain sanctions: Evaluating Hawaii's HOPE. National Criminal Justice Reference Service. NCJ 229023.

Hemphill, J. F., Hare, R. D., & Wong, S. (1998). Psychopathy and recidivism: A review. *Leg Criminol Psychol, 3*(1), 139–170.

Hirono, N., Mega, M. S., Dinov, I. D., Mishkin, F., & Cummings, J. L. (2000). Left frontotemporal hypoperfusion is associated with aggression in patients with dementia. *Arch Neurology, 57*(6), 861–866.

Institute of Medicine and National Research Council. (2011). *The science of adolescent risk-taking: Workshop summary.* Washington, DC: National Academies Press.

James, D. J., & Glaze, L. E. (2006). *Mental health problems of prison and jail inmates.* Washington, DC: US Department of Justice, Office of Justice Programs, Bureau of Justice Statistics.

Jones, A., Laurens, K., Herba, C., Barker, G., & Viding, E. (2009). Amygdala hypoactivity to fearful faces in boys with conduct problems and callous-unemotional traits. *Am J Psychiatry, 166*(1), 95–102.

Karberg, J. C., & James, D. J. (2005). Substance dependence, abuse, and treatment of jail inmates, 2002. National Criminal Justice Reference Service. NCJ 209588.

Kaufman, J., Yang, B. Z., Douglas-Palumberi, H., Houshyar, S., Lipschitz, D., Krystal, J. H., ... Gelernter, J. (2004). Social supports and serotonin transporter gene moderate depression in maltreated children. *Proc Natl Acad Sci USA, 1010*, 17316–17321.

Kendler, K. S., Kuhn, J. W., Vittum, J., Prescott, C. A., & Riley, B. (2005). The interaction of stressful life events and a serotonin transporter polymorphism in the prediction of episodes of major depression: A replication. *Arch Gen Psychiatry, 62*, 529–535.

Kiehl, K. A. (2006). A cognitive neuroscience perspective on psychopathy: Evidence for paralimbic system dysfunction. *Psych Res, 12*(2), 107–128.

Kiehl, K. A., Smith, A. M., Hare, R. D., Mendrek, A., Forster, B. B., Brink, J., & Liddle, P. F. (2001). Limbic abnormalities in affective processing by criminal psychopaths as revealed by functional magnetic resonance imaging. *Biol Psychiatry, 50*(9), 677–684.

Koenigs, M., Baskin-Sommers, A., Zeier, J., & Newman, J. P. (2011). Investigating the neural correlates of psychopathy: A critical review. *Molecular Psych, 16*(8), 792–799.

Kohnke, M. D., Batra, A., Kolb, W., Kohnke, A. M., Lutz, U., Schick, S., & Gaertner, I. (2005). Association of the dopamine transporter gene with alcoholism. *Alcohol, 40*, 339–342.

Kuruoglu, A.Ç., Arikan, Z.,Vural, G., Karatas, M., Araç, M., & Isik, E. (1996). Single photon emission computerized tomography in chronic alcoholism: Antisocial personality disorder may be associated with decreased frontal perfusion. *British J Psych, 169*, 348–354.

Lai, C. K., & Arthur, D. G. (2003). Wandering behaviour in people with dementia. *J Advanced Nursing, 44*(2), 173–182.

Large, M., & Nielssen, O. (2008). Evidence for a relationship between the duration of untreated psychosis and the proportion of psychotic homicides prior to treatment. *Soc Psychiatry Psychiatr Epidemiol, 43*, 37–44.

Large, M. M., & Nielssen, O. (2011). Violence in first-episode psychosis: A systematic review and meta-analysis. *Schizophr Res, 125*(2), 209–220.

Lenroot, R. K., & Giedd, J. N. (2006). Brain development in children and adolescents: Insights from anatomical magnetic resonance imaging. *Neurosci Biobehav Rev, 30*, 718–729.

Mason, D. A., & Frick, P. J. (1994). The heritability of antisocial behavior: A meta-analysis of twin and adoption studies. *J Psychopathol Behav Assessment, 16*, 301–323.

Marsh, A., Finger, E., Mitchell, D., Reid, M., Sims, C., Kosson, D., ... Blair, R. (2008). Reduced amygdala response to fearful expressions in children and adolescents with callous-unemotional traits and disruptive behavior disorders. *Am J Psychiatry, 165*(6), 712–720.

McEllistrem, J. E. (2004). Affective and predatory violence: A bimodal classification system of human aggression and violence. *Aggress Violent Behav, 10*, 1–30.

Meyer-Lindenberg, A., Buckholtz, J. W., Kolachana, B., Hariri, A. R., Pezawas, L., Blasi, G., ... Weinberger, D. R. (2006). Neural mechanisms of genetic risk for impulsivity and violence in humans. *Proc Natl Acad Sci USA, 103*, 6269–6274.

McGurk, S. R., Twamley, E. W., Sitzer, D. I., McHugo, G. J., & Mueser, K. T. (2007). A meta-analysis of cognitive remediation in schizophrenia. *Am J Psychiatry, 164*(12), 1791.

Moffitt, T. E. (2005). The new look of behavioral genetics in developmental psychopathology: Gene–environment interplay in antisocial behaviors. *Psychol Bull, 131*, 533–554.

Müller, J. L., Sommer, M., Döhnel, K., Weber, T., Schmidt-Wilcke, T., & Hajak, G. (2008). Disturbed prefrontal and temporal brain function during emotion and cognition interaction in criminal psychopathy. *Behav Sci Law, 26*(1), 131–150.

National Center on Addiction and Substance Abuse (CASA). (2010). Behind bars II: Substance abuse and America's prison population. Institute of Education Sciences. ED509000.

NIELSSEN, O., & LARGE, M. (2010). Rates of homicide during the first-episode of psychosis and after treatment: A systematic review and meta-analysis. *Schizophr Bull, 36,* 702–712.

PAPEZ, J. W. (1937). A proposed mechanism of emotion. *Arch Neurol Psychiatry, 38*(4), 725.

PATRICK, C. J. (1994). Emotion and psychopathy: Startling new insights. *Psychophysiology, 31*(4), 319–330.

PESSIGLIONE, M., SEYMOUR, B., FLANDIN, G., DOLAN, R. J., & FRITH, C. D. (2006). Dopamine-dependent prediction errors underpin reward-seeking behavior in humans. *Nature, 442,* 1042–1045.

PORTER, S., BIRT, A. R., & BOER, D. P. (2001). Investigation of the criminal and conditional release profiles of Canadian federal offenders as a function of psychopathy and age. *Law Hum Behav, 25*(6), 647.

PORTER, S., BRINKE, L., & WILSON, K. (2009). Crime profiles and conditional release performance of psychopathic and non-psychopathic sexual offenders. *Leg Criminol Psychol, 14*(1), 109–118.

RAINE, A., BUCHSBAUM, M., & LACASSE, L. (1997). Brain abnormalities in murderers indicated by positron emission tomography. *Biol Psych, 42*(6), 495–508.

RAINE, A., BUCHSBAUM, M. S., STANLEY, J., LOTTENBERG, S., ABEL, L., & STODDARD, J. (1994). Selective reductions in prefrontal glucose metabolism in murderers. *Biol Psychiatry, 36*(6), 365–373.

RAINE, A., MELOY, J. R., BIHRLE, S., STODDARD, J., LACASSE, L., & BUCHSBAUM, M. S. (1998). Reduced prefrontal and increased subcortical brain functioning assessed using positron emission tomography in predatory and affective murderers. *Behav Sci Law, 16*(3), 319–332.

RAINE, A., STODDARD, J., BIHRLE, S., & BUCHSBAUM, M. (1998). Prefrontal glucose deficits in murderers lacking psychosocial deprivation. *Neuropsychiatry, Neuropsychology, Behav Neurol, 11*(1), 1–7.

REYNA, V., & FARLEY, F. (2006). Risk and rationality in adolescent decision-making: Implications for theory, practice, and public policy. *Psychol Sci Publ Interest, 35,* 1.

RICE, M. E., & HARRIS, G. T. (1997). Cross-validation and extension of the Violence Risk Appraisal Guide for child molesters and rapists. *Law Hum Behav, 21*(2), 231–241.

RILLING, J. K., GLENN, A. L., JAIRAM, M. R., PAGNONI, G., GOLDSMITH, D. R., ELFENBEIN, H. A., & LILIENFELD, S. O. (2007). Neural correlates of social cooperation and noncooperation as a function of psychopathy. *Biol Psychiatry, 61*(11), 1260–1271.

RITCHIE, T., & NOBLE, E. P. (2003). Association of seven polymorphisms of the D2 dopamine receptor gene with brain receptor-binding characteristics. *Neurochem Res, 28,* 73–82.

SALEKIN, R. T., ROGERS, R., & SEWELL, K. W. (1996). A review and meta-analysis of the Psychopathy Checklist and Psychopathy Checklist-Revised: Predictive validity of dangerousness. *Clin Psychol Sci Pract, 3*(3), 203–215.

SCHMITT, L. A., FOX, N. A., & HAMER, D. H. (2007). Evidence for a gene-gene interaction in predicting children's behavior problems: Association of serotonin transporter short and dopamine receptor D4 long genotypes with internalizing and externalizing behaviors in typically developing 7-year-olds. *Dev Psychopathol, 19,* 1105–1116.

SILVA, J. A., DERECHO, D. V., LEONG, G. B., WEINSTOCK, R., & FERRARI, M. M. (2001). A classification of psychological factors leading to violent behavior in posttraumatic stress disorder. *J Forensic Sci, 46*(2), 309–316.

SPEAR, L. P. (2000). The adolescent brain and age related behavioral manifestations. *Neurosci Biobehav Rev, 24,* 417–463.

STEINBERG, L. (2004). Risk taking in adolescence: What changes, and why? *Ann NY Acad Sci, 1021*(1), 51–58.

STEINBERG, L. (2007). Risk taking in adolescence: New perspectives from brain and behavioral sciences. *Curr Dir Psychol Sci, 16,* 55.

SULLIVAN, E. V., LIM, K. O., MATHALON, D., MARSH, L., BEAL, D. M., HARRIS, D., & PFEFFERBAUM, A. (1998). A profile of cortical gray matter volume deficits characteristic of schizophrenia. *Cereb Cortex, 8*(2), 117–124.

TAMMINGA, C. A., THAKER, G. K., BUCHANAN, R., KIRKPATRICK, B., ALPHS, L. D., CHASE, T. N., & CARPENTER, W. T. (1992). Limbic system abnormalities identified in schizophrenia using positron emission tomography with fluorodeoxyglucose and neocortical alterations with deficit syndrome. *Arch Gen Psychiatry, 49*(7), 522.

VAN ELST, L. T., WOERMANN, F. G., LEMIEUX, L., THOMPSON, P. J., & TRIMBLE, M. R. (2000). Affective aggression in patients with temporal lobe epilepsy: A quantitative MRI study of the amygdala. *Brain, 123*(2), 234–243.

VEIT, R., FLOR, H., ERB, M., HERMANN, C., LOTZE, M., GRODD, W., & BIRBAUMER, N. (2002). Brain circuits involved in emotional learning in antisocial behavior and social phobia in humans. *Neurosci Lett, 328*(3), 233–236.

VIDING, E., BLAIR, R. J. R., MOFFITT, T. E., & PLOMIN, R. (2005). Evidence for substantial genetic risk for psychopathy in 7-year-olds. *J Child Psychol Psychiatry, 46*(6), 592–597.

VOLKOW, N. D., FOWLER, J. S., & WANG, G. J. (1999). Imaging studies on the role of dopamine in cocaine reinforcement and addiction in humans. *J Psychopharmacol, 13,* 337–345.

WITT, K., VAN DORN, R., & FAZEL, S. (2013). Risk factors for violence in psychosis: Systematic review and meta-regression analysis of 110 studies. *PLoS ONE, 8*(2), e55942.

WONG, M. T., FENWICK, P. B., LUMSDEN, J., FENTON, G. W., MAISEY, M. N., LEWIS, P., & BADAWI, R. (1997). Positron emission tomography in male violent offenders with schizophrenia. *Psych Res: Neuroimaging, 68*(2), 111–123.

YANG, Y., & RAINE, A. (2009). Prefrontal structural and functional brain imaging findings in antisocial, violent, and psychopathic individuals: A meta-analysis. *Psychiatry Res, 174*(2), 81.

YANG, Y., RAINE, A., COLLETTI, P., TOGA, A. W., & NARR, K. L. (2010). Morphological alterations in the prefrontal cortex and the amygdala in unsuccessful psychopaths. *J Abnorm Psychol, 119*(3), 546.

YANG, Y., RAINE, A., COLLETTI, P., TOGA, A. W., & NARR, K. L. (2011). Abnormal structural correlates of response perseveration in individuals with psychopathy. *J Neuropsychiatry Clin Neurosci, 23*(1), 107–110.

YOUNG, S. E., SMOLEN, A., CORLEY, R. P., KRAUTER, K. S., DEFRIES, J. C., CROWLEY, T. J., & HEWITT, J. K. (2002). Dopamine transporter polymorphism associated with externalizing behavior problems in children. *Am J Med Genet, 114,* 144–149.

ZIPURSKY, R. B., LIM, K. O., SULLIVAN, E. V., BROWN, B. W., & PFEFFERBAUM, A. (1992). Widespread cerebral gray matter volume deficits in schizophrenia. *Arch Gen Psychiatry, 49*(3), 195.

90 Neuropharmacology and Society

ANJAN CHATTERJEE

ABSTRACT Advances in our understanding of large-scale neurocognitive and affective systems, along with their chemical underpinnings, are contributing to better treatments for neuropsychiatric diseases and greater options to ameliorate their symptoms. These clinical interventions can also spill over into nonclinical settings to enhance healthy cognitive abilities and modulate normal affective states. Decisions to use enhancements are complicated by limits in our knowledge of their efficacy and the range of adverse effects that they might incur. However, these limits and the relative modesty of their benefits have not inhibited their use among certain populations, such as athletes and students. The ethical concerns that arise from widespread use of neuroenhancements relate to safety, justice in their availability, possible erosion of character, and threats to individual autonomy. Societal norms, informed but not dictated by scientific evidence, are likely to determine the extent of and acceptance of nonclinical uses of pharmacological interventions within different sectors of society.

Neuropharmacology has moved from the bench, through the bedside, and into playing fields, classrooms, recital halls, and even ivory towers (Chatterjee, 2008; Sahakian & Morein-Zamir, 2007). Should we celebrate these societal trends, or should we be alarmed? Our ability to treat diseases, modify their course, or at least alleviate symptoms has been improving. In the wake of these clinical advances, do we face unintended societal consequences of manipulating health?

"Cosmetic neurology" refers to the use of neurologic interventions to enhance movement, mood, and mentation in healthy people (Chatterjee, 2004, 2006). These interventions are typically developed to treat disease before their use is considered in healthy people. In what follows, I review drug interventions currently available for enhancement. I discuss what we know and limits of that knowledge. I then outline the ethical concerns that surround cosmetic neurology, including special concerns that apply to children and adolescents.

Enhancements

Enhancements apply to three general domains: motor systems, cognition, and mood and affect. This chapter will focus on pharmacologic cognitive and affective enhancements, although other interventions are also available (such as brain stimulation; Hamilton, Messing, & Chatterjee, 2011).

COGNITION Most pharmacologic cognitive treatments target catecholamine and cholinergic systems. Catecholamine effects on neuronal plasticity may help cognitive training (Repantis, Schlattmann, Laisney, & Heuser, 2010). The observation that amphetamines improve speech therapy in aphasic patients (Walker-Batson et al., 2001) gave rise to the hypothesis that they might enhance language abilities in people without aphasia. It turns out that amphetamines can facilitate novel vocabulary learning (Breitenstein et al., 2004; Whiting, Chenery, Chalk, Darnell, & Copland, 2007) and speed up information processing (Fillmore, Kelly, & Martin, 2005) in healthy subjects. Biomarkers (Hamidovic, Dlugos, Palmer, & Wit, 2010; Volkow et al., 2008) might predict individual responses to these drugs and help target their enhancement use. Methylphenidate is also used widely to improve attention, concentration, spatial working memory, and planning (Mintzer & Griffiths, 2007; Weber & Lutschg, 2002; Zeeuws, Deroost, & Soetens, 2010). Students commonly use amphetamines and their analogs (McCabe, Knight, Teter, & Wechsler, 2005), despite the fact that these drugs sometimes impair performance (Babcock & Byrne, 2000; Diller, 1996) and the actual empirical data in support of their effects are modest (Smith & Farah, 2011). Newer, nonaddictive drugs such as atomoxetine (a selective norepinephrine-reuptake inhibitor) may improve executive control (Chamberlain et al., 2009). Other stimulants, like modafinil, improve arousal and selective attention in demanding situations (Marchant et al., 2008) and ameliorate cognitive deficits associated with sleep deprivation (Lagarde, Batejat, Van Beers, Sarafian, & Pradella, 1995).

Cholinesterase inhibitors are used to improve attention and memory (see Repantis, Laisney, & Heuser, 2010, for a review). These medications were developed to treat memory deficits in Alzheimer's disease but can be used in the healthy. Yesavage and colleagues (2001) found that commercial pilots taking 5mg of donepezil performed better than pilots on placebo when facing demanding flight-simulation tasks. These drugs may also improve semantic processing (FitzGerald et al., 2008) and memory (Grön et al., 2005; Zaninotto et al., 2009) and mitigate the effects of sleep deprivation (Chuah et al., 2009).

Mood and Affect By some estimates, between 9.5 and 20% of Americans are depressed (National Institute of Mental Health, 2003). Horowitz and Wakefield (2007) argue that the epidemic of depression is explained in part by the reclassification of normal experiences of sadness as depression. Here, we see a familiar pattern of therapy extending to enhancement. When the boundary between a clinical condition and a healthy state is graded, drug use expands to borderline cases. It is a short step for the drugs to then be considered in the healthy. Consistent with this pattern, selective serotonin-reuptake inhibitor use in the healthy can selectively dampen negative affect (Knutson et al., 1998) and increase affiliative behavior in social settings (Tse & Bond, 2002).

We might be able to modulate our emotional states in more subtle ways. For example, oxytocin and vasopressin might be used to induce trust and promote affiliative behavior (Insel, 2010). The possibility of such manipulations raises questions about what constitutes improvement. Debates about whether these kinds of drugs can manipulate morality itself are beginning to surface (Douglas, 2008; Harris, 2011; Persson & Savulescu, 2008)

Drugs can modulate the memory of emotional events (Cahill, 2003; Strawn & Geracotti, 2008). Epinephrine consolidates emotional memories, and beta-blockers dampen them. Subjects given propranolol remember emotionally arousing stories as if they were emotionally neutral (Cahill, Prins, Weber, & McGaugh, 1994). Propranolol also enhances the memory of events surrounding emotionally charged events that are otherwise suppressed (Strange, Hurlemann, & Dolan, 2003). In one study, patients in an emergency room given propranolol after a traumatic event had fewer post-traumatic stress disorder symptoms when assessed one month later (Pitman et al., 2002). Benzodiazepines can lessen the emotional impact of memories (Brignell, Rosenthal, & Curran, 2006). These examples demonstrate the difficulty of distinguishing between therapy and enhancement. How should such preventive measures be regarded? We might regard the dampening of emotional memories in veterans suffering from post-traumatic stress disorder as therapy. But emotional stresses of events that occur in everybody's lives such as romantic rejections, failures at work, death of loved ones might also be dampened. These nonclinical interventions fall under the rubric of cosmetic neurology.

Ethical dilemmas

If drugs can make us smarter and happier, surely this development is a good thing. The reasons to be cautious about the promise of a pharmacological utopia involve concerns about safety, authenticity, justice, and autonomy (Chatterjee, 2006).

Safety Virtually all medications have potential side effects that range from minor inconveniences to severe disability or death. For example, amphetamines have FDA black box warnings pointing to the risk of addiction and serous cardiac side effects, including sudden death. Recent large-scale studies do not find greater cardiovascular side effects of stimulants (Cooper et al., 2011; Habel, 2011) and mitigate this concern. However, physicians tend to be concerned about the safety of enhancements (Chatterjee, 2009), given what they see as their professional roles. They are even suspicious of safety claims made by pharmaceutical companies (Banjo, Nadler, & Reiner, 2010) and think enhancements should only be made available if they are safe (Hotze, Shah, Anderson, & Wynia, 2011).

A subtler version of the safety concern is that of trade-offs rather than side effects. Would cognitive enhancement in one cognitive process detract from others? For example, medications that enhanced attention and concentration might limit imagination and creativity (but see Farah, Haimm, Sankoorikal, & Chatterjee, 2009). Another potential trade-off is that enhancing long-term memory could impair working memory; enhancing consolidation of long-term memories could disrupt the flexibility of those memories to respond to a changed environment and alter behavior (Schermer, Bolt, de Jongh, & Olivier, 2009).

Finally, short- and long-term effects of enhancements might be different. Most pharmacologic studies are conducted in relatively short clinical trials. An underlying concern is that chronic use of such medications might have unpredictable and even detrimental effects. Whitaker (2010) observes that since the advent of neuropsychiatric medications in the 1950s, the natural history of disorders like schizophrenia, depression, and bipolar disorder has worsened rather than improved. Most of these medications work by increasing synaptic concentrations of neurotransmitters such as acetylcholine or norepinephrine. As a result of flooding of these neurotransmitters within the synaptic cleft, one might expect down-regulation of postsynaptic receptors and up-regulation of presynaptic reuptake mechanisms. Synaptic homeostasis may mitigate the chronic effects of pharmacologic interventions and potentially have long-term adverse consequences.

Authenticity The concern about individual authenticity takes two general forms, one about eroding character and the other about altering individuals. The concern about erosion of character draws on a "no pain, no gain" belief (Chatterjee, 2008). Many people believe that struggling with pain builds character, and eliminating that pain undermines good character. Easy benefits without effort cheapen us (Kass, 2003).

While the concerns about authenticity run deep, they are mitigated by several factors. Which pains are worth their hypothetical gains? We live in homes with central heat and air, eat food prepared by others, travel vast distances in short times, and take Tylenol for headaches and H2 blockers for heartburn. Perhaps these conveniences have eroded our collective character and cheapened us. But few choose to turn back.

A fundamental concern is that chemically changing the brain threatens our notion of ourselves. The central intuition is that such interventions threaten essential characteristics of what it means to be human (President's Council on Bioethics, 2003). For example, does selectively dampening the impact of our painful memories change who we are? This is a difficult issue, given that there is little consensus on the essence of human nature (Fukayama, 2002; Kolber, 2011; Wolpe, 2002). However, the search and desire for an authentic self probably drives both the desire for and the worry about consequences of enhancement (Elliott, 2011).

Justice Who gets to use enhancements? Insurance companies or the state (in the United States) are unlikely to pay for nonclinical interventions. Only those who can afford to pay privately would get enhancements. A common counter to the worry of widening inequities is that enhancement is not a zero-sum game. With general improvements, benefits will trickle down even to those at the bottom of a material hierarchy. However, this argument assumes that people's sense of well-being is determined by an absolute level of quality, rather than a recognition of one's relative place. However, beyond worries about basic subsistence, well-being is mostly affected by expectations and relative positions in society (Frank, 1987).

One might argue that the critical issue is access and not availability (Caplan, 2003). If access to enhancements were open to all, then differences might even be minimized. This argument has logical merit, but skirts the issue in practice (in the United States). We tacitly accept wide disparities in modifiers of cognition, as demonstrated by the presence of inequities in education, nutrition, and shelter. The access concern might be less relevant to new, nonpharmacological interventions like transcranial direct current stimulation. These cheap and relatively safe interventions may have cosmetic uses (Hamilton et al., 2011) and do not need physicians to serve as gatekeepers.

Autonomy Matters of choice can become coercive. Coercion takes two forms. One is an implicit pressure to maintain or better one's position in some perceived social order. Such pressures increase in "winner-take-all" environments in which more people compete for fewer and bigger prizes (Frank & Cook, 1995). Many professionals work 60, 80, or more than 100 hours a week without regard to their health. Emergency department residents use zolpidem and modafinil to regulate sleep and enhance their effectiveness (McBeth et al., 2009). Athletes take steroids to compete at the highest levels, and children at competitive preparatory schools take methylphenidate in epidemic proportions (Hall, 2003). To not take enhancements might mean being left behind. Students frequently cite academic assignments or grades as reasons to take amphetamines (Arria, O'Grady, Caldeira, Vincent, & Wish, 2008; DeSantis, Webb, & Noar, 2008). In a US survey from 2005, nonmedical uses of stimulants were highest in competitive colleges (McCabe et al., 2005). Similar practices are evident in Europe (Schermer et al., 2009).

A second form of coercion is an explicit demand for superior performance. Soldiers have been encouraged to take enhancements for the greater good (Russo, Stetz, & Stetz, 2013). Might this logic extend to civilians? Yesavage and colleagues' (2001) findings that pilots taking donepezil performed better in emergencies than those on placebo could have wide implications. If these results are reliable and meaningful, should pilots be required to take such medications? Should medical students and post-call residents take stimulants to attenuate deficits of sustained attention brought on by sleep deprivation (Webb, Thomas, & Valasek, 2010)? Could hospital administrators or patients require this practice?

Children and Adolescents Concerns about the consequences of coercion take on special force in the context of limited autonomy that applies to children and adolescents. These young people face similar pressures as adults to use enhancements (Singh & Kelleher, 2013). Many children, especially in affluent environments, have demanding social schedules, sports commitments, and other extracurricular activities added to burdensome levels of schoolwork. In this pressured environment, the demand for enhancements has risen over the last few years (Johnston et al., 2006). In 2005, 7.4% of eighth graders reported trying amphetamines without medical instruction (Johnston et al., 2006). Physicians frequently write prescriptions for psychotropic drugs, especially stimulants and antidepressants, to young people without a clear diagnosis of a mental illness (Thomas, Conrad, Casler, & Goodman, 2006).

From 2002 to 2010, physicians wrote fewer prescriptions for antibiotics among adolescents, but increased prescriptions for stimulants in this age group by 46% (Chai et al., 2012).

Estimates of nonprescription use of stimulants were below 0.5% until 1995 across the age range from high school to adults. Since the mid-1990s, 2.5% of high school students report nonprescription use of stimulants (Smith & Farah, 2011). Data from the Monitoring the Future Survey suggest that young people use different psychotropic prescription drugs, including tranquillizers, painkillers, stimulants, and hypnotics, for nonmedical purposes (Johnston et al., 2006). By the twelfth grade, in 2012, the lifetime prevalence of the use of amphetamines was 12.0% and of sedatives was 6.9%. These drugs are used recreationally and to enhance performance (Friedman, 2006; Teter, McCabe, Cranford, Boyd, & Guthrie, 2005). University chat sites and listservs make prescription drugs readily available for nonmedical use (Talbot, 2009). In 2012, 45.4% of twelfth graders thought it was very easy or fairly easy to get amphetamines, and 28.7% thought it was very easy or fairly easy to get sedatives (Johnston et al., 2013).

Ethical concerns for the use of enhancements in children are amplified when considering safety, authenticity, and autonomy. The American Academy of Neurology recently issued guidance opposing the prescription of cognitive enhancements for children (Graf et al., 2013). The long-term biological impact of enhancements on the developing nervous system is unknown (Kim et al., 2009). Another concern is that use of enhancements could erode the development of children's character, notions of authenticity, and sense of personal responsibility. Stimulants can alter reward circuitry (Kim et al., 2009) and change behavior, motivation, attention, and interaction with others. Decisions involving long-term risks may be particularly challenging for younger children because they involve calculation of future risk-benefit ratios (Singh et al., 2010). Parents can be sources of coercion, driven by performance pressures or goals to produce highly successful children even at the expense of the child's physical or mental health. Schools can add to such pressures. Teachers often suggest to parents that a child might benefit from stimulant treatment (Sax, 2003).

In summary, we face a paradox when it comes to enhancements in children and adolescents. Their limited autonomy amplifies other concerns of safety (unknown long-term consequences), coercion (parental, school pressure), and character (altering developing brains). However, the use and acceptance of enhancements continues to be on the rise among adolescents and young adults.

Limits to our knowledge

Any discussion of the advantages and disadvantages of cosmetic neurology relies on adequate information about their efficacy and adverse effects. Unfortunately, our knowledge is limited and may continue to be so for the foreseeable future.

Placebo Effects A pervasive issue with ascertaining the specific effects of enhancing drugs is determining the extent of placebo effects (Kirsch, 2009; Rutherford, Mori, Sneed, Pimontel, & Roose, 2012). The belief that drugs are helpful often contributes to their demonstrable effects (Benedetti, 2008). For example, participants may feel better about their performance on stimulants even when there is no measurable improvement (Ilieva, Boland, & Farah, 2013). At a societal level, the greater the general belief that enhancements work, regardless of whether that belief is generated by the media or advertising or what peers say, the more likely people will feel positive effects of these medications. Such beliefs, if widespread, are likely to influence public policy regardless of the modesty of scientific support.

Bias Several impediments limit our ability to conduct research into the effects of enhancing medications in healthy people. Most funding agencies do not support such research, making systematic progress difficult. The lack of funding and regulatory burdens prevents large multicenter randomized controlled trials. Accumulated data are often biased by fields with few researchers and small true effect sizes that might not be meaningful (Ioannidis, 2005). Limited power is an endemic problem in studies of enhancements. These studies typically enroll small numbers of participants. Often the drugs are given once, and if repeated only for relatively short durations. Because studies that show significant effects are more likely to be published, well-designed negative studies are not accounted for in any systematic manner. This publication bias is common in neuroscience (Button et al., 2013). As such, reviews of this literature and various meta-analyses may overestimate the effects of enhancement medications.

Future considerations

When ethical discussions of pharmacologic enhancements began in earnest over a decade ago (Chatterjee, 2004; Farah et al., 2004; Savulescu, 2005; Wolpe, 2002), the general assumption was that the armamentarium of applicable drugs would grow. Such growth, in the near future, appears unlikely. Development of novel neuropsychiatric drugs is declining (Hyman, 2012; Insel,

2012). Truly innovative drugs are not evident in the pipeline. For now, discussion about pharmacologic enhancements will largely focus on drugs currently available.

How should we, as a society, respond to cosmetic neurology? Some version of the practice is inevitable (Chatterjee, 2007). Strict prohibition of the use of enhancements is unlikely to be effective. This approach simply moves the market for such medications underground and would probably inhibit thoughtful discussion about the actual use of these medications. Some advocate for policies to maximize benefits and minimize harm by supporting fairness, protecting individuals from coercion, and minimizing enhancement-related socioeconomic disparities (Appel, 2008; Greely et al., 2008).

Cultural norms about the use of enhancements have not coalesced into a consensus. Physicians, who would do the prescribing, may need to think beyond traditional disease-treatment models of care (Bostrom, 2008; Chatterjee, 2004; Ravelingien, Braeckman, Crevits, De Ridder, & Mortier, 2009; Synofzik, 2009). Physicians are typically pragmatic (Ott, Lenk, Miller, Buhler, & Biller-Andorno, 2012) about prescribing enhancements, but also ambivalent, viewing the practice as alleviating suffering while being wary of exaggerating social inequities (Hotze et al., 2011). Approaching enhancement as a public health issue may advance the discussion (Outram & Racine, 2011). Physician organizations will help structure discussions. For example, the American Academy of Neurology has published guidance specifying that prescribing enhancements might be permissible for adults (Larriviere et al., 2009), but not for children (Graf et al., 2013).

Research and clinical neuroscientists along with ethicists can lay the groundwork for broader public discussions. As of now, there is little agreement among academics (Boot, Partridge, & Hall, 2012; Forlini & Racine, 2009; Heinz, Kipke, Heimann, & Wiesing, 2012). Some regard the ethical concerns as exaggerated (Partridge, Bll, Luck, Yeates, & Hall, 2011). It remains unclear where these issues will settle, and if and when we will arrive at consensus. However, such discussions are critical as the effects of neuropharmacology on society continue to be pervasively felt and our attitudes evolve.

REFERENCES

Appel, J. M. (2008). When the boss turns pusher: A proposal for employee protections in the age of cosmetic neurology. *J Med Ethics, 34*(8), 616–618.

Arria, A., O'Grady, K., Caldeira, K., Vincent, K., & Wish, E. (2008). Nonmedical use of prescription stimulants and analgesics: Associations with social and academic behaviors among college students. *Pharmacotherapy, 38*(4), 1045–1060.

Babcock, Q., & Byrne, T. (2000). Student perceptions of methylphenidate abuse at a public liberal arts college. *J Am Coll Health, 49*(3), 143–145.

Banjo, O., Nadler, R., & Reiner, P. (2010). Physician attitudes towards pharmacological cognitive enhancement: Safety concerns are paramount. *PLoS ONE, 5*(12), e14322.

Benedetti, F. (2008). Mechanisms of placebo and placebo-related effects across diseases and treatments. *Annu Rev Pharmacol Toxicol, 48*(1), 33–60.

Boot, B. P., Partridge, B., & Hall, W. (2012). Better evidence for safety and efficacy is needed before neurologists prescribe drugs for neuroenhancement to healthy people. *Neurocase, 18*(3), 181–184.

Bostrom, N. (2008). Drugs can be used to treat more than disease. *Nature, 451*(7178), 520.

Breitenstein, C., Wailke, S., Bushuven, S., Kamping, S., Zwitserlood, P., Ringelstein, E. B., & Knecht, S. (2004). D-Amphetamine boosts language learning independent of its cardiovascular and motor arousing effects. *Neuropsychopharmacology, 29*(9), 1704–1714.

Brignell, C. M., Rosenthal, J., & Curran, H. V. (2006). Pharmacological manipulations of arousal and memory for emotional material: Effects of a single dose of methylphenidate or lorazepam. *J Psychopharmacol, 21*(7), 673–683.

Button, K. S., Ioannidis, J. P. A., Mokrysz, C., Nosek, B. A., Flint, J., Robinson, E. S. J., & Munafo, M. R. (2013). Power failure: Why small sample size undermines the reliability of neuroscience. *Nat Rev Neurosci, 14*(5), 365–376.

Cahill, L. (2003). Similar neural mechansims for emotion-induced memory impairment and enhancement. *Proc Natl Acad Sci USA, 100,* 13123–13124.

Cahill, L., Prins, B., Weber, M., & McGaugh, J. (1994). Beta-adrenergic activation and memory for emotional events. *Nature, 371,* 702–704.

Caplan, A. (2003). Is better best? *Sci Am, 289,* 104–105.

Chai, G., Governale, L., McMahon, A. W., Trinidad, J. P., Staffa, J., & Murphy, D. (2012). Trends of outpatient prescription drug utilization in US children, 2002–2010. *Pediatrics, 133*(3), 375–385.

Chamberlain, S. R., Hampshire, A., Muller, U., Rubia, K., sel Campo, N., Craig, K., ... Sahakian, B. J. (2009). Atomoxetine modulates right inferior frontal activation during inhibitory control: A pharmacological functional magnetic resonance imaging study. *Biol Psychiatry, 65*(7), 550–555.

Chatterjee, A. (2004). Cosmetic neurology: The controversy over enhancing movement, mentation and mood. *Neurology, 63,* 968–974.

Chatterjee, A. (2006). The promise and predicament of cosmetic neurology. *J Med Ethics, 32,* 110–113.

Chatterjee, A. (2007). Cosmetic neurology and cosmetic surgery: Parallels, predictions and challenges. *Camb Q Healthc Ethics, 16,* 129–137.

Chatterjee, A. (2008). Framing pains, pills, and professors. *Expositions, 2*(2), 139–146.

Chatterjee, A. (2009). A medical view of potential adverse effects. *Nature, 457*(7229), 532–533.

Chuah, L. Y. M., Chong, D. L., Chen, A. K., Rekshan, W. R., 3rd, Tan, J.-C., Zheng, H., & Chee, M. W. L. (2009). Donepezil improves episodic memory in young individuals vulnerable to the effects of sleep deprivation. *Sleep, 32*(8), 999.

Cooper, W. O., Habel, L. A., Sox, C. M., Chan, K. A., Arbogast, P. G., Cheetham, T. C., … Ray, W. A. (2011). ADHD Drugs and serious cardiovascular events in children and young adults. *N Engl J Med, 365*(20), 1896–1904.

DeSantis, A., Webb, E., & Noar, S. (2008). Illicit use of prescription ADHD medications on a college campus: A multimethodological approach. *J Am Coll Health, 57*(3), 315–324.

Diller, L. (1996). The run on Ritalin: Attention deficit disorder and stimulant treatment in the 1990s. *Hastings Cent Rep, 26,* 12–14.

Douglas, T. (2008). Moral enhancement. *J Appl Philos, 25*(3), 228–245.

Elliott, C. (2011). Enhancement technologies and the modern self. *J Med Philos, 36*(4), 364–374.

Farah, M. J., Haimm, C., Sankoorikal, G., & Chatterjee, A. (2009). When we enhance cognition with Adderall, do we sacrifice creativity? A preliminary study. *Psychopharmacology, 202,* 541–547.

Farah, M. J., Illes, J., Cook-Deegan, R., Gardner, H., Kandel, E., King, P., … Wolpe, P. (2004). Neurocognitive enhancement: What can we do and what should we do? *Nat Rev Neurosci, 5,* 421–425.

Fillmore, M. T., Kelly, T. H., & Martin, C. A. (2005). Effects of d-amphetamine in human models of information processing and inhibitory control. *Drug Alcohol Depend, 77*(2), 151–159.

FitzGerald, D. B., Crucian, G. P., Mielke, J. B., Shenal, B. V., Burks, D., Womack, K. B., … Heilman, K. M. (2008). Effects of donepezil on verbal memory after semantic processing in healthy older adults. *Cogn Behav Neurol, 21*(2), 57–64.

Forlini, C., & Racine, E. (2009). Disagreements with implications: Diverging discourses on the ethics of non-medical use of methylphenidate for performance enhancement. *BMC Med Ethics, 10*(1), 1–13.

Frank, R. (1987). *Choosing the right pond.* New York, NY: Oxford University Press.

Frank, R., & Cook, P. (1995). *The winner-take-all strategy.* New York, NY: Free Press.

Friedman, R. A. (2006). The changing face of teenage drug abuse—the trend toward prescription drugs. *N Engl J Med, 354*(14), 1448–1450.

Fukayama, F. (2002). *Our posthuman future.* New York, NY: Farrar, Straus & Giroux.

Graf, W. D., Nagel, S. K., Epstein, L. G., Millert, G., Nass, R., & Larriviere, D. (2013). Pediatric neuroenhancement: Ethical, legal, social, and neurodevelopmental implications. *Neurology, 80,* 1251–1260.

Greely, H., Sahakian, B., Harris, J., Kessler, R. C., Gazzaniga, M., Campbell, P., & Farah, M. J. (2008). Towards responsible use of cognitive-enhancing drugs by the healthy. *Nature, 456*(7223), 702–705.

Grön, G., Kirstein, M., Thielscher, A., Riepe, M., & Spitzer, M. (2005). Cholinergic enhancement of episodic memory in healthy young adults. *Psychopharmacology, 182*(1), 170–179.

Habel, L. A. (2011). ADHD medications and risk of serious cardiovascular events in young and middle-aged adults. *JAMA, 306*(24), 2673.

Hall, S. (2003). The quest for a smart pill. *Sci Am, 289,* 54–65.

Hamidovic, A., Dlugos, A., Palmer, A. A., & de Wit, H. (2010). Polymorphisms in dopamine transporter (SLC6A3) are associated with stimulant effects of d-amphetamine: An exploratory pharmacogenetic study using healthy volunteers. *Behav Genet, 40*(2), 255–261.

Hamilton, R., Messing, S., & Chatterjee, A. (2011). Rethinking the thinking cap: Ethics of neural enhancement using noninvasive brain stimulation. *Neurology, 76*(2), 187–193.

Harris, J. (2011). Moral enhancement and freedom. *Bioethics, 25*(2), 102–111.

Heinz, A., Kipke, R., Heimann, H., & Wiesing, U. (2012). Cognitive neuroenhancement: False assumptions in the ethical debate. *J Med Ethics, 38*(6), 372–375.

Horowitz, A. V., & Wakefield, J. C. (2007). *The loss of sadness: How psychiatry transformed normal sorrow into depressive disorder.* New York, NY: Oxford University Press.

Hotze, T. D., Shah, K., Anderson, E. E., & Wynia, M. K. (2011). "Doctor, would you prescribe a pill to help me … ?" A national survey of physicians on using medicine for human enhancement. *Am J Bioeth, 11*(1), 3–13.

Hyman, S. E. (2012). Revolution stalled. *Sci Transl Med, 4*(155), 155cm111.

Ilieva, I., Boland, J., & Farah, M. J. (2013). Objective and subjective cognitive enhancing effects of mixed amphetamine salts in healthy people. *Neuropharmacology, 64,* 496–505.

Insel, T. R. (2010). The challenge of translation in social neuroscience: A review of oxytocin, vasopressin, and affiliative behavior. *Neuron, 65*(6), 768–779.

Insel, T. R. (2012). Next-generation treatments for mental disorders. *Sci Transl Med, 4*(155), 155ps119.

Ioannidis, J. P. A. (2005). Why most published research findings are false. *PLoS Med, 2*(8), e124.

Johnston, L. D., O'Malley, P. M., Bachman, J. G., Schulenberg, J. E., & National Institute on Drug Abuse. (2006). Monitoring the future: National survey results on drug use, 1975–2006. National Institutes of Health, 07-6205.

Johnston, L. D., O'Malley, P. M., Bachman, J. G., Schulenberg, J. E., & National Institute on Drug Abuse. (2013). *Monitoring the future national results on adolescent drug use: Overview of key findings, 2012.* Ann Arbor: Institute for Social Research, The Unversity of Michigan.

Kass, L. (2003, October 16). The pursuit of biohappiness. *Washington Post,* A25.

Kim, Y., Teylan, M. A., Baron, M., Sands, A., Nairn, A. C., & Greengard, P. (2009). Methylphenidate-induced dendritic spine formation and ΔFosB expression in nucleus accumbens. *Proc Natl Acad Sci USA, 106*(8), 2915–2920.

Kirsch, I. (2009). *The emperor's new drugs: Exploding the antidepressant myth.* New York, NY: Random House.

Knutson, B., Wolkowitz, O., Cole, S., Chan, T., Moore, E., Johnson, R., … Reus, V. (1998). Selective alteration of personality and social behavior by serotonergic intervention. *Am J Psychiatry, 155,* 373–379.

Kolber, A. (2011). Neuroethics: Give memory-altering drugs a chance. *Nature, 476,* 275–276.

Lagarde, D., Batejat, D., van Beers, P., Sarafian, D., & Pradella, S. (1995). Interest of modafinil, a new psychostimulant, during a sixty-hour sleep deprivation experiment. *Fund Clin Pharmacol, 9,* 1–9.

Larriviere, D., Williams, M. A., Rizzo, M., Bonnie, R. J., & AAN Ethics, Law and Humanities Committee. (2009).

Responding to requests from adult patients for neuroenhancements: Guidance of the Ethics, Law and Humanities Committee. *Neurology, 73*(17), 1406–1412.

Marchant, N. L., Kamel, F., Echlin, K., Grice, J., Lewis, M., & Rusted, J. M. (2008). Modafinil improves rapid shifts of attention. *Psychopharmacology, 202*(1–3), 487–495.

McBeth, B. D., McNamara, R. M., Ankel, F. K., Mason, E. J., Ling, L. J., Flottemesch, T. J., & Asplin, B. R. (2009). Modafinil and zolpidem use by emergency medicine residents. *Acad Emerg Med, 16*(12), 1311–1317.

McCabe, S. E., Knight, J. R., Teter, C. J., & Wechsler, H. (2005). Non-medical use of prescription stimulants among US college students: Prevalence and correlates from a national survey. *Addiction, 100*(1), 96–106.

Mintzer, M., & Griffiths, R. (2007). A triazolam/amphetamine dose–effect interaction study: Dissociation of effects on memory versus arousal. *Psychopharmacology, 192*(3), 425–440.

National Institute of Mental Health. (2003). *The numbers count: Mental disorders in America.* NIH, 01-4584.

Ott, R., Lenk, C., Miller, N., Buhler, R. N., & Biller-Andorno, N. (2012). Neuroenhancement: Perspectives of Swiss psychiatrists and general practitioners. *Swiss Med Wkly, 142,* w13707.

Outram, S. M., & Racine, E. (2011). Developing public health approaches to cognitive enhancement: An analysis of current reports. *Public Health Ethics, 4*(1), 93–105.

Partridge, B. J., Bell, S. K., Lucke, J. C., Yeates, S., & Hall, W. D. (2011). Smart drugs "as common as coffee": Media hype about neuroenhancement. *PLoS ONE, 6*(11), e28416.

Persson, I., & Savulescu, J. (2008). The perils of cognitive enhancement and the urgent imperative to enhance the moral character of humanity. *J Appl Philos, 25*(3), 162–177.

Pitman, R., Sanders, K., Zusman, R., Healy, A., Cheema, F., Lasko, N., ... Orr, S. (2002). Pilot study of secondary prevention of posttraumatic stress disorder with propranolol. *Biol Psychiatry, 51,* 189–192.

President's Council on Bioethics. (2003). *Beyond therapy: Biotechnology and the pursuit of happiness.* New York, NY: Harper Perennial.

Ravelingien, A., Braeckman, J., Crevits, L., de Ridder, D., & Mortier, E. (2009). "Cosmetic neurology" and the moral complicity argument. *Neuroethics, 2*(3), 151–162.

Repantis, D., Laisney, O., & Heuser, I. (2010). Acetylcholinesterase inhibitors and memantine for neuroenhancement in healthy individuals: A systematic review. *Pharmacol Res, 61*(6), 473–481.

Repantis, D., Schlattmann, P., Laisney, O., & Heuser, I. (2010). Modafinil and methylphenidate for neuroenhancement in healthy individuals: A systematic review. *Pharmacol Res, 62*(3), 187–206.

Russo, M. B., Stetz, M. C., & Stetz, T. A. (2013). Brain enhancement in the military. In A. Chatterjee & M. J. Farah (Eds.), *Neuroethics in practice: Medicine, mind, and society* (pp. 35–45). New York, NY: Oxford University Press.

Rutherford, B. R., Mori, S., Sneed, J. R., Pimontel, M. A., & Roose, S. P. (2012). Contribution of spontaneous improvement to placebo response in depression: A meta-analytic review. *J Psychiatr Res, 46*(6), 697–702.

Sahakian, B., & Morein-Zamir, S. (2007). Professor's little helper. *Nature, 450,* 1157–1159.

Savulescu, J. (2005). New breeds of humans: The moral obligation to enhance. *Reprod Biomed Online, 10*(Suppl. 1), 36–39.

Sax, L. (2003). Who first suggests the diagnosis of attention-deficit/hyperactivity disorder? *Ann Fam Med, 1*(3), 171–174.

Schermer, M., Bolt, I., de Jongh, R., & Olivier, B. (2009). The future of psychopharmacological enhancements: Expectations and policies. *Neuroethics, 2*(2), 75–87.

Singh, I., & Kelleher, K. J. (2013). Brain enhancement and children. In A. Chatterjee & M. J. Farah (Eds.), *Neuroethics in practice: Medicine, mind, and society* (pp. 16–34). New York, NY: Oxford University Press.

Singh, I., Kendall, T., Taylor, C., Mears, A., Hollis, C., Batty, M., & Keenan, S. (2010). Young people's experience of ADHD and stimulant medication: A qualitative study for the NICE guideline. *Child Adolesc Ment Health, 15*(4), 186–192.

Smith, E. M., & Farah, M. J. (2011). Are prescription stimulants "smart pills"? The epidemiology and cognitive neuroscience of prescription stimulant use by normal healthy individuals. *Psychol Bull, 137*(5), 717–741.

Strange, B., Hurlemann, R., & Dolan, R. (2003). An emotion-induced retrograde amnesia in humans is amygdala- and ß-adrenergic-dependent. *Proc Natl Acad Sci USA, 100,* 13626–13631.

Strawn, J. R., & Geracotti, T. D. (2008). Noradrenergic dysfunction and the psychopharmacology of posttraumatic stress disorder. *Depress Anxiety, 25,* 260–271.

Synofzik, M. (2009). Ethically justified, clinically applicable criteria for physician decision-making in psychopharmacological enhancement. *Neuroethics, 2*(2), 89–102.

Talbot, M. (2009, April 27). Brain gain. The underground world of "neuroenhancing" drugs. *The New Yorker.* Retrieved from www.newyorker.com/reporting/2009/04/27/090427fa_fact_talbot

Teter, C. J., McCabe, S. E., Cranford, J. A., Boyd, C. J., & Guthrie, S. K. (2005). Prevalence and motives for illicit use of prescription stimulants in an undergraduate student sample. *J Am Coll Health, 53*(6), 253–262.

Thomas, C. P., Conrad, P., Casler, R., & Goodman, E. (2006). Trends in the use of psychotropic medications among adolescents, 1994 to 2001. *Psychiatr Serv, 57*(1), 63–69.

Tse, W., & Bond, A. (2002). Serotonergic intervention affects both social dominance and afiliative behavior. *Psychopharmacology, 161,* 373–379.

Volkow, N. D., Fowler, J. S., Wang, G.-J., Telang, F., Logan, J., Wong, C., ... Swanson, J. M. (2008). Methylphenidate decreased the amount of glucose needed by the brain to perform a cognitive task. *PLoS ONE, 3*(4), e2017.

Walker-Batson, D., Curtis, S., Natarajan, R., Ford, J., Dronkers, N., Salmeron, E., ... Unwin, D. (2001). A double-blind, placebo-controlled study of the use of amphetamine in the treatment of aphasia. *Stroke, 32,* 2093–2098.

Webb, J. R., Thomas, J. W., & Valasek, M. A. (2010). Contemplating cognitive enhancement in medical students and residents. *Perspect Biol Med, 53*(2), 200–214.

Weber, P., & Lutschg, J. (2002). Methylphenidate treatment. *Pediatr Neurol, 26,* 261–266.

Whitaker, R. (2010). *Anatomy of an epidemic: Magic bullets, psychiatric drugs, and the astonishing rise of mental illness in America.* New York, NY: Crown Publishing Group.

Whiting, E., Chenery, H., Chalk, J., Darnell, R., & Copland, D. (2007). Dexamphetamine enhances explicit new word learning for novel objects. *Int J Neuropsychopharmacol, 10*(6), 805–816.

Wolpe, P. (2002). Treatment, enhancement, and the ethics of neurotherapeutics. *Brain Cogn, 50,* 387–395.

Yesavage, J., Mumenthaler, M., Taylor, J., Friedman, L., O'Hara, R., Sheikh, J., ... Whitehouse, P. (2001). Donezepil and flight simulator performance: Effects on retention of complex skills. *Neurology, 59,* 123–125.

Zaninotto, A. L. C., Bueno, O. F. A., Pradella-Hallinan, M., Tufik, S., Rusted, J., Stough, C., & Pompéia, S. (2009). Acute cognitive effects of donepezil in young, healthy volunteers. *Hum Psychopharmacol, 24*(6), 453–464.

Zeeuws, I., Deroost, N., & Soetens, E. (2010). Effect of an acute d-amphetamine administration on context information memory in healthy volunteers: Evidence from a source memory task. *Hum Psychopharmacol, 25*(4), 326–334.

91 Neuroeconomics

SCOTT A. HUETTEL

ABSTRACT Neuroeconomics is an emerging hybrid discipline that applies cognitive neuroscience methods to understand the brain mechanisms that shape decision making. It has grown out of interactions between neuroscientists and scholars in economics, decision science, and other social sciences—with considerable cross-talk between these diverse areas of research. A core research theme has been to understand the neural basis of subjective value: how value is constructed in the brain, how other decision variables (e.g., probability, time) influence value signals, and how signals from different sorts of rewards are compared in a common currency. Other key topics include risk and uncertainty, temporal discounting, social influences on decision making, and information integration. As neuroeconomics has grown as an academic discipline, there has been parallel growth in neuromarketing as an applied discipline. While neuromarketing shows promise as an application of cognitive neuroscience to commercial problems, it also raises some ethical and practical challenges. And, as neuroeconomics itself matures as a discipline, it will be important to build connections back to other emerging areas within cognitive neuroscience, so that research advances in the study of decision making can influence and be influenced by new theories of brain function.

Over the past decade, there has been remarkable progress in the neuroscientific study of decision making. Much of the impetus for that progress has come from outside cognitive neuroscience, as researchers interested in decision-making behavior apply the tools of neuroscience to long-standing problems. The resulting collision of neuroscience methods and social science questions—often involving collaborative research among researchers in neuroscience, economics, marketing, psychology, and other disciplines—has become known as *neuroeconomics*. Broadly defined, neuroeconomic research seeks to identify the neural mechanisms that underlie decision making, to understand how those mechanisms are shaped by contextual factors, and to create models that use brain structure and function to predict decisions. Despite its relatively brief history, neuroeconomics has developed a coherent research community and patterns of scholarship (Levallois, Clithero, Wouters, Smidts, & Huettel, 2012)—leading to descriptions of it as a "nascent" discipline with a character distinct from both neuroscience and economics.

No single event marks the birth of neuroeconomics. Economic research on decision making can be readily traced back to the seventeenth and eighteenth centuries, when scholars interested in improving their odds while gambling developed statistical methods for understanding probability and value (Bernoulli, 1738). Such research led to the development of formal, normative models of decision making—often referred to as *rational choice models*—that came to dominate economic scholarship. But, beginning in the mid-twentieth century, new research incorporated cognitive bounds on rationality, the imperfect knowledge of real agents in markets, and psychological factors like heuristics and biases (Kahneman & Tversky, 1979; Simon, 1955; Tversky & Kahneman, 1974). These new approaches led to recognition that actual human decision-making behavior often diverged from the predictions of formal models, in turn sparking the new discipline of behavioral economics. Concurrently, the burgeoning field of neuroscience now considered neural processes that contributed to decision making. In particular, studies of the dopamine system revealed the central role for that neurotransmitter in motivation and value (Berridge & Robinson, 1998; Olds & Milner, 1954). The use of formal computational models later showed that the firing of dopamine neurons was reference-dependent (i.e., firing rates tracked deviations in reward from an expected reference point), which connected basic neurophysiology to a key assumption of behavioral economics (Schultz, Dayan, & Montague, 1997). And, in 1999, physiologists demonstrated that neurons in parietal cortex were sensitive to the expected value of potential actions (Platt & Glimcher, 1999)—ushering in neuroeconomic studies of value signals throughout the brain.

Today, neuroeconomic research spans a wide range of experimental techniques and modeling approaches (Platt & Huettel, 2008; Rangel, Camerer, & Montague, 2008). Studies of human and nonhuman animals are each well-represented, with considerable intellectual crossover between species. Functional magnetic resonance imaging (fMRI) remains the most common technique, but seminal research has been conducted using single-unit recording (Padoa-Schioppa & Assad, 2006), electroencephalography (Gehring & Willoughby, 2002), transcranial magnetic stimulation (Knoch, Pascual-Leone, Meyer, Treyer, & Fehr, 2006), hormone administration (Kosfeld, Heinrichs, Zak, Fischbacher, & Fehr, 2005), lesion analysis (Camille et al., 2004), magnetoencephalography (Hunt et al., 2012), and other approaches. Moreover, concepts from economics (and other social sciences) shape neuroeconomic modeling, as seen in topics like uncertainty and temporal discounting. Because of this breadth of methodological and theoretical span, relatively few individuals have been cross-trained within neuroeconomics itself—meaning that research disproportionately involves collaborations. Examination of the social network of neuroeconomics, as identified through a comprehensive survey of researchers in the field, revealed a striking feature of that collaboration: the research groups that were relatively balanced between neuroscientists and social scientists tended to occupy more central and influential places in the network (Levallois et al., 2012).

Research methods in neuroeconomics

Neuroeconomic research shares many methodological similarities with cognitive neuroscience research. Key goals include mapping distinct functions onto brain regions and systems, describing how systems in the brain interact to shape complex functions, and understanding how brain function leads to behavior. Yet there are clear differences in both methods and experimental conventions—and those differences carry implications for the inferences that can be drawn from neuroscience data.

Broadly considered, neuroeconomic experiments can be described as involving three basic steps. First, researchers simplify a real-world decision problem into something amenable for laboratory manipulation, such as choices between probabilistic outcomes (or "gambles") and certain outcomes. Such a task fails to capture the complexity of real-world risky decisions like financial investments—and in many ways the difference between laboratory paradigms and real-world behavior is greater for research in neuroeconomics than for cognitive neuroscience, generally. Second, participants make decisions that involve meaningful incentives (e.g., monetary payments) while brain function is measured or manipulated. Third, analysis methods seek to identify some brain signal associated with a decision variable that predicts choice. Those decision variables can be objective properties of a stimulus (e.g., magnitude of reward) or estimated internal states like subjective value (Levy & Glimcher, 2012).

Experiments in neuroeconomics also typically adopt methodological conventions from economics, which can differ from those in psychology and cognitive neuroscience. The use of *incentive-compatible* tasks is a hallmark of research in economics, which often uses complex procedures to ensure that participants report their true preferences (Hertwig & Ortmann, 2001). Even though direct comparisons have indicated that people often report their preferences accurately in the absence of incentives (Carlsson & Martinsson, 2001), incentive compatibility remains the cardinal constraint upon experimental design. Additionally, most neuroeconomic experiments avoid any substantive *deception* of research participants; this convention provides the clearest distinction between research in neuroeconomics and in the cognate discipline of social neuroscience. The use of deception, it is argued, could change experimental participants' decisions (e.g., through suspicions that a partner in a game might be an experimental confederate). Finally, neuroeconomic experiments seek to minimize external factors that motivate decisions. As an example, if an experimenter observes participants while they choose whether to donate money to a charity, then those donations might be influenced by factors other than altruism (e.g., the desire to please the experimenter). Implementing these conventions—and those of economic research, more generally—does restrict the range of paradigms available to neuroeconomic researchers. But that cost is thought to be outweighed by the benefits of increased participant motivation and increased fidelity of decision making.

Key research topics

Like cognitive neuroscience more generally, neuroeconomics cannot be readily partitioned into a small set of discrete research topics. Even seemingly well-circumscribed topics—say, the neural basis for strategic interaction in games—involve complex decision-making behavior that resists simple categorization. Nevertheless, there are several large themes that have pervaded neuroeconomic research since its outset, and thus provide illustrative examples of progress in the field.

VALUE The most central topic for neuroeconomic research has been subjective *value*. The earliest studies presented participants with highly desirable rewards—such as juice for monkeys or money for humans—and then measured brain signals associated with the anticipation and receipt of those rewards. The dopamine system provided a natural target for such studies. Dopamine neurons in the ventral tegmental area track not simply rewards themselves, but *information* about rewards (Schultz, 2004; Schultz et al., 1997). Specifically, these neurons increase their firing rate when a reward is greater than expected or when a cue signals an increase in the future expectation of reward, but decrease their firing rate when a reward is less than expected or when a cue signals a diminishment of future rewards. This signal, which tracks the difference between what was expected and what was received, is now called the *reward-prediction error.* Neuroimaging studies examined brain responses in key projection targets of those dopamine neurons, notably the ventral striatum and the orbitofrontal cortex, revealing that both the anticipation and receipt of rewards reliably modulate activation in those regions. Many subsequent studies have used variants of a monetary incentive delay task—which involves playing a simple reaction-time game for the opportunity to win (or lose) money—to evoke robust activation in the ventral striatum (Knutson, Adams, Fong, & Hommer, 2001; Knutson & Cooper, 2005). Of note, there now exists good meta-analytic specificity for reward-related activation in these dopamine target regions. Activation in the ventral striatum, for example, provides a reliable signal for reward-related processing (Ariely & Berns, 2010), as evident in reverse-inference maps generated by the meta-analytic tool NeuroSynth (neurosynth.org).

The brain's ability to make decisions about an arbitrarily large set of potential rewards poses challenges for understanding decision making. Early work speculated that decision making requires the conversion of different rewards into a common currency (Montague & Berns, 2002), a modality-independent scale of subjective value (cf. economic "utility") on which any potential reward could be valued. Evidence for such a common-currency signal came both from single-unit (Padoa-Schioppa & Assad, 2006) and functional neuroimaging studies (Plassmann, O'Doherty, & Rangel, 2007), each of which revealed that the orbitofrontal cortex (OFC) tracked the willingness to pay for some good (e.g., juice, food). More recently, there have been direct tests of the common-currency hypothesis that used brain responses in medial OFC (sometimes called ventromedial prefrontal cortex) to predict subsequent value-based decisions (Levy & Glimcher, 2011; Smith et al., 2010). Collectively, these results suggest a reconception of the earlier links between OFC and behavioral control, such that processing in OFC reflects value signals (including emotional value) rather than control signals that inhibit undesirable behavior (Winecoff et al., 2013).

UNCERTAINTY Understanding how the brain processes uncertainty represents a second major thrust of neuroeconomics. The most commonly considered form of uncertainty is *risk,* as when a decision maker chooses between one option that leads to a certain reward and another option with known probabilities of better or worse rewards. Influential early work identified risk-related signals in the insular cortex, perhaps reflecting the aversive properties of risky options (Paulus, Rogalsky, Simmons, Feinstein, & Stein, 2003). More recent conceptions indicate that risk processing occurs in a distributed fashion, with regions of dorsolateral and dorsomedial prefrontal cortex (dlPFC, dmPFC) also making important contributions (Mohr, Biele, & Heekeren, 2010). Another form of uncertainty, *ambiguity,* describes decisions when the probabilities of rewards are unknown, as if one were playing a raffle with an unknown number of tickets (Ellsberg, 1961). Decisions involving ambiguity evoke neural processing distinct from decisions involving only risk (Hsu, Bhatt, Adolphs, Tranel, & Camerer, 2005; Huettel, Stowe, Gordon, Warner, & Platt, 2006); in particular, regions in posterior lateral prefrontal cortex are engaged when people face an uncertain outcome and they know they do not have all of the available information (Bach, Seymour, & Dolan, 2009). Like other decision variables, uncertainty influences the subjective value of a potential outcome (e.g., increasing uncertainty leads to decreased value) and the brain systems that track that value (Levy, Snell, Nelson, Rustichini, & Glimcher, 2010).

TEMPORAL DISCOUNTING Everyday decisions often involve a consideration of future consequences. When deciding to save for retirement, people forego smaller rewards now in order to obtain larger rewards at a later date. Such a trade-off lies at the heart of neuroeconomic research on *temporal discounting.* An early and very influential study (McClure, Laibson, Loewenstein, & Cohen, 2004) posited that intertemporal decisions arise from the competition between two competing neural systems: a relatively impulsive system involving value-related regions (e.g., ventral striatum and OFC) and a relatively patient system involving control-related regions (e.g., dlPFC). Such a result was consistent with the long-standing tradition in decision science of dual-system models (Kahneman, 2011). Subsequent research led to a more nuanced perspective on temporal discounting. The regions previously associated with impulsive choice map well to core regions in the brain's value network—and these regions were shown to not delay itself, but the subjective value of rewards regardless of delay (Kable & Glimcher, 2007). Yet other evidence supports links between the dlPFC and patient intertemporal choices; application of low-frequency repetitive transcranial magnetic stimulation to the left dlPFC was show to increase the likelihood that participants choose smaller, sooner rewards (Figner et al., 2010). There is a growing consensus, accordingly, that the brain contains a core system for processing value—not two interacting systems—but that factors like the exertion of self-control can shape processing in that value system.

Most research on temporal discounting has examined decisions by human participants about monetary rewards. Temporal discounting is ubiquitous, however, across species and other sorts of rewards (Hayden & Platt, 2007; Monterosso, Piray, & Luo, 2012). Varying the reward modality—as when comparing human participants' discounting for money and juice—tends to alter the rate of discounting but not the basic form nor the underlying neural mechanisms. Moreover, experimental manipulations that call attention to the needs of one's future self can minimize the rate of temporal discounting, potentially through engagement of processes of self-cognition (Loewenstein, 1988; Weber et al., 2007). There exists at least one notable exception to temporal discounting: when people make decisions about aversive rewards (e.g., electric shocks), they are often willing to receive a larger shock sooner rather than wait for a smaller shock. This effect has been linked to the psychological state of dread evoked by the anticipation of electrical shock, such that the actual waiting period itself carries negative value (Berns et al., 2006).

Social Influences Decisions are not always made by individuals in isolation. A variety of social influences can shape choices: benefits to others, peer pressure, consequences for our self-identity, and strategic considerations associated with social competition. Borrowing from economics, early neuroeconomic studies of social interactions examined two-player games like the prisoner's dilemma and trust games, finding that cooperation and signals of intended cooperation engage reward-related brain regions (King-Casas et al., 2005; Rilling et al., 2002). Conversely, maladaptive social behavior (e.g., cheating) reliably modulates regions associated with aversive responses, like insular cortex (Sanfey, Rilling, Aronson, Nystrom, & Cohen, 2003). In addition, stimuli that have value because of their social meaning (e.g., attractive faces, bodies) have been used as rewards in a wide variety of experiments.

While it is clear social influences do shape many decision processes in a manner similar to other rewards, an important area of research has been to identify ways in which social decisions differ from other sorts of decisions. One intriguing tool has been the administration of the social hormone oxytocin, which plays a role in affiliation and social bonding. When people or nonhuman primates receive this hormone in advance of playing economic games (in double-blind, placebo-controlled studies), they become more likely to trust others (Kosfeld et al., 2005), particularly ingroup members (De Dreu et al., 2010). Another approach has been to identify patterns of brain function that seem to be uniquely social; that is, engaged under conditions of social interaction, but not under nonsocial decisions with similar statistical properties. Several such candidates for specifically social decision making have been identified recently. Within the dmPFC, single-unit studies indicate distinct regions supporting learning from social cues and from nonsocial rewards, with both regions feeding into ventromedial prefrontal cortex and presumably influencing its representations of value (Behrens, Hunt, Woolrich, & Rushworth, 2008). Another candidate is the temporoparietal junction, a region that has been frequently associated with social cognition (Saxe & Kanwisher, 2003) and whose structural (Morishima, Schunk, Bruhin, Ruff, & Fehr, 2012) and functional (Tankersley, Stowe, & Huettel, 2007) properties have been linked to prosocial behaviors like altruism. Information contained within the temporoparietal junction, as measured from fMRI data using machine-learning techniques, has been shown to predict choices specifically during social decision making tasks (Carter, Bowling, Reeck, & Huettel, 2012). This specificity may reflect the role for this region in establishing a social context for behavior (Carter & Huettel, 2013).

Information Integration As computational models of decision making become better integrated with neuroscience methods, there has been increasing interest in how the brain balances potentially competing goals, such as learning about the properties of an uncertain environment (i.e., *exploring*) and obtaining maximal rewards from that environment (i.e., *exploiting*). Direct comparisons of exploration/exploitation have suggested that exploratory processes depend on the anterior dlPFC (Daw, O'Doherty, Dayan, Seymour, & Dolan, 2006), a result that dovetails nicely with cognitive neuroscience studies of higher-order executive function (Koechlin & Hyafil, 2007). More broadly, the application of computational models has been useful for understanding information integration—often through application of drift-diffusion models (Ratcliff & McKoon, 2008)—both in patterns of neuronal activity (Roitman & Shadlen, 2002) and in measures of eye-tracking and of choice behavior (Krajbich, Armel, & Rangel, 2010; Krajbich, Lu, Camerer, & Rangel, 2012). This line of research holds promise for connecting basic properties of neuronal function to observable choice behavior.

Neuromarketing

As neuroeconomics has progressed, there has been concomitant interest in applying neuroscience methods to better understand consumer choices, both through academic marketing research and through industry-sponsored *neuromarketing* (Venkatraman, Clithero, Fitzsimons, & Huettel, 2012). Like many other applications of neuroscience (e.g., lie detection in legal cases), the idea of *neuro*marketing seems to raise ethical issues—a concern sparked by breathless news coverage in the popular media. Many of those ethical issues rest on misinformation about the goals of neuromarketing, especially since all neuromarketing is necessarily indirect: neural data is collected from a relatively small focus group to generate recommendations for the broader population. Direct use of neuroscience to shape the brain function of consumers remains closer to science fiction than science practice.

At its core, neuromarketing seeks to gain information about consumer behavior that is not readily available using other means. The outputs of neuromarketing research are no different from those obtained using standard marketing approaches (e.g., redesigned packaging, television commercials)—what differs is simply the data being used to make those recommendations. All major techniques within human cognitive neuroscience are being used within current neuromarketing practice, although the relative prevalence of those techniques differs from that in the cognitive neuroscience community. There is proportionally more use of scalp electroencephalography, measurement of peripheral physiology (e.g., heart rate, skin conductance), and eye-tracking—in large part because these techniques have better temporal resolution and much lower capital costs than fMRI or other neuroimaging techniques.

Neuromarketing can, in principle, provide information about the decision process that would be difficult to obtain using the standard tools of marketing. A common claim from neuromarketing firms is that they can assess covert information that individuals are unable or unwilling to report. Consider the marketing goal of optimizing a 30-second television commercial for a given audience. Traditional techniques involve collecting individuals' reactions to that commercial after its presentation—perhaps many minutes later, if the commercial were embedded in a mock television show. That delay poses challenges for understanding which specific images or messages drive consumer reactions. By collecting neuroscience data during the presentation of the commercial, neuromarketers can track some biomarker that predicts attitudes or subsequent intent to purchase, without requiring any overt response from the individuals surveyed. Moreover, cognitive neuroscience data can provide some information not accessible through traditional methods; for example, signals associated with aversion or disgust that may not be reported in traditional focus groups. There are also some intriguing examples of how measures collected using neuroscience can outperform self-reports in predicting large-scale marketing success (Berns & Moore, 2012; Falk, Berkman, & Lieberman, 2012).

Yet, despite the promise of neuromarketing, there are clear limitations. The most striking difference between academic neuroeconomics and industry neuromarketing can be seen in methodological transparency. Whereas academic researchers describe their methods in detail in publications—and freely adopt new paradigms and analysis techniques developed in their colleagues' laboratories—neuromarketing firms are incentivized to be proprietary. Such firms exist in a very competitive market with considerable prior art; for example, every year, hundreds of peer-reviewed publications apply cognitive neuroscience methods to questions of decision making. This pushes firms to conceal their own approaches, both so that they can obtain and retain a competitive advantage and so that their claims about capabilities are difficult to validate independently. Nor is transparency desirable to the customers of neuromarketing firms, since they likewise want to obtain a competitive advantage through these methods —and they often wish to avoid the undesirable connotations that could accompany neuromarketing.

Without external evaluation of neuromarketing practices, it can be difficult to assess whether a given firm is appropriately rigorous in its methods and appropriately cautious in its interpretations. Skepticism is justifiable, in many cases, given the breadth of the claims of neuromarketing proponents; these claims often go well beyond what would be considered practical, even for cutting-edge research in cognitive neuroscience. There are, however, some laudable attempts to subject neuromarketing claims to a form of peer review. As an example, the Advertising Research Foundation and its industry partners have commissioned independent evaluations of neuromarketing by experts in cognitive neuroscience. These evaluations have included assessments of neuromarketing reports for quality of the methods and whether claims are appropriately derived from the data, as well as recent evaluations of neuromarketing data themselves. As neuromarketing matures, a key contributor to its success will be reduced barriers between academic and industry research, so that new findings flow freely in both directions. This model would be more akin to that in other fields where there is substantial industry-funded research and development (e.g., computer science), and thus would reflect an important new frontier for cognitive neuroscience.

Challenges and future directions

Neuroeconomics has grown rapidly over the past decade. Through continual interactions among scholars from a variety of fields—neuroscience, psychology, economics, and business—it has generated both a considerable breadth of research and findings and a deep, integrated research community. Such breadth and depth are rare among emerging disciplines; for example, analysis of citation patterns indicates that neuroeconomic research covers a broader range of the scientific landscape than other, similar research areas (Levallois et al., 2012). Yet rapid growth carries costs—and how neuroeconomics faces those challenges will determine its future viability and direction.

A primary challenge lies in the integration with cognitive neuroscience. The focus on decision variables and relatively simple conceptions of processing has led to major advances so far, but progress will be accelerated as concepts from cognitive neuroscience enter the neuroeconomics lexicon. One developing connection comes from work on self-control, for which high-profile studies have identified links between prefrontal control regions and value signals (Hare, Camerer, & Rangel, 2009). As paradigms within neuroeconomics become increasingly complex, then the growing specification of control systems within cognitive neuroscience could provide new insights. For example, hierarchical models of prefrontal cortex function (Badre & D'Esposito, 2009) could provide an overarching theory that integrates control processes and decision variables (Coutlee & Huettel, 2012). The other direction of integration—neuroeconomics feeding back into cognitive neuroscience—seems more clearly successful, in that manipulations of reward properties have become commonplace within cognitive neuroscience. Broadly considered, this may be part of an ongoing shift away from a focused *neuroeconomics* to a more general *decision neuroscience* (Levallois et al., 2012).

A second challenge is much more practical: Can neuroeconomics develop into a truly integrated discipline, given the constraints of training and hiring within academic science? As mentioned earlier, most current interdisciplinarity in neuroeconomics comes from collaboration—as, say, an economist and cognitive neuroscientist collaborate on an experiment. Increasing numbers of junior scholars are being trained in neuroeconomics laboratories, often gaining both expertise in cognitive neuroscience methods as well as a theoretical or experiment framework from economics or another decision science. Not yet clear is whether such cross-trained individuals will be advantaged or disadvantaged as they take positions and seek funding from more traditionally organized institutions. These issues recapitulate some of the same structural issues that arose for cognitive neuroscience as it grew as a discipline. One encouraging development has been the increased willingness of new disciplines to hire and promote individuals doing cognitive neuroscience research. In particular, some marketing programs within business schools have been developing neuroscience to complement their existing focus on consumer behavior; in an irony, the growth of neuroeconomic research within marketing has been generally greater than that within economics.

Finally, there are ongoing debates about the value of neuroscience for questions specific to the social sciences, particularly economics (Gul & Pesendorfer, 2008). Some of those debates arise from overly strict bounds on the nature of social science inquiry, along with an underestimation of the contributions of neuroscience to a larger research enterprise (Clithero, Tankersley, & Huettel, 2008). Nevertheless, there are real and substantive criticisms whose concerns have not abated even as neuroeconomics makes clear scientific advances. Answering those criticisms may require some direct case studies demonstrating how some key theory in economics can be tested using neuroscience data, at least in part. Such direct tests may not be obvious when they are published, but instead will gain in acceptance over time (cf. the rise of behavioral economics). But, more likely, the increased communication between disciplines promoted by neuroeconomics will lead to clearer research questions—and, in turn, an improved understanding of which questions neuroscience can address and which it cannot within the social sciences.

ACKNOWLEDGMENTS The author thanks Adina Roskies, Walter Sinnott-Armstrong, and Vinod Venkatraman for comments on this chapter.

REFERENCES

Ariely, D., & Berns, G. S. (2010). Neuromarketing: The hope and hype of neuroimaging in business. *Nat Rev Neurosci, 11*(4), 284–292.

Bach, D. R., Seymour, B., & Dolan, R. J. (2009). Neural activity associated with the passive prediction of ambiguity and risk for aversive events. *J Neurosci, 29*(6), 1648–1656.

Badre, D., & D'Esposito, M. (2009). Is the rostro-caudal axis of the frontal lobe hierarchical? *Nat Rev Neurosci, 10*(9), 659–669.

Behrens, T. E., Hunt, L. T., Woolrich, M. W., & Rushworth, M. F. (2008). Associative learning of social value. *Nature, 456*(7219), 245–249.

Bernoulli, D. (1738). Specimen theoriae novae de mensura sortis. *Commentarii Academiae Scientarum Imperialis Petropolitanae, 5*, 175–192.

Berns, G. S., Chappelow, J., Cekic, M., Zink, C. F., Pagnoni, G., & Martin-Skurski, M. E. (2006). Neurobiological substrates of dread. *Science, 312*(5774), 754–758.

Berns, G. S., & Moore, S. E. (2012). A neural predictor of cultural popularity. *J Consum Psychol, 22*(1), 154–160.

Berridge, K. C., & Robinson, T. E. (1998). What is the role of dopamine in reward: Hedonic impact, reward learning, or incentive salience? *Brain Res Reviews, 28*(3), 309–369.

Camille, N., Coricelli, G., Sallet, J., Pradat-Diehl, P., Duhamel, J. R., & Sirigu, A. (2004). The involvement of the orbitofrontal cortex in the experience of regret. *Science, 304*, 1167–1170.

Carlsson, F., & Martinsson, P. (2001). Do hypothetical and actual marginal willingness to pay differ in choice experiments? Application to the valuation of the environment. *J Environ Econ Manag, 41*(2), 179–192.

Carter, R. M., Bowling, D. L., Reeck, C., & Huettel, S. A. (2012). A distinct role of the temporal-parietal junction in predicting socially guided decisions. *Science, 337*(6090), 109–111.

Carter, R. M., & Huettel, S. A. (2013). A nexus model of the temporal-parietal junction. *Trends Cogn Sci, 17*(7), 328–336.

Clithero, J. A., Tankersley, D., & Huettel, S. A. (2008). Foundations of neuroeconomics: From philosophy to practice. *PLoS Biol, 6*(11), e298.

Coutlee, C. G., & Huettel, S. A. (2012). The functional neuroanatomy of decision making: Prefrontal control of thought and action. *Brain Res, 1428*, 3–12.

Daw, N. D., O'Doherty, J. P., Dayan, P., Seymour, B., & Dolan, R. J. (2006). Cortical substrates for exploratory decisions in humans. *Nature, 441*, 876–879.

De Dreu, C. K. W., Greer, L. L., Handgraaf, M. J. J., Shalvi, S., van Kleef, G. A., Baas, M., … Feith, S. W. W. (2010). The neuropeptide oxytocin regulates parochial altruism in intergroup conflict among humans. *Science, 328*(5984), 1408–1411.

Ellsberg, D. (1961). Risk, ambiguity, and the savage axioms. *Q J Econ, 75*, 643–669.

Falk, E. B., Berkman, E. T., & Lieberman, M. D. (2012). From neural responses to population behavior: Neural focus group predicts population-level media effects. *Psychol Sci, 23*(5), 439–445.

Figner, B., Knoch, D., Johnson, E. J., Krosch, A. R., Lisanby, S. H., Fehr, E., & Weber, E. U. (2010). Lateral prefrontal cortex and self-control in intertemporal choice. *Nat Neurosci, 13*(5), 538–539.

Gehring, W. J., & Willoughby, A. R. (2002). The medial frontal cortex and the rapid processing of monetary gains and losses. *Science, 295*(5563), 2279–2282.

Gul, F., & Pesendorfer, W. (2008). The case for mindless economics. In A. Caplin & A. Schotter (Eds.), *The foundations of positive and normative economics* (pp. 3–42). Oxford, UK: Oxford University Press.

Hare, T. A., Camerer, C. F., & Rangel, A. (2009). Self-control in decision-making involves modulation of the vmPFC valuation system. *Science, 324*(5927), 646–648.

Hayden, B. Y., & Platt, M. L. (2007). Temporal discounting predicts risk sensitivity in rhesus macaques. *Curr Biol, 17*(1), 49–53.

Hertwig, R., & Ortmann, A. (2001). Experimental practices in economics: A methodological challenge for psychologists? *Behav Brain Sci, 24*(3), 383–403; discussion 403–351.

Hsu, M., Bhatt, M., Adolphs, R., Tranel, D., & Camerer, C. F. (2005). Neural systems responding to degrees of uncertainty in human decision-making. *Science, 310*(5754), 1680–1683.

Huettel, S. A., Stowe, C. J., Gordon, E. M., Warner, B. T., & Platt, M. L. (2006). Neural signatures of economic preferences for risk and ambiguity. *Neuron, 49*(5), 765–775.

Hunt, L. T., Kolling, N., Soltani, A., Woolrich, M. W., Rushworth, M. F., & Behrens, T. E. (2012). Mechanisms underlying cortical activity during value-guided choice. *Nat Neurosci, 15*(3), 470–476.

Kable, J. W., & Glimcher, P. W. (2007). The neural correlates of subjective value during intertemporal choice. *Nat Neurosci, 10*(12), 1625–1633.

Kahneman, D. (2011). *Thinking, fast and slow.* New York, NY: Farrar, Straus & Giroux.

Kahneman, D., & Tversky, A. (1979). Prospect theory: An analysis of decision under risk. *Econometrica, 47*(2), 263–291.

King-Casas, B., Tomlin, D., Anen, C., Camerer, C. F., Quartz, S. R., & Montague, P. R. (2005). Getting to know you: Reputation and trust in a two-person economic exchange. *Science, 308*(5718), 78–83.

Knoch, D., Pascual-Leone, A., Meyer, K., Treyer, V., & Fehr, E. (2006). Diminishing reciprocal fairness by disrupting the right prefrontal cortex. *Science, 314*(5800), 829–832.

Knutson, B., Adams, C. M., Fong, G. W., & Hommer, D. (2001). Anticipation of increasing monetary reward selectively recruits nucleus accumbens. *J Neurosci, 21*(16), RC159.

Knutson, B., & Cooper, J. C. (2005). Functional magnetic resonance imaging of reward prediction. *Curr Opin Neurol, 18*(4), 411–417.

Koechlin, E., & Hyafil, A. (2007). Anterior prefrontal function and the limits of human decision-making. *Science, 318*(5850), 594–598.

Kosfeld, M., Heinrichs, M., Zak, P. J., Fischbacher, U., & Fehr, E. (2005). Oxytocin increases trust in humans. *Nature, 435*(7042), 673–676.

Krajbich, I., Armel, C., & Rangel, A. (2010). Visual fixations and the computation and comparison of value in simple choice. *Nat Neurosci, 13*(10), 1292–1298.

Krajbich, I., Lu, D., Camerer, C., & Rangel, A. (2012). The attentional drift-diffusion model extends to simple purchasing decisions. *Front Psychol, 3*, 193.

Levallois, C., Clithero, J. A., Wouters, P., Smidts, A., & Huettel, S. A. (2012). Translating upwards: Linking the neural and social sciences via neuroeconomics. *Nat Rev Neurosci, 13*(11), 789–797.

Levy, D. J., & Glimcher, P. W. (2011). Comparing apples and oranges: Using reward-specific and reward-general subjective value representation in the brain. *J Neurosci, 31*(41), 14693–14707.

Levy, D. J., & Glimcher, P. W. (2012). The root of all value: A neural common currency for choice. *Curr Opin Neurobiol, 22*(6), 1027–1038.

Levy, I., Snell, J., Nelson, A. J., Rustichini, A., & Glimcher, P. W. (2010). Neural representation of subjective value under risk and ambiguity. *J Neurophysiol, 103*(2), 1036–1047.

Loewenstein, G. F. (1988). Frames of mind in intertemporal choice. *Manage Sci, 34*(2), 200–214.

McClure, S. M., Laibson, D. I., Loewenstein, G., & Cohen, J. D. (2004). Separate neural systems value immediate and delayed monetary rewards. *Science, 306*(5695), 503–507.

Mohr, P. N., Biele, G., & Heekeren, H. R. (2010). Neural processing of risk. *J Neurosci, 30*(19), 6613–6619.

Montague, P. R., & Berns, G. S. (2002). Neural economics and the biological substrates of valuation. *Neuron, 36*(2), 265–284.

Monterosso, J., Piray, P., & Luo, S. (2012). Neuroeconomics and the study of addiction. *Biol Psychiatry, 72*(2), 107–112.

Morishima, Y., Schunk, D., Bruhin, A., Ruff, C. C., & Fehr, E. (2012). Linking brain structure and activation in temporoparietal junction to explain the neurobiology of human altruism. *Neuron, 75*(1), 73–79.

Olds, J., & Milner, P. (1954). Positive reinforcement produced by electrical stimulation of septal area and other regions of rat brain. *J Comp Physiol Psychol, 47*(6), 419–427.

Padoa-Schioppa, C., & Assad, J. A. (2006). Neurons in the orbitofrontal cortex encode economic value. *Nature, 441*(7090), 223–226.

Paulus, M. P., Rogalsky, C., Simmons, A., Feinstein, J. S., & Stein, M. B. (2003). Increased activation in the right insula during risk-taking decision making is related to harm avoidance and neuroticism. *NeuroImage, 19*(4), 1439–1448.

Plassmann, H., O'Doherty, J., & Rangel, A. (2007). Orbitofrontal cortex encodes willingness to pay in everyday economic transactions. *J Neurosci, 27*(37), 9984–9988.

Platt, M. L., & Glimcher, P. W. (1999). Neural correlates of decision variables in parietal cortex. *Nature, 400*(6741), 233–238.

Platt, M. L., & Huettel, S. A. (2008). Risky business: The neuroeconomics of decision making under uncertainty. *Nat Neurosci, 11*(4), 398–403.

Rangel, A., Camerer, C., & Montague, P. R. (2008). A framework for studying the neurobiology of value-based decision making. *Nat Rev Neurosci, 9*(7), 545–556.

Ratcliff, R., & McKoon, G. (2008). The diffusion decision model: Theory and data for two-choice decision tasks. *Neural Comput, 20*(4), 873–922.

Rilling, J., Gutman, D., Zeh, T., Pagnoni, G., Berns, G., & Kilts, C. (2002). A neural basis for social cooperation. *Neuron, 35*(2), 395–405.

Roitman, J. D., & Shadlen, M. N. (2002). Response of neurons in the lateral intraparietal area during a combined visual discrimination reaction time task. *J Neurosci, 22*(21), 9475–9489.

Sanfey, A. G., Rilling, J. K., Aronson, J. A., Nystrom, L. E., & Cohen, J. D. (2003). The neural basis of economic decision-making in the Ultimatum Game. *Science, 300*(5626), 1755–1758.

Saxe, R., & Kanwisher, N. (2003). People thinking about thinking people. The role of the temporo-parietal junction in "theory of mind." *NeuroImage, 19*(4), 1835–1842.

Schultz, W. (2004). Neural coding of basic reward terms of animal learning theory, game theory, microeconomics and behavioural ecology. *Curr Opin Neurobiol, 14*(2), 139–147.

Schultz, W., Dayan, P., & Montague, P. R. (1997). A neural substrate of prediction and reward. *Science, 275*(5306), 1593–1599.

Simon, H. A. (1955). A behavioral model of rational choice. *Q J Econ, 69*(1), 99–118.

Smith, D. V., Hayden, B. Y., Truong, T. K., Song, A. W., Platt, M. L., & Huettel, S. A. (2010). Distinct value signals in anterior and posterior ventromedial prefrontal cortex. *J Neurosci, 30*(7), 2490–2495.

Tankersley, D., Stowe, C. J., & Huettel, S. A. (2007). Altruism is associated with an increased neural response to agency. *Nat Neurosci, 10*(2), 150–151.

Tversky, A., & Kahneman, D. (1974). Judgment under uncertainty: Heuristics and biases. *Science, 185,* 1124–1131.

Venkatraman, V., Clithero, J. A., Fitzsimons, G. J., & Huettel, S. A. (2012). New scanner data for brand marketers: How neuroscience can help better understand differences in brand preferences. *J Consum Psychol, 22*(1), 143–153.

Weber, E. U., Johnson, E. J., Milch, K. F., Chang, H., Brodscholl, J. C., & Goldstein, D. G. (2007). Asymmetric discounting in intertemporal choice: A query-theory account. *Psychol Sci, 18*(6), 516–523.

Winecoff, A., Clithero, J. A., Carter, R. M., Bergman, S. R., Wang, L. H., & Huettel, S. A. (2013). Ventromedial prefrontal cortex encodes emotional value. *J Neurosci, 33*(27), 11032–11039.

92 The Significance of Cognitive Neuroscience: Findings, Applications, and Challenges

ANAT ARZI, SNIGDHA BANERJEE, JUSTIN C. COX, DEAN D'SOUZA, FELIPE DE BRIGARD, BRADLEY B. DOLL, JACQUELINE FAIRLEY, STEPHEN M. FLEMING, SIBYLLE C. HERHOLZ, DANIELLE R. KING, LAURA A. LIBBY, JOHN C. MYERS, MAITAL NETA, DAVID PITCHER, JONATHAN D. POWER, OLGA RASS, MAUREEN RITCHEY, EDUARDO ROSALES JUBAL, ASHLEY ROYSTON, DYLAN D. WAGNER, WEI-CHUN WANG, JILL D. WARING, JAMAL WILLIAMS, AND SUZANNE WOOD

The field of cognitive neuroscience has been lauded for its potential to integrate disparate domains such as neurobiology and philosophy, and to provide new insights into human behavior. Although initially cognitive neuroscience was seen as perhaps no more than the intersection between neuropsychology and cognitive science, it has since become an independent area of inquiry, with a membership easily rivaling those of its parent disciplines. Given its origin as a multidisciplinary science, it should come as little surprise that the current state of the field is best characterized by expansion into areas of research that were formerly considered off-limits to neuroscientific study. For instance, since the early 2000s, topics such as emotion, decision making, and social cognition have risen to prominence both in terms of the number of papers published and also in terms of the number of pages they occupy in the current edition of this book. This latest edition also includes a new section on "neuroscience and society," reflecting the increasing role of cognitive neuroscience research in informing social policy.

However, the story of cognitive neuroscience's success isn't solely one of increasing breadth, but also one of rapid adoption of new methods and technologies. As is evident from the preceding chapters, these advances have paved the way for significant progress in our understanding of the human brain and human behavior. In this chapter, we chart the advances that the field has made in recent years. Some of these advances have been theoretical in nature, reflecting recent paradigm shifts in how we view the roles of individual brain regions and large-scale brain networks. Other advances have been primarily methodological: for instance, the five editions of this book have borne witness to the rapid development of functional imaging methods, spurred on by enormous advances in technology. For example, developments in both scanner technology and computational power have allowed for the collection of finer resolution data at a much faster rate. Early neuroimaging studies typically acquired slices as large as seven millimeters (Buckner et al., 1996; Tootell et al., 1995). In contrast, current functional MRI (fMRI) scanners are capable of collecting scans that are only one millimeter thick, and whole-brain coverage is typically achieved in about two seconds. The increasing size and complexity of modern-day data sets has been paralleled by increases in computational power, allowing scientists to develop new methods to store, catalog, and mine large data sets for new discoveries, leading to entirely new fields of study such as neuroinformatics (see chapter 83 by Marcus, this volume). New research initiatives such as the Human Connectome Project (www.humanconnectomeproject.org), which are amassing an enormous collection of neuroimaging data, are already capitalizing on these resources. Of course, these technological changes carry their own theoretical impact; for instance, increasing emphasis on neuroimaging has led to new efforts to understand neural function and dysfunction by studying the active, behaving brain.

Today, cognitive neuroscience should be viewed not only as a field with great potential, but also of great achievements. Here we highlight many of these achievements, as well as our vision of the future of cognitive neuroscience.

Applications of cognitive neuroscience

An important achievement of cognitive neuroscience today is its capacity to translate basic research results into the realm of clinical research and intervention. Indeed, the balance of cognitive neuroscience studies has recently shifted toward an increase in clinical research, sometimes at the expense of research on foundational questions in basic neuroscience. While the health implications of neuroscience research should inform and often guide research questions, this should not preclude the pursuit of basic research. To best lay the foundation for future discoveries of clinical importance within cognitive neuroscience research, it is imperative that basic research be encouraged and funded.

Examples of the societal impact of cognitive neuroscience research follow; all are grounded in basic research. Studies of individual differences, from the level of the gene to the level of brain connectivity, are increasingly informing truly personalized medication and treatment choices. The brain, once thought to be immutable after development, is now known to be alterable through behavioral training. While these findings are exciting, it is critical to ensure that they are clearly presented to a broad audience in order to produce the greatest benefit to society. In particular, education practices and policies could be strengthened by such research findings.

Individual Differences Traditional cognitive neuroscience imaging studies focus on mean group analysis of physiological fluctuations across samples, the results of which are assumed to generalize to wider populations. However, this approach may provide uninformative results, neglecting salient aspects of large interindividual variants relevant to identifying cognitive brain networks. Evidence for the vital impact of interindividual differences can be seen in imaging studies on motor behavior and decision making that link variations in anatomical brain connections to behavioral and cognitive outcomes (Kanai & Rees, 2011). These studies show promise in illuminating the contribution of interindividual variants on brain circuitry and plasticity.

However, tracking the time course of structural plasticity in interindividual imaging studies is a primary issue (Kanai & Rees, 2011). Network science (NS), the study of complex networks and topologies using mathematics, offers tools to track structural plasticity. More specifically, NS economy of brain networks traces wiring and rewiring on various time scales, allowing measurements of plasticity to be linked to fluctuations in cognitive states (Bullmore & Sporns, 2012). Merging NS techniques with studies of interindividual variation will advance brain network research in cognitive neuroscience and could lead to the development of imaging-based metrics to assist in clinical treatment of cognitive disorders.

Cognitive Intervention, Rehabilitation, and Optimization Better insight into individual differences in structural and functional brain characteristics, individual differences in genetic makeup, and the relation of these differences to behavioral variation may open new opportunities for personalized approaches in cognitive neuroscience applications, as can be seen in the following examples. In psychiatry, individual differences in gene expression can influence responses to psychoactive drugs, mediated by differences in brain metabolism and neural networks (Costa & Silva, 2012; Gvozdic, Brandl, Taylor, & Muller, 2012). Characterization of a brain injury patient's affected and nonaffected brain networks using imaging methods can help guide treatment selections, optimizing neurological rehabilitation (Ham & Sharp, 2012). In the context of training, individual white matter integrity within the corpus callosum predicts training response in aging adults (Wolf et al., 2012), while hippocampal structural characteristics and functional connectivity partly predict response to mathematics tutoring in school-age children (Supekar et al., 2013). More detailed genotyping and phenotyping of individuals before administration of pharmacological and nonpharmacological interventions can improve health care, much as knowledge of individual characteristics can inform therapeutic decisions about customized intervention strategies. Such knowledge will not only improve individual health care outcomes, but may also improve clinical research, for example via targeted patient selection for clinical trials.

Reorganization and optimization of neural networks can be achieved through physical exercise (i.e., experience-dependent neural plasticity) and cognitive training (i.e., learning; experience-dependent cognitive plasticity). Structural, functional, and chemical plasticity accompany behavioral changes following training, including in neurogenesis and synaptogenesis; promotion of neurotrophic activity and neurotransmitter efficiency; recovery of function; and reduced cognitive decline and psychiatric symptoms (Barbour, Edenfield, & Blumenthal, 2007; Dresler et al., 2013; Will, Galani, Kelche, & Rosenzweig, 2004). Improved performance and neural activation during cognitive tasks have been suggested to occur by engagement of previously underactive brain systems or compensation from other neural regions (Kelly, Foxe, & Garavan, 2006). Based on these findings, we propose that exercise increases neural efficiency and facilitates the neural context for learning. Therefore, combining exercise and cognitive training within a study may produce a synergistic effect with lasting neurocognitive outcomes.

Technological advancement may provide novel training opportunities. The use of video games for "brain training" has recently become popular, captivating the public interest. Some findings have suggested that these training programs do not lead to generalized benefits in a normal adult population (Owen et al., 2010), while others show promise in promoting maintenance of information (Jaeggi, Buschkuehl, Jonides, & Shah, 2011), sustained attention (Dye, Green, & Bavelier, 2009), executive control (Strobach, Frensch, & Schubert, 2012), reasoning (Strenziok et al., 2013), and in remediating symptoms in patients and other special populations (Vinogradov, Fisher, & de Villers-Sidani, 2012). Future work will show whether evidence-based game development aimed at improving cognition could provide insight for incorporating neurocognitive training into our daily lives. Biofeedback and neurofeedback technologies may be the future in optimizing training by providing real-time feedback on performance and psychological/physiological state (e.g., attention, arousal, stress). Additionally, transcranial direct current stimulation and transcranial magnetic stimulation hold promise in modulating cognition. These tools may be used to enhance cognition in healthy aging (Dresler et al., 2013) and to achieve functional recovery in impaired states such as addiction (Sokhadze, Cannon, & Trudeau, 2008) and psychiatric disorders (Mizenberg & Carter, 2012). Moreover, transcranial direct current stimulation, transcranial magnetic stimulation, and other technologies may act as elements of neuroprosthetic systems for locomotion, environmental control, or communication in cases of neural insult or congenital conditions (Lehembre et al., 2012; Nicolas-Alonso & Gomez-Gil, 2012). Finally, they may facilitate collaboration in a multiuser environment (Pope & Stevens, 2012). Therefore, these technologies can be used as adjunctive interventions in illness and can boost cognitive performance in healthy individuals or groups.

Neuroscience and Education One primary role of neuroscience is to inform education practices through advocacy and community outreach. Research on marginalized populations, such as those with financial disadvantages or mental illnesses, has revealed the negative impact of inadequate or ineffective education and its downstream consequences. As the intricacies of cognitive development are uncovered, we find that a better understanding of brain systems is fundamental in supporting both pedagogical theory and practice. Neural systems display degrees of plasticity that vary throughout development, and this knowledge can be used to target at-risk groups with neural training programs (Neville et al., 2013). Institutions such as the Economic and Social Research Council and the Society for Neuroscience recognize that we are now equipped to provide better training for teachers by bolstering their understanding of research on learning, memory, attention, and social behavior (e.g., Blakemore & Choudhury, 2006; Evans, Saffran, & Robe-Torres, 2009; Stevens, Sanders, & Neville, 2006). Teachers who are informed by such research may, in turn, promote more adaptive curricula fashioned to address the strengths and weaknesses of students from various environments and genetic backgrounds.

Theoretical neuroscience

Mapping Brain to Cognition Twenty-five years ago, cognitive neuroscience was conceptualized as the scientific study of the neural substrates of cognition. Its initial research program was envisioned in terms of finding the brain mechanisms responsible for the production of cognitive processes and functions. However, as the field has developed, the research objectives have evolved. The astonishing technical and methodological advances that have taken place in the last 10 years have revealed that the mapping of cognitive functions onto brain mechanisms is substantially more complicated than originally thought. On the one hand, the same neural mechanisms appear to be implicated in a number of ostensibly distinct cognitive functions; on the other hand, these same cognitive processes appear to be supported by the interactions amongst numerous, seemingly disparate brain areas. As a result, the field of cognitive neuroscience is moving toward a more complex understanding of the neural substrates of cognition, one that does not assume simplistic one-to-one mapping between any single brain region and a specific cognitive domain.

An example of this theoretical transition comes from recent work studying the hippocampus, a brain structure that has traditionally been considered to be critically and selectively involved in conscious, episodic, autobiographical memory. However, emerging neuroimaging, neurophysiological, and neuropsychological evidence demonstrates that the hippocampus is also involved in a range of other processes, including simulating possible future events and counterfactual thinking (Schacter et al., 2012), discriminating complex visuospatial stimuli (Lee et al., 2012), and learning and retrieving associative relationships without conscious awareness (Hannula & Greene, 2012). A similar story can be told about many other brain regions. Indeed, the engagement of a particular brain region during apparently distinct psychological functions seems to be the norm rather than the exception, suggesting that brain function cannot easily be classified by psychological taxonomy (Anderson, 2010).

Mapping Brain Networks Just as it has become clear that no individual brain region performs a single cognitive function, a wealth of recent work also suggests that cognition is supported by the dynamics of large-scale networks of brain regions. This shift in perspective has been facilitated largely by the widespread adoption of studies using functional connectivity MRI (fc-MRI) to characterize the spontaneous activity of the brain at rest in the absence of an explicit experimental task. Spontaneous activity is believed to reflect not only anatomical constraints but also Hebbian sculpting by co-activation (Lewis et al., 2009); thus, resting-state fc-MRI provides a window into statistical histories of functional coupling. Therefore, a brain region's resting functional connectivity profile can inform questions about cognition and provide additional correlates of behavioral or physiological measures. Although much still remains to be learned about how resting-state functional connectivity relates to cognition, studies thus far have yielded promising insights.

The growing use of resting-state fMRI (and other measures of connectivity) in cognitive neuroscience reflects a broader trend across many scientific disciplines to approach data sets from the perspective of NS. The human brain is a network with several levels of organization, so an NS approach is relevant to most neuroscientists, regardless of whether they study cognition at the scale of microcircuits or large-scale brain systems. At the large-scale level, efforts are underway to describe the domain-general functional organization of the cortex, cerebellum, and subcortical tissues, and to relate this organization to patterns of co-activation seen within and across particular cognitive domains. These initial efforts have resulted in coarse maps of human brain organization (Power et al., 2011; Yeo et al., 2011) that reflect and inform decades of functional neuroimaging. Studies of specific nodes in functional networks have begun to identify properties that correlate with cognitive abilities such as intelligence (Cole, Yarkoni, Repovs, Anticevic, & Braver, 2012).

At the microcircuit level, interactions among small clusters of neurons has revealed mechanistic principles of neural computation, which seem to be ubiquitous throughout the brain and powerful enough to implement complex features of neural machinery. Interneurons can configure networks with different properties, depending on whether they exert inhibitory or excitatory connections within their local circuits (Lee et al., 2012; Wang, 2002). The study of network properties and dynamics at much finer scales will become more common in the near future as the dissemination and refinement of techniques for controlling and quantifying neural activity with fine temporal and spatial scales improves (e.g., via optogenetics and the CLARITY process; see Chung & Deisseroth, 2013, etc.).

Mapping Networks Across Time Recent advances have also elucidated some of the mechanisms by which neural and cognitive processes emerge and develop across time. Developmental cognitive neuroscience is important because it can help us understand individual differences, inform educational practices, and pave the way for tailoring remediation techniques for atypically developing children. However, until recently, most neuroscientific accounts have disappointed developmental theorists by relegating developmental processes to brain maturation. The maturational or "predetermined epigenesist" approach to development cannot account for the complex and dynamic ("probabilistic") interactions that happen within and between all levels of organization across time, from genes to the external environment (Gottlieb, 1992). Nor can they account for the dynamics of change in genetic and environmentally induced disorders, nor answer questions such as whether an early, basic-level deficit might be followed by compensation or compounding of effects. Take, for example, the dyadic interaction between a child and her mother. If the mother were told that her child had a neurodevelopmental disorder (e.g., autism), would the mother interact with her child differently? If yes, then the child's responses would reflexively change. Indeed, Karmiloff-Smith and colleagues (2012) have observed that some parents find it difficult to allow their atypically developing child to freely roam about and learn from their environment as a typically developing child would. This may result in a less richly explored environment, which in turn would constrain brain, motor, and sociocognitive development.

Individual differences in mother-child interactions are known to constrain cognitive development even in the case of typical development (Karmiloff-Smith et al., 2010, 2012), raising questions not contemplated in the past. As a consequence of such empirical findings, researchers are abandoning the idea that brain and cognitive development are yoked to some predetermined maturational process. Old, static questions regarding the "age" at which a certain "brain module" comes online, where the modules are located, and which modules are "impaired" or "intact" are giving way to new, dynamic questions concerning the emergence of and changes in neural circuits and cognitive functions over developmental time, which domains interact across developmental trajectories, and which aspects of our dynamic environments interact with and alter ontogenesis.

Challenges and conclusion

The cognitive neurosciences have progressed markedly since the last edition of this volume (Aminoff et al., 2009). Borne along in part by advancing new and refined methodologies, our understanding of the relationship between brain and behavior has advanced considerably. This progress paves the way for future breakthroughs, but also presents new challenges. Here, we mention a few noteworthy developments and the advances that they may foreshadow, as well as some potential pitfalls the field may face as it continues to develop.

New Directions in Cognitive Neuroscience The chapters included in this book delineate the current state of the cognitive neurosciences, and hint at the headway that may be gained in the future. Recent findings in neuroimaging have shown that it is possible to decode a person's conscious experience based only on his or her brain activity (Horikawa, Tamaki, Miyawaki, & Kamitani, 2013; Nishimoto et al., 2011). This ability to detect the presence of certain cognitive states, during both wake and sleep, may enable reconstruction of dreams and have interesting implications, such as legal implications for lie detection.

Some of the latest developments in brain-machine interface (BMI) technologies enable the restoration of body mobility in individuals suffering from motor deficits (e.g., paraplegics; Lebedev & Nicolelis, 2011; Nirenberg & Pandarinath, 2012; Wang et al., 2013) and the restoration to near-normal vision in the blind (Nirenberg & Pandarinath, 2012). Integration of cognitive neuroscience and engineering may enable whole-body BMI and sensory substitutions (Reich, Maidenbaum, & Amedi, 2012), improving quality of life across many domains.

Recent developments on olfaction research—including theoretical models positing geometrical relationship among odorants (Haddad, Lapid, Harel, & Sobel, 2008)—have revealed connections between odor molecules and their corresponding neural and perceptual responses. Such theories may allow us to determine what sensation a given odorant will have on our olfactory systems or to sense odorants outside the range of normal human sensation, laying the foundation for the development of an electronic nose that could detect diseases (Wilson & Baietto, 2011), allowing for "photographing" an odor (e.g., for categorization, reconstruction, or later comparison), and enabling the reconstruction of odor experience through BMI (e.g., for those with anosmia). The cognitive neuroscience of odor-space research is still relatively nascent (e.g., as compared with vision neuroscience), so it is reasonable to expect great continuing advances in this area in the years to come.

Previously, progress in areas such as developmental and clinical cognitive neurosciences was slow, because well-established techniques in healthy adult research (e.g., fMRI) are often unsuitable for research with difficult-to-test infants or children and clinical populations. Research in children and special populations (e.g., patients with sensory processing disorder, schizophrenia, fragile X) is now possible with the introduction of new, lightweight, comfortable, and quickly and easily positioned functional near-infrared spectroscopy (which measures the hemodynamic response to cortical neural activation; see, e.g., Lloyd-Fox, Blasi, & Elwell, 2010); quick-application electroencephalographic "hairnets" and high-impedance electroencephalographic systems; and head-mounted and fixation-responsive eye trackers. These powerful new tools (and numerous others) can reveal how neural and cognitive processes become specialized over developmental time, recover from insult, or respond to medical intervention, having considerable implications for basic research as well as health care and education.

The recent development of optogenetic tools for precise online control of neural activity (see chapter 82 by Zalocusky and Deisseroth, this volume) has already begun to deliver substantial insight, providing causal evidence for theories of learning (Steinberg et al., 2013) and memory (Ramirez et al., 2013). This technique offers great promise to advance our understanding of neuronal signaling as well as to provide better treatment in the clinic (for example, supplanting beneficial but imprecise deep-brain stimulation in Parkinson's patients, and enhancing the viability of BMI prostheses for patients with brain injury or amputated limbs).

From Science to Society A deluge of media coverage has accompanied the headway made in the cognitive neurosciences, highlighting the challenge of how to best disseminate neuroscience knowledge. There remains a large gap between the empirical evidence and the public perception in both our understanding of the brain and our applications of that knowledge (Eagleman, 2013; Racine, Waldman, Rosenberg, & Illes, 2010). For example, the concept of "brain training" has recently experienced an increase in both popularity and criticism (Cook, 2013), and despite a paucity of data supporting the idea that improvements in cognitive-training tasks can transfer to a quantitative increase in general intelligence (e.g., Owen et al., 2010), companies implying as much have proved to be incredibly popular. However, evidence is building that some types of training can enhance durable and transferable cognitive performance in several domains (see, e.g., Jaeggi et al., 2011; Strenziok et al., 2013), drawing attention to the need for further research and the fact that greater efforts must be made to clearly communicate both the promise and the limitations of neuroscience research.

This issue extends well beyond the popular press and has nontrivial implications. In criminal law, similar uncertainty exists between what neuroscience *can* tell us and what neuroscience is *expected* to tell us. There is a growing appreciation of research that questions the validity of eyewitness testimonies (Schacter & Loftus, 2013; see the section XI introduction by Sinnott-Armstrong and Roskies, this volume) and that highlights the neural substrates of the "criminal mind" (e.g., Farisco & Petrini, 2012; see chapter 89 by Gaudet, Anderson, and Kiehl, this volume). Notably, recent advances in fMRI and optogenetics have respectively led to claims of "mindreading" (Stahl, 2009) and "total recall" (Hornyak, 2013) from the media, when in fact the utility of such methods to the legal system remains limited for the foreseeable future. Moving forward, the field must continue to encourage public dissemination of neuroscience research without overstating the implications of our work (Eagleman, 2013; Racine et al., 2010).

Just as new technologies have changed the way that scientists acquire, store, and share data, they have also offered new means for scientists to develop studies and communicate their findings with each other and with the general public. For example, blogs and tools like Twitter allow scientists to rapidly respond to new findings and papers in a form of post-publication peer review, and growing support for study preregistration (Chambers & Munafo, 2013) may enhance the collaborative basis and quality of studies during their formation. As academic journals have moved almost entirely into the online sector, the barriers to disseminating new findings are diminishing, and there is increasing emphasis on establishing new mechanisms for rapid, open-access publishing (Kriegeskorte, Walther, & Deca, 2012). Although these developments affect the scientific community as a whole, given the widespread public interest in the brain sciences, it behooves cognitive neuroscientists to embrace technologies that will enable us to share our knowledge with the general public.

The proliferation of literature and methods (as illustrated by the breadth of the preceding chapters) exposes another challenge for the future of the field. With maturation, cognitive neuroscience risks the fractionation of its subdisciplines into independent fields, diluting its interdisciplinary strengths. Emerging technologies that facilitate the interpretation of these vast literatures (e.g., Yarkoni et al., 2011) and their data will be of increasing importance as the field progresses.

We look forward to the developments over the next five years, and expect the next edition of this volume will be as rich with progress and promise for the future of cognitive neuroscience as this one.

REFERENCES

Aminoff, E. M., Balslev, D., Borroni, P., Bryan, R. E., Chua, E. F., Cloutier, J., ... Yamada, M. (2009). The landscape of cognitive neuroscience: Challenges, rewards, and new perspectives. In M. Gazzaniga (Ed.), *The cognitive neurosciences* (4th ed., pp. 1255–1292). Cambridge, MA: MIT Press.

Anderson, M. (2010). Neural reuse: A fundamental organizational principle of the brain. *Behav Brain Sci, 33,* 245–313.

Barbour, K. A., Edenfield, T. M., & Blumenthal, J. A. (2007). Exercise as a treatment for depression and other psychiatric disorders: A review. *J Cardiopulm Rehabil Prev, 27*(6), 359–367.

Blakemore, S. J., & Choudhury, S. (2006). Development of the adolescent brain: Implications for executive function and social cognition. *J Child Psychol Psychiatry, 47*(3–4), 296–312.

Buckner, R. L., Bandettini, P. A., O'Craven, K. M., Savoy, R. L., Petersen, S. E., Raichle, M. E., & Rosen, B. R. (1996). Detection of cortical activation during averaged single trials of a cognitive task using functional magnetic resonance imaging. *Proc Natl Acad Sci USA, 93,* 14878–14883.

Bullmore, E., & Sporns, O. (2012). The economy of brain network organization. *Nat Rev Neurosci, 13,* 336–349.

Chambers, C., & Munafo, M. (2013, June 5). Trust in science would be improved by study pre-registration. *The Guardian.*

Chung, K., & Deisseroth, K. (2013). CLARITY for mapping the nervous system. *Nat Methods, 10,* 508–513.

Cole, M. W., Yarkoni, T., Repovs, G., Anticevic, A., & Braver, T. S. (2012). Global connectivity of prefrontal cortex predicts cognitive control and intelligence. *J Neurosci, 32*(26), 8988–8999.

Cook, G. (2013, April 5). Brain games are bogus. *The New Yorker.*

Costa, E., & Silva, J. A. (2012). Personalized medicine in psychiatry: New technologies and approaches. *Metabolism, 62*(Suppl. 1), S40-S44.

Dresler, M., Sandberg, A., Ohla, K., Bublitz, C., Trenado, C., Mroczko-Wąsowicz, A., ... Repantis, D. (2013). Non-pharmacological cognitive enhancement. *Neuropharmacology, 64*, 529–543.

Dye, M. W. G., Green, C. S., & Bavelier, D. (2009). The development of attention skills in action video game players. *Neuropsychologia, 47*(8–9), 1780–1789.

Eagleman, D. M. (2013). Why public dissemination of science matters: A manifesto. *J Neurosci, 33*(30), 12147–12149.

Evans, J. L., Saffran, J. R., & Robe-Torres, K. (2009). Statistical learning in children with specific language impairment. *J Speech Lang Hear Res, 52*, 321–335.

Farisco, M., & Petrini, C. (2012). The impact of neuroscience and genetics on the law: A recent Italian case. *Neuroethics, 5*(3), 317–319.

Gottlieb, G. (1992). *Individual development and evolution: The genesis of novel behavior.* New York, NY: Oxford University Press.

Gvozdic, K., Brandl, E. J., Taylor, D. L., & Muller, D. J. (2012). Genetics and personalized medicine in antidepressant treatment. *Curr Pharm Des, 18*, 5853–5878.

Haddad, R., Lapid, H., Harel, D., & Sobel, N. (2008). Measuring smells. *Curr Opin Neurobiol, 18*(4), 438–444.

Ham, T. E., & Sharp, D. J. (2012). How can investigation of network function inform rehabilitation after traumatic brain injury? *Curr Opin Neurol, 25*(6), 662–669.

Hannula, D. E., & Greene, A. J. (2012). The hippocampus reevaluated in unconscious learning and memory: At a tipping point? *Front Hum Neurosci, 6*(80), 1–20.

Horikawa, T., Tamaki, M., Miyawaki, Y., & Kamitani, Y. (2013). Neural decoding of visual imagery during sleep. *Science, 340*(6132), 639–642.

Hornyak, T. (2013, July 26). False memories created in mice: "Total Recall" can't be far. CNET.

Jaeggi, S. M., Buschkuehl, M., Jonides, J., & Shah, P. (2011). Short- and long-term benefits of cognitive training. *Proc Natl Acad Sci USA, 108*(25), 10081–10086.

Kanai, R., & Rees, G. (2011). The structural basis of interindividual differences in human behavior and cognition. *Nat Rev Neurosci, 12*, 231–242.

Karmiloff-Smith, A., Aschersleben, G., de Schonen, S., Elsabbagh, M., Hohenberger, A., & Serres, J. (2010). Constraints on the timing of infant cognitive change: Domain-specific or domain-general? *Eur J Dev Sci, 4*(1), 31–45.

Karmiloff-Smith, A., D'Souza, D., Dekker, T. M., van Herwegen, J., Xu, F., Rodic, M., & Ansari, D. (2012). Genetic and environmental vulnerabilities in children with neurodevelopmental disorders. *Proc Natl Acad Sci USA, 109*, 17261–17265.

Kelly, C., Foxe, J. J., & Garavan, H. (2006). Patterns of normal human brain plasticity after practice and their implications for neurorehabilitation. *Arch Phys Med Rehabil, 87*(12), 20–29.

Kriegeskorte, N., Walther, A., & Deca, D. (2012). An emerging consensus for open evaluation: 18 visions for the future of scientific publishing. *Front Comput Neurosci, 6*(94), 1–5.

Lebedev, M. A., & Nicolelis, M. A. (2011). Toward a whole body neuroprosthetic. *Prog Brain Res, 194*, 47–60.

Lee, S. H., Kwan, A. C., Zhang, S., Phoumthipphavong, V., Flannery, J. G., Masmanidis, S. C., ... Dan, Y. (2012). Activation of specific interneurons improves V1 feature selectivity and visual perception. *Nature, 488*, 379–383.

Lehembre, R., Gosseries, O., Lugo, Z., Jedidi, Z., Chatelle, C., Sadzot, B., ... Noirhomme, Q. (2012). Electrophysiological investigations of brain function in coma, vegetative and minimally conscious patients. *Arch Ital Biol, 150*(2–3), 122–139.

Lewis, C. M., Baldassarre, A., Committeri, G., Romani, G. L., & Corbetta, M. (2009). Learning sculpts the spontaneous activity of the resting human brain. *Proc Natl Acad Sci USA, 106*(41), 17558–17563.

Lloyd-Fox, S., Blasi, A., & Elwell, C. E. (2010). Illuminating the developing brain: The past, present and future of functional near infrared spectroscopy. *Neurosci Biobehav Rev, 34*(3), 269–284.

Mizenberg, M. J., & Carter, C. S. (2012). Developing treatments for impaired cognition in schizophrenia. *Trends Cogn Sci, 16*(1), 35–42.

Neville, H. J., Stevens, C., Pakulak, E., Bell, T. A., Fanning, J., Klein, S., & Isbell, E. (2013). Family-based training program improves brain function, cognition and behavior in lower socioeconomic status preschoolers. *Proc Natl Acad Sci USA, 110*(29), 12138–12143.

Nicolas-Alonso, L. F., & Gomez-Gil, J. (2012). Brain computer interfaces, a review. *Sensors, 12*, 1211–1279.

Nirenberg, S., & Pandarinath, C. (2012). Retinal prosthetic strategy with the capacity to restore normal vision. *Proc Natl Acad Sci USA, 109*(37), 15012–15017.

Nishimoto, S., Vu, A. T., Naselaris, T., Benjamini, Y., Yu, B., & Gallant, J. L. (2011). Reconstructing visual experiences from brain activity evoked by natural movies. *Curr Biol, 21*(19), 1641–1646.

Owen, A. M., Hampshire, A., Grahn, J. A., Stenton, R., Dajani, S., Burns, A. S., ... Ballard, C. G. (2010). Putting brain training to the test. *Nature, 465*, 775–778.

Pope, A. T., & Stevens, C. L. (2012). Interpersonal biocybernetics: Connecting through social psychophysiology. *Proc. ACM International Conference on Multimodal Interaction* (pp. 561–566).

Power, J. D., Cohen, A. L., Nelson, S. M., Wig, G. S., Barnes, K. A., Church, J. A., ... Petersen, S. E. (2011). Functional network organization of the human brain. *Neuron, 72*(4), 665–678.

Racine, E., Waldman, S., Rosenberg, J., & Illes, J. (2010). Contemporary neuroscience in the media. *Soc Sci Med, 71*(4), 725–733.

Ramirez, S., Liu, X., Lin, P., Suh, J., Pignatelli, M., Redondo, R. L., ... Tonegawa, S. (2013). Creating a false memory in the hippocampus. *Science, 341*(6144), 387–391.

Reich, L., Maidenbaum, S., & Amedi, A. (2012). The brain as a flexible task machine: Implications for visual rehabilitation using noninvasive vs. invasive approaches. *Curr Opin Neurol, 25*(1), 86–95.

Schacter, D. L., Addis, D. R., Hassabis, D., Martin, V. C., Spreng, R. N., & Szpunar, K. K. (2012). The future of memory: Remembering, imagining, and the brain. *Neuron, 7*, 677–694.

Schacter, D. L., & Loftus, E. F. (2013). Memory and law: What can cognitive neuroscience contribute? *Nat Neurosci, 16*, 119–123.

Sokhadze, T. M., Cannon, R. L., & Trudeau, D. L. (2008). EEG biofeedback as a treatment for substance use disorders: Review, rating of efficacy, and recommendations for further research. *Appl Psychophysiol Biofeedback, 33*(1), 1–28.

Stahl, L. (2009, June 26). How technology may soon "read" your mind. *60 Minutes*.

Steinberg, E. E., Keiflin, R., Boivin, J. R., Witten, I. B., Deisseroth, K., & Janak, P. H. (2013). A causal link between prediction errors, dopamine neurons and learning. *Nat Neurosci, 16*, 966–973.

Stevens, C., Sanders, L., & Neville, H. (2006). Neurophysiological evidence for selective auditory attention deficits in children with specific language impairment. *Brain Res, 1111*, 143–152.

Strenziok, M., Parasuraman, R., Clarke, E., Cisler, D. S., Thompson, J. C., & Greenwood, P. M. (2013). Neurocognitive enhancement in older adults: Comparison of three cognitive training tasks to test a hypothesis of training transfer in brain connectivity. *NeuroImage, 85*(Pt. 3), 1027–1039.

Strobach, T., Frensch, P. A., & Schubert, T. (2012). Video game practice optimizes executive control skills in dual-task and task switching situations. *Acta Psychol (Amst), 140*(1), 12–24.

Supekar, K., Swigart, A. G., Tenison, C., Jolles, D. D., Rosenberg-Lee, M., Fuchs, L., & Menon, V. (2013). Neural predictors of individual differences in response to math tutoring in primary-grade school children. *Proc Natl Acad Sci USA, 110*(20), 8230–8235.

Tootell, R. B. H., Reppas, J. B., Kwong, K. K., Malach, R., Born, R. T., Brady, T. J., ... Belliveau, J. W. (1995). Functional analysis of human MT and related visual cortical areas using magnetic resonance imaging. *J Neurosci, 15*(4), 3215–3230.

Vinogradov, S., Fisher, M., & de Villers-Sidani, E. (2012). Cognitive training for impaired neural systems in neuropsychiatric illness. *Neuropsychopharmacology, 37*(1), 43–76.

Wang, W., Collinger, J. L., Degenhart, A. D., Tyler-Kabara, E. C., Schwartz, A. B., Moran, D. W., ... Boninger, M. L. (2013). An electrocorticographic brain interface in an individual with tetraplegia. *PLoS ONE, 8*(2).

Wang, X. J. (2002). Probabilistic decision making by slow reverberation in cortical circuits. *Neuron, 36*(5), 955–968.

Will, B., Galani, R., Kelche, C., & Rosenzweig, M. R. (2004). Recovery from brain injury in animals: Relative efficacy of environmental enrichment, physical exercise or formal training (1990–2002). *Prog Neurobiol, 72*(3), 167–182.

Wilson, A. D., & Baietto, M. (2011). Advances in electronic-nose technologies developed for biomedical applications. *Sensors, 11*(1), 1105–1176.

Wolf, D., Fischer, F. U., Fesenbeckh, J., Yakushev, I., Lelieveld, I. M., Scheurich, A., ... Fellgiebel, A. (2012). Structural integrity of the corpus callosum predicts long-term transfer of fluid intelligence-related training gains in normal aging. *Hum Brain Mapp*.

Yarkoni, T., Poldrack, R. A., Nichols, T. E., Van Essen, D. C., & Wager, T. D. (2011). Large-scale automated synthesis of human functional neuroimaging data. *Nat Methods, 8*, 665–670.

Yeo, B. T. T., Krienen, F. M., Sepulcre, J., Sabuncu, M. R., Lashkari, D., Hollinshead, M., ... Buckner, R. L. (2011). The organization of the human cerebral cortex estimated by intrinsic functional connectivity. *J Neurophysiol, 106*(3), 1125–1165.

CONTRIBUTORS

EDDY ALBARRAN Howard Hughes Medical Institute and Department of Neurobiology, Stanford University, Stanford, California

NALINI AMBADY Department of Psychology, Stanford University, Stanford, California

KATRIN AMUNTS Institute of Neuroscience and Medicine, Research Centre Juelich, Juelich, Germany

ADAM K. ANDERSON Department of Human Development, Cornell University, Ithaca, New York

NATHANIEL E. ANDERSON Department of Psychology and Neuroscience, Baylor University, Waco, Texas

ANAT ARZI Department of Neurobiology, Weizmann Institute of Science, Rehovot, Israel

YANIV ASSAF Department of Neurobiology, The George S. Wise Faculty of Life Sciences, Sagol School of Neuroscience, Tel Aviv University, Tel Aviv, Israel

RENÉE BAILLARGEON Department of Psychology, University of Illinois, Champaign, Illinois

ANTONELLO BALDASSARE Department of Neurology, Washington University School of Medicine, Saint Louis, Missouri; Department of Neurosciences and Imaging, University of Chieti, Chieti, Italy

SNIGDHA BANERJEE The Cognitive Neurophysiology Laboratory, Children's Evaluation and Rehabilitation Center, Departments of Pediatrics and Neuroscience, Albert Einstein College of Medicine, New York, New York; Program in Cognitive Neuroscience, Department of Psychology, City College of the City University of New York, New York, New York

DANIELLE S. BASSETT Department of Bioengineering, University of Pennsylvania, Philadelphia, Pennsylvania; Sage Center for the Study of the Mind, University of California, Santa Barbara, Santa Barbara, California

AMY BASTIAN Department of Neuroscience, Johns Hopkins University, Baltimore, Maryland; The Kennedy Krieger Institute, Baltimore, Maryland

AARON BATISTA Department of Bioengineering, Center for the Neural Basis of Cognition, University of Pittsburgh, Pittsburgh, Pennsylvania

MICHAEL S. BEAUCHAMP Departments of Neurosurgery and Neuroscience, Baylor College of Medicine, Houston, Texas

ANNABELLE M. BELCHER National Institute on Drug Abuse, Intramural Research Program, National Institutes of Health, Bethesda, Maryland

LIN BIAN Department of Psychology, University of Illinois, Champaign, Illinois

SARAH-JAYNE BLAKEMORE Institute of Cognitive Neuroscience, University College London, London, United Kingdom

OLAF BLANKE Center for Neuroprosthetics and Brain Mind Institute, School of Life Sciences, Ecole Polytechnique Fédérale de Lausanne, Lausanne, Switzerland

JENNIFER BLAZE Department of Psychology, University of Delaware, Newark, Delaware

HAL BLUMENFELD Departments of Neurology, Neurobiology, and Neurosurgery, Yale University, New Haven, Connecticut

ALI BORJI University of Southern California, Los Angeles, California

ANDREEA C. BOSTAN Center for the Neural Basis of Cognition, Systems Neuroscience Institute and the Department of Neurobiology, University of Pittsburgh, Pittsburgh, Pennsylvania

CHARNESE BOWES Department of Neurosurgery, Stanford University School of Medicine, Stanford, California

MIRJANA BOZIC Department of Psychology, University of Cambridge, Cambridge, United Kingdom

JOHANNES BURGE Center for Perceptual Systems, University of Texas at Austin, Austin, Texas

ALICIA CALLEJAS Department of Neurology, Washington University School of Medicine, Saint Louis, Missouri

BEATRICE H. CAPESTANY Department of Psychology and Neuroscience, Duke University, Durham, North Carolina

BJ CASEY Sackler Institute for Developmental Psychobiology, Weill Cornell Medical College, New York, New York

MARCO CATANI NatBrainLab, Department of Forensic and Neurodevelopmental Sciences, King's College London, London, United Kingdom

HANAH A. CHAPMAN Department of Psychology, Brooklyn College, New York, New York

ANJAN CHATTERJEE Department of Neurology and the Center for Cognitive Neuroscience, University of Pennsylvania, Philadelphia, Pennsylvania

B. K. H. CHAU Department of Experimental Psychology, University of Oxford, Oxford, United Kingdom

MAN CHEN Department of Physics and Astronomy, Rice University, Houston, Texas

MORTEN H. CHRISTIANSEN Department of Psychology, Cornell University, Ithaca, New York

TYLER CLUFF Centre for Neuroscience Studies, Queen's University, Kingston, Ontario

MARLENE R. COHEN Department of Neuroscience and Center for the Neural Basis of Cognition, University of Pittsburgh, Pittsburgh, Pennsylvania

MAURIZIO CORBETTA Departments of Neurology, Radiology, and Anatomy and Neurobiology, Washington University School of Medicine, Saint Louis, Missouri; Department of Neurosciences and Imaging, University of Chieti, Chieti, Italy

JUSTIN C. COX Department of Psychology, Washington University in Saint Louis, Saint Louis, Missouri

FRÉDÉRIC CREVECOEUR Centre for Neuroscience Studies, Queen's University, Kingston, Ontario

EVELINE A. CRONE Department of Psychology, Leiden University, Leiden, The Netherlands

ANTHONY D. D'ANTONA Center for Perceptual Systems, University of Texas at Austin, Austin, Texas

Dean D'Souza Centre for Brain and Cognitive Development, Department of Psychological Sciences, Birkbeck, University of London, London, United Kingdom

Lila Davachi Department of Psychology, Center for Neural Science, New York University, New York, New York

Nathaniel D. Daw Center for Neural Science, Department of Psychology, New York University, New York, New York

Felipe De Brigard Department of Philosophy, Center for Cognitive Neuroscience, and Duke Institute for Brain Sciences, Duke University, Durham, North Carolina

Michael W. Deem Department of Bioengineering, Rice University, Houston, Texas

Karl Deisseroth Howard Hughes Medical Institute, Departments of Bioengineering and Psychiatry, Stanford University, Stanford, California

Athena Demertzi Cyclotron Research Center, University of Liège, Liège, Belgium

James J. DiCarlo Department of Brain and Cognitive Sciences, Massachusetts Institute of Technology, Cambridge, Massachusetts

Jörn Diedrichsen Institute of Cognitive Neuroscience, University College London, London, United Kingdom

Ian G. Dobbins Washington University in Saint Louis, Saint Louis, Missouri

Bradley B. Doll Center for Neural Science, New York University, New York, New York; Department of Psychology, Columbia University, New York, New York

Rodney J. Douglas Institute of Neuroinformatics, University of Zürich and Eidgenössische Technische Hochschule Zürich, Zürich, Switzerland

Richard P. Dum Center for the Neural Basis of Cognition, Systems Neuroscience Institute and the Department of Neurobiology, University of Pittsburgh, Pittsburgh, Pennsylvania

Karen Emmorey Laboratory for Language and Cognitive Neuroscience, San Diego State University, San Diego, California

Adrienne Fairhall Department of Physiology and Biophysics, University of Washington, Seattle, Washington

Jacqueline Fairley Sleep Medicine Program, Department of Neurology, School of Medicine, Emory University, Atlanta, Georgia

Kara D. Federmeier Department of Psychology, University of Illinois, Champaign, Illinois

Sergi Ferré National Institute on Drug Abuse, Intramural Research Program, National Institutes of Health, Bethesda, Maryland

Stephen M. Fleming Center for Neural Science, New York University, New York, New York; Department of Experimental Psychology, University of Oxford, Oxford, United Kingdom.

Jonathan B. Freeman Dartmouth College, Hanover, New Hampshire

Scott H. Frey Departments of Neurology, Psychiatry, and Physical Medicine and Rehabilitation, University of Missouri, Columbia, Missouri

Pascal Fries Donders Institute for Brain, Cognition and Behaviour, Radboud University Nijmegen, Nijmegen, The Netherlands

Lyn M. Gaudet Center for Law, Science, and Innovation, Arizona State University, Tempe, Arizona

Dylan G. Gee Department of Psychology, University of California, Los Angeles, Los Angeles, California

Wilson S. Geisler Center for Perceptual Systems and Department of Psychology, University of Texas at Austin, Austin, Texas

Asif A. Ghazanfar Neuroscience Institute, Princeton University, Princeton, New Jersey

Jay N. Giedd Child Psychiatry Branch, National Institute of Mental Health, Bethesda, Maryland

Paul Glimcher Center for Neuroeconomics, New York University, New York, New York

Anne-Lise Goddings Institute of Child Health and Institute of Cognitive Neuroscience, University College London, London, United Kingdom

Rainer Goebel Department of Cognitive Neuroscience, Faculty of Psychology and Neuroscience, Maastricht University, Maastricht, The Netherlands; Department of Neuroimaging and Neuromodeling, Netherlands Institute for Neuroscience, Royal Netherlands Academy of Arts and Sciences, The Netherlands

Manuel Gomez-Ramirez Krieger Mind/Brain Institute, Johns Hopkins University, Baltimore, Maryland

Alan M. Gordon Department of Psychology, Stanford University, Stanford, California

Scott T. Grafton Department of Psychological and Brain Sciences, University of California, Santa Barbara, Santa Barbara, California

Joshua D. Greene Department of Psychology, Harvard University, Cambridge, Massachusetts

Kalanit Grill-Spector Department of Psychology and Neuroscience Institute, Stanford University, Stanford, California

Patrick Haggard Institute of Cognitive Neuroscience, University College London, London, United Kingdom

Peter Hagoort Donders Institute for Brain, Cognition and Behaviour, Radboud University Nijmegen, Nijmegen, The Netherlands

Lasana T. Harris Department of Psychology and Neuroscience and Center for Cognitive Neuroscience, Duke University, Durham, North Carolina

Catherine A. Hartley Sackler Institute of Developmental Psychobiology, Weill Cornell Medical College, New York, New York

Michael E. Hasselmo Center for Memory and Brain, Department of Psychology and Graduate Program for Neuroscience, Boston University, Boston, Massachusetts

Todd F. Heatherton Department of Psychological and Brain Sciences, Dartmouth College, Hanover, New Hampshire

Sibylle C. Herholz German Center for Neurodegenerative Diseases, Bonn, Germany

Steven A. Hillyard University of California, San Diego, La Jolla, California

Steven Hsiao Solomon H. Snyder Department of Neuroscience, Krieger Mind/Brain Institute, Johns Hopkins University, Baltimore, Maryland

Scott A. Huettel Department of Psychology and Neuroscience, Duke Center for Interdisciplinary Decision Science, Duke University, Durham, North Carolina

Laurent Itti University of Southern California, Los Angeles, California

Kyong-sun Jin Department of Psychology, University of Illinois, Champaign, Illinois

Eduardo Rosales Jubal Institute for Microscopic Anatomy and Neurobiology, Johannes Gutenberg University of Mainz, Mainz, Germany; Department of Neurophysiology, Max Planck Institute for Brain Research, Frankfurt am Main, Germany

Jon H. Kaas Department of Psychology, Vanderbilt University, Nashville, Tennessee

Michael J. Kahana Department of Psychology and Computational Memory Lab, University of Pennsylvania, Philadelphia, Pennsylvania

Sabine Kastner Department of Psychology, Princeton University, Princeton, New Jersey

Daniel Kersten Department of Brain and Cognitive Engineering, Korea University, Seoul, South Korea

Kent A. Kiehl Departments of Psychology, Neuroscience, and Law, University of New Mexico, Albuquerque, New Mexico

Danielle R. King Miller Memory Lab, Department of Psychological and Brain Sciences, University of California, Santa Barbara, Santa Barbara, California

Tilman J. Kispersky Biology Department and Volen Center, Brandeis University, Waltham, Massachusetts

Christof Koch Allen Institute for Brain Science, Seattle, Washington; California Institute of Technology, Pasadena, California

N. Kolling Department of Experimental Psychology, University of Oxford, Oxford, United Kingdom

Timothy R. Koscik Department of Psychology, University of Toronto, Toronto, Ontario

Sid Kouider Brain and Consciousness Group, Institute for Cognitive Studies, École Normale Supérieure, Paris, France; Centre National de la Recherche Scientifique, Paris, France

Gabriel Kreiman Department of Neurology and Ophthalmology, Children's Hospital Boston, Harvard Medical School, Boston, Massachusetts; Center for Brain Science, Harvard University, Cambridge, Massachusetts

Leah Krubitzer Laboratory of Evolutionary Neurobiology, Center for Neuroscience, University of California, Davis, Davis, California

Marta Kutas Departments of Cognitive Science and Neuroscience and Center for Research in Language, University of California, San Diego, La Jolla, California

Ayelet N. Landau Ernst Strüngmann Institute for Neuroscience in Cooperation with Max Planck Society, Frankfurt am Main, Germany

Steven Laureys Cyclotron Research Center, University of Liège, Liège, Belgium

Victoria K. Lee Department of Psychology and Neuroscience, Duke University, Durham, North Carolina

Roger N. Lemon Sobell Department of Motor Neuroscience and Movement Disorders, Institute of Neurology, University College London, London, United Kingdom

Jill K. Leutgeb Neurobiology Section and Center for Neural Circuits and Behavior, Division of Biological Sciences, University of California, San Diego, La Jolla, California

Stefan Leutgeb Neurobiology Section and Center for Neural Circuits and Behavior, Division of Biological Sciences, University of California, San Diego, La Jolla, California

Stephen C. Levinson Donders Institute for Brain, Cognition and Behaviour, Radboud University Nijmegen, Nijmegen, The Netherlands

Jarrod A. Lewis-Peacock Department of Psychology, University of Texas at Austin, Austin, Texas

Laura A. Libby Department of Psychology, University of California, Davis, Davis, California

Steven J. Luck University of California, Davis, Davis, California

Mary-Ellen Lynall Academic Neurosciences Foundation Programme, University of Cambridge, Cambridge, United Kingdom

Eleanor A. Maguire Wellcome Trust Centre for Neuroimaging, Institute of Neurology, University College London, London, United Kingdom

Jeremy R. Manning Princeton Neuroscience Institute and Department of Computer Science, Princeton University, Princeton, New Jersey

Uri Maoz Division of Biology, California Institute of Technology, Pasadena, California

Daniel Marcus Neuroinformatics Research Group and Neuroimaging Informatics and Analysis Center, Washington University in Saint Louis, Saint Louis, Missouri

Eve Marder Biology Department and Volen Center, Brandeis University, Waltham, Massachusetts

William D. Marslen-Wilson Department of Psychology, University of Cambridge, Cambridge, United Kingdom

Kevan A.C. Martin Institute of Neuroinformatics, University of Zürich and Eidgenössische Technische Hochschule Zürich, Zürich, Switzerland

Julio C. Martinez-Trujillo Department of Physiology, McGill University, Montréal, Québec

Marcello Massimini Department of Biomedical and Clinical Sciences, University of Milan, Milan, Italy

Markus Meister Division of Biology and Biological Engineering, California Institute of Technology, Pasadena, California

Peter Mende-Siedlecki Department of Psychology, Princeton University, Princeton, New Jersey

Kyle A. Meyer Department of Neurobiology and Kavli Institute for Neuroscience, Yale School of Medicine, Yale University, New Haven, Connecticut

Melchi M. Michel Department of Psychology, Rutgers University, New Brunswick, New Jersey

Michael B. Miller University of California, Santa Barbara, Santa Barbara, California

Kathryn L. Mills Child Psychiatry Branch, National Institute of Mental Health, Bethesda, Maryland

F. Gerard Moeller Departments of Psychiatry and Pharmacology and Toxicology, School of Medicine, Virginia Commonwealth University, Richmond, Virginia

Tirin Moore Howard Hughes Medical Institute and Department of Neurobiology, Stanford University, Stanford, California

Justin M. Moscarello Center for Neural Science, New York University, New York, New York

Sinéad L. Mullally Wellcome Trust for Neuroimaging, Institute of Neurology, University College London, London, United Kingdom

Ralph-Axel Müller Department of Psychology, San Diego State University, San Diego, California

John C. Myers Department of Psychology and the Institute of Intelligent Systems, University of Memphis, Memphis, Tennessee

Maital Neta Department of Neurology, Washington University School of Medicine in Saint Louis, Saint Louis, Missouri

F.-X. Neubert Department of Experimental Psychology, University of Oxford, Oxford, United Kingdom

Helen Neville Psychology Department, University of Oregon, Eugene, Oregon

Kenneth A. Norman Department of Psychology and Princeton Neuroscience Institute, Princeton University, Princeton, New Jersey

Behrad Noudoost Howard Hughes Medical Institute and Department of Neurobiology, Stanford University, Stanford, California

Timothy O'Leary Biology Department and Volen Center, Brandeis University, Waltham, Massachusetts

Kevin N. Ochsner Department of Psychology, Columbia University, New York, New York

Bruno A. Olshausen Helen Wills Neuroscience Institute, School of Optometry, and Redwood Center for Theoretical Neuroscience, University of California, Berkeley, Berkeley, California

Asli Özyürek The Max Planck Institute for Psycholinguistics, Radboud University Nijmegen, Nijmegen, The Netherlands

Benjamin Pasquereau Department of Neurobiology, University of Pittsburgh, Pittsburgh, Pennsylvania

Elizabeth A. Phelps Center for Neural Science, New York University, New York, New York

David Pitcher Laboratory of Brain and Cognition, National Institute for Mental Health, Bethesda, Maryland

Xaq Pitkow Department of Neuroscience, Baylor College of Medicine, Houston, Texas

David Poeppel Department of Psychology, New York University, New York, New York

Jonathan D. Power Department of Neurology, Washington University School of Medicine in Saint Louis, Saint Louis, Missouri

Alison Preston Departments of Psychology and Neuroscience, University of Texas at Austin, Austin, Texas

Todd M. Preuss Yerkes National Primate Research Center, Emory University, Atlanta, Georgia

Gregory J. Quirk Department of Psychiatry, School of Medicine, University of Puerto Rico, San Juan, Puerto Rico

Lenny Ramsey Division of Biomedical and Biological Sciences, Washington University in Saint Louis, Saint Louis, Missouri

Charan Ranganath Center for Neuroscience, University of California, Davis, Davis, California

Olga Rass Behavioral Pharmacology Research Unit, Department of Psychiatry and Behavioral Sciences, Johns Hopkins University School of Medicine, Baltimore, Maryland

Gregg H. Recanzone Center for Neuroscience, University of California, Davis, Davis, California

Fred Rieke Department of Physiology and Biophysics, Howard Hughes Medical Institute, University of Washington, Seattle, Washington

James K. Rilling Department of Anthropology, Emory University, Atlanta, Georgia

Maureen Ritchey Center for Neuroscience, University of California, Davis, Davis, California

Jason S. Robert School of Life Sciences, Arizona State University, Tempe, Arizona

Alard Roebroeck Department of Cognitive Neuroscience, Faculty of Psychology and Neuroscience, Maastricht University, Maastricht, The Netherlands

Adina L. Roskies Department of Philosophy, Dartmouth College, Hanover, New Hampshire

Eric D. Roth Department of Psychology, University of Delaware, Newark, Delaware

Tania L. Roth Department of Psychology, University of Delaware, Newark, Delaware

Ashley Royston Department of Psychology and Center for Mind and Brain, University of California, Davis, Davis, California

Douglas A. Ruff Department of Neuroscience and Center for the Neural Basis of Cognition, University of Pittsburgh, Pittsburgh, Pennsylvania

Matthew F. S. Rushworth Department of Experimental Psychology, University of Oxford, Oxford, United Kingdom

Nicole C. Rust Department of Psychology, University of Pennsylvania, Philadelphia, Pennsylvania

Jérôme Sackur Brain and Consciousness Group, Institute for Cognitive Studies, École Normale Supérieure, Paris, France; Centre National de la Recherche Scientifique, Paris, France

Elad Schneidman Department of Neurobiology, Weizmann Institute of Science, Rehovot, Israel

U. Schüffelgen Department of Experimental Psychology, University of Oxford, Oxford, United Kingdom

Stephen H. Scott Department of Biomedical and Molecular Sciences, Department of Medicine, Queen's University, Kingston, Ontario

John T. Serences Department of Psychology, University of California, San Diego, La Jolla, California

Nenad Šestan Department of Neurobiology and Kavli Institute for Neuroscience, Yale School of Medicine, Yale University, New Haven, Connecticut

Peipei Setoh Department of Psychology, University of Illinois, Champaign, Illinois

Daphna Shohamy Department of Psychology, Columbia University, New York, New York

Gordon L. Shulman Department of Neurology, Washington University School of Medicine in Saint Louis, Saint Louis, Missouri

Walter Sinnott-Armstrong Philosophy Department and Kenan Institute for Ethics, Duke University, Durham, North Carolina

Stephanie Sloane Department of Psychology, University of Illinois, Champaign, Illinois

Leah H. Somerville Department of Psychology, Harvard University, Cambridge, Massachusetts

André M. M. Sousa Department of Neurobiology and Kavli Institute for Neuroscience, Yale School of Medicine, Yale University, New Haven, Connecticut

Olaf Sporns Department of Psychological and Brain Sciences, Indiana University, Bloomington, Indiana

Chantal E. Stern Center for Memory and Brain, Department of Psychology and Graduate Program for Neuroscience, Boston University, Boston, Massachusetts

Courtney Stevens Department of Psychology, Willamette University, Salem, Oregon

Dietrich Stout Department of Anthropology, Emory University, Atlanta, Georgia

Peter L. Strick Center for the Neural Basis of Cognition, Systems Neuroscience Institute, and the Department of Neurobiology, University of Pittsburgh, Pittsburgh, Pennsylvania

Alexander Todorov Department of Psychology, Princeton University, Princeton, New Jersey

Giulio Tononi Department of Psychiatry, University of Wisconsin, Madison, Wisconsin

Stefan Treue German Primate Center, Department of Cognitive Neurosciences, Göttingen Graduate Graduate School for Neurosciences, Biophysics, and Molecular Biosciences Göttingen, Germany

Naotsugu Tsuchiya School of Psychological Sciences, Monash University, Melbourne, Australia; Japan Science and Technology Agency, Tokyo, Japan

Robert S. Turner Department of Neurobiology, University of Pittsburgh, Pittsburgh, Pennsylvania

Lorraine K. Tyler Department of Psychology, University of Cambridge, Cambridge, United Kingdom

Melina R. Uncapher Department of Psychology, Stanford University, Stanford, California

Thomas P. Urbach Department of Cognitive Science, University of California, San Diego, La Jolla, California

Nora D. Volkow National Institute on Drug Abuse, Intramural Research Program, National Institutes of Health, Bethesda, Maryland

Anthony D. Wagner Department of Psychology and Neuroscience Program, Stanford University, Stanford, California

Dylan D. Wagner Department of Psychological and Brain Sciences, Dartmouth College, Hanover, New Hampshire

B. A. Wandell Department of Psychology and Center for Cognitive and Neurobiological Imaging, Stanford University, Stanford, California

Wei-chun Wang Department of Psychology, University of California, Davis, Davis, California

Jill D. Waring Department of Psychiatry and Behavioral Sciences, Stanford University School of Medicine, Stanford, California; Department of Psychiatry, Palo Alto Veterans Affairs Healthcare System, Palo Alto, California

Kevin S. Weiner Department of Psychology and Neuroscience Institute, Stanford University, Stanford, California

Paul J. Whalen Department of Psychological and Brain Sciences, Dartmouth College, Hanover, New Hampshire

Nicole White Department of Psychology, University of Toronto, Toronto, Ontario

Jamal Williams Princeton Computational Memory Lab, Department of Psychology, Princeton University, Princeton, New Jersey

Rachel I. Wilson Department of Neurobiology, Harvard Medical School, Boston, Massachusetts; Howard Hughes Medical Institute, Chevy Chase, Maryland

Steven P. Wise Olschefskie Institute for the Neurobiology of Knowledge, Potomac, Maryland

Jeremy Wolfe Visual Attention Lab, Department of Surgery, Brigham and Women's Hospital, Boston, Massachusetts

Thilo Womelsdorf Centre for Vision Research, Department of Biology, York University, North York, Ontario

Suzanne Wood Department of Psychology, Columbia University, New York, New York

Gideon Yaffe Department of Philosophy and Yale Law School, Yale University, New Haven, Connecticut

Daniel Yoshor Departments of Neurosurgery and Neuroscience, Baylor College of Medicine, Houston, Texas

Alan Yuille Department of Brain and Cognitive Engineering, Korea University, Seoul, South Korea

Kelly A. Zalocusky Howard Hughes Medical Institute, Departments of Bioengineering and Psychiatry, Stanford University, Stanford, California

INDEX

B

C

F

G

H

I

J

K

L

M

N

O

Q

R

S

T